D0138715

TEXTBOOK OF BIOCHEMISTRY WITH CLINICAL CORRELATIONS

Fifth Edition

TEXTBOOK OF BIOCHEMISTRY WITH CLINICAL CORRELATIONS

Fifth Edition

EDITED BY

Thomas M. Devlin, Ph.D.

Professor Emeritus
Department of Biochemistry
School of Medicine
MCP-Hahnemann University
Philadelphia, Pennsylvania

WILEY-LISS

A JOHN WILEY & SONS, INC., PUBLICATION

Cover Illustration: The front and back covers contain four different views of the crystallographic structure of the 70S ribosome of Thermus thermophilus at 5.5 angstrom resolution, connected by a strand of messenger RNA. The ribosomes are illustrated as catalyzing protein synthesis, as indicated by the peptide chains (red) exiting from the individual ribosomes. See page 243 for a detailed description of the structure of ribosomes. Reprinted with permission from Yusupov, M. M., Yusupova, G. Z., Baucom, A., Lieberman, K., Earnest, T. N., Cate, J. H. D., and Noller, H. F. Crystal Structure of the Ribosome at 5.5 Å Resolution. Science 292: 883, 2001. Copyrighted 2001 by the American Association for the Advancement of Science. A very special thanks to Drs. A. Baucom and H. F. Noller, University of California at Santa Cruz, who generously supplied the ribosome figures.

This book is printed on acid-free paper. ♾

Published simultaneously in Canada.

For ordering and customer service, call 1-800-CALL-WILEY.

Library of Congress Cataloging-in-Publication Data:
Library of Congress Cataloging-in-Publication Data is available. 0-471-411361

Printed in the United States of America.

10 9 8 7 6 5 4 3 2

To Marjorie,
my best friend, companion, and devoted loving wife.

CONTRIBUTORS

Carol N. Angstadt, Ph.D.
Professor Emerita
Department of Arts and Sciences
MCP Hahnemann University
490 S. Old Middletown Road
Media, PA 19063

Email: angstadtc@drexel.edu

William Awad, JR., M.D., Ph.D.
Professor
Departments of Medicine and
of Biochemistry
University of Miami School of Medicine
PO Box 016960
Miami, FL 33101

Email: wawad@med.miami.edu

Diana S. Beattie, Ph.D.
Professor and Chair
Department of Biochemistry
West Virginia University School
of Medicine
PO Box 9142
Morgantown, WV 26506-9142

Email: dbeattie@hsc.wva.edu

Stephen G. Chaney, Ph.D.
Professor
Departments of Biochemistry and Biophysics,
and of Nutrition
School of Medicine CB# 7260
University of North Carolina at Chapel Hill
Mary Ellen Jones Building
Chapel Hill, NC 27599-7260

Email: stephen_chaney@med.unc.edu

Marguerite W. Coomes, Ph.D.
Associate Professor
Department of Biochemistry
and Molecular Biology
Howard University College of Medicine
520 W Street, N.W.
Washington, DC 20059-0001

Email: mcoomes@fac.howard.edu

Joseph G. Cory, Ph.D.
Professor and Chair
600 Moye Blvd
Department of Biochemistry
Brody School of Medicine
East Carolina University
Greenville, NC 27858-4354

Email: coryjo@mail.ecu.edu

David W. Crabb, M.D.
John B. Hickam Professor and Chair
Department of Medicine
Indiana University School of Medicine
Emerson Hall 317
545 Barnhill Drive
Indianapolis, IN 46202-5124

Email: dcrabb@iupui.edu

Thomas M. Devlin, Ph.D.
Professor Emeritus
Department of Biochemistry
School of Medicine
MCP Hahnemann University
159 Greenville Court
Berwyn, PA 19312-2071

Email: tdevlin@drexel.edu

John E. Donelson, Ph.D.
Professor and Head
Department of Biochemistry
University of Iowa College of Medicine
Bowen Science Building
Iowa City, IA 52242-0001

Email: john-donelson@uiowa.edu

Howard J. Edenberg, Ph.D.
Chancellor's Professor
Department of Biochemistry and Molecular
Biology
Indiana University School of Medicine
635 Barnhill Drive, Med. Sci. 4063
Indianapolis, IN 46202-5122

Email: edenberg@iupui.edu

Robert H. Glew, Ph.D.
Professor
Department of Biochemistry
and Molecular Biology
Basic Medical Science Building, Room 249
School of Medicine
University of New Mexico
915 Camino de Salud NE
Albuquerque, NM 87131-5221

Email: rglew@salud.unm.edu

Dohn G. Glitz, Ph.D.
Professor Emeritus
Department of Biological Chemistry
UCLA School of Medicine
11260 Barnett Valley Road
Sebastopol, CA 95472

Email: dglitz@biochem.medsch.ucla.edu

Robert A. Harris, Ph.D.
Distinguished Professor
Showalter Professor and Chair
Department of Biochemistry
and Molecular Biology
Indiana University School of Medicine
635 Barnhill Drive
Indianapolis, IN 46202-5122

Email: raharris@iupui.edu

Ulrich Hopfer, M.D., Ph.D.
Professor
Department of Physiology and Biophysics
Case Western Reserve University
10900 Euclid Ave.
Cleveland, OH 44106-4970

Email: uxh@po.cwru.edu

Michael N. Liebman, Ph.D.
Director of Computational Biology
University of Pennsylvania Cancer Center
Abramson Family Cancer Research Institute
511 BRB II/III
University of Pennsylvania School
of Medicine
421 Curie Boulevard
Philadelphia, PA 19104

Email: liebmanm@mail.med.upenn.edu

Gerald Litwack, Ph.D.
Professor and Chair
Department of Biochemistry and Molecular
Pharmacology
Jefferson Medical College
Thomas Jefferson University
233 South 10th Street
Philadelphia, PA 19107

Email: gerry.litwack@mail.tju.edu

Bettie Sue Siler Masters, Ph.D.
Robert A. Welch Professor of Chemistry
Department of Biochemistry
University of Texas Health Science Center
at San Antonio
7703 Floyd Curl Drive
San Antonio, TX 78229-3900

Email: masters@uthscsa.edu

J. Denis McGarry, Ph.D.
Professor
Departments of Internal Medicine
and of Biochemistry
University of Texas Southwestern Medical
Center at Dallas
Bldg. G5, Room 210
5323 Harry Hines Blvd
Dallas, TX 75235-9135

Email: dmcgar@mednet.swmed.edu

Richard T. Okita, Ph.D.
Professor and Associate Chair
Department of Pharmaceutical Science
Washington State University
PO Box 646534
Pullman, WA 99164-6534

Email: okitar@wsu.edu

Francis J. Schmidt, Ph.D.
Professor
Department of Biochemistry
School of Medicine
Univ. of Missouri-Columbia
M121 Medical Sciences
Columbia, MO 65212-0001

Email: schmidtf@missouri.edu

Thomas J. Schmidt, Ph.D.
Professor
Department of Physiology and Biophysics
University of Iowa College of Medicine
5-610 Bowen Science Building
Iowa City, IA 52242-1109

Email: thomas-schmidt@uiowa.edu

Richard M. Schultz, Ph.D.
Professor
Division of Molecular and Cellular
Biochemistry
Department of Cell Biology, Neurobiology,
and Anatomy
Stritch School of Medicine
Loyola University of Chicago
2160 South First Avenue
Maywood, IL 60153

Email: rschult@lumc.edu

Nancy B. Schwartz, Ph.D.
Professor
Departments of Pediatrics and of
Biochemistry and Molecular Biology
University of Chicago, MC 5058
5841 S. Maryland Ave.
Chicago, IL 60637-1463

Email: n-schwartz@uchicago.edu

Thomas E. Smith, Ph.D.
Professor
Department of Biochemistry
and Molecular Biology
College of Medicine
Howard University
520 W Street, N.W.
Washington, DC 20059

Email: tsmith@fac.howard.edu

Gerald Soslau, Ph.D.
Professor
Department of Biochemistry
School of Medicine, M.S. 344
MCP Hahnemann University
245 North 15th Street
Philadelphia, PA 19102-1192

Email: Gerald.Soslau@drexel.edu

Daniel L. Weeks, Ph.D.
Professor
Department of Biochemistry
University of Iowa College of Medicine
Bowen Science Building
Iowa City, IA 52242

Email: daniel-weeks@uiowa.edu

Stephen A. Woski, Ph.D.
Associate Professor
Department of Chemistry
University of Alabama
Box 870336
Tuscaloosa, AL 35487-0336

Email: swoski@bama.ua.edu

J. Lyndal York, Ph.D.
Professor
Department of Biochemistry and Molecular Biology
College of Medicine
University of Arkansas for Medical Science
4301 W. Markham St.
Little Rock, AR 72205-7199

Email: yorklyndal@uams.edu

PREFACE

The purposes of the fifth edition of the Textbook of Biochemistry With Clinical Correlations are: to present a clear and precise discussion of the biochemistry of mammalian cells and where appropriate prokaryotic and other eukaryotic cells; to relate the biochemical events at the cellular level to the physiological processes occurring in the whole animal; and to cite examples of abnormal biochemical processes in human disease. These remain unchanged from the earlier editions. The scope and depth of presentation should fulfill the requirements of most biochemistry courses. Topics for inclusion were selected to cover the essential areas of both biochemistry and physiological chemistry for upper-level undergraduate, graduate-level, and especially professional school courses in biochemistry. Since the application of biochemistry is so important to human medicine, the text continues to have an overriding emphasis on the biochemistry of mammalian cells. Information from biochemical investigations of prokaryotes and other eukaryotes is presented, however, when these studies are the primary source of knowledge about the topic. The textbook is organized and written such that any sequence of topics considered most appropriate by an instructor can be presented.

The rapid advances in knowledge in the biological sciences, particularly due to the techniques of molecular biology, and the continued evolution of biochemistry courses, required a critical rethinking of the sequence of topics and content of each chapter in the previous edition. The editor and contributors sought input from biochemistry instructors in the review, and no part of the previous edition was excluded from the evaluation. The outcome was a decision to change the sequence of the material but to maintain the division of material in the same chapters, except for combining into one chapter the presentation of the structures of nucleic acids. The chapter on Gas Transport and pH Regulation was deleted because very few biochemistry courses include this topic. Every chapter was revised and updated with inclusion of significant new information and deletion of some material.

The content of the fifth edition is divided into five sections, in which related topics are grouped together. **Part I, Structure of Macromolecules**, contains an introductory chapter on cell structure, followed by chapters on nucleic acid and protein structure. **Part II, Transmission of Information**, describes the synthesis of the major cellular macromolecules, that is, DNA, RNA, and protein. A chapter on biotechnology is included because information from this area has had such a significant impact on the development of our current knowledge base. Part II concludes with a chapter on the Regulation of Gene Expression in which mechanisms of both prokaryotes and eukaryotes are presented. **Part III, Functions of Proteins**, opens with a presentation of the structure-function relationship of four major families of proteins. This is followed by a discussion of enzymes, including a separate chapter on the cytochromes P450. A presentation of membrane structure and function concludes Part III. **Part IV, Metabolic Pathways and Their Control**, starts with a discussion of bioenergetics followed by separate chapters on the synthesis and degradation of carbohydrates, lipids, amino acids, and purine and pyrimidine nucleotides. A chapter on the integration of these pathways in humans completes this part. **Part V, Physiological Processes**, covers those areas unique to mammalian cells and tissues beginning with two chapters on hormones that emphasize their biochemical functions as messengers, and a chapter on molecular cell biology containing discussions of four major signal transducing systems. The textbook concludes with presentations of iron and heme metabolism, digestion and absorption of basic nutritional constituents, and principles of human nutrition.

The **illustrations** were reviewed and updated where appropriate, and many new figures were added including a number of protein structures published recently. The adage "A picture is worth a thousand words" is appropriate and the reader is encouraged to study the illustrations because they are meant to clarify confusing aspects of a topic.

In each chapter the relevancy of the topic to human life processes is presented in **Clinical Correlations**, which describe the aberrant biochemistry of disease states. In the past few years the genetic and biochemical bases of many diseases have been documented, thus a number of new correlations have been included. There has been no attempt to review all of the major diseases but rather to present examples of disease processes where the ramifications of deviant biochemical processes are well established. References are included in the correlations to facilitate exploration of the topic in more detail. In some cases similar clinical problems are presented in different chapters, but each from a different perspective. All pertinent biochemical information is presented in the main text, and an understanding of the material does not require a reading of the correlations. In some cases, clinical conditions are discussed as part of the primary text because of the significance of the medical condition to an understanding of the biochemical process.

Each chapter concludes with a set of **Questions and Answers**. The multiple-choice format has been retained because they are valuable to students for self-assessment of their knowledge and they are the type used in national examinations. New questions have been added that include clinical vignettes, as well as problem solving questions. Brief annotated answers are given.

The appendix, **Review of Organic Chemistry**, is designed as a ready reference for the nomenclature and structures of organic molecules encountered in biochemistry and is not intended as a comprehensive review of organic chemistry. The material is presented in the Appendix rather than at the beginning of chapters dealing with the different biologically important molecules. The reader should become familiar with the content of the Appendix and then use it as a ready reference when reading related sections in the main text.

We still believe that a **multi-contributor textbook** is the best approach to achieve an accurate and current presentation of biochemistry. Each contributor is involved actively in teaching biochemistry in a medical or graduate school and has an active research interest in the field in which he or she has written. Thus, each has the perspective of the classroom instructor, with the experience to select the topics and determine the emphasis required for students in a course of biochemistry. Every contributor, however, brings to the book an individual approach, leading to some differences in presentation. However, every chapter was edited to have a consistent writing style and to eliminate unnecessary repetitions and redundancies. Presentations of some topics, such as the structure of DNA binding proteins, are included in two different chapters in order to make the individual discussions complete and self contained; in these cases the individual contributors approach the topic from different perspectives. The repetition should facilitate the learning process.

The contributors prepared their chapters for a **teaching book**. The textbook is not intended as a compendium of biochemical facts or a review of the current literature, but each chapter, however, contains sufficient detail on the subject to make it useful as a resource. Contributor were requested not to reference specific researchers; our apologies to those many scientists who have made outstanding research contributions to the field of biochemistry. Each chapter contains a **Bibliography** that serves as an entry point to the research literature.

In any project one person must accept the responsibility for the final product. The decisions concerning the selection of topics and format, reviewing the drafts, and responsibility for the final checking of the book were entirely mine. I welcome comments, criticisms, and suggestions from the students, faculty, and professionals who use this textbook. It is our hope that this work will be of value to those embarking on the exciting experience of learning biochemistry for the first time as well as those returning to a topic in which the information is expanding so rapidly.

THOMAS M. DEVLIN

ACKNOWLEDGMENTS

This project would never have been accomplished without the encouragement and participation of many people. My personal and very deep appreciation goes to each of the contributors for accepting the challenge of preparing the chapters, sharing their ideas and making recommendations to improve the book, accepting so readily suggestions to modify their contributions, and cooperating throughout the period of preparation. To each I extend my sincerest thanks for a job well done.

The contributors received the support of associates and students in the preparation of their chapters, and, for fear of omitting someone, it was decided not to acknowledge individuals by name. To everyone who gave time unselfishly and shared in the objective and critical evaluation of the text, we extend a sincere thank you. In addition, every contributor has been influenced by former teachers and colleagues, various reference resources, and, of course, the research literature of biochemistry. We are very indebted to these many sources of inspiration.

I am deeply grateful to Dr. Frank Vella, Professor of Biochemistry at the University of Saskatchewan, Canada, who assisted me in editing the text. Dr. Vella is a distinguished biochemist who has made a major personal effort to improve the teaching of biochemistry throughout the world. He read every chapter in draft form and made significant suggestions for clarifying and improving the presentation. I extend to him my appreciation and thanks for his participation and friendship.

A special note of appreciation to Dr. Carol Angstadt, my friend and colleague, who reviewed many of the chapters and gave me many valuable suggestions. Our thanks to Dr. Harry F. Noller, University of California at Santa Cruz, for generously giving us permission to use his figures of the structure of the ribosome on the cover.

I extend my sincerest appreciation and thanks to the members of the staff of the STM Division of John Wiley & Sons who participated in the preparation of this edition. My deepest gratitude to Dr. Darla Henderson, Editor, Chemistry and Biochemistry, who conscientiously guided the planning of this edition and made many valuable suggestions. She has been a constant support. Many thanks to Amy Romano, Editorial Program Assistant, who handled efficiently the administrative details and my special requests. My appreciation to Janet Bailey, Executive Publisher for her unqualified support of the project. I am in debt to Kristin Cooke Fasano, Associate Managing Editor, who patiently and meticulously oversaw the production. Kristin kept me well informed, managed the many details, acted promptly to my suggestions and concerns, and kept us on schedule. It has been a pleasure to work with an efficient, knowledgeable and conscientious professional, and a very pleasant individual; to her I extend my heartfelt thanks. A special recognition to John Sollami, Senior Managing Editor and friend, who guided the production of the 4th edition and was an enthusiastic supporter of the 5th edition; I wish John the very best in his new position.

My sincerest thanks and gratitude are extended to Dean Gonzalez, Illustration Manager; Dean oversaw the preparation and revision of the artwork, many times making corrections himself to expedite production. My appreciation to Camille Pecoul Carter, Director, Books Production and Manufacturing, for her continuing support. Thanks to J.C. Morgan and the staff at Precision Graphics who prepared the illustrations. The outstanding design of the book is the work of Laura Ierardi, Designer, to whom I extend a very special thanks. My appreciation to Christina Della Bartolomea, Copy-editor, and Alexandra Nickerson, Indexer with Coughlin Indexing; both did an excellent job. Three individuals deserve special recognition for their efforts in the development of the book for presentation on the Web; my gratitude is extended to Kimi Sugueno, Senior manager, Online Book Production, Colleen Finley, Web Development Manager, and Eileen Dolan, Director Interscience Development. No book is successful without the activities of a Marketing Department; special recognition and my thanks to Greg Giblin, Director of Marketing, and Adam Kirszner, Marketing Manager, and their colleagues at Wiley for their ideas and efforts.

Finally, a very special note of appreciation to my supportive and considerate wife, Marjorie, who had the foresight many years ago to encourage me to undertake the preparation of a Textbook, who supported me during the days of intensive work, and who created an environment in which I could devote the many hours required for the preparation of this textbook. To her my deepest and sincerest thank you.

Thomas M. Devlin

CONTENTS IN BRIEF

CONTENTS

PART I

STRUCTURE OF MACROMOLECULES

Living cells are complex structures with a variety of functions. They synthesize numerous small and large molecules, including very large molecular weight structures, termed macromolecules, using only a limited number of different building blocks. The major macromolecules are deoxyribonucleic acid (DNA), ribonucleic acid (RNA), and protein. The figure above represents the binding of DNA to specific proteins and is an example of the interaction of two different types of macromolecules; details are presented in Figure 2.48. Part I focuses first on the cell and its components, then on the structure of nucleic acids and of proteins. Many diseases are due to changes in the structure of these macromolecules.

Figure reprinted from R.D. Kornberg and Y. Lorch, Cell 98:285, 1999, with permission from the authors and Elsevier Science.

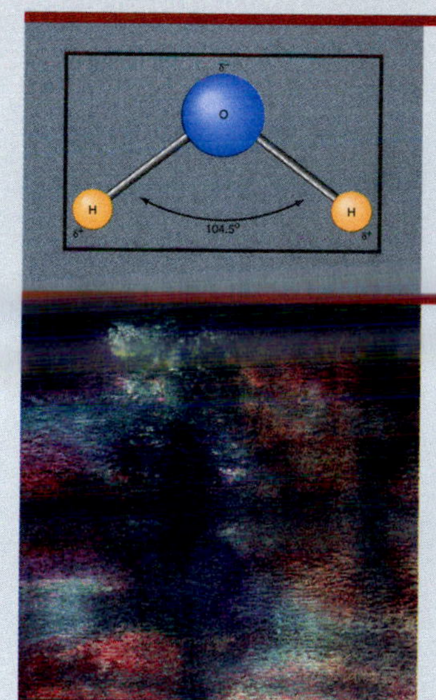

1 EUKARYOTIC CELL STRUCTURE

Thomas M. Devlin

1.1 | OVERVIEW: CELLS AND CELLULAR COMPARTMENTS

Over three and a half billion years ago, under conditions not entirely clear and in a time span difficult to comprehend, the elements carbon, hydrogen, oxygen, nitrogen, sulfur, and phosphorus formed simple chemical compounds. These combined, dispersed, and recombined to form a variety of larger molecules until a structure was achieved that was capable of replicating itself. With continued evolution and formation of ever more complex molecules, the environment around some of these self-replicating molecules became enclosed by a lipid membrane. This development gave these primordial structures the ability to control their own environment to some extent. A form of life had evolved and a unit of three-dimensional space—a cell—had been established. With the passing of time a diversity of cells evolved, and their structure and chemistry became more complex. They could extract nutrients from the environment, convert these nutrients to sources of energy or to complex molecules, control chemical processes that they catalyzed, and carry out cellular replication. Thus the vast diversity of life observed today began. The cell is the basic unit of life in all living organisms, from the smallest single-celled bacterium to the most complex multicellular animal.

All cells have a limiting outer membrane, the **plasma membrane**, that delineates the space occupied and separates a variable and potentially hostile environment outside from a relatively constant milieu within. The plasma membrane and its components are the link to the outside, controlling the movement of substances into and out of the cell and serving as means of communication.

Living cells are divided into two major classes: **prokaryotes** that do not have a nucleus or internal membrane structures and **eukaryotes** that have a defined nucleus and intracellular organelles surrounded by membranes. Prokaryotes, which include eubacteria and archaebacteria, are usually unicellular (Figure 1.1*a*) but in some cases form colonies or filaments. They have a variety of shapes and sizes, usually with a volume 1/1000 to 1/10,000 that of eukaryotic cells. The plasma membrane is often invaginated.

Deoxyribonucleic acid (DNA) of prokaryotes is often segregated into a discrete mass, the **nucleoid** region, that is not surrounded by a membrane or envelope. Even without defined membrane compartments, the intracellular milieu of prokaryotes is organized into functional compartments. Eukaryotic cells, which include those of yeasts, fungi, plants, and animals, have a well-defined membrane that surrounds a central **nucleus** and a variety of intracellular structures and organelles (Figure 1.1*b*). Intracellular membrane systems establish distinct subcellular compartments, as described in Section 1.4, that permit a unique degree of subcellular organization. By compartmentalization different chemical reactions that require different environments can occur simultaneously. In addition, many reactions occur in or on specific membranes that create additional environments for diverse cellular functions.

Besides structural distinctions between prokaryotic and eukaryotic cells (Figure 1.1) there are significant differences in chemical composition and biochemical activities. As an example, eukaryotes but not prokaryotes contain histones, a highly conserved class of proteins in all eukaryotes that complex with DNA. There are differences in the ribonucleic acid–protein complexes involved in biosynthesis of proteins and in enzyme content between these cell types. The many similarities, however, are equally striking. Emphasis throughout this book is on the chemistry of eukaryotes, particularly mammalian cells, but much of our knowledge of the biochemistry of living cells has come from studies of prokaryotic and nonmammalian eukaryotic cells. The basic chemical components and fundamental chemical reactions of all living cells are very similar. The universality of many biochemical phenomena, however, permits many extrapolations from prokaryotes to humans.

The intracellular environment, and particularly the water it contains, places constraints on many cell activities. Thus it is appropriate to review some of the chemical and physical characteristics of this environment. The concluding section

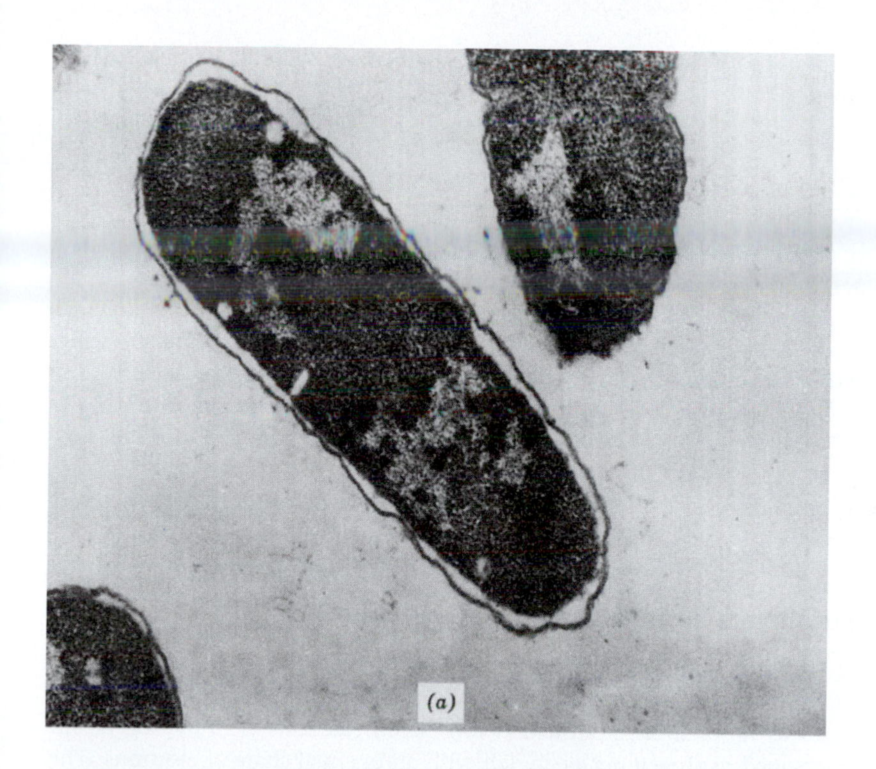

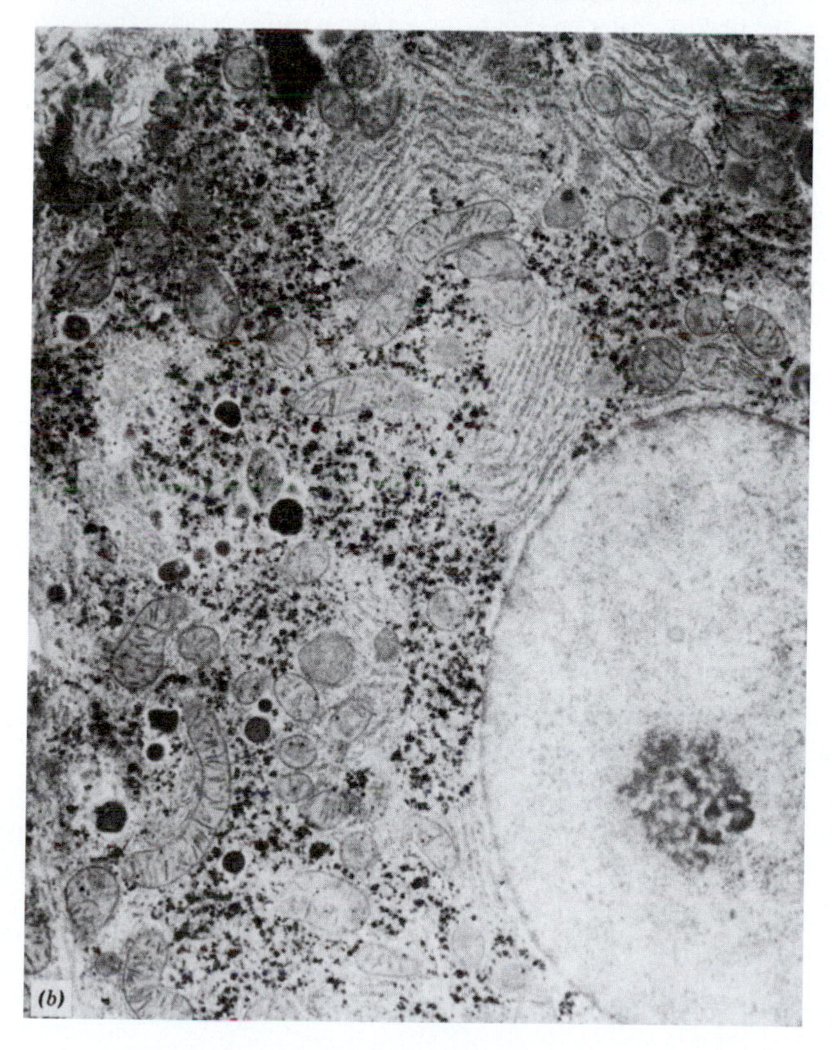

FIGURE 1.1
Cellular organization of prokaryotic and eukaryotic cells.
(*a*) Electron micrograph of *Escherichia coli,* a representative prokaryote; approximate magnification ×30,000. There is little apparent intracellular organization and no cytoplasmic organelles. Chromatin is condensed in a nuclear zone but not surrounded by a membrane. Prokaryotic cells are much smaller than eukaryotic cells. (*b*) Electron micrograph of a thin section of a liver cell (rat hepatocyte), a representative eukaryotic cell; approximate magnification ×7500. Note the distinct nuclear membrane, different membrane-bound organelles or vesicles, and extensive membrane systems. Various membranes create a variety of intracellular compartments.
Photograph (a) generously supplied by Dr. M. E. Bayer, Fox Chase Cancer Institute, Philadelphia, PA; photograph (b) reprinted with permission of Dr. K. R. Porter, from Porter, K. R. and Bonneville, M. A. In: Fine Structure of Cells and Tissues, *Philadelphia: Lea & Febiger, 1972.*

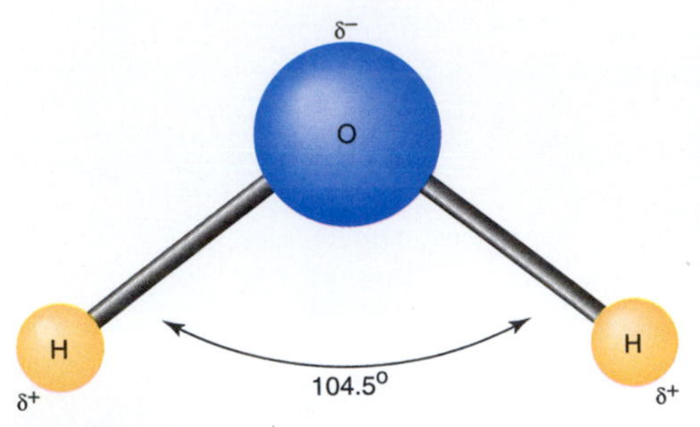

FIGURE 1.2
Structure of a water molecule.
The H—O—H bond angle is 104.5°. Both hydrogen atoms carry a partial positive charge and the oxygen a partial negative charge, creating a dipole.

of this chapter outlines the activities and roles of subcellular mammalian compartments.

1.2 | CELLULAR ENVIRONMENT: WATER AND SOLUTES

All living cells contain essentially the same inorganic ions, small organic molecules, and types of macromolecules. The general classes of cellular components are presented in Table 1.1. Intracellular concentrations of inorganic ions in mammalian cells are essentially similar but very different from the extracellular milieu (see p. 15). **Microenvironments,** created within eukaryotic cells by organelles as well as around macromolecules and membranes, lead to variations in concentration of components throughout a cell. **Water** is the one common component of all environments. It is the solvent in which substances required for the cell's existence are dissolved or suspended. Life as we know it exists because of the unique physicochemical properties of water.

Hydrogen Bonds Form Between Water Molecules

Two hydrogen atoms share their electrons with an unshared pair of electrons of an oxygen atom to form a water molecule. Water is a polar molecule because the oxygen nucleus has a stronger attraction for shared electrons than hydrogen, and positively charged hydrogen nuclei are left with an unequal share of electrons. This creates a partial positive charge on each hydrogen and a partial negative charge on oxygen. The bond angle between hydrogens and oxygen is 104.5°, making the molecule electrically asymmetric and producing an electric dipole (Figure 1.2).

Water molecules interact with each other because positively charged hydrogen atoms on one molecule are attracted to a negatively charged oxygen atom on another, with formation of a weak bond between two molecules (Figure 1.3*a*). This bond, indicated by a dashed line, is a **hydrogen bond.** Recent studies, however, indicate that the bond between two water molecules is partially covalent. A detailed discussion of noncovalent interactions, including electrostatic, van der Waals, and hydrophobic between molecules, is presented on page 135. Five molecules of water can form a tetrahedral structure (Figure 1.3*b*), with each oxygen sharing its electrons with four hydrogen atoms and each hydrogen with another oxygen. A tetrahedral lattice structure is formed in ice and creates its crystalline structure. In the transition from ice to liquid water, some hydrogen bonds are broken. Hydrogen bonds are relatively weak compared to covalent bonds but their large number between molecules in liquid water is the reason for the stability of water. Liquid water has a variable structure due to hydrogen bonding but it is a rapidly changing structure as hydrogen bonds break and new bonds form. The half-life of hydrogen

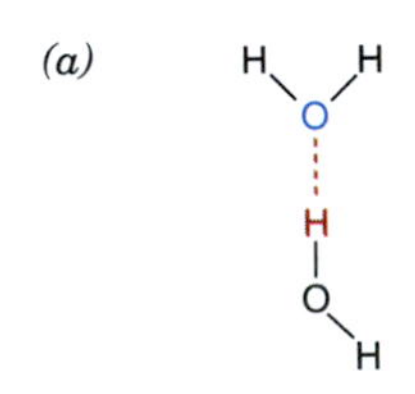

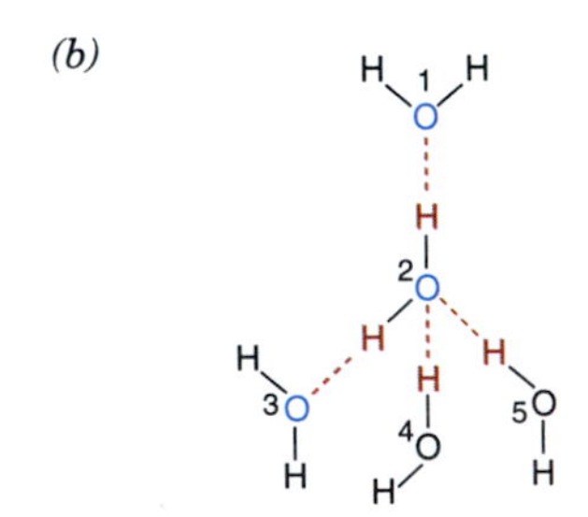

FIGURE 1.3
(*a*) Hydrogen bonding, indicated by dashed lines, between two water molecules. (*b*) Tetrahedral hydrogen bonding of five water molecules. Water molecules 1, 2, and 3 are in the plane of the page, molecule 4 is below, and molecule 5 is above.

TABLE 1.1 Chemical Components of Biological Cells

Component	*Range of Molecular Weights*
H_2O	18
Inorganic ions	23–100
Na^+, K^+, Cl^-, SO_4^{2-},	
HCO_3^- Ca^{2+}, Mg^{2+}, etc.	
Small organic molecules	100–1200
Carbohydrates, amino acids,	
lipids, nucleotides, peptides	
Macromolecules	50,000–1,000,000,000
Proteins, polysaccharides,	
nucleic acids	

bonds in water is less than 1×10^{-10}s. Even at 100°C liquid water contains a significant number of hydrogen bonds, which accounts for its high heat of vaporization. In the transformation from the liquid to vapor state, hydrogen bonds are disrupted. Many models for the structure of liquid water have been proposed, but none adequately explains all its properties.

The structure of pure water is altered when water molecules hydrogen bond to different chemical structures. The water environment in cells is not uniform. As an example, water molecules are more ordered along the surface of a membrane because of the amphiphilic nature of lipid-containing membranes. This creates a microenvironment that can alter the activities of substances present in this milieu. A similar change can occur in the structure of water when it is present within protein and nucleic acid molecules and stabilizes these macromolecules.

Hydrogen bonding also occurs between other molecules and within a molecule wherever electronegative oxygen or nitrogen atoms come into close proximity to hydrogen covalently bonded to another electronegative atom. Representative hydrogen bonds are presented in Figure 1.4. Intramolecular hydrogen bonding occurs extensively in large macromolecules such as proteins and nucleic acids and is partially responsible for their structural stability.

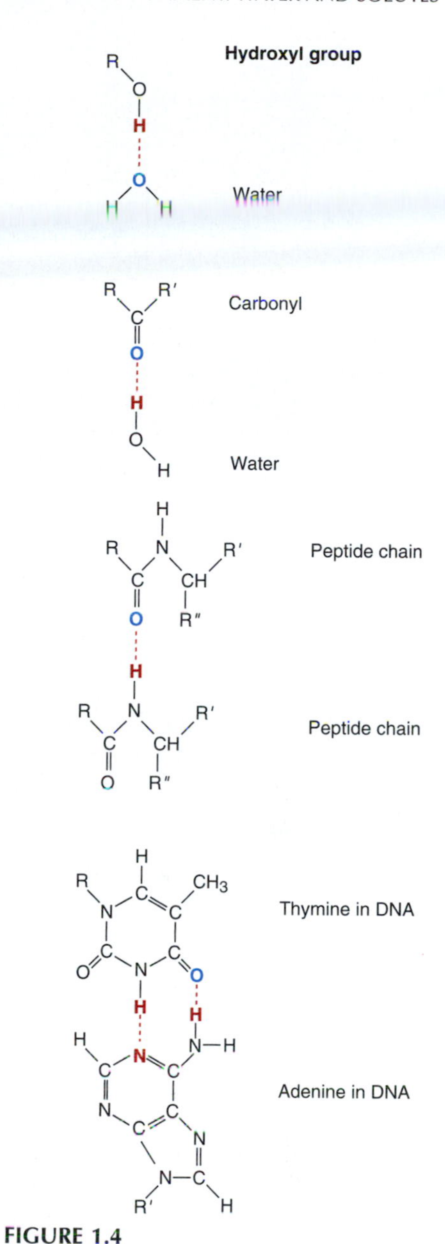

FIGURE 1.4
Representative hydrogen bonds of importance in biological systems.

Water Has Unique Solvent Properties

The polar nature and ability to form hydrogen bonds are the basis for the unique **solvent properties** of water. Polar molecules are readily dispersed in water. **Salts,** in which a crystal lattice is held together by attraction of positive and negative groups, dissolve in water because electrostatic forces in the crystal can be overcome by attraction of charges to the dipole of water. NaCl is an example where electrostatic attraction of individual $Na^+$and Cl^- ions is overcome by interaction of Na^+ with the negative charge on oxygen atoms of water, and Cl^- with positive charges on the hydrogen atoms. Thus a shell of water surrounds individual ions. The number of weak charge–charge interactions between water and $Na^+$and Cl^- ions is sufficient to maintain separation of the charged ions.

Many organic molecules contain nonionic but weakly polar groups and are soluble in water because of the attraction of these groups to molecules of water. Sugars and alcohols are readily soluble in water for this reason. **Amphipathic molecules,** compounds that contain both polar and nonpolar groups, disperse in water if attraction of the polar group for water can overcome hydrophobic interactions of nonpolar portions of the molecules. Very hydrophobic molecules, such as compounds that contain long hydrocarbon chains, however, do not readily disperse as single molecules in water but interact with one another to exclude the polar water molecules.

Some Molecules Dissociate to Form Cations and Anions

Substances that dissociate in water into a **cation** (positively charged ion) and an **anion** (negatively charged ion) are classified as **electrolytes** because these ions facilitate conductance of an electrical current through water. Sugars or alcohols are classified as **nonelectrolytes** because they dissolve readily in water but do not carry a charge or dissociate into charged species.

Salts of alkali metals (e.g., Li, Na, and K) dissociate completely when dissolved in water at low concentrations but not necessarily at high concentrations. It is customary, however, to consider such compounds and salts of organic acids, for example, sodium lactate, to be dissociated totally in biological systems because their concentrations are low. The dissociated anion of an organic acid, for example, lactate ion, reacts to a limited extent with a proton from water to form undissociated acid (Figure 1.5). If a solution has been prepared that contains several different salts (e.g., NaCl, K_2SO_4, and Na lactate) these molecules do not exist as such in solution, only the dissociated ions (e.g., Na^+, K^+, $SO_4{}^{2-}$, and lactate$^-$). Many acids,

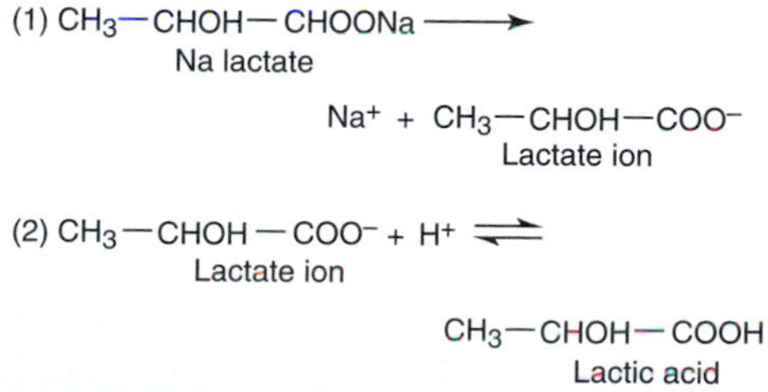

(1) CH_3—CHOH—CHOONa ⟶ Na^+ + CH_3—CHOH—COO^-
Na lactate → Lactate ion

(2) CH_3—CHOH—COO^- + H^+ ⇌ CH_3—CHOH—COOH
Lactate ion → Lactic acid

FIGURE 1.5
Reactions that occur when sodium lactate is dissolved in water.

however, when dissolved in water do not dissociate totally but rather establish an equilibrium between undissociated and dissociated components. Thus lactic acid, an important metabolic intermediate, partially dissociates into a lactate anion and a proton (H^+) as follows:

$$CH_3\text{—}CHOH\text{—}COOH \rightleftharpoons CH_3\text{—}CHOH\text{—}COO^- + H^+$$

A dynamic equilibrium is established between the components in which the products of the reaction reform the undissociated reactant while other molecules dissociate. The degree of dissociation of such an electrolyte depends on the affinity of the anion for a H^+. There will be more dissociation if the weak dipole forces of water that interact with the anion and cation are stronger than the electrostatic forces between anion and H^+. Such compounds on a molar basis have a lower capacity to carry an electrical charge in comparison to those that dissociate totally; they are termed **weak electrolytes.**

Weak Electrolytes Dissociate Partially

In partial dissociation of a weak electrolyte, represented by HA, the concentration of various species can be determined from the equilibrium equation

$$K'_{eq} = \frac{[H^+][A^-]}{[HA]} \tag{1.1}$$

K'_{eq} is a physical constant, A^- represents the dissociated anion, and square brackets indicate concentration of each component in units such as moles per liter (mol L^{-1} or M) or millimoles per liter (mmol L^{-1} or mM). The **activity** of each species rather than concentration should be employed in the equilibrium equation but since most compounds of interest in biological systems are present in low concentration, the value for activity approaches that of concentration. The equilibrium constant, however, is indicated as $\boldsymbol{K'_{eq}}$ to indicate that it is an apparent constant based on concentrations. The extent of dissociation of an acid increases with increasing temperatures. From the dissociation equation, it is apparent that K'_{eq} will be a small number if the degree of dissociation of a substance is small (large denominator in Eq. 1.1) but large if the degree of dissociation is large (small denominator). Obviously, for compounds that dissociate totally, a K'_{eq} cannot be determined because at equilibrium there is no remaining undissociated solute.

Water Is a Weak Electrolyte

Water dissociates as follows:

$$HOH \rightleftharpoons H^+ + OH^-$$

A proton that dissociates interacts with oxygen of another water molecule to form the **hydronium ion,** H_3O^+. It is a generally accepted practice, and one that will be employed in this book, to present the proton as H^+ rather than H_3O^+, recognizing that the latter is the actual chemical species. At 25°C the value of K'_{eq} for dissociation of water is very small and is about 1.8×10^{-16}:

$$K'_{eq} = 1.8 \times 10^{-16} = \frac{[H^+][OH^-]}{[H_2O]} \tag{1.2}$$

With such a small K'_{eq} an extremely small number of water molecules actually dissociate relative to the number of undissociated molecules. Thus the concentration of water, which is 55.5 M, is essentially unchanged by the dissociation. Equation 1.2 can be rewritten as follows:

$$K'_{eq} \times [H_2O] = [H^+][OH^-] \tag{1.3}$$

$K'_{eq} \times [55.5]$ is a constant and is termed the **ion product of water.** Its value at 25°C is 1×10^{-14}. In pure water the concentration of H^+ equals OH^-, and by substituting $[H^+]$ for $[OH^-]$ in Eq. 1.3, $[H^+]$ is 1×10^{-7} M. Similarly, $[OH^-]$ is also 1×10^{-7} M. The equilibrium reaction of H_2O, H^+, and OH^- always exists in solutions regardless of the presence of dissolved substances. If dissolved material alters the H^+ or OH^- concentration, as occurs on addition of an acid or a base, a concomitant change in the other ion must occur in order to satisfy the equilibrium relationship of water. Using the equation for ion product, $[H^+]$ or $[OH^-]$ can be calculated if the concentration of one ion is known. The importance of hydrogen ions in biological systems will become apparent in discussions of enzyme activity (see p. 431) and metabolism.

For convenience $[H^+]$ is usually expressed in terms of **pH**, calculated as follows:

$$\mathbf{pH} = \log \frac{1}{[H^+]} \qquad (1.4)$$

In pure water $[H^+]$ and $[OH^-]$ are both 1×10^{-7} M, and pH = 7.0. $[OH^-]$ can be expressed as the pOH. For the equation describing dissociation of water, $1 \times 10^{-14} = [H^+][OH^-]$, taking negative logarithms of both sides, the equation becomes $14 = \text{pH} + \text{pOH}$. Table 1.2 presents the relationship between pH and $[H^+]$.

pH values of different biological fluids are presented in Table 1.3. In blood plasma, $[H^+]$ is 0.00000004 M or a pH of 7.4. Other cations are between 0.001 and 0.10 M, well over 10,000 times higher than $[H^+]$. An increase in hydrogen ion to 0.0000001 M (pH 7.0) or a decrease to 0.00000002 M (pH 7.8) of blood leads to serious medical consequences and are life threatening.

TABLE 1.2 Relationships Between $[H^+]$ and pH and $[OH^-]$ and pOH

$[H^+]$ (*M*)	*pH*	$[OH^-]$ (*M*)	*pOH*
1.0	0	1×10^{-14}	14
0.1 (1×10^{-1})	1	1×10^{-13}	13
1×10^{-2}	2	1×10^{-12}	12
1×10^{-3}	3	1×10^{-11}	11
1×10^{-4}	4	1×10^{-10}	10
1×10^{-5}	5	1×10^{-9}	9
1×10^{-6}	6	1×10^{-8}	8
1×10^{-7}	7	1×10^{-7}	7
1×10^{-8}	8	1×10^{-6}	6
1×10^{-9}	9	1×10^{-5}	5
1×10^{-10}	10	1×10^{-4}	4
1×10^{-11}	11	1×10^{-3}	3
1×10^{-12}	12	1×10^{-2}	2
1×10^{-13}	13	0.1 (1×10^{-1})	1
1×10^{-14}	14	1.0	0

Many Biologically Important Molecules Are Acids or Bases

The definitions of an acid and a base proposed by Lowry and Brønsted are most convenient in considering biological systems. An **acid** is a **proton donor** and a **base** is a **proton acceptor.** Hydrochloric acid (HCl) and sulfuric acid (H_2SO_4) are **strong acids** because they dissociate totally, releasing H^+. OH^- ion is a base. Addition of either an acid or a base to water will lead to establishment of a new equilibrium of $OH^- + H^+ \rightleftharpoons H_2O$. When a strong acid and OH^- are combined, H^+ from the acid and OH^- interact nearly totally and neutralize nearly totally each other.

Anions produced when strong acids dissociate totally, such as Cl^- from HCl, are not bases because they do not associate with protons in solution. Most organic acids (HA) when dissolved in water dissociate only partially, establishing an equilibrium between HA (proton donor), an anion of the acid (A^-), and a proton as follows:

$$HA \rightleftharpoons A^- + H^+$$

Organic acids are usually **weak acids** because they only partially dissociate. The anion formed in this dissociation is a base because it can accept a proton to reform the acid. A weak acid and its base (anion) formed on dissociation are referred to as a **conjugate pair**; examples of some biologically important conjugate pairs are presented in Table 1.4.

Ammonium ion (NH_4^+) is an acid because it dissociates to yield H^+ and uncharged ammonia (NH_3), a conjugate base. Phosphoric acid (H_3PO_4) is an acid and PO_4^{3-} is a base, but $H_2PO_4^-$ and HPO_4^{2-} are either a base or an acid depending on whether the phosphate group is accepting or donating a proton.

The tendency of a **conjugate acid** to release H^+ can be evaluated from the K'_{eq} (Eq. 1.2). The smaller the value of K'_{eq}, the less the tendency to give up a proton and the weaker the acid; the larger the value of K'_{eq}, the greater the tendency to dissociate a proton and the stronger the acid. Water is a very weak acid with a K'_{eq} of 1.8×10^{-16} at 25°C.

A convenient method of stating the K'_{eq} is in the form of **p*K*′** as

$$\mathbf{p}K' = \log \frac{1}{K'_{eq}} \qquad (1.5)$$

TABLE 1.3 pH of Some Biological Fluids

Fluid	*pH*
Blood plasma	7.4
Interstitial fluid	7.4
Intracellular fluid	
Cytosol (liver)	6.9
Lysosomal matrix	Below 5.0
Gastric juice	1.5–3.0
Pancreatic juice	7.8–8.0
Human milk	7.4
Saliva	6.4–7.0
Urine	5.0–8.0

TABLE 1.4 Some Conjugate Acid–Base Pairs of Importance in Biological Systems

Proton Donor (Acid)		*Proton Acceptor (Base)*
$CH_3—CHOH—COOH$ (lactic acid)	$\rightleftharpoons$	$H^+ + CH_3—CHOH—COO^-$ (lactate)
$CH_3—CO—COOH$ (pyruvic acid)	$\rightleftharpoons$	$H^+ + CH_3—CO—COO^-$ (pyruvate)
$HOOC—CH_2—CH_2—COOH$ (succinic acid)	$\rightleftharpoons$	$2H^+ + {}^-OOC—CH_2—CH_2—COO^-$ (Succinate)
${}^+H_3NCH_2—COOH$ (glycine)	$\rightleftharpoons$	$H^+ + {}^+H_3N—CH_2—COO^-$ (glycinate)
H_3PO_4	$\rightleftharpoons$	$H^+ + H_2PO_4^-$
$H_2PO_4^-$	$\rightleftharpoons$	$H^+ + HPO_4^{2-}$
HPO_4^{2-}	$\rightleftharpoons$	$H^+ + PO_4^{3-}$
Glucose 6-PO_3H^-	$\rightleftharpoons$	H^+ + glucose 6-PO_3^{2-}
H_2CO_3	$\rightleftharpoons$	$H^+ + HCO_3^-$
NH_4^+	$\rightleftharpoons$	$H^+ + NH_3$
H_2O	$\rightleftharpoons$	$H^+ + OH^-$

Note the similarity of this definition with that of pH; as with pH and $[H^+]$, the relationship between pK' and K'_{eq} is an inverse one, and the smaller the K'_{eq} the larger the pK'. Representative values of K'_{eq} and pK' for conjugate acids of importance in biological systems are presented in Table 1.5.

A special case of a weak acid important in medicine is carbonic acid (H_2CO_3). **Carbon dioxide** when dissolved in water is involved in the following equilibrium reactions:

$$CO_2 + H_2O \overset{K'_2}{\rightleftharpoons} H_2CO_3 \overset{K'_1}{\rightleftharpoons} H^+ + HCO_3^-$$

TABLE 1.5 Apparent Dissociation Constant and pK' of Some Compounds of Importance in Biochemistry

Compound	*Structures*	K'_{eq} (M)	pK'
Acetic acid	$CH_3—COOH$	1.74×10^{-5}	4.76
Alanine	$CH_3—CH(NH_3^+)—COOH$	4.57×10^{-3}	2.34 (COOH)
		2.04×10^{-10}	9.69 (NH_3^+)
Citric acid	$HOOC—CH_2—COH(COOH)—CH_2—COOH$	8.12×10^{-4}	3.09
		1.77×10^{-5}	3.74
		3.89×10^{-6}	5.41
Glutamic acid	$HOOC—CH_2—CH_2—CH(NH_3^+)—COOH$	6.45×10^{-3}	2.19 (COOH)
		5.62×10^{-5}	4.25 (COOH)
		2.14×10^{-10}	9.67 (NH_3^+)
Glycine	$CH_2(NH_3^+)—COOH$	4.57×10^{-3}	2.34 (COOH)
		2.51×10^{-10}	9.60 (NH_3^+)
Lactic acid	$CH_3—CHOH—COOH$	1.38×10^{-4}	3.86
Pyruvic acid	$CH_3—CO—COOH$	3.16×10^{-3}	2.50
Succinic acid	$HOOC—CH_2—CH_2—COOH$	6.46×10^{-5}	4.19
		3.31×10^{-6}	5.48
Glucose 6-PO_3H^-	$C_{12}H_{11}O_5\ PO_3H^-$	7.76×10^{-7}	6.11
	H_3PO_4	1×10^{-2}	2.0
	$H_2PO_4^-$	2.0×10^{-7}	6.7
	HPO_4^{2-}	3.4×10^{-13}	12.5
	H_2CO_3	1.70×10^{-4}	3.77
	NH_4^+	5.62×10^{-10}	9.25
	H_2O	1.8×10^{-16}	15.74

Carbonic acid (H_2CO_3) has a pK_1' of 3.77, which is comparable to organic acids such as lactic acid. The equilibrium equation for this reaction is

$$K_1' = \frac{[H^+][HCO_3^-]}{[H_2CO_3]} \tag{1.6}$$

H_2CO_3 is, however, in equilibrium with dissolved CO_2 and the equilibrium equation for this reaction is

$$K_2' = \frac{[H_2CO_3]}{[CO_2][H_2O]} \tag{1.7}$$

Solving Eq. 1.7 for H_2CO_3 and substituting for the H_2CO_3 in Eq. 1.6, the two equilibrium reactions can be combined into one equation:

$$K_1' = \frac{[H^+][HCO_3^-]}{K_2'[CO_2][H_2O]} \tag{1.8}$$

Rearranging to combine constants, including the concentration of H_2O, simplifies the equation and yields a new combined constant, K_3', as follows:

$$K_1'K_2'[H_2O] = K_3' = \frac{[H^+][HCO_3^-]}{[CO_2]} \tag{1.9}$$

It is common practice in medicine to refer to dissolved CO_2 as the conjugate acid; it is the acid anhydride of H_2CO_3. The term K_3' has a value of 7.95×10^{-7} and $pK_3' = 6.1$. If the aqueous system is in contact with an air phase, dissolved CO_2 will also be in equilibrium with CO_2 in the air phase. Decrease or increase of one component—that is, CO_2 (air), CO_2 (dissolved), H_2CO_3, H^+, or HCO_3^-—causes a change in every component. The CO_2–HCO_3^- system is extremely important for maintaining pH homeostasis in animals.

Henderson–Hasselbalch Equation Defines the Relationship Between pH and Concentrations of Conjugate Acid and Base

A change in concentration of any component in an equilibrium reaction necessitates a concomitant change in every component. For example, an increase in $[H^+]$ decreases the concentration of **conjugate base** (e.g., lactate ion) with an equivalent increase in the **conjugate acid** (e.g., lactic acid). This relationship is conveniently expressed by rearranging the equilibrium equation and solving for H^+, as shown for the following dissociation:

$$\text{conjugate acid} \rightleftharpoons \text{conjugate base} + H^+$$

$$K_{eq}' = \frac{[H^+][\text{conjugate base}]}{[\text{conjugate acid}]} \tag{1.10}$$

Rearranging Eq. 1.10 by dividing through by both $[H^+]$ and K_{eq}' leads to

$$\frac{1}{[H^+]} = \frac{1}{[K_{eq}']} \cdot \frac{[\text{conjugate base}]}{[\text{conjugate acid}]} \tag{1.11}$$

Taking the logarithm of both sides gives

$$\log \frac{1}{[H^+]} = \log \frac{1}{K_{eq}'} + \log \frac{[\text{conjugate base}]}{[\text{conjugate acid}]} \tag{1.12}$$

Since $pH = \log(1/[H^+])$ and $pK' = \log(1/K_{eq}')$, Eq. 1.12 becomes

$$\mathbf{pH = pK' + \log \frac{[conjugate\ base]}{[conjugate\ acid]}} \tag{1.13}$$

Equation 1.13, developed by Henderson and Hasselbalch, is a convenient way of viewing the relationship between pH of a solution and relative amounts of conjugate

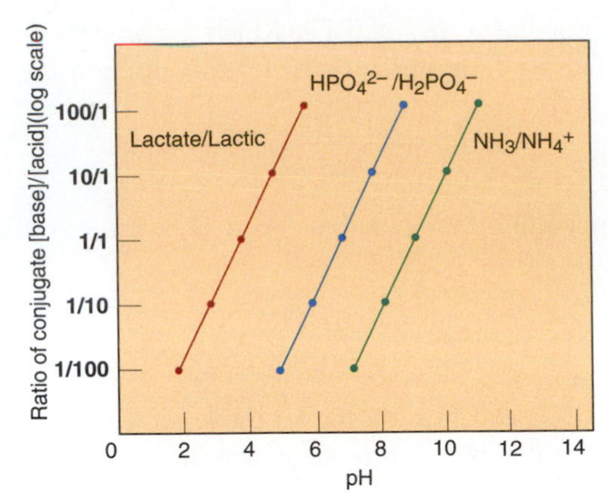

FIGURE 1.6
Ratio of conjugate [base]/[acid] as a function of the pH.
When the ratio of [base]/[acid] is 1, pH equals pK' of weak acid.

base and acid present. Analysis of Eq. 1.13 demonstrates that when the ratio of [base]/[acid] is 1:1, pH equals the pK' of the acid because log 1 = 0, and pH = pK'. If pH is one unit less than pK', the [base]/[acid] ratio is 1:10, and if pH is one unit above pK', the [base]/[acid] ratio is 10:1. Figure 1.6 is a plot of ratios of conjugate base to conjugate acid versus pH of several weak acids; note that ratios are presented on a logarithmic scale.

Buffering Is Important to Control pH

When NaOH is added to a solution of a weak acid such as lactic acid, the ratio of [conjugate base]/[conjugate acid] changes. NaOH dissociates totally and the OH^- formed is neutralized by existing H^+ to form H_2O. The decrease in $[H^+]$ causes further dissociation of weak acid to comply with requirements of its equilibrium reaction. The amount of weak acid dissociated will be so nearly equal to the amount of OH^- added that it is considered to be equal. Thus the decrease in amount of conjugate acid is equal to the amount of conjugate base that is formed. The events are represented in the titration curves of two weak acids presented in Figure 1.7. When 0.5 equiv of OH^- is added, 50% of the weak acid is dissociated and the [acid]/[base] ratio is 1.0; pH at this point is equal to pK' of the acid. Shapes of individual titration curves are similar but displaced due to differences in pK' values. There is a rather steep rise in pH when only 0.1 equiv of OH^- is added but between 0.1 and 0.9 equiv of added OH^-, the pH change is only ~2. Thus a large amount of OH^- is added with a relatively small change in pH. This is called **buffering** and is defined as the ability of a solution to resist a change in pH when an acid or base is added. If weak acid is not present, a small amount of OH^- would lead to a high pH because there would be no source of H^+ to neutralize the OH^-.

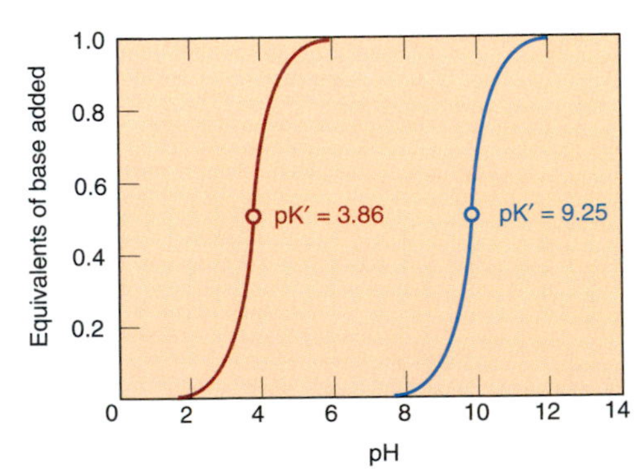

FIGURE 1.7
Acid–base titration curves for lactic acid (pK' 3.86) and NH_4^+; (pK' 9.25).
At pH equal to respective pK' values, there will be an equal amount of acid and base for each conjugate pair.

The best buffering range for a conjugate pair is in the pH range near the pK' of the weak acid. Starting from a pH one unit below to a pH one unit above pK', about 82% of a weak acid in solution will dissociate, and therefore an amount of base equivalent to about 82% of original acid can be neutralized with a change in pH of 2. Maximum buffering ranges for conjugate pairs are considered to be between 1 pH unit below and 1 pH unit above pK'. The conjugate pair lactate ion/lactic acid with pK' = 3.86 is an effective buffer between pH 3 and pH 5, with no significant buffering capacity at pH = 7.0. The $HPO_4^{2-}/H_2PO_4^-$ pair with pK' = 6.7 is an effective buffer at pH = 7.0. Thus at the pH of the cell's cytosol (~7.0), the lactate ion/lactic acid pair is not an effective buffer but $HPO_4^{2-}/H_2PO_4^-$ is.

Buffering capacity also depends on concentrations of conjugate acid and base. The higher the concentration of conjugate base, the more added H^+ with which it can react. The more conjugate acid the more added OH^- can be neutralized by dissociation of the acid. A case in point is blood plasma at pH 7.4. The p*K′* for $HPO_4^{2-}/H_2PO_4^-$ of 6.7 would suggest that this conjugate pair would be an effective buffer; the concentration of the phosphate pair, however, is low compared to that of the HCO_3^-/CO_2 system with a p*K′* of 6.1, which is present at a 20-fold higher concentration and accounts for most of the buffering capacity. Both p*K′* and concentration of a conjugate pair must be taken into account when considering buffering capacity. Most organic acids are relatively unimportant as buffers in cellular fluids because their p*K′* values are more than several pH units lower than the pH of the cell, and their concentrations are too low in comparison to buffer systems such as $HPO_4^{2-}/H_2PO_4^-$ and HCO_3^-/CO_2.

Figure 1.8 presents some typical problems using the Henderson–Hasselbalch equation. Control of pH and buffering in humans cannot be over emphasized; Clinical Correlation 1.1 is a representative problem in clinical practice.

1. Calculate the ratio of $HPO_4^{2-}/H_2PO_4^-$ (p*K*=6.7) at pH 5.7, 6.7, and 8.7.

Solution:

$$\text{pH} = \text{p}K + \log \frac{[HPO_4^{2-}]}{[H_2PO_4^-]}$$

5.7 = 6.7 + log of ratio; rearranging

5.7 – 6.7 = –1 = log of ratio

The antilog of –1 = 0.1 or 1/10. Thus, $HPO_4^{2-}/H_2PO_4^-$ = 1/10 at pH 5.7. Using the same procedure, the ratio at pH 6.7 = 1/1 and at pH 8.7 = 100/1.

2. If the pH of blood is 7.1 and the HCO_3^- concentration is 8 mM, what is the concentration of CO_2 in blood (p*K′* for HCO_3^-/CO_2 = 6.1)?

Solution:

$$\text{pH} = \text{p}K + \log \frac{[HCO_3^-]}{[CO_2]}$$

7.1 = 6.1 + log (8 mM / $[CO_2]$); rearranging

7.1 – 6.1 = 1 = log (8 mM / $[CO_2]$).

The antilog of 1 = 10. Thus, 10 = 8 mM / $[CO_2]$, or $[CO_2]$ = 8 mM/10 = 0.8 mM.

3. At a normal blood pH of 7.4, the sum of $[HCO_3^-] + [CO_2]$ = 25.2 mM. What is the concentration of HCO_3^- and CO_2 (p*K′* for HCO_3^-/CO_2 = 6.1)?

Solution:

$$\text{pH} = \text{p}K + \log \frac{[HCO_3^-]}{[CO_2]}$$

7.4 = 6.1 + log ($[HCO_3^-]$ / $[CO_2]$); rearranging

7.4 – 6.1 = 1.3 = log ($[HCO_3^-]$ / $[CO_2]$).

The antilog of 1.3 is 20. Thus $[HCO_3^-]/[CO_2]$ = 20. Given $[HCO_3^-] + [CO_2]$ = 25.2, solve these two equations for $[CO_2]$ by rearranging the first equation:

$[HCO_3^-] = 20\,[CO_2]$.

Substituting in the second equation,

$20\,[CO_2] + [CO_2] = 25.2$

or

CO_2 = 1.2 mM

Then substituting for CO_2, 1.2 + $[HCO_3^-]$ = 25.2, and solving, $[HCO_3^-]$ = 24 mM.

FIGURE 1.8
Typical problems of pH and buffering.

CLINICAL CORRELATION 1.1
Blood Bicarbonate Concentration in Metabolic Acidosis

In normal adults blood buffers maintain blood pH at about 7.40; if pH should drop below 7.35, the condition is referred to as an acidosis. A blood pH near 7.0 could lead to serious consequences and possibly death. Thus in acidosis, particularly that caused by a metabolic change, it is important to monitor the acid–base parameters of a patient's blood. Values of interest to a clinician include the blood pH and HCO_3^- and CO_2 concentrations. Normal values for these are pH = 7.40, $[HCO_3^-]$ = 24.0 mM, and $[CO_2]$ = 1.20 mM. If the blood values of a patient with a metabolic acidosis were pH = 7.03 and $[CO_2]$ = 1.10 mM, what is the patient's blood $[HCO_3^-]$ and how much of the normal $[HCO_3^-]$ has been used in buffering the acid causing the condition?

1. The Henderson–Hasselbalch equation is

$$\text{pH} = \text{p}K' + \log([HCO_3^-]/[CO_2])$$

The p*K′* value for $[HCO_3^-]/[CO_2]$ is 6.10.

2. Substitute the given values in the equation.

$$7.03 = 6.10 + \log([HCO_3^-]/1.10\text{ mM})$$

or

$$7.03 - 6.10 = 0.93 = \log([HCO_3^-]/1.10\text{ mM})$$

The antilog of 0.93 is 8.5; thus

$$8.5 = [HCO_3^-]/1.10\text{ mM}$$

or

$$[HCO_3^-] = 9.4\text{ mM}$$

3. Since the normal value of $[HCO_3^-]$ is 24 mM, there has been a decrease of 14.6 mmol of HCO_3^- per liter of blood in this patient. If much more HCO_3^- is lost, a point would be reached when this important buffer would be unavailable to buffer any more acid in the blood and the pH would drop rapidly.

1.3 | ORGANIZATION AND COMPOSITION OF EUKARYOTIC CELLS

Eukaryotic cells contain well-defined cellular organelles such as a nucleus, mitochondria, lysosomes, and peroxisomes, each delineated by a membrane (Figure 1.9). Membranes also form a tubule-like network throughout the cell, the endoplasmic reticulum and Golgi complex, enclosing an interconnecting space or cisternae, respectively. Specific functions and activities of these organelles are described in Section 1.4.

The lipid nature of **cellular membranes** prevents rapid movement of many molecules, including water, across them. Specific mechanisms for translocation of

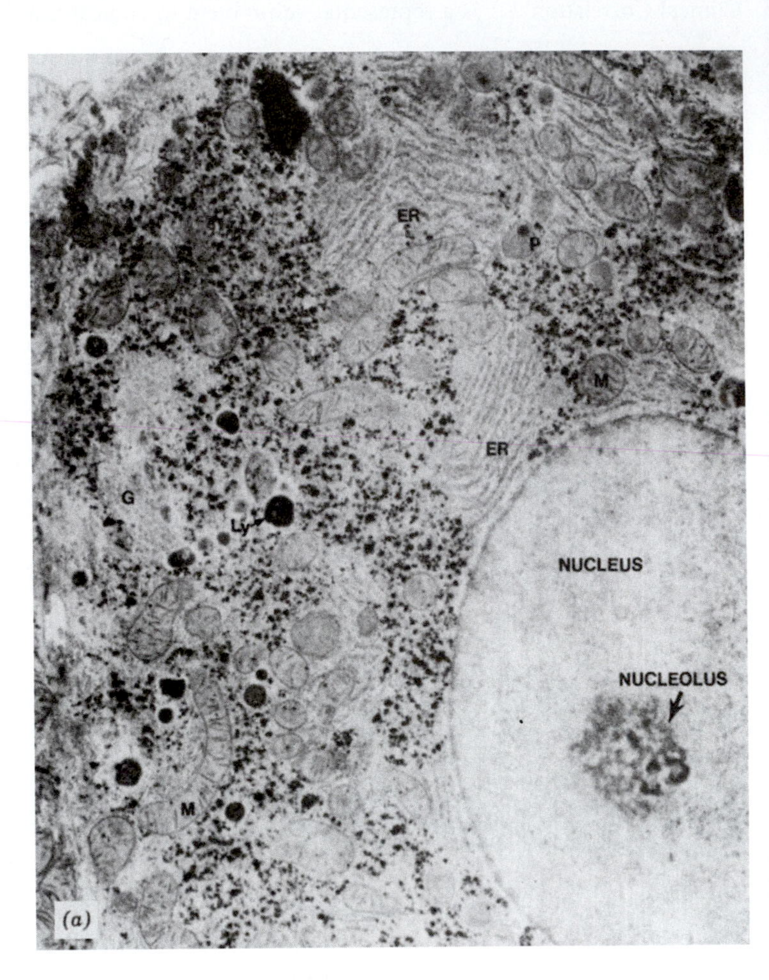

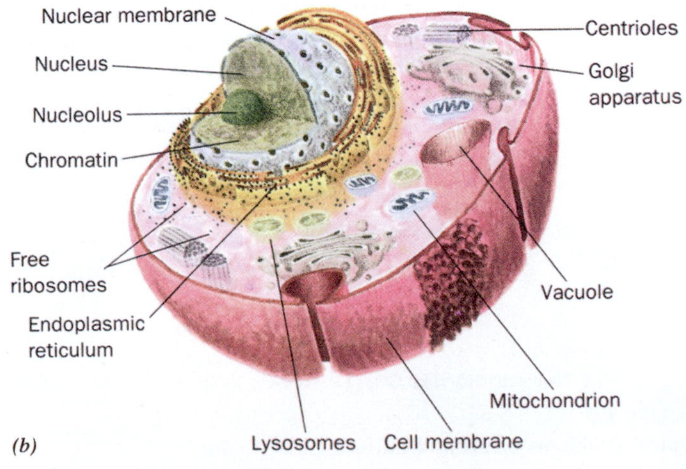

FIGURE 1.9
(*a*) Electron micrograph of a rat liver cell labeled to indicate the major structural components of eukaryotic cells and (*b*) a schematic drawing of an animal cell. Note the number and variety of subcellular organelles and the network of interconnecting membranes enclosing channels, that is, cisternae. All eukaryotic cells are not as complex in their appearance, but most contain the major structures shown in the figure. ER, endoplasmic reticulum; G, Golgi zone; Ly, lysosomes; P, peroxisomes; M, mitochondria.
Photograph (a) reprinted with permission of Dr. K. R. Porter from Porter, K. R. and Bonneville, M. A. In: Fine Structure of Cells and Tissues. *Philadelphia: Lea & Febiger, 1972; and schematic (b) from Voet, D. and Voet, J. G.* Biochemistry, *2nd ed. New York: Wiley, 1995. Reprinted by permission of John Wiley & Sons, Inc.*

large or small, charged or uncharged molecules allow membranes to modulate concentrations of substances in various compartments. Thus the cytosol and fluid compartment of organelles have a distinguishing composition of inorganic ions, organic molecules, proteins, and nucleic acids. Partitioning of activities and components in membrane-enclosed compartments and organelles has a number of advantages for the economy of the cell. These include the sequestering of substrates and cofactors where they are required, and adjustment of pH and ionic composition for maximum activity of biological processes.

The activities and composition of cellular structures and organelles are determined on intact cells by histochemical, immunological, and fluorescent staining methods. Continuous observations in real time of cellular events in intact viable cells are possible. An example are studies of changes of pH and ionic calcium concentration in the cytosol by use of ion-specific indicators. Individual organelles, membranes, and components of the cytosol are isolated and analyzed after disruption of the plasma membrane. Techniques for disrupting membranes include use of detergents, osmotic shock, or homogenization of tissues where shearing forces break down the plasma membrane. In appropriate isolation media, cell organelles and membrane systems are separated by centrifugation because of differences in size and density. These techniques permit isolation of cellular fractions from most mammalian tissues. In addition, components of organelles, for example, mitochondria and peroxisomes, are isolated after disruption of the organelle membrane.

In many instances isolated structures and cellular fractions appear to retain the chemical and biochemical characteristics of the structure *in situ*. But biological membrane systems are very sensitive structures, subject to damage even under very mild conditions. Alterations that occur during isolation lead to changes in composition. Damage to a membrane alters its permeability properties to allow transfer of substances that would normally be excluded by the membrane barrier. In addition, many proteins are loosely associated with membranes and dissociate easily (see p. 502).

Not unexpectedly, there are differences in structure, composition, and activities of cells from different tissues due to their diverse functions. The major biochemical activities, however, are fairly constant from tissue to tissue. The differences between cell types are usually in distinctive specialized activities.

Chemical Composition of Cells

Each cellular compartment has an aqueous fluid or **matrix** that contains various ions, small molecular weight organic molecules, and a variety of proteins, and in some cases nucleic acids. The localization of most enzymes and nucleic acids in cells is known, but the exact ionic composition of the matrix of specific organelles is still uncertain. Each presumably has a distinctly different ionic composition and pH. The ionic composition of intracellular fluid, considered to represent the cytosol primarily, compared to blood plasma is presented in Figure 1.10. $\mathbf{Na^+}$ is the major extracellular cation, with a concentration of ~140 meq L^{-1} (mM); very little Na^+ is present in intracellular fluid. $\mathbf{K^+}$ is the major intracellular cation. $\mathbf{Mg^{2+}}$ is present in both extra- and intracellular compartments at concentrations much lower than Na^+ and K^+. The major extracellular anions are $\mathbf{Cl^-}$ and $\mathbf{HCO_3^-}$ with smaller amounts of phosphate and sulfate. Most proteins have a negative charge at pH 7.4 (see Chapter 3), being anions at the pH of tissue fluids. Major intracellular anions are inorganic phosphate, organic phosphates, and proteins. Other inorganic and organic anions and cations are present in concentrations well below the meq L^{-1} level. Except for very small differences created by membranes and leading to development of membrane potentials, *total cation concentration equals total anion concentration in the different fluids.*

Intracellular concentrations of most small molecular weight organic molecules, such as sugars, organic acids, amino acids, and phosphorylated intermediates, are in the range of 0.01–1.0 meq L^{-1} but can be significantly lower. Coenzymes,

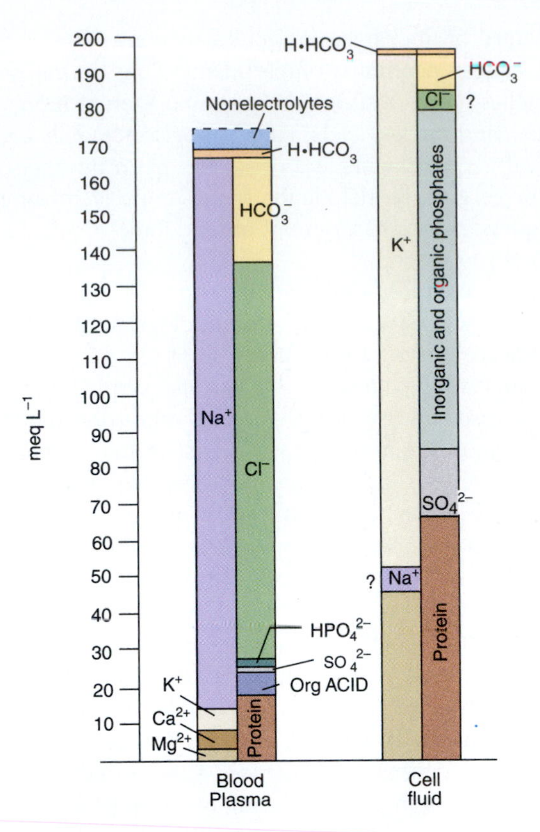

FIGURE 1.10
Major chemical constituents of blood plasma and cell fluid.
Height of left half of each column indicates total concentration of cations; that of right half, concentrations of anions. Both are expressed in milliequivalents per liter (meq L^{-1}) of fluid. Note that chloride and sodium values in cell fluid are questioned. It is probable that, at least in muscle, the cytosol contains some sodium but no chloride.
Adapted from Gregersen, M. I. In P. Bard (Ed.), Medical Physiology, *11th ed. St. Louis, MO: Mosby, 1961, p. 307.*

organic molecules required for the activity of some enzymes, are in the same range of concentration. Substrates for enzymes are present in relatively low concentration in contrast to inorganic ions but localization in a specific organelle or cellular microenvironment can increase their concentrations significantly.

It is not very meaningful to determine the molar concentration of individual proteins in cells. In many cases they are localized with specific structures or in combination with other proteins to create functional units. It is in a restricted compartment that individual proteins carry out their role, whether structural, catalytic, or regulatory.

1.4 | FUNCTIONAL ROLE OF SUBCELLULAR ORGANELLES AND MEMBRANE SYSTEMS

The cellular localization of each metabolic pathway or process will be described in later chapters. Some entire pathways are located in a single cellular compartment but many are divided between two locations, with the pathway intermediates diffusing or being translocated from one compartment to another. Organelles have very specific functions and their specific enzymatic activities serve as markers for the organelle during isolation. The following describes briefly some major roles of eukaryotic cell structures and is intended to indicate the complexity and organization of cells. A summary of functions and division of labor within eukaryotic cells is presented in Table 1.6 and the structures are identified in Figure 1.9.

TABLE 1.6 Summary of Eukaryotic Cell Compartments and Their Major Functions

Compartment	*Major Functions*
Plasma membrane	Transport of ions and molecules Recognition Receptors for small and large molecules Cell morphology and movement
Nucleus	DNA synthesis and repair RNA synthesis
Nucleolus	RNA processing and ribosome synthesis
Endoplasmic reticulum	Membrane synthesis Synthesis of proteins and lipids for some organelles and for export Lipid and steroid synthesis Detoxification reactions
Golgi apparatus	Modification and sorting of proteins for incorporation into membranes and organelles, and for export Export of proteins
Mitochondria	Production of ATP Cellular respiration Oxidation of carbohydrates and lipids Urea and heme synthesis
Lysosomes	Cellular digestion: hydrolysis of proteins, carbohydrates, lipids, and nucleic acids
Peroxisomes	Lipid oxidation Oxidative reactions involving O_2 Utilization of H_2O_2
Microtubules, intermediate filaments, and microfilaments	Cell cytoskeleton Cell morphology Cell motility Intracellular movements
Cytosol	Metabolism of carbohydrates, lipids, amino acids, and nucleotides Protein synthesis

Plasma Membrane Is the Limiting Boundary of a Cell

The outer surface of a **plasma membrane** is in contact with a variable external environment and the inner surface with a relatively constant environment provided by the cell's cytoplasm. As will be discussed in Chapter 12, the two sides of mammalian plasma membranes have different chemical compositions and functions. The outer surface makes adhesive interactions with the extracellular matrix and other cells through **integrins,** transmembrane proteins. Integrins bind to the cytoskeleton inside the cells. These proteins also have a role in signal transduction across the membrane. Through cytoskeletal elements the plasma membrane is involved in determining cell shape and movement. The lipid nature of membranes excludes many substances, including proteins and other macromolecules, but specific transport mechanisms or pores permit uptake of selective ions and organic molecules. Through the plasma membrane cells communicate with other cells; the membrane contains many specific protein receptors for binding extracellular signals, such as hormones (see Chapter 21) and neurotransmitters released by other cells. Plasma membranes of eukaryotes have different proteins in distinct **membrane domains,** and their protein composition varies between cell types. Details of membrane structure and biochemistry are presented in Chapter 12.

Nucleus Is Site of DNA and RNA Synthesis

Early microscopists described the interior of eukaryotic cells as containing a **nucleus,** a large membrane-bound compartment, and **cytoplasm.** The nucleus is surrounded by two membranes, termed the **nuclear envelope,** with the outer membrane being continuous with membranes of the endoplasmic reticulum. The nuclear envelope contains numerous pores about 90 Å in diameter that permit controlled movement of particles and molecules between the nuclear matrix and cytoplasm. Some macromolecules such as RNA are actively transported across the nuclear envelope utilizing the pores. The nucleus contains a major subcompartment, seen clearly in electron micrographs, the **nucleolus. Deoxyribonucleic acid (DNA),** the repository of genetic information, is located in the nucleus as a DNA–protein complex, **chromatin,** that is organized into chromosomes. The nucleus contains the biochemical processes involved in replication of DNA before mitosis and in repair of DNA that has been damaged (see Chapter 4). Transcription of genetic information in DNA into a form that can be translated into cell proteins (see Chapter 5) involves synthesis and processing into a variety of forms of ribonucleic acid (RNA) in the nucleus. The processing of ribosomal RNA and formation of ribosomal subunits occurs in the nucleolus.

Endoplasmic Reticulum Has a Role in Protein Synthesis and Many Synthetic Pathways

The cytoplasm of eukaryotic cells contains a network of interconnecting membranes that thread from the nuclear envelope to the plasma membrane. This extensive structure, termed **endoplasmic reticulum (ER),** consists of membranes with a smooth (**smooth ER** or CER) appearance in some areas and rough (**rough ER** or RER) appearance in other places. The ER encloses a cellular compartment, the ER lumen, where newly synthesized proteins for export or incorporation into organelles and membranes are modified. The rough appearance is due to the presence of **ribonucleoprotein particles,** that is **ribosomes,** attached on its cytosolic side. During cell fractionation the endoplasmic reticulum network is disrupted and the membrane reseals to form small vesicles, referred to as **microsomes,** that can be isolated by differential centrifugation. Microsomes, as such, do not occur in cells.

A major function of ribosomes on **rough endoplasmic reticulum** is biosynthesis of proteins for incorporation into membranes and cellular organelles, and for export to the outside of the cell. **Smooth endoplasmic reticulum** is involved in lipid synthesis and contains an important class of enzymes termed **cytochromes P450** that catalyze hydroxylation of a variety of endogenous and exogenous compounds. These enzymes are important in biosynthesis of steroid hormones and removal of toxic substances (see Chapter 11). Endoplasmic reticulum and Golgi apparatus are involved in formation of other cellular organelles such as lysosomes and peroxisomes.

Golgi Complex Is Involved in Secretion of Proteins

Golgi complex is a network of flattened smooth membrane sacs—**cisternae**—and vesicles responsible for secretion from the cell of proteins such as hormones, blood plasma proteins, and digestive enzymes. It works in consort with endoplasmic reticulum where proteins for certain destinations are synthesized. Enzymes in Golgi membranes catalyze transfer of carbohydrate units to proteins to form glycoproteins, a process that is important in determining the proteins' eventual destination. Golgi complex is the major site of new membrane synthesis and is responsible for formation of lysosomes and peroxisomes. Membrane vesicles shuttle proteins between the cisternae; vesicles are pinched off of one cisterna and fuse with another with the aid of a family of proteins, termed **SNARE** proteins. Vesicles originating from Golgi complex transport proteins to the plasma membrane for secretion.

Mitochondria Supply Most of the Cell's Need for ATP

In electron micrographs of cells, **mitochondria** appear as spheres, rods, or filamentous bodies that are usually about 0.5–1 μm in diameter and up to 7 μm in length. The internal matrix, **mitosol,** is surrounded by two membranes, distinctly different in appearance and biochemical function. The inner membrane convolutes into the matrix to form **cristae** and contains numerous small spheres attached by stalks on the inner surface. The structure of cristae varies from tubular to lamellar depending on the tissue and conformational state of mitochondria. Outer and inner membranes are very different in their function and enzyme activities. Components of the electron transport system and synthesis of ATP in oxidative phosphorylation are part of the inner membrane and are described in detail in Chapter 13. Metabolic pathways for oxidation of carbohydrates, fatty acids, and amino acids, and some reactions in biosynthesis of urea and heme are located in the mitosol. The outer membrane contains pores that permit access to larger molecules but the inner membrane is highly selective and contains a variety of transmembrane transport systems. Mitochondria have a key role in aging and cytochrome *c*, a component of the mitochondrial electron transport system, is an initiator of apoptosis or programmed cell death (see p. 23.)

Mitochondria have a role in their own replication. They contain several copies of a circular DNA (**mtDNA**), with genetic information for 13 mitochondrial proteins, and some RNAs. The DNA is similar in size (16,569 base pairs) to bacterial chromosomes. The presence of an independent "genome" and the similarity to bacteria have led to the widely accepted hypothesis that these organelles were bacteria that formed a symbiotic relationship with eukaryotic cells some three billion years ago. The inheritance of mitochondria is by maternal transmission and by evaluating variations in mtDNA in individuals it has been possible to study the global movement of humans. Mitochondria also have unique RNAs (see p. 85) and enzymes required for protein synthesis. Most mitochondrial proteins, however, are derived from genes present in nuclear DNA and are synthesized on free ribosomes in the cytosol and then imported into the organelle. Over 150 genetic diseases of mitochondrial activities have been described; some are based on mutations of nuclear DNA coding for mitochondrial proteins and others are due to mutation of mitochondrial DNA (see Clin. Corr. 1.2).

CLINICAL CORRELATION 1.2

Mitochondrial Diseases

The first disease (Luft's disease) specifically involving mitochondrial energy transduction was reported in 1962. A 30-year-old patient had general weakness, excessive perspiration, a high caloric intake without increase in body weight, and an excessively elevated basal metabolic rate (a measure of oxygen utilization). It was demonstrated that the patient had a defect in the mechanism that controls mitochondrial oxygen utilization (see Chapter 13). Since that time over 150 genetic abnormalities of mitochondria have been identified that lead to alterations in enzymes, ribonucleic acids, electron transport components, and membrane transport systems. Mutations of mtDNA as well as of nuclear DNA lead to mitochondrial genetic diseases. The first disease to be identified as due to a mutation of mtDNA was Leber's hereditary optic neuropathy, which leads to sudden blindness in early adulthood. Many of the diseases involve skeletal muscle and the central nervous system. Mitochondrial DNA damage may occur due to free radicals (superoxides) formed in the mitochondria. Thus age-related degenerative diseases, such as Parkinson's disease, Alzheimer's disease, and cardiomyopathies, may have a component of mitochondrial damage. For details of specific diseases see Clinical Correlations 6.5 (p. 254), 13.2 (p. 557), 13.4 (p. 558), 13.5 (p. 589), and 13.6 (p. 589).

Luft, R. The development of mitochondrial medicine. *Proc. Natl. Acad. Sci. USA* 91: 8731,1944; Chalmers, R. M. and Schapira, A. H. V. Clinical, biochemical and molecular genetic features of Leber's hereditary optic neuropathy. *Biochim. Biophys. Acta* 1410: 147, 1999; Wallace, D. C. Mitochondrial DNA in aging and disease. *Sci. Am.* 280:40, 1997; and Wallace, D. C. Mitochondrial diseases in man and mouse. *Science* 283:1482, 1999.

TABLE 1.7 Representative Lysosomal Enzymes and Their Substrates

Type of Substrate and Enzyme	*Specific Substrate*
POLYSACCHARIDE-HYDROLYZING ENZYMES	
α-Glucosidase	Glycogen
α-Flucosidase	Membrane fucose
β-Galactosidase	Galactosides
α-Mannosidase	Mannosides
β-Glucuronidase	Glucuronides
Hyaluronidase	Hyaluronic acid and chondroitin sulfates
Arylsulfatase	Organic sulfates
Lysozyme	Bacterial cell walls
PROTEIN-HYDROLYZING ENZYMES	
Cathepsins	Proteins
Collagenase	Collagen
Elastase	Elastin
Peptidases	Peptides
NUCLEIC ACID-HYDROLYZING ENZYMES	
Ribonuclease	RNA
Deoxyribonuclease	DNA
LIPID-HYDROLYZING ENZYMES	
Lipases	Triacylglycerol and cholesterol esters
Esterase	Fatty acid esters
Phospholipase	Phospholipids
PHOSPHATASES	
Phosphatase	Phosphomonoesters
Phosphodiesterase	Phosphodiesters
SULFATASES	
	Heparan sulfate
	Dermatan sulfate

Lysosomes Are Required for Intracellular Digestion

Lysosomes are responsible for **intracellular digestion** of extracellular and intracellular substances. With a single limiting membrane, they maintain an acidic matrix pH of ~5. Encapsulated in these organelles is a class of glycoprotein enzymes—hydrolases—that catalyze hydrolytic cleavage of carbon–oxygen, carbon–nitrogen, carbon–sulfur, and oxygen–phosphorus bonds in proteins, lipids, carbohydrates, and nucleic acids. A partial list of lysosomal enzymes is presented in Table 1.7. As in gastrointestinal digestion, lysosomal hydrolases split complex molecules into simple low molecular weight compounds that can be reutilized and are most active at acidic pH values. The relationship between pH and enzyme activity is discussed on page 451.

The enzyme content of lysosomes varies in different tissues and depends on specific tissue needs. The lysosomal membrane is impermeable to small and large molecules and specific protein mediators, receptors, and transporters are required for translocation of substances across it. Isolated lysosomes *in vitro* do not catalyze hydrolysis of added substrates until the lysosomal membrane is disrupted; their enzymatic activity is termed **"latent."** Membrane disruption within cells can lead to cellular digestion. Various pathological conditions have been attributed to release of lysosomal enzymes, including arthritis, allergic responses, several muscle diseases, and drug-induced tissue destruction (see Clin. Corr. 1.3).

Lysosomes are involved in digestion of intra- and extracellular substances that must be removed. Through **endocytosis,** material from the exterior is taken into cells and encapsulated in membrane-bound vesicles (Figure 1.11). The process occurs by several different mechanisms; one involves several cytosolic proteins, including clathrin (see p. 941), to form coated pits, which bud off segments of membrane. The plasma membrane invaginates around formed foreign substances such as microorganisms by **phagocytosis** or extracellular fluid that contains suspended material by **pinocytosis.** Vesicles containing external material fuse with lysosomes to form **secondary lysosomes** or **digestive vacuoles,** containing material to be digested and lysosomal hydrolases to carry out the digestion. They are identified microscopically by their size and often by the presence of partially digested structures. **Primary lysosomes** are those in which the enzymes are not involved in the digestive process.

Cellular proteins, nucleic acids, and lipid as well as structures such as mitochondria are in a dynamic state of synthesis and degradation. Lysosomes are responsible for hydrolysis of these cellular components, referred to as **autophagy.** Substances destined to be degraded are identified and taken up by lysosomes or are first encapsulated within membrane vesicles that fuse with a lysosome. An outline of the process is presented in Figure 1.11.

CLINICAL CORRELATION 1.3

Lysosomal Enzymes and Gout

Catabolism of purines, nitrogen-containing heterocyclic compounds found in nucleic acids, leads to formation of uric acid, which is excreted in urine normally (see p. 840 for details). Gout is an abnormality in which uric acid is produced in excess, leading to an increase of uric acid in blood and deposition of urate crystals in joints. Clinical manifestations include inflammation, pain, swelling, and increased warmth of some joints, particularly the big toe. Uric acid is rather insoluble and some of the clinical symptoms of gout can be attributed to damage done by urate crystals. Crystals are phagocytosed by cells in the joint and accumulate in digestive vacuoles that contain lysosomal enzymes. Crystals cause physical damage to the vacuoles, releasing lysosomal hydrolyases into the cytosol. Even though the pH optimum of lysosomal enzymes is lower than the pH of the cytosol, they have some hydrolytic activity at the higher pH, which causes digestion of cellular components, release of substances from the cell, and cellular autolysis.

Weissmann, G. Crystals, lysosomes and gout. *Adv. Intern. Med.* 19:239, 1974; and Burt, H. M., Kalkman, P. H., and Mauldin, D. Membranolytic effects of crystalline monosodium urate monohydrate. *J. Rheumatol.* 10:440,1983.

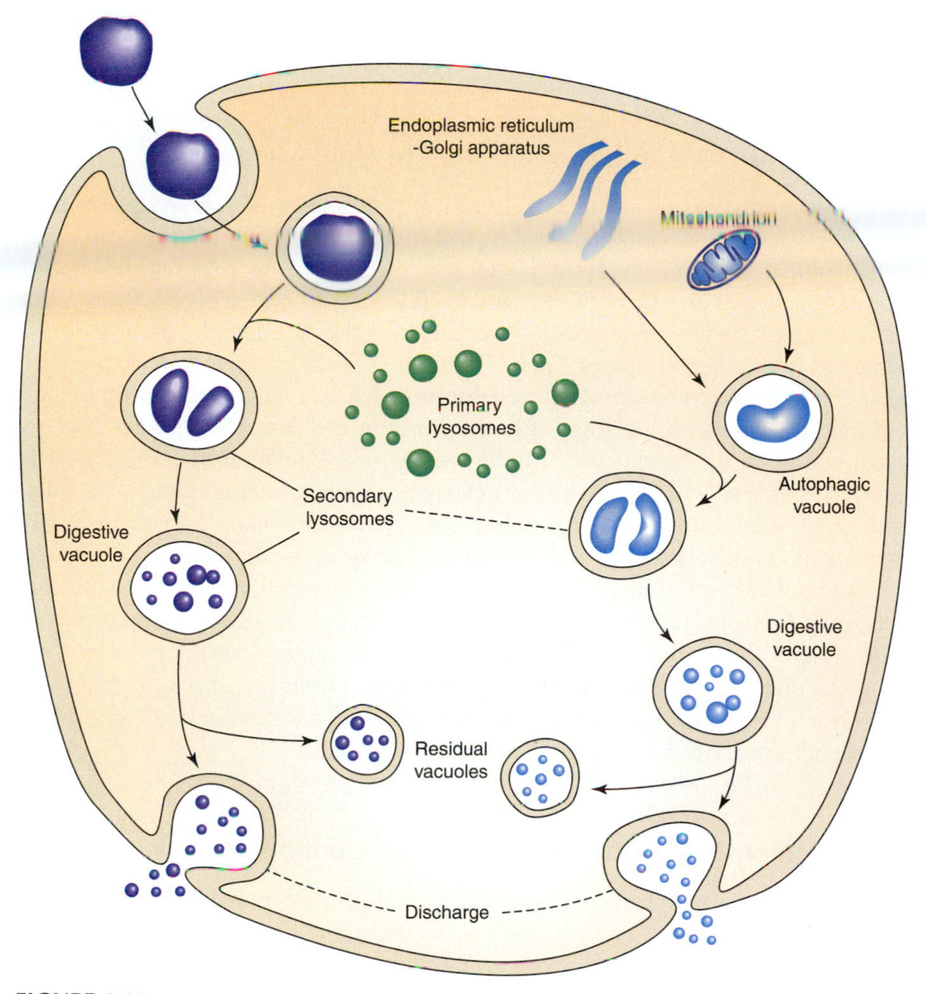

FIGURE 1.11
Diagrammatic representation of the role of lysosomes in intracellular digestion of substances internalized by phagocytosis (heterophagy) and of cellular components (autophagy).
In both processes substances to be digested are enclosed in a membrane vesicle, followed by fusing with a primary lysosome to form a secondary lysosome.

Products of lysosomal digestion are released from lysosomes and are reutilized by the cell. Indigestible material accumulates in vesicles referred to as residual bodies whose contents are removed from the cell by **exocytosis.** Some residual bodies contain high concentrations of a pigmented substance that is chemically heterogeneous and contains polyunsaturated fatty acids and proteins, termed **lipofuscin,** or "**age pigment,**" or "**wear and tear pigment,**" that accumulate in cells. Lipofuscin has been observed particularly in postmitotic neurons and muscle cells and has been implicated in the aging process.

Under controlled conditions some lysosomal enzymes are normally secreted from the cell for digestion of extracellular material in connective tissue and the prostate gland. In a number of genetic diseases individual lysosomal enzymes are missing and lead to accumulation of the substrate of the missing enzyme. Lysosomes of affected cells become enlarged with undigested material, which interferes with normal cell processes. The diseases are referred to as "**lysosomal storage diseases**" (see p. 762); see Clinical Correlation 1.4 for a discussion of a deficiency of lysosomal lipase. A disease of lysosomes is **I cell disease** in which the cellular mechanism for directing lysosomal enzymes into lysosomes during their synthesis

CLINICAL CORRELATION 1.4
Lysosomal Acid Lipase Deficiency

Human lysosomal acid lipase (hLAL) hydrolyzes triacylglycerol to free fatty acids and glycerol, and cholesteryl esters to free cholesterol and fatty acids. It is a critical enzyme in cholesterol metabolism, serving to make available free cholesterol for the needs of cells. Cholesteryl ester storage disease (CESD) and Wolman's disease are two distinct phenotypic forms of a genetic deficiency of hLAL; both are rare autosomal recessive diseases. Cholesteryl ester storage disease is usually diagnosed in adulthood and is evidenced by hypercholesterolemia, hepatomegaly, and early onset of severe atherosclerosis. Affected individuals express hLAL activity but at a very low level (<5% of normal). Apparently this level is sufficient to hydrolyze triacylglycerol but not cholesteryl esters. Analysis of the gene for the enzyme indicates that a mutation has occurred on both alleles at a splice junction (see p. 225) leading to a catalytically defective and unstable enzyme.

Wolman's disease is manifested in infants and is usually fatal by age one. There is no detectable activity of hLAL. Both triacylglycerol and cholesteryl esters accumulate in tissues. Gene analysis has demonstrated that a homozygous mutation leads to a gene product that cannot be translated into an active enzyme.

Hegele, R. A., Little, J. A., Vezina, C., et al. Hepatic lipase deficiency. Clinical, biochemical and molecular genetic characteristics. *Atherosclerosis and Thrombosis* 13:720, 1993; Lohse, P., Maas, S., Lohse, P., Elleder, M., Kirk, J. M., Besley, G. T., and Seidel, D. Compound heterozygosity for a Wolman mutation is frequent among patients with cholesteryl ester storage disease. *J. Lipid Res.* 41:23, 2000; and Anderson, R. A., Bryson, G. M., and Parks, J. S. Lysosomal acid lipase mutations that determine phenotype in Wolman and cholesteryl ester storage disease. *Mol. Genet. Metab.* 68:333, 1999.

$$(1)\ 2H_2O_2 \longrightarrow 2H_2O + O_2$$

$$(2)\ RH_2 + H_2O_2 \longrightarrow R + 2H_2O$$

FIGURE 1.12
Reactions catalyzed by catalase.

is defective and the hydrolytic enzymes are exported out of the cell, damaging the extracellular matrix. (See Clin. Corr. 6.8, p. 263.)

Peroxisomes Have an Important Role In Lipid Metabolism

Most eukaryotic cells of mammalian origin and those of protozoa and plants have organelles, designated **microbodies** or **peroxisomes,** the latter because they were first identified by their ability to produce or utilize **hydrogen peroxide** (H_2O_2). They are small (0.3–1.5 *o*m in diameter), spherical or oval in shape, with a fine network of tubules in their matrix. Over 50 enzymes have been identified in peroxisomes from different tissues. Their number in a cell fluctuates depending on cellular conditions. Several classes of **xenobiotics,** that is, foreign substances, including aspirin, lead to proliferation of peroxisomes in the liver. Peroxisomes have an essential role in the breakdown of lipids, particularly oxidation of very-long-chain fatty acids (see p. 722), and synthesis of glycerolipids, glycerol ether lipids (plasmalogens), and isoprenoids (see Chapter 16). They also contain enzymes that oxidize D-amino acids, uric acid, and various 2-hydroxy acids using molecular O_2 with formation of H_2O_2. **Catalase,** an enzyme present in peroxisomes, catalyzes the conversion of H_2O_2 to water and oxygen, and the oxidation of various compounds by H_2O_2 (Figure 1.12). By having both peroxide-producing and peroxide-utilizing enzymes in one compartment, cells protect themselves from the toxicity of H_2O_2.

Over 25 known diseases that involve peroxisomes are considered together as disorders of peroxisome biogenesis (see Clin. Corr. 1.5).

CLINICAL CORRELATION 1.5
Peroxisome Biogenesis Disorders (PBDs)

Peroxisomes are responsible for a number of important metabolic reactions, including synthesis of glycerol ethers, shortening very-long-chain fatty acids so that mitochondria can completely oxidize them, and oxidation of the side chain of cholesterol needed for bile acid synthesis. Peroxisome biogenesis disorders (PBDs) are more than 25 genetically and phenotypically related disorders that involve enzymatic activities of peroxisomes. They are rare autosomal recessive diseases characterized by abnormalities of the liver, kidney, brain, and skeletal system. A number of biochemical abnormalities have been described in PBD patients including decreased levels of glycerol ether lipids (plasmalogens), and increased levels of very-long-chain fatty acids (C_{24} and C_{26}) and cholestanoic acid derivatives (precursors of bile acids). The most severe condition is Zellweger syndrome, a condition due to the absence of functional peroxisomes; death frequently occurs by age 6 months. In this condition, the genetic defect is in the mechanism for importing enzymes into the matrix of peroxisomes. Specific mutations for some PBD conditions have been determined and include donor splice and missense mutations (see p. 192); subsets of the disease include absence of a single metabolic enzyme and defects in membrane transport components. In some instances the disease can be diagnosed prenatally by assay of peroxisomal enzymes or fatty acids in cells of amniotic fluid.

Wanders, R. J., Schutgens, R. B., and Barth, P. G. Peroxisomal disorders: a review. *J. Neuropathol. Exp. Neurol.* 54:726,1995; Fitzpatrick, D. R. Zellweger syndrome and associated phenotypes. *J. Med. Genet.* 33:863, 1996; and Warren, D. S., Wolfe, B. D., and Gould, S. J. Phenotype–genotype relationships in PEX10-deficient peroxisome biogenesis disorder patients. *Hum. Mutat.* 15:509, 2000.

Cytoskeleton Organizes the Intracellular Contents

Eukaryotic cells contain microtubules, intermediate filaments, and actin filaments (microfilaments) as parts of a cytoskeletal network. The **cytoskeleton** has a role in maintenance of cellular morphology, intracellular transport, cell motility, and cell division. **Microtubules** are multimeres of **tubulin,** a protein, which rapidly assembles and disassembles depending on the needs of cells. Actin and myosin form very important cellular filaments in striated muscle, where they are responsible for muscular contraction (see Chapter 23). Three mechanochemical proteins—**myosin, dynein,** and **kinesin**—convert chemical energy into mechanical energy for movement of cellular components. These **molecular motors** are associated with the cytoskeleton. Dynein is involved in ciliary and flagellar movement, whereas kinesin is a driving force for the movement of vesicles and organelles along microtubules especially in neuronal axons.

Cytosol Contains Soluble Cellular Components

The least complex in structure, but not in chemistry, is the organelle-free cell sap, or **cytosol.** It is here that many reactions and pathways of metabolism occur. Although there is no apparent structure to the cytosol, the high protein content precludes it from being a truly homogeneous mixture of soluble components and it is believed that there are functional compartments throughout the cytosol. Many reactions are localized in selected areas where substrate availability is more favorable. The actual physicochemical state of the cytosol is poorly understood. A major role of cytosol is to support synthesis of proteins on both free ribosomes often in a polysome form and those bound to endoplasmic reticulum by supplying required intermediates.

Some enzymes occur as soluble proteins when the cytosol is isolated, but in the intact cell many of them may be loosely attached to membrane structures or to cytoskeletal components.

1.5 | INTEGRATION AND CONTROL OF CELLULAR FUNCTIONS

A eukaryotic cell is a complex structure that maintains an intracellular environment to permit a myriad of complex reactions and functions to occur as efficiently as possible. Cells of multicellular organisms also participate in maintaining the well-being of the whole organism by exerting influences on each other to maintain all tissue and cellular activities in balance. Intracellular processes and metabolic pathways are tightly controlled and integrated in order to achieve this balance. Very few functions operate totally independently; changes in one function can exert an influence, positive or negative, on other pathways. As will be described throughout this textbook, controls of function occur at many levels from the expression of a gene to alter the concentration of an enzyme or a protein effector to changes in substrate or coenzyme levels to alter the rate of a specific enzyme reaction. The integration of many cellular processes is controlled by proteins that serve as activators or inhibitors. Both positive and negative controls are required to maintain cellular homeostasis. Many cellular processes are programmed to occur under specific conditions; as an example, cell division in normal cells does not continue unchecked, but rather occurs only when the processes required for cell division are activated. Then and only then is there an orderly and integrated series of reactions culminating in the division of a cell into two daughter cells. A fascinating process is **apoptosis**, programmed cell death, also referred to as cell suicide. This highly regulated process occurs in cells of all mammalian tissues but individual steps in the process vary from tissue to tissue. Clinical Correlation 1.6 presents an overview of apoptosis and some of the clinical conditions where deviations in this pathway have been suggested as the cause of the disease. As emphasized throughout this textbook, many diseases are due to a failure in specific control mechanisms. As one continues to comprehend the complexity of biological cells, one becomes amazed that there are

CLINICAL CORRELATION 1.6
Apoptosis: Programmed Cell Death

Eukaryotic cells have genetically regulated mechanisms for programmed cell death, termed apoptosis. Programmed cell death is important during embryogenesis and throughout adult life and occurs when a cell has fulfilled its biological function. It is distinct from necrotic death of a cell caused by injury due to radiation or anoxia. Initiation of apoptosis has three phases: an initiation signal, activation of a cascade of reactions involving protein factors, and activation of specific proteolytic enzymes.

Different cells have highly specific receptors on the plasma membrane, termed death receptors, that bind specific proteins instructing the cell to initiate apoptosis. Various proteins such as tumor necrosis factor (TNF) have been identified as "death signals." The specificity of receptors and of death signals determines which cells will undergo apoptosis. Death receptors are transmembrane proteins with intracellular domains, termed death domains. When the receptor is activated by binding of the extracellular signal, the death domain binds specific proteins and promotes a cascade of protein–protein interactions. Proteases, termed caspase 8 and 9, are activated, which then activate other caspases. The caspase family of enzymes are present as proenzymes (inactive) and are activated by the cleavage to three subunits, two of which will form an active heterodimer. They catalyze hydrolysis of very specific cellular proteins that leads to disassembly of the cell.

Apoptosis is also caused by intracellular stress, such as a disruption of the mitochondrial membrane. This mechanism involves the release of cytochrome *c* from mitochondria, which along with another protein factor, Apaf-1, activates caspase 9 and initiates the cell death cascade. The cascade is controlled by pro- and antiapoptotic proteins, such as Bcl-2/Bax. A change in the balance between these factors can lead either to premature cell death or to unchecked cell division.

Dysfunctions in the apoptotic pathway have been implicated in a variety of diseases including autoimmune, inflammatory, malignant, and neurodegenerative diseases, and in viral infections such as HIV that attack the immune system. An understanding of the different pathways of apoptosis may lead to therapeutic approaches for these diseases.

Randhawa, B. A. S., Apoptosis. *Mol. Pathol.* 53:55, 2000; Konopleva, M., Zhao, S., Xie, Z., Segall, H., et.al. Apoptosis. Molecules and mechanisms. *Adv. Exp. Med. Biol.* 457:217, 1999; and Wei, M. C., Zong, W., Cheng, E. H., Lindsten, T., et.al. Proapoptoxic BAX and BAK: A requisite gateway to mitochondrial dysfunction and death. *Science* 292:727, 2001.

not many more errors occurring and many more individuals with abnormal conditions. Thus as we proceed to study the separate chemical components and activities of cells in subsequent chapters, it is important to keep in mind the concurrent and surrounding activities, constraints, and influence of the environment. Only by bringing together all the parts and activities of a cell, that is, reassembling the puzzle, will we appreciate the wonder of living cells.

BIBLIOGRAPHY

Water and Electrolytes

Dick, D. A. T. *Cell Water.* Washington, DC: Butterworths, 1966.

Eisenberg, D. and Kauzmann, W. *The Structures and Properties of Water.* Fairlawn, NJ: Oxford University Press, 1969.

Morris, J. G. *A Biologist's Physical Chemistry.* Reading, MA: Addison-Wesley, 1968.

Westof, E. *Water and Biological Macromolecules.* Boca Raton, FL: CRC Press, 1993.

Wiggins, P. M. Role of water in some biological processes. *Microbiol. Rev.* 54:432, 1990.

Cell Structure

Alberts, B., Bray, D., Lewis, J., Raff, M., Roberts, K., and Watson, J. D. *Molecular Biology of the Cell,* 3rd ed. New York: Garland Publishing, 1994.

Becker, W. M. and Deamer, D. W. *The World of the Living Cell,* 2nd ed. Redwood City, CA: Benjamin, 1991.

DeDuve, C. *Guided Tour of the Living Cell,* Vols. 1 and 2. New York: Scientific American Books, 1984.

Dingle, J. T., Dean, R. T., and Sly, W. S. (Eds.). *Lysosomes in Biology and Pathology.* Amsterdam: Elsevier (a serial publication covering all aspects of lysosomes).

Holtzman, E. and Novikoff, A. B. *Cells and Organelles,* 3rd ed. New York: Holt, Rinehart & Winston, 1984.

Hoppert, M. and Mayer, F. Principles of macromolecular organization and cell function in bacteria and archaea. *Cell Biochem. Biophys.* 31:247, 1999.

Porter, K. R. and Bonneville, M. A. *Fine Structure of Cells and Tissues.* Philadelphia: Lea & Febiger, 1972.

Vale, R. D. Intracellular transport using microtubule-based motors. *Annu. Rev. Cell. Biol.* 3:347, 1987.

Cell Organelles

Attardi, G. and Chomyn, A. *Mitochondrial Biogenesis and Genetics.* San Diego, CA: Academic Press, 1995.

Fujiki, Y. Peroxisome biogenesis and peroxisome biogenesis disorders. *Fed. Eur. Biochem. Lett.* 476: 42, 2000.

Holtzman, E. *Lysosomes.* New York: Plenum Press, 1989.

Kiberstis, P. A. Mitochondria make a comeback. *Science* 283:1475, 1999. This is a series of articles describing recent developments in our knowledge about mitochondria.

Latruffe, N. and Bugaut, M. *Peroxisomes.* New York: Springer-Verlag, 1994.

Lestienne, P. (Ed.). *Mitochondrial Diseases: Models and Methods.* Heidelberg: Springer, 1999.

Moser, H. W. Molecular genetics of peroxisomal disorders. *Frontiers Biosci.* 5:D298, 2000.

Pavelka, M. *Functional Morphology of the Golgi Apparatus.* New York: Springer-Verlag, 1987.

Preston. T. M., King, C. A., and Hyams, J. S. *The Cytoskeleton and Cell Motility.* New York: Chapman and Hall, 1990.

Rothman, J. E. and Orci, L. Budding vesicles in living cells. *Sci. Am.* 274:70, 1996.

Scheffler, I. E. *Mitochondria.* New York: Wiley, 1999.

Strauss, P. R. and Wilson, S. H. *The Eukaryotic Nucleus: Molecular Biochemistry and Macromolecular Assemblies.* Caldwell, NJ: Telford Press, 1990.

Tyler, D. D. *Mitochondrion in Health and Disease.* New York: VCH,1992.

QUESTIONS C. N. ANGSTADT

Multiple Choice Questions

1. Prokaryotic cells, but not eukaryotic cells, have:
   A. endoplasmic reticulum.
   B. histones.
   C. nucleoid.
   D. a nucleus.
   E. a plasma membrane.

2. Factors responsible for a water molecule being a dipole include:
   A. the similarity in electron affinity of hydrogen and oxygen.
   B. the tetrahedral structure of liquid water.
   C. the magnitude of the H—O—H bond angle.
   D. the ability of water to hydrogen bond to various chemical structures.
   E. the difference in bond strength between hydrogen bonds and covalent bonds.

3. Hydrogen bonds can be expected to form only between electronegative atoms such as oxygen or nitrogen and a hydrogen atom bonded to:
   A. carbon.
   B. an electronegative atom.
   C. hydrogen.
   D. iodine.
   E. sulfur.

4. The ion product of water:
   A. is independent of temperature.
   B. has a numerical value of 1×10^{-14} at 25°C.
   C. is the equilibrium constant for the reaction HOH $\rightleftharpoons H^+ + OH^-$.
   D. requires that $[H^+]$ and $[OH^-]$ always be identical.
   E. is an approximation that fails to take into account the presence of the hydronium ion, H_3O^+.

5. Which of the following is both a Brønsted acid and a Brønsted base in water?
   A. $H_2PO_4^-$
   B. H_2CO_3
   C. NH_3
   D. NH_4^+
   E. Cl^-

6. Biological membranes are associated with all of the following EXCEPT:
   A. prevention of free diffusion of ionic solutes.
   B. release of proteins when damaged.
   C. specific systems for the transport of uncharged molecules.
   D. sites for biochemical reactions.
   E. free movement of proteins and nucleic acids across the membrane.

7. Analysis of the composition of the major fluid compartments of the body shows that:
   A. the major blood plasma cation is K^+.
   B. the major cell fluid cation is Na^+.
   C. one of the major intracellular anions is Cl^-.
   D. one of the major intracellular anions is phosphate.
   E. plasma and cell fluid are all very similar in ionic composition.

Refer to the following for Questions 8–10:

A. peroxisome
B. nucleus
C. cytoskeleton
D. endoplasmic reticulum
E. Golgi complex

8. Oxidizes very-long-chain fatty acids.
9. Connected to plasma membrane.
10. Transfers carbohydrate precursors to proteins during glycoprotein synthesis.

Questions 11 and 12: A patient with Luft's disease presented with general weakness, excessive perspiration, a high caloric intake without increase in body weight, and an excessively elevated basal metabolic rate. Luft's disease was the first disease involving a defect in mitochondria to be described. It is a defect in the mechanism that controls oxygen utilization in mitochondria.

11. Mitochondria are associated with all of the following EXCEPT:
    A. ATP synthesis.
    B. DNA synthesis.
    C. protein synthesis.
    D. hydrolysis of various macromolecules at low pH.
    E. apoptosis.

12. Which of the following is/are characteristics of mitochondria?
    A. The inner membrane forms cristae and contains small spheres attached by stalks on the inner surface.
    B. The outer membrane is highly selective in its permeability.
    C. Components of the electron transport system are located in the mitosol.
    D. Mitochondrial DNA is similar to nuclear DNA in size and shape.
    E. All of the above are correct.

Questions 13 and 14: Gout is a condition in which excessive production of uric acid leads to deposition of urate crystals in joints. Clinical manifestations include inflammation, pain, and swelling of joints, especially the joint of the big toe. Crystals are phagocytosed by cells in the joint and accumulate in digestive vacuoles that contain lysosomal enzymes. Crystals cause physical damage to the vacuoles, releasing lysosomal enzymes into the cytosol.

13. Lysosomal enzymes:
    A. are hydrolases.
    B. usually operate at acidic pH.
    C. are normally isolated from their substrates by the impermeable lysosomal membrane.
    D. can lead to cellular digestion if the lysosomal membrane is disrupted by certain disease states or drug damage.
    E. All of the above are correct.

14. Individual lysosomal enzymes are missing in a number of genetic diseases referred to as lysosomal storage diseases. In these diseases:
    A. a defect is an inability to direct enzymes to the lysosome after synthesis.
    B. lysosomes of affected cells become enlarged with undigested materials.
    C. lipofuscin, or "wear and tear pigment," accumulates in cells.
    D. any material taken into the lysosome will accumulate.
    E. residual bodies contain the products of digestion.

Problems

15. If a weak acid is 91% neutralized at pH 5.7, what is the pK' of the acid?
16. Metabolic alkalosis is a condition in which the blood pH is higher than its normal value of 7.40 and can be caused, among other reasons, by an increase in $[HCO_3^-]$. If a patient has the following blood values, pH = 7.45, $[CO_2]$ = 1.25 mM, what is the $[HCO_3^-]$? The pK' for the bicarbonate/carbonic acid system is 6.1. How does the calculated value of bicarbonate compare with the normal value?

ANSWERS

1. **C** Prokaryotic DNA is organized into a structure that also contains RNA and protein, called nucleoid. A, B, and D are found in eukaryotic cells, and E is an element of both prokaryotic and eukaryotic cells.
2. **C** Water is a polar molecule because the bonding electrons are attracted more strongly to oxygen than to hydrogen. The bond angle gives rise to asymmetry of the charge distribution; if water were linear, it would not be a dipole. A: Hydrogen and oxygen have very different electron affinities. B and D are consequences of water's structure, not factors responsible for it.
3. **B** Only hydrogen atoms bonded to one of the electronegative elements (O, N, F) can form hydrogen bonds. A hydrogen participating in hydrogen bonding must have an electronegative element on both sides of it.
4. **B** The constant is a function of temperature and is numerically equal to the equilibrium constant for the dissociation of water divided by the molar concentration of water. D: $[H^+] = [OH^-]$ in pure water, but not in solutions of solutes that contribute H^+ or OH^-.
5. **A** $H_2PO_4^-$ can donate a proton to become HPO_4^{2-}. It can also accept a proton to become H_3PO_4. B and D are Brønsted acids; C is a Brønsted base. The Cl^- ion in water is neither.
6. **E** These molecules are too large to cross the membrane freely unless the membrane is damaged. A: Ionic solutes do not cross the

lipid membrane readily. C: Most substances require transport across the membrane. D: Different membranes have different reactions.

7. **D** Phosphate and protein are the major intracellular anions. A, B, E: Plasma and cell fluid are strikingly different. The Na^+ ion is the major cation of plasma. C: Most chloride is extracellular.
8. **A** Fatty acid oxidation occurs in the mitochondria, but the oxidation of very-long-chain fatty acids involves the peroxisomes.
9. **B** This describes the cytoskeleton.
10. **E** Lipids, too, are attached covalently to certain proteins in the Golgi complex.
11. **D** This is a lysosomal function. A: This is a major role of mitochondria. B, C: Mitochondria synthesize some of their own proteins. E: Cytochrome *c*, part of the mitochondrial electron transport system, is an initiator of programmed cell death.
12. **A** This is very different from the outer membrane. B: The inner membrane is highly selective, the outer membrane relatively permeable. C: The electron transport system is in the inner membrane. D: Mitochondrial DNA (mtDNA) is small and circular.
13. **E** A: There are hydrolases for all classes of macromolecules. B, D: Lysosomal pH is usually about 5 but the enzymes retain some activity at higher pH. C: Substrates and enzymes are brought together by phagocytosis or pinocytosis.
14. **B** Engorgement interferes with normal cell processes. A: This is I cell disease and affects all of the lysosomal enzymes, which are exported out of the cell. C: This is a normal process. Lipofuscin is a chemically heterogeneous pigmented substance and not caused by a missing lysosomal enzyme. D: Each lysosomal storage disease affects a single enzyme so only substrates of that enzyme would be affected. E: Residual bodies contain indigestible material.
15. If a weak acid is 91% neutralized, 91 parts are present as conjugate base and 9 parts remain as the undissociated acid. Thus the conjugate base/acid ratio is approximately 10:1. Substituting this ratio into the Henderson–Hasselbalch equation gives $5.7 = pK' + \log(10/1)$. Solving the equation for pK' gives an answer of 4.7. The acid could be β-hydroxybutyric acid, an important physiological acid, which has this pK'.
16. One need simply substitute the given values into the Henderson–Hasselbalch equation:

$$7.45 = 6.1 + \log(x/1.25)$$
$$1.35 = \log(x/1.25)$$

The antilog of 1.35 is 22.39. Therefore

$$[HCO_3^-] = 22.39 \times 1.25 = 27.98 \text{ mM}$$

Normal $[HCO_3^-] = 24.0$ mM so there is an increase of 3.98 mmol (per liter) of HCO_3^-.

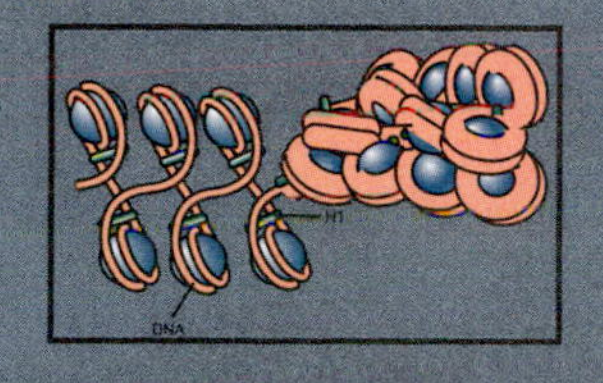

2

DNA AND RNA: COMPOSITION AND STRUCTURE

Stephen A. Woski
and Francis J. Schmidt

2.1 | DNA OVERVIEW

One of the hallmarks of life is its ability to reproduce. The information that makes each individual life form unique must be preserved and then passed on to progeny. All life on earth uses **nucleic acids** for storage of genetic information. With the exception of some viruses, the biomolecule utilized for information storage is **deoxyribonucleic acid (DNA).** This molecule is remarkably well suited for its task because of its chemical stability and its ability to encode vast amounts of information using a simple four-letter code.

Central Dogma of Molecular Biology

Because of its role as the repository of genetic information, DNA can arguably be considered the most important biomolecule in living systems. However, DNA is only a part of the core architecture of life. Two other types of biomolecule, **ribonucleic acid (RNA)** and protein, play equally important roles. The interrelationship of these three classes of molecules constitutes the **"central dogma of molecular biology"** (Figure 2.1), which holds that DNA stores information that controls all cellular processes. Much of the structure and biochemistry of cells is due to the properties of their constituent proteins. These properties are determined by the sequence of DNA, which directs synthesis of proteins. Information, however, cannot flow directly from DNA to protein, but depends on ribonucleic acid (RNA) to transport the information. Genetic information is transmitted from DNA to RNA by **transcription.** The sequence of RNA is then **translated** into a protein sequence at the ribosome. DNA also plays an essential role in heredity by serving as the template for its **replication.** Several discoveries have begun to blur the distinct roles of each of these biomolecules outlined in the "central dogma." For example, it has been demonstrated that RNA can act as a catalyst in biochemical reactions (**ribozymes**) and as a template for DNA synthesis (**reverse transcription**). In addition, RNA constitutes the active component of ribosomes, the protein synthesis machinery of the cell. DNA and RNA interact closely with proteins in many of their functions, and it has been suggested that proteins can store inheritable information. Nevertheless, the central role that nucleic acids play in life is undeniable.

DNA Can Transform Cells

These universally accepted principles were rejected outright not long ago. In fact, prior to the 1950s, the general view was that nucleic acids were substances of some-

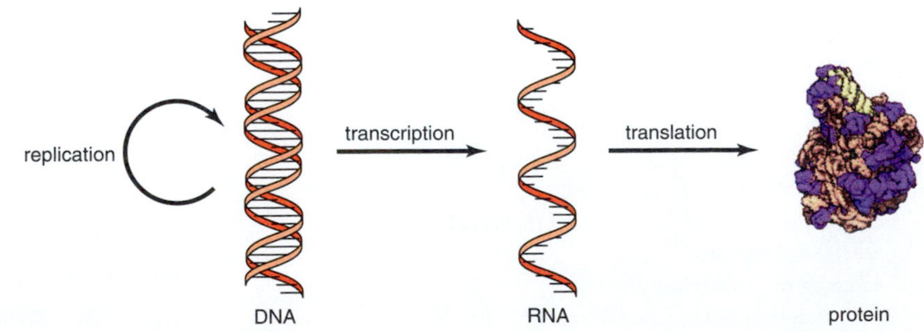

FIGURE 2.1
Central dogma of molecular biology.

CLINICAL CORRELATION 2.1

DNA Vaccines

Traditional procedures of vaccination have used purified components of an infectious organism, dead or attenuated intact cells or viruses, to provide individuals with active immunity by eliciting production of specific antibodies. Many have been successful in providing protection against diseases such as polio, smallpox, whooping cough, typhoid fever, and diphtheria.

Recently, a new approach to immunization has been explored that uses "naked" DNA. The source of this DNA is typically a bacterial plasmid that is free of bacterial proteins. This DNA can enter a number of cell types and can be expressed by host cells. Sequences encoding antigenic portions of a pathogen are encoded in this DNA, and protein products are synthesized using cellular transcription and translation machinery. In this respect, DNA vaccines act much like viruses. However, these DNAs contain only a very limited set of genetic information and cannot become infectious. The mechanisms of uptake and induction of the immune response are not yet clear. However, promising results have been observed against a number of pathogens including viruses, bacteria, and parasitic microorganisms. This approach holds great promise for development of effective vaccines against a number of intractable diseases including HIV/AIDS, tuberculosis, and malaria.

Lewis, P. J. and Babiuk, L. A. DNA vaccines: a review. *Adv. Virus Res.* 54:129, 1999.

what limited cellular importance. The first evidence that DNA is the genetic material was found during the 1920s, but definitive demonstration of DNA's role was not accomplished until 1944. The key experiments involved two strains of pneumococcus, a bacterium that causes a form of pneumonia. When cultured, one strain formed smooth colonies and the other formed rough colonies; these were labeled the S- and R-forms, respectively. In addition to their differing appearance, the S-form was virulent while the R-form was nonvirulent. These two forms are genetically distinct and cannot interconvert spontaneously. Treatment of R-form bacteria with pure DNA extracted from the S-form resulted in its **transformation** into the S-form. Furthermore, the transformation was permanently inheritable by subsequent generations. It was thus demonstrated that DNA was the transforming agent, as well as the material responsible for transmitting genetic information from one generation to the next. Clinical Correlation 2.1 describes vaccines based on transformation of mammalian cells with DNA.

Information Capacity of DNA Is Enormous

A striking characteristic of DNA is its ability to encode an enormous quantity of biological information. For example, a human cell contains information for the synthesis of about 50,000–100,000 proteins. This information is stored in the cell nucleus, a package roughly 0.00001 meter in diameter. Despite this compactness, information in DNA is readily accessed and duplicated on command. The ability to store large amounts of information on molecules and to access it readily is still far beyond modern information technology. The capacity of nucleic acids to maintain and transmit the archived information efficiently arises directly from their chemical structure.

2.2 | STRUCTURAL COMPONENTS OF NUCLEIC ACIDS: BASES, NUCLEOSIDES, AND NUCLEOTIDES

Nucleic acids are linear polymers consisting of repeating **nucleotide** units. Each nucleotide consists of three components: a phosphate ester, a pentose sugar, and a heterocyclic base. In RNA, the sugar is **D-ribose;** whereas DNA utilizes a closely related sugar, **2-deoxy-D-ribose.** In either case, the base is attached to the 1-position of the sugar through a β-*N*-glycosidic bond. Two classes of major bases are found in nucleic acids: the **purines** and **pyrimidines** (Figure 2.2). The major purines are **guanine** and **adenine,** which are found in both DNA and RNA and are attached to the sugar at N-9. There are three major pyrimidine bases: **cytosine, uracil,** and

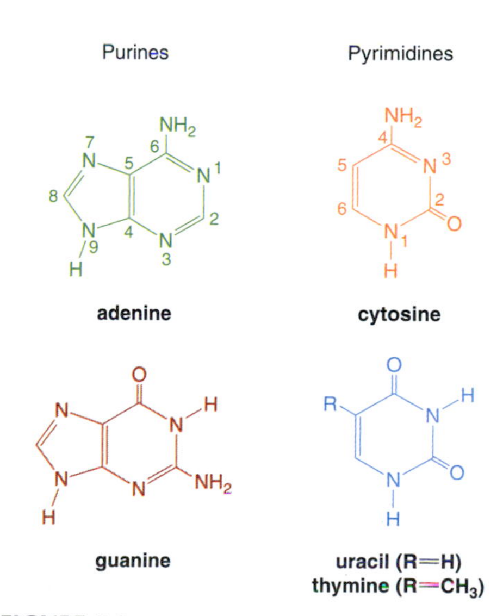

FIGURE 2.2
Major purine and pyrimidine bases found in DNA and RNA.

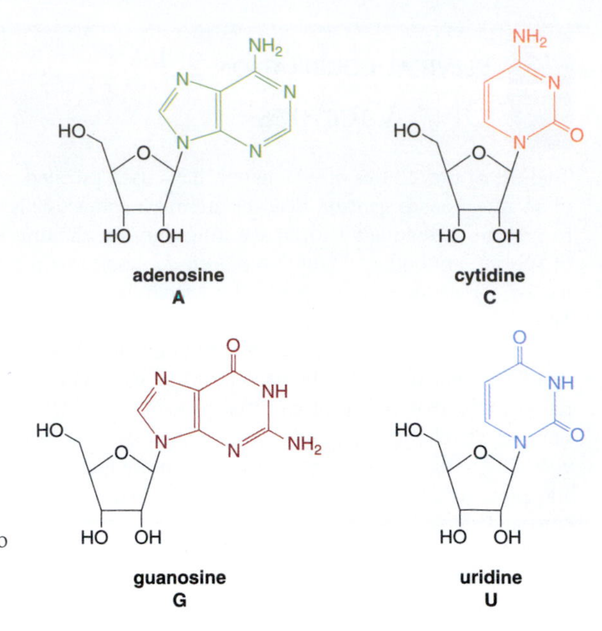

FIGURE 2.3
Structures of ribonucleosides.
Shown are one-letter abbreviations for each compound. These abbreviations are also used for the corresponding bases and nucleotides and, in some instances, for the deoxyribonucleosides

thymine. Cytosine is present in both DNA and RNA. However, uracil is generally found only in RNA, and thymine is generally found only in DNA. Each pyrimidine is linked to the sugar through the N-1 position. A base glycosylated with either pentose sugar is referred to as a **nucleoside.** Nucleosides that contain ribose are designated as **ribonucleosides** (Figure 2.3), whereas those with deoxyribose are **deoxyribonucleosides** (Figure 2.4). Four ribonucleosides are commonly found in RNA—**adenosine (A), guanosine (G), cytidine (C),** and **uridine (U)**—and four deoxyribonucleosides in DNA—**deoxyadenosine (dA), deoxyguanosine (dG), deoxycytidine (dC),** and **deoxythymidine (dT).** In addition to these nucleosides, more than 80 minor nucleosides have been found in naturally occurring nucleic acids.

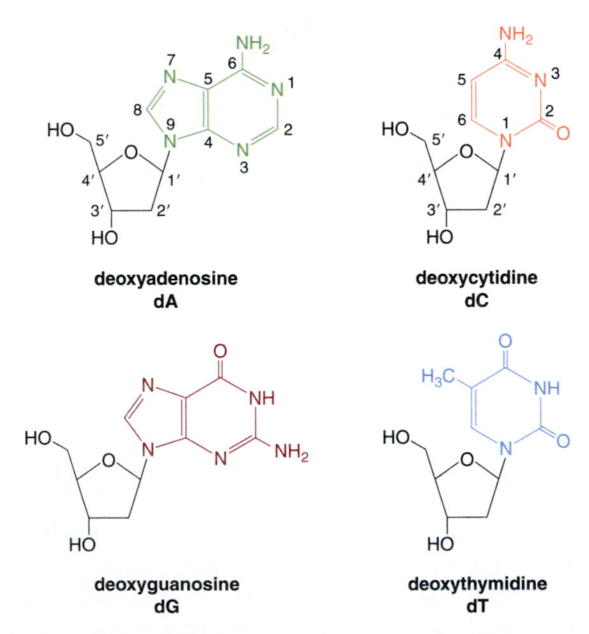

FIGURE 2.4
Structures of deoxyribonucleosides.
Presence of 2-deoxyribose is abbreviated by "d" preceding the one-letter notation. Note that some sources use thymidine (T) interchangeably with deoxythymidine (dT).

Nucleotides are phosphate esters of nucleosides (Figure 2.5). Any of the hydroxyl groups on the sugars of nucleosides can be phosphorylated, but the bases generally are not. Nucleotides that contain a phosphate monoester are called nucleoside monophosphates. For example, the 5′-nucleotide of deoxycytidine is named deoxycytidine 5′-monophosphate (5′-dCMP). More than one phosphate can be linked together by an anhydride linkage, resulting in the corresponding di-, tri-, and tetraphosphate esters. The 5′-triphosphate of guanosine would be guanosine 5′-triphosphate (GTP). Phosphate diesters are also possible, including the important "second messengers" adenosine-3′,5′-cyclic monophosphate (cAMP) and guanosine-3′,5′-cyclic monophosphate (cGMP).

deoxycytidine 5′-monophosphate
dCMP

guanosine 5′-triphosphate
GTP

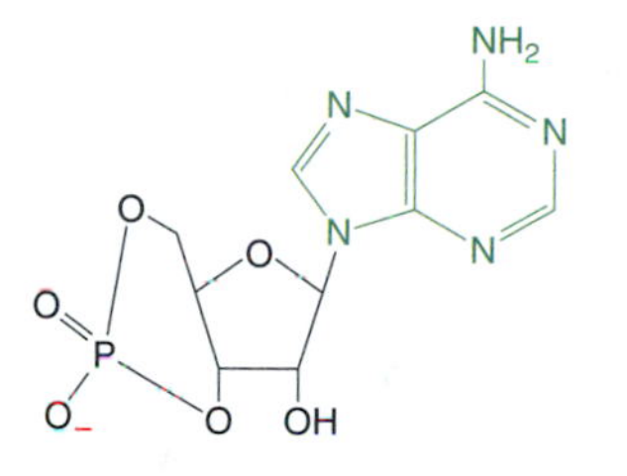

adenosine 3′,5′-cyclic monophosphate
cAMP

FIGURE 2.5
Structures of some representative nucleotides.

Physical Properties of Nucleosides and Nucleotides

To varying degrees, nucleobases, nucleosides, and nucleotides are all soluble in water over a wide range of pH values. Because of the hydrophilic nature of the sugars, nucleosides are more soluble than the corresponding bases. However, at biologically relevant pH values (pH 5–9), nucleosides are neutral species. At low pH values, the amino-containing nucleobases (G, C, A) are protonated. The sites of protonation are not the exocyclic NH_2 groups but are ring nitrogens. Protonation of these sites avoids disruption of the aromatic character of the heterocyclic system and maintains conjugation of the nitrogen lone pair of the sp^2-hybridized amino groups with the heterocycle. The pentose hydroxyl groups are quite weak bases, only becoming significantly deprotonated at very high pH values (>12).

A further increase in aqueous solubility is seen upon phosphorylation of the nucleosides. The pK_a of phosphate mono- and diesters is ~1; thus phosphates carry a negative charge at physiological pH values. The second pK_a of a phosphate monoester is ~6.5; thus an equilibrium exists between the monoanion and the dianion at neutral pH values. Nucleotides with di- and triphosphates carry multiple negative charges. The presence of charged phosphate groups on nucleotides provides sites for electrostatic interactions with positively charged sites on proteins and metal ions.

Nucleosides and nucleotides are stable over a wide range of pH values. Under strongly basic conditions, hydrolysis of phosphate esters occurs. However, the rates of these hydrolyses are reduced because of electrostatic repulsion between the negatively charged phosphates and the negatively charged hydroxide ions. *N*-Glycosidic bonds of nucleosides and nucleotides are stable under these conditions but under acidic conditions they are considerably more labile. In addition, protonation of purines bases (G and A) at elevated temperatures results in rapid scission of the bond between the sugar and the base. Pyrimidine (C, T, U) nucleosides and nucleotides are much more resistant to acid treatment. Conditions that result in breakage of the glycosidic bond (e.g., 60% perchloric acid at 100 °C) also lead to destruction of the sugar.

Molecules that contain purine or pyrimidine bases strongly absorb ultraviolet (UV) light. Purines and purine-containing nucleosides and nucleotides have higher **molar extinction coefficients** (absorb light more strongly) than pyrimidine derivatives. The wavelength of UV light at which maximum absorption occurs varies with the bases but is usually near 260 nm. Because proteins and other cellular components typically do not absorb strongly in this region, a strong absorbance at 260 nm is suggestive of the presence of nucleobases. Because of the high extinction coefficients of purine and pyrimidine bases and their high concentration in nucleic acids, the absorbance at 260 nm can be used to accurately quantitate the amount of nucleic acids (DNA or RNA) present in a sample. Further analysis and separation of nucleobases, nucleosides, nucleotides, and nucleic acids can be accomplished by use of a variety of chromatographic techniques, including high-performance liquid chromatography (HPLC), thin layer chromatography (TLC), paper chromatography, and electrophoresis.

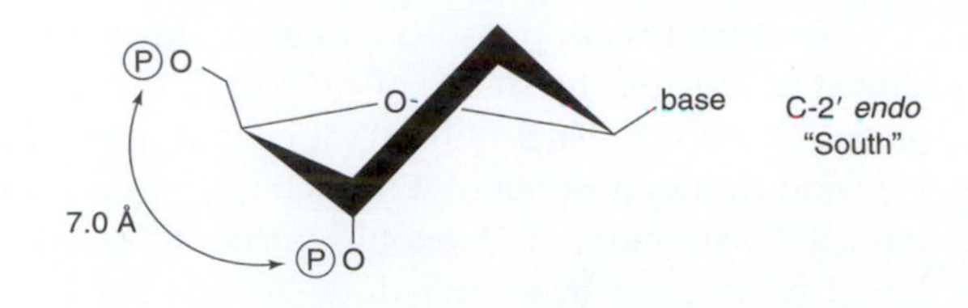

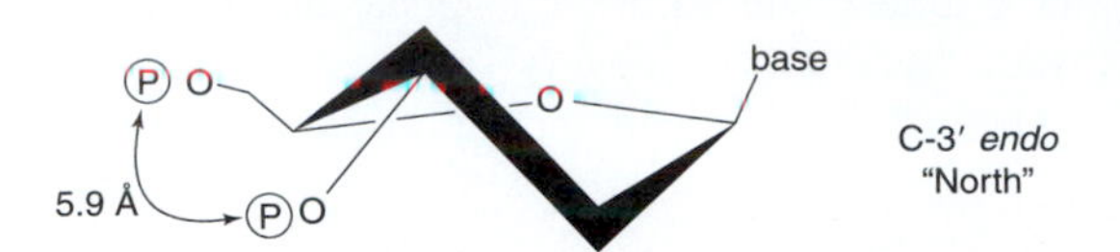

FIGURE 2.6
Preferred conformations of pentose sugars.
The two conformations produce variations in relative orientation of the base (with respect to the sugar) and in the distance between 3′- and 5′-phosphate groups (P). Ultimately, these differences affect overall conformation of the double-helical complex.

Structural Properties of Nucleosides and Nucleotides

A striking feature of nucleosides and nucleotides is the considerable number of possible conformations that they can adopt. Unlike six-membered rings (cyclohexanes) that have relatively high barriers to conformational change, five-membered rings are highly flexible. The five ring atoms in a pentose sugar are not coplanar because this conformation results in a number of unfavorable through-space interactions between groups bound to the ring. To alleviate this, one or two of the atoms of the ring twist out of plane. In cyclopentane rings, there are several "envelope" and "half-chair" conformations that rapidly interconvert. Because the substitution pattern of the pentose ring in nucleosides is not symmetric, two conformations are preferred (Figure 2.6). These two modes of **sugar puckering** are defined by the displacement of the 2′- and 3′-carbons "above" the plane of the C-1′–O-4′–C-4′ atoms. By convention, "above" is the direction in which the base and C-5′ project from the ring, and this is the ***endo*** face of the pentose. If C-2′ is displaced above the pentose ring, the conformation is called **C-2′ *endo*.** When viewed edge on, the pentose carbon atoms in this arrangement resemble the letter S; thus this conformation is sometimes called the S-conformation or the South-conformation. In the second common pucker C-3′ is displaced toward the *endo* face and is called **C-3′ *endo*.** The pentose carbons trace the letter N, and this conformation is sometimes called the N-conformation or North-conformation. Notably, the groups attached to the sugar have very different orientations in each of these conformations. For example, 5′- and 3′-phosphate groups are much farther apart in the C-2′ *endo* pucker than in the C-3′ *endo* pucker. The orientation of the glycosidic bond also changes significantly in the two conformations.

C-2′ *endo* and C-3′ *endo* conformations are generally in rapid equilibrium. The presence of an electronegative substituent at the 2′-position of the pentose favors the C-3′ *endo* conformation. Thus ribonucleosides that comprise RNA prefer this sugar pucker. In the case of RNA, factors such as hydrogen bonds between the 2′-OH group and the O-4′ atom of the neighboring residue also shift the equilibrium toward the C-3′ *endo* conformation. However, the 2′-deoxynucleosides of DNA contain a hydrogen in place of the 2′-OH group, and the C-2′ *endo* conformation is preferred.

The bases of nucleosides are planar. While free rotation around the glycosidic bond is possible, there are two predominant orientations of the base with respect to the sugar (Figure 2.7). In purines, the ***anti*** **conformation** places H-8 over the sugar, whereas the ***syn*** **conformation** positions this atom away from the sugar and the bulk of the bicyclic purine over the sugar. In pyrimidines, the H-6 atom is above the pentose ring in the *anti* conformation, and the larger O-2 atom is above the ring in the *syn* glycosidic conformation. Pyrimidines, therefore, show a large preference for the less sterically hindered *anti* conformation. Purines rapidly interconvert between the two conformations but favor the *anti* orientation. However, guanine 5′-nucleotides are exceptions. In these cases, favorable interactions between the 2-NH_2 group and the 5′-phosphate group stabilize the *syn* conformation. 2′-De-

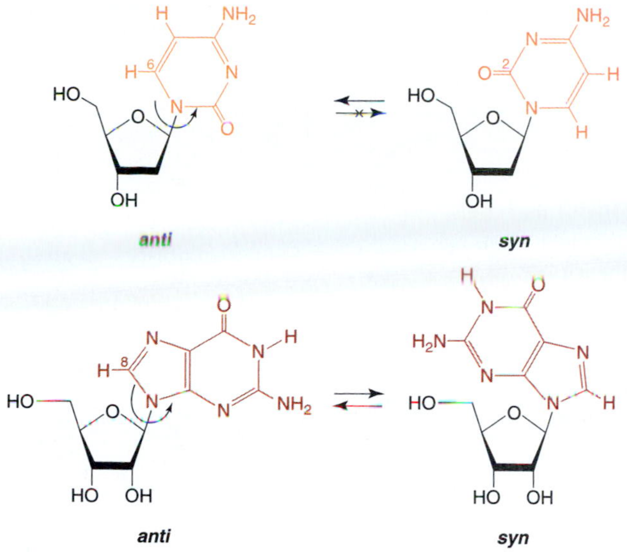

FIGURE 2.7
Glycosidic conformations of purines and pyrimidines.
In pyrimidines, steric clashes between the sugar and the O-2 of the base strongly disfavor the *syn* conformation. In purines, the *anti* and *syn* conformations readily interconvert, with *anti* being more stable in most cases. The *syn* conformation is stabilized in guanosine 5′-phosphates because of favorable interactions between the 2-NH_2 group and the phosphate oxygens.

oxyguanosine 5′-monophosphate (dGMP), for example, prefers the *syn* glycosidic conformation. This conformation also has been observed in double-stranded DNAs with sequences of alternating Gs and Cs. The *syn* conformation of the G residues in these DNAs results in formation of an unusual left-handed helix (see p. 47).

2.3 | STRUCTURE OF DNA

Polynucleotide Structure

Nucleic acids are strands of nucleosides linked by **phosphodiester bonds** (Figure 2.8). The length of these strands varies considerably, ranging from two residues to thousands and even millions of residues. Typically, strands of nucleic acids con-

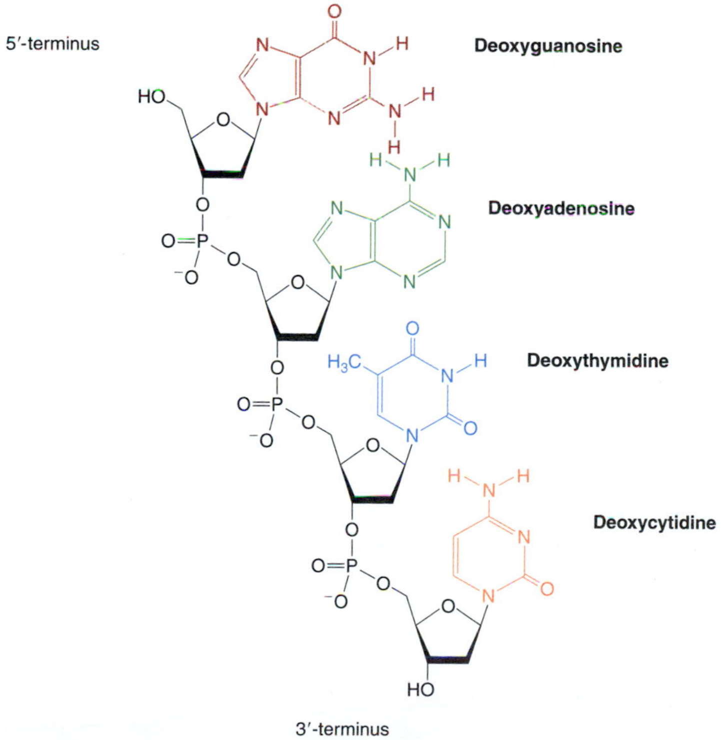

FIGURE 2.8
Structure of a DNA polynucleotide segment.
Shown is a tetranucleotide. Generally, nucleic acids less than 50 nucleotides long are referred to as oligonucleotides. Longer nucleic acids are called polynucleotides.

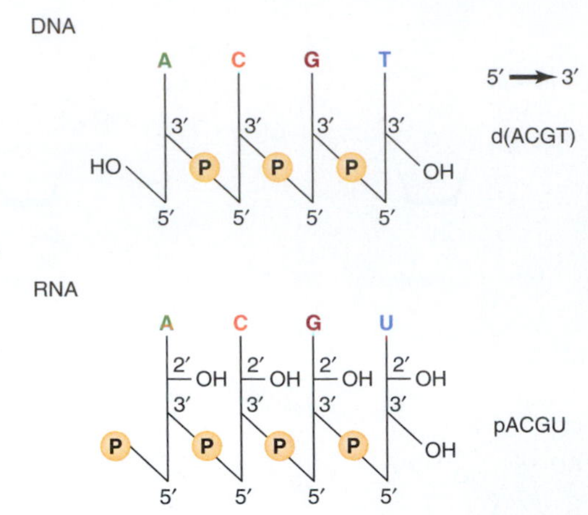

FIGURE 2.9
Shorthand notations for structure of oligonucleotides.
The convention used in writing the structure of an oligo- or polynucleotide is a perpendicular bar representing the sugar moiety, with the 5′-OH position of the sugar located at the bottom of the bar and the 3′-OH (and 2′-OH, if present) at a midway position. Bars joining 3′- and 5′-positions represent the 3′,5′-phosphodiester bond, and the P on the left or right side of the perpendicular bar represents a 5′-phosphate or 3′-phosphate ester, respectively. The base is represented by its initial placed at the top of the bar. An alternative shorthand form is to use the one-letter initials for the bases written in the 5′ → 3′ direction from left to right. Internal phosphodiester groups are assumed, and terminal phosphates are denoted with a "p." Oligonucleotide sequences containing deoxyribose sugars are preceded by a "d."

taining ≤50 nucleotides are called **oligonucleotides,** whereas those that are longer are called **polynucleotides.** The phosphodiester links the 5′-hydroxyl group of one residue to the 3′-hydroxyl group of the next. Linkages between two 5′-OHs or two 3′-OHs are not seen in naturally occurring DNA or RNA. The **directionality** of the phosphodiester bond means that linear oligo- and polynucleotides have ends that are not structurally equivalent. One end of a polynucleotide must terminate in a 5′-OH; the other terminates with a 3′-OH. These ends are referred to as the **5′-terminus** and **3′-terminus,** respectively. In many polynucleotides, one or both termini are chemically modified with moieties such as phosphate groups or amino acids. Circular polynucleotides have no free termini and are formed by joining the 5′-terminus of a linear polynucleotide with its own 3′-terminus through a phosphodiester bond.

In both linear and circular polynucleotides, the sugar–phosphodiester backbone is highly uniform and the chemical diversity of the molecules arises from the bases. More specifically, the **sequence** of bases on a polynucleotide provides the molecule with a unique chemical identity (Figure 2.9). This arrangement is analogous to that in polypeptides and proteins (see Chapter 3), where variations in chemical structure primarily arise from the identity of the amino acid side chains.

Polynucleotide Conformations

The bases are, to a large extent, responsible for the conformations of polynucleotides. The edges of the bases contain nitrogen and oxygen atoms (—NH_2, =N—, and =O groups) that can interact with other polar groups or surrounding water molecules. The faces of the rings, however, cannot participate in such interactions and tend to avoid contact with water. Instead they interact with one another, to produce a stacked conformation. **Base stacking** reduces the hydrophobic surface area that must be solvated by polar water molecules. Release of these water molecules into the bulk solvent is entropically favorable. The stacked arrangement of bases is further stabilized by favorable electronic interactions (van der Waals forces). Like other aromatic systems, the bases possess highly delocalized **π-orbitals** above and below the planes of their rings. Electron density in these orbitals can be polarized by nearby dipoles. The resulting induced dipole can then interact with the polarizing group in an energetically favorable fashion. These weak interactions are further enhanced by attractive **London dispersion forces** (induced dipole–induced dipole). Because the strength of these electronic interactions is very dependent upon distance, no empty space remains between the stacked bases.

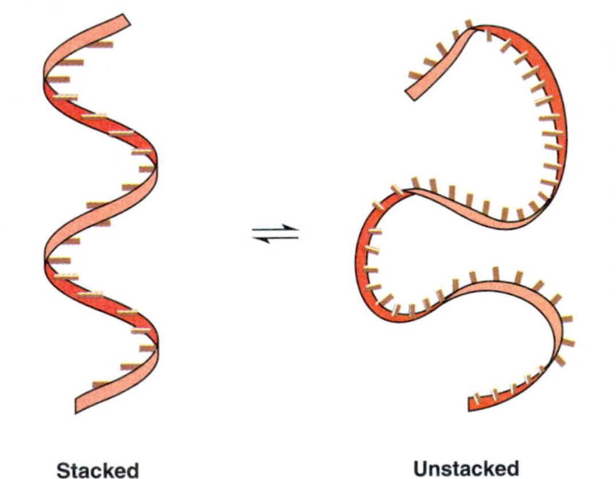

FIGURE 2.10
Stacked and random coil conformations of a single-stranded polynucleotide.
The helical band represents the phosphate backbone of the polynucleotide. Bases are shown in a side view as thick lines in tight contact with their neighbors, above and below each base. Unstacking of bases leads to a more flexible structure with bases oriented randomly.

Polynucleotides adopt conformations that maximize favorable stacking interactions between neighboring bases. The constraints imposed by the structure of the sugar–phosphodiester backbone favor helical structures (Figure 2.10). The overall stability of these helical structures is dependent on factors that include sequence-

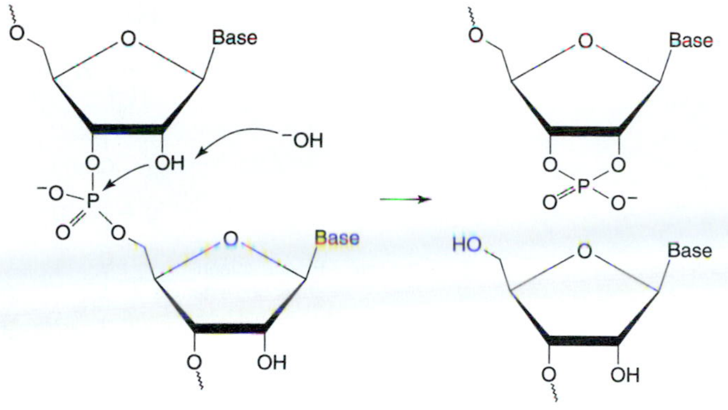

FIGURE 2.11
Hydrolysis of RNA is accelerated by participation of the 2′-OH group.
Intramolecular nucleophilic attack by the 2′-OH group of ribose greatly increases the rate of hydrolytic cleavage of the phosphodiester backbone, especially under basic conditions. The resulting 2′,3′-cyclic phosphodiester can subsequently react with water or hydroxide to a mixture of 2′- and 3′-phosphate monoesters.

dependent base stacking, pH, salt concentration, and temperature. For example, some synthetic polynucleotides, including poly(C) and poly(A), form right-handed helices in which the bases are highly stacked. Many other polynucleotides remain largely disordered in solution, a conformation called a **random coil.** Large polynucleotides with random sequences have both helical and disordered regions depending on the local sequence. The characteristics of the solution also play a role in determining whether polynucleotides adopt stacked conformations. For example, high concentrations of cations, especially some divalent ions such as Mg^{2+}, stabilize the helical conformation of polynucleotides by shielding the charges of the phosphodiester groups in the backbone. Without this shielding, the negative charges on the phosphates would destabilize the helix by electrostatic repulsion.

Stability of the Polynucleotide Backbone

Polynucleotides are relatively stable in aqueous solutions near neutral pH. It has been estimated that the half-life for hydrolysis of phosphodiester linkages in DNA is about 200 million years. This high stability makes DNA suitable for the long-term storage of genetic information. In contrast, RNA is much more prone to hydrolysis. The presence of the 2′-OH group provides an internal nucleophile for transesterification of the 3′,5′-phosphodiester linkage (Figure 2.11). The result is scission of the polynucleotide backbone, leaving a 2′,3′-cyclic phosphate on one fragment and a free 5′-OH group on the other. The greater lability of RNA toward hydrolysis makes it less suitable as a genetic material than DNA.

While the DNA backbone is relatively stable to hydrolysis, numerous enzymes catalyze phosphodiester scission. These enzymes, called nucleases, are characterized on the basis of the types of polynucleotides they hydrolyze and the specific bonds that are broken (Figure 2.12). For example, **exonucleases** cleave the last nucleotide residue at either the 5′- or 3′-terminus of a polynucleotide. Stepwise removal of individual nucleotides from one end of a polynucleotide can result in its complete degradation. Nucleases can sever one of the two nonequivalent bonds in the phosphodiester linkage, leaving either a free 5′-OH group and a 3′-phosphate or a 5′-phosphate and a 3′-OH. For example, treatment of an oligodeoxyribonucleotide with snake venom phosphodiesterase yields deoxyribonucleoside 5′-phosphates. In contrast, treatment with a spleen phosphodiesterase produces deoxyribonucleoside 3′-phosphates.

Other nucleases that cleave phosphodiester bonds located in the interior of polynucleotides are designated as **endonucleases.** These enzymes do not require a free terminus; therefore they can cleave circular polynucleotides. Endonucleases

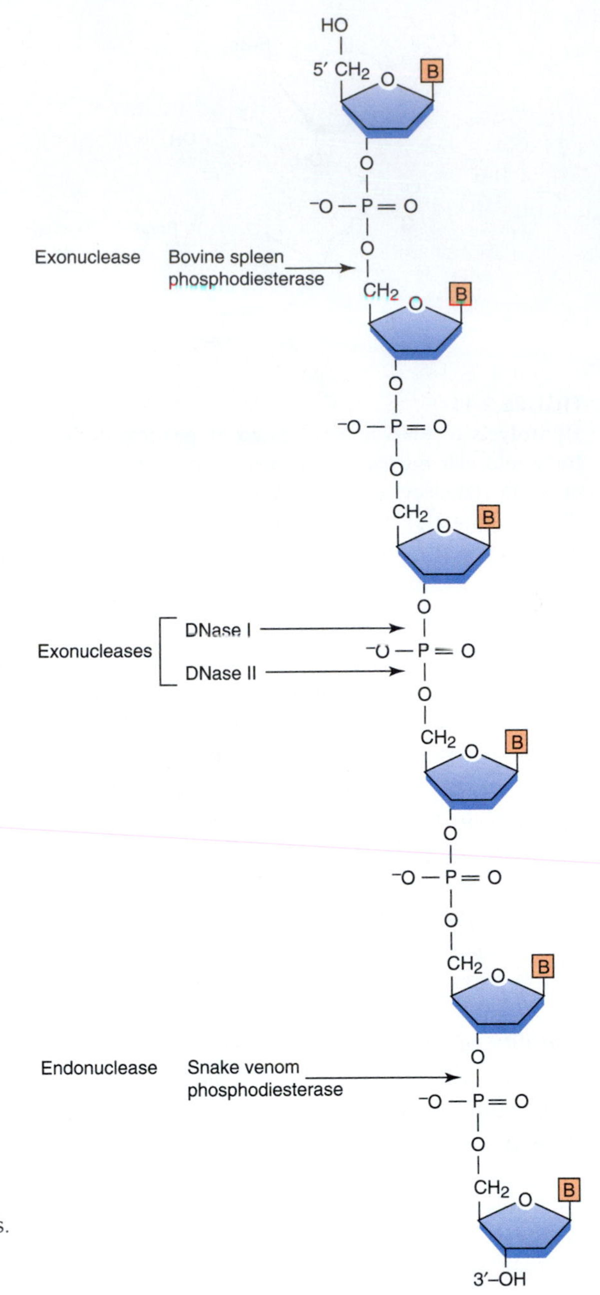

FIGURE 2.12
Specificities of nucleases.
Exonucleases remove nucleotide residues from either end of a polynucleotide, depending on their specificity. Endonucleases hydrolyze interior phosphodiester bonds. Both endo- and exonucleases typically show specificity regarding which oxygen–phosphorus bond is hydrolyzed.

such as DNase I and DNase II hydrolyze DNA with little sequence selectivity. Other enzymes, the **restriction endonucleases,** recognize and cleave very specific sequences. Restriction endonucleases have been particularly useful in the development of methodologies for sequencing of DNA polynucleotides and provide the basis for recombinant DNA techniques. Nucleases also exhibit specificity with respect to the overall structure of polynucleotides. For instance, some nucleases act on both single- or double-stranded polynucleotides, whereas others discriminate between these two structures. Some nucleases exclusively act on either DNA or RNA, whereas other nucleases have activity on only one type of polynucleotide.

Double-Helical DNA

With the recognition in the early and mid-20th century that DNA served as the carrier of genetic information, work to establish the structural basis for information storage intensified. However, several misconceptions slowed progress. The first of these was the (erroneous) experimental observation that DNA contained equal amounts of each of the four nucleosides (deoxyadenosine, deoxyguanosine, deoxycytidine, and deoxythymidine). This led to the proposal that DNA consisted of a cyclic tetranucleotide containing one of each of the nucleosides (Figure 2.13). A glaring problem with this structure was that DNA was found to have molecular masses thousands of times larger than that of a tetranucleotide. It was not obvious how the monotonous structure of repeating tetranucleotides could possess the complexity to convey an enormous number of hereditary traits.

Synthetic and X-ray diffraction experiments led to the acceptance of the linear polynucleotide structure for DNA. However, the three-dimensional structure of DNA still remained a mystery. Three key pieces of information were necessary for deduction of the structure of DNA. The first was the determination that DNA did not contain equal amounts of the four nucleosides, but contained variable amounts in different organisms. However, a relationship that held true is that the abundance of deoxyadenosine always equaled that of thymidine and the abundance of deoxyguanosine always equaled that of deoxycytidine. This led to consideration of structures with the bases specifically paired together (dA with dT and dG with dC). The second was that X-ray diffraction data suggested that DNA contained **double-helical** structures, and symmetry suggested that the two polynucleotide strands were oriented **antiparallel** to each other. The final piece in the puzzle was the suggestion that the bases were in the keto and amino **tautomeric forms** rather than enolic and imino forms. Tautomers are isomers of a molecule that differ only in the position of a hydrogen atom. Each of the four nucleoside bases has two or more possible tautomeric forms that are in equilibrium (Figure 2.14). The incorrect assignment of the predominant tautomeric structures meant that researchers were attempting to form base pairs using in-

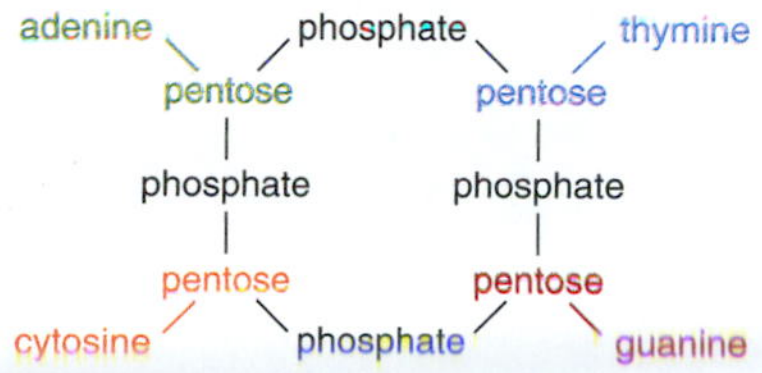

FIGURE 2.13
Tetranucleotide structure for DNA proposed in the 1930s.

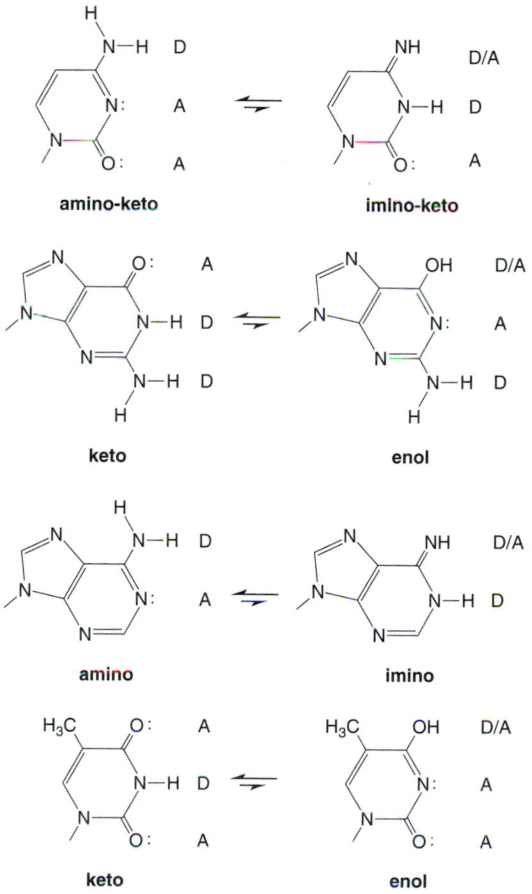

FIGURE 2.14
Tautomeric forms of bases.
The pattern of hydrogen bonding donating (D) and accepting (A) groups changes depending on the tautomer of the base that is present. Shown are predominant tautomeric forms (left) and one of the alternate forms (right) for each base. The hydrogen bonding pattern for each tautomer is also shown; ambiguity at the —OH and =NH positions of the minor tautomers arises from rotation or isomerism of these groups. The equilibrium ratios of the predominant form to all others is greater than 99:1.

correct hydrogen bonding patterns. With the correct tautomers, a model for DNA rapidly fell into place.

The structure that Watson and Crick proposed in 1953 for **double-helical DNA** was quite attractive because of its simplicity and symmetry (Figure 2.15). Moreover, it explained all available data for DNA structure and immediately led to hypotheses regarding the mechanism for storage of genetic information. The Watson–Crick double helix can be visualized as resulting from interwinding of two right-handed helical polynucleotide strands around a common axis. The two strands achieve contact through hydrogen bonds, formed at the hydrophilic edges of the bases. The N—H groups of the bases are good hydrogen-bond donors, and the electron pairs on the sp^2-hybridized oxygens of the C═O groups and nitrogens of the ═N— groups are good hydrogen-bond acceptors. Pairing occurs when an acceptor and a donor are in a position to form a hydrogen bond. These bonds extend between purine residues in one strand and pyrimidine residues in the other, and the matching of hydrogen-bond donors and acceptors results in two types of **base pairs:** adenine–thymine and guanine–cytosine (Figure 2.16). A direct consequence of these hydrogen bonding specificities is that **double-stranded DNA (dsDNA)** must contain ratios of nucleosides that agree with experimental observations (dA = dT and dG = dC). Finally, the geometries of the dA/dT and dG/dC base pairs result in similar C-1′–C-1′ distances (~10.6 Å) and glycosidic bond orientations. This **structural isomorphism** means that any of the four possible base pairs (dA-dT, dT-dA, dG-dC, dC-dG) can be placed into the double helix without changing the structure of the backbone.

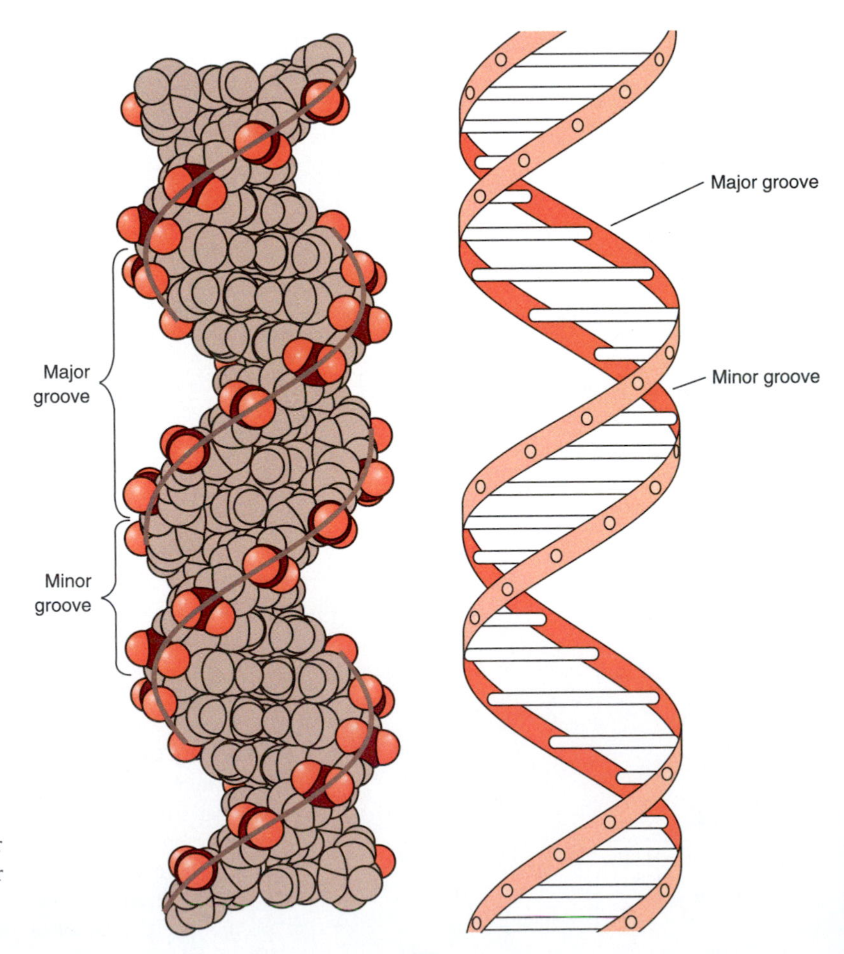

FIGURE 2.15
The Watson–Crick model of DNA.
On the left is a space-filling model of DNA; on the right is an idealized ribbon model. Bases are stacked in the interior of the helix, whereas the hydrophilic sugar–phosphodiester backbone is located on the exterior.
Redrawn from Rich, A. J. Biomol. Struct. Dyn. *1:1, 1983.*

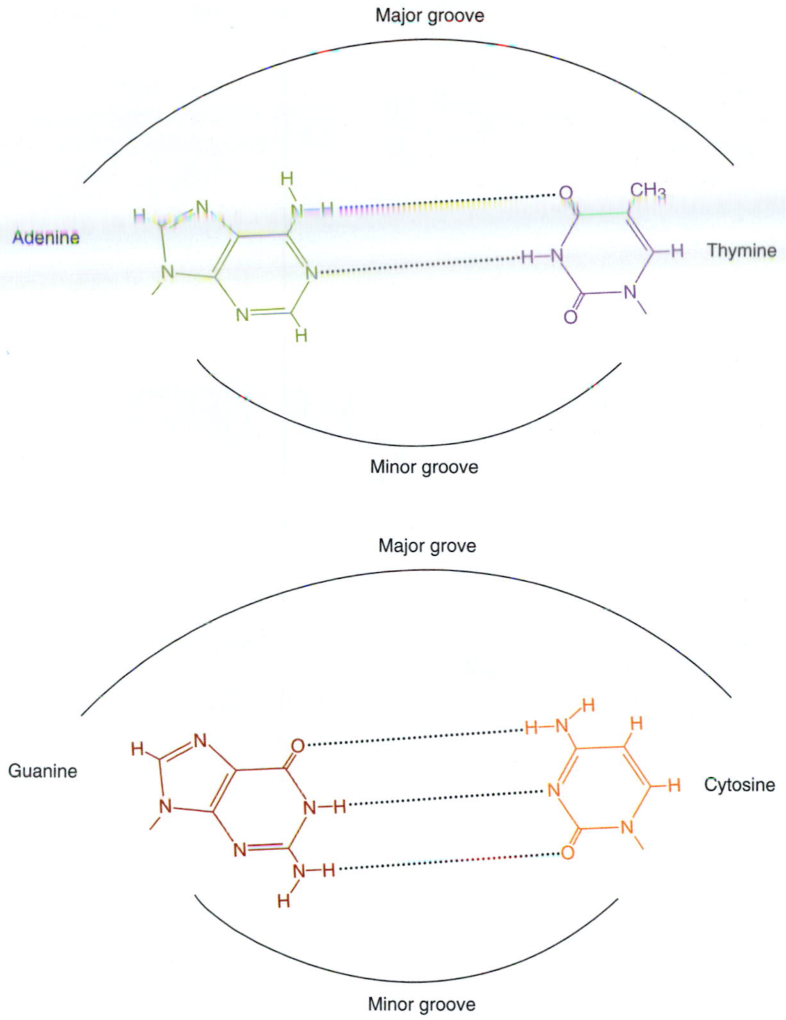

FIGURE 2.16
Watson–Crick base pairs.
Selective base pairs are formed between adenine and thymine and between guanine and cytosine.

This relationship between bases in the double helix is described as **complementarity.** Bases are complementary because every base of one strand is matched by a complementary hydrogen bonding base on the other strand (Figure 2.17). For instance, for each adenine projecting toward the common axis of the double helix, a thymine must be projected from the opposite chain so as to fill exactly the space between strands by hydrogen bonding with adenine. Neither cytosine nor guanine fits precisely in the available space across from adenine in a manner that allows formation of hydrogen bonds across strands. These **hydrogen bonding specificities** ensure that the entire base sequence of one strand is complementary to that of the other strand.

The exterior of the double helix consists of the sugar–phosphate backbones of the two antiparallel polynucleotide strands. Polynucleotides are asymmetric structures with an intrinsic sense of polarity built into them. The two strands are aligned in opposite directions; if two adjacent bases in the same strand, for example, thymine and cytosine, are connected in the $5' \rightarrow 3'$ direction, their complementary bases adenine and guanine will be linked in the $3' \rightarrow 5'$ direction. This antiparallel alignment produces a stable association between strands to the exclusion of the alternate parallel arrangement.

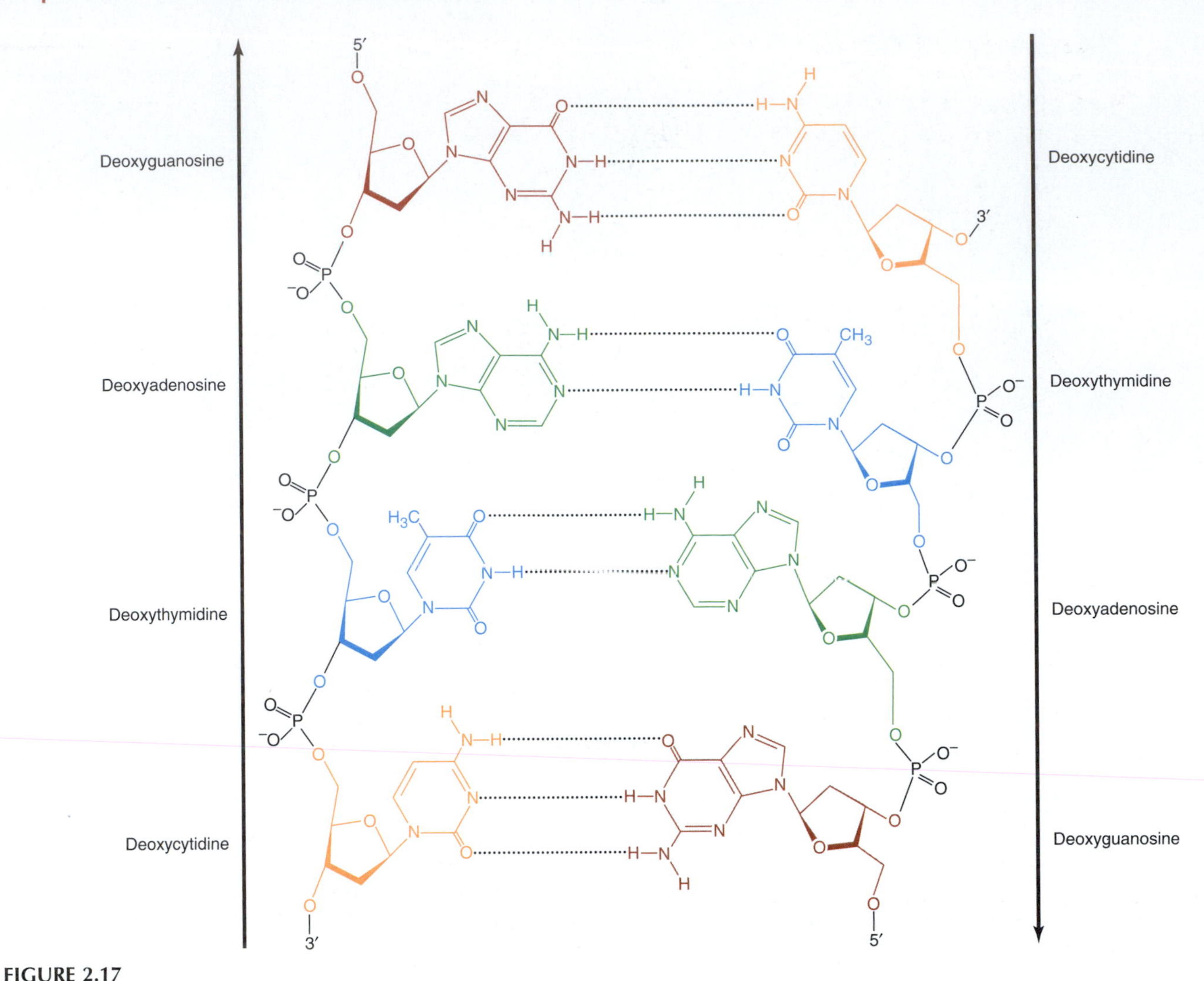

FIGURE 2.17

Formation of hydrogen bonds between complementary bases in double-stranded DNA. Interaction between polynucleotide strands is highly selective. Complementarity depends not only on the geometric factors that allow the proper fitting between the complementary bases of the two strands, but also on the formation of specific hydrogen bonds between complementary bases. Note the antiparallel orientation of the strands of a double-stranded DNA. Geometry of the helices does not prevent a parallel alignment, but such an arrangement is not found in DNA.

Interwinding of the two antiparallel strands produces a structure that has two distinct helical grooves between the sugar–phosphate backbones (Figure 2.15). One of the grooves is much wider (**major groove**) than the other (**minor groove**); this disparity arises from the geometry of the base pairs. The glycosidic bonds between the bases and the backbone pentose are not arranged directly opposite to each other but are displaced toward the minor groove. Significantly, the nucleotide sequence of DNA can be discerned without dissociating the double helix by looking inside these grooves. Because each of the four bases has its own orientation with respect to the rest of the helix, each base always displays the same atoms into the grooves. These atoms then constitute an important means of sequence-specific recognition of DNA by other molecules such as proteins. For example, N-7 of purines is always displayed in the major groove and can serve as a hydrogen-bond acceptor in interactions with donor groups on proteins (Figure 2.16). Similarly, the exocyclic 2-NH_2 group of guanine always projects into the minor groove and can form a steric blockade to the binding of small molecules.

Factors that Stabilize Double-Helical DNA

Factors that stabilize the helical structures of single-stranded polynucleotides—that is, stacking interactions—are also instrumental in stabilizing the double helix. The separation between the hydrophobic core of the stacked bases and the hydrophilic exterior of the charged sugar–phosphate groups is even more pronounced in the double helix than in single-stranded helices. The stacking tendency of single-stranded polynucleotides can be viewed as resulting from a tendency of the bases to reduce their contact with water. The double-stranded helix is a more favorable arrangement, essentially removing the bases from the aqueous environment while permitting the hydrophilic phosphate backbone to be highly solvated by water.

Stacking interactions, a combination of hydrophobic forces and van der Waals interactions, are estimated to generate 4–15 kcal mol^{-1} of stabilization energy for each adjacent pair of stacked bases (Table 2.1). Additional stabilization of the double helix results from extensive networks of cooperative hydrogen bonds. Typically, hydrogen bonds are relatively weak (3–7 kcal mol^{-1}) and are even weaker in DNA (2–3 kcal mol^{-1}) because of geometric constraints within the double helix. Cumulatively, hydrogen bonds provide substantial energies of stabilization for the double helix. However, because hydrogen bonds in the base pairs merely replace those between the bases and water in single-stranded DNA, they do not confer any selective stabilization of the double helix over the single-stranded polynucleotide. Yet, in contrast to stacking forces, hydrogen bonds are highly directional and are able to provide a discriminatory function for choosing between correct and incorrect base pairs. Because of their directionality, hydrogen bonds tend to orient the bases in a way that favors stacking. Therefore the contribution of hydrogen bonds is indirectly vital for the stability of the double helix.

The relative importance of stacking interactions versus hydrogen bonding in stabilizing the double helix was not always appreciated. Indeed, hydrogen bonds have been considered to be the "glue" that holds the two strands together. However, experiments with reagents that reduce the stability of the double helix (**denaturants**) illustrate the greater relative importance of stacking interactions (Table 2.2). These results show that the destabilizing effect of a reagent is not related to its ability to break hydrogen bonds but is determined by the solubility of the free bases in the reagent. As the reagent becomes a better solvent for the bases, the driving force for stacking diminishes, and the double helix is destabilized.

Ionic forces also have an effect on stability and conformation of the double helix. Phosphodiester groups are fully ionized at physiological pH; thus the exterior of the double helix carries two negative charges per base pair. The interstrand elec-

TABLE 2.1 Base-Pair Stacking Energies

Dinucleotide Base Pairs	*Stacking Energies (kcal mol^{-1} per stacked pair)*[a]
(GC) · (GC)	− 14.59
(AC) · (GT)	− 10.51
(TC) · (GA)	− 9.81
(CG) · (CG)	− 9.69
(GG) · (CC)	− 8.26
(AT) · (AT)	− 6.57
(TG) · (CA)	− 6.57
(AG) · (CT)	− 6.78
(AA) · (TT)	− 5.37
(TA) · (TA)	− 3.82

[a] Data from Ornstein, R. L., Reim. R., Breen, D. L., and Mc Elroy, R. D. *Biopolymers* 17:2341, 1978.

TABLE 2.2 Effects of Various Reagents on the Stability of the Double Helix[a]

Reagent	*Adenine Solubility $\times 10^{-3}$ (in 1 M reagent)*	*Molarity Producing 50% Denaturation*
Ethylurea	22.5	0.60
Propionamide	22.5	0.62
Ethanol	17.7	1.2
Urea	17.7	1.0
Methanol	15.9	3.5
Formamide	15.4	1.9

Source: Data from Levine, L., Gordon, J., and Jencks, W. P. *Biochemistry* 25168, 1963.

[a] The destabilizing effect of the reagents listed on the double helix is independent of the ability of these reagents to break hydrogen bonds. Rather, the destabilizing effect is determined by the solubility of adenine. Similar results would be expected if the solubilities of the other bases were examined.

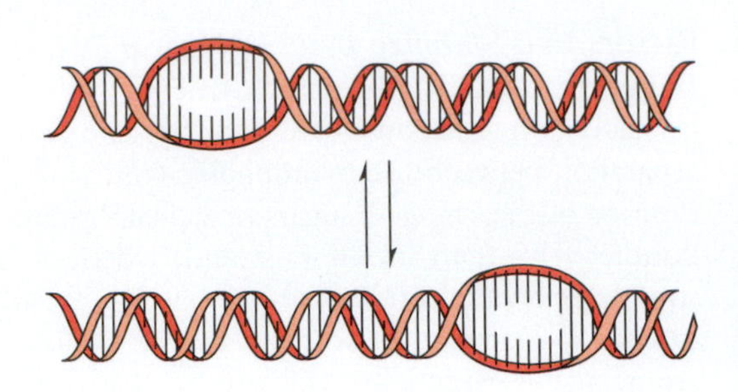

FIGURE 2.18
Migration of bubbles through double-helical DNA.
DNA contains short sections of open-strandedness that can "move" along the helix.

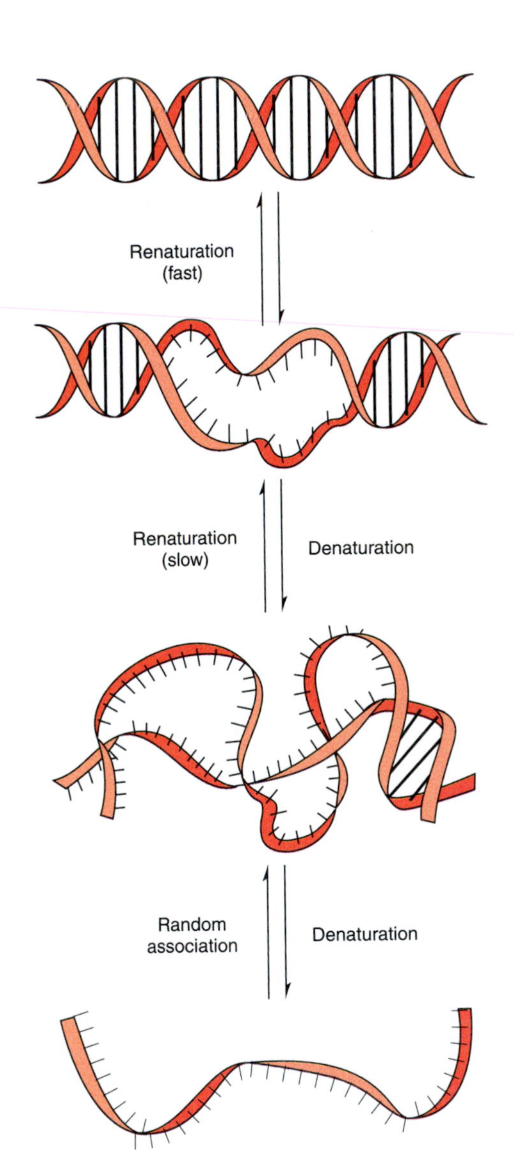

FIGURE 2.19
Denaturation of DNA.
At high temperatures the double-stranded structure of DNA is completely disrupted, with eventual separation of strands and formation of single-stranded open coils. Denaturation also occurs at extreme pH ranges or at extreme ionic strengths.

trostatic repulsion between negatively charged phosphates is destabilizing and tends to separate the complementary strands. In distilled water, DNA strands separate at room temperature. However, the presence of cations such as Mg^{2+}, spermine^{4+} (a tetraamine), and the basic side chains of proteins can shield the phosphate groups and decrease repulsive forces.

Denaturation and Renaturation

The double helix is disrupted during almost every important biological process in which DNA participates, including DNA replication, transcription, repair, and recombination. The forces that hold the two strands together are adequate for providing stability but weak enough to allow facile strand separation. The double helix is stabilized relative to the single strands by about 1 kcal per base pair so that a relatively minor perturbation can produce a local disruption in the double helix, provided that only a short section of the DNA is involved. These base pairs can then close up again, releasing free energy that can then cause the adjacent base pairs to unwind. In this manner minor disruptions of the double helix can migrate along its length (Figure 2.18). Thus, at any particular moment, the large majority of bases remain hydrogen bonded, but all bases can pass through the single-stranded state, a short stretch at a time. The ability of the DNA double helix to "breathe" is an essential prerequisite for its biological functions.

Separation of DNA strands can be studied by increasing the temperature of a solution. At relatively low temperatures a few base pairs are disrupted, creating one or more **"open-stranded bubbles"** (Figure 2.19). These bubbles form initially in sections that contain relatively higher proportions of adenine–thymine pairs because of the lower stacking energies of dimers of such pairs. As the temperature is raised, the size of the bubbles increases, and eventually the thermal motion of the polynucleotides overcomes the forces that stabilize the double helix. At even higher temperatures, the strands separate physically and acquire a random-coil conformation. The process is most appropriately described as a helix-to-coil transition, but is commonly called denaturation or melting.

Denaturation is accompanied by several physical changes, including a buoyant density increase, a reduction in viscosity, a change in ability to rotate polarized light, and changes in absorbance of UV light. Changes in UV absorption are frequently used to follow the process of denaturation experimentally (Figure 2.20). Because of the strong absorbance of the purine and pyrimidine bases, DNA absorbs strongly in the UV region with a maximum near 260 nm. However, the absorbance of individual bases is reduced by electronic interactions that arise from base stacking. The total absorbance of the stacked bases may be reduced by as much as 40% compared to an unstacked state. This reduction in the extinction coefficient of the aggregate bases is termed **hypochromicity.** As the ordered structure of the double helix is disrupted as temperature increases, stacking interactions decrease gradually. Therefore a totally disordered polynucleotide approaches an absorbance not very different from the sum of the absorbances of its purine and pyrimidine constituents.

Measurement of the absorbance of a DNA complex at 260 nm while slowly increasing the temperature provides a means to observe denaturation. In a thermal denaturation experiment monitored using absorption spectroscopy, the polynucleotide absorbance typically changes very slowly at first, then rapidly rises to a maximum value (Figure 2.20). Before the rise, the DNA is double stranded. In the rising section of the curve an increasing number of base pairs is interrupted. Complete strand separation occurs at a critical temperature corresponding to the upper plateau of the curve. However, if the temperature is decreased before the complete separation of the strands, the native structure is completely restored. The midpoint te mperature of this transition, the T_m, is characteristic of the base content of DNA under standard conditions of concentration and ionic strength. The higher the guanine–cytosine content, the higher the transition temperature between double-stranded helix and single strands. This difference in T_m values is attributed to increased stability of guanine–cytosine pairs, which arises from more favorable stacking interactions.

DNA can also be denatured at pH values >11.3 as the N—H groups on the bases become deprotonated, preventing them from participating in hydrogen bonding. Alkaline denaturation is often used in preference to heat denaturation to prevent breakage of phosphodiester bonds that can occur at a high temperature or low pH. Denaturation can also be induced at low ionic strengths, because of enhanced interstrand repulsion between negatively charged phosphates, and by various denaturing agents (compounds that can effectively hydrogen bond to the bases while disrupting hydrophobic stacking interactions). A complete denaturation curve similar to that shown in Figure 2.20 is obtained at a relatively low constant temperature by variation of the concentration of an added denaturant such as urea.

Complementary DNA strands, separated by denaturation, can reform a double helix if appropriately treated. This is called **renaturation** or **reannealing.** If denaturation is not complete and a few bases remain hydrogen bonded between the two strands, the helix-to-coil transition is rapidly reversible. Annealing is possible even after complementary strands have been completely separated. Under these conditions the renaturation process depends on two complementary DNA strands meeting in an exact manner that can lead to reformation of the original structure. Not surprisingly, this is a slow, concentration-dependent process. As renaturation begins, some of the hydrogen bonds formed are extended between short tracts of polynucleotides that may have been distant in the original native structure. These randomly base-paired structures are short-lived because the bases that surround the short complementary segments cannot pair and lead to the formation of a stable fully hydrogen-bonded structure. Once the correct bases begin to pair by chance, the double helix is rapidly reformed over the entire DNA molecule.

Sudden onset of denaturation/renaturation reveals the "all-or-none" nature of the helix-to-coil transition. The renaturation process described above requires formation of a short double-helical region to start formation of the double helix (Figure 2.21). This begins with formation of a single base pair that is rather unstable. However, formation of a second neighboring base pair is enhanced because the process is now less entropically disfavored. As new base pairs begin to form a stacked structure, formation of subsequent base pairs is further facilitated. Unpaired nucleotides of each strand begin to stack on the growing helix, if they are positioned optimally for base pairing. After formation of a double helix with four to five base

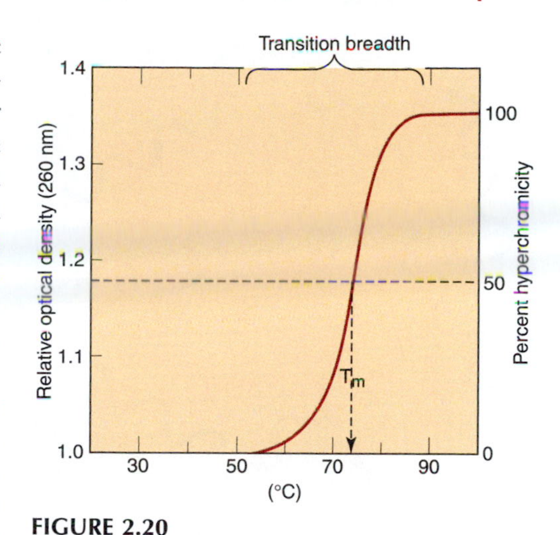

FIGURE 2.20
Temperature–optical density profile for DNA. When DNA is heated, the absorbance at 260 nm increases with rising temperature. A graph in which absorbance versus temperature is plotted is called a "melting curve." Relative optical density is the ratio of the absorbance at any temperature to that at low temperatures. The temperature at which one-half of the maximum optical density is reached is the midpoint temperature (T_m).

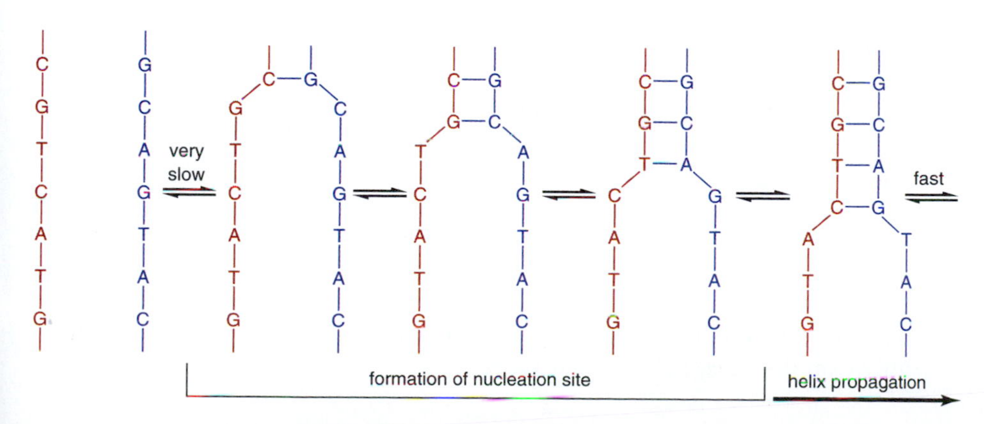

FIGURE 2.21
Cooperativity of renaturation/denaturation of DNA double helices.
During renaturation, formation of the first base pair is very slow. Annealing of neighboring pairs is facilitated, especially after formation of a 3–5 bp "nucleation site." Denaturation follows a similar course, but order of steps is reversed.
Redrawn from Saenger, W. Principles of Nucleic Acid Structure. *New York: Springer-Verlag, 1984, p. 141.*

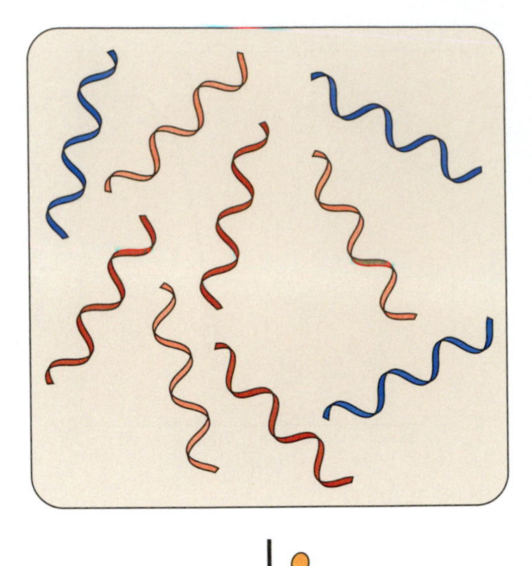

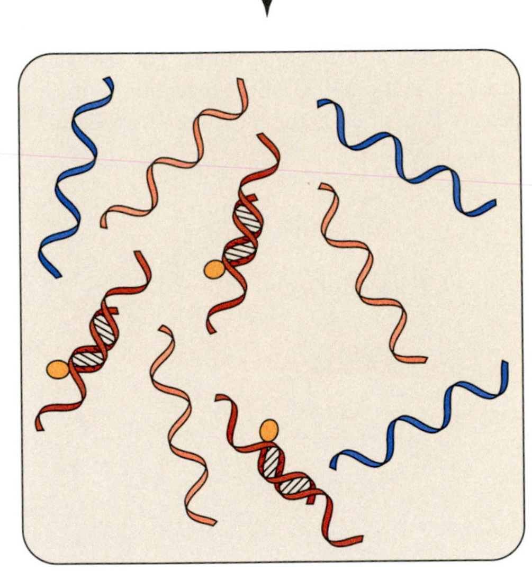

FIGURE 2.22
General scheme for hybridization experiments.
A mixture of denatured DNAs is treated with a DNA probe bearing a label. The probe can hybridize with those DNAs with complementary sequences. Detection of the double-helical complexes allows for detection and quantitation of DNA that contains the sequence of interest. Specific applications often feature steps to separate and immobilize the different DNAs in the mixture.

pairs, a stable double helix is formed, and the remainder of the complex will rapidly and spontaneously "zip up." After formation of the initial base pair, stacking and base pairing are not independent events but are influenced by the neighboring pairs. Such a process is called **cooperative.** The presence of a short double helix serves as a nucleation site for annealing by facilitating formation of subsequent base pairs. Denaturation is the same process in reverse: a bubble serves as an initiation site for unstacking of the bases and rapid unzipping of the helix.

Hybridization

Self-association of complementary polynucleotide strands has provided the basis for development of **hybridization.** This technique depends on association between two polynucleotide chains, which may be of the same or of different origin or length, provided that a base complementarity exists between these chains. Hybridization occurs not only between DNA chains but also between complementary RNA chains as well as DNA–RNA combinations. A number of techniques have been developed for detecting and quantitating the hybridized complex and for measuring the rates of hybridization. These techniques are important basic tools of contemporary molecular biology and are being used for (1) determining whether a certain sequence occurs more than once in the DNA of a particular organism, (2) demonstrating a genetic or evolutionary relatedness between different organisms, (3) determining the number of genes transcribed in a particular mRNA, and (4) determining the location of any given DNA sequence by annealing with a complementary polynucleotide, called a **probe,** that is appropriately tagged for easy detection of the hybrid. In a typical experiment, DNA to be tested for hybridization is denatured and immobilized by binding to a suitable insoluble matrix. Labeled **DNA probes** are then allowed to hybridize to complementary sequences bound to the matrix (Figure 2.22). Probes are short single-stranded RNA or DNA oligonucleotides that are complementary to specific sequences of interest in genomic DNA. Under the proper conditions probes interact only with the segment of interest, indicating whether it is present in a particular sample of DNA. Probe molecules are generally at least 15–20 nucleotides long. Appropriate labels include radioactive elements, fluorescent chromophores, and biotin. Because the double-helical complex containing the hybridized probe molecules is usually bound to an insoluble matrix, unhybridized probes can be washed away. Detection of bound labels allows direct quantitation of the sequence of interest. Determination of the maximum amount of DNA that can be hybridized can establish homologies between DNA of different species since the base sequences in each organism are unique. On this basis, annealing can be used to compare the degree to which DNAs isolated from different species are related to one another. The observed homologies serve as indices of evolutionary relatedness and have been particularly useful for defining phylogenies in prokaryotes. Hybridization studies between DNA and RNA have also provided very useful information about the biological role of DNA, particularly the mechanism of transcription. Arrays of probes are useful for definitive and rapid diagnosis of genetic disorders, infectious disease, and cancer as described briefly in Clinical Correlation 2.2.

Conformations of Double-Helical DNA

The early X-ray diffraction studies showed that there was more than one conformation of DNA (Figure 2.23). One of these, **A-DNA,** was found under conditions of low humidity and high salt concentration. Adding organic solvents such as ethanol reduced the "humidity" of these aqueous solutions. A second distinct form, **B-DNA,** appeared under conditions of high humidity and low salt concentration and was the basis of the Watson–Crick structure. Eleven distinct conformations of double-helical DNA have been described. These conformations vary in orientation of the bases relative to the helix and to each other as well as in other geometric parameters of the double helix. One form, **Z-DNA,** incorporates a left-handed helix rather than the usual right-handed variety.

CLINICAL CORRELATION 2.2

Diagnostic Use of DNA Arrays in Medicine and Genetics

With pending completion of the Human Genome Project, a wealth of genetic information is rapidly becoming available. Application of this knowledge to medicine requires development of new techniques to monitor gene expression and rapidly evaluate genes for mutations and other sequence variations. Oligonucleotide arrays have shown great promise for these applications. Such arrays consist of a number of gene-specific oligonucleotide probes immobilized at specific sites on a solid matrix (chip). Arrays can contain thousands of unique probe molecules, each fixed within an "address." Gene chips can then be treated with labeled target nucleic acids (DNA or RNA) derived from cells of an organism. Hybridization of the targets with complementary probe sequences allows for immobilization of the label at specific sites on the chip. In this way, the presence of specific sequences can be determined and the amount of labeled target hybridized to a site can be quantitated, allowing for determination of amount of each target in a sample.

Use of DNA arrays to analyze sequences has led to detection of mutations in human cells. For example, high-density DNA arrays with thousands of oligonucleotide probes have been used to screen for mutations that lead to ataxia telangiectasia, a recessive genetic disease characterized by neurological disorders, recurrent respiratory infections, and dilated blood vessels in the skin and eyes. Similar studies have examined mutations in the hereditary breast and ovarian cancer gene BRCA1 and other genetic markers for disease. In the future, such techniques may be developed into diagnostics for rapid screening of genomic DNA for disease-associated mutations.

Freeman, W. M., Robertson, D. J., and Vrana, K. E. Fundamentals of DNA hybridization arrays for gene expression analysis. *BioTechniques* 29:1042, 2000; and Hacia, J. G., Brody, L. C., Chee, M. S., Fodor, S. P., and Collins, F. S. Detection of heterozygous mutations in *BRCA1* using high density oligonucleotide arrays and two-colour fluorescence analysis. *Nat. Genet.* 14:441, 1996.

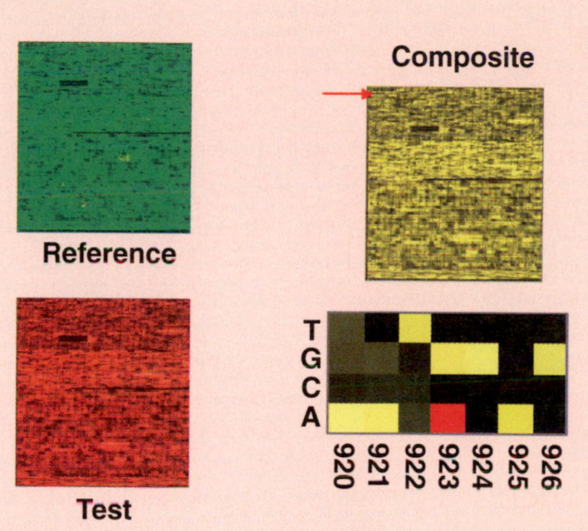

High density DNA arrays for mutation analysis.
Arrays for the ataxia telangiectasia (A-T) gene are interrogated separately with a reference (unmutated) sample and a test sample. The composite image is made by combination of the images from the two experiments. At sites where the test and reference DNAs have identical sequences, the composite image is yellow (red + green). In this sample, the person has a mutation at position 923, which appears red. This mutation, the replacement of a G with an A, causes premature termination of protein synthesis. Because this sample also has a yellow spot at position 923, the mutation has occurred on only one copy of the gene (e.g., it is heterozygous). In recessive disorders such as A-T, people with heterozygous mutations are carriers, but will not develop the disease themselves.

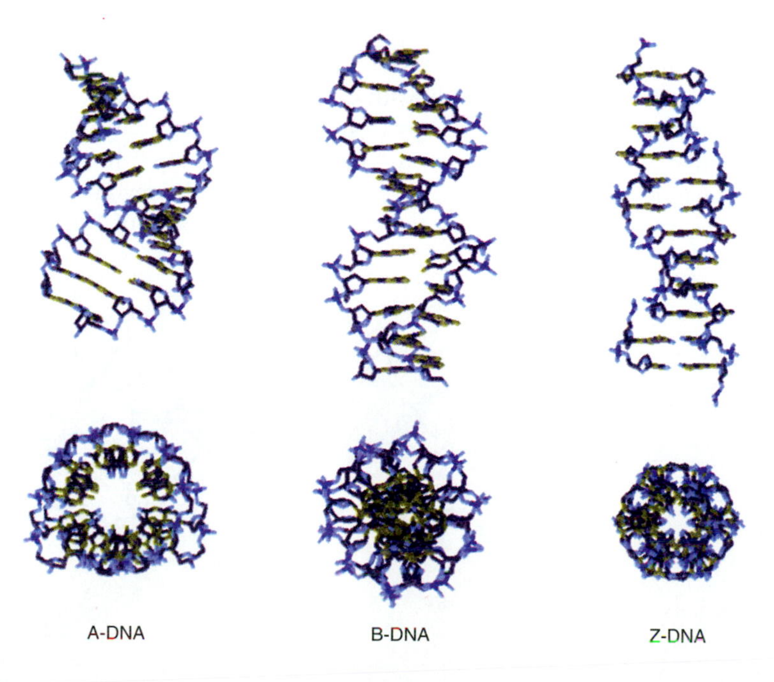

FIGURE 2.23
The varied geometries of double-helical DNA.
Depending on conditions and base sequence, the double helix can acquire various forms of distinct geometries. There are three main families of DNA conformations: A, B, and Z. The right-handed forms, B-DNA and A-DNA, differ in sugar pucker, which leads to differing helical structures. The A-form is underwound compared to the B-form, and the resulting helix is shorter and fatter. The Z-DNA structure is a left-handed helix with a zigzagging backbone. The sugar puckers and glycosidic conformations alternate from one residue to the next, producing a local reversal in the chain direction.

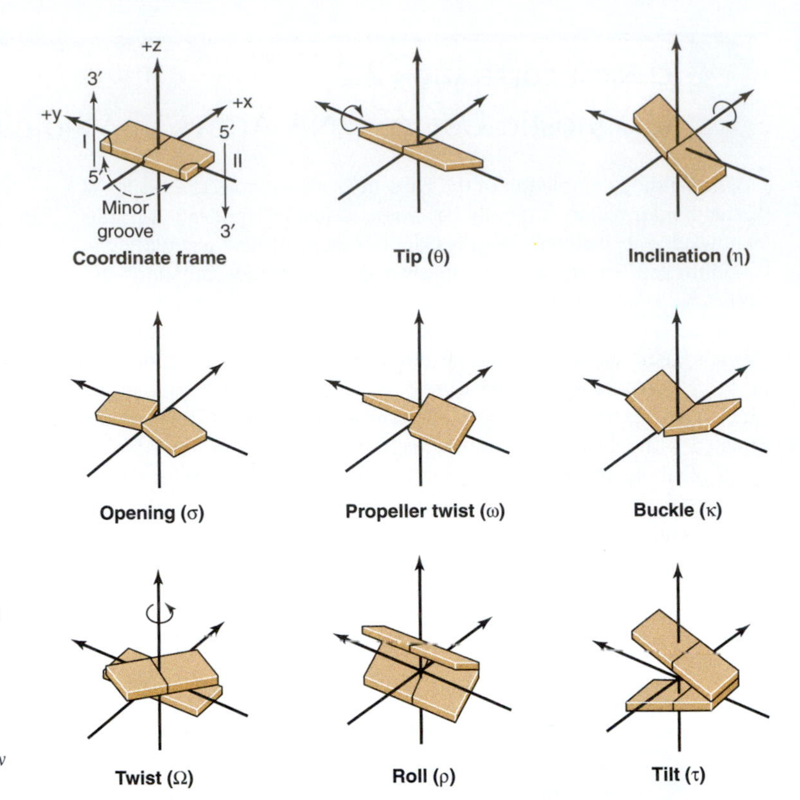

FIGURE 2.24
Parameters used to describe the local orientation of a base pair in a double helix.
Perturbations from the ideal Watson–Crick base pair can involve motions of the individual bases relative to each other (middle row) or the coordinated motion of the whole base pair with respect to the coordinate system (top row) or a neighboring base pair (bottom row).
Redrawn from Hecht, S.M. (Ed.). Bioorganic Chemistry: Nucleic Acids. *New York, Oxford University Press, 1996, p. 109.*

The **structural polymorphism** of double-helical DNA depends on the base composition and on physical conditions. The local structure of DNA is sufficiently flexible to allow for changes in conformation that maximize stacking while minimizing unfavorable steric interactions. Variation in shapes and in electronic properties of the bases leads to a sequence-dependent variability in stacking. Indeed, calculations of stacking energies of various pairs of stacked bases show such differences. The overlap between bases changes according to sequence and conformation. The stacking preferences of bases can favor one conformation over others. For this reason, consecutive guanines on one strand favor A-DNA-like conformations. Repulsive steric interactions between two bases or between bases and the backbone sugars can also cause changes in the geometry of the base pairs that lead to changes in conformation. Differences in orientation between planes of H-bonded bases may produce double helix variants with different base tilt, roll, twist, or propeller twist (Figure 2.24).

The solution conditions also play a key role in determining the favored conformation. Water molecules interact differently with double helices in different conformations. For example, the phosphate groups in B-DNA are more accessible to water molecules than in A-DNA. Also, polar groups on the bases are better hydrated in a B-DNA conformation. In fact, in AT-rich sequences, an ordered array of water molecules, the spine of hydration, occupies the narrow minor groove of B-DNA

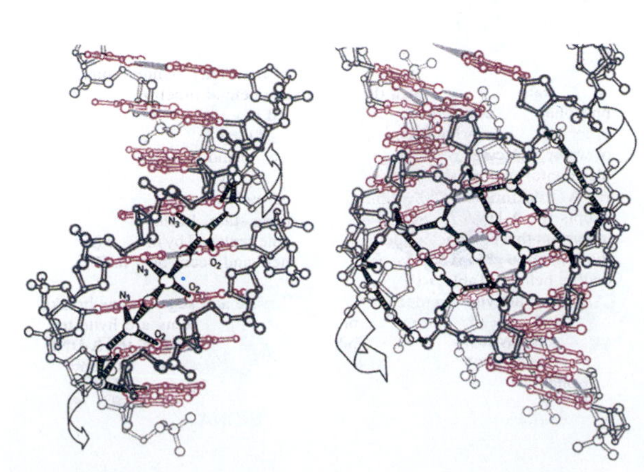

FIGURE 2.25
Hydration of the grooves of DNA.
(*a*) An organized spine of hydration fills the minor groove of B-DNA.
(*b*) The phosphates lining the major groove are spanned by a network of waters in A-DNA.
Reprinted with permission from Saenger, W. Principles of Nucleic Acid Structure. *New York: Springer-Verlag, 1984, p. 379.*

(Figure 2.25*a*). Upon reduction in the "humidity," the available water molecules solvate the highly polar phosphate groups in preference to the bases. The major groove narrows, allowing water molecules to bridge the phosphates (Figure 2.25*b*). This network of waters stabilizes the A-DNA conformation. The different conformations of DNA can be grouped into three families: A-DNA, B-DNA, and Z-DNA. The parameters for these conformations, listed in Table 2.3, have been determined by X-ray diffraction methods. While these experiments provide very accurate information about molecular geometry and dimensions of crystalline samples, they give only approximate dimensions for the macromolecule in solution. It must be emphasized that the average overall structure of DNA in living organisms is believed to be B-DNA-like. *Notably, this conformation, unlike the A- and Z-forms, is highly flexible.* In native B-DNA, considerable local variation in conformation of individual nucleotides may occur. Such variations may be important in regulation of gene expression, since they can influence the extent of DNA binding with various types of regulatory proteins.

DNA conformations in the B-family feature base pairs that are nearly perpendicular to the helical axis, which passes through the base pairs. The major and minor grooves are roughly the same depth, and the minor groove is relatively narrow. Sugar puckers are generally C-2′ *endo*, and the bases occupy the *anti* glycosidic conformation. The helix is relatively long and thin, with approximately 10 base pairs per helical turn. The rise per residue is 3.4 Å, the approximate thickness of the bases.

In contrast, the A-DNA structure is shorter and thicker. There are about 11 base pairs per helical turn with a vertical rise of 2.56 Å per residue. In order to accommodate the thickness of the bases, the base pair is tilted 20° from the plane perpendicular to the helical axis. The helical axis is displaced to the major groove side of the base pairs. This results in a very deep major groove and a shallow minor groove. Displacement of the axis also forms a hole ~3 Å in diameter that runs through the center of the helix. While the glycosidic conformation remains *anti* for all bases, the sugar pucker changes to C-3′ *endo*. This change in pucker brings the phosphates closer together in A-DNA compared to the B-conformation (5.9 Å vs. 7.0 Å). Low humidity favors the A-DNA structure that exposes more hydrophobic surface to the solvent than B-DNA.

Z-DNA is a radically different conformation for double-stranded DNA. The most striking feature is that it forms a left-handed double helix. It is generally observed in sequences of alternating purines and pyrimidines, particularly $d(GC)_n$. The designation "Z" was chosen because the phosphodiester backbone assumes a "zigzag" arrangement rather than the smooth conformation that characterizes other

TABLE 2.3 Structural Features of A-, B-, and Z-DNA

Features	*A-DNA*	*B-DNA*	*Z-DNA*
Helix rotation	Right-handed	Right-handed	Left-handed
Base pairs per turn (crystal)	10.7	9.7	12
Base pairs per turn (fiber)	11	10	—
Base pairs per turn (solution)	—	10.5	—
Pitch per turn of helix	24.6 Å	33.2 Å	45.6 Å
Proportions	Short-end broad	Longer and thinner	Elongated and thin
Helix packing diameter	25.5 Å	23.7 Å	18.4 Å
Rise per base pair (crystal)	2.3 Å	3.3 Å	3.7
Rise per base pair (fiber)	2.6	3.4 Å	—
Base-pair tilt	+19°	−1.2° (but varies)	−9°
Propeller twist	+18°	+16°	0°
Helix axis rotation	Major groove	Through base pairs	Minor groove
Sugar ring conformation (crystal)	C-3′ *endo*	Variable	Alternating
Sugar ring conformation (fiber)	C-3′ *endo*	C-2′ *endo*	—
Glycosyl bond conformation	anti	anti	*anti* at C, *syn* at G

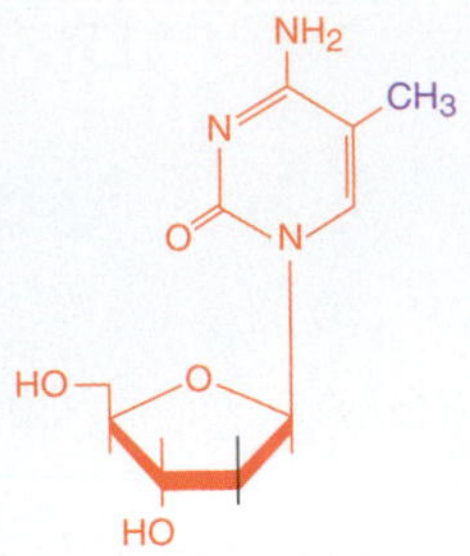

FIGURE 2.26
Structure of 5-methyldeoxycytidine.

double-stranded forms. This zigzag arises from a change to the *syn* glycosidic conformation of the purines, while the pyrimidines remain in the *anti* orientation. The alternating conformations result in a local reversal of the chain direction. The sugar puckers of the purines and pyrimidines also alternate, with the former in the C-3′ *endo* conformation and the latter in the C-2′ *endo*. The Z-DNA structure is longer and much thinner than that of B-DNA and completes one turn in 12 base pairs. The minor groove is very deep and contains the helical axis. The base pairs are displaced so far into the major groove that a distinct channel no longer exists. Therefore the conformation of Z-DNA may be viewed as the result of the major groove of B-DNA having "popped out" in order to form the outer surface of Z-DNA. This change places the stacked bases on the outer part of Z-DNA rather than in their conventional positions in the interior of the double helix.

The biological function of Z-DNA is not known. Some evidence exists that suggests that Z-DNA influences gene expression and regulation. Apparently small stretches of DNA with the potential of forming Z-DNA are more commonly found at the 5′-ends of genes, regions that regulate transcriptional activities. These stretches consist of alternating purines and pyrimidines that favor formation of the Z-conformation. The possible function of Z-DNA in gene regulation is also supported by the fact that methylation of either guanine residues in C-8 and N-7 positions or cytosine residues in C5 position stabilizes the Z-form. Sequences that are not strictly alternating purines and pyrimidines may also acquire the Z-conformation as a result of methylation. For instance, the hexanucleotide m5CGATm5CG (m5C is 5-methyldeoxycytidine, Figure 2.26) forms Z-DNA. On this basis it might be expected that *in vivo* methylation of cytosine also induces a B → Z transition in cellular DNA. The suggestion that Z-DNA may have a role in gene regulation is supported by modifications in methylation patterns that accompany the process of gene expression. However, Z-DNA has not yet been detected in DNA *in vivo*.

Noncanonical DNA Structures

Conformational variants of DNA—that is, A-, B-, and Z-DNA—are associated mainly with variation in the conformation of the nucleotide constituents of DNA. It is now recognized that DNA is not a straight, monotonous, and uniform structure. Instead, DNA forms unusual structures such as cruciforms or triple-stranded arrangements and bends as it interacts with certain proteins. Such variations in DNA conformation appear to be an important recurring theme in the process of molecular recognition of DNA by proteins and enzymes. Variations in DNA structure or conformation are favored by specific DNA sequence motifs. These include DNA sequences with symmetry elements such as inverted repeats, palindromes, mirror repeats, and direct repeats (Figure 2.27), as well as homopurine–homopyrimidine sequences, phased A tracts, and G-rich regions. AT-rich sequences, which are prone to easy strand separation, exist near the origins of DNA replication. The human genome is rich in homopurine–homopyrimidine sequences and alternating purine–pyrimidine tracts. DNA bending, slipped DNA, cruciform formation, triplex DNA, and quadruplex arrangements are among the structures reviewed in this section.

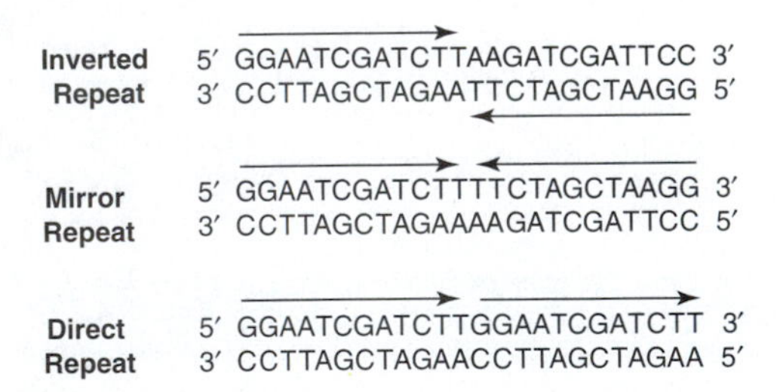

FIGURE 2.27
Symmetry elements of DNA sequences.
Three types of symmetry elements for double-stranded DNA sequences are shown. Arrows illustrate the special relationship of these elements in each one of these sequences. In inverted repeats, also referred to as palindromes, each single DNA strand is self-complementary within the inverted region that contains the symmetry elements. A mirror repeat is characterized by the presence of identical base pairs equidistant from a center of symmetry within the DNA segment. Direct repeats are regions of DNA in which a particular sequence is repeated. The repeats need not be adjacent to one another.

Bent DNA

DNA sequences with runs of four to six adenines phased by 10-bp spacers produce bent conformations. Interestingly, NMR spectroscopic evidence suggests that the A-tracts are unusually rigid and straight. Mixed-sequence DNA may tend to writhe or twist, but these curves would normally cancel out over long stretches of DNA. However, if straight A-tracts were placed in phase with these curves, a strongly bent conformation would arise. Alternative models where the bent conformation arises from the collapse of the helix into the minor groove of the A-tract or from an un-

CLINICAL CORRELATION 2.3

Antitumor Antibiotics that Change the Shape of DNA

The local three-dimensional structure of DNA is important in interactions with proteins involved in repair, transcription, recombination, and chromatin condensation. Recently, it has been proposed that antibiotics can induce formation of DNA structures that can recruit these proteins with cytotoxic results. The best studied example of this phenomenon is the antitumor drug cisplation, a tetracoordinate platinum complex [*cis*-$Pt(NH_2)_2Cl_2$]. Cisplatin is used alone or in combination with other antitumor agents to treat a variety of tumors including testicular, ovarian, bone, and lung cancers. This platinum complex forms inter- and intrastrand cross-links in double-stranded DNA with the latter adduct comprising 90% of DNA lesions. These bonds arise from displacement of chloride ligands on platinum by N-7 atoms of two neighboring guanines. Structural studies on intrastrand cross-linked DNA adduct show that the double helix is strongly bent toward the major groove.

From Zamble, D. B. and Lippard, S. J. In: B. Lippard (Ed.), Cisplatin: Chemistry and Biochemistry of a Leading Anticancer Drug. *New York: Wiley-VCH 1999, p. 74.*

Bent structures of the DNA–cisplatin adduct are specifically recognized by several DNA-binding proteins that include nucleotide excision repair (NER) proteins and high-mobility-group proteins such as HMG-1. It has been proposed that the cytotoxicity of cisplatin adducts is a complicated process mediated by specific interactions with these proteins. Cellular processes such as transcription and apoptosis (programmed cell death) are affected by formation of cisplatin–DNA adducts. The lesions themselves and the adduct–protein complexes are likely to interfere with transcription. NER proteins are recruited to repair the lesion, but excision repair is prone to introduction of DNA strand breaks. Accumulation of these breaks will ultimately induce apoptosis as the DNA becomes too damaged to function. Similar mechanisms have also been proposed to account for cytotoxicity of other DNA-binding drugs such as ditercalinium. This bifunctional molecule forms noncovalent adducts with DNA that are also highly bent. Cytotoxicity is thought to arise from induction of abortive repair pathways that lead to DNA strand breaks.

Interactions of the cisplatin–DNA adduct with HMG proteins may also contribute to its cytotoxicity. Binding of HMG proteins may incorrectly signal that the damaged region of DNA is transcriptionally active and prevent condensation into folded chromatin structures. These complexes might also perpetuate the lesion by shielding the DNA–cisplatin adduct from repair.

Zamble, D. B. and Lippard, S. J. The response of cellular proteins to cisplatin-damaged DNA. In: B. Lippert (Ed.), *Cisplatin: Chemistry and Biochemistry of a Leading Anticancer Drug.* New York: Wiley-VCH, 1999, pp. 73–134; and Lambert, B., Segal-Bendirdjian, E., Esnault, C., Le Pecq, J.-B., Roques, B. P., Jones, B., and Yeung, A. T. Recognition by the DNA repair system of DNA structural alterations induced by reversible drug–DNA interactions. *Anti-Cancer Drug Des.* 5:43, 1990.

usual structure at the junction between the A-tract and general-sequence DNA cannot yet be discounted. DNA bending appears to be a fundamental element in the interaction between DNA sequences and proteins that catalyze central processes, such as replication, transcription, and site-specific recombination. Bending induced by interactions of DNA with enzymes and other proteins, such as histones, does not require the exacting nucleotide sequence conditions that are needed for bending of protein-free DNA. Bending also occurs because of photochemical damage or mispairing of bases and serves as a recognition signal for initiation of DNA repair. Antitumor antibiotics that produce bent structures are discussed in Clinical Correlation 2.3.

Cruciform DNA

Inverted repeats are quite widespread within the human genome and are often found near putative control regions of genes or at origins of DNA replication. It has been speculated that inverted repeats may function as molecular switches for replication and transcription. Disruption of hydrogen bonds between the complementary strands and formation of intrastrand hydrogen bonds within the region of an

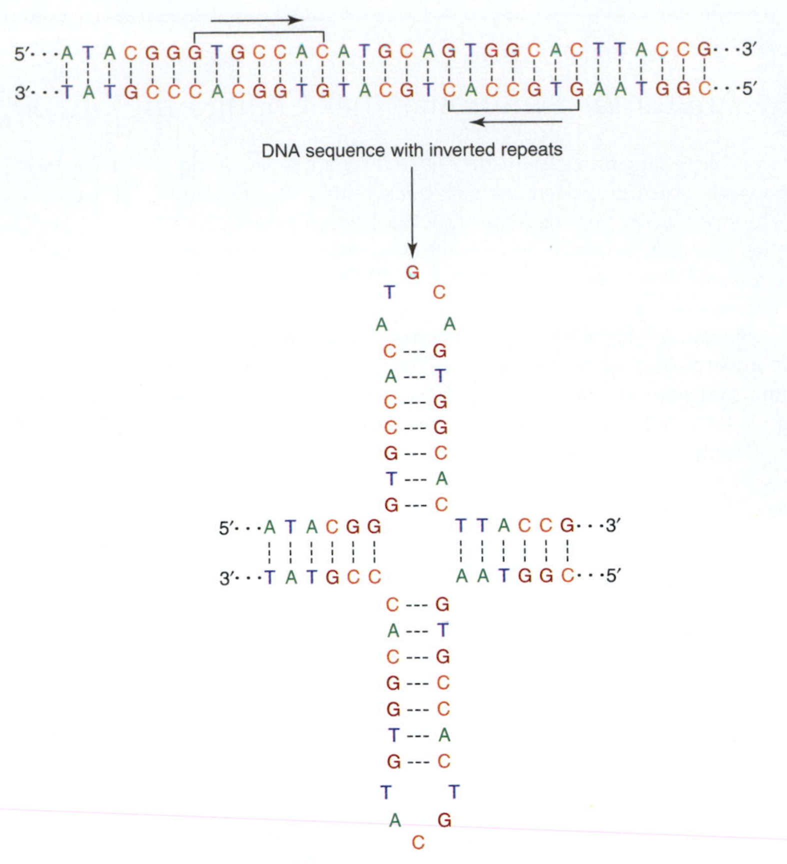

FIGURE 2.28
Formation of cruciform structures in DNA.
The existence of inverted repeats in double-stranded DNA is a necessary but not a sufficient condition for the formation of cruciform structures. In relaxed DNA, cruciforms are not likely to form because the linear DNA accommodates more hydrogen-bonded stacked base pairs than the cruciform structure, making the formation of the latter thermodynamically unfavored. Unwinding is followed by intrastrand hydrogen bond formation between the two symmetrical parts of the repeat to produce the cruciform structure. Formation of cruciform structures is not favored over DNA regions that consist of mirror repeats because such cruciforms would be constructed from parallel rather than antiparallel DNA strands. Instead, certain mirror repeats tend to form triple helices.

inverted repeat produce a cruciform structure (Figure 2.28). The loops generated by cruciform formation require the unpairing of 3–4 bases at the end of the "hairpin." Depending on the sequence, these structures may be only slightly destabilizing because residues in the loop can remain stacked at the end of the helix. Overall, a biological function for cruciforms is only circumstantial and has not been established.

Triple-Stranded DNA

Work in the 1950s and 1960s with polynucleotides showed that some combinations of sequences formed triple-stranded complexes rather than the expected double helices. For example, poly(dA) and poly(dT) formed complexes with a 1:2 stoichiometry. An important clue to the structure of these complexes was the fact that at least one strand of the complex must contain a homopurine sequence. Even when participating in Watson–Crick base pairing, purines possess two potential hydrogen bonding sites in the major groove: N-7 and O-6 for guanine, N-7 and 6-NH_2 for adenine. Nucleoside bases in the major groove with appropriate patterns

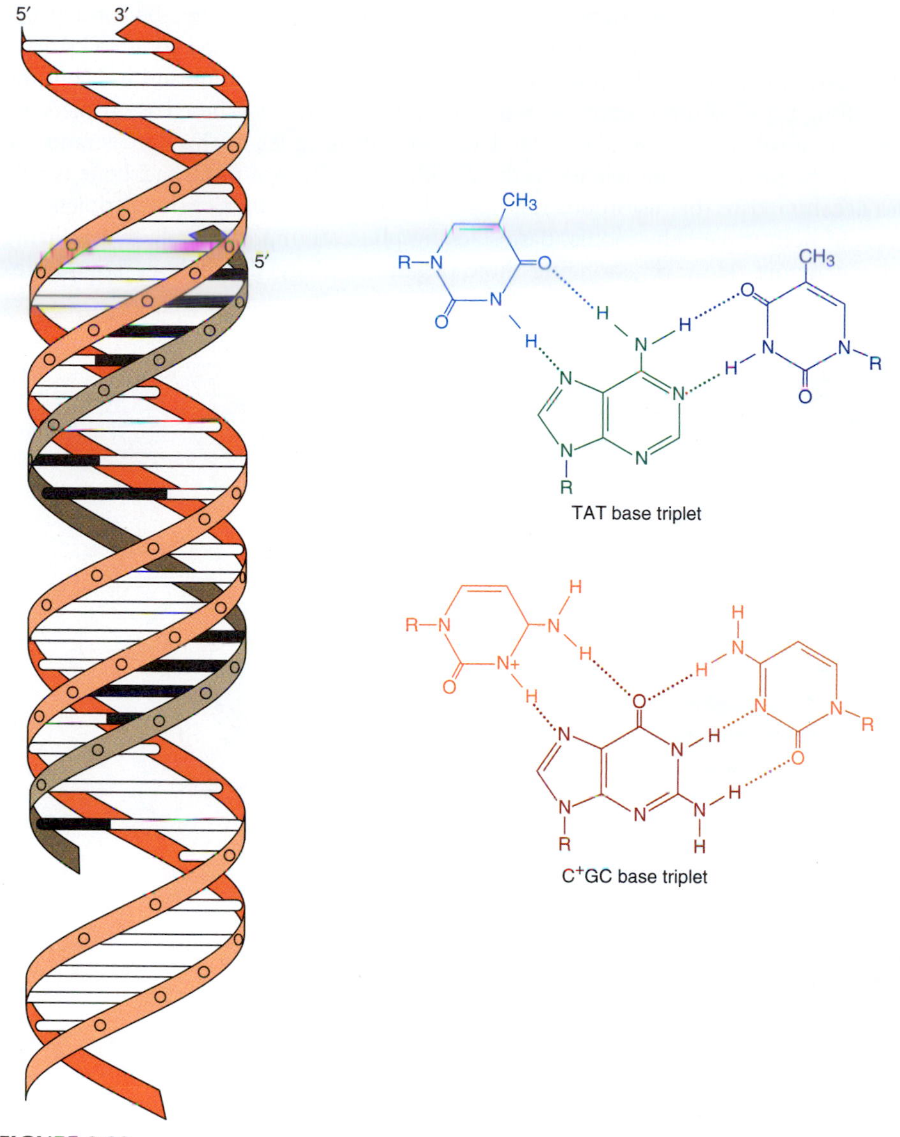

FIGURE 2.29
Hoogsteen triple helix.
Triple helices can be formed by formation of Hoogsteen hydrogen bonds between the homopurine strand of a double helix and a parallel homopyrimidine strand. The resulting isomorphous base triplets, TAT and C^+GC, provide for sequence selective binding.

of hydrogen-bond donors and acceptors can form selective **Hoogsteen** (Figure 2.29) or **reversed-Hoogsteen** (Figure 2.30) base triplets. For example, a thymine can selectively form two Hoogsteen hydrogen bonds to the adenine of an A-T pair. Likewise, a protonated cytosine can form two Hoogsteen H bonds with the guanine of a G-C pair, resulting in a base triplet isomorphous to the T-A-T. The pK_a of cytidine is approximately 4.5, and triple helices containing C-G-C triplets show a strong dependence on pH. However, the templating effect of the triplet raises the apparent pK_a of the cytosine, making it easier to protonate this residue, even in solutions that are only mildly acidic (pH 6). When binding to a homopurine-homopyrimidine (pu · py) sequence of DNA, the third strand orients itself parallel to the purine-rich strand and binds with high sequence selectivity (Figure 2.29). The unique Hoogsteen hydrogen bonding patterns of guanine and adenine provide for specificity similar to the Watson–Crick pairs. Reversed-Hoogsteen hydrogen bonding is also possible if the orientation of the third strand is reversed (i.e., antiparallel to the purine strand, Figure 2.30). Selective triplets can be formed between an adenine and the A-T pair as well as between a guanine and a G-C pair. A reversed-Hoogsteen triplet could also be formed between thymine and A-T, but the resulting triplet is not isomorphous to the pu·pu·py triplets. These base triplets are not isomorphous; that is, the sugar 1′-carbons of residues in the third strand are not in the same location relative to the double helix. However, the backbone is able to accommodate the distortions that result from incorporation of these triplets.

In both triple helix motifs, base stacking plays an important role in stabilizing the structure. However, bringing three negatively charged backbone strands to-

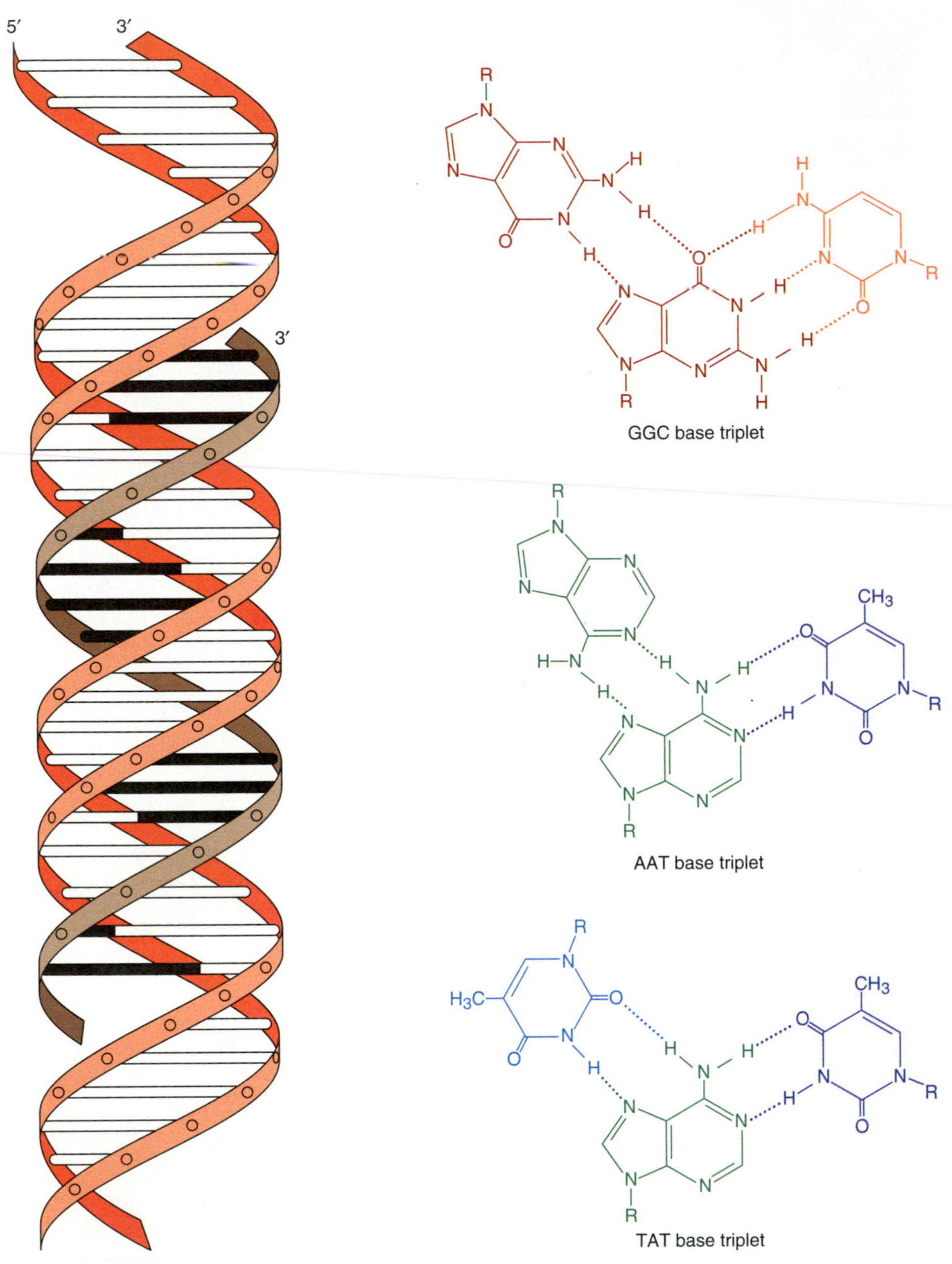

FIGURE 2.30
Reversed Hoogsteen triple helix.
Triple helices can be formed by antiparallel binding of an oligonucleotide to the homopurine strand of a Watson–Crick double helix. Reversed Hoogsteen hydrogen bonding produces three possible triplets: GGC, AAT, and TAT. These triplets are not isomorphous because the sugars (represented by R) are positioned differently with respect to the Watson–Crick base pair.

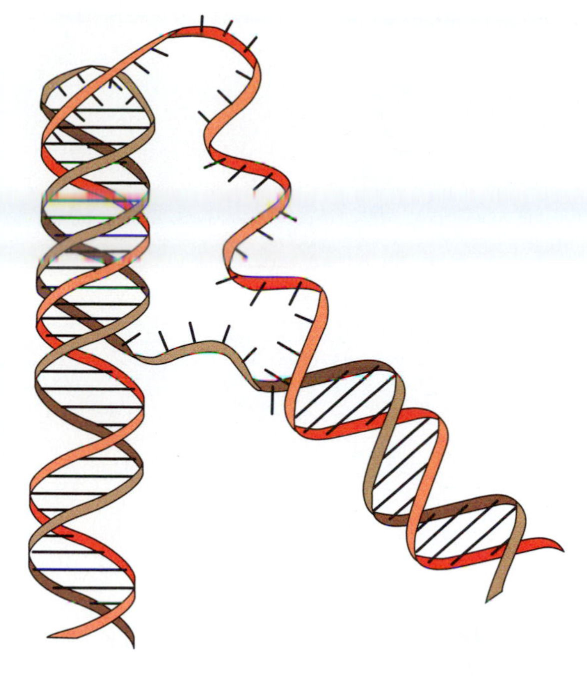

FIGURE 2.31
Intramolecular triple helices—H-DNA.
Polypurine–polypyrimidine regions of DNA with a mirror repeat symmetry can form an intramolecular triple helix in which the third strand lays in the major groove, whereas its complementary strand acquires a single-stranded conformation.
Redrawn from Sinden, R. R. DNA Structure and Function. *New York: Academic Press, 1994.*

gether in a triple helix results in a significant increase in electrostatic repulsion. Thus the triple-helical complex is less stable than the associated Watson–Crick double helix. Furthermore, triple helix formation is highly salt dependent. In particular, the presence of Mg^{2+} and other multivalent cations stabilizes the triple helix by shielding the phosphate charges.

Intramolecular triple helices can be formed by disruption of double-helical DNA with polypurine sequences in mirror repeats. A mirror repeat is a region such as AGGGGA that has the same base sequence when read in either direction from a central point. Refolding generates a triple-stranded region and a single-stranded loop in a structure called **H-DNA** (Figure 2.31). Even though the formation of H-DNA is thermodynamically unfavorable, primarily because of a loss of stacking interactions in the single-stranded loop, intramolecular triple helices have been detected in cellular DNA when under superhelical stress. Apparently, DNA supercoiling provides the energy to drive the unwinding of DNA that is necessary for the formation of the triple helix. Triple-strand formation produces a relaxation of negative supercoils. Also, the binding of proteins to the single-stranded DNA may further stabilize the H-DNA structure and prevent degradation of the loop by nucleases.

Many sequences in eukaryotic genomes have the potential to form triple-stranded DNA structures. Such regions occur with much higher frequency than expected from probability considerations alone. Polypurine tracts over 25 nucleotides long constitute as much as 0.5% of some eukaryotic genomes. These potential triple helical regions are especially common near sequences involved in gene regulation. Because of this fact, it has been proposed that H-DNA may play a role in control of transcription (synthesis of RNA from DNA). The abilities of triple-helical DNAs to interfere with transcription have also led to efforts to use intermolecular triple helices to control artificially RNA synthesis and subsequent protein synthesis. The therapeutic potential of oligonucleotides capable of forming triplex DNA with segments of DNA having Hoogsteen base pairing potential is discussed in Clinical Correlation 2.4. A multiplicity of other potential biological tasks has been proposed for triple-helical DNA, including possible roles in initiation and termination of replication and recombination. However, there is not definitive evidence to prove a biological role for triple-helical DNA.

CLINICAL CORRELATION 2.4
Hereditary Persistence of Fetal Hemoglobin

Hereditary persistence of fetal hemoglobin (HPFH) is a group of conditions in which fetal hemoglobin synthesis is not turned off with development but continues into adulthood. The homozygous form of the disease is extremely uncommon, being characterized by changes in red blood cells similar to those found in heterozygous β-thalassemia. HPFH, in either the homozygous or heterozygous state, is associated with mild clinical or hematologic abnormalities. Mild musculoskeletal pains may occur infrequently, but HPFH patients are generally asymptomatic.

The disease results from failure in control of transcription from human Gγ- and Aγ-globin genes. Affected chromosomes fail to switch from γ- to β-chain synthesis. Expression of these genes appears to be affected substantially by formation of an intramolecular DNA triplex structure located about 200 bp upstream from the initiation site for transcription of genes, specifically between positions −194 and −215.

Hemoglobin genes of patients contain mutations in positions −195, −196, −198, and −202. Mutations at −202 involve changes from C to G and C to T, at −198 from T to C, at −196 from C to T, and at −195 from C to G. These mutations influence the stability of the intramolecular DNA triple helix.

In general, the presence of polypurine–polypyrimidine sequences sufficiently long to form intramolecular triple helices tends to repress transcription, while short polypurine–polypyrimidine segments that are unable to induce triple helix formation have no effect on transcription. In the case of HPFH, a remarkable correspondence is noted between base changes that destabilize formation of the triple helix and presence of the genetic disease.

Ulrich, M. J., Gray, W. J., and Ley, T. J. An intramolecular DNA triplex is disrupted by point mutations associated with hereditary persistence of fetal hemoglobin. *J. Biol. Chem.* 267:18649, 1992; and Bacolla, A., Ulrich, M. J., Larson, J. E., Ley, T. J., and Wells, R. D. An intramolecular triplex in the human gamma-globin 5′-flanking region is altered by point mutations associated with hereditary persistence of fetal hemoglobin. *J. Biol. Chem.* 270:24556, 1995.

Four-Stranded DNA

Guanine nucleotides and highly G-rich polynucleotides form novel tetrameric structures called **G-quartets.** These structures feature a planar array of guanines connected by Hoogsteen hydrogen bonds (Figure 2.32). The quartets can stack upon each other to form a multilayered structure. A variety of four-stranded structures is possible, differing in the relative orientations of the strands. When four separate polynucleotides interact, a parallel arrangement of all four strands is possible (Figure 2.33). There are also three possible isomeric antiparallel arrangements; these are preferred where the **tetraplex** involves intramolecular folding when one polynucleotide supplies two or more strands of the complex. All of the structures are stabilized by the presence of metal cations, especially sodium and potassium. These cations interact with guanine oxygens (O-6), binding in the hole formed in the center of the quartet structure. Metal ions can be positioned in the quartet plane or between two adjacent planes. In the latter arrangement, eight oxygens are available to coordinate the ion. The selective stabilization by Na^+ and K^+ compared to Li^+ or Cs^+ is due to the abilities of the ions to fit into the cavity formed by the quartet structure: lithium ions are too small and cesium ions are too large to bind effectively. Uniquely, these ions are completely buried within the polynucleotide, and only those ions on the ends of a tetraplex structure can remain coordinated to any water molecules. In addition to the entropically favored release of water molecules from the metal ions, the ions reduce the overall amount of negative charge that accumulates from the four-stranded structure.

While G-tetraplexes have been observed by X-ray diffraction and NMR spectroscopy, their existence *in vivo* has not been proved. However, there are biological processes where G-rich polynucleotides are important, and a possible role for G-tetraplexes has been proposed. The ends of eukaryotic chromosomes (**telomeres**) contain repetitive G-rich sequences. Human telomeres contain 800–2400 copies of

FIGURE 2.32
The structure of a G-quartet.
The four coplanar guanines form a tetrameric structure by formation of Hoogsteen hydrogen bonds. The cavity in the center of the quartet can accommodate a sodium or potassium ion with coordination by the four O-6 oxygens.

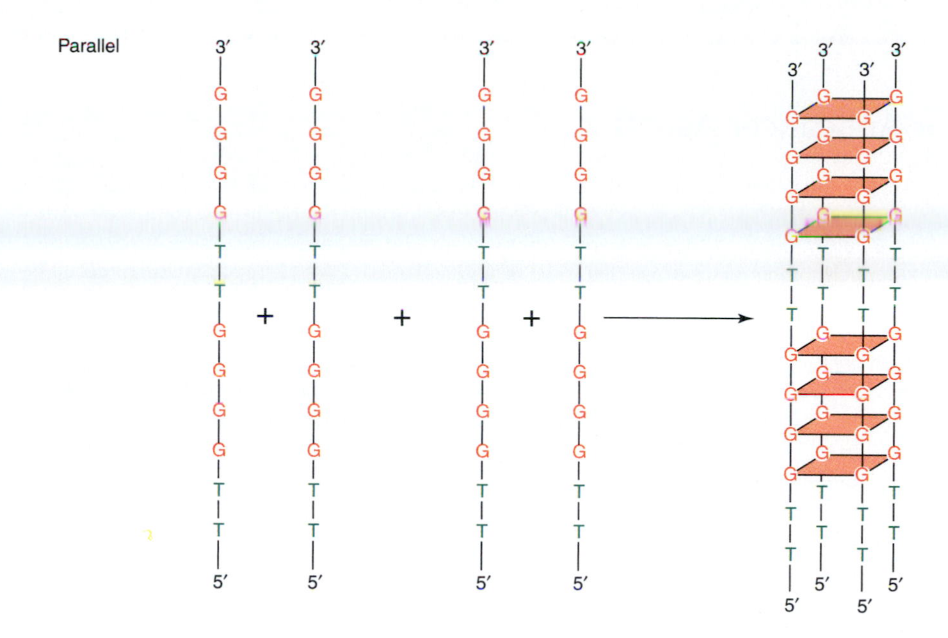

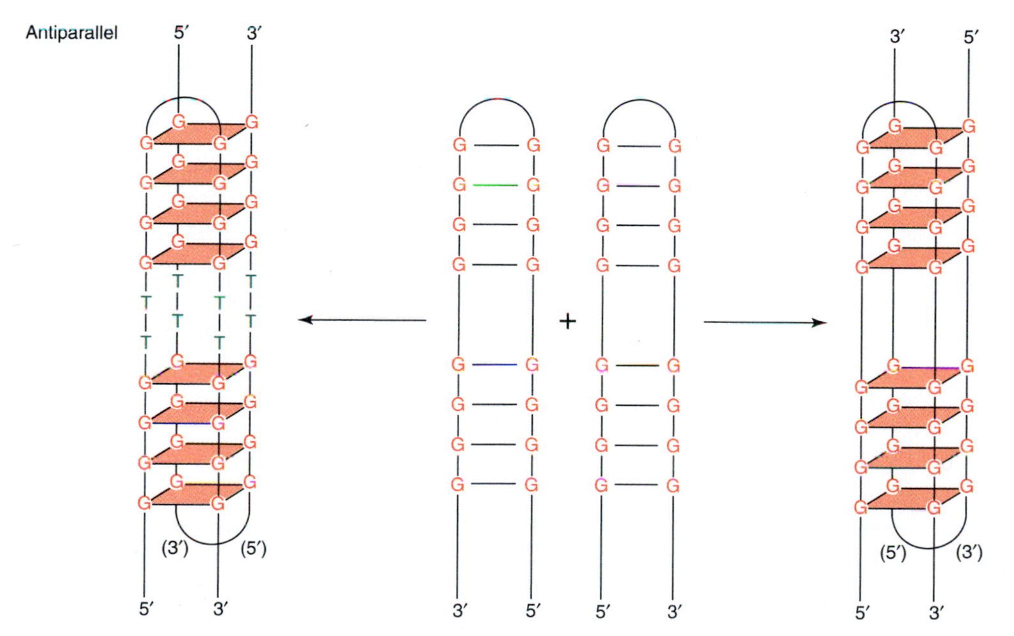

FIGURE 2.33
G-quadruplex DNA.
Four-stranded structures can arise from stacking of G-quartets. Quadruplex structures in which all four strands are parallel can form from four single-stranded G-rich tracts of polynucleotides. Several isomeric quadruplexes can form some strands oriented antiparallel to the others. This arrangement frequently occurs when one polynucleotide provides two or more strands of the quadruplex structure and may be formed by the G-rich sequences of telomeric DNA.
Redrawn from Sinden, R. R. DNA Structure and Function. *New York: Academic Press, 1994.*

the hexameric repeat sequence $(TTdAdGdGdG)_n$ and terminate in a single-stranded overhang that is roughly 150 nucleotides long. *In vitro* studies indicate that oligonucleotides with this sequence can form tetraplex structures. It is not known what role, if any, tetraplexes play in telomere functions. However, telomeres are attracting attention as targets for new anticancer chemotherapies (see Clin. Corr. 2.5). In addition, G-tetraplexes have been implicated in recombination of immunoglobulin genes and in dimerization of double-stranded genomic RNA of the human immunodeficiency virus (HIV).

Investigation of the potential quadruplex structures of the guanine-rich sequences of telomeres has led to the identification of another four-stranded DNA structure involving the complementary cytosine-rich sequences. X-ray diffraction and nuclear magnetic resonance (NMR) experiments have shown that C-rich

CLINICAL CORRELATION 2.5
Telomerase as a Target for Anticancer Agents

Telomeres, the ends of linear eukaryotic chromosomes, are critical for maintaining the stability of the genome. Telomeres are progressively shortened during each cycle of cell division. Upon reaching a critical length, programmed cell death (apoptosis) occurs. Ribonucleoprotein telomerase acts to maintain or lengthen the telomeres but is not active in normal somatic cells. Telomerase activity has been detected in most tumor cell lines and may be responsible for their immortalization, and increased telomerase activity in tumors can be correlated to poorer clinical prognoses. Thus this enzyme is an attractive target for anticancer chemotherapies.

Two approaches are being examined for selective inhibition of telomerase. The first involves targeting of the RNA-containing portion of the enzyme. This RNA serves as the template for extension of the telomeric repeat sequence. Modified nucleic acids such as 2′-*O*-methyl RNAs and peptide nucleic acids have been shown to bind to telomeric RNA in immortal human cells and inhibit telomerase activity. This treatment ultimately resulted in cell death. These nucleic acid derivatives were chosen because of their resistance to nuclease degradation and high affinities for forming double-helical complexes with RNA. A second approach toward telomerase inhibition involves use of drugs that bind to G-quadruplex DNA. Large aromatic molecules such as porphyrins and anthraquinones selectively bind and stabilize G-quadruplex DNA structures. Although *in vivo* existence of G-quadruplexes has not been demonstrated, some of these compounds have been shown to inhibit telomerase activity. Further studies will be necessary to determine antitumor activities of these compounds.

Herbert, B.-S., Pitts, A. E., Baker, S. I., Hamilton, S. E., Wright, W. E., Shay, J. W., and Corey, D. R. Inhibition of human telomerase in immortal human cells leads to progressive telomere shortening and cell death. *Proc. Natl. Acad. Sci. USA* 96:14276, 1999; and Sun, D., Thompson, B., Cathers, B. E., Salazar, M., Kerwin, S. M., Trent, J. O., Jenkins, T. C., Neidle, S., and Hurley, L. H. Inhibition of human telomerase by a G-quadruplex-interactive compound. *J. Med. Chem.* 40:2113, 1997.

oligodeoxyribonucleotides can form a completely novel type of four-stranded complex called **i-DNA** (Figure 2.34). This structure is composed of two pairs of duplexes that mutually intercalate to form a four-stranded structure. The duplex region is composed of two parallel strands joined by C-C base pairs. Instead of stacking with neighboring base pairs, each C-C pair stacks on C-C pairs from a second duplex, forming an interdigitated four-stranded complex. The duplexes are oriented antiparallel to each other. The structure is stabilized by protonation of one of the cytosine bases at N-3. This allows for the formation of three hydrogen bonds in the C-C base pair, and the positive charges partly counteract the interstrand phosphate–phosphate repulsions. Although C-rich telomeric polynucleotide sequences have been found to form i-DNA complexes, their considerably higher stability at pH 5–6 casts doubt on their physiological relevance. However, the highly unusual architecture demonstrates the structural variability that is possible in nucleic acids.

Slipped DNA

DNA regions with direct repeat symmetry can form structures known as slipped, mispaired DNA (SMP-DNA). Their formation involves the unwinding of the double helix and realignment and subsequent pairing of one copy of the direct repeat with an adjacent copy on the other strand. This realignment generates two single-

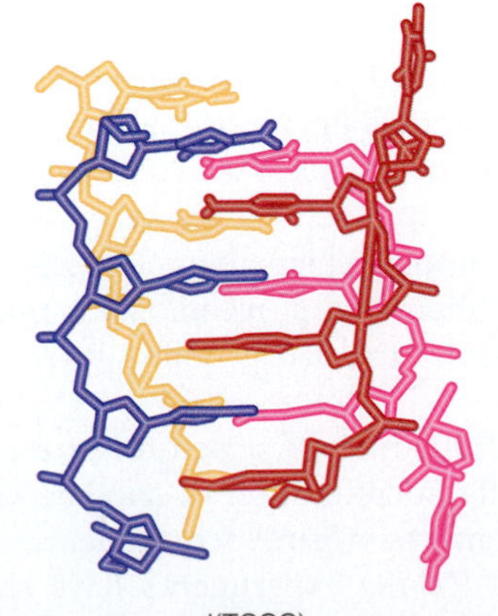

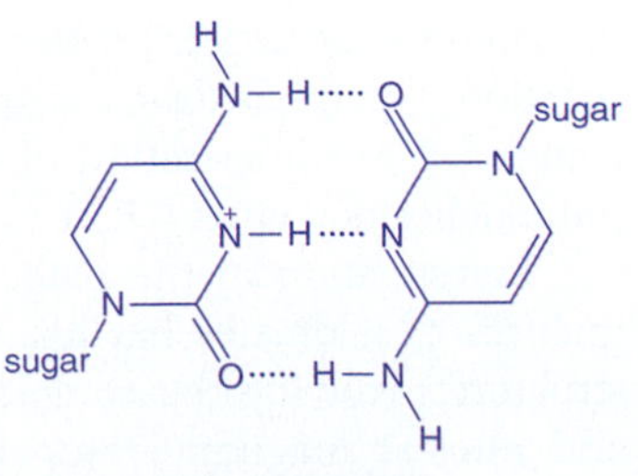

FIGURE 2.34
i-DNA.
Another four-stranded DNA that may be related to telomeric DNA is that formed by C-rich strands. i-DNA consists of two interdigitated pairs of duplexes containing C-C base pairs. The favorable protonation of one of the cytosines in each pair explains the greater stability of this structure at acidic pH values.
Redrawn from Feigon, J. Curr. Biol. *3:611, 1993.*

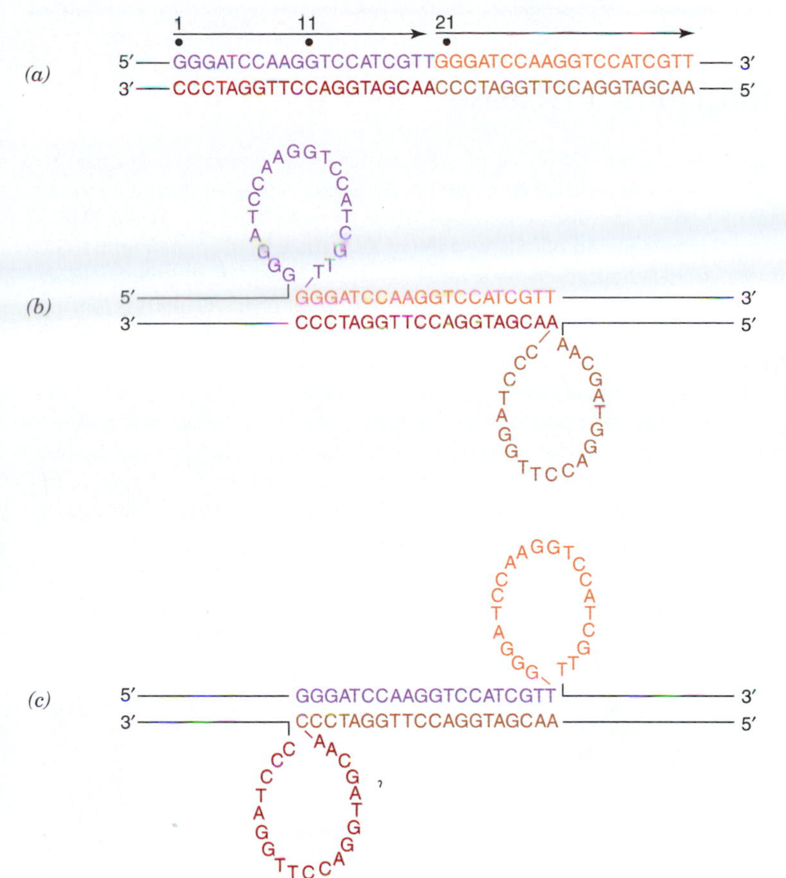

FIGURE 2.35
Slipped, mispaired DNA.
Presence of two adjacent tandem repeats (*a*) can give rise to one of two isomers of slipped, mispaired DNA. In one of these isomers (*b*) the second copy of the direct repeat in top strand pairs with first copy of repeat on bottom strand. Pairing of first copy of direct repeat in top strand with second copy of direct repeat in bottom strand produces second isomer (*c*). A pair of single-stranded loops is generated in both isomers.

stranded loops (Figure 2.35). Two isomeric structures of an SMP-DNA are possible. One generates a loop that consists of the 5′-direct repeat in both strands, and the other produces loops of the 3′-direct repeat. Although SMP-DNA has not yet been identified *in vivo,* genetic evidence suggests that this type of DNA is undoubtedly involved in spontaneous frameshift mutagenesis that is manifested as base addition or deletion occurring within runs of single bases (Figure 2.36). Deletions and du-

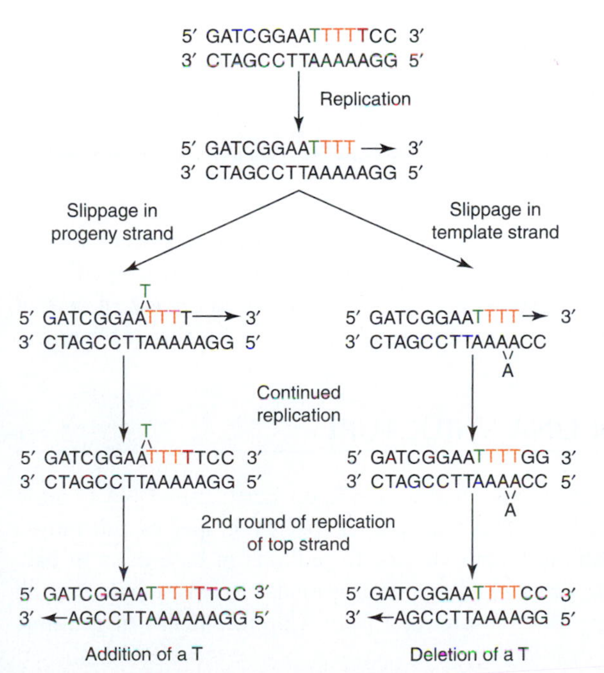

FIGURE 2.36
Frameshift mutagenesis by DNA slippage.
DNA replication within a run of a single base can produce a single-base frameshift. In the example shown here, a run of five As is replicated and, depending on whether a slippage occurs in progeny strand or template strand, a T may be added or deleted from the DNA.

CLINICAL CORRELATION 2.6
Expansion of DNA Triple Repeats and Human Disease

The presence of reiterated three-base-pair DNA sequences has been noted in a number of human genetic diseases including fragile X syndrome, myotonic dystrophy, X-linked spinal and bulbar muscular atrophy (Kennedy's syndrome), spinocerebellar ataxia, and Huntington's disease. These diseases are associated with expansion of certain triplet nucleotide repeats that appear to be overrepresented in the human genome. Repeats can be present in different locations near or within the associated gene. In all cases, expansion of the triplet interferes with normal functioning of the protein. In many cases, a loss of protein function occurs, but in some cases, the gain of a deleterious function occurs. Fragile X syndrome, a leading cause of mental retardation, is characterized by expansion of a GCC triplet. Diseases associated with expansion of triplets are characterized by an increase in severity of the disease with each successive generation, which is known as anticipation. Normally, about 30 copies of this triplet are present on the 5′-side of a gene associated with the disease, the *FMR-1* gene. The site of the repeat is expanded to as many as 300 copies in males that carry fragile X gene mutations but have no symptoms of the disease. Offspring of male carriers who express the disease can have a remarkable expansion of the triplet repeat, up to thousands of copies. The disease develops when normal expression of the *FMR-1* gene is turned off. Methylation of CpG dinucleotides present in CGG triplets appears to be associated with shutting off the *FMR-1* gene.

Triplet expansion may result from slipped mispairing during DNA synthesis. Because of massive amplification that characterizes the diseases associated with triplet expansion, repeated or multiple slippage would have to be involved to explain the high degree of expansion. One possible mechanism for expansion involves slippage of nascent DNA during lagging strand synthesis. This process may be aided by formation of a stable hairpin structure by the slipped loop. Repetition of this process leads to accumulation of large numbers of triplet repeats that result in disease.

Timchenko, L. T. and Caskey, C. T. Triplet repeat disorders: discussion of molecular mechanisms. *Cell. Mol. Life Sci.* 55:1432, 1999. Figure redrawn from Wells, R. D. *J. Biol. Chem.* 271:2875, 1996.

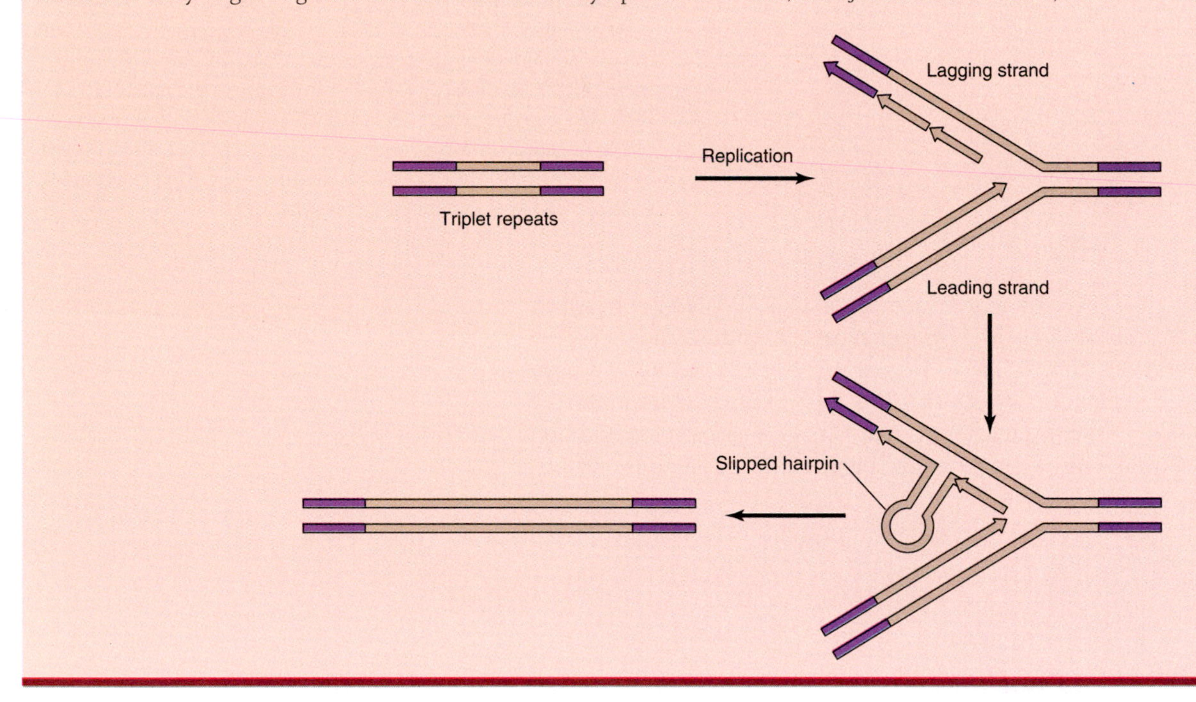

plications of DNA segments that are longer than a single base can occur during DNA replication between direct repeats, causing slipped-looped structures. Duplication of certain simple triplet repeats that are implicated as the basis of several human genetic diseases (see Clin. Corr. 2.6) may also occur by this mechanism.

2.4 | HIGHER ORDER DNA STRUCTURE

With the exception of RNA-containing viruses, all life on Earth uses DNA to store genetic information. The length of DNA varies from species to species and ranges from a few thousand base pairs for small viruses, to millions of base pairs in bacteria, and to billions of base pairs in plants and animals. In all organisms, the **contour length** (length of the DNA assuming a B-form double helix) of genomic

DNA is usually much larger than the size of the cell containing the DNA. For example, a medium-sized virus, such as λ phage, contains 4.8×10^4 bp of DNA, which has a length of 16.5 μm. However, the length of the viral particle is only 0.19 μm. The common bacterium *Escherichia coli* is approximately 2 μm long and possesses a single chromosome with 4.6×10^6 bp and a contour length of 1.5 mm. The size of the DNA of higher cells is even larger. A single diploid human cell contains 6×10^9 bp of chromosomal DNA packaged in 46 chromosomes. The contour length of this DNA approaches 2 meters, all of which is contained within a nucleus, about 10 μm in diameter. It is clear that the DNA of all organisms must be exquisitely packaged.

Genomic DNA May Be Linear or Circular

With the exception of a few small bacteriophages such as ϕX174 that can acquire a single-stranded form, most DNAs exist as double-helical complexes. Depending on the source of the DNA, the complexes can be linear or circular. For example, DNAs of several small viruses are **linear double-stranded helices.** Some DNAs have naturally occurring interior single-stranded breaks. Breaks found in natural bacteriophage DNA result mostly from broken phosphodiester bonds, although occasionally a deoxyribonucleoside residue may be missing. DNA of T5 phage consists of one intact strand and a complementary strand, which is really four well-defined complementary fragments ordered perfectly along the intact strand. A similar regularity in the points of strand breaks is noted, for example, in *Pseudomonas aeruginosa* phage B3, but generally interior breaks seem to be randomly distributed. The double-helical structure is maintained because the breaks in one strand are generally in different locations from breaks in the complementary strand. Chromosomes in higher organisms can also be linear. For example, each of the 46 chromosomes in a diploid human cell is a linear DNA and each is comprised of two very long DNA strands in a double-helical complex.

Circular DNA results from the formation of phosphodiester bonds between the 3′- and 5′-termini of linear polynucleotides. The circular nature of the single-stranded phage ϕX174 DNA was suspected from studies that showed that no ends were available for reactions with exonucleases. Sedimentation studies also revealed that enzymatic cleavage at an interior nucleotide residue yielded only one polynucleotide. These results were later confirmed by direct observation with electron microscopy.

Double-helical DNAs can also be circularized by joining each of the polynucleotide strands through phosphodiester bonds. This process has been observed for the linear DNA of λ phage. Before entering *E. coli,* the host cell, the phage DNA is linear and double-stranded with single-stranded overhangs on the 5′-terminus of each strand. These overhangs are approximately 20 nucleotides in length and have complementary sequences. Upon infection of the host, circularization occurs by hybridization of the ends to each other. Formation of phosphodiester bonds between the 3′- and 5′-ends of each strand by the enzyme **DNA ligase** produces a covalently closed circle (Figure 2.37). Obviously, the strands of a circular DNA cannot be irreversibly separated by denaturation because they exist as intertwined closed circles. The absence of 3′- or 5′-termini provides an evolutionary advantage because it endows the circular DNA with complete resistance toward exonucleases, which improves the longevity of DNA.

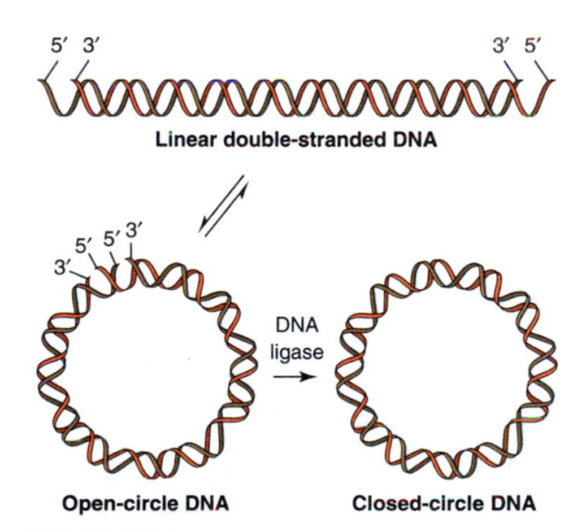

FIGURE 2.37
Circularization of λ DNA.
DNA of bacteriophage λ exists in linear and circular forms, which are interconvertible. Circularization of λ DNA is possible because the 5′-overhangs of the linear form are complementary sequences.

DNA from a number of sources occurs naturally in a circular form. Many viruses have circular DNA at some point in their life cycle. Most (if not all) DNA in bacteria exists as closed circles. This includes the bacterial chromosomes as well as smaller extrachromosomal DNA called **plasmids.** Plasmid DNAs are generally a few thousand base pairs long and encode accessory genes such as antibiotic resistance genes. Plasmids are maintained and replicated separately from chromosomal DNA. A bacterium can contain hundreds of copies of a given plasmid. Other organisms, such as yeasts, also carry circular plasmids. Mitochondria and chloroplasts in eukaryotic cells contain circular DNAs. These DNAs are typically similar in size to bacterial chromosomes ($200–1500 \times 10^3$ bp) and encode for unique proteins used by the organelles. The presence of an independent "genome" and a marked similarity to cyanobacteria have lead to the hypothesis that these organelles were derived from bacteria that formed a symbiotic relationship with eukaryotic cells billions of years ago.

Circular DNA Is Superhelical

Circular double-stranded DNA is formed by ligating together the free termini of a linear DNA. If no other manipulations are introduced, the resulting circular DNA will be **relaxed;** that is, it will have the thermodynamically favored structure of B-DNA. Surprisingly, relaxed DNA has greatly reduced activity in a number of crucial biological processes including replication, translation, and recombination. The biologically active form of DNA is **superhelical,** a topologically strained isomer created by either underwinding or overwinding the double helix. Before ligation of the two ends of a linear B-form DNA to form a relaxed circle, the double helix contains about 10.5 bp per complete turn. If the DNA is untwisted before sealing the circle, the resulting structure will be strained (Figure 2.38). The untwisting of the helix reduces the total number of helical turns present in the circular structure. There are two ways that the underwinding of the helix can be accommodated. One is to disrupt the base pairing over a small region of the DNA, producing a pair of single-stranded loops in a relaxed circular structure. Loss of base stacking in this structure is energetically unfavorable. However, if all base stacking is maintained, the underwinding generates a torsional strain in the backbone of the double helix. To relieve this strain the entire circular DNA can twist in a direction opposite to the one in which it was initially rotated. This results in the formation of a coiled coil, better known as a **superhelix.** Underwound DNA is said to be negatively supercoiled, and

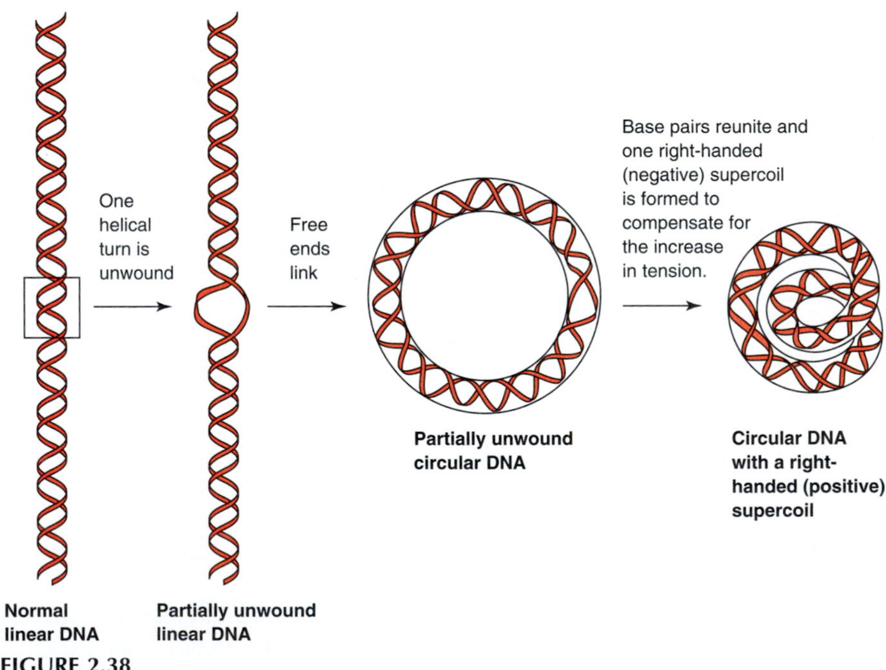

FIGURE 2.38
Negative DNA supercoiling.
Right-handed supercoils (negatively supercoiled DNA) are formed if relaxed DNA is partially unwound. Unwinding may lead to a disruption of hydrogen bonds or alternatively produce negative supercoils. The negative supercoils are formed to compensate for the increase in tension that is generated when disrupted base pairs are reformed.
Redrawn from Darnell, J., Lodish, H., and Baltimore, D. Molecular Cell Biology. *New York: Freeman, 1986.*

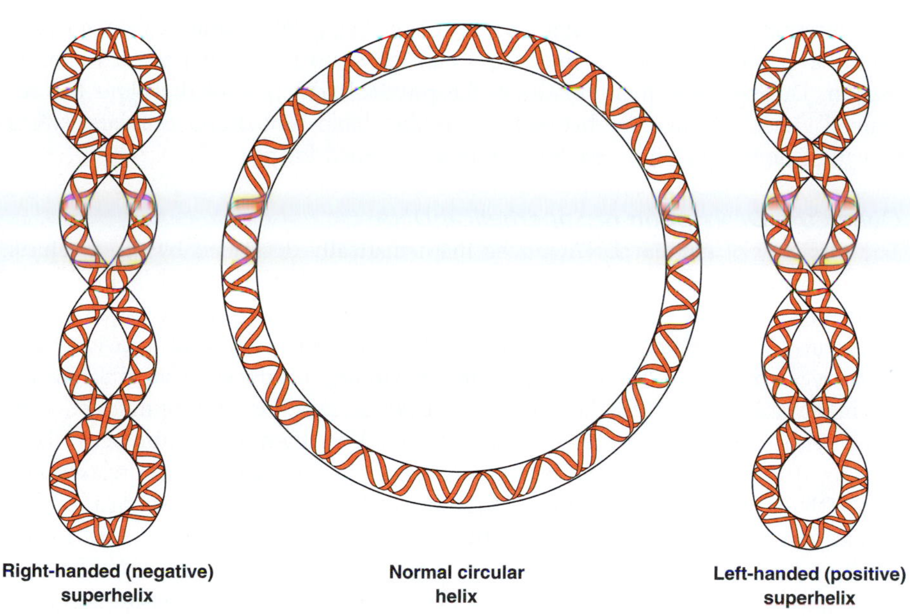

FIGURE 2.39
Relaxed and supercoiled DNA.
Relaxed DNA can be converted to either right- or left-handed superhelical DNA. Right-handed DNA (negatively supercoiled DNA) is the form normally present in cells. Left-handed DNA may also be transiently generated as DNA is subjected to enzymatically catalyzed transformations (replication, recombination, etc.) and it is also present in certain bacterial species. The distinctly different patterns of folding for right- and left-handed DNA superhelices are apparent in this representation.
Redrawn from Dornell, J., Lodish, H., and Baltimore, D. Molecular Cell Biology. *New York: Freeman, 1986.*

the resulting superhelix is right-handed (Figure 2.39). Similarly, overwound DNA is positively supercoiled and forms a left-handed superhelix. In general, a dynamically imposed compromise is reached between hydrogen bond disruption and supercoiling. In practice, however, this means that supercoiled DNA may consist of superhelical structures with an enhanced tendency to generate regions with disrupted hydrogen bonding (bubbles).

In circular DNA that is initially relaxed, the transient strand unwinding would tend to introduce compensating supertwists. However, if DNA is superhelical to begin with, the density of the superhelix will tend to fluctuate with the "breathing" of the helix. All naturally occurring DNA molecules are underwound and have a deficit of helical turns. They exist as negative superhelices with a superhelical density that remains remarkably constant among different DNAs. Normally one negative twist is found for every 20 turns of the helix. Superhelical stress from an overwound helix can be relieved by positive supercoiling or by unwinding of a left-handed double helix. Positive supercoils can be generated by specialized enzymes, the topoisomerases, and may be present transiently *in vivo* but are rarely found in cellular DNA.

Positive and negative supercoils can, in principle, coexist temporarily within a DNA molecule. If equal numbers of negative and positive supercoils are present, the DNA molecule must still be considered "relaxed" because it may return to a relaxed state without the breaking of phosphodiester bonds. A rubber band, which in its normal unstrained form might be visualized as a circular relaxed structure (without supercoils), can be used as such a model (Figure 2.40). Grasping this band firmly at opposite sides and twisting one side of the band generates a structure characterized by two topological domains, with twisting of opposite handedness, that are clearly visible when the two sides are pulled apart. If the opposite sides are brought back close together, each domain becomes supertwisted; that is, each domain generates a supercoil. This requires an input of energy since the supertwisted state does

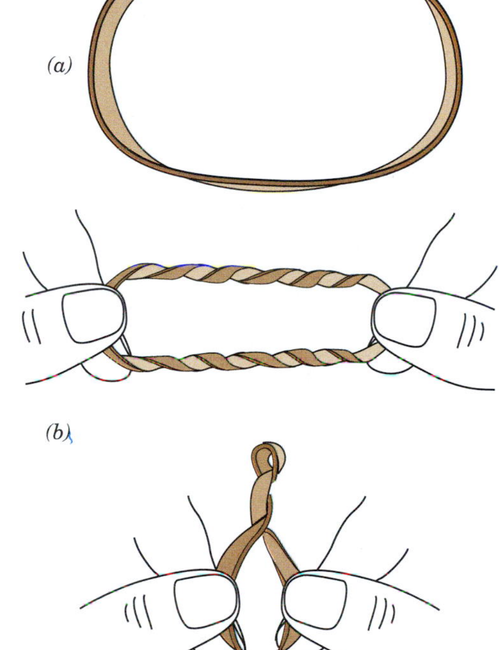

FIGURE 2.40
Superhelical model for DNA.
A rubber band represents the topological properties of double-stranded circular DNA. The relaxed form of the band, shown in (*a*), has been twisted to generate two distinct domains, separated by the pair of "thumb–forefinger anchors," as shown in (*b*). Left-handed (counterclockwise) turns have been introduced into the upper section of the band, with compensating right-handed (clockwise) turns present into the bottom section. When the "anchors" are brought into close proximity with each other as shown in (*c*), the upper section that contained the left-handed turns forms a right-handed superhelix. The bottom section produces a left-handed superhelix. Clearly, superhelicity is not the property of a DNA molecule as a whole but rather a property of specific DNA domains.
Redrawn from Sinden, R. R. and Wells, R. D. Curr. Opin. Biotech. *3:612, 1992.*

not represent the low-energy state of the rubber band. When the band is released from the grasp that restrains rotation, it can return to its original relaxed conformation. During these manipulations, the physical structure of the band has remained intact. A difference between the rubber band model and cellular DNA is that the latter exists almost exclusively in supercoiled form.

Geometric Description of Superhelical DNA

The topology of circular DNA can be mathematically described by three values: **linking number, twisting number,** and **writhing number.** The linking number, L, is an integer number defined as the number of times one polynucleotide strand winds around the other. In circular DNAs, this number is constant, provided that no covalent bonds are broken. Right-handed winding produces positive values for L. The number of double-helical turns in a DNA defines the twisting number, T. In a B-DNA, the twisting number can be determined by dividing the number of base pairs by 10.5, the average number of base pairs per helical turn. In a relaxed circular DNA in the B-conformation, one polynucleotide strand twists around the other every 10.5 bp. Thus, in this situation, the linking number is equal to the twisting number ($L = T$). However, this is not always the case. In DNA that has been underwound or overwound before closure into a circular structure, the linking number is changed. Local unwinding or overwinding (if possible) of the double helix relieves the strain and is accompanied by a change in the twisting number such that $\Delta L = \Delta T$. However, as was discussed above, the DNA prefers to incorporate supercoils to relieve the strain while maintaining the B-form structure. The number of supercoils defines the writhing number, W. One supercoil can compensate for each change in the linking number; therefore, if the twisting is not changed, $\Delta L = \Delta W$. However, the circular DNA can compensate for changes in linking by incorporating both changes in twisting and writhing. This relationship can be described by the equation

$$\Delta L = \Delta T + \Delta W$$

For any circular DNA, the linking number will equal the sum of the twisting and writhing numbers,

$$L = T + W$$

Thus DNAs with a specific linking number can acquire various different topological conformations, and different types of superhelices may be formed. However, all conformations with the same linking number are interconvertible without breaking any covalent bonds. Therefore linking number is a constant for any covalently closed circular DNA.

Various forms of supercoiled DNAs can be described using L, T, and W numbers. The mental exercise shown in Figure 2.41 illustrates how these numbers apply. It should be recalled that the turns of the typical double helix are right handed. Therefore, if a hypothetical linear DNA duplex that is 10 turns long ($L = 10$ and $T = 10$) is unwound by, say, one turn, the resulting structure will have the following characteristics: $L = 9$ and $T = 9$. A potentially equivalent structure can be formed if ends of the same hypothetical DNA are secured so that they cannot rotate and the molecule is looped in a counterclockwise manner. Since in this case untwisting is not permitted to occur, the number of helical turns remains unchanged; that is, $T = 10$. However, as a result of "looping" operations, linking number is now reduced by 1; that is, $L = 9$. The structure resulting from this deliberate introduction of a loop is visibly superhelical. Furthermore, the application of the equation that relates values of L, T, and W indicates that W must be equal to -1; that is, the structure is a negative superhelix with one supercoil.

The two structures described above, [$L = 9, T = 9, W = 0$] and [$L = 9, T = 10, W = -1$], have the same linking number and are interconvertible without the disruption of any phosphodiester bonds. The potential equivalence of these two types of structure becomes more apparent when ends of polynucleotides in each struc-

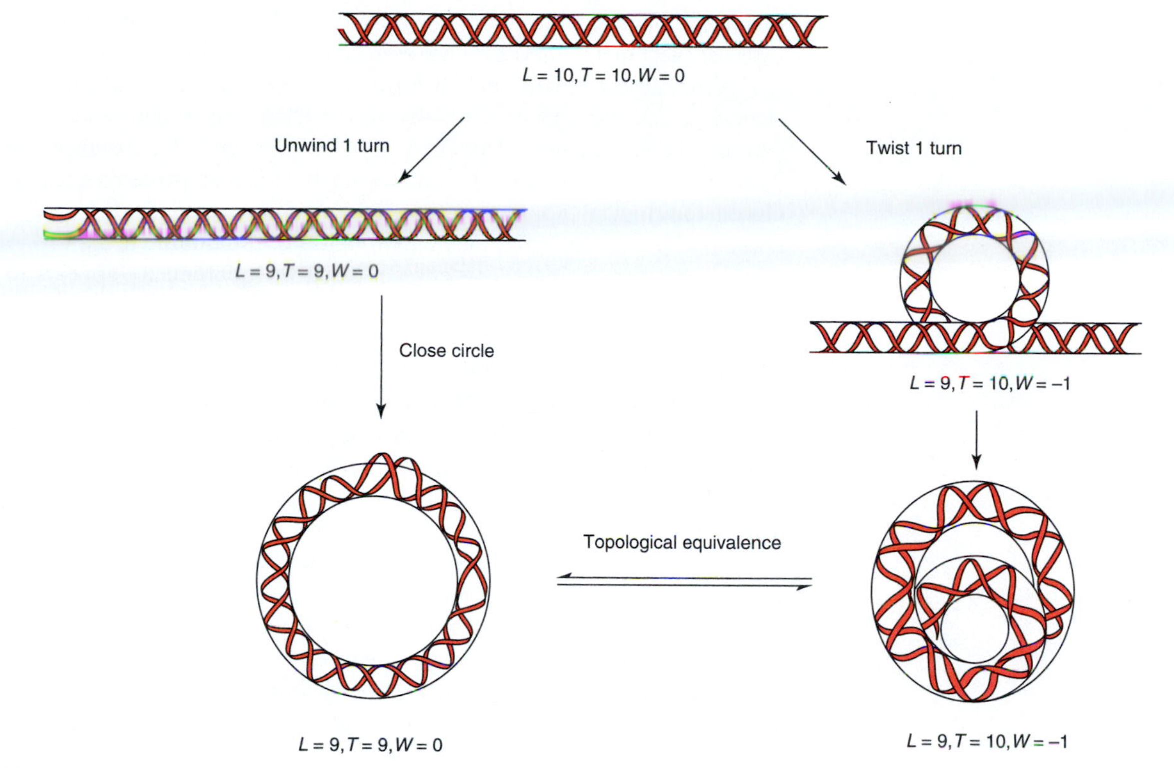

FIGURE 2.41
Relationship of twisting and writhing.
An accurate representation of superhelical DNA structures can be made, using the number of helical turns or twists, T, and the number of supercoils or writhing number, W, along with a third parameter, the linking number, L, as defined in the text. The figure shows ways of introducing one supercoil into a DNA segment of 10 duplex turns and the parameters of the resulting superhelices.
Redrawn with permission from Cantor, C. R. and Schimmel, P. R. Biophysical Chemistry, *Part III. San Francisco: Freeman, 1980. Copyright © 1980.*

ture are joined into a circle without strands being allowed to rotate. Circularization produces an interwound circular structure (a number 8-shaped structure referred to as a **plectonemic** coil) or a doughnut-shaped superhelical arrangement referred to as a **toroidal** turn, both of which are freely interconvertible. An interwound turn, shown in Figure 2.42, can be produced by unfolding a toroidal turn along an axis that is distinct from the supercoil axis.

Real superhelical DNA exists as an equilibrium among many forms that have the same linking number but different numbers of helical turns and supertwists. Although linking number is a constant and an integer, the number of twists can change in positive and negative increments, which are compensated by negative and

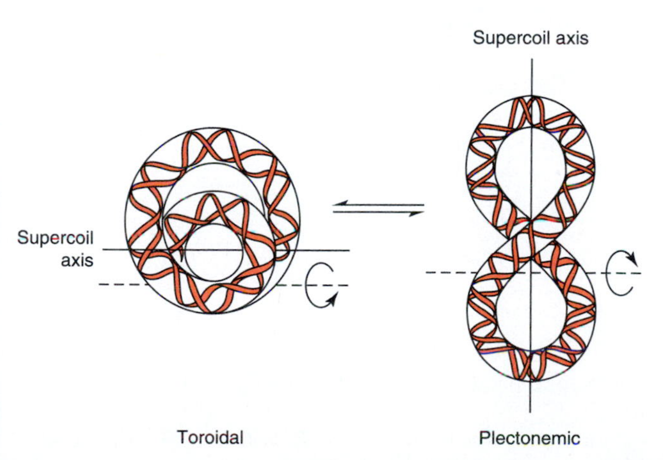

FIGURE 2.42
Equilibrium between two equivalent supercoiled forms of DNA.
Forms shown are freely interconvertible by unfolding the doughnut-shaped toroidal form along an axis parallel to the supercoil axis or by folding the number 8-shaped interwound (plectonemic) form along an axis perpendicular to the supercoil axis. The two forms have the same W, T, and L numbers.
Redrawn with permission from Cantor, C. R. and Schimmel, P. R. Biophysical Chemistry, *Part III. San Francisco: Freeman, 1980. Copyright ©1980.*

positive changes in the writhing number. DNA supercoils are distributed as mixtures of plectonemic and toroidal coils and untwisted regions of the double helix.

Although the closed-circular form of DNA is ideal for acquiring a superhelical structure, any segment of double-stranded DNA that is immobilized at both ends can also be superhelical. The DNA of eukaryotic cells, for instance, can acquire a superhelical form because its anchoring by nuclear proteins creates numerous closed **topological domains.** A topological domain is defined as a DNA segment contained in a manner that restrains rotation of the double helix. Overall, whether DNA is circular or linear, existence of negative superhelicity appears to be an important feature. Supercoiling promotes packaging of DNA within the confines of the cell by facilitating formation of compact structures. For instance, while the length of DNA in each human chromosome is of the order of centimeters, condensed mitotic chromosomes are only a few nanometers long. Negative superhelicity may also be instrumental in facilitating the process of localized DNA strand separation during DNA repair, synthesis, and recombination.

Topoisomerases

Topoisomerases regulate the formation of superhelices. These enzymes change the linking number, *L*, of DNA. Topoisomerases act by catalyzing the concerted breakage and rejoining of DNA strands, which produces a DNA that is more or less superhelical than the original. Topoisomerases are classified into type I, which break only one strand, and type II, which break both strands simultaneously. **Topoisomerases I** act by making a transient single-strand break in a supercoiled DNA duplex, which changes the linking number by increments of one and results in relaxation of the supercoiled DNA (Figure 2.43).

Topoisomerases II act by binding to a DNA molecule in a manner that generates two supercoiled loops, as shown in step 1 of Figure 2.44. Because one loop is positive and the other negative and there is no disruption of phosphodiester bonds, the overall linking number remains unchanged. Subsequently, the enzyme nicks both strands and passes one DNA segment through this break before resealing it. This manipulation inverts the sign of the positive supercoil, resulting in the introduction of two negative supercoils and changing the linking number in increments of two. Energy for the topoisomerase II reaction comes from the hydrolysis of ATP, which is used to restore the enzyme's conformation. During the reaction, topoisomerases remain bound to DNA by a covalent bond between a tyrosyl residue and a phosphoryl group at the incision site (a 5′-phosphotyrosine bond). This

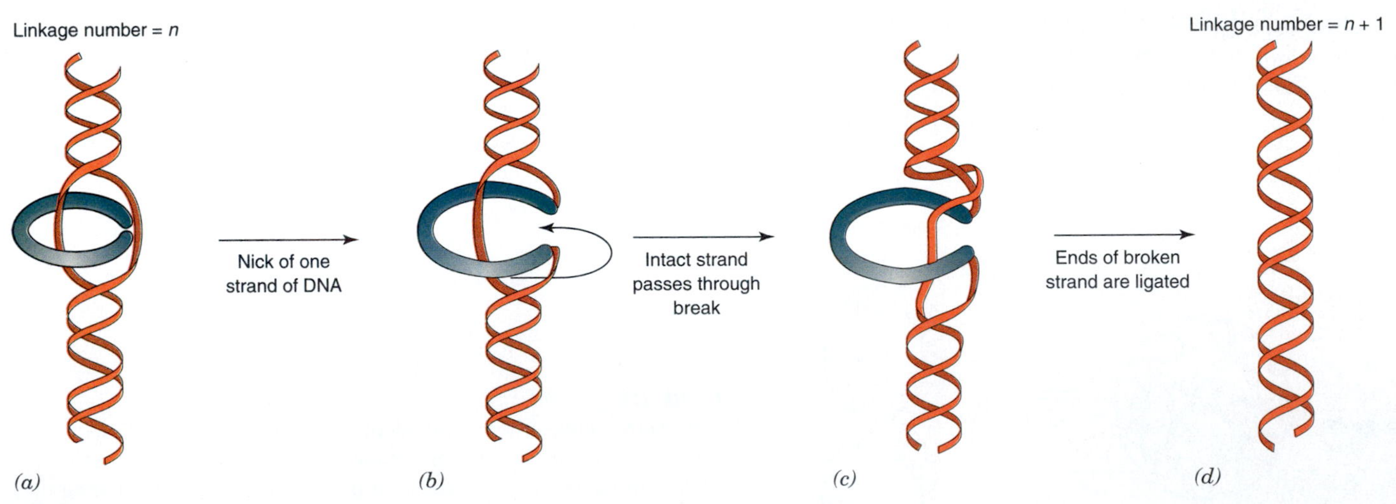

FIGURE 2.43
Mechanism of action of topoisomerases I.
Topoisomerases I relax DNA by (*a*) binding to it and locally separating the complementary strands; then (*b*) nicking one strand and binding to the newly generated termini; and (*c*) passing the intact strand through the gap generated by the nick and closing the gap by restoring the phosphodiester bond. This gives rise to a relaxed structure (*d*).
Redrawn from Dean, F., Kaasnow, M. A., Otter, R., Matzuk, M. M., and Spengler, S. J. Cold Spring Harbor Symp. Quant. Biol. 47:769, 1983.

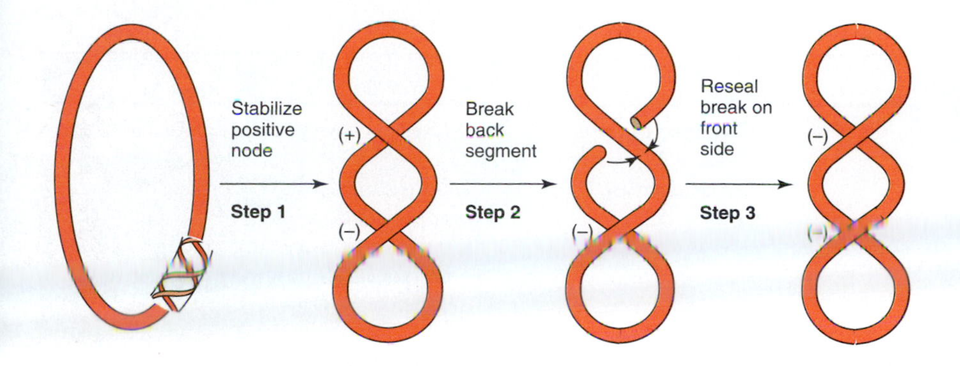

FIGURE 2.44
Mechanism of action of topoisomerases II. Topoisomerases II (and gyrase) change the linking number of DNA by binding to a DNA molecule and passing one DNA segment through a reversible break formed at a different segment of the same DNA molecule. The mechanism of action of gyrase is illustrated above using as an example the conversion of a relaxed DNA molecule to a molecule that contains first two supercoils, one positive and one negative (Step 1). Passage of a DNA segment through the positive supercoil shown on the rightmost part of the figure (Step 3) changes the linking number, producing a molecule that contains two negative supercoils.
Redrawn with permission from Brown, P. O. and Cozzarelli, N. R. Science 206:1081, 1979. Copyright © 1979 by the American Association for the Advancement of Science.

enzyme–polynucleotide bond conserves the energy of the interrupted phosphodiester bond for the subsequent repair of the nick.

A subset of type II topoisomerases, the **gyrases**, are the only enzymes that introduce negative supercoils into relaxed DNA. These enzymes occur only in bacteria; analogous eukaryotic topoisomerases have not been discovered. Eukaryotes apparently use the wrapping of DNA around chromosomal proteins for introduction of negative supercoils into relaxed DNA. Gyrase adds negative supercoils to DNA at a rate of about 100 per minute. Regulation of supercoiling in cells requires involvement of both topoisomerase II/gyrase and topoisomerase I activities. The balance between these two opposing enzymatic activities keeps DNA at a precisely regulated cellular level of superhelicity. The ATP to ADP ratio may play a role in this process, since this ratio influences gyrase activity. Compounds that inhibit topoisomerases and gyrases are also effective antibacterial and antitumor agents (see Clin. Corr. 2.7).

CLINICAL CORRELATION 2.7

Topoisomerases in Treatment of Disease

Topoisomerases are emerging as important targets of antimicrobial and antineoplastic agents. These agents share a common principal mechanism of action by interfering with the enzyme-catalyzed rejoining of DNA strands, in effect inhibiting one of two substeps in the action of topoisomerases. Therefore topoisomerase drugs do not inhibit overall activity of the enzyme, as is the case with most enzyme-targeting drugs. Instead, they trap the immediate complex between topoisomerase and DNA. This may result in degradation of DNA, introduction of mutations, or inhibition of translation and replication.

In treatment of cancers, both topoisomerases I and II can be targeted with therapeutic results. Camptothecin and its derivatives modify the function of topoisomerase I. An excellent correlation has been noted between antitumor activity of various camptothecin derivatives on murine leukemia and their interference with topoisomerase activity. Camptothecins may cause potentially lethal lesions in cells in the form of drug-stabilized covalent DNA cleavage complexes. Subsequent DNA replication may be a prerequisite for cell toxicity. The therapeutic efficacy of camptothecin derivatives may be improved by increased levels of topoisomerase I in some tumor cells such as advanced colon cancers. Studies with two other potent antineoplastic agents—amsacrine and etoposide—that act selectively on topoisomerases II indicate that these clinically useful drugs stabilize covalent topoisomerase II–DNA cleavage complexes by interfering with the enzyme-mediated DNA religation reaction. Indirect evidence also suggests that these drugs may stimulate formation of these complexes. Contrary to observations regarding the importance of DNA replication in expression of the cytotoxic effect of drugs that target topoisomerase I, topoisomerase II–mediated DNA breaks can exert their cytotoxic effect in the absence of ongoing DNA synthesis. Instead, lethal lesions induced by topoisomerase II-targeted drugs may be dependent on recombinations and mutations at sites of formation of drug-induced topoisomerase II–DNA complexes. Many anticancer agents including anthracyclines (including adriamycin and doxorubicin), synthetic intercalators, ellipticines, and podophyllotoxins exert their therapeutic effects on topoisomerases II. Hematologic neoplasms, such as lymphoid and nonlymphoid leukemias, high-grade non-Hodgkin's lymphomas, and Hodgkin's disease, are treated mostly with combinations of one or more topoisomerase II inhibitors with or without additional cytotoxic agents.

Some topoisomerase inhibitors are also powerful antibacterial agents. The quinolones are a large class of antibiotics that selectively inhibit gyrase and/or topoisomerase IV. Some common antibiotics in this class include nalidixic acid and norfloxacin. These antibiotics bind to the gyrase–DNA complex, trapping an intermediate state. Because these enzymes are unique to bacteria, these drugs show high selectivity in their cytotoxic action. Quinolone antibiotics generally show their greatest activity against gram-negative bacteria such as *Enterobacteriaceae, Neisseria,* and *Haemophilus,* but recent analogs show improved activity against gram-positive and anaerobic bacteria.

Potmesil, M. and Kohn, K. W. (Eds.). *DNA Topoisomerases in Cancer.* New York: Oxford University Press, 1991; and Hooper, D. C. Clinical applications of quinolones. *Biochem. Biophys. Acta* 1400:45, 1998.

TABLE 2.4 Properties of DNA Topoisomerases

Enzyme	*Type*[a]	ΔL	*Activities*
E. coli topoisomerase I (*top A*)[b]	I	Increase L $\Delta L = 1$	Relaxes negatively supercoiled DNA
Eukaryotic topoisomerase I from yeast (*top 1*)	I	Increase or decrease L $\Delta L = \pm 1$	Relaxes either positively or negatively supercoiled DNA
E. coli topoisomerase II or DNA gyrase (*gyrA*, *gyrB*)	II	Increase or decrease L $\Delta L = \pm 2$	Introduces negative supercoiling to DNA; relaxes either positively or negatively supercoiled DNA
E. coli topoisomerase IV (*parC*, *parE*)	II	Increase L $\Delta L = +2$	DNA relaxing activity; it cannot introduce negative supercoils
Eukaryotic topoisomerase II from yeast (*top 2*)	II	Increase or decrease L $\Delta L = \pm 2$	Relaxes positively or negatively supercoiled DNA
E. coli topoisomerase III (*top B*)	I	Increase L $\Delta L = +1$	Relaxes negatively supercoiled DNA decatenation activity
Eukaryotic topoisomerase III (*top 3*)	I	Increase L $\Delta L = +1$	Specific activity on DNA with single-stranded heteroduplex

[a] Type I topoisomerases use Mg^{2+} as cofactor but do not use ATP. Type II topoisomerases require Mg^{2+} plus ATP.

[b] The name of the gene coding for the topoisomerase is shown in parentheses.

Topoisomerases also catalyze other topological isomerizations (Table 2.4). Bacterial type III topoisomerases have type I topoisomerase properties; that is, they can relax supercoils without the requirement of an energy source, such as ATP hydrolysis. These topoisomerases may specialize in the resolution of interlocking circular DNA products (catenates) that are generated just prior to the completion of DNA replication. An unusual class of topoisomerases, **reverse gyrases,** has been isolated from various species of archaebacteria. Remarkably, these gyrases introduce positive supercoils into DNA. Positive supercoiling may protect DNA from the denaturing conditions of high temperature and acidity under which these bacteria exist.

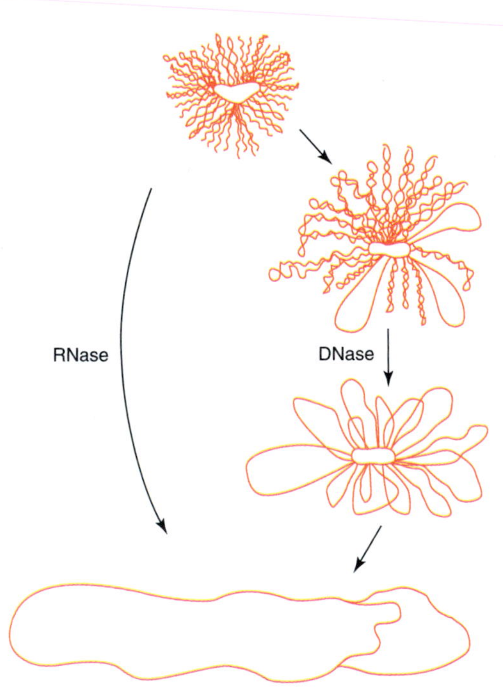

FIGURE 2.45
Bacterial chromosomes are packaged in nucleoids.
The circular chromosome of a bacterium is compacted into about 40–50 loops of supercoiled DNA organized by a central RNA–protein scaffold. DNase relaxes the structure progressively by opening individual loops, one at a time. RNase completely unfolds the chromosome in a single step by disrupting the nucleoid core.
Redrawn from Worcel, A. and Burgi, E. J. Mol. Biol. 71:127, 1972.

Packaging of Prokaryotic DNA

In prokaryotic cells DNA is organized as a single chromosome that contains a double-stranded circular supercoil. In an average bacterium such as *E. coli,* the contour length of the DNA is some 80 times larger than the diameter of the cell. Thus DNA must be packaged in a highly condensed form to fit inside the cell. Bacterial chromosomes are organized into compacted structures, called **nucleoids,** by interaction of **HU** and **H-NS** proteins and participation of various cations, **polyamines** (such as spermine, spermidine, putrescine, and cadaverine), RNA, and nonhistone proteins (Figure 2.45). HU (molecular mass 18 kDa) exists as a heterodimer of two nearly identical subunits (HU-1 and HU-2). Upon binding to DNA, HU changes the shape and supercoiling of the double helix. *In vitro,* HU binding compacts DNA and restrains supercoils in a concentration-dependent manner. At an equimolar ratio, the interaction of DNA with HU prevents topoisomerases from relaxing negatively supercoiled DNA in the DNA–HU complex. It also means that HU can introduce supercoils in relaxed DNA. Higher concentrations of HU do not result in the restraining of additional supercoils. From the effects of HU on DNA supercoiling and other evidence, it appears that HU bends DNA sharply into a tight circle. HU is primarily responsible for formation of a beaded nucleoid structure seen in prokaryotes. H-NS, another abundant small protein, may be involved in chromosomal organization either directly or indirectly through interaction with the HU proteins. In *E. coli,* the nucleoid consists of a single supercoiled DNA molecule organized into about 40 loops, each of approximately 10^5 bp of DNA, that merge into a scaffold rich in protein and RNA. In prokaryotic scaffolds, the loops are maintained by interaction between DNA and RNA, rather than DNA–protein interactions only, as in eukaryotes. As a result of formation of nucleoids, which have diameters of 2 μm, the *E. coli* genome can easily be fit into the cell. Bacterial chromosomes are dynamic structures that bind and dissociate fairly rapidly. This may reflect the need for rapid DNA synthesis, cell division, and transcription that characterize bacterial cells.

Organization of Eukaryotic Chromatin

DNA in eukaryotic cells is associated with various proteins to form **chromatin.** In nondividing (interphase) cells, chromatin is amorphous and dispersed throughout the nucleus. Just prior to cell division (metaphase), chromatin becomes organized into compact structures called **chromosomes.** Each chromosome is characterized by a **centromere,** which is the site for attachment to proteins that link the chromosome to the mitotic spindle. **Telomeres** define the termini of linear chromosomes. Chromosomes also contain sequences required for initiation of DNA replication (**origin of replication**).

The enormous length of the genome of most eukaryotes necessitates the division of genetic information into numerous independent domains, that is, chromosomes. The number of chromosomes observed is species specific; human cells contain 46 chromosomes (chromatids) organized into 23 pairs. The average DNA length of chromosomes is 1.3×10^8 bp or approximately 5 cm. Each human chromosome consists of an intact DNA molecule varying in size from 263×10^6 bp for chromosome 1 to less than 50×10^6 bp for chromosome 23. The chromosomal organization that makes it possible for DNA to fit within a cell nucleus with a diameter of approximately 10 μm requires a "condensation ratio" of more than five orders of magnitude. During metaphase the DNA molecule is very tightly wound. For example, human chromosome 16 is 2.5 μm long, whereas the DNA molecule is 3.7 cm in each of the two chromatids, giving a condensation ratio of $1.5 \times 10^4{:}1$. The parceling of DNA in 46 chromosomes provides for a further increase in the condensation ratio to 105:1.

The early stages of DNA packing that lead to formation of **30-nm fibers** have been studied extensively. Less is known about later stages of DNA compaction in which looped domains of the 30-nm fiber are organized into scaffolds and chromatid coils. At each stage of packing, shown in this model, DNA is condensed severalfold (Figure 2.46). The cumulative effect of successive folding stages provides the large condensation ratio necessary for packing of DNA within the nucleus. The first stage of organization is formation of a "beads-on-a-string" structure consisting

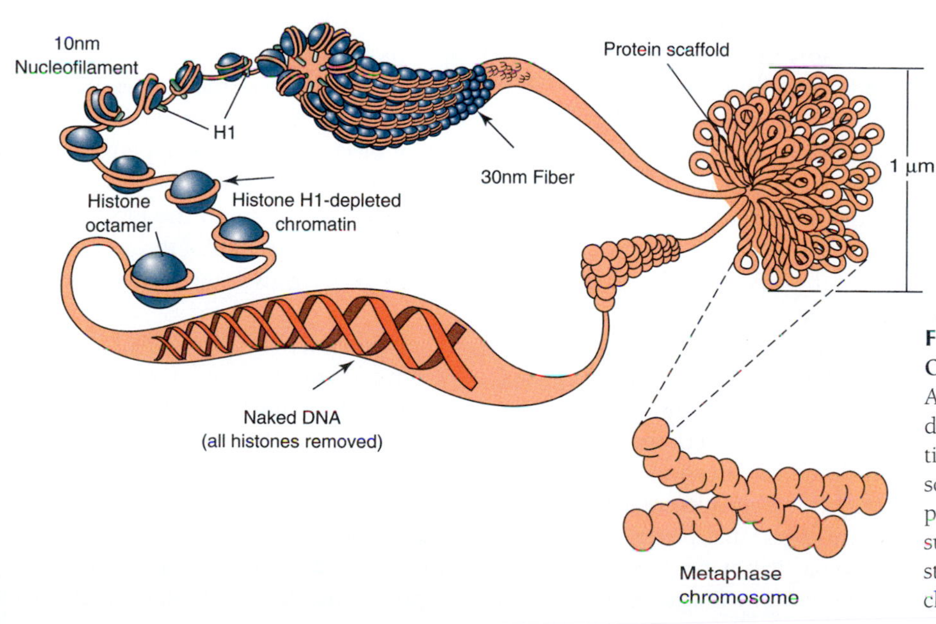

FIGURE 2.46
Organization of DNA into chromosomes. A speculative drawing showing the stepwise condensation of DNA into chromatin. The DNA initially wraps around the histones of the nucleosome core. Condensation with histone H1 produces the 10-nm nucleofilament, which is subsequently packaged into a twisted, looped structure attached to a protein scaffold within the chromosome.

of DNA associated with a class of highly basic proteins known as **histones.** These bind tightly to DNA, forming very stable complexes. The "beads-on-a-string" arrangement is seen in chromatin that has been treated under conditions of low ionic strength and examined under the electron microscope. The "string" is free DNA, and the "beads" contain DNA coiled around histones.

Histones, regardless of their source, consist of five types of polypeptides of different size and composition. The most "conserved" histones are **H4** and **H3,** which differ very little, even between extremely diverse species; histones H4 from peas and cows are very similar, differing by only two amino acids, although these species diverged more than a billion years ago. The **H2A** and **H2B** are less highly conserved but still exhibit substantial evolutionary stability, especially within their nonbasic portions. **H1** is quite distinct from the other histones, being larger, more basic, and by far the most tissue-specific and species-specific histones. Vertebrates contain an additional histone, **H5,** which has a function similar to H1. As a result of their unusually high content of the basic amino acids lysine and arginine, histones are highly polycationic and interact with the polyanionic phosphate backbone of DNA to produce uncharged nucleoproteins. All five histones are characterized by a central nonpolar domain, which forms a globular structure, and N-terminal and C-terminal regions that contain most of the basic amino acids. The basic N-terminal regions of H2A, H2B, H3, and H4 are the major sites of interaction with DNA. A heterogeneous group of proteins with high species and organ specificity is also present in chromatin. These proteins, grouped together as nonhistone proteins, consist of several hundred members, most of which are present in trace amounts. Many nonhistone proteins are associated with various chromosome functions, such as replication, gene expression, and chromosome organization.

Nucleosomes and Polynucleosomes

Histones interact with DNA to form a periodic "beads-on-a-string" structure, called a polynucleosome, in which an elementary unit, a **nucleosome,** is regularly repeated (Figure 2.47). Each nucleosome is a disk-shaped structure about 11 nm in diameter and 6 nm in height that consists of a DNA segment and an octameric histone cluster composed of two molecules each of H2A, H2B, H3, and H4 histones. Each cluster consists of a tetramer consisting of $(H3)_2$-$(H4)_2$ with an H2A-H2B dimer stacked on each face in the disk. The DNA is wrapped around the octamer as a negative toroidal superhelix with the central $(H3)_2$–$(H4)_2$ core interacting with the central 70–80 bp of the surrounding DNA (Figure 2.48). Histones are in contact with the minor groove of DNA and leave the major groove available for interaction with proteins that regulate gene expression and other DNA functions.

The structure of the nucleosome core explains why eukaryotic cells lack topoisomerases that can underwind DNA. It appears that negative superhelicity is, instead, introduced into eukaryotic cells as a result of DNA forming a toroidal coil around the histone core (Figure 2.49). Such wrapping requires the removal of approximately one helical turn in DNA. When subjected to such wrapping, relaxed DNA generates a negative toroidal supercoil within the region wrapped around the histone core and a compensating positive supercoil elsewhere in the molecule, so as to maintain a constant linking number. Subsequent relaxation of the positive supercoil by eukaryotic topoisomerases leaves one net negative supercoil within the nucleosomal region.

Overall the nucleosome core consists of approximately 146 bp of DNA wrapped around the histone octamer complex. Polynucleosomes consist of numerous nucleosomes joined by **"linker" DNA,** about 20–90 bp long. This DNA is associated with histone H1 that locks the coiled DNA in place; the resulting complex is called a **chromatosome** (Figure 2.48). The periodicity of nucleosome distribution along the polynucleosome structure has been determined by controlled digestion by a nuclease that preferentially attacks linker DNA. The distribution of these nucleosomes is not random with respect to the DNA base sequence. DNA does not bend uniformly but rather bends gently and then more sharply around the histone

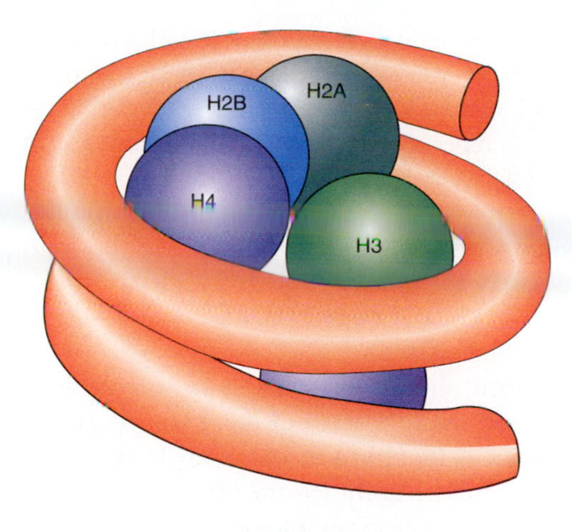

Nucleosome core
(1 3/4 turn particle)

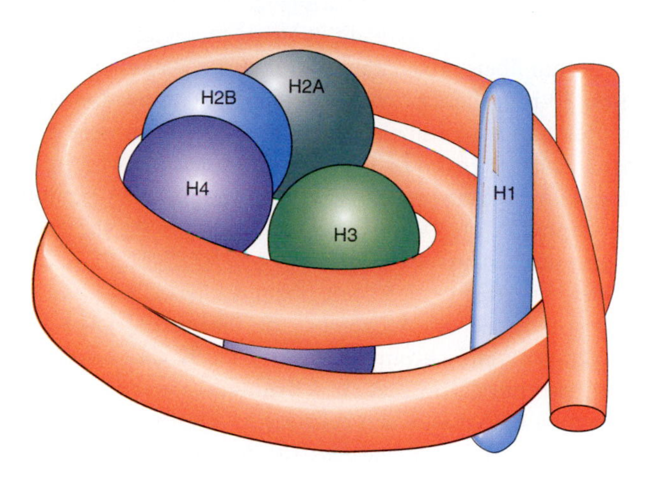

Chromatosome
(Two-turn particle +H1)

FIGURE 2.47
Models of the structure of the nucleosome complex.
The nucleosome core consists of approximately 146 bp of DNA corresponding to $1\frac{3}{4}$ superhelical turns wound around a histone octamer. The chromatosome consists of about 166 bp of DNA (two superhelical turns). The H1 histone is retained by this particle and is associated with the linker DNA, as shown.

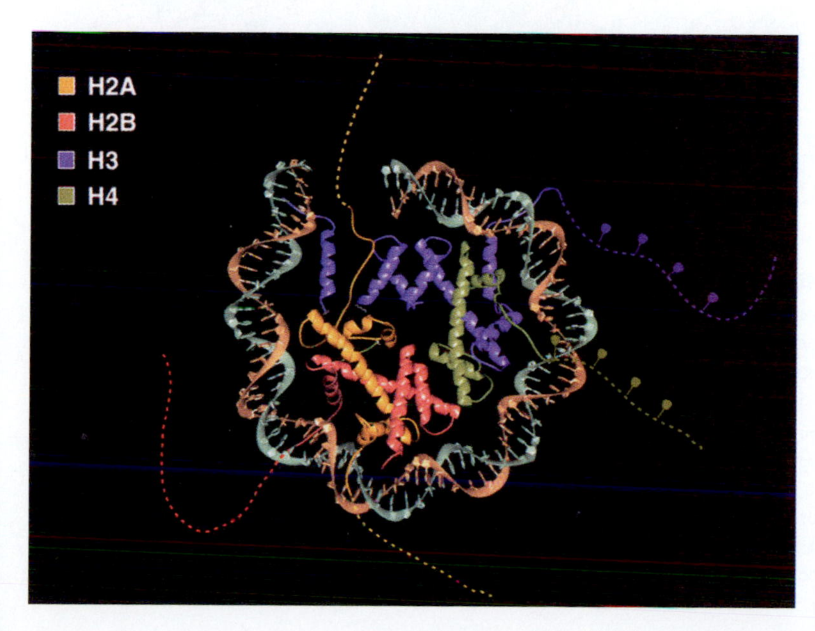

FIGURE 2.48
X-ray structure of the nucleosome core.
The disk of the nucleosome core has been split in half to show one turn of DNA wrapped around four histone molecules. The dashed lines represent unstructured parts of the histone tails.
Reprinted from Kornberg, R. D. and Lorch, Y. Cell *98:285, 1999,* with permission from Elsevier Science.

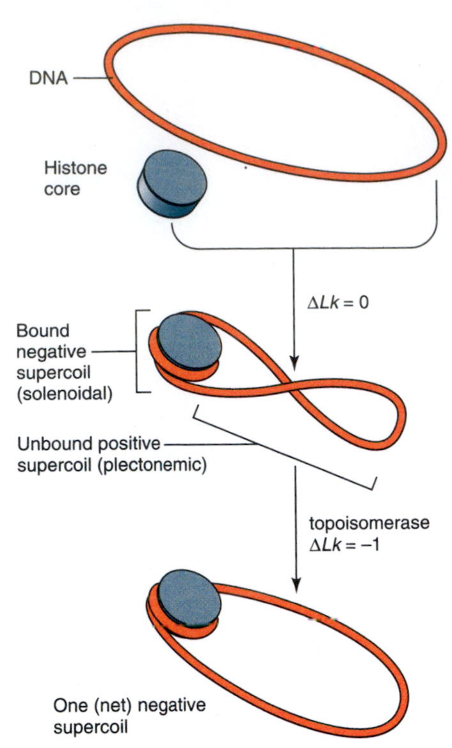

FIGURE 2.49
Generation of negative supercoiling in eukaryotic DNA.
The binding of a histone octamer to a relaxed, closed-domain DNA forces the DNA to wrap around the octamer, generating a negative supercoil. In the absence of any strand breaks, the domain remains intact and a compensating positive supercoil must be generated elsewhere within the domain. The action of a eukaryotic type I topoisomerase subsequently relaxes the positive supercoil, leaving the closed domain with one net negative supercoil.

octamers. This suggests that DNA binding is sequence dependent and that positioning of nucleosomes may be influenced by the DNA sequence. In fact, nucleosomes tend to associate preferentially with certain DNA regions. DNA tracts that resist binding, such as long A tracts or G-C repeats, are not usually associated with nucleosomes. In contrast, certain bent DNA regions, for instance, periodically phased A tracts, associate strongly with histones. Nucleosome core particles migrate over the polynucleosome DNA sequence. This mobility allows access to the DNA by polymerases and other proteins necessary for transcription and replication.

Polynucleosome Packing into Higher Structures

Wrapping of DNA around histones to form nucleosomes results in a tenfold reduction in the apparent length of DNA and the formation of the so-called **10-nm fiber** (which is actually 11 nm wide), which is the diameter of the nucleosomes. In chromosomes isolated by very gentle methods, both 10-nm fibers and thicker 30-nm fibers (in fact, 34 nm wide) can be seen in electron micrographs. The **30-nm fibers** dissociate into 10-nm fibers by treatment at low ionic strength. The 30-nm fibers appear to form by condensation of 10-nm fibers into a solenoid arrangement involving six to seven chromatosomes per solenoid turn (Figure 2.50). Histone H1 molecules may also bind to one another cooperatively, bringing the neighboring nucleosomes closer together in 30-nm fibers. Formation of polynucleosomes and their subsequent condensation into 30-nm fibers provides for a DNA compaction ratio that may be as high as two orders of magnitude. The 30-nm fibers form only over selected regions of DNA that are characterized by the absence of binding with other sequence-specific nonhistone DNA-binding proteins. The presence of DNA-binding proteins and the effects on formation of 30-nm fibers may depend on the transcriptional status of the regions of DNA involved.

Models of the higher levels of packing of 30-nm fibers are based on indirect evidence obtained from two specialized chromosomes: the **lampbrush chromosomes** of vertebrate oocytes and **polytene chromosomes** of fruit fly giant secretory cells. These chromosomes are exceptional in that they maintain precisely defined higher-order structures in interphase, that is, when cells are in a resting (nondividing) state. By extrapolation, the structural features of interphase lampbrush

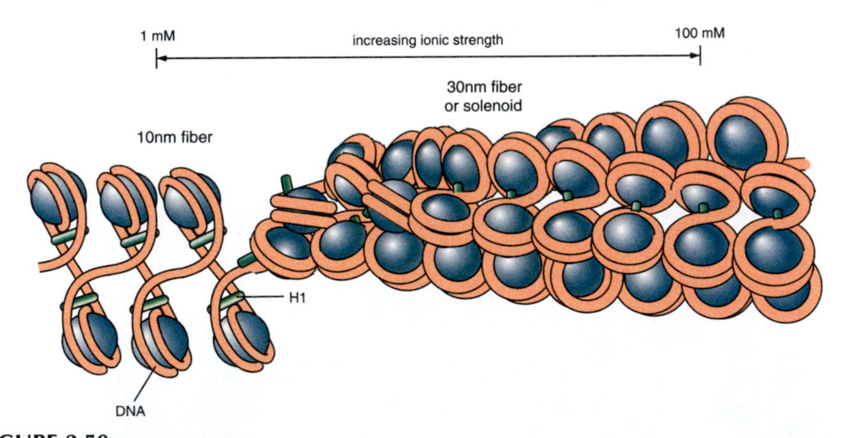

FIGURE 2.50
Nucleofilament structure.
Nucleofilament has the "beads-on-a-string" appearance, which corresponds to an extended polynucleosome chain. Histone H1 is attached to the "linker" regions between nucleosomes, but in the resulting structure H1 molecules, associated to adjacent nucleosomes, are located close to one another. At higher salt concentrations, polynucleosomes are transformed into the higher order structure of the 30-nm fiber. It has been proposed that at higher ionic strengths the nucleofilament forms a very compact helical structure or a helical solenoid, as illustrated in the upper part of the figure. H1 histones appear to interact strongly with one another in this structure. In fact, the organization of the 10-nm nucleofilament into the 30-nm coil or solenoid requires the presence of H1.
Adapted from Kornberg, R. D. and Klug, A. The Nucleosome. San Diego, CA: Academic Press, 1989.

chromosomes have led to the proposal that chromosomes in general are organized as a series of looped, condensed domains of 30-nm fibers of variable size for different organisms. It is estimated that these loops may contain from 5000 to 120,000 bp with an average of about 20,000. Thus the haploid human genome would contain about 60,000 loops, which is close to the estimated number of genes (30,000–60,000). It appears likely that each loop contains one or a few linked genes. The domains are bound to a nuclear scaffold consisting of H1 histone and several nonhistone proteins, including two major scaffold proteins Sc1 (a topoisomerase II) and Sc2. The loops are fixed at their bases and can therefore accumulate supercoils. Specific AT-rich regions of DNA known as **SARs (scaffold attachment regions)** are preferentially associated with the scaffold. SARs also contain topoisomerase II binding sites. The presence of type II topoisomerase at the base of closed topological domains, which define the scaffold loops, suggests that changes in supercoiling within these domains are biologically important. Formation of looped domains may account for as much as an additional 200-fold condensation in the length of DNA and an overall packing ratio of more than four orders of magnitude. Each loop can be coiled and then supercoiled into 0.4 μm of a 30-nm fiber. Since a chromatid is about 1 μm in diameter, packing of the 30-nm fiber into a chromatid would require just one more order of folding.

The next level of chromosomal organization probably involves the packing of loops. The packing may be achieved by arranging the loops of the 30-nm fiber in tightly stacked helical coils. **Chromatids** of metaphase chromosomes may consist of helically packed loops of 30-nm fibers. Changes in packing, and therefore the transition between the various forms of chromatin, seem to be partially controlled by covalent modification of core histones. Histones H3 and H4 undergo cell-cycle-dependent reversible acetylation on the ε-amino group of lysine by two different enzymes, a histone acetylase and a histone deacylase. Acetylation appears to affect the negative superhelical tension within domains and, in certain instances, the binding of transcription factors. The hydroxyl group of the N-terminal serine residue in H4 is subject to phosphorylation catalyzed by a kinase. Acetylation and phosphorylation change the charge of the N-terminal region of histone H4 from +5 to −2. The resulting negative charge causes the core histones to bind less tightly to DNA, promoting the unraveling of 30-nm fibers and decondensation of chromatin. Finally, phosphorylation of terminal H1 correlates with chromatin condensation into metaphase chromosomes. This may result from modulation of interaction between phosphorylated–dephosphorylated H1 and the histone octamer. The change from compact to decondensed chromatin is promoted by binding of proteins, known as HMG proteins (high-mobility-group proteins), which interact preferentially with transcriptionally active decondensed chromatin, that is, 10-nm fiber. Control of eukaryotic transcription and replication involves both histone and nonhistone proteins. While dissociation of histones from chromosomal DNA may be a prerequisite for transcription, nonhistone proteins provide more finely tuned transcription controls. Whatever the details of control may be, chromosomal regions actively synthesizing RNA are less condensed than inactive regions. Active genes must be packaged in a way that makes them accessible to regulatory proteins while permanently repressed genes must remain inaccessible. Packaging may also determine accessibility of DNA to DNA-damaging agents. Lastly, nonhistone proteins control gene expression during differentiation and development and may serve as sites for binding of hormones and other regulatory molecules.

2.5 | DNA SEQUENCE AND FUNCTION

The size and average base composition of DNA vary widely between species. The property that makes the DNA of a species unique is the nucleotide sequence. Until recently, direct determination of nucleotide sequences in genomic DNA was an intimidating undertaking. The technology developed in connection with the Human

Genome Project has accelerated the rate at which DNA sequences are determined. Soon, complete sequences for a number of species will be available for analysis.

Restriction Endonucleases and Palindromes

One key event that enabled development of methods to sequence genomic DNA was the discovery of **restriction endonucleases.** These enzymes cleave double-stranded DNA chains at a specific sequence by making two cuts, one in each strand of dsDNA (Figure 2.51). Bacteria developed restriction enzymes as a defense against infection by phages. Cleavage exposes the viral DNA to eventual degradation by nonspecific bacterial exonucleases. Bacterial DNA can be protected from cleavage by sequence-specific methylation. The recognition sites for restriction methylases correspond to those of the endonucleases. Methylation of specific bases within the recognition site prevents cleavage by the cognate nuclease. The most common sites for base methylation are the 5-position of cytosine and the 6-NH_2 group of adenine. Notably, base methylation may be critical in gene regulation in higher organisms.

Many hundreds of restriction endonucleases have been purified and the list of new restriction enzymes is growing. With few exceptions, they have been found to recognize sequences four to six nucleotides long. "Rare cutters," endonucleases with unusually large recognition sites, are valuable because of the relative infrequency of cleavage of very large DNAs such as eukaryotic chromosomes. Not I, for example, has an eight-nucleotide recognition sequence. The recognized sequences are completely symmetrical inverted repeats, known as **palindromes.** The order of the bases is the same when the two complementary strands of the palindrome are each read $5' \rightarrow 3'$. For example, in the case of the restriction enzyme EcoRI, isolated from *E. coli*, the sequence of bases is 5′-GAATTC-3′. Restriction endonucleases are classified into three categories. Types I and III make cuts at sites remote from the recognition site. Type II specifically cleaves DNA within the recognition sequence.

Restriction endonucleases recognize specific sequences that occur along large DNAs with relatively low frequencies and fragment DNA very selectively. For example, a typical bacterial DNA, which may contain about 3×10^6 bp, will be cleaved by a restriction endonuclease into a few hundred fragments. A small virus or plasmid may have few or no cutting sites at all for a particular restriction endonuclease (Figure 2.52). The practical significance is that a particular restriction enzyme generates a unique family of fragments for any given DNA molecule. This unique fragmentation pattern is called a restriction digest. Availability of restriction enzymes for fragmenting large DNA sequences and development of gel electrophoresis techniques for separating DNA fragments have made determination of sequences a simple matter. These sequencing techniques are described in Chapter 7.

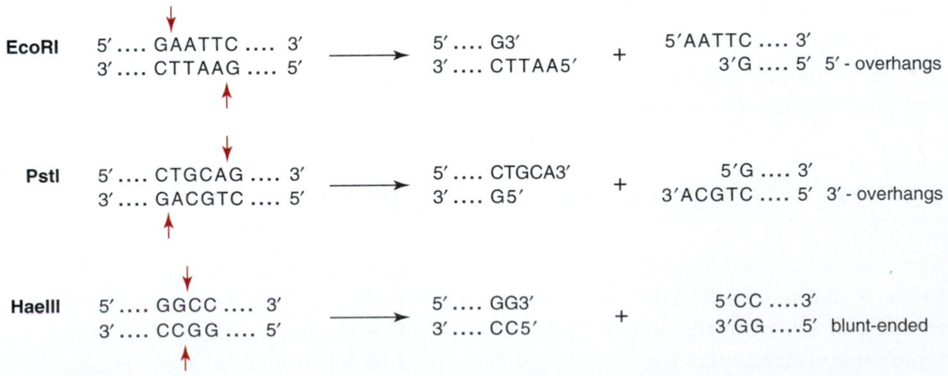

EcoRI 5′.... GAATTC .... 3′ / 3′.... CTTAAG .... 5′ → 5′.... G3′ / 3′.... CTTAA5′ + 5′AATTC .... 3′ / 3′G .... 5′ 5′-overhangs

PstI 5′.... CTGCAG .... 3′ / 3′.... GACGTC .... 5′ → 5′.... CTGCA3′ / 3′.... G5′ + 5′G .... 3′ / 3′ACGTC .... 5′ 3′-overhangs

HaeIII 5′.... GGCC .... 3′ / 3′.... CCGG .... 5′ → 5′.... GG3′ / 3′.... CC5′ + 5′CC ....3′ / 3′GG ....5′ blunt-ended

FIGURE 2.51
Types of products generated by type II restriction endonucleases.
Enzymes exemplified by EcoRI and PstI nick on both sides of the center of symmetry of the palindrome, generating single-stranded overhangs. Many commonly used enzymes generate 5′-overhangs, although some produce ends with 3′-overhangs as shown for PstI. Other restriction nucleases cut across the center of symmetry of the recognition sequence, to produce flush or blunt ends, as exemplified by HaeIII.

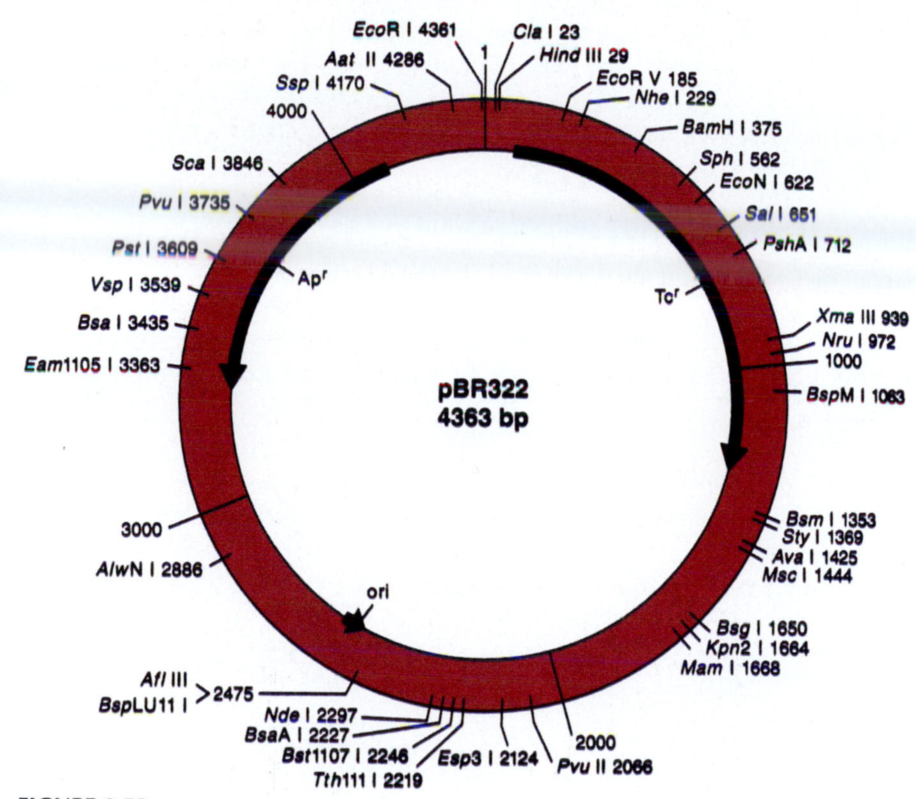

FIGURE 2.52
Restriction map for plasmid DNA.
Bacterial plasmid pBR322 is a 4363-bp circular DNA containing genes for resistance to the antibiotics ampicillin (Ap^r) and tetracycline (Tc^r). The locations of unique DNA sequences that are recognized by a number of restriction endonuclease are marked. Other known endonucleases will cleave this plasmid many times or not at all. For example, the recognition sequence for the enzyme Bg lI appears in three locations on this plasmid, whereas Bg lII does not cleave this plasmid at all.

Most Prokaryotic DNA Codes for Specific Proteins

In prokaryotes a large percentage of total chromosomal DNA codes for specific proteins. The entire *E. coli* genome has been sequenced and consists of about 4.6×10^6 bp of DNA and contains ~4200 genes. Not all of the genes code for expressible, functional proteins. For example, 80 genes code for tRNA molecules, and some may not encode functional molecules at all. In an overall sense, *E. coli* DNA is densely packed with sequence information, and there is little repetition of information in the genome. As much as 1% of the *E. coli* genome is composed of multiple copies of short repetitive sequences known as **repeated extragenic palindromic (REP) elements.** These are present at sites of DNA interaction with functional proteins, for example, in the region of initiation of DNA synthesis (referred to as **OriC**). At OriC, REP elements have a consensus sequence of 34 nucleotides and bind topoisomerase II. REP elements with the sequence GCTGGTGG (Chi sites) bind the enzyme RecBCD, initiating DNA recombination. Chi sites are regularly spaced at intervals separated by about 4000 bp. Genetic information is even more densely organized in smaller organisms, such as bacteriophages, where the primary sequence of DNA reveals that structural genes—nucleotide sequences coding for protein—do not always have distinct physical locations. Rather, they frequently overlap with one another, as illustrated by the partial sequence of phage **ϕX174** (Figure 2.53). This overlap provides for the efficient and economic utilization of the limited DNA present and may control the order of gene expression.

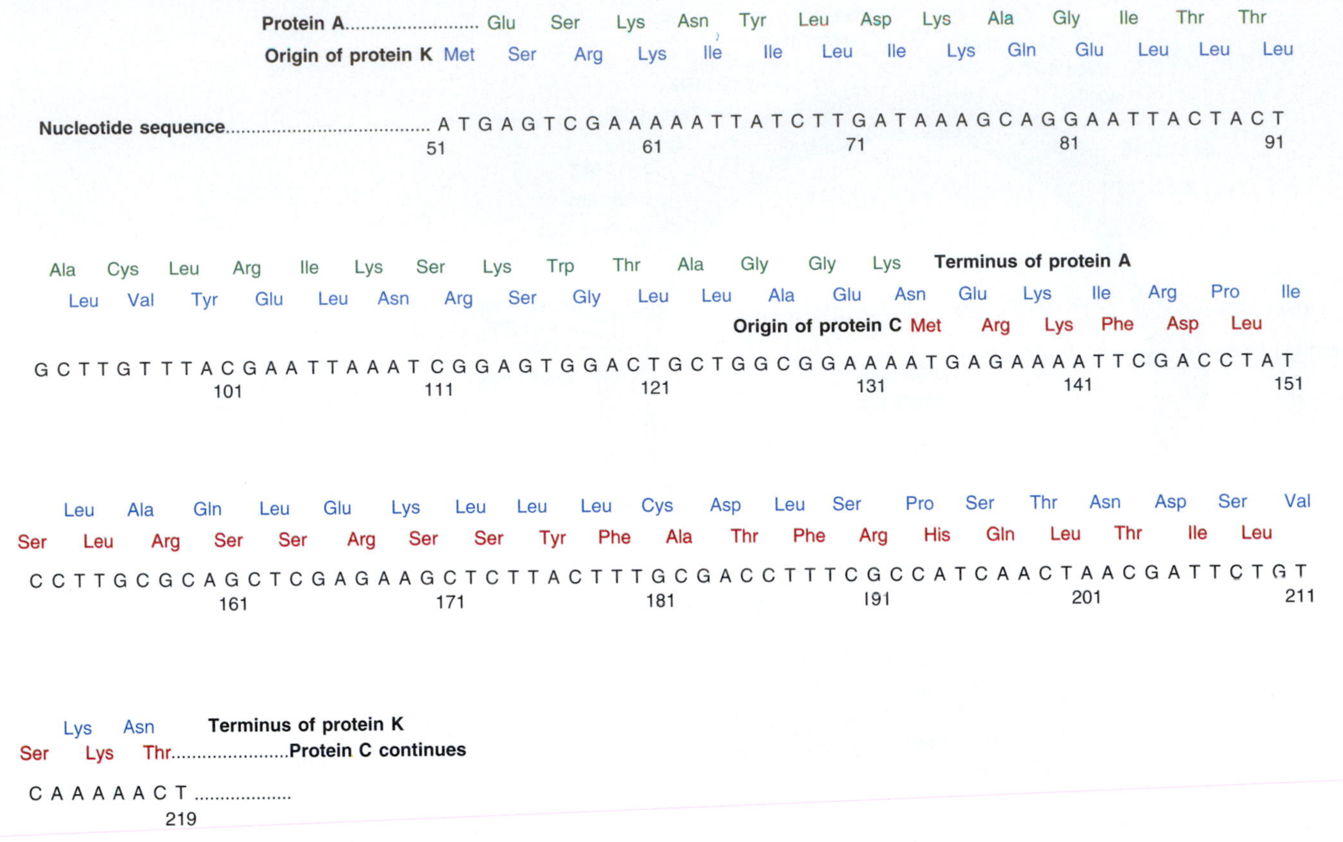

FIGURE 2.53

Partial nucleotide sequences of contiguous and overlapping genes of phage ϕX174. Complete nucleotide sequence of ϕX174 is known, but only the sequence starting with nucleotide 51 and continuing to nucleotide 219 is shown in this figure. This sequence codes for the complete sequence of one of the proteins of ϕX174, protein K. A part of the same sequence, nucleotide 51 to nucleotide 133, codes for part of the nucleotide sequence of another protein, protein A. The sequence coding for protein K, which starts with nucleotide 133, also codes for part of a third protein, protein C. Similar overlaps are noted in other genes of ϕX174.

Adapted with permission from Smith, M. Am. Sci. 67:61, 1979. Journal of Sigma Xi, The Scientific Research Society.

Only a Small Percentage of Eukaryotic DNA Codes for Functional Genes

Eukaryotes have much larger genomes than prokaryotes, from about 1.5×10^7 bp for yeast to about 1.15×10^{11} bp for the haploid genome of the lily *Fritillaria assyrinca.* The latter contains sufficient DNA to code for nearly 3,000,000 genes. Prior to February 2001, it was thought that the human genome contained between 70,000 and 120,000 genes. Initial results from the Human Genome Project, however, suggest that the human genome codes as few as 30,000 genes. As a result, genetic information need not be as densely packed as in bacteria. A typical mammalian DNA, with only seven times as many genes as that of *E. coli,* contains 500 times more DNA than *E. coli.* Clearly, genes for specific proteins and sequences that control gene expression cannot account for the entire DNA in eukaryotic cells. In fact, only 2–4% of DNA in a mammalian cell may suffice for all its genes. Some of the remaining DNA, such as DNA in centromeres and telomeres, has well-defined function. No specific function can be assigned to the majority of noncoding DNA and it has been referred to as "junk DNA." However, there is increasing evidence that junk DNA may have a vital role in regulation of gene expression during development.

Nucleotide sequences indicate that eukaryotic genes do not overlap and are spaced, on the average, 40,000 bp apart. However, some eukaryotic genes may be closer together in regions that contain genes expressed in a tightly coordinated manner (gene families). Most eukaryotic genes are interrupted by noncoding intervening nucleotide sequences, called **introns** (Figure 2.54). The sequences in the gene

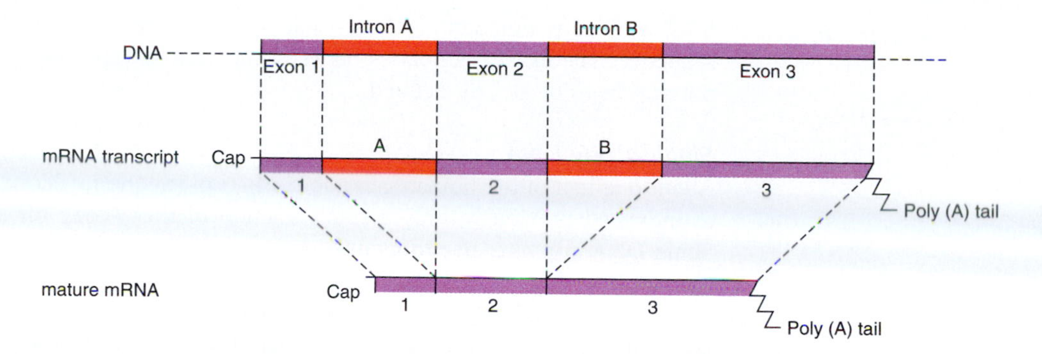

FIGURE 2.54
Schematic presentation of a eukaryotic gene.
The top horizontal line represents a part of the DNA genome of a eukaryote; the bottom line represents the mRNA produced by it. In this example the DNA consists of two introns and three exons. The intron sequences are transcribed but are removed during splicing, which produces a mature mRNA.
Redrawn from Crick, F. Science *204:264, 1979. Copyright © 1979 by the American Association for the Advancement of Science.*

that are expressed, either in the final RNA product (mature RNA) or as a protein, are termed **exons.** The introns are removed during the processing of the RNA transcript, and the remaining exons are ligated together. This tailoring of the original transcript is referred to as **splicing.** The sequence and size of introns vary greatly among species, but generally they are very large. Added together they may be five to ten times the length of the exons they separate. Most genes are interrupted by introns at least once, whereas others are interrupted repeatedly. Some genes, however, such as the gene for human **interferon-α,** contain no introns.

Introns are common in genes of vertebrates and flowering plants but infrequent in genes of other species. The biological role of introns is not clear. Their presence in eukaryotes may represent a stage in the evolution of genes, since introns are rare in prokaryotes and much less common in lower eukaryotes, such as yeasts. Introns in eukaryotic genes may have arisen relatively recently in evolution as a result of migration of mobile DNA elements (transposons, see p. 345) from other parts of the genome and their insertion into protein-coding genes. Through mutation, these inserts may have subsequently lost their mobility.

Repeated Sequences

While repetition of particular DNA sequences in prokaryotes is very limited, the DNA of eukaryotes contains nucleotide sequences that are repeated anywhere from a few times, for certain coding genes, to millions of times per genome for certain simple, relatively short, sequences. Depending on the species, repetitive DNA may constitute 3–80% of total DNA. In mammalian genomes, including the human genome, 25–35% of the DNA is repetitive. Sequences can be classified as single copy, moderately reiterated, and highly reiterated. The content of single-copy DNA varies among eukaryotes.

A distinction between the terms "**reiterated**" and "**repetitive**" in describing a DNA sequence should be made. "Reiterated" is used to describe a unique DNA sequence, usually several hundred nucleotides long, present in multiple copies in a genome. A sequence is termed "repetitive" if a certain, usually short, nucleotide sequence is repeated many times over the DNA sequence.

Most highly reiterated sequences have a characteristic base composition and can be isolated by shearing the DNA into segments of a few hundred nucleotides long and separating the fragments by density gradient centrifugation. These fragments are termed **satellite DNA** because after centrifugation they appear as satellites of the band of bulk DNA. Other highly reiterated sequences, which cannot be isolated by centrifugation, can be identified by their property of rapid reannealing. Some of the highly reiterated sequences can also be isolated by digestion of total

DNA with restriction endonucleases that cleave at specific sites within the reiterated sequence. The exact boundaries between the various types of reiterated DNAs do not appear to be strictly defined.

Single-Copy DNA

About one-half of the human genome consists of unique nucleotide sequences but, as indicated previously, only a small fraction of these code for specific proteins. Some DNA contains **pseudogenes,** sequences of DNA that have significant nucleotide homology to a functional gene but contain mutations that prevent gene expression. These genes, which may be present in a frequency as high as one pseudogene for every four functional genes, significantly increase the size of eukaryotic genomes without contributing to their expressible genetic content. Additional single-copy DNA sequences serve as introns and as control regions that flank genes.

Moderately Reiterated DNA

This class of DNA consists of copies of identical or closely related sequences that are reiterated from a few to a thousand times. These sequences are relatively long, varying between a hundred and many thousand nucleotides before the same polynucleotide sequence is repeated. About 20% of mouse DNA of sequences a few hundred base pairs long are repeated more than a thousand times. About 15% of the human genome consists of moderately reiterated DNA. Normally, single-copy and moderately reiterated sequences occur on a chromosome in an orderly pattern known as the **interspersion pattern,** which consists of alternating blocks of single-copy DNA and moderately reiterated DNA. Moderately repetitive sequences are further classified as short and long interspersed repeats that are present at 1000–100,000 copies per genome. **Long interspersed repeats** consist of sequences several thousand nucleotides long; there can be up to 1000 copies of these repeats per genome. These repeats are flanked on either side of the sequence by sequences that are direct repeats. One example of a **short interspersed repeat** is the Alu family that constitutes a substantial portion (about 5%) of the human genome. Alu sequences are approximately 300 bp long and are repeated over 500,000 times. The structures of the short interspersed repeats, including the Alu family, are reminiscent of transposons. The function of the Alu family remains to be established.

Highly Reiterated DNA

Single-copy and moderately repetitive sequences together normally account for more than 80% of the total nucleotide content of the eukaryotic genomes. The remaining DNA consists of sequences reiterated thousands or millions of times and are typically shorter than 20 nucleotides. About 10% of mouse DNA consists of 10-bp repeats that are reiterated millions of times. Because of the manner in which they are constructed, **highly reiterated DNAs** are also referred to as simple sequence DNA. Simple sequences are typically present in the DNA of most, if not all, eukaryotes. In some species, one major simple sequence is present, while in others, several simple sequences are repeated up to one million times. Some considerably longer repeat units for simple sequence DNA have been identified. For instance, in the genome of the African green monkey a 172-bp segment is highly reiterated and contains a few sequence repetitions within the segment. Because of its characteristic composition, simple sequence DNA can often be isolated as satellite DNA. Satellite DNA found in the centromeres of higher eukaryotes consists of thousands of tandem copies of one or a few short sequences. Satellite sequences are only 5–10 bp long. Simple sequence (satellite) DNA is a constituent of telomeres where it has a well-defined role in DNA replication.

Inverted Repeat DNA

Inverted repeats are a structural motif of DNA. Short inverted repeats, consisting of up to six nucleotides, for example, the palindromic sequence GAATTC, occur by chance about once for every 3000 nucleotides. Such short repeats cannot form a stable "hairpin" structure as is formed by longer palindromic sequences. Inverted repeat sequences that are long enough to form stable "hairpins" are not likely to occur by chance and should be classified as a separate class of eukaryotic sequence. In human DNA about two million inverted repeats are present, with an average length of about 200 bp; however, inverted sequences longer than 1000 bp have been detected. Most inverted repeat sequences are repeated 1000 or more times per cell.

2.6 | RNA OVERVIEW

RNA molecules play different roles in cellular information transfer (Table 2.5). Most RNA in a cell is ribosomal RNA (rRNA). Ribosomes are large complexes of proteins and RNA; they carry out translation. Messenger RNAs (mRNAs) serve as templates for the synthesis of protein; they carry information from the DNA to the cellular protein synthetic machinery. Transfer RNAs (tRNAs) transfer specific amino acids from soluble amino acid pools to ribosomes and ensure their proper alignment prior to peptide bond formation. Other RNAs in the cell are important in RNA processing and in protein export from the cell.

TABLE 2.5 Characteristics of Cellular RNAs

Type of RNA	*Abbreviation*	*Function*	*Size and Sedimentation Coefficient*	*Site of Synthesis*	*Structural Features*
Messenger RNA Cytosolic	mRNA	Transfer of genetic information from nucleus to cytoplasm, or from gene to ribosome	Depends on size of protein 1000–10,000 nucleotides	Nucleoplasm	Blocked 5′-end; poly (A) tail on 3′-end; nontranslated sequences before and after coding regions; few base pairs and methylations
Mitochondrial	mt mRNA		9S–40S	Mitochondria	
Transfer RNA Cytosolic	tRNA	Transfer of amino acids to mRNA–ribosome complex and correct sequence insertion	65–110 nucleotides 4S	Nucleoplasm	Highly base paired; many modified nucleotides; common specific structure
Mitochondrial	mt tRNA		3.2S–4S	Mitochondria	
Ribosomal RNA Cytosolic	rRNA	Structural framework for ribosomes	28S, 5400 nucleotides 18S, 2100 nucleotides 5.8S, 158 nucleotides 5S, 120 nucleotides	Nucleolus Nucleolus Nucleolus Nucleoplasm	5.8S and 5S highly base paired; 28S and 18S have some base paired regions and some methylated nucleotides
Mitochondrial	mt rRNA		16S, 1650 nucleotides 12S, 1100 nucleotides	Mitochondria	
Heterogeneous nuclear RNA	hnRNA	Some are precursors to mRNA and other RNAs	Extremely variable 30S–100S	Nucleoplasm	mRNA precursors may have blocked 5′-ends and 3′-poly(A) tails; many have base paired loops
Small nuclear RNA	snRNA	Structural and regulatory RNAs in chromatin	100–300 nucleotides	Nucleoplasm	
Small cytosolic RNA [7S(L) RNA]	scRNA	Selection of proteins for export	129 nucleotides	Cytosol and rough endoplasmic reticulum	Associated with proteins as part of signal recognition particle

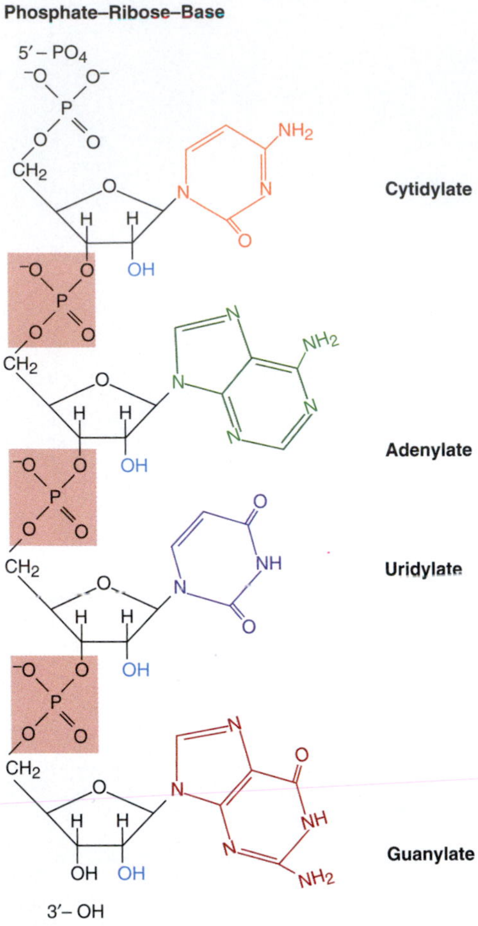

Phosphate–Ribose–Base

5′–PO_4

Cytidylate

Adenylate

Uridylate

Guanylate

3′–OH

FIGURE 2.55
Structure of 3′,5′-phosphodiester bonds between ribonucleotides forming a single strand of RNA.
Phosphate joins the 3′-OH group of one ribose with the 5′-OH group of the next ribose. This linkage produces a polyribonucleotide having a sugar–phosphate "backbone." Purine and pyrimidine bases extend away from the axis of the backbone and may pair with complementary bases to form double-helical base-paired regions.

2.7 | STRUCTURE OF RNA

RNA Is a Polymer of Ribonucleoside 5′-Monophosphates

RNA is an unbranched linear polymer of ribonucleoside monophosphates. The purines found in RNA are adenine and guanine; the pyrimidines are cytosine and uracil. Except for uracil, which replaces thymine, these are the same bases found in DNA. A, C, G, and U nucleotides are incorporated into RNA during transcription. Many RNAs also contain **modified nucleotides,** which are produced by processing. Modified nucleotides are especially characteristic of stable RNA species (i.e., tRNA and rRNA); however, some methylated nucleotides are also present in eukaryotic mRNA. For the most part, modified nucleotides in RNA have "fine tuning" rather than indispensable roles in the cell.

The 3′,5′-phosphodiester bonds of RNA form a chain or backbone from which the bases extend (Figure 2.55). Eukaryotic RNAs vary from approximately 65 nucleotides long to more than 200,000 nucleotides long. RNA sequences are complementary to the base sequences of specific portions of only one strand of DNA. Thus, unlike the base composition of DNA, molar ratios of (A + U) and (G + C) in RNA are not equal. Cellular RNA is linear and single stranded, but double-stranded RNA is present in some viral genomes.

Chemically, RNA is similar to DNA. Both contain negatively charged phosphodiester bonds, and the bases are very similar chemically. The chemical differences between DNA and RNA are largely due to two factors. First, RNA contains ribose rather than 2′-deoxyribose as the nucleotide sugar component, and, second, RNAs are generally single stranded rather than double stranded.

The 2′-hydroxyl group makes the phosphodiester bonds of an RNA molecule more susceptible to chemical hydrolysis than those of DNA. RNA phosphodiester bonds rapidly hydrolyze in alkaline solution, whereas the chemically similar phosphodiester bonds of DNA chains are quite stable. The chemical instability of RNA is reflected in its metabolic instability. Some RNAs, such as bacterial mRNA, are synthesized, used, and degraded within minutes. Others, such as human rRNA, are more stable metabolically, with a lifetime measured in days. Nevertheless, even the most stable RNAs are much less stable than is DNA.

Secondary Structure of RNA Involves Intramolecular Base Pairing

Since RNA molecules are single stranded, they do not usually form extensive double helices. Rather, the secondary structure of an RNA molecule results from relatively short regions of **intramolecular base pairing,** that is, base pairs formed from complementary sequences in the same molecule. Even the nonpaired sequences of single-stranded RNAs contain considerable helical structure (Figure 2.56). This hel-

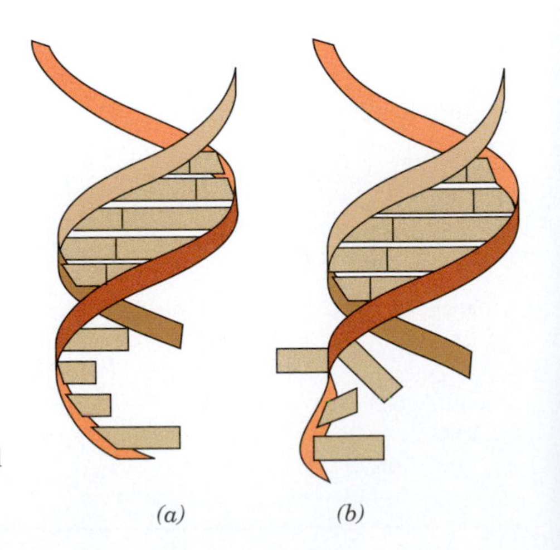

FIGURE 2.56
Helical structure of RNA.
Models indicating a helical structure due to (*a*) base stacking in the CCA terminus of tRNA and (*b*) lack of an ordered helix when no stacking occurs in this non-base-paired region.
Redrawn from Sprinzl, M. and Cramer, F. Prog. Nucl. Res. Mol. Biol. *22:9, 1979.*

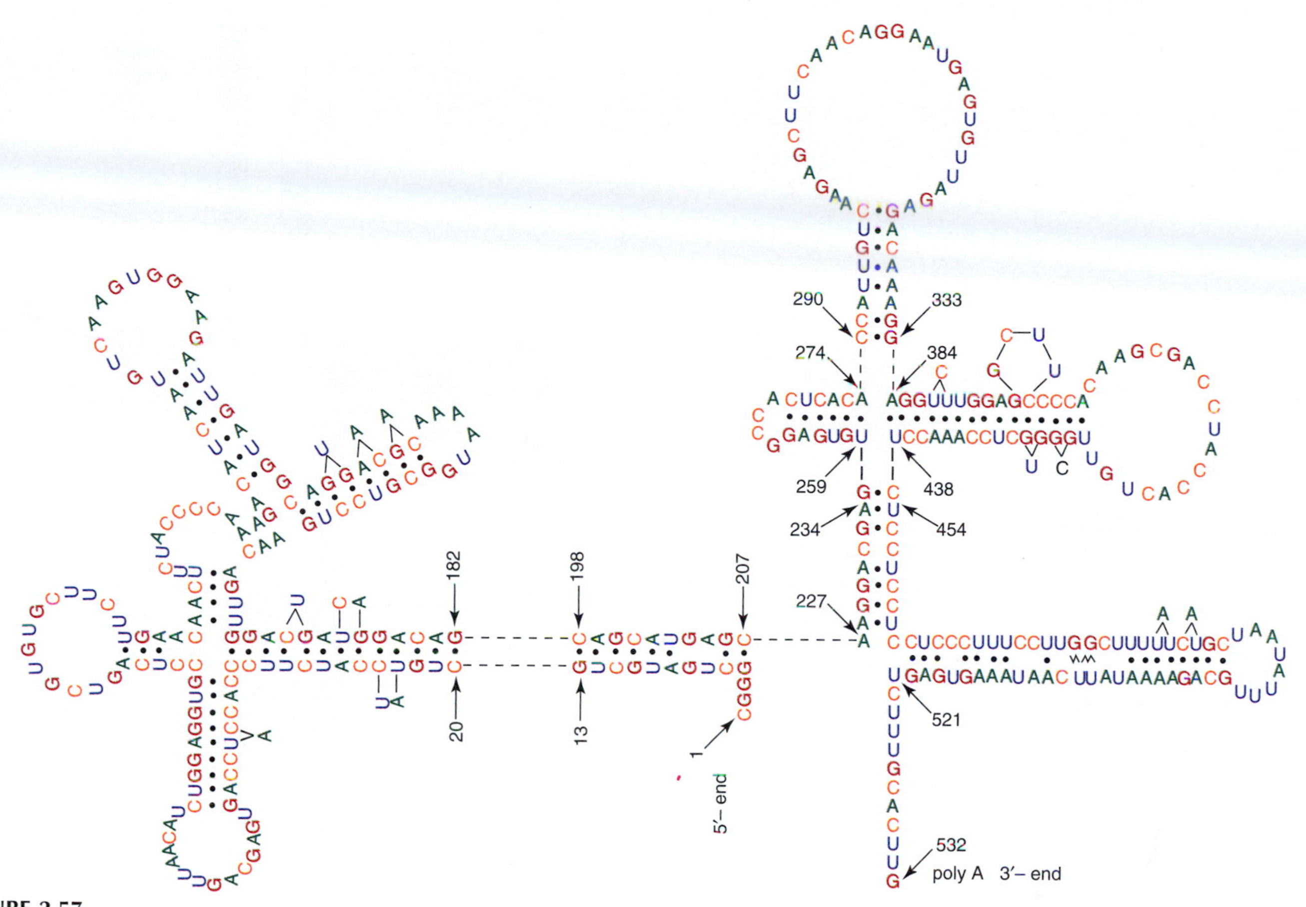

FIGURE 2.57
Proposed base-pairing regions in the mRNA for mouse immunoglobulin light chain. Base-paired structures shown have free energies of at least −5 kcal. Note the variance in loop size and length of paired regions.
Redrawn from Hamlyn, P. H., Browniee, G. G., Cheng, C. C., Gait, M. J., and Milstein, C. Cell 15:1067, 1978.

ical structure results from the strong base-stacking forces between A, G, and C. RNA double-helical structures generally are of the "A-type" with 11 nucleotides per turn in the double helix.

Double-helical stem–loop regions in RNA are often called "**hairpins.**" There are considerable variations in the fine structural details of "hairpin" structures, including the length of base-paired regions and the size and number of unpaired loops (Figure 2.57). Transfer RNAs are excellent examples of base stacking and hydrogen bonding in a single-chain RNA molecule (Figure 2.58). About 60% of bases are paired in four double-helical stems. In addition, the "unpaired" regions have the capability to form base pairs with free bases in the same or other looped regions, thereby contributing to the molecule's tertiary structure. The anticodon region in tRNA is an unpaired, base-stacked loop of seven nucleotides. The partial helix caused by base stacking in this loop binds, by base pairing, to a complementary codon in mRNA so that translation can occur.

RNA Molecules Have Tertiary Structures

Structures of functional RNA molecules are more complex than the base-stacked and hydrogen-bonded helices mentioned above. *In vivo* RNAs are dynamic molecules that change their conformation during synthesis, processing, and function-

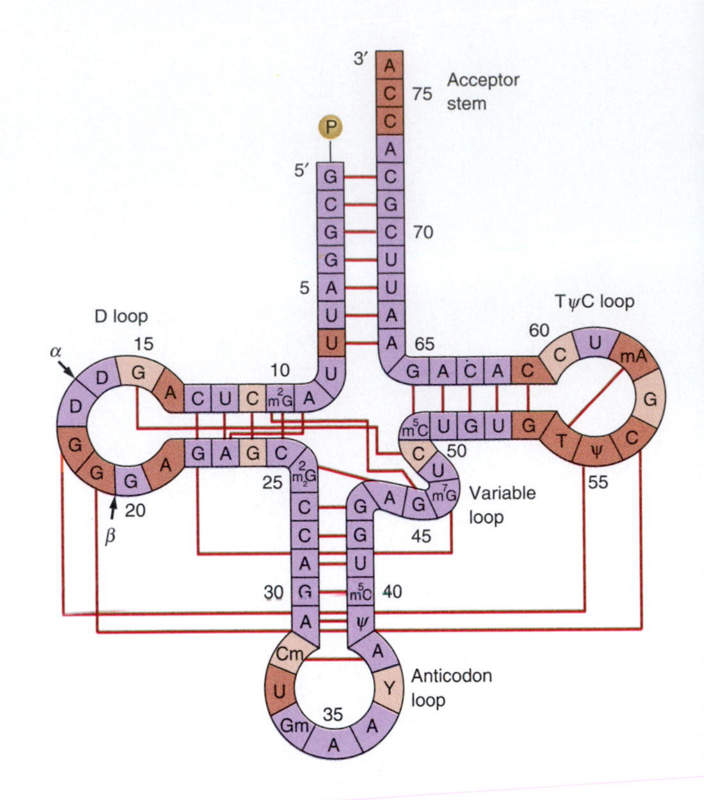

FIGURE 2.58
Cloverleaf structure of tRNA.
Cloverleaf diagram of the two-dimensional structure and nucleotide sequence of yeast $tRNA^{Phe}$. Red lines connecting nucleotides indicate hydrogen-bonded bases. Rose squares indicate constant nucleotides; tan squares indicate a constant purine or pyrimidine. Insertion of nucleotides in the D loop occurs at positions α and β for different tRNAs.
Redrawn with permission from Quigley, G. J. and Rich, A. Science 194:797, 1976. Copyright © 1976 by the American Association for the Advancement of Science.

ing. Proteins associated with RNA molecules often lend stability to the RNA structure. In fact, cellular RNA functions as RNA–protein complexes, rather than as free RNA molecules. The **tertiary structure** of RNA molecules results from base stacking and hydrogen bonding between different parts of the molecule. tRNA provides a number of examples. In solution, tRNA is folded into a compact "L-shaped" conformation (Figure 2.59), stabilized by Watson–Crick base pairing and

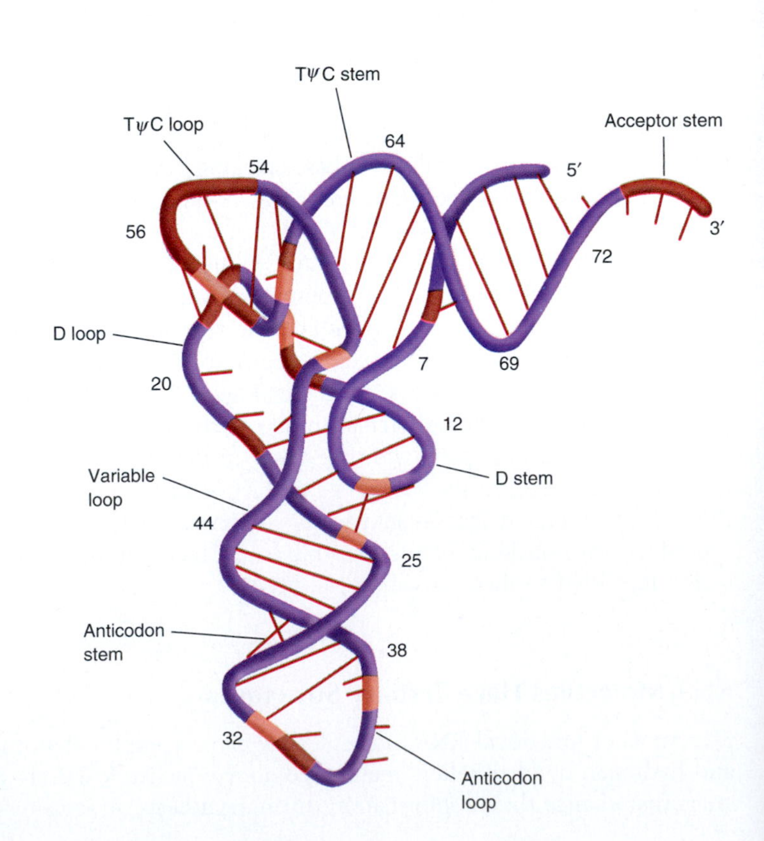

FIGURE 2.59
Tertiary structure on tRNA.
Tertiary folding of the cloverleaf structure of $tRNA^{Phe}$. Hydrogen bonds are indicated by cross rungs. Compare the presentation with Figure 2.58.
Redrawn with permission from Quigley, G. J. and Rich, A. Science 194:797, 1976. Copyright © 1976 by the American Association for the Advancement of Science.

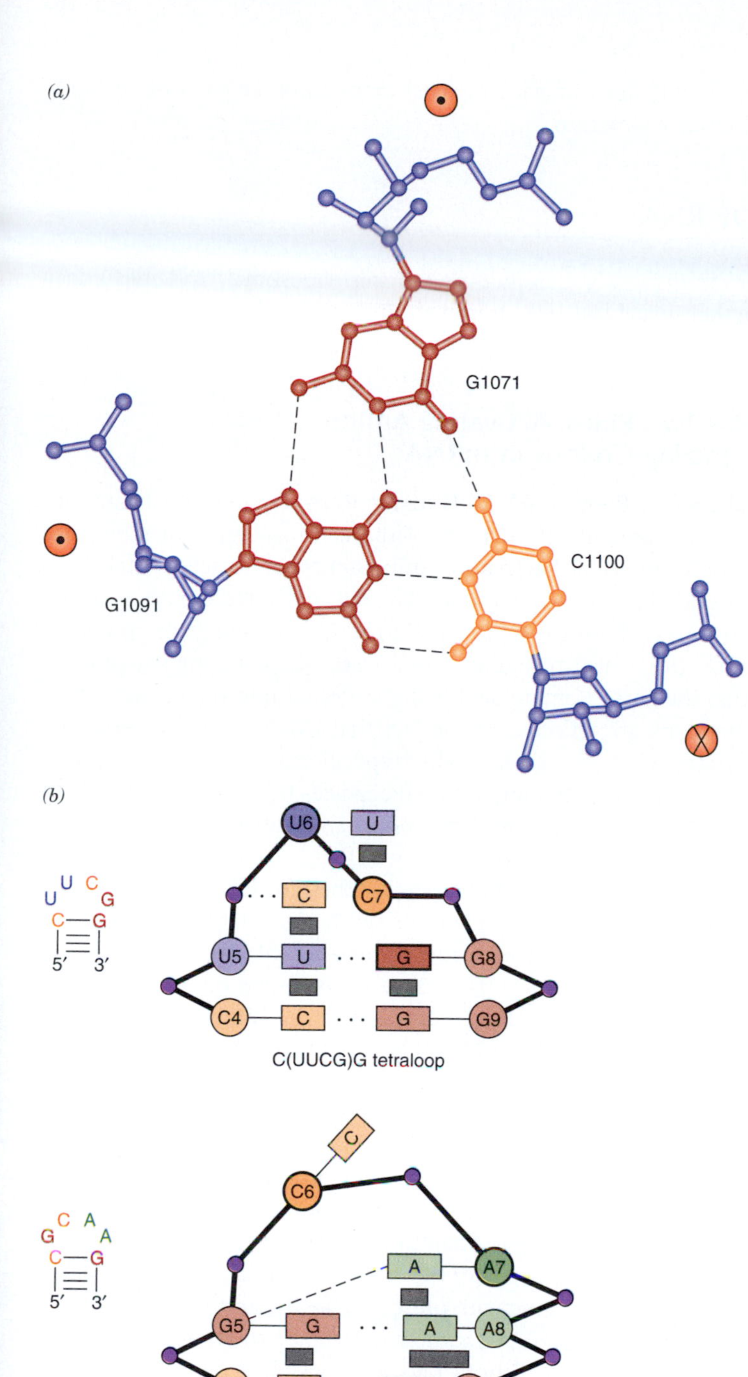

FIGURE 2.60
Hydrogen bonding in tRNA.
(*a*) A base triplet from 23S ribosomal RNA. In this case, the third base (G1071) forms hydrogen bonds with the "top" or "Hoogsteen" face of both G1091 and C1100. (*b*) Schematic diagram showing that two RNA tetraloops have similar structures. Both structures are stabilized by stacking of the loop bases, by non-Watson–Crick hydrogen bonds between bases and by hydrogen bonding between sugar, phosphate, and ribose residues.
Part (a) redrawn from Westhof, E. and Fritsch, V. Structure *8:R55, 2000; part (b) redrawn from Wyatt, J. and Tinoco, I. Jr. In: R. F. Gesteland and J. F. Atkins, (Eds.),* The RNA World. *Cold Spring Harbor, NY: Cold Spring Harbor Press, 1993, p. 471.*

base interactions involving more than two nucleotides. This is true for other RNAs. Figure 2.60 shows a well-defined base triplet from bacterial ribosomal RNA. Bases can donate hydrogen atoms to bond with the phosphodiester backbone. The 2′-OH of the ribose is also an important donor and acceptor of hydrogens. All these interactions contribute to the folded shape of an RNA molecule.

Other tertiary features were deduced from the three-dimensional structure of RNAs in solution or in crystal form. The term **"tetraloop"** refers to specific sequences (e.g., UUCG or GAGA) that are often found at the closed end of RNA hairpins (Figure 2.60). These tetraloops impart extra stability to RNA hairpins by forming internal hydrogen bonds between bases and sugars. Tetraloops can hydrogen bond with regions of the double helix or with internal loops in other parts of an RNA, thereby stabilizing tertiary structure.

2.8 | TYPES OF RNA

RNA molecules are traditionally classified as transfer, ribosomal, and messenger RNAs according to their usual function; however, RNA molecules perform a variety of other functions in a cell.

Transfer RNA Has Two Roles: Activating Amino Acids and Recognizing Codons in mRNA

About 15% of total cellular RNA is **tRNA. Transfer RNA** is an essential intermediary between DNA and protein information in the central dogma of molecular biology. Binding to tRNA molecules **activates amino acids** for protein synthesis so that formation of peptide bonds is energetically favored. Activated amino acids are transported to ribosomes where they are transferred to growing peptide chains (hence tRNA's name). tRNA also recognizes nucleotide sequence information in mRNA to ensure that the correct amino acid is incorporated into the growing peptide chain. Both functions are reflected in the fact that tRNAs have two essential parts, the 3′-OH terminal CCA sequence, to which specific amino acids are attached enzymatically, and the **anticodon triplet,** which recognizes mRNA sequence. Three nucleotides in the mRNA form a codon. The codon and anticodon triplets form base pairs during peptide bond formation.

Each tRNA can transfer only one amino acid. Although only 20 amino acids are used in protein synthesis, free-living organisms synthesize a larger set of tRNAs. This is a consequence of the genetic code being redundant (having more than one codon per amino acid). For example, the genomic sequence of the bacterium *Haemophilus influenzae* contains genes for 54 tRNA species. Mitochondria synthesize a much smaller number of tRNAs. tRNAs that accept the same amino acid are called **isoacceptors.** A tRNA that accepts phenylalanine would be written as $tRNA^{Phe}$, whereas one accepting tyrosine would be written $tRNA^{Tyr}$.

tRNAs are 65–110 nucleotides long, corresponding to a molecular weight range of 22,000–37,000. The sequences of all tRNA molecules (thousands are known) can be arranged into a common **cloverleaf secondary structure** by complementary Watson–Crick base pairing to form three stem and loop structures. The anticodon triplet sequence is at one "leaf" of the cloverleaf while the CCA acceptor stem is at the "stem." This arrangement is preserved in the tertiary structure of $tRNA^{Phe}$ (see Figure 2.59), where the anticodon and acceptor stems are at each end of the L-shaped molecule. Additional, non-Watson–Crick, hydrogen bonds form the tertiary structure of the L-shaped molecule. tRNAs contain **modified nucleotides,** for example, 7-methyguanine at position 46 in Figure 2.58. Modified nucleotides affect tRNA structure and stability but are not required for their basic functioning. For example, a modified base next to the anticodon makes codon recognition more efficient but a tRNA without this modification can still be read correctly by the ribosome.

Many structural features are common to all tRNA cloverleafs. Seven base pairs form the amino acid acceptor stem, which terminates with the nucleotide triplet CCA. This CCA triplet is not base paired. The dihydrouracil or D stem has three or four base pairs, while the anticodon and T stems have five base pairs each. Both the anticodon loop and T loop contain seven nucleotides. Differences in the number of nucleotides in different tRNAs are accounted for by the variable loop. Most tRNAs have small variable loops of 4–5 nucleotides, while others have larger loops of 13–21 nucleotides. The positions of some nucleotides are constant in all tRNAs (see the dark orange boxes in Figure 2.58).

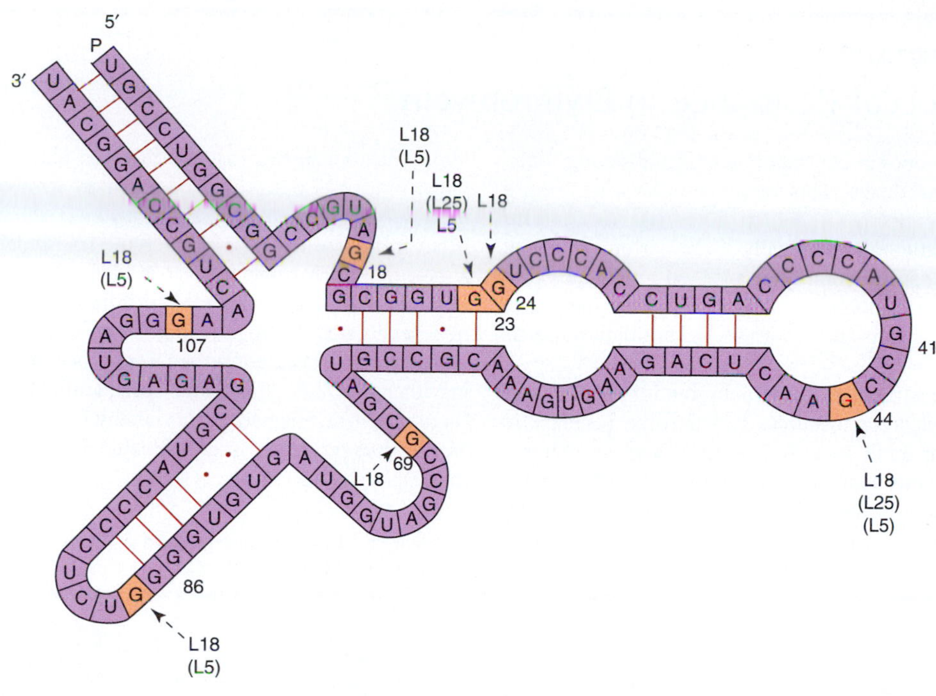

FIGURE 2.61
Secondary, base-paired, structure proposed for 5S rRNA.
Arrows indicate regions protected by proteins in the large ribosomal subunit.
Combined information from Fox, G. E. and Woese, C. R. Nature 256:505, 1975; and R. A. Garrett and P. N. Gray.

Ribosomal RNA Is Part of the Protein Synthesis Apparatus

Protein synthesis takes place on ribosomes. In eukaryotes these complex assemblies are composed of four RNA molecules, representing about two-thirds of the particle mass, and 82 proteins. The smaller subunit, the **40S particle,** contains one **18S RNA** and 33 proteins. The larger subunit, the **60S particle,** contains **28S, 5.8S,** and **5S** rRNAs and 49 proteins. The total assembly forms the **80S ribosome.** Prokaryotic ribosomes are smaller; the **30S** subunit contains a single **16S rRNA** and 21 proteins while the larger 50S subunit contains **5S** and **23S rRNAs** and 34 proteins (see p. 242 for a discussion of the structure of ribosomes).

rRNA accounts for 80% of cellular RNA and is metabolically stable. This stability is required for repeated functioning of the ribosome and is enhanced by association with ribosomal proteins. Eukaryotic 28S (4718 nucleotides), 18S (1874 nucleotides), and 5.8S (160 nucleotides) rRNAs are synthesized in the nucleolar region of the nucleus. The 5S rRNA (120 nucleotides) is not transcribed in the nucleolus but rather from separate genes within the nucleoplasm (Figure 2.61). The three larger rRNAs are synthesized in one long polynucleotide chain, which is then processed to yield the individual molecules. Nucleotide modifications in rRNA are primarily **methylations** on the 2′-position of ribose, to yield 2′-*O*-methylribose. Methylation of rRNA has been directly related to bacterial antibiotic resistance in a pathogenic species (see Clin. Corr. 2.8). A small number of N^6-dimethyladenines are present in 18S rRNA. The 28S rRNA has about 45 methyl groups and the 18S rRNA has 30 methyl groups.

Ribosomal RNA is the catalytic component of the ribosome. The X-ray crystal structure of the large subunit of a bacterial ribosome species shows that the active site for peptide bond synthesis is composed of 23S RNA only—any potential catalytic amino acid side chains are too far away to participate directly in the reaction.

CLINICAL CORRELATION 2.8

Staphylococcal Resistance to Erythromycin

Bacteria exposed to antibiotics in clinical or agricultural settings often develop resistance to the drugs. This resistance can arise from a mutation in the target cell's DNA, which gives rise to resistant descendants. An alternative and clinically more serious mode of resistance arises when plasmids coding for antibiotic resistance proliferate through the bacterial population. These plasmids may carry multiple resistance determinants and render several antibiotics useless at the same time.

Erythromycin inhibits protein synthesis by binding to the large ribosomal subunit. *Staphylococcus aureus* can become resistant to erythromycin and similar antibiotics as a result of a plasmid-borne RNA methylase that converts a single adenosine in 23S rRNA to N^6-dimethyladenosine. Since the same ribosomal site binds lincomycin and clindamycin the plasmid causes cross-resistance to these antibiotics as well. Synthesis of the methylase is induced by erythromycin.

A microorganism that produces an antibiotic must also be immune to it or else it would be killed by its own toxic product. The producer of erythromycin, *Streptomyces erythreus,* itself possesses an rRNA methylase that acts at the same ribosomal site as the one from *S. aureus.* Which came first? It is likely that many of the resistance genes in target organisms evolved from those of producer organisms. In several cases, DNA sequences from resistance genes of the same specificity have been conserved in producer and target organisms. We may therefore look on plasmid-borne antibiotic resistance as a case of "natural genetic engineering," whereby DNA from one organism (e.g., *Streptomyces* producer) is appropriated and expressed in another (e.g., the *Staphylococcus* target).

Cundliffe, E. How antibiotic-producing microorganisms avoid suicide. *Annu. Rev. Microbiol.* 43:207, 1989.

Messenger RNAs Carry the Information for the Primary Structure of Proteins

mRNAs are the direct carriers of genetic information from genome to ribosomes. Each eukaryotic mRNA is **monocistronic,** that is, contains information for only one polypeptide chain. In prokaryotes, mRNA species are often **polycistronic,** encoding more than one protein. A cell's phenotype and functional state are related directly to its mRNA content. In the cytoplasm, mRNAs have relatively short life spans. Some are synthesized and stored in an inactive or dormant state in the cytoplasm, ready for a quick response when protein synthesis is required. In various animals, immediately upon fertilization the egg undergoes rapid protein synthesis in the absence of transcription, indicating that preformed mRNA and ribosomes are present in the egg cytoplasm.

Eukaryotic mRNAs have unique structural features that are not found in rRNA or tRNA (Figure 2.62). Since the information within mRNA lies in the linear sequence of the nucleotides, the integrity of this sequence is extremely important. Any loss or change of nucleotides could alter the structure of the protein being

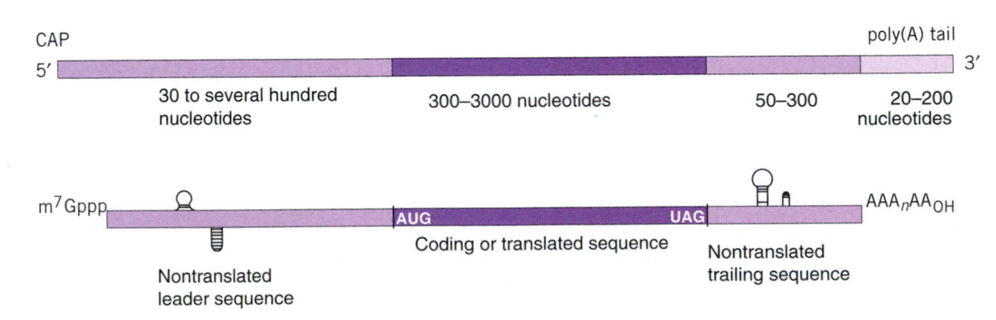

FIGURE 2.62
General structure for a eukaryotic mRNA.
There is a "blocked" 5′-terminus, cap, followed by a nontranslated leader containing a promoter sequence. Coding region usually begins with the initiator codon AUG and continues to the translation termination sequence UAG, UAA, or UGA. This is followed by the nontranslated trailer and a poly(A) tail on the 3′-end.

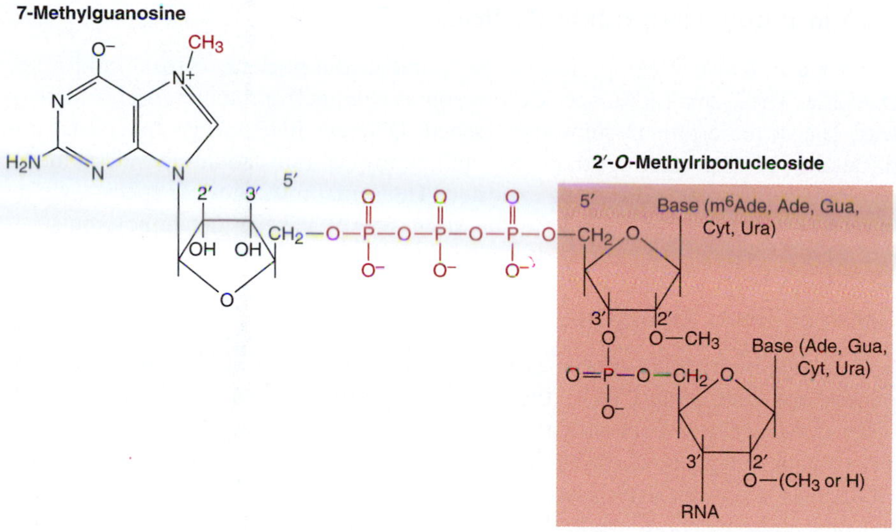

FIGURE 2.63
Diagram of "cap" structure or blocked 5′-terminus in mRNA.
The 7-methylguanosine is inverted to form a 5′-phosphate to 5′-phosphate linkage with the first nucleotide of the mRNA. This nucleotide is often a methylated purine.

translated. Translation of mRNA on the ribosomes must also begin and end at specific sequences. In eukaryotes, a **"cap" structure** at the 5′-end of an mRNA specifies where translation should begin. The caps of eukaryotic mRNAs contain an inverted phospodiester bond, a methylated base attached to the mRNA via a **5′-phosphate–5′-phosphate** rather than the usual 3′,5′-phosphodiester linkage. The cap is attached to the first transcribed nucleotide, usually a purine, methylated on the 2′-OH of the ribose (Figure 2.63). The cap is followed by a nontranslated or **"leader" sequence** to the 5′-side of the coding region. Following the leader sequence are the **initiation codon,** most often AUG, and the translatable coding region of the molecule. At the end of the coding sequence is a **termination sequence** signaling termination of polypeptide formation and release from the ribosome. A second nontranslated or **"trailer"** sequence follows, terminated by a string of 20–200 adenine nucleotides, called a **poly(A) tail,** which makes up the 3′-terminus of the mRNA.

The 5′-cap has a positive effect on initiation of message translation. During the initiation of translation of a mRNA, the cap structure is recognized by a single ribosomal protein, an initiation factor (see p. 245). The poly(A) sequence affects the stability of the mRNA; degradation of an mRNA often begins with shortening of its poly(A) tail.

Mitochondria Contain Unique RNA Species

Mitochondria (mt) have their own protein-synthesizing apparatus, including ribosomes, tRNAs, and mRNAs. Mitochondrial rRNAs, 12S, and 16S are transcribed from the mitochondrial DNA (**mtDNA**) as are 22 specific tRNAs and 13 mRNAs, most of which encode proteins of the electron transport chain and ATP synthase. Note that there are fewer mitochondrial tRNAs than prokaryotic or cytosolic tRNA species; there is generally only one mitochondrial tRNA species per amino acid. The mtRNAs account for 4% of the total cellular RNA. They are transcribed by a mitochondria-specific RNA polymerase and are processed from a pair of RNA precursors that contain tRNA and mRNA sequences. Expression of the nuclear and mitochondrial genomes is tightly coordinated. Most of the aminoacylating enzymes for the mitochondrial tRNAs and all of the mitochondrial ribosomal proteins are specified by nuclear genes, translated in the cytosol and transported into mitochondria. The modified bases in mitochondrial tRNA species are synthesized by enzymes encoded in nuclear DNA.

RNA in Ribonucleoprotein Particles

Other small, stable RNA species can be found in the nucleus, cytosol, and mitochondria. These small RNA species function as ribonucleoprotein particles (RNPs), with one or more protein subunits attached. Different RNP species have been implicated in RNA processing, transport, and control of translation, as well as in the recognition of proteins due to be exported. The actual roles of these species, where known, are described more fully in the discussion of specific metabolic events.

Catalytic RNA: Ribozymes

RNA can be catalytic. In several cases, the RNA component of a ribonucleoprotein particle has been shown to be the catalytically active subunit of the enzyme. In other cases, catalytic reactions can be carried out *in vitro* by RNA in the absence of any protein. Enzymes whose RNA subunits carry out catalytic reactions are called **ribozymes.** There are five classes of ribozyme. Three of these RNA species carry out self-processing reactions while the others, **ribonuclease P (RNase P)** and rRNA, are true catalysts that act on separate substrates.

In the ciliated protozoan *Tetrahymena thermophila,* an intron in the rRNA precursor is removed by a multistep reaction (Figure 2.64). A guanine nucleoside or nucleotide reacts with the intron–exon phosphodiester linkage to displace the donor exon from the intron. This reaction, a transesterification, is promoted by the folded intron itself. The free donor exon then similarly attacks the exon–intron phosphodiester bond at the acceptor end of the intron. Introns of this type **(Group I introns)** have been found in a variety of genes in fungal mitochondria, in bacteria, and in the bacteriophage T4. Although these introns are not true enzymes *in vivo* as they only work for one reaction cycle, they can be made to carry out catalytic reactions under specialized conditions.

Group II self-splicing introns are found in the mitochondrial RNA precursors of yeasts and other fungi. Even though these RNAs self-splice, there are many parallels between this reaction and the removal of introns from mRNA precursors during processing (see Chapter 5).

Group III of **self-cleaving RNAs** is found in the genomic RNAs of several plant viruses. These RNAs self-cleave during generation of single genomic RNA molecules from large multimeric precursors. The three-dimensional structure of the **hammerhead ribozyme,** a member of this third class, has been determined (Figure 2.65). Although the catalytic cycle is not completely determined, the amino group of a cytosine base plays an essential role in the cleavage reaction. The phosphate of the cleaved bond is left at the 3′ hydroxyl position of the RNA product. A self-cleaving RNA is found in a small satellite virus, hepatitis delta virus, that is implicated in severe cases of human infectious hepatitis. All of the above self-processing RNAs can be made to act as true catalysts (i.e., exhibiting multiple turnover) *in vitro* and *in vivo.*

The fourth type of ribozyme, **ribonuclease P,** contains both a protein and an RNA component. It acts as a true enzyme in cells, cleaving tRNA precursors to generate the mature 5′-end of NA molecules. RNase P recognizes constant structures associated with tRNA precursors (e.g., the acceptor stem and CCA sequence) rather than using extensive base pairing to bind the substrate RNA to the ribozyme. The product of cleavage contains a 5′ phosphate in contrast to the products of hammerhead and similar RNAs. In all of these events the structure of the catalytic RNA is essential for catalysis.

Finally, and most recently, the X-ray crystallographic structure of a bacterial ribosome reveals that it, too, is a ribozyme. The active site of the ribosome is on the 50S subunit. There are no protein functional groups close enough to catalyze peptide bond formation, leading to the conclusion that some part of the RNA chain serves as the catalyst for peptide bond formation.

Discovery of RNA catalysis has greatly altered our concepts of biochemical evolution and the range of allowable cellular chemistry. First, we now recognize that

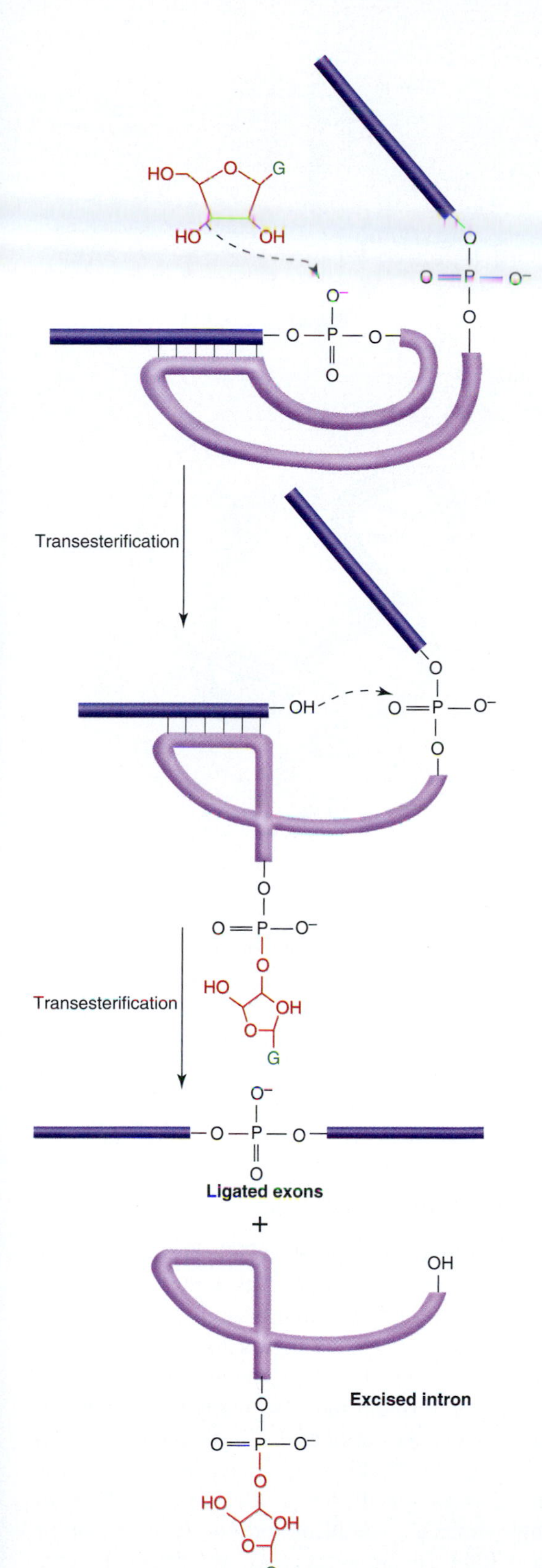

FIGURE 2.64
Mechanism of self-splicing of the rRNA precursor of *Tetrahymena.*
Two exons of rRNA are denoted by dark blue. Catalytic functions reside in the intron, which is purple. This splicing function requires an added guanine nucleoside or nucleotide.
Redrawn from Cech, T. R. JAMA 260:308, 1988.

RNA can serve as both a catalyst and a carrier of genetic information. This has raised the possibility that the earliest living organisms were based entirely on RNA and that DNA and proteins evolved later. This model is sometimes referred to as the **"RNA world."** Second, we know that many viruses, including human pathogens, use RNA genetic information; some of these RNAs have been shown to be catalytic. Thus catalytic RNA presents opportunities for the discovery of RNA-based phar-

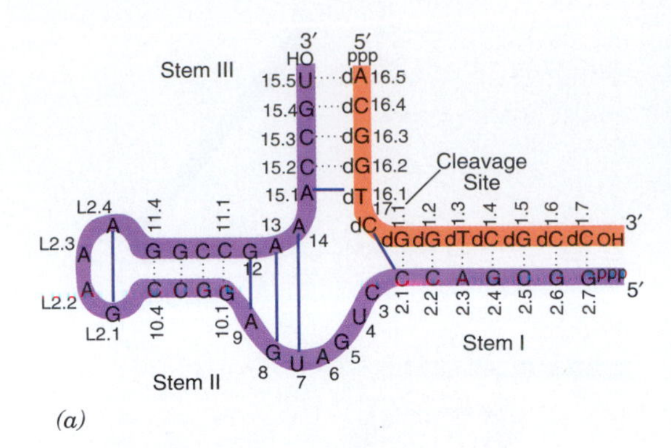

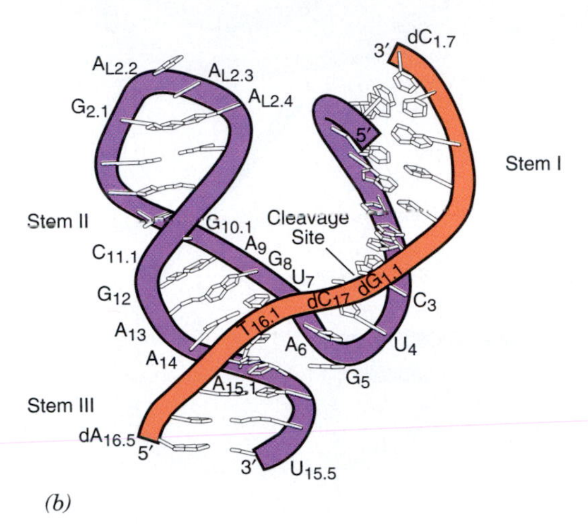

FIGURE 2.65
"Hammerhead" structure of viral RNA.
(*a*)"Hammerhead" structure of a self-cleaving viral RNA. This artificial molecule is formed by the base pairing of two separate RNAs. Cleavage of the RNA sequence at the site indicated by the arrow in the top strand requires its base pairing with the sequence at the bottom of the molecule. The boxed nucleotides are a consensus sequence found in self-cleaving RNA viral RNAs. (*b*) Three-dimensional folding of the hammerhead catalytic RNA. Star indicates the position of the cleaved bond while M indicates a binding site for a metal ion. Helices II and III stack to form an apparently continuous helix while non-Watson–Crick interactions position the noncomplementary bases in the hammerhead into a "uridine turn" structure identical to that found in tRNA.
Part (a) redrawn from Sampson, J. R., Sullivan, F. X., Behlen, L. S., DiRenzo, A. B., and Uhlenbeck, O.C. Cold Spring Harbor Symp. Quant. Biol. 52:267, 1987; part (b) redrawn from Pley, H. W., Flaherty, K. M., and McKay, D. B. Nature 372:68, 1994.

maceuticals. Third, many of the information-processing events in mRNA splicing require RNA components. These RNAs may also be acting as catalysts. Finally, ribozymes can be made to specifically recognize and cleave various cellular mRNAs. When introduced into cultured cells, these ribozymes are efficient agents to prevent gene expression. This is only a research tool at present, but it has been a useful strategy for determining the function of genes whose metabolic activity can only be guessed at from genomic sequence information.

RNAs Can Form Binding Sites for Other Molecules

Consideration of the RNA world has led to a new type of "combinatorial biochemistry" based on the large number of potential sequences (4^N) that could be made if A,C,G, or U were inserted randomly in each of *N* positions in a nucleic acid. A set of chemically synthesized, randomized, nucleic acid molecules 25 nucleotides long would contain $4^{25} = 10^{15}$ potential members. Individual molecules within this large collection of RNAs would be expected to fold into a similarly large collection of shapes. The large number of molecular shapes implies that some member of this collection should be capable of strong, specific binding to any ligand, much as Group I introns bind guanine nucleotides specifically. These binding RNAs are termed "**aptamers.**" Though a single molecule would be too rare to study within the original population, the aptamer RNA can be selected and preferentially replicated *in vitro*. In one case, for example, an RNA capable of distinguishing **theophylline** from **caffeine** was selected from a complex population (Figure 2.66). Theophylline is used in the treatment of chronic asthma but the level must be carefully controlled to avoid side effects. The monitoring of theophylline by conventional antibody-based clinical chemistry is difficult because caffeine and theophylline differ only by a single methyl group. Therefore anti-theophylline antibodies show considerable cross-reaction with caffeine. RNA aptamers have been found that

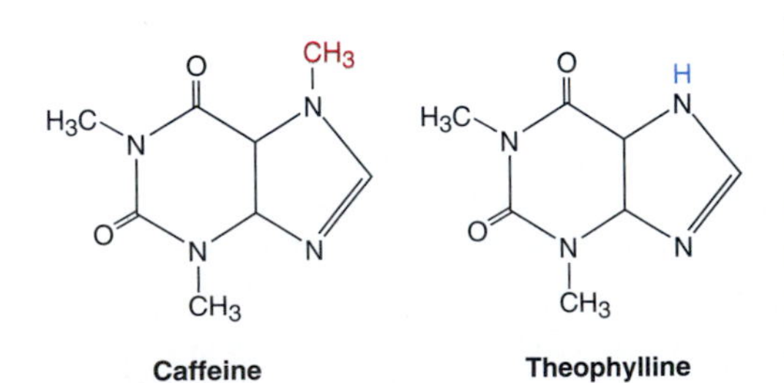

FIGURE 2.66
Structures of caffeine and theophylline. Although these compounds differ only by a single methyl group, a specific synthetic RNA can bind to theophylline 10,000-fold more tightly than to caffeine.

bind theophylline 10,000-fold more tightly than caffeine. Other extensions of the technology have used selection procedures to identify new, synthetic ribozymes and potential therapeutic RNAs.

BIBLIOGRAPHY

DNA

Bates, A. D. *DNA Topology.* Oxford, England: IRL Press, 1993.

Blackburn, G. M. and Gait, M. J. *Nucleic Acids in Chemistry and Biology,* 2nd ed. New York: Oxford University Press, 1996.

Bouffler, S., Silver, A., and Cox, R. The role of DNA repeats and associated secondary structures in genomic instability and neoplasia. *Bioessays* 15:409, 1993.

Cech, T. R. Life at the end of the chromosome: telomeres and telomerase. *Angew. Chem. Int. Ed.* 39:34, 2000.

Crothers, D. M. and Shakked, Z. DNA bending by adenine–thymine tracts. In: S. Neidle (Ed.), *Oxford Handbook of Nucleic Acid Structure.* New York: Oxford University Press, 1999, p. 455.

De Bruijn, F. J., Lupski, J. R., and Weinstock, G. M. (Eds.). *Bacterial Genomes: Physical Structure and Analysis.* New York: Chapman and Hall, 1998.

De Vries, J. E. and de Waal Malefyt, R. *Therapeutic Applications of Oligonucleotides.* New York: Demos Vermande, 1995.

Dickerson, R. E. DNA structure from A to Z. *Methods Enzymol.* 211:67, 1992.

Liu, M. A., Hilleman, M. R., and Kurth, R. (Eds). *DNA Vaccines: A New Era in Vaccinology.* New York: New York Academy of Sciences, 1995.

Feigon, J. A new DNA quadruplex. *Curr. Biol.* 3:611, 1993.

Hecht, S. M. (Ed.). *Bioorganic Chemistry: Nucleic Acids.* New York: Oxford University Press , 1996.

Kornberg, R. D. and Lorch, Y. Twenty-five years of the nucleosome, fundamental particle of the eukaryote chromosome. *Cell* 98:285, 1999.

Laughlin, G., Murchie, A. I. H., Norman, D. G., Moore, M. H., Moody, P. C. E., Lilley, D. M. J., and Luisi, B. The high-resolution crystal structure of a parallel-stranded guanine tetraplex. *Science* 265:520, 1994.

Leach, D. R. Long DNA palindromes, cruciform structures, genetic instability and secondary structure repair. *Bioessays* 16:893, 1994.

Lee, M. S. and Garrard, W. T. Positive DNA supercoiling generates a chromatin conformation characteristic of highly active genes. *Proc. Natl. Acad. Sci. USA* 88:9675, 1991.

Murchie, A. I. H. and Lilley, D. M. J. Supercoiled DNA and cruciform structures. *Methods Enzymol.* 211:158, 1992.

Praseuth, D., Guieysse, A. L., and Helene, C. Triple helix formation and the antigene strategy for sequence-specific control of gene expression. *Biochim. Biophys. Acta* 1489:181, 1999.

Rich, A. DNA comes in many forms. *Gene* 135:99, 1993.

Rich, A. Speculation on the biological roles of left-handed Z-DNA. In: S. S. Wallace, B. van Houten, and Y. W. Kow (Eds.), *DNA Damage: Effects on DNA Structure and Protein Recognition.* New York: New York Academy of Sciences, 1994, p. 1.

Saenger, W. *Principles of Nucleic Acid Structure.* New York: Springer-Verlag, 1984.

Tenover, F. C. *DNA Probes for Infectious Diseases.* Boca Raton, FL: CRC Press, 1989.

Ussery, D. W. and Sinden, R. R. Environmental influences on the *in vivo* level of intramolecular triplex DNA in *Escherichia coli. Biochemistry* 32:6206, 1993.

Van Holde, K. and Zlatanova, J. Unusual DNA structures, chromatin and transcription. *Bioessays* 16:59, 1994.

Wells, R. D. Unusual DNA structures. J. Biol. Chem. 263:1095, 1988.

Williamson, J. R. G-quartet structures in telomeric DNA. *Annu. Rev. Biophys. Biomol. Struct.* 23:703, 1994.

Yakubovskaya, E. A. and Gabibov, A. G. Topoisomerases: mechanisms of DNA topological alterations. *Mol. Biol.* 33:318, 1999.

RNA

Cundliffe, E. How antibiotic-producing microorganisms avoid suicide. *Annu. Rev. Microbiol.* 43:207, 1989.

Dahlberg, A. E. The ribosome in action. *Science* 292:868, 2001.

Gesteland, R. F., Cech, T. and Atkins, J. F. (Eds.). *The RNA World,* 2nd ed. Cold Spring Harbor, NY: Cold Spring Harbor Laboratory Press, 1998.

Gold, L., Polisky, B., Uhlenbeck, O., and Yarus, M. Diversity of oligonucleotide functions. *Annu. Rev. Biochem.* 64:763, 1995.

Ibba, M., Becker, H. D., Stathopoulos, C., Tumbula, D. L., and Söll, D. The adaptor hypothesis revisited. *Trends Biochem. Sci.* 25:311, 2000

Lai, M. M. C. The molecular biology of hepatitis delta virus. *Annu. Rev. Biochem.* 64:259, 1995.

Pace, N. R. and Brown, J. W. Evolutionary perspective on the structure and function of ribonuclease P, a ribozyme. *J. Bacteriol.* 177:1919, 1995.

Söll, D. and Rajbhandary, U. L. (Eds.). *tRNA: Structure, Biosynthesis and Function.* Washington DC: American Society for Microbiology, 1994.

Westhof, E. and Fritsch, V. RNA folding: beyond Watson–Crick pairs. *Structure* 8:R55, 2000.

QUESTIONS | C. N. ANGSTADT

Multiple Choice Questions

1. A polynucleotide is a polymer in which:
   A. the two ends are structurally equivalent.
   B. the monomeric units are joined together by phosphodiester bonds.
   C. there are at least 20 different kinds of monomers that can be used.
   D. the monomeric units are not separated by hydrolysis.
   E. purine and pyrimidine bases are the repeating units.

2. The best definition of an endonuclease is an enzyme that hydrolyzes:
   A. a nucleotide from only the 3′-end of an oligonucleotide.
   B. a nucleotide from either terminal of an oligonucleotide.
   C. a phosphodiester bond located in the interior of a polynucleotide.
   D. a bond only in a specific sequence of nucleotides.
   E. a bond that is distal (d) to the base that occupies the 5′-position of the bond.

3. A palindrome is a sequence of nucleotides in DNA that:
   A. is highly reiterated.
   B. is part of the introns of eukaryotic genes.
   C. is a structural gene.
   D. has local symmetry and may serve as a recognition site for various proteins.
   E. has the information necessary to confer antibiotic resistance in bacteria.

4. In a DNA double helix:
   A. the individual strands are not helical.
   B. hydrogen bonds form between a purine and a pyrimidine base on the same strand.
   C. adenine on one strand is hydrogen bonded to thymine on the opposite strand.
   D. phosphodiester bonds are oriented toward the interior of the helix.
   E. the outside of the helix is uncharged.

5. The A helix of DNA differs from the B helix in all of the following EXCEPT:
   A. appearance of the major and minor grooves.
   B. pitch of the base pairs relative to the helix axis.
   C. thickness of the helix.
   D. tilt of the bases.
   E. polarity of the strands.

6. The Z-DNA helix:
   A. has fewer base pairs per turn than the B-DNA.
   B. is favored by an alternating GC sequence.
   C. tends to be found at the 3′-end of genes.
   D. is inhibited by methylation of the bases.
   E. is a permanent conformation of DNA.

7. An interspersion pattern in DNA consists of:
   A. highly repetitive DNA sequences.
   B. the portion of DNA composed of single-copy DNA.
   C. Alu sequences.
   D. alternating blocks of single-copy DNA and moderately repetitive DNA.
   E. alternating blocks of short interspersed repeats and long interspersed repeats.

8. A nucleosome:
   A. is a regularly repeating structure of DNA and histone proteins.
   B. has a core of DNA with proteins wrapped around the outside.
   C. uses only one type of histone per nucleosome.
   D. is separated from a second nucleosome by nonhistone proteins.
   E. has histones in contact with the major groove of the DNA.

9. RNA:
   A. incorporates both modified and unmodified purine and pyrimidine bases during transcription.
   B. does not exhibit any double-helical structure.
   C. structures exhibit base stacking and hydrogen-bonded base pairing.
   D. usually contains about 65–100 nucleotides.
   E. does not exhibit Watson–Crick base pairing.

10. Ribozymes:
   A. are any ribonucleoprotein particles.
   B. are enzymes whose catalytic function resides in RNA subunits.
   C. carry out self-processing reactions but cannot be considered true catalysts.
   D. require a protein cofactor to form a peptide bond.
   E. function only in the processing of mRNA.

Questions 11 and 12: Nearly every process in which DNA participates requires that the DNA interact with proteins. Interaction with proteins for repair, transcription, recombination, and chromatin condensation, for example, depend on the local three-dimensional structure of DNA. One antitumor drug, cisplatin, which is used to treat tumors in testicular, ovarian, bone, and lung cancers is a tetracoordinate platinum complex. It forms intrastrand cross-links with DNA, causing the double helix to strongly bend toward the major groove. Cellular processes such as transcription and programed cell death are affected.

11. Bent DNA:
   A. occurs only in the presence of external agents like the antitumor drugs.
   B. may be a fundamental element in the interaction between DNA sequences and proteins for such processes as transcription and site-specific recombination.
   C. occurs primarily in the presence of triple-stranded DNA.
   D. requires the presence of inverted repeats.
   E. occurs only in DNA which is in the Z-form.

12. Bent DNA is not the only unusual conformation of DNA. Another form of DNA is a triple-stranded complex. Triple-stranded DNA:
   A. generally occurs in DNA in regions that play no role in transcription.
   B. involves Hoogsteen hydrogen bonding.
   C. is characterized by the presence of a string of alternating purine–pyrimidine bases.
   D. forms only intermolecularly.
   E. assumes a cruciform conformation.

Questions 13 and 14: Telomeres are guanine-rich repetitive sequences at the ends of linear eukaryotic chromosomes. They are progressively shortened during each cycle of cell division until a critical length is reached and programmed cell death occurs. The ribonucleoprotein, telomerase, maintains the length of telomeres for a finite number of cell divisions and then is turned off, but most tumor cells maintain telomerase activity. The higher the tumor telomerase activity, the poorer the clinical prognosis is. A current chemotherapeutic approach being studied is selective inhibition of telomerase. Certain drugs selectively bind to G-quadruplex DNA structure in telomerase and have been shown to inhibit telomerase activity. These drugs are now being studied for antitumor activity.

13. G-quadruplex DNA:
   A. is a stacked structure of guanine tetramers.
   B. is stabilized by metal cations.
   C. involves bases held together by Hoogsteen hydrogen bonds.
   D. can form either parallel or antiparallel arrangements.
   E. all of the above are correct.

14. Another approach to cancer treatment is drugs that inhibit topoisomerases. Topoisomerases:
   A. regulate the level of superhelicity of DNA in cells.
   B. always break only one strand of DNA.
   C. can create but not remove supercoils.
   D. must hydrolyze ATP for their action.
   E. of the subclass gyrases, introduce negative superhelices in eukaryotic DNA.

Problems

15. The figure shown below represents conformations of DNA (A, B, C) at different temperatures. Which conformation has the highest relative optical density at 260 nm? Which section, D or E, has the higher content of guanine and cytosine? Would increasing section

D relative to section E have any effect on T_m of the "melting" curve?

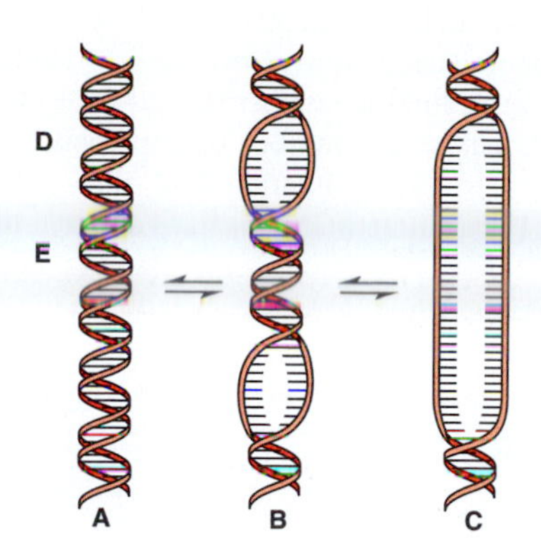

16. Linear DNA has 15 turns ($T = 15$) and no supercoils and will close to become a circular DNA. In situation A, the DNA is unwound 2 turns before closing. In situation B, the ends of the DNA are secured so they cannot rotate, the molecule is looped in a counterclockwise direction once, and the ends are linked to form circular DNA. Are A and B topologically equivalent—that is, are they interconvertible without breaking any covalent bonds?

ANSWERS

1. **B** The sugar phosphates, not the bases, are joined by phosphodiester bonds. A: The structure of a polynucleotide possesses an intrinsic sense of direction that does not depend on whether a 3′-OH or 5′-OH terminal is esterified. C, E: There are only four different monomers, and the repeating unit is the nucleoside monophosphate. D: Nucleases hydrolyze the bonds.
2. **C** Both A and B describe exonucleases. D does refer to an endonuclease but only to a specific type, a restriction endonuclease, and is therefore not a definition of the general type. E: Both endo- and exonucleases show specificity toward the bond hydrolyzed and so this is not a definition of an endonuclease.
3. **D** A palindrome, by definition, reads the same forward and backward. Short palindromic segments of DNA are recognized by a variety of proteins such as restriction endonucleases. A: This is not likely since it would be incompatible with specific recognition. B is possible but has not been shown. C: Genes are thousands of base pairs in length, whereas palindromes are short segments. E: This would not be likely because palindromes are too short.
4. **C** This results in complementarity of the strands. A: Single strands are right-handed helices. B: Bases in a single strand interact through the hydrophobic faces of the rings. D, E: The phosphate groups are negative and on the exterior of the helix.
5. **E** The two strands are always antiparallel. A: The A helix has a narrower and deeper major groove and a wider and shallower minor groove than the B helix. B, D: Bases in the B helix are almost perpendicular to the helix axis while those in A are tilted. C: The B helix is thinner than the A helix.
6. **B** The alternating purine–pyrimidine sequence is important. A: Z-DNA is longer and thinner than the B-form because it has 12 bp per turn instead of 10. C: It is more likely to be found at the 5′-end, consistent with one of its proposed roles in transcriptional regulation. D: Methylation favors the Z-form in which the methyl is protected from water. E: B $\rightarrow$ Z transition is influenced by such things as methylation and rotation of the G to the *syn* conformation.
7. **D** A, B: These are two of several kinds of DNA but do not constitute patterns. C, E: Alu is a type of short interspersed repeat. Short and long interspersed repeats are the two classes of moderately repetitive DNA.
8. **A** The "beads-on-a-string" structure is called a polynucleosome. B: Histones form the core with DNA on the outside. C: All five types of histones are present—four in the core and one outside. D: The linker regions are DNA. E: Proteins that regulate gene expression and other activities bind to the major groove. Histones bind to the minor groove.
9. **C** Stacking stabilizes the single-strand helix. Folded portions of the structure have hydrogen-bonded base pairing. A: Only the four bases A, G, U, and C are incorporated during transcription. B: Although single stranded, RNA exhibits considerable secondary and tertiary structure. D: Only tRNA would be this small; sizes can range to more than 6000 nucleotides. E: This occurs in the intrachain helical regions.
10. **B** Ribozymes are a very specific type of particle. A: See previous comment. C: One of the four classes, RNase P, catalyzes a cleavage reaction. D: X-ray crystallography has shown there is no amino acid chain sufficiently close to catalyze peptide bond formation. E: Ribozymes have been implicated in the processing of ribosomal and tRNAs.
11. **B** The bent DNA may be a recognition site for specific proteins to bind. A, D: Bent DNA occurs naturally either in runs of four to six adenosines separated by spacers or is induced by interactions of DNA with certain proteins. C: Bent DNA is double stranded. E: Z-DNA does not have a major groove and the bending is toward the major groove.
12. **B** Hoogsteen bonding in TAT and GGC triplets is responsible for holding the third strand in the major groove. A: They are found frequently in regions involved in gene regulation. C: The required sequence is a homopurine string. D: They can also form intramolecularly by unfolding and refolding of the DNA. E: A cruciform is an alternate conformation of DNA but does not involve a third strand.
13. **E** A, C: The guanines are held together by Hoogsteen bonds and are called G-quartets. B: Sodium and potassium are most effective because they fit into the center of the quartet and bind to guanine oxygens. D: Parallel structures occur when four separate polynucleotides interact and antiparallel structures are favored when one polynucleotide provides two or more strands.
14. **A** Topoisomerase I relaxes DNA and topoisomerase II generates supercoils. Balance between the two opposing activities maintains the proper level of superhelicity. B: Topoisomerase I breaks one strand but topoisomerase II nicks both strands. C: Topoisomerase I removes supercoils. D: Topoisomerase II uses ATP

but topoisomerase I does not. E: Gyrases are found only in bacteria.

15. C has the highest relative optical density and also represents the conformation at the highest temperature. Optical density increases as stacking of bases decreases. Denaturation increases with increasing temperature. E has a higher G-C content since it is less denatured than D at the intermediate temperature. T_m is the midpoint of the transition between double-stranded and separated DNA. Increasing the extent of D would increase the A-T content relative to G-C and thus lower the temperature required to separate the strands.

16. To be topologically equivalent, the two conformations must have the same linking number (L). $L = T + W$ (writhing number as indicated by number of supercoils). In situation A, $L = 13 + 0 = 13$. In situation B, $L = 15 - 1 = 14$. A and B are not equivalent because they have different linking numbers.

3

PROTEINS I: COMPOSITION AND STRUCTURE

Richard M. Schultz and Michael N. Liebman

3.1 | FUNCTIONAL ROLES OF PROTEINS IN HUMANS

Proteins perform a surprising variety of essential functions in mammalian organisms. Dynamic functions include catalysis of chemical transformations, transport, metabolic control, and contraction. In their structural functions, proteins provide the matrix for bone and connective tissue, giving structure and form to the human organism.

An important class of dynamic proteins are the enzymes that catalyze chemical reactions, converting a substrate to a product at the enzyme's active site. Almost all of the thousands of chemical reactions in living organisms require a specific enzyme catalyst to ensure that reactions occur at a rate compatible with life. The character of any cell is based on its particular chemistry, which is determined by its specific enzyme composition. Numerous genetic traits are expressed through synthesis of enzymes, which catalyze reactions that establish the phenotype. Many genetic diseases result from altered levels of enzyme production or specific alterations to their amino acid sequence. Transport is another major function of proteins. Examples discussed in greater detail in this text are hemoglobin and myoglobin, which transport oxygen in blood and in muscle, respectively. Transferrin transports iron in blood. Other transport proteins bind and carry steroid hormones in blood from their site of synthesis to their site of action. Many drugs and toxic compounds are transported bound to proteins. Proteins participate in contractile mechanisms. Myosin and actin function in muscle contraction.

Proteins have a protective role through a combination of dynamic functions. Immunoglobulins and interferon are proteins that protect the body against bacterial or viral infection. Fibrin stops the loss of blood on injury to the vascular system.

Many hormones are proteins or peptides. Protein hormones include insulin, thyrotropin, somatotropin (growth hormone), prolactin, luteinizing hormone, and follicle-stimulating hormone. Many polypeptide hormones have a low molecular weight ($<$5000) and are referred to as peptides. In general, the term **protein** is used for molecules composed of over 50 amino acids and **peptide** is used for those of less than 50 amino acids. Important peptide hormones include adrenocorticotropic hormone, antidiuretic hormone, glucagon, and calcitonin.

Proteins control and regulate gene transcription and translation. These include histones that are closely associated with DNA, repressor and enhancer transcription factors that control gene transcription, and proteins that form a part of the heteronuclear RNA particles and ribosomes.

Structural proteins function in "brick-and-mortar" roles. They include collagen and elastin, which form the matrix of bone and ligaments and provide structural strength and elasticity to organs and the vascular system. α-Keratin is present in hair and other epidermal tissue.

An understanding of normal functioning and pathology of the mammalian organism requires a clear understanding of the properties of proteins.

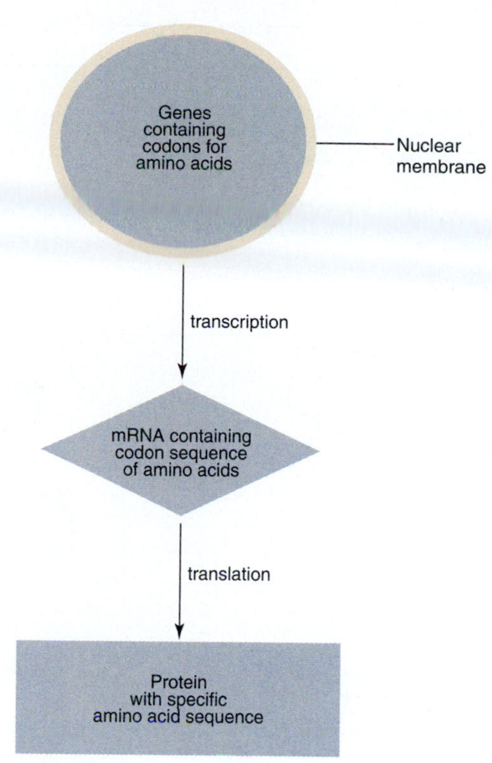

FIGURE 3.1
Genetic information is transcribed from a DNA sequence into mRNA and then translated to the amino acid sequence of a protein.

3.2 | AMINO ACID COMPOSITION OF PROTEINS

Proteins Are Polymers of α-Amino Acids

It is notable that all the different types of proteins are synthesized as polymers of only 20 amino acids. These **common amino acids** are defined as those for which at least one codon exists in the genetic code. Transcription and translation of the DNA code result in polymerization of amino acids into a specific linear sequence characteristic of a protein (Figure 3.1). Proteins may also contain **derived amino acids,** which are usually formed by an enzymatic reaction on a common amino acid after that amino acid has been incorporated into a protein structure. Examples of derived amino acids are cystine (see p. 100), desmosine, and isodesmosine found in elastin, hydroxyproline and hydroxylysine in collagen, and γ-carboxyglutamate in prothrombin.

Common Amino Acids Have a General Structure

Common amino acids have the general structure depicted in Figure 3.2. They contain in common a central *alpha* (α)-carbon atom to which a carboxylic acid group, an amino group, and a hydrogen atom are covalently bonded. In addition, the α-carbon atom is bound to a specific chemical group, designated R and called the side chain, that uniquely defines each of the 20 common amino acids. Figure 3.2 depicts the ionized form of a common amino acid in solution at pH 7. The α-amino group is protonated and in its ammonium ion form; the carboxylic acid group is in its unprotonated or carboxylate ion form.

$$\overset{+}{N}H_3-\underset{R}{\overset{COO^-}{C_\alpha}}-H$$

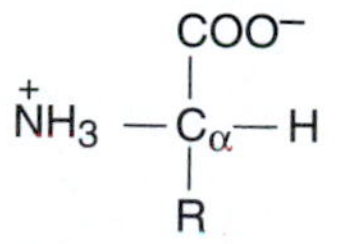

FIGURE 3.2
General structure of the common amino acids.

Side Chains Define Chemical Nature and Structures of Different Amino Acids

Structures of the common amino acids are shown in Figure 3.3. Alkyl amino acids have alkyl group side chains and include glycine, alanine, valine, leucine, and isoleucine. **Glycine** has the simplest structure, with R = H. **Alanine** contains a methyl (CH_3—) group. **Valine** has an isopropyl R group (Figure 3.4). The leucine and isoleucine R groups are butyl groups that are structural isomers of each other. In **leucine** the branching in the isobutyl side chain occurs on the *gamma* (γ)-carbon and in **isoleucine** it is branched at the *beta* (β)-carbon of the amino acid.

The aromatic amino acids are phenylalanine, tyrosine, and tryptophan. **Phenylalanine** contains a benzene ring, **tyrosine** a phenol group, and **tryptophan** the heterocyclic structure, indole. In each case the aromatic moiety is attached to the α-carbon through a methylene (—CH_2—) carbon (Figure 3.3).

Sulfur-containing amino acids are cysteine and methionine. The **cysteine** side chain group is a thiolmethyl ($HSCH_2$—). In **methionine** the side chain is a methyl ethyl thiol ether ($CH_3SCH_2CH_2$—).

The two hydroxy (alcohol)-containing amino acids are serine and threonine. The **serine** side chain is a hydroxymethyl ($HOCH_2$—). In **threonine** an ethanol

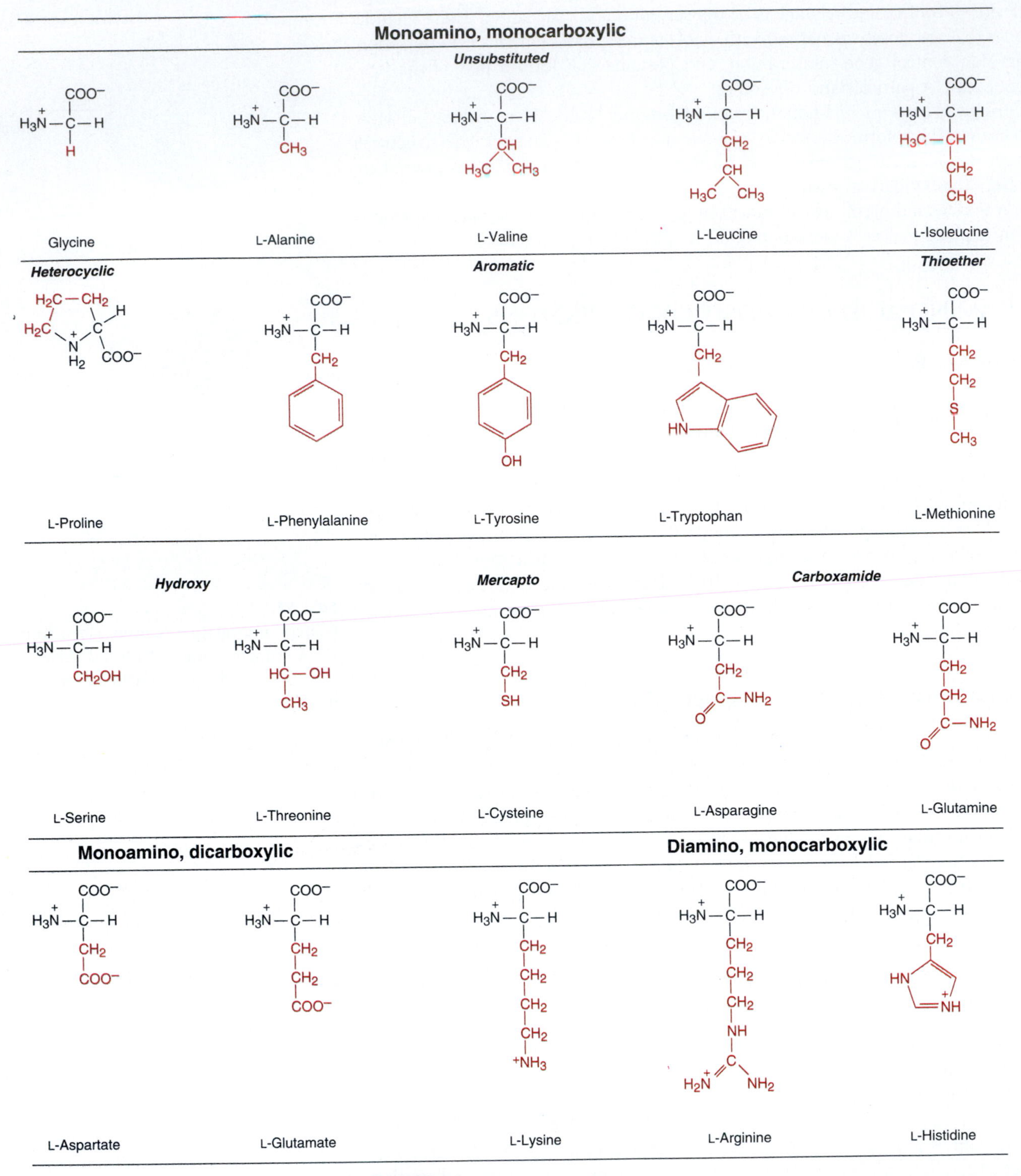

FIGURE 3.3
Structures of the common amino acids.
Charge forms are those present at pH 7.0.

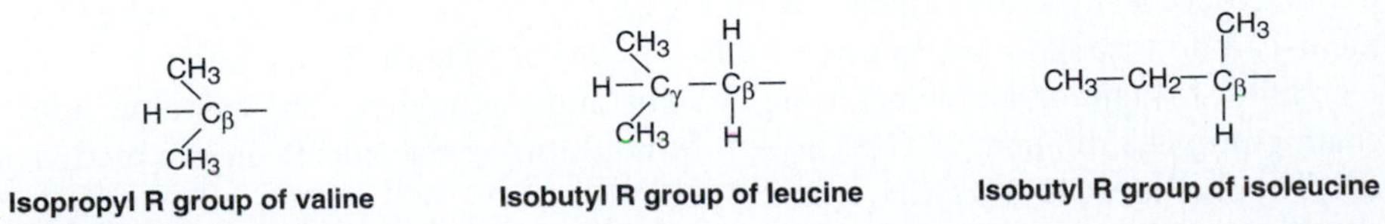

Isopropyl R group of valine

Isobutyl R group of leucine

Isobutyl R group of isoleucine

FIGURE 3.4
Alkyl side chains of valine, leucine, and isoleucine.

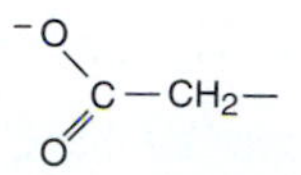

^{-}O—C(=O)—CH_2—

Aspartate R group

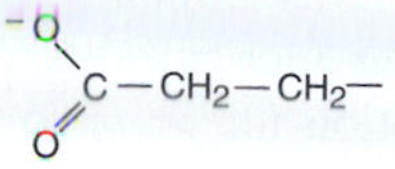

^{-}O—C(=O)—CH_2—CH_2—

Glutamate R group

FIGURE 3.5
Side chains of aspartate and glutamate.

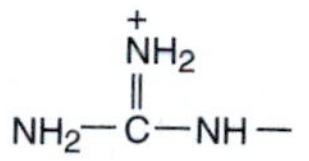

structure is connected to the α-carbon through the carbon containing the hydroxyl substituent, resulting in a secondary alcohol structure (CH_3—CHOH—CH_α—).

Proline is unique in that it incorporates the α-amino group in its side chain and is more accurately classified as an α-imino acid, since its α-amine is a secondary amine with its α-nitrogen having two covalent bonds to carbon (to the α-carbon and side chain carbon), rather than a primary amine. Incorporation of the α-amino nitrogen into a five-membered ring constrains the rotational freedom around the —N_α—C_α— bond in proline to a specific rotational angle, which limits participation of proline in polypeptide chain conformations.

The amino acids discussed so far contain side chains that are uncharged at physiological pH. The **dicarboxylic monoamino acids** contain a carboxylic group in their side chain. **Aspartate** contains a carboxylic acid group separated by a methylene carbon (—CH_2—) from the α-carbon (Figure 3.5). In **glutamate** (Figure 3.5), the carboxylic acid group is separated by two methylene (—CH_2—CH_2—) carbon atoms from the α-carbon (Figure 3.2). At physiological pH, side chain carboxylic acid groups are unprotonated and negatively charged. **Dibasic monocarboxylic** acids include lysine, arginine, and histidine (Figure 3.3). In these structures, the R group contains one or two nitrogen atoms that act as a base by binding a proton. The **lysine** side chain is an *N*-butyl amine. In **arginine,** the side chain contains a guanidino group (Figure 3.6) separated from the α-carbon by three methylene carbon atoms. Both the guanidino group of arginine and the ε-amino group of lysine are protonated at physiological pH (~7) and in their positively charged form. In **histidine** the side chain contains a five-membered heterocyclic structure, the imidazole (Figure 3.6). The pK'_a of the imidazole group is approximately 6.0 in water; physiological solutions contain relatively high concentrations of both basic (imidazole) and acidic (imidazolium) forms of the histidine side chain (see Section 3.3).

Glutamine and **asparagine** are structural analogs of glutamic acid and aspartic acid with their side chain carboxylic acid groups amidated. Unique DNA codons exist for glutamine and asparagine separate from those for glutamic acid and aspartic acid. The amide side chains of glutamine and asparagine cannot be protonated and are uncharged at physiological pH.

To represent sequences of amino acids in proteins, three-letter and one-letter abbreviations for the common amino acids have been established (Table 3.1). These abbreviations are universally accepted and will be used throughout this book. The three-letter abbreviations of aspartic acid (Asp) and glutamic acid (Glu) should not be confused with those for asparagine (Asn) and glutamine (Gln). In experimental determination of the amino acids of a protein by chemical procedures, one cannot easily differentiate between Asn and Asp, or between Gln and Glu, because the amide groups in Asn and Gln are hydrolyzed and generate Asp and Glu (see Section 3.9). The symbols of Asx for Asp or Asn, and Glx for Glu or Gln indicate this ambiguity. A similar scheme is used with the one-letter abbreviations for Asp or Asn, and Glu or Gln.

NH_2—C(=$\overset{+}{N}H_2$)—NH—

Guanidinium group (charged form) of arginine

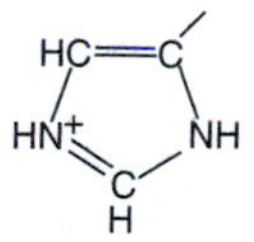

Imidazolium group of histidine

FIGURE 3.6
Guanidinium and imidazolium groups of arginine and histidine.

TABLE 3.1 Abbreviations for the Amino Acids

Amino Acid	*Abbreviation*	
	Three Letter	*One Letter*
Alanine	Ala	A
Arginine	Arg	R
Asparagine	Asn	N
Aspartate	Asp	D
Asparagine or aspartate	Asx	B
Cysteine	Cys	C
Glycine	Gly	G
Glutamine	Gln	Q
Glutamate	Glu	E
Glutamine or glutamate	Glx	Z
Histidine	His	H
Isoleucine	Ile	I
Leucine	Leu	L
Lysine	Lys	K
Methionine	Met	M
Phenylalanine	Phe	F
Proline	Pro	P
Serine	Ser	S
Threonine	Thr	T
Tryptophan	Trp	W
Tyrosine	Tyr	Y
Valine	Val	V

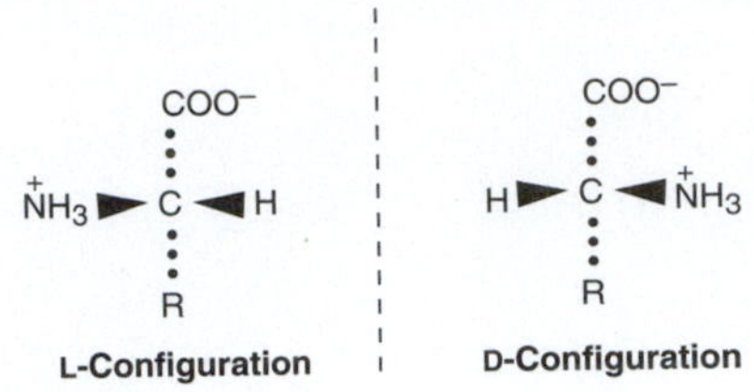

L-Configuration D-Configuration

FIGURE 3.7
Absolute configuration of an amino acid.

Amino Acids Have an Asymmetric Center

The common amino acids in Figure 3.2 have four substituents (R, H, COO^-, NH_3^+) covalently bonded to an α-carbon atom. A carbon atom with four different substituents in a **tetrahedral configuration** is asymmetric and exists in two enantiomeric forms. Thus all amino acids exhibit optical isomerism except glycine, in which R = H and thus two of the four substituents on the α-carbon atom are hydrogen. The absolute configuration for an amino acid is depicted in Figure 3.7 using the Fischer projection to show the position in space of the tetrahedrally arranged α-carbon substituents. The α-COO^- group is directed up and behind the plane of the page, and the R group is directed down and behind the plane of the page. The α-H and α-NH_3^+ groups are directed toward the reader. An amino acid held in this way projects its α-NH_3^+ group either to the left or right of the α-carbon atom. By convention, if the α-NH_3^+ is projected to the left, the amino acid has an **L absolute configuration.** Its optical enantiomer, with α-NH_3^+ projected toward the right, has a **D absolute configuration.** In mammalian proteins only amino acids of L configuration are found. The L and D designations refer to the ability to rotate polarized light to the left (L, levo) or right (D, dextro) from its plane of polarization. As the amino acids in proteins are asymmetric, the proteins also exhibit asymmetric properties.

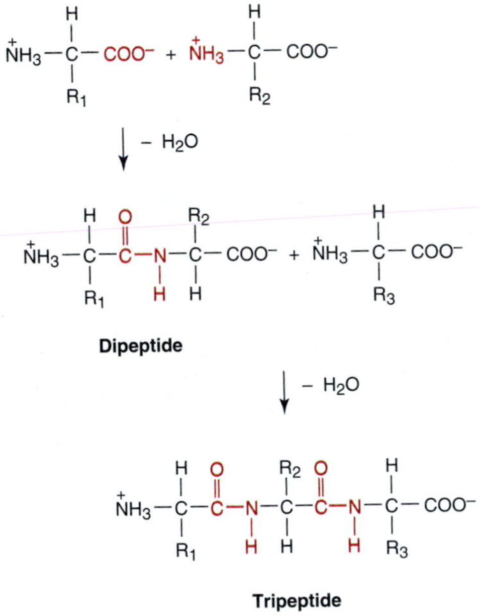

Dipeptide

Tripeptide

FIGURE 3.8
Peptide bond formation.

Amino Acids Are Polymerized into Peptides and Proteins

Polymerization of the 20 common amino acids into polypeptide chains in cells is catalyzed by enzymes and is associated with the ribosomes (see Chapter 6). Chemically, this polymerization is a dehydration reaction (Figure 3.8). The α-carboxyl group of one amino acid with side chain R_1 forms a covalent **peptide bond** with the α-amino group of another amino acid with side chain R_2 by elimination of a molecule of water. The **dipeptide** (two amino acid residues joined by a single peptide bond) can then form a second peptide bond through its terminal carboxylic acid group and the α-amino of a third amino acid (R_3), to generate a tripeptide (Figure 3.8). Repetition of this process generates a **polypeptide** or protein of specific amino acid sequence (R_1-R_2-R_3-R_4-$\cdots R_n$). This is the **primary structure** of the protein, and it is predetermined by the nucleotide sequence of its gene (see Chapter 6). It is the unique primary structure that enables a polypeptide chain to fold into a specific three-dimensional structure that gives the protein its chemical and physiological properties.

A peptide bond can be represented as two **resonance isomers** (Figure 3.9). In structure I, a double bond is located between the carbonyl carbon and carbonyl oxygen (C′=O), and the carbonyl carbon to nitrogen (C′—N) linkage is a single bond. In structure II, the carbon to oxygen bond (C′—O—) is a single bond and the bond located between the carbon and nitrogen is a double bond (C′=N). In structure II there is a negative charge on the oxygen and a positive charge on the nitrogen. Actual peptide bonds are a resonance hybrid of these two electron isomer structures, the carbon to nitrogen bond having a 50% double-bond character. The presence of this hybrid bond is supported by spectroscopic measurements and X-ray diffraction studies, the latter showing that the carbonyl carbon to nitrogen bond length (1.33 Å) is approximately half-way between that of a C′—N single bond (~1.45 Å) and a C′=N double bond (~1.25 Å).

A consequence of this partial double-bond character is that, as for normal double-bond structures, rotation does not occur about the carbonyl carbon to nitrogen of a peptide bond at physiological temperatures. All the atoms attached to C′ and N lie in a common plane.

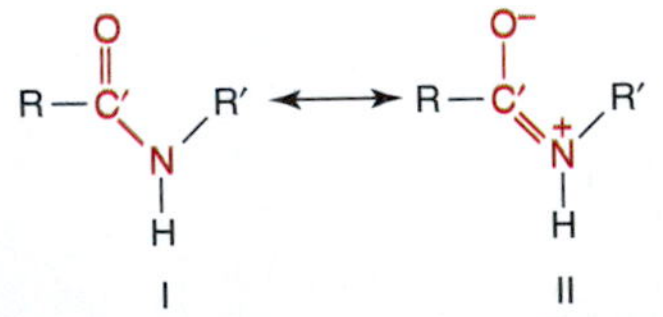

FIGURE 3.9
Electronic isomer structures of a peptide bond.

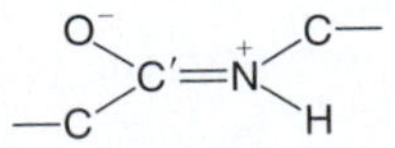

Thus a polypeptide chain is a polymer that consists of peptide-bond planes interconnected at the α-carbon atoms. The α-carbon interconnects peptide bonds through single bonds that allow rotation of adjacent peptide planes with respect to each other. Each **amino acid residue** contributes one α-carbon, two single bonds, and a peptide bond to the polypeptide chain (Figure 3.10). The term residue refers to the atoms contributed by an amino acid to a polypeptide chain including the atoms of the side chain.

The peptide bond in Figure 3.11*a* shows *a* ***trans*** **configuration** of its oxygen (O) and hydrogen (H) atoms. This is the most stable configuration with the two side chains (R and R′) also being in *trans*. The ***cis*** **configuration** (Figure 3.11*b*) brings the two side chain groups to the same side of the C′=N bond, where unfavorable repulsive steric forces occur between the R groups. Accordingly, *trans*-peptide bonds occur in proteins except where there are proline residues. In proline the side chain includes the α-amino group, and the *cis*- and *trans*-peptide bonds with the proline α-imino group have near equal energies. The configuration of the peptide bond with proline depends on the specific forces generated by the unique folded three-dimensional structure of the protein molecule.

One of the largest polypeptides in humans is that of apolipoprotein B-100, which contains 4536 amino acid residues. Chain length alone, however, does not determine the function of a polypeptide. Many small peptides of less than ten amino

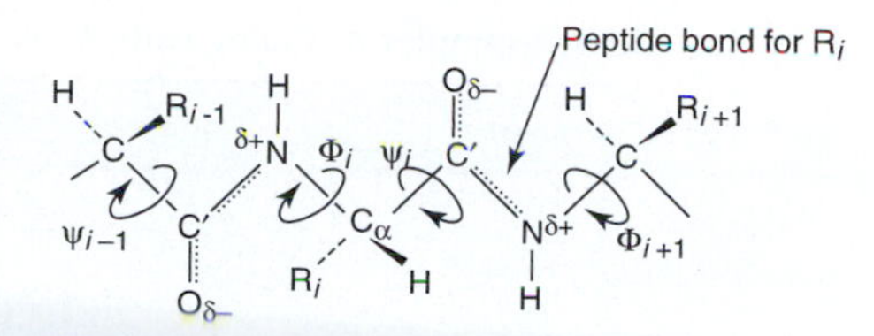

FIGURE 3.10
Amino acid residue within a polypeptide chain.
Each residue of a polypeptide contributes two single bonds and one peptide bond to the chain. The single bonds are those between the C_α and carbonyl C′ atoms, and the C_α and N atoms. See p. 113 for definition of ϕ and ψ.

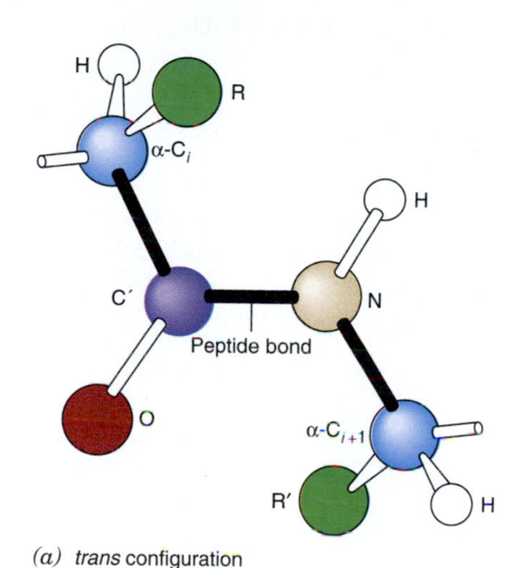

(*a*) *trans* configuration

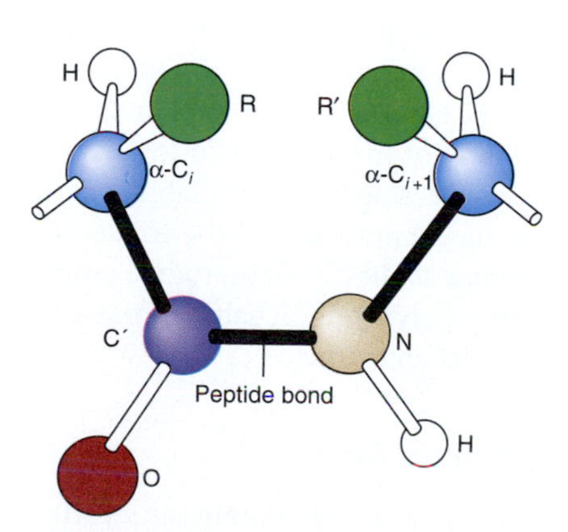

(*b*) *cis* configuration

FIGURE 3.11
(*a*) *Trans*-peptide bond, and (*b*) the rare *cis*-peptide bond. C′—N have a partial double-bond character.

TABLE 3.2 Some Examples of Biologically Active Peptides

Amino Acid Sequence	Name	Function
1 3 pyroGlu-His-Pro(NH_2)[a]	Thyrotropin-releasing hormone	Secreted by hypothalamus; causes anterior pituitary gland to release thyrotropic hormone
1 9 H-Cys-Tyr-Phe-Gln-Asn-Cys-Pro-Arg-Gly(NH_2)[b,d] (S——S bridge between Cys-1 and Cys-6)	Vasopressin (antidiuretic hormone)	Secreted by posterior pituitary gland; causes kidney to retain water from urine
1 5 H-Tyr-Gly-Gly-Phe-Met-OH	Methionine enkephalin	Opiate-like peptide found in brain that inhibits sense of pain
1 10 pyroGlu-Gly-Pro-Trp-Leu-Glu-Glu-Glu-Glu- 11 17 Ala-Tyr-Gly-Trp-Met-Asp-Phe(NH_2)[c,d] (Tyr: SO_3)	Little gastrin (human)	Hormone secreted by mucosal cells in stomach; causes parietal cells of stomach to secrete acid
1 10 H-His-Ser-Gln-Gly-Thr-Phe-Thr-Ser-Asp-Tyr- 11 20 Ser-Lys-Tyr-Leu-Asp-Ser-Arg-Arg-Ala-Gln- 21 29 Asp-Phe-Val-Gln-Trp-Leu-Met-Asn-Thr-OH	Glucagon (bovine)	Pancreatic hormone involved in regulating glucose metabolism
1 8 H-Asp-Arg-Val-Tyr-Ile-His-Pro-Phe-OH	Angiotensin II (horse)	Pressor or hypertensive peptide; also stimulates release of aldosterone from adrenal cortex
1 9 H-Arg-Pro-Pro-Gly-Phe-Ser-Pro-Phe-Arg-OH	Plasma bradykinin (bovine)	Vasodilator peptide
1 10 H-Arg-Pro-Lys-Pro-Gln-Phe-Phe-Gly-Leu-Met(NH_2)[d]	Substance P	Neurotransmitter

[a] The NH_2-terminal Glu is in the pyro form in which its γ-COOH is covalently joined to its α-NH_2 via amide linkage; the COOH-terminal amino acid is amidated and thus also not free.

[b] Cysteine-1 and cysteine-6 are joined to form a disulfide bond structure within the nonapeptide.

[c] The Tyr 12 is sulfonated on its phenolic side chain OH.

[d] The COOH-terminal amino acid is amidated.

acids perform important biochemical and physiological functions in humans (Table 3.2). Primary structures are written in a standard convention and sequentially numbered from their NH_2-terminal end to their COOH-terminal end, consistent with the order of addition of the amino acid to the chain during biosynthesis. Accordingly, for thyrotropin-releasing hormone (Table 3.2) the glutamic acid residue on the left is the NH_2-terminal amino acid of the tripeptide or amino acid residue 1 in the sequence. The proline is the COOH-terminal amino acid and is residue 3. The defined direction of the polypeptide chain is from Glu to Pro (NH_2-terminal amino acid to COOH-terminal amino acid).

Two cysteines —(−2H)→ Cystine

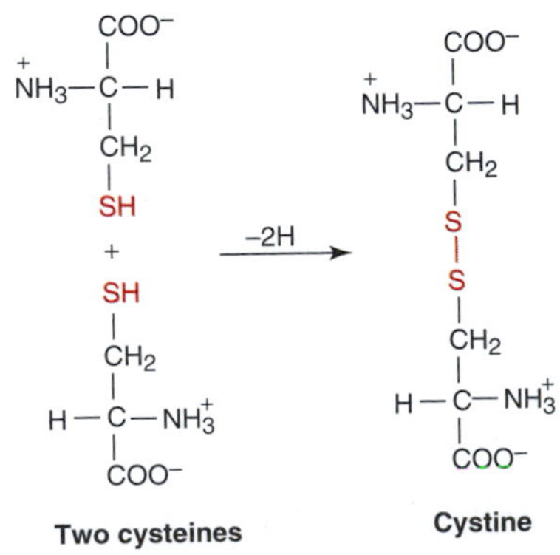

FIGURE 3.12
Cystine bond formation.

Cystine Is a Derived Amino Acid

A derived amino acid found in many proteins is **cystine.** It is formed by the oxidation of two cysteine thiol side chains to form a disulfide covalent bond (Figure 3.12). Within proteins, disulfide links of cystine formed from cysteines, separated from each other in a polypeptide chain (intrachain) or between two polypeptide chains (interchain), have an important role in stabilizing the folded conformation of proteins.

3.3 | CHARGE AND CHEMICAL PROPERTIES OF AMINO ACIDS AND PROTEINS

Ionizable Groups of Amino Acids and Proteins Are Critical for Biological Function

Ionizable groups common to proteins and amino acids are shown in Table 3.3. The acid forms are on the left of the equilibrium sign and the base forms on the right side. In forming its conjugate base, the acid form releases a proton. In reverse, the base form associates with a proton to form the respective acid. The proton dissociation of an acid is characterized by an acid **dissociation constant (K'_a)** and its pK'_a value: $pK'_a = \log_{10}(1/K'_a)$. Table 3.3 shows a range of pK'_a values for each acid group, as the actual pK'_a depends on the environment in which an acid group is placed. For example, when a positive-charged ammonium group ($—NH_3^+$) is placed near a negative-charged group within a protein, the negative charge stabilizes the positively charged acid form, making it more difficult to dissociate its proton. The pK'_a of the $—NH_3^+$ will have a higher value than normal for an ammonium group in the absence of a nearby negative charge stabilization. Factors other than charge that affect the pK'_a include polarity of the environment, absence or presence of water, and the potential for hydrogen bond formation. Also, acid groups (α-COOH or α-NH_3^+) at the ends of polypeptides typically have a lower pK'_a value than the same types of acid groups in the side chains (Table 3.4). The amino acids whose R groups contain nitrogen atoms (Lys and Arg) are the **basic amino acids,** since their side chains have relatively high pK'_a values and function as bases at physiological pH. They are usually in their acid form and positively charged at physiological pH. Amino acids whose side chains contain a carboxylic acid group have relatively low pK'_a values and are **acidic amino acids** and are predominantly in their unprotonated forms and are negatively charged at physiological

TABLE 3.3 Characteristic pK'_a Values for the Common Acid Groups in Proteins

Where Acid Group Is Found	*Acid Form*		*Base Form*	*Approximate pK_a Range for Group*
NH_2-terminal residue polypeptide chains, lysine	$R—NH_3^+$ Ammonium	⇌	$R—NH_2 + H^+$ Amine	7.6–10.6
COOH-terminal residue polypeptide chains, glutamate, aspartate	R—COOH Carboxylic acid	⇌	$R—COO^- + H^+$ Carboxylate	3.0–5.5
Arginine	$R—NH—\overset{+}{C}{\cdots}NH_2$ with NH_2 on C Guanidinium	⇌	$R—NH—C{=}NH + H^+$ with NH_2 on C Guanidino	11.5–12.5
Cysteine	R—SH Thiol	⇌	$R—S^- + H^+$ Thiolate	8.0–9.0
Histidine	R—C═CH, HN, +NH, C, H (ring) Imidazolium	⇌	R—C═CH, HN, N (ring) $+ H^+$ Imidazole	6.0–7.0
Tyrosine	R—(benzene ring)—OH	⇌	R—(benzene ring)—$O^- + H^+$	9.5–10.5

TABLE 3.4 pK'_a of Side Chain and Terminal Acid Groups in Ribonuclease

	$—NH_3^+$	—COOH
Side chain	Lysines ≃ 10.2	Glu and Asp ≃ 4.6
Chain end	N-terminal = 7.8	C-terminal = 3.8

pH. Proteins in which the ratio (ΣLys + ΣArg)/(ΣGlu + ΣAsp) is greater than 1 are referred to as **basic proteins.** Proteins in which the above ratio is less than 1 are referred to as **acidic proteins.**

Ionic Form of an Amino Acid or Protein Can Be Determined at a Given pH

$$pH = pK_a + \log \frac{[\text{conjugate base}]}{[\text{conjugate acid}]}$$

or

$$pH - pK_a = \log \frac{[\text{conjugate base}]}{[\text{conjugate acid}]}$$

FIGURE 3.13
Henderson–Hasselbalch equation.
(For a more detailed discussion of this equation, see p. 11)

From the pK'_a value for each acid group in an amino acid or protein and the **Henderson–Hasselbalch equation** (Figure 3.13), the ionic form of the molecule can be calculated at a given pH. This is an important relationship as it shows the change in ionization state and charge of a molecule with pH. Physiological activities of a molecule differ with changes in pH and ionization state. For example, an enzyme may require a histidine imidazole in its base form for catalytic activity. If the pK'_a of this histidine 6.0, at pH 6.0 one-half of the enzyme molecules are in the active base (imidazole) form and one-half in the inactive acid (imidazolium) form. Accordingly, the enzyme exhibits 50% of its potential activity. At pH 7.0, the pH is one unit above the imidazolium pK'_a and the ratio of [imidazole]/[imidazolium] is 10:1 (Table 3.5). Based on this ratio, the enzyme exhibits $[10/(10 + 1)] \times 100\% = 91\%$ of its maximum potential activity. Thus a change in pH has a dramatic effect on the enzyme's activity. Most protein activities demonstrate similar pH dependency due to their acid and base group(s).

Titration of a Monoamino Monocarboxylic Acid: Determination of the Isoelectric pH

An understanding of a protein's acid and base forms and their relation to charge is made more clear by following the titration of the ionizable groups for a simple amino acid. In Figure 3.14, leucine has an α-COOH with $pK'_a = 2.4$ and an $\alpha\text{-}NH_3^+$ group with $pK'_a = 9.6$. At pH 1.0 the predominant ionic form (form I) has a charge of +1 and migrates toward the cathode in an electrical field. Addition of 0.5 equivalent of base half-titrates the α-COOH group; that is, the ratio of $[COO^-]/[COOH]$ will equal 1. The Henderson–Hasselbalch equation, with the second term on the right side of the equation $\log_{10}[(\text{base})/(\text{acid})] = \log_{10}[1] = 0$

TABLE 3.5 Relationship Between the Difference of pH and Acid pK'_a and the Ratio of the Concentrations of Base to Its Conjugate Acid

$pH - pK'_a$ (Difference Between pH and pK'_a)	Ratio of Concentration of Base to Conjugate Acid
0	1
1	10
2	100
3	1000
−1	0.1
−2	0.01
−3	0.001

	I		II		III
	$\overset{+}{N}H_3$—C(H)(COOH)—CH_2—C(H)(CH_3)—CH_3	$\underset{+H^+}{\overset{+OH^-}{\rightleftharpoons}}$	$\overset{+}{N}H_3$—C(H)(COO^-)—CH_2—C(H)(CH_3)—CH_3	$\underset{+H^+}{\overset{+OH^-}{\rightleftharpoons}}$	NH_2—C(H)(COO^-)—CH_2—C(H)(CH_3)—CH_3
Charge	+1		0		−1
	pH < 2.4		2.4 < pH < 9.6		9.6 < pH

FIGURE 3.14
Ionic forms of leucine.

at a ratio of conjugate base to acid of 1:1, shows that the pH (when the α-COOH is half-titrated) is directly equal to the $pK_{a(\alpha COOH)}$ (Figure 3.15).

Addition of 1 equivalent of base completely titrates the α-COOH but leaves the α-NH_3^+ group intact. In the resulting form (II), the negatively charged α-COO^- and positively charged α-NH_3^+ cancel each other and the net charge is zero. Form II is thus the **zwitterion** form, that is, the ionic form in which the total of positive charges equals the total of negative charges. As the net charge on a zwitterion molecule is zero, it will not migrate toward either the cathode or anode in an electric field. Further addition of 0.5 equivalent of base to the zwitterion form of leucine (total base added is 1.5 equivalents) will half-titrate the α-NH_3^+ group. At this point in the titration, the ratio of $[NH_2]/[NH_3^+] = 1$, and the pH is equal to the value of the pK'_a for the α-NH_3^+ group (Figure 3.15). Addition of a further 0.5 equivalent of base (total of 2 full equivalents of base added; Figure 3.15) completely titrates the α-NH_3^+ group to its base form (α-NH_2). pH becomes greater than 11, and the predominant molecular species has a negative charge of −1 (form III).

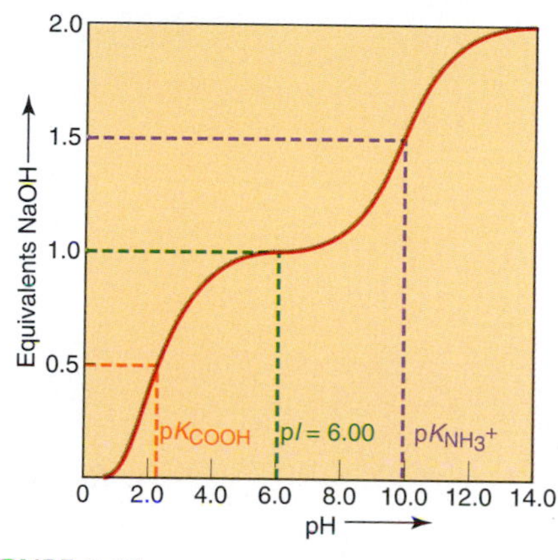

FIGURE 3.15
Titration curve of leucine.

It is useful to calculate the pH at which an amino acid is electrically neutral and in its zwitterion form. This pH is known as the **isoelectric pH** for the molecule, represented as **p*I*.** The p*I* value is a constant of a compound at a particular ionic strength and temperature. For monoamino and monocarboxylic molecules, such as leucine, p*I* is calculated as the average of the two pK'_a values that regulate the boundaries of the zwitterion form. For leucine the p*I* is calculated as follows:

$$pI = \frac{pK'_a COOH + pK'_a NH_3^+}{2} = \frac{2.4 + 9.6}{2} = 6.0$$

At pH > 6.0, leucine has a partial negative charge that increases at high pH to a full negative charge of −1 (form III) (Figure 3.14). At pH < 6, it has a partial positive charge until at a very low pH it has a charge of +1 (form I) (Figure 3.14). The partial charge at any pH can be calculated from the Henderson–Hasselbalch equation or from extrapolation from the titration curve of Figure 3.15.

Titration of a Monoamino Dicarboxylic Acid

A more complicated example of the relationship between molecular charge and pH is provided by glutamic acid. As shown in Figures 3.16 and 3.17, in glutamic acid

$\overset{+}{N}H_3$—C(H)(COOH)—CH_2—CH_2—COOH	$\underset{+H^+}{\overset{+OH^-}{\rightleftharpoons}}$	$\overset{+}{N}H_3$—C(H)(COO^-)—CH_2—CH_2—COOH	$\underset{+H^+}{\overset{+OH^-}{\rightleftharpoons}}$	$\overset{+}{N}H_3$—C(H)(COO^-)—CH_2—CH_2—COO^-	$\underset{+H^+}{\overset{+OH^-}{\rightleftharpoons}}$	NH_2—C(H)(COO^-)—CH_2—CH_2—COO^-
Charge +1		0		−1		−2
pH < 2.2		2.2 < pH < 4.3		4.3 < pH < 9.7		9.7 < pH

FIGURE 3.16
Ionic forms of glutamic acid.

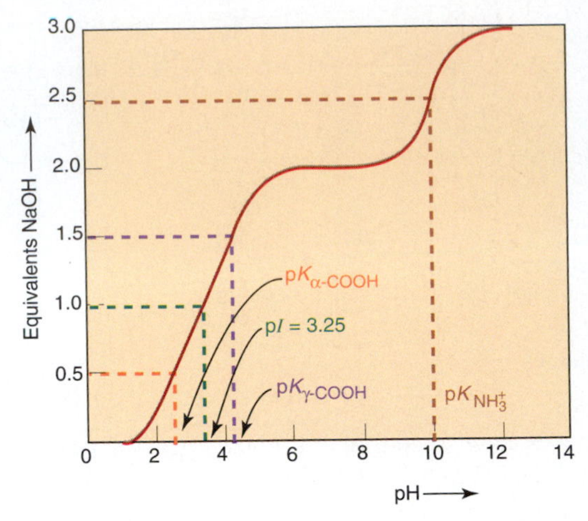

FIGURE 3.17
Titration curve of glutamic acid.

the α-COOH pK'_a = 2.2, the γ-COOH pK'_a = 4.3, and the α-NH_3^+ pK'_a = 9.7. The zwitterion form is generated after 1.0 equivalent of base is added to the low-pH form, and the isoelectric pH (p*I*) is calculated from the average of the two pK'_a values that control the boundaries of the zwitterion form:

$$pI = \frac{2.2 + 4.3}{2} = 3.25$$

Accordingly, at values above pH 3.25 the molecule assumes a net negative charge until at high pH it has a net charge of −2. At pH < 3.25 glutamic acid is positively charged, and at extremely low pH it has a net positive charge of +1.

TABLE 3.6 p*I* Values for Some Representative Proteins

Protein	*pI*
Pepsin	~1
Human serum albumin	5.9
α_1-Lipoprotein	5.5
Fibrinogen	5.8
Hemoglobin A	7.1
Ribonuclease	7.8
Cytochrome *c*	10.0
Thymohistone	10.6

General Relationship Between Charge Properties of Amino Acids and Proteins and pH

Analysis of charge forms of other common amino acids shows that the relationship between pH and charge for leucine and glutamate is generally true. At a pH less than p*I*, the amino acid is positively charged. At a pH greater than p*I*, the amino acid is negatively charged. The degree of positive or negative charge is a function of the magnitude of the difference between pH and p*I*. As proteins are complex polyelectrolytes that contain many ionizable acid groups that regulate the boundaries of their zwitterion forms, calculation of a protein's isoelectric pH from its acid pK'_a values utilizing the Henderson–Hasselbalch relationship would be difficult. Accordingly, p*I* values of proteins are always experimentally measured by determining the pH value at which the protein does not move in an electrical field. p*I* values for some representative proteins are given in Table 3.6.

As with amino acids, at a pH greater than its p*I*, a protein has a net negative charge. At a pH less than its p*I*, a protein has a net positive charge (Figure 3.18). The magnitude of the net charge of a protein increases as the difference between pH and p*I* increases. An example is human plasma albumin with 585 amino acid

pH > p*I*, then protein charge negative
pH < p*I*, then protein charge positive

FIGURE 3.18
Relationship between solution pH, protein p*I*, and protein charge.

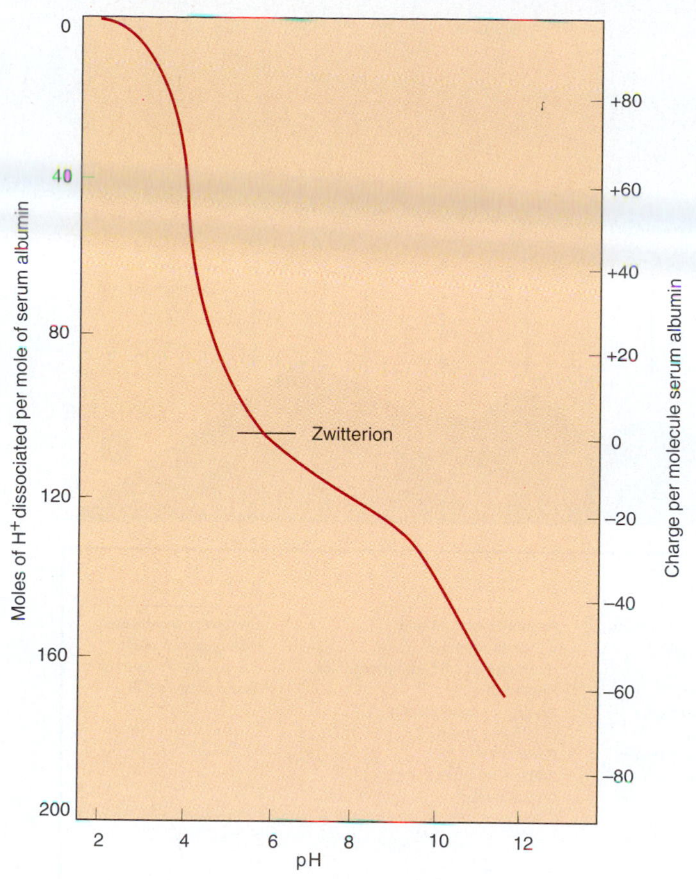

FIGURE 3.19
Titration curve of human serum albumin at 25°C and an ionic strength of 0.150.
Redrawn from Tanford, C. J. Am. Chem. Soc. 72:441, 1950.

residues of which there are 61 glutamates, 36 aspartates, 57 lysines, 24 arginines, and 16 histidines. The titration curve for this complex molecule is shown in Figure 3.19. Albumin's pI = 5.9, at which pH its net charge is zero. At pH 7.5 the imidazolium groups of histidines are partially titrated and albumin has a negative charge of −10. At pH 8.6 the net charge is approximately −20. At pH 11 the net charge is approximately −60. On the acid side of the pI value, at pH 3, the approximate net charge of albumin is +60.

Amino Acids and Proteins Can Be Separated Based on p*I* Values

The techniques of electrophoresis, isoelectric focusing, and ion-exchange chromatography separate and characterize biological molecules on the basis of their pI (see p. 140). In clinical medicine, electrophoretic separation of plasma proteins has led to their classification based on their relative electrophoretic mobility. The separation is commonly carried out at pH 8.6, which is higher than the pI values of the major **plasma proteins.** Accordingly, the proteins are negatively charged and move toward the anode at a rate dependent on their net charge. Major fractions observed in order of their migration are those of albumin, α_1-, α_2-, and β-globulins, fibrinogen, and γ_1- and γ_2-globulins (Figure 3.20). Some of these fractions represent tens to hundreds of different proteins that migrate similarly at pH 8.6. However, certain proteins predominate in each peak and variation in their relative

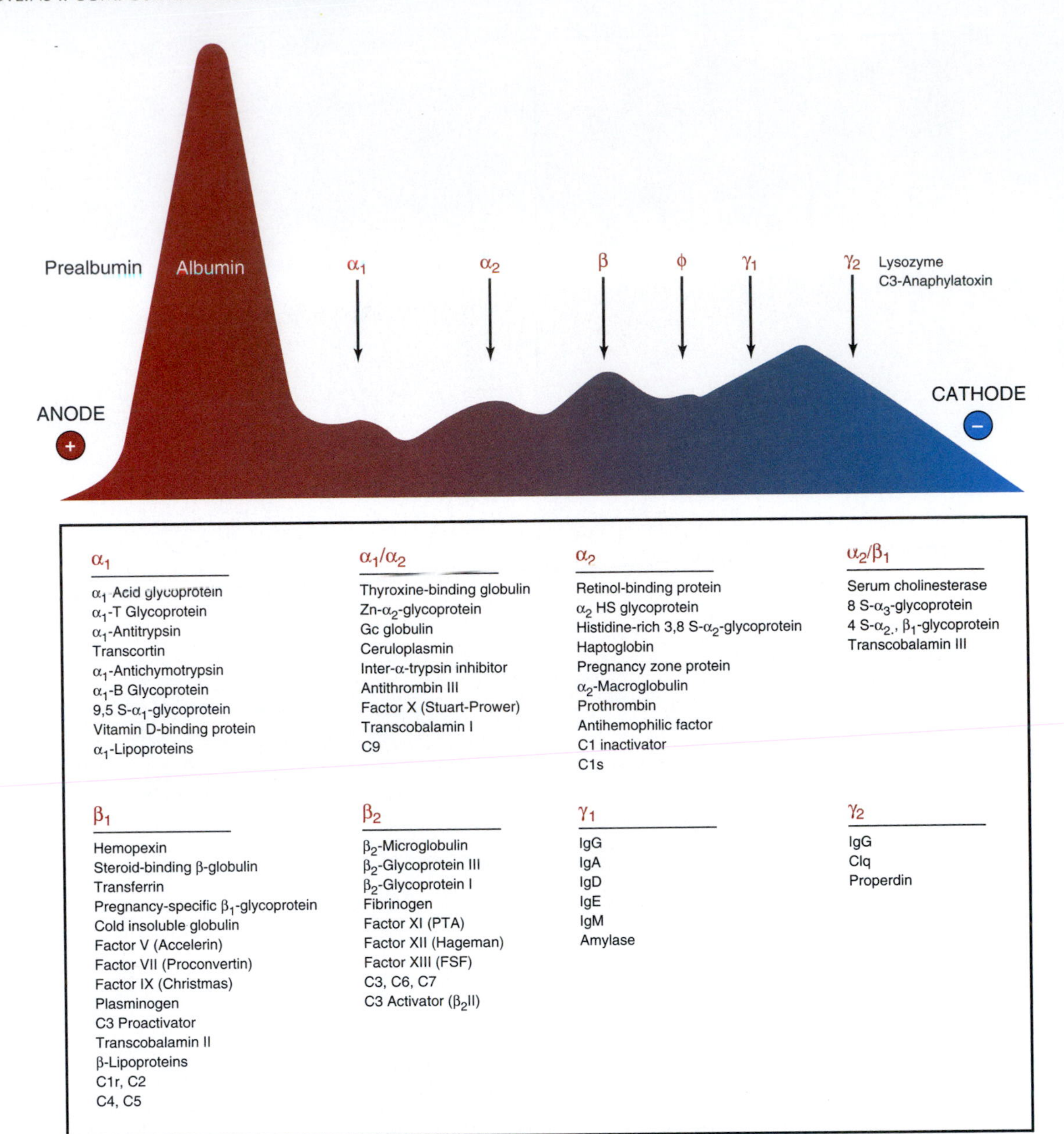

FIGURE 3.20
Electrophoresis pattern for plasma proteins at pH 8.6.
Plot shows order of migration along horizontal axis with proteins of highest mobility closest to anode. Height of band along vertical axis shows protein concentration. Different major proteins are designated underneath their electrophoretic mobility peaks.

Reprinted with permission from Heide, K., Haupt, H., and Schwick, H. G. In: F. W. Putnam (Ed.), The Plasma Proteins, 2nd ed., Vol. III. New York: Academic Press, 1977, p. 545.

amounts is characteristic of certain diseases (Figures 3.20 and 3.21; see Clin. Corr. 3.1).

Amino Acid Side Chains Have Polar or Apolar Properties

The **hydrophobicity** of amino acid side chains is critical for the folding of a protein to its native structure and for stability of the folded protein. Figure 3.22 plots the values of relative hydrophobicity of the common amino acids on the basis of their tendency to partition themselves in a mixture of water and a nonpolar solvent. The scale is based on a value of zero for glycine. Side chains that preferentially dissolve in the nonpolar solvent relative to glycine show a positive (+) hy-

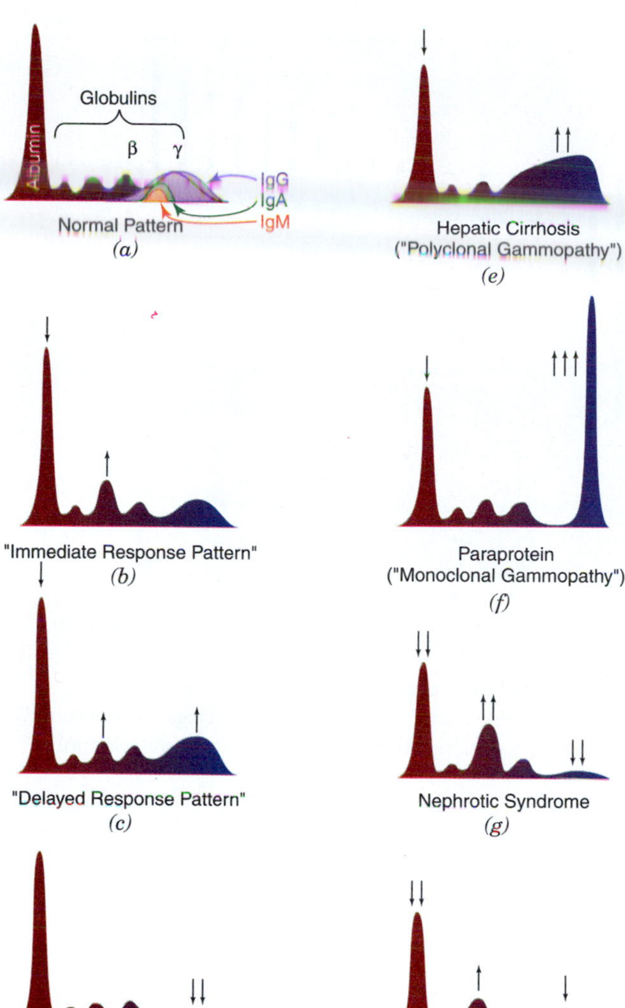

FIGURE 3.21
Examples of electrophoretic mobility patterns observed for a normal individual and patients with abnormal concentrations of serum proteins, analyzed by agarose gel electrophoresis.
Redrawn from McPherson, R. A. In: J. B. Henry (Ed.), Clinical Diagnosis and Management, *17th ed. Philadelphia: Saunders, 1984.*

CLINICAL CORRELATION 3.1
Plasma Proteins in Diagnosis of Disease

Electrophoretic analysis of the plasma proteins is commonly used in diagnosis of disease. Electrophoresis of plasma buffered at pH 8.6 separates the major plasma proteins as they migrate to the anode in the electric field into bands or peaks, based on their charge differences (see text). Examples of abnormal electrophoresis patterns are shown in Figure 3.21. An "immediate response" that occurs with stress or inflammation caused by infection, injury, or surgical trauma is shown in pattern (*b*) in which haptoglobins in the α_2 mobility band are selectively increased. A "late response" shown in pattern (*c*) is correlated with infection and shows an increase in the γ-globulin peaks due to an increase in immunoglobulins. An example of a hypo-γ-globulinemia due to an immunosuppressive disease is shown in pattern (*d*). In hepatic cirrhosis there is a broad elevation of the γ-globulins with reduction of albumin, as in pattern (*e*). Monoclonal gammopathies are due to the clonal synthesis of a unique immunoglobulin and give rise to a sharp γ-globulin band, as in pattern (*f*). Nephrotic syndrome shows a selective loss of lower molecular weight proteins from plasma, as in pattern (*g*). The pattern shows a decrease in albumin (65 kDa), but a retention of the bands composed of the higher molecular weight proteins α_2-macroglobulin (725 kDa) and β-lipoproteins (2000 kDa) in the α_2-band. Pattern (*h*) is from a patient with a protein-losing enteropathy. The slight increase in the α_2-band in pattern (*h*) is due to an immediate or late response from a stressful stimulus, as previously observed in patterns (*b*) and (*c*).

Ritzmann, S. E. and Daniels, J. C. Serum protein electrophoresis and total serum proteins. In: S. E. Ritzmann and J. C. Daniels (Eds.), *Serum Protein Abnormalities, Diagnostic and Clinical Aspects.* Boston: Little, Brown and Co., 1975, pp. 3–25; and McPherson, R. A. Specific proteins. In: J. B. Henry (Ed.), *Clinical Diagnosis and Management by Laboratory Methods,* 17th ed. Philadelphia: Saunders, 1984, p. 204.

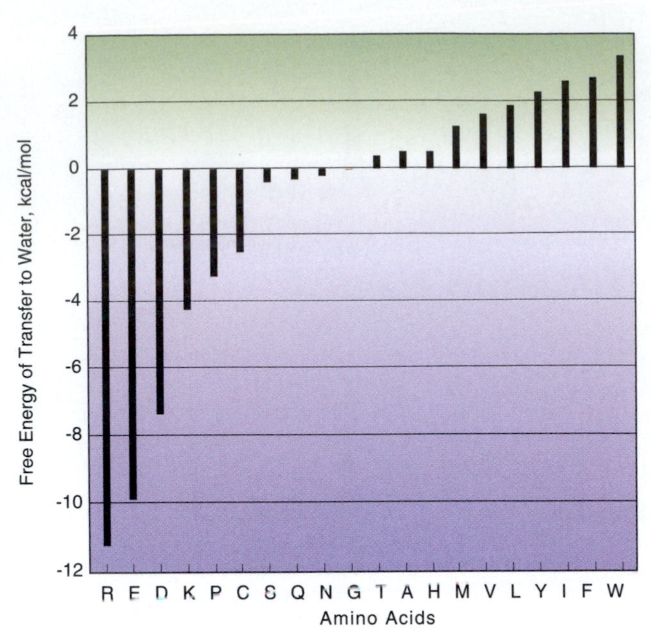

FIGURE 3.22
Relative hydrophobicity of amino acid side chains.
Based on partition of amino acid between organic solvent and water. Negative values indicate preference for water and positive values preference for nonpolar solvent (ethanol or dioxane) relative to glycine (see text).
Based on data from Von Heijne, G. and Blomberg, C. Eur. J. Biochem. *97:175, 1979; and from Nozaki, Y. and Tanford, C.* J. Biol. Chem. *246:2211, 1971.*

drophobicity value, the more positive the greater the preference for the nonpolar solvent. Most hydrophobic are those amino acids found buried in folded protein structures away from the water solvent that interacts with the surface of a soluble protein. However, the general correlation is not perfect due to the amphoteric nature of many of the hydrophobic amino acids that place the more polar portions of their side chain structure near the surface to interact with water on the outside. In addition, contrary to expectation, not all hydrophobic side chains are in a buried position in a folded three-dimensional structure of a globular protein. When on the surface, the hydrophobic groups are generally dispersed among the polar side chains. When clustering of nonpolar side chains occurs on the surface, it is usually associated with a function, such as to provide a site for binding of substrate molecules through hydrophobic interactions.

Most charged side chains occur on the surface of soluble globular proteins where they are stabilized by favorable energetic interactions with water. The rare positioning of a charged side chain in the interior usually implies an important functional role for that "buried" charge within the nonpolar interior in stabilizing protein conformation or participation in catalysis.

Amino Acids Undergo a Variety of Chemical Reactions

Amino acids in proteins react with a variety of reagents that may be used to investigate the function of specific side chains. Some common chemical reactions are presented in Table 3.7. Reagents that modify the acid side chain have been synthesized to bind to specific sites in a folded protein's structure, like the substrate-binding site. The strategy is to model the structural features of the enzyme's natural substrate into the modifying reagent. The reagent binds to the active site like a natural substrate and reacts with a specific side chain. This identifies the modified amino acid as being located in the substrate-binding site and helps identify its role in catalysis.

TABLE 3.7 Some Chemical Reactions of the Amino Acids

Reactive Group	*Reagent or Reaction*	*Product*
Amine (—NH_2) groups	Ninhydrin	Blue-colored product that absorbs at 540 nm[a]
	Fluorescamine	Product that fluoresces
Carboxylic acid groups	Alcohols	Ester products
	Amines	Amide products
	Carbodiimide	Activates for reaction with nucleophiles
—NH_2 of Lys	2,3,6-Trinitrobenzene sulfonate	Product that absorbs at 367 nm
	Anhydrides	Acetylate amines
	Aldehydes	Form Shiff base adducts
Guanidino group of Arg	Sakaguchi reaction	Pink-red product that can be used to assay Arg
Phenol of Tyr	I_2	Iodination of positions ortho to hydroxyl group on aromatic ring
	Acetic anhydride	Acetylation of —OH
S atom of Met side chain	CH_3I	Methyl sulfonium product
	$[O^-]$ or H_2O_2	Methionine sulfoxide or methionine sulfone
—SH of Cys	Iodoacetate	Carboxymethyl thiol ether
	N-Ethylmaleimide	Addition product with S
	Organic mercurials	Mercurial adducts
	Performic acid	Cysteic acid (—SO_3H)
	Dithionitrobenzoic acid	Yellow product that can be used to quantitate —SH groups
Imidazole of His and phenol of Tyr	Pauly's reagent	Yellow to reddish product

[a] Proline imino group reacts with ninhydrin to form product that absorbs light at 440 nm (yellow color).

3.4 | PRIMARY STRUCTURE OF PROTEINS

The **primary structure** (amino acid sequence) of a protein is required for understanding of its structure, its mechanism of action, and its relationship to other proteins with similar physiological roles. The primary structure of **insulin** illustrates the value of this knowledge for understanding a protein's biosynthesis and physiological forms. Insulin is produced in pancreatic islet cells as a single-chain, inactive precursor, **proinsulin**, with the primary structure shown in Figure 3.23. The polypeptide chain contains 86 amino-acids and 3 intrachain cystine disulfide bonds. It is transformed into biologically active insulin by proteolytic cleavage of the primary structure prior to its secretion from islet cells. Proinsulin is cleaved by proteases present in the islet cells that cleave two peptide bonds in proinsulin between residues 30 and 31 and 65 and 66. This releases a 35-amino-acid segment (the **C-peptide**) and insulin, which consists of two polypeptide chains (A and B) of 21 amino acids and 30 amino acids, respectively, covalently joined by the same disulfide bonds present in proinsulin (Figure 3.23). The C-peptide is further processed in the pancreatic islet cells by proteases that hydrolyze a dipeptide from the COOH-terminal and a dipeptide from the NH_2-terminal of the C-peptide. The modified C-peptide is secreted into the blood with insulin.

Besides giving information on the pathway for formation of active insulin, the role of particular amino acids in insulin is indicated by comparison of the sequence

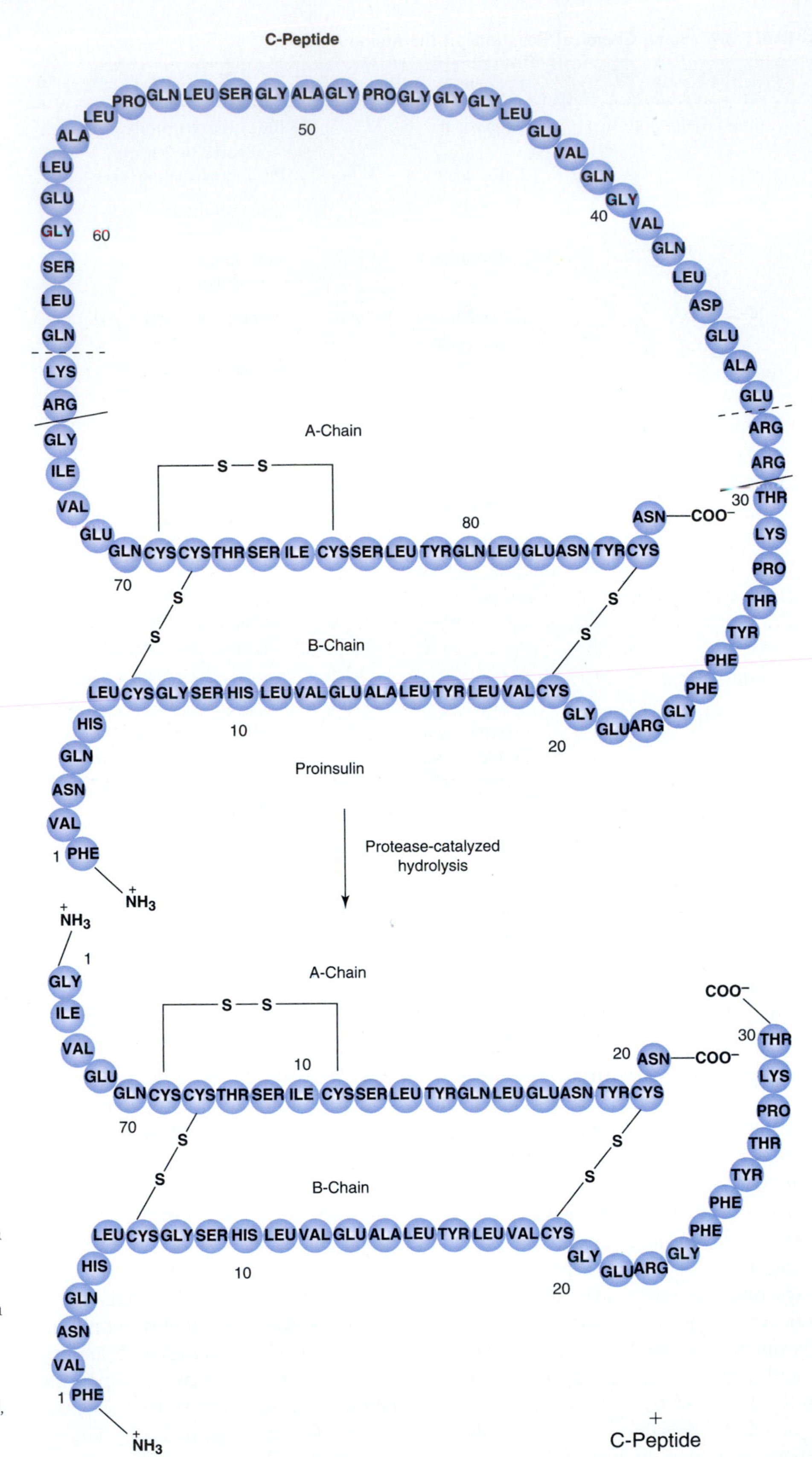

FIGURE 3.23
Primary structures of human proinsulin, insulin, and C-peptide.
In proinsulin, the B-chain peptide extends from Phe at position 1 to Thr at position 30, the C-peptide from Arg at position 31 to Arg at position 65, and the A-peptide from Gly at position 66 to Asn at position 86. Cystine bonds from positions 7 to 72, 19 to 85, and 71 to 76 are found in proinsulin.
Redrawn from Bell, G. I., Swain, W. F., Pictet, R., Cordell, B., Goodman, H. M., and Rutter, W. J., Nature *282:525, 1979.*

TABLE 3.8 Variation in Positions A8, A9, A10, and B30 of Insulin

Species	*A8*	*A9*	*A10*	*B30*
Human	Thr	Ser	Ile	Thr
Cow	Ala	Ser	Val	Ala
Pig	Thr	Ser	Ile	Ala
Sheep	Ala	Gly	Val	Ala
Horse	Thr	Gly	Ile	Ala
Dog	Thr	Ser	Ile	Ala
Chicken[a]	His	Asn	Thr	Ala
Duck[a]	Glu	Asn	Pro	Thr

[a] Positions 1 and 2 of B chain are both Ala in chicken and duck; whereas in the other species in the table, position 1 is Phe and position 2 is Val in B chain.

of insulins from different animal species. The aligned primary structures show a residue identity in most amino acid positions, except for residues 8, 9, and 10 of the A chain and residue 30 of the B chain. Amino acids in these positions vary widely in different animal insulins (Table 3.8) and apparently do not affect the biological properties (see Clin. Corr. 3.2). Other residues are rarely substituted, suggesting that they have an essential role in insulin function.

Sequence comparison is commonly used to predict the similarity in structure and function between proteins. These comparisons typically require aligning of sequences to maximize the number of identical residues and minimize the number of insertions or deletions required to achieve this alignment. Two sequences are termed **homologous** when their sequences are highly alignable. In its correct usage homology refers only to proteins that have evolved from the same gene. **Analogy** is used to describe sequences from proteins that are structurally similar but for which no evolutionary relationship has been demonstrated. Substitution of an amino acid by another amino acid of similar polarity (i.e., Val for Ile in position 10 of insulin) is called a **conservative** substitution and is commonly observed in sequences of the same protein from different animal species. If a particular amino acid is regularly found at the same position, then these are designated **invariant residues** and it can be assumed that these residues have an essential role in the structure or function of the protein. In contrast, a **nonconservative** substitution involves replacement of an amino acid by another of different polarity. This may produce severe changes in the

CLINICAL CORRELATION 3.2

Differences in Primary Structure of Insulins Used in Treatment of Diabetes Mellitus

Both pig (porcine) and cow (bovine) insulins are commonly used in the treatment of human diabetics. Because of the differences in amino acid sequence from the human insulin, some diabetic individuals will have an initial allergic response to the injected insulin as their immunological system recognizes the insulin as foreign, or develop an insulin resistance due to a high anti-insulin antibody titer at a later stage in treatment. However, the number of diabetics who have a deleterious immunological response to pig and cow insulins is small; the great majority of human diabetics can utilize the nonhuman insulins without immunological complication. The compatibility of cow and pig insulins in humans is due to the small number and the conservative nature of the changes between the amino acid sequences of the insulins. These changes do not significantly perturb the three-dimensional structure of the insulins from that of human insulin. Pig insulin is usually more acceptable than cow insulin in insulin-reactive individuals because it is more similar in sequence to human insulin (see Table 3.8). Human insulin is now available for clinical use. It can be made using genetically engineered bacteria or by modifying pig insulin.

Brogdon, R. N. and Heel, R. C. Human insulin: a review of its biological activity, pharmacokinetics, and therapeutic use. *Drugs* 34:350, 1987.

CLINICAL CORRELATION 3.3

A Nonconservative Mutation Occurs in Sickle Cell Anemia

Hemoglobin S (HbS) is a variant form of the normal adult hemoglobin (HbA_1) in which a nonconservative substitution occurs in the sixth position of the β-globin chain of the HbA_1. Whereas in HbA_1 this position is taken by a glutamic acid residue, in HbS the position is occupied by a valine. Consequently, in HbS a polar side chain group on the molecule's outside surface has been replaced with a nonpolar hydrophobic side chain (a nonconservative mutation). Through hydrophobic interactions with this nonpolar valine, HbS in its deoxy conformation polymerizes with other molecules of deoxy-HbS, leading to a precipitation of the hemoglobin within the red blood cell. This precipitation makes the red blood cell assume a sickle shape that results in a high rate of hemolysis and a lack of elasticity during circulation through the small capillaries, which become clogged by the abnormal shaped cells.

Only individuals homozygous for HbS exhibit the disease. Individuals heterozygous for HbS have approximately 50% HbA_1 and 50% HbS in their red blood cells and do not exhibit symptoms of the sickle cell anemia disease except under extreme conditions of hypoxia.

Individuals heterozygous for HbS have a resistance to the malaria parasite, which spends a part of its life cycle in red blood cells. This is a factor selecting for the HbS gene in malarial regions of the world and is the reason for the high frequency of this lethal gene in the human genetic pool. Approximately 10% of American blacks are heterozygous for HbS, and 0.4% of American blacks are homozygous for HbS and exhibit sickle cell anemia. HbS is detected by gel electrophoresis. Because it lacks a glutamate, it is less acidic than HbA. HbS, therefore, does not migrate as rapidly toward the anode as does HbA. It is also possible to diagnose sickle cell anemia by recombinant DNA techniques.

Bunn, H. F. Pathogenesis and treatment of sickle cell disease. *N Engl. J. Med.* 337:762, 1997: and Embury, S. H. The clinical pathophysiology of sickle-cell disease. *Annu. Rev. Med.* 37:361, 1986.

properties of the resultant protein or occur in regions that are apparently unimportant functionally (see Clin. Corr. 3.3). Polarity is only one physical property of amino acids that determines whether a substitution will significantly alter the protein's function. Other important properties are the volume and surface area of the residue.

3.5 | HIGHER LEVELS OF PROTEIN ORGANIZATION

Primary structure of a protein refers to the covalent structure of a protein. It includes amino acid sequence and location of disulfide (cystine) bonds. Higher levels of protein organization refer to noncovalently generated conformational properties of the primary structure. These higher levels of protein conformation and organization are defined as the secondary, tertiary, and quaternary structures. **Secondary structure** refers to the local three-dimensional folding of the polypeptide chain in the protein. The polypeptide chain in this context is the covalently interconnected atoms of the peptide bonds and α-carbon linkages that sequentially link the amino acid residues of the protein. Side chains are not considered at the level of secondary structure. **Tertiary structure** refers to the three-dimensional structure of the polypeptide. It includes the conformational relationships in space of the side chains and the geometric relationship between distant regions of the polypeptide chain. **Quaternary structure** refers to the noncovalent association of discrete polypeptide subunits into a multisubunit protein. Not all proteins have a quaternary structure.

Proteins generally assume unique secondary, tertiary, and quaternary conformations termed the **native conformation.** Folding of the primary structure into the native conformation occurs, in most cases, spontaneously through noncovalent interactions. This conformation is the one of lowest total Gibbs free energy kinetically accessible to the polypeptide(s) for the particular conditions of ionic strength, pH, and temperature in which the folding occurs. Chaperone proteins (see p. 133) may facilitate the rate of protein folding.

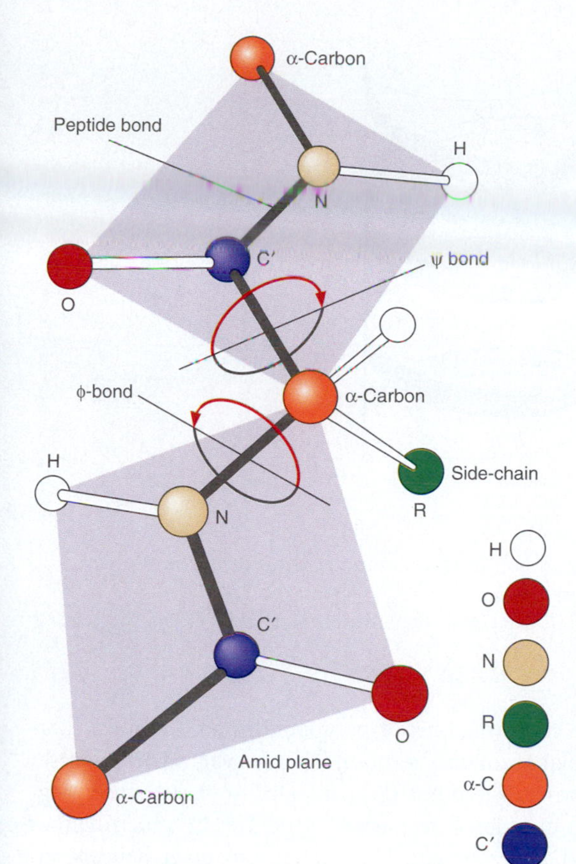

FIGURE 3.24
Polypeptide chain showing ϕ, ψ, and peptide bonds for residue R_i within a polypeptide chain.
Redrawn with permission from Dickerson, R. E. and Geis, I. The Structure and Action of Proteins. *Menlo Park, CA: Benjamin, 1969, p. 25.*

Proteins Have a Secondary Structure

The conformation of a polypeptide chain may be described by the rotational angles of covalent bonds that contribute to the chain. These bonds are contributed by each of the amino acids between (1) the nitrogen and α-carbon and (2) the α-carbon and the carbonyl carbon atoms. The first is designated the **phi (ϕ) bond** and the second the **psi (ψ) bond** for an amino acid residue in a polypeptide chain (Figure 3.24). The third bond contributed by each amino acid to the polypeptide chain is the peptide bond. As previously discussed, due to the partial double-bond character of the C′$\cdots$N bonds, there is a barrier to free rotation about this peptide bond.

Regular secondary structure occurs in segments of a polypeptide chain in which all ϕ bond angles are equal, and all ψ bond angles are equal. The rotational angles for ϕ and ψ bonds for common regular secondary structures are given in Table 3.9.

TABLE 3.9 Helix Parameters of Regular Secondary Structures

Structure	*Approximate Bond Angles (°)* ϕ	ψ	*Residues per turn, n*	*Helix Pitch,[a] p (Å)*
Right-handed α-helix [3.6_{13}-helix)	−57	−47	3.6	5.4
3_{10}-helix	+49	−26	3.0	6.0
Parallel β-strand	−119	+113	2.0	6.4
Antiparallel β-strand	−139	+135	2.0	6.8
Polyproline type II[b]	−78	+149	3.0	9.4

[a] Distance between repeating turns on a line drawn parallel to helix axis.
[b] Helix type found for polypeptide chains of collagen.

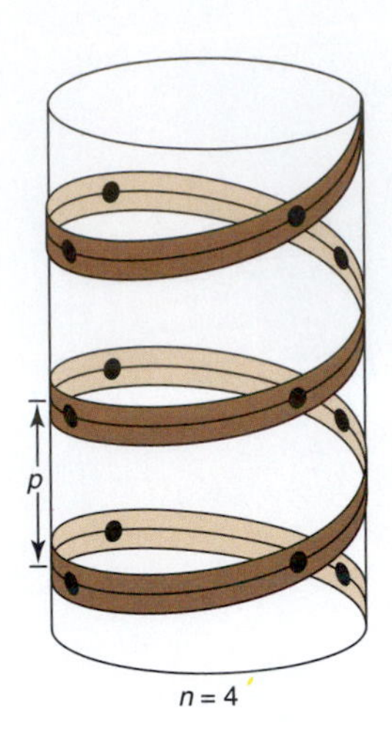

FIGURE 3.25
Helix pitch (p) for a helix with n = 4.
Each circle on a line represents an α-carbon from an amino acid residue. The rise per residue would be p/n (see equation in text).
Redrawn with permission from Dickerson, R. E. and Geis, I. The Structure and Action of Proteins. *Menlo Park, CA: Benjamin, 1969, p. 26.*

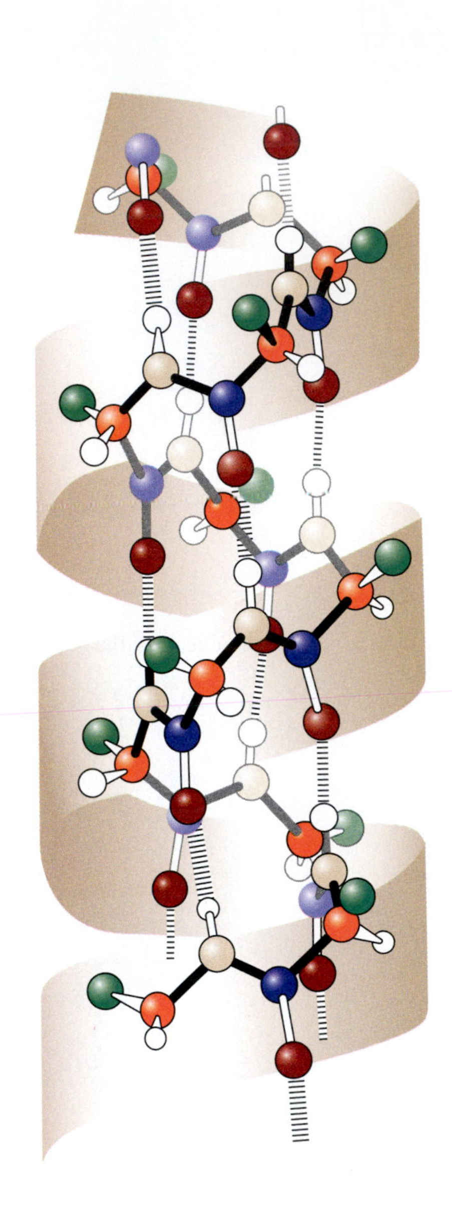

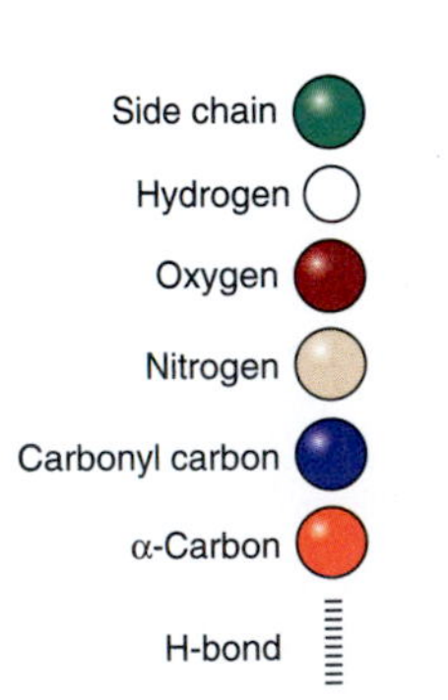

FIGURE 3.26
An α-helix.
Redrawn with permission, based on figure from Pauling, L. The Nature of the Chemical Bond, *3rd ed. Ithaca, NY: Cornell University Press, 1960.*

The α-helix and β-structure conformations of polypeptides are the most thermodynamically stable of the regular secondary structures. However, a particular sequence may form regular conformations other than α-helix or β-structure. Segments of polypeptides may have unordered secondary structure, in which neither the ϕ bond angles nor the ψ bond angles are equal. Proline interrupts α-helical conformation since its pyrrolidine side chain sterically interacts with the residue preceding it and prevents formation of an α-helical structure.

Helical structures of polypeptide chains are characterized by the number of amino acid residues per turn of helix (n) and the distance between α-carbon atoms of adjacent amino acids measured parallel to the axis of the helix (d). The **helix pitch** (p), defined as the product of $n \times d$, then measures the distance between repeating turns of the helix on a line drawn parallel to the helix axis (Figure 3.25):

$$p = n \times d$$

α-Helical Structure

An amino acid sequence in an α-helical conformation is shown in Figure 3.26. Characteristic are 3.6 amino acid residues per 360° turn (n = 3.6). The peptide bond planes in the α-helix are parallel to the axis of the helix. In this geometry each peptide forms two hydrogen bonds, one to the peptide bond of the fourth residue above and the second to the peptide bond of the fourth amino acid below in the primary structure. Other parameters, such as the pitch (p), are given in Table 3.9. In the hydrogen bonds between the peptide groups, the distance between the hydrogen-donor atom and the hydrogen-acceptor atom is 2.9 Å. Also, the donor, acceptor, and hydrogen atoms are approximately collinear, in that they approximate a straight line. This is an optimum geometry and distance for maximum hydrogen-bond strength (see Section 3.7).

The side chains in an α-helix are on the outside of the spiral structure generated by the polypeptide chain. Due to the characteristic 3.6 residues per turn, the first and every third and fourth R group of the amino acid sequence in the helix come close to the other. Helices often present separable polar and nonpolar faces as determined by their amino acid sequences, which place polar or nonpolar side chains three or four residues apart. This gives rise to unique functional characteristics of the helix. However, if every third or fourth side chain that come close together have the same charge sign or are branched at their β-carbon (valine and isoleucine), their unfavorable ionic or steric interactions destabilize the helix

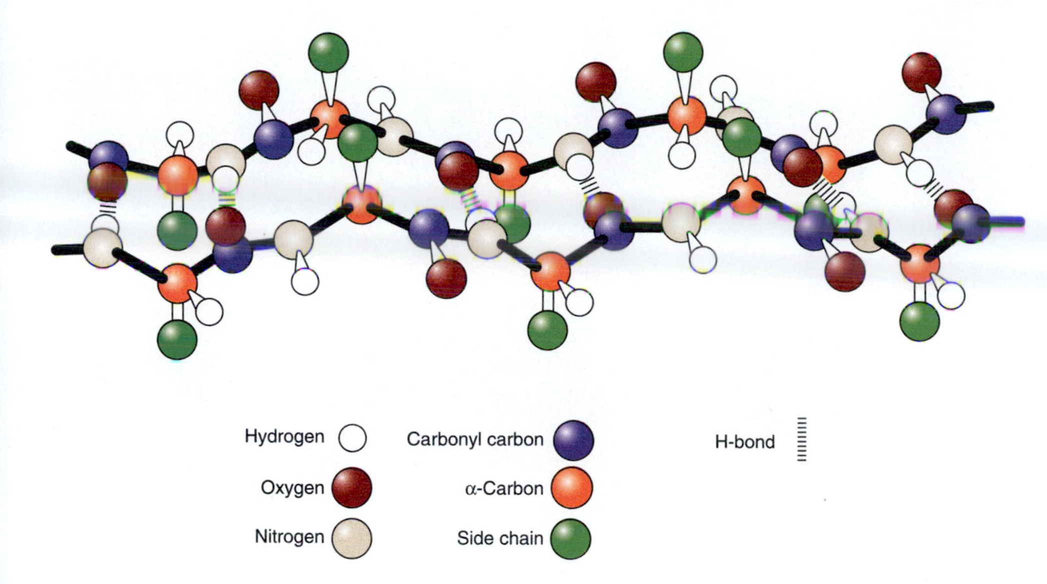

FIGURE 3.27
Two polypeptide chains in a β-structure conformation.
Additional polypeptide chains may be added to generate more extended structure.
Redrawn with permission from Fersht, A. Enzyme Structure and Mechanism. *San Francisco: Freeman, 1977, p. 10.*

structure. The α-helix may theoretically form its spiral in either a left-handed or right-handed sense, giving the helix asymmetric properties and correlated optical activity. The structure shows a right-handed α-helix; this is more stable than the left-handed helix.

β-Structure

A polypeptide chain in a β-strand conformation (Figure 3.27) is hydrogen bonded to another similar region aligned in a parallel or an antiparallel direction to generate the **β-structure** (Figure 3.28). Hydrogen-bonded β strands appear like a pleated sheet (Figure 3.29). The side chains project above and below the pleated sheet-like structure.

Structural Motifs and Protein Folds

Certain arrangements of secondary structure elements can occur in different folded protein structures. These combinations may be rather simple and are referred to as

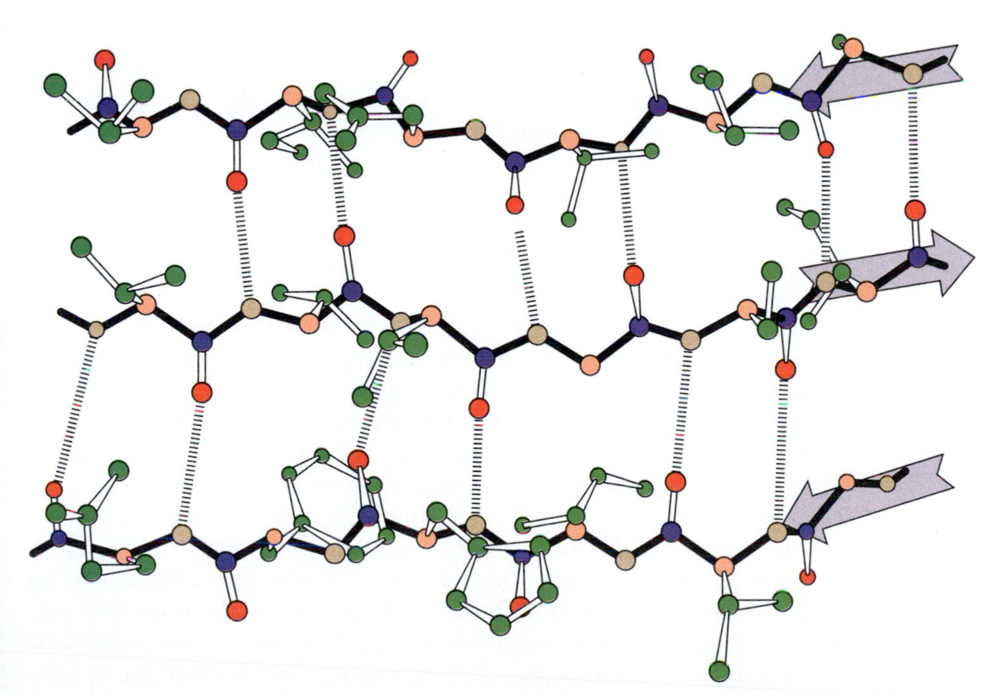

FIGURE 3.28
Example of antiparallel β-structure (residues 93–98, 28–33, and 16–21 of Cu, Zn superoxide dismutase).
Dashed line shows hydrogen bonds between carbonyl oxygen atoms and peptide nitrogen atoms; arrows show direction of polypeptide chains from N-terminal to C-terminal. In the characteristic antiparallel β-structure, pairs of closely spaced interchain hydrogen bonds alternate with widely spaced hydrogen bond pairs.
Redrawn with permission from Richardson, J. S. Adv. Protein Chem. *34:168, 1981.*

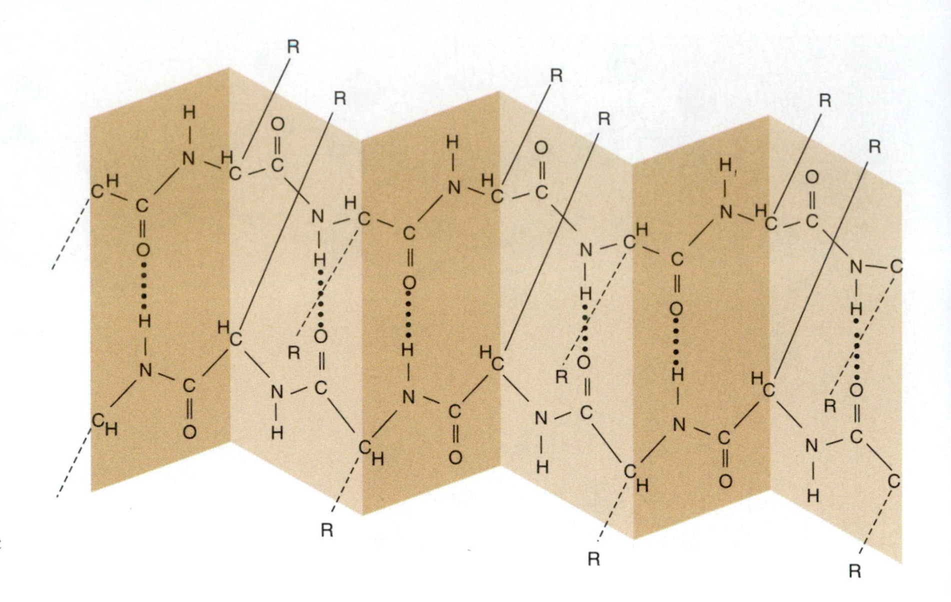

FIGURE 3.29
β-Pleated sheet structure between two polypeptide chains.
Additional polypeptide chains may be added above and below to generate a more extended structure.

structural motifs. They include the **helix–turn–helix** motif found in many DNA-binding proteins, the **β-strand–turn–β-strand** motif found in proteins with antiparallel β-structure, and the alternating **β-strand–turn–helix–turn–β-strand** motif found in many α,β-proteins. In these motifs a turn refers to small segments of the polypeptide (approximately 3 or 4 residues) of nonregular secondary structure that interconnect regions of regular secondary structure, whereas a loop is a larger segment of interconnecting polypeptide chain. Combinations of motifs or more complex orderings of secondary structure elements may form a **fold,** which comprises the secondary structure pattern of a domain.

Proteins Form Three-Dimensional Structures: Tertiary Structure

The **tertiary structure** of a polypeptide depicts the location of each of its atoms in space. It includes the geometric relationship between distant segments of primary structure and the positional relationship of the side chains with one another. An example of a protein's tertiary structure, that of trypsin, is shown in Figure 3.30. In Figure 3.30*a* the ribbon structure shows the conformation of polypeptide chains and the overall pattern of chain folding (fold structure). The tertiary structure depicted in Figure 3.30*b* shows the position of side chains. Active site catalytic side chains are shown in yellow, which include the hydroxymethyl group of serine (residue 177 in the sequence), the imidazole of histidine (residue 40), and the side chain carboxylate of aspartate (residue 85). Although these catalytic residues are widely separated in the primary structure, the tertiary structure brings them together in space to form the catalytic site. In Figure 3.30*c* a space-filling model shows C, N, and O atoms represented by balls of radius proportional to their van der Waals radius.

The tertiary structure of trypsin conforms to the general rules for folded proteins (see Section 3.3). Hydrophobic side chains are generally in the interior, away from the water interface. Ionized side chains occur on the outside, where they are stabilized by water of solvation. Within the protein structure (not shown) are buried water molecules exhibiting specific arrangements. A large number of water molecules form a solvation shell around the outside of the protein.

A long polypeptide strand often folds into multiple compact semi-independent regions or **domains**, each having a characteristic compact geometry with a hydrophobic core and polar surface. They typically contain 100–150 contiguous amino acids. The domains of a **multidomain protein** may be connected by a segment of the polypeptide chain lacking regular secondary structure. Alternatively, the dense spherical folded regions are separated by a cleft or less dense region of terti-

(a)

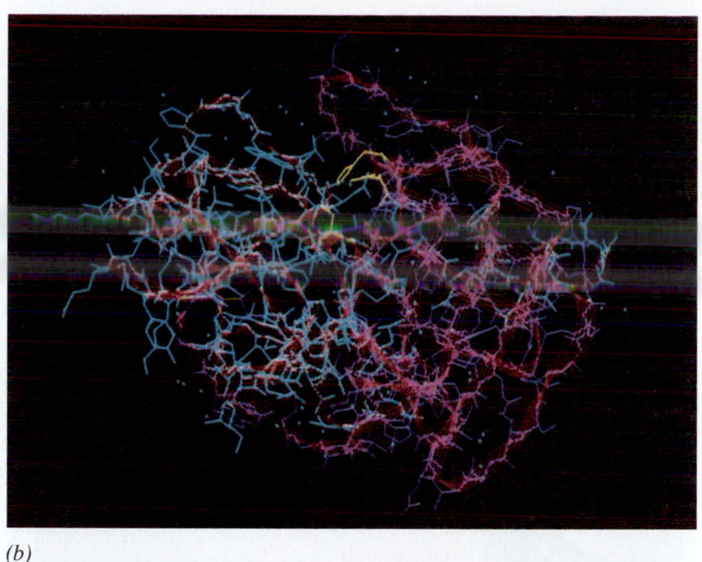

(b)

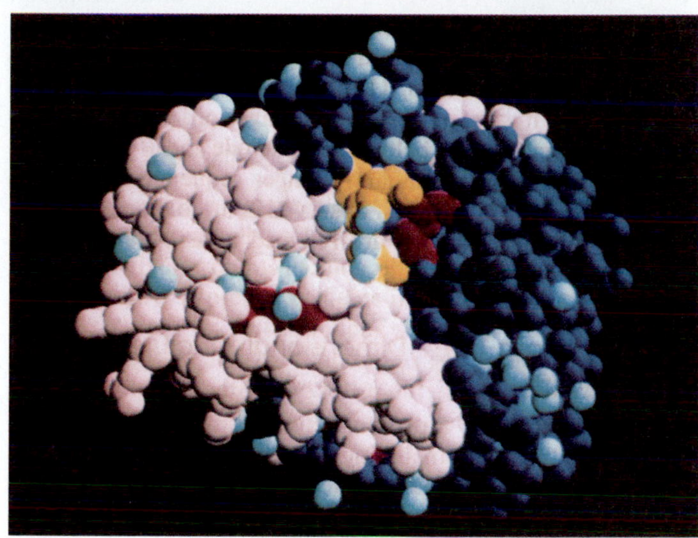

(c)

FIGURE 3.30
Tertiary structure of trypsin.
(*a*) Ribbon structure outlines conformation of the polypeptide chain. (*b*) Structure shows side chains including active site residues (in yellow) with outline of polypeptide chain (ribbon) superimposed. (*c*) Space-filling structure in which each atom is depicted as the size of its van der Waals radius. Hydrogen atoms are not shown. Different domains are shown in dark blue and white. Active site residues are in yellow and intrachain disulfide bonds of cystine in red. Light blue spheres represent water molecules associated with the protein. This structure shows the density of packing within interior of the protein.

ary structure (Figure 3.31). There are two domains in the trypsin molecule with a cleft between the domains that includes the substrate-binding catalytic site of the protein. An active site within an interdomain interface is an attribute of many enzymes. Different domains within a protein can move with respect to each other. Hexokinase (Figure 3.32), which catalyzes phosphorylation of glucose by adenosine triphosphate (ATP), has a glucose-binding site in a region between two domains. When glucose binds in the active site, the surrounding domains move to enclose the substrate to trap it for phosphorylation (Figure 3.32). In enzymes with more than one substrate or allosteric effector site (see p. 436), the different sites may be located within different domains. In multifunctional proteins, each domain performs a different task.

Bioinformatics Relates Structure and Function of Protein Gene Products

Bioinformatics is a computationally based research area that focuses on integration and analysis of complex biological data with computer algorithms. A major emphasis of bioinformatics has been to identify patterns within nucleic acid or amino acid sequences which are signatures of structural features or motifs as well as signatures of the protein family or class to which the gene product belongs. These homology-searching algorithms are used to identify and classify gene products in terms of the enzyme or protein family to which they "belong" on the basis of sequence similarity, and when the structure is known homology searching may be based on structural similarity. The scoring for sequence similarity uses criteria from that of absolute identity of sequence to similarity based on physicochemical properties such as hydrophobicity and size. The algorithms allow for insertion or deletion of segments of polypeptide chain or structure to give the maximum possible overlap between two proteins.

Protein domains are classified by class, fold, and family. The **class** of the protein is determined by the predominant type of secondary structure present in the protein. Some examples are mainly α-helix, mainly β-strand, and approximately equal amounts of α-helix and β-strand as classes for proteins. The **fold** classification is determined by the arrangement of secondary structure elements within the domain. The **family** classification is determined by the sequence identity between proteins. Proteins that are members of the same family have a common evolutionary relationship; that is, they are derived from the same primordial gene. Proteins of the same family have the same fold and often have similar functions.

Of interest to clinical medicine is the finding of limited mutations in the amino acid sequence, which can give rise to significant alterations in function. Biochemi-

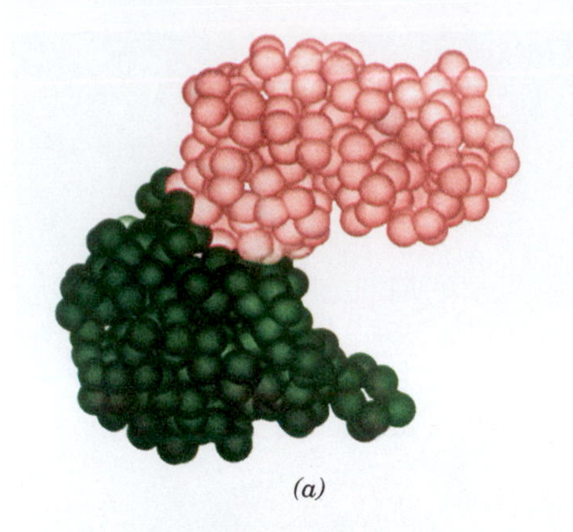

(a)

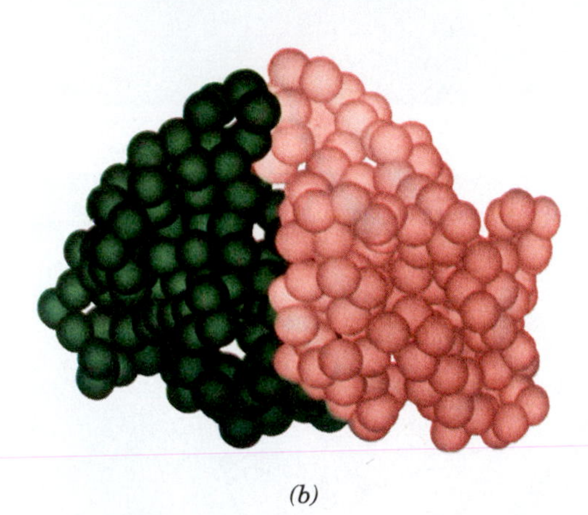

(b)

FIGURE 3.31
Globular domains within proteins.
(a) Phosphoglycerate kinase has two domains with a relatively narrow neck in between.
(b) Elastase has two tightly associated domains separated by a narrow cleft. Each sphere in the space-filling drawing represents the α-carbon position for an amino acid within the protein structure.
Reprinted with permission from Richardson, J. S. Adv. Protein Chem. 34:168, 1981.

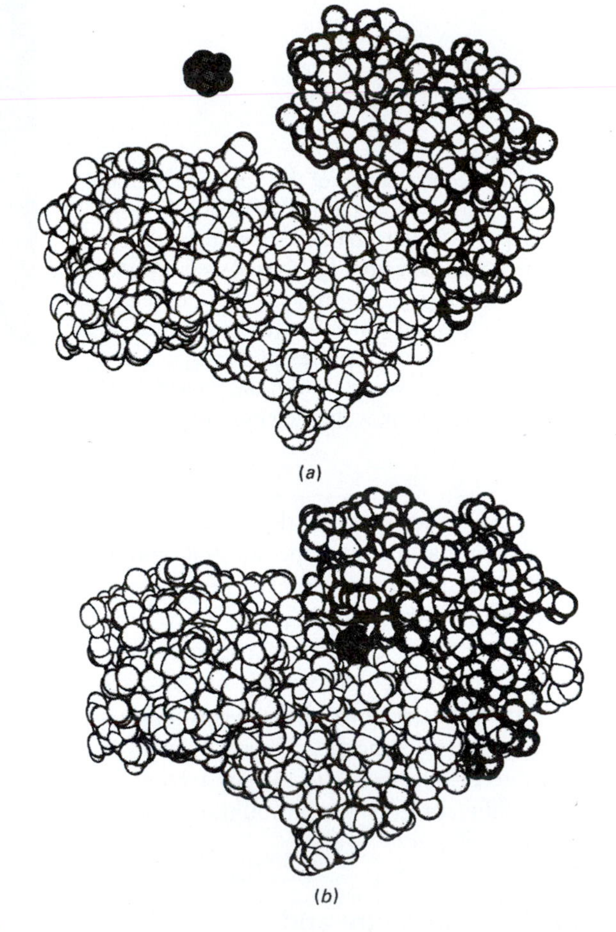

FIGURE 3.32
Drawings of *(a)* unliganded form of hexokinase and free glucose and *(b)* conformation of hexokinase with glucose bound.
In this space-filling drawing each circle represents a van der Waals radius of an atom in the structure. Glucose is black, and each domain is differently shaded.
Reprinted with permission from Bennett, W. S. and Huber, R. CRC Rev. Biochem. 15:291, 1984. Copyright © 1984 CRC Press, Inc., Boca Raton, FL.

cal measurements show that, at the molecular level, changes in even one amino acid can be significant. Perturbations on amino acid mutation have been observed in the native protein conformation, the number or states of conformational flexibility, the energetics or dynamical motion of the molecule, and the selectivity of proteins toward substrates and inhibitors. Examples of this include hemoglobin for which there are extensive catalogs of single-site mutations. Some hemoglobin mutations present significant clinical symptoms, such as sickle cell anemia, where the substitution of a single amino acid enables the hemoglobin molecules to form aggregates that precipitate within the red blood cell (see Clin. Corr. 3.3). Certain positions in the amino acid sequence are observed to be variant across diverse populations. These sequence positions, when they involve single changes in the base codon for that amino acid, are termed **single nucleotide polymorphisms** (SNPs) and give an understanding of the differences in response by individuals to the same disease or therapeutic treatment.

Homologous Fold Structures Are Often Formed from Nonhomologous Amino Acid Sequences

Although each native structure is unique, comparison of tertiary structures of different proteins derived by X-ray crystallography shows that similar arrangements of secondary structure motifs are often observed in the fold structures of domains. Folds of similar structure from proteins unrelated by function, sequence, or evolution are designated **superfolds.** The superfold structures form because of the thermodynamic stability of these secondary structure arrangements or the kinetic accessibility of the arrangements.

A common **all-α secondary structure** arrangement is found in the enzyme lysozyme and is designated the globin fold (Figure 3.33). Other examples of the globin fold are found in myoglobin and the subunits of hemoglobin, whose structures are discussed in Chapter 9. These all-α structures have seven or eight sections of α-helix joined by smaller segments of polypeptide chains that allow the helices to fold back upon themselves to form a characteristic globular shape. As this fold is generated from proteins from different sequence homology families, the globin fold is a superfold. Another common superfold is the **α,β-domain structure** shown by triose phosphate isomerase (Figure 3.34) in which the strands (designated by arrows) are wound into a central **β-barrel** with each β-strand in the interior interconnected by α-helical regions of the polypeptide chain on the outside of the fold. A similar fold structure is found in pyruvate kinase (Figure 3.34). A different type of **α,β-domain superfold** structure is seen in lactate dehydrogenase and

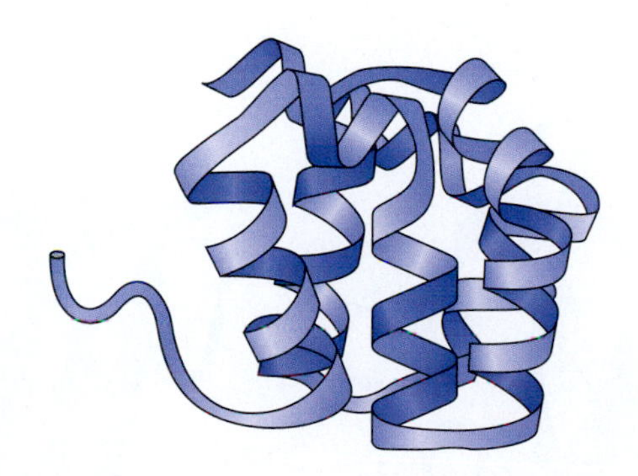

FIGURE 3.33
An example of an all α-domain globin fold in the enzyme lysozyme.
In this drawing and those that follow (Figures 3.34–3.36), only the outline of the polypeptide chain is shown. β-Structure strands are shown by arrows with direction of arrow showing N → C terminal direction of chain; lightning bolts represent disulfide bonds, and circles represent metal ion cofactors (when present). All α domain is the globin fold.
Redrawn with permission from Richardson, J. S. Adv. Protein Chem. *34:168, 1981.*

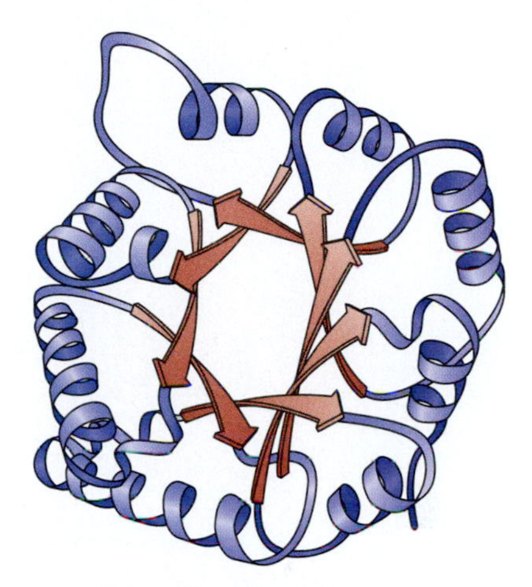

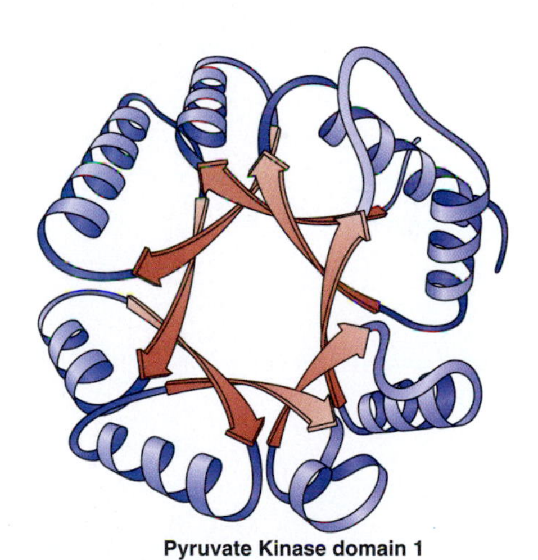

FIGURE 3.34
Example of an α,β-domain fold in triose phosphate isomerase.
See legend to Figure 3.33. In this commonly formed superfold the β-strands form a β-barrel in the center of the domain while α-helix segments are on the outside of the domain. β-Strands are in parallel directions. Regions of α-helix alternate with β-strands within the polypeptide chain.
Redrawn with permission from Richardson, J. S. Adv. Protein Chem. *34:168, 1981.*

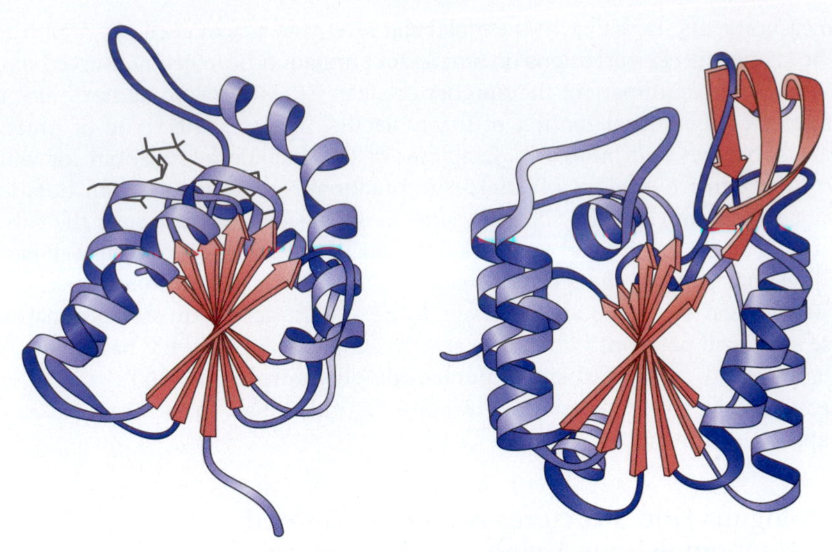

FIGURE 3.35
Example of an α,β-domain fold in which β-structure strands are in the form of a classical twisted β-sheet of lactate dehydrogenase and phosphoglycerate kinase.
See legend to Figure 3.33. As in the previous α,β-domain fold, regions of α-helix alternate with regions of β-strand within polypeptide chain. β-Sheet structure is on the inside while the α-helical segments are on the outside. β-Strands are in parallel within the β-structure.
Redrawn with permission from Richardson, J. S. Adv. Protein Chem. 34:168, 1981.

phosphoglycerate kinase (Figure 3.35). In these the interior polypeptide sections participate in a **twisted-sheet β-structure.** Again, the β-strand segments are joined by α-helix regions positioned on the outside of the molecule to give a characteristic fold pattern. An **all-β-domain** superfold structure is present in Cu, Zn superoxide dismutase, in which the antiparallel β-sheet forms a "Greek key" β-barrel (Figure 3.36). A similar pattern occurs in each of the domains of the immunoglobulins, discussed in Chapter 9. **Concanavalin A** (Figure 3.36) shows an all-β-domain structure in which the antiparallel β-strands form a β-barrel fold called a "jellyroll." Proteins that are not water soluble may contain different fold patterns (see Sections 3.6).

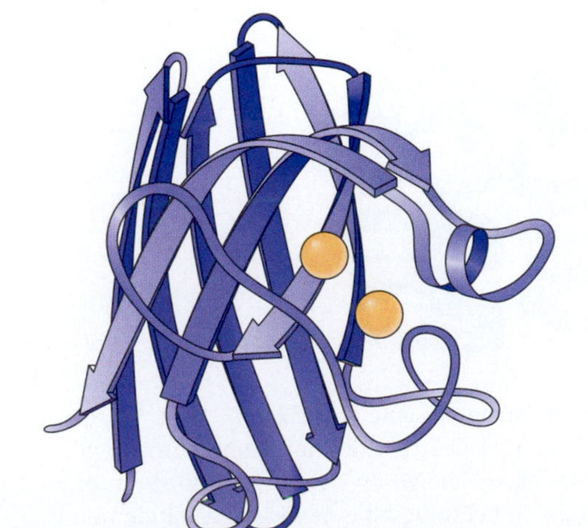

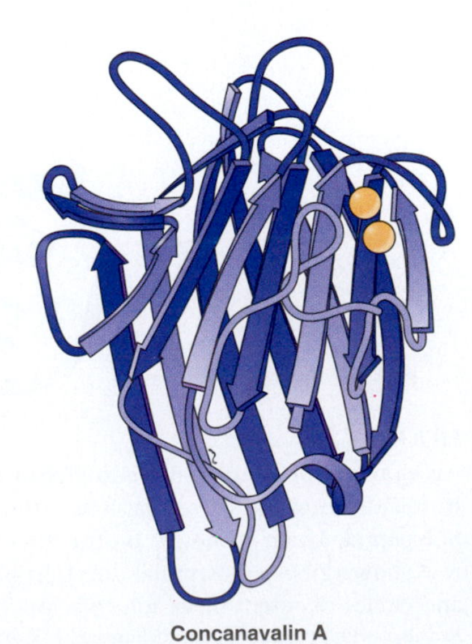

FIGURE 3.36
Examples of all β-domain folds (see legend to Figure 3.33).
β-strands are mostly antiparallel in all β-domain folds.
Redrawn with permission from Richardson, J. S. Adv. Protein Chem. 34:168, 1981.

A Quaternary Structure Occurs When Several Polypeptide Chains Form a Specific Noncovalent Association

Quaternary structure refers to the arrangement of polypeptide chains in a multichain protein. The subunits in a quaternary structure must be in noncovalent association. α-Chymotrypsin contains three polypeptides covalently joined together by interchain disulfide bonds into a single covalent unit and therefore does not have a quaternary structure. Myoglobin consists of one polypeptide and has no quaternary structure. However, hemoglobin A contains four polypeptide subunits ($\alpha_2\beta_2$) held together noncovalently in a specific conformation as required for its function (see Chapter 9). Thus hemoglobin has a quaternary structure. Aspartate transcarbamylase (see p. 840) has a quaternary structure comprised of 12 polypeptide subunits. The poliovirus protein coat contains 60 polypeptide subunits, and the tobacco mosaic virus protein has 2120 subunits held together noncovalently.

3.6 | OTHER TYPES OF PROTEINS

The characteristics of protein structure, as discussed above, are based on observations on globular and water-soluble proteins. Other proteins, such as the fibrous proteins, are nonglobular and have a low water solubility; lipoproteins and glycoproteins contain nonprotein components and may or may not be water soluble.

Fibrous Proteins Include Collagen, Elastin, α-Keratin, and Tropomyosin

Globular proteins have a spheroidal shape, variable molecular weights, relatively high water solubility, and a variety of functional roles as catalysts, transporters, and regulators of metabolic pathways and gene expression. In contrast, **fibrous proteins** characteristically contain larger amounts of regular secondary structure, have a long cylindrical (rod-like) shape, have a low solubility in water, and have a structural rather than a dynamic role in the cell or organism. Examples of fibrous proteins are collagen, elastin, α-keratin, and tropomyosin.

Distribution of Collagen in Humans

Collagen is present in all tissues and organs where it provides the framework that gives the tissues their form and structural strength. Its importance is shown by its high concentration in all organs; the percentage of collagen by weight for some representative human tissues and organs is liver 4%, lung 10%, aorta 12–24%, cartilage 50%, cornea 64%, whole cortical bone 23% , and skin 74% (see Clin. Corr. 3.4).

CLINICAL CORRELATION 3.4

Symptoms of Diseases of Abnormal Collagen Synthesis

Collagen is present in virtually all tissues and is the most abundant protein in the human. Certain organs depend heavily on normal collagen structure to function physiologically. Abnormal collagen synthesis or structure causes dysfunction of cardiovascular organs (aortic and arterial aneurysms and heart valve malfunction), bone (fragility and easy fracturing), skin (poor healing and unusual distensibility), joints (hypermobility and arthritis), and eyes (dislocation of the lens). Examples of diseases caused by abnormal collagen synthesis include Ehlers–Danlos syndrome, osteogenesis imperfecta, and scurvy. These diseases may result from abnormal collagen genes, abnormal posttranslational modification of collagen, or deficiency of cofactors needed by the enzymes that carry out posttranslational modification of collagen.

Byers, P. H. Disorders of collagen biosynthesis and structure. In: C. R. Scriver, A. L. Beaudet, W. S. Sly, and D. Valle (Eds.), *The Metabolic and Molecular Bases of Inherited Disease,* 7th ed. New York: McGraw-Hill, 1995, Chap. 134.

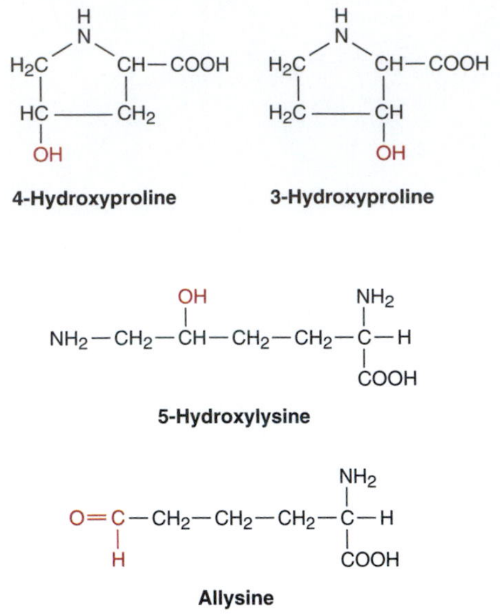

4-Hydroxyproline

3-Hydroxyproline

5-Hydroxylysine

Allysine

FIGURE 3.37
Derived amino acids found in collagen. Carbohydrate is attached to 5-OH in 5-hydroxylysine by a type III glycosidic linkage (see Figure 3.45).

Amino Acid Composition of Collagen

The amino acid compositions of type I skin collagen and of the globular proteins ribonuclease and hemoglobin are given in Table 3.10. Skin collagen is rich in **glycine** (33% of its amino acids), **proline** (13%), and the derived amino acids **4-hydroxyproline** (9%) and **5-hydroxylysine** (0.6%) (Figure 3.37). Hydroxyproline is unique to collagens, being formed enzymatically from prolines within a collagen polypeptide chain. The enzyme-catalyzed hydroxylation of proline requires ascorbic acid (vitamin C); thus in vitamin C deficiency **(scurvy)** there is poor synthesis of new collagen. Most of the hydroxyprolines in a collagen have the hydroxyl group in the 4-position (γ carbon) of the proline structure, although a small amount of 3-hydroxyproline is also formed (Table 3.10). Collagens are glycoproteins with carbohydrate covalently joined to the derived amino acid, 5-hydroxylysine, by an *O*-glycosidic bond through the δ-carbon hydroxyl group. Formation of 5-hydroxylysine and addition of the carbohydrate occur after polypeptide chain formation but prior to the folding of the collagen chains into their unique supercoiled structure.

Amino Acid Sequence of Collagen

Each mature collagen or tropocollagen molecule contains three polypeptide chains. The various collagen polypeptides have their own genes. Some types of collagen contain three polypeptide chains with identical amino acid sequences. In others such as type I (Table 3.11), two of the chains are identical while the sequence of the third chain is slightly different. In type I collagen, the identical polypeptides are

TABLE 3.10 Comparison of the Amino Acid Content of Human Skin Collagen (Type I) and Mature Elastin with That of Two Typical Globular Proteins[a]

Amino Acid	*Collagen (Human Skin)*	*Elastin (Mammalian)*	*Ribonuclease (Bovine)*	*Hemoglobin (Human)*
COMMON AMINO ACIDS		PERCENT OF TOTAL		
Ala	11	22	8	9
Arg	5	0.9	5	3
Asn			8	3
Asp	5	1	15	10
Cys	0	0	0	1
Glu	7	2	12	6
Gln			6	1
Gly	33	31	2	4
His	0.5	0.1	4	9
Ile	1	2	3	0
Leu	2	6	2	14
Lys	3	0.8	11	10
Met	0.6	0.2	4	1
Phe	1	3	4	7
Pro	13	11	4	5
Ser	4	1	11	4
Thr	2	1	9	5
Trp	2	1	9	2
Tyr	0.3	2	8	3
Val	2	12	8	10
DERIVED AMINO ACIDS				
Cystine	0	0	7	0
3-Hydroxyproline	0.1		0	0
4-Hydroxyproline	9	1	0	0
5-Hydroxylysine	0.6	0	0	0
Desmosine and isodesmosine	0	1	0	0

[a] Boxed numbers emphasize important differences in amino acid composition between the fibrous proteins (collagen and elastin) and typical globular proteins.

designated $\alpha1(I)$ chains and the third nonidentical chain, $\alpha2(I)$. Type V collagen contains three different chains, designated $\alpha1(V)$, $\alpha2(V)$, and $\alpha3(V)$. Different types of collagen differ in their physical properties due to differences in the amino acid sequence among chains, even though there are large regions of homologous sequence among the different chain types. Collagen has covalently attached carbohydrate and the collagen types differ in their carbohydrate component. Table 3.11 describes some characteristics of collagen types I–VI; additional collagen types (designated up through type XIX) have been reported.

The amino acid sequence of the chains of collagens is unusual. In long segments of all the collagen types glycine is repeated as the third residue and proline or hydroxyproline also occurs three residues apart in these same regions. Accordingly, the amino acid sequences Gly-Pro-Y and Gly-X-Hyp (where X and Y are any of the amino acids) are repeated in tandem several hundred times. In type I collagen, the triplet sequences are reiterated over 200 times, encompassing over 600 of approximately 1000 amino acids in a polypeptide.

Structure of Collagen

Polypeptides that contain only proline can be synthesized in the laboratory. These polyproline chains assume a regular secondary structure in aqueous solution, a tightly twisted extended helix with three residues per turn ($n = 3$). This helix with all *trans*-peptide bonds is designated the **polyproline type II** helix (see Figure 3.11 for differences between *cis*- and *trans*-peptide bonds). The polyproline helix has the same characteristics as the helix found in collagen chains in regions of the primary structure that contain a proline or hydroxyproline at approximately every third position. This indicates that the thermodynamic forces leading to formation of the collagen helix structure are due to the properties of proline. In proline, the ϕ angle

TABLE 3.11 Classification of Collagen Types

Type	*Chain Designations*	*Tissue Found*	*Characteristics*
I	$[\alpha1(I)]_2\alpha2(I)$	Bone, skin, tendon, scar tissue, heart valve, intestinal, and uterine wall	Low carbohydrate; <10 hydroxylysines per chain; two types of polypeptide chains
II	$[\alpha1(II)]_3$	Cartilage, vitreous	10% carbohydrate; >20 hydroxylysines per chain
III	$[\alpha1(III)]_3$	Blood vessels, newborn skin, scar tissue, intestinal, and uterine wall	Low carbohydrate; high hydroxyproline and Gly; contains Cys
IV	$[\alpha1(IV)]_3$ $[\alpha2(IV)]_3$	Basement membrane, lens capsule	High 3-hydroxyproline; >40 hydroxylysines per chain; low Ala and Arg; contains Cys; high carbohydrate (15%)
V	$[\alpha1(V)]_2\alpha2(V)$ $[\alpha1(V)]_3$ $\alpha1(V)\alpha2(V)\alpha3(V)$	Cell surfaces or exocytoskeleton; widely distributed in low amounts	High carbohydrate, relatively high glycine, and hydroxylysine
VI	—	Aortic intima, placenta, kidney, and skin in low amounts	Relatively large globular domains in telopeptide region; high Cys and Tyr; molecular weight relatively low (~160,000); equimolar amounts of hydroxylysine and hydroxyproline

contributed to the polypeptide chain is part of the five-member cyclic side chain. This constrains the C_α—N bond to an angle compatible with the polyproline helix.

In polyproline type II helix, the plane of each peptide bond is perpendicular to the helix. In this geometry the carbonyl groups point toward neighboring chains and are correctly oriented to form strong interchain hydrogen bonds with other polyproline helical molecules. This contrasts with the α-helix, in which the plane containing the atoms of the peptide bond is parallel to the α-helix axis and the peptide bonds form only intrachain hydrogen bonds with peptide bonds in the same polypeptide chain. The three chains of a collagen molecule, where each of the chains is in a polyproline type II helix conformation, are wound about each other in a defined way to form a **superhelical structure** (Figure 3.38). The three-chain superhelix has a characteristic rise (d) and pitch (p) as does the single-chain helix. The collagen superhelix forms because glycines have a side chain of low steric bulk (R = H). As the polyproline type II helix has three residues per turn (n = 3) and glycine is at every third position, the glycines in each of the polypeptide chains are aligned along one side of the helix, forming an **apolar edge** of the chain. The

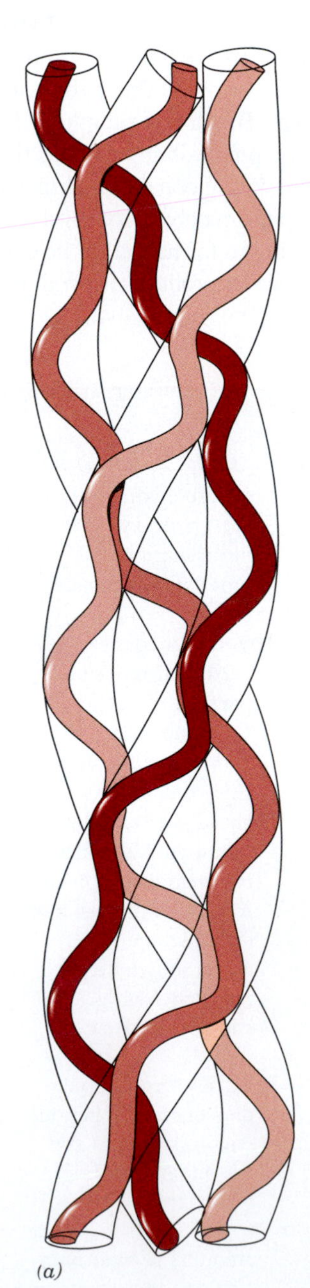

(a)

(b)

FIGURE 3.38
Diagram of collagen demonstrating necessity for glycine in every third residue to allow different chains to be in close proximity in the structure. (*a*) Ribbon model for supercoiled structure of collagen with each individual chain in a polyproline type II helix. (*b*) More detailed model of supercoiled conformation. All α-carbon atoms are numbered and proposed hydrogen bonds are shown by dashed lines.

Redrawn with permission from Dickerson, R. E. and Geis, I. The Structure and Actions of Proteins. *Menlo Park, CA: Benjamin, 1969, pp. 41, 42.*

glycine edges from the three polypeptide chains associate noncovalently in a close arrangement, held together by hydrophobic interactions, to form the superhelix structure of collagen. A larger side chain than that of glycine would impede the adjacent chains from coming together in the superhelix structure (Figure 3.38).

In collagen the superhelix may extend for long stretches of the sequence, especially in type I collagen where only the COOH-terminal and NH_2-terminal segments (known as the **telopeptides**) are not in a superhelical conformation. The type I collagen molecule is 3000 Å long and only 15 Å wide, a very long cylindrical structure. In other types, superhelical regions may be broken periodically by globular regions.

Formation of Covalent Cross-links in Collagen

An extracellular enzyme acts on secreted collagen molecules (see p. 269) to convert some ε-amino groups of lysine side chains to δ-aldehydes (Figure 3.39). The resulting amino acid, containing an aldehydic R group, is the derived amino acid **allysine**. The newly formed aldehyde side chain spontaneously undergoes nucleophilic addition reactions with nonmodified lysine ε-amino groups and with the δ-carbon atoms of other allysine aldehydic groups to form linking covalent bonds (Figure 3.39). These covalent linkages can be between chains within the superhelical structure or between adjacent superhelical collagen molecules in a collagen fibril.

Elastin Is a Fibrous Protein with Allysine-Generated Cross-links

Elastin gives tissues and organs the capacity to stretch without tearing. It is classified as a fibrous protein because of its structural function and relative insolubility in water; it is abundant in ligaments, lungs, walls of arteries, and skin. Elastin does not contain repeating sequences of Gly-Pro-Y or Gly-X-Hyp and does not fold into a polyproline helix or a superhelix. It lacks a regular secondary structure but contains an unordered coiled structure in which amino acid residues within the folded structure are highly mobile. This kinetically free, though extensively cross-linked structure, gives the protein a rubber-like elasticity. As in collagen, allysines form cross-links in elastin. An extracellular **lysine amino oxidase** converts lysine side chains to allysines, being specific for lysines in the sequence -Lys-Ala-Ala-Lys- and -Lys-Ala-Ala-Ala-Lys-. Three allysines and an unmodified lysine in these sequences,

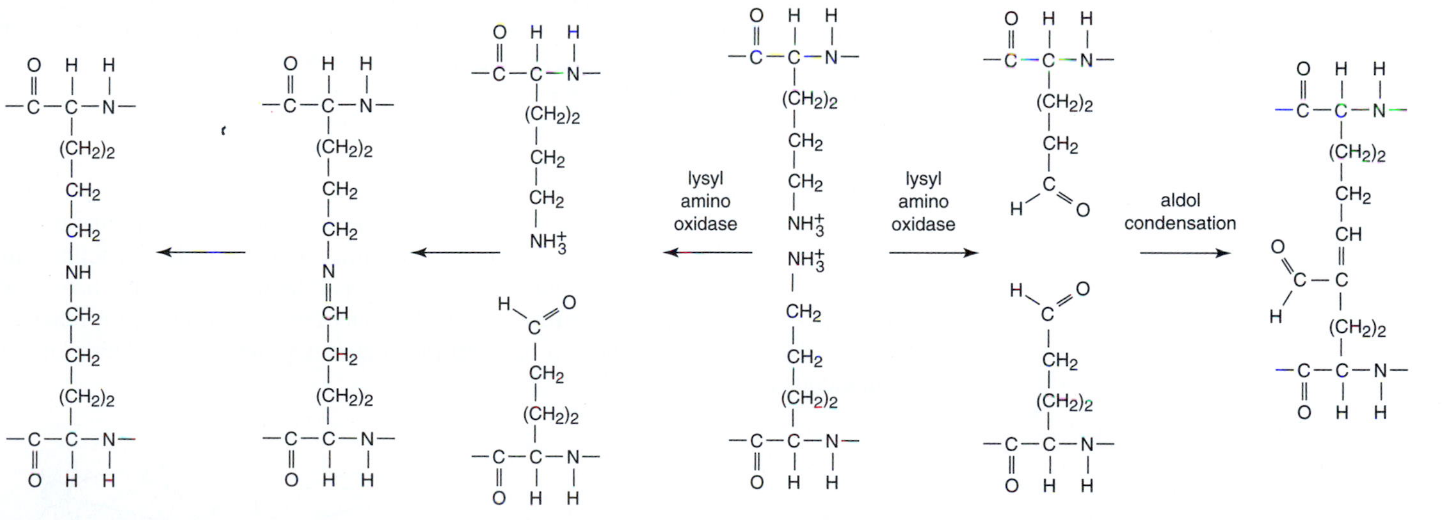

Lysinonorleucine (a cross-link in collagen or elastin) ← Schiff base ← Aldehyde derivative (allysyl) ← lysyl amino oxidase — Lysyl residues in collagen — lysyl amino oxidase → Aldehyde derivatives (allysyls) — aldol condensation → Aldol cross-link in collagen

FIGURE 3.39
Covalent cross-links formed in collagen through allysine intermediates.
Formation of allysines is catalyzed by lysyl amino oxidase.

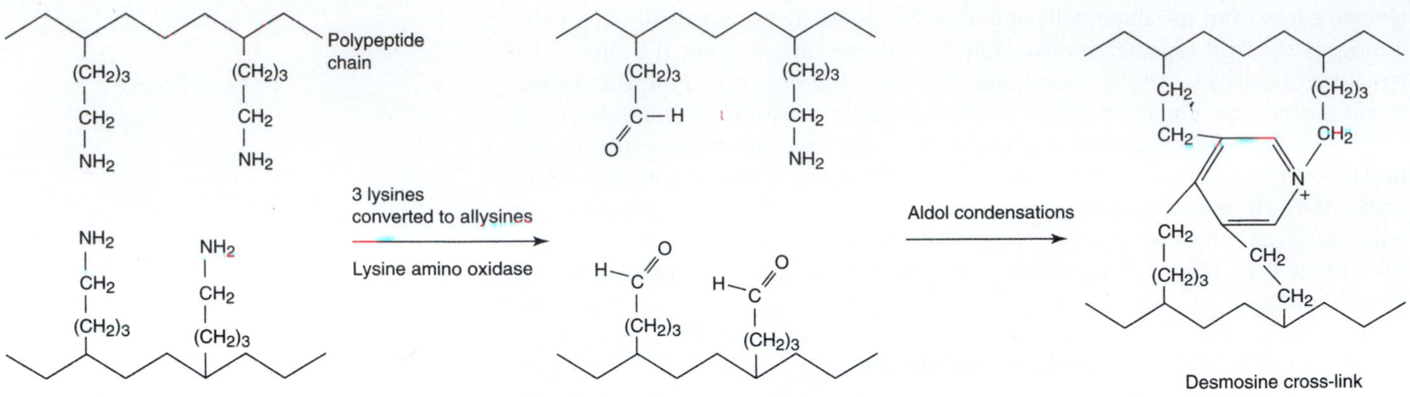

Polypeptide chain

3 lysines converted to allysines

Lysine amino oxidase

Aldol condensations

Desmosine cross-link

FIGURE 3.40
Formation of desmosine covalent cross-link in elastin from lysine and allysines.
Polypeptide chain drawn schematically with intersections of lines representing placement of α-carbons.

from different regions in the polypeptide chains, react to form the heterocyclic structure of **desmosine** or **isodesmosine.** The desmosines covalently cross-link the polypeptide chains in elastin fibers (Figure 3.40).

α-Keratin and Tropomyosin

α-Keratin and **tropomyosin** are fibrous proteins in which each polypeptide has an α-helical conformation. α-Keratin is found in the epidermal layer of skin, in nails, and in hair. Tropomyosin is a component of the thin filament in muscle tissue. The sequences in both proteins shows tandem repetition of seven residue segments, in which the first and fourth amino acids have hydrophobic side chains and the fifth and seventh polar side chains. The reiteration of hydrophobic and polar side chains in seven amino acid segments is symbolically represented by the formulation $(a\text{-}b\text{-}c\text{-}d\text{-}e\text{-}f\text{-}g)_i$, where a and d are hydrophobic amino acids, and e and g are polar or ionized side chain groups. Since a seven-amino-acid segment represents two complete turns of an α-helix ($n = 3.6$), the apolar residues at a and d align to form an apolar edge along one side of the α-helix (Figure 3.41). This apolar edge in α-keratin interacts with polypeptide apolar edges of other α-keratin chains to form a superhelical structure containing two or three polypeptide chains. Each polypeptide also contains a polar edge, due to residues e and g, that interacts with water on the outside of the superhelix and also stabilizes the superhelical structure. Similarly, two tropomyosin polypeptides in α-helical conformation wind around each other to form a tropomyosin superhelical structure.

Thus collagen, α-keratin, and tropomyosin molecules are multistrand structures where polypeptide chains have a highly regular secondary structure (polyproline type II helix or α-helix) and form a rod-shaped supercoiled conformation. In turn, the supercoiled molecules are aligned into multimolecular fibrils stabilized in some cases by covalent cross-links. The amino acid sequences of the chains are repetitive, generating edges on the cylindrical surfaces of each of the chains that stabilize a hydrophobic interaction between the chains required for generation of the supercoiled conformation.

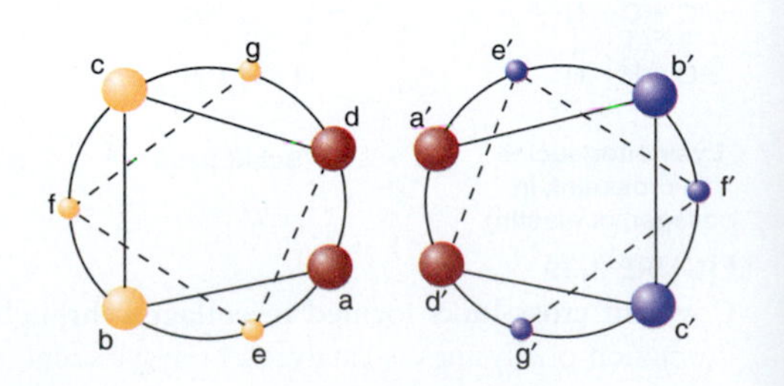

FIGURE 3.41
Interaction of an apolar edge of two chains in α-helical conformation as α-keratin and tropomyosin.
Interaction of apolar d-a and a′-d residues of two α-helices aligned parallel in an NH_2-terminal (top) to COOH-terminal direction is presented.
Redrawn from McLachlan, A. D. and Stewart, M. J. Mol. Biol. 98:293, 1975.

CLINICAL CORRELATION 3.5
Hyperlipidemias

Hyperlipidemias are disorders of the rates of synthesis or clearance of lipoproteins from the bloodstream. Usually they are detected by measuring plasma triacylglycerol and cholesterol and are classified on the basis of which class of lipoproteins is elevated.

Type I hyperlipidemia is due to accumulation of chylomicrons. Two genetic forms are known: lipoprotein lipase deficiency and ApoC-II deficiency. ApoC-II is required by lipoprotein lipase for full activity. Patients with type I hyperlipidemia have exceedingly high plasma triacylglycerol concentrations (over 1000 mg dL^{-1}) and suffer from eruptive xanthomas (triacylglycerol deposits in the skin) and pancreatitis.

Type II hyperlipidemia is characterized by elevated LDL levels. Most cases are due to genetic defects in the synthesis, processing, or function of the LDL receptor. Heterozygotes have elevated LDL levels, hence the trait is dominantly expressed. Homozygous patients have very high LDL levels and may suffer myocardial infarctions before age 20.

Type III hyperlipidemia is due to abnormalities of ApoE, which interfere with the uptake of chylomicron and VLDL remnants. Hypothyroidism can produce a very similar hyperlipidemia. These patients have an increased risk of atherosclerosis.

Type IV hyperlipidemia is the commonest abnormality. The VLDL levels are increased, often due to obesity, alcohol abuse, or diabetes. Familial forms are also known but the molecular defect is unknown.

Type V hyperlipidemia is, like type I, associated with high chylomicron triacylglycerol levels, pancreatitis, and eruptive xanthomas.

Hypercholesterolemia also occurs in certain types of liver disease in which biliary excretion of cholesterol is reduced. An abnormal lipoprotein called lipoprotein X accumulates. This disorder is not associated with increased cardiovascular disease from atherosclerosis.

Havel, R. J. and Kane, J. P. Introduction: structure and metabolism of plasma lipoproteins. In: C. R. Scriver, A. L. Beaudet, W. S. Sly, and D. Valle (Eds.), *The Metabolic and Molecular Bases of Inherited Disease,* 7th ed. New York: McGraw-Hill, 1995, Chap. 56; and Goldstein, J. L., Hobbs, H. H., and Brown, M. S. Familial hypercholesterolemia. In: C. R. Scriver, A. L. Beaudet, W. S. Sly, and D. Valle (Eds.), *The Metabolic and Molecular Bases of Inherited Disease,* 7th ed. New York: McGraw-Hill, 1995, Chap. 62.

Plasma Lipoproteins Are Complexes of Lipids with Proteins

Lipoproteins are multicomponent complexes of proteins and lipids that form distinct molecular aggregates with an approximate stoichiometry between protein and lipid components. Each type of lipoprotein has a characteristic molecular mass, size, composition, density, and physiological role. The protein and lipid in each complex are held together by noncovalent forces.

Plasma lipoproteins have been extensively characterized. Changes in their relative concentrations are predictive of atherosclerosis, a major human disease (see Clin. Corr. 3.5). Their roles in blood include transport of lipids from tissue to tissue and participating in lipid metabolism (see p. 745). Four classes of plasma lipoproteins exist in normal fasting humans (Table 3.12); in the postabsorptive period a fifth type, chylomicrons, is also present. The classes are identified by their density, as determined by ultracentrifugation and by electrophoresis (Figure 3.42). The protein components of a lipoprotein particle are the **apolipoproteins.** Each type of lipoprotein has a characteristic apolipoprotein composition, the different apolipoproteins often being present in a set ratio. The most prominent apolipoprotein in **high density lipoproteins (HDLs)** is apolipoprotein A-I **(apoA-I)**

TABLE 3.12 Hydrated Density Classes of Plasma Lipoproteins

Lipoprotein Fraction	*Density ($g\ mL^{-1}$)*	*Flotation Rate, S_f (Svedberg units)*	*Molecular Weight (daltons)*	*Particle Diameter (Å)*
HDL	1.063–1.210		HDL_2, 4×10^5	70–130
			HDL_3, 2×10^5	50–100
LDL (or LDL_2)	1.019–1.063	0–12	2×10^6	200–280
IDL (or LDL_1)	1.006–1.019	12–20	4.5×10^6	250
VLDL	0.95–1.006	20–400	5×10^6–10^7	250–750
Chylomicrons	<0.95	>400	10^9–10^{10}	10^3–10^4

Source: Data from Soutar, A. K. and Myant, N. B. In: R. E. Offord (Ed.), *Chemistry of Macromolecules,* IIB. Baltimore, MD: University Park Press, 1979.

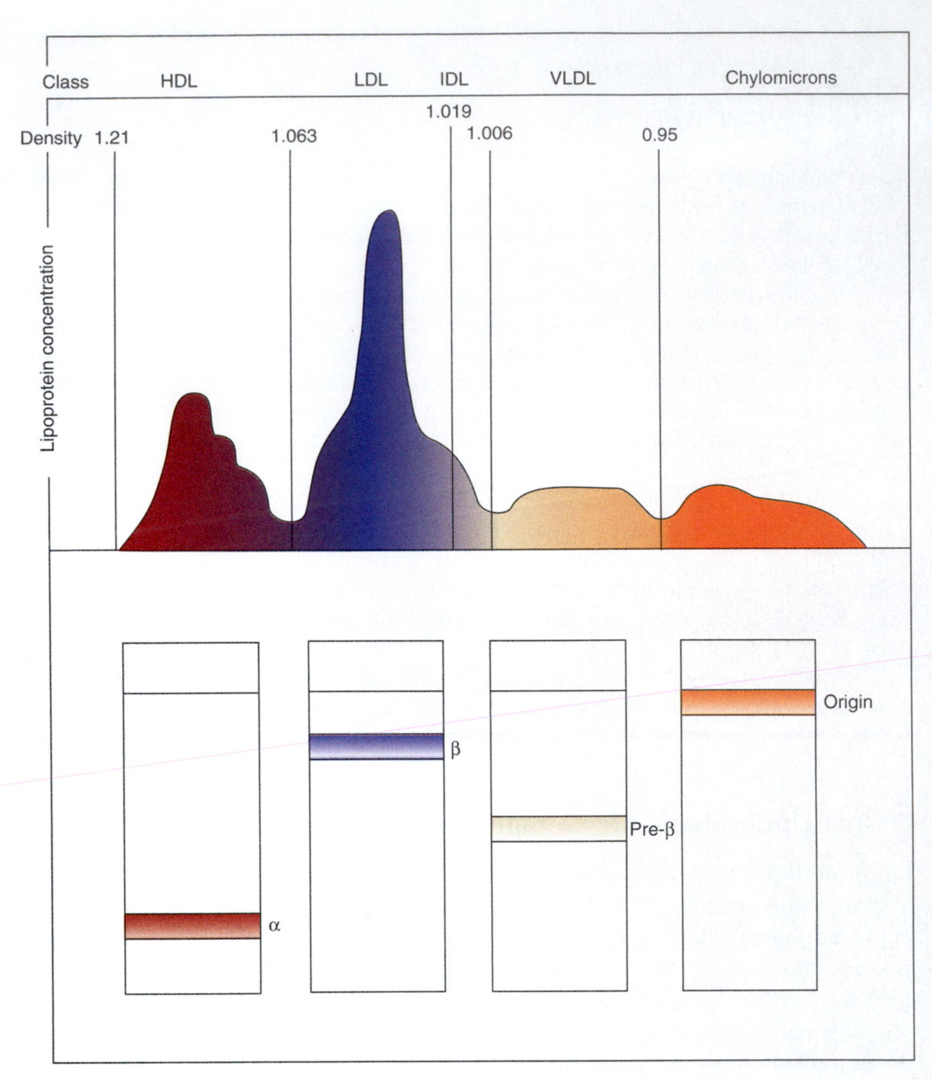

FIGURE 3.42
Correspondence of plasma lipoprotein density classes with electrophoretic mobility in a plasma electrophoresis.
In upper diagram an ultracentrifugation Schlieren pattern is shown. At bottom, electrophoresis on a paper support shows mobilities of major plasma lipoprotein classes with respect to α- and β-globulin bands.
Reprinted with permission from Soutar, A. K. and Myant, N. B. In: R. E. Offord (Ed.), Chemistry of Macromolecules, IIB. *Baltimore, MD: University Park Press, 1979.*

CLINICAL CORRELATION 3.6
Hypolipoproteinemias

Abetalipoproteinemia is a genetic disease that is characterized by absence of chylomicrons, VLDLs, and LDLs due to an inability to synthesize apolipoprotein B-100. Patients show accumulation of lipid droplets in small intestinal cells, malabsorption of fat, acanthocytosis (spiny shaped red cells), and neurological disease (retinitis pigmentosa, ataxia, and retardation).

Tangier disease, an α-lipoprotein deficiency, is a rare autosomal recessive disease in which the HDL level is 1–5% of its normal value. Clinical features are due to the accumulation of cholesterol in the lymphoreticular system, which may lead to hepatomegaly and splenomegaly. In this disease the plasma cholesterol and phospholipids are greatly reduced.

Deficiency of the enzyme lecithin:cholesterol acyltransferase is a rare disease that results in the production of lipoprotein X (see Clin. Corr. 3.5). Also characteristic of this disease is the decrease in the α-lipoprotein and pre-β-lipoprotein bands, with an increase in the β-lipoprotein (lipoprotein X) in electrophoresis.

Kane, J. P. and Havel, R. J. Disorders of the biogenesis and secretion of lipoproteins containing the β apolipoproteins. In: C. R. Scriver, A. L. Beaudet, W. S. Sly, and D. Valle (Eds.), *The Metabolic and Molecular Bases of Inherited Disease,* 7th ed. New York: McGraw-Hill, 1995, Chap.57; and Assmann, G., von Eckardstein, A., and Brewer, H. B. Jr. Familial high density lipoprotein deficiency: Tangier disease. In: C. R. Scriver, A. L. Beaudet, W. S. Sly, and D. Valle (Eds.), *The Metabolic and Molecular Bases of Inherited Disease,* 7th ed. New York: McGraw-Hill, 1995, Chap. 64.

(Table 3.13). **Low density lipoproteins (LDLs)** contain **apoB,** which is also present in **intermediate density lipoproteins (IDLs)** and **very low density lipoproteins (VLDLs).** The **apoC** family is also present in high concentrations in IDLs and VLDLs. Each apolipoprotein class (A, B, etc.) is genetically and structurally distinct (see Clin. Corr. 3.6). The molecular weights of the apolipoproteins vary from 6 kDa (apoC-I) to 550 kDa for apoB-100. This latter is one of the longest single-chain polypeptides known (4536 amino acids).

A model for a VLDL particle is shown in Figure 3.43. On the inside are neutral lipids such as cholesterol esters and triacylglycerols. Surrounding this inner core of neutral lipids, in a shell ~20 Å thick, reside the proteins and the charged amphoteric lipids such as unesterified cholesterol and phosphatidylcholines (see p. 729). Amphoteric lipids and proteins in the outer shell place their hydrophobic apolar regions toward the inside of the particle and their charged groups toward the outside, where they interact with each other and with water. This spherical model with a hydrophobic inner core of neutral lipids and amphoteric lipids and proteins in the

TABLE 3.13 Apolipoproteins of Human Plasma Lipoproteins (Values in Percentage of Total Protein Present)[a]

Apolipoprotein	*HDL_2*	*HDL_3*	*LDL*	*IDL*	*VLDL*	*Chylomicrons*
ApoA-I	85	70–75	Trace	0	0–3	0–3
ApoA-II	5	20	Trace	0	0–0.5	0–1.5
ApoD	0	1–2			0	1
ApoB	0–2	0	95–100	50–60	40–50	20–22
ApoC-I	1–2	1–2	0–5	<1	5	5–10
ApoC-II	1	1	0.5	2.5	10	15
ApoC-III	2–3	2–3	0–5	17	20–25	40
ApoE	Trace	0–5	0	15–20	5–10	5
ApoF	Trace	Trace				
ApoG	Trace	Trace				

Source: Data from Soutar, A. K, and Myant, N. B. In: R. E. Offord (Ed.), *Chemistry of Macromolecules,* IIB. Baltimore, MD: University Park Press, 1979; Kostner, G. M. *Adv. Lipid Res.* 20:1, 1983.

[a] Values show variability from different laboratories.

outer shell applies to all plasma lipoproteins. As the diameter of a particle decreases, the molecules in the outer shell make up a greater percentage of the total molecules in the particle. The smaller HDL particles would therefore be predicted to have a higher percentage of surface proteins and amphoteric lipids than the larger VLDL particles. In fact, HDL particles are 45% protein and 55% lipid, while the larger VLDL particles are only 10% protein with 90% lipid (Table 3.14).

The apolipoproteins, with the exception of apoB, have a high α-helical content when in association with lipid. The helical regions have amphipathic properties. Every third or fourth amino acid in the helix is charged and forms a polar edge that associates with the polar heads of phospholipids and the aqueous environment. The opposite side of the helix has hydrophobic side chains that associate with the nonpolar neutral lipid core of the lipoprotein particle. The α-helical structure of part of apoC-I is shown in Figure 3.44. ApoB appears to have both α-helical and β-structural regions embedded in the lipid outer core. The 4536-residue polypeptide of apoB-100 surrounds the circumference of the LDL particle like a belt weaving in and out of the monolayer phospholipid outer shell (Figure 3.43). One apoB molecule associates per LDL particle.

Glycoproteins Contain Covalently Bound Carbohydrate

Glycoproteins participate in many normal and disease-related functions of clinical relevance. Many plasma membrane proteins are glycoproteins. Some are antigens

TABLE 3.14 Chemical Composition of the Different Plasma Lipoprotein Classes

Lipoprotein Class	*Total Protein (%)*	*Total Lipid (%)*	*Percent Composition of Lipid Fraction*			
			Phospholipids	*Esterified Cholesterol*	*Unesterified Cholesterol*	*Triacylglycerols*
HDL_2[a]	40–45	55	35	12	4	5
HDL_3[a]	50–55	50	20–25	12	3–4	3
LDL	20–25	75–80	15–20	35–40	7–10	7–10
IDL	15–20	80–85	22	22	8	30
VLDL	5–10	90–95	15–20	10–15	5–10	50–65
Chylomicrons	1.5–2.5	97–99	7–9	3–5	1–3	84–89

Source: Data from Soutar, A. K. and Myant, N. B. In: R. E. Offord (Ed.), *Chemistry of Macromolecules,* IIB. Baltimore, MD: University Park Press, 1979.

[a] Subclasses of HDL.

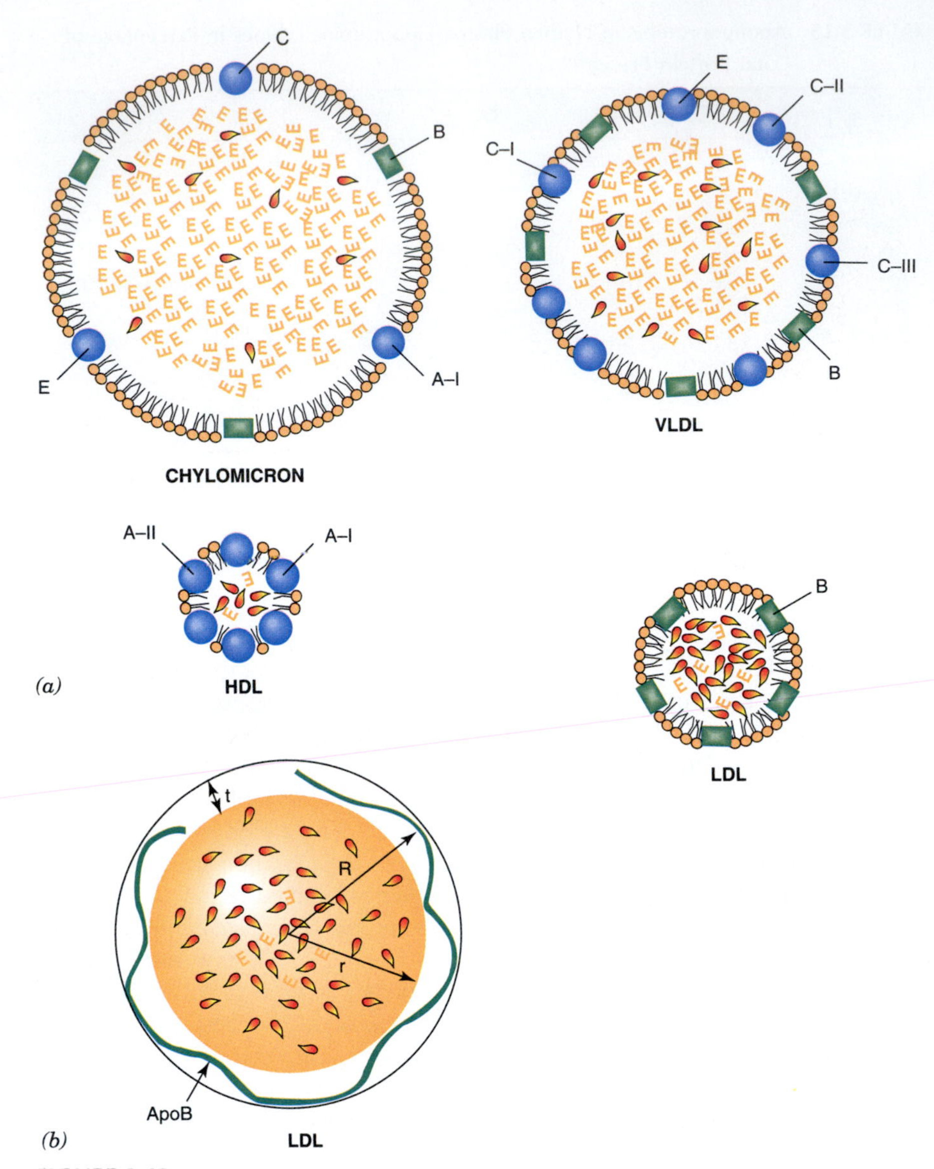

FIGURE 3.43
Generalized structure of plasma lipoproteins.
(*a*) Spherical particle model consisting of a core of triacylglycerol (yellow E's) and cholesterol esters (orange drops) with a shell ~20 Å thick of apolipoproteins (lettered), phospholipids, and unesterified cholesterol. Apolipoproteins are embedded with their hydrophobic edges oriented toward core and their hydrophilic edges toward the outside.
(*b*) LDL particle showing ApoB-100 imbedded in outer shell of particle.
Part (a) from Segrest, J. P., et al. Adv. Protein Chem. *45:303, 1994; part (b) from Schumaker, V. N., et al.* Protein Chem. *45:205, 1994.*

such as those that determine the **blood antigen system** (A, B, O) and the histocompatibility and transplantation determinants of an individual. Immunoglobulin antigenic sites, and viral and hormone receptor binding sites on plasma membranes are often glycoproteins. These carbohydrates in membranes provide a surface code for identification by other cells and for contact inhibition in the regulation of cell growth. Changes in membrane glycoproteins have been correlated with tumorigenesis and malignant transformation in cancer. Most plasma proteins, except albumin, are glycoproteins including blood-clotting proteins, immunoglobulins, and many of the complement proteins. Some protein hormones, such as follicle-stimulating hormone (FSH) and thyroid-stimulating hormone (TSH), are glycoproteins. The structural proteins collagen, laminin, and fibronectin contain carbohy-

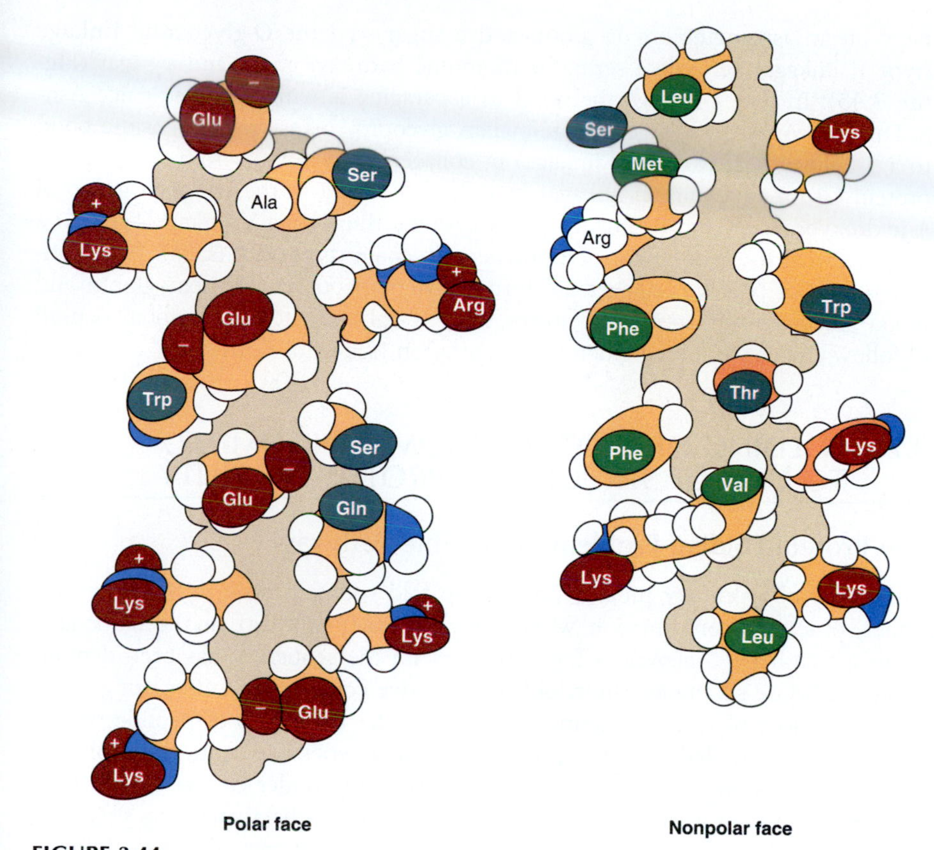

FIGURE 3.44
Illustration showing side chains of a helical segment of apolipoprotein C-1 between residues 32 and 53.
The polar face shows ionizable acid residues in the center and basic residues at the edge. On the other side of the helix, hydrophobic residues form a nonpolar longitudinal face.
Redrawn with permission from Sparrow, J. T. and Gotto, A. M. Jr. CRC Crit. Rev. Biochem. *13:87, 1983. Copyright © 1983 CRC Press, Inc., Boca Raton, FL.*

drate, as do proteins of mucous secretions that perform a role in lubrication and protection of epithelial tissue.

The percentage of carbohydrate in glycoproteins is variable. IgG contains small amounts of carbohydrate (4%); glycophorin of human red blood cell membranes is 60% carbohydrate and human gastric glycoprotein is 82% carbohydrate. The carbohydrate can be distributed evenly along the polypeptide chain or concentrated in defined regions. In proteins of plasma membrane only the extracellular portion has carbohydrate covalently attached. The carbohydrate attached at one point or at multiple points along a polypeptide chain usually contains less than 15 sugar residues and in some cases only one sugar residue. Glycoproteins with the same function from different animal species often have homologous amino acid sequences but variable carbohydrate structures. Heterogeneity in carbohydrate content can occur in the same protein within a single organism. For example, pancreatic ribonuclease A and B forms have an identical amino acid sequence but differ in their carbohydrate composition.

Addition of complex carbohydrate units occurs in a series of enzyme-catalyzed reactions as the polypeptide chain is transported through the endoplasmic reticulum and Golgi network (see Chapter 6). Immature nonfunctional glycoproteins sometimes contain intermediate stages of carbohydrate additions.

Types of Carbohydrate–Protein Covalent Linkages

Different types of covalent linkages join the sugar moieties to protein in glycoproteins. The two most common are the **N-glycosidic linkage** (type I linkage)

CLINICAL CORRELATION 3.7

Glycosylated Hemoglobin, HbA_{1c}

A glycosylated hemoglobin, designated HbA_{1c}, is formed spontaneously in red blood cells by combination of the NH_2-terminal amino groups of the hemoglobin β chain and glucose. The aldehyde group of the glucose first forms a Schiff base with the NH_2-terminal amino group,

$$-N{=}\underset{H}{C}-\overset{OH}{\underset{H}{C}}-$$

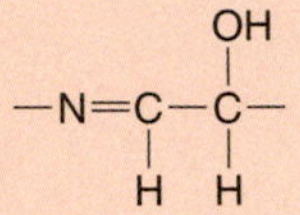

which then rearranges to a more stable amino ketone linkage,

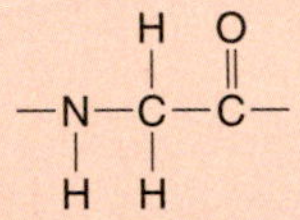

$$-\underset{H}{N}-\overset{H}{\underset{H}{C}}-\overset{O}{\overset{\|}{C}}-$$

by a spontaneous (nonenzymatic) reaction known as the Amadori rearrangement. The concentration of HbA_{1c} is dependent on the concentration of glucose in the blood and the duration of hyperglycemia. In prolonged hyperglycemia the concentration may rise to 12% or more of the total hemoglobin. Patients with diabetes mellitus have high concentrations of blood glucose and therefore high amounts of HbA_{1c}. The changes in the concentration of HbA_{1c} in diabetic patients can be used to follow the effectiveness of treatment for the diabetes.

Bunn, H. F. Evaluation of glycosylated hemoglobin in diabetic patients. *Diabetes* 30:613, 1980; and Brown, S. B. and Bowes, M. A. Glycosylated haemoglobins and their role in management of diabetes mellitus. *Biochem. Educ.* 13:2, 1985.

between an asparagine amide group and a sugar, and the **O-glycosidic linkage** (type II linkage) between a serine or threonine hydroxyl group and a sugar (Figure 3.45). In type I linkage the bond to asparagine is within the sequence Asn-X-Thr(Ser). Another linkage in mammalian glycoproteins is an *O*-glycosidic bond to a 5-hydroxylysine (type III linkage) in collagens and in the serum complement protein C1q. Less common linkages include attachment to the hydroxyl group of 4-hydroxyproline (type IV linkage), to a cysteine thiol (type V linkage), and to a NH_2-terminal α-amino group of a polypeptide chain (type VI linkage). High concentrations of type VI linkages are formed spontaneously with hemoglobin and blood glucose in uncontrolled diabetes. Assay of glycosylated hemoglobin is used to follow changes in blood glucose concentration (see Clin. Corr. 3.7).

3.7 | FOLDING OF PROTEINS FROM RANDOMIZED TO UNIQUE STRUCTURES: PROTEIN STABILITY

The Protein Folding Problem: A Possible Pathway

The ability of a primary protein structure to fold spontaneously to its native secondary or tertiary conformation, without any information other than the amino acid sequence and the noncovalent forces that act on the sequence, has been demonstrated. RNAase spontaneously refolds to its native conformation after being denatured with loss of native structure but without the hydrolysis of peptide bonds. Such observations led to the hypothesis that a polypeptide sequence promotes spontaneous folding to its unique active conformation under correct solvent conditions and in the presence of prosthetic groups that may be a part of its structure. As described below chaperone proteins may facilitate the rate of protein folding. Quaternary structures also assemble spontaneously, after the tertiary structure of the subunits has formed.

It may appear surprising that a protein folds into a single unique conformation given all the possible rotational conformations available around single bonds in its primary structure. For example, the α-chain of hemoglobin contains 141 amino acids and there are at least four single bonds per amino acid residue around which free rotation can occur. If each bond about which free rotation occurs has two or more stable rotamer conformations accessible to it, then there are a minimum of 4^{141} or 5×10^{86} possible conformations for the α-chain amino acid sequence.

The conformation of a protein is the one of lowest Gibbs free energy accessible to the amino acid sequence within a physiological time frame. Thus folding is under thermodynamic and kinetic control. Although detailed knowledge of *de novo*

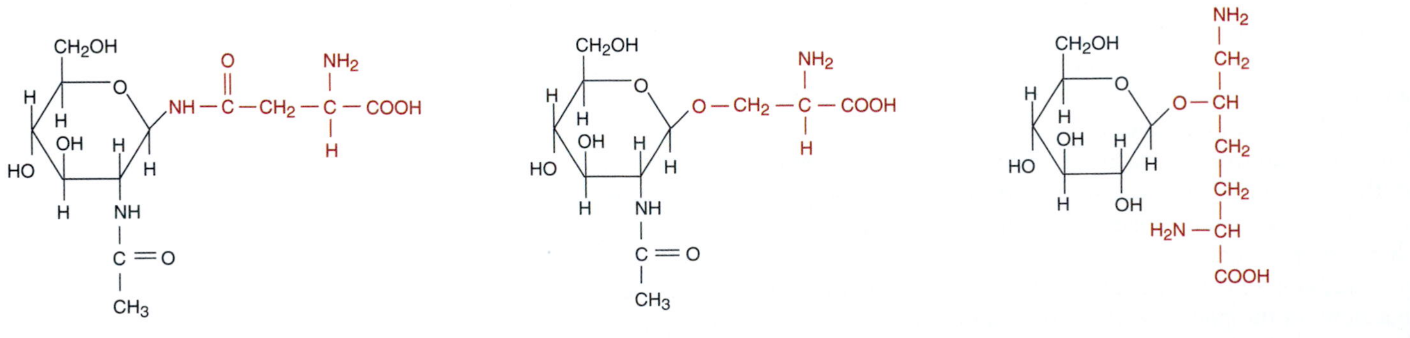

Type I *N*-Glycosyl linkage to asparagine

Type II *O*-Glycosyl linkage to serine

Type III *O*-Glycosyl linkage to 5-hydroxylysine

FIGURE 3.45
Examples of glycosidic linkages to amino acids in proteins.
Type I is an *N*-glycosidic linkage through an amide nitrogen of Asn; type II is an *O*-glycosidic linkage through OH of Ser or Thr; and type III is an *O*-glycosidic linkage to 5-OH of 5-hydroxylysine. Amino acids, red and sugars, black.

folding of a polypeptide is at present an unattainable goal, the involvement of certain processes appears reasonable. There is evidence that folding is initiated by short-range noncovalent interactions between a side chain and its nearest neighbors that form secondary structures in small regions of the polypeptide. Particular side chains have a propensity to promote the formation of α-helices, β-structure, and sharp turns or bends (β-turns) in the polypeptide. The interaction of a side chain with its nearest neighbors in the polypeptide determines the secondary structure into which that section of the polypeptide strand folds. Sections of polypeptide, called **initiation sites,** thus spontaneously assume a secondary structure. The partially folded structures then condense with each other to form a **molten-globule** state. This is a condensed intermediate on the folding pathway that contains much of the secondary structure elements of the native structure, but a large number of incorrect tertiary structure interactions. Segments of secondary structure in the molten-globule state are highly mobile relative to one another, and the molten-globule structure is in rapid equilibrium with the fully unfolded denatured state. The correct medium- and long-range interactions between different initiation sites are found by rearrangements within the molten-globule and the low free energy, native tertiary structure for the polypeptide chain is formed. With formation of the native tertiary structure, the correct disulfide bonds (cystine) are formed. The rate-determining step for folding and unfolding of the native conformation lies between the molten-globule and the native structure. For some proteins there may exist two or more thermodynamically stable folded conformations of low Gibbs free energy.

Clinical Correlation 3.8 contains a discussion of protein folding and prions, infectious agents.

Chaperone Proteins May Assist the Protein Folding Process

Cells contain proteins that facilitate the folding process. These include *cis-trans*-prolyl isomerases, protein disulfide isomerases, and chaperone proteins. ***cis-trans*-Prolyl isomerases** interconvert *cis*- and *trans*-peptide bonds of proline residues and increase the rate of folding. This allows the correct prolyl peptide bond conformation to form for each proline as required by the folded native structure. **Protein disulfide isomerases** catalyze the breakage and formation of disulfide cystine linkages so incorrect linkages are not stabilized and the correct arrangement of cystine linkages for the folded conformation is rapidly achieved.

Chaperone proteins were discovered as **heat shock proteins (hsps)**, a family of proteins whose synthesis is increased at elevated temperatures. The chaperones do not change the final outcome of the folding process but prevent protein aggregation prior to completion of folding and prevent formation of metastable dead-end or nonproductive intermediates. They increase the rate of the folding process by limiting the number of unproductive folding pathways available to a polypeptide. Chaperones of the hsp 70-kDa family bind to polypeptides as they are synthesized on the ribosomes, shielding the hydrophobic surfaces that would normally be exposed to solvent. This protects the protein from aggregation until the full chain is synthesized and folding can occur. Some proteins, however, cannot complete their folding process while in the presence of hsp70 chaperones and are delivered to the hsp60 family (GroEL in *Escherichia coli*) of chaperone proteins, also called **chaperonins.** The chaperonins form long cylindrical multisubunit quaternary structures that bind unfolded polypeptides in their molten-globule state within their central hydrophobic cavity. Chaperonins have an ATPase activity and hydrolyze ATP as they facilitate folding. The folding process in *E. coli* is presented in Figure 3.46. Chaperone proteins are also required for refolding of proteins after they cross cellular membranes. A system of chaperones facilitates protein transport into mitochondria and into and through the endoplasmic reticulum. Proteins cross the lipid bilayer of the mitochondrial and endoplasmic reticulum membranes in an unfolded conformation, and local chaperones are required to facilitate their folding.

CLINICAL CORRELATION 3.8

Proteins as Infectious Agents: Human Transmissible Spongiform Encephalopathies (TSEs)

Proteins that act as infectious agents in the absence of DNA or RNA are known as prion proteins. Creutzfeldt–Jakob disease (CJD) is the most common of the prion diseases. The disease is characterized by ataxia, dementia, and paralysis and is almost always fatal. Pathological examination of the brain shows amyloid plaques and spongiform (vacuolous) degeneration of the brain. The disease can be generated by an inherited mutation (familial disease), by sporadic mutation, or by an infectious process. The infectious form was first observed among the Fore people of New Guinea who exhibited a transmissible spongiform encephalopathy called Kuru, in which the prion proteins were transmitted through a ritual cannibalism. More recently, bovine spongiform encephalopathy or mad cow disease has been transmitted to humans through the ingestion of meat from cattle infected by bovine spongiform encephalopathy. While there is strong evidence that the infective agent is a protein, the direct infection of an animal with pure recombinant prion protein has not been demonstrated.

The disease is believed to be caused by a conformational change in the prion protein, PrP, from its normal soluble cellular conformation, PrP^c, to a toxic conformation, PrP^{Sc}. The PrP^{Sc} conformation polymerizes into insoluble amyloid fibers, which cause the neurotoxic pathologies. The soluble PrP^c domain fold is composed of three α-helical and two small β-strand segments. The conversion to the PrP^{Sc} form is characterized by the conversion of two of the α-helical segments into β-strand conformations. The PrP^{Sc} conformation polymerizes into amyloid fibers through β-strand interactions between monomers. Formation of the amyloid polymer is irreversible. An unknown factor, designated protein X (perhaps a chaperone protein), facilitates the conversion of the α-helical to β-structure conformation and/or promotes the polymerization of the β-structure conformation into the amyloid fiber. Two mechanisms are suggested for amyloid plaque formation initiated by an infective prion protein. In the Nucleation–Polymerization Mechanism (a), the PrP^c conformation (green) is in rapid equilibrium with the PrP^{Sc} conformation (brown) (step 1), but the polymerization into an amyloid fiber is slow (step 2) in the absence of an initiator molecule. The introduction of a PrP^{Sc} conformation protein or fragment (infective prion, red) nucleates the polymerization process (step 3). The process is propagated by fragments from the newly generated polymer (step 5).

In the Template-Directed Mechanism (b), the infective prion, PrP^{Sc}, serves as a template to promote the change in conformation of endogenous PrP^c to PrP^{Sc} (steps 1 and 2). Once converted to the PrP^{Sc}, the PrP^{Sc} conformations rapidly polymerize into insoluble fibrils. Familial forms of Creutzfeldt–Jakob disease are induced by mutant PrP proteins that have a higher tendency to spontaneously form the PrP^{Sc} conformation.

Horwich, A. L. and Weissman, J. S. Deadly conformations—protein misfolding in prion disease. *Cell* 89:499, 1997; and Prusiner, S. B. Prion diseases and the BSE crises. *Science* 278:245, 1997.

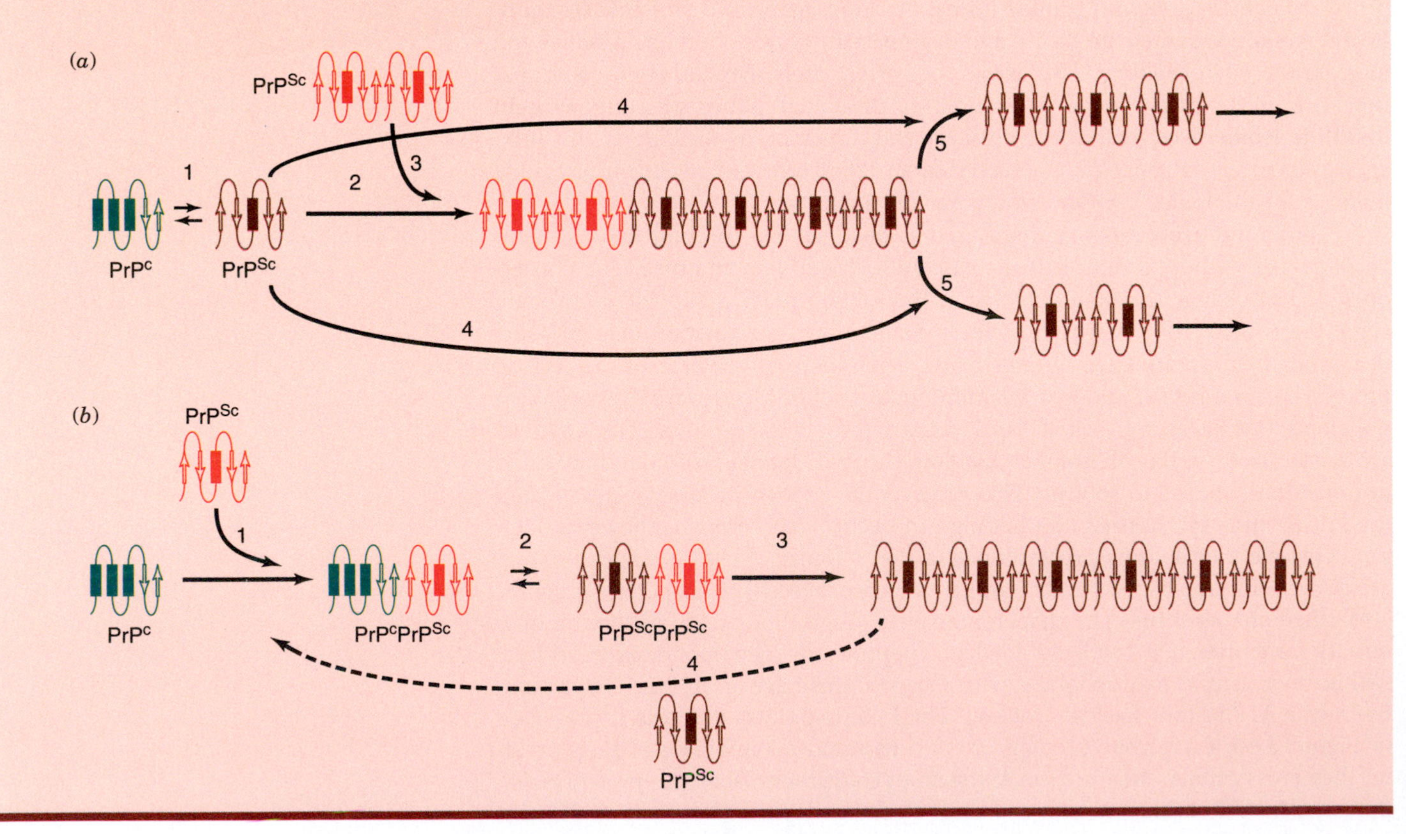

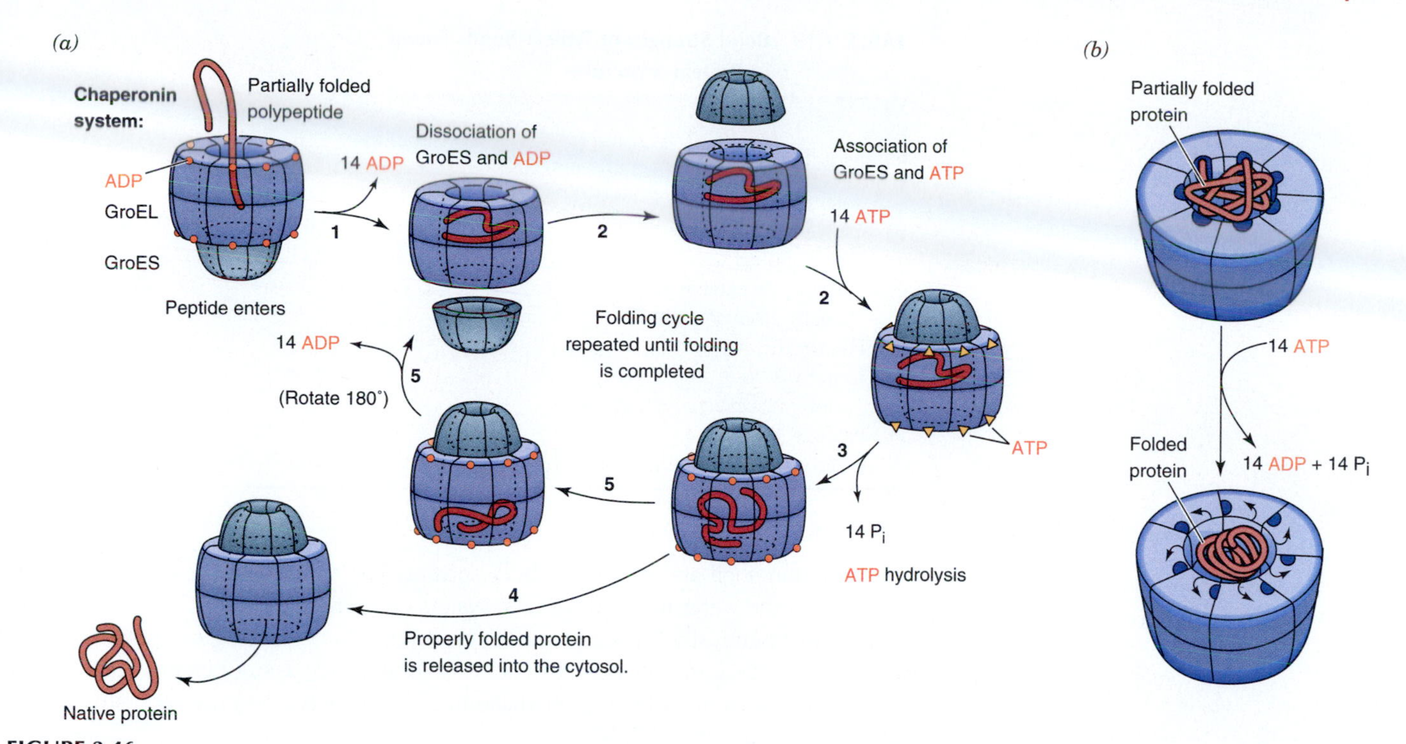

FIGURE 3.46
Chaperonin-directed protein folding in *E. coli*.
(*a*) A proposed reaction cycle of *E. coli* chaperonins GroEL and GroES in protein folding. (1) GroEL is a complex of 14 subunits, each with ADP attached. An associated ring of 7 GroES subunits binds an unfolded polypeptide in its central cavity and ADP and GroES subunits are released. (2) Each GroEL subunit binds an ATP, weakening interaction between unfolded polypeptide and GroEL. GroES is rebound on the opposite face of GroEL. (3) The 14 ATPs are simultaneously hydrolyzed, releasing the bound polypeptide inside GroEL. The polypeptide, which enters in its molten-globule state, folds in a protected microenvironment, preventing aggregation with other partially folded polypeptides. (4) The polypeptide is released from GroEL after folding into its native conformation. (5) If the polypeptide fails to attain its native fold, it remains bound to GroEL and reenters the reaction cycle at step 2. In the diagram GroEL turns over by 180°. GroES binds but does not hydrolyze ATP and facilitates binding of ATP to GroEL. It coordinates simultaneous hydrolysis of ATP and prevents escape of a partially folded polypeptide from the GroEL cavity. (*b*) A model for ATP-dependent release of an unfolded polypeptide from its multiple attachment sites in GroEL. ATP binding and hydrolysis mask the hydrophobic sites of GroEL subunits (darker areas) that bind to unfolded polypeptide, thus permitting it to fold in an isolated environment.
Adapted from Hartl, R.-U., Hlodan, R., and Langer, T. Trends Biochem. Sci. 19: 23, 1994. Figure reproduced from Voet, D. and Voet, J. Biochemistry, 2nd ed. New York: Wiley, 1995. Reprinted by permission of John Wiley & Sons, Inc.

Noncovalent Forces Lead to Protein Folding and Contribute to a Protein's Stability

Noncovalent forces cause a polypeptide to fold into a unique conformation and then stabilize the native structure against denaturation. Noncovalent forces are weak bonding forces with strengths of 1–7 kcal mol^{-1} (4–29 kJ mol^{-1}). This may be compared to the strength of covalent bonds that have a bonding strength of at least 50 kcal mol^{-1} (Table 3.15). Even though individually weak, the large number of individually weak noncovalent contacts within a protein provide a large energy factor that promotes protein folding.

Hydrophobic Interaction Forces

The most important noncovalent forces that cause a randomized polypeptide conformation to fold into its native structure are **hydrophobic interaction forces.** Their strength is not due to any intrinsic attraction between nonpolar groups, but rather to the properties of the water that surrounds the nonpolar groups. A nonpolar molecule or a region of a protein molecule dissolved in water induces a solvation shell of water in which water molecules are highly ordered. When two nonpolar side chains come together on folding of a polypeptide, the surface area exposed to solvent is reduced and some of the highly ordered water molecules in

TABLE 3.15 Bond Strength of Typical Bonds Found in Protein Structures

Bond Type	*Bond Strength (kcal mol^{-1})*
Covalent	>50
Noncovalent	0.6–7
Hydrophobic (i.e., two benzyl side chain groups of Phe)	2–3
Hydrogen	1–7
Ionic (low dielectric environment)	1–6
van der Waals	<1
Average energy of kinetic motion (37°C)	0.6

the solvation shell are released to bulk solvent. The entropy of the system (i.e., net disorder of the water molecules in the system) is increased. The increase in entropy is thermodynamically favorable and is the driving force causing nonpolar moieties to come together in aqueous solvent. A favorable free energy change of −2 kcal mol^{-1} for association of two phenylalanine side chain groups in water is due to this gain in entropy (Figure 3.47).

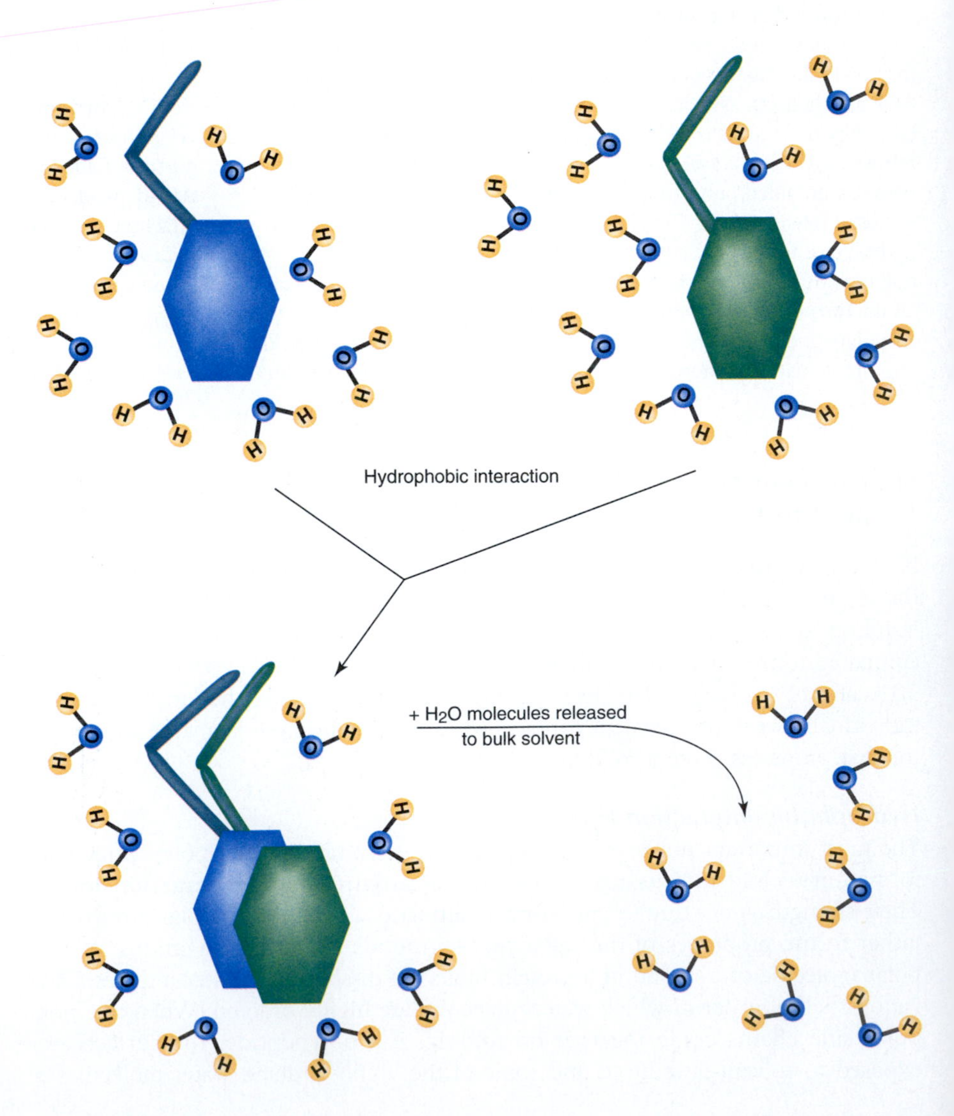

FIGURE 3.47
Formation of hydrophobic interaction between two phenylalanine side chain groups.

In transition from random to regular secondary conformation such as an α-helix or β-structure, approximately one-third of the water of solvation about the unfolded polypeptide is lost to bulk solvent. This approximates 0.5–0.9 kcal mol^{-1} for each peptide residue. Another one-third of the original solvation shell is lost when a polypeptide that has folded into a secondary structure folds into a tertiary structure. The tertiary folding brings different segments of folded polypeptide chains into close proximity with the release of water of solvation between the polypeptide chains.

Hydrogen Bonds

Hydrogen bonds are formed when a hydrogen atom covalently bonded to an electronegative atom is shared with a second electronegative atom. The atom to which the hydrogen atom is bonded is designated the **hydrogen-donor atom.** The atom with which the hydrogen atom is shared is the **hydrogen-acceptor atom.** Typical hydrogen bonds found in proteins are shown in Figure 3.48. α-Helical and β-structure conformations are extensively hydrogen bonded.

The strength of a hydrogen bond depends on the distance between donor and acceptor atoms. High bonding energies occur when this distance is between 2.7 and 3.1 Å. Of lesser importance, but not negligible, to bonding strength is the dependence of hydrogen-bond strength on geometry. Bonds of higher energy are geometrically collinear, with donor, hydrogen, and acceptor atoms lying in a straight line. The dielectric constant of the medium around the hydrogen bond may also be reflected in the bonding strength. Typical hydrogen bond strengths in proteins are 1–7 kcal mol^{-1}. Although they contribute to thermodynamic stability of a protein's conformation, their formation may not be a major driving force for folding. This is because peptide bonds and other hydrogen bonding groups in proteins form hydrogen bonds to the water solvent in the denatured state, and these bonds must be broken before the polypeptide folds. The energy required to break the hydrogen bonds to water must be subtracted from the energy gained from formation of new hydrogen bonds between atoms in the folded protein in calculating the net contribution of hydrogen bonding forces to the folding.

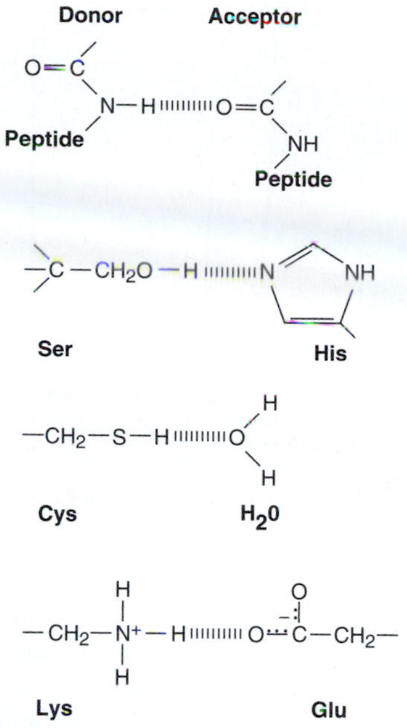

FIGURE 3.48
Some common hydrogen bonds found in proteins.

Electrostatic Interactions

Electrostatic interactions (also referred to as **ionic** or **salt linkages**) between charged groups are important in the stabilization of protein structure and in binding of charged ligands and substrates to proteins. Electrostatic forces are repulsive or attractive depending on whether the interacting charges are of the same or opposite sign. The strength of an electrostatic force (E_{el}) is directly dependent on the charge (Z) of each ion and inversely dependent on the dielectric constant (D) of the solvent and the distance between the charges (r_{ab}) (Figure 3.49).

Water has a high dielectric constant ($D = 80$), and interactions in water are relatively weak compared to charge interactions in the interior of a protein where the dielectric constant is low. However, most charged groups of proteins remain on the surface where they do not interact with other charged groups from the protein because of the high dielectric constant of water, but are stabilized by hydrogen bonding and polar interactions to the water. These water interactions generate the dominant forces that place most charged groups of a protein on the outside of the folded structures.

$$E_{el} \approx \frac{Z_A \cdot Z_B \cdot \varepsilon^2}{D \cdot r_{ab}}$$

FIGURE 3.49
Strength of electrostatic interactions.

Van der Waals–London Dispersion Forces

Van der Waals and London dispersion forces are the weakest of the noncovalent forces. They have an attractive term (A) inversely dependent on the 6th power of the distance between two interacting atoms (r_{ab}), and a repulsive term (B) inversely dependent on the 12th power of r_{ab} (Figure 3.50). The A term contributes at its optimum distance an attractive force of less than 1 kcal mol^{-1} per atomic interaction due to the induction of complementary partial charges or dipoles in the electron density of adjacent atoms when the electron orbitals of the two atoms ap-

$$E_{VDW} = -\frac{A}{r_{ab}^{6}} + \frac{B}{r_{ab}^{12}}$$

FIGURE 3.50
Strength of van der Waals interactions.

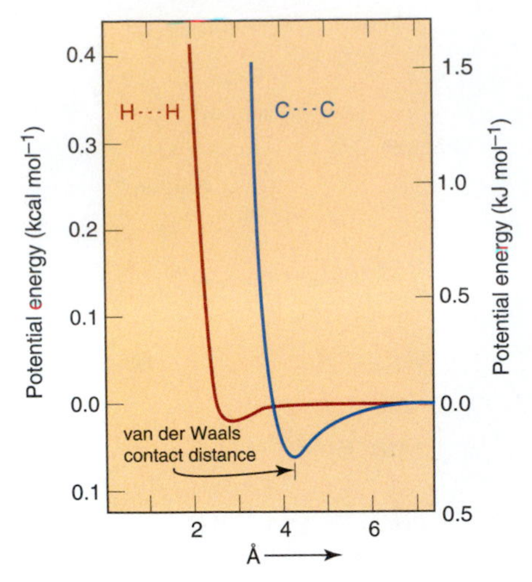

FIGURE 3.51
Van der Waals–London dispersion interaction energies between two hydrogen atoms and two (tetrahedral) carbon atoms.
Negative energies are favorable and positive energies unfavorable.
Redrawn from Fersht, A. Enzyme Structure and Mechanism. *San Francisco: Freeman, 1977, p. 228.*

proach to a close distance. As the atoms come even closer, however, the repulsive component (term *B*) of the van der Waals force predominates as the electron orbitals of the adjacent atoms begin to overlap. The repulsive force is commonly called **steric hindrance.**

The distance of maximum favorable interaction between two atoms is the **van der Waals contact distance,** which is the sum of van der Waals radii for the two atoms (Figure 3.51). The van der Waals radii for atoms found in proteins are given in Table 3.16.

The van der Waals repulsive forces between atoms of a peptide bond are weakest at the specific ϕ and ψ angles compatible with the α-helix and β-strand structures. Thus van der Waals forces are critical for secondary structure formation in proteins. In folding into a tertiary structure, the number of weak van der Waals interactions that occur are in the thousands. Thus the total contribution of van der Waals–London dispersion forces to the stability of a folded structure is substantial, even though a single interaction between any two atoms is less than 1 kcal mol^{-1}.

A special type of interaction (π-electron–π-electron) occurs when two aromatic rings approach each other with electrons of their aromatic rings favorably interacting (Figure 3.52). This interaction can result in attractive forces of up to 6 kcal mol^{-1}. A number of π–π aromatic interactions occur in a typical folded protein, contributing to the stability of the folded structure.

Denaturation of Proteins Leads to Loss of Native Structure

Denaturation occurs when a protein loses its native secondary, tertiary, and/or quaternary structure. The primary structure is not necessarily broken by denaturation. The **denatured state** is always correlated with loss of a protein's function. This, however, is not necessarily synonymous with denaturation since small conformational changes can lead to loss of function. A change in conformation of a single side chain in the active site of an enzyme or a change in protonation of a side chain can result in loss of activity but does not lead to a complete loss of the native protein structure.

Although conformational differences between denatured and native structures may be substantial, the free energy difference between such structures can be as low as the free energy of three or four noncovalent bonds. Thus loss of a single hydrogen bond or electrostatic or hydrophobic interaction can lead to destabilization of a folded structure. A change in stability of a noncovalent bond can, in turn, be

TABLE 3.16 Covalent Bond Radii and van der Waals Radii for Selected Atoms

Atom	*Covalent Radius (Å)*	*van der Waals Radius (Å)*[a]
Carbon (tetrahedral)	0.77	2.0
Carbon (aromatic)	0.69 along═bond	1.70
	0.73 along—bond	
Carbon (amide)	0.72 to amide N	1.50
	0.67 to oxygen	
	0.75 to chain C	
Hydrogen	0.33	1.0
Oxygen (—O—)	0.66	1.35
Oxygen (═O)	0.57	1.35
Nitrogen (amide)	0.60 to amide C	1.45
	0.70 to hydrogen bond H	
	0.70 to chain C	
Sulfur, diagonal	1.04	1.70

Source: Fasman, G. D. (Ed.), *CRC Handbook of Biochemistry and Molecular Biology*, 3rd ed., Sect. D, Vol. II. Boca Raton, FL: CRC Press, 1976, p. 221.

[a] The van der Waals contact distance is the sum of the two van der Waals radii for the two atoms in proximity.

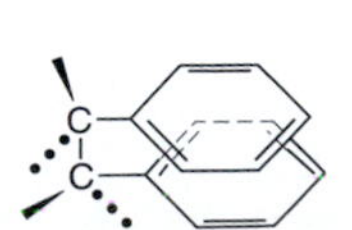

FIGURE 3.52
π-Electron–π-electron interactions between two aromatic rings.

caused by a change in pH, ionic strength, or temperature. Binding of prosthetic groups, cofactors, and substrates also affects stability of the native conformation.

The statement that breaking of a single noncovalent bond in a protein can cause denaturation apparently conflicts with the observation that an amino acid sequence can often be extensively varied without loss of a protein's structure. The key to the resolution of this apparent conflict is the word "essential." Many noncovalent interactions are not essential to the structural stability of the native conformation of a protein. However, substitution or modification of an essential amino acid that provides a critical noncovalent interaction without a compensating stabilizing interaction dramatically affects the stability of a native protein structure.

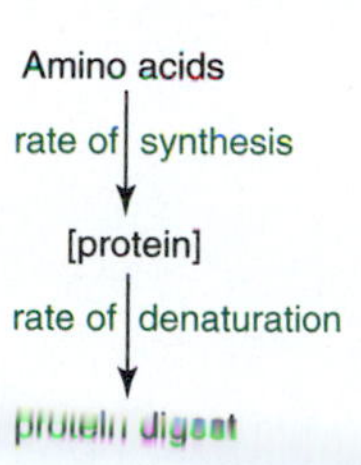

FIGURE 3.53
Steady-state concentration of a protein is due to its rates of synthesis and denaturation.

The cellular concentration of a protein is controlled by its rate of synthesis and degradation (Figure 3.53). Under many circumstances the denaturation of a protein is the rate-controlling step in its degradation. Cellular enzymes and organelles that digest proteins "recognize" denatured protein conformations and eliminate them rapidly. In experimental situations, protein denaturation occurs on addition of urea or detergents (sodium dodecyl sulfate or guanidine hydrochloride) that weaken hydrophobic bonding in proteins. These reagents stabilize the denatured state and shift the equilibrium toward the denatured form of the protein. Addition of strong base, acid, or organic solvent, or heating to temperatures above 60°C are also common ways to denature a protein.

3.8 | DYNAMIC ASPECTS OF PROTEIN STRUCTURE

While high-resolution X-ray diffraction yields atomic coordinates for each atom in a protein structure, experimental evidence from NMR, fluorescence spectroscopy, and the temperature dependence of the X-ray diffraction reveals that the atoms in a folded protein molecule have a fluid-like dynamic motion and do not exist in a single static position. Rather than an exact location, the atomic coordinates obtained by X-ray diffraction represent the time-averaged position for each atom. The time for position averaging is the duration of data collection, which may be several days. Thus the active conformation may differ from the average conformation. An X-ray structure also shows small "defects" in packing of the folded structure, indicating the existence of "holes" in the structure that will allow the protein space for flexibility. The concept that each atom in a protein is in constant motion such as molecules within a fluid, although constrained by its covalent bonds and the secondary and tertiary structure, is an important aspect of protein structure.

Calculations of theoretical **molecular dynamics** describe the changes in coordinates of atoms in a folded protein structure and in position of regions of the structure due to summation of the movements of atoms in that region. The dynamic motion computation is based on the solving of Newton's equations of motion simultaneously for all the atoms of the protein and the solvent that interacts with the protein. The energy functions used in the equation include representations of covalent and noncovalent bonding energies due to electrostatic forces, hydrogen bonding, and van der Waals forces. Individual atoms are randomly assigned a velocity from a theoretical distribution and Newton's equations are used to "relax" the system at a given "temperature." The calculation is a computationally intensive activity, even when limited to less than several hundred picoseconds (1 ps = 10^{-12} s) of protein dynamic time, and frequently requires supercomputers. These calculations indicate that the average atom in a protein oscillates over a distance of 0.7 Å on the picosecond scale. Some atoms or groups of atoms move smaller or larger distances than this calculated average (Figure 3.54).

Net movement of any segment of a polypeptide over time represents the sum of forces due to rapid atomic oscillations, and the local jiggling and elastic movements of covalently attached groups of atoms. These movements within the closely packed interior of a protein frequently allow the planar aromatic rings of buried tyrosines to flip. The small-amplitude fluctuations provide the "lubricant" for large motions in proteins such as domain motions and quaternary structure changes, like

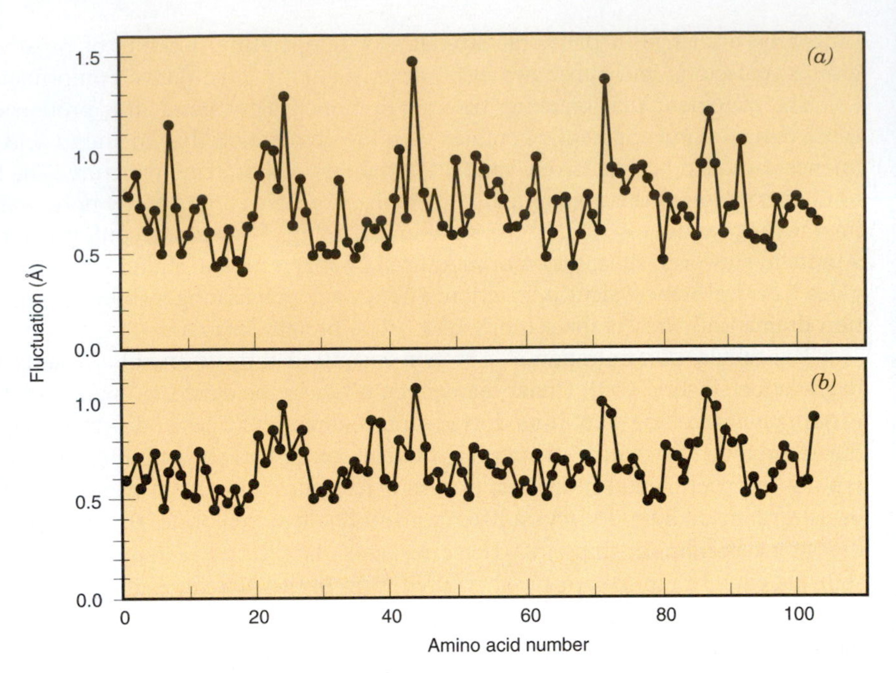

FIGURE 3.54
Fluctuation of structure of cytochrome *c*.
(*a*) Calculated fluctuation on a picosecond time scale of α-carbons within each amino acid residue in folded structure of cytochrome *c* and (*b*) experimentally observed fluctuation of each α-carbon of the amino acid residues determined from the temperature dependency of the X-ray diffraction pattern for the protein. Cytochrome *c* has 103 amino acid residues. The *x*-axis plots amino acid residues in cytochrome *c* from 1 to 103, and the *y*-axis the fluctuation distances in angstroms.
Redrawn from Karplus, M. and McCammon, J. A. Annu. Rev. Biochem. *53:263, 1983.*

those observed in hemoglobin on O_2 binding (see Chapter 9). The dynamic behavior of proteins is implicated in conformational changes induced by a substrate, an inhibitor, or a drug on binding to an enzyme and receptor, generation of allosteric effects in hemoglobins, electron transfer in cytochromes, and in the formation of supramolecular assemblies such as viruses. The movements may also have a functional role in the protein's action.

3.9 | METHODS FOR CHARACTERIZATION, PURIFICATION, AND STUDY OF PROTEIN STRUCTURE AND ORGANIZATION

Separation of Proteins on Basis of Charge

In **electrophoresis**, protein dissolved in a buffer solution at a particular pH is placed in an electric field. Depending on the relationship of the buffer pH to the p*I* of the protein, the protein moves toward the cathode (−) or the anode (+) or remains stationary (pH = p*I*). Supports such as polymer gels (e.g., polyacrylamide), starch, or paper are used. The inert supports are saturated with buffer solution, a protein sample is placed on the support, an electric field is applied across the support, and the charged proteins migrate in the support toward the oppositely charged pole.

An electrophoresis technique of extremely high resolution is **isoelectric focusing**, in which mixtures of polyamino–polycarboxylic acid ampholytes with a defined range of p*I* values are used to establish a pH gradient across the applied electric field. A charged protein migrates through the pH gradient in the electric field until it reaches a pH region in the gradient equal to its p*I* value. At this point the protein becomes stationary and may be visualized (Figure 3.55). Proteins that differ by as little as 0.0025 in p*I* value are separated on the appropriate pH gradient.

Ion-exchange column chromatography is used for preparative separation of proteins by charge.

Ion-exchange resins consist of insoluble materials (agarose, polyacrylamide, cellulose, and glass) that contain charged groups (Figure 3.56). Negatively charged resins bind cations strongly and are **cation-exchange resins.** Positively charged

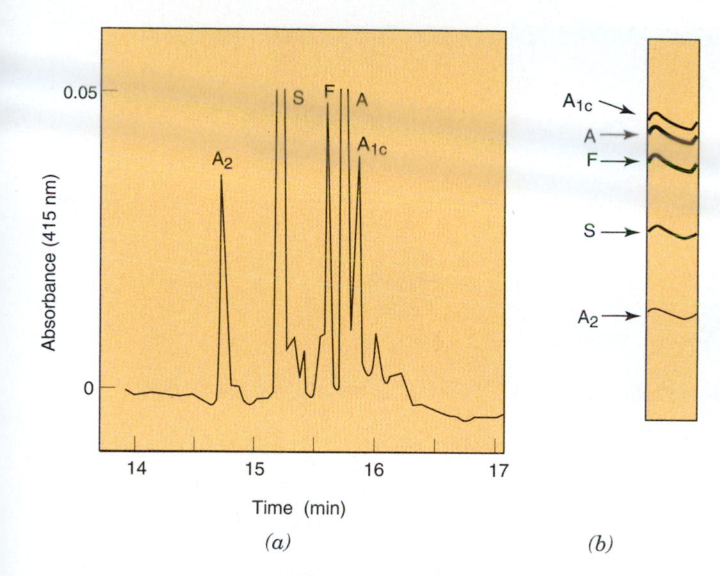

FIGURE 3.55
Isoelectric focusing of hemoglobins from patient heterozygous for HbS and β-thalassemia.
Figure shows separation by isoelectric focusing of HbA_{1c} (HbA glycosylated on NH_2-end; see Clin. Corr. 3.7), normal adult HbA, fetal HbF, sickle cell HbS (see Clin. Corr. 3.3), and minor adult HbA_2. *(a)* Isoelectric focusing carried out by capillary electrophoresis with ampholyte pH range between 6.7 and 7.7 and detection of bands at 415 nm. *(b)* Isoelectric focusing carried out on gel with Pharmacia Phast System; ampholyte pH range is between 6.7 and 7.7.
From Molteni, S., Frischknecht, H., and Thormann, W. Electrophoresis *15:22, 1994 (Figure 4, parts A and B).*

resins bind anions strongly and are **anion-exchange resins.** The degree of retardation of a protein (or an amino acid) by a resin depends on the magnitude of the charge on the protein at the particular pH of the experiment. Molecules with the same charge as the resin are eluted first in a single band, followed by those with an opposite charge to that of the resin, in an order based on the protein's charge density (Figure 3.57). When it is difficult to remove a molecule from the resin because of the strength of the attractive interaction between the bound molecule and resin, systematic changes in pH or in ionic strength are used to weaken the interaction. For example, an increasing pH gradient through a cation-exchange resin reduces the difference between the solution pH and the p*I* of the bound protein. This decrease between pH and p*I* reduces the magnitude of the net charge on the protein and decreases the strength of the charge interaction between the protein and the resin. An increasing gradient of ionic strength also decreases the charge interactions and elutes tightly bound electrolytes from the resin.

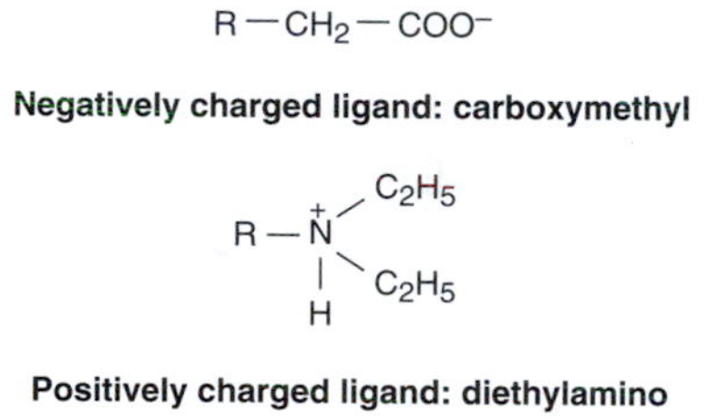

$R-CH_2-COO^-$

Negatively charged ligand: carboxymethyl

$R-\overset{+}{N}H(C_2H_5)_2$

Positively charged ligand: diethylamino

FIGURE 3.56
Two examples of charged ligands used in ion-exchange chromatography.

Capillary Electrophoresis

Electrophoresis within a fused silica capillary tube has a high separation efficiency, utilizes very small samples, and requires only several minutes for an assay. A long capillary tube is filled with the electrophoretic medium, a sample is injected in a narrow band near the anode end of the tube, and the molecules of the sample are separated by their mobility toward the negatively charged pole. The fused silica wall of the capillary has a negative charge and an immobile cationic layer is fixed to it. An adjacent diffuse layer of cations moves toward the cathode in the applied electric field and causes a flow of solvent toward the cathode. This electro-osmotic flow creates a "current" that carries analyte molecules toward the cathode, irrespective of the analyte's charge (Figure 3.58). Molecules with a high positive charge-to-mass ratio move with the current and have the highest mobility, followed by neutral molecules. Anionic molecules are repelled by the cathode and will move against the electro-osmotic flow. However, the electro-osmotic current toward the cathode overcomes any negative migration, and anions also migrate toward the cathode but at a slower rate than the cationic or neutral molecules.

In zone electrophoresis, separations are made in the presence of a single buffer. Capillary electrophoresis may be performed in the presence of ampholytes to separate proteins by isoelectric focusing, in the presence of a porous gel to separate proteins by molecular weight, or in the presence of a micellar component to separate by hydrophobicity. Detectors that utilize UV light, fluorescence, Raman spectroscopy, electrochemical detection, or mass spectroscopy make the capillary method sensitive and versatile.

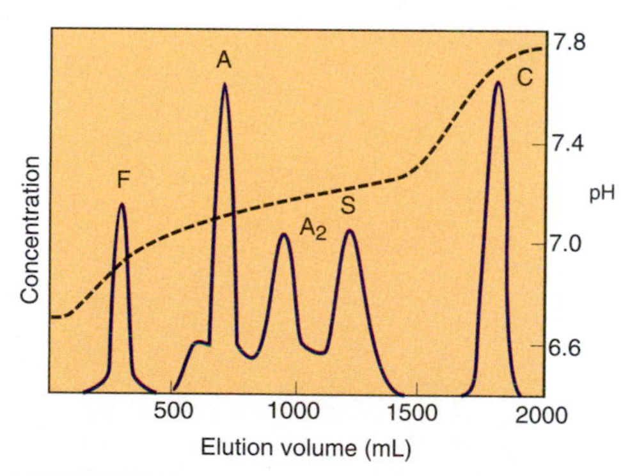

FIGURE 3.57
Example of ion-exchange chromatography.
Elution diagram of an artificial mixture of hemoglobins F, A_1, A_2, S, and C on carboxymethyl–Sephadex C-50.
From Dozy, A. M. and Juisman, T. H. J. J. Chromatogr. *40:62, 1969.*

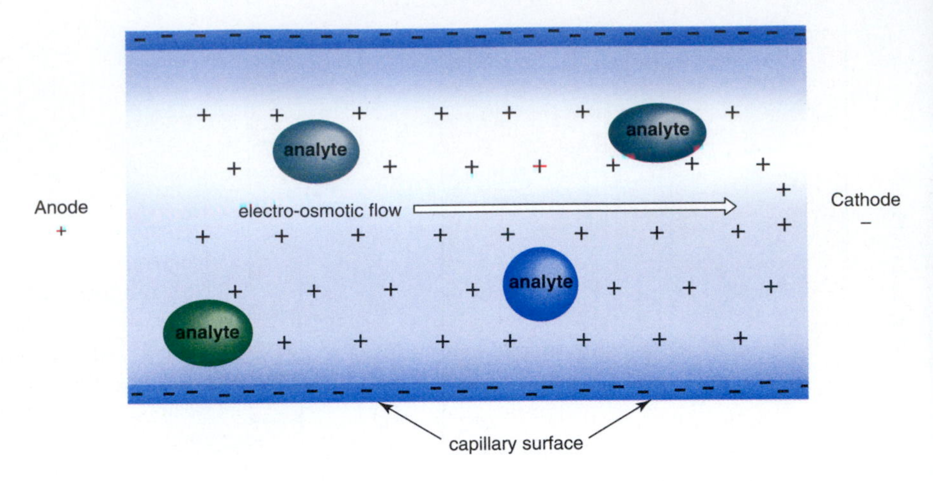

FIGURE 3.58
Generation of electro-osmotic flow toward cathode in capillary electrophoresis.

Separation of Proteins Based on Molecular Mass or Size

Ultracentrifugation: Definition of Svedberg Coefficient

A protein subjected to centrifugal force moves in the direction of the force at a velocity dependent on its mass. The rate of movement is measured with an appropriate optical detection system, and from the rate the sedimentation coefficient is calculated in **Svedberg units** (units of 10^{-13} s). In the equation (Figure 3.59), v is the measured velocity of protein movement, ω the angular velocity of the centrifuge rotor, and r the distance from the center of the tube in which the protein is placed to the center of rotation. Sedimentation coefficients between 1 and 200 Svedberg units (S) have been found for proteins (Table 3.17). Equations have been derived to relate the sedimentation coefficient to the molecular mass of a protein. One of the more simple equations is shown in Figure 3.60, in which R is the gas constant, T the temperature, s the sedimentation coefficient, D the diffusion coefficient of the protein, $\bar{v}$ the partial specific volume of the protein, and p the density of solvent. The quantities D and $\bar{v}$ must be measured in independent experiments. The equation assumes a spheroidal geometry for the protein. Because this assumption may not be true and independent measurements of D and $\bar{v}$ are difficult, usually only the sedimentation coefficient for a molecule is reported. A protein's sedimentation coefficient is a qualitative measurement of molecular mass.

$$s = \frac{v}{\omega^2 r}$$

FIGURE 3.59
Equation for calculation of the Svedberg coefficient.

$$\text{Molecular weight} = \frac{RTs}{D(1 - \bar{v}\rho)}$$

FIGURE 3.60
An equation relating the Svedberg coefficient to molecular weight.

Molecular Exclusion Chromatography

A porous gel in the form of small insoluble beads is commonly used to separate proteins by size in column chromatography. Small proteins penetrate the pores of

TABLE 3.17 Svedberg Coefficients for Some Plasma Proteins of Different Molecular Weights

Protein	s_{20}, $\times 10^{-13}$ ($cm\ s^{-1}\ dyn^{-1}$)[a]	*Molecular Weight*
Lysozyme	2.19	15,000–16,000
Albumin	4.6	69,000
Immunoglobulin G	6.6–7.2	153,000
Fibrinogen	7.63	341,000
C1q (factor of complement)	11.1	410,000
α_2-Macroglobulin	19.6	820,000
Immunoglobulin M	18–20	1,000,000
Factor VIII of blood coagulation	23.7	1,120,000

Source: Fasman, G. D. (Ed.). *CRC Handbook of Biochemistry and Molecular Biology*, 3rd ed., Sect. A, Vol. II. Boca Raton, FL: CRC Press, 1976, p. 242.

[a] s_{20}, $\times 10^{-13}$ is sedimentation coefficient in Svedberg units, referred to water at 20°C, and extrapolated to zero concentration of protein.

the gel and have a larger solvent volume through which to travel in the column than large proteins, which are sterically excluded from the pores. Accordingly, a protein mixture is separated by size. The larger proteins are eluted first, followed by the smaller proteins, which are retarded by their accessibility to a larger solvent volume (Figure 3.61). As with ultracentrifugation, an assumption is made as to the geometry of an unknown protein in the determination of molecular mass. Elongated nonspheroidal proteins give anomalous molecular masses when analyzed using a standard curve determined with proteins of spheroidal geometry.

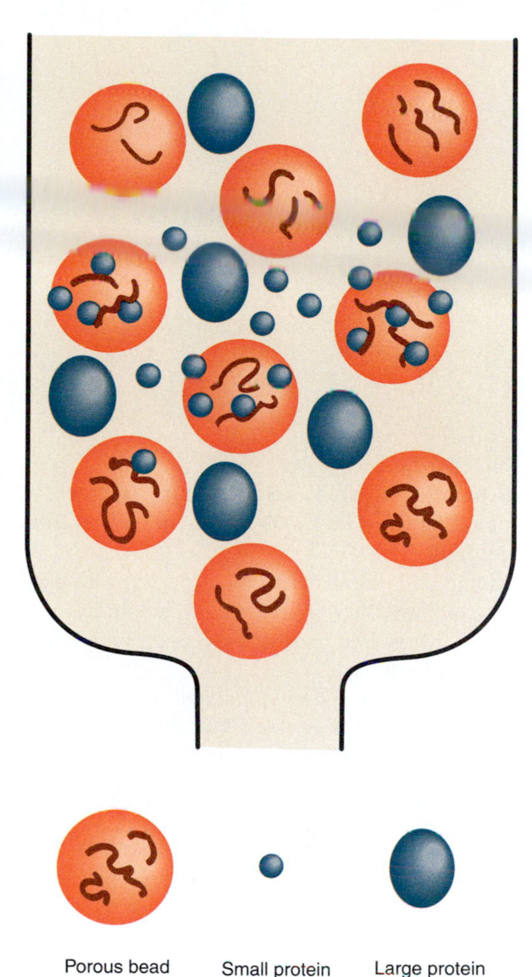

FIGURE 3.61
Molecular exclusion chromatography.
A small protein can enter the porous gel particles and will be retarded on the column with respect to a larger protein that cannot enter the porous gel particles.

Polyacrylamide Gel Electrophoresis in the Presence of a Detergent

If a charged detergent is added to a protein electrophoresis assay and electrophoresis occurs through a sieving support, separation of proteins is based on protein size and not charge. A common detergent is **sodium dodecyl sulfate (SDS)** and a common sieving support is **cross-linked polyacrylamide.** The dodecyl sulfates are amphiphilic C12 alkyl sulfates, which stabilize a denatured protein by forming a charged micellar SDS solvation shell around its polypeptide chain. The inherent charge of the polypeptide chain is obliterated by the negatively charged SDS molecules, and each protein–SDS solubilized aggregate has an identical charge per unit volume. Negatively charged particles move through the polyacrylamide gel toward the anode. Polyacrylamide acts as a molecular sieve and the protein–micelle complexes are separated by size; proteins of larger mass are retarded. A single band in an SDS polyacrylamide electrophoresis is often taken to demonstrate the purity of a protein. The conformation of the native structure is not a factor in the calculation of molecular mass, which is determined by comparison to known standards that are similarly denatured. The detergent dissociates quaternary structure and releases a protein's constituent subunits. Only the molecular mass of subunits of a protein are determined by this method.

HPLC Chromatographic Techniques Separate Amino Acids, Peptides, and Proteins

In **high-performance liquid chromatography** (HPLC), a liquid solvent containing a mixture of molecules to be identified is passed through a column densely packed with a small-diameter insoluble bead-like resin. In column chromatography, the smaller and more tightly packed the resin beads, the greater the resolution of the separation technique. In HPLC, the resin is so tightly packed that the liquid must be pumped through the column at high pressure. Therefore HPLC uses precise high-pressure pumps with metal plumbing and columns rather than glass and plastics used in gravity chromatography. Resin beads are coated with charged groups to separate compounds by ion exchange or with hydrophobic groups to retard hydrophobic nonpolar molecules. In hydrophobic chromatography, tightly associated nonpolar compounds are eluted from the hydrophobic beads in aqueous solvents containing various percentages of an organic reagent. The higher the percentage of organic solvent in the effluent, the faster the nonpolar component is eluted from the hydrophobic resin. This latter type of chromatography over nonpolar resin beads is called **reverse-phase HPLC** (Figure 3.62). HPLC separations have extremely high resolution and reproducibility.

Affinity Chromatography

Proteins have a high affinity for their substrates, prosthetic groups, membrane receptors, specific noncovalent inhibitors, and antibodies made against them. These high-affinity compounds can be covalently attached to an insoluble resin and the modified resin used to purify its conjugate protein in column chromatography. In a mixture of proteins eluted through the resin, the one of interest is selectively retarded.

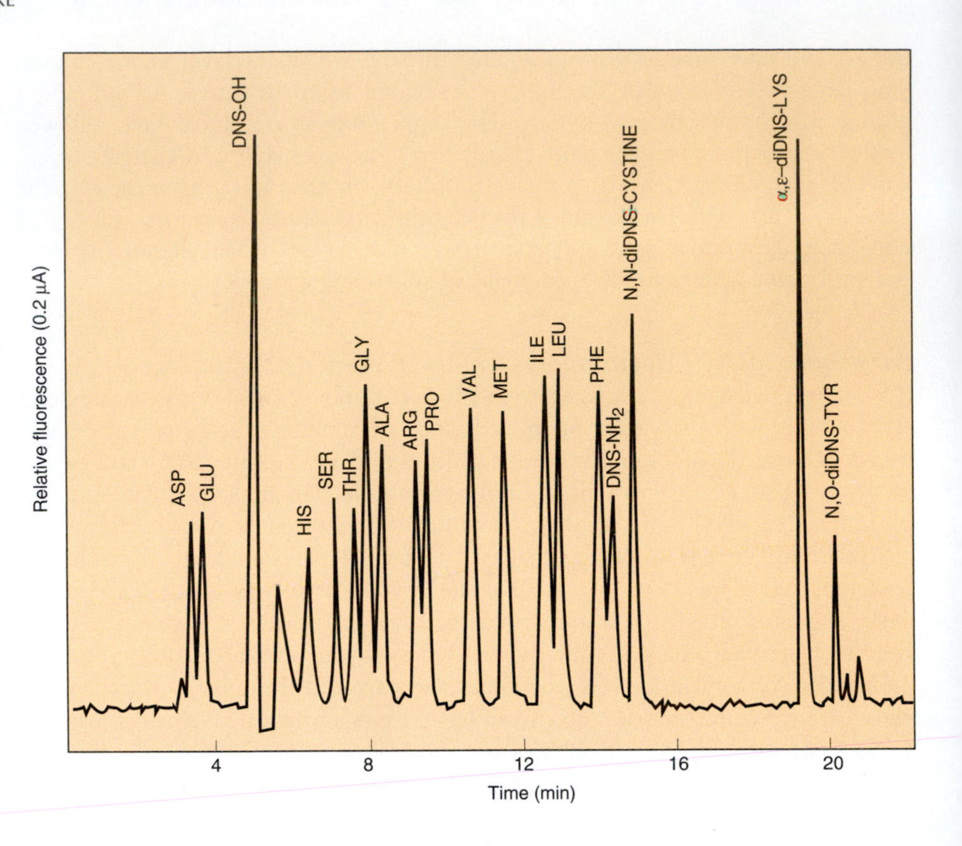

FIGURE 3.62
Separation of amino acids utilizing reverse-phase HPLC. The *x*-axis is time of elution from column.
Amino acids are derivatized by reaction with dansyl chloride (DNS) so that they emit a fluorescence that is used to assay them as they are eluted from the column.
Reprinted with permission from Hunkapiller, M. W., Strickler, J. E., and Wilson, K. J. Science 226:304, 1984. Copyright © 1984 by American Association for the Advancement of Science.

General Approach to Protein Purification

A protein must be purified prior to a meaningful determination of its chemical composition, structure, and function. As living cells contain thousands of genetically distinct proteins, the purification of a single protein from the other cellular constituents may be difficult. The first task in the purification of a protein is the development of a simple assay for the protein. Whether it utilizes the rate of transformation of substrate to product, antibody–antigen reaction, or a physiological response in an animal assay system, a protein assay must give a quantitative measure of activity per unit of protein concentration. This quantity is known as the sample's **specific activity.** The purpose of a purification procedure is to increase a sample's specific activity to the value expected for the pure protein. A typical protocol for purification of a soluble cellular protein involves disruption of the cell membranes, followed by differential centrifugation in a density gradient to isolate the protein from subcellular particles and high molecular weight aggregates. Further purification may utilize selective precipitation by inorganic salts (salting out) or by organic solvent. Final purification includes a combination of techniques that separate based on molecular charge, molecular size, and/or affinity.

Proteomic Techniques Determine All the Proteins Expressed in a Cell or Tissue in a Single Assay

The number of unique genes in the human genome is estimated to be between 30,000 and 80,000. Given that multiple protein products may be produced from a single gene, the number of unique proteins present in the human probably exceeds 250,000. **Proteomics** is the science of determining exactly which proteins are produced in a cell or tissue, under a specific set of conditions.

Any cell or tissue may express thousands of different proteins simultaneously. In order to understand the properties of a cell or tissue it is of interest to determine the type of proteins expressed and how the pattern of expressed proteins change with development, differentiation, and disease. Techniques have been developed to assay expressed mRNAs in cells and tissues in a single assay by hybridizing to DNA microarrays. More recently, techniques have been developed to analyze the active gene products, the proteins expressed by cells and tissues by use of **2-D electrophoresis.** In this technique, the proteins are first extracted from cells or tissues and then spotted onto a polyacrylamide gel in an electrophoresis apparatus. The proteins are separated in the first direction based on their differences in isoelectric pH (p*I*, see Isoelectric Focusing). The gel is then turned in a second direction and sodium dodecyl sulfate (SDS) is added to the buffer. The proteins are separated in the second direction based on differences in their molecular mass (see Gel Electrophoresis in the Presence of Detergent). The resulting gel is stained for protein and the intensity of each of the thousands of protein spots measured in order to determine whether a particular protein is expressed and its concentration (Figure 3.63).

Determining the identity of each of the protein spots in the 2-D gel is not a trivial task. The 2-D gel pattern, if carried out under a standard condition, may be compared to patterns obtained by reference laboratories who have determined the identity of the majority of spots in the 2-D pattern from a particular cell type. These reference 2-D patterns are available over the Internet. More definitively, spots may be extracted from the gel and the protein partially hydrolyzed into smaller peptide fragments by proteolytic enzyme digestion (e.g., trypsin or chymotrypsin), and the peptide fragments may then be subjected to mass spectroscopy. The mass spectroscopy rapidly determines the amino acid sequence of many of the small frag-

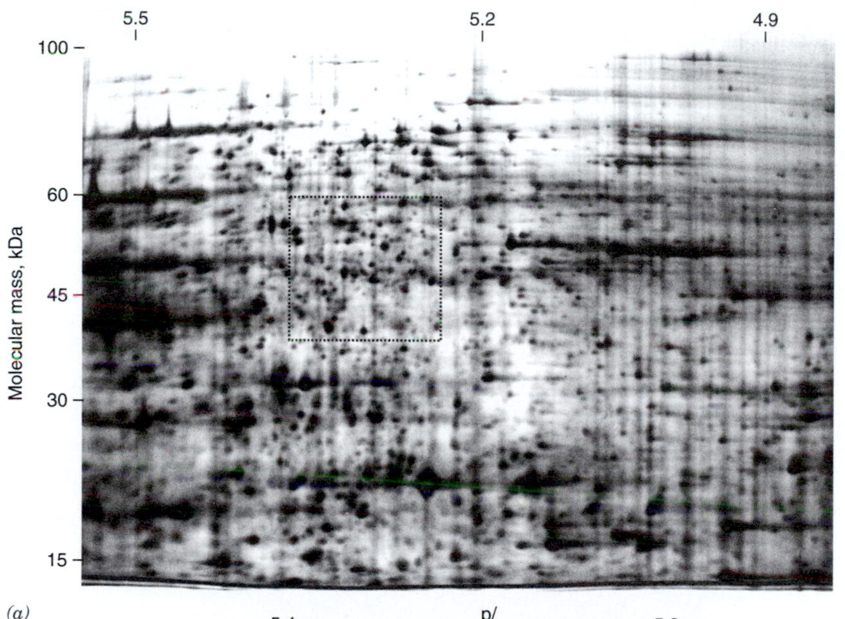

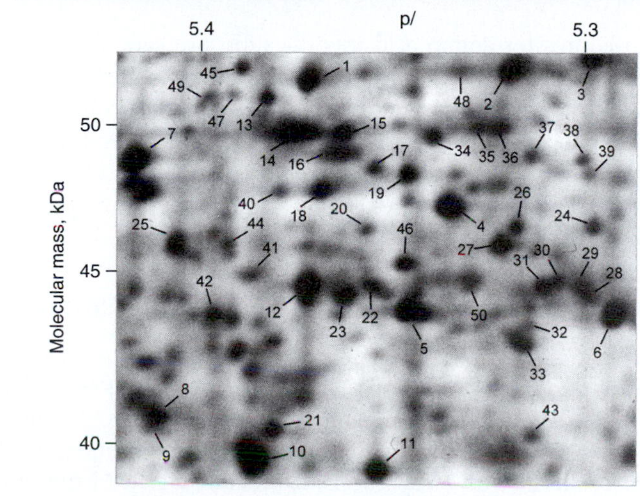

FIGURE 3.63
Two-dimensional (2-D) display of expressed proteins from cultured cells.
(*a*) Soluble proteins from cellular extract (500 μg) loaded on the gel and separated by isoelectric focusing (between pH 4.9 and 5.5) in the horizontal direction and by molecular mass in the vertical direction (electrophoresed in the presence of SDS detergent). More than 1500 proteins are observed in the gel by silver staining. (*b*) A region from the gel expanded to show detail. Numbered proteins were analyzed by protease hydrolysis and mass spectrometry to determine its partial amino acid sequence, leading to the protein's identification.
Reproduced with permission from Gygi, S. P., Corthals, G. L., Zhang, Y., Rochon, Y., and Aebersold, R. Proc. Natl. Acad. Sci. USA *97: 9390, 2000.*

CLINICAL CORRELATION 3.9

Use of Amino Acid Analysis in Diagnosis of Disease

There are a number of clinical disorders in which high concentrations of amino acids are found in plasma or urine. An abnormally high concentration in urine is called an aminoaciduria. Phenylketonuria is a metabolic defect in which patients lack sufficient amounts of the enzyme phenylalanine hydroxylase, which catalyzes the transformation of phenylalanine to tyrosine. As a result, large concentrations of phenylalanine, phenylpyruvate, and phenyllactate accumulate in the plasma and urine. Phenylketonuria occurs clinically in the first few weeks after birth, and if the infant is not placed on a special diet, severe mental retardation will occur (see Clin. Corr. 18.5). Cystinuria is a genetically transmitted defect in the membrane transport system for cystine and the basic amino acids (lysine, arginine, and the derived amino acid ornithine) in epithelial cells. Large amounts of these amino acids are excreted in urine. Other symptoms of this disease may arise from the formation of renal stones composed of cystine precipitated within the kidney (see Clin. Corr. 5.3). Hartnup disease is a genetically transmitted defect in epithelial cell transport of neutral-type amino acids (monoamino monocarboxylic acids), and high concentrations of these amino acids are found in the urine. The physical symptoms of the disease are primarily caused by a deficiency of tryptophan. These symptoms may include a pellagra-like rash (nicotinamide is partly derived from tryptophan precursors) and cerebellar ataxia (irregular and jerky muscular movements) due to the toxic effects of indole derived from the bacterial degradation of unabsorbed tryptophan present in large amounts in the gut (see Clin. Corr. 26.2). Fanconi's syndrome is a generalized aminoaciduria associated with hypophosphatemia and a high excretion of glucose. Abnormal reabsorption of amino acids, phosphate, and glucose by the tubular cells is the underlying defect.

ments. This technique is called peptide **mass fingerprinting.** Utilizing these sequences to search protein sequence or gene sequence databases leads to the identification of the protein extracted from the 2-D gel. Robotic instruments exist to perform each of the steps in protein spot extraction and identification. In this way, the thousands of expressed proteins may be identified.

The technique currently fails to identify low abundance proteins in cells or tissues. Certain types of proteins are difficult to analyze due to low solubility, low molecular charge, or very low molecular mass. For example, integral membrane proteins are highly hydrophobic and not soluble in the standard isoelectric focusing solvents.

Determination of Amino Acid Composition of a Protein

Determination of the amino acid composition is an essential component in the study of a protein's structure and physiological properties. A protein is hydrolyzed to its constituent amino acids by heating the protein at 110°C in 6 N HCl for 18–36 h, in a sealed tube under vacuum to prevent degradation of oxidation-sensitive amino acid side chains by oxygen in air. Tryptophan is destroyed in this method and alternative procedures are used for its analysis. Asparagine and glutamine side chain amides are hydrolyzed to aspartate and glutamate and free ammonia; thus they are included within the glutamic acid and aspartic acid content in the analysis.

Common procedures for amino acid identification use cation-exchange chromatography or reverse-phase HPLC to separate the amino acids, which are then reacted with ninhydrin, fluorescamine, dansyl chloride, or similar chromophoric or fluorophoric reagents for quantitation (Figure 3.62). With some types of derivatization, amino acids are identified at concentrations as low as 0.5×10^{-12} mol (pmol). Analysis of the amino acid composition of physiological fluids (i.e., blood and urine) is utilized in diagnosis of disease (see Clin. Corr. 3.9).

Techniques to Determine Amino Acid Sequence of a Protein

The ability to clone genes has led to determination of the amino acid sequence of a protein as derived from their DNA or messenger RNA sequences (see Chapter 7). This is a much faster method for obtaining an amino acid sequence. Sequencing of a protein, however, is required for the determination of modifications to the protein structure that occur after its biosynthesis, to identify a part of the protein sequence in order that its gene can be cloned, and to identify a protein as the product of a particular gene (see Chapter 6). Determination of primary structure of a protein requires a purified protein and it is necessary to determine the number of chains in the protein. Individual chains are purified by the same techniques used in purification of the whole protein. If disulfide bonds covalently join the chains, these bonds have to be broken (Figure 3.64).

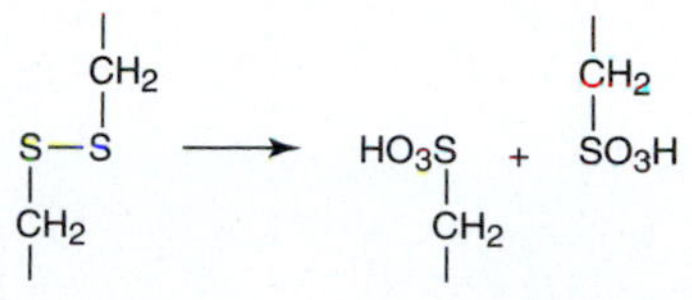

Polypeptide chains are most commonly sequenced by the **Edman reaction** or **mass spectroscopy.** In the Edman reaction the polypeptide chain is reacted with phenylisothiocyanate, which reacts with the NH_2-terminal amino acid. Acidic conditions catalyze intramolecular cyclization that cleaves the NH_2-terminal amino acid as a phenylthiohydantoin derivative (Figure 3.65). This NH_2-terminal amino acid derivative may be separated chromatographically and identified against standards. The remaining polypeptide chain is isolated, and the Edman reaction is repeated to identify the next NH_2-terminal amino acid. Theoretically this can be repeated until the sequence of the entire polypeptide chain is determined but under favorable conditions can only be carried out for 30 or 40 amino acids into the polypeptide chain. At this point in the analysis, impurities generated from incomplete reactions in the reaction series make further **Edman cycles** unfeasible. Polypeptide chains that contain more than 30 or 40 amino acids are hydrolyzed into smaller fragments and sequenced in sections. For sequencing by mass spectroscopy it is also necessary to break long polypeptide chains into smaller fragments prior to analysis.

Enzymatic and chemical methods are used to break polypeptide chains into smaller fragments (Figure 3.66). The enzyme **trypsin** preferentially catalyzes hydrolysis of the peptide bond on the COOH-terminal side of the basic amino acid residues of lysine and arginine within polypeptide chains. **Chymotrypsin** hydrolyzes peptide bonds on the COOH-terminal side of residues with large apolar side chains. Other enzymes cleave polypeptide chains on the COOH-terminal side of glutamic and aspartic acid. **Cyanogen bromide** specifically cleaves peptide bonds on the COOH-terminal side of methionine residues within polypeptide chains (Figure 3.66). A large polypeptide is subjected to **partial hydrolysis** by a specific cleaving reagent, the segments are separated, and the amino acid sequence of each is determined by the Edman reaction or mass spectroscopy. To place the sequenced peptides correctly into the complete sequence of the original polypeptide, a sample of the original polypeptide is subjected to a second partial hydrolysis by a different hydrolytic reagent from that used initially. This generates overlapping sequences to the first set of sequences, leading to the complete sequence (Figure 3.67).

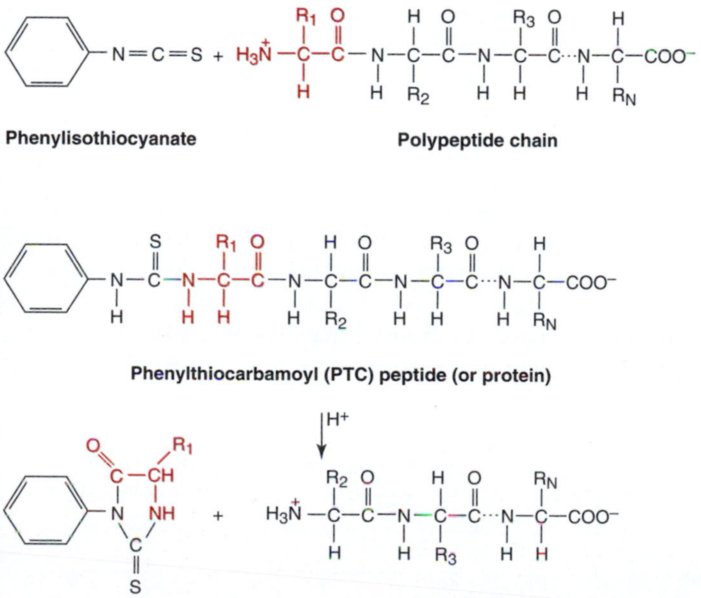

Cystine bond → Two cysteic acids

FIGURE 3.64
Breaking of disulfide bonds by oxidation to cysteic acids.

Phenylisothiocyanate + Polypeptide chain

Phenylthiocarbamoyl (PTC) peptide (or protein)

H^+

Phenylthiohydantoin + Polypeptide chain (minus NH_2-terminal amino acid)

FIGURE 3.65
Edman reaction.

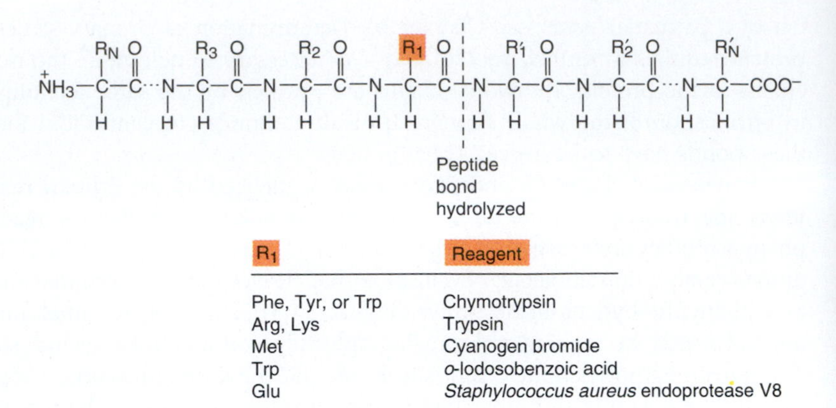

Peptide bond hydrolyzed
R1
Reagent
Phe, Tyr, or Trp
Chymotrypsin
Arg, Lys
Trypsin
Met
Cyanogen bromide
Trp
o-Iodosobenzoic acid
Glu
Staphylococcus aureus endoprotease V8

FIGURE 3.66
Specificity of some polypeptide cleaving reagents.

X-Ray Diffraction Techniques Are Used to Determine the Three-Dimensional Structure of Proteins

X-ray diffraction enables determination of the three-dimensional structure of proteins at near atomic resolution. The approach requires formation of a protein crystal, which contains solvent and is thus a concentrated solution, for use as the target. Our understanding of the detailed components of protein structure derived from experiments performed in this crystalline state correlate well with other physical measurements of protein structure in solution such as those made using NMR spectroscopy (see p. 152).

Generation of protein crystals can be the most time-consuming aspect of the process. A significant factor in experimental and computational handling of protein crystals, in contrast with most small molecule crystals, stems from the content of the protein crystalline material. Proteins exhibit molecular dimensions an order of magnitude greater than small molecules, and the packing of large protein molecules into the crystal lattice generates a crystal with large "holes" or solvent channels. A protein crystal typically contains 40–60% solvent and may be considered a concentrated solution rather than the hard crystalline solid associated with most small molecules. Presence of solvent and unoccupied volume in the crystal permits the infusion of inhibitors and substrates into the protein molecules in the "crystalline state" and a **dynamic flexibility** within regions of the protein structure. This flexibility may be seen as "disorder" in the X-ray diffraction experiment. Disorder is used to describe the situation in which the observed electron density can be fitted by more than a single local conformation. Two explanations for the disorder exist and must be distinguished. The first involves the presence of two or more static molecular conformations, which are present in a stoichiometric relationship. The second involves the actual dynamic range of motion exhibited by atoms or groups of atoms in localized regions of the molecule. These explanations can be distinguished by lowering the temperature of the crystal to a point where dynamic disorder is "frozen out"; in contrast, the static disorder is not temperature dependent and persists. Analysis of dynamic disorder by its temperature dependency using X-ray diffraction determinations is an important method for studying protein dynamics (see Section 3.8).

Crystallization techniques have so advanced that crystals are obtainable even from less abundant proteins. Interesting structures have been reported for proteins

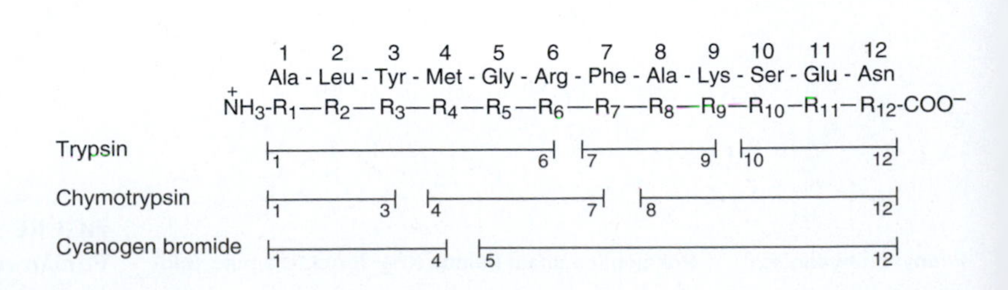

FIGURE 3.67
Ordering of peptide fragments from overlapping sequences produced by specific proteolysis of a peptide.

in which specific amino acid residues have been substituted, for antibody–antigen complexes, and for viral products such as the protease required for the infection of the human immunodeficiency virus (HIV) that causes acquired immune deficiency syndrome (AIDS). Many structures have been solved by X-ray diffraction and the details are stored in a database called the **Protein Data Bank**, which is readily accessible.

Diffraction of X-rays by a crystal occurs with incident radiation of a characteristic wavelength (e.g., copper, K_{α} = 1.54 Å). The X-ray beam is diffracted by the electrons that surround the atomic nuclei in the crystal, with an intensity proportional to the number of electrons around the nucleus. Thus the technique establishes the **electron distribution** of the molecule and infers the nuclear distribution. Actual positions of atomic nuclei can be determined directly by diffraction with **neutron beam radiation,** an interesting but very expensive technique as it requires a source of neutrons (nuclear reactor or particle accelerator). With the highest resolution now available for X-ray diffraction determination of protein structure, the electron diffraction from C, N, O, and S atoms can be observed. Diffraction from hydrogen atoms is not observed due to the low density of electrons—that is, a single electron around a hydrogen nucleus. Detectors of the diffracted beam, typically photographic film or electronic area detectors, permit the recording of the amplitude (intensity) of radiation diffracted in a defined orientation. However, the data do not give information about phases of the radiation, which are essential to the solution of a protein's structure. Determination of the **phase angles** historically required the placement of heavy atoms (such as iodine, mercury, or lead) in the protein molecule. Modern procedures can often solve the phase problem without use of heavy atoms.

It is convenient to consider an analogy between X-ray crystallography and light microscopy to understand the processes involved in carrying out the three-dimensional structure determination. In light microscopy, incident radiation is reflected by an object under study and the reflected beam is recondensed by the objective lens to produce an image of the object. The analogy is appropriate to incident X-rays with the notable exception that no known material exists that can serve as an objective lens for X-ray radiation. To replace the objective lens, amplitude and phase angle measurements of the diffracted radiation are mathematically reconstructed by **Fourier synthesis** to yield a three-dimensional **electron-density map** of the diffracted object. Initially a few hundred reflections are obtained to construct a low-resolution electron-density map. For example, for one of the first protein crystallographic structures, 400 reflections were used to obtain a 6-Å map of myoglobin. At this level of resolution it is possible to locate clearly the molecule within the unit cell of the crystal and study the overall packing of the subunits in a protein with a quaternary structure. A trace of the polypeptide chain of an individual protein molecule is made with difficulty. Utilizing the low-resolution structure as a base, further reflections may be used to obtain higher-resolution maps. For myoglobin, where 400 reflections were utilized to obtain the 6-Å map, 10,000 reflections were needed for a 2-Å map, and 17,000 reflections for an extremely high-resolution 1.4-Å map. Many of these steps are now partially automated using computer graphics. A two-dimensional slice through a three-dimensional electron-density map of trypsinogen is shown in Figure 3.68. The known primary structure of the protein is fitted to the electron-density pattern by **refinement,** the computer-intensive process of aligning a protein's amino acid sequence to the electron-density map until the best fit is obtained.

Whereas X-ray diffraction has provided extensive knowledge on protein structure, such a structure provides incomplete evidence of a protein's mechanism of action. The X-ray determined structure is an average structure of a molecule whose atoms are normally undergoing rapid fluctuations in solution (see Section 3.8). Thus the average structure determined by X-ray diffraction may not be the active structure of a particular protein in solution. A second important consideration is that it currently takes at least a day to collect data in order to determine a structure. On this time scale, the structures of reactive enzyme–substrate complexes, in-

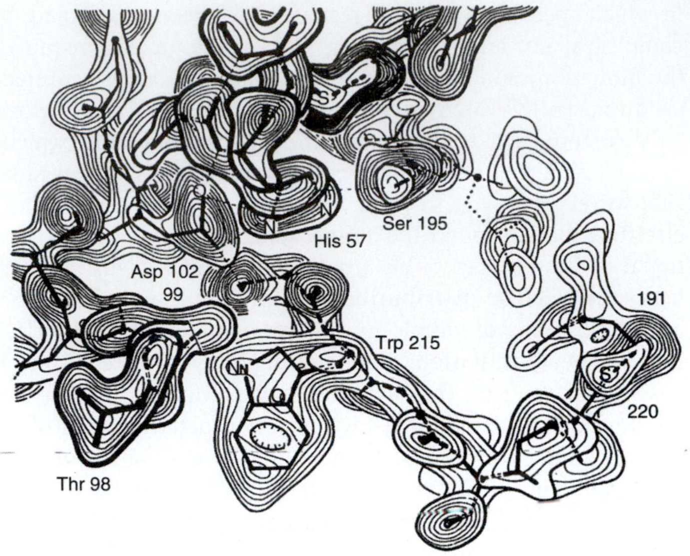

FIGURE 3.68
Electron-density map at 1.9-Å resolution of active site region of proenzyme form of trypsin.
Active site amino acid residues are fitted onto density map.
Reprinted with permission from Kossiakoff, A. A., Chambers, J. L., Kay, L. M., and Stroud, R. M. Biochemistry 16:654, 1977. Copyright © 1977 by American Chemical Society.

termediates, and reaction transition states of an enzyme can not be observed. Rather, these structures must be inferred from the static pictures of an inactive form of the protein or from complexes with inactive analogs of the substrates of the protein (Figure 3.69). Newer methods for X-ray diffraction using synchrotron radiation to generate an X-ray beam at least 10,000-times brighter than that of standard X-ray generators will enable collection of diffraction data to solve a protein structure on a millisecond time scale. Application of the latter X-ray techniques will enable scientists to determine short-lived structures and solve mechanistic and dynamic structural questions not addressable by the standard technology.

Various Spectroscopic Methods Are Employed in Evaluating Protein Structure and Function

Ultraviolet Light Spectroscopy

Side chains of tyrosine, phenylalanine, and tryptophan, as well as peptide bonds in proteins, absorb ultraviolet (UV) light. The efficiency of light energy absorption for

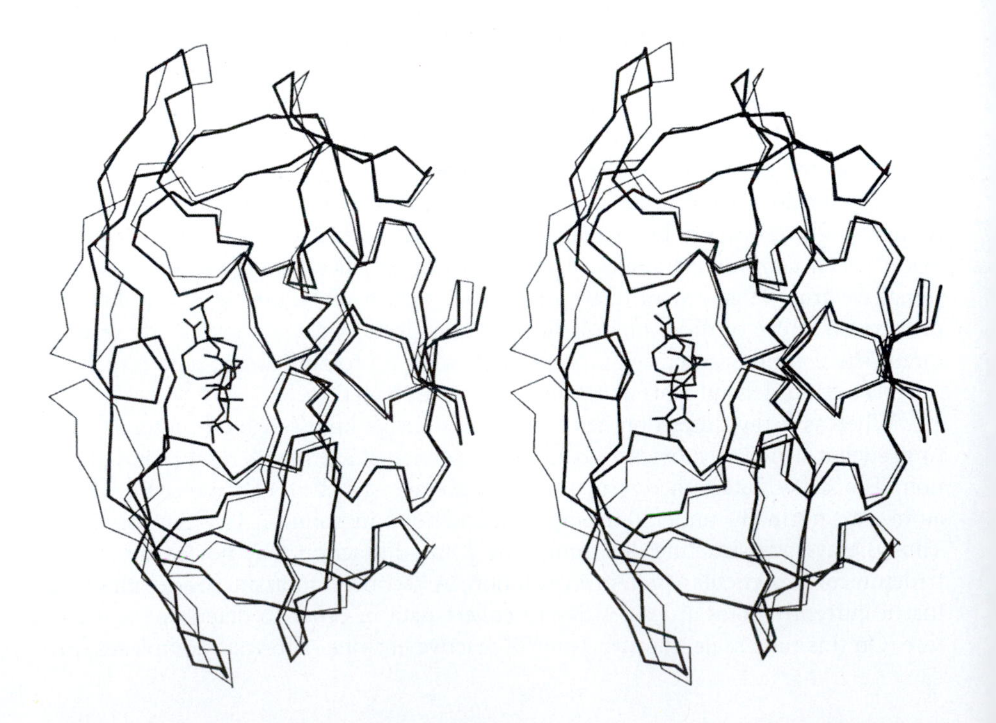

FIGURE 3.69
Stereo tracing of superimposed α-carbon backbone structure of HIV protease with inhibitor bound (thick lines) and the native structure of HIV protease without inhibitor bound (thin lines).
Redrawn with permission from Miller, M., Schneider, J., Sathyanarayana, B. K., Toth, M. V., Marshall, G. R., Clawson, L., Selk, L., Kent, S. B. H., and Wlodawer, A. Science 246:1149, 1989. Copyright © 1989 by the American Association for the Advancement of Science.

each **chromophore** is related to its **molar extinction coefficient** (ε). A typical protein spectrum is shown in Figure 3.70. Absorbance between 260 and 300 nm is primarily due to side chains of phenylalanine, tyrosine, and tryptophan (Figure 3.71). When the tyrosine side chain is ionized at high pH (the tyrosine R group has a $pK'_a \approx 10$), the absorbance for tyrosine is shifted to a higher wavelength (red shifted) and its molar absorptivity is increased (Figure 3.71). Peptide bonds absorb in the far-UV (180–230 nm). A peptide bond in α-helix conformation interacts with the electrons of other peptide bonds above and below it in the spiral conformation to create an **exciton system** in which electrons are delocalized. The result is a shift of the absorption maximum from that of an isolated peptide bond to either a lower or higher wavelength (Figure 3.72). Thus UV spectroscopy can be used to study changes in a protein's secondary and tertiary structures. As a protein is denatured (helix unfolded), differences are observed in the absorption characteristics of the peptide bonds due to the disruption of the exciton system. In addition, the absorption maximum for an aromatic chromophore appears at a lower wavelength in an aqueous environment than in a nonpolar environment.

The **molar absorbancy** of a chromophoric substrate often changes on binding to a protein and can be used to measure its binding constant. Changes in chromophore extinction coefficients during enzyme catalysis of a chemical reaction are used to obtain the kinetic parameters for the reaction.

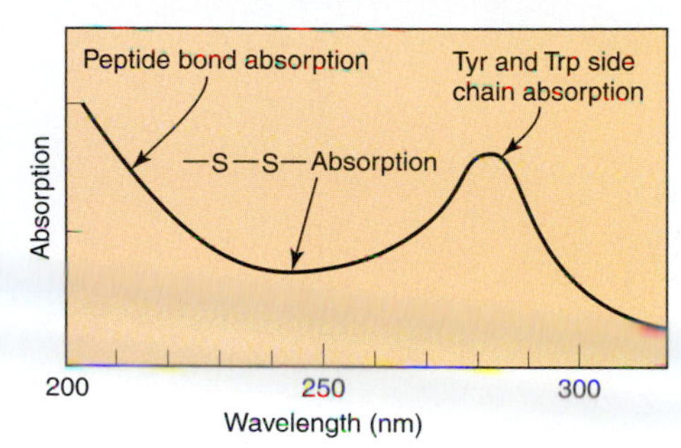

FIGURE 3.70
Ultraviolet absorption spectrum of the globular protein α-chymotrypsin.

Fluorescence Spectroscopy

The energy of an excited electron produced by light absorption is lost by various mechanisms and most commonly as thermal energy in a collision process. In some chromophores the excitation energy is dissipated by fluorescence. The **fluorescent emission** is always at a longer wavelength of light (lower energy) than the absorption wavelength of the fluorophore. Higher vibrational energy levels, formed in the excited electron state during the excitation event, are lost prior to the fluorescent event (Figure 3.73). If a chromophoric molecule is present that absorbs light energy emitted by the **fluorophore**, the emitted fluorescence is not observed. Rather, the fluorescence energy is transferred to the absorbing molecule. The **acceptor molecule**, in turn, either emits its own characteristic fluorescence or loses its excitation energy by an alternative process. If the acceptor molecule loses its excitation energy by a nonfluorescent process, it is acting as a **quencher** of the **donor molecule's** fluorescence. The efficiency of the **excitation transfer** is dependent on the distance and orientation between donor and acceptor molecules.

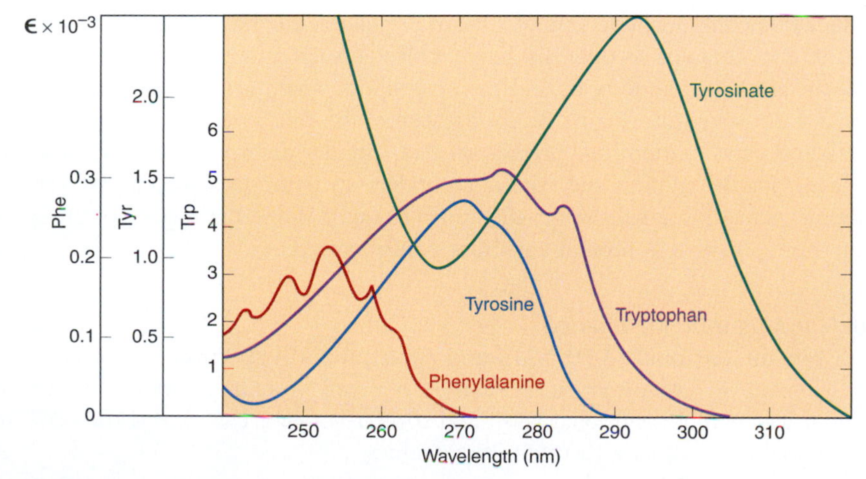

FIGURE 3.71
Ultraviolet absorption for the aromatic chromophores in Phe, Tyr, Trp, and tyrosinate. Note differences in extinction coefficients on left axis for the different chromophores.
Redrawn from d'Albis, A. and Gratzer, W. B. In: A. T. Bull, J. R. Lagmado, J. O. Thomas, and K. F. Tipton (Eds.), Companion to Biochemistry. *London: Longmans, 1974, p. 170.*

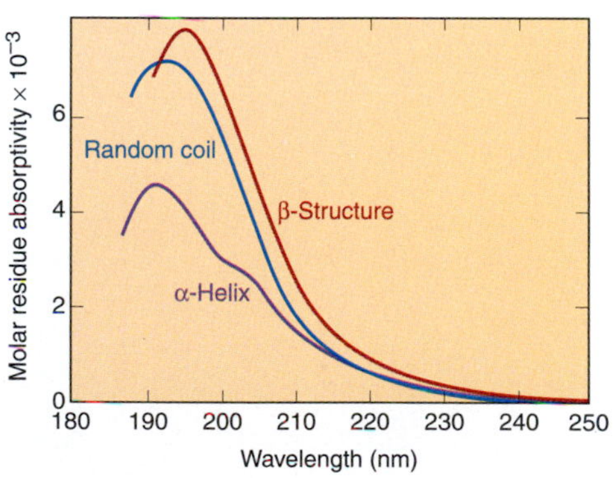

FIGURE 3.72
Ultraviolet absorption of the peptide bonds of a polypeptide chain in α-helix, random-coil, and antiparallel β-structure conformations.
Redrawn from d'Albis, A. and Gratzer, W. B. In: A. T. Bull, J. R. Lagmado, J. O. Thomas, and K. F. Tipton (Eds.), Companion to Biochemistry. *London: Longmans, 1970, p. 175.*

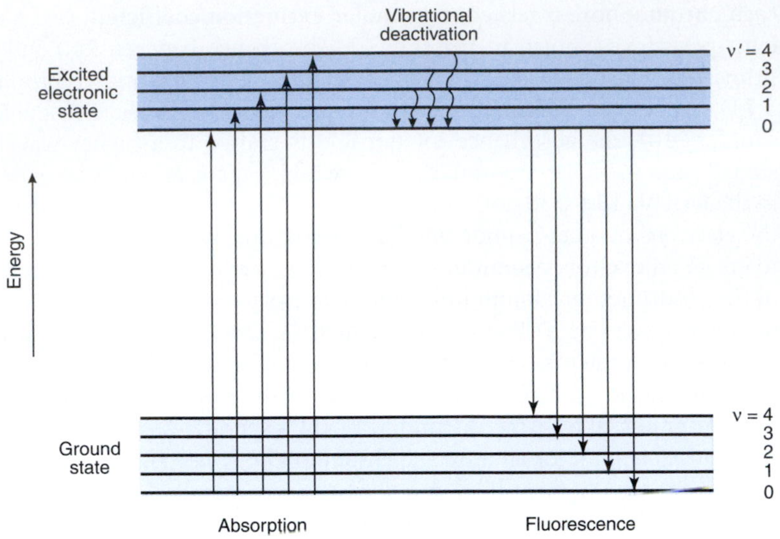

Vibrational deactivation
Excited electronic state
Energy
Ground state
v′ = 4 3 2 1 0
v = 4 3 2 1 0
Absorption
Fluorescence

FIGURE 3.73
Absorption and fluorescence electronic transitions.
Excitation is from zero vibrational level in ground state to various higher vibrational levels in the excited state. Fluorescence is from zero vibrational level in excited electronic state to various vibrational levels in the ground state.
Redrawn from d'Albis, A. and Gratzer, W. B. In: A. T. Bull, J. R. Lagmado, J. O. Thomas, and K. F. Tipton (Eds.), Companion to Biochemistry. *London: Longmans, 1970, p. 166.*

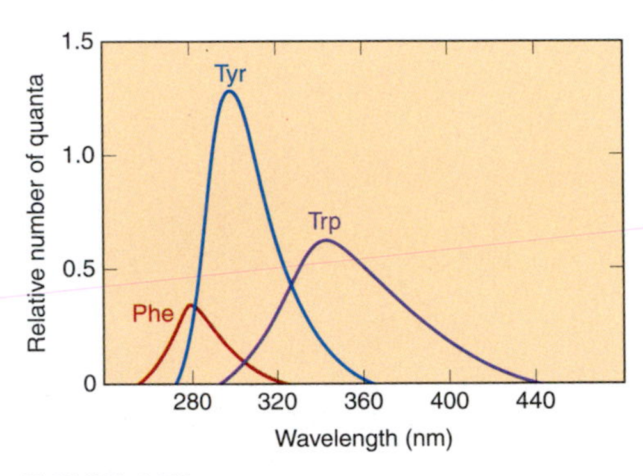

FIGURE 3.74
Characteristic fluorescence of aromatic groups in proteins.
Redrawn from d'Albis, A. and Gratzer, W. B. In: A. T. Bull, J. R. Lagmado, J. O. Thomas, and K. F. Tipton (Eds.), Companion to Biochemistry. *London: Longmans, 1970, p. 478.*

Fluorescence emission spectra for phenylalanine, tyrosine, and tryptophan side chains are shown in Figure 3.74. The emission wavelengths for phenylalanine overlap with the absorption wavelengths for tyrosine. In turn, the emission wavelengths for tyrosine overlap with the absorption wavelengths for tryptophan. Because of these overlaps in emission and absorption wavelengths, primarily only tryptophan fluorescence is observed in proteins that contain all of these amino acids. **Excitation energy transfers** occur over distances up to 80 Å, which are typical diameter distances in folded globular proteins. On protein denaturation, the distances between donor and acceptor groups become greater and decrease the efficiency of energy transfer to tryptophan. Accordingly, an increase in fluorescence due to tyrosines and/or phenylalanines is observed on denaturation of a protein. Since excitation transfer processes in proteins are distance and orientation dependent, the fluorescence yield is dependent on the conformation of the protein. The greatest sensitivity of this analysis occurs in its ability to detect changes due to solvent or binding interactions rather than establish absolute structure.

Optical Rotatory Dispersion and Circular Dichroism Spectroscopy

Optical rotation is caused by differences in the refractive index and **circular dichroism** (CD) by differences in light absorption between the clockwise and counterclockwise component vectors of a beam of polarized light as it travels through a solution that contains an optically active molecule such as an L-amino acid. In proteins the aromatic amino acids and the polypeptide chain generate an optical rotation and CD spectrum (Figure 3.75). Because of the differences between α-helical, β-structure, and random polypeptide spectra, circular dichroism has been a sensitive assay for the amount and type of secondary structure in a protein. Newer developments in vibrational circular dichroism examine the CD in regions of the spectrum more sensitive to protein backbone conformation.

Nuclear Magnetic Resonance

With **two-dimensional (2-D) NMR** and powerful NMR spectrometers it is possible to obtain the conformation in solution of small proteins of about 150 amino acids or less. Multidimensional NMR and triple resonance can extend the NMR to solve protein structures with up to 250 residues.

Conventional NMR techniques involve use of radiofrequency (rf) radiation to study the environment of atomic nuclei that are magnetic. The requirement for magnetic nuclei is absolute and based on an unpaired spin state in the nucleus. Thus naturally abundant carbon (^{12}C), nitrogen (^{14}N), and oxygen (^{16}O) do not absorb, while ^{13}C, ^{15}N, and ^{17}O do absorb. The absorption bands in an NMR spectrum are characterized by (1) a position or chemical shift value, reported as the frequency difference between that observed for a specific absorption band and that for a standard reference material; (2) the intensity of the peak or integrated area, which is proportional to the total number of absorbing nuclei; (3) the half-height peak width, which reflects the degree of motion in solution of the absorbing species; and (4) the coupling constant, which measures the extent of direct interaction or influence of neighboring nuclei on the absorbing nuclei. These four measurements enable the determination of the identity and number of nearest-neighbor groups that can affect the response of absorbing species through bonded interactions. They give no information on through-space (nonbonded) interaction due to the three-dimensional structure of the protein. To determine through-space interactions and protein tertiary structure requires the use of **nuclear Overhauser effects (NOEs)** and the application of the two-dimensional technique.

The major difference between two-dimensional and one-dimensional (1-D) NMR is the addition of a second time delay rf pulse. The technique requires identification in the spectrum of proton absorbance from a particular position in the protein structure. A maximum distance of about 5 Å is the limit for which these through-space interactions can be observed. Upon the generation of distance information for interresidue pairs through the protein structure, three-dimensional protein conformations consistent with the spectra are generated. In this calculation, a distance matrix is constructed containing ranges of distances (minimum and maximum) for as many interresidue interactions as may be measured. Possible structures are generated from the data consistent with the constraints imposed by the NMR spectra. Computational refinements of the initially calculated structures can be made to optimize covalent bond distances and bond angles. The method generates a family of structures, the variability showing either the imprecision of the technique or the dynamic "disorder" of the folded structure (Figure 3.76). Such computations based on NMR experiments have yielded structures for proteins that do not significantly differ from the time-averaged structure observed with X-ray diffraction methods.

Other enhancements of NMR, applicable to determination of protein structure, include the ability to synthesize proteins that contain isotopically enriched (e.g., containing ^{13}C or ^{15}N) amino acids, and development of paramagnetic shift reagents to study localized environments on paramagnetic resonances, such as the lanthanide ion reporting groups.

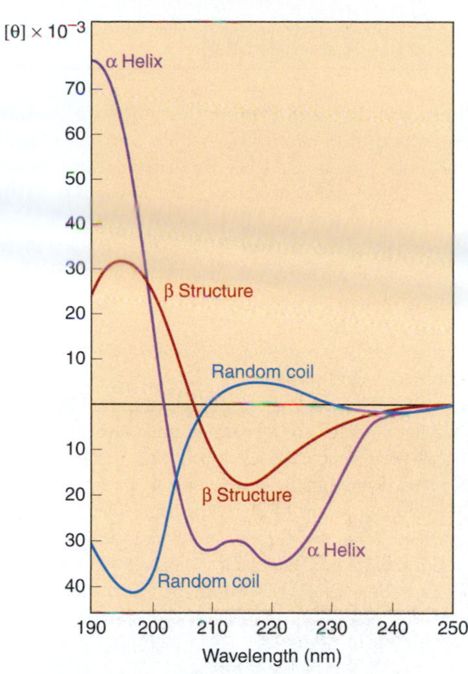

FIGURE 3.75
Circular dichroism spectra for polypeptide chains in α-helix, β-structure, and random-coil conformations.
Redrawn from d'Albis, A. and Gratzer, W. B. In: A. T. Bull, J. R. Lagmado, J. O. Thomas, and K. F. Tipton (Eds.), Companion to Biochemistry. *London: Longmans, 1970, p. 190.*

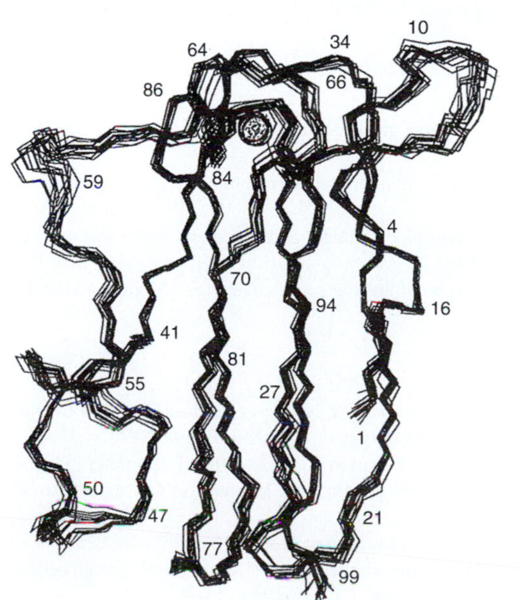

FIGURE 3.76
NMR structure of the protein plastocyanin from the French bean.
Structure shows superposition of eight structures of the polypeptide backbone for the protein, calculated from constraints of NMR spectrum.
From Moore, J. M., Lepre, C. A., Gippert, G. P., Chazin, W. J., Case, D. A., and Wright, P. E. J. Mol. Biol. *221:533, 1991. Figure generously supplied by P. E. Wright.*

BIBLIOGRAPHY

Bioinformatics and Proteomics Software Portals

http://www3.ncbi.nlm.nih.gov/Entrez/ National Institutes of Health Entrez site accesses protein sequence databases and the Protein Data Bank of protein three-dimensional structures. Contains tool (Cn3D) for visualization of protein structures.

http://www.expasy.ch/tools/ The ExPASy (Expert Protein Analysis System) proteomics site of the Swiss Institute of Bioinformatics (SIB) contains tools for protein study and links to other sites and databases for the study and analysis of proteins.

http://www.sdsc.edu/restools/ A compendium of electronic and Internet accessible tools and resources.

http://archive.uwcm.ac.uk/uwcm/mg/hgmd0.html A human gene mutation database. Collection of data on mutations in human proteins.

http://www.biochem.ucl.ac.uk/bsm/cath. A structural classification of proteins.

http://www.usm.maine.edu/~rhodes/SPVTut/index.html/ A tutorial for the Swiss PDB viewer by Gale Rhodes, University of Southern Maine. The Swiss PDB viewer allows the viewing of molecular structures of proteins downloaded from the Protein Data Bank. The Swiss PDV is available from the ExPASy site. This is an alternative viewer to the Cn3D viewer obtained from the Entrez site.

http://www.umass.edu/microbio/chime/explorer/index.htm/ A third type of protein structure viewer called protein explorer. This site also contains links to other protein sites of interest.

Commercial Website Portals (Charge Fee for Use)

http://www.bionavigator.com/ A bioinformatics workspace with tools for protein analysis.

http://www.doubletwist.com/ Bioinformatic tools and access to databases.

Physical and Structural Properties of Proteins

Bryson, J. W., Betz, S. F., Lu, H. S., Suich, D. H., Zhou, H. X., O'Neil, K. T., and DeGrado, W. F. Protein design: a hierarchic approach. *Science* 270:935, 1995.

Chothia, C. Principles that determine the structure of proteins. *Annu. Rev. Biochem.* 53:537, 1984.

Doolittle, R. F. The multiplicity of domains in proteins. *Annu. Rev. Biochem.* 64:287, 1995.

Eisenberg, D. and McLachlan, A. D. Solvation energy in protein folding and binding. *Nature* 319:199, 1986.

Fasman, G. D. Protein conformational prediction. *Trends Biochem. Sci.* 14:295, 1989.

Finkelstein, A. V., Gutun, A. M., and Badretdinov, A. Ya. Why are the same protein folds used to perform different functions? *FEBS Lett.* 325:23, 1993.

Orengo, C. A., Jones, D. T., and Thornton, J. M. Protein superfamilies and domain superfolds. *Nature* 372:631, 1994.

Richardson, J. S. The anatomy and taxonomy of protein structure. *Adv. Protein Chem.* 34:168, 1981.

Rose, G. D. and Wolfenden, R. Hydrogen bonding, hydrophobicity, packing, and protein folding. *Annu. Rev. Biophys. Biomol. Struct.* 22:381, 1993.

Srinivasan, R. and Rose, G. D. A physical basis for protein secondary structure. *Proc. Natl. Acad. Sci. USA* 96:14258, 1999.

Protein Folding

Anfinsen, C. B. and Scheraga, H. Experimental and theoretical aspects of protein folding. *Adv. Protein Chem.* 29:205, 1975.

Baldwin, R. L. and Rose, G. D. Is protein folding hierarchic? I. Local structure and peptide folding. *Trends Biochem. Sci.* 24:26, 1999.

Baldwin, R. L. and Rose, G. D. Is protein folding hierarchic? II. Folding intermediates and transition states. *Trends Biochem. Sci.* 24:77, 1999.

Bukau, B. and Horwich, A. L. The Hsp70 and Hsp60 chaperone machines. *Cell* 92:351, 1998.

Bychkova, V. E. and Ptitsyn, O. B. The molten globule state is involved in genetic disease? *FEBS Lett.* 359:6, 1995.

Chen, S., Roseman, A. M., Hunter, A. S., Wood, S. P., Burston, S. G., Ranson, N. A., Clarke, A. R. and Saibil, H. R. Location of a folding protein and shape changes in GroEL–GroES complexes imaged by cryo-electron microscopy. *Nature* 371:261, 1994.

Dinner, A. R., Sali, A., Smith, L. J., Dobson, C. M., and Karplus, M. Understanding protein folding via free-energy surfaces from theory and experiment. *Trends Biochem. Sci.* 25:331, 2000.

Dobson, C. M., Evans, P. A., and Radford, S. E. Understanding how proteins fold: the lysozyme story so far. *Trends Biochem. Sci.* 19:31, 1994.

Feltham, J. L. and Gierasch, L. M. GroEL–substrate interactions: molding the fold, or folding the mold? *Cell* 100:193, 2000.

Jaenicke, R. Protein folding: local structures, domains, subunits and assemblies. *Biochemistry* 30:3147, 1991.

Lins, L. and Brasseur, R. The hydrophobic effect in protein folding. *FASEB J.* 9:535, 1995.

Ptitsyn, O. B. How the molten globule became. *Trends Biochem. Sci.* 20:376, 1995.

Shakhnovich, E., Abkevich, V., and Ptitsyn, O. Conserved residues and the mechanism of protein folding. *Nature* 379:96, 1996.

Wolynes, P. G., Onuchic, J. N., and Thirumalai, D. Navigating the folding routes. *Science* 267:1619, 1995.

Proteomics and Bioinformatics

Eisenberg, D., Marcotte, E. M., Xenarios, I., and Yeates, T. O. Protein function in the postgenomic era. *Nature* 405:823, 2000.

Pandey, A. and Mann, M. Proteomics to study genes and genomes. *Nature* 405:837, 2000.

Techniques for the Study of Proteins

Bax, A. and Grzesiek, S. Methodological advances in protein NMR. *Acc. Chem. Res.* 26:131, 1993.

Reif, O. W., Lausch, R., and Fritag, R. High-performance capillary electrophoresis of human serum and plasma proteins. *Adv. Chromatogr.* 34:1, 1994.

Rhodes, G. *Crystallography Made Crystal Clear,* 2nd ed. San Diego,CA: Academic Press, 2000.

Dynamics in Folded Proteins

Daggett, V. and Levitt, M. Realistic simulations of native-protein dynamics in solution and beyond. *Annu. Rev. Biophys. Biomol. Struct.* 22:353, 1993.

Joseph, D., Petsko, G. A., and Karplus, M. Anatomy of a conformational change: hinged lid motion of the triosephosphate isomerase loop. *Science* 249:1425, 1990.

Karplus, M. and McCammon, J. A. Dynamics of proteins: elements and function. *Annu. Rev. Biochem.* 53:263, 1983.

Karplus, M. and McCammon, J. A. The dynamics of proteins. *Sci. Am.* 254:42, 1986.

Glycoproteins

Brockhausen, I. Clinical aspects of glycoprotein biosynthesis. *Crit. Rev. Clin. Lab. Sci.* 30:65, 1993.

Drickamer, K. and Taylor, M. E. Evolving views of protein glycosylation. *Trends Biochem. Sci.* 23:321, 1998.

Lis., H. and Sharon, N. Protein glycosylation. Structural and functional aspects. *Eur. J. Biochem.* 218:1, 1993.

Paulson, J. C. Glycoproteins: what are the sugar chains for? *Trends Biochem. Sci.* 14:272, 1989.

Lipoproteins

Gotto, A. M. Jr. Plasma lipoproteins. In: A. Neuberger and L. L. M. van Deenen (Eds.), *New Comprehensive Biochemistry.* Amsterdam: Elsevier, 1987.

Myers, G. L., Cooper, G. R., and Sampson, E. J. Traditional lipoprotein profile: clinical utility, performance requirement, and standardization. *Atherosclerosis* 108:S157, 1994.

Schumaker, V. N., Phillips, M. L., and Chatterton, J. E. Apolipoprotein B and low-density lipoprotein structure: implications for biosynthesis of triglyceride-rich lipoproteins. *Adv. Protein Chem.* 45:205, 1994.

Segrest, J. P., Garber, D. W., Brouillette, C. G., Harvey, S. C., and Anantharamaiah, G. M. The amphipathic α helix: a multifunctional structural motif in plasma apolipoproteins. *Adv. Protein Chem.* 45:303, 1994.

Collagen

Brodsky, B. and Shah, N. K. The triple-helix motif in proteins. *FASEB J.* 9:1537, 1995.

Kuivaniemi, H., Tromp, G., and Prockop, D. J. Mutations in collagen genes: causes of rare and some common diseases in humans. *FASEB J.* 5:2052, 1991.

Prockop, D. J. and Kivirikko, K. I. Collagens: molecular biology, diseases, and potentials for therapy. *Annu. Rev. Biochem.* 64:403, 1995.

QUESTIONS | C. N. ANGSTADT

Multiple Choice Questions

Refer to the drawing for Questions 1 and 2.

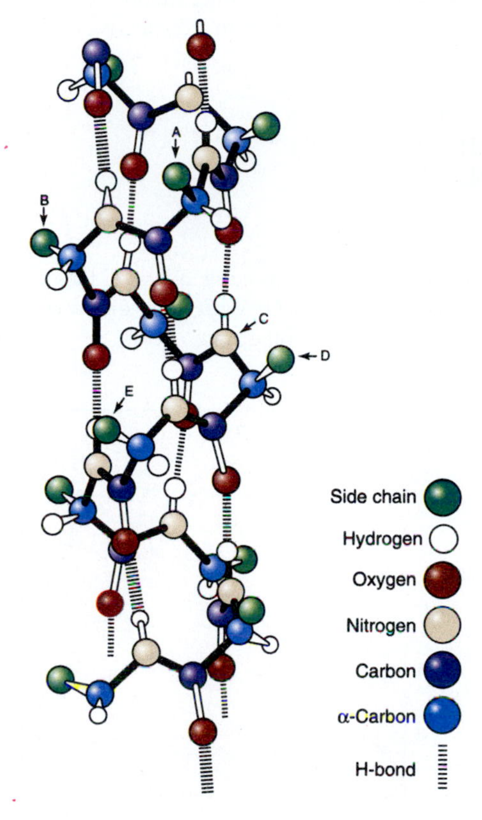

1. When group E contains a negatively charged carboxyl function, the structure is destabilized by:
   A. alanine at position A.
   B. arginine at position B.
   C. serine at position B.
   D. proline at position D.
   E. aspartate at position D.

2. The properties of atom C are essential to which stabilizing force in the structure?
   A. hydrogen bonding
   B. steric effects
   C. ionic attraction
   D. disulfide bridge
   E. none of the above

Refer to the following for Questions 3 and 4:

A. disulfide bond formation
B. hydrogen bonding
C. hydrophobic interaction
D. electrostatic interactions
E. van der Waals forces

3. Which interaction is NOT formed when protein subunits combine to form a quaternary structure?
4. Which is driven by the entropy of water?

5. In collagen:
   A. intrachain hydrogen bonding stabilizes the native structure.
   B. three chains with polyproline type II helical conformation can wind about one another to form a superhelix because of the structure of glycine.
   C. the ϕ angles contributed by proline are free to rotate, but the ψ angles are constrained by the ring.
   D. regions of superhelicity comprise the entire structure except for the N- and C-termini.
   E. cross-links between triple helices form after an intracellular enzyme converts some of the lysine to allysine.

6. Chaperone proteins:
   A. all require ATP to exert their effect.
   B. cleave incorrect disulfide bonds, allowing correct ones to subsequently form.
   C. guide the folding of polypeptide chains into patterns that would be thermodynamically unstable without the presence of chaperones.
   D. of the hsp70 class are involved in transport of proteins across mitochondrial and endoplasmic reticulum membranes.
   E. act only on fully synthesized polypeptide chains, unfolding incorrect structures so they can refold correctly.

7. Proteins may be separated according to size by:
   A. isoelectric focusing.
   B. polyacrylamide gel electrophoresis.
   C. ion-exchange chromatography.
   D. molecular exclusion chromatography.
   E. reverse-phase HPLC.

8. Changes in protein conformation can be detected rapidly by:
   A. ultraviolet absorbance spectroscopy.
   B. fluorescence emission spectroscopy.
   C. optical rotatory dispersion.
   D. circular dichroism.
   E. all of the above.

9. All lipoprotein particles in the blood have the same general architecture, which includes:
   A. a neutral core of triacylglycerols and cholesteryl esters.
   B. amphipathic lipids oriented with their polar head groups at the surface and their hydrophobic chains oriented toward the core.

C. most surface apoproteins containing amphipathic helices.
D. unesterified cholesterol associated with the outer shell.
E. all of the above.

10. Glycoproteins:
A. are found in cells but not in plasma.
B. if located in a plasma membrane, typically have the carbohydrate portion on the cytosolic side.
C. may have the carbohydrate portion covalently linked to the protein at an asparagine residue.
D. that have the carbohydrate portion linked to a hydroxyl group in the protein always have the linkage to hydroxylysine.
E. of a given type always have identical carbohydrate chains.

Questions 11 and 12: More than 300 variants of human hemoglobin have been discovered, about 95% of these resulting from a single amino acid substitution. Many of these show no clinical effects while others cause serious abnormalities. One of the best studied variants is HbS, which causes sickle cell disease. In HbS, glutamate in the sixth position of the β-chain of HbA (normal adult hemoglobin) has been replaced by a valine. This replacement causes HbS, in its deoxy form, to polymerize with other HbS molecules. The result is a change in red cell structure to a sickle shape, which leads to clogging of capillaries. Some other hemoglobin variants are Hiroshima (His 146 $\rightarrow$ Asp), Riverdale–Bronx (Gly 24 $\rightarrow$ Asp), and $M_{\text{Hyde Park}}$ (His 92 $\rightarrow$ Tyr).

11. Which of the following statements is correct?
A. The substitution in HbS is conservative.
B. All four hemoglobins have nonconservative substitutions.
C. HbS has a nonconservative substitution but the other three have conservative substitutions.
D. The substitution in $M_{\text{Hyde Park}}$ would not be likely to cause a change in structure.
E. All amino acid substitutions in proteins cause changes in the protein's function.

12. HbS polymerizes with other molecules because:
A. the structure shifts to bury valine in the interior of the protein.
B. the C-terminal end of the chain has been modified.
C. a positive charge has been introduced on the surface to attract negative charges on other molecules.
D. the introduction of a hydrophobic group on the surface is attracted to other hydrophobic groups on other HbS molecules.
E. valine has a strong tendency to hydrogen bond.

Questions 13 and 14: There are a number of pathological conditions known as hyperlipidemias that result from abnormalities in the rate of synthesis or the rate of clearance of lipoproteins in the blood. They are usually characterized by elevated levels of cholesterol and/or triacylglycerols in the blood. Type I is characterized by very high plasma triacylglycerol levels (>1000 g dL^{-1}) because of an accumulation of chylomicrons. Type II (sometimes called familial hypercholesterolemia) is characterized by elevated cholesterol, specifically in the form of LDL. Another abnormality of lipoproteins is hypolipoproteinemia in which lipoproteins are not formed because of the inability to make a particular apoprotein.

13. If the serum of a patient with type I hyperlipidemia were centrifuged, the lipid band would be found:
A. at the top of the tube.
B. just below the top of the tube.
C. below the top but still in the upper half of the tube.
D. about the middle of the tube.
E. near the bottom of the tube.

14. Abetalipoproteinemia is a disease in which chylomicrons, VLDLs, and LDLs are absent from the blood. Based on the fact that it is present in significant (>20%) concentration in all three of these particles, the missing protein in this disease is:
A. ApoA-I.
B. ApoB-100.
C. ApoC-II.
D. ApoD.
E. ApoE.

Problems

15. The figure shows the titration curve of one of the common amino acids. What is the amino acid? What point on the curve represents the p*I* (isoelectric point) of the amino acid?

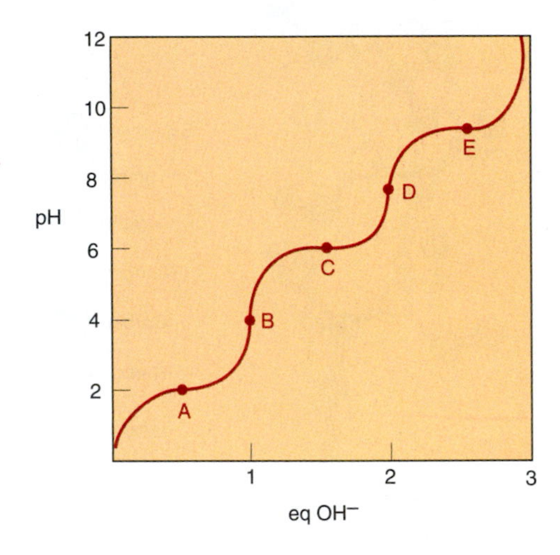

16. After proper purification, the Edman reaction was used to sequence a dodecapeptide. The following data were obtained. The C-terminal amino acid is isoleucine; N-terminal amino acid is methionine; peptide fragments are Ala-Ala-Ile, Leu-Arg-Lys-Lys-Glu-Lys-Glu-Ala, Met-Gly-Leu, and Met-Phe-Pro-Met. What is the sequence of this peptide?

ANSWERS

1. **D** Proline destabilizes the α-helix conformation and is usually not found in α-helix structures. A: Alanine has a small side chain. B: Arginine is neutral and should not destabilize the helix. E: Like charges in the third or fourth position in either direction from the designated position destabilize the helix due to charge repulsion. Thus aspartate at position D is harmless, whereas glutamate at position A or B would destabilize.
2. **A** Atom C is an amide nitrogen. The attached hydrogen atom participates in hydrogen bonding. Hydrogen bonds contribute to the stability of the structure.

3. **A** Quaternary structure is stabilized exclusively by noncovalent interactions. Disulfide bonds are covalent.
4. **C** Hydrophobic groups in contact with water result in formation of a relatively highly ordered solvation shell of water around the group. If the hydrophobic groups come together, eliminating the bound water, the water becomes more random, a favorable process.
5. **B** The close contacts in the interior of the triple helix are possible only when the R group of the amino acid at that position is very small, that is, hydrogen. A: The hydrogen bonding in collagen is interchain. C: The ϕ angle is part of the proline ring and is not free to rotate. D: Although the statement is true of type I collagen, the superhelical regions in other collagen types may be broken by regions of globular domains. E: The conversion and cross-linking are extracellular.
6. **D** The proteins cross the membrane in an unfolded state and refold once they cross the membrane. A: The hsp60 family of chaperones is ATP-linked, but the hsp70 family is not. B: Disulfide isomerases catalyze this reaction. C: The final product is thermodynamically stable; chaperones merely prevent unfavorable intermediate interactions. E: Hsp70 chaperones react with nascent polypeptide chains as they are synthesized by the ribosome. The protein may then be delivered to a hsp60 chaperone for facilitation of final folding.
7. **D** Another method that separates on the basis of size is SDS PAGE. A–C separate molecules on the basis of charge. E: Reverse-phase HPLC effects separations on the basis of polarity.
8. **E** A: Peptide bond absorption (180–230 nm) in the α-helical conformation differs from that in other conformations. B: Excitation transfers become less efficient as donor and acceptor groups become further apart, as in denaturation. C, D: These effects of optically active chromophores upon polarized light are sensitive to environment; in addition, the peptide bond itself becomes part of an optically active system when it forms an asymmetric structure like the α-helix.
9. **E** All lipoproteins share these characteristics. C: The polar face interacts with water and the hydrophobic face is oriented toward the core. D: Cholesterol's hydroxyl group is sufficiently polar that it orients toward the outer shell.
10. **C** This is the N-linked type and the asparagine must be in the sequence Asn-X-Thr(Ser). A: Most plasma proteins, except albumin, are glycoproteins—for example, the blood determinants and immunoglobulins. B: The carbohydrate is on the outside of the cell because its main functions include cell–cell recognition and formation of cellular receptors that bind extracellular proteins. D: This link is in collagen only. Other O-linked carbohydrates have the linkage to serine or threonine. E: Carbohydrate structure is not determined by genes and is variable.
11. **B** A nonconservative substitution is one in which the polarity changes significantly. A: Negative charge to hydrophobic group is definitely nonconservative. C: Hiroshima is positive to negative, Riverdale–Bronx is slightly hydrophobic to positive, and Hyde Park is positive to very hydrophobic. D: This mutation has been shown to lead to the loss of the ligand bond to the iron heme and the loss of the heme from the heme-binding site in the beta chain. E: Changes in function depend on many factors, including whether the substitution changes the structure and whether it is in a critical position. All of the variants indicated in this problem happen to be pathological.
12. **D** These have sometimes been referred to as "sticky ends" since valine is strongly hydrophobic and attracts other hydrophobic groups. A: Glutamate's original position is on the exterior and this is where the valine remains. B: Chains are numbered from the N-terminal so position 6 is near the N-terminal. C: There is no positive charge. E: Valine cannot hydrogen bond.
13. **A** Chylomicrons have a density less than water so float on the surface. B: VLDLs might be found here. C: This band would be LDL. D: Probably none of the lipoproteins remain at this point. E: HDLs are found near the bottom of the tube because they are the most dense.
14. **B** ApoB-100 is the protein recognized by LDL receptors. Chylomicrons have the smallest amount but it is still at least 20%. A: This is the major protein of HDL. C: ApoC-II is present in all three particles to activate lipoprotein lipase but accounts for only 2–15% of total protein. D: This is present only in chylomicrons and then only at 1%. E: This is found in all three particles but at 5–20% level.
15. Look carefully at the labels on the axes in doing your analysis. Points A, C, and E represent the p*K* values of three dissociable groups, the points where each of the groups is half-titrated. They have values of ~2, ~6, and ~9.5 so the amino acid must be histidine. At point B, the charge is $-1 + 1 + 1 = +1$. At D the charge is $-1 + 0 + 1 = 0$. Point D must be the isoelectric point.
16. You must have the correct N- and C-terminal amino acids and the correct total number (12) of amino acids in the peptide. Keeping these criteria in mind and overlapping the fragments you should get the sequence: Met-Phe-Pro-Met-Gly-Leu-Arg-Lys-Glu-Ala-Ala-Ile. Note that the Lys-Glu fragment doesn't give any additional information for the sequencing.

PART II

TRANSMISSION OF INFORMATION

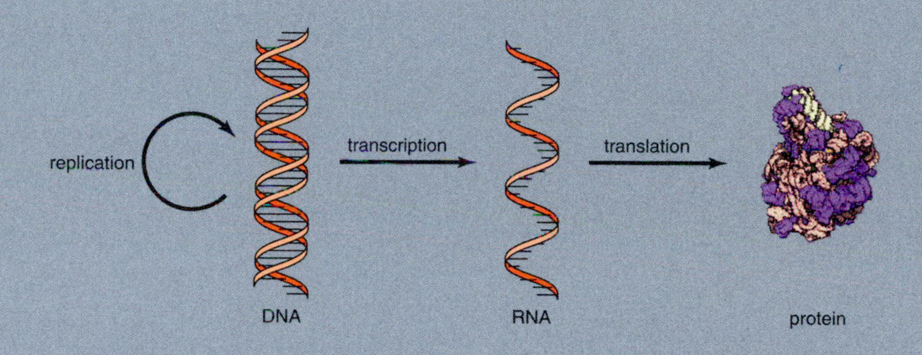

Transmission of information is an absolute requirement of cells. The information encoded in DNA must be replicated into new DNA with a high degree of fidelity for transmission to daughter cells. Encoded in DNA is the information to synthesize proteins. The information is transcribed into various forms of RNA. One form of RNA serves as a template to translate the information into a specific sequence of amino acids of a protein. Other forms of RNA function in the mechanism of protein synthesis. The following chapters describe the processes of replication, transcription, and translation, how they are controlled, and how humans are able to manipulate these processes to alter the genetic make up of cells.

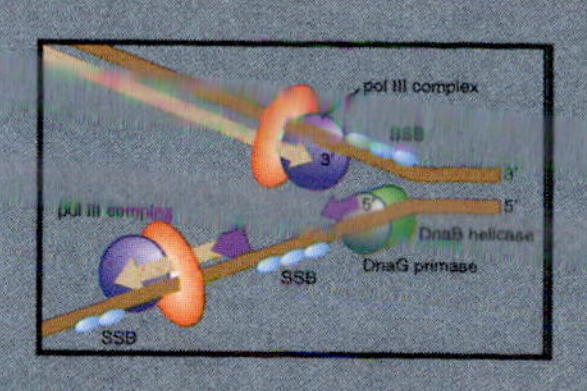

4

DNA REPLICATION, RECOMBINATION, AND REPAIR

Howard J. Edenberg

4.1 | COMMON FEATURES OF REPLICATION, RECOMBINATION, AND REPAIR

In addressing "the three R's" of DNA—replication, recombination, and repair—it is useful to consider some unifying similarities among them. Because these processes act on the same substrate, DNA chains, they share fundamental chemical mechanisms of making and breaking phosphodiester bonds. The chemistry of DNA is largely that of these phosphodiester bonds. The theme of a nucleophilic attack on a **phosphodiester bond,** leading to a switch in the partners across the bond, recurs in all three processes. Other key features of DNA that influence these processes are the double-stranded nature of cellular DNA and complementarity of two strands. Thus information is present in two copies, and each strand can serve as a template for synthesis or repair of the opposite strand. The antiparallel nature of the two strands is critical in all processes that act on DNA. Because replication, recombination, and repair are critical for all cells, their fundamental mechanisms are conserved throughout evolution. Similarities among replication, repair, and recombination often lead to use of the same enzyme in different processes. Requirements of accuracy and regulation, however, sometimes lead to specialization of a group of enzymes for a particular task. Recognizing similarities and differences will make understanding the processes easier.

As the carrier of genetic information, DNA in a cell must be duplicated, maintained, and passed down to daughter cells accurately. The scale of the problem is large. The common bacterium *Escherichia coli* (***E. coli***) has a genome containing 4.6 million base pairs (bp). Individual human **chromosomes** each contain a single DNA molecule; sizes range from 34 million base pairs for the smallest chromosome to 263 million base pairs for the largest. The human genome consists of two copies of 23 chromosomes, totaling approximately 6×10^9 bp. Overall accuracy of replication is extremely high. During a single cycle of replication, approximately 1 error is introduced per billion base pairs. But even with this extraordinary accuracy, a handful of errors is introduced into an average human cell at each cell division. This results in mutations, heritable changes in the DNA sequence. Mutations are generally harmful; they can affect cell viability or trigger the uncontrolled growth characteristic of cancer. There are situations in which the change of a single base pair in the 6×10^9 bp human genome can cause serious disease. Eukaryotes have the additional task of ensuring that all parts of the genome are replicated only once during each cell cycle.

Accurate replication is not enough. DNA in a cell is subject to continuing attack by both chemical and physical agents. These include water, activated oxygen species that are by-products of normal cellular metabolism, chemicals in foods and in the environment, and radiation. There are systems that recognize and repair DNA damage and thereby help maintain the integrity of the genome. In spite of all this care, accumulation of mutations can lead to cell death, aging, and cancer.

4.2 | DNA REPLICATION

The Basics

Replication leads to doubling of the DNA, preserving the genetic information carried as the sequence of bases. Replication requires a **template** to provide sequence information. The basic mechanism of DNA replication was obvious as soon as the complementary, double-stranded structure of DNA was recognized by Watson and Crick in 1953. Because each strand is the complement of the other, with an A in one strand always paired with a T in the other and likewise for C and G, the two strands can be separated and each can be used as a template for synthesis of a new complement. This is called **semiconservative replication,** because half of the

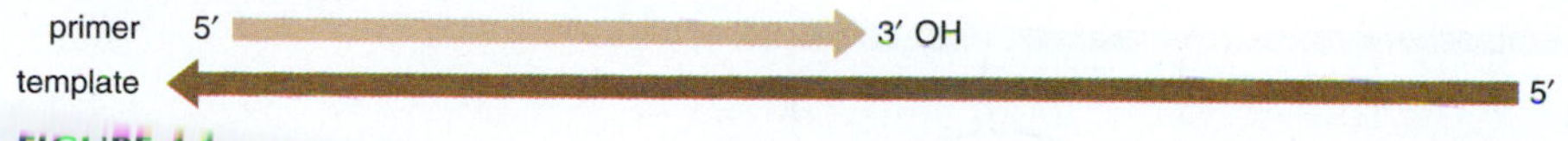

FIGURE 4.1
Template and primer.
Substrate for addition of nucleotides is a primer (light tan) hydrogen-bonded to a template (dark tan). Primer provides a free 3′-hydroxyl residue to which nucleotides can be added. The two strands run antiparallel.

parental DNA molecule (one strand) is conserved in each new double helix, paired with a newly synthesized complementary strand.

A template alone, however, is not sufficient for DNA synthesis. **DNA polymerases,** the enzymes that catalyze addition of mononucleotides, require **primers.** DNA polymerases do not join the first two nucleotides to start a strand. Rather, they add nucleotides to an existing primer that provides a 3′-hydroxyl residue (Figure 4.1, Table 4.1). The requirement for a primer is not due to the chemistry of phosphodiester bond formation, a fact demonstrated by the ability of **RNA polymerases** to start polynucleotide chains *de novo* (without primers). It probably evolved to increase the accuracy of the process, because proofreading of the first few nucleotides is not likely to be effective. Marking these nucleotides, by making the primer RNA, allows their subsequent identification and removal, after which the gap can be resynthesized more accurately.

Chemistry of Chain Elongation

The chemistry of DNA replication, repair, and recombination is largely that of the **phosphodiester bonds** that link neighboring nucleotides along each DNA chain. Figure 4.2 depicts formation of a phosphodiester bond between a short chain of DNA (a **primer**) and an incoming nucleotide. This process is repeated for each nucleotide added to the growing chain.

Addition of a mononucleotide to a growing chain is not a spontaneous process, because the decrease in entropy is large. For this reason, nucleotide precursors of DNA are 5′-deoxyribonucleoside triphosphates (5′-dNTPs). The phosphodiester bond connecting the first (α) phosphate (attached to the 5′ carbon of the sugar) and the outer two (β and γ) phosphates undergoes nucleophilic attack by the 3′-OH of the growing DNA chain (Figure 4.2). The bond is not hydrolyzed, but rather **transesterified:** the phosphodiester bond that had joined the α-phosphate to the β-phosphate now joins the α-phosphate to the 3′ end of the growing chain. The terminal two phosphates are released as inorganic pyrophosphate, which is hydrolyzed by phosphodiesterases present in cells. Cleavage of inorganic pyrophosphate renders the reaction essentially irreversible. Much of the chemistry of replication, repair, and recombination involves transesterification reactions like this.

DNA Polymerases

DNA polymerases catalyze addition of nucleotides during chain elongation. DNA polymerases require both templates and primers, and use 5′-dNTP precursors, as shown in Figure 4.2. Cells have multiple DNA polymerases that carry out specialized

TABLE 4.1 Requirements of DNA Replication

Template	Provides sequence information
Primer	Provides free 3′-OH to which nucleotides are added
Precursors	5′-Deoxynucleoside triphosphates (5′-dNTPs)
Enzymes	DNA polymerases, sliding clamps, helicases, primases, single-stranded DNA binding proteins, nucleases, ligases

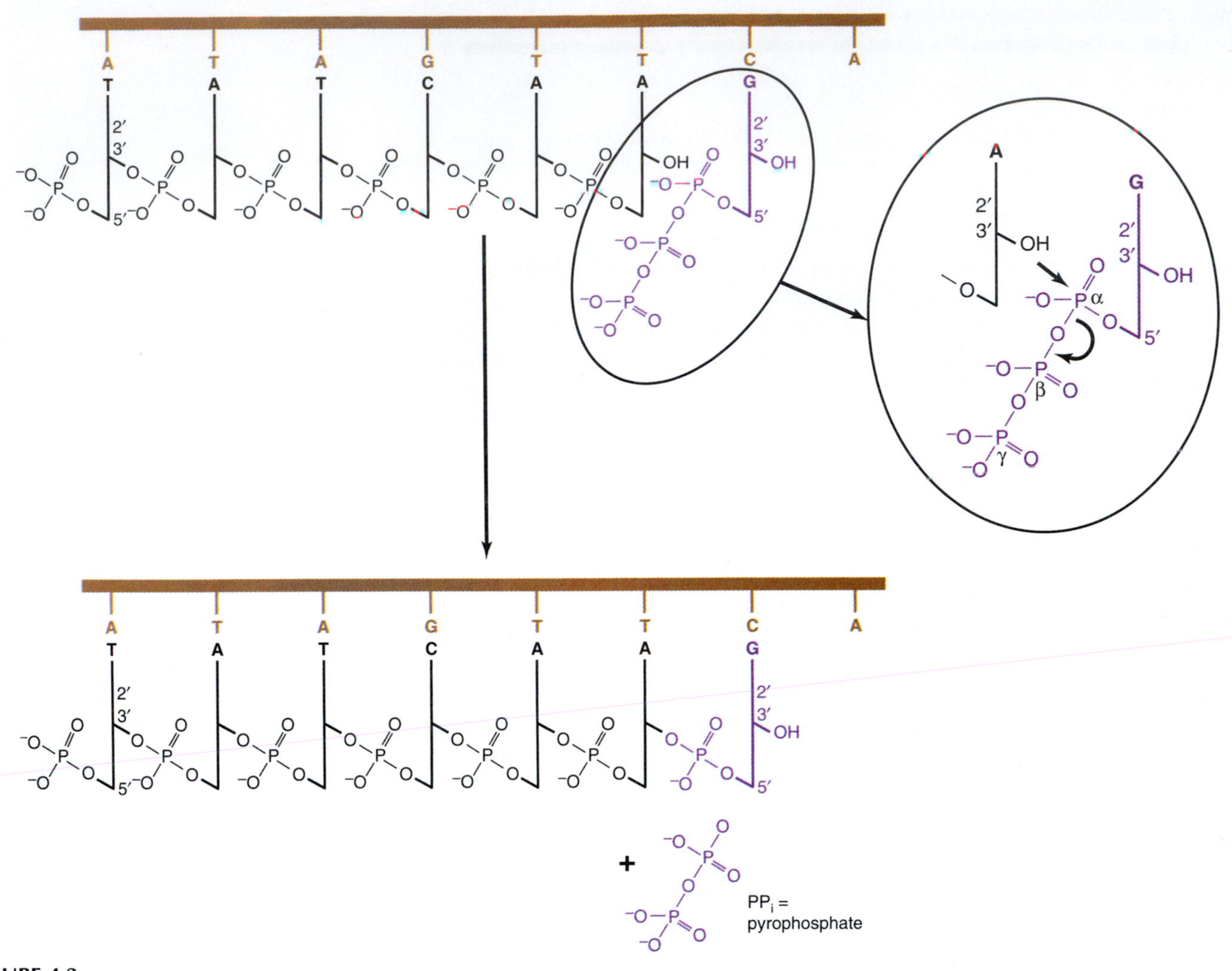

FIGURE 4.2
Formation of a phosphodiester bond.
An incoming nucleoside triphosphate that forms correct hydrogen bonds with template in the active site of a DNA polymerase is depicted in purple. 3′-OH of primer attacks innermost (α) phosphate of nucleotide, displacing a pyrophosphate, as shown in inset. This results in the elongation of primer by one nucleotide. This cycle is repeated as long as template is available.

functions (Table 4.2). DNA polymerases help to ensure accuracy in two ways: initial selection of the proper nucleotide to add and enzymatic proofreading.

The initial selection is based on the fit of incoming nucleotide into the enzyme's active site. An incoming nucleotide that makes the proper hydrogen bonds with the nucleotide on the template strand can be added to the growing chain. An incoming nucleotide that does not make the correct hydrogen bonds is not aligned properly for catalysis. This leads to accurate synthesis with error rates in the range of 10^{-4}. Although good for many processes, this is not nearly as accurate as required for replication.

The second way DNA polymerases increase accuracy is by a **proofreading** step. This is carried out by a 3′ to 5′ **exonucleolytic activity** that removes mispaired nucleotides from the 3′ end of the chain. This activity is integral to many replicative DNA polymerases, but in other cases it is carried out by an associated protein in the replication complex. A primer with a properly base-paired terminal nucleotide is a good substrate for addition of the next nucleotide, and a poor substrate for the 3′ to 5′ exonuclease (Figure 4.3). A primer with a mismatched terminal nucleotide is a poor substrate for addition of a nucleotide and a good substrate for 3′ to 5′ exonuclease. This combination results in removal of virtually all mismatched nucleotides before the next nucleotide is added. It also leads to removal of some

TABLE 4.2 DNA Polymerases

Enzyme	*Function*
Escherichia coli	
Pol I	Completion of Okazaki fragment, DNA repair
Pol II	Damage bypass, DNA repair
Pol III	Major replication polymerase
Pol IV	Damage bypass
Pol V	Damage bypass
Eukaryotic	
Pol α	Priming replication
Pol β	Base excision repair
Pol γ	Mitochondrial DNA replication
Pol δ	Major replication polymerase
Pol ε	(Unclear; replication and repair)
Pol ξ	Damage bypass
Pol η	Damage bypass
Pol ι	Damage bypass

properly incorporated nucleotides, but that is a price that must be paid for the increased accuracy provided by proofreading. The ability to discriminate between properly and incorrectly incorporated nucleotides is compromised when the growing chain is very short, because the strength of base pairing between a short chain and template is weak. This is the most likely explanation for the evolution of DNA polymerases unable to start chains *de novo*.

DNA polymerases did not evolve ever-increasing accuracy of proofreading. Some mutated DNA polymerases are more error-prone and lead to a higher mutation rate, but other mutated DNA polymerases lead to fewer errors. This is paradoxical if one assumes that there is an evolutionary advantage to ever-increasing accuracy. However, there are reasons to expect that beyond a certain level, higher accuracy is not advantageous. First, extremely accurate proofreading is energetically costly. To be sure that virtually all misincorporated bases are removed, a polymerase must also remove a large number of correctly incorporated bases that have moved slightly with respect to the template due to normal thermal fluctuations. This would slow replication and utilize additional energy in the form of dNTPs. Second, the ability of DNA repair systems to deal with residual errors reduces the advantages of ever more accurate polymerases. It has also been argued that a low level of muta-

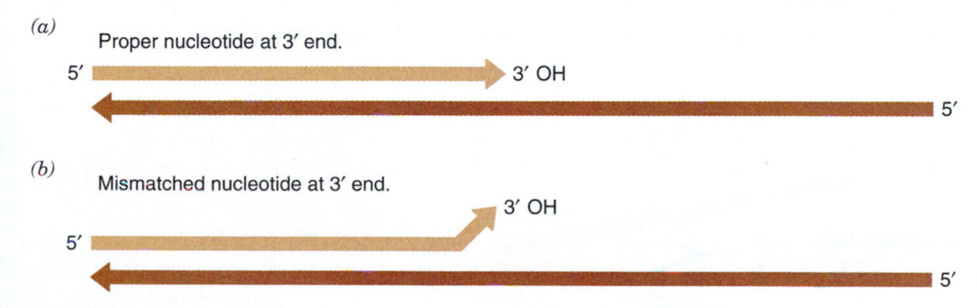

FIGURE 4.3
Proofreading.
(*a*) When the proper nucleotide is added to the 3′ end of primer, the result is a good substrate for addition of a nucleotide, but a poor substrate for 3′ to 5′ exonuclease. (*b*) On rare occasions when an incorrect nucleotide is added to primer, the resulting primer with a mismatch at the end is a poor substrate for addition of a nucleotide, but a good substrate for 3′ to 5′ proofreading exonuclease that removes the mismatched nucleotide. This leaves primer shorter but ready for a correct nucleotide to be incorporated.

tions provides the raw material for evolution, producing a population with some genetic variation; such populations are better able to survive changing conditions.

Despite their very high accuracy, polymerases can incorporate nucleotide analogs. Nucleoside analogs are often used in chemotherapy to kill rapidly growing cancer cells or viruses. Analogs that are phosphorylated to nucleotides can be incorporated into DNA, where they can inhibit further synthesis or lead to a high level of mutation. Differences in the ability of bacterial or viral polymerases to incorporate nucleotide analogs can provide a therapeutic window, allowing physicians to target infected cells.

Replication Fork Movement

Separation of parental strands that allows each to serve as a template for a new complementary strand creates a structure called a **replication fork** (Figure 4.4). Recall that the paired strands of DNA run in antiparallel directions. Thus one daughter strand is oriented with the 3′ end toward the fork and the other strand has its 5′ end oriented toward the fork. This complicates the process of replication. Both prokaryotes and eukaryotes solve problems of replication in fundamentally similar ways. The same types of enzymatic activities are employed, but the enzymes associate into different complexes. We will first address what happens to the DNA, then the protein machinery involved.

The Polarity Problem

Antiparallel strands at a replication fork present an immediate problem for the process of replication (Figure 4.5*a*). The strand with its 3′-OH oriented toward the fork can be elongated simply by sequential addition of new nucleotides to this end; this strand is called the **leading** or **continuous** strand. The other strand presents the problem: no DNA polymerase catalyzes addition of nucleotides to the 5′ end of a growing chain. Yet when viewed on a larger scale, replication appears to proceed along both strands in the direction of fork movement. This problem is solved by synthesizing one strand as a series of short pieces, each made in the normal 5′ to 3′ direction, and later joining them (Figure 4.5*b*). This is called **semidiscontinuous** synthesis, and the strand made in pieces is called the **discontinuous** or **retrograde** (backward going) strand. The small pieces from which the retrograde strand is made are called **Okazaki fragments** in honor of Reiji Okazaki, who first demonstrated this process. Okazaki fragments in human cells average about 130–200 nucleotides (nt) in length. In *E. coli*, they are about ten times larger.

Priming

DNA polymerases require **primers.** The leading (continuous) strand needs only a single priming event, which occurs during replication initiation (below). But each Okazaki fragment of the retrograde strand requires a separate primer. These primers

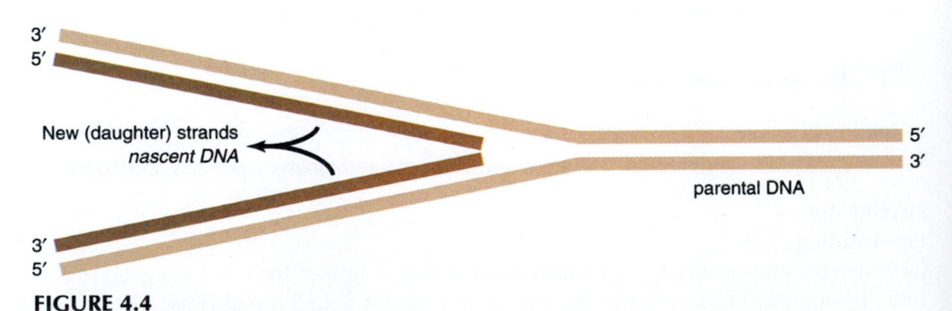

FIGURE 4.4
Semiconservative DNA synthesis.
DNA synthesis occurs at a replication fork. Parental strands separate, and each serves as template for synthesis of a new (daughter) strand. Result is a pair of duplexes each containing one old and one new strand.

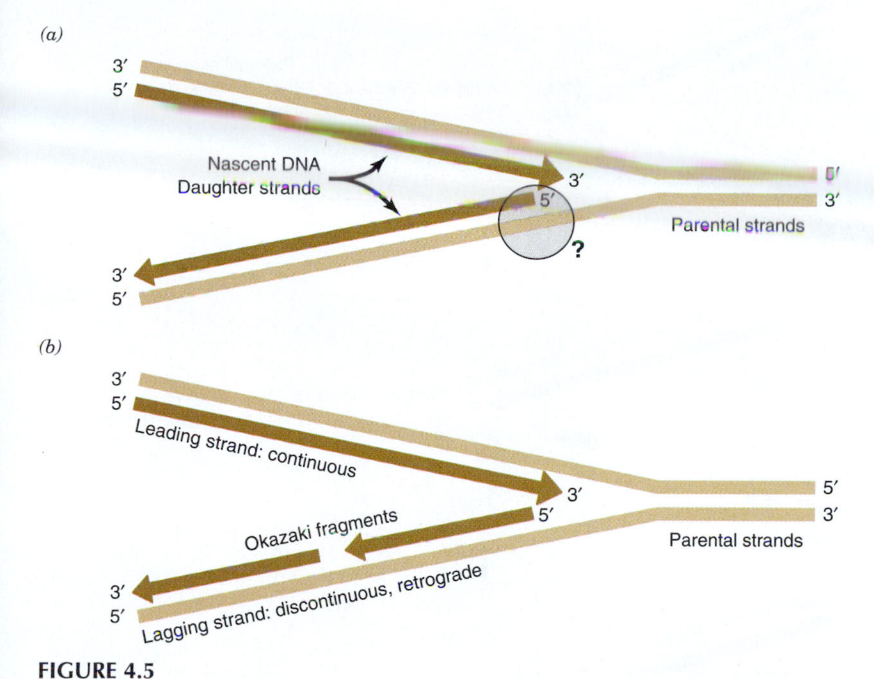

FIGURE 4.5
The polarity problem.
(*a*) The problem: The replication fork moves with both sides replicated nearly simultaneously. DNA strands are synthesized only from the 5′ end toward the 3′ end; no polymerase synthesizes 3′ to 5′. But the DNA strands are antiparallel. How can the strand with its 5′ end toward the replication fork be elongated? (*b*) The solution: Semidiscontinuous DNA synthesis. Both strands are elongated at their 3′ ends. The one with its 3′ end toward the fork can grow continuously; it is called the leading strand. The one with its 5′ end toward the fork is made as a series of short pieces, called Okazaki fragments, each synthesized in the 5′ to 3′ direction and later joined. This is called the lagging, discontinuous, or retrograde strand.

are short stretches of RNA, synthesized by a special enzyme called a **primase.** In eukaryotic cells, the RNA primers are about 8–10 nucleotides long. They provide a free 3′-OH to which the first deoxyribonucleotide can be added covalently (Figure 4.6).

Synthesis

DNA chains grow by repeated cycles of addition of nucleotides to the 3′-OH ends of the chains by the mechanism shown in Figure 4.2. This reaction is catalyzed by a DNA polymerase.

Primer Removal

Newly synthesized strands of DNA do not contain short stretches of RNA where the primers were. The RNA primers are removed from the 5′ ends of Okazaki fragments by enzymes with **RNaseH (RNA hybridase)** activity. RNaseH catalyzes hydrolysis of an RNA chain hydrogen bonded to a DNA chain (i.e., an RNA–DNA hybrid). All of the ribonucleotides are removed, leaving only DNA.

Gap Filling

Removal of RNA primers leaves a **gap.** Synthesis must fill the gap with deoxyribonucleotides, leaving only a **nick.** Note the terminology: a gap means that at least one nucleotide is missing. A nick is an interruption in the phosphodiester backbone with no missing nucleotides. The gap is filled with deoxyribonucleotides by a DNA polymerase, using the more recently synthesized Okazaki fragment as primer. The Okazaki fragment provides a secure primer–template combination that allows accurate proofreading.

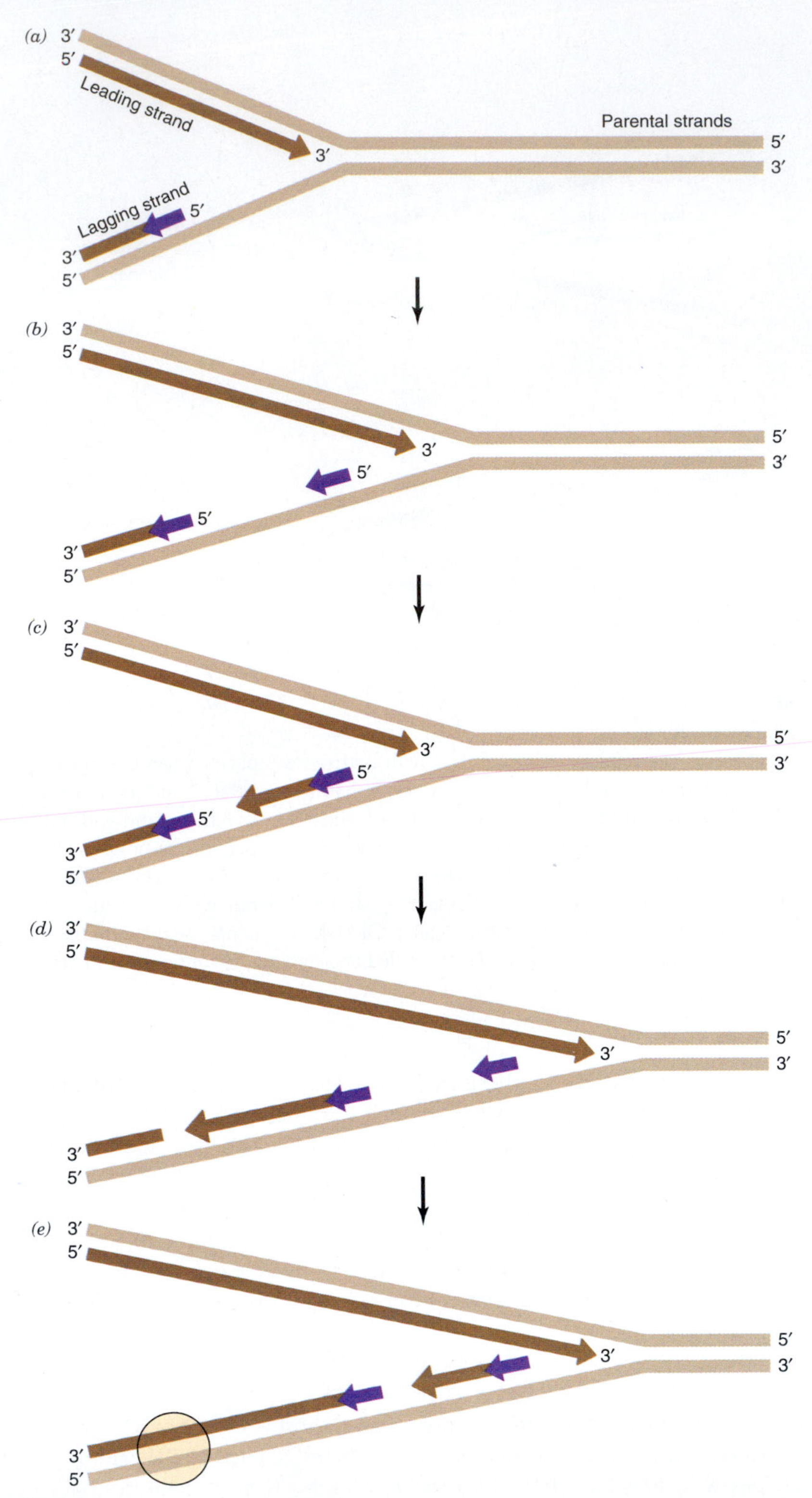

FIGURE 4.6

DNA synthesis at a replication fork.

(*a*) The leading strand is elongated at its 3′ end by repeated cycles of nucleotide addition as shown in Figure 4.2. A previously made Okazaki fragment with a short RNA primer is shown on the other side. (*b*) Using the template for the lagging strand, primase synthesizes an RNA primer (purple). (*c*) A DNA polymerase covalently extends the 3′ end of the RNA primer, incorporating deoxyribonucleotides (tan) as shown in Figure 4.2. (*d*) As the growing Okazaki fragment approaches a previously synthesized fragment, the RNA primer on the older fragment is removed, and the gap is filled by a DNA polymerase that elongates the Okazaki fragment. Meanwhile, primase has synthesized a new primer as the fork opens further. (*e*) When the gap has been filled and only a nick remains, a DNA ligase seals the two fragments together (within circle). The cycle will be repeated, with the new primer elongated, the old primer removed, the gap filled, and the nick sealed.

Ligation

The remaining nick is sealed by a **DNA ligase.** Sealing of a nick requires formation of a new phosphodiester bond. It therefore requires energy, supplied by coupling the reaction to the splitting of ATP or NAD^+. In the first step, an AMP residue is attached to the enzyme (Figure 4.7). This is transferred to the 3′ end of the nick and then serves as a good leaving group that is displaced by the 3′-OH at the other side of the nick.

Separating Parental Strands

In order to serve as templates, parental DNA strands must be separated. This requires considerable input of energy. Melting double-stranded DNA into two single strands normally occurs only at elevated temperatures, usually over 90°C. To separate the parental strands at physiological temperatures, cells use enzymes called **helicases.** Helicases bind to single-stranded DNA and move along it in a fixed direction, with each step requiring hydrolysis of ATP. This "pushes apart" the parental DNA. In the absence of additional proteins, parental strands would quickly reanneal behind the helicase, because the complementary strands are close and in proper register. This reannealing is prevented by **single-stranded DNA binding proteins (SSBs).** SSBs bind to single-stranded DNA, keep the strands apart, reduce potential secondary structure (hairpins that might impede polymerization), and align the template strands for rapid DNA synthesis. SSBs are important not only in replication but also in recombination and repair.

Untwisting Parental Strands and Topological Problems

The above discussion and diagrams were simplified by depicting strands as straight lines. They omitted any reference to the need to untwist parental strands, which wrap around each other approximately once every 10.5 bp (Figure 4.8). Cells don't have the option of ignoring this twisting. In the bacterium *E. coli,* for example, the genome can replicate in approximately 40 minutes. To untwist the 4.6×10^6 bp DNA completely in 40 minutes requires rotation at approximately 12,000 rpm! Clearly, the entire chromosome, compactly folded within the cell and carrying large replication (and transcription) complexes, cannot simply rotate without whipping the contents.

Another problem caused by the double-helical nature of DNA is **topological:** in circular DNA (such as the *E. coli* chromosome, mitochondrial DNA, many viruses, and plasmids) or long linear DNA whose ends are not free to rotate around each other (such as in our chromosomes, bound at intervals to the nuclear matrix), the number of times one strand wraps around the other, called the **linking number**, is fixed. The linking number cannot be altered without breaking at least one of the two strands. In DNA molecules, the wrapping is of two types, the "Watson–Crick" **twisting** of one strand around the other approximately every 10.5 bp and the coiling of the double helix on its axis (**writhing** or **supercoiling**). Topologically, both are equivalent and can be interconverted; it is only the total number of links, the linking number, that must remain constant as long as both strands are intact.

One way to allow untwisting and reduction in linking number would be to nick a parental strand ahead of the replication fork. This allows one strand to rotate around the other. A nick presents a serious danger, however, because if the replication fork reached the nick it would be transformed into a double-strand break in DNA. A double-strand break presents a serious threat to genome stability; if unrepaired, it is lethal. Enzymes, called **topoisomerases,** solve the unlinking problem by catalyzing changes in the linking number that allow the untwisting and eventual separation of parental strands. Topoisomerases act by forming a transient break in the DNA backbone and then resealing it. This break is not formed by hydrolysis of the sugar–phosphate bond, but by a transesterification reaction that creates a phosphate–enzyme bond as a transient intermediate. Rejoining of the backbone phosphodiester bond displaces the enzyme (Figure 4.9). Thus there is no net loss or creation of phosphodiester bonds, just a switch in partners across the bond.

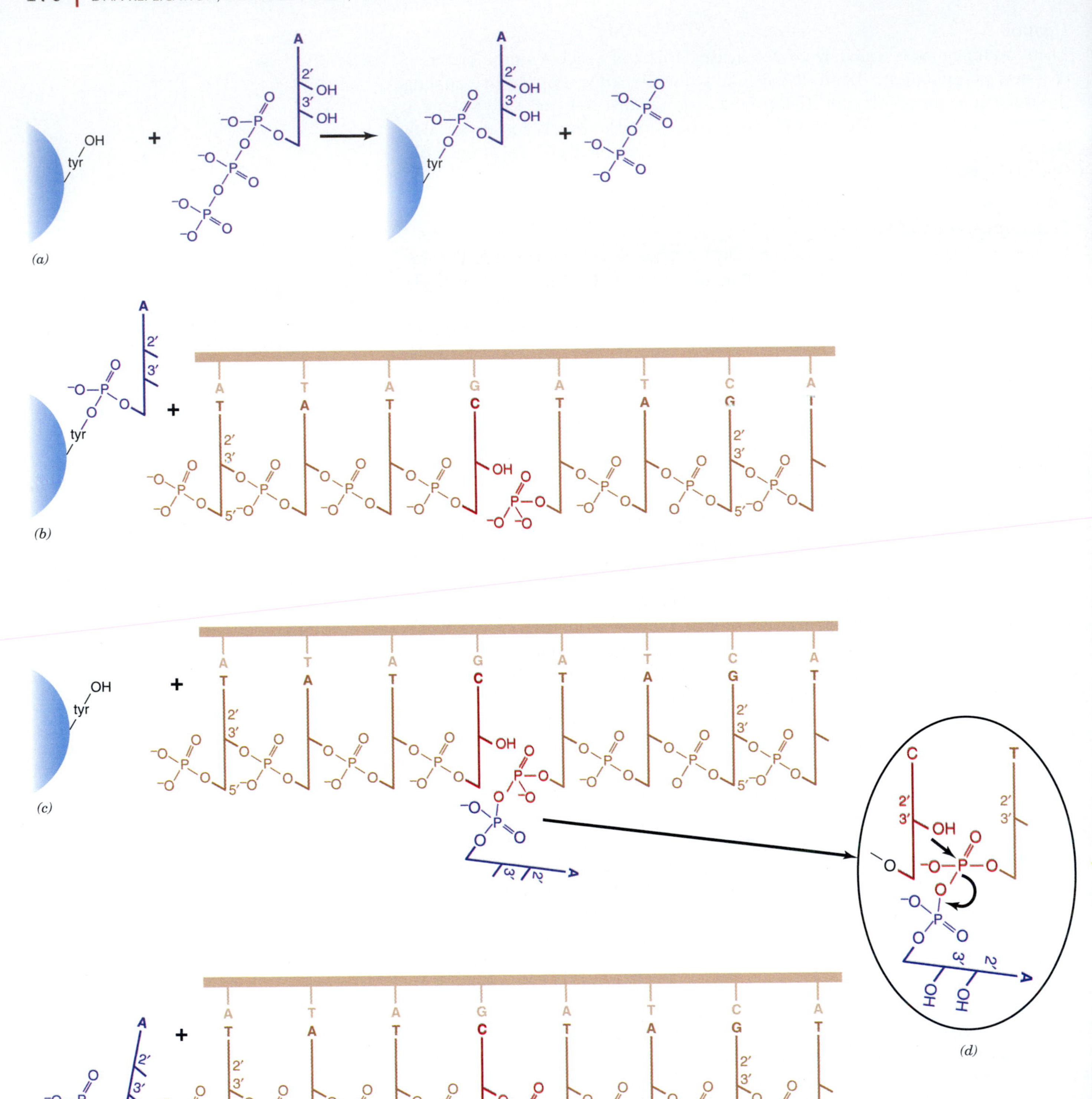

FIGURE 4.7
DNA ligase mechanism.
(*a*) DNA ligase first catalyzes the addition of an AMP residue to a tyrosine on the enzyme. The AMP donor can be ATP (split to AMP + pyrophosphate, as shown here) or NAD^+ (split to AMP + NMN). (*b*) Ligase–AMP complex binds to a nicked DNA duplex (broken backbone shown in red). (*c*) Ligase transfers the AMP residue onto the 5′ phosphate at a nick. (*d*) Inset: 3′-OH at the other side of a nick attacks the phosphodiester bond closest to 5′ carbon to form a new phosphodiester bond in the DNA backbone and release AMP. This attack is similar to that which occurs during DNA synthesis (Figure 4.2), but with an AMP as the group that leaves, rather than a pyrophosphate. (*e*) Result is a resealed phosphodiester backbone, at a net "cost" of an ATP split into an AMP + PP_i.

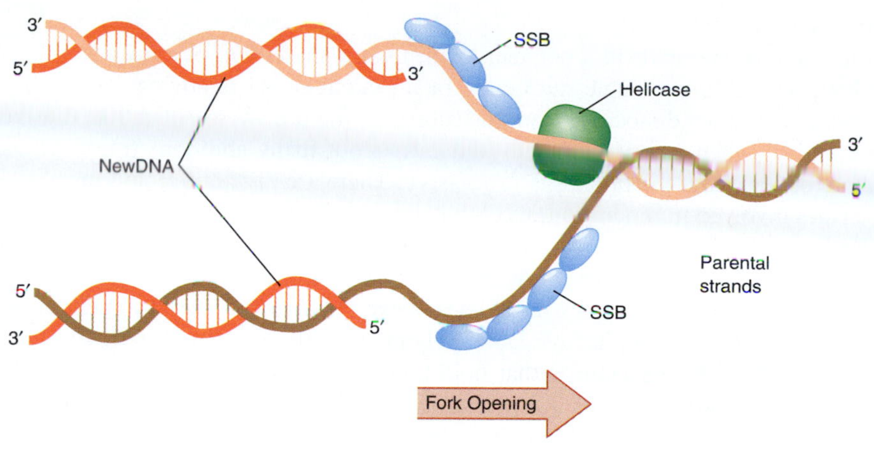

FIGURE 4.8
Separation of parental strands.
A helicase is needed to separate parental strands under physiological conditions. Single-stranded DNA binding proteins (SSB) are needed to keep the strands apart. But parental strands twist around each other approximately every 10.5 bases, and must untwist to separate. This creates topological problems.

This means that "nicking" and "resealing" can occur without the need for coupling with a high-energy intermediate (e.g., hydrolysis of ATP or NAD^+). It also avoids the presence of free nicks in DNA.

There are two main classes of topoisomerases: Type I topoisomerases make a transient break in one strand (forming a protein–DNA bond) and allow the other strand to pass through. This changes the linking number in steps of one. Type II topoisomerases make transient breaks in both strands (slightly staggered) and allow a double helix to pass through, changing the linking number in steps of two. Both play important roles in DNA replication and are important targets of chemotherapy.

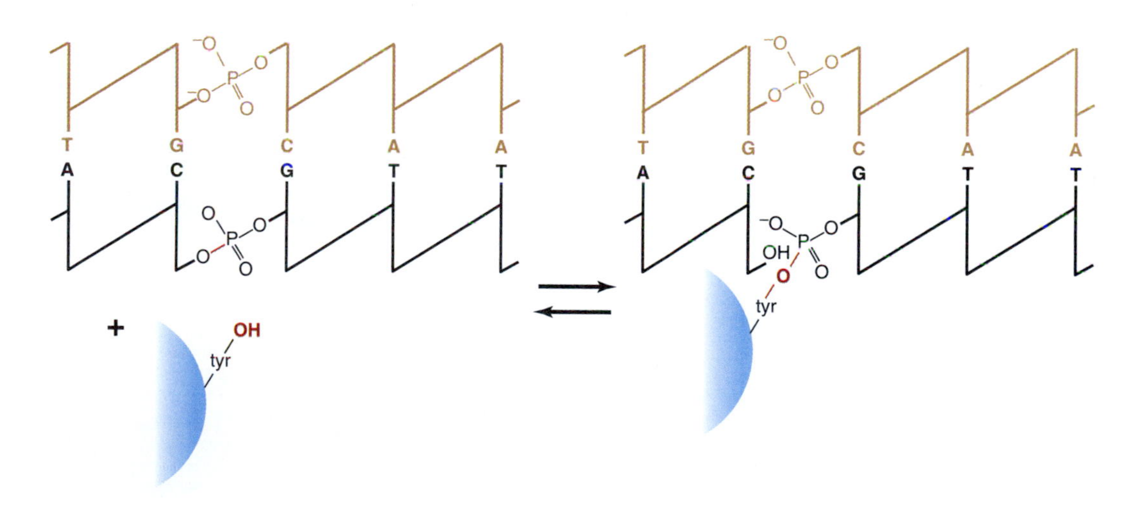

FIGURE 4.9
Topoisomerase mechanism.
Topoisomerase I (mammalian nicking–closing enzyme is depicted here) catalyzes a transesterification reaction that results in a phosphodiester bond between 3′ carbon of one residue on the DNA chain and a tyrosine residue on the topoisomerase (right panel). This breaks the continuity of the sugar–phosphate backbone and releases the other portion of DNA chain to rotate around (or pass through) its partner, which allows changes in linking number. Then the 5′-OH that was left at the other side of the interruption attacks the phosphodiester bond between nucleotide and protein, reforming DNA backbone and freeing protein (left panel). Note that because this is a series of transesterification reactions, no high-energy cofactor is needed.

Processivity

When an enzyme binds to a polymer, carries out a single step of a reaction (e.g., a polymerase adding a single nucleotide or an exonuclease removing a single nucleotide), and then dissociates from the substrate, the process is called **distributive.** In contrast, when an enzyme binds and carries out many additions (or excisions) before dissociating, the process is **processive.** There are distributive and processive enzymes involved in replication, recombination, and repair. Because it takes a finite time for an enzyme to dissociate from a template and then reassociate, distributive enzymes tend to work more slowly than processive enzymes.

For rapid synthesis of new DNA, processive polymerases are advantageous. To increase processivity, replicative DNA polymerases are associated with accessory proteins called **sliding clamps** that hold them in contact with the growing DNA chain. These sliding clamps are multimers that bind together around the DNA double helix, forming a doughnut-shaped molecule with the DNA threaded through the hole (Figure 4.10). By binding to a sliding clamp, DNA polymerase is made more processive, which increases the speed of synthesis and its accuracy.

Choreography in Three Dimensions

The complex choreography described above occurs within a very confined space: a small bacterial cell or the nucleus of a eukaryotic cell. This small volume is filled with many other molecules carrying out their functions. The idea of two huge replication complexes moving along DNA at the required speeds at each replication fork, and having to circle around it to accommodate the helical structure of DNA, is untenable. It is much easier for the very thin, cylindrical DNA to rotate around its own axis and move through a large replication complex. By assembling the polymerases working on the leading and lagging strands together, DNA can feed through this

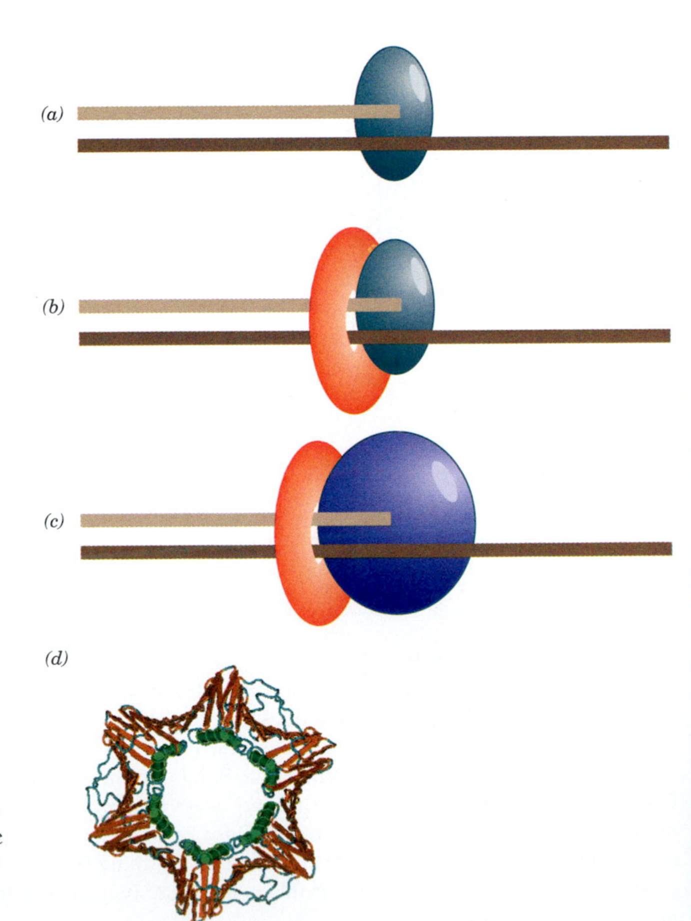

FIGURE 4.10
Sliding clamps.
(*a*) A clamp-loading protein binds to DNA. (*b*) The clamp loader assembles the sliding clamp from its subunits. (*c*) DNA polymerase associates with the assembled clamp and becomes processive. (*d*) Structure of sliding clamp in mammalian cells, a trimer of PCNA subunits. DNA passes freely through a large hole in center of complex.

large complex, called a **replisome,** as both strands are synthesized (Figure 4.11). The continuous strand feeds straight through the complex; the retrograde strand binds to allow synthesis of an Okazaki fragment and is then released to allow the finishing steps (primer removal, gap filling, and ligation) while the template further along the molecule binds to allow synthesis of the next Okazaki fragment.

Prokaryotic Enzymes

Enzymes that carry out the movement of a replication fork in *E. coli* are shown in Figure 4.12. The main replicative DNA polymerase in *E. coli* is **DNA polymerase III** (pol III). Pol III synthesizes the continuous strand and most of the discontinuous (retrograde) strand. Pol III is a large complex of subunits that carry out several functions. The core polymerase contains an α subunit that catalyzes phosphodiester bond formation, an ε subunit that functions as the 3′ to 5′ proofreading exonuclease, and a θ subunit. The sliding clamp consists of two β subunits that are assembled onto the DNA by the γ complex (the clamp loader) in a step that requires the hydrolysis of ATP. The two β subunits form a ring around the double helix that can slide along it, but not dissociate until other factors disassemble the complex. The complex of pol III and sliding clamp proteins is highly processive and can synthesize DNA essentially indefinitely, until it runs out of template or hits certain types of damage in DNA. On the lagging strand, the complex releases the DNA when it encounters the previously made Okazaki fragment. Two molecules of pol III are held together by the τ subunit and synthesize leading and lagging strands in one large complex.

On the lagging strand the DNA primase and DnaG form a complex with the DnaB helicase (DnaB). DnaB causes the complex to travel along the template for the discontinuous strand, pushing the parental strands apart. This movement requires ATP hydrolysis. The single-stranded DNA that results from helicase action is coated with **single-stranded DNA binding (SSB) protein** that prevents reannealing of parental strands and also prevents hairpins and other secondary structures from forming in the single-stranded DNA. Approximately once per 1000–2000 bp the

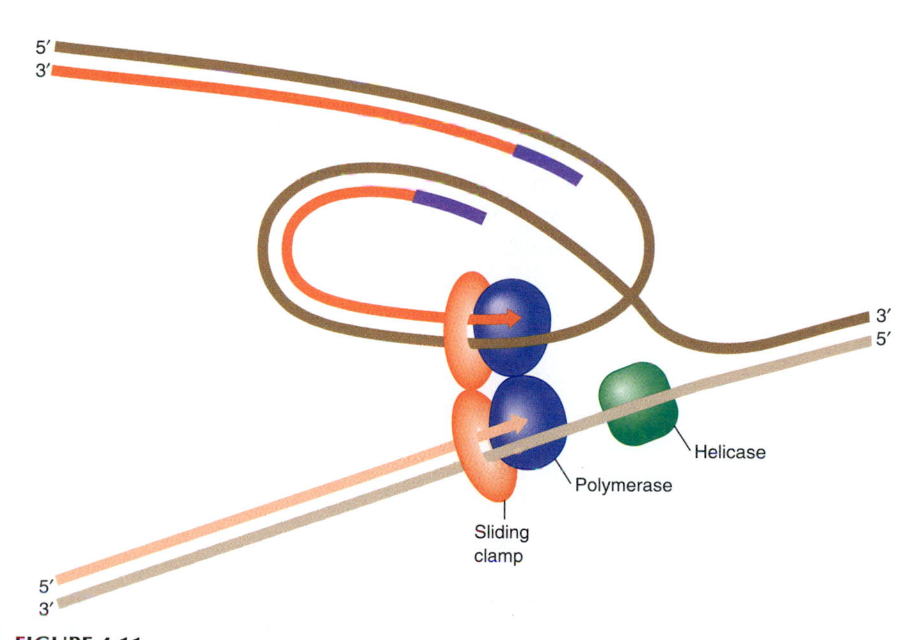

FIGURE 4.11
Replication of both strands in a replisome complex.
By binding two replicative polymerases together and looping the discontinuous strand so that it can pass through the complex, both strands can be made in one place. DNA feeds through the complex, rotating as it passes through; the large protein complexes do not have to rotate around the DNA.

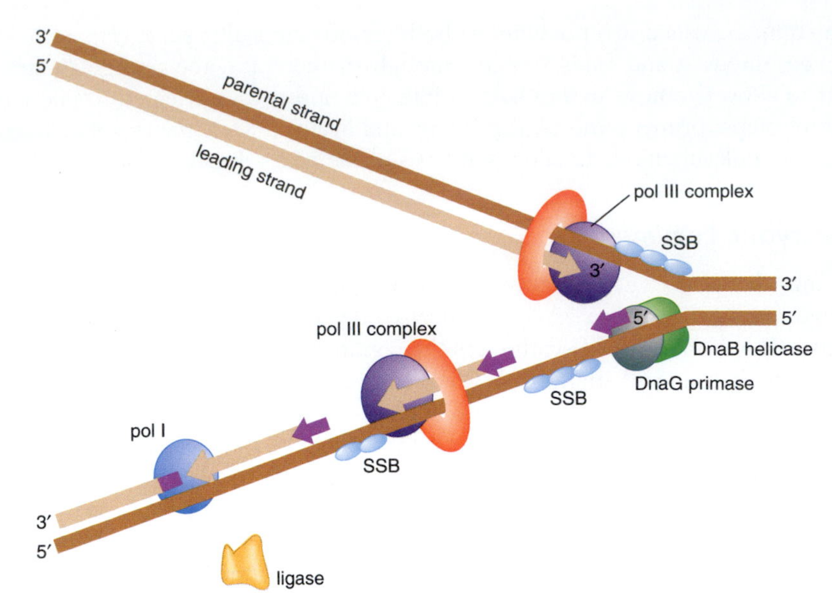

FIGURE 4.12
Replication enzymes in *E. coli.*
The continuous strand is shown at the top, being elongated by pol III associated with the sliding clamp. DNA helicase/primase complex (DnaB/DnaG) moves along the template for the lagging strand; it pauses to allow DnaG to synthesize an RNA primer every 1000–2000 bp. Pol III elongates the primer until it reaches the primer of the previously synthesized Okazaki fragment. At that time, pol III releases DNA, and pol I binds to the end of the Okazaki fragment. Pol I removes RNA primer and fills the gap simultaneously in a process called nick translation. When there is a remaining nick with deoxyribonucleotides on both sides, DNA ligase can seal the nick.

helicase/primase complex stops and the DnaG primase synthesizes a short RNA primer. The RNA primer is elongated into an Okazaki fragment of about 1000–2000 bp by DNA polymerase III, associated with a sliding clamp. Elongation stops when the polymerase complex encounters the RNA primer from the previously synthesized Okazaki fragment, and the DNA dissociates from the polymerase III complex.

In *E. coli,* RNA primer removal and the gap filling are both catalyzed by a single enzyme, DNA polymerase I (pol I). Pol I has several distinct functions in a single polypeptide chain. It contains a 5′ to 3′ exonuclease that removes the RNA primer from the 5′ end of the previously synthesized Okazaki fragment, thereby functioning as an RNase H. It contains a DNA polymerase activity that catalyzes addition of deoxyribonucleotides to the 3′ end of the more recently made Okazaki fragment until the gap created by removal of the primer is filled. An intrinsic 3′ to 5′ proofreading exonuclease increases accuracy of gap filling. Pol I coordinates primer removal and gap filling by excising a ribonucleotide from the primer and elongating the newer Okazaki fragment with a deoxyribonucleotide. It repeats these steps until the primer is removed. Pol I is capable of continuing down the DNA, removing deoxyribonucleotides and replacing them, an activity called "nick translation." This would be a wasteful procedure, but it is limited by the fact that pol I acts in a distributive manner. When the gap has been filled and only a nick remains, dissociation of pol I allows **DNA ligase** to bind to the nick and catalyze formation of a phosphodiester bond, sealing the Okazaki fragment to the growing chain.

As noted earlier, the circular *E. coli* chromosome is under **topological constraint.** Yet the parental double helix must be unwound completely for replication and chromosomal segregation to occur. Replication requires removal of positive supercoils, because the linking number of the parental strands must be reduced from a large positive number (they are twisted around each other once per 10.5 bp) to zero. **Topoisomerases** are therefore crucial for DNA replication. *Escherichia coli* has both type I and type II topoisomerases. The *E. coli* type I topoisomerase is called the omega protein (ω). Omega acts unidirectionally, removing only negative supercoils. Therefore ω is not sufficient to allow replication to occur. **DNA gyrase,** a type II topoisomerase, is essential for DNA replication. DNA gyrase acts as a "power swivel" to remove positive supercoils or introduce negative supercoils—the direction is identical. Gyrase is a heterotetramer with two "swivelase" subunits encoded by *gyrA* and two ATPase subunits encoded by *gyrB*. The swivelase subunits catalyze transesterification reactions that break and reform the phosphodiester backbone. Hydrolysis of ATP is coupled with gyrase action not to form new phosphodiester bonds but rather to trigger the conformational changes that allow a double helix to pass through the transient gap resulting from the double-strand break, resulting in unidirectional reduction in linking number. Antibiotics that target one or the other subunit of DNA gyrase rapidly stop *E. coli* replication. *Escherichia coli* has a second type II topoisomerase activity, called topo IV, that is important in chromosome segregation into daughter cells (see Clin. Corr. 4.1).

CLINICAL CORRELATION 4.1

Topoisomerases as Drug Targets

Antibiotics that target either subunit of DNA gyrase rapidly stop *E. coli* replication, because preventing the reduction in linking number of the parental strands prevents the strands from untwisting. There are two modes of targeting topoisomerases. Topoisomerase inhibitors prevent catalytic activity. Coumermycin A1 and novobiocin are inhibitors that target the ATPase subunits encoded by *gyrB*. Topoisomerase poisons freeze the covalent DNA–protein links; these complexes are lethal if converted into double-strand breaks during replication. Naladixic acid is a topoisomerase poison that targets the swivelase subunits encoded by *gyrA*. Ciprofloxacin, another topoisomerase poison, is one of the most effective oral antibiotics in clinical use today.

Eukaryotic Enzymes

In eukaryotes, the same types of enzymatic activities are required, because replication follows essentially the same pathway. Differences between bacteria and human cells in the structure of the enzymes are used in antibacterial therapy to target pathogen replication and spare the host cells.

The continuous strand at a replication fork is synthesized by DNA polymerase δ, associated with the sliding clamp called **PCNA** (proliferating cell nuclear antigen). PCNA was first detected as an antigen in the nuclei of replicating cells, hence its name. Three subunits of PCNA are assembled to form a ring around DNA, to which DNA polymerase δ attaches (Figure 4.10). Assembly of this ring requires a **clamp-loading factor** called replication factor C (RFC).

The situation on the lagging strand is slightly more complicated than in *E. coli* (Figure 4.13). The helicase activity in eukaryotes separates the parental strands, and a single-stranded DNA binding protein called **replication protein A (RPA)** binds to the exposed single strands. In eukaryotes, the primase forms a complex with DNA polymerase α that initiates Okazaki fragment synthesis. This **pol α/primase** complex synthesizes an approximately 10-bp RNA primer and then switches from primase to DNA polymerase activity and elongates the primer with approximately 15–30 deoxyribonucleotides. The product of this dual reaction is a short stretch of DNA attached to the RNA primer. Once the Okazaki fragment has reached this length, the pol α/primase complex dissociates from DNA. RFC binds to this elongated primer and serves as a **clamp loader** to assemble the PCNA sliding clamp. Then pol δ binds to the PCNA and completes the Okazaki fragment to a final length of about 130–200 bp. The pol δ/PCNA complex releases the DNA when it encounters the 5′ end of the previously synthesized Okazaki fragment. It must then reattach to the next partly synthesized Okazaki fragment. As in *E. coli*, two molecules of the main replicative polymerase, pol δ in this case, are held together in a "replication factory" in which both strands are synthesized (Figure 4.10). A related polymerase, pol ε, is essential for viability, but its role in replication is unknown.

Primer removal is carried out in two steps by a pair of enzymes, RNase H and FEN1. **Rnase H** degrades the RNA primer, leaving a single ribonucleotide attached to the end of the Okazaki fragment. **FEN1 (flap endonuclease 1)** is required to complete the removal of the last ribonucleotide. FEN1 acts by peeling back one or a few nucleotides to form a small "flap" and then cleaving the phosphodiester bond at the angle (an endonucleolytic cleavage) to release the flap. An important aspect of this reaction is that if there is a mismatch within the first few nucleotides of the Okazaki fragment, as a result of misincorporation by pol α, the mismatch would destabilize the 5′ end and create a larger flap that could be excised by FEN1. This

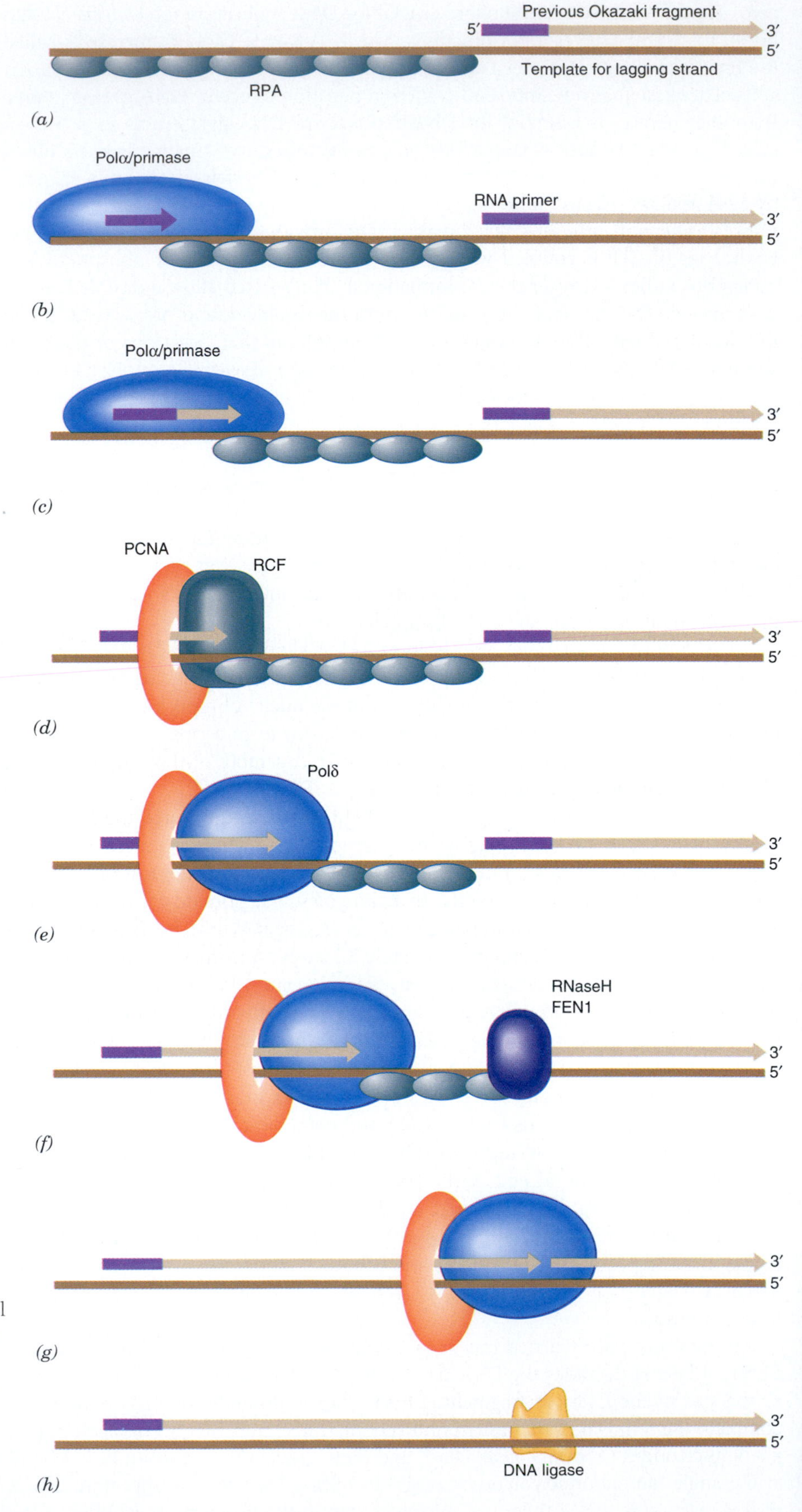

FIGURE 4.13
Enzymes replicating the lagging strand in eukaryotes (leading strand not shown). (*a*) RPA, a single-stranded DNA binding protein, binds to single-stranded template that has been opened up by the progressing replication fork. A previous Okazaki fragment is shown at right, with RNA primer in purple. (*b*) Primase activity of pol α/primase complex initiates synthesis of an RNA primer. (*c*) After RNA primer reaches 10 nucleotides, the pol α activity of the complex takes over and elongates the primer with about 20–30 deoxyribonucleotides. Then pol α/primase dissociates. (*d*) RFC binds to the end of this partly completed Okazaki fragment and catalyzes assembly of sliding clamp from three molecules of PCNA. (*e*) Pol δ complex binds to the sliding clamp and elongates the Okazaki fragment. (*f*) As the replication complex approaches RNA primer, the primer is degraded by the combined action of RNase H and FEN1. (*g*) The gap is filled by continued elongation of the Okazaki fragment. (*h*) The remaining nick is sealed by DNA ligase.

increases accuracy of the replication process by removing some of the errors introduced by pol α. The gap that remains is filled by pol δ, extending the 3′ end of the more recent Okazaki fragment. The remaining nick is sealed by DNA ligase.

Eukaryotic DNA is packaged into **nucleosomes** that contain approximately 200 bp of DNA. Dissociation of nucleosomes is required for replication and probably limits the rate of DNA synthesis. When a single nucleosome is dissociated, about 200 bp of parental DNA is available to be separated, and primer synthesis can occur somewhere in the exposed single-stranded DNA. This would explain the limited size of the Okazaki fragments.

Humans have both type I and type II topoisomerases. The human type I topoisomerase, called nicking-closing enzyme, is capable of removing both positive and negative supercoils and functions during DNA replication and transcription. Human type II topoisomerase is not a gyrase, in that it does not introduce negative supercoiling, but is critical at the termination step and for segregation of chromosomes, and appears to play a role in attaching DNA to special sites in the nuclear matrix during interphase. Cancer chemotherapy often targets topoisomerases, using poisons that lead to double-strand breaks during replication. Rapidly replicating cells are more sensitive to these drugs than are quiescent cells.

Origins of Replication

In the previous discussion, we addressed progression of a replication fork. How does a fork get started? That is a key issue, because replication is regulated by determining whether to start a cycle of replication. Once started, cells generally proceed through the entire replication process. Failure to replicate the entire genome would lead to chromosome instability and loss of genetic information. If a portion of the chromosome's DNA is not replicated, the two progeny chromosomes cannot separate properly, or the chromosome could be broken during the attempt. Broken chromosomes are unstable and trigger recombination, which causes further problems. The lack of key genes in a cell missing a chromosome due to mis-segregation or excessive expression of other genes in a cell with an extra copy of a chromosome can be disastrous. In humans, loss of one of a pair of chromosomes in cells of a fetus usually leads to miscarriage. Presence of an extra chromosome is almost always lethal, except trisomy of a few small chromosomes, which leads to serious birth defects. Down's syndrome is caused by trisomy of chromosome 21.

Replication begins from specific sites called **origins of replication (ori).** Known origins of replication contain multiple, short, repeated sequences that bind specific proteins, and AT-rich regions at which the initial separation of parental strands occurs. The *E. coli* chromosome has a single origin of replication, *oriC* (*o*rigin of *c*hromosomal replication), a region of approximately 245 bp. There are thousands of origins in eukaryotic cells (Figure 4.14). In yeast the *oris* are termed the **autonomously replicating sequences (ARS).** In humans, specific sequences that serve as *oris* have not been identified.

Initiation Process

In *E. coli, oriC* is bound by an **initiator protein** called **DnaA. DnaC** then associates and acts like a "matchmaker" to allow **DnaB,** the helicase, to bind and begin separating the parental strands to create a replication fork. At the *ori,* a replication bubble is formed and a pair of replication forks are established that move away from the *ori,* one in each direction. Thus replication is **bidirectional.**

In eukaryotes, *oris* contain special sequences at which a large complex of proteins called the **origin recognition complex (ORC)** assembles to initiate the formation of the replication forks (Figure 4.15). Assembly of an ORC at an origin is not sufficient for initiation to occur. A second complex called **MCM (minichromosome maintenance proteins),** which has a weak helicase activity, must bind and

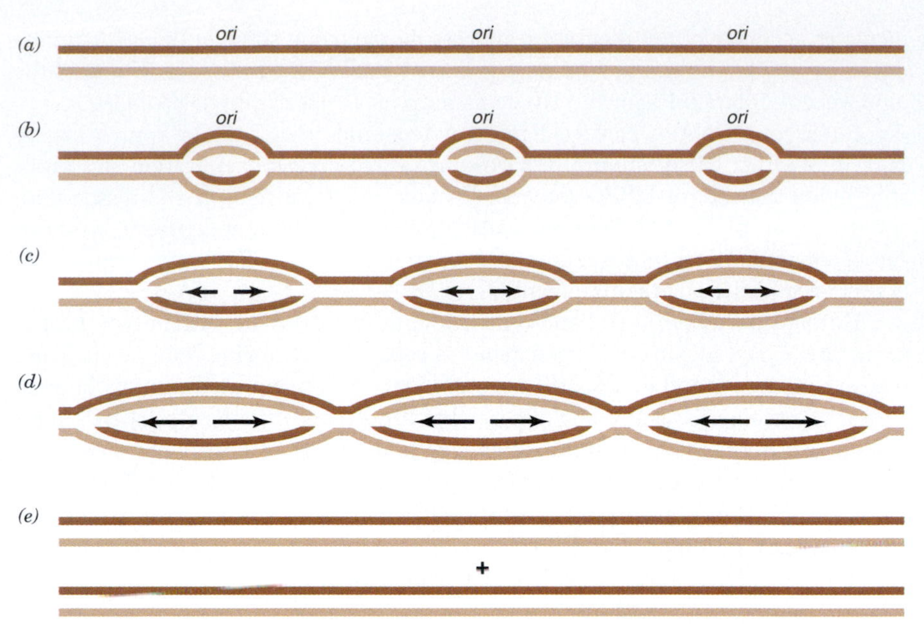

FIGURE 4.14

Tandem replicons and bidirectional replication in eukaryotes.
(*a*) There are multiple origins of replication (*ori*) tandemly arrayed along each eukaryotic chromosome. In humans they are spaced at approximately 50,000–100,000 bp intervals. (*b*) Initiation occurs at each origin (*ori*). Adjacent clusters of *oris* tend to function together. (*c*) Two replication forks are established at each *ori*, so replication is bidirectional from each. The structures are called replication bubbles. (*d*) Replication bubbles enlarge as replication continues, until they are in close proximity. At that stage, termination of replication joins adjacent bubbles and unlinks parental DNA. Topoisomerase II is essential to termination and segregation. (*e*) The resulting duplicated chromosomes can then segregate.

be activated. These activation steps are regulated by cyclins and cyclin-dependent protein kinases.

Once the ORC/MCM complex at an *ori* is activated, it catalyzes the initial separation of parental strands to form a small **replication bubble** (Figure 4.15). SSBs bind to hold the separated strands apart. **Helicases** are loaded on the replication bubble to allow replication fork unwinding. The initiation of DNA synthesis uses most of the same mechanisms as the movement of a replication fork. A primase initiates synthesis of an RNA primer, which is elongated by a DNA polymerase in the same way as on the retrograde strand at a replication fork. The polymerase elongating this first strand can continue synthesis and forms the leading strand of the fork on one side of the replication bubble. On that half of the bubble, synthesis of the first Okazaki fragment on the retrograde strand is initiated by the same mechanism as is used in moving the replication fork. However, as this first Okazaki fragment is elongated, it will not run into a previous fragment and continues to grow in the 5′ to 3′ direction, and becomes the leading strand of the opposite fork. As it progresses, Okazaki fragments are synthesized on the opposite side. The result is a pair of replication forks diverging from the origin.

Cell Cycle

In eukaryotes, there is a distinct and regulated pattern of activities that constitutes a **cell cycle** (Figure 4.16). The most striking morphological feature of the cell cycle is mitosis, the condensation, alignment, and separation of chromosomes followed by division of a cell into two occurring in the **M-phase.** DNA replication takes place during the **S-phase** (synthesis). **G1-phase** occurs between mitosis and S-phase, and **G2-phase** between S-phase and mitosis.

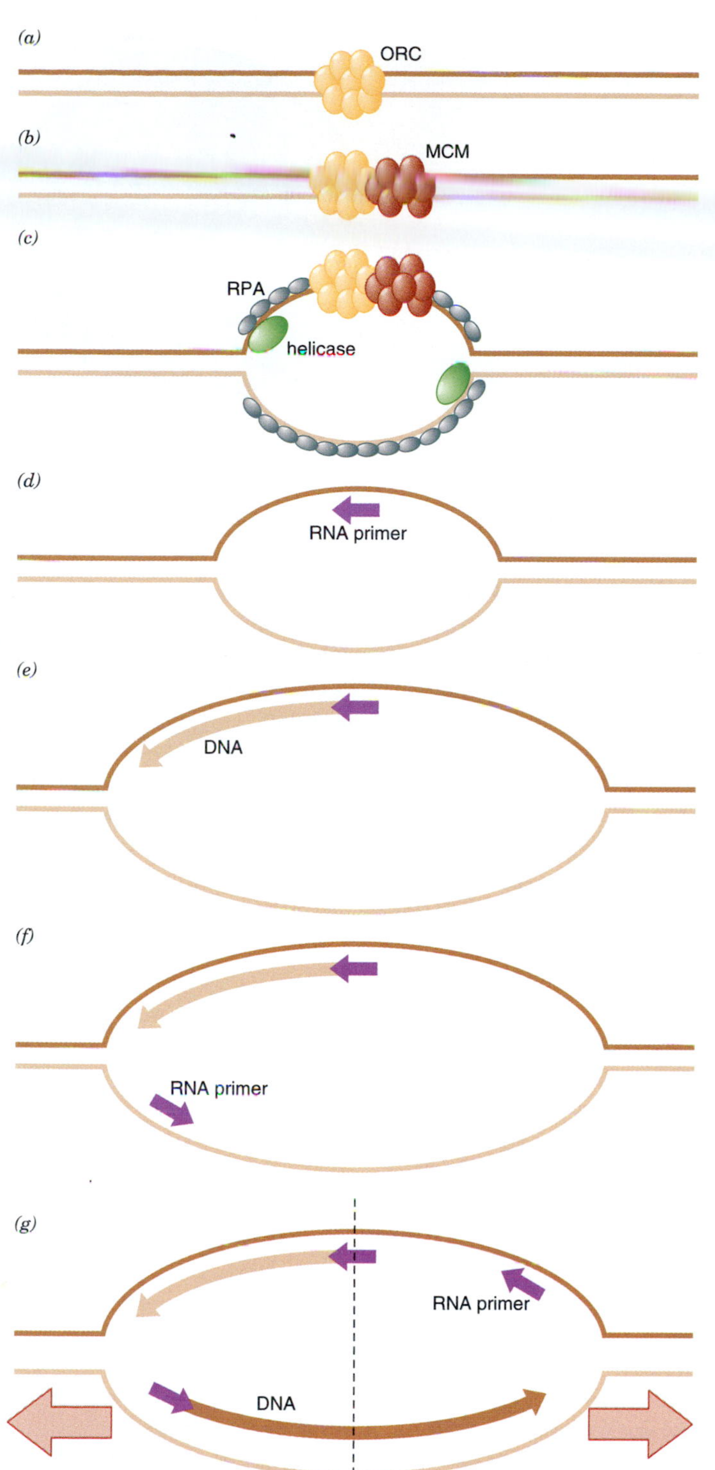

FIGURE 4.15
Initiation of replication at a eukaryotic replication origin. (*a*) The origin recognition complex (ORC) binds to an *ori*. (*b*) An MCM complex binds to this; cdc6 is important in this assembly. (*c*) After an appropriate signal, regulated to a specific part of the cell cycle, initiator complex is activated and helicase activity opens parental strands to form a very small bubble. SSB (RPA) binds to exposed single strands, helicases are loaded onto DNA, and the bubble is enlarged. (*d*) Pol α/primase synthesizes the first RNA primer and short DNA. (In this and subsequent panels, the focus is on processes occurring on DNA, and proteins are not shown.) (*e*) RNA primer is elongated by a DNA polymerase δ, incorporating deoxyribonucleotides. To this stage, the process is like that occurring on the discontinuous strand. However, the elongating strand will not encounter a previous Okazaki fragment; it can continue elongating, becoming the leading strand on the leftward-moving replication fork. (*f*) An RNA primer is synthesized on the discontinuous side of this replication fork, as in normal fork progression. (*g*) This primer is elongated as previously described. However, the elongating strand will not encounter a previous Okazaki fragment; it can continue elongating, becoming the leading strand on the rightward-moving replication fork. An RNA primer can be synthesized on this fork by the normal mechanism. The result is two replication forks diverging from the origin. The process is symmetrical around the axis indicated by the dotted line, with mirror-image forks diverging from *ori*.

For cell viability, it is important that DNA be completely replicated before mitosis. Cells have regulatory mechanisms that ensure coordination of the events of the cell cycle. DNA damage and blockage of replication forks causes slowing of S-phase progression and arrest of the cell cycle either at the G1/S boundary or the G2/M boundary. These mechanisms also induce synthesis of proteins that aid in DNA repair. In yeast, pol ε mutants are deficient in induction of damage response genes and checkpoint control. PCNA may play an important role in coordination of replication and repair functions. The interaction of PCNA with $p21^{cip1}$, an

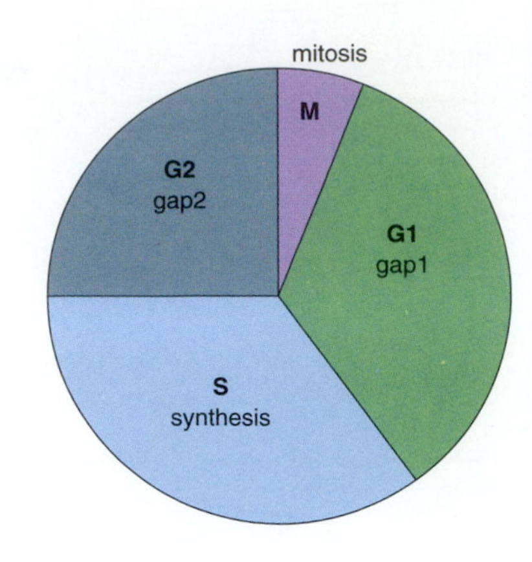

FIGURE 4.16
Cell cycle.
In eukaryotic cells, DNA replication and mitosis are separated by two gaps, G1 and G2.

inhibitor of cyclin-dependent kinase, appears to inhibit replication but not repair; it may thus allow repair to proceed before replication continues (see Clin. Corr. 4.2).

In eukaryotes, initiation of replication at *oris* must be coordinated to ensure that all regions of all chromosomes are replicated in each cell cycle, and that no region is replicated more than once. This is, obviously, a difficult challenge. Although the details of the process are not yet clear, some general principles can be stated. The current model is referred to as "**licensing.**" A **licensing factor** must be bound to the origins to allow initiation of replication at that origin. This occurs before the start of S-phase and involves the MCM protein complex that binds to the origin recognition complex, ORC. This binding is facilitated by "**matchmaker**" proteins, including Cdc6. Assembly of this prereplication complex only occurs during G1-phase, before DNA synthesis occurs. Once initiation of replication has occurred, the licensing factor is inactivated by dissociation of Cdc6 and the MCM protein complex. For that origin to be used again, a new licensing factor must bind, but the assembly of the complex cannot occur until the next G1-phase. This prevents premature reinitiation during a single cell cycle. Cyclin-dependent kinase activities regulate these processes.

Different regions of the genome are replicated at different times during the S-phase of the cell cycle. Typically, regions that are actively transcribed are replicated early during S-phase. Conversely, heterochromatic regions are typically replicated late in S-phase.

CLINICAL CORRELATION 4.2

Cancer and the Cell Cycle

Cancer is defined as the excessive division of cells. It is recognized pathologically by a higher fraction of cells actively in the cell cycle than is expected for the normal tissue from which it arose. This includes a higher fraction of cells in mitosis, recognizable by microscopy, and a higher fraction of cells in S-phase.

Most drugs used for cancer chemotherapy attack DNA replication or enzymes involved in the process. Cells that are actively dividing are more sensitive to these drugs than quiescent cells. In addition to cancer cells, the targets of therapy, there are rapidly dividing cells in the intestinal tract, bone marrow, and hair follicles. The sensitivity of these normal but rapidly dividing cells explains many of the limiting side effects of chemotherapy.

Many kinds of drugs are used to disrupt replication. Antimetabolites interfere with synthesis of precursors of 5′-dNTP. Methotrexate inhibits dihydrofolate reductase, needed to maintain reduced tetrahydrofolate required for nucleotide synthesis. 5-Fluorouracil inhibits thymidylate synthase; when metabolized into its triphosphate it can be incorporated into DNA and lead to strand breaks. Vinca alkaloids inhibit microtubule assembly and thereby interfere with mitosis and chromosome segregation. Topoisomerase inhibitors slow or stop the process of replication by preventing untwisting of parental strands. Topoisomerase poisons inhibit resealing of the phosphodiester bond, leaving covalent protein–DNA junctions that are converted into strand breaks.

Many widely used chemotherapeutic drugs are alkylating agents, often ones that have two functional groups and thereby create both intrastrand and interstrand cross-links in DNA. Cyclophosphamide, busulfan, and nitrosoureas are alkylating agents. Cisplatin, a key part of the combined chemotherapy for testicular cancer, cross-links DNA. Alkylating agents are not only cytotoxic but also mutagenic, sometimes resulting in secondary cancers such as leukemias.

Termination of Replication in Circular Genomes

Termination of replication of a circular genome generally occurs 180° away from the origin. Two converging replication forks meet, and the last portion of the genome is synthesized. Topological unlinking of the two new chromosomes must occur. This is a key function of type II topoisomerases. In some small viruses like SV40, termination occurs wherever replication forks meet; there is no special sequence involved. The *E. coli* genome contains special termination sequences that constrain termination within a region, by preventing forks from proceeding past the region.

Termination of Replication in Linear Genomes

Human cells, and eukaryotic cells in general, have linear chromosomes. There are special difficulties in replicating the ends of linear chromosomes. What exactly is the problem in replicating the ends of linear chromosomes? Although the continuous strand can theoretically be synthesized to the very end of its template, the retrograde strand cannot. There is no place to synthesize a primer to which the nucleotides opposite the end of the template can be added (Figure 4.17*a*). Even in the unlikely event that a primase could start at the very end of the template, removal of the RNA would leave a short gap. While this may not be a problem in a single generation, over many cycles of replication chromosome ends would be shortened until essential genes become lost and the cell dies. Therefore a key problem is to prevent continued loss of DNA at the ends of chromosomes.

A second problem that eukaryotes face is that the ends of DNA molecules tend to trigger recombination. To avoid both problems, the ends of eukaryotic linear chromosomes are special structures called **telomeres,** which contain many repeats of a six-nucleotide, G-rich repeated sequence. Human telomeres contain thousands of the repeat TTAGGG. The 3′ end of the chromosome extends about 18 nucleotides beyond the 5′ end (Figure 4.17*b*), leaving about three repeats as an overhang. The overhanging 3′ end folds back on itself, forming non-Watson–Crick G-G hydrogen bonds, and binds proteins that define its length and protect it from excessive recombination.

Telomerase

Telomeres are maintained by **telomerase,** enzymes that add new six-nucleotide repeats to the 3′ ends of the telomeres. Telomerases are ribonucleoprotein complexes containing a small RNA that serves as a template for addition of a new six-nucleotide repeat (Figure 4.18). A telomerase binds to the end of the 3′ strand, with part of the small RNA hydrogen bonded to the last few nucleotides. A six-nucleotide

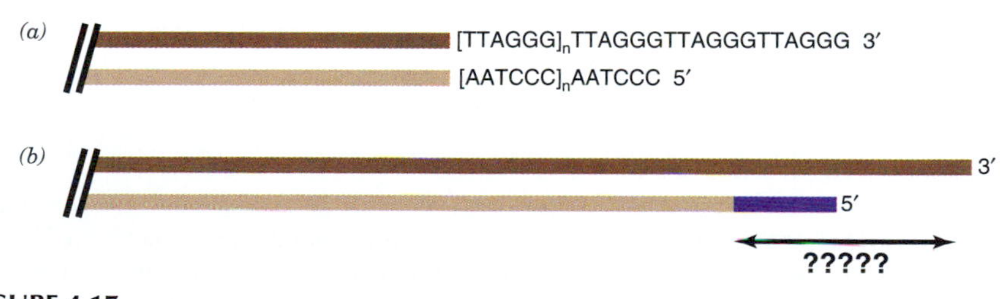

FIGURE 4.17
Human telomeres.
(*a*) A telomere consists of many tandem repeats of a 6-nt sequence, with the G-rich strand extending beyond the C-rich strand by about 12–18 nt. (*b*) The telomere replication problem. The 3′ end of one parental strand, the template for the discontinuous daughter strand, is shown in dark tan; the daughter strand is shown in light tan, and an RNA primer is in purple. The problem is that there is no place to synthesize a primer that would allow the daughter strand to be completed (region shown with ????), so the daughter strand will be shorter than the parental strand. Removal of the last RNA primer makes it even shorter.

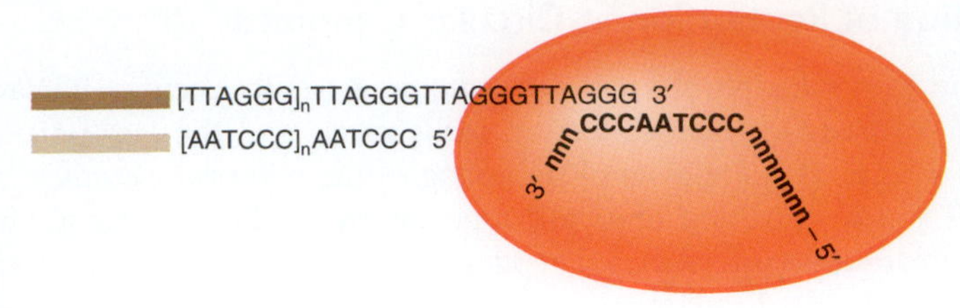

FIGURE 4.18
Telomerase.
Telomerase is an enzyme that catalyzes the addition of new 6-nt telomere repeats to the 3′ end of a DNA chain. Telomerase is a ribonucleoprotein complex with a short RNA strand as an integral part. Telomerase RNA serves as the template for the reaction, while the protein component functions as a reverse transcriptase, synthesizing DNA using the RNA template. After a 6-nt repeat is added, the enzyme can dissociate and bind again, and add additional 6-nt repeats.

repeat is synthesized, using the RNA as a template. Then the telomerase can dissociate and reassociate to add another hexamer. There is a balance between shortening of the lagging strands due to the problems in replication (Figure 4.19) and addition of six-nucleotide repeat units (Figure 4.20).

Telomeres do not have to remain exactly the same length; some shortening is not a problem, because the repeats do not encode proteins. Telomeres undergo cycles of shortening and addition of new repeats to the 3′ end by telomerase. Although the length of the telomere does not remain constant, progressive shorten-

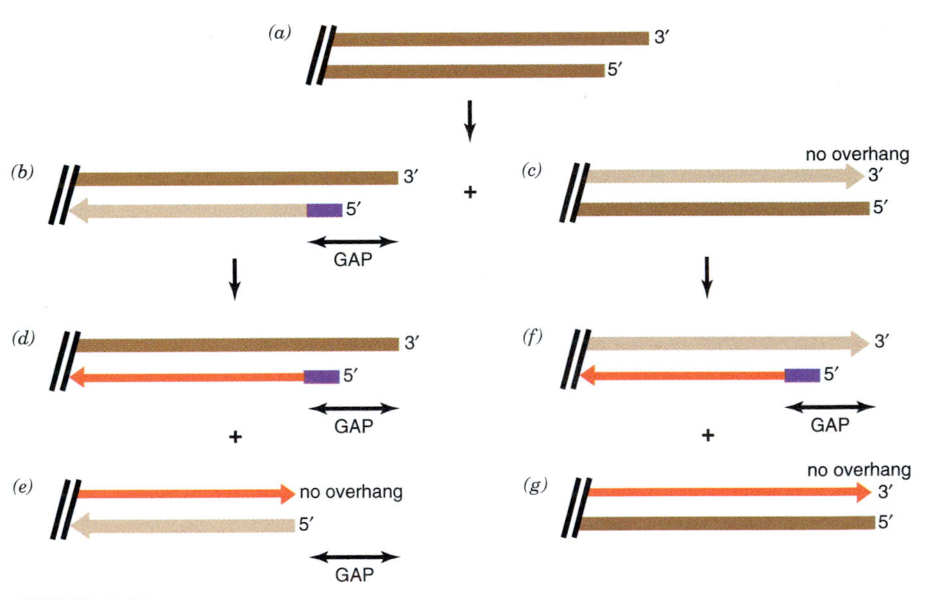

FIGURE 4.19
Replication in the absence of telomerase.
(*a*) Chromosome end, with 3′ strand overhanging. (*b*, *c*) Products of the first round of replication. (*b*) Normal process of replication cannot finish the discontinuous strand. (*c*) The other parental strand can serve as a template to its very end, so there is no shortening, but no 3′ overhang is left. (*d*, *e*) Products of a second round of replication of the molecule in (*b*). In the next generation, the original strand from A has the same problem as shown in (*b*). (*e*) The shorter daughter strand in (*b*) can be completed to the end of the DNA (primer, shown as a thick line, has been removed by RNase H plus FEN1). The result is a shorter chromosome with flush ends. (*f*, *g*) Products of a second round of replication of the molecule in (*c*). (*f*) Template for the discontinuous strand (newly synthesized strand in (*c*) has the same problems as that in (*b*); the result is further shortening. (*g*) The other parental strand can serve as a template to its very end, so there is no shortening, but also no overlap. As this cycle is repeated, there is progressive shortening of one strand and loss of 3′ overhang.

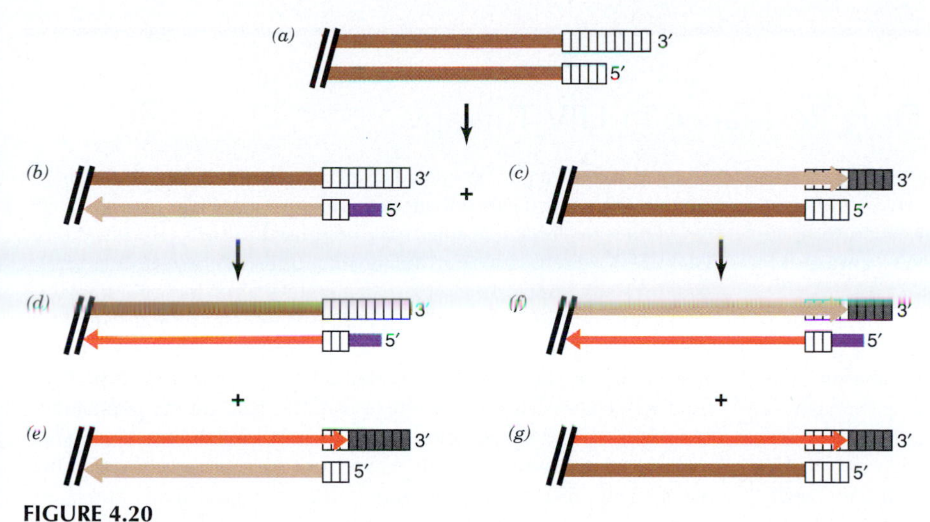

FIGURE 4.20
Replication with telomerase.
(*a*) Chromosome end, with overhang and telomeric repeats shown as boxes. (*b, c*) Products of first round of replication. (*b*) The normal process of replication cannot finish the discontinuous strand. The region that is not completed is made of noncoding telomeric repeats. (*c*) The other parental strand can serve as a template to its very end, so there is no shortening, but also no overlap. However, telomerase can add new subunits to the 3′ end (dark boxes), reestablishing overlap. (*d, e*) Products of a second round of replication of the molecule in (*b*). (*d*) In the next generation, the original strand from A has the same problem as shown in (*b*). (*e*) The shorter daughter strand in (*c*) can be completed to the end of DNA (primer, shown as a thick line, has been removed by Rnase H plus FEN1). The result is a shorter chromosome with flush ends. However, telomerase can add new subunits to the 3′ end (dark boxes), lengthening the strand and reestablishing overlap. Thus there is no progressive shortening. (*f, g*) Products of a second round of replication of the molecule in (*c*). (*f*) The template for the discontinuous strand [newly synthesized strand in (*c*)] has the same problems as that in (*b*); result is further shortening. (*g*) The other parental strand can serve as a template to its very end, so there is no shortening, but also no overlap. However, telomerase can add new subunits to the 3′ end (dark boxes), reestablishing overlap. As this cycle is repeated, telomerase prevents progressive shortening and preserves 3′ overhang.

ing is avoided by addition of repeats. Telomerases also reestablish the 3′ overhang characteristic of telomeres.

Cells that have differentiated and no longer divide, or will divide only a limited number of times, do not express telomerase. Telomerase expression is generally reactivated in tumor cells, which allows them to continue division indefinitely without chromosome shortening. This makes telomerase an attractive target for cancer chemotherapy. It should be noted that inactivating telomerase in a tumor would not lead to a rapid halt in tumor growth; any effect would be delayed by many cell cycles.

Replication of RNA Genomes

Some viruses have an RNA genome. Such genomes are replicated with much lower accuracy and can accumulate variations relatively rapidly. A particularly relevant example of this is the **human immunodeficiency virus (HIV)** that causes **AIDS** (acquired immune deficiency syndrome; see Clin. Corr. 4.3).

Rapid generation of variants due to the reduced accuracy of replicating its RNA genome leads to the generation of a collection of viruses within an individual. A small fraction of these variants is likely to be resistant to any single drug being used to treat the infection. The fraction can grow in the presence of the therapeutic agent and is thus selected for, leading to loss of efficacy of the drug. Combination therapies are designed to reduce the probability that a virus will be

CLINICAL CORRELATION 4.3

Nucleoside Analogs and Drug Resistance in HIV Therapy

AIDS (acquired immune deficiency syndrome) is caused by infection with a retrovirus, HIV (human immunodeficiency virus). A key step in the life cycle of this virus is synthesis, catalyzed by a reverse transcriptase, of a DNA copy of the viral RNA genome. Reverse transcriptase is a major target of chemotherapy.

AZT (3′-azido-2′,3′-dideoxythymidine; Zidovudine) was the first drug approved for treatment of HIV infection. It is a nucleoside analog with an azido group on the sugar. It can be phosphorylated into the triphosphate form and competes with dTTP for incorporation into the reverse transcript. Once incorporated, it terminates the growing chain of the transcript because the azido group on the 3′ carbon of the sugar is not a substrate for nucleotide addition. AZT is much less efficient at competing with the more accurate cellular DNA polymerases, providing a therapeutic opportunity in which the effect is primarily upon viral replication. Nevertheless, side effects include toxicity to bone marrow, which contains rapidly dividing cells, and myopathy that might be related to toxicity to mitochondria, which contain their own DNA polymerase, pol γ.

Other nucleoside analogs have been used to treat HIV infection. DDI (2′,3′-dideoxyinosine; didanosine) and dideoxycytidine (zalcitabine) also function as chain terminators after incorporation of their phosphorylated derivatives by the HIV reverse transcriptase. Reverse transcriptases do not carry out proofreading; thus their error rate is much higher than that of cellular DNA polymerases. This high error rate complicates the treatment of AIDS, because the population of viruses carried by any one patient contains many mutants. Some of these mutants are likely to be resistant to any given therapeutic agent. Thus many drugs initially reduce the viral load, but later become ineffective due to the selective growth of viruses in which the drug target is mutated. Combination therapy, with multiple drugs that target different viral proteins, is an attempt to circumvent this problem.

simultaneously resistant to all drugs in the combination. Current combination therapies target both the reverse transcriptase that synthesizes DNA copies of the RNA genome (to allow its integration into the genome of cells) and HIV protease.

DNA Sequencing

The human genome project is an effort to determine the complete nucleotide sequence of the human genome, as well as the genomes of several important model organisms: *E. coli*, *Saccharomyces cerevisiae* (bakers' yeast), *Caenorhabditis elegans* (a nematode that has widely been used in studies of development and aging), *Drosophila melanogaster* (the fruit fly), and *Mus musculus* (mouse). Other goals are to develop physical and genetic maps and tools for sequencing and for analyzing sequences, and to examine the ethical, legal, and social implications of the availability of these sequences. By the summer of 2000, most of these goals had been met or exceeded, and about 90% of the human genome sequence was available in unfinished form, as patches of sequence not yet assembled, with gaps between the patches (see Clin. Corr. 4.4).

DNA sequencing is based on two simple aspects of DNA synthesis: the fact that DNA polymerases require primers, and the chemistry of the addition of a nucleotide to a growing DNA chain (see Figure 4.2). The key idea is to synthesize a nested set of DNA strands all starting at the 5′ end of the primer and stopping after adding one of the four nucleotides. Then one separates the nested set by length, using electrophoresis. From the pattern of lengths one can read the sequence.

Fragments of DNA to be sequenced are cloned into a plasmid vector at a specific site. A synthetic oligonucleotide complementary to the plasmid sequence, with its 3′ end near the point at which the fragment was inserted, is used as a universal primer (Figure 4.21). In the version of sequencing used for the human genome project, specific termination occurs after addition of **dideoxyribonucleotides** (ddNTPs; Figure 4.21), each of which is labeled with a different fluorescent dye. ddNTPs can be added to the growing chain just like any normal dNTP, but once added there is no longer a free 3′-OH at the end of the chain, so the next base cannot be added. This results in specific termination of chain elongation after ddA, ddC, ddG, or ddT is incorporated. A mixture of ddNTPs and dNTPs is used, the ratio being adjusted so that the ddNTP is incorporated only about 0.5% of the time at each site. This produces a nested set of fragments each with the same known

CLINICAL CORRELATION 4.4
The Human Genome Project

The Human Genome Project is a large international effort aimed at determining the sequence of the entire human genome of 3,200,000,000 bp. But it is much more. Other aims include the creation of physical and genetic maps of the genome, sequencing the genomes of model organisms, developing technology for sequencing and informatics for understanding the sequence, training of scientists in use of these data, and exploring the ethical, legal, and social implications of the data.

The genetic and physical maps are already of great importance in research directed toward finding genes that contribute to the risk for diseases. Parallel sequencing of model organisms allows comparative studies that illuminate the potential functions of genes that are discovered. Model organisms originally targeted were *E. coli, Saccharomyces cerevisiae* (bakers' yeast), the nematode *Caenorhabditis elegans,* the fruit fly *Drosophila melanogaster,* and the mouse. These are key experimental organisms about which much is known. Yeast is a simple eukaryote and has been an organism of choice for studies of eukaryotic replication, repair, and recombination as well as cell cycle regulation. Many of the results described in this chapter come from studies on yeast. The nematode *C. elegans* is a relatively simple animal, for which the developmental fate of every cell has been mapped. Drosophila is more complex, with complex behaviors, and it has been important in genetic and developmental studies during nearly the entire 20th century. The mouse is a mammal with many similarities to us in development, physiology, and behavior. The ability to genetically engineer gene changes in mice allows us to create animal models of human disease for study. The project is well ahead of the published schedule and has expanded to include other model organisms and a search for sequence variation (polymorphisms; see Clin. Corr. 4.5). Knowing the sequence of the human genome (and that of other organisms) is already having a major impact on science and medicine. But it is only the first step. Much work remains to enable us to understand the vast amounts of data generated. These data also raise issues about privacy and discrimination that need to be addressed carefully.

Collins, F. S., Patrinos, A., Jordan, E., Chakravarti, A., Gesteland, R., Walters, L., and members of the DOE and NIH planning groups. New goals for the U.S. Human Genome Project: 1998–2003. *Science* 282:682, 1998.

end and the other end base-specific (Figure 4.21). These DNA fragments are separated by electrophoresis with a resolution of a single nucleotide. The different fluorescent labels are detected by automated instruments, and the sequence is read (Figure 4.21).

Data available from the human genome project will have tremendous implications for understanding how cells and organisms work, and for medicine. Sequence differences between individuals affect their risk for diseases, and their metabolism of drugs and responses to them. Knowledge of individual differences in DNA

TGACTATAGCTCAATAGTCAG-----
GATATCGAGTTAT*
GATATCGAGTTATC*
GATATCGAGTTATCA*
GATATCGAGTTATCAG*
GATATCGAGTTATCAGT*
GATATCGAGTTATCAGTC*
5′ -primer ⟶

(a)

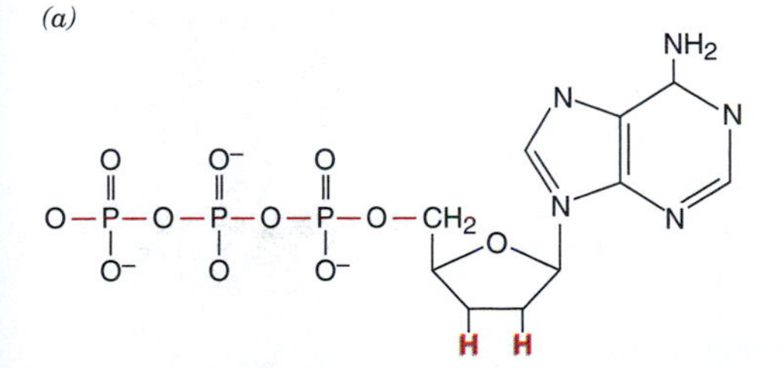

(b)

(c)

FIGURE 4.21
DNA sequencing.
(*a*) Nested fragments. DNA is synthesized by extending an oligonucleotide primer. A fraction of the chains is terminated after addition of each ddNTP (* indicates that ddNTP is fluorescent and can be detected). The result is a nested set of fragments all with the same 5′ end (5′ end of primer), and with base-specific 3′ ends. (*b*) 2′,3′-Dideoxyribonucleotide: ddATP. The 5′ triphosphate of a ddNTP is identical to that in a normal dNTP, so a ddNTP can be added to the growing DNA chain as depicted in Figure 4.2. However, there is no 3′-OH to which a new nucleotide can be added, so the chain terminates after addition of a ddNTP. (*c*) Electrophoretic separation of the nested set of fragments and detection of the specific fluorescent dye at each length (a different dye for each nucleotide) allows automated reading of the sequence from bottom (smallest fragment) to top.

CLINICAL CORRELATION 4.5

Genomic Medicine

Individuals differ in the sequence of DNA in their genome in subtle ways. About once per 1000 bp, there is a difference between individuals, called a polymorphism. Most polymorphisms do not affect cell functions, but others contribute to the many differences that we see. Some affect susceptibility to specific diseases, some affect metabolism of drugs and other compounds (pharmacokinetics), and still others affect response to these drugs (pharmacodynamics).

Along with completion of the human genome sequence, there is a major effort to find and catalog polymorphisms. As more is learned about the effects of these sequence differences on disease and treatment, medicine will become more tailored to the individual. Technology already exists to rapidly determine which polymorphisms are present at specific sites in the genome of an individual. When these technologies are made cheaper and more routine, determining the alleles present in key genes will become a routine part of diagnosis and planning of treatment. For example, there are significant differences between individuals in the rate at which they metabolize drugs. If a standard dose is given based on body weight, it may be too low to be effective in those who metabolize the drug rapidly, but may be toxic to those who metabolize it very slowly. Determination of the relevant polymorphisms and routine testing could lead to more effective therapies with fewer side effects.

sequence will enable physicians to better detect diseases at an early stage and to tailor treatments to individual patients (see Clin. Corr. 4.5).

4.3 | RECOMBINATION

Recombination is the exchange of genetic information. There are two basic types of recombination, **homologous recombination** and **nonhomologous recombination.** Homologous recombination (also called **general recombination**) occurs between identical or nearly identical sequences, for example, between paternal and maternal chromosomes of a pair. Chromosomes are not passed down intact from generation to generation (Figure 4.22); rather, the chromosome you inherit from your father contains portions from both of his parents. This is a normal part of the process of chromosome alignment and segregation necessary to ensure that each germ cell gets a single haploid set of chromosomes during meiosis. This recombination is reciprocal, transferring part of one chromosome to the other and vice versa. It shuffles the combination of genes before they are passed to the next generation, generating genetic diversity. The probability that a recombination event will occur between any two points on a chromosome is roughly proportional to the physical distance between them. This is the basis for genetic mapping. A 1% frequency of recombination between two genes or markers is defined as a genetic distance of 1 **centimorgan (cM).** In humans, 1 cM is approximately 1,000,000 base pairs along the chromosome.

Recombination is also important for DNA replication and repair. There are three major models for homologous recombination. They are related in many ways and have a common intermediate called a **Holliday junction.** The simplest is the Holliday model and understanding its key points will make it easy to see the distinctions of the others.

The Holliday Model

The key features of the **Holliday model** are (1) homology, (2) symmetry of both breaks and strand invasion, and (3) presence of a four-stranded "Holliday junction" as a key intermediate (Figure 4.23). In this model, recombination is initiated by single-strand breaks at homologous positions in the two aligned DNA duplexes. The strands of each duplex partly unwind, and each "invades" the opposite duplex. That is, a portion of one strand from each duplex base pairs with the opposite duplex in a reciprocal, symmetrical fashion. These strands can be joined covalently to the opposite duplex, creating a joint molecule in which two strands cross between the DNA molecules. **Branch migration** then occurs. Branch migration involves simultaneous unwinding and rewinding of the two duplexes, such that the total number of hydrogen bonds remains constant but the position of the crossover moves. Because the number of hydrogen bonds remains constant, this requires little energy.

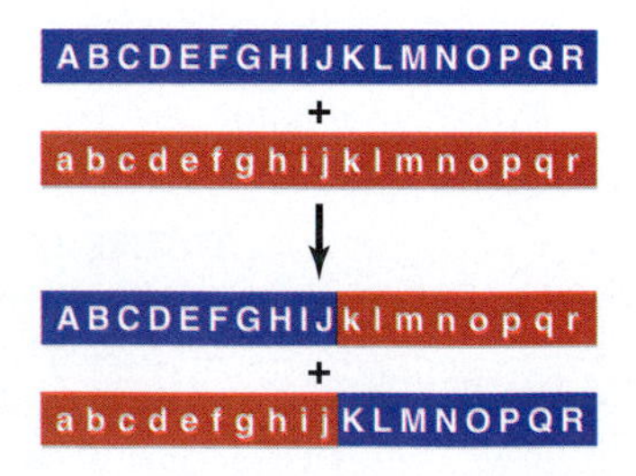

FIGURE 4.22
Homologous recombination.
(*a*) Paternal chromosome is shown in blue with alleles shown in capital letters. Homologous maternal chromosome is shown in red with alleles shown in lowercase letters. (*b*) After homologous recombination (between genes j and k) both chromosomes contain DNA from both parents. There has been an equal exchange between them.

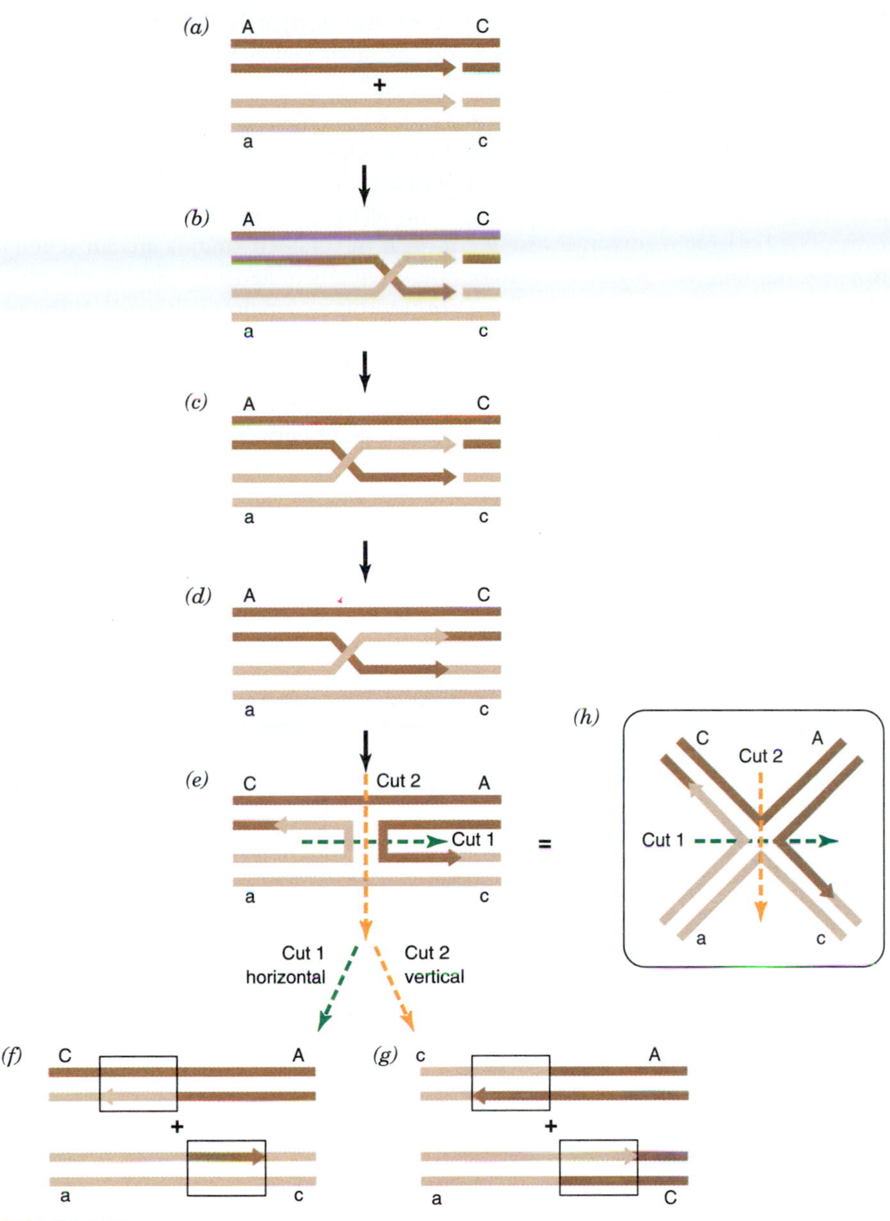

FIGURE 4.23
Holliday model of homologous recombination.
(*a*) Synapsis (pairing): two homologous DNA molecules come together properly aligned. (A,a and C,c represent different alleles of a gene or marker on either side of crossover.) Single-strand breaks are made at homologous positions in strands of the same polarity in each. (*b*) Strands partly unwind and reciprocally "invade" (base pair with) the opposite molecule, forming a structure in which two intact strands are joined by two crossed strands. This is a symmetrical and reciprocal event. Regions in which one strand was from one duplex and the other from the homologous duplex are called heteroduplexes. (*c*) "Branch migration" occurs; there is simultaneous winding and unwinding so there is no net change in the amount of base pairing, but the position of the crossover point moves, resulting in more heteroduplex DNA. (*d*) DNA ligase seals nicks in the DNA. Further branch migration can occur. All four strands are held together at one crossover point. (*e*) Rotation of molecule forms a "Holliday intermediate" shown in inset (*h*). (*f*, *g*) Resolution: molecules are separated by a pair of symmetrical cuts in either of two directions (arrows). The ends are resealed by ligase. Direction of cuts determines whether flanking regions are exchanged. (*f*) If crossed strands are cut (cut 1, horizontal) flanking genes or markers remain as they were (AC/ac). You would not detect recombination between these loci. (*g*) If intact parental strands are cut (cut 2, vertical) flanking genes or markers are exchanged (Ac/aC). This would be detected as a recombination event. Note that there is a region of heteroduplex in either case (boxed); mismatch repair in this region can lead to gene conversion.

Branch migration creates a region of **heteroduplex,** that is, a region where one strand comes from one original duplex and the other comes from the other duplex. In humans, where on average two individuals differ by approximately one base pair per thousand, a region of heteroduplex is likely to contain at least one mismatched base pair. The two DNA duplexes, joined by a single crossover point, can rotate as shown in Figure 4.23 to create a four-stranded Holliday junction. The resolution of a Holliday junction into two duplexes can occur in either of two ways, depicted as vertical or horizontal cuts in Figure 4.23*f–h*. If the crossed strands are cut (cut 1), markers flanking the region of heteroduplex remain linked together in the same phase as originally; on a gross level, the chromosomes remain unchanged. If the intact strands are cut (cut 2), the flanking regions are exchanged between the two chromosomes, creating a new chromosome with a part from one parent and another from the other. The heteroduplex may undergo mismatch repair (see below), converting one allele into another, a process called **gene conversion.**

The Meselson–Radding Model

Once you understand the Holliday model, it is easy to understand the variant proposed by Meselson and Radding. It is initiated by a single nick and a single invading strand, rather than a pair of nicks and reciprocal invasion (Figure 4.24). The

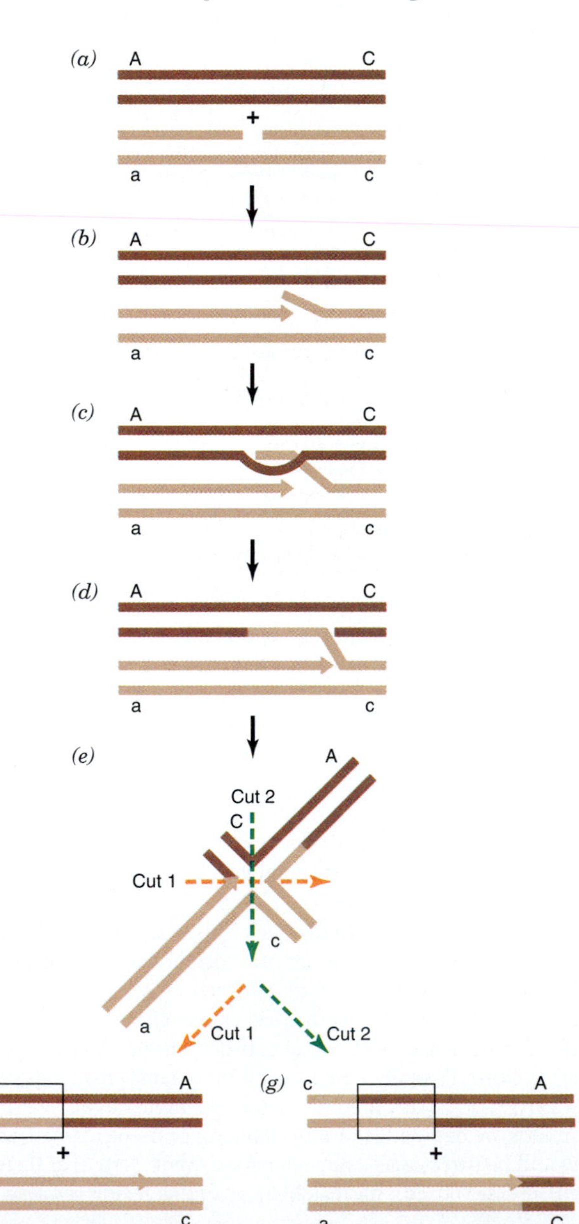

FIGURE 4.24
Meselson–Radding model of homologous recombination. This model is similar to the Holliday model, but initiation is by a single nick and a single invading strand. (*a*) Synapsis (pairing): two homologous DNA molecules come together properly aligned. A single-strand break is made in one strand. (*b*) The 3′-OH end is elongated by a DNA polymerase, displacing the 5′ end. (*c*) The free 5′ end invades the homologous duplex, forming a "D-loop." (*d*) The D-loop is degraded, and repair synthesis and resealing by DNA ligase lead to a molecule in which one strand crosses between two duplexes. Branch migration occurs as one strand is elongated further (not shown). The elongating strand crosses over and is joined to the opposite duplex, leaving all four strands joined at a single crossover point, as in Figure 4.23. (*e*) Strands rotate to form a Holliday junction exactly as in Figure 4.23*e*, *h*. Resolution of the Holliday junction can occur in two ways, as shown in Figure 4.23. Direction of cuts determines whether flanking regions are exchanged. (*f*) If crossed strands are cut (cut 1, horizontal) flanking markers remain as they were (AC/ac). You would not detect recombination between these markers. (*g*) If intact parental strands are cut (cut 2, vertical) flanking markers are exchanged (Ac/aC). This could be detected as a recombination event. Note that there is a region of heteroduplex in either case (boxed), but it is not reciprocal, it is only on one duplex; mismatch repair in this region can lead to gene conversion.

nicked strand serves as a template–primer combination for a DNA polymerase, and this elongation displaces part of the strand. The displaced strand invades the opposite, aligned duplex, forming a structure called a D-loop. The single-stranded portion of the D-loop is degraded, and repair synthesis and resealing result in crossing over of one strand. Branch migration occurs. The remaining nick is sealed and the strands can swivel to form a Holliday junction identical to that described previously. The Holliday junction can be resolved exactly as described for the Holliday model, resulting in an exchange of flanking markers (cut 2) or the original configuration of markers (cut 1). The key difference in the **Meselson–Radding model** is the asymmetrical heteroduplex that results from the invasion by one strand from one of the two duplexes. Mismatch repair can lead to gene conversion, as above.

The Double-Strand Break Model

The third model is the double-strand break model. In this, recombination is initiated by a double-strand break in one duplex (Figure 4.25). The ends are resected, resulting in loss of genetic information from one of the two duplexes. There are alternative forms of this model. In one, the two single-stranded segments of the broken duplex both invade the intact duplex, forming a joint molecule that has two Holliday junctions. The broken ends can be repaired by a polymerase, using the intact strand as template, and a DNA ligase. Branch migration occurs as in the previous models. An alternative model has been suggested in which the 3′ end of only one of the two sides of the break invades the other strand and establishes a D-loop that allows copying of the intact duplex. This can result in a small replication bubble with synthesis on both strands for a short distance, and resolution when the bubble encounters the other side of the break. This would account for extensive gene conversion.

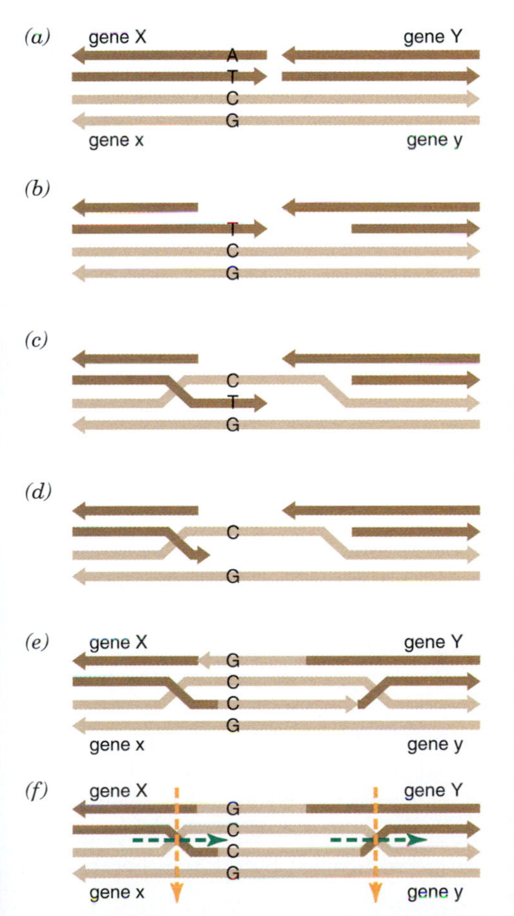

FIGURE 4.25
Double-strand break model of homologous recombination.
This model is similar in many ways to the two models discussed previously, but initiation is by a double-strand break in one of the two DNAs. (*a*) Initiation is by a double-strand break in one DNA molecule. Two aligned duplexes are shown, with flanking genes with different alleles (X,x;Y,y). Also shown is a position at which one duplex has an AT base pair and the other has a GC base pair. (*b*) DNA is resected from the break by an exonuclease that leaves 3′-OH overhangs. This resection may remove one of the nucleotides at the site that differs between strands. (*c*) One or both of these overhangs can invade the intact homologous duplex. (*d*) Where there is a mismatch, the free end can be resected further by enzymes of mismatch repair. (*e*) The free end can be extended by a DNA polymerase, resulting in gene conversion as the template is from one of the two duplexes. Remaining nicks are repaired, forming two Holliday junctions. (*f*) Both Holliday junctions are then resolved. Because there are two junctions and two ways to resolve each (comparable to Figures 4.22 and 4.23), there are four possible results. Two of the four possible ways to resolve the Holliday junctions (vertical/vertical or horizontal/horizontal cuts) leave flanking markers as they were. The other two ways to resolve the Holliday junctions lead to exchange of flanking markers. Resolution can also be carried out by topoisomerase.

Resolution of the two Holliday junctions is by a pair of cuts. Because there are two possible directions in which to resolve each junction, there are four possible outcomes (Figure 4.25). Two leave flanking markers together, and two lead to the exchange of flanking markers. Resolution by a type II topoisomerase can also occur. Because one duplex has lost DNA in the region of the break, there is a region in which gene conversion is essential to allow closure of the gap.

Key Enzymes of Recombination

RecA

The key *E. coli* protein in the process of recombination is **RecA,** which binds in a cooperative manner to single-stranded DNA and forms a helical coil. This structure facilitates pairing of the single strand with a homologous duplex DNA to form a three-stranded D-loop structure. Strand switching can occur, leading to heteroduplex DNA, as discussed. Eukaryotes have homologous proteins, the yeast version of which is called Rad51.

Other Proteins

RecBCD is an *E. coli* complex with multiple activities; its function in recombination is to bind at a double-strand break and process it into a substrate for recombination. RecBCD does this by chewing back the DNA and, upon encountering a special sequence called a **X** site, leaving a protruding single strand with 3′-OH onto which RecA is loaded. **RuvA** and **RuvB** form a complex with DNA helicase activity that catalyzes extensive branch migration. RuvC is an endonuclease that binds to this complex and catalyzes the resolution of Holliday junctions.

DNA-PK

In mammalian cells, a heterodimeric molecule called Ku binds avidly to DNA ends. **Ku** is a **DNA-dependent ATPase** with helicase activity, so it can move along the DNA. Defects in Ku lead to X-ray sensitivity and to severe combined immunodeficiency disease (SCID), because Ku plays a role in V(D)J recombination, the site-specific cleavage and rejoining required to assemble functional immunoglobulin genes. Ku associates with a large catalytic subunit to form **DNA-PK,** a DNA-dependent serine/threonine protein kinase. It is thought to bring the two free ends of the break together and stimulate the Ku ATPase activity that unwinds the ends until very short regions of "**microhomology**" (2–6 bp) can pair. The unpaired flaps could be removed by FEN1, the gaps filled by DNA polymerase, and the nicks ligated by DNA ligase.

Nonhomologous Recombination

Site-Specific Recombination

In site-specific recombination, specific enzymes catalyze the integration of a sequence into particular sites in the DNA. Integration of the bacteriophage λ is the best studied example of this. The **λ-integrase** catalyzes specific nicking of λ DNA and of a special sequence in the *E. coli* chromosome, and also catalyzes the resealing involved. This is a reversible process, and the integrated λ sequence can be excised from the chromosome. There are important examples of site-specific recombination that control processes such as mating type conversion in yeast and phase change (alterations in major surface protein) in trypanosomes.

Transposition

Transposition is the movement of specific pieces of DNA in the genome. Transposition resembles site-specific recombination in being catalyzed by special enzymes. Some pieces of DNA, called **transposons,** have two key features that enable them to move nearly anywhere into a target chromosome. They encode **transposase** enzymes and have **insertion sequences** that are recognized by the transposase. The transposase catalyzes the process of excision from one place in the genome and insertion into another, or the replicative insertion into a new place without losing the original piece of DNA. Other pieces of DNA contain insertion sequences but do not encode their own transposase; these can move when a transposase encoded elsewhere is expressed.

Transposons can integrate into other chromosomes and can move to new places within a chromosome. A significant fraction of the human genome has resulted from accumulation of transposons and insertion sequences. **LINEs (long interspersed elements** repeated DNA of which the L1 element is most common) are transposons present in about 50,000 copies in the human genome. Alu sequences are **SINEs** (**short interspersed elements** of repeated DNA) present in about 500,000 copies per haploid human genome; they do not encode a reverse transcriptase. Some mutations that cause disease are due to insertion of a transposon into a gene.

Retrotransposons are sequences that were transcribed into RNA and then reverse transcribed and integrated back into a random place in the genome. Retroviruses make DNA copies of their RNA genome (using **reverse transcriptase,** an enzyme that makes a DNA copy of an RNA) and insert them into the host chromosome by a transposition event catalyzed by a viral-encoded **integrase.**

Nonhomologous (Illegitimate) Recombination

In illegitimate recombination no homology is needed, nor are there special sequences. When genes are introduced into mammalian cells, they usually integrate randomly into the chromosome by nonhomologous recombination. This process probably uses the enzymes involved in double-strand break repair/recombination. Integration of a fragment of DNA can disrupt genes and cause mutations or dysregulation. This is a major limitation to gene therapy at present (see Clin. Corr. 4.6).

CLINICAL CORRELATION 4.6

Gene Therapy

Gene therapy is the introduction of new or altered genes into cells to correct a genetic defect or treat a disease. As we learn more about specific diseases, the possibilities for genetic intervention increase. For some diseases, one could theoretically cure the disease by replacing a defective gene with a normal one. For others, genes could be engineered either to supply missing proteins (e.g., insulin for diabetics) or to circumvent the problems caused by the disease.

Several major limitations impede progress. One is the difficulty of introducing genes into the relevant cells; this is being addressed by more work on vectors for gene introduction and on stem cell biology. Another is the problem created by the tendency of introduced genes to be incorporated randomly into the genome.

DNA introduced into mammalian cells is most frequently integrated into the genome by nonhomologous (illegitimate) recombination, probably as a result of double-strand break repair processes. This random insertion can result in the creation of new mutations by the insertion of the transgene into an inappropriate location. It can also create difficulties in sustaining gene expression over time when the gene is inserted into an inappropriate region of chromatin, affecting its regulation.

4.4 | REPAIR

Although we have emphasized the accuracy of DNA replication, it is not 100%. Even at an error rate of 10^{-9}, a handful of errors are introduced during each round of replication. Also, DNA in cells is constantly being altered by cellular constituents, including active oxygen species that are by-products of metabolism. Many environmental agents attack and modify DNA. Thus maintenance of the genetic information requires constant repair of DNA damage.

All free-living organisms have mechanisms to repair damage to DNA. In this section, we will discuss some of the most important lesions in DNA and the mechanisms by which they are repaired. Some lesions are repaired directly, but most are removed from DNA as part of the repair process. A key feature in most repair processes is the double-stranded nature of DNA, which contains the genetic information in duplicate. This allows restoration of the correct sequence on the damaged strand using the information on the other strand.

DNA Damage

In aqueous solution at 37°C, spontaneous **deamination** occurs of some C, A, and G bases in DNA. C deaminates to form U (Figure 4.26), A to hypoxanthine, and G to xanthine. Spontaneous **depurination** due to cleavage of the glycosyl bond connecting purines to the backbone (leaving the backbone of the DNA intact; Figure 4.26) occurs at a substantial rate. It has been estimated that between 2000 and 10,000 purines are lost per mammalian cell in 24 hours. These depurinated sites are called abasic (lacking a base) or **AP sites** (originally meaning apurinic, lacking a purine, but since generalized to lacking any base).

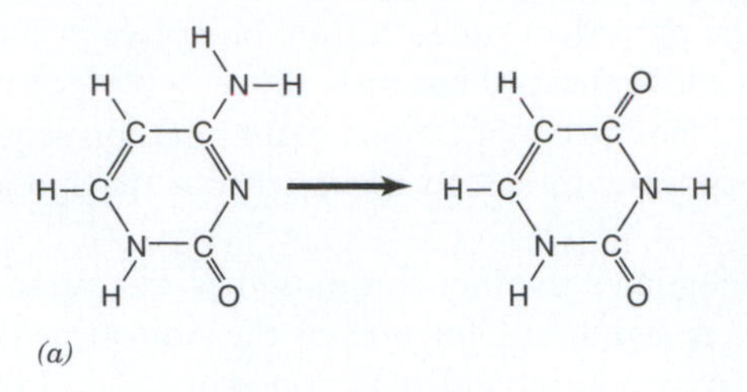

(a)

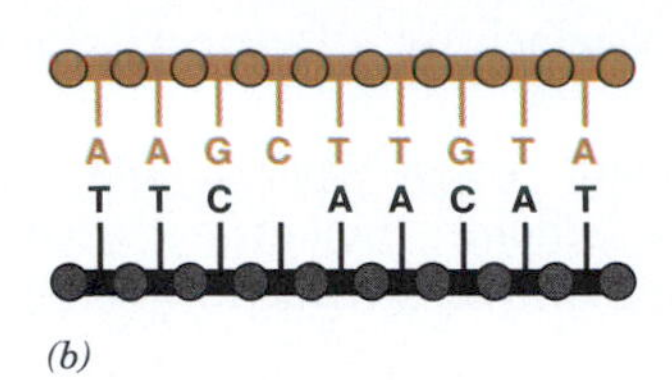

(b)

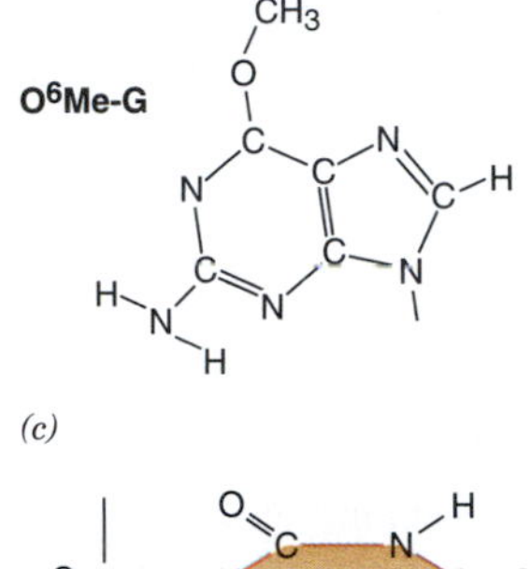

(c)

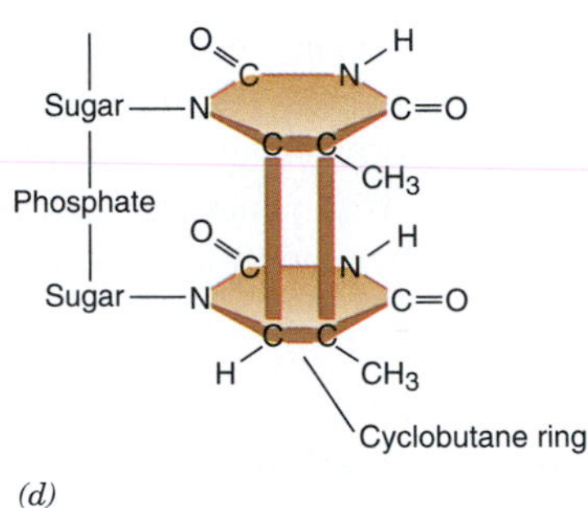

(d)

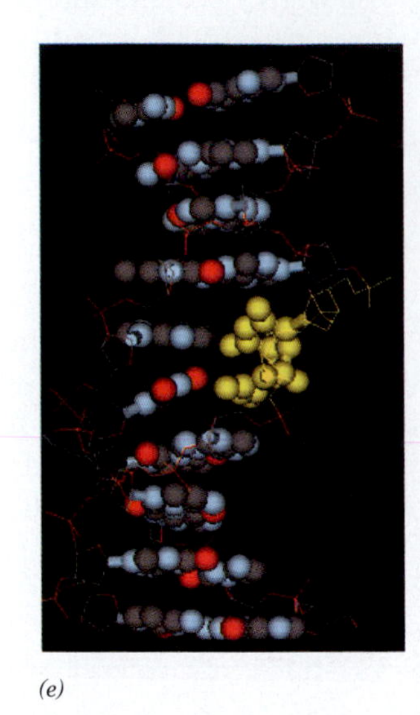

(e)

FIGURE 4.26
DNA damage.
(*a*) Oxidative deamination converts C to U. (*b*) AP site. Depurination (or, less frequently, depyrimidination) is cleavage of glycosyl bond between sugar and base. This leaves an abasic or AP site but does not break sugar–phosphate backbone. (*c*) O^6-methyl guanosine is a highly mutagenic lesion. (*d*) Ultraviolet light leads to cross-linking of adjacent pyrimidines along one strand of DNA. A *cis–syn* cyclobutane thymine dimer is shown in schematic. (*e*) A cyclobutane dimer is shown within double-stranded DNA; the backbone is shown as a stick figure to reveal distortion of cross-linked bases (yellow) and complementary bases on other strand.

Bases are oxidized at a substantial rate by reactive oxygen species that are byproducts of metabolism, creating many other altered bases, for example, 8-hydroxyguanine (8-oxo G). Oxidative damage tends to increase with age. *S*-adenosyl methionine, a carrier of methyl groups in normal metabolism, occasionally methylates a base in DNA. Products of lipid peroxidation can form adducts with bases, particularly G residues. Alkylating agents add methyl or other alkyl groups to the bases.

Ultraviolet radiation (from sunlight or tanning lamps) covalently links adjacent pyrimidines along one strand of the DNA to form *cis–syn* cyclobutane **pyrimidine dimers** (Figure 4.26) and other photoproducts. Ionizing radiation, including X-rays and radioactive decay, creates strand breaks and reactive oxygen species that damage DNA.

Carcinogens and mutagens attack DNA. Some are direct acting, others are **procarcinogens.** Procarcinogens in their native form do not damage DNA, but they can be activated by metabolic processes (e.g., oxidized by cytochrome P450s) into carcinogens that damage DNA. This process is called **metabolic activation. Benzo[*a*]pyrene** from coal tar is an extremely potent procarcinogen that requires oxidation to the epoxide to attack DNA. Many chemotherapeutic agents attack DNA. Some cause bases to become alkylated, others cross-link the two strands.

Mutations

Mutations are inherited changes in the DNA sequence. They can result from replication errors, from damage to the DNA, or from errors during repair of damage. Mutations that are changes of a single base pair are called **point mutations. Tran-**

sitions are point mutations in which one purine is substituted for another (e.g., A for G or G for A) or one pyrimidine is substituted for another (e.g., T for C or C for T). Deamination of C, if unrepaired, would lead to a transition. The frequency of transitions is increased by base analogs, including 2-amino purine. **Transversions** are point mutations in which a purine is substituted for a pyrimidine, or vice versa (e.g., A for C). **Missense mutations** are point mutations that change a single base pair in a codon such that the codon now encodes a different amino acid (Figure 4.27). **Nonsense mutations** are point mutations that change a single base pair in a codon to a stop codon that terminates translation; they usually have more severe effects than missense mutations.

Insertions or deletions of one or more base pairs can lead to **frameshifts** (if the number of base pairs is not a multiple of 3), disrupting the coding of a protein. Frameshifts not only change the amino acids encoded beyond the point of the insertion or deletion, but also generally lead to premature termination (or more rarely elongation) of the encoded polypeptide chain. This occurs when stop codons are generated or removed by the frameshift. Some chemicals, including **acridines** and **proflavin,** intercalate into the DNA; that is, they insert between adjacent base pairs. This usually leads to insertions or deletions of a single base pair. Some mutations also result from large-scale changes including the insertion of transposons.

Recently, a different type of mutation has been discovered, a **triplet expansion.** This is a great increase in the number of repeating triplets. Repeating triplets are a type of simple sequence repeat polymorphism (also called a microsatellite) that is frequent in the genome. There is generally a range of repeat numbers in normal individuals, since these sites are highly polymorphic. When repeats get significantly

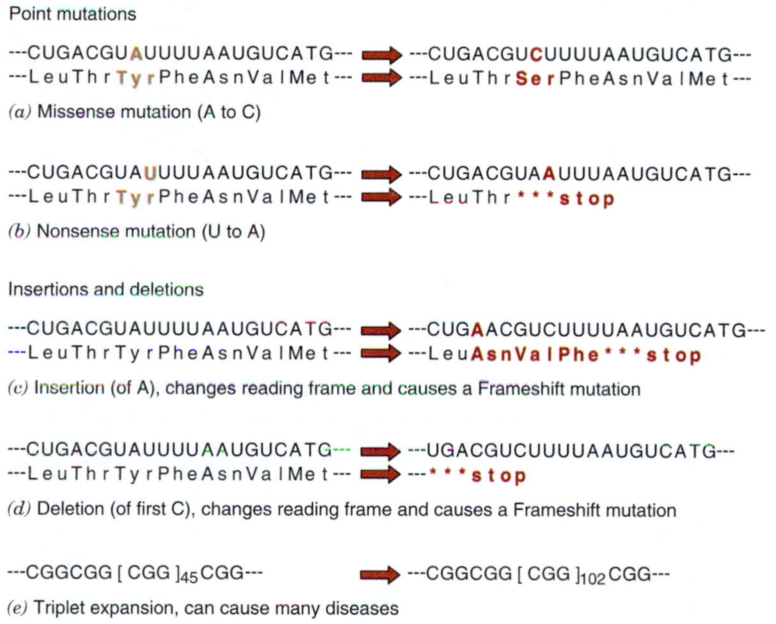

Point mutations

---CUGACGUAUUUUAAUGUCATG--- → ---CUGACGUCUUUUAAUGUCATG---
---LeuThrTyrPheAsnValMet--- → ---LeuThrSerPheAsnValMet---

(a) Missense mutation (A to C)

---CUGACGUAUUUUAAUGUCATG--- → ---CUGACGUAAUUUAAUGUCATG---
---LeuThrTyrPheAsnValMet--- → ---LeuThr***stop

(b) Nonsense mutation (U to A)

Insertions and deletions

---CUGACGUAUUUUAAUGUCATG--- → ---CUGAACGUCUUUUAAUGUCATG---
---LeuThrTyrPheAsnValMet--- → ---LeuAsnValPhe***stop

(c) Insertion (of A), changes reading frame and causes a Frameshift mutation

---CUGACGUAUUUUAAUGUCATG--- → ---UGACGUCUUUUAAUGUCATG---
---LeuThrTyrPheAsnValMet--- → ---***stop

(d) Deletion (of first C), changes reading frame and causes a Frameshift mutation

---CGGCGG $[\,CGG\,]_{45}$ CGG--- → ---CGGCGG $[\,CGG\,]_{102}$ CGG---

(e) Triplet expansion, can cause many diseases

FIGURE 4.27
Mutations.
Several different types of mutations are shown. *(a)* Missense mutation changes a single amino acid in encoded polypeptide. *(b)* Nonsense mutation changes a codon for an amino acid into a stop codon, terminating synthesis of encoded polypeptide. *(c)* Insertion of a single nucleotide changes the reading frame of all codons beyond the point of insertion; this usually leads to formation of a stop codon that terminates synthesis. *(d)* Deletion of a single nucleotide changes the reading frame of all codons beyond the point of insertion; this usually leads to formation of a stop codon that terminates synthesis. *(e)* Triplet expansion is a newly discovered type of mutation that leads to a great increase in the number of triplet repeats. Triplet expansion causes many diseases, including Huntington's disease and fragile X disease, by adding long stretches of a single amino acid to encoded polypeptide or by disrupting regulation of gene.

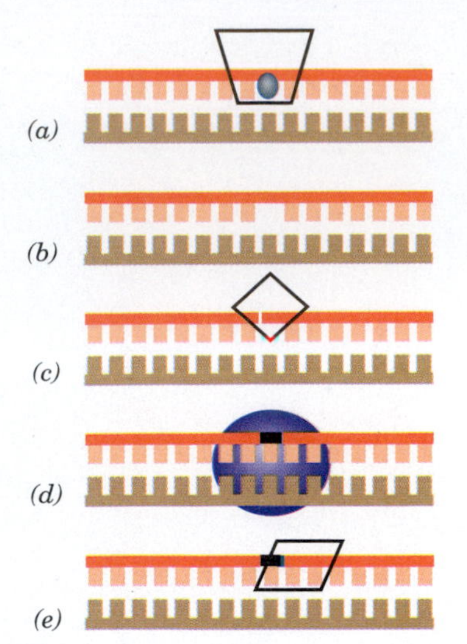

FIGURE 4.28
Base excision repair.
(*a*) The damaged site (blue dot) is recognized by a DNA glycosylase. (*b*) The base is removed by cleavage of a glycosyl bond that connects it to deoxyribose. The sugar–phosphate backbone is not broken. This leaves an AP site. (*c*) The sugar–phosphate backbone is cleaved 5′ to the abasic site by an AP endonuclease. There is also a cleavage at the 3′ side of this site to remove the sugar residue; this can be done by some AP endonucleases or an AP lyase activity. (*d*) Single nucleotide gap is filled by a DNA polymerase. (*e*) Remaining nick is sealed by a DNA ligase.

longer than normal, they may form a premutation that does not have a detectable phenotype but frequently leads to a further, dramatic expansion in the next generation. These much longer than normal repeats can cause disease by inserting a long stretch of the same amino acid residue into the encoded polypeptide, as in **Huntington's disease.** They can also affect regulation of a gene, as in **fragile X syndrome.**

Excision Repair

Excision repair is a general mechanism that can repair many different kinds of damage. The general pathway is the same, but there are differences in particular enzymes used to deal with different damages. The key defining characteristic of excision repair is removal of the damaged nucleotide(s), leaving a gap in the DNA, followed by resynthesis using the genetic information on the opposite strand, and then ligation to restore continuity of the DNA. There are two major modes of excision repair: base excision repair (BER) and nucleotide excision repair (NER) (Table 4.3).

Base Excision Repair

Base excision repair is an essential process that repairs many different types of damaged bases, including methylated, deaminated (e.g., U resulting from deamination of C), and oxidized bases, and abasic (AP) sites. The initial step is removal of a single damaged base from the DNA backbone by a **glycosylase** that cuts the *N*-glycosyl bond between the sugar and the base. This step does not break the sugar–phosphate backbone of the DNA (Figure 4.28). It leaves an abasic deoxyribose in the backbone, an AP site, that must be removed. Two different activities are required to remove this sugar: an **AP endonuclease** that cleaves the phosphodiester bond at the 5′ side but leaves the sugar still attached to the next nucleotide, and an **AP lyase** that cuts 3′ to the AP site to remove the sugar. The resulting single nucleotide gap has a free 3′-hydroxyl. The gap is filled and ligated by a DNA polymerase and DNA ligase, respectively.

There are many different glycosylases, at least eight in humans, each of which recognizes certain types of damaged bases. One of these, **uracil DNA glycosylase (UNG),** is specialized to remove U from DNA. U is not a normal constituent of DNA but is formed when C is deaminated, and can occasionally be incorporated into DNA during replication. Frequent deamination of C to U could lead to a high rate of mutation; therefore it is not surprising that a special repair mechanism to remove U evolved. But many C residues in mammalian DNA, in the sequence CG, are methylated at the 5 position; this is important in the regulation of gene expression. Deamination of these 5-methyl C residues leaves a T, rather than a U, in the DNA (cf. Figure 4.26). These T residues are not recognized by uracil DNA glyco-

TABLE 4.3 Excision Repair

Base excision repair	Nucleotide excision repair
Basic steps • Recognize damage • Remove damage by excising part of one strand • Resynthesize to fill gap; genetic information from other strand used • Ligate to restore continuity of DNA	
• Glycosylase removes base, leaves backbone intact • AP endonuclease cuts backbone, AP lyase removes sugar	• Double excision removes oligonucleotide (12–13 nt in *E. coli*, 27–29 nt in humans)
• DNA polymerase fills gap • DNA ligase seals nick	• DNA polymerase fills gap • DNA ligase seals nick

sylase. There are glycosylases that recognize the resulting T:G mismatch and preferentially remove the T. Sometimes, however, the mismatch repair system removes the G instead of the T, creating a mutation. Over millions of years, this has led to a much lower than expected frequency of CG dinucleotides in mammalian DNA.

The most important AP endonuclease in humans is APE1. After cleaving the backbone at the 5′ side of the AP site to leave a free 3′-OH, APE1 recruits DNA polymerase β (pol β) to the site. Pol β has AP lyase activity that removes the abasic sugar–phosphate remaining at the site of the strand break, leaving a single-nucleotide gap with a free 3′-OH. Pol β also has DNA polymerase activity that can fill the gap. For many types of damage, the gap is filled with only one or two nucleotides; this is called **short patch repair.** In cases where AP lyase cannot readily remove the sugar–phosphate, strand displacement can create a flap of several nucleotides and endonuclease FEN1 can cleave it. The somewhat larger gap is filled by a polymerase such as pol β, or possibly pol δ (or pol ε) stimulated by PCNA. In either case, the remaining nick is sealed by a DNA ligase. Protein–protein interactions coordinate this process, keeping the nick and the intermediates protected until the next enzyme has been recruited to the site.

Nucleotide Excision Repair

Nucleotide excision repair acts on a wide variety of damage, typically involving large adducts or distortion of the double helical structure of DNA. The best studied are pyrimidine dimers, the major products of ultraviolet damage. Chemical adducts with carcinogens such as benzo[*a*]pyrene and **aflatoxin** and with chemotherapeutic agents such as **cisplatin** are removed by this pathway, as are mismatched bases and small loops in DNA. The damage is removed by an enzyme complex that cuts several nucleotides away on both sides of the damaged base(s), so the damage is released as part of an oligonucleotide (Figure 4.29).

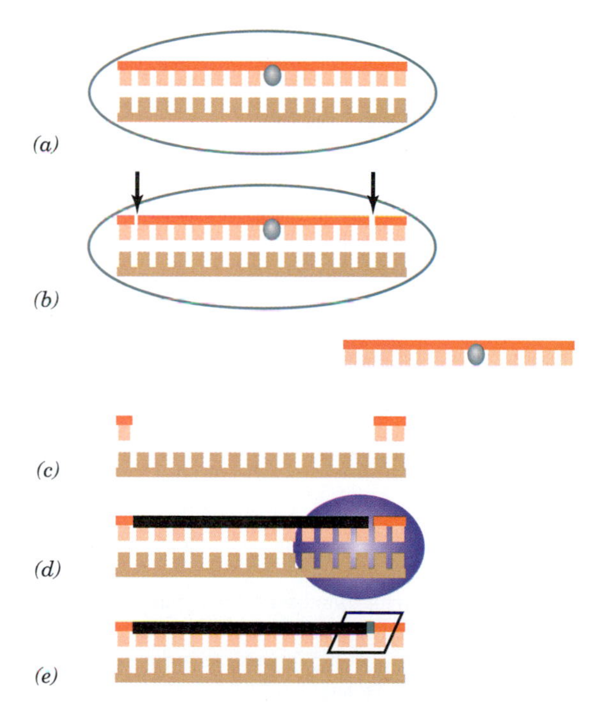

FIGURE 4.29
Nucleotide excision repair.
(*a*) The damaged base (blue dot) is recognized by a DNA repair complex. (*b*) The segment around the damage is excised by an enzyme complex that makes two nicks in the same strand, one on either side of the damage. (*c*) An oligonucleotide (27–29 nt in humans, 12–13 nt in *E. coli*) is released. (*d*) Resynthesis: a DNA polymerase fills the gap, using the opposite strand as a template. (*e*) DNA ligase seals the remaining nick.

In *E. coli,* a damage excision complex containing UvrA, UvrB, and UvrC recognizes the damage and makes two endonucleolytic cuts, one 3–5 nucleotides 3′ to the lesion and the other 8 nucleotides 5′ to the lesion. This excises the lesion as part of an oligonucleotide 12–13 nucleotides long. Damage recognition by UvrA appears to start the process, and opening of the strands by UvrB (a helicase) requires energy. UvrC binds and catalyzes the dual cleavage, with one cut on either side of the damage. A DNA polymerase fills in the gap, followed by ligation of the nick.

In humans, several large enzyme complexes are involved in damage recognition, opening of DNA, and the two cleavages. The rare, autosomal recessive disease **xeroderma pigmentosum** (XP), which renders patients photosensitive and susceptible to skin cancers, results from mutation in any of seven genes involved in excision repair or one involved in postreplication repair (see Clin. Corr. 4.7). XPC is part of a complex that binds tightly to lesions that distort DNA structure. XPA is a key factor involved in damage recognition and assembly of the excision complex; mutations in the *XPA* gene render individuals essentially unable to remove UV-induced pyrimidine dimers. The single-stranded DNA binding protein RPA is required for excision repair. XPA recruits a large complex called **TFII-H,** a general transcription factor involved in initiation of transcription. TFII-H contains XPB and XPD subunits, both of which are ATPases with limited helicase activity. The DNA is partially unwound by helicase activity of TFII-H, creating a bubble of about 25 bp. Then an XPF/ERCC1 heterodimer associates and makes an endonucleolytic cut on the damaged strand approximately 22–24 nucleotides 5′ to the lesion, and XPG makes an endonucleolytic cleavage approximately 5 nucleotides 3′ to the lesion. These two cuts liberate a lesion-containing oligonucleotide approximately 27–29 nucleotides in length. The gap in DNA is bound by RPA and filled by a replicative polymerase (pol δ or pol ε) stimulated by PCNA. The remaining nick is sealed by a DNA ligase, probably LIG1. These latter steps resemble completion of an Okazaki fragment.

Coupling of Repair and Transcription

Many lesions block transcription in the same manner that they block replication. Thus cells face an immediate problem even if they are not synthesizing DNA: they must be able to make key proteins in order to survive. A mechanism called **transcription-coupled repair** directs repair to the template strand of transcribed regions, so that lesions there are repaired faster than lesions elsewhere. The survival value of this is obvious. Presence of an RNA polymerase blocked by a lesion may prevent normal damage recognition. Transcription-coupled repair is a form of excision repair triggered by a halted RNA polymerase complex. In eukaryotic cells, a complex containing CSA and CSB (mutations that cause Cockayne syndrome, another DNA repair deficiency disease) recognizes the stalled RNA polymerase, caus-

CLINICAL CORRELATION 4.7
Xeroderma Pigmentosum

Xeroderma pigmentosum (XP) was the first disease recognized to be caused by defective DNA repair. Patients are photosensitive and highly susceptible to skin cancers in sun-exposed areas of the body. Many patients also have neurological problems. XP is a rare, autosomal recessive disease.

XP can be caused by defects in eight different genes. Defects in seven of these lead to defects in the initial incision step in nucleotide excision repair. This reflects the complexity of the repair process. Cells defective in any of these seven genes cannot carry out the first step, incision, at cyclobutane pyrimidine dimers. These dimers are the most common damage introduced by exposure to ultraviolet light, hence the extreme photosensitivity that is characteristic of this disease. An additional class, called XP-V (for variant), is not detectably deficient in excision repair but has a defect in postreplication repair after ultraviolet damage.

XPA is critical for the recognition of pyrimidine dimers and interacts with other repair proteins. XPB and XPD encode proteins that function as subunits of TFII-H, a general transcription factor that is also required for transcription-coupled DNA repair. TFII-H can function as a helicase, opening a region around the dimer. ERCC1–XPF is a complex that binds to TFII-H and incises at the 5′ side of the dimer. XPG incises at the 3′ side.

ing it to back up away from the lesion, and to recruit repair proteins. For many kinds of damage, the nucleotide excision repair pathway is used. This involves XPA, TFII-H, and other repair factors, except that XPC is not required for damage recognition. Recent evidence suggests that transcription-coupled repair also acts on oxidative lesions repaired by base excision repair. Once repair is complete, the RNA polymerase can resume synthesis.

Mismatch Repair

Mismatch repair is a specialized form of nucleotide excision repair that removes replication errors. Mismatches are not like DNA damage: there is no damaged or modified base present, just the wrong one of the four bases. Thus recognition of mismatches relies on the distortion of the double helical structure. A major difference between repair of DNA damage and repair of mismatches is in the choice of which base to excise. Enzymes can recognize damaged bases specifically and remove the damaged base, either individually in base excision repair or as part of an oligonucleotide in nucleotide excision repair. But when a mismatch is recognized, both bases are normal: which one should be excised? Clearly, excising the newly synthesized base would preserve the genetic information, whereas excising the base on the parental strand would permanently alter the DNA, producing a mutation. Randomly choosing one would lead to mutations half of the time, which is unacceptably high. Therefore the challenge is to recognize the newly synthesized strand and remove the mismatched base on that strand.

The brief period during which newly synthesized DNA is not methylated provides time within which that strand can be recognized. DNA in most organisms (but not all) is methylated at specific positions. In bacteria such as *E. coli,* primary methylation is of A in a GATC sequence. In mammalian DNAs, the primary methylation is of C in a CG sequence. Both GATC and CG are palindromic sequences, meaning that the opposite strand has the same sequence when read from 5′ to 3′, and both strands are methylated. This methylation does not affect base pairing. It does allow recognition of the methylated position and affects gene regulation. These bases are incorporated into the DNA in unmodified form and later methylated by a maintenance methylase. The **maintenance methylase** recognizes **hemimethylated** (methylated on only one strand, the parental strand) sequences in DNA and methylates the appropriate base. But there is a short time during which the DNA is only hemimethylated, and it is then that the *E. coli* mismatch repair system can recognize and specifically remove the newly synthesized strand in the region of a mismatch. Although a eukaryotic cell has a similar mismatch repair system, the manner in which it recognizes the new strand has not been determined; hemimethylated CG sequences or transient single-strand breaks from unligated Okazaki fragments may play roles.

The mismatch repair system is best studied in *E. coli*. Mismatches are recognized by a MutS homodimer, to which MutL binds (Figure 4.30). The two MutS monomers create a loop that contains the mismatch. The unmethylated, newly synthesized strand is nicked near the hemimethylated site. The nicked strand is excised by helicase II and either a 5′ to 3′ exonuclease such as RecJ or exo VII or a 3′ to 5′ exonuclease such as exo I. This excision can occur in either direction and is usually several hundred nucleotides in length, extending past the mismatch. The long gap that is created is filled by the action of DNA polymerase, adding nucleotides to the 3′ end of the nicked strand. When the gap has been filled, it is sealed by DNA ligase.

Humans have a similar mismatch repair system. Several complexes recognize mismatches. MutSα recognizes base–base mismatches and short insertions/deletions, and MutSβ recognizes the short insertions/deletions; both contain MSH2. A human MutLα complex is required for the incision step. The enzymes responsible for excision of the nicked strand to a point beyond the mismatch have not been identified. As in *E. coli,* the excised region is hundreds of nucleotides long. PCNA, the sliding clamp, is required for mismatch repair. In addition to its role as a processivity factor for the DNA polymerase, it appears to function during the excision

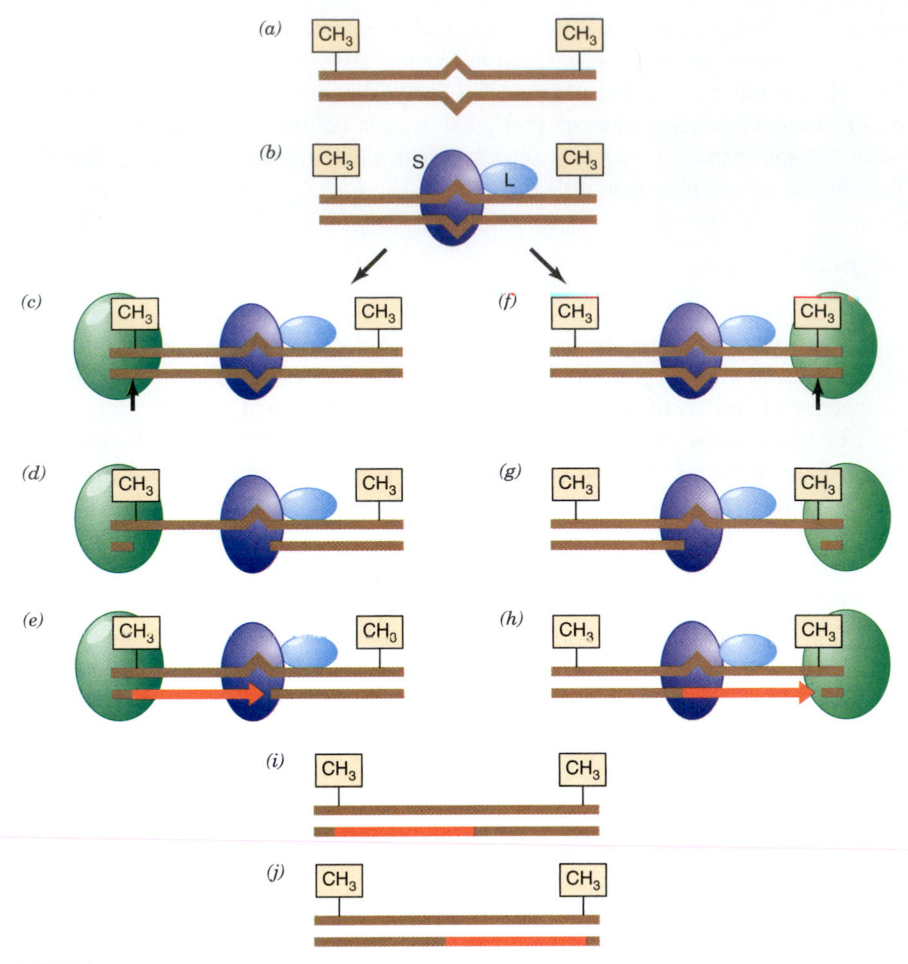

FIGURE 4.30
Mismatch repair in *E. coli*.
(*a*) A mismatch in newly replicated DNA. The parental strand is methylated, but for the first few minutes after it is synthesized, the new strand is not yet methylated. (*b*) MutS and MutL bind to the mismatch. ATP is hydrolyzed as they extrude a short loop of DNA. (*c, f*) MutH binds at a hemimethylated site (which could be to either side of the mismatch: left and right sides). (*d, g*) DNA of the unmethylated strand is nicked and then excised, extending back past the site of mismatch. This can be in either direction (*d* and *g*). (*e, h*) The resulting long gap is filled by a DNA polymerase, adding nucleotides to the 3′ end of the strand. (*i, j*) Remaining nick is sealed by a DNA ligase. Hemimethylated sites are methylated (not shown).

step. The resynthesis of the large gap requires DNA pol δ (or pol ε) and PCNA. The nick that remains is sealed by DNA ligase.

Defects in mismatch repair have been shown to cause **hereditary nonpolyposis colon cancer** (HNPCC) (see Clin. Corr. 4.8). Mismatch repair plays other roles in cells. It can lead to gene conversion during recombination by removing one strand of a heteroduplex and resynthesizing it using the other strand as a template. Mismatch repair plays a paradoxical role in resistance of cells to cytotoxic agents such as methylating agents (*N*-nitrosourea) and cisplatin, agents used in cancer chemotherapy. Resistant cells are often deficient in mismatch repair. This suggests that recognition of such damage by mismatch repair proteins (probably MutSα) can trigger cell death.

Enzymatic Demethylation

In addition to the normally methylated bases in DNA, some bases become alkylated inappropriately, by carcinogens or by interaction with the normal methyl carriers in

CLINICAL CORRELATION 4.8
Mismatch Repair and Cancer

Defects in mismatch repair cause hereditary nonpolyposis colon cancer (HNPCC) and probably are important in other cancers. This was initially suggested by the finding that some colon tumors showed frequent mutations in microsatellites (short repeating sequences, particularly mono- and dinucleotides), a phenotype called microsatellite instability.

HNPCC is an autosomal dominant disease. A gene contributing to the risk for HNPCC was mapped to a region that contains a mismatch repair gene. This led to studies that demonstrated the primary defect to be a mutation in one of the mismatch repair genes. Defects in *MLH1*, *MSH2*, or *PMS2* lead to HNPCC.

Why does the inheritance of a single defective allele in a mismatch repair gene lead to HNPCC? Cells lining the intestinal tract actively divide throughout a person's life. There is a high probability that at some time in one's life, a somatic mutation will occur in one allele of a mismatch repair gene in at least one colon cell. Cells in which one of the alleles at any of these loci is inactive can still carry out mismatch repair, so this does not generally lead to a tumor. However, in an individual who inherits one inactive allele, a somatic mutation that inactivates the single active copy leads to defective mismatch repair. Mutation rates can be elevated 100- to 1000-fold in cells with inactivated mismatch repair. The result is the accumulation of mutations that eventually lead to tumor formation through mutations in growth-regulating genes.

cells. Alkylation on positions that affect base pairing is highly mutagenic. For example, O^6-methyl guanine base pairs with T rather than with C (Figure 4.31). Replication of this base would lead to a mutation. Cells have glycosylases that recognize inappropriately methylated bases and trigger base excision repair, but cells also have special proteins that recognize O^6-methyl guanine in DNA and directly remove the methyl group, leaving the DNA intact. **O^6-methyl guanine methyl transferase** (MGMT) recognizes this specific lesion in DNA and transfers the methyl group from the guanine to a cysteine on the protein itself (Figure 4.31). A single molecule of MGMT can transfer only one methyl group; methylation of the cysteine renders the protein inactive for further transfers, and the inactivated protein is degraded through the ubiquitin proteolytic pathway. The cost to the cell is high. It must replicate a stretch of DNA to maintain the gene for this enzyme, transcribe the gene and process the transcript, and synthesize the protein. All of these steps require considerable energy and material input. And all of this energy and material produces the protein that removes a single methyl group from the DNA. This reemphasizes how important it is to maintain the integrity of the genome, even at great metabolic cost.

Photoreactivation

Photoreactivation is a specific mechanism for repair of cyclobutane pyrimidine dimers, the major lesions produced by ultraviolet irradiation. Photoreactivation is a direct reversal of the damage. A specialized enzyme called **photolyase** binds to the

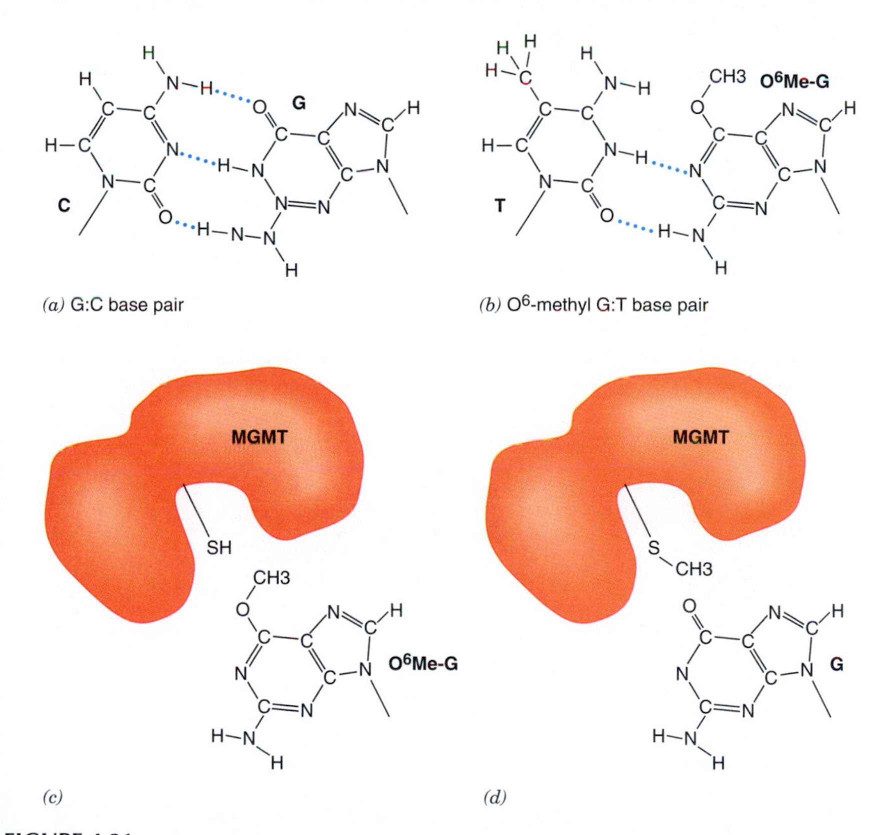

FIGURE 4.31
Enzymatic demethylation.
(*a*) Normal G:C base pair. (*b*) O^6-methyl G:T base pair. This base pairing would allow incorporation of a T opposite G during DNA replication and would lead to a mutation in the next generation. (*c*) MGMT (methyl guanine methyl transferase) can bind to O^6-methyl G. (*d*) MGMT transfers a methyl group from the O^6 of guanine to a sulfhydryl group on the protein, leaving an intact G in DNA. The methylated protein is inactive.

damage (Figure 4.32). Upon absorbing light, it catalyzes the reversal of the bonds between the adjacent pyrimidines, leaving the DNA exactly as it was before the dimer formed. This is a direct reversal of the damage. Photoreactivation, while of great importance in most organisms, is not significant in mammals.

Lesions Can Block Replication

The accuracy and proofreading abilities of DNA polymerases generally prevent them from inserting nucleotides opposite damaged bases. Thus when a replication complex reaches a lesion in the template strand, it halts. This causes a delay in replication and must be accommodated for replication to be completed. There are three general ways for this to occur: the complex can either disassemble or back up to allow repair of the lesion, skip past the lesion and resume synthesis beyond it (leaving a gap in the daughter strand), or bypass the lesion. All three of these occur, and all present some problems.

Daughter-Strand Gap Repair

Many lesions block replication. If replication restarts beyond the lesion, a gap is left in the DNA. This situation is easiest to picture on the retrograde strand where it can occur simply by dissociation of the polymerase from the site of the lesion and reattachment to synthesize the next Okazaki fragment by the normal replication mechanism (Figure 4.33). Gaps in DNA are potentially lethal. Also, a lesion opposite a gap is not a substrate for excision repair, which requires an intact strand opposite the lesion to provide coding information necessary for repair. **Daughter-strand gap repair** (postreplication recombination) does not remove the lesion, but does repair the gap that was caused by the lesion. This protects the cell from the potential damage that gaps can cause and also provides a substrate for excision repair to act on the lesion itself.

The mechanism of daughter-strand gap repair relies on recombination (Figure 4.33). Single-stranded DNA such as that found opposite a gap is recombinogenic; that is, it stimulates recombination. Recombination between the two duplexes allows transfer of a piece of the parental strand into the gap (the parental strand on one side of the fork has the same polarity as the daughter strand on the other side). This leaves a gap in the parental strand that donated the patch, but because it has a normal strand of DNA opposite, that gap can readily be filled by DNA polymerase and sealed by a DNA ligase. The result is that the gap is repaired, although the lesion remains. As noted above, in this configuration the lesion is a substrate for excision repair.

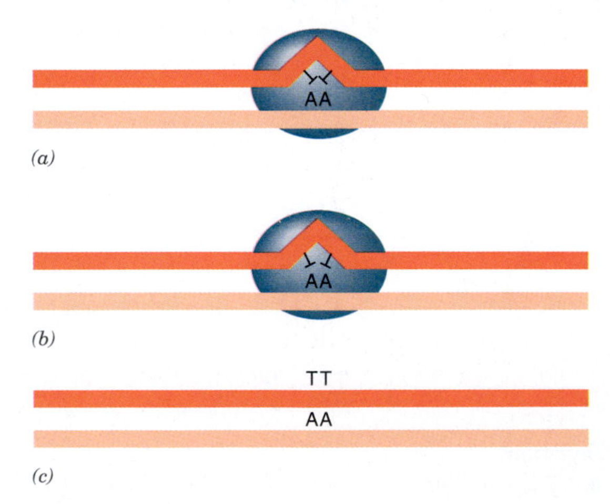

FIGURE 4.32
Photoreactivation.
(*a*) Photolyase binds to a cyclobutane pyrimidine dimer; light is not needed for this step. (*b*) The complex absorbs light, which results in cleavage of bonds linking adjacent pyrimidines. (*c*) The enzyme then dissociates from the DNA.

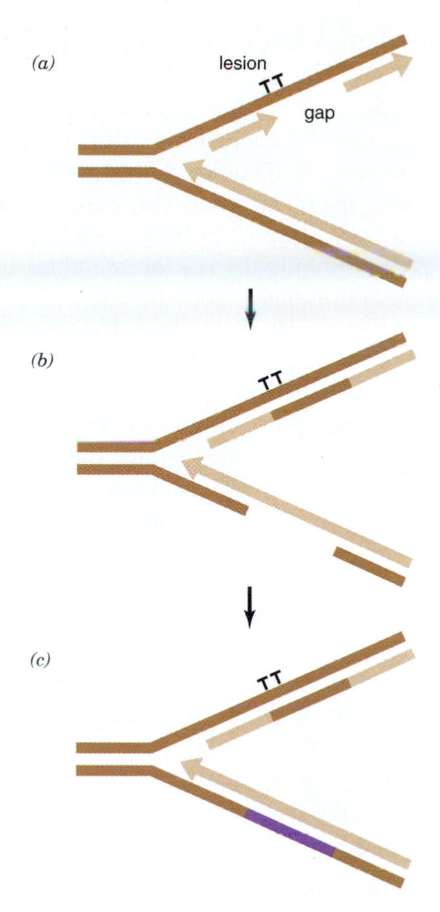

FIGURE 4.33
Daughter-strand gap repair (recombination repair).
(*a*) Gaps are left in newly replicated DNA where replication forks are halted by lesions. Because a lesion does not have intact DNA opposite, it is not a substrate for excision repair. (*b*) Recombination allows isopolar parental strand to fill gap in daughter strand. This leaves a gap in parental strand. (*c*) Gap in parental strand can be filled by a DNA polymerase, because there is an intact template opposite it. The result is repair of gap in the daughter strand, but the lesion remains. Note that the lesion can now be repaired by excision repair, because it is now opposite an intact strand.

It is easy for a replication complex that has stopped synthesis of the retrograde strand at a lesion to resume synthesis at the next Okazaki fragment, using the normal mechanisms for synthesis of the discontinuous strand. Reinitiating the continuous strand beyond a block is more difficult, because initiations are normally tightly controlled at replication origins. Blocked forks can be restarted through recombination with extensive branch migration and then resolution, which can recreate the DNA structure of a replication fork beyond the site of the lesion. This still leaves the task of reassembling the replication complex in an origin-independent manner, the details of which are not yet clear.

When a replication fork encounters a single-strand break, that break is converted into a double-strand break. Double-strand breaks are potentially lethal. The major pathway for cell survival involves double-strand break repair, by mechanisms similar to the double-strand break model of recombination. It has been estimated that about ten such breaks are formed during a single cycle of replication in mammalian cells, showing the importance of double-strand break repair for cell survival.

Bypass Synthesis

An alternative to forming gaps when lesions are encountered in the template strand is to continue synthesis through the lesion, a process called **bypass synthesis** (Figure 4.34). A very accurate, proofreading polymerase cannot do this readily.

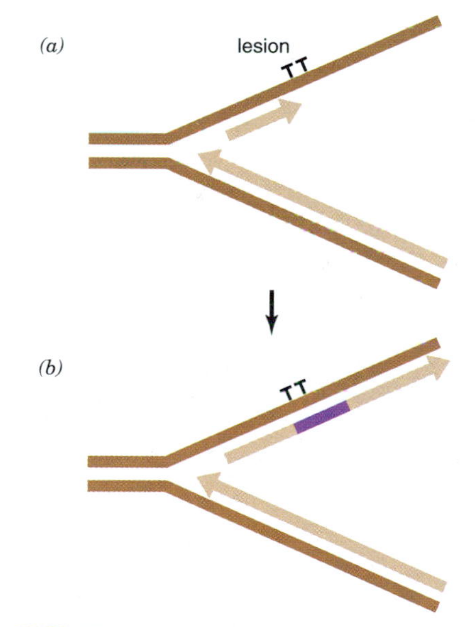

FIGURE 4.34
Translesion synthesis.
(*a*) Damage to the template generally halts replicative DNA polymerases, since they cannot add a nucleotide opposite a noncoding or strand-distorting lesion. (*b*) Specialized DNA polymerases can insert nucleotides opposite lesions in the template and thereby allow bypass of damage and synthesis of nongapped daughter strands. To do this, proofreading must be relaxed, which means that errors (potential mutations) are introduced at a high frequency.

There is a family of bypass DNA polymerases with reduced fidelity that carry out this function (see Table 4.2). The reduced accuracy greatly increases the probability that the wrong nucleotide will be incorporated. The harmful consequences of this are evidently less than those of a long (or permanent) block to replication.

In *E. coli*, DNA polymerases IV (DinB) and V ($UmuD'_2C$ complex) are induced as part of the SOS response described below. This ensures that the inaccurate synthesis mechanism is functioning only when there is sufficient damage to the DNA to trigger the SOS response, limiting the risk of mutations to situations in which the cell's viability is seriously threatened. Both are much less accurate than the normal replicative DNA polymerases III and I. They have low processivity, generally synthesizing fewer than 10 nucleotides before dissociating from the DNA. This serves to limit the extent of mutations introduced.

Eukaryotes also have bypass polymerases (see Table 4.2). DNA polymerase η predominantly incorporates A opposite the thymine dimers introduced by UV irradiation, which reduces the probability of mutagenesis at those lesions. However, polymerase η lacks intrinsic proofreading capability and incorporates one mismatch per 20–400 nucleotides synthesized, suggesting that its activity must be tightly controlled. A second bypass polymerase is DNA polymerase ι, also very inaccurate, which incorporates nucleotides opposite other highly distorting lesions or lesions such as abasic sites that cannot normally serve as templates. These polymerases appear to function with a third bypass enzyme, DNA polymerase ζ, which doesn't itself add nucleotides opposite lesions but can extend the structures formed by the other bypass polymerases until the normal replicative polymerase can take over. These are distributive enzymes, so the region in which mismatches are likely to be introduced is typically small, limiting the introduction of mutations.

Double-Strand Break Repair

Double-strand breaks (DSBs) in DNA are potentially lethal events. They stimulate genetic recombination that can lead to chromosomal translocations and, if unrepaired, lead to broken chromosomes and cell death. Double-strand breaks are caused by many agents, particularly ionizing radiation, reactive oxygen species, and many chemotherapeutic agents that generate oxidative free radicals (bleomycin) or are topoisomerase poisons. DSBs can be repaired by homologous recombination (described earlier; see Figure 4.25) or by nonhomologous end-joining. In yeast, homologous recombination predominates; in mammalian cells, nonhomologous end-joining predominates.

Regulation of Repair: The SOS Regulon

In *E. coli*, damage to DNA triggers the **SOS response,** a coordinated change in gene expression that aids in recovery from damage. The SOS response is a coordinated induction of a large number of genes whose transcription is at least partly regulated by a common repressor, LexA. Among these genes are *uvrA*, *uvrB*, *uvrC*, *recA* and *lexA* itself. (There is a low, constitutive level of many of the genes in the SOS regulon, because repression by LexA is not complete.) A group of operons that are controlled by a common repressor is called a **regulon.** The SOS response is triggered by long stretches of single-stranded DNA such as those in gaps left by blocked replication (Figure 4.35). A low level of RecA protein is normally present in cells and can cooperatively polymerize along single-stranded DNA to form a nucleoprotein filament. In this state, it is "activated" and binds a protein called LexA, increasing the rate at which LexA is cleaved into inactive fragments. Thus the genes normally repressed by LexA are turned on. RecA participates in daughter-strand gap repair, as noted earlier. The other damage-inducible genes (*din* genes) produce proteins that aid in cell recovery and inhibit initiation of new replication forks and cell division until repair is completed (Figure 4.35).

Autoregulation of *lexA* is based on a simple feedback loop and effectively allows a rapid but transient response to DNA damage. When LexA protein is cleaved,

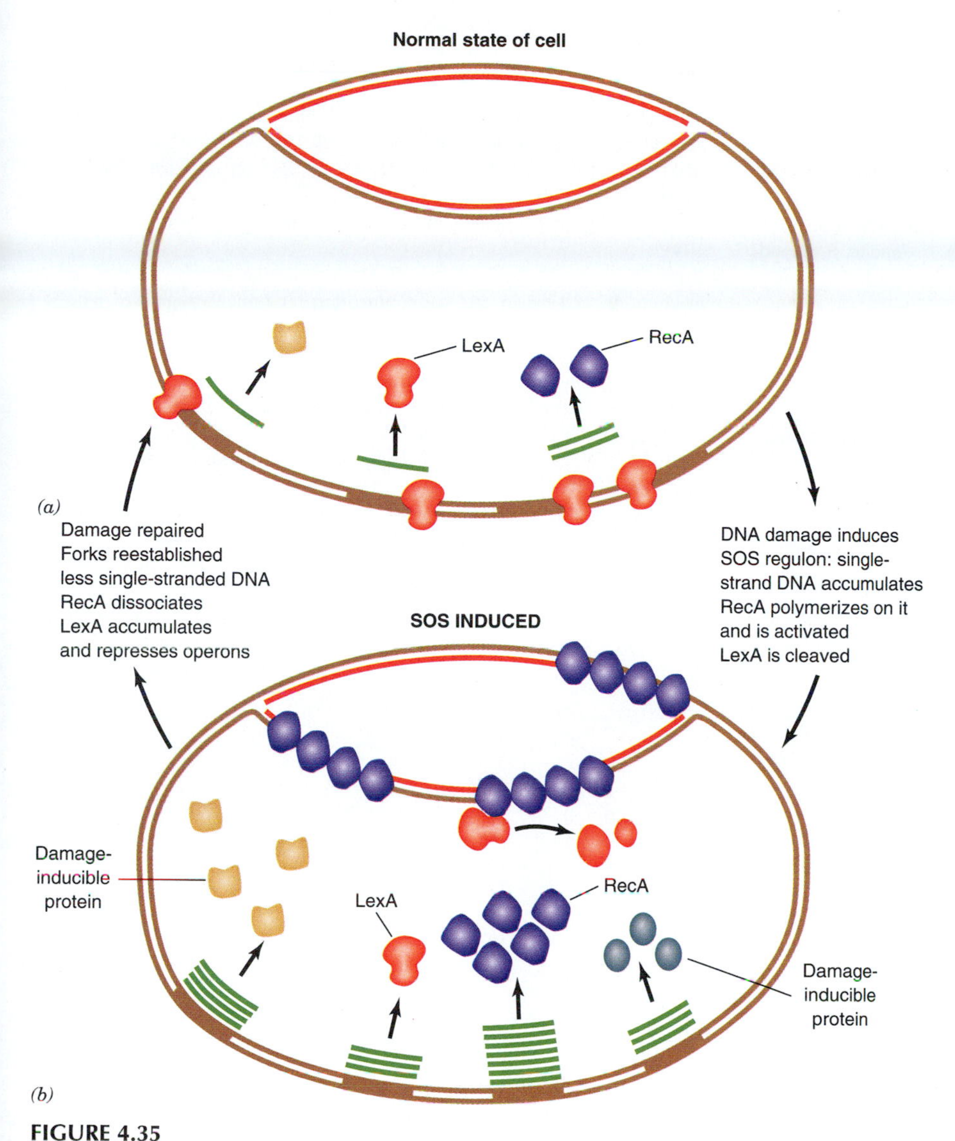

FIGURE 4.35
Regulation of DNA repair and recovery in *E. coli:* the SOS response.
(*a*) In undamaged cells, a group of operons that constitute the SOS regulon are regulated by a common repressor, called LexA. This regulon includes the *lexA* gene itself (autoregulation) and the *recA* gene, along with many genes encoding enzymes that act to repair damage. Most of these genes are largely repressed by binding of LexA to operator sites in their regulatory region, although there is residual transcription of some of them, including *lexA* and *recA*. (*b*) Damage blocks replication forks and thereby leaves single-stranded DNA. RecA binds to single-stranded DNA and is "activated." In that form, it aids in the cleavage of LexA. The cleaved fragments of LexA cannot bind to operators, so the entire set of operons is induced. Much more RecA is made, along with other proteins including LexA. As long as the damage remains and RecA remains activated, LexA continues to be cleaved so the operons remain on. When damage is repaired, RecA is "deactivated" and no longer cleaves LexA, so active LexA builds up and shuts off regulon.

among the many genes of the SOS regulon that are induced is *lexA* itself, so the amount of LexA produced increases significantly. However, while DNA damage is still present and RecA protein is still activated, this LexA protein is cleaved and cannot repress the many genes. As the damage is repaired, the amount of activated RecA drops and therefore cleavage of LexA (whose synthesis was also increased, because it is autoregulated by binding of LexA to the *lexA* promoter) decreases. The increasing levels of LexA in the cell bind to the operators of the *din* genes and gradually shut down the SOS response.

BIBLIOGRAPHY

Baynton, K. and Fuchs, R. P. P. Lesions in DNA: hurdles for polymerases. *Trends Biochem. Sci.* 25:75, 2000.

Berger, J. M., Gamblin, S. J., Harrison, S. C., and Wang, J. C. Structure and mechanism of DNA topoisomerase II. *Nature* 379:225, 1996.

Blow, J. J. and Tada, S. A new check on issuing the license. *Nature* 404:560, 2000.

Cox, M. M., Goodman, M. F., Kreuzer, K. N., Sherratt, D. J., Sandler, S. J., and Marians, K. J. The importance of repairing stalled replication forks. *Nature* 404:37, 2000.

Evans, A. R., Limp-Foster, M., and Kelley, M. R. Going ape over Ref-1. *Mutat. Res.* 461:83, 2000.

Flores-Rozas, H. and Kolodner, R. D. Links between replication, recombination, and genome instability in eukaryotes. *Trends Biochem. Sci.* 25:196, 2000.

Friedberg, E. C., Walker, G. C., and Siede, W. *DNA Repair and Mutagenesis.* Washington, D.C.: American Society of Microbiology, 1995.

Haber, J. E. DNA recombination: the replication connection. *Trends Biochem. Sci.* 25: 271, 2000.

Hanawalt, P. C. Transcription-coupled repair and human disease. *Science* 266:1957, 1994.

Hanawalt, P. C. The bases for Cockayne syndrome. *Nature* 405:415, 2000.

Hubscher, U., Nasheuer, H.-P., and Syvaoja, J. E. Eukaryotic DNA polymerases, a growing family. *Trends Biochem. Sci.* 25:143, 2000.

Human genome project. http://www.nhgri.nih.gov/HGP/.

Human genome sequencing. http://www.ncbi.nlm.nih.gov/genome/seq/.

Johnson, R. E., Washington, M. T., Haracska, L., Prakash, S., and Prakash, L. Eukaryotic polymerases ι and ζ act sequentially to bypass DNA lesions. *Nature* 406:1015, 2000.

Kowalczykowski, S. C. Initiation of genetic recombination and recombination-dependent replication. *Trends Biochem. Sci.* 25:156, 2000.

Lindahl, T., and Wood, R. D. Quality control by DNA repair. *Science* 286:1897, 1999.

Modrich, P. Mismatch repair, genetic stability, and cancer. *Science* 266:1959, 1994.

Naktinis, V., Turner, J., and O'Donnell, M. A molecular switch in a replication machine defined by an internal competition for protein rings. *Cell* 84:137, 1996.

Sancar, A. Excision repair in mammalian cells. *J. Biol. Chem.* 270:15915, 1995.

Stillman, B. Cell cycle control of DNA replication. *Science* 274:1659, 1996.

Stahl F. Meiotic recombination in yeast: coronation of the double-strand-break repair model. *Cell* 87:965, 1996.

Von Hippel, P. H. and Jing, D. H. Bit players in the trombone orchestra. *Science* 287:2435, 2000.

Waga, S. and Stillman, B. Anatomy of a DNA replication fork revealed by reconstitution of SV40 replication in vitro. *Nature* 369:207, 1994.

Watson, J. D. and Crick, F. H. C. Genetical implications of the structure of deoxyribonucleic acid. *Nature* 171:964, 1953.

Yu, Z., Chen, J., Ford, B. N., Brackley, M. E., and Glickman, B. W. Human DNA repair systems: an overview. *Environ. Mol. Mutagen.* 33:3, 1999.

Zakian, V. A. Telomeres: beginning to understand the end. *Science* 270:1601, 1995.

QUESTIONS | C. N. ANGSTADT

Multiple Choice Questions

1. Which of the following statements about *E. coli* DNA polymerases is correct?
   A. All polymerases have both 3′ to 5′ and 5′ to 3′ exonuclease activity.
   B. The primary role of polymerase III is in DNA repair.
   C. Polymerases I and III require both a primer and a template.
   D. Polymerase I tends to remain bound to the template until a large number of nucleotides have been added.
   E. The specificity of the polymerase reaction is inherent in the nature of the polymerases.

2. Both strands of DNA serve as templates concurrently in:
   A. replication.
   B. excision repair.
   C. mismatch repair.
   D. transcription-coupled repair.
   E. all of the above.

3. Replication:
   A. is semiconservative.
   B. requires only proteins with DNA polymerase activity.
   C. uses 5′ to 3′ polymerase activity to synthesize one strand and 3′ to 5′ polymerase activity to synthesize the complementary strand.
   D. requires a primer in eukaryotes but not in prokaryotes.
   E. must begin with an excision step.

4. The discontinuous nature of DNA synthesis:
   A. requires that DNA polymerase III (or appropriate eukaryotic enzyme) dissociate from the template when it reaches the end of each single-stranded region.
   B. is necessary only because synthesis is bidirectional from the initiation point.
   C. leads to the formation of Okazaki fragments.
   D. means that synthesis occurs on the second strand of DNA only after synthesis on the first strand is completed.
   E. means that both 3′ to 5′ and 5′ to 3′ polymerases are used.

5. All of the following are factors in the unwinding and separation of DNA strands for replication EXCEPT:
   A. the tendency of negative superhelices to partially unwind.
   B. destabilization of complementary base pairs by helicases.
   C. the action of topoisomerases.
   D. the enzymatic activity of SSB proteins.
   E. energy in the form of ATP.

6. In eukaryotic DNA replication:
   A. only one replisome forms because there is a single origin of replication.

B. the Okazaki fragments are 1000–2000 nucleotides in length.
C. helicase dissociates from DNA as soon as the initiation bubble forms.
D. at least one DNA polymerase has a 3′ to 5′ exonuclease activity.
E. the process occurs throughout the cell cycle.

7. All of the following statements about telomerase are correct EXCEPT:
 A. the RNA component acts as a template for the synthesis of a segment of DNA.
 B. it adds telomeres to the 5′ ends of the DNA strands.
 C. it provides a mechanism for replicating the ends of linear chromosomes in most eukaryotes.
 D. it recognizes a G-rich single strand of DNA.
 E. it is a reverse transcriptase.

8. A transition mutation:
 A. occurs when a purine is substituted for a pyrimidine, or vice versa.
 B. results from the insertion of one or two bases or base analogs into the DNA chain.
 C. decreases in frequency in the presence of base analogs such as 2-amino purine.
 D. results from the substitution of one purine for another or of one pyrimidine for another.
 E. always is a missense mutation.

9. Homologous recombination:
 A. occurs only between two segments from the same DNA molecule.
 B. requires that a specific DNA sequence be present.
 C. requires that one of the duplexes undergoing recombination be nicked in both strands.
 D. may result in strand exchange by branch migration.
 E. is catalyzed by transposases.

10. All of the following are true about transpositions EXCEPT:
 A. transposons move from one location to a different one within a chromosome.
 B. both the donor and target sites must be homologous.
 C. transposons have insertion sequences that are recognized by transposases.
 D. the transposon may either be excised and moved or be replicated with the replicated piece moving.
 E. transposition may either activate or inactivate a gene.

Questions 11 and 12: Retroviruses, like HIV (human immunodeficiency virus), which causes AIDS (acquired immune deficiency syndrome), have their genetic information in the form of RNA. However, reverse transcriptase synthesizes a DNA copy of the viral genome. One of the major drugs used in treating AIDS is AZT, an analog of deoxythymidine, which has an azido group at the 3′ position of the sugar. It can be phosphorylated and competes with dTTP for incorporation into the reverse transcript. Once incorporated, the presence of the analog terminates chain elongation.

11. The growing chain is terminated because:
 A. the analog cannot hydrogen bond to RNA.
 B. the presence of the AZT analog inhibits the proofreading ability of reverse transcriptase.
 C. elongation requires a 3′-OH to attack the phosphate of the incoming nucleotide and the analog of AZT does not have a free 3′-OH.
 D. the analog causes distortion of the growing chain, which inhibits reverse transcriptase.
 E. dTTP can no longer be added to the growing chain.

12. There is a therapeutic window in which the effect is primarily on viral replication because AZT is much less effective at competing with dTTP for incorporation by cellular DNA polymerases. This may be related to the proofreading ability of DNA polymerases. Proofreading activity to maintain the fidelity of DNA synthesis:
 A. occurs after the synthesis has been completed.
 B. is a function of 3′ to 5′ exonuclease activity intrinsic to or associated with DNA polymerases.
 C. requires the presence of an enzyme separate from the DNA polymerases.
 D. occurs in prokaryotes but not eukaryotes.
 E. is independent of the polymerase activity in prokaryotes.

Questions 13 and 14: Patients with the rare genetic disease xeroderma pigmentosum (XP) are very sensitive to light and are highly susceptible to skin cancers in areas of the body exposed to the sun. Many patients also have neurological problems. Although the condition is rare, study of such patients has enhanced our knowledge of DNA repair because XP is caused by defective DNA repair—specifically nucleotide excision repair. (There is a variant, XP-V, that is deficient in postreplication repair.)

13. All of the following are true about nucleotide excision repair EXCEPT:
 A. removal of the damaged bases occurs on only one strand of the DNA.
 B. it removes thymine dimers generated by UV light.
 C. it involves the activity of an excision nuclease, which is an endonuclease.
 D. it requires polymerase I (*E. coli*) and ligase.
 E. only the damaged nucleotides are removed.

14. The defect in XP is in nucleotide excision repair. Another type of DNA repair is base excision repair. Base excision repair:
 A. is used only for bases that have been deaminated.
 B. uses enzymes called DNA glycosylases to generate an abasic sugar site.
 C. removes about 10–15 nucleotides.
 D. does not require an endonuclease.
 E. recognizes a bulky lesion.

Problems

15. Mismatch repair removes replication errors by excising the incorrect bases. There is no DNA damage or modified bases present—only a normal, but incorrect, base. How does the cell distinguish the newly synthesized strand to be removed from the correct parental DNA strand, which must be preserved?

16. In the coding strand of DNA for the alpha gene of normal hemoglobin (HbA), the three bases that correspond to position 142 of the mRNA synthesized are TAA and the alpha chain has 141 amino acids. In the coding strand of the gene for the alpha chain of hemoglobin constant spring, the three bases in the same position as above are CAA and the alpha chain produced contains 172 amino acids. Explain the mutation that has occurred. Is the mutation a frameshift or a point mutation?

ANSWERS

1. **C** The primer is the initial 3′ terminus of an existing strand and the template is the free portion of the complementary strand. A: Pol III has 3′ to 5′ activity but not the other. B: Polymerase III functions in synthesis. D: Polymerase I has low processivity because it dissociates after only a few nucleotides add. E: Specificity is a function of complementary hydrogen bonding between the base being added and the template.
2. **A** This allows for the synthesis of two identical DNA molecules. B, C: In both of these the damaged segment is removed so both strands are not available. D: This is an excision repair.
3. **A** The new DNA has one parent and one new strand. B, D: Replication requires a primer, usually synthesized by a primase. Ligases, helicases, and other proteins are required as well. C: Replication involves Okazaki fragments because synthesis occurs only in the 5′ → 3′ direction. E: Excision is the recognition step for DNA repair.
4. **C** These are the segments of DNA built upon the primer. A: DNA polymerase remains bound to the template and slides over the next primer to continue synthesis. B, E. This mechanism compensates for the inability to synthesize 3′ to 5′ and would be necessary even if synthesis were unidirectional. D: Both strands are synthesized concurrently.
5. **D** SSB proteins have a very important role in stabilizing the single strands after separation but they are not enzymes. A: This is especially true in regions of high AT pairs. B, E: This helps in the original unwinding at the expense of ATP. C: Topoisomerases nick and reseal one of the strands to prevent the introduction of an increasing number of positive supercoils.
6. **D** Polymerase α shows this activity that provides proofreading during synthesis. δ and ε have this for proofreading. A: There are multiple initiation sites. B: This is the size of prokaryotic Okazaki fragments; eukaryotic fragments are about 200 nucleotides long. C: Helicase activity is also necessary for the continuation of synthesis, that is, the opening of the forks. E: Replication is confined to the S-phase.
7. **B** Telomeres are at the 3′ end of each strand so that the 5′ ends can be replicated. A, C: Telomerase both positions itself at the 3′ end of the DNA and provides the template for extending that end. D: This is a characteristic of the 3′ end. E: It is using an RNA template to synthesize DNA.
8. **D** This is the definition—for example, A for G or T for C. B: This is a frameshift and transitions are point mutations. C: The frequency of mutation increases. E: It could be a missense mutation if the change coded for a different amino acid or a nonsense mutation if the code was changed to a stop signal.
9. **D** This is just one of the events in this complex process. A, B: It may occur between two distinct DNA molecules; the requirement is that the two sequences be homologous but not that they be specific sequences. C: The nicks are usually on a single strand as depicted by the Holliday and Meselson–Radding models. E: These are the enzymes of transpositional site-specific recombination.
10. **B** Only the donor site requires a specific nucleotide sequence; homology is not required. A: This is the definition. C: This is one of the key features. D: Both types of event are catalyzed by transposases. E: Insertion into the middle of a gene would inactivate it; insertion of a promoter next to a gene may activate it.
11. **C** The chemistry of nucleotide formation requires a free 3′-OH at the end of the chain for attachment of the next nucleotide. A: The presence of the azido group does not affect hydrogen bonding. B: Reverse transcriptase does not have proofreading properties. D: This does not happen. E: The AZT analog competes with dTPP; it does not eliminate it.
12. **B** This activity removes a newly added base if there is a mismatch with the template. A: This is called repair. C: Most polymerases are multifunctional enzymes and have proofreading ability. Sometimes another protein associated with the complex is used. D: Not all eukaryotic polymerases have 3′ → 5′ exonuclease activity but some do. E: The polymerase active site seems to be the one that detects the mismatch and directs the 3′terminus to the proofreading site.
13. **E** The cuts are made several nucleotides on either side of the damaged bases. A: The uncut strand serves as the template for repair. B: Thymine dimers are only one cause of bulky lesions. C, D: The excision nuclease is a complex of proteins needed to unwind the DNA and remove the lesion. The polymerase and ligase fill in the gap (pol δ or ε in eukaryotes).
14. **B** These catalyze the first step of the process. A: Methylated and other chemically modified bases can also be removed. C, E: These are characteristics of a different repair system. D: The abasic sugar–phosphate must be removed.
15. The process has been best studied in *E. coli*. DNA is methylated on the A of a GATC sequence. Since this is a palindrome, both strands are methylated. Methylation occurs after the unmethylated bases are incorporated into DNA. During synthesis, the DNA will be hemimethylated for a short period of time until the methylase recognizes the new strand and adds the methyl group. The mismatch repair system recognizes the hemimethylated state and can remove the mismatch on the unmethylated (new) strand. Eukaryotes have a similar repair system although the details of recognizing the new strand are not yet known.
16. The coding strand of the gene has the same sequence as the mRNA (except U replaces T in the RNA). In HbA, the codon at position 142 of mRNA is a stop codon so the last amino acid added is 141. In Hb Constant Spring, a point mutation has mutated the DNA so that the mRNA codon at 142 now codes for an amino acid instead of stop. Translation continues until a stop codon appears at position 173 (so 172 amino acids). This could be a transition mutation—pyrimidine for pyrimidine.

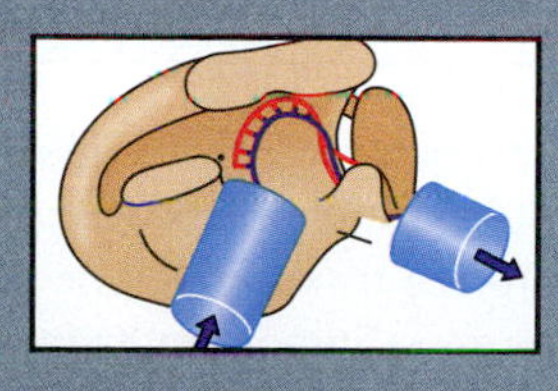

5

RNA: TRANSCRIPTION AND RNA PROCESSING

Francis J. Schmidt

5.1 | OVERVIEW

Synthesis of an RNA molecule involves copying one strand of a template sequence using Watson–Crick base pairing between nucleotides of the template (usually DNA) and the nucleotides that are being incorporated into the **transcript.** The initiation of transcription by RNA polymerase is perhaps the most important event in the control of gene expression. Transcription initiation requires specialized DNA sequences, called **promoters,** which signal where RNA synthesis should begin. The recognition of promoter sequences involves molecular contacts between DNA and protein **factors.** Factors bind both to RNA polymerase enzyme and to DNA bases through hydrogen bonds and other contacts. During elongation, **RNA processing** reactions remove, add, or modify nucleotides in the primary transcript. These processing reactions occur cotranscriptionally. In other words, they occur on parts of the transcript while downstream sequences are still being transcribed.

5.2 | MECHANISMS OF TRANSCRIPTION

The Initial Process of RNA Synthesis Is Transcription

Transcription is the process by which RNA chains are made from DNA templates. Transcription reactions take the following form:

$$\text{DNA template} + n(\text{NTP}) \longrightarrow \text{pppN(pN)}_{n-1} + (n-1)\text{PP}_\text{i} + \text{DNA template}$$

Enzymes that catalyze this reaction are the RNA polymerases; it is important to recognize that, like DNA polymerases, they are absolutely template dependent. Unlike DNA polymerases, however, **RNA polymerases** do not require a primer molecule. The energetics favoring the RNA polymerase reaction are twofold. First, the 5′ α-nucleotide phosphate of the ribonucleoside triphosphate is converted from a phosphate anhydride to a phosphodiester bond with a change in free energy ($\Delta G'$) of approximately 3 kcal (12.5 kJ) mol^{-1} under standard conditions. Second, the released pyrophosphate, PP_i, is cleaved to two phosphates by pyrophosphatase so that its concentration is low and phosphodiester bond formation is more favored relative to standard conditions (see p. 541).

Since a DNA template is required for RNA synthesis, eukaryotic transcription takes place in the nucleus or mitochondrial matrix. Within the nucleus, the nucleolus is the site of rRNA synthesis, whereas mRNA and tRNA are synthesized in the nucleoplasm. Prokaryotic transcription is accomplished on the cell's DNA, which is located in a relatively small region of the cell. Transcription of plasmid DNAs, of course, occurs independent of the chromosome.

Structural changes occur in DNA during its transcription. In polytene chromosomes of *Drosophila,* transcriptionally active genes are visualized in the light microscope as puffs distinct from the condensed, inactive chromatin. The nucleosome patterns of active genes are disrupted so that active chromatin is more accessible to chemical reagents or enzymes. In both prokaryotes and eukaryotes, the DNA double helix is transiently opened (unwound) as the transcription complex proceeds down the DNA.

These openings and **unwindings** are necessary because DNA is a double helix. Imagine what would happen if the RNA chain were copied off DNA without this process. The transcription complex and the growing end of the RNA chain would have to wind around the double helix once every 10 base pairs as they travel from the beginning of the gene to its end. Such a process would wrap the newly synthesized RNA chain around the DNA double helix. Local opening and unwinding of the DNA solves this problem before it occurs by allowing transcription to proceed on a single face or side of the DNA. In addition, the opening of DNA base pairs during transcription allows Watson–Crick base pairing between template DNA and the bases in the newly synthesized RNA.

The process of transcription is generally divided into three parts: **initiation** refers to the recognition of a specific DNA sequence by RNA polymerase and the beginning of the bond formation process; **elongation** is the actual synthesis of the RNA chain, which is followed by chain **termination** and **release.**

DNA Sequence Information Signals RNA Synthesis

Most DNA sequences in the human genome are never transcribed into RNA. Even in bacteria, where almost all the DNA specifies gene products, not all genes are expressed at any one time. RNA polymerase molecules must therefore recognize where the genes are and ignore a vast excess of useless information.

Initiation of transcription starts with recognition of a promoter sequence in the DNA template. Promoters are located a short distance upstream of the nucleotide where transcription starts. Different promoters within an organism contain similar sequences. An example is shown in Figure 5.1, where several different bacterial promoter sequences are aligned to show the conservation of sequence information among them. **Conserved sequences** are found in both prokaryotic and eukaryotic promoters. DNA sequences that stimulate transcription but are located further away from the initiation site are called **enhancers.** Enhancers stimulate the synthesis of some prokaryotic RNAs and of most, if not all, eukaryotic mRNAs. Enhancer sequences can stimulate transcription whether they are located at the beginning, in the middle, or at the end of a gene. An enhancer sequence must be on the same DNA molecule (chromosome) as is the transcribed gene (this relationship is termed *cis*-acting in genetics) but can function in either orientation and at variable distance from their respective promoters. Enhancers work by binding specific protein factors, called activators. When an activator binds to an enhancer it causes a structural change (often a looping or bending) in the DNA template that allows the interaction of the activator with other factors or with RNA polymerase. This interaction facilitates transcription by "recruiting" RNA polymerase to form an initiation complex.

RNA Polymerase Catalyzes the Transcription Process

RNA polymerases synthesize RNA in the 5′ → 3′ direction using a DNA template, being similar to the template-dependent DNA polymerases discussed in Chapter 4.

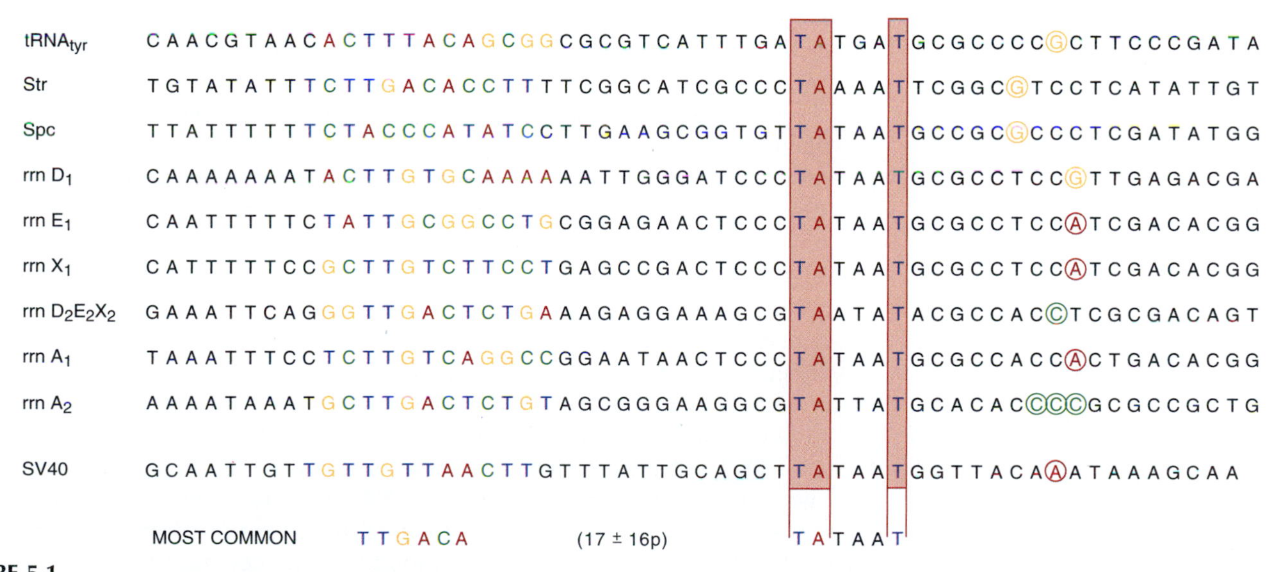

FIGURE 5.1

Determination of a consensus sequence for prokaryotic promoters. A portion of the data set used for identification of consensus sequence for *E. coli* promoter activity. The −10 region (sometimes called the Pribnow box) is shaded in red and the −35 region is shaded in blue. Note that none of the individual promoters has the entire consensus sequence.

Modified from Rosenberg, M. and Court, D. Annu. Rev. Genet. *13:319, 1979.*

TABLE 5.1 Comparative Properties of Some RNA Polymerases

	Nuclear				
	I(A)	*II(B)*	*III(C)*	*Mitochondrial*	*E., coli*
High MW subunits[a]	195–197	240–214	155	65	160 (β')
	117–126	140	138		150 (β)
Low MW subunits	61–51	41–34	89		86 (σ)
	49–44	29–25	70		40 (α)
	29–25	27–20	53		10 (ω)
	19–16.5	19.5	49		
		19	41		
		16.5	32		
			29		
			19		
Variable forms	2–3 types	3–4 types	2–4 types	1	1
Specialization	Nucleolar; rRNA	mRNA Viral RNA	tRNA 5S rRNA	All mtRNA	None
Inhibition by α-amanitin	Insensitive (>1 mg mL^{-1})	Very sensitive ($10^{-9} - 10^{-8}$ M)	Sensitive ($10^{-5} - 10^{-4}$ M)	Insensitive, but sensitive to rifampicin	Rifampicin sensitive

[a] Molecular weight $\times 10^{-3}$.

Unlike DNA polymerases, however, RNA polymerases initiate polymerization at promoter sequences without the need for a DNA or RNA primer.

Prokaryotic and eukaryotic RNA polymerases are large multisubunit enzymes whose mechanisms are only partially understood. RNA polymerase from *Escherichia coli* consists of five subunits with an aggregate molecular weight (MW) of over 500,000 (Table 5.1). Two α subunits, one β subunit, and one β' subunit constitute the **core enzyme,** which is capable of faithful transcription but not of specific (i.e., promoter-initiated) RNA synthesis. Addition of a fifth protein subunit, designated σ, forms the **holoenzyme** that is capable of specific RNA synthesis *in vitro* and *in vivo.* That σ is involved in the specific recognition of promoters has been borne out by a variety of biochemical studies. Specific **σ factors** can recognize different classes of genes. Thus a specific σ factor recognizes promoters for genes that are induced as a result of heat shock. In sporulating bacteria, specific σ factors recognize genes induced during sporulation. Some bacteriophage (viruses that infect bacteria) synthesize σ factors that allow the appropriation of the cell's RNA polymerase for transcription of the viral DNA.

The common prokaryotic RNA polymerases are inhibited by the antibiotic **rifampicin** (used in treating tuberculosis), which binds to the β subunit (see Clin. Corr. 5.1). Eukaryotic nuclear RNA polymerases are distinguished because they are inhibited differentially by **α-amanitin,** which is synthesized by the poisonous mushroom *Amanita phalloides.* Three nuclear RNA polymerase classes can be distinguished since very low concentrations of α-amanitin inhibit the synthesis of mRNA and some small nuclear RNAs (snRNAs); higher concentrations inhibit the synthesis of tRNA and other small nuclear RNAs, called snRNAs. rRNA synthesis is not inhibited much even by very high concentrations of the toxin. The purified enzymes are differentially inhibited by α-amanitin, so it is possible to conclude that mRNA synthesis is the function of RNA polymerase II, the most sensitive of the purified RNA polymerase forms. Synthesis of tRNA, 5S rRNA, and some snRNAs is carried out by RNA polymerase III. rRNA genes are transcribed by RNA polymerase I, which is concentrated in the nucleolus. (The numbers refer to the order of elution of the enzymes from a chromatography column.) Each enzyme has a highly complex structure (Table 5.1). An RNA polymerase in mitochondria is responsible for synthesis of mitochondrial mRNA, tRNA, and rRNA species. This enzyme, like bacterial RNA polymerase, is inhibited by rifampicin (see Clin. Corr. 5.1).

CLINICAL CORRELATION 5.1
Antibiotics and Toxins that Target RNA Polymerase

RNA polymerase is an essential enzyme for life since transcription is the first step of gene expression. No RNA polymerase means no other enzymes. Two natural products point out this principle; in both cases inhibition of RNA polymerase leads to death of the organism. The "death cap" or "destroying angel" mushroom, *Amanita phalloides,* is highly poisonous and still causes several deaths each year despite widespread warnings to amateur mushroom hunters (it is reputed to taste delicious, incidentally). The most lethal toxin, α-amanitin, inhibits the largest subunit of eukaryotic RNA polymerase II, thereby inhibiting mRNA synthesis. The poisoning starts with relatively mild, gastrointestinal symptoms, followed about 48 h later by massive liver failure as essential mRNAs and their proteins are degraded but not replaced by newly synthesized molecules. The only therapy is supportive including liver transplantation; but this is clearly a desperate measure of unproven efficacy.

More benign (at least from the point of view of our own species) is the action of the antibiotic rifampicin to inhibit the RNA polymerases of a variety of bacteria, most notably in the treatment of tuberculosis. *Mycobacterium tuberculosis,* the causative agent, is insensitive to many commonly used antibiotics but it is sensitive to rifampicin, the product of a soil streptomycetes. Since mammalian RNA polymerase differs from the prokaryotic variety, inhibition of the latter enzyme is possible without great toxicity to the host. This implies a good therapeutic index for the drug, that is, the ability to treat a disease without causing undue harm to the patient. Together with improved public health measures, antibiotic therapy with rifampicin and isoniazid (an antimetabolite) has greatly reduced the morbidity due to tuberculosis in industrialized countries. Unfortunately, the disease is still endemic in impoverished populations in the United States and in other countries. In increasing numbers immunocompromised individuals, especially AIDS patients, have active tuberculosis.

Mitchel, D. H. Amanita mushroom poisoning. *Annu. Rev. Med.* 31:51, 1980; Gilman, A. G., Rall, T. W., Nies, A. S., and Taylor, P. (Eds.). *The Pharmacological Basis of Therapeutics,* 8th ed. New York: Pergamon Press, 1990, pp. 129–130; DeCock, K. M., Soro, B., Colibaly, I. M., and Lucas, S. B. Tuberculosis and HIV infection in sub-Saharan Africa. *JAMA* 268:1581, 1992.

Steps of Transcription in Prokaryotes

Chromosomal DNA is usually transcribed in only one direction. This is illustrated as follows:

$$\begin{aligned}\text{DNA: } 5' &\longrightarrow 3'\\ \text{DNA: } 3' &\longrightarrow 5'\\ \textbf{RNA: } 5' &\longrightarrow 3' \cdots\end{aligned}$$

The DNA strand that serves as the template for RNA synthesis is complementary to the RNA transcript. Conventionally, the template strand is usually the "bottom" strand of a double-stranded DNA as written. The other strand, the "top" strand, has the same direction as the transcript when read in the 5′ to 3′ direction; this strand is sometimes called the coding strand. When only a single DNA sequence is given in this book, the coding strand is represented. Its sequence can be converted to the RNA transcript of a gene by simply substituting U (uracil) for T (thymine) bases.

Promoter Recognition

Prokaryotic transcription begins with binding of RNA polymerase to a gene's promoter (Figures 5.2 and 5.3). RNA polymerase holoenzyme binds to one face of the DNA extending 45 bp or so upstream and 10 bp downstream from the RNA initiation site. Two short sequences in this region are highly conserved (Figure 5.1). One sequence that is located about 10 bp upstream from the transcription start is the **consensus sequence** (sometimes called a "**−10**" or "**Pribnow" box**):

T*A*TAAT*

The positions marked with an asterisk are the most conserved; indeed, the last T residue is always found in *E. coli* promoters.

A second consensus sequence is located upstream from the Pribnow or "−10" box. This "**−35 sequence**"

T*T*G*ACA

is centered about 35 bp upstream from the transcription start; the nucleotides with asterisks are most conserved. The spacing between the "−35" and "−10" sequences is crucial with 17 bp being highly conserved. As shown in Figure 5.1, the TTGACA and TATAAT sequences are asymmetrical; that is, they do not have the same sequence if the complementary sequence is read. Thus the promoter sequence itself determines that transcription will proceed in only one direction.

What difference do these consensus sequences make to a gene? Measurements of RNA polymerase binding affinity and initiation efficiency to various promoter sequences show that the most active promoters fit the consensus sequences most closely. Statistical measurements of promoter homology to the consensus sequence conform closely to the measured "strength" of a promoter, that is, its kinetic ability to initiate transcription with purified RNA polymerase.

Bases flanking the "−35" and "−10" sequences, bases near the transcription start, and bases located near the "−16" position are weakly conserved. In some of these weakly conserved regions, RNA polymerase may require that a particular nucleotide not be present or that local variations in DNA helical structure be present. Promoters for *E. coli* heat shock genes have different consensus sequences at the "−35" and "−10" regions. This is consistent with their being recognized by a different σ factor.

An RNA transcript usually starts with a purine riboside triphosphate; that is, pppG ⋯ or pppA ⋯, but pyrimidine starts are also known (Figure 5.1). The position of transcription initiation differs slightly among various promoters, but usually is 5–8 bp downstream from the invariant T of the Pribnow box.

Start of Synthesis

Two kinetically distinct steps are required for RNA polymerase to initiate synthesis of an RNA transcript (Figure 5.2*a*, *b*). In the first step, described above, RNA polymerase holoenzyme binds relatively weakly to the promoter DNA to form a **"closed complex."** In the second step, the holoenzyme forms a more tightly bound **"open complex,"** characterized by a local opening of about 10 bp of the DNA double helix. Since the consensus Pribnow box is A-T rich, it can facilitate this local unwinding: its base pairs are more easily disrupted during opening. As discussed in Chapter 2, the opening 10 bp of DNA is topologically equivalent to the relaxation of a single negative supercoil. As might be predicted from this observation, the activity of some promoters depends on the superhelical state of the DNA template. Some promoters are more active on highly supercoiled DNA while others are more active when the superhelical density of the template is lower. The unwound DNA binds the initiating triphosphate and RNA polymerase then forms the first phosphodiester bond. The enzyme translocates to the next position (this is the rifampicin-inhibited step) and continues synthesis. At or soon after the initial bond formation, σ factor is released and the enzyme enters an elongation mode. Other RNA polymerase molecules can now bind to promoter so that the gene can be transcribed many times (Figure 5.3).

Elongation

RNA polymerase continues the binding–bond formation–translocation cycle at an average rate of about 40 nucleotides per second. However, many examples are known where RNA polymerase pauses or slows down at particular sequences, usually inverted repeats (palindrome sequence of nucleotides). As discussed below, these pauses can bring about transcription termination.

As RNA polymerase continues down the double helix, it continues to separate the two strands of the DNA template. As seen in Figure 5.2 *c*, *d*, this process allows the template strand of the DNA to base pair with the growing RNA chain. Thus a single mechanism of information transfer (Watson–Crick base pairing) serves several processes: DNA replication, DNA repair, and transcription of genetic information into RNA. (As will be seen in Chapter 6, base pairing is essential for translation as well.) The process of unwinding and restoring the DNA double helix is aided by DNA topoisomerases I and II, which are components of the transcription complex.

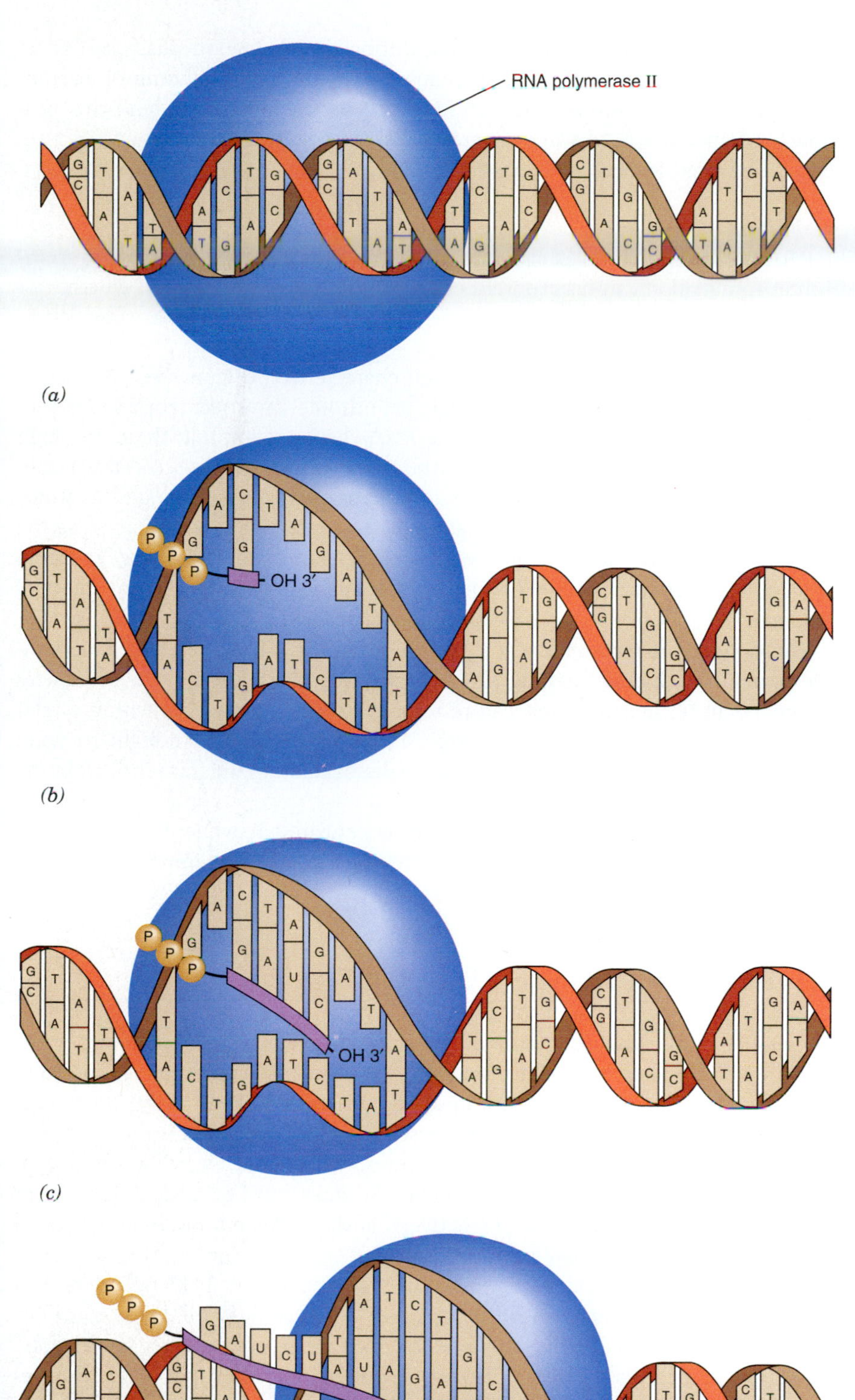

FIGURE 5.2
Early events in prokaryotic transcription.
(*a*) Recognition: RNA polymerase with "sigma" factor binds to a DNA promoter region in a "closed" conformation. (*b*) Initiation: The complex is converted to an "open" conformation and the first nucleoside triphosphate aligns with DNA. (*c*) The first phosphodiester bond is formed and the "sigma" factor is released. (*d*) Elongation: Synthesis of nascent RNA proceeds with movement of RNA polymerase along the DNA. The double helix reforms.

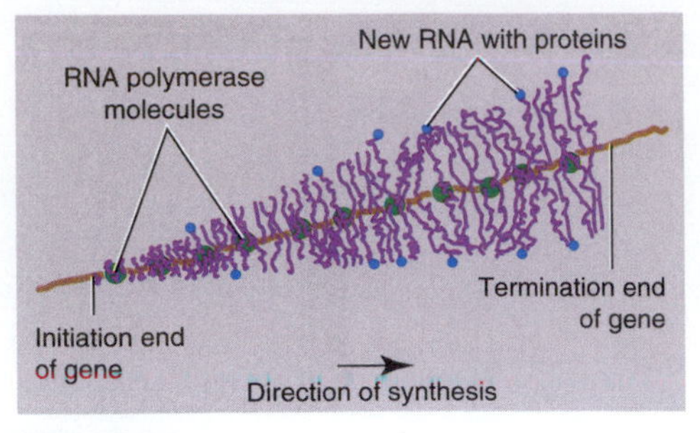

FIGURE 5.3
Simultaneous transcription of a gene by many RNA polymerases, depicting increasing length of nascent RNA molecules.
Courtesy of Dr. O. L. Miller, University of Virginia. Reproduced with permission from Miller, O. L. and Beatty, B. R. J. Cell Physiol. *74:225, 1969.*

Changes in the transcription complex during the elongation phase can affect subsequent termination events. These changes depend on the binding of another cellular protein (NusA protein) to core RNA polymerase. Failure to bind this protein factor sometimes results in an increased frequency of termination and, consequently, a reduced level of gene expression.

Termination

The RNA polymerase complex also recognizes the ends of genes (Figure 5.4). Transcription termination can occur in either of two modes, depending on whether or not it is dependent on the protein factor ρ (rho). Terminators are thus classified as ρ-independent or ρ-dependent.

Rho-Independent terminators are well characterized (Figure 5.4): A consensus-type sequence is involved: a G-C rich palindrome (inverted repeat) that precedes a sequence of 6–7 U residues in the RNA chain. As a result, the RNA chain forms a stem and loop structure ahead of the polyU residues. This secondary structure of the stem and loop is crucial for termination since base change mutations that disrupt pairing also reduce termination. The most efficient terminators are the most G-C rich and therefore lead to the most stable terminator structures. The stem and loop left after termination stabilizes prokaryotic mRNA against nucleolytic degradation.

Rho-dependent terminators are less well-defined. The Rho factor is a hexameric protein that has an essential RNA-dependent ATPase activity. The sequences of ρ-dependent termination sites feature regularly spaced C residues within a relatively unstructured part of the transcript. The nascent RNA is thought to wrap around ρ factor while ATP hydrolysis leads to dissociation of the transcript from the template.

Prokaryotic ribosomes usually attach to nascent mRNA while it is being transcribed. This coupling between transcription and translation is important in gene control by attenuation (see Chapter 8).

5.3 | TRANSCRIPTION IN EUKARYOTES

Initiation of eukaryotic transcription differs substantially from its prokaryotic counterpart. While the definition of a promoter is the same—DNA sequence information that specifies the start of transcription—the molecular events required for transcription initiation are more complex. First, chromatin containing the promoter sequence must be made accessible to the transcription machinery. Second, **transcription factors** distinct from RNA polymerase must bind to DNA sequences in the promoter region for a gene to be active. Third, **enhancers** bind other protein factors (activators) to stimulate transcription.

Eukaryotic transcription factors bind to DNA and recruit RNA polymerase to the promoter. This contrasts with the action of bacterial σ factors, which do not bind DNA without first binding to the RNA polymerase core enzyme. Eukaryotic RNA polymerase consists of three distinct enzyme forms, each specific form capable of transcribing different classes of cellular RNA. By contrast, all prokaryotic genes are transcribed by a single form of core RNA polymerase, although different σ factors may be involved in initiation of different genes.

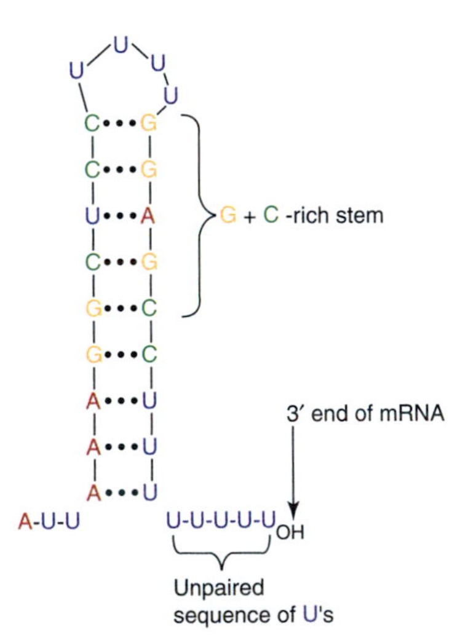

FIGURE 5.4
Stem–loop structure of RNA transcript that determines e-independent transcriptional termination.
Note the two components of the structure: the (G + C)-rich stem and loop, followed by a sequence of U residues.

The Nature of Active Chromatin

The structural organization of eukaryotic chromosomes is discussed in Chapters 4 and 8. Although chromatin is organized into **nucleosomes** whether or not it is capable of being transcribed, an active gene has a generally "looser" conformation than transcriptionally inactive chromatin. This difference is most striking in the promoter sequences, parts of which are not organized into nucleosomes at all (Figure 5.5). The lack of nucleosomes is manifested experimentally by the enhanced

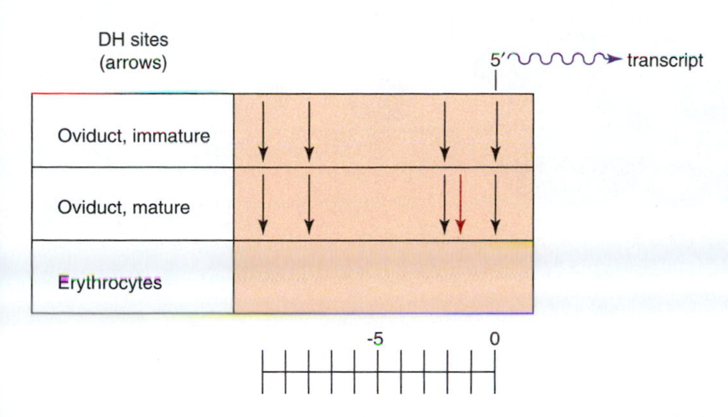

FIGURE 5.5
DNase-hypersensitive (DH) sites upstream of the promoter for the chick lysozyme gene, a typical eukaryotic transcriptional unit. Hypersensitive sites, that is, sequences around the lysozyme gene where nucleosomes are not bound to DNA, are indicated by arrows. Note that some hypersensitive sites are found in the lysozyme promoter whether the oviduct is synthesizing or not synthesizing lysozyme; the synthesis of lysozyme is accompanied by the opening up of a new hypersensitive site in mature oviduct. In contrast, no hypersensitive sites are present in nucleated erythrocytes that never synthesize lysozyme.
Adapted from Elgin, S. C. R. J. Biol. Chem. *263:1925, 1988.*

sensitivity of promoter sequences to external reagents that cleave DNA, such as the enzyme DNase I. This enhanced accessibility of promoter sequences (termed **DNase I hypersensitivity**) ensures that transcription factors can bind to appropriate regulatory sequences. Although the transcribed sequences of a gene may be organized into nucleosomes, the nucleosomes are less tightly bound than those in an inactive gene. This looser structure is due to acetylation of the histones by histone acetyltransferase, a reaction that transfers acetyl groups from acetyl-CoA to histones, especially to the N-terminus of histone H4 and to H3. Finally, DNA may be transcriptionally inactivated by methylation (see Clin. Corr. 5.2). The overall theme is one of partially unfolded chromatin being necessary but not sufficient for transcription.

Transcription Activation Operates by Recruitment of RNA Polymerase

Eukaryotic protein factors, regardless of the sequence they are bound to, operate in a fundamentally different way than *E. coli* σ factor. Rather than first forming part of a protein complex and then seeking out the relevant DNA sequence, the factors bind to a specific site (sequence) on DNA and then bind to RNA polymerase (with or without involvement of intermediary factors). This mechanism is termed "recruitment." Recruitment is a minor means of gene activation in prokaryotes and the major mechanism in eukaryotes.

CLINICAL CORRELATION 5.2
Fragile X Syndrome: A Chromatin Disease?

Fragile X syndrome is the single most common form of inherited mental retardation, affecting 1/1250 males and 1/2000 females. A variety of anatomical and neurological symptoms result from inactivation of the *FMR1* gene, located on the X chromosome. The genetics of the syndrome are complex due to the molecular mechanism of the fragile X mutation.

The fragile X condition results from expansion of a trinucleotide repeat sequence, CGG, found at the 5′ untranslated region of the *FRM1* gene. Normally, this repeat is present in 30 copies, although normal individuals can have up to 200 copies of the repeat. In individuals with fragile X syndrome, the *FMR1* gene contains many more copies, from 200 to thousands, of the CGG repeat. The complex genetics of the disease result from the potential of the CGG repeat sequence to expand from generation to generation.

The presence of an abnormally high number of CGG repeats induces extensive DNA methylation of the entire promoter region of *FMR1*. Methylated DNA is transcriptionally inactive, so FMR1 mRNA is not synthesized. Absence of FMR1 protein leads to the pathology of the disease.

FMR1 protein normally is located in the cytoplasm in all tissues of the early fetus and, later, especially in fetal brain. Its sequence has some characteristics of an RNA-binding protein. One hypothesis is that the protein aids in the translation of brain-specific mRNAs during development.

Warren, S. L. and Nelson, D. L. Advances in molecular analysis of fragile X syndrome. *JAMA* 271:536, 1994; and Caskey, C. T. Triple repeat mutations in human disease. *Science* 256:784, 1992.

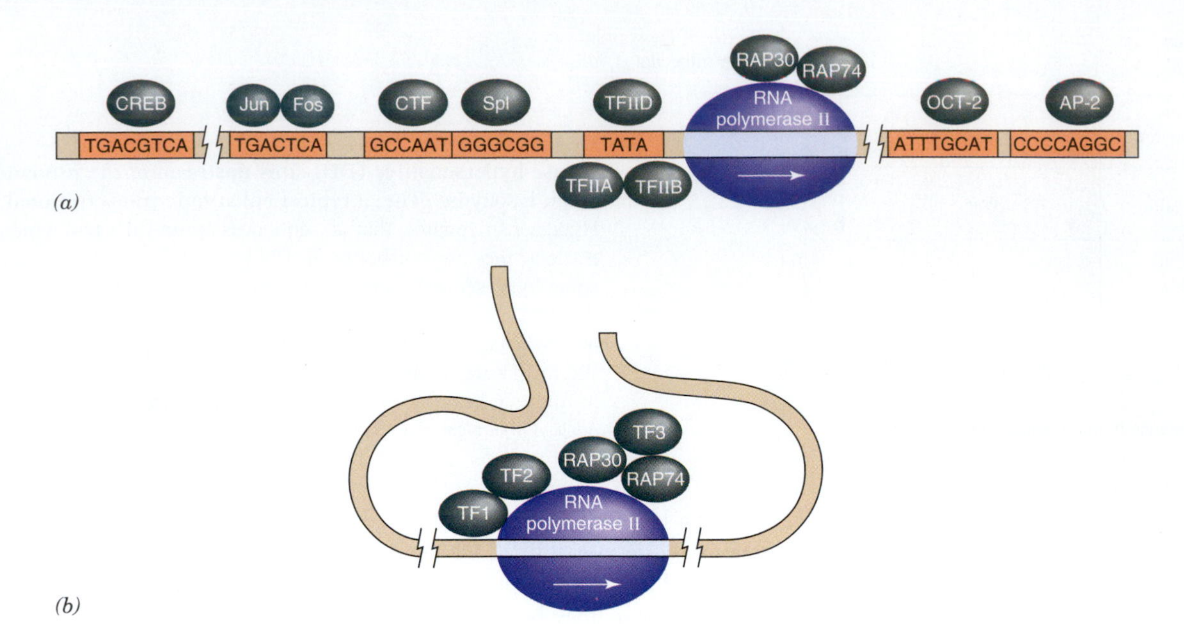

FIGURE 5.6
Interaction of transcription factors with promoters.
A large number of transcriptional factors interact with eukaryotic promoter regions. (*a*) Hypothetical array of factors that interact with specific DNA sequences near the promoter, including TFIID, which binds to the TATA box and the Jun and Fos proteins, which are protooncogenes (see Clin. Corr. 5 .3). The figure is not meant to imply that all of the DNA-binding factors bind to the promoter simultaneously. (*b*) One way in which the DNA-binding factors are hypothesized to bind to each other and to RNA polymerase. Although this model is not completely proved, it is known that proteins that bind to distant DNA sequences make protein—protein contacts with each other.
Reprinted with permission from Mitchell, P. J. and Tjian, R. Science 245:371, 1989.

Enhancers

Enhancers increase expression of a gene about 100-fold. Enhancer-binding transcription factors are called activators. Activator proteins have two domains, one of which binds to the enhancer sequence, and the other binds to other protein factors or to RNA polymerase. The most accepted model for these effects is that chromatin forms a "loop" that allows the enhancer and the promoter to be close together in space, even though they are separated by a relatively long sequence of DNA (Figure 5.6).

Transcription by RNA Polymerase II

RNA polymerase II is responsible for synthesis of mRNA in the nucleus. Several common themes have emerged from research on a large number of genes (Figure 5.6). (1) The DNA sequences that control transcription are complex; a single gene may be controlled by as many as six or eight DNA sequence elements in addition to the promoter (RNA polymerase binding region) itself. Controlling sequence elements function in combination to give a finely tuned pattern of control. (2) The effect of controlling sequences on transcription is mediated by binding of proteins to each sequence element. These transcription factors recognize the nucleotide sequence of the appropriate controlling sequence element. (3) Bound transcription factors bind with each other and recruit RNA polymerase. The DNA-binding and activation activities of the factors reside in separate domains of the proteins. (4) RNA polymerase II is modified during the transcription reaction. The modified polymerase recruits other nuclear components, including RNA processing enzymes, during the elongation phase of transcription.

Promoters for mRNA Synthesis

In contrast to prokaryotic RNA polymerase, which recognizes only a single promoter sequence, RNA polymerase II can initiate transcription by recognizing several classes of **consensus sequences** upstream from the mRNA start site. The first and most prominent of these, sometimes called the **TATA box,** has the sequence

TATA(A/T)(A/T) A

where the nucleotides in parentheses can be either an A or a T.

The TATA box is centered about 25 bp upstream from the transcription unit. Experiments in which it was deleted suggest that it is required for efficient transcription, although some promoters may lack it entirely.

A second region of homology that contains the **CAAT box** is located further upstream and has the sequence

GG(T/C) CAATCT

This sequence is not as highly conserved as the TATA box, and some active promoters may not possess it. Other sequences, described in Figure 5.6, may also promote transcription. CAAT and TATA boxes, as well as other sequences shown in Figure 5.6, do not contact RNA polymerase II directly. Rather, they require the binding of specific **transcription factors** to function. Note how protein factors bind not only to their recognition sequences but also to each other and to RNA polymerase, itself a very large and complex enzyme. Despite complexities of the detailed interactions, the principles elaborated above account for the known mechanisms of all **class II transcription factors,** that is, those factors that promote transcription by RNA polymerase II. Mutated forms of several of these transcription factors are products of oncogenes (see Clin. Corr. 5.3).

CLINICAL CORRELATION 5.3

Involvement of Transcriptional Factors in Carcinogenesis

Conversion of a normally well-regulated cell into a cancerous one requires a number of independent steps whose end result is a transformed cell capable of uncontrolled growth and metastasis. Insights into this process have come from recombinant DNA studies of the genes, termed oncogenes, whose mutated or overexpressed products contribute to carcinogenesis. Oncogenes were first identified as products of DNA or RNA tumor viruses but they are also present in normal cells. The normal, nonmutated cellular analogs of oncogenes are termed protooncogenes. Their products are components of the many pathways that regulate growth and differentiation of a normal cell; mutation into an oncogenic form involves a change that makes the regulatory product less responsive to normal control.

Some protooncogene products are involved in transduction of hormonal signals or recognition of cellular growth factors and act cytoplasmically. Other protooncogenes have a nuclear site of action; their gene products are often associated with the transcriptional apparatus and they are synthesized in response to growth stimuli. It is easy to visualize how overproduction or permanent activation of such a positive transcription factor could aid the transformation of a cell to malignancy: genes normally transcribed at a low or controlled level would be overexpressed by such a deranged control mechanism.

A more subtle genetic effect predisposing to cancer is exemplified by the human tumor suppressor protein p53. This protein is the product of a dominant oncogene. A single copy of the mutant gene causes Li–Fraumeni syndrome, an inherited condition predisposing to carcinomas of the breast and adrenal cortex, sarcomas, leukemia, and brain tumors.

Somatic mutations in p53 can be identified in about half of all human cancers. Mutations represent a loss of function, affecting either the stability or DNA-binding ability of p53. Thus wild-type p53 functions as a tumor suppressor. The wild-type protein helps to control the checkpoint between the G1 and S phases of the cell cycle, activates DNA repair, and, in other circumstances, leads to programmed cell death (apoptosis). Thus the biochemical actions of p53 serve to keep cell growth regulated, maintain the information content of the genome, and, finally, eliminate damaged cells. All of these functions would counteract neoplastic transformation of a cell.

These varied roles are a function of the action of p53 as a transcription factor, inhibiting some genes and activating others. For example, p53 inhibits transcription of genes with TATA sequences, perhaps by binding to the complex formed between transcription factors and the TATA sequence. Alternatively, p53 is a site-specific DNA-binding protein and promotes transcription of some other genes, for example, those for DNA repair.

The three-dimensional structure of p53 has been determined. Mutations found in p53 from tumors affect the DNA-binding domain of the protein. For example, nearly 20% of all mutated residues involve mutations at two positions in p53. The crystal structure of the protein–DNA complex shows that these two amino acids, both arginines, form hydrogen bonds with DNA. Arginine 248 forms hydrogen bonds in the minor groove of the DNA helix with a thymine oxygen and with a ring nitrogen of adenine. Mutation disrupts this H-bonded network and therefore the ability of p53 to regulate transcription.

Weinberg, R. A. Oncogenes, antioncogenes, and the molecular basis of multistep carcinogenesis. *Cancer Res.* 49:3713, 1989; Cho, Y., Gorina, S., Jeffrey, P. D., and Pavletich, N. P. Crystal structure of a p53 tumor suppressor–DNA complex: understanding tumorigenic mutations. *Science* 265:346, 1994; Friend, S. p43: a glimpse at the puppet behind the shadow play. *Science* 265:334, 1994; and Harris, C.C. and Hollstein, M. Clinical implications of the p53 tumor-suppressor gene. *N. Engl. J. Med.* 329:1318, 1993.

Transcription by RNA Polymerase I

Recall that rRNA genes are located in the **nucleolus.** Ribosomal RNAs are synthesized in a specific nuclear body, the nucleolus, and the rRNA genes are located on a specific chromosomal region termed the **"nucleolar organizer."** Each transcriptional unit contains sequences for 28S, 5.8S, and 18S rRNAs, in that order. Several hundred copies of each transcriptional unit occur tandemly (one after another) in the chromosome. Transcriptional units are separated by spacer sequences. Spacer sequences include sequences that specify binding of RNA polymerase I and of **class I transcription factors,** which promote RNA polymerase I activity. The repeat units contain a copy of each RNA sequence (28S, 5.8S, and 18S) and are separated from each other by nontranscribed spacer regions. Figure 5.7 is a diagram of this arrangement. Each repeat unit is transcribed as a unit, yielding a primary transcript that contains one copy each of the 28S, 5.8S, and 18S sequences, ensuring synthesis of equimolar amounts of these three RNAs. The primary transcript is then processed by ribonucleases and modifying enzymes to the three mature rRNA species. Termination of transcription occurs within the nontranscribed spacer region before RNA polymerase I reaches the promoter of the next repeat unit.

The promoter recognized by RNA polymerase I is located within the nontranscribed spacer, from about positions −40 to +10 and from −150 to −110. A transcription factor binds to the promoter and thereby directs recognition of the promoter sequence by RNA polymerase I. In addition, an enhancer element is located about 250 bp upstream from the promoter in human ribosomal DNA. The size of nontranscribed spacer varies considerably among organisms, as does the position of the enhancer element.

Transcription of rRNA can be very rapid; this reflects the fact that synthesis of ribosomes is rate-limiting for cell growth. Phosphorylation of RNA polymerase I may activate especially rapid transcription of rRNA, for example, during embryonic growth or liver regeneration. In other situations, when growth is not so rapid, only some of the rDNA repeats are available for transcription.

Transcription by RNA Polymerase III

The themes elaborated above for transcription of class I and class II promoters hold for transcription of 5S RNA and tRNA by RNA polymerase III. Transcription factors bind to DNA and direct the action of RNA polymerase. One unusual feature of RNA polymerase III action in the transcription of 5S RNA is the location of the factor-binding sequence; it can be located within the DNA sequence encoding the RNA. The DNA in the region that would normally be thought of as a promoter, that is, the sequence immediately 5′ to the transcribed region of the gene, has no specific sequence and can be substituted by other sequences without a substantial effect on transcription. Figure 5.8 diagrams this unusual sequence arrangement. In other cases, for ex-

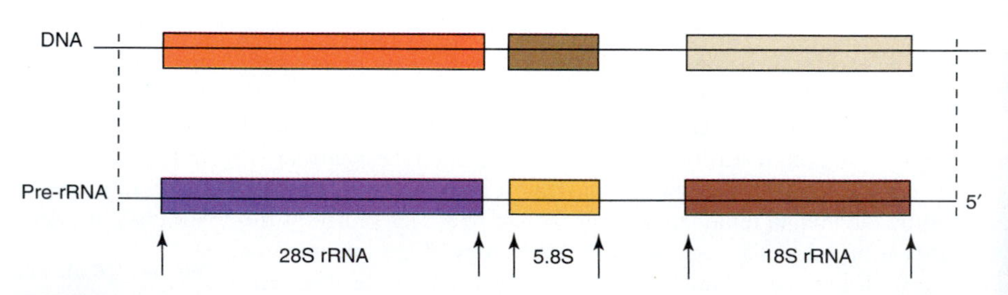

FIGURE 5.7
Structure of a rRNA transcription unit.
Ribosomal RNA genes are arranged with many copies one after another. Each copy is transcribed separately and each transcript is processed into three separate RNA species. Promoter and enhancer sequences are located in the nontranscribed regions of the tandemly repeated sequences.

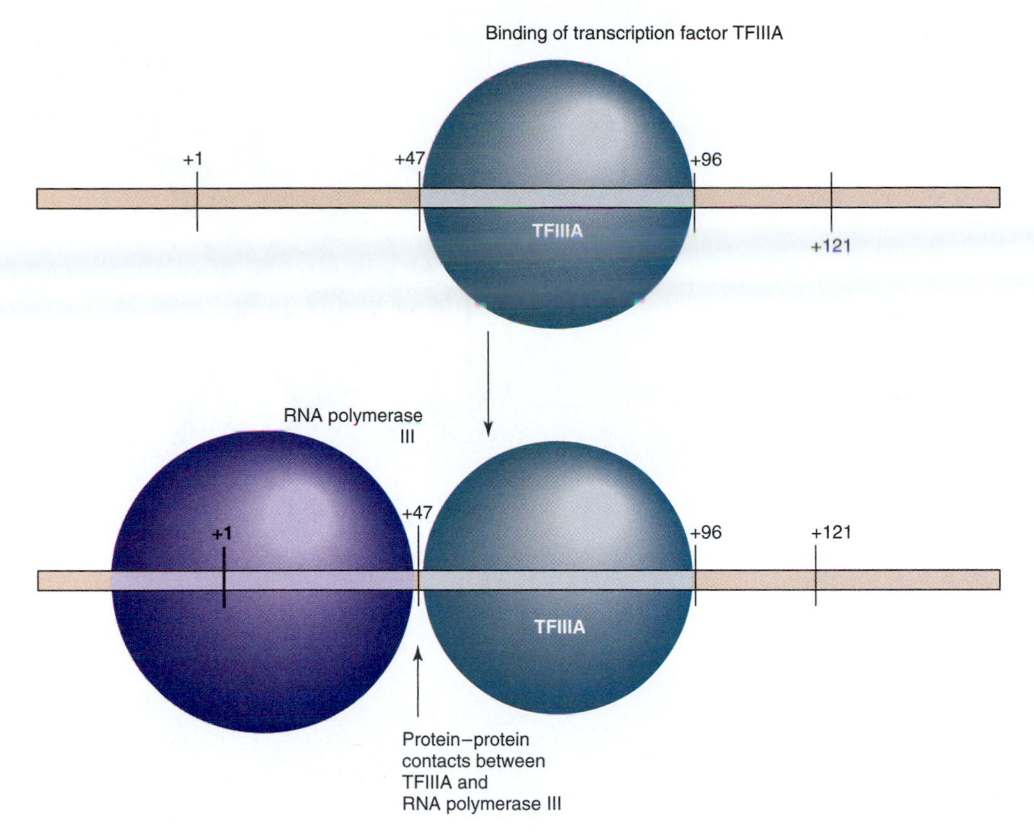

FIGURE 5.8
Transcription factor for a class III eukaryotic gene.
Transcription factor TFIIIA binds to a sequence located within the *Xenopus* gene for 5S rRNA. The RNA polymerase III then binds to the factor and initiates transcription of the 5S sequence. No specific sequence in the DNA is required other than the factor-binding sequence.

ample, tRNA transcription, the factor-binding sequence is located more conventionally at the 5′ region of the gene, that is, preceding the transcribed sequences.

The Common Enzymatic Basis for RNA Polymerase Action

Our understanding of the molecular basis of **RNA polymerase** action has increased recently, as the three-dimensional structures of a σ factor, of bacterial core RNA polymerase, and of a eukaryotic RNA polymerase II have been determined by X-ray crystallography. Despite all the differences in subunit composition, size, and mechanism, these enzymes seem to interact with their DNA templates and nucleotide substrates in a similar fashion.

σ Factor is an oblong protein, composed of a bundle of α-helical residues, packed into an open "V" shape (Figure 5.9*a*). One arm of the "V" contains critical residues for promoter recognition and core polymerase binding. One side of this arm contains an α-helix that binds to the "-10" sequence and the other face binds to core polymerase through hydrophobic interactions. The bacterial core enzyme is shaped rather like a computer mouse or the claw of a crab (Figure 5.9*b*). DNA comes in at one end of the molecule, makes a 90° bend and exits from the polymerase. The double-helical structure of DNA is opened up as it makes the bend. Nucleotides come in and RNA comes out of two other, separate channels in the molecule. A flexible structure in the β subunit acts like a flap to hold the DNA in the channel.

The three-dimensional structure of eukaryotic RNA polymerase II shows a channel in the enzyme through which the DNA template passes, a pore for entry of the ribonucleotide triphosphates to enter, and another pore for the RNA product to exit through. The DNA template also must bend as it opens up to accommodate the hybrid formed by the template strand and the growing RNA chain. Thus, very

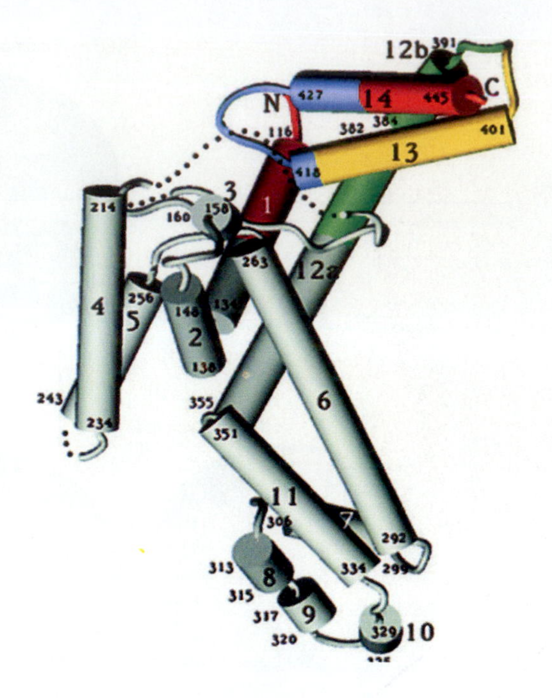

(a)

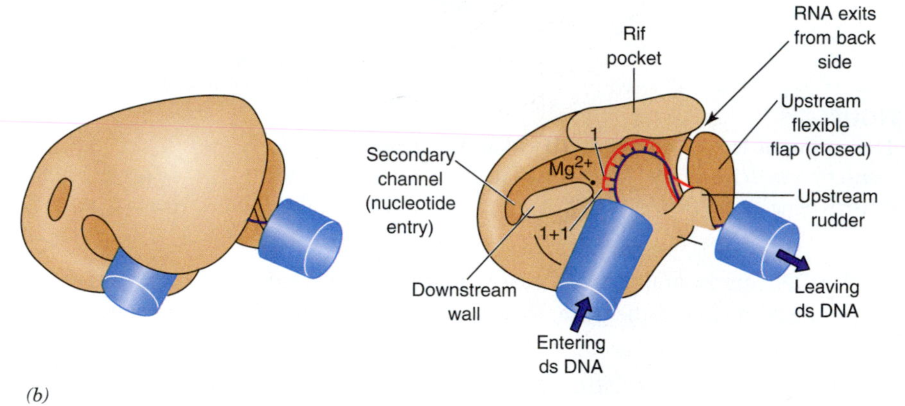

(b)

FIGURE 5.9

Structures of some important parts of bacterial RNA polymerase.
(*a*) Schematic diagram of the major bacterial σ subunit. Yellow- and green-colored helices (shown as cylinders) contact the core structure to form holoenzyme, while the blue region recognizes the promoter sequence. (*b*) A diagram of the action of RNA polymerase core enzyme during elongation. Left: The overall enzyme structure looks like a clamshell or crab claw. The "clamp" region at the top of the enzyme holds the DNA and the enzyme together. Right: A cutaway view of the enzyme showing the abrupt bending of DNA at the site of phosphodiester bond synthesis.
Part (a) from Malhotra, A. Severinova, E., and Darst, S. A. Cell 87: 127, 1996; part (b) from Zhang, G., Campbell, E. A., Minaklin, L., et al. Cell 98: 811, 1999. Reprinted with permission of the authors and Elsevier Science.

different in subunit composition, all RNA polymerases carry out the same processes: threading the DNA template through a channel, bringing in nucleotides to the active site through a pore in the enzyme, and extruding the RNA product through another pore.

5.4 | RNA PROCESSING

RNA copies of DNA sequences must be modified to mature, functional, molecules in prokaryotes and eukaryotes. The reactions of **RNA processing** can include removal of extra nucleotides, base modification, addition of nucleotides, and separation of different RNA sequences by the action of specific nucleases. Finally, in eukaryotes, RNAs must also be exported from the nucleus.

Transfer RNA Is Modified by Cleavage, Additions, and Base Modification

Cleavage

The primary transcript of a tRNA gene contains extra nucleotide sequences both 5′ and 3′ to the tRNA sequence. In some cases these primary transcripts contain **introns** in the anticodon region of the tRNA also. Processing reactions occur in a closely defined but not necessarily rigid temporal order. First, the primary transcript is trimmed in a relatively nonspecific manner to yield a precursor molecule with shorter 5′ and 3′ extensions. Then ribonuclease P, a ribozyme (see p. 86), removes the 5′ extension by endonucleolytic cleavage. The 3′-end is trimmed exonucleolytically, followed by synthesis of the CCA-terminus. Synthesis of the modified nucleotides occurs in any order relative to the nucleolytic trimming. Intron removal is dictated by the secondary structure of the precursor (see Figure 5.10) and is carried out by a soluble, two-component enzyme system; one enzyme removes the intron and the other reseals the nucleotide chain.

Additions

Each functional tRNA has the sequence CCA at its 3′-terminus. This sequence is essential for tRNA to accept amino acids. In most instances the sequence is added sequentially by the enzyme **tRNA nucleotidyltransferase.** Nucleotidyltransferase uses ATP and CTP as substrates and always incorporates them into tRNA at a ratio of 2C/1A. The CCA ends are found on both cytoplasmic and mitochondrial tRNAs.

Modified Nucleosides

Transfer RNA nucleotides are the most highly modified of all nucleic acids. More than 60 different modifications to the bases and ribose, requiring well over 100 different enzymatic reactions, have been found in tRNA. Many are simple, one-step methylations, but others involve multistep synthesis. Two derivatives, **pseudouridine** and **queuosine** (7-4, 5-*cis*-dihydroxy-1-cyclopenten-3-ylamino methyl-7-deazaguanosine), actually require severing of the β-glycosidic bond of the altered nucleotide. One enzyme or set of enzymes produces a single site-specific modification in more than one species of tRNA. Separate enzymes or sets of enzymes produce the same modifications at more than one location in tRNA. In other words, most modification enzymes are site or nucleotide sequence specific, not tRNA specific. Most modifications are completed before the tRNA precursors have been cleaved to mature tRNA size.

Ribosomal RNA Processing Releases the Various RNAs from a Longer Precursor

The primary product of rRNA transcription is a long RNA, termed 45S RNA, which contains the sequences of 28S, 5.8S, and 18S rRNAs. Processing of 45S RNA occurs in the nucleolus. Like the processing of mRNA precursors (see below), processing of the rRNA precursors is carried out by large multisubunit ribonucleoprotein assemblies. At least three RNA species are required for processing. These all function as small nucleolar ribonucleoprotein complexes. Processing of the rRNAs follows a sequential order (Figure 5.11).

Processing of pre-rRNA in prokaryotes also involves cleavage of high molecular weight precursors to smaller molecules. Some bases are modified by methylation on the ring nitrogens of the bases rather than the ribose and by the formation of pseudouridine. These reactions are specified by **small nucleolar ribonucleoprotein particles (snoRNPs)** in the nucleolus. Each snoRNP contains a guide RNA that base-pairs with a short sequence in the rRNA transcript and thereby specifies the site of modification.

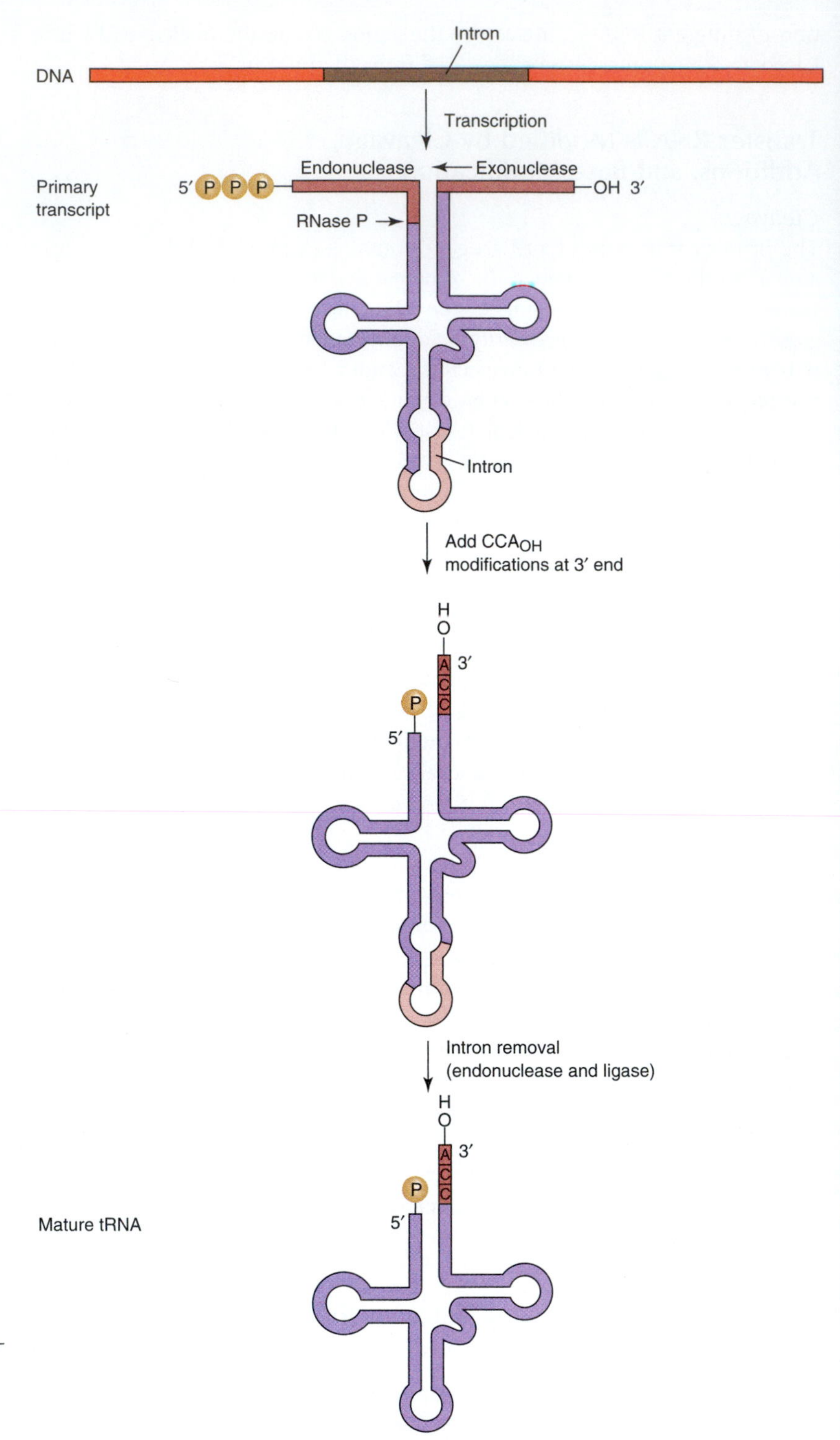

FIGURE 5.10
Scheme for processing a eukaryotic tRNA.
Primary transcript is cleaved by RNase P and a 3′-exonuclease and the terminal CCA is synthesized by tRNA nucleotidyltransferase before the intron is removed, if necessary.

Messenger RNA Processing Requires Maintenance of the Coding Sequence

Most eukaryotic mRNAs have distinctive structural features added in the nucleus by enzyme systems other than RNA polymerase. These include the 3′-terminal poly(A) tail, methylated internal nucleotides, and the cap 5′-terminus. Cytoplasmic mRNAs are shorter than their primary transcripts, which can contain additional terminal and internal sequences. Noncoding sequences present within pre-mRNA, but not present in mature mRNAs, are called **intervening sequences** or **introns.** The expressed or retained sequences are called **exons.** The general pattern for mRNA

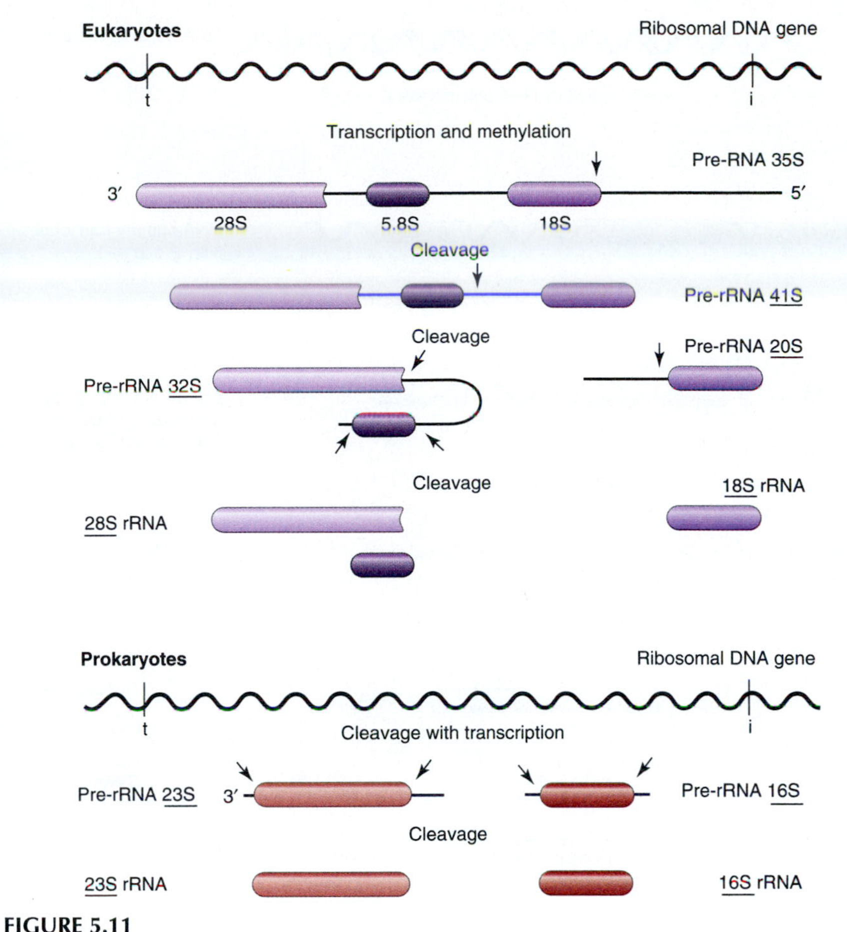

FIGURE 5.11
Schemes for transcription and processing of rRNAs.
Redrawn from Perry, R. Annu. Rev. Biochem. 45:611, 1976. Copyright © 1976 by Annual Reviews, Inc.

processing is depicted in Figure 5.12. Incompletely processed mRNAs make up a large part of the heterogeneous nuclear RNAs (HnRNAs).

Processing of eukaryotic pre-mRNA involves a number of molecular reactions, all of which must be carried out with exact fidelity. This is most clear in removal of introns from an mRNA transcript. An extra nucleotide in the coding sequence of mature mRNA would cause the reading frame of that message to be shifted and the resulting protein will almost certainly be nonfunctional. Indeed, mutations that interfere with intron removal are a major cause of human genetic diseases, for example, **β-thalassemia** (see Clin. Corr. 5.4). The task for cells becomes even more daunting since some important human genes consist of over 90% intron sequences. The complex reactions to remove introns are accomplished by multicomponent enzyme systems that act in the nucleus; after these reactions are completed the mRNA is exported to the cytoplasm where it interacts with ribosomes to initiate translation. Processing is carried out on the transcript while it is still being elongated by RNA polymerase—an arrangement that is called cotranscriptional.

RNA Polymerase II Recruits Processing Enzymes During Transcription in Eukaryotes

The large subunit of RNA polymerase II contains a C-terminal domain that functions to couple transcription and processing. When RNA polymerase II is in an initiation mode, the C-terminal domain is not modified. When the transcript is about 25 nucleotides long, the C-terminal domain is extensively phosphorylated. Phosphorylated C-terminal domain recruits the capping enzyme, splicing, and

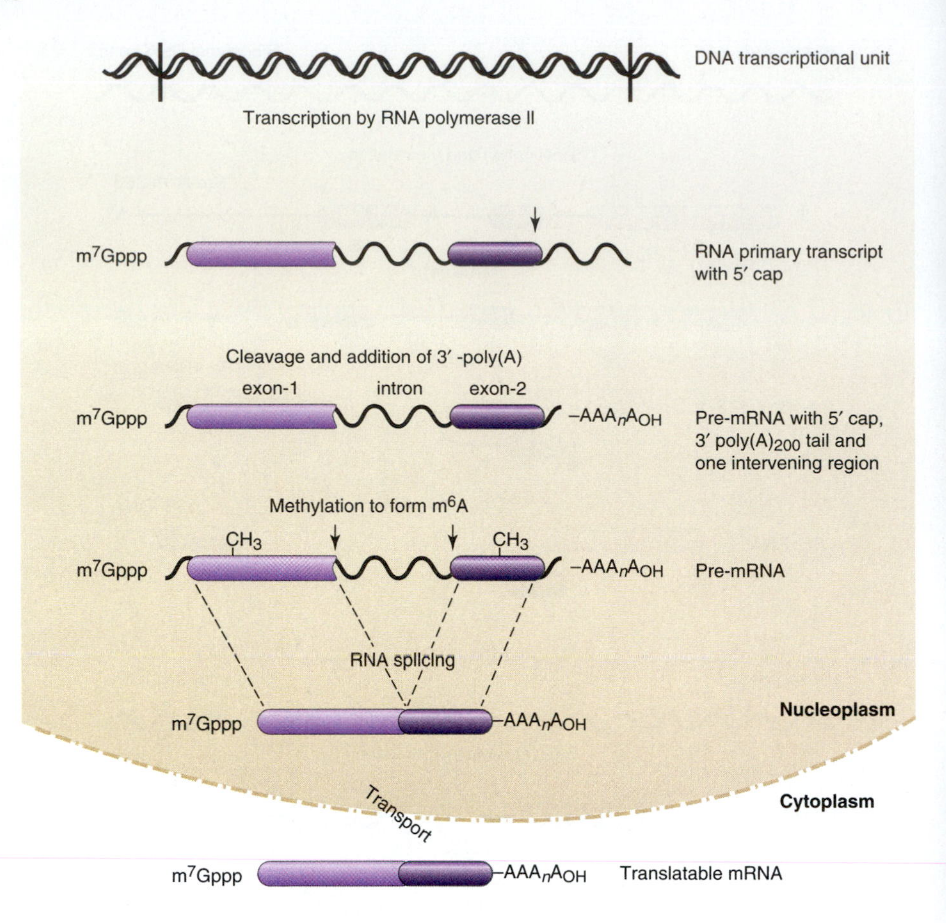

FIGURE 5.12
Scheme for processing mRNA.
Points for initiation and termination of transcription are indicated on the DNA. Arrows indicate cleavage points. Many proteins associated with the RNA and tertiary conformations are not shown.

CLINICAL CORRELATION 5.4
Thalassemia Due to Defects in Messenger RNA Synthesis

Thalassemias are genetic defects in the coordinated synthesis of α- and β-globin peptide chains; a deficiency of β-globin chains is termed β-thalassemia while a deficiency of α-globin chains is termed α-thalassemia. Patients suffering from either of these conditions present with anemia at about 6 months of age as HbF synthesis ceases and HbA synthesis becomes predominant. The severity of symptoms leads to the classification of the disease into either thalassemia major, where a severe deficiency of globin synthesis occurs, or thalassemia minor, representing a less severe imbalance. Occasionally an intermediate form is seen. Therapy for thalassemia major involves frequent blood transfusions, leading to a risk of complications from iron overload. Unless chelation therapy is successful, the deposition of iron in peripheral tissues, termed hemosiderosis, can lead to death before adulthood. Carriers of the disease usually have thalassemia minor, involving mild anemia. Ethnographically, the disease is common in persons of Mediterranean, Arabian, and East Asian descent. As in sickle cell anemia (HbS) and glucose 6-phosphate dehydrogenase deficiency, the abnormality of the carriers' erythrocytes affords some protection from malaria. Maps of the regions where one or another of these diseases is frequent in the native population superimpose over the areas of the world where malaria is endemic.

α-Thalassemia is usually due to a genetic deletion, which can occur because the α-globin genes are duplicated; unequal crossing over between adjacent α alleles has apparently led to the loss of one or more loci. In contrast, β-thalassemia can result from a wide variety of mutations. Known events include mutations leading to frameshifts in the β-globin coding sequence, as well as mutations leading to premature termination of peptide synthesis. Many β-thalassemias result from mutations affecting the biosynthesis of β-globin mRNA. Genetic defects are known that affect the promoter of the gene, leading to inefficient transcription. Other mutations result in aberrant processing of the nascent transcript, either during splicing out of its two introns from the transcript or during polyadenylation of the mRNA precursor. Examples where the molecular defect illustrates a general principle of mRNA synthesis are discussed in the text.

Orkin, S. H. Disorders of hemoglobin synthesis: the thalassemias. In: G. Stamatoyannopoulis, A. W. Nienhuis, P. Leder, and P. W. Majerus (Eds.), *The Molecular Basis of Blood Diseases.* Philadelphia: Saunders, 1987; and Weatherall, D. J., Clegg, J. B., Higgs, D. R., and Wood, W. G. The hemoglobinopathies. In: C. R. Scriver, A. L. Beaudet, W. S. Sly, and D. Valle (Eds.), *The Metabolic and Molecular Bases of Inherited Disease,* 7th ed. New York: McGraw-Hill, 1995.

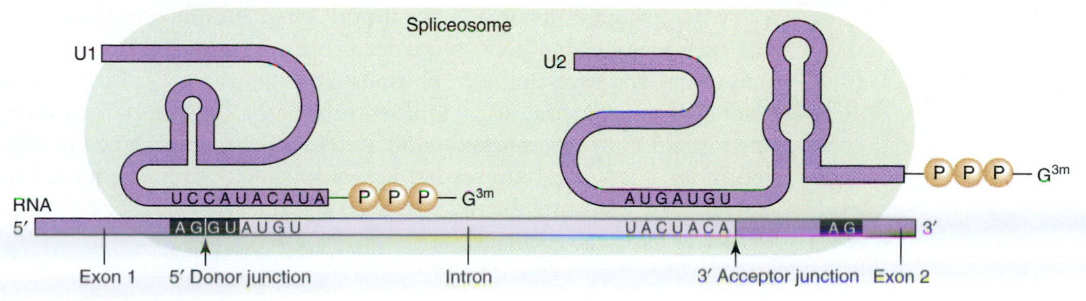

FIGURE 5.13
Mechanism of splice junction recognition.
Recognition of the 5′ splice junction involves base pairing between the intron–exon junction and the U1 RNA snRNP. This base pairing targets the intron for removal.
Adapted from Sharp, P. A. JAMA 260:3035, 1988.

polyadenylation complexes to the transcript in turn. This means that all these modifications happen during transcription.

Capping

As the transcription complex moves along the DNA, the **capping** enzyme complex modifies the 5′-end of the nascent mRNA. Capping involves the synthesis of a 5′ to 5′ triphosphate bond and a guanylyl residue is added to the transcript. This structure is further modified by methylation.

Removal of Introns from mRNA Precursors

As pre-mRNA is extruded from the RNA polymerase complex, it is rapidly bound by **small nuclear ribonucleoproteins, snRNPs** ("snurps"), which carry out the dual steps of RNA splicing: (1) breakage of the intron at the 5′ donor site and (2) joining the upstream and downstream exon sequences together. All **introns** begin with a GU sequence and end with AG; these are termed the **donor** and **acceptor intron–exon junctions,** respectively. Not all GU or AG sequences are spliced out of RNA, however. How does the cell know which GU sequences are in introns (and therefore must be removed) and which are destined to remain in mature mRNA? This discrimination is accomplished by formation of base pairs between **U1 RNA** and the sequence of the mRNA precursor surrounding the donor GU sequence (Figure 5.13) (see Clin. Corr. 5.5). Another snRNP, containing **U2 RNA,** recognizes important sequences at the 3′ acceptor end of the intron. Still other snRNP species, among them U5 and U6, then bind to the RNA precursor, forming a large complex termed a **spliceosome** (by analogy with the large ribonucleoprotein assembly involved in protein synthesis, the ribosome). The spliceosome uses ATP energy to

CLINICAL CORRELATION 5.5

Autoimmunity in Connective Tissue Disease

Humoral antibodies in sera of patients with various connective tissue diseases recognize a variety of ribonucleoprotein complexes. Patients with systemic lupus erythematosus exhibit a serum antibody activity designated Sm, and those with mixed connective tissue disease exhibit an antibody designated RNP. Each antibody recognizes a distinct site on the same RNA–protein complex, U1 RNP, that is involved in mRNA processing in mammalian cells. The U1–RNP complex contains U1 RNA, a 165-nucleotide sequence highly conserved among eukaryotes, that at its 5′-terminus includes a sequence complementary to intron–exon splice junctions. Addition of this antibody to *in vitro* splicing assays inhibits splicing, presumably by removal of the U1 RNP from the reaction. Sera from patients with other connective tissue diseases recognize different nuclear antigens, nucleolar proteins, and/or chromosomal centromeres. Sera of patients with myositis have been shown to recognize cytoplasmic antigens such as aminoacyl-tRNA synthetases. Although humoral antibodies have been reported to enter cells that via Fc receptors, there is no evidence that this is part of the mechanism of autoimmune disease.

carry out accurate removal of the intron. First, the phosphodiester bond between the exon and the donor GU sequence is broken, leaving a free 3′-OH group at the end of the first exon and a 5′ phosphate on the donor G of the intron. This pG is then used to form an unusual linkage with the 2′-OH group of an adenosine within the intron to form a branched or lariat RNA structure, as shown in Figure 5.14. After the lariat is formed, the second step of splicing occurs. The phosphodiester bond immediately following the AG is cleaved and the two exon sequences are ligated together.

Polyadenylation

RNA polymerase continues transcribing the gene until a **polyadenylation signal sequence is** reached (Figure 5.15). This sequence, which has the consensus AAUAAA, appears in the mature mRNA but does not form part of its coding region. Rather, it signals cleavage of the nascent mRNA precursor about 20 or so nucleotides downstream. The poly(A) sequence is then added by polymerase to the free 3′-end produced by this cleavage. The C-terminal domain of RNA polymerase II recruits the cleavage and **polyadenylation** complex to the transcript. Along with this event, RNA polymerase terminates transcription although the detailed mechanism is not clear in all cases. The end result of processing is a fully functional coding mRNA, all introns removed, and ready to direct protein synthesis.

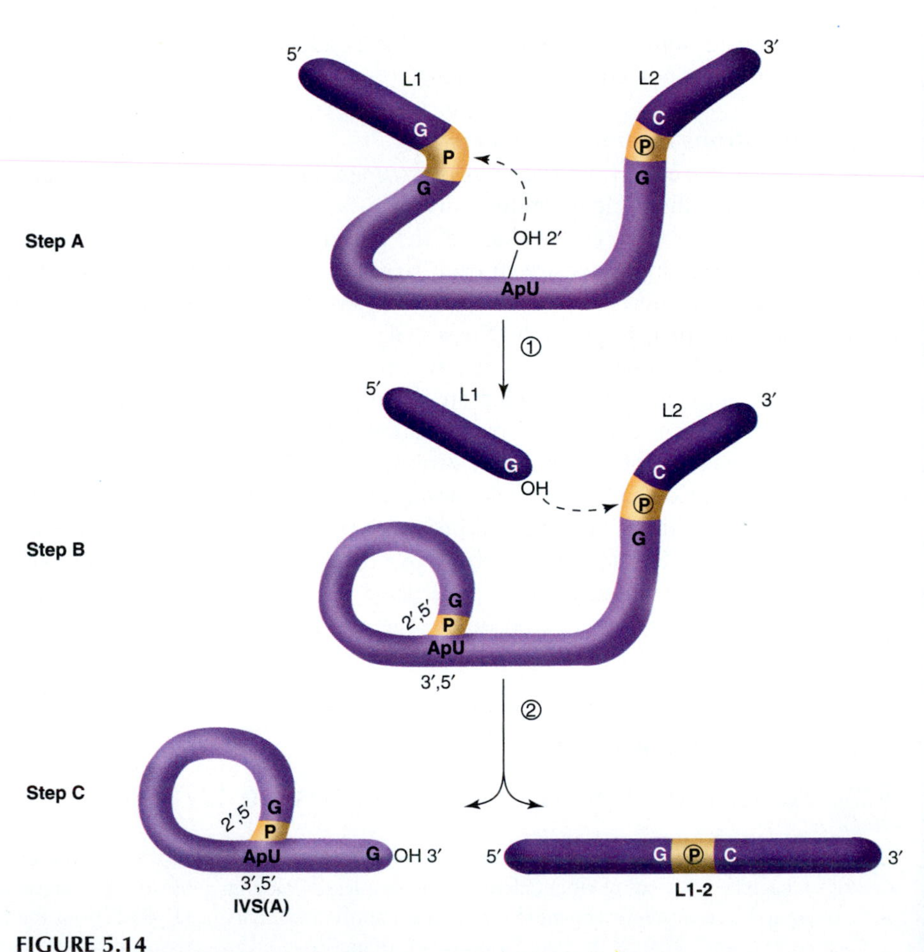

FIGURE 5.14
Proposed scheme for mRNA splicing to include the lariat structure.
A messenger RNA is depicted with two exons (in dark blue) and an intervening intron (in light blue). A 2′-OH group of the intron sequence reacts with 5′-phosphate of the intron's 5′-terminal nucleotide producing a 2′–5′ linkage and the lariat structure. Simultaneously, the exon 1–intron phosphodiester bond is broken leaving a 3′-OH terminus on this exon free to react with 5′-phosphate of the exon 2, displacing the intron and creating the spliced mRNA. The released intron lariat is subsequently digested by cellular nucleases.

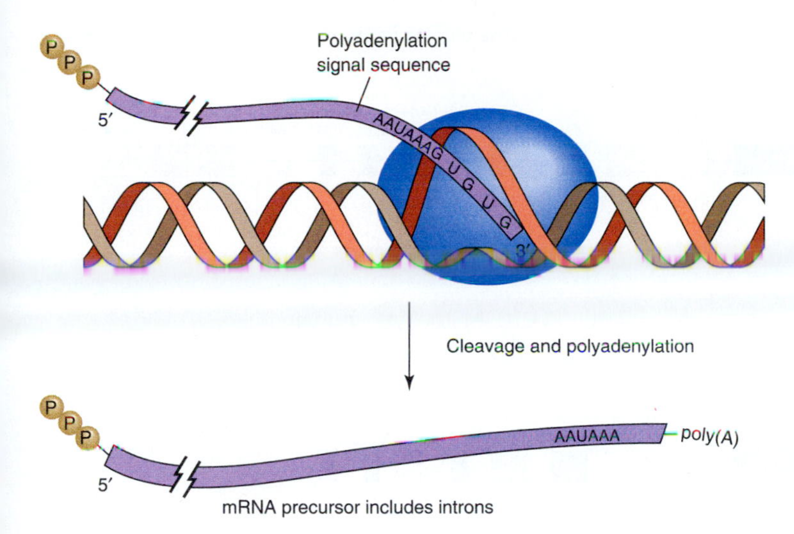

FIGURE 5.15
Cleavage and polyadenylation of eukaryotic mRNA precursors.
The 3′-termini of eukaryotic mRNA species are derived by processing. The sequence AAUAAA in mRNA specifies cleavage of mRNA precursor. The free 3′-OH end of the mRNA is a primer for poly(A) synthesis.
Adapted from Proudfoot, N. J. Trends Biochem. Sci. *14:105, 1989.*

Mutations in Splicing Signals Cause Human Diseases

Messenger RNA splicing is an intricate process dependent on many molecular events. If these events are not performed with precision, functional mRNA is not produced. This is illustrated in the human **thalassemias,** which affect the balanced synthesis of α- and β-globin chains (see Clin. Corr. 5.4). Some mutations that lead to β-thalassemia interfere with splicing of β-globin mRNA precursors. For example, we know that all intron sequences begin with the dinucleotide GU. Mutation of the G in this sequence to an A means that the splicing machinery will no longer recognize this dinucleotide as a donor site. Splicing will "skip over" the correct exon–intron junction. This could lead to extra sequences appearing in the β-globin mRNA that would normally be spliced out, or alternatively, sequences could be deleted from the mRNA product (Figure 5.16). In either event, functional β-globin will be synthesized in reduced amounts and the anemia characteristic of the disease will result.

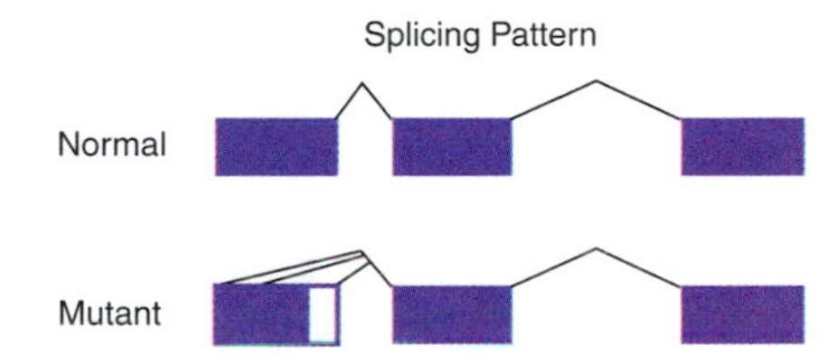

FIGURE 5.16
Nucleotide change at an intron–exon junction of human β-globin gene, which leads to aberrant splicing and β-thalassemia.
This figure shows the splicing pattern of a mutated transcript containing a change of G-U to A-U at the first two nucleotides of the first intron. Loss of this invariant sequence means that the correct splice junction cannot be used; therefore transcript sequences that base pair with the U1 snRNA less well than the correct sequence junction are used as splice donors. The diagonal lines indicate the portions spliced together in mutant transcripts. Note that some of the mutant mRNA precursor molecules are spliced so that portions of the first intron (denoted as a white box) appear in the processed product. In other instances the donor junction lies within the first exon and portions of the first exon are deleted. In no case is wild-type globin mRNA produced.
Adapted from Orkin, S. H. In: G. Stamatoyannopoulis, et al. (Eds.), The Molecular Basis of Blood Diseases. *Philadelphia: Saunders, 1987.*

Alternate pre-mRNA Splicing Can Lead to Multiple Proteins Being Made from a Single DNA Coding Sequence

The existence of intron sequences is paradoxical. Introns must be removed precisely so that the mRNA can accurately encode a protein. As we have seen above, a single base mutation can drastically interfere with splicing and cause a serious disease. Furthermore, the presence of intron sequences in a gene means that its overall sequence is much larger than is required to encode its protein product. A large gene is a target for more mutagenic events than is a small one. Indeed, common human genetic diseases like **Duchenne's muscular dystrophy** occur in genes that encompass millions of base pairs of DNA information. Why has nature not removed introns completely over the long time scale of eukaryotic evolution? There is not a clear answer to questions of this type but some introns do have beneficial effects.

Tropomyosin proteins are essential components of the contractile apparatus in the three types of muscle (see p. 1023) and each contains a specific tropomyosin. This diversity arises from a single gene that is transcribed into a primary transcript. The transcript is then processed as diagramed in Figure 5.17. All cells that contain tropomyosin make the same primary transcript but each cell type processes this transcript in a characteristic fashion. The resulting mRNA species are translated to yield the tropomyosins characteristic of each cell type. About 40 examples are well documented of tissue-specific splicing. Thus the existence of introns supplies the organism with still another method of generating protein diversity. Perhaps not surprisingly, RNA polymerase II and/or the gene-specific transcription factors are involved in this splice site choice.

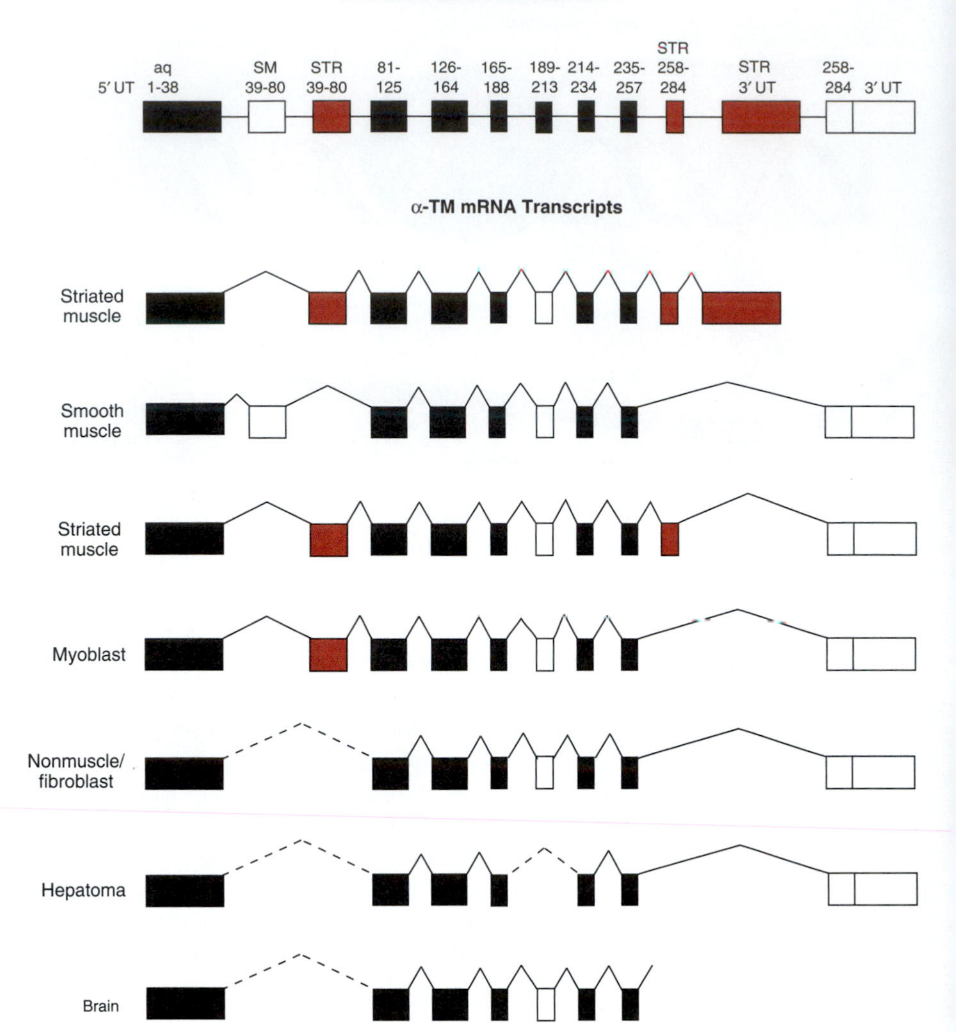

FIGURE 5.17
Alternate splicing of tropomyosin gene transcripts results in a family of tissue-specific tropomyosin proteins.
Redrawn from Breitbart, R. E., Andreadis, A., and Nadal-Ginard, B. Annu. Rev. Biochem. *56:467, 1986.*

5.5 | RNA EXPORT FROM THE NUCLEUS

Messenger, transfer, and ribosomal RNAs function in the cytoplasm. Movement out of the nucleus is through the **nuclear pore.** The nuclear pore assembly is a cylinder that spans the membrane. Within the cylinder lie two stacked rings and a central "plug." Small ions can move through the pore between the plug and the rings but RNAs are too big to pass through that space. RNA export occurs through a channel in the plug, a process that requires that the RNAs be bound to proteins to form ribonucleoprotein particles. The movement of ribosomal RNAs occurs after they are assembled into ribosomes in the nucleolus. Transfer RNAs may be bound to aminoacyl-tRNA synthetases, since they can be aminoacylated in the nucleus prior to export. Messenger RNAs are exported in complexes that contain a cap-binding protein at the 5′-end and RNA-binding proteins along the rest of the sequence. The process requires metabolic energy in the form of GTP, specific signal sequences in the protein, and a GTPase.

5.6 | TRANSCRIPTION-COUPLED DNA REPAIR

A chemical lesion in DNA affects gene expression only if the altered sequence is transcribed. Sequences that aren't expressed will not lead to an observed mutation. It makes sense, therefore, that the repair of DNA would be targeted to sequences

CLINICAL CORRELATION 5.6

Cockayne Syndrome: Transcription Coupled DNA Repair

Cockayne syndrome (CS) is a complex autosomal recessive disorder caused by a mutation in one of two genes. CS patients have developmental and neurological disorders, skeletal and retinal abnormalities, and a "bird-like" facial dysmorphy. Death usually results by the age of 20 years and is caused by neural degeneration. Most CS patients also are photosensitive and predisposed to skin cancer.

Photosensitivity points to a defect in DNA repair. For example, xeroderma pigmentosum (XP) results from mutations in a number of components of the DNA repair pathway. CS also is a DNA repair deficiency. Surprisingly, one of the two genes responsible for the syndrome is a subunit of RNA polymerase II. The protein encoded by the Cockayne syndrome B gene enhances the rate of elongation by RNA polymerase II.

How can an RNA polymerase deficiency cause a problem with DNA repair? The answer is that CS patients are deficient in transcription-coupled DNA repair. Transcription-coupled repair occurs when RNA polymerase is stalled by encountering an altered base (e.g., a thymine photodimer). Transcription is halted, the partial transcript is degraded, and the template DNA strand is repaired. Apparently, enhancement of transcription by the CSB protein also stimulates transcription-coupled DNA repair.

If this were all to the story, Cockayne's syndrome would be a variant of xeroderma pigmentosum; however, XP patients are developmentally and neurologically normal while still being photosensitive. What causes the other features of CS? It is likely that these other symptoms are due to a primary deficiency in transcription elongation caused by the mutationally altered elongation factor CSB. This idea expands our understanding of the relationship between mutation and disease. Usually, genetic diseases are caused by a mutation in biochemical processes that are outside the central information pathways of cells. This makes sense because overall inhibition of DNA, RNA, or protein synthesis would be lethal at an early stage of development. The widespread defects of CS must be due to the mutation affecting the transcription of some genes more than of others.

Citterio, E., Vermeulen, W., and Hoeijmakers, J. H. J. Transcriptional healing. *Cell* 101:447, 2000; Selby, C. P. and Sancar, A. Cockayne syndrome group B protein enhances elongation by RNA polymerase II. *Proc. Natl. Acad. Sci. U. S. A* 94:11205, 1997; and van Gool, A. J., van der Horst, G. T. J., Citterio, E., and Hoeijmakers, J. H. J. Cockayne syndrome: defective repair of transcription? *EMBO J.* 16:4155, 1997.

that are actively being transcribed. This coupling of transcription and repair occurs because DNA lesions (such as **cyclobutane dimers,** see Chapter 4) that occur on the template strand cause RNA polymerase II to stall. Stalled RNA polymerase provides a signal to the excision-repair systems to remove the altered sequences and resynthesize the correct one. Obviously, this cannot occur with the bulky RNA polymerase bound to the DNA, so the stalled transcript is sacrificed. A deficiency in transcription-coupled repair is a hallmark of **Cockayne syndrome** (see Clin. Corr. 5.6).

5.7 | NUCLEASES AND RNA TURNOVER

The different roles of RNA and DNA in genetic expression are reflected in their metabolic fates. A cell's information store (DNA) must be preserved, thus the multiple DNA repair and editing systems in the nucleus. Although individual stretches of nucleotides in DNA may turn over, the molecule as a whole is metabolically inert when not replicating. The various RNA molecules, on the other hand, are individually dispensable and can be replaced by newly synthesized species of the same specificity. It is no surprise that RNA repair systems are not known. Instead, defective RNAs are removed from cells by degradation into nucleotides, which then are reused for new RNA species.

This is clearest for mRNA species, which are classified as unstable. However, even the so-called stable RNAs turn over; for example, the half-life of tRNA species in liver is about 5 days. A fairly long half-life for a mammalian mRNA is 30 h.

Removal of RNAs from the cytoplasm is accomplished by cellular **ribonucleases.** Messenger RNAs are initially degraded in the cytoplasm. The rates vary for different mRNA species, raising the possibility of control by differential degradation.

Two examples of the role of RNA stability in gene control illustrate how the stability of mRNA influences gene expression. **Tubulin** is the major component of the microtubules of the cytoskeleton of many cell types. When there is an excess of tubulin in the cell, the monomeric protein binds to and promotes degradation of tubulin mRNA, thereby reducing tubulin synthesis. A second example is provided by **herpes simplex viruses (HSVs)**, the agent causing cold sores and some genital infections. An early event in the establishment of HSV infection is the ability of the virus to destabilize all cellular mRNA molecules, thereby reducing the competition for free ribosomes. Thus viral proteins are more efficiently translated.

Nucleases are of several types and specificities. The most useful distinction is between **exonucleases**, which degrade RNA from the 5′- or 3′-end, and **endonucleases**, which cleave phosphodiester bonds within a molecule. Products of RNase action contain either 3′- or 5′-terminal phosphates, and both endo- and exonucleases can be further characterized by the position (5′ or 3′) at which the monophosphate created by the cleavage is located.

The structure of RNA also affects nuclease action. Most ribonucleases are less efficient on regions of highly ordered RNA structure. Thus tRNAs are preferentially cleaved in unpaired regions of the sequence. On the other hand, many RNases involved in RNA processing require the substrate to have a defined three-dimensional structure for recognition.

BIBLIOGRAPHY

Bentley, D. Coupling RNA polymerase II transcription with pre-mRNA processing. *Curr. Opin. Cell Biol.* 11:347, 1999.

Berk, A. J. Activation of RNA polymerase II transcription. *Curr. Opin. Cell Biol.* 11:330, 1999.

Carmo-Fonseca, M., Mendes-Soares, L., and Campos, I. To be or not to be in the nucleolus. *Nature Cell Biol.* 2:E107, 2000.

Caskey, C. T. Triple repeat mutations in human disease. *Science* 256:784, 1992.

Citterio, E., Vermeulen, W., and Hoeijmakers, J. H. J. Transcriptional healing. *Cell* 101:447, 2000.

Conaway, J. W. and Conaway, R. C. Transcription elongation and human disease. *Annu. Rev. Biochem.* 68:301, 1999.

Daneholt, B. A look at messenger RNP moving through the nuclear pore. *Cell* 88:585, 1997.

Friend, S. A glimpse at the puppet behind the shadow play. *Science* 265:334, 1994.

Gesteland, R. F., Cech, T., and Atkins, J. F., (Eds.). *The RNA World,* 2nd ed. Cold Spring Harbor, NY: Cold Spring Harbor Laboratory Press, 1998.

Izaurralde, E., Kann, M., Panté, N., Sodeik, B., and Hohn, T. Viruses, microorganisms and scientists meet at the nuclear pore. *EMBO J.* 18:289, 1999.

Koleske, A. J. and Young, R. A. The RNA polymerase II holoenzyme and its implications for gene regulation. *Trends Biochem. Sci.* 20:113, 1995.

Larson, D. E., Zahradka, P., and Sells, B. H. Control points in eucaryotic ribosome biogenesis. *Biochem. Cell. Biol.* 69:5, 1991.

Lewin, B. *Genes VII.* New York: Oxford University Press, 2000.

Mooney, R. A. and Landick, R. RNA polymerase unveiled. *Cell* 98:687, 1999.

Orkin, S. H. Disorders of hemoglobin synthesis: the thalassemias. In: G. Stamatoyannopoulis, A. W. Nienhuis, P. Leder, and P. W. Majerus (Eds.), *The Molecular Basis of Blood Diseases.* Philadelphia: Saunders, 1987, pp. 106–126.

Paule, M. R. and White, R. J. Survey and summary: transcription by RNA polymerases I and III. *Nucleic Acids Res.* 28: 1283, 2000.

Proudfoot, N. Connecting transcription to messenger RNA processing. *Trends Biochem. Sci.* 25:290, 2000.

Ptashne, M., and Gann, A. Transcriptional activation by recruitment. *Nature* 386:569, 1997.

Rosenberg, M. and Court, D. Regulatory sequences involved in the promotion and termination of RNA transcription. *Annu. Rev. Genet.* 12:319, 1979.

Smith, C. W. J. and Valcárcel, J. Alternative pre-mRNA splicing: the logic of combinatorial control. *Trends Biochem. Sci.* 25:381, 2000.

Weinberg, R. A. Oncogenes, antioncogenes and the molecular basis of multistep carcinogenesis. *Cancer Res.* 49:3713, 1989.

Weatherall, D. J., Clegg, J. B., Higgs, D. R., and Wood, W. G. The hemoglobinopathies. In: C. R. Scriver, A. L. Beaudet, W. S. Sly, and D. Valle (Eds.), *The Metabolic and Molecular Bases of Inherited Disease,* 7th ed. New York: McGraw-Hill, 1995.

Weinstein, L. B., and Steitz, J. A. Guided tours: from precursor snoRNA to functional snRNP. *Curr. Opin. Cell Biol.* 11:378, 1999.

Wise, J. A. Guides to the heart of the spliceosome. *Science* 262:1978, 1993.

QUESTIONS | C. N. ANGSTADT

Multiple Choice Questions

Refer to the following for Questions 1–3:

A. HnRNA
B. mRNA
C. rRNA
D. snRNA
E. tRNA

1. Has the highest percentage of modified bases of any RNA.

2. Stable RNA representing the largest percentage by weight of cellular RNA.

3. Contains both a 7-methylguanosine triphosphate cap and a polyadenylate segment.

4. In eukaryotic transcription:
   A. RNA polymerase does not require a template.
   B. all RNA is synthesized in the nucleolus.
   C. consensus sequences are the only known promoter elements.
   D. phosphodiester bond formation is favored, in part, because it is accomplished by pyrophosphate hydrolysis.
   E. RNA polymerase requires a primer.

5. The sigma (σ) subunit of prokaryotic RNA polymerase:
   A. is part of the core enzyme.
   B. binds the antibiotic rifampicin.
   C. is inhibited by α-amanitin.
   D. must be present for transcription to occur.
   E. specifically recognizes promoter sites.

6. Termination of a prokaryotic transcript:
   A. is a random process.
   B. requires the presence of the rho subunit of the holoenzyme.
   C. does not require rho factor if the end of the gene contains a G-C rich palindrome.
   D. is most efficient if there is an A-T rich segment at the end of the gene.
   E. requires an ATPase in addition to rho factor.

7. Eukaryotic transcription:
   A. is independent of the presence of consensus sequences upstream from the start of transcription.
   B. may involve a promoter located within the region transcribed rather than upstream.
   C. requires a separate promoter region for each of the three ribosomal RNAs transcribed.
   D. requires that the entire gene be in the nucleosome form of chromatin.
   E. is affected by enhancer sequences only if they are adjacent to the promoter.

8. All of the following are correct about a primary transcript in eukaryotes EXCEPT it:
   A. is usually longer than the functional RNA.
   B. may contain nucleotide sequences that are not present in functional RNA.
   C. will contain no modified bases.
   D. could contain information for more than one RNA molecule.
   E. contains a TATA box.

9. The processing of transfer RNA involves all of the following EXCEPT:
   A. addition of a methylated guanosine at the 5′-end.
   B. cleavage of extra bases from both the 3′- and 5′-ends.
   C. nucleotide sequence-specific methylation of bases.
   D. addition of the sequence CCA by a nucleotidyl transferase.
   E. sometimes, removal of intron from the anticodon region.

10. In the cellular turnover of RNA:
    A. repair is more active than degradation.
    B. regions of extensive base pairing are more susceptible to cleavage.
    C. endonucleases may cleave the molecule starting at either the 5′- or 3′-end.
    D. the products are nucleotides with a phosphate at either the 3′- or 5′-OH group.
    E. all species except rRNA are cleaved.

Questions 11 and 12: Fragile X syndrome is a common form of inherited mental retardation. The mutation in the disease allows the increase of a CGC repeat in a particular gene from a normal of about 30 repeats to 200–1000 repeats. This repeat is normally found in the 5′ untranslated region of a gene for a protein called FMR1. This protein might be involved in the translation of brain-specific mRNAs during brain development. The consequence of the very large number of CGC repeats in the DNA is extensive methylation of the entire promoter region of the *FMR1* gene.

11. Methylation of bases in DNA usually:
    A. facilitates the binding of transcription factors to the DNA.
    B. makes a difference in activity only if it occurs in an enhancer region.
    C. prevents chromatin from unwinding.
    D. inactivates DNA for transcription.
    E. results, ultimately, in increased production of the product of whatever gene is methylated.

12. Transcription of eukaryotic genes requires not only the presence of a promoter but also, usually, the presence of enhancers. An enhancer:
    A. is a consensus sequence in DNA located where RNA polymerase first binds.
    B. may be located in various places in different genes.
    C. may be on either strand of DNA in the region of the gene.
    D. functions by binding RNA polymerase.
    E. stimulates transcription in both prokaryotes and eukaryotes.

Questions 13 and 14: The synthesis of normal adult hemoglobin (HbA) requires the coordinated synthesis of the two types of protein subunits, α-globin and β-globin. β-Thalassemia is a genetic disease leading to a deficiency of β-globin chains. The result is an inability of the blood to deliver oxygen properly. β-Thalassemia can result from a wide variety of mutations.

13. All of the following could lead to lack of production of β-globin or to nonfunctional β-globin EXCEPT:
    A. a frame-shift mutation leading to premature termination of protein synthesis.
    B. mutation in the promoter region of the β-globin gene.
    C. mutation toward the 3′-end of the β-globin gene that codes for the polyadenylation site.
    D. mutation in the middle of an intron that is not at an A base.
    E. all of the above could lead to this result.

14. One known mutation leading to β-thalassemia occurs at a splice junction. Which of the following statements about removing introns (cleavage/splicing) is/are correct?
    A. Small nuclear ribonucleoproteins (snRNPs) are necessary for removing introns.
    B. The consensus sequences at the 5′- and 3′-ends of introns are identical.
    C. Removal of an intron does not require metabolic energy.
    D. The exon at one end of an intron must always be joined to the exon at the other end of the intron.
    E. The nucleoside at the end of the intron first released forms a bond with a 3′-OH group on one of the nucleotides within the intron.

Problems

15. Using a schematic representation of a prokaryotic gene such as the one shown below, indicate what regions would most likely be involved in the following events:

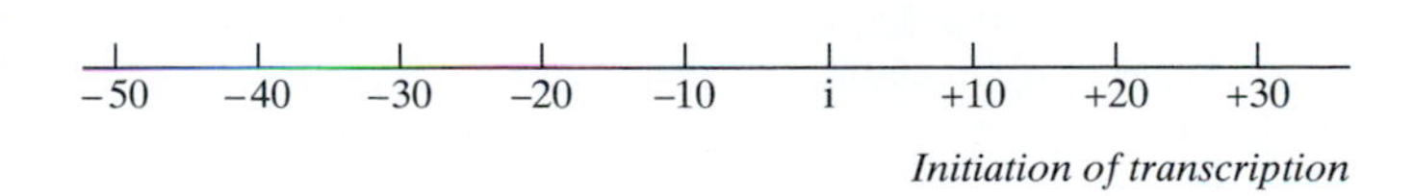

    a. RNA polymerase forms an “open complex.”
    b. RNA polymerase first binds to DNA.
    c. Core enzyme is sufficient for catalysis.
    d. Sigma (σ) factor would probably be released.

16. (a) How could you experimentally determine whether a purified preparation of an RNA polymerase is from a prokaryotic or eukaryotic source? (b) A purified preparation of RNA polymerase is sensitive to inhibition by α-amanitin at a concentration of 10^{-8} M. The synthesis of what type or types of RNA is inhibited?

ANSWERS

1. **E** Modified bases seem to be very important in the three-dimensional structure of tRNA.
2. **C** Stability of rRNA is necessary for repeated functioning of ribosomes.
3. **B** These are important additions during processing that yield a functional eukaryotic mRNA.
4. **D** This is an important mechanism for driving reactions. A,B: Transcription is directed by the genetic code, generating rRNA precursors in the nucleolus and mRNA and tRNA precursors in nucleoplasm. C: Eukaryotic transcription may have internal promoter regions as well as enhancers. E: This is a difference from DNA polymerase.
5. **E** A, D, E: Sigma subunit is required for correct initiation and dissociates from the core enzyme after the first bonds have been formed. Core enzyme can transcribe but cannot correctly initiate transcription. B, C: Rifampicin binds to the β–subunit, and α-amanitin is an inhibitor of eukaryotic polymerases.
6. **C** Rho-independent termination involves secondary structure, which is stabilized by high G-C content. A, B, E: There is a rho-dependent as well as a rho-independent process. Rho is a separate protein from RNA polymerase and appears to possess ATPase activity. D: G-C rich region is required.
7. **B** RNA polymerase III uses an internal promoter. A: RNA polymerase II activity involves the TATA and CAAT boxes. C: RNA polymerase I produces one transcript, which is later processed to yield three rRNAs. D: Parts of the promoter are not in a nucleosome. E: Enhancers may be as much as 1000 bp away.
8. **E** The TATA box is part of the promoter, which is not transcribed. A–D: Modification of bases, cleavage, and splicing are all important events in posttranscriptional processing to form functional molecules.
9. **A** Capping is a feature of mRNA. B: The primary transcript is longer than the functional molecule. C: The same modifications, catalyzed by a certain (set of) enzyme(s), occur at more than one location. D: This is a posttranscriptional modification.
10. **D** The location of the —OH depends on which side of the phosphodiester bond is split. A: There are no RNA repair processes. B: Most degradative enzymes are less efficient on an ordered structure. C: An endonuclease cleaves an interior phosphodiester bond. E: Even rRNA turns over although it is more stable than the other species.
11. **D** Methylation–demethylation of DNA is one form of control, with the methylated DNA being inactive for transcription. A, E: These would indicate that methylation activates DNA. B: Altering the enhancer region could make a difference but alteration of the promoter region certainly would. C: Unfolding of chromatin in this region is necessary for transcription but methylation of DNA is not directly involved in the process.
12. **B** B, C: Enhancer sequences seem to work whether they are at the beginning or end of the gene, but they must be on the same DNA strand as the transcribed gene. A: RNA polymerase first binds at the promoter. D: They seem to function by binding proteins, which themselves bind RNA polymerase.
13. **D** Mutation at either terminus of the intron would lead to a splicing error but a mutation in the middle of the intron should not. The exception would be if the specific mutation site were at the adenosine, which forms the branch point. A–C: These are all correct.
14. **A** snRNPs are responsible both for cleaving the intron at the 5′-end and joining the exons together. B: The sequences at the two ends of an intron are different—GU at the 5′-end and AG at the 3′-end. C: The spliceosome requires ATP to remove the intron. D: Nonconsecutive exons can be joined together, leading to multiple proteins from one gene. E: The lariat forms at a 2′-OH. All the 3′-OHs are already occupied in forming the chain.
15. (a) The Pribnow box with its high A-T content facilitates opening and is in the region −10. (b) The −35 region has a consensus sequence recognized by the RNA polymerase. (c) Certainly from +20 on (or maybe even closer to the start) is into the elongation phase, where core enzyme is all that is required. (d) Sigma factor leaves at or shortly after the start of transcription so between *i* and +10 would be the region.

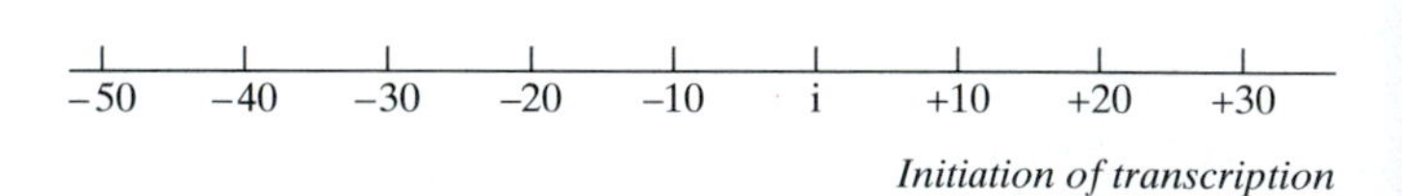

16. (a) Test the preparation for inhibition by α-amanitin at a concentration of about 2 mg mL^{-1}. At that concentration, all forms of eukaryotic RNA polymerase are inhibited but prokaryotic enzyme is not. If the preparation is not inhibited by the amanitin, confirm that it is inhibited by rifampicin. It will not be possible to distinguish eukaryotic mitochondrial RNA polymerase from prokaryotic enzyme because they have the same sensitivity. (b) RNA polymerase II is sensitive to α-amanitin at this concentration so mRNA synthesis would be inhibited.

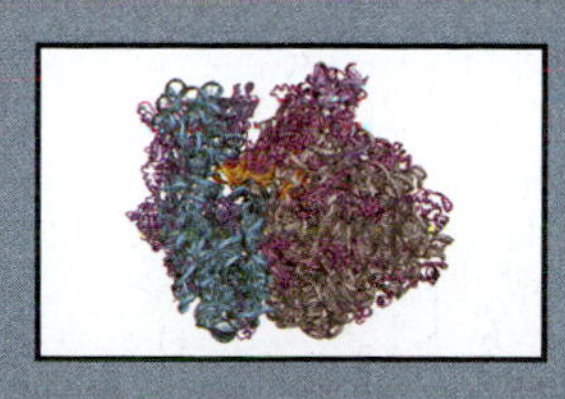

6

PROTEIN SYNTHESIS: TRANSLATION AND POSTTRANSLATIONAL MODIFICATIONS

Dohn Glitz

6.1 | OVERVIEW

Genetic information is stored and transmitted in the four-letter alphabet and language of DNA and ultimately expressed in the 20-letter alphabet and language of proteins. Protein biosynthesis is also called **translation** because it involves the biochemical translation of information between these languages. Translation has many requirements that center on functional RNA species: an informational messenger RNA molecule that is exported from the nucleus, several "bilingual" transfer RNA species that read the message, and RNA-rich ribosomes that serve as catalytic and organizational centers for protein synthesis. Further requirements include a variety of protein "factors" and an energy source. Polypeptides are formed by sequential addition of amino acids in the specific order determined by information carried in the nucleotide sequence of the mRNA. Correct folding into a three-dimensional structure may be assisted by a protein chaperone. A protein is often matured or processed by a variety of modifications that may target it to a specific intracellular location or to secretion from the cell, or might modulate its activity or function. These complex processes are carried out with considerable speed and precision, and regulated both globally and for specific proteins. Finally, if a protein is mistranslated, becomes nonfunctional, or is no longer needed, it is degraded and its amino acids are catabolized or recycled into new proteins.

Cells vary in their need and ability to synthesize proteins. Terminally differentiated red blood cells lack the components of the biosynthetic apparatus, cannot synthesize proteins, and have a life span of only 120 days. Other cells need to maintain concentrations of enzymes and other proteins and, to remain viable, carry out limited protein synthesis. Growing and dividing cells must synthesize much larger amounts of protein, and some cells also synthesize proteins for export. For example, liver cells synthesize many enzymes needed in multiple metabolic pathways plus proteins for export including serum albumin, the major protein of blood plasma or serum. Liver cells are protein factories that are particularly rich in the machinery for translation.

6.2 | COMPONENTS OF THE TRANSLATIONAL APPARATUS

Messenger RNA Transmits Information Present in DNA

Genetic information is inherited and stored in the nucleotide sequences of DNA. Selective expression of this information requires its transcription into mRNA that carries specific and precise messages from the nuclear "data bank" to the cytoplasmic sites of protein synthesis. In eukaryotes the messengers, mRNAs, are usually synthesized as significantly larger precursor molecules that are processed prior to their organized export from the nucleus (see p. 222). Eukaryotic mRNA has several identifying characteristics. It is almost always **monocistronic,** that is, encoding a single polypeptide. The 5′-end is capped with **7-methylguanosine** linked through a 5′-triphosphate bridge to the 5′-end of the messenger sequence (see p. 225). A 5′-nontranslated region, which may be short or up to a few hundred nucleotides in length, separates the cap from the translational **initiation signal.** Usually, but not always, this is the first **AUG** sequence encountered as the message is read 5′ → 3′. Uninterrupted sequences that specify a unique polypeptide sequence follow the initiation signal until a specific translation termination signal is reached. This is followed by a 3′-untranslated sequence, usually about 100 nucleotides in length, before the mRNA terminates in a 100–200-nucleotide long polyadenylate tail.

Prokaryotic mRNA differs from eukaryotic mRNA in that the 5′-end is not capped but retains a terminal triphosphate from initiation of its synthesis by RNA polymerase. Also, most messengers are **polycistronic,** that is, encoding several polypeptides, and including more than one initiation AUG sequence. A ribosome-positioning sequence is located about 10 nucleotides upstream of a valid AUG initiation signal. An untranslated sequence follows the termination signal but there is no polyadenylate tail.

Transfer RNA Acts as a Bilingual Translator Molecule

All tRNA molecules have several common structural characteristics including the 3′-terminal–CCA sequence to which an amino acid is bound, a conserved cloverleaf secondary structure, and an L-shaped three-dimensional structure (see p. 80). But each of the many molecular species has a unique nucleotide sequence, giving it individual characteristics that allow great specificity in interactions with mRNA and with the aminoacyl-tRNA synthetase that couples one specific amino acid to it.

The Genetic Code Uses a Four-Letter Alphabet of Nucleotides

Information in cells is stored as linear sequences of nucleotides in DNA, in a manner that is analogous to the linear sequence of alphabet letters in the words you are now reading. The DNA language uses a **four-letter alphabet** comprised of two purines, A and G (adenine and guanine) and two pyrimidines, C and T (cytosine and thymine). In mRNA the information is transcribed into a similar four-letter alphabet, but U (uracil) replaces T. The language of RNA is thus a dialect of the genetic language of DNA. Genetic information is **expressed** predominantly in the form of proteins that derive their properties from their linear sequence of amino acids. Thus, during protein biosynthesis, the four-letter language of nucleic acids is **translated** into the 20-letter language of proteins. Implicit in the analogy to language is the directionality of these sequences. By convention, nucleic acid sequences are written in a 5′ → 3′ direction, and protein sequences from the amino terminus to the carboxy terminus. These directions correspond in their reading and biosynthetic senses.

Codons in mRNA Are Three-Letter Words

A 1:1 correspondence of bases to amino acids would only permit mRNA to encode four amino acids, while a 2:1 correspondence would encode $4^2 = 16$ amino acids. Neither is sufficient since 20 amino acids occur in most proteins. The actual three-letter genetic code has $4^3 = 64$ permutations or words; this is also sufficient to encode start and stop signals, equivalent to punctuation. The three-base words are called **codons** and are customarily shown in the form of Table 6.1. Two amino acids are designated by single codons: methionine by AUG and tryptophan by UGG. The rest are designated by two, three, four, or six codons. Multiple codons for a single amino acid represent **degeneracy** in the code. The genetic code is nearly **universal.** The same code words are used in all living organisms, prokaryotic and eukaryotic. An exception to universality occurs in mitochondria, in which a few codons have a different meaning than in the cytosol of the same organism (Table 6.2).

Punctuation

Four codons function partly or totally as punctuation, signaling the start and stop of protein synthesis. The **start signal, AUG,** also specifies methionine. An AUG at an appropriate site in mRNA signifies methionine as the initial, amino-terminal

TABLE 6.1 The Genetic Code[a]

5′ Base	U	C	A	G	3′ Base
U	UUU Phe	UCU	UAU Tyr	UGU Cys	U
	UUC	UCC Ser	UAC	UGC	C
	UUA Leu	UCA	UAA Stop	UGA Stop	A
	UUG	UCG	UAG	UGG Trp	G
C	CUU	CCU	CAU His	CGU	U
	CUC Leu	CCC Pro	CAC	CGC Arg	C
	CUA	CCA	CAA Gln	CGA	A
	CUG	CCG	CAG	CGG	G
A	AUU	ACU	AAU Asn	AGU Ser	U
	AUC Ile	ACC Thr	AAC	AGC	C
	AUA	ACA	AAA Lys	AGA Arg	A
	AUG Met	ACG	AAG	AGG	G
G	GUU	GCU	GAU Asp	GGU	U
	GUC Val	GCC Ala	GAC	GGC Gly	C
	GUA	GCA	GAA Glu	GGA	A
	GUG	GCG	GAG	GGG	G

[a] The genetic code comprises 64 codons, which are permutations of four bases taken in threes. Note the importance of sequence: three bases, each used once per triplet codon, given six permutations: ACG, AGC, GAC, GCA, CAG, and CGA, for threonine, serine, aspartate, alanine, glutamine, and arginine, respectively.

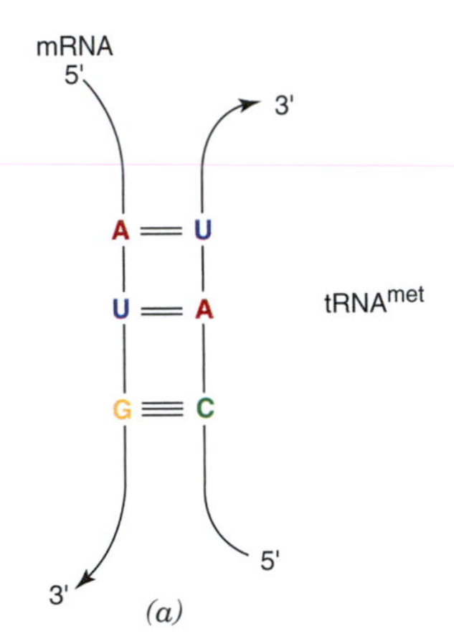

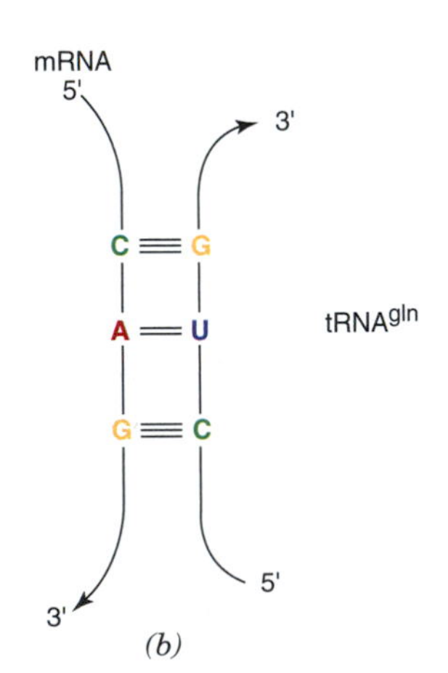

FIGURE 6.1
Codon–anticodon interactions.
Shown are interactions between the AUG (methionine) codon and its CAU anticodon (*a*) and the CAG (glutamine) codon and a CUG anticodon (*b*). Note that base pairing of mRNA with tRNA is antiparallel.

residue. AUG codons elsewhere in the message specify methionine residues within the protein. Three codons, UAG, UAA, and UGA, are **stop signals** that specify no amino acid and are known as **termination codons** or, less appropriately, as **nonsense codons.**

Codon–Anticodon Interactions Permit the Reading of mRNA

Translation of the codons of mRNA involves their direct interaction with complementary **anticodon sequences** in tRNA. Each tRNA species carries a unique amino acid, and each has a specific three-base anticodon sequence. Like base pairing in DNA, codon–anticodon base pairing is **antiparallel,** as shown in Figure 6.1, and codons are read in a **sequential, nonoverlapping reading frame.** Anticodon and amino acid-acceptor sites are located at opposite extremes of all tRNA molecules (see p. 82). The tRNA both conceptually and physically bridges the gap between the nucleotide sequence of the ribosome-bound mRNA and the site of protein assembly on the ribosome.

Since 61 codons designate 20 amino acids, it might seem necessary to have 61 different tRNA species. This is *not* the case. Variances from standard base pairing are common in codon–anticodon interactions. Many amino acids can be carried by more than one tRNA species, and degenerate codons can be read by more than one tRNA (but always one carrying the correct amino acid). Much of this complexity is explained by the **wobble hypothesis,** which permits less stringent base pairing be-

TABLE 6.2 Nonuniversal Codon Usage in Mammalian Mitochondria

Codon	*Usual Code*	*Mitochondrial Code*
UGA	Termination	Tryptophan
AUA	Isoleucine	Methionine
AGA	Arginine	Termination
AGG	Arginine	Termination

tween the first position of the anticodon and the **degenerate** (third) position of the codon. A second modulator of codon–anticodon interactions is the presence of **modified nucleotides** at or beside the first nucleotide of the anticodon in many tRNA species. A frequent anticodon nucleotide is inosinic acid (I), the nucleotide of hypoxanthine, which base pairs with U, C, or A. Wobble base-pairing rules are shown in Table 6.3.

If the wobble rules are followed, the 61 nonpunctuation codons could be read by as few as 31 tRNA molecules, but most cells have 50 or more tRNA species. Some codons are read more efficiently by one anticodon than another. Not all codons are used equally, some being used very rarely. Examination of many mRNA sequences has allowed construction of "codon usage" tables that show that different organisms preferentially use different codons to generate similar polypeptide sequences.

TABLE 6.3 Wobble Base-Pairing Rules

3′ Codon Base	*5′ Anticodon Bases Possible*
A	U or I
C	G or I
G	C or U
U	A or G or I

"Breaking" the Genetic Code

The genetic code (Table 6.1) was determined before methods were developed to sequence natural mRNA. These code-breaking experiments used simple artificial mRNAs or chemically synthesized trinucleotide codons and provided insight into how proteins are synthesized.

Polynucleotide phosphorylase catalyzes the template-independent and readily reversible reaction

$$x\text{NDP} \rightleftharpoons \text{polynucleotide } (\text{pN})_x + x\text{P}_i$$

where NDP is any nucleoside 5′-diphosphate or a mixture of two or more. If the nucleoside diphosphate is UDP, a polymer of U, designated poly(U), is formed. *In vitro* protein synthesis with poly(U) as mRNA generates the "protein" polyphenylalanine. Similarly, poly(A) encodes polylysine and poly(C) polyproline. An mRNA with a random sequence of only U and C produces polypeptides that contain not only proline and phenylalanine as predicted, but also serine (from UCU and UCC) and leucine (from CUU and CUC). Degeneracy in the code and the complexity of products made experiments with random sequence mRNAs difficult to interpret, so synthetic messengers of defined sequence were transcribed from simple repeating synthetic DNAs. Thus poly(AU), transcribed from a repeating poly(dAT), produces only a repeating copolymer of Ile-Tyr-Ile-Tyr, read from successive triplets AUA UAU AUA UAU and so on. A synthetic poly(CUG) has possible codons CUG for Leu, UGC for Cys, and GCU for Ala, each repeating itself once the **reading frame** has been selected. Since selection of the initiation codon is random in these *in vitro* experiments, three different homopolypeptides are produced: polyleucine, polycysteine, and polyalanine. A perfect poly(CUCG) produces a polypeptide with the sequence (-Leu-Ala-Arg-Ser-) whatever the initiation point. These relationships are summarized in Table 6.4; they show codons to be triplets read in exact sequence, without overlap or omission. Other experiments used chemically synthesized trinucleotide codons as minimal messages. No proteins were made, but ribosome binding of only one amino acid (as a tRNA conjugate) was stimulated by a given codon. The meaning of each possible codon was later verified by determination of mRNA sequences.

TABLE 6.4 Polypeptide Products of Synthetic mRNAs[a]

mRNA	*Codon Sequence*	*Products*
—$(AU)_n$—	—AUA UAU AUA UAU —	—$(\text{Ile-Tyr})_{n/3}$—
—$(CUG)_n$—	—CUG CUG CUG CUG —	—Leu_n—
	—UGC UGC UGC UGC —	—Cys_n—
	—GCU GCU GCU GCU —	—Ala_n—
—$(CUCG)_n$—	CUC GCU CGC UCG	—$(\text{Leu-Ala-Arg-Ser})_{n/3}$—

[a] The horizontal brackets accent the reading frame.

CLINICAL CORRELATION 6.1

Missense Mutations: Hemoglobin

Clinically the most important missense mutation that affects hemoglobin structure is the change from A to U in either the GAA or GAG codon for glutamate to give a GUA or GUG codon for valine in the sixth position of the β chain of hemoglobin. An estimated 1 of 10 African-Americans are carriers of this mutation, which in its homozygous state is the basis for sickle-cell disease, the most common of all hemoglobinopathies. The second most common hemoglobinopathy is hemoglobin C disease, in which a change from G to A in either the GAA or GAG codon for glutamate results in an AAA or AAG codon for lysine in the sixth position of the β chain. Hundreds of other hemoglobin missense mutations are now known (see Clin. Corr. 2.4 and 9.6). A recent advance in therapy of sickle cell anemia uses hydroxyurea treatment to stimulate synthesis of γ-chains and thus increase fetal hemoglobin production in affected adults. This decreases the tendency of the HbS in erythrocytes to form linear multimers that result in the cell shape distortion, that is, sickling, when the oxygen tension decreases. The hemoglobinopathies are also major targets for gene therapy.

Charache, S., Terrin, M. L., Moore, R. D., Dover, G. J., et al. Effect of hydroxyurea on the frequency of painful crises in sickle cell anemia. *N. Engl. J. Med.* 332:1317, 1995; and Persons, D. A. and Nienhuis, A. W. Gene therapy for the hemoglobin disorders: past, present, and future. *Proc. Natl. Acad. Sci. USA* 97:5022, 2000.

Mutations

An understanding of the genetic code and how it is read provides a basis for understanding the nature of mutations. A mutation is simply a change in a gene. **Point mutations** involve a change in a single base pair in the DNA, and thus a single base in the corresponding mRNA. If this change occurs in the third position of a degenerate codon, there may be no change in the amino acid specified (e.g., UCC to UCA still codes for serine). Such **silent mutations** are detected by gene sequence determination. They are commonly seen in comparison of genes for similar proteins, for example, hemoglobins from different species. **Missense mutations** arise from a base change that causes incorporation of a different amino acid in the encoded protein (see Clin. Corr. 6.1). Point mutations can also form or destroy a termination codon and thus change the length of a protein. Formation of a termination codon from one that encodes an amino acid (see Clin. Corr. 6.2) is often called a **nonsense mutation;** it results in premature termination and a truncated protein. Mutation of a termination codon to one for an amino acid allows the message to be "read through" until another stop codon is encountered. The result is a larger than normal protein. This phenomenon is the basis of several disorders (see Table 6.5 and Clin. Corr. 6.3).

CLINICAL CORRELATION 6.2

Disorders of Terminator Codons

In hemoglobin McKees Rocks the UAU or UAC codon normally designating tyrosine in position 145 of the β chain has mutated to the terminator codon UAA or UAG. This results in shortening of the β chain from its normal 146 residues to 144 residues. This change gives the hemoglobin molecule an unusually high oxygen affinity since the normal C-terminal sequence involved in binding 2,3-bisphosphoglycerate is modified. The response to decreased oxygen delivery is secretion of erythropoietin by the kidney and increased red blood cell production that produces a polycythemic phenotype.

Another illness that results from a terminator mutation is a variety of β-thalassemia. Thalassemias are a group of disorders characterized at the molecular level by an imbalance in the stoichiometry of α- and β-globin synthesis. In β^0-thalassemia no β-globin is synthesized. As a result α-globin, unable to associate with β-globin to form hemoglobin, accumulates and precipitates in erythroid cells. The precipitation damages cell membranes, causing hemolytic anemia and stimulation of erythropoiesis. One variety of β^0-thalassemia, common in Southeast Asia, results from a terminator mutation at codon 17 of the β-globin; the normal codon AAG that designates a lysyl residue at β-17 becomes the stop codon UAG. In contrast to hemoglobin McKees Rocks, in which the terminator mutation occurs late in the β-globin message, the mutation occurs so early in the mRNA of β^0-thalassemia that no useful β-globin sequence can be synthesized, and β-globin is absent. This leads to anemia and aggregation of unused α-globin in the red cell precursors. β-Globin mRNA concentrations are depressed, probably because premature termination of translation leads to instability of the mRNA.

Winslow, R. M., Swenberg, M. L., Gross, E., Chervenick, P. A., Buchman, R. R., and Anderson, W. F. Hemoglobin McKees Rocks ($\alpha_2\,\beta_2\,^{145\ \mathrm{Tyr}\rightarrow\mathrm{term}}$). A human nonsense mutation leading to a shortened β chain. *J. Clin. Invest.* 57:772, 1976; and Chang, J. C. and Kan, Y. W. β-Thalassemia: a nonsense mutation in man. *Proc. Natl. Acad. Sci. USA* 76:2886, 1979.

TABLE 6.5 "Read Through" Mutations in Termination Codons Produce Abnormally Long α-Globin Chains

Hemoglobin	*α-Codon 142*	*Amino Acid 142*	*α-Globin Length (Residues)*
A	UAA		141
Constant Spring	CAA	Glutamine	172
Icaria	AAA	Lysine	172
Seal Rock	GAA	Glutamate	172
Koya Dora	UCA	Serine	172

Insertion or deletion of a single nucleotide within the coding region of a gene results in a **frameshift mutation.** The reading frame is altered at that point and subsequent codons are read in the new context until a termination codon is reached. Table 6.6 illustrates this phenomenon with the mutant hemoglobin Wayne. The significance of reading frame selection is underscored by a phenomenon in some very small viruses in which the size of the virus limits the amount of DNA it can hold. As compensation, a single segment of DNA encodes different polypeptides that are translated using different reading frames. An example is the tumor-causing simian virus SV40 (Figure 6.2). Other viruses rely on frameshifting during translation to generate different proteins from a single message (see Clin. Corr. 6.4).

Aminoacylation of Transfer RNA Activates Amino Acids for Protein Synthesis

In order to be incorporated into proteins, amino acids must first be "activated" by linkage to their appropriate tRNA carriers. This two-step process is catalyzed by a family of **aminoacyl-tRNA synthetases,** each of which is specific for a single amino acid and its appropriate tRNA species. The reactions are normally written as follows:

$$\mathrm{H{-}N(H){-}CH(R){-}C({=}O){-}OH + ATP + E \rightleftharpoons [H{-}N(H){-}CH(R){-}C({=}O)\sim AMP \cdot E] + PP_i} \qquad (1)$$

$$\mathrm{[H{-}N(H){-}CH(R){-}C({=}O)\sim AMP \cdot E] + tRNA \rightleftharpoons H{-}N(H){-}CH(R){-}C({=}O){-}tRNA + AMP + E} \qquad (2)$$

Sum:

$$\mathrm{H{-}N(H){-}CH(R){-}C({=}O){-}OH + ATP + tRNA \rightleftharpoons H{-}N(H){-}CH(R){-}C({=}O){-}tRNA + AMP + PP_i}$$

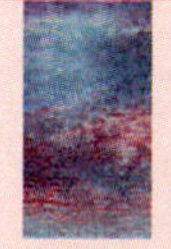

CLINICAL CORRELATION 6.3
α-Thalassemia

There are two expressed α-globin genes on each chromosome 17. Many instances of α-thalassemia arise from deletion of two, three, or all four α-globin genes. The clinical severity increases with the number of genes deleted. In contrast, the disorders summarized in Table 6.5 are forms of α-thalassemia that arise from abnormally long α-globin molecules, which replace normal α-globin, and are present only in small amounts as a result of a decreased rate of synthesis or more likely from an increased rate of breakdown of the abnormal α-globin. The normal stop codon, UAA, for α-globin mutates to any of four sense codons with resultant placement of four different amino acids at position 142. Normal α-globin is only 141 residues long, but the four abnormal α-globins are 172 residues long because a triplet of nucleotides in the normally untranslated region of the mRNA becomes a terminator codon in the abnormal position 173. Elongated globin chains can also result from frameshift mutations or insertions.

Weatherall, D. J. and Clegg, J. B. The α-chain termination mutants and their relationship to the α-thalassemias. *Philos. Trans. R. Soc. London B* 271:411, 1975.

TABLE 6.6 A Frameshift Mutation Results in Production of Hemoglobin Wayne[a]

Position	*137*	*138*	*139*	*140*	*141*	*142*	*143*	*144*	*145*	*146*	*147*
Normal α-globin amino acid sequence	- Thr	- Ser	- Lys	- Tyr	- Arg						
Normal α-globin codon sequence	- ACP	- UC(U)	- AAA	- UAC	- CGU	- [UAA]	- GCU	- GGA	- GCC	- UCG	- GUA
Wayne α-globin codon sequence	- ACP	- UCA	- AAU	- ACC	- GUU	- AAG	- CUG	- GAG	- CCU	- CGG	- [UAG]
Wayne α-globin amino acid sequence	- Thr	- Ser	- Asn	- Thr	- Val	- Lys	- Leu	- Glu	- Pro	- Arg	

[a] The base deletion causing the frameshift is encircled. The stop codons are boxed.
P = A, G, U, or C.

Brackets surrounding the aminoacyl-AMP–enzyme complex indicate that it is a transient, enzyme-bound intermediate. The aminoacyl-adenylate, a mixed acid anhydride with carboxyl and phosphoryl components, is a "high-energy" intermediate. The aminoacyl ester linkage in tRNA is lower in energy than in aminoacyl-adenylate, but still higher than in the carboxyl group of the free amino acid. Pyrophosphatases cleave the pyrophosphate that is produced and the equilibrium is strongly shifted toward formation of aminoacyl-tRNA. From the viewpoint of precision in translation, the amino acid, which had only its side chain (R group) to distinguish it, becomes linked to a large, complex, and easily recognized carrier.

Specificity and Fidelity of Aminoacylation Reactions

Almost all cells contain 20 different aminoacyl-tRNA synthetases, each specific for one amino acid, and at most a small family of carrier tRNAs for that amino acid. Since codon–anticodon interactions define the amino acid to be incorporated, an incorrect amino acid carried by the tRNA will cause an error in the protein. Correct selection of both tRNA and amino acid by the synthetase is therefore central to fidelity in protein synthesis.

Aminoacyl-tRNA synthetases share a common mechanism and many are physically associated with one another in the cell. Nevertheless, they are a diverse group of proteins that may contain one, two, or four identical subunits or pairs of dis-

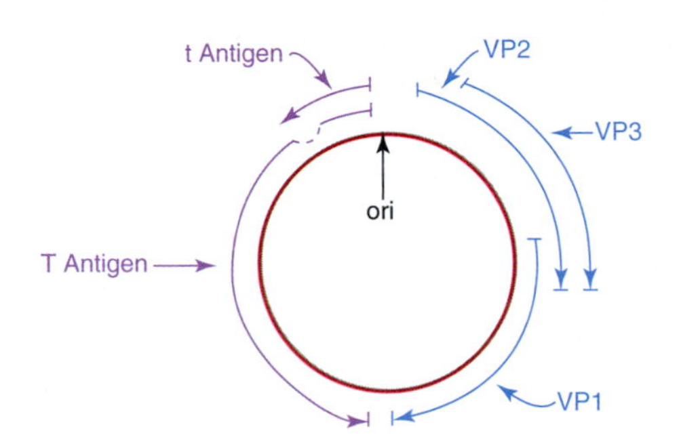

FIGURE 6.2
Genome of simian virus 40 (SV40).
DNA of SV40, shown in red, is a double-stranded circle of slightly more than 5000 base pairs that encodes all information needed by the virus for its survival and replication within a host cell. It is an example of extremely efficient use of the information-coding potential of a small genome. Proteins VP1, VP2, and VP3 are structural proteins of the virus; VP2 and VP3 are translated from different initiation points to the same carboxyl terminus. VP1 is translated in a different reading frame so that its amino-terminal section overlaps the VP2 and VP3 genes but its amino acid sequence in the overlapping segment is different from that of VP2 and VP3. Two additional proteins, the large T and small t tumor antigens, which promote transformation of infected cells, have identical amino-terminal sequences. The carboxyl-terminal segment of small t protein is encoded by a segment of mRNA that is spliced out of the large T message, and the carboxyl-terminal sequence of large T is encoded by DNA that follows termination of small t. This occurs through differential processing of a common mRNA precursor. The single site of origin of DNA replication (ori) is outside all coding regions of the genome.

CLINICAL CORRELATION 6.4

Programmed Frameshifting in the Biosynthesis of HIV Proteins

Maintaining the reading frame during translation is central to the accuracy and fidelity of translation. However, many retroviruses, including HIV, the human immunodeficiency virus that causes AIDS, take advantage of mRNA slippage and a change in reading frame to generate different proteins from the same message. A single mRNA, the *gag-pol* messenger, encodes two polyproteins that overlap by about 200 nucleotides and are in different reading frames. The gag polyprotein is translated from the initiation codon to an in-frame termination codon near the *gag-pol* junction; gag polyprotein is then cleaved to generate several structural proteins of the virus. However, about 5% of the time a one-nucleotide frameshift occurs within the overlapping segment of mRNA and the termination codon is bypassed because it is no longer in the reading frame. Instead a gag-pol fusion polyprotein is produced; proteolytic cleavage of the pol polyprotein produces the viral reverse transcriptase and other proteins needed in virus reproduction.

Jacks, T., Power, M. D., Masiarz, F. R., Luciw, P. A., and Varmus, H. E. Characterization of ribosomal frameshifting in HIV-1 gag-pol expression. *Nature* 331:280, 1988.

similar subunits. Detailed studies indicate that separate structural domains are involved in aminoacyl-adenylate formation, tRNA recognition, and, if it occurs, subunit interactions. In spite of their structural diversity, each enzyme is capable of almost error-free formation of correct aminoacyl-tRNA combinations.

Selection of a correct amino acid requires great discrimination on the part of some synthetases. While some amino acids are easily recognized by their bulk (e.g., tryptophan), or lack of bulk (glycine), or by positive or negative charges on the side chains (e.g., lysine and glutamate), others are much more difficult to discriminate. Recognition of valine rather than threonine or isoleucine by the valyl-tRNA synthetase is difficult since the side chains differ by either an added hydroxyl or single methylene group. The amino acid-recognition and -activation sites of each enzyme have great specificity, but sometimes misrecognition does occur. An additional **proofreading** or **editing** step increases discrimination. This most often occurs through hydrolysis of the aminoacyl-adenylate intermediate, with the release of amino acid and AMP. Valyl-tRNA synthetase efficiently hydrolyzes threonyl-adenylate and it hydrolyzes isoleucyl-adenylate in the presence of bound (but not aminoacylated) $tRNA^{val}$. In other cases a misacylated tRNA is recognized and deacylated. Valyl-, phenylalanyl-, and isoleucyl-tRNA synthetases deacylate tRNAs that have been mischarged with threonine, tyrosine, and valine, respectively. This proofreading is analogous to editing of misincorporated nucleotides by the $3' \rightarrow 5'$ exonuclease activity of DNA polymerases (see p. 164). Editing is performed by many but not all aminoacyl-tRNA synthetases. The net result is an average level of misacylation of one in 10^4 to 10^5.

Each synthetase must correctly recognize one to several tRNA species that correctly serve to carry the same amino acid, while rejecting incorrect tRNA species. Given the complexity of tRNA molecules this should be simpler than selection of a single amino acid. However, recall the conformational similarity and common sequence elements of all tRNAs (see p. 82). Different synthetases recognize different elements of tRNA structure. Usually multiple structural elements contribute to recognition, but many are not absolute determinants. One logical element of tRNA recognition by the synthetase is the anticodon, specific to one amino acid. In the case of $tRNA^{met}$, changing the anticodon also alters recognition by the synthetase. In other instances this is only partly true, and sometimes the anticodon is not a recognition determinant. Consider, for example, **suppressor mutations** that "suppress" the expression of classes of chain termination (nonsense) mutations. A point mutation in a glutamine (CAG) codon produces a termination (UAG) codon that causes the premature termination of the encoded protein. A second suppressor mutation in the anticodon of a $tRNA^{tyr}$, in which the normal GUA anticodon is changed to CUA, allows "read through" of the termination codon. The initial mutation is suppressed as a nearly normal protein is made, with the affected glutamine replaced by tyrosine. Aminoacylation of the mutant $tRNA^{tyr}$ with tyrosine shows that in this

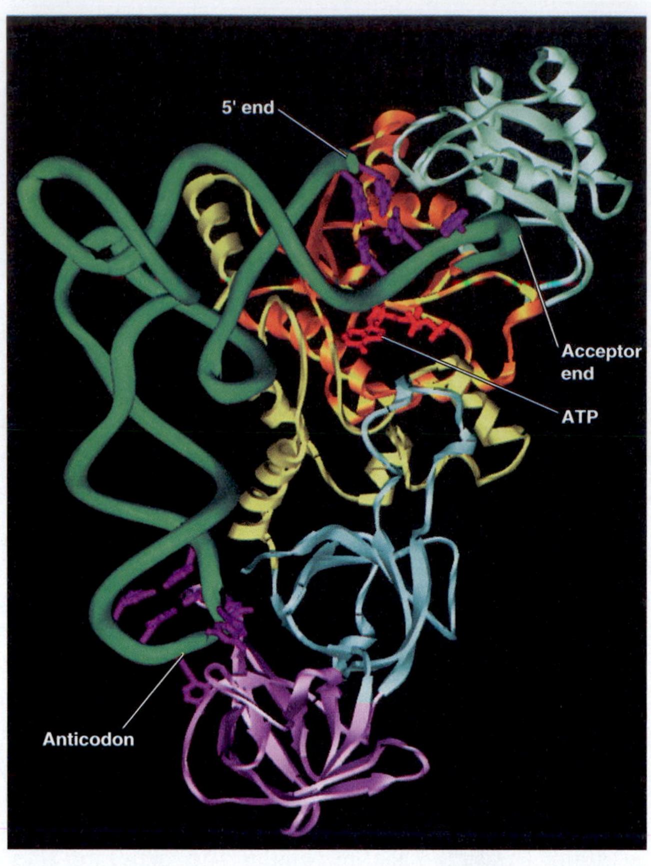

FIGURE 6.3
Interaction of a tRNA with its cognate aminoacyl-tRNA synthetase.
The sugar–phosphate backbone of *E. coli* glutaminyl-tRNA is shown in green and the peptide backbone of the glutamine $tRNA^{glx}$ synthetase in multiple colors. Note strong interactions of the synthetase with partially unwound acceptor stem and anticodon loop of the tRNA, and placement of ATP, shown in red, within a few angstroms of the 3′-end of tRNA. Space-filling models of the enzyme and tRNA would show both molecules as solid objects with several sites of direct contact.
Adapted from Perona, J., Rould, M., and Steitz, T. Biochemistry *32:8758, 1993.*

case the anticodon does not determine synthetase specificity. In *E. coli* $tRNA^{ala}$, the primary recognition characteristic is a G_3-U_{70} base pair in the acceptor stem; even if no other changes in the $tRNA^{ala}$ occur, any variation at this position destroys its acceptor ability with alanine-$tRNA^{ala}$ synthetase. Other tRNA-identification features include additional elements of the acceptor stem, and sometimes parts of the variable loop or the D-stem/loop. Different tRNA segments may be recognized for aminoacylation and proofreading. The X-ray structure of glutaminyl synthetase–tRNA complex shown in Figure 6.3 shows binding at the concave tRNA surface, which is typical and compatible with the biochemical observations.

Ribosomes Are Workbenches for Protein Biosynthesis

Proteins are assembled on **ribosomes** that are complex **ribonucleoprotein** particles made up of two dissimilar subunits, each of which contains RNA and many proteins. With one exception each protein is present in a single copy per ribosome, as is each RNA species. Ribosome architecture and function have been conserved in evolution, and similarities between ribosomes and subunits from different sources are more important than the differences. The composition of major ribosome types is shown in Table 6.7.

Ribosomes and their subunits can be crystallized and X-ray structural determination is well advanced. Figure 6.4 shows the structure of a prokaryotic ribo-

TABLE 6.7 Ribosome Classification and Composition

Ribosome Source	*Monomer Size*	*Subunits*	
		Small	*Large*
Eukaryotes			
Cytosol	80S	40S:	60S:
		34 proteins	50 proteins
		18S RNA	28S, 5.8S, 5S RNAs
Mitochondria			
Animals	55S–60S	30S–35S:	40–45S:
		12S RNA	16S RNA
		70–100 proteins	
Higher plants	77S–80S	40S:	60S:
		19S RNA	25S, 5S RNAs
		70–75 proteins	
Chloroplasts	70S	30S:	50S:
		20–24 proteins	34–38 proteins
		16S RNA	23S, 5S, 4.5S RNAs
Prokaryotes			
Escherichia coli	70S	30S:	50S:
		21 proteins	34 proteins
		16S RNA	23S, 5S RNAs

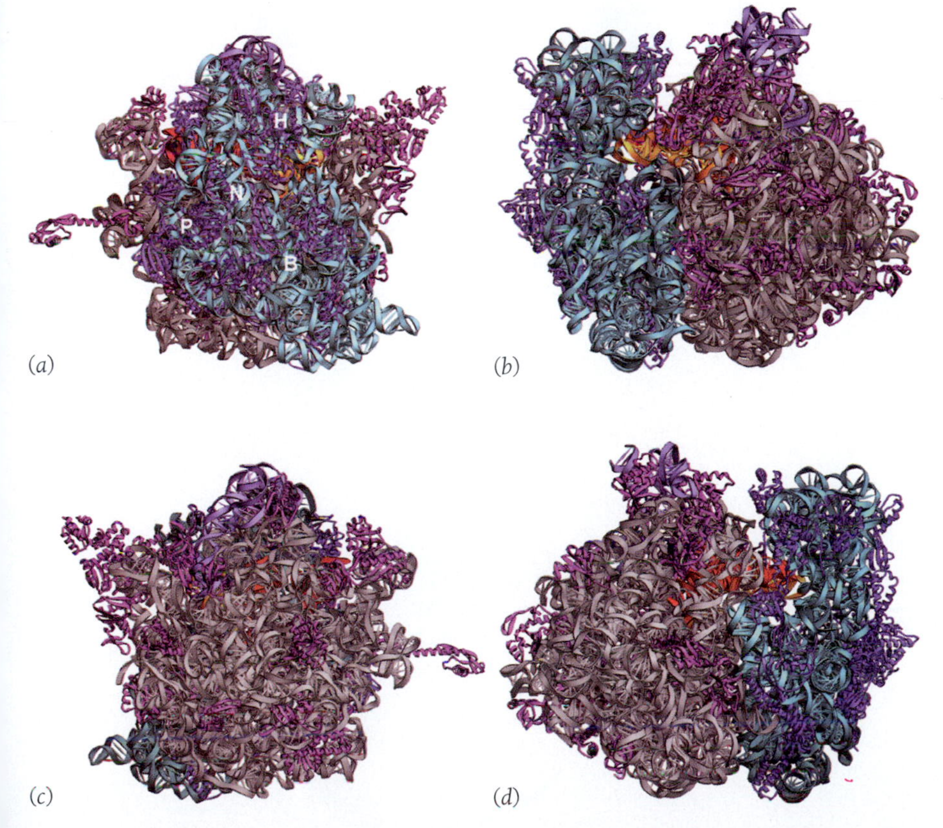

FIGURE 6.4
Crystallographic structure of a 70S ribosome. The structure of the *Thermus thermophilus* ribosome is shown at 5.5 Å resolution. Successive views are rotated 90° around the vertical axis. (*a*) The small subunit lies atop the large subunit. Small subunit features include the head (H), connected by a neck (N) to the body (B), and the platform (P) which projects toward the large subunit. The 16S RNA is colored cyan and small subunit proteins are dark blue. (*b*) The large subunit is at the right; 23S RNA is grey, 5S RNA is light blue, and large subunit proteins are magenta. A-site tRNA (gold) spans the subunits. (*c*) The large subunit lies on top with the stalk protruding to the left. (*d*) The large subunit is at the left, and elements of E- and A-side tRNAs are visible in the subunit interface.
Reprinted with permission from Yusupov, M.M., Yusupova, G. Z., Baucom, A., Lieberman, K., Earnest, T.N., Cate, J.H.D., and Noller, H.F. Science *292: 883, 2001. Copyright © 2001 by American Association for the Advancement of Science. Figure generously supplied by Drs. A. Baucom and H. Noller. Also see Figure 6.5.*

some at 5.5 Å resolution. Each subunit includes an RNA core, folded into a specific three-dimensional structure, upon which proteins are positioned through protein–RNA and protein–protein interactions.

Helical RNA elements and some individual protein molecules can be distinguished in the X-ray structures. Additional evidence, obtained using a variety of techniques, allows further understanding of the structure of this large complex. Many experiments were made possible because prokaryotic ribosomes can **self-assemble;** that is, the native structures can be reconstituted from mixtures of purified individual proteins and RNAs. **Reconstitution** of subunits from mixtures in which a single component is omitted or modified can show if, for example, a given

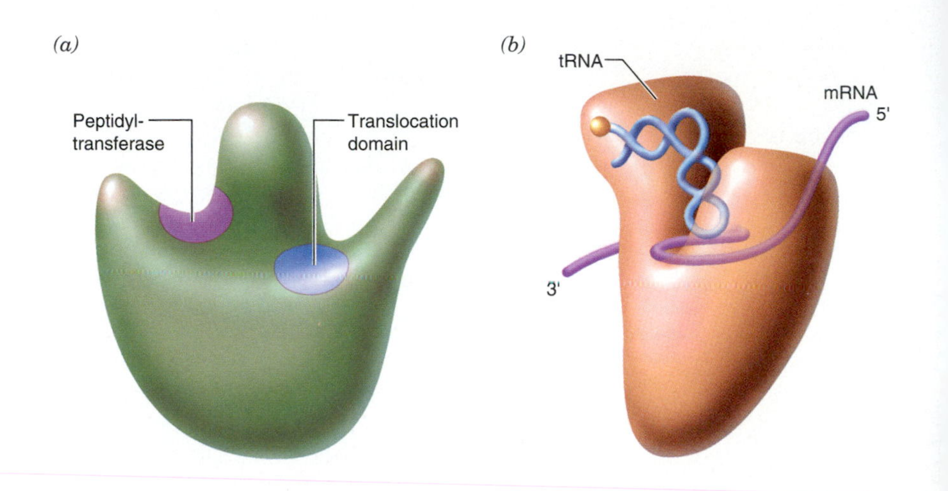

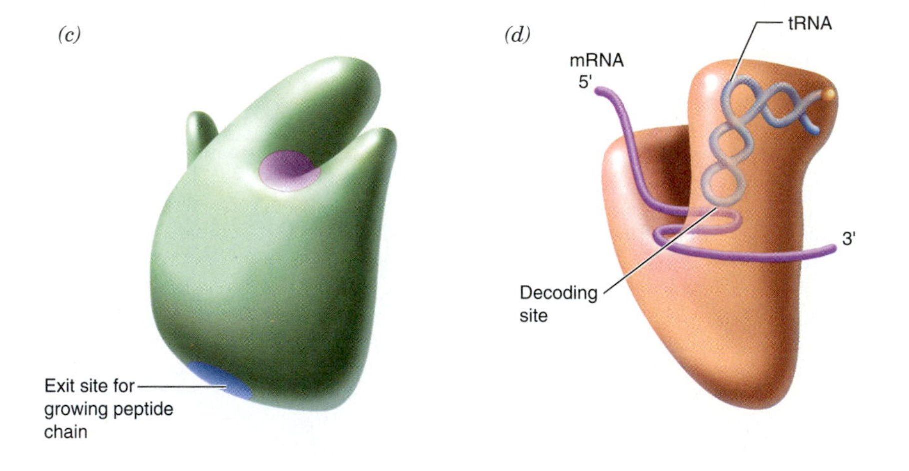

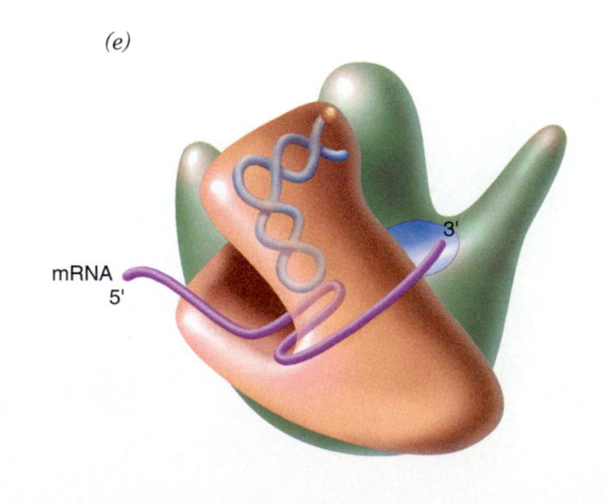

FIGURE 6.5
Ribosome structure and functional sites. Top row shows the faces of each subunit that interact in the functional ribosome. In (*a*) the large subunit is shown; note that sites of peptide bond formation and of binding of the elongation factors are on opposite sides of the central protuberance. The arm-like structure is formed by protein L7/L12 (or its eukaryotic equivalent). In (*b*) the small subunit platform protrudes toward the reader. mRNAs and tRNAs interact in a decoding site, deep in the cleft between the platform and the subunit body. The orientation of mRNA and tRNA is depicted, although their interaction in the decoding site is obscured by the platform. In (*c*) the large subunit is rotated 90° and the arm projects into the page. The exit site near the base of the subunit is where newly synthesized protein emerges from the ribosome. This area is in contact with membranes in the bound ribosomes of rough endoplasmic reticulum. The site of peptide bond formation, the peptidyltransferase center, is distant from the exit site; the growing peptide passes through a groove or tunnel in the ribosome to reach the exit site. In (*d*) the small subunit has been rotated 90°: the platform projects toward the face of the large subunit, showing the cleft. In (*e*) subunits have been brought together to show their relative orientation in the ribosome. Aminoacyl-tRNA bound by the small subunit is oriented with the acceptor end near peptidyltransferase while translocation domain (where EF-1 and EF-2 bind) is near the decoding region and the area in which mRNA enters the complex.

protein is important for assembly of the subunit or for some specific function. Total reconstitution of subunits from eukaryotes has not been achieved but conclusions about how ribosomes function are fully applicable to eukaryotic systems.

Ribosomes and their subunits can be visualized by electron microscopy. Locations of most ribosomal proteins, some elements of the RNA, and functional sites on each subunit were shown by electron microscopy of subunits in complexes with antibodies against a single ribosomal component. The antibody molecule acts as a pointer to the site on the ribosome. Further structural information was obtained from chemical cross-linking, which identifies near neighbors within the structure, and from neutron diffraction measurements, which quantitate the distances between pairs of proteins. Sequence comparisons and chemical and enzymatic probes give information about RNA conformation. Correlation of structural data with functional measurements in protein synthesis have allowed development of models, such as that in Figure 6.5, that link ribosome morphology to function in translation.

Ribosomes are organized in two additional ways. First, several ribosomes often translate a single mRNA molecule simultaneously. Purified mRNA-linked **polysomes** can be visualized by electron microscopy (Figure 6.6). Second, in eukaryotic cells some ribosomes occur free in the cytosol, but many are bound to membranes of the rough endoplasmic reticulum. In general, **free ribosomes** synthesize proteins that remain within the cytosol or become targeted to the nucleus, mitochondria, or some other organelle. **Membrane-bound ribosomes** synthesize proteins that will be secreted from the cell or sequestered and function in other cellular membranes or vesicles. In cell homogenates, membrane fragments with bound ribosomes constitute the **microsome** fraction; detergents that disrupt membranes release these ribosomes.

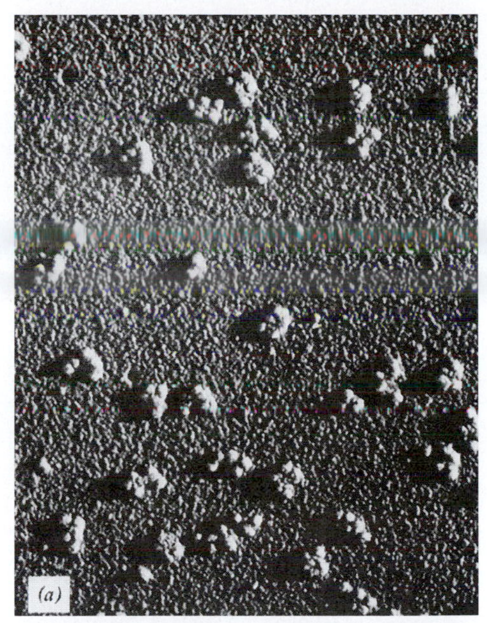

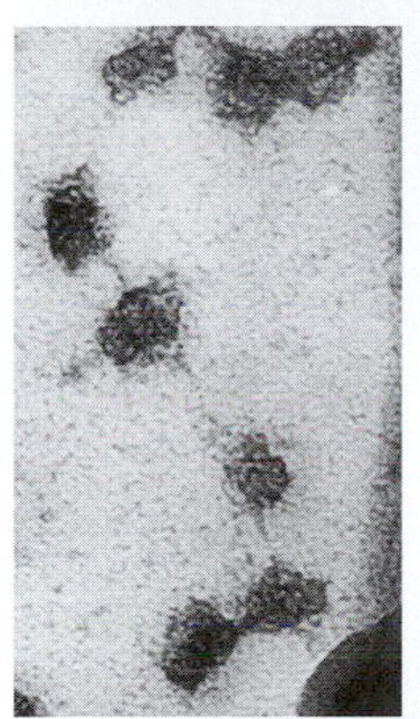

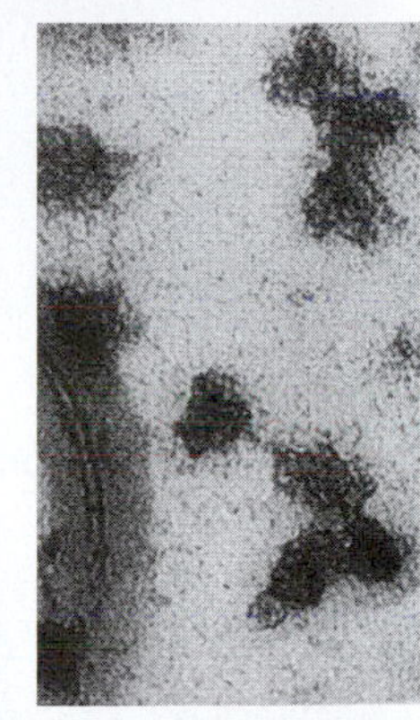

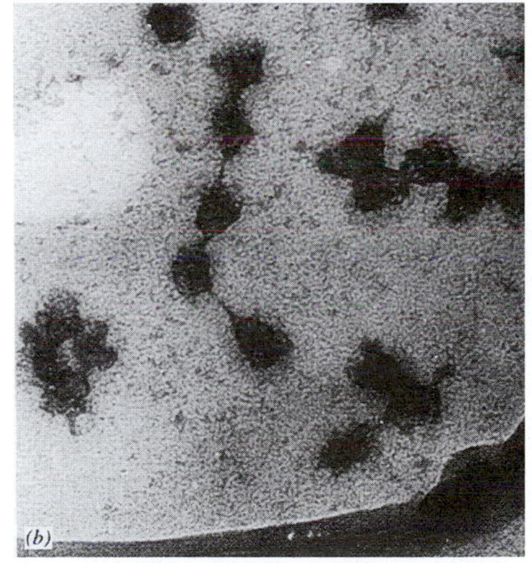

FIGURE 6.6
Electron micrographs of polysomes.
(*a*) Reticulocyte polyribosomes shadowed with platinum are seen in clusters of three to six ribosomes, a number consistent with the size of mRNA for a globin chain. (*b*) Uranyl acetate staining and visualization at a higher magnification shows polysomes in which parts of the mRNA are visible.
Courtesy of Dr. Alex Rich, MIT.

6.3 | PROTEIN BIOSYNTHESIS

Translation Is Directional and Collinear with mRNA

In the English language words are read from left to right and not the reverse. Similarly, mRNA sequences are written (and transcribed) $5' \rightarrow 3'$, and during the translation process they are read in the same direction. Amino acid sequences are both written and synthesized from the amino-terminal residue to the carboxy terminus. A ribosome remains bound to an mRNA molecule and moves along the length of the mRNA until it is fully read. Comparison of mRNA sequences with sequences of the proteins they encode shows a perfect, collinear, nonoverlapping, and gap-free correspondence of the mRNA coding sequence and that of the polypeptide synthesized. (For a rare exception, see Clin. Corr. 6.4.) In fact, it is common to deduce the sequence of a protein solely from the nucleotide sequence of its mRNA or the DNA of the gene encoding it. However, the deduced sequence may differ from the genuine protein because of posttranslational modifications.

A story can be analyzed in terms of its beginning, its development or middle section, and its ending. Protein biosynthesis will be described in a similar framework: initiation of the process, elongation during which the great bulk of the protein is formed, and termination of synthesis and release of the completed polypeptide. We will then examine posttranslational modifications that a protein may undergo.

Initiation of Protein Synthesis Is a Complex Process

Initiation requires bringing together a small (40S) ribosomal subunit, a mRNA, and a tRNA complex of the amino-terminal amino acid, all in a proper orientation. This is followed by association of the large (60S) subunit to form a completed initiation complex on an 80S ribosome. This process requires a complex group of proteins, known as **initiation factors,** that act only in initiation. These are not ribosomal proteins, although many bind transiently to ribosomes. There are many eukaryotic initiation factors and the specific functions of some remain unclear; prokaryotic protein synthesis provides a less complex model for comparison.

The initiation of translation is shown in Figure 6.7. As a first step, eukaryotic **initiation factor 2a (eIF-2a)** binds to GTP and an **initiator** species of $tRNA^{met}$, designated **Met-$tRNA_i^{met}$**, to form a ternary complex. No other aminoacyl-tRNA, including Met-$tRNA_e^{met}$, which participates only in elongation steps of protein synthesis, can replace the initiation-specific Met-$tRNA_i^{met}$ in this step. Prokaryotes also utilize a specific initiator tRNA whose methionine is modified by formylation of its amino group. Only **fMet-$tRNA_i^{met}$**, is recognized by prokaryotic IF-2.

The second step in initiation requires 40S ribosomal subunits associated with a very complex protein, eIF-3. Mammalian eIF-3 contains eight different polypeptides and has a mass of 600–650 kDa; it binds to the 40S subunit surface that will contact the 60S subunit and physically blocks subunit association. Hence eIF-3 is also called a ribosome **anti-association factor**, as is eIF-6, which binds to 60S subunits. A complex that includes eIF-2a–Met-$tRNA_i^{met}$–GTP, eIF-3–40S, and additional protein factors now forms and binds mRNA, which is also in a complex with several proteins. Factor **eIF-4g** serves as the core of the mRNA complex; it binds mRNA, protein **eIF-4e**, also called the **cap binding protein, eIF-4a,** which binds the mRNA and prevents secondary structure formation, and a polyA-binding protein that loops the 3′-end of the messenger near to the 5′-cap. A complex that includes several protein factors, the correctly oriented 40S subunit, the initiator tRNA, and the message is thus formed. A final **preinitiation complex** is generated as the message is "scanned" from the capped end until an initiation AUG sequence is reached. Usually this is the first AUG triplet in the mRNA, but sometimes the surrounding nucleotide sequence or secondary structure is not appropriate for initiation and a later AUG is selected.

Formation of a complete initiation complex proceeds with involvement of a 60S subunit and an additional factor, eIF-5. Protein **eIF-5** first interacts with the preinitiation complex; GTP is hydrolyzed to GDP and P_i, and eIF-2a–GDP, eIF-3, and other factors are released. The 40S–Met-$tRNA_i^{met}$–mRNA complex interacts with a 60S subunit and initiation factor eIF-4d to generate an 80S ribosome with the mRNA and initiator tRNA correctly positioned on the ribosome. The eIF-2a–GDP that is released interacts with the **guanine nucleotide exchange factor eIF-2b** and GTP to regenerate eIF-2a–GTP for another round of initiation.

Prokaryotes use fewer nonribosomal factors to form a similar initiation complex. Their 30S subunits complexed with a simpler IF-3 can first bind either mRNA or a ternary complex of IF-2, fMet-$tRNA_i^{met}$, and GTP. Orientation of the mRNA relies in part on base pairing between a pyrimidine-rich sequence of eight nucleotides in 16S rRNA and a purine-rich sequence about 10 nucleotides upstream of the initiator AUG codon. Complementarity between rRNA and the message-positioning sequence of an mRNA may include several mismatches but, as a first approximation, greater complementarity leads to more efficient initiation. It is interesting that eukaryotes do not utilize an RNA–RNA base-pairing mechanism, but instead use many protein factors to position mRNA correctly. A third initiation factor, IF-1, also participates in formation of the preinitiation complex. A 50S subunit is now bound; in the process, GTP is hydrolyzed to GDP and P_i, and the initiation factors are released.

Elongation Is the Stepwise Formation of Peptide Bonds

Protein synthesis now occurs by stepwise elongation to form a polypeptide chain. At each step ribosomal **peptidyltransferase** transfers the growing peptide (or in the first step the initiating methionine residue) from its carrier tRNA to the α-amino group of the amino acid residue of the aminoacyl-tRNA specified by the next codon. Efficiency and fidelity are enhanced by nonribosomal protein **elongation factors** that utilize the energy released by GTP hydrolysis to ensure selection of the proper aminoacyl-tRNA species and to move the mRNA and associated tRNAs through the **decoding region** of the ribosome. Elongation is illustrated in Figure 6.8.

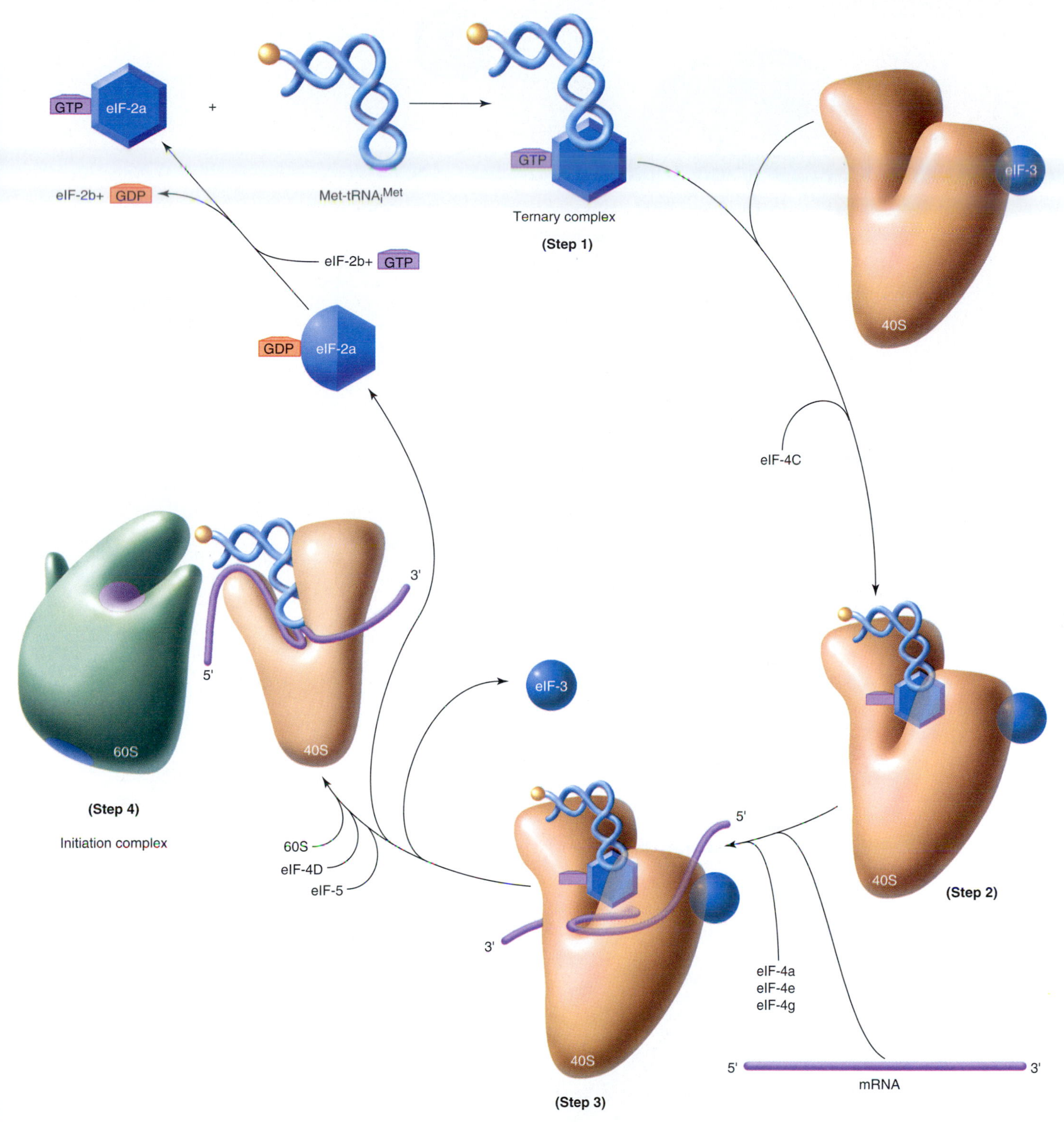

FIGURE 6.7
Initiation of translation in eukaryotes.
Details are given in text. A ternary complex (Step 1) first combines with a small ribosomal subunit to place the initiator tRNA (Step 2). Interaction with mRNA then forms a preinitiation complex (Step 3). Large subunit binding completes formation of the initiation complex (Step 4). The different shape of eIF-2a in complexes with GTP and GDP indicates that conformational change in the protein occurs upon hydrolysis of triphosphate. After elongation has begun, additional small subunits will complex with the same mRNA as polysomes are formed.

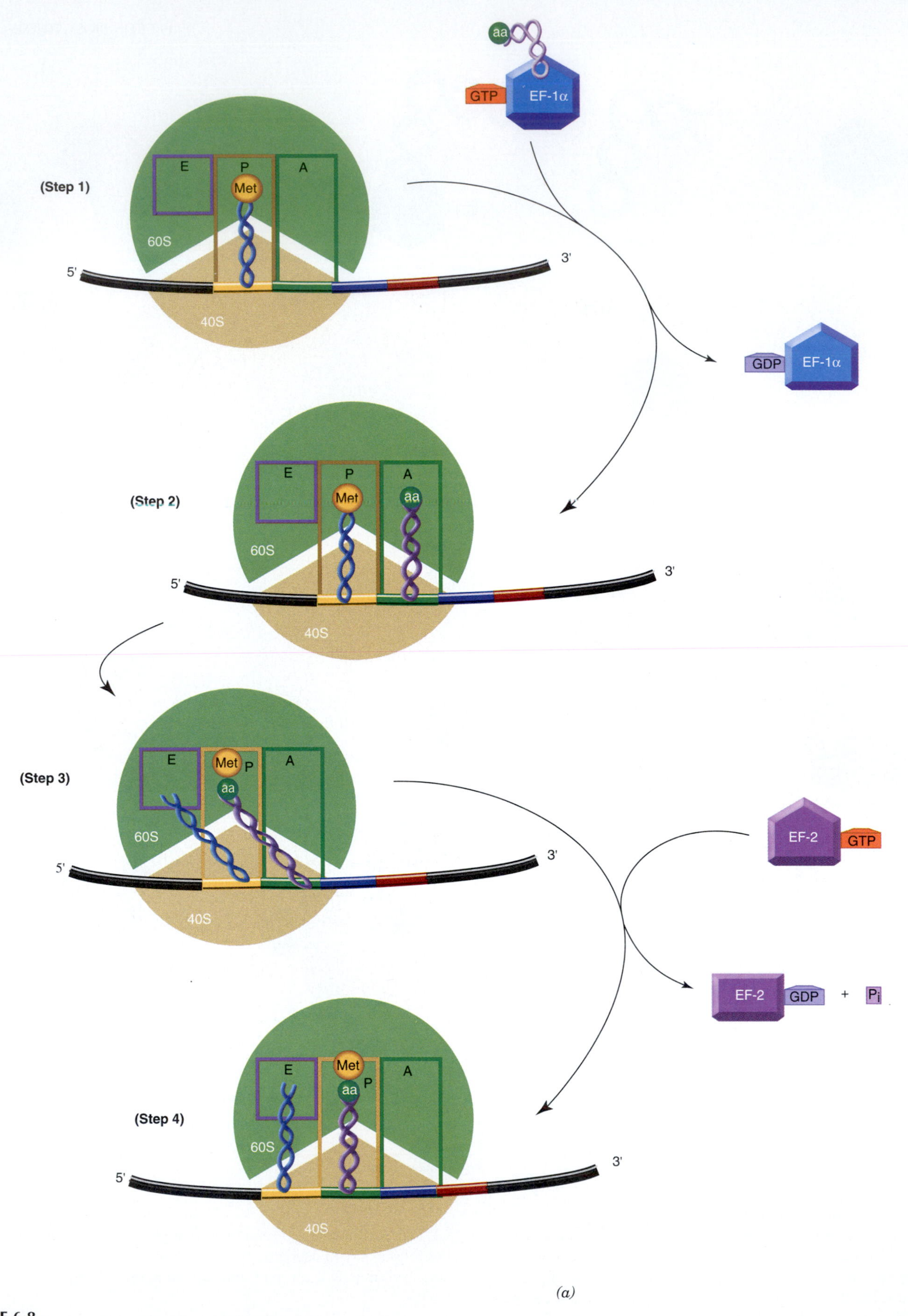

FIGURE 6.8

Elongation steps in eukaryotic protein synthesis.

(*a*) The first cycle of elongation is shown. Step 1: Completed initiation complex with methionyl-$tRNA_i^{met}$ in 80S P site. Step 2: EF-1α has placed an aminoacyl-tRNA in the A site. GTP hydrolysis results in a conformational change in EF-1α. Step 3: The first peptide bond has

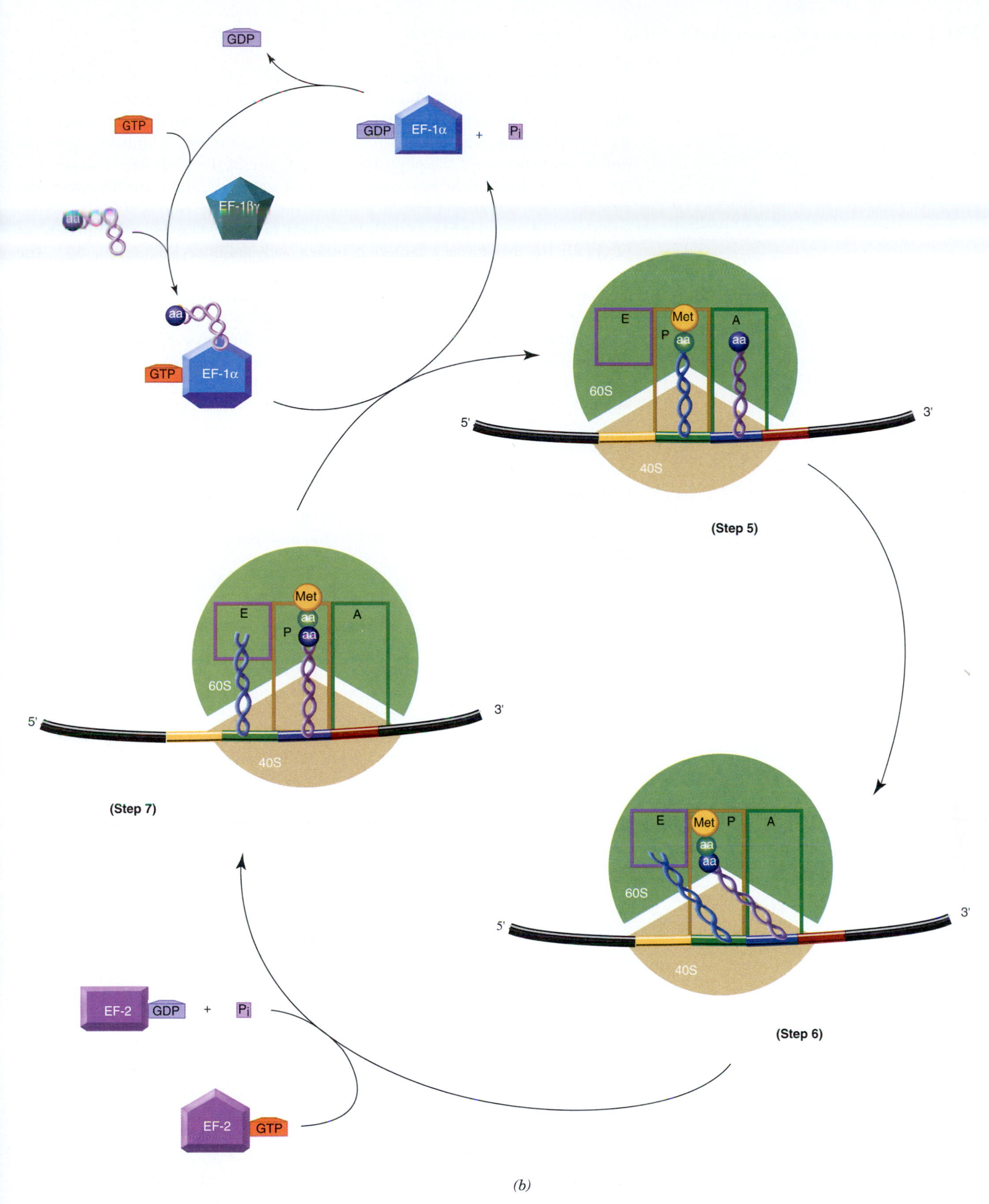

(*b*)

been formed, new peptidyl-tRNA occupies a hybrid (A/P) site on ribosome, and the acceptor stem of deacylated $tRNA_i^{met}$ is in the large subunit E site. Step 4: The mRNA–peptidyl-tRNA complex has been fully translocated to the P site while deacylated initiator tRNA is moved to the E site. (*b*) Further rounds of elongation. Step 5: Binding of aminoacyl-tRNA causes concomitant release of deacylated tRNA from the E site. Step 6: Formation of next peptide bond again results in the new peptidyl-RNA occupying a hybrid A/P site on ribosome. Step 7: Translocation moves mRNA and new peptidyl-tRNA in register into P site. Additional amino acids are added by successive repetitions of the cycle. For further details see text.

During translation up to three different tRNA molecules may be bound at specific sites that span both ribosomal subunits. The initiating methionyl-tRNA is placed in position so that its methionyl residue may be transferred (or donated) to the free α-amino group of the incoming aminoacyl-tRNA; it thus occupies the **donor site,** also called the **peptidyl site** or **P site** of the ribosome. The aminoacyl-tRNA specified by the next codon of the message is bound at the **acceptor site,** also called the **aminoacyl site** or **A site** of the ribosome. Selection of the correct aminoacyl-tRNA is enhanced by **elongation factor 1 (EF-1);** a component of EF-1, EF-1α, first forms a ternary complex with aminoacyl-tRNA and GTP. The EF-1α– aminoacyl-tRNA– GTP complex binds to the ribosome and if codon–anticodon interactions are correct the aminoacyl-tRNA is placed at the A site, GTP is hydrolyzed to GDP and P_i, and the **EF-1α–**GDP complex dissociates. The initiating methionyl-tRNA and the incoming aminoacyl-tRNA are now juxtaposed on the ribosome. Their anticodons are paired with successive codons of the mRNA in the decoding region of the small subunit, and their amino acids are beside one another at the peptidyltransferase site of the large subunit. Peptide bond formation now occurs. Peptidyltransferase catalyzes attack of the α-amino group of aminoacyl-tRNA onto the carbonyl carbon of methionyl-tRNA. The result is transfer of methionine to the amino group of the aminoacyl-tRNA, which then occupies a "hybrid" position on the ribosome. The anticodon remains in the 40S A site, while the acceptor end and the attached peptide are in the 60S P site. The anticodon of the deacylated tRNA remains in the 40S P site, and its acceptor end is located in the 60S **exit** or **E site.**

The mRNA and the dipeptidyl-tRNA at the 40S A site must now be repositioned to permit another elongation cycle to begin. This is done by **elongation factor 2 (EF-2),** also called **translocase.** EF-2 moves the mRNA and dipeptidyl-tRNA, in codon–anticodon register, from the 40S A site to the P site. In the process GTP is hydrolyzed to GDP plus P_i, providing energy for the movement, and the A site is fully vacated. As the dipeptidyl-tRNA moves to the P site, the deacylated donor (methionine) tRNA also moves to the E site, which only exists on the 60S subunit. The ribosome can now enter a new cycle. The next aminoacyl-tRNA specified by the mRNA is delivered by **EF-1α** to the A site and the deacylated tRNA in the E site is probably released. Peptide transfer again occurs. Successive cycles of binding of aminoacyl-tRNA, peptide bond formation, and translocation result in the stepwise elongation of the polypeptide toward its eventual carboxyl terminus. Note that whatever the length of the growing chain, peptide bond formation always occurs through attack of the α-amino group of the incoming aminoacyl-tRNA on the peptide carboxyl-tRNA linkage; hence the geometric arrangement of the reacting molecules at the peptidyltransferase site remains constant.

Peptide bond formation does not require any additional energy source such as ATP or GTP. The energy of the methionyl (or peptidyl) ester linkage to tRNA drives the reaction toward peptide bond formation; recall that ATP is used to form each aminoacyl-tRNA and that these reactions are reversible. Isolated 60S subunits can catalyze peptidyltransferase activity, and nonribosomal factors are not involved in the reaction. Peptidyltransferase has never been dissociated from the large subunit or characterized as a specific ribosomal protein. A growing body of evidence indicates that this is because the large subunit is a complex **ribozyme** in which peptide bond formation is an **RNA-catalyzed** reaction. Experiments with isolated, conformationally stable large subunit RNA from a thermophilic bacterium suggest that the rRNA is the catalytic component of peptidyltransferase while the proteins serve to stabilize RNA folding, but this hypothesis was controversial and difficult to prove. However, the crystallographic structure of the large subunit shows the transferase region to be composed entirely of RNA, with the nearest amino acid side chain about 25 Å away. Moreover, mutagenesis of a single adenine residue in the RNA results in the loss of transferase activity. Hence the central reaction in the formation of proteins is almost certainly catalyzed by RNA. It is speculated that the primordial ribosome was a naked RNA molecule.

As determined from prokaryotic models, the role of GTP in the action of EF-1α and EF-2 probably relates to conformational changes in these proteins. Crystallographic studies have shown that a large rearrangement of domains with movements of several angstroms occurs upon GTP hydrolysis in **EF-Tu**, the prokaryotic equivalent of EF-1α. Both EF-1α and EF-2 bind ribosomes tightly as GTP complexes, while GDP complexes dissociate from the ribosome more easily. Viewed another way, GTP stabilizes a protein conformation that confers upon EF-1α high affinity toward aminoacyl-tRNA and the ribosome, while GDP stabilizes a conformation with lower affinity for aminoacyl-tRNA and ribosome, thus allowing tRNA delivery and factor dissociation. Restoration of the higher affinity GTP-associated conformation of EF-1α requires participation of **EF-1$\beta\gamma$** (Figure 6.9). This protein displaces GDP from EF-1α, forming an EF-1α–EF-1$\beta\gamma$ complex. GTP then displaces EF-1$\beta\gamma$, forming an EF-1α–GTP complex that can successively bind an aminoacyl-tRNA and then a ribosome. Prokaryotes use a similar mechanism in which EF-Tu binds GTP and aminoacyl-tRNA and **EF-Ts** displaces GDP and helps recycle the carrier molecule. Prokaryotes also utilize a GTP-dependent translocase, like EF-2 but called **EF-G** or **G factor.**

Termination of Polypeptide Synthesis Requires a Stop Codon

A chain-terminating UAG, UAA, or UGA codon in the A site does not promote binding of any tRNA species. Instead, another complex nonribosomal protein, **release factor (eRF)**, binds the ribosome as an eRF–GTP complex (Figure 6.10). The peptide-tRNA ester linkage is cleaved through the action of peptidyltransferase, acting here as a hydrolase, and the completed polypeptide is released from its carrier tRNA and the ribosome. Dissociation of eRF from the ribosome requires hydrolysis of the bound GTP and frees the ribosome to dissociate into subunits and then reenter the protein synthesis cycle at the initiation stage. Structural analysis of release factors shows a strong physical resemblance to tRNA; human eRF is similar in size and shape to tRNA, and amino acids at the end of protein domain recognize and interact with a termination codon. This phenomenon of "**molecular mimicry**" is also apparent in the structure of bacterial EF-G, in which a protein domain mimics the anticodon stem of tRNA and functions in elongation by displacing the tRNA from the A to the P subsite on the small subunit.

Translation Has Significant Energy Cost

There is considerable use of energy in synthesis of a polypeptide. Amino acid activation converts an ATP to AMP and pyrophosphate, which is normally hydrolyzed to P_i; the net cost is two high-energy phosphates. Two more high-energy bonds are hydrolyzed in the actions of EF-1α and EF-2, for a total of four per peptide bond formed. Posttranslational modifications may add to the energy cost, and of course energy is needed for biosynthesis of the multi-use mRNA, tRNAs, ribosomes, and protein factors, but these costs are distributed more generally between the proteins formed during their lifetime.

Protein Synthesis in Mitochondria Differs Slightly

Many characteristics of mitochondria suggest that they are descendants of aerobic prokaryotes that invaded and set up a symbiotic relationship within a eukaryotic cell. Some of their independence and prokaryotic character are retained. Human mitochondria have a circular DNA genome of 16,569 base pairs that encodes 13 proteins, 22 tRNA species, and 2 mitochondrion-specific rRNA species. Their independent apparatus for protein synthesis includes RNA polymerase, aminoacyl-tRNA synthetases, tRNAs, and ribosomes. Although the course of protein biosynthesis in mitochondria is like that in the cytosol, some details are different. The synthetic components, tRNAs, aminoacyl-tRNA synthetases, and ribosomes, are unique

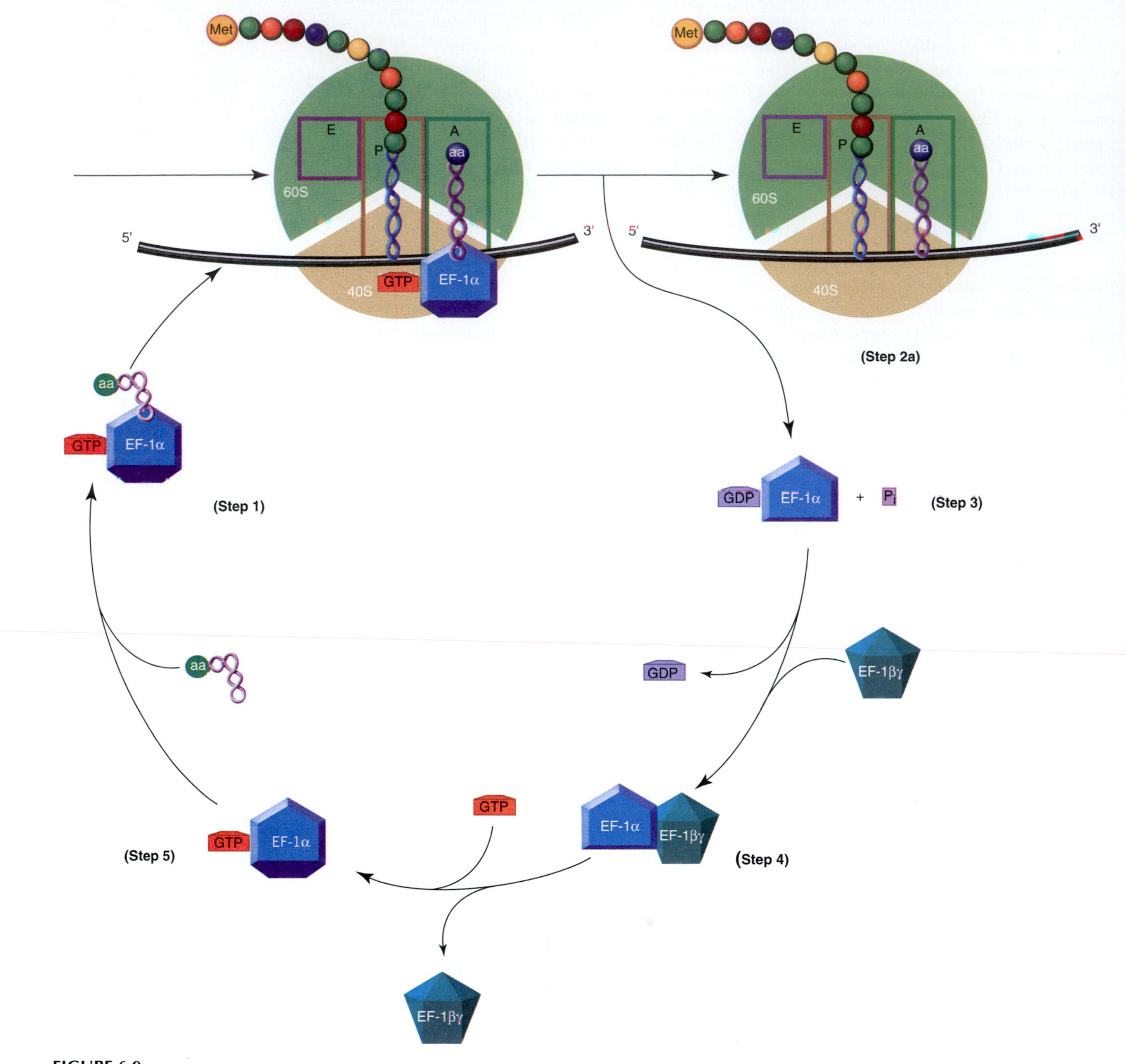

FIGURE 6.9
EF1 in the elongation cycle.
Step 1: EF-1α–GTP–aminoacyl-tRNA complex binds the ribosome. Step 2: Aminoacyl-tRNA is placed on the ribosome (2a) with concomitant hydrolysis of GTP and a change in conformation of EF-1α (3) that reduces its affinity for tRNA and the ribosome. Step 4: GDP is displaced from EF-1α by EF-1$\beta\gamma$. Binding of GTP then displaces EF-1$\beta\gamma$ (Step 5) and allows binding of an aminoacyl-tRNA by EF-1α in its higher affinity conformation (1). In prokaryotes a similar cycle exists; EF-Tu functions as carrier of aminoacyl-tRNA and EF-Ts is a guanine nucleotide exchange factor.

to the mitochondrion. The number of tRNA species is small and the genetic code is slightly different (Table 6.2). Mitochondrial ribosomes are smaller and the rRNAs are shorter than those of either the eukaryotic cytosol or of prokaryotes (Table 6.7). An initiator Met-$tRNA_i^{met}$ is modified by a **transformylase** that uses N^{10}-formyl H_4-folate to produce **fMet-$tRNA_i^{met}$**. Most mitochondrial proteins are encoded in nuclear DNA and synthesized in the cytosol, but mitochondrial protein synthesis is

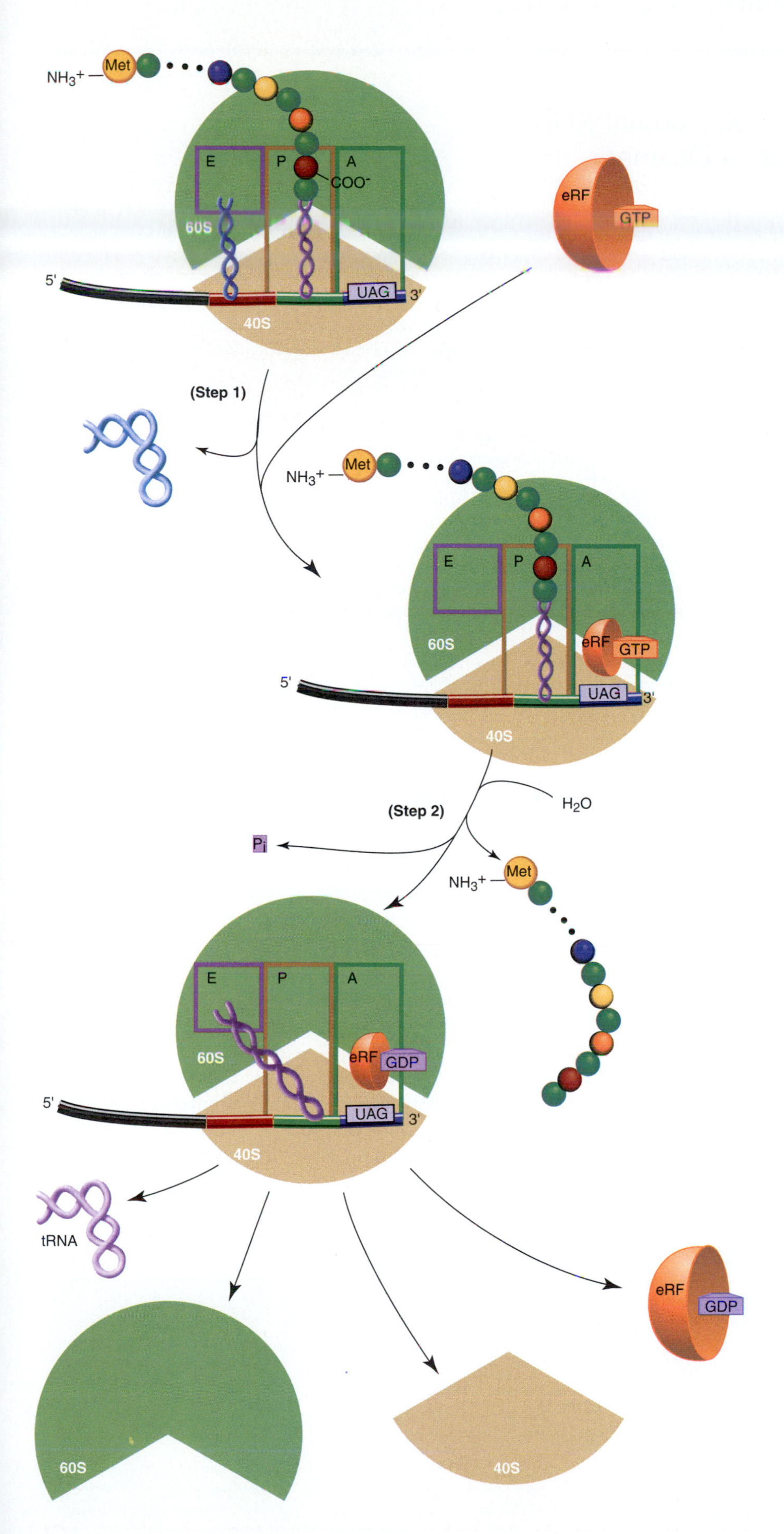

FIGURE 6.10
Termination of protein biosynthesis. When a termination codon (UAG, UAA, or UGA) in mRNA occupies the A site, binding of a release factor–GTP complex occurs (Step 1) and deacylated tRNA is released. Step 2: Peptidyltransferase functions as a hydrolase; protein is released by hydrolysis of the ester bond linking it to tRNA, and the acceptor end of deacylated tRNA is probably displaced. GTP is hydrolyzed to GDP and P_i, presumably altering the conformation of the release factor. Dissociated components of the complex can enter additional rounds of protein synthesis.

CLINICAL CORRELATION 6.5

Mutation in Mitochondrial Ribosomal RNA Results in Antibiotic-Induced Deafness

In some regions of China a significant percentage of irreversible cases of deafness have been linked to use of normally safe and effective amounts of aminoglycoside antibiotics such as streptomycin and gentamicin. The unusual sensitivity to aminoglycosides is transmitted only through women. This maternal transmission suggests a mitochondrial locus, since sperm do not contribute mitochondria to the zygote. Aminoglycosides are normally targeted to bacterial ribosomes, so the mitochondrial ribosome is a logical place to look for a mutation site.

A single A →G point mutation at nucleotide 1555 of the gene on mitochondrial DNA for large subunit rRNA has been identified in families with this susceptibility to aminoglycosides. The mutation site is in a highly conserved region of the rRNA sequence that is involved in aminoglycoside binding; some mutations in the same region confer resistance to the antibiotics, and the RNA region is part of the ribosomal A site. It is hypothesized that the mutation makes the region more "prokaryote-like," increasing its affinity for aminoglycosides and the ability of the antibiotic to interfere in protein synthesis in mitochondria. Proteins synthesized in mitochondria are needed to form the enzyme complexes of the oxidative phosphorylation system, so affected cells are starved of ATP. Aminoglycosides accumulate in the cochlea, making this a particularly sensitive target and leading to sensorineural deafness. A simple missense mutation in the gene *(DPP)* that encodes a protein involved in intermembrane transport in mitochondria also results in deafness.

Fischel-Ghodsian, N., Prezant, T., Bu, X., and Öztas, S. Mitochondrial ribosomal RNA gene mutation in a patient with sporadic aminoglycoside ototoxicity. *Am. J. Otolaryngol.* 14:399, 1993; Prezant, T., et al. Mitochondrial ribosomal RNA mutation associated with both antibiotic-induced and non-syndromic deafness. *Nature Genet.* 4:289, 1993; and Trannebjaerg, L., Hamel, B. C., Gabreels, F. J., Renier, W. O., and Van Ghelue, M. A *de novo* missense mutation in a critical domain of the X-linked DPP gene causes the typical deafness–dystonia–optic atrophy syndrome. *Eur. J. Hum. Genet.* 8:464, 2000.

clearly important (see Clin. Corr. 6.5). Cells must coordinate protein synthesis within mitochondria with the cytosolic synthesis of proteins destined for import into mitochondria.

Some Antibiotics and Toxins Inhibit Protein Biosynthesis

Protein biosynthesis is central to the continuing life and reproduction of cells. An organism can gain a biological advantage by interfering with the ability of its competitors to synthesize proteins, and many antibiotics and toxins function in this way. Some are selective for prokaryotic rather than eukaryotic protein synthesis and so are extremely useful in clinical practice. Examples of antibiotic action are listed in Table 6.8.

Several mechanisms of interfering in ribosome subunit–tRNA interactions are utilized by different antibiotics. **Streptomycin** binds the small subunit of prokaryotic ribosomes, interferes with the initiation of protein synthesis, and causes misreading of mRNA. Mutations in a ribosomal protein or the rRNA of the small subunit can confer resistance to or even dependence upon streptomycin. Streptomycin alters the interactions of tRNA with the ribosomal subunit and mRNA, probably by affecting subunit flexibility and conformation. Other aminoglycoside antibiotics, such as the **neomycins** and **gentamicins**, also cause mistranslation; they interact

TABLE 6.8 Some Inhibitors of Protein Biosynthesis

Inhibitor	*Processes Affected*	*Site of Action*
Streptomycin	Initiation, elongation	Prokaryotes: 30S subunit
Neomycins	Translation	Prokaryotes: multiple sites
Tetracyclines	Aminoacyl-tRNA binding	30S or 40S subunits
Puromycin	Peptide transfer	70S or 80S ribosomes
Erythromycin	Translocation	Prokaryotes: 50S subunit
Fusidic acid	Translocation	Prokaryotes: EF-G
Cycloheximide	Elongation	Eukaryotes: 80S ribosomes
Ricin	Multiple	Eukaryotes: 60S subunit

with the small ribosomal subunit, but at sites that differ from that for streptomycin. **Kasugamycin** inhibits the initiation of translation; sensitivity to kasugamycin depends on base methylation that normally occurs on two adjacent adenines of small subunit rRNA. **Tetracyclines** bind to ribosomes and interfere in aminoacyl-tRNA binding.

Other antibiotics interfere with elongation. **Puromycin** (Figure 6.11) resembles an aminoacyl-tRNA; it binds at the large subunit A site and acts as an acceptor in the peptidyltransferase reaction. However, since its aminoacyl derivative is not in an ester linkage to the nucleoside, it cannot serve as a peptide donor. Instead puromycin prematurely terminates translation, leading to release of peptidyl-puromycin. **Chloramphenicol** inhibits peptidyltransferase by binding at the transferase center; no transfer occurs, and peptidyl-tRNA remains associated with the ribosome.

The translocation step is also a potential target. **Erythromycin**, a macrolide antibiotic, interferes with translocation on prokaryotic ribosomes. Eukaryotic translocation is inhibited by **diphtheria toxin**, a protein produced by *Corynebacterium diphtheriae;* the toxin binds at the cell membrane and a subunit enters the cytoplasm and catalyzes the ADP-ribosylation and inactivation of EF-2, as represented in the reaction

$$\underset{\text{(active)}}{\text{EF-2}} + \text{NAD}^+ \rightleftharpoons \underset{\text{(inactive)}}{\text{ADP-ribosyl EF-2}} + \text{nicotinamide} + \text{H}^+$$

ADP-ribose is attached to EF-2 at a **diphthamide** residue that is a posttranslationally modified form of histidine. Posttranslational events are discussed in the next section.

A third group of toxins attack the rRNA. **Ricin** (from castor beans) and related toxins are *N*-glycosidases that cleave a single adenine from the large subunit rRNA backbone. The ribosome is inactivated by this apparently minor damage. A fungal toxin, **α-sarcin,** cleaves large subunit rRNA at a single site and similarly inactivates the ribosome. Some *E. coli* strains make extracellular toxins that affect other bacteria. One of these, **colicin E3,** is a ribonuclease that cleaves 16S RNA near the mRNA-binding sequence and decoding region; it thus inactivates the small subunit and halts protein synthesis in competitors of the colicin-producing cell.

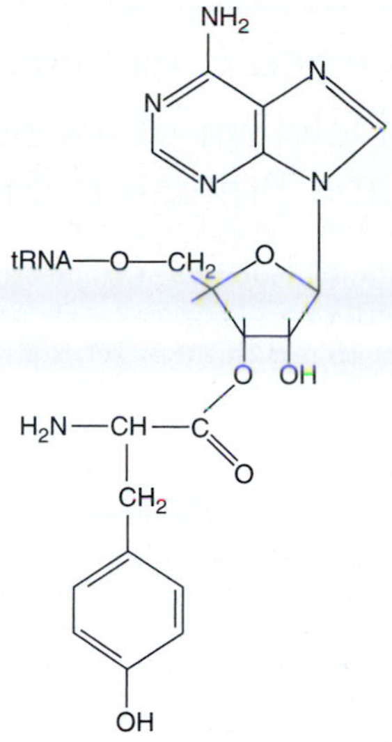

3' end of tyrosyl-tRNA

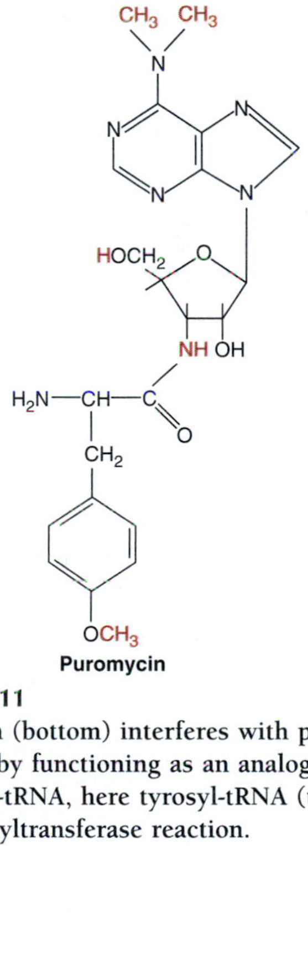

Puromycin

FIGURE 6.11
Puromycin (bottom) interferes with protein synthesis by functioning as an analog of aminoacyl-tRNA, here tyrosyl-tRNA (top) in the peptidyltransferase reaction.

6.4 | PROTEIN MATURATION: FOLDING, MODIFICATION, SECRETION, AND TARGETING

Some proteins emerge from the ribosome ready to function, while others undergo a variety of **posttranslational modifications.** These alterations may result in conversion to a functional form, direction to a specific subcellular compartment, secretion from the cell, or an alteration in activity or stability. Information that determines the posttranslational fate of a protein resides in its structure. That is, the amino acid sequence and conformation of the polypeptide determine whether a protein will be a substrate for a modifying enzyme and/or identify it for direction to a subcellular or extracellular location.

Chaperones Aid in Protein Folding

Many proteins can independently and spontaneously generate their correct three-dimensional conformation. For example, fully denatured pancreatic ribonuclease can, under appropriate conditions, refold and generate correct disulfide bridges and full activity. Bacterial ribosomal subunits can reassemble from denatured RNA and protein components. In cells folding of some proteins may be unaided, but many proteins attain their correct conformation only with assistance of one or more of a diverse group of protein **chaperones.** A subgroup of chaperones, known as **chaperonins,** are multicomponent tubular structures that require ATP. Chaperones

CLINICAL CORRELATION 6.6

Deletion of a Codon, Incorrect Posttranslational Modification, and Premature Protein Degradation: Cystic Fibrosis

Cystic fibrosis is the most common autosomal recessive disease in Caucasians, with a frequency of about 1 per 2000. The CF gene is 230 kb in length and includes 27 exons that encode a protein of 1480 amino acids. The protein known as the cystic fibrosis transmembrane conductance regulator or CFTR is a member of a family of ATP-dependent transport proteins and it includes two membrane-spanning domains, two nucleotide-binding domains that interact with ATP, and one regulatory domain that includes several phosphorylation sites. CFTR functions as a cyclic AMP-regulated chloride channel. CF epithelia are characterized by defective electrolyte transport. The organs most strongly affected include the lungs, pancreas, and liver, and the most life-threatening effects involve thick mucous secretions that lead to chronic obstructive lung disease and persistent infections of lungs.

In about 70% of affected individuals the problem is traced to a deletion of three nucleotides that results in deletion of a single amino acid, phenylalanine 508, normally located in ATP-binding domain 1 on the cytoplasmic side of the plasma membrane. As with several other CF mutations, the Phe 508-deletion protein does not fold correctly in the ER and is not properly glycosylated or transported to the cell surface. Instead, it is returned to the cytoplasm to be degraded within proteasomes.

Ward, C., Omura, S., and Kopito, R. Degradation of CFTR by the ubiquitin–proteasome pathway. *Cell* 83:121, 1995; and Plemper, R. K. and Wolf, D. H. Retrograde protein translocation: eradication of secretory proteins in health and disease. *Trends Biochem. Sci.* 24:266, 1999.

reversibly bind hydrophobic regions of unfolded proteins and folding intermediates. They can stabilize intermediates, maintain proteins in an unfolded state to allow passage through membranes, help unfold misfolded segments, prevent formation of incorrect intermediates, and prevent inappropriate interactions with other proteins. All of these activities can assist a bound protein in attaining its functional, usually most compact conformation. Failure to fold correctly usually leads to rapid protein degradation (see Clin. Corr. 6.6), and accumulation of misfolded proteins can result in protein aggregation and serious diseases (see Clin. Corr. 6.7).

Proteins for Export Follow the Secretory Pathway

Proteins destined for export are synthesized on membrane-bound ribosomes of the **rough endoplasmic reticulum (ER)** (Figure 6.12). Initiation and elongation begin on free cytosolic ribosomes. Proteins of the secretory pathway have a hydrophobic **signal peptide,** usually at or near the amino terminus. There is no unique signal peptide sequence, but its characteristics include a positively charged N-terminal region and a core of 8–12 hydrophobic amino acids followed by a more polar C-terminal segment that eventually serves as a cleavage site for excision of the signal peptide.

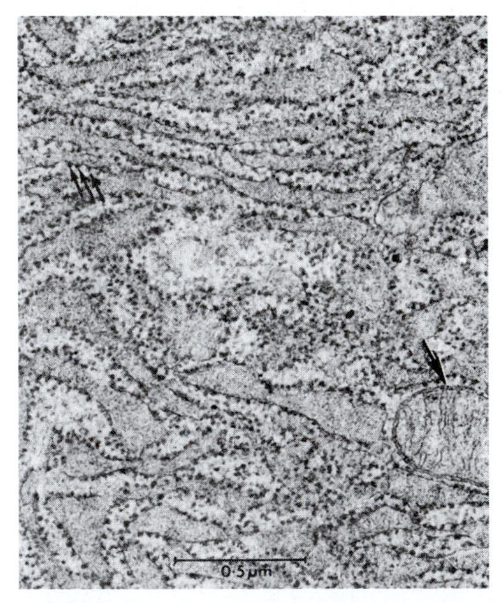

FIGURE 6.12
Rough endoplasmic reticulum of a plasma cell.
Three parallel arrows indicate three ribosomes among the many attached to the extensive membranes. Single arrow indicates a mitochondrion for comparison.
Courtesy of Dr. U. Jarlfors, University of Miami.

As the signal peptide of 15–30 amino acids emerges from the ribosome it binds a **signal recognition particle (SRP)** (see Figure 6.13). The SRP is an elongated particle made up of six different proteins and a small (7S) RNA molecule. Binding of signal peptide in a hydrophobic pocket of the SRP with the positively charged N-terminal segment in contact with SRP RNA orients the ribosome and halts protein synthesis. The complex moves to the ER where the SRP recognizes and binds to an SRP receptor or **docking protein,** localized at the cytosolic surface of the ER membrane. The ribosome is transferred to a **translocon,** a **ribosome receptor** that crosses the membrane and serves as a passageway through the membrane. Both SRP and docking protein are released and can act to direct other ribosomes to the ER, the translational block caused by SRP binding is relieved, and the translocon passageway is unplugged and broadened to allow passage of the nascent protein. The hydrophobic signal sequence, probably complexed by a receptor protein, is inserted into the membrane, further anchoring the ribosome to the ER. Translation and extrusion into or through the membrane are now coupled; even very hydrophilic or ionic segments are directed through the membrane into the ER lumen. A cleavage site on the protein is hydrolyzed by **signal peptidase,** an integral membrane protein located at the luminal surface of the ER. The protein can now fold into a three-

CLINICAL CORRELATION 6.7

Protein Misfolding and Aggregation: Creutzfeldt–Jacob Disease, Mad Cow Disease, Alzheimer's Disease, and Huntington's Disease

Misfolded proteins are usually removed by proteolytic degradation, as occurs in the case of mutant CFTR protein in cystic fibrosis (see Clin. Corr. 6.6). Several neurological diseases illustrate the result of cellular accumulation of aggregates of misfolded proteins or their partial-degradation products.

Prions (*pro*teinaceous *in*fectious agents) are devoid of nucleic acid but show some of the characteristics of viral or microbial pathogens. The first prion-caused disease identified was scrapie, a disorder of sheep. Human disorders include kuru, Creutzfeldt–Jacob disease (CJD), and a form of CJD arising from bovine spongiform encephalopathy (BSE), popularly known as the "mad cow disease" in cattle. These diseases vary in the course and rate of development, can be inherited, spontaneous, or result from infection, but all cause neurodegeneration and eventual death. Autopsy shows spongiform degeneration of brain and the intracellular accumulation of rod-shaped deposits of amyloid plaque.

Disease results from a conformational transition in a normal, cellular 254-residue-long prion protein, designated PrP^{C}. The protein is found mostly on the outer surface of neurons, but its function is still unclear. Structural studies show it to be rich in α-helix and devoid of β-sheet; it is soluble in mild detergent solutions and has little tendency to form large aggregates. The infectious form of prion protein, designated PrP^{Sc}, is rich in β-sheet and is much less soluble. A protease-resistant fragment is particularly likely to aggregate and form plaque, and a number of point mutations PrP^{C} are associated with formation of infectious PrP^{Sc}. Infection occurs by consumption of food, especially neural tissue, that contains PrP^{Sc}. Ritual consumption of human brain by the Fore in New Guinea spread kuru, and inclusion of infected offal in animal feeds helped spread scrapie and BSE. Humans have acquired the CJD variation by eating BSE-infected beef. PrP^{Sc} acts by altering the conformation of normal cellular PrP^{C}; although the mechanism is somewhat controversial, PrP^{Sc} appears to function with or as a chaperone, fostering adoption of the PrP^{Sc} conformation in existing or newly synthesized PrP^{C}. The protease-resistant form and products of its partial proteolysis accumulate and form the aggregates characteristic of the disease (but not necessarily the cause of its symptoms).

Alzheimer's disease is characterized by progressive memory loss and cognition, and it is most common in elderly people. Alzheimer's disease is not caused by a prion or other external infectious agent, but it too is characterized by accumulation of intraneuronal and extracellular bundles and filaments that form plaques. The major component of the plaques is β-amyloid, a 39–43 amino acid peptide derived from a larger amyloid precursor protein. Other components of the plaque are highly ubiquitinated but protease resistant in compacted and tangled fiber form. Huntington's disease is a neurodegenerative disorder that is also associated with accumulation of protein aggregates. In this case, a variant of a cellular protein called huntingtin is involved; a CAG codon within the protein is expanded in affected individuals, resulting in polyglutamine repeats of up to 180 residues within huntingtin. Protein or peptides derived from the protein accumulate in inclusions in the nucleus. Several other neurodegenerative disorders are also linked to polyglutamine tract expansion and protein deposition.

Baldwin, M. A., Cohen, F. A., and Prusiner, S. P. Prion protein isoforms, a convergence of biological and structural investigations. *J. Biol. Chem.* 270:19197, 1995; Kaytor, M. D. and Warren, S. T. Aberrant protein deposition and neurological disease. *J. Biol. Chem.* 274:37507, 1999; Selkoe, D. J. Amyloid β-protein and the genetics of Alzheimer's disease. *J. Biol. Chem.* 271:18295, 1996; and Weissmann, C. Molecular genetics of transmissible spongiform encephalopathies. *J. Biol. Chem.* 274:3, 1999.

dimensional conformation, disulfide bonds can form, and components of multisubunit proteins may assemble. Other steps may include proteolytic processing and glycosylation that occur within the ER lumen and during transit of the protein through the Golgi apparatus and into secretory vesicles.

Glycosylation of Proteins Occurs in the Endoplasmic Reticulum and Golgi Apparatus

Glycosylation of proteins to form glycoproteins (see p. 129) is important for two reasons. Glycosylation alters the properties of proteins, changing their stability, solubility, and physical bulk. In addition, carbohydrates of glycoproteins act as recognition signals that are central to aspects of protein targeting and for cellular recognition of proteins and other cells. Glycosylation can involve addition of a few carbohydrate residues or the formation of large branched oligosaccharide chains. Sites and types of glycosylation are determined by amino acids and sequences in the protein and by availability within the cell of enzymes and substrates to carry out the glycosylation reactions.

Glycosylation involves many **glycosyltransferases**, classes of which are summarized in Table 6.9. Up to 100 different enzymes each carry out a similar basic reaction in which a sugar is transferred from an activated donor substrate to an

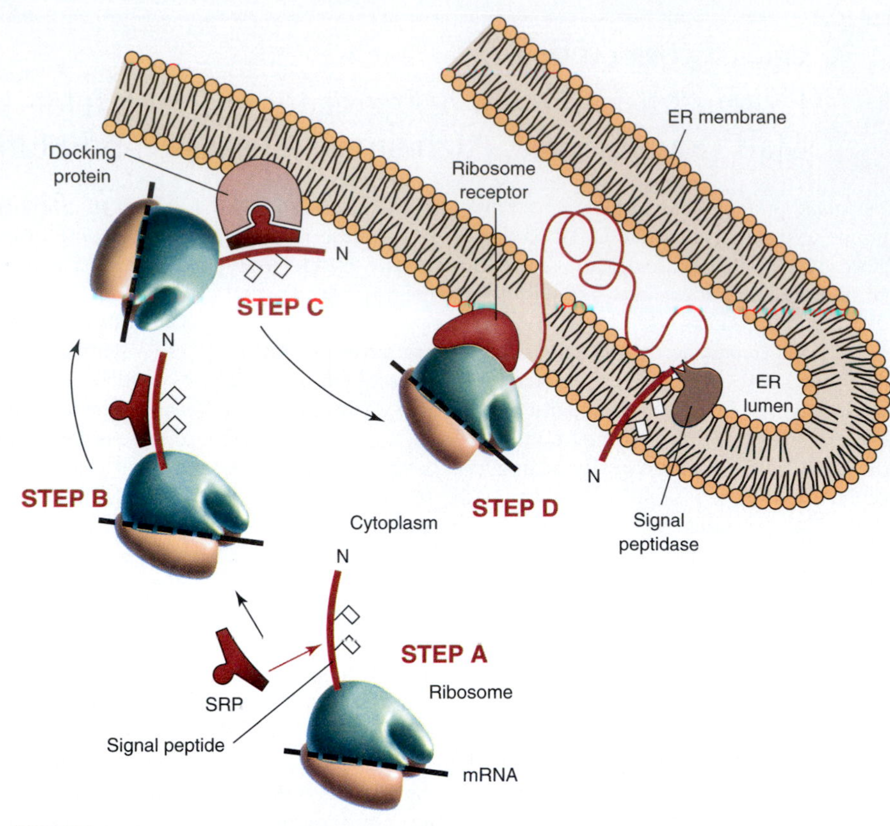

FIGURE 6.13
Secretory pathway: signal peptide recognition.
Step A: A hydrophobic signal peptide emerges from the exit site of a free ribosome in the cytosol. Step B: Signal recognition particle (SRP) binds the peptide and elongation is temporarily halted. Step C: The ribosome moves to ER membrane where docking protein binds SRP. Step D: Ribosome is transferred to a translocon, elongation is resumed, and newly synthesized protein is extruded through the membrane into the ER lumen.

appropriate acceptor, usually another sugar residue that is part of an oligosaccharide under construction. The enzymes show three kinds of specificity: for the **monosaccharide** that is transferred, for structure and sequence of the **acceptor** molecule, and for the site and configuration of the **anomeric linkage** formed.

One class of glycoproteins has sugars linked through the amide nitrogen of asparagine residues in the process of **N-linked glycosylation.** (The antibiotic **tunicamycin** prevents *N*-glycosylation). Formation of *N*-linked oligosaccharides begins in the ER lumen and continues after transport of the protein to the Golgi ap-

TABLE 6.9 Glycosyltransferases in Eukaryotic Cells

Sugar Transferred	*Abbreviation*	*Donors*	*Glycosyltransferase*
Mannose	Man	GDP-Man Dolichol-Man	Mannosyltransferase
Galactose	Gal	UDP-Gal	Galactosyltransferase
Glucose	Glc	UDP-Glc Dolichol-Glc	Glucosyltransferase
Fucose	Fuc	GDP-Fuc	Fucosyltransferase
N-Acetylgalactosamine	GalNAc	UDP-GalNac	*N*-acetylgalactosaminyltransferase
N-Acetylglucosamine	GlcNAc	UDP-GlcNAc	*N*-acetylglucosaminyltransferase
N-Acetylneuraminic acid (or sialic acid)	NANA or NeuNAc SA	CMP-NANA CMP-SA	*N*-Acetylneuraminyltransferase (sialyltransferase)

paratus. A specific sequence, Asn-X-Thr (or Ser) in which X may be any amino acid except proline or aspartic acid, is required. Not all Asn-X-Thr/Ser sequences are glycosylated because some may be unavailable because of the protein's conformation.

Biosynthesis of *N*-linked oligosaccharides uses a lipid-linked intermediate (Figure 6.14). **Dolichol phosphate** at the cytoplasmic surface of the ER membrane serves as glycosyl acceptor of *N*-acetylglucosamine. Stepwise glycosylation and

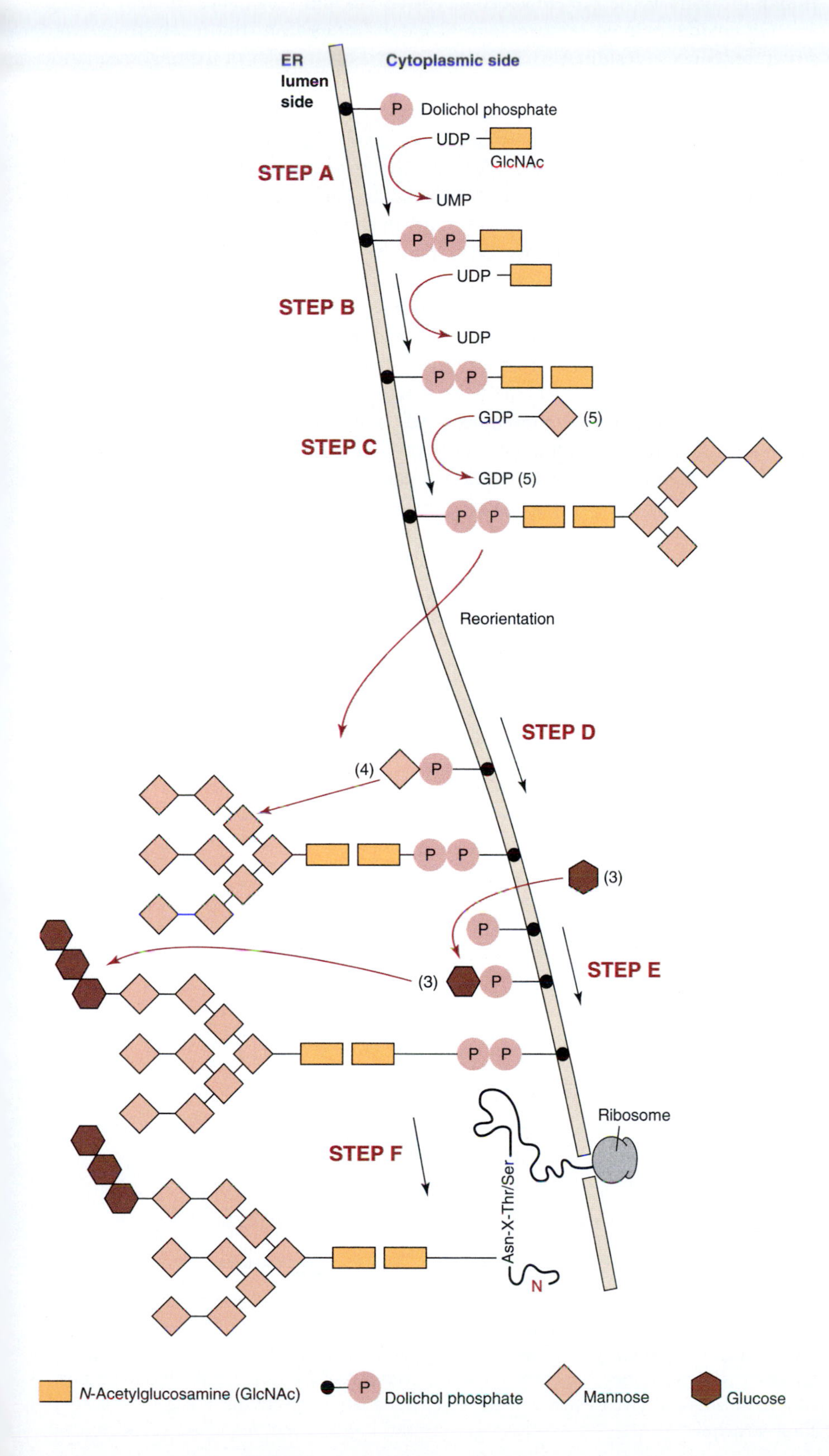

FIGURE 6.14
Biosynthesis of *N*-linked oligosaccharides at the surface of the endoplasmic reticulum. Step A: Synthesis begins on the cytoplasmic face of the ER membrane with transfer of *N*-acetylglucosamine phosphate to a dolichol acceptor. Step B: Formation of the first glycosidic bond occurs upon transfer of a second residue of *N*-acetylglucosamine. Step C: Five residues of mannose (from a GDP mannose carrier) are added sequentially. The lipid-linked oligosaccharide is then reoriented to the luminal face of the membrane. Step D: Additional mannose and (Step E) glucose residues are transferred from dolichol-linked intermediates. Dolichol sugars are generated from cytosolic nucleoside diphosphate sugars. Step F: The completed oligosaccharide is transferred to a protein in the process of being synthesized at the membrane surface; signal peptide may have already been cleaved at this point.

formation of a branched $(Man)_5(GlcNAc)_2$-pyrophosphoryl-dolichol on the cytosolic side of the membrane follows. This intermediate is then **reoriented** to the luminal surface of the ER membrane, and four additional mannose and then three glucose residues are sequentially added to complete the structure. The complete oligosaccharide is then transferred from its dolichol carrier to an asparagine residue of the polypeptide as it emerges into the ER lumen. Thus *N*-glycosylation is **cotranslational;** it occurs as the protein is being synthesized and it can affect protein folding.

Glycosidases now remove some sugar residues from the newly transferred structure. Glucose residues, which were required for transfer of the oligosaccharide from the dolichol carrier, are sequentially removed, as is one mannose. These alterations mark the fully folded glycoprotein for transport to the Golgi apparatus where further trimming by glycosidases may occur. Additional sugars may also be added by a variety of glycosyltransferases. The resulting *N*-linked oligosaccharides are diverse, but two classes are distinguishable. Each has a common core region ($GlcNAc_2Man_3$) linked to asparagine and originating from the dolichol-linked intermediate. The **high-mannose type** includes mannose residues in a variety of linkages and shows less processing from the dolichol-linked intermediate. The **complex type** is more highly processed and diverse, with a larger variety of sugars and linkages. Examples of mature oligosaccharides are shown in Figure 6.15.

The second major class of glycoproteins have sugars bound through serine or threonine hydroxyl groups (see p. 131). There is no defined amino acid sequence in which the serine or threonine must occur, but only residues whose side chains are in an appropriate environment on the protein surface serve as acceptors for the GalNAc-transferase that attaches *N*-acetyl galactosamine to the protein. ***O*-linked**

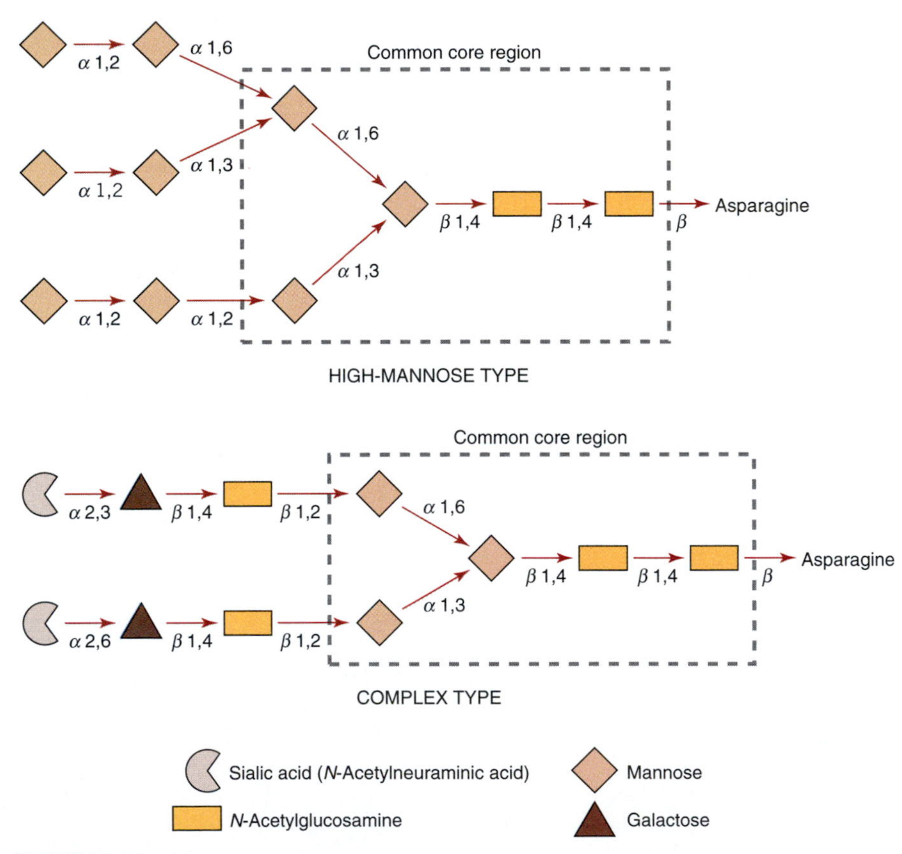

FIGURE 6.15
Structure of *N*-linked oligosaccharides.
Basic structures of both types of *N*-linked oligosaccharides are shown. In each case the structure is derived from that of the initial dolichol-linked oligosaccharide through action of glycosidases and glycosyltransferases. Note the variety of glycosidic linkages involved in these structures.

glycosylation occurs only after the protein has reached the Golgi apparatus, hence *O*-glycosylation is posttranslational and occurs only on fully folded proteins. Sequential addition of sugars to the GalNAc acceptor follows. The structures synthesized depend on the types and amounts of glycosyltransferases in a given cell. If an acceptor is a substrate for more than one transferase, the amount of each transferase controls the competition between them. Some oligosaccharides may be formed that are not acceptors for any glycosyltransferase present; hence no further growth of the chain occurs. Other structures may be excellent acceptors that continue to grow until completed by one of a number of nonacceptor termination sequences. These processes can lead to many different oligosaccharide structures on otherwise identical proteins, so **heterogeneity** in glycoproteins is common. Examples are shown in Figure 6.16.

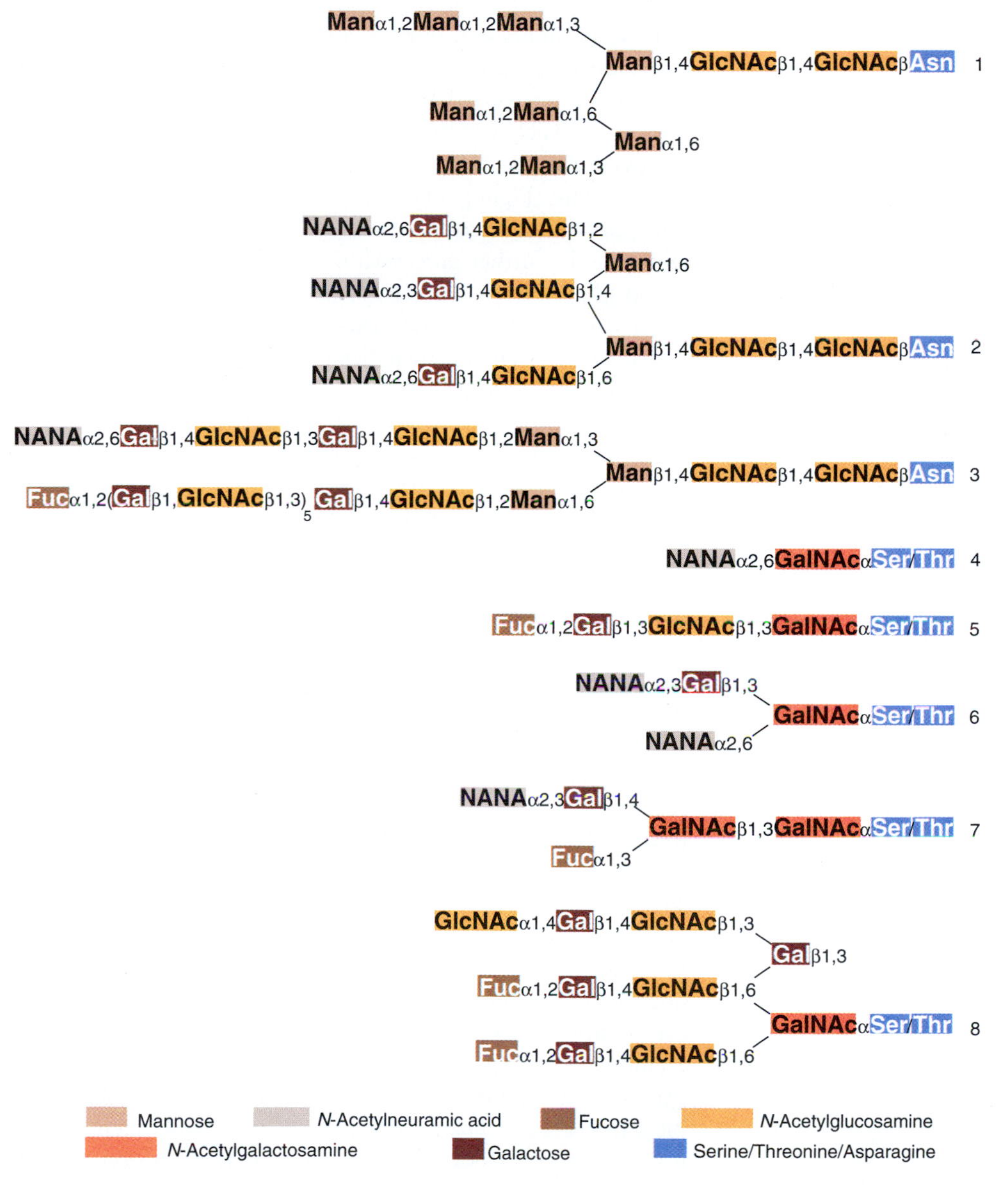

FIGURE 6.16
Examples of oligosaccharide structure.
Structures 1–3 are typical *N*-linked oligosaccharides of high-mannose (1) and complex types (2, 3); note the common core structure from the protein asparagine residue through the first branch point. Structures 4–8 are common *O*-linked oligosaccharides that may be quite simple or highly complex. Note that although the core structure (GalNAc-Ser/Thr) is unlike that of *N*-linked oligosaccharides, the termini can be quite similar (e.g., structures 2 and 6, 3, and 7). Abbreviations: Man, mannose; Gal, galactose; Fuc, fucose; GlcNAc, *N*-acetylglucosamine; GalNAc, *N*-acetylgalactosamine; NANA, *N*-acetylneuraminic acid (sialic acid).
Adapted from Paulson, J. Trends Biochem. Sci. *14:272, 1989.*

6.5 | MEMBRANE AND ORGANELLE TARGETING

Protein transport from the ER to and through the Golgi apparatus and beyond uses carrier vesicles. Only correctly folded proteins are recognized as **cargo** for transport, and chaperones in the ER assist folding and foster correct disulfide formation. Misfolded or damaged proteins are exported to the cytosol for degradation. Sorting of proteins for their ultimate destinations occurs in conjunction with their glycosylation and proteolytic trimming as they pass through the *cis*, medial, and *trans* elements of the Golgi apparatus. Families of vesicle and receptor **SNARE proteins** provide specificity for membrane targeting and fusion.

Sorting of Proteins Targeted for Lysosomes Occurs in the Secretory Pathway

The best understood sorting process is targeting of specific glycoproteins to **lysosomes.** In the *cis* Golgi some aspect of tertiary structure causes lysosomal proteins to be recognized by a glycosyltransferase that attaches *N*-acetylglucosamine phosphate (GlcNAc-P) to high-mannose type oligosaccharides. A glycosidase then removes the GlcNAc, forming an oligosaccharide that contains **mannose 6-phosphate** (Figure 6.17) that is responsible for compartmentation and vesicular transport of these proteins to lysosomes. Other oligosaccharide chains on the proteins may be further processed to form complex type structures, but the mannose 6-phosphate determines the lysosomal destination of these proteins. Patients with **I-cell disease** lack the GlcNAc-P glycosyltransferase and cannot correctly mark lysosomal enzymes for their destination. Thus the enzymes are secreted from the cell (see Clin. Corr. 6.8).

Other sorting signals are reasonably well understood. Soluble proteins are retained in the ER lumen in response to a C-terminal **KDEL** (Lys-Asp-Glu-Leu) sequence, and a different sequence in an exposed C-terminus signals retention in the ER membrane. Transmembrane domains have been identified that result in retention in the Golgi. Polypeptide-specific glycosylation and sulfation of some glycoprotein hormones in the anterior pituitary mediates their sorting into storage granules. Polysialic acid modification of a neural cell adhesion protein appears to be both specific to the protein and regulated developmentally. Many other sorting signals must still be deciphered to explain fully how the Golgi apparatus directs proteins to its own subcompartments, various storage and secretory granules, and specific elements of the plasma membrane.

The secretory pathway directs proteins to lysosomes, the plasma membrane, or secretion outside the cell. Proteins of the ER and Golgi apparatus are targeted through partial use of the pathway. For example, localization of proteins on either side of or spanning the ER membrane can utilize the signal recognition particle in slightly different ways (Figure 6.18). If the signal sequence is downstream from the amino terminus of the protein the amino end may not be inserted into the membrane and remains on the cytoplasmic surface. Internal hydrophobic anchoring sequences within a protein can allow much of the sequence to either remain on the cytoplasmic surface or to be retained, anchored on the luminal surface of the ER membrane. Multiple anchoring sequences in a single polypeptide can cause it to span the membrane several times and thus be largely buried in it. Such hydrophobic sequences are separated by polar loops whose orientation is determined by positively charged flanking residues that predominate on the cytoplasmic side of the membrane.

Import of Proteins by Mitochondria Requires Specific Signals

Mitochondria provide a particularly complex targeting problem since specific proteins are located in the mitochondrial matrix, inner or outer membrane, or inter-

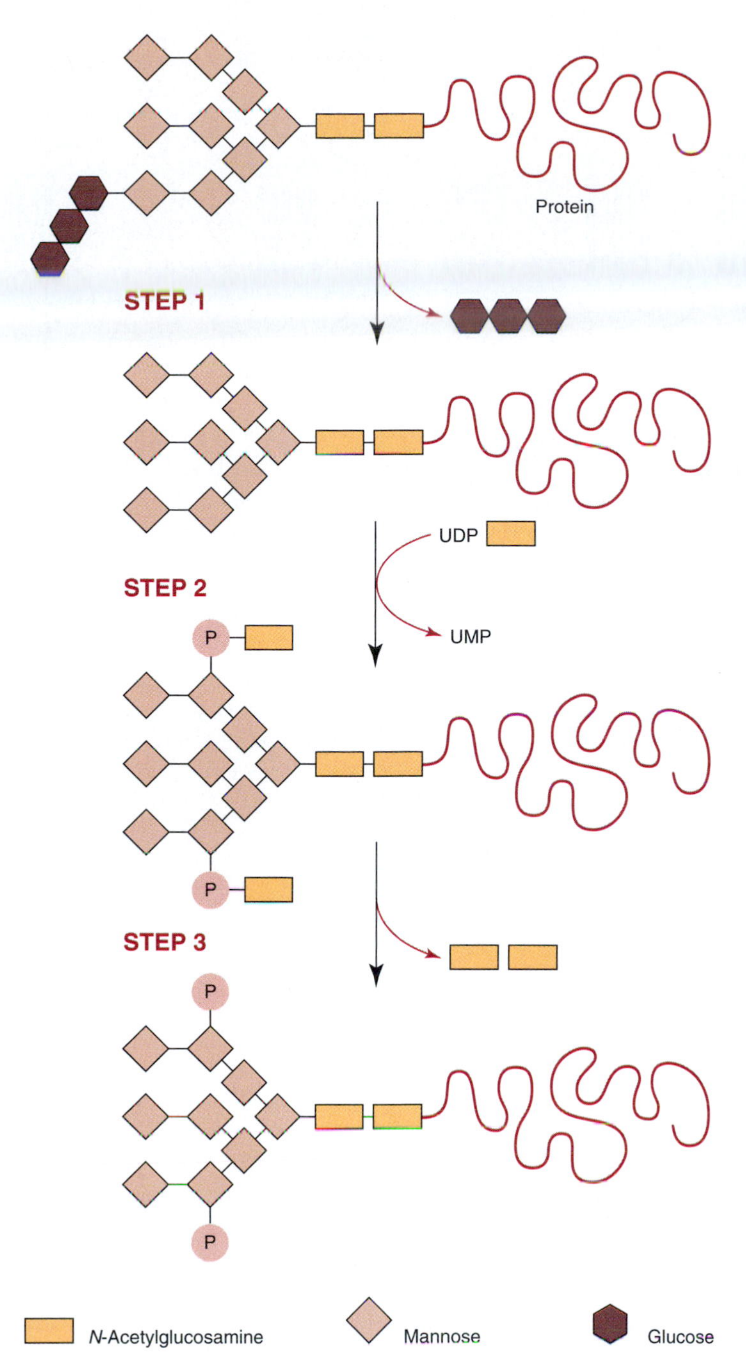

FIGURE 6.17
Targeting of enzymes to lysosomes.
Completed and folded *N*-linked glycoprotein is released from the ER membrane and prior to transport to the Golgi apparatus glycosidases that remove glucose residues (Step 1). A mannose residue may also be removed. Step 2: In the Golgi apparatus a glycosyltransferase links one or sometimes two *N*-acetylglucosamine phosphate residues to the oligosaccharide. Step 3: A glycosidase removes *N*-acetyglucosamine, leaving one or two mannose 6-phosphate residues on the oligosaccharide. The protein is then recognized by a mannose 6-phosphate receptor and directed to vesicles that are targeted to lysosomes.
Adapted from Kornfeld, R. and Kornfeld, S. Annu. Rev. Biochem. *54:631, 1985.*

CLINICAL CORRELATION 6.8
I-Cell Disease

I-cell disease (mucolipidosis II) and pseudo-Hurler polydystrophy (mucolipidosis III) are related diseases that arise from defects in lysosomal enzyme targeting because of a deficiency in the enzyme that transfers N-acetylglucosamine phosphate to the high-mannose type oligosaccharides of proteins destined for the lysosome. Fibroblasts *in vitro* from affected individuals show dense inclusion bodies (hence I-cells) and are defective in multiple lysosomal enzymes that are found secreted into the medium. Patients have abnormally high levels of lysosomal enzymes in their sera and other body fluids. The disease is characterized by severe psychomotor retardation, many skeletal abnormalities, coarse facial features, and restricted joint movement. Symptoms are usually observable at birth and progress until death, usually by age 8. Pseudo-Hurler polydystrophy is a much milder form of the disease. Onset is usually delayed until the age of 2–4 years, the disease progresses more slowly, and patients survive into adulthood. Prenatal diagnosis of both diseases is possible, but there is as yet no definitive treatment.

Kornfeld, S. Trafficking of lysosomal enzymes in normal and disease states. *J. Clin. Invest.* 77: 1, 1986. Shields, D. and Arvan, P. Disease models provide insights into post-Golgi protein trafficking, localization and processing. *Curr. Opin. Cell Biol.* 11:489, 1999. For a comprehensive review see Kornfeld, S. and Sly, W. S. I-cell disease and pseudo-Hurler polydystrophy: disorders of lysosomal enzyme phosphorylation and localization. In: C. R. Scriver, A. L. Beaudet, W. S. Sly, and D. Valle (Eds.), *The Metabolic and Molecular Bases of Inherited Disease,* 8th ed. New York: McGraw-Hill, 2001, p. 3469.

membrane space. Most of these proteins are synthesized in the cytosol on free ribosomes, transported (unfolded) with the aid of chaperones to the mitochondrion, and imported. Most are synthesized as larger preproteins; N-terminal **presequences** mark the protein for the mitochondrion and for a specific subcompartment. The **mitochondrial matrix targeting signal** is not a specific sequence, but rather a

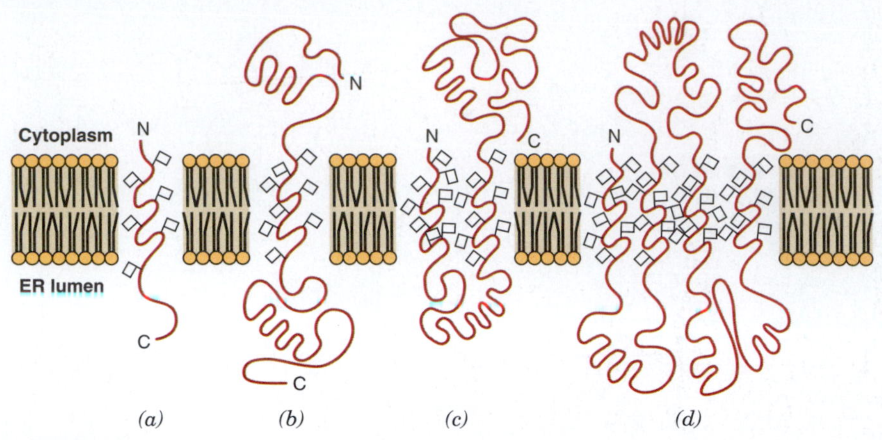

FIGURE 6.18
Topology of proteins at membranes of endoplasmic reticulum.
Proteins are shown in several orientations with respect to membrane. In (*a*) the protein is anchored to luminal surface of membrane by an uncleaved signal peptide with several hydrophobic residues. In (*b*) the signal sequence is not near the N-terminus; a domain of the protein was synthesized before emergence of signal peptide. Insertion of internal signal sequence, followed by completion of translation, resulted in a protein with a cytoplasmic N-terminal domain, a membrane-spanning central segment, and a C-terminal domain in the ER lumen. Diagram (*c*) shows a protein with the opposite orientation: an N-terminal signal sequence, which might also have been cleaved by signal peptidase, resulted in extrusion of a segment of protein into the ER lumen. A second hydrophobic anchoring sequence remained membrane associated and prevented passage of rest of protein through the membrane, thus allowing formation of a C-terminal cytoplasmic domain. In (*d*) several internal signal and anchoring sequences allow various segments of the protein to be oriented on each side of membrane.

positively charged amphiphilic α-helix that is recognized by a mitochondrial **receptor.** The protein is translocated across both membranes and into the mitochondrial matrix in an energy-dependent reaction. Passage occurs at adhesion sites where inner and outer membranes are close together, and several intermembrane proteins are involved. (See Clin. Corr. 6.5.) Proteases remove the matrix targeting signal but may leave other sequences that further sort the protein within the mitochondrion. For example, in response to a hydrophobic signal sequence a clipped precursor of cytochrome b_2 is moved back across the inner membrane where further proteolysis frees it in the intermembrane space. In contrast, cytochrome *c* apoprotein (without heme) binds at the outer membrane and is passed into the intermembrane space. There it acquires its heme and undergoes a conformational change that prevents return to the cytosol. Outer membrane localization can utilize the matrix targeting mechanism to translocate part of the protein, but a large apolar sequence blocks full transfer and leaves a membrane-bound protein with a C-terminal domain on the mitochondrial surface.

Targeting to Other Organelles Requires Specific Signals

Nuclei must import many proteins involved in their own structure and for DNA replication, transcription, and ribosome biogenesis. Nuclear pores permit passage of small proteins, but larger proteins are targeted by **nuclear localization signals** that include clusters of basic amino acids. Some nuclear proteins may be retained in the nucleus by forming complexes within the organelle. **Peroxisomes** contain a limited array of enzymes. One targeting signal is a carboxy-terminal tripeptide, Ser-Lys-Leu (SKL). An N-terminal nonapeptide targeting signal also functions, and others may exist.

A different targeting problem exists for proteins that reside in more than one subcellular compartment. Sometimes gene duplication and divergence have resulted in different targeting signals on closely related mature polypeptides. **Alternative transcription initiation** sites or pre-mRNA splicing can generate different messages from a single gene. An example of the latter is a calcium/calmodulin-dependent protein kinase; **alternatively spliced mRNAs** differ with respect to an internal segment that encodes a nuclear localization signal. Without this segment, the protein remains in the cytosol. **Alternative translation initiation** sites lead to two forms of rat liver fumarase, one of which includes a mitochondrial targeting sequence while the other does not and remains in the cytosol. A suboptimal localization signal can lead to inefficient targeting and a dual location, as in the partial secretion of an inhibitor of the plasminogen activator. Finally, some proteins contain more than one targeting signal, which must compete with each other.

6.6 | FURTHER POSTTRANSLATIONAL PROTEIN MODIFICATIONS

Several additional maturation events may modify newly synthesized polypeptides to help generate their final, functional forms. Many of these events are very common, while others are restricted to one or a few known instances.

Partial Proteolysis Activates Insulin and Zymogens

Partial proteolysis of proteins is a common maturation step. Sequences can be removed from either end or from within the protein. Proteolysis in the ER and Golgi apparatus helps to mature the protein hormone insulin (Figure 6.19). **Preproinsulin** encoded by mRNA is inserted into the ER lumen. Signal peptidase cleaves the

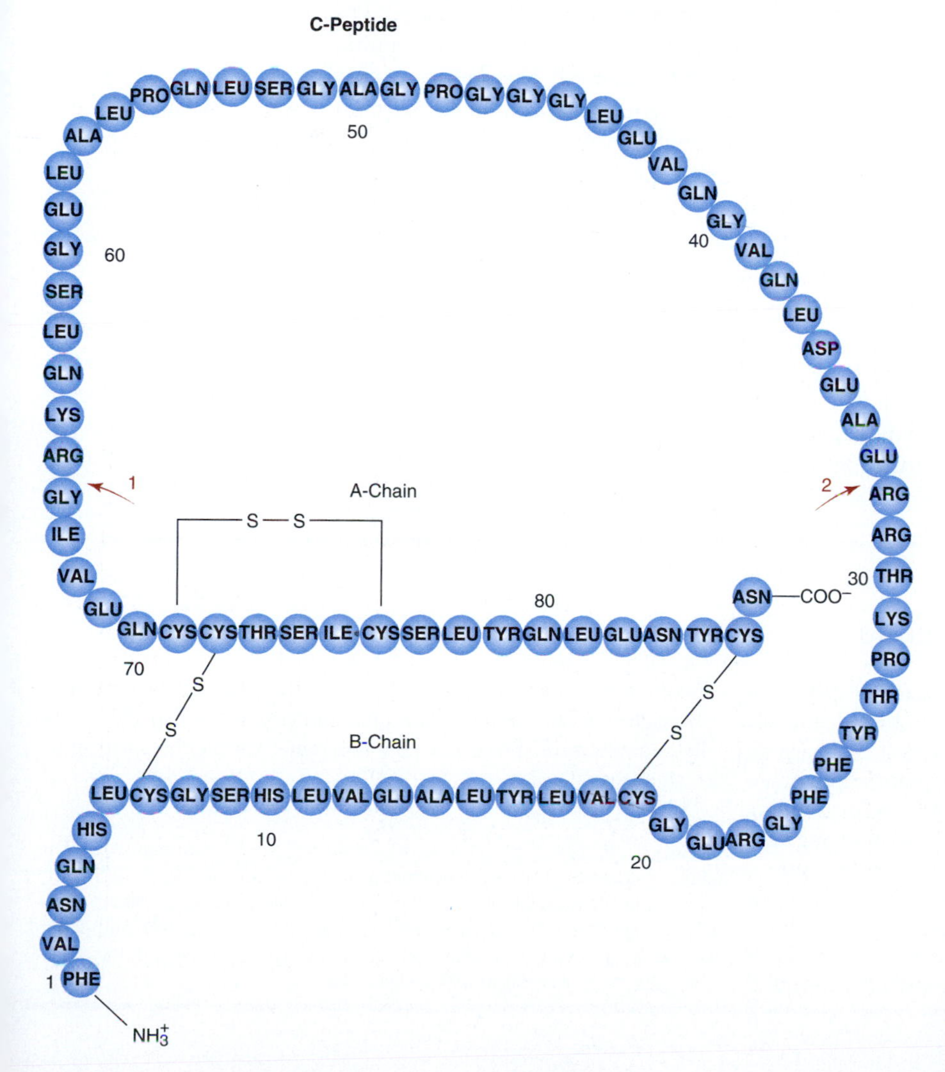

FIGURE 6.19
Maturation of human proinsulin.
After cleavage at the two sites indicated by arrows, arginine residues 31, 32, and 65 and lysine residue 64 are removed to produce insulin and C-peptide.
Redrawn from Bell, G. I., Swain, W. F., Pictet, R., Cordell, B., Goodman, H. M., and Rutter, W. J. Nature *282:525, 1979.*

signal peptide to generate **proinsulin**, which folds to form the correct disulfide linkages. Proinsulin is transported to the Golgi apparatus, where it is packaged into secretory granules. An internal **connecting peptide (C peptide)** is removed by proteolysis, and mature **insulin** is secreted. In familial hyperproinsulinemia processing is incomplete (see Clin. Corr. 6.9). This pathway for insulin biosynthesis has advantages over synthesis and joining of two separate polypeptides. First, it ensures production of equal amounts of A and B chains without coordination of two translational activities. Second, proinsulin folds into a three-dimensional structure in which the cysteine residues are placed for correct disulfide bond formation. Reduced and denatured proinsulin refolds correctly to form proinsulin, while renaturation of reduced and denatured insulin is inefficient, and incorrect disulfide linkages are also formed.

Precursor protein cleavage is a common means of enzyme activation. Digestive proteases are classic examples of this phenomenon (see p. 1098). Inactive zymogen precursors are packaged in storage granules and activated by proteolysis upon secretion. Thus **trypsinogen** is cleaved to give an amino-terminal peptide plus trypsin, and chymotrypsinogen is cleaved to form **chymotrypsin** and two peptides.

Amino Acids Can Be Modified After Incorporation into Proteins

Only 20 amino acids are encoded genetically and incorporated during translation. **Posttranslational modification** of proteins, however, leads to formation of 100 or more different amino acid derivatives in proteins. Modification may be permanent or highly reversible. The amounts of modified amino acids may be small, but they often play a major functional role in proteins. Examples are listed in Table 6.10.

Protein amino termini are frequently modified. Protein synthesis is initiated using methionine, but in the majority of cases the amino-terminal residue is not methionine; proteolysis has occurred. The amino terminus is then frequently altered. Acetylation is common. The α subunits of signal transducing G-proteins (these are central to cell–cell communication; see p. 925) are derivatized with myristic or palmitic acid. Amino-terminal glutamine residues spontaneously cyclize; one possible result is the stabilization of the protein. Amino-terminal sequences are occasionally lengthened by the addition of an amino acid (see Section 6.8 on protein degradation).

Posttranslational disulfide bond formation is catalyzed by a **disulfide isomerase.** The cystine-containing protein is conformationally stabilized. Disulfide formation can prevent unfolding of proteins and their passage across membranes, so it also becomes a means of localization. As seen in the case of insulin, disulfide bonds can covalently link separate polypeptides and be necessary for biological function. Cysteine modification also occurs; as an example, γ subunits of the het-

CLINICAL CORRELATION 6.9
Familial Hyperproinsulinemia

Familial hyperproinsulinemia, an autosomal dominant condition, results in approximately equal amounts of insulin and an abnormally processed proinsulin being released into the circulation. Although affected individuals have high concentrations of proinsulin in their blood, they are apparently normal in terms of glucose metabolism, being neither diabetic nor hypoglycemic. The defect was originally thought to result from a deficiency of one of the proteases that process proinsulin. Three enzymes process proinsulin: endopeptidases that cleave the Arg 31–Arg 32 and Lys 64–Arg 65 peptide bonds, and a carboxypeptidase. In several families the defect is the substitution of Arg65 by His or Leu, which prevents cleavage between the C-peptide and the A chain of insulin, resulting in secretion of a partially processed proinsulin. In one family a point mutation (His 10 → Asp 10) causes the hyperproinsulinemia, but how this mutation interferes with processing is not known.

Steiner, D. F., Tager, H. S., Naujo, K., Chan, S. J., and Rubenstein, A. H. Familial syndromes of hyperproinsulinemia with mild diabetes. In: C. R. Scriver, A. L. Beaudet, W. S. Sly, and D. Valle (Eds.), *The Metabolic and Molecular Bases of Inherited Disease*, 7th ed. New York: McGraw-Hill, 1995, pp. 897; and Zhou, A., Webb, G., Zhou, X., and Steiner, D. F. Proteolytic processing in the secretory pathway. *J. Biol. Chem.* 274:20745, 1999.

TABLE 6.10 Modified Amino Acids in Proteins[a]

Amino Acid	*Modifications Found*
Amino terminus	Formylation, acetylation, aminoacylation, myristoylation, glycosylation
Carboxyl terminus	Methylation, glycosyl-phosphatidylinositol anchor formation, ADP-ribosylation
Arginine	*N*-Methylation, ADP-ribosylation
Asparagine	*N*-Glycosylation, *N*-methylation, deamidation
Aspartic acid	Methylation, phosphorylation, hydroxylation
Cysteine	Cystine formation, selenocysteine formation, palmitoylation, linkage to heme, *S*-glycosylation, prenylation, ADP-ribosylation
Glutamic acid	Methylation, γ-carboxylation, ADP-ribosylation
Glutamine	Deamidation, cross-linking, pyroglutamate formation
Histidine	Methylation, phosphorylation, diphthamide formation, ADP-ribosylation
Lysine	*N*-acetylation, *N*-methylation, oxidation, hydroxylation, cross-linking, ubiquitination, allysine formation, biotinylation
Methionine	Sulfoxide formation
Phenylalanine	β-Hydroxylation and glycosylation
Proline	Hydroxylation, glycosylation
Serine	Phosphorylation, glycosylation, acetylation
Threonine	Phosphorylation, glycosylation, methylation
Tryptophan	β-Hydroxylation, dione formation
Tyrosine	Phosphorylation, iodination, adenylation, sulfonylation, hydroxylation

Source: Adapted from Krishna, R. G. and Wold, F. Post-translational modification of proteins. In: A. Meister (Ed.), *Advances in Enzymology,* Vol. 67. New York: Wiley-Interscience, 1993, pp. 265–298.
[a]The listing is not comprehensive and some of the modifications are very rare. Note that no derivatives of alanine, glycine, isoleucine, and valine have been identified in proteins.

erotrimeric G-proteins are modified by thioester linkage of an isoprenoid to a cysteine at or near the carboxy terminus. Multiple sulfatase deficiency arises from reduced ability to carry out a posttranslational modification (see Clin. Corr. 6.10).

Methylation of lysine ε-amino groups occurs in histones and may modulate their interactions with DNA. A fraction of the H2A histone is also modified through

CLINICAL CORRELATION 6.10

Absence of Posttranslational Modification: Multiple Sulfatase Deficiency

A variety of biological molecules are sulfated; examples include glycosaminoglycans, steroids, and glycolipids. Ineffective sulfation of the glycosaminoglycans chondroitin sulfate and keratan sulfate (see p. 684) of cartilage results in major skeletal deformities.

Degradation of sulfated molecules depends on the activity of several related sulfatases, most of which are located in lysosomes. Multiple sulfatase deficiency is a rare lysosomal storage disorder that combines features of metachromatic leukodystrophy and mucopolysaccharidosis. Affected individuals develop slowly and from their second year of life lose the abilities to stand, sit, or speak; physical deformities and neurological deficiencies develop and death before the age of 10 years is usual. Biochemically, multiple sulfatase deficiency is characterized by severe lack of all the sulfatases. In contrast, deficiencies in individual sulfatases are also known, and several distinct diseases are linked to single-enzyme defects.

Multiple sulfatase deficiency arises from a defect in a posttranslational modification that is common to all sulfatase enzymes and is necessary for their enzymatic activity. In each case a cysteine residue of the enzyme is normally converted to 2-amino-3-oxopropionic acid; the —CH_2SH side chain of cysteine becomes a —CHO (aldehyde) group, which may react with amino or hydroxyl groups of the enzyme, a cofactor, and so on. Fibroblasts from individuals with multiple sulfatase deficiency catalyze this modification with significantly lowered efficiency, and the unmodified sulfatases are catalytically inactive.

Schmidt, B., Selmer, T., Ingendoh, A., and von Figura, K. A novel amino acid modification in sulfatases that is deficient in multiple sulfatase deficiency. *Cell* 82:271, 1995.

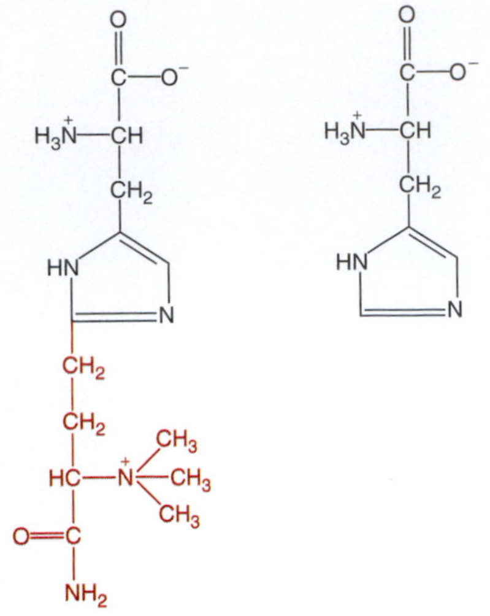

FIGURE 6.20
Diphthamide (left) is a posttranslational modification of a specific residue of histidine (right) in EF-2.

isopeptide linkage of a small protein, ubiquitin, from its C-terminal glycine to a lysine ε-amino group on histone. A role in DNA interactions has been postulated. Biotin is also linked to a few proteins through amide linkages to lysine.

Serine and threonine hydroxyl groups are major sites of glycosylation and of reversible phosphorylation by protein kinases and protein phosphatases. A classic example of phosphorylation of a serine residue is glycogen phosphorylase, which is modified by phosphorylase kinase (see p. 652). Tyrosine kinase activity is a property of many growth factor receptors in which growth factor binding stimulates cell division, which stimulates the proliferation of specific cell types. Oncogenes, responsible in part for cell transformation and proliferation of tumor cells, often have tyrosine kinase activity and show strong homology with normal growth factor receptors. Dozens of other examples exist; together the protein kinases and protein phosphatases control the activity of many proteins that are central to normal and abnormal cellular development.

ADP-ribosylation of EF-2 at a modified histidine residue represents a doubling of posttranslational modifications. First, a specific EF-2 histidine residue is modified to generate the diphthamide derivative (Figure 6.20) of the functional protein. ADP-ribosylation of the diphthamide by diphtheria toxin then inhibits EF-2 activity. Physiological ADP-ribosylation, not mediated by bacterial toxins, involves modification of arginine and cysteine residues. Formation of γ-carboxyglutamate from glutamic acid residues occurs in several blood clotting proteins including prothrombin and factors VII, IX, and X. The γ-carboxyglutamate residues chelate calcium ion, which is required for normal blood clotting (see p. 1045). In each case the modification requires vitamin K and can be blocked by coumarin derivatives, which antagonize vitamin K. As a result the rate of coagulation is greatly decreased. Coumarins are used in therapy to control clotting.

Collagen Biosynthesis Requires Many Posttranslational Modifications

Collagen, the most abundant family of related proteins in the human, is a fibrous protein that provides the structural framework for tissues and organs. It undergoes a wide variety of posttranslational modifications that directly affect its structure and function, and defects in its modification result in serious diseases. Collagen is an excellent example of the importance of posttranslational modification.

Different species of collagen, designated types I, II, III, IV, and so on (see p. 123 for details of structure) are encoded on several chromosomes and expressed in different tissues. Their amino acid sequences differ, but a repeating sequence **Gly-X-Y** that is about 1000 residues long predominates. Every third residue is glycine, about one-third of the X positions are occupied by proline, and a similar number of Y positions are 4-hydroxyproline, a posttranslationally modified form of proline. Proline and hydroxyproline residues impart considerable rigidity to the structure, which exists as a polyproline type II helix (Figure 6.21). Each collagen polypeptide is designated an **α chain**; a collagen molecule includes three α chains intertwined in a collagen **triple helix** in which the glycine residues occupy the center of the structure (see p. 124).

Procollagen Formation in the Endoplasmic Reticulum and Golgi Apparatus

Collagen α chain synthesis starts in the cytosol, where the amino-terminal signal sequences bind signal recognition particles. Precursor forms, designated for example, prepro α1(I), are extruded into the ER lumen and the signal peptides are cleaved. Hydroxylation of proline and lysine residues occurs **cotranslationally**, before assembly of a triple helix. Prolyl 4-hydroxylase requires an -X-Pro-Gly- sequence (hence 4-hydroxyproline is found only at Y positions in the -Gly-X-Y- sequence). Also present in the ER is a prolyl 3-hydroxylase, which modifies a smaller number of proline residues, and a lysyl hydroxylase that modifies some of the Y position lysine residues. These hydroxylases require Fe^{2+} and ascorbic acid (vitamin C); the

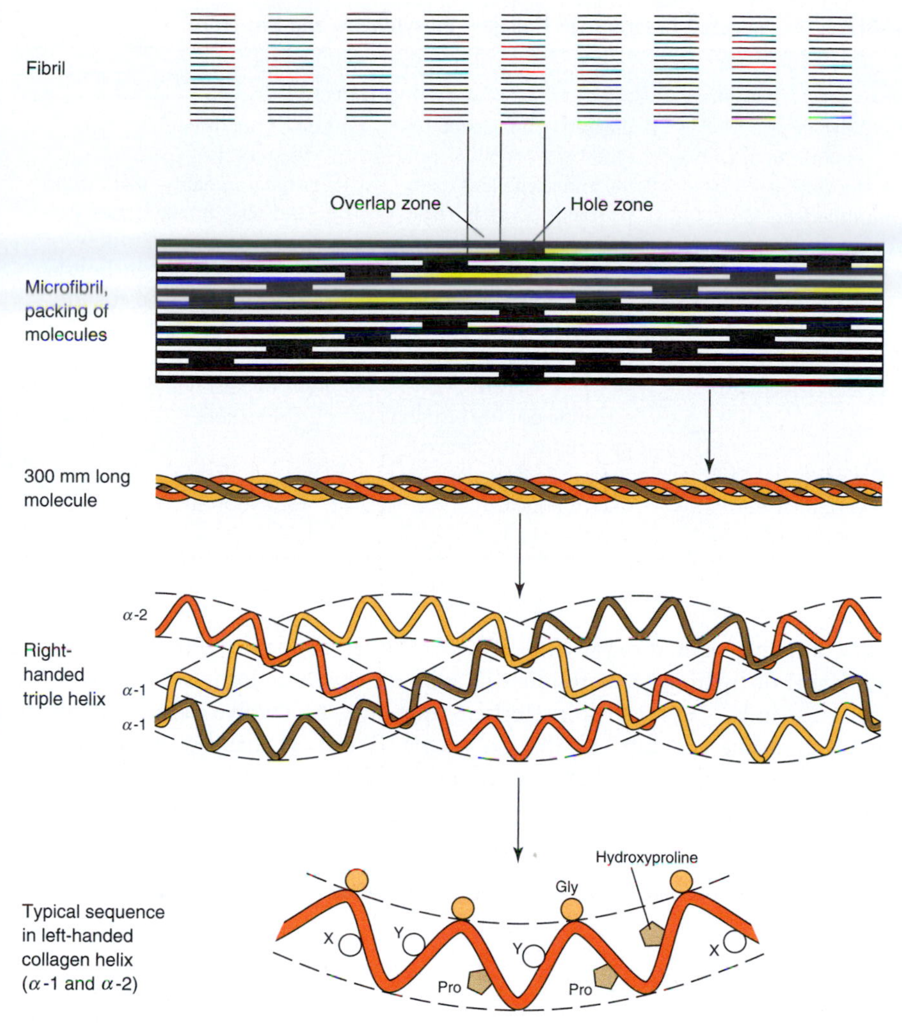

FIGURE 6.21
Collagen structure, illustrating (bottom to top) the regularity of primary sequence, the left-handed α helix, the right-handed triple helix, the 300-nm molecule, and the organization of molecules in a typical fibril, within which collagen molecules are cross-linked.

extent of modification depends on the specific α-chain type. Proline hydroxylation stabilizes collagen and lysine hydroxylation provides sites for interchain cross-linking and for glycosylation by specific glycosyltransferases of the ER. Asparagine residues are also glycosylated at this point, eventually leading to high-mannose type oligosaccharides.

Triple helix assembly occurs after the polypeptide chains have been completed. Carboxy-terminal globular proprotein domains fold and disulfide bonds are formed. Interaction of these domains initiates winding of the triple helix from the carboxyl toward the amino terminus. The completed triple helix, with globular **proprotein** domains at each end, moves to the Golgi apparatus where oligosaccharides are processed and matured. Sometimes tyrosine residues are modified by sulfation and some serines are phosphorylated. The completed procollagen is then released from the cell via secretory vesicles.

Collagen Maturation

Conversion of procollagen to collagen occurs extracellularly. The amino- and carboxyl-terminal propeptides are cleaved by separate proteases. Concurrently, the triple helices assemble into fibrils and the collagen is stabilized by extensive cross-linking. Lysyl oxidase converts some lysine or hydroxylysine to the reactive alde-

TABLE 6.11 Selected Disorders in Collagen Biosynthesis and Structure

Disorder	*Collagen Defect*	*Clinical Manifestations*
Osteogenesis imperfecta 1	Decreased synthesis of type I	Long bone fractures prior to puberty
Osteogenesis imperfecta 2	Point mutations and exon rearrangements in triple helical regions	Perinatal lethality; malformed and soft, fragile bones
Ehlers–Danlos IV	Poor secretion, premature degradation of type III	Translucent skin, easy bruising, arterial and colon rupture
Ehlers–Danlos VI	Decreased hydroxylysine in types I and III	Hyperextensive skin, joint hypermobility
Ehlers–Danlos VII	Type I procollagen accumulation: N-terminal propeptide not cleaved	Joint hypermobility and dislocation
Cutis laxa (occipital horn syndrome)	Decreased hydroxylysine due to poor Cu distribution	Lax, soft skin; occipital horn formation in adolescents

hydes, allysine, or hydroxyallysine. These residues condense with each other or with lysine or hydroxylysine residues in adjacent chains to form **Schiff's base** and aldol cross-links. Less well-characterized reactions can involve other residues including histidine and can link three α chains. Defects at many of these steps are known. Some of the best characterized are listed in Table 6.11 and described in Clinical Correlation 6.11.

CLINICAL CORRELATION 6.11

Defects in Collagen Synthesis

Ehlers–Danlos Syndrome, Type IV

Ehlers–Danlos syndrome is a group of at least ten disorders that are clinically, genetically, and biochemically distinguishable but that share manifestations of structural weaknesses in connective tissue. The usual problems are with fragility and hyperextensibility of skin and hypermobility of the joints. The weaknesses result from defects in collagen structure. For example, type IV Ehlers–Danlos syndrome is caused by defects in type III collagen (see p. 123), which is particularly important in skin, arteries, and hollow organs. Characteristics include thin, translucent skin through which veins are easily seen, marked bruising, and sometimes an appearance of aging in the hands and skin. Clinical problems arise from arterial rupture, intestinal perforation, and rupture of the uterus during pregnancy or labor. Surgical repair is difficult because of tissue fragility. The basic defects in type IV Ehlers–Danlos are due to changes in the primary structure of type III chains. These arise from point mutations that result in replacement of glycine residues and thus disruption of the collagen triple helix, and from exon-skipping, which shortens the polypeptide and can result in inefficient secretion and decreased thermal stability of the collagen and in abnormal formation of type III collagen fibrils. In some cases type III collagen is accumulated in the rough ER, overmodified, and degraded very slowly.

Superti-Furga, A., Gugler, E., Gitzelmann, R., and Steinmann, B. Ehlers–Danlos syndrome type IV: a multi-exon deletion in one of the two COL 3A1 alleles affecting structure, stability, and processing of type III procollagen. *J. Biol. Chem.* 263:6226, 1988.

Osteogenesis Imperfecta

Osteogenesis imperfecta is a group of at least four clinically, genetically, and biochemically distinguishable disorders characterized by multiple fractures and resultant bone deformities. Several variants result from mutations producing modified α(I) chains. In the clearest example a deletion mutation causes absence of 84 amino acids in the $\alpha 1$(I) chain. The shortened $\alpha 1$(I) chains are synthesized, because the mutation leaves the reading frame in register. The short $\alpha 1$(I) chains associate with normal $\alpha 1$(I) and $\alpha 2$(I) chains, thereby preventing normal collagen triple helix formation, with resultant degradation of all of the chains, a phenomenon aptly named "protein suicide." Three-fourths of all the collagen molecules formed have at least one short (defective) $\alpha 1$(I) chain, an amplification of the effect of a heterozygous gene defect. Other forms of osteogenesis imperfecta result from point mutations that substitute another amino acid for a glycine. Since glycine has to fit into the interior of the collagen triple helix, these substitutions destabilize that helix.

Barsh, G. S., Roush, C. L., Bonadio, J., Byers, P. H., and Gelinas, R. E. Intron mediated recombination causes an α(I) collagen deletion in a lethal form of osteogenesis imperfecta. *Proc. Natl. Acad. Sci. USA* 82:2870, 1985.

Scurvy and Hydroxyproline Synthesis

Most animals, but not humans, can synthesize ascorbic acid (vitamin C). Scurvy results from dietary deficiency of ascorbic acid. Among other problems, ascorbic acid deficiency causes decreased hydroxyproline synthesis because prolyl hydroxylase requires ascorbic acid.

Hydroxyproline provides additional hydrogen-bonding atoms that stabilize the collagen triple helix. Collagen containing insufficient hydroxyproline loses temperature stability and is significantly less stable than normal collagen at body temperature. The resultant clinical manifestations are distinctive and understandable: suppression of the orderly growth process of bone in children, poor wound healing, and increased capillary fragility with resultant hemorrhage, particularly in the skin. Severe ascorbic acid deficiency leads secondarily to a decreased rate of procollagen synthesis.

Peterofsky, B. Ascorbate requirement for hydroxylation and secretion of procollagen: relationship to inhibition of collagen synthesis in scurvy. *Am. J. Clin. Med.* 54:1135S, 1991.

Deficiency of Lysyl Hydroxylase

In type VI Ehlers–Danlos syndrome lysyl hydroxylase is deficient. As a result type I and III (see p. 123) collagens in skin are synthesized with decreased hydroxylysine content, and subsequent cross-linking of collagen fibrils is less stable. Some cross-linking between lysine and allysine occurs but these are not as stable and do not mature as readily as do hydroxylysine-containing cross-links. In addition, carbohydrate is transferred to the hydroxylysine residues but the function of this carbohydrate is unknown. The clinical features include marked hyperextensibility of the skin and joints, poor wound healing, and musculoskeletal deformities. Some patients with this form of Ehlers–Danlos syndrome have a mutant form of lysyl hydroxylase with a higher Michaelis constant for ascorbic acid than the normal enzyme. Accordingly, they respond to high doses of ascorbic acid.

Pinnell, S. R., Krane, S. M., Kenzora, J. E., and Glimcher, M. J. A heritable disorder of connective tissue: hydroxylysine-deficient collagen disease. *N. Engl. J. Med.* 286:1013, 1972.

Ehlers–Danlos Syndrome, Type VII

In Ehlers–Danlos syndrome, type VII, skin bruises easily and is hyperextensible, but the major manifestations are dislocations of major joints, such as hips and knees. Laxity of ligaments is caused by incomplete removal of the amino-terminal propeptide of the procollagen chains. One variant of the disease results from deficiency of procollagen *N*-protease. A similar deficiency occurs in the autosomal recessive disease called dermatosparaxis of cattle, sheep, and cats, in which skin fragility is so extreme as to be lethal. In other variants the proα1(I) and proα2(I) chains lack amino acids at the cleavage site because of skipping of one exon in the genes. This prevents normal cleavage by procollagen *N*-protease.

Cole, W. G., Chan, D., Chambers, G. W., Walker, I. D., and Bateman, J. F. Deletion of 24 amino acids from the proα(I) chain of type I procollagen in a patient with the Ehlers–Danlos syndrome type VII. *J. Biol. Chem.* 261:5496, 1986.

Occipital Horn Syndrome

In type IX Ehlers–Danlos syndrome and in Menkes' kinky hair syndrome there is thought to be a deficiency in lysyl oxidase activity. In type IX Ehlers–Danlos syndrome there are consequent cross-linking defects manifested in lax, soft skin and in the appearance during adolescence of bony occipital horns. Copper-deficient animals have deficient cross-linking of elastin and collagen, apparently because of the requirement for cuprous ion by lysyl oxidase. In Menkes' kinky hair syndrome there is a defect in intracellular copper transport that results in low activity of lysyl oxidase, and in occipital horn syndrome there is also a defect in intracellular copper distribution. A woman taking high doses of the copper-chelating drug, *d*-penicillamine, gave birth to an infant with an acquired Ehlers–Danlos-like syndrome, which subsequently cleared. Side effects of *d*-penicillamine therapy include poor wound healing and hyperextensible skin.

Peltonen, L., Kuivarnieme, H., Palotie, H., Horn, N., Kaitila, I., and Kivirikko, K. I. Alterations of copper and collagen metabolism in the Menkes syndrome and a new subtype of Ehlers–Danlos syndrome. *Biochemistry* 22:6156, 1983.

For a detailed overview of collagen disorders see: Byers, P. H. Disorders of collagen biosynthesis and structure. In: C. R. Scriver, A. L. Beaudet, W. S. Sly, and D. Valle (Eds.), *The Metabolic and Molecular Bases of Inherited Disease*, 8th ed., New York: McGraw-Hill, 2001, p. 5241.

6.7 | REGULATION OF TRANSLATION

The amount of protein in cells is regulated at the stages of transcription, translation, and degradation. Formation of functioning proteins has significant consequences for cells and translation requires considerable energy. It is thus necessary that translation be carefully controlled. This is done both globally and for specific proteins. The most effective and common mechanisms for regulation are at the initiation stage.

The best understood means of overall regulation of translation involves reversible **phosphorylation** of eIF-2a. Under conditions that include nutrient starvation, heat shock, and viral infection, eIF-2a is phosphorylated. Phosphorylated eIF-2a–GDP binds tightly to eIF-2b, the guanine nucleotide exchange factor, which is present in limiting amounts. Since eIF-2b is unavailable for nucleotide exchange, no eIF-2a–GTP is available for initiation. Phosphorylation can be catalyzed by a **heme-regulated inhibitor kinase,** which, in the absence of heme, is activated by autophosphorylation. This kinase is present in many cells, but best studied in reticulocytes that synthesize hemoglobin. Deficiencies in energy supply or any heme precursor activate the kinase. A related **double-stranded RNA-dependent kinase** is autophosphorylated and activated in response to binding of dsRNA that results from many viral infections. Production of this kinase is also induced by interferon. In contrast, initiation factor eIF-4e is activated by phosphorylation in response to, for example, growth factors, and inactivated by a protein phosphatase following, for example, viral infection. These effects are mediated by eIF-4g, which is also activated by phosphorylation; they may be most important in the translation of mRNAs with highly structured leader sequences that most need to be unwound by eIF-4a to reveal a translational start site.

Regulation of translation of specific genes also occurs. A clear example is the regulation by iron of synthesis of the iron-binding protein, ferritin. In the absence of iron a repressor protein binds to the **iron responsive element** (IRE), a stem–loop structure in the 5′ leader sequence of ferritin mRNA. This mRNA is sequestered for future use. δ-Aminolevulinic acid synthase, an enzyme of heme biosynthesis, is also regulated by a 5′-IRE in its mRNA. In contrast, more ferritin receptor mRNA is needed if iron is limited; it has IREs in its 3′-untranslated region. Binding of the repressor protein stabilizes the mRNA and prolongs its useful lifetime. Many growth-regulated mRNAs, including those for ribosomal proteins, have a polypyrimidine tract in their leader sequence. A polypyrimidine-binding protein helps regulate their translation.

6.8 | PROTEIN DEGRADATION AND TURNOVER

Proteins have finite lifetimes. They are subject to environmental damage such as oxidation, proteolysis, conformational denaturation, or other irreversible modifications. Errors in translation and folding lead to nonfunctional proteins and proteolytic processing generates nonfunctional peptides. Equally important, cells must change their protein complements in order to respond to different needs and situations. In each case "trash disposal" is needed. Specific proteins have very different lifetimes. Cells of the eye lens are not replaced and their proteins are not recycled. Hemoglobin in red blood cells lasts the life of these cells, about 120 days. Other proteins have lifetimes measured in days, hours, or even minutes. Some blood clotting proteins survive only a few days, so hemophiliacs are only protected for a short period by transfusions or injections of required factors. Diabetics require insulin injections regularly since the hormone is metabolized. Metabolic enzymes vary quantitatively depending on need, for example, the concentration of urea cycle enzymes change in response to diet. Most amino acids produced by protein degradation are recycled to synthesize new proteins but some will be metabolized and their degradation products will be excreted. In all cases, proteolysis first reduces the proteins in question to peptides and eventually amino acids. Several proteolytic systems accomplish this end.

FIGURE 6.22
Model of the proteasome.
A 20S central segment is made up of four stacked heptameric rings of two types. The core is hollow and includes 12–15 different polypeptides; several proteases with different specificities are localized within the rings. V-shaped segments at each end cap the cylinder and are responsible for ATP-dependent substrate recognition, unfolding, and translocation into the proteolytic core. The upper cap structure is also in contact with central segment but it is shown displaced from it in order to illustrate the hollow core of the cylinder.
Adapted from Rubin, D. and Finley, D. Curr. Biol. *5: 854, 1995; and Peters, J.-M.* Trends Biochem. Sci. *19:377, 1994.*

ATP-Dependent Proteolysis Occurs in Proteasomes

One well-described proteolytic pathway uses **proteasomes.** A proteasome is a dumbbell-shaped structure made up of about 28 polypeptides; a proteolytically active cylindrical core is capped at each end by V-shaped complexes (Figure 6.22). The cap acts in recognizing and unfolding polypeptides and transporting them to the proteolytic core in an ATP-dependent mechanism.

Targeting to proteasomes normally requires **ubiquitin,** a highly conserved 76 amino acid protein. One function of ubiquitin is to mark proteins for degradation. Ubiquitin has other roles; as an example, linkage of ubiquitin to histones H2A and H2B is unrelated to turnover since the proteins are stable, but modification may affect chromatin structure or transcription.

The ubiquitin-dependent proteolytic cycle is shown in Figure 6.23. Ubiquitin is activated by enzyme E1 to form a thioester; ATP is required and a transient AMP–ubiquitin complex is involved. The ubiquitin is then passed to enzyme E2, and finally via one of a group of E3 enzymes it is coupled to a targeted protein. Linkage of ubiquitin is through isopeptide bonds between ε-amino groups of lysine residues of the protein and the carboxyl-terminal glycine residues of ubiquitin. Several ubiquitin molecules may be attached to the protein and to each other. Proteasome proteases then degrade the tagged protein but release intact ubiquitin for reuse in further degradation cycles.

Ubiquitin-dependent proteolysis plays a major role in the regulation of cellular events. **Cyclins** and receptor **protein tyrosine kinases** are involved in control of cell division. Ubiquitin-dependent destruction of a cyclin allows cells to pass from the M-phase into G1, while degradation of receptors prevents signal transduction and halts cell proliferation. Other proteins degraded by ubiquitin-dependent proteolysis include transcription factors, the p53 tumor suppressor and other oncoproteins, a protein kinase, and immune system and other cell surface receptors.

Damaged, defective, and mutant proteins are rapidly degraded via the ubiquitin pathway. In **cystic fibrosis** a mutation that results in deletion of one amino acid greatly alters the stability of a protein (see Clin. Corr. 6.6). It is not always clear how native proteins are identified for degradation. Selectivity occurs at the level of the E3 enzyme; both conformation and amino acid sequence are important. Destabilizing PEST sequences (rich in Pro, Glu, Ser, and Thr) have been identified in several short-lived proteins. Another determinant is simply the identity of the amino-terminal amino acid. Otherwise identical proteins with different amino-terminal residues are degraded at widely differing rates, and the lifetime of a protein can be modified by incorporation of an N-terminal destabilizing residue from an aminoacyl-tRNA donor. Like proteasomes, the complex *E. coli* proteases Lon and Clp and similar enzymes in other microorganisms and in mitochondria require ATP hydrolysis for their action, but ubiquitin is absent in prokaryotes and the means of identification of proteins for degradation is still obscure.

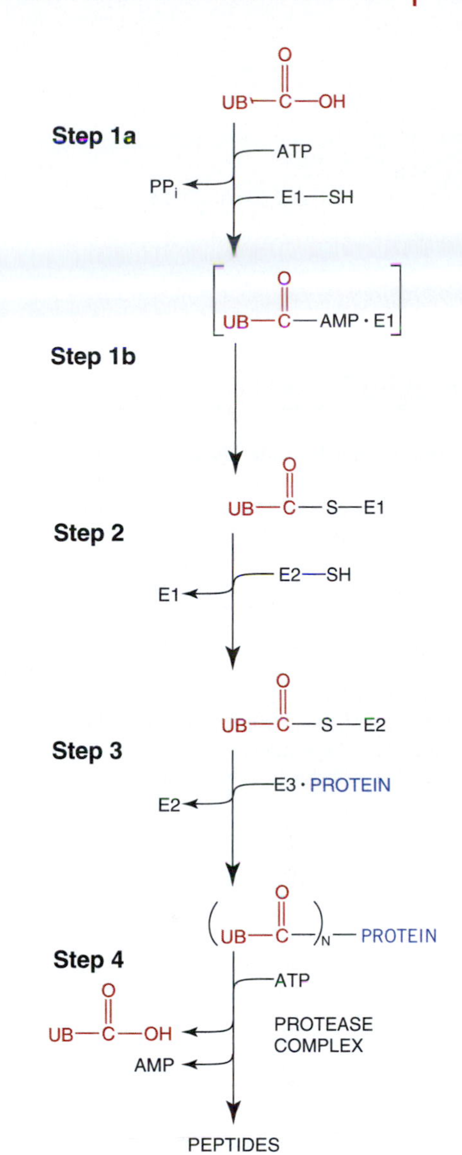

FIGURE 6.23
ATP and ubiquitin-dependent protein degradation.
Ubiquitin is activated in a two-step reaction involving formation of a transient mixed anhydride of AMP and the carboxy terminus of ubiquitin (Step 1a), followed by generation of a thioester with enzyme E1 (Step 1b). Enzyme E2 can now form a thioester with ubiquitin (Step 2) and serve as a donor in E3-catalyzed transfer of ubiquitin to a targeted protein (Step 3). Several more ubiquitin molecules are usually attached to lysine residues of the ubiquitin and/or other lysines of the targeted protein at this stage. Ubiquitinylated protein is now degraded by ATP-dependent proteolysis (Step 4); ubiquitin is not degraded and can reenter the process at Step 1.

Intracellular Digestion of Some Proteins Occurs in Lysosomes

Additional proteolytic pathways are also important. Intracellular digestion of proteins from the extracellular environment occurs within protease-rich **lysosomes.** Material that is impermeable to the plasma membrane is imported by endocytosis. In **pinocytosis** large particles, molecular aggregates, or other molecules in the extracellular fluid are ingested by engulfment. Macrophages ingest bacteria and dead cells by this mechanism. **Receptor-mediated endocytosis** uses cell surface receptors to bind specific molecules at pits in the cell surface that are coated internally with the multisubunit protein **clathrin.** Invagination of the plasma membrane and the receptors forms intracellular coated vesicles, one fate of which is fusion with a lysosome and degradation of the contents. Some intracellular protein turnover may also occur within lysosomes, and under some conditions significant amounts of cellular material can be mobilized via lysosomes. For example, serum starvation of fibroblasts in culture or starvation of rats leads to the lysosomal degradation of a subpopulation of cellular proteins. Recognition of a specific peptide sequence is involved, again indicating that the lifetime of a protein is ultimately encoded in its sequence.

Many Other Proteolytic Systems Exist

Calcium-dependent proteases, also called **calpains,** are present in most cells. Activators and inhibitors of these enzymes are also present, and calpains are logical candidates for enzymes involved in protein turnover. However, their role in these processes is not quantitatively established. Golgi and ER proteases degrade peptide fragments that arise during maturation of proteins in the secretory pathway. They could be involved in turnover of ER proteins, although proteins are also known to be exported from the ER for degradation by proteasomes. **Apoptosis,** programmed cell death, follows upon activation of proteases known as **caspases.** It is likely that other uncharacterized mechanisms exist in both the cytosol and in the mitochondria. Digestive proteases such as pepsin, trypsin, chymotrypsin, and elastase hydrolyze dietary protein and have no part in intracellular protein turnover, but the amino acids they help generate contribute to the metabolic pool used in translation. It is likely that protein degradation will turn out to be as complex and important a problem as protein biosynthesis.

BIBLIOGRAPHY

Ribosomes and Transfer RNA

Ban, N., Nissen, P., Hansen, J., Moore, P. B., and Steitz, T. A. The complete atomic structure of the large ribosomal subunit at 2.4 Å resolution. *Science* 289:905, 2000.

Cate, J. H., Yusupov, M. M., Yusupova, G. Z., Earnest, T. N., and Noller, H. F. X-ray crystal structures of 70S ribosome structural complexes. *Science* 285:2095, 1999.

Filonenko, V. V. and Deutscher, M. P. Evidence for similar structural organization of the multienzyme aminoacyl-tRNA complex *in vivo* and *in vitro*. *J. Biol. Chem.* 269:17375, 1994.

Hale, S. P., Auld, D. S., Schmidt, E., and Schimmel, P. Discrete determinants in transfer RNA for editing and aminoacylation. *Science* 276:1250, 1997.

Herold, M. and Nierhaus, K. H. Incorporation of six additional proteins to complete the assembly map of the 50S subunit from *Escherichia coli* ribosomes. *J. Biol. Chem.* 262:8826, 1987.

Hong, K. W., Ibba, M., Weygand-Durasevic, I., Rogers, M. J., Thomann, H. U., and Söll, D. Transfer RNA-dependent cognate amino acid recognition by an aminoacyl-tRNA synthetase. *EMBO.* 15:1983, 1996.

Hou, Y.-M., Francklyn, C., and Schimmel, P. Molecular dissection of a transfer RNA and the basis for its identity. *Trends Biochem. Sci.* 14:233, 1989.

Muth, G. W., Ortoleva-Donnelly, L., and Strobel, S. A. A single adenosine with a neutral pK_a in the ribosomal peptidyl transferase center. *Science* 289:947, 2000.

Nissen, P., Hansen, J., Ban, N., Moore, P. B., and Steitz, T. A. The structural basis of ribosome activity in peptide bond synthesis. *Science* 289:920, 2000.

Nomura, M. The role of RNA and protein in ribosome function: a review of early reconstitution studies and prospects for future studies. *Cold Spring Harbor Symp. Quant. Biol.* 52:653, 1987.

Perona, J., Rould, M., and Steitz, T. Structural basis for transfer RNA aminoacylation by *Escherichia coli* glutaminyl-tRNA synthetase. *Biochemistry* 32:8758, 1993.

Silvian, L. F., Wang, J., and Steitz, T. A. Insights into editing from an Ile-tRNA synthetase structure with tRNA and mupirocin. *Science* 285:1074, 1999.

Protein Biosynthesis and its Regulation

Agarwal, R. K., Heagle, A. B., Penczek, P., Grassucci, R. A., and Frank, J. EF-G-dependent GTP hydrolysis induces translocation accompanied by large conformational changes in the 70S ribosome. *Nature Struct. Biol.* 6:643, 1999.

Barrell, B., et al. Different pattern of codon recognition by mammalian mitochondrial tRNAs. *Proc. Natl. Acad. Sci. USA* 77:3164, 1980.

Berchtold, H., Reshetnikova, L., Reiser, C. O., Shirmer, N. K., Sprinzl, M., and Hilgenfeld, R. Crystal structure of active elongation factor T_u reveals major domain rearrangements. *Nature* 365:126, 1993.

Burkhardt, N., Jünemann, R., Spahn, C. M. T., and Nierhaus, K. M. Ribosomal tRNA binding sites: three site models of translation. *Crit. Rev. Biochem. Mol. Biol.* 33:95, 1998.

Gnirke, A., Geigenmuller, U., Rheinberger, H., and Nierhaus, K. The allosteric three-site model for the ribosomal elongation cycle. *J. Biol. Chem.* 264:7291, 1989.

Green, R. Ribosomal translocation: EF-G turns the crank. *Curr. Biol.* 10:R369, 2000.

Merrick, W. C. Eukaryotic protein synthesis: an *in vitro* analysis. *Biochimie* 76:822, 1994.

Nakamura, Y., Ito, K., and Ehrenberg, M. Mimicry grasps reality in translation termination. *Cell* 101:349, 2000.

Rhoads, R. E. Regulation of eukaryotic protein biosynthesis by initiation factors. *J. Biol. Chem.* 268:3017, 1993.

Samuel, C. E. The eIF2a protein kinases as regulators of protein synthesis in eukaryotes from yeasts to humans. *J. Biol. Chem.* 268:7603, 1993.

Song, H., et al. The crystal structure of human eukaryotic release factor eRF1—mechanism of stop codon recognition and peptidyl-tRNA hydrolysis. *Cell* 100:311, 2000.

Ziff, E. B. Transcription and RNA processing by the DNA tumour viruses. *Nature* 287:491, 1980.

Protein Targeting and Posttranslational Modification

Batey, R. T., Rambo, R. P., Lucast, L., Rha, B., and Doudna, J. A. Crystal structure of the ribonucleoprotein core of the signal recognition particle. *Science* 287:1232, 2000.

Bukau, B., Deuerling, E., Pfund, C., and Craig, E. A. Getting newly synthesized proteins into shape. *Cell* 101:119, 2000.

Corsi, A. K. and Schekman, R. Mechanism of polypeptide translocation into the endoplasmic reticulum. *J. Biol. Chem.* 271:30299, 1996.

Dahms, N. M., Lobel, P., and Kornfeld, S. Mannose 6-phosphate receptors and lysosomal enzyme targeting. *J. Biol. Chem.* 264:12115, 1989.

Danpure, C. J. How can products of a single gene be localized to more than one subcellular compartment? *Trends Cell Biol.* 5:230, 1995.

Dever, T. E. Translation initiation: adept at adapting. *Trends Biochem. Sci.* 24:398, 1999.

Dobson, C. M. Protein misfolding, evolution and disease. *Trends Biochem. Sci.* 24: 329, 1999.

Ellgaard, L., Molinari, M., and Helinius, A. Setting the standards: quality control in the secretory pathway. *Science* 286:1882, 1999.

Feldman, D. E. and Frydman, J. Protein folding *in vivo*: the importance of molecular chaperones. *Curr. Opin. Struct. Biol.* 10:26, 2000.

Gilbert, H. F. Protein disulfide isomerase and assisted protein folding. *J. Biol. Chem.* 272:29399, 1997.

Koehler, C. M., Merchant, S., and Schatz, G. How proteins travel across the mitochondrial intermembrane space. *Trends Biochem. Sci.* 24:428, 1999.

Krishna, R. G. and Wold, F. Post-translational modification of proteins. *Adv. Enzymol.* 67:265, 1993.

Kuhn, K. The classical collagens. In: R. Mayne and R. E. Burgeson (Eds.), *Structure and Function of Collagen Types.* Orlando, FL: Academic, 1987, p.1.

Paulson, J. C. and Colley, K. J. Glycosyltransferases: structure, localization, and control of cell type-specific glycosylation. *J. Biol. Chem.* 264:17615, 1989.

Petrescu, S. M., Branza-Nichita, N., Negroiu, G., Petrescu, A. J., and Dwek, R. A. Tyrosinase and glycoprotein folding: role of chaperones that recognize glycans. *Biochemistry* 39:5229, 2000.

Rudd, P. M., et al. The effects of variable glycosylation on the functional activities of ribonuclease, plasminogen and tissue plasminogen activator. *Biochem. Biophys. Acta* 1248:1, 1995.

Sachs, A. B. and Varani, G. Eukaryotic translation initiation: there are (at least) two sides to every story. *Nature Struct. Biol.* 7:356, 2000.

Schimöller, F., Simon, I., and Pfeffer, S. R. Rab GTPases, directors of vesicle docking. *J. Biol. Chem.* 273:22161, 1998.

Stroud, R. M. and Walter, P. Signal sequence recognition and protein targeting. *Curr. Opin. Struct. Biol.* 9:754, 1999.

Subramani, S. Protein translocation into peroxisomes. *J. Biol. Chem.* 271:32483, 1996.

Tokalidis, K. and Schatz, G. Biogenesis of mitochondrial inner membrane proteins. *J. Biol. Chem.* 274:35285, 1999.

Wedegaertner, P. B., Wilson, P. T., and Bourne, H. R. Lipid modification of trimeric G proteins. *J. Biol. Chem.* 270:503, 1995.

Wickner, S., Maurizi, M. R., and Gottesman, S. Posttranslational quality control: folding, refolding, and degrading proteins. *Science* 286:1888, 1999.

Wolin, S. From the elephant to *E. coli*: SRP-dependent protein targeting. *Cell* 77:787, 1994.

Zhou, A., Webb, G., Zhu, X., and Steiner, D. F. Proteolytic processing in the secretory pathway. *J. Biol. Chem.* 274:20745, 1999.

Protein Turnover and Proteasomes

Ciechanover, A. The ubiquitin–proteasome proteolytic pathway. *Cell* 79:13, 1994.

DeMartino, G. N. and Slaughter, C. A. The proteasome, a novel protease regulated by multiple mechanisms. *J. Biol. Chem.* 274:22123, 1999.

Joazeiro, C. A. P., Wing, S. S., Huang, H., Leverson, J. D., Hunter, T., and Liu, Y. C. The tyrosine kinase negative regulator c-Cbl as a RING-type, E2-dependent ubiquitin–protein ligase. *Science* 286:309, 1999.

Kessel, M., et al. Homology in structural organization between *E. coli* C1pAP protease and the eukaryotic 26S proteasome. *J. Mol. Biol.* 250:587, 1995.

Löwe, J., Stock, D., Jap, B., Zwickl, P., Baumeister, W., and Huber, R. Crystal structure of the 20S proteasome from the archaeon *T. acidophilum* at 3.4 Å resolution. *Science* 268:533, 1995.

Miller, M. H., Finger, A., Schweiger, M., and Wolf, D. H. ER degradation of a misfolded luminal protein by the cytosolic ubiquitin–proteasome pathway. *Science* 273:1725, 1996.

Rock, K., et al. Inhibitors of the proteasome block the degradation of most cell proteins and the generation of peptides presented on MHC class II molecules. *Cell* 78:761, 1994.

Rogers, S., Wells, R., and Rechsteiner, M. Amino acid sequences common to rapidly degraded proteins: the PEST hypothesis. *Science* 234:364, 1986.

Wolf, B. B. and Green, D. R. Suicidal tendencies: apoptotic cell death by caspase family proteinases. *J. Biol. Chem.* 274:20049, 1999.

QUESTIONS | C. N. ANGSTADT

Multiple Choice Questions

1. Degeneracy of the genetic code denotes the existence of:
   A. multiple codons for a single amino acid.
   B. codons consisting of only two bases.
   C. base triplets that do not code for any amino acid.
   D. different protein synthesis systems in which a given triplet codes for different amino acids.
   E. codons that include one or more of the "unusual" bases.

2. In the formation of an aminoacyl-tRNA:
   A. ADP and P_i are products of the reaction.
   B. aminoacyl-adenylate appears in solution as a free intermediate.
   C. the aminoacyl-tRNA synthetase is believed to recognize and hydrolyze incorrect aminoacyl-tRNAs it may have produced.
   D. there is a separate aminoacyl-tRNA synthetase for every amino acid appearing in the final, functional protein.
   E. there is a separate aminoacyl-tRNA synthetase for every tRNA species.

3. During initiation of protein synthesis:
   A. methionyl-tRNA appears at the A site of the 80S initiation complex.
   B. eIF-3 and the 40S ribosomal subunit participate in forming a preinitiation complex.
   C. eIF-2 is phosphorylated by GTP.
   D. the same methionyl-tRNA is used as is used during elongation.
   E. a complex consisting of mRNA, the 60S ribosomal subunit, and certain initiation factors is formed.

4. Requirements for eukaryotic protein synthesis include all of the following EXCEPT:
   A. mRNA.
   B. ribosomes.
   C. GTP.
   D. 20 different amino acids in the form of aminoacyl-tRNAs.
   E. fMet-$tRNA_i^{met}$.

5. During the elongation stage of eukaryotic protein synthesis:
   A. the incoming aminoacyl-tRNA binds to the P site.
   B. a new peptide bond is synthesized by peptidyltransferase site of the large ribosomal subunit in a GTP-requiring reaction.
   C. the peptide, still bound to a tRNA molecule, is translocated to a different site on the ribosome.
   D. streptomycin can cause premature release of the incomplete peptide.
   E. peptide bond formation occurs by the attack of the carboxyl group of the incoming aminoacyl-tRNA on the amino group of the growing peptide chain.

6. Termination of protein synthesis:
   A. requires a stop codon to be located at the P site of the large ribosomal subunit.
   B. occurs when a nonribosomal protein release factor binds to the ribosome.
   C. requires the action of a nonribosomal hydrolase to release the peptide.
   D. does not require energy.
   E. coincides with the degradation of the ribosomes.

7. Formation of mature insulin includes all of the following EXCEPT:
   A. removal of a signal peptide.
   B. folding into a three-dimensional structure.
   C. disulfide bond formation.
   D. removal of a peptide from an internal region.
   E. γ-carboxylation of glutamate residues.

8. 4-Hydroxylation of specific prolyl residues during collagen synthesis requires all of the following EXCEPT:
   A. Fe^{2+}.
   B. a specific amino acid sequence.
   C. ascorbic acid.
   D. succinate.
   E. individual α chains, not yet assembled into a triple helix.

9. Chaperones:
   A. are always required to direct the folding of proteins.
   B. when bound to protein, increase the rate of protein degradation.
   C. usually bind to strongly hydrophilic regions of unfolded proteins.
   D. sometimes maintain proteins in an unfolded state to allow passage through membranes.
   E. do not require energy (ATP) for action.

10. Diphtheria toxin:
   A. acts catalytically.
   B. releases incomplete polypeptide chains from the ribosome.
   C. activates translocase.
   D. prevents release factor from recognizing termination signals.
   E. attacks the RNA of the large subunit.

Questions 11 and 12: Cystic fibrosis is a fairly frequent (1 per 2000) genetic disease of Caucasians. The CF gene codes for a protein called the cystic fibrosis transmembrane conductance regulator (CFTR), which functions as a cAMP-regulated chloride channel. The protein has two membrane-spanning domains, two domains that interact with ATP, and one regulatory domain. The most common defect is a three-nucleotide deletion in the gene for a protein of one of the ATP-binding domains. The result is a protein that does not fold correctly in the endoplasmic reticulum, is not properly glycosylated, and is not transported to the cell surface. Rather, it is degraded in the cytosol within proteasomes.

11. The particular mutation described above most likely leads to a protein that:
   A. must be longer than it should be.
   B. must be several amino acids shorter than it should be.
   C. would always have a different amino acid sequence from the point of mutation to the end of the protein compared to the unmutated situation.
   D. has a substitution of one amino acid for another compared to the protein from the unmutated gene.
   E. has the deletion of one amino acid compared to the protein from the unmutated gene.

12. Targeting a protein to be degraded within proteasomes usually requires ubiquitin. In the functions of ubiquitin all of the following are true EXCEPT:
   A. ATP is required for activation of ubiquitin.
   B. a peptide bond forms between the carboxyl terminal of ubiquitin and an ε-amino group of a lysine.
   C. linkage of a protein to ubiquitin does not always mark it for degradation.
   D. the identity of the N-terminal amino acid is one determinant of selection for degradation.
   E. ATP is required by the enzyme that transfers the ubiquitin to the protein to be degraded.

Questions 13 and 14: Collagen, the most abundant type of protein in the human body, is unusual in its amino acid composition and requires a wide variety of posttranslational modifications to convert it to a functional molecule. Because of the complexity of collagen synthesis, there are many diseases caused by defects in the process. These diseases result in structural weaknesses in connective tissue. Ehlers–Danlos syndrome is a collection of ten different collagen diseases. Type VII Ehlers–Danlos syndrome results from a deficiency of procollagen *N*-protease, resulting in the incomplete removal of the amino-terminal peptide of the procollagen chains.

13. During collagen synthesis, events that occur extracellularly include all of the following EXCEPT:
   A. modification of prolyl residues.
   B. amino-terminal peptide cleavage.
   C. carboxyl-terminal peptide cleavage
   D. modification of lysyl residues.
   E. covalent cross-linking.

14. Since much of procollagen formation occurs in the endoplasmic reticulum and Golgi apparatus, signal peptide is required. All of the following statements about targeting a protein for the ER are true EXCEPT:
   A. the signal peptide usually has a positively charged N-terminus and a stretch of hydrophobic amino acids.
   B. as the signal peptide emerges from a free ribosome, it binds to a signal recognition particle (SRP).
   C. the signal peptide is usually cleaved from the protein before the protein is inserted into the ER membrane.
   D. docking protein is actually an SRP receptor and serves to bind the SRP to the ER.
   E. SRP and docking protein do not enter the ER lumen but are recycled.

Problems

15. I-cell (inclusion bodies) disease results from a defect in the enzyme that transfers *N*-acetylglucosamine phosphate to proteins containing high-mannose type oligosaccharides. What enzymes would be deficient in I-cell patients? How does the disease name correspond to the problem in these patients?

16. What is the significance of the following structures? A positively charged amphiphilic α-helix, a cluster of lysine and arginine residues, and a carboxy-terminal Ser-Lys-Leu (SKL) sequence?

ANSWERS

1. **A** This is the definition of degeneracy. B and E are not known to occur, although sometimes tRNA reads only the first two bases of a triplet (wobble), and sometimes unusual bases occur in anticodons. C denotes the stop (nonsense) codons. D is a deviation from universality of the code, as found in mitochondria.
2. **C** Bonds between a tRNA and an incorrect smaller amino acid may form but are rapidly hydrolyzed. A, B: ATP and the amino acid react to form an enzyme-bound aminoacyl-adenylate; PP_i is released into the medium. D: Some amino acids, such as hydroxyproline and hydroxylysine, arise by co- or posttranslational modification. E: An aminoacyl-tRNA synthetase may recognize any of several tRNAs specific for a given amino acid.
3. **B** This then binds the mRNA. A: Methionyl-$tRNA_i^{met}$ appears at the P site. C: Phosphorylation of eIF-2 inhibits initiation. D: Methionyl-$tRNA_c^{met}$ is used internally. E: mRNA associates first with the 40S subunit.

4. **E** fMet-$tRNA_i^{met}$ is involved in initiation of protein synthesis in prokaryotes. A–D are all required.
5. **C** This is necessary to free the A site for the next incoming tRNA. A: The incoming aminoacyl-tRNA binds to the A site. B: Peptide bond formation requires no energy source other than the aminoacyl-tRNA. D: Streptomycin inhibits formation of the prokaryotic 70S initiation complex (analogous to the eukaryotic 80S complex) and causes misreading of the genetic code when the initiation complex is already formed. E: The electron pair of the amino group carries out a nucleophilic attack on the carbonyl carbon.
6. **B** Human eRF is similar in size and shape to a tRNA. A: The stop codon is at the A site. C: Peptidyltransferase catalyzes the hydrolysis. D: eRF binds as an eRF–GTP complex. E: The subunits dissociate after termination and are recycled.
7. **E** γ-Carboxylation is of special importance in several blood clotting proteins but not in insulin formation. A: Preproinsulin is inserted into the ER. B: All proteins, except fibrous ones, have to fold into a three-dimensional structure. C: Proinsulin folds and forms disulfide bonds before the chain is cleaved. D: This is called the C-peptide.
8. **D** This is not required. A, C: Prolyl hydroxylase requires both Fe^{2+} and ascorbic acid. B: The sequence is -S-Pro-Gly-. E: Hydroxylation is a cotranslational event.
9. **D** This is only one of many functions chaperones serve. A: Many proteins spontaneously fold correctly. B: Misfolded proteins are recognized and rapidly degraded. C: Chaperones bind to hydrophobic regions of unfolded proteins. E: Chaperonins require ATP.
10. **A** This toxin catalyzes the formation of an ADP-ribosyl derivative of translocase (EF-2), which irreversibly inactivates the translocase.
11. **E** Deletion of three nucleotides is deletion of an entire codon so that a particular amino acid doesn't appear. A: The only way this would happen is if the particular three nucleotides deleted were a stop codon. B: The new protein is one amino acid shorter but not several. C: This is an example of a frameshift mutation that arises from a one- or two- nucleotide deletion. The only way this could happen with three nucleotides would be for the three to span two codons, which apparently is not the usual process. D: This arises from a point mutation.
12. **E** ATP is required in the ubiquitin activation and the protease steps but not here. B, D: These are both correct. C: Linkage to histones does not result in their degradation.
13. **A** Hydroxylation of prolines and some lysines is cotranslational so intracellular. B and C require that the chains be complete. D, E: These are involved in cross-linking, which means that the three chains must be assembled.
14. **C** Signal peptidase is located on the luminal surface of the ER. A: These are common features along with a polar segment that signal peptidase recognizes. B, D, E: These are all essential features of the process.
15. Multiple lysosomal enzymes are severely deficient in the lysosomes and are actually found in abnormally high levels in sera and other body fluids. Mannose 6-phosphate is the signal to target enzymes to lysosomes. The first step is addition of *N*-acetylglucosamine phosphate to high-mannose oligosaccharides. Subsequent cleavage, leaving the phosphate behind, produces the mannose 6-phosphate signal. Lack of lysosomal enzymes means that undegraded products accumulate in lysosomes—thus forming inclusion bodies.
16. These structures are all targeting signals to send proteins to particular subcellular particles. The three indicated are targets for mitochondria, nucleus, and peroxisomes, respectively.

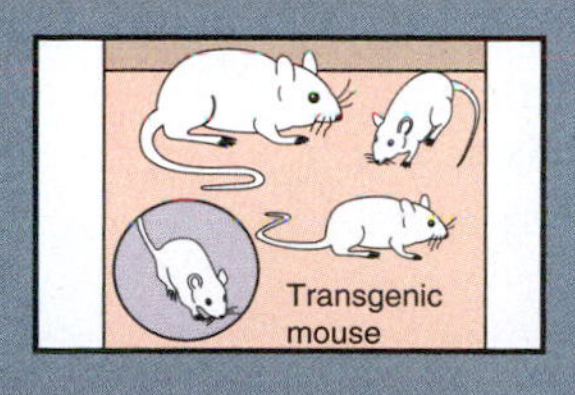

7

RECOMBINANT DNA AND BIOTECHNOLOGY

Gerald Soslau

CLINICAL CORRELATIONS

7.1 | OVERVIEW

By 1970, the stage was set for modern molecular biology based on the studies of numerous scientists in the previous 30 years during which ignorance of what biochemical entity orchestrated the replication of life forms with such fidelity gave way to a state where sequencing and manipulation of the expression of genes would be feasible. The relentless march toward a full understanding of gene regulation under normal and pathological conditions has moved with increasing rapidity since then. Deoxyribonucleic acid, composed of only four different nucleotides covalently linked by a sugar–phosphate backbone, is deceptively complex. Complexity is conferred on the DNA molecule by the nonrandom sequence of its bases, multiple conformations that exist in equilibrium in the biological environment, and specific proteins that recognize and associate with selected regions. Biochemical knowledge of the cellular processes and their macromolecular components had established several facts required for the surge forward. It was clear that gene expression was highly regulated. Enzymes involved in DNA replication and RNA transcription had been purified and their function in the biosynthesis defined. The genetic code had been broken. Genetic maps of prokaryotic chromosomes had been established based on gene linkage studies with thousands of different mutants. Finally, RNA species could be purified, enzymatically hydrolyzed into discrete pieces, and laboriously sequenced. However, further progress in the understanding of gene regulation would require techniques to cut DNA selectively into homogeneous pieces. Even small, highly purified viral DNA genomes were too complex to decipher. The thought of tackling the human genome with more than 3×10^9 base pairs was all the more onerous.

Identification, purification, and characterization of restriction endonucleases that faithfully cleave DNA molecules at specific sequences permitted the development of recombinant DNA methodologies. Development of DNA sequencing opened the previously tightly locked molecular biology gates to the secrets held within the organization of the diverse biological genomes. Genes could finally be sequenced, but perhaps more importantly so could the flanking regions that regulate their expression. Sequencing regulatory regions of numerous genes defined consensus sequences such as those found in promoters, enhancers, and many binding sites for regulatory proteins (see Chapter 8). Each gene contains an upstream promoter where a DNA-dependent RNA polymerase binds prior to the initiation of transcription. While some DNA regulatory sites lie just upstream of the transcription initiation site, other regulatory regions are hundreds to thousands of bases removed and still others are downstream.

This chapter presents many of the sophisticated techniques, developed in the past 30 years, that allow for the dissection of complex genomes into defined fragments with the complete analysis of the nucleotide sequence and function of these DNA fragments. The modification and manipulation of genes, that is, genetic engineering, facilitates the introduction and expression of genes in both prokaryotic and eukaryotic cells. Many methodological approaches in genetic engineering have been greatly simplified by the employment of a method that rapidly amplifies selected regions of DNA, the polymerase chain reaction (PCR). Proteins for experimental and clinical uses are readily produced by these procedures and it is anticipated that in the not too distant future these methods will allow for the rapid increase of treatment modalities of genetic diseases with gene replacement therapy. Current and potential uses of recombinant DNA technologies are also described. The significance to our society of the advancements in the understanding of genetic macromolecules and their manipulation cannot be overstated.

7.2 | POLYMERASE CHAIN REACTION

The ability to analyze fully the sequence and function of a selected region of DNA requires relatively large amounts of a purified preparation of the DNA segment. The rapid production of large quantities of a specific DNA sequence took a leap forward with the development of the **polymerase chain reaction (PCR).** The PCR requires two nucleotide oligomers that hybridize to the complementary DNA strands in a region of interest. The oligomers serve as primers for a DNA polymerase that extends each strand. Repeated cycling of the PCR yields large amounts of each DNA molecule of interest in a matter of hours as opposed to days and weeks required for cloning techniques.

The PCR amplification of a specific DNA sequence can be accomplished with a purified DNA sample or a small region within a complex mixture of DNA. The principles of the reaction are shown in Figure 7.1. The nucleotide sequence of the DNA to be amplified must be known or it must be cloned in a vector (see p. 305) where the sequence of the flanking DNA has been established. The product of PCR is a double-stranded DNA (dsDNA) molecule and the reaction is completed in each cycle when all of the template molecules have been copied. In order to initiate a new round of replication the sample is heated to melt the dsDNA and, in the presence of excess oligonucleotide primers, cooled to permit hybridization of the single-stranded template with free oligomers. A new cycle of DNA replication will initiate in the presence of DNA polymerase and all four deoxyribonucleoside triphosphates (dNTPs). Heating to about 90°C as required for melting DNA inactivates most DNA polymerases, but a heat stable polymerase, termed Taq DNA polymerase isolated from *Thermus aquaticus,* is now employed, obviating the need for fresh polymerase after each cycle. This has permitted the automation of PCR with each DNA molecule capable of being amplified one million-fold.

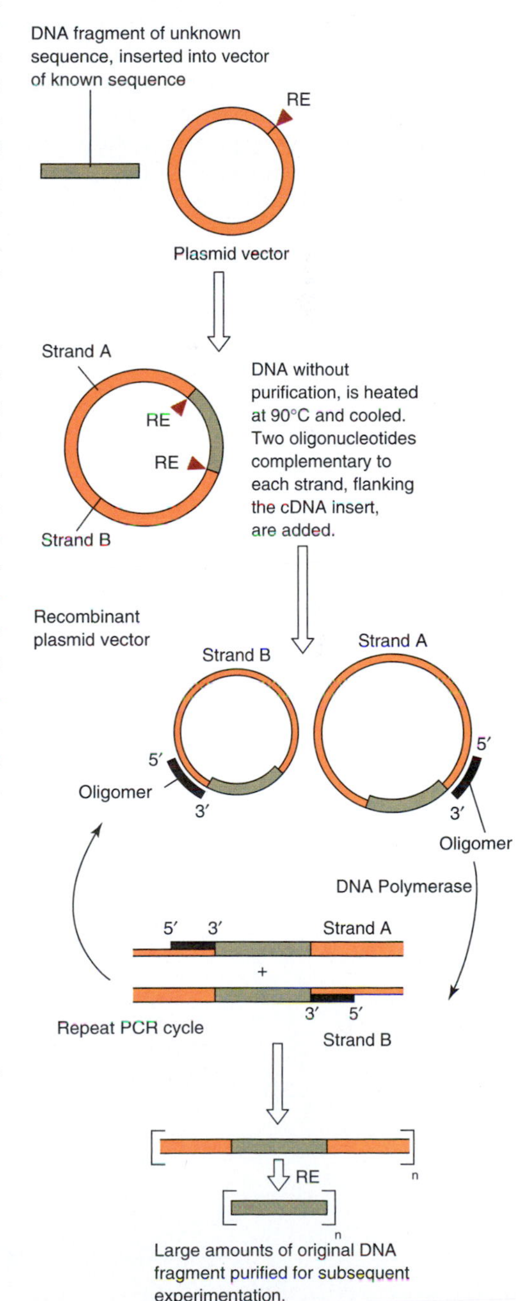

FIGURE 7.1
Polymerase chain reaction (PCR).
A DNA fragment of unknown sequence is inserted into a vector of known sequence by normal recombinant methodologies. The recombinant DNA of interest does not need to be purified. The DNA is heated to 90°C to dissociate the double strands and cooled in the presence of excess amounts of two different complementary oligomers that hybridize to the known vector DNA sequences flanking the foreign DNA insert. Only recombinant single-stranded DNA species can serve as templates for DNA replication, yielding double-stranded DNA fragments of foreign DNA bounded by the oligomer DNA sequences. The heating–replication cycle is repeated many times to rapidly produce large amounts of the original foreign DNA. The DNA fragment of interest can be purified from the polymerase chain reaction mixture by cleaving it with the original restriction endonuclease (RE), electrophoresing the DNA mixture through an agarose gel, and eluting the band of interest from the gel.

CLINICAL CORRELATION 7.1

Polymerase Chain Reaction

Polymerase Chain Reaction and Screening for Human Immunodeficiency Virus

The use of the polymerase chain reaction (PCR) to amplify minute quantities of DNA has revolutionized the ability to detect and analyze DNA species. With PCR it is possible to synthesize sufficient DNA for analysis. Conventional methods for detection and identification of the human immunodeficiency virus (HIV), such as Southern blot–DNA hybridization and antigen analysis, are labor intensive and expensive and have low sensitivity. An infected individual, with no sign of AIDS (acquired immune deficiency syndrome), may test false negative for HIV by these procedures. Early detection of HIV infections in these individuals is crucial to initiate treatment and/or monitor the progression of their disease. In addition, a sensitive method is required to be certain that blood contributed by donors does not contain HIV. PCR amplification of potential HIV DNA sequences within DNA isolated from an individual's white blood cells permits the identification of viral infections prior to appearance of antibodies, the so called seronegative state. Current methods are too costly to apply this testing to large-scale screening of donor blood samples. PCR can also be used to increase the sensitivity to detect and characterize DNA sequences of any other human infectious pathogen.

Kwok, S. and Sninsky, J. J. Application of PCR to the detection of human infectious diseases. In: H. A. Erlich (Ed.), *PCR Technology. Principles and Applications for DNA Amplification.* New York: Stockton Press, 1989, p. 235.

The Nested Polymerase Chain Reaction to Detect Microchimerism

Donor leukocytes transferred to patients receiving blood transfusions have been shown to survive in the recipient's peripheral blood. The significance, if any, of this microchimeric population of leukocytes remains to be resolved. One of the best ways to detect donor-derived cells in the recipient's blood is to use PCR detection of polymorphisms in the HLA-DR region of the major histocompatibility complex (MHC). A nested PCR assay was at least 100-fold more sensitive than a standard PCR assay. However, because of the increased sensitivity, nonspecific products may appear due to mispriming events that are generally associated with pseudogenes. As such it is essential to establish a baseline pattern with pretransfusion blood samples. Once potential false positives could be excluded with these baseline patterns, the detection of donor leukocytes is greatly enhanced.

Carter, A. S., Cerundolo, L., Bunce, M., Koo, D. D. H., Welsh, K. I., Morris, P. J., and Fuggle, S. V. Nested polymerase chain reaction with sequence-specific primers typing for HLA-A, -B, and -C alleles: detection of microchimerism in DR-matched individuals. *Blood* 94:1471, 1999.

When the DNA to be amplified is present in very low concentrations relative to the total DNA in the sample, it is possible to amplify the DNA region of interest along with other spurious sequences. In this situation the specificity of the amplification reaction can be enhanced by **nested PCR.** After conducting the first PCR with one set of primers for 10–20 cycles a small aliquot is removed for a second PCR. However, the second PCR is conducted with a new set of primers that are complementary to the template DNA just downstream of the first set of primers, or "nested" between the original set of primers. This process amplifies the DNA region of interest twice with a greatly enhanced specificity.

PCR has many applications in gene diagnosis, forensic investigations, in which only a drop of dried blood or a single hair is available, and evolutionary studies with preserved biological material. Use of PCR for screening for human immunodeficiency virus and nested PCR to detect microchimerism are presented in Clinical Correlation 7.1.

7.3 | RESTRICTION ENDONUCLEASE AND RESTRICTION MAPS

Restriction Endonucleases Selectively Hydrolyze DNA to Generate Restriction Maps

Nature possesses a diverse set of tools, the **restriction endonucleases,** capable of selectively dissecting DNA molecules of many sizes and origins into smaller fragments. These enzymes confer some protection on bacteria against invading viruses, that is, bacteriophages. Bacterial DNA sequences normally recognized by a restriction endonuclease are protected from cleavage in host cells by methylation of bases within the palindrome. The unmethylated viral DNA is recognized as foreign and is hydrolyzed. Numerous type II restriction endonucleases, differing in specificity, have been identified and purified; many are now commercially available (see p. 72 for discussion of restriction endonuclease activities).

Restriction endonucleases permit construction of a new type of genetic map, a **restriction map,** in which the site of enzyme cleavage within the DNA is identified. Purified DNA species that contain susceptible sequences are subjected to restriction endonuclease cleavage. By regulating the time of exposure of the purified DNA molecules to restriction endonuclease cleavage, a population of DNA fragments of different sizes that are partially to fully hydrolyzed can be generated. Fully hydrolyzed samples are ones where all restriction sites have been cleaved. Separation of these enzyme-generated fragments by agarose gel electrophoresis allows for the construction of restriction maps; an example of this procedure with circular DNA is presented in Figure 7.2. Analysis of a DNA completely hydrolyzed by a restriction endonuclease establishes how many sites the restriction endonuclease recognizes within the molecule and what size fragments are generated. The size distribution of composite fragments generated by the partial enzymatic cleavage of the DNA molecules demonstrates linkage of all potential fragments. The sequential use of different restriction endonucleases has permitted a detailed restriction map of numerous circular DNA species including bacterial plasmids, viruses, and mitochondrial DNA. The method is also equally amenable to linear DNA fragments that have been purified to homogeneity.

Restriction Maps Permit Routine Preparation of Defined Segments of DNA

Restriction maps may yield little information as to the genes or regulatory elements within the various DNA fragments. They have been used to demonstrate sequence

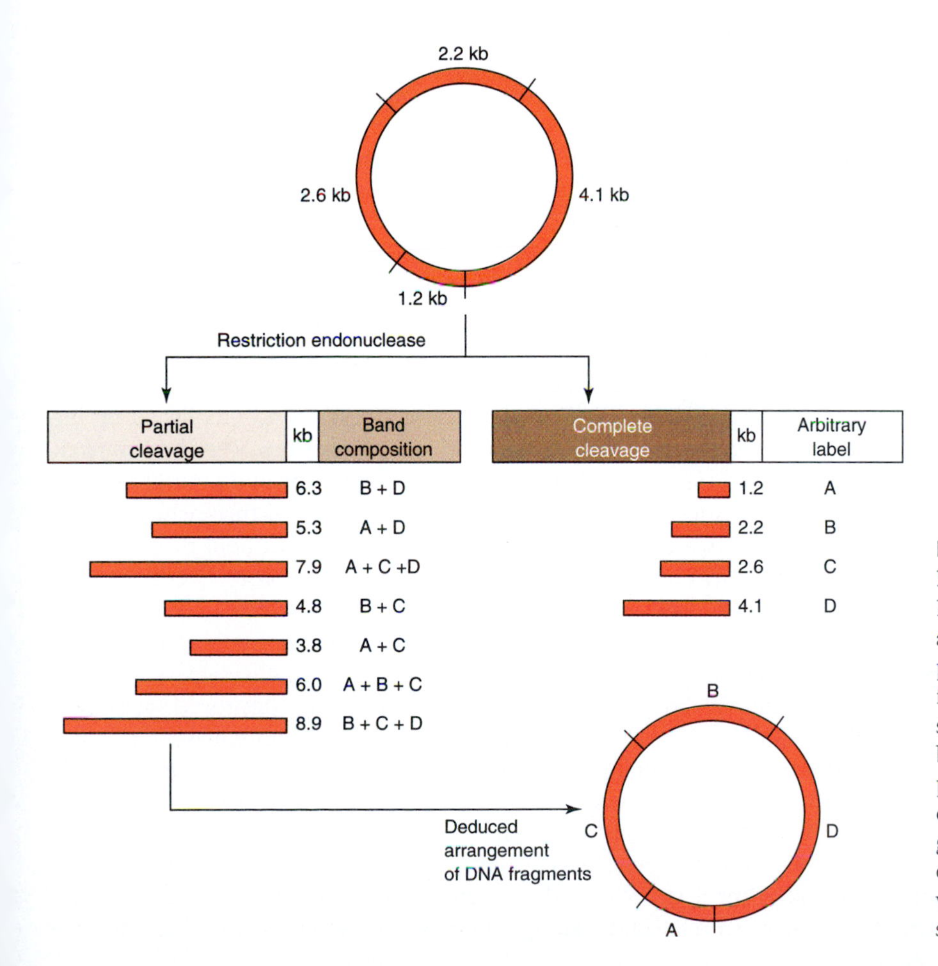

FIGURE 7.2
Restriction endonuclease mapping of DNA. Purified DNA is subjected to restriction endonuclease digestion for varying times, which generates partially to fully cleaved DNA fragments. The DNA fragments are separated by agarose gel electrophoresis and stained with ethidium bromide. The DNA bands are visualized with a UV light source and photographed. The size of the DNA fragments is determined by the relative migration through the gel as compared to coelectrophoresed DNA standards. The relative arrangement of each fragment within the DNA molecule can be deduced from the size of the incompletely hydrolyzed fragments.

CLINICAL CORRELATION 7.2

Restriction Mapping and Evolution

In the past, evolutionary studies of species have depended solely on anatomical changes observed in fossil records and on carbon dating. More recently, these studies are being supported by the molecular analysis of the sequence and size of selected genes or whole DNA molecules. Evolutionary alterations of a selected DNA molecule from different species can rapidly be assessed by restriction endonuclease mapping. Generation of restriction endonuclease maps requires a pure preparation of DNA. Mammalian mitochondria contain a covalently closed circular DNA molecule of 16,569 bp that can rapidly be purified from cells. The mitochondrial DNA (mtDNA) can be employed directly for the study of evolutionary changes in DNA without the need of cloning a specific gene.

Mitochondrial DNA has been purified from the Guinea baboon, rhesus macaque, guenon, and human and cleaved with 11 different restriction endonucleases. Restriction maps were constructed for each species. The maps were all aligned relative to the direction and the nucleotide site where DNA replication is initiated. A comparison of restriction endonuclease sites allowed for the calculation of the degree of divergence in nucleotide sequence between species. The rate of base substitution (calculated from the degree of divergence versus the time of divergence) has been about tenfold greater than changes in the nuclear genome. This high rate of mutation of the readily purified mtDNA molecule makes it an excellent model to study evolutionary relationships between species.

The analysis of mtDNA can also be used to assess the migration patterns of people who settled in diverse geographic regions but possess the same restriction-enzyme-generated patterns. Native Americans have been grouped into four major haplogroups based on their mtDNA restriction patterns. These groups appear to have migrated from Asia and Siberia some 18,000 years ago, just before the major ice age period.

Brown, W. M., George, M. Jr., and Wilson, A. C. Rapid evolution of animal mitochondrial DNA. *Proc. Natl. Acad. Sci. USA* 76:1967, 1979; Schurr, J. G. Mitochondrial DNA and the people of the new world. *Am. Sci.* 88:246, 2000.

diversity of organelle DNA, such as mitochondrial DNA, within species (see Clin. Corr. 7.2). They can also be used to detect deletion mutations where a defined DNA fragment from the parental strain migrates as a smaller fragment in the mutated strain. Most importantly, the enzymatic microscissors used to generate restriction maps cut DNA into defined homogeneous fragments that can readily be purified. These maps are crucial for cloning and for sequencing of genes and their flanking DNA regions.

7.4 | DNA SEQUENCING

To explore the complexities of regulation of gene expression and to seek the basis for genetic diseases, techniques were necessary to determine the exact sequence of bases in DNA. In the late 1970s two different sequencing techniques were developed, one by A. Maxam and W. Gilbert, the chemical cleavage approach, and the other by F. Sanger, the enzymatic approach. Both procedures employ labeling of a terminal nucleotide, followed by the separation and detection of generated oligonucleotides.

Chemical Cleavage Method: Maxam–Gilbert Procedure

This procedure requires (1) labeling of the terminal nucleotide, (2) selective hydrolysis of the phosphodiester bond for each nucleotide separately to produce fragments with 1, 2, 3, or more bases, (3) quantitative separation of the hydrolyzed fragments, and (4) a qualitative determination of the label added in step 1. The following describes one approach of the **Maxam–Gilbert procedure.** The overall approach is presented in Figure 7.3.

One end of each strand of DNA is selectively radiolabeled with ^{32}P. This is accomplished when a purified double-helical DNA fragment contains restriction endonuclease sites on either side of the region to be sequenced. Hydrolysis of the DNA with two different restriction endonucleases then results in different staggered ends, each with a different base in the first position of the single-stranded region. Labeling of the 3′-end of each strand is accomplished with the addition of the next nucleotide as directed by the corresponding base sequence on the complementary DNA strand. A fragment of *Escherichia coli* DNA polymerase I, termed the **Klenow fragment,** catalyzes this reaction. The Klenow fragment, produced by partial proteolysis of the polymerase holoenzyme, lacks $5' \rightarrow 3'$ exonuclease activity but retains the $3' \rightarrow 5'$ exonuclease and polymerase activities. Each strand can therefore be selectively labeled in separate experiments. The complementary unlabeled strand will not be detectable when analyzing the sequence of the labeled strand.

Hydrolysis of the labeled DNA into different lengths is accomplished by first selectively destroying one or two bases of the four nucleotides. The procedure exposes the phosphodiester bond that connects adjoining bases and permits selective cleavage of the DNA at the altered base. In separate chemical treatments, samples of labeled DNA are treated to alter purines and pyrimidines without disrupting the sugar–phosphate backbone; a method is not currently available to alter adenine or thymine specifically. Conditions for base modification are selected such that only one or a few bases are destroyed randomly within any one molecule. The four separate DNA samples are then reacted with **piperidine,** which chemically breaks the sugar–phosphate backbone at sites where a base has been destroyed, generating fragments of different sizes. Since labeling is specific at the end while the chemical alteration of the base is random and not complete, some of the fragments will be end labeled. For example, wherever a cytosine residue had been randomly destroyed in the appropriate reaction tube, a break will be introduced into the DNA fragment. The series of chemically generated, end-labeled DNA fragments from each of the four tubes are electrophoresed through a polyacrylamide gel. Bases destroyed near the end-labeled nucleotide will generate fragments that migrate faster through

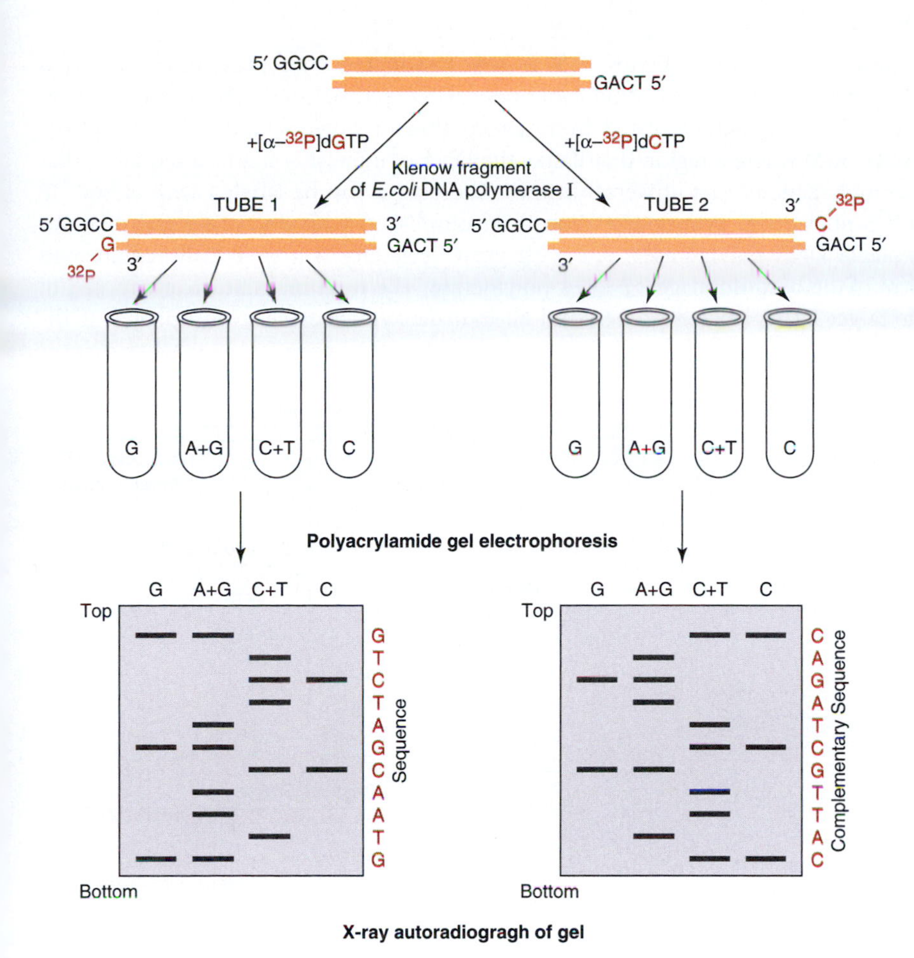

FIGURE 7.3
Maxam–Gilbert chemical method to sequence DNA.
A double-stranded DNA fragment to be sequenced is obtained by restriction endonuclease cleavage and purified. Both strands are sequenced by selectively labeling the ends of each DNA strand. One strand of DNA is end-labeled with [^{32}P]-dGTP in reaction tube 1, while the other is end-labeled with [^{32}P]-dCTP in reaction tube 2. The end-labeled DNA is then subdivided into four fractions where the different bases are destroyed chemically at random positions within the single-stranded DNA molecule. The less selective chemical destruction of adenine simultaneously destroys G and the destruction of thymine destroys the C bases. The single-stranded DNA is cleaved at the sites of the destroyed bases. This generates end-labeled fragments of all possible lengths corresponding to the distance from the end to the sites of base destruction. Labeled DNA fragments are separated according to size by electrophoresis. The DNA sequence can then be determined from the electrophoretic patterns detected on autoradiograms.

the gel, as low molecular weight species, while fragments derived from bases destroyed further from the end will migrate through the gel more slowly as higher molecular weight molecules. The gel is then exposed to X-ray film, which detects the ^{32}P and the radioactively labeled bands within the gel can be visualized. The sequence can be read manually or directly by automated methods from the X-ray autoradiograph from the bottom (smaller fragments) to the top of the film (larger fragments). Sequencing the complementary strand checks the correctness of the sequence.

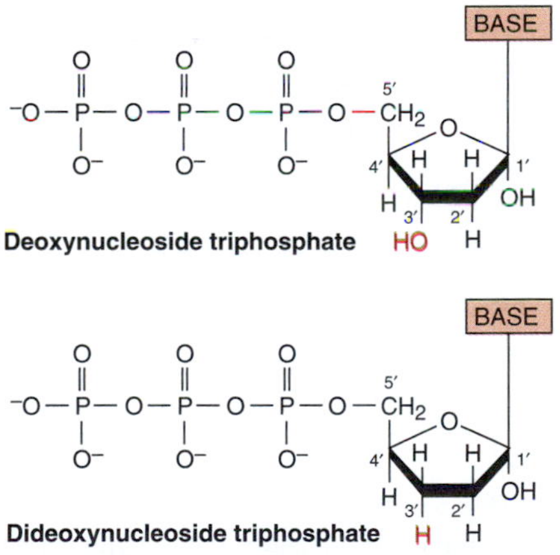

Interrupted Enzymatic Cleavage Method: Sanger Procedure

The **Sanger procedure** of DNA sequencing is based on the random termination of a DNA chain during enzymatic synthesis. The technique is possible because the dideoxynucleotide analog of each of the four normal nucleotides (Figure 7.4) can be incorporated into a growing DNA chain by DNA polymerase. The ribose of the **dideoxynucleotide triphosphate (ddNTP)** has the OH group at both the 2′- and 3′-positions replaced with a proton, whereas dNTP has only a single OH group replaced by a proton at the 2′-position. Thus the ddNTP incorporated into the growing chain is unable to form a phosphodiester bond with another dNTP because the 3′-position of the ribose does not contain a OH group. The growing DNA molecule can be terminated at random points, from the first nucleotide incorporated to the last, by including in the reaction system both the normal nucleotide and the ddNTP (e.g., dATP and ddATP) at concentrations such that the two nucleotides compete for incorporation.

Identification of DNA fragments requires labeling of the 5′-end of the DNA molecules or the incorporation of labeled nucleotides during synthesis. The

BASE

Deoxynucleoside triphosphate

BASE

Dideoxynucleoside triphosphate

FIGURE 7.4
Structure of deoxynucleoside triphosphate and dideoxynucleoside triphosphate.
The 3′-OH group is lacking on the ribose component of the dideoxynucleoside triphosphate (ddNTP). This molecule can be incorporated into a growing DNA molecule through a phosphodiester bond with its 5′-phosphates. Once incorporated the ddNTP blocks further synthesis of the DNA molecule since it lacks the 3′-OH acceptor group for an incoming nucleotide.

technique, outlined in Figure 7.5, is best conducted with pure single-stranded DNA; however, denatured double-stranded DNA can be used. The DNA to be sequenced is frequently isolated from a recombinant single-stranded bacteriophage (see p. 305) where a region that flanks the DNA of interest contains a sequence that is complementary to a universal primer. The primer can be labeled with either ^{32}P or ^{35}S nucleotide. Primer extension is accomplished with one of several different available DNA polymerases; one with great versatility is a genetically engineered form of the bacteriophage T7 DNA polymerase. The reaction mixture, composed of the target DNA, labeled primer, and all four deoxynucleoside triphosphates, is di-

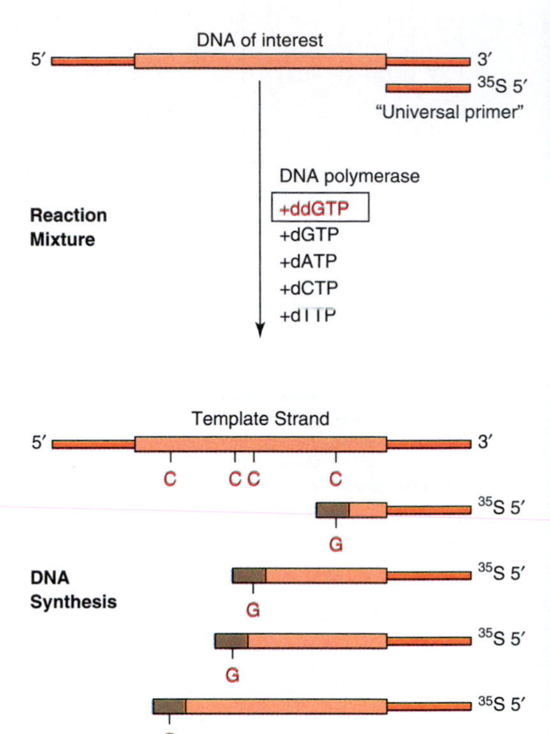

(a) Recombinant M13 bacteriophage

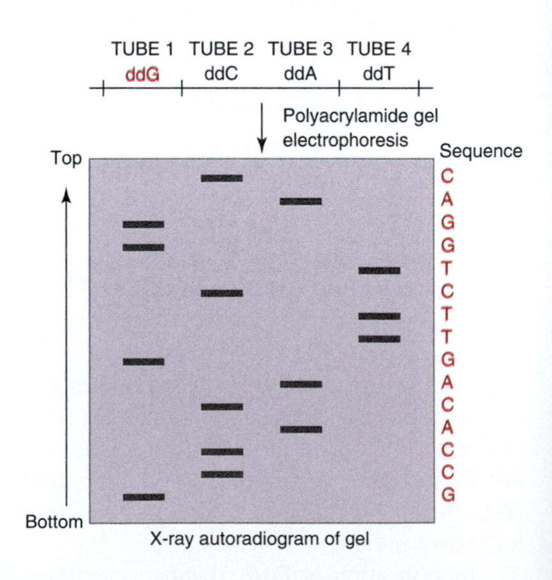

(b) Polyacrylamide gel electrophoresis of reaction mixture

FIGURE 7.5
Sanger dideoxynucleoside triphosphate method to sequence DNA.
The DNA region of interest is inserted into a bacteriophage DNA molecule. Replicating bacteriophage produce a single-stranded recombinant DNA molecule that is readily purified. The known sequence of the bacteriophage DNA downstream of the DNA insert serves as a hybridization site for an end-labeled oligomer with a complementary sequence, a universal primer. Extension of this primer is catalyzed with a DNA polymerase in the presence of all four deoxynucleoside triphosphates plus one dideoxynucleoside triphosphate, for example, ddGTP. Synthesis stops whenever a dideoxynucleoside triphosphate is incorporated into the growing molecule. Note that the dideoxynucleotide competes for incorporation with the deoxynucleotide. This generates end-labeled DNA fragments of all possible lengths that are separated by electrophoresis. The DNA sequence can then be determined from the electrophoretic patterns.

CLINICAL CORRELATION 7.3

Direct Sequencing of DNA for Diagnosis of Genetic Disorders

The X-linked recessive hemorrhagic disorder, hemophilia B, is caused by a coagulation factor IX deficiency. The factor IX gene has been cloned and sequenced and contains 8 exons spanning 34 kb that encode a glycoprotein secreted by the liver. Over 300 mutations of the gene have been discovered of which about 85% are single base substitutions and the rest are complete or partial gene deletions. Several methods have been employed to identify carriers of a defective gene copy and for prenatal diagnoses. Unfortunately, these methods were costly, time consuming, and all too often inaccurate. Direct sequencing of PCR-amplified genomic DNA has been employed to circumvent these diagnostic shortcomings. Between 0.1 and 1 μg of genomic DNA can readily be isolated from patient blood samples and each factor IX exon can be PCR amplified with appropriate primers. Amplified DNA can be used for direct sequencing to determine if a mutation in the gene exists that would be diagnostic of one of the forms of hemophilia B. For example, a patient with a moderate hemophilia B (London 6) had an A to G transition at position 10442 that led to a substitution of Asp64 by Gly.

Green, P. M., Bentley, D. R., Mibashan, R. S., Nilsson, I. M., and Gianelli, F. Molecular pathology of hemophilia B. *EMBO J.* 8:1067, 1989.

vided into four tubes, each containing a different dideoxynucleoside triphosphate. ddNTPs are randomly incorporated during the enzymatic synthesis of DNA and cause chain termination. Since the ddNTP is present in the reaction tube at a low level, relative to the corresponding dNTP, termination of DNA synthesis occurs randomly at all possible complementary sites to the DNA template. This yields DNA molecules of varying sizes, labeled at the 5′-end, that can be separated by polyacrylamide gel electrophoresis. The labeled species are detected by X-ray autoradiography and the sequence is read.

Initially, this method required a single-stranded DNA template, production of a specific complementary oligonucleotide primer, and the need for a relatively pure preparation of the Klenow fragment of *E. coli* DNA polymerase I. These difficulties have been overcome and modifications have simplified the approach. The Sanger method can rapidly sequence as many as 400 bases while the Maxam–Gilbert method is limited to about 250 bases.

The PCR and the Sanger methods can be combined for **direct sequencing** of small DNA regions of interest. The double-stranded PCR product is employed directly as template. Conditions are set such that one strand of melted DNA (template) anneals with the primer in preference to reannealing of template with its complementary strand. Sequencing then follows the standard dideoxy chain termination reaction (typically with Sequenase in lieu of the Klenow polymerase) with synthesis of random length chains occurring as extensions of the PCR primer. This method has been successfully employed for diagnosis of genetic disorders (see Clin. Corr. 7.3).

7.5 | RECOMBINANT DNA AND CLONING

DNA from Different Sources Can Be Ligated to Form a New DNA Species: Recombinant DNA

The ability to hydrolyze a population of DNA molecules selectively with a battery of restriction endonucleases led to the technique for joining two different DNA molecules termed **recombinant DNA.** This technique combined with techniques for replication, separation, and identification permits the production of large quantities of purified DNA fragments. The combined techniques, referred to as recombinant DNA technologies, allow removal of a piece of DNA out of a larger complex molecule, such as the genome of a virus or human, and amplification of the DNA fragment. Recombinant DNAs have been prepared that combine DNA fragments from bacteria with fragments from humans, viruses with viruses, and so on. The ability to join two different pieces of DNA together at specific sites within the molecules is achieved with two enzymes, a restriction endonuclease and a DNA ligase. There

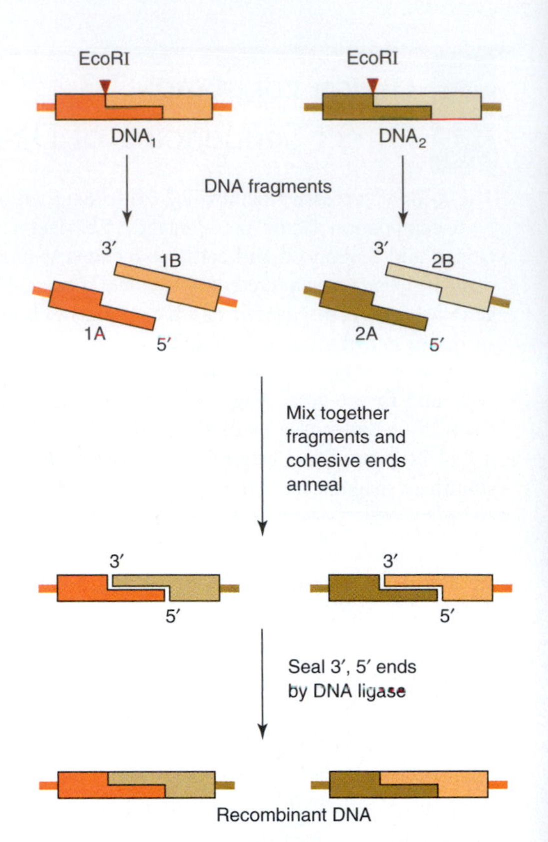

FIGURE 7.6
Formation of recombinant DNA from restriction-endonuclease-generated fragments that contain cohesive ends.
Many restriction endonucleases hydrolyze DNA in a staggered fashion, yielding fragments with single-stranded regions at their 5′- and 3′-ends. DNA fragments generated from different molecules with the same restriction endonuclease have complementary single-stranded ends that can be annealed and covalently linked together with a DNA ligase. All different combinations are possible in a mixture. When two DNA fragments of different origin combine, a recombinant DNA molecule results.

are a number of different restriction endonucleases, varying in their nucleotide sequence specificity, that can be used (Section 7.3). Some hydrolyze the two strands of DNA in a staggered fashion to produce **"sticky or cohesive" ends** (Figure 7.6) while others cut both strands symmetrically to produce a **blunt end.** A specific restriction enzyme cuts DNA at exactly the same sequence site regardless of source of DNA (bacteria, plant, mammal, etc.). A DNA molecule may have one, several, hundreds, thousands, or no recognition sites for a particular restriction endonuclease. The staggered cut results in DNA fragments with single-stranded ends. When different DNA fragments generated by the same restriction endonuclease are mixed, their single-stranded ends can hybridize, that is, anneal together. In the presence of DNA ligase the two fragments are connected covalently, producing a recombinant DNA molecule.

The DNA fragments produced from a restriction endonuclease that form blunt ends can also be ligated but with much lower efficiency. The efficiency can be increased by adding enzymatically a poly(dA) tail to one species of DNA and a poly(dT) tail to the ends of the second species of DNA. The fragments with complementary tails can be annealed and ligated in the same manner as fragments with single-stranded ends.

Recombinant DNA Vectors Are Produced in Significant Quantities by Cloning

Synthesis of a recombinant DNA opens the way for production of significant quantities of interesting DNA fragments. By incorporating a recombinant DNA into a cell that allows replication of recombinant DNA, amplification of DNA of interest can be achieved. A carrier DNA, termed a **cloning vector,** is employed. Bacterial plasmids are ideally suited as recombinant DNA vectors.

Many bacteria contain a single circular chromosome of approximately 4×10^6 base pairs (bp) and minicircular DNA molecules called **plasmids.** Plasmids are usually composed of only a few thousand base pairs and are rarely associated with the

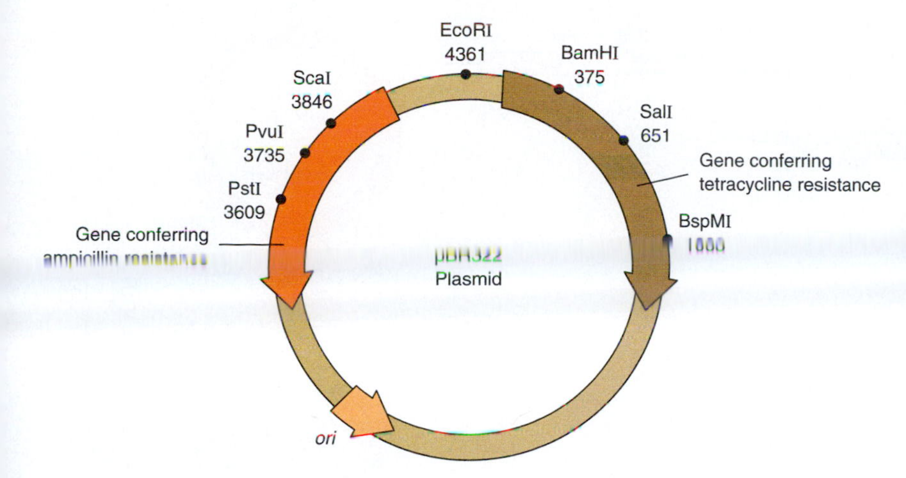

FIGURE 7.7
pBR322 Plasmid constructed in the laboratory to contain features that facilitate cloning of foreign DNA fragments
By convention the numbering of the nucleotides begins with the first T in the unique EcoRI recognition sequence (GAATTC) and the positions on the map refer to the 5′-base of the various restriction endonuclease-recognition sequences. Only a few of the unique restriction sites within the antibiotic resistance genes and none of the numerous sites where an enzyme cuts more than once within the plasmid are shown.

large chromosomal molecule. Genes within a plasmid have various functions that include the ability to confer antibiotic resistance to the bacterium, an attribute useful in selecting specific colonies of the bacteria. Plasmids replicate independently of the replication of the main bacterial chromosome. One type of plasmid, the **relaxed-control** plasmids, may be present in tens to hundreds of copies per bacterium, and its replication depends solely on host enzymes that have long half-lives. Therefore replication of **"relaxed" plasmids** can occur in the presence of a protein synthesis inhibitor. Bacteria can accumulate several thousand plasmid copies per cell under these conditions. Other types are subject to **stringent control** and their replication depends on continued synthesis of plasmid-encoded proteins. These plasmids replicate at about the same rate as the large bacterial chromosome, and only a low number of copies occur per cell. The relaxed plasmid type is routinely used for production of recombinant DNA.

The first practical recombinant DNA that could be cloned involved the ligation of foreign DNA with the vector, *E. coli* **plasmid pSC101,** which contains a single EcoRI restriction endonuclease site and a gene that encodes for a protein that confers antibiotic resistance to the bacteria. This plasmid contains an origin of replication and associated DNA regulatory sequences that are together referred to as a **replicon.** This vector suffers from several limiting factors. The single restriction endonuclease site limits the DNA fragments that can be cloned to those generated by EcoRI and the one antibiotic resistance selectable marker reduces the convenience in selection; in addition it replicates poorly.

Plasmid vectors with broad versatility have been constructed by recombinant DNA technology. The desirable features of a plasmid vector include a relatively low molecular weight (3–5 kb) to accommodate larger fragments, several different restriction sites useful in cloning a variety of fragments, multiple selectable markers to aid in selecting bacteria with recombinant DNA molecules, and a high rate of replication. The first plasmid constructed (Figure 7.7) to satisfy these requirements was **pBR322** and this plasmid has been used for the subsequent generation of many newer vectors in use today. Most vectors in use currently contain an inserted DNA sequence termed a **polylinker, restriction site bank** or **polycloning site,** which contains numerous restriction endonuclease sites unique to the plasmid.

Directional Cloning: DNA Inserted into Vector DNA in a Specific Direction

Directional cloning reduces the number of variable "recombinants" and enhances the probability of selection of the desired recombinant. Insertion of foreign DNA, with a defined polarity, into a plasmid vector without the plasmid resealing itself can be accomplished by use of two restriction endonuclease to cleave the plasmids

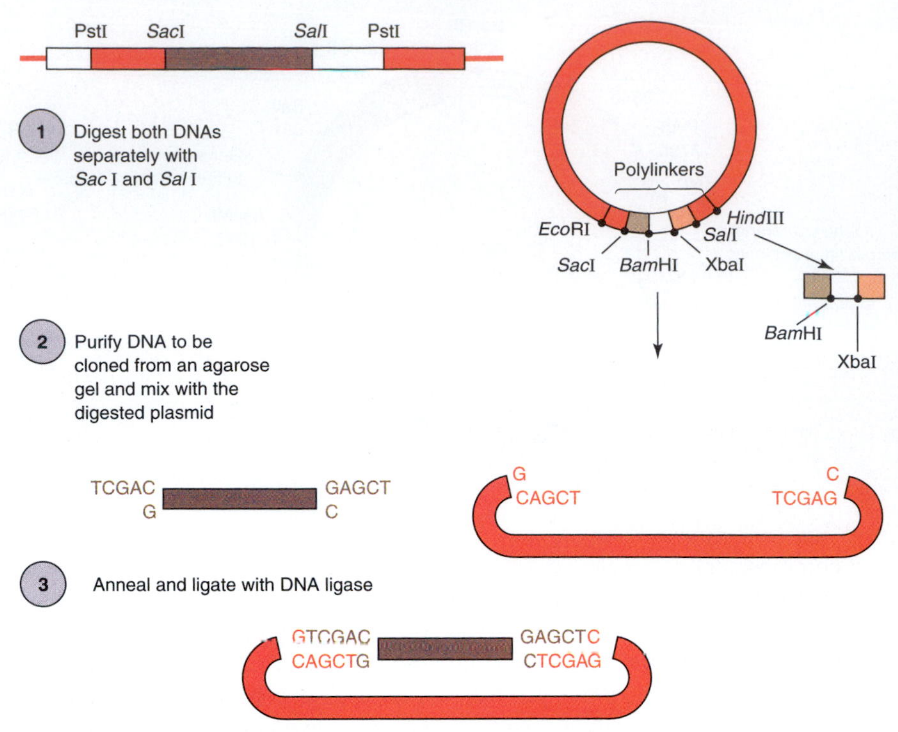

FIGURE 7.8
Directional cloning of foreign DNA into vectors with a specified orientation. Insertion of a foreign DNA fragment into a vector with a specified orientation requires two different annealing sequences at each end of the fragment and the corresponding complementary sequence at the two ends generated in the vector. A polylinker with numerous unique restriction endonuclease sites within the vector facilitates directional cloning. Knowledge of the restriction map for the DNA of interest allows for the selection of appropriate restriction endonucleases to generate specific DNA fragments that can be cloned in a vector.

(Figure 7.8); vectors with polylinkers are ideally suited for this purpose. The use of two enzymes yields DNA fragments and linearized plasmids with different "sticky" ends. Under these conditions the plasmid is unable to reanneal with itself. In addition, the foreign DNA can be inserted into the vector in only one orientation. This is extremely important when one clones a potentially functional gene downstream from the promoter-regulatory elements in expression vectors (see p. 305).

Bacteria Transformed with Recombinant DNA

The artificial introduction of DNA into bacteria is referred to as **transformation.** It is accomplished by briefly exposing the cells to divalent cations that make them transiently permeable to small DNA molecules. Recombinant plasmid molecules containing foreign DNA can be introduced into bacteria where they would replicate normally.

Selection of Transformed Bacteria

Once a plasmid has been introduced into a bacterium, both can replicate. Methods are available to select those bacteria that carry the recombinant DNA molecules. In the recombinant process some bacteria may not be transformed or may be transformed with a vector not carrying foreign DNA; in preparation some vectors may reanneal without inclusion of the foreign DNA. In some experimental conditions DNA fragments can be generated that are readily purified. Such fragments can be generated from small, highly purified DNA species, for example, some DNA viruses. More typically, however, a single restriction endonuclease will generate hundreds to hundreds of thousands of DNA fragments depending on the size and complexity of DNA being studied. Individual fragments cannot be isolated from these samples to be individually incorporated into the plasmid. Methods have therefore been developed to select bacteria that contain the desired DNA.

Restriction endonucleases do not necessarily hydrolyze DNA into fragments that contain intact genes. If a fragment contains an entire gene it may not contain the required flanking regulatory sequences, such as the promoter region. If the foreign gene is of mammalian origin, its regulatory sequences would not be recognized by the bacterial synthetic machinery. The primary gene transcript (pre-mRNA) may contain introns that cannot be processed by the bacteria.

Recombinant DNA Molecules in a Gene Library

When a complex mixture of thousands of different genes, located on different chromosomes, as in the human genome, is subjected to cleavage by a single restriction endonuclease, thousands of DNA fragments are generated. These DNA fragments are annealed with a plasmid vector that has been cleaved to a linear molecule by the same restriction endonuclease. By adjusting the ratio of plasmid to foreign DNA the probability of joining at least one copy of each DNA fragment within a cyclized recombinant-plasmid DNA approaches one. Usually, only one of the many DNA fragments is inserted into each plasmid vector. Bacteria are transformed with the recombinant molecules such that only one plasmid is taken up by a single bacterium. Each recombinant molecule can be replicated within the bacterium, which produces progeny, each carrying multiple copies of the recombinant DNA. The total bacterial population will contain fragments of DNA that may represent the entire human genome. This is termed a **gene library.** As in any library containing thousands of volumes, a selection system must be available to retrieve the book or gene of interest.

Plasmids are commonly employed to clone DNA fragments generated from DNA of limited size and complexity, such as viruses, and to subclone large DNA fragments previously cloned in other vectors. Genomic DNA fragments are usually cloned from vectors capable of carrying larger foreign DNA fragments than plasmids (see p. 307).

PCR Circumvents the Need to Clone DNA

Cloning and amplification of a DNA fragment can be employed in subcloning, mutagenesis, and sequencing studies. The PCR has, in many instances, replaced the need to amplify recombinant DNA in a replicating biological system, greatly reducing the time and preparative steps required. It is not necessary to know the sequence of the DNA insert (up to 6 kb) to amplify it by the PCR since the sequence of the vector DNA flanking the insert is known.

PCR may completely circumvent the need to clone the DNA of interest. For instance, a gene previously cloned and sequenced can readily be analyzed in DNA of a patient for detection of mutations within this gene by a **multiplex PCR strategy.** DNA is isolated from a patient's blood cells and multiple pairs of oligonucleotide primers are synthesized to amplify the entire gene or selected regions within the gene (Figure 7.9). Analysis of the amplified DNA fragments by agarose gel electrophoresis allows detection of any potential deletion mutation as compared to the normal gene products. Direct sequencing of multiple PCR products can be employed to detect point mutations in the patient's gene. Multiplex PCR has been

Step 1

Amplification of the DNA region of interest by the PCR

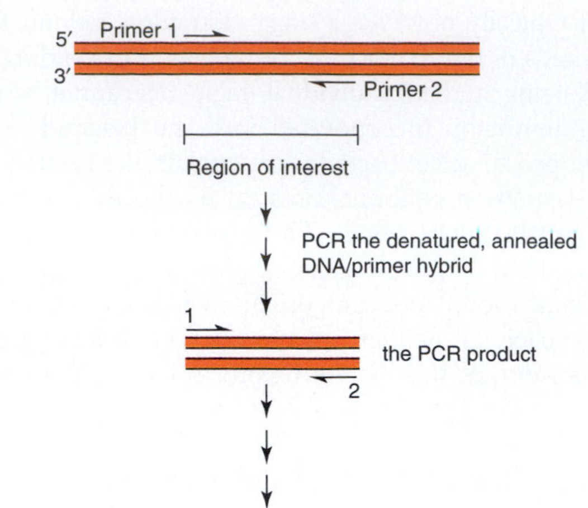

Step 2a

Employ multiple primers (1′ – 6′) simultaneously with the isolated PCR product to yield secondary PCR products of varying sizes

1′ 3′ 5′
a b c
2′ 4′ 6′
Region a Region b Region c
PCR products a,b,c

Step 2b

Repeat steps 1 and 2a with products from patient material containing a potentially deleted region b

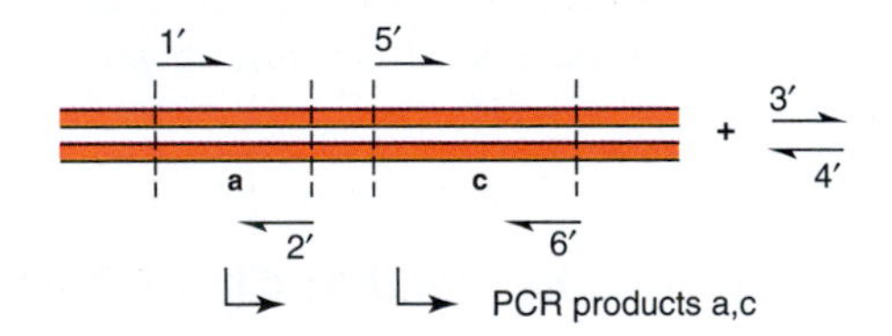

Step 3

Separate and detect the PCR products from control and patient samples by agarose gel electrophoresis and ethidium bromide staining to demonstrate the deletion of region b in the patient DNA

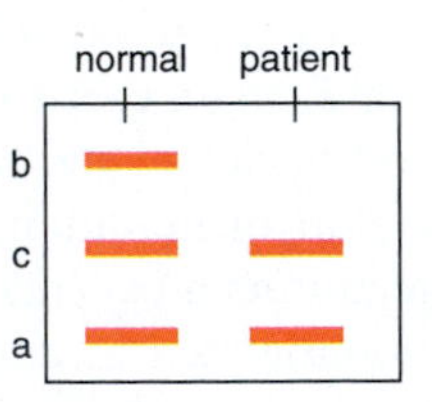

FIGURE 7.9

Multiplex PCR strategy to analyze a DNA region of interest for mutated alterations.

A region of DNA within a complex DNA molecule, derived from any source, can be amplified by the PCR with specific primers that are complementary to sequences flanking the DNA region of interest (Step 1). After multiple PCR cycles the amplified DNA (PCR product) can then be used as a template simultaneously for multiple pairs of primers (Step 2a) that are complementary throughout the DNA (here they cover three segments—a, b, and c). This procedure requires prior knowledge of the sequence of the normal DNA/gene. Step 2a is repeated for DNA derived from a patient with potential mutation(s) in the DNA region of interest (Step 2b). The amplified DNA products from the multiplex PCR step (Steps 2a and 2b) are then analyzed by agarose gel electrophoresis to ascertain if the patient sample contains a mutation (Step 3).

CLINICAL CORRELATION 7.4

Multiplex PCR Analysis of the *HGPTase* Gene Defects in Lesch–Nyhan Syndrome

Lesch–Nyhan syndrome, as described in Clinical Correlation 19.2 results from a deficiency in hypoxanthine–guanine phosphoribosyltransferase (HGPRTase) activity. Several variant forms of HGPRTase defects have been detected. Multiplex PCR amplification of the *HGPRT* gene locus has been employed to analyze this gene in cells derived from Lesch–Nyhan patients and results account for the variability of the HGPRTase. The gene, comprised of 9 exons, can be multiplex amplified using 16 different primers in a single PCR reaction. The products can be separated by agarose gel electrophoresis. Analysis of the *HGPRT* gene locus by multiplex amplification of DNA derived from cells of several patients detected great variations in deletions of different exons to total absence of the exons.

Rossiter, B. J. F., et al. In: M. J. McPherson, P. Quirke, and G. R. Taylor (Eds.), *PCR. A Practical Approach,* Vol. 1. Oxford, England: Oxford University Press, 1994, p. 67.

used to detect defects in the *HGPRTase* gene in Lesch–Nyhan patients (see Clin. Corr. 7.4).

7.6 | SELECTION OF SPECIFIC CLONED DNA IN LIBRARIES

Selection of Transformed Bacteria by Loss of Antibiotic Resistance

When a single transformed bacterium carrying a recombinant DNA multiplies, its progeny are all genetically the same and all will carry copies of the same recombinant plasmid. The foreign DNA has been amplified and is derived from a single cloned DNA fragment. The problem is how to identify the colony that contains the desired plasmid in a field of thousands to millions of different bacterial colonies. The plasmid construct **pBR322** and its descendants carry two genes that confer antibiotic resistance. Within these antibiotic-resistant genes are DNA sequences sensitive to restriction endonuclease. When a fragment of foreign DNA is inserted into a restriction site within a gene for antibiotic resistance, the gene becomes nonfunctional. Bacteria carrying this recombinant plasmid are sensitive to the antibiotic (Figure 7.10). The second antibiotic resistance gene within the plasmid, however, remains intact and the bacteria will be resistant to this antibiotic. This technique of **insertional inactivation** of plasmid gene products permits selection of bacteria that carry recombinant plasmids.

pBR322 contains genes that confer resistance to **ampicillin**(amp^r) and **tetracycline**(tet^r). A gene library with cellular DNA fragments inserted within the tet^r gene can be selected and screened in two stages (Figure 7.10). First, the bacteria are grown in an ampicillin-containing growth medium. Bacteria that are not transformed by a plasmid (they lack a normal or recombinant plasmid) during the construction of the gene library will not grow in the presence of the antibiotic and this population of bacteria will be eliminated. This, however, does not indicate which of the remaining viable bacteria carry a recombinant plasmid vector rather than a plasmid with no DNA insert. The second step is to identify bacteria carrying recombinant vectors with nonfunctional tet^r genes, which are therefore sensitive to tetracycline.

Bacteria insensitive to ampicillin are plated and grown on agar plates containing ampicillin (Figure 7.10). Replica plates can be made by touching the colonies on the original agar plate with a filter and then touching additional sterile plates with the filter. All the plates will contain portions of each original colony at identifiable positions on the plates. The replica plate can contain tetracycline, which does not support the growth of bacteria harboring recombinant plasmids with their tet^r gene disrupted. Comparison of replica plates with and without tetracycline indicates which colonies on the original ampicillin plate contain recombinant plasmids. Thus individual colonies containing the recombinant DNA can be selected, cultured, and analyzed.

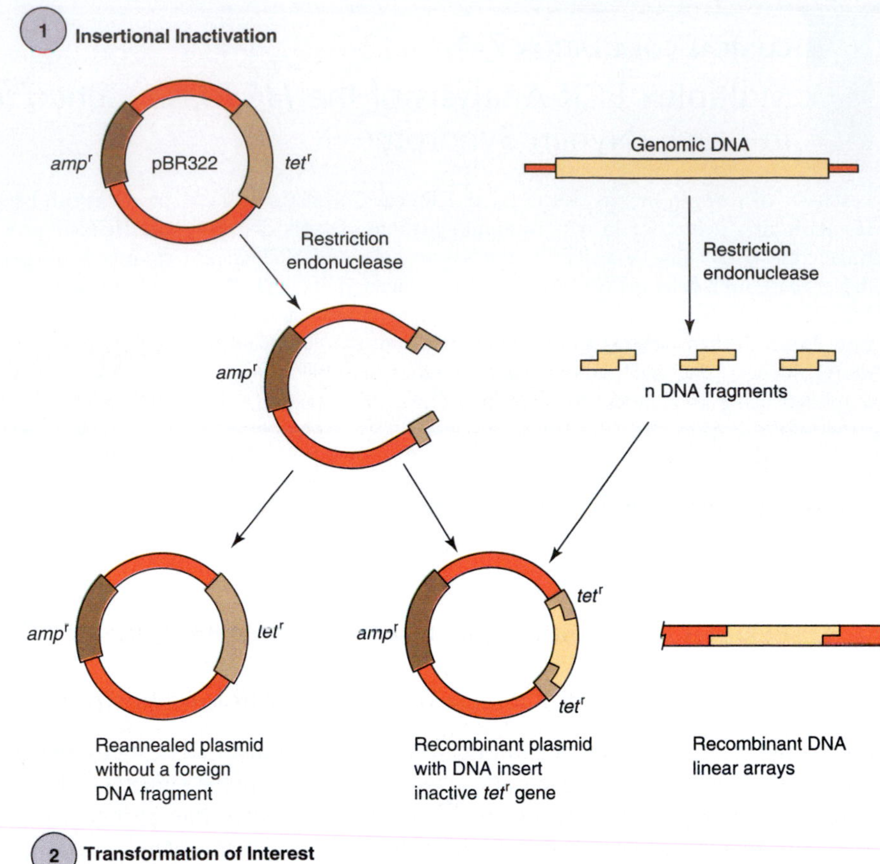

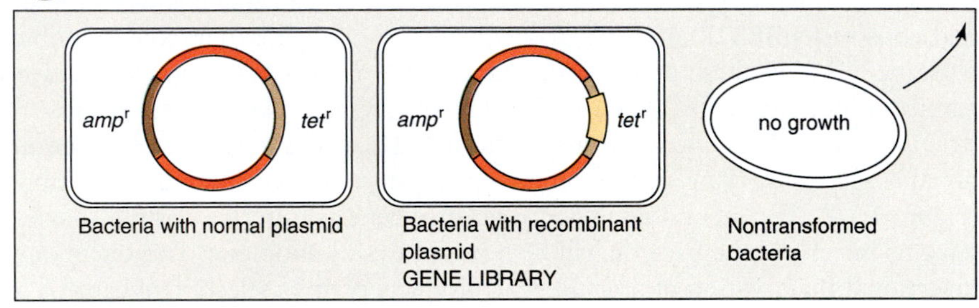

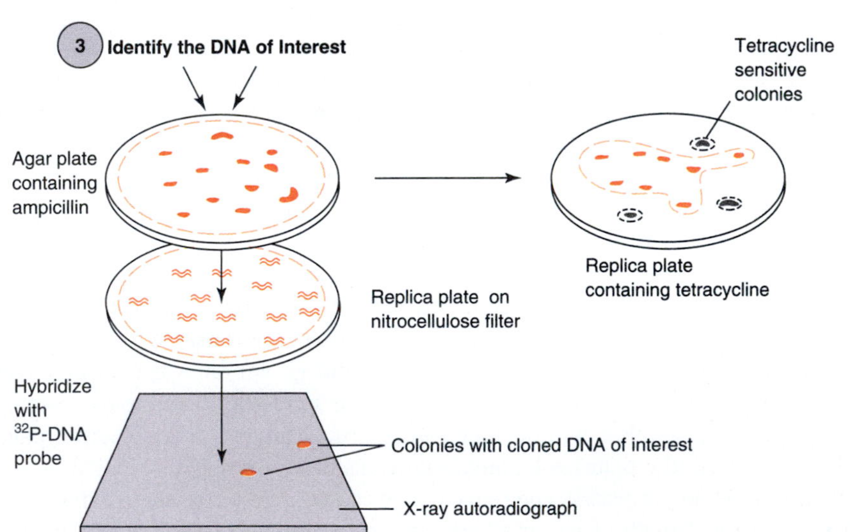

FIGURE 7.10
Insertional inactivation of recombinant plasmids and detection of transformed bacteria carrying a cloned DNA of interest.
When the insertion of a foreign DNA fragment into a vector disrupts a functional gene sequence the resulting recombinant DNA does not express the gene. The gene that codes for antibiotic resistance to tetracycline (*tet*^{r}) is destroyed by DNA insertion while the ampicillin resistance gene (*amp*^{r}) remains functional. Destruction of one antibiotic resistance gene and the retention of a second antibiotic resistance gene allows for the detection of bacterial colonies carrying the foreign DNA of interest within the replicating recombinant vector.

Either DNA or RNA probes (see pp. 44 and 296) can be utilized to identify the DNA of interest. Ampicillin-resistant bacterial colonies on agar can be replica plated onto a nitrocellulose filter and adhering cells from each colony can be lysed with NaOH (Figure 7.10). DNA within the lysed bacteria is also denatured by the NaOH and becomes firmly bound to the filter. A labeled DNA or RNA probe that is complementary to the DNA of interest can be hybridized to the nitrocellulose-bound DNA. The filter is tested by X-ray autoradiography. Any colony that carries the cloned DNA of interest will appear as a developed signal on the X-ray film. These spots indicate the colonies on the original agar plate that can then be grown in a large-scale culture for further manipulation.

Cloned and amplified DNA fragments usually do not contain a complete gene and are not expressed. The DNA inserts, however, can readily be purified for sequencing or used as probes to detect genes within a mixture of genomic DNA, to assay transcription levels of mRNA, and to detect disease-producing mutations.

α-Complementation for Selecting Bacteria Carrying Recombinant Plasmids

Other selection techniques can identify bacteria carrying recombinant DNA molecules. Vectors have been constructed (the pUC series) such that selected bacteria transformed with these vectors, carrying foreign DNA inserts, can be identified visually (Figure 7.11). The pUC plasmids contain the regulatory sequences and a part of the 5′-end coding sequence (N-terminal 146 amino acids) for the β-galactosidase gene (*lacZ* gene) of the *lac* operon (see p. 332). The translated N-terminal fragment of β-galactosidase, containing 146 amino acids, is an inactive polypeptide. Mutant *E. coli* that code for the missing inactive carboxy-terminal portion of β-galactosidase can be transformed with the pUC plasmids. The translation of the host cell and plasmid portions of the β-galactosidase in response

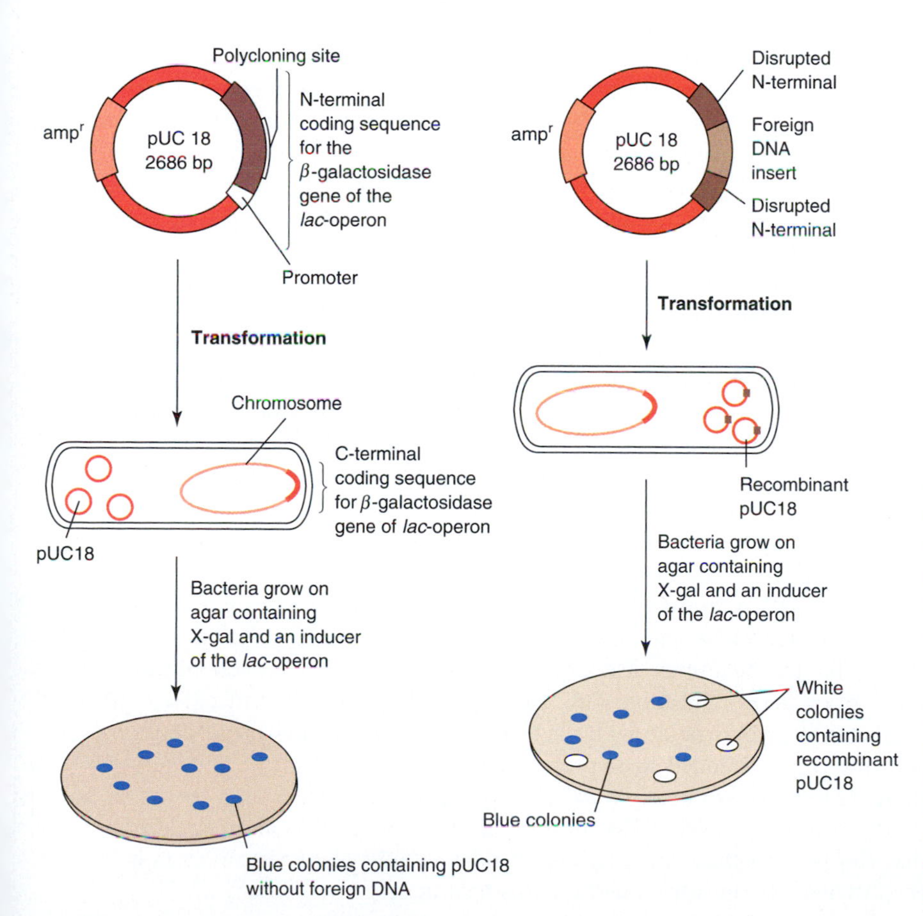

FIGURE 7.11
α-Complementation for the detection of transformed bacteria.
A constructed vector (pUC 18) expresses the N-terminal coding sequence for the enzyme β-galactosidase of the *lac* operon. Bacterial mutants coding for the C-terminal portion of β-galactosidase are transformed with pUC 18. These transformed bacteria grown in the presence of a special substrate for the intact enzyme (X-gal) result in blue colonies because they contain the enzyme to react with substrate. The functional N-terminal and C-terminal coding sequences for the enzyme complement each other to yield a functional enzyme. If, however, a foreign DNA fragment insert disrupts the pUC 18 N-terminal coding sequence for β-galactosidase, bacteria transformed with this recombinant molecule will not produce a functional enzyme. Bacterial colonies carrying these recombinant vectors can then be detected visually as white colonies.

to an inducer, **isopropylthio-β-D-galactoside,** complement each other, yielding an active enzyme. The process is referred to as **α-complementation.** When these transformed bacteria are grown in the presence of a chromogenic substrate (5-bromo-4-chloro-3-indolyl-β-D-galactoside [X-gal]) for β-galactosidase they form blue colonies. If, however, a foreign DNA fragment has been inserted into the base sequence for the N-terminal portion of β-galactosidase, the active enzyme cannot be formed. Bacteria transformed with these recombinant plasmids and grown on X-gal will yield white colonies and can be selected visually from nontransformed blue colonies.

7.7 | TECHNIQUES FOR DETECTION AND IDENTIFICATION OF NUCLEIC ACIDS AND DNA-BINDING PROTEINS

Nucleic Acids as Probes for Specific DNA or RNA Sequences

Selection of bacteria that harbor recombinant DNA of interest, analysis of mRNA expressed in a cell, or identification of DNA sequences within a genome require sensitive and specific detection methods. DNA and RNA **probes** meet these requirements. These probes contain nucleotide sequences complementary to the target nucleic acid and will hybridize with the nucleic acid of interest. The degree of complementarity of a probe with the DNA under investigation determines the tightness of binding of the probe. The probe does not need to contain the entire complementary sequence of the DNA. The probe, RNA or DNA, can be labeled, usually with ^{32}P. Nonradioactive labels are also employed that depend on enzyme substrates coupled to nucleotides, which when incorporated into the nucleic acid can be detected by an enzyme catalyzed reaction.

Labeled probes can be produced by **nick translation** of double-stranded DNA. Nick translation (Figure 7.12) involves the random enzymatic hydrolysis of a phosphodiester bond in the backbone of a DNA strand by DNase I; the breaks in the DNA backbone are referred to as nicks. A second enzyme, *E. coli* DNA polymerase I, with its $5' \rightarrow 3'$ exonucleolytic activity and its DNA polymerase activity, creates single-strand gaps by hydrolyzing nucleotides from the 5′-side of the nick and then filling in the gaps with its polymerase activity. The polymerase reaction is usually carried out in the presence of an α-^{32}P-labeled deoxynucleoside triphosphate and three unlabeled deoxynucleoside triphosphates. The DNA employed in this method is usually purified and is derived from cloned DNA, viral DNA, or cDNA.

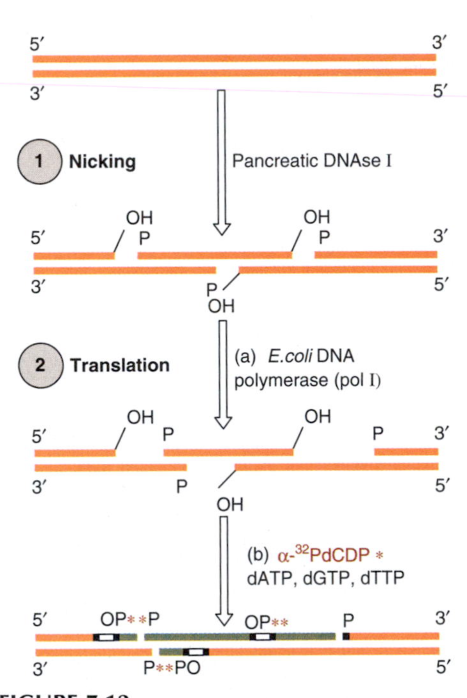

FIGURE 7.12
Nick translation to label DNA probes.
Purified DNA molecules can be labeled radioactively and used to detect, by hybridization, the presence of complementary RNA or DNA in experimental samples. (1) Nicking step introduces random single-stranded breaks in DNA. (2) Translation step: *E. coli* DNA polymerase (pol I) has both (*a*) $5' \rightarrow 3'$ exonuclease activity that removes nucleotides from the 5′-end of the nick; and (*b*) polymerase activity that simultaneously fills in the single-stranded gap with radioactively labeled nucleotides using the 3′-end as a primer.

Another method to label DNA probes, **random primer labeling of DNA,** has distinct advantages over the nick translation method. The random primer method typically requires only 25 ng of DNA as opposed to 1–2 μg of DNA for nick translation and results in labeled probes with a specific activity ($>10^9$ cmp μg^{-1}) approximately ten times higher. This method generally produces longer labeled **DNA probes.** The double-stranded probe is melted and hybridized with a mixture of random hexanucleotides containing all possible sequences (ACTCGG, ACTCGA, ACTCGC, etc.). Hybridized hexanucleotides serve as primers for DNA synthesis with a DNA polymerase, such as the Klenow enzyme, in the presence of one or more radioactively labeled deoxynucleoside triphosphates.

Labeled **RNA probes** have advantages over DNA probes. For one, relatively large amounts of RNA can be transcribed from a template, which may be available in very limited quantities. A double-stranded DNA (dsDNA) probe must be denatured prior to hybridization with the target DNA and rehybridization with itself competes for hybridization with the DNA of interest. No similar competition occurs with the single-stranded RNA probes that hybridize with complementary DNA or RNA molecules. Synthesis of an RNA probe requires DNA as a template. To be transcribed the template DNA must be covalently linked to an upstream promoter that can be recognized by a DNA-dependent RNA polymerase. Vectors have been constructed that are well suited for this technique.

A labeled DNA or RNA probe can be hybridized to nitrocellulose-bound nucleic acids and identified by the detection of the labeled probe. Nucleic acids of interest can be transferred to nitrocellulose from bacterial colonies grown on agar or from agarose gels where the nucleic acid species have been electrophoretically separated by size.

Southern Blot Technique for Identifying DNA Fragments

A technique to transfer DNA species separated by agarose gel electrophoresis to a filter for analysis was developed in the 1970s, and it is an indispensable tool. The method, developed by E. M. Southern, is referred to as the **Southern blot technique** (Figure 7.13). A DNA mixture of discrete restriction endonuclease-generated fragments from any source and complexity can be separated according

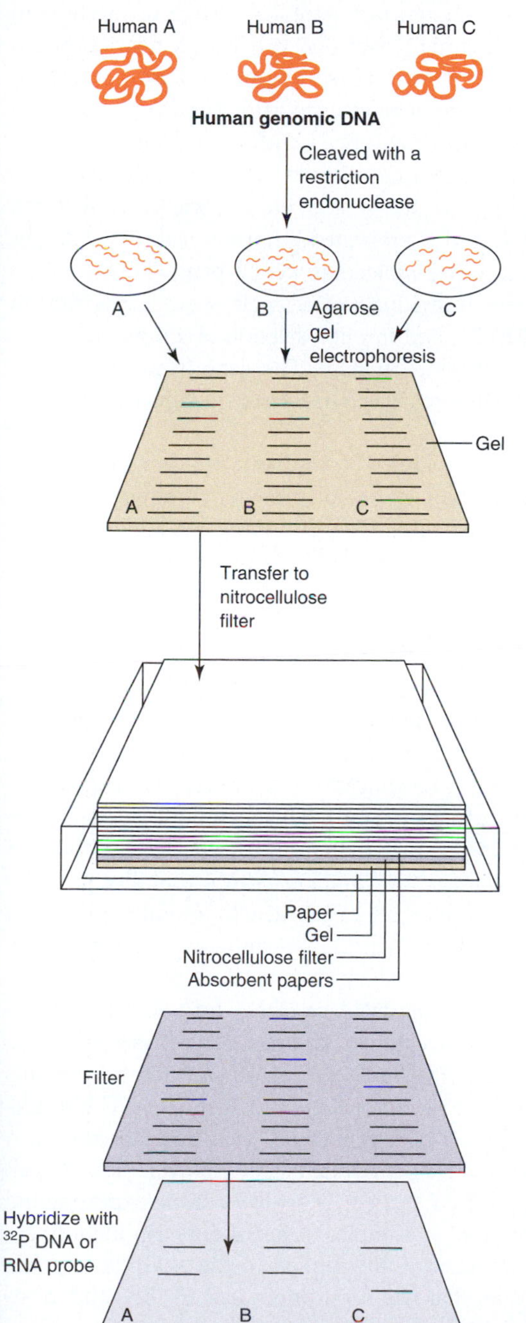

FIGURE 7.13
Southern blot to transfer DNA from agarose gels to nitrocellulose.
Transfer of DNA to nitrocellulose, as single-stranded molecules, allows for the detection of specific DNA sequences within a complex mixture of DNA. Hybridization with nick-translation- labeled probes can demonstrate if a DNA sequence of interest is present in the same or different regions of the genome.

to size by electrophoresis through an agarose gel. The DNA is denatured by soaking the gel in alkali. The gel is then placed on absorbent paper and a nitrocellulose filter placed directly on top of the gel. Several layers of absorbent paper are placed on top of the nitrocellulose filter. The absorbent paper under the gel is kept wet with a concentrated salt solution that is pulled up through the gel, the nitrocellulose, and into the absorbent paper layers above by capillary action. The DNA is eluted from the gel by the upward movement of the high salt solution onto the nitrocellulose filter directly above, where it becomes bound. The position of the DNA bound to the nitrocellulose filter is the same as that present in the agarose gel. In its single-stranded membrane-bound form, the DNA can be analyzed with labeled probes.

The Southern blot technique is invaluable in analytical procedures for detection and determination of the number of copies of particular sequences in complex genomic DNA, for confirmation of DNA cloning results, and for demonstration of polymorphic DNA in the human genome that correspond to pathological states. An example of the use of Southern blots is shown in Figure 7.13. Here whole human genomic DNA, isolated from three individuals, was digested with a restriction endonuclease generating thousands of fragments. These fragments were distributed throughout the agarose gel according to size in an electric field. The DNA was transferred (blotted) to a nitrocellulose filter and hybridized with a ^{32}P-labeled DNA or RNA probe that represents a portion of a gene of interest. The probe detected two bands in all three individuals, indicating that the gene of interest was cleaved at one site within its sequence. Individuals A and B presented a normal pattern while patient C had one normal band and one lower molecular weight band. This is an example of altered DNA within different individuals of a single species, **restriction fragment length polymorphism (RFLP),** and implies deletion of a segment of the gene that may be associated with a pathological state. The gene from this patient can be cloned, sequenced, and fully analyzed to characterize the altered nature of the DNA (see Clin. Corr. 7.5).

Other techniques that employ the principles of Southern blot are the transfer of RNA (Northern blots), as described below, and of proteins (Western blots) to nitrocellulose filters or nylon membranes.

Single-Strand Conformation Polymorphism

Southern blot analysis and detection of base changes in DNA from different individuals by RFLP analysis are dependent on alteration of a restriction endonuclease site. Often a base substitution, deletion, or insertion does not occur within a restriction endonuclease site. However, these modifications can readily be detected by **single-strand conformation polymorphism (SSCP).** This technique takes advantage of the fact that single-stranded DNA, smaller than 400 bases long, subjected to electrophoresis through a polyacrylamide gel migrates with a mobility partially dependent on its conformation. A single base alteration usually modifies the DNA conformation sufficiently to be detected as a mobility shift on electrophoresis through a nondenaturing polyacrylamide gel.

The analysis of a small region of genomic DNA or cDNA for SSCP can be accomplished by PCR amplification of the region of interest. Sense and antisense oligonucleotide primers are synthesized that flank the region of interest, and this DNA is amplified by PCR in the presence of radiolabeled nucleotide(s). The resulting purified radiolabeled double-stranded PCR product is then heat denatured in 80% formamide and immediately loaded onto a nondenaturing polyacrylamide gel. The mobilities of control products are compared to samples from experimental studies or patients. Detection of mutations in samples from patients can identify genetic lesions. These procedures were successfully applied to the analysis of genes associated with the long QT syndrome that has been implicated in the sudden in-

CLINICAL CORRELATION 7.5

Restriction Fragment Length Polymorphisms Determine the Clonal Origin of Tumors

It is generally assumed that most tumors are monoclonal in origin; that is, a rare event alters a single somatic cell genome such that the cells grow abnormally into a tumor mass with all-daughter cells carrying the identically altered genome. Proof that a tumor is of monoclonal origin versus polyclonal in origin can help to distinguish hyperplasia (increased production and growth of normal cells) from neoplasia (growth of new or tumor cells). The detection of restriction fragment length polymorphisms (RFLPs) of Southern blotted DNA samples allows one to define the clonal origin of human tumors. If tumor cells were collectively derived from different parental cells they should contain a mixture of DNA markers characteristic of each cell of origin. However, an identical DNA marker in all tumor cells would indicate a monoclonal origin. The analysis is limited to females where one can take advantage of the fact that each cell carries only one active X chromosome of either paternal or maternal origin with the second X chromosome being inactivated. Activation occurs randomly during embryogenesis and is faithfully maintained in all-daughter cells with one-half the cells carrying an activated maternal X chromosome and the other one-half an activated paternal X chromosome.

Analysis of the clonal nature of a human tumor depends on the fact that activation of an X chromosome involves changes in the methylation of selected cytosine (C) residues within the DNA molecule. Several restriction endonucleases, such as HhaI, which cleaves DNA at GCGC sites, do not cleave DNA at their recognition sequences if a C is methylated within this site. Therefore the methylated state (activated versus inactivated) of the X chromosome can be probed with restriction endonucleases. Also, the paternal X chromosome can be distinguished from the maternal X chromosome in a significant number of individuals by differences in the electrophoretic migration of restriction-endonuclease-generated fragments derived from selected regions of the chromosome. These fragments are identified on a Southern blot by hybridization with a DNA probe that is complementary to this region of the X chromosome. An X-linked gene that is amenable to these studies is the hypoxanthine–guanine phosphoribosyltransferase (*HGPRTase*) gene. The *HGPRTase* gene consistently has two BamHI restriction endonuclease sites (B1 and B3 in figure attached), but in many individuals a third site (B2) is also present (see figure).

The presence of site B2 in only one parental X chromosome HGPRT allows for the detection of restriction-enzyme-generated polymorphisms. Therefore a female cell may carry one X chromosome with the *HGPRT* gene possessing two BamHI sites (results in a single detectable DNA fragment of 24 kb) or three BamHI sites (results in a single detectable DNA fragment of 12 kb). This figure depicts the expected results for the analysis of tumor cell DNA to determine its monoclonal or polyclonal origin. As expected, three human tumors examined by this method were shown to be of monoclonal origin.

Vogelstein, B., Fearon, E. R., Hamilton, S. R., and Feinberg, A. B. Use of restriction fragment length polymorphism to determine the clonal origin of tumors. *Science* 227:642, 1985.

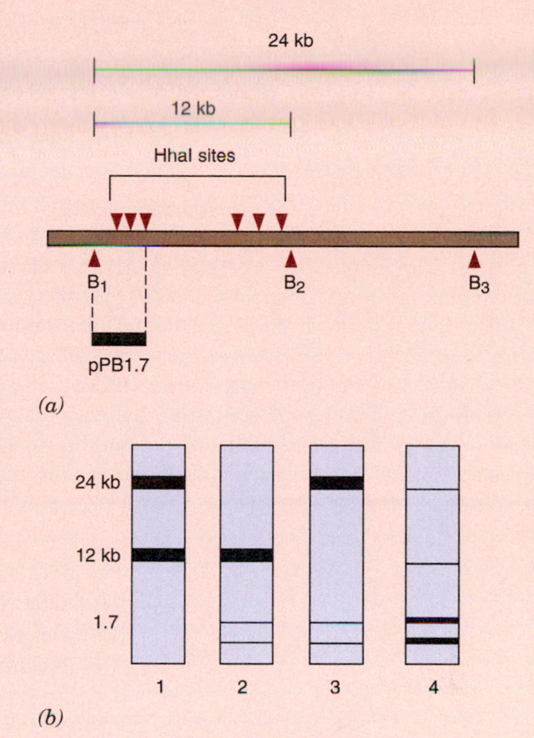

Analysis of genomic DNA to determine the clonal origin of tumors.
(*a*) The X chromosome-linked hypoxanthine–guanine phosphoribosyltransferase (*HGPRTase*) gene contains two invariant BamHI restriction endonuclease sites (B1 and B3) while in some individuals a third site, B2, is also present. The *HGPRTase* gene also contains several HhaI sites, however, all of these sites, except H1, are usually methylated in the active X chromosome. Therefore only the H1 site would be available for cleavage by HhaI in the active X chromosome. A cloned, labeled probe, pPB1.7, is employed to determine which form of the *HGPRTase* gene is present in a tumor and if it is present on an active X chromosome. (*b*) Restriction endonuclease patterns predicted for monoclonal versus polyclonal tumors are as follows: (1) Cleaved with BamHI alone, 24-kb fragment derived from *HGPRTase* gene containing only B1 and B3 sites and 12-kb fragment derived from *HGPRTase* gene containing extra B2 site. Pattern is characteristic for heterozygous individual. (2) Cleaved with BamHI plus HhaI. Monoclonal tumor with the 12 kb derived from an active X chromosome (methylated). (3) Cleaved with BamHI plus HhaI; monoclonal tumor with the 24-kb fragment derived from an active X chromosome (methylated). (4) Cleaved with BamHI plus HhaI. Polyclonal tumor. All tumors studied displayed patterns as in lane 2 or lane 3.

CLINICAL CORRELATION 7.6

Single-Strand Conformational Polymorphism for Detection of Spontaneous Mutations that May Lead to SIDS

Sudden infant death syndrome (SIDS) is a major cause of death during the first year of life in the United States. Prospective study of more then 34,000 newborns who were monitored by electrocardiography indicated a strong correlation for increased risk of SIDS associated with a prolonged QT interval in their heart EKG. Based on this study, it was decided to look for a mutation in one or more of the genes known to be related to the long QT syndrome in a 44-day-old infant that presented cyanotic, apneic, and pulseless to a hospital emergency room. The child's arrhythmia with a prolonged QT interval was stabilized with multiple electrical dc shocks followed by drug treatment. Peripheral blood lymphocytes from the infant and his parents were employed to prepare genomic DNA. One gene associated with the long QT syndrome was found to contain the substitution of AAC for TCC at position 2971 to 2972 (protein associated with the sodium channel) by single-strand conformational polymorphism (SSCP) analysis in the child's but not the parent's DNA sample. The mutation results in the replacement of a serine residue with an asparagine in a highly conserved region of the protein that is presumed to participate in the function of the sodium channel. The mutation was not detected in 200 control subjects. The conclusion of this case study was that the child had a spontaneous mutation in a gene that was associated with a prolonged QT interval and that this contributed to a SIDS-like event. After treatment and monitoring, the child was symptom free at nearly five years of age. This study points to the potential value of neonatal electrocardiographic screening to reduce infant mortality due to SIDS.

Schwartz, P. J., Priori, S. G., Dumaine, R., Napolitano, C., Antzelevitch, C., Stramba-Badiale, M., Richard, T. A., Berji, M. R., and Bloise, R. A molecular link between the sudden infant death syndrome and the long-QT syndrome. *N. Engl. J. Med.* 343:262, 2000.

fant death syndrome (SIDS) (see Clin. Corr. 7.6). The method depends on prior knowledge of the sequence of the gene/gene fragment of interest, while analysis by RFLP requires only restriction map analysis of DNA.

Methods to Detect mRNA

The analysis of the presence, absence, or quantity of mRNA species in tissue or total RNA preparations is often critical for our understanding of gene regulation of cell growth and tissue differentiation. One can assess timing, level, and site within a tissue of gene expression through the analysis of specific mRNA species. Several different techniques are available to detect the presence and/or the quantity of a specific mRNA species in a total RNA preparation: (1) **Northern blot analysis** requires the electrophoretic separation of RNA species, by size, on an agarose gel followed by the transfer and cross-linking to a membrane in a fashion similar to the Southern blot technique. The fixed RNA species are then hybridized with a labeled probe specific for the mRNA species of interest. If the probe is radiolabeled, hybrids can be detected by autoradiography. Nonradioactive methods are also available. Northern blot analysis allows ready determination of the size of the mRNA and identification of potential alternatively spliced transcripts and/or the presence of multigene family transcripts. (2) **RT-PCR** (see Section 7.8), unlike the Northern blot, allows one to quantitate the amount of a rare (theoretically as low as one copy) or abundant mRNA species present in a RNA sample. Quantitation is based on comparison of the relative amount of a PCR product of an internal control transcript to the level detected for a specific mRNA in the same RNA sample. Typical internal control transcripts include GAPDH (glyceraldehyde 3-phosphate dehydrogenase) and β-actin. (3) The **nuclease protection assay** (NPA) does not give any information about the size of the mRNA of interest; however, it is ideally suited for the simultaneous analysis of multiple mRNA species that may be present in an RNA sample. Radiolabeled or nonisotopic probe(s) are hybridized with the RNA sample in solution. Hybridization in solution is more efficient than hybridization to membrane-fixed RNA, as used in Northern blots. Any single-stranded unhybridized probe and RNA are then hydrolyzed with nucleases after hybridization. The mixture of hybridized species are then separated through a low percentage (6%) acrylamide gel. The size of each hybridized species is determined by the size of the probe employed. NPA is commonly used to detect the presence or absence of mRNA transcripts in different tissues (Figure 7.14). (4) ***In situ* hybridization** is the only method of the four described in this section to define which cells in a sample of formalin-fixed tissue expresses a specific gene. In this method RNA is neither isolated nor subjected to electrophoretic separations. Here tissues of interest are fixed, thin sectioned, and mounted onto microscope slides. Proteins are digested with proteinase K to increase accessibility for the labeled probe to hybridize with the cellular mRNA of interest. This method would support findings of the nuclease protection assays and would also define which cell types within the tissues are expressing the mRNA of interest.

Analysis of Sequence-Specific DNA-Binding Proteins

It is established that regulatory proteins bind to specific DNA sequences that flank genes that up- or downregulate gene expression. The **electrophoretic mobility shift assay** (EMSA) or gel retardation method has been employed extensively to analyze sequence-specific DNA-binding proteins along with the DNA sequence required for binding. Proteins with potential DNA-binding characteristics can be prepared from whole cell extracts, nuclear extracts, purified protein preparations, or recombinant proteins from genetically engineered expression systems. The DNA probe is radiolabeled. The DNA employed may be DNA fragments, double-stranded synthetic oligonucleotides, or cloned DNA containing a known protein-binding site(s) or potential protein-binding site(s). It is critical that the DNA be double-stranded since single-stranded DNA would bind nonspecific single-strand binding proteins (with cell or nuclear extracts) that would interfere with the interpretation

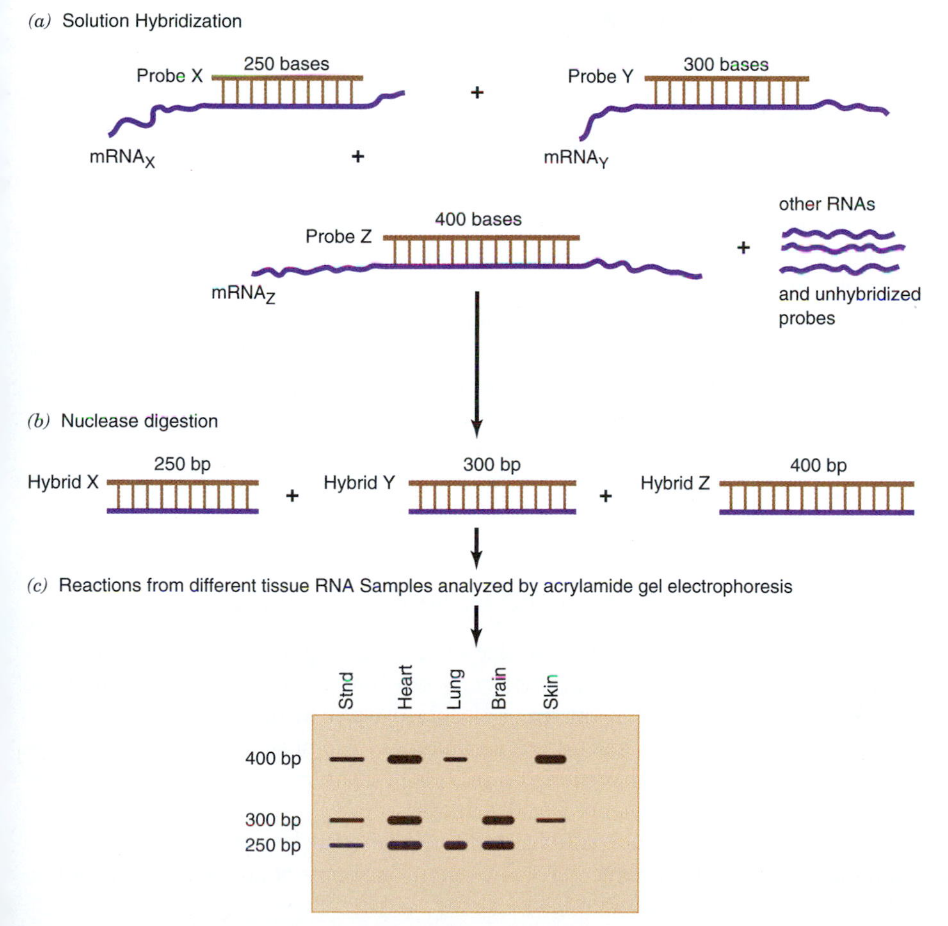

FIGURE 7.14
Nuclease protection assay.
Total cellular mRNA can be isolated from different tissues. Single-stranded DNA probes that are complementary to known sequences of different gene transcripts (mRNA$_x$, mRNA$_y$, mRNA$_z$) are hybridized with the RNA mixture. Nuclease digestion with a ribonuclease will hydrolyze single-stranded RNA regions of mRNA not hybridized with the DNA probe and all nonhybridized RNA species. Only the nuclease protected DNA–RNA hybrids will remain for analysis by acrylamide gel electrophoresis. Differential expression of genes in specific tissues is then readily observed.

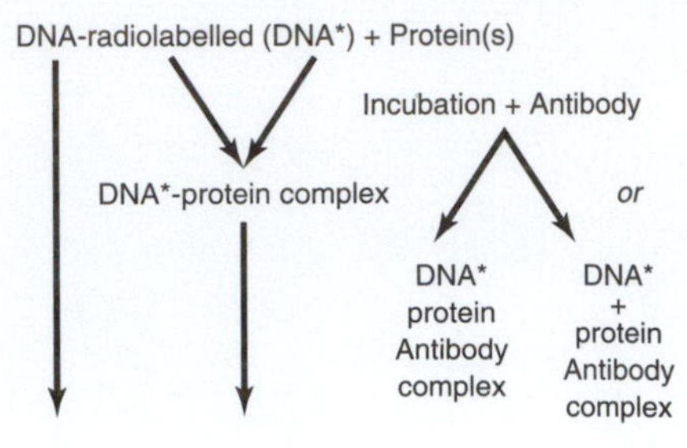

FIGURE 7.15
Electrophoretic mobility shift assay.
Purified DNA migrates through a gel when subjected to an electric field based on its charge and molecular weight. If a protein(s) is added to this purified DNA that complexes with the DNA, the total apparent molecular weight of the DNA is increased. This would slow its migration through the gel—a mobility shift. The addition of an antibody that reacts with the protein complexed with the DNA, but does not interfere with the DNA–protein interaction, will further increase the size of the DNA–protein complex, slowing its migration. If the antibody reacts with the DNA-binding region of the protein, a DNA–protein complex will not form. In both cases the antibody helps to identify the DNA-binding protein.

of the results. The purified labeled DNA probe and protein sample are preincubated to form a stable protein–DNA complex prior to the electrophoretic analysis of the sample.

Figure 7.15 schematically depicts the expected results where the radiolabeled DNA complexes with a protein, resulting in a retarded movement through the gel relative to the unreacted DNA. If an antibody to the DNA-binding protein of interest is added to the preincubation tube, one of two possible reactions can occur. If the antibody binds to an epitope that does not impede protein binding to the DNA the complex size is increased resulting in a **supershift** (greater retardation) in the gel migration (Figure 7.15). Conversely, if the antibody blocks the protein's DNA-binding site a protein–antibody complex will form and the labeled DNA probe will be unmodified in its migration through the gel. In either case the result with the antibody would confirm the identity of the DNA-binding protein.

7.8 | COMPLEMENTARY DNA AND COMPLEMENTARY DNA LIBRARIES

Insertion of specific functional eukaryotic genes into vectors that can be expressed in a prokaryotic cell could produce large amounts of "genetically engineered" proteins with significant medical, agricultural, and experimental potential. Hormones and enzymes, including insulin, erythropoietin, thrombopoietin, interleukins, interferons, and tissue plasminogen activator, are currently produced by these methods. Unfortunately, it is not possible to clone functional genes from genomic DNA, except in rare instances. One reason for this is that most genes within the mammalian genome yield transcripts that contain introns that must be spliced out of the primary mRNA transcript. Prokaryotic systems cannot splice out the introns to yield functional mRNA transcripts. This problem can be circumvented by synthesizing **complementary DNA (cDNA)** from functional eukaryotic mRNA.

mRNA as a Template for DNA Synthesis Using Reverse Transcriptase

Messenger RNA can be reverse transcribed to cDNA and the cDNA inserted into a vector for amplification, identification, and expression. Mammalian cells normally contain 10,000–30,000 different species of mRNA molecules at any time during the cell cycle. In some cases, however, a specific mRNA species may approach 90% of the total mRNA, such as mRNA for globin in reticulocytes. Many mRNAs are normally present at only a few (1–14) copies per cell. A **cDNA library** can be constructed from the total cellular mRNA but if only a few copies per cell of mRNA of interest are present, the cDNA may be very difficult to identify. Methods that enrich the population of mRNAs or their corresponding cDNAs permit reduction in number of different cDNA species within a cDNA library and greatly enhance the probability of identifying the clone of interest.

Desired mRNA Can Be Enriched by Separation Techniques

Messenger RNA can be separated by size by gel electrophoresis or centrifugation. Utilization of mRNA in a specific molecular size range will enrich severalfold an mRNA of interest. Knowledge of the molecular weight of the protein encoded by the gene of interest gives a clue to the approximate size of the mRNA transcript or its cDNA; variability in the predicted size, however, arises from differences in the length of the untranslated regions of the mRNAs.

Enrichment of a specific mRNA molecule can also be accomplished by immunological procedures, but this requires the availability of antibodies against the

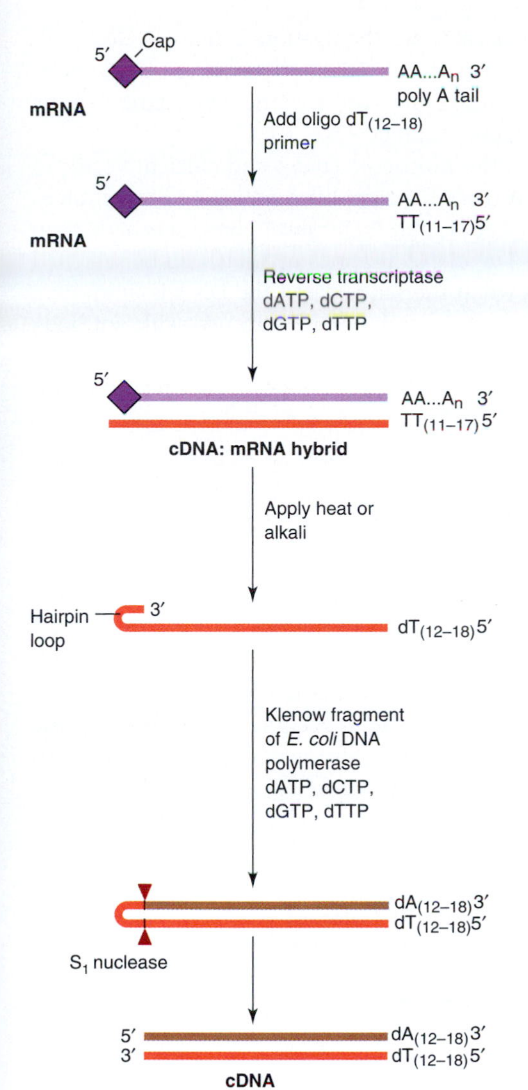

FIGURE 7.16
Synthesis of cDNA from mRNA.
The 3′ poly(A) tail of mRNA is hybridized with an oligomer of dT [oligo(dT)12–18] that serves as a primer for reverse transcriptase, which catalyzes the synthesis of the complementary DNA (cDNA) strand in the presence of all four deoxynucleotide triphosphates (dNTPs). The resulting cDNA–mRNA hybrid is separated into single-stranded cDNA by melting with heat or hydrolyzing the mRNA with alkali. The 3′-end of the cDNA molecule forms a hairpin loop that serves as a primer for the synthesis of the second DNA strand catalyzed by the Klenow fragment of *E. coli* DNA polymerase. The single-stranded unpaired DNA loop is hydrolyzed by S_1 nuclease to yield a double-stranded DNA molecule.

encoded protein by the gene of interest. Antibodies added to an *in vitro* protein synthesis mixture react with the growing polypeptide chain associated with polysomes and precipitate it. The mRNA can be purified from the immunoprecipitated polysomal fraction.

Complementary DNA Synthesis

An mRNA mixture is used as a template to synthesize a complementary strand of DNA using RNA-dependent DNA polymerase, **reverse transcriptase** (Figure 7.16). A primer is required; advantage is taken of the poly(A) tail at the 3′-terminus of eukaryotic mRNA. An oligo(dT) with 12–18 bases is employed as the primer that hybridizes with the poly(A) sequence. After cDNA synthesis, the hybrid is denatured or the mRNA hydrolyzed in alkali in order to obtain the single-stranded cDNA. The 3′-termini of single-stranded cDNAs form a hairpin loop that serves as a primer for the synthesis of the second strand of the cDNA. Either the Klenow fragment or a reverse transcriptase can be used for this step. The resulting double-stranded cDNA contains a single-stranded loop that is selectively recognized and digested by S1 nuclease. The ends of the cDNA must be modified prior to cloning in a vector. One method involves incubating blunt-ended cDNA molecules with linker molecules

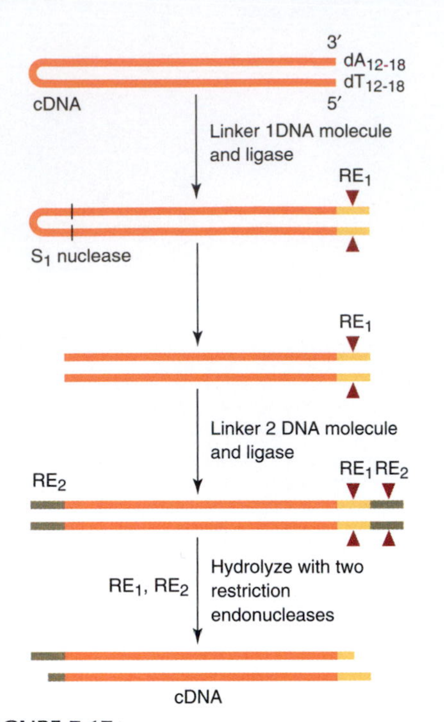

FIGURE 7.17
Modification of cDNA for cloning.
The procedure begins with double-stranded DNA that contains a hairpin loop. A linker DNA that contains a restriction endonuclease site (RE_1) is added to the free end of the cDNA by blunt-end ligation. The single-stranded hairpin loop is next hydrolyzed with S_1 nuclease. A second linker with a different restriction endonuclease site within RE_2 is blunt-end ligated to the newly created free cDNA. The second linker will probably bind to both ends but will not interfere with the first restriction endonuclease site. The modified DNA is hydrolyzed with the two restriction endonucleases and can be inserted into a plasmid or bacteriophage DNA by directional cloning.

and a bacteriophage T4 DNA ligase that catalyzes the ligation of blunt-ended molecules (Figure 7.17). The synthetic linker molecules contain restriction endonuclease sites that can now be cleaved with the appropriate enzyme for insertion of the cDNA into a vector cleaved with the same endonuclease.

Bacteriophage DNA (see p. 305) is the most convenient and efficient vector to create cDNA libraries because they can readily be amplified and stored indefinitely. Two bacteriophage vectors, **λgt10** and **λgt11**, and their newer constructs have been employed to produce cDNA libraries. The cDNA libraries in λgt10 can be screened only with labeled nucleic acid probes when the sequence of the DNA is unknown, whereas those in λgt11, an expression vector, can also be screened with antibody for the production of the protein or antigen of interest. If the sequence of the desired cDNA is known, then PCR can be used to screen the recombinant bacteriophage.

Total Cellular RNA as a Template for DNA Synthesis Using RT-PCR

Alternative methods to construct cDNA libraries employ a **reverse transcriptase–PCR (RT-PCR) technique** and obviate the need to purify mRNA. One such strategy is depicted in Figure 7.18 and begins with the reverse transcriptase production of a DNA–mRNA hybrid. The method then adds a dG homopolymer tail to the 3′-end catalyzed by terminal transferase and the subsequent hydrolysis of the mRNA. PCR primers are synthesized to hybridize with the dG, dA tails and terminate with two different restriction endonuclease sequences. The resulting PCR-

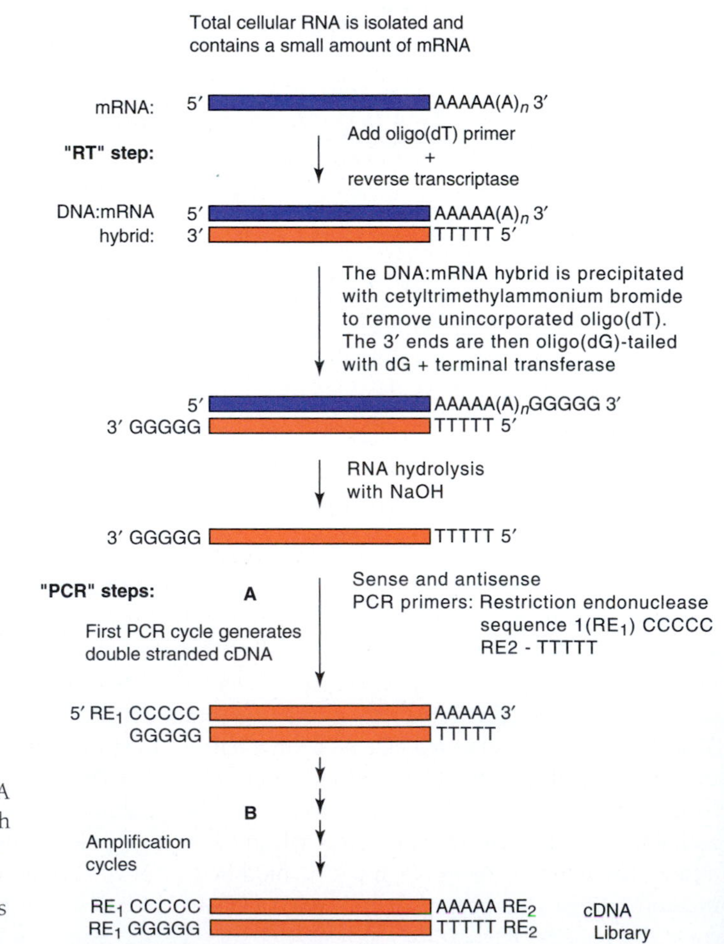

FIGURE 7.18
Generation of cDNA by reverse transcriptase–PCR (RT-PCR).
Total cellular RNA or mRNA can be used to generate cDNA by RT-PCR. The mRNA with an oligo(rA) tail is reverse transcribed with an oligo(dT) primer. An oligo(dG) tail is added to the 3′-ends of the RNA and DNA strands and the RNA strand is subsequently hydrolyzed with NaOH. Sense and antisense primers, modified with restriction site sequences, are then employed to amplify the cDNA by the PCR. The products can be hydrolyzed with the specific restriction endonucleases (RE_1 and RE_2) for cloning and subsequent studies.

amplified cDNA can then be hydrolyzed with two different restriction endonucleases for directional cloning (see p. 287) into an appropriate vector.

7.9 | BACTERIOPHAGE, COSMID, AND YEAST CLONING VECTORS

Detection of noncoding sequences in most eukaryotic genes and distant regulatory regions that flank the genes necessitated cloning strategies to package larger DNA fragments than can be cloned in **plasmids.** Plasmids can accommodate foreign DNA inserts with a maximum length in the range of 5–10 kb (kilobases). Portions of recombinant DNA fragments larger than this are randomly deleted during replication of the plasmid within the bacterium. Thus alternate vectors have been developed.

Bacteriophages as Cloning Vectors

Bacteriophage λ (λ phage), a virus that infects and replicates in bacteria, is an ideal vector for DNA inserts of approximately 15 kb long. The λ phage selectively infects bacteria and replicates by a lytic or nonlytic (lysogenic) pathway. The λ phage contains a self-complementary 12-base single-stranded tail (cohesive termini) at both ends of its 50-kb double-stranded DNA. On infection of the bacteria the cohesive termini (cos sites) of a single λ phage DNA self-anneal and the ends are covalently linked with the host cell DNA ligase. The circular DNA serves as a template for transcription and replication. λ Phage, with restriction endonuclease-generated fragments representing a cell's whole genomic DNA inserted into it, are used to infect bacteria. Recombinant bacteriophages, released from the lysed cells, are collected and constitute a genomic library in λ phage. The phage library can be screened more rapidly than a plasmid library due to the increased size of the DNA inserts.

Numerous λ phage vectors have been constructed for different cloning strategies. For the sake of simplicity only a generic λ phage vector will be described here. Cloning large fragments of DNA in λ phage takes advantage of the fact that a 15–25-kb segment of the phage DNA can be replaced without impairing its replication in *E. coli* (Figure 7.19). Packaging of phage DNA into the virus particle is constrained by its total length, which must be approximately 50 kb. The linear λ phage DNA can be digested with a specific restriction endonuclease that generate small terminal fragments with their cos sites (arms), which are separated from the larger intervening fragments. Cellular genomic DNA is partially digested by appropriate restriction enzymes to permit annealing and ligation with the phage arms in order to randomly generate fragments that can be properly packaged into phage particles. The DNA fragments smaller or larger than 15–25 kb can hybridize with the cos arms but are excluded from being packaged into infectious bacteriophage particles. All the information required for phage infection and replication in bacteria is carried within the cos arms. The recombinant phage DNA is mixed with λ phage proteins *in vitro,* which assemble into infectious virions. The infectious recombinant λ phage particles are then propagated in an appropriate *E. coli* strain to yield a λ phage library. Many different *E. coli* strains have been genetically altered to sustain replication of specific recombinant virions.

Screening Bacteriophage Libraries

A **bacteriophage library** can be screened by plating the virus on a continuous layer of bacteria (a bacterial lawn) grown on agar plates (Figure 7.20). Individual phage will infect, replicate, and lyse one cell. The progeny virions will then infect and subsequently lyse bacteria immediately adjacent to the site of the first

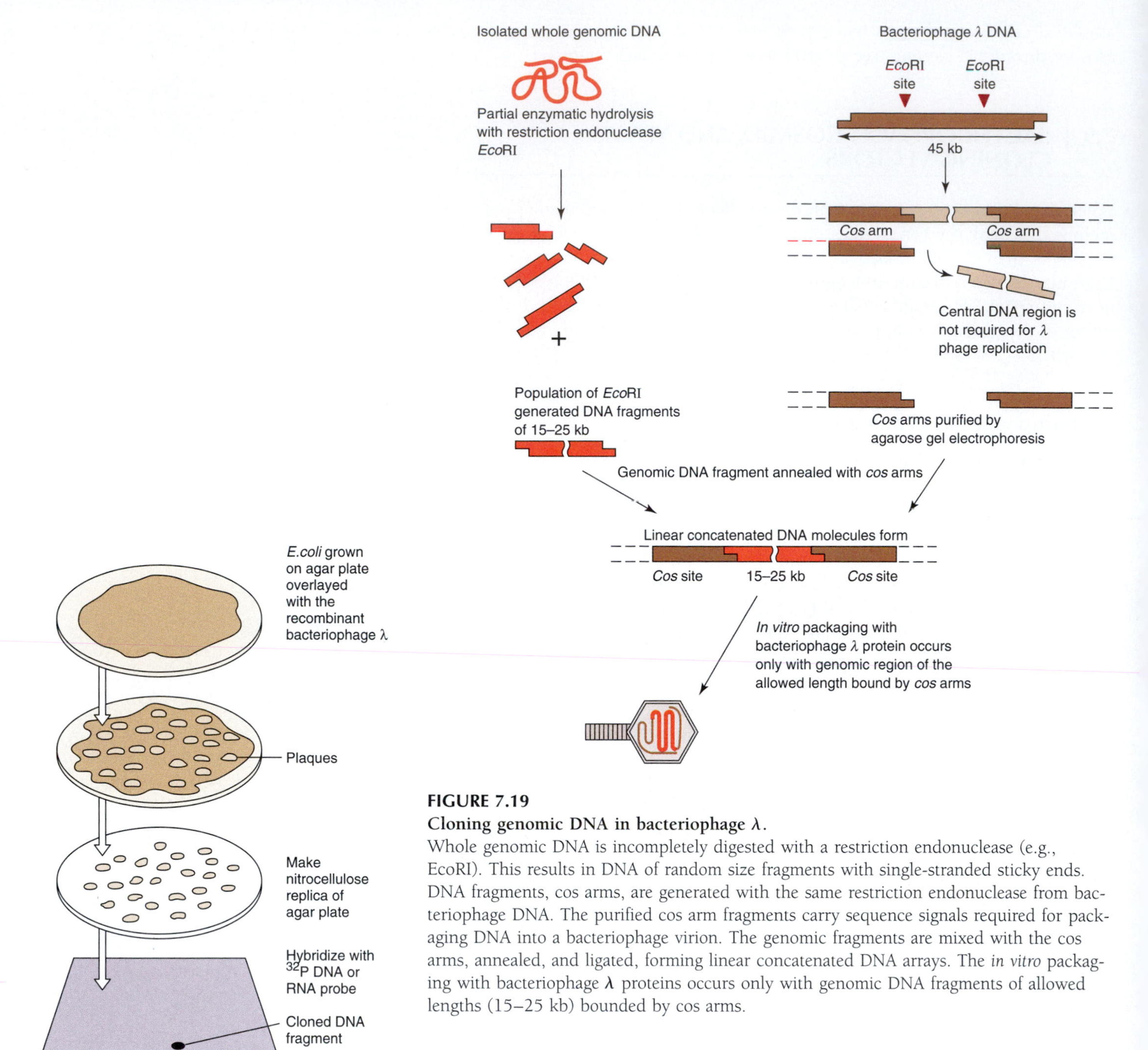

FIGURE 7.19
Cloning genomic DNA in bacteriophage λ.
Whole genomic DNA is incompletely digested with a restriction endonuclease (e.g., EcoRI). This results in DNA of random size fragments with single-stranded sticky ends. DNA fragments, cos arms, are generated with the same restriction endonuclease from bacteriophage DNA. The purified cos arm fragments carry sequence signals required for packaging DNA into a bacteriophage virion. The genomic fragments are mixed with the cos arms, annealed, and ligated, forming linear concatenated DNA arrays. The *in vitro* packaging with bacteriophage λ proteins occurs only with genomic DNA fragments of allowed lengths (15–25 kb) bounded by cos arms.

FIGURE 7.20
Screening genomic libraries in bacteriophage λ.
Competent *E. coli* are grown to confluence on an agar plate and then overlayed with the recombinant bacteriophage. Plaques develop where bacteria are infected and subsequently lysed by the phage λ. Replicas of the plate can be made by touching the plate with a nitrocellulose filter. The DNA is denatured and fixed to the nitrocellulose with NaOH. The fixed DNA is hybridized with a ^{32}P-labeled probe and exposed to X-ray film. The autoradiograph identifies the plaque(s) with recombinant DNA of interest.

infected cell, creating a clear region or plaque in the opaque bacterial field. Phage, within each plaque, can be picked up on a nitrocellulose filter (as for replica plating) and the DNA fixed to the filter with NaOH. The location of cloned DNA fragments of interest is determined by hybridizing the filter-bound DNA with a labeled DNA or RNA probe followed by autoradiography. The bacteriophages in the plaque corresponding to the labeled filter-bound hybrid are picked up and amplified in bacteria for further analysis. PCR can also be employed if the full or partial sequence is known of the desired cDNA. Here one takes portions of several plaques and if the DNA of interest is present in one or more, a region can be amplified with an appropriate primer pair by PCR. The PCR product is then detected by gel electrophoresis. Complementary DNA libraries in bacteriophages are also constructed that contain the phage cos arms. If the cDNA is recombined with phage DNA that permits expression of the gene, such as λgt11, then plaques can be screened immunologically with antibodies specific for the antigen of interest.

Cloning DNA Fragments into Cosmid and Yeast Artificial Chromosome Vectors

Even though λ phage is the most commonly used vector to construct genomic DNA libraries, the length of many genes exceeds the maximum size of DNA that can be inserted between the phage arms. A **cosmid vector** can accommodate foreign DNA inserts of approximately 45 kb. **Bacterial artificial chromosomes (BACs)** and **yeast artificial chromosomes (YACs)** have been developed to clone DNA fragments of 100–200-kb and 200–500-kb lengths, respectively. While cosmid and yeast artificial chromosome vectors are difficult to work with, their libraries permit the cloning of large genes with their flanking regulatory sequences, as well as families of genes or contiguous genes.

Cosmid vectors are a cross between plasmid and bacteriophage vectors. Cosmids contain an antibiotic resistance gene for selection of recombinant DNA molecules, an origin of replication for propagation in bacteria, and a cos site for packaging of recombinant molecules in bacteriophage particles. The bacteriophage with recombinant cosmid DNA can infect *E. coli* and inject its DNA into the cell. Cosmid vectors contain only approximately 5 kb of the 50-kb bacteriophage DNA and, therefore, cannot direct its replication and assembly of new infectious phage particles. Instead, the recombinant cosmid DNA circularizes and replicates as a large plasmid. Bacterial colonies with recombinants of interest can be selected and amplified by methods similar to those described for plasmids.

Standard cloning procedures and some novel methods are employed to construct YACs. Very large foreign DNA fragments are joined to yeast DNA sequences, one that functions as telomeres (distal extremities of chromosome arms), and another that functions as a centromere and as an origin of replication. The recombinant YAC DNA is introduced into the yeast by transformation. The YAC constructs are designed so that yeast transformed with recombinant chromosomes grow as visually distinguishable colonies. This facilitates the selection and analysis of cloned DNA fragments.

7.10 | TECHNIQUES TO ANALYZE LONG STRETCHES OF DNA

Subcloning Permits Definition of Large Segments of DNA

Complete analysis of functional elements in a cloned DNA fragment requires sequencing of the entire molecule. Current techniques can sequence 200–400 bases in a DNA fragment, yet cloned DNA inserts are frequently much larger. Restriction maps of the initial DNA clone are essential for cleaving the DNA into smaller pieces to be recloned, or **subcloned,** for further analysis. The sequence of each of the small subcloned DNA fragments can be determined. Overlapping regions of the subcloned DNA properly align and confirm the entire sequence of the original DNA clone.

Sequencing can often be accomplished without subcloning. Antisense primers can be synthesized that are complementary to the initially sequenced 3′-ends of the cloned DNA. This process is repeated until the full length of the cloned DNA has been sequenced. This method obviates the need to prepare subclones but requires synthesis/purchase of numerous primers. On the other hand, the subcloned DNA is always inserted back into the same region of the plasmid. Therefore one set of primers complementary to the plasmid DNA sequences flanking the inserted DNA can be used for all of the sequencing reactions with subcloned DNA.

Chromosome Walking Is a Technique to Define Gene Arrangement in Long Stretches of DNA

Knowledge of how genes and their regulatory elements are arranged in a chromosome should lead to an understanding of how sets of genes may be coordinately regulated. It is difficult to clone DNA fragments large enough to identify contiguous genes. The combination of several techniques allows for analysis of very long stretches of DNA (50–100 kb). The method, **chromosome walking,** is possible because λ phage or cosmid libraries contain partially cleaved genomic DNA cut at specific restriction endonuclease sites. The cloned fragments will contain overlapping sequences with other cloned fragments. Overlapping regions are identified by restriction mapping, subcloning, screening λ phage or cosmid libraries, and sequencing procedures.

The procedure of chromosome walking is shown in Figure 7.21. Initially the λ phage library is screened for a sequence of interest with a DNA or RNA probe.

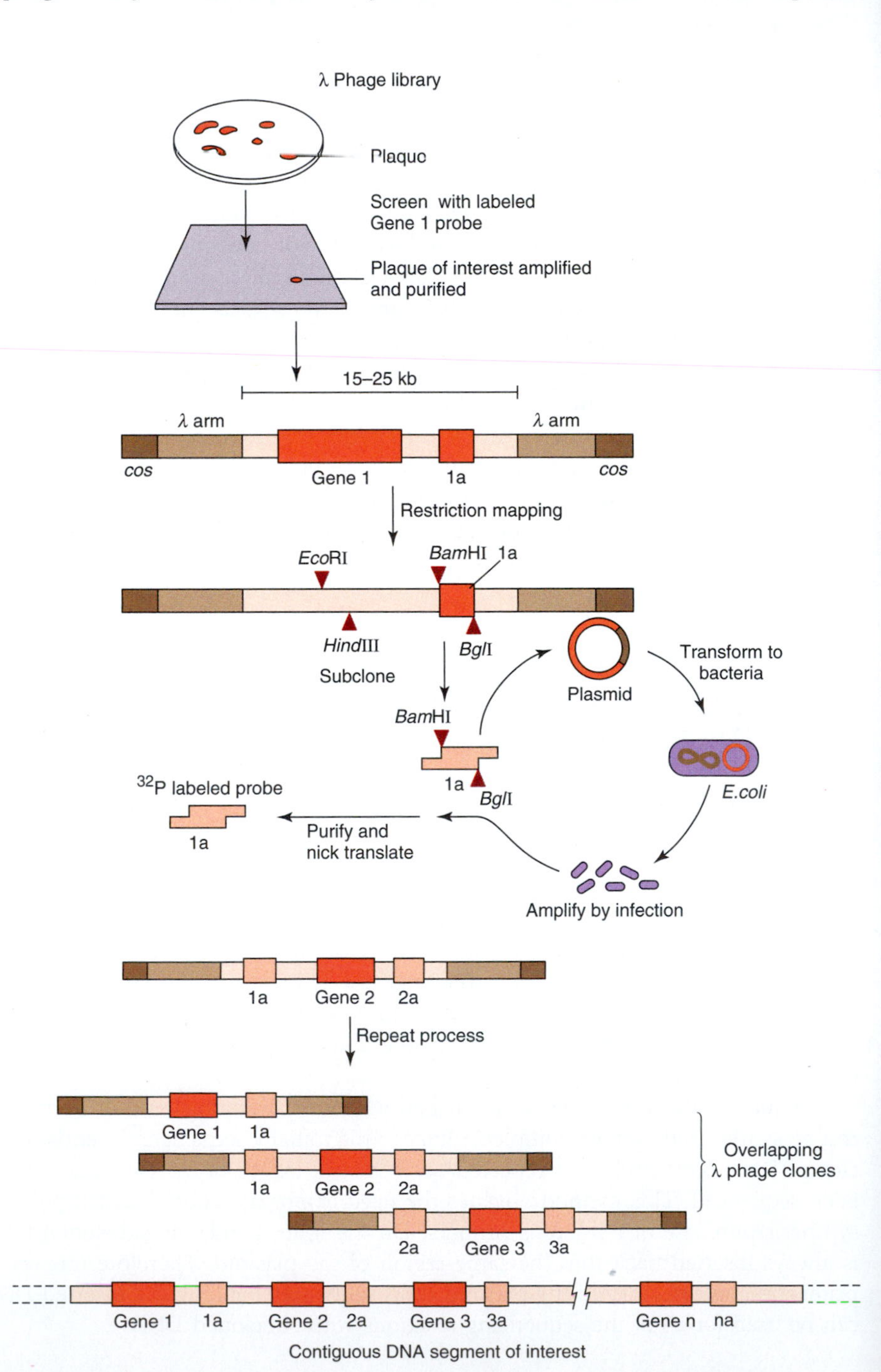

FIGURE 7.21
Chromosome walking to analyze contiguous DNA segments in a genome.
Initially, a DNA fragment is labeled by nick translation to screen a library for recombinant λ phage carrying a gene of interest. The amplified DNA is mapped with a battery of restriction endonucleases to select a new region (1a) within the original cloned DNA that can be recloned (subcloned). The subcloned DNA (1a) is used to identify other DNA fragments within the original library that would overlap the initially amplified DNA region. The process can be repeated many times to identify contiguous DNA regions upstream and downstream of the initial DNA (gene 1) of interest.

The cloned DNA is restriction mapped and a small segment is subcloned in a plasmid, amplified, purified, and labeled by nick translation. This labeled probe is then used to rescreen the λ phage library for complementary sequences, which are then cloned. The newly identified overlapping cloned DNA is then treated in the same fashion as the initial DNA clone to search for other overlapping sequences. In higher eukaryotic genomes caution must be taken so that subcloned DNA does not contain a sequence common to the large numbers of repeating DNA sequences. If a subcloned DNA probe contains a repeat sequence it hybridizes to numerous bacteriophage plaques and prevents identification of a specific overlapping clone.

7.11 | EXPRESSION VECTORS AND FUSION PROTEINS

Recombinant DNA methodology described to this point has dealt primarily with screening, amplification, and purification of cloned DNA species. An important goal of recombinant DNA studies, as stated earlier, is to have a foreign gene expressed in bacteria with the product in a biologically active form. Sequencing the DNA of many bacterial genes and their flanking regions has identified the spatial arrangement of regulatory sequences required for expression of genes. A promoter and other regulatory elements upstream of the gene are required to transcribe a gene (see p. 332). The mRNA transcript of a recombinant eukaryotic gene, however, is not translated in a bacterial system because it lacks the bacterial recognition sequence, the Shine–Dalgarno sequence, required to properly orient it with a functional bacterial ribosome. Vectors that facilitate the functional transcription of DNA inserts, termed **expression vectors,** have been constructed such that a foreign gene can be inserted into the vector downstream of a regulated promoter but within a bacterial gene, commonly the *lacZ* gene. The mRNA transcript of the recombinant DNA contains the *lacZ* Shine–Dalgarno sequence, codons for a portion of the 3′-end of the *lacZ* gene protein, followed by the codons of the complete foreign gene of interest. The protein product is a **fusion protein** that contains a few N-terminal amino acids of the *lacZ* gene protein and the complete amino acid sequence of the foreign gene product.

Foreign Genes Expressed in Bacteria Allow Synthesis of Their Encoded Proteins

Many plasmid and bacteriophage vectors have been constructed to permit the expression of eukaryotic genes in bacterial cells. Rapidly replicating bacteria can serve as a biological factory to produce large amounts of specific proteins, which have research, clinical, and commercial value. As an example, recombinant technologies have produced human protein hormones that serve as replacement or supplemental hormones in patients with aberrant or missing hormone production. Figure 7.22 depicts a generalized plasmid vector for the expression of a mammalian gene. Recall that the inserted foreign gene must be in the form of cDNA from its corresponding mRNA since the bacterial system cannot remove the introns in the pre-mRNA transcript. The DNA must be inserted in register with the codons of the 3′-terminal codons of the bacterial protein when creating a fusion protein. That is, insertion must occur after a triplet codon of the bacterial protein and at the beginning of a triplet codon of the eukaryotic gene protein to ensure proper translation. Finally, the foreign gene must be inserted in the proper orientation relative to the promoter to yield a functional transcript. This can be achieved by directional cloning.

Eukaryotic proteins synthesized within bacteria are often unstable and are degraded by intracellular proteases. Fusion protein products, however, are usually stable. The fusion protein amino acids encoded by the prokaryotic genome may be cleaved from the purified protein of interest by enzymatic or chemical procedures. An alternative cloning strategy to circumvent the intracellular instability of some proteins is to produce a foreign protein that is secreted. This requires cloning the

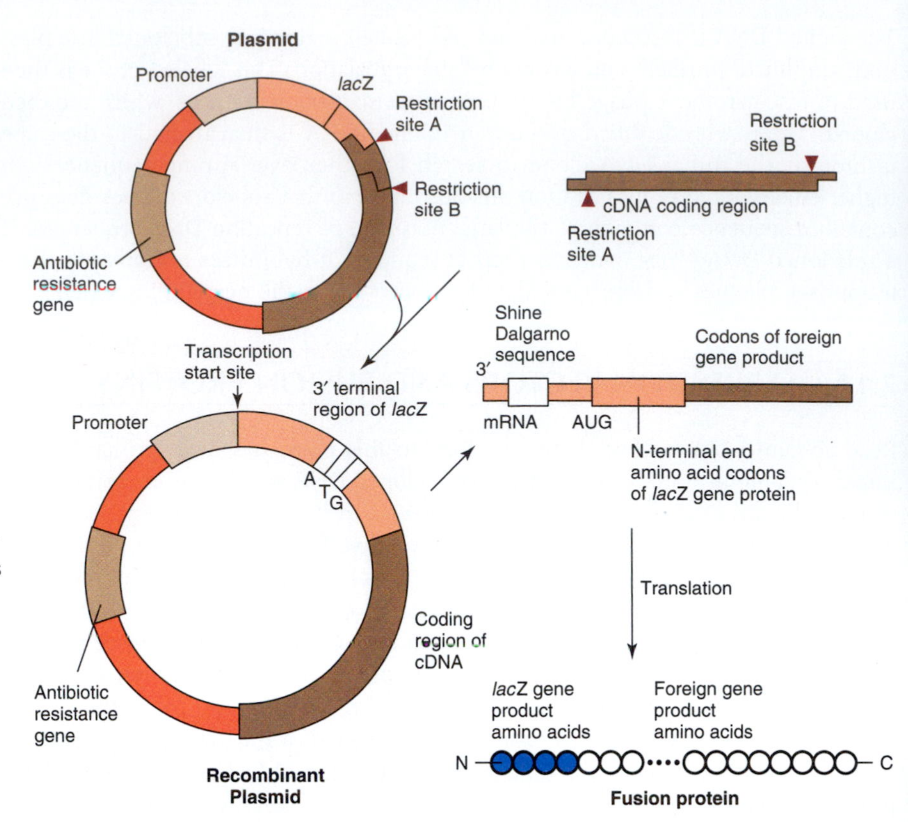

FIGURE 7.22
Construction of a bacterial expression vector. A cDNA coding region of a protein of interest is inserted downstream of bacterial regulatory sequences (promoter, P) for the *lac*Z gene, the coding sequence for the mRNA Shine–Dalgarno sequence, the AUG codon, and a few codons for the N-terminal amino acids of the *lac*Z protein. The mRNA produced from this expression vector directs synthesis in the bacterium of a foreign protein with a few of its N-terminal amino acids of bacterial protein origin (a fusion protein).

foreign gene in a vector such that the fusion protein synthesized contains a signal peptide that can be recognized by the bacterial signal peptidase that properly processes the protein for secretion.

7.12 | EXPRESSION VECTORS IN EUKARYOTIC CELLS

Mammalian genetic diseases result from missing or defective intracellular proteins. To utilize recombinant techniques to treat these diseases, vectors have to be constructed that can be incorporated into mammalian cells. In addition, these vectors have to be selective for the tissue or cells containing the aberrant protein. Numerous vectors permit the expression of foreign DNA genes in mammalian cells grown in tissue culture. These vectors have been used extensively for elucidation of the posttranslational processing and synthesis of proteins in cultured eukaryotic cells. Unfortunately, the goal to selectively express genes in specific tissues or at specific developmental stages within an animal has met with very limited success.

Several types of **expression vectors** have been developed that allow replication, transcription, and translation of foreign genes in eukaryotic cells grown *in vitro,* including both RNA and DNA viral vectors that contain a foreign DNA insert. These viral vectors can infect and then replicate in a host cell. Experimentally constructed vectors that contain essential DNA elements, usually derived from a viral genome, permit expression of foreign gene inserts. **Shuttle vectors** contain both bacterial and eukaryotic replication signals, thus permitting replication of the vector in both bacteria and mammalian cells. A shuttle vector allows a gene to be cloned and purified in large quantities from a bacterial system and then the same recombinant vector can be expressed in a mammalian cell. Some expression vectors become integrated into the host cell genome while others remain as extrachromosomal entities (episomes) with stable expression of their recombinant gene in the daughter cells. Other expression vectors remain as episomal DNA, permitting only transient expression of their foreign gene prior to cell death.

Foreign DNA, such as viral expression vectors, may be introduced into cultured eukaryotic cells by **transfection,** a process that is analogous to transformation of DNA into bacterial cells. The most commonly employed transfection methods involve the formation of a complex of DNA with calcium phosphate or diethylaminoethyl (DEAE)-dextran, which is then taken up by the cell by endocytosis. The DNA is subsequently transferred from the cytoplasm to the nucleus, where it is replicated and expressed. The details of the mechanism of transfection are not known. Both methods are employed to establish transiently expressed vectors while the calcium phosphate procedure is also used for permanently expressed foreign genes. Typically, 10–20% of the cells in culture can be transfected by these procedures.

DNA Elements Required for Expression of Vectors in Mammalian Cells

Expression of recombinant genes in mammalian cells requires the presence of DNA controlling elements within the vector that are not necessary in the bacterial system. To be expressed in a eukaryotic cell the cloned gene is inserted in the vector in the proper orientation relative to control elements, including a promoter, polyadenylation signals, and an enhancer sequence. Expression may be improved by the inclusion of an intron. Some or all of these DNA elements may be present in the recombinant gene if whole genomic DNA is used for cloning. A particular cloned fragment generated by restriction endonuclease cleavage, however, may not contain the required controlling elements. A cDNA would not possess these required DNA elements. It is therefore necessary that the expression vector to be used in mammalian cells be constructed such that it contains all of the required controlling elements.

An expression vector can be constructed by insertion of required DNA controlling elements into the vector by recombinant technologies. Enhancer and promoter elements, engineered into an expression vector, should be recognized by a broad spectrum of cells in culture for the greatest applicability of the vector. Controlling elements derived from viruses with a broad host range are used for this purpose and are usually derived from the **papovavirus, simian virus 40 (SV40), Rous sarcoma virus,** or the **human cytomegalovirus.**

The vector must replicate so as to increase the number of copies within each cell or to maintain copies in daughter cells. The vector is, therefore, constructed to contain DNA sequences derived from a virus and referred to as the origin of replication (ori) that promote its replication in the eukaryotic cell. Specific protein factors, encoded by genes engineered into the vector or previously introduced into the host genome, recognize and interact with the ori sequence to initiate DNA replication.

Selection of Transfected Eukaryotic Cells by Utilizing Mutant Cells that Require Specific Nutrients

It is important to have a means of growing the transfected cells selectively since they often represent 10–20% of the cell population. As was the case for the bacterial plasmid, a gene can be incorporated into the vector that encodes an enzyme that confers resistance to a drug or confers selective growth capability to the cells carrying the vector. Constructing vectors that express both a selectable marker and a foreign gene is difficult. **Cotransfection** circumvents this problem. Two different vectors are efficiently taken up by those cells capable of being transfected. In most cases greater than 90% of transfected cells carry both vectors, one with the selectable marker and the second carrying the gene of interest.

Two commonly employed selectable markers are the thymidine kinase (*tk*) and the dihydrofolate reductase genes. The *tk* gene product, thymidine kinase, is ex-

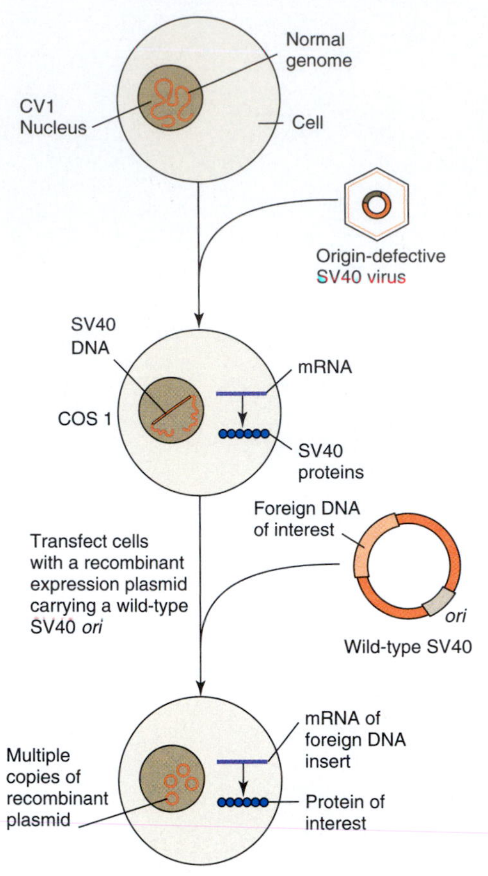

FIGURE 7.23
Expression of foreign genes in eukaryotic COS cells.
CV1, an established tissue culture cell line of simian origin, can be infected and supports the lytic replication of the simian DNA virus, SV40. Cells are infected with an origin (ori)-defective mutant of SV40 whose DNA permanently integrates into the host CV1 cell genome. The defective viral DNA continuously codes for proteins that can associate with a normal SV40 ori to regulate replication. Due to its defective ori, the integrated viral DNA will not produce viruses. The SV40 proteins synthesized in the permanently altered CV1 cell line, COS-1, can, however, induce the replication of recombinant plasmids carrying a wild-type SV40 ori to a high copy number (as high as 10^5 molecules per cell). The foreign protein synthesized in the transfected cells may be detected immunologically or enzymatically.

pressed in most mammalian cells and participates in the salvage pathway of thymidine. Several mutant cell lines have been isolated that lack a functional thymidine kinase gene (tk^-) and in growth medium containing hypoxanthine, aminopterin, and thymidine these cells will not survive. Only those tk^- mutant cells cotransfected with a vector carrying a *tk* gene, usually of herpes simplex virus origin, will grow in the medium. In most instances, these cells have been cotransfected with the gene of interest.

The dihydrofolate reductase gene (*dhfr*) is required to maintain cellular concentrations of tetrahydrofolate for nucleotide biosynthesis (see p. 855). Cells lacking this enzyme will only survive in media containing thymidine, glycine, and purines. Mutant cells ($dhfr^-$), which are transfected with the *dhfr* gene, can therefore be selectively grown in a medium lacking these supplements. Expression of foreign genes in mutant cells, cotransfected with selectable markers, is limited to cell types with the required gene defect that can be isolated. Normal cells, however, transfected with a vector carrying the *dhfr* gene, are also resistant to methotrexate, an inhibitor of dihydrofolate reductase, and these cells can be selected for by growth in methotrexate.

Another approach for selecting nonmutated cells involves use of a bacterial gene coding for aminoglycoside 3′-phosphotransferase (APH) for cotransfection. Cells that express APH are resistant to aminoglycoside antibiotics such as neomycin and kanamycin, which inhibit protein synthesis in both prokaryotes and eukaryotes. Vectors that carry an *APH* gene can be used as a selectable marker in both bacterial and mammalian cells.

Foreign Genes Expressed in Virus-Transformed Eukaryotic Cells

Figure 7.23 depicts the transient expression of a transfected gene in COS cells, a commonly used system to express foreign eukaryotic genes. The COS cells are permanently cultured simian cells, transformed with an origin-defective SV40 genome. The defective viral genome has integrated into the host cell genome and constantly expresses viral proteins. Infectious viruses that are normally lytic to infected cells are not produced because the viral origin of replication is defective. The SV40 proteins expressed by the transformed COS cell will recognize and interact with a normal SV40 ori carried in a transfected vector. These SV40 proteins will therefore promote the repeated replication of the vector. A transfected vector containing both an SV40 ori and a gene of interest may reach a copy number in excess of 10^5 molecules/cell. Transfected COS cells die after 3–4 days possibly due to a toxic overload of the episomal vector DNA.

7.13 | SITE-DIRECTED MUTAGENESIS

By mutating selected regions or single nucleotides within cloned DNA it is possible to define the role of DNA sequences in gene regulation and amino acid sequences in protein function. **Site-directed mutagenesis** is the controlled alteration of selected regions of a DNA molecule. It may involve insertion or deletion of selected DNA sequences or the replacement of a specific nucleotide with a different base. A variety of chemical methods mutate DNA *in vitro* and *in vivo* usually at random sites within the molecule.

Role of DNA Flanking Regions Evaluated by Deletion and Insertion Mutations

Site-directed mutagenesis can be carried out in various regions of a DNA sequence including the gene itself or the flanking regions. Figure 7.24 depicts a simple deletion mutation strategy where the sequence of interest is selectively cleaved with restriction endonuclease, the specific sequences are removed, and the altered recom-

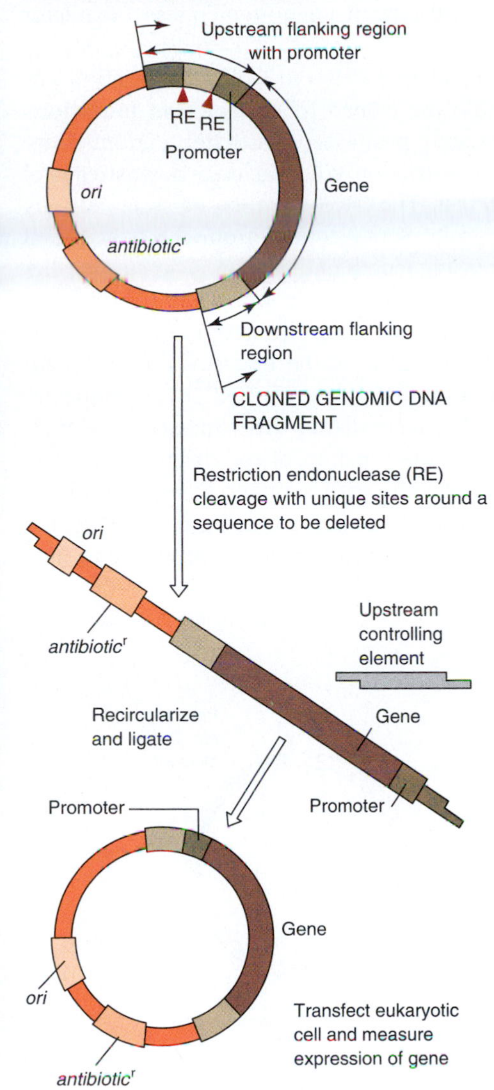

FIGURE 7.24
Use of expression vectors to study DNA regulatory sequences.
The gene of interest with upstream and/or downstream DNA flanking regions is inserted and cloned in an expression vector, and the baseline gene expression is determined in an appropriate cell. Defined regions of potential regulatory sequences can be removed by restriction endonuclease cleavage and the truncated recombinant DNA vector can be recircularized, ligated, and transfected into an appropriate host cell. The level of gene expression in the absence of the potential regulator is determined and compared to controls to ascertain the regulatory role of the deleted flanking DNA sequence.

binant vector is recircularized with DNA ligase. The role of the deleted sequence can be determined by comparing the level of expression (translation) of the gene product, measured immunologically or enzymatically, to the unaltered recombinant expression vector. A similar technique is used to insert new sequences at the site of cleavage. Deletion of a DNA sequence within the flanking region of a cloned gene can help to define its regulatory role in gene expression. The presence or absence of a regulatory sequence may not be sufficient to evaluate its role in controlling expression. The spatial arrangement of regulatory elements relative to one another, to the gene, and to its promoter may be important in the regulation of gene expression (see Chapter 8).

Analysis of potential regulatory sequences is conveniently conducted by inserting the sequence of interest upstream of a reporter gene in an expression vector. A reporter gene, usually of prokaryotic origin, encodes a gene product that can readily be distinguished from proteins normally present in the nontransfected cell and for which there is a convenient and rapid assay. A commonly used reporter gene is the **chloramphenicol acetyltransferase** (*CAT*) gene of bacteria. The gene product catalyzes the acetylation and inactivation of chloramphenicol, an inhibitor of protein synthesis in prokaryotic cells. The ability of a regulatory element to enhance or suppress expression of the *CAT* gene can be determined by assaying the level of acetylation of chloramphenicol in extracts prepared from transfected cells.

The regulatory element can be mutated prior to insertion into the vector carrying the reporter gene to determine its spatial and sequence requirements as a regulator of gene expression.

A difficulty in analysis of regulatory elements is the lack of restriction endonuclease sites at useful positions within the cloned DNA. **Deletion mutations** can be made, in the absence of appropriately positioned restriction endonuclease sites, by linearizing cloned DNA with a restriction endonuclease downstream of the potential regulatory sequence of interest. The DNA can then be systematically truncated with an exonuclease that hydrolyzes nucleotides from the free end of both strands of the linearized DNA. Increasing length of digestion generates smaller DNA fragments. Figure 7.25 demonstrates how larger deletion mutations (yielding smaller fragments) can be tested for functional activity. The enzymatic hydrolysis of the double strand of DNA occurs at both ends of the linearized recombinant vector, destroying the original restriction endonuclease site (RE_2). A unique restriction endonuclease site is reestablished to recircularize the truncated DNA molecule for further manipulations to evaluate the function of the deleted sequence. This is accomplished by ligating the blunt ends with a linker DNA, a synthetic oligonucleotide containing one or more restriction endonuclease sites. The ligated linkers are cut with the appropriate enzyme, permitting recircularization and ligation of the DNA.

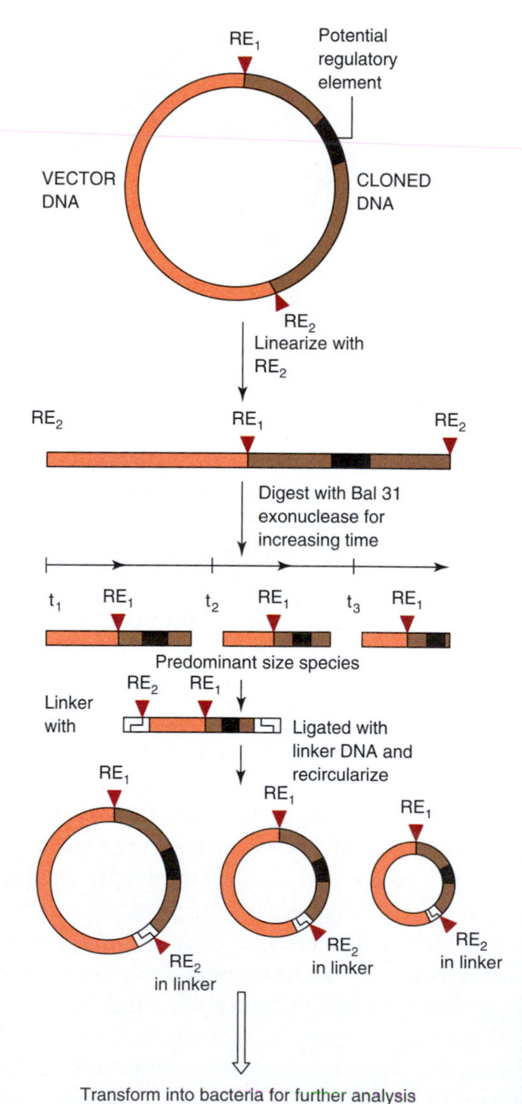

FIGURE 7.25
Enzymatic modification of potential DNA regulatory sequences.
A purified recombinant DNA molecule with a suspected gene regulatory element within flanking DNA regions is cleaved with a restriction endonuclease (RE_2). The linearized recombinant DNA is digested for varying time periods with the exonuclease, Bal31, reducing the size of the DNA flanking the potential regulatory element. The resulting recombinant DNA molecules of varying reduced sizes have small DNA oligomers (linkers) containing a restriction endonuclease sequence for RE_2 ligated to their ends. The linker-modified DNA is hydrolyzed with RE_2 to create complementary single-stranded sticky ends that permit recircularization of the recombinant vectors. The potential regulatory element, bounded by various reduced-sized flanking DNA sequences, can be amplified, purified, sequenced, and inserted upstream of a competent gene in an expression vector. Modification of the expression of the gene in an appropriate transfected cell can then be monitored to evaluate the role of the potential regulatory element placed at varying distances from the gene.

CLINICAL CORRELATION 7.7
Site-Directed Mutagenesis of HSV I gD

The structural and functional roles of a carbohydrate moiety covalently linked to a protein can be studied by site-directed mutagenesis. The gene for a glycoprotein in which an asparagine residue(s) is normally glycosylated (N-linked) must first be cloned. The herpes simplex virus, type I (HSV I) glycoprotein D (gD) may contain up to three N-linked carbohydrate groups. The envelope-bound HSV I gD appears to play a central role in virus adsorption and penetration. Carbohydrate groups may play a role in these processes.

The cloned *HSV I gD* gene has been modified by site-directed mutagenesis to alter codons for the asparagine residue at the three potential glycosylation sites. These mutated genes, cloned within an expression vector, were transfected into eukaryotic cells (COS-1), where the gD protein was transiently expressed. The mutated HSV I gD, lacking one or all of its normal carbohydrate groups, can be analyzed with a variety of available monoclonal anti-gD antibodies to determine if immunological epitopes (specific sites on a protein recognized by an antibody) have been altered. Altered epitopes would indicate that the missing carbohydrate moiety is directly associated with the normal recognition site or play a role in the protein's native conformation. An altered protein conformation can impact on immunogenicity (e.g., for vaccines) and protein processing (movement of the protein from the endoplasmic reticulum, where it is synthesized, to the membrane where it is normally bound). Mutations at two of the glycosylation sites altered the native conformation of the protein such that it was less reactive with selected monoclonal antibodies. Alteration at a third site had no apparent effect on protein structure and loss of the carbohydrate chain at all three sites did not prevent normal processing of the protein.

Sodora, D. L., Cohen, G. H., and Eisenberg, R. J. Influence of asparagine-linked oligosaccharides on antigenicity, processing, and cell surface expression of herpes simplex virus type I glycoprotein D. *J. Virol.* 63:5184, 1989.

Site-Directed Mutagenesis of a Single Nucleotide

The procedures discussed previously can elucidate the functional roles of small to large DNA sequences. Frequently, however, one wants to evaluate the role of a single nucleotide at selected sites within the DNA molecule. A single base change permits evaluation of the role of specific amino acids in a protein (see Clin. Corr. 7.7). This method also allows one to create or destroy a restriction endonuclease site at specific locations within a DNA sequence. The **site-directed mutagenesis** of a specific nucleotide is a multistep process that begins with cloning the normal type gene in a bacteriophage (Figure 7.26). The M13 series of recombinant bacteriophage vectors are commonly employed for these studies. The M13 is a filamentous bacteriophage that specifically infects male *E. coli* that express sex pili encoded for by a plasmid (F factor). The M13 bacteriophage contains DNA in a single-stranded or replicative form, which is replicated to double-stranded DNA within an infected cell. The double-stranded form of the DNA is isolated from infected cells and used for cloning the gene to be mutated. The plaques of interest can be visually identified by α-complementation (see p. 295).

The M13 carrying the cloned gene of interest is used to infect susceptible *E. coli*. The progeny bacteriophages are released into the growth medium and contain single-stranded DNA. An oligonucleotide (18–30 nucleotides long) is synthesized that is complementary to a region of interest except for the nucleotide to be mutated. This oligomer, with one mismatched base, hybridizes to the single-stranded gene cloned in the M13 DNA and serves as a primer. Primer extension is accomplished by bacteriophage T4 DNA polymerase and the resulting double-stranded DNA can be transformed into susceptible *E. coli*, where the mutated DNA strand serves as a template to replicate new (+) strands now carrying the mutated nucleotide.

The bacteriophage plaques that contain mutated DNA are screened by hybridization with a labeled probe of the original oligonucleotide. By adjusting the wash temperature of the hybridized probe only the perfectly matched hybrid remains complexed while the wild-type DNA oligomer with mismatched nucleotide will dissociate. The M13 that carries the mutated gene is then replicated in bacteria, the DNA is purified, and the mutated region of the gene is sequenced to confirm the identity of the mutation. Many modifications have been developed to improve the efficiency of site-directed mutagenesis of a single nucleotide including a method to selectively replicate the mutated strand. The M13 bacteriophage, repli-

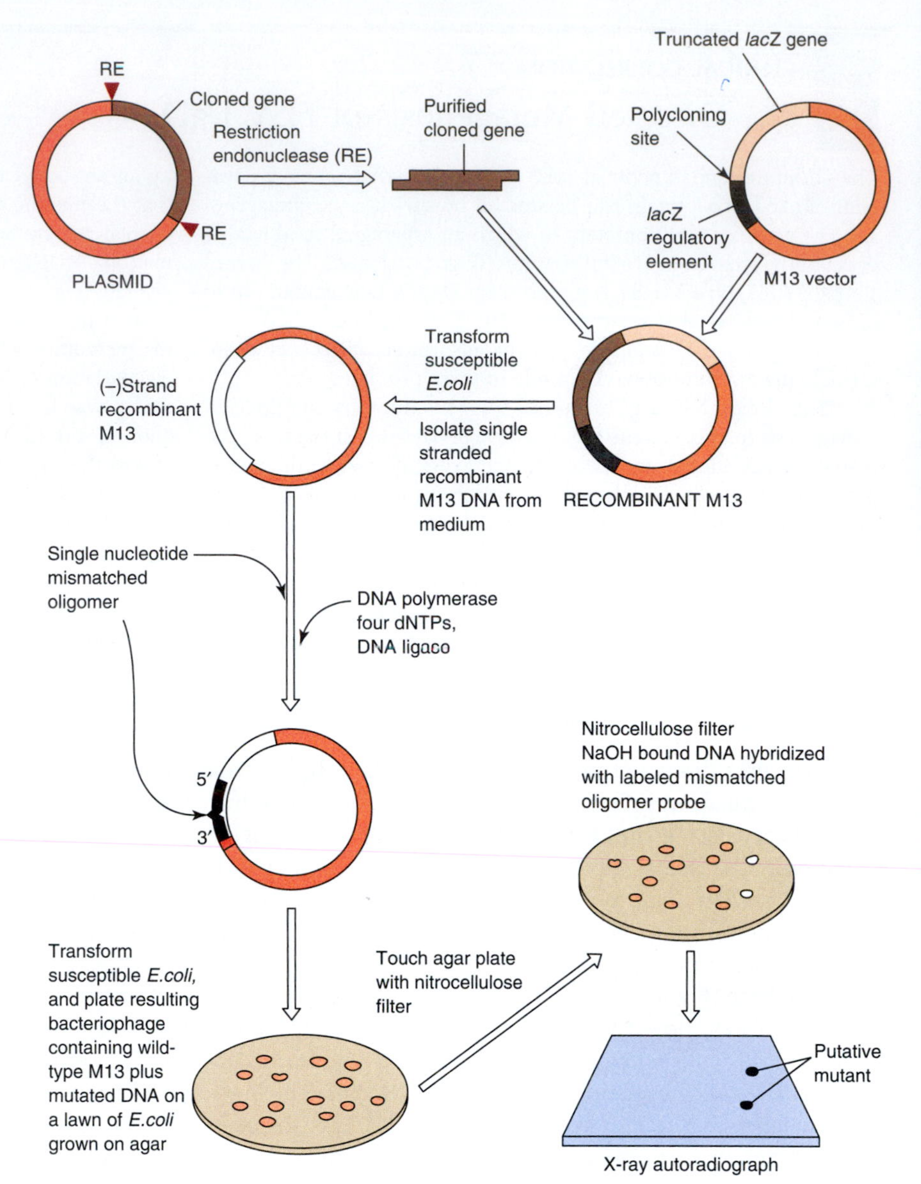

FIGURE 7.26
Site-directed mutagenesis of a single nucleotide and detection of the mutated DNA.
The figure is a simplified overview of the method. This process involves the insertion of an amplified pure DNA fragment into a modified bacteriophage vector, M13. Susceptible *E. coli,* transformed with the recombinant M13 DNA, synthesize the (+) strand DNA packaged within the virion bacteriophage proteins. The bacteriophages are isolated from the growth medium and the single-stranded recombinant M13 DNA is purified. The recombinant M13 DNA serves as a template for DNA replication in the presence of DNA polymerase, deoxynucleoside triphosphates (dNTPs), DNA ligase, and a special primer. The DNA primer (mismatched oligomer) is synthesized to be exactly complementary to a region of the DNA (gene) of interest except for the one base intended to be altered (mutated). The newly synthesized M13 DNA, therefore, contains a specifically mutated base, which, when reintroduced into susceptible *E. coli,* will be faithfully replicated. The transformed *E. coli* are grown on agar plates with replicas of the resulting colonies picked up on a nitrocellulose filter. DNA associated with each colony is denatured and fixed to the filter with NaOH and the filter bound DNA is hybridized with a ^{32}P-labeled mismatched DNA oligomer probe. The putative mutants are then identified by exposing the filter to X-ray film.

cated in a mutant *E. coli,* incorporates some uracil residues into its DNA in place of thymine due to a metabolic defect in the synthesis of dTTP from dUTP and the lack of an enzyme that normally removes uracil residues from DNA. The purified single-stranded M13 uracil-containing DNA is hybridized with a complementary oligomer containing a mismatched base at the nucleotide to be mutated. The oligomer serves as the primer for DNA replication *in vitro* with the template (+) strand containing uracils and the new (−) strand containing thymines. When this double-stranded M13 DNA is transformed into a wild-type *E. coli,* the uracil-containing strand is destroyed and the mutated (−) strand serves as the template for the progeny bacteriophages, most of which will carry the mutation of interest.

The PCR can also be employed for site-directed mutagenesis. Strategies have been developed to incorporate a mismatched base into one of the oligonucleotides that primes the PCR. Some of these procedures employ M13 bacteriophage and follow the principles described in Figure 7.26. A variation of these PCR methods, **inverse PCR mutagenesis,** has been applied to small recombinant plasmids (4–5 kb) (Figure 7.27). The method is very rapid with 50–100% of the generated colonies containing the mutant sequence. The two primers are synthesized so that they anneal back-to-back with one primer carrying the mismatched base.

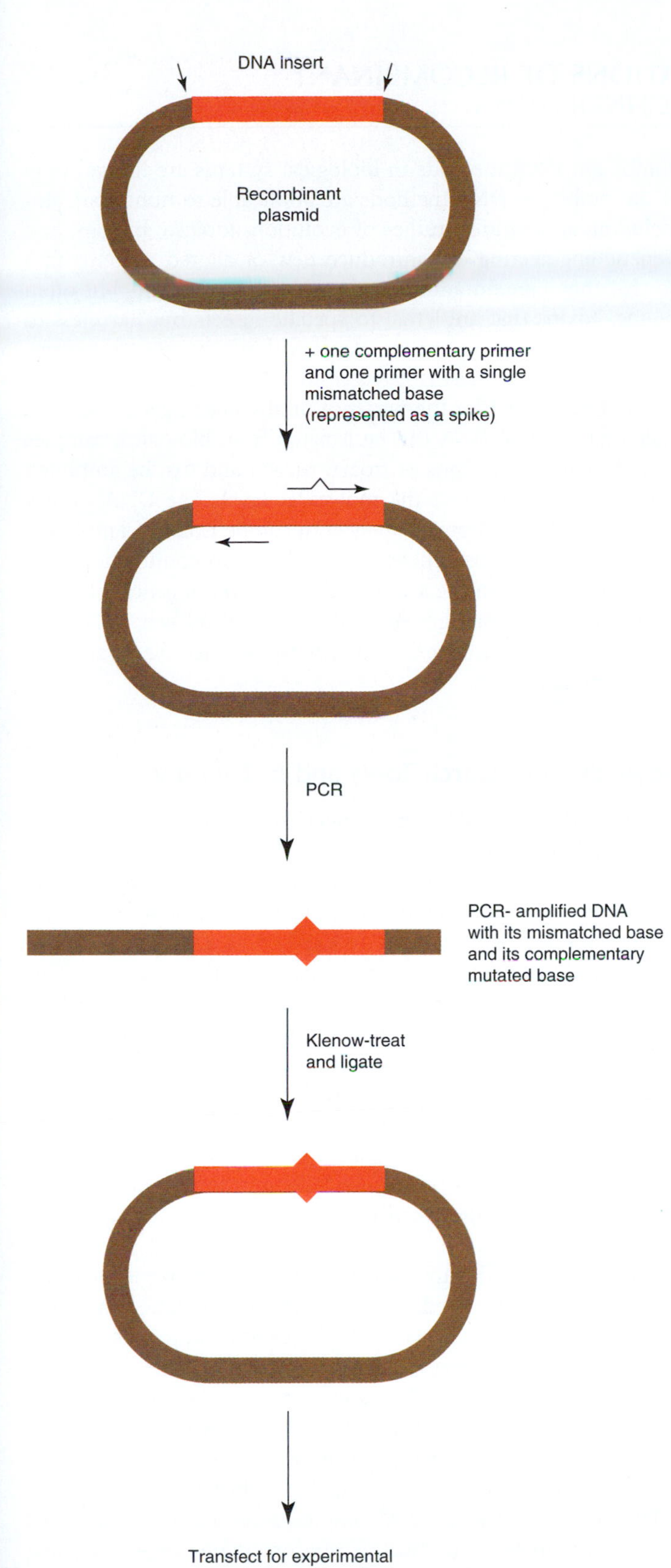

FIGURE 7.27
Inverse PCR mutagenesis.
A single base can be mutated in recombinant DNA plasmids by inverse PCR. Two primers are synthesized with their antiparallel 5′-ends complementary to adjacent bases on the two strands of DNA. One of the two primers carries a specific mismatched base that is faithfully copied during the PCR amplification steps, yielding ultimately a recombinant plasmid with a single mutated base.

7.14 | APPLICATIONS OF RECOMBINANT DNA TECHNOLOGIES

Practical uses of recombinant DNA methods in biological systems are limited only by one's imagination. Recombinant DNA methods are applicable to numerous biological disciplines including agriculture, studies of evolution, forensic biology, and clinical medicine. Genetic engineering can introduce new or altered proteins into crops (e.g., corn), so that they contain amino acids essential to humans but often lacking in plant proteins. Toxins that are lethal to specific insects but harmless to humans can be introduced into crops to protect plants without the use of environmentally destructive pesticides. The DNA isolated from cells in the amniotic fluid of a pregnant woman can be analyzed for the presence or absence of genetic defects in the fetus. Minuscule quantities of DNA can be isolated from biological samples that have been preserved in ancient tar pits or frozen tundra and can be amplified and sequenced for evolutionary studies at the molecular level. The DNA from a single hair, a drop of blood, or sperm from a rape victim can be isolated, amplified, and mapped to aid in identifying felons. Current technologies in conjunction with future invented methods should permit the selective introduction of genes into cells with defective or absent genes. Developing methodologies are also likely to become available to introduce nucleic acid sequences into specific cells to selectively turn off the expression of detrimental genes.

Antisense Nucleic Acids as Research Tools and in Therapy

Antisense nucleic acids (RNA or DNA) have been introduced to study the intracellular expression and function of specific proteins. Natural and synthetic antisense nucleic acids that are complementary to mRNAs do hybridize within the cell and inactivate the mRNA and block translation. The introduction of antisense nucleic acids into cells has opened new avenues to explore how proteins, whose expression has been selectively repressed in a cell, function within that cell. This method also holds great promise in control of viral infections. Antisense technology and site-directed mutagenesis are part of an approach termed **reverse genetics.** Reverse genetics (from gene to phenotype) selectively modifies a gene to evaluate its function, as opposed to classical genetics, which depends on the isolation and analysis of cells carrying random mutations that can be identified. A second use of the term reverse genetics refers to the mapping and ultimate cloning of a human gene associated with a disease where no prior knowledge of the molecular agents causing the disease exists. The use of the term "reverse genetics" in this latter case is likely to be modified.

Antisense RNA can be introduced into a cell by common cloning techniques. In Figure 7.28 one method is demonstrated. A gene of interest is cloned in an expression vector in the wrong orientation. That is, the sense or coding strand that is normally inserted into the expression vector downstream of a promoter is intentionally inserted in the opposite direction. This now places the complementary or antisense strand of the DNA under the control of the promoter with expression or transcription yielding antisense RNA. Transfection of cells with the antisense expression vector introduces antisense RNA that is capable of hybridizing with normal cellular mRNA. The mRNA–antisense RNA complex is not translated due to a number of reasons such as its inability to bind to ribosomes, blockage of normal processing, and rapid enzymatic degradation.

DNA oligonucleotides have been synthesized that are complementary to the known sequence of mRNA for a selected gene. Introduction of specific DNA oligomers to cells in culture have inhibited viral infection including infection by the human immunodeficiency virus (HIV). It is conceivable that one day bone marrow cells will be removed from AIDS patients and antisense HIV nucleic acids will be introduced into these cells in culture. These "protected" cells can then be reintro-

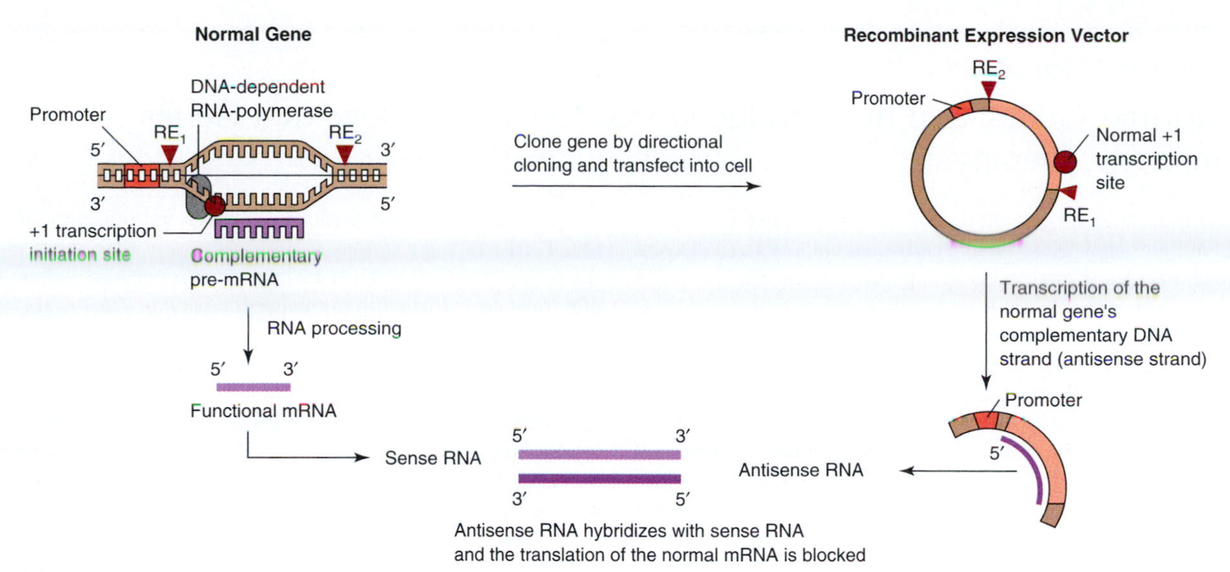

FIGURE 7.28
Production of antisense RNA.
A gene, or a portion of it, is inserted into a vector by directional cloning downstream of a promoter and in the reverse orientation to that normally found in the cell of origin. Transfection of this recombinant DNA into the parental cell that carries the normal gene results in the transcription of RNA (antisense RNA) from the cloned reversed-polarity DNA along with a normal cellular mRNA (sense RNA) transcript. The two antiparallel complementary RNAs hybridize within the cell, resulting in blocked expression (translation) of the normal mRNA transcript.

duced into the AIDS patient's bone marrow (autologous bone marrow transplantation) and replace those cells normally destroyed by the virus. Experimental progress is also being made with antisense nucleic acids that can regulate the expression of oncogenes, genes involved in cancer formation. Harnessing antisense technologies holds great promise for the treatment of human diseases.

Normal Genes Can Be Introduced into Cells with a Defective Gene: Gene Therapy

It is sometimes desirable for the transfected recombinant DNA to replicate to high copy numbers independent of the cell cycle. In other situations it is preferable for only one or few copies to integrate into the host genome with replication being regulated by the cell cycle. Individuals who possess a defective gene resulting in a debilitating or fatal condition could theoretically be treated by supplying their cells with a normal gene. **Gene therapy** is in its infancy; however, the successful transfer of a normal gene to humans has been accomplished employing retroviral vectors (see Clin. Corr. 7.8). The success of gene transfer depends, in part, on the integration of the gene into the host genome. This is directed by the retroviral integration system. Integration, however, is normally a random event that could result in deleterious consequences. Exciting studies are in progress that indicate that the viral integration machinery can be selectively tethered to specific target sequences within the host DNA by protein–protein interactions to obviate these potential problems.

Recombinant DNA in Agriculture Has Significant Commercial Impact

Perhaps the greatest gain to all humanity would be the practical use of recombinant technology to improve our agricultural crops. Genes must be identified and isolated that code for properties such as higher crop yield, rapid plant growth, resistance to adverse conditions such as arid or cold periods, and plant size. New genes, not common to plants, may be engineered into plants that confer resistance to insects,

CLINICAL CORRELATION 7.8

Normal Genes Can Be Introduced into Cells with Defective Genes in Gene Therapy

More than 4000 different genetic diseases are known, many of which are debilitating or fatal. Most are currently incurable. With the advent of new technologies in molecular biology, the clinical application of gene transfer and gene therapy is becoming a reality. Diseases that result from a deficiency in adenosine deaminase (ADA) or a mutation in the gene that encodes a subunit of several cytokines, γ_c, are but two of many genetic diseases that may readily be cured by gene therapy.

ADA is important in purine salvage, catalyzing the conversion of adenosine to inosine or deoxyadenosine to deoxyinosine. It is a protein of 363 amino acids with highest activity in thymus and other lymphoid tissues. A defect in the *ADA* gene is inherited as an autosomal recessive disorder. Over 30 mutations are associated with the disease. ADA deficiency causes a severe combined immunodeficiency disease (SCID), by an unknown mechanism. These immune-compromised children usually die in the first few years of life from overwhelming infections. The first authorized gene therapy in humans began on September 14, 1990 with the treatment of a four-year-old girl with ADA deficiency. The patient's peripheral blood T cells were expanded in tissue culture with appropriate growth factors. The *ADA* gene was introduced within these cells by retroviral mediated gene transfer. A modified retrovirus was constructed to contain the human *ADA* gene such that it would be expressed in human cells without virus replication. (These viruses that cannot replicate are first propagated in a cell line that contains a helper virus to produce "infectious" viruses. The "infectious" viruses with foreign genetic information can now infect and transfer information to cells without helper virus functions and therefore cannot replicate.) The transfer of the *ADA* gene to the patient's T cells was mediated by retroviral infection. The modified T cells carrying a normal ADA gene were then reintroduced to the patient by autologous transfusion. Levels of ADA as low as 10% of normal are sufficient to normalize the patient. The patient, now ten years later at age 13, is alive and well and has maintained circulating gene-corrected T cells at a level of 20–25% of total T cells.

The X-linked inherited disorder, SCID-X1, results in the blocked differentiation of T and natural killer (NK) lymphocytes. SCID-X1 results from a mutation in the cytokine subunit, γ_c, common to interleukin-2, -4, -7, -9, and -15 receptors. The methodology of treatment for an 8- and an 11-month-old patient was similar to that employed for the ADA-deficient SCID patients, which tends to be a less severe disease than SCID-X1. Unlike the earlier studies that employed mature T cells, the SCID-X1 studies transduced $CD34^+$ stem cells with a retroviral vector carrying the normal γ_c cDNA. This resulted in a much higher level of gene transduction than observed in earlier studies. Both infants had their immune systems normalized sufficiently after 3 months to leave the protective isolation of the hospital.

Blaese, R. M. Progress toward gene therapy. *Clin. Immunol Immunopathol.* 61:574, 1991; Mitani, K., Wakamiya, M., and Caskey, C.T., Long-term expression of retroviral-transduced adenosine deaminase in human primitive hematopoietic progenitors. *Hum. Gene Ther.* 4:9, 1993; Anderson, W. F. The best of times, the worst of times. *Science* 288:627, 2000; Cavazzana-Calvo, M., Hacein-Bey, S., de Saint Basile, G., Gross, F., Yvon, E., Nusbaum, P., Selz, F., Hue, C., Certain, S., Casanova, J.-L., Bousso, P., Le Deist, F., and Fischer, A. Gene therapy of human severe combined immunodeficiency (SCID)-X1 disease. *Science* 288:669, 2000.

fungi, or bacteria. Finally, genes encoding existing structural proteins can be modified to contain essential amino acids not normally present in the plant, without modifying the function of the protein. The potential to produce plants with new genetic properties depends on the ability to introduce genes into plant cells that can differentiate into whole plants.

New genetic information carried in crown gall plasmids can be introduced into plants infected with soil bacteria known as agrobacteria. Agrobacteria naturally contain a crown gall or Ti (tumor-inducing) plasmid whose genes integrate into an infected cell's chromosome. The plasmid genes direct the host plant cell to produce new amino acid species that are required for bacterial growth. A crown gall, or tumor mass of undifferentiated plant cells, develops at the site of bacterial infection. New genes can be engineered into the Ti plasmid, and the recombinant plasmid introduced into plant cells on infection with the agrobacteria. Transformed plant cells can then be grown in culture and under proper conditions can be induced to redifferentiate into whole plants. Every cell would contain the new genetic information and would represent a transgenic plant.

Some limitations in producing plants with improved genetic properties must be overcome before significant advances in our world food supply can be realized. Clearly, proper genes must yet be identified and isolated for desired characteristics. Also, important crops such as corn and wheat cannot be transformed by Ti plasmids; therefore other vectors must be identified. However, significant success has been achieved in recent years in designing crop plants with resistance to insects and viruses. Of equal importance is the very recent genetic engineering feat of inserting a foreign gene into pea plants that now produce a protein that inhibits the feeding of weevil larvae on the pea seeds. Peas and other legume seeds will be able to be stored without the need of protective chemical fumigants (currently Brazilian farmers lose 20–40% of their stored beans to pests).

7.15 | MOLECULAR TECHNIQUES APPLIED TO THE WHOLE ANIMAL

Many *in vitro* and intracellular systems have been described to this point that facilitate the purification, sequencing, and modification of genes or their cDNAs and DNA flanking regions of genes with potential regulatory functions. Methods have been developed and improved on in the 1990s that allow introduction of foreign genes into the whole animal or deletion of a gene from the whole animal genome so that the function of the gene product can be ascertained *in vivo*. Finally, methods are now available to clone the whole animal. While there are some clear bioethical issues and hurdles that must be addressed when using some of these methodologies, the methods themselves hold great promise for future gene therapy.

Transgenic Animals

Recombinant DNA methods allow production of large amounts of foreign gene products in bacteria and cultured cells. They also facilitate the evaluation of the role of a specific gene product in cell structure or function. In order to investigate the role of a selected gene product in growth and development of a whole animal the gene must be introduced into the fertilized egg. Foreign genes can be inserted into the genome of a fertilized egg. Animals that develop from a fertilized egg with an inserted foreign gene carry that gene in every cell and are referred to as **transgenic animals.**

The most commonly employed method to create transgenic animals is outlined in Figure 7.29. The gene of interest is usually a cloned recombinant DNA molecule, which includes its own promoter or is cloned in a construct with a different promoter that can be selectively regulated. Multiple copies of the foreign gene are microinjected into the pronucleus of the fertilized egg. The foreign DNA inserts randomly within the chromosomal DNA. If the insert disrupts a critical cellular gene the embryo will die. Usually, nonlethal mutagenic events result from the insertion of foreign DNA into the chromosome.

Transgenic animals are currently being used to study several different aspects of the foreign gene, including the analysis of DNA regulatory elements, expression of proteins during differentiation, tissue specificity, and the potential role of oncogene products on growth, differentiation, and the induction of tumorigenesis. Eventually, it is expected that these and related technologies will allow for methods to replace defective genes in the developing embryo (see Clin. Corr. 7.9).

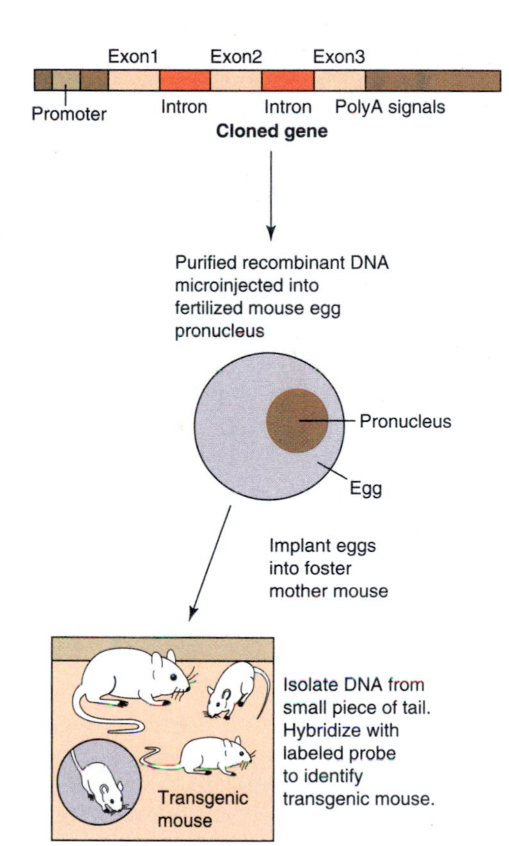

FIGURE 7.29
Production of transgenic animals.
Cloned, amplified, and purified functional genes are microinjected into several fertilized mouse egg pronuclei *in vitro*. The eggs are implanted into a foster mother. DNA is isolated from a small piece of each offspring pup's tail and hybridized with a labeled probe to identify animals carrying the foreign gene (transgenic mouse). The transgenic mice can be mated to establish a new strain of mice. Cell lines can also be established from tissues of transgenic mice to study gene regulation and the structure/function of the foreign gene product.

Knockout Mice

The creation of an animal with a selected gene "destroyed" in every cell allows researchers to define the biological role of the gene if its loss is nonlethal. The basic principles behind creating a **null** or **knockout mouse** involve inactivating a recombinant purified gene, introducing this altered gene into an embryonic stem (ES) cell such that it replaces the normal gene, injecting the modified ES cell into a developing blastocyst, implanting the blastocyst into a foster mother, and selecting offspring lacking a normal gene. A common approach to inactivate a selected gene is to insert an antibiotic resistance gene, known as *neo,* into the cloned gene of interest. This both inactivates the gene of interest and allows one to identify/select ES cells that have been successfully transfected with the altered gene. The normal gene in a very low percentage of ES cells may be replaced by the *neo*-disrupted recombinant gene by a process known as **homologous recombination.** The nucleotide sequences of the altered/disrupted gene align with the homologous se-

CLINICAL CORRELATION 7.9
Transgenic Animal Models

Transgenic animal models hold promise to correct genetic diseases early in fetal development. These animals are used to study the regulation of expression and function of specific gene products in a whole animal and have the potential for creating new breeds of commercially valuable animals. Transgenic mice have been developed from fertilized mouse eggs with rat growth hormone (*GH*) genes microinjected into their male pronuclei (see p. 321). The rat *GH* gene, fused to the mouse metallothionein-I (MT-I) promoter region, was purified from the plasmid in which it had been cloned. Approximately 600 copies of the promoter–gene complex were introduced into each egg, which was then inserted into the reproductive tract of a foster mother mouse. The resulting transgenic mouse was shown to carry the rat *GH* gene within its genome by hybridizing a labeled DNA probe to mouse DNA that had been purified from a slice of the tail, restriction endonuclease digested, electrophoresed, and Southern blotted. The diet of the animals was supplemented with $ZnSO_4$ at 33 days postparturition. The $ZnSO_4$ presumably can activate the mouse MT-I promoter to initiate transcription of the rat *GH* gene. The continuous overexpression of rat *GH* in some transgenic animals produced mice nearly twice the size of littermates that did not carry the rat *GH* gene. A transgenic mouse transmitted the rat *GH* gene to one-half of its offspring, indicating that the gene stably integrated into the germ cell genome and that new breeds of animals can be created.

Palmiter, R. D., et al. Dramatic growth of mice that develop from eggs microinjected with metallothionein-growth hormone fusion genes. *Nature* 300:611, 1982.

quence of the normal gene and swap places, or may insert within the normal gene. In either case the net result is to destroy/knock out the normal ES cellular gene. The selection of altered ES cells, microinjection into a mouse blastocyst, birth of chimeric offspring, and the ultimate breeding of offspring that is homozygous for the altered gene (Figure 7.30) is an extremely labor-intensive process. The process,

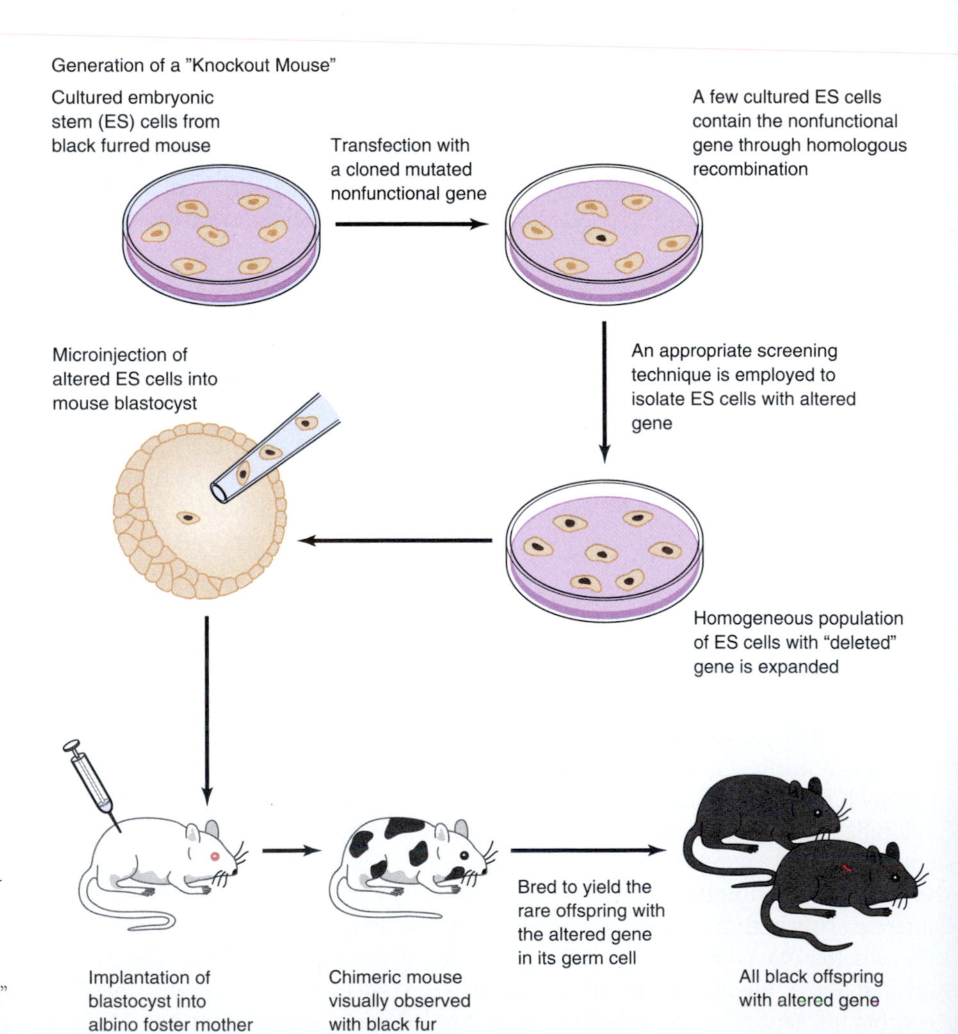

FIGURE 7.30
Generation of a knockout mouse.
Cultured embryonic cells can be manipulated through recombinant technologies to carry a defective or "deleted" gene. These altered cells can be introduced into a blastocyst that is then implanted into a foster mother. Offspring can then be selected that have the nonfunctional "deleted" gene in all of their cells—the knockout animal.

CLINICAL CORRELATION 7.10

Knockout Mice to Define a Role for the $P2Y_1$ Purinoceptor

Platelets play a central role in hemostasis and thrombosis. An important physiologic agonist that induces platelet aggregation is ADP. Early studies in the 1970s established the dogma that ADP, bound to a single G-protein-coupled receptor, induced aggregation while extracellular ATP, acting at the same receptor site, competitively inhibits the reaction. Sometimes rules, based on limited evidence, become firmly entrenched and take years to be reworked. In the 1990s it became clear that platelets, in fact, possess at least three different purinoceptors with different binding specificities for ATP versus ADP. An understanding of how these potentially interactive receptors function is very important clinically. ATP and ADP are released into the cardiovascular system/blood under a variety of normal and pathological conditions. The balance between the agonist action of ADP and the action of ATP may play a significant role in the regulation of platelet responses. Defining the mechanisms of action may lead to the production of new drugs to combat thrombosis. The recent development of a knockout (null) mouse model allowed for the assessment of the role of one of these purine receptors, $P2Y_1$.

While the $P2Y_1$ receptor is found in many tissues other than platelets, mice whose *$P2Y_1$* gene was knocked out had no apparent abnormalities. However, normal levels of ADP failed to induce platelet aggregation in samples derived from these animals, while high concentrations of ADP did work. When these platelets aggregated in response to high levels of agonist, it was not accompanied by the usual change in the shape of platelets. It was found the these $P2Y_1$-null mice were resistant to thromboembolism and this receptor might be a good target for antithrombotic drugs.

Soslau, G. and Youngprapakorn, D. A possible dual physiological role of extracellular ATP in the modulation of platelet aggregation. *Biochim. Biophys. Acta* 1355:131, 1997; and Leon, C., Hechler, G., Freund, M., Eckly, A., Vial C., Ohlmann, P., Dierich, A., LeMeus, M., Cazenave, J.-P., and Gachet, C. Defective platelet aggregation and increased resistance to thrombosis in purinergic $P2Y_1$ receptor-null mice. *J. Clin. Invest.* 104:1731, 1999.

to date, has been applied in numerous experiments to evaluate the role of selected genes such as the $P2Y_1$ **purinoceptor** (see Clin. Corr. 7.10).

Dolly, a Lamb Cloned from an Adult Cell

The **cloning** of animals raises numerous ethical questions such as: Is it humanity's purview to select the genetic composition of animals and to propagate these animals? Clearly there are many experimental advantages to developing animals with identical genetic information. However, when the talk shifts to employing whole animal cloning techniques to humans the ethical issues reach a level that most consider unacceptable.

Cloning viable offspring from adult mammalian cells is very labor intensive. The method employed to produce "Dolly," the cloned lamb, began with the culturing of mammary gland cells derived from a six-year-old Finn Dorset ewe. Cells were arrested in G_0 by being maintained in culture medium with a low serum content. The quiescent diploid donor cells were then employed for the transfer of their nuclei to anucleated oocytes derived from a Scottish Blackface ewe. The artificially "fertilized" eggs were allowed to develop to the morula or blastocyst stage in culture and then transferred to a recipient ewe who carried the fetus to term. Only a small percentage of implanted embryos survive to term. However, the methodology to clone offspring from adult somatic cells has been successfully extended to cattle, mice, and monkeys.

7.16 | CONCLUDING REMARKS

The commitment to sequence the entire human genome by the early 2000s helped spark a flurry of sequencing activities and advanced new ways to speed up the process. By 1996 the eukaryotic yeast genome, which consists of approximately 14×10^6 bp distributed among 16 chromosomes, had been sequenced. The genome of the nematode **Caenorhabditis elegans** was sequenced by 1998. The *C. elegans* genome contains approximately 18,000 genes as compared to the 6000 genes of yeast. In 2000 a multidisciplinary group at Celera Genomics of Rockville, Maryland adopted a new approach to sequence essentially the whole 180×10^6 bp genome of the fruit fly *Drosophila melanogaster.* They employed a whole genome **shotgun sequencing strategy** as opposed to the deliberate, chromosome-by-chromosome approach used by the Human Genome Project. Essentially the method involves the breakage of the whole genome into small pieces, sequencing the pieces by a rapid automated procedure, and assembling the sequenced fragments into the proper order assisted by overlapping sequences and a supercomputer. The process was also supported by previously cloned sequences. The determination of a genomic sequence is only a part of the genome puzzle. The start and stop sites of each gene must be determined along with what they do—a process referred to as **annotation.** The feat of sequencing and annotation of the *Drosophila* genome by the Celera Genomics group and their colleagues was all the more impressive when one realizes it was accomplished in less than one year. The Celera Genomics group looked to apply their shotgun strategy to speed up the sequencing of the human genome. Several human chromosomes had already been mapped and thousands of cDNA clones had been sequenced or were ready to be sequenced in early 2000, which would ultimately provide landmarks of the huge human genetic map. These procedures initially resulted in the complete or nearly complete sequencing of five chromosomes, chromosome 5, 16, 19, 21, and 22. On June 26, 2000 President Bill Clinton, in the presence of J. Craig Venter (President of Celera Genomics) and Francis S. Collins (director of the National Human Genome Research Institute), announced that 97% of the entire human genome had been sequenced, well ahead of schedule. Eukaryotic genomes usually contain permanently condensed or heterochromatic regions around the centromeres. These regions usually contain repetitive DNA that cannot be cloned and sequenced by any of the current sequencing methods. As such, no eukaryotic genome sequence will be "complete." It may be decades or more before the complete determination of all the genes and regulatory elements is accomplished. Sequencing of the human genome is only the beginning of our quest to understand the workings of our genetic makeup. The new challenge is being dealt with in the burgeoning field of **proteomics** (see p. 144), which is directed at the identification of the proteins encoded by the newly sequenced genomes.

The delineation of all of the human genes and their regulatory sequences should greatly enhance our understanding of many genetic diseases. This knowledge should also open new avenues to regulate and/or cure these diseases. Hundreds of clinical trials in gene therapy involving many patients have been initiated. Genetic diseases now identified and to be identified should eventually be curable by gene replacement therapy when the technical roadblocks are surmounted. A major hurdle is the production of safe, efficient vectors that will not induce a lethal toxic response, as occurred in 1999 with a young patient being treated for a deficiency of the liver enzyme ornithine transcarbamylase. If one looks at the enormous advances made in molecular biology in just the past three decades, it is reasonable to believe the "when" will not be that far off. The old challenge confronting scientists was how to sequence the human genome; the new challenge is how to effectively manipulate that knowledge to benefit humankind.

BIBLIOGRAPHY

Adams, D. M., Celniker, S. E., Holt, R. A., plus 192 collaborators. The genome sequence of *Drosophila melanogaster. Science* 287:2185, 2000.

Brown, W. M., George, M. Jr., and Wilson, A. C. Rapid evolution of animal mitochondrial DNA. *Proc. Natl. Acad. Sci. USA* 76:1967, 1979.

Bushman, F. Targeting retroviral integration. *Science* 267:1443, 1995.

Capecchi, M. R. Altering the genome by homologous recombination. *Science* 244:1288, 1989.

C. elegans Sequencing Consortium. Genome sequence of the nematode *C. elegans:* a platform for investigating biology. *Science* 282:2012, 1998.

Battey, J. F., Davis, L. G., and Kuehl, W. M. *Basic Methods in Molecular Biology,* 2nd ed. New York: Appleton, 1994.

Erlich, H. A. (Ed.). *PCR Technology. Principles and Applications for DNA Amplification.* New York: Stockton Press, 1989.

Feinberg, A. and Vogelstein, B. Addendum: a technique for radiolabeling DNA restriction endonuclease fragments to high specific activity. *Anal. Biochem.* 137:266, 1984.

Fields, S. Proteomics; Proteomics in genomeland. *Science* 291:1221, 2001.

Goffeau, A., et al. Life with 6,000 genes. *Science* 274:546, 1996.

International Human Genome Sequencing Consortium. Initial sequencing of the human genome. *Nature* 409:860, 2001.

Jaenisch, R. Transgenic animals. *Science* 240:1468, 1988.

Kerr, L. D. Electrophoretic mobility shift assay. *Methods Enzymol.* 254:619, 1995.

Kreeger, K.Y. Influential consortium's cDNA clones praised as genome research time-saver. *Scientist* 9:1, 1995.
Kunkel, T. A. Rapid and efficient site-specific mutagenesis without phenotypic selection. *Proc. Natl. Acad. Sci. USA* 82:488, 1985.
McPherson, M. J., Quirke, P., and Taylor, G. R. (Eds.). *PCR. A Practical Approach*, Vol. 1. Oxford, England: Oxford University Press, 1994.
Marshall, E. Gene therapy's growing pains. *Science* 269:1050, 1995.
Maxam, A. M. and Gilbert, W. A new method of sequencing DNA. *Proc. Natl. Acad. Sci. USA* 74:560, 1977.
Mulligan, R. C. The basic science of gene therapy. *Science* 260:926, 1993.
Palmiter, R. D., et al. Dramatic growth of mice that develop from eggs microinjected with metallothionein-growth hormone fusion genes. *Nature* 300:611, 1982.
Rigby, P. W. J., Dieckmann, M., Rhodes, C., and Berg, P. Labelled deoxyribonucleic acid to high specific activity *in vitro* by nick translation with DNA polymerase I. *J. Mol. Biol.* 113:237, 1977.
Sambrook, J. and Russell, D. W. *Molecular Cloning. A Laboratory Manual,* 3rd ed. Cold Spring Harbor, NY: Cold Spring Harbor Laboratory Press, 2001.
Sanger, F., Nicklen, S., and Coulson, A. R. DNA sequencing with chain-terminating inhibitors. *Proc. Natl. Acad. Sci. USA* 74:5463, 1977.
Southern, E. M. Detection of specific sequences among DNA fragments separated by gel electrophoresis. *J. Mol. Biol.* 98:503, 1975.
Venter, J. C., Adams, M. D., Myers, E. W., Li, P. W., et al. The sequence of the human genome. *Science* 291:1304, 2001.
Vogelstein, B., Fearon, E. R., Hamilton, S. R., and Feinberg, A. P. Use of restriction fragment length polymorphism to determine the clonal origin of human tumors. *Science* 227:642, 1985.
Watson, J. D., Tooze, J., and Kurtz, D. T. *Recombinant DNA: A Short Course*. San Francisco: Scientific American Books, Freeman, 1983.
Weintraub, H. M. Antisense RNA and DNA. *Sci. Am.* 262:40, 1990.
Wilmut, I., Schnieke, A. E., McWhir, J., Kind, A. J., and Campbell, K. H. S. Viable offspring derived from fetal and adult mammalian cells. *Nature* 385:810, 1997.
Zhang, Y. and Yunis, J. J. Improved blood RNA extraction micro-technique for RT-PCR. *BioTechniques* 18:788, 1998.

QUESTIONS | C. N. ANGSTADT

Multiple Choice Questions

1. Development of recombinant DNA methodologies is based on discovery of:
   A. the polymerase chain reaction (PCR).
   B. restriction endonucleases.
   C. plasmids.
   D. complementary DNA (cDNA).
   E. yeast artificial chromosomes (YACs).

2. Construction of a restriction map of DNA requires all of the following EXCEPT:
   A. partial hydrolysis of DNA.
   B. complete hydrolysis of DNA.
   C. electrophoretic separation of fragments on a gel.
   D. staining of an electrophoretic gel to locate DNA.
   E. cyclic heating and cooling of the reaction mixture.

3. In the Maxam–Gilbert method of DNA sequencing:
   A. cleavage of the DNA backbone occurs randomly at only some of the sites where the base had been destroyed.
   B. all nucleotides produced during cleavage of the DNA backbone are detected by radioautography.
   C. electrophoretic separation of DNA fragments is due to differences in both size and charge.
   D. the sequence of bands in the four lanes of the autoradiogram contains the base sequence information.
   E. dideoxynucleoside triphosphates are used.

4. The difference between the Sanger and the Maxam–Gilbert methods of DNA sequencing is:
   A. that the Maxam–Gilbert method must label the 5′-end, while the Sanger method can label either end of the DNA.
   B. only the Maxam–Gilbert method involves electrophoresing a mixture of fragments of different sizes.
   C. the Sanger method employs DNA cleavage, while the Maxam–Gilbert method employs interrupted DNA synthesis.
   D. only the Maxam–Gilbert method uses radioautography to detect fragments in which one of the termini is radioactively labeled.
   E. the Sanger method can sequence a longer segment of DNA than the Maxam–Gilbert method can.

5. Preparation of recombinant DNA requires:
   A. restriction endonucleases that cut in a staggered fashion.
   B. restriction endonucleases that cleave to yield blunt-ended fragments.
   C. poly(dT).
   D. DNA ligase.
   E. cDNA.

6. In the selection of colonies of bacteria that carry cloned DNA in plasmids, such as pBR322, and that contain two antibiotic resistance genes:
   A. one antibiotic resistance gene is nonfunctional in the desired bacterial colonies.
   B. untransformed bacteria are antibiotic resistant.
   C. both antibiotic resistance genes are functional in the desired bacterial colonies.
   D. radiolabeled DNA or RNA probes play a role.
   E. none of the above.

7. A technique for defining gene arrangement in very long stretches of DNA (50–100 kb) is:
   A. RFLP.
   B. chromosome walking.
   C. nick translation.
   D. Southern blotting.
   E. SSCP.

Refer to the following for Questions 8–10:

A. Antisense nucleic acid
B. Polymerase chain reaction
C. Site-directed mutagenesis
D. Shuttle vector
E. Transfection

8. Contains both bacterial and eukaryotic replication signals.
9. Complementary to mRNA and will hybridize to it, blocking translation.
10. Oligomer with one mismatched base is used as a primer.

Questions 11 and 12: In this country, a major cause of death of babies during the first year is sudden infant death syndrome (SIDS). One study showed a strong correlation for an increased risk of SIDS with a prolonged QT interval in their heart rate. In one child studied, a gene associated with the long QT syndrome had a substitution of AAC for TCC. This gene codes for a protein associated with the sodium channel. The mutation was detected by single-strand conformation polymorphism (SSCP).

11. Which of the following statements about SSCP is/are correct?
    A. The electrophoretic mobility on polyacrylamide gel of small, single-stranded DNA during the SSCP technique depends partly on the secondary conformation.
    B. There must be a restriction endonuclease site in the region studied for SSCP to work.
    C. Radiolabeling is not used in this technique.
    D. It is not necessary to know the sequence of the DNA to be studied.
    E. All of the above are correct.

12. In order to get enough DNA to analyze in many techniques, including SSCP, DNA is amplified by a polymerase chain reaction (PCR). The essential property of the DNA polymerase employed in PCR is that it:
    A. does not require a primer.
    B. is unusually active.
    C. is thermostable.
    D. replicates double-stranded DNA.
    E. can replicate both eukaryotic and prokaryotic DNA.

Questions 13 and 14: The purpose of gene therapy is to introduce a normal gene into cells containing a defective gene. The first authorized human gene therapy was given to a four-year-old girl with adenosine deaminase deficiency (ADA). ADA children have a severe combined immunodeficiency disease (SCID) and usually die within the first few years from overwhelming infections. A modified retrovirus was constructed to contain the human ADA gene, which could be expressed in human cells but the virus could not replicate. The child's isolated T cells were "infected" with the retrovirus to transfer the normal ADA gene to them. The modified T cells were reintroduced into the patient. The girl is now 13 years old and doing well.

13. Expression of recombinant genes in mammalian cells:
    A. will not occur if the gene contains an intron.
    B. occurs most efficiently if cDNA is used in the vector.
    C. does not require directional insertion of the gene into the vector.
    D. requires that the vector have enhancer and promoter elements engineered into the vector.
    E. does not require that the vector have an origin of replication (ori).

14. Another clinical use for recombinant DNA technology is to have rapidly replicating bacteria produce large amounts of specific proteins (e.g., hormones). Expression of a eukaryotic gene in prokaryotes involves:
    A. a Shine–Delgarno (SD) sequence in mRNA.
    B. absence of introns.
    C. regulatory elements upstream of the gene.
    D. a fusion protein.
    E. all of the above.

Problems

15. The X-ray autoradiograph of one strand of a fragment of DNA sequenced by the Maxam–Gilbert method is shown below. The 5′-end of the nucleotide was labeled with ^{32}P. What is the sequence of the fragment? Construct the autoradiograph pattern of the complementary sequence.

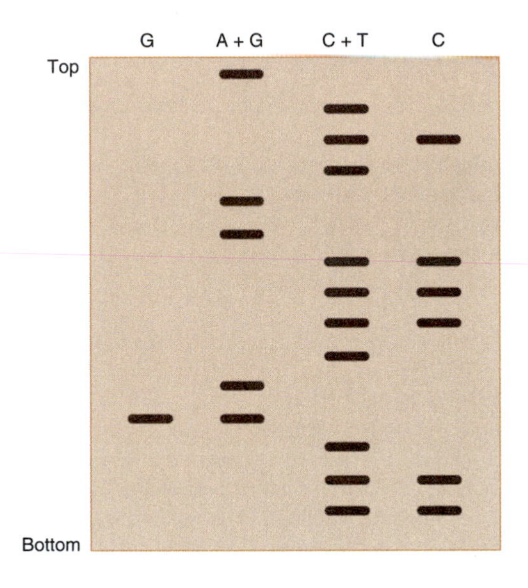

16. What would be the best vector to use to carry a segment of DNA that is 250 kb in size? Would any other type of vector work? Why?

ANSWERS

1. **B** The ability to cleave DNA predictably at specific sites is essential to recombinant DNA technology. A, C, E. These techniques or structures are involved in recombinant DNA technology but came later.
2. **E** Cyclic heating and cooling is part of the PCR process, not of restriction mapping. A, B: Restriction mapping involves all degrees of hydrolysis. Partial hydrolysis gives fragments of varying sizes, and complete hydrolysis gives the smallest possible fragments. C, D: Fragments are electrophoretically separated by size on agarose gel, which is stained to reveal the DNA.
3. **D** The relative positions of G are given by the bands in the lane corresponding to the destruction of G; of A by the bands in the AG lane that are not duplicated in the G lane; of C by the bands in the C lane; of T by the bands in the CT lane that are not duplicated in the C lane. A: Cleavage occurs at all such sites. Limited destruction of the bases is random. B: Only the nucleotides that contain the labeled 5′-terminal are detected. Other nucleotides are produced but are not detected by the method, and do not contribute information to the analysis. C: Although charge is, of course, required in order to produce movement of a particle in a field, the separation of these fragments is not due to charge differences, but to size differences, with the smallest fragments migrating farthest. E: This is part of the Sanger method.
4. **E** The Sanger method can sequence about 400 bases while the Maxam–Gilbert method is limited to about 250 bases. A: The Sanger method involves a labeled 5′-end. With the Maxam–

Gilbert method either end could be labeled. B: Both methods do this. C: This statement reverses the methodologies. D: They both use radioautography to detect fragments in which one of the termini is radioactively labeled.

5. **D** DNA ligase covalently connects fragments held together by interaction of cohesive ends. A: This is the most desirable type of restriction endonuclease to use, but it is not essential. B: Restriction nucleases that make blunt cuts can also be used if necessary. C: This is used in conjunction with poly (dA) if restriction endonucleases that make blunt cuts are employed, but it is not essential to all of recombinant DNA preparation. E: cDNA is only one type that can be used.
6. **A** The foreign DNA is inserted into one antibiotic resistance gene, thus destroying it. B: Resistance is due to the plasmids. C: See the comment for A above. D: Radiolabeling detects the DNA of interest, not the colonies that contain cloned DNA.
7. **B** A: Restriction fragment length polymorphism (RFLP) is a characteristic of DNA, not a technique. C: Nick translation is used to label DNA during chromosome walking. D: Southern blotting is a method for analyzing DNA. E: Single-strand conformation polymorphism (SSCP) is a method for detecting base changes in DNA that do not alter restriction endonuclease sites.
8. **D** This allows a gene to be cloned and purified in large quantities in bacteria and then the same vector can be expressed in a mammalian cell.
9. **A** This technology is part of an approach called reverse genetics.
10. **C** This allows evaluation of the role of a specific amino acid in a protein and can also be used to create or destroy a restriction endonuclease site at specific locations in DNA.
11. **A** A single base substitution usually modifies the conformation enough to shift the mobility as detected by SSCP. B: SSCP is the method of choice if there is no restriction endonuclease site. C: DNA is amplified by PCR in the presence of radiolabeled nucleotides. There has to be a detection method for the bands. D: SSCP requires prior knowledge of the sequence.
12. **C** PCR requires cycling between low temperatures, where hybridization of template DNA and oligomeric primers occurs, and high temperatures, where DNA melts. The Taq DNA polymerase, isolated from a thermophilic organism discovered in a hot spring on federal land, is stable at high temperatures and makes the cycling possible with no addition of fresh polymerase after each cycle.
13. **D** Eukaryotic systems require controlling elements, which are not necessary in bacterial systems. A: Expression may be improved if an intron is present. B: cDNA does not possess the required controlling elements. C: The gene must be inserted in the proper orientation relative to the control elements. E: The vector must replicate so it needs the sequence to promote replication.
14. **E** A: The SD sequence is necessary for the bacterial ribosome to recognize the mRNA. B: Bacteria do not have the intracellular machinery to remove introns from mRNA. C: Appropriate regulatory elements are necessary to allow the DNA to be transcribed. D: A fusion protein may be a product of the reaction.
15. The sequence is C C T G A T C C C A A T C T A reading 5′ to 3′. The autoradiograph of the complementary strand should look like:

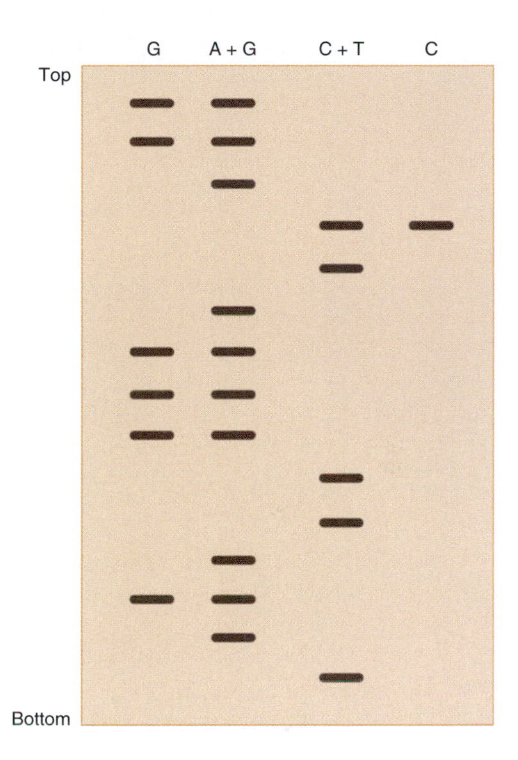

16. YACs (yeast artificial chromosomes) would be the best vector since they can accept DNA between 200 and 500 kb. Cosmids accept the next largest piece but it is only 45 kb. Plasmids and bacteriophages accept much smaller pieces.

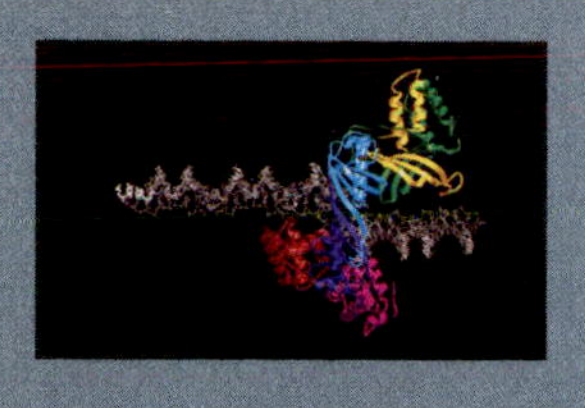

8

REGULATION OF GENE EXPRESSION

Daniel L. Weeks and
John E. Donelson

8.1 | OVERVIEW

To survive, a living cell must be able to respond to changes in its environment. One way in which cells adjust to changes is to alter expression of specific genes, which, in turn, affects the number of the corresponding protein molecules in the cell. This chapter will focus on some of the molecular mechanisms that determine when a given gene is expressed and to what extent. The attempt to understand how expression of genes is regulated is one of the most active areas of biochemical research today.

It makes sense for a cell to vary the amount of a given gene product available under different conditions. For example, the bacterium *Escherichia coli* (*E. coli*) contains genes for about 4300 different proteins, but it does not need to synthesize all of these proteins at the same time. Therefore it regulates the number of molecules of these proteins that are made. The classic illustration of this phenomenon is the regulation of the number of **β-galactosidase** molecules in the cell. This enzyme converts the disaccharide lactose to the two monosaccharides, glucose and galactose. When *E. coli* grows in a medium containing glucose as the carbon source, β-galactosidase is not required and only about five molecules of the enzyme are present in the cell. When lactose is the sole carbon source, however, 5000 or more molecules of β-galactosidase occur in the cell. Clearly, the bacteria respond to the need to metabolize lactose by increasing the synthesis of β-galactosidase molecules. If lactose is removed from the medium, the synthesis of this enzyme stops as rapidly as it began.

The complexity of eukaryotic cells means that they have even more extensive mechanisms of gene regulation than do prokaryotic cells. The differentiated cells of higher organisms have a much more complicated physical structure and often a more specialized biological function that is determined, again, by the expression of their genes. For example, **insulin** is synthesized in β cells of the pancreas and not in kidney cells even though the nuclei of all cells of the body contain the insulin genes. Molecular regulatory mechanisms facilitate the expression of insulin in pancreas and prevent its synthesis in kidney and other cells. In addition, during development of the organism the appearance or disappearance of proteins in specific cell types is tightly controlled with respect to timing and sequence of the developmental events.

As expected from the differences in complexities, far more is understood about the regulation of genes in prokaryotes than in eukaryotes. However, studies on the control of gene expression in prokaryotes often provide exciting new ideas that can be tested in eukaryotic systems. Sometimes, discoveries about eukaryotic gene structure and regulation alter the interpretation of data on the control of prokaryotic genes.

In this chapter several of the best studied examples of gene regulation in bacteria will be discussed, followed by some illustrations of the many protein transcription factors that regulate expression of genes in the human genome.

8.2 | UNIT OF TRANSCRIPTION IN BACTERIA: THE OPERON

The single *E. coli* chromosome is a circular double-stranded DNA molecule of 4.6 million base pairs whose entire sequence has been determined. Most of the approximately 4300 *E. coli* genes are not distributed randomly throughout this DNA; instead, the genes that code for the enzymes of a specific metabolic pathway are clustered in one region of the DNA. In addition, genes for associated structural proteins, such as the 70 or so proteins that comprise the **ribosome**, are frequently adjacent to one another. Members of a set of clustered genes are usually coordinately regulated; they are transcribed together to form a **"polycistronic" mRNA** species that contains the coding sequences for several proteins. The term **operon** describes the complete regulatory unit of a set of clustered genes. An operon includes the adjacent **structural genes** that code for the related enzymes or associated proteins, a **regulatory gene** or genes that code for regulator protein(s), and **control elements** that are sites on the DNA near the structural genes at which regulator proteins act. Figure 8.1 shows a partial genetic map of the *E. coli* genome that gives locations of structural genes of a few of the many *E. coli* operons.

When transcription of the structural genes of an operon increases in response to the presence of a specific substrate in the medium, the effect is known as **induction.** The increase in transcription of the β-galactosidase gene when lactose is the sole carbon source is an example of induction. Bacteria also respond to nutritional changes by quickly turning off the synthesis of enzymes that are no longer needed. As will be described below, *E. coli* synthesizes the amino acid tryptophan as the end product of a specific biosynthetic pathway. However, if tryptophan is supplied in the medium, the bacteria do not need to make it themselves, and the

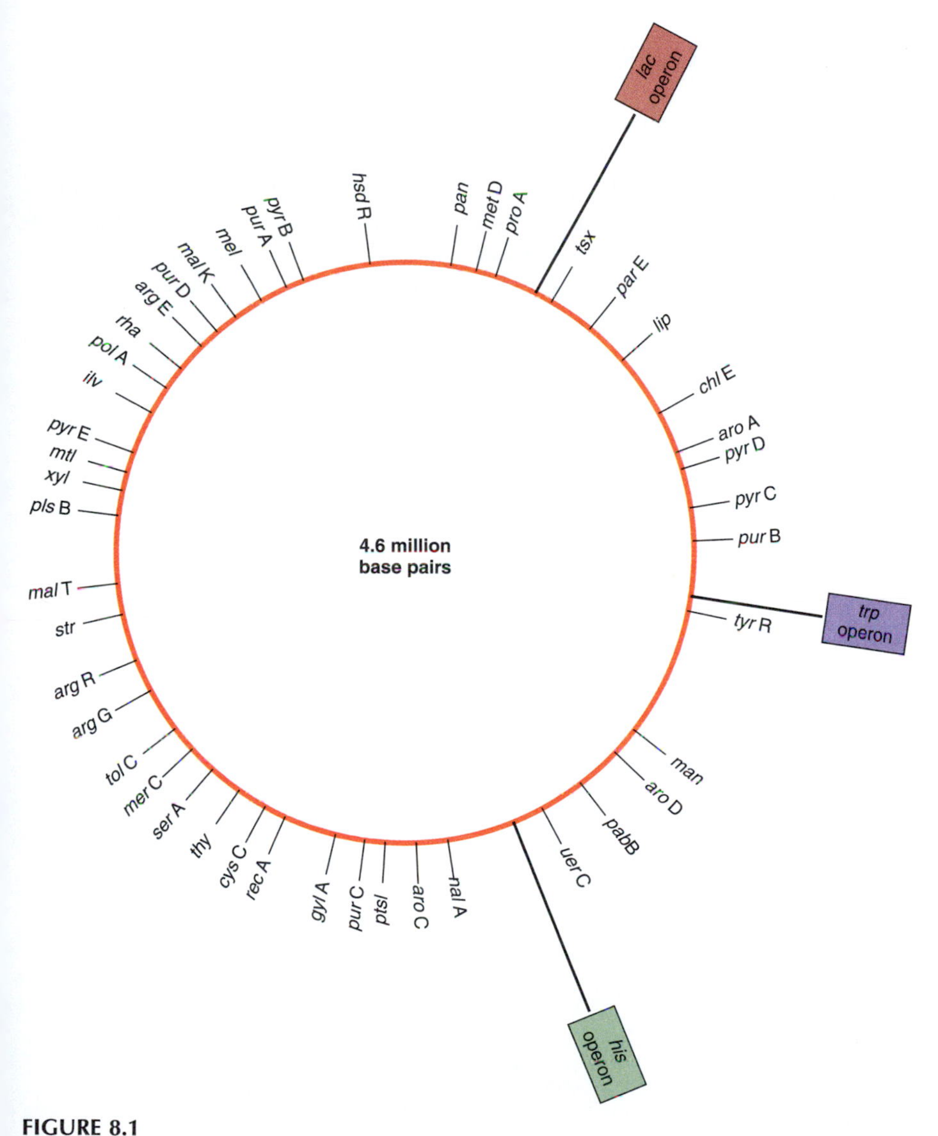

FIGURE 8.1
Partial genetic map of *E. coli*.
Locations of a few of the 4300 genes in the circular *E. coli* genome are shown. Three operons discussed in this chapter are indicated.
Reproduced with permission from Stent, G. S. and Calendar, R. Molecular Genetics, An Introductory Narrative. San Francisco: Freeman, 1978, p. 289; modified from Bachmann, B. J., Low, K. B., and Taylor, A. L. Bacteriol. Rev. 40:116, 1976.

synthesis of the enzymes for this metabolic pathway is stopped. This process is called **repression.** It permits the bacteria to avoid using their energy to make unnecessary and even harmful proteins.

Induction and repression are manifestations of the same phenomenon. In one case the bacterium changes its enzyme composition to utilize a specific substrate in the medium; in the other it reduces the number of enzyme molecules so that it does not overproduce a specific metabolic product. The signal for each type of regulation is the small molecule that is a substrate for the metabolic pathway or a product of the pathway, respectively. These small molecules are called **inducers** when they stimulate induction and **corepressors** when they cause repression.

Section 8.3 will describe in detail the lactose operon, an extensively studied set of inducible genes. Section 8.4 will present the tryptophan operon, an example of a repressible operon. Sections 8.5 and 8.6 will briefly describe other operons and gene systems in which physical movement of the genes themselves within the DNA (i.e., gene rearrangements) plays a role in their regulation.

8.3 | LACTOSE OPERON OF *E. COLI*

The **lactose operon** contains three adjacent structural genes as shown in Figure 8.2. *Lac*Z codes for the enzyme β-galactosidase, which is composed of four identical subunits of 1021 amino acids. *Lac*Y codes for a permease, which is a 275-amino-acid protein that occurs in the cell membrane and participates in the transport of sugars, including lactose, across the membrane. The third gene, *lac*A, codes for β-galactoside transacetylase, a 275-amino-acid enzyme that transfers an acetyl group from acetyl-CoA to β-galactosides. Of these three proteins, only β-galactosidase participates in a known metabolic pathway. However, the permease is clearly important since it is involved in transporting lactose into the cell. The acetylation reaction may be associated with detoxification and excretion reactions of nonmetabolized analogs of β-galactosides.

Mutations in *lac*Z or *lac*Y that destroy the function of β-galactosidase or the permease prevent cells from cleaving lactose or acquiring it from the medium, respectively. Mutations in *lac*A that destroy transacetylase activity do not seem to have an identifiable effect on cell growth and division. Perhaps there are other related enzymes in the cell that serve as backups for this enzyme, or perhaps it has an unknown function that is required only under certain conditions.

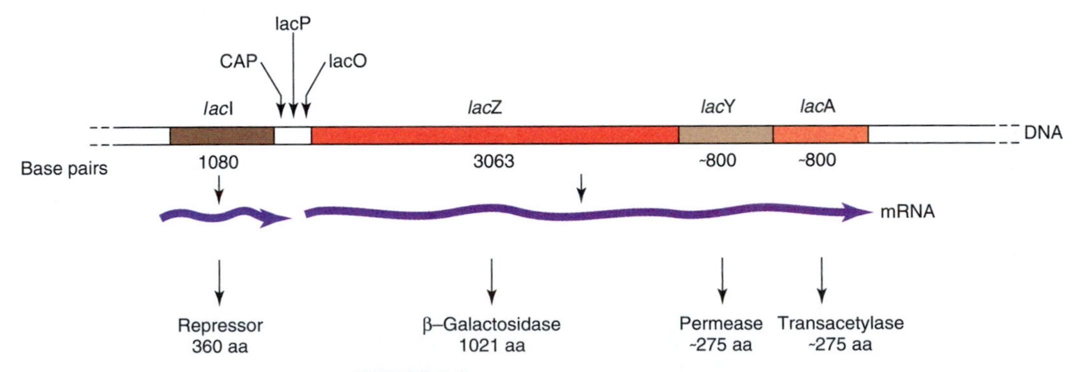

FIGURE 8.2
Lactose operon of *E. coli*.
Lactose operon is composed of *lac*I gene, which codes for a repressor, control elements of CAP, *lac*P, and *lac*O, and three structural genes, *lac*Z, *lac*Y, and *lac*A, which code for β-galactosidase, a permease, and a transacetylase, respectively. The *lac*I gene is transcribed from its own promoter. The three structural genes are transcribed from the promoter, *lac*P, to form a polycistronic mRNA from which the three proteins are translated.

A single mRNA species containing the coding sequences of all three structural genes is transcribed from a **promoter,** called *lac*P, located just upstream from the *lac*Z gene. Induction of these three genes occurs during initiation of their transcription. Without the inducer, transcription of the gene cluster occurs only at a very low level. In the presence of the inducer, transcription begins at *lac*P and proceeds through all three genes to a transcription terminator located slightly beyond the end of *lac*A. Therefore the genes are **coordinately expressed;** all three are transcribed in unison or none is transcribed.

The presence of three coding sequences on the same mRNA molecule might suggest that the relative amounts of the three proteins are always the same under varying conditions of induction. An inducer that causes a high rate of transcription will result in a high level of all three proteins; an inducer that stimulates only a little transcription of the operon will result in a low level of the proteins. The inducer can be thought of as a molecular switch that influences synthesis of the single mRNA species for all three genes. However, it turns out that the number of molecules of each protein encoded by the same operon is often different, but this does not reflect differences in transcription; it reflects differences in rates of translation of the coding sequences or in degradation of the proteins themselves.

The mRNA induced by lactose is very unstable; it is degraded with a half-life of about 3 min. Therefore expression of the operon can be altered very quickly. Transcription ceases as soon as inducer is no longer present, existing mRNA molecules disappear within a few minutes, and cells stop making the proteins.

Repressor of the Lactose Operon Is a Diffusible Protein

The regulatory gene of the lactose operon, *lac*I, codes for a protein whose only function is to control the initiation of transcription of the three *lac* structural genes. This regulator protein is called the lactose **repressor.** The *lac*I gene is located just in front of the controlling elements for the *lac*ZYA gene cluster. However, it is not obligatory that a regulatory gene be physically close to the gene cluster it regulates. In some of the other operons it is not. Transcription of *lac*I is not regulated; instead, this single gene is always transcribed from its own promoter at a low rate that is relatively independent of the cell's status. Therefore affinity of the *lac*I promoter for **RNA polymerase** seems to be the only factor involved in initiation of its transcription.

The lactose repressor is initially synthesized as a monomer of 360 amino acids and four monomers associate to form a homotetramer, the active form of the repressor. Usually there are about 10 tetramers per cell. The repressor has a strong affinity for a specific DNA sequence that lies between *lac*P and the start of *lac*Z. This sequence is called the **operator** and is designated *lac*O. The operator overlaps the promoter somewhat so that presence of repressor bound to the operator physically prevents RNA polymerase from binding to the promoter and initiating transcription.

In addition to recognizing and binding to the *lac*O sequence, the repressor has a strong affinity for inducer molecules of the lactose operon. Each monomer has a binding site for an inducer molecule. Binding of inducer to the monomers causes a **conformational change** in the repressor that greatly lowers its affinity for the operator sequence (Figure 8.3). The result is that repressor no longer binds to the operator so that RNA polymerase, in turn, can begin transcription from the promoter. A repressor molecule that is already bound to the operator when the inducer becomes available can still bind to inducer so that the repressor–inducer complex immediately disassociates from the operator.

Study of the lactose operon was greatly facilitated by the discovery that some small molecules fortuitously serve as inducers but are not metabolized by β-galactosidase. Isopropylthiogalactoside (**IPTG**) is one of several thiogalactosides with this property. They are called **gratuitous inducers.** They bind to inducer sites on the repressor molecule causing the conformational change but are not cleaved by the induced β-galactosidase. Therefore they affect the system without themselves being

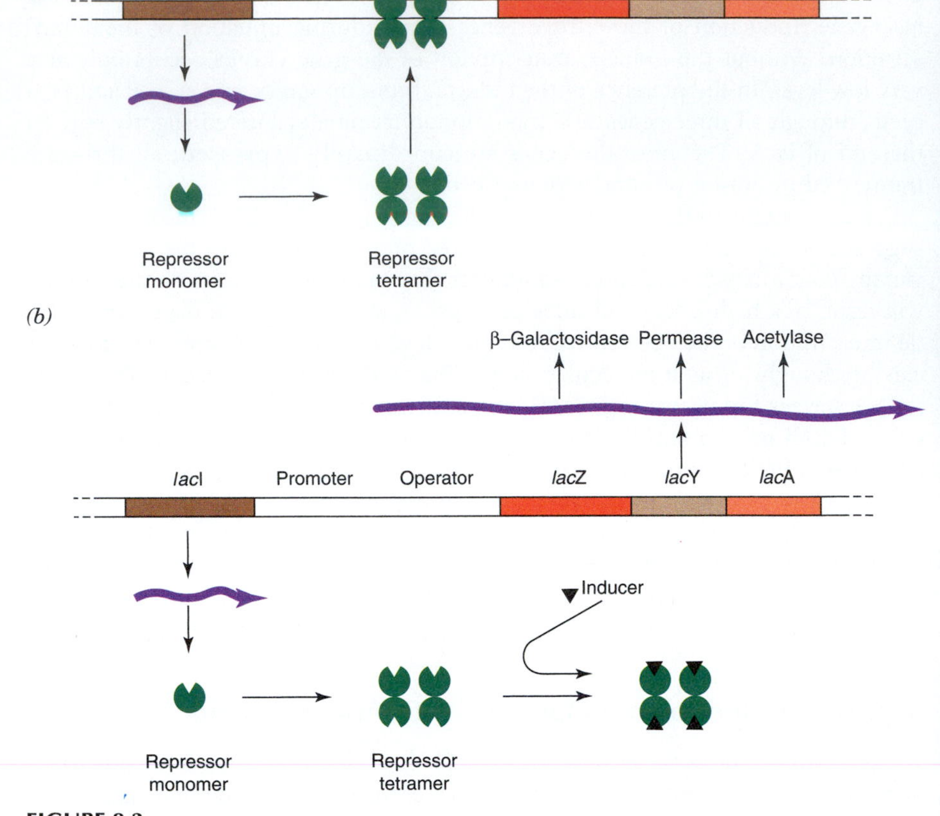

FIGURE 8.3
Control of *lac* operon.
(*a*) Repressor tetramer binds to operator and prevents transcription of structural genes.
(*b*) Inducer binds to repressor tetramer, which prevents repressor from binding to operator. Transcription of three structural genes can occur from the promoter.

altered (metabolized) by it. If it were not possible to manipulate the system with these gratuitous inducers experimentally, it would have been much more difficult to reach our current understanding of the lactose operon in particular and bacterial gene regulation in general.

The product of the *lac*I gene, the repressor protein, acts in trans; that is, it is a diffusible product that moves through the cell to its site of action. Therefore mutations in the *lac*I gene can exert an effect on the expression of other genes located far away. *Lac*I mutations can be of several types. One class of mutations changes or deletes amino acids of the repressor that are located in the binding site for the inducer. These changes interfere with interaction between the inducer and the repressor but do not affect the affinity of repressor for the operator. Therefore the repressor is always bound to the operator, even in the presence of inducer, and the *lac*ZYA genes are never transcribed above a very low basal level. Another class of *lac*I mutations changes the amino acids in the operator binding site of the repressor. Most of these mutations lessen the affinity of the repressor for the operator. This means that the repressor does not bind to the operator and the *lac*ZYA genes are always being transcribed. These mutations are called **repressor-constitutive mutations** because the *lac* genes are permanently turned on. Interestingly, a few rare *lac*I mutants actually increase the affinity of repressor for the operator over that of wild-type repressor. In these cases inducer molecules can bind to repressor, but they are less effective in releasing repressor from the operator.

Repressor-constitutive mutants illustrate the features of a negative control system. An active repressor, in the absence of an inducer, shuts off expression of the *lac* structural genes. An inactive repressor results in the constitutive, unregulated, expression of these genes. It is possible, using recombinant DNA techniques described in Chapter 7, to introduce into constitutive *lac*I mutant cells a recombinant plasmid containing the wild-type *lac*I gene (but not the rest of the lactose operon). Therefore these cells have one wild-type and one mutant *lac*I gene and synthesize both active and inactive repressor molecules. Under these conditions, normal wild-type regulation of the lactose operon occurs. In genetic terms, the wild-type induction is dominant over the mutant constitutivity. This property is the main feature of a negative control system.

Operator Sequence of the Lactose Operon Is Contiguous on DNA with a Promoter Sequence and Three Structural Genes

The control elements upstream of the structural genes of the lactose operon are the operator and the promoter. The operator was identified, like the *lac*I gene, by mutations that affected the transcription of the *lac*ZYA region. Some of these mutations also result in the constitutive synthesis of *lac* mRNA; that is, they are **operator-constitutive mutations.** In these cases the operator DNA sequence has undergone one or more base pair changes so that the repressor no longer binds as tightly to the sequence. Thus the repressor is less effective in preventing RNA polymerase from initiating transcription.

In contrast to mutations in the *lac*I gene that affect the diffusible repressor, mutations in the operator do not affect a diffusible product. They exert their influence on transcription of only the three *lac* genes that lie immediately downstream of the operator on the same DNA molecule. Thus if a second lactose operon is introduced into a bacterium on a recombinant plasmid, the operator of one operon does not influence action on the other operon. Therefore an operon with a wild-type operator will be repressed under the usual conditions, whereas in the same engineered bacterium a second operon that has an operator-constitutive mutation will be continuously transcribed.

Operator mutations are frequently referred to as **cis-dominant** to emphasize that these mutations affect only adjacent genes on the same DNA molecule and that they are not influenced by the presence in the cell of other copies of the unmutated sequence. Cis-dominant mutations occur in DNA sequences that are recognized by proteins rather than in DNA sequences that code for the diffusible proteins. **Trans-dominant** mutations occur in genes that specify the diffusible products. Therefore cis-dominant mutations also occur in promoter and transcription termination sequences, whereas trans-dominant mutations also occur in the genes for the subunit proteins of RNA polymerase, the ribosomes, and so on.

Figure 8.4 shows the sequence of both the *lac* operator and promoter. The operator sequence has an axis of **dyad symmetry.** The sequence of the upper strand on the left side of the operator is nearly identical to the lower strand on the right

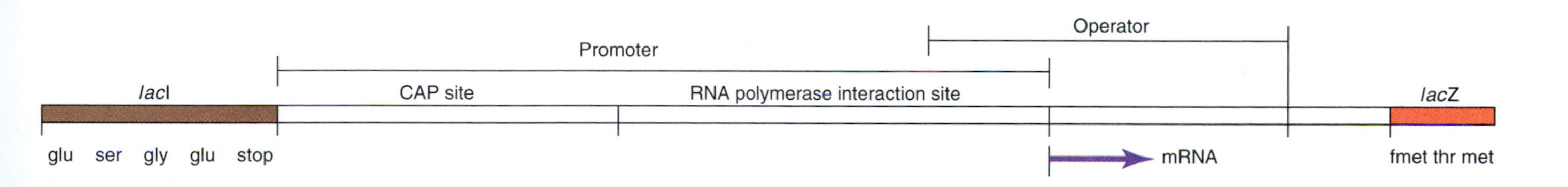

5′ GGAAAGCGGGCAGTGAGCGCAACGCAATTAATGTGAGTTAGCTCACTCATTAGGCACCCCAGGCTTTACACTTTATGCTTCCGGCTCGTATGTTGTGTGGAATTGTGAGCGGATAACAATTTCACACAGGAAACAGCTATGACCATG 3′
3′ CCTTTCGCCCGTCACTCGCGTTGCGTTAATTACACTCAATCGAGTGAGTAATCCGTGGGGTCCGAAATGTGAAATACGAAGGCCGAGCATACAACACACCTTAACACTCGCCTATTGTTAAAGTGTGTCCTTTGTCGATACTGGTAC 5′

FIGURE 8.4
Nucleotide sequence of control elements of lactose operon.
The end of *lac*I gene (coding for the lactose repressor) and beginning of *lac*Z gene (coding for β-galactosidase) are also shown. Lines above and below the sequence indicate symmetrical sequences within the CAP site and operator.

side; only three differences occur between these inverted DNA repeats. This symmetry in the DNA recognition sequence reflects symmetry in the tetrameric repressor and facilitates tight binding of the subunits of the repressor to the operator. A common feature of many protein-binding or recognition sites on double-stranded DNA, including most recognition sites for restriction enzymes, is a dyad symmetry in the nucleotide sequence.

The 30 bp that constitute the *lac* operator are an extremely small fraction of the total *E. coli* genome of 4.6 million bp and occupy an even smaller fraction of the total volume of the cell. Therefore it would seem that the approximately 10 tetrameric repressors in a cell might have trouble finding the *lac* operator if they just randomly diffuse about the cell. However, at least two factors confine the repressor to a much smaller space than the entire volume of the cell. First, the repressor gene is located very close to the *lac* operator. This means that the repressor does not have far to diffuse if its translation begins before its mRNA is fully synthesized. Second, and much more importantly, the repressor possesses a low general affinity for all DNA sequences. When the inducer binds to the repressor, its affinity for the operator is reduced about a 1000-fold, but its low affinity for random DNA sequences is unaltered. Therefore all of the lactose repressors of the cell probably spend the majority of the time in loose association with the DNA. When the binding of the inducer releases a repressor molecule from the operator, it quickly reassociates with another nearby region of the DNA. Therefore induction redistributes the repressor on the DNA rather than generating freely diffusing repressor molecules. This confines the repressor to a smaller volume within the cell.

How does lactose enter a lactose-repressed cell in the first place if the *lac*Y gene product, the permease, is repressed, yet is required for lactose transport across the cell membrane? The answer is that even in the fully repressed state, there is a very low basal level of transcription of the *lac* operon that provides five or six molecules of the permease per cell. Perhaps this is just enough to get a few molecules of lactose inside the cell and begin the process.

An even more curious observation is that, in fact, lactose is not the natural inducer of the lactose operon as we would expect. When the repressor is isolated from fully induced cells, the small molecule bound to each repressor monomer is **allolactose,** not lactose.

Allolactose, like lactose, is composed of galactose and glucose, but the linkage between the two sugars is different. It turns out that a side reaction of β-galactosidase (which normally breaks down lactose to galactose and glucose) converts these two products to allolactose. Therefore it appears that a few molecules of lactose are taken up and converted by β-galactosidase to allolactose, which then binds to the repressor and induces the operon. Further confirmation that lactose itself is not the real inducer comes from experiments indicating that binding of lactose to purified repressor slightly increases the repressor's affinity for the operator. Therefore, in the induced state, a small amount of allolactose must be present to overcome this "anti-inducer" effect of the lactose substrate.

Promoter Sequence of Lactose Operon Contains Recognition Sites for RNA Polymerase and a Regulator Protein

Immediately upstream of the *lac* operator sequence is the promoter sequence. This sequence contains the recognition sites for RNA polymerase and **catabolite activator protein (CAP;** also called **cAMP receptor protein** or **CRP).** (Figure 8.4). The site at which RNA polymerase binds to DNA to initiate transcription has been identified by genetic and biochemical approaches. Point mutations in this region frequently affect the affinity with which RNA polymerase will bind the DNA. Deletions (or insertions) that extend into this region also dramatically affect the binding of RNA polymerase to the DNA. The end points of the sequence to which RNA polymerase binds were identified by DNase protection experiments. Purified RNA polymerase was bound to the *lac* promoter region cloned in a bacteriophage DNA or a plasmid, and this protein–DNA complex was digested with DNase I. The DNA segment protected from degradation by DNase was recovered and its sequence determined. The ends of this protected segment varied slightly with different DNA molecules but corresponded closely to the boundaries of the RNA polymerase binding site shown in Figure 8.4.

The sequence of the RNA polymerase interaction site is not composed of symmetrical elements like those described for the operator sequence. This is not surprising since RNA polymerase must associate with the DNA in an asymmetrical fashion for RNA synthesis to be initiated in only one direction from the binding site. However, that portion of the promoter sequence recognized by CAP does contain some symmetry. A **DNA–protein interaction** at this region enhances transcription of the lactose operon and is described next.

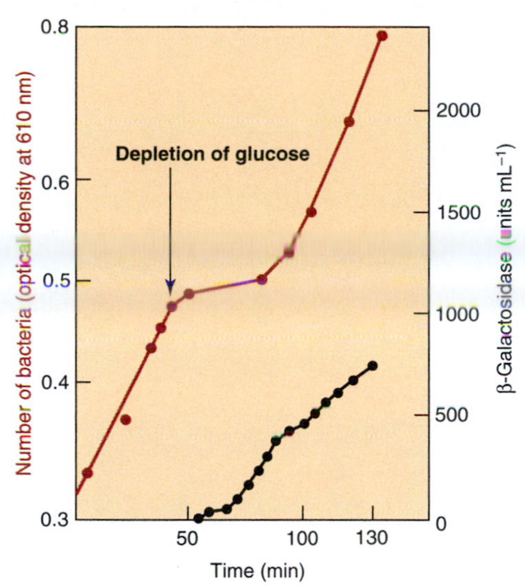

FIGURE 8.5
Lack of synthesis of β-galactosidase in *E. coli* when glucose is present.
Bacteria are growing in a medium containing initially 0.4 mg mL^{-1} of glucose and 2 mg mL^{-1} lactose. Left-hand ordinate indicates optical density of growing culture, an indicator of the number of bacterial cells. Right-hand ordinate indicates units of β-galactosidase per milliliter. Note that appearance of β-galactosidase is delayed until the glucose is depleted.
Redrawn from Epstein, W., Naono, S., and Gros, F. Biochem. Biophys. Res. Commun. *24:588, 1966.*

Catabolite Activator Protein Binds at a Site on the Lactose Promoter

Escherichia coli prefers to use glucose instead of other sugars as a carbon source. For example, if the concentrations of glucose and lactose in the medium are the same, the bacteria will selectively metabolize glucose and not utilize lactose. This phenomenon is illustrated in Figure 8.5, which shows that the appearance of β-galactosidase, the *lac*Z product, is delayed until all of the glucose in the medium is depleted. Only then is lactose used as the carbon source. This delay indicates that glucose interferes with the induction of the lactose operon, called **catabolite repression** since it occurs during the catabolism of glucose. An identical effect is exerted on a number of other inducible operons, including the arabinose and galactose operons, which code for enzymes involved in the utilization of other sugars as energy sources. It is probably a general coordinating system for turning off synthesis of unwanted enzymes whenever the preferred substrate, glucose, is present.

Catabolite repression occurs because the presence of glucose deactivates adenylyl cyclase, which synthesizes **cyclic AMP (cAMP)**, causing a lower intracellular concentration of cAMP. cAMP binds to **catabolite activator protein (CAP),** an allosteric protein that, when combined with cAMP, is capable of binding to the CAP regulatory site at the promoter of the lactose (and other) operons (Figure 8.6). The

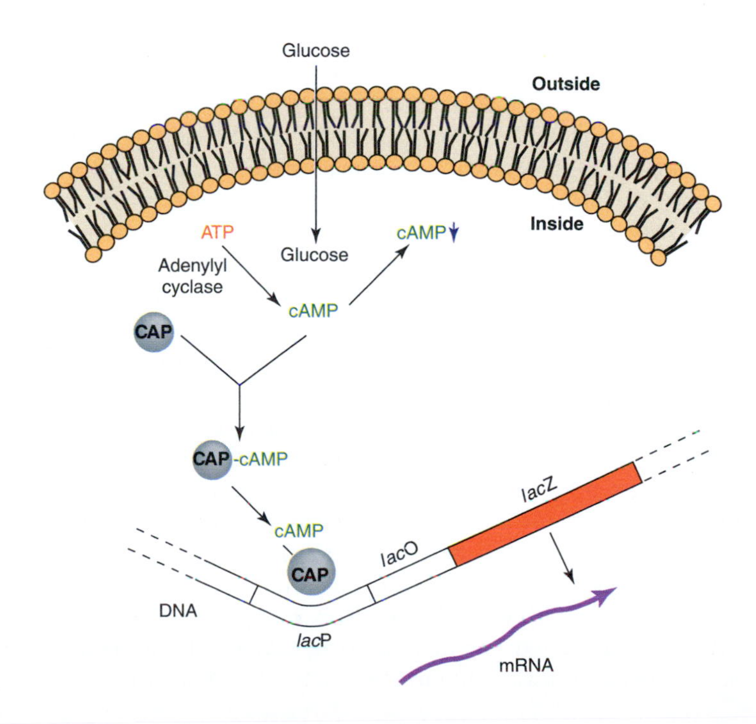

FIGURE 8.6
Control of *lac*P by cAMP.
A CAP–cAMP complex binds to CAP site and enhances transcription at *lac*P. Catabolite repression occurs when glucose lowers intracellular concentration of cAMP. This reduces the amount of the CAP–cAMP complex and decreases transcription from *lac*P and from promoters of several other operons.

CAP–cAMP complex exerts positive control on the transcription of these operons. Its binding to the CAP site on the DNA double helix causes the helix itself to undergo a dramatic bend, or kink, of about 90° at the site of the binding. This bend in DNA and the interaction between the CAP–cAMP complex and *E. coli* RNA polymerase activate the initiation of transcription at the promoter. Alternatively, if the CAP site is not occupied, RNA polymerase has more difficulty binding to the promoter, and transcription of the operon occurs much less efficiently. Thus, when glucose is present, the cAMP level drops, the CAP–cAMP complex does not form, and the positive influence on RNA polymerase does not occur. Conversely, when glucose is absent, the cAMP level is high, the CAP–cAMP complex binds to the CAP site, and transcription is enhanced.

Thus the lactose operon demonstrates how bacteria can coordinate both a general response to a metabolic condition (the need to use a sugar as an energy or carbon source) and a specific response to that condition (the need to utilize lactose as the sugar). Other operons in *E. coli* illustrate how bacteria respond when confronted with other metabolic situations.

8.4 | TRYPTOPHAN OPERON OF *E. COLI*

Another issue faced by bacteria is the need for the proper amount and relative balance of the 20 amino acids that are the constituents of proteins. Thus the cell must be able to sense whether a given amino acid is available and whether its intracellular concentration is sufficient. **Tryptophan,** for example, is required for the synthesis of all proteins that contain tryptophan. Therefore if tryptophan is not present in sufficient quantity in the medium, the bacterial cell has to make it. In contrast, lactose is not absolutely required for the cell's growth; many other sugars can substitute for it, and, in fact, as we saw in the previous section, the bacterium prefers to use some of these other sugars for the carbon source. As a result, synthesis of biosynthetic enzymes for tryptophan is regulated differently than synthesis of the proteins encoded by the lactose operon.

Tryptophan Operon Is Controlled by a Repressor Protein

In *E. coli* tryptophan is synthesized from chorismic acid in a five-step pathway that is catalyzed by three different enzymes as shown in Figure 8.7. The **tryptophan operon** contains five structural genes that code for these three enzymes (two of which contain two different subunits). Upstream from this gene cluster is a promoter where transcription begins and an operator to which binds a repressor protein encoded by the unlinked *trp*R gene. Transcription of the lactose operon is generally "turned off" unless it is induced by a small molecule inducer. The tryptophan operon, on the other hand, is always "turned on" unless it is repressed by a small molecule **corepressor** (a term used to distinguish it from the repressor protein). Hence the lactose operon is inducible, whereas the tryptophan operon is repressible. When the tryptophan operon is being actively transcribed, it is said to be **derepressed,** since the tryptophan repressor is not preventing RNA polymerase from binding. This is mechanistically the same as an induced lactose operon in which the lactose repressor is not interfering with RNA polymerase.

The tryptophan biosynthetic pathway is regulated by mechanisms that affect both the synthesis and the activity of the enzymes that catalyze the pathway. For example, anthranilate synthetase, which catalyzes the first step of the pathway, is a two-subunit protein encoded by the *trp*E and *trp*D genes of the tryptophan operon. The number of molecules of this enzyme that is present in the cell is determined by the transcriptional regulation of the tryptophan operon. However, the activity of the enzyme is regulated by **feedback inhibition.** This is a common short-term means of regulating the first committed step in a metabolic pathway. In this case, tryptophan, the end product of the pathway, can bind to an allosteric site on the anthranilate syn-

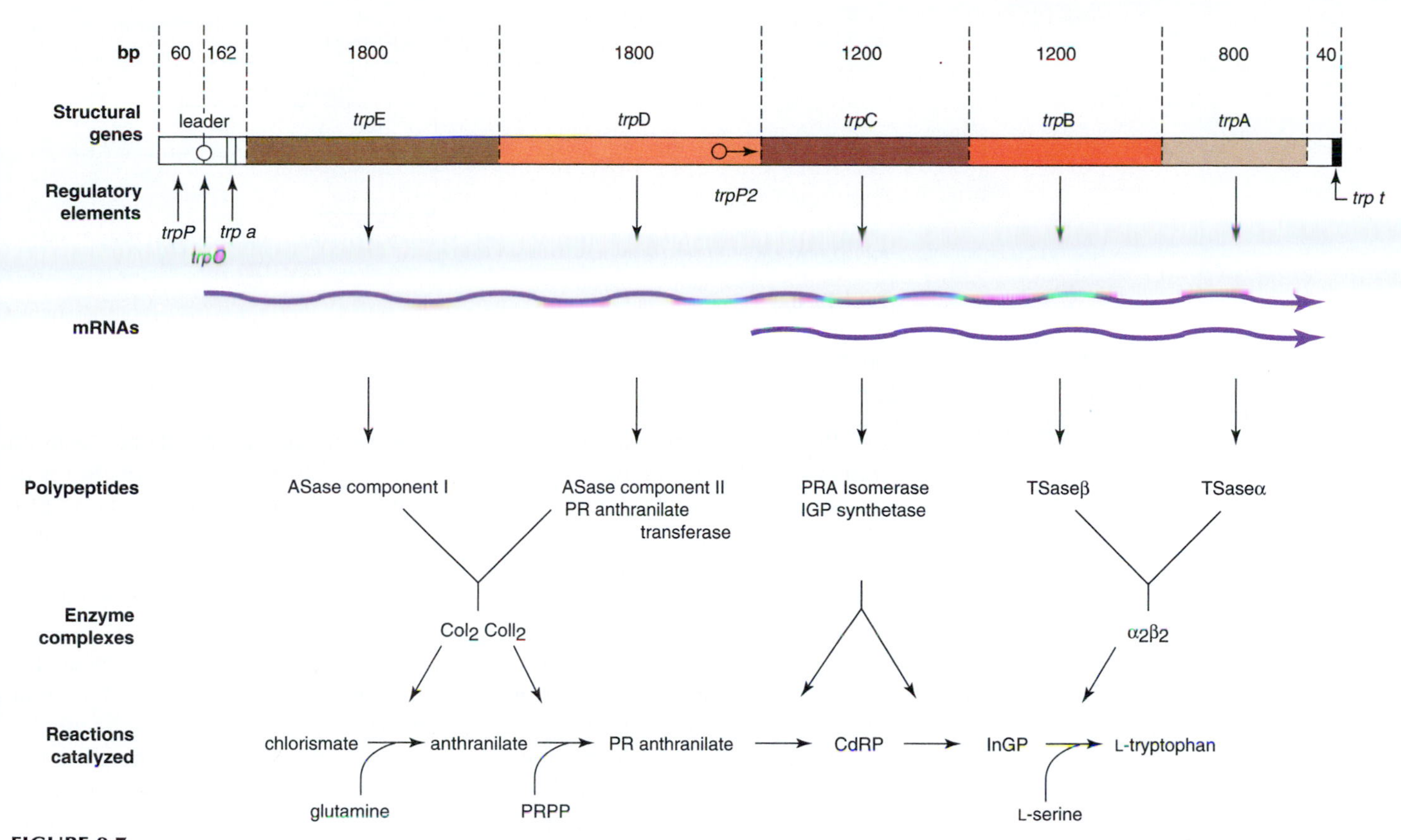

FIGURE 8.7
Genes of tryptophan operon of *E. coli*.
Regulatory elements are the primary promoter (*trp*P), operator (*trp*O), attenuator (*trp* a), secondary internal promoter (*trp*P2), and terminator (*trp* t). Direction of mRNA synthesis is indicated on the wavy lines representing mRNAs. CoI_2 and $CoII_2$ signify components I and II, respectively, of the anthranilate synthetase (ASase) complex; PR-anthranilate is *N*-5′-phosphoribosyl-anthranilate; CdRP is 1-(*o*-carboxy-phenylamino)-1-deoxyribulose-5-phosphate; InGP is indole-3-glycerol phosphate; PRPP is 5-phosphoribosyl-1-pyrophosphate; and TSase is tryptophan synthetase.
Redrawn from Platt, T. In: J. H. Miller and W. Reznikoff (Eds.), The Operon. *Cold Spring Harbor, NY: Cold Spring Harbor Laboratory Press, 1978, p. 263.*

thetase and interfere with its catalytic activity at another site. Thus, as the concentration of tryptophan builds up, it binds to anthranilate synthetase and immediately decreases its activity on the substrate, chorismic acid. In addition, tryptophan also acts as a corepressor to shut down the synthesis of new enzyme molecules from the tryptophan operon. Therefore feedback inhibition is a short-term control that has an immediate effect on the pathway, whereas repression takes a little longer but has the more permanent effect of reducing the number of enzyme molecules.

The **tryptophan repressor** is a homodimer of two identical subunits of 108 amino acids each. Under normal conditions about 20 molecules of the repressor dimer are present. The repressor by itself does not bind to the *trp* operator. It must be complexed with two molecules of tryptophan in order to bind to the operator and therefore only acts *in vivo* in the presence of tryptophan. This is opposite to the lactose repressor, which binds to its operator only in the absence of its inducer. Interestingly, the tryptophan repressor also regulates transcription of *trp*R, its own gene. As the tryptophan repressor accumulates in cells, the repressor–tryptophan complex binds to a region upstream of this gene, turning off its transcription and maintaining the equilibrium of 20 repressors per cell. Another difference from the lactose operon is that the *trp* operator occurs entirely within the *trp* promoter rather than adjacent to it, as shown in Figure 8.8. The operator sequence is a region of dyad symmetry, and the mechanism for preventing transcription is the same as in the lactose operon. Binding of the repressor–corepressor complex to the operator physically blocks binding of RNA polymerase to the promoter.

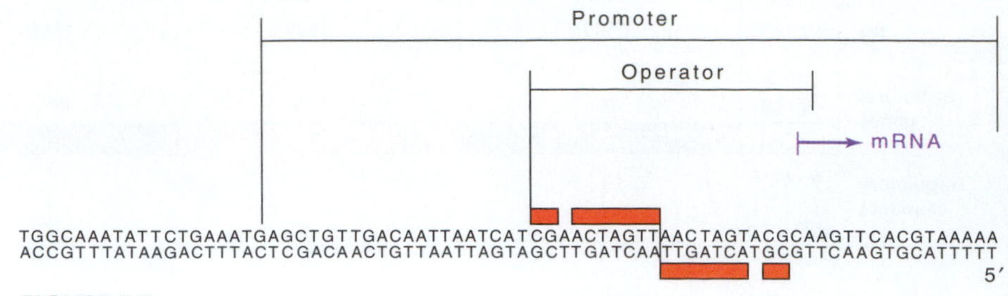

FIGURE 8.8
Nucleotide sequence of control elements of tryptophan operon.
Boxes above and below sequence indicate symmetrical sequences within operator.

Repression results in about a 70-fold decrease in the rate of transcription initiation at the *trp* promoter. (In contrast, the basal level of lactose operon gene products is about 1000-fold lower than the induced level.) However, the tryptophan operon contains additional regulatory elements that impose further control on the extent of its transcription. One control site is a secondary promoter, designated *trp*P2, which is located within the coding sequence of the *trp*D gene (shown in Figure 8.7). This promoter is not regulated by the tryptophan repressor. Transcription from it occurs constitutively at a relatively low rate and is terminated at the same location as transcription from the regulated promoter for the whole operon, *trp*P. The transcription product from *trp*P2 is an mRNA that contains the coding sequences for *trp*CBA, the last three genes of the operon. Therefore two polycistronic mRNAs are derived from the tryptophan operon, one containing all five structural genes and one possessing only the last three genes. Under conditions of maximum repression the basal level of mRNA coding sequence for the last three genes is about five times higher than the basal mRNA level for the first two genes.

The reason for a second internal promoter is not clear. Perhaps the best explanation comes from the observation that three of the five proteins do not contain tryptophan; only the *trp*B and *trp*C genes contain the single codon that specifies tryptophan. Therefore, under extreme tryptophan starvation, these two proteins would not be synthesized, which would prevent the pathway from being activated. However, since both of these genes lie downstream of the unregulated second promoter, their protein products will always be present at the basal level necessary to maintain the pathway.

Tryptophan Operon Has a Second Control Site: The Attenuator Region

Another important control element of the tryptophan operon not present in the lactose operon is the **attenuator** region (Figure 8.9). This lies within the 162 nucleotides between the start of transcription from *trp*P and the initiator codon of the *trp*E gene. Its existence was first deduced by the identification of mutations that mapped in this region and increased transcription of all five structural genes. Within the 162 nucleotides, also called the **leader sequence,** are 14 adjacent codons that begin with a methionine codon and end with an in-phase termination codon. These codons are preceded by a canonical ribosome-binding site and could potentially specify a 14-residue leader peptide. This peptide has never been detected in bacterial cells, perhaps because it is degraded very rapidly. The ribosome-binding site does function properly when its corresponding DNA sequence is ligated upstream of a structural gene using recombinant DNA techniques.

The attenuator region provides RNA polymerase with a second chance to regulate transcription of the tryptophan operon, based on whether the cell needs the tryptophan-biosynthesizing enzymes. In the presence of tryptophan, transcription begins at the promoter but is prematurely terminated at the end of the attenuator

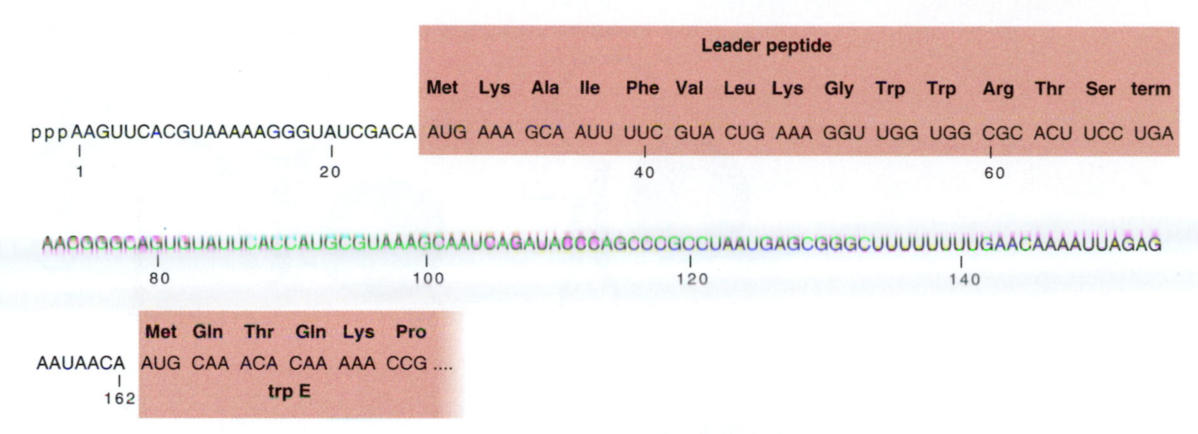

FIGURE 8.9
Nucleotide sequence of leader RNA from tryptophan operon. The 14 amino acids of the putative leader peptide are indicated over their codons.

Redrawn with permission from Oxender, D. L., Zurawski, G., and Yanofsky, C. Proc. Natl. Acad. Sci. USA *76:5524, 1979.*

region to produce a short 140-nucleotide transcript. In the absence of tryptophan, the attenuator region has no effect on transcription, and the entire polycistronic mRNA of the five structural genes is synthesized. Therefore, at both operator and attenuator, tryptophan exerts the same general influence. At the operator it participates in repressing transcription, and at the attenuator it participates in stopping transcription by those RNA polymerases that have escaped repression. It has been estimated that attenuation has about a 10-fold effect on transcription of the tryptophan structural genes. When multiplied by the 70-fold effect of derepression at the operator, about a 700-fold range exists in the level at which the tryptophan operon can be transcribed.

The mechanism by which transcription is terminated at the attenuator site is a marvelous example of the cooperative, linked interaction between transcription and translation that can occur in bacteria to achieve the desired amount of a given mRNA. The tryptophan leader peptide (Figure 8.9) of 14 residues contains two adjacent tryptophans in positions 10 and 11. This is unusual because tryptophan is a relatively rare amino acid in *E. coli*. These two tryptophan codons also provided an early clue that $tRNA^{trp}$ is involved in attenuation. If the tryptophan in the cell is low, the amount of the charged tryptophanyl-$tRNA^{trp}$ will also be low and the ribosomes may be unable to translate through the two tryptophan codons of the leader peptide region. Therefore they will stall at this place in the leader RNA sequence.

The RNA sequence of the attenuator region can adopt several possible secondary structures (Figure 8.10). The position of the ribosome within the leader peptide-coding sequence determines the secondary structure that will form. This secondary structure, in turn, is recognized (or sensed) by the RNA polymerase that has just transcribed through the leader peptide-coding region and is now located a small distance downstream. The RNA secondary structure that forms when a ribosome is not stalled at the tryptophan codons is a termination signal for the RNA polymerase. Under these conditions the cell does not need to make tryptophan, and transcription stops after the synthesis of a 140-nucleotide transcript, which is quickly degraded. On the other hand, the secondary structure that results when the ribosomes are stalled at the tryptophan codons is not recognized as a termination signal, and transcription continues into the *trp*E gene.

Transcription Attenuation Is a General Mechanism of Control in Operons for Amino Acid Biosynthesis

Transcription attenuation does not occur in eukaryotic organisms because transcription (in the nucleus) and translation (in the cytoplasm) take place in different

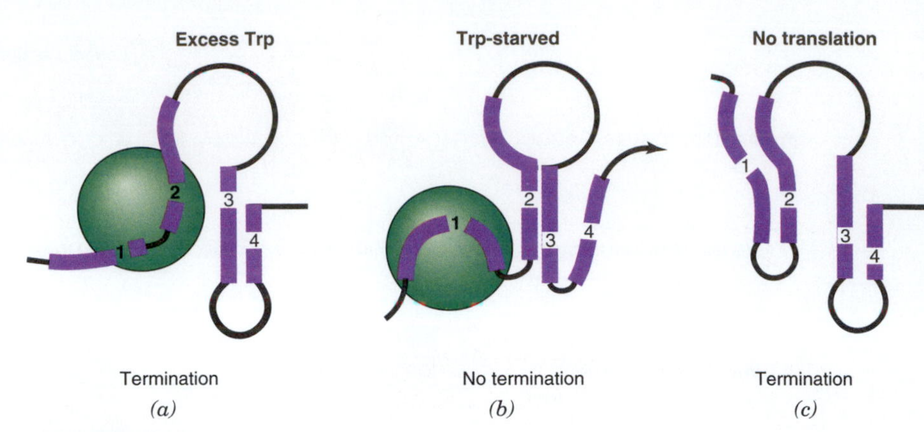

FIGURE 8.10

Schematic model for attenuation in the tryptophan operon of *E. coli*.
(*a*) Under conditions of excess tryptophan, the ribosome (green sphere) translating the newly transcribed leader RNA synthesizes the complete leader peptide. During this synthesis the ribosome binds to regions 1 and 2 of the RNA and prevents formation of stem and loop 1–2 or 2–3. Stem and loop 3–4 will be free to form and signal the RNA polymerase molecule (not shown) to terminate transcription. (*b*) Under conditions of tryptophan starvation, tryptophanyl-tRNA^{Trp} will be limiting, and the ribosome will stall at the adjacent tryptophan codons at the beginning of region 1 in the leader peptide-coding region. Because region 1 is bound to the ribosomes, stem and loop 2–3 will form, excluding formation of stem and loop 3–4, which is required as the signal for transcription termination. Therefore RNA polymerase continues transcription into the structural genes. (*c*) Under conditions in which the leader peptide is not translated, stem and loop 1–2 forms, preventing formation of stem and loop 2–3, and thereby permitting formation of stem and loop 3–4. This signals transcription termination.
Reprinted with permission from Oxender, D. L., Zurawski, G., and Yanofsky, C. Proc. Natl. Acad. Sci. USA *76:5524, 1979.*

cellular compartments separated by the nuclear membrane. Thus a direct link between the processes of transcription and translation cannot be utilized as a form of gene regulation in eukaryotes. Attenuation is a common phenomenon in bacterial gene expression and occurs in at least six other operons that code for enzymes catalyzing amino acid biosynthetic pathways. Figure 8.11 shows the **leader peptide** sequences specified by each of these operons. In each, the leader peptide contains several codons for the amino acid product of the biosynthetic pathway. The most extreme case is the 16-residue leader peptide encoded in the histidine operon that contains seven contiguous histidines. Starvation for histidine results in a decrease in the amount of histidinyl-tRNA^{his} and a dramatic increase in transcription of the *his* operon. The nucleotide sequence of the attenuator region suggests that ribosome

Operon	Leader peptide sequence	Regulatory amino acids
his	Met-Thr-Arg-Val-Gln-Phe-Lys-His-His-His-His-His-His-His-Pro-Asp	His
*phe*A	Met-Lys-His-Ile-Pro-Phe-Phe-Phe-Ala-Phe-Phe-Phe-Thr-Phe-Pro	Phe
thr	Met-Lys-Arg-Ile-Ser-Thr-Thr-Ile-Thr-Thr-Thr-Ile-Thr-Ile-Thr-Thr-Gly-Asn-Gly-Ala-Gly	Thr Ile
leu	Met-Ser-His-Ile-Val-Arg-Phe-Thr-Gly-Leu-Leu-Leu-Leu-Asn-Ala-Phe-Ile-Val-Arg-Gly-Arg-Pro-Val-Gly-Gly-Ile-Gln-His	Leu
ilv	Met-Thr-Ala-Leu-Leu-Arg-Val-Ile-Ser-Leu-Val-Val-Ile-Ser-Val-Val-Val-Ile-Ile-Ile-Pro-Pro-Cys-Gly-Ala-Ala-Leu-Gly-Arg-Gly-Lys-Ala	Leu, Val, Ile

FIGURE 8.11

Leader peptide sequences specified by biosynthetic operons of *E. coli*.

All of the leader peptide sequences contain multiple copies of amino acid(s) synthesized by enzymes coded for by that operon.

stalling at the histidine codons also influences the formation of alternate hairpin loops, one of which resembles a termination hairpin followed by several U residues. In contrast to the *trp* operon, transcription of the *his* operon is regulated primarily by attenuation; it does not possess an operator that is recognized by a repressor protein. Instead, the ribosome acts rather like a positive regulator protein, similar to the cAMP–CAP complex discussed with the *lac* operon. If the ribosome is bound to (i.e., stalled at) the attenuator site, then transcription of the downstream structural genes is enhanced. If the ribosome is not bound, then transcription of these genes is greatly reduced.

Transcription of some of the other operons shown in Figure 8.11 can be attenuated by more than one amino acid. For example, the ***thr* operon** is attenuated by threonine or isoleucine; the ***ilv* operon** by leucine, valine, or isoleucine. This effect can be explained in each case by stalling of the ribosome at the corresponding codon, which, in turn, interferes with the formation of a termination hairpin. It is possible that in the longer leader peptides, stalling at more than one codon is necessary to achieve maximal transcription through the attenuation region.

8.5 | OTHER BACTERIAL OPERONS

Synthesis of Ribosomal Proteins Is Regulated in a Coordinated Manner

Many bacterial operons possess the same general regulatory mechanisms as the *lac*, *trp*, and *his* operons, as discussed in Section 8.4. However, each operon has evolved its own distinctive quirks. One interesting example concerns the structural genes for the 70 or more proteins that comprise a ribosome (Figure 8.12). Each ribosome contains one copy of each **ribosomal protein** (except for protein L7–L12, which is probably present in four copies). Therefore all 70 proteins are required in equimolar amounts, and it makes sense that their synthesis is regulated in a coordinated fashion. Six different operons, containing about one-half of the ribosomal protein genes, occur in two major gene clusters. One cluster contains four adjacent operons (*Spc*, S10, *str*, and *a*), and the other cluster has two operons (L11 and *rif*) located elsewhere in the *E. coli* chromosome. There is no obvious pattern in distribution of the genes among these operons. Some operons contain genes for proteins of just one ribosomal subunit; others code for proteins of both subunits. In addition to structural genes for ribosomal proteins, these operons also contain genes for other (related) proteins. For example, the *str* operon contains genes for soluble **translation elongation factors,** EF-Tu and EF-G, as well as genes for some proteins in the 30S ribosomal subunit. The *a* operon contains genes for proteins of 30S and 50S ribosomal subunits plus a gene for the α subunit of RNA polymerase. The *rif* operon has genes for the β and β' subunits of RNA polymerase and genes for ribosomal proteins.

Operon	Regulator protein	Proteins specified by the operon
Spc	S8	L14-L24-L5-S14-S8-L6-L18-S5-L15-L30
S10	L4	S10-L3-L2-L4-L23-S19-L22-S3-S17-L16-L29
str	S7	S12-S7-EF•G-EF•Tu
α	S4	S13-S11-S4-α-L17
L11	L1	L11-L1
rif	L10	L10-L7-β-β′

FIGURE 8.12
Operons containing genes for ribosomal proteins *E. coli*.
Genes for the protein components of the small (S) and large (L) ribosomal subunits of *E. coli* are clustered on several operons. Some of these operons also contain genes for RNA polymerase subunits α, β, and β', and protein synthesis factors EF-G and EF-Tu. At least one of the protein products of each operon usually regulates expression of that operon (see text).

Common to the six ribosomal operons is that their expression is regulated by one of their own structural gene products; that is, they are **self-regulated.** The precise mechanism for this self-regulation varies considerably with each operon and is not yet completely understood. In some cases, regulation occurs at the level of translation, not transcription as discussed for the *lac* and *trp* operons. After the polycistronic mRNA is made, the "regulatory" ribosomal protein binds to this mRNA and determines which regions, if any, are translated. In general, the ribosomal protein that regulates expression of its own operon, or part of its own operon, is a protein that is associated with one of the ribosomal RNAs (rRNAs) in the intact ribosome. This ribosomal protein has a high affinity for the rRNA and a lower affinity for one or more regions of its own mRNA. Therefore a competition between the rRNA and the operon's mRNA for binding with the ribosomal protein occurs. As the ribosomal protein accumulates to a higher level than the free rRNA, it binds to its own mRNA and prevents initiation of protein synthesis at one or more of the coding sequences on this mRNA (Figure 8.13). As more ribosomes are formed, the excess of this particular ribosomal protein is used up and translation of its mRNA can begin again.

Stringent Response Controls Synthesis of rRNAs and tRNAs

Bacteria respond in several ways to emergency situations, that is, times of **extreme general stress.** One of these situations is when the bacterium does not have a sufficient pool of amino acids to maintain protein synthesis. Under these conditions the cell invokes the **stringent response,** a mechanism that reduces the synthesis of the rRNAs and tRNAs about 20-fold. This response places many of the activities

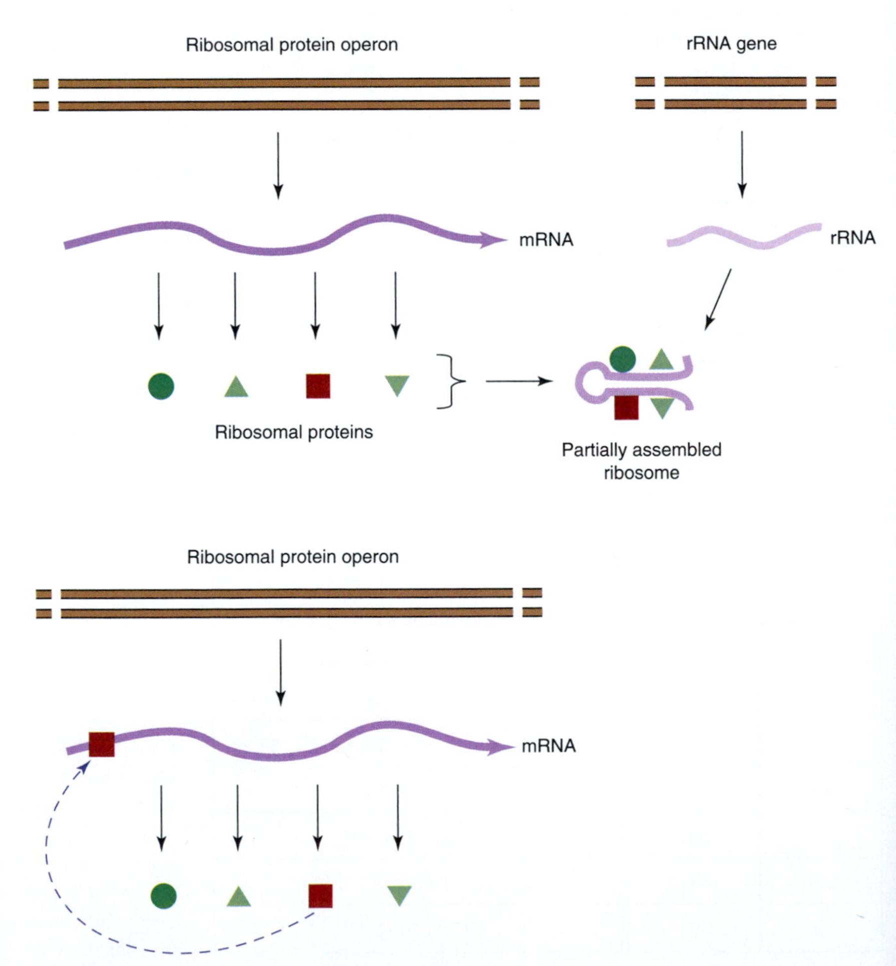

FIGURE 8.13
Self-regulation of ribosomal protein synthesis. If free rRNA is not available for assembly of new ribosomal subunits, individual ribosomal proteins bind to polycistronic mRNA from their own operon, blocking further translation.

within the cell on hold until conditions improve. The mRNAs are less affected, but there is also about a threefold decrease in their synthesis.

The stringent response is triggered by the presence of an uncharged tRNA in the A site of the ribosome. This situation occurs when the concentration of the corresponding charged tRNA is very low. The first result, of course, is that further peptide elongation by the ribosome stops. This event causes a protein called the **stringent factor**, the product of the *rel*A gene, to synthesize **guanosine tetraphosphate** (ppGpp) and **guanosine pentaphosphate** (pppGpp), from ATP and GTP or GDP as shown in Figure 8.14.

Stringent factor is loosely associated with some, but not all, ribosomes of the cell. Perhaps a conformational change in the ribosome is induced by the occupation of the A site by an uncharged tRNA, which, in turn, activates the associated stringent factor. The exact functions of ppGpp and pppGpp are not known. However, they seem to inhibit transcription initiation of the rRNA and tRNA genes. In addition, they affect transcription of some operons more than others.

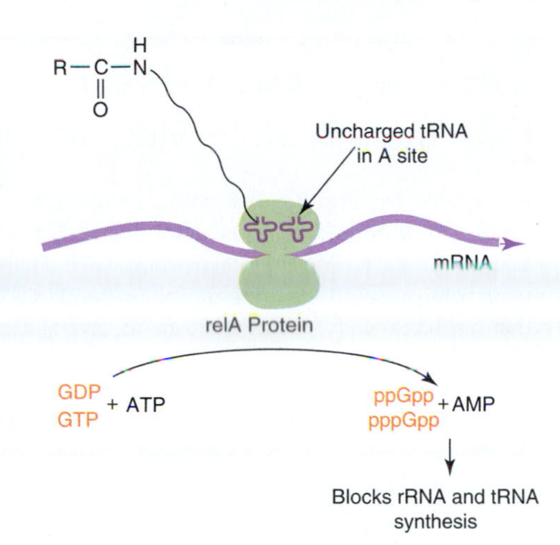

FIGURE 8.14
Stringent control of protein synthesis in *E. coli*.
During extreme amino acid starvation, an uncharged tRNA in the A site of the ribosome activates the relA protein to synthesize ppGpp and pppGpp, which are involved in decreasing transcription of the genes coding for rRNAs and tRNAs.

8.6 | BACTERIAL TRANSPOSONS

Transposons Are Mobile Segments of DNA

So far we have discussed regulation of bacterial genes with fixed locations in the chromosome. Their positions relative to the neighboring genes do not change. The vast majority of bacterial genes are of this type. In fact, the genetic maps of *E. coli* and *Salmonella typhimurium* are quite similar, indicating the lack of much evolutionary movement of most genes within bacterial chromosomes. A class of bacterial genes exists, however, in which newly duplicated gene copies "jump" from one genomic site to another with a frequency of about 10^{-7} per generation, the same rate as for spontaneous point mutations. These mobile segments of DNA containing these genes are called **transposable elements** or **transposons** (Figure 8.15). Regulation of genes within bacterial transposons has several interesting features; some control the presence and transposition of the transposon itself, whereas others, usually antibiotic resistance genes, can provide the bacterium with a selective advantage against other bacteria.

Transposons were first detected as rare insertions of foreign DNA into structural genes in bacterial operons. Usually, these insertions interfere with expression of the structural gene into which they have inserted and all downstream genes of the operon. This is not surprising since they can potentially destroy the translation reading frame, introduce transcription termination signals, affect the mRNA stability, and so on. Many transposons and the sites into which they insert have been isolated using recombinant DNA techniques and have been extensively characterized.

Transposons vary tremendously in length. Some consist of a few thousand base pairs and contain only two or three genes; others are many thousands of base pairs long, containing several genes. Sometimes, small transposons can occur within a large transposon. All active transposons contain at least one gene that codes for a **transposase**, an enzyme required for the transposition or "jumping" event. Often they contain genes that code for resistance to antibiotics or heavy metals. Most transpositions involve generation of an additional copy of the transposon and insertion of this copy into another location. The original transposon copy is the same after the duplication as before; that is, the donor copy is unaffected by insertion of its duplicate into the recipient site. Transposons contain short inverted **terminal repeat sequences** that are essential for the insertion mechanism and are often used to define the two boundaries of a transposon. The multiple target sites into which most transposons can insert seem to be fairly random in sequence; other transposons have a propensity for insertion at specific "hot spots." The duplicated transposon can be located in a different DNA molecule than its donor. Frequently, transposons are found on plasmids that pass from one bacterial strain to another and are the

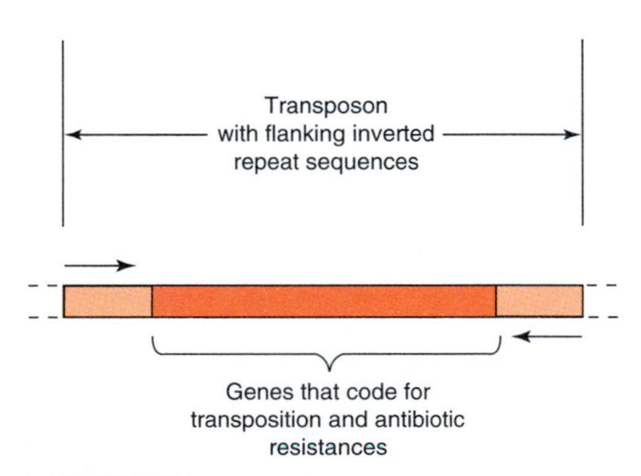

FIGURE 8.15
General structure of transposons.
Transposons are relatively rare mobile segments of DNA that contain genes coding for their own rearrangement and (usually) genes that specify resistance to various antibiotics.

CLINICAL CORRELATION 8.1

Transmissible Multiple Drug Resistance

Pathogenic bacteria are becoming increasingly resistant to a large number of antibiotics. This is viewed with alarm by many physicians. Many cases have been documented in which a bacterial strain in a patient being treated with one antibiotic suddenly became resistant to that antibiotic and, simultaneously, to several other antibiotics even though the bacterial strain had never previously been exposed to these other antibiotics. This occurs when the bacteria suddenly acquire from another bacterial strain a plasmid that contains several different transposons, each containing one or more antibiotic resistance genes. Examples include the genes encoding β-lactamase, which inactivates penicillins and cephalosporins, chloramphenicol acetyltransferase, which inactivates chloramphenicol, and phosphotransferases that modify aminoglycosides such as neomycin and gentamicin.

Neu, H. C. The crisis in antibiotic resistance. *Science* 257:1064, 1992.

source of a suddenly acquired resistance to one or more antibiotics by a bacterium (see Clin. Corr. 8.1).

As for bacterial operons, each transposon or set of transposons has its own distinctive characteristics. The well-characterized transposon *Tn*3 is discussed as an example of their general properties.

Transposon *Tn*3 Contains Three Structural Genes

Transposon *Tn*3 contains 4957 bp including 38 bp at one end that occur as an inverted repeat at the other end (Figure 8.16). Three genes are present in *Tn*3. One gene codes for the enzyme **β-lactamase,** which hydrolyzes ampicillin and renders the cell resistant to this antibiotic. The other two genes, *tnp*A and *tnp*R, code for a **transposase** and a **repressor protein,** respectively. The transposase contains 1021 amino acids and binds to single-stranded DNA. It recognizes the repetitive ends of the transposon and participates in the cleavage of the recipient site into which the new transposon copy inserts. The *tnp*R gene product is a protein of 185 amino acids. In its role as a repressor it controls transcription of both the transposase gene and its own gene. The *tnp*A and *tnp*R genes are transcribed divergently from a 163-bp control region located between the two genes that is recognized by the repressor. The *tnp*R product also participates in the recombination process that results in the insertion of the new transposon but does not affect transcription of the ampicillin resistance gene.

Mutations in the transposase gene generally decrease the frequency of *Tn*3 transposition. Mutations that destroy the repressor function of the *tnp*R product cause an increased frequency of transposition. These mutations derepress the *tnp*A

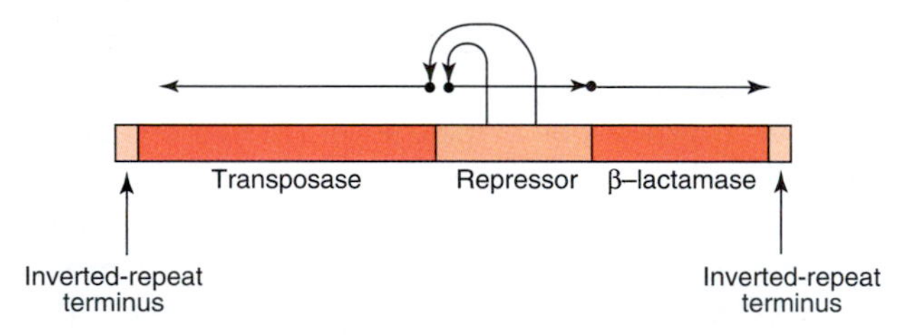

FIGURE 8.16
Functional components of the transposon *Tn*3.
Genetic and DNA sequence analyses show there are at least four kinds of regions: the inverted repeat termini; a gene for the enzyme β-lactamase, which confers resistance to ampicillin and related antibiotics; a gene encoding an enzyme required for transposition (transposase); and a gene for a repressor protein that controls transcription of the genes for the transposase and the repressor itself. Horizontal arrows indicate direction in which DNA of various regions is transcribed.
Redrawn from Cohen, S. N. and Shapiro, J. A. Sci. Am. 242:40, 1980. W. H. Freeman and Company, Copyright © 1980.

gene, resulting in more molecules of the transposase, which enhances the formation of more duplicated transposon copies. They also derepress the *tnp*R gene, but since the repressor is inactive, this has no effect on the system.

The transposons located on bacterial plasmids are of increasing importance in the clinical use of antibiotics. Bacterial plasmids that have not been altered for experimental use usually contain genes that facilitate their transfer from one bacterium to another. As the plasmids transfer (e.g., between different infecting bacterial strains), their transposons containing **antibiotic resistance genes** are moved into new bacterial strains. Once inside a new bacterium, the transposon can be duplicated onto the chromosome and become permanently established in that cell's lineage. The result is that more and more pathogenic bacterial strains have become resistant to an increasing number of antibiotics.

8.7 | GENE EXPRESSION IN EUKARYOTES

As in prokaryotes, gene transcription in eukaryotic organisms is regulated to provide the appropriate response to biological needs. However, especially in multicellular organisms, determination of developmental cell fate and expression of specialized genes by specific cell types are also regulated transcriptionally. Some genes (so-called housekeeping genes) are expressed in most cells, other genes are activated upon demand, and still other genes are rendered permanently inactive in all but a few cell types. In eukaryotic cells the nuclear membrane serves as a barrier that selectively allows some proteins access to DNA while keeping others in the cytosol. Thus one of the differences between the regulation of gene expression in humans and most prokaryotic organisms is the specialized activation and inactivation of some genes in differentiated cells.

While in bacteria one RNA polymerase is responsible for transcription of all RNAs (tRNA, rRNA, and mRNA), in eukaryotic organisms the task is divided among three different RNA polymerases (see p. 210). **RNA polymerase I** transcribes the rRNA genes, **RNA polymerase II** the protein-encoding genes whose transcripts become the mRNAs, and **RNA polymerase III** the genes for tRNAs and for most other small RNAs. Although some principles for eukaryotic gene activation and control are applicable for all three RNA polymerases, in this section the focus will be on transcription by RNA polymerase II. RNA polymerase II is comprised of at least ten different subunits, ranging in size from 10 to 220 kDa. Some of the subunits found in RNA polymerase II are also part of the RNA polymerase I and III complexes, while others are unique to RNA polymerase II. The largest subunit of RNA polymerase II has, depending on the species, as many as 52 repeats of the amino acid motif PTSPSYS in its C-terminal region (or its CTD for C-terminal domain). One of the distinguishing features of these repeats is that the threonines (T), serines (S), and tyrosine (Y) can be phosphorylated.

To better understand the regulation of transcription in eukaryotes it is useful to recall the organization of DNA into **chromatin** and the role of modification of DNA, specifically **methylation** of the DNA bases, on gene activation. In addition, we'll consider how RNA polymerase II is positioned at the appropriate spot in the promoter of a gene to transcribe that gene by the formation of a preinitiation complex that involves the assembly of **general transcription factors (TFs)** with RNA polymerase II. Next, we'll look at how specific gene activity can be regulated through the use of **enhancers, transcription factor binding sites,** and **RNA polymerase assembly sites.** Finally, we will discuss the activation of transcription by specific transcription factors, some of their general characteristics, and how they are regulated.

Eukaryotic DNA Is Bound by Histones to Form Chromatin

Segments of eukaryotic DNA are wrapped around octamers of **histone proteins** that contain two molecules each of histone H2A, H2B, H3, and H4 to form

nucleosomes (see p. 68). The association of the histones with the DNA is by electrostatic interactions between a large number of positively charged lysine residues and the negatively charged phosphodiester backbone of DNA. In most cells histone H1 or H5 binds when the octamer–DNA association is established. The length of DNA associated with each octamer has been estimated by its differential accessibility of DNA to various nucleases. About 200 bp of DNA make up a single nucleosomal unit, with about 130–160 bp in direct contact with the octamer core. Some of the remaining DNA binds histones H1 and H5 and the rest is the **linker** DNA between nucleosomes. Although the histone octamer–DNA interaction is not sequence-specific, some sequence-dependent patterns of association have been noted. When, for instance, the double-stranded DNA helix bends because of the presence of an A-T rich region, the minor groove of the helix faces on the inside of a nucleosome core particle. On the contrary, in G-C rich regions the minor groove faces away from the histone octamer core. Despite these tendencies, there is not, as yet, any clear way to predict which specific sequences within a DNA duplex region will end up with the major and minor grooves on the outside of the nucleosome complex. This positioning is important because assembly of the **transcription initiation complex** and binding of other proteins that influence gene expression need access to the grooves in the DNA duplex. Sequence-specific interactions occur via hydrogen bonds formed between the edges of the paired bases of the DNA that face the major or minor groove. When DNA is wrapped around a histone octamer, the specific DNA sequences that transcription factors bind to may be occluded. It does not take the complete removal of DNA from the nucleosome to allow access to specific sequences. By rolling the DNA helix so that there is, for instance, a 5-bp shift in the association of the DNA with the histone octamer, different major and minor groove contacts can be made available on the outer surface of the nucleosome (Figure 8.17).

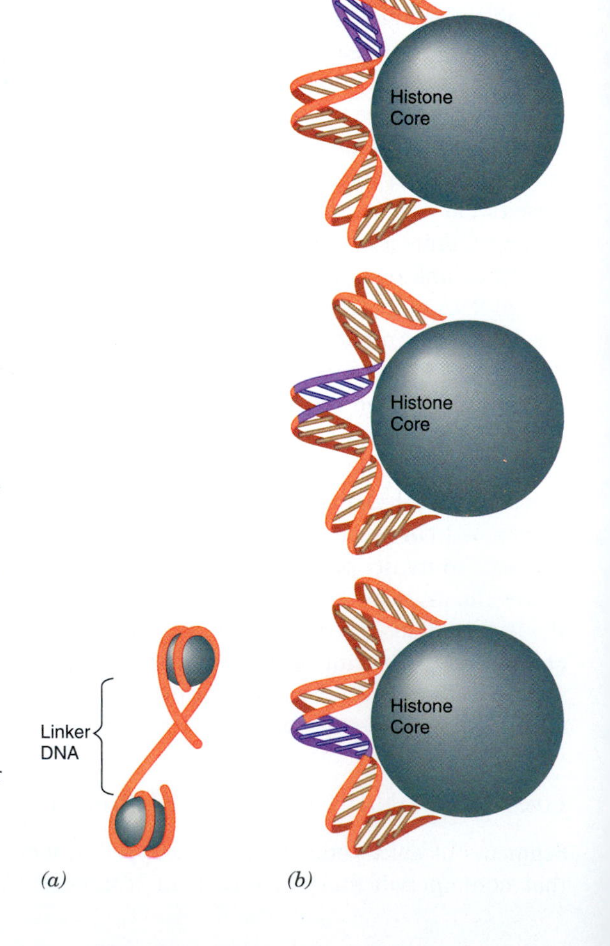

FIGURE 8.17
Access to specific sequences in major and minor grooves of the DNA depends on nucleosome positioning.
(*a*) DNA wrapped around the core histone octamer of two molecules each of histone H2A, H2B, H3, and H4 is called the nucleosome core particle. Under most conditions histone H1 or H5 (not shown) associates with DNA where the wrap around the histone octamer begins. DNA between nucleosome core particles is called linker DNA. When exposed to agents like micrococcal nuclease, linker regions are more accessible than DNA wrapped around the histone core particle. (*b*) Within the nucleosome, the side of the helix facing away from the histone core is accessible, but the side facing in is not. If we rotate the helix with respect to the histone core of a nucleosome and follow the position of the same five base pairs, access to the major and minor grooves for other interactions is dependent on whether the base pair is on the inside or the outside of the wrapped nucleosome core particle.
Redrawn from Wolffe, A. P. Chromatin: Structure and Function, *3rd ed. New York: Academic Press, 1998.*

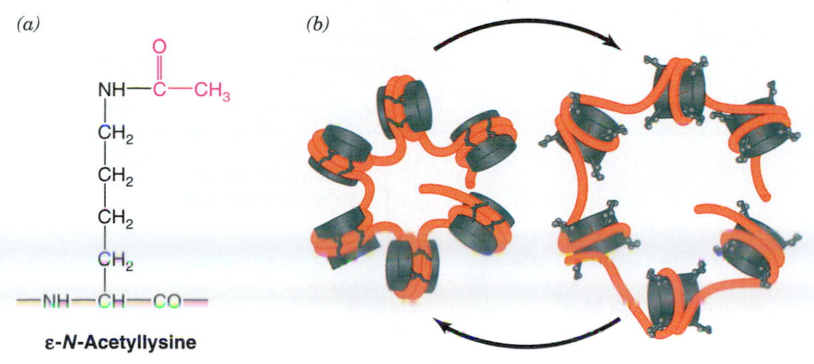

FIGURE 8.18
Strength of histone DNA association is modified by acetylation of lysine residues in the N-terminus of histone proteins.
Histones that make up the core octamer have a number of lysine residues that, by virtue of their positively charged side group, can promote interaction with the negatively charge phosphodiester linkages of DNA. The modification of the lysines by acetylation replaces the positive charge with a neutral acetyl group (shown in red in (a)) and weakens the electrostatic interaction between the octamer core and the DNA. This process is reversible (b), and the acetylation and deacetylation of histones provides a way to loosen or tighten chromatin structure.
Redrawn from Wolffe, A. P. Sci. Med. *6:28, 1999.*

Genes that are not transcribed within a particular cell form highly condensed heterochromatin. In contrast, transcriptionally active regions of DNA have a less condensed, more open structure. At least part of this difference is due to post-translational modification of histones. The **acetylation** of the ε-amino group of the lysine residues near the N-termini of histones reduces the positive charge carried by the histones and thus weakens the electrostatic attraction between the histones and the DNA. In general, the acetylation of histones leads to the activation of gene expression, while the deacetylation of histones reverses the effect. Acetylation of histones, and the subsequent repositioning of the nucleosome by the destabilization of histone DNA interactions, is important in providing sequence-specific access to DNA (Figure 8.18).

The relative position of a nucleosome can also be influenced by the **SWI/SNF complex,** first identified and best characterized in the yeast *S. cerevisiae* and named after mating type **sw**itching and **s**ucrose **n**on**f**ermenting strains. This complex of about ten proteins interacts with the C-terminal domain of the large subunit of RNA polymerase II. Its function is to disrupt nucleosomal arrays in an ATP-dependent manner. The result is to open regions of DNA for interaction with transcription factors and thus facilitate gene activation. There are fewer SWI/SNF complexes in the cell than genes being transcribed, suggesting that the SWI/SNF complex acts in a catalytic manner (see Clin. Corr. 8.2).

CLINICAL CORRELATION 8.2
Rubinstein–Taybi Syndrome

Regulation of histone acetylation influences activation and inactivation of gene expression. Two major enzymes responsible for acetylation of histones in mammalian cells are the acetyltransferases p300 and CBP. p300 was named for its molecular mass and CBP stands for CREB binding protein, with CREB being a transcription factor that is a *c*AMP *r*egulatory-*e*lement *b*inding protein. These acetyltransferases allow activation of appropriate genes in association with CREB and other transcription factors, leading to the opening of chromatin structure through the weakening of histone–DNA interactions.

Rubinstein–Taybi syndrome, characterized by mental retardation and other developmental abnormalities, is caused by sequence mutations in the *CBP* gene. Alterations identified include point mutations, small deletions, and rearrangements. More severe developmental defects correspond to more substantial mutations. The complete loss of the *CBP* gene product is probably lethal.

Petrij, F., Giles, R. H., Dauwerse, H. G., et al. Rubinstein–Taybi syndrome caused by mutations in the transcriptional co-activator CBP. *Nature* 376:348, 1995; and Wolffe, A. P. The cancer–chromatin connection. *Sci. Med.* 6:28, 1999.

Methylation of DNA Correlates with Inactivation of Genes

Methylation in human DNA centers on the formation of 5-methylcytosine (Figure 8.19) from cytosine in the sequence CG on both strands. About 70% of CG sequences in human DNA are methylated. Methylation is implicated in the **imprinting** of genomic DNA. In DNA imprinting the different methylation patterns of DNA inherited from the sperm or egg correlate with choice of allelic expression. Methylation patterns are conserved after replication by the action of hemimethylase (which methylates only one of the two strands containing the CG), although a *de novo* methylase must also exist. At present only the hemimethylase has been identified.

A terminology often encountered in the DNA methylation literature is **"CG islands"** or **"CG-rich regions."** The sequence CG is underrepresented in the genome. The underrepresentation is likely due to the accumulation, over many generations, of C to T transitions caused by the deamination of 5-methylcytosine. Maintenance

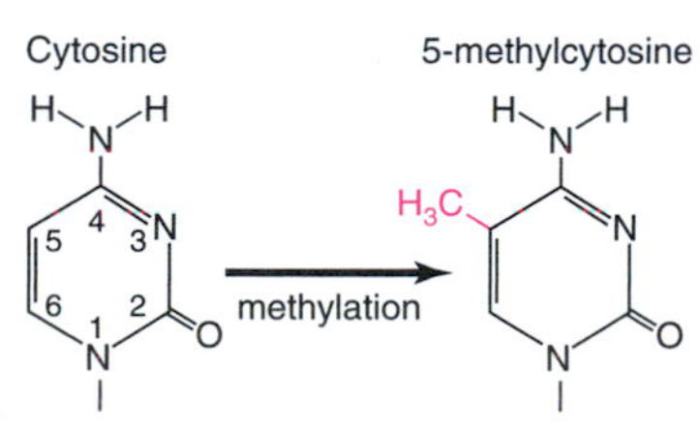

FIGURE 8.19
The most common methylated base in humans is 5-methyl cytosine.

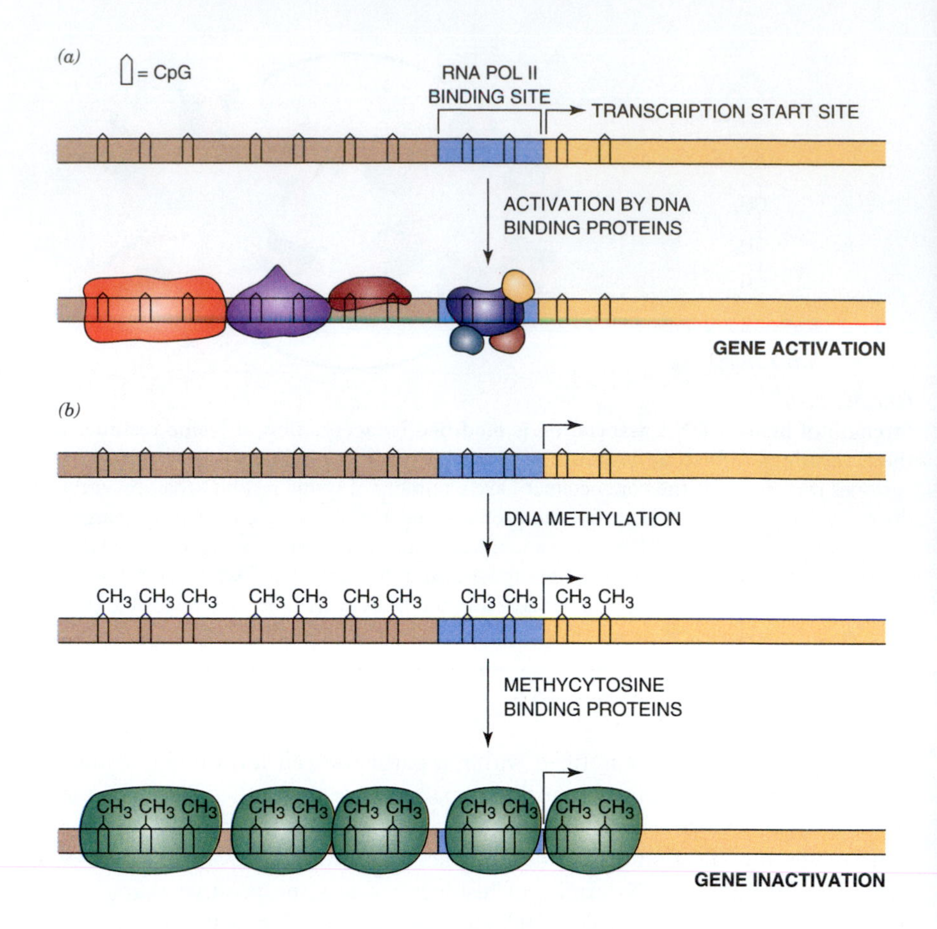

FIGURE 8.20
Methylation of DNA leads to altered gene activity.
In order for most genes to be maximally transcribed, transcription factors recognize and bind to specific sequences of DNA in the promoter region. Their interaction with the DNA and the general transcription factors in the RNA polymerase II initiation complex leads to the expression of a gene. Methylation of DNA, specifically the formation of 5-methyl cytosine, provides a new target for protein–DNA interaction. The association of 5-methyl cytosine DNA-binding proteins with methylated DNA may block the ability of other transcription factors to find and bind to DNA. This inhibition is usually not through the sequence-specific competition for DNA binding, but rather by steric hindrance.
Redrawn from Alberts, B., Bray, D., Lewis, J., Raff, M., Roberts, K., and Watson, J. Molecular Biology of the Cell. *New York: Garland Publishing, 1994.*

of CGs are presumably due to selective pressure to maintain a regulatory region, and CG islands are most commonly found in gene promoter regions.

Methylation of DNA often correlates with lack of transcriptional activity. This is thought to be mediated by proteins that recognize and bind methylated DNA, which prevents the binding of transcription factors. Methylation of DNA correlates with the deacetylation of histones, providing two different means of repression of transcription at a specific location (Figure 8.20).

Hypermethylation of DNA is a common feature in cells of cancerous tissue. There is increasing evidence that methylation occurs in genes encoding proteins that would direct abnormally dividing cells into programmed cell death (apoptosis). Conversely, in mammalian totipotent and pluripotent cells, such as a fertilized egg and cells of the very early embryo, DNA undergoes fairly global demethylation.

8.8 | PREINITIATION COMPLEX IN EUKARYOTES: TRANSCRIPTION FACTORS, RNA POLYMERASE II, AND DNA

Unlike bacterial RNA polymerase, eukaryotic RNA polymerase II does not undergo sequence-specific binding to eukaryotic promoters. Rather, an initiation complex is formed through the initial contact of the promoter with the **general transcription factor TFIID.** TFIID is one of at least six general transcription factors **(TFIIA, TFIIB**, TFIID, **TFIIE, TFIIF**, and **TFIIH)** that are required for basal transcription by RNA polymerase II. The nomenclature of these transcription factors (TF) reflects the fact that they are involved in the assembly of the RNA polymerase II preinitiation complex (Table 8.1).

TFIID is a multisubunit complex that contains the protein TBP (TATA *b*inding *p*rotein) and a number of different TAFs (TBP-*a*ssociated *f*actors). TFIID's contact with the DNA is via the binding of TBP in the minor groove of the DNA at a

TABLE 8.1 General Transcription Factors Found in Eukaryotes

Factor	*Number of Subunits*	*Mass (kDa)*
TFIID		
TBP	1	38
TAFs	12	15–250
TFIIA	3	12, 19, 35
TFIIB	1	15
TFIIE	2	34, 57
TFIIF	2	30, 74
TFIIH	9	35–89

Source: Roeder, R. G. *Trends Biochem. Sci.* 21: 329, 1996.

consensus sequence called the TATA box, which is about 27 bp upstream of the transcription start site. The TATA box must be accessible to TFIID, so, as mentioned above, it cannot be on the inner face of a nucleosome core particle if the preinitiation complex is to form. Interaction of TBP with the DNA causes a large distortion in the DNA duplex (Figure 8.21). TBP was originally thought to have an irreplaceable role in the transcription of all RNA polymerase genes, but the recent identification of TLF (TBP-*l*ike *f*actor) suggests that for some genes an alternate protein serves the same function. Binding to the TATA box by TBP directs assembly of the preinitiation complex by ordered addition of several general transcription factors and RNA polymerase II (Figure 8.22). As mentioned earlier,

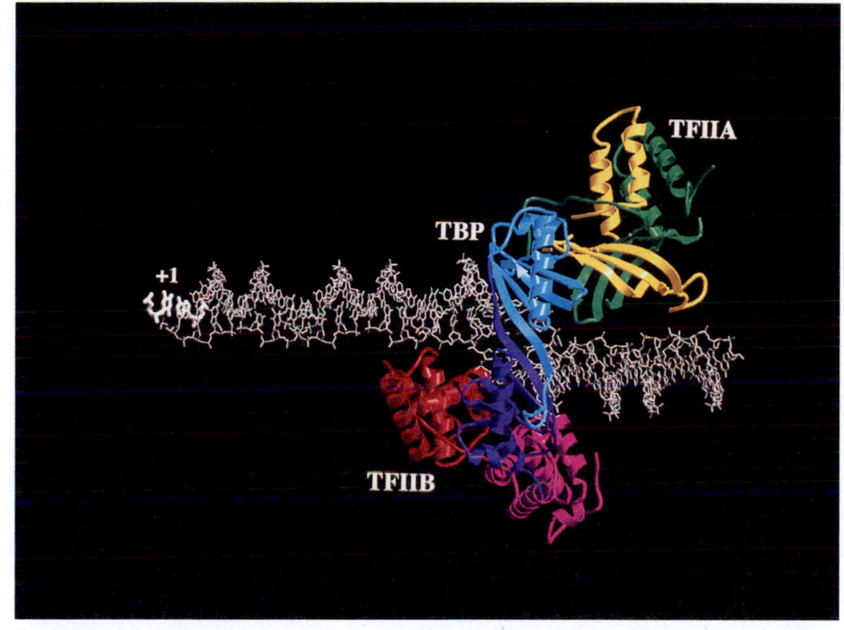

FIGURE 8.21
The TATA binding protein (TBP) has been cocrystallized with DNA.
The first step in forming the transcription complex that will allow RNA polymerase II-mediated gene transcription is the association of the general transcription factor TFIID with DNA. In most cases, the interaction is mediated through one protein of the TFIID–protein complex, TBP, which binds to DNA through contacts with the sequence TATA. The cocrystallization of TBP with DNA allows a closer look at the interaction that occurs and indicates that the binding of TBP introduces a significant bend in the DNA. Included in the figure are two other general transcription factors, TFIIA and TFIIB, neither of which are involved in the initial contact with DNA. TFIIA binds to TBP and stabilizes the TBP–DNA interaction. TFIIB binds to TBP and leads to the recruitment of RNA polymerase II to the growing complex and is involved in identification of the transcription start site. See Figure 9.27, p. 391, for another view of the TATA-binding protein.
Reproduced from Voet, D., Voet, J., and Pratt, C. W. Fundamentals of Biochemistry. *New York: Wiley, 1999. Reprinted by permission of John Wiley and Sons Inc. Figure courtesy of S. K. Burley and D. B. Nikolova, Rockefeller University.*

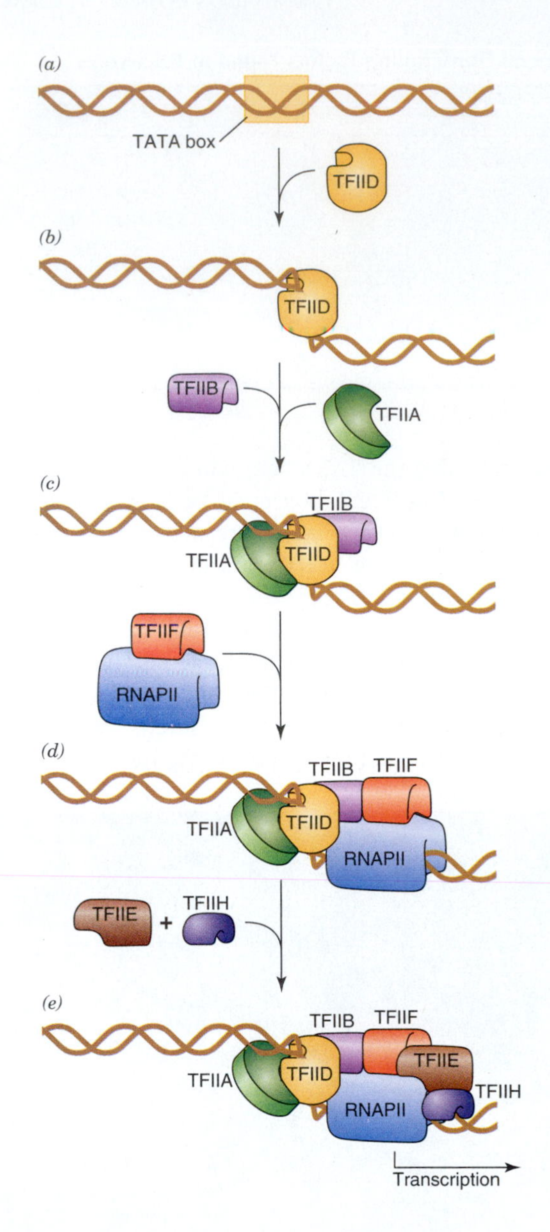

FIGURE 8.22

Formation of the initiation complex for genes transcribed with RNA polymerase II is ordered.

Many of the current models of initiation complex assembly include the following steps. (*a*) The promoter region of a gene with a TATA box before TFIID binding. (*b*) TFIID, which includes TBP, binds at the TATA box and bends the DNA. (*c*) The DNA–TFIID complex serves as a coordination site for binding other general transcription factors. The TFIID complex associates with TFIIA and TFIIB. TFIIA serves to stabilize the TFIID–DNA interaction, and TFIIB provides an appropriate interaction site for the binding of RNA polymerase II. (*d*) TFIIF and RNA polymerase II join the complex. At this stage the carboy-terminal domain (CTD) of RNA polymerase II is unphosphorylated. TFIIF is thought to destabilize nonspecific RNA polymerase II–DNA interactions, thus targeting the RNA polymerase II to the growing initiation complex. (*e*) Two additional general transcription factors join the complex. TFIIE recruits TFIIH to the complex. TFIIH has helicase, ATPase, and kinase activity. TFIIH is though to have roles in transient opening of the DNA to provide a single-stranded template for RNA polymerase to copy, and also in the phosphorylation of the CTD of RNA polymerase II, which is required for the polymerase to leave the initiation complex and facilitate transcription.

Redrawn from Voet, D., Voet, J., and Pratt, C. W. Fundamentals of Biochemistry. *New York: Wiley, 1999.*

the large subunit of RNA polymerase II has a C-terminal domain (CTD) containing the repeating amino acid sequence PTSPSYS. The entry of RNA polymerase II into the preinitiation complex only occurs when its CTD is not phosphorylated. However, the movement of RNA polymerase II away from the assembly site is correlated with the phosphorylation of CTD. Thus the assembly of the preinitiation complex does not ensure transcription; further signaling, such as phosphorylation, must occur for transcription to begin.

The formation of the preinitiation complex allows the ATP-dependent **helicase** activity associated with TFIIH to open the two strands of the DNA, providing a template for the subsequent elongation phase of RNA synthesis. TFIID then remains behind, still bound to the TATA box, to promote the assembly of additional preinitiation complexes; TFIIF continues along with RNA polymerase II during elongation phase but the other general transcription factors dissociate from the elongation complex.

Eukaryotic Promoter and Other Sequences that Influence Transcription

Promoters of eukaryotic genes transcribed by RNA polymerase II are operationally defined as those sequences that influence the initiation of gene transcription. The hallmark of eukaryotic promoters is the utilization of multiple transcription factor binding sites to regulate gene activity. In general, these binding sites are relatively close to the **TATA box** that marks the site of assembly of the preinitiation complex. Other consensus sequences are often found in eukaryotic promoters, including the **CAAT** and **GC** boxes (the GC box refers to a consensus sequence of GGGCGG) (Figure 8.23). The exact position of the CAAT box and the exact positions, orientation and number of GC boxes varies with the promoter. These sequences (like the TATA box) are present in most but not all promoters. The CAAT box serves as a binding site for several different transcription factors, including NF1. The placement of multiple GC boxes provides multiple sites where the transcription factor SP1 can bind. The presence of a CAAT box is usually taken as an indicator of a strong promoter, and the GC box is characteristic of many housekeeping genes.

The binding of transcription factors to other sequences in promoters is influenced either by the presence of a signaling molecule, such as hormones associated with steroid response transcription factors, or by phosphorylation of amino acid side groups. DNA sequences that bind to such transcription factors are often referred to as **response elements.** In addition, binding sites in promoters of some genes are used by transcription factors unique to specific tissues or that appear at developmentally specific times. The relative effectiveness of the regulatory sequences found in promoters is dependent on their orientation or directionality and may be diminished by altering their distance from the TATA box.

An additional regulatory sequence associated with many eukaryotic cellular and viral genomes is an **enhancer.** Enhancer elements also provide a binding site for specific proteins that regulate transcription. However, they differ from promoter sequences in two important respects. First, enhancers can be located many thousands of base pairs away from the site where the preinitiation complex is assembled

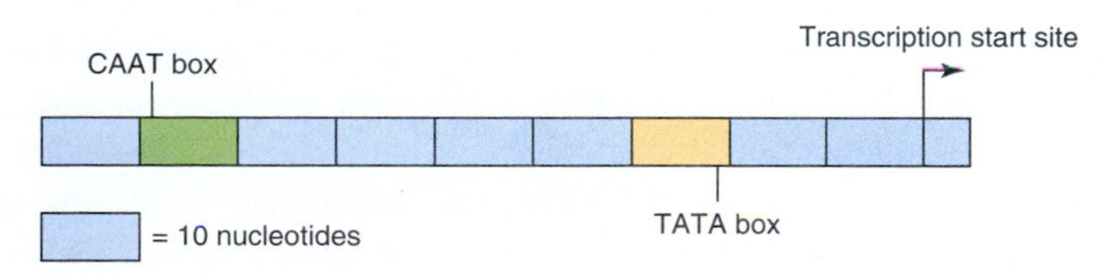

FIGURE 8.23
DNA sequences in genes transcribed in eukaryotes typically contain a TATA box and may contain a CAAT box, GC-rich regions (GC boxes), and multiple transcription factor binding sites not shown.

and can occur either upstream or downstream of the transcription start site. Second, the enhancer element can act in either orientation. The existence of enhancer elements illustrates the spatial constraints imposed on transcription factors that bind to DNA and coordinate interaction with the preinitiation complex, affecting initiation of transcription. Transcription factors that bind to response elements close to the transcription initiation site may act directly with the initiation complex or through the use of proteins that make a bridge to the initiator complex. It is quite common for proteins influencing transcription to serve this bridging function. However, in the case of enhancer elements, the linear distance between the enhancer and the preinitiation complex may be very large. The transcription factors bound to enhancers are brought together with the preinitiation complex by looping of the DNA, that is, the bending of the DNA to bring elements well separated on the linear DNA into close proximity. This flexibility also helps to explain how the enhancer can act at a distance, and why the orientation of the enhancer sequence does not matter. As long as the protein binds to the enhancer, the proper alignment of the protein with the preinitiation complex can be adjusted as the DNA loops back.

Modular Design of Eukaryotic Transcription Factors

Eukaryotic transcription factors have multiple domains that carry out specific interactions. These domains include DNA recognition domains that participate in site-specific binding and activation domains that contact general transcription factors, RNA polymerase II, or other regulators of transcription. Other domains in many transcription factors are dimerization domains that promote the formation of homodimers or heterodimers with another monomeric transcription factor and protein interaction domains that allow association with other proteins such as histone acetylase. In addition, many transcription factors have domains that bind to coactivators. Examples of coactivators include steroid hormones and cAMP that, when bound, change the ability of the transcription factor to either bind to DNA or serve as an activator. Although their overall amino acid sequence and composition uniquely identify each transcription factor, the domains involved in each activity can be grouped into a few characteristic motifs.

Common Motifs in Proteins that Interact with DNA and Regulate Transcription

Several characteristic **amino acid motifs** are commonly found in transcription factors (trans-acting regulators). The motifs include helix–turn–helix (HTH), zinc finger, helix–loop–helix (HLH), and basic region–leucine zipper (bZIP) domains (see p. 387). These motifs account for about 80% of known sequence-specific binding proteins.

Examples of **helix–turn–helix** (HTH) motif proteins include the lactose repressor of *E. coli* (see Section 8.3) and a group of developmentally important transcription factors called **homeodomain proteins.** The homeodomain is the DNA recognition portion of these transcription factors, and received its name because it was characteristic of proteins required for segment specification in fruit fly development. Mutations in the genes encoding these proteins lead to homeosis, where, for instance the normal developmental identity of antennae may be altered, resulting in legs forming where the antennae should be. Homeodomain proteins have turned out to be key regulators in mammalian development as well. The HTH motif is about 20 amino-acids and is thus often a relatively small part of a much larger protein. It is the domain within the protein that allows sequence-specific interaction with DNA. Within the motif the first seven amino-acids form one helix followed by a four-amino-acid turn and then a nine-amino-acid helix. The nine-amino-acid helix is the "recognition helix" that binds in a sequence-specific manner in the major groove, while hydrophobic interactions between the seven-residue helix and one side of the nine-residue helix stabilize the structure. The rest of the protein may have allosteric sites that can be occupied by regulators of structure or regions that allow other protein–protein interactions (Figure 8.24 and p. 387).

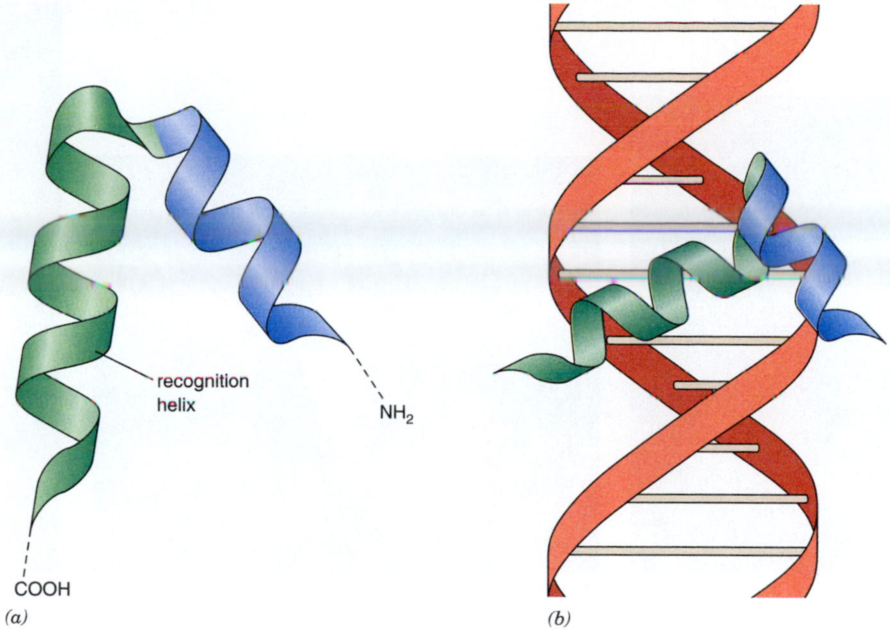

FIGURE 8.24
Helix–turn–helix proteins use one helix to bind in the major groove while the other supports that binding through hydrophobic interaction.
(*a*) HTH domain of a protein is typically about 20 amino acids. These 20 amino acids form two individual α-helices, joined by a nonhelical turn. (*b*) The dimensions of the α-helix of the protein allow it to fit into the major groove of the DNA helix. The helix (green) that interacts directly with the DNA (the recognition helix) includes about nine amino acids with side chains capable of forming hydrogen bonds with specific bases that are exposed in the major groove. The second helix (blue) in the HTH motif does not directly interact with DNA, but stabilizes the binding of the recognition helix through hydrophobic interactions. Thus both helices include amino acids, like valine or leucine, that allow these hydrophobic interactions to occur.
Redrawn from Alberts, B., Bray, D., Lewis, J., Raff, M., Roberts, K., and Watson, J. Molecular Biology of the Cell. *New York: Garland Publishing, 1994.*

Examples of **zinc finger** proteins include the general transcription factor TFIIIA that binds to the promoters of RNA polymerase III-transcribed genes, Sp1 (which gives 10- to 20-fold stimulation of all genes with GC boxes), Gal4 in yeast, and the steroid hormone receptor superfamily. Different subclasses of zinc finger proteins are defined by the specific amino acids that coordinate Zn binding. For instance, in TFIIIA two Cys and two His coordinate Zn binding and are in the C2H2 class, whereas the steroid hormone receptor transcription factors use four Cys for each Zn and are of the C_x class. The zinc finger motif binds in the major grove of the DNA in a sequence-specific manner, mediated by an α-helix formed on one side of the finger region. As was seen with HLH proteins, the ability of an α-helix to fit in the major groove with amino acid side chains forming hydrogen bonds with the bases in the major groove is the basis for sequence-specific binding (See p. 388).

The two classes of zinc finger proteins have binding sites characteristic of the way each class positions itself on DNA. For TFIIIA, sequential zinc fingers follow the major groove, each forming hydrogen bonds with specific bases. Zinc finger proteins of the C_x class, such as the steroid hormone receptor transcription factors, use one zinc finger to bind to DNA and a second in a manner reminiscent of the HTH proteins. The second zinc finger stabilizes the binding of the first finger by a protein–protein interaction. The steroid hormone receptor transcription factors associate with DNA as dimers, and the binding site in the DNA reflects this by having two palindromic "half-sites" spaced to accommodate the two DNA-binding fingers formed by dimer formation (Figure 8.25).

Examples of basic region–**leucine zipper** (bZIP) proteins include fos, jun, and CREB. The bZIP proteins are named for the periodic repeat of leucine residues in an α-helix. These leucines form hydrophobic interactions with a second protein in which a similar helix allows the formation of a dimer. Thus the leucine zipper refers

(a)

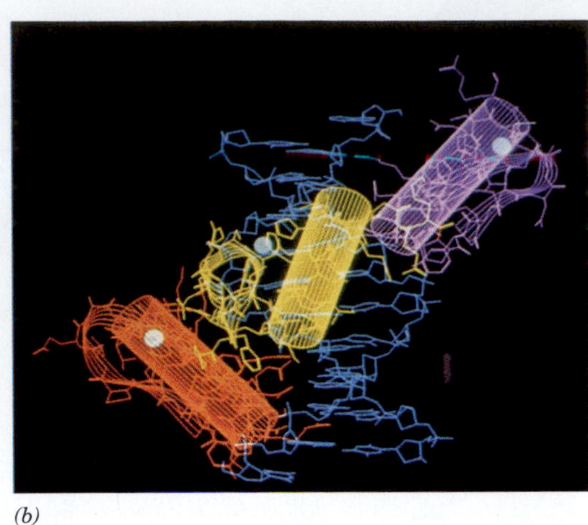

(b)

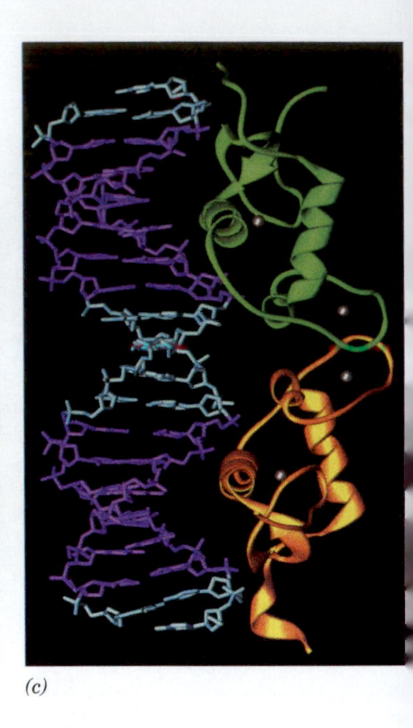

(c)

FIGURE 8.25

Two different zinc finger motifs are found in transcription factors. Although both zinc finger proteins use the coordinate binding of a Zn molecule to assume their final structure, there are recognizable differences in the two major motifs that have been identified. (*a*) Both classes take advantage of the formation of α-helical structure to form domains that bind in the major groove of DNA. Either two cysteines and two histidines or four or more cysteines may coordinate binding of Zn. (*b*) The C2H2 class includes proteins that may have many zinc finger domains. Each α-helix from a zinc finger has the potential to bind in a sequence-specific manner to sites along the major groove. The result may be a procession of protein–DNA interactions, each dependent on the particular array of amino acid side chains found in each α-helical domain of the zinc finger protein. (*c*) Cx class of zinc finger proteins commonly has two zinc finger domains. The α-helix from one zinc finger binds the DNA in the major groove, while the α-helix in the other zinc finger supports that interaction by hydrophobic interactions with the domain binding to the DNA. The Cx class of zinc finger proteins normally binds DNA by forming dimers. Shown is the interaction of the glucocorticoid receptor dimer, with each monomer contacting the DNA. Zn molecules are indicated by spheres.
Reproduced from Voet, D. and Voet, J. G., Biochemistry, 2nd ed., New York: Wiley, 1995. Reprinted by permission of John Wiley and Sons Inc. Part (b) courtesy of C. Pabo, MIT.

to the protein–protein interaction domain. The dimer formed can be a homodimer or a heterodimer (Figure 8.26 and see p. 389.).

The DNA contact surface of the protein is a basic region, defined by the presence of arginine and lysine. This basic region assumes the conformation of two α-helices with a small break that allows the helices to follow the major groove of the DNA. The basic amino acids serve to stabilize the DNA–protein association through electrostatic interactions between the positively charged amino acids and the negatively charged DNA backbone. The binding site on the DNA is different when the bZIP protein acts as a homodimer than when it is a heterodimer. Homodimers bind to a site that has dyad symmetry, whereas this symmetry is not found when heterodimers bind.

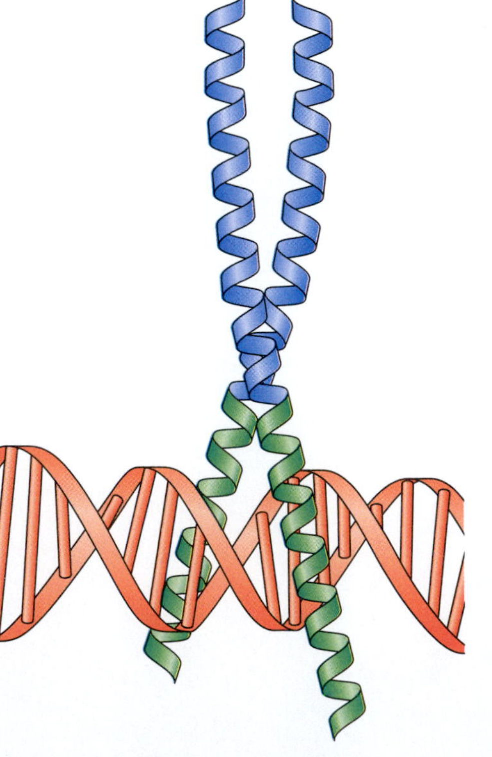

FIGURE 8.26

Leucine zipper proteins bind to DNA as dimers.
Leucine zipper proteins form dimers by virtue of the periodic placement of leucines along an α-helix domain. These leucines form a hydrophobic face that will interact with the hydrophobic face of a protein with an α-helix with similarly placed leucines. The protein-protein interaction domains are indicated in blue. The α-helical regions may continue beyond the protein–protein interaction domain, allowing binding to the major groove of DNA (green). For homodimers, the DNA-binding site is characterized by two recognizable and symmetric half sites.
Modified from Alberts, B., Bray, D., Lewis, J., Raff, M., Roberts, K., and Watson, J. Molecular Biology of the Cell. *New York: Garland Publishing, 1994.*

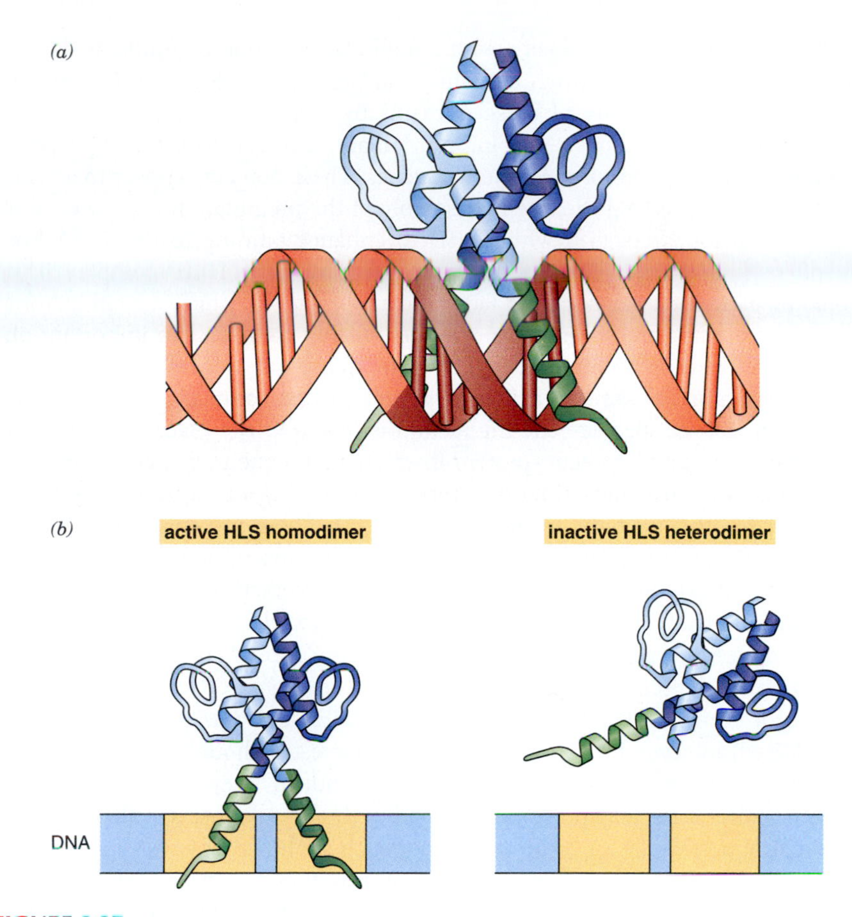

FIGURE 8.27
Transcription factor dimer formation is mediated through helix–loop–helix interactions. The helix–loop–helix motif brings together two monomers to form a dimer that binds to DNA. (*a*) Each monomer has two helices joined by a loop. One helix (blue) is used for protein–protein interaction; the other (green) is used to bind the major groove of DNA. Thus the dimer consists of a four-helix bundle. If the dimer is formed by two identical monomers, then the DNA-binding sites are expected to be very similar or identical; however, if the monomers are different proteins (forming heterodimers), then the DNA-binding sites may be unrelated. (*b*) When transcription factors bind as dimers, the presence of a truncated monomer can prevent DNA binding even in the presence of full-length monomers. For example, if the protein dimerization helix is made without the DNA-binding domain, the dimerization with a full-length monomer produces a product unable to bind effectively to DNA.
Modified from Alberts, B., Bray, D., Lewis, J., Raff, M., Roberts, K., and Watson, J., Molecular Biology of the Cell. *New York: Garland Publishing, 1994.*

The **helix–loop–helix** class of transcription factors includes myoD, myc, and max. Two amphipathic α-helical segments separated by an intervening loop characterize helix–loop–helix proteins. The helices are not responsible for DNA binding, as in the zinc finger proteins, but for dimerization with another protein. As was described for the bZIP proteins, the dimers formed can be homo- or heterodimers. The DNA-binding domains extend from helices that are involved in the protein-protein interaction that forms the dimer. (Figure 8.27).

The structures of helix-turn-helix, zinc fingers, bZIP, and helix-loop-helix proteins are described in detail in Secion 9.4, DNA Binding Proteins (p. 387).

8.9 | REGULATION OF EUKARYOTIC GENE EXPRESSION

As indicated above, a relatively small region of a protein may be dedicated to sequence-specific binding of DNA, while other domains are involved in protein–protein or ligand interactions. Several characteristic activation domains have been identified, including acidic domains with a high concentration of amino acids with acidic side chains, glutamine-rich domains, and proline-rich domains. Experimentally, overproduction of any of these domains by recombinant techniques, even without their corresponding DNA-binding domain, can lead to the inappropriate activation of transcription from a variety of genes. These domains appear to activate transcription by increasing the rate of assembly of the preinitiation complex. Some act through direct interaction with TFIID to enhance binding to the TATA box, while others interact with TFIIB or TAFs that are part of the TFIID complex. Thus the binding of multiple transcription factors to a promoter may have a combinatorial effect on the binding and assembly of the preinitiation complex. An important observation is that the activation domains of many transcription factors mediate their regulatory effect on the same proteins in the preinitiation complex. Thus, for many transcription factors, specific-gene interaction is linked to the placement of DNA-binding sites and not to protein–protein interactions unique to a specific gene.

A second way that transcription factors can regulate gene expression is by the recruitment of other proteins to the promoter area. These proteins may not bind DNA but can form a regulatory bridge between the transcription factor bound to DNA and the initiation complex. Alternatively, the interacting protein may bring chromatin modification enzymes, such as histone acetylase (e.g., the CBP/p300 complex) to specific genes. Acetylation of histones loosens chromatin structure. Some regulated transcription factors that recruit the CBP/p300 complex, and the molecules or events to which they respond, are CREB and cAMP, SPEBP and cholesterol, NF-κB and cytokines, and p53 and growth arrest. Although few transcription factors that reduce gene expression have been identified, they may act in several ways. A negative factor may preemptively bind to the DNA, either specifically to the same site as a positive factor or more globally, such as the negative effect of methylated DNA-binding proteins. Alternatively, negative factors may inhibit binding or assembly of the preinitiation complex. Finally, negative factors may bind to a positive transcription factor to prevent its binding to DNA or to speed its degradation.

Regulating the Regulators

Transcription factors themselves are regulated in a variety of ways. Probably the simplest examples are those that are only synthesized by certain cells, or at specific times in development. The presence or absence of the transcription factor determines activity. Some factors are regulated by cofactor binding, which may either inhibit or stimulate their binding to DNA. For transcription factors that bind as dimers or multimers, the formation of nonproductive homo- or heterodimer complexes can alter DNA binding and protein–protein interactions that allow gene activation. Posttranslational modification, such as protein processing or phosphorylation, may affect not only the ability of transcription factors to bind to DNA or other proteins, but also their ability to move from the cytoplasm into the nucleus (see Clin. Corr. 8.3 and 8.4).

CLINICAL CORRELATION 8.3

Tamoxifen and Targeting of Estrogen Receptor

Tamoxifen, a drug used to treat breast cancer, binds to the estrogen receptor (ER) in place of estrogen. Since estrogen activates the estrogen receptor, tamoxifen reduces the transcription from genes that are regulated by the estrogen receptor, thus reducing the growth of breast cancer cells. There is a complication, however, in that sometimes tamoxifen treatment increases the risk for uterine cancer. The apparent cause lies in the presence of two estrogen receptor subtypes (α and β). Both subtypes exist in breast tissue but estrogen receptor-α predominates in uterine tissue. When tamoxifen is bound to estrogen receptor-α, it leads to activation rather than inhibition of transcription in conjunction with transcription factors fos and jun.

Paech, K., Webb, P., Kuiper, G., Nilsson, S., Gustafsson, J.-A., Kushner, P. J., and Scanlan, T. S. Differential ligand activation of estrogen receptors ERα and ERβ at AP1 sites. *Science* 277:1508, 1997.

Activation of Transcription of the LDL Receptor Gene Illustrates Many Features Found in Eukaryotic Gene Regulation

The description of the regulation of expression of specific eukaryotic genes can rapidly turn into an alphabet soup of transcription factors and regulators of transcription factors that makes sense to researchers studying that gene but is confusing to most others. However, many features of transcription factors described above are illustrated by the transcriptional control of the gene for the **low density lipoprotein (LDL) receptor.** This gene is transcribed in response to the lack of cellular cholesterol. Increased transcription of the gene leads to an increased amount of the LDL receptor protein and enhanced uptake of LDLs and their cholesterol in the blood (see p. 745).

The promoter of the LDL receptor gene has a TATA box and several regions that are binding sites for known transcription factors (Figure 8.28). The TATA box provides the site for TFIID binding and the formation of the preinitiation complex with RNA polymerase II. There are also three consensus sites at which the zinc finger containing transcription factor Sp1 can bind (recall that SP1 binds to GC boxes in promoters). The modular Sp1 protein has, in addition to the zinc finger motif, a glutamine-rich activation domain, which is thought to assist in the recruitment of TFIID to the TATA box. This recruitment uses an additional factor called the *c*ofactor *r*equired for *Sp*1 activation (CRSP). However, the binding of Sp1 is not sufficient to activate transcription of the gene when the gene is assembled into chromatin. Activation also requires the participation of a second transcription factor called **SREBP-1a** (**s**terol **r**esponsive **e**lement-**b**inding **p**rotein 1a). SREBP-1a is a helix–loop–helix–leucine zipper transcription factor that binds to the sterol response element sequence that lies between the Sp1 binding sites in the LDL receptor gene.

The transport of SREB-1a into the nucleus affords the cell the capacity to respond to the concentration of cholesterol. SREBP-1a contains two domains that loop through the membrane of the endoplasmic reticulum (ER), leaving the domains capable of serving as a transcription factor exposed to the cytoplasm but tethered to the ER. In order for SREBP-1a to move to the nucleus it must be cleaved by two proteolytic steps, the first of which is carried out by a protease found in an active form in a post-ER compartment called the cis-Golgi. Transport of SREBP-1a to the cis-Golgi depends on interaction with **S**REBP **c**leavage **a**ctivating **p**rotein **(SCAP)**, which also is partially embedded in the ER membrane. SCAP has a cholesterol-sensing region and when membrane cholesterol levels are low, the SCAP–SREBP-1a complex is transported to the cis-Golgi, where proteolysis of

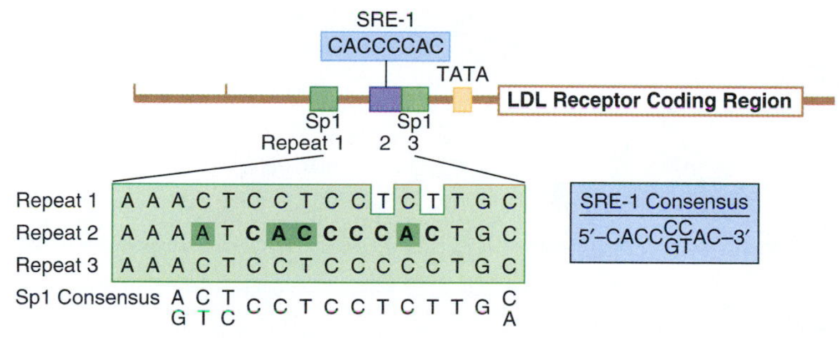

FIGURE 8.28
A schematic of the LDL receptor gene promoter.
The LDL receptor gene has a structure typical of many eukaryotic genes with a number of different transcription factor binding sites. Shown here are the principal sites that are involved in the regulation of the LDL receptor gene in response to cholesterol levels. These sites include the TATA box, just upstream from the transcription initiation site, several GC boxes (Sp1 binding sites), and the steroid response element (SRE) where SREBP binds.
Modified from Goldstein, J. L. and Brown, M. S. Nature *343:425, 1990.*

CLINICAL CORRELATION 8.4
Transcription Factors and Cardiovascular Disease

Among the genes recently identified as causing human cardiovascular disease are two genes encoding transcription factors. One gene, NKX2-5, encodes a homeodomain containing protein. Homeodomain proteins have been recognized as regulators of gene expression during embryonic development, and NKX2-5 regulates genes involved in heart formation. Mutations in a single allele of NKX2-5 leads to defects in the atrial septum, A/V valve, and the conduction system. The null mutation, so far only identified in mouse, is embryonic lethal. The other gene, Tbx is responsible for Holt–Oram syndrome. Holt–Oram syndrome causes holes between the atria and sometimes the ventricles as well as hand and arm defects.

Barinaga, M. Tracking down mutations that can stop the heart. *Science* 128:32, 1998; Schott, J. J., Benson, D. W., Basson, C. T., Pease, W., et al. Congenital heart disease caused by mutations in the transcription factor NKX2-5. *Science* 281:108, 1998; and Li, Q. Y., Newbury-Ecob R.A., Terrett, J.A., Wilson, D. I., et al. Holt–Oram syndrome is caused by mutations in TBX5, a member of the Brachyury (T) gene family. *Nature Genet.* 15:21, 1997.

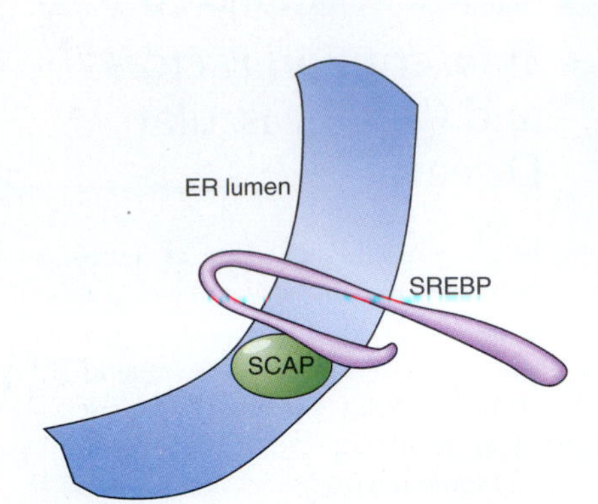

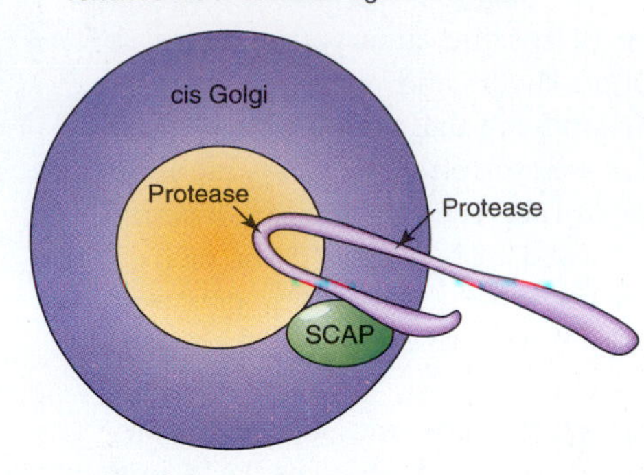

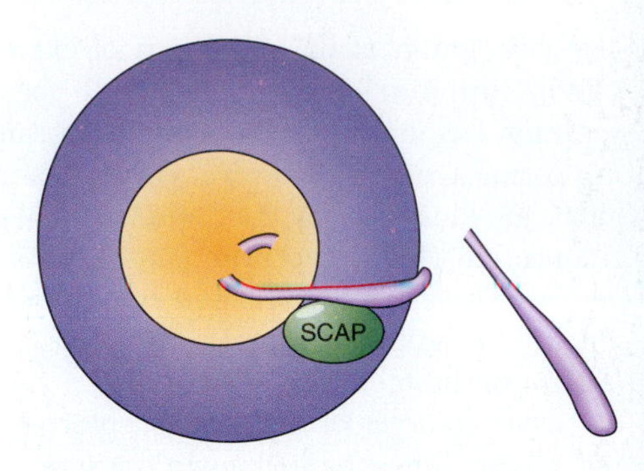

FIGURE 8.29
SREBP is released from membrane-bound precursor by protease action.
(*a*) The steroid response element binding protein (SREBP) is synthesized as a precursor protein that must be proteolytically processed before it acts as a transcription factor. The precursor is tethered to the membrane of the ER by two membrane-spanning regions. It is in close association with SREBP cleavage activating protein (SCAP) but remains unprocessed when cholesterol levels in the ER membrane are normal. (*b*) SCAP senses low levels of cholesterol and moves with SREBP to the cis-Golgi, where two different proteases cleave (SREBP), (*c*) allowing the cytoplasmic domain to move to the nucleus. In this way the SREBP does not activate the LDL receptor gene unless cholesterol levels are low.
Adapted from Brown, M. S., Ye, J., Rawson, R. B., and Goldstein, J. L. Cell *100:391, 2000.*

SREBP-1a occurs, releasing the portion of the protein that can act as a transcription factor to the nucleus (Figure 8.29). Once in the nucleus, it can bind to the steroid response element (SRE) on the LDL promoter where it recruits a histone acetyltransferase called CBP and other proteins to the promoter area. Then, along with the action of Sp1 and CRSP, the LDL receptor gene becomes transcriptionally active (Figure 8.30).

This example illustrates the composite nature of eukaryotic gene regulation, in which the activation of transcription factors, coordination of multiple transcription factors, and recruitment of chromatin remodeling enzymes, all work in concert to regulate gene expression.

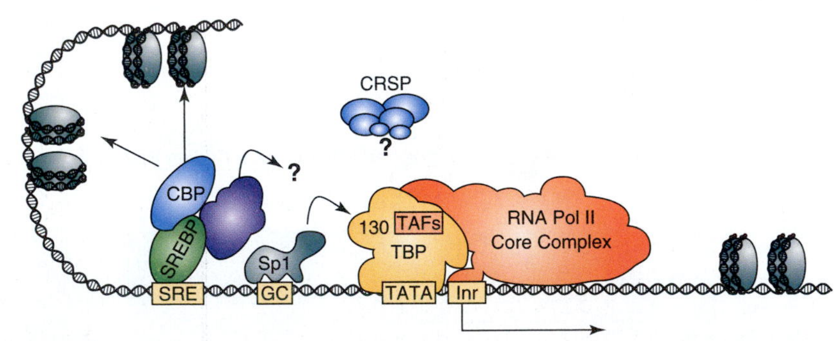

FIGURE 8.30
LDL receptor gene is activated through coordinate effect of several transcription factors.
Activation of the LDL receptor gene requires participation of several different factors. Once SREBP, Sp1, and the cofactor required for Sp1 activation (CRSP) bind, the histone acetyltransferase, CBP, is recruited to the promoter region. A combination of factors provides the positive signal needed to enhance binding of the initiation complex to the TATA box, and through recruitment of CBP also affects nearby chromatin structure.
Naar, A. M., Ryu, S., and Tijian, R. Cold Spring Harbor Symp. Quant. Biol. *LXIII:189, 1998.*

BIBLIOGRAPHY

Prokaryotic Gene Expression

Blattner, F. R., Plunkett, G., Bloch, C. A., Perna, N. T., et al. The complete genome sequence of *Escherichia coli* K-12. *Science* 277:1453, 1997.

Cohen, S. N. and Shapiro, J. A. Transposable genetic elements. *Sci. Am.* 242:40, 1980.

Miller, J. H. The *lac* gene: its role in *lac* operon control and its use as a genetic system. In: J. H. Miller and W. S. Resnikoff (Eds.), *The Operon*. Cold Spring Harbor, NY: Cold Spring Harbor Laboratory Press, 1978, p. 31.

Platt, T. Regulation of gene expression in the tryptophan operon of *Escherichia coli*. In: J. H. Miller and W. S. Resnikoff (Eds.), *The Operon*. Cold Spring Harbor, NY: Cold Spring Harbor Laboratory Press, 1978, p. 263.

Schultz, S.C., Shields, G. C., and Steitz, T. A. Crystal structure of a CAP–DNA complex: the DNA is bent by 90 degrees. *Science* 253:1001, 1991.

Eukaryotic Gene Expression

Alberts, B., Bray, D., Lewis, J., Raff, M., Roberts, K., and Watson, J. *Molecular Biology of the Cell*. New York: Garland Publishing, 1994.

Brown, M. S., Ye, J., Rawson, R. B., and Goldstein, J. L. Regulated intramembrane proteolysis: a control mechanism conserved from bacteria to humans. *Cell* 100:391, 2000.

Brown, C. E., Lechner, T., Howe, L., and Workman, J. L. The many HATs of transcription coactivators. *Trends Biochem. Sci.* 25:15, 2000.

Burley, S. K. and Roeder, R. Biochemistry and structural biology of transcription factor IID (TFIID). *Annu. Rev. Biochem.* 65:769, 1996.

Cramer, P. Bushnell, D. A., and Kornberg, R. D. Structural basis of transcription: RNA polymerase II at 2.8 Å resolution. Science 292:1863, 2001.

Dahmus, M. E. Reversible phosphorylation of the C-terminal domain of RNA polymerase II. *J. Biol. Chem.* 271:19009, 1996.

Dantonel, J.-C., Wurtz, J.-M., Poch, J. M., Moras, D., and Tora, L. The TBP-like factor: an alternative transcription factor in Metazoa. *Trends Biochem. Sci.* 24: 335, 1999.

Goldstein, J. L. and Brown, M. S. Regulation of the mevalonate pathway. *Nature* 343:425, 1990.

Harrison, S. C. A structural taxonomy of DNA-binding domains. *Nature* 353: 715, 1991.

Kornberg, R. D. Mechanism and regulation of yeast RNA polymerase II transcription. *Cold Spring Harbor Symp. Quant. Biol.* LXIII: 229, 1998.

Latchman, D. S. *Eukaryotic Transcription Factors*, 3rd ed. New York: Academic Press, 1998.

Lemon, B. and Tjian, R. Orchestrated response: a symphony of transcription factors for gene control. *Genes Dev.* 14:2551, 2000.

Lewin, B. *Genes VII*. New York: Oxford University Press, 2000.

Miller, J., McLachlan, A. D., and Klug, A. Repetitive zinc-binding domains in the protein transcription factorIIA from *Xenopus* oocytes. *EMBO J.* 4:1609, 1985.

Naar, A. M., Ryu, S., and Tijian, R. Cofactor requirements for transcriptional activation by Sp1. *Cold Spring Harbor Symp. Quant. Biol.* LXIII:189, 1998.

Ng, H. and Bird, A. Histone deacetylases: silencers for hire. *Trends Biochem. Sci.* 25:121, 2000.

Orphanides, G., Lagrange, T., and Reinberg, D. The general transcription factors of RNA polymerase II. *Genes Dev.* 10:2657, 1996.

Pabo, C. O. and Sauer, R. T. Transcription factors: structural families and principles of DNA recognition. *Annu. Rev. Biochem.* 61:1053, 1992.

Pruss, D., Hayes, J. J., and Wolffe, A. Nucleosomal anatomy—where are the histones? *BioEssays* 17:161, 1995.

Roeder, R. G. Role of general and gene-specific cofactors in the regulation of eukaryotic transcription. *Cold Spring Harbor Symp. Quant. Biol.* LXIII:201, 1998.

Stewart, S. and Crabtree, G. Regulating the regulators. *Nature* 408:46, 2000.

Wolffe, A. P. The cancer–chromatin connection. *Sci. Med.* 6:28, 1999.

Wolffe, A. P. *Chromatin: Structure and Function*, 3rd ed. New York: Academic Press, 1998.

Woychik, N. and Young, R. RNA polymerase II: subunit structure and function. *Trends Biochem. Sci.* 15:347, 1990.

Zawel, L. and Reinberg, D. Initiation of transcription by RNA polymerase II: a multi-step process. *Curr. Opin. Cell Biol.* 4:488, 1992.

QUESTIONS | C. N. ANGSTADT

Multiple Choice Questions

1. Full expression of the *lac* operon requires:
   A. lactose and cAMP.
   B. allolactose and cAMP
   C. lactose alone.
   D. allolactose alone.
   E. *lac* corepressor.

2. In an operon:
   A. each gene of the operon can be transcribed independently to achieve different levels of expression required by the cell.
   B. control may be exerted via induction or via repression.
   C. operator and promoter may be trans to the genes they regulate.
   D. the structural genes are either not expressed at all or they are fully expressed.
   E. control of gene expression consists exclusively of induction and repression.

3. The *E. coli lac*ZYA region will be upregulated if:
   A. there is a defect in binding of the inducer to the product of the *lac*I gene.
   B. glucose and lactose are both present in the growth medium, but there is a defect in the cell's ability to bind the CAP protein.
   C. glucose and lactose are both readily available in the growth medium.
   D. the operator has mutated so it can no longer bind repressor.
   E. the *lac* corepressor is not present.

4. All of the following describe an operon EXCEPT:
   A. control mechanism for eukaryotic genes.
   B. includes structural genes.
   C. expected to code for polycistronic mRNA.
   D. contains control sequences such as an operator.
   E. can have multiple promoters.

Refer to the following for Questions 5–8:

A. repression
B. corepression
C. attenuation
D. stringent response

5. Associated with guanosine tetraphosphate and guanosine pentaphosphate.
6. Inhibition of transcription by a complex of a protein and a small molecule.
7. Involves a leader peptide containing several occurrences of the same amino acid.

8. The primary regulatory mechanism for the *his* operon.
9. In eukaryotic transcription by RNA polymerase II, formation of a preinitiation complex:
   A. begins with the binding of a protein (TBP) to the TATA box of the promoter.
   B. involves the ordered addition of several transcription factors and the RNA polymerase.
   C. allows an ATP-dependent opening of the two strands of DNA.
   D. requires that the C-terminal domain (CTD) of RNA polymerase II not be phosphorylated at this point.
   E. all of the above are correct.

10. Enhancers:
    A. are sequences in the promoter that bind to hormone–transcription factor complexes.
    B. are more effective the closer they are to the TATA box.
    C. are DNA sequences that bind transcription factors but may be thousands of base pairs away from the site of the preinitiation complex assembly.
    D. must be upstream of the site of the preinitiation complex assembly.
    E. bind transcription factors that act directly with the initiation complex.

Questions 11 and 12: The problem of pathogenic bacteria becoming resistant to a large number of antibiotics is a serious public health concern. A bacterial strain in a patient being treated with one antibiotic may suddenly become resistant not only to that antibiotic but to others as well, even though it has not been exposed to the other antibiotics. This occurs when the bacteria acquire a plasmid from another strain that contains several different transposons.

11. All of the following phrases describe transposons EXCEPT:
    A. a means for the incorporation of antibiotic resistant genes into the bacterial chromosome.
    B. contain short inverted terminal repeat sequences.
    C. code for an enzyme that synthesizes guanosine tetraphosphate and guanosine pentaphosphate, which inhibit further transposition.
    D. include at least one gene that codes for a transposase.
    E. contain varying numbers of genes, from two to several.

12. In the operation of transposons:
    A. the typical mode of action is the physical removal of the transposon from its original site and relocation to a different site.
    B. a duplicated transposon must be inserted into the same DNA molecule as the original.
    C. all transposons are approximately the same size.
    D. the insertion sites must be in a consensus sequence.
    E. the transposase may recognize the repetitive ends of the transposon and participate in the cleavage of the recipient site.

Questions 13 and 14: Genes that are present in a region of DNA that is in the highly condensed heterochromatin form cannot be transcribed. To be transcribed, this region of DNA must change to a more open structure, which may occur by acetylation of histones. One of the enzymes responsible for acetylation of histones is an acetyltransferase called CBP, CREB binding protein. CREB is a transcription factor. Rubinstein–Taybi syndrome is caused by mutations of the *CBP* gene. Rubinstein–Taybi patients are mentally retarded and have other developmental abnormalities, the severity of which depends on the extent of mutation.

13. In chromatin:
    A. a nucleosome consists of four molecules of histones surrounding a DNA core.
    B. DNA positioned so that the major and minor grooves are on the outside of the nucleosome is more accessible for transcription than if the grooves face the interior.
    C. DNA must be completely removed from the nucleosome structure to be available for transcription.
    D. linker DNA is the only DNA capable of binding transcription factors.
    E. the histone octamer consists of eight different kinds of histone proteins.

14. Acetylation of histones can lead to a more open DNA structure by:
    A. weakening the electrostatic attraction between histones and DNA.
    B. causing histones to interact with the C-terminal domain (CTD) of RNA polymerase.
    C. causing electrostatic repulsion between histones and DNA.
    D. facilitating methylation of DNA.
    E. attracting transcription factors to DNA.

Problems

15. What will be the status of transcription of the *lac* operon in (a) the presence of glucose and (b) the absence of glucose, in each case with lactose present, if there is a mutation that produces an inactive adenyl cyclase?
16. In an operon for synthesis of an amino acid which is controlled wholly or in part by attenuation, why does the presence of the amino acid prevent transcription of the whole operon while the absence of the amino acid permits it?

ANSWERS

1. **B** The true inducer is allolactose, which is usually formed from lactose by the action of β-galactosidase. A: See above. C, D: In addition to the sugar binding to the repressor, cAMP must bind to the CAP protein, and the cAMP–CAP complex serves as a positive control of transcription. E: The *lac* operon does not involve corepression.
2. **B** Induction and repression are among the mechanisms used to control operons. A: In an operon the structural genes are under coordinate control. C: The operator and promoter are elements of the same strand of DNA as the operon they control; they are not diffusible. D: Typically, regulation of operators is somewhat leaky; some gene product is produced even in the repressed state. E: Another mechanism for regulation of an operon is attenuation.
3. **D** If the operator is unable to bind repressor, the rate of transcription is greater than the basal level. A: The product of the *lacI* gene is the repressor protein. When this protein binds an inducer, it changes its conformation, no longer binds to the operator site of DNA, and transcription occurs at an increased rate. Failure to bind an inducer prevents this sequence. B, C: In

the presence of glucose, catabolite repression occurs. Glucose lowers the intracellular level of cAMP. The catabolite activator protein (CAP) then cannot complex with cAMP, so there is no CAP–cAMP complex to activate transcription. E: The *lac* operon does not involve corepression.

4. **A** Operons are prokaryotic mechanisms. B–D: An operon is the complete regulatory unit of a set of clustered genes, including the structural genes (which are transcribed together to form a polycistronic mRNA), regulatory genes, and control elements, such as the operator. E: An operon may have more than one promoter, as does the tryptophan operon of *E. coli*.
5. **D** The exact functions of these species are not yet known, but their production is very rapid after the onset of amino acid starvation.
6. **B** Tryptophan, for the tryptophan operon, is called a corepressor to distinguish it from the repressor protein to which it must bind.
7. **C** Synthesis of the leader peptide depends strongly on the availability of this amino acid, since it must be incorporated several times. When it is insufficiently available, the ribosome stalls, and the secondary structure that forms is different from the hairpin loop signaling termination, so transcription continues.
8. **C** In this operon the stalled ribosome stimulates transcription of downstream structural genes.
9. **E** A: TBP is part of the general transcription factor TFIID. B: Many proteins are involved. C: The helicase activity is associated with one of the transcription factors (TFIIH). D: The CTD must be in the unphosphorylated form for RNA polymerase to enter the preinitiation complex. Later it must be phosphorylated to move away from the site of assembly.
10. **C** Transcription factors bound to enhancers are brought together with the preinitiation complex by looping of the DNA to bring distant portions together. A, B, E: These are characteristics of response elements such as the steroid response element. D: Enhancers can act in either orientation, again because the DNA loops back on itself.
11. **C** These guanosine phosphates are synthesized by the product of the *rel*A gene; they inhibit initiation of transcription of the rRNA and tRNA genes, shutting off protein synthesis in general. This is the stringent response.
12. **E** This has been demonstrated for transposon *Tn*3. A: Most transpositions involve generation of an additional copy of the transposon, which is then inserted somewhere else in DNA. B: The copy could be inserted into a different DNA molecule. C: Length varies considerably, depending on the number of genes incorporated into the transposon. D: Most target sites seem to be fairly random in sequence although some transposons tend to insert at specific "hot spots."
13. **B** Assembly of the transcription initiation complex and other proteins involved in gene expression need access to the grooves of DNA. A, E: The core of the nucleosome is an octamer of two molecules each of four different histones. C: A 5-bp shift in the association of DNA with the histone core is sufficient to change the orientation of the grooves and allow access. D: It seems to be the DNA associated with the nucleosome that is involved in binding transcription factors and other proteins.
14. **A** Acetylation of an ϵ-amino group of lysines near the N-terminus of histones changes a positive charge to a neutral species so there is less attraction to the negative phosphates of DNA. B: The entity that does this is a complex of proteins called the SWI/SNF complex. C: The change is from positive to neutral, not negative, so there is no electrostatic repulsion. D: Methylation of DNA does affect transcription but is a separate phenomenon from acetylation of histones. E: Opening of the DNA structure by acetylation may make it possible for transcription factors to bind but acetylated histones do not attract transcription factors.
15. Normally, glucose lowers the intracellular level of cAMP. There is then little or no CAP–cAMP complex to activate transcription. The lack of positive control means transcription of the operon is low. The same situation would exist in this mutation. Normally, in the absence of glucose, cAMP is high, positive control is exerted, and the *lac* operon is transcribed. In this mutation, cAMP cannot form since the enzyme for its formation is defective. There is no positive control and the *lac* operon will not be transcribed effectively even though there is no glucose and lactose is present.
16. Such operons will code for a leader peptide that has one or more codons for the amino acid in question. Once the RNA for the leader peptide has been synthesized, it can form different secondary structures depending on whether or not the ribosome is stalled at this region. If the ribosome does not stall because there is enough amino acid (and thus charged tRNA), the secondary structure is a termination signal and synthesis stops. If the ribosome stalls because of insufficient charged tRNA, the secondary structure that forms is not recognized as a termination signal and transcription through the operon continues.

PART III

FUNCTIONS OF PROTEINS

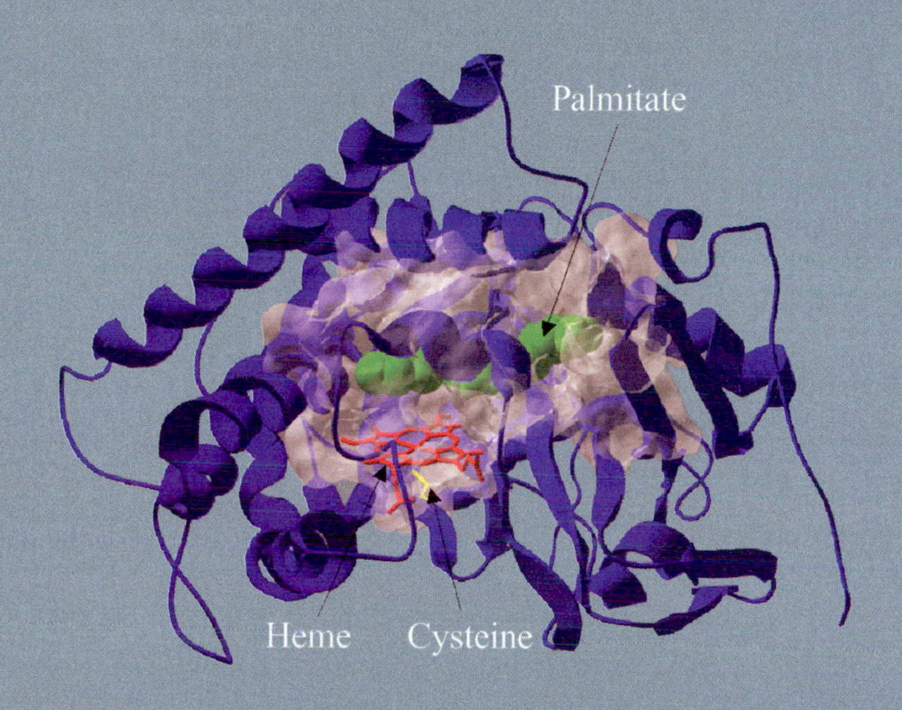

The phenotypic expressions of genes are the different proteins synthesized by cells. Functions of proteins include structural, catalytic, messenger, regulatory, and transport. The figure above is a representive of an important family of proteins, the cytochromes P450; a description of the protein is presented in Figure 11.2. The next four chapters describe the structure and function of different classes of proteins. The activities of specific proteins are described in later chapters as they relate to the various functions of cells. Structural modification of a protein due to genetic changes or directly by exogenous factors can produce disease conditions.

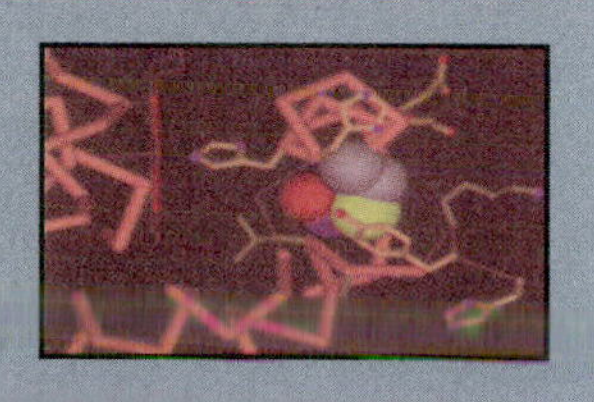

9

PROTEINS II: STRUCTURE–FUNCTION RELATIONSHIPS IN PROTEIN FAMILIES

Richard M. Schultz and Michael N. Liebman

9.1 | OVERVIEW

The fundamentals of protein architecture including the physical and chemical properties of the amino acids, hierarchical organization of primary, secondary, tertiary, and quaternary structure, and energetic forces that hold these molecules together and provide the flexibility observed in their dynamic motion were presented in Chapter 3. Computational and experimental tools were introduced that enable the analysis of high-resolution structural features and their conformational response to perturbations, which may be a simple alteration of the solution environment or aspects of their interactions with other molecules that define their biological function. The concept that structure and function are interrelated was introduced through examples of conservation of structure with function.

In this chapter we examine the specific relationships between structure and function in four protein functional groups: immunoglobulins, serine proteases, DNA-binding proteins, and hemoglobins. We pursue this study through the examination of the variability in amino acid sequence, structural organization, and biological function. The significance of the structure–function relationship can best be appreciated through observation of the range of such variations within specific **protein families.**

The **immunoglobulin** family provides examples of multidomain architecture that supports recognition and binding to foreign molecules and leads to their sequestration. Diversity among family members is the source of specific recognition of molecular and individual binding capabilities.

Serine proteases provide examples of a family of enzymes that appear to have diverged to perform unique physiological functions, frequently within highly organized enzyme cascade processes. Their inherent similarities in the catalytic mechanism and three-dimensional structure are a common link.

DNA-binding proteins are multifamily proteins that bind to regulatory sites in DNA and regulate expression of a specific subset of genes, an amazing feat as the mammalian genome contains approximately 30,000 unique genes. These proteins contain structural motifs that allow them to selectively bind the regulatory sites of specific genes.

The **hemoglobin** family offers examples of a highly fine-tuned protein that simultaneously performs multiple highly regulated physiological functions. The proteins can accommodate small substitutions or mutations, many of which have been studied as to their clinical implications, and still retain their physiological functions. This family reveals the potential diversity of amino acid sequence substitutions that can be tolerated and allow the protein to function in an acceptable physiological manner.

9.2 | ANTIBODY MOLECULES: THE IMMUNOGLOBULIN SUPERFAMILY OF PROTEINS

Antibody molecules are produced in response to invasion by foreign compounds that can be proteins, carbohydrate polymers, and nucleic acids. An antibody molecule noncovalently associates with the foreign substance, initiating a process by which the foreign substance can be eliminated from the organism.

Molecules that induce antibody production are **antigens** and may contain multiple antigenic determinants, small regions of the antigen molecule that elicit the production of a specific antibody to which the antigen binds. In proteins, an antigenic determinant may comprise only six or seven amino acids.

A **hapten** is a small molecule that cannot alone elicit production of specific antibodies but when covalently attached to a larger molecule it acts as an antigenic determinant and induces antibody synthesis. Whereas hapten molecules need attachment to a larger molecule to elicit antibody synthesis, when detached from their carrier, they will retain the ability to bind strongly to antibody.

It is estimated that each human can potentially produce about 1×10^8 different antibody structures. All antibodies, however, have a similar structure. The determination of the structure has been accomplished from studies of immunoglobulin amino acid sequence and X-ray diffraction of the antibody molecule alone or in complex with antigen.

Structural studies of proteins require pure homogeneous preparations. Such samples of antibodies are extremely difficult to isolate from plasma because of the wide diversity of antibody molecules present. Homogeneous antibodies can be obtained, however, by the monoclonal hybridoma technique in which mouse myeloma cells are fused with mouse antibody-producing B lymphocytes to construct immortalized hybridoma cells that express a single antibody.

Antibody (Immunoglobulin) Molecules Contain Four Polypeptide Chains

Antibody molecules are **glycoproteins** with four polypeptide chains, two identical copies of each of two nonidentical polypeptide chains. Two light chains (L) of identical sequence combine with two identical heavy chains (H) to form the structure $(LH)_2$. In the most common immunoglobulin type, IgG, the H chains have approximately 440 amino acids (50 kDa). The smaller L polypeptide chains contain about 220 amino acids (25 kDa). The four chains are covalently interconnected by disulfide bonds (Figures 9.1 and 9.2). Each H chain is associated with an L chain such that the NH_2-terminal ends of both chains are near each other. Since the L chain is half the size of the H chain, only the NH_2-terminal half of the H chain is associated with the L chain.

In the other classes of immunoglobulins (Table 9.1) the H chains are slightly longer than those of the IgG class. A variable amount of carbohydrate (2–12%, depending on immunoglobulin class) is attached to the H chain.

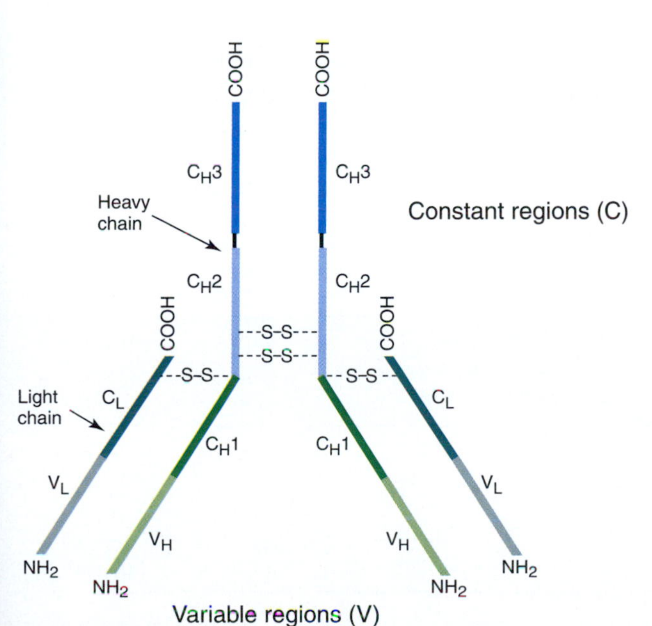

FIGURE 9.1
Linear representation of IgG antibody molecule.
Two H chains and two L chains are cooriented in their COOH-terminal to NH_2-terminal directions, as shown. Interchain disulfide bonds link heavy (H) chains and light (L) chains to the H chains. Domains of the constant (C) region of the H chain are C_H1, C_H2, and C_H3. The constant region of the L chain is designated C_L, and variable (V) regions are V_H and V_L of H and L chains, respectively.
Adopted from Burton, D. R. In: F. Calabi and M. S. Neuberger (Eds.), Molecular Genetics of Immunoglobulin. *Amsterdam: Elsevier, 1987, pp. 1–50.*

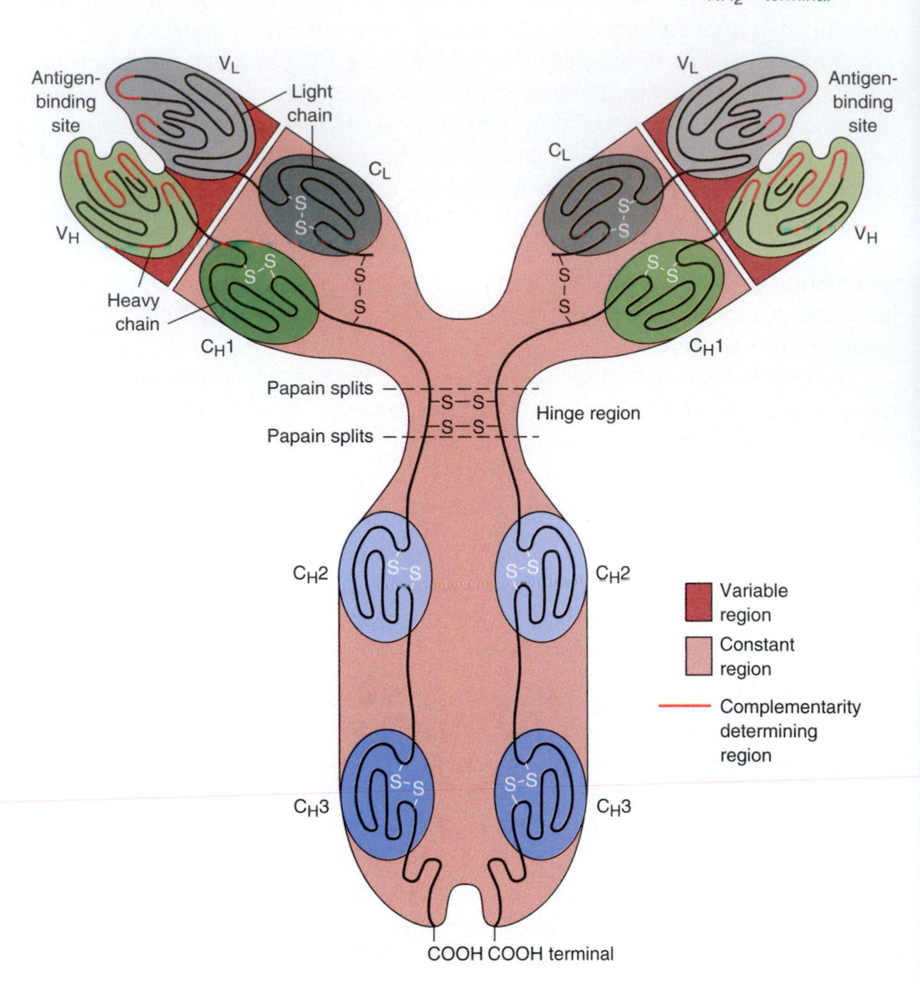

FIGURE 9.2
Diagrammatic structure for IgG.
Light chains (L) are divided into domains V_L (variable amino acid sequence) and C_L (constant amino acid sequence). Heavy chains (H) are divided into domains V_H (variable amino acid sequence) and C_H1, C_H2, and C_H3. Antigen-binding sites are V_H–V_L. "Hinge" polypeptides interconnect domains. Positions of inter- and intrachain cystine bonds are shown.
From Cantor, C. R. and Schimmel, P. R. Biophysical Chemistry, Part I. San Francisco: Freeman, 1980. Reprinted with permission of Mr. Irving Geis, New York.

Constant and Variable Regions of Primary Structure

Comparison of amino acid sequences of antibody molecules elicited by different antigens shows regions of sequence homology and other regions of sequence variability. In particular, sequences of the NH_2-terminal half of L chains and the NH_2-terminal quarter of H chains are highly variable. These NH_2-terminal segments are the **variable (V) regions** and designated V_H and V_L domains of H and L chains, respectively. Within these V domains certain segments are "hypervariable." Three **hypervariable regions** of between 5 and 7 residues in the V_L domain and three or four hypervariable regions of between 6 and 17 residues in the V_H domain are commonly found. The hypervariable sequences are also termed the **complementarity-**

TABLE 9.1 Immunoglobulin Classes

Classes of Immunoglobulin	*Approximate Molecular Mass*	*H Chain Isotype*	*Carbohydrate by Weight (%)*	*Concentration in Serum (mg 100 mL^{-1})*
IgG	150,000	γ, 53,000	2–3	600–1800
IgA	170,000–720,000[a]	α, 64,000	7–12	90–420
IgD	160,000	δ, 58,000		0.3–40
IgE	190,000	ε, 75,000	10–12	0.01–0.10
IgM	950,000[a]	μ, 70,000	10–12	50–190

[a] Forms polymer structures of basic structural unit.

CLINICAL CORRELATION 9.1

The Complement Proteins

At least 11 distinct complement proteins exist in plasma (see p. 1032 for their designations). They are activated by IgG or IgM antibody binding to antigens on the outer surface of invading bacterial cells, protozoa, or tumor cells. After the immunoglobulin-binding event, the 11 complement proteins are sequentially activated and associate with the cell membrane to cause a lysis of the membrane and death of the target cell.

Many complement proteins are precursors of proteolytic enzymes present in a nonactive form. Upon their activation, they will, in turn, activate a succeeding protein of the pathway by the hydrolysis of a specific peptide bond in the second protein, leading to a cascade phenomenon. Activation by specific proteolysis (i.e., hydrolysis of a specific peptide bond) is an important general method for activating extracellular enzymes. For example, the enzymes that catalyze blood clot formation, induce fibrinolysis of blood clots, and digest dietary proteins in the gut are all activated by a specific proteolysis catalyzed by a second enzyme (see pp. 1034, 1098).

Upon association to a cellular antigen the exposure of a complement-binding site in the antibody's F_c region occurs and causes the binding of the C1 complement proteins, which are a protein complex composed of three individual proteins: C1q, C1r, and C1s. C1r and C1s undergo a conformational change and become active enzymes on the cell surface. The activated C1 complex (C1a) hydrolyzes a peptide bond in complement proteins C2 and C4, which then also associate on the cell surface. The now active C2–C4 complex has a proteolytic activity that hydrolyzes a peptide bond in complement protein C3. Activated C3 protein binds to the cell surface and the activated C2–C4–C3 complex activates protein C5. Activated protein C5 will associate with complement proteins C6, C7, C8, and six molecules of complement protein C9. This multiprotein complex binds to the cell surface and initiates membrane lysis.

The mechanism is a cascade in which amplification of the trigger event occurs. In summary, activated C1 can activate a number of molecules of C4–C2–C3, and each activated C4–C2–C3 complex can, in turn, activate many molecules of C5 to C9. The reactions of the classical complement pathway are summarized below, where "a" and "b" designate the proteolytically modified proteins and a line above a protein indicates an enzyme activity.

$$\text{IgG or IgM} \xrightarrow{\text{C1q,C1r,C1s}} \overline{\text{C1a}} \xrightarrow{\text{C2,C4}}$$
$$\text{C4b} \cdot \overline{\text{C2a}} \xrightarrow{\text{C3}}$$
$$\text{C4b} \cdot \overline{\text{C2a}} \cdot \text{C3b} \xrightarrow{\text{C5,C6,C7,C8,C9}}$$

There is an "alternative pathway" for C3 complement activation, initiated by aggregates of IgA or by bacterial polysaccharide in the absence of immunoglobulin binding to cell membrane antigens. This pathway involves the proteins properdin, C3 proactivator convertase, and C3 proactivator.

A major role of the complement systems is to generate opsonins—an old term for proteins that stimulate phagocytosis by neutrophils and macrophages. The major opsonin is C3b; macrophages have specific receptors for this protein. Patients with inherited deficiency of C3 are subject to repeated bacterial infections.

Colten, H. R. and Rosen, F. S. Complement deficiencies. *Annu. Rev. Immunol.* 10:809, 1992; and Morgan, B. P. Physiology and pathophysiology of complement: progress and trends. *Crit. Rev. Clin. Lab. Sci.* 32:265, 1995.

determining regions (**CDRs**) as they form the antigen-binding site complementary to the topology of the antigen.

In contrast, the COOH-terminal three-quarters of H chains and the COOH-terminal half of L chains are homologous in sequence with other H or L chains of the same class. These **constant (C) regions** are designated C_H and C_L in the H and L chains, respectively. The C_H regions determine the antibody class, provide for binding of complement proteins (see Clin. Corr. 9.1), and are the site necessary for antibodies to cross the placental membrane. The V regions determine the antigen specificity of the molecule.

Immunoglobulins in a Single Class Contain Common Homologous Regions

Differences in sequence of the C_H regions between immunoglobulin classes are responsible for the characteristics of each class. In some classes, the C_H sequence promotes the polymerization of molecules of the basic structure $(LH)_2$. Thus antibodies of the IgA class are often covalently linked dimeric structures $[(LH)_2]_2$. Similarly, IgM molecules are pentamers $[(LH)_2]_5$. The different H chains, designated τ, α, δ, ε, and μ, occur in IgG, IgA, IgM, IgD, and IgE classes, respectively (Table 9.1; see Clin. Corr. 9.2). Two types of L chain sequences are synthesized, designated lambda (λ) and kappa (κ) chains, either of which is found combined with the five classes of H chains.

CLINICAL CORRELATION 9.2

Functions of Different Antibody Classes

The IgA class of immunoglobulins is found primarily in the mucosal secretions (bronchial, nasal, and intestinal mucous secretions, tears, milk, and colostrum). These immunoglobulins are the initial defense against invading viral and bacterial pathogens prior to their entry into plasma or other internal spaces.

The IgM class is found primarily in plasma. They are the first antibodies elicited in significant quantity on exposure of a foreign antigen. IgM antibodies promote phagocytosis of microorganisms by macrophage and polymorphonuclear leukocytes and are also potent activators of complement (see Clin. Corr. 9.1). IgM antibodies occur in many external secretions but at levels lower than those of IgA.

The IgG class occurs in high concentration in plasma. Their response to foreign antigens takes a longer period of time than that of IgM. At maximum concentration they are present in significantly higher concentration than IgM. Like IgM, IgG antibodies promote phagocytosis in plasma and activate complement.

The normal biological functions of the IgD and IgE classes are not known; however, the IgE play an important role in allergic responses such as anaphylactic shock, hay fever, and asthma.

Immunoglobulin deficiency usually causes increased susceptibility to infection. X-linked agammaglobulinemia and common variable immunodeficiency are examples. The commonest disorder is selective IgA deficiency, which results in recurrent infections of sinuses and the respiratory tract.

Rosen, F. S., Cooper, M. D., and Wedgewood, R. J. P. The primary immunodeficiencies. *N. Engl. J. Med.* 311:235 (Part 1); 300 (Part II), 1984.

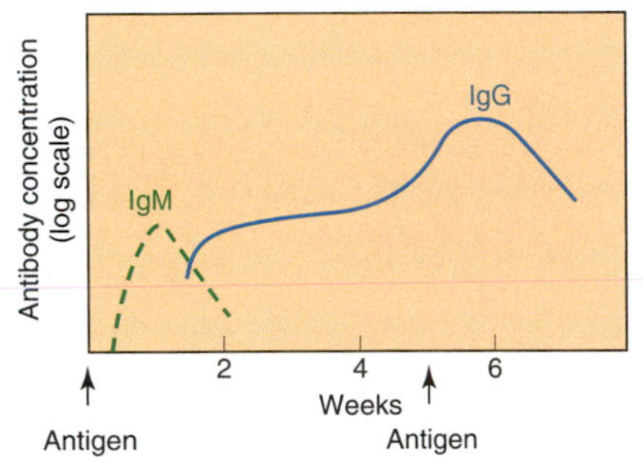

FIGURE 9.3
Time course of specific antibody IgM and IgG response to added antigen.
Redrawn from Stryer, L. Biochemistry. *San Francisco: Freeman, 1988, p. 890.*

IgG is the major immunoglobulin in plasma. Biosynthesis of a specific IgG in significant concentrations takes about 10 days after exposure to a new antigen (see Clin. Corr. 9.3). In the absence of an initially high concentration of IgG to a specific antigen, antibodies of the IgM class, which are synthesized at a faster rate, associate with the antigen and serve as the first line of defense until large quantities of IgG are produced (Figure 9.3; see Clin. Corr. 9.3).

Repeating Amino Acid Sequences and Homologous Three-Dimensional Domains Occur Within an Antibody

Within each of the polypeptide chains of an antibody molecule is a repeating pattern of amino acid sequences. For the IgG class, the repetitive pattern contains segments of approximately 110 amino acids within both L and H chains. This homology is far from exact, but clearly a number of amino acids match identically following alignment of 110 amino acid segments. Other residues are matched in the sequence by having similar nonpolar or polar side chains. As the H chains are about 440 amino acids in length, the repetition of the homologous sequence occurs four

CLINICAL CORRELATION 9.3

Immunization

An immunizing vaccine consists of killed bacterial cells, inactivated viruses, killed parasites, a nonvirulent form of live bacterium, a denatured bacterial toxin, or recombinant protein. The introduction of a vaccine into a human can lead to protection against virulent forms of microorganisms or toxic agents that contain the same antigen. Antigens in nonvirulent material not only cause the differentiation of lymphoid cells so that they produce antibody toward the foreign antigen but also cause differentiation of some lymphoid cells into memory cells. Memory cells do not secrete antibody but place antibodies to the antigen onto their outer surface, where they act as future sensors for the antigen. These memory cells are like a longstanding radar for the potentially virulent antigen. On reintroduction of the antigen at a later time, the binding of the antigen to the cell surface antibody in the memory cells stimulates the memory cell to divide into antibody-producing cells as well as new memory cells. This reduces the time required for antibody production on introduction of an antigen and increases the concentration of antigen-specific antibody produced. This is the basis for the protection provided by immunization.

Vaccines recently introduced for adults include pneumococcal vaccine (to prevent pneumonia due to *Diplococcus pneumoniae*), hepatitis B vaccine, and influenza vaccine. The composition of the latter changes each year to account for antigenic variation in the influenza virus.

Flexner, C. New approaches to vaccination. *Adv. Pharmacol.* 21:51, 1990; and Sparling, P. F., Elkins, C., Wyrick, P. B., and Cohen, M. S. Vaccines for bacterial sexually transmitted infections: a realistic goal? *Proc. Natl. Acad. Sci. USA* 91:2456, 1994.

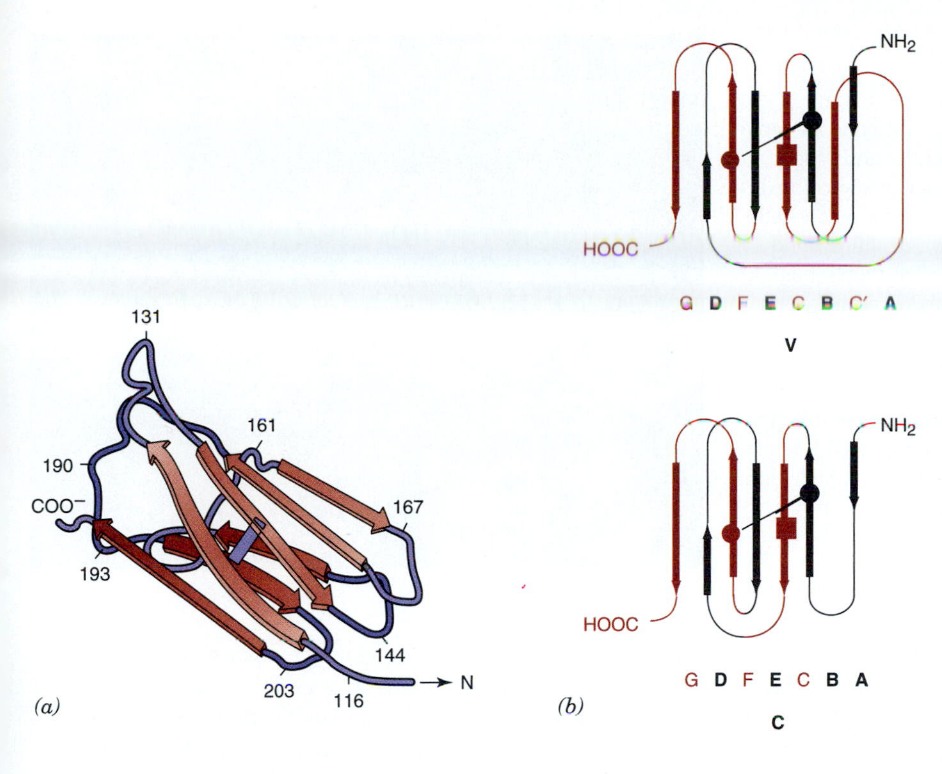

FIGURE 9.4
Immunoglobulin fold.
(*a*) Schematic diagram of folding of a C_L fold, showing β-pleated sheet structure. Arrows show strands of β-sheet and bar (light blue) shows position of cystine bond. Pink arrows are for β-strands in plane above and dark red arrows are β-strands in plane below. (*b*) Diagrammatic outline of arrangement (topology) of β-strands in immunoglobulin fold motif. Examples are for IgG variable (upper diagram) and constant (lower diagram) regions. Thick arrows indicate β-strands and thin lines loops that connect the β-strands. Circles indicate cysteines that form intrafold disulfide bond. Squares show positions of tryptophan residues that are an invariant component of the core of the immunoglobulin fold. Boldface black letters indicate strands that form one plane of the sheet, while other letters form the parallel plane behind the first plane.
Part (a) reprinted with permission from Edmundson, A. B., Ely, K. R., Abola, E. E., Schiffer, M., and Pavagiotopoulos, N. Biochemistry *14:3953, 1975. Copyright ©1975 by American Chemical Society. Part (b) redrawn from Calabi, F. In: F. Calabi and M. S. Neuberger (Eds.),* Molecular Genetics of Immunoglobulin. *Amsterdam: Elsevier, 1987, p. 203.*

times along an immunoglobulin H chain. Based on this sequence pattern, the chain is divided into one V_H region and three C_H regions (designated C_H1, C_H2, and C_H3) (see Figures 9.1 and 9.2). The L chain of about 220 amino acids is divided into one V_L region and one C_L region. Each repeat contains an intrachain disulfide bond linking two cysteines (Figure 9.2).

The 110 amino acid segments form separate structural domains of similar tertiary structure as shown by X-ray diffraction studies. Each 110 amino acid segment has a unique but similar arrangement of antiparallel β-strands, an arrangement of secondary elements known as an **immunoglobulin fold** (Figure 9.4). This fold consists of 7 to 9 polypeptide strands that form two antiparallel β-sheets aligned face-to-face. Globular domains result from the strong interaction between two immunoglobulin folds from two separate chains (Figure 9.5). The associations are

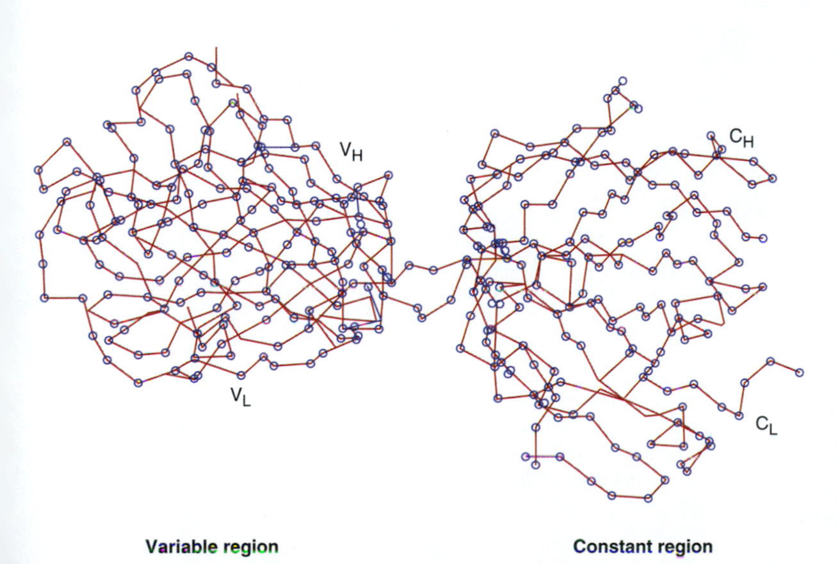

FIGURE 9.5
α-Carbon (○) structure of F_{ab} fragment of IgG KOL showing V_L–V_H and C_L–C_H1 domains interconnected by the hinge polypeptides.
Redrawn with permission from Huber, R., Deisenhofer, J., Coleman, P. M., Matsushima, M., and Palm, W. In: The Immune System, *27th Mosbach Colloquium. Berlin: Springer-Verlag, 1976, p. 26.*

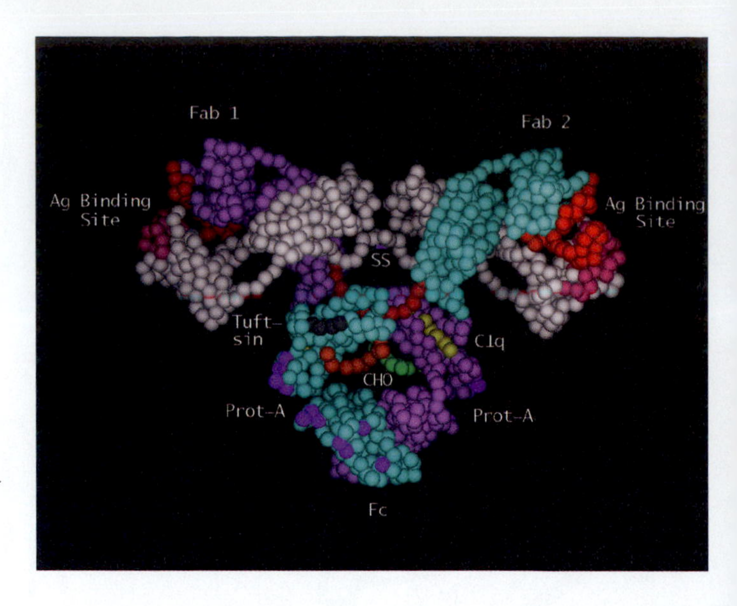

FIGURE 9.6
Model of an IgG antibody molecule.
Only the α-carbons of the structure appear. The two L chains are represented by light gray spheres and H chains by lavender and blue spheres. Carbohydrates attached to the two C_H2 domains are colored light green and orange. The CDR of the V_H–V_L domains are dark red in the H chains and pink in L chains. The interchain disulfide bond between the L and H chains is a magenta ball-and-stick representation (partially hidden). The heptapeptide hinge between C_H1 and C_H2 domains, connecting the F_{ab} and F_c units, are dark red. The center of the C1q complement site in the C_H2 domains is yellow, the protein A docking sites at the junction of C_H2 and C_H3 are magenta, and the tuftsin-binding site in C_H2 is gray. Tuftsin is a natural tetrapeptide that induces phagocytosis by macrophages and may be transported bound to an immunoglobulin. Protein A is a bacterial protein with a high affinity to immunoglobulins. See Figure 9.8 for definition of Fab and Fc.
Photograph generously supplied by Dr. Allen B. Edmundson, from Guddat, L. W., Shan, L., Fan, Z.-C., et al. FASEB J. *9:101, 1995.*

between folds V_L and V_H, and C_L and C_H1 in the H and L chains. In the C-terminal half of the H chains, the two chains associate to generate domains C_H2–C_H2 and C_H3–C_H3 (Figure 9.2). A "hinge" polypeptide sequence interconnects the two V_H–C_H1 folds with the C_H2–C_H2 folds of the H chains in the antibody structure (Figure 9.2). Thus the antibody structure exhibits six domains, each domain due to the association of two immunoglobulin folds (Figures 9.2 and 9.6). The NH_2-terminal V_L–V_H domains contain a shallow crevice in the center of a hydrophobic core that binds the antigen. Hypervariable sequences in the V domain crevices form loops that come close together and are the complementarity binding site (complementary-determining regions, CDRs) for the antigen (see Figures 9.6 and 9.7). The sequences of the hypervariable loops give a unique three-dimensional conformation for each antibody that makes it specific to its antigenic determinant. Small changes

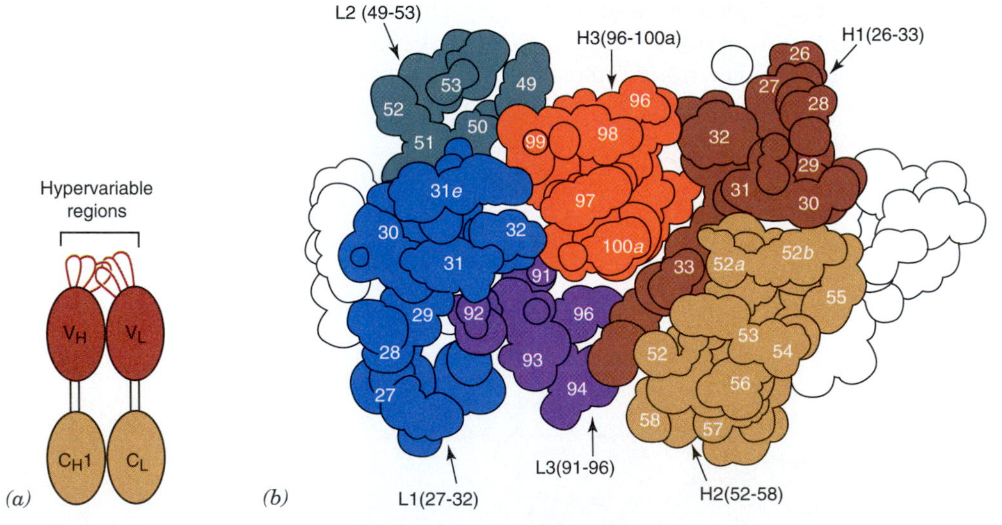

FIGURE 9.7
Hypervariable loops in immunoglobin.
(*a*) Schematic diagram showing hypervariable loops (CDRs) in V_L–V_H domain that form the antigen-binding site. (*b*) A cut through an antigen-binding site showing contributions of different CDRs using CPK space-filling models of the atoms. Numbers refer to amino acid residue numbers in the sequence of the hypervariable loop regions of the light (L) and heavy (H) chains.

Part (a) redrawn from Branden, C. and Tooze, J. Introduction to Protein Structure. *New York: Garland Publishing, 1991, p. 187. Part (b) redrawn from Branden, C. and Tooze, J.* Introduction to Protein Structure. *New York: Garland Publishing, 1991, p. 189, and attributed to Chothia, C. and Lesk, A. J.* Mol. Biol. *196:914, 1987.*

in conformation of the CDRs occur on binding of antigen to V_L–V_H domains, indicating that antigen binding induces an optimal complementary fit to the variable CDR site. Antigen binding may also induce conformational changes between the V_L–V_H domains and the other domains to activate effector sites, such as the binding of complement to a site within the C_H2–C_H2 domain (see C1q site in Figure 9.6). The strength of association between antibody and antigen is due to noncovalent forces (see p. 135). Complementarity of the structures of the antigenic determinant and antigen-binding site results in extremely high equilibrium affinity constants, between 10^5 and 10^{10} M^{-1} (strength of 7–14 kcal mol^{-1}) for this noncovalent association.

There Are Two Antigen-Binding Sites per Antibody Molecule

The NH_2-terminal variable (V) domains of each pair of L and H chains (V_L–V_H) comprise an antigen-binding site; thus there are two antigen-binding sites per antibody molecule. The existence of an antigen-binding site in each LH pair is demonstrated by treating antibody molecules with the proteolytic enzyme papain, which hydrolyzes a peptide bond in the hinge peptide of each H chain (see Figures 9.2, 9.6, and 9.8). The antibody molecule is cleaved into three fragments. Two are identical, each consisting of the NH_2-terminal half of the H chain (V_H–C_H1) associated with the full L chain (Figure 9.8). Each fragment binds antigen with a similar affinity to that of the intact antibody molecule and is designated an $\mathbf{F_{ab}}$ (antigen binding) **fragment.** The other product from the papain hydrolysis is the COOH-terminal half of the H chains (C_H2–C_H3) joined together by a disulfide bond. This is the $\mathbf{F_c}$ (crystallizable) **fragment,** which exhibits no binding affinity for the antigen. The L chain can be dissociated from its H chain fragment within the F_{ab} fragment by reduction of disulfide bonds, which eliminates antigen binding. Accordingly, each antigen-binding site must be formed from components of both the L chain (V_L) and the H chain (V_H) acting together.

In summary, the major features of antibody structure and antibody–antigen interactions include the following: (1) The polypeptide chains fold into multiple folds, each having an immunoglobulin fold structure. (2) Two immunoglobulin folds on separate chains associate to form the six domains of the basic immunoglobulin structure. The V_L and V_H associate to form the two NH_2-terminal domains that bind to antigen. (3) The antigen-binding site of the V_L–V_H domains is generated by hypervariable loops (CDRs), which form a continuous surface with a topology complementary to the antigenic determinant. (4) The strong interactions between antigen and antibody CDRs are noncovalent and include van der Waals, hydrogen bonding, and hydrophobic interactions. Ionic salt bridges participate in

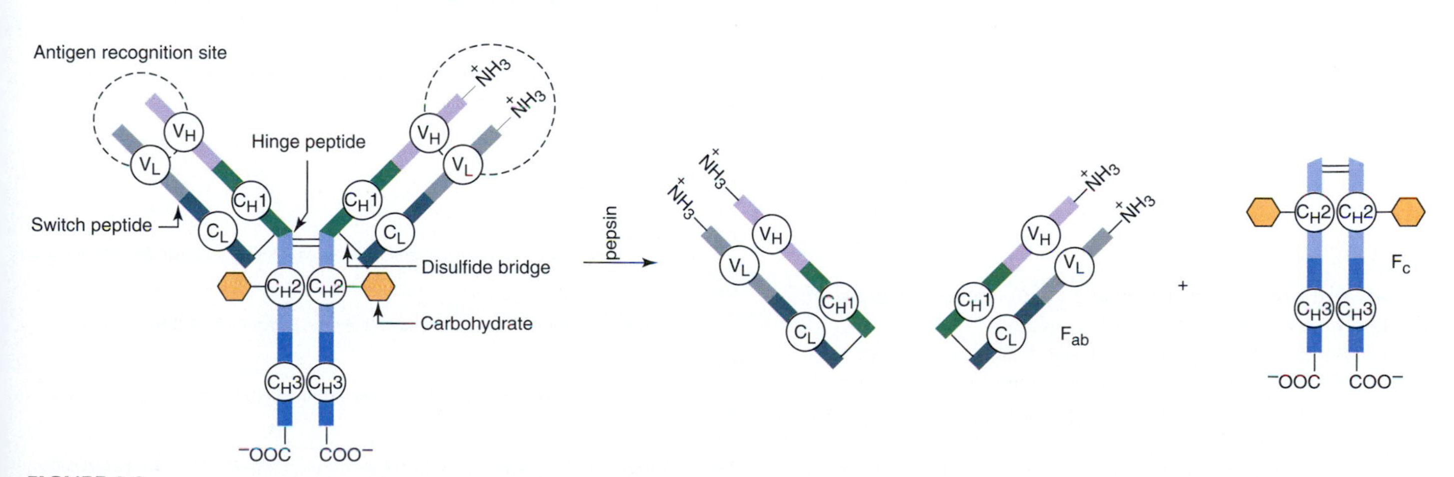

FIGURE 9.8
Hydrolysis of IgG into two F_{ab} fragments and one F_c fragment by papain, a proteolytic enzyme.

antigen–antibody associations to a much lesser extent. (5) Small conformational changes occur in the V_L–V_H domain upon binding of antigen, indicating an "induced-fit" mechanism in binding of antigen to antibody. (6) The binding of antigen to the V_L–V_H domains induces conformational changes between binding and distant domains of the antibody. These allosteric movements alter the binding affinity of effector sites in the constant domains such as that for binding of complement protein C1q to the C_H2–C_H2 domain (see Clin. Corr. 9.1).

Genetics of the Immunoglobulins

The V and C regions of the L and H chains are specified by distinct genes. There are four unique genes that code for the C regions of the H chain in the IgG antibody class. Each gene codes for a complete constant region, thus coding for all the amino acids of the H chain except for the V_H region sequence. These four genes are known as gamma (γ) genes—γ_1, γ_2, γ_3, and γ_4—that give rise to **IgG isotypes** IgG_1, IgG_2, IgG_3, and IgG_4. Figure 9.9 presents the sequences of three γ-gene proteins. There is a 95% homology in amino acid sequence among the genes.

It is likely that a primordial gene coded for a single immunoglobulin fold sequence of approximately 110 amino acids, and **gene duplication** events resulted in

```
Constant Region C1:
Cγ1: AlaSerThrLysGlyProSerValPheProLeuAlaProSerSerLysSerThrSerGlyGlyThrAlaAlaLeuGly
Cγ2                                          C     R           E  S
Cγ4

Cγ1  CysLeuValLysAspTyrPheProGluProValThrValSerTrpAsnSerGlyAlaLeuThrSerGlyValHisThr
Cγ2
Cγ4

Cγ1  PheProAlaValLeuGlnSerSerGlyLeuTyrSerLeuSerSerValValThrValProSerSerSerLeuGly
Cγ2                                                                         N  F
Cγ4

Cγ1  ThrGlnThrTyrIleCysAsnValAsnHisLysProSerAsnThrLysValAspLysLysVal
Cγ2              T           D                                 T
Cγ4      K                                                     R

Hinge Region H:
Cγ1  GluProLysSerCysAspLysThrHisThrCysProProCysPro
Cγ2      R     C     V  E  C  P  P        -  -  -
Cγ4      S     Y  G  P  P        S        -  -  -

Constant Region C2:
Cγ1  AlaPro    GluLeuLeuGlyGlyProSerValPheLeuPheProProLysProLysAspThrLeuMetIleSerArg
Cγ2                 -  P  V  A
Cγ4        E   F     G

Cγ1  ThrProGluValThrCysValValValAspValSerHisGluAspProGluValLysPheAsnTrpTyrValAspGly
Cγ2                                                        Q
Cγ4                                      Q

Cγ1  ValGluValHisAsnAlaLysThrLysProArgGluGluGlnTyrAsnSerThrThrArgValValSerValLeuThr
Cγ2                                             F           F
Cγ4                                                         Y

Cγ1  ValLeuHisGlnAspTrpLeuAsnGlyLysGluTyrLysCysLysValSerAsnLysAlaLeuProAlaProIleGlu
Cγ2      V                                                      G
Cγ4                                                                      S  S

Cγ1  LysThrIleSerLysAlaLys
Cγ2                 T
Cγ4

Constant Region C3:
Cγ1  GlyGlnProArgGluProGlnValTyrThrLeuProProSerArgAspGluLeuThrLysAsnGlnValSerLeuThr
Cγ2                                                 E     M
Cγ4                                             Q

Cγ1  CysLeuValLysGlyPheTyrProSerAspIleAlaValGluTrpGluSerAsnGlyGlnProGluAsnAsnTyrLys
Cγ2
Cγ4

Cγ1  ThrThrProProValLeuAspSerAspGlySerPhePheLeuTyrSerLysLeuThrValAspLysSerArgTrpGln
Cγ2               M
Cγ4                                                   R

Cγ1  GlnGlyAsnValPheSerCysSerValMetHisGluAlaLeuHisAsnHisTyrThrGlnLysSerLeuSerLeuSer
Cγ2
Cγ4  E

Cγ1  ProGlyLysStop
Cγ2
Cγ4  L
```

FIGURE 9.9
Amino acid sequence of the heavy chain constant regions of the IgG heavy chain γ_1, γ_2, and γ_4 genes.
Domains of constant region C_H1, hinge region H, constant region C_H2, and constant region C_H3 are presented. Sequence for γ_1 is complete and differences in γ_2 and γ_4 from γ_1 sequence are shown using single-letter amino acid abbreviations. Dashed line (—) indicates absence of an amino acid in position correlated with γ_1, in order to better align sequences to show maximum homology.
Sequence of γ_1 chain from Ellison, J. W., Berson, B. J., and Hood, L. E. Nucleic Acids Res. *10:4071, 1982; and sequences of the γ_2 and γ_4 genes from Ellison, J. and Hood, L.* Proc. Natl. Acad. Sci. USA *79:1984, 1982.*

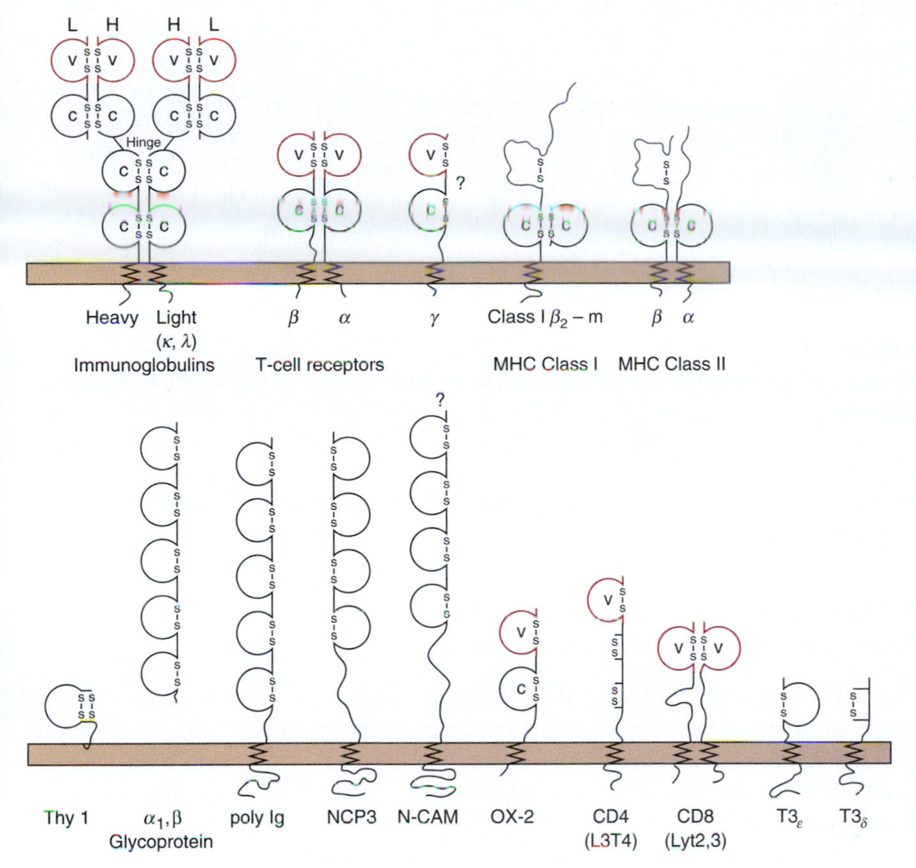

FIGURE 9.10
Diagrammatic representation of immunoglobulin domain structures from different proteins of immunoglobulin gene superfamily.
Proteins presented include heavy and light chains of immunoglobulins, T-cell receptors, major histocompatibility complex (MHC) Class I and Class II proteins, T-cell accessory proteins involved in Class I (CD8) and Class II (CD4) MHC recognition and possible ion channel formation, a receptor responsible for transporting certain classes of immunoglobulin across mucosal membranes (poly-Ig), β_2-microglobulin, which associates with Class I molecules, a human plasma protein with unknown function (α_1/β-glycoprotein), two molecules of unknown function with a tissue distribution that includes lymphocytes and neurons (Thy-1, OX-2), and two brain-specific molecules, neuronal cell-adhesion molecule (N-CAM) and neurocytoplasmic protein 3 (NCP3).
Redrawn from Hunkapiller, T. and Hood, L. Nature 323:15, 1986.

the three repeating units within the same IgG H γ gene. Mutations modified the individual sequences so that an exact correspondence in sequence no longer exists. However, each immunoglobulin fold has a similar length and folding pattern stabilized by a cystine linkage. Later in evolution further gene duplications led to the multiple genes (γ_1, γ_2, γ_3, and γ_4) that code for the constant regions of the IgG class sequences.

Immunoglobulin Fold Is a Tertiary Structure Found in a Large Family of Proteins with Different Functional Roles

The immunoglobulin fold motif is present in many nonimmunological proteins that exhibit widely different functions. Based on their structural homology they are grouped into a **protein superfamily** (Figure 9.10). For example, the Class I major histocompatibility complex proteins are in this superfamily; they have immunoglobulin folds that consist of two stacked antiparallel β-sheets enclosing an internal space filled mainly by hydrophobic amino acids. Two cysteines in the structure form a disulfide bond linking the facing β-sheets. Transcription factors NF-κB and p53 also contain an immunoglobulin fold motif. It can be speculated that gene duplication during evolution led to distribution of the structural fold in the functionally diverse proteins.

9.3 | PROTEINS WITH A COMMON CATALYTIC MECHANISM: SERINE PROTEASES

Serine proteases are a family of enzymes that use a single uniquely activated serine residue in their substrate-binding site to catalytically hydrolyze peptide bonds. This serine can be characterized by the irreversible reaction of its side chain hydroxyl group with diisopropylfluorophosphate (DFP) (Figure 9.11). Of all the ser-

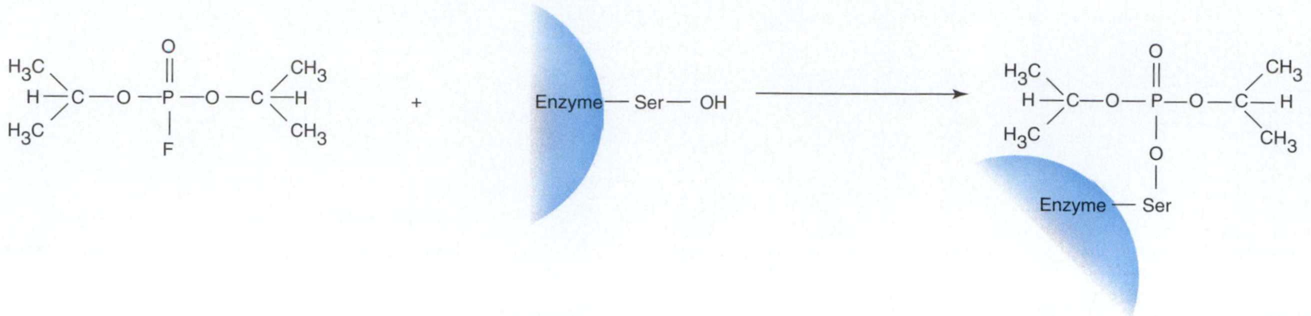

FIGURE 9.11
Reaction of diisopropylfluorophosphate (DFP) with the active-site serine in a serine protease.

ines in the protein, DFP reacts only with the catalytically active serine to form a phosphate ester.

Proteolytic Fnzymes Are Classified Based on Their Catalytic Mechanism

Proteolytic enzymes are classified according to their catalytic mechanism. Besides serine proteases, other classes utilize cysteine (**cysteine proteases**), aspartate (**aspartate proteases**), or metal ions (**metallo proteases**) to perform their catalytic function. Proteases that hydrolyze peptide bonds within a polypeptide chain are **endopeptidases** and those that cleave the peptide bond of either the COOH- or NH_2-terminal amino acid are classified as **exopeptidases.**

Serine proteases often activate other serine proteases from their inactive precursor form, termed a **zymogen,** by the catalytic cleavage of a specific peptide bond. Serine proteases participate in carefully controlled physiological processes such as blood coagulation (see Clin. Corr. 9.4), fibrinolysis, complement activation (see

CLINICAL CORRELATION 9.4

Fibrin Formation in a Myocardial Infarct and the Action of Recombinant Tissue Plasminogen Activator (rt-PA)

Coagulation is an enzyme cascade process in which inactive serine proteases (zymogens) are catalytically activated by other serine proteases in a stepwise manner (the coagulation pathway is described in Chapter 23). These multiple activation events generate catalytic products with a dramatic amplification of the initial signal of the pathway. The end product of the coagulation pathway is a cross-linked fibrin clot. The zymogen of the serine protease components of coagulation include factor II (prothrombin), factor VII (proconvertin), factor IX (Christmas factor), factor X (Stuart factor), factor XI (plasma thromboplastin antecedent), and factor XII (Hageman factor). The roman numeral designation was assigned in the order of their discovery and not from their order of action within the pathway. Upon activation of their zymogen forms, the activated enzymes are noted with the suffix "a." Thus prothrombin is denoted as factor II, and the activated enzyme, thrombin, is factor IIa.

The main function of coagulation is to maintain the integrity of the closed circulatory system after blood vessel injury. The process, however, can be dangerously activated in a myocardial infarction and decrease blood flow to heart muscle. About 1.5 million individuals suffer heart attacks each year, resulting in 600,000 deaths. A fibrinolysis pathway also exists in blood to degrade fibrin clots. This pathway also utilizes zymogen factors that are activated to serine proteases. The end reaction is the activation of plasmin, a serine protease. Plasmin acts directly on fibrin to catalyze the degradation of the fibrin clot. Tissue plasminogen activator (t-PA) is one of the plasminogen activators that activates plasminogen to form plasmin. Recombinant t-PA (rt-PA) is produced by gene cloning technology (see Chapter 7). Clinical studies show that the administration of rt-PA shortly after a myocardial infarct significantly enhances recovery. Other plasminogen activators such as urokinase and streptokinase are also effective.

The GUSTO investigators (authors). An international randomized trial comparing four thrombolytic strategies for acute myocardial infarction. *N. Engl. J. Med.* 329:673, 1993; International Study Group (authors). In hospital mortality and clinical course of 20,891 patients with suspected acute myocardial infarction randomized between alteplase and streptokinase with or without heparin. *Lancet* 336:71, 1990; and Gillis, J. C., Wagstaff, A. J., and Goa, K. L. Alteplase. A reappraisal of its pharmacological properties and therapeutic use in acute myocardial infarction. *Drugs* 50:102, 1995.

TABLE 9.2 Some Serine Proteases and Their Biochemical and Physiological Roles

Protease	*Action*	*Possible Disease Due to Deficiency or Malfunction*
Plasma kallikrein Factor XIIa Factor XIa Factor IXa Factor VIIa Factor Xa Factor IIa (thrombin) Activated protein C	Coagulation (see Clin. Corr. 9.4)	Cerebral infarction (stroke), coronary infarction, thrombosis, bleeding disorders
Factor C1r Factor C1s Factor D Factor B C3 convertase	Complement (see Clin. Corr. 9.1)	Inflammation, rheumatoid arthritis, autoimmune disease
Trypsin Chymotrypsin Elastase (pancreatic) Enteropeptidase	Digestion	Pancreatitis
Urokinase plasminogen activator Tissue plasminogen activator Plasmin	Fibrinolysis, cell migration, embryogenesis, menstruation	Clotting disorders, tumor metastasis (see Clin. Corr. 9.5)
Tissue kallikreins	Hormone activation	
Acrosin	Fertilization	Infertility
α-Subunit of nerve growth factor γ-Subunit of nerve growth factor	Growth factor activation	
Granulocyte elastase Cathepsin G Mast cell chymases Mast cell tryptases	Extracellular protein and peptide degradation, mast cell function	Inflammation, allergic response

Clin. Corr. 9.1), fertilization, and hormone production (Table 9.2). The protein activations catalyzed by serine proteases are examples of “limited proteolysis” because only one or two specific peptide bonds of the hundreds in a protein substrate are hydrolyzed. Under denaturing conditions, however, these same enzymes hydrolyze multiple peptide bonds and lead to digestion of peptides, proteins, and even self-digestion (autolysis). Several diseases, such as emphysema, arthritis, thrombosis, cancer metastasis (see Clin. Corr. 9.5), and some forms of hemophilia, are thought to result from the lack of regulation of specific serine proteases.

Serine Proteases Exhibit Remarkable Specificity for the Site of Peptide Bond Hydrolysis

Many serine proteases exhibit preference for hydrolysis of peptide bonds adjacent to a particular type of amino acid. Thus trypsin preferentially cleaves following basic amino acids such as arginine and lysine, and chymotrypsin cleaves peptide bonds following large hydrophobic amino acid residues such as tryptophan, phenylalanine, tyrosine, and leucine. Elastase cleaves peptide bonds following small hydrophobic residues such as alanine. A serine protease may be called trypsin-like if it prefers to cleave peptide bonds of lysine and arginine, chymotrypsin-like if it prefers large hydrophobic amino acids, and elastase-like if it prefers amino acids

CLINICAL CORRELATION 9.5

Involvement of Serine Proteases in Tumor Cell Metastasis

The serine protease urokinase is believed to be required for the metastasis of cancer cells. Metastasis is the process by which a cancer cell leaves a primary tumor and migrates through the blood or lymph systems to a new tissue or organ, where a secondary tumor grows. Increased synthesis of urokinase has been correlated with an increased ability to metastasize in many cancers. Urokinase activates plasminogen to form plasmin. Plasminogen is ubiquitously located in extracellular space and its activation to plasmin can cause the catalytic degradation of the proteins in the extracellular matrix through which the metastasizing tumor cells migrate. Plasmin can also activate procollagenase to collagenase, promoting the degradation of collagen in the basement membrane surrounding the capillaries and lymph system. This promotion of proteolytic degradative activity by the urokinase secreted by tumor cells allows the tumor cells to invade the target tissue and form secondary tumor sites.

Dano, K., Andreasen, P. A., Grondahl-Hansen, J., Kristensen, P., Nielsen, L. S., and Skriver, L. Plasminogen activators, tissue degradation and cancer. *Adv. Cancer Res.* 44:139, 1985; Yu, H. and Schultz, R. M. Relationship between secreted urokinase plasminogen activator activity and metastatic potential in murine B16 cells transfected with human urokinase sense and antisense genes. *Cancer Res.* 50:7623, 1990; and Fazioli, F. and Blasi, F. Urokinase-type plasminogen activator and its receptor: new targets for anti-metastatic therapy? *Trends Pharmacol. Sci.* 15:25, 1994.

with small side chain groups like alanine. The specificity for a certain type of amino acid only indicates its relative preference. Trypsin can also cleave peptide bonds following hydrophobic amino acids, but at a much slower rate than for the basic amino acids. Thus specificity for hydrolysis of a peptide bond of a particular type may not be absolute, but may be more accurately described as a range of most likely targets. Each of the identical amino acid hydrolysis sites within a protein substrate is not equally susceptible. Trypsin hydrolyzes each of the arginine peptide bonds in a particular protein at a different catalytic rate, and some may require a conformational change to make them accessible.

Detailed studies of the specificity of serine proteases for a particular peptide bond have been performed with synthetic substrates with fewer than 10 amino acids (Table 9.3). Because these are significantly smaller than the natural ones, they interact only with the catalytic site (primary binding site S_1, see below) and are said to be **active-site directed.** Studies with small substrates and inhibitors indicate that the site of hydrolysis is flanked by approximately four amino acid residues in both directions that bind to the enzyme and impact on the reactivity of the bond hydrolyzed. The two amino acids in the substrate that contribute the hydrolyzable bond are designated $\mathbf{P_1—P_1'}$. Thus in trypsin-like substrates P_1 will be lysine or arginine and in chymotrypsin-like substrates P_1 will be a hydrophobic amino acid.

TABLE 9.3 Reactivity of α-Chymotrypsin and Elastase Toward Substrates of Various Structures

Structure	*Variation of Side Chain Group in S_1 Site (Chymotrypsin)*	*Relative Reactivity*[a]
Glycyl	H—	1
Leucyl	$(H_3C)_2CH—CH_2—$	1.6×10^4
Methionyl	$CH_3—S—CH_2—CH_2—$	2.4×10^4
Phenylalanyl	$C_6H_5—CH_2—$	4.3×10^6
Hexahydrophenylalanyl	$C_6H_{11}—CH_2—$ (S)	8.2×10^6
Tyrosyl	$HO—C_6H_4—CH_2—$	3.7×10^7
Tryptophanyl	indolyl (N—H)—$CH_2—$	4.3×10^7
Variation in chain length (elastase hydrolysis of Ala N-terminal amide)[b]		
$Ac\text{-}Ala\text{-}NH_2$		
$Ac\text{-}Pro\text{-}Ala\text{-}NH_2$		1
$Ac\text{-}Ala\text{-}Pro\text{-}Ala\text{-}NH_2$		1.4×10^1
$Ac\text{-}Pro\text{-}Ala\text{-}Pro\text{-}Ala\text{-}NH_2$		4.2×10^3
Ac-Ala-Pro-Ala-Pro-Ala-		4.4×10^5
NH_2		2.7×10^5

[a] Calculated from values of k_{cat}/K_m found for *N*-acetyl amino acid methyl esters in chymotrypsin substrates.

[b] Calculated from values of k_{cat}/K_m in Thompson, R. C. and Blout, E. R. *Biochemistry* 12:57, 1973.

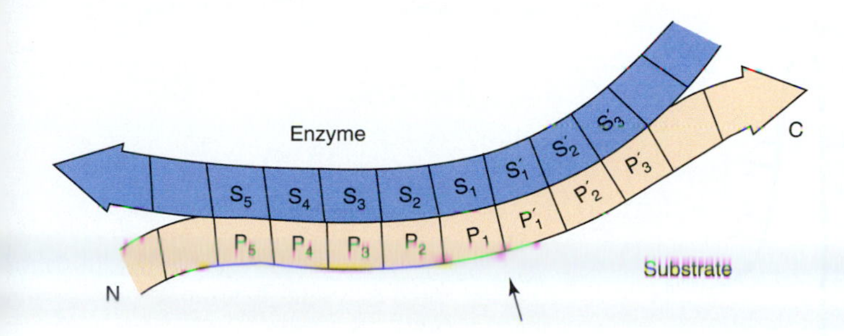

FIGURE 9.12
Schematic diagram of binding of a polypeptide substrate to binding site in a proteolytic enzyme.
P_5–P_5' are amino acid residues in the substrate that bind to subsites S_5–S_5' in the enzyme with peptide hydrolysis occurring between P_1–P_1' (arrow). NH_2-terminal direction of substrate polypeptide chain is indicated by N, and COOH-terminal direction by C.
Redrawn from Polgar, L. In: A. Neuberger and K. Brocklehurst (Eds.), Hydrolytic Enzymes. *Amsterdam: Elsevier, 1987, p. 174.*

The other interacting residues in the substrate are labeled P_4—P_2 on the NH_2-terminal side of P_1—P_1' and the COOH-terminal residues to the scissile bond are substrate residues P_2'—P_4'. Thus the residues in the substrate that interact with the **extended active site** in the serine protease will be P_4—P_3—P_2—P_1—P_1'—P_2'—P_3'—P_4'. The complementary regions in the enzyme that bind the amino acid residues in the substrate are designated **S_4–S_4'** (Figure 9.12). It is the **secondary interactions** with the substrate, outside S_1–S_1', that ultimately determine a protease's specificity toward a particular protein substrate. Thus the serine protease in coagulation, factor Xa, only cleaves a particular arginine in prothrombin, activating it to thrombin. It is the secondary interaction that allows factor Xa to recognize the particular arginine in the structure of prothrombin to be cleaved. The interaction of the substrate residues $P_4 \cdots P_4'$ with the enzyme-binding subsites $S_4 \cdots S_4'$ are due to noncovalent interactions. The substrate interacts with the enzyme-binding site to extend a β-sheet structure between the polypeptide chain in the enzyme and the polypeptide chain of the substrate, which places the scissile peptide bond of the substrate into S_1—S_1' (Figure 9.13).

Serine Proteases Are Synthesized in a Zymogen Form

Serine proteases are synthesized in an inactive zymogen form, that requires limited proteolysis to produce the active enzyme. Those for coagulation are synthesized in

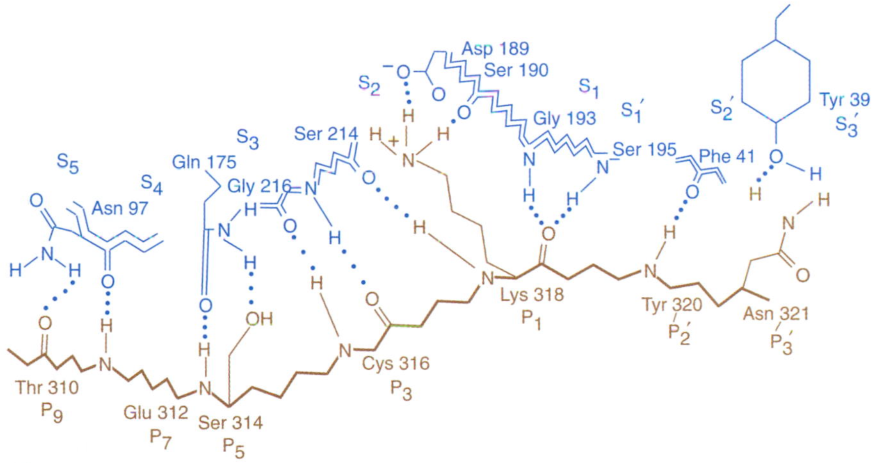

FIGURE 9.13
Schematic drawing of binding of pancreatic trypsin inhibitor to trypsinogen based on X-ray diffraction data.
Binding-site region of trypsinogen in the complex assumes a conformation like that of active trypsin with inhibitor, which is believed to bind in a similar manner to a substrate in the active enzyme-binding site. One cannot obtain X-ray structures of a natural enzyme–substrate complex because substrate is used up at a rate faster than the time of the X-ray diffraction experiment (see p. 148). Note that inhibitor has an extended conformation so that amino acids P_9, P_7, P_5, P_3, P_1–P_3' interact with binding subsites S_5–S_3'. Potentially hydrolyzable bond in inhibitor is between P_1–P_1'.
Redrawn from Bolognesi, M., Gatti, B., Menegatti, E., Guarneri, M., Papamokos, E., and Huber, R. J. Mol. Biol. *162:839, 1983.*

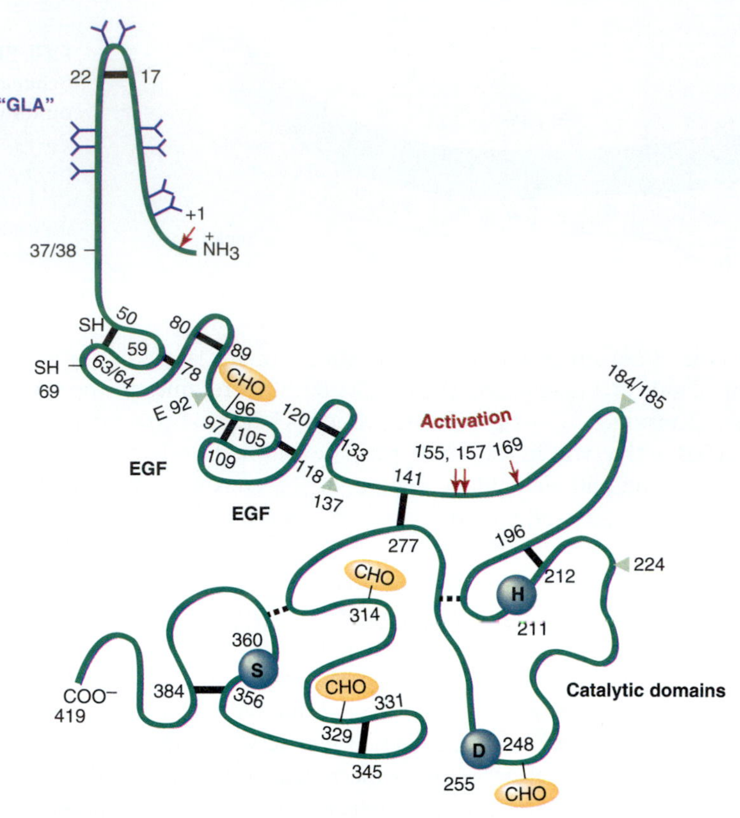

FIGURE 9.14
Schematic of domain structure for protein C showing multidomain structure. "GLA" refers to the γ-carboxyglutamic residues (indicated by tree structures) in the NH_2-terminal domain, disulfide bridges are indicated by thick bars, EGF indicates positions of epidermal growth factor-like domains, and CHO indicates positions where sugar residues are joined to the polypeptide chain. Proteolytic cleavage sites leading to catalytic activation are shown by arrows. Amino acid sequence is numbered from NH_2-terminal end, and catalytic sites of serine, histidine, and aspartate are shown in the catalytic domains by circled one letter abbreviations S, H, and D, respectively.
Redrawn from Long, G. L. J. Cell. Biochem. 33:185, 1987.

liver cells and are secreted into the blood for subsequent activation by other serine proteases following vascular injury. Zymogen forms are usually designated by the suffix *-ogen* to the enzyme name; the zymogen form of trypsin is termed trypsin*ogen* and for chymotrypsin is termed chymotrypsin*ogen.* In some cases the zymogen form is referred to as a **proenzyme;** the zymogen form of thrombin is prothrombin.

Several plasma serine proteases have zymogen forms that contain **multiple nonsimilar domains.** Protein C, involved in a fibrinolysis pathway in blood, has four distinct domains (Figure 9.14). The NH_2-terminal domain contains the derived amino acid **γ-carboxyglutamic acid** (Figure 9.15), which is enzymatically formed by carboxylation of glutamic acid residues in a vitamin K-dependent reaction (see pp. 1047 and 1146). The γ-carboxyglutamates chelate Ca^{2+} and form part of a binding site to membranes. The COOH-terminal segment contains the catalytic domains. Activation of these zymogens requires specific proteolysis outside the catalytic domains (Figure 9.14) and is controlled by the binding through the nine γ-carboxyglutamate residues at the NH_2-terminal end through calcium ions to a membrane site that promotes activation.

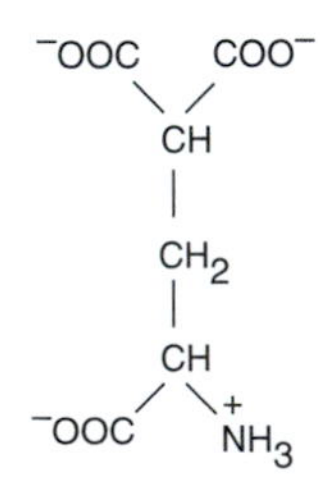

FIGURE 9.15
Structure of the derived amino acid γ-carboxyglutamic acid (abbreviation Gla), found in NH_2-terminal domain of many clotting proteins.

There Are Specific Protein Inhibitors of Serine Proteases

Evolutionary selection of this enzyme family for participation in physiological processes requires a parallel evolution of control factors. Specific proteins inhibit

TABLE 9.5 Invariant Sequences Found Around the Catalytically Essential Serine (S) and Histidine (H)

Enzyme	*Sequence (Identical Residues to Chymotrypsin Are in Bold*																									
	Residues Around Catalytically Essential Histidine																									
Chymotrypsin A	F	**H**	**F**	**C**	**G**	**G**	**S**	**L**	**I**	**N**	**E**	**N**	**W**	**V**	**V**	**T**	**A**	**A**	**H***	**C**	G	V	T	T	S	D
Trypsin	Y	**H**	**F**	**C**	**G**	**G**	**S**	**L**	**I**	**N**	S	Q	**W**	**V**	**V**	S	**A**	**A**	**H**	**C**	Y	K	S	G	I	Q
Pancreatic elastase	A	**H**	T	**C**	**G**	**G**	T	**L**	**I**	R	Q	**N**	**W**	**V**	M	**T**	**A**	**A**	**H**	**C**	V	D	R	E	L	T
Thrombin	E	L	L	**C**	**G**	A	**S**	**L**	**I**	S	D	R	**W**	**V**	L	**T**	**A**	**A**	**H**	**C**	L	L	Y	P	P	W
Factor X	E	G	**F**	**C**	**G**	**G**	T	I	L	**N**	**E**	F	Y	**V**	L	**T**	**A**	**A**	**H**	**C**	L	H	Q	A	K	R
Plasmin	M	**H**	**F**	**C**	**G**	**G**	T	**L**	**I**	S	P	E	**W**	**V**	L	**T**	**A**	**A**	**H**	**C**	L	E	K	S	P	R
Plasma kallikrein	S	F	Q	**C**	**G**	**G**	V	**L**	V	**N**	P	K	**W**	**V**	L	**T**	**A**	**A**	**H**	**C**	K	N	D	N	Y	E
Streptomyces trypsin	–	–	–	**C**	**G**	**G**	A	**L**	Y	A	Q	D	I	**V**	L	**T**	**A**	**A**	**H**	**C**	V	S	G	S	G	N
Subtilisin	V	G	G	A	S	F	V	A	G	E	A	Y	N	T	D	G	N	G	**H**	G	T	H	V	A	G	T
	Residues Around Catalytically Essential Serine																									
Chymotrypsin A	**C**	**A**	**G**	–	–	–	A	S	**G**	**V**	–	–	**S**	**S**	**C**	**M**	**G**	**D**	**S***	**G**	**G**	**P**	**L**	**V**		
Trypsin	**C**	**A**	**G**	Y	–	–	L	E	**G**	G	K	–	D	**S**	**C**	Q	***G***	***D***	***S***	***G***	***G***	***P***	V	**V**		
Pancreatic elastase	**C**	**A**	**G**	–	–	–	G	N	**G**	**V**	R	–	**S**	G	**C**	Q	**G**	**D**	**S**	**G**	**G**	**P**	**L**	**H**		
Thrombin	**C**	**A**	**G**	Y	K	P	G	E	**G**	K	R	G	D	A	**C**	E	**G**	**D**	**S**	**G**	**G**	**P**	F	**V**		
Factor X	**C**	**A**	**G**	Y	–	–	D	T	**Q**	P	E	–	D	A	**C**	Q	**G**	**D**	**S**	**G**	**G**	**P**	H	**V**		
Plasmin	**C**	**A**	**G**	H	–	–	L	A	**G**	G	T	–	D	**S**	**C**	Q	**G**	**D**	**S**	**G**	**G**	**P**	**L**	**V**		
Plasma kallikrein	**C**	**A**	**G**	Y	–	–	L	P	**G**	G	K	–	D	T	**C**	**M**	**G**	**D**	**S**	**G**	**G**	**P**	**L**	I		
Streptomyces trypsin	**C**	**A**	**G**	Y	–	P	D	T	**G**	G	V	–	D	T	**C**	Q	**G**	**D**	**S**	**G**	**G**	**P**	M	F		
Subtilisin	A	G	V	Y	S	T	Y	P	T	N	T	Y	A	T	L	N	**G**	T	**S**	M	A	S	P	H		

Source: Barrett, A. J. In: A. J. Barrett and G. Salvesen (Eds.), *Proteinase Inhibitors*. Amsterdam: Elsevier, 1986, p. 7.

and functional homology with chymotrypsin. A separate family of serine proteases initially discovered in bacteria has no structural homology to the mammalian chymotrypsin family. The serine protease subtilisin, isolated from *Bacillus subtilis*, hydrolyzes peptide bonds and contains an activated serine with a histidine and aspartate in its active site but the active site arises from structural regions of the protein that bear no sequence or structural homology with the chymotrypsin serine proteases. This serine protease is an example of **convergent evolution** of an enzyme catalytic mechanism. Apparently a completely different gene evolved with the same catalytic mechanism utilizing an active-site serine. The primary and tertiary structures are also different from that of the trypsin- and chymotrypsin-like structures.

Tertiary Structures of Trypsin Family of Serine Proteases Are Similar

Ser 195 in chymotrypsin reacts with **diisopropylfluorophosphate (DFP)**, with a 1:1 enzyme/DFP stoichiometry that inhibits the enzyme. The three-dimensional structure of chymotrypsin reveals that the Ser 195 is situated within an internal pocket, with access to the solvent interface. His 57 and Asp 102 are oriented so that they participate with the Ser 195 in the catalytic mechanism (see p. 447).

Structure determinations by X-ray crystallography have been carried out on many members of this family of proteins (Table 9.6). Structural data are available for catalytically active enzyme forms, zymogens, the same enzyme in multiple species, enzyme–inhibitor complexes, and a particular enzyme at different temperatures and in different solvents. The most complete analysis has been that of trypsin. Its X-ray diffraction analysis has yielded a three-dimensional structure at better than 1.7-Å resolution, which can resolve atoms at a separation of 1.3 Å such as the C═O on of the carbonyl group (1.2 Å). This resolution, however, is not uniform over the entire trypsin structure. Different regions vary in their tendency to be localized in space during the time course of the X-ray diffraction experiment, and for some atoms in the structure their exact positions cannot be as precisely defined as for

TABLE 9.6 Serine Protease Structures Determined by X-Ray Crystallography

Enzyme	*Species Source*	*Inhibitors Present*	*Resolution (Å)*
Chymotrypsin[a]	Bovine	Yes[b]	1.67[c]
Chymotrypsinogen	Bovine	No	2.5
Elastase	Porcine	Yes	2.5
Kallikrein	Porcine	Yes	2.05
Proteinase A	*S. griseus*	No	1.5
Proteinase B	*S. griseus*	Yes	1.8
Proteinase II	Rat	No	1.9
Trypsin[a]	Bovine	Yes[b]	1.4[c]
Trypsinogen[a]	Bovine	Yes[b]	1.65[c]

[a] Structure of this enzyme molecule independently determined by two or more investigators.

[b] Structure obtained with no inhibitor present (native structure) and with inhibitors. Inhibitors used include low molecular weight inhibitors (i.e., benzamidine, DFP, and tosyl) and protein inhibitors (i.e., bovine pancreatic trypsin inhibitor).

[c] Highest resolution for this molecule of the multiple determinations.

others. The structural disorder is especially apparent in surface residues not in contact with neighboring molecules. Rapid methods for X-ray data acquisition (see p. 148) further support this observation of dynamic fluctuation. Trypsin is globular and consists of two distinct domains of approximately equal size (Figure 9.17), which do not penetrate one another. There is little α-helix, except in the COOH-terminal region. The structure is predominantly β-structure, with each of the domains in a "deformed" β-barrel. Loop regions protrude from the barrel ends, being almost symmetrically presented by each of the two folded domains. These loop structures combine to form a surface region of the enzyme that extends outward, above the catalytic site, and are structurally and functionally similar to the CDRs of immunoglobulins.

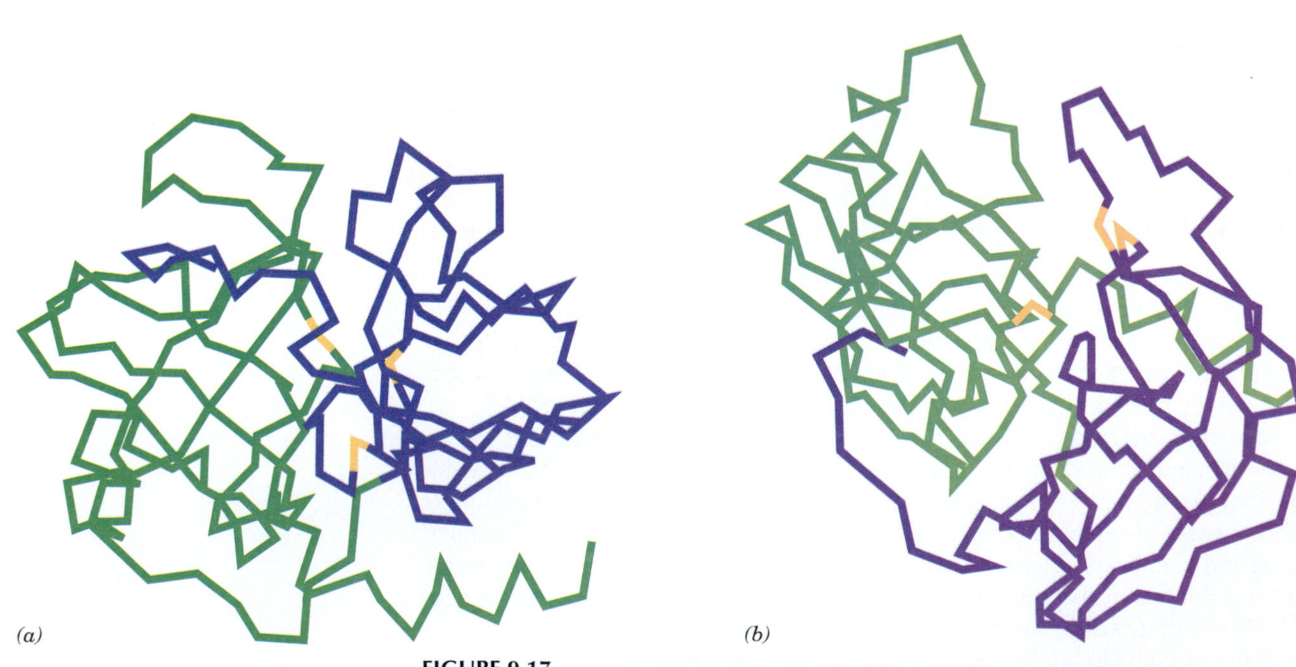

FIGURE 9.17
Two views of the structure of trypsin showing tertiary structure of two domains. Active-site serine, histidine, and aspartate are indicated in yellow.

TABLE 9.7 Structural Superposition of Selected Trypsin Family Serine Proteases and the Resultant Amino Acid Sequence Comparison

Comparison	*Number of Amino Acids in Sequence*		*Number of Structurally Equivalent Residues*	*Number of Chemically Identical Residues*
	Protease 1	*Protease 2*		
Trypsin–elastase	223	240	188	81
Trypsin–chymotrypsin	223	241	185	93
Trypsin–mast cell protease	223	224	188	69
Trypsin–prekallikrein	223	232	194	84
Trypsin–*S. griseus* protease	223	180	121	25

Alignment of three-dimensional structures can be performed on serine proteases using a mathematical function that compares structural equivalence and allows for insertion and deletion of amino acids. The data of Table 9.7 contrast the extent of structural superimposability with the homology of sequences brought into coincidence by the structural superposition. This table shows the total number of amino acids and the number that are statistically identical in each structure. Topologically equivalent amino acids have the same relationship in three-dimensional space to the point where they cannot be distinguished from one another by X-ray diffraction. The last column presents the number of amino acids that are chemically identical. In these structural alignments the regions of greatest difference appear to be localized to the CDR-like loop regions, which extend from the β-barrel domains to form the surface region out from the catalytic site. The effect of altering the amino acids in these loops is to alter the **macromolecular binding specificity** of the protease. It is the structure of the loop in factor Xa, for example, that allows it to bind specifically to prothrombin. Serpins interact with different proteases based on their affinity for the loop structures. Bacterial proteases related to the eukaryotic serine protease family contain the same two domains but lack most of the loop structures. This agrees with their lack of a requirement for complex interactions and the observation that bacterial proteases are not synthesized in a zymogen form.

Thus the serine protease family constitutes a structurally related series of proteins that use a catalytically active serine. During evolution, the basic two-domain structure and the catalytically essential residues have been maintained, but the region of the secondary interactions (loop regions) have changed to give the different members their specificities toward substrates, activators, and inhibitors, characteristic of their physiological functions.

9.4 | DNA-BINDING PROTEINS

Regulatory sites in DNA bind proteins that control gene expression (see p. 350). These sites contain a nucleotide sequence that binds regulatory proteins known as transcription factors. The specific DNA sequence, or **transcription factor binding element,** is usually less than 10 nucleotides long. Noncovalent interactions between the protein and DNA allow the protein to recognize the nucleotide sequence and bind. This is highly selective as the human genome has about 30,000 genes, each with its own regulatory sequences. While there are huge gaps in knowledge of how proteins regulate gene expression, some common structural motifs of DNA-binding proteins are apparent.

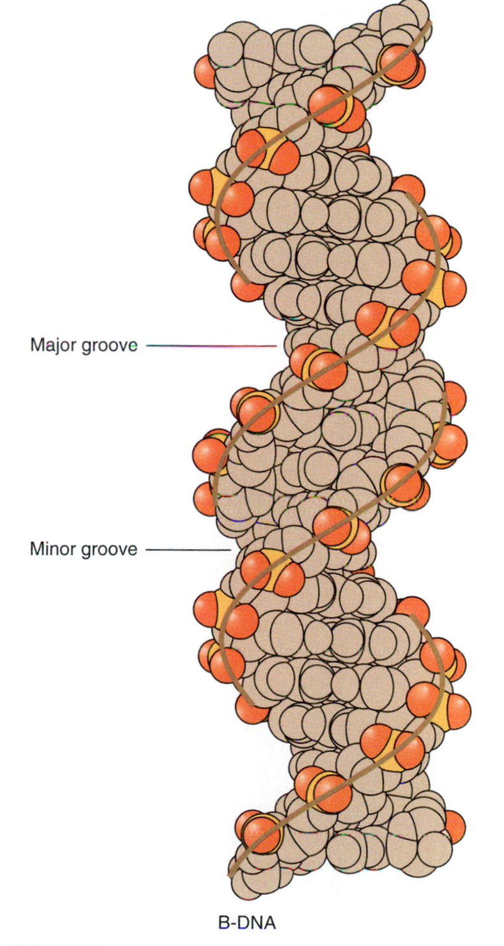

FIGURE 9.18
Space-filling model of DNA in B conformation showing major and minor grooves.
Reprinted with permission from Rich, A. J. Biomol. Struct. Dyn. 1:1, 1983.

Three Major Structural Motifs of DNA-Binding Proteins

Along the double helix of a DNA molecule in its most common form (B form) are two grooves, the major and minor grooves (Figure 9.18) (see p. 38), to which the proteins must associate. A common structural motif found in many DNA-binding proteins is the **helix–turn–helix (HTH).** A HTH places one of its α-helices, designated the **recognition helix**, across the major groove where side chain residues of the helix form specific noncovalent interactions with the base sequence of the target DNA sequence. The interaction often induces distortion of the B-DNA-binding site that better accommodates the interactions with protein. Nonspecific interactions are made between the protein and sugar–phosphate backbone of DNA. Most HTH proteins bind as dimers, which then form two helix–turn–helix motifs per active regulatory protein. X-ray structures show the two helix–turn–helix motifs protruding from the structure of each monomer domain binding at two adjacent turns of the major groove in the DNA, the dimer dual-binding interaction making a much stronger and more specific protein–DNA interaction than possible for a single monomer protein–DNA interaction. The dimer proteins are symmetrical and the corresponding DNA binding sequences in the regulatory element are palindromes (see p. 72; Figures 9.19–9.21).

The **zinc finger** motif of some DNA-binding proteins contains a repeating motif of a Zn^{2+} atom bonded to two cysteine and two histidine side chains (Figure 9.22). In some cases the histidines are substituted by cysteines. The primary structure for the motif contains two close cysteines separated by about 12 amino acids from a second pair of Zn^{2+} liganding amino acids (histidine or cysteine). The three-dimensional structure of a common zinc finger structure has been deduced by ^{1}H-NMR (Figure 9.23). The motif contains an α-helix segment that can bind within

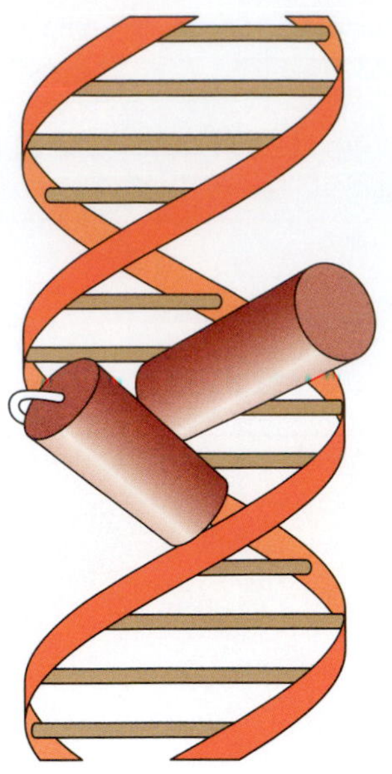

FIGURE 9.19
Binding of a helix–turn–helix motif into the major groove of B-DNA.
The recognition helix lies across the major groove.
Redrawn from Schleif, R. Science 241:241, 1988.

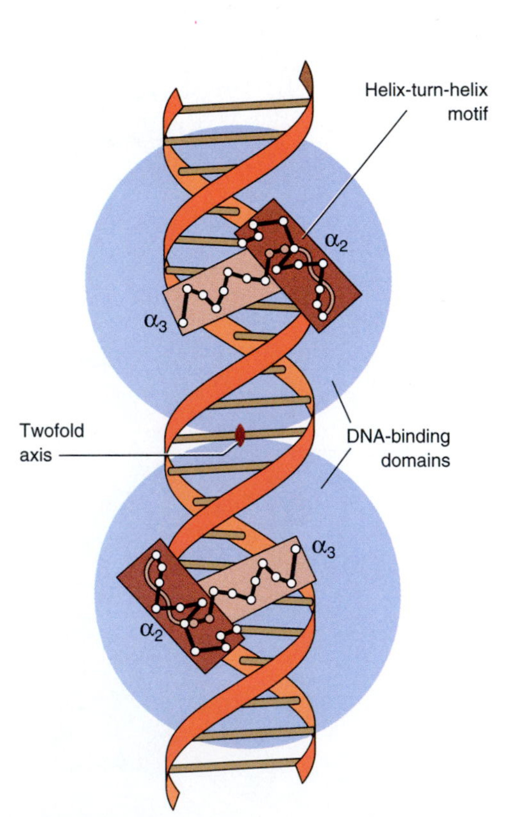

FIGURE 9.20
Association of two helix–turn–helix motifs from a DNA-binding protein (dimer) into adjacent turns of the major grooves of B-DNA.
Redrawn from Brennan, R. G. and Matthews, B. W. Trends Biochem. Sci. 14:287, 1989.

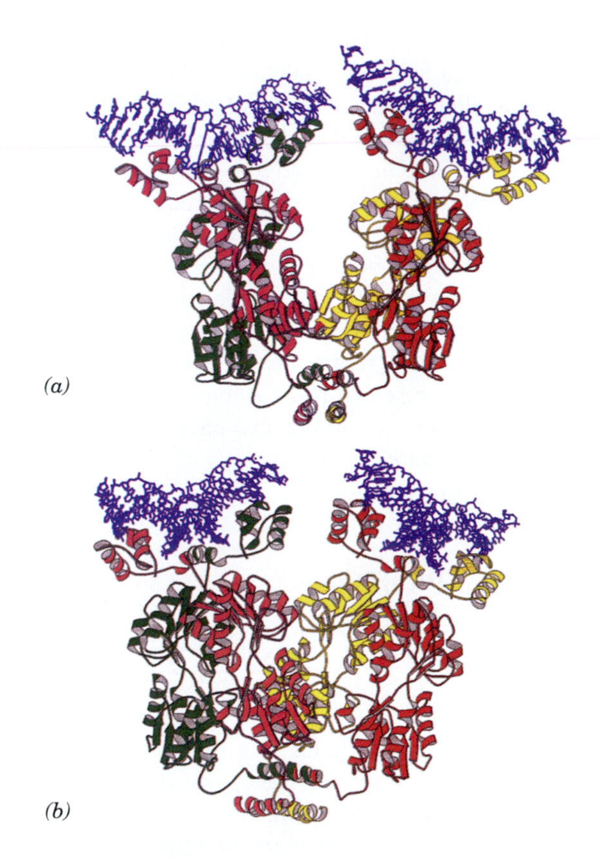

FIGURE 9.21
X-ray crystallographic structure of helix–turn–helix motif *lac* repressor protein in association with target DNA.
(*a*) Repressor is a tetramer protein with individual monomers colored green and red (left), red and yellow (right). The DNA targets are colored purple (top). Recognition helices are shown to interact in adjacent major grooves of target DNAs. Each dimer in tetramer interacts with a discrete (separated) target consensus sequence present in DNA. (*b*) A different view of the same tetramer.
Reprinted with permission from Lewis, M., Chang, G., Horton, N. C., Kercher, M. A., Pace, H. C., Schumacher, M. A., Brennan, R. G., and Lu, P. Science 271:1247, 1996. Copyright © 1996 by the American Association for the Advancement of Science.

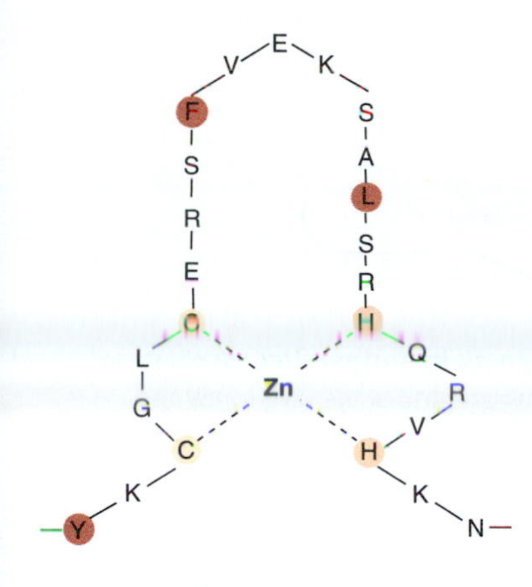

FIGURE 9.22
Primary sequence of a zinc finger motif found in DNA-binding protein Xfin from *Xenopus*.
Invariant and highly conserved amino acids in structure are circled in red.
Redrawn from Lee, M. S., Gippert, G. P., Soman, K. V., Case, D. A., and Wright, P. E. Science 245:635, 1989.

the major groove at its target site in DNA and makes specific interactions with the nucleotide sequence (see p. 354).

A third structural motif found in some of the DNA-binding proteins is the **leucine zipper.** Leucine zippers are formed from a region of α-helix that contains at least four leucines, each leucine separated by six amino acids from one another (i.e., Leu-X_6-Leu-X_6-Leu-X_6-Leu, where X is any common amino acid). With 3.6 residues per turn of the α-helix, the leucines align on one edge of the helix, with a leucine at every second turn of the helix (Figure 9.24). The leucine-rich helix forms a hydrophobic interaction with a second leucine helix on another subunit, to "zipper" the two together to form a dimer (Figures 9.24 and 9.25). The leucine-zipper motif does not directly interact with the DNA, as do the zinc finger or helix–turn–helix motifs. However, if a dimer is not formed the protein will not strongly bind to DNA. The DNA-binding region is adjacent to the zipper motif toward the NH_2-terminal end in the primary structures. The DNA-binding region contains a high concentration of basic amino acids, arginine and lysine. This evolutionary conserved basic region directly interacts with the negatively charged DNA. The positive charges of the arginine and lysine side chains are drawn to the negatively charged DNA phosphate groups. Due to the DNA-binding region of basic amino acids just NH_2-terminal to the leucine-zipper region of each of the polypeptide chains, the structure is designated a "bZIP" structure, where b is for the **b**asic amino acid region and ZIP for the adjacent **zip**per motif (see p. 355).

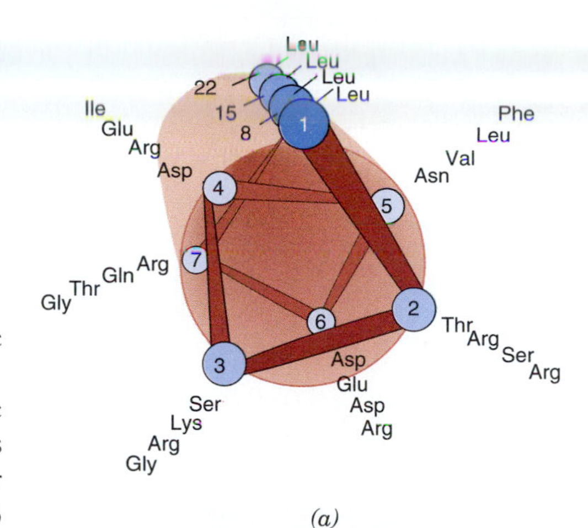

(a)

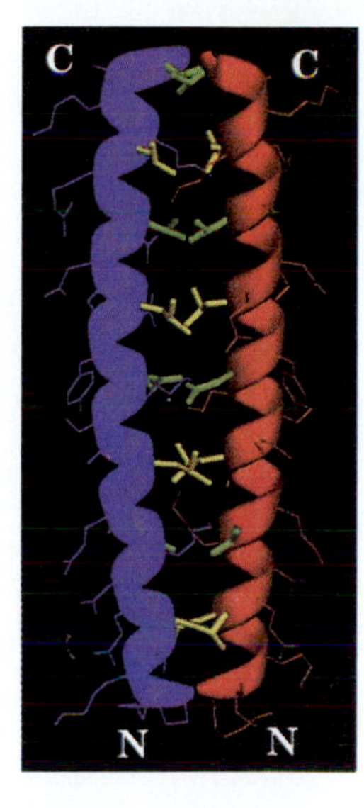

(b)

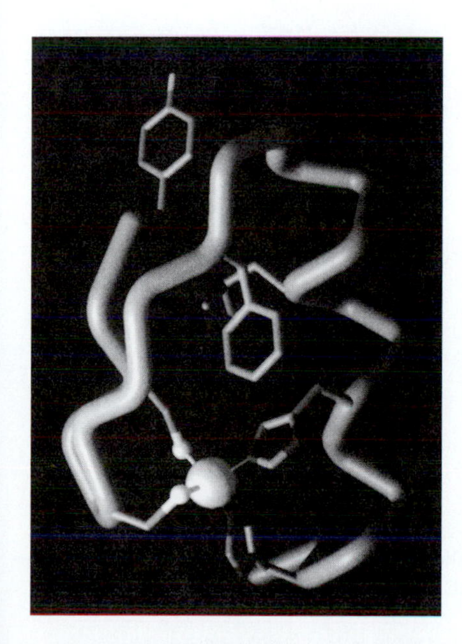

FIGURE 9.23
Three-dimensional structure obtained by ¹H-NMR of zinc finger motif from *Xenopus* protein Xfin (sequence shown in Figure 9.22).
Superposition of 37 possible structures derived from calculations based on the ¹H-NMR. NH_2-terminal is at upper left and COOH-terminal is at bottom right. Zinc is sphere at the bottom with Cys residues to the left and His residues to the right.
Photograph provided by Michael Pique and Peter E. Wright, Department of Molecular Biology, Research Institute of Scripps Clinic, La Jolla, California.

FIGURE 9.24
Leucine-zipper motif of DNA-binding proteins.
(*a*) Helical wheel analysis of the leucine-zipper motif in DNA-binding protein. The amino acid sequence in the wheel analysis is displayed end-to-end down the axis of a schematic α-helix structure. The leucines (Leu) are observed in alignment along one edge of the helix (residues 1, 8, 15, and 22 in the sequence). (*b*) The X-ray structure, in side view, in which the helices are presented in ribbon form and side chains in stick form. Contacting leucine residues in yellow and green.
Part (a) redrawn from Landschulz, W. H., Johnson, P. F., and McKnight, S. L. Science 240:1759, 1988; part (b) reprinted from Voet, D. and Voet, J. Biochemistry, 2nd ed. New York: Wiley, 1995 and based on an X-ray structure by Peter Kim, MIT, and Tom Alber, University of Utah School of Medicine. Reprinted by permission of John Wiley & Sons, Inc.

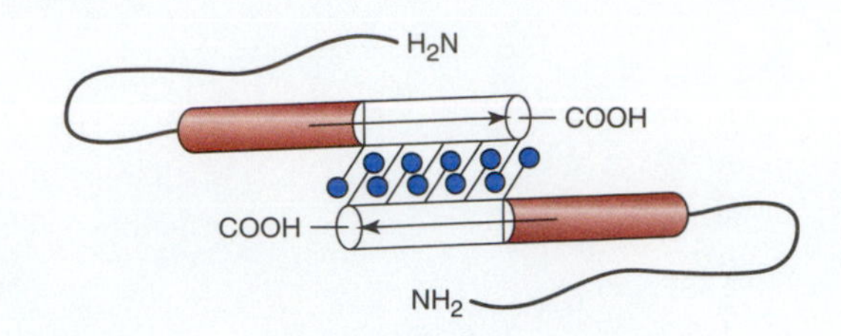

FIGURE 9.25
Schematic diagram of two proteins with leucine zippers in antiparallel association. DNA-binding domains containing a high content of basic amino acids (arginines and lysines) are shown in pink.
Redrawn from Landschulz, W. H., Johnson, P. F., and McKnight, S. L. Science *240:1759, 1988.*

The yeast transcription factor GCN4 contains the leucine-zipper (bZIP) motif. Figure 9.26 shows its association with its palindromic DNA recognition element. It is a dimer of two continuous α-helical subunits joined by a leucine-zipper interface. The α-helices cross at this interface and then diverge with their two N-terminal ends separated to pass directly through different sides of the same major groove of the DNA target site (Figure 9.26). Amazingly, there are no bends or kinks in the linear helical structure of each subunit of the dimer.

Many regulatory proteins that contain the leucine-zipper motif have been shown to be oncogene products (Myc, Jun, and Fos). Fos forms a heterodimer with Jun through a leucine-zipper interaction, and the Fos/Jun dimers bind to gene regulatory sites. If these regulatory proteins are mutated or produced in an unregulated manner, the cell may be transformed to a cancer cell.

DNA-Binding Proteins Utilize a Variety of Strategies for Interaction with DNA

The helix–turn–helix motif was the first motif to be identified for interaction with DNA. X-ray studies of protein–DNA complexes show a great variety of other mechanisms for protein–DNA association. The TATA box-binding protein (TBP) associ-

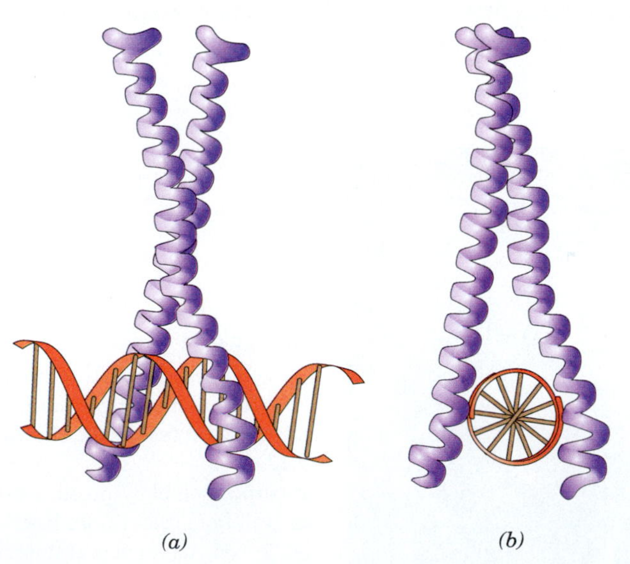

FIGURE 9.26
Structure of the bZIP GCN4–DNA complex.
(*a*) bZIP protein is a dimer (polypeptide chains colored blue) with each monomer joined by a leucine-zipper motif. NH_2-termini diverge to allow the basic region of the sequence to interact in the major groove of DNA target site (DNA colored orange). (*b*) Same interaction viewed down the DNA axis.
Redrawn from Ellenberger, T. E., Brandl, C. J., Struhl, K., and Harrison, S. C. Cell *71:1223, 1992.*

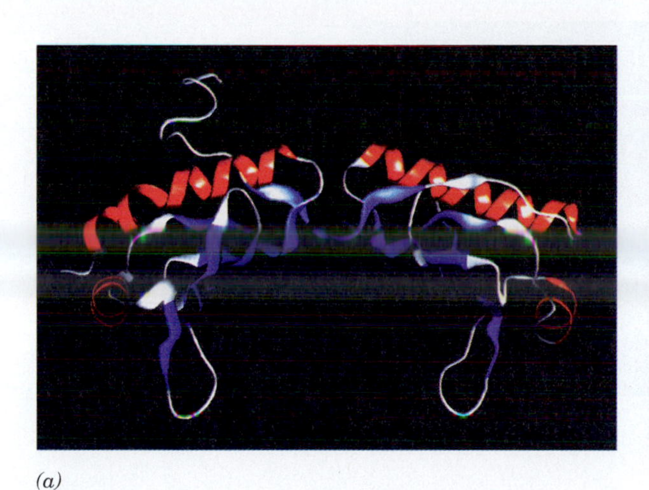

(a)

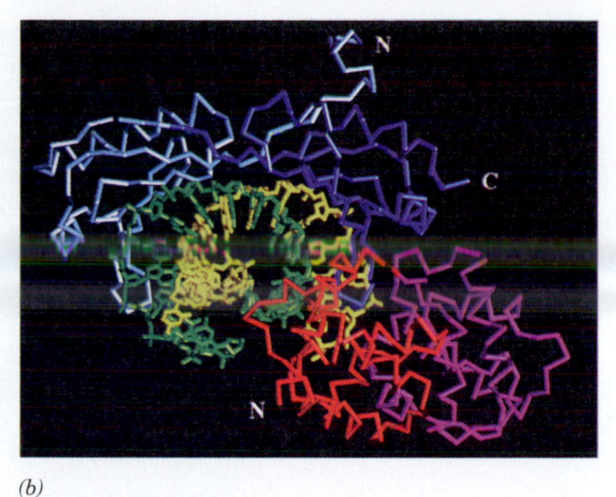

(b)

FIGURE 9.27

Structures of TATA-binding protein (TBP)–DNA binary and TBP–TFIIB–DNA ternary complexes. (*a*) Computer model generated from X-ray structure of the "saddle-like" structure of TBP; α-helices and β-strands are shown in red and blue, respectively, with the remainder in white. (*b*) TBP–TFIIB–DNA complex. Proteins are depicted as α-carbon traces while the DNA is shown as an atomic stick model. TFIIB first repeat is colored red and the second repeat magenta. One domain of TBP is light blue while the second is dark blue. DNA-coding strand is colored green and noncoding strand is in yellow. N- and C-termini of TBP and TFIIB are labeled when visible. See Figure 8.21, p. 351, for another view of the TATA-binding protein.

Courtesy of S. K. Burley. Reprinted with permission from (a) Nikolov, D. B., Hu, S.-H., Lin, J., Gasch, A., Hoffmann, A., Horikoshi, M., Chua, N.-H., Roeder, R. G., and Burley, S. K. Nature *360:40, 1992; and (b) Nikolov, D. B., Chen, H., Halay, E. D., Usheva, A. A., Hisatake, K., Lee, D. K., Roeder, R. G., and Burley, S. K.* Nature *377:119, 1995. Copyright © 1992 and 1995 by Macmillan Magazines Limited.*

ates with the TATA sequence of gene promoters. Association of TBP with the TATA sequence forms the foundation for a large protein complex that initiates gene transcription by RNA polymerase. The X-ray structure of the C-terminal domain of the TBP bound to a TATA sequence shows that TBP contains two domains, each composed of a curved antiparallel β-sheet with a concave surface. The two-domain structure forms the shape of a "saddle" that sits over the DNA double helix. The concave surface of the "saddle" distorts the B-DNA structure and partially unwinds the DNA helix. This distortion, in turn, produces a wide open, though shallow, minor groove that interacts extensively with the underportion of the TBP saddle (Figure 9.27*a*). One critical component of the initiation complex for RNA transcription is TFIIB. An X-ray structure shows TFIIB associates with one of the "stirrups" of the TBP "saddle" in the TATA sequence complex (Figure 9.27*b*).

The p53 protein is a transcription factor that, on sensing damaged DNA, up-regulates the expression of genes that inhibit cell division, to give the cell time to repair the damaged DNA. Alternatively, it can instruct the cell to undergo apoptosis (programmed cell death) if the DNA damage is too extensive for repair. This transcription factor is a key tumor suppressor protein, as its absence allows mutated DNA in a cell to accumulate, leading to its transformation to a cancer cell. Mutant or inactive forms of p53 are found in the majority of human cancers. The DNA-binding domain of p53 consists of two sheets of antiparallel β-strands like an immunoglobulin fold. This central fold provides the scaffolding for the **loop–sheet–helix motif** and for the two large loops (15 and 32 residues) that interact with the DNA. The α-helix (designated H2) of the loop–sheet–helix motif fits into a major groove with loop 1 (L1), while loop 3 (L3) interacts strongly with the adjacent minor groove (Figure 9.28*a*). Figure 9.28*b* shows the residues commonly found mutated in human cancers. Many are in residues that interact directly with DNA, such as Arg 248, which is a part of loop 3. Others are in residues within the domain core required for protein stability. p53 binds DNA as a tetramer (Figure 9.28*c*).

NF-κB transcription factors are ubiquitous members of the Rel family. They regulate a variety of genes, especially genes with roles in cellular defense mechanisms against infection and in differentiation. The NF-κB p50 protein has two domains interconnected by a 10 amino acid linker region (Figure 9.29*a*). Each

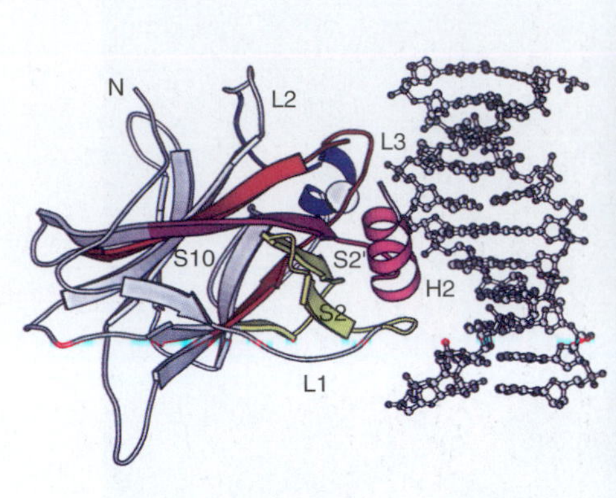

(a)

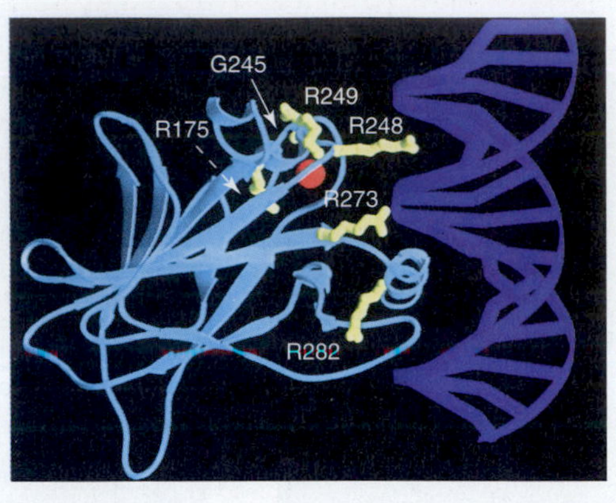

(b)

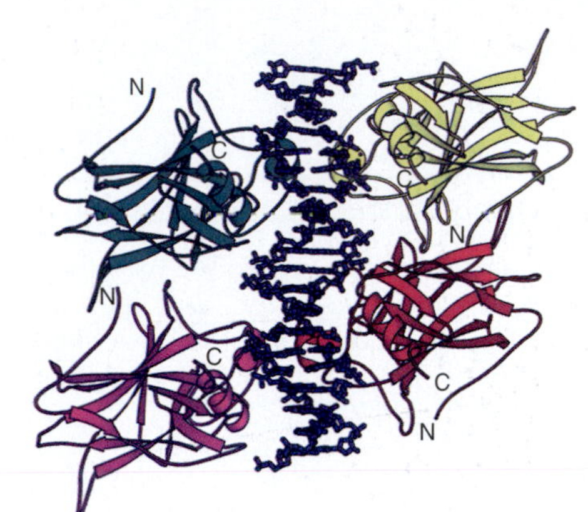

(c)

FIGURE 9.28
Structure of p53–DNA complex.
(*a*) Structure of monomer p53 core domain complexed with DNA. β-Strands (S), α-helices (H), loops (L), and zinc atom (sphere) are lettered and numbered. Helix (H2), loop 1 (L1), and loop 3 (L3) associate in major and minor grooves of target DNA. (*b*) Frequently mutated amino acid side chains commonly found in human cancers are colored yellow. Zinc atom is colored orange. (*c*). Structure of tetramer p53 in association with DNA. Each monomer of tetramer binds to a discrete consensus binding site in the target DNA as shown in (*a*). Four core domains of the tetramer are colored yellow, dark green, red, and lavender, and DNA is colored blue.
Reprinted with permission from Cho, Y., Gorina, S., Jeffrey, P. D., and Pavletich, N. P. Science 265:346, 1994. Copyright © 1994 by the American Association for the Advancement of Science. Figures courtesy of Dr. N. P. Pavletich, Memorial Sloan-Kettering.

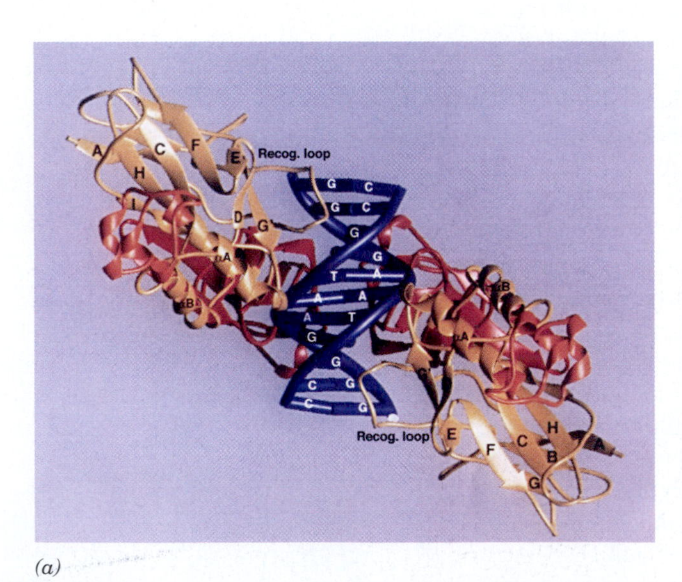

(a)

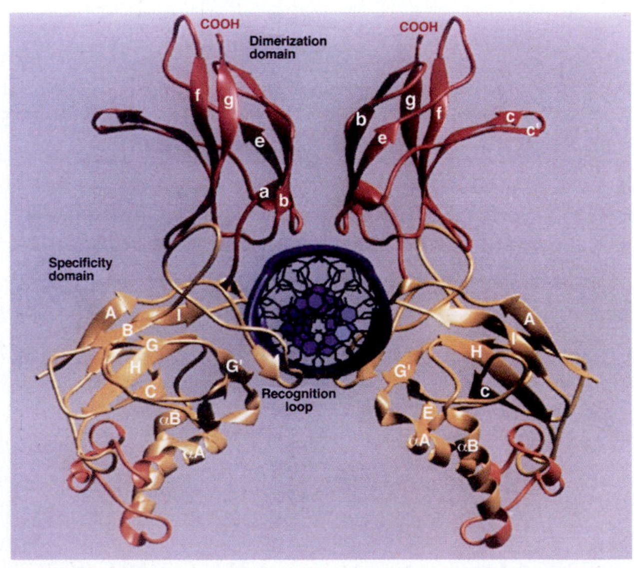

(b)

FIGURE 9.29
Structure of the NF-κB p50 homodimer to DNA.
Only residues 43 through 352 of both subunits are shown in structures. NF-κB p50 protein binds as a dimer. In each monomer, the N-terminal domain is colored yellow and the C-terminal domain is colored red. (*a*) View along DNA axis. (*b*) Alternative view of same complex.

Reprinted with permission from Müller, C. W., Rey, F. A., Sodeoka, M., Verdine, G. L., and Harrison, S. C. Nature 373:311, 1995. Copyright © 1995 by Macmillan Magazines Limited.

domain contains a β-barrel core with antiparallel strands that have structural homology to the immunoglobulin fold. The C-terminal domains provide the dimer interface, where the immunoglobulin folds back together to form the subunit interface. Both N-terminal and C-terminal domains, as well as the loop that connects them, bind to the DNA surface, contributing 10 loops (5 from each subunit in the dimer) that fill the entire major groove in the target DNA (Figure 9.29). N-terminal domains also have an α-helical segment that forms a strong interaction in the minor groove near the center of the target element. In contrast to many other DNA-binding proteins, the NF-κB p50 dimer does not make contact with two separated sites on the DNA target. Rather, the contacts from one monomer combine with those of the second monomer to form a continuous interaction through the single binding site in the DNA.

The structures of helix-turn-helix, zinc fingers, bZIP, and helix-loop-helix proteins are also described in Section 8.8, Preinitiation Complex in Eukaryotes: Transcription Factors, RNA Polymerase II, and DNA (p. 354).

9.5 | HEMOGLOBIN AND MYOGLOBIN

Hemoglobins are globular proteins, present in high concentrations in red blood cells, that bind oxygen in the lungs and transport it to cells. They also transport CO_2 and H^+ from the tissues to the lungs. Hemoglobins also carry and release nitric oxide (NO) in the blood vessels of the tissues. NO is a potent vasodilator and inhibitor of platelet aggregation. The molecular structure and chemistry of the hemoglobins perform these multiple complex and interrelated physiological roles beautifully.

Human Hemoglobin Occurs in Several Forms

A **hemoglobin** molecule consists of four polypeptide chains, two each of two different amino acid sequences. The major form of human adult hemoglobin, **HbA_1**, consists of two α chains and two β chains ($\alpha_2 \beta_2$). The α polypeptide has 141 and the β polypeptide has 146 amino acids. Other forms of hemoglobin predominate in the blood of the human fetus and early embryo (Figure 9.30). The fetal form (**HbF**) contains the same α chains as HbA_1, but a second type of chain (γ chain) occurs in the tetramer molecule and differs in amino acid sequence from that of the β chain of adult HbA_1 (Table 9.8). Two other forms appear in the first months after conception (embryonic) in which the α chains are substituted by *zeta* (ζ) chains of different amino acid sequence and the ε chains serve as the β chains. A minor adult hemoglobin, **HbA_2**, comprises about 2% of normal adult hemoglobin and contains

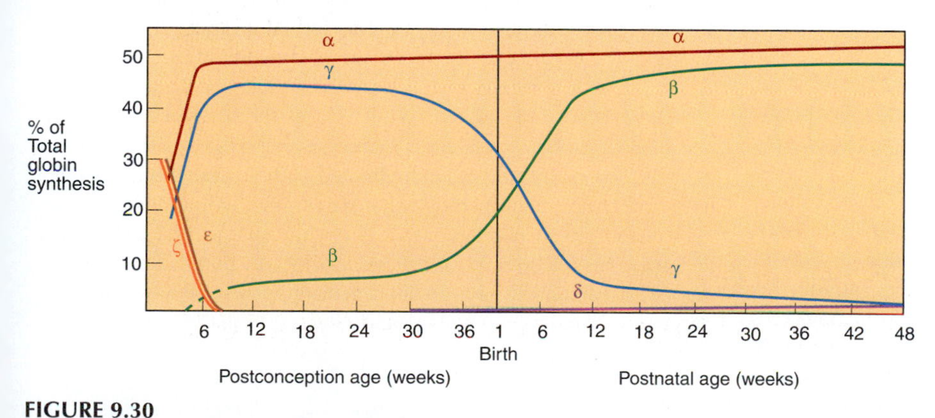

FIGURE 9.30
Changes in globin chain production during development.
Redrawn from Nienhuis, A. W. and Maniatis, T. In: G. Stamatoyannopoulos, A. W. Nienhuis, P. Leder, and P. W. Majerus (Eds.), The Molecular Basis of Blood Diseases. *Philadelphia: Saunders, 1987, p. 68, in which reference of Weatherall, D. J. and Clegg, J. B.,* The Thalassemia Syndromes, *3rd ed. Oxford: Blackwell Scientific Publications, 1981, is acknowledged.*

TABLE 9.8 Chains of Human Hemoglobin

Developmental Stage	*Symbol*	*Chain Composition*
Adult	HbA_1	$\alpha_2\beta_2$
Adult	HbA_2	$\alpha_2\delta_2$
Fetus	HbF	$\alpha_2\gamma_2$
Embryo	Hb Gower-1	$\zeta_2\varepsilon_2$
Embryo	Hb Portland	$\zeta_2\gamma_2$

two α chains and two chains designated delta (δ) (Table 9.8). A discussion of abnormal hemoglobins is presented in Clinical Correlation 9.6.

Myoglobin: A Single Polypeptide with One O_2-Binding Site

Myoglobin (Mb) is an O_2-carrying protein that binds and releases O_2 with changes in the O_2 concentration in the sarcoplasm of skeletal muscle cells. In contrast to hemoglobin, which has four chains and four O_2-binding sites, myoglobin contains only a single polypeptide chain and one O_2-binding site. Myoglobin is a model for what occurs when a single protomer molecule acts alone without the interactions exhibited among the four O_2-binding sites in the more complex tetramer molecule of hemoglobin. The binding of O_2 in hemoglobin reveals that the four sites of O_2-binding do not act independently but act in a cooperative manner.

CLINICAL CORRELATION 9.6

Hemoglobinopathies

There are over 400 documented different mutant human hemoglobins. Mutations cause instability in hemoglobin structure, increased or decreased oxygen affinity, or an increase in the rate of oxidation of the heme ferrous iron (Fe^{2+}) to the ferric oxidation state (Fe^{3+}). A nonfunctional ferric hemoglobin is designated methemoglobin and is symbolized HbM. Unstable hemoglobins arise by the substitution of proline for an amino acid within an α-helical region of the globin fold. Prolines do not participate in α-helical structure and the breaking of an α-helical segment can cause an unstable hemoglobin (examples are $Hb_{Saki}^{\beta Leu14\rightarrow Pro}$ and $Hb_{Genova}^{\beta 28Leu\rightarrow Pro}$). Other unstable hemoglobins are generated by the substitution of an amino acid side chain by one that is either too large or small to make appropriate contacts, or by placement of a charged or polar group on the inside of a domain. Unstable hemoglobins easily denature and precipitate within the erythrocyte, forming Heinz bodies, which damage the erythrocyte membrane. Patients with unstable hemoglobins may develop anemia, reticulocytosis, splenomegaly, and urobilinuria. A number of mutant hemoglobins have an increased oxygen affinity (lower P_{50}). An interesting example is $Hb_{Cowtown}^{\beta His146\rightarrow Leu}$, in which the histidine that dissociates 50% of the Bohr effect protons is lost. This mutation impedes the regulation of oxygen dissociation by hydrogen ion concentration and destabilizes the T conformation relative to the R conformation, causing an increase in oxygen affinity. Hemoglobin mutations that impair BPG binding also increase oxygen affinity. Hemoglobinopathies of increased oxygen affinity are also often characterized by hemolytic anemia and Heinz body formation. These hemoglobins cannot dissociate their oxygen to the tissues as needed.

Mutant hemoglobins that form methemoglobin include $HbM_{Iwate}^{\alpha 87His\rightarrow Tyr}$ and $HbM_{Hyde\ Park}^{\beta 92His\rightarrow Tyr}$, where the proximal histidines F8 in the α and β chains, respectively, are mutated. In $HbM_{Boston}^{\alpha 58His\rightarrow Tyr}$ and $HbM_{Saskatoon}^{\beta 63His\rightarrow Tyr}$ the distal histidines E7 are mutated. Mutations of amino acids that pack against the heme or form the oxygen-binding site often lead to methemoglobinemia. Patients with high concentrations of methemoglobin show a cyanosis (bluish color of the skin).

Two of the most prevalent mutations found in hemoglobin occur at the same amino acid position, the β6Glu. When this glutamate is replaced by valine, the result is $HbS^{\beta 6Glu\rightarrow Val}$, whereas replacement of the glutamate by lysine yields $HbC^{\beta 6Glu\rightarrow Lys}$. Homozygotes with HbS express sickle cell anemia in which the hemoglobin molecules precipitate as tactoids or long arrays. This formation produces the sickle shape of the erythrocyte characteristic of the disease (see Clin. Corr. 3.3). HbC differs from HbS in that it forms a different aggregate structure, blunt-ended crystalloids. This reduces the survival time for the red blood cells but exhibits less hemolysis than HbS and this form of hemoglobinopathy exhibits more limited pathological effects. As both HbS and HbC mutant hemoglobins are commonly found among certain black African populations, it is not unusual to find individuals heterozygous for both mutant genes among this population. Individuals with HbSC will have an intermediate anemia between that observed for homozygotes of HbS and HbC.

Dickerson, R. E. and Geis, I. *Hemoglobin: Structure, Function, Evolution, and Pathology.* Menlo Park, CA: Benjamin-Cummings, 1983; and Arcasoy, M. O. and Gallagher, P. G. Molecular diagnosis of hemoglobinopathies and other red blood cell disorders. *Semin. Hematol.* 36:328, 1999.

Heme Prosthetic Group Is at the Site of O_2 Binding

The four polypeptides of the globin subunits in hemoglobin and the one of myoglobin each contain a heme prosthetic group. A **prosthetic group** is a nonpolypeptide moiety that forms a functional part of a protein. Without its prosthetic group, a protein is designated an **apoprotein.** With its prosthetic group it is a **holoprotein.**

Heme is **protoporphyrin IX** (see p. 1063) with an iron atom in its center (Figure 9.31). The iron atom is in the ferrous (2^+ charge) oxidation state in functional hemoglobin and myoglobin. The ferrous atom in the heme can form five or six ligand bonds, depending on whether or not O_2 is bound to the molecule. Four bonds are to the pyrrole nitrogen atoms of the porphyrin. Since all pyrrole rings of porphyrin lie in a common plane, the four ligand bonds from the porphyrin to the iron atom will have a tendency to lie in the plane of the porphyrin ring. The fifth and the potentially sixth ligand bonds to the ferrous atom are directed along an axis perpendicular to the plane of the porphyrin ring (Figure 9.32). The fifth coordinate bond of the ferrous atom is to a nitrogen of a histidine imidazole. This is designated the **proximal histidine** in hemoglobin and myoglobin structures (Figures 9.32 and 9.33). O_2 forms a sixth coordinate bond to the ferrous atom when bound to hemoglobin and myoglobin. In this bonded position the O_2 is placed between the ferrous atom to which it is liganded and a second histidine imidazole, designated the **distal histidine.** In deoxyhemoglobin, the sixth coordination position of the ferrous atom is unoccupied. The heme is positioned within a hydrophobic pocket of each globin subunit, where approximately 80 interactions are provided by about 18 residues to the heme. Most of these noncovalent interactions are between apolar side chains of amino acids and the apolar regions of the porphyrin. As discussed in Chapter 3, the driving force for these interactions is the expulsion of water of solvation on association of the hydrophobic heme with the apolar amino acid side

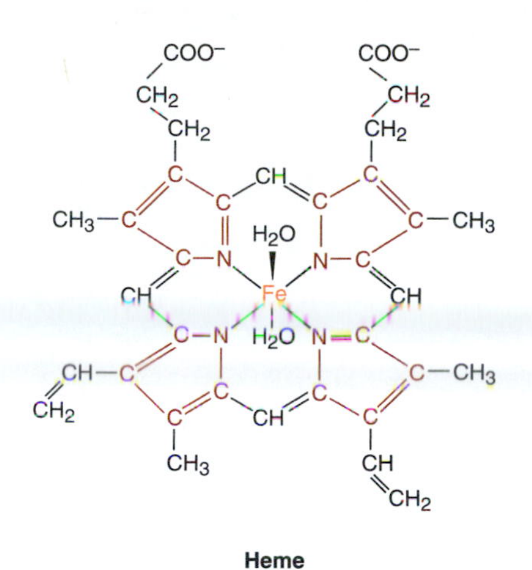

FIGURE 9.31
Structure of heme.

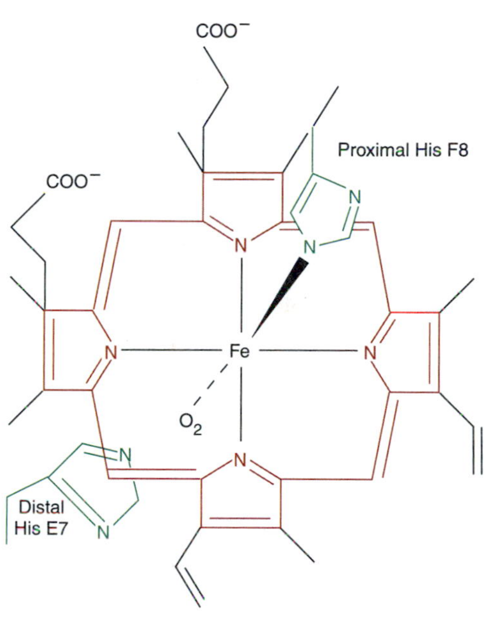

FIGURE 9.32
Ligand bonds to ferrous atom in oxyhemoglobin.

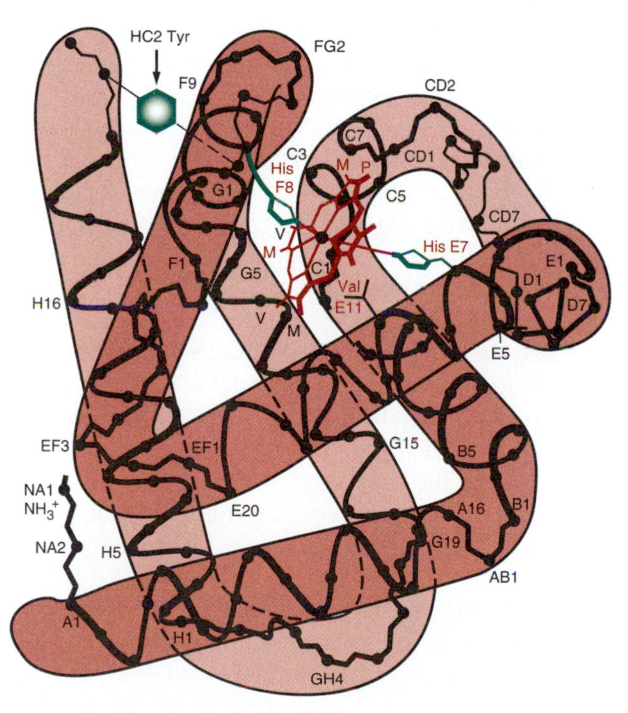

FIGURE 9.33
Secondary and tertiary structure characteristics of globins of hemoglobin.
Proximal His F8, distal His E7, and Val E11 side chains are shown. Other amino acids of polypeptide chain are represented by α-carbon positions only; the letters M, V, and P refer to the methyl, vinyl, and propionate side chains of the heme.
Reprinted with permission from Perutz, M. Br. Med. Bull. 32:195, 1976.

chains in the heme pocket. In myoglobin, additional noncovalent interactions are made between the negatively charged propionate groups of the heme and positively charged arginine and histidine side chains of the globin. However, in hemoglobin chains a difference in the amino acid sequence in this region of the heme-binding site leads to stabilization of the porphyrin propionates by interaction with an uncharged histidine imidazole and with water molecules of solvent toward the outer surface of the molecule.

X-Ray Crystallography Has Assisted in Defining the Structure of Hemoglobin and Myoglobin

The structures of deoxy and oxy forms of hemoglobin and myoglobin have been resolved by X-ray crystallography. In fact, sperm whale myoglobin was the first globular protein whose full three-dimensional structure was determined by this technique. This was followed by the X-ray structure of the more complex horse hemoglobin. These structures show that each globin in the hemoglobins and the single subunit of myoglobin are composed of multiple α-helical regions connected by turns of the polypeptide chain that allow the protein to fold into a spheroidal shape characteristic of the globin fold (Figure 9.33; see also Figure 3.33). The cooperative associations of O_2, discussed below, are based on the X-ray structures of oxyhemoglobin, deoxyhemoglobin, and a variety of hemoglobin derivatives.

Primary, Secondary, and Tertiary Structures of Myoglobin and Individual Hemoglobin Chains

The amino acid sequences of the globin of myoglobin of 23 different animal species have been determined. All myoglobins contain 153 amino acids in their polypeptide chains, of which 83 are invariant. Only 15 of these invariant residues in the myoglobin sequence are identical to the invariant residues of the sequenced globins of mammalian hemoglobin. However, the changes are, in the great majority of cases, conservative and preserve the general physical properties of the residues (Table 9.9). Since myoglobin functions as a monomer, many of its surface positions interact with water and prevent another molecule of myoglobin from associating. In contrast, surface residues of the individual subunits in hemoglobin are designed to provide hydrogen bonds and nonpolar contacts with other subunits in the hemoglobin quaternary structure. The proximal and distal histidines are preserved in the sequences of all the polypeptide chains. Other invariant residues are in the hydrophobic heme pocket and form essential nonpolar contacts with the heme that stabilize the heme–protein complex.

While there is surprising variability in sequences among the different polypeptide chains, to a first approximation the secondary and tertiary structures are almost identical (Figure 9.34). Significant differences in physiological properties between α, β, γ, and δ chains of hemoglobins and the polypeptide chain of myoglobin are due to rather small specific changes in their structures. The similarity in tertiary structure, resulting from widely varied amino acid sequences, shows that the same fold structure for a polypeptide chain can be arrived at by many different sequences.

Approximately 70% of the residues participate in the α-helical secondary structures, generating seven helical segments in the α chain and eight in the β chain. These latter eight helical regions are commonly labeled A–H, starting from the first (A) helix at the NH_2-terminal end. The interhelical regions are designated as AB, BC, CD, . . . , GH, respectively. The nonhelical region between the NH_2-terminal end and the A helix is designated the NA region; and the region between the COOH-terminal end and the H helix is designated the HC region (Figure 9.33 and Table 9.9). This system allows discussion of particular residues that have similar functional and structural roles in hemoglobin and myoglobin.

TABLE 9.9 Amino Acid Sequences of Human Hemoglobin Chains and of Sperm Whale Myoglobin[a]

	NA 1	2	3	A 1	2	3	4	5	6	7	8	9	10	11	12	13	14	15	A 16	AB 1	B 1	2	3	4	5	6
MYOGLOBIN	Val	...	Leu	Ser	Glu	Gly	Glu	Trp	Gln	Leu	Val	Leu	His	Val	**Trp**	Ala	**Lys**	**Val**	Glu	Ala	Asp	Val	Ala	Gly	His	**Gly**
HEMOGLOBIN Horse α	Val	...	Leu	Ser	Ala	Ala	Asp	**Lys**	Thr	Asn	Val	Lys	Ala	Ala	**Trp**	Ser	**Lys**	**Val**	Gly	Gly	His	Ala	Gly	Glu	Tyr	**Gly**
β	Val	Gln	Leu	Ser	Gly	Glu	Glu	**Lys**	Ala	Ala	Val	Leu	Ala	Leu	**Trp**	Asp	**Lys**	**Val**	Asn	...	...	Glu	Glu	Glu	Val	**Gly**
Human α	Val	...	Leu	Ser	Pro	Ala	Asp	**Lys**	Thr	Asn	Val	Lys	Ala	Ala	**Trp**	Gly	**Lys**	**Val**	Gly	Ala	His	Ala	Gly	Glu	Tyr	**Gly**
β	Val	His	Leu	Thr	Pro	Glu	Glu	**Lys**	Ser	Ala	Val	Thr	Ala	Leu	**Trp**	Gly	**Lys**	**Val**	Asn	...	...	Val	Asp	Glu	Val	**Gly**
γ	Gly	His	Phe	Thr	Glu	Glu	Asp	**Lys**	Ala	Thr	Ilu	Thr	Ser	Leu	**Trp**	Gly	**Lys**	**Val**	Asn	...	...	Val	Glu	Asp	Ala	**Gly**
δ	Val	His	Leu	Thr	Pro	Glu	Glu	**Lys**	Thr	Ala	Val	Asn	Ala	Leu	**Trp**	Gly	**Lys**	**Val**	Asn	...	...	Val	Asp	Ala	Val	**Gly**

	7	8	9	10	11	12	13	14	15	16	C 1	2	3	4	5	6	7	CD 1	2	3	4	5	6	7	8	D 1
MYOGLOBIN	Gln	Asp	Ilu	**Leu**	Ilu	**Arg**	Leu	Phe	Lys	Ser	His	**Pro**	Glu	**Thr**	Leu	Glu	Lys	**Phe**	Asp	Arg	**Phe**	Lys	His	**Leu**	Lys	Thr
HEMOGLOBIN Horse α	Ala	**Glu**	Ala	**Leu**	Glu	**Arg**	Met	Phe	Leu	Gly	Phe	**Pro**	Thr	**Thr**	Lys	Thr	Tyr	**Phe**	Pro	His	**Phe**	...	Asp	**Leu**	**Ser**	His
β	Gly	**Glu**	Ala	**Leu**	Gly	**Arg**	Leu	Leu	Val	Val	Tyr	**Pro**	Trp	**Thr**	Gln	Arg	Phe	**Phe**	Asp	Ser	**Phe**	Gly	Asp	**Leu**	**Ser**	Gly
Human α	Ala	**Glu**	Ala	**Leu**	Glu	**Arg**	Met	Phe	Leu	Ser	Phe	**Pro**	Thr	**Thr**	Lys	Thr	Tyr	**Phe**	Pro	His	**Phe**	...	Asp	**Leu**	**Ser**	His
β	Gly	**Glu**	Ala	**Leu**	Gly	**Arg**	Leu	Leu	Val	Val	Tyr	**Pro**	Trp	**Thr**	Gln	Arg	Phe	**Phe**	Glu	Ser	**Phe**	Gly	Asp	**Leu**	**Ser**	Thr
γ	Gly	**Glu**	Thr	**Leu**	Gly	**Arg**	Leu	Leu	Val	Val	Tyr	**Pro**	Trp	**Thr**	Gln	Arg	Phe	**Phe**	Asp	Ser	**Phe**	Gly	Asn	**Leu**	**Ser**	Ser
δ	Gly	**Glu**	Ala	**Leu**	Gly	**Arg**	Leu	Leu	Val	Val	Tyr	**Pro**	Trp	**Thr**	Gln	Arg	Phe	**Phe**	Glu	Ser	**Phe**	Gly	Asp	**Leu**	**Ser**	Ser

	2	3	4	5	6	7	E 1	2	3	4	5	6	7	8	9	10	11	12	13	14	E 15	16	17	18	19	20
MYOGLOBIN	Glu	Ala	Glu	Met	Lys	Ala	Ser	Glu	Asp	Leu	**Lys**	Lys	**His**	**Gly**	Val	Thr	**Val**	Leu	Thr	Ala	Leu	Gly	Ala	Ilu	Leu	Lys
HEMOGLOBIN Horse α	...	...	...	...	...	**Gly**	Ser	Ala	Gln	**Val**	**Lys**	Ala	**His**	**Gly**	**Lys**	**Lys**	**Val**	Ala	Asp	Gly	Leu	Thr	Leu	Ala	Val	Gly
β	Pro	Asp	Ala	Val	Met	**Gly**	Asn	Pro	Lys	**Val**	**Lys**	Ala	**His**	**Gly**	**Lys**	**Lys**	**Val**	Leu	His	Ser	Phe	Gly	Glu	Gly	Val	His
Human α	...	...	...	...	...	**Gly**	Ser	Ala	Gln	**Val**	**Lys**	Gly	**His**	**Gly**	**Lys**	**Lys**	**Val**	Ala	Asp	Ala	Leu	Thr	Asn	Ala	Val	Ala
β	Pro	Asp	Ala	Val	Met	**Gly**	Asn	Pro	Lys	**Val**	**Lys**	Ala	**His**	**Gly**	**Lys**	**Lys**	**Val**	Leu	Gly	Ala	Phe	Ser	Asp	Gly	Leu	Ala
γ	Ala	Der	Ala	Ilu	Met	**Gly**	Asn	Pro	Lys	**Val**	**Lys**	Ala	**His**	**Gly**	**Lys**	**Lys**	**Val**	Leu	Thr	Ser	Leu	Gly	Asp	Ala	Ilu	Lys
δ	Pro	Asp	Ala	Val	Met	**Gly**	Asn	Pro	Lys	**Val**	**Lys**	Ala	**His**	**Gly**	**Lys**	**Lys**	**Val**	Leu	Gly	Ala	Phe	Ser	Asp	Gly	Leu	Ala

	EF 1	2	3	4	5	6	7	8	F 1	2	3	4	F 5	6	7	8	9	FG 1	2	3	4	5	G 1	2	3	4
MYOGLOBIN	Lys	Lys	Gly	His	His	Glu	Ala	Glu	Leu	Lys	Pro	**Leu**	Ala	Gln	Ser	**His**	Ala	Thr	**Lys**	His	Lys	Ilu	Pro	Ilu	Lys	Tyr
HEMOGLOBIN Horse α	**His**	Leu	**Asp**	Asp	Leu	Pro	Gly	Ala	Leu	Ser	Asp	**Leu**	**Ser**	Asn	**Leu**	**His**	Ala	His	**Lys**	**Leu**	Arg	**Val**	**Asp**	**Pro**	Val	**Asn**
β	**His**	Leu	**Asp**	Asn	Leu	Lys	Gly	Thr	Phe	Ala	Ala	**Leu**	**Ser**	Glu	**Leu**	**His**	Cys	Asp	**Lys**	**Leu**	His	**Val**	**Asp**	**Pro**	Glu	**Asn**
Human α	**His**	Val	**Asp**	Asp	Met	Pro	Asn	Ala	Leu	Ser	Ala	**Leu**	**Ser**	Asp	**Leu**	**His**	Ala	His	**Lys**	**Leu**	Arg	**Val**	**Asp**	**Pro**	Val	**Asn**
β	**His**	Leu	**Asp**	Asn	Leu	Lys	Gly	Thr	Phe	Ala	Thr	**Leu**	**Ser**	Glu	**Leu**	**His**	Cys	Asp	**Lys**	**Leu**	His	**Val**	**Asp**	**Pro**	Glu	**Asn**
γ	**His**	Leu	**Asp**	Asp	Leu	Lys	Gly	Thr	Phe	Ala	Gln	**Leu**	**Ser**	Glu	**Leu**	**His**	Cys	Asp	**Lys**	**Leu**	His	**Val**	**Asp**	**Pro**	Glu	**Asn**
δ	**His**	Leu	**Asn**	Asp	Leu	Lys	Gly	Thr	Phe	Ser	Gln	**Leu**	**Ser**	Glu	**Leu**	**His**	Cys	Asp	**Lys**	**Leu**	His	**Val**	**Asp**	**Pro**	Glu	**Asn**

	5	6	7	8	G 9	10	11	12	13	14	15	16	17	18	19	GH 1	2	3	4	5	6	H 1	2	H 3	4	5
MYOGLOBIN	Leu	Glu	Phe	Ilu	Ser	Glu	Ala	Ilu	Ilu	His	Val	Leu	His	Ser	Arg	His	Pro	Gly	Asn	**Phe**	Gly	Ala	Asp	Ala	Gln	Gly
HEMOGLOBIN Horse α	**Phe**	Lys	**Leu**	**Leu**	Ser	His	Cys	**Leu**	Leu	Ser	Thr	Leu	**Ala**	Val	His	Leu	Pro	Asn	Asp	**Phe**	**Thr**	**Pro**	Ala	Val	His	**Ala**
β	**Phe**	Arg	**Leu**	**Leu**	Gly	Asn	Val	**Leu**	Ala	Leu	Val	Val	**Ala**	Arg	His	Phe	Gly	Lys	Asp	**Phe**	**Thr**	**Pro**	Glu	Leu	Gln	**Ala**
Human α	**Phe**	Lys	**Leu**	**Leu**	Ser	His	Cys	**Leu**	Leu	Val	Thr	Leu	**Ala**	Ala	His	Leu	Pro	Ala	Glu	**Phe**	**Thr**	**Pro**	Ala	Val	His	**Ala**
β	**Phe**	Srg	**Leu**	**Leu**	Gly	Asn	Val	**Leu**	Val	Cys	Val	Leu	**Ala**	His	His	Phe	Gly	Lys	Glu	**Phe**	**Thr**	**Pro**	Pro	Val	Gln	**Ala**
γ	**Phe**	Lys	**Leu**	**Leu**	Gly	Asn	Val	**Leu**	Val	Thr	Val	Leu	**Ala**	Ilu	His	Phe	Gly	Lys	Glu	**Phe**	**Thr**	**Pro**	Glu	Val	Gln	**Ala**
δ	**Phe**	Arg	**Leu**	**Leu**	Gly	Asn	Val	**Leu**	Val	Cys	Val	Leu	**Ala**	Arg	Asn	Phe	Gly	Lys	Glu	**Phe**	**Thr**	**Pro**	Gln	Met	Gln	**Ala**

	6	7	8	9	10	11	12	13	14	15	16	17	18	19	20	H 21	22	23	2	HC 1	2	3	4	5
MYOGLOBIN	Ala	Met	Asn	**Lys**	Ala	Leu	Glu	Leu	Phe	Arg	Lys	Asp	Ilu	Ala	Ala	Lys	**Tyr**	Lys	Glu	Leu	Gly	Tyr	Gln	Gly
HEMOGLOBIN Horse α	Ser	Leu	Asp	**Lys**	Phe	Leu	Ser	Ser	**Val**	Ser	Thr	Val	**Leu**	Thr	Ser	Lys	**Tyr**	Arg						
β	Ser	Tyr	Gln	**Lys**	Val	Val	Ala	Gly	**Val**	Ala	Asn	Ala	**Leu**	Ala	His	Lys	**Tyr**	His						
Human α	Ser	Leu	Asp	**Lys**	Phe	Leu	Ala	Ser	**Val**	Ser	Thr	Val	**Leu**	Thr	Ser	Lys	**Tyr**	Arg						
β	Ala	Tyr	Gln	**Lys**	Val	Val	Ala	Gly	**Val**	Ala	Asn	Ala	**Leu**	Ala	His	Lys	**Tyr**	His						
γ	Ser	Trp	Gln	**Lys**	Met	Val	Thr	Gly	**Val**	Ala	Ser	Ala	**Leu**	Ser	Ser	Arg	**Tyr**	His						
δ	Ala	Tyr	Gln	**Lys**	Val	Val	Ala	Gly	**Val**	Ala	Asn	Ala	**Leu**	Ala	His	Lys	**Tyr**	His						

Source: Based on diagram in Dickerson, R. E. and Geis, I. *The Structure and Function of Proteins.* New York: Harper & Row, 1969, p. 52.

[a] Residues that are identical are enclosed in box. A, B, C, . . . designate different helices of tertiary structure (see text).

A Simple Equilibrium Defines O_2 Binding to Myoglobin

The association of oxygen to myoglobin is characterized by a simple equilibrium constant (Eqs. 9.1 and 9.2). In Eq. 9.2 $[MbO_2]$ is the solution concentration of oxymyoglobin, [Mb] is that of deoxymyoglobin, and $[O_2]$ is the concentration of oxygen, in moles per liter. The equilibrium constant, K_{eq}, has the units of moles per liter. As for any true equilibrium constant, the value of K_{eq} is dependent on pH, ionic strength, and temperature.

$$\mathrm{Mb} + \mathrm{O_2} \overset{K_{eq}}{\rightleftharpoons} \mathrm{MbO_2} \tag{9.1}$$

$$K_{eq} = \frac{[\mathrm{Mb}][\mathrm{O_2}]}{[\mathrm{MbO_2}]} \tag{9.2}$$

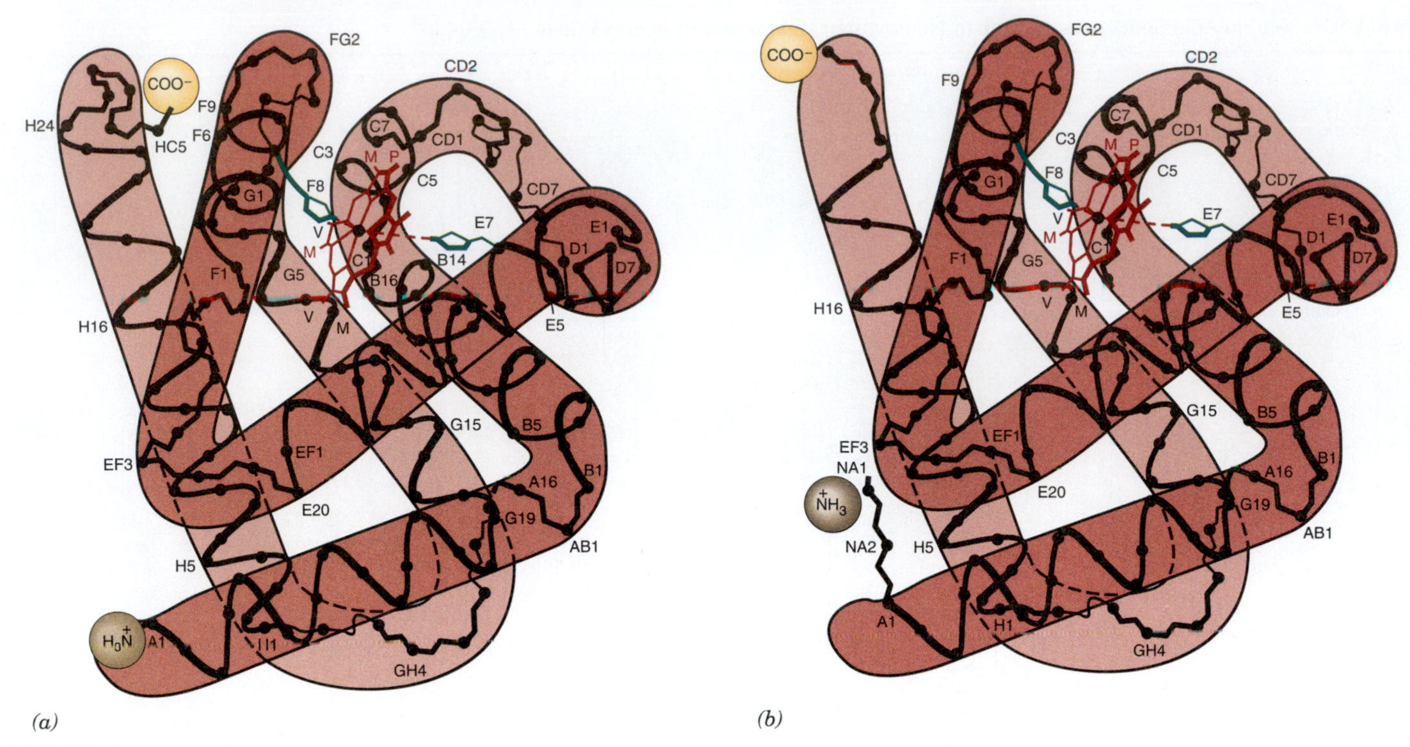

FIGURE 9.34
Comparison of conformation of (*a*) myoglobin and (*b*) β chain of HbA_1.
Overall structures are very similar, except at NH_2-terminal and COOH-terminal ends.

Reprinted with permission from Fersht, A. Enzyme Structure and Mechanism. *San Francisco: Freeman, 1977, pp. 12, 13.*

Since oxygen is a gas, it is more convenient to express its concentration as the pressure of oxygen in torr (1 torr equals the pressure of 1 mm Hg at 0°C and standard gravity). In Eq. 9.3 this conversion of units has been made: P_{50}, the equilibrium constant, and pO_2, the concentration of oxygen, being expressed in torr.

$$P_{50} = \frac{[\text{Mb}] \cdot pO_2}{[\text{MbO}_2]} \tag{9.3}$$

An oxygen-saturation curve characterizes the properties of an oxygen-binding protein. In this plot the fraction of oxygen-binding sites in solution that contain oxygen (Y, Eq. 9.4) is plotted on the ordinate versus pO_2 (oxygen concentration) on the abscissa. The Y value is simply defined for myoglobin by Eq. 9.5. Substitution into Eq. 9.5 of the value of $[\text{MbO}_2]$ obtained from Eq. 9.3, and then dividing through by [Mb], results in Eq. 9.6, which shows the dependence of Y on the value of the equilibrium constant, P_{50}, and the oxygen concentration. It is seen from Eqs. 9.3 and 9.6 that the value of P_{50} is equal to the oxygen concentration, pO_2, when $Y = 0.5$ (50% of the available sites occupied)—hence the designation of the equilibrium constant by the subscript 50.

$$Y = \frac{\text{number of binding sites occupied}}{\text{total number of binding sites in solution}} \tag{9.4}$$

$$Y = \frac{[\text{MbO}_2]}{[\text{Mb}] + [\text{MbO}_2]} \tag{9.5}$$

$$Y = \frac{pO_2}{P_{50} + pO_2} \tag{9.6}$$

A plot of Eq. 9.6 for Y versus pO_2 generates an oxygen-saturation curve for myoglobin and has the form of a rectangular hyperbola (Figure 9.35).

A simple algebraic manipulation of Eq. 9.6 leads to Eq. 9.7. Taking the logarithm of both sides of Eq. 9.7 results in Eq. 9.8, the **Hill equation.** A plot of log $([Y/(1-Y)]$ versus log pO_2, according to Eq. 9.8, yields a straight line with a slope equal to 1 for myoglobin (Figure 9.36). This is the Hill plot, and the slope (n_H) is the **Hill coefficient** (see Eq. 9.9).

$$\frac{Y}{1-Y} = \frac{pO_2}{P_{50}} \quad (9.7)$$

$$\log\frac{Y}{1-Y} = \log pO_2 - \log P_{50} \quad (9.8)$$

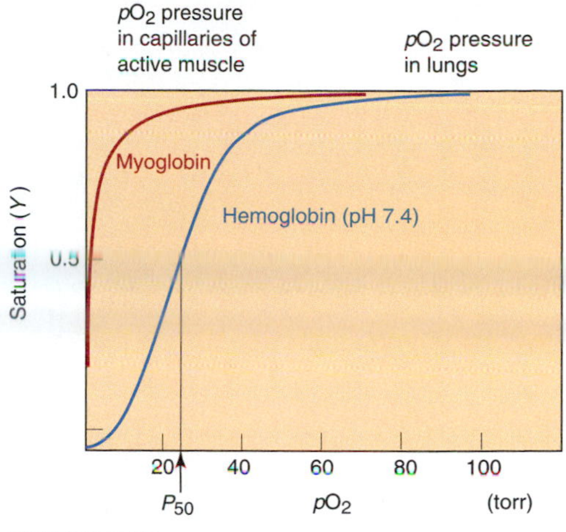

FIGURE 9.35
Oxygen-binding curves for myoglobin and hemoglobin.

Binding of O_2 to Hemoglobin Involves Cooperativity Between Hemoglobin Subunits

Whereas myoglobin has a single O_2-binding site per molecule, hemoglobins, with four monomeric subunits, have four heme-binding sites for O_2. Binding of the four O_2 to hemoglobin manifests as **positively cooperative**, since binding of the first O_2 to deoxyhemoglobin facilitates the binding of O_2 to the other subunits in the molecule. Conversely, dissociation of the first O_2 from fully oxygenated hemoglobin, $Hb(O_2)_4$, will make the dissociation of O_2 from the other subunits of the tetramer easier.

Because of cooperativity in association and dissociation of oxygen, the oxygen-saturation curve for hemoglobin differs from that for myoglobin. A plot of Y versus pO_2 for hemoglobin is a sigmoidal line, indicating cooperativity in oxygen association (Figure 9.35). A plot of the Hill equation (Eq. 9.9) gives a value of the slope (n_H) equal to 2.8 (Figure 9.36).

$$\log\frac{Y}{1-Y} = n_H \log pO_2 - \text{constant} \quad (9.9)$$

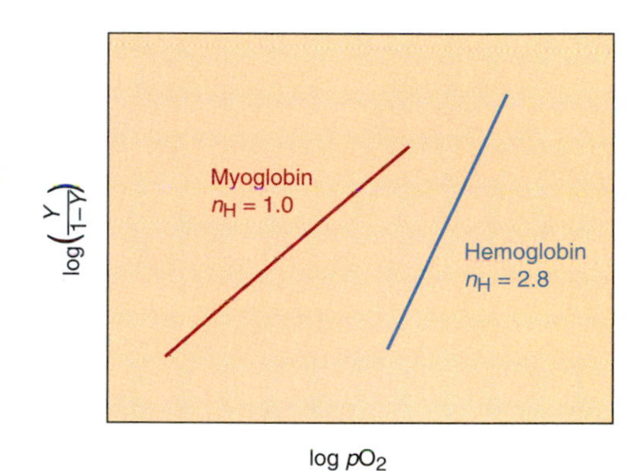

FIGURE 9.36
Hill plots for myoglobin and hemoglobin A_1.

The meaning of the Hill coefficient to cooperative O_2 association can be evaluated quantitatively as presented in Table 9.10. A parameter known as the **cooperativity index**, R_x, is calculated, which shows the fold-change of pO_2 required to change Y from a value of $Y = 0.1$ (10% of sites filled) to a value of $Y = 0.9$ (90% of sites filled) for designated Hill coefficient values found experimentally. For myoglobin, $n_H = 1$, and an 81-fold change in oxygen concentration is required to change from $Y = 0.1$ to $Y = 0.9$. For hemoglobin, where positive cooperativity is observed, $n_H =$ 2.8 and only a 4.8-fold change in oxygen concentration is required to change the fractional saturation from 0.1 to 0.9.

TABLE 9.10 Relationship Between Hill Coefficient (n_H) and Cooperativity Index (R_x)

n_H	R_x	*Observation*
0.5	6560	Negative substrate cooperativity
0.6	1520	
0.7	533	
0.8	243	
0.9	132	
1.0	81.0	Noncooperativity
1.5	18.7	
2.0	9.0	
2.8	4.8	Positive substrate cooperativity
3.5	3.5	
6.0	2.1	
10.0	1.6	
20.0	1.3	

Source: Based on Table 7.1 in Cornish-Bowden, A. *Principles of Enzyme Kinetics*. London: Butterworths Scientific Publishers, 1976.

Molecular Mechanism of Cooperativity in O_2 Binding

X-ray diffraction data on deoxyhemoglobin show that the ferrous atoms actually sit out of the plane of their porphyrins by about 0.4–0.6 Å. This occurs for two reasons. The electronic configuration of the five-coordinated ferrous atom in deoxyhemoglobin has a slightly larger radius than the distance from the center of the porphyrin to each of the pyrrole nitrogen atoms. Accordingly, the iron can be placed in the center of the porphyrin only with some distortion of the porphyrin conformation. Probably more important is that if the iron atom sits in the plane of the porphyrin, the proximal His F8 imidazole will interact unfavorably with atoms of the porphyrin. The strength of this unfavorable steric interaction is due, in part, to conformational constraints on the His F8 and the porphyrin in the deoxyhemoglobin conformation that forces the approach of the His F8 toward the porphyrin to a particular path (Figure 9.37). These constraints become less significant in oxyhemoglobin.

With the iron atom out of the plane of the porphyrin, the conformation is unstrained and energetically favored for the five-coordinate ferrous atom. When O_2

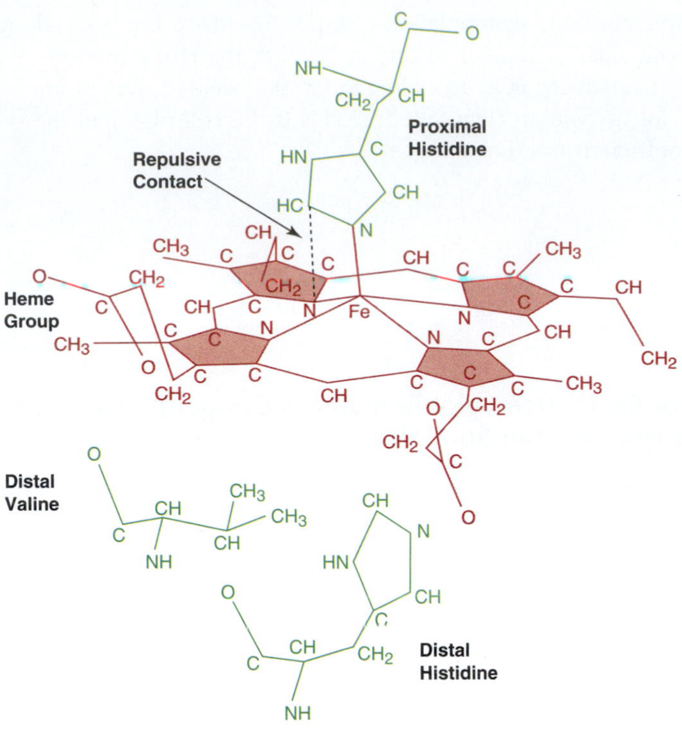

FIGURE 9.37
Steric hindrance between proximal histidine and porphyrin in deoxyhemoglobin.
Redrawn from Perutz, M. Sci. Am. *239:92, 1978. Copyright © 1978 by Scientific American, Inc. All rights reserved.*

binds the sixth coordinate position of the iron, however, this conformation becomes strained. A more energetically favorable conformation for the O_2 liganded iron is one in which the iron atom is within the plane of the porphyrin structure.

On binding of O_2 to a ferrous atom the favorable free energy of bond formation overcomes the repulsive interaction between His F8 and porphyrin, and the ferrous atom moves into the plane of the porphyrin ring. This is the most thermodynamically stable position for the six-bonded iron atom; one axial ligand is on either side of the plane of the porphyrin ring, and the steric repulsion of one of the axial ligands with the porphyrin is balanced by the repulsion of the second axial ligand on the opposite side when the ferrous atom is in the center. If the iron atom is displaced from the center, the steric interactions of the two axial ligands with the porphyrin in the deoxy conformation are unbalanced, and the stability of the unbalanced structure will be lower than that of the equidistant conformation. Also, the radius of the iron atom with six ligands is reduced so that it can just fit into the center of the porphyrin without distortion of the porphyrin conformation.

Since steric repulsion between porphyrin and His F8 in the deoxy conformation must be overcome on O_2 association, binding of the first O_2 is characterized by a relatively low affinity constant. However, when O_2 association occurs to the first heme in deoxyhemoglobin, the change in position of the iron atom from above the plane of the porphyrin into the center of the porphyrin triggers a conformational change in the whole molecule. The change in conformation results in a greater affinity of O_2 to the other heme sites after the first O_2 has bound.

The binding of O_2 pulls the Fe^{2+} atom into the porphyrin plane and moves the proximal His F8 toward the porphyrin and with it the F helix of which the His F8 is a part. Movement of the F helix, in turn, moves the FG corner of its subunit, destabilizing the FG noncovalent interaction with the C helix of the adjacent subunit at an α_1–β_2 or α_2–β_1 subunit interface (Figures 9.38–9.40).

The FG to C intersubunit contacts act as a "switch," because they exist in two different arrangements with different modes of contact between the FG corner of

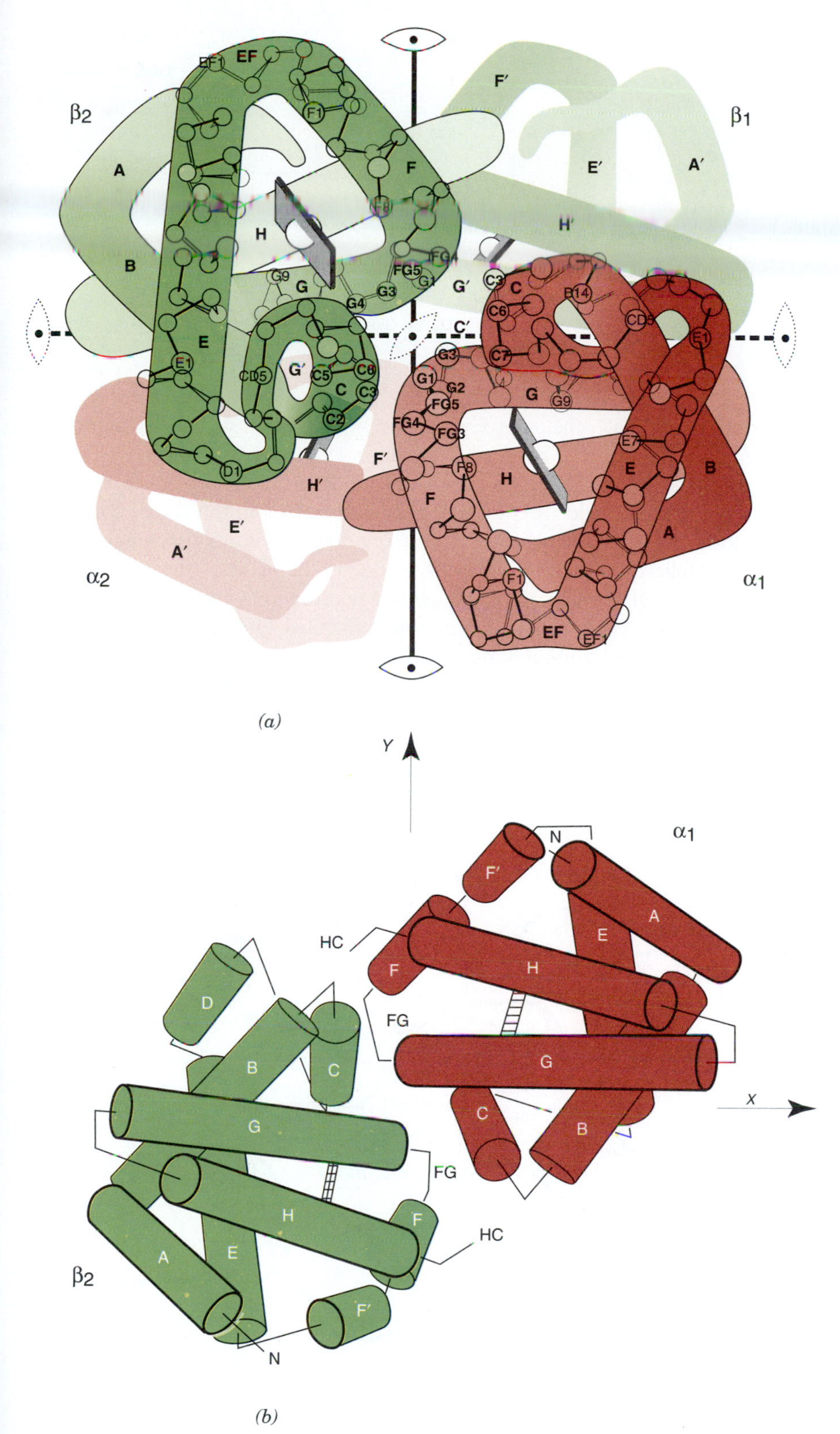

FIGURE 9.38
Quaternary structure of hemoglobin showing FG corner C helix interactions across α_1–β_2 interface.
(*a*) α_1–β_2 interface contacts between FG corners and C helix are shown. (*b*) Cylinder representation of α_1 and β_2 subunits in hemoglobin showing α_1 and β_2 interface contacts between FG corner and C helix, viewed from opposite side of *xy* plane from (*a*).
Part (a) reprinted with permission from Dickerson, R. E. and Geis, I. The Structure and Action of Proteins. Menlo Park, CA: Benjamin, 1969, p. 56. Part (b) reprinted with permission from Baldwin, J. and Chothia, C. J. Mol. Biol. 129:175, 1979. Copyright © 1979 by Academic Press, Inc. (London) Ltd.

one subunit and the C helix of the adjacent subunit. The switch in noncovalent interactions between the two positions involves a relative movement of FG and C in adjacent subunits of about 6 Å. In the second position of the "switch," the tertiary conformation of the subunits participating in the FG to C intersubunit contact is less constrained and the adjacent subunit changes to a new tertiary conformation (oxy conformation) even without bound O_2. This oxy conformation allows the His

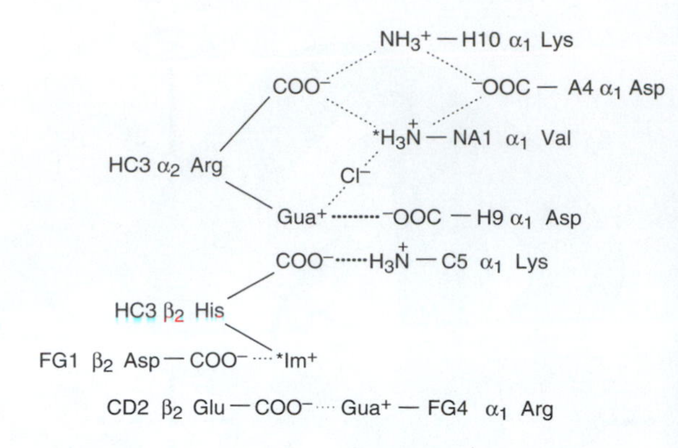

FIGURE 9.39
Salt bridges between subunits in deoxyhemoglobin that are broken in oxyhemoglobin.
Im^+ is imidazolium; Gua^+ is guanidinium; starred residues account for approximately 60% of alkaline Bohr effect.
Redrawn from Perutz, M. Br. Med. Bull. 32:195, 1976.

F8 residues to approach their porphyrins on O_2 association with a less significant steric repulsion than in the deoxy conformation (Figure 9.40). Thus O_2 binds to the empty hemes in the less constrained oxy conformation more easily than to a subunit conformation held by the quaternary interactions in the deoxy conformation. In addition, ionic interactions stabilize the deoxy conformation (Figure 9.39) and are destabilized on the binding of O_2 to one of the hemes.

The deoxy conformation of hemoglobin is referred to as the "tense" or **T conformational state.** The oxyhemoglobin conformational form is referred to as the "relaxed" or **R conformational state.** The allosteric mechanism describes how initial binding of the O_2 to one of the heme subunits of the tetrameric molecule pushes the molecular conformation from the T to R conformational state. The affinity constant for O_2 is greater in the R state hemes than the T state by a factor of 150–300, depending on the solution conditions.

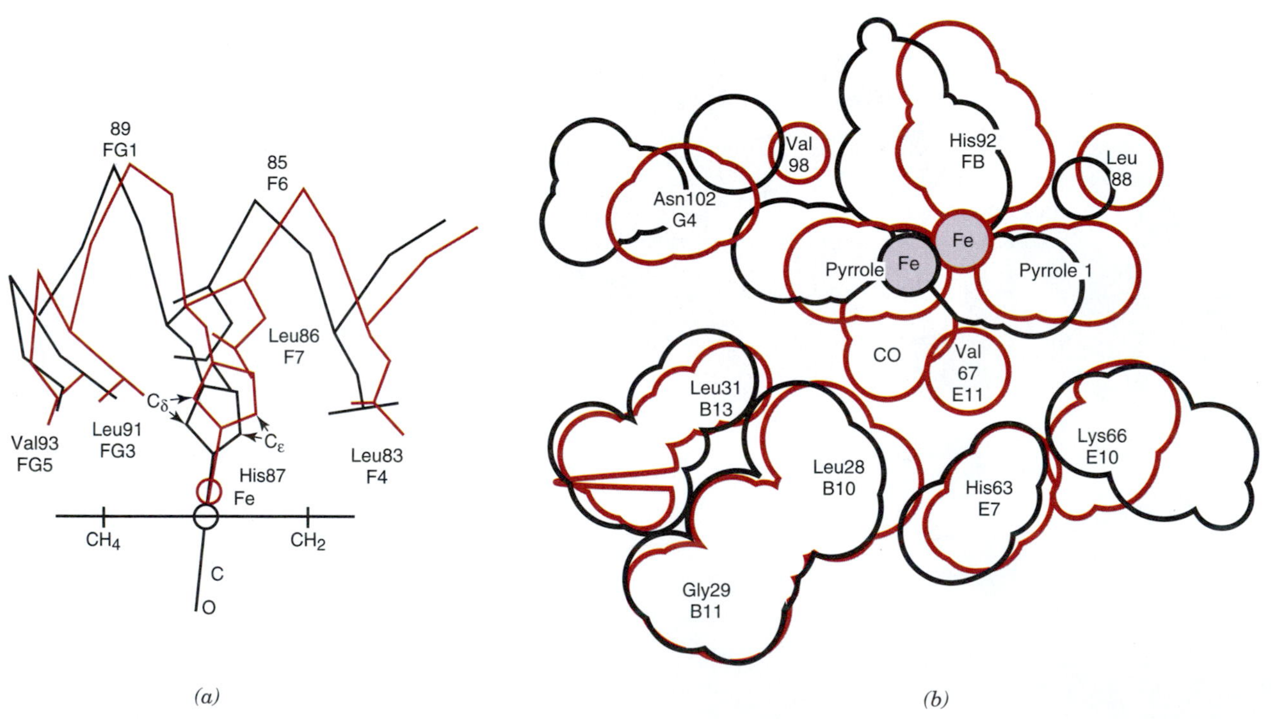

FIGURE 9.40
Stick and space-filling diagrams drawn by computer graphics showing movements of residues in heme environment on transition from deoxyhemoglobin into oxyhemoglobin.
(*a*) Black line outlines position of polypeptide chain and His F8 in carbon monoxide hemoglobin, a model for oxyhemoglobin. Red line outlines the positions of the same regions for deoxyhemoglobin. Position of iron atom is shown by circle. Movements are for an α subunit. (*b*) Similar movements of amino acid residue position in a β subunit using space-filling diagram are shown. Residue labels centered in density for the deoxyconformation.
Redrawn with permission from Baldwin, J. and Chothia, C. J. Mol. Biol. 129:175, 1979. Copyright © 1979 by Academic Press, Inc. (London) Ltd.

Hemoglobin Facilitates Transport of CO_2 and NO

The survival of cells relies on the ability of hemoglobin in red blood cells to both deliver O_2 for cellular metabolism and to facilitate the transport of the end product of cellular metabolism, CO_2, away from the cells to the lung. In addition, hemoglobin binds the vasodilator, **NO**, and delivers the NO to the blood vessel walls in the tissues. The T to R conformational equilibrium of the hemoglobin molecule controls the delivery of O_2, CO_2, and NO by hemoglobin to their appropriate delivery sites. Furthermore, the T $\leftrightharpoons$ R equilibrium is regulated by the hydrogen ion concentration, which links the delivery of O_2 and NO to the site of the tissues and the transport of CO_2 away from the tissues by hemoglobin.

Decrease in pK_a of Acid Groups with Change in Conformation from T to R Leads to Dissociation of Protons

Equation 9.10 shows the release of protons as the **T conformation** is converted to the **R conformation.** The release of protons as the T conformation changes to the R conformation at blood pH is called the alkaline **Bohr effect** and is of physiological importance.

$$\underset{\mathrm{T}}{\mathrm{Hb}} + 4\mathrm{O}_2 \rightleftharpoons \underset{\mathrm{R}}{\mathrm{Hb(O_2)_4}} + n\mathrm{H}^+ \qquad \textbf{(9.10)}$$

The value of n in Eq. 9.10 (equivalents of protons released when one molecule of deoxy-Hb is converted to oxy-Hb) will vary between 1.2 and 2.7 depending on solution conditions and the concentration of chloride ion and **2,3-bisphosphoglycerate.** According to the law of mass action, increasing the hydrogen ion concentration on the right side of the equilibrium sign of Eq. 9.10 forces the equilibrium to the left, toward increasing concentrations of deoxy-Hb and free O_2, in order to reestablish the equilibrium ratio. Cells actively carrying out metabolism release carbonic acid (hydrated CO_2) and lactic acid (see p. 599), increasing the acidity of their environment. The lowered blood pH (increase in H^+ concentration) will force oxy-Hb to deoxy-Hb and the delivery of O_2 to the cells. This mechanism is a feedback loop, where the cellular metabolic waste (CO_2) induces increases in the concentration of a key substrate (O_2) required for the continuation of cellular metabolism. The proton concentration variation links the two processes of O_2 delivery and CO_2 disposal by the hemoglobin molecule.

Why do protons dissociate from hemoglobin as the T conformation converts to the R conformation? Protons dissociate when acid side chain groups of the hemoglobin molecule are more acidic in the R conformation than in the T conformation. This requires the pK_a of these groups to be lower in the R conformation than in the T conformation.

The imidazolium group of His 146(β) is the major contributor to proton dissociation on conversion of the T to R conformation. In the T conformation, the positive charge on the His 146(β) side-chain imidazolium is in an ion-pair with the negative charge of the carboxylate side chain of the Asp 94(β) (Figure 9.41). The negatively charged carboxylate stabilizes the positively charged imidazolium. Thus the T conformation His 146(β) imidazolium in ion-pair with the Asp 94(β) carboxylate has a higher than normal pK_a (approximately 7.7). The conversion of hemoglobin conformation to the R conformation breaks the ion pair between Asp 94 and His 146 and places the two residues at new positions in the R conformation, where the oppositely charged groups no longer strongly interact. Consequently, in the R conformation, the His 146(β) imidazolium reverts to a more normal pK_a of approximately 7.3. The change in the His 146(β) pK_a to a more acidic value on conversion of Hb from the T to R conformation results in the dissociation of protons at the pH of blood (~7.4). Approximately 50% of the protons that dissociate come from the His 146(β) imidazolium. The other dissociated Bohr effect protons come from other acid groups of the protein that similarly

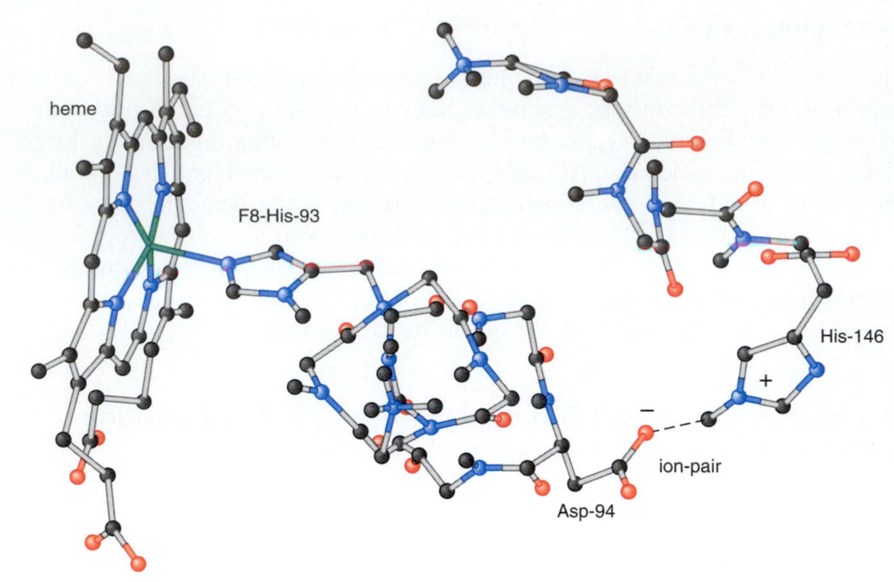

FIGURE 9.41
Ion pair between the His 146(β) imidazolium and the Asp 94(β) carboxylate side chain groups in the deoxy (T) conformation of hemoglobin.
A partial structure of the β chain showing the backbone polypeptide chain for amino acids 87 through 95 and 142 through 146 of the β chain in the deoxy conformation. Only the side chains of Asp 94, His 146, and the F8 His 93 liganded to the β subunit heme are shown. Oxygen atoms are colored red, nitrogen atoms are blue, and carbon and hydrogen atoms are black. The hydrogen bond (broken line) is shown between the positively charged imidazolium N-H of His 146(β) and negatively charged carboxylate oxygen of Asp 94(β).
Drawing made with the Swiss-PdbViewer using PDB structure 1A3N.

change their pK_a from higher to lower values as the conformation changes from T to R.

In Figure 9.42 is shown the plot of the fractional saturation of O_2-binding sites in hemoglobin versus O_2 concentration for Hb at different values of pH. At more acidic pH (higher [H^+]), hemoglobin more easily dissociates its O_2 (Eq. 9.10) and the curve is perturbed to the right of the curve at normal values of pH. Accordingly, the P_{50} value at more acidic pH is higher than the value at normal pH reflecting a poorer affinity of hemoglobin for O_2. In the opposite pH direction, at higher than normal pH (lower [H^+]), the hemoglobin association–dissociation curve is perturbed to the left of the normal pH curve and the O_2-Hb dissociation equilibrium has a lower P_{50} value, reflecting a higher affinity of O_2 to hemoglobin.

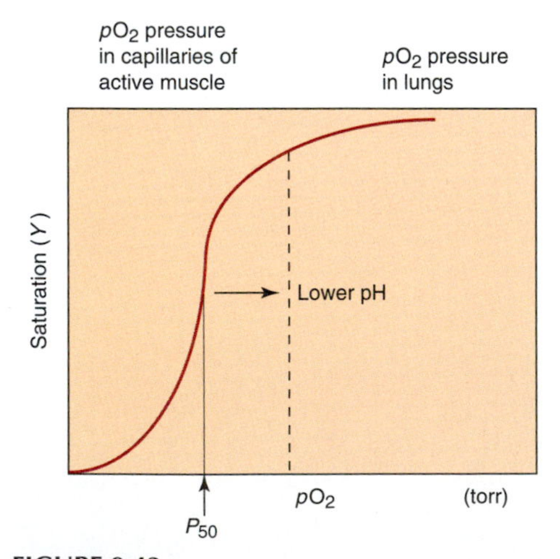

FIGURE 9.42
Change in oxygen–hemoglobin saturation curve to higher P_{50} value with decrease in pH (increase in [H^+]).

Bohr Effect Protons Link CO_2 Transport from Tissues to Lung with O_2 Transport from Lung to Tissues

Metabolizing cells utilize O_2 and produce CO_2. The CO_2 produced by the cells diffuses into the surrounding blood and enters the red blood cell. There the CO_2 is rapidly converted by the enzyme **carbonic anhydrase** to carbonic acid. The conversion of CO_2 to carbonic acid is a simple hydration reaction (Eq. 9.11).

$$CO_2 + H_2O \rightleftharpoons H_2CO_3 \tag{9.11}$$

$$H_2CO_3 \rightleftharpoons HCO_3^- + H^+ \tag{9.12}$$

In turn, the carbonic acid spontaneously dissociates to HCO_3^- and a H^+ (Eq. 9.12). The HCO_3^- diffuses out of the red blood cell and is carried to the lung by the plasma (Figure 9.43). The transport of CO_2 from the tissues to the lung as bicarbonate (HCO_3^-) is known as **isohydric transport.** Approximately 70–80% of cellular produced CO_2 is transported to the lung by this mechanism.

The hydrogen ion that dissociates from the carbonic acid is a major contributor to the increased acidity of the blood around rapidly metabolizing tissue. Fortunately, the pH of the blood is prevented from becoming too acidic due to the binding of protons by hemoglobin, according to Eq. 9.10. The binding of protons to hemoglobin also forces $Hb(O_2)_4$ to dissociate its O_2 by the equilibrium of Eq. 9.10. The O_2 then diffuses out of the red blood cell to the cells. Hemoglobin thus participates in the regulation of blood pH by binding the hydrogen ions released from carbonic acid. The protons are bound to deoxy-Hb (H^+Hb) by acid/base side chain groups, such as the imidazolium side chains of the two His 146(β) of the hemoglobin molecule. The deoxy H^+Hb is transported to the lungs in the red blood cell.

In the lungs, the high pO_2 levels promote the conversion of the deoxy (T conformation) Hb to oxy (R conformation) Hb. On conversion to the oxy-Hb, protons dissociate (Eq. 9.10). The protons then combine with HCO_3^- molecules, which diffuse back into the red blood cells from the plasma, to reform carbonic acid (H_2CO_3, reverse of Eq. 9.12). The enzyme carbonic anhydrase, which catalyzed the hydration of CO_2 to H_2CO_3 at the tissues, now catalyzes the conversion of the H_2CO_3 formed in the lung to CO_2 and H_2O (reverse of Eq. 9.11). The CO_2 diffuses out of the blood into the lung alveoli and is expired into air (Figure 9.43).

A second mechanism of CO_2 transport, accounting for 15–20% of the CO_2 transported from tissues to lungs, is **carbamino-hemoglobin.** This transport mechanism makes use of the spontaneous reaction of CO_2 with the NH_2-terminal amino groups of Hb polypeptide chains to form carbamino-Hb (Eq. 9.13). This reaction also produces a H^+.

$$HbNH_2 + CO_2 \rightleftharpoons HbNHCO_2^- + H^+ \qquad \textbf{(9.13)}$$

(a) Red Blood Cell in Capillaries of Tissues

CO_2 from cells of tissues → CO_2 + H_2O (carbonic anhydrase) → H_2CO_3 + $Hb(O_2)_4$ → HCO_3^- + H^+Hb + $4O_2$ → O_2 to cells of tissues; HCO_3^- → plasma

(b) Red Blood Cell in Capillaries of Lung

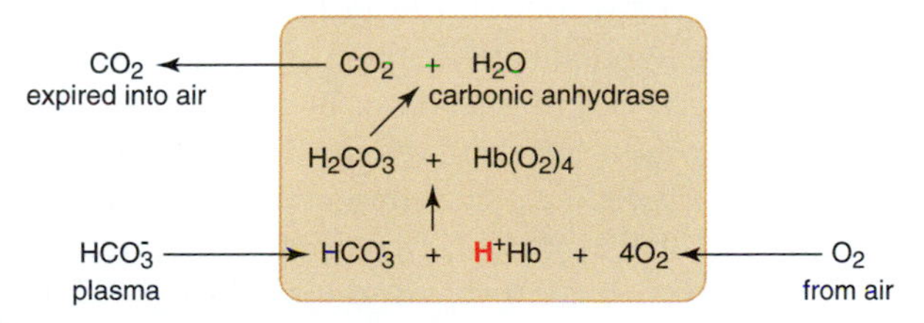

FIGURE 9.43
The isohydric transport of CO_2 as bicarbonate.
(a) Reactions in the red blood cell at the tissues. CO_2 diffuses into red blood cell from tissues and is converted into carbonic acid by carbonic anhydrase. The carbonic acid spontaneously dissociates into H^+ and HCO_3^- (Eq. 9.12). The proton that dissociates from carbonic acid (boldface H^+) binds to deoxy-Hb forcing the O_2-Hb equilibrium from oxyhemoglobin to deoxy-hemoglobin (Eq. 9.10) with the dissociation of O_2, which diffuses out of the red blood cell to the tissues. The HCO_3^- diffuses out of the red blood cell and is transported in plasma to the lung (outside of cell). *(b)* Reactions in the red blood at the level of the lung. In the lung, the high O_2 pressure forces the reactions in the opposite direction. Reactions are the reverse of those in the capillaries.

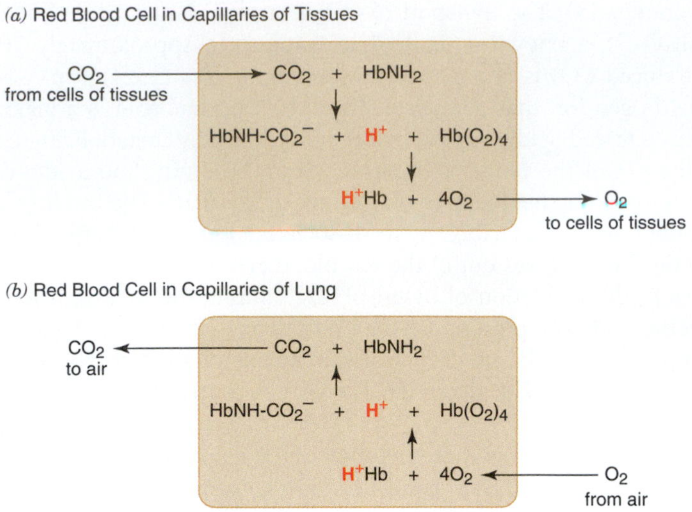

(*a*) Red Blood Cell in Capillaries of Tissues

CO_2 from cells of tissues → CO_2 + $HbNH_2$ → $HbNH\text{-}CO_2^-$ + H^+ + $Hb(O_2)_4$ → H^+Hb + $4O_2$ → O_2 to cells of tissues

(*b*) Red Blood Cell in Capillaries of Lung

O_2 from air → H^+Hb + $4O_2$ → $HbNH\text{-}CO_2^-$ + H^+ + $Hb(O_2)_4$ → CO_2 + $HbNH_2$ → CO_2 to air

FIGURE 9.44
Transport of CO_2 as carbamino-hemoglobin.
(*a*) Reactions in the red blood cell at the tissues. CO_2 diffuses into red blood cell and reacts with the NH_2-terminal amino group of hemoglobin chains to form carbamino-hemoglobin. The reaction releases a proton (boldface H^+), which promotes the dissociation of O_2 from hemoglobin. The O_2 diffuses out of the red blood cell to the tissues. (*b*) Reactions in the red blood cell at the level of the lung. In the lung, the high O_2 pressure forces the reactions in the opposite direction, leading to the expiration of CO_2 from the lung. Reactions are the reverse of those in the capillaries.

The increase in H^+ due to carbamino-Hb formation promotes the dissociation and delivery of O_2 to the actively metabolizing cells (Eq. 9.10). In the lung, the high pO_2 concentration generates the formation of oxy-Hb with dissociation of H^+ (Eq. 9.10). The free H^+ promotes the dissociation of the carbamino group from Hb to reform CO_2 (reverse of Eq. 9.13), which is expired from the lung (Figure 9.44).

2,3-Bisphosphoglycerate (BPG) in Red Blood Cells Modulates Oxygen Dissociation from Hemoglobin

An important modulator of the hemoglobin equilibrium is **2,3-bisphosphoglycerate (BPG)** (Figure 9.45). An alternative chemical name for BPG, often found in the older literature, is **2,3-diphosphoglycerate,** abbreviated **DPG.** BPG is an intermediate in a minor pathway for glucose metabolism and is present in small amounts in all cells. However, in the red blood cell this metabolic pathway is highly active, and BPG concentrations are approximately equimolar to that of hemoglobin. BPG has an affinity to the deoxy (T) conformation of hemoglobin and not the oxy (R) conformation. Binding of BPG to Hb stabilizes the T conformation, and therefore the presence of BPG will increase the concentration of the T conformation relative to the R conformation. As BPG binds tightly only to the T conformation, a BPG molecule dissociates as deoxy-Hb is converted to oxy-Hb (Eq. 9.14).

$$H^+BPG \cdot Hb + 4O_2 \rightleftharpoons Hb(O_2)_4 + BPG + nH^+ \qquad \textbf{(9.14)}$$

Increased concentrations of BPG will force the equilibrium to the left, in order to reestablish the equilibrium ratio of reactants and products. In contrast, lower than normal concentrations of BPG increase the concentrations of the oxy-Hb relative to deoxy-Hb and free O_2. High BPG concentrations perturb the oxygen–Hb dissociation plot to the right with a corresponding increase in the P_{50} value (Figure 9.46). At lower than normal concentrations of BPG, the P_{50} value is lower than for normal Hb and the oxygen saturation plot will be to the left of the oxygen–Hb dissociation plot at normal concentrations of BPG.

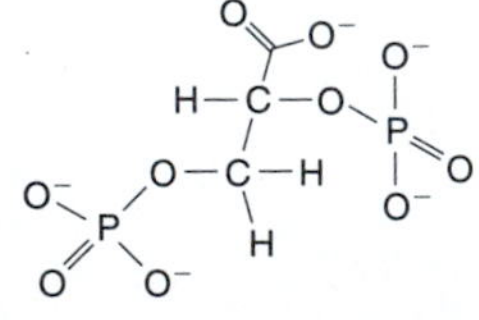

FIGURE 9.45
Structure of 2,3-bisphosphoglycerate (BPG). Molecule has a charge of −5 at pH 7.4.

BPG binds within a pocket in the hemoglobin molecule formed at the β_1-subunit–β_2-subunit interface. This site contains eight positively charged groups [2 × His 143(β), 2 × Lys 82(β), 2 × His 2(β), 2 × NH_2-terminal ammonium from amino acid 1(β)] (Figure 9.47). The BPG molecule contains a charge of −5 (Figure 9.45) and is strongly attracted to the positive charges of the β–β interface binding site. In the R conformation of hemoglobin, the size of the binding pocket is decreased, resulting in an inability of the BPG molecule to easily fit into the binding pocket.

Conditions that cause hypoxia (deficiency of oxygen) such as anemia, smoking, and high altitude will increase BPG levels in the red blood cells. In turn, conditions leading to hyperoxia will result in lower levels of BPG. Changes in red blood cell levels of BPG are not immediate, but rather occur over hours and days to compensate for chronic changes in the pO_2 levels.

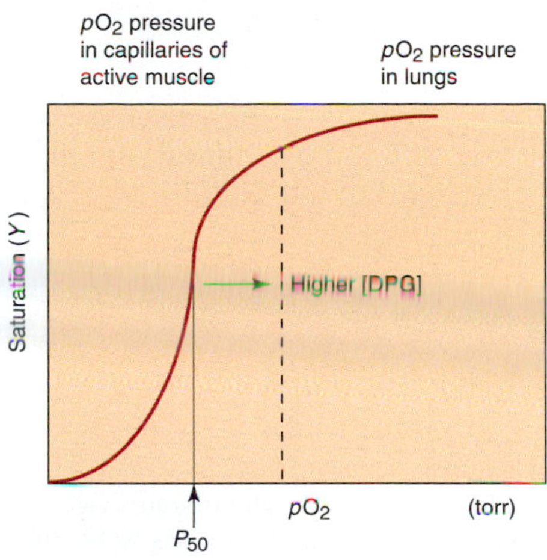

FIGURE 9.46
Change in oxygen–hemoglobin saturation curve to higher P_{50} value with increase in BPG concentration.

Hemoglobin Delivers Nitric Oxide (NO) to the Capillary Wall of Tissues Where It Promotes O_2 Delivery

Hemoglobin reversibly binds **nitric oxide** (NO), a potent vasodilator, which has a very short lifetime in blood. By binding NO, hemoglobin sequesters the NO molecule from rapid destruction. Hb dissociates NO by transferring the bound NO to

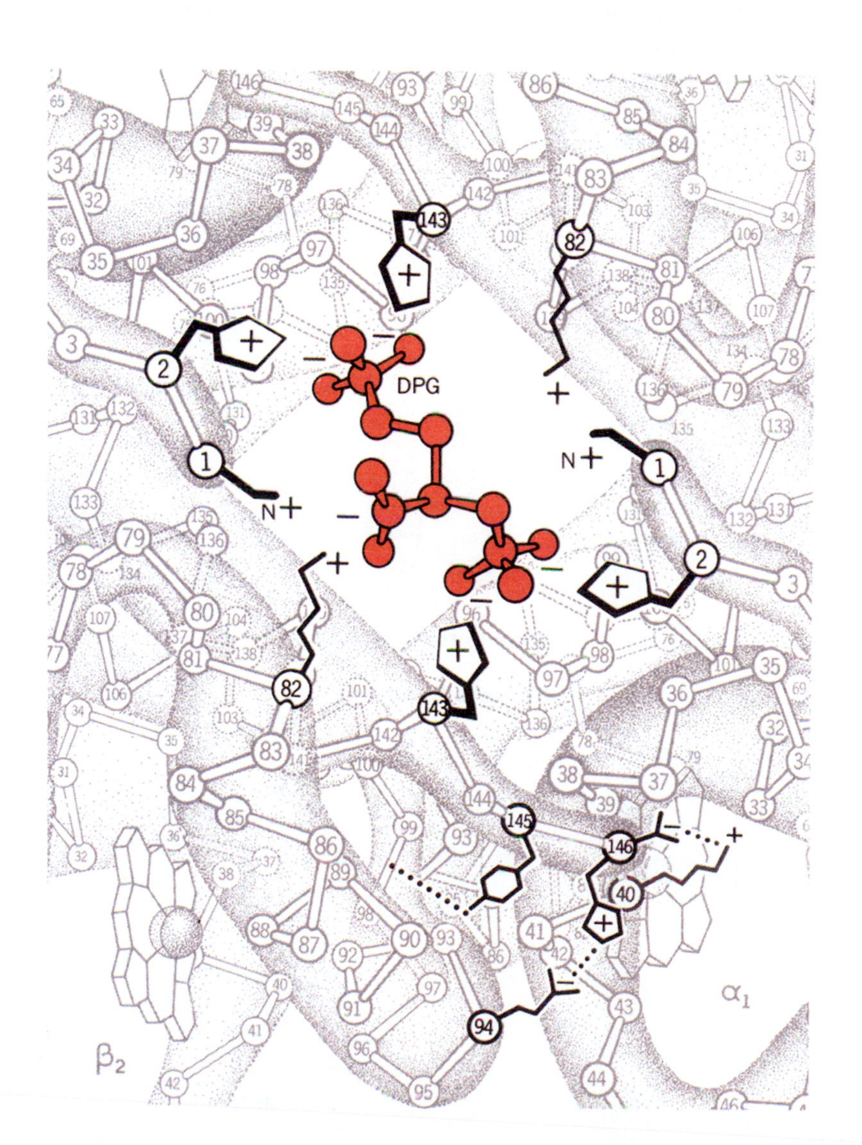

FIGURE 9.47
2,3-Bisphosphoglycerate-binding site at the β–β interface of deoxy-hemoglobin.
Shown are the positively charged side chains of two Val 1(β) amino-terminus ammonium, His 2(β) imidazolium, Lys 82(β) ε-ammonium, and His 143(β) imidazolium. The negatively charged 2,3-bisphosphoglycerate (designated as DPG) binds in the middle of the ring of positively charged groups.
Reprinted with permission from Dickerson, R. E. and Geis, I. Hemoglobin: Structure, Function, Evolution, and Pathology. Menlo Park, CA: Benjamin-Cummings, 1983. Illustration by Irving Geis, Geis Archives Trust. Rights owned by Howard Hughes Medical Institute.

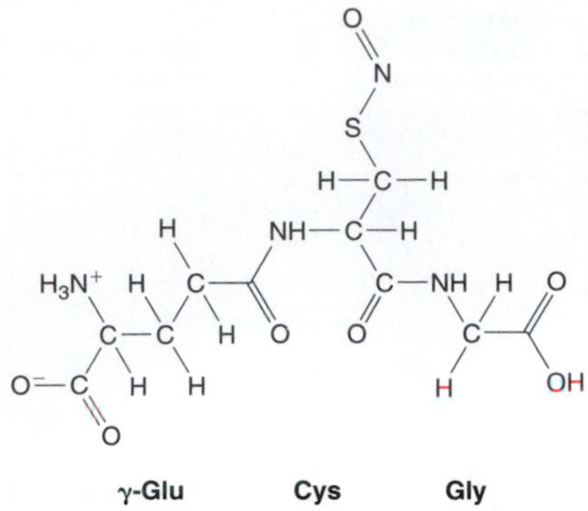

γ-Glu Cys Gly

FIGURE 9.48
The X-S-NO transporter glutathione (γ-glutamyl-cysteinyl-glycine) transports NO bound to the sulfhydryl side chain of its cysteine.

small sulfhydryl (X-SH) molecules as hemoglobin changes from the oxy (R) conformation to the deoxy (T) conformation. A common X-SH molecule is glutathione (Figure 9.48). The released NO in the form of X-S-NO also stabilizes NO against rapid degradation and allows efficient delivery of a bioactive NO equivalent (X-S-NO) to the NO receptors in cells of the blood vessel wall, promoting the relaxation of the vascular wall. This facilitates the transfer of gases between the blood and the cells of the tissues. The molar concentration of NO in blood is 1/70th the concentration of Hb. However, the low concentration of the NO is physiologically significant due to the potent biological activity of the NO (X-S-NO).

NO is first captured by the Fe^{2+} at the heme sites and, secondly, transferred to the sulfhydryl (S atom) of the β-chain cysteine 93 (βCys^{93}) residues. The heme iron preferentially binds NO when in the T conformation. The NO is transferred from the heme-binding sites to βCys^{93} when hemoglobin is in the R conformation. Then when the R conformation changes again to the T conformation to deliver O_2 to the tissues, the βCys^{93}-S-NO transfers the NO to small S-XH molecules such as glutathione (Figures 9.48 and 9.49). The net effect is the catalysis by hemoglobin of the conversion of unstable free NO to stable X-S-NO, in order to deliver the potent vasodilator to blood vessel targets, where needed.

Figure 9.50 shows the region of the hemoglobin protein that contains the modified βCys^{93}-S-NO. In the T conformation the S-NO side chain of the βCys^{93} is on the outside of the molecule and accessible to glutathione molecules. This allows *trans*-nitrosylation between the βCys^{93}-S-NO and X-SH molecules in solution. In contrast, in the R conformation the βCys^{93}-SH is pointed toward the heme iron and is inaccessible to the outside solvent. In the buried R-conformation position, the distance is optimal for transfer of NO from its iron heme-binding site to the βCys^{93}-SH. Thus the conformational changes in the hemoglobin molecule, induced by changes in oxygen pressure between the lungs and tissues, regulates NO uptake and release from the βCys^{93}. The net effect is delivery of bioactive NO to the capillaries of the tissues under low oxygen tension.

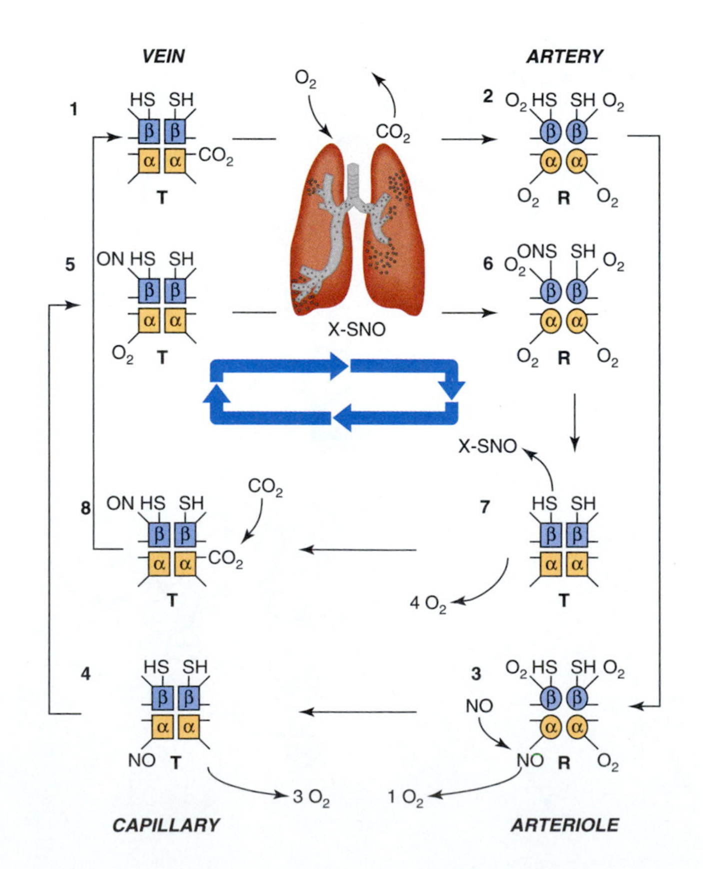

FIGURE 9.49
Binding and release of NO by hemoglobin during the respiratory cycle.
The model shows the binding and dissociation of NO, O_2, and CO_2 as a hemoglobin molecule makes two complete cycles through the circulation. The first cycle involves intermediates 1–4 and the second cycle intermediates 5–8. The T and R conformations are shown and the SH groups are from the side chain of βCys^{93}. The NO is either directly bound to a heme iron or to the βCys^{93} SH. The key steps in the NO transport are its initial binding to a heme in intermediate 3 and transfer from a β-subunit heme to βCys^{93} in intermediate 6 (R conformation), and its transfer to a small molecular thiol X-SH in intermediate 7 (T conformation) when hemoglobin is converted from R to T. The hemoglobin molecule depicted may be only 1 in 1000 hemoglobin molecules that are cycling, due to the low relative molar concentration of NO in blood.
Redrawn from Gross, S. S. and Lane, P. Proc. Natl. Acad. Sci. USA *96:9967, 1999.*

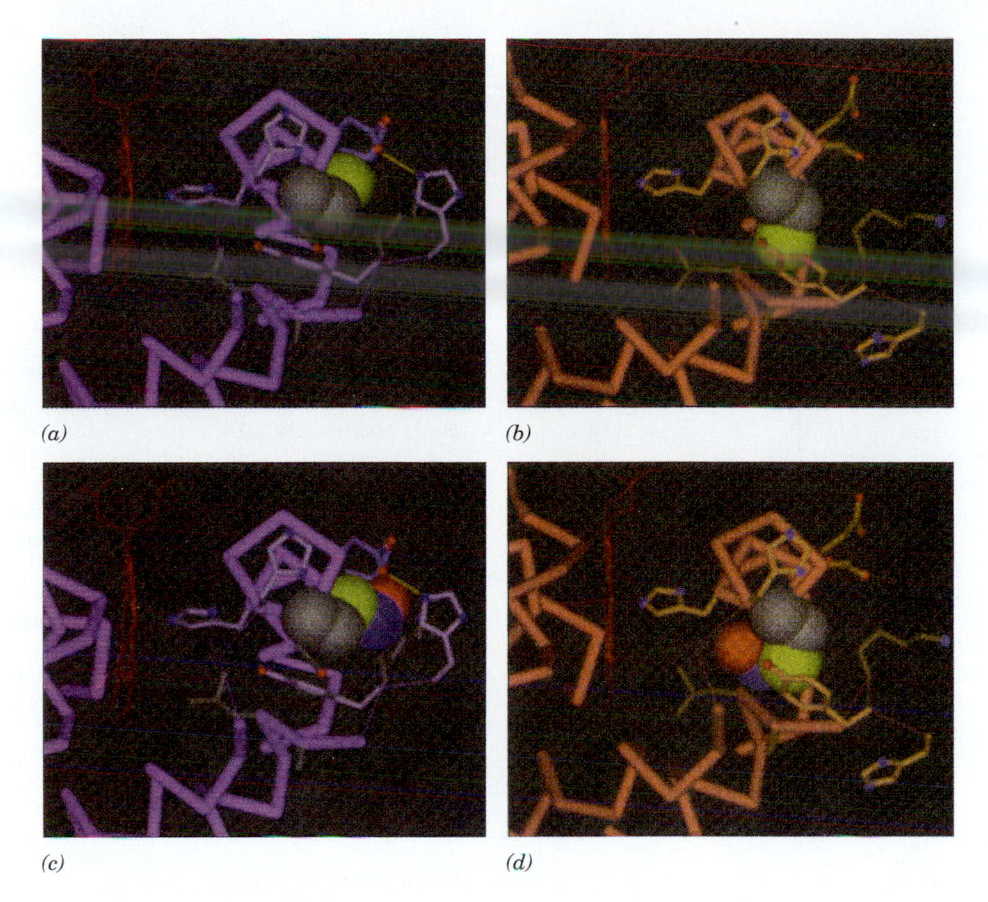

FIGURE 9.50
Structure of βCys[93] and βCys[93]-S-NO in the T and R conformations. In all structures the heme (red) edge is pointed toward reader and is shown to be bonded to proximal histidine imidazole. The two carbons (grey) and sulfur (yellow-green) of the Cys 93 side chain are shown with space-filling models. In panels (*c*) and (*d*), NO is bonded to the sulfur of βCys[93] with the N atom colored blue and the O atom colored red. (*a*) The deoxy (T) conformation with the βCys[93] side chain (—CH_2—SH) on the surface of the molecule away from the heme. The cysteine side chain is prevented from entering the heme-binding site by the βHis[146]-imidazolium to βAsp[94]-carboxylate hydrogen-bonded ion pair on the upper right and behind the βCys[93]-SH side chain. (*b*) The oxy(R) conformation with the βCys[93] side chain pointed toward the heme away from the solvent on the outside of the molecule. The βHis[146] to βAsp[94] salt bridge is broken in the R conformation, which allows the folding of the βCys-SH toward the heme pocket. (*c*) Model of βCys[93]-SNO in the deoxy (T) conformation. SNO is positioned on the outside, accessible to reaction with X-SH small molecules in the solvent. As in (*a*), the cysteine side chain is prevented from entering heme site by the βHis[146]–βAsp[94] ion pair. (*d*) Model of βCys[93]-SNO in oxy (R) conformation. SNO is buried near the heme and away from the outside solvent. This conformation facilitates the transfer of NO from the heme iron to the cysteine SH and prevents the reaction of the βCys[93]-SNO with X-SH molecules in the solvent.

Reprinted with permission from Stamler, J. S., Jia, L., Eu, J. P., McMahon, T. J., Demchenko, I. T., Bonaventura, J., Gernert, K., and Piantadosi, C. A. Science *276:2034, 1997.*

BIBLIOGRAPHY

Immunoglobulins

Alzari, P. M., Lascombe, M. B., and Poljak, R. J. Structure of antibodies. *Annu. Rev. Immunol.* 6:555, 1988.

Barclay, A. N. Ig-like domains: evolution from simple interaction molecules to sophisticated antigen recognition. *Proc. Natl. Acad. Sci. USA* 96:14672, 1999.

Chothia, C., Lesk, A. M., Tramontano, A., Levitt, M., Smith-Gill, S. J., Air, G., Sheriff, S., Padlan, E. A., Davies, D., Tulip, W. R., Colman, P. M., Spinelli, S., Alzari, P. M., and Poljak, R. J. Conformations of immunoglobulin hypervariable regions. *Nature* 342:877, 1989.

Davies, D. R., Padlan, E. A., and Sheriff, S. Antibody–antigen complexes. *Acc. Chem. Res.* 26:421, 1993.

Guddat, L. W., Shan, L., Fan, Z.-C., Andersen, K. N., Rosauer, R., Linthicum, D. S., and Edmundson, A. B. Intramolecular signaling upon complexation. *FASEB J.* 9:101, 1995.

Hunkapiller, T. and Hood, L. Diversity of the immunoglobulin gene superfamily. *Adv. Immunol.* 44:1, 1989.

Padlan, E. A. Anatomy of the antibody molecule. *Mol. Immunol.* 31:169, 1994.

Rini, J. M., Schultze-Gahmen, U., and Wilson, I. A. Structural evidence for induced fit as a mechanism for antibody–antigen recognition. *Science* 255:959, 1992.

Serine Proteases

Birk, Y. Proteinase inhibitors. In: A. Neuberger and K. Brocklehurst (Eds.), *Hydrolytic Enzymes.* Amsterdam: Elsevier, 1987, p. 257.

Liebman, M. N. Structural organization in the serine proteases. *Enzyme* 36:115, 1986.

Neurath, H. Proteolytic processing and physiological regulation. *Trends Biochem. Sci.* 14:268, 1989.

Perona, J. J. and Craik, C. S. Structural basis of substrate specificity in the serine proteases. *Protein Sci.* 4:337, 1995.

Polgar, L. Structure and function of serine proteases. In: A. Neuberger and K. Brocklehurst (Eds.), *Hydrolytic Enzymes,* Series in New Comprehensive Biochemistry, Vol. 16. Amsterdam: Elsevier, 1987, 159.

DNA-Binding Proteins

Bewley, C. A., Gronenborn, A. M., and Clore, G. M. Minor groove-binding architectural proteins: structure, function, and DNA recognition. *Annu. Rev. Biophys. Biomol. Struct.* 27:105, 1998.

Cho, Y., Gorina, S., Jeffrey, P. D., and Pavletich, N. P. Crystal structure of a p53 tumor suppressor–DNA complex: understanding tumorigenic mutations. *Science* 265:346, 1994.

Choo, Y. and Klug, A. Physical basis of a protein–DNA recognition code. *Curr. Opin. Struct. Biol.* 7:117, 1997.

Ghosh, G., Van Duyne, G., Ghosh, S., and Sigler, P. B. Structure of NF-κB p50 homodimer bound to a κB site. *Nature* 373:303, 1995.

Landschulz, W. H., Johnson, P. F., and McKnight, S. L. The leucine zipper: a hypothetical structure common to a new class of DNA binding proteins. *Science* 240:1759, 1988.

Lee, M. S., Gippert, G. P., Soman, K. V., Case, D. A., and Wright, P. E. Three-dimensional solution structure of a single zinc finger DNA-binding domain. *Science* 245:635, 1989.

Nikolov, D. B., Chen, H., Halay, E. D., Usheva, A. A., Hisatake, K., Lee, D. K., Roeder R. G., and Burley, S. K. Crystal structure of a TFIIB-TBP-TATA-element ternary complex. *Nature* 377:119, 1995.

Pavletich, N. P. and Pabo, C. O. Zinc finger–DNA recognition: crystal structure of a Zif-268-DNA complex at 2.1 Å. *Science* 252:809, 1991.

Hemoglobin

Website of interest: The Globin Gene Server database of sequence alignments and experimental results for the β-like globin gene cluster of mammals. http://globin.cse.psu.edu/

Baldwin, J. and Chothia, C. Haemoglobin: the structural changes related to ligand binding and its allosteric mechanism. *J. Mol. Biol.* 129:175, 1979.

Dickerson, R. E. and Geis, I. *Hemoglobin: Structure, Function, Evolution and Pathology.* Menlo Park, CA: Benjamin-Cummings, 1983

Gross, S. S. and Lane, P. Physiological reactions of nitric oxide and hemoglobin: a radical rethink. *Proc. Natl. Acad. Sci. USA* 96:9967, 1999.

Hsia, C. C. W. Mechanisms of disease: respiratory function of hemoglobin. *N. Engl. J. Med.* 338:239, 1998.

Perutz, M. Hemoglobin structure and respiratory transport. *Sci. Am.* 239:92, 1978.

Perutz, M. F., Fermi, G., and Shih, T.-B. Structure of deoxy cowtown [His HC3(146)beta to Leu]: origin of the alkaline Bohr effect and electrostatic-interactions in hemoglobin. *Proc. Natl. Acad. Sci. USA* 81:4781, 1984.

Perutz, M. F., Wilkinson, A. J., Paoli, M., and Dodson, G. G. The stereochemical mechanism of the cooperative effects in hemoglobin revisited. *Annu. Rev. Biophys. Biomol. Struct.* 27:1, 1998.

Veeramachaneni, N. K., Harken, A. H., and Cairns, C. B. Clinical implications of hemoglobin as a nitric oxide carrier. *Arch. Surg.* 134:434, 1999.

QUESTIONS | C. N. ANGSTADT

Multiple Choice Questions

1. Haptens:
   A. can function as antigens.
   B. strongly bind to antibodies specific for them.
   C. may be macromolecules.
   D. never act as antigenic determinants.
   E. can directly elicit the production of specific antibodies.

2. In the three-dimensional structure of immunoglobulins:
   A. β-sheets align edge to edge.
   B. in each chain (H and L) the C and V regions fold onto one another, forming C–V associations.
   C. C_L–V_L associations form the complementary sites for binding antigens.
   D. free —SH groups are preserved to function in forming tight covalent bonds to antigens.
   E. hinge domains connect globular domains.

3. The active sites of all serine proteases contain which of the following amino acid residues?
   A. asparagine
   B. γ-carboxyglutamate
   C. histidine
   D. lysine or arginine
   E. threonine

Refer to the following for Questions 4–7:

A. helix–turn–helix
B. leucine zipper
C. zinc finger
D. two-domain "saddle" of β-pleated sheets
E. immunoglobulin fold

4. Not a DNA-binding motif.
5. Motif in TATA-binding protein (TBP).
6. Contains a single α-helix.
7. Two antiparallel β-sheets aligned face-to-face.

8. Hemoglobin and myoglobin both have all of the following characteristics EXCEPT:
   A. consist of subunits designed to provide hydrogen bonds to and nonpolar interaction with other subunits.
   B. highly α helical.
   C. bind one molecule of heme per globin chain.
   D. bind heme in a hydrophobic pocket.
   E. can bind one O_2 per heme.

9. All of the following are believed to contribute to the stability of the deoxy or T conformation of hemoglobin EXCEPT:
   A. the larger ionic radius of the six-coordinated ferrous ion as compared to the five-coordinated ion.
   B. unstrained steric interaction of His F8 with the porphyrin ring when iron is above the plane.
   C. interactions between the FG corner of one subunit and the C helix of the adjacent subunit.
   D. ionic interactions.

10. When hemoglobin is converted from the deoxy form to oxyhemoglobin:
    A. it becomes more acidic and releases protons.
    B. carbamino formation is promoted.
    C. binding of BPG is favored.
    D. bound NO is transferred to glutathione.
    E. all of the above are correct.

Questions 11 and 12: Immunoglobulins (antibody molecules) are a family of proteins that are produced in response to invasion by foreign substances, initiating a process by which the foreign substance can be eliminated from the organism. Binding of antibodies to antigens on the cell membrane of the invading organism, like bacteria, activates the complement cascade. Activation of the complement system leads to the production of proteins that stimulate phagocytosis by neutrophils and macrophages. An individual with either an immunoglobulin deficiency or a deficiency of one of the complement proteins is susceptible to recurrent infections.

11. In immunoglobulins all of the following are true EXCEPT:
    A. there are four polypeptide chains.
    B. there are two copies of each type of chain.
    C. all chains are linked by disulfide bonds.
    D. carbohydrate is covalently bound to the protein.
    E. immunoglobulin class is determined by the C_L regions.

12. IgG:
    A. is found primarily in mucosal secretions.
    B. is the first antibody elicited when a foreign antigen is introduced into the host's plasma.
    C. has the highest molecular weight of all the immunoglobulins.
    D. contains carbohydrate covalently attached to the H chain.
    E. plays an important role in allergic responses.

Questions 13 and 14: When a blood vessel is injured, the coagulation process is initiated to prevent loss of blood and maintain the integrity of the circulatory system. Coagulation is a process in which zymogens are converted to active serine proteases in a stepwise, cascade process with the final result the production of a cross-linked fibrin clot. In a myocardial infarction, the process can be activated to such an extent that blood flow to the heart muscle is decreased. The fibrinolysis pathway to dissolve fibrin clots also involves activating zymogens to active serine proteases, the final step being the activation of plasminogen to plasmin, which acts directly on the fibrin clot. One of the current treatments for myocardial infarctions is rapid administration of t-PA (tissue plasminogen activator); actually recombinant t-PA, produced by gene cloning technology, is used.

13. Serine proteases:
    A. hydrolyze peptide bonds involving the carboxyl groups of serine residues.
    B. are characterized by having several active sites per molecule, each containing a serine residue.
    C. are inactivated by reacting with one molecule of diisopropylfluorophosphate per molecule of protein.
    D. are exopeptidases.
    E. recognize only the amino acids that contribute to the bond to be broken.

14. All of the following are characteristic of serine proteases as a class EXCEPT:
    A. only one serine residue is catalytically active.
    B. natural protein substrates and inhibitors bind very tightly to the protease.
    C. the genes that code for them are analogously organized.
    D. catalytic units exhibit two structural domains of dramatically different size.
    E. conversion of zymogen to active enzyme usually involves one or more hydrolytic reactions.

Problems

15. One of the adaptations to high altitude is an increase in the concentration of BPG. What effect does this have on a saturation versus pO_2 curve? Why does increasing [BPG] increase the delivery of O_2 to tissues?

16. How did papain hydrolysis of an immunoglobulin molecule confirm that V_L–V_H comprises the antigenic site of an immunoglobulin?

17. Both myoglobin and hemoglobin consist of globin (protein) bound to a heme prosthetic group. One heme group binds one O_2. Why is the oxygen saturation curve (saturation versus pO_2) of myoglobin a rectangular hyperbola while that of hemoglobin is sigmoidal?

ANSWERS

1. **B** Haptens are small molecules and cannot alone elicit antibody production; thus they are not antigens. They can act as antigenic determinants if covalently bound to a larger molecule, and free haptens may bind strongly to the antibodies thereby produced.
2. **E** See Figure 9.2. A: The β-sheets align face-to-face. B: The V and C regions are adjacent to each other. C: The complementarity regions are the variable regions of both the heavy and light chains [V(H)–V(L)]. D: Antigen binding is noncovalent.
3. **C** Histidine participates in the catalytic mechanism. A: Aspartate, not asparagine, is involved. B: γ-Carboxyglutamate is essential to some of the serine proteases, but it is not at the active site. D: These are the substrate specificities of the trypsin-like proteases. E: This is not involved.
4. **B** The leucine zipper binds two subunits in a head-to-head manner but does not itself interact with DNA.
5. **D** The saddle fits over the DNA double helix at the TATA box.
6. **C** The zinc finger has an α-helix and a β-sheet joined by a loop.
7. **E** This structure is found in immunoglobulins but also in other proteins—for example, certain transcription factors.
8. **A** Hemoglobin has four chains and four oxygen-binding sites, whereas myoglobin has one chain and one oxygen-binding site. Each oxygen-binding site is a heme.
9. **A** Six-coordinated ferrous ion has a smaller ionic radius than the five-coordinated species and just fits into the center of the porphyrin ring without distortion.
10. **A** Positively charged histidine is no longer stabilized by close proximity to an aspartate carboxylate group. This is the Bohr effect. B: Increased H^+ favors dissociation of carbamino groups, which is what happens in the lung. C: BPG does not bind effectively to oxyhemoglobin because the binding pocket is too small. D: In oxyhemoglobin, NO is bound to cysteine and it gets transferred to glutathione when O_2 is released (R form reverts to T form).
11. **E** The C_H regions determine class. A: There are two copies of each of two types of polypeptide chain.
12. **D** All immunoglobulins are glycoproteins. A: IgA is associated with mucosal secretions. B: IgM arises first; IgG takes longer to appear but eventually is present in higher concentration. C: IgM has the highest molecular weight. E: IgE plays an important role in allergic responses.

13. **C** This is the distinguishing characteristic of the serine proteases, and of the serine hydrolases in general. A: They have various specificities. B: There is only one active site per molecule. D: They are all endopeptidases. E: An "extended active site" containing the hydrolyzable bond and about four amino acids on either side is responsible for specificity.
14. **D** The domains are of about equal size.
15. Binding of BPG to hemoglobin stabilizes the T conformation, so for any given amount of O_2, less O_2 will be bound to hemoglobin. This shifts the saturation curve to the right and increases P_{50}. (Verify this for yourself by drawing the curve.) The equilibrium shown illustrates that increasing BPG causes the release of O_2, which must happen in order for O_2 to enter tissues.

$$H^+BPG \cdot Hb + 4O_2 \rightleftharpoons Hb(O_2)_4 + BPG + nH^+ \qquad \mathbf{(9.14)}$$

16. Papain hydrolyzes at the hinge region giving two identical products, F_{ab}, which consist of the N-terminal portion of the H chain plus the whole L chain. These are variable domains, V_L and V_H. The third product, F_c, is the C-terminal portion of the two heavy chains joined together covalently. F_{ab} binds antigen with the same affinity as the whole molecule, while F_c has no binding affinity for antigen. If F_{ab} is separated into two chains (one L and one part of H), neither chain alone binds antigen. Therefore the antigen-binding site requires both V_L and V_H working together.
17. Myoglobin is a monomer with one globin and one heme binding one O_2. O_2 binding is a simple equilibrium with a Hill coefficient of 1. Hemoglobin is a tetramer with four O_2 bound to the four hemes. Hemoglobin without oxygen is in the T conformation in which binding of O_2 is difficult. Thus the curve starts with a slow increase as pO_2 increases. Binding of the first O_2 shifts the conformation of the whole molecule to favor the R form, which binds O_2 more readily. Thus the slope rises steeply giving it a sigmoidal shape. This is an example of cooperativity. The Hill coefficient for hemoglobin is 2.8.

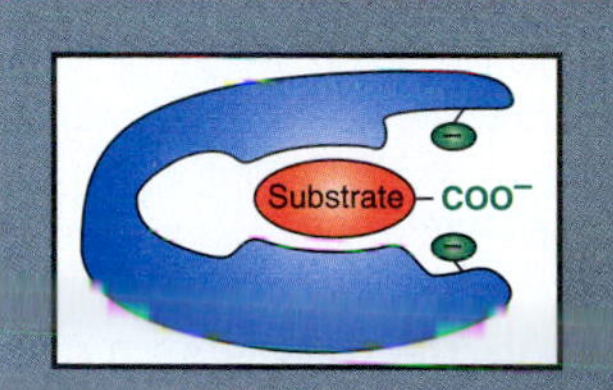

10

ENZYMES: CLASSIFICATION, KINETICS, AND CONTROL

J. Lyndal York

10.1 | GENERAL CONCEPTS

Enzymes are proteins that function in the acceleration of chemical reactions in biological systems. Many reactions required for the living cell would not proceed fast enough at the pH and temperature of the body without enzymes. The term defining the speed of a chemical reaction, whether catalyzed or uncatalyzed, is **rate** or **velocity.** Rate is the change in amount (moles) of starting materials or products per unit time. Velocity is the change in concentration of starting material or product per unit time. Enzymes increase the rate by acting as catalysts. A **catalyst** increases the rate of a chemical reaction but is not itself changed in the process. An enzyme may become temporarily covalently bound to a molecule being transformed during intermediate stages of the reaction but at the end of the reaction the enzyme will again be in its original form as the product is released.

Two important characteristics of enzyme catalysts are that the enzyme is not changed as a result of catalysis and the enzyme does not change the equilibrium constant of the reaction but simply increases the rate at which the reaction approaches equilibrium. As will be discussed later, it accomplishes the rate increase by lowering the barrier to reaction; that is, it lowers the energy of activation. Therefore a catalyst increases the rate but does not change the thermodynamic properties of the system with which it is interacting.

Several terms need to be defined before we enter into a discussion of the mechanism of enzyme action. An **apoenzyme** is the protein part of an enzyme without any cofactors or prosthetic groups that may be required for the enzyme to be functional. The apoenzyme is catalytically inactive. Not all enzymes require cofactors or prosthetic groups. **Cofactors** are small organic or inorganic molecules that an apoenzyme requires for its activity. For example, in lysine oxidase copper is loosely bound but is required for the enzyme to be active. A prosthetic group is similar to a cofactor but is tightly bound to an apoenzyme. For example, in the cytochromes, the heme prosthetic group is very tightly bound and requires strong acids to disassociate it from the apocytochrome. Addition of a cofactor or prosthetic group to the apoprotein yields the **holoenzyme,** that is, the active enzyme. The molecule acted upon by the enzyme to form product is the **substrate.** Since most reactions are reversible, the products of the forward reaction become substrates for the reverse reaction.

Enzymes have a great deal of specificity. For example, glucose oxidase will oxidize glucose but not galactose. The specificity resides in a particular region on the enzyme surface, the **substrate-binding site,** a particular arrangement of amino acid side chains in the polypeptide that is specially formulated to bind a specific substrate. Some enzymes have broad specificity; glucose, mannose, and fructose are phosphorylated by hexokinase, whereas glucokinase is specific for glucose. The substrate-binding site may contain the active site. In some cases, however, the active site may not be within the substrate-binding site but may be contiguous to it in the primary sequence. In other instances the active-site residues lie in distant regions of the primary sequence but are brought adjacent to the substrate-binding site by folding in the tertiary structure. The **active site** contains the machinery, in the form of particular amino acid side chains, involved in catalyzing the reaction.

Some enzymes have variants called **isoenzymes (isozymes)** that catalyze the same chemical reaction. Isoenzymes are electrophoretically distinguishable because of mutations in one or more amino acids in noncritical areas of the protein.

Some enzymes have a region of the molecule, the **allosteric** site, that is not at the active site or substrate-binding site but is a unique site where small molecules bind and effect a change in the substrate-binding site or the activity occurring in the active site. The binding of a specific small molecule at the allosteric site causes a change in the conformation of the enzyme. This can cause the active site to become either more active or less active by increasing or decreasing the affinity of the binding site for substrate. Such interactions regulate the enzyme's activity and are discussed in detail on page 436.

10.2 | CLASSIFICATION OF ENZYMES

The International Union of Biochemistry and Molecular Biology (IUBMB) has established a system whereby all enzymes are classified into six major classes, each subdivided into subclasses that are further subdivided. In naming an enzyme, the substrates are stated first, followed by the reaction type to which the ending -ase is affixed. For example, alcohol dehydrogenase is alcohol:NAD^+ oxidoreductase because it catalyzes an oxidation–reduction reaction and the electron donor is an alcohol and the acceptor is NAD^+. Many common names persist but are not very informative. For example, "aldolase" does not tell much about the substrates, although it does identify the reaction type. Trivial names recognized by the IUBMB and that are in common usage will be used in this chapter. Table 10.1 summarizes the six major classes and subclasses of enzymes.

TABLE 10.1 Summary of the Enzyme Classes and Major Subclasses

1. Oxidoreductases	2. Transferases
Dehydrogenases	Transaldolase and transketolase
Oxidases	Acyl, methyl, glucosyl, and phosphoryltransferases
Reductases	Kinases
Peroxidases	Phosphomutases
Catalase	
Oxygenases	
Hydroxylases	
3. Hydrolases	4. Lyases
Esterases	Decarboxylases
Glycosidases	Aldolases
Peptidases	Hydratases
Phosphatases	Dehydratases
Thiolases	Synthases
Phospholipases	Lyases
Amidases	
Deaminases	
Ribonucleases	
5. Isomerases	6. Ligases
Racemases	Synthetases
Epimerases	Carboxylases
Isomerases	
Mutases (not all)	

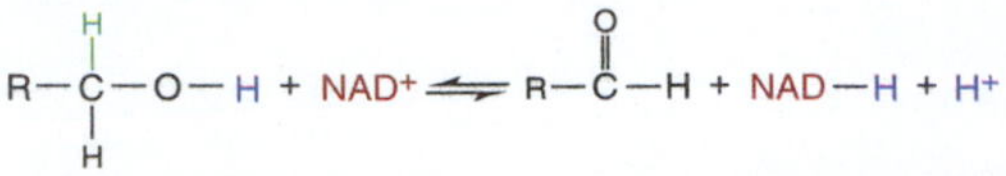

$$R-\underset{H}{\overset{H}{C}}-O-H + NAD^+ \rightleftharpoons R-\overset{O}{\overset{\|}{C}}-H + NAD-H + H^+$$

FIGURE 10.1
Oxidation of ethanol by alcohol dehydrogenase.

Class 1: Oxidoreductases

These enzymes catalyze oxidation–reduction reactions. For example, alcohol:NAD^+ oxidoreductase (a **dehydrogenase**—specifically, alcohol dehydrogenase) catalyzes the oxidation of an alcohol to an aldehyde. Two electrons and two hydrogen atoms are removed from the alcohol to yield an aldehyde. The two electrons in the carbon–hydrogen bond of the alcohol are transferred to the NAD^+, which is reduced (Figure 10.1). NAD^+, whose structure is presented in Figure 10.18, is a cofactor that mediates many biological oxidation–reduction reactions. The redox site in NAD^+ is shown in Figure 10.19. In addition to the alcohol and aldehyde functional groups, dehydrogenases also act on the following functional groups as electron donors: $—CH_2—CH_2—$, $—CH_2—NH_2$, and $—CH{=}NH$, as well as the coenzymes NADH, NADPH, FADH, and FMNH (p. 426).

There are subclasses to the oxidoreductases. **Oxidases** transfer two electrons from the donor to oxygen, yielding hydrogen peroxide (H_2O_2). Usually the flow of electrons from substrate to product is mediated by flavin coenzymes or metals. For example, D-amino acid oxidase catalyzes the overall reaction

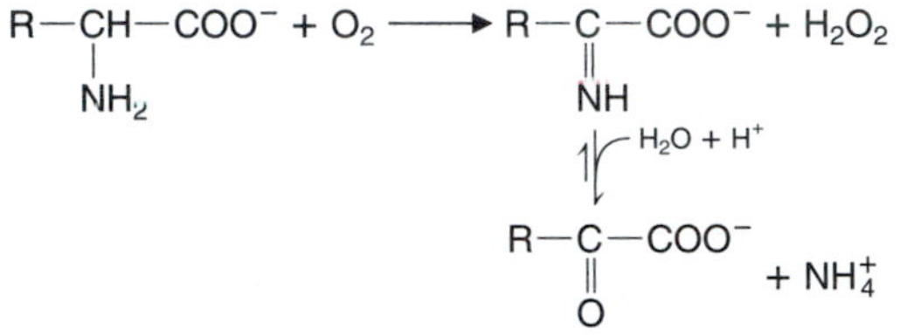

$$R-\underset{NH_2}{\underset{|}{CH}}-COO^- + O_2 \longrightarrow R-\underset{NH}{\underset{\|}{C}}-COO^- + H_2O_2$$

$$\xrightarrow{H_2O + H^+} R-\underset{O}{\underset{\|}{C}}-COO^- + NH_4^+$$

The initial oxidation of the amino acid by FAD and the subsequent reoxidation of the $FADH_2$ by oxygen is schematically shown in Figure 10.2. **Cytochrome oxidase** is a little unusual in that it produces H_2O rather than H_2O_2 from the reduction of oxygen. **Oxygenases** catalyze the incorporation of oxygen into a substrate. With **dioxygenases** both atoms of O_2 are incorporated in a single pro-

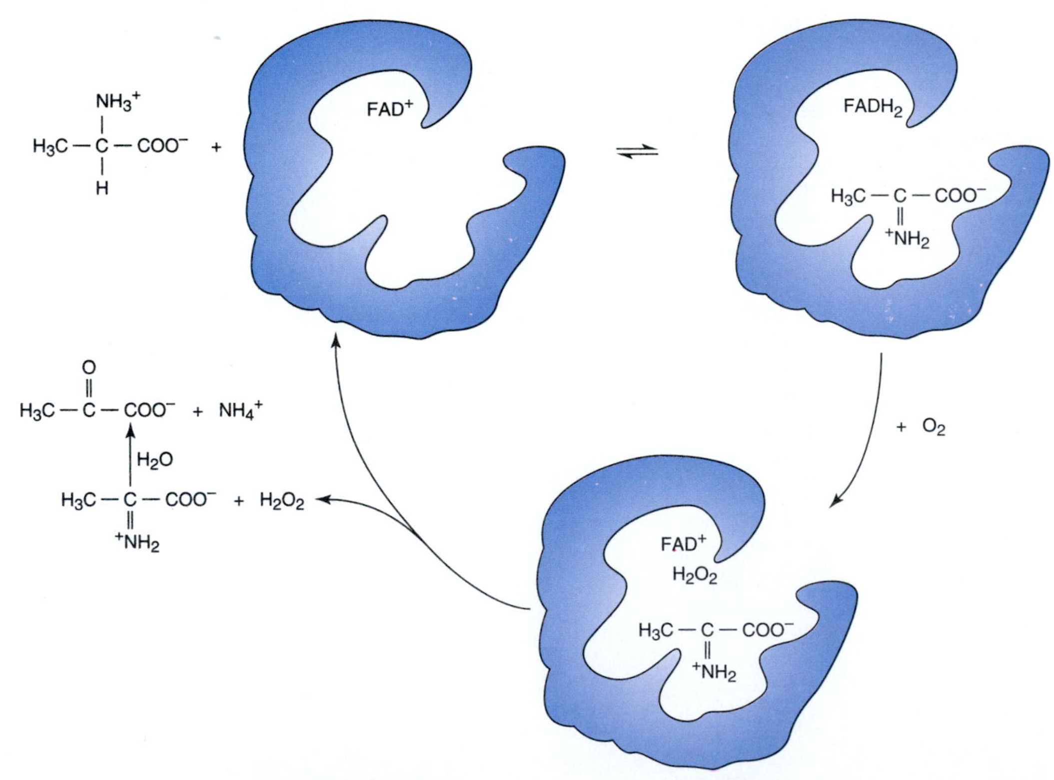

FIGURE 10.2
Oxidation of alanine by D-amino acid oxidase.

duct, whereas with the **monooxygenases** a single oxygen atom is incorporated as a hydroxyl group and the other oxygen atom is reduced to water by electrons from the substrate or from a second substrate that is not oxygenated. Catechol oxygenase catalyzes the dioxygenase reaction (Figure 10.3); steroid hydroxylase illustrates a monooxygenase (mixed function oxygenase) reaction where one oxygen atom goes to water and the other to the steroid (Figure 10.4). **Peroxidases** utilize H_2O_2 rather than oxygen as the oxidant. NADH peroxidase catalyzes the reaction

$$\text{NADH} + \text{H}^+ + \text{H}_2\text{O}_2 \rightleftharpoons \text{NAD}^+ + 2\text{H}_2\text{O}$$

Catalase is unique in that H_2O_2 serves as both donor and acceptor. Catalase functions in the cell to detoxify H_2O_2:

$$\text{H}_2\text{O}_2 + \text{H}_2\text{O}_2 \rightleftharpoons \text{O}_2 + 2\text{H}_2\text{O}$$

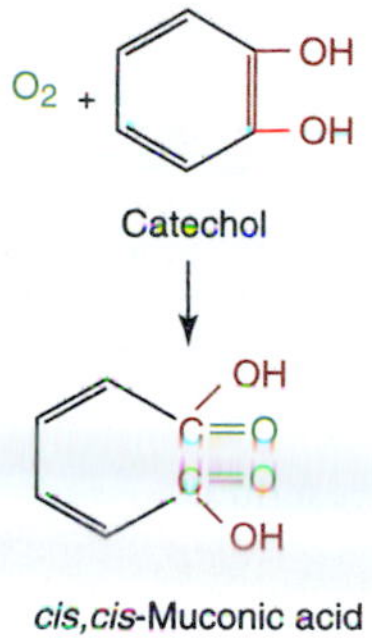

O_2 + Catechol → *cis,cis*-Muconic acid

FIGURE 10.3
Oxygenation of catechol by an oxygenase.

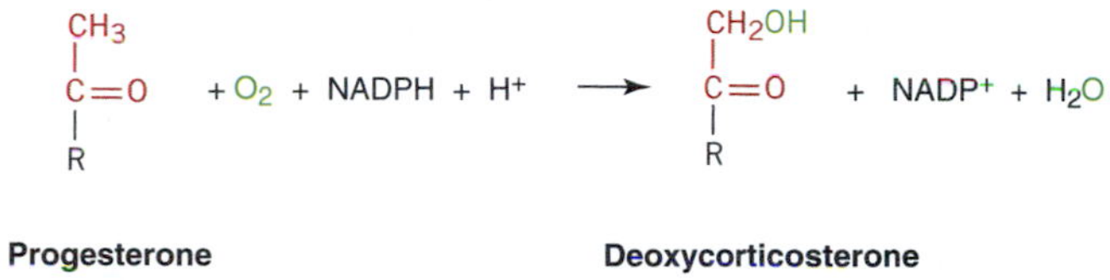

Progesterone (CH_3—C=O—R) + O_2 + NADPH + H^+ → Deoxycorticosterone (CH_2OH—C=O—R) + $NADP^+$ + H_2O

FIGURE 10.4
Hydroxylation of progesterone by a monooxygenase.

Class 2: Transferases

These enzymes transfer functional groups between donors and acceptors. The amino, acyl, phosphate, one-carbon, and glycosyl groups are the major moieties that are transferred. **Aminotransferases (transaminases)** transfer an amino group from one amino acid to an α-keto acid acceptor, resulting in the formation of a new amino acid and a new keto acid (Figure 10.5).

Kinases are enzymes that catalyze the transfer of the chemically labile γ phosphoryl group from ATP or other nucleoside triphosphate to alcohol or amino group acceptors. These acceptors may be small organic molecules such as glucose or macromolecules such as proteins. For example, protein kinase A transfers phosphate from ATP to the serine hydroxyl of specific enzymes. The phosphorylated form of the protein may be more or less active than the nonphosphorylated species, depending on the species.

In general the group to be transferred must be activated, that is, made chemically labile before transfer can occur. This usually is accomplished by making an ester with ATP as follows:

$$\text{ATP} + \text{R—OH} \rightleftharpoons \text{R—O—ADP} + \text{P}_i$$

The "R" group can now be transferred to an acceptor to form a new compound.

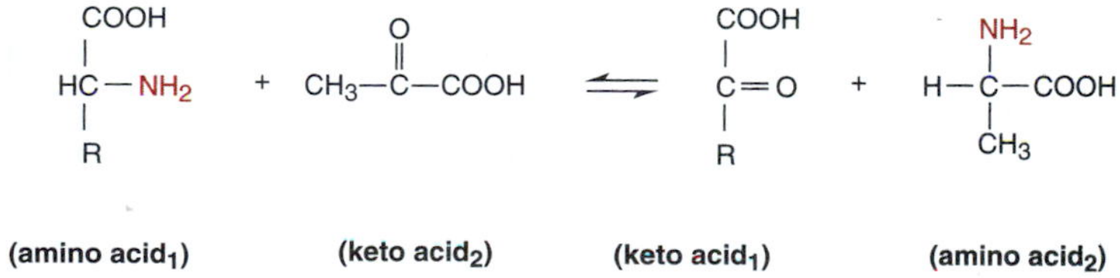

(amino acid$_1$) HC(COOH)(R)—NH_2 + (keto acid$_2$) CH_3—C(=O)—COOH ⇌ (keto acid$_1$) COOH—C=O—R + (amino acid$_2$) H—C(NH_2)(CH_3)—COOH

FIGURE 10.5
Example of a reaction catalyzed by an aminotransferase.

Class 3: Hydrolases

This group of enzymes can be considered as a special class of the transferases in which the donor group is transferred to water. The generalized reaction involves the hydrolytic cleavage of C—O, C—N, O—P, and C—S bonds. The cleavage of a peptide bond is a good example of this reaction:

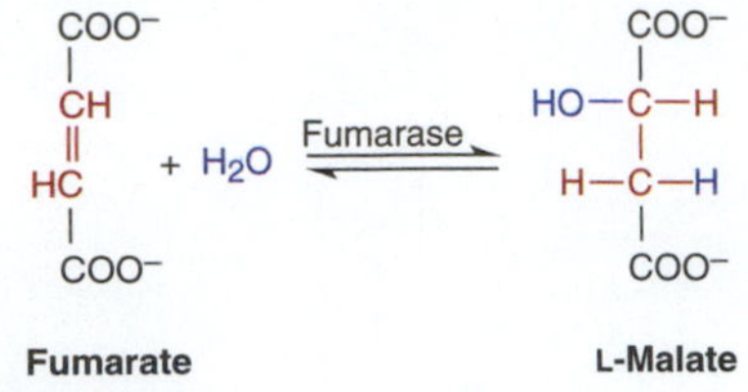

Fumarate   L-Malate

FIGURE 10.7
The fumarase reaction.

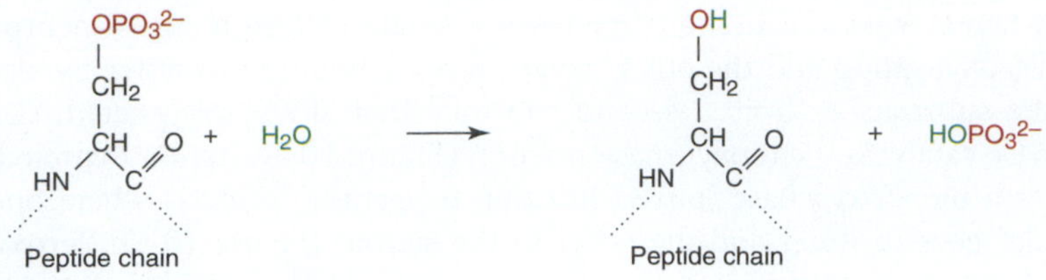

Peptide chain   Peptide chain

FIGURE 10.6
Hydrolysis of a phosphorylated protein by a phosphatase.

$$R_1{-}\overset{O}{\overset{\|}{C}}{-}NH{-}R_2 + H_2O \longrightarrow R_1{-}\overset{O}{\overset{\|}{C}}{-}O^- + H_3\overset{+}{N}{-}R_2$$

Proteolytic enzymes are a special class of hydrolases called peptidases.

Phosphatases are enzymes that replace a phosphate group with a hydroxyl group from water (Figure 10.6). In the case of protein phosphatases, their action is to negate the effects of the protein kinases.

D-Xylulose 5-phosphate   D-Ribulose 5-phosphate

D-Lactic acid   L-Lactic acid

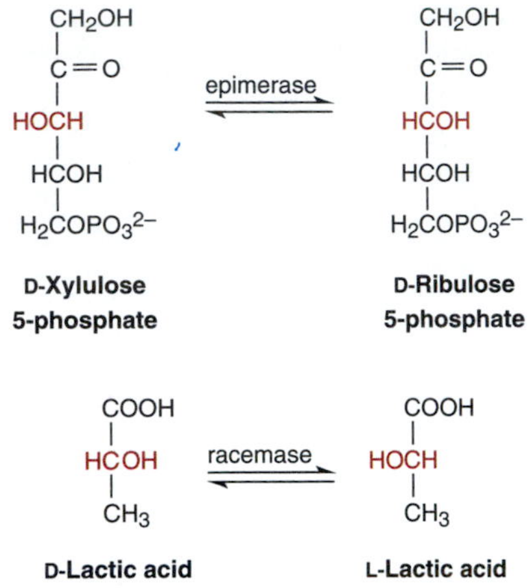

FIGURE 10.8
Examples of reactions catalyzed by an epimerase and a racemase.

Class 4: Lyases

Lyases add or remove the elements of water, ammonia, or carbon dioxide. **Decarboxylases** remove the element of CO_2 from α- or β-keto acids or amino acids:

$$R{-}\overset{O}{\overset{\|}{C}}{-}\overset{O}{\overset{\|}{C}}{-}O^- + H^+ \longrightarrow R{-}\overset{O}{\overset{\|}{C}}{-}H + CO_2$$

Dehydratases remove H_2O in a dehydration reaction. Fumarase converts fumarate to malate (Figure 10.7).

2-Phosphoglycerate

phosphoglyceromutase

3-Phosphoglycerate

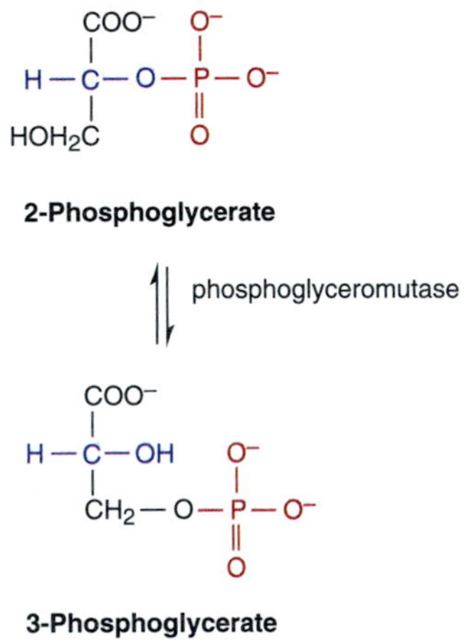

FIGURE 10.9
Interconversion of the 2- and 3-phosphoglycerates.

Class 5: Isomerases

This very heterogeneous group of enzymes catalyze isomerizations of several types. These include cis–trans and aldose–ketose interconversions. **Isomerases** that catalyze inversion at asymmetric carbon atoms are either epimerases or racemases (Figure 10.8). **Mutases** involve intramolecular transfer of a group such as a phosphoryl. The transfer may be direct but can involve a phosphorylated enzyme as an intermediate. Phosphoglycerate mutase catalyzes conversion of 2-phosphoglycerate to 3-phosphoglycerate (Figure 10.9).

Class 6: Ligases

Since to ligate means to bind, these enzymes are involved in synthetic reactions where two molecules are joined at the expense of a "high-energy phosphate bond" of ATP. The term **synthetase** is reserved for this particular group of enzymes. The formation of aminoacyl tRNAs, acyl coenzyme A, and glutamine and the addition of CO_2 to pyruvate are reactions catalyzed by ligases. Pyruvate carboxylase is a good example of a ligase enzyme (Figure 10.10). The substrates bicarbonate and pyruvate are ligated to form a four-carbon (C4) α-keto acid.

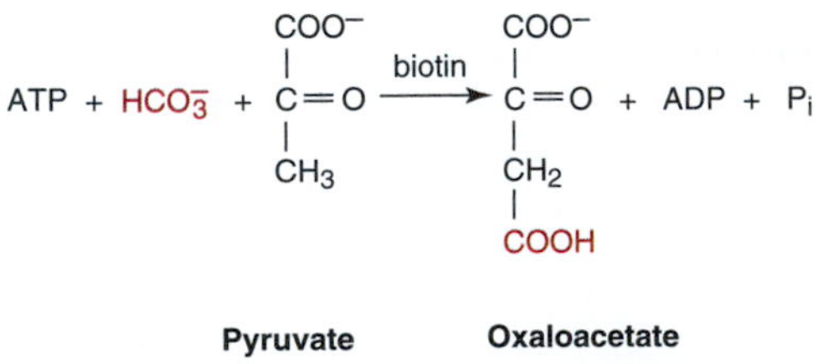

Pyruvate   Oxaloacetate

FIGURE 10.10
Pyruvate carboxylase reaction.

10.3 | KINETICS

Kinetics Studies the Rate of Change of Reactants to Products

Since enzymes affect the rate of chemical reactions, it is important to understand basic chemical kinetics and how kinetic principles apply to enzyme-catalyzed reactions. **Kinetics** is a study of the rate of change of reactants to products. **Velocity** is expressed in terms of change in the concentration of substrate or product per unit time, whereas **rate** refers to changes in total quantity (moles or grams) per unit time. Biochemists tend to use these terms interchangeably.

The velocity of a reaction A → P is determined from its progress curve or velocity profile. The progress curve can be determined by following the disappearance of reactants or the appearance of product at several different times. In Figure 10.11 product appearance is plotted against time. The slope of tangents to the progress curve yields the instantaneous velocity at that point in time. The **initial velocity** is the change in reactant or product concentration during the first few seconds of the reaction and is determined as the slope of the line through the linear phase of the reaction—extrapolated to zero time (Figure 10.11). Note that the velocity changes constantly as the reaction proceeds to equilibrium, where it becomes zero. Mathematically, the velocity is expressed as

$$\text{Velocity} = v = -\frac{d[\text{A}]}{dt} = \frac{d[\text{P}]}{dt} \tag{10.1}$$

and represents the change in concentration of reactants or products per unit time.

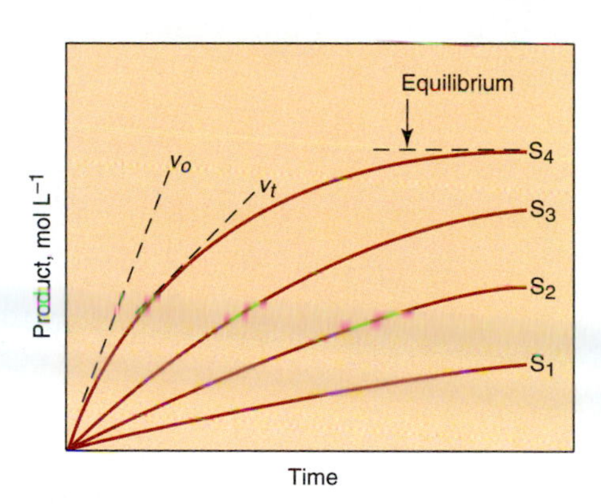

FIGURE 10.11
Progress curves for an enzyme-catalyzed reaction.
The initial velocity (v_0) of the reaction is determined from the slope of the progress curve at the beginning of the reaction. The initial velocity increases with increasing substrate concentration (S_1–S_4) but reaches a limiting value characteristic of each enzyme. The velocity at any time, t, is denoted as v_t.

Rate Equation

Determination of the velocity of a reaction reveals nothing about the stoichiometry of the reactants and products or about the reaction mechanism. An equation is needed that relates the experimentally determined initial velocity to the concentration of reactants. This is the velocity or rate equation. In the reaction A → P, the velocity equation is

$$\frac{-d[\text{A}]}{dt} = v = k[\text{A}]^n \tag{10.2}$$

Thus the observed initial velocity depends on the starting concentration of A to the nth power multiplied by a proportionality constant (k). The latter is known as the **rate constant.** The exponent n is usually an integer from 1 to 3 that is required to satisfy the mathematical identity of the velocity expression.

Characterization of Reactions Based on Order

Another term useful in describing a reaction is the **order of reaction.** Empirically the order is determined as the sum of the exponents on each concentration term in the rate expression. In the case under discussion the reaction is **first order**, since the velocity depends on the concentration of A to the first power, $v = k[\text{A}]^1$. In the reaction A + B → C, if the order with respect to A and B is 1, that is, $v = k[\text{A}]^1[\text{B}]^1$, overall the reaction is second order. Note that the order of reaction is independent of the stoichiometry of the reaction; that is, if the reaction were third order, the rate expression could be either $v = k[\text{A}][\text{B}]^2$ or $v = k[\text{A}]^2[\text{B}]$, depending on the order in A and B.

Since the velocity of the reaction is constantly changing as the reactant concentration changes, first-order reaction conditions would not be ideal for assaying an enzyme-catalyzed reaction because one would have two variables, the changing substrate concentration and the unknown enzyme concentration.

If the differential first-order rate expression Eq. 10.2 is integrated, one obtains

$$k_1 \cdot t = 2.3 \log\left(\frac{[\text{A}]}{[\text{A}] - [\text{P}]}\right) \tag{10.3}$$

where [A] is the initial reactant concentration and [P] is the concentration of product formed at time t. The first-order rate constant k_1 has the units of reciprocal time. If the data shown in Figure 10.11 were replotted as log[P] versus time for any one of the substrate concentrations, a straight line would be obtained whose slope is equal to $k_1/2.303$. The rate constant k_1 should not be confused with the rate or velocity of the reaction.

Many biological processes proceed under first-order conditions. The clearance of many drugs from the blood by peripheral tissues is a first-order process. A specialized form of the rate equation can be used in these cases. If we define $t_{1/2}$ as the time required for the concentration of the reactants or the blood level of a drug to be reduced by one-half the initial value, then Eq. 10.3 reduces to

$$k_1 \cdot t_{1/2} = 2.3 \log\left(\frac{1}{1-\frac{1}{2}}\right) = 2.3 \log 2 = 0.69 \qquad (10.4)$$

or

$$t_{1/2} = \frac{0.69}{k_1} \qquad (10.5)$$

Note that $t_{1/2}$ is not one-half the time required for the reaction to be completed. The term $t_{1/2}$ is referred to as the **half-life** of the reaction.

Many second-order reactions that involve water or any one of the reactants in large excess can be treated as pseudo-first-order reactions. In the hydrolysis of an ester,

$$R-\overset{O}{\overset{\|}{C}}-O-CH_3 + H_2O \rightleftharpoons R-\overset{O}{\overset{\|}{C}}-OH + CH_3OH$$

the second-order rate expression is

$$\text{velocity} = v = k_2[\text{ester}]^1[H_2O]^1 \qquad (10.6)$$

but since water is in abundance (55.5 M) compared to the ester (10^{-2}–10^{-3} M), the system obeys the first-order rate law Eq. 10.2, and the reaction appears to proceed as if it were a first-order reaction. Reactions in the cell that involve hydration, dehydration, or hydrolysis are pseudo-first-order.

The rate expression for the **zero-order** reaction is $v = k_0$. Note that there is no concentration term for reactants; therefore the addition of more reactant does not augment the rate. The disappearance of reactant or the appearance of product proceeds at a constant velocity irrespective of reactant concentration. The units of the rate constant are concentration per unit time. Zero-order reaction conditions only occur in catalyzed reactions where the concentration of reactants is large enough to saturate all the catalytic sites. Under these conditions the catalyst is operating at maximum velocity, and all catalytic sites are filled; therefore addition of more reactant cannot increase the rate.

Reversibility of Reactions

Although most chemical reactions are reversible, some directionality is imposed on particular steps in a metabolic pathway by rapid removal of end product by subsequent reactions in the pathway. Many ligase reactions involving the nucleoside triphosphates result in release of pyrophosphate (PP_i). These reactions are rendered irreversible by the hydrolysis of the pyrophosphate to 2 moles of inorganic phosphate, P_i. Schematically,

$$A + B + ATP \rightarrow A{-}B + AMP + PP_i$$
$$PP_i + H_2O \rightarrow 2P_i$$

Conversion of the "high-energy" pyrophosphate to inorganic phosphate imposes irreversibility on the system by virtue of the thermodynamic stability of the products.

For reactions that are reversible, the equilibrium constant for

$$A + B \rightarrow C$$

is

$$K_{eq} = \frac{[C]}{[A][B]} \qquad (10.7)$$

and can also be expressed in terms of rate constants of the forward and reverse reactions:

$$A + B \underset{k_2}{\overset{k_1}{\rightleftharpoons}} C$$

where

$$\frac{k_1}{k_2} = K_{eq} \tag{10.8}$$

Equation 10.7 shows the relationship between thermodynamic and kinetic quantities. The term K_{eq} is a thermodynamic expression of the state of the system, while k_1 and k_2 are kinetic expressions that are related to the speed at which that state is reached.

Enzymes Show Saturation Kinetics

Terminology

Enzyme activity is usually expressed as micromoles (μmol) of substrate converted to product per minute under specified assay conditions. One standard unit of enzyme activity (U) is that activity that catalyzes transformation of 1 μmol min^{-1} of substrate. **Specific activity** of an enzyme preparation is defined as the number of enzyme units per milligram of protein (μmol min^{-1} mg of protein^{-1} or U/mg of protein). This expression, however, does not indicate whether the sample tested contains only the enzyme protein; during enzyme purification the value will increase as contaminating protein is removed. The catalytic constant, or **turnover number,** of an enzyme is equal to the units of activity per mole of enzyme (μmol min^{-1} mol of enzyme^{-1}). Where the enzyme has more than one catalytic center, the catalytic constant is often given on the basis of the particle weight of the subunit rather than the molecular weight of the entire protein.

The catalytic constant or turnover number allows direct comparison of relative catalytic ability between enzymes. For example, the constants for catalase and α-amylase are 5×10^6 and 1.9×10^4, respectively, indicating that catalase is about 2500 times more active than amylase. **Maximum velocity, V_{max},** is the velocity obtained under conditions of substrate saturation of the enzyme under specified conditions of pH, temperature, and ionic strength. V_{max} is a constant for a given concentration of enzyme.

Interaction of Enzyme and Substrate

The initial velocity of an enzyme-catalyzed reaction is dependent on the concentration of substrate (S) (Figure 10.11). As concentration of substrate increases (S_1–S_4), initial velocity increases until the enzyme is completely saturated with substrate. If initial velocities obtained at given substrate concentrations are plotted (Figure 10.12), a rectangular hyperbola is obtained like that obtained for binding of oxygen to myoglobin as a function of increasing oxygen pressure. In general, a rectangular hyperbola is obtained for any process that involves interaction or binding of reactants or other substances at a specific but limited number of sites. The

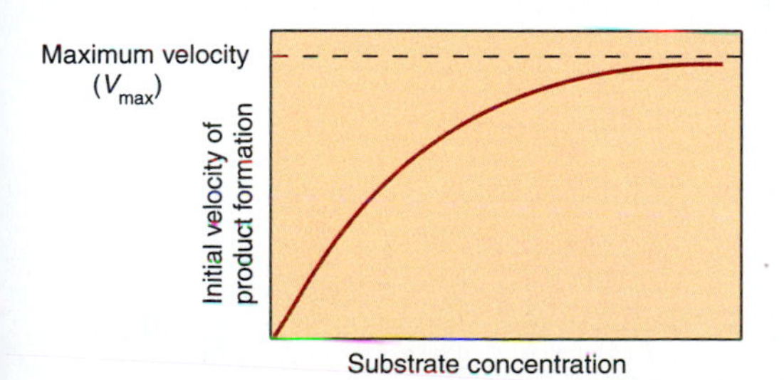

FIGURE 10.12
Plot of velocity versus substrate concentration for an enzyme-catalyzed reaction. Initial velocities are plotted against the substrate concentration at which they were determined. The curve is a rectangular hyperbola, which asymptotically approaches the maximum velocity possible with a given amount of enzyme.

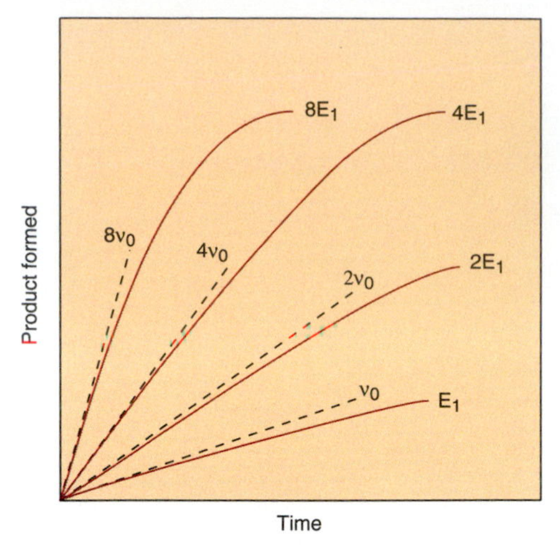

FIGURE 10.13
Progress curves at variable concentrations of enzyme and saturating concentrations of substrate.
The initial velocity (v_0) doubles as the enzyme concentration doubles. Since the substrate concentrations are the same, the final equilibrium concentrations of product will be identical in each case; however, equilibrium will be reached at a slower rate in those assays containing small amounts of enzyme.

velocity of the reaction reaches a maximum at the point at which all the available sites are saturated. The curve in Figure 10.12 is referred to as the substrate saturation curve of an enzyme-catalyzed reaction and reflects the fact that an enzyme has a specific binding site for the substrate. Enzyme (E) and substrate must interact in some way if the substrate is to be converted to products. Initially there is formation of a complex between the enzyme and substrate:

$$\mathrm{E} + \mathrm{S} \underset{k_2}{\overset{k_1}{\rightleftharpoons}} \mathrm{ES} \tag{10.9}$$

The **rate constant** for formation of the ES complex is defined as k_1, and the rate constant for disassociation of the ES complex is defined as k_2. So far, we have described only an equilibrium binding of enzyme and substrate. The chemical event in which bonds are made or broken occurs in the ES complex. The conversion of substrate to products (P) then occurs from the ES complex with a rate constant k_3. Therefore Eq. 10.9 is transformed to

$$\mathrm{E} + \mathrm{S} \underset{k_2}{\overset{k_1}{\rightleftharpoons}} \mathrm{ES} \xrightarrow{k_3} \mathrm{E} + \mathrm{P} \tag{10.10}$$

Equation 10.10 is a general statement of the mechanism of enzyme action. The equilibrium between E and S can be expressed as an affinity constant, K_a, only if the rate of the chemical phase of the reaction, k_3, is small compared to k_2; then $K_a = k_1/k_2$. Earlier we used K_{eq} to describe chemical reactions. In enzymology the association or affinity constant K_a is preferred.

The **initial velocity,** v_0, of an enzyme-catalyzed reaction is dependent on amount of substrate present and on enzyme concentration. Figure 10.13 shows progress curves for increasing concentrations of enzyme, where there is enough substrate initially to saturate the enzyme at all levels. The initial velocity doubles as the concentration of enzyme doubles. At the lower concentrations of enzyme, equilibrium is reached more slowly than at higher concentrations, but the final equilibrium position is the same.

From this discussion, we can conclude that the velocity of an enzyme reaction is dependent on both substrate and enzyme concentrations.

Formulation of the Michaelis–Menten Equation

In the discussion of chemical kinetics, rate equations were developed so that velocity of the reaction could be expressed in terms of substrate concentration. This approach also holds for enzyme-catalyzed reactions, where the goal is to develop a relationship that will allow the velocity of a reaction to be correlated with the amount of enzyme. First, a rate equation must be developed that relates the velocity of the reaction to the substrate concentration.

Development of this rate equation, known as the **Michaelis–Menten** equation, requires three basic assumptions based on Eq. 10.10. The first is that the ES complex is in a **steady state;** that is, during the initial phases of the reaction, the concentration of the ES complex remains constant, even though many molecules of substrate are converted to products via the ES complex. The second assumption is that under saturating conditions all of the enzyme is converted to ES complex, and none is free. This occurs when the substrate concentration is high. The third assumption is that if all the enzyme is in the ES complex, then the rate of formation of products will be maximal; that is,

$$V_{\text{max}} = k_3[\mathrm{ES}] \tag{10.11}$$

If we then write the steady-state expression for formation and breakdown of the ES complex as

$$K_{\text{m}} = \frac{k_2 + k_3}{k_1} \tag{10.12}$$

the rate expression obtained by algebraic manipulation becomes

$$\text{velocity} = v_0 = \frac{V_{max} \cdot [S]}{K_m + [S]} \quad (10.13)$$

where v_0 is the initial velocity at a substrate concentration [S]. Equation 10.13 is the Michaelis–Menten equation. This is the fundamental equation of enzymology that relates the initial velocity of an enzyme-catalyzed reaction to the substrate concentration. The two constants in this rate equation, V_{max} and K_m, are unique to each enzyme under specific conditions of pH and temperature with a given concentration of enzyme. For enzymes in which $k_3 \ll k_2$, K_m becomes the reciprocal of the enzyme-substrate binding constant, and V_{max} reflects the catalytic phase of the enzyme mechanism as suggested by Eq. 10.11. Thus, in this model, activity of the enzyme can be separated into two phases: binding of substrate followed by chemical modification of the substrate. This biphasic nature of the enzyme mechanism is reinforced in the clinical example discussed in Clinical Correlation 10.1.

Significance of K_m

The concept of K_m may appear to have no physiological or clinical relevance. The truth is quite the contrary. As discussed in Section 10.9 (p. 453), all valid enzyme assays performed in the clinical laboratory are based on knowledge of K_m values for each substrate. In terms of the physiological control of glucose and phosphate metabolism, two hexokinases have evolved, one with a high K_m and one with a low K_m for glucose. Together, they contribute to maintaining steady-state levels of blood glucose and phosphate, as discussed on page 615.

In general, K_m values are near the concentrations of substrate found in cells. Perhaps enzymes have evolved substrate-binding sites with affinities comparable to *in vivo* levels of their substrates. Occasionally, mutation of an enzyme-binding site occurs, or an isoenzyme with an altered K_m is expressed. Either of these events can result in an abnormal physiology. An interesting example (see Clin. Corr. 10.2) is the expression of only the atypical form of aldehyde dehydrogenase in people of Asiatic origin.

Note that if one allows the initial velocity, v_0, to be equal to $\frac{1}{2} V_{max}$ in Eq. 10.13, K_m will be equal to [S]:

$$\tfrac{1}{2}V_{max} = \frac{V_{max} \cdot [S]}{K_m + [S]}$$

$$K_m + [S] = \frac{2V_{max} \cdot [S]}{V_{max}}$$

$$K_m = [S]$$

CLINICAL CORRELATION 10.1

A Case of Gout Demonstrates Two Phases in the Mechanism of Enzyme Action

The two phases of the Michaelis–Menten model of enzyme action, binding followed by modification of substrate, are illustrated by studies on a family with gout. The patient excreted three times the normal amount of uric acid per day and had markedly increased levels of 5′-phosphoribosyl-α-pyrophosphate (PRPP) in his red blood cells. PRPP is an intermediate in the biosynthesis of AMP and GMP, which are converted to ATP and GTP. Uric acid arises directly from degradation of AMP and GMP. Assays *in vitro* revealed that the patient's red cell PRPP synthetase activity was increased threefold. The pH optimum and the K_m of the enzyme for ATP and ribose 5-phosphate were normal, but V_{max} was increased threefold! This increase was not due to an increase in the amount of enzyme; immunologic testing with an antibody specific to the enzyme revealed similar quantities of the enzyme protein as in normal red cells. This finding demonstrates that the binding of substrate as reflected by K_m and the subsequent chemical event in catalysis, which is reflected in V_{max}, are separate phases of the overall catalytic process. This situation holds only for those enzyme mechanisms in which $k_3 \gg k_2$.

Becker, M. A., Kostel, P. J., Meyer, L. J., and Seegmiller, J. E. Human phosphoribosylpyrophosphate synthetase: increased enzyme specific activity in a family with gout and excessive purine synthesis. *Proc. Natl. Acad. Sci. USA* 70:2749, 1973.

CLINICAL CORRELATION 10.2

Physiological Effect of Changes in Enzyme K_m Values

The unusual sensitivity of Asians to alcoholic beverages has a biochemical basis. In some Japanese and Chinese, much less ethanol is required to produce vasodilation that results in facial flushing and rapid heart rate than is required to achieve the same effect in Europeans. The physiological effects are due to acetaldehyde generated by liver alcohol dehydrogenase:

$$CH_3CH_2OH + NAD^+ \rightleftharpoons CH_3CHO + H^+ + NADH$$

Acetaldehyde is normally removed by a mitochondrial aldehyde dehydrogenase that converts it to acetate. In some Asians, the normal form of the mitochondrial aldehyde dehydrogenase, with a low K_m for acetaldehyde, is missing. These individuals have only the cytosolic high K_m (lower affinity) form of the enzyme, which leads to a high steady-state level of acetaldehyde in the blood after alcohol consumption. This accounts for the increased sensitivity to alcohol.

Crabb, D. W., Edenberg, H. J., Bosron, W. F., and Li, T.-K. Genotypes for aldehyde dehydrogenase deficiency and alcohol sensitivity: The ALDH2(2) allele is dominant. *J. Clin. Invest.* 83:314, 1989.

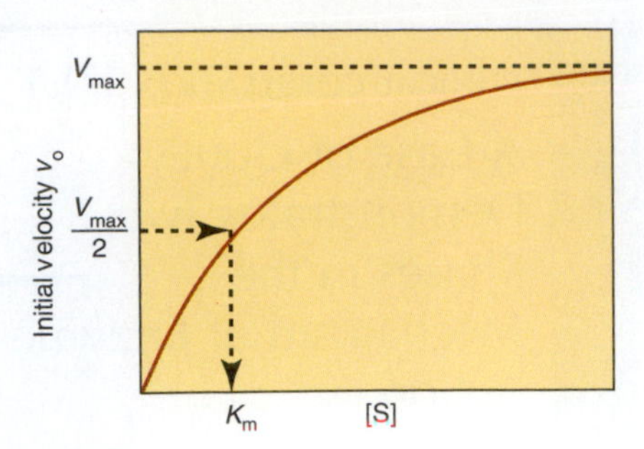

FIGURE 10.14
Graphic estimation of K_m for the v_0 versus [S] plot.
K_m is the substrate concentration at which the enzyme has half-maximal activity.

Thus, from a substrate saturation curve, the numerical value of K_m can be derived by graphical analysis (Figure 10.14). Here the K_m is equal to the substrate concentration that gives one-half the maximum velocity.

Linear Form of the Michaelis–Menten Equation

In practice the determination of K_m from the substrate saturation curve is not very accurate, because V_{max} is approached asymptotically. If one takes the reciprocal of Eq. 10.13 and separates the variables into a format consistent with the equation of a straight line ($y = mx + b$), then

$$\frac{1}{v_0} = \frac{K_m}{V_{max}} \times \frac{1}{[S]} + \frac{1}{V_{max}}$$

A plot of the reciprocal of the initial velocity versus the reciprocal of the initial substrate concentration yields a line whose slope is K_m/V_{max} and whose y-intercept is $1/V_{max}$. Such a plot is shown in Figure 10.15. It is often easier to obtain the K_m from the intercept on the x-axis, which is $-1/K_m$.

This linear form of the Michaelis–Menten equation is the **Lineweaver–Burk** or **double–reciprocal plot.** Its advantage is that statistically significant values of K_m and V_{max} can be obtained directly with six to eight data points.

An Enzyme Catalyzes Both Forward and Reverse Directions of a Reversible Reaction

As indicated previously, enzymes do not alter the equilibrium constant of a reaction; consequently, in the reaction

$$S \underset{k_2}{\overset{k_1}{\rightleftharpoons}} P$$

the direction of flow of material, either in the forward or the reverse direction, will depend on the concentration of S relative to P and the equilibrium constant of the reaction. Since enzymes catalyze the forward and reverse reactions, a problem may arise if product has an affinity for the enzyme that is similar to that of substrate. In this case the product can easily rebind to the active site of the enzyme and will compete with the substrate for that site. **Product inhibition** occurs when the product progressively inhibits the reaction as concentration of product increases. The Lineweaver–Burk plot will not be linear in those cases where the enzyme is susceptible to product inhibition. If the subsequent enzyme in a metabolic pathway removes the product rapidly, then product inhibition should not occur.

Product inhibition in a metabolic pathway provides a limited means of controlling or modulating flux of substrates through the pathway. As the end product of a pathway increases, each intermediate will also increase by mass action. If one or more enzymes are particularly sensitive to product inhibition, output of end product of the pathway will be suppressed. Reversibility of a pathway or a particular enzyme-catalyzed reaction is dependent on the rate of product removal. If the end product is quickly removed, then the pathway may be physiologically unidirectional.

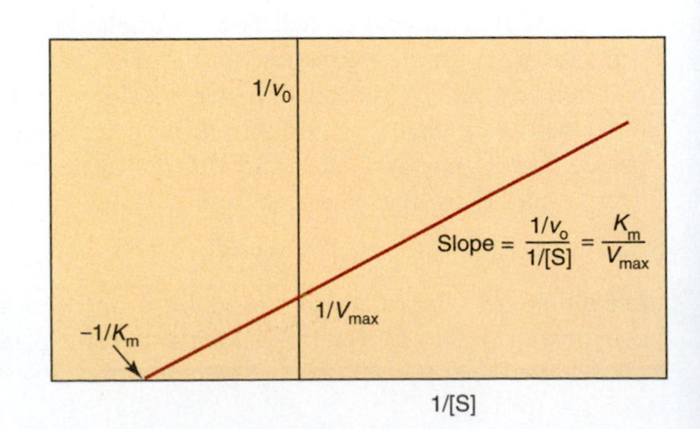

FIGURE 10.15
Determination of K_m and V_{max} from the Lineweaver–Burk double-reciprocal plot.
Plots of the reciprocal of the initial velocity versus the reciprocal of the substrate concentration used to determine the initial velocity yield a line whose x-intercept is $-1/K_m$.

Multisubstrate Reactions Follow Either a Ping–Pong or Sequential Mechanism

Most enzymes utilize more than one substrate, or act upon one substrate plus a coenzyme and generate one or more products. In any case, a K_m must be determined for each substrate and coenzyme involved in the reaction when establishing an enzyme assay.

Mechanistically, enzyme reactions are divided into two major categories, ping–pong or sequential. There are many variations on these major mechanisms. The **ping–pong mechanism** can be represented as follows:

$$E + A \longrightarrow EA \longrightarrow \overset{\overset{P_1}{\uparrow}}{E'} \longrightarrow \overset{\overset{B}{\downarrow}}{E'B} \longrightarrow P_2 + E$$

where substrate A reacts with E to produce product P_1, which is released before the second substrate B binds to the modified enzyme E′. Substrate B is then converted to product P_2 and the enzyme is regenerated. A good example of this mechanism is the transaminase-catalyzed reaction (see p. 782) in which the α-amino group of amino $acid_1$ is transferred to the enzyme and the newly formed α-keto $acid_1$ is released, as the first product, followed by the binding of the acceptor α-keto $acid_2$ and release of amino $acid_2$. This reaction is outlined in Figure 10.16.

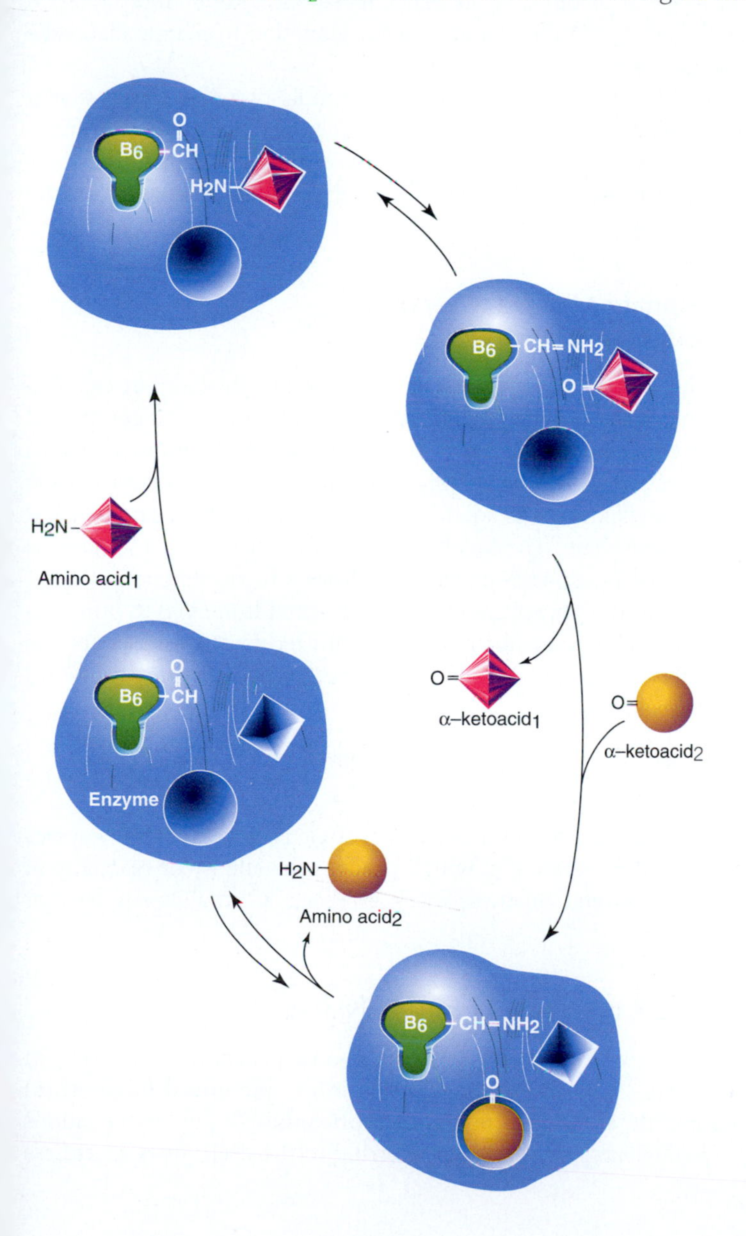

FIGURE 10.16
Schematic representation of the transaminase reaction mechanism: an example of a ping–pong mechanism.
Enzyme-bound pyridoxal phosphate (vitamin B_6 coenzyme) accepts the α-amino group from the first amino acid (AA_1), which is then released from the enzyme as an α-keto acid. The acceptor α-keto acid (AA_2) is then bound to the enzyme, and the bound amino group is transferred to it, forming a new amino acid, which is then released from the enzyme. The terms "oxy" and "keto" are used interchangeably.

CLINICAL CORRELATION 10.3

Mutation of a Coenzyme-Binding Site Results in Clinical Disease

Cystathioninuria is a genetic disease in which γ-cystathionase is either deficient or inactive. Cystathionase catalyzes the reaction

$$\text{cystathionine} \leftrightarrow \text{cysteine} + \alpha\text{-ketobutyrate}$$

Deficiency of the enzyme leads to accumulation of cystathionine in the plasma. Since cystathionase is a pyridoxal phosphate-dependent enzyme, vitamin B_6 was administered to patients whose fibroblasts contained material that cross-reacted with antibody against cystathionase. Many responded to B_6 therapy with a fall in plasma levels of cystathionine. These patients produce the apoenzyme that reacted with the antibody. In one patient the enzyme activity was undetectable in fibroblast homogenates but increased to 31% of normal with the addition of 1 mM of pyridoxal phosphate to the assay mixture. It is thought that the K_m for pyridoxal phosphate binding to the enzyme was increased because of a mutation in the binding site. Activity is partially restored by increasing the concentration of coenzyme. Apparently these patients require a higher steady-state concentration of coenzyme to maintain γ-cystathionase activity.

Pascal, T. A., Gaull, G. E., Beratis, N. G., Gillam, B. M., Tallan, H. H., and Hirschhorn, K. Vitamin B_6-responsive and unresponsive cystathionuria: two variant molecular forms. *Science* 190:1209, 1975.

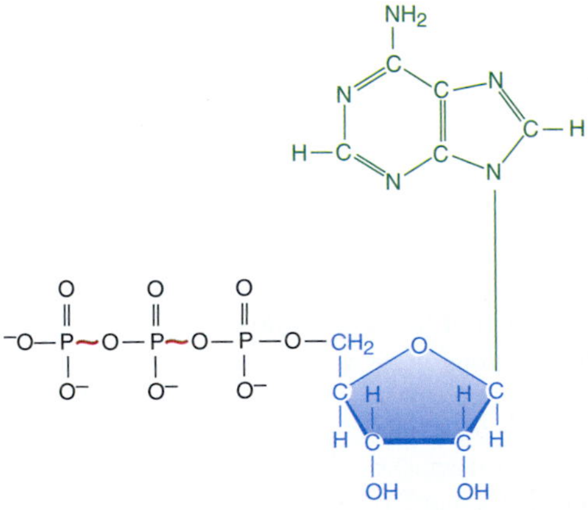

In a **sequential mechanism,** if the two substrates A and B can bind in any order, it is a random mechanism; if binding of A is required before B can be bound, then it is an ordered mechanism. In either case the reaction is bimolecular; that is, both A and B must be bound before reaction occurs. Examples of these mechanisms are found among the dehydrogenases in which the second substrate is a coenzyme (NAD^+, FAD, etc.) (see p. 427). Release of products may or may not be ordered in either case.

10.4 | COENZYMES: STRUCTURE AND FUNCTION

Coenzymes are small organic molecules, often derivatives of vitamins, that function with the enzyme in the catalytic process. Often the coenzyme has an affinity for the enzyme that is similar to that of the substrate; consequently, the coenzyme can be considered to be a second substrate. In some cases, the coenzyme is covalently bound to the apoenzyme and functions at or near the active site in catalysis. In other enzymes the role of the coenzyme falls between these two extremes.

Several coenzymes are derived from the B vitamins. Vitamin B_6, pyridoxine, requires little modification to form the active coenzyme, pyridoxal phosphate (see p. 1151). Clinical Correlation 10.3 points out the importance of the coenzyme-binding site and how alterations in this site cause metabolic dysfunction.

In contrast to vitamin B_6, niacin requires major alteration in mammalian cells to form a coenzyme (see p. 850).

The structures and functions of the coenzymes of only two B vitamins, niacin and riboflavin, and of ATP are discussed in this chapter. The structures and functions of coenzyme A (CoA) (see p. 852), thiamine (see p. 1148), biotin, and vitamin B_{12} are included in those chapters dealing with enzymes dependent on the given coenzyme for activity.

Adenosine Triphosphate May Be a Second Substrate or a Modulator of Activity

Adenosine triphosphate (ATP) often functions as a second substrate but can also serve as a cofactor in modulation of the activity of specific enzymes. This compound is central in biochemistry (Figure 10.17) and it is synthesized *de novo* in all mammalian cells. The nitrogenous heterocyclic ring is adenine. The combination of the base, adenine, plus ribose is known as adenosine; hence ATP is adenosine that has at the 5′-hydroxyl a triphosphate. The biochemically functional end is the reactive triphosphate. The terminal phosphate–oxygen bond has a high free energy of hydrolysis, which means that the phosphate can be transferred from ATP to other acceptor groups. For example, as a cosubstrate ATP is utilized by the kinases for the transfer of the terminal phosphate to various acceptors. A typical example is the reaction catalyzed by **glucokinase:**

$$\text{glucose} + \text{ATP} \rightarrow \text{glucose 6-phosphate} + \text{ADP}$$

where ADP is adenosine diphosphate.

ATP also serves as a modulator of the activity of some enzymes. These enzymes have binding sites for ATP, occupancy of which changes the affinity or reactivity of the enzyme toward its substrates. In these cases, ATP acts as an allosteric effector (see p. 436).

FIGURE 10.17
Structure of adenosine triphosphate (ATP).

NAD and NADP Are Coenzyme Forms of Niacin

Niacin is pyridine-3-carboxylic acid. It is converted to two coenzymes involved in oxidoreductase reactions. They are **NAD (nicotinamide adenine dinucleotide)** and **NADP (nicotinamide adenine dinucleotide phosphate).** The abbreviations NAD and NADP are convenient to use when referring to the coenzymes regardless of their state of oxidation or reduction. NAD^+ and $NADP^+$ represent the oxidized forms, and NADH and NADPH represent the reduced forms. Some dehydrogenases are specific for NADP and others for NAD; some function with either coenzyme. This arrangement allows for specificity and control over dehydrogenases that reside in the same subcellular compartment.

NAD^+ consists of adenosine and *N*-ribosyl-nicotinamide linked by a pyrophosphate linkage between the 5′-OH groups of the two ribosyl moieties (Figure 10.18). NADP differs structurally from NAD in having an additional phosphate esterified to the 2′-OH group of the adenosine moiety. Both coenzymes function as intermediates in transfer of two electrons between an electron donor and an acceptor. The donor and acceptor need not be involved in the same metabolic pathway. Thus the reduced form of these nucleotides acts as a common "pool" of electrons that arise from many oxidative reactions and can be used for various reductive reactions. The adenine, ribose, and pyrophosphate components of NAD are involved in binding of NAD to the enzyme. Enzymes requiring NADP do not have a conserved aspartate residue present in the NAD-binding site. If the aspartate were present, a charge–charge interaction between the negative charged aspartate and the 2′-phosphate of NADP would prevent binding. Since there is no negatively charged phosphate on the 2′-OH in NAD, there is discrimination between NAD and NADP binding. The nicotinamide reversibly accepts and donates two electrons at a time. It is the active center of the coenzyme. In oxidation of deuterated ethanol by alcohol dehydrogenase, NAD^+ accepts two electrons and the deuterium from the ethanol. The other hydrogen is released as a H^+ (Figure 10.19).

The binding of NAD^+ to the enzyme surface confers a chemically recognizable "top side" and "bottom side" to the planar nicotinamide. The former is known as the **A** face and the latter as the **B** face. In the case of alcohol dehydrogenase, the proton or deuterium ion that serves as a tracer is added to the A face. Other dehydrogenases utilize the B face. This particular effect demonstrates how enzymes can induce **stereospecificity** in chemical reactions by virtue of the asymmetric binding of coenzymes and substrates.

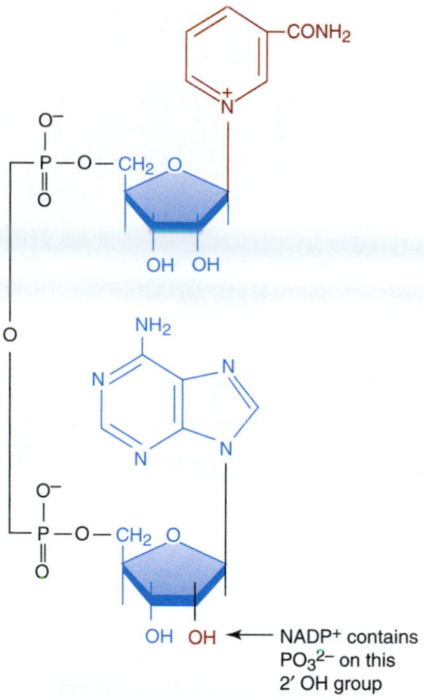

CONH2
N
O⁻
P—O—CH2 O
O
OH OH
O
NH2
N
N
N
N
O⁻
P—O—CH2 O
O
OH OH ← NADP+ contains PO3 2− on this 2′ OH group

FIGURE 10.18
Nicotinamide adenine dinucleotide (NAD^+).

FMN and FAD Are Coenzyme Forms of Riboflavin

The two coenzyme forms of riboflavin are **FMN (flavin mononucleotide)** and **FAD (flavin adenine dinucleotide).** The vitamin riboflavin consists of the heterocyclic ring, isoalloxazine (flavin), connected through N-10 to the alcohol ribitol (Figure 10.20). FMN has a phosphate esterified to the 5′-OH group of ribitol. FAD is structurally analogous to NAD in having adenosine linked by a pyrophosphate linkage to a heterocyclic ring, in this case riboflavin (Figure 10.21). Both FAD and FMN function in oxidoreduction reactions by accepting and donating $2e^-$ in the isoalloxazine ring. A typical example of FAD participation in an enzyme reaction is the oxidation of succinate to fumarate by succinate dehydrogenase (see p. 556) (Figure 10.22). In some cases, these coenzymes are $1e^-$ acceptors, which lead to flavin semiquinone formation (a free radical).

Flavin coenzymes tend to be bound much tighter to their apoenzymes than the niacin coenzymes and often function as prosthetic groups rather than as coenzymes.

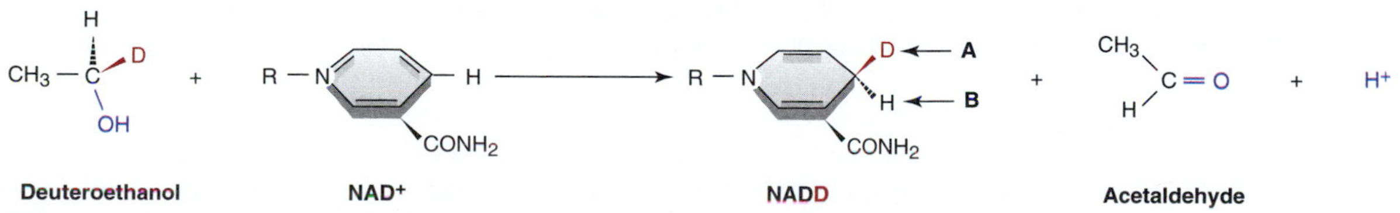

H
CH3—C—D
OH
+ R—N —H → R—N D ← A H ← B + CH3 C=O H + H+
CONH2
CONH2
Deuteroethanol
NAD+
NADD
Acetaldehyde

FIGURE 10.19
Stereo-specific transfer of deuterium from deuterated ethanol to NAD^+.

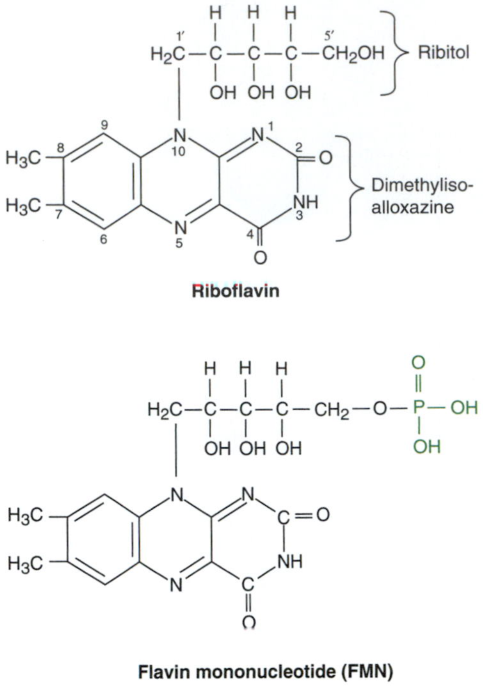

Riboflavin

Flavin mononucleotide (FMN)

FIGURE 10.20
Riboflavin and flavin mononucleotide.

Flavin adenine dinucleotide (FAD)

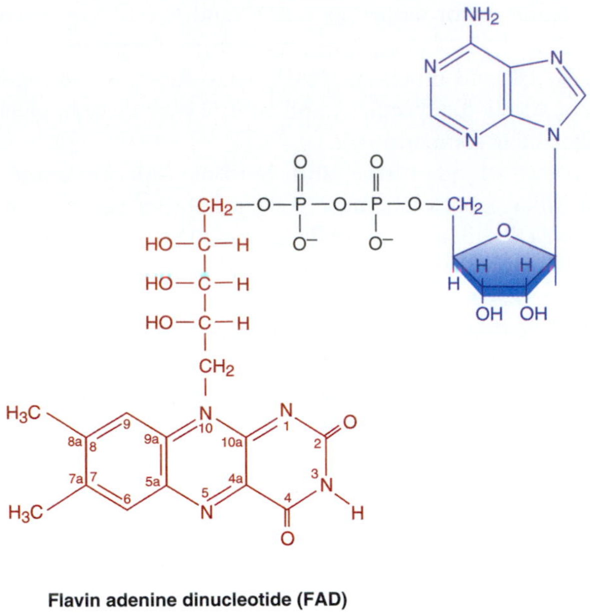

FIGURE 10.21
Flavin adenine dinucleotide (FAD).

Succinic acid + FAD ⇌ Fumaric acid + $FADH_2$

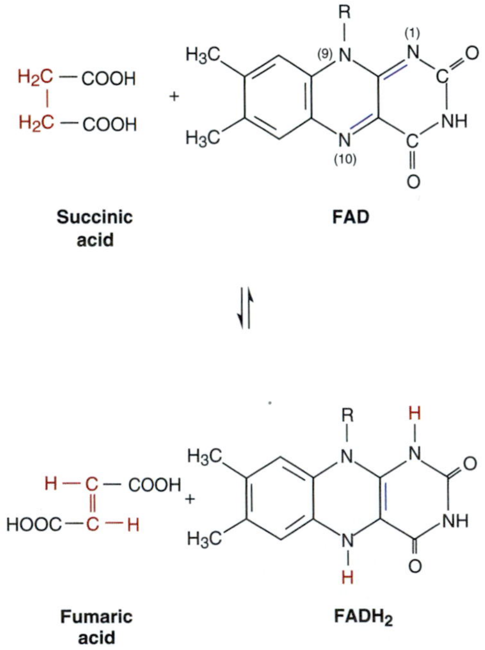

FIGURE 10.22
FAD is the coenzyme in the succinic dehydrogenase reaction.

Metal Cofactors Have Various Functions

Metals are not coenzymes in the same sense as FAD, FMN, NAD^+, and $NADP^+$ but are required as cofactors in approximately two-thirds of all enzymes. Metals participate in enzyme reactions by acting as Lewis acids and by various modes of chelate formation. **Chelates** are organometallic coordination complexes. A good example of a chelate is the complex between iron and protoporphyrin IX to form a heme (see p. 1063). Metals that act as Lewis acid catalysts are found among the transition metals like Zn, Fe, Mn, and Cu, which have empty *d* electron orbitals that act as electron sinks. The alkaline earth metals such as K and Na do not possess this ability. A good example of a metal functioning as a Lewis acid is found in **carbonic anhydrase**, a zinc enzyme that catalyzes the reaction

$$CO_2 + H_2O \rightleftharpoons H_2CO_3$$

The first step (Figure 10.23) can be visualized as the *in situ* generation of a proton and a hydroxyl group from water binding to the zinc (Lewis acid function of zinc). The proton and hydroxyl group are then added to the carbon dioxide and carbonic acid is released. Actually, the reactions presented in sequence may occur in a concerted fashion, that is, all at one time.

Metals can also promote catalysis either by binding substrate and promoting electrophilic catalysis at the site of bond cleavage or by stabilizing intermediates in the reaction pathway. In the case of **carboxypeptidase** and **thermolysin,** zinc proteases with identical active sites, the zinc functions to generate a hydroxyl group from water, and then to stabilize the transition state resulting from attack of the hydroxyl on the peptide bond. Figure 10.24 depicts the generation of the active-site hydroxyl by zinc. As shown, Glu 270 functions as a base in plucking the proton from water. Stabilization of the tetrahedral transition state by zinc is shown in Figure 10.25. The positive zinc provides a counterion to stabilize the negative oxygens on the tetrahedral carbon.

Role of the Metal as a Structural Element

The function of a metal as a Lewis acid in carbonic anhydrase and carboxypeptidase requires chelate formation. Various modes of chelation occur between metal, enzyme, and substrate that are structural in nature, but in which no acid catalysis occurs.

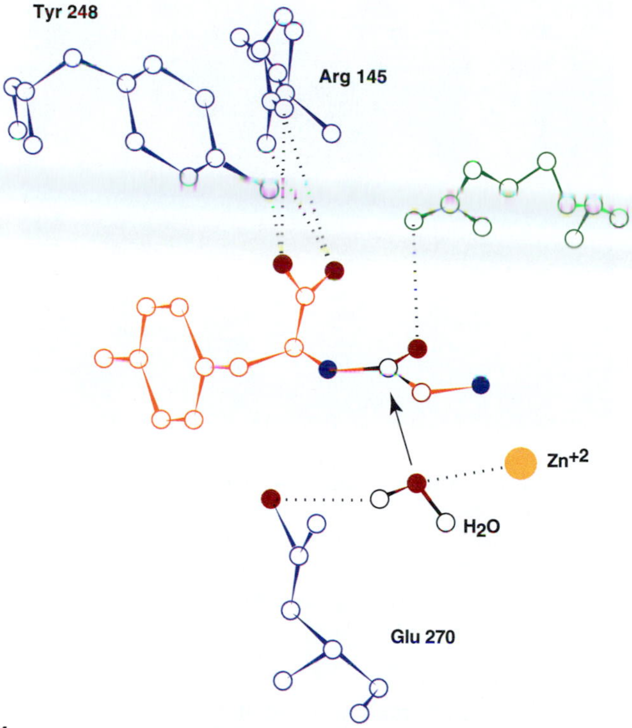

FIGURE 10.24
Zinc in the mechanism of reaction of carboxypeptidase A.
Enzyme-bound zinc generates a hydroxyl nucleophile from bound water, which attacks the carbonyl of the peptide bond as indicated by the arrow. Glu 270 assists by pulling the proton from the zinc-bound water.
Redrawn from Lipscomb, W. N. Robert A. Welch Found. Conf. Chem. Res. 15:140,1971.

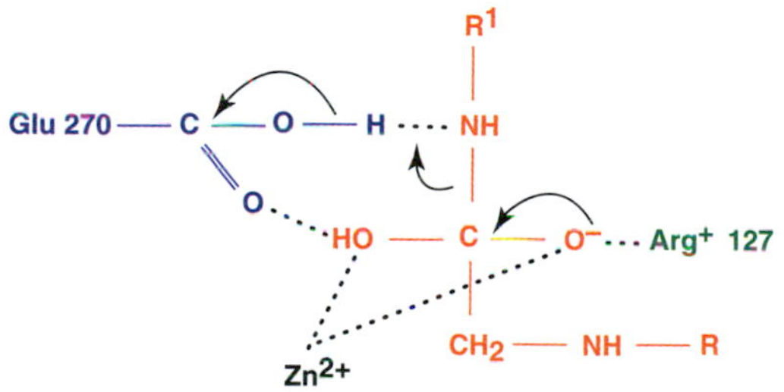

FIGURE 10.25
Stabilization of the transition state of the tetrahedral intermediate by zinc.
Positive charge on the zinc stabilizes the negative charge that develops on the oxygens of the tetrahedral carbon in the transition state. The tetrahedral intermediate then collapses as indicated by the arrows, resulting in breakage of the peptide bond.

In several kinases, creatine kinase being the best example, the true substrate is not ATP but Mg^{2+}–ATP (Figure 10.26). In this case, Mg^{2+} does not interact directly with the enzyme. It may serve to neutralize the negative charge density on ATP and facilitate binding to the enzyme. Ternary complexes of this conformation are known as "substrate-bridged" complexes and can be schematically represented as Enz–S–M. A hypothetical scheme for the binding of Mg^{2+}–ATP and glucose in the active site of hexokinase is presented in Figure 10.27. All kinases except muscle pyruvate kinase and phosphoenolpyruvate carboxykinase are substrate-bridged complexes.

In pyruvate kinase, Mg^{2+} chelates ATP to the enzyme. Absence of the metal cofactor results in failure of ATP to bind to the enzyme. Enzymes of this class are "metal-bridged" ternary complexes, Enz–M–S. **Metalloenzymes** are of this type

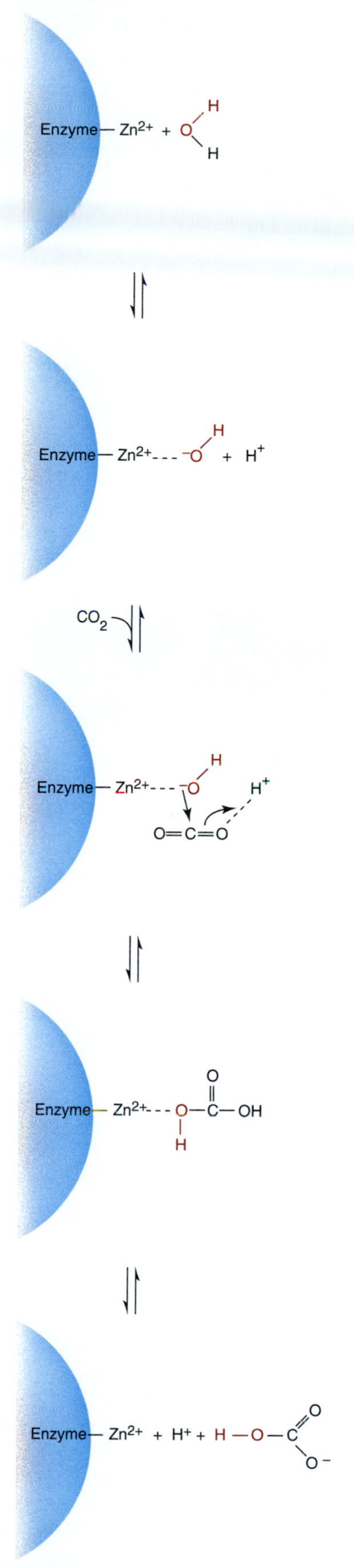

FIGURE 10.23
Zinc functions as a Lewis acid in carbonic anhydrase.

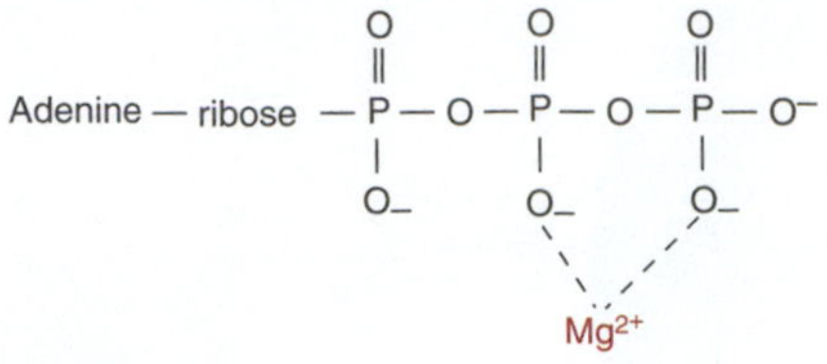

Adenine — ribose — P — O — P — O — P — O$^-$ (each P with =O above and O$^-$ below; Mg^{2+} bridging the two terminal O$^-$)

FIGURE 10.26
Structure of Mg^{2+}–ATP.

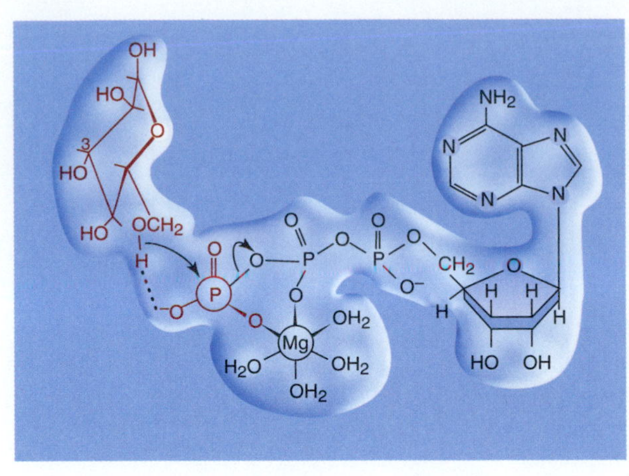

FIGURE 10.27
Role of Mg^{2+} as a substrate-bridged complex in the active site of the kinases. In hexokinase the terminal phosphate of ATP is transferred to glucose, yielding glucose 6-phosphate. Mg^{2+} coordinates with the ATP to form the true substrate and may labilize the terminal P—O bond of ATP to facilitate transfer of the phosphate to glucose. There are specific binding sites (light blue) on the enzyme for glucose (upper left in red) as well as the adenine and ribose moieties of ATP (black).

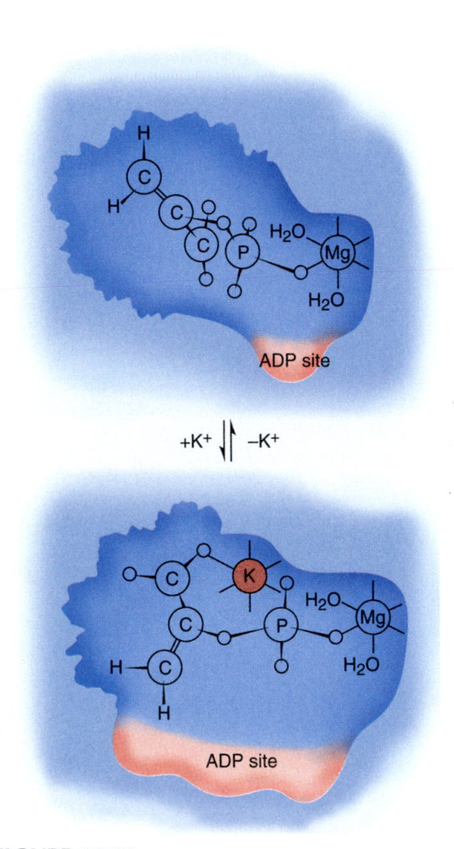

FIGURE 10.28
Model of the role of K^+ in the active site of pyruvate kinase.
Pyruvate kinase catalyzes the reaction: phosphoenolpyruvate + ADP ⟶ ATP + pyruvate. Initial binding of K^+ induces conformational changes in the kinase, which result in increased affinity for phosphoenolpyruvate. In addition, K^+ orients the phosphoenolpyruvate in the correct position for transfer of its phosphate to ADP (not shown), the second substrate. Mg^{2+} coordinates the substrate to the enzyme active site.
Modified with permission from Mildvan, A. S. Annu. Rev. Biochem. 43:365, 1971. Copyright © 1974 by Annual Reviews, Inc.

and contain a tightly bound transition metal such as Zn^{2+} or Fe^{2+}. Several enzymes that catalyze enolization and elimination reactions are metal-bridged complexes.

In addition to the role of binding enzyme and substrate, metals may also bind directly to the enzyme to stabilize it in the active conformation or perhaps to induce the formation of a binding site or active site. Not only do the strongly chelated metals like Mn^{2+} play a role in this regard, but the weakly bound alkali metals (Na^+ or K^+) are also important. In pyruvate kinase, K^+ induces an initial conformational change, which is necessary, but not sufficient, for ternary complex formation. Upon substrate binding, K^+ induces a second conformational change to the catalytically active ternary complex as indicated in Figure 10.28. Thus Na^+ and K^+ stabilize the active conformation of the enzyme but are passive in catalysis.

Role of Metals in Oxidation and Reduction

Iron–sulfur proteins, often referred to as nonheme iron proteins, are a unique class of metalloenzymes in which the active center consists of one or more clusters of sulfur-bridged iron chelates. The structures are presented on page XXX. In some cases the sulfur comes only from cysteine and in others from both cysteine and free ionic sulfur. The free sulfur is released as hydrogen sulfide upon acidification. These nonheme iron enzymes have reasonably low reducing potentials (E_0') and function in electron-transfer reactions (see p. 567).

Cytochromes are heme iron proteins that function as cosubstrates for their respective reductases (see p. 571). Iron in hemes of cytochromes undergoes reversible 1e$^-$ transfers. Heme is bound to the apoprotein by coordination of an amino acid side chain to iron of heme. Thus the metal serves not only a structural role but also participates in the chemical event.

Metals, specifically copper and iron, also have a role in activation of molecular oxygen. Copper is an active participant in several oxidases and hydroxylases. For example, **dopamine β-hydroxylase** catalyzes the introduction of one oxygen atom from O_2 into dopamine to form norepinephrine (Figure 10.29). The active enzyme contains one atom of cuprous ion that reacts with oxygen to form an activated oxygen–copper complex. The copper–hydroperoxide complex shown in Figure 10.29 is thought to be converted to a copper(II)–O$^-$ species that serves as the "active oxygen" in the hydroxylation of DOPA. In other metalloenzymes other species of "active oxygen" are generated and used for hydroxylation.

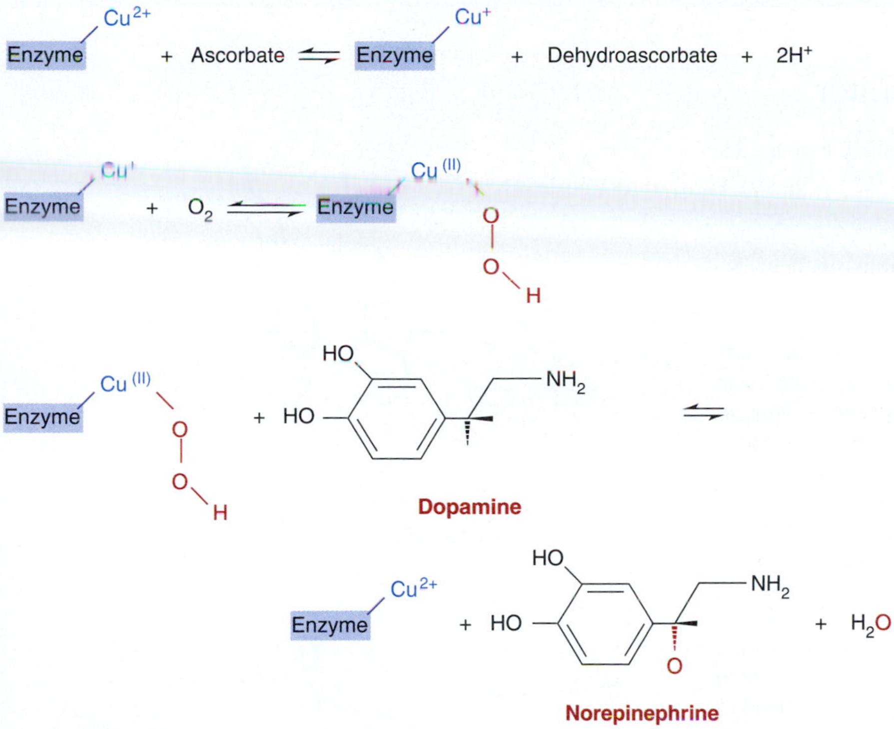

Enzyme–Cu^{2+} + Ascorbate ⇌ Enzyme–Cu^{+} + Dehydroascorbate + $2H^+$

Enzyme–Cu^{+} + O_2 ⇌ Enzyme–Cu (II)–O–O–H

Enzyme–Cu (II)–O–O–H + Dopamine ⇌ Enzyme–Cu^{2+} + Norepinephrine + H_2O

FIGURE 10.29
Role of copper in activation of molecular oxygen by dopamine hydroxylase.
The normal cupric form of the enzyme is not reactive with oxygen but on reduction by the cosubstrate, ascorbate, generates a reactive enzyme–copper bound oxygen species that then reacts with dopamine to form norepinephrine and an inactive cupric enzyme.

10.5 | INHIBITION OF ENZYMES

Mention was made of product inhibition of enzyme activity and how an entire pathway can be controlled or modulated by this mechanism (see p. 425). In addition to inhibition by the immediate product, products of other enzymes can also inhibit or activate a particular enzyme. Much of current drug therapy is based on inhibition of specific enzymes by a substrate analog. There are three major classes of inhibitors: competitive, noncompetitive, and uncompetitive.

Competitive Inhibition May Be Reversed by Increased Substrate

Competitive inhibitors are inhibitors whose action can be reversed by increasing amounts of substrate. Competitive inhibitors are structurally similar to the substrate and bind at the substrate-binding site, thus competing with the substrate for the enzyme. Once bound, the enzyme cannot convert the inhibitor to products. Increasing substrate concentrations will displace the reversibly bound inhibitor by the law of mass action. For example, in the succinate dehydrogenase reaction, malonate is structurally similar to succinate and is a competitive inhibitor (Figure 10.30).

Clinical Correlation 10.4 illustrates how very subtle differences in the active site of two similar enzymes can be utilized in the design of an inhibitor that will differentially inhibit the two enzymes.

Since substrate and inhibitor compete for the same binding site, the K_m for the substrate shows an apparent increase in the presence of inhibitor. This can be seen in a double-reciprocal plot as a shift in the x-intercept ($-1/K_m$) and in the slope of the line (K_m/V_{max}). If we first establish the velocity at several levels of substrate

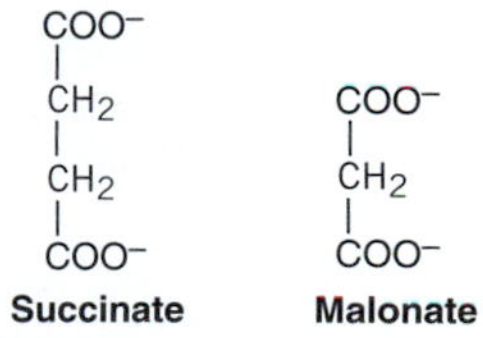

Succinate: COO^-–CH_2–CH_2–COO^-

Malonate: COO^-–CH_2–COO^-

FIGURE 10.30
Substrate and inhibitor of succinate dehydrogenase.

CLINICAL CORRELATION 10.4
Design of a Selective Inhibitor

Prostaglandins are a very important class of paracrine hormones derived from long chain unsaturated fatty acids (p. 766). One of the early steps in their synthesis involves an enzyme, cyclooxygenase (COX). Among the many physiological effects of the prostaglandins are the pain and inflammation associated with arthritis. Traditionally, arthritis has been treated with aspirin and other nonsteroidal anti-inflammatory drugs that have been shown to work by inhibiting the cyclooxygenases (p. 768), thus decreasing the levels of "bad" prostaglandins causing the symptoms. A serious problem with these drugs is the development of gastrointestinal bleeding and perforation caused by a concomitant decrease in the "good" prostaglandins, which are protective of the gastrointestinal mucosa.

Recently, it has been observed that there are two different cyclooxygenases, COX-1 and COX-2 (see p. 768) that have different tissue distributions. The COX-2 enzyme is inducible by mediators from the arthritic condition; whereas COX-1 constitutively makes the mucosa-sparing prostaglandins. The only significant difference in the active sites of the two enzymes is that COX-2 has a valine rather than isoleucine at residue 523. The smaller valine allows for preferential binding of custom-designed inhibitors, resulting in selective inhibition of COX-2. Such a custom designed drug, Vioxx™ (in green) is shown modeled into the COX-1 active site.

Key amino acid side chains forming the substrate-binding pocket are shown in red and the heme, which participates in prostaglandin synthesis, is shown in blue. It has been shown that aspirin acetylates Ser 530 in both COX-1 and -2 whereas Vioxx™ selectively binds to and inhibits COX-2, thus preserving the production of "good" prostaglandins by COX-1. The phenyl group of aspirin occupies the identical location in the active site as the phenyl group of Vioxx. Initially Vioxx-like drugs appear to be competitive inhibitors, but with COX-2 and not with COX-1, showing a time-dependent irreversible inhibition that does not involve a covalent linkage. Vioxx came on the market in 2000 and has made a tremendous difference in the quality of life for arthritic patients.

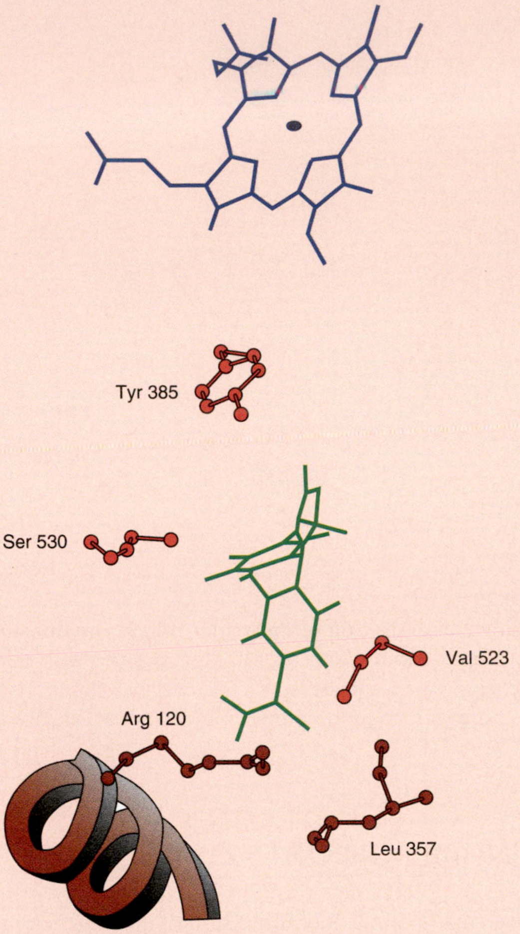

Tyr 385
Ser 530
Val 523
Arg 120
Leu 357

Gierse, J. K., McDonald, J. J., Hauser, S. D., Rangwala, S. H., Koboldt, C. M., and Seibert K. A single amino acid difference between cyclooxygenase-1 (COX-1) and -2 (COX-2) reverses the selectivity of COX-2 specific inhibitors. *J. Biol. Chem.* 271:15810, 1996.

and then repeat the experiment with a given but constant amount of inhibitor at various substrate levels, two different straight lines will be obtained (Figure 10.31). V_{max} does not change; hence the intercept on the y-axis remains the same. In the presence of inhibitor, the x-intercept is no longer the negative reciprocal of the true K_m, but of an apparent value, $K_{m,app}$ where

$$K_{m,app} = K_m \cdot \left(1 + \frac{[I]}{K_I}\right)$$

Thus the inhibitor constant, K_I, can be determined from the concentration of inhibitor used and the K_m, which was obtained from the x-intercept of the line obtained in the absence of inhibitor.

Noncompetitive Inhibitors Do Not Prevent Substrate from Binding

A **noncompetitive inhibitor** binds at a site other than the substrate-binding site. Inhibition is not reversed by increasing concentration of substrate. Both binary (EI)

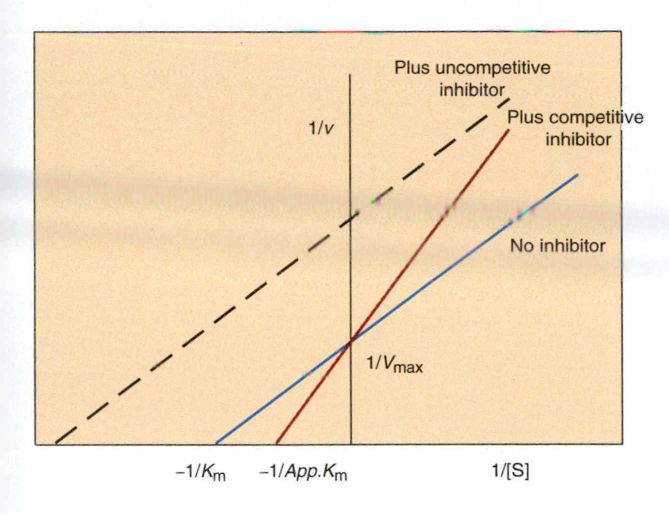

FIGURE 10.31
Double-reciprocal plots for competitive and uncompetitive inhibition. A competitive inhibitor binds at the substrate-binding site and effectively increases the K_m for the substrate. An uncompetitive inhibitor causes an equivalent shift in both V_{max} and K_m, resulting in a line parallel to that given by the uninhibited enzyme.

and ternary (EIS) complexes form, which are catalytically inactive and are therefore dead-end complexes. A noncompetitive inhibitor behaves as though it were removing active enzyme from the solution, resulting in a decrease in V_{max}. This is seen graphically in the double-reciprocal plot (Figure 10.32), where K_m does not change but V_{max} does change. Inhibition can often be reversed by exhaustive dialysis of the inhibited enzyme provided that the inhibitor has not reacted covalently with the enzyme as discussed under irreversible inhibitors.

An **uncompetitive inhibitor** binds only with the ES form of the enzyme in the case of a one-substrate enzyme. The result is an apparent equivalent change in K_m and V_{max}, which is reflected in the double-reciprocal plot as a line parallel to that of the uninhibited enzyme (Figure 10.31). In the case of multisubstrate enzymes the interpretation is complex and will not be considered further.

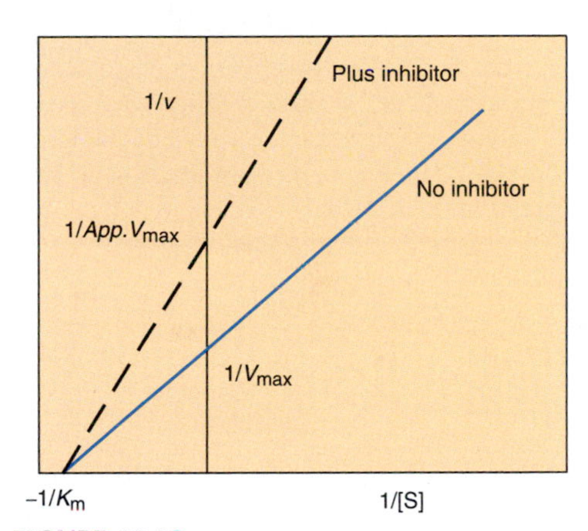

FIGURE 10.32
Double-reciprocal plot for an enzyme subject to reversible noncompetitive inhibition. A noncompetitive inhibitor binds at a site other than the substrate-binding site; therefore the effective K_m does not change, but the apparent V_{max} decreases.

Irreversible Inhibition Involves Covalent Modification of an Enzyme Site

When covalent modification occurs at the binding site or the active site, inhibition will not be reversed by dialysis unless the linkage is chemically labile like that of an ester or thioester. The active-site thiol group in glyceraldehyde-3-phosphate dehydrogenase reacts with ***p*-chloromercuribenzoate** to form a mercuribenzoate adduct of the enzyme (Figure 10.33). Such adducts are not reversed by dialysis or by addition of substrate. Double-reciprocal plots show the characteristic pattern for noncompetitive inhibition (Figure 10.32). Many pesticides function by alkylating the active site serine of acetylcholine esterase, thereby inhibiting nerve function. However, the esterase may be reactivated by a strong nucleophile as indicated in Clinical Correlation 10.5.

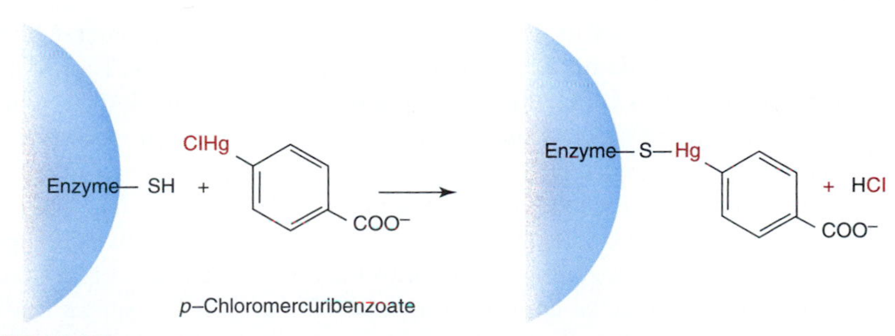

FIGURE 10.33
Enzyme inhibition by covalent modification of an active-site cysteine.

CLINICAL CORRELATION 10.5
A Case of Poisoning

Emergency room personnel encounter many instances of pesticide poisoning and must be equipped to recognize and treat these cases. Many of the common insecticides are organophosphate compounds that irreversibly inhibit the action of acetylcholine esterase, AChE, in the postsynaptic fibers of the cholinergic neurons (p. 998) by forming stable phosphate esters with a specific serine in the active site of the esterase. Inhibition of AChE prevents the hydrolysis of acetylcholine in the synapse, resulting in constant stimulation of the end organs of these neurons. The most prominent effects of pesticide poisoning in humans are paralysis of the respiratory muscles and pulmonary edema. If given early enough, a drug like Pralidoxime can displace the alkylphosphate from the pesticide bound to the active- site serine and regenerate an active AchE:

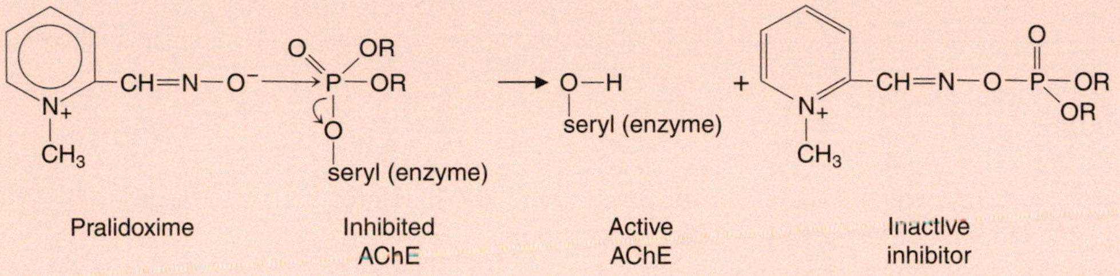

Pralidoxime + Inhibited AChE → Active AChE + Inactive inhibitor

In the disease myasthenia gravis, poisoning the AChE is good medical practice. In this condition there is a functional decrease in the amount of postsynaptic acetylcholine receptor. One way to overcome this deficit is to increase levels of acetylcholine by treating the patient with an AChE inhibitor.

Main, R. In E. Hodgson and F. E. Guthrie (Eds.,) *Introduction to Biochemical Toxicology*. New York: Elsevier, 1980, pp. 193–223.

Many Drugs Are Enzyme Inhibitors

Most modern drug therapy is based on the concepts of enzyme inhibition that were described in the previous section. Drugs are designed to inhibit a specific enzyme in a metabolic pathway. This application is most easily appreciated with antiviral, antibacterial, and antitumor drugs that are administered to the patient under conditions of limited toxicity. Such toxicity is often unavoidable because, with the exception of cell wall biosynthesis in bacteria, there are few critical metabolic pathways that are unique to tumors, viruses, or bacteria. Hence drugs that kill these organisms will often kill host cells. The one characteristic that can be taken advantage of is the comparatively short generation time of the undesirable organisms. They are much more sensitive to antimetabolites and in particular those that inhibit enzymes involved in replication. **Antimetabolites** are compounds with some structural difference from the natural substrate and are inhibitors. In subsequent chapters, numerous examples of antimetabolites will be described. A few representative examples will be presented to illustrate the concept.

Sulfa Drugs

Modern chemotherapy had its beginning in compounds of the general formula R—SO_2—NHR′. **Sulfanilamide,** the simplest member of the class, is an antibacterial agent because of its competition with ***p*-aminobenzoic acid** (PABA), which is required for bacterial growth. Structures of these compounds are shown in Figure 10.34.

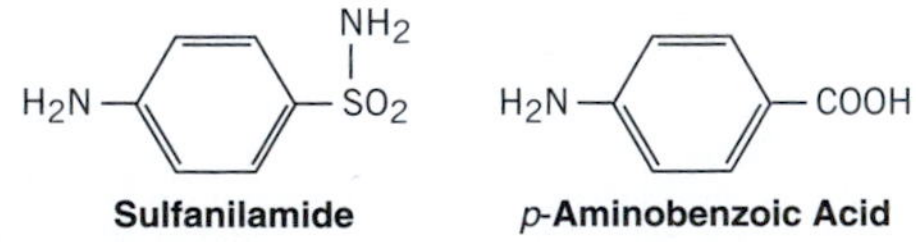

Sulfanilamide   *p*-Aminobenzoic Acid

FIGURE 10.34
Structure of *p*-aminobenzoic acid and sulfanilamide, a competitive inhibitor.

Bacteria cannot absorb folic acid, a required vitamin for the host, but must synthesize it. Since sulfanilamide is a structural analog of *p*-aminobenzoate, the bacterial dihydropteroate synthetase is tricked into making an intermediate, containing sulfanilamide, that cannot be converted to folate. Figure 10.35*b* shows the fully reduced or coenzyme form of folate. Thus the bacterium is starved of the required

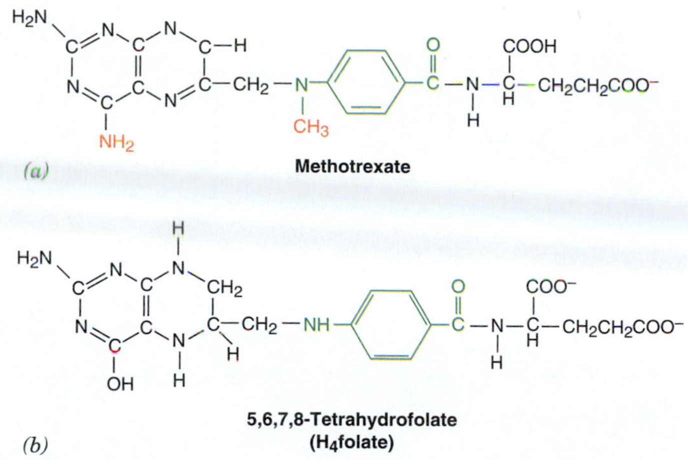

FIGURE 10.35
(*a*) Methotrexate (4-amino-N^{10}-methyl folic acid) and (*b*) tetrahydrofolic acid. Contribution from *p*-aminobenzoate is shown in green.

folate and cannot grow or divide. Since humans obtain folate from dietary sources, the sulfanilamide is not harmful at the doses that kill bacteria.

Methotrexate

Biosynthesis of purines and pyrimidines, heterocyclic bases required for synthesis of RNA and DNA, requires **folic acid,** which serves as a coenzyme in transfer of one-carbon units from various amino acid donors (see p. 794). **Methotrexate** (Figure 10.35*a*), a structural analog of folate, has been used with great success in childhood leukemia. Its mechanism of action is based on competition with dihydrofolate for **dihydrofolate reductase.** It binds 1000-fold more strongly than the natural substrate and is a powerful competitive inhibitor of the enzyme. The synthesis of thymidine monophosphate stops in the presence of methotrexate because of failure of the one-carbon transfer reaction. Since cell division depends on thymidine monophosphate as well as the other nucleotides, the leukemia cell cannot multiply. One problem is that rapidly dividing human cells such as those in bone marrow and intestinal mucosa are sensitive to the drug for the same reasons. Also, prolonged usage leads to amplification of the gene for dihydrofolate reductase, with increased levels of the enzyme and preferential growth of methotrexate-resistant cells.

Nonclassical Antimetabolites

A nonclassical antimetabolite is a substrate for an enzyme that upon action of the enzyme generates a highly reactive species. This species forms a covalent adduct with an amino acid side chain at the active site, leading to irreversible inactivation of the enzyme. An example of this type of inhibitor is the drug Omeprazole that is used in the treatment of excess stomach acidity. These inhibitors are referred to as **suicide substrates** and are very specific.

Another group of inhibitors contains a reactive functional group that is located outside the active site. For example, the compound shown in Figure 10.36 is an irreversible inhibitor of dihydrofolate reductase because it is specifically bound to the reductase and the reactive sulfonyl fluoride is positioned to react with a serine hydroxyl group in the substrate-binding site. Covalent binding of this substrate analog to the enzyme blocks binding of the normal substrate and thus inhibits the enzyme.

Other Antimetabolites

Two other analogs of the purines and pyrimidines will be mentioned to emphasize the structural similarity of chemotherapeutic agents to normal substrates.

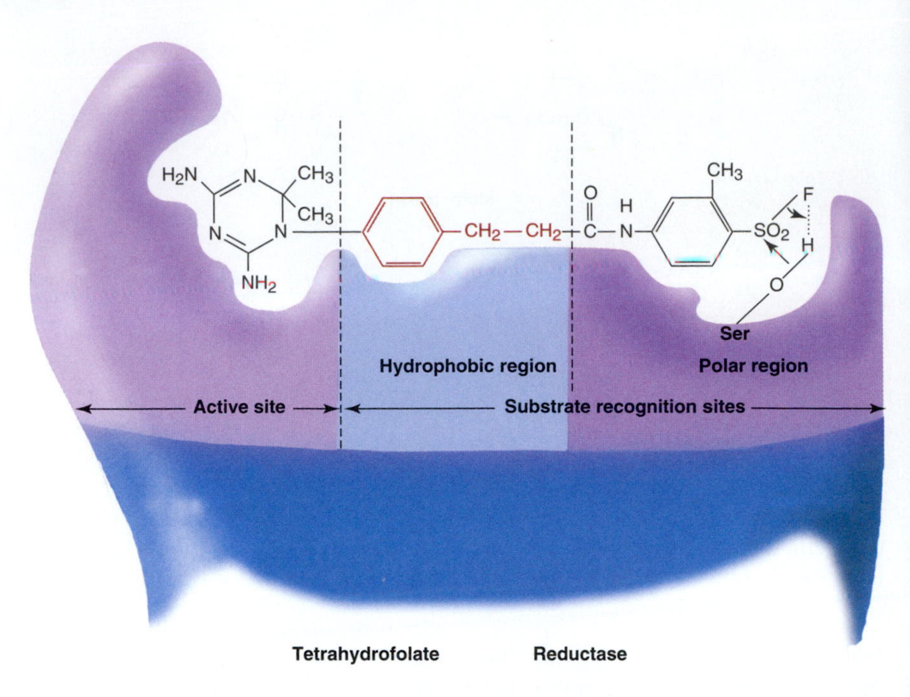

FIGURE 10.36
Site-directed inactivation of tetrahydrofolate reductase.
The irreversible inhibitor, a substituted dihydrotriazine, structurally resembles dihydrofolate and binds specifically to the dihydrofolate site on dihydrofolate reductase. The triazine portion of the inhibitor resembles the pterin moiety and therefore binds to the active site. The ethylbenzene group (in red) binds to the hydrophobic site normally occupied by the *p*-aminobenzoyl group. The reactive end of the inhibitor contains a reactive sulfonyl fluoride that forms a covalent linkage with a serine hydroxyl on the enzyme surface. Thus this inhibitor irreversibly inhibits the enzyme by blocking access of dihydrofolate to the active site.

Fluorouracil (Figure 10.37) is a thymine analog in which the ring-bound methyl is substituted by fluorine. The deoxynucleotide of this compound is an irreversible inhibitor of thymidylate synthetase. **6-Mercaptopurine** (Figure 10.37) is an analog of hypoxanthine, adenine, and guanine, which is converted to a 6-mercaptopurine nucleotide in cells. This nucleotide is a broad spectrum antimetabolite because of its competition in reactions involving adenine and guanine nucleotides. 6-Mercaptopurine is very effective in the treatment of childhood leukemias. The antimetabolites discussed relate to purine and pyrimidine metabolism but the general concepts can be applied to any enzyme or metabolic pathway.

10.6 | ALLOSTERIC CONTROL OF ENZYME ACTIVITY

Allosteric Effectors Bind at Sites Different from Substrate-Binding Sites

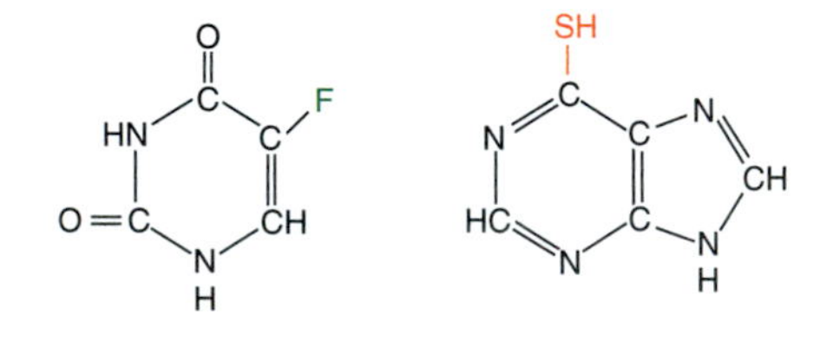

FIGURE 10.37
Structures of two antimetabolites.

Although the substrate-binding and active sites of an enzyme are well-defined structures, the activity of many enzymes can be modulated by ligands acting in ways other than as competitive or noncompetitive inhibitors. A **ligand** is any molecule that is bound to a macromolecule; the term is not limited to small organic molecules, such as ATP, but includes low molecular weight proteins. Ligands can be activators, inhibitors, or even the substrates of enzymes. Those ligands that change enzymatic activity, but are unchanged as a result of enzyme action, are referred to as effectors, modifiers, or modulators. Most of the enzymes subject to modulation by ligands are rate-determining enzymes in metabolic pathways. To appreciate the mechanisms by which metabolic pathways are controlled, the principles governing the allosteric and cooperative behavior of individual enzymes must be understood.

Enzymes that respond to modulators have additional site(s) known as allosteric site(s). **Allosteric** is derived from the Greek root *allo,* meaning "the other." An allosteric site is a unique region of the enzyme quite different from the substrate-binding site. The existence of allosteric sites is illustrated in Clinical Correlation 10.6. The ligands that bind at the allosteric site are called allosteric effectors or modulators. Binding of an allosteric effector causes a conformational change of the enzyme so that the affinity for the substrate or other ligands also changes. Positive (+) allosteric effectors increase the enzyme affinity for substrate or other ligand. The reverse is true for negative (−) allosteric effectors. The allosteric site at which the positive effector binds is referred to as an activator site; the negative effector binds at an inhibitory site.

Allosteric enzymes are divided into two classes based on the effect of the allosteric effector on the K_m and V_{max}. In the **K class** the effector alters the K_m but not V_{max}, whereas in the **V class** the effector alters V_{max} but not K_m. K class enzymes give double-reciprocal plots like those of competitive inhibitors (Figure 10.31) and V class enzymes give double reciprocal plots like those of noncompetitive inhibitors (Figure 10.32). The terms competitive and noncompetitive are inappropriate for allosteric enzyme systems because the mechanism of the effect of an allosteric inhibitor on a V or K enzyme is different from that of a simple competitive or noncompetitive inhibitor. For example, in the K class, the negative effector binding at an allosteric site affects the affinity of the substrate-binding site for the substrate, whereas in simple competitive inhibition the inhibitor competes directly with substrate for the site. In V class enzymes, positive and negative allosteric modifiers increase or decrease the rate of breakdown of the ES complex to products. The catalytic rate constant, k_3, is affected and not the substrate-binding constant. There are a few enzymes in which both K_m and V_{max} are affected.

In theory, a monomeric enzyme can undergo an allosteric transition in response to a modulating ligand. In practice, only two monomeric allosteric enzymes are known, ribonucleoside diphosphate reductase and pyruvate-UDP-*N*-acetylglucosamine transferase. Most allosteric enzymes are **oligomeric;** that is, they consist of several subunits. Identical subunits are designated as **protomers,** and each protomer may consist of one or more polypeptide chains. As a consequence of the oligomeric nature of allosteric enzymes, binding of ligand to one protomer can affect the binding of ligands on the other protomers in the oligomer. Such ligand effects are referred to as **homotropic interactions.** Transmission of the homotropic effects between protomers is one aspect of cooperativity, considered later. Substrate influencing substrate, activator influencing activator, or inhibitor influencing inhibitor binding are homotropic interactions. Homotropic interactions are almost always positive.

A **heterotropic interaction** is the effect of one ligand on the binding of a different ligand. For example, the effect of a negative effector on the binding of substrate or on binding of an allosteric activator are heterotropic interactions. Heterotropic interactions can be positive or negative and can occur in monomeric allosteric enzymes. Heterotropic and homotropic effects in oligomeric enzymes are mediated by cooperativity between subunits.

Based on the foregoing descriptions of allosteric enzymes, two models are pictured in Figure 10.38. In (*a*) a monomeric enzyme is shown, and in panel (*b*) an oligomeric enzyme consisting of two protomers is visualized. In both models heterotropic interactions can occur between the activator and substrate sites. In model (*b*), homotropic interactions can occur between the activator sites or between the substrate sites.

Allosteric Enzymes Exhibit Sigmoidal Kinetics

As a consequence of interaction between substrate site, activator site, and inhibitor site, a characteristic sigmoid or S-shaped curve is obtained in v_0 versus [S] plots of allosteric enzymes, as shown in Figure 10.39 (curve A). Negative allosteric effectors move the curve toward higher substrate concentrations and enhance the sigmoidicity of the curve. If we use $\frac{1}{2}V_{max}$ as a guideline, Figure 10.39 shows that a higher concentration of substrate would be required to achieve $\frac{1}{2}V_{max}$ in the presence of a negative effector (curve C) than required in the absence of negative effector (curve A). In the presence of a positive modulator (curve B), $\frac{1}{2}V_{max}$ can be reached at a lower substrate concentration than is required in the absence of the positive modulator (curve A). Positive modulators shift the v_0 versus [S] plots toward the hyperbolic plots observed in Michaelis–Menten kinetics.

CLINICAL CORRELATION 10.6
A Case of Gout Demonstrates the Difference Between an Allosteric and Substrate-Binding Site

The realization that allosteric inhibitory sites are separate from allosteric activator sites as well as from the substrate-binding and the catalytic sites is illustrated by a study of a gouty patient whose red blood cell PRPP level was increased (see Clin. Corr. 10.1). It was found that the patient's PRPP synthetase had normal K_m and V_{max} values, and sensitivity to activation by phosphate. The increased PRPP levels and hyperuricemia arose because the end products of the pathway (ATP, GTP) were not able to inhibit the synthetase through the allosteric inhibitory site (I). It was suggested that a mutation in the inhibitory site or in the coupling mechanism between the inhibitory and catalytic site led to failure of the feedback control mechanism.

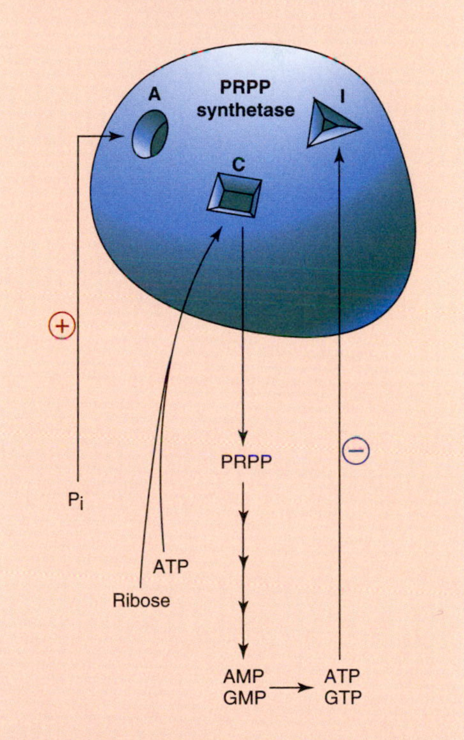

Sperling, O., Persky-Brosh, S., Boen, P., and DeVries, A. Human erythrocyte phosphoribosylpyrophosphate synthetase mutationally altered in regulatory properties. *Biochem. Med.* 7:389, 1973.

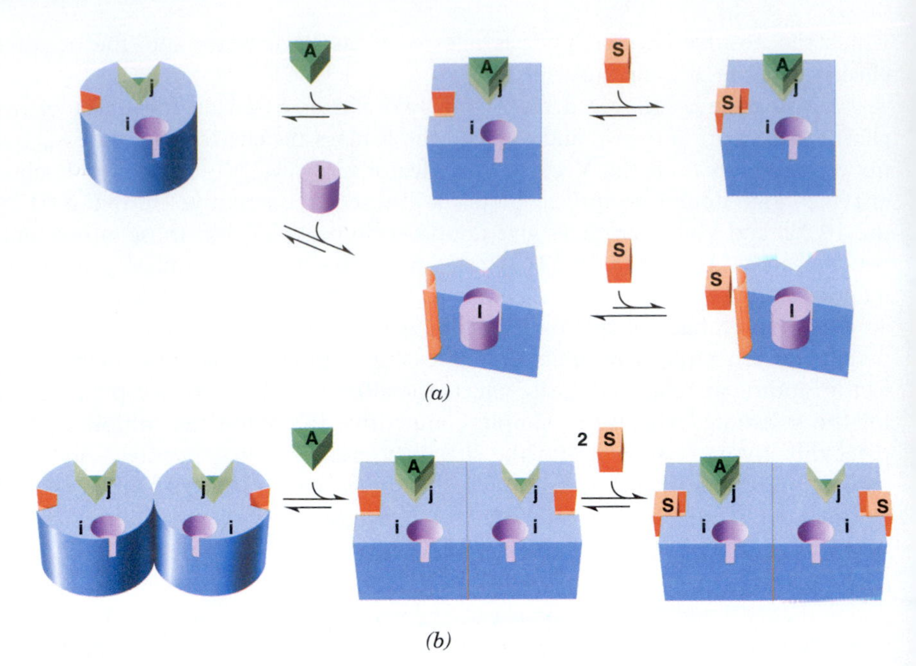

FIGURE 10.38
Models of allosteric enzyme systems.
(*a*) Model of a monomeric enzyme. Binding of a positive allosteric effector, A (green), to the activator site, j, induces a new conformation to the enzyme, one that has a greater affinity for the substrate. Binding of a negative allosteric effector (purple) to the inhibitor site, i, results in an enzyme conformation having a decreased affinity for substrate (orange). (*b*) A model of a polymeric allosteric enzyme. Binding of the positive allosteric effector, A, at the j site causes an allosteric change in the conformation of the protomer to which the effector binds. This change in the conformation is transmitted to the second protomer through cooperative protomer–protomer interactions. The affinity for the substrate is increased in both protomers. A negative effector decreases the affinity for substrate of both protomers.

The rates of allosteric-controlled enzymes can be finely controlled by small fluctuations in the level of substrate; often the *in vivo* concentration of substrate corresponds with the sharply rising segment of the sigmoid v_0 versus [S] plot; thus large changes in enzyme activity are effected by small changes in substrate concentration (see Figure 10.39). It is also possible to "turn an enzyme off" with small amounts of a negative allosteric effector by having the apparent K_m shifted to values well above the *in vivo* level of substrate. Note that at a given *in vivo* concentration of substrate the initial velocity, v_0, is decreased in the presence of a negative effector (compare curves A and C).

Cooperativity Explains Interaction Between Ligand Sites in an Oligomeric Protein

Since allosteric enzymes are usually oligomeric with sigmoid v_0 versus [S] plots, the concept of cooperativity was proposed to explain the interaction between ligand sites in oligomeric enzymes. **Cooperativity** is the influence that the binding of a ligand to one protomer has on the binding of ligand to another protomer in an oligomeric protein. It should be emphasized that kinetic mechanisms other than co-

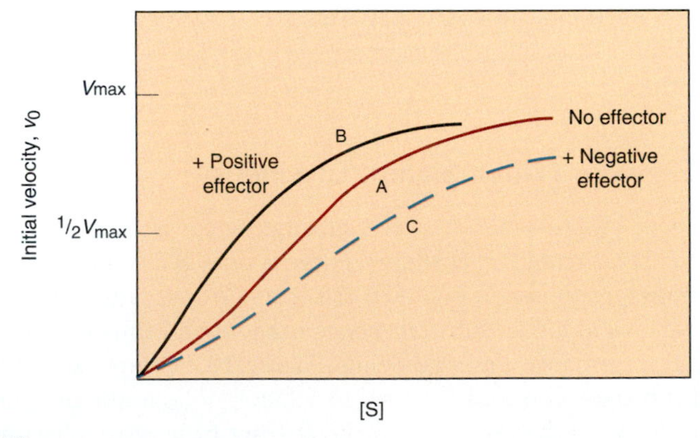

FIGURE 10.39
Kinetic profile of a K class allosteric enzyme.
The enzyme shows sigmoid v_0 versus [S] plots. Negative effectors shift the curve to the right, resulting in an increase in K_m. Positive effectors shift the curve to the left and effectively lower the apparent K_m. The V_{max} is not changed.

operativity can also produce sigmoid v_0 versus [S] plots; consequently, sigmoidicity is not diagnostic of cooperativity in a v_0 versus [S] plot. The relationship between allosterism and cooperativity has frequently been confused. Conformational change occurring in a given protomer in response to ligand binding at an allosteric site is an allosteric effect. Cooperativity generally involves a change in conformation of an effector-bound protomer that in turn transforms an adjacent protomer into a new conformation with an altered affinity for the effector ligand or for a second ligand. The conformation change may be induced by an allosteric effector or it may be induced by substrate, as it is in the case of hemoglobin where the oxygen-binding site on each protomer corresponds to the substrate site on an enzyme rather than to an allosteric site. Therefore the oxygen-induced conformational change in the hemoglobin protomers is technically not an allosteric effect, although some authors describe it as such. It is a homotropic cooperative interaction. Those who consider the oxygen-induced changes in hemoglobin to be "allosteric" are using the term in a much broader sense than the original definition allows; however, "allosteric" is now used by many to describe any ligand-induced change in the tertiary structure of a protomer.

An allosteric effect can occur in the absence of any cooperativity. For example, in alcohol dehydrogenase, conformational changes occur independently in each of the protomers upon the addition of positive allosteric effectors. The active site of each protomer is completely independent of the other and there is no cooperativity between protomers; that is, induced conformational changes in one protomer are not transmitted to adjacent protomers.

To describe experimentally observed ligand saturation curves mathematically, several models of cooperativity have been proposed. The two most prominent are the concerted model and the sequential induced-fit model. Although the concerted model is rather restrictive, most of the nomenclature associated with allosterism and cooperativity arose from it. The **concerted model** proposes that the enzyme exists in only two states, the T (tense or taut) and the R (relaxed) (Figure 10.40*a*). The T and R states are in equilibrium. Activators and substrates favor the R state and shift the preexisting equilibrium toward the R state by the law of mass action. Inhibitors

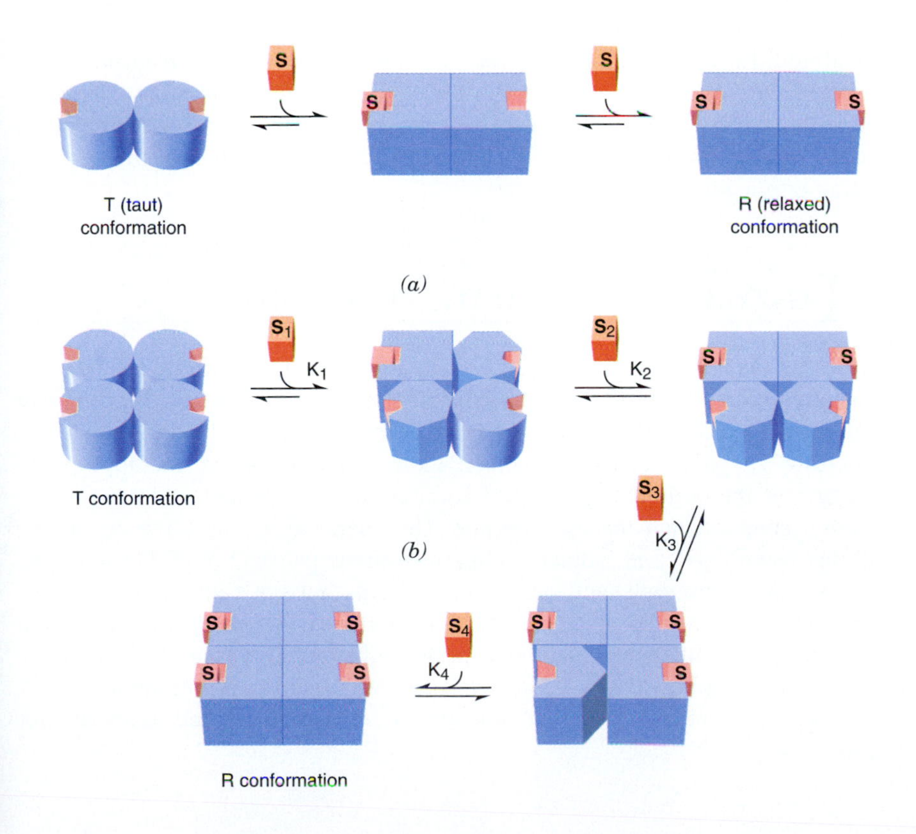

FIGURE 10.40
Models of cooperativity.
(*a*) The concerted model. The enzyme exists in only two states, the T (tense or taut) and R (relaxed) conformations. Substrates and activators have a greater affinity for the R state and inhibitors for the T state. Ligands shift the equilibrium between the T and R states. (*b*) The sequential induced-fit model. Binding of a ligand to any one subunit induces a conformational change in that subunit. This conformational change is transmitted partially to adjoining subunits through subunit–subunit interaction. Thus the effect of the first ligand bound is transmitted cooperatively and sequentially to the other subunits (protomers) in the oligomer, resulting in a sequential increase or decrease in ligand affinity of the other protomers. The cooperativity may be either positive or negative, depending on the ligand.

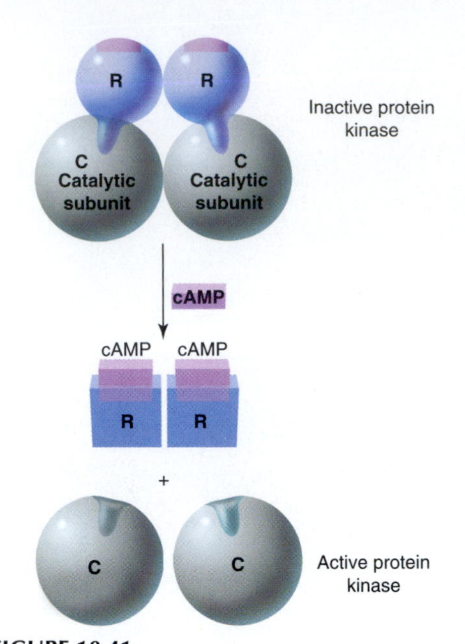

FIGURE 10.41
Model of allosteric enzyme with separate catalytic (C) and regulatory (R) subunits.
The regulatory subunit of protein kinase A contains a pseudosubstrate region in its primary sequence that binds to the substrate site of the catalytic subunit. In the presence of cAMP the conformation of the R subunit changes so that the pseudosubstrate region can no longer bind, resulting in release of active C subunits.

favor the T state. A conformational change in one protomer causes a corresponding change in all protomers. No hybrid states occur. Although this model accounts for the kinetic behavior of many enzymes, it cannot account for negative cooperativity.

The **sequential model** proposes that ligand binding induces a conformational change in a protomer. A corresponding conformational change is then partially induced in an adjacent protomer contiguous with the protomer containing the bound ligand. The effect of ligand binding is sequentially transmitted through the oligomer, producing increased or decreased affinity for the ligand by contiguous protomers (Figure 10.40*b*). In this model numerous hybrid states occur, giving rise to cooperativity and sigmoid v_0 versus [S] plots. Both positive and negative cooperativity can be accommodated by the model. A **positive modulator** induces a conformation in the protomer, which has an increased affinity for the substrate. A **negative modulator** induces a different conformation in the protomer, one that has a decreased affinity for substrate. Both effects are cooperatively transmitted to adjacent protomers. For the V class enzymes the effect is on the catalytic event (k_3) rather than on K_m.

Regulatory Subunits Modulate the Activity of Catalytic Subunits

In the foregoing an allosteric site was considered to reside on the same protomer as the catalytic site and all protomers were considered to be identical. In several very important enzymes a distinct **regulatory subunit** exists. These regulatory subunits have no catalytic function, but their binding with the catalytic protomer modulates the activity of the catalytic subunit through an induced conformational change. One strategy for regulation by regulatory subunits is outlined in Figure 10.41 for the **protein kinase A** (PKA) complex. Each regulatory subunit (R) has a segment of its primary sequence that is a pseudosubstrate for the catalytic subunit (C). In the absence of cAMP, the R subunit binds to the C subunit at its active site through the pseudosubstrate sequence, which inhibits the protein kinase activity. When cellular cAMP levels rise, cAMP binds to a site on the R subunits, causing a conformational change. This removes the pseudosubstrate sequence from the active site of the C subunit. The C subunits are released and can accept other protein substrates containing the pseudosubstrate sequence.

Calmodulin (see p. 525), a 17-kDa Ca^{2+}-binding protein, is a regulatory subunit for enzymes using Ca^{2+} as a modulator of their activity. Binding of calcium to calmodulin induces a conformational change in calmodulin, allowing it to bind to the Ca^{2+} enzyme. This binding induces a conformational change in the enzyme, restoring enzymatic activity.

10.7 | ENZYME SPECIFICITY: THE ACTIVE SITE

Enzymes are the most specific catalysts known, as regards the substrate and the type of reaction undergone by substrate. Specificity resides in the **substrate-binding site** on the enzyme surface. The tertiary structure of the enzyme is folded in such a way as to create a region that has the correct molecular dimensions, the appropriate topology, and the optimal alignment of counterionic groups and hydrophobic regions to accommodate a specific substrate. The tolerances in the active site are so small that usually only one isomer of a diastereomeric pair will bind. For example, D-amino acid oxidase will bind only D-amino acids but not L-amino acids. Some enzymes show absolute specificity for substrate. Others have broader specificity and will accept several different analogs of a specific substrate. For example, hexokinase catalyzes the phosphorylation of glucose, mannose, fructose, glucosamine, and 2-deoxyglucose, but at different rates. Glucokinase, on the other hand, is specific for glucose.

The specificity of the reaction rests in the active site and the amino acids that participate in the bond-making and bond-breaking phase of catalysis (see p. 443).

FIGURE 10.42
Lock-and-key model of the enzyme-binding site.
The enzyme contains a negative impression of the molecular features of the substrate, thus allowing specificity of the enzyme for a particular substrate. Specific ion pair formation can contribute to recognition of the substrate.

Complementarity of Substrate and Enzyme Explains Substrate Specificity

Various models have been proposed to explain the substrate specificity of enzymes. The first proposal was the "**lock-and-key**" **model** (Figure 10.42), in which a negative impression of the substrate is considered to exist on the enzyme surface. Substrate fits in this binding site just as a key fits into the proper lock or a hand into the proper sized glove. Hydrogen and ionic bonding and hydrophobic interactions contribute in binding substrate to the binding site. This model gives a rigid picture of the enzyme and cannot account for the effects of allosteric ligands.

A more flexible model of the binding site is provided by the **induced-fit model** in which the binding and active sites are not fully preformed. The essential elements of the binding site are present to the extent that the correct substrate can position itself properly. Interaction of substrate with enzyme induces a conformational change in the enzyme, resulting in the formation of a stronger binding site and the repositioning of the appropriate amino acids to form the active site. A schematic of the induced-fit model is shown in Figure 10.43*a*. Figure 10.44 depicts a specific example of induced fit: the glucose-induced conformational change in hexokinase that results in formation of the catalytic site. The smaller lobe, shown in green, of the hexokinase twists and moves some 12 Å to close around the glucose and bring the active-site residues into proximity with the glucose substrate. There is very little change in conformation of the larger blue lobe.

Induced fit combined with substrate strain accounts for more experimental observations concerning enzyme action than other models. In this model (Figure 10.43*b*), substrate is "strained" toward product formation by an induced conformational transition of the enzyme. A good example of enzyme-induced **substrate strain** is that of lysozyme (Figure 10.45) in which the conformation of the sugar residue "D" at which bond breaking occurs is strained from the stable chair to the unstable half-chair conformation upon binding. These conformations of glucose are shown in Figure 10.46. The concept of substrate strain explains the role of the enzyme in increasing the rate of reaction (see p. 445).

Asymmetry of the Binding Site

Not only are enzymes able to distinguish between isomers of the substrate, but they are able to distinguish between two equivalent atoms in a symmetrical

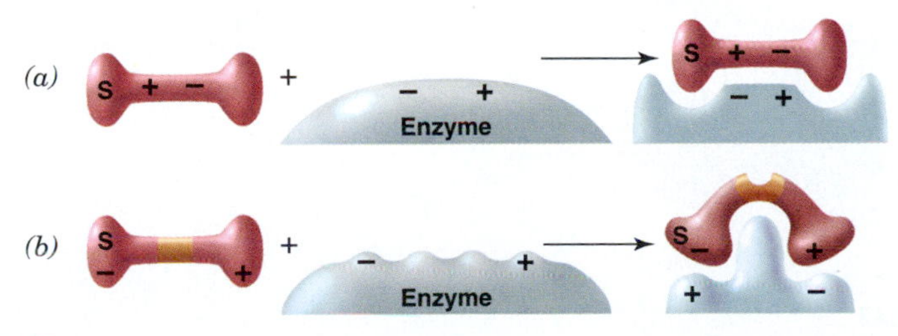

FIGURE 10.43
Models for induced fit and substrate strain.
(*a*) Approach of substrate to the enzyme induces the formation of the active site. (*b*) Substrate strain, induced by substrate binding to the enzyme, contorts normal bond angles and "activates" the substrate.
Reprinted with permission from Koshland, D. Annu. Rev. Biochem. *37:374, 1968. Copyright © 1968 by Annual Reviews, Inc.*

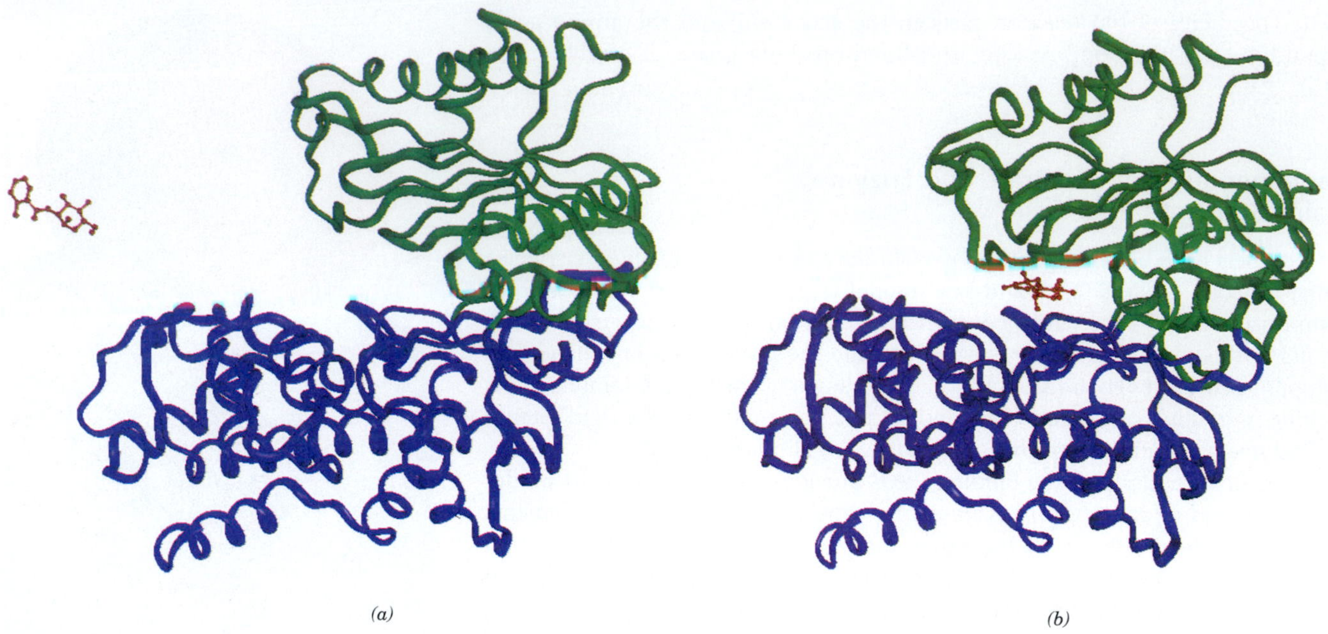

FIGURE 10.44
Glucose-induced conformational change of hexokinase.
(*a*) Hexokinase without glucose. (*b*) Hexokinase with glucose. The three-cord ribbon traces the peptide backbone of hexokinase.

Drawn from PDB files 1HKG and 2YHX; Bennett, W. S. Jr. and Steitz, T. A. J. Mol. Biol. *140:211, 1980.*

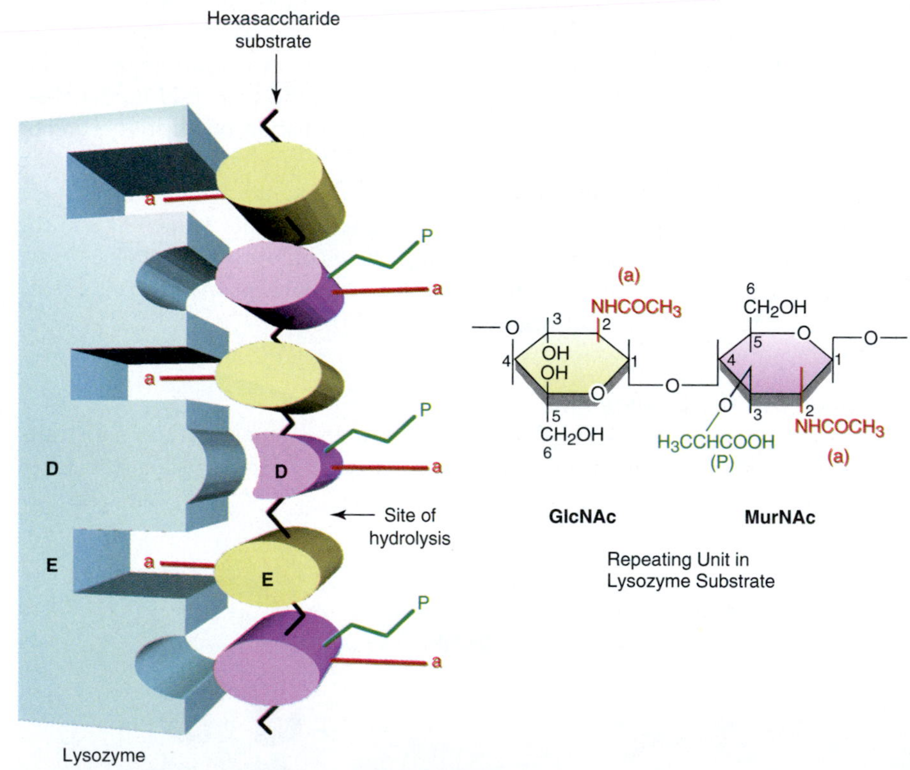

FIGURE 10.45
Hexasaccharide binding at active site of lysozyme.
In the model substrate pictured, the ovals represent individual pyranose rings of the repeating units of the lysozyme substrate shown to the right. Ring D is strained by the enzyme to the half-chair conformation and hydrolysis occurs between the D and E rings. Six subsites on the enzyme bind the substrate. Alternate sites are specific for acetamido groups (a) but are unable to accept the lactyl (P) side chains, which occur on the *N*-acetylmuramic acid residues. Thus the substrate can bind to the enzyme in only one orientation.
Redrawn based on model proposed by Imoto, T., et al. In: P. Boyer (Ed.), The Enzymes, 3rd ed., Vol. 7. *New York: Academic Press, 1972, p. 713.*

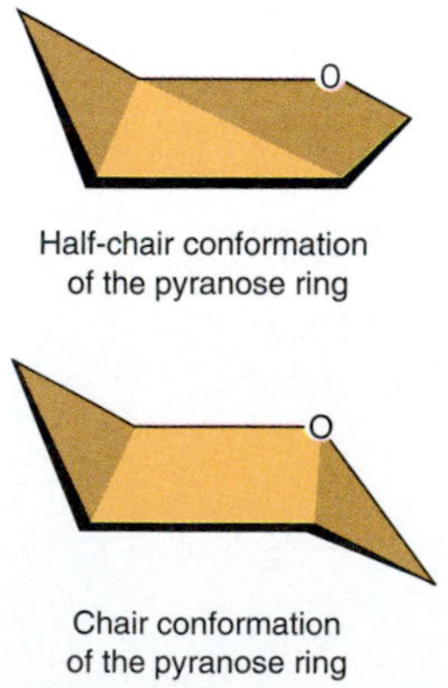

O
Half-chair conformation of the pyranose ring

O
Chair conformation of the pyranose ring

FIGURE 10.46
Two possible conformations of glucose.

molecule. For example, glycerol kinase distinguishes between configurations of H and OH on C-2 in the symmetric substrate glycerol, so that only the asymmetric product L-glycerol 3-phosphate is formed. These prochiral substrates have two identical substituents and two additional but dissimilar groups on the same carbon ($C_{aa'bd}$).

$$\begin{array}{llll} (a) & CH_2OH & & CH_2OH \\ & CHOH\ (b,d) & \xrightarrow[ATP]{\text{glycerol kinase}} & HOCH \quad + ADP \\ (a') & CH_2OH & & CH_2OPO_3^{2-} \end{array}$$

Glycerol — L-Glycerol 3-phosphate

Prochiral substrates possess no optical activity but can be converted to chiral compounds, that is, ones that possess an asymmetric center. The explanation for this enigma is provided if the enzyme binds the two dissimilar groups at specific sites and only one of the two similar substituents is able to bind at the active site (Figure 10.47). Thus the enzyme is able to recognize only one specific orientation of the symmetrical molecule. Asymmetry is produced in the product by modification of one side of the bound substrate. A minimum of three different binding sites on the enzyme surface is required to distinguish between identical groups on a prochiral substrate.

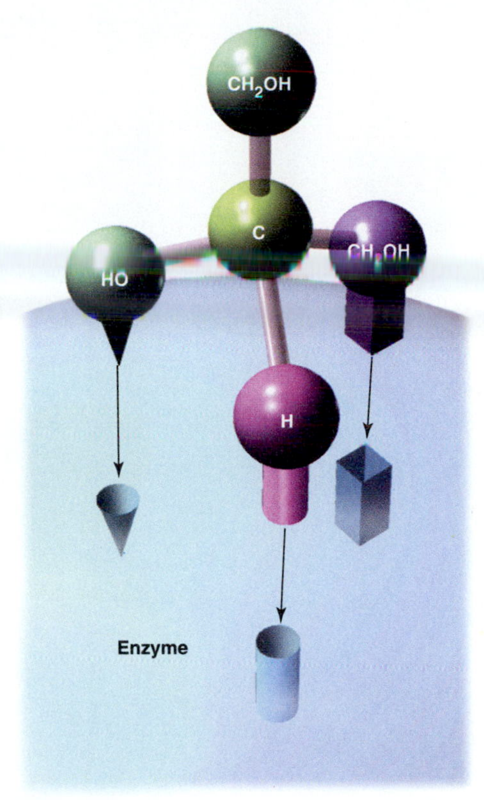

FIGURE 10.47
Three-point attachment of a symmetrical substrate to an asymmetric substrate-binding site. Glycerol kinase by virtue of dissimilar binding sites for the —H and —OH groups of glycerol binds only the α'-hydroxymethyl group to the active site. One stereoisomer results from the kinase reaction, L-glycerol 3-phosphate.

10.8 | MECHANISM OF CATALYSIS

All chemical reactions have a potential energy barrier that must be overcome before reactants can be converted to products. In the gas phase the reactant molecules can be given enough kinetic energy by heating them so that collisions result in product formation. The same is true with solutions. However, a well-controlled body temperature of 37°C does not allow temperature to be increased to accelerate the reaction, and 37°C is not warm enough to provide the reaction rates required for fast-moving species of animals. Enzymes employ other means of overcoming the barrier to reaction.

Diagrams for catalyzed and noncatalyzed reactions are shown in Figure 10.48. The energy barrier represented by the uncatalyzed curve in Figure 10.48 is a measure of the **activation energy,** E_a, required for the reaction to occur. The reaction coordinate is simply the pathway in terms of bond stretching between reactants and products. At the apex of the energy barrier is the activated complex known as the **transition state,** Ts, that represents the reactants in their activated state. In this state reactants are in an intermediate stage along the reaction pathway and cannot be identified as starting material or products. For example, in the hydrolysis of ethyl acetate,

$$CH_3-\overset{\overset{O}{\|}}{C}-O-CH_2-CH_3 \xrightarrow{H_2O} CH_3-CH_2-OH + CH_3-\overset{\overset{O}{\|}}{C}-OH$$

the Ts might look like

$$\left[\begin{array}{c} O^- \\ | \\ CH_3-C\cdots O-CH_2-CH_3 \\ \vdots \\ H\diagup O\cdots H \end{array}\right]$$

The transition state complex can break down to products or go back to reactants. The Ts is not an intermediate and cannot be isolated! In the case of the enzyme-catalyzed reaction (Figure 10.48) the energy of the reactants and products is no different than in the uncatalyzed reaction. Enzymes do not change the thermodynamics of the system but they do change the pathway for reaching the final state.

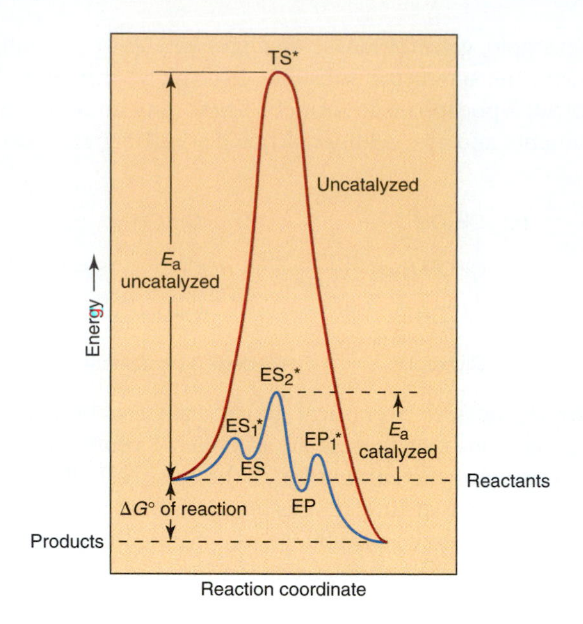

FIGURE 10.48
Energy diagrams for catalyzed versus noncatalyzed reactions.
The overall energy difference between reactants and products is the same in catalyzed and noncatalyzed reactions. The enzyme-catalyzed reaction proceeds at a faster rate because the energy of activation is lowered.

As noted on the energy diagram, there may be several plateaus or valleys on the energy contour for an enzyme reaction. At these points metastable intermediates exist. An important point is that each valley may be reached with the heat input available in 37°C. The enzyme allows the energy barrier to be scaled in increments. The Michaelis–Menten ES complex is not the transition state but may be found in one of the valleys because in the ES complex substrates are properly oriented and may be "strained." The bonds to be broken lie further along the reaction coordinate.

If our concepts of the transition state are correct, one would expect that compounds designed to resemble closely the transition state would bind more tightly to the enzyme than the natural substrate. This has proved to be the case. In such substrate analogs one finds affinities 10^2–10^5 times greater than those for substrate. These compounds are called **transition state analogs** and are potent enzyme inhibitors. Previously, lysozyme was discussed in terms of substrate strain, and mention was made of the conversion of sugar ring D from a chair to a strained half-chair conformation. Synthesis of a transition state analog in the form of the δ-lactone of tetra-*N*-acetylchitotetrose (Figure 10.49), which has a distorted half-chair conformation, followed by binding studies, showed that this transition state analog was bound 6000 times tighter than the normal substrate.

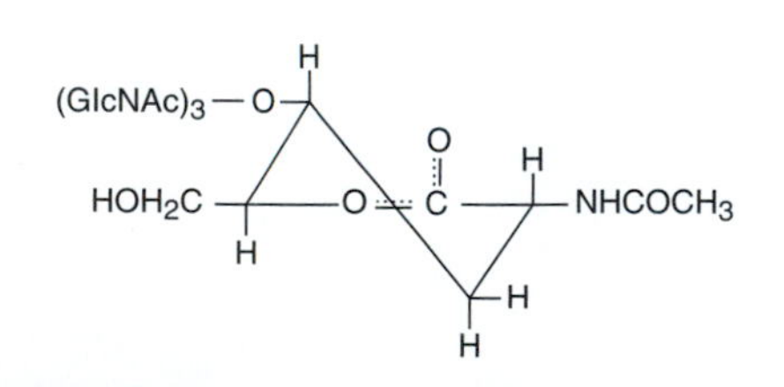

FIGURE 10.49
A transition state analog (tetra-*N*-acetylchitotetrose-γ-lactone) of ring D of the substrate for lysozyme.

Enzymes Decrease Activation Energy

On the average, enzymes can enhance the rates of reaction by a factor of 10^9–10^{12} times that of the noncatalyzed reaction. Orotidine 5′-phosphate decarboxylase is the most efficient enzyme known with a rate enhancement of 10^{17}-fold. In general, enzymatic rate enhancement can be accounted for by the following mechanisms: acid–base catalysis, substrate strain (transition state stabilization), covalent catalysis, ground state destabilization, and entropy effects. A given enzyme may utilize one or more of these mechanisms to achieve the tremendous rate enhancements observed.

Acid–Base Catalysis

Free protons and hydroxide ions are not encountered in most enzyme reactions and then only in some metal-dependent enzymes (see p. 429). A **general acid** or base is a compound that is weakly ionizable. In the physiological pH range, the protonated form of histidine is the most important general acid and its conjugate base is an important general base (Figure 10.50). Other acids are the thiol—SH of cysteine, tyrosine—OH, and the ε-amino group of lysine. Other bases are carboxylic acid anions and the conjugate bases of the general acids.

Ribonuclease (RNase) exemplifies the role of acid and base catalysis at the enzyme active site. RNase cleaves an RNA chain at the 3′-phosphodiester linkage of pyrimidine nucleotides with an obligatory formation of a cyclic 2′,3′-phosphoribose on a pyrimidine nucleotide as intermediate. In the mechanism outlined in Figure 10.51, His 119 acts as a general acid to protonate the phosphodiester bridge, whereas His 12 acts as a base in generating an alkoxide on the ribose-3′-hydroxyl. The latter then attacks the phosphate group, forming a cyclic phosphate and breakage of the RNA chain at this locus. The cyclic phosphate is then cleaved in phase 2 by a reversal of the reactions in phase 1, but with water replacing the leaving group. The active-site histidines revert to their original protonated state.

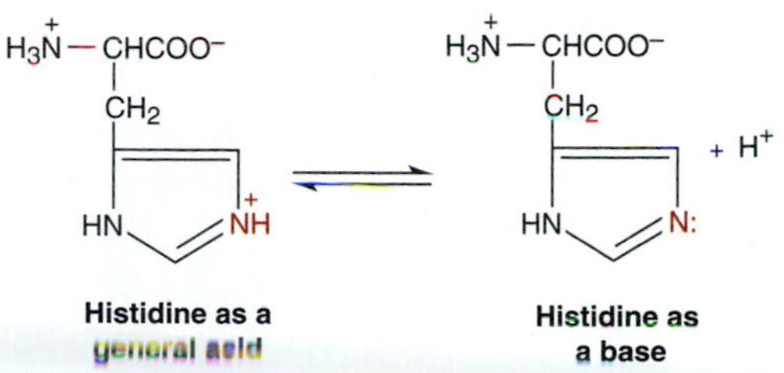

Histidine as a general acid

Histidine as a base

FIGURE 10.50
Acid and base forms of histidine.

Substrate Strain

Previous discussion of this topic related to induced fit of enzymes to substrate. Binding of substrate to a preformed site on the enzyme can induce strain in the substrate. Irrespective of the mechanism of strain induction, the energy level of the substrate is raised, and the bond lengths and angles of the substrate more closely resemble those found in the transition state.

A combination of substrate strain and acid–base catalysis is observed in the action of lysozyme (Figure 10.52). X-ray evidence shows that ring D of the hexasaccharide substrate is strained to the half-chair conformation upon binding to isozyme. General acid catalysis by active-site glutamic acid promotes the unstable half-chair

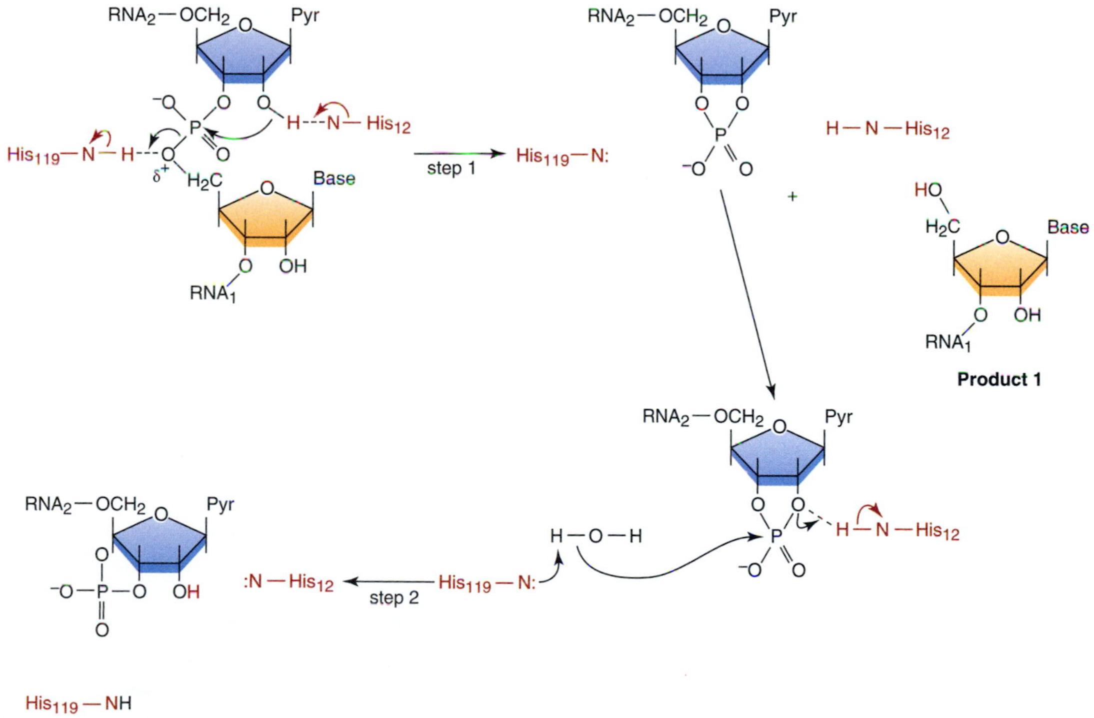

FIGURE 10.51
Role of acid and base catalysis in the active site of ribonuclease. RNase cleaves the phosphodiester bond in a pyrimidine locus in RNA. Histidine residues 12 and 119, respectively, at the ribonuclease active site function as acid and base catalysts in enhancing the formation of an intermediate 2′,3′-cyclic phosphate and release of a shorter fragment of RNA (product 1). These same histidines then play a reverse role in the hydrolysis of the cyclic phosphate and release of the other fragment of RNA (product 2) that ends in a pyrimidine nucleoside 3′-phosphate. As a result of the formation of product 2, the active site of the enzyme is regenerated.

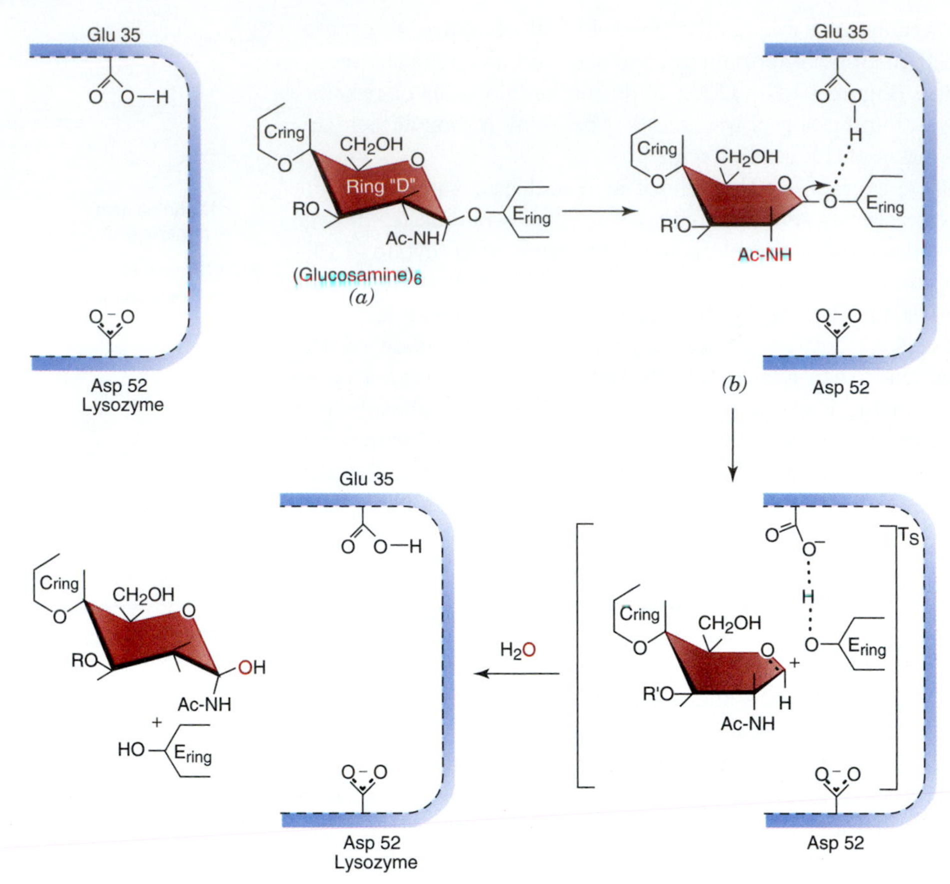

FIGURE 10.52
Mechanism for lysozyme action: substrate strain.
Binding of the stable chair (*a*) conformation of the substrate to the enzyme generates the strained half-chair conformation (*b*) in the ES complex. In the transition state, acid-catalyzed hydrolysis of the glycosidic linkage by an active-site glutamic acid residue generates a carbonium ion on the D ring, which relieves the strain generated in the initial ES complex and results in collapse of the transition state to products.

into the transition state. The oxycarbonium ion formed in the transition state is stabilized by the negatively charged aspartate. Breakage of the glycosidic linkage between rings D and E relieves the strained transition state by allowing ring D to return to the stable chair conformation.

Covalent Catalysis

In **covalent catalysis,** the attack of a nucleophilic (negatively charged) or electrophilic (positively charged) group in the enzyme active site upon the substrate results in covalent binding of the substrate to the enzyme as an intermediate in the reaction sequence. Enzyme-bound coenzymes often form covalent bonds with the substrate. For example, in the transaminases, the amino acid substrate forms a Schiff base with enzyme-bound pyridoxal phosphate (see p. 783). In all cases of covalent catalysis, the enzyme- or coenzyme-bound substrate is more labile than the original substrate. The enzyme–substrate adduct represents one of the valleys on the energy profile (Figure 10.48).

Serine proteases, such as trypsin, chymotrypsin, and thrombin, are good representatives of the covalent catalytic mechanism (see p. 377). Acylated enzyme has been isolated in the case of chymotrypsin. Covalent catalysis is assisted by acid–base catalysis in these particular enzymes (Figure 10.53). In chymotrypsin the attacking nucleophile is Ser 195, which is not dissociated at pH 7.4 and a mechanism for ionizing this very basic group is required. It is now thought that in the anhydrous milieu of the active site, Ser 195 and His 57 have similar p*K* values and that the negative charge on Asp 102 stabilizes the transfer of the proton from the OH of Ser 195 to N3 of His 57 (Figure 10.53). The resulting serine alkoxide attacks the car-

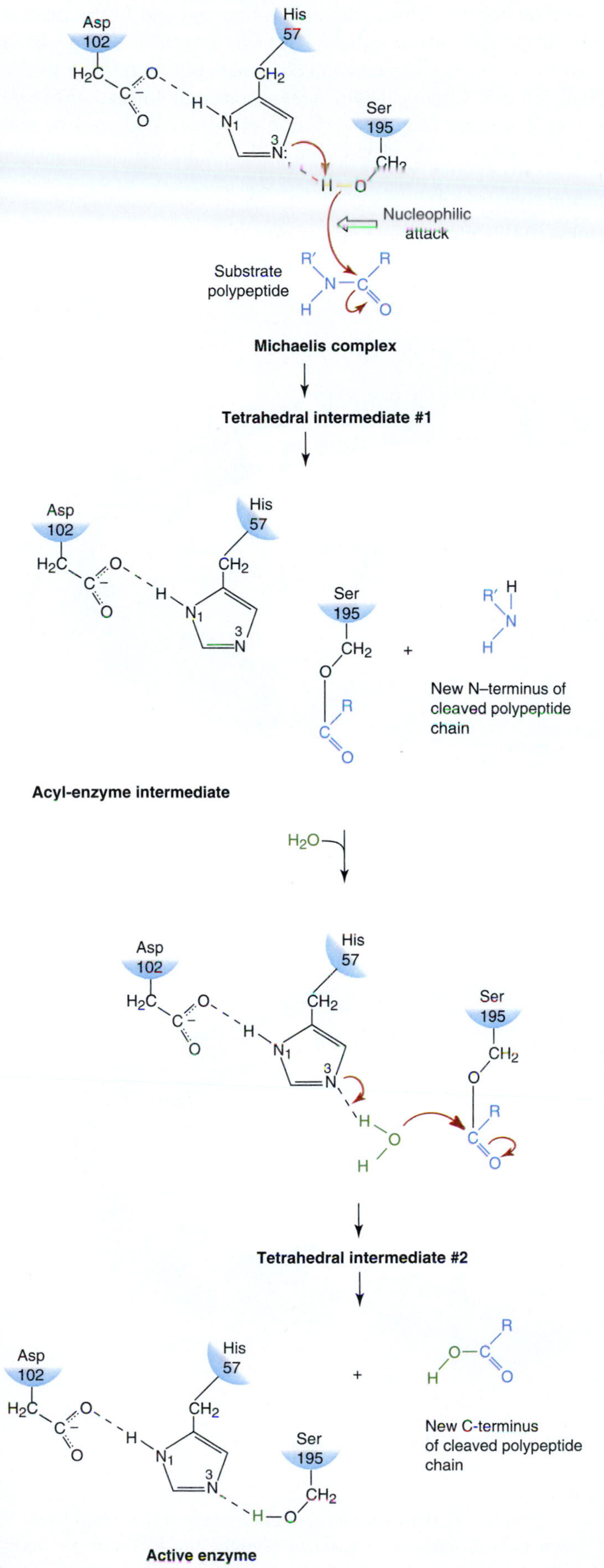

Asp 102
His 57
Ser 195
Nucleophilic attack
Substrate polypeptide
Michaelis complex
Tetrahedral intermediate #1

New N–terminus of cleaved polypeptide chain
Acyl-enzyme intermediate

H_2O

Tetrahedral intermediate #2

New C-terminus of cleaved polypeptide chain
Active enzyme

FIGURE 10.53
Covalent catalysis in active site of chymotrypsin.
Through acid-catalyzed nucleophilic attack, as shown by red arrows, the stable amide linkage of the peptide substrate is converted into an unstable acylated enzyme through Ser 195 of the enzyme. The latter is hydrolyzed in the rate-determining step. The new amino-terminal peptide, shown in blue, is released concomitant with formation of the acylated enzyme.

bonyl carbon of the peptide bond, releasing the amino-terminal end of the protein and forming an **acylated enzyme intermediate** (through Ser 195). The acylated enzyme is then cleaved by reversal of the reaction sequence, but with water as the nucleophile rather than Ser 195. Chemical evidence indicates the formation of two tetrahedral intermediates, one preceding the formation of the acylated enzyme and one following the attack of water on the acyl-enzyme (Figure 10.54).

Transition State Stabilization

The previously mentioned effects promote the substrate to enter the transition state. Since the active site binds the transition state with a much greater affinity

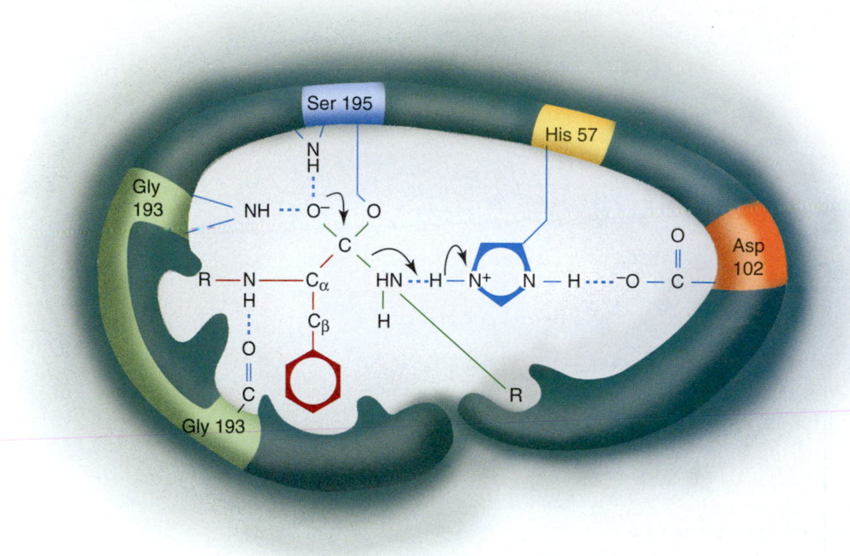

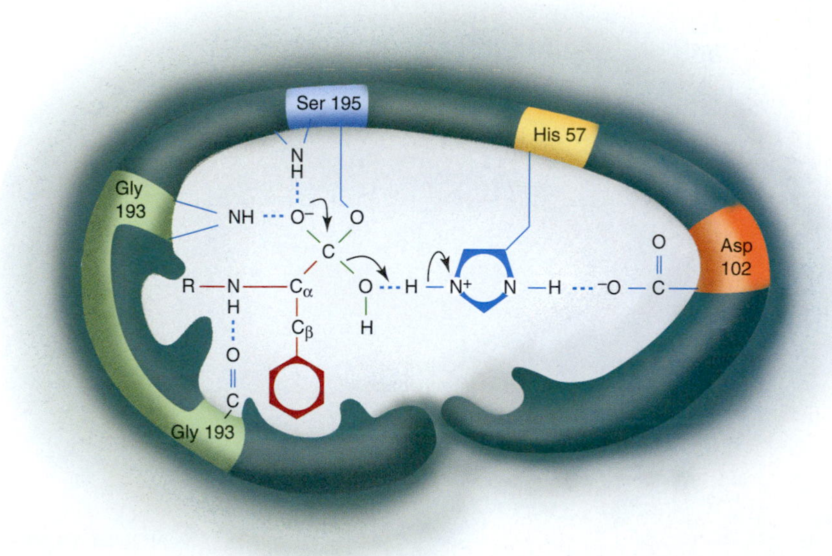

FIGURE 10.54
Tetrahedral intermediates.
(*a*) Model of tetrahedral intermediate #1 that precedes formation of the acyl-enzyme intermediate. (*b*) Model of tetrahedral intermediate #2 resulting from the attack of water on acyl enzyme intermediate.

than the substrate, that small fraction of substrate molecules existing in a transition state geometry will be converted to products quickly. Thus, by mass action, all the substrate can rapidly be converted to products. Any factor that increases the population of substrate molecules resembling the transition state will contribute to catalysis.

Entropy Effect

Entropy is a thermodynamic term, ΔS, which defines the extent of disorder in a system. At equilibrium, entropy is maximal. For example, in solution two reactants A and B exist in many different orientations. The chances of A and B coming together with the correct geometric orientation and with enough energy to react is small at 37°C and in dilute solution. However, if an enzyme with two high-affinity binding sites for A and B is introduced into the dilute solution of these reactants, as suggested in Figure 10.55, A and B will be bound to the enzyme in the correct orientation for the reaction to occur. They will be bound with the correct stoichiometry, and the effective concentration of the reactants will be increased on the enzyme surface, all of which will contribute to an increased rate of reaction.

When correctly positioned and bound on the enzyme surface, the substrates may be "strained" toward the transition state. At this point the substrates have been "set up" for acid–base and/or covalent catalysis. Proper orientation and the nearness of the substrate with respect to the catalytic groups, which has been dubbed the **"proximity effect,"** contribute 10^3–10^5-fold to the rate enhancement observed with enzymes. It is estimated that the decrease in entropy contributes a factor of 10^3 to the rate enhancement.

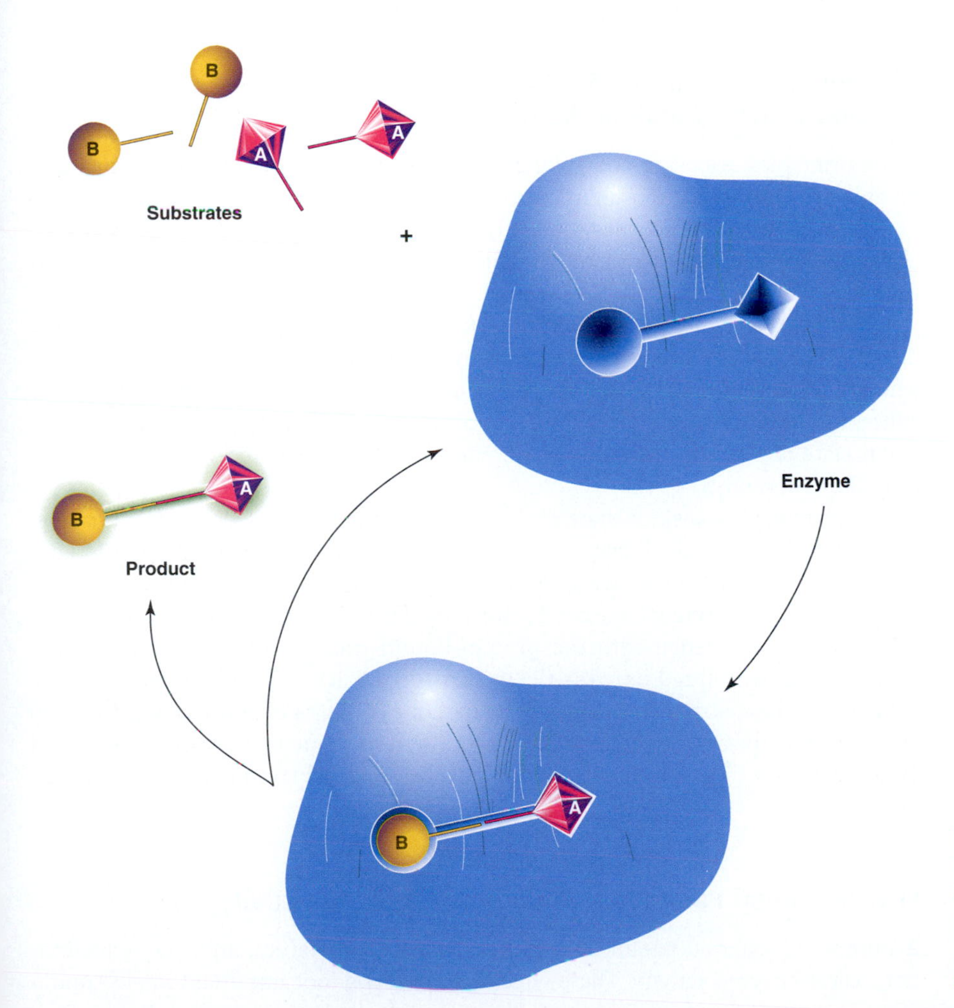

FIGURE 10.55
Role of the enzyme in enhancing reaction rate by decreasing entropy.
Substrates in dilute solution are concentrated and oriented on the enzyme surface so as to enhance the rate of the reaction.

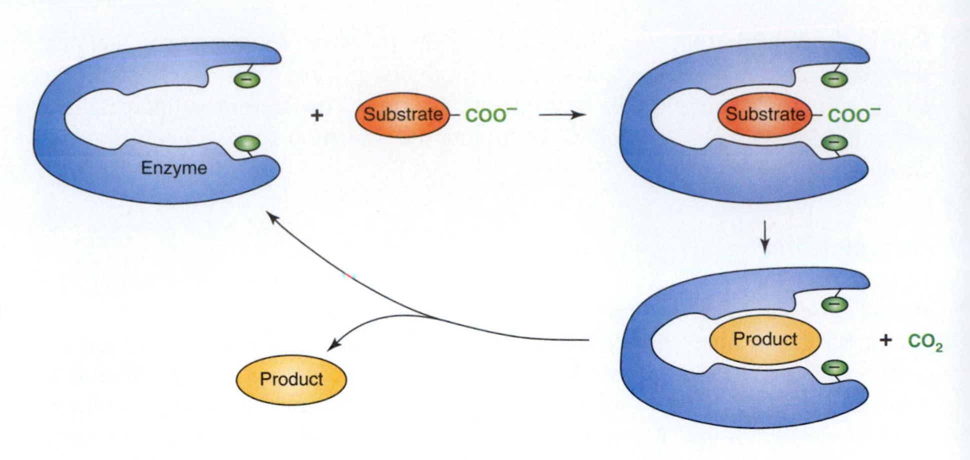

FIGURE 10.56
Model for the role of the enzyme in raising the ground state energy of the ES complex. This destabilization of the ground state results in a lowering of the energy of activation. The charge–charge repulsion, destabilized ground state, that develops on binding the substrate is relieved by decarboxylation of the substrate.

Ground State Destabilization

In addition to lowering the E_a by the mechanisms just discussed, the energy barrier also can be lowered by raising the energy of the **ground state.** Enzymes can accomplish this destabilization of the ground state by selecting conformations of substrates and enzyme active-site residues that approach the bonding distances and angles found in the transition state. The formation of these near attack conformers is enthalpy driven. The ground state can also be destabilized by bringing like charges between enzyme and substrate together so that chemical alteration of the substrate will relieve the unfavorable energetics as suggested in Figure 10.56, where the release of the carboxyl group as CO_2 relieves the unfavorable charge interaction of substrate and enzyme.

Abzymes Are Artificially Synthesized Antibodies with Catalytic Activity

If the principles discussed above for enzyme catalysis are correct, then one should be able to design an artificial enzyme. This feat has been accomplished by the use of several different approaches, but only the synthesis of antibodies that have catalytic activity will be considered in this discussion. These antibodies are called **abzymes.**

Design of abzymes is based on two principles. The first principle is the ability of the immune system to recognize any arrangement of atoms in the foreign antigen and to make a binding site on the resulting immunoglobulin that is exquisitely suited to binding that antigen. The second principle is that strong binding of transition state-like substrates reduces the energy barrier along the reaction pathway (see discussion on p. 444).

In abzymes a transition state analog serves as the hapten. For a lipase abzyme, a racemic phosphonate (Figure 10.57) serves as a hapten. Two enantiomeric fatty acid ester substrates are shown in Figure 10.57*b*,*c*. See page 448 for the transition state structure expected for ester hydrolysis. Among many antibodies produced by rabbits on challenge with the protein-bound transition state analog (Figure 10.57*a*), one hydrolyzed only the (*R*) isomer (Figure 10.57*b*) and another only the (*S*) isomer. These abzymes enhanced the rate of hydrolysis of substrates (*a*) and (*b*) 10^3–10^5-fold above the background rate in a stereospecific manner. Acceleration of 10^6-fold, which is close to the enzymatic rate, has been achieved in another esterase-like system.

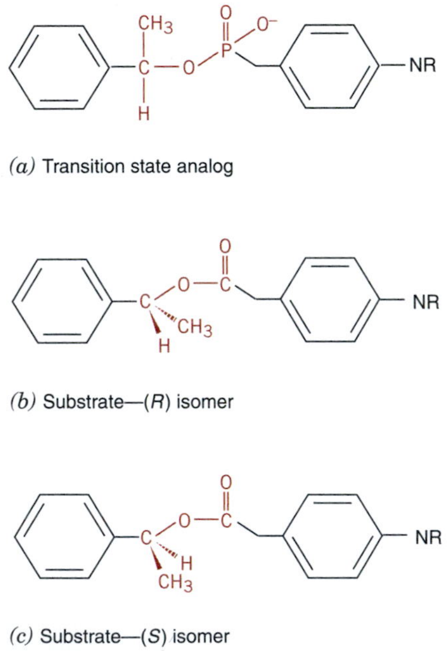

(*a*) Transition state analog

(*b*) Substrate—(*R*) isomer

(*c*) Substrate—(*S*) isomer

FIGURE 10.57
Hapten and substrate for a catalytic antibody (abzyme).
Phosphonate (*a*) is the transition state analog used as the hapten to generate antibodies with lipase-like catalytic activity. Specific abzymes can be generated for either the (*R*) isomer (*b*) or the (*S*) isomer (*c*) of methyl benzyl esters.

Environmental Parameters Influence Catalytic Activity

A number of external parameters, including pH, temperature, and salt concentration, affect enzyme activity. These effects are probably not important *in vivo* under

CLINICAL CORRELATION 10.7

Thermal Lability of Glucose 6-Phosphate Dehydrogenase Results in Hemolytic Anemia

In red cells, glucose 6-phosphate (G6PD) is an important enzyme in the red cell for the maintenance of the membrane integrity. A deficiency or inactivity of this enzyme leads to a hemolytic anemia. In other cases, a variant enzyme is present that normally has sufficient activity to maintain the membrane but fails under conditions of oxidative stress. A mutation of this enzyme leads to a protein with normal kinetic constants but a decreased thermal stability. This condition is especially critical to the red cell, since it is devoid of protein-synthesizing capacity and cannot renew enzymes as they denature. The end result is a greatly decreased lifetime for those red cells that have an unstable G6PD. These red cells are also susceptible to drug-induced hemolysis. See Clinical Correlation 15.1, p. 668.

Luzzato, L. and Mehta, A. Glucose-6-phosphate dehydrogenase deficiency. In: C. R. Scriver, A. L. Beaudet, W. S. Sly, and D. Valle (Eds.), *The Metabolic and Molecular Bases of Inherited Disease,* 7th ed. New York: McGraw-Hill, 1995, p. 3369.

normal conditions but are very important in setting up enzyme assays *in vitro* to measure enzyme activity in samples of a patient's plasma or tissue.

Temperature

Plots of velocity versus temperature for most enzymes reveal a bell-shaped curve with an optimum between 40°C and 45°C for mammalian enzymes, as indicated in Figure 10.58. Above this temperature, heat denaturation of the enzyme occurs. Between 0°C and 40°C, most enzymes show a twofold increase in activity for every 10°C rise. Under conditions of hypothermia, most enzyme reactions are depressed, which accounts for the decreased oxygen demand of living organisms at low temperature. Mutation of an enzyme to a thermolabile form can have serious consequences (see Clin. Corr. 10.7).

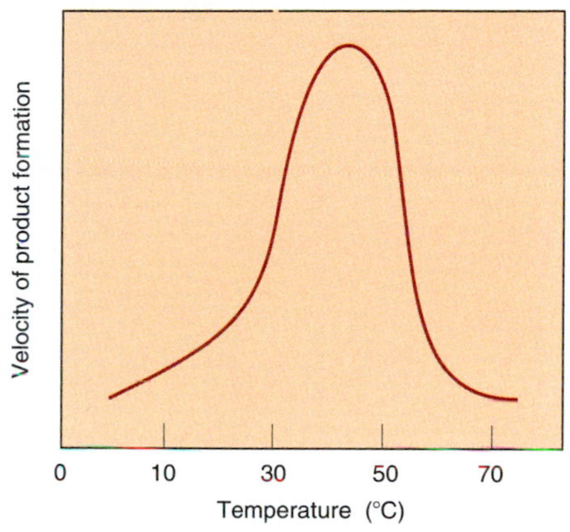

FIGURE 10.58
Temperature dependence of a typical mammalian enzyme.
To the left of the optimum, the rate is low because the environmental temperature is too low to provide enough kinetic energy to overcome the energy of activation. To the right of the optimum, the enzyme is inactivated by heat denaturation.

pH

Nearly all enzymes show a bell-shaped pH–velocity profile, but the maximum **(pH optimum)** varies greatly with different enzymes. Alkaline and acid phosphatases with very different pH optima are both found in humans, as shown in Figure 10.59. The bell-shaped curve and its position on the *x*-axis are dependent on the particular ionized state of the substrate that will be optimally bound to the enzyme. This in turn is related to the ionization of specific amino acid residues that constitute the substrate-binding site. In addition, amino acid residues involved in catalyzing the reaction must be in the correct charge state to be functional. For example, if aspartic acid is involved in catalyzing the reaction, the pH optimum may be in the region of 4.5 at which the α-carboxyl of aspartate ionizes; whereas if the ε-amino of lysine is the catalytic group, the pH optimum may be around 9.5, the pK_a of the ε-amino group. Studies of the pH dependence of enzymes are useful for suggesting which amino acid(s) may be operative in catalysis.

Clinical Correlation 10.8 points out the effect of a mutation leading to a change in the pH optimum of a physiologically important enzyme. Such a mutated enzyme may function on the shoulder of the pH-rate profile, but not be optimally active, even under normal physiological conditions.

When an abnormal condition such as alkalosis (observed in severe vomiting) or acidosis (observed in pneumonia and often in surgery) occurs, the enzyme activity may decrease because the pH is inappropriate. Thus, under normal conditions, the enzyme may be active enough to meet normal requirements, but under stress conditions the enzyme may be less active.

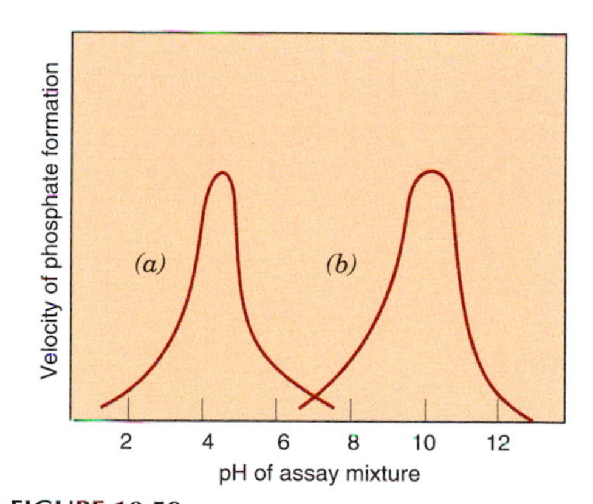

FIGURE 10.59
The pH dependence of (*a*) acid and (*b*) alkaline phosphatase reactions.
In each case the optimum represents the ideal ionic state for binding of enzyme and substrate and the correct ionic state for the amino acid side chains involved in the catalytic event.

CLINICAL CORRELATION 10.8

Alcohol Dehydrogenase Isoenzymes with Different pH Optima

In addition to the change in aldehyde dehydrogenase isoenzyme composition in some Asians (see Clin. Corr. 10.2), different alcohol dehydrogenase isoenzymes are also observed. Alcohol dehydrogenase (ADH) is encoded by three genes, which produce three different polypeptides: α, β, and γ. Three alleles are found for the β gene that differ in a single nucleotide base, which causes substitutions for arginine. The substitutions are shown below:

	Residue 47	*Residue 369*
β_1	Arg	Arg
β_2	His	Arg
β_3	Arg	Cys

The liver β_3 form has ADH activity with a pH optimum near 7, compared with 10 for β_1, and 8.5 for β_2. The rate-determining step in alcohol dehydrogenase is the release of NADH. NADH is held on the enzyme by ionic bonds between the phosphates of the coenzyme and the arginines at positions 47 and 369. In the β_1 isozyme this ionic interaction is not broken until the pH is quite alkaline and the guanidinium group of arginine starts to dissociate H^+. Substitution of amino acids with lower pK values, as in β_2 and β_3, weakens the interaction and lowers the pH optimum. Since the release of NADH is facilitated, the V_{max} values for β_2 and β_3 are also higher than for β_1.

Burnell, J. C., Carr, L. G., Dwulet, F. E., Edenberg, H. J., Li, T.-K., and Bosron, W. F. The human β_3 alcohol dehydrogenase subunit differs from β_1 by a Cys-for Arg-369 substitution which decreases NAD(H) binding. *Biochem. Biophys. Res. Commun.* 146:1227, 1987.

10.9 | CLINICAL APPLICATIONS OF ENZYMES

The principles of enzymology outlined in previous sections are applied in the clinical laboratory in measurement of plasma or tissue enzyme activities and concentrations of substrates in patients.

The rationale for measuring plasma enzyme activities is based on the premise that changes in activities reflect changes that have occurred in a specific tissue or organ. Plasma enzymes are of two types: (1) one type is present in the highest concentration, is specific to plasma, and has a functional role in plasma; and (2) the second is normally present at very low levels and plays no functional role in the plasma. The former includes the enzymes associated with blood coagulation (e.g., thrombin), fibrin dissolution (plasmin), and processing of chylomicrons (lipoprotein lipase). In disease of tissues and organs, the nonplasma-specific enzymes are most important. Normally, the plasma levels of these enzymes are low to absent. A disease process may cause changes in cell membrane permeability or increased cell death, resulting in release of intracellular enzymes into the plasma. When permeability changes, those enzymes of lower molecular weight will appear in the plasma first and the greater the concentration gradient between intra- and extracellular levels, the more rapidly the enzyme diffuses out. Cytosolic enzymes will appear in the plasma before mitochondrial enzymes, and the greater the quantity of tissue damaged, the larger the increase in plasma level. The nonplasma-specific enzymes will be cleared from the plasma at varying rates, which depend on the stability of the enzyme and its uptake by the reticuloendothelial system.

In the diagnosis of specific organ involvement in a disease process it would be ideal if enzymes unique to each organ could be identified. This is unlikely because the metabolic processes of various organs are very similar. Alcohol dehydrogenase of the liver and acid phosphatase of the prostate are useful for specific identification of disease in these organs. Other than these two examples, there are few enzymes that are tissue or organ specific. However, the ratio of various enzymes does vary from tissue to tissue. This fact, combined with a study of the kinetics of appearance and disappearance of particular enzymes in plasma, allows a diagnosis of specific organ involvement to be made. Figure 10.60 illustrates the time dependence of the plasma activities of enzymes released from the myocardium following a heart attack. Such profiles allow one to establish when the attack

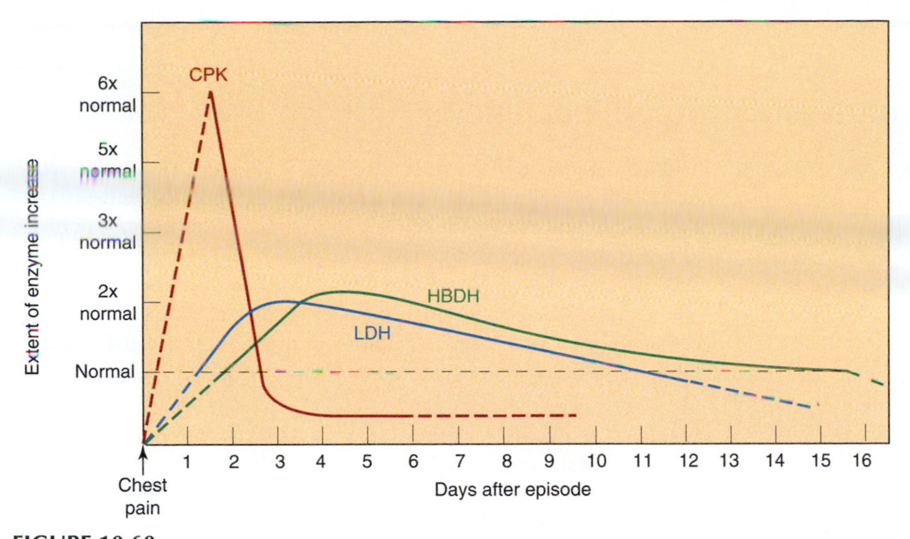

FIGURE 10.60
Kinetics of release of cardiac enzymes into serum following a myocardial infarction. CPK, creatine phosphokinase; LDH, lactic dehydrogenase; HBDH, α-hydroxybutyric dehydrogenase. Such kinetic profiles allow one to determine where the patient is with respect to the infarct and recovery. Note: CPK rises sharply but briefly; HBDH rises slowly but persists. *Reprinted with permission from Coodley, E. L.* Diagnostic Enzymes. *Philadelphia: Lea & Febiger, 1970, p. 61.*

occurred and whether treatment is effective. Clinical Correlation 10.9 demonstrates how diagnosis of a specific enzyme defect led to a rational clinical treatment that restored the patient to health.

Studies of the kinetics of appearance and disappearance of plasma enzymes require a valid enzyme assay. A good assay is based on temperature and pH control, as well as saturating levels of all substrates, cosubstrates, and cofactors. To accomplish the latter, the K_m must be known for those particular conditions of pH, ionic strength, and so on, that are to be used in the assay. Recall that K_m is the substrate concentration at half-maximal velocity ($\frac{1}{2}V_{max}$). To assure that the system is saturated, substrate concentration is generally increased five- to tenfold over the K_m. With saturation of the enzyme with substrate, the reaction is zero order. This fact is emphasized in Figure 10.61. Under zero-order conditions, changes in velocity are proportional to enzyme concentration alone. Under first-order conditions, the velocity is dependent on both the substrate and enzyme concentrations. Clinical Correlation 10.10 demonstrates the importance of determining if the assay conditions accurately reflect the amount of enzyme actually present. Clinical laboratory assay conditions are optimized for the properties of the normal enzyme and may not correctly measure levels of mutated enzyme. pH dependence and/or the K_m for substrate and cofactors may drastically change in a mutated enzyme. Under optimal conditions a valid enzyme assay reflects a linear dependence of velocity and amount of enzyme. This can be tested by determining if the velocity of the reaction doubles when the plasma sample volume is doubled, while keeping the total volume of the assay constant (Figure 10.62).

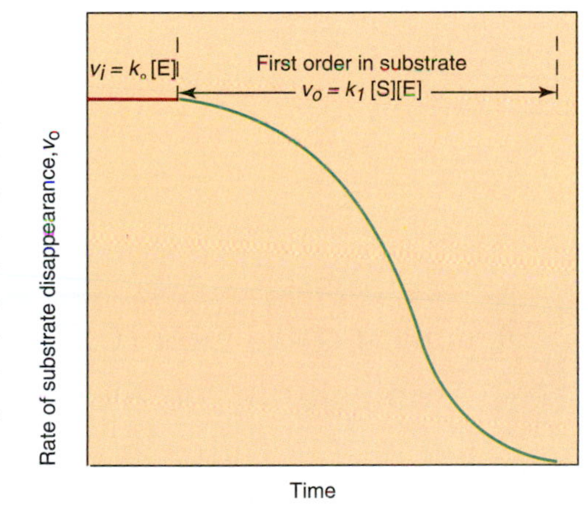

FIGURE 10.61
Relation of substrate concentration to order of the reaction.
When the enzyme is completely saturated, the kinetics are zero order with respect to substrate and are first order in the enzyme; that is, the rate depends only on enzyme concentration. When the substrate concentration falls below saturating levels, the kinetics are first order in both substrate and enzyme and are therefore second order; that is, the observed rate is dependent on both enzyme and substrate.

Coupled Assays Utilize the Optical Properties of NAD, NADP, or FAD

Enzymes that employ the coenzymes NAD^+, $NADP^+$, and FAD are easily measured because of the optical properties of NADH, NADPH, and FAD. The absorption spectra of NADH and FAD in the ultraviolet and visible light regions are shown in Figure 10.63. Oxidized FAD absorbs strongly at 450 nm, while NADH has maximal absorption at 340 nm. The concentrations of both FAD and NADH are related to their absorption of light at the respective absorption maximum by the **Beer–Lambert relation**

$$A = \varepsilon \cdot c \cdot l$$

CLINICAL CORRELATION 10.9
Identification and Treatment of an Enzyme Deficiency

Enzyme deficiencies usually lead to increased accumulation of specific intermediary metabolites in plasma and hence in urine. Recognition of the intermediates that accumulate in biological fluids is useful in pinpointing possible enzyme defects. After the enzyme deficiency is established, metabolites that normally occur in the pathway but are distal to the block may be supplied exogenously in order to overcome the metabolic effects of the enzyme deficiency.

In hereditary orotic aciduria there is a double enzyme deficiency in the pyrimidine biosynthetic pathway, leading to accumulation of orotic acid. Both orotate phosphoribosyltransferase and orotidine 5′-phosphate decarboxylase are deficient, causing decreased *in vivo* levels of CTP and TTP. The two activities are deficient because they reside in separate domains of a bifunctional polypeptide of 480 amino acids. dCTP and dTTP, which arise from CTP and TTP, are required for cell division. In these enzyme deficiency diseases the patients are pale, weak, and fail to thrive. Administration of the missing pyrimidines as uridine or cytidine promotes growth and general well-being and also decreases orotic acid excretion. The latter occurs because the TTP and CTP formed from the supplied uridine and cytidine repress carbamoyl-phosphate synthetase, the committed step, by feedback inhibition, resulting in a decrease in orotate production.

Webster, D. R., Becroft, D. M. O., and Suttie, D. P. Hereditary orotic aciduria and other diseases of pyrimidine metabolism. In C. R. Scriver, A. L. Beaudet, W. S. Sly, and D. Valle (Eds.), *The Metabolic and Molecular Bases of Inherited Disease*, 7th ed. New York: McGraw-Hill, 1995, p. 1799.

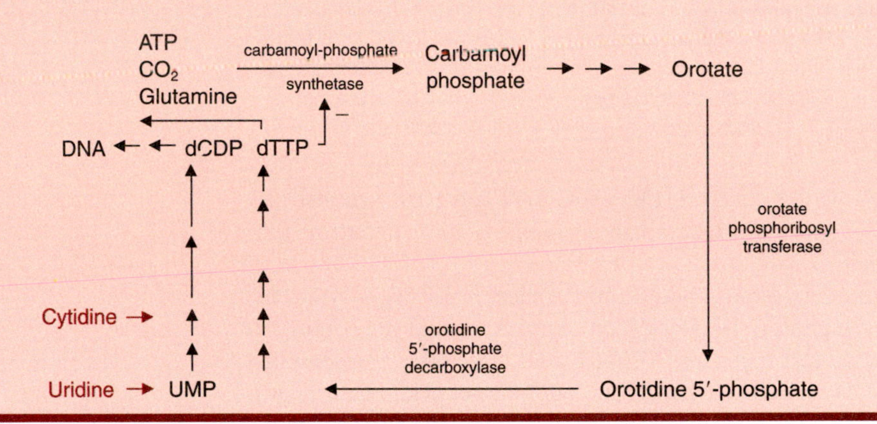

where l is the pathlength of the spectrophotometer cell in centimeters (usually 1 cm), ε is absorbance of a molar solution of the substance being measured at a specific wavelength of light, A is absorbance, and c is concentration. Absorbance is the log of transmittance (I_0/I). I_0 is the faction of the light transmitted in the absence of the sample and I in the presence of the sample. The term ε is a constant that varies from substance to substance; its value can be found in a handbook of

CLINICAL CORRELATION 10.10
Ambiguity in the Assay of Mutated Enzymes

Structural gene mutations leading to production of enzymes with increases or decreases in K_m are frequently observed. A case in point is a patient with hyperuricemia and gout, whose red blood cell hypoxanthine–guanine–phosphoribosyltransferase (HGPRT) showed little activity in assays *in vitro*. This enzyme is involved in the salvage of purine bases and catalyzes the reaction

$$\text{hypoxanthine} + \text{PRPP} \rightarrow \text{inosine monophosphate} + \text{PP}_i$$

where PRPP is phosphoribosylpyrophosphate.

The absence of HGPRT activity results in a severe neurological disorder known as Lesch–Nyhan syndrome (see p. 836), yet this patient did not have the clinical signs of this disorder. Immunological testing with a specific antibody to the enzyme revealed as much cross-reacting material in the patient's red blood cells as in normal controls. The enzyme was therefore being synthesized but was inactive in the assay *in vitro*. Increasing the substrate concentration in the assay restored full activity in the patient's red cell hemolysate. This anomaly is explained as a mutation in the substrate-binding site of HGPRT, leading to an increased K_m. Neither the substrate concentration in the assay nor in the red blood cells was high enough to bind to the enzyme. This case reinforces the point that an accurate enzyme determination is dependent on zero-order kinetics, that is, the enzyme being saturated with substrate.

Sorenson, L. and Benke, P. J. Biochemical evidence for a distinct type of primary gout. *Nature* 213:1122, 1967.

biochemistry. In an optically clear solution, the concentration c can be calculated after determination of the absorbance A and substituting into the Beer–Lambert equation.

Many enzymes do not employ NAD^+ or FAD but do generate products that can be utilized by a NAD^+- or FAD-linked enzyme. For example, glucokinase catalyzes the reaction

$$\text{glucose} + \text{ATP} \rightleftharpoons \text{glucose 6-phosphate} + \text{ADP}$$

ADP and glucose 6-phosphate (G6P) are difficult to measure directly; however, glucose 6-phosphate dehydrogenase catalyzes the reaction

$$\text{glucose 6-phosphate} + \text{NADP}^+ \rightleftharpoons \text{6-phosphogluconolactone} + \text{NADPH} + \text{H}^+$$

Thus by adding an excess of G6P dehydrogenase and $NADP^+$ to the assay mixture, the velocity of production of G6P by glucokinase is proportional to the rate of reduction of $NADP^+$, which can be measured directly in the spectrophotometer.

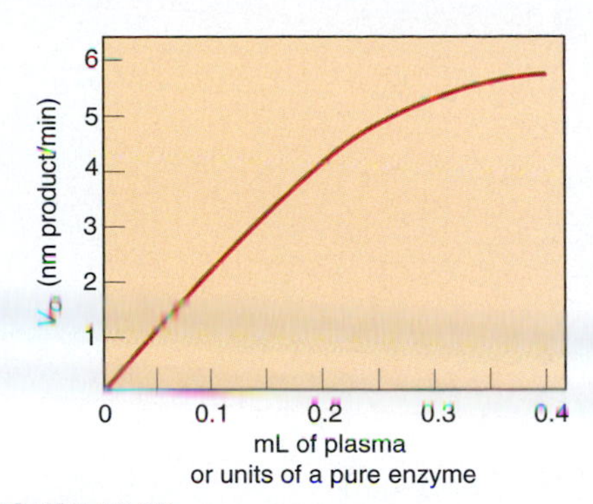

FIGURE 10.62
Assessing the validity of an enzyme assay.
The line shows what is to be expected for any reaction where the concentration of substrate is held constant and the amount of enzyme is increased. In this example linearity between initial velocity observed and amount of enzyme, whether pure or in a plasma sample, is only observed up to 0.2 mL of plasma or 0.2 units of pure enzyme. If more than 0.2 mL is used, the actual amount of enzyme in the sample would be underestimated.

Clinical Analyzers Use Immobilized Enzymes as Reagents

Enzymes are used as chemical reagents in desktop clinical analyzers in offices or for screening purposes in shopping centers and malls. For example, screening tests for cholesterol and triacylglycerols can be completed in a few minutes using 10 mL of plasma. The active components in the assay system are cholesterol oxidase for the cholesterol determination and lipase for the triacylglycerols. The enzymes are immobilized in a bilayer along with the necessary buffer salts, cofactors or cosubstrates, and indicator reagents. The ingredients are arranged in a multilayered vehicle the size and thickness of a 35-mm slide. The plasma sample provides the substrate and water necessary to activate the system. In the case of cholesterol oxidase, hydrogen peroxide is a product that subsequently oxidizes a colorless dye to a colored product that is measured by reflectance spectroscopy. Peroxidase is included in the reagents to catalyze the latter reaction.

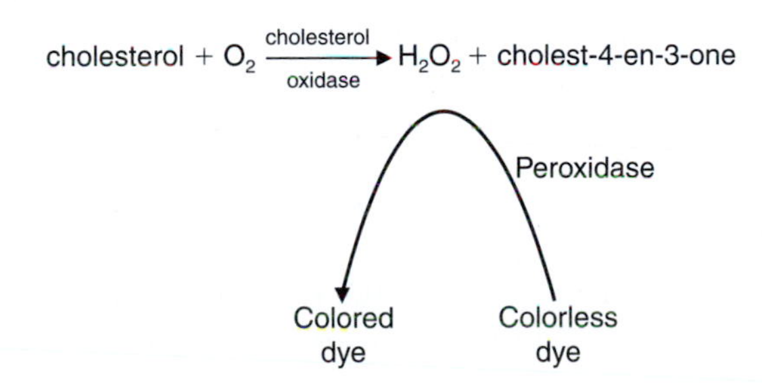

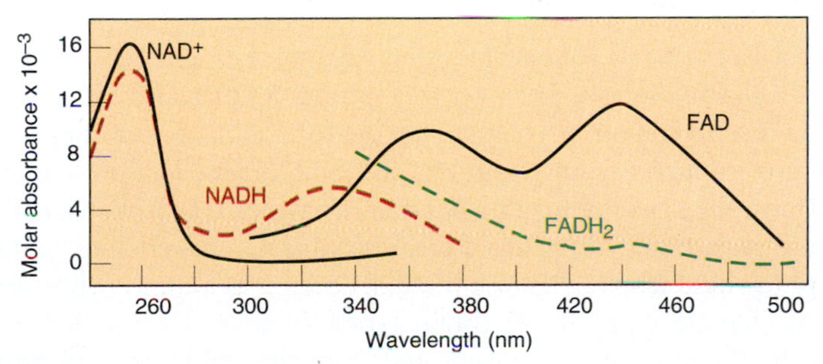

FIGURE 10.63
Absorption spectra of niacin and flavin coenzymes.
The reduced form of NAD (NADH) absorbs strongly at 340 nm. The oxidized form of flavin coenzymes absorbs strongly at 450 nm. Thus one can follow the rate of reduction of NAD^+ by observing the increase in the absorbance at 340 nm and the formation of $FADH_2$ by following the decrease in absorbance at 450 nm.

Each slide packet is constructed to measure a specific substance or enzyme and is stored in the cold for use as needed. In many cases the slide packet contains several enzymes in a coupled assay system that eventually generates a reduced nucleotide or a colored dye that can be measured spectroscopically. This technology has been made possible, in part, by the fact that the enzymes involved are stabilized when bound to immobilized matrices and are stored in the dry state or in the presence of a stabilizing solvent such as glycerol.

Enzyme-Linked Immunoassays Employ Enzymes as Indicators

Modern clinical chemistry has benefitted from the marriage of enzyme chemistry and immunology. Antibodies specific to a protein antigen are coupled to an indicator enzyme such as horseradish peroxidase to generate a very specific and sensitive assay. After binding of the peroxidase-coupled antibody to the antigen, the peroxidase is used to generate a colored product that is measurable and whose concentration is related to the amount of antigen in a sample. Because of the catalytic nature of the enzyme the system greatly amplifies the signal. This assay has been given the acronym **ELISA** for enzyme-linked immunoadsorbent assay.

Application of these principles is demonstrated by an assay for **human immunodeficiency virus (HIV)** coat protein antigens. This virus can lead to development of **acquired immunodeficiency syndrome** (AIDS). Antibodies are prepared in a rabbit against HIV coat proteins. In addition, a reporter antibody is prepared in a goat against rabbit IgG directed against the HIV protein. To this goat anti-rabbit IgG is linked the enzyme, horseradish peroxidase. The test for the virus is performed by incubating patient serum in a polystyrene dish that binds the proteins in the serum sample. Any free protein-binding sites remaining on the dish after incubation with patient serum are then covered by incubating with a nonspecific protein like bovine serum albumin. Next, the rabbit IgG antibody against the HIV protein is incubated in the dish, during which time the IgG attaches to any HIV coat proteins that are attached to the polystyrene dish. All unbound rabbit IgG is washed out with buffer. The goat anti-rabbit IgG–peroxidase is now placed in the dish where it binds to any rabbit IgG attached to the dish via the HIV viral coat protein. Unattached antibody–peroxidase is washed out. Peroxidase substrates are added and the amount of color developed in a given time period is a measurement of the amount of HIV coat protein present in a given volume of patient plasma when compared against a standard curve. This procedure is schematically diagramed in Figure 10.64. This assay amplifies the signal because of the catalytic nature of the reporter group, the enzyme peroxidase. Such amplified enzyme assays allow the measurement of remarkably small amounts of antigens.

Measurement of Isozymes Is Used Diagnostically

Isozymes (or isoenzymes) are enzymes that catalyze the same reaction but migrate differently on electrophoresis. Their physical properties may also differ, but not necessarily. The most common mechanism for the formation of isozymes involves the arrangement of subunits arising from two different genetic loci in different combinations to form the active polymeric enzyme. Isozymes that have wide clinical application are lactate dehydrogenase, creatine kinase, and alkaline phosphatase. **Creatine phosphokinase** (CPK) (see p. 1028) occurs as a dimer with two types of subunits, M (muscle type) and B (brain type). In brain both subunits are electrophoretically the same and are designated B. In skeletal muscle the subunits are both of the M type. The isozyme containing both M and B type subunits (MB) is found only in the myocardium. Other tissues contain variable amounts of the MM and BB isozymes. The isozymes are numbered beginning with the species migrating the fastest to the anode on electrophoresis—thus, CPK_1 (BB), CPK_2 (MB), and CPK_3 (MM).

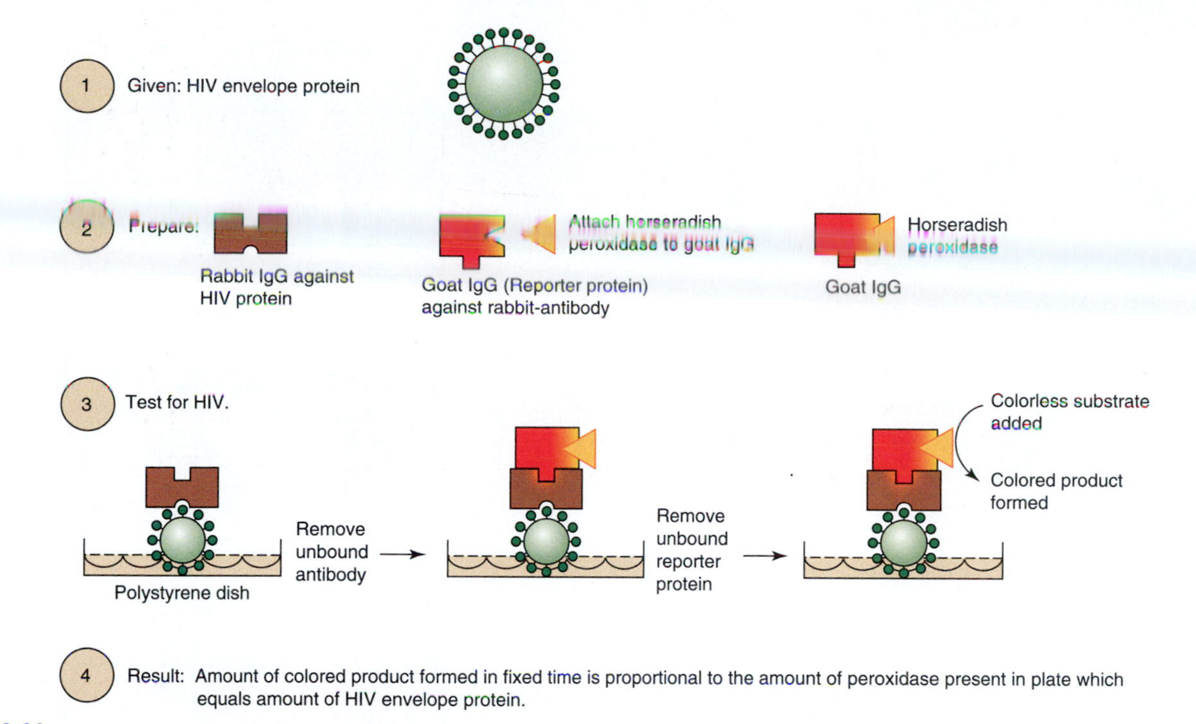

FIGURE 10.64
Schematic of ELISA (enzyme-linked immunoadsorbent assay) for detecting the human immunodeficiency virus (HIV) envelope proteins.

Lactate dehydrogenase is a tetrameric enzyme containing only two distinct subunits: those designated H for heart (myocardium) and M for muscle. These two subunits are combined in five different ways. The lactate dehydrogenase isozymes, subunit compositions, and major locations are as follows:

Type	*Composition*	*Location*
LDH_1	HHHH	Myocardium and RBC
LDH_2	HHHM	Myocardium and RBC
LDH_3	HHMM	Brain and Kidney
LDH_4	HMMM	
LDH_5	MMMM	Liver and Skeletal muscle

To illustrate how kinetic analyses of plasma enzyme activities are useful in medicine, activities of some CPK and LDH isozymes are plotted in Figure 10.65 as a function of time after infarction. After damage to heart tissue the cellular breakup releases CPK_2 into the blood within the first 6–18 h after an infarct, but LDH release lags behind the appearance of CPK_2 by 1–2 days. Normally, the activity of the LDH_2 isozyme is higher than that of LDH_1; however, in the case of infarction the activity of LDH_1 becomes greater than LDH_2, at about the time CPK_2 levels are back to baseline (48–60 h). Figure 10.66 shows the fluctuations of all five LDH isozymes after an infarct. The increased ratio of LDH_2 and LDH_1 can be seen in the 24-h tracing. The LDH isozyme "switch" coupled with increased CPK_2 is diagnostic of myocardial infarct (MI) in virtually 100% of the cases. Increased activity of LDH_5 is an indicator of liver congestion. Thus secondary complications of heart failure can be monitored.

The electrophoresis method for determining cardiac enzymes is too slow and insensitive to be of value in the emergency room situation. ELISAs based on monoclonal antibodies to CPK_2 are both quick (30 min) and sensitive enough to detect CPK_2 in the serum within an hour or so of a heart attack.

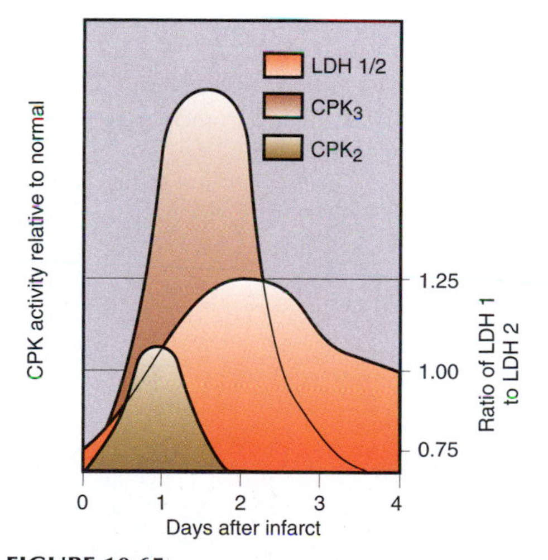

FIGURE 10.65
Characteristic changes in serum CPK and LDH isozymes following a myocardial infarction.
CPK_2 (MB) isozyme increases to a maximum within 1 day of the infarction. CPK_3 lags behind CPK_2 by about 1 day. Total LDH level increases more slowly. The increase of LDH_1 and LDH_2 within 12–24 h coupled with an increase in CPK_2 is diagnostic of myocardial infarction.

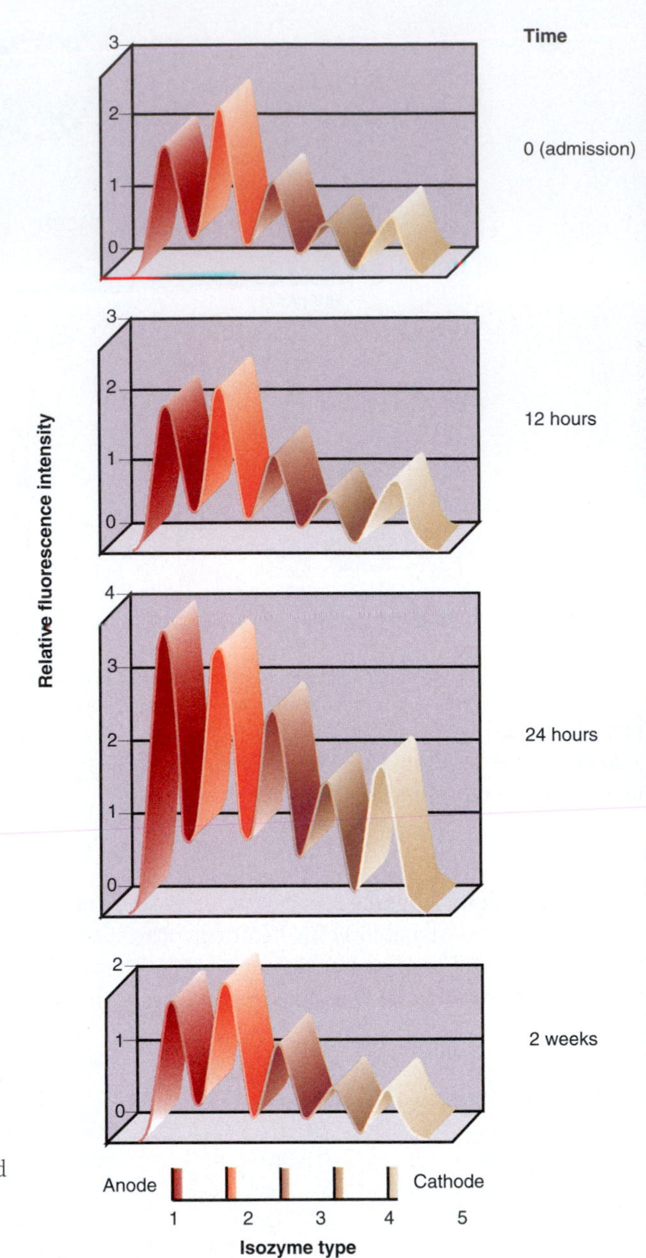

FIGURE 10.66
Tracings of densitometer scans of LDH isozymes at time intervals following a myocardial infarction.
Total LDH increases and LDH_1 becomes greater than LDH_2 between 12 and 24 h. Increase in LDH_5 is diagnostic of a secondary congestive liver involvement. Note the Y-axis scales are not identical. After electrophoresis on agarose gels, the LDH activity is assayed by measuring the fluorescence of the NADH formed in LDH-catalyzed reaction.
Courtesy of Dr. A. T. Gajda, Clinical Laboratories, The University of Arkansas for Medical Science.

Some Enzymes Are Used as Therapeutic Agents

In a few cases enzymes have been used as drugs in the therapy of specific medical problems. **Streptokinase**, an enzyme mixture prepared from a streptococcus, is useful in clearing blood clots that occur in myocardial infarcts and in the lower extremities. It activates the fibrinolytic proenzyme **plasminogen** that is normally present in plasma. The activated enzyme is plasmin.

Plasmin is a serine protease that cleaves the insoluble fibrin in blood clots into several soluble components (see p. 1045). Another serine protease, human tissue plasminogen activator, t-PA, is being commercially produced by bioengineered *Escherichia coli (E. coli)* for use in dissolving blood clots in patients suffering myocardial infarction (see p. 1045). t-PA also functions by activating the patient's plasminogen.

Asparaginase therapy is used for some types of adult leukemia. Tumor cells have a requirement for asparagine and must scavenge it from the host's plasma. Intravenous (i.v.) administration of asparaginase lowers the host's plasma level of asparagine, which results in depressing the viability of the leukemia cells.

Most enzymes have a short half-life in blood; consequently, unreasonably large amounts of enzyme are required to maintain therapeutic levels. Work is in progress to enhance enzyme stability by coupling enzymes to solid matrices and implanting these materials in areas that are well perfused. In the future, **enzyme replacement** in individuals that are genetically deficient in a particular enzyme may become feasible.

Enzymes Linked to Insoluble Matrices Are Used as Chemical Reactors

Specific enzymes linked to insoluble matrices are used in the pharmaceutical industry as highly specific chemical reactors. For example, immobilized β-galactosidase is used to decrease the lactose content of milk for lactose-intolerant people. In production of prednisolone, immobilized steroid 11-β-hydroxylase and a δ-1,2-dehydrogenase convert a cheap precursor to prednisolone in a rapid, stereospecific, and economical manner.

10.10 | REGULATION OF ENZYME ACTIVITY

Our discussion up to this point has centered on the chemical and physical characteristics of individual enzymes, but we must be concerned with the physiological integration of many enzymes into a metabolic pathway and the interrelationship of the products of one pathway with the metabolic activity of other pathways. Control of a pathway occurs through modulation of the activity of one or more key enzymes in the pathway. One of the key enzymes is the **rate-limiting enzyme**, which is the enzyme with the lowest V_{max}. It usually occurs early in the pathway.

Another is that catalyzing the **committed step** of the pathway, the first irreversible reaction that is unique to a metabolic pathway. The rate-limiting enzyme is not necessarily the enzyme associated with the committed step. Specific examples of these regulatory enzymes will be pointed out in discussions of metabolic pathways.

The activity of the enzyme associated with the committed step or with the rate-limiting step can be regulated in a number of ways. First, the absolute amount of the enzyme can be regulated by change in *de novo* synthesis of the enzyme. Second, the activity of the enzyme can be modulated by activators, by inhibitors, and by covalent modification through mechanisms previously discussed. Finally, the activity of a pathway can be regulated by physically partitioning the pathway from its initial substrate and by controlling access of the substrate to the enzymes of the pathway. This is referred to as **compartmentation.**

Anabolic and catabolic pathways are usually segregated into different organelles in order to maximize the cellular economy. There would be no point to oxidation of fatty acids occurring at the same time and in the same compartment as biosynthesis of fatty acids. If such occurred, a futile cycle would exist. By maintaining fatty acid biosynthesis in the cytoplasm and oxidation in the mitochondria, control can be exerted by regulating transport of common intermediates across the mitochondrial membrane. Table 16.4 (p. 717) contains a compilation of some of the metabolic pathways and their intracellular distribution.

As indicated earlier, the velocity of any reaction is dependent on the amount of enzyme present. Many rate-controlling enzymes are present in very low concentrations. More enzyme may be synthesized or existing rates of synthesis repressed through hormonally instituted activation of the mechanisms controlling gene expression. In some instances substrate can repress the synthesis of enzyme. For example, glucose represses the *de novo* synthesis of pyruvate carboxykinase, which is the rate-limiting enzyme in the conversion of pyruvate to glucose. If there is plenty of glucose available there is no point in synthesizing glucose. This effect of glucose may be mediated by insulin and is not direct feedback inhibition.

Many rate-controlling enzymes have relatively short half-lives; for example, that of pyruvate carboxykinase is 5 h. Teleologically this is reasonable because it provides a mechanism for effecting much larger fluctuations in the activity of a pathway than would be possible by inhibition or activation of existing levels of enzyme.

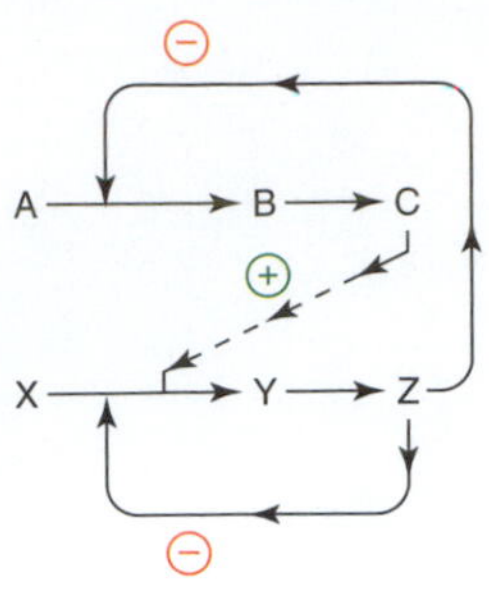

FIGURE 10.67
Model of feedback inhibition and cross-regulation.
Open bar indicates inhibition and broken line indicates activation. Product Z cross-regulates production of C by its inhibitory effect on the enzyme responsible for the conversion of A to B in the A → B pathway. C in turn cross-regulates the production of Z. The product Z inhibits its own formation by feedback inhibition of the conversion of X to Y.

Short-term regulation occurs through modification of the activity of existing enzyme. For example, when the cellular concentration of deoxyribonucleotides builds up such that the cell has sufficient amounts for synthesis of DNA, the key enzyme of the synthetic pathway is inhibited by the end products, resulting in shutdown of the pathway. This is referred to as **feedback inhibition.**

The inhibition may take the form of competitive inhibition or allosteric inhibition. In any case, the apparent K_m may be raised above the *in vivo* levels of substrate, and the reaction ceases or decreases in velocity.

In addition to feedback within the pathway, feedback on other pathways also occurs. This is referred to as cross-regulation. In **cross-regulation** a product of one pathway serves as an inhibitor or activator of an enzyme occurring early in another pathway as depicted in Figure 10.67. A good example, considered in detail on p. 844, is the cross-regulation of the production of the four deoxyribonucleotides for DNA synthesis. This is important since all four must be present in similar amounts for DNA synthesis.

An example of reversible covalent modification is glycogen phosphorylase, in which the interconvertible active and inactive forms are phosphorylated and dephosphorylated proteins, respectively. Protein kinases and protein phosphatases are also regulated by phosphorylation and dephosphorylation. Other examples of reversible covalent modification include acetylation–deacetylation, adenylylation–deadenylylation, uridylylation–deuridylylation, and methylation–demethylation. The phosphorylation–dephosphorylation scheme is most common.

BIBLIOGRAPHY

Blackburn, G. M., Kang, A. S., Kingsbury, G. A., and Burton, D. R. Review of abzymes. *Biochem. J.* 262:381, 1989.

Bugg, T. *An Introduction to Enzyme and Coenzyme Chemistry.* London: Blackwell Press, 1997.

Cornish-Bowden, A. *Fundamentals of Enzyme Kinetics.* London: Portland Press, 1995.

Fersht, A. *Enzyme Structure and Mechanism in Protein Science: A Guide to Enzyme Catalysis and Protein Folding.* New York: Freeman, 1999.

Kyte, J. *Mechanism in Protein Chemistry.* New York: Garland Publishing, 1995.

Keffer, J. H. Myocardial markers of injury. *Am. J. Clin. Pathol.* 105:305, 1996.

Knowles, J. R. and Alberty, W. J. Evolution of enzyme function and the development of catalytic efficiency. *Biochemistry* 15:5631, 1976.

Kraut, J. How do enzymes work? *Science* 242:533, 1988.

Lerner, R. A., Benkovic, S. J., and Schultz. P. G. At the crossroads of chemistry and immunology: catalytic antibodies. *Science* 252:659, 1991.

QUESTIONS | C. N. ANGSTADT

Multiple Choice Questions

1. In all enzymes the active site:
   A. contains the substrate-binding site.
   B. is contiguous with the substrate-binding site in the primary sequence.
   C. lies in a region of the primary sequence distant from the substrate-binding site.
   D. contains a metal ion as a prosthetic group.
   E. contains the amino acid side chains involved in catalyzing the reaction.

2. Which of the following types of oxidoreductase enzymes usually form hydrogen peroxide (H_2O_2) as one of its products?
   A. dehydrogenases
   B. oxidases
   C. oxygenases
   D. peroxidases
   E. none of the above

3. Although enzymic catalysis is reversible, a given reaction may appear irreversible:
   A. if the products are thermodynamically far more stable than the reactants.
   B. under equilibrium conditions.
   C. if a product accumulates.
   D. at high enzyme concentrations.
   E. at high temperatures.

4. Metal cations may do all of the following EXCEPT:
   A. donate electron pairs to functional groups found in the primary structure of the enzyme protein.
   B. serve as Lewis acids in enzymes.
   C. participate in oxidation–reduction processes.
   D. stabilize the active conformation of an enzyme.
   E. form chelates with the substrate, with the chelate being the true substrate.

5. Enzymes may be specific with respect to all of the following EXCEPT:

A. chemical identity of the substrate.
B. the atomic mass of the elements in the reactive group (e.g., ^{12}C but not ^{14}C).
C. optical activity of product formed from a symmetrical substrate.
D. type of reaction catalyzed.
E. which of a pair of optical isomers will react.

6. All of the following can be chemically isolated EXCEPT:
A. enzymes.
B. enzyme–substrate complexes.
C. enzyme–inhibitor complexes.
D. enzyme–substrate covalent intermediates.
E. transition states.

7. Which of the following necessarily results in formation of an enzyme–substrate intermediate?
A. substrate strain
B. acid–base catalysis
C. entropy effects
D. allosteric regulation
E. covalent catalysis

8. In the reaction sequence below the best point for controlling production of Compound 6 is reaction:

Cpd 3
$\Updownarrow$ B
Cpd 1 $\xrightarrow{A}$ Cpd 2 $\overset{C}{\rightleftharpoons}$ Cpd 4 $\xrightarrow{D}$ Cpd 5 $\overset{E}{\rightleftharpoons}$ Cpd 6

A. A.
B. B.
C. C.
D. D.
E. E.

9. If the plasma activity of an intracellular enzyme is abnormally high all of the following may be a valid explanation EXCEPT:
A. the rate of removal of the enzyme from plasma may be depressed.
B. tissue damage may have occurred.
C. the enzyme may have been activated.
D. determination of the isozyme distribution may yield useful information.
E. the rate of synthesis of the enzyme may have increased.

10. Types of physiological regulation of enzyme activity include all of the following EXCEPT:
A. covalent modification.
B. changes in rate of synthesis of the enzyme.
C. allosteric activation.
D. suicide inhibition.
E. competitive inhibition.

Questions 11 and 12: A man of Japanese ancestry found himself to be experiencing severe flushing and a very rapid heart rate after consuming one alcoholic beverage. His companion, a Caucasian male, did not have the same symptoms even though he had finished his second drink. These physiological effects are related to the presence of acetaldehyde (CH_3CHO) generated from the alcohol. Acetaldehyde is normally removed by the reaction of mitochondrial aldehyde dehydrogenase, which catalyzes the reaction

$$CH_3CHO + NAD^+ \rightleftharpoons CH_3COO^- + NADH + H^+$$

11. Acetaldehyde dehydrogenase is:
A. an oxidoreductase.
B. a transferase.
C. a hydrolase.
D. a lyase.
E. a ligase.

12. The explanation for the difference in physiological effects is that the Japanese man was missing the normal mitochondrial aldehyde dehydrogenase and had only a cytosolic isomer. The cytosolic isomer:
A. does not react with CH_3CHO.
B. activates the enzyme that produces CH_3CHO.
C. differs from the mitochondrial enzyme in that it has a higher K_m for CH_3CHO.
D. would be expected to have a greater affinity for the substrate than the mitochondrial enzyme.
E. produces a low steady-state level of acetaldehyde following alcohol consumption.

Questions 13 and 14: A research technician who is working with organophosphate compounds is required to have a weekly blood test for acetylcholine esterase activity. Typically, esterase activity remains relatively constant for some time and then abruptly drops to zero. If this happens, the technician must immediately stop working with the organophosphate compounds. The organophosphate compounds form stable esters with a critical serine hydroxyl group in the esterase.

13. In this type of enzyme, serine transfers a proton to a histidine residue. Serine is acting as a:
A. specific acid.
B. general acid.
C. specific base.
D. general base.
E. a transition stabilization catalyst.

14. Histidine accepts a proton in the above mechanism and, in a subsequent step, donates the proton to another species. Such a histidine would be most likely to have a pH–activity profile resembling which curve on the following drawing?

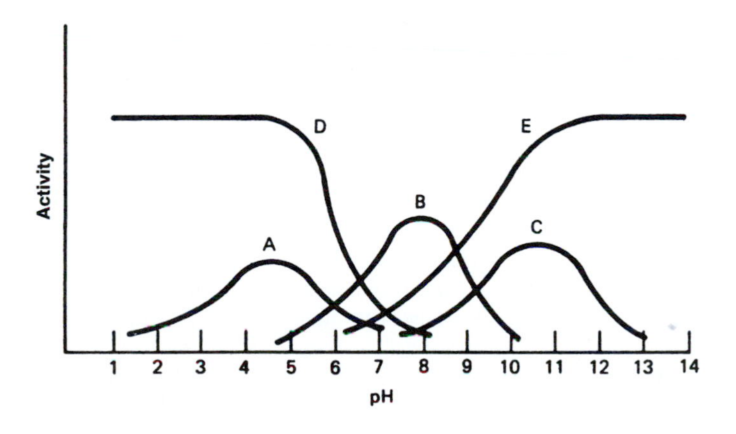

A. Curve A
B. Curve B
C. Curve C
D. Curve D
E. Curve E

Problems

15. An experiment measuring velocity versus substrate concentration was run, first in the absence of substance A, and then in the presence of substance A. The following data were obtained:

[S] *(μM)*	*Velocity in Absence of A* *(μmol min⁻¹)*	*Velocity in Presence of A* *(μmol min⁻¹)*
2.5	0.32	0.20
3.3	0.40	0.26
5.0	0.52	0.36
10.0	0.69	0.56

Is substance A an activator or an inhibitor? If it is an inhibitor, what kind of inhibitor is it?

16. For the experiment in Question 15, calculate the K_m and V_{max} both in the absence and in the presence of substance A. Are these results consistent with your answer for Question 15?

ANSWERS

1. **E** The active site contains all of the machinery, including the amino acid side chains, involved in catalyzing the reaction. A–D are all possible, but none is necessarily true.
2. **B** Most oxidases produce H_2O_2 as a product of the transfer of two electrons from the donor to oxygen. A: Dehydrogenases produce reduced pyridine or flavin coenzymes. C: Oxygenases add one or both atoms of oxygen to the substrate. D: Peroxidases degrade H_2O_2.
3. **A** Stable products do not react in the reverse direction at an appreciable rate. B: At equilibrium the forward and reverse reactions proceed at identical rates. C: Product accumulation would tend to reverse the reaction. D: Enzymes merely catalyze reactions and do not affect the equilibrium of the reaction. E: Temperature affects the rates of reactions and may also affect the position of the equilibrium, but does not interconvert reversible and irreversible reactions.
4. **A** Metal cations are electron deficient and may accept electron pairs, serving as Lewis acids, but they do not donate electrons to other functional groups. C: On the contrary, they sometimes accept electron pairs from groups in amino acid side chains. D: In doing so they may become chelated, which may stabilize the appropriate structure. E: Sometimes they are chelated by the substrate, with the chelate being the true substrate.
5. **B** Enzymes do not distinguish among different nuclides of an element, although the rate of reaction of a heavier nuclide might be less than that of a lighter one. A, D: Enzymes are specific for the substrate and the type of reaction. C, E: The asymmetry of the binding site generally permits only one of a pair of optical isomers to react, and only one optical isomer is generated when a symmetric substrate yields an asymmetric product.
6. **E** The transition state is not an intermediate and cannot be isolated. Rather, it can be thought of as a state in which old bonds are partly broken and new bonds partly formed. A–D: All the other species can be isolated under suitable experimental conditions.
7. **E** Only in covalent catalysis is a covalent bond between enzyme and a portion of the substrate involved. A–D: All enzyme-catalyzed reactions involve an enzyme–substrate complex but this is different than an intermediate, which requires formation of a covalent bond. There is always at least one transition state involved.
8. **D** Reaction D is irreversible; if it were not controlled, Cpd 5 might build up to toxic levels. A: Control of reaction A would control production of both Cpds 3 and 6. B: Reaction B is not on the direct route. C, E: Reactions C and E are freely reversible, so do not need to be controlled.
9. **E** Since appearance of intracellular enzymes in plasma arises from leakage, typically from damaged or destroyed cells, changes in their rates of synthesis within the cell would not be expected to affect plasma concentration. A: This would lead to elevated levels. B: Intracellular enzymes may appear in abnormal amounts when tissues are damaged. C: Laboratory assays depend on all factors except amount of enzyme being constant. D: Different tissues have characteristic distributions of isozymes.
10. **D** Suicide inhibitors are sometimes used as drugs but would be inappropriate as a regulatory mechanism because the inhibition is irreversible. A: Covalent modification includes zymogen activation and phospho–dephospho protein conversions. B: Enzyme levels may be controlled, often by hormones. C: Allosteric activation is common. E: End products of a reaction or reaction sequence may inhibit their own formation by competitive inhibition.
11. **A** Aldehyde to acid conversion is an oxidation. This is of the subclass dehydrogenases as indicated by the presence of the NAD^+.
12. **C** The lower affinity (higher K_m) makes it more difficult for the enzyme to remove CH_3CHO. A: Isozymes, by definition, catalyze the same reaction. B: It is highly unlikely that one enzyme would activate another enzyme. D: With a greater affinity (lower K_m) the reaction would be occurring at a rapid rate, thus removing CH_3CHO. E: These effects are due to a high steady-state concentration.
13. **B** General acids are weakly ionized at physiological pH. The particular environment of the serine in the protein facilitates its ability to transfer a proton and thus act as a nucleophile. A, C: Specific acids and bases do not exist at physiological pH. D: Bases accept protons. E: There is no indication that serine does this.
14. **B** A group must be in the correct ionization state to act catalytically. For a histidyl group to serve as a general acid and a general base, the pH must be compatible with both ionization states of histidine. Since the p*K* of the histidyl side chain is about 6.8, the maximum activity is likely to be near that pH. A, C: These peaks are out of the normal physiological range. D, E: Both of these curves indicate that the species is acting as only an acid or a base but not both.
15. The best way to handle the data is to take the reciprocals of both [S] and *v* and construct a Lineweaver–Burk plot. You

should find that the two curves cross the y-axis at the same point but the curve in the presence of A crosses the x-axis closer to the origin. This pattern indicates that A is a competitive inhibitor.

16.

	$-1/K_m$	K_m	$1/V_{max}$	V_{max}
Absence of A	−0.14	7.1	0.8	1.25
Presence of A	−0.08	12.5	0.8	1.25

With a competitive inhibitor, V_{max} remains constant (be sure you understand why) but the apparent K_m is larger. It takes more substrate to reach a given velocity because the substrate has to compete with the inhibitor.

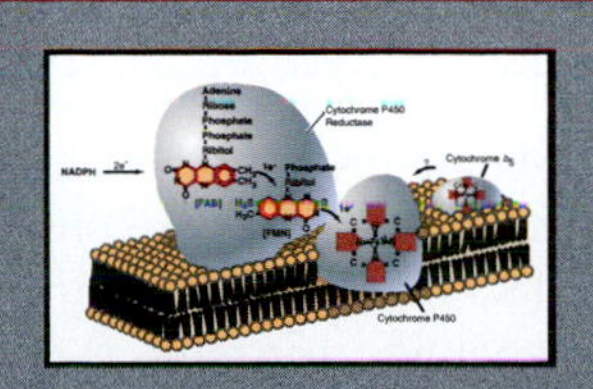

11

THE CYTOCHROMES P450 AND NITRIC OXIDE SYNTHASES

Richard T. Okita
and Bettie Sue Siler Masters

11.1 | OVERVIEW

The term **cytochrome P450** refers to a unique family of **heme proteins** present in bacteria, fungi, insects, plants, fish, mammals, and primates, which can be considered to be universal oxygenases (oxygen-utilizing enzymes) because of the variety of reactions they catalyze and the structurally diverse compounds that serve as substrates. Cytochrome P450 proteins are encoded by a gene superfamily that contains hundreds of genes. Substrates for this enzyme system include endogenously synthesized compounds, such as cholesterol, steroid hormones, and fatty acids, as well as exogenous compounds, such as drugs, food additives, components of cigarette smoke, pesticides, and chemicals used for industry that enter the body by ingestion of foods, inhalation, or absorption through the skin.

The cytochrome P450 system has far reaching implications in medicine. It is involved in (1) inactivation or activation of therapeutic agents; (2) conversion of chemicals to highly reactive molecules, which may produce unwanted cellular damage, cell death, or mutations; (3) production of steroid hormones; (4) metabolism of fatty acids, prostaglandins, leukotrienes, and retinoids; and (5) enzyme inhibition or induction that results in drug–drug interactions and adverse effects. Cytochrome P450 proteins are cysteine thiolate-bound heme proteins that share catalytic mechanisms with thromboxane, prostacyclin, and allene oxide synthases and nitric oxide synthases. This chapter will address these heme–cysteine thiolate proteins by discussing the cytochromes P450 in detail and will provide an introduction to the various nitric oxide synthases.

11.2 | CYTOCHROME P450: GENERAL CLASSIFICATION AND OVERALL REACTION DESCRIPTION

Designation of a particular protein as a cytochrome P450 originated from its spectral properties before its catalytic function was known. This group of proteins has a unique absorbance spectrum that is obtained by adding a reducing agent, such as sodium dithionite, to a suspension of endoplasmic reticulum vesicles (microsomes) followed by bubbling of carbon monoxide (CO) gas into the solution. CO is bound to the reduced form of the heme protein and produces an absorbance spectrum with a peak at approximately 450 nm (Figure 11.1); thus the name P450 for a pigment with an absorbance at 450 nm. CO binds to the heme iron, in lieu of oxygen, with a much higher binding affinity and thereby is a potent inhibitor of its function. Specific forms of cytochromes P450 differ in their maximum absorbance wavelengths, with a range between 446 and 452 nm.

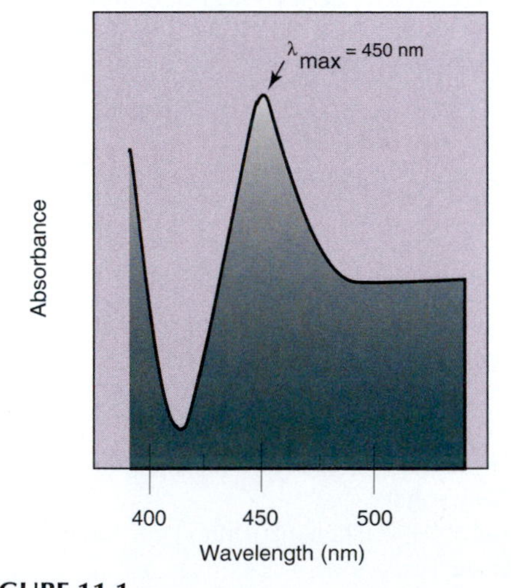

FIGURE 11.1
Absorbance spectrum of the carbon monoxide-bound cytochrome P450.
The reduced-form of this hemeprotein binds carbon monoxide to produce a maximum absorbance at approximately 450 nm. Hence this cytochrome was designated P450.

The general reaction catalyzed by a cytochrome P450 is written as follows:

$$\text{NADPH} + \text{H}^+ + \text{O}_2 + \text{SH} \rightarrow \text{NADP}^+ + \text{H}_2\text{O} + \text{SOH}$$

where the substrate (S) may be a steroid, fatty acid, drug, or other chemical that has an alkane, alkene, aromatic ring, or heterocyclic ring substituent that can serve as a site for oxygenation. The reaction is a monooxygenation and the enzyme is a **monooxygenase** due to the incorporation of only one of the two oxygen atoms into the substrate. In mammalian cells, cytochromes P450 serve as terminal electron acceptors in **electron transport systems,** which are present either in the endoplasmic reticulum or inner mitochondrial membrane. The cytochrome P450 protein contains a single iron **protoporphyrin IX** (heme) prosthetic group (see p. 1063), which binds oxygen and contains binding sites for substrates. Heme iron of cytochromes P450 is bound to the four pyrrole nitrogen atoms of the porphyrin ring and two axial ligands, one of which is a sulfhydryl group from a cysteine residue located toward the carboxyl end of the cytochrome P450 molecule (Figure 11.2). Heme iron may exist in two different spin states: (1) hexa-coordinated low-spin state or (2) penta-coordinated high-spin state. Low- and high-spin states are de-

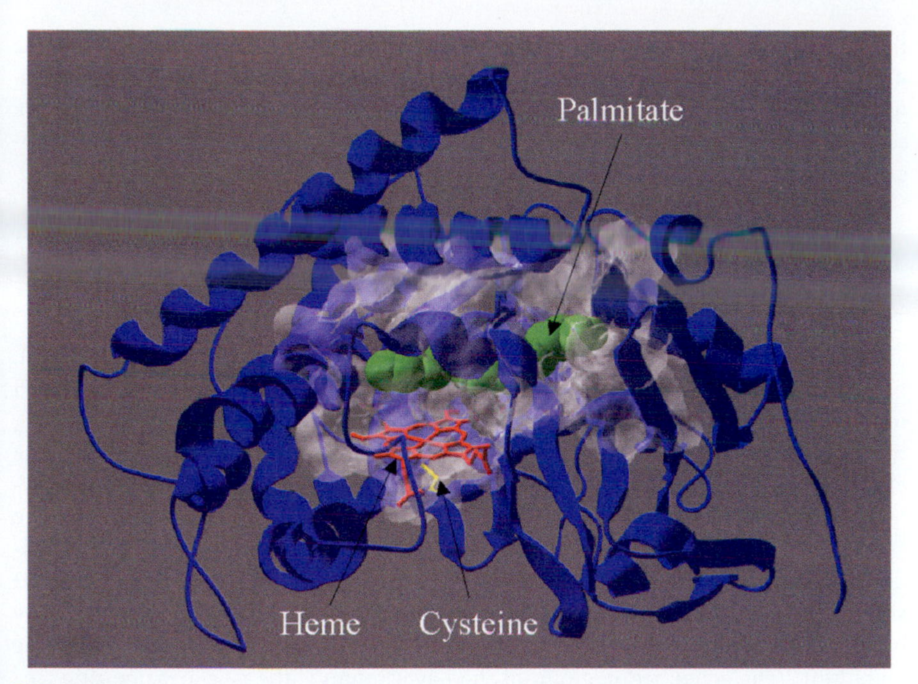

FIGURE 11.2
Model of the active site of P450 from *Bacillus megaterium* (known as BM3 or CYP102).
The model shows the protoporphyrin IX prosthetic group (red) with the cysteine thiolate ligand (yellow) attached to the heme iron. The substrate palmitate (in green) is in the active site of the P450. The white area represents the substrate channel of the enzyme.
Generated by Dawn Harris from Dr. Tom Poulos's P450BM3 (CYP102) structure using Glaxo Wellcome Experimental Research's Swiss Pdb Viewer. Guex, N. and Peitsch, M. C. Electrophoresis *18:2714, 1997.*

scriptions of the *d*-electronic shells around the iron atom. When a cytochrome P450 molecule binds a substrate, there is a conformational change of the protein structure surrounding the heme iron, resulting in a more positive reduction potential (−170 mV) than in the absence of substrate (−270 mV). This accelerates the rate at which the P450 may be reduced by electrons donated from NADPH through the flavoprotein enzyme NADPH–cytochrome P450 reductase (Figure 11.3). In order for **hydroxylation (monooxygenation)** to occur, heme iron must be reduced from

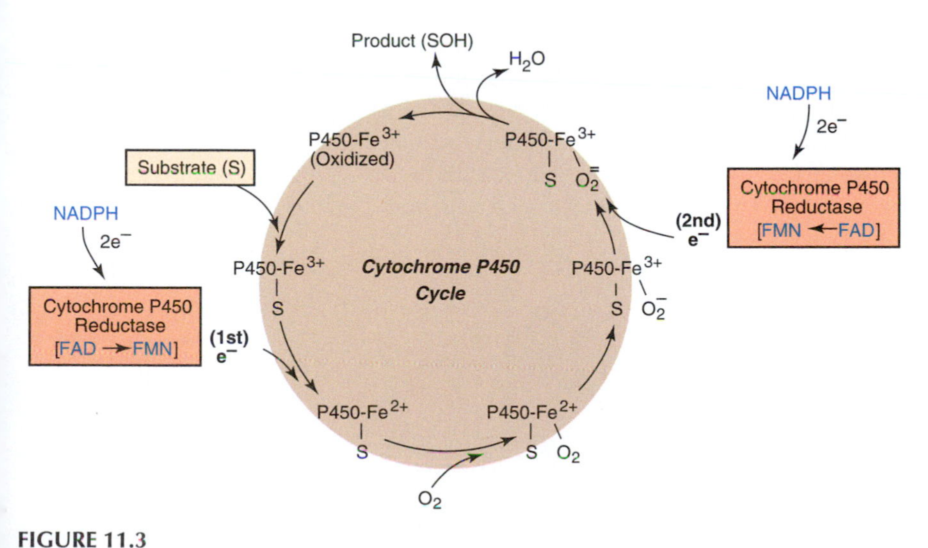

FIGURE 11.3
Sequence of reactions at cytochrome P450.
Diagram shows the binding of substrate, transfer of the first and second electrons from NADPH–cytochrome P450 reductase, and binding of O_2.

the ferric (Fe^{3+}) to its ferrous (Fe^{2+}) state so that oxygen may bind to the heme iron. Two electrons are required for the monooxygenation reaction. They are transferred to the P450 molecule individually, the first to allow oxygen binding and the second to cleave the oxygen molecule to generate the active oxygen species for insertion of a hydroxyl group into the substrate.

11.3 | CYTOCHROME P450 ELECTRON TRANSPORT SYSTEMS

The two electrons required for the monooxygenase reaction are provided by NADPH, but a basic mechanistic problem is the facilitation of electron transfer from NADPH to the cytochrome P450. NADH and NADPH are two electron donors (see p. 427), but cytochrome P450, with its single heme prosthetic group, can only accept one electron at a time. Thus a protein that transfers electrons from NADPH to the cytochrome P450 molecule must have the capacity to accept two electrons but serve as a one-electron donor. This problem is solved by the presence of a NADPH-dependent flavoprotein reductase, which accepts two electrons from NADPH simultaneously but transfers the electrons individually, either to an intermediate iron–sulfur protein (mitochondria) or directly to cytochrome P450 (endoplasmic reticulum). The active redox group of the flavoprotein is the isoalloxazine ring (see p. 427). The isoalloxazine nucleus is uniquely suited to perform this chemical task since it can exist in fully oxidized and one- and two-electron reduced states (Figure 11.4). The transfer of electrons from NADPH to cytochrome P450 is accomplished by two distinct electron transport systems that reside almost exclusively in either mitochondria or endoplasmic reticulum

NADPH–Cytochrome P450 Reductase Is the Flavoprotein Electron Donor in the Endoplasmic Reticulum

Most mammalian cytochrome P450 proteins are found in the endoplasmic reticulum in hepatocytes, in renal cells, and in cells of the respiratory tract. In the endoplasmic reticulum, NADPH donates electrons to a flavoprotein called **NADPH–cytochrome P450 reductase.** The human enzyme has a molecular mass of 76,558 Da and contains both flavin adenine dinucleotide (FAD) and flavin mononucleotide (FMN) as prosthetic groups. Until the recent characterization of nitric oxide synthases, it was the only mammalian flavoprotein known to contain both FAD and FMN. A significant number of residues at the amino end of the molecule are hydrophobic and this segment of the protein is embedded in the endoplasmic reticulum membrane (Figure 11.5). The FAD moiety serves as the entry point for electrons from NADPH, and FMN serves as the exit point, transferring electrons individually to cytochrome P450. Because a single flavin molecule may exist as one- or two-electron-reduced forms and two flavin molecules are bound per reductase molecule, the enzyme may receive electrons from NADPH and store them between

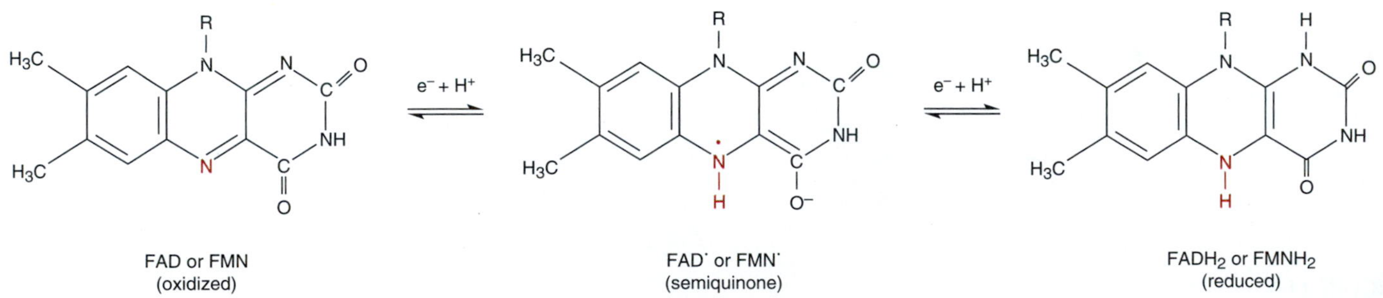

FAD or FMN (oxidized) $\underset{}{\overset{e^- + H^+}{\rightleftharpoons}}$ FAD˙ or FMN˙ (semiquinone) $\underset{}{\overset{e^- + H^+}{\rightleftharpoons}}$ $FADH_2$ or $FMNH_2$ (reduced)

FIGURE 11.4
Isoalloxazine ring of FMN or FAD in its oxidized, semiquinone ($1e^-$ reduced form), or fully reduced ($2e^-$ reduced) state.

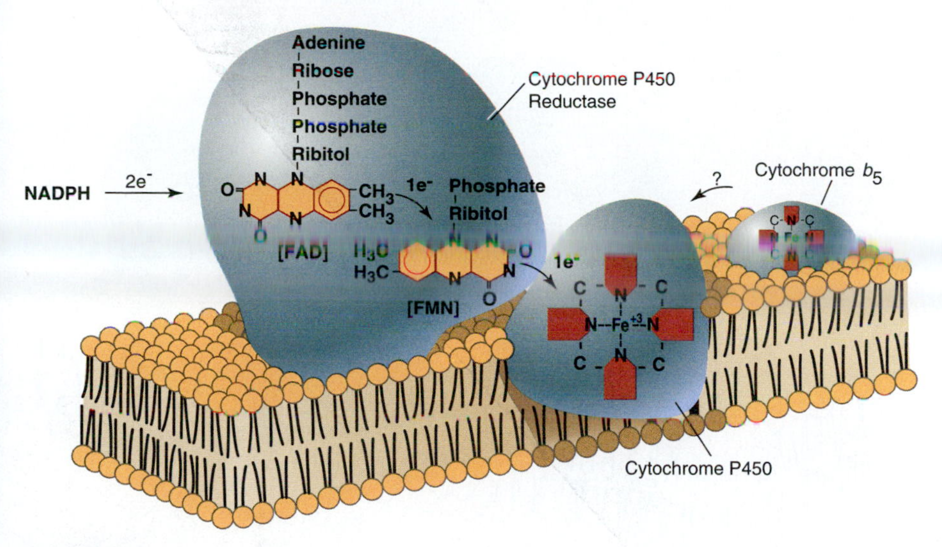

FIGURE 11.5
Components of the endoplasmic reticulum (microsomal) cytochrome P450 system. NADPH–cytochrome P450 reductase is bound by its hydrophobic tail to the membrane, whereas cytochrome P450 is deeply embedded in the membrane. Also shown is cytochrome b_5, which may participate in selected cytochrome P450-mediated reactions.

the two flavin molecules before transferring them individually to the cytochrome P450.

In certain reactions catalyzed by the microsomal cytochrome P450, the transfer of the second electron may not be directly from NADPH–cytochrome P450 reductase, but may occur from **cytochrome b_5,** a small heme protein of molecular mass of 15,199 Da. Cytochrome b_5 is reduced either by NADPH–cytochrome P450 reductase or another microsome-bound flavoprotein, **NADH–cytochrome b_5 reductase.** It is not known why reactions catalyzed by specific cytochromes P450 apparently require cytochrome b_5 for optimal enzymatic activity.

NADPH–Adrenodoxin Reductase Is the Flavoprotein Electron Donor in Mitochondria

In mitochondria, a flavoprotein reductase also acts as the electron acceptor from NADPH. This protein is referred to as **NADPH–adrenodoxin reductase** because its characteristics were described for the **flavoprotein** first isolated from the adrenal gland. This protein contains only FAD and the human NADPH–adrenodoxin reductase has a mass of 50,709 Da. Adrenodoxin reductase is only weakly associated with the inner mitochondrial membrane, unlike the integration of the NADPH–cytochrome P450 reductase in the endoplasmic reticulum. Adrenodoxin reductase cannot directly transfer either the first or second electron to heme iron of cytochrome P450 (Figure 11.6). A second protein, called **adrenodoxin** (molecular mass 12,500 Da) serves as an electron carrier between the adrenodoxin reductase and **mitochondrial cytochrome P450.** The adrenodoxin molecule is also weakly associated with the inner mitochondrial membrane through its interaction with the membrane-bound cytochrome P450. Adrenodoxin contains two **iron–sulfur clusters,** which serve as redox centers for this molecule and function as an electron shuttle between the adrenodoxin reductase and the mitochondrial cytochromes P450. One adrenodoxin molecule receives an electron from its mitochondrial flavoprotein reductase and interacts with a second adrenodoxin, which then transfers its electron to the cytochrome P450 (Figure 11.6). Components of the mitochondrial cytochrome P450 system are synthesized in the cytosol as larger molecular weight precursors, transported into mitochondria, and processed by proteases into smaller molecular weight, mature proteins.

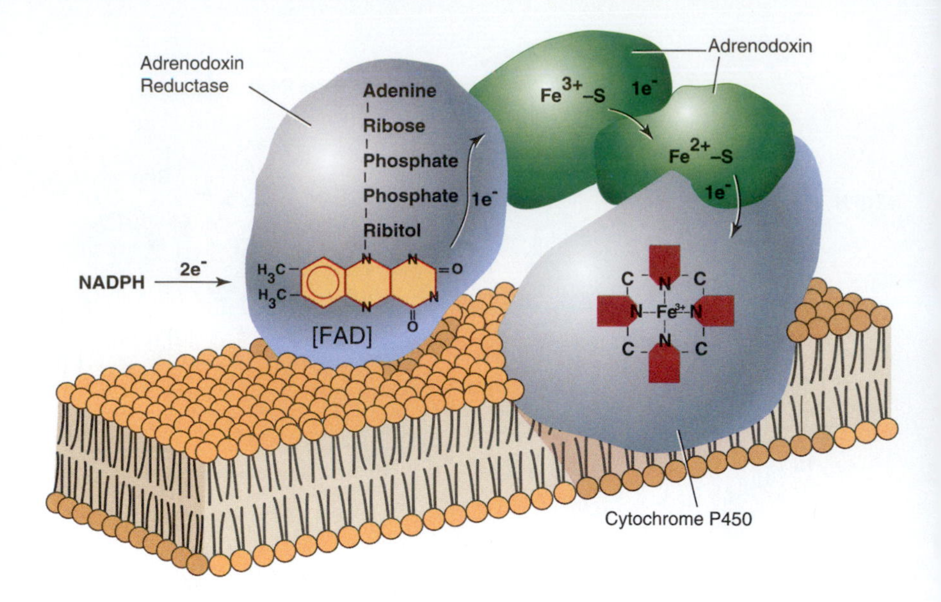

FIGURE 11.6
Components of mitochondrial cytochrome P450 system.
Cytochrome P450 is an integral protein of the inner mitochondrial membrane. NADPH–adrenodoxin reductase and adrenodoxin (ADR) are peripheral proteins and are not embedded in the membrane.

11.4 | CYTOCHROME P450: NOMENCLATURE AND TERMINOLOGY

Within the **superfamily** of cytochrome P450 proteins are several related cytochrome P450 proteins within a species or in different species that share, to varying degrees of similarity, common amino acid sequences. As researchers began to isolate different cytochrome P450 forms from various species including mouse, rat, rabbit, and humans, it became apparent that it was difficult to discuss the numerous P450 proteins because a uniform system of nomenclature was lacking. Thus a classification system was developed in which the cytochromes P450 are identified according to their amino acid sequences into various families and subfamilies, which are designated by numbers and letters, respectively. Cytochrome P450 forms in the same family may metabolize different substrates but are assigned to the same family because of the similarities in amino acid sequences. It was decided that a **cytochrome P450 family** will contain proteins that share at least 40% sequence homology at the amino acid level and is identified by an Arabic number. Cytochrome P450 proteins, assigned to the same family, that share at least 55% sequence homology are further classified into a **subfamily** by assignment of capital letters. Thus cytochrome P450 forms that are designated 2A, 2B, 2C represent cytochrome P450 proteins that share at least 40% sequence homology to be classified as members of the 2 family. Those cytochrome P450 proteins that share 55–99% homology are then further grouped into subfamilies that are identified by a capital letter. Cytochrome P450 forms in the same subfamily may be from different species. Finally, to identify individual cytochrome P450 proteins in a subfamily, the protein is assigned a second number, for example, 2A1, 2A2, or 2A3. The term **CYP**, which represents the first two letters of cytochrome and the first letter in P450, has been designated as the preface for a cytochrome P450 gene or protein. Thus cytochromes P450 1A2 and P450 2D6 are abbreviated CYP1A2 and CYP2D6 in this nomenclature system. At one time the term CYP was used to designate the gene whereas P450 referred to the protein, but today it is common to use the designations P450 or CYP interchangeably to identify either a gene or a protein. Table 11.1 lists several human P450 forms.

With certain cytochrome P450 forms, the nomenclature rules have been modified and there is only a single identifier, for example, CYP2E1, to describe the proteins that are in a particular CYP subfamily despite the fact that distinct gene sequences are present for these cytochrome P450 proteins in different species. The explanation is that, although there are distinct amino acid sequences in various species, the reaction or reactions catalyzed in this subfamily are similar and they share common substrates across species. In addition, there are cytochrome P450

TABLE 11.1 Major Human Cytochrome P450 Forms

CYP1	*CYP2*	*CYP3*	*CYP4*	*CYP11*	*CYP17*	*CYP19*	*CYP21*	*CYP26*	*CYP51*
1A1	2A6	3A4	4A11	11A1	17[a]	19[a]	21A2	26A1	51[a]
1A2	2A7	3A5	4B1	11B1				26B1	
1B1	2A13	3A7	4F2	11B2					
	2B6		4F3						
	2C8		4F8						
	2C9								
	2C18								
	2C19								
	2D6								
	2E1								
	2J2								

[a]P450 forms in these subfamilies are designated by number only, because of the existence of only one gene product.

enzymes in a single cytochrome P450 family that are identified by only a single number, for example, CYP17 or CYP19. These cytochrome P450 proteins catalyze a distinct reaction with specific steroid substrates and the number designates the specific carbon atom in the steroid molecule that is hydroxylated.

Although cytochromes P450 are enzymes, the terms isoenzyme or isozymes are generally not used to describe these related proteins. The terms "form" or "isoform" have been used to describe related cytochrome P450 proteins.

11.5 | SUBSTRATE SPECIFICITY OF CYTOCHROMES P450: PHYSIOLOGICAL FUNCTIONS

Cytochromes P450 metabolize a variety of lipophilic compounds of endogenous or exogenous origin. They may catalyze **hydroxylation** of the carbon atom of a methyl group, the methylene carbon of an alkane, hydroxylation of an aromatic ring to form a phenol, or addition of an oxygen atom across a double bond to form an **epoxide.** Reactions catalyzed by cytochrome P450 forms are shown in Figure 11.7. Cytochrome P450 proteins also catalyze **dealkylation** reactions in which alkyl groups attached to oxygen, nitrogen, or sulfur atoms are removed. **Oxidation** of nitrogen, sulfur, and phosphorus atoms and **dehalogenation** reactions are also catalyzed by cytochrome P450 forms.

It is important to characterize the substrates metabolized by individual cytochrome P450 forms. This is essential for researchers who are characterizing the human cytochrome P450 forms because of the need to determine how therapeutic drugs are metabolized and whether cytochrome P450 forms can be inhibited by different compounds. The Food and Drug Administration's (FDA) drug approval process requires that extensive information be submitted on the metabolism of drugs before they are finally approved for general human use. Recombinant DNA techniques have made it possible to determine the substrate profiles of human cytochrome P450 proteins by expressing the cDNA for the particular protein in an appropriate cellular expression system. This has been achieved in bacterial, insect, yeast, and mammalian cell systems and permits the unequivocal determination of substrate specificity. The assumption is that knowledge of the nucleotide sequence of the expressed gene leaves little doubt as to the source of enzyme activity expressed in those cells. The ability to express recombinant cytochrome P450 forms has permitted extensive *in vitro* studies to characterize the substrate specificity of a given cytochrome P450 and its inhibition by various compounds. *In vitro* studies with expressed human recombinant cytochrome P450 proteins will identify which cytochrome P450 form or forms catalyze the metabolism of a test drug, which product or products are formed, whether the test drug inhibits the metabolism of other drugs, or if other drugs may reduce or stimulate the metabolism of the test drug.

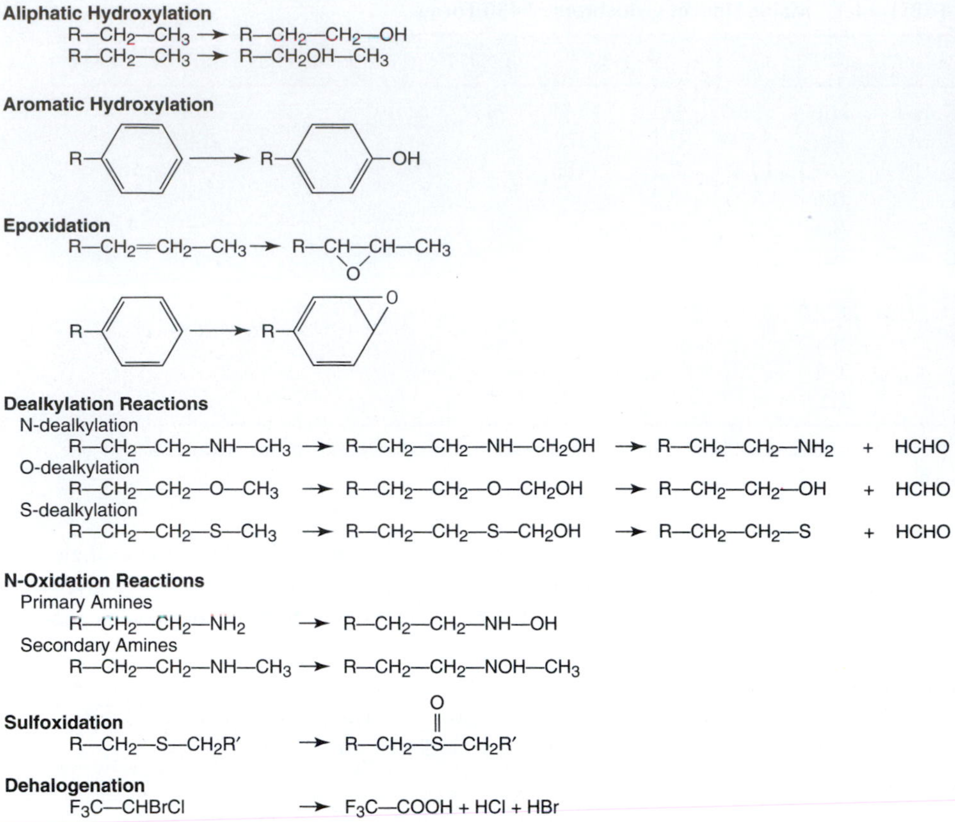

Aliphatic Hydroxylation

R—CH_2—CH_3 → R—CH_2—CH_2—OH

R—CH_2—CH_3 → R—CH_2OH—CH_3

Aromatic Hydroxylation

R—(benzene ring) → R—(benzene ring)—OH

Epoxidation

R—CH_2=CH_2—CH_3 → R—CH—CH—CH_3 (with O bridging the two CH)

R—(benzene ring) → R—(ring epoxide, O)

Dealkylation Reactions

N-dealkylation

R—CH_2—CH_2—NH—CH_3 → R—CH_2—CH_2—NH—CH_2OH → R—CH_2—CH_2—NH_2 + HCHO

O-dealkylation

R—CH_2—CH_2—O—CH_3 → R—CH_2—CH_2—O—CH_2OH → R—CH_2—CH_2—OH + HCHO

S-dealkylation

R—CH_2—CH_2—S—CH_3 → R—CH_2—CH_2—S—CH_2OH → R—CH_2—CH_2—S + HCHO

N-Oxidation Reactions

Primary Amines

R—CH_2—CH_2—NH_2 → R—CH_2—CH_2—NH—OH

Secondary Amines

R—CH_2—CH_2—NH—CH_3 → R—CH_2—CH_2—NOH—CH_3

Sulfoxidation

R—CH_2—S—CH_2R' → R—CH_2—S(=O)—CH_2R'

Dehalogenation

F_3C—CHBrCl → F_3C—COOH + HCl + HBr

FIGURE 11.7
Common reactions catalyzed by cytochromes P450.

Cytochromes P450 Participate in the Synthesis of Steroid Hormones and Oxygenation of Endogenous Compounds

The importance of cytochrome P450-catalyzed reactions is illustrated by the synthesis of **steroid hormones** from cholesterol in the **adrenal cortex** and sex organs. Mitochondrial and endoplasmic reticulum cytochromes P450 are required to convert cholesterol to aldosterone and cortisol in adrenal cortex, **testosterone** in testes, and **estradiol** in ovaries. Cytochrome P450 forms are responsible for several steps in the adrenal synthesis of **aldosterone,** the mineralocorticoid responsible for regulating salt and water balance, and **cortisol,** the glucocorticoid that governs protein, carbohydrate, and lipid metabolism. In addition, adrenal cytochromes P450 catalyze the synthesis of small quantities of the androgen, **androstenedione,** a regulator of secondary sex characteristics and a precursor of both estrogens and testosterone. Figure 11.8 presents a summary of these pathways.

In adrenal mitochondria, a cytochrome P450 (CYP11A1) catalyzes the **side chain cleavage** reaction converting cholesterol to pregnenolone, a committed step in steroid biosynthesis. The removal of isocaproic aldehyde is the result of sequential hydroxylation at carbon atoms 22 and 20 to produce 22-hydroxycholesterol and then 20,22-dihydroxycholesterol, respectively (Figure 11.9). An additional cytochrome P450-catalyzed step is necessary to cleave the carbon–carbon bond between C-20 and C-22 to produce pregnenolone, a 21-carbon steroid. This reaction sequence requires three NADPH and three O_2 molecules and is catalyzed by a single cytochrome P450 enzyme, CYP11A1. Two intermediates are produced and remain in the active site until the final product, **pregnenolone,** is formed. After pregnenolone is produced in mitochondria, it moves to the cytosol where it is metabolized by 3β-hydroxysteroid dehydrogenase/$\Delta^{4,5}$-isomerase to **progesterone.** Progesterone is metabolized to **deoxycorticosterone (DOC)** by an endoplasmic reticulum cytochrome P450 (CYP21), which catalyzes the 21-hydroxylation reaction. DOC is hydroxylated at carbon atom 11 by a mitochondrial cytochrome P450,

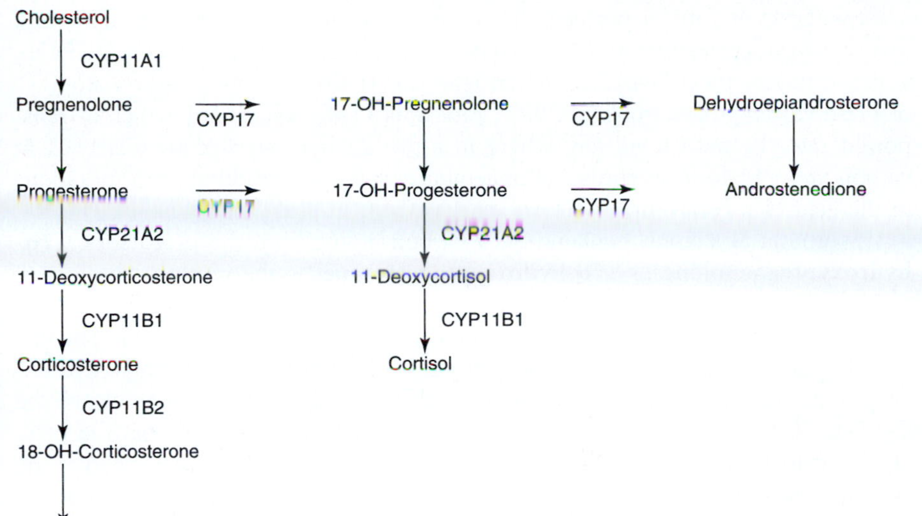

FIGURE 11.8
Steroid hormone synthesis in the adrenal gland.
The reactions catalyzed by cytochromes P450 (CYP) are indicated.

CYP11B1, to form **corticosterone** and then hydroxylated by another mitochondrial cytochrome P450 enzyme, CYP11B2, at carbon atom 18 to form the mineralocorticoid aldosterone (see p. 967). Although it was stated above that cytochrome P450 forms that hydroxylate steroids are given CYP designations that identify the carbon atom that is hydroxylated, there are exceptions to this process as noted for CYP11A1 and CYP11B2, which catalyze hydroxylation reactions at carbon positions other than carbon atom 11.

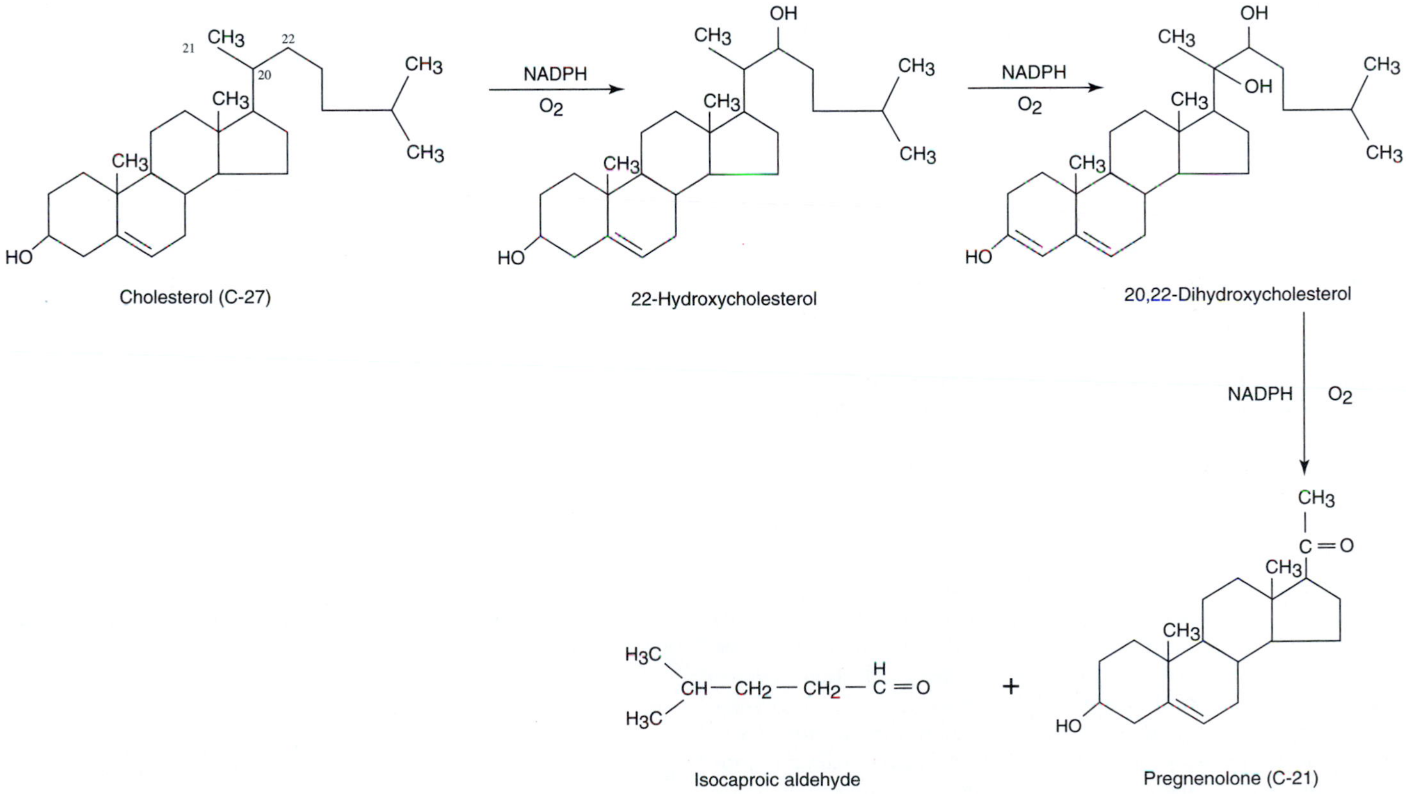

Cholesterol (C-27) → (NADPH, O_2) → 22-Hydroxycholesterol → (NADPH, O_2) → 20,22-Dihydroxycholesterol → (NADPH, O_2) → Isocaproic aldehyde + Pregnenolone (C-21)

FIGURE 11.9
Side chain cleavage reaction of cholesterol.
Three sequential reactions are catalyzed by cytochrome P450 to produce pregnenolone and isocaproic aldehyde.

Synthesis of cortisol proceeds from either pregnenolone or progesterone and involves a cytochrome P450 (CYP17), an endoplasmic reticulum cytochrome P450, which catalyzes the 17α-hydroxylation reaction. Hydroxylation at carbon atom 21 of 17α-hydroxyprogesterone by CYP21 produces **11-deoxycortisol,** which is transported into the mitochondrion, where it is further hydroxylated by CYP11B1 at carbon atom 11 to form cortisol. Congenital deficiencies resulting from mutations in the gene that codes for CYP21 are discussed in Clinical Correlation 11.1.

Synthesis of steroids containing 19 carbon atoms from the precursors, **17α-hydroxypregnenolone** or **17α-hydroxyprogesterone,** is the result of the loss of the acetyl group at carbon atom 17. This reaction is also catalyzed by CYP17, identified as the same cytochrome P450 that hydroxylates steroids at C-17. The factors that determine whether this cytochrome P450 performs only a single hydroxylation step to produce the 17-hydroxy product or proceeds further to cleave the C-17–C-20 bond has not been determined, although it was shown that the presence of cytochrome b_5 in gonadal tissues caused the CYP17 enzyme to catalyze the cleavage reaction. The products that are formed after removal of the acetyl group are **dehydroepiandrosterone (DHEA)** from 17α-hydroxypregnenolone or **androstenedione** from 17α-hydroxyprogesterone. DHEA may be dehydrogenated at the 3-OH group to androstenedione, a potent androgenic steroid that serves as the immediate precursor of testosterone.

Another physiologically important reaction catalyzed by a cytochrome P450 is the synthesis of estrogens from androgens, a reaction called **aromatization** because an aromatic ring is produced in ring A of the steroid product. In this complex reaction, not unlike the side-chain cleavage of cholesterol, multiple hydroxylation reactions are catalyzed by a single cytochrome P450 enzyme, to form the aromatic ring with removal of the methyl group at C-19. Figure 11.10 outlines the aromatization reaction. Two sequential hydroxylation reactions at methyl C-19 result in the introduction of an aldehyde group. The final step may involve a peroxidative attack at C-19 with the loss of the methyl group and elimination of the hydrogen atom to produce the aromatic ring. All three hydroxylation reactions are catalyzed by the same cytochrome P450 and the enzyme is called **aromatase** or $P450_{arom}$, a member of the CYP19 subfamily. The complexity of steroid hormone

CLINICAL CORRELATION 11.1

Congenital Adrenal Hyperplasia: Deficiency of CYP21A2

The adrenal cortex is a major site of steroid hormone production during fetal and adult life. The adrenal gland is metabolically more active in fetal life and may produce 100–200 mg of steroids per day in comparison to the 20–30 mg produced per day in the nonstressed adult adrenal gland. Enzyme deficiencies have been reported at all steps of cortisol production. Diseases associated with insufficient cortisol production are referred to as congenital adrenal hyperplasias (CAHs). The enzyme deficiency that is most common in CAHs is a defective 21-hydroxylase enzyme or CYP21A2, resulting in the failure to metabolize 17α-hydroxyprogesterone to 11-deoxycortisol and its conversion to cortisol. This causes an increase in ACTH secretion, the pituitary hormone that regulates adrenal cortex production of cortisol. Prolonged periods of elevated ACTH levels cause adrenal hyperplasia and an increased production of the androgenic hormones, dehydroepiandrosterone (DHEA) and androstenedione. Clinical problems arise because the additional production of androgenic steroids causes virilization in females, precocious sex organ development in prepubertal males, or diseases related to salt imbalance because of decreased levels of aldosterone. Clinical consequences of severe 21-hydroxylase deficiency may be recognizable at birth, particularly in females, because the excessive buildup of androgenic steroids may cause obvious irregular development of their genitalia. In male newborns, a deficiency in 21-hydroxylase activity may be overlooked, because male genitalia will appear normal, but there will be precocious masculinization and physical development. In late onset CAH, individuals are born without obvious signs of prenatal exposure to excessive androgen levels, and clinical symptoms may vary considerably from early development of pubic hair, early fusion of epiphyseal growth plates causing premature cessation of growth, or male baldness patterns in females.

Donohoue, P. A., Parker, K., and Migeon, C. J. Congenital adrenal hyperplasia. In: C. S. Scriver, A. L. Beaudet, W. S. Sly, and D. Valle (Eds.), *The Metabolic and Molecular Bases of Inherited Disease,* 7th ed. Vol. II, New York; McGraw-Hill, 1995, Chap. 94, p. 2929; and White, P. C. and Speiser, P. W. Congenital adrenal hyperplasia due to 21-hydroxylase deficiency. *Endocr. Rev.* 21:245, 2000.

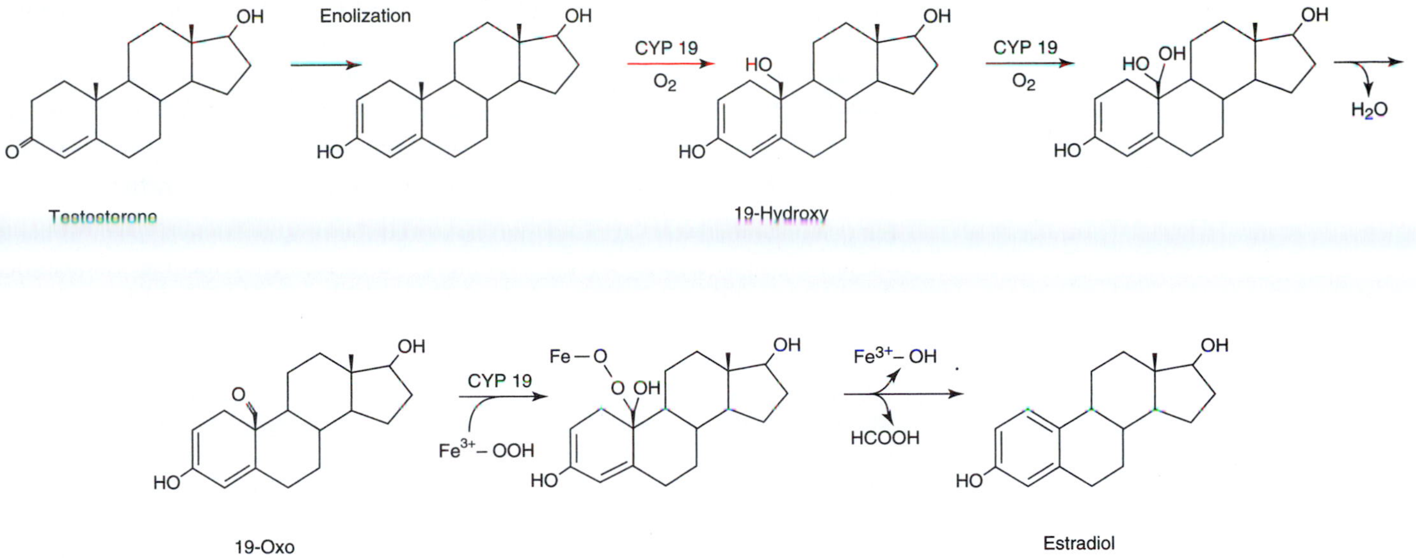

FIGURE 11.10
Sequence of reactions leading to the aromatization of androgens to estrogens.
Adapted from Graham-Lorence, S., Amarneh, B., White R. E., Peterson, J. A., and Simpson E. R. A three dimensional model of aromatase cytochrome P450. Protein Sci. *4:1065, 1995.*

production during pregnancy and the role of P450 forms are illustrated in Clinical Correlation 11.2.

Cytochrome P450 forms also metabolize other endogenous compounds and are important in regulating their activity. For example, CYP27B1 catalyzes the 1-hydroxylation of **25-hydroxy-vitamin D_3** to produce the 1,25-dihydroxy metabolite, which is the active form of this important hormone (see p. 1141). **Leukotriene** B_4 is a polymorphonuclear leukocyte chemoattractant and is hydroxylated by CYP4F3 to the less active 20-hydroxy-leukotriene B_4. **Arachidonic acid** is a

CLINICAL CORRELATION 11.2

Steroid Hormone Production During Pregnancy

Cytochrome P450 forms play a major role in estrogen synthesis as described in the text. During pregnancy, a unique interaction among cytochrome P450 forms in different organs is needed in order to synthesize the large quantities that are required. Steroid hormone production increases dramatically during pregnancy and, at term, the pregnant woman produces 15–20 mg of estradiol, 50–100 mg of estriol, and approximately 250 mg of progesterone per 24-h period. The amount of estrogens synthesized during pregnancy far exceeds the amount synthesized by nonpregnant women. For example, the pregnant woman at the end of gestation produces 1000 times more estrogen than premenopausal women per day.

The corpus luteum of the ovary is the major site for estrogen production in the first few weeks of pregnancy, but at approximately 4 weeks of gestation, the placenta begins synthesizing and secreting progesterone and estrogens. After 8 weeks of gestation, the placenta becomes the dominant source for the synthesis of progesterone. An interesting difference between the steroid hydroxylating systems in the placenta and the ovary is that the human placenta lacks the cytochrome P450 (CYP17) that catalyzes both the 17α-hydroxylation reaction and the cleavage of the C17–C20 bond. (see p. 963, for details of synthesis of steroid hormones). Thus the placenta cannot synthesize estrogens from cholesterol. The placenta catalyzes the side chain cleavage reaction to form pregnenolone from cholesterol and oxidizes pregnenolone to progesterone but releases this hormone into the maternal circulation. How then does the placenta produce estrogens if it cannot synthesize DHEA or androstenedione from progesterone? This is accomplished in the fetal adrenal gland, a highly active steroidogenic organ during fetal life, which catalyzes the synthesis of DHEA from cholesterol and releases it into the fetal circulation. A large proportion of fetal DHEA is metabolized by the fetal adrenal gland and liver to 16α-hydroxy-DHEA, which is converted by CYP19 (or CYP_{arom}) in the placenta to the estrogen estriol. This is an elegant demonstration of the cooperativity of the cytochrome P450-mediated hydroxylating systems in the fetal and maternal organ systems leading to the progressive formation of estrogens during the gestational development of the human fetus.

Cunningham, F. G., MacDonald, P. C., Gant, N. F., Leveno, K. J., Gilstrap, L. C, Hankins, G. D. V., and Clark, S. L. The placental hormones. In: *Williams Obstetrics*, 20th ed. East Norwalk, CT: Appleton and Lange, 1997, Chap. 6, p. 125.

substrate for CYP2J2, CYP2B6, and CYP4A11 and is metabolized to various epoxides or hydroxy and dihydroxy derivatives, which may have important regulatory functions (see p. 766). **Retinoic acid** is metabolized by CYP26A1 to the 4-hydroxy derivative, which may have greater activity than retinoic acid in certain organs.

Cytochromes P450 Oxidize Exogenous Lipophilic Substrates

Exogenous substrates, often referred to as **xenobiotics,** meaning "foreign to life," are important substrates for cytochrome P450 forms. They include therapeutic drugs, chemicals used in the workplace, industrial by-products that become environmental contaminants, and food additives. Cytochromes P450 oxidize a variety of xenobiotics, particularly lipophilic compounds. Addition of a hydroxyl group makes the compound more polar and, therefore, more soluble in the aqueous environment of the cell. Many exogenous compounds are highly lipophilic and will accumulate within cells over time, and interfere with cellular function. Examples of xenobiotics that are oxidized by cytochrome P450 are presented in Tables 11.2 and 11.3. In many cases, the action of the cytochromes P450 leads to a compound with reduced pharmacological activity, which can readily be excreted in the urine or feces. It is estimated that CYP3A4 metabolizes approximately 50% of therapeutic drugs, CYP2D6 approximately 20%, CYP2C9 and CYP2C19 approximately 15%, with CYP1A2, CYP2A6, CYP2B6, and other P450 forms the remaining 15% of drugs. It should be noted that because P450 forms have broad substrate specificities, a compound may be metabolized *in vivo* by more than one cytochrome P450 form and at different sites of the molecule as shown in Figure 11.11 for the compound diazepam. If more than one P450 metabolizes a substrate, the products that are formed in individuals may differ because there may be different amounts of the proteins expressed and cytochrome P450 forms may differ in their K_m and V_{max} values.

Xenobiotics can also be modified by a variety of conjugating enzyme systems, which form products that are nontoxic and readily eliminated from the body. A list of enzymes that metabolize xenobiotics is presented in Table 11.3. Many of these enzymes, including those catalyzing conjugation reactions occur primarily in the liver.

One example of a xenobiotic whose metabolism has received considerable attention is **benzo[*a*]pyrene**, a common environmental contaminant produced from

TABLE 11.2 Types of Reactions Catalyzed by Cytochromes P450

Reaction	*Example*
Aliphatic hydroxylation	Hydroxylation of ibuprofen at its alkyl side chain
Aromatic hydroxylation	Hydroxylation of S-warfarin
N-hydroxylation	Hydroxylation of aminoglutethimide to its *N*-hydroxy derivative
Epoxidation	Metabolism of aflatoxin B_1 to its epoxide derivative, a liver carcinogen
Oxygen and nitrogen dealkylation	*O*-demethylation of codeine to morphine, its active analgesic agent
	N-demethylation of diazepam (Valium) to nordazepam
Oxidative deamination	Removal of the amine group of amphetamine to form phenylacetone
Phosphorus oxidation	Oxidation of malathion, an insecticide, to malaoxon, its active agent
Sulfur oxidation	Oxidation of cimetidine (Tagamet) to its sulfoxide derivative
Dehalogenation	Dechlorination of the solvent carbon tetrachloride (CCl_4) to form the radical $CCl_3\bullet$
Dehydrogenation	Metabolism of ethanol (CH_3CH_2OH) to acetaldehyde (CH_3CHO)

TABLE 11.3 Examples of Xenobiotic-Metabolizing Reactions and Enzymes

Type of Reaction	*General Enzyme Name*	*General Enzymatic Reaction*
Oxidation	Dehydrogenases	Oxidation of alcohol or aldehyde groups
	Flavin-containing monooxygenases (FMO)	Oxidation of primary, secondary, and tertiary amines, nitrones, thioethers, thiocarbamates and phosphines.
	Monoamine oxidases	Removal of an amine by oxidative deamination
	Cytochrome P450	See Table 11.2 and Figure 11.7 for reactions catalyzed by P450 enzymes
Reduction	Aldehyde and ketone reductases	Reduction of carbonyl group to a hydroxyl group
	Azoreductases	Reduction of azo group to form primary amines
	Quinone reductases	Reduction of quinones to hydroquinones
Hydrolysis	Epoxide hydrolases	Addition of water to arene oxide or epoxide groups to form vicinal hydroxy groups
	Carboxylesterases	Addition of water to an ester bond
	Amidases	Addition of water to an amide bond
Conjugation	*N*-acetyltransferases	Addition of CoA from acetyl-CoA to aromatic amines or compounds containing hydrazines
	UDP–glucuronosyl transferases	Addition of glucuronic acid from uridine-5′-diphosphate (UDP)–glucuronic acid to carboxyl, phenolic, aliphatic hydroxyl, aromatic or aliphatic amines, or sulfhydryl groups
	Sulfotransferases	Addition of sulfate group from phosphoadenosyl phosphosulfate (PAPS) to phenolic groups, aliphatic alcohols, or aromatic amines
	Methyltransferases	Addition of methyl group from *S*-adenosyl methionine to compounds containing phenolic, catechol, aliphatic or aromatic amines, or sulfhydryl groups
	Glutathione *S*-transferase	Conjugation of compounds containing electrophilic heteroatoms, nitrenium ions, carbonium ions, free radicals, epoxide, or arene oxide groups

the burning of coal, from the combustion of materials in tobacco products, from food barbecued on charcoal briquettes, and from industrial processing. Benzo[*a*]pyrene, a weak carcinogen, is metabolized to a potent carcinogen in animals and to a mutagen in bacteria, prompting considerable work in identifying the enzymes involved in this process. Benzo[*a*]pyrene is metabolized by P450 1A1, 1A2, and 1B1. Benzo[*a*]pyrene also binds to the **aryl hydrocarbon receptor (AhR)** and induces these P450 forms, thus increasing its own metabolism. The product found to be the ultimate **carcinogen** is **benzo[*a*]pyrene-7,8-dihydrodiol-9,10-epoxide,** whose formation is illustrated in Figure 11.12. The reactions involve a cytochrome P450-catalyzed epoxidation at the 7,8 position, hydrolysis by **epoxide hydrolase** to the vicinal hydroxy derivative, benzo[*a*]pyrene-7,8-dihydrodiol, and a second epoxidation reaction to form benzo[*a*]pyrene-7,8-dihydrodiol-9,10-epoxide. This latter metabolite has been shown to form guanine adducts, which are capable of disrupting gene function leading to mutations. The action of benzo[*a*]pyrene-7,8-dihydrodiol-9,10-epoxide on the *p53* gene was a critical step in

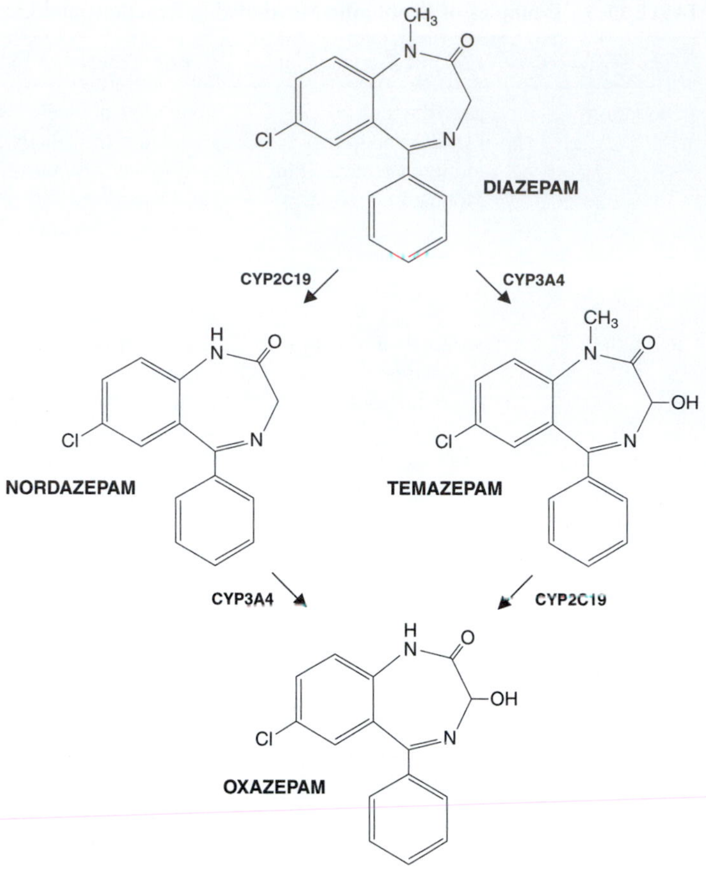

FIGURE 11.11
Metabolism of diazepam by cytochromes P450 3A4 and 2C19. Diazepam may be metabolized by more than one P450. The metabolites that are formed from a compound will be affected by the amount of P450 form that is expressed in tissues.

understanding the mechanism by which benzo[*a*]pyrene, a compound that is present in **cigarette smoke**, is carcinogenic in humans. Elucidation of the metabolic pathway of benzo[*a*]pyrene played a key role in understanding how it exerts its toxic effects on cells.

The metabolism of the common analgesic agent **acetaminophen** is another example of the conversion of a xenobiotic to a toxic metabolite. The metabolism of acetaminophen demonstrates the complex roles that P450 and other drug-

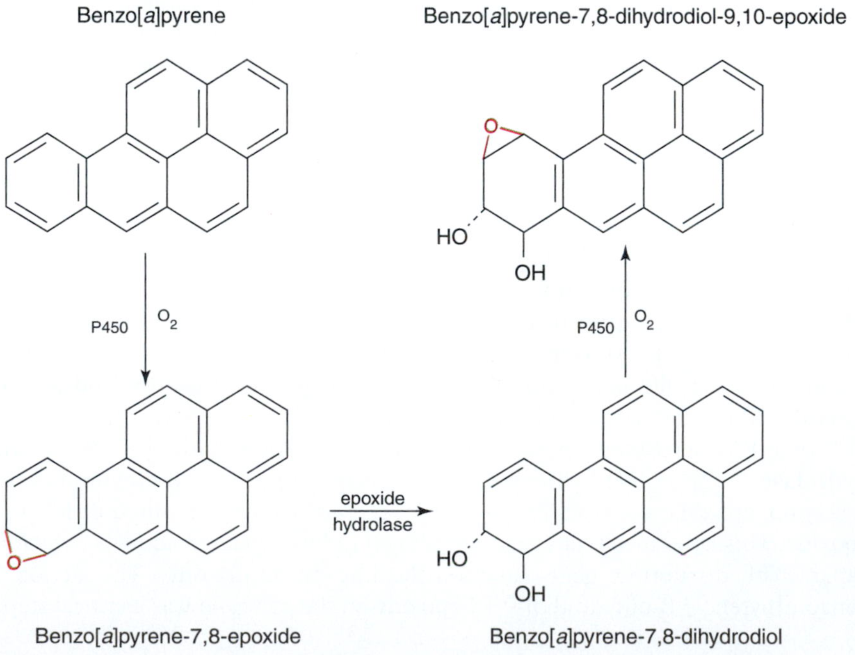

FIGURE 11.12
Metabolism of benzopyrene by cytochrome P450 and epoxide hydrolase to form benzopyrene-7,8-dihydrodiol-9,10-epoxide.

metabolizing enzymes play in drug-induced toxicities. Normally, acetaminophen is metabolized primarily by sulfation and glucuronidation pathways to polar, inactive conjugates that are readily excreted (Figure 11.13). Acetaminophen is also metabolized by CYP2E1 to a highly reactive intermediate, *N*-acetyl-*p*-benzoquinoneimine (NAPQI). The amount of acetaminophen that is metabolized by CYP2E1 is normally low in comparison to sulfation and glucuronidation. When small amounts of NAPQI are formed, they are rapidly conjugated by glutathione (see p. 818) to a nontoxic metabolite. However, if glutathione levels are depleted, NAPQI will escape glutathione conjugation and react with liver cell components to cause cell damage. Normally, the levels of CYP2E1 that are present are low in comparison to other P450 forms and the production of NAPQI, when normal doses of acetaminophen are taken, is small. However, when individuals take large quantities of acetaminophen, there will be an increase in the production of NAPQI that may cause severe liver damage. Another factor in acetaminophen-induced liver toxicity is that the consumption of alcoholic beverages induces CYP2E1, which will increase the production of NAPQI. However, **alcohol** is also a substrate for this cytochrome P450 and will inhibit the metabolism of other CYP2E1 substrates. The extent of liver damage depends on the timing and amount of acetaminophen taken. A description of the effects of alcohol on acetaminophen metabolism is presented in Clinical Correlation 11.3.

As exemplified by benzo[*a*]pyrene and acetaminophen, the P450 system can produce metabolites that are toxic. Formation of toxic compounds by the cytochrome P450, however, does not mean that cell damage or a tumor will occur. Other cellular processes will determine the extent of cellular damage, for example,

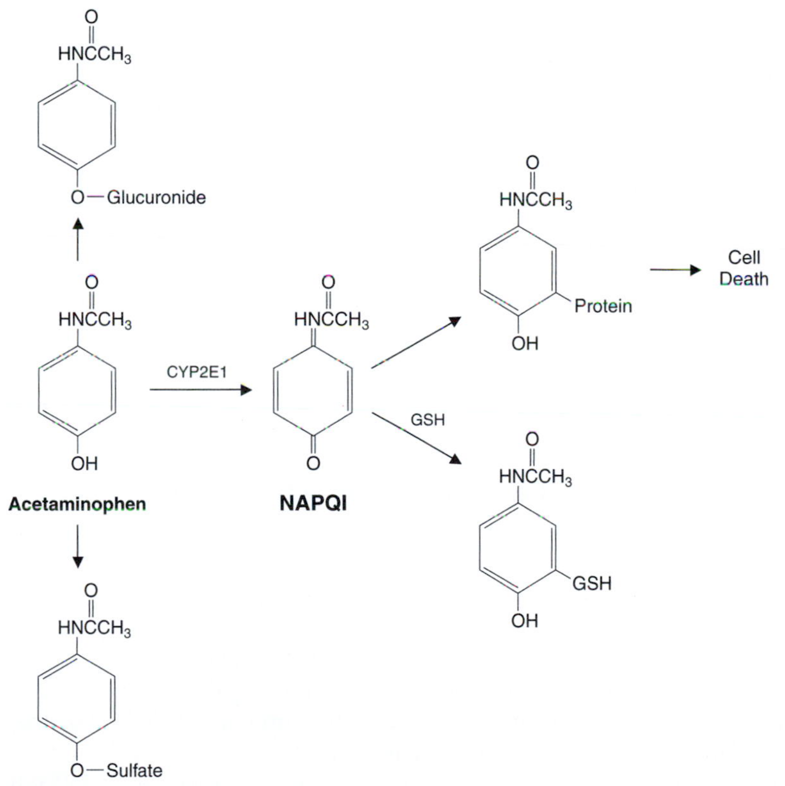

FIGURE 11.13
Acetaminophen may be metabolized by the phase II enzymes, sulfotransferase and glucuronsyltransferase, and by CYP2E1.
Metabolism by CYP2E1 will lead to the formation of *N*-acetyl-*p*-benzoquinoneimine (NAPQI). When NAPQI formation is low, NAPQI can be conjugated by glutathione (GSH) to a nontoxic metabolite; however, if NAPQI is formed in excessive amounts, it may escape glutathione conjugation and react with liver cell components to cause cell damage.

CLINICAL CORRELATION 11.3

Role of Cytochrome P450 2E1 in Acetaminophen-Induced Liver Toxicity

Acetaminophen-induced liver toxicity involves multiple factors. The pathway of acetaminophen metabolism is shown in Figure 11.13. One occurrence that has a profound effect on acetaminophen metabolism is the consumption of alcohol, which is both an inducer and a substrate for CYP2E1. In studies in which individuals consume alcohol that is the equivalent of drinking a 750 mL bottle of wine, six 12-oz. cans of beer, or 9 oz. of 80 proof liquor over a 6–7-h period, there is a 22% increase in CYP2E1-mediated metabolism of acetaminophen. This increase in CYP2E1-mediated acetaminophen metabolism was only observed when acetaminophen was given several hours after alcohol was administered. The timing between acetaminophen administration and the last consumption of alcohol may be very critical in the development of acetaminophen-induced liver injury. In a person drinking alcohol, the metabolism of acetaminophen by CYP2E1 is delayed because ethanol is also a substrate for CYP2E1 and competes with other CYP2E1 substrates, such as acetaminophen. In chronic alcohol consumption, a high CYP2E1 level is induced and produces greater amounts of *N*-acetyl-*p*-benzoquinoneimine (NAPQI); however, the alcohol consumed may actually protect drinkers from acetaminophen-induced liver injury, if they take the acetaminophen at the same time as the alcohol or shortly thereafter. If they take the acetaminophen several hours after they have stopped drinking, then increased levels of CYP2E1 may metabolize acetaminophen in the absence of alcohol, and there will be a greater risk of liver damage.

The length of time and amount of alcohol consumed daily affect the amount of CYP2E1 expressed in hepatocytes. The amount of acetaminophen consumed is a major determining factor as normally recommended doses do not usually generate sufficient NAPQI to produce liver cell damage. Higher doses of drug increase the risk of liver cell injury as more drug will be metabolized through the CYP2E1 pathway. Both the sulfation and glucuronidation pathways are essential in determining the amount of acetaminophen that will be metabolized by CYP2E1. An additional factor is the presence of glutathione, which conjugates with NAPQI to protect the liver from this reactive metabolite.

Brunner, L. J., McGuinness, M. E., Meyer, M. M., and Munar, M. Y. Acute acetaminophen toxicity during chronic alcohol use. *U.S. Pharmacist* Sept:HS11-HS19, 1999. This article may be accessed from the website at http://www.uspharmacist.com/; and Thummel, K. E., Slattery, J. T., Ro, H., Chien, J. Y., Nelson, S. D., Lown, K. E., and Watkins, P. B. Ethanol and production of the hepatotoxic metabolite of acetaminophen in healthy adults. *Clin. Pharmacol. Ther.* 67:591, 2000; and Slattery, J. T., Nelson, S. D., and Thummel, K. E. The complex interaction between ethanol and acetaminophen. *Clin. Pharmacol. Ther.* 60:241, 1996.

conjugation of the benzo[*a*]pyrene-7,8-dihydrodiol-9,10-epoxide by **glutathione transferases** was shown to play a key protective role in inhibiting DNA adduct formation and reducing its carcinogenic activity. The role of various detoxification enzyme systems including conjugation reactions such as glucuronidation and sulfation, the status of the immune system, nutritional state, genetic predisposition, and exposure to various environmental chemicals may play a key part in determining the role of P450 forms in cell damage and toxicity. One may ask: Why should the mammalian system possess an enzyme that would create highly toxic compounds? The purpose of the cytochrome P450 system is to add or expose functional groups making the molecule more polar and/or more susceptible to metabolism by other detoxification enzyme systems and the creation of toxic metabolites is an unexpected consequence of this process. The need to eliminate xenobiotics is more important than if these chemicals were permitted to accumulate in cells.

11.6 | INHIBITORS OF CYTOCHROMES P450

Due to the many known forms of cytochrome P450, it is of interest to examine the metabolic roles of these various enzymes in the organs in which they function. Cytochrome P450 inhibition has become a topic of clinical importance, because there is interest in the design of specific inhibitors to prevent the metabolism of endogenous and exogenous compounds. There is also significant interest in P450 inhibition because altering the metabolism of a drug from its predicted rate may cause drug accumulation and the onset of adverse effects.

An example of a therapeutic benefit for selective cytochrome P450 inhibition is the use of anti-fungal azole inhibitors (**ketoconazole**, **itraconazole**, and **fluconazole**) to control yeast infections. **Yeast infections** are associated with AIDS patients, in cancer patients undergoing chemotherapy, and in patients who are treated with immunosuppressive drugs. CYP51 catalyzes the demethylation of sterol molecules and in yeasts this reaction is involved in the synthesis of ergosterol, an important membrane lipid. A related P450 form, also identified as CYP51, is present in humans and catalyzes the demethylation of lanosterol, a reaction that is essential in cholesterol synthesis for the formation of steroid hormones. Although the human and yeast CYP51 proteins are similar, there are sufficient differences between mammalian and yeast P450 forms that permit azole antifungal agents to be effective inhibitors of ergosterol formation.

Because of the broad substrate specificity of P450 forms, some chemicals that are substrates for P450 can also serve as competitive or noncompetitive inhibitors of these proteins. This inhibition of P450 forms by various drugs is significant because a reduction in the metabolic activity of P450 proteins may create problems *in vivo* for individuals who take a combination of drugs. The observation that a drug may inhibit a cytochrome P450 enzyme *in vitro* using recombinant forms has helped to identify potential drug–drug interactions. For certain drugs, whose elimination from the body is dependent on its metabolism, *in vivo* inhibition will produce accumulations of the parent drug that may lead to the development of adverse effects. In certain cases, the severity of the adverse effects has led to the withdrawal of the drug for further human use (see Clin. Corr. 11.4). It should be noted that, although a drug inhibits a cytochrome P450-mediated reaction *in vitro,* it does not necessarily mean that the inhibitory effect will be observed *in vivo,* because the drug concentration that is needed to inhibit a P450 *in vitro* may not be attainable *in vivo.*

CLINICAL CORRELATION 11.4

Consequences of P450 Inhibition: Drug–Drug Interactions and Adverse Effects

The role that cytochrome P450 plays in drug metabolism and the serious consequences of drug–drug interactions were clearly shown for two drugs, terfenadine (Seldane) and cisapride (Propulsid), when their metabolism was inhibited by other drugs. A small percentage of their users experienced severe adverse effects that forced their manufacturers and the Food and Drug Administration (FDA) to issue warnings that the drugs should not be taken with other drugs that inhibit CYP3A4. The seriousness of these adverse effects forced the FDA to have both drugs removed from the market.

The FDA approved terfenadine in 1985 to treat seasonal allergies. Terfenadine is rapidly metabolized in the liver by CYP3A4 to fexofenadine, resulting in low levels of the parent drug soon after drug ingestion. This rapid metabolism does not reduce the therapeutic effects of terfenadine because the effects are associated with fexofenadine. Because a number of other drugs are substrates for or serve as inhibitors of CYP3A4, the metabolism of terfenadine is potentially inhibitable. Individuals who took terfenadine with the macrolide antibiotic erythromycin or the antifungal agent ketoconazole were found to have significantly elevated plasma levels of terfenadine since these drugs bind to CYP3A4 and are strong inhibitors of this P450 form. In a small percentage of individuals who used terfenadine, serious cardiac problems developed as a result of this rise in plasma levels of terfenadine, because the parent drug caused alteration in cardiac potassium channels and increased the risk of a rare ventricular tachycardia, called torsades de pointes. There were reports of individuals who died from cardiac problems that developed after taking terfenadine with erythromycin or ketoconazole. The FDA issued warnings that terfenadine should not be coadministered with erythromycin or ketoconazole, or to individuals who had reduced liver function and could not metabolize drugs normally. However, because of the continued reports of serious drug interactions with terfenadine, the FDA reevaluated this drug and took the unusual action in 1998 of proposing to withdraw its approval. The fact that the therapeutic properties of terfenadine were associated with its metabolite led the pharmaceutical manufacturer to test fexofenadine as a new medication and seek FDA approval for this metabolite, now marketed as Allegra.

Cisapride was approved in 1993 for patients who suffer from symptoms of nighttime heartburn as a result of gastroesophageal reflux disease or GERD. The elimination of this drug from the body is dependent on its metabolism by CYP3A4. When administered alone or with other drugs, that do not inhibit CYP3A4, the parent drug does not accumulate in the plasma. However, when cisapride is taken with certain drugs that are substrates or inhibitors of CYP3A4, its metabolism is reduced, resulting in an accumulation of the parent compound with subsequent administrations of the drug. In some individuals, increased cisapride levels cause cardiac arrhythmias. By the end of 1999, heart rhythm abnormalities were reported in 341 patients taking cisapride, resulting in 80 deaths. Accordingly, the FDA issued strict warnings to doctors and pharmacists about using cisapride with drugs that were substrates or inhibitors of CYP3A4 and, as a result, the manufacturer of cisapride decided to stop marketing this drug in the United States after 2000.

Terfenadine: proposal to withdraw approval of two new drug applications and one abbreviated new drug application. *Fed. Reg.* 62:1889, 1997. (This document may be accessed from the internet site for the *Federal Register* at http://www.access.gpo.gov/su_docs/aces/aces140.html); and Desta, Z., Soukhova, N., Mahal, S. K., and Flockhart, D. A. Interaction of cisapride with the human cytochrome P450 system: metabolism and inhibition studies. *Drug Metab. Dispos.* 28:789, 2000.

Mechanism-based inhibitors, so-called **suicide substrates**, bear strong resemblance to a substrate of an enzyme but during catalytic turnover form an irreversible inhibition product with the enzyme prosthetic group or protein. Such inhibitors would be highly specific for a particular cytochrome P450 form. This represents a possible tactical approach to drug design. There is an interest in the design of mechanism-based inhibitors of CYP19, the cytochrome P450 that catalyzes estradiol formation from androstenedione. Certain tumors are dependent on estrogen for their growth, such as **breast tumors**, and the prevention of estrogen synthesis in tissues without inhibiting production of other steroid hormones can serve as a selective chemotherapeutic tool to reduce and eliminate tumors.

11.7 | REGULATION OF CYTOCHROME P450 EXPRESSION

Over 1000 cytochrome P450 genes, coding for different proteins and catalyzing the oxygenation of a variety of endogenous and exogenous substrates, have now been characterized in mammals, insects, bacteria, fungi, and plants. There remain other members of this gene superfamily for which sequences have not yet been determined. Within a single species it is estimated that there may be 20–60 different cytochromes P450 present in different organs and tissues. Some P450 proteins may be present only at certain stages of life, such as during fetal development. Others may be present in low amounts but increase when the animal is exposed to chemicals termed inducers.

Numerous cytochrome P450 forms have emerged due to gene duplication and mutational events occurring in the last 5–50 million years. The different forms of cytochrome P450 among various prokaryotic and eukaryotic species have likely arisen from the selective pressure of environmental influences, such as dietary habits or exposure to environmental agents. An examination of the phylogenetic tree, generated by comparing amino acid sequences and assuming a constant evolutionary change rate, leads to the conclusion that the earliest cytochromes P450 evolved to metabolize cholesterol and fatty acids. Therefore they may have played a role in the maintenance of membrane integrity in early eukaryotes.

Induction of Cytochrome P450

Induction of various cytochromes P450 by both endogenous and exogenous compounds has been known since the mid-1960s. **Induction** is the increase in amount of an enzyme in response to a chemical agent, called an inducer. The mechanisms of induction of cytochromes P450 have been demonstrated to be at either the transcriptional or posttranscriptional level (e.g., stabilization of mRNA or a decrease in protein degradation). It is not possible to predict the mode of induction based on the inducing compound. An example of the complexity of the induction process occurs with rat CYP2E1. The amount of this cytochrome P450 in liver is increased as a result of treatment with small organic molecules, such as ethanol, acetone, or pyrazole, or under fasting or diabetic conditions. Administration of these small organic compounds increases the amount of the CYP2E1 protein without affecting the levels of mRNA for the protein. While the mechanism is not completely understood, pyrazole may stabilize this specific P450 from proteolytic degradation. However, in diabetic rats the sixfold induction of CYP2E1 protein is accompanied by a tenfold increase in mRNA without involving an increase in gene transcription, suggesting stabilization of the mRNA.

The role of specific cytosolic receptor proteins has been indicated for some of the inducing agents. One of the most extensively studied is the cytosolic receptor called the **aryl hydrocarbon** (or Ah) **receptor**, which interacts with 2,3,7,8-tetrachlorodibenzo-*p*-dioxin (TCDD) and causes induction of CYP1A1, CYP1A2, and CYP1B1 forms. Polycyclic aromatic hydrocarbons are ligands for the Ah receptor, producing a ligand–receptor complex that is translocated into the nucleus and binds to specific response elements in the upstream regulatory regions of cytochrome P450 genes. A second protein called the **Ah receptor nuclear translocator (Arnt)** protein was found to interact with the ligand-bound Ah receptor. The Arnt protein serves as a heterodimeric receptor partner that is essential for the ligand–Ah receptor complex to recognize and bind its specific DNA response element. Figure 11.14 illustrates the interaction of the ligand with its receptor and the binding of the heterodimer receptor partner. Utilizing cytochrome P450 gene transfection and expression vector technology, it has been possible to express those portions of the P450 DNA representing the RNA polymerase II promoter region and the upstream DNA sequences in conjunction with another gene coding for an enzyme that is not expressed in eukaryotes. In an assay of the prokaryotic enzyme activity, for example, chloramphenicol acetyltransferase (CAT) in the expression system, it is possible to determine which specific nucleotide sequences of DNA are involved in regulating these genes.

Another receptor system of P450 genes is the **pregnane X receptor (PXR)**, which regulates the *CYP3A* genes. Pregnane refers to 21-carbon steroids such as the progesterone metabolite 5α-pregnane-3,20-dione. This compound was identified as a ligand for PXR and is a potent inducer of the human *CYP3A4* gene. In addition to C_{21} steroids, other structurally dissimilar compounds such as **carbamazepine** (a drug used for the treatment of convulsions) and **rifampin** (an antituberculosis drug), as well as **St. John's wort** (a nonprescription herbal agent used to treat mood disorders) are ligands for PXR. PXR, like the Ah receptor, also requires a receptor partner to recognize specific response elements in the 5′-flanking region of *CYP3A* genes. This protein is the **retinoid X receptor (RXR),** which forms a heterodimeric complex (PXR–RXR) that initiates transcription of *CYP3A* and other genes after this complex recognizes its response element. The induction of the *CYP3A4* gene is important because approximately 50% of prescription drugs are metabolized by this cytochrome P450 form.

Another inducer of cytochrome P450 genes is **phenobarbital**, which increases the transcription rate of several cytochrome P450 forms in different species. A

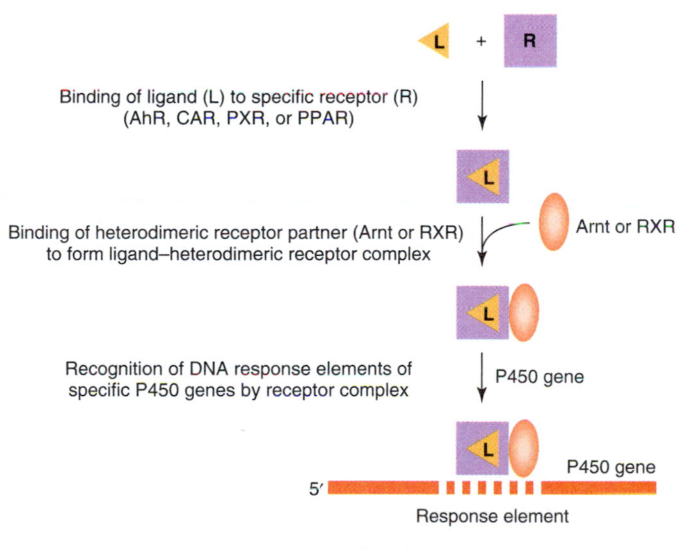

FIGURE 11.14
Flowchart demonstrating the interaction of a ligand with its receptor and receptor partner to form the heterodimeric receptor complex to initiate induction of cytochrome P450 forms.

protein called the **constitutive androstane receptor (CAR)** was found to mediate the induction of *CYP2B* genes by phenobarbital. CAR also forms a heterodimeric complex with **RXR** and interacts with a specific response element that is present in the 5′-flanking region of *CYP2B* genes. It was found that two endogenous androstane steroids, androstanol and androstenol, are ligands for CAR and when either ligand is bound to CAR, they prevent the interaction of the CAR–RXR receptor complex with the protein **SRC-1.** SRC-1 is a coactivator of nuclear receptors and is required for transcriptional activation. Administration of phenobarbital or other phenobarbital-like inducers displaces the androstane ligands, permitting SRC-1 binding and the transcription of the *CYP2B* gene.

Another nuclear receptor is the **peroxisome proliferator activated receptor alpha (PPARα),** which forms a heterodimeric complex with RXR, permitting recognition of response elements for *CYP4A* genes in rat, mouse, and rabbit. The human liver CYP4A11 protein has not been reported to be inducible. A wide variety of exogenous and endogenous compounds have been reported to serve as ligands for the PPARα including fatty acids, leukotrienes, hypolipidemic drugs, and phthalate plasticizers.

Because cytochrome P450 forms metabolize a number of therapeutic agents, their induction increases the rates at which therapeutic drugs are metabolized and may have profound effects on therapeutic outcomes. Certain medications may become ineffective and others may not be activated. Clinical consequences that relate to induction of drug-metabolizing enzymes are presented in Clinical Correlation 11.5.

Genetic Polymorphisms of Cytochrome P450

In addition to exposure to various inducing agents that may change the cytochrome P450 expression pattern in the liver and other organs, individuals may differ in their metabolism rates for particular drugs because of unique cytochrome P450 genes or alleles they possess. These genetic polymorphisms cause the expression of a cytochrome P450 protein that is unable to metabolize a drug at a sufficient rate, thereby producing significantly elevated drug levels. **Genetic polymorphisms** are characterized as being present in over 1% of a population. There may be greater expression of specific allelic variants in ethnic populations;

CLINICAL CORRELATION 11.5
Consequences of Induction of P450 Enzymes

Induction of specific P450 forms may decrease the therapeutic effects of drugs, because increases in liver P450 levels increase rates of metabolism and faster inactivation and/or excretion of drugs. Induction of the P450 system may have adverse effects due to the increased formation of toxic metabolites.

A drug–drug interaction that is caused by P450 induction is the enhanced elimination of oral contraceptives that may occur when a woman is treated with rifampicin, an antituberculosis drug that is a potent CYP3A4 inducer. Induction of hepatic CYP3A4 will enhance the metabolism of ethinyl estradiol, an active contraceptive agent. The accelerated turnover results in lower levels of the contraceptive drug, which increases the risk of pregnancy. Treatment with the anticonvulsant drugs phenytoin and carbamezepine may also induce CYP3A4 and reduce the contraceptive potency of ethinyl estradiol. It is imperative that a woman taking ethinyl estradiol, as an oral contraceptive simultaneously with a CYP3A4 inducer, increase the dose of the contraceptive or use an alternative contraceptive method or drug.

The widely used herbal agent St. John's wort, which can be purchased without a prescription, induces CYP3A4 activity. This has raised concerns that individuals who take St. John's wort may not receive the therapeutic benefits of drugs that are CYP3A4 substrates. A reduction in plasma levels of oral contraceptives, HIV antiprotease drugs, the immunosuppressive drug cyclosporin, or certain statin drugs, used to lower cholesterol levels, can occur because each of these drugs is a substrate for CYP3A4. Individuals may take St. John's wort without knowing its effects upon the efficacy of other drugs they are taking.

Roby, C. A., Anderson, G. D., Kantor, E., Dryer, D. A., and Burstein, A. H. St. John's wort: effect on CYP3A4 activity. *Clin. Pharmacol. Ther.* 67:451, 2000; and Shader, R. I. and Oesterheld, J. R. Contraceptive effectiveness: cytochromes and induction. *J. Clin. Psychopharmacol.* 20:119, 2000.

therefore specific ethnic groups may experience more frequent adverse effects with certain drugs. The variant alleles may code for proteins that contain an altered substrate docking site or an altered binding site for the NADPH–cytochrome P450 reductase, which reduces electron transfer. In some cases, the variant allele is caused by a single nucleotide substitution or a frameshift or missense mutation, whereas in other cases, the mutation leads to an early termination codon that codes for a truncated or shortened protein that is inactive. Such **allelic variants** result in individuals who are **poor metabolizers.** These subjects may be at risk for a dose-dependent toxicity if the unmetabolized form of the drug is pharmacologically active and the drug accumulates in the body with repeated dosing (see Clin. Corr. 11.6). It is estimated that 40% of the different human cytochrome P450 forms that metabolize xenobiotics are polymorphic. In the case of CYP2D6, 65 different alleles have been identified in humans. It should be noted that the existence of allelic variants for a cytochrome P450 gene does not mean that the expressed protein will be nonfunctional or demonstrate less metabolic turnover. In some cases, the amino acid substitution will not affect either the substrate-binding site or the docking of the NADPH cytochrome P450 reductase with the specific cytochrome P450.

Polymorphisms in which more than one active gene is expressed have also been reported and are due to gene duplication of a cytochrome P450 gene such that two or more copies of a gene are transcribed, leading to greater amounts of active protein and an accelerated rate of metabolism for substrates for this form. These individuals are extensive or **rapid metabolizers.**

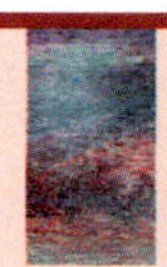

CLINICAL CORRELATION 11.6

Genetic Polymorphisms of P450 Enzymes

Multiple alleles have been demonstrated for approximately 40% of human P450 genes. These genetic polymorphisms may produce defective P450 proteins. A description of the polymorphic CYP21 and its importance in the metabolism of an endogenous compound was given in Clinical Correlations 11.1. The absence of a P450 form or the expression of nonfunctional protein may cause an individual many problems due to the accumulation of therapeutic drugs that cannot be metabolized. The discovery of an individual who suffered exaggerated hypotensive effects when administered the antihypertensive drug debrisoquine led to the characterization of individuals who inefficiently metabolized substrates catalyzed by CYP2D6. Approximately 5–10% of the Caucasian, 8% of the African and African-American, and 1% of the Asian populations are deficient for the catalytically active CYP2D6 form. In addition to debrisoquine, other drugs that are metabolized by CYP2D6 are sparteine, amitriptyline, dextromethorphan, and codeine. In the case of codeine, CYP2D6 catalyzes the *O*-demethylation of codeine to morphine. Approximately 10% of the dose of codeine is metabolized to morphine in individuals who have a normal CYP2D6. Individuals who lack the normal gene for CYP2D6 are unable to catalyze this reaction and do not achieve the analgesic effects associated with codeine.

Another genetic polymorphism has been demonstrated in individuals who are poor metabolizers of the drug mephenytoin. This drug is used in the treatment of epilepsy. Poor metabolizers of this drug suffer greater sedative effects at normal dosages. The 4-hydroxylation of the S-enantiomer of mephenytoin is carried out by CYP2C19. Approximately 14–22% of the Asian, 4–7% of the African and African-American, and 3% of the Caucasian populations are reported to be lacking the active form of the *CYP2C19* gene.

Polymorphisms associated with CYP2C9 have been found in less than 0.003% of the African or African-American, 0.08% of the Asian, and 0.36% of the Caucasian populations. The absence of a functional CYP2C9 may have serious consequences in an individual who is prescribed S-warfarin, an orally administered drug used to inhibit blood coagulation. This drug is used for extensive periods in patients who have suffered a heart attack or a stroke, to prevent the reoccurrence of clots that may precipitate another life-threatening episode. It is very important that plasma levels of the drug be maintained in a specific range because excessive amounts of the drug will cause uncontrolled bleeding and, possibly, death. Warfarin is dependent on CYP2C9 for its metabolism and elimination. A substitution of an isoleucine for a leucine at residue 359 in this P450 will cause a substantial loss of enzymatic activity. This allelic variant is denoted as CYP2C9*3. If a 5-mg dose of warfarin were prescribed each day to an individual lacking a functional CYP2C9, uncontrollable bleeding could result from a simple cut. Individuals deficient in CYP2C9 may require only 0.5–1 mg of warfarin per week in contrast to the 4–5 mg per day prescribed for an individual who possesses the normally active form of this P450 enzyme.

Ingelman-Sundberg., M., Oscarson, M., and McLellan, R. A. Polymorphic human cytochrome P450 enzymes: an opportunity for individualized drug treatment. *Trends Pharmacol. Sci* 20:342, 1999.

11.8 | OTHER HEMOPROTEIN AND FLAVOPROTEIN-MEDIATED OXYGENATIONS: SOLUBLE CYTOCHROMES P450 AND THE NITRIC OXIDE SYNTHASES

The yeast CYP51 was described earlier with regard to its sensitivity to azole antifungal agents to reduce yeast infections. **Bacterial cytochrome P450** proteins because of their unique structures have proved to be an important research tool in providing basic structural information on different cytochromes P450 including mammalian forms.

A *Bacillus megaterium* cytochrome P450, CYP102, has undergone extensive analysis and its unique structure has permitted one of the first detailed examinations at the atomic level of the substrate-binding and heme-containing domains of cytochrome P450 proteins. The cytochrome P450 in *Bacillus megaterium* is similar to the mammalian microsomal enzymes in that FAD, FMN, and heme prosthetic groups are required for cytochrome P450 activity. However, its structure is unique because these prosthetic groups are incorporated into a single polypeptide chain in which the reductase and heme-containing peptides are fused into a single large protein. Our basic understanding of the structural properties of a particular cytochrome P450, such as protein folding pattern and composition of β-sheets and α-helices, was developed from bacterial cytochromes P450, because they are soluble and more readily crystallized. Unlike mammalian cytochrome P450 proteins, which are integral membrane proteins and difficult to crystallize, CYP102 could be cleaved into distinct heme and flavin proteins, which were crystallized and studied by X-ray crystallography.

Another bacterial cytochrome P450, the CYP51 from *Mycobacterium tuberculosis,* was recently characterized and was found to be a soluble protein that was not fused with the FAD/FMN domain of the NADPH–cytochrome P450 reductase as in CYP102. This soluble CYP51 has been crystallized and its three-dimensional structure has been determined.

Although the CYP102 protein is unique among bacterial cytochrome P450 forms because its reductase and heme-containing domains are fused into a single soluble protein, there is another group of mammalian proteins that resemble CYP102 in its molecular and biochemical properties. These mammalian proteins are the **nitric oxide synthases** (NOS), which have received considerable attention in the last decade. The NOS proteins are the first mammalian proteins to be identified to contain FAD, FMN, and heme prosthetic groups in a single polypeptide chain. The NOS proteins function in a variety of important physiological processes and the significance of these enzymes in mammalian biology was recognized by the awarding of the 1998 Nobel Prize to Drs. Ferid Murad, Louis J. Ignarro, and Robert F. Furchgott for their elucidation of the physiological functions of nitric oxide (NO) in health and disease.

Three Nitric Oxide Synthases Display Diverse Physiological Functions

Release of nitric oxide from therapeutic drugs has been used as a treatment for angina pectoris since 1867, when Sir Thomas Lauder Brunton reported the use of nitroglycerin and amyl nitrate in his patients. However, it was not known until the 1980s that **nitric oxide**, or NO•, was the active agent in the dilation of blood vessels. The demonstration that this free radical diatomic gas was the primary endogenous **vasodilator** released by the vascular endothelium led to the search for an enzymatic source of NO•. The source of NO• is the guanidino group of the naturally occurring amino acid L-arginine. The reaction catalyzed by the enzymes responsible for the conversion of L-arginine to L-citrulline and NO• is

$$\text{L-arginine} + \text{NADPH} \longrightarrow \text{L-citrulline} + \text{NO}\bullet + \text{NADP}^+$$

Three genes have been identified for the isoforms of NOS. The respective enzymes have been designated as neuronal (NOS-I), macrophage or induced (NOS-II), or endothelial (NOS-III). Any tissue or cell may contain more than one isoform of nitric oxide synthase, thus contributing to the production of NO• under various physiological circumstances. Studies of the macrophage-type of nitric oxide synthase led to the conclusion that, upon treatment of animals with **cytokines** or **lipopolysaccharide**, the increase in production of NO• was due to this isoform, since it is quantitatively the major source of NO•. Subsequently, L-arginine was shown to be the precursor of NO• in both endothelial and neuronal cells. Production of NO• is necessary for the maintenance of **vascular tone, platelet aggregation, neural transmission**, and bacterial and/or tumor **cytotoxicity** (see Clin. Corr. 11.7).

Structural Aspects of Nitric Oxide Synthases

Although the written reaction does not reveal the overall stoichiometry, it is representative of a monooxygenation reaction and the mechanism is similar to that catalyzed by cytochromes P450. The oxygen atoms incorporated into both L-citrulline and NO• are derived from atmospheric oxygen. It was originally assumed that the oxygenation was occurring through the mediation of **tetrahydrobiopterin** (H_4B), a

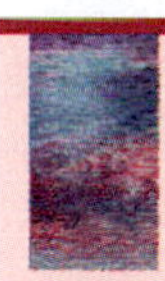

CLINICAL CORRELATION 11.7

Clinical Aspects of Nitric Oxide Production

Although NO• is essential in tumoricidal and bactericidal functions of macrophages, overproduction of NO• (from the inducible isoform of nitric oxide synthase, iNOS or NOS-II) has been implicated in septic/cytokine-induced circulatory shock in humans through the activation of guanylate cyclase. This mechanism is responsible for profound hypotension in postoperative patients whose recovery is complicated by bacterial infections that produce endotoxins. Hypotension in these patients is often refractory to treatment with conventional vasoconstrictor drugs. Therapeutic intervention by NOS inhibitors is being examined in inflammatory diseases of the gastrointestinal tract, such as pancreatitis and ulcerative colitis, and in arthritis.

The endothelial isoform of nitric oxide synthase, eNOS or NOS-III, is thought to play a critical role in maintaining a basic vasotonus in hemodynamic regulation such that an imbalance in the production of NO• could result in hypertension, thrombosis, or atherosclerosis. On the other hand, direct application of NO• gas may also be beneficial in the treatment of pulmonary hypertension. Following reports of the administration of inhaled NO• in the laboratory and to adult patients with primary pulmonary hypertension, its clinical utility was determined. In selected populations of patients, for example, hypoxic children and adults, inhaled NO• has been shown to improve arterial oxygenation and reduce pulmonary arterial hypertension. It has been used in newborn patients with hypoxic respiratory failure to reduce the need for extracorporeal membrane oxygenation, which is both expensive and invasive. Inhaled NO• has also been used to treat acute respiratory distress syndrome (ARDS), resulting in NO•-induced pulmonary vasodilation, decreased pulmonary arterial pressure (MPAP), and increased cardiac index.

Recent experiments with mice in which the gene for nNOS or NOS-I has been deleted have produced animals with distended stomachs due to constriction of the pyloric sphincter. nNOS is located in the neurons that innervate the myenteric plexis. This work has unexpectedly produced a model for the clinical disease infantile hypertrophic pyloric stenosis. These nNOS-deficient mice are resistant to brain damage as a result of ischemic injury, which produces vascular strokes, suggesting that nNOS is involved in this injurious process. While the direct connection to human disease has not yet been made, in this instance, it presents a paradigm, which can now be examined in clinical and pathological settings. Such a paradigm must include the continuum of effects from physiological to pathophysiological events that NO• produces as a signaling molecule. As the NO• concentration increases, there is a gradient of consequences ranging from the activation of guanylate cyclase, action as an antioxidant, inactivation of enzymes, induction of stress proteins, apoptosis, and, finally, DNA damage. The existence of this gradient suggests that specific inhibition of the isoforms of nitric oxide synthase is an ideal goal for therapeutic intervention in various circumstances. The development of potent, specific inhibitors of the isoforms of nitric oxide synthase is an active area of research being pursued collaboratively by investigators in academia and the pharmaceutical industry.

Hurford, W. E., Steudel, W., and Zapol, W. Clinical therapy with inhaled nitric oxide in respiratory diseases. In: Louis J. Ignarro (Ed.), *Nitric Oxide Biology and Pathobiology.* San Diego, CA: Academic Press, 2000, Chap. 56, p.931: and Kim, Y.-M., Tzeng, E., and Billiar, T. M. Role of NO and nitrogen intermediates in regulation of cell functions. In: M. S. Goligorsky and S. S. Gross (Eds.), *Nitric Oxide and the Kidney. Physiology and Pathophysiology.* New York: Chapman and Hall, 1997, Chap. 2, p. 22.

required cofactor for the overall reaction, analogous to the phenylalanine hydroxylase reaction (see p. 797). The discovery that heme (iron protoporphyrin IX) is a functional prosthetic group associated with all three isoforms of NOS has directed subsequent studies to include interactions between the flavoprotein and hemoprotein domains of these enzymes. These complex proteins must now be understood from the standpoint of the roles of the flavins, heme, and H_4B, under the control of Ca^{2+}/calmodulin in the case of the neuronal (NOS-I) and endothelial (NOS-III) isoforms. Figure 11.15 shows the overall structural organization of neuronal NOS (NOS-I).

The flow of electrons may occur in an analogous fashion to that of cytochrome P450-mediated systems. The electron donor is NADPH, which donates two electrons to **FAD,** which, in turn, reduces **FMN.** The FMN reduces the heme iron prosthetic group to its ferrous form to which oxygen can now bind for the oxygenation of the substrate, L-arginine. The overall reaction is inhibited by carbon monoxide and enzyme activity is totally dependent on bound calmodulin, which requires high concentrations of Ca^{2+} for the neuronal and endothelial isoforms. **Calmodulin** (see p. 525) is involved in the control of electron flow between the flavin prosthetic groups and between the exit flavin, FMN, and the heme prosthetic group in the oxygenase module.

Similar to the situation observed with the cytochrome P450-mediated hydroxylation systems, there can be leakage of electrons to molecular O_2 to form the one-electron-reduced, **superoxide anion,** under conditions in which electron flow to the heme does not occur ("uncoupled"). This can occur in the absence of H_4B, for example, in the case of endothelial nitric oxide synthase. The addition of Ca^{2+}/calmodulin to the nitric oxide synthase isoforms in the absence of L-arginine and/or H_4B increases electron flux, which is dissipated in the form of superoxide anion production. The formation of reduced oxygen species, such as superoxide anion, has been implicated in **ischemia/reperfusion injury** under circumstances in which tissues are first depleted of oxygen and then reperfused with oxygen-rich blood. This type of injury occurs in **cerebral thrombosis** and in tissues that have been removed for **transplantation.**

The analogy between the systems synthesizing nitric oxide and the cytochrome P450-mediated systems is remarkable but the differences are significant and the oxygenase module probably represents an example of convergent evolution with the cytochromes P450. The three-dimensional structures of mammalian representatives of both the cytochromes P450 and the heme domains of nitric oxide synthases have been determined.

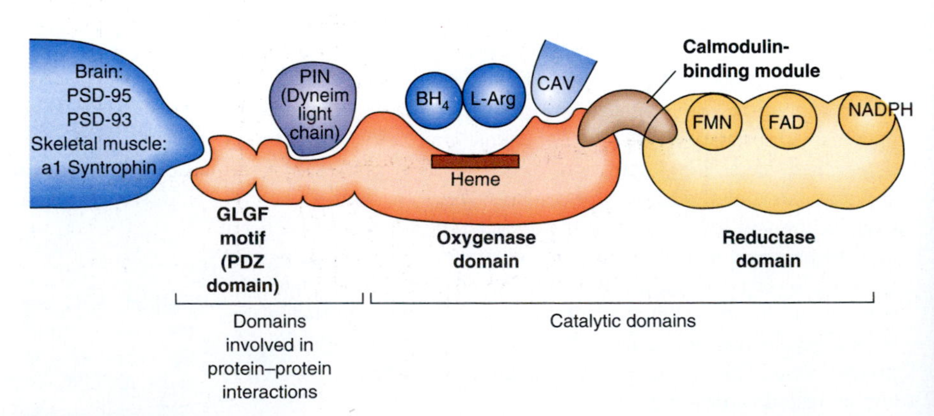

FIGURE 11.15
Modular structure of neuronal nitric oxide synthase showing approximate locations of prosthetic groups and cofactors.

BIBLIOGRAPHY

General References on Drug Metabolism and Cytochrome P450

Cupp, M. and Tracy, T. S. Cytochrome P450: new nomenclature and clinical implications. *Am. Fam. Physician* 57:107, 1998.

Estabrook, R. W., Cooper, D. Y., and Rosenthal, O. The light reversible carbon monoxide inhibition of the steroid C21-hydroxylase system of the adrenal cortex. *Biochem. Z.* 338:741, 1963.

Gibson, G. G. and Skett, P. *Introduction to Drug Metabolism*, 2nd ed. London: Academic, 1994.

Lewis, D. F. V. *Cytochromes P450. Structure, Function and Mechanism.* Bristol, PA: Taylor and Francis, 1996.

Nelson, D. R. Cytochrome P450 and individuality of species. *Arch. Biochem. Biophys.* 369:1, 1999.

Ortiz de Montellano, P. R. *Cytochrome P450. Structure, Mechanism, and Biochemistry.* New York: Plenum Press, 1995.

Parkinson, A. Biotransformation of xenobiotics. In C. D. Klassen (Ed.), *Casarett and Doull's Toxicology. The Basic Science of Poisons.* New York: McGraw Hill, 1996, Chap. 6, p. 113.

Substrate Specificity of Cytochromes P450: Physiological Functions

Cunningham, F. G., MacDonald, P. C., Gant, N. F., Leveno, K. J., Gilstrap, L.C., Hankins, G. D. V., and Clark, S. L. The placental hormones. In: *Williams Obstetrics,* 20th ed. Stamford, CT: Appleton and Lange, 1997, Chap. 6, p. 125.

Donohoue, P. A., Parker, K., and Migeon, C. J. Congenital adrenal hyperplasia. In: C. R. Scriver, A. L. Beaudet, W. S. Sly, and D. Valle (Eds.), *The Metabolic and Molecular Bases of Inherited Disease,* 7th ed., Vol II. New York: McGraw-Hill, 1995, Chap. 94, p. 2929.

Graham, S. E. and Peterson, J. A. How similar are P450s and what can their differences teach us? *Arch. Biochem. Biophys.* 369:24, 1999.

Graham-Lorence, S., Amarneh, B., White, R. E., Peterson, J. A., and Simpson, E. R. A three-dimensional model of aromatase cytochrome P450. *Protein Sci.* 4:1065, 1995.

Masters, B. S. S., Muerhoff, A. S., and Okita, R. T. Enzymology of extrahepatic cytochromes P450. In: F. P. Guengerich (Ed.), *Mammalian Cytochromes P450.* Boca Raton, FL: CRC Press, 1987, Chap. 3, p. 107.

Cytochrome P450 Inhibition

Lin, J. H. and Lu, A. Y. H. Role of pharmacokinetics and metabolism in drug discovery and development. *Pharmacol. Rev.* 49:403, 1997.

Thummel, K. E. and Wilkinson, G. R. *In vitro* and *in vivo* drug interactions involving human CYP3A. *Annu. Rev. Pharmacol. Toxicol.* 38:389, 1998.

Regulation of Cytochrome P450 Expression

Denison, M. S. and Whitlock, J. P. Xenobiotic-inducible transcription of cytochrome P450 genes. *J. Biol. Chem.* 270:18175, 1995.

Ingelman-Sundberg, M., Oscarson, M., and McLellan, R. A. Polymorphic human cytochrome P450 enzymes: an opportunity for individualized drug treatment. *Trends Pharmacol. Sci.* 20:342, 1999.

Kalow, W. *Pharmacogenetics of Drug Metabolism.* New York: Pergamon Press, 1992.

Kalow, W. Pharmacogenetics in biological perspective. *Pharmacol. Rev.* 49:369, 1997.

Meyer, U. A. and Zanger, U. M. Molecular mechanisms of genetic polymorphisms of drug metabolism. *Annu. Rev. Pharmacol. Toxicol.* 37:269, 1997.

Waxman, D. J. P450 gene induction by structurally diverse xenochemicals: central role of nuclear receptors CAR, RXR, PPAR. *Arch. Biochem. Biophys.* 369:11, 1999.

Biochemistry and Physiology of Nitric Oxide Formation

Bredt, D. S. and Snyder, S. H. Nitric oxide: a physiologic messenger molecule. *Annu. Rev. Biochem.* 63:175, 1994.

Garthwaite, J. and Boulton, C. L. Nitric oxide signalling in the central nervous system. *Annu. Rev. Physiol.* 57:683, 1995.

Griffith, O. W. and Stuehr, D. J. Nitric oxide synthases—properties and catalytic mechanism. *Annu. Rev. Physiol.* 57:707, 1995.

Ignarro, L. J., Buga, G. M., Wood, K. S., Byrns, R. E., and Chaudhuri, G. Endothelium-derived relaxing factor produced and released from artery and vein is nitric oxide. *Proc. Natl. Acad. Sci. USA* 84:9265, 1987.

Khan, M. T. and Furchgott, R. F. Additional evidence that endothelium-derived relaxing factor is nitric oxide. In: M. J. Rand and C. Raper (Eds.), *Pharmacology.* Amsterdam: Elsevier, 1987, p. 341.

Marletta, M. A. Approaches toward selective inhibition of nitric oxide synthase. *J. Med. Chem.* 37:1899, 1994.

Masters, B. S. S. Structural variations to accommodate functional themes of the isoforms of nitric oxide synthases. In: L. J. Ignarro (Ed.), *Nitric Oxide, Biology and Pathobiology.* San Diego, CA: Academic Press, 2000, p. 91.

Palmer, R. M. J., Ferrige, A. G., and Moncada, S. Nitric oxide release acccounts for the biological activity of endothelium-derived relaxing factor. *Nature* 327:524, 1987.

QUESTIONS C. N. ANGSTADT

Multiple Choice Questions

1. All of the following are correct about a molecule designated as a cytochrome P450 EXCEPT:
   A. it contains a heme as a prosthetic group.
   B. it catalyzes the hydroxylation of a hydrophobic substrate.
   C. it may accept electrons from a substance such as NADPH.
   D. it undergoes a change in the heme iron upon binding a substrate.
   E. it comes from the same gene family as all other molecules designated as cytochromes P450.

2. Known roles for cytochromes P450 include all of the following EXCEPT:
   A. synthesis of steroid hormones.
   B. conversion of some chemicals to mutagens.
   C. hydroxylation of an amino acid.
   D. inactivation of some hydrophobic drugs.
   E. metabolism of fatty acid derivatives.

3. Flavoproteins are usually intermediates in the transfer of electrons from NADPH to cytochrome P450 because:
   A. NADPH cannot enter the membrane.
   B. flavoproteins can accept two electrons from NADPH and donate them one at a time to cytochrome P450.
   C. they have a more negative reduction potential than NADPH and so accept electrons more readily.
   D. as proteins, they can bind to cytochrome P450 while the non-protein NADPH cannot.
   E. they contain iron–sulfur centers.

4. NADPH–cytochrome P450 reductase:
   A. uses both FAD and FMN as prosthetic groups.
   B. is found in mitochondria.
   C. requires an iron–sulfur center for activity.
   D. always passes its electrons to cytochrome b_5.
   E. can use NADH as readily as NADPH.

5. NADPH–adrenodoxin reductase:
   A. is located in the endoplasmic reticulum.
   B. passes its electrons to a protein with iron–sulfur centers.
   C. has a stretch of hydrophobic amino acid residues at the N-terminal end.
   D. reacts directly with cytochrome P450.
   E. reacts directly with cytochrome b_5.

6. Cytochrome P450 systems are able to oxidize:
   A. $—CH_2—$ groups.
   B. benzene rings.
   C. nitrogen atoms in an organic compound.
   D. sulfur atoms in an organic compound.
   E. all of the above.

7. In the conversion of cholesterol to steroid hormones in the adrenal gland:
   A. all of the cytochrome P450 oxidations occur in the endoplasmic reticulum.
   B. all of the cytochrome P450 oxidations occur in the mitochondria.
   C. side chain cleavage of cholesterol to pregnenolone is one of the cytochrome P450 systems that uses adrenodoxin reductase.
   D. cytochrome P450 is necessary for the formation of aldosterone and cortisol but not for the formation of the androgens and estrogens.
   E. aromatization of the first ring of the steroid does not use cytochrome P450 because it involves removal of a methyl group, not a hydroxylation.

8. Benzopyrene, a xenobiotic produced by combustion of a variety of substances:
   A. induces the synthesis of cytochrome P450.
   B. undergoes epoxidation by a cytochrome P450.
   C. is converted to a potent carcinogen in animals by cytochrome P450.
   D. would be rendered more water-soluble after the action of cytochrome P450.
   E. all of the above.

9. Genetic polymorphism in genes for cytochromes P450:
   A. could cause an individual to poorly metabolize certain drugs.
   B. could cause an individual to metabolize certain drugs more rapidly than normal.
   C. could lead to a specific ethnic group experiencing more frequent adverse effects with certain drugs.
   D. all of the above are correct.
   E. none of the above is correct.

Questions 10 and 11: Because herbal remedies are not prescription drugs, many patients fail to inform their physicians that they are using such products. A patient was told by his physician that his blood cholesterol levels were not responding to the statin drug he was taking although the drug is usually very effective. Further investigation elicited the fact that the patient was also taking St. John's wort, a widely used herbal agent, to improve his mood. St. John's wort induces a cytochrome P450 (CYP3A4) activity. Statins are one of the many drugs that are metabolized by CYP3A4.

10. The induction of cytochromes P450:
    A. occurs only by exogenous compounds.
    B. occurs only at the transcriptional level.
    C. necessarily results from increased transcription of the appropriate mRNA.
    D. necessitates the formation of an inducer–receptor protein complex.
    E. may occur by posttranscriptional processes.

11. Statins could be considered xenobiotics (exogenous substances metabolized by the body). Many xenobiotics are oxidized by cytochromes P450 in order to:
    A. make them carcinogenic.
    B. increase their solubility in an aqueous environment.
    C. enhance their deposition in adipose tissue.
    D. increase their pharmacological activity.
    E. all of the above.

Questions 12 and 13: Some patients show profound hypotension after abdominal surgery or abdominal trauma complicated by bacterial infections that produce endotoxins. Such hypotension is often refractory to treatment with conventional vasoconstrictor drugs. This hypotension is thought to be caused by an overproduction of nitric oxide by the induced form of nitric oxide synthase (NOS). Administration of NOS inhibitors specific to this form might be an appropriate treatment for such patients.

12. Nitric oxide:
    A. is formed spontaneously by a reduction of NO_2.
    B. is synthesized only in macrophages.
    C. is synthesized from arginine.
    D. acts as a potent vasoconstrictor.
    E. has three isoforms.

13. Nitric oxide synthase:
    A. catalyzes a dioxygenase reaction.
    B. is similar mechanistically to phenylalanine hydroxylase since it requires tetrahydrobiopterin.
    C. accepts electrons from NADH.
    D. uses a flow of electrons from NADPH to FAD to FMN to heme iron.
    E. is inhibited by Ca^{2+}.

Problems

14. Phenobarbital is a potent inducer of cytochrome P450. Warfarin, an anticoagulant, is a substrate for cytochrome P450 with the result that the drug is metabolized and cleared from the body more rapidly than normal. If phenobarbital is added to the therapeutic regimen of a patient, with no change in the dosage of warfarin, what would the expected consequence be? What would happen if the warfarin dosage were adjusted for a proper response, and then phenobarbital were withdrawn without adjusting the warfarin dosage?
15. Acetaminophen is primarily metabolized by sulfation and glucuronidation to compounds that are excreted. It can also be metabolized by a cytochrome P450 (CYP2E1) to a highly reactive intermediate (NAPQI), which can damage liver cells. Alcohol is an inducer and substrate for CYP2E1. If acetaminophen is taken several hours after drinking alcohol, liver toxicity of acetaminophen is greatly enhanced. If acetaminophen and alcohol are consumed simultaneously, this enhancement of toxicity is not observed. Why?

ANSWERS

1. **E** Several gene families are known. The number after CYP designates the family. A: All cytochromes are heme proteins. B: The types of substrates are hydrophobic. It is classified as a monooxygenase. C: This is the usual electron donor. D: The change from hexa to penta coordinated gives the compound a more positive reduction potential.
2. **C** Cytochromes P450 are not the only hydroxylases and other types are active with amino acids.
3. **B** Heme can accept only one electron at a time while NADPH always donates two at a time. A: NADPH passes only electrons; it does not have to enter the membrane. C: If this were true, the flow of electrons would not occur in the way it does. D: Protein–protein binding is not known to play a role here. E: Iron–sulfur centers play a role in some, but not all, systems. Flavoproteins are not the only system with iron–sulfur centers.
4. **A** This enzyme is one of two mammalian proteins known to do so. B: This is in the endoplasmic reticulum. C: Some reductases do so but not this one. D: Only certain reactions catalyzed by the enzyme do. E: There are NADH-dependent reductases but they are different enzymes.
5. **B** The iron–sulfur protein is adrenodoxin, which passes the electron to cytochrome P450. A, C: This is a mitochondrial enzyme. D, E: See B.
6. **E** These are all legitimate reaction types.
7. **C** This is a mitochondrial process. A, B: Hormone synthesis involves a series of reactions that move back and forth between mitochondria and endoplasmic reticulum. D, E: Removal of side chains frequently begins with oxidation reactions.
8. **E** A: It is not uncommon for xenobiotics to induce synthesis of something that will enhance their own metabolism. B, C: Epoxidation is the first step in the conversion of this compound to one that is carcinogenic—again, a common occurrence. D: Benzopyrene, with its four fused benzene rings, is highly hydrophobic; introducing oxygens increases water solubility.
9. **D** A: This occurs if the allelic variants code for inactive or less active proteins. B: If there is gene duplication, an individual may have larger amounts of active protein. C: There may be greater expression of specific allelic variants in specific ethnic populations.
10. **E** There may be a stabilization of mRNA (as seen in diabetic rats) or decrease in the degradation of the protein, which may be a mechanism for pyrazole. A: Both endogenous and exogenous substances can induce cytochromes P450. B, C: Transcriptional modification is only one of the mechanisms of induction (see E). D: This has been shown with induction by some compounds, but with others, like phenobarbital, this is not so.
11. **B** The types of xenobiotics oxidized by cytochrome P450 are usually highly lipophilic but must be excreted in the aqueous urine or bile. A: This may happen but is certainly not the purpose. C: They do that prior to oxidation. D: Oxidation tends to reduce pharmacological activity.
12. **C** The other product is citrulline. A: It is formed from arginine. B: One of the isoforms of NO synthase has been found in macrophages but neuronal and endothelial isoforms also exist. The macrophage isoform is inducible. D: Nitric oxide is a vasodilator, which is the basis for the use of nitroglycerin in angina pectoris. E: Three isoforms of nitric oxide synthase have been identified but nitric oxide is a specific compound.
13. **D** This is the second mammalian enzyme known to use both FAD and FMN. A: The reaction is a monooxygenation. F: BH_4 is required but the action of the enzyme is similar to a cytochrome P450-mediated system. C: The donor is NADPH. E: The system requires Ca^{2+}-calmodulin, at least the neuronal and endothelial isoforms.
14. If warfarin is metabolized and cleared more rapidly by cytochrome P450, its therapeutic efficiency is decreased. Therefore, at the same dosage, it will be less effective as an anticoagulant and there is an increased possibility of clot formation. If the phenobarbital is withdrawn, the cytochrome P450 levels will decrease with time to the uninduced level. With no change in the warfarin level, eventually there will be an increased possibility of hemorrhaging.
15. The toxic effects of acetaminophen are due to the production of NAPQI generated by CYP2E1. In the presence of alcohol, CYP2E1 levels are increased (induced). When acetaminophen and alcohol are consumed together, they compete for the enzyme and the metabolism of acetaminophen by this pathway is slowed. If alcohol is consumed alone, it induces the enzyme. When acetaminophen is later consumed, its metabolism by this pathway is more rapid (there is no competition with alcohol) and more NAPQI is formed. Obviously the levels of acetaminophen consumed is also a factor. Small amounts may be able to be handled by the normal conjugation pathways in a nontoxic fashion.

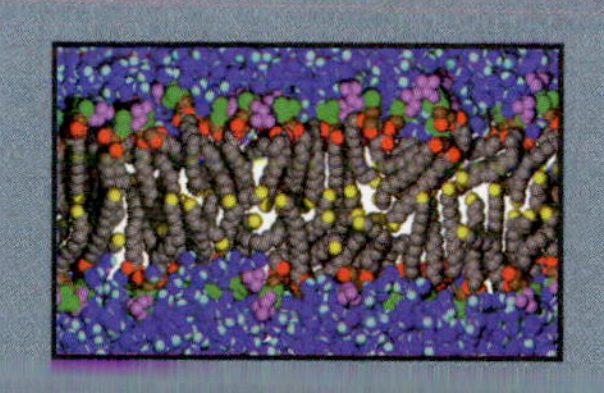

12

BIOLOGICAL MEMBRANES: STRUCTURE AND MEMBRANE TRANSPORT

Thomas M. Devlin

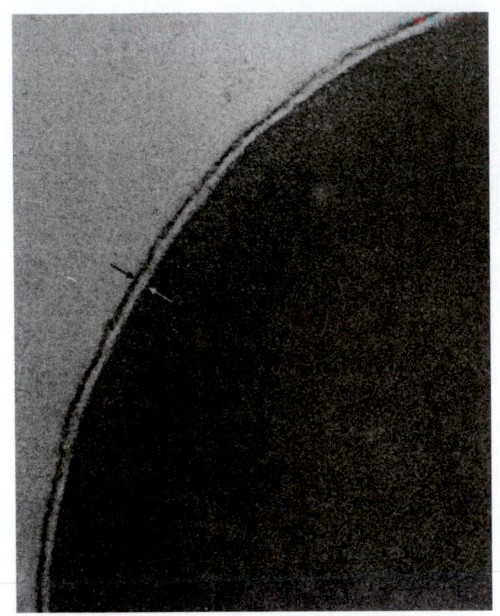

FIGURE 12.1
Electron micrograph of the erythrocyte plasma membrane showing the trilaminar appearance.
A clear space separates the two electron-dense lines. Electron microscopy has demonstrated that the inner dense line is frequently thicker than the outer line. Magnification about × 150,000.
Courtesy of J. D. Robertson, Duke University, Durham, North Carolina.

12.1 | OVERVIEW

Biological membranes from eukaryotic or prokaryotic cells have the same classes of chemical components, are similar in structural organization, and have many properties in common. There are major differences, however, in specific lipid, protein, and carbohydrate components but not in physicochemical interaction of these molecules. Membranes have a trilaminar appearance when viewed by electron microscopy (Figure 12.1), with two dark bands on each side of a light band. The overall width of most mammalian membranes is 7–10 nm but some are significantly thinner. Intracellular membranes are usually thinner than plasma membranes. Many do not appear symmetrical, with an inner dense layer often thicker than an outer dense layer. There is a chemical asymmetry of membranes. With development of sophisticated techniques for preparation of tissue samples and staining, including negative staining and freeze fracturing, surfaces of membranes have been viewed; at the molecular level surfaces are not smooth but dotted with protruding globular-shaped masses.

Membranes are very dynamic structures and undergo movement that permits cells and subcellular structures to adjust their shape and to change position. Chemical components of membranes, that is, lipids and protein, are ideally suited for this dynamic role. Membranes are an organized sea of lipid in a fluid state, in which both proteins and lipids are able to move and interact.

Cellular membranes control the composition of space that they enclose by excluding a variety of molecules and by selective transport systems allowing movement of specific molecules from one side to the other. These transporters are proteins. By controlling translocation of substrates, cofactors, and ions, the concentrations of substances in cellular compartments are modulated, thereby exerting an influence on metabolic pathways. Hormones, and growth and metabolic regulators bind to specific **protein receptors** on plasma membranes (Chapter 21) and information of their presence is transmitted from the membrane receptors to the appropriate metabolic pathway by a series of intracellular intermediates, termed **"second messengers."** Plasma membranes of cells have a role in cell–cell recognition, in maintenance of the cell shape, and in cell locomotion.

The discussion that follows is directed at the chemistry and transport functions of membranes, primarily of mammalian cells, but the observations and activities described are applicable to all biological membranes.

12.2 | CHEMICAL COMPOSITION OF MEMBRANES

Lipids and proteins are the two major components of all membranes. The amount of each varies greatly in different membranes (Figure 12.2). Protein ranges from about 20% in the myelin sheath to over 70% in the inner membrane of the mitochondria. Proteins are responsible for the dynamic function of membranes and, thus, individual membranes have very specific complements of proteins. Intracellular membranes have a high percentage of protein because of the large number of enzymic activities of these membranes. Membranes also contain a small amount of various polysaccharides in the form of glycoprotein and glycolipid, but contain no free carbohydrate.

Lipids Are Major Components of Membranes

The three major lipid components of eukaryotic membranes are glycerophospholipids, sphingolipids, and cholesterol. Glycerophospholipids and sphingomyelin, a sphingolipid containing phosphate, are classified as phospholipids. Bacteria and blue-green algae contain glycerolipids in which a carbohydrate is attached directly to the glycerol. Individual membranes also contain small quantities of other lipids, such as triacylglycerol and diol derivatives (see the Appendix (p. 1175) for structures), as well as lipid covalently linked to proteins.

Glycerophospholipids Are the Most Abundant Lipids of Membranes

Glycerophospholipids (phosphoglycerides) have a glycerol molecule with a phosphate esterified at the α-carbon (Figure 12.3) and two long-chain fatty acids esterified to the remaining carbon atoms (Figure 12.4). Glycerol does not have an asymmetric carbon, but the α-carbon atoms are not stereochemically identical. Esterification of a phosphate to an α-carbon makes the molecule asymmetric. Naturally occurring glycerophospholipids are designated by the stereospecific numbering system (*sn*)(Figure 12.3) discussed on p. 728.

1,2-Diacylglycerol 3-phosphate, or **phosphatidic acid**, is the parent compound of a series of glycerophospholipids, where different hydroxyl-containing compounds, some of which are polar, are esterified to the phosphate. The major

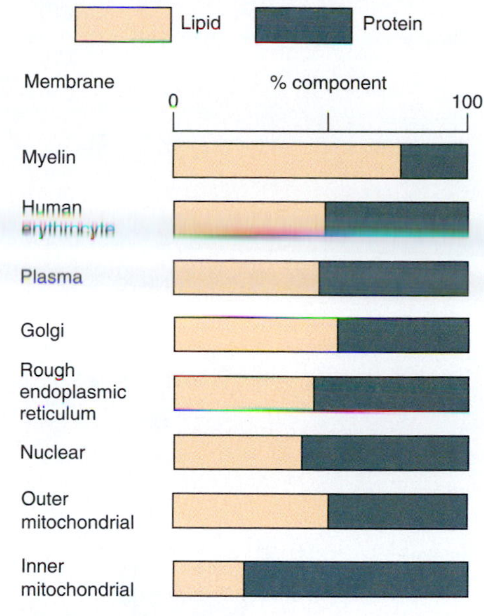

FIGURE 12.2
Representative values for the percentage of lipid and protein in various cellular membranes.
Values are for rat liver, except for the myelin and human erythrocyte plasma membrane. Values for liver from other species, including human, indicate a similar pattern.

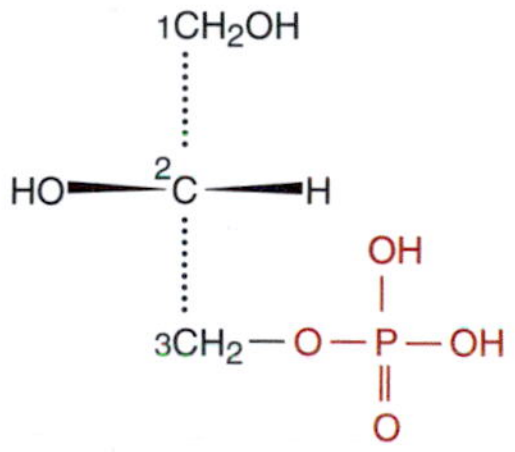

FIGURE 12.3
Stereochemical configuration of L-glycerol 3-phosphate (*sn*-glycerol 3-phosphate).
The H and OH attached to C-2 are above and C-1 and C-3 are below the plane of the page.

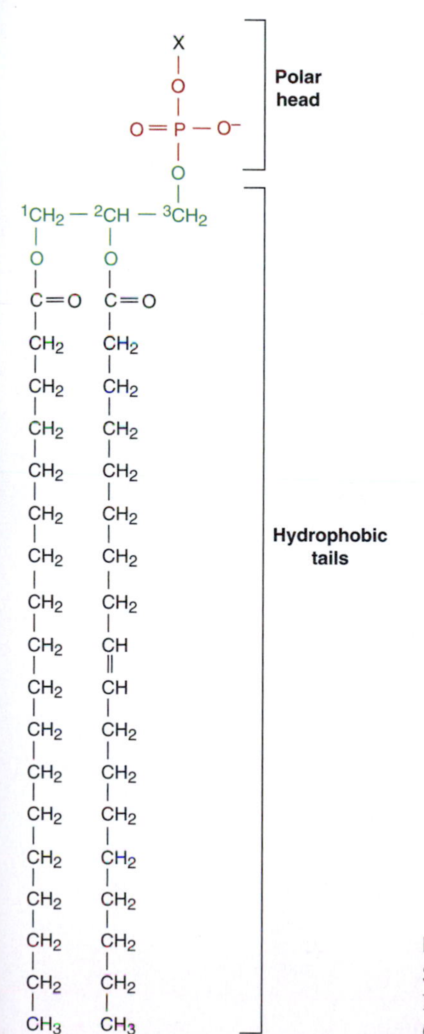

FIGURE 12.4
Structure of glycerophospholipid.
Long-chain fatty acids are esterified at C-1 and C-2 of the L-glycerol 3-phosphate. X can be a H (phosphatidic acid) or one of several alcohols presented in Figure 12.5.

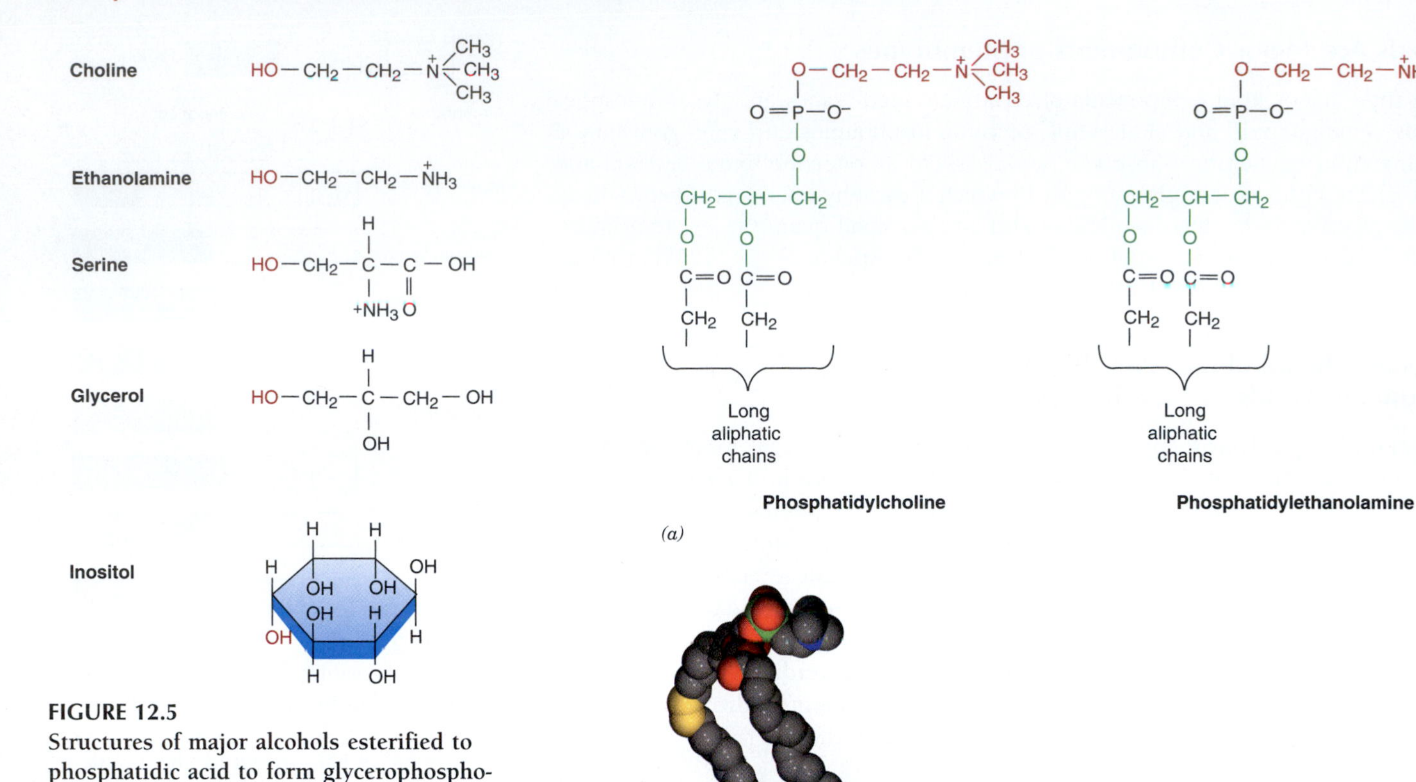

FIGURE 12.5
Structures of major alcohols esterified to phosphatidic acid to form glycerophospholipids.

FIGURE 12.6
Common glycerophospholipids.
(*a*) Structures of the two most common glycerophospholipids: phosphatidylcholine and phosphatidylethanolamine. (*b*) Space-filling model of phosphatidylcholine C gray, vinyl C yellow, O red, N blue, and P green. H not shown.
Figure (b) generously supplied by Richard Pastor and Richard Venable, FDA, Bethesda, Maryland.

compounds attached by a phosphodiester bridge to glycerol are choline, ethanolamine, serine, glycerol, and inositol (Figure 12.5). **Phosphatidylethanolamine (ethanolamine glycerophospholipid** or the trivial name **cephalin**) and **phosphatidylcholine (choline glycerophospholipid** or **lecithin)** are the most common glycerophospholipids in membranes (Figure 12.6). **Phosphatidylglycerol phosphoglyceride** (Figure 12.7) **(diphosphatidylglycerol** or **cardiolipin)** contains

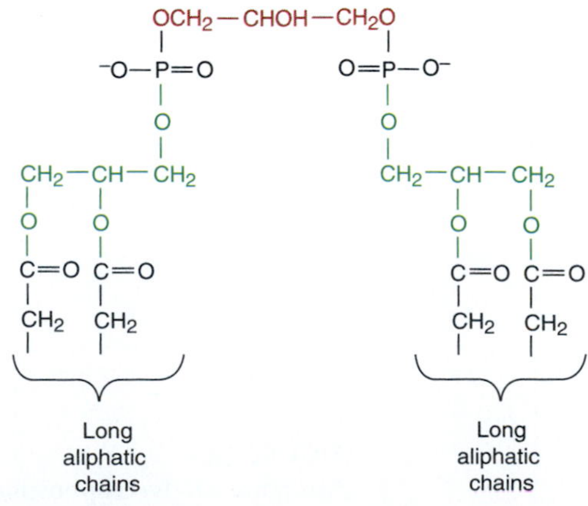

FIGURE 12.7
Phosphatidylglycerol phosphoglyceride (cardiolipin).

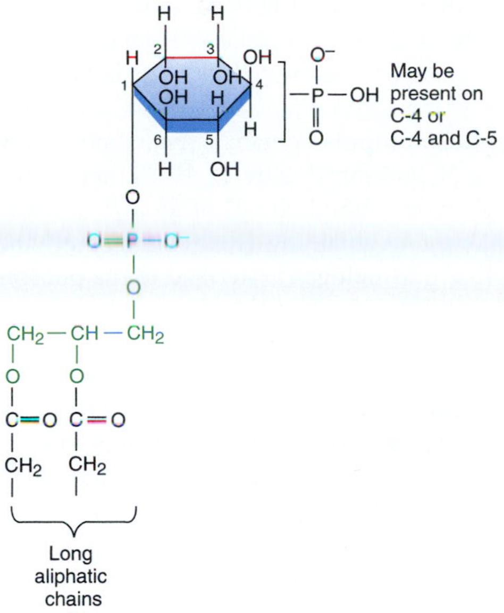

FIGURE 12.8
Phosphatidylinositol.
Phosphate groups are also found on C-4 or C-4 and C-5 of the inositol. The additional phosphate groups increase the charge on the polar head of this glycerophospholipid.

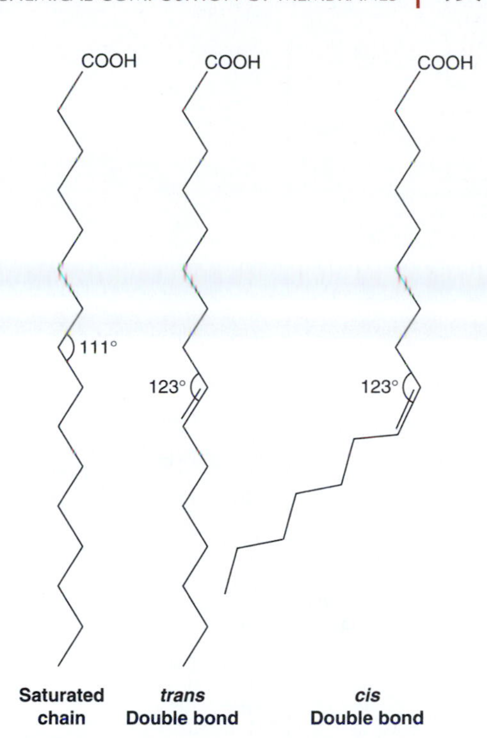

FIGURE 12.9
Conformation of fatty acyl groups in phospholipids.
Saturated (palmitic) and unsaturated fatty acids (palmitoleic) with *trans* double bonds are straight chains in their minimum energy conformation, whereas a chain (palmitoleic) with a *cis* double bond has a bend. The *trans* double bond is rare in naturally occurring fatty acids.

two phosphatidic acids linked by a glycerol and is found nearly exclusively in mitochondrial inner membranes and bacterial membranes.

Inositol, a hexahydroxy alcohol, is esterified to phosphate in **phosphatidylinositol** (Figure 12.8). 4-Phospho- and **4,5-bisphosphoinositol glycerophospholipids** (Figure 12.8) are present in plasma membranes; the latter is the source of inositol trisphosphate and diacylglycerol, referred to as "second messengers," that are involved in the action of some hormones (see p. 930).

Glycerophospholipids contain two fatty acyl groups esterified to carbon atoms 1 and 2 of glycerol; some of the major fatty acids found in glycerophospholipids are presented in Table 12.1. A saturated fatty acid is usually found on C-1 of the glycerol and an unsaturated fatty acid on C-2. Designation of the different classes of glycerophospholipids does not specify which fatty acids they contain. Phosphatidylcholine usually contains palmitic or stearic acid in the *sn*-1 position and a C18 unsaturated fatty acid, oleic, linoleic, or linolenic, on the *sn*-2 carbon. Phosphatidylethanolamine contains palmitic or oleic acid on *sn*-1 but a long-chain polyunsaturated fatty acid, such as arachidonic, on the *sn*-2 position.

A saturated fatty acid is a straight chain, as is a fatty acid with an unsaturation in the *trans* position. *Cis* double bonds, which occur in most naturally occurring fatty acids, create a kink in their hydrocarbon chain (Figure 12.9). A straight chain diagram, as shown in Figures 12.4 and 12.9, does not adequately represent the

TABLE 12.1 Major Fatty Acids in Glycerophospholipids

Common Name	*Systematic Name*	*Structural Formula*
Myristic acid	*n*-Tetradecanoic	$CH_3—(CH_2)_{12}—COOH$
Palmitic acid	*n*-Hexadecanoic	$CH_3—(CH_2)_{14}—COOH$
Palmitoleic acid	*cis*-9-Hexadecenoic	$CH_3—(CH_2)_5—CH{=}CH—(CH_2)_7—COOH$
Stearic acid	*n*-Octadecanoic	$CH_3—(CH_2)_{16}—COOH$
Oleic acid	*cis*-9-Octadecenoic acid	$CH_3—(CH_2)_7—CH{=}CH—(CH_2)_7—COOH$
Linoleic acid	*cis,cis*-9,12-Octadecadienoic	$CH_3—(CH_2)_3—(CH_2—CH{=}CH)_2—(CH_2)_7—COOH$
Linolenic acid	*cis,cis,cis*-9,12,15-Octadecatrienoic	$CH_3—(CH_2—CH{=}CH)_3—(CH_2)_7—COOH$
Arachidonic acid	*cis,cis,cis,cis*-5,8,11,14-Icosatetraenoic	$CH_3—(CH_2)_3—(CH_2—CH{=}CH)_4—(CH_2)_3—COOH$

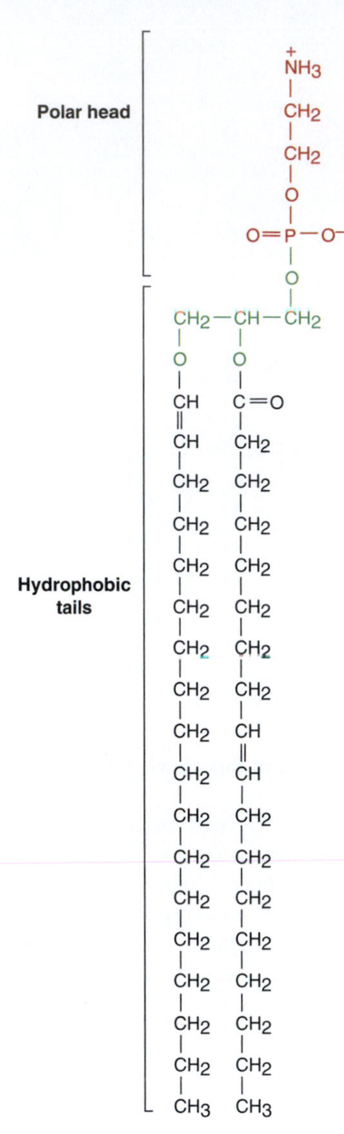

FIGURE 12.10
Ethanolamine plasmalogen.
Note the ether linkage of the aliphatic chain on C-1 of glycerol.

chemical configuration of a long-chain fatty acid. Actually, there is a high degree of coiling of the hydrocarbon chain in a glycerophospholipid that is disrupted by a double bond. The presence of unsaturated fatty acids has a marked effect on the physicochemical state of the membrane (see p. 509).

Glycerol ether phospholipids contain a long aliphatic chain in ether linkage to the glycerol at the *sn*-1 position (Figure 12.10). Ether phospholipids contain an alkyl group (alkyl acyl glycerophospholipid) or an α,β-unsaturated ether, termed a **plasmalogen.** Plasmalogens containing ethanolamine **(ethanolamine plasmalogen)** or choline **(choline plasmalogen)** esterified to the phosphate are abundant in nervous tissue and heart but not in liver. In human hearts more than 50% of the ethanolamine glycerophospholipids are plasmalogens. High levels of ether-linked lipids in plasma membranes of very metastatic cancer cells have been reported, suggesting that they may have a role in the invasive properties of these cells.

Glycerophospholipids are **amphipathic,** containing both a polar end, or head group, due to the charged phosphate and substitutions on the phosphate and a nonpolar tail due to hydrophobic hydrocarbon chains of the fatty acyl groups. Polar groups are negatively charged at pH 7.0 due to ionization of phosphate (p$K \sim 2$) and carry the charge of groups esterified to phosphate (Table 12.2). Choline and ethanolamine glycerophospholipids are zwitterions at pH 7.0, as they carry a negative charge on phosphate and a positive charge on nitrogen. Phosphatidylserine has a positive charge on the α-amino group of serine and two negative charges, one on phosphate and one on the carboxyl group of serine, with a net charge of -1. In contrast, glycerophospholipids containing inositol and glycerol have only a single negative charge on phosphate. The 4-phosphoinositol and 4,5-bisphosphoinositol derivatives are very polar compounds with negative charges on phosphate.

Each type of cell membrane has a distinctive fatty acid composition in its glycerophospholipids. Fatty acyl groups of the same tissue in a variety of species are very similar. In addition, the fatty acid content of the glycerophospholipids can vary, depending on the physiological or pathophysiological state of the tissue.

Sphingolipids Are Present in Membranes

The amino alcohols **sphingosine** (D-4-sphingenine) and **dihydrosphingosine** (Figure 12.11) are the basis for the **sphingolipids. Ceramides** have a saturated or unsaturated long-chain fatty acyl group in amide linkage on the amino group of sphingosine (Figure 12.12). With two nonpolar tails a ceramide is similar in structure to diacylglycerol. Various substitutions are found on the hydroxyl group at position 1. The **sphingomyelin** series has phosphorylcholine esterified to the 1-OH (Figure 12.13) and is the most abundant sphingolipid in mammalian tissues. The similarity of the sphingomyelin structure to that of choline glycerophospholipids is apparent, and they have many properties in common. Note that the sphingomyelins

TABLE 12.2 Predominant Charge on Glycerophospholipids and Sphingomyelin at pH 7.0

Lipid	*Phosphate Group*	*Base*	*Net Charge*
Phosphatidylcholine	− 1	+ 1	0
Phosphatidylethanolamine	− 1	+ 1	0
Phosphatidylserine	− 1	+ 1, − 1	− 1
Phosphatidylglycerol	− 1	0	− 1
Diphosphatidylglycerol (cardiolipin)	− 2	0	− 2
Phosphatidylinositol	− 1	0	− 1
Sphingomyelin	− 1	+ 1	0

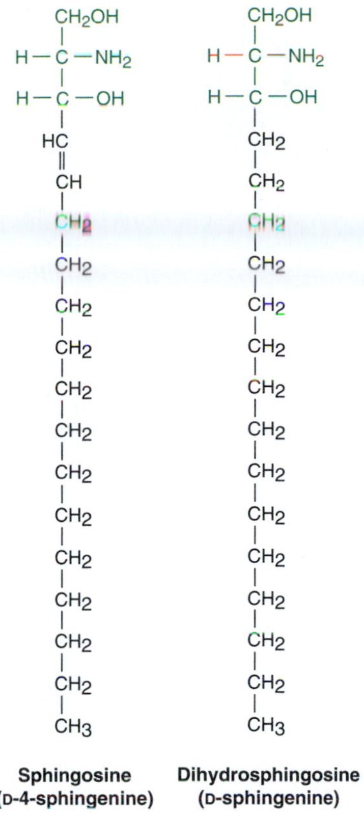

Sphingosine (D-4-sphingenine)
Dihydrosphingosine (D-sphingenine)

FIGURE 12.11
Structures of sphingosine and dihydrosphingosine.

are amphipathic compounds with a charged phosphate. Sphingomyelins and glycerophospholipids are classified as phospholipids. Sphingomyelin of myelin contains predominantly the longer chain fatty acids, with carbon lengths of 24; as with glycerophospholipids, there is a specific fatty acid composition of the sphingomyelin, depending on the tissue.

Glycosphingolipids have a sugar attached by a β-glycosidic linkage to the 1-OH group of sphingosine in a ceramide; they do not contain phosphate and are electrically neutral. A subgroup is the **cerebrosides,** which contain either a glucose **(glucocerebrosides)** or galactose **(galactocerebrosides)** attached to ceramide. Figure 12.14 presents the structure of **phrenosine,** a galactocerebroside with a 2-OH C24 fatty acid. Galactocerebrosides are predominantly in brain and nervous tissue, whereas the small quantities of cerebrosides in nonneural tissues usually contain glucose. Galactocerebrosides containing a sulfate group esterified on the 3 position of the sugar are classified as sulfatides (Figure 12.15). Cerebrosides and sulfatides usually contain fatty acids with 22–26 carbon atoms.

In place of monosaccharides, neutral glycosphingolipids often have 2 (dihexosides), 3 (trihexosides), or 4 (tetrahexosides) sugar residues attached to the 1-OH group of sphingosine. Diglucose, digalactose, **N-acetylglucosamine,** and **N-acetyldigalactosamine** are the usual sugars.

Gangliosides, the most complex glycosphingolipids, contain an oligosaccharide group with one or more residues of **N-acetylneuraminic acid (sialic acid);** they are amphipathic compounds with a negative charge at pH 7.0. The gangliosides represent 5–8% of the total lipids in brain. Some 20 different types have been identified differing in the number and relative position of the hexose and sialic acid residues. A detailed description of the nomenclature and structures of gangliosides is presented on p. 760. The carbohydrate moiety of gangliosides extends beyond the surface of the membrane, is involved in cell–cell recognition, and serves as binding site for hormones and bacteria toxins such as cholera toxin.

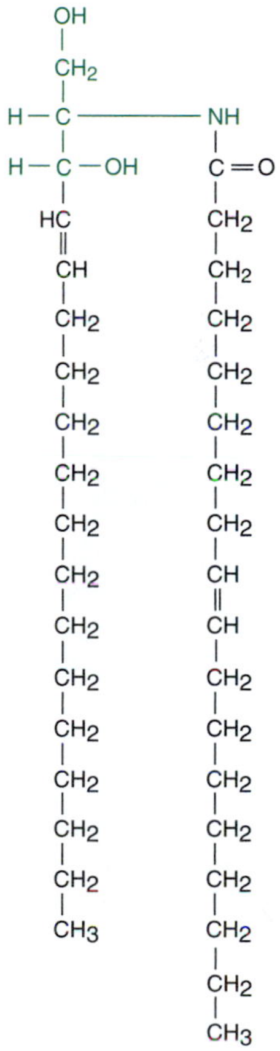

FIGURE 12.12
Structure of a ceramide.

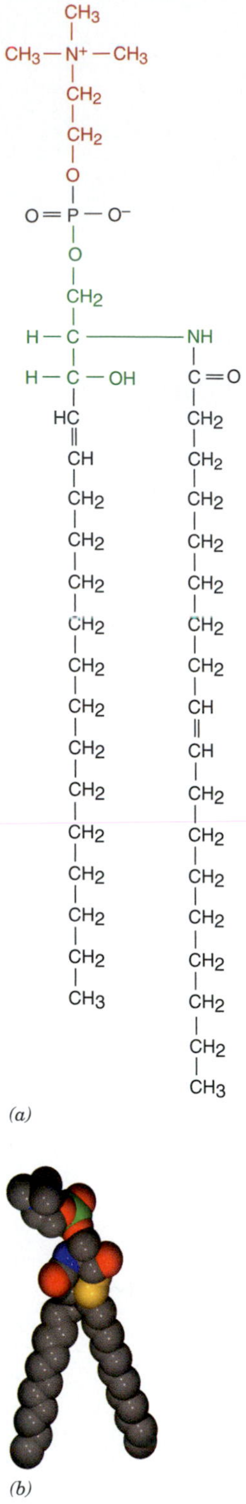

(a)

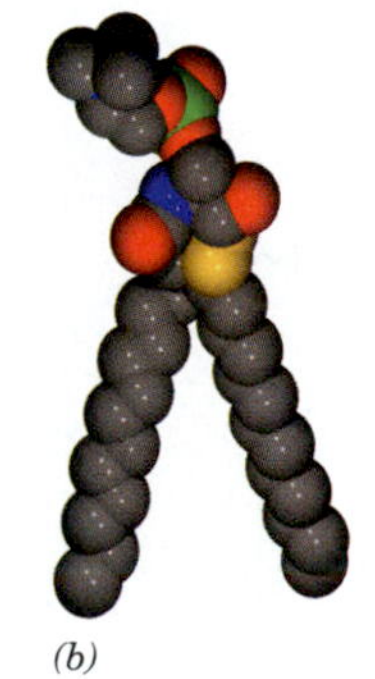

(b)

FIGURE 12.13
Sphingomyelin.
(*a*) Structure of a choline-containing sphingomyelin. (*b*) Space-filling model with C gray, vinyl C yellow, O red, N blue, and P green. H not shown.
Figure (b) generously supplied by Richard Pastor and Richard Venable, FDA, Bethesda, Maryland.

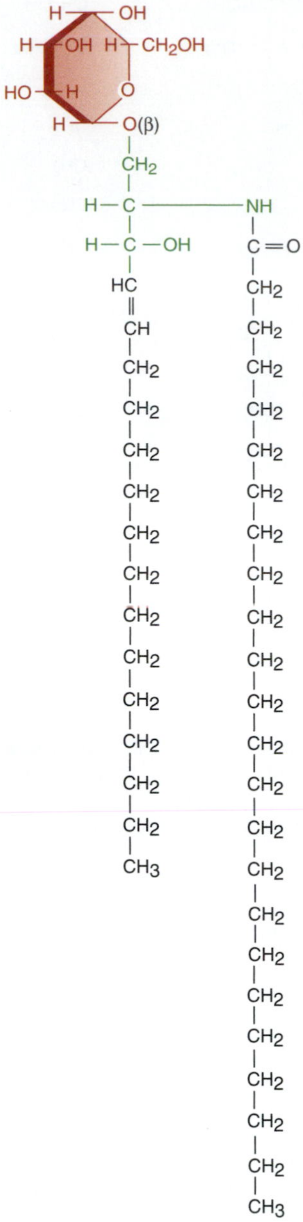

FIGURE 12.14
Structure of a galactocerebroside containing a C24 fatty acid.

Cholesterol Is an Important Component of Plasma Membranes

Cholesterol is the third major lipid in membranes. With four fused rings and a C8 branched hydrocarbon chain attached to the D ring at position 17, cholesterol is a compact, rigid, hydrophobic molecule (Figure 12.16). It also has a polar hydroxyl group at C-3. Cholesterol alters the fluidity of membranes and participates in controlling the microstructure of plasma membranes (see p. 510).

Lipid Composition Varies Between Membranes

Each tissue and various cell membranes have a distinctive composition of lipids (Figure 12.17). The same intracellular membranes of a specific tissue, for example, liver, in different species have very similar classes of lipids. The plasma membrane exhibits the greatest variation in percentage composition because the amount

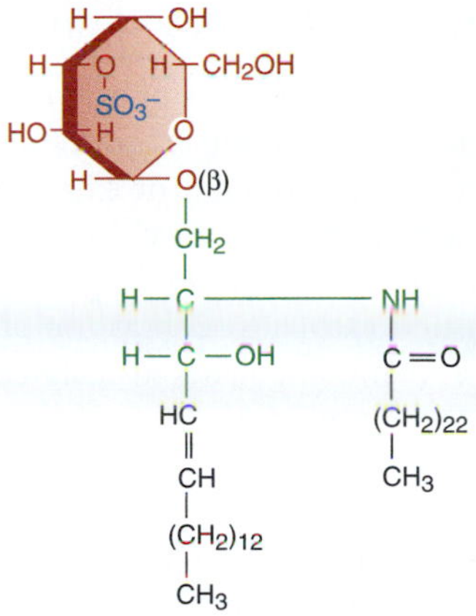

FIGURE 12.15
Structure of a sulfatide.

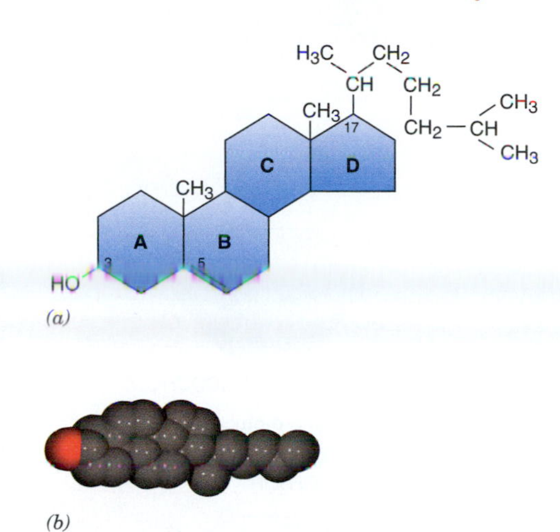

FIGURE 12.16
Cholesterol.
(*a*) Structure of cholesterol. (*b*) Space-filling model with C gray, vinyl C yellow, O red, N blue, and P green. H not shown.
Figure (b) generously supplied by Richard Pastor and Richard Venable, FDA, Bethesda, Maryland.

of cholesterol is affected by the nutritional state of the animal. Myelin membranes of neuronal axons are rich in sphingolipids, with a high proportion of glycosphingolipids. Intracellular membranes primarily contain glycerophospholipids with little sphingolipids. The membrane lipid composition of mitochondria, nuclei, and rough endoplasmic reticulum are similar, with Golgi membrane being somewhere between other intracellular membranes and the plasma membrane. The amount of cardiolipin is high in the inner mitochondrial membrane and low in the outer membrane but essentially none in other membranes. Choline-containing lipids, phosphatidylcholine and sphingomyelin, are predominant, followed by ethanolamine glycerophospholipid. The constancy of composition of various membranes indicates the relationship between lipids and the specific functions of individual membranes.

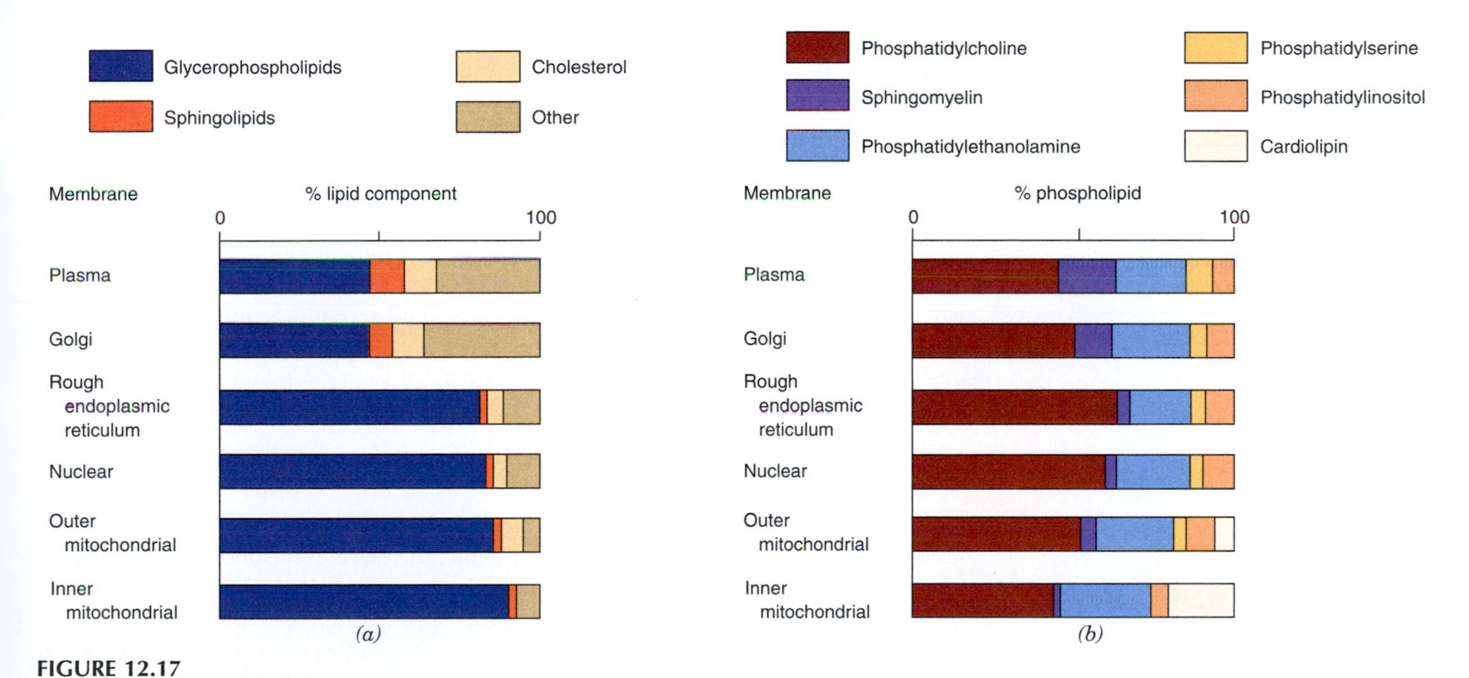

FIGURE 12.17
Lipid composition of cellular membranes isolated from rat liver.
(*a*) Amount of major lipid components as percentage of total lipid. The area labeled "Other" includes mono-, di-, and triacylglycerol, fatty acids, and cholesterol esters.
(*b*) Phospholipid composition as percentage of total phospholipid.
Values from Harrison, R. and Lunt, G. G. Biological Membranes. *New York: Wiley, 1975.*

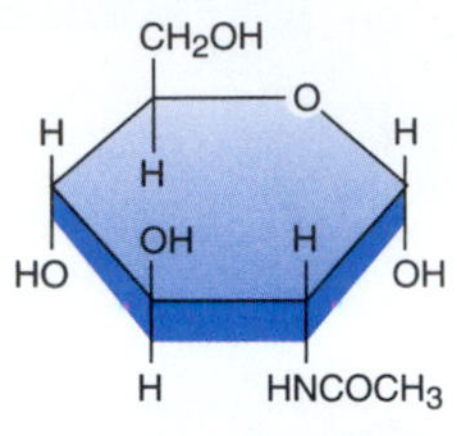

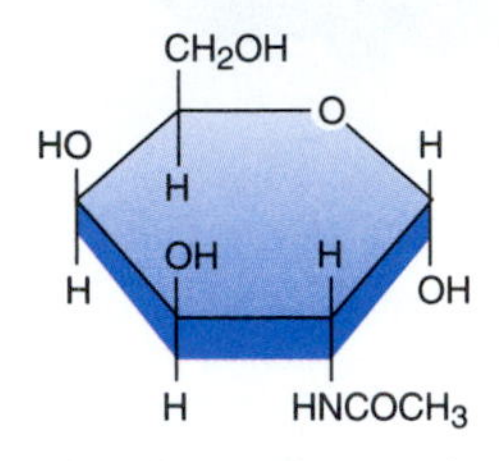

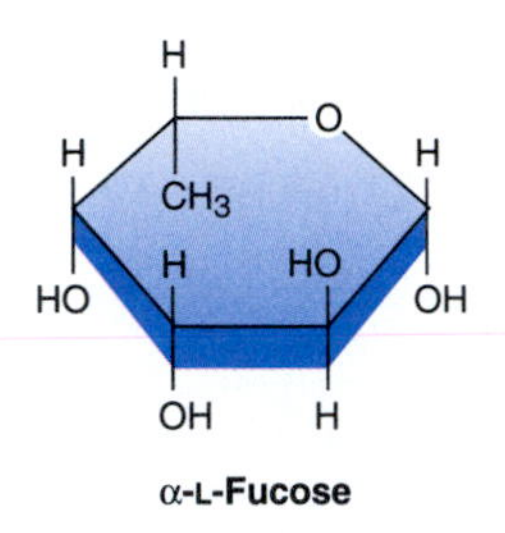

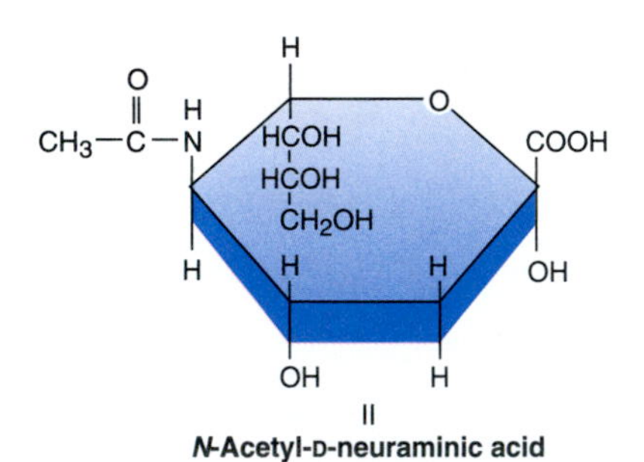

FIGURE 12.18
Structures of some membrane carbohydrates.

Membrane Proteins Are Classified Based on Their Ease of Removal

Membrane proteins are classified on the basis of ease of their removal from isolated membrane fractions. **Peripheral** (or **extrinsic**) **proteins** are released by treatment with salt solutions of different ionic strength or extremes of pH. Their names imply a physical location on the membrane surface. Isolated peripheral proteins, many of which are enzymes, are typical water-soluble proteins. **Integral** (or **intrinsic**) **proteins** require rather drastic treatment, such as use of detergents or organic solvents, to be released from a membrane. They usually contain tightly bound lipid, which if removed leads to denaturation of the protein and loss of biological function. Integral proteins contain sequences of hydrophobic amino acids, which create hydrophobic domains that interact with the hydrophobic hydrocarbons of the lipids and stabilize the protein–lipid complex. Removal of integral proteins leads to disruption of the membrane, whereas peripheral proteins can be removed with little or no change in the integrity of the membrane.

Proteolipids are hydrophobic lipoproteins soluble in chloroform and methanol but insoluble in water. They are present in many membranes but particularly in myelin, where they represent about 50% of the protein component. An example is **lipophilin,** a major lipoprotein of brain myelin that contains over 65% hydrophobic amino acids and covalently bound fatty acids.

Glycoproteins are another class of integral membrane proteins. Plasma membranes contain a number of different glycoproteins, each with its own unique carbohydrate content.

The complexity, variety, and interaction of membrane glycoproteins with lipids are being resolved. Many membrane proteins are enzymes located within or on specific cellular membranes. Membrane proteins have a role in transmembrane movement of molecules and as receptors for the binding of hormones and growth factors. In many cells, such as neurons and erythrocytes, membrane proteins have a structural role to maintain the shape of the cell.

Membrane Carbohydrates Are Present as Glycoproteins or Glycolipids

Carbohydrates are present in membranes as oligosaccharides covalently attached to proteins to form glycoproteins and to a lesser amount to lipids to form glycolipids. Their sugars include glucose, galactose, mannose, fucose, *N*-acetylgalactosamine, *N*-acetylglucosamine, and sialic acid (see Figure 12.18 and the Appendix for structures). Structures of glycoproteins and glycolipids are presented on pages 679 and 756, respectively. The carbohydrate is on the exterior side of the plasma membrane or the luminal side of the endoplasmic reticulum. Roles for membrane carbohydrates include cell–cell recognition, adhesion, and receptor action.

12.3 | MICELLES AND LIPOSOMES

Lipids Form Vesicular Structures

The basic structural characteristic of membranes is derived from the physicochemical properties of the major lipid components, the glycerophospholipids and sphingolipids. These amphipathic compounds contain a hydrophilic head and a hydrophobic tail (Figure 12.19*a*), which at appropriate concentrations interact in aqueous systems *in vitro* to form spheres, termed **micelles** (Figure 12.19*b*). Charged polar head groups will be on the outside of the sphere while hydrophobic tails interact to exclude water. The specific concentration of lipid required for micelle formation is referred to as the critical micelle concentration. Micelles that contain a single lipid or a mixture of lipids can be prepared. Formation of micelles also depends on the temperature of the system and, if a mixture of lipids is used, on the ratio of concentrations of the different lipids in the mixture (see p. 1107). The micelle structure is very stable because of hydrophobic interaction of hydrocarbon chains and attraction of polar groups to water. Micelles are important in the intestinal digestion and absorption of lipids (see p. 1108).

Liposome Membranes Are Similar to Biological Membranes

Depending on conditions, amphipathic lipids such as glycerophospholipids form a bimolecular structure that consists of two layers of lipid. The polar head groups will be at the interface between the aqueous medium and the lipid, and the hydrophobic tails will interact to form an environment that excludes water (Figure 12.19*c*). This bilayer conformation is the basic lipid structure of all biological membranes.

Lipid bilayers are extremely stable structures, being held together by hydrophobic forces of the hydrocarbon chains and ionic interactions of charged head groups with water. Interactions of the hydrocarbon chains lead to the smallest possible area for water to be in contact with the chains, and water is essentially excluded from the interior of the bilayer. If disrupted, bilayers self-seal because hydrophobic groups will seek to establish a structure in which there is minimal contact of the hydrocarbon chains with water, a condition that is most favorable thermodynamically. A lipid bilayer will close in on itself, forming a spherical vesicle separating the external environment from an internal aqueous compartment. Such vesicles are termed **liposomes.** Because individual lipid–lipid interactions have low energies of activation, lipids in a bilayer have a circumscribed mobility, breaking and forming interactions with surrounding molecules but not readily escaping from the lipid bilayer (Figure 12.19*d*). Self-assembly of amphipathic lipids into bilayers is an important property and is involved in formation of cell membranes.

Individual phospholipid molecules exchange places with neighboring molecules in a bilayer, leading to rapid lateral diffusion in the plane of the membrane (Figure 12.20). There is rotation around the carbon–carbon bonds in fatty acyl chains; in fact, there is a greater degree of rotation nearer the methyl end, leading to greater motion at the center than at the peripheral region of the lipid bilayer. Individual lipid molecules, however, do not migrate readily from one monolayer to the other, a transverse movement termed flip-flop, because of thermodynamic constraints on movement of a charged head group through the lipophilic core. Thus

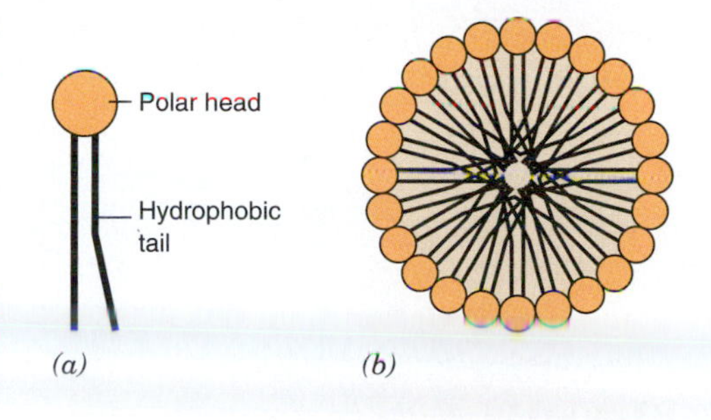

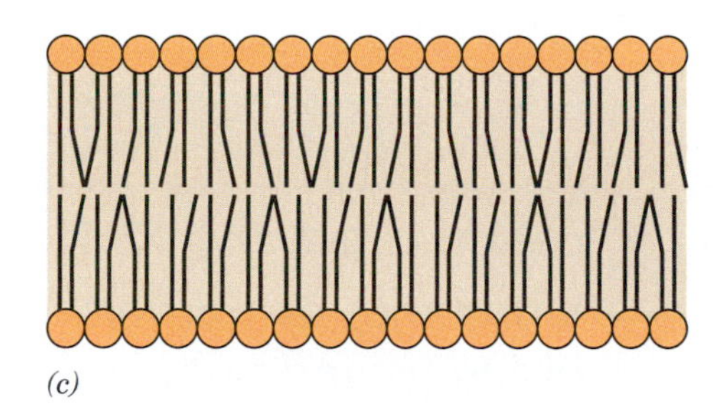

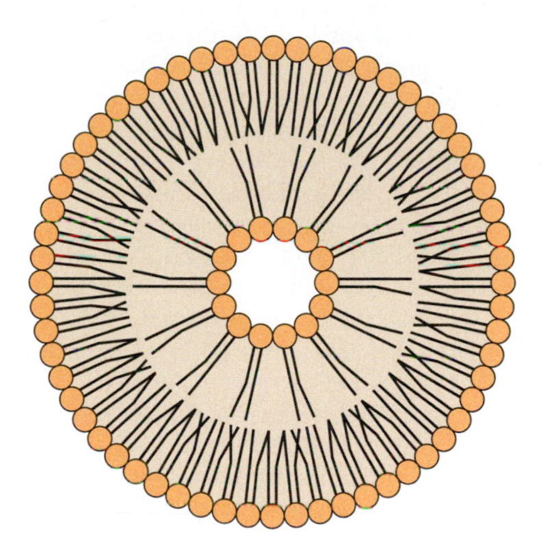

FIGURE 12.19
Interactions of phospholipids in an aqueous medium.
(*a*) Representation of an amphipathic lipid. (*b*) Cross-sectional view of the structure of a micelle. (*c*) Cross-sectional view of the structure of lipid bilayer. (*d*) Cross section of a liposome. Each structure has an inherent stability due to the hydrocarbon chains and the attraction of the polar head groups to water.

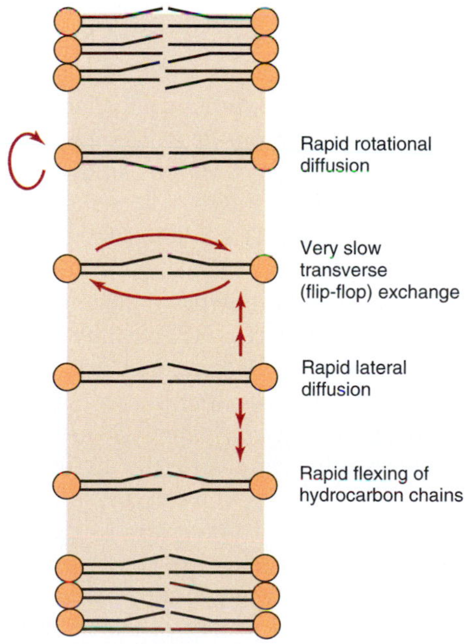

FIGURE 12.20
Mobility of lipid components in membranes.

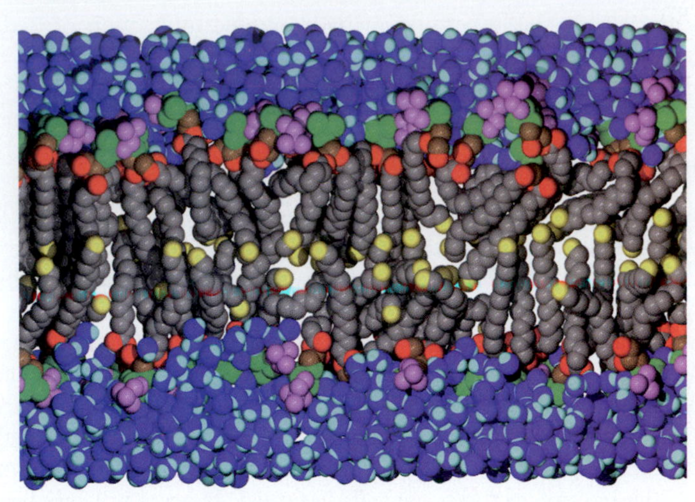

FIGURE 12.21
Model of a lipid bilayer stopped at a moment in time.
An artificial bilayer consisting of dipalmitoyl phosphatidylcholine surrounded by water as modeled by computer. Atom colors are chain C gray (except terminal methyl C yellow and glycerol C brown), ester O red, P and O green, choline C and N pale violet, water O dark blue, and water H light blue. Lipid H have been omitted.
Figure generously supplied by Richard Pastor and Richard Venable, FDA, Bethesda, Maryland.

lipid bilayers have an inherent stability and fluidity in which individual molecules move rapidly in their own monolayer but do not readily exchange with an adjoining monolayer. Lipid components distributed themselves randomly in the monolayers in artificial membranes composed of different lipids.

Interaction of lipids in membranes is very different from that illustrated in Figure 12.19. The interior of membrane bilayers is very fluid and in constant motion. Acyl chains of phospholipids have a random motion, intercalating with chains in the opposing surface. The viscosity increases significantly closer to lipid head groups. A computer representation of the arrangement of phospholipids in a bilayer at a moment in time is presented in Figure 12.21. Several layers of ordered water molecules cover the surface and influence the environment of a membrane.

Artificial membrane systems have been studied extensively in an effort to understand the properties of biological membranes. A variety of techniques are available to prepare liposomes, using synthetic phospholipids and lipids extracted from natural membranes. Depending on the procedure, unilamellar and multilamellar (vesicles within vesicles) vesicles of various sizes (20-nm to 1-μm diameter) can be prepared. Figure 12.19*d* represents a liposome structure. The interior of the vesicle is an aqueous environment, and it is possible to prepare liposomes with different substances entrapped. Both the external and internal environments of liposomes can be manipulated and properties—including ability to exclude molecules, interaction with various substances, and stability under different conditions—of these synthetic membranes have been studied. Na^+, K^+, Cl^-, and most polar molecules do not readily diffuse across lipid bilayers of liposomes. Lipid-soluble nonpolar substances such as triacylglycerol and undissociated organic acids readily diffuse into the bilayers and remain in the hydrophobic environment of the hydrocarbon chains. Membrane-bound enzymes and proteins involved in translocation of ions have been isolated from various tissues and incorporated into the membrane of liposomes for evaluation of their function. With liposomes it is easier to manipulate various parameters of membrane systems and, thus, study various activities free of interfering reactions present in cell membranes. Liposomes are used in delivery of drugs in humans (see Clin. Corr. 12.1) and have been tested as vectors in gene therapy.

CLINICAL CORRELATION 12.1
Liposomes as Carriers of Drugs and Enzymes

A major obstacle in the use of many drugs is lack of tissue specificity in the action of the drug. Administration of drugs orally or intravenously leads to a drug acting on many tissues and not exclusively on a target organ, resulting in toxic side effects. An example is the commonly observed suppression of bone marrow cells by anticancer drugs. Some drugs are metabolized rapidly and their period of effectiveness is relatively short. Liposomes have been prepared with drugs, enzymes, and DNA encapsulated inside and used as carriers for these substances to target organs. Liposomes prepared from purified phospholipids and cholesterol are nontoxic and biodegradable. Alteration of surface charge enhances drug incorporation and release. Attempts have been made to prepare liposomes for interaction at a specific target organ. Antibiotic, antineoplastic, antimalarial, antiviral, antifungal, and anti-inflammatory agents have been found to be effective when administered in liposomes. Some drugs have a longer period of effectiveness when administered encapsulated in liposomes. It may be possible to prepare liposomes with a high degree of tissue specificity so that drug administration and perhaps even enzyme replacement can be carried out with this technique.

Ranade, V. V. Drug delivery systems. 1. Site-specific drug delivery using liposomes as carriers. *J. Clin. Pharmacol.* 29:685, 1989; Caplen, N. J., Gao, X., Hayes, P., et al. Gene therapy for cystic fibrosis in humans by liposome-mediated DNA transfer: the production of resources and the regulatory process. *Gene Ther.* 1:139, 1994; Desormeaux, A. and Bergeron, M. G. Liposomes in drug delivery systems: a strategic approach for the treatment of HIV infection. *J. Drug Targeting* 6:1, 1998: and Gregoriadis, G. Engineering liposomes for drug delivery: progress and problems. *Trends Biotechnol.* 13:527, 1995.

12.4 | STRUCTURE OF BIOLOGICAL MEMBRANES

Fluid Mosaic Model of Biological Membranes

Knowledge of membrane structure has evolved based on evidence from physicochemical, biochemical, and electron microscopic investigations. All biological membranes have a bilayer arrangement of lipids, as in liposomes. Amphipathic lipids and cholesterol are oriented so that their hydrophobic portions interact to minimize their contact with water or other polar groups, and polar head groups of lipids are at the interface with the aqueous environment. J. D. Davson and J. Danielli in 1935 proposed this model, which was refined by J. D. Robertson. A major question with the earlier models was how to explain the interaction of membrane proteins with the lipid bilayer. In the early 1970s, S. J. Singer and G. L. Nicolson proposed the **fluid mosaic model** for membranes in which some proteins (intrinsic) are actually immersed in the lipid bilayer while others (extrinsic) are loosely attached to the surface. It was suggested that some proteins span the lipid bilayer and are in contact with the aqueous environment on both sides. Both lipids and proteins were proposed to diffuse laterally. Figure 12.22 is a current representation of a biological membrane. The characteristics of the lipid bilayer explain many of the observed properties of the cellular membrane, including fluidity, flexibility that permits changes of shape and form, ability to self-seal, and impermeability. The model continues to undergo modification and refinement. Thus lipids in a membrane are now considered to assume nonlamellar structural domains, and long acyl chains in one layer can interdigitate with those of the opposite layer.

Integral Membrane Proteins Are Immersed in Lipid Bilayer

The development of techniques for isolation of **integral membrane proteins,** for determination of their primary structure, and for identification of their specific functional domains has led to an understanding of the structural relationship

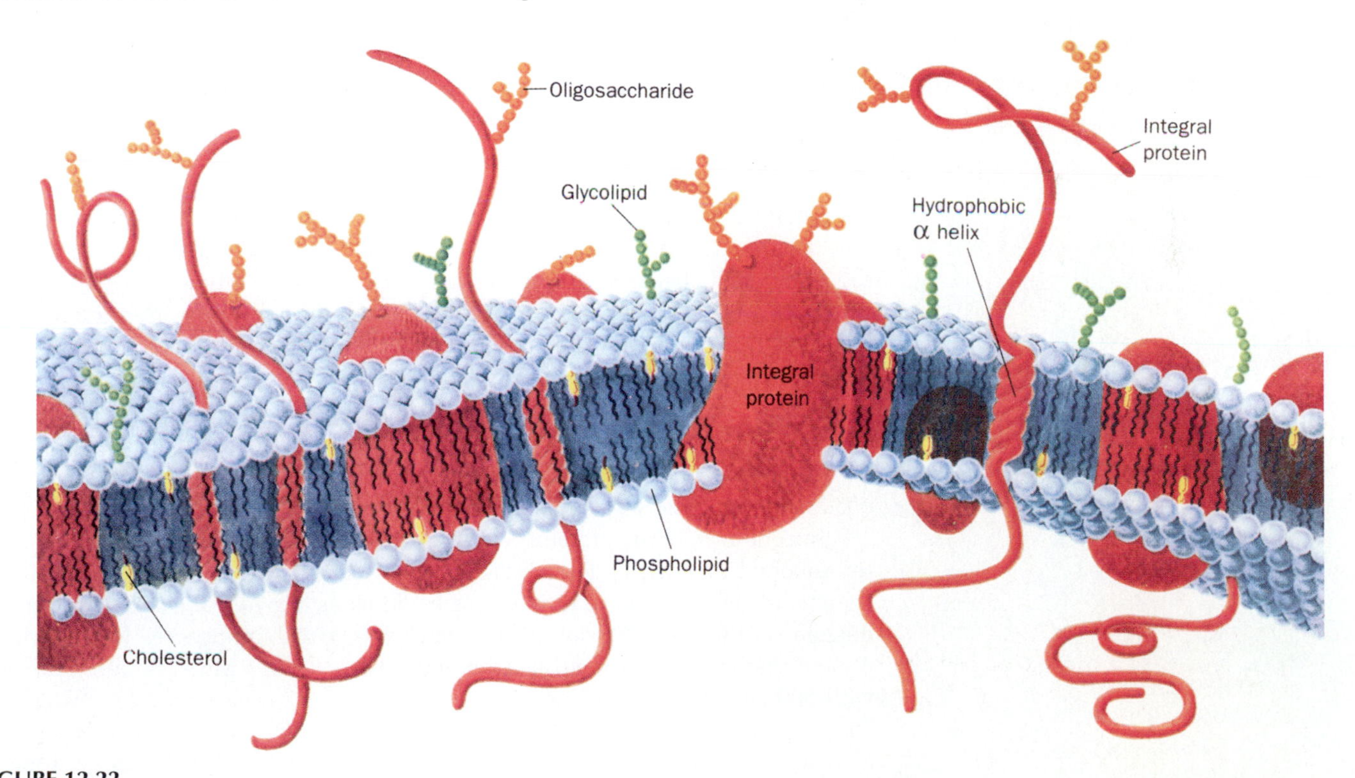

FIGURE 12.22
Fluid mosaic model of biological membranes.
Figure reproduced from Voet, D. and Voet, J. Biochemistry, 2nd ed. New York: Wiley, 1995. Reprinted by permission of John Wiley & Sons, Inc.

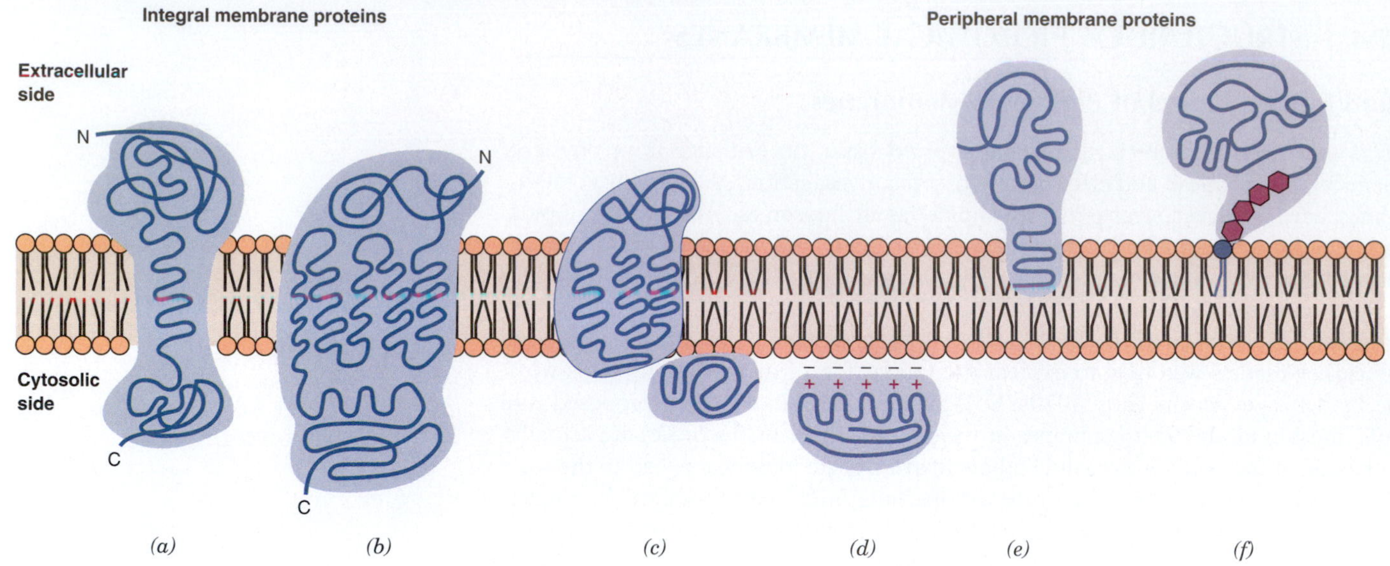

FIGURE 12.23
Interactions of membrane proteins with the lipid bilayer.
Diagram illustrates the multiple types of binding of proteins in or to the lipid bilayer: (*a*) a single transmembrane segment; (*b*) multiple transmembrane segments; (*c*) bound to an integral protein; (*d*) bound electrostatically to the lipid bilayer; (*e*) attached by a short terminal hydrophobic sequence of amino acids; and (*f*) attached by covalently bound lipid.

between the hydrophobic lipid bilayer and membrane proteins. Figure 12.23 illustrates various ways of attachment of proteins to a biological membrane. Some integral membrane proteins (see p. 502) span the membrane, whereas others may be immersed only partially in the bilayer. Measurements of the hydrophobicity of the amino acid residues and partial proteolytic digestion of proteins have revealed sequences of amino acids that are embedded in the membrane. Some proteins contain an α-helical structure consisting primarily of hydrophobic amino acids (such as leucine, isoleucine, valine, and phenylalanine), which is the transmembrane sequence. This is illustrated in Figure 12.23*a*. An example is **glycophorin,** a polypeptide of 131 residues present in the plasma membrane of human erythrocytes; amino acid residues 73–91 are the transmembrane sequence and are predominantly hydrophobic. Glycophorin has three domains: a sequence exterior to the cell containing the amino-terminal end, the transmembrane sequence, and a sequence extending into the cell with the carboxyl-terminal segment. In other transmembrane proteins the amino acid chain loops back and forth across the membrane (Figure 12.23*b*). In some cases there are 12 loops snaking across the lipid bilayer. Often multiple α helices spanning the membrane are organized to form a tubular structure. The **anion channel** of human erythrocytes, a glycopeptide with 926 amino acids responsible for the exchange of Cl^- and HCO_3^- across the membrane, has a tubular structure (p. 520). Secondary and tertiary structures of proteins are critical in the topography of the protein in the membrane. Some proteins in membranes form a quaternary structure with multiple subunits.

Integral membrane proteins may contain specific domains, for ligand binding, for catalytic activity, and for attachment of carbohydrate or lipid. The anion channel of the erythrocyte has two major domains: a hydrophilic amino-terminal domain on the cytosolic side of the membrane with binding sites for ankyrin, a protein that anchors the cytoskeleton and other cytosolic proteins, and a domain with 509 amino acids that traverses the membrane 12 times and mediates the exchange of Cl^- and HCO_3^-. Glycophorin contains 60% by weight carbohydrate, all of which is attached to the extracellular domain of the protein. Integral membrane proteins have a defined, rather than a random, orientation in the membrane. Specific structural orientation of both lipids and proteins demonstrates another important aspect of membrane structure: biological membranes are asymmetric, with each surface having specific characteristics. The orientation of proteins is determined during synthesis of the membrane or replacement of the protein; the bulkiness of the proteins, as well as thermodynamic restrictions, prevents transverse (flip-flop) movement.

Many enzymes that are integral membrane proteins require a membrane lipid for activity. Examples are D-β-hydroxybutyrate dehydrogenase, located in the inner mitochondrial membrane that requires phosphatidylcholine, and Ca^{2+}-ATPase of skeletal muscle sarcoplasmic reticulum, which has a specific binding site for lipid. Cholesterol has been implicated in the activity of various membrane ion pumps, including Na^+/K^+-exchanging ATPase (see p. 522), Ca^{2+}-transporting ATPases (see p. 523), and acetylcholine receptors. Some modulating effects of lipids may reflect a change in ordering and fluidity of the membrane but the lipid may also have a direct influence on the activity.

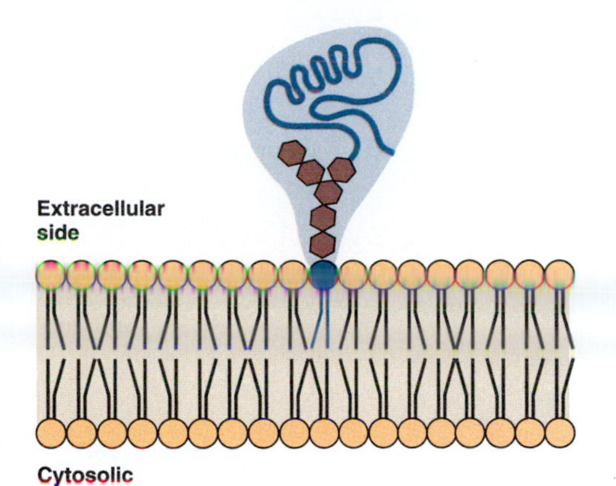

FIGURE 12.24
Attachment of a protein to a membrane by a glycosyl phosphatidylinositol anchor.

Peripheral Membrane Proteins Have Various Modes of Attachment

Peripheral membrane proteins are loosely attached to membranes and if removed do not disrupt lipid bilayers. Some bind to integral membrane proteins, such as **ankyrin** binding to the anion channel in erythrocytes (Figure 12.23*c*). Negatively charged phospholipids of membranes interact with positively charged regions of proteins, allowing electrostatic binding (Figure 12.23*d*). Some peripheral proteins have sequences of hydrophobic amino acids at one end of the peptide chain that serve as an anchor in the membrane lipid (Figure 12.23*e*); cytochrome b_5 is attached to the endoplasmic reticulum by such an anchor.

Phosphatidylinositol also anchors proteins to membranes (Figures 12.23*f* and 12.24). A **glycan**, consisting of ethanolamine, phosphate, mannose, mannose, mannose, and glucosamine, is covalently bound to the carboxyl terminal of the protein. This glycan has been conserved throughout evolution because it is found in different species attached to carboxyl-terminal residues of various membrane-bound proteins. Additional carbohydrate can be attached to the last mannose. The glucosamine of the glycan is covalently linked to phosphatidylinositol. The fatty acids of this glycerophospholipid are inserted into the lipid membrane, thus anchoring the protein. These molecules are referred to as **glycosyl phosphatidylinositol (GPI) anchors.** Over 50 proteins are so attached in this manner including enzymes, antigens, and cell adhesion proteins; a partial list is presented in Table 12.3. Fatty acyl groups of phosphatidylinositol are apparently specific for different proteins. To date, proteins found to be attached by a GPI anchor are on the external surface of plasma membranes. The significance of this form of anchoring has yet to be determined but it may be important for localization of the protein on a membrane, control of function of the protein, and controlled release of the protein from the membrane. A specific phosphatidylinositol-specific **phospholipase C** catalyzes the hydrolysis of the phosphate–inositol bond, leading to release of the protein.

TABLE 12.3 Proteins that Have a Glycosyl Phosphatidylinositol Anchor

Alkaline phosphatase
5'-Nucleotidase
Acetylcholinesterase
Trehalase
Renal dipeptidase
Lipoprotein lipase
Carcinoembryonic antigen
Neural cell adhesion molecule
Scrapie prion protein
Oligodendrocyte–myelin protein

Source: Low, M.G. Glycosyl-phosphatidylinositol: a versatile anchor for cell surface proteins. *FASEB J.* 3:1600, 1989.

Myristic and palmitic acid can also be covalently linked to proteins and serve to anchor proteins by insertion of the acyl chain into the lipid bilayer (Figure 12.23*f*). Myristic acid (C14) is attached by an amide linkage to an amino-terminal glycine, and palmitic acid (C16) is most often attached by a thioester linkage to cysteine or by a hydroxy ester bond to serine or threonine.

Although membrane models suggest that some proteins are randomly distributed throughout and on the membrane, there is a high degree of functional organization with definite restrictions on the localization of some proteins. As an example, proteins participating in the electron transport chain in the inner membrane of mitochondria function in consort and are organized into functional units both laterally and transversely. The location of specific proteins on the surface of plasma membranes is also controlled. Cells lining the lumen of kidney nephrons have specific plasma membrane enzymes on the luminal surface but not on the contraluminal surface of cells; enzymes restricted to a particular region of the membrane are located to meet specific functions of these cells. Thus there is a high degree of molecular organization of biological membranes that is not apparent from diagrammatic models.

Human Erythrocytes Are Ideal for Study of Membrane Structure

The structure of the plasma membrane of the human erythrocyte has been investigated extensively because of the ease with which the membrane can be purified from other cellular components. Figure 12.25 is a representation of the interaction of some of the many proteins in this membrane.

Lipids Are Distributed in an Asymmetric Manner in Membranes

There is an asymmetric distribution of lipid components across biological membranes. Each layer of the bilayer has a different composition with respect to individual glycerophospholipids and sphingolipids. The asymmetric distribution of lipids in the human erythrocyte membrane is shown in Figure 12.26. Sphingomyelin is predominantly in the outer layer, whereas phosphatidylethanolamine is predominantly in the inner lipid layer. In contrast, cholesterol is equally distributed on both layers.

Uncatalyzed transverse movement from one side to the other (i.e., flip-flop) of glycerophospholipids and sphingolipids is slow. Asymmetry of lipids is maintained by specific membrane proteins, termed **lipid transporters,** which promote the

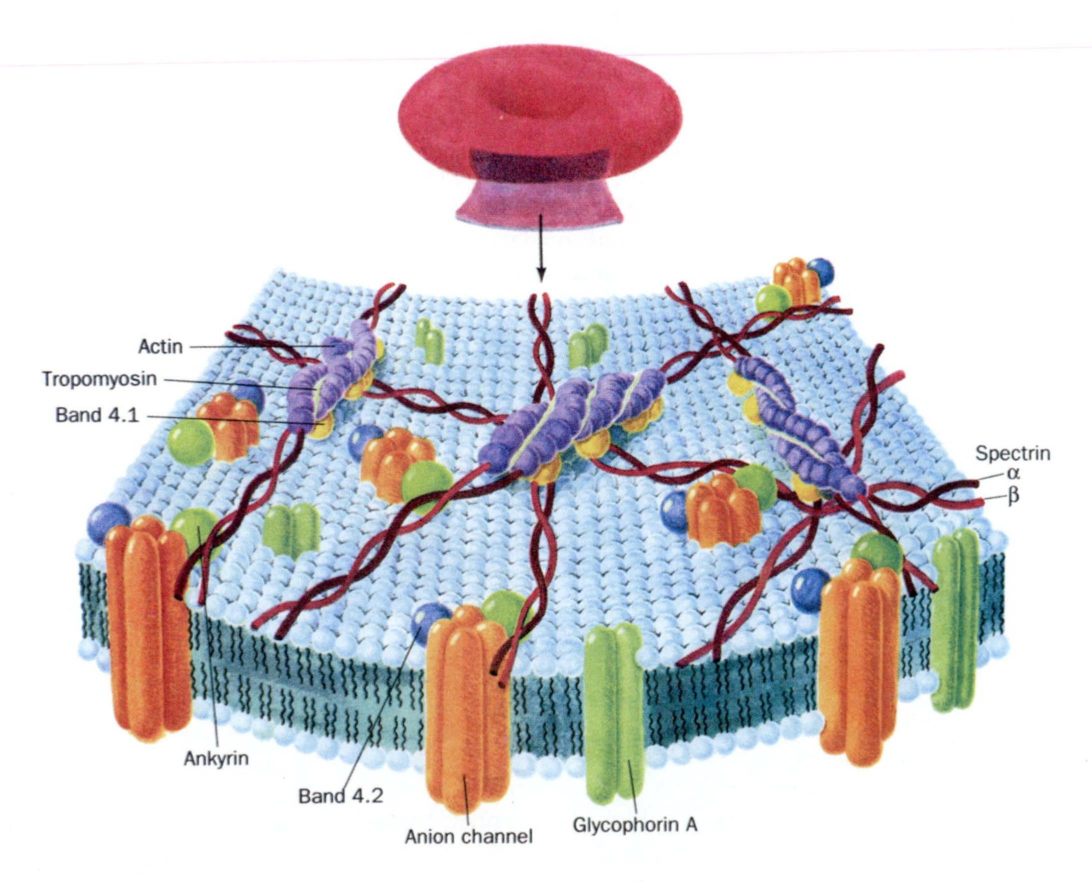

FIGURE 12.25
Schematic diagram of the erythrocyte membrane.
Diagram indicates the relationship of four membrane-associated proteins with the lipid bilayer. Glycophorin is a glycoprotein that contains 131 amino acids but whose function is unknown. Band 3, so designated because of its mobility in electrophoresis, contains over 900 amino acids and is involved in interacting with ankyrin and possibly in the facilitated diffusion of Cl^- and HCO_3^-. Ankyrin and spectrin are part of the cytoskeleton and are peripheral membrane proteins. Ankyrin binds to band 3 and spectrin is anchored to the membrane by ankyrin. *Figure reproduced from Voet, D. and Voet, J.* Biochemistry, *2nd ed. New York: Wiley, 1995.* Reprinted by permission of John Wiley & Sons, Inc.

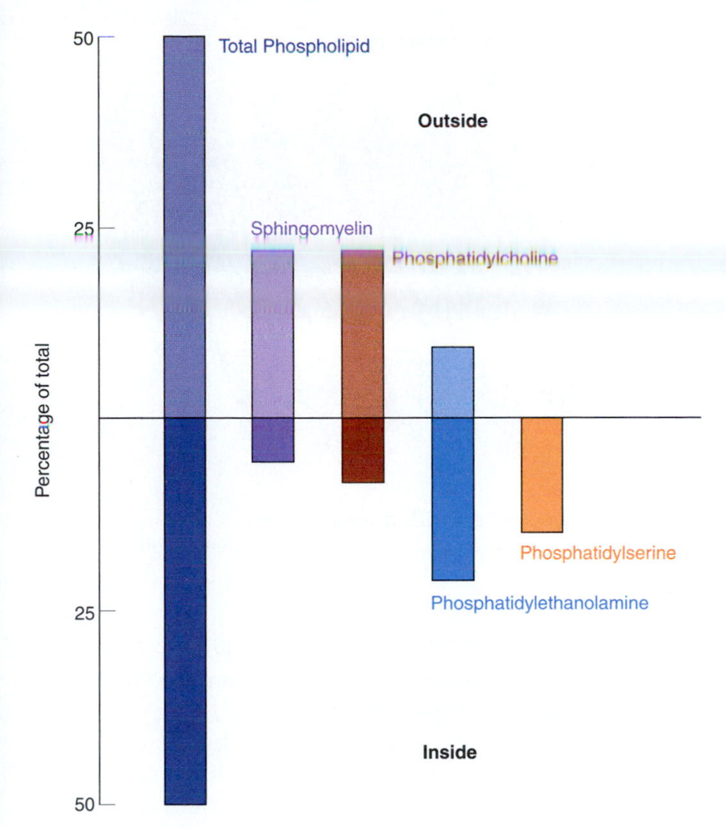

FIGURE 12.26
Distribution of phospholipids between inner and outer layers of the human erythrocyte membrane.
Values are percentage of each phospholipid in the membrane.
Redrawn from Verkeij, A. J., Zwaal, R. F. A., Roelofsen, B., Comfurius, P., Kastelijn, D., and Van Deenan, L. L. M. Biochim. Biophys. Acta *323:178, 1973.*

transverse movement of specific lipids from one layer to the other. Three transport proteins for movement of phospholipids between membrane layers have been identified. An ATP-dependent transporter specific for phosphatidylserine and phosphatidylethanolamine catalyzes the inward transport of the aminoglycerolipids. The protein, which has been given the trivial name "**flippase,**" maintains the low concentration of these phospholipids in the outer layer of the plasma membrane. An outward-directed transporter, termed "**floppase,**" is nonspecific for phospholipids and may be energy dependent. Floppase may be the P-glycoprotein present in plasma membranes and responsible for multidrug resistance of some cancer cells (see p. 858). A third lipid transporter, termed "**scramblase,**" mixes phospholipids between the two layers; it is nonspecific with respect to phospholipids and its activity is stimulated by increases in intracellular Ca^{2+}. Scramblase apparently has an important role in the cell-mediated coagulation cascade (see p. 1031) and recognition of injured cells for removal by the reticuloendothelial system. In both responses it is proposed that movement of phosphatidylserine into the outer layer initiates the response. The asymmetric partitioning of phospholipids can be controlled by these lipid transporters.

Proteins and Lipids Diffuse in Membranes

Interactions among different lipids and between lipids and proteins are very complex and dynamic. There is a fluidity in the lipid portion of membranes so that both lipids and proteins move. The degree of fluidity is dependent on the temperature and membrane composition. At low temperatures, lipids are in a gel–crystalline state, with lipids restricted in their mobility. As temperature is increased, there is a phase transition into a liquid–crystalline state, with an increase in fluidity (Figure 12.27). With liposomes prepared from a single pure phospholipid, the phase transition temperature, T_m, is rather precise; but with liposomes prepared from a mixture of lipids, T_m becomes less precise because individual clusters of lipids may be in either the gel–crystalline or the liquid–crystalline state. T_m is not precise for

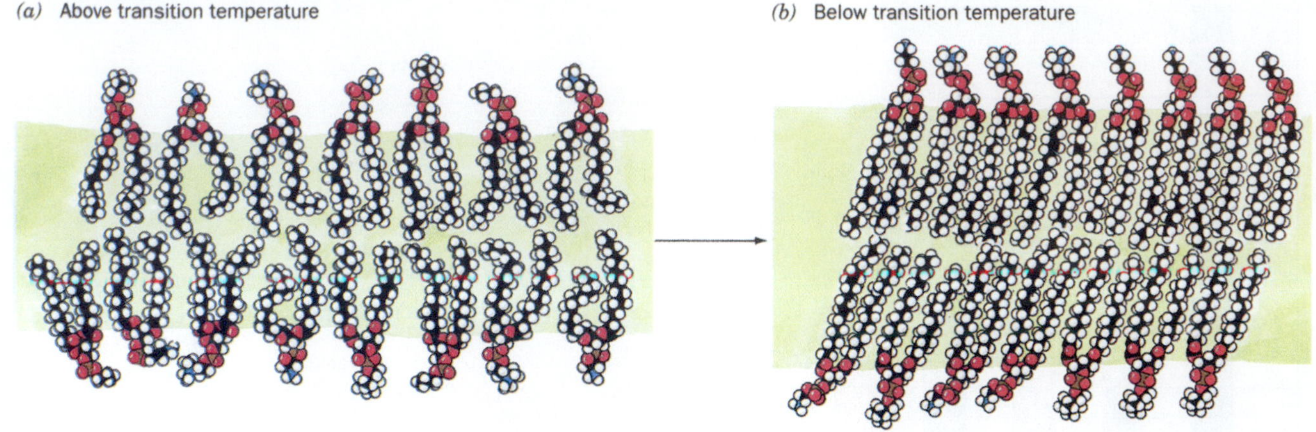

FIGURE 12.27
Structure of lipid bilayer above and below transition temperature.
Figure reproduced from Voet, D. and Voet, J. Biochemistry, *2nd ed. New York: Wiley, 1995.* Reprinted by permission of John Wiley & Sons, Inc.

biological membranes because of their heterogeneous composition. Interactions between lipids and proteins lead to variations in the gel–liquid state throughout the membrane and differences in fluidity in different areas.

The specific composition of individual biological membranes leads to differences in fluidity. Glycerophospholipids containing short-chain fatty acids increase the fluidity as does an increase in unsaturation of the fatty acyl groups. *Cis* double bonds in unsaturated fatty acids of phospholipids lead to kinks in the hydrocarbon chain, prevent the tight packing of the chains, and create pockets in the hydrophobic areas. It is assumed that these spaces, which are mobile due to the mobility of the hydrocarbon chains, are filled with water molecules and small ions. Cholesterol with its flat stiff ring structure reduces the coiling of the fatty acid chain and decreases fluidity. Consideration has been given to the potential clinical significance of high blood cholesterol on the fluidity of cell membranes (see Clin. Corr. 12.2). Ca^{2+} ion decreases the fluidity of membranes because of its interaction with the negatively charged phospholipids, reducing repulsion between polar groups and increasing packing of lipid molecules. This ion causes aggregation of lipids into clusters, reducing membrane fluidity.

CLINICAL CORRELATION 12.2
Abnormalities of Cell Membrane Fluidity in Disease States

Membrane fluidity can control the activity of membrane-bound enzymes and functions such as phagocytosis and cell growth. A major factor in controlling the fluidity of the plasma membrane in higher organisms and mammals is the presence of cholesterol. With increasing cholesterol content the lipid bilayers become less fluid on their outer surface but more fluid in the hydrophobic core. Erythrocyte membranes of individuals with spur cell anemia have an increased cholesterol content and a spiny shape, and the cells are destroyed prematurely in the spleen. This condition occurs in severe liver disease such as alcoholic cirrhosis of the liver. Cholesterol content is increased 25–65%, and the fluidity of the membrane is decreased. Erythrocyte membranes require a high degree of fluidity for their function and any decrease would have serious effects on the cell's ability to pass through the capillaries. Increased plasma membrane cholesterol in other cells leads to an increase in intracellular membrane cholesterol, which also affects their fluidity. The intoxicating effect of ethanol on the nervous system is probably due to modification of membrane fluidity and alteration of membrane receptors and ion channels. Individuals with abetalipoproteinemia have an increase in sphingomyelin content and a decrease in phosphatidylcholine in cellular membranes, thus causing a decrease in fluidity. The ramifications of these changes in fluidity are not completely understood, but as techniques for the measurement and evaluation of cellular membrane fluidity improve, some of the pathological manifestations in disease states will be explained on the basis of changes in membrane structure and function.

Cooper, R. A. Abnormalities of cell membrane fluidity in the pathogenesis of disease. *N. Engl. J. Med.* 297:371, 1977.

Fluidity at different levels within the membrane also varies. The hydrocarbon chains of the lipids have a motion, which produces fluidity in the hydrophobic core. The central area of a bilayer is occupied by ends of the hydrocarbon chains and is more fluid than areas closer to the two surfaces, where there are more constraints due to stiffer portions of the hydrocarbon chains. Cholesterol makes membranes more rigid toward the periphery because it does not reach into their central core.

Individual lipids and proteins move rapidly in a lateral motion along the surface of membranes. Electrostatic interactions of polar head groups, hydrophobic interactions of cholesterol with selected phospholipids or glycolipids, and protein–lipid interactions, however, lead to constraints on movement. There are lipid domains in which lipids move together as a unit. Glycerophospholipids (G) and cholesterol (C) cluster together in distinct microdomains, **G,C-lipid rafts,** and move as a unit. In some cells these G,C-rafts along with specific glycosyl phosphatidylinositol-anchored proteins (see p. 507) are involved in formation of **caveolae,** flask-like invaginations of the plasma membrane. These structures are involved in cell signaling and may be responsible for the separation at distinct poles of specific proteins in epithelial cells.

Movement of integral membrane proteins in the membrane has been demonstrated by fusion of human and rat cells. When antigenic membrane proteins on cells of these species were labeled with different antibody markers, the markers indicated the localization of the proteins on the membrane. Immediately following fusion of the cells, proteins on the membranes of the human and rat cells were segregated in different hemispheres of the new cell, but within 40 minutes the two groups of proteins were evenly distributed over the membrane of the new cell. These observations may apply only to some membrane proteins because in most cells protein movements are restricted by other membrane proteins, matrix proteins, or cellular structural elements such as microtubules or microfilaments to which they may be attached.

Evidence is accumulating that the fluidity of cellular membranes can change in response to changes in diet or physiological state. Fatty acid and cholesterol content of membranes is modified by a variety of factors. In addition, pharmacological agents may have a direct effect on membrane fluidity. **Anesthetics** that induce sleep and muscular relaxation may have their action because of their effect on membrane fluidity of specific cells. A number of structurally unrelated compounds induce anesthesia, but their common feature is lipid solubility. Anesthetics increase membrane fluidity *in vitro*.

Thus cellular membranes are in a constantly changing state, with not only movement of proteins and lipids laterally on the membrane but with molecules moving into and out of the membrane. The membrane creates a number of microenvironments, from the hydrophobic portion of the core of the membrane to the interface with the surrounding environments. It is difficult to express in words or pictures the very fluid and dynamic state, in that neither captures the time-dependent changes that occur in the structure of biological membranes. Figure 12.28 attempts to illustrate the structural and movement aspects of cellular membranes.

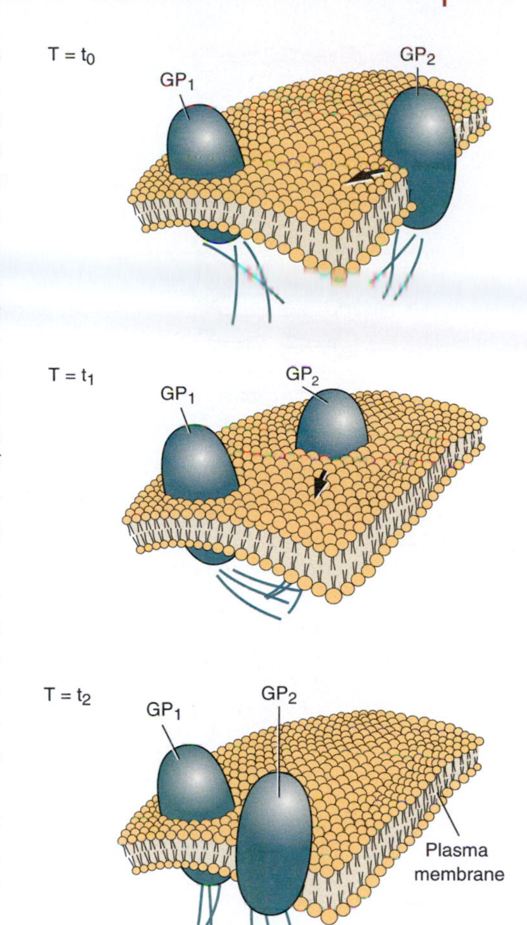

FIGURE 12.28
Modified version of the fluid mosaic model of biological membranes to indicate the mobility of membrane proteins.
t_0, t_1, and t_2 represent successive points in time. Some integral proteins (GP_2) are free to diffuse laterally in the plane of the membrane directed by the cytoskeletal components, whereas others (GP_1) may be restricted in their mobility.

12.5 | MOVEMENT OF MOLECULES THROUGH MEMBRANES

The lipid nature of biological membranes severely restricts the type of molecules that diffuse readily from one side to another. Inorganic ions and charged organic molecules do not diffuse at a significant rate because of their attraction to water molecules and exclusion of charged species by the hydrophobic environment of lipid membranes. The size of large molecules such as proteins and nucleic acids precludes significant diffusion. Diffusion rates of most nutrients and inorganic ions is too slow to accommodate a cell's requirements for the substance. To overcome this,

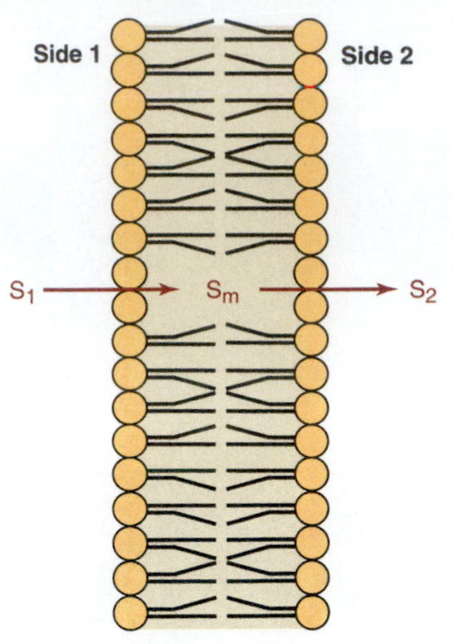

FIGURE 12.29
Diffusion of a solute molecule through a membrane.
S_1 and S_2 are the solute on each side of the membrane, and S_m is solute in the membrane.

translocation of water, ions, nutrients, and even waste products across membranes involve a variety of specialized transporters and channels. Several hundred transporters have been identified, many by use of recombinant technology.

The basic mechanisms by which molecules cross cellular membranes is presented in the following sections with examples of the processes. In later chapters the roles of some transporters in cellular homeostasis are discussed. Many inherited diseases are due to changes in transport mechanisms.

Some Molecules Diffuse Through Membranes

Diffusion of a substance through a membrane involves three major steps: (1) solute leaves the aqueous environment on one side and enters the membrane; (2) solute traverses the membrane; and (3) solute leaves the membrane to enter a new environment on the opposite side (Figure 12.29). Each step involves an equilibrium of solute between two states. Thermodynamic and kinetic constraints control the concentration equilibrium of a substance on two sides of a membrane and the rate at which it attains equilibrium. Diffusion of gases such as O_2, N_2, CO_2, and NO occurs rapidly and depends entirely on the concentration gradient. Water diffuses through biological membranes under osmotic forces, presumably via gaps in the hydrophobic environment created by random movement of fatty acyl chains of lipids. For diffusion of a solute with strong interaction with water molecules, the shell of water surrounding the solute must be stripped away before it enters the lipid milieu and then regained on leaving the membrane. Distribution of hydrophobic substances between the aqueous phase and lipid membrane will depend on the degree of lipid solubility of the substance; very lipid-soluble materials will dissolve readily in the membrane.

The rate of diffusion of a lipophilic substance is directly proportional to its lipid solubility and diffusion coefficient in lipids; the latter is a function of the size and shape of the substance.

Direction of movement of solutes by diffusion is always from a higher to a lower concentration and the rate is described by **Fick's first law of diffusion:**

$$J = -D\left(\frac{\delta c}{\delta x}\right)$$

where J is the amount of substance moved per time, D is the diffusion coefficient, and $\delta c/\delta x$ is the chemical gradient of substance. As the concentration of solute on one side of the membrane is increased, there will be an increasing *initial rate* of diffusion as illustrated in Figure 12.30. A net movement of molecules from one side to another will continue until the concentration in each is at chemical equilibrium. A continued exchange of solute molecules from one side to another occurs after equilibrium is attained but no net accumulation on either side can occur because this would recreate a concentration gradient if it occurred.

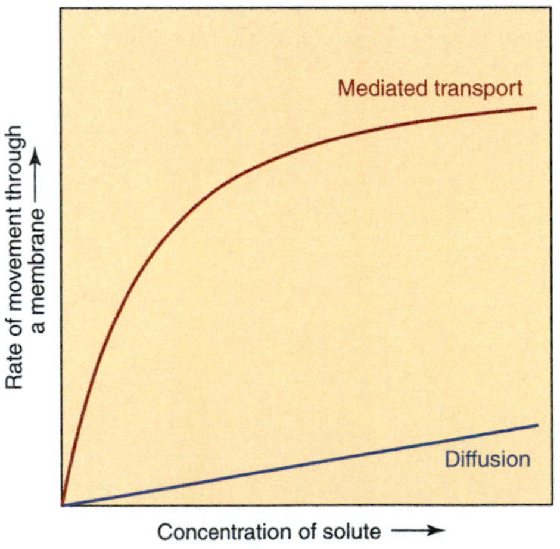

FIGURE 12.30
Kinetics of movement of a solute molecule through a membrane.
Initial rate of diffusion is directly proportional to the concentration of solute. In mediated transport, rate will reach a V_{max} when carrier is saturated.

Movement of Molecules Across Membranes Can Be Facilitated

Membranes have translocation mechanisms with a high degree of specificity for various substances. They involve intrinsic membrane proteins and are classified on the basis of their mechanism of substrate translocation and the energetics of the system. A classification of **transport systems** is presented in Table 12.4. Each will be discussed in more detail in subsequent sections but for now it is important to distinguish the three main types.

Membrane Channels

Membranes of most cells contain specific **channels,** in some cases referred to as **pores,** which permit rapid movement of specific molecules or ions from one side of a membrane to the other. The tertiary and quaternary structures of these intrinsic

TABLE 12.4 Classification of Membrane Translocation Systems

Type	*Class*	*Example*
Channel	1. Voltage regulated	Na^+ channel
	2. Agonist regulated	Acetylcholine receptor
	3. cAMP regulated	Cl^- channel
	4. Other	Pressure sensitive
		Stretch sensitive
		Heat sensitive
Transporter	1. Facilitated diffusion	Glucose transporter
	2. Active mediated	
	a. Primary-redox coupled	Respiratory chain linked
	Primary-ATPases	Na^+, K^+–ATPase
	ATP-binding cassette	Multidrug resistance protein transporter
	b. Secondary	Na^+-dependent glucose transport
Group translocation		Amino acid translocation

membrane proteins create an aqueous hole that permits diffusion of substances in both directions through the membrane. Substances move only in the direction of a lower concentration and while the channel is open many molecules move to the other side of the membrane. Channel proteins do not bind the molecules or ions to be transported, but can be inhibited. The channels have some degree of specificity, however, based on the size and charge of the substance. Flow through channels is regulated by various mechanisms (Table 12.4) that open and shut the passageway.

Transporters

Transporters actually translocate the molecule or ion across the membrane by binding and physically moving the substance. The activity can be evaluated in the same kinetic terms as an enzyme-catalyzed reaction except that no chemical reaction occurs. Transporters have specificity for the substance to be transported, referred to as the substrate, have defined reaction kinetics, and can be inhibited by both competitive and noncompetitive inhibitors. Some transporters only move substrates down their concentration gradient (referred to as passive transport or facilitated diffusion), while others move substrate against its concentration gradient (active transport) and require expenditure of some form of energy. With both channels and transporters the molecule is unchanged following translocation across the membrane.

A major difference between channels and transporters is the rate of substrate translocation; for a channel, rates in the range of 10^7 ions $s^{-1}$are usual, whereas with a transporter the rate is in the range of 10^2–10^3 molecules s^{-1}. The activity of all translocation systems can be modulated, permitting cells and tissues to control the movement of substances across membranes. Drugs for specific channels and transporters have been developed to control these processes.

Group Translocation

Group translocation involves not only movement of a substance across a membrane but also a chemical modification of the substrate during the process. One mechanism of uptake of sugars by bacteria involves transport and then phosphorylation of the sugar before release into the cytosol. In some mammalian tissues uptake of amino acids involves a group translocation mechanism.

Membrane Transport Systems Have Common Characteristics

The proteins or protein complexes involved in transport systems have been designated by a variety of names, including **transporter, translocase, translocator,**

TABLE 12.5 Characteristics of Membrane Transporters

Facilitated Diffusion	*Active Mediated*
1. Saturation kinetics	1. Saturation kinetics
2. Specificity for solute transported	2. Specificity for solute transported
3. Can be inhibited	3. Can be inhibited
4. Solute moves down concentration gradient	4. Solute can move against concentration gradient
5. No expenditure of energy	5. Requires coupled input of energy

Recognition:	$S_1 + T_1 \rightleftharpoons S-T_1$
Transport:	$S-T_1 \rightleftharpoons S-T_2$
Release:	$S-T_2 \rightleftharpoons T_2 + S_2$
Recovery:	$T_2 \rightleftharpoons T_1$

FIGURE 12.31
Reactions involved in mediated transport across a biological membrane.
S_1 and S_2 are solutes on sides 1 and 2 of the membrane, respectively; T_1 and T_2 are binding sites on the transporter on sides 1 and 2, respectively.

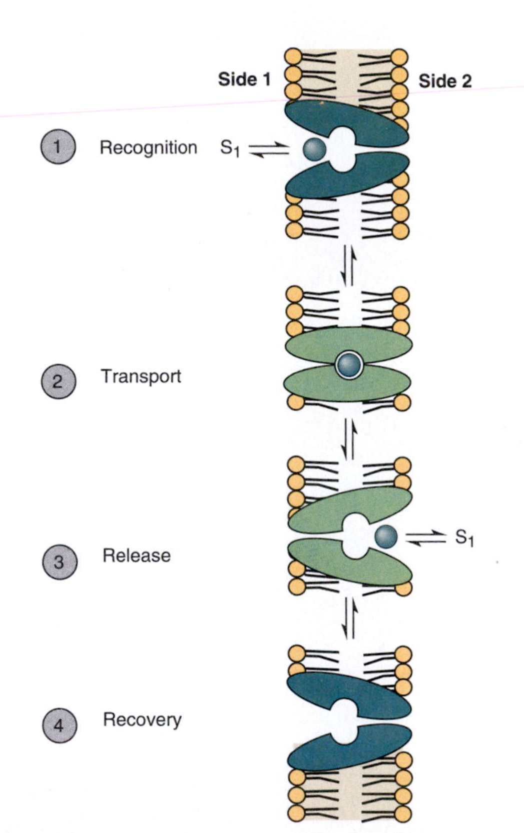

FIGURE 12.32
Model for a mediated transport system in a biological membrane.
Model is based on the concept of specific sites for binding of substrate and a conformational change in the transporter to move bound solute a short distance but into the environment of the other side of the membrane. Once moved, solute is released from the transporter.

permease, and **pump**, or termed **transporter system, translocation mechanism,** and **mediated transport system.** These designations are used interchangeably, but for convenience we will use transporter when referring to the proteins involved in translocation. Transporters have a number of characteristics in common. They facilitate the movement of a molecule or molecules through the lipid bilayer at a rate that is significantly faster than can be accounted for by simple diffusion. If S_1 is the solute on side 1 and S_2 on side 2, then the transporter promotes the establishment of an equilibrium as follows:

$$[S_1] \rightleftharpoons [S_2]$$

where the brackets represent the concentration of solute. If the transporter (T) is included in the equilibrium the reaction is

$$[S_1] + T \rightleftharpoons [S - T] \rightleftharpoons [S_2] + T$$

If no energy is used, the concentration on both sides of the membrane will be equal at equilibrium, but if there is an expenditure of energy, a concentration gradient can be established. Note the similarity of the role of a transporter to that of an enzyme; in both cases the protein increases the rate but does not determine the final equilibrium.

Table 12.5 lists major characteristics of membrane transport systems. As presented in Figure 12.30, transporters demonstrate **saturation kinetics;** as the concentration of the substance to be translocated increases, the initial rate of transport increases but reaches a maximum when the substance saturates the protein transporter. Simple diffusion does not demonstrate saturation kinetics. Constants such as V_{max} and K_m can be calculated for transporters. As with enzymes, transporters can catalyze movement of a solute in both directions across the membrane depending on the $\Delta G'$ for the reaction.

Most transporters have a high degree of structural and stereo specificity for the substance transported. An example is mediated transport of D-glucose in erythrocytes, where the K_m for D-galactose is 10 times larger and for L-glucose 1000 times larger than for D-glucose. The transporter has essentially no activity with D-fructose or disaccharides. Structural analogs of the substrate inhibit competitively and reagents that react with specific groups on proteins are noncompetitive inhibitors.

Four Common Steps in the Transport of Solute Molecules

We need to expand the equation above and consider four aspects of mediated transport (Figure 12.31). These are (1) recognition by the transporter of appropriate solute from a variety of solutes in the aqueous environment, (2) translocation of solute across the membrane, (3) release of solute by the transporter, and (4) recovery of transporter to its original condition to accept another solute molecule.

The first step, **recognition** of a specific substrate by the transporter, is explained on the same basis as that described for recognition of a substrate by an enzyme (see p. 441). Specific binding sites on the protein recognize the correct structure of solute to be translocated. The second step, **translocation,** occurs when the bound substrate induces a conformational change in the protein (Figure 12.32).

This moves the substrate a short distance, perhaps only 2 or 3 Å into the environment of the opposite side of the membrane. In this manner, it is not necessary for the transporter to move the molecule the entire distance across the membrane. Step 3, **release** of the substrate, occurs if the concentration of substrate is lower in the new environment than on the initial side of binding. Without a change in affinity of protein for substrate (K_{eq}), there would be a shift in the equilibrium and release of a portion of the solute. For transporters that move a solute against a concentration gradient, release of the solute at the higher concentration requires a decrease in affinity for the solute by the transporter. A change in conformation of transporter can lead to a decrease in affinity. In group translocation (p. 819) the solute is chemically altered while attached to the transporter and the modified molecule has a lower affinity for the transporter. Finally, in **recovery** (step 4), the transporter returns to its original conformation upon release of substrate.

The discussion above has centered on the movement of a single solute molecule by a transporter. There are systems that move two molecules simultaneously in one direction (symport mechanisms), two molecules in opposite directions (antiport mechanism), as well as a single molecule in one direction (uniport mechanism) (Figure 12.33). When a charged substance, such as K^+, is translocated and no ion of the opposite charge is moved, a charge separation occurs across the membrane. This mechanism is termed electrogenic and leads to development of a membrane potential. If an oppositely charged ion is moved to balance the charge, the mechanism is called neutral or electrically silent.

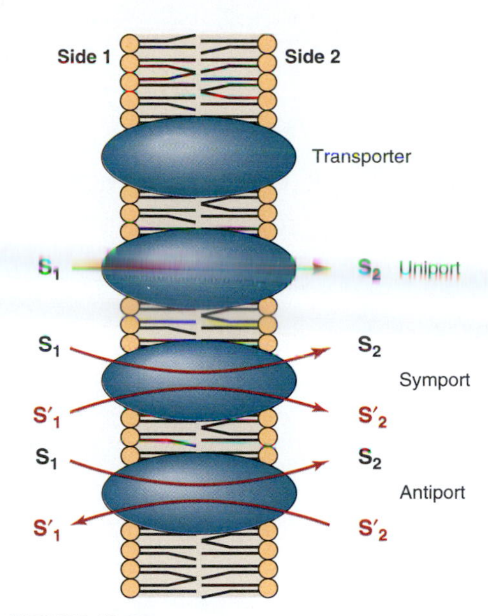

FIGURE 12.33
Uniport, symport, and antiport mechanisms for translocation of substances.
S and S′ represent different molecules; subscripts 1 and 2 indicate S molecules on different sides of membrane.

Energetics of Membrane Transport Systems

The change in free energy when an uncharged molecule moves from concentration C_1 to concentration C_2 on the other side of a membrane is given by Eq. 12.1:

$$\Delta G' = 2.3RT \log\left(\frac{C_2}{C_1}\right) \tag{12.1}$$

When $\Delta G'$ is negative—that is, there is release of free energy—movement of solute will occur without the need for a driving force. When $\Delta G'$ is positive, as would be the case if C_2 is larger than C_1, then there needs to be an input of energy to drive the transport. For a charged molecule (e.g., Na^+) both the electrical potential and concentrations of solute are involved in calculating the change in free energy, as indicated in Eq. 12.2:

$$\Delta G' = 2.3RT \log\left(\frac{C_2}{C_1}\right) + Z\mathcal{F}\Psi \tag{12.2}$$

where Z is the charge of the species moving, $\mathcal{F}$ is the Faraday constant (23.062 kcal V^{-1} mol^{-1}), and Ψ is the difference in electrical potential in volts across the membrane. The electrical component is the membrane potential and $\Delta G'$ is the electrochemical potential.

A **facilitated transport** system is one in which $\Delta G'$ is negative and the movement of solute occurs spontaneously. When $\Delta G'$ is positive, coupled input of energy from some source is required for movement of the solute and the process is called active transport. **Active transport** is driven by either hydrolysis of ATP to ADP or utilization of an electrochemical gradient of Na^+ or H^+ across the membrane. In the first the chemical energy released on hydrolysis of a pyrophosphate bond drives the reaction, whereas in the latter an electrochemical gradient is dissipated to transport the solute.

Transport systems that maintain very large concentration gradients are present in various membranes. An example is the plasma membrane transport system that maintains the Na^+ and K^+ gradients. One of the most striking examples of an active transport system is that present in the parietal cells of gastric glands, which are responsible for secretion of HCl into the lumen of the stomach (see p. 1096). The pH of plasma is about 7.4 (4×10^{-8} M H^+), and the luminal pH of the stomach can reach 0.8 (0.15 M H^+). The cells transport H^+ against a concentration gradient of $1 \times 10^{6.6}$. Assuming there is no electrical component, the energy for H^+ secretion under these conditions can be calculated from Eq. 12.1 and is 9.1 kcal mol^{-1} of HCl.

12.6 | CHANNELS AND PORES

Channels and Pores in Membranes Function Differently

Membrane channels and pores are intrinsic membrane proteins and are differentiated on the basis of their degree of specificity for molecules crossing the membrane. Channels are selective for specific inorganic cations and anions, whereas pores are not selective, permitting inorganic and organic molecules to pass through the membrane. The **Na^+ channel** of plasma membranes of eukaryotic cells, for example, permits movement of Na^+ at a rate more than ten times greater than that for K^+. This difference between channels and pores is due to differences in size of the aqueous area created in the protein structure as well as amino acid residues lining the channel area. Over 100 families of channel-forming proteins have been identified. A common motif is a structure formed by amphipathic α helices of associated protein subunits or from domains within a single polypeptide chain creating a central aqueous space as pictured in Figure 12.34. Exceptions to the α-helical structure are the porins (see below) of Gram-negative bacteria, which have a β-sheet structure lining the central pore.

Opening and Closing of Channels Are Controlled

Opening and closing of membrane channels involve a conformational change in the channel protein. Those controlled by changes in the **transmembrane potential** are referred to as voltage-gated channels; a specific sensor detects membrane potential changes and transfers its energy to the channel to control its gate. In the case of the Na^+ channel, depolarization of the membrane leads to an opening of the channel. Voltage-gated channels for Na^+, K^+, and Ca^{2+} are present in the plasma membrane of most cells; Clinical Correlation 23.1 (p. 997) describes changes in voltage-gated channels in myotonic muscle disorders. Mitochondria have a voltage-dependent channel for anions. Binding of a specific agent, termed an **agonist,** is another mechanism to control opening of a channel. Binding of acetylcholine opens the channel in the **nicotinic–acetylcholine receptor**, allowing the flow of Na^+ into the cell. This is important to neuronal electrical signal transmission (see p. 998). In addition, some channels are controlled by cAMP (see p. 928); Clinical Correlation 12.3 describes the modification of the Cl^- channel in cystic fibrosis. These forms of control for opening channels are very fast, permitting bursts of ion flow through the membrane at rates of over 10^7 ions s^{-1}, which is near their diffusion rate in solution. This rate is necessary because these channels are involved in nerve conduction and muscle contraction. The surface of a nerve terminal can contain several hundred different channel molecules, including voltage-dependent Ca^{2+} channels, K^+ channels, Ca^{2+}-gated K^+ channels, Cl^- channels, ligand-gated channels, and stretch-activated channels. A number of pharmacological agents are used therapeutically to modulate channels.

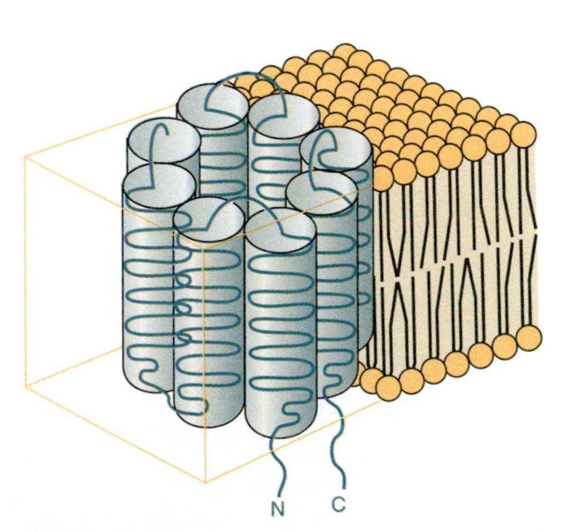

FIGURE 12.34
Arrangement of protein subunits or domains to form a membrane channel.

Sodium and Chloride Channels

Voltage-sensitive **Na^+ channels** mediate rapid increase in intracellular Na^+ following depolarization of the plasma membrane in nerve and muscle cells. The channel consists of a single large glycopolypeptide and several smaller glycoproteins. Genes for some of the peptides of Na^+ channels have been cloned and the amino acid sequences determined. There are four repeat homology units in the channel, each with six transmembrane α helices. A model for this transporter is presented in Figure 12.35*a* and a possible arrangement of the helices in the membrane as viewed looking down at a membrane is presented in Figure 12.35*b*. One transmembrane segment, labeled as SIV, has a positively charged amino acid at every third position and may serve as a voltage sensor. A mechanical shift of this region due to a change in the membrane potential may lead to a conformational change in the protein, resulting in the opening of the channel. The channel size created by the protein, however, cannot totally explain the specificity for Na^+.

Chloride channels participate in regulating cell volume, stabilizing membrane potential, controlling signal transduction, and regulating transepitheal transport. Ligand-gated, voltage-gated, ATP-binding cassette types (see p. 525) have been identified. Some are activated by cyclic AMP (cAMP) and Ca^{2+}. The cystic fibrosis transmembrane regulator is a Cl^- channel (see Clin. Corr. 12.3).

CLINICAL CORRELATION 12.3
Cystic Fibrosis and the Cl^- Channel

Cystic fibrosis (CF), an autosomal recessive disease, is the commonest, fatal, inherited disease of Caucasians, occurring with a frequency of 1 in 2000 live births. It is a multiorgan disease, with a principal manifestation being pulmonary obstruction; thick mucous secretions obstruct the small airways, allowing recurrent bacterial infections. Exocrine pancreatic dysfunction occurs early and leads to steatorrhea (fatty stool) in CF patients; see p. 1104 for a discussion of the role of the pancreas in fat digestion and absorption. CF patients have reduced Cl^- permeability, which impairs fluid and electrolyte secretion, leading to luminal dehydration. Diagnosis of CF is confirmed by a significant increase of Cl^- content of sweat of affected individuals in comparison to normal individuals.

The gene responsible for CF was identified in 1989 and over 400 mutations leading to CF have been found. The most common mutation (about 70%) leads to a deletion of a single phenylalanine at position 508 on the protein, but missense, nonsense, frameshift, and splice-junction mutations (see p. 192) have also been reported. Many mutations lead to a change in protein folding. The CF gene product is the **c**ystic **f**ibrosis **t**ransmembrane conductance **r**egulator (CFTR). CFTR is a cAMP-dependent Cl^- channel and may itself regulate other ion channels; it is expressed in epithelial tissues. Phosphorylation of a cytoplasmic regulatory domain by protein kinase A activates the channel. CFTR is a polypeptide of 1480 amino acids with structural homology to the superfamily of ATP-binding cassette (ABC) transporters. The gene has been cloned (see p. 287) and a major effort is under way to treat the disease by gene therapy, using both viral and nonviral vectors including liposomes (see Clin. Corr. 12.1).

Alton, E. W. and Geddes, D. M. Gene therapy for cystic fibrosis; a clinical perspective. *Gene Ther.* 2:88, 1995; Frizzell R. A. Functions of the cystic fibrosis transmembrane conductance regulator protein. *Am. J. Respir. Crit. Care Med.* 151:54, 1995; Naren, A. P., Cormet-Boyaka, E., Fu, J., et al. CFTR chloride channel regulation by an interdomain interaction. *Science* 286:544, 1999; and Wagner, J. A., Chao, A. C., and Gardner, P. Molecular strategies for therapy of cystic fibrosis. *Annu. Rev. Pharmacol. Toxicol.* 35:257, 1995.

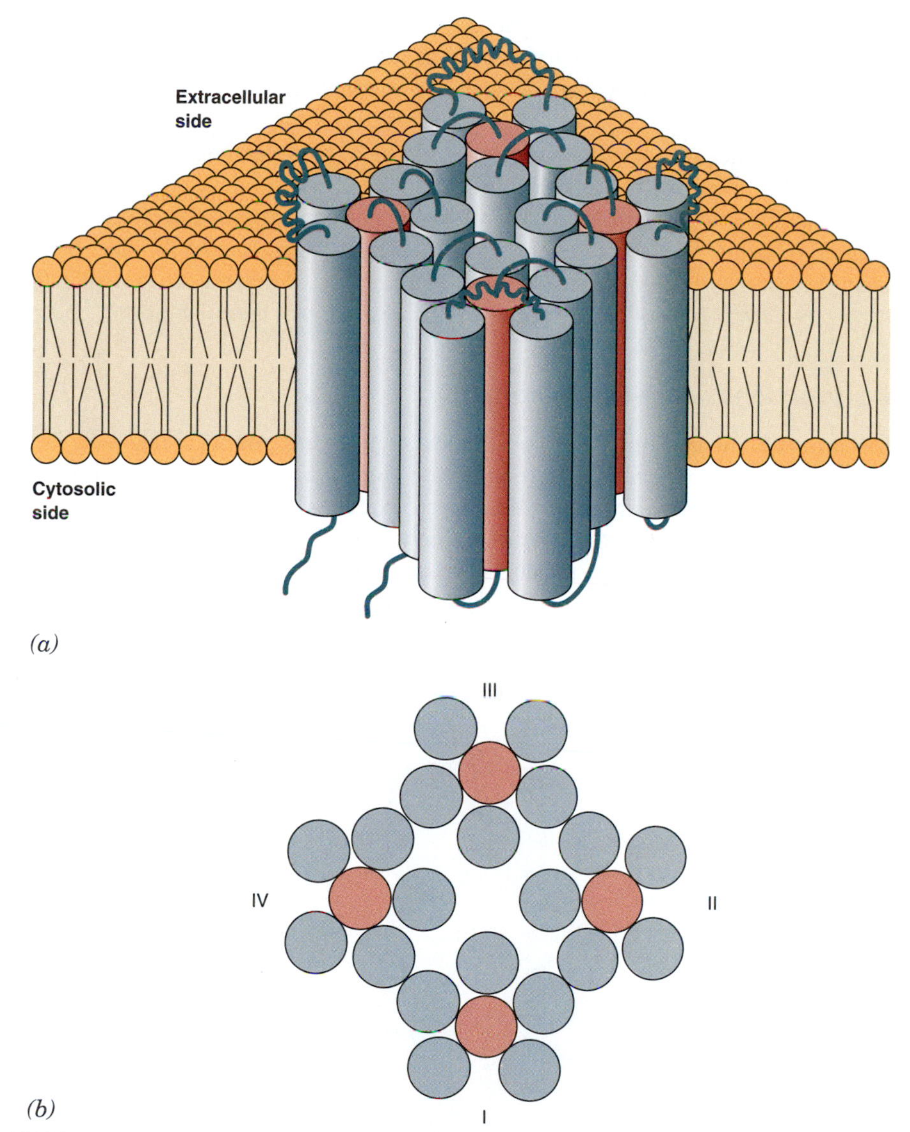

FIGURE 12.35
Possible model of the Na^+ channel.
(*a*) The single peptide consists of four repeating units with each unit folding into six transmembrane helices. (*b*) Proposed arrangement of the transmembrane sequence as viewed down on the membrane.
Redrawn from Noda, M., et al. Nature *320:188, 1986.*

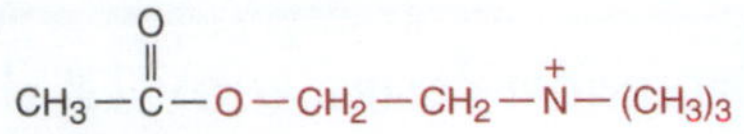

$$CH_3-\overset{\overset{\Large O}{\|}}{C}-O-CH_2-CH_2-\overset{+}{N}-(CH_3)_3$$

FIGURE 12.36
Structure of acetylcholine.

Nicotinic–Acetylcholine Channel (nAChR)

The **nicotinic–acetylcholine** channel, also referred to as the **acetylcholine receptor,** is an example of a chemically regulated channel, in which binding of acetylcholine (Figure 12.36) opens the channel. The dual name is used to differentiate this receptor from other acetylcholine receptors, which function in a different manner. Acetylcholine, a neurotransmitter, is released at the neuromuscular junction by a neuron when electrically excited. The acetylcholine diffuses to the skeletal muscle membrane where it interacts with the acetylcholine receptor, opening the channel and allowing selective cations to move across the membrane (see p. 998). The change in transmembrane potential leads to a series of events culminating in muscle contraction. The nicotinic–acetylcholine receptor consists of five polypeptide subunits, with two α subunits and one each of β, γ, and δ; each α subunit is glycosylated and two others contain covalently bound lipid. Phosphorylation of the α subunit is required for activity. The channel opens when acetylcholine molecules bind to each of the two α subunits, causing a conformational change in the protein; closure of the channel occurs within a millisecond due to hydrolysis of acetylcholine to acetate and choline, and their release from the protein. In the open conformation, cations and small nonelectrolytes flow through the channel but not anions; negatively charged amino acid residues in the channel are sufficient to repel negatively charge ions.

The nicotinic–acetylcholine receptor is inhibited by several deadly neurotoxins including ***d*-tubocurarine**, the active ingredient of curare, and several toxins from snakes including **α-bungarotoxin, erabutoxin,** and **cobratoxin. Succinylcholine,** a muscle relaxant, opens the channel, leading to depolarization of the membrane; succinylcholine is used in surgical procedures because its activity is reversible due to the rapid hydrolysis of the compound after cessation of administration.

Gap Junctions and Nuclear Pores

Plasma membrane **gap junctions** and **nuclear membrane pores** are relatively large openings created by specific proteins. Gap junctions are channels lined by proteins spanning two plasma membranes that create aqueous connections between two cells. They mediate the electrical coupling between cells, and the exchange of ions and metabolites but not large molecular weight compounds such as proteins. The

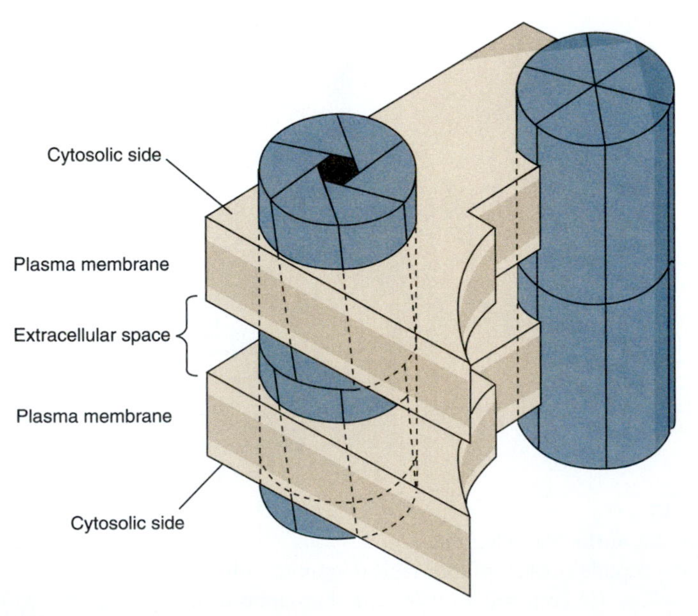

FIGURE 12.37
Model for a channel in the gap junction.

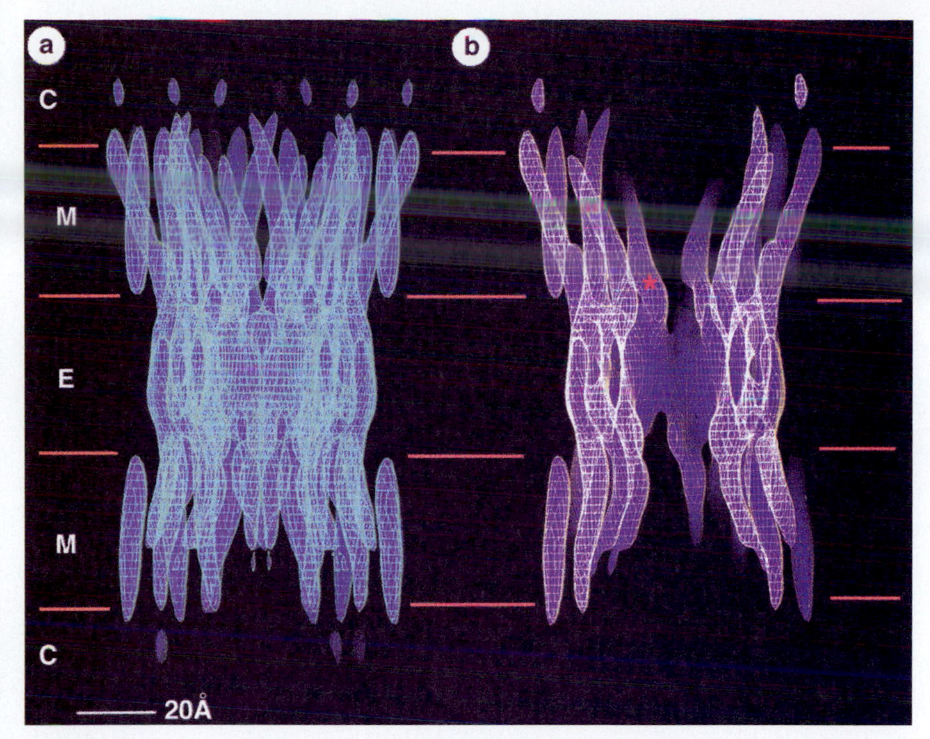

FIGURE 12.38
Molecular organization of recombinant cardiac gap junction.
Model developed from electron crystallography. (*a*) A full side view is shown and (*b*) the density has been cropped to show the channel interior. M, membrane bilayer; E, extracellular space; and C, cytoplasmic space.
Reprinted with permission from Unger, V. M., Kumar, N. M., Gilula, N. B., and Yeager, M. Science *283:1176, 1999. Copyright© 1999 by American Association for the Advancement of Science.*

diameter of the opening ranges from 12 to 20 Å. Oligomers of the gap junction polypeptide (32 kDa), referred to as **connexin,** form the channel. Twelve subunits, six in each cell, form a hexameric structure in each membrane as shown in Figure 12.37. The molecular organization of a recombinant gap junction is presented in Figure 12.38. The channels are normally open but increases in cytosolic Ca^{2+}, a change in metabolism, a drop in transmembrane potential, or acidification of the cytosol cause closure. When the channel is open the subunits appear to be slightly tilted but when closed they appear to be more nearly parallel to a perpendicular to the membrane, suggesting that subunits slide over each other.

Like gap junctions, nuclear pores span two membranes, creating aqueous channels in the nuclear envelope. Pores are about 90 Å in diameter and permit the movement of large macromolecules. They are very complex structures and participate in the movement of RNA from the nucleus to the cytosol. The plasma membranes of Gram-negative bacteria contain protein pores, termed **porins.** Over 40 different porins have been isolated and they range in size from 28 to 48 kDa. In contrast to most mammalian channels, these transmembrane segments are β sheets not α helices and exist in the membrane as trimers. Porins are water-filled transmembrane channels and range in diameter from 6 to 23 Å with some degree of selectivity for inorganic ions; some, however, permit the uptake of sugars.

12.7 | PASSIVE MEDIATED TRANSPORT SYSTEMS

Passive mediated transport, also referred to as **facilitated diffusion,** is a mechanism for translocation of solutes through cell membranes without expenditure of metabolic energy (see Table 12.5, p. 514). As with nonmediated diffusion the di-

TABLE 12.6 Major Isoforms of Glucose Transporter

Isoform	*Tissue*	*Cellular Localization*	*Comments*
GLUT1	Muscle Heart Blood–brain barrier Glia cells Placenta	Plasma membrane	Activity in muscle increased by insulin, hypoxia, and diet
GLUT2	Liver Pancreas Intestine Kidney	Plasma membrane	
GLUT3	Neuron Kidney Placenta	Plasma membrane	
GLUT4	Muscle Adipose Heart Blastocysts	Plasma membrane	Activity in muscle and adipose tissue stimulated by insulin, and in muscle by hypoxia and diet
GLUT5	Muscle Spermatozoa	Sarcolemmal vesicles	Fructose in substrate

rection of flow is always from a higher to a lower concentration. The distinguishing differences between the process of nonmediated diffusion and passive mediated transport are the demonstration of saturation kinetics, a structural specificity for the class of molecule moving across the membrane, and specific inhibition of solute movement in the latter process.

Phloretin

2,4,6-Trihydroxyacetosphenone

FIGURE 12.39
Inhibitors of passive mediated transport of D-glucose in erythrocytes.

Glucose Transport Is Facilitated

Eight members of a superfamily of membrane proteins that mediate D-glucose transport have been reported in mammalian cells. The proteins are products of unique genes that are expressed in a tissue-specific manner. The **glucose transporters** are designated as GLUT1, GLUT2, and so on (Table 12.6). All have 12 hydrophobic segments considered to be the transmembrane regions and there is significant amino acid similarity between the transporters. Most are in the plasma membrane where direction of movement of glucose is usually out to in. GLUT2, however, may be responsible for glucose export from liver cells. The transporters catalyze a uniport mechanism and are most active with D-glucose. D-Galactose, D-mannose, D-arabinose, and several other D-sugars as well as glycerol are translocated by some isoforms. GLUT5 of sarcolemmal membranes of skeletal muscle transports fructose preferentially. The affinity of the erythrocyte transporter is highest with glucose with a K_m of ~6.2 mM, whereas for other sugars the affinity is much lower. With isolated erythrocytes, glucose moves into or out of the erythrocyte, depending on the direction of the concentration gradient, demonstrating the reversibility of the system. GLUT4 is an insulin-responsive transporter. Activity of some GLUTs in muscle is stimulated by exercise and hypoxia; under these conditions there is an increase in glucose utilization by muscle. Several sugar analogs as well as **phoretin** and 2,4,6-trihydroxyacetophenone (Figure 12.39) are competitive inhibitors.

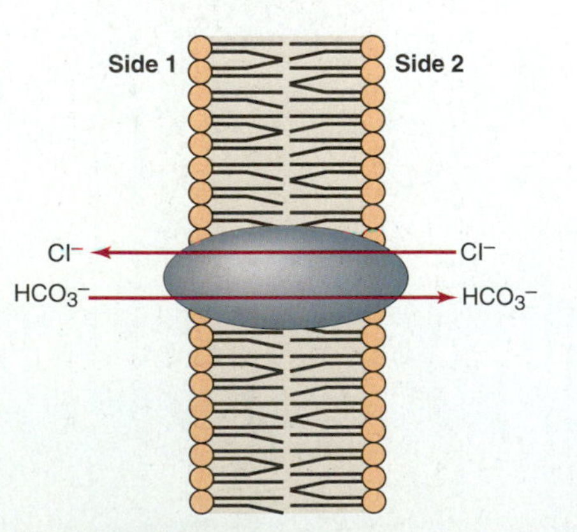

FIGURE 12.40
Passive anion antiport mechanism for movement of Cl^- and HCO_3^- across the erythrocyte plasma membrane.

Cl^- and HCO_3^+ Are Transported by an Antiport Mechanism

An anion transporter in erythrocytes and kidneys involves the antiport movement of Cl^- and HCO_3^- (Figure 12.40). The transporter is referred to as the Na^+-independent **Cl^-–HCO_3^- exchanger** or **anion exchange protein** and, in erythrocytes, **band 3,** the latter because of its position in SDS polyacrylamide gel electrophoresis of erythrocyte membrane proteins. The direction of ion flow is reversible and depends on the concentration gradients of the ions across the membrane. The transporter is important in adjusting the erythrocyte HCO_3^- concentration in arterial and venous blood.

The kidney transporter is a truncated form of the erythrocyte protein and is responsible for base (HCO_3^-) efflux to balance the ATP-driven H^+ efflux.

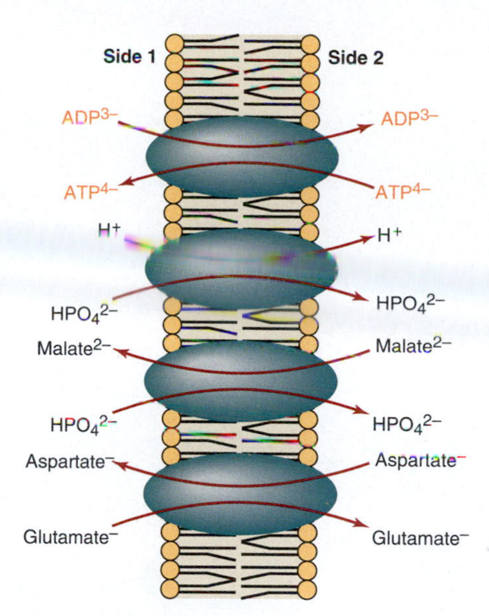

FIGURE 12.41
Representative anion transport systems in liver mitochondria.
Note that each is an antiport mechanism. Mitochondrial transport systems are discussed in Chapter 13.

Mitochondria Contain a Number of Transport Systems

The inner mitochondrial membrane contains antiport systems for exchange of anions between cytosol and mitochondrial matrix, including (1) a transporter for exchange of ADP and ATP, (2) a symport transporter for phosphate and H^+, (3) a dicarboxylate carrier that catalyzes an exchange of malate for phosphate, and (4) a translocator for exchange of aspartate and glutamate (Figure 12.41). The relationship of these transporters and energy coupling are discussed on p. 583. In the absence of an input of energy these transporters mediate a passive exchange of metabolites down their concentration gradient to achieve a thermodynamic equilibrium. As an antiport mechanism, a concentration gradient of one compound can drive the movement of the other solute. In several cases, the transporter mediates the antiport movement of an equal number of electrical charges on the substrates; in such movement the mitochondrial membrane potential influences the equilibrium. ADP–ATP and phosphate transporters, as well as an **uncoupling protein** (see p. 586) that translocates H^+, have significant amino acid homology and are presumably derived from a common ancestor. It has been suggested that each subunit has six transmembrane α helices. The uncoupling protein, found in mitochondria of brown adipose tissue, is involved in generation of heat.

The **ATP–ADP transporter** is very specific for ATP and ADP but does not transport AMP or other nucleotides. The protein represents about 12% of the total protein in heart mitochondria. It is very hydrophobic and can exist in two conformations. **Atractyloside** and **bongkrekic** acid (Figure 12.42) are specific inhibitors,

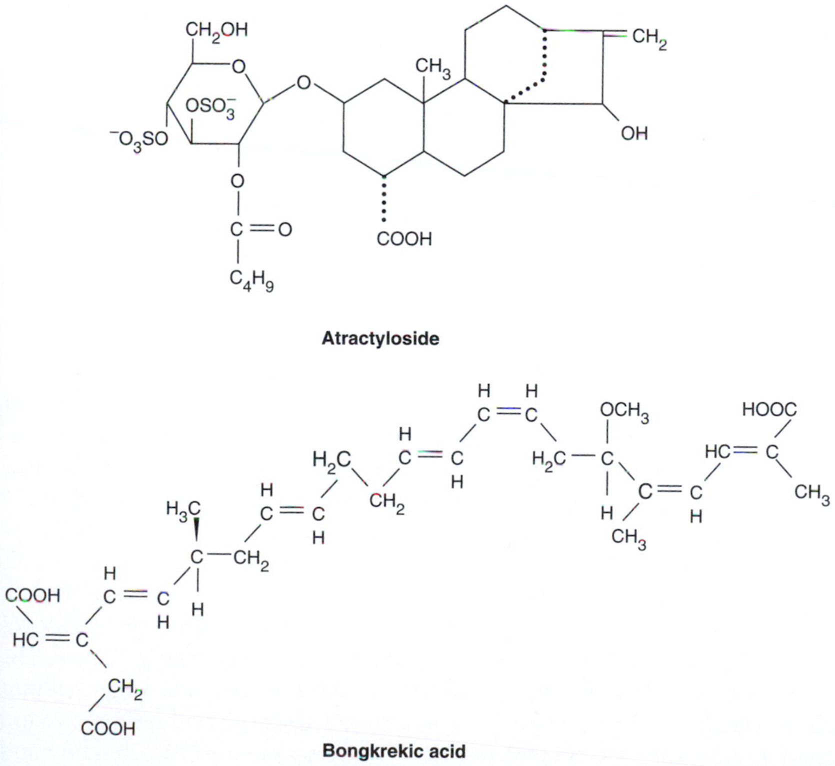

Atractyloside

Bongkrekic acid

FIGURE 12.42
Structure of two inhibitors of the ATP–ADP transport system of liver mitochondria.

each apparently reacting with a different conformation of the protein. The mitochondrial membrane potential can drive the movement of the nucleotides by this translocator, but in the absence of the membrane potential it functions as a passive mediated transporter.

12.8 | ACTIVE MEDIATED TRANSPORT SYSTEMS

Active mediated transporters have the same three characteristics as facilitated transport systems, that is, saturation kinetics, substrate specificity, and inhibitability (see Table 12.5). They also require utilization of energy to translocate solutes. Active transporters are classified as **primary active transporters** if they require ATP as the energy source, or **secondary active transporters** if a transmembrane chemical gradient of Na^+ or H^+ is required for translocation (Figure 12.43). Transporters utilizing ATP are also referred to as an ATPase, an enzyme activity, because during translocation ATP is hydrolyzed to ADP and phosphate.

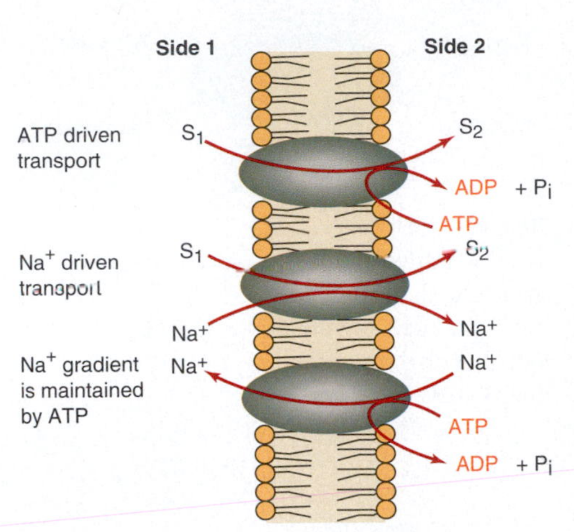

FIGURE 12.43
Involvement of metabolic energy (ATP) in active mediated transport systems.
Chemical energy released on hydrolysis of ATP to ADP and inorganic phosphate is used to drive active transport of various substances, including Na^+. The transmembrane concentration gradient of Na^+ is also used for active transport of substances.

Primary active transporters are classified as **P type transporters** if the protein is phosphorylated and dephosphorylated during the transport activity. V, for **vacuole,** type are responsible for acidification (proton pumps) of the interior of lysosomes, endosomes, Golgi vesicles, and secretory vesicles. **F type transporters** are involved in ATP synthesis and are present in mitochondria and chloroplasts (see p. 579).

Secondary active transporters use the transmembrane electrochemical gradient of Na^+ or H^+ to drive transport of sugars and amino acids. Maintenance of a Na^+ gradient across cell membranes is achieved by expenditure of ATP (Figure 12.43) by the activity of a primary active transporter. Inhibition of ATP synthesis leads to dissipation of the Na^+ gradient, which in turn causes a cessation of secondary active transport activity.

Translocation of Na^+ and K^+ Is by Primary Active Transport

All mammalian cells contain a Na^+–K^+ transporter, of type P, which utilizes ATP to drive translocation of Na^+ and K^+. Information about this transporter developed along two paths: (1) from studies of a membrane enzyme that required Na^+ and K^+ to catalyze ATP hydrolysis; the activity was termed Na^+/K^+–ATPase; and (2) from measurements of Na^+ and K^+ movements across intact plasma membranes by a protein referred to as the Na^+/K^+–pump or just Na^+ pump. The two activities are catalyzed by the same protein, the **Na^+/K^+-exchanging ATPase;** it is frequently referred to just as **Na^+/K^+– ATPase** or the trivial name, **Na^+ pump.**

All Plasma Membranes Contain a Na^+/K^+-Exchanging ATPase

All plasma membranes catalyze the reaction

$$ATP \xrightarrow[Mg^{2+}]{Na^+ + K^+} ADP + P_i$$

by an ATPase with an absolute requirement for both Na^+ and K^+ ions, and for Mg^{2+}, a cofactor for ATP-requiring reactions. ATPase activities in plasma membranes correlate with the Na^+/K^+-transport activity. Excitable tissues, such as muscle and nerve, and cells actively involved in movement of Na^+ ion, such as those in the salivary gland and kidney cortex, have high activities of Na^+/K^+-exchanging ATPase. The transporter is an integral membrane protein; it is an oligomer with two α subunits of about 110 kDa each and two β subunits of about 55 kDa each. There are several isoforms of both subunits. The smallest subunits are glycoproteins. Figure 12.44 is a schematic diagram of the Na^+/K^+-exchanging ATPase. The activity has a requirement for phospholipids. During transport the larger subunit is phosphorylated and dephosphorylated on a specific aspartic acid residue to form a β-aspartyl phosphate. Phosphorylation of the protein requires Na^+ binding and

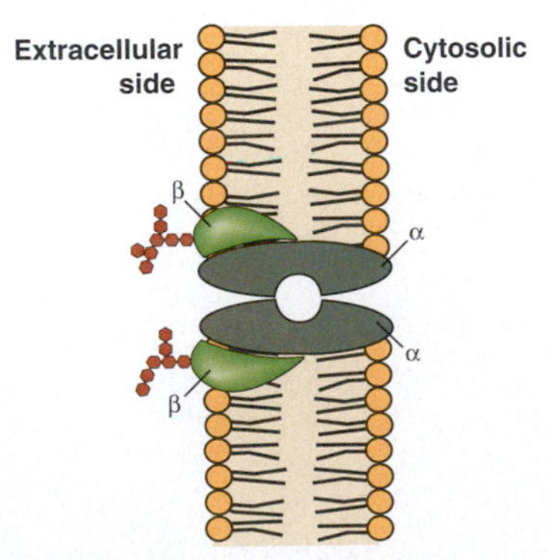

FIGURE 12.44
Schematic drawing of the Na^+/K^+-exchanging ATPase of plasma membranes.

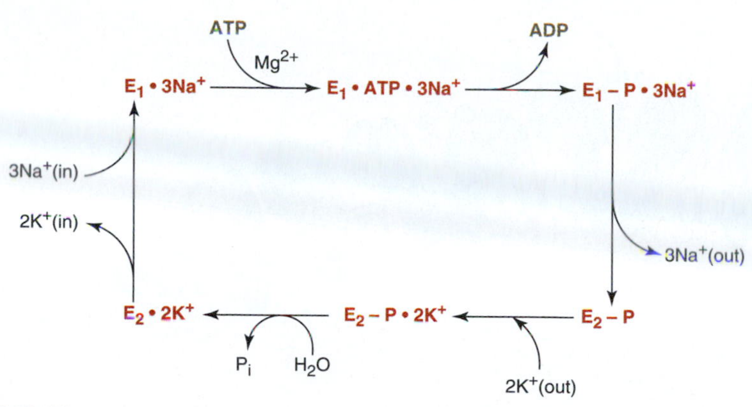

FIGURE 12.45
Proposed sequence of reactions and intermediates in hydrolysis of ATP by the Na^+/K^+-exchanging ATPase.
E_1 and E_2 are different conformations of the enzyme. Phosphorylation of the enzyme requires Na^+ and Mg^{2+} and dephosphorylation involves K^+.

Mg^{2+} but not K^+, whereas dephosphorylation requires K^+ binding but not Na^+ or Mg^{2+}. Reactions for the enzyme are presented in Figure 12.45. The isolated enzyme has an absolute requirement for Na^+, but K^+ can be replaced with $NH_4{}^+$ or Rb^+. The protein complex has two distinguishable conformations and thus is classified as an E_1–E_2 type transporter. Various isoforms of the transporter are regulated by different mechanisms including substrate concentrations, the cytoskeleton, circulating inhibitors, and a variety of hormones. Cardiotonic steroids such as **digitalis** are inhibitors of the Na^+/K^+–ATPase and increase the force of contraction of heart muscle by altering the excitability of the tissue. **Ouabain** (Figure 12.46) is one of the most active inhibitors of Na^+/K^+-exchanging ATPase; it binds on the smaller subunit of the enzyme complex on the external surface of the membrane and at some distance from the ATP-binding site of the larger monomer on the inside surface.

Movement of Na^+ and K^+ is an electrogenic antiport process, with three Na^+ ions moving out and two K^+ ions into the cell. This leads to an increase in external positive charge and is part of the mechanism for the maintenance of the transmembrane potential in cells. ATP hydrolysis by the transporter occurs only if Na^+ and K^+ are translocated. This transporter is responsible for maintaining the high K^+ and low Na^+ concentrations in a mammalian cell (see p. 15).

A model for movement of Na^+ and K^+ is presented in Figure 12.47. The protein goes through conformational changes during which the Na^+ and K^+ are moved short distances. During this conformational transition a change in the affinity of the binding site for the cations occurs such that there is a decrease in affinity constants, resulting in the release of the cation into a milieu where the concentration is higher than that from which it was transported.

As an indication of the importance of this enzyme, it has been estimated that Na^+/K^+-exchanging ATPase uses about 60–70% of the ATP synthesized by nerve and muscle cells and may utilize about 35% of ATP generated in a resting individual.

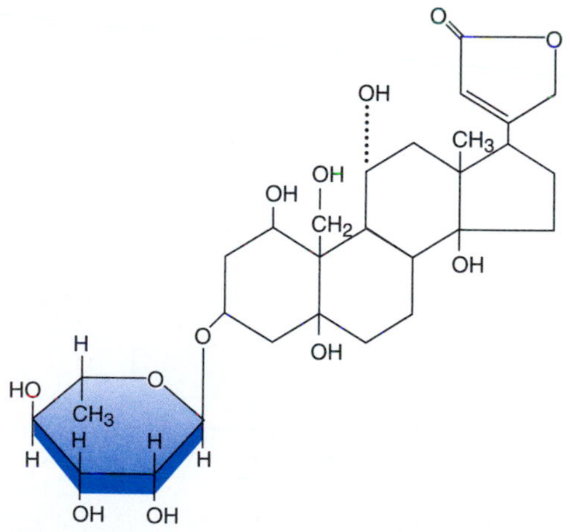

FIGURE 12.46
Structure of ouabain, a cardiotonic steroid, a potent inhibitor of the Na^+/K^+-exchanging ATPase.

Ca^{2+} Translocation Is Another Example of a Primary Active Transport System

Ca^{2+} is an important intracellular messenger that regulates cellular processes as varied as muscle contraction and carbohydrate metabolism. The signal initiated by some hormones, the primary messenger to direct cells to alter their function, is transmitted by changes in cytosolic Ca^{2+}; for this reason Ca^{2+} is referred to as a second messenger. Cytosolic Ca^{2+} concentration is in the range of 0.10 μM, over

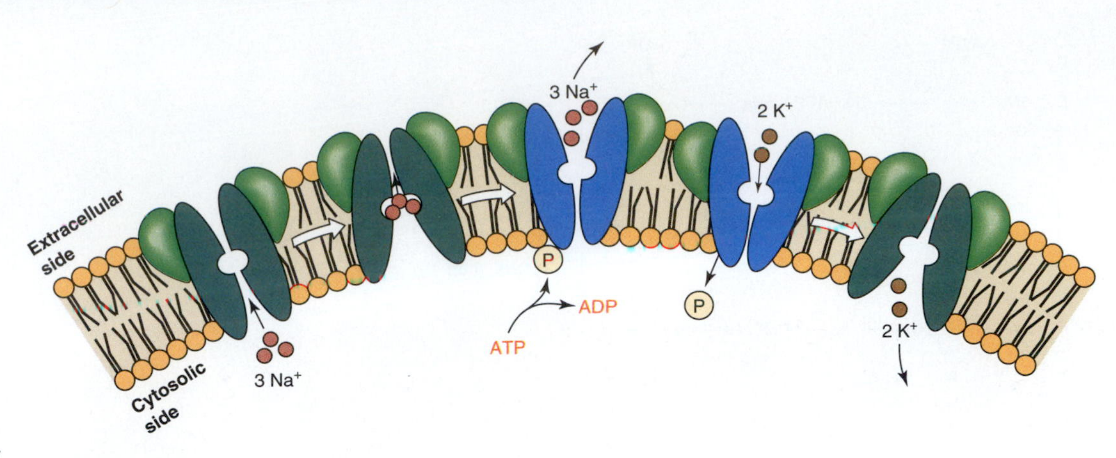

FIGURE 12.47
Hypothetical model for translocation of Na^+ and K^+ across plasma membrane by the Na^+/K^+-exchanging ATPase.
(1) Transporter in conformation 1 picks up Na^+. (2) Transporter in conformation 2 translocates and releases Na^+. (3) Transporter in conformation 2 picks up K^+. (4) Transporter in conformation 1 translocates and releases K^+.

10,000 times lower than that of extracellular Ca^{2+}. Intracellular Ca^{2+} concentrations can be increased rapidly (1) by transient opening of Ca^{2+} channels in the plasma membrane, permitting flow of Ca^{2+} down the large concentration gradient, or (2) by release from stores of Ca^{2+} in endoplasmic or sarcoplasmic reticulum. To reestablish low cytosolic levels, Ca^{2+} is actively transported out of cells across the plasma membrane or into the endoplasmic or sarcoplasmic reticulum by a **Ca^{2+}-transporting ATPase.** The various Ca^{2+} transporters are an E_1–E_2 type, having two conformations, and are structurally and mechanistically similar to Na^+/K^+-exchanging ATPase (Figure 12.48).

Ca^{2+}–ATPase of sarcoplasmic reticulum of muscle, which is involved in the contraction–relaxation cycles of muscle, represents 80% of the integral membrane protein and occupies one-third of the surface area (see p. 1025). The protein has ten membrane-spanning helices and is phosphorylated on an aspartyl residue during Ca^{2+} translocation. Two Ca^{2+} ions are translocated for each ATP hydrolyzed and it can move Ca^{2+} against a very large concentration gradient.

The Ca^{2+} transporter of plasma membranes has properties similar to the enzyme of sarcoplasmic reticulum. In eukaryotic cells, the transporter is regulated by cytosolic Ca^{2+} levels through a calcium-binding protein termed **calmodulin.** As cellular Ca^{2+} levels increase, Ca^{2+} is bound to calmodulin, which has a dissociation constant of ~1 μM. The Ca^{2+}–calmodulin complex binds to the Ca^{2+} transporter, leading to an increased rate in Ca^{2+} transport. This rate is increased by lowering the K_m for Ca^{2+} of the transporter from about 20 to 0.5 μM. Increased activity reduces cytosolic Ca^{2+} to its normal resting level (~0.10 μM) at which concentration the Ca^{2+}–calmodulin complex dissociates and the rate of the Ca^{2+} transporter returns to a lower value. Thus the Ca^{2+}–calmodulin complex exerts fine control on the Ca^{2+} transporter. Calmodulin is one of several Ca^{2+}-binding proteins, including **parvalbumin** and **troponin C,** all of which have very similar structures. The Ca^{2+}–calmodulin complex is also involved in control of other cellular processes, which are affected by Ca^{2+}. The protein (17 kDa) has the shape of a dumbbell with two globular ends connected by a seven-turn α helix; it contains four Ca^{2+}-binding sites, two of high affinity on one lobe and two of low affinity on the other. It is believed that binding of Ca^{2+} to the lower affinity binding sites causes a conformational change in the protein, revealing a hydrophobic area that can interact with a protein that it controls. Each Ca^{2+}-binding site consists of a helix–loop–helix structural motif (Figure 12.49) and Ca^{2+} is bound in the loop connecting the helices. A similar structure is found in other Ca^{2+}-binding proteins. The motif is referred to as the **EF hand,** based on studies with parvalbumin where the Ca^{2+} is bound between helices E and F of the protein.

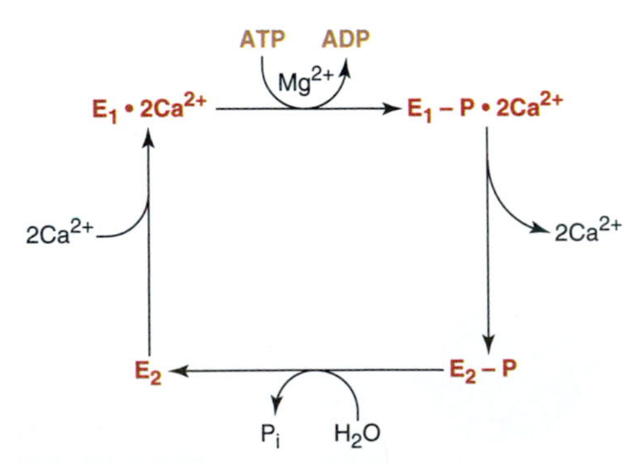

FIGURE 12.48
Proposed sequence of reactions and intermediates in translocating Ca^{2+} by Ca^{2+}-transporting ATPase.
E_1 and E_2 are different conformations of enzyme.

ATP-Binding Cassette Transport Proteins

Both eukaryotes and prokaryotes have a number of remarkable ATP-dependent transporters that belong to a superfamily of proteins termed **ATP-binding cassette (ABC) transport proteins.** The ATP-binding site of all these transporters share extensive sequence homology and a conserved domain organization: thus the name, ATP-binding cassette, to indicate that the genetic information for the common highly conserved binding site was inserted into the genes for the transmembrane segment of the transporters during evolution, as a cassette is inserted into a cassette player. The ABC domain is the driving force for transport. These transporters translocate a variety of substances, including ions, heavy metals, sugars, drugs, amino acids, peptides, and proteins. Over 30 ABC transporters, grouped into eight subfamilies, have been identified. Many have a broad range of substrate specificity. They all consist of a membrane-spanning domain that recognizes the substrate and an ATP-binding domain.

Depending on the subclass of transporter, translocation occurs by several different mechanisms, including channels, transporters, and receptors. Little is known about the mechanism for coupling the energy of ATP hydrolysis with movement of substrate. One subfamily of ABC transporters in humans is coded in DNA by the **multidrug resistance (MDR) family of genes.** The gene products are glycoproteins, termed **P-glycoproteins,** which transport a wide range of hydrophobic drugs and various analogs of naturally occurring lipids such as phosphatidylcholine. The **multidrug resistance associated proteins (MRPs)** have a tandem repeat of six transmembrane helices each with an ATP-binding domain. Overexpression of *MDR* genes leads to increased resistance to a variety of drugs in many tumor cells. With increased numbers of transporters, chemotherapeutic agents are rapidly transported out of the cells and do not attain effective intracellular concentrations. P-glycoproteins are responsible for extrusion from cells of a variety of xenobiotics. Other P-glycoprotein transporters are responsible for efflux of cholesterol from cells and for transport of bile acids across the canalicular plasma membrane of hepatocytes (see p. 1083). Another ABC transporter is the **sulfonylurea receptor,** which has an important role in the regulation of glucose-induced insulin secretion.

The **cystic fibrosis transmembrane conductance regulator (CFTR)** is a unique member of the ABC transporter family in that it is both a conductance regulator and a Cl^- channel. The channel mediates transepithelial salt and liquid movement in the apical plasma membrane of epithelial cells. A dysfunction of CFTR is the cause of the genetic disease cystic fibrosis. CFTR consists of five domains, two membrane-spanning domains, two nucleotide-binding domains, and a regulatory domain. Figure 12.50 is a representation of the CFTR. Phosphorylation of the

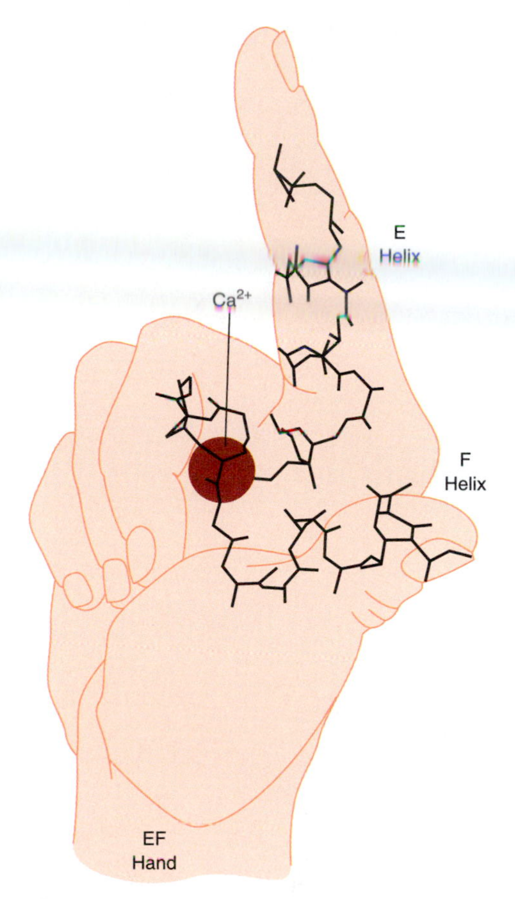

FIGURE 12.49
Binding site for Ca^{2+} in calmodulin.
Calmodulin contains four Ca^{2+}-binding sites, each with a helix–loop–helix motif. Ca^{2+} ion is bound in the loop that connects two helices. This motif occurs in various Ca^{2+}-binding proteins and is referred to as the EF hand.

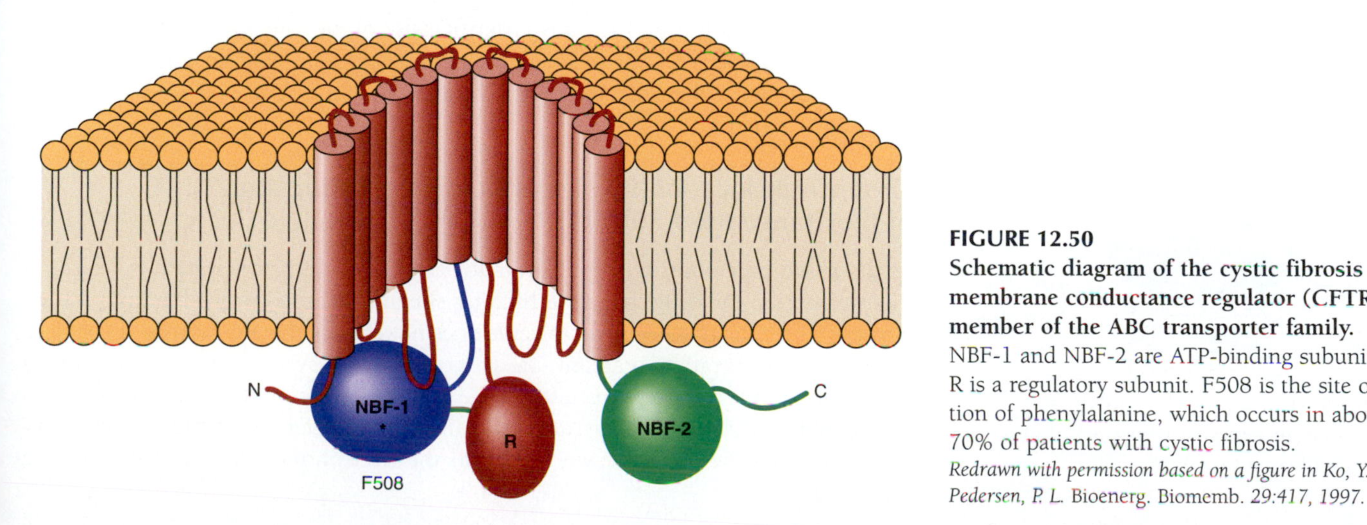

FIGURE 12.50
Schematic diagram of the cystic fibrosis transmembrane conductance regulator (CFTR), a member of the ABC transporter family.
NBF-1 and NBF-2 are ATP-binding subunits and R is a regulatory subunit. F508 is the site of deletion of phenylalanine, which occurs in about 70% of patients with cystic fibrosis.
Redrawn with permission based on a figure in Ko, Y. H. and Pedersen, P. L. Bioenerg. Biomemb. *29:417, 1997.*

regulatory domain regulates channel activity and ATP hydrolysis by the ABC domain controls the gating of the channel. A deletion of phenylalanine residue F508 in the ATP-binding subunit NBF-1 (also referred to as NBD-1) is the cause of about 70% of the cases of cystic fibrosis. This deletion prevents the proper folding of NBF-1. Other genetic diseases attributed to defects in specific ABC transport protein genes include adrenoleukodystrophy and Stargardt's disease, a hereditary degenerative disease of the macula lutea leading to rapid loss of visual acuity.

Na^+-Dependent Transport of Glucose and Amino Acids Are Secondary Active Transport Systems

The Na^+ electrochemical gradient across the plasma membrane (see p. 1095) is an energy source for active symport movement with Na^+ of sugars, amino acids, and Ca^{2+}. The general mechanism is presented in Figure 12.51 for the **sodium/glucose cotransporter (SGLT).** The diagram represents transport of D-glucose driven by the movement of Na^+. In the process two Na^+ are moving by passive facilitated transport down the electrochemical gradient and glucose can be carried along against its concentration gradient. In the transport the Na^+gradient is dissipated and unless it is continuously regenerated, transport of glucose will cease. The Na^+ gradient is maintained by the Na^+/K^+-exchanging ATPase described above and also represented in Figure 12.51. Thus metabolic energy in the form of ATP is indirectly involved in cotransport because it is required to maintain the Na^+ ion gradient. A decrease in cellular ATP concentrations or inhibition of the Na^+/K^+-exchanging ATPase will cause a decrease in the Na^+ gradient and prevent glucose uptake by this mechanism. Over 35 members of the *SGLT* gene family have been identified in animal cells, yeast, and bacteria. Some isoforms are expressed in proximal renal tubules and small intestinal epithelium (see p. 1103 for a discussion of SGLT in sugar absorption). All SGLTs have a common core structure of 13 transmembrane helices; different isoforms have additional transmembrane spans. Na^+/glucose cotransporters catalyze a uniport movement of Na^+ and water in the absence of glucose.

Amino acids are translocated by luminal epithelial cells of intestines by **Na^+/amino acid cotransporters** by a symport mechanism. At least seven different cotransporters have been identified for different classes of amino acids (see p. 1099 for details). The Na^+ gradient across the plasma membrane is also utilized to drive the transport of other ions, including a symport mechanism for uptake of Cl^- and an antiport mechanism for the excretion of Ca^{2+}. Symport movement of molecules utilizing the Na^+ gradient involves cooperative interaction of Na^+ ions and another molecule translocated on the protein. A conformational change of the protein occurs following association of ligands, which moves them the necessary distance in the transporter to bring them into contact with the cytosolic environment. Dissociation of Na^+ ion from the transporter because of the low intracellular Na^+ concentration leads to a return of the protein to its original conformation and a decrease in affinity and release of the other ligand.

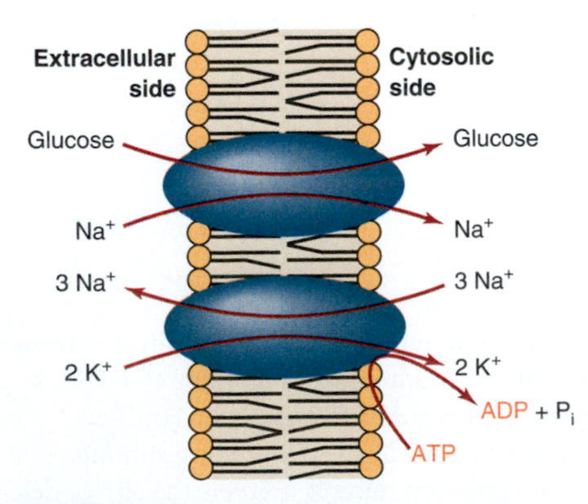

FIGURE 12.51
Na^+-dependent symport transport of glucose across the plasma membrane.

Group Translocation Involves Chemical Modification of the Transported Substrate

As discussed previously, a major hurdle for any active transport system is release of the transported molecule from the binding site after translocation. If affinity of the transporter for the translocated molecule does not change there cannot be movement against a concentration gradient. In active transport systems described above a change in affinity for the substrate by the transporter occurs by a conformational change of the protein. An alternate mechanism for release of a substrate is by chemically changing the substrate while still on the transporter. The process is termed

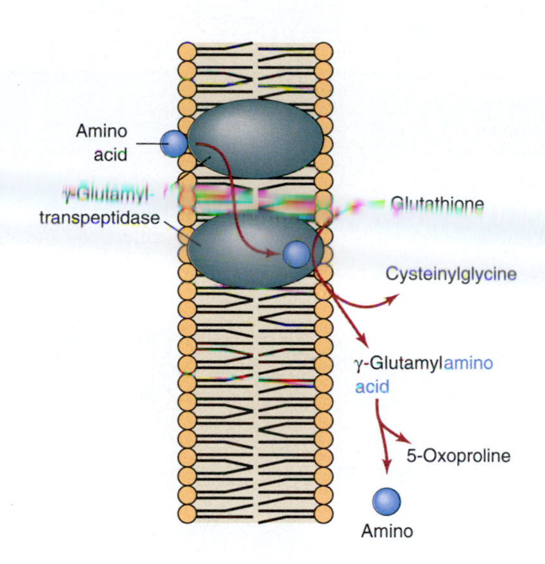

FIGURE 12.52
γ-Glutamyl cycle.
Represented are key reactions involved in group translocation of amino acids across liver cell plasma membranes. Continued uptake of amino acids requires constant resynthesis of glutathione via a series of ATP-requiring reactions described in Chapter 18, page 819.

group translocation. The modified substrate bound to the transporter has a lower affinity for the transporter and will be released. In the **γ-glutamyl cycle** for transport of amino acids across plasma membranes of some tissues the original substrate is chemically changed and released into the cell as a different molecule (Figure 12.52). **γ-Glutamyltranspeptidase,** a membrane-bound enzyme, catalyzes transfer of a γ-glutamyl residue from **glutathione** (Figure 12.53; see p. 1242) to an amino acid being translocated forming a dipeptide. The dipeptide is released from the translocator because of its low affinity and subsequently hydrolyzed, yielding the translocated amino acid and oxoproline.

All amino acids except proline can be transported by group translocation. The energy for transport comes from hydrolysis of a peptide bond in glutathione. For the system to continue, glutathione must be resynthesized, which requires the expenditure of three ATP molecules (see p. 1242). Group translocation is an expensive energetic mechanism for transport of amino acids. The pathway is present in many tissues but some doubt has been raised about its physiological significance in that individuals who genetically lack γ-glutamyltranspeptidase have no apparent difficulty in amino acid transport.

A group translocation mechanism for uptake of sugars is found in bacteria. This pathway involves phosphorylation of the sugar, using phosphoenolpyruvate as the phosphate donor. The mechanism is referred to as the **phosphoenolpyruvate-dependent phosphotransferase system (PTS).**

Summary of Transport Systems

The foregoing has presented the major mechanisms for movement of molecules across cellular membranes, particularly plasma membranes. Mammalian cells have a wide variety of different transport systems and in many instances different mechanisms to translocate the same substrate. Table 12.7 summarizes characteristics of some major transport systems found in mammalian cells. The same membranes can have a variety of different transporters. In many instances the rate of transport of a substance across a membrane is controlled by the number of transporters

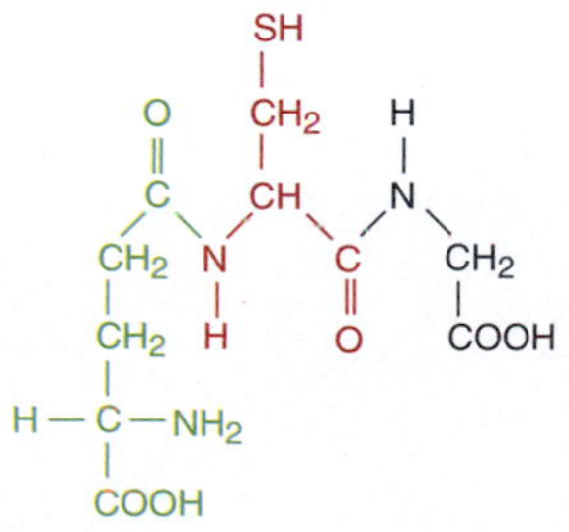

FIGURE 12.53
Glutathione (γ-glutamylcysteinylglycine).

TABLE 12.7 Major Transport Systems in Mammalian Cells[a]

Substance Transported	*Mechanism of Transport*	*Tissues*
Sugars		
Glucose	Passive	Widespread
	Active symport with Na^+	Small intestines and renal tubular cells
Fructose	Passive	Intestines and liver
Amino acids		
Amino acid-specific transporters	Active symport with Na^+	Intestines, kidney, and liver
All amino acids except proline	Active group translocation	Liver
Specific amino acids	Passive	Small intestine
Other organic molecules		
ATP–ADP	Antiport transport of nucleotides; can be active transport	Mitochondria
Ascorbic acid	Active symport with Na^+	Widespread
Biotin	Active symport with Na^+	Liver
Cholesterol	Active: ABC transporter	Widespread
Cholic acid deoxycholic acid, and taurocholic acid	Active symport with Na^+ and ABC transporter	Intestines and liver
Dicarboxylic acids	Active symport with Na^+	Kidney
Folate	Active	Widespread
Various drugs	Active: ABC transporter	Widespread
Lactate and monocarboxylic acids	Active symport with H^+	Widespread
Neurotransmitters (e.g., γ-amino butyric acid, norepinephrine, glutamate, dopamine)	Active symport with Na^+	Brain
Organic anions (e.g., malate, α-ketoglutarate, glutamate)	Antiport with counterorganic anion	Mitochondria
Peptides (2 to 4 amino acids)	Active symport with H^+	Intestines
Urea	Passive	Erythrocytes and kidney
Inorganic ions		
H^+	Active	Mitochondria
H^+	Active; vacuolar ATPase	Widespread; lysosomes, endosomes, and Golgi complex
Na^+	Passive	Distal renal tubular cells
Na^+, H^+	Active antiport	Proximal renal tubular cells and small intestines
Na^+, K^+	Active: ATP driven	Plasma membrane of all cells
Na^+, HPO_4^{2-}	Active cotransport	Kidney
Ca^{2+}	Active : ATP driven	Plasma membrane and endoplasmic (sarcoplasmic) reticulum
Ca^{2+}, Na^+	Active antiport	Widespread
H^+, K^+	Active antiport	Parietal cells of gastric mucosa secreting H^+
Cl^-, HCO_3^-	Passive antiport	Erythrocytes and kidneys

[a]The transport systems are only indicative of the variety of transporters known; others responsible for a variety of substances have been proposed. Most systems have been studied in only a few tissues and their localization may be more extensive than indicated. ABC, ATP-binding cassette.

in a membrane at a given time. Many genetic diseases are attributable to defects in transport systems (see Clin. Corr. 12.4). We have emphasized the role of Na^+ in transport, but it is important to note the important role of the proton gradient in translocation in mitochondria (see p. 577). Transport systems analogous to those in eukaryotes are found in prokaryotes.

CLINICAL CORRELATION 12.4

Diseases Due to Loss of Membrane Transport Systems

A number of pathological conditions are due to an alteration in the transport systems for specific cellular components. Some of these are discussed in the appropriate sections describing the metabolism of the intermediates. Individuals have been observed with a decrease in glucose uptake from the intestinal tract due to a loss of the specific sodium-coupled glucose–galactose transporter. Fructose malabsorption syndromes have been observed, which are due to an alteration in the activity of the transport system for fructose. In Hartnup's disease there is a decrease in the transport of neutral amino acids in the epithelial cells of the intestine and renal tubules.

In cystinuria, renal reabsorption of cystine and the basic amino acids lysine and arginine is abnormal, resulting in formation of cystine kidney stones. In hypophosphatemic, vitamin D resistant rickets, renal absorption of phosphate is abnormal. Little is known concerning possible changes of transport activities in tissues such as muscle, liver, and brain but it is likely that pathological states due to the loss of specific transport mechanisms in these tissues will be discovered.

Evans, L., Grasset, E., Heyman, M., et al. Congenital selective malabsorption of glucose and galactose. *J. Pediatr. Gastroenterol. Nutr.* 4:878, 1985; and Mueckler, M. Facilitative glucose transporters. *Eur. J. Biochem.* 219:713, 1994.

12.9 | IONOPHORES

An interesting class of antibiotics of bacterial origin facilitates the movement of monovalent and divalent inorganic ions across biological and synthetic membranes. These molecules, called **ionophores,** are relatively small molecular weight compounds (up to several thousand daltons). Ionophores are divided into two major groups. (1) **Mobile carriers** are ionophores that readily diffuse in a membrane and can carry an ion across a membrane. (2) **Channel formers** are ionophores that create a channel that transverses the membrane and through which ions can diffuse. Some major ionophores are listed in Table 12.8.

Each ionophore has a definite ion specificity. Valinomycin (Figure 12.54) has an affinity for K^+ 1000 times greater than for Na^+, and A23187 (Figure 12.55) an affinity for Ca^{2+} 10 times greater than for Mg^{2+}. Several of the diffusion type ionophores have a cyclic structure and the ion coordinates to oxygen atoms in the core of the structure. The periphery of the molecule consists of hydrophobic groups. When an ion is chelated by the ionophore, its water shell is stripped away and the ion is encompassed by a hydrophobic shell. Thus an ionophore–ion complex freely diffuses across the membrane. Since interaction of ion and ionophore is an equilibrium reaction, a steady-state concentration of the ion will develop on both sides of the membrane.

Valinomycin transports K^+ by an electrogenic uniport mechanism that creates an electrochemical gradient across a membrane as it carries a positively charged K^+. **Nigericin** is an electrically neutral antiporter; its carboxyl group when dissociated binds a positive ion, such as K^+, leading to a neutral complex that moves across a membrane. On diffusion back through the membrane it transports a proton; overall K^+ exchanges for H^+. These mechanisms are presented in Figure 12.56.

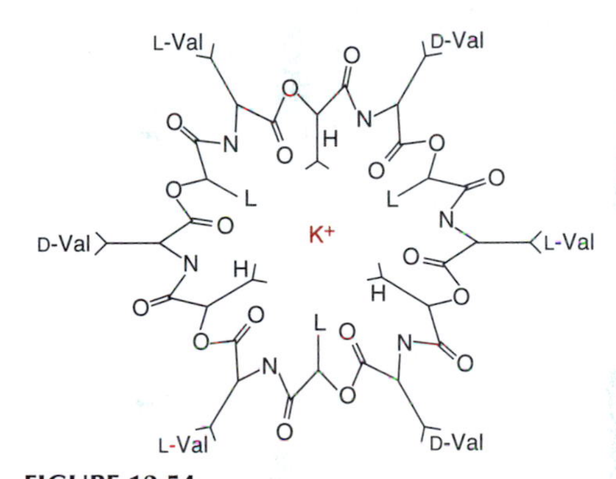

FIGURE 12.54
Structure of valinomycin–K^+ complex.
Abbreviations: D-Val, D-valine; L-Val, L-valine; L, L-lactate; and H, D-hydroxyisovalerate.

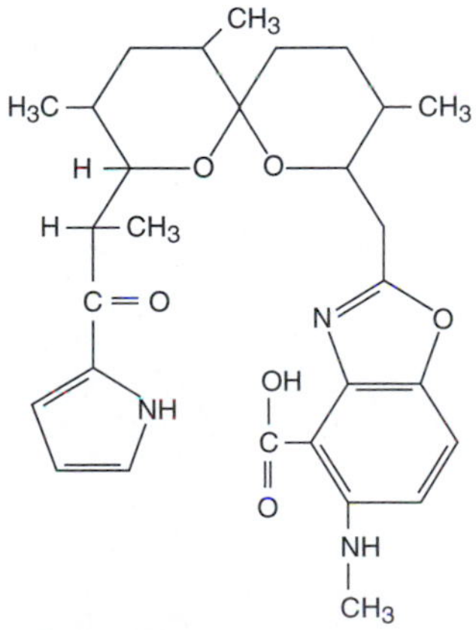

FIGURE 12.55
Structure of A23187, a Ca^{2+} ionophore.

TABLE 12.8 Major Ionophores

Compound	*Major Cations Transported*	*Action*
Valinomycin	K^+ or Rb^+	Uniport, electrogenic
Nonactin	NH_4^+, K^+	Uniport, electrogenic
A23187	Ca^{2+}/2 H^+	Antiport, electroneutral
Nigericin	K^+/H^+	Antiport, electroneutral
Monensin	Na^+/H^+	Antiport, electroneutral
Gramicidin	H^+, Na^+, K^+, Rb^+	Forms channels
Alamethicin	K^+, Rb^+	Forms channels

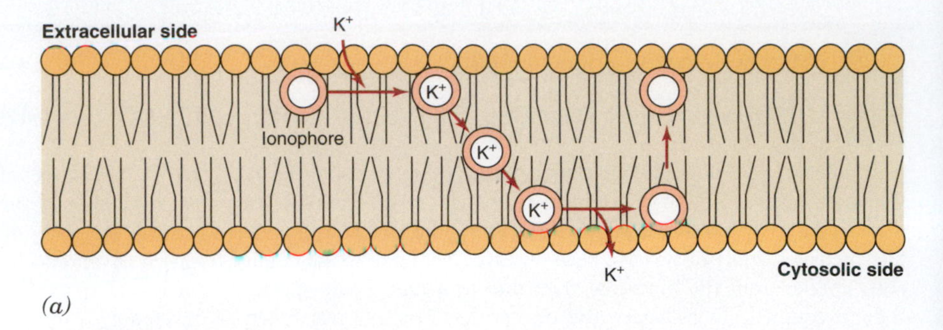

(a)

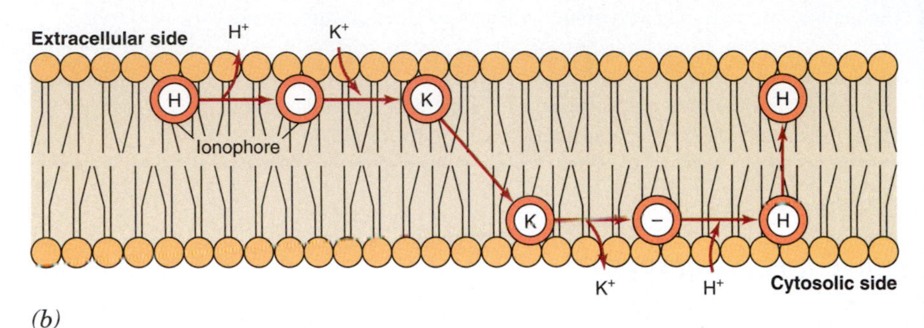

(b)

FIGURE 12.56
Proposed mechanism for ionophoretic activities of valinomycin and nigericin.
(*a*) Transport by valinomycin. (*b*) Transport by nigericin. The valinomycin–K^+ complex is positively charged and translocation of K^+ is electrogenic, leading to creation of a charge separation across membrane. Nigericin translocates K^+ in exchange for a H^+ across membrane and the mechanism is electrically neutral.
Diagram adopted from Pressman, B. C. Annu. Rev. Biochem. 45:501, 1976.

The polypeptide **gramicidin A** forms a channel when two molecules in a membrane hydrogen bond at their N-terminals to form a dimer that transverses the membrane (Figure 12.57). Ions can diffuse through the aqueous channel that is formed. Association and dissociation of monomers control the rate of ion flux. Polar peptide groups line the channel and hydrophobic groups are on the periphery of the channel interacting with the lipid membrane. This type of ionophore has a low ion selectivity.

Ionophores are valuable experimental tools in studies of ion translocation in biological membranes and for manipulation of the ionic compositions of cells.

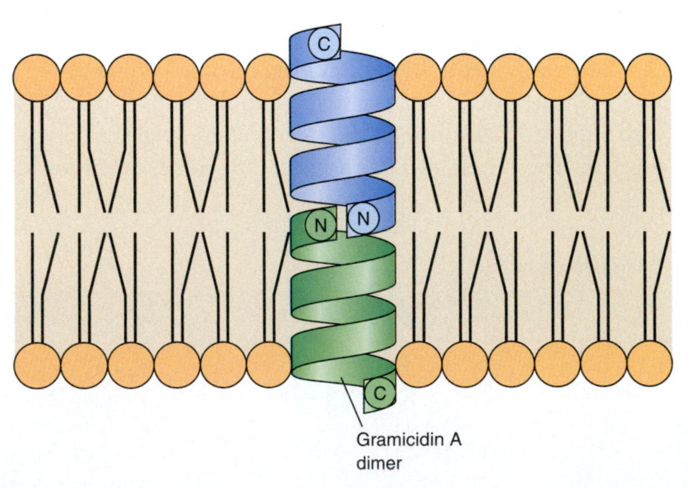

FIGURE 12.57
Action of gramicidin A.
Two molecules of gramicidin A form a dimer by hydrogen bonding at the amino ends of the peptides, creating a channel in a membrane.

BIBLIOGRAPHY

General

Dobler, M. *Ionophores and Their Structure*. New York: Wiley-Interscience, 1981.

Jones, M. N. *Micelles, Monolayers, and Biomembranes*. New York: Wiley-Liss, 1995.

Lipowsky, R. and Sackmann, E. *Structure and Dynamics of Membranes*. New York: Elsevier Science, 1995.

Papa, S. and Tager, J. M. *Biochemistry of Cell Membranes*. Basel: Birkhauser Verlag, 1995.

Vance, D. E. and Vance, J. E. (Eds.). *Biochemistry of Lipids, Lipoproteins, and Membranes*. Amsterdam: Elsevier, 1996.

Watts, A. *Protein–Lipid Interactions*. New York: Elsevier, 1993.

Yeagle, P. *The Membranes of Cells*. San Diego: Academic Press, 1993.

Membrane Structure

Daleke, L. D. and Lyles, J. V. Identification and purification of aminophospholipid flippase. *Biochim. Biophys. Acta* 1486:108, 2000.

Devaux, P. F. Static and dynamic lipid asymmetry in cell membranes. *Biochemistry* 30:1163, 1991.

Gregoriadis, G. Engineering liposomes for drug delivery: progress and problems. *Trends Biotechnol.* 13:527, 1995.

Low, M. and Saltiel, A. R. Structural and functional roles of glycosylphosphatidyl inositol in membranes. *Science* 239:268, 1988.

McMurchie, E. J. Dietary lipids and the regulation of membrane fluidity and function. In: *Physiological Regulation of Membrane Fluidity*. New York: Liss, 1988, p. 189.

Pastor, R. W. Molecular dynamics and Monte Carlo simulations of lipid bilayers. *Curr. Opin. Struct. Biol.* 4:486, 1994.

Somerharju, P., Virtanen, J. A., and Cheng, K. W. Lateral organization of membrane lipids. *Biochim. Biophys. A* 1440:32, 1999.

Tien, H. T. and Ottova-Leitmannova, A. (Eds.). *Membrane Biophysics: As Viewed from Experimental Bilayer Lipid Membranes (Planar Lipid Bilayers and Spherical Liposomes)*. Amsterdam: Elsevier, 2000.

Transport Processes

Agutter, P. S. Intracellular structure and nucleocytoplasmic transport. *Int. Rev. Cytol.* 162B:183, 1995.

Borgnia, B. M., Nelson, S., Engel, A., and Agre, P. Cellular and molecular biology of the aquaporin water channels. *Annu. Rev. Biochem.* 68:425, 1999.

Carafoli, E. and Brini, M. Calcium pumps: structural basis for and mechanism of calcium transmembrane transport. *Curr. Opin. Chem. Biol.* 4:152, 2000.

Catterall, W. A. Structure and function of voltage-gated ion channels. *Annu. Rev. Biochem.* 64:493, 1995.

Chin, D. and Means, A. R. Calmodulin: a prototypical calcium sensor. *Trends Cell Biol.* 10:322, 2000.

Davis, L. I. The nuclear pore complex. *Annu Rev. Biochem.* 64:865, 1995.

Gould, G. W. and Holman, G. D. The glucose transporter family: structure, function and tissue specific expression. *Biochem. J.* 295:329, 1993.

Hediger, M. A., Kanai, U., You, G., and Nussberger S. Mammalian ion-coupled solute transporters. *J. Physiol. (London)* 482:75, 1995.

Holland, I. B. and Blight, M. A. ABC-ATPases, adaptable energy generators fueling transmembrane movement of a variety of molecules in organisms from bacteria to humans. *J. Mol. Biol.* 293:381, 1999.

Kakuda, D. K. and MacLeod C. L. Na^+-independent transport (uniport) of amino acids and glucose in mammalian cells. *J. Exp. Biol.* 196:93, 1994.

Katz, A. M. Cardiac ion channels. *N. Engl. J. Med.* 328:1244, 1993.

Kurachi, Y., Jan, L. Y., and Lazdunski, M. (Eds.). *Potassium Ion Channels: Molecular Structure, Function, and Diseases*. San Diego: Academic Press, 1999.

Leo, T. W. and Clarke, D. M. Molecular dissection of the human multidrug resistance P-glycoprotein. *Biochem. Cell. Biol.* 77:11, 1999.

MacLennan, D. H., Rice, W. J., and Green, N. M. The mechanism of Ca^{2+} transport by sarco (endo)plasmic reticulum Ca^{2+}-ATPases. *J. Biol. Chem.* 272:28815, 1997.

Marsh, D. Peptide models for membrane channels. *Biochem. J.* 315:345, 1996.

McGiven, J. D. and Pastor-Anglada, M. Regulatory and molecular aspects of mammalian amino acid transport. *Biochem. J.* 299:321, 1994.

Pederson, P. and Carafoli, E. Ion motive ATPases. I. Ubiquity, properties, and significance to cell function, and II. Energy coupling and work output. *Trends Biochem. Sci.* 12:146, 186, 1987.

Peracchia, C. (Ed.). *Gap Junctions: Molecular Basis of Cell Communication in Health and Disease*. San Diego: Academic Press, 2000.

Poole, R. C. and Halestrap, A. P. Transport of lactate and other monocarboxylates across mammalian plasma membranes. *Am. J. Physiol.* 264:C761, 1993.

Rudy, B. and Seeburg, P. *Molecular and Functional Diversity of Ion Channels and Receptors*. Proceedings of the N.Y. Academy of Sciences, Vol. 868, 1999.

Scarborough, G. A. Structure and function of the P-type ATPases. *Curr. Opin. Cell Biol.* 11:517, 1999.

Schultz, S. G. (Ed.). *Molecular Biology of Membrane Transport Disorders*. New York: Plenum Press, 1996.

Van Winkle, L. J. *Biomembrane Transport*. San Diego: Academic Press, 1999.

QUESTIONS | C. N. ANGSTADT

Multiple Choice Questions

1. The glycerophospholipids and sphingolipids of membranes:
   A. all contain phosphorus.
   B. are all amphipathic.
   C. all have individual charges but are zwitterions (no net charge).
   D. are present in membranes in equal quantities.
   E. all contain one or more monosaccharides.

2. According to the fluid mosaic model of a membrane:
   A. proteins are always completely embedded in the lipid bilayer.
   B. transverse movement (flip-flop) of a protein in the membrane is thermodynamically favorable.
   C. the transmembrane domain has largely hydrophobic amino acids.
   D. proteins are distributed symmetrically in the membrane.
   E. peripheral proteins are attached to the membrane only by noncovalent forces.

3. Characteristics of a mediated transport system include:
   A. nonspecific binding of solute to transporter.
   B. release of the transporter from the membrane following transport.
   C. a rate of transport directly proportional to the concentration of solute.
   D. release of the solute only if the concentration on the new side is lower than that on the original side.
   E. a mechanism for translocating the solute from one side of the membrane to the other.

4. Which of the following require(s) a transporter that specifically binds a solute?
   A. active mediated transport
   B. gap junction
   C. membrane channel
   D. simple diffusion
   E. all of the above

5. Which of the following can transport a solute against its concentration gradient?
   A. active mediated transport
   B. passive mediated transport

C. both of the above systems
D. neither of the above systems

6. The transport system that maintains the Na^+ and K^+ gradients across the plasma membrane of cells:
   A. involves an enzyme that is an ATPase.
   B. is a symport system.
   C. moves Na^+ either into or out of the cell.
   D. is an electrically neutral system.
   E. in the membrane, hydrolyzes ATP independently of the movement of Na^+ and K^+.

7. A mediated transport system would be expected to:
   A. show a continuously increasing initial rate of transport with increasing substrate concentration.
   B. exhibit structural and/or stereospecificity for the substance transported.
   C. be slower than that of a simple diffusion system.
   D. establish a concentration gradient across the membrane.
   E. exist only in plasma membranes.

8. The translocation of Ca^{2+} across a membrane:
   A. is a passive mediated transport.
   B. is an example of a symport system.
   C. involves the phosphorylation of a serine residue by ATP.
   D. may be regulated by the binding of a Ca^{2+}–calmodulin complex to the transporter.
   E. maintains $[Ca^{2+}]$ very much higher in the cell than in extracellular fluid.

9. The group translocation type of transport system:
   A. does not require metabolic energy.
   B. involves the transport of two different solute molecules simultaneously.
   C. has been demonstrated for fatty acids.
   D. results in the alteration of the substrate molecule during the transport process.
   E. uses ATP to maintain a concentration gradient.

10. All of the following are correct about an ionophore EXCEPT it:
   A. requires the input of metabolic energy for mediated transport of an ion.
   B. may diffuse back and forth across a membrane.
   C. may form a channel across a membrane through which an ion may diffuse.
   D. may catalyze electrogenic mediated transport of an ion.
   E. will have specificity for the ion it moves.

Questions 11 and 12: Two of the major problems encountered with the oral or intravenous administration of drugs are the lack of tissue specificity in the action of the drug and the rapid metabolism, and therefore limited period of effectiveness, of some drugs. One attempt to circumvent these problems is the use of liposomes to encapsulate the drugs. Some drugs have a longer period of effectiveness when administered this way. Research is ongoing to prepare liposomes with a high degree of tissue specificity. Liposomes are also very useful as a research tool to study the properties of biological membranes since they have a similar structure and properties. Much of our understanding of biological membranes has been obtained using these artificial membranes.

11. Cell membranes typically:
   A. are about 90% phospholipid.
   B. have both integral and peripheral proteins.
   C. contain cholesteryl esters.
   D. contain free carbohydrate such as glucose.
   E. contain large amounts of triacylglycerols.

12. Which of the following statements concerning membranes is/are correct?
   A. The major lipid of inner mitochondrial membrane is sphingomyelin.
   B. Because of transverse (flip-flop) movement, the lipid compositions of the two layers of the membrane equilibrate.
   C. As demonstrated with liposomes, the membrane is most fluid at the surfaces.
   D. An increase in the cholesterol content of a membrane decreases the fluidity of the membrane.
   E. All of the above are correct.

Questions 13 and 14: Cystic fibrosis is a relatively common (1 in 2000 live births) genetic disease of Caucasians. Although it affects many organs, pulmonary obstruction is a major problem. CF patients have reduced Cl^- permeability and the disease can be diagnosed by elevated $[Cl^-]$ in sweat. Genetic mutations lead to defects in cystic fibrosis transmembrane conductance regulator (CFTR) protein, which is a cAMP-dependent Cl^- channel. CFTR shows structural homology with the superfamily of ATP-binding cassette (ABC) transporters.

13. Membrane channels:
   A. have a large aqueous area in the protein structure so are not very selective.
   B. commonly contain amphipathic α helices.
   C. are opened or closed only as a result of a change in the transmembrane potential.
   D. are the same as gap junctions.
   E. allow substrates to flow only from outside to inside the cell.

14. ATP-binding cassette (ABC) transporters:
   A. all have both a membrane-spanning domain that recognizes the substrate and an ATP-binding domain.
   B. all effect translocation by forming channels.
   C. are found only in eukaryotes.
   D. all have two functions—forming a channel and conductance regulation.
   E. are all P-glycoproteins.

Problems

15. Draw a curve illustrating the rate of movement of a solute through a membrane as a function of the concentration of the solute for (a) O_2 and (b) uptake of glucose into the erythrocyte.

16. Using the figure shown below, identify each of the types of transport systems illustrated by the letters A through E.

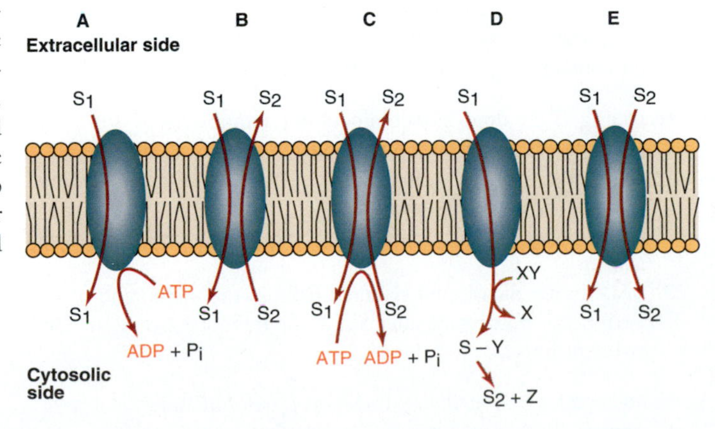

ANSWERS

1. **B** This property is essential to their function in membranes. A: Sphingolipids, except for sphingomyelins, do not. C: Cerebrosides are not charged; some glycerophospholipids and sphingomyelins are neutral; others have net negative charges. D: Glycerophospholipids are usually present in much larger quantities than sphingolipids. E: Only sphingolipids, except for sphingomyelin, contain carbohydrate.
2. **C** Hydrophobic domains will be in the interior; hydrophilic domains will be at either surface of the membrane. A: Proteins may also be on the surface. B: Transverse motion of proteins is very limited. D: Proteins are distributed asymmetrically. E: Glycans bind covalently to an amino acid as part of GPI anchors.
3. **E** This is essential to having the solute transported across the membrane. A: Specificity of binding is an integral part of the process. B: Recovery of the transporter to its original condition is one of the characteristics of mediated transport. C: Only at low concentrations of solute; transporters show saturation kinetics. D: Active transport, movement against a gradient, is also mediated transport.
4. **A** Specific binding by the transporter is a characteristic of mediated systems. B–D: These do not require a transporter.
5. **A** Transportation against a gradient requires the input of energy.
6. **A** The Na^+–K^+ transporter is the Na^+/K^+–ATPase. It is an antiport, vectorial (Na^+ out), electrogenic ($3Na^+$, $2K^+$) system. E: ATP hydrolysis is not useless.
7. **B** A, B: Mediated transport systems show saturation kinetics and substrate specificity. C: The purpose of the transporter is to aid the transport of water-soluble substances across the lipid membrane. D: This would be true only if the system were an active one. E: Mediated systems are also present in other membranes, for example, mitochondrial membrane.
8. **D** This occurs with the eukaryotic plasma membrane. A, B: Ca^{2+} translocation is an active uniport. C: Like Na^+/K^+–ATPase, phosphorylation occurs on an aspartyl residue. E: Extracellular Ca^{2+} is about 10,000 times higher than intracellular.
9. **D** In some eukaryotic cells, amino acids are transported by group translocation in which they are converted to a γ-glutamyl amino acid during transport. A, E: It is an active system with the ATP used to resynthesize the intermediate, glutathione. B, C: The system transports a single amino acid at a time.
10. **A** Ionophores transport by passive mediated mechanisms. B, C: These are the two major types of ionophores. D: Valinomycin transports K^+ by a uniport mechanism. There are also antiport systems that are electroneutral. E: For example, valinomycin has an affinity for K^+ 1000 times greater than for Na^+.
11. **B** Some proteins are embedded in the membrane while others are at the surface. A: This is more than the total lipid. C: Cholesterol in membranes is unesterified. D: All carbohydrate in membranes is in the form of glycoproteins and glycolipids. E: This is a minor component, if present.
12. **D** Cholesterol is a rigid structure. Its orientation, with the hydroxyl group near the surface and the rings projecting into the hydrophobic core, restricts the movement of the acyl chains of the lipids. A: Phosphatidylcholine is the major lipid of most membranes. B: Moving a charged head group through the lipophilic core of the membrane is thermodynamically unfavorable. The two layers of the membrane do not equilibrate. C: The area of greatest fluidity is the center of the membrane where the methyl ends of the acyl chains are located.
13. **B** The helices typically form the channel with hydrophilic side chains lining the channel and hydrophobic ones facing the lipid core of the membrane. A: This describes a pore; channels are quite specific. C: Voltage-gated channels, like that for Na^+, are controlled this way but others, like the nicotinic acetylcholine channel, are chemically regulated. D: Clusters of membrane channels work together to form a gap junction. E: Substances may move in either direction as dictated by the concentration gradient.
14. **A** These are the common features. B: They operate by a variety of mechanisms, including transporters and receptors. C: Both prokaryotes and eukaryotes have them. D: CFTR is unusual in having both of these functions. E: Some, such as those coded for by the *MDR* gene family or the transporter for cholesterol out of cells, are P-glycoproteins, but others are not.
15.

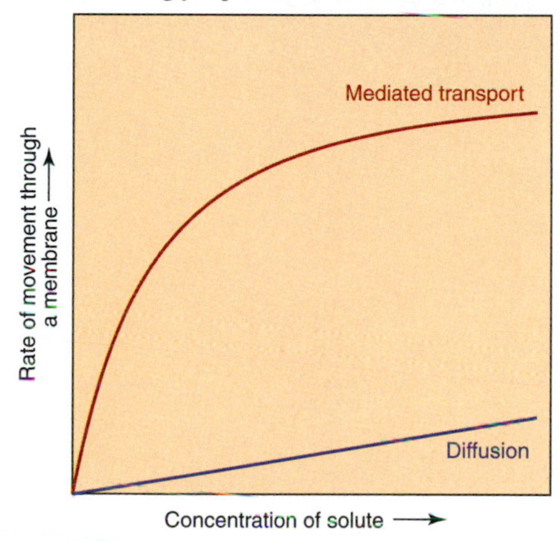

(a) O_2 is a gas and crosses membranes by simple diffusion and, thus, rate increases with concentration as illustrated by the lower curve on the above figure.

(b) Uptake of glucose by erythrocytes is a passive mediated transport so should show saturation kinetics as illustrated by the upper curve in the figure.

16. All of the systems illustrated represent mediated transport. A: An active uniport. B: A passive antiport; for example, Cl^-–HCO_3^-. C: An active antiport; for example, Na^+/K^+–ATPase. D: A group translocation representing a change in S_1 during transport. E: A symport system; in this case, S_1 could be glucose and S_2 could be Na^+.

PART IV

METABOLIC PATHWAYS AND THEIR CONTROL

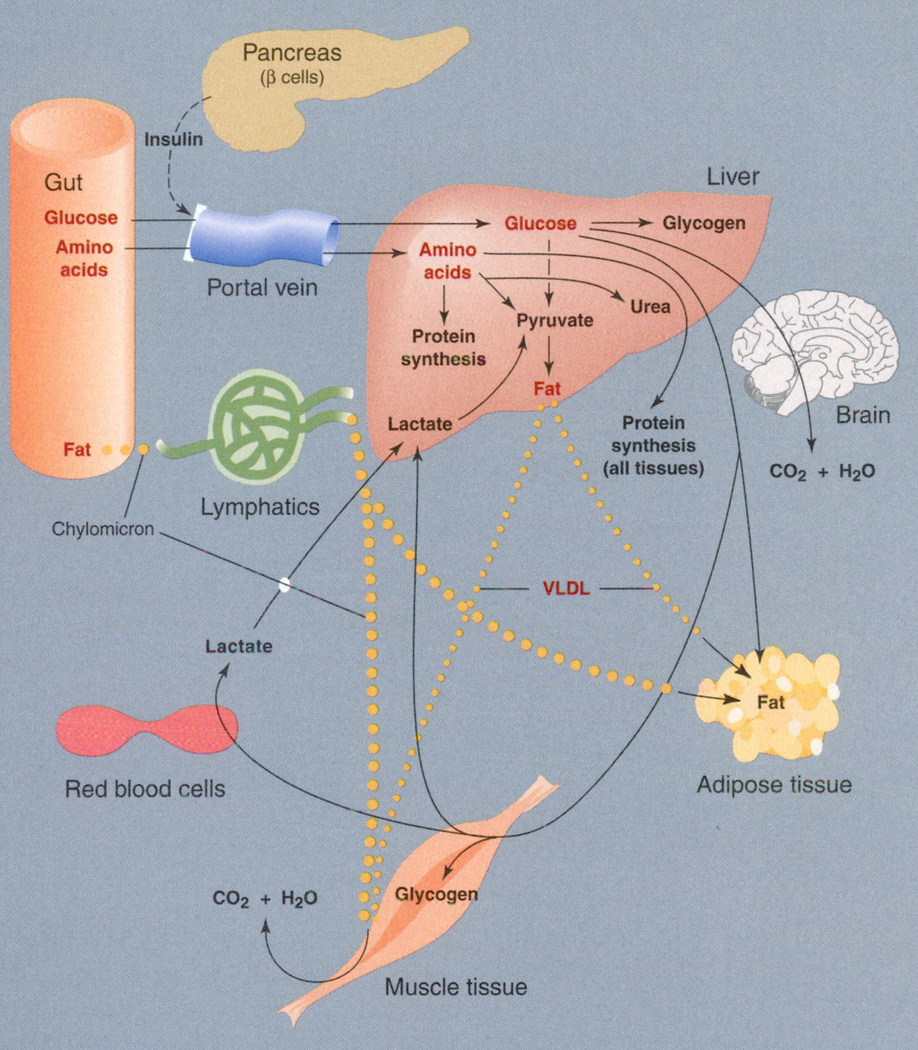

Cells transform chemical energy in food and light energy in the case of plants into a form that can be utilized for their various functions. They metabolize carbohydrates, lipids, amino acids, and nucleotides, and synthesize other organic molecules for a variety of purposes. These metabolic pathways involve numerous chemical interconversions catalyzed by enzymes which work together in a coordinated and controlled manner. In higher organisms metabolic processes can involve cells in different tissues. This is represented in the figure above; details of these interrelationships are presented in Figure 20.2. A discussion of energy transduction begins this Part, followed by descriptions of the major metabolic pathways. The concluding chapter presents the integration of these pathways under various physiological and pathophysiological conditions.

13

BIOENERGETICS AND OXIDATIVE METABOLISM

Diana S. Beattie

13.1 | ENERGY-PRODUCING AND ENERGY-UTILIZING SYSTEMS

Living cells are composed of a complex, intricately regulated system of energy-producing and energy-utilizing chemical reactions called metabolism. Metabolism consists of two contrasting processes, **catabolism** and **anabolism**, which represent the sum of chemical changes that convert foodstuffs into usable forms of energy and into complex biological molecules. Catabolism involves degradation of ingested foodstuffs or stored fuels such as carbohydrate, lipid, and protein into either usable or storable forms of energy. These reactions generally result in conversion of large complex molecules to smaller molecules (ultimately CO_2 and H_2O), and in mammals often require consumption of O_2. Energy-utilizing reactions perform various necessary and, in many instances, tissue-specific, cellular functions, for example, nerve impulse conduction, muscle contraction, growth, and cell division. Catabolic reactions are generally exergonic with the released energy generally trapped in the formation of ATP. The oxidative reactions of catabolism also result in transfer of reducing equivalents to the coenzymes NAD^+ and $NADP^+$ to form NADH and NADPH and a proton (H^+). Anabolic pathways are involved in biosynthesis of large, complex molecules from smaller precursors and require expenditure of energy either in the form of ATP or using reducing equivalents stored in NADPH (Figure 13.1).

ATP Links Energy-Producing and Energy-Utilizing Systems

The relationship between energy-producing and energy-utilizing functions of cells is illustrated in Figure 13.1. Energy may be derived from oxidation of metabolic

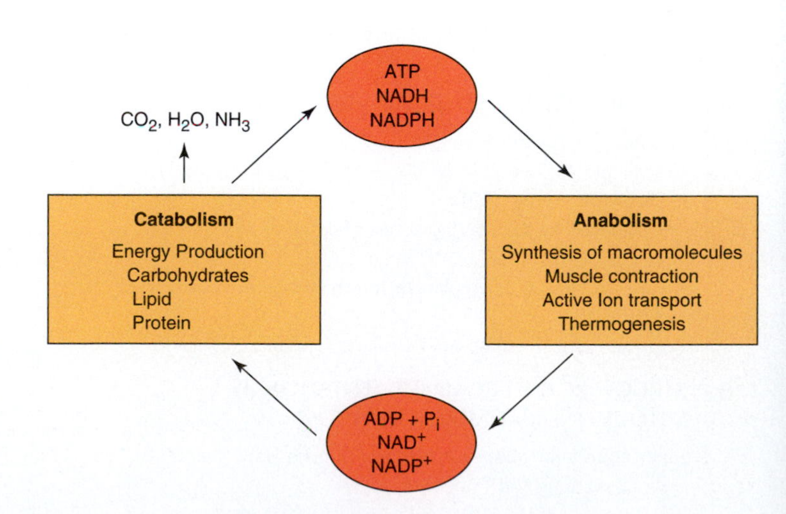

FIGURE 13.1
Energy relationships between energy production (catabolism) and energy utilization (anabolism).
Oxidative breakdown of foodstuffs is an exergonic process releasing free energy and reducing power that are trapped as ATP and NADPH, respectively. Anabolic processes are endergonic and use chemical energy stored as ATP and NADPH.

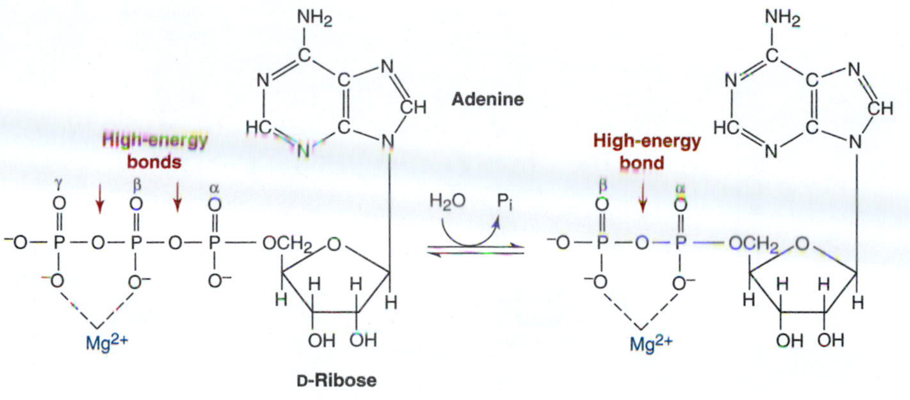

FIGURE 13.2
Structures of ATP and ADP complexed with Mg^{2+}.
High-energy bonds are highlighted.

fuels presented to the organism usually in the form of carbohydrate, lipid, and protein. The proportion of each fuel utilized as an energy source depends on the tissue and the dietary and hormonal state of the organism. For example, mature erythrocytes and adult brain in the fed state use only carbohydrate as a source of energy, whereas the liver of a diabetic or fasted individual metabolizes primarily lipid to meet the energy demands. Energy may be consumed during performance of various energy-linked (work) functions, some of which are indicated in Figure 13.1. Note that liver and pancreas are primarily involved in biosynthetic and secretory work functions, whereas the primary function of cardiac and skeletal muscle involves converting metabolic energy into mechanical energy during muscle contraction.

The essential link between energy-producing and energy-utilizing pathways is the nucleoside triphosphate, **adenosine 5′-triphosphate (ATP)** (Figure 13.2). ATP is a purine (adenine) nucleotide in which adenine is attached in a glycosidic linkage to D-ribose. Three phosphoryl groups are esterified to the 5′ position of the ribose moiety in **phosphoanhydride** bonds. The two terminal phosphoryl groups (i.e., β and γ) are designated as energy-rich or **high-energy bonds.** Synthesizing ATP as a result of a catabolic process or consuming ATP in an energy-linked cellular function involves formation and either hydrolysis or transfer of the terminal phosphate group of ATP. The physiological form of this nucleotide is chelated with a divalent metal cation such as magnesium.

NAD^+ and NADPH in Catabolism and Anabolism

Many catabolic processes are oxidative in nature, because the carbons in the substrates, carbohydrates, fats, and proteins, are in a partially or highly reduced state (Figure 13.3). **Reducing equivalents** are released from substrates in the form of hydride ions (a proton containing two electrons $H:^-$), which are transferred from the substrates to **nicotinamide adenine dinucleotide (NAD^+)** by enzymes called dehydrogenases with formation of NADH (Figure 13.4). The NADH is then transported into mitochondria where reducing equivalents are transferred in a series of reactions by the electron transport chain to O_2 as the ultimate electron acceptor. The oxidative reactions in mitochondria are exergonic, producing energy used for synthesis of ATP in a process called **oxidative phosphorylation.** The reductive and oxidative reactions in the NAD^+–NADH cycle play a central role in conversion of the chemical energy of carbon compounds in foodstuffs into chemical energy of phosphoric anhydride bonds of ATP. This process, called energy transduction, will be discussed in detail later in this chapter.

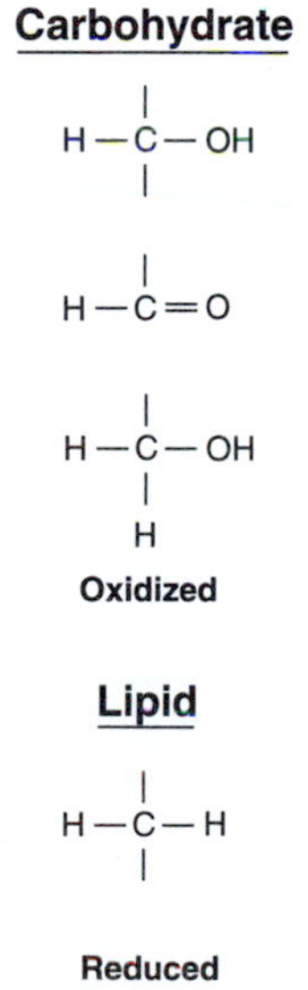

FIGURE 13.3
Oxidation states of typical carbon atoms in carbohydrates and lipids.

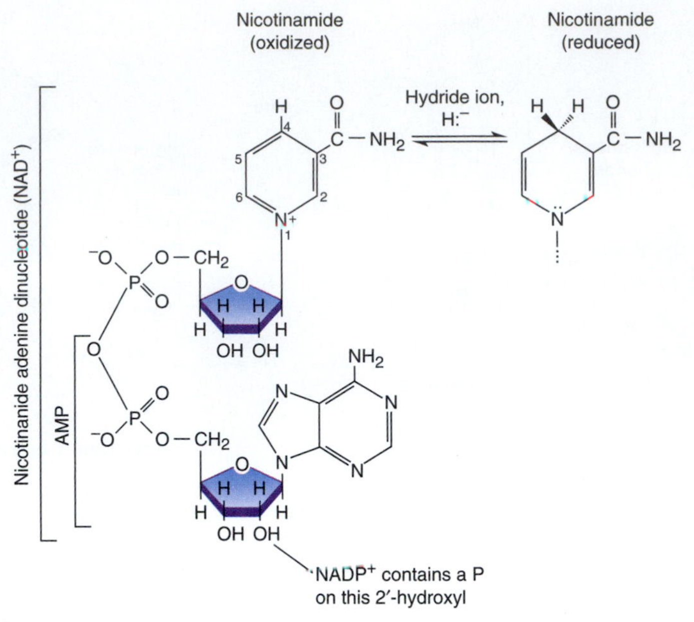

FIGURE 13.4
Structure of nicotinamide adenine dinucleotide (NAD).
Hydride ion (H:$^-$, a proton with two electrons) transfer to NAD$^+$ forms NADH.

Anabolism in contrast to catabolism is largely a reductive process as small more highly oxidized molecules are converted into large complex molecules (Figure 13.1). The reducing power used in biosynthesis of highly reduced compounds, such as fatty acids, is provided by NADPH, a 2′ phosphorylated NAD (see p. 426), produced by transfer of electrons to NADP$^+$ (Figure 13.5).

13.2 | THERMODYNAMIC RELATIONSHIPS AND ENERGY-RICH COMPONENTS

Living cells can interconvert different forms of energy and also may exchange energy with their surroundings. Hence it is necessary to review the principles of **thermodynamics,** which govern reactions of this type. Knowledge of these principles will facilitate a perception of how energy-producing and energy-utilizing metabolic reactions are permitted to occur within the same cell and how an organism is able to accomplish various work functions.

The **first law of thermodynamics** states that energy can neither be created nor destroyed. This law of energy conservation stipulates that, although energy may be converted from one form to another, the total energy in a system must remain constant. For example, chemical energy available in a metabolic fuel such as glucose is converted in the process of glycolysis to another form of chemical energy, ATP. In skeletal muscle chemical energy involved in energy-rich phosphate bonds of ATP may be converted to mechanical energy during the process of muscle contraction. The energy involved in an osmotic electropotential gradient of protons across the mitochondrial membrane may be converted to chemical energy using the proton gradient to drive ATP synthesis.

To discuss the **second law of thermodynamics** the term **entropy** must be defined. Entropy, designated by S, is a measure or indicator of the degree of disorder or randomness in a system. Entropy is viewed as the energy in a system that is unavailable to perform useful work. All processes, whether chemical or biological, tend to progress toward a situation of maximum entropy. Hence living systems that

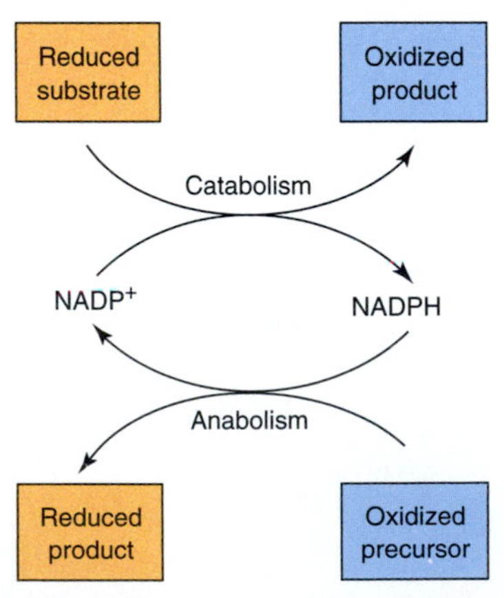

FIGURE 13.5
Transfer of reducing equivalents during catabolism and anabolism using NADPH and NADH.

are highly ordered are never at equilibrium with their surrounding as equilibrium in a system will result when the randomness or disorder (entropy) is at a maximum. In biological systems, however, it is nearly impossible to quantitate entropy changes as such systems are rarely at equilibrium. For simplicity and because of its inherent utility in these considerations, a quantity termed **free energy** is employed.

Free Energy Is the Energy Available for Useful Work

The free energy (denoted by G or Gibbs free energy) of a system is that portion of the total energy in a system that is available for useful work. It is further defined by

$$\Delta G = \Delta H - T\,\Delta S$$

In this expression for a system proceeding toward equilibrium at a constant temperature and pressure, ΔG is the change in free energy, ΔH is the change in enthalpy or the heat content, T is the absolute temperature, and ΔS is the change in entropy of the system. If ΔG is equal to zero, the process is at equilibrium and there is no net flow in either direction of the reaction. Furthermore, any process that exhibits a negative ΔG (free-energy change) will proceed spontaneously toward equilibrium in the direction written, in part, due to an increase in entropy or disorder in the system. Such a reaction releases energy and is called an **exergonic reaction.** A process that exhibits a positive ΔG will proceed spontaneously in the reverse direction as written. Energy from some other source must be applied to this process to allow it to proceed toward equilibrium. This process is termed an **endergonic reaction.** It should be noted that the sign and value of ΔG do not predict how fast the reaction will go. The rate of a given reaction depends on the free energy of activation but not on the magnitude of ΔG. In addition, the change in free energy in a biochemical process is the same regardless of the pathway or mechanism employed to attain the final state. The change in free energy for a chemical reaction is related to the equilibrium constant of that reaction. For example, an enzymatic reaction may be described as

$$\mathrm{A} + \mathrm{B} \rightleftharpoons \mathrm{C} + \mathrm{D}$$

and an expression for the equilibrium constant may be written as

$$K_{eq} = \frac{[\mathrm{C}][\mathrm{D}]}{[\mathrm{A}][\mathrm{B}]}$$

Under standard conditions, when reactants and products are initially present at 1 M concentrations, at 1 atm pressure and 1 M $[\mathrm{H}^+]$ or pH 0, the standard free-energy change is defined as $\Delta G°$. Biochemists have modified this expression and define the standard free energy at pH 7.0 ($[\mathrm{H}^+] = 10^{-7}$ M), where biological reactions generally occur. Under these conditions the change in free energy is expressed as $\Delta G°'$ and K'_{eq}. Since the value of ΔG is zero at equilibrium, the following relationship is defined:

$$\Delta G°' = -RT \ln K'_{eq}$$

R is the gas constant, which is 1.987 cal mol^{-1} K^{-1} or 8.134 J mol^{-1} K^{-1}, depending on whether the resultant free-energy change is expressed in calories (cal) or joules (J) per mole; and T is the absolute temperature in kelvin units (K).

Hence, if the **equilibrium constant** for a reaction is determined, the standard free energy change ($\Delta G°'$) for that reaction also can be calculated. The relationship between $\Delta G°'$ and K'_{eq} is illustrated in Table 13.1. When the equilibrium constant of a reaction is less than unity, the reaction is endergonic, and $\Delta G°'$ is positive. When the equilibrium constant is greater than unity, the reaction is exergonic, and $\Delta G°'$ is negative.

As discussed above, $\Delta G°'$of a reaction defines work available in a reaction when substrates and products are present at 1 M concentrations. In cells this situation

TABLE 13.1 Tabulation of Values of K_{eq} and $\Delta G°$

K_{eq}	$\Delta G°'$ (*kcal mol*$^{-1}$)	$\Delta G°'$ (*kJ mol*$^{-1}$)
10^{-4}	5.46	22.8
10^{-3}	4.09	17.1
10^{-2}	2.73	11.4
10^{-1}	1.36	5.7
1	0	0
10	−1.36	−5.7
10^2	−2.73	−11.4
10^3	−4.09	−17.1
10^4	−5.46	−22.8

does not occur, as compounds are rarely present at 1 M concentration. Hence an expression related to actual intracellular concentrations of substrates and products can provide insight into work available in a reaction. The expression for ΔG at any concentration of substrate or product includes the energy change for a 1 M concentration of substrate and product to reach equilibrium ($\Delta G°'$) and the energy change to reach a 1 M concentration of substrates and products:

$$\Delta G = \Delta G°' + RT \ln \left(\frac{[C][D]}{[A][B]} \right)$$

For example, in a muscle cell the concentration of ATP = 8.1 mM, [ADP] = 0.93 mM, and P_i = 8.1 mM. If $\Delta G°'$ for the reaction ATP + HOH $\rightleftharpoons$ ADP + P_i is 7.7 kcal mol^{-1} at 37°C, pH 7.4, then the overall ΔG for the reaction is

$$\Delta G = \Delta G°' + RT \ln \left(\frac{[\text{ADP}][\text{P}_i]}{[\text{ATP}]} \right)$$

$$\Delta G = \Delta G°' + RT \ln [0.93 \times 10^{-3}]$$

$$\Delta G = -7.7 \text{ kcal mol}^{-1} + (-4.2 \text{ kcal mol}^{-1}) = -12 \text{ kcal mol}^{-1}$$

These calculations demonstrate that considerably more free energy is available for work in a muscle cell than indicated by the value of $\Delta G°'$. Moreover, synthesis of ATP in muscle cells under these conditions, the reverse reaction, would require +12 kcal mol^{-1} of energy.

In energy-producing and energy-utilizing metabolic pathways in cellular systems, free-energy changes characteristic of individual enzymatic reactions in a metabolic pathway are additive; for example,

$$\text{A} \rightarrow \text{B} \rightarrow \text{C} \rightarrow \text{D}$$

$$\Delta G°'_{\text{A–D}} = \Delta G°'_{\text{A} \rightarrow \text{B}} + \Delta G°'_{\text{B} \rightarrow \text{C}} + \Delta G°'_{\text{C} \rightarrow \text{D}}$$

Although any given enzymatic reaction in a sequence may have a characteristic positive free-energy change, as long as the sum of all free-energy changes is negative, the pathway will proceed. Another way of expressing this principle is that enzymatic reactions with positive free-energy changes may be coupled to or driven by reactions with negative free-energy changes associated with them. In a metabolic pathway such as glycolysis, various individual reactions either have positive $\Delta G°'$ values or $\Delta G°'$ values close to zero. On the other hand, there are other reactions that have large and negative $\Delta G°'$ values, which drive the entire pathway. The crucial consideration is that the sum of $\Delta G°'$ values for individual reactions in a pathway must be negative in order for such a metabolic sequence to be thermodynamically feasible. Also, as for all chemical reactions, individual enzymatic reactions in a metabolic pathway or the pathway as a whole would be facilitated if the concentrations of the reactants (substrates) of the reaction exceed the concentrations of products of the reaction.

Caloric Value of Dietary Substances

During complete stepwise oxidation of glucose, a primary metabolic fuel in cells, a large quantity of energy is available. The free energy released during oxidation of glucose in a functioning cell is illustrated in the following equation:

$$C_6H_{12}O_6 + 6\ O_2 \rightarrow 6\ CO_2 + 6\ H_2O; \qquad \Delta G^{\circ\prime} = -686{,}000 \text{ cal mol}^{-1}$$

When this process occurs under aerobic conditions in most cells, it is possible to conserve less than one-half of this "available" energy as 38 molecules of ATP. The $\Delta G^{\circ\prime}$ and **caloric values** for oxidation of other metabolic fuels are listed in Table 13.2. Carbohydrates and proteins (amino acids) have a caloric value of 3–4 kcal g^{-1}, while lipid (i.e., palmitate, a long-chain fatty acid, or a triacylglycerol) exhibits a caloric value nearly three times greater. The reason that more energy is derived from lipid than from carbohydrate or protein relates to the average oxidation state of carbon atoms in these substances. Carbon atoms in carbohydrates are considerably more oxidized (or less reduced) than those in lipids (Figure 13.3). Hence during sequential breakdown of lipid nearly three times as many reducing equivalents (a reducing equivalent is defined as a proton plus an electron, i.e., $H^+ + e^-$) can be extracted than from carbohydrate.

Compounds Are Classified on the Basis of Energy Released on Hydrolysis of Specific Groups

The two terminal phosphoryl groups of ATP are referred to as **high-energy bonds,** indicating that free energy of hydrolysis of an energy-rich phosphoanhydride bond is much greater than would be obtained for a simple phosphate ester. High energy is not synonymous with stability of the chemical bond in question, nor does it refer to the energy required to break such bonds. The concept of high-energy compounds implies that products of hydrolytic cleavage of the energy-rich bond are in more stable forms than the original compound. As a rule, simple phosphate esters (low-energy compounds) exhibit negative $\Delta G^{\circ\prime}$ values of hydrolysis in the range 1–3 kcal mol^{-1}, whereas high-energy bonds have negative $\Delta G^{\circ\prime}$ values in the range 5–15 kcal mol^{-1}. Phosphate esters such as glucose 6-phosphate and glycerol 3-phosphate are examples of low-energy compounds. Table 13.3 lists various types of energy-rich compounds with approximate values for their $\Delta G^{\circ\prime}$ values of hydrolysis.

There are various reasons why certain compounds or bonding arrangements are energy rich. First, products of hydrolysis of an energy-rich bond may exist in more **resonance forms** than the precursor molecule. The more possible resonance forms in which a molecule can exist stabilize that molecule. The resonance forms for inorganic phosphate (P_i) can be written as indicated in Figure 13.6. Fewer resonance forms can be written for ATP or pyrophosphate (PP_i) than for phosphate (P_i).

Second, many high-energy bonding arrangements have groups of similar electrostatic charges located in close proximity to each other in such compounds. Because like charges repel one another, hydrolysis of energy-rich bonds alleviates this

TABLE 13.2 Free-Energy Changes and Caloric Values Associated with the Total Metabolism of Various Metabolic Fuels

Compound	*Molecular Weight*	*$\Delta G^{\circ\prime}$ (kcal mol^{-1})*	*Caloric Value (kcal g^{-1})*
Glucose	180	−686	3.81
Lactate	90	−326	3.62
Palmitate	256	−2380	9.30
Tripalmitin	809	−7510	9.30
Glycine	75	−234	3.12

TABLE 13.3 Examples of Energy-Rich Compounds

Type of Bond	*$\Delta G^{\circ\prime}$ of Hydrolysis (kcal mol^{-1})*	*$\Delta G^{\circ\prime}$ of Hydrolysis (kJ mol^{-1})*	*Example*
Phosphoric acid anhydrides	−7.3	−35.7	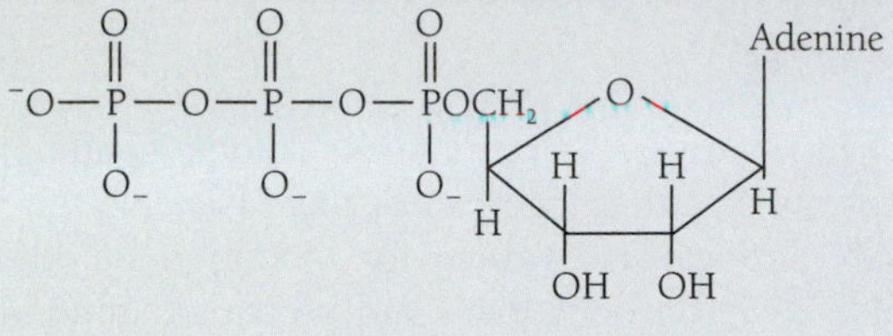 ATP
	−11.9	−50.4	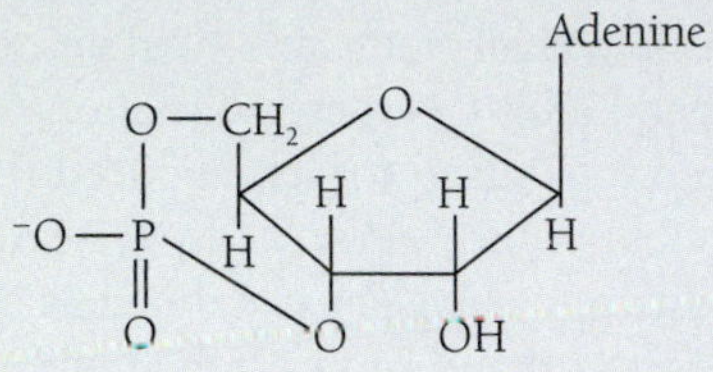 3′,5′ cyclic AMP
Phosphoric–carboxylic acid anhydrides	−10.1	−49.6	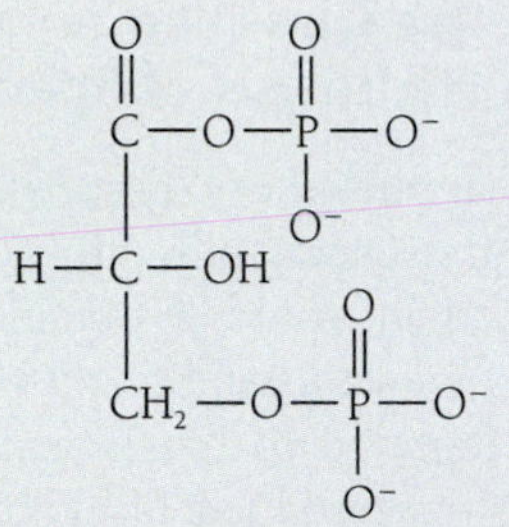 1,3-Bisphosphoglycerate
	−10.3	−43.3	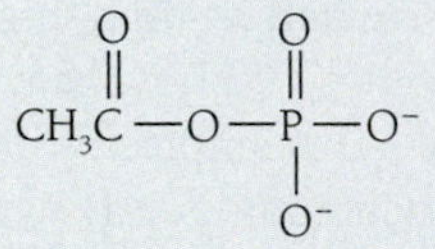 Acetyl phosphate
Phosphoguanidines	−10.3	−43.3	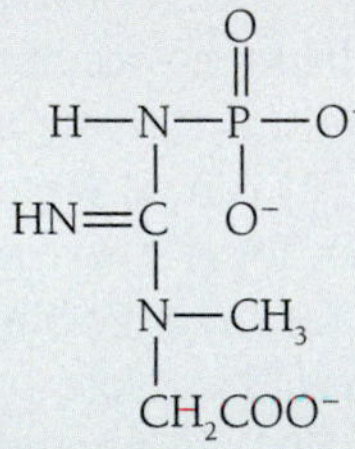 Creatine phosphate
Enol phosphates	−14.8	−62.2	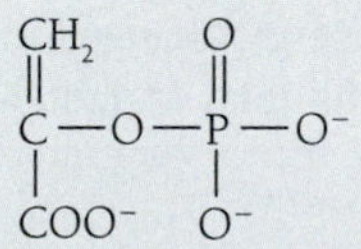 Phosphoenolpyruvate
Thiol esters	−7.7	−31.5	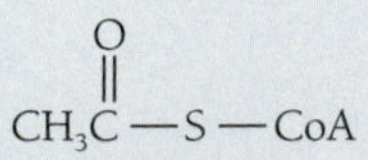 Acetyl CoA

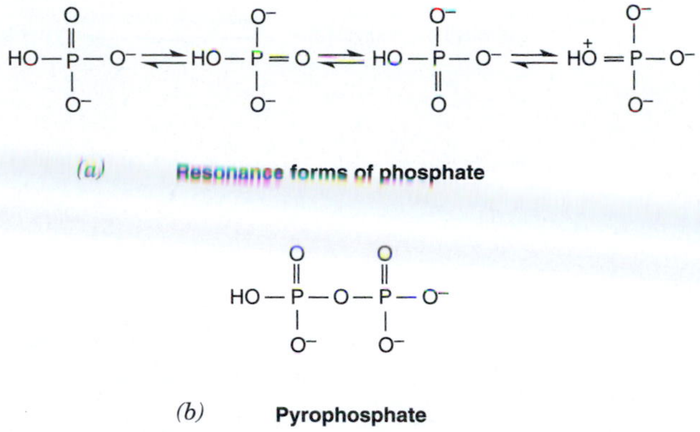

$$HO-\overset{O}{\underset{O^-}{\overset{\|}{\underset{|}{P}}}}-O^- \rightleftharpoons HO-\overset{O^-}{\underset{O^-}{\overset{|}{\underset{|}{P}}}}=O \rightleftharpoons HO-\overset{O^-}{\underset{O}{\overset{|}{\underset{\|}{P}}}}-O^- \rightleftharpoons H\overset{+}{O}=\overset{O^-}{\underset{O^-}{\overset{|}{\underset{|}{P}}}}-O^-$$

(*a*) Resonance forms of phosphate

$$HO-\overset{O}{\underset{O^-}{\overset{\|}{\underset{|}{P}}}}-O-\overset{O}{\underset{O^-}{\overset{\|}{\underset{|}{P}}}}-O^-$$

(*b*) Pyrophosphate

FIGURE 13.6
(*a*) Resonance forms of phosphate. (*b*) Structure of pyrophosphate.

situation and, again, lends stability to products of hydrolysis. Third, hydrolysis of certain energy-rich bonds results in formation of an unstable compound, which may isomerize spontaneously to form a more stable compound. Hydrolysis of phosphoe*nol*pyruvate is an example of this type of compound (Figure 13.7). The $\Delta G^{\circ\prime}$ for isomerization is considerable, and the final product, in this case pyruvate, is much more stable. Finally, if a product of hydrolysis of a high-energy bond is an undissociated acid, dissociation of the proton and its subsequent buffering may contribute to the overall $\Delta G^{\circ\prime}$ of the hydrolytic reaction. In general, any property or process that lends stability to products of hydrolysis tends to confer a high-energy character to that compound. The high-energy character of **3′,5′-cyclic adenosine monophosphate (cAMP)** has been attributed to the fact that the phosphoanhydride bonding character in this compound is strained as it bridges the 3′ and 5′ positions on ribose. The energy-rich character of thiol ester compounds such as **acetyl CoA** or succinyl CoA results from the relatively acidic character of the thiol group. Hence the thioester bond of acetyl CoA is nearly equivalent in energy terms to a phosphoanhydride bond rather than to a simple thioester.

Free-Energy Changes Can Be Determined in Coupled Enzyme Reactions

The $\Delta G^{\circ\prime}$ value of hydrolysis of the terminal phosphate of ATP is difficult to determine simply because the K_{eq} of the hydrolytic reaction is far to the right.

$$\text{ATP} + \text{HOH} \rightleftharpoons \text{ADP} + \text{P}_i + \text{H}^+$$

However, $\Delta G^{\circ\prime}$ of hydrolysis of ATP is determined indirectly because of the additive nature of free-energy changes discussed above. Hence free energy of hydrolysis of ATP is determined by adding $\Delta G^{\circ\prime}$ of an ATP-utilizing reaction such as hexokinase to $\Delta G^{\circ\prime}$ of a reaction that cleaves phosphate from the product of the hexokinase reaction, glucose 6-phosphate (G6P), as indicated below:

$$\text{glucose} + \text{ATP} \xrightleftharpoons{\text{hexokinase}} \text{G6P} + \text{ADP} + \text{H}^+ \qquad \Delta G^{\circ\prime} = -4.0 \text{ kcal mol}^{-1}$$

$$\text{G6P} + \text{HOH} \xrightleftharpoons{\text{glucose 6-phosphatase}} \text{glucose} + \text{P}_i \qquad \Delta G^{\circ\prime} = -3.3 \text{ kcal mol}^{-1}$$

$$\text{ATP} + \text{HOH} \rightleftharpoons \text{ADP} + \text{P}_i + \text{H}^+ \qquad \Delta G^{\circ\prime} = -7.3 \text{ kcal mol}^{-1}$$

Free energies of hydrolysis for other energy-rich compounds are determined in a similar fashion.

High-Energy Bond Energies of Various Groups Can Be Transferred from One Compound to Another

Energy-rich compounds can transfer various groups from the parent (donor) compound to an acceptor compound in a thermodynamically feasible fashion as long

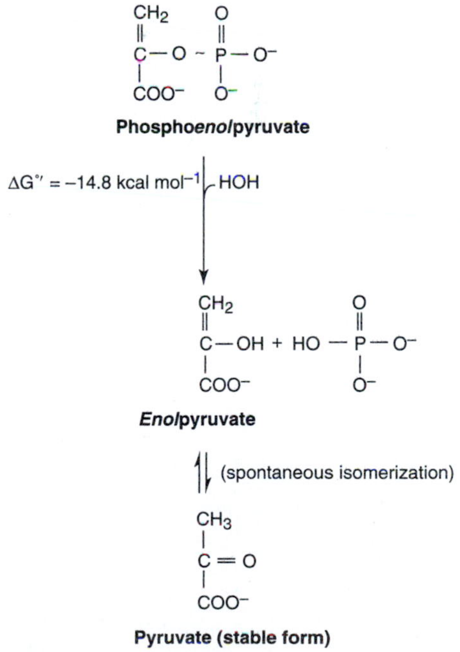

$CH_2{=}C(COO^-)-O\sim PO_3^{2-}$

Phospho*enol*pyruvate

$\Delta G^{\circ\prime} = -14.8 \text{ kcal mol}^{-1}$ ↓ HOH

$CH_2{=}C(COO^-)-OH + HO-PO_2(O^-)-O^-$

*Enol*pyruvate

⇅ (spontaneous isomerization)

$CH_3-C(=O)-COO^-$

Pyruvate (stable form)

FIGURE 13.7
Hydrolysis of phospho*enol*pyruvate indicating the free energy released.

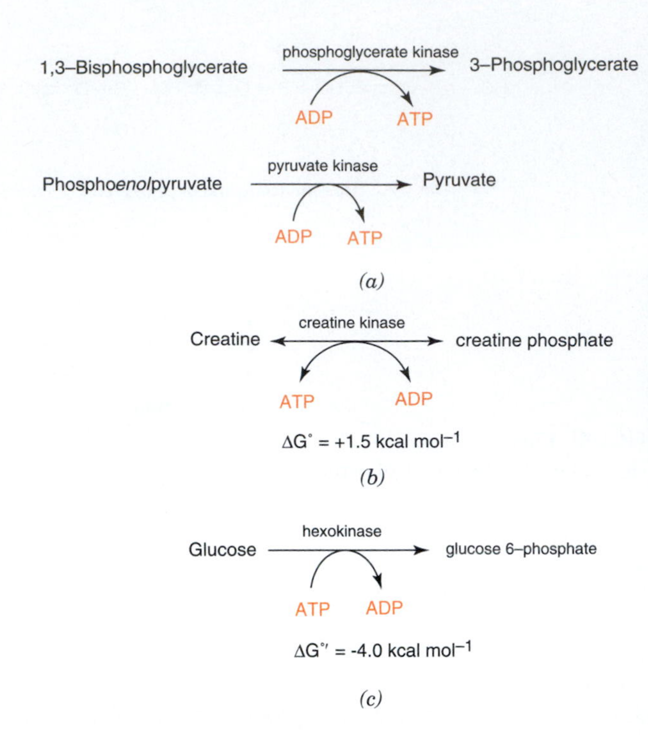

FIGURE 13.8
Examples of reactions involved in transfer of "high-energy" phosphate.

as an appropriate enzyme is present to facilitate transfer. The energy-rich intermediates in the glycolytic pathway such as 1,3-bisphosphoglycerate and phospho*enol*pyruvate can transfer their high-energy phosphate moieties to ATP in the phosphoglycerate kinase and pyruvate kinase reactions, respectively (Figure 13.8*a*). The $\Delta G^{\circ\prime}$ values of these two reactions are -4.5 and -7.5 kcal mol^{-1}, respectively, and hence transfer of "high-energy" phosphate is thermodynamically possible, and ATP synthesis is the result. ATP can transfer its terminal high-energy phosphoryl groups to form compounds of relatively similar high-energy character [i.e., creatine phosphate in the creatine kinase reaction (Figure 13.8*b*)] or compounds of considerably lower energy, such as glucose 6-phosphate formed in the hexokinase reaction (Figure 13.8*c*).

Thus phosphate or other transferable groups can be transferred from compounds that contain energy-rich bonding arrangements to compounds that have bonding characteristics of a lower energy in thermodynamically permissible enzymatic reactions. This principle is a major premise of interaction between energy-producing and energy-utilizing metabolic pathways in living cells.

Although adenine nucleotides are mainly involved in energy generation or conservation, various nucleoside triphosphates, including ATP, are involved in transferring energy during biosynthetic processes. As indicated in Figure 13.9, the guanine nucleotide GTP serves as the source of energy in gluconeogenesis and protein synthesis, whereas UTP (uracil) and CTP (cytosine) are utilized in glycogen

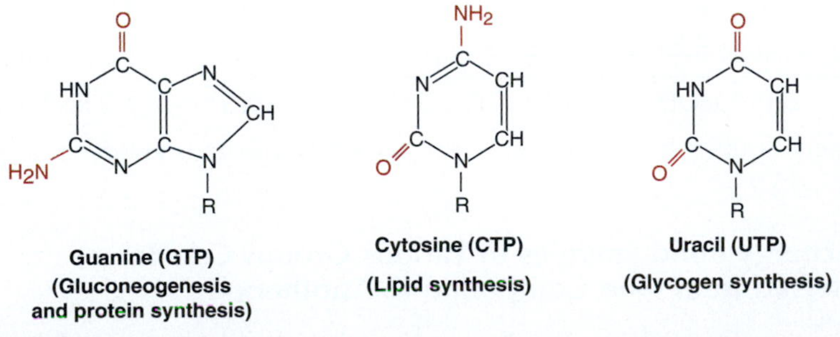

Guanine (GTP)
(Gluconeogenesis and protein synthesis)

Cytosine (CTP)
(Lipid synthesis)

Uracil (UTP)
(Glycogen synthesis)

FIGURE 13.9
Structures of purine and pyrimidine bases involved in various biosynthetic pathways.

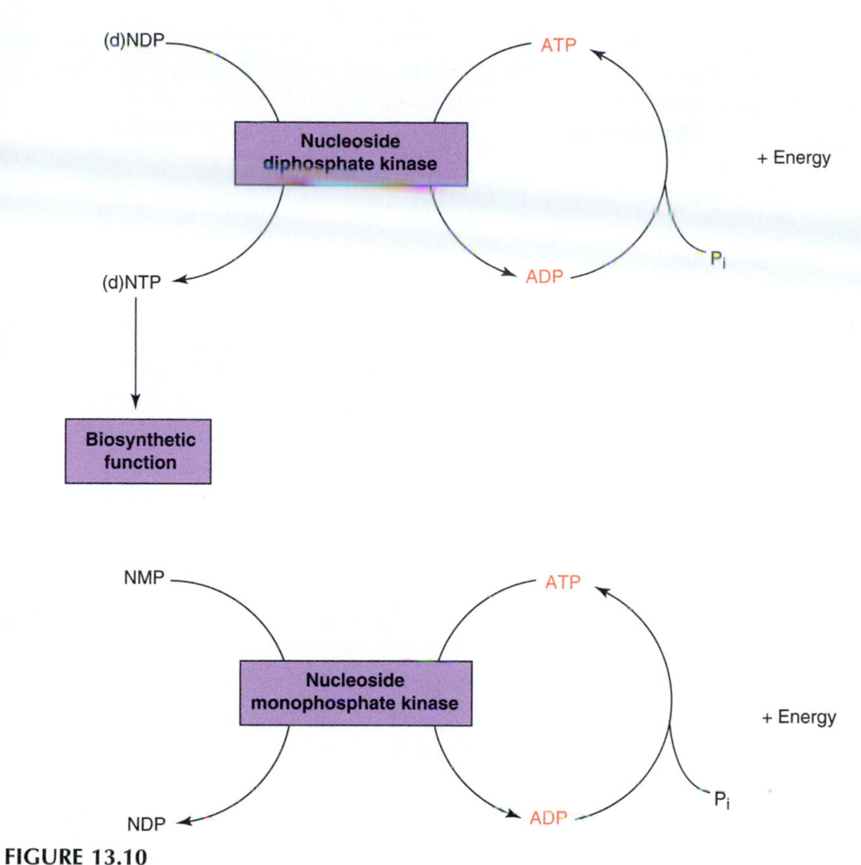

FIGURE 13.10
Nucleoside diphosphate kinase and nucleoside monophosphate kinase reactions. N represents any purine or pyrimidine base; (d) indicates a deoxyribonucleotide.

2ADP
Adenylate kinase (myokinase)
ATP + AMP

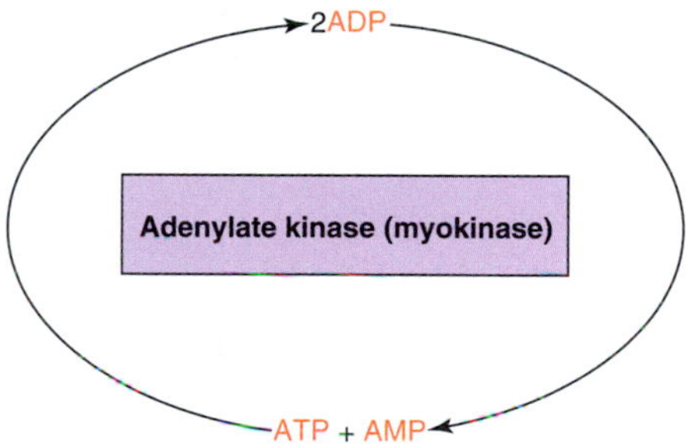

FIGURE 13.11
Adenylate kinase (myokinase) reaction.

and lipid synthesis, respectively. The energy in the terminal phosphate bonds of ATP may be transferred to the other nucleotides, using either the nucleoside diphosphate kinase or the nucleoside monophosphate kinase reactions illustrated in Figure 13.10. Two nucleoside diphosphates are converted to a nucleoside triphosphate and a nucleoside monophosphate in various nucleoside monophosphate kinase reactions, such as the adenylate kinase reaction (Figure 13.11). The significance of these types of enzymes is that the terminal energy-rich phosphate bonds of ATP may be transferred to appropriate nucleotides and utilized in a variety of biosynthetic processes.

13.3 | SOURCES AND FATES OF ACETYL COENZYME A

The acetate group that serves as the source of fuel for the **tricarboxylic acid (TCA)** cycle is derived from the major energy-generating metabolic pathways of cells. These pathways include oxidation of long-chain fatty acids by β-oxidation, breakdown of ingested or stored carbohydrate by **glycolysis**, oxidation of **ketone bodies, acetoacetate** and **β-hydroxybutyrate,** and oxidation of ethanol as well as oxidative breakdown of certain amino acids (Figure 13.12). All of these processes eventually result in production of the two-carbon unit **acetyl coenzyme A** (CoA). As indicated in Figure 13.13, coenzyme A, abbreviated either as CoA or CoASH, is a complex molecule composed of β-mercaptoethylamine, the vitamin **pantothenic acid,** and the adenine nucleotide, adenosine 3′-phosphate 5′-diphosphate. In cells, coenzyme A exists as the reduced thiol (CoASH), which forms high-energy thioester bonds with acyl groups and is involved in a variety of acyl group transfer reactions. As such, CoA alternately serves as acceptor, then donor, of the acyl group. Various metabolic pathways involve only acyl CoA derivatives, for example, β-oxidation of

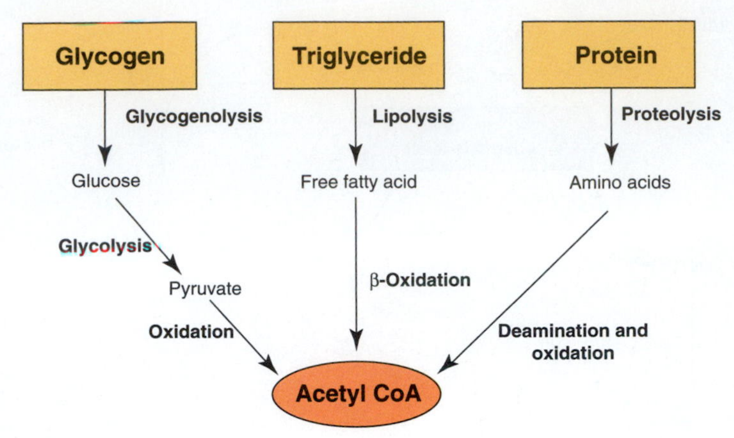

FIGURE 13.12
General precursors of acetyl CoA.
Carbohydrates, lipids, and proteins are broken down to form acetyl CoA.

fatty acids and branched-chain amino acid degradation. Because CoA is a large, hydrophilic molecule, derivatives of CoA such as acetyl CoA, are not freely transported across cellular membranes. This property has necessitated evolution of certain transport or shuttle mechanisms by which various intermediates or groups are transferred across membranes. Such acyltransferase reactions for acetyl groups and long-chain acyl groups will be discussed in Chapter 16. Since the thiol ester linkage in acyl CoA derivatives is an energy-rich bond, these compounds can serve as effective donors of acyl groups in acyltransferase reactions. Also, to synthesize an acyl CoA derivative a high-energy bond of ATP must be expended, such as in the **acetate thiokinase** reaction,

$$\text{acetate} + \text{CoASH} + \text{ATP} \xrightarrow{\text{acetate kinase}} \text{acetyl CoA} + \text{AMP} + \text{PP}_i$$

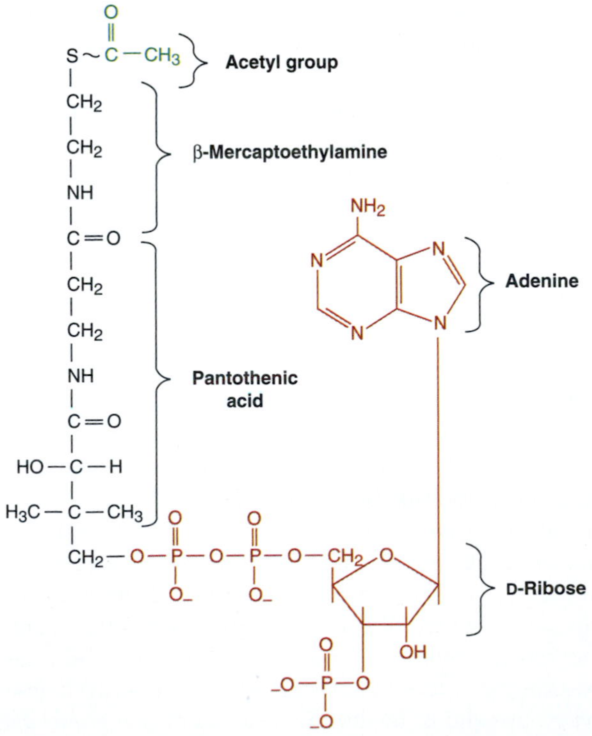

FIGURE 13.13
Structure of acetyl CoA.

Metabolic Sources and Fates of Pyruvate

During aerobic glycolysis (see p. 599), glucose or other monosaccharides are converted to pyruvate, the end product of this cytosolic pathway. Pyruvate, formed in degradation of amino acids such as alanine or serine, has a number of fates depending on the tissue and its metabolic state. The fates of pyruvate and types of reactions in which it participates are indicated in Figure 13.14. The oxidative decarboxylation of pyruvate in the **pyruvate dehydrogenase** reaction is discussed next; see p. 782 for a discussion of other reactions involving pyruvate.

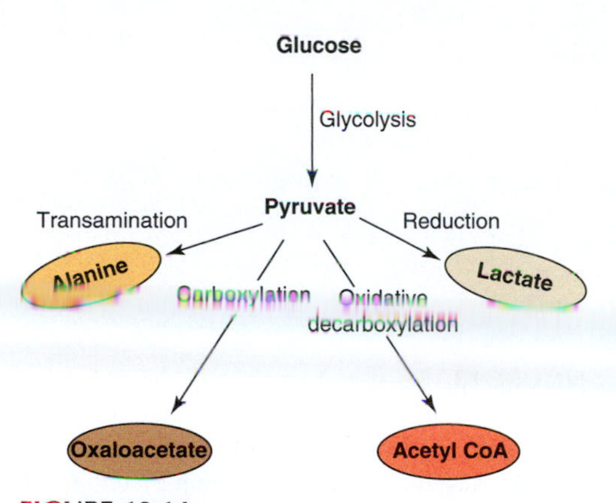

FIGURE 13.14
Metabolic fates of pyruvate.
Pyruvate is at a crossroads of metabolism. Pyruvate can be converted to lactate, alanine, oxaloacetate, or acetyl CoA depending on the needs of the cell.

Pyruvate Dehydrogenase Is a Multienzyme Complex

Pyruvate is converted to acetyl CoA by the pyruvate dehydrogenase multienzyme complex:

$$\text{pyruvate} + \text{NAD}^+ + \text{CoASH} \xrightarrow{\text{pyruvate dehydrogenase}} \text{acetyl CoA} + \text{CO}_2 + \text{NADH} + \text{H}^+; \quad \Delta G^{\circ\prime} = -8 \text{ kcal mol}^{-1}$$

The mechanism of this reaction is more complex than might be inferred from the overall stoichiometry. Three catalytic cofactors, **thiamine pyrophosphate** (TPP), **lipoamide,** and **flavin adenine dinucleotide** (FAD), are bound to subunit proteins of the complex in addition to the stoichiometric cofactors, CoASH and NAD^+. The pyruvate dehydrogenase reaction has a $\Delta G^{\circ\prime}$ of -8 kcal mol^{-1} and hence is irreversible under physiological conditions. The mammalian pyruvate dehydrogenase complex consists of three different types of catalytic subunits associated in a multienzyme complex with molecular weights of the complex from kidney, heart, or liver ranging from 7 to 8.5×10^6. The catalytic subunits with their associated catalytic cofactors are presented in Table 13.4.

The structure of the pyruvate dehydrogenase complex derived from *Escherichia coli* with a particle weight of 4.6×10^6 has been visualized by electron microscopy. These electron micrographs of the bacterial enzyme (Figure 13.15) indicate that transacetylase, which consists of 24 identical polypeptide chains (MW = 64,500), forms the cube-like core of the complex (white spheres in model shown in Figure 13.15). Twelve pyruvate dehydrogenase dimers (black spheres; MW = 90,500) are distributed symmetrically on the 12 edges of the transacetylase cube. Six dihydrolipoyl dehydrogenase dimers (gray spheres; MW = 56,000) are distributed on the six faces of the cube. The arrangement of these subunits in a structured multienzyme complex provides for greater efficiency in the overall reaction of the pyruvate dehydrogenase complex as the intermediates of the reaction are tightly bound to the enzyme subunits and are not released into solution.

The functional group of TPP participates in the formation of a covalent intermediate (Figure 13.16). Lipoic acid forms an amide bond with a lysine on the transacetylase and is called lipoamide, while FAD is tightly bound to the

TABLE 13.4 Pyruvate Dehydrogenase Complex of Mammals

Enzyme	*Number of Chains*	*Prosthetic Group*	*Reaction Catalyzed*
Pyruvate dehydrogenase	20 or 30	TPP	Oxidative decarboxylation of pyruvate
Dihydrolipoyl transacetylase	60	Lipoamide	Transfer of the acetyl group to CoA
Dihydrolipoyl dehydrogenase	6	FAD	Regeneration of the oxidized form of lipoamide and transfer of electrons to NAD^+

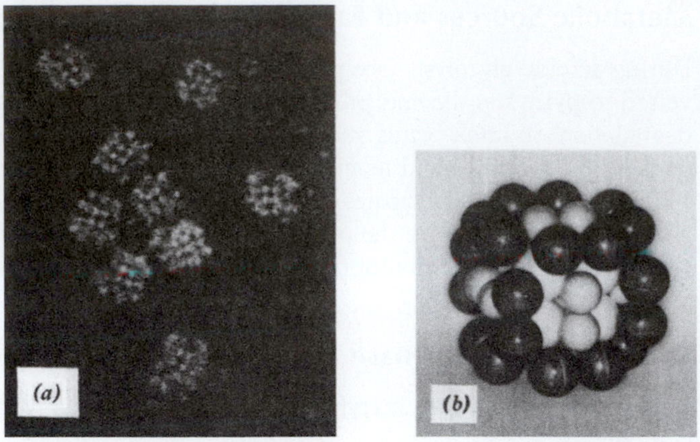

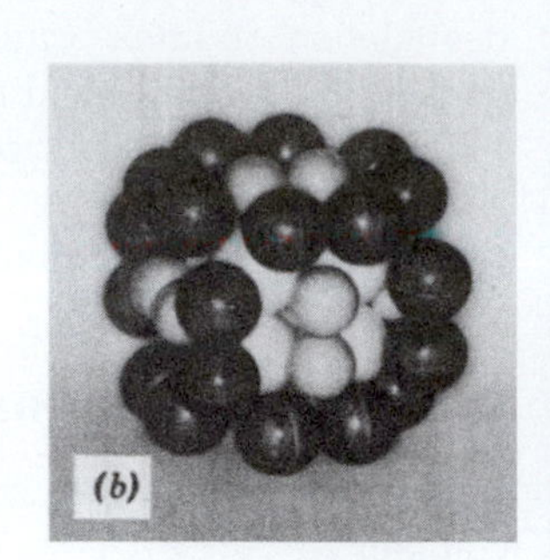

FIGURE 13.15
Pyruvate dehydrogenase complex from *E. coli*.
(*a*) Electron micrograph. (*b*) Molecular model. The enzyme complex was negatively stained with phosphotungstate (× 200,000).
Courtesy of Dr. Lester J. Reed, University of Texas, Austin.

dihydrolipoyl dehydrogenase subunit. The structures of NAD^+ and CoASH are presented in Figure 13.4 and 13.13, respectively. The mechanism of the reaction is illustrated in Figure 13.17.

Pyruvate Dehydrogenase Is Strictly Regulated

Two types of regulation of the pyruvate dehydrogenase complex have been characterized. First, two products of the pyruvate dehydrogenase reaction, acetyl CoA and NADH, inhibit the complex in a competitive fashion, as feedback inhibitors. The second and most important regulation of the pyruvate dehydrogenase

R = —H

R = —CH—CH₃ (OH) **(hydroxyethyl)**

Thiamine pyrophosphate

Lipoamide (oxidized)

(reduced/acetylated)

Flavin adenine dinucleotide (oxidized)

(Reduced)

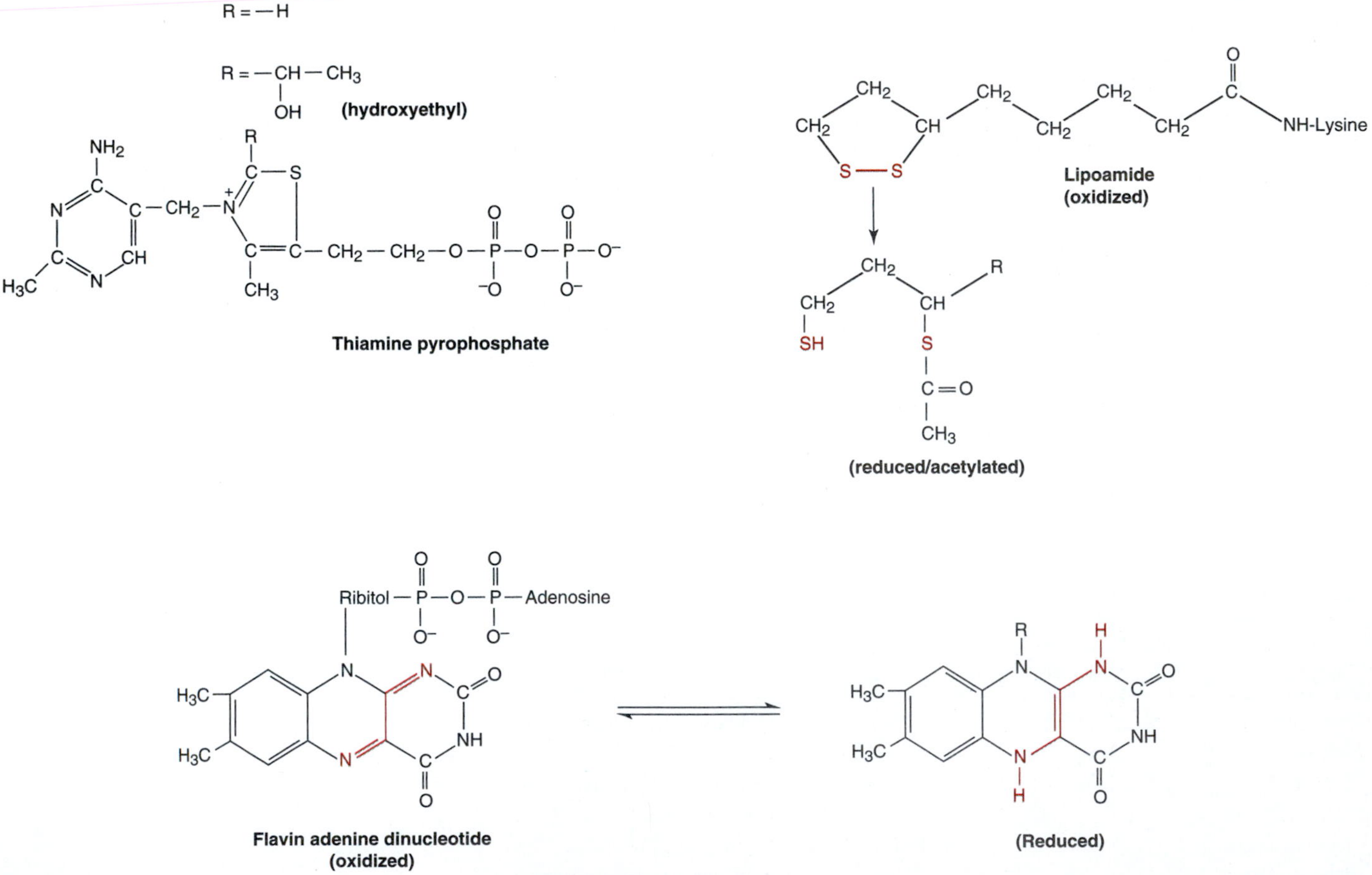

FIGURE 13.16
Structures of coenzymes involved in the pyruvate dehydrogenase reaction.
See Figure 13.4 for structure of NAD^+ and Figure 13.13 for structure of CoA.

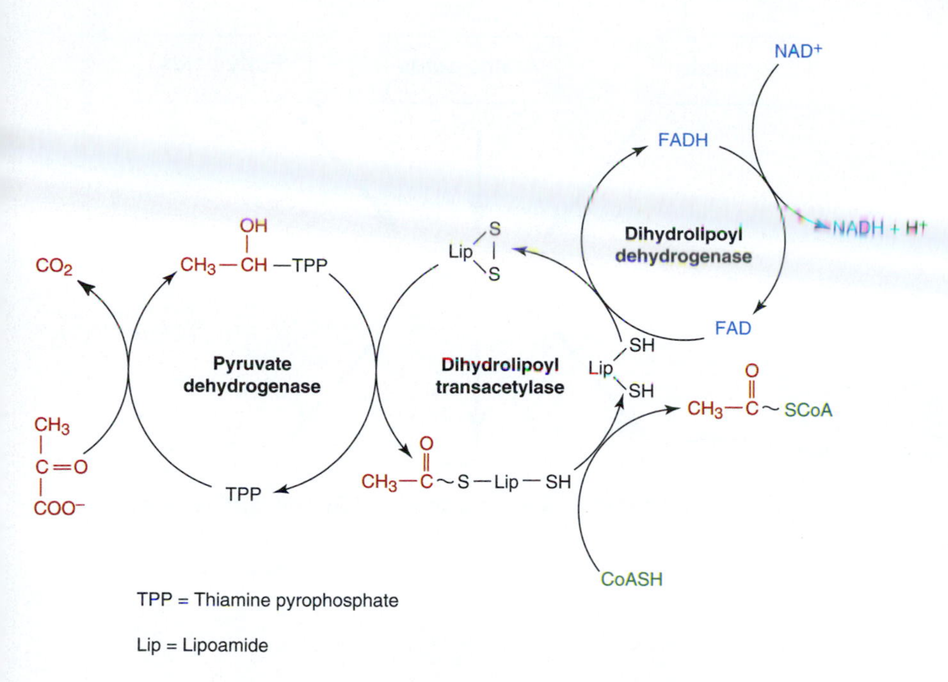

FIGURE 13.17
Mechanism of the pyruvate dehydrogenase multienzyme complex.
Pyruvate dehydrogenase catalyzes oxidative decarboxylation of pyruvate and transfer of the acetyl group to lipoamide. Dihydrolipoyl transacetylase transfers the acetyl group from lipoamide to coenzyme A. Dihydrolipoyl dehydrogenase oxidizes reduced lipoamide.

complex involves regulation by phosphorylation and dephosphorylation. Two forms of the complex have been described: (1) an active, dephosphorylated complex and (2) an inactive, phosphorylated complex. Inactivation of the complex is accomplished by a Mg^{2+}–ATP-dependent **protein kinase,** which is tightly bound to the enzyme complex. Reactivation is accomplished by a **phosphoprotein phosphatase,** which is also bound to the complex and dephosphorylates the complex in a Mg^{2+}- and Ca^{2+}-dependent reaction. The differential regulation of pyruvate dehydrogenase kinase and phosphatase is the key to the overall regulation of the pyruvate dehydrogenase complex. Essential features of this complex regulatory system are illustrated in Figure 13.18. Acetyl CoA and NADH inhibit the dephospho (active)

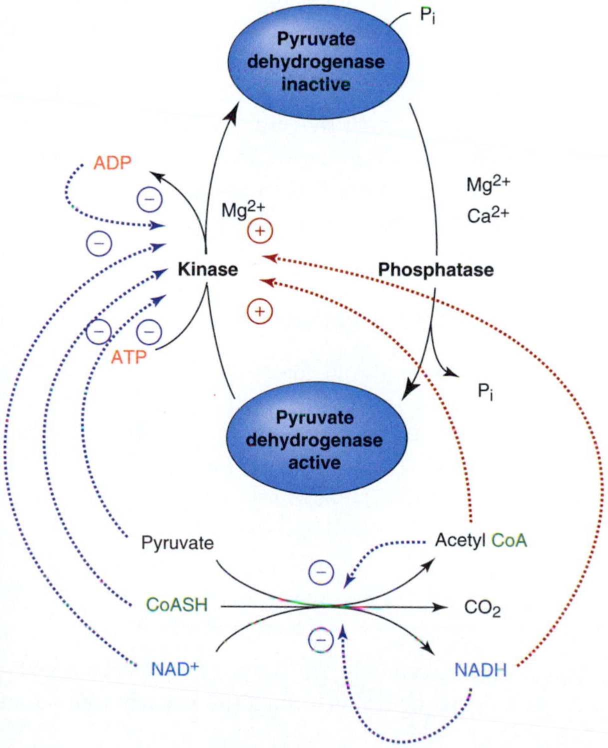

FIGURE 13.18
Regulation of the pyruvate dehydrogenase multienzyme complex.

CLINICAL CORRELATION 13.1

Pyruvate Dehydrogenase Deficiency

A variety of disorders in pyruvate metabolism have been detected in children. Some involve deficiency of the catalytic or regulatory subunits of the pyruvate dehydrogenase multienzyme complex. Children diagnosed with pyruvate dehydrogenase deficiency usually exhibit elevated serum levels of lactate, pyruvate, and alanine, which produce a chronic lactic acidosis. They frequently exhibit severe neurological defects, and in most situations this type of enzymatic defect results in death. The diagnosis of pyruvate dehydrogenase deficiency is usually made by assaying the enzyme complex and/or its various enzymatic subunits in cultures of skin fibroblasts taken from the patient. In certain instances patients respond to dietary management in which a ketogenic diet is administered and carbohydrates are minimized. Patients may be in shock from lactic acidosis because decreased delivery of O_2 to tissues inhibits pyruvate dehydrogenase and increases anaerobic metabolism. Patients with this condition have been treated with dichloroacetate, an inhibitor of the kinase subunit of pyruvate dehydrogenase. Inhibition of the kinase, which causes inhibition of the enzyme, will therefore activate the enzyme complex.

Patel, M. S. and Harris, R. A. Mammalian α-keto acid dehydrogenase complexes: gene regulation and genetic defects. *FASEB J.* 9:1164, 1995.

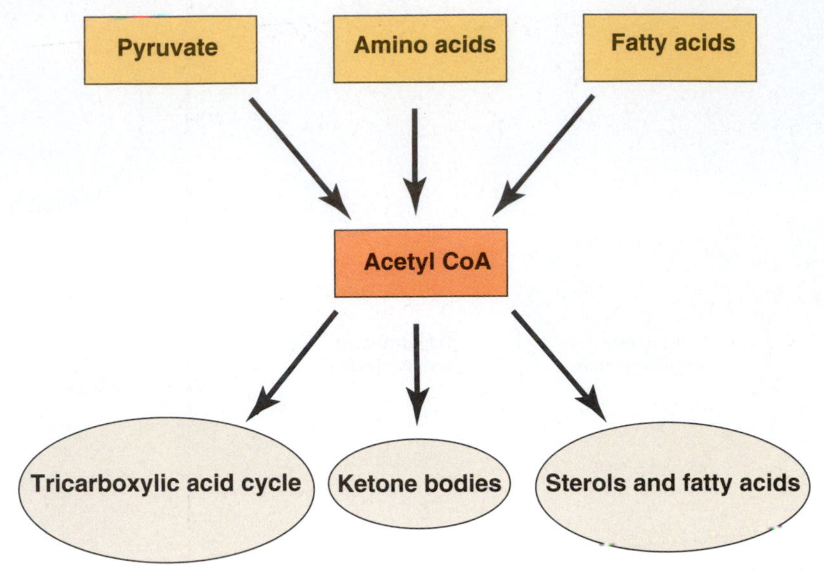

FIGURE 13.19
Sources and fates of acetyl CoA.

form of the enzyme. These two compounds also stimulate the protein kinase reaction, leading to an interconversion of the complex to its inactive form. In addition, free CoASH and NAD^+ inhibit the protein kinase. Hence any increase of the mitochondrial $NADH/NAD^+$ or acetyl CoA/CoASH ratio with concomitant lowering of NAD^+ and CoASH will result in an inactivation of pyruvate dehydrogenase by the stimulated kinase reaction. In addition, pyruvate, the substrate of the enzyme, is a potent inhibitor of the protein kinase, and therefore in the presence of elevated tissue pyruvate levels the kinase will be inhibited and the complex maximally active. The activity of the pyruvate dehydrogenase complex is stimulated by Ca^{2+} that acts as a potent activator of the protein phosphatase reaction. These effects of Ca^{2+} may play an important role in skeletal muscle where the release of Ca^{2+} during contraction will activate the protein phosphatase of pyruvate dehydrogenase, thus stimulating the oxidation of pyruvate by this enzyme and hence energy production. Finally, insulin administration activates pyruvate dehydrogenase in adipose tissue, and catecholamines, such as epinephrine, activate pyruvate dehydrogenase in cardiac tissue. The mechanisms of these hormonal effects are not well understood, but alterations of intracellular distribution of Ca^{2+}, such that the phosphoprotein phosphatase reaction is stimulated in the mitochondrial matrix, may be involved. These hormonal effects are not mediated directly by alterations in tissue cAMP levels, because pyruvate dehydrogenase protein kinase and phosphatase are cAMP-independent or insensitive (see Clin. Corr. 13.1).

Acetyl CoA Is Used by Several Different Pathways

Fates of acetyl CoA generated in the mitochondrial matrix include (1) complete oxidation of the acetyl group in the TCA for energy generation; (2) in liver, conversion of excess acetyl CoA into ketone bodies, acetoacetate, and β-hydroxybutyrate; and (3) transfer of acetyl units as citrate to the cytosol with subsequent biosynthesis of long-chain fatty acids (see p. 698) and sterols (Figure 13.19) (see p. 742).

13.4 | TRICARBOXYLIC ACID CYCLE

Acetyl CoA produced in energy-generating catabolic pathways of most cells is completely oxidized to CO_2 in a cyclic series of reactions termed the **tricarboxylic acid**

(TCA) cycle. This metabolic cycle is also commonly referred to as the **citric acid cycle** or the **Krebs cycle** after Sir Hans Krebs who postulated its essential features in 1937. The primary location of enzymes of the TCA cycle is in mitochondria although isozymes of some enzymes are found in cytosol. A mitochondrial location for the TCA cycle is appropriate as the pyruvate dehydrogenase multienzyme complex and the fatty acid β-oxidation sequence, the two primary sources of acetyl CoA, are also located in mitochondria. Four reactions, involved in oxidation of acetyl CoA by the TCA cycle, transfer electrons to either NAD^+ or FAD. The resulting NADH + H^+ or $FADH_2$ are subsequently oxidized by the mitochondrial **electron transport chain** (also referred to as the **electron transfer chain** or **respiratory chain**) to generate energy that is used to form ATP in a process called oxidative phosphorylation, discussed on page 577. The enzymes of the electron transport chain and those involved in ATP synthesis are also exclusively localized in mitochondria. Figure 13.20 is an overview of reactions of the TCA cycle. In the first step of the TCA cycle, the acetyl moiety of acetyl CoA is condensed with oxaloacetate (a four-carbon dicarboxylic acid) to form citrate (a six-carbon tricarboxylic acid). After rearrangement of the carbons of citrate, two oxidative decarboxylation reactions result in formation of two molecules of CO_2, two molecules of NADH + H^+ and succinate (a four-carbon dicarboxylic acid) with production of the high-energy bond as GTP. In the next steps of the cycle two more oxidations occur with production of another NADH + H^+ plus one $FADH_2$ and regeneration of oxaloacetate.

In summary, the substrate of the TCA cycle is the two-carbon unit acetyl CoA and the products of a complete turn of the cycle are two CO_2, one high-energy phosphate bond (as GTP), and four reducing equivalents (i.e., three NADH and one $FADH_2$). The NADH and $FADH_2$ are subsequently oxidized by the electron transport chain with the production of 9 ATP (see p. 578 for yields of ATP during oxidation of the coenzymes).

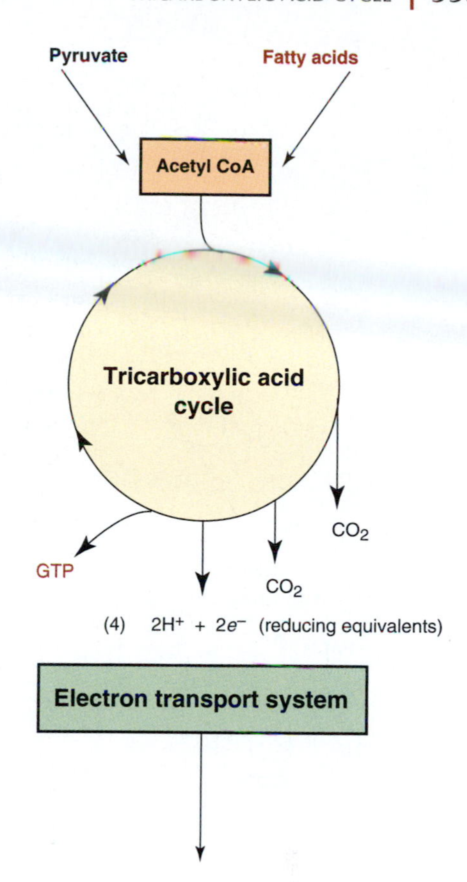

FIGURE 13.20
General description of synthesis of ATP within mitochondria.
Acetyl CoA produced by oxidation of pyruvate and fatty acids is metabolized by the tricarboxylic acid cycle to reducing equivalents, which are oxidized by the electron transport system. Energy released during the oxidative process is used to drive synthesis of ATP.

Reactions of Tricarboxylic Acid Cycle

The individual enzymatic reactions of the TCA cycle are presented in Figure 13.21. The initial step of the cycle is catalyzed by **citrate synthase**, an enzyme localized in the matrix. This highly exergonic reaction commits acetyl groups to **citrate** formation and complete oxidation in the Krebs cycle. As shown below citrate synthase involves condensation of an acetyl moiety and the α-keto carbon of the dicarboxylic acid **oxaloacetate.** The citroyl-SCoA intermediate is not released but remains bound to the catalytic site on citrate synthase.

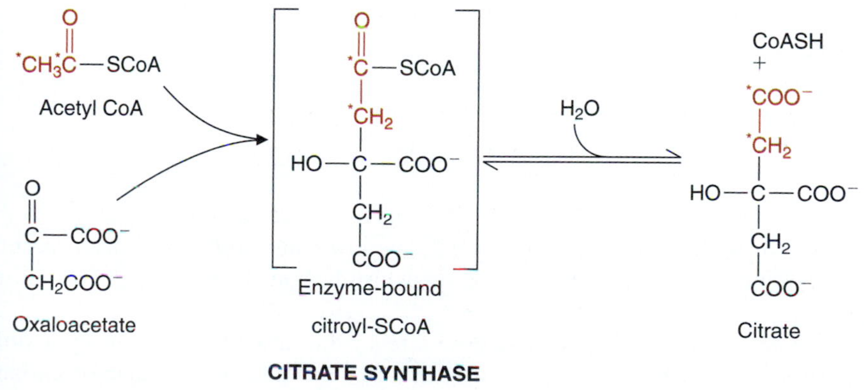

Acetyl CoA ($^*CH_3^*C(=O)$—SCoA) + Oxaloacetate ($O=C(COO^-)CH_2COO^-$) → [Enzyme-bound citroyl-SCoA] —H_2O→ Citrate + CoASH

CITRATE SYNTHASE

The equilibrium of this reaction is far toward citrate formation with a $\Delta G^{\circ\prime}$ near -9 kcal mol^{-1}. It should be noted that the intramitochondrial concentration of oxaloacetate is very low (less than 1 μM); however, the strong, negative

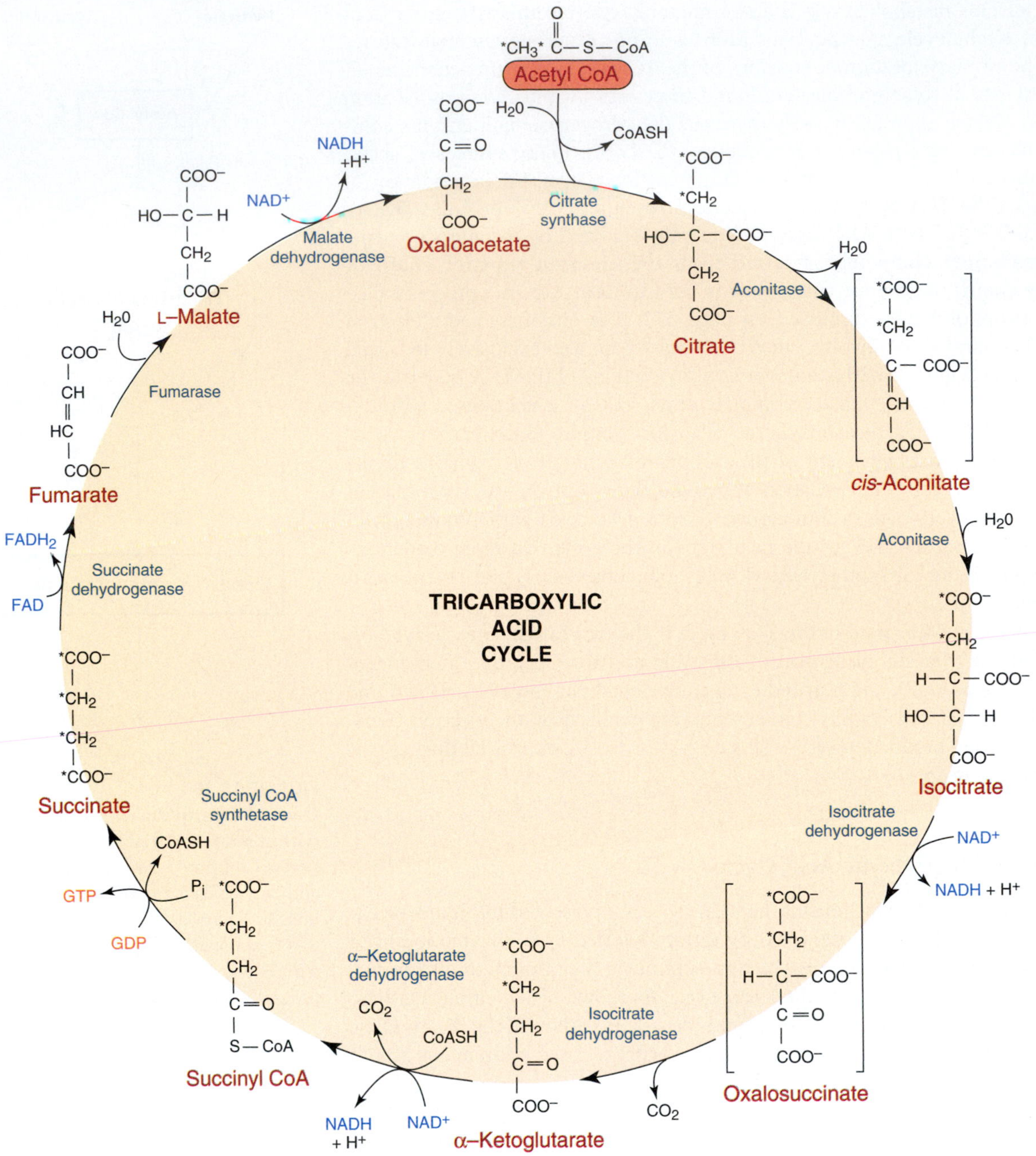

FIGURE 13.21
Tricarboxylic acid cycle.
Asterisked carbons indicate fate of the carbons of the acetyl group.

$\Delta G^{\circ\prime}$ serves to drive the reaction forward. The low concentrations of oxaloacetate, which are below the K_m of the reaction, may also be a major factor controlling this reaction.

Citrate is converted to **isocitrate** by a reversible reaction catalyzed by **aconitase** in which the hydroxyl group of citrate is transferred to an adjacent carbon with a C—H bond. The hydroxyl group is thus located next to the carboxyl group of isocitrate where oxidative decarboxylation can occur. Conversion of citrate to isocitrate occurs on the surface of the aconitase enzyme without release of the proposed intermediate ***cis*-aconitate.** Aconitase contains a nonheme iron–sulfur cluster that is involved in the catalytic mechanism. The overall equilibrium of the reaction favors the formation of citrate.

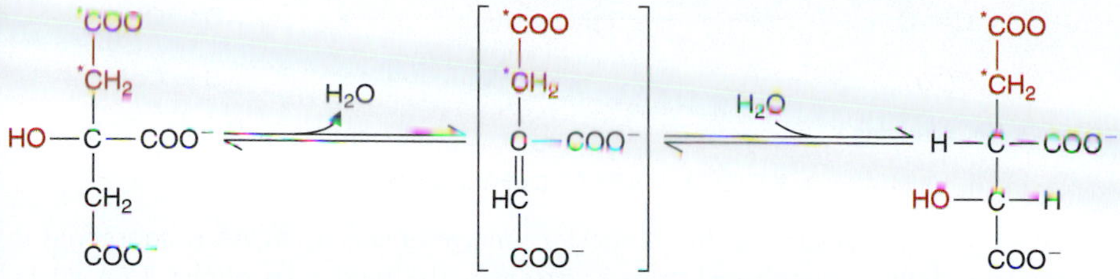

Citrate — *cis*-Aconitate — Isocitrate

ACONITASE

Fluoroacetate is a potent inhibitor of the cycle, although it does not inhibit directly any of the enzymes of the cycle. The mechanism of inhibition is explained by the observation that fluoroacetate is converted in two steps to fluorocitrate, which is a potent inhibitor of aconitase. Fluoroacetyl CoA is formed by action of acetyl CoA synthetase. Citrate synthase then converts fluoroacetyl CoA to fluorocitrate. Fluoroacetate is lethal in small doses and has been used as a rat poison; the LD_{50}, the lethal dose for 50% of animals consuming it, is 0.2 mg per kilogram of body weight.

In the next step of the TCA cycle, **isocitrate dehydrogenase** converts isocitrate to α-ketoglutarate in an oxidative decarboxylation reaction with concomitant reduction of NAD^+ to $NADH + H^+$. The isocitrate dehydrogenase present in mammalian mitochondria requires NAD^+ as the acceptor of reducing equivalents, has a molecular weight of 380,000, and consists of eight identical subunits. The reaction requires a divalent metal cation (e.g., Mn^{2+} or Mg^{2+}) in decarboxylation of the β position of oxalosuccinate. The equilibrium of this reaction lies strongly toward α-ketoglutarate formation with a $\Delta G^{\circ\prime}$ of nearly -5 kcal mol^{-1}.

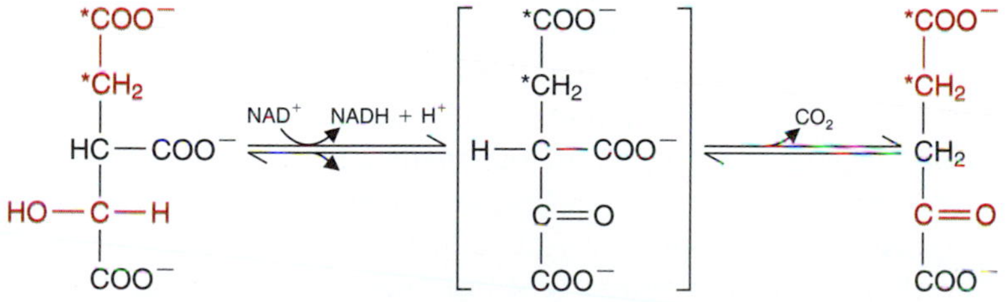

Isocitrate — Oxalosuccinate — α-Ketoglutarate

ISOCITRATE DEHYDROGENASE

Mitochondria also have an isocitrate dehydrogenase that requires $NADP^+$. The $NADP^+$-linked enzyme is also found in the cytosol, where it is involved in providing reducing equivalents for cytosolic reductive processes.

Conversion of α-ketoglutarate to succinyl CoA is catalyzed by the **α-ketoglutarate dehydrogenase** multienzyme complex, which is nearly identical to the pyruvate dehydrogenase complex in terms of the individual reactions catalyzed and its structural features. Again, **thiamine pyrophosphate, lipoic acid,** CoASH, **FAD,** and NAD^+ participate in the catalytic mechanism. The multienzyme complex consists of α-ketoglutarate dehydrogenase, **dihydrolipoyl transsuccinylase,** and dihydrolipoyl dehydrogenase as three catalytic subunits. The equilibrium of the α-ketoglutarate dehydrogenase reaction lies strongly toward succinyl CoA formation with a $\Delta G^{\circ\prime}$ of -8 kcal mol^{-1}. In this reaction the second molecule of CO_2 and the second reducing equivalent (i.e., $NADH + H^+$) of the TCA cycle are produced. Another product of this reaction, **succinyl CoA,** is an energy-rich thiol ester compound similar to acetyl CoA.

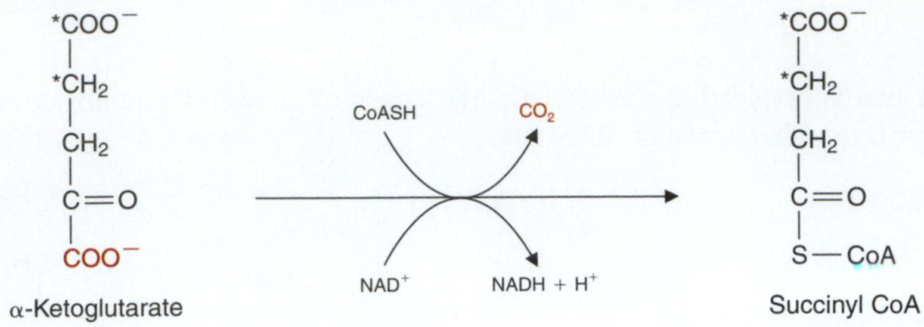

α-KETOGLUTARATE DEHYDROGENASE

The energy-rich character of the thiol ester linkage of succinyl CoA is conserved in a **substrate-level phosphorylation** reaction in the next step of the TCA cycle. **Succinyl-CoA synthetase** (or **succinate thiokinase**) converts succinyl CoA to succinate and in mammalian tissues results in the phosphorylation of GDP to GTP.

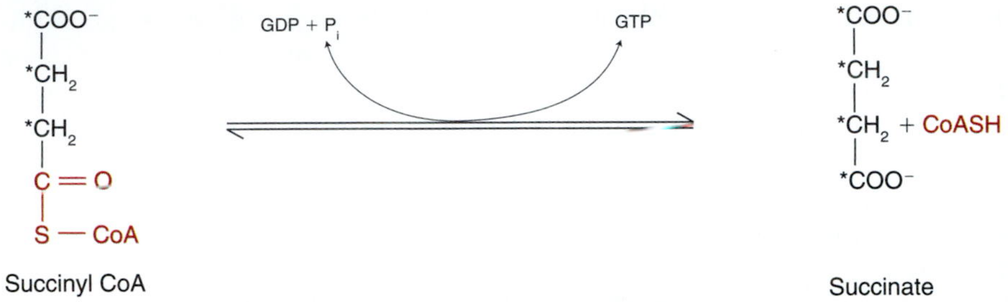

SUCCINYL CoA SYNTHETASE

This reaction is freely reversible with a $\Delta G^{\circ\prime} = -0.7$ kcal mol^{-1} and the catalytic mechanism involves an enzyme–succinyl phosphate intermediate:

$$\text{succinyl CoA} + P_i + \text{enz} \rightleftharpoons \text{enz–succinyl phosphate} + \text{CoASH}$$
$$\text{enz–succinyl phosphate} \rightleftharpoons \text{enz–phosphate} + \text{succinate}$$
$$\text{enz–phosphate} + \text{GDP} \rightleftharpoons \text{enz} + \text{GTP}$$

The enzyme is phosphorylated on the 3 position of a histidine residue during the reaction, which conserves the energy of the thioester for formation of GTP. The presence of the nucleoside diphosphate kinase provides the mechanism for the transfer of the γ-phosphate of GTP to ADP to generate ATP (Figure 13.10).

Succinate is oxidized to fumarate by **succinate dehydrogenase,** a complex enzyme tightly bound to the inner mitochondrial membrane. Succinate dehydrogenase is composed of four subunits, a 70,000 MW subunit contains the substrate-binding site (FAD covalently bound to a histidine residue), and a 30,000 MW subunit that contains three **iron–sulfur centers (nonheme iron),** plus two small hydrophobic proteins. This enzyme is a typical flavoprotein in which electrons and protons are transferred from the substrate, succinate, through covalently bound FAD and the iron–sulfur centers in which the nonheme iron undergoes oxidation and reduction. The electrons are eventually transferred to coenzyme Q for further transport through the electron transfer chain as will be discussed in Section 13.6 (see p. 564). Succinate dehydrogenase is strongly inhibited by malonate and oxaloacetate and is activated by ATP, P_i, and succinate. Malonate inhibits succinate dehydrogenase competitively with respect to succinate due to the very close structural similarity between malonate and succinate (Figure 13.22).

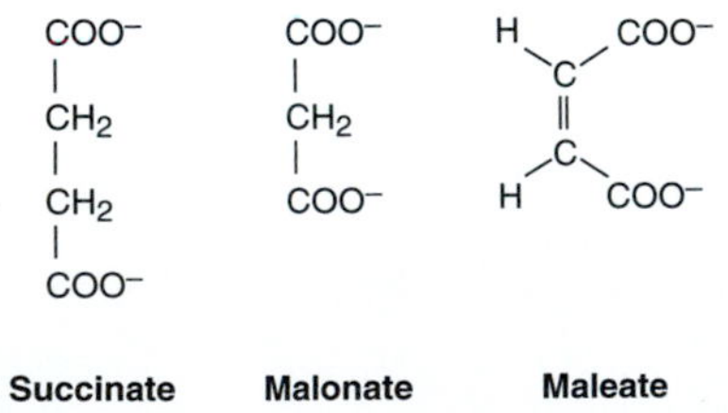

FIGURE 13.22
Structures of succinate, a TCA cycle intermediate; malonate, an inhibitor of succinate dehydrogenase and the cycle; and maleate, a compound not involved in the cycle.

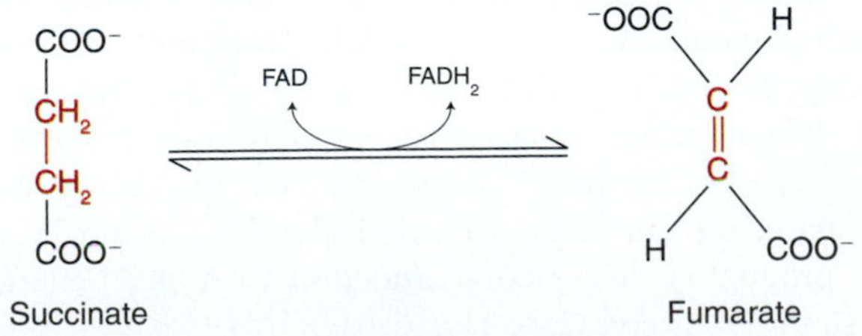

SUCCINATE DEHYDROGENASE

Fumarate is then hydrated to form L-malate by fumarase. **Fumarase** is a tetramer (MW 200,000) and is stereospecific for the *trans* form of substrate (the *cis* form, maleate, is not a substrate; Figure 13.22). The reaction is freely reversible under physiological conditions. Clinical Correlation 13.2 describes a genetic deficiency of fumarase.

$$\text{Fumarate} + H_2O \rightleftharpoons \text{L-Malate}$$

Fumarate: $^{-}OOC{-}CH{=}CH{-}COO^{-}$ (trans); L-Malate: $COO^{-}{-}HOCH{-}CH_2{-}COO^{-}$

FUMARASE

The final reaction in the TCA cycle is catalyzed by **malate dehydrogenase** in which the reducing equivalents are transferred to NAD^+ to form of NADH + H^+. The equilibrium of the malate dehydrogenase reaction lies far toward L-malate formation with a $\Delta G^{\circ\prime} = +7.0$ kcal mol^{-1} in the reaction as written below. This endergonic reaction is pulled in the forward direction by the actions of citrate synthase and other reactions of the cycle, which remove oxaloacetate. In addition, the NADH produced in the three NAD^+-linked dehydrogenases in the TCA cycle is oxidized rapidly to NAD^+ by the respiratory chain, another factor favoring the forward direction of malate dehydrogenase.

$$\text{L-Malate} + NAD^+ \rightleftharpoons \text{Oxaloacetate} + NADH + H^+$$

L-Malate: $COO^{-}{-}HOCH{-}CH_2{-}COO^{-}$; Oxaloacetate: $COO^{-}{-}C{=}O{-}CH_2{-}COO^{-}$

MALATE DEHYDROGENASE

Conversion of Acetyl Group of Acetyl CoA to CO_2 and H_2O Conserves Energy

In summary, the TCA cycle (Figure 13.20) serves as a terminal oxidative pathway for most metabolic fuels. Two-carbon moieties in the form of acetyl CoA are oxidized completely to CO_2 and H_2O. During this process 4 reducing equivalents (3 as NADH + H^+ and 1 as $FADH_2$) are produced, which are used subsequently for energy generation. Oxidation of each NADH + H^+ results in the formation of 2.5 ATP in oxidative phosphorylation, while oxidation of $FADH_2$ formed in the succinate

CLINICAL CORRELATION 13.2
Fumarase Deficiency

Deficiency of enzymes of the TCA cycle is rare, indicating the importance of this pathway for survival. Several cases, however, are on record in which there is a severe deficiency of fumarase in both mitochondria and cytosol of tissues (e.g., blood lymphocytes). The condition is characterized by severe neurological impairment, encephalomyopathy, and dystonia developing soon after birth. Urine contains abnormal amounts of fumarate and elevated levels of succinate, α-ketoglutarate, citrate, and malate. Both mitochondrial and cytosolic isozymes of fumarase are derived from a single gene. In the patients described with fumarase deficiency, both parents had half-normal levels of enzyme activity but were clinically normal, as expected for an autosomal recessive disorder. The first description of a mutation in the gene for fumarase reported that glutamate at residue 319 was replaced by glutamine.

Bourgeron, T., Chretien, D., Poggi-Bach, J., et al. Mutation of the fumarase gene in two siblings with progressive encephalopathy and fumarase deficiency. *J. Clin. Invest.* 93:2514, 1994.

dehydrogenase reaction yields 1.5 ATP. Also, a high-energy bond is formed as GTP in the succinyl-CoA synthetase reaction. Hence the net yield of ATP or its equivalent (i.e., GTP) for the complete oxidation of an acetyl group in the Krebs cycle is 10.

Tricarboxylic Acid Cycle Serves as a Source of Biosynthetic Intermediates

The discussion of the TCA cycle thus far has concentrated on its catabolic role in the oxidative breakdown of acetate to CO_2 and H_2O with the formation of reduced coenzymes and the synthesis of ATP. In general, the TCA cycle serves as the final steps in the breakdown of foodstuffs; however, as summarized in Figure 13.23, the four-, five-, and six-carbon compounds generated in the reactions of the TCA cycle are important intermediates in **biosynthetic processes.** Succinyl CoA, oxaloacetate, α-ketoglutarate, malate, and citrate are all precursors in the biosynthesis of important cellular compounds.

Transamination converts α-ketoglutarate to the amino acid glutamate, which can leave mitochondria and serve as a precursor for several other amino acids. In nervous tissue, α-ketoglutarate is converted to the neurotransmitters, glutamate and γ-aminobutyric acid (GABA). Glutamate is also produced from α-ketoglutarate by the action of the mitochondrial enzyme glutamate dehydrogenase in the presence of NADH or NADPH and ammonia. The amino group incorporated into glutamate can then be transferred to a variety of amino acids by different transaminases. These enzymes and the relevance of the incorporation or release of ammonia into or from α-keto acids are discussed in Chapter 18. The four-carbon compound, succinyl CoA, represents a metabolic branch point in that intermediates may enter or exit the TCA cycle at this point (Figure 13.24). Succinyl CoA may be formed either from α-ketoglutarate in the cycle or from methylmalonyl CoA in the final steps of breakdown of odd-chain length fatty acids or the branched-chain amino acids valine and isoleucine. Metabolic fates of succinyl CoA include its conversion to succinate in the succinyl-CoA synthetase reaction of the Krebs cycle and its condensation with glycine to form δ-aminolevulinate by δ-aminolevulinate synthase, the initial reaction in porphyrin biosynthesis (see p. 1065).

Oxaloacetate is transaminated to aspartate, a key intermediate in biosynthesis of other amino acids. Oxaloacetate is converted to **phosphoenolpyruvate (PEP)**, a key intermediate in synthesis of glucose by **gluconeogenesis** (see p. 631). Oxaloacetate cannot cross the mitochondrial membrane and hence it is first converted to malate, which is transported on a specific carrier out of mitochondria where it is oxidized to regenerate oxaloacetate, which is then converted to PEP (see p. 631).

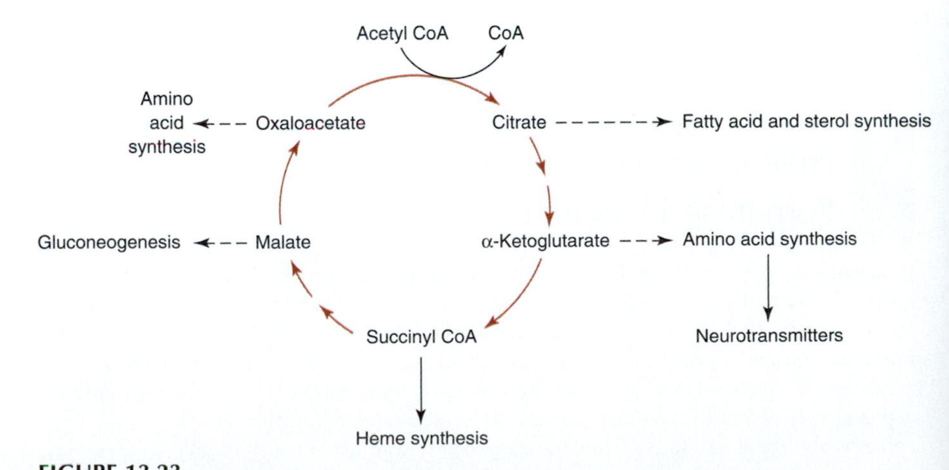

FIGURE 13.23
The TCA cycle serves as a source of precursors for amino acid, fatty acid, and glucose synthesis.

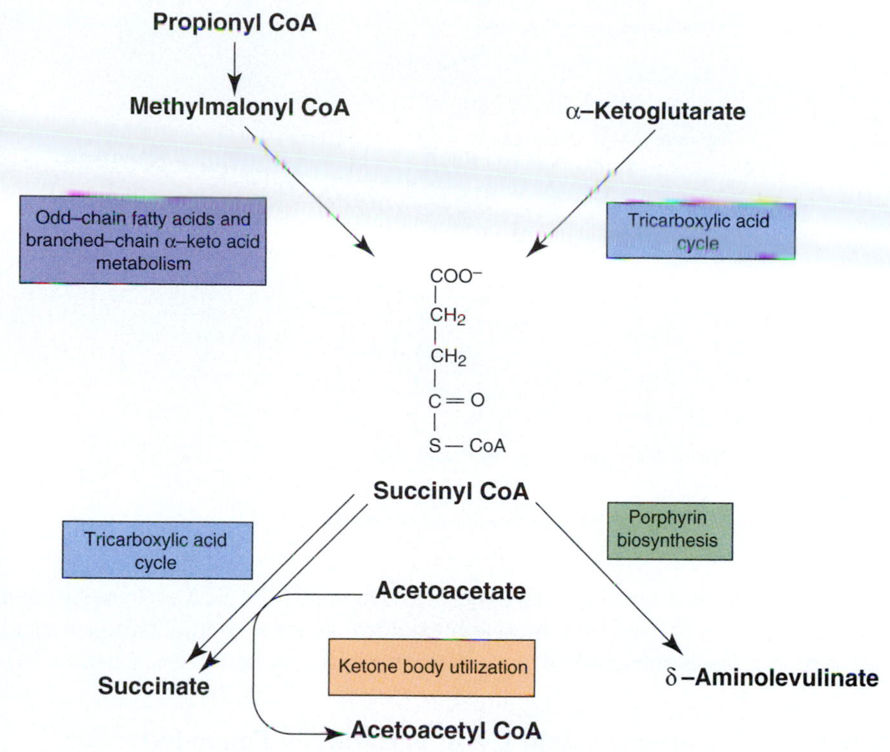

FIGURE 13.24
Sources and fates of succinyl CoA.

Finally, citrate is exported from mitochondria into the cytosol. **Citrate lyase** converts it to oxaloacetate and acetyl CoA, a precursor for biosynthesis of long-chain fatty acids and sterols. The oxaloacetate produced in this reaction is rapidly reduced to malate, which is converted by action of **malic enzyme** to form pyruvate and NADPH, a source of reducing equivalents for biosynthetic processes in cytosol. In addition, citrate serves as a regulatory effector of other metabolic pathways (see Chapters 14 and 16).

Anaplerotic Reactions Replenish Intermediates of Tricarboxylic Acid Cycle

In its role in catabolism, the TCA cycle oxidizes acetyl CoA with release of two acetate carbons as CO_2 with no concomitant loss of carbons from the four-carbon dicarboxylic acids, succinate, fumarate, malate, and oxaloacetate. Indeed, oxaloacetate, the acceptor of the acetate group, is regenerated during the cycle. However, metabolic pathways in all tissues remove intermediates of the TCA cycle for biosynthetic pathways. Hence, in order to maintain a functional TCA cycle, a source of four-carbon acids is required to replenish the resultant loss of oxaloacetate. The reactions that supply four- or five-carbon intermediates to the cycle are called **anaplerotic** (meaning "filling up") **reactions** (Figure 13.25). The most important is catalyzed by **pyruvate carboxylase,** which catalyzes conversion of pyruvate and CO_2 to oxaloacetate (Figure 13.26). The enzyme contains a covalently bound cofactor, biotin, which binds CO_2 in the presence of ATP and Mg^{2+} ions. The CO_2 group is then transferred to pyruvate as a carboxyl group (see p. 731). Levels of pyruvate carboxylase are high in both liver and nervous tissues, because these tissues have a constant efflux of intermediates used for gluconeogenesis in liver and neurotransmitter synthesis in nervous tissues.

Amino acids also serve as sources of four-carbon intermediates. Glutamate is converted to α-ketoglutarate in mitochondria by action of glutamate dehydrogenase. Aspartate is converted to oxaloacetate by transamination, while valine and

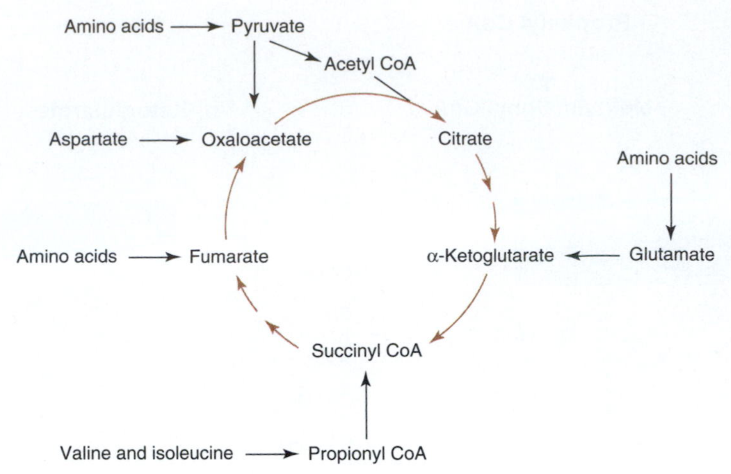

FIGURE 13.25
Anaplerotic reactions replenish intermediates of the TCA cycle.

isoleucine are broken down to propionyl CoA that enters the TCA cycle as succinyl CoA. Amino acids derived from muscle breakdown become an important source of four-carbon acids as substrates for gluconeogenesis during fasting (see Chapter 14).

Activity of Tricarboxylic Acid Cycle Is Carefully Regulated

Various factors regulate of the activity of the TCA cycle. First, the supply of acetyl units, whether derived from pyruvate (i.e., carbohydrate) or fatty acids, is a crucial factor in determining the rate of the cycle. Regulation of pyruvate dehydrogenase has an important effect on the cycle. Similarly, any control exerted on transport of fatty acids into mitochondria or rate of β-oxidation of fatty acids would serve as an effective determinant of the cycle activity. Second, because the dehydrogenases of the cycle are dependent on a continuous supply of both NAD^+ and FAD, their activities are very stringently controlled by the respiratory chain, which is responsible for oxidizing NADH and $FADH_2$. As discussed in Section 13.7 (see p. 577), the activity of the respiratory chain is coupled obligatorily to generation of ATP in reactions of oxidative phosphorylation, a process called **respiratory control.** Consequently, activity of the TCA cycle is very much dependent on rate of ATP synthesis (and hence rate of electron transport), which is strongly affected by availability of ADP, phosphate, and O_2. Hence an inhibitory agent or metabolic condition that interrupts supply of O_2, the continuous supply of ADP, or the source of reducing equivalents (e.g., substrate for the cycle) would result in decreased activity of the TCA cycle. In general, these control mechanisms of the TCA cycle are considered to function as a coarse control of the cycle. There are a variety of postulated effector-mediated regulatory interactions between various intermediates or nucleotides and individual enzymes of the cycle, which may serve to exert a fine control on the activity of the cycle. Illustrations of these interactions are shown in Figure 13.27. It should be noted that the physiological relevance of many of these types of individual regulatory interactions has not been established rigorously in intact metabolic systems.

Purified citrate synthase is inhibited by ATP, NADH, succinyl CoA, and long-chain acyl CoA derivatives; however, these effects have not been demonstrated in intact metabolic systems under physiological conditions. The most probable means for regulating the citrate synthase reaction is the availability of its two substrates, acetyl CoA and oxaloacetate. As discussed above, very low concentrations of oxaloacetate (lower than the K_m for oxaloacetate on citrate synthase) are present in mitochondria.

The NAD^+-linked isocitrate dehydrogenase, often considered to be the key regulatory enzyme of the TCA cycle, is stimulated by ADP and in some cases AMP and

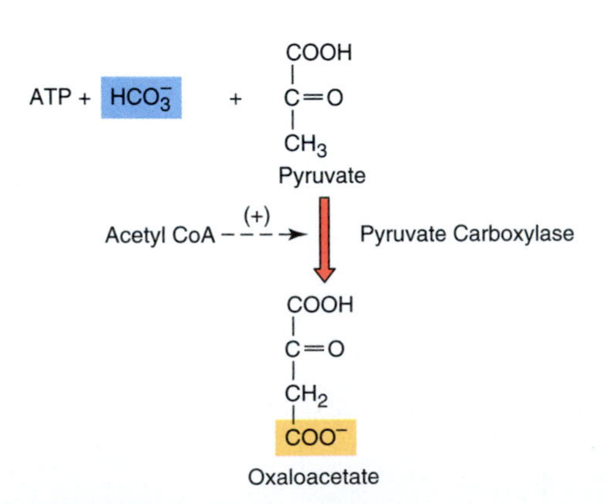

FIGURE 13.26
Pyruvate carboxylase reaction.

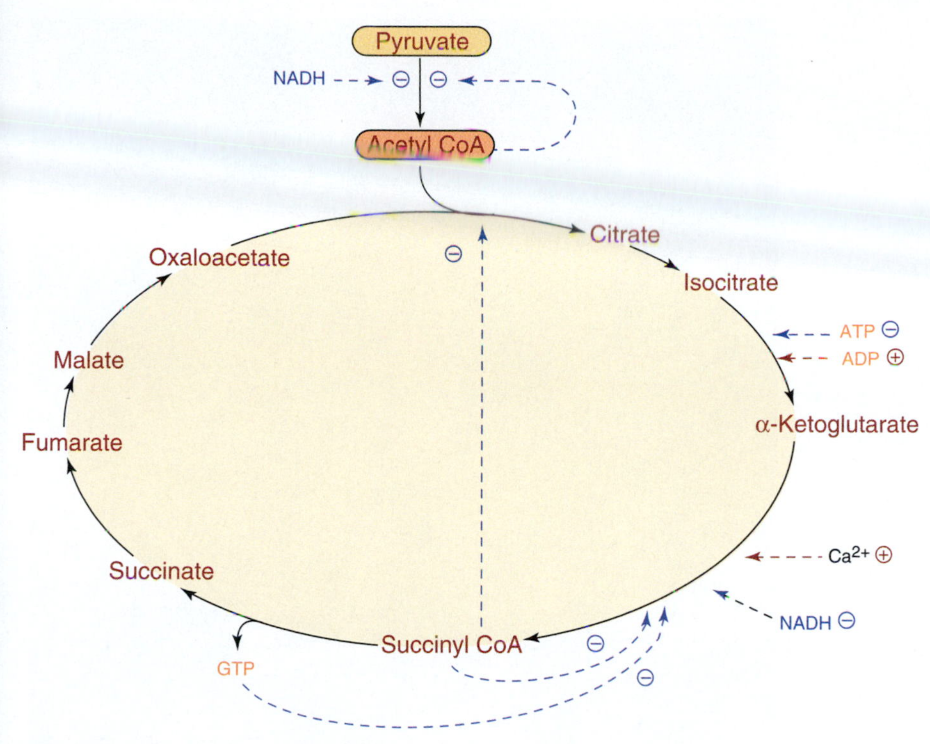

FIGURE 13.27
Representative examples of regulatory interactions in the TCA cycle.

is inhibited by ATP and NADH. Hence, under high-energy conditions (i.e., high ATP/ADP + P_i and high $NADH/NAD^+$ ratios), the activity of NAD^+-linked isocitrate dehydrogenase of the TCA cycle is inhibited. By contrast, during periods of low energy the activity of this enzyme and consequently the complete TCA cycle is stimulated. Respiratory control by the electron transport chain coupled to ATP synthesis thus acts to regulate the TCA cycle at the NAD^+-linked isocitrate dehydrogenase step by affecting levels of ADP and NAD^+ necessary to activate or inhibit the enzyme.

The other irreversible enzyme of the TCA cycle, the α-ketoglutarate dehydrogenase complex, is inhibited by both ATP and GTP, NADH, and succinyl CoA, while Ca^{2+} has been shown to activate the complex in certain tissues. Unlike the pyruvate dehydrogenase complex, α-ketoglutarate dehydrogenase is not regulated by a protein kinase-mediated phosphorylation.

13.5 | STRUCTURE AND COMPARTMENTATION BY MITOCHONDRIAL MEMBRANES

The final steps in breakdown of carbohydrates and fatty acids are located in **mitochondria** where energy released during oxidation of NADH and $FADH_2$ is transduced into chemical energy of ATP. This process is called oxidative phosphorylation and mitochondria are often called the **"powerhouse of the cell."** The role of a given tissue in aerobic metabolic functions and its need for energy based on its physiological function are reflected in the number and activity of its mitochondria (Figure 13.28). Cardiac muscle is highly aerobic, needing a constant supply of ATP. It has been estimated that about one-half of the cytoplasmic volume of cardiac cells is composed of mitochondria, which contain numerous infolding of the inner membrane called **cristae** and, consequently, a high concentration of the enzyme complexes of the electron transport chain. The liver is also highly dependent on aerobic metabolic processes for its various functions, and it has been estimated that

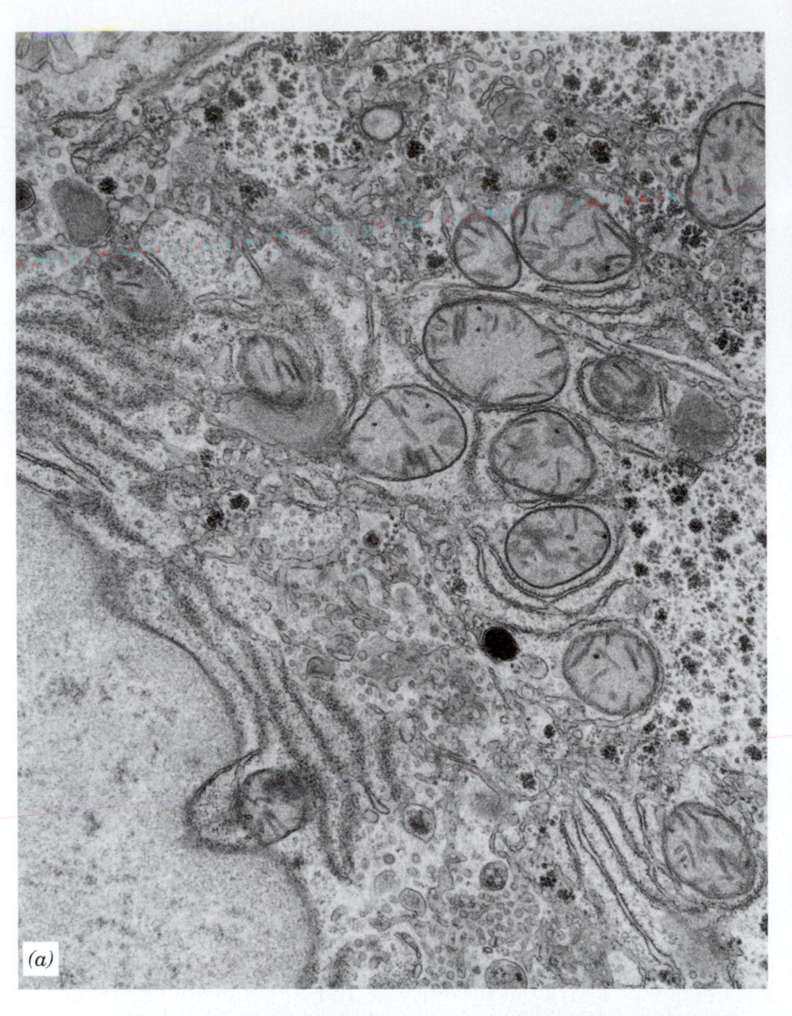

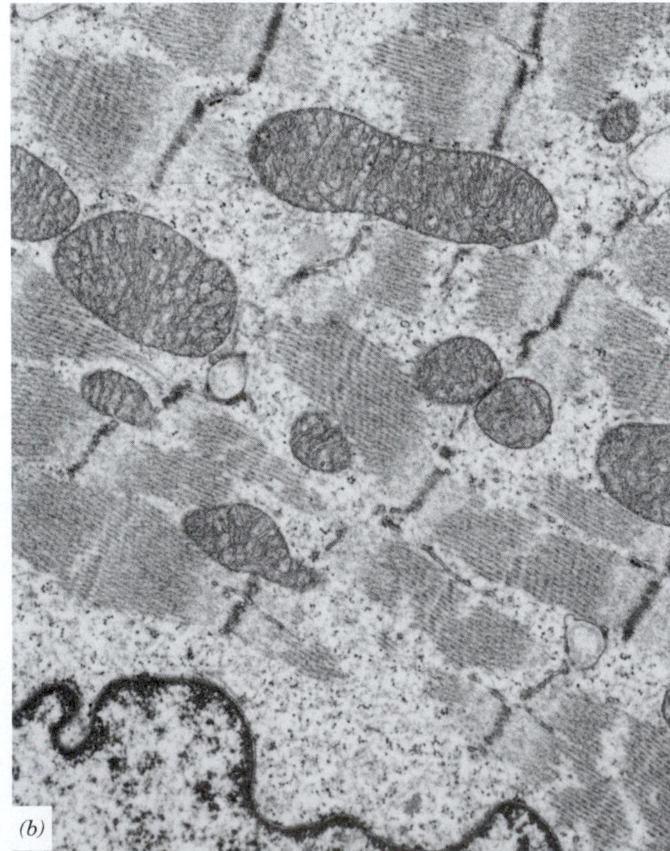

FIGURE 13.28

Electron micrographs of mitochondria (*a*) in hepatocytes from rat liver (× 39,600) and (*b*) in muscle fibers from rabbit heart (×39,600).

Courtesy of Dr. W. B. Winborn, Department of Anatomy, The University of Texas Health Science Center at San Antonio, and the Electron Microscopy Laboratory, Department of Pathology, The University of Texas Health Science Center at San Antonio.

mammalian hepatocytes contain between 800 and 2000 mitochondria. By contrast, erythrocytes have no mitochondria and do not possess the capacity to generate energy using O_2 as a terminal electron acceptor and instead obtain energy from glycolysis. Mitochondria exist in a variety of different shapes, depending on the cell type from which they are derived. As is seen in Figure 13.28, mitochondria from liver are nearly spherical in shape, whereas those found in cardiac muscle are oblong or cylindrical and contain more numerous cristae than do liver mitochondria.

Inner and Outer Mitochondrial Membranes Have Different Compositions and Functions

Mitochondria have two membranes, an **outer membrane** and a structurally and functionally complex **inner membrane** (Figure 13.29); the space between is the **intermembrane space.** Several enzymes involved in transfer of the high-energy bond of ATP, such as adenylate kinase, creatine kinase, and nucleoside diphosphate kinase, are located in the intermembrane space (Table 13.5). The outer membrane is composed of about 30–40% lipid and 60–70% protein, with relatively few enzymatic or transport functions. The outer membrane is rich in a protein called **porin,** a membrane-spanning protein consisting of β-sheets, which forms a channel to permit the movement of molecules with molecular weights up to 10,000 through the membrane. In addition, monoamine oxidase and kynurenine hydroxylase, enzymes of importance in nervous tissues for removal of neurotransmitters, are located on the outer surface of the outer membrane.

The inner membrane, a complex membrane, consists of 80% protein and is rich in unsaturated fatty acids. In addition, **cardiolipin** and diphosphatidylglycerol are present in high concentrations in this membrane. The enzyme complexes involved in electron transport and oxidative phosphorylation are located in the inner membrane as well as various dehydrogenases and several transport systems, which are involved in transferring substrates, metabolic intermediates, and adenine nucleotides between cytosol and matrix. The inner membrane appears to be invaginated forming folds termed **cristae,** increasing the surface area (Figure 13.29). The space inside the inner membrane, the **matrix,** contains the enzymes of the TCA cycle with the exception of succinate dehydrogenase, which is bound to the inner membrane, and enzymes for fatty acid oxidation. In addition, mitochondrial DNA (mtDNA), ribosomes, and proteins necessary for transcription of mtDNA and translation of mRNA are located in the matrix.

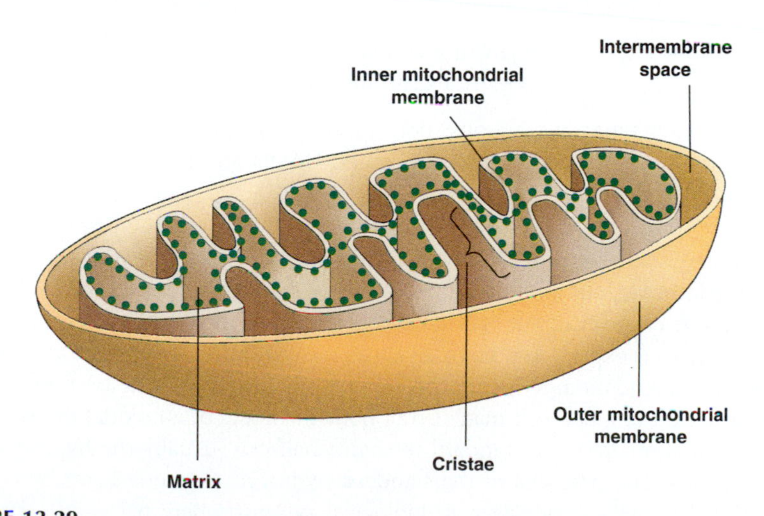

FIGURE 13.29
Diagram of various submitochondrial compartments.
Green spheres represent localization of F_1 portion of ATP synthase on the inner mitochondrial membrane.

TABLE 13.5 Enzymatic Composition of the Various Mitochondrial Subcompartments

Outer Membrane	*Intermembrane Space*	*Inner Membrane*	*Matrix*
Monoamine oxidase	Adenylate kinase	Succinate dehydrogenase	Pyruvate dehydrogenase
Kynurenine hydroxylase	Nucleoside diphosphate kinase	F_1F_o ATP synthase	Citrate synthase
Nucleoside diphosphate kinase	Creatine kinase	NADH dehydrogenase	Isocitrate dehydrogenase
Phospholipase A		β-Hydroxybutyrate dehydrogenase	α-Ketoglutarate dehydrogenase
Fatty acyl-CoA synthetases		Cytochromes b, c_1, c, a, a_3	Aconitase
NADH: cytochrome-c reductase (rotenone-insensitive)		Carnitine: acyl-CoA transferase	Fumarase
			Succinyl-CoA synthetase
Choline phosphotransferase		Adenine nucleotide translocase	Malate dehydrogenase
		Mono-, di-, and tricarboxylate translocase	Fatty acid β-oxidation system
			Glutamate dehydrogenase
		Glutamate–aspartate translocase	Glutamate–oxaloacetate transaminase
		Glycerol 3-phosphate dehydrogenase	Ornithine transcarbamoylase
			Carbamoyl phosphate synthetase I
			Heme synthesis enzymes

13.6 | ELECTRON TRANSPORT CHAIN

During reactions involved in fatty acid oxidation and the TCA cycle, reducing equivalents are derived from sequential breakdown and oxidation of the substrates. The reducing equivalents are transferred to NADH and $FADH_2$, which are subsequently oxidized by the **electron transport chain,** a system of electron carriers located in the inner membrane. In the presence of O_2, this system converts reducing equivalents into utilizable energy, as ATP, by the process of oxidative phosphorylation. The complete oxidation of NADH and $FADH_2$ by the electron transport chain results in production of approximately 2.5 and 1.5 mol of ATP per mole of reducing equivalent transferred to O_2, respectively.

Oxidation–Reduction Reactions

The mitochondrial electron transport system is little more than a sequence of linked oxidation–reduction reactions. Oxidation–reduction reactions occur when there is a transfer of electrons from a suitable electron donor **(reductant)** to a suitable electron acceptor **(oxidant).** In some oxidation–reduction reactions only electrons are transferred from reductant to oxidant (i.e., electron transfer between cytochromes),

$$\text{cytochrome } c\ (Fe^{2+}) + \text{cytochrome } a\ (Fe^{3+}) \rightleftharpoons \text{cytochrome } c\ (Fe^{3+}) + \text{cytochrome } a\ (Fe^{2+})$$

whereas in other types of reactions, both electrons and protons (hydrogen atoms) are transferred (e.g., electron transfer between NADH and FAD):

$$NADH + H^+ + FAD \rightleftharpoons NAD^+ + FADH_2$$

Oxidized and reduced forms of compounds or groups operating in oxidation–reduction reactions are referred to as **redox couples** or pairs. The facility with which a given electron donor (reductant) gives up its electrons to an electron acceptor (oxidant) is expressed quantitatively as the **oxidation–reduction potential** of the system. An oxidation–reduction potential is measured in volts as an **electromotive force** (emf) of a half-cell made up of both members of an oxidation–reduction couple when compared to a standard reference half-cell (usually the hydrogen electrode reaction). The potential of the standard hydrogen electrode is set by convention at 0.0 V at pH 0.0; however, in biological systems where pH is 7.0, the reference hydrogen potential becomes −0.42 V. The oxidation–reduction potentials for a variety of important biochemical reactions are tabulated in Table 13.6. To interpret

TABLE 13.6 Standard Oxidation–Reduction Potentials for Various Biochemical Reactions

Oxidation–Reduction System	*Standard Oxidation–Reduction Potential E_0' (V)*
Acetate + $2H^+$ + $2e^-$ ⇌ acetaldehyde	−0.60
$2H^+$ + $2e^-$ ⇌ H_2	−0.42
Acetoacetate + $2H^+$ + $2e^-$ ⇌ β-hydroxybutyrate	−0.35
NAD^+ + $2H^+$ + $2e^-$ ⇌ NADH + H^+	−0.32
Acetaldehyde + $2H^+$ + $2e^-$ ⇌ ethanol	−0.20
Pyruvate + $2H^+$ + $2e^-$ ⇌ lactate	−0.19
Oxaloacetate + $2H^+$ + $2e^-$ ⇌ malate	−0.17
Coenzyme Q_{ox} + $2e^-$ ⇌ coenzyme Q_{red}	+0.10
Cytochrome *b* (Fe^{3+}) + e^- ⇌ cytochrome *b* (Fe^{2+})	+0.12
Cytochrome *c* (Fe^{3+}) + e^- ⇌ cytochrome *c* (Fe^{2+})	+0.22
Cytochrome *a* (Fe^{3+}) + e^- ⇌ cytochrome *a* (Fe^{2+})	+0.29
$\frac{1}{2}O_2$ + $2H^+$ + $2e^-$ ⇌ H_2O	+0.82

the data in the table, remember that the reductant of an oxidation–reduction pair with a large negative potential will give up its electrons more readily than redox pairs with smaller negative or positive redox potentials. Compounds with large negative potentials are considered to be strong reducing agents. By contrast, a strong oxidant (e.g., characterized by a large positive potential) has a very high affinity for electrons and will act to oxidize compounds with more negative standard potentials.

The **Nernst equation** characterizes the relationship between standard oxidation–reduction potential of a particular redox pair (E_0'), observed potential (E), and ratio of concentrations of oxidant and reductant in the system:

$$E = E_0' + 2.3\ RTn\mathcal{F} \log\left(\frac{[\text{oxidant}]}{[\text{reductant}]}\right)$$

where E is observed potential when all reactants are present at a concentration of 1 M. E_0' is the standard potential at pH 7.0. R is the gas constant of 8.3 J deg^{-1} mol^{-1}, T is absolute temperature in kelvin units (K), n is number of electrons being transferred, and $\mathcal{F}$ is the Faraday constant of 96,500 J V^{-1}.

When an observed potential is equal to the standard potential, a potential is defined that is referred to as the midpoint potential where concentration of oxidant is equal to that of reductant. Knowing standard oxidation–reduction potentials of a diverse variety of biochemical reactions allows one to predict direction of electron flow or transfer when more than one redox pair is linked together by the appropriate enzyme that causes a reaction to occur. For example, Table 13.6 shows that the NAD^+–NADH pair has a standard potential of −0.32 V, and the pyruvate–lactate pair has a standard potential of −0.19V. This means that electrons will flow from NAD^+–NADH to pyruvate–lactate as long as lactate dehydrogenase is present as indicated below:

$$\text{pyruvate} + \text{NADH} + H^+ \rightleftharpoons \text{lactate} + NAD^+$$

In the electron transfer system, electrons or reducing equivalents are produced in NAD^+- and FAD-linked dehydrogenase reactions, which have standard potentials at or close to that of NAD^+–NADH. The electrons are subsequently transferred through the electron transfer chain, which has as its terminal acceptor the O_2–water couple with a standard redox potential of +0.82 V.

Free-Energy Changes in Redox Reactions

Oxidation–reduction potential differences between two redox pairs are similar to free-energy changes in chemical reactions, in that both quantities depend on concentration of reactants and products of the reaction and the following relationship exists:

$$\Delta G^{o\prime} = -n\mathcal{F}\,\Delta E_0'$$

Using this expression, the free-energy change for electron transfer reactions can be calculated if the potential difference between two oxidation–reduction pairs is known. Hence, for the mitochondrial electron transfer process in which electrons are transferred between the NAD^+–NADH couple ($E_0' = -0.32$ V) and the $\frac{1}{2}O_2$–H_2O couple ($E_0' = +0.82$ V), the free-energy change for this process can be calculated:

$$\Delta G^{o\prime} = -n\mathcal{F}\,\Delta E_0' = -2 \times 96.5 \text{ kJ V}^{-1} \times 1.14 \text{ V}$$
$$\Delta G^{o\prime} = -219 \text{ kJ mol}^{-1}$$

where 96.5 is the Faraday constant in kJ V^{-1} and n is number of electrons transferred; for example, in the case of NADH $\rightarrow O_2$, $n = 2$. The free energy available from the potential span between NADH and O_2 in the electron transfer chain is capable of generating more than enough energy to synthesize three molecules of ATP per two reducing equivalents or two electrons transported to O_2. In addition, because of the negative sign of the free energy available in the electron transfer, this process is exergonic and will proceed provided that the necessary enzymes are present.

Mitochondrial Electron Transport Is a Multicomponent System

The final steps in the overall oxidation of foodstuffs—carbohydrates, fats, and amino acids—results in formation of NADH and $FADH_2$ in the matrix. The electron transport chain oxidizes these reduced cofactors by transferring electrons in a series of steps to O_2, the terminal electron acceptor, while capturing the free energy of the oxidation–reduction reactions to drive the synthesis of ATP. During removal of electrons from the coenzymes, protons are also removed and pumped from the matrix across the inner membrane to form an electrochemical gradient, which provides energy for synthesis of ATP (Figure 13.30). The various electron carriers involved in transfer of electrons from NADH to O_2 have standard redox potentials that span the range from that of the most electronegative electron donor NADH, with a standard redox potential of −0.32 V, to the most electropositive electron acceptor O_2 with a standard redox potential of +0.82 V (Table 13.6). While electrons are removed from reactants with more negative reduction potentials and transferred to electron acceptors with more positive reduction potentials, the mitochondrial

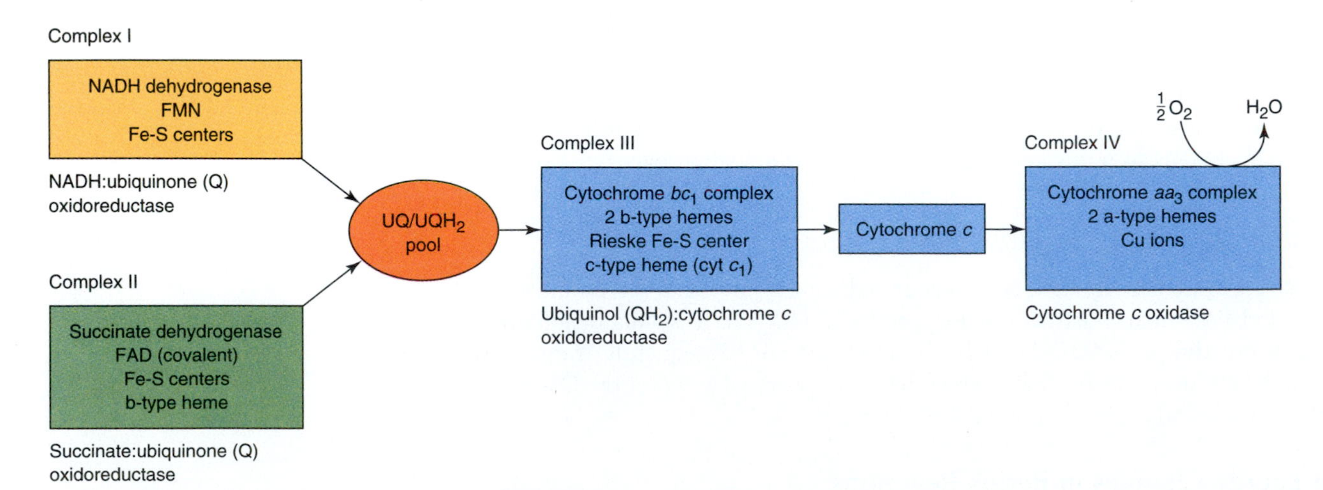

FIGURE 13.30
Overview of the complexes and pathways of electron transfer in mitochondrial electron transport chain.

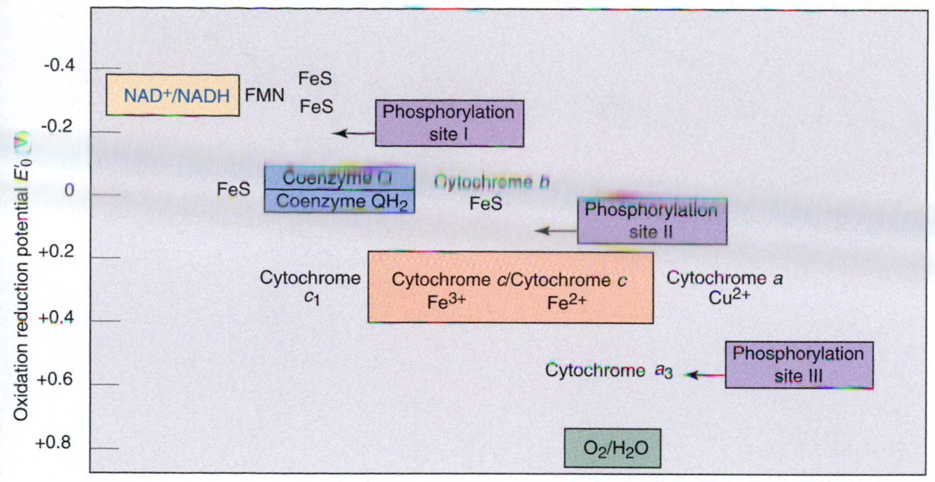

FIGURE 13.31
Oxidation–reduction potentials of the mitochondrial electron transport chain carriers.

electron carriers do not exist in a linear arrangement. Instead, the electron carriers are grouped into four large multisubunit enzyme complexes (complexes I–IV) that catalyze different partial reactions of the electron transport chain.

The four complexes of the electron transport chain include: **complex I**, NADH–ubiquinone oxidoreductase, that catalyzes the transfer of electrons from NADH to **ubiquinone (UQ)** or also referred to as **coenzyme Q (CoQ); complex II**, succinate–ubiquinone oxidoreductase or succinate dehydrogenase, that transfers electrons from succinate to coenzyme Q; **complex III**, the **cytochrome** bc_1 **complex**, ubiquinol–cytochrome *c* reductase, that transfers electrons from ubiquinol (reduced form of ubiquinone abbreviated as **$CoQH_2$** or **UQH_2**) to **cytochrome** ***c***; and **complex IV**, cytochrome *c* oxidase, that transfers electrons from cytochrome *c* to O_2 (Figure 13.30). Another multiprotein complex, the **ATP synthase**, also called **complex V**, uses energy of the electrochemical gradient produced during electron transfer for synthesis of ATP (Figure 13.30). As shown in Figure 13.31, the standard reduction potentials of electron carriers in the four complexes span the range of redox potentials from NADH to O_2. These four complexes consist of several different electron carriers including **flavoproteins** that contain tightly bound FMN or FAD and can transfer one or two electrons, the heme-containing proteins, **cytochromes** (cytochromes b, c_1, c, a, and a_3), that transfer one electron from Fe^{2+} of heme, **iron–sulfur proteins** containing bound inorganic Fe and S that transfer one electron, and **copper** in complex IV (cytochrome *c* oxidase) that is involved in one electron transfer. UQ participates in either one or two electron transfer reactions. In the following section, the structure and function of the four complexes will be described in detail.

Complex I: NADH–Ubiquinone Oxidoreductase

Complex I is the most complicated and, in mammals, consists of at least 40 different polypeptides with a molecular mass of approximately 1,000,000 Da. Complex I, often called NADH dehydrogenase, oxidizes NADH and transfers electrons to ubiquinone (coenzyme Q). The first step in oxidation of NADH involves transfer of two electrons and two protons from NADH to **FMN**, flavin mononucleotide (Figure 13.32), a cofactor tightly bound to one polypeptide subunit of complex I. Two electrons and two protons are added across the double bond of riboflavin, the reactive part of FMN, also known as vitamin B_2.

$$NADH + H^+ + FMN \rightleftharpoons NAD^+ + FMNH_2$$

The electrons are subsequently transferred from FMN via a series of FeS centers, both 2Fe2S and 4Fe4S (Figure 13.33) to ubiquinone. One important role of FMN

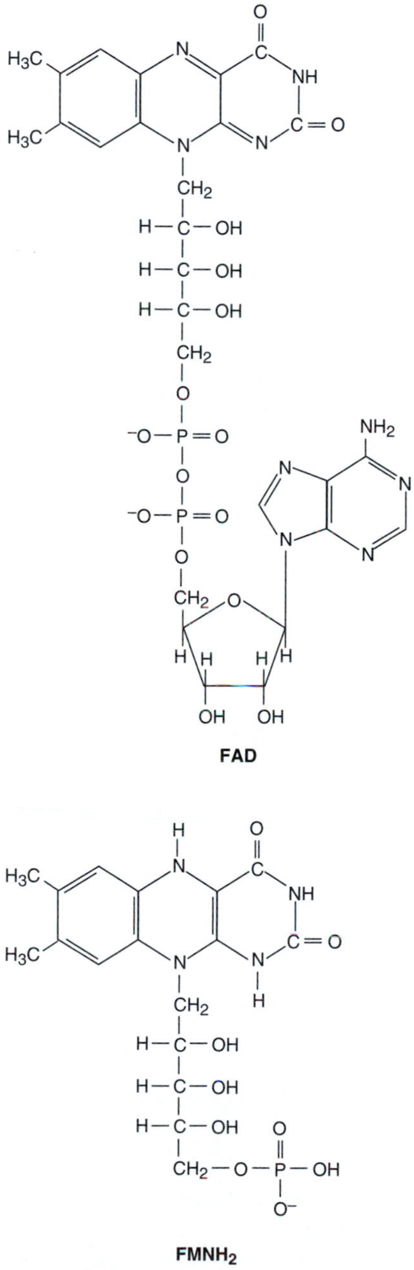

FAD

$FMNH_2$

FIGURE 13.32
Structures of flavin adenine dinucleotide (FAD) and flavin mononucleotide ($FMNH_2$).

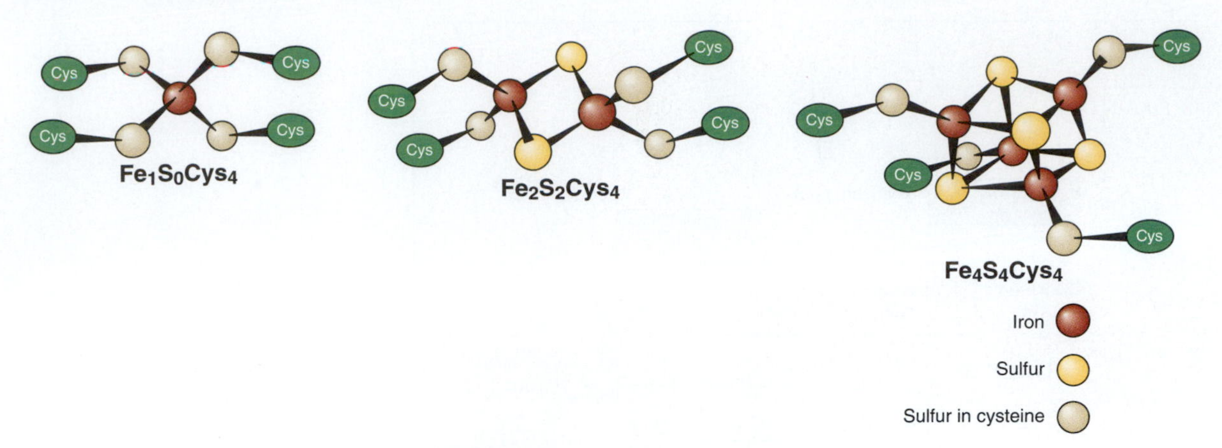

FIGURE 13.33
Structures of iron–sulfur centers.
Yellow, inorganic sulfur; gray, sulfur in cysteine; and red, iron.

in complex I is to serve as both a two-electron acceptor from NADH and as a one-electron donor to the FeS centers. The ability of FMN to act as both a one- and two-electron carrier results from existence of a stable semiquinone form of FMN. Ubiquinone, the terminal electron acceptor of complex I, can also act as a one- or two-electron acceptor due to the presence of a stable semiquinone intermediate (Figure 13.34). In addition to its hydrophilic quinone portion, ubiquinone also has a long hydrophobic side chain consisting of 10 isoprene units, which is buried in the membrane lipid bilayer. Ubiquinone and ubiquinol are both freely diffusible in the membrane and can act to transfer electrons from complexes I and II to complex III, ubiquinol–cytochrome *c* oxidoreductase.

During oxidation of one NADH and transfer of two electrons to ubiquinone by complex I, four protons are also pumped across the mitochondrial membrane from the matrix side (**N** for negative face) to the cytosolic side (**P** for positive face). The energy released during oxidative reactions occurring in complex I is conserved by concomitant pumping of protons across the membrane. Little is known of the mechanism by which protons are transferred across the membrane during electron transfer in complex I. Figure 13.35 provides a schematic representation of the events occurring during electron transfer and proton pumping through complex I.

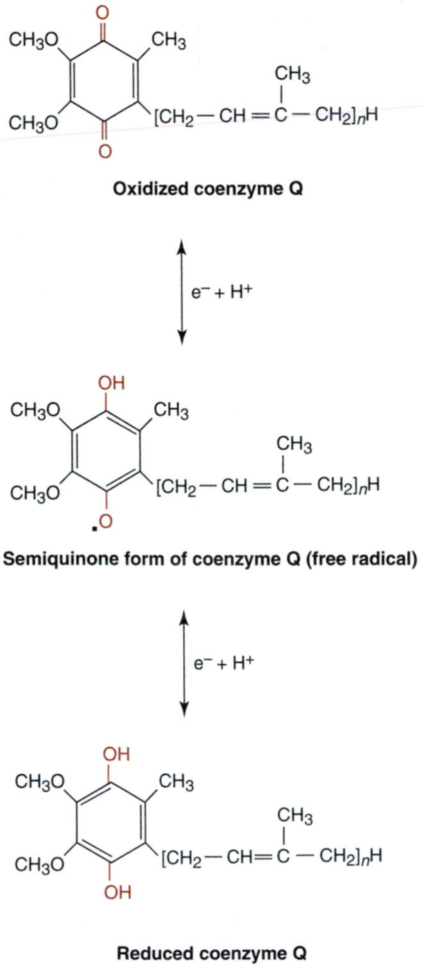

Oxidized coenzyme Q

$e^- + H^+$

Semiquinone form of coenzyme Q (free radical)

$e^- + H^+$

Reduced coenzyme Q

FIGURE 13.34
Oxidation–reduction of ubiquinone (coenzyme Q).
Note that ubiquinone can accept one electron at a time, forming an intermediate semiquinone.

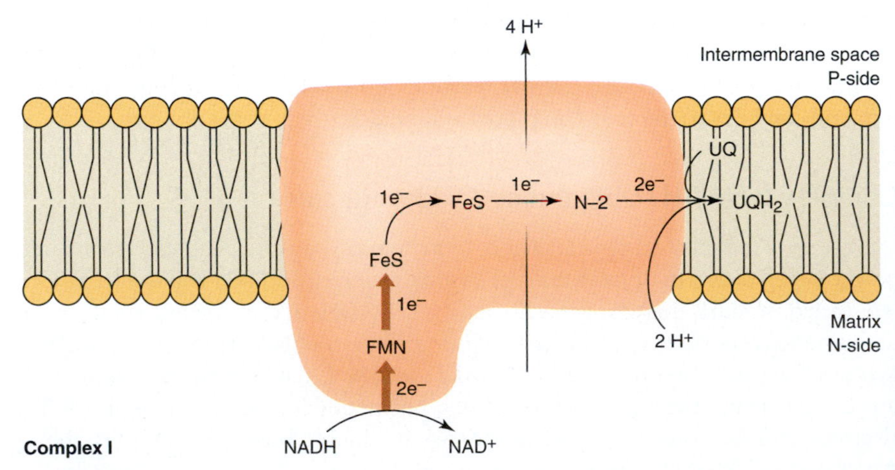

FIGURE 13.35
Model of complex I.
The presence of both the membrane and peripheral domains are indicated. Electrons are transferred from NADH to ubiquinone (UQ) in the membrane via FMN and several FeS centers.

Complex II: Succinate–Ubiquinone Oxidoreductase

Complex II, better known as **succinate dehydrogenase,** catalyzes oxidation of succinate. The complex is composed of four subunits: a 70,000-MW subunit that contains FAD covalently bound to a histidine residue, a 30,000-MW subunit that contains three iron–sulfur centers, and two small hydrophobic proteins. During oxidation of succinate to fumarate, two electrons and two protons are first transferred to FAD (Figure 13.36). The $FADH_2$, thus formed, transfers electrons to ubiquinone via FeS centers of complex II in the following two reactions:

$$\text{succinate} \longrightarrow \text{fumarate} + 2\ \text{H}^+ + 2\ \text{e}^-$$

$$\text{UQ} + 2\ \text{H}^+ + 2\ \text{e}^- \longrightarrow \text{UQH}_2$$

$$\text{(overall)} \qquad \text{succinate} + \text{UQ} \longrightarrow \text{fumarate} + \text{UQH}_2$$

$$\Delta E_0' = 0.029\ \text{V}; \qquad \Delta G^{\circ\prime} = -5.6\ \text{kJ mol}^{-1}$$

The small amount of free energy liberated during oxidation of succinate and transfer of electrons to ubiquinone is insufficient for proton pumping across the mitochondrial membrane and hence no gain in free energy is accomplished by the reactions of complex II. Figure 13.36 provides a schematic representation of events occurring during electron transfer through complex II.

Other Mitochondrial Flavoprotein Dehydrogenases

Other dehydrogenases located in mitochondria feed electrons into the electron transport chain at the level of ubiquinone. Glycerol 3-phosphate, formed by release of glycerol from triacylglycerols or by reduction of dihydroxyacetone phosphate produced during glycolysis, is oxidized by **glycerol 3-phosphate dehydrogenase** (Figure 13.36). This flavoprotein contains a single polypeptide chain and is localized on the outer face of the inner membrane where it transfers electrons directly to ubiquinone in the mitochondrial membrane. The importance of glycerol 3-phosphate dehydrogenase in shuttling of reducing equivalents from NADH in the cytosol to the mitochondrial electron transport chain will be discussed in Section 13.8 (see p. 583).

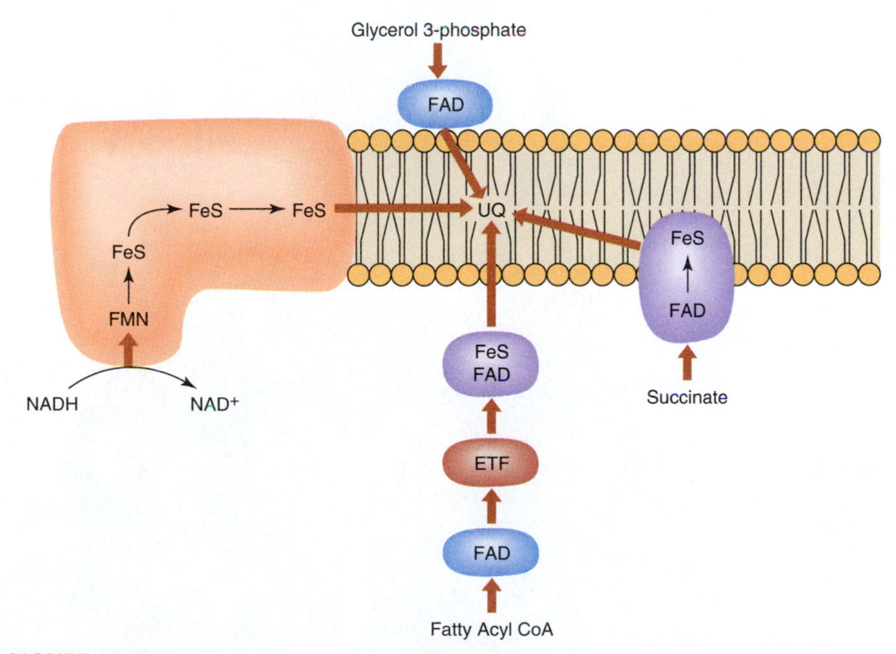

FIGURE 13.36
Reduction of ubiquinone (UQ) in the mitochondrial inner membrane by the flavoproteins NADH, succinate, glycerol 3-phosphate, and fatty acyl-CoA dehydrogenases.

The second dehydrogenase of importance in intermediary metabolism is **acyl-CoA dehydrogenase,** a flavoprotein that catalyzes the first step in β-oxidation of fatty acids. This enzyme transfers electrons from fatty acyl-CoA to FAD on the dehydrogenase, which are then transferred to **e**lectron **t**ransferring **f**lavoprotein **(ETF).** Electrons are then transferred from ETF to **ETF–ubiquinone oxidoreductase** that transfers electrons directly to ubiquinone in the inner membrane. Figure 13.36 illustrates the reduction of the ubiquinone pool by all of these flavoprotein dehydrogenases including complex I, complex II, glycerol 3-phosphate dehydrogenase, and ETF–ubiquinone oxidoreductase. Ubiquinol produced by these dehydrogenases is oxidized by complex III.

Complex III: Ubiquinol–Cytochrome *c* Oxidoreductase

Complex III, also called the cytochrome bc_1 complex, catalyzes transfer of electrons from ubiquinol to cytochrome *c* coupled to the translocation of protons across the inner mitochondrial membrane. In mammals this multisubunit enzyme complex consists of 11 subunits of which three have prosthetic groups that serve as redox centers, **cytochrome *b*** and **cytochrome c_1,** which each contain a heme group, and the **Rieske iron–sulfur protein,** which contains a 2Fe2S cluster. The recent resolution of the complete structure of complex III by X-ray crystallography has provided valuable structural information that has proved useful in explaining previous biochemical observations and in suggesting novel insights into the function of this complex (Figure 13.37). In the crystal structure, the dimeric complex with a molecular mass of 250,000 Da for each monomer, is pear-shaped with a large domain protruding 75 Å into the mitochondrial matrix and a smaller domain containing the head groups of the Rieske iron–sulfur protein and cytochrome c_1 protruding into the intermembrane space. The transmembrane domain of complex III consists of eight α-helices of the hydrophobic protein, cytochrome *b*, plus membrane-anchoring helices of the iron–sulfur protein, cytochrome c_1, and other subunits of the complex.

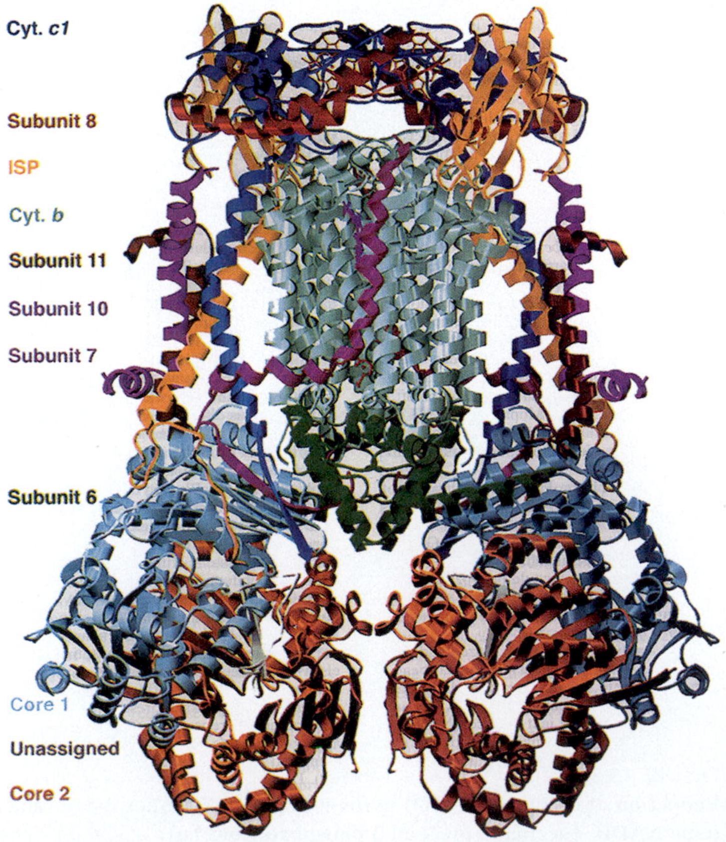

FIGURE 13.37
Model of the crystal structure of the dimeric cytochrome bc_1 complex.
The α helices of cytochrome *b* (pale green) form the transmembrane domain of the complex. The enzyme protrudes 75 Å into the matrix and 38 Å into the intermembrane space. Colors identifying the subunits are shown on the left.
Reprinted with permission from Kim, H., Xia, D., Yu, C.-A., Xia, J.-Z., Kachurin, A. M., Zhang, L., Yu, L., and Deisenhofer, J. Proc. Natl. Acad. Sci. USA *95:8026, 1998. Copyright 1998, National Academy of Sciences, U.S.A. Figure generously supplied by Dr. J. Deisenhofer.*

Cyt. *c1*
Subunit 8
ISP
Cyt. *b*
Subunit 11
Subunit 10
Subunit 7
Subunit 6
Core 1
Unassigned
Core 2

Cytochromes

The cytochromes are a class of proteins that contain an iron-containing heme group tightly bound to the protein (see p. 1063). Unlike the heme group in hemoglobin or myoglobin in which heme iron remains in the Fe^{2+} state, iron in heme of a cytochrome is alternately oxidized (Fe^{3+}) or reduced (Fe^{2+}) as it functions in the electron transport chain. The cytochromes of mammalian mitochondria are designated as *a*, *b*, and *c* on the basis of the α band of their absorption spectrum and the type of heme group attached to the protein (Figure 13.38). The absorption band of the different cytochromes and their standard redox potentials depend on the structure of the heme and its environment in the protein. Cytochrome *b* and other ***b*-type cytochromes** contain iron–protoporphyrin IX (Figure 13.38), the same heme found in hemoglobin and myoglobin. The heme of cytochrome *b*, however, is buried in the membrane and cannot bind O_2. The ***c*-type cytochromes** contain heme *c* that is covalently bound to cysteine residues of the protein via thioether linkages involving vinyl side chains of protoporphyrin IX. The ***a*-type cytochromes** contain heme *a* that is a modified form of protoporphyrin IX (see p. 1063) in which a

Heme *a*

Heme *b*

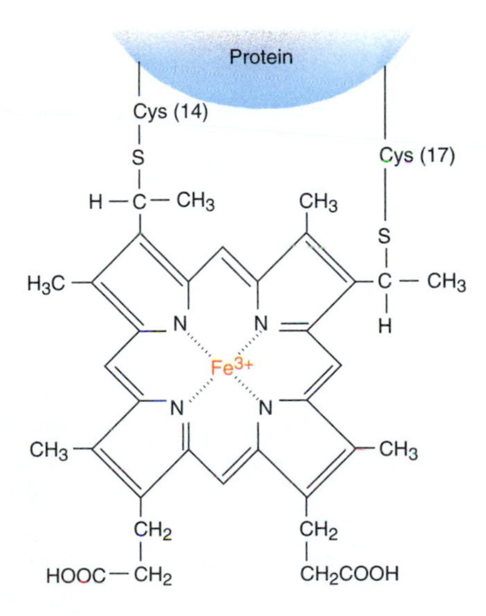

Heme *c*

FIGURE 13.38
Structures of heme *a*, heme *b*, and heme *c*.

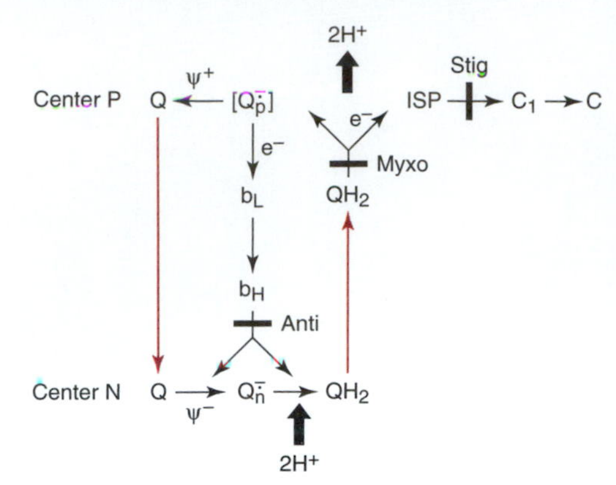

FIGURE 13.39
The Q cycle.
Q cycle explains electron transfer through cytochrome bc_1 complex. QH_2 is reduced ubiquinol, $[Q^-_p]$ is the semiquinone formed at center Q_O and Q^-_n is the semiquinone formed at center Q_I. ISP is iron–sulfur protein. Sites of inhibitors myxothiazol (Myxo), stigmatellin (Stig), and antimycin (Anti) are indicated.

formyl group and an isoprenoid side chain have been added to the molecule. Two forms of cytochrome *a* are present in cytochrome *c* oxidase, complex IV of the electron transport chain.

Q-Cycle Mechanism for Electron Transfer and Proton Pumping in Complex III

Two heme *b* groups with different redox potentials, a high-potential heme of +0.50 V and a low-potential heme of −0.100 V, are bound to cytochrome *b* in the cytochrome bc_1 complex, or complex III. Complex III also contains one *c*-type cytochrome, known as **cytochrome c_1.** The transfer of electrons through complex III is best explained by the **Q-cycle mechanism** in which four protons are translocated across the mitochondrial membrane per two electrons transferred from ubiquinol to cytochrome *c* (Figure 13.39). For electron transfer to continue according to the Q cycle, two separate ubiquinone or ubiquinol-binding sites are required in the bc_1 complex. A ubiquinol oxidizing site (Q_O) involving the low-potential heme *b* **(heme b_L)** is located at the positive (P) side of the membrane, where inhibitors such as myxothiazol bind. A ubiquinone-reducing site (Q_i), involving the high-potential heme *b* (heme b_H) is located at the negative (N) side of the membrane, where inhibitors such as antimycin bind. The oxidation of ubiquinol at the Q_O site results in the transfer of one electron to the 2Fe2S cluster of the iron–sulfur protein, which subsequently is oxidized by transfer of an electron to heme of cytochrome c_1. The strongly reducing ubisemiquinone anion formed during ubiquinol oxidation in the Q_O site immediately reduces low-potential heme of cytochrome b_L that rapidly transfers an electron to high-potential heme of cytochrome b_H at the Q_i site. The reduced cytochrome b_H is then oxidized by transfer of an electron to ubiquinone at the Q_i site to form a stable ubisemiquinone. To complete the Q cycle, a second molecule of ubiquinol is oxidized at the Q_O site and transfers one electron to the iron–sulfur protein and the second electron to heme *b* with eventual reduction of ubisemiquinone at the Q_i site to form ubiquinol. Overall, the Q cycle provides a mechanism to explain the observation that four protons are pumped across the inner membrane during transfer of two electrons from ubiquinol to cytochrome *c*. Two protons are taken up on the matrix side of the inner membrane to reduce the ubiquinone at the Q_i site, while four protons are released on the cytosolic side of the membrane during transfer of two electrons to the iron–sulfur protein and cytochrome c_1.

Proposed Movement of the Iron–Sulfur Protein During Electron Transfer in Complex III

Recent crystallographic studies have indicated that the **iron–sulfur protein** can exist in several different conformations in complex III depending on both crystal form and presence of specific inhibitors. In one conformation the head group of the iron–sulfur protein containing the **2Fe2S cluster** is located next to one of the extramembranous loops of cytochrome *b* that form the Q_O site. In a second conformation, the 2Fe2S cluster of the head group is close to its electron acceptor, the heme of cytochrome c_1, while in other conformations the 2Fe2S cluster is located between cytochrome *b* and cytochrome c_1. Thus this head group of the iron–sulfur protein may move during electron transfer through complex III. Several experimental approaches have provided biochemical evidence for this suggestion. A model to explain electron transfer through the bc_1 complex suggests that when ubiquinol binds in the Q_O site and is deprotonated, the head group of the iron–sulfur protein moves closer to cytochrome *b* and assumes the "*b*" state. The deprotonated ubiquinol bound at the Q_O site is oxidized then by a concerted mechanism in which one electron is transferred to the 2Fe2S cluster of iron–sulfur protein and the second electron is immediately transferred to the low-potential heme at the Q_O site. Ubiquinone is released from the Q_O-binding pocket followed by subsequent movement of reduced iron–sulfur protein to the "c_1" state close to the heme of cytochrome c_1 (Figure 13.40).

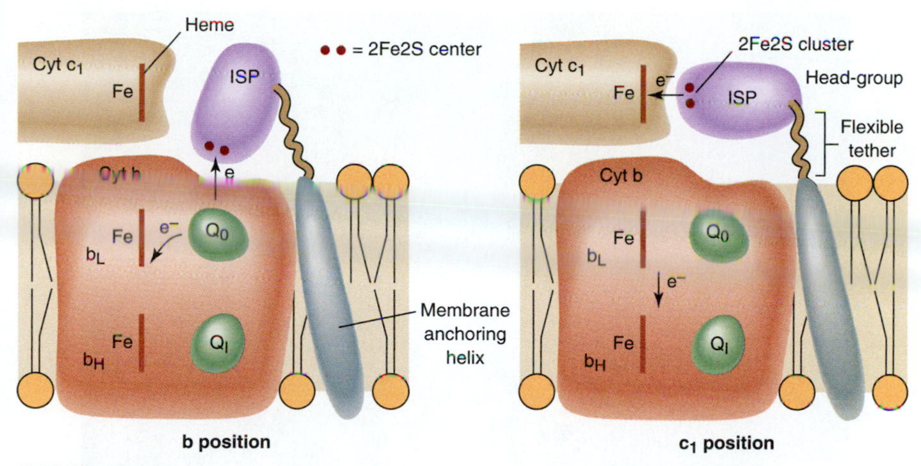

FIGURE 13.40
Proposed movement of iron–sulfur protein during electron transport through cytochrome bc_1 complex.
In "b" position, the iron–sulfur cluster (2Fe2S) is located on the head group of iron–sulfur protein (ISP), which is docked on cytochrome *b* near Q_O site. In "c_1" position, the head group of ISP is docked such that the 2Fe2S cluster is close to the heme of cytochrome c_1.

Cytochrome c Is a Mobile Carrier of Electrons

Electrons are transferred through complex III to **cytochrome *c***, a small hydrophilic protein with a molecular mass of 13,000 Da. Cytochrome *c* is a globular protein with the planar heme group located in the middle of the protein surrounded by hydrophobic residues and covalently bound to two conserved cysteine residues on the protein through vinyl ether linkages (Figure 13.38). The iron in the porphyrin ring is coordinated to nitrogen of a histidine and a sulfur atom of a methionine, thus preventing the interaction of the heme with O_2 (Figure 13.41).

Cytochrome *c*, like ubiquinone, functions as a mobile carrier in the electron transport chain. The protein is held loosely to the outer face of the inner membrane by electrostatic forces where it binds to cytochrome c_1 of complex III and accepts electrons. The reduced cytochrome *c* then apparently moves along the membrane surface where it interacts with subunit II of cytochrome *c* oxidase, again through electrostatic linkages, and donates electrons to the Cu_A site.

Complex IV: Cytochrome *c* Oxidase

Complex IV catalyzes transfer of electrons from cytochrome *c* to O_2, the terminal electron acceptor, to form water coupled to translocation of protons across the membrane. This multisubunit enzyme complex in mammalian mitochondria consists of 13 subunits with a molecular mass of 200,000 Da and contains as redox components two cytochromes, ***a*** and ***a_3***, and two copper centers, known as **Cu_A** and **Cu_B**. A simpler cytochrome *c* oxidase containing only three or four subunits but catalyzing similar electron transfer and proton pumping reactions as the more complex mammalian enzyme is present in bacterial membranes. Thus only three subunits of complex IV appear to be essential for a functional enzyme. The three subunits present in bacterial cytochrome *c* oxidases are homologous to the three largest subunits of cytochrome *c* oxidase, which are encoded in mitochondrial DNA (mtDNA). The remaining subunits of mammalian complex IV are encoded in nuclear DNA and may function either as regulatory subunits or in assembly of the enzyme. In addition, identification of tissue-specific isozymes of certain nuclear-encoded subunits of complex IV has provided tentative evidence of a regulatory role for these subunits.

The crystal structures of both a bacterial cytochrome *c* oxidase and a complex IV isolated from beef heart mitochondria have been solved (Figure 13.42).

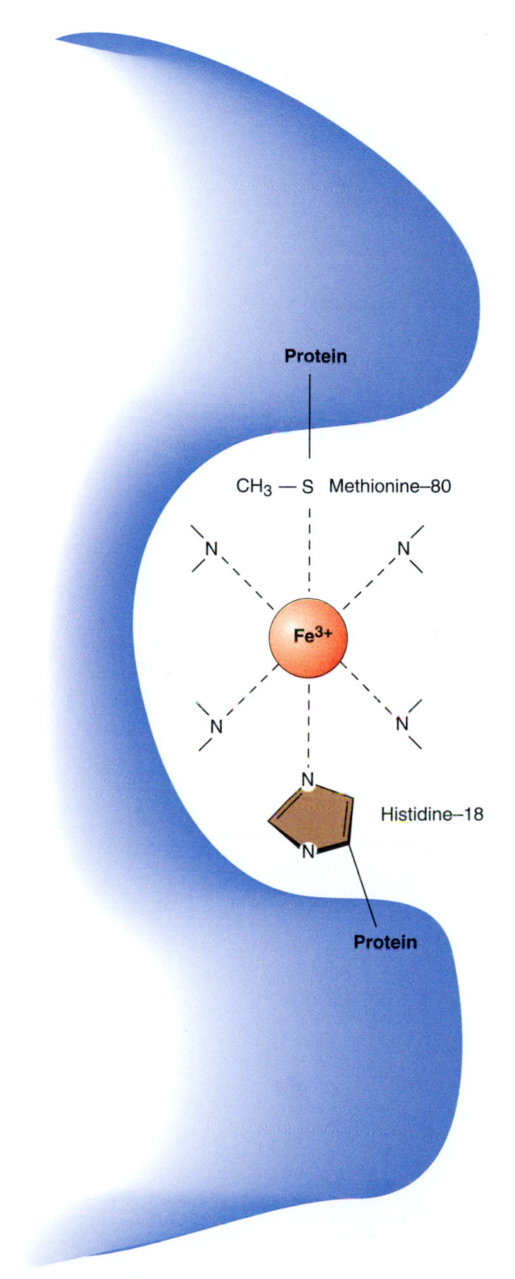

FIGURE 13.41
Six coordination positions of cytochrome *c*.

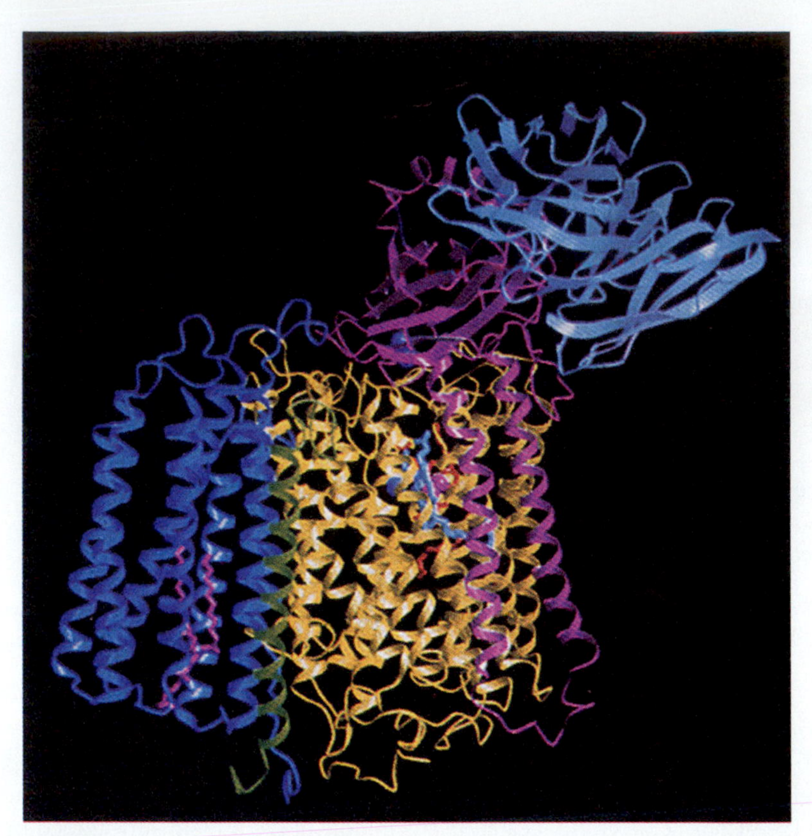

FIGURE 13.42
Model of crystal structure of cytochrome *c* oxidase from bacterium *Paracoccus denitrificans.*
Subunit I is yellow, subunit II is purple, and subunit III is blue with an embedded phospholipid in pink. The antibody fragment used to drive crystallization is cyan.
Reprinted with permission from Iwata, S., Ostermeier, C., Ludwig, B., Michel, H., et al. Nature *376:660, 1995. Copyright 1995, Macmillan Magazines Limited. Figure generously supplied by Professor S. Iwata.*

Subunit I, the largest polypeptide of the complex, contains 12 transmembrane helices but lacks any significant extramembranous domains. Two heme groups, *a* and a_3, are bound to subunit I such that the iron in the protoporphyrin ring is coordinated by nitrogen atoms of conserved histidine residues. The plane of both heme *a* and a_3 lies perpendicular to the membrane. In addition, subunit I contains a copper atom **(Cu_B)** that with heme a_3 forms a binuclear center involved in transfer of electrons from heme *a* to O_2 (Figure 13.43). Subunit II of cytochrome *c* oxidase has a large domain protruding from the cytosolic face of the inner membrane, where reduced cytochrome *c* binds, and contains two atoms of copper bound through sulfhydryl groups to two cysteine moieties (called **Cu_A**). Subunit III contains seven transmembrane helices with negligible extramembranous domains but does not have any redox carriers. Subunits II and III are localized on opposite sides of subunit I; the role of subunit III is unclear.

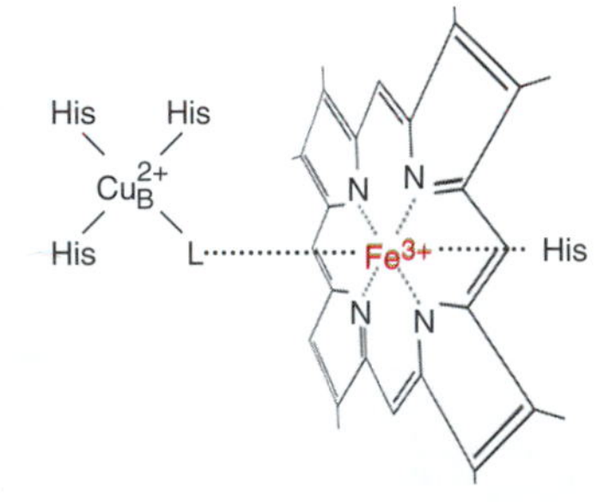

FIGURE 13.43
Binuclear center of cytochrome *c* oxidase indicated heme a_3 and Cu_B.

Pathways of Electron Transfer Through Complex IV

Electrons are transferred from reduced cytochrome *c* to the Cu_A site on subunit II and then to heme *a* on subunit I of complex IV (Figure 13.44). The Cu_A and heme *a* are localized within 1.5 Å of each other, permitting rapid electron transfer to occur. Electrons are then transferred to the binuclear center consisting of Cu_B and heme a_3 where final transfer of electrons to O_2 occurs. Initially, two electrons are transferred to an O_2 tightly bound to the binuclear center to form a peroxy derivative of oxygen (O_2^{2-}). Two additional electrons are transferred to the binuclear center with concomitant uptake of four protons from the matrix to form water. Since each of the redox carriers present in complex IV is a one-electron carrier and the reduction of O_2 to water requires four electrons, the reactions catalyzed by this enzyme have evolved to prevent the release of partially reduced toxic oxygen intermediates such a superoxide, hydrogen peroxide, or hydroxyl radicals (see Section 13.10, p. 590). Each of the intermediates formed in the reduction of O_2 remains tightly bound to the binuclear center and is thus prevented from dissociating from the enzyme until water is produced.

Complex IV also Pumps Protons

As discussed above, the reduction of O_2 by four electrons to form water by complex IV also involves the uptake of four protons from the mitochondrial matrix. The pathway for proton uptake is unclear, but experimental evidence has suggested that charged amino acids located on subunit I are involved in forming a channel for proton movement. In addition to the protons required for water formation, cytochrome *c* oxidase also pumps additional protons across the membrane to contribute to the electrochemical gradient. In the overall stoichiometry for proton pumping in cytochrome *c* oxidase during transfer of two electrons to O_2, two protons are released to the cytosol for every four protons taken up from the matrix (Figure 13.44).

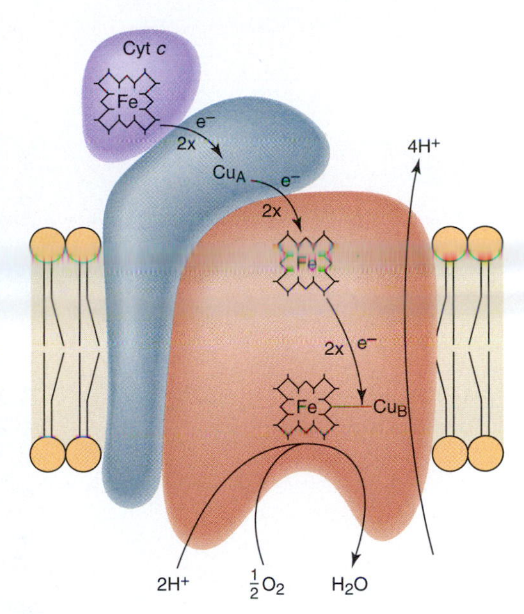

FIGURE 13.44
Pathways of electron transfer through cytochrome ***c*** oxidase. Cytochrome ***c*** binds on the surface of subunit II and transfers electrons to Cu_A. Electrons are transferred from Cu_A to heme *a* and then to the binuclear center (heme a_3 and Cu_B) where oxygen is reduced to water.

Overview of Electron Transport Chain Including Inhibitors

A dynamic picture of the electron transport chain has developed with our increasing knowledge of the detailed chemistry of the different respiratory chain complexes (Figure 13.45). Each of the four complexes exists independently in the inner membrane and is freely mobile. Complexes I and II, plus the other flavoprotein dehydrogenases, diffuse in the membrane and transfer electrons to the ubiquinone pool in the membrane. The reduced ubiquinol also freely diffuses in the membrane where it interacts with and is oxidized by complex III. The electrons are transferred from complex III to cytochrome *c*, which moves along the membrane to complex IV where it is oxidized and the electrons are eventually transferred to O_2. During electron transport from NADH or reduced flavoproteins down the electron transport chain to O_2, a total of ten protons are pumped across the inner mitochondrial membrane from the matrix to the intermembrane space. Four protons are pumped during transfer of two electrons through complexes I and III and two protons during transfer of two electrons through complex IV. The electrochemical gradient, thus formed, provides a source of potential energy that is used to drive the synthesis of ATP by the ATP synthase as will be discussed in Section 13.7 (p. 577).

Figure 13.45 also indicates the sites where specific inhibitors bind to the complexes of the electron transport chain and block electron flow. For example, **rotenone,** a commonly used insecticide, binds stoichiometrically to complex I and prevents the reduction of ubiquinone. Other agents, such as **piericidin, Amytal,** and other barbiturates, also inhibit complex I by preventing the transfer of electrons from the iron–sulfur centers to ubiquinone. Complex II is inhibited by **carboxin** and **thenoyltrifluoroacetone** as well as by **malonate** that acts as a competitive inhibitor with the substrate succinate. **Antimycin** inhibits electron transfer through complex III, the bc_1 complex, by binding to the Q_i site and blocking the transfer of electrons from the high-potential heme b_H to ubiquinone. Other antibiotics, such as **myxothiazol** and **stigmatellin,** inhibit electron transfer through complex III by binding to the Q_O site and blocking the transfer of electrons from ubiquinol to the 2Fe2S center of the iron–sulfur protein. Complex IV is inhibited by **cyanide** (CN^-), **azide** (N_3^-), and **carbon monoxide** (CO). Cyanide and azide bind tightly to the oxidized form of heme a_3 (Fe^{3+}) and prevent the transfer of electrons from heme *a* to the binuclear center. By contrast, carbon monoxide binds to the reduced form of heme a_3 (Fe^{2+}) competitively with O_2 and prevents electron transfer to O_2. Hence inhibition of mitochondrial electron transport results in impairment of the energy-generating function of oxidative phosphorylation leading to the death of the organism (see Clin. Corr. 13.3 for a discussion of cyanide toxicity).

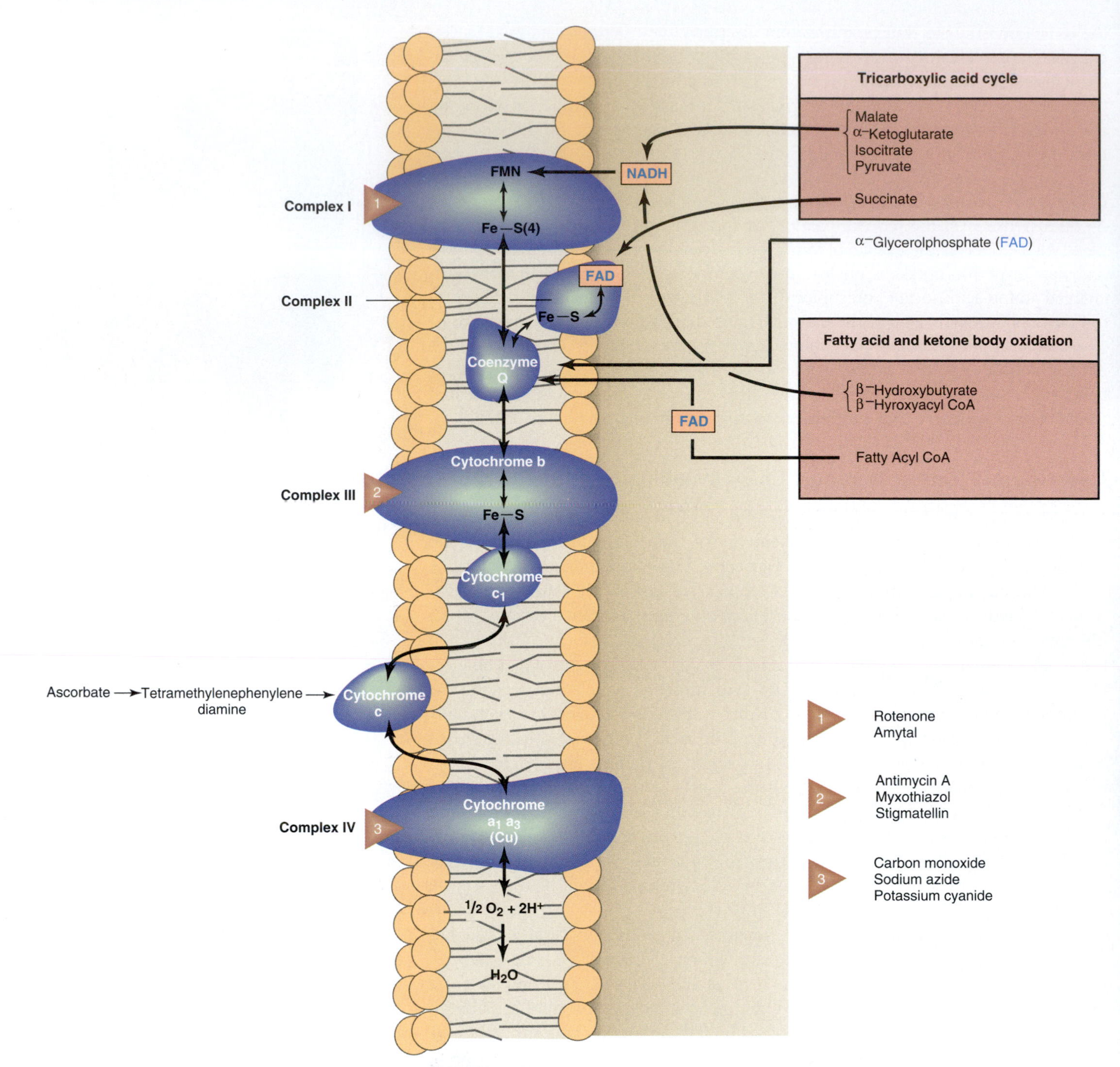

FIGURE 13.45
Overview of mitochondrial electron transport chain indicating pathways of electron transfer and binding sites for specific inhibitors, rotenone, amytal, antimycin A, myxothiazol, stigmatellin, carbon monoxide, sodium azide, and potassium cyanide.

Each of the four complexes can be isolated and studied independently of the other complexes. The electron transfer and proton pumping reactions of each individual complex occurs in the isolated complexes where the mechanisms involved can be more easily investigated. The polypeptides present in each of the complexes are tightly associated in stoichiometric relationships to form these multiprotein enzyme complexes, which catalyze the partial reactions of the electron transport chain. Recent successes in determining the crystal structures of complexes III and IV and the ATP synthase have provided a greater understanding of their structures and functions.

CLINICAL CORRELATION 13.3

Cyanide Poisoning

Inhalation of hydrogen cyanide gas or ingestion of potassium cyanide causes a rapid and extensive inhibition of the mitochondrial electron transport chain at the cytochrome oxidase step. Cyanide is one of the most potent and rapidly acting poisons known. It binds to the Fe^{3+} in the heme a_3 of cytochrome c oxidase that catalyzes the terminal step in the electron transport chain. Cyanide thus prevents the binding of oxygen to the binuclear center and the role of oxygen as the final electron acceptor. Mitochondrial respiration and energy production cease, and cell death occurs rapidly. Death due to cyanide poisoning occurs from tissue asphyxia, most notably of the central nervous system. An antidote to cyanide poisoning, if the poisoning is diagnosed rapidly, is the administration of various nitrites that convert oxyhemoglobin to methemoglobin by oxidizing Fe^{2+} of hemoglobin to Fe^{3+}. Methemoglobin (Fe^{3+}) competes with cytochrome a_3 (Fe^{3+}) for cyanide, forming a methemoglobin–cyanide complex. Administration of thiosulfate causes the cyanide to react with the enzyme rhodanese, forming the nontoxic thiocyanate.

Holland, M. A. and Kozlowski, L. M. Clinical features and management of cyanide poisoning. *Clin. Pharmacol.* 5:737, 1986.

13.7 | OXIDATIVE PHOSPHORYLATION

During the transfer of electrons from NADH and other respiratory substrates to O_2 via the mitochondrial electron transport chain, protons are translocated across the inner membrane to establish a **proton and charge gradient** (Figure 13.46). The energy released during the electron transfer reactions, which are exergonic, is used to pump protons across the inner mitochondrial membrane. This pumping of protons across the membrane results in a difference in proton concentration on the two sides of the membrane such that the intermembrane space becomes more acidic and the matrix space becomes more alkaline. Simultaneously, the external face of the inner membrane becomes more positively charged and the matrix face becomes more negatively charged due to the electrical potential energy resulting from the transfer of a positively charged ion (H^+) across the membrane without a compensating negatively charged ion.

During the transfer of two electrons from NADH to O_2, approximately 10 protons are pumped across the membrane to establish the **electrochemical gradient.** The total free energy obtained by the translocation of protons and charge distribution across the membrane can be calculated by the following equation in which Z is the absolute value of the charge, $\mathcal{F}$ is the Faraday constant, and $\Delta\psi$ is the membrane potential:

$$\Delta G = 2.3RT\ \Delta\text{pH} + Z\mathcal{F}\ \Delta\psi$$

In actively respiring mitochondria, the observed change in pH across the membrane is 0.75–1.0 pH units and the observed change in membrane potential is 0.15–2.0 V. Hence the ΔG calculated for respiring mitochondria is roughly 200 kJ for the transfer of 10 H^+ across the membrane during the transfer of electrons from NADH to O_2. The ΔG for this reaction can also be calculated from the difference in standard redox potentials of the electron donor and electron acceptor. The calculated ΔG for the electron pairs NADH and O_2 is 219 kJ mol^{-1} (Section 13.6, see p. 566), suggesting that the energy of the electron transfer reactions is efficiently captured in the electrochemical potential. The energy stored in the electrochemical gradient, often called the **protonmotive force,** is used to drive the synthesis of ATP by the movement of protons down the electrochemical gradient through the ATP synthase. This process, called **chemiosmosis,** was proposed originally by Professor Peter Mitchell who won the Nobel Prize in 1978. The detailed mechanisms by which the movement of protons through the ATP synthase is coupled to the chemical synthesis of ATP will be discussed later in this section.

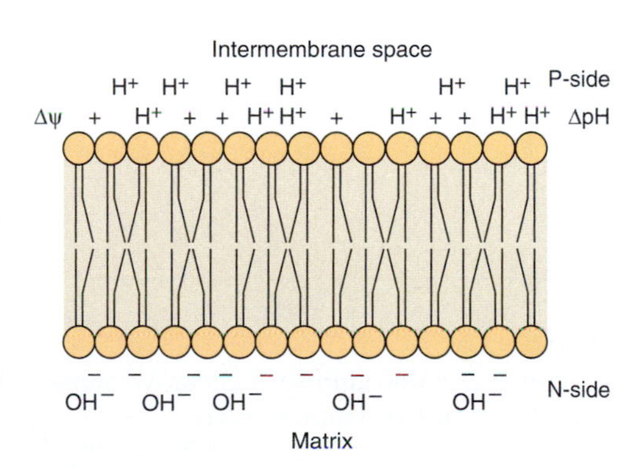

FIGURE 13.46
The electrochemical gradient consists of a gradient of charges ($\Delta\Psi$) and proton concentration (ΔpH) across inner mitochondrial membrane.

Coupling of ATP Synthesis and Electron Transport

The rate of ATP utilization, reflecting the need of a cell for ATP, regulates the rate of ATP synthesis in mitochondria, which in turn regulates the rate of electron trans-

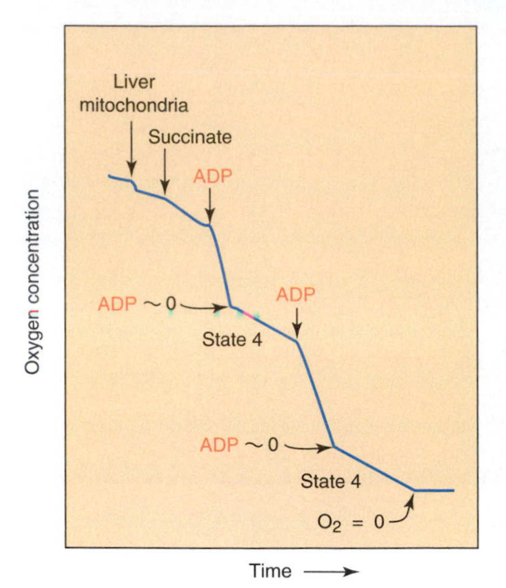

FIGURE 13.47
Demonstration of coupling of electron transport to oxidative phosphorylation in a suspension of liver mitochondria.
Addition of ADP stimulates rate of electron transfer measured as oxygen uptake defined as respiratory control.

fer. The coupling of ATP synthesis to electron transport is achieved by the electrochemical gradient as illustrated by the experiment shown in Figure 13.47. The rate of mitochondrial electron transport monitored by measuring the rate of O_2 consumption by a suspension of liver mitochondria can occur at a rapid rate only following the addition of an electron donor (succinate in this experiment) and ADP (a phosphate acceptor) plus phosphate (P_i). Following conversion of all the externally added ADP to ATP, the rate of electron transport returns to the rate observed prior to addition of ADP. Hence the rate of electron transport, or respiration, is tightly coupled to ATP synthesis. The chemiosmotic model can readily explain this relationship, termed respiratory control. When the energy needs of the cell are low, ATP will accumulate and the proton gradient will not be used for ATP synthesis. The magnitude of the proton gradient will increase until the energy required to pump protons across the membrane from the matrix against the existing electrical gradient equals the energy released during the transfer of electrons from NADH to O_2. At this point, electron transport will cease as equilibrium has been obtained. In cells using ATP, ADP will accumulate, leading to a stimulation of the ATP synthase. As ATP is synthesized, the magnitude of the proton gradient will decrease as protons move through the ATP synthase to provide the energy for ATP synthesis. As a result, the proton back-pressure on the electron transport chain will decrease. The increased rate of electron transport through the chain will stimulate the oxidation of NADH, resulting in the formation of NAD^+. The increased concentrations of NAD^+ coupled with the increased concentrations of ADP in cells actively using ATP act to stimulate the reactions of the TCA cycle and fatty acid oxidation. In this way, the need for ATP in a cell acts to regulate in a coordinated fashion the rate of electron flow through the respiratory chain as well as the reactions of the TCA cycle and fatty acid oxidation.

P/O Ratios for Mitochondrial Electron Transport and Oxidative Phosphorylation

The **P/O ratio** (phosphate incorporated into ATP to atoms of O_2 utilized) is a measure of the number of ATP molecules formed during the transfer of two electrons through a segment of the electron transport chain. Classically, the P/O ratio had been thought to be a whole number, 3 with transfer of two electrons from NADH-linked substrates to O_2, 2 with succinate to O_2, and 1 with reduced cytochrome *c* to O_2. These P/O ratios had suggested that one ATP could be produced during electron transfer through each of the proton-pumping complexes I, III, and IV. Questions about actual P/O ratios, however, have arisen with the determination that ten protons are pumped across the mitochondrial membrane during the transfer of two electrons from NADH to O_2 and that four protons are required for the synthesis (and translocation) of one ATP molecule. These proton stoichiometries result in a calculated P/O ratio of 2.5. Indeed, recent experimental determinations of the P/O ratio have provided values of approximately 2.5 with NADH-linked substrates and 1.5 with succinate.

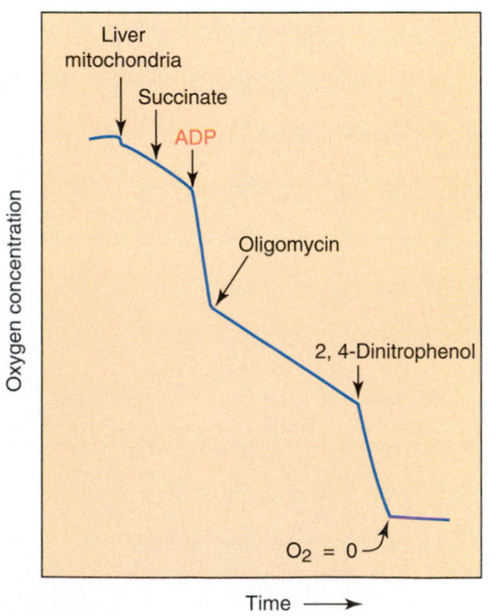

FIGURE 13.48
Inhibition and uncoupling of oxidative phosphorylation in liver mitochondria.
Stimulation of oxygen uptake by ADP is inhibited by oligomycin, which blocks proton movements through F_0 of ATP synthase. Addition of the uncoupler, dinitrophenol, relieves inhibition by oligomycin and stimulates rate of oxygen uptake.

Effects of Uncouplers and Inhibitors of the Electron Transport–Oxidative Phosphorylation System

As illustrated in Figure 13.48 addition of an inhibitor of the ATP synthase, **oligomycin,** to liver mitochondria actively respiring in the presence of ADP results in an inhibition of the rate of O_2 uptake. Oligomycin blocks the synthesis of ATP by preventing the movement of protons through the ATP synthase. Since ATP synthesis and electron flow are tightly coupled, respiration or electron transport decreases to the rate observed in the absence of ADP.

The mandatory coupling between electron transport and ATP synthesis can be "uncoupled" by certain conditions and chemical reagents such as **2,4-dinitrophenol** or **carbonylcyanide-p-trifluoromethoxy phenylhydrazone.** After addition of an **uncoupler** to a system in which both O_2 consumption and ATP have been inhibited by oligomycin as in Figure 13.48, a rapid increase in the rate of O_2 consump-

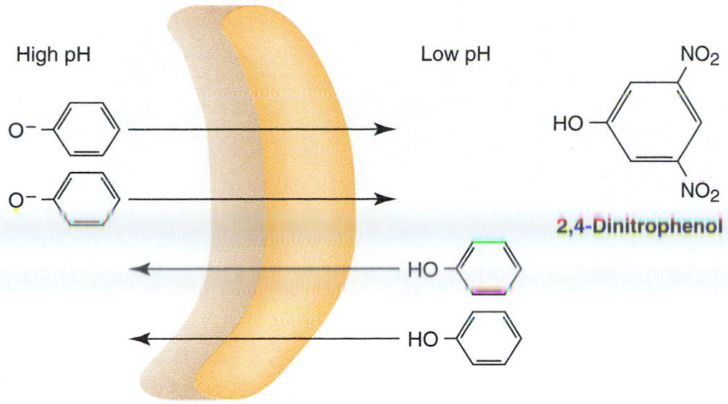

High pH
Low pH
O^-
O^-
HO
HO
NO_2
HO
NO_2
2,4-Dinitrophenol

FIGURE 13.49
Action of the uncoupler, dinitrophenol, a proton ionophore that equilibrates pH across the inner mitochondrial membrane.
Dinitrophenol is a weak acid that picks up a proton from the intermembrane space (P-side of membrane) that has a high proton concentration and carries it across the membrane to the matrix (N-side of membrane) where the proton dissociates due to the low proton concentration there.

tion is observed. Because respiration or electron transport is now uncoupled from ATP synthesis, electron transport may continue but without ATP synthesis. Uncouplers are hydrophobic weak acids that are protonated in the intermembrane space where a higher concentration of protons results from active electron transfer. These protonated uncouplers due to their lipophilic nature rapidly diffuse across the membrane into the mitochondrial matrix where they are deprotonated since the matrix has a lower concentration of protons there. In this way the proton gradient can be completely dissipated as shown in Figure 13.49.

ATP Synthase

ATP synthase, or complex V, a multiprotein enzyme found in the inner mitochondrial membrane of mammals, yeast, and fungi and in the cytoplasmic membrane of bacteria, catalyzes ATP synthesis using the energy of the proton gradient. The flow of protons through the ATP synthase down the electrochemical gradient from the positive intermembrane space to the negative matrix provides the energy for ATP synthesis in all of these organisms. ATP synthase consists of two domains: $\mathbf{F_1}$, a peripheral enzyme complex originally observed in electron micrographs as small particles bound to the inner mitochondrial membrane (Figure 13.50), and $\mathbf{F_0}$, an integral membrane protein complex. The F_1 domain of the ATP synthase contains the binding sites for ATP and ADP and is involved in the catalytic reactions of ATP synthesis. The F_0 domain is embedded in the membrane and provides a channel for the translocation of protons across the membrane. The removal of F_1 from the mitochondrial inner membrane by gentle agitation leaves an intact electron transport chain capable of electron transfer without formation of a proton gradient. The protons pumped across the membrane during electron transfer flow back into the matrix through the F_0 domain in the absence of F_1. Adding the F_1 domain back to the stripped membranes again permits the formation of a proton gradient as F_1 reconstitutes with F_0 and acts as a plug to block the flow of protons through F_0. In addition, the ability to synthesize ATP is restored in the stripped membranes reconstituted with F_1. Isolated F_1 is an ATPase that catalyzes the hydrolysis of ATP, the reverse reaction to ATP synthesis. Consequently, the ATP synthase was originally called the $\mathbf{F_1/F_0}$ **ATPase.** The entire ATP synthase, containing both F_1 and F_0, can be isolated and, when incorporated into artificial membrane vesicles, is able to synthesize ATP when an electrochemical gradient is established across the membrane. Incorporation of purified F_0 into an artificial membrane renders the membrane permeable to protons.

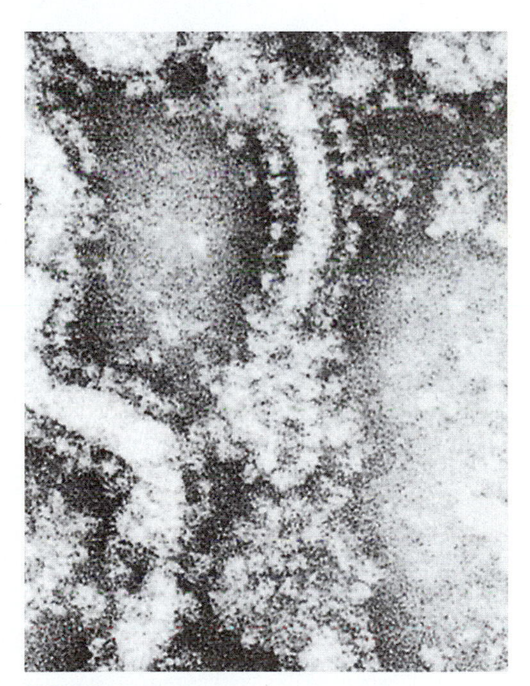

FIGURE 13.50
Electron micrograph of mitochondrial F_1courtesy of Dr. Parsons. Reprinted with permission from Parsons, D. F. Science 140:985, 1963. Copyright 1963, American Association for the Advancement of Science.

TABLE 13.7 Subunits of F_1F_0-ATP Synthase from *Escherichia coli*

Complex	*Protein Subunit*	*Mass (kDa)*	*Stoichiometry*
F_1	α	55	3
	β	52	3
	γ	30	1
	δ	15	1
	ε	5.6	1
F_0	a	30	1
	b	17	2
	c	8	9–12

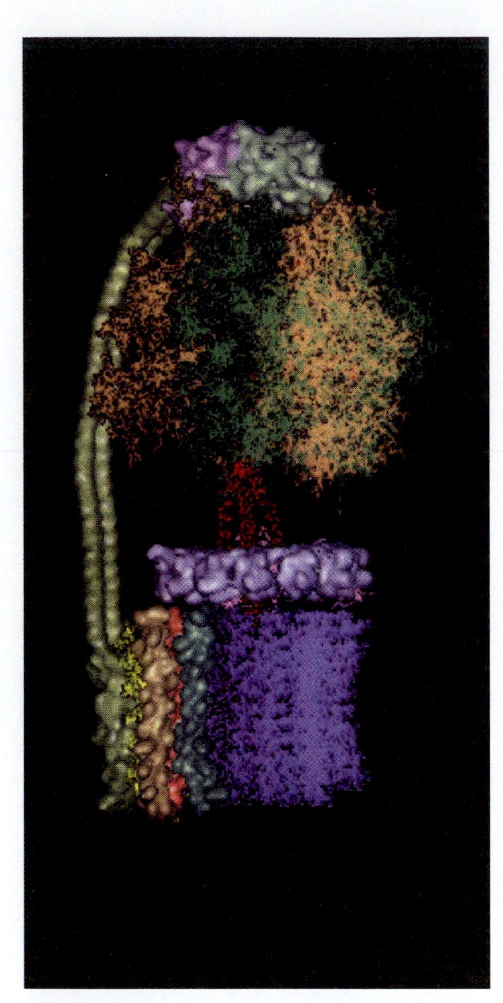

FIGURE 13.51
A model for mitochondrial F_1F_0-ATP synthase, a rotating molecular motor.
ATP synthesis occurs on F_1 domain, while F_0 domain contains a proton channel. The a, b, α, β, and δ subunits constitute the stator while the c, γ, and ε subunits provide the rotor. Protons flow through the structure causing the rotor to turn, resulting in conformational changes in the β subunits where ATP is synthesized.
Figure generously supplied by Drs. Peter L. Pedersen, Young Hee Ko, and Sangjin Hong, The Johns Hopkins University School of Medicine. Reprinted with permission from Pedersen, P. L., Ko, Y. H., and Hong, S. J. Bioenerg. Biomembr. 32:325, 2000.

The ATP synthase is a multicomponent complex having a molecular weight of 480,000–500,000 (Table 13.7 and Figure 13.51). The F_1 domain consists of five nonidentical subunits (α, β, γ, δ, and ε) with a subunit stoichiometry of α_3, β_3, γ, δ, ε and a molecular weight of 350,000–380,000. Binding sites for ATP and ADP have been identified on both the α and β subunits. The catalytic sites are on the β subunits, while the function of the nucleotides bound to the α subunits is unknown. The γ subunit forms the central core of F_1, while the δ subunit may be involved in the attachment of the F_1 domain to the membrane. The F_0 domain of the *E. coli* enzyme consists of three nonidentical hydrophobic subunits termed a, b, and c that are present in the apparent stoichiometry of a_1, b_2, c_{9-12}. The c subunits each contain an essential charged amino acid (aspartate 61 in *E. coli*) that is involved in proton pumping. Each individual c subunit consists of two α helices bent to form a hairpin with aspartate residue 61 located in the middle of the membrane. Mutating this aspartate to an asparagine abolishes proton pumping. Similar mutations of charged amino acids have also implicated the a subunit of F_0 in proton movements, while the b subunits appear to act to attach the F_1 domain to F_0. The F_0 domain of ATP synthase found in mitochondria contains subunits homologous to subunits a, b, and c of the *E. coli* enzyme; however, additional subunits that appear to be integral parts of F_0 are also present.

Synthesis of ATP on the Surface of F_1

Insights into the mechanism of ATP synthesis on the catalytic surface of F_1 were derived from isotope exchange experiments, which revealed the fact that when stoichiometric amounts of ADP, ATP, and inorganic phosphate were present with isolated F_1 the reaction was essentially in equilibrium with a $\Delta G^{\circ\prime}$ close to zero. This exchange reaction

$$\text{enz:ADP} + P_i \rightleftharpoons \text{enz:ATP}$$

proceeds readily on the surface of the enzyme even in the absence of a proton gradient. It was concluded that the synthesis of ATP tightly bound to F_1 does not require energy because of the nature of the binding of the substrates to conserved residues of the β subunit of F_1. The movement of protons through the ATP synthase was required for the release of ATP from F_1. Proton movement through the channel in F_0 thus must affect the binding site for ATP on the β subunit, resulting in the release of ATP. It was proposed that the energy released during the movement of protons across the membrane caused a conformational change in the ATP synthase resulting in the release of the tightly bound ATP. The **binding-change mechanism** suggests that the three β subunits of the ATP synthase adopt different conformations that change during catalysis with only one β subunit acting as the catalytic site. As depicted in Figure 13.52, one subunit has an open conformation (O) with a low affinity for ligands and is empty. A second subunit has a loose conformation (L) with a low affinity for ligands and is inactive, while the third subunit has a tight conformation (T) that has a high affinity for ligands and is active in catal-

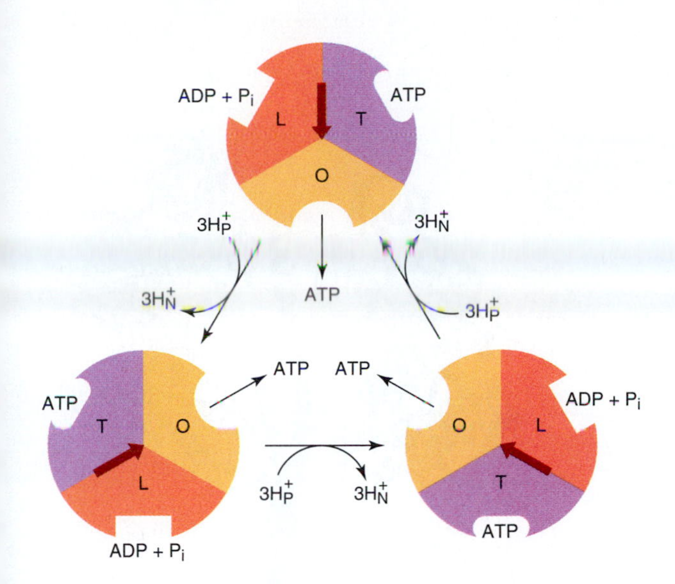

FIGURE 13.52
The binding-change model for ATP synthesis by ATP synthase.

ysis. According to this model, the synthesis of ATP occurs on the surface of the β subunit in the T conformation. During catalysis, ADP and P_i, the substrates of the enzyme bind to the β subunit in the L conformation. The energy provided by passage of protons through F_0 to F_1 results in the following conformational changes in the different β subunits: the L site changes to the T conformation with the resultant synthesis of ATP, the O site changes to the L conformation and binds ADP and P_i, and the T site changes to the O conformation with the release of ATP. According to this model, the energy released by electron transfer is conserved as a proton gradient that drives conformational changes in the ATP synthase, resulting in the binding of substrates, the synthesis of ATP on the enzyme, and the release of the product ATP.

Structural Studies of the ATP Synthase

Resolution of the crystal structure of F_1 has provided dramatic visualization of the conformations of the different β subunits that provide evidence for the binding-change model described above. In this crystal structure, alternating α and β subunits form the knob of F_1 with the single γ subunit forming a central shaft through the center of F_1 (Figure 13.53*a*, *b*). Each β subunit had a different conformation depending on the presence of substrate. Thus F_1 crystallized in the presence of ADP and a nonhydrolyzable analog of ATP revealed binding of ATP analog to one β subunit, binding of ADP to a second β subunit, and an empty third β subunit (Figure 13.53*c*).

The model for ATP synthesis that has developed is that protons flow through the membrane by first binding to conserved acidic amino acid present in the c subunit of F_0, causing the ring of c subunits attached to γ and ε subunits to rotate (Figure 13.51). The movement of the γ subunit causes conformational changes in β subunits as the γ subunit associates sequentially with each β subunit in turn. The a and b subunits of F_0 plus the δ subunit of F_1 form the "stator" to hold α and β subunits in position while γ and c subunits form the moving rotor. The binding-change model predicts that the γ subunit should move in one direction during ATP synthesis and in the opposite direction during ATP hydrolysis, when the energy released during breakdown of ATP is converted into a proton gradient. The rotation of the γ subunit in a single F_1 subunit was demonstrated by attaching a fluorescent actin polymer to the γ subunit of an F_1 in which $\alpha_3\beta_3$ subunits were fixed to a microscope slide. Rotation of the fluorescent γ subunit was observed upon addition of ATP. Similar experiments were performed using the entire F_1/F_0 complex in

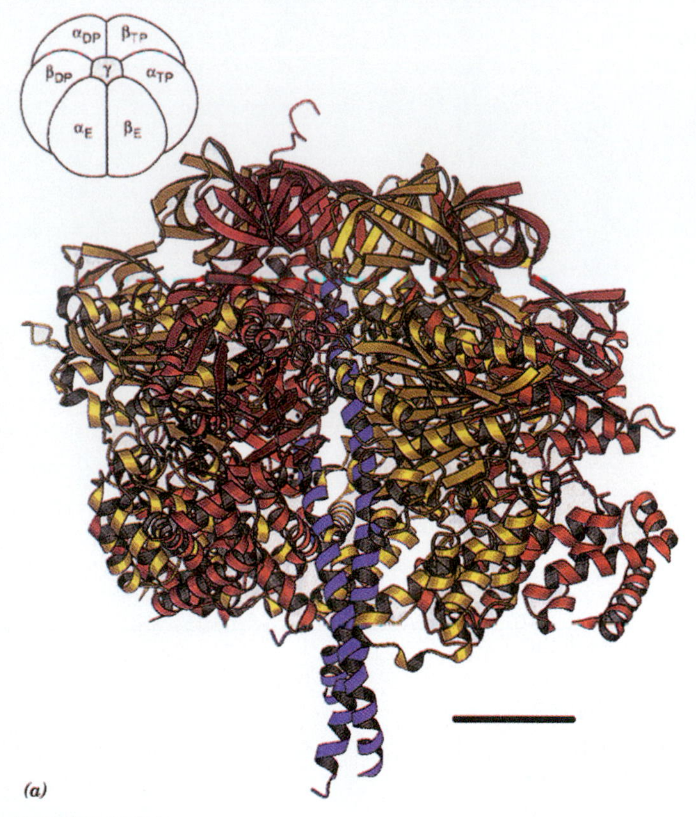

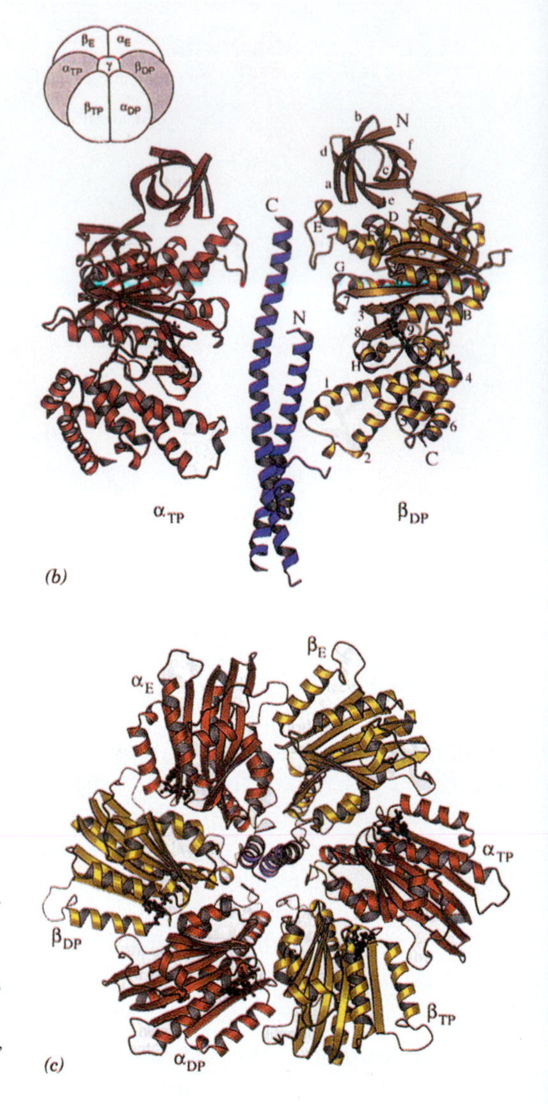

FIGURE 13.53
Mitochondrial ATP synthase complex.
(*a*) Side view of F_1complex structure deduced from the crystal structure. Three α (red) and three β (yellow) subunits alternate around a central shaft, the γ subunit (blue). (*b*) Side view of F_1 subunit in which two α and β subunits have been removed to reveal the central γ subunit. Subunits are colored as indicated for part (*a*). (*c*) Top view of F_1 complex shows alternating α and β subunits surrounding central γ subunit.
Reprinted with permission from Abrahams, J. P., Leslie, A. G. W., Lutter, R., and Walker, J. E. Nature *370:621, 1994. Copyright 1994, Macmillan Magazines Limited.*

which the complex of c subunits along with the γ subunit rotated as indicated by the fluorescent actin attached to one c subunit (Figure 13.54). Under both experimental conditions, the rotor movement was not continuous but occurred in discrete steps of approximately 120°, which is consistent with stepwise movement of the γ subunit from one β subunit to another. ATP synthase is the smallest known **molecular motor.**

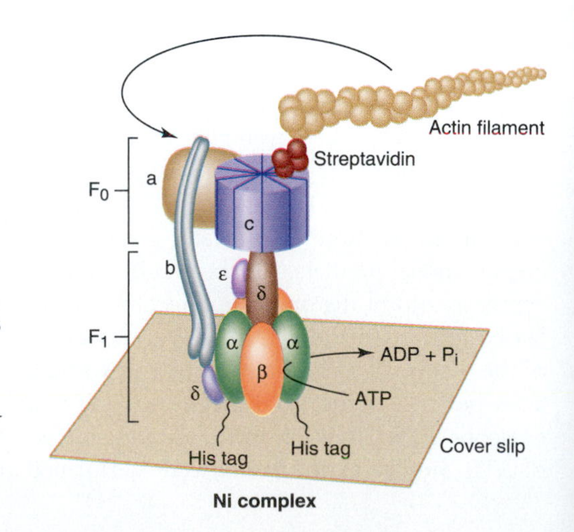

FIGURE 13.54
Experimental evidence for rotation of γ and c subunits.
F_1 domain is attached to a nickel-coated coverslip by histidine residues genetically engineered at the N terminus of the α subunits. Biotin, covalently attached to c subunits, binds very tightly to the protein streptavidin that is covalently attached to an actin filament containing a fluorescent probe. Addition of ATP that is hydrolyzed by the ATPase of F_1 causes actin filament to rotate in one direction, proving that c subunit of F_0 rotates. Earlier experiments in which actin filament was attached to γ subunit provided evidence that γ subunit can also rotate. Presumably, both γ and c subunits rotate as a unit.
Redrawn from Sambongi, Y., Iko, Y., Tanabe, M., Omote, H. et al. Science *286:1722, 1999.*

13.8 | MITOCHONDRIAL INNER MEMBRANES CONTAIN SUBSTRATE TRANSPORT SYSTEMS

Whereas the outer membrane presents little or no permeability barrier to substrate or nucleotide molecules of interest in energy metabolism, the inner membrane limits the types of substrates, intermediates, and nucleotides that can diffuse from cytosol into the matrix. Various transport systems have been described in mitochondria (Figure 13.55), some of which have been thoroughly characterized. These transport systems facilitate selective movement of various substrates and intermediates back and forth across the inner mitochondrial membrane. Through these transporters, various substrates can be accumulated in the matrix, since the transporters can move the substrate against a concentration gradient.

Transport of Adenine Nucleotides and Phosphate

Continued synthesis of ATP in the mitochondrial matrix requires that cytosolic ADP formed during energy-consuming reactions be transported back across the inner membrane into the matrix where it is converted to ATP. Similarly, newly synthesized ATP must be transported back across the inner membrane into the cytosol to meet the energy needs of the cell. This exchange of adenine nucleotides, which are very highly charged hydrophilic molecules, is catalyzed by a very specific **adenine nucleotide translocase** located in the inner membrane (Figure 13.56). The adenine nucleotide translocase, a homodimer consisting of two subunits with molecular weights of 30,000, catalyzes a 1 : 1 exchange of ATP for ADP. The presence of one nucleotide-binding site on the transporter suggests that the enzyme alternately faces the matrix or the inner membrane space during the transport process. Newly synthesized ATP is bound to the translocase in the matrix, which then changes its conformation to face the cytosol where the ATP is released

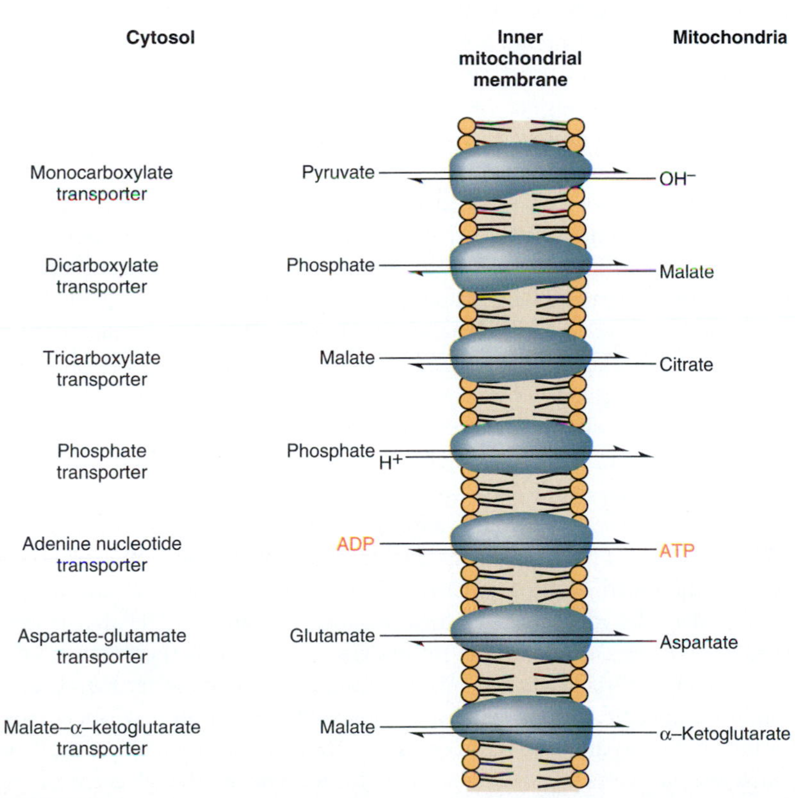

FIGURE 13.55
Mitochondrial metabolite transporters.

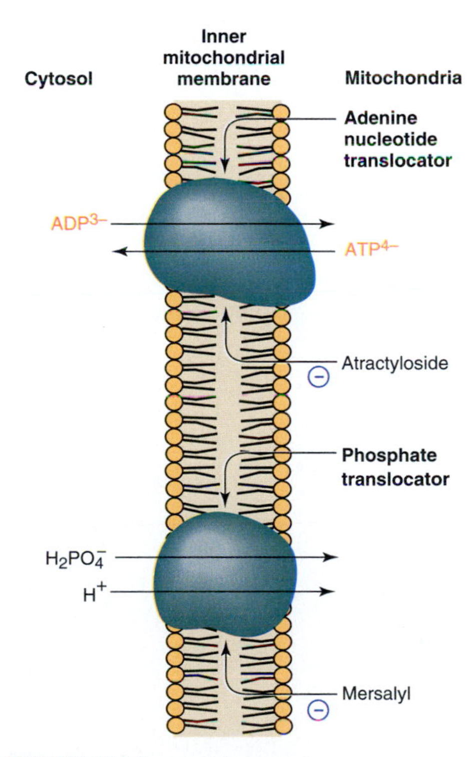

FIGURE 13.56
Adenine nucleotide and phosphate translocators.

in exchange for an ADP. The translocase then changes its conformation again to bring the nucleotide-binding site containing ADP back to face the matrix. The translocase favors outward movement of ATP and inward movement of ADP despite observations that both nucleotides bound equally well to the binding site. The explanation for this observation suggests that, at pH 7, ADP has three negative charges and ATP has four. Hence the even exchange of one ATP for an ADP results in net outward movement of one negative charge, which is equivalent to import of one proton. The membrane potential established during electron transfer is positive outside, which would favor outward transport of more negatively charged ATP over that of ADP. The adenine nucleotide translocase is present in high concentrations, up to 14% of total protein, in the inner membrane. Hence it is unlikely that transport of adenine nucleotides across the inner mitochondrial membrane is ever rate limiting.

A second transporter essential for oxidative phosphorylation is the **phosphate translocase,** which transports cytosolic phosphate into the matrix along with a proton (Figure 13.56). This symport also requires the proton gradient as phosphate and protons are transported into the matrix in a 1:1 ratio. This transport of ADP and phosphate requires a significant fraction of the energy present in the electrochemical gradient produced during electron transfer. Thus the protonmotive force provides energy for ATP synthesis by ATP synthase as well as for uptake of the two required substrates.

Substrate Shuttles Transport Reducing Equivalents Across the Inner Mitochondrial Membrane

The nucleotides involved in cellular oxidation–reduction reactions (e.g., NAD^+, NADH, $NADP^+$, NADPH, FAD, and $FADH_2$) and CoA and its derivatives are not permeable to the inner mitochondrial membrane. For example, to transport reducing equivalents (e.g., protons and electrons) from cytosol to matrix or the reverse, "substrate shuttle mechanisms" involving reciprocal transfer of reduced and oxidized members of various oxidation–reduction couples are used to accomplish net transfer of reducing equivalents across the membrane.

Two examples of how this transfer of reducing equivalents from cytosol to matrix occurs are shown in Figure 13.57. The malate–aspartate shuttle and α-glycerol phosphate shuttle are employed in various tissues to translocate reducing equivalents from cytosol to matrix, where they are oxidized to yield energy. The operation of such substrate shuttles requires that appropriate enzymes are localized on the correct side of the membrane and that appropriate transporters or translocases be present on/in the mitochondrial inner membrane. In the **glycerol phosphate shuttle,** two different glycerol phosphate dehydrogenases, one located in the cytosol and one on the outer face of the inner mitochondrial membrane are involved. NADH produced in the cytosol is used to reduce dihydroxyacetone phosphate to glycerol 3-phosphate catalyzed by the cytosolic glycerol phosphate dehydrogenase. The glycerol 3-phosphate in turn is oxidized by glycerol phosphate dehydrogenase, an FAD-containing enzyme, in mitochondria to produce dihydroxyacetone phosphate and $FADH_2$ that is oxidized by the electron transport chain.

The second shuttle, the **malate–aspartate shuttle,** operates on the same principle. NADH in the cytosol reduces oxaloacetate to malate, which enters mitochondria on the dicarboxylic acid carrier. This malate is readily oxidized by mitochondrial malate dehydrogenase to form oxaloacetate and NADH that is oxidized by the electron transport chain. The oxaloacetate produced is then converted to aspartate by mitochondrial aspartate aminotransferase since the inner membrane does not contain a specific transporter for oxaloacetate. Aspartate can then cross the membrane via the aspartate–glutamate carrier to the cytosol where cytosolic aspartatetransferase converts it to oxaloacetate. In contrast to the glycerol phosphate shuttle, the malate–aspartate shuttle is reversible and serves as a mechanism to bring reducing equivalents out of mitochondrial matrix to cytosol.

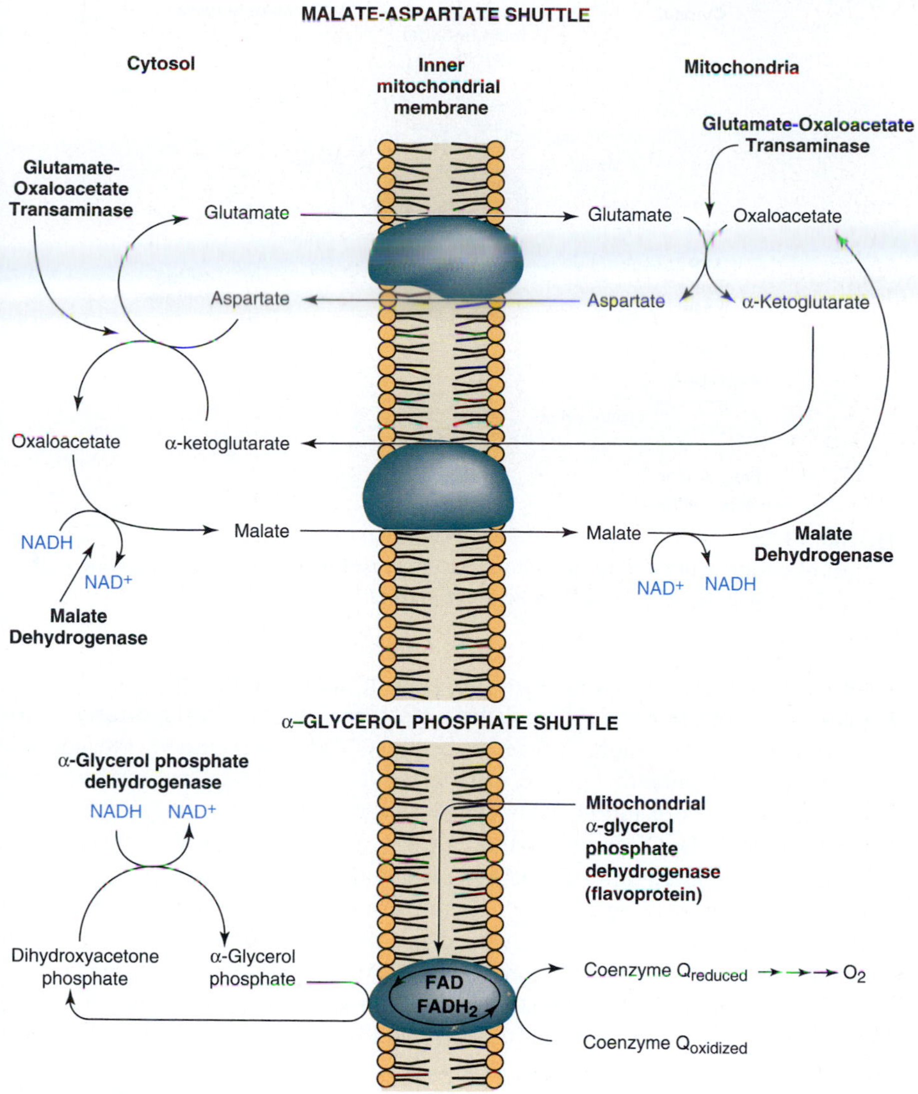

FIGURE 13.57
Transport shuttles for reducing equivalents.

Acetyl Units Are Transported by Citrate

The inner mitochondrial membrane is impermeable to acetyl CoA but acetyl groups are transferred from the mitochondrial compartment to cytosol, where acetyl moieties are required for fatty acid or sterol biosynthesis (Figure 13.58). Intramitochondrial acetyl CoA is converted to citrate by citrate synthase of the TCA cycle. Subsequently, citrate is exported to cytosol by a tricarboxylate transporter in exchange for a dicarboxylate such as malate. Cytosolic citrate is then cleaved to acetyl CoA and oxaloacetate at the expense of an ATP by **ATP: citrate lyase** (see p. 702). Substrate shuttle mechanisms in liver are involved in movement of appropriate substrates and intermediates in both directions across the inner membrane during periods of active gluconeogenesis and ureagenesis (see p. 635).

Mitochondria Have a Specific Calcium Transporter

Mitochondria in most mammalian tissues possess a transport system for translocating Ca^{2+} across the inner membrane. The distribution/redistribution of cellular Ca^{2+} pools within cells is critical for different cell functions, such as muscle contraction, neural transmission, secretion, and hormone action. Distinct pools of Ca^{2+}

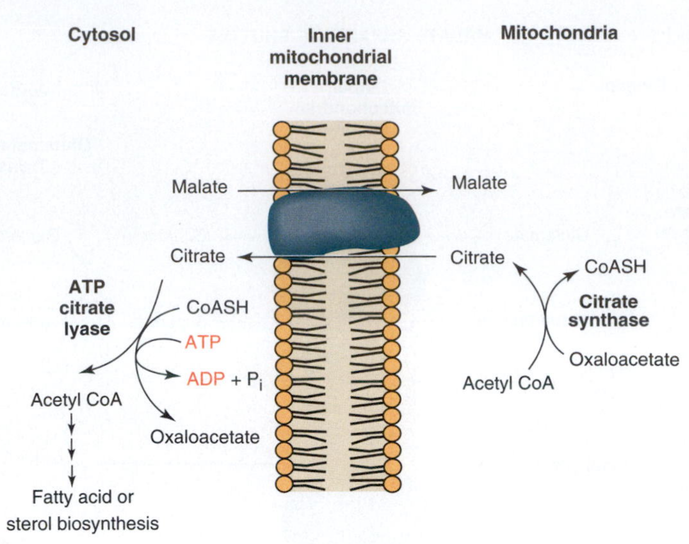

FIGURE 13.58
Export of citrate generated in mitochondria to cytosol where it serves as a source of acetyl CoA for biosynthesis of fatty acids or sterols.

have been detected in highest concentration in endoplasmic reticulum (or sarcoplasmic reticulum) but Ca^{2+} pools have also been detected in the cytosol, mitochondria, nuclei, and Golgi. Some intracellular Ca^{2+} is bound to nucleotides, metabolites, or membrane ligands, while a portion of intracellular Ca^{2+} is free in solution. Estimates of intracellular cytosolic Ca^{2+} are in the range of 10^{-7} M, whereas extracellular Ca^{2+} is at least four orders of magnitude greater. Mitochondria take up Ca^{2+} by a uniporter in the inner membrane using the energy of the electrochemical gradient (Figure 13.59). Recent studies using confocal microscopy of living cells have provided convincing evidence that mitochondria may take part in regulation of intracellular Ca^{2+} concentration. Mitochondria are localized in cells in close proximity to both endoplasmic reticulum and sarcoplasmic reticulum. The binding of hormones to cell membranes results in release of inositol 3-phosphate (IP_3) from phosphatidyl inositol, which acts to cause release of Ca^{2+} from endoplasmic reticulum (see p. 928). The resulting transient microdomains of high Ca^{2+} concentrations within cells may be modulated by uptake into nearby mitochondria. In mitochondria Ca^{2+} acts to regulate Ca^{2+}-sensitive dehydrogenases of the TCA cycle, while also acting to remove Ca^{2+} from local areas of the cytosol. Hence sequestering and release of an intracellular store of Ca^{2+} influences intracellular Ca^{2+} pools and various cell functions. One consequence of uptake of high concentrations of Ca^{2+} into mitochondria is the opening of a pore on the outer membrane leading to release of cytochrome *c* and consequent activation of cell death by apoptosis or necrosis.

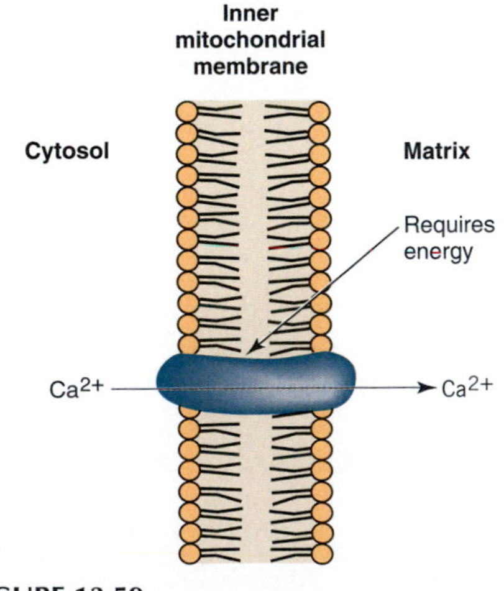

FIGURE 13.59
Mitochondrial calcium carrier.

Uncoupling Proteins

Brown adipose tissue plays a major role in nonshivering thermogenesis in newborns, in hibernating animals, and in experimental animals in diet-induced thermogenesis. The primary agent involved in cold-induced thermogenesis in brown fat is the **uncoupling protein**, UCP-1, which is localized exclusively in the inner membrane of brown adipose tissue. UCP-1 functions to carry protons back across the inner mitochondrial membrane and thus acts to uncouple ATP synthesis from electron transport (Figure 13.60). Thermogenesis results from activation of sympathetic nerves in brain responding to cold exposure with the resultant release of norepinephrine that binds to β-adrenergic receptors on cell membranes of brown fat cells. The binding of norepinephrine to the β-adrenergic receptors causes release of cAMP

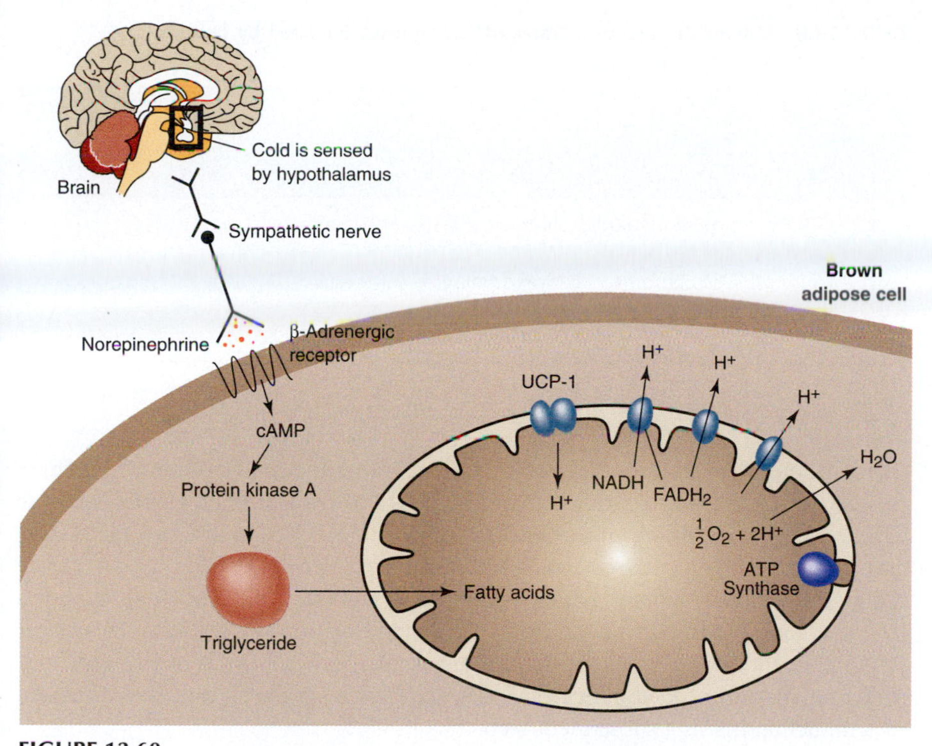

FIGURE 13.60
Activation of UCP-1 by cold adaptation.
Cold stimulates release of norepinephrine from sympathetic nerve cells. The norepinephrine binds to the β-adrenergic receptor resulting in activation of a lipase with production of free fatty acids, which activate proton conducting protein, UCP-1.

and the activation of protein kinase A, which results in stimulation of lipolysis. The production of free fatty acids during lipolysis activates UCP-1 for transport of protons back across the membrane (Figure 13.60). The stimulation of proton transport by free fatty acids is believed to result from release of a proton from the carboxyl group of the free fatty acid. The UCP-1 is a member of the mitochondrial transporter family with close structural similarities to the adenine nucleotide translocator but with a specific pore for the transport of protons into the matrix. Chronic cold-induced stimulation of the β-adrenergic receptor by norepinephrine results in increased transcription of the *UCP-1* gene, stimulation of mitochondrial biogenesis, and eventual hyperplasia of brown adipose tissue. In large mammals such as dogs, cats, and primates including humans, who do not hibernate, discrete deposits of brown fat are present at birth, but become sparse during later development. Recently, four other uncoupling proteins, UCP-2, UCP-3, UCP-4, and UCP-5 with amino acid sequences similar to that of UCP-1 have been discovered in tissues other than brown adipose tissue. The presence of uncoupling proteins in tissues such as skeletal muscle has prompted investigations into the possible role of these proteins in regulation of energy expenditure and perhaps obesity. Development of potential pharmacological agents that might affect uncoupling proteins has been suggested as a possible treatment for obesity.

13.9 | MITOCHONDRIAL GENES AND MITOCHONDRIAL DISEASES

Mitochondria contain their own genome, a circular double-stranded DNA that contains structural genes for 13 proteins of the electron transport chain including seven subunits of NADH:ubiquinone oxidoreductase (complex I), one subunit

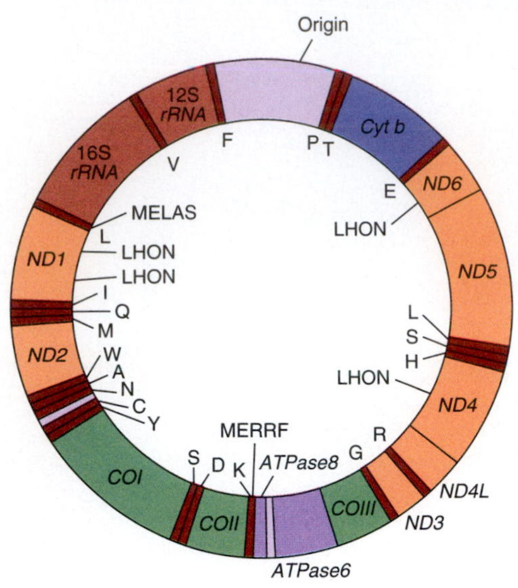

FIGURE 13.61
Map of genes on mitochondrial DNA.

TABLE 13.8 Subunits of Electron Transport Complexes Encoded by Human Mitochondrial DNA

Complex		Total Number of Subunits	Number of Subunits Encoded by Mitochondrial DNA
I	NADH–ubiquinone oxidoreductase	>35	7
II	Succinate dehydrogenase	4	0
III	Ubiquinol–cytochrome *c* oxidoreductase	11	1
IV	Cytochrome *c* oxidase	13	3
V	ATP synthase	12	2

(cytochrome *b*) of ubiquinol:cytochrome *c* oxidoreductase (complex III), three subunits of cytochrome *c* oxidase (complex IV), and two subunits of ATP synthase (Table 13.8). In addition, mitochondrial DNA (mtDNA) contains genes encoding two ribosomal RNAs (rRNAs) plus all necessary transfer RNAs (tRNAs), which are required for protein synthesis in mitochondria (Figure 13.61). While mitochondria have the ability to transcribe their own DNA and translate the resulting messenger RNAs, mitochondria are not self-replicating organelles. Over 90% of all mitochondrial proteins are encoded in nuclear DNA, synthesized in the cytosol, and imported into mitochondria in a subsequent step.

Mitochondrial defects have been implicated in a number of degenerative diseases of aging including Parkinson's and Alzheimer's diseases. Several diseases result from point mutations in mtDNA involving either tRNAs or one of the structural genes. Other mitochondrial diseases result from deletions of large portions of mtDNA. One hallmark of all mitochondrial diseases is that these diseases are invariably maternally inherited, because essentially all mitochondria present in a fertilized ovum are derived from the egg. Generally, mitochondrial diseases result in decreased activity of the electron transport chain and consequently pyruvate and fatty acids accumulate, leading to a lactate acidosis and accumulation of triglycerides. In addition, the rate of ATP synthesis is decreased, resulting in muscle weakness and exercise intolerance. See Clinical Correlations 13.4, 13.5, and 13.6 for a discussion of diseases resulting from three different types of mutations in mtDNA.

CLINICAL CORRELATION 13.4
Leber's Hereditary Optic Neuropathy

The first mitochondrial disease to be elucidated at the molecular level is maternally inherited Leber's hereditary optic neuropathy (LHON), which affects the central nervous system, including the optic nerves, causing sudden-onset blindness in early adulthood due to the death of the optic nerve. In nearly all families, LHON results from single base changes in the mitochondrial genes for three subunits of complex I (ND1, ND4, and ND6), resulting in a lowered activity of NADH–ubiquinone oxidoreductase (complex I).

The severity of the diseases resulting from mutations in mitochondrial DNA (mtDNA) depends on the content of mutated mtDNA present in a given cell or tissue. The presence of hundreds or thousands of mitochondria in each cell allows for different percentages of mutant mtDNA to exist in a tissue as a consequence of the random distribution of mutant mtDNA to the daughter cells at cell division. The more mutant mtDNA in a tissue the more the deleterious effects on mitochondrial functions, especially energy production. An example is the LHON that results from a point mutation in the gene for ND6 in which a valine is substituted for a conserved alanine. Patients with a lower percentage of mutant mtDNA develop the sudden-onset blindness in early adulthood and other symptoms typical of LHON. Patients with a higher percentage of mutant mtDNA carrying the identical mutation develop dystonia, a severe disease characterized by early onset of generalized movement disorder, impaired speech, and mental retardation. This indicates the difficulties of making generalizations about diseases resulting from specific mutations in mtDNA.

Chalmers, R. M. and Schapira, A. H. V. Clinical, biochemical and molecular genetic features of Leber's hereditary optic neuropathy. *Biochim. Biophys. Acta* 1410:147, 1999.

CLINICAL CORRELATION 13.5
Mitochondrial Myopathies Due to Mutations in tRNA genes

Point mutations in genes encoding mitochondrial tRNAs result in two of the most common mitochondrial diseases characterized by abnormalities of the central nervous system as well as mitochondrial myopathy with ragged-red fibers, an association known as mitochondrial encephalomyopathy. A mutation in the tRNA gene for lysine causes myoclonic epilepsy and ragged red fibers (MERRF). Symptoms include myoclonus and ataxia with generalized seizures and myopathy. The skeletal muscles of patients with MERRF have abnormally shaped mitochondria that contain paracrystalline structures called ragged red fibers (see figure) and decreased cytochrome *c* oxidase activity.

Mutation in the mitochondrial gene for the tRNA for leucine results in the most complex of all mtDNA defects, the common MELAS mutation resulting in mitochondrial encephalopathy, lactic acidosis, and stroke-like activity. The skeletal muscles of patients with MELAS have ragged red fibers but retain cytochrome *c* oxidase activity. The severity of symptoms observed in patients with the mutation in the $tRNA^{Leu}$ varies with the percentage of mitochondrial DNAs containing the mutant tRNA gene. Patients with >85% mutant genes present with the more severe central nervous symptoms described above, while patients with 5–30% mutant genes often present with maternally inherited diabetes mellitus and deafness. The biochemical consequence of both of these tRNA mutations is impaired mitochondrial protein synthesis leading to decreased activities of complex I and cytochrome *c* oxidase. Explaining the different phenotypes of affected individuals with similar mutations is a challenge facing the mitochondrial research community.

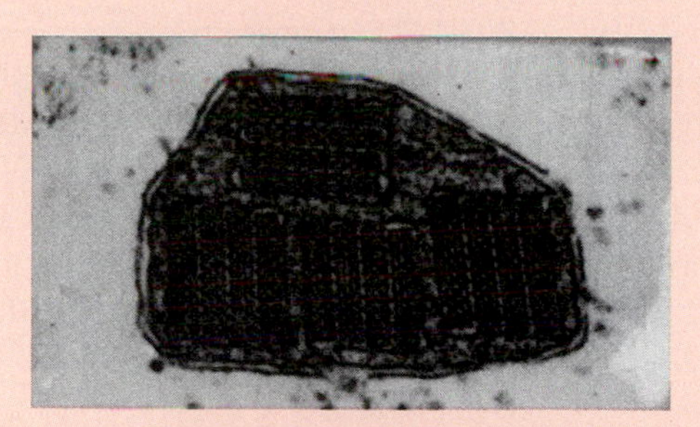

Example of paracrystalline inclusions in mitochondria from muscles of ocular myopathic patients (× 36,000).
Courtesy of Dr. D. N. Landon, Institute of Neurology, University of London.

Wallace, D. C. Mitochondrial diseases in man and mouse. *Science* 283:1482, 1999.

Other mutations in mitochondrial genes lead to progressive muscular weakness, retinitus pigmentosa (loss of retinal response), hearing loss, and ataxia (irregular muscular action), as well as enlargement and deterioration of heart muscle. The deleterious effects of aging may also result from mutations that are proposed to accumulate in mtDNA throughout the life of the individual resulting from constant exposure to DNA-damaging agents such as oxygen radicals.

CLINICAL CORRELATION 13.6
Exercise Intolerance in Patients with Mutations in Cytochrome *b*

In 1993, the first report of a mutation in cytochrome *b* resulting in lowered activity of the cytochrome bc_1 complex was reported in a 25-year-old man who presented with exercise intolerance. The mutation in the cytochrome *b* gene involved the substitution of an aspartate residue for a conserved glycine at position 290. Subsequently, patients with similar symptoms of progressive exercise intolerance and proximal weakness with lowered bc_1 complex activity were shown to have mutations in the cytochrome *b* gene in which a glutamate was substituted for a conserved glycine at position 339, and a serine for a conserved glycine at position 34. More recently, a patient with severe hypertrophic cardiomyopathy was shown to have a mutation in the cytochrome *b* gene in which a glutamate was substituted for a conserved glycine at position 166. The glycine to aspartate and glutamate mutations were located in the cytochrome *b* protein close to the Q_O site for ubiquinol oxidation, while the glycine to serine mutation was located near the Q_i site of ubiquinone reduction. All these mutations of cytochrome *b* involved a guanine to adenine transition in the mtDNA suggesting that the mutation might have resulted from oxidative damage. Moreover, in each of the missense mutations a conserved glycine was replaced by a larger charged molecule, which may alter significantly the structure of cytochrome *b*, leading to the observed lowered catalytic activity of the bc_1 complex. Additional nonsense mutations resulting in the synthesis of truncated cytochrome *b* have been identified in the cytochrome *b* gene as well as mutations involving deletions of 4 to 24 base pairs of mtDNA. These nonsense and deletion mutations often lead to severe exercise intolerance, lactic acidosis in the resting state, and occasionally myglobinuria resulting from the decreased activity of the bc_1 complex. In contrast to the majority of mutations in mtDNA, the mutations identified in the cytochrome *b* gene are not maternally inherited. Moreover, most of these mutations have only been expressed in muscle tissues, suggesting that the mutations identified in the cytochrome *b* gene are somatic and occur during germ-layer differentiation of myogenic stem cells.

Andreu, A. L., et al. Exercise intolerance due to mutations in the cytochrome *b* gene of mitochondrial DNA. *N. Engl. J. Med.* 341:1037, 1999.

O_2
Oxygen
$1e^-$
O_2^-
Superoxide
$1e^- + 2H^+$
H_2O_2
Hydrogen Peroxide
$1e^- + 1H^+$
H_2O + $OH^\bullet$
Hydroxyl Radical
$1e^- + 1H^+$
H_2O
Water

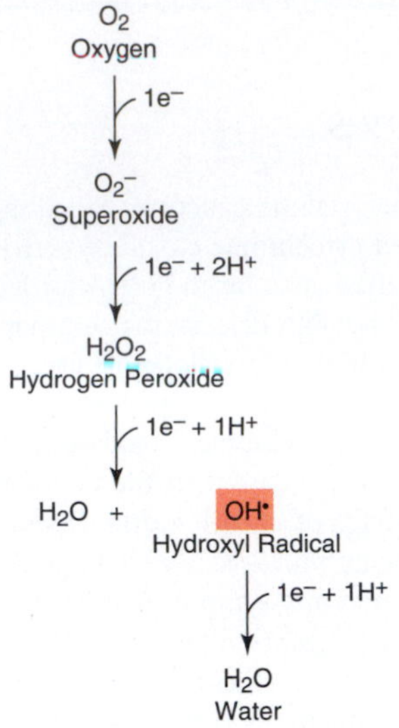

FIGURE 13.62
One-electron steps in reduction of oxygen leading to formation of reactive oxygen species superoxide, hydrogen peroxide, and hydroxyl radical.

Fenton Reaction

$$Fe^{2+} + H_2O_2 \longrightarrow Fe^{3+} + OH^\bullet + OH^-$$

Haber–Weiss Reaction

$$O_2^- + H_2O_2 \xrightarrow{H^+} O_2 + H_2O + OH^\bullet$$

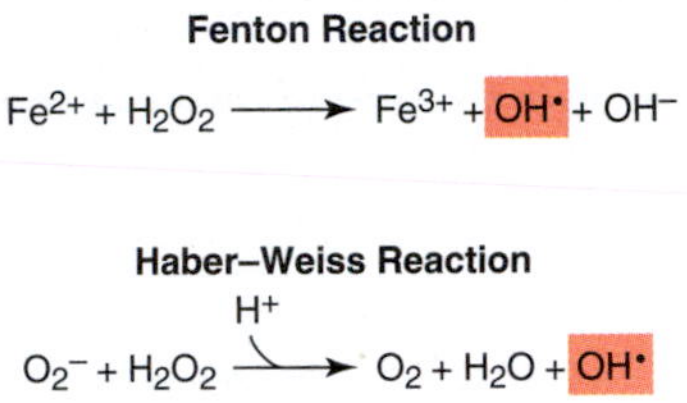

FIGURE 13.63
The Fenton and Haber–Weiss reactions for formation of toxic hydroxyl radical.

13.10 | REACTIVE OXYGEN SPECIES (ROS)

Oxygen is essential to life. The majority of intracellular oxidation of substrates results in transfer of two-electrons to appropriate acceptors such as NAD^+ or FAD, which are subsequently oxidized by the electron transport chain. The terminal step in which O_2 is reduced to water is catalyzed by cytochrome *c* oxidase, which tightly binds O_2 to the binuclear center where stepwise reduction of O_2 occurs without release of intermediates in the oxidation process (see Section 13.6, see p. 574). The electronic structure of O_2, however, favors its reduction by addition of one electron at a time leading to the generation of **oxygen radicals** that can cause cellular damage. A radical is defined as a molecule with a highly reactive unpaired electron in an outer orbital, which can initiate chain reactions by removal of an electron from another molecule to complete its own orbital. The stepwise transfer of electrons to O_2 results in formation of the following intermediates as indicated in Figure 13.62: namely, **superoxide anion** (O_2^-), the partially reduced **hydrogen peroxide** (H_2O_2), and **hydroxyl free radical** (OH•). Of these intermediates in reduction of O_2 to water, the hydroxyl radical is undoubtedly the most dangerous free radical as it is involved in reactions such as lipid peroxidation and generation of other toxic radicals. Hydrogen peroxide itself is not a free radical but is converted by the Fenton or Haber–Weiss reactions to the hydroxyl radical in the presence of Fe^{2+} or Cu^+ prevalent in cells (Figure 13.63).

Production of Reactive Oxygen Species

While oxidative processes in cells generally result in transfer of electrons to O_2 to form water without release of intermediates, a small number of oxygen radicals are inevitably formed due to leakage in electron transfer reactions. The major intracellular source of oxygen radicals is the mitochondrial electron transport chain where superoxide is produced by transfer of one electron to O_2 from the stable semiquinone produced during reduction of ubiquinone by complexes I and II of the electron transport chain (Figure 13.64). The oxygen radicals produced in mitochondria include superoxide, hydrogen peroxide, and hydroxyl radical. Toxic oxygen species are also produced in peroxisomes; they oxidize fatty acids and other compounds by transfer of two electrons from a $FADH_2$ to O_2 with formation of hydrogen perox-

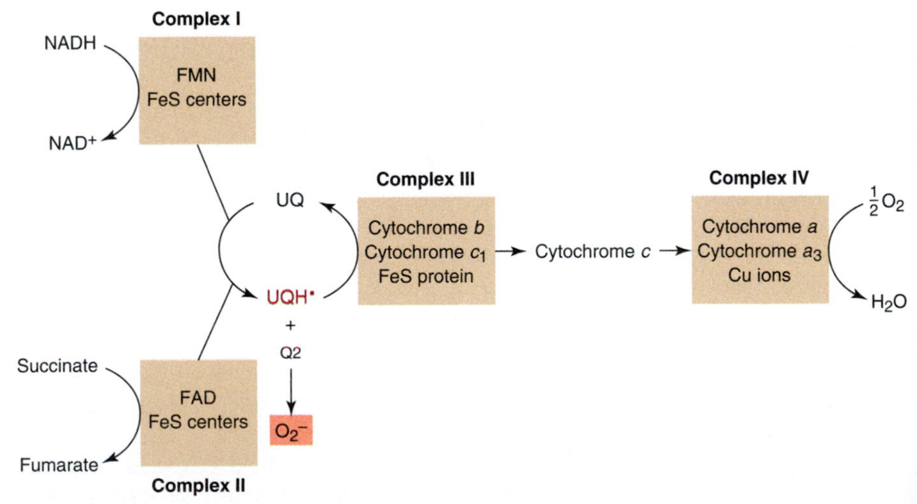

FIGURE 13.64
Generation of superoxide anions by mitochondrial electron transfer chain.
Stable semiquinone formed during the two-electron reduction of ubiquinone by the iron–sulfur centers of both complexes I and II can transfer an electron to oxygen to form the superoxide anion. By contrast, the binuclear center of cytochrome *c* oxidase prevents release of intermediates in the reduction of oxygen.

ide, which is readily converted to hydroxyl radical (Figure 13.63). The cytochrome P450 system localized in endoplasmic reticulum can also produce oxygen radicals.

Oxygen radicals are also produced in cells during processes such as inflammation due to bacterial infection. To combat microbial infections, phagocytes produce and release toxic oxygen radicals to kill invading bacteria in a process known as respiratory burst. The phagocytes (Figure 13.65) then engulf killed bacteria. In an acute infection, production of oxygen radicals and killing of bacteria are efficient processes; however, in prolonged infections, phagocytes tend to die, releasing toxic oxygen radicals that affect surrounding cells.

In addition, cosmic radiation, ingestion of chemicals and drugs, as well as smog can lead to formation of reactive oxygen species. Damage due to reactive oxygen species often occurs during perfusion of tissues with solutions containing high concentrations of O_2; this procedure is used with patients who have suffered an ischemic episode in which localized O_2 levels are lowered due to blockage of an artery (see Clin. Corr. 13.7).

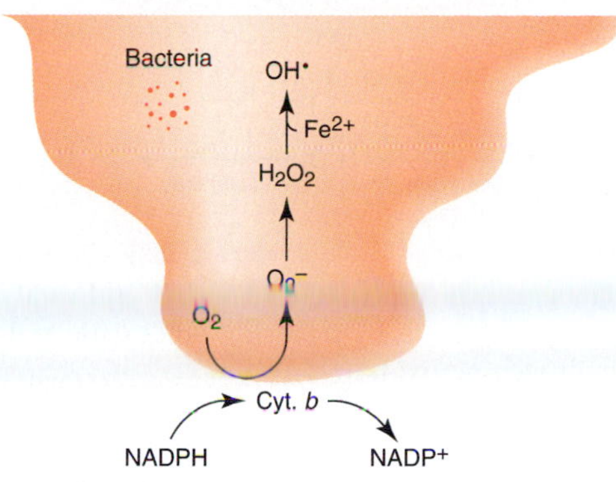

FIGURE 13.65
Respiratory burst in phagocytes.
An electron transfer chain involving a unique cytochrome *b* transfers electrons from NADPH to oxygen with formation of superoxide anion. Superoxide is converted to the hydroxyl radical that kills bacteria subsequently engulfed by phagocytes.

Damage Caused by Reactive Oxygen Species

Reactive oxygen species react with and cause damage to all major classes of macromolecules in cells. The phospholipids present in plasma and organelle membranes are subject to **lipid peroxidation,** a free radical chain reaction initiated by removal of hydrogen from a polyunsaturated fatty acid by hydroxyl radical. The resulting lipid radicals then react with O_2 to form lipid peroxy radicals and lipid peroxide along with malondialdehyde, which is water soluble and can be detected in blood. An example of the effects of lipid peroxidation in humans is the brown spots commonly observed on hands of the elderly. These "age"spots contain the pigment lipofuscin, suggested to be a mixture of cross-linked lipids and products of lipid peroxidation, which accumulate in individuals over the course of a lifetime. One significant consequence of lipid peroxidation is increased membrane permeability leading to an influx of Ca^{2+} and other ions with subsequent swelling of the cell. Similar increases in permeability of organelle membranes may also result in maldistribution of ions and result in intracellular damage. For example, accumulation of excessive amounts of Ca^{2+} in mitochondria may trigger apoptosis.

CLINICAL CORRELATION 13.7

Ischemia/Reperfusion Injury

The occlusion of a major coronary artery during myocardial infarction results in ischemia or lowered oxygen supply. Consequently, the mitochondrial electron transport–oxidative phosphorylation sequence is inhibited, resulting in a decline of intracellular levels of ATP and creatine phosphate. As cellular ATP levels diminish, anaerobic glycolysis is activated in an attempt to maintain normal cellular functions. Glycogen levels are rapidly depleted and lactic acid levels in the cytosol increase, lowering the intracellular pH. Despite the direct consequences of ischemia, the damage to the affected tissues appears to become more severe when oxygen is reintroduced (or reperfused) and oxygen radicals including superoxide (O_2^-), hydrogen peroxide, and hydroxyl radical (OH•) are formed. One recent study observed the formation of oxygen radicals including peroxynitrite ($ONOO^-$) produced from nitric oxide (NO) and superoxide during acute reperfusion of the ischemic heart. Peroxynitrite has been shown to contribute to the poor recovery of mechanical function in ischemic hearts, while antioxidants such as glutathione that scavenge the peroxynitrite radicals protect against damage to mechanical function. Reperfusion of the ischemic heart may also lead to the activation of leukocytes, which act as mediators of the inflammatory process leading to further tissue injury.

Myocardial ischemia/reperfusion represents a clinical problem associated with thrombolysis, angioplasty, and coronary bypass surgery. Injury to the myocardium due to ischemia/reperfusion include cardiac contractile dysfunction, arrhythmias, and irreversible myocyte damage. Ischemia may also arise during surgery especially during transplantation of tissues. The postulated role of oxygen radicals in ischemia/reperfusion injury is strengthened by the observations that antioxidants protect against reperfusion injury to ischemic tissue. Currently, active investigations of methods to protect against reperfusion injury are underway using animal models in which the effects of administering antioxidants and preventing the generation of oxygen radicals are being explored. The increased use of invasive procedures used in clinical medicine indicates the importance of developing methods to protect against ischemia/reperfusion injury.

Cheung, P. Y., Wang, W., and Schulz, R. Glutathione protects against myocardial ischemia–reperfusion injury by detoxifying peroxynitrite. *J. Mol. Cell. Cardiol.* 32:1669, 2000.

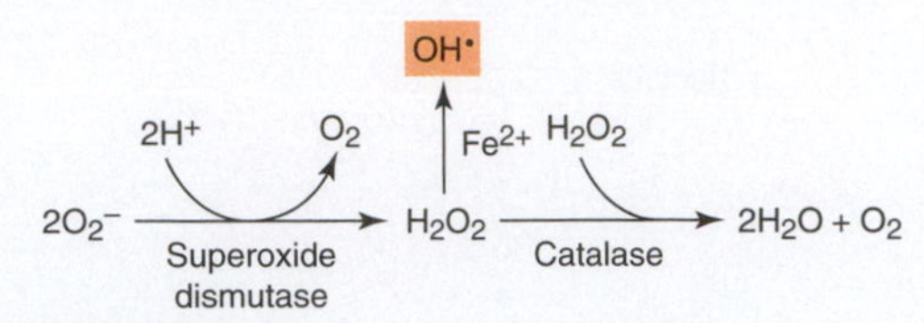

FIGURE 13.66
The enzymes superoxide dismutase and catalase protect cells by removing superoxide and hydrogen peroxide.

Proline, histidine, arginine, cysteine, and methionine are susceptible to attack by hydroxyl radicals with subsequent fragmentation of proteins, cross-linking, and aggregation. Proteins that have been damaged due to attack by oxygen radicals may be subjected to increased rates of digestion by intracellular proteases.

Undoubtedly, the most important consequence of oxygen radicals is damage to both mitochondrial and nuclear DNA resulting in mutations. The nonspecific binding of ferrous ions (Fe^{2+}) to DNA may result in localized formation of hydroxyl radicals that attack individual bases and cause strand breaks. Mitochondrial DNA is more susceptible to damage by oxygen radicals, since the electron transport chain is a major source of toxic oxygen radicals. Nuclear DNA is protected from permanent damage by a protective coat of histones as well as by active and efficient mechanisms for DNA repair. Damage to mtDNA will generally result in mutations that affect energy production. The symptoms in affected individual will be manifest in energy-requiring processes such as muscle contraction. An example of the consequences of a somatic mutation in the mitochondrial gene for cytochrome *b* that may have been caused by oxygen radicals is presented in Clinical Correlation 13.7.

Cellular Defenses Against Reactive Oxygen Species

Cells that live in an aerobic environment have developed multiple ways to remove reactive oxygen species and thus protect themselves against deleterious effects of these radicals. Mammals have three different isozymes of **superoxide dismutase** that catalyzes conversion of superoxide to hydrogen peroxide (Figure 13.66). The cytosolic form of superoxide dismutase contains Cu/Zn at its active site, as does the extracellular form of the enzyme; however, a unique mitochondrial form of superoxide dismutase exists with Mn at its active site. The importance of superoxide removal in cells is emphasized by the presence of these three forms of superoxide dismutase in different cell compartments. Hydrogen peroxide is removed by catalase, a heme-containing enzyme present in highest concentration in peroxisomes and to a lesser extent in mitochondria and cytosol (Figure 13.66).

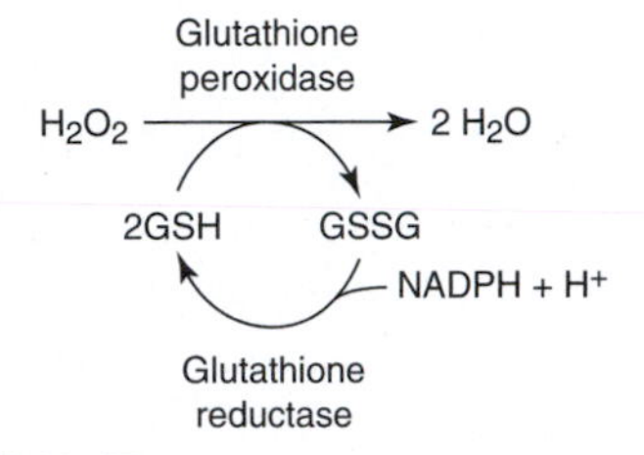

FIGURE 13.67
Glutathione peroxidase removes hydrogen peroxide as well as lipid peroxides.
Electrons are transferred to hydrogen peroxide from the sulfhydryl groups of reduced glutathione (GSH) with formation of oxidized glutathione (GSSG). Glutathione reductase then reduces GSSG to GSH with NADPH as the reducing agent.

A major mechanism for protecting against the damage caused by oxygen radicals is **glutathione peroxidase,** which catalyzes reduction of both hydrogen peroxide and lipid peroxides (Figure 13.67). This selenium-containing enzyme uses sulfhydryl groups of glutathione (GSH) as a hydrogen donor with formation of the oxidized disulfide form of glutathione (GSSG). **Glutathione reductase** converts the disulfide form of glutathione back to the sulfhydryl form using NADPH produced in the pentose phosphate pathway as an electron donor. Protection against reactive oxygen species may also be gained from ingestion of oxygen scavengers such as vitamins C and E and β-carotene.

BIBLIOGRAPHY

Energy-Producing and Energy-Utilizing Systems

Atkinson, D. E. *Cellular Energy Metabolism and Its Regulation.* New York: Academic Press, 1977.

Bock, R. M. Adenosine nucleotides and properties of pyrophosphate compounds. In: P. D. Boyer, H. Lardy, and K. Myrback (Eds.), *The Enzymes,* 2nd ed., Vol. 2. New York: Academic Press, 1960.

Hanson, R. W. The role of ATP in metabolism. *Biochem. Educ.* 17:86, 1989.

Lipmann, F. Metabolic generation and utilization of phosphate bond energy. *Adv. Enzymol.* 1:99, 1941.

Sources and Fates of Acetyl Coenzyme A

Behal, R. H., Buxton, D. B., Robertson, J. G., and Olson, M. S. Regulation of the pyruvate dehydrogenase multienzyme complex. *Annu. Rev. Nutr.* 13:497, 1993.

Reed, L. J. and Hackert, M. L. Structure–function relationships in dihydrolipoamide acyl transferases. *J. Biol. Chem.* 265:8971, 1990.

The Tricarboxylic Acid Cycle

Hansford, R. G. Control of mitochondrial substrate oxidation. *Curr. Top. Bioenerg.* 10:217, 1980.

Krebs, H. A. The history of the tricarboxylic acid cycle. *Perspect. Biol. Med.* 14:154, 1970.

Kornberg, H. L. Tricarboxylic acid cycles. *BioEssays* 7:236, 1987.

Hajnoczky, G., Csordas, G., Krishnamurthy, R., and Szalai, G. Mitochondrial calcium signaling driven by the IP_3 receptor. *J. Bioenerg. Biomembr.* 32:15, 2000.

Ovadi, J. and Srere, P. A. Macromolecular compartmentation and channeling. *Int. Rev. Cytol.* 192:255, 2000.

Patel, M. S. and Harris, R. A. Mammalian α-keto acid dehydrogenase complexes: gene regulation and genetic defects. *FASEB J.* 9:1164, 1995.

Williamson, J. R. and Copper, R. V. Regulation of the citric acid cycle in mammalian systems. *FEBS Lett.* 117 (Suppl.): K73, 1980.

Structure and Compartmentation of Mitochondrial Membranes

Ernster, L. (Ed.). *Bioenergetics*. Amsterdam: Elsevier, 1984.

Tzagoloff, A. *Mitochondria*. New York: Plenum Press, 1982.

Electron Transport

Babcock, G. T. and Wickstrom, M. Oxygen activation and the conservation of energy in cell respiration. *Nature* 356:301, 1992.

Berry, E. A., Huang, L. S., Zhang, Z., and Kim, S. H. Structure of the avian mitochondrial cytochrome bc_1 complex. *J. Bioenerg. Biomembr.* 31:177, 1999.

Calhoun, M. W., Thomas, J. W., and Gennis, R. B. The cytochrome oxidase superfamily of redox-driven proton pumps. *Trends Biochem. Sci.* 19:325, 1994.

Iwata, M., Bjorkman, J., and Iwata, S. Conformational change of the Rieske [2Fe2S] protein in cytochrome bc_1 complex. *J. Bioenerg. Biomembr.* 31:169, 1999.

Trumpower, B. L. and Gennis, R. B. Energy transduction by cytochrome complexes in mitochondrial and bacterial respiration. *Annu. Rev. Biochem.* 63:675, 1994.

Xia, D., et al. Crystal structure of the cytochrome bc_1 complex from bovine heart mitochondria. *Science* 277:60, 1997.

Oxidative Phosphorylation

Boyer, P. D. The ATP synthase—a splendid molecular machine. *Annu. Rev. Biochem.* 66:717, 1997.

Cramer, W. A. and Knaff, B. A. *Energy Transduction in Biological Membranes. A Textbook of Bioenergetics*. New York: Springer-Verlag, 1990.

Mitchell, P. Vectorial chemistry and the molecular mechanism of chemiosmotic coupling: power transmission by proticity. *Biochem. Soc. Trans.* 4:398, 1976.

Sambongi, Y., et al. Mechanical rotation of the c subunit oligomer in ATP synthase (F_0F_1): direct observation. *Science* 286:1722, 1999.

Saraste, M. Oxidative phosphorylation at the fin de siècle. *Science* 283:1488, 1999.

Stock, D., Leslie, A. G. W., and Walker, J. E. Molecular architecture of the rotary motor in ATP synthase. *Science* 286:1700, 1999.

Mitochondrial Transport Systems

Klingenberg, M. Structure–function of the ADP/ATP carrier. *Biochem. Soc. Trans.* 20:547, 1992

Klingenberg, M., Winkler, E., and Huang, S. ADP/ATP carrier and uncoupling protein. *Methods Enzymol.* 260:369, 1995.

LaNoue, K. F. and Schoolwerth, A. C. Metabolite transport in mitochondria. *Annu. Rev. Biochem.* 48:871, 1979.

Mitochondrial Genes and Mitochondrial Diseases

DiMauro, S., Bonilla, E., Davidson, M., Hiran, M. L., and Schon, E. A. Mitochondria in neuromuscular disorders. *Biochim. Biophys. Acta* 1366:199, 1998.

Morgan-Hughes, J. A. and Hanna, M. G. Mitochondrial encephalomyopathies: the enigma of genotype versus phenotype. *Biochim. Biophys. Acta* 1410:125, 1999.

Wallace, D. C. Mitochondrial diseases in mice and man. *Science* 283:1482, 1999.

Reactive Oxygen Species

Fridovich, I. Superoxide radical and superoxide dismutases. *Annu. Rev. Biochem.* 64:97, 1995.

Hansford, R., Tsuchiya, N., and Pepe, S. Mitochondria in heart ischaemia and aging. *Biochem. Soc. Symp.* 66:141, 1999.

Sohal, R. S. Mitochondria generate superoxide anion radicals and hydrogen peroxide. *FASEB J.* 11:1269, 1997.

QUESTIONS

C. N. Angstadt

Multiple Choice Questions

1. A bond may be "high energy" for any of the following reasons EXCEPT:
   A. products of its cleavage are more resonance stabilized than the original compound.
   B. the bond is unusually stable, requiring a large energy input to cleave it.
   C. electrostatic repulsion is relieved when the bond is cleaved.
   D. a cleavage product may be unstable, tautomerizing to a more stable form.
   E. the bond may be strained.

2. At which of the following enzyme-catalyzed steps of the tricarboxylic acid cycle does net incorporation of the elements of water into an intermediate of the cycle occur?
   A. aconitase
   B. citrate synthase
   C. malate dehydrogenase
   D. succinate dehydrogenase
   E. succinyl-CoA synthase

3. All of the following tricarboxylic acid cycle intermediates may be added or removed by other metabolic pathways EXCEPT:
   A. citrate
   B. fumarate
   C. isocitrate
   D. α-ketoglutarate
   E. oxaloacetate

4. Regulation of tricarboxylic acid cycle activity *in vivo* may involve the concentration of all of the following EXCEPT:
   A. acetyl CoA.
   B. ADP.
   C. ATP.
   D. CoA
   E. oxygen.

5. The mitochondrial membrane contains a transporter for:
   A. NADH.
   B. acetyl CoA.
   C. GTP.
   D. ATP.
   E. NADPH.

6. If rotenone is added to the mitochondrial electron transport chain:
   A. the P/O ratio of NADH is reduced from 3:1 to 2:1.
   B. the rate of NADH oxidation is diminished to two-thirds of its initial value.
   C. succinate oxidation remains normal.
   D. oxidative phosphorylation is uncoupled at site I.
   E. electron flow is inhibited at site II.

7. If cyanide is added to tightly coupled mitochondria that are actively oxidizing succinate:
   A. subsequent addition of 2,4-dinitrophenol will cause ATP hydrolysis.
   B. subsequent addition of 2,4-dinitrophenol will restore succinate oxidation.

C. electron flow will cease, but ATP synthesis will continue.
D. electron flow will cease, but ATP synthesis can be restored by subsequent addition of 2,4-dinitrophenol.
E. subsequent addition of 2,4-dinitrophenol and the phosphorylation inhibitor, oligomycin, will cause ATP hydrolysis.

8. The heme iron of which of the following is bound to the protein by only one coordination linkage?
A. cytochrome *a*
B. cytochrome a_3
C. cytochrome *b*
D. cytochrome *c*
E. none of the above

9. Copper is an essential component, participating in the transfer of electrons, of:
A. complex I.
B. complex II.
C. complex III.
D. complex IV.
E. all of the above.

10. The chemiosmotic hypothesis involves all of the following EXCEPT:
A. a membrane impermeable to protons.
B. electron transport by the respiratory chain pumps protons out of the mitochondrion.
C. proton flow into the mitochondria depends on the presence of ADP and P_i.
D. ATPase activity is reversible.
E. only proton transport is strictly regulated; other positively charged ions can diffuse freely across the mitochondrial membrane.

Questions 11 and 12: A child presented with severe neurological defects. Blood tests indicated elevated serum levels of lactate, pyruvate, and alanine. Cultures of skin fibroblasts showed deficient activity of the pyruvate dehydrogenase complex. Further study might indicate which subunit of the complex is defective, although the metabolic effects are essentially the same regardless of which subunit is defective.

11. The active form of pyruvate dehydrogenase is favored by the influence of all of the following on pyruvate dehydrogenase kinase EXCEPT:
A. low $[Ca^{2+}]$.
B. low acetyl CoA/CoASH.
C. high [pyruvate].
D. low $NADH/NAD^+$.

12. Suppose the specific defect were a mutant pyruvate dehydrogenase (the first catalytic subunit) with poor binding of its prosthetic group. In this type of defect, sometimes greatly increasing the dietary precursor of the prosthetic group is helpful. In this case, increasing which of the following might be helpful?
A. lipoic acid
B. niacin (for NAD)
C. pantothenic acid (for CoA)
D. riboflavin (for FAD)
E. thiamine (for TPP)

Questions 13 and 14: When a major coronary artery is blocked, ischemia (lowered oxygen supply) results, inhibiting mitochondrial electron transport and oxidative phosphorylation. The ischemia causes damage to affected tissues. Reperfusion (introducing oxygen), however, appears to cause even more tissue damage as oxygen radicals are formed. Some hope for minimizing the reperfusion damage lies in the administration of antioxidants.

13. All of the following statements are correct EXCEPT:
A. reactive oxygen species (oxygen radicals) result when there is a concerted addition of four electrons at a time to O_2.
B. superoxide anion (O_2^-) and hydroxyl radical (OH•) are two forms of reactive oxygen.
C. superoxide dismutase is a naturally occurring enzyme that protects against damage by converting O_2^- to H_2O_2.
D. reactive oxygen species damage phospholipids, proteins, and nucleic acids.
E. glutathione protects against H_2O_2 by reducing it to water.

14. All of the following result from ischemia EXCEPT:
A. decrease in intracellular ATP.
B. decrease in intracellular creatine phosphate.
C. decrease in $NADH/NAD^+$.
D. lactic acidosis.
E. depletion of cellular glycogen.

Problems

15. For the reaction A $\rightleftharpoons$ B, $\Delta G^{\circ\prime} = -7.1$ kcal mol^{-1}. At 37°C, $-2.303\ RT = -1.42$ kcal mol^{-1}. What is the equilibrium ratio of B/A?

16. Using pyruvate, labeled with ^{14}C in its keto group, via the pyruvate dehydrogenase reaction and the TCA cycle, where would the carbon label be at the end of one turn of the TCA cycle? Where would the carbon label be at the end of the second turn of the cycle?

ANSWERS

1. **B** High energy does not refer to a high energy of formation (bond stability). A "high-energy" bond is so designated because it has a high free energy of hydrolysis. This could arise for reasons A, C, D, or E.
2. **B** Water is required to hydrolyze the thioester bond of acetyl CoA. A: Aconitase removes water, then adds it back. C, D: The dehydrogenases remove two protons and two electrons. E. Here the thioester undergoes phosphorolysis, not hydrolysis; the phosphate is subsequently transferred from the intermediate succinyl phosphate to GDP.
3. **C** A: Citrate is transported out of the mitochondria to be used as a source of cytoplasmic acetyl CoA. B: Fumarate is produced during phenylalanine and tyrosine degradation. D: This can be formed from glutamate. E: Oxaloacetate is produced by pyruvate carboxylase and is used in gluconeogenesis. Clearly most of the tricarboxylic acid cycle intermediates play multiple roles in the body.
4. **D** CoA is not a regulator, though there is a reciprocal relationship between CoA and acetyl CoA in the short term. A is the substrate. B activates isocitrate dehydrogenase, and C inhibits it. E:

The cycle requires oxygen to oxidize NADH and ADP to be converted to ATP (respiratory control).

5. **D** ATP and ADP are transported in opposite directions. A, B: Reducing equivalents from NADH are shuttled across the membrane, as is the acetyl group of acetyl CoA, but NADH and acetyl CoA themselves cannot cross. C: Of the nucleotides, only ATP and ADP are transported. E: Like NADH, NADPH does not cross the membrane.
6. **C** Rotenone inhibits at the level of NADH dehydrogenase (site I), preventing all electron flow and all ATP synthesis from NADH. Flavin-linked dehydrogenases feed in electrons below site I and are unaffected by site I inhibitors.
7. **A** Cyanide inhibits electron transport at site III, blocking electron flow throughout the system. In coupled mitochondria, ATP synthesis ceases too. Addition of an uncoupler permits the mitochondrial ATPase (which is normally driven in the synthetic direction) to operate, and it catalyzes the favorable ATP hydrolysis reaction unless it is inhibited by a phosphorylation inhibitor such as oligomycin.
8. **B** Fe^{2+} has six coordination positions. In heme, four are filled by the porphyrin ring. In cytochromes *a*, *b*, and *c* the other two are filled by the protein. But in cytochrome a_3, one position must be left vacant to provide an oxygen-binding site.
9. **D** The copper is directly involved in cytochrome a_3 reacting with O_2. A–C: These complexes contain Fe–S protein clusters but do not contain copper.
10. **E** If the charge separation could be dissipated by free diffusion of other ions, the energy would be lost, and no ATP could be synthesized.
11. **A** High Ca^{2+} favors the active dehydrogenase but by activating the phosphatase. B–D: NADH and acetyl CoA activate pyruvate dehydrogenase kinase, thus inactivating pyruvate dehydrogenase. Pyruvate inhibits the kinase, favoring the active dehydrogenase.
12. **E** TPP is the cofactor for the first reaction, which decarboxylates pyruvate. A, C: These are cofactors for the dihydrolipoyl transacetylase. B, D: These are cofactors for the dihydrolipoyl dehydrogenase.
13. **A** Four-electron transfer to oxygen produces water. Oxygen radicals occur when electrons are added stepwise to O_2. B: H_2O_2 is also included as reactive oxygen because it can produce OH•. C: The H_2O_2 is then degraded by catalase. D: All of these are subject to oxidation with deleterious effects. E: Glutathione peroxidase catalyzes this reaction.
14. **C** Inhibition of electron transport because of low O_2 prevents reoxidation of NADH and, therefore, the NADH/NAD^+ increases. A, B: ATP decreases because oxidative phosphorylation is inhibited. This will shift the creatine kinase reaction toward ATP. D: Anaerobic glycolysis is stimulated to provide ATP, leading to increased lactate. E: Glycogen is the substrate for anaerobic glycolysis.
15. $\Delta G^{\circ\prime} = -RT \ln K'_{eq} = -2.303\ RT \log K'_{eq}$. Substitution gives $\log K'_{eq} = 5$. K'_{eq} then is 100,000 so the B/A = 100,000/1.
16. The labeled keto carbon of pyruvate becomes the labeled carboxyl carbon of acetyl CoA. After condensation with oxaloacetate, the first carboxyl group of citrate is labeled. This label is retained through subsequent reactions to succinate. However, succinate is a symmetrical compound to the enzyme so, in effect, both carboxyl groups of succinate are labeled. This means that the oxaloacetate regenerated is labeled in both carboxyl groups at the end of one turn (actually half the molecules are labeled in one carboxyl and half in the other but this can't be distinguished experimentally). Note that CO_2 is not labeled. In the second turn, the same carboxyl labeled acetyl CoA is added but this time to labeled oxaloacetate. Both carboxyl groups of the oxaloacetate are released as CO_2 so it will be labeled, as will the regenerated oxaloacetate.

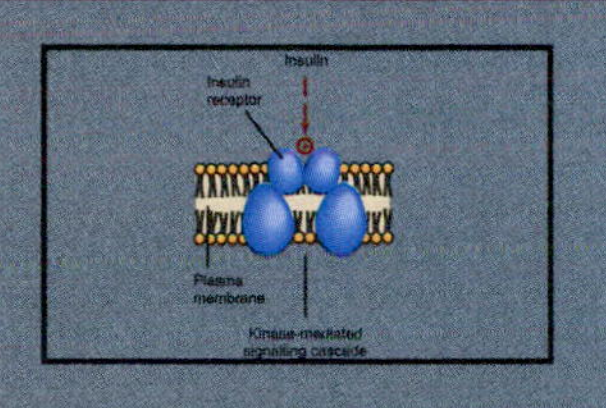

14

CARBOHYDRATE METABOLISM I: MAJOR METABOLIC PATHWAYS AND THEIR CONTROL

Robert A. Harris

14.1 | OVERVIEW

The major pathways of carbohydrate metabolism either begin or end with glucose (Figure 14.1). This chapter describes the utilization of glucose as a source of energy, formation of glucose from noncarbohydrate precursors, storage of glucose in the form of glycogen for later use, and release of glucose from glycogen for use by cells. An understanding of the pathways and their regulation is necessary because of the important role played by glucose in the body. Glucose is the major form in which carbohydrate absorbed from the intestinal tract is presented to cells of the body. Glucose is the only fuel used to any significant extent by a few specialized cells, and the major fuel used by the brain. Indeed, glucose is so important to these specialized cells and the brain that several of the major tissues of the body have evolved a working relationship that assures a continuous supply of this essential substrate. Glucose metabolism is defective in two very common metabolic diseases, obesity and diabetes, which contribute to the development of a number of major medical problems, including atherosclerosis, hypertension, small vessel disease, kidney disease, and blindness.

The discussion begins with **glycolysis,** a pathway used by all cells of the body to extract part of the chemical energy inherent in the glucose molecule. This pathway also converts glucose to pyruvate and sets the stage for complete oxidation of glucose to CO_2 and H_2O. The *de novo* synthesis of glucose, that is, **gluconeogenesis,** is a function of the liver and kidneys and can be conveniently discussed following glycolysis because it makes use of some of the same enzymes used in the glycolytic pathway, although the reactions catalyzed are in the opposite direction. In contrast to glycolysis, which produces ATP, gluconeogenesis requires ATP and is therefore an energy-requiring process. The consequence is that only some of the enzyme-catalyzed steps can be common to both the glycolytic and gluconeogenic pathways. Indeed, additional enzymes including some in mitochondria become involved to make the overall process of gluconeogenesis exergonic. How regulation is exerted at key enzyme-catalyzed steps will be stressed throughout the chapter. This will be particularly true for glycogen synthesis **(glycogenesis)** and degradation **(glycogenolysis).** Many cells store glycogen for the purpose of having glucose

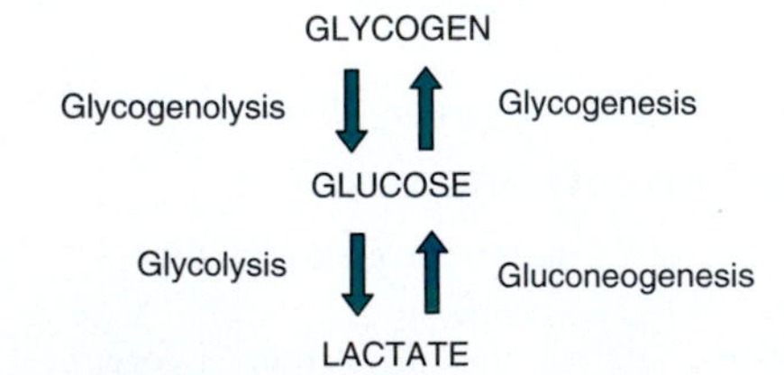

FIGURE 14.1
Relationship of glucose to major pathways of carbohydrate metabolism.

available for later use. The liver is less selfish, storing glycogen not for its own need, but for maintenance of blood glucose in order to ensure that other tissues, especially the brain, have an adequate supply of this important substrate. Regulation of the synthesis and degradation of glycogen is a model for our understanding of how hormones work and how other metabolic pathways may be regulated. This subject contributes to our understanding of the diabetic condition, starvation, and how tissues of the body respond to stress, severe trauma, and injury. The Appendix (see p. 1172) presents the nomenclature and chemistry of the carbohydrates.

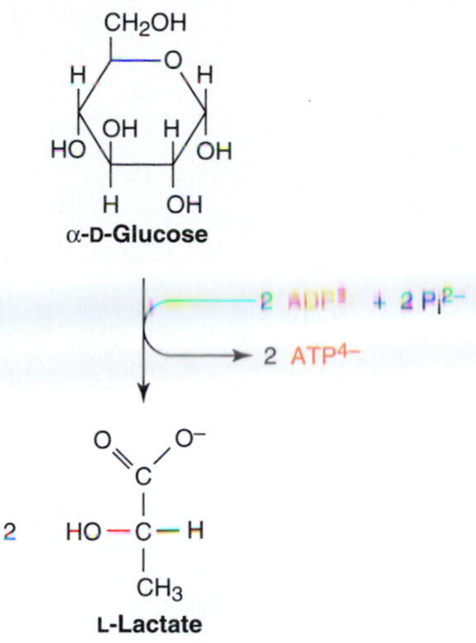

14.2 | GLYCOLYSIS

Glycolysis Occurs in All Human Cells

The **Embden–Meyerhof** or **glycolytic pathway** represents an ancient process, possessed by all cells of the human body, in which anaerobic degradation of glucose to lactate occurs. This is one example of **anaerobic fermentation,** a term used to refer to pathways by which organisms extract chemical energy from high-energy fuels in the absence of molecular oxygen. For many tissues glycolysis is an emergency energy-yielding pathway, capable of producing 2 mol of ATP from 1 mol of glucose in the absence of molecular oxygen (Figure 14.2). Thus when the oxygen supply to a tissue is shut off, ATP levels can still be maintained by glycolysis for at least a short period of time. Many examples could be given, but the capacity to use glycolysis as a source of energy is particularly important to the human being at birth. With the exception of the brain, circulation of blood decreases to most parts of the body of the neonate during delivery. The brain is not normally deprived of oxygen during delivery, but other tissues must depend on glycolysis for their supply of ATP until circulation returns to normal and oxygen becomes available again. This conserves oxygen for use by the **brain,** illustrating one of many mechanisms that have evolved to assure survival of brain tissue in times of stress. Oxygen is not necessary for glycolysis, and the presence of oxygen can indirectly suppress glycolysis, a phenomenon called the **Pasteur effect,** which is considered later. Nevertheless, glycolysis can and does occur in cells with an abundant supply of molecular oxygen. Provided cells also contain mitochondria, the end product of glycolysis in the presence of oxygen is pyruvate rather than lactate. Pyruvate can then be completely oxidized to CO_2 and H_2O by enzymes housed within the mitochondria. Glycolysis therefore sets the stage for aerobic oxidation of carbohydrate. The overall process of glycolysis plus the subsequent mitochondrial oxidation of pyruvate to CO_2 and H_2O has the following equation:

$$\text{D-glucose} + 6\,O_2 + 32\,ADP^{3-} + 32\,P_i^{2-} + 32\,H^+ \rightarrow 6\,CO_2 + 6\,H_2O + 32\,ATP^{4-} (C_6H_{12}O_6)$$

Much more ATP is produced in complete oxidation of glucose to CO_2 and H_2O than in the conversion of glucose to lactate. This has important consequences, which are considered in detail later. For glucose to be completely oxidized to CO_2 and H_2O, it must first be converted to pyruvate by glycolysis (Figure 14.3). The im-

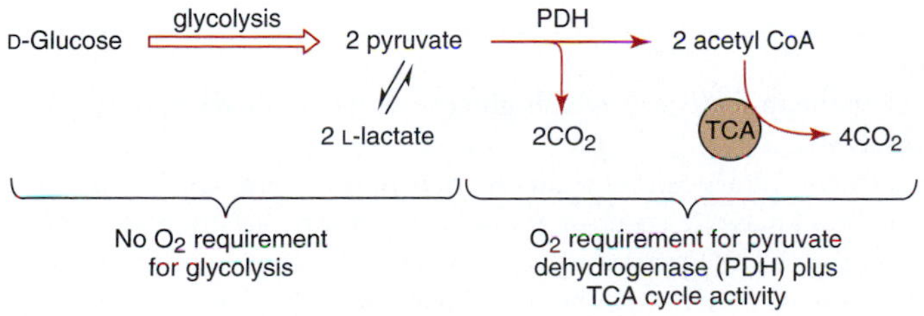

FIGURE 14.2
Overall balanced equation for the sum of reactions of glycolytic pathway.

FIGURE 14.3
Glycolysis is a preparatory pathway for aerobic metabolism of glucose.
TCA refers to the tricarboxylic acid cycle.

portance of glycolysis as a preparatory pathway is best exemplified by the brain. This tissue has an absolute need for glucose and processes most of it via glycolysis. Pyruvate produced is then oxidized to CO_2 and H_2O in mitochondria. An adult human brain uses approximately 120 g of glucose each day in order to meet its need for ATP. In contrast, glycolysis with lactate as the end product is the major mechanism of ATP production in a number of other tissues. Red blood cells lack mitochondria and therefore are unable to convert pyruvate to CO_2 and H_2O. The cornea, lens, and regions of the retina have a limited blood supply and also lack mitochondria (because mitochondria would absorb and scatter light) and depend on glycolysis as the major mechanism for ATP production. Kidney medulla, testis, leukocytes, and white muscle fibers are almost totally dependent on glycolysis as a source of ATP because these tissues have relatively few mitochondria. Tissues dependent primarily on glycolysis for ATP production consume about 40 g of glucose per day in a normal adult.

Starch is the storage form of glucose in plants and contains **α-1,4-glycosidic linkages** along with **α-1,6-glycosidic branches. Glycogen** is the storage form of glucose in animal tissues and contains the same type of glycosidic linkages and branches. Exogenous glycogen refers to that which we eat and digest; endogenous glycogen is that synthesized or stored in our tissues. Exogenous starch or glycogen is hydrolyzed in the intestinal tract with the production of glucose, whereas stored glycogen endogenous to our tissues is converted to glucose or glucose 6-phosphate by enzymes present within the cells. Disaccharides that serve as important sources of glucose in our diet include milk sugar (lactose) and grocery store sugar (sucrose). Hydrolysis of these sugars by enzymes of the brush border of the intestinal tract is discussed on page 1102. Glucose can be used as a source of energy for cells of the intestinal tract. However, these cells do not depend on glucose to any great extent; most of their energy requirement is met by glutamine catabolism (p. 784). Most of the glucose passes through the cells of the intestinal tract into the portal vein blood and then the general circulation to be used by other tissues. Liver is the first major tissue to have an opportunity to remove glucose from the portal vein blood. When blood glucose is high, the liver removes glucose for the glucose-consuming processes of glycogenesis and glycolysis. When blood glucose is low, the liver supplies the blood with glucose by the glucose-producing processes of glycogenolysis and gluconeogenesis. The liver is also the first organ exposed to the blood flowing from the pancreas and therefore "senses" the highest concentrations of the hormones (**glucagon** and **insulin**) released from this endocrine tissue. These important hormonal regulators of blood glucose levels have effects on key enzyme-catalyzed steps in the liver.

Glucose Is Metabolized Differently in Various Cells

After penetrating the plasma membrane by facilitated transport by way of **GLUT-1 (*glucose transporter isoform 1*)**, glucose is metabolized mainly by glycolysis in **red blood cells** (Figure 14.4*a*). Since red blood cells lack mitochondria, the end product of glycolysis is lactic acid, which is released into the blood plasma. Glucose used by the **pentose phosphate pathway** in red blood cells provides NADPH

FIGURE 14.4
Overviews of the major ways in which glucose is metabolized within cells of selected tissues of the body.
(*a*) Glucose transport into a cell by a glucose transport protein (GLUT); (*b*) glucose phosphorylation by hexokinase; (*c*) pentose phosphate pathway; (*d*) glycolysis; (*e*) lactic acid transport out of the cell; (*f*) pyruvate decarboxylation by pyruvate dehydrogenase; (*g*) TCA cycle; (*h*) glycogenesis; (*i*) glycogenolysis; (*j*) lipogenesis; (*k*) gluconeogenesis; (*l*) hydrolysis of glucose 6-phosphate and release of glucose from the cell into the blood; (*m*) formation of glucuronides (drug and bilirubin detoxification by conjugation) by the glucuronic acid pathway.

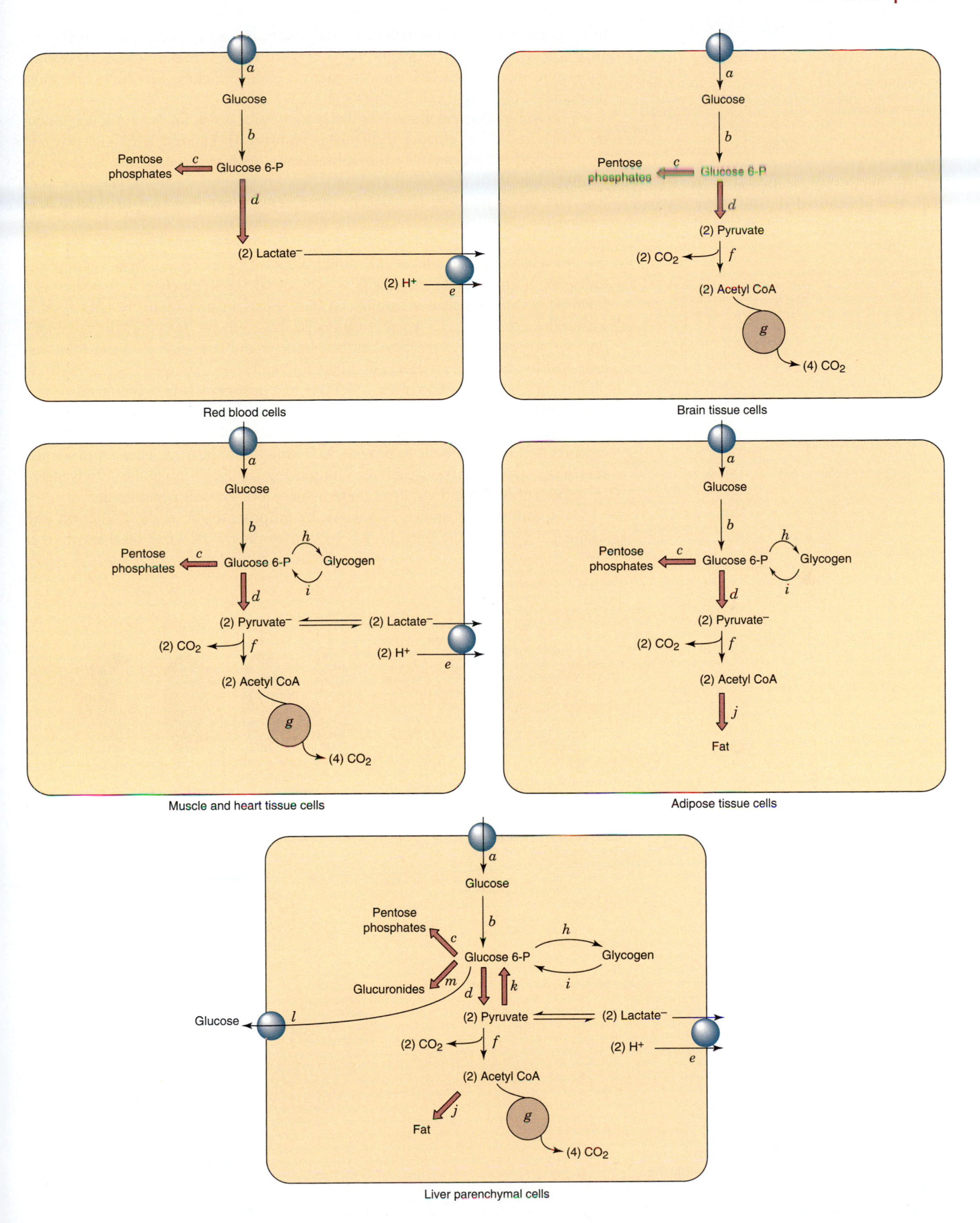
a
Glucose
b
Pentose phosphates
c
Glucose 6-P
d
(2) Lactate⁻
(2) H⁺
e
Red blood cells
a
Glucose
b
Pentose phosphates
c
Glucose 6-P
d
(2) Pyruvate
(2) CO_2
f
(2) Acetyl CoA
g
(4) CO_2
Brain tissue cells
a
Glucose
b
Pentose phosphates
c
Glucose 6-P
h
Glycogen
i
d
(2) Pyruvate⁻
(2) Lactate⁻
(2) CO_2
f
(2) H⁺
e
(2) Acetyl CoA
g
(4) CO_2
Muscle and heart tissue cells
a
Glucose
b
Pentose phosphates
c
Glucose 6-P
h
Glycogen
i
d
(2) Pyruvate⁻
(2) CO_2
f
(2) Acetyl CoA
j
Fat
Adipose tissue cells
a
Glucose
b
Pentose phosphates
c
Glucose 6-P
h
Glycogen
Glucuronides
m
d
k
i
Glucose
l
(2) Pyruvate
(2) Lactate⁻
(2) CO_2
f
(2) H⁺
e
(2) Acetyl CoA
j
Fat
g
(4) CO_2
Liver parenchymal cells

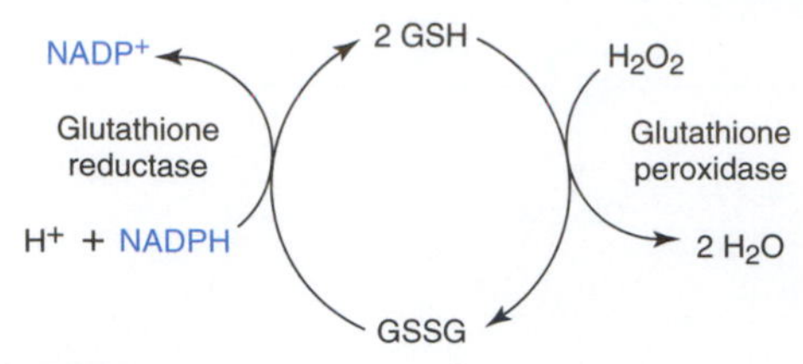

FIGURE 14.5
Destruction of H_2O_2 is dependent on reduction of oxidized glutathione by NADPH generated by pentose phosphate pathway.

to keep glutathione in the reduced state, which has an important role in the destruction of organic peroxides and H_2O_2 (Figure 14.5). Peroxides cause irreversible damage to membranes, DNA, and numerous other cellular components and must be removed to prevent cell death (see p. 22).

The **brain** takes up glucose by facilitated transport in an insulin-independent manner by **GLUT-3 (glucose transporter isoform 3)** (Figure 14.4*b*). Glycolysis in the brain yields pyruvate, which is subsequently oxidized to CO_2 and H_2O by the combination of the pyruvate dehydrogenase complex and the TCA cycle. The pentose phosphate pathway is active in these cells, generating part of the NADPH needed for reductive synthesis and the maintenance of glutathione in the reduced state.

Muscle and **heart cells** readily utilize glucose (Figure 14.4*c*). Insulin stimulates transport of glucose into these cells by way of **GLUT-4 (glucose transporter isoform 4).** In the absence of insulin, GLUT-4 exists in membrane vesicles located within the cytosol of cells where it cannot facilitate glucose transport (Figure 14.6). Binding of insulin to its receptor on the plasma membrane initiates a signaling cascade that promotes translocation and fusion of GLUT-4-containing vesicles with the plasma membrane, thereby placing GLUT-4 in a position where it can facilitate glucose transport. Glucose taken into muscle and heart cells can be utilized by glycolysis to give pyruvate, which is used by the combination of the pyruvate dehydrogenase complex and the TCA cycle to provide ATP. Muscle and heart, in contrast to the tissues just considered, are capable of synthesizing significant quantities of glycogen, an important fuel that the cells of these tissues store for later consumption.

Like muscle, the uptake of glucose by **adipose tissue** is dependent on and stimulated by insulin (Figure 14.4*d*). Insulin works by the same mechanism; that

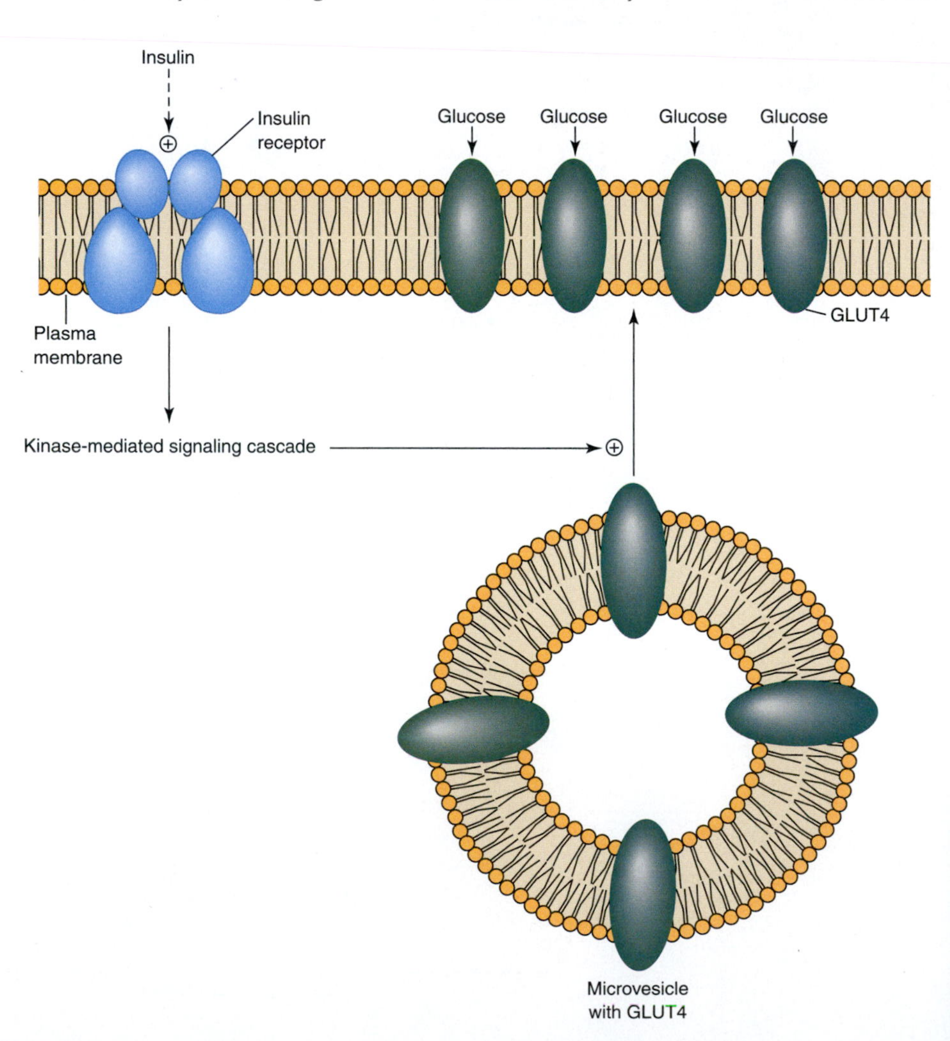

FIGURE 14.6
Insulin stimulates glucose uptake by adipose tissue and muscle by increasing the number of glucose transporters (GLUT-4) in the plasma membrane.

is, it initiates a signaling cascade that culminates in fusion of intracellular GLUT-4 vesicles with the plasma membrane of adipocytes where GLUT-4 functions to bring glucose into the cell (Figure 14.6). Pyruvate, as in other cells, is generated by glycolysis and is oxidized by the pyruvate dehydrogenase complex to give acetyl CoA, which is used primarily for *de novo* fatty acid synthesis. The pentose phosphate pathway is important because it produces the NADPH necessary for the reductive steps of fatty acid synthesis. Adipose tissue has the capacity for glycogenesis and glycogenolysis, but these processes are very limited in this tissue relative to that of muscle, heart, and liver.

Liver has the greatest number of ways to utilize glucose (Figure 14.4*e*). Uptake of glucose occurs independent of insulin by means of **GLUT-2,** a low-affinity, high-capacity glucose transporter. Glucose is used rather extensively by the pentose phosphate pathway for the production of NADPH, which is needed for reductive synthesis, maintenance of reduced glutathione, and numerous reactions catalyzed by endoplasmic reticulum enzyme systems. A quantitatively less important but nevertheless vital function of the pentose phosphate pathway (see p. 666) is the provision of ribose phosphate, required for the synthesis of nucleotides such as ATP and those in DNA and RNA. Glucose is also used for glycogen synthesis, with glycogen storage being a particularly important feature of the liver. Glucose can also be used in the **glucuronic acid pathway,** which is important in drug and **bilirubin detoxification** (see Chapters 11, p. 477 and 24, p. 1073). The liver is also capable of glycolysis, the pyruvate produced being used as a source of acetyl CoA for complete oxidation by the TCA cycle and for the synthesis of fat by the process of *de novo* fatty acid synthesis. In contrast to the other tissues, the liver is unique in that it has the capacity to convert three-carbon precursors (lactate, pyruvate, glycerol, and alanine) into glucose by the process of gluconeogenesis to meet the glucose need of other cells.

14.3 | THE GLYCOLYTIC PATHWAY

Glucose is combustible and will burn in a test tube to yield heat and light but, of course, no ATP. Cells use some 30 steps to take glucose to CO_2 and H_2O, a seemingly inefficient process, since it can be done in a single step in a test tube. However, side reactions and some of the actual steps used by the cell to "burn" glucose to CO_2 and H_2O lead to the conservation of a significant amount of energy in the form of ATP. In other words, ATP is produced by the controlled "burning" of glucose in the cell, glycolysis representing only the first few steps, shown in Figure 14.7, in the overall process.

Glycolysis Occurs in Three Stages

Glycolysis can conveniently be pictured as occurring in three major stages (Figure 14.7).

Priming stage:

$$\text{D-glucose} + 2\ \text{ATP}^{4-} \rightarrow \text{D-fructose 1,6-bisphosphate}^{4-} + 2\ \text{ADP}^{3-} + 2\ \text{H}^+$$

Splitting stage:

$$\text{D-fructose 1,6-bisphosphate}^{4-} \rightarrow 2\ \text{D-glyceraldehyde 3-phosphate}^{2-}$$

Oxidoreduction–phosphorylation stage:

$$2\ \text{D-glyceraldehyde 3-phosphate}^{2-} + 4\ \text{ADP}^{3-} + 2\ \text{P}_i^{2-} + 2\ \text{H}^+ \rightarrow 2\ \text{L-lactate}^- + 4\ \text{ATP}^{4-} + 2\ \text{H}_2\text{O}$$

Sum:

$$\text{D-glucose} + 2\ \text{ADP}^{3-} + 2\ \text{P}_i^{2-} \rightarrow 2\ \text{L-lactate}^- + 2\ \text{ATP}^{4-} + 2\ \text{H}_2\text{O}$$

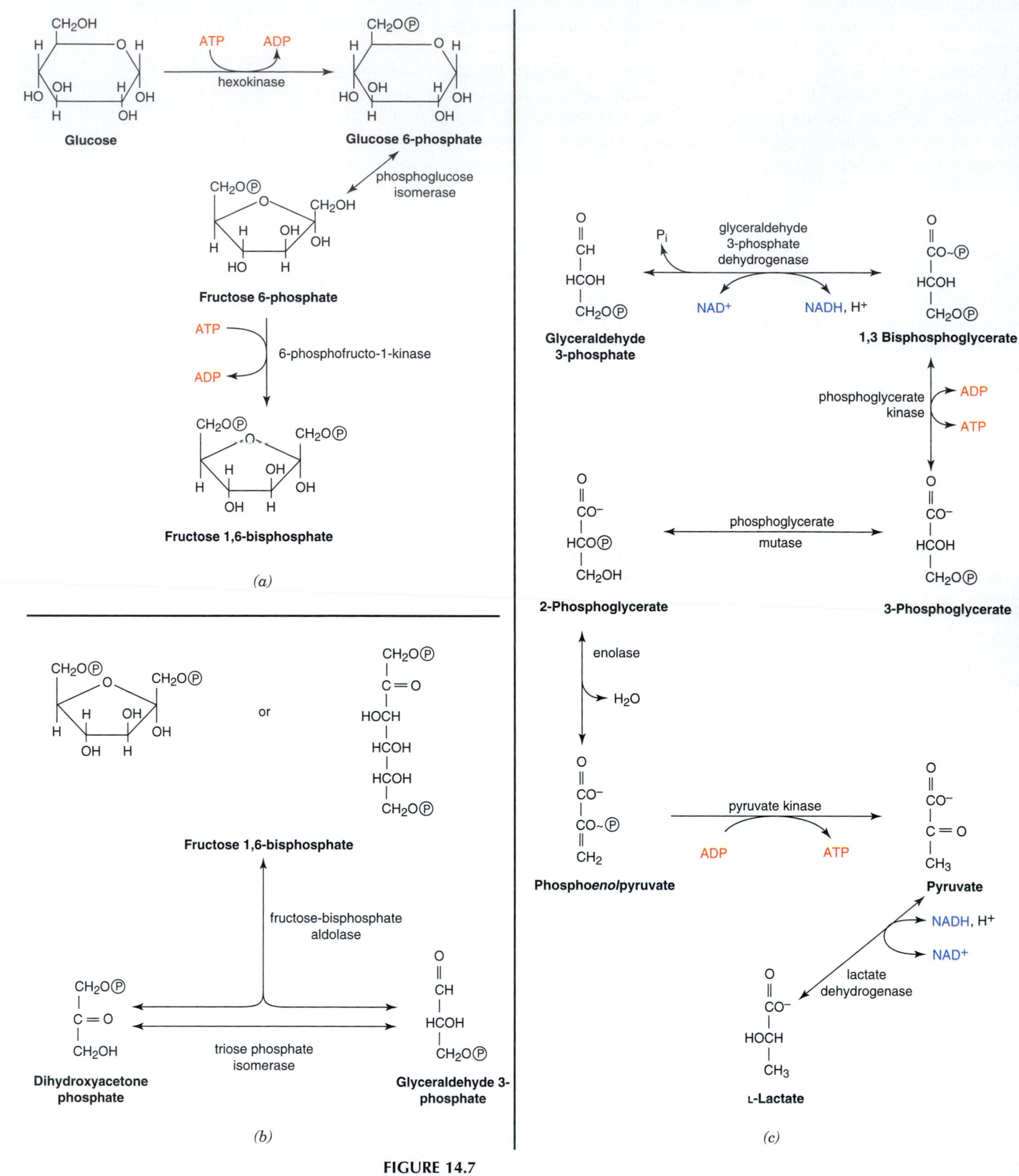

FIGURE 14.7

The glycolytic pathway, divided into its three stages.

The symbol P refers to phosphoryl group PO_3^{2-}; ~indicates a high-energy phosphate bond. (*a*) Priming stage, (*b*) splitting stage, and (*c*) oxidoreduction–phosphorylation stage.

The **priming stage** involves input of two molecules of ATP to convert glucose into a molecule of fructose 1,6-bisphosphate. ATP is therefore "invested" in the priming stage of glycolysis. However, ATP beyond this investment is subsequently gained from the glycolytic process. The **splitting stage** "splits" the six-carbon molecule fructose 1,6-bisphosphate into two molecules of glyceraldehyde 3-phosphate. In the **oxidoreduction–phosphorylation stage** two molecules of glyceraldehyde 3-phosphate are converted into two molecules of lactate with the production of four molecules of ATP. The overall process of glycolysis therefore generates two molecules of lactate and two molecules of ATP at the expense of one molecule of glucose.

Stage One Primes the Glucose Molecule

Hexokinase catalyzes the first step of glycolysis (see Figure 14.7*a* and Step 1). Although this reaction consumes ATP, it gets glycolysis off to a good start by trapping glucose as glucose 6-phosphate (G6P) within the cytosol of the cell where all of the glycolytic enzymes are located. Phosphate esters are charged, hydrophilic compounds that do not readily penetrate cell membranes. The phosphorylation of glucose with ATP is a thermodynamically favorable reaction, requiring the use of one high-energy phosphate bond. It is irreversible under cellular conditions. It is not, however, a way to synthesize ATP or to hydrolyze G6P to give glucose by the reverse reaction. Hydrolysis of G6P is accomplished by a different reaction, catalyzed by **glucose 6-phosphatase:**

$$\text{glucose 6-phosphate}^{2-} + H_2O \rightarrow \text{glucose} + P_i^{2-}$$

This reaction is thermodynamically favorable in the direction written and cannot be used in cells for the synthesis of G6P from glucose. (A common mistake is to note that ATP and ADP are involved in the hexokinase reaction and then assume that they are also involved in the glucose 6-phosphatase reaction.) Glucose 6-phosphatase is an important enzyme in liver, functioning to produce free glucose from G6P in the last step of both gluconeogenesis and glycogenolysis; it has no role in glycolysis.

The next reaction is a readily reversible step of the glycolytic pathway, catalyzed by **phosphoglucose isomerase** (Step 2). This step is not subject to regulation and, since it is readily reversible, functions in both glycolysis and gluconeogenesis.

6-Phosphofructo-1-kinase (or **phosphofructokinase-1**) catalyzes the next reaction, an ATP-dependent phosphorylation of fructose 6-phosphate (F6P) to give fructose 1,6-bisphosphate (FBP) (Step 3). This is a favorite enzyme of many students of biochemistry, being subject to regulation by several effectors and often considered the most important regulatory enzyme of the glycolytic pathway. The reaction is irreversible under intracellular conditions; that is, it represents a way to produce FBP but not a way to produce ATP or F6P by the reverse reaction. This reaction utilizes the second ATP needed to "prime" glucose, thereby completing the first stage of glycolysis.

Stage Two Is Splitting of a Phosphorylated Intermediate

Fructose 1,6-bisphosphate aldolase catalyzes the cleavage of fructose 1,6-bisphosphate into a molecule each of dihydroxyacetone phosphate and glyceraldehyde 3-phosphate (GAP) (see Figure 14.7*b;* Step 4). This is a reversible reaction, the enzyme being called aldolase because the overall reaction is a variant of an aldol cleavage in one direction and an aldol condensation in the other. **Triose phosphate isomerase** then catalyzes the reversible interconversion of dihydroxyacetone phosphate and GAP to complete the splitting stage of glycolysis (Step 5). With the transformation of dihydroxyacetone phosphate (DHAP) into GAP, one molecule of glucose is converted into two molecules of GAP.

CH_2OH (ring) + ATP^{4-} $\xrightarrow{Mg^{3+}}$ $CH_2OPO_4^{2}$ (ring) + ADP^{3-} + H^+

α-D-Glucose → α-D-Glucose 6-phosphate

Step 1

$CH_2OPO_3^{2-}$ (ring)

α-D-Glucose 6-phosphate

⇅

CH_2OH – C=O – HOCH – HCOH – HCOH – $CH_2OPO_3^{2-}$

α-D-Fructose 6-phosphate

Step 2

CH_2OH – C=O – HOCH – HCOH – HCOH – $CH_2OPO_3^{2-}$ + ATP^{4-}

D-Fructose 6-phosphate

↓ Mg^{2+}

$CH_2OPO_3^{2-}$ – C=O – HOCH – HCOH – HCOH – $CH_2OPO_3^{2-}$ + ADP^{3-} + H^+

D-Fructose 1,6-bisphosphate

Step 3

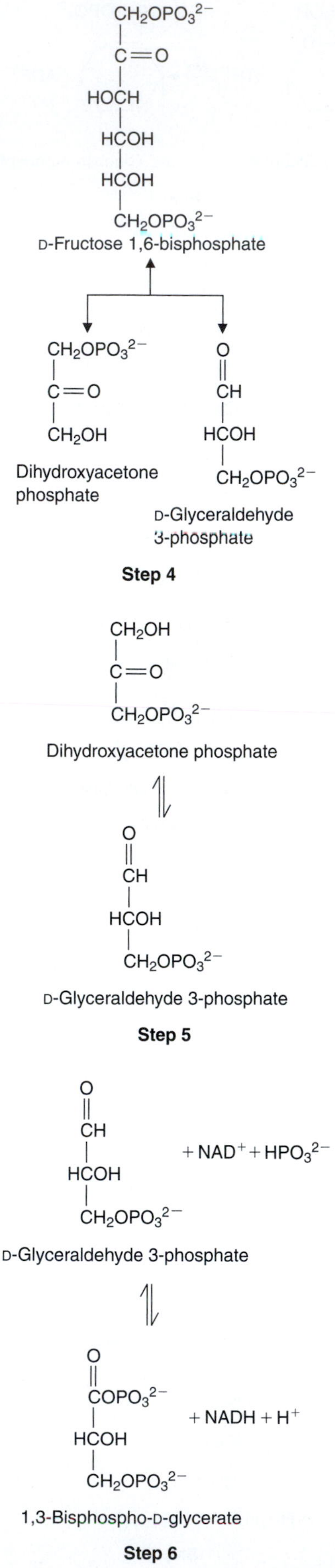

$CH_2OPO_3^{2-}$
|
C=O
|
HOCH
|
HCOH
|
HCOH
|
$CH_2OPO_3^{2-}$
D-Fructose 1,6-bisphosphate

$CH_2OPO_3^{2-}$ | C=O | CH_2OH — Dihydroxyacetone phosphate

O || CH | HCOH | $CH_2OPO_3^{2-}$ — D-Glyceraldehyde 3-phosphate

Step 4

CH_2OH
|
C=O
|
$CH_2OPO_3^{2-}$
Dihydroxyacetone phosphate

⇅

O
||
CH
|
HCOH
|
$CH_2OPO_3^{2-}$
D-Glyceraldehyde 3-phosphate

Step 5

O
||
CH
|
HCOH $+ NAD^+ + HPO_3^{2-}$
|
$CH_2OPO_3^{2-}$
D-Glyceraldehyde 3-phosphate

⇅

O
||
$COPO_3^{2-}$
|
HCOH $+ NADH + H^+$
|
$CH_2OPO_3^{2-}$
1,3-Bisphospho-D-glycerate

Step 6

Stage Three Involves Oxidoreduction Reactions and Synthesis of ATP

The first reaction of the last stage of glycolysis (Figure 14.7*c*) is catalyzed by **glyceraldehyde 3-phosphate dehydrogenase** (Step 6). This reaction is of considerable interest because of what is accomplished in a single enzyme-catalyzed step. An aldehyde (glyceraldehyde 3-phosphate) is oxidized to a carboxylic acid with the reduction of NAD^+ to NADH. In addition to NADH, the reaction produces 1,3-bisphosphoglycerate, a mixed anhydride of a carboxylic acid and phosphoric acid. 1,3-Bisphosphoglycerate has a large negative free energy of hydrolysis, enabling it to participate in a subsequent reaction that yields ATP. The overall reaction catalyzed by glyceraldehyde 3-phosphate dehydrogenase can be visualized as the coupling of a very favorable exergonic reaction with an unfavorable endergonic reaction. The exergonic reaction can be thought of as being composed of a half-reaction in which an aldehyde is oxidized to a carboxylic acid, which is then coupled with a half-reaction in which NAD^+ is reduced to NADH:

$$R-\overset{O}{\overset{\|}{C}}H + H_2O \longrightarrow R-\overset{O}{\overset{\|}{C}}OH + 2H^+ + 2e^-$$

$$NAD^+ + 2H^+ + 2e^- \longrightarrow NADH + H^+$$

The overall reaction (sum of the half-reactions) is quite exergonic, with the aldehyde being oxidized to a carboxylic acid and NAD^+ being reduced to NADH:

$$R-\overset{O}{\overset{\|}{C}}H + NAD^+ + H_2O \longrightarrow R-\overset{O}{\overset{\|}{C}}OH + NADH + H^+, \quad \Delta G^{\circ\prime} = -10.3 \text{ kcal mol}^{-1}$$

A second endergonic component of the reaction corresponds to the formation of a mixed anhydride between the carboxylic acid and phosphoric acid:

$$R-\overset{O}{\overset{\|}{C}}OH + HPO_3^{2-} \longrightarrow R-\overset{O}{\overset{\|}{C}}-OPO_3^{2-} + H_2O, \quad \Delta G^{\circ\prime} = +11.8 \text{ kcal mol}^{-1}$$

The overall reaction involves coupling of the endergonic and exergonic components to give an overall standard free energy change of $+1.5$ kcal mol^{-1}.

$$\text{Sum: } R-\overset{O}{\overset{\|}{C}}H + NAD^+ + HPO_3^{2-} \longrightarrow$$

$$R-\overset{O}{\overset{\|}{C}}OPO_3^{2-} + NADH + H^+, \quad \Delta G^{\circ\prime} = +1.5 \text{ kcal mol}^{-1}$$

The reaction is freely reversible in cells and is used in both the glycolytic and gluconeogenic pathways. The proposed mechanism for the enzyme-catalyzed reaction is shown in Figure 14.8. Glyceraldehyde 3-phosphate reacts with a sulfhydryl group of a cysteine residue of the enzyme to generate a **thiohemiacetal.** An internal oxidation–reduction reaction occurs in which bound NAD^+ is reduced to NADH and the thiohemiacetal is oxidized to give a high-energy thiol ester. Exogenous NAD^+ then replaces the bound NADH and the high-energy thiol ester reacts with P_i to form the mixed anhydride and regenerate the free sulfhydryl group. The mixed anhydride dissociates from the enzyme. Note that a free carboxylic acid group (—COOH) is not generated during the reaction. Instead, the enzyme generates a carboxyl group in the form of high-energy thiol ester, which is converted by reaction with P_i into another high-energy compound, a mixed anhydride of carboxylic and phosphoric acids.

The reaction catalyzed by glyceraldehyde 3-phosphate dehydrogenase requires NAD^+ and produces NADH. Since the cytosol has only a limited amount of NAD^+, continuous glycolytic activity can only occur if NADH is oxidized back to NAD^+, otherwise glycolysis will stop for want of NAD^+. The options that cells have for regeneration of NAD^+ from NADH are described in a later section (see p. 609).

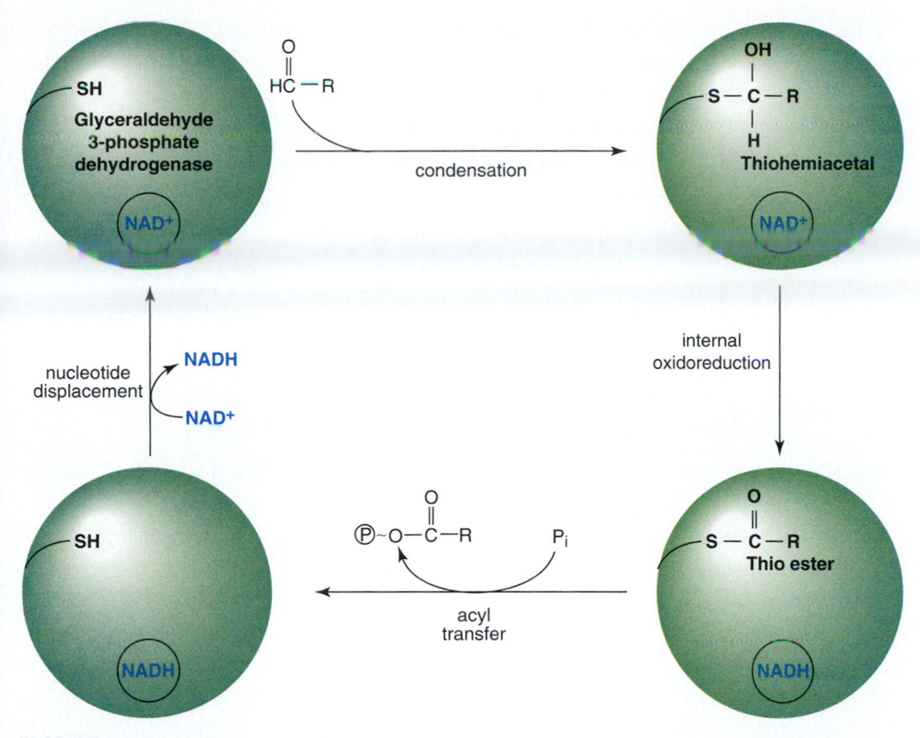

FIGURE 14.8
Mechanism of action of glyceraldehyde 3-phosphate dehydrogenase.
Large circle represents enzyme; small circle binding site for NAD^+; RCOH, glyceraldehyde 3-phosphate; —SH, the sulfhydryl group of the cysteine residue located at the active site; and ~P, the high-energy phosphate bond of 1,3-bisphosphoglycerate.

The next reaction, catalyzed by **phosphoglycerate kinase,** produces ATP from the high-energy compound 1,3-bisphosphoglycerate (Figure 14.7*c*; Step 7). This is the first site of ATP production in the glycolytic pathway. Because two ATP molecules were invested for each glucose molecule in the priming stage, and because two molecules of 1,3-bisphosphoglycerate are produced from each glucose, all of the ATP "invested" in the priming stage is recovered in this step of glycolysis. Since ATP production occurs in the forward direction and ATP utilization in the reverse direction, it may seem surprising that this reaction is freely reversible and *can* be used in both the glycolytic and gluconeogenic pathways. The reaction provides a means for the generation of ATP in the glycolytic pathway but, when needed for glucose synthesis, can also be used in the reverse direction for the synthesis of 1,3 bisphosphoglycerate at the expense of ATP. The glyceraldehyde 3-phosphate dehydrogenase–phosphoglycerate kinase system is an example of substrate-level phosphorylation, a term used for a process in which a substrate participates in an enzyme-catalyzed reaction that yields ATP or GTP. **Substrate-level phosphorylation** stands in contrast to mitochondrial oxidative phosphorylation (see p. 577). Note, however, that the combination of the reactions catalyzed by glyceraldehyde 3-phosphate dehydrogenase and phosphoglycerate kinase accomplishes the coupling of an oxidation (an aldehyde is oxidized to a carboxylic acid) to a phosphorylation.

Phosphoglycerate mutase converts 3-phosphoglycerate to 2-phosphoglycerate (Step 8). This is a freely reversible reaction in which **2,3-bisphosphoglycerate** (or **2,3-diphosphoglycerate**) functions as an obligatory intermediate at the active site of the enzyme (E):

E-phosphate + 3-phosphoglycerate ⇌ E + 2,3-bisphosphoglycerate
E + 2,3-bisphosphoglycerate ⇌ E-phosphate + 2-phosphoglycerate
Sum: 3-phosphoglycerate ⇌ 2-phosphoglycerate

O‖$COPO_3^{2-}$ | HCOH | $CH_2OPO_3^{2-}$ + ADP^{3-}

1,3-Bisphospho-D-glycerate

⇌ Mg^{2+}

O‖CO^- | HCOH | $CH_2OPO_3^{2-}$ + ATP^{4-}

3-Phospho-D-glycerate

Step 7

O‖CO^- | HCOH | $CH_2OPO_3^{2-}$

3-Phospho-D-glycerate

⇌

O‖CO^- | $HCOPO_3^{2-}$ | CH_2OH

2-Phospho-D-glycerate

Step 8

The involvement of 2,3-bisphosphoglycerate as an intermediate creates an absolute requirement for the presence of a catalytic amount of this compound in cells. This can be appreciated by noting that E-P in this reaction cannot be generated without 2,3-bisphosphoglycerate. Cells synthesize 2,3-bisphosphoglycerate, independent of the reaction catalyzed by phosphoglycerate mutase, by a reaction catalyzed by 2,3-bisphosphoglycerate mutase:

$$\text{1,3-Bisphospho-D-glycerate } (\text{O}{=}\text{COPO}_3^{2-}\text{–HCOH–CH}_2\text{OPO}_3^{2-}) \longrightarrow \text{2,3-Bisphospho-D-glycerate } (\text{O}{=}\text{CO}^-\text{–HCOPO}_3^{2-}\text{–CH}_2\text{OPO}_3^{2-}) + \text{H}^+$$

The latter enzyme is unusual in that it is **catalytically bifunctional,** functioning first as a mutase for the formation of 2,3-bisphosphoglycerate and then as a phosphatase that hydrolyzes 2,3-bisphosphoglycerate to 3-phosphoglycerate and P_i. All cells contain minute quantities of 2,3-bisphosphoglycerate to produce the phosphorylated form of newly synthesized phosphoglycerate mutase. The amounts needed are small because phosphorylation needs to occur only once, the phosphorylated enzyme being regenerated by reaction with the substrate during each reaction cycle. In contrast to all others, red blood cells contain very high concentrations of 2,3-bisphosphoglycerate where this compound serves as a negative allosteric effector for regulation of oxygen binding with hemoglobin (see p. 406). From 15% to 25% of the glucose converted to lactate in red blood cells goes by way of the steps of the **"BPG shunt"** (Figure 14.9). No net production of ATP occurs when a molecule of glucose is converted to two molecules of lactate by way of this shunt around the phosphoglycerate kinase step.

Enolase catalyzes elimination of water from 2-phosphoglycerate to form phospho*enol*pyruvate (PEP) in the next reaction (Step 9; Figure 14.7*c*). This is a remarkable reaction from the standpoint that a high-energy phosphate compound is generated from one of markedly lower energy level. The standard free energy change ($\Delta G^{\circ\prime}$) for the hydrolysis of phospho*enol*pyruvate is -14.8 kcal mol^{-1}, a much greater value than 2-phosphoglycerate (-4.2 kcal mol^{-1}). Although the reaction catalyzed by enolase is freely reversible, a large change in the distribution of energy occurs within the molecule as a consequence of its action. Although the free energy levels of PEP and 2-phosphoglycerate are not markedly different, the free energy levels of their products of hydrolysis (pyruvate and glycerate, respectively) are quite different. Since $\Delta G^{\circ\prime} = G^{\circ\prime}_{\text{products}} - G^{\circ\prime}_{\text{substrates}}$, this accounts for the marked differences in the standard free energy of hydrolysis of these two compounds.

$$\text{2-Phospho-D-glycerate } (\text{O}{=}\text{CO}^-\text{–HCOPO}_3^{2-}\text{–CH}_2\text{OH}) \rightleftharpoons \text{Phospho}enol\text{pyruvate } (\text{O}{=}\text{CO}^-\text{–C(OPO}_3^{2-}){=}\text{CH}_2) + \text{H}_2\text{O}$$

Step 9

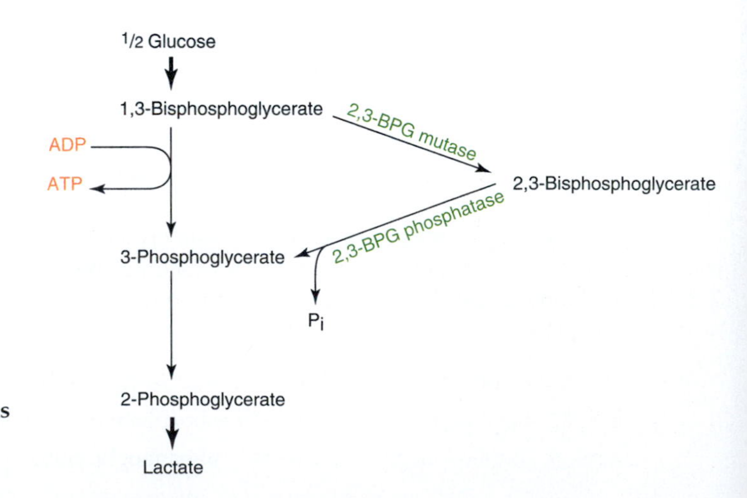

FIGURE 14.9
The 2,3-bisphosphoglycerate (2,3-BPG) shunt consists of reactions catalyzed by the bifunctional enzyme, 2,3-BPG mutase/ phosphatase.

Pyruvate kinase (Step 10; Figure 14.7*c*) accomplishes another **substrate-level phosphorylation:** that is, the synthesis of ATP with the conversion of the high-energy compound PEP into pyruvate. The reaction is not reversible under intracellular conditions. It constitutes a way to synthesize ATP but is not reversible under conditions that exist in cells and cannot be used for the synthesis of PEP when needed for glucose synthesis.

The last step of the glycolytic pathway is an oxidoreduction reaction catalyzed by **lactate dehydrogenase** (Step 11; Figure 14.7*c*). Pyruvate is reduced to give L-lactate and NADH is oxidized to NAD^+ by this freely reversible reaction. This is the only reaction of the body that produces L-lactate. It also is the only reaction by which the body can utilize L-lactate.

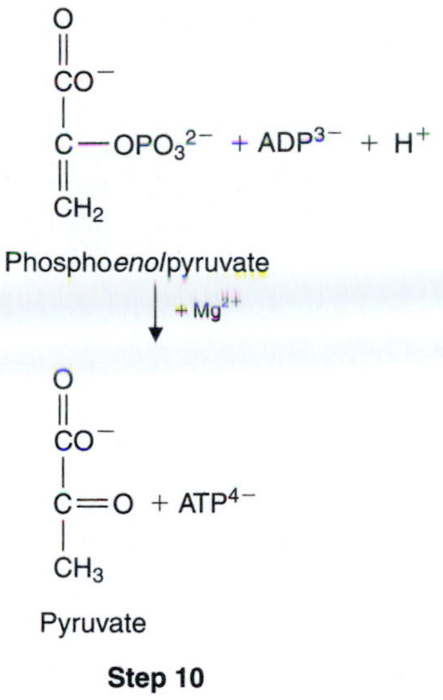

O=C(O⁻)–C(–OPO_3^{2-})=CH_2 + ADP^{3-} + H^+

Phospho*enol*pyruvate

↓ + Mg^{2+}

O=C(O⁻)–C(=O)–CH_3 + ATP^{4-}

Pyruvate

Step 10

O=C(O⁻)–C(=O)–CH_3 + NADH + H^+

Pyruvate

⇅

O=C(O⁻)–HOCH–CH_3 + NAD^+

L-Lactate

Step 11

ATP Yield and Balanced Equation for Anaerobic Glycolysis

Conversion of one molecule of glucose to two molecules of lactate by glycolysis results in the net formation of two molecules of ATP. Two molecules of ATP are used in the priming stage to set glucose up so that it can be cleaved. However, subsequent steps then yield four molecules of ATP so that the overall net production of ATP by the glycolytic pathway is two molecules of ATP:

$$\text{D-glucose} + 2\ ADP^{3-} + 2\ P_i^{2-} \rightarrow 2\ \text{L-lactate}^- + 2\ ATP^{4-} + 2\ H_2O$$

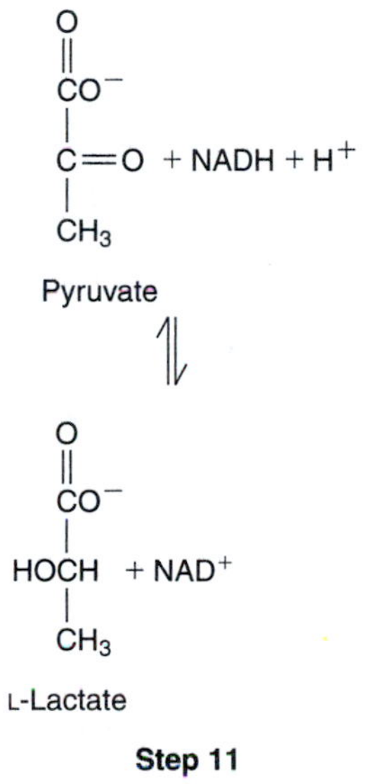

Biological cells have only a limited amount of ADP and P_i. Flux through the glycolytic pathway is therefore dependent on an adequate supply of these substrates. If the ATP is not utilized for performance of work, glycolysis will stop for want of ADP and/or P_i. Consequently, the ATP generated has to be used, that is, turned over, in normal work-related processes in order for glycolysis to occur. The equation for the use of ATP for any work-related process is simply

$$ATP^{4-} + H_2O \rightarrow ADP^{3-} + P_i^{2-} + H^+ + \text{"work"}$$

When this equation is doubled and added to that given above for glycolysis, excluding the work accomplished since this is understood to be necessary for ATP turnover, the overall balanced equation becomes

$$\text{D-glucose} \rightarrow 2\ \text{lactate}^- + 2\ H^+$$

The equation illustrates an extremely important point—anaerobic glycolysis generates acid, a problem created by metabolism for cells since intracellular pH must be maintained near neutrality for optimum enzyme activities.

NADH Generated During Glycolysis Has to Be Oxidized Back to NAD^+: Role of Lactate Dehydrogenase

NADH and NAD^+ do not appear in the balanced sum equation for the steps of anaerobic glycolysis. A perfect coupling exists between the generation of NADH and its utilization in glycolysis (Figure 14.7*c*). Two molecules of NADH are generated at the level of glyceraldehyde 3-phosphate dehydrogenase and two molecules of NADH are utilized by lactate dehydrogenase during conversion of one molecule of glucose into two molecules of lactate. NAD^+, a soluble cytosolic molecule, is available in only limited amounts and must be regenerated from NADH for glycolysis to continue unabated. The two reactions involved are

$$\text{D-glyceraldehyde 3-phosphate} + NAD^+ + P_i \rightarrow \text{1,3-bisphospho-D glycerate} + NADH + H^+$$

$$\text{pyruvate} + NADH + H^+ \rightarrow \text{L-lactate} + NAD^+$$

Coupling of these reactions gives the sum reaction:

$$\text{D-glyceraldehyde 3-phosphate} + \text{pyruvate} + P_i \rightarrow \text{1,3-bisphosphoglycerate} + \text{L-lactate}$$

Perfect coupling of reducing equivalents by these two reactions has to occur under conditions of anaerobiosis or in cells that lack mitochondria for glycolysis to occur. With the availability of oxygen and mitochondria, reducing equivalents in the form of NADH generated at the level of glyceraldehyde 3-phosphate dehydrogenase can be shuttled into the mitochondria for the synthesis of ATP, leaving pyruvate rather than lactate as the end product of glycolysis. Two shuttle systems are known to exist for the transport of reducing equivalents from the cytosolic space to the mitochondrial matrix space (mitosol). The mitochondrial inner membrane is not permeable to NADH (see p. 584).

NADH Generated During Glycolysis Can Be Reoxidized Via Substrate Shuttle Systems

The **glycerol phosphate shuttle** is shown in Figure 14.10*a* and the **malate–aspartate shuttle** in Figure 14.10*b* (also see p. 584). Tissues with cells that contain mitochondria have the capability of shuttling reducing equivalents from the cy-

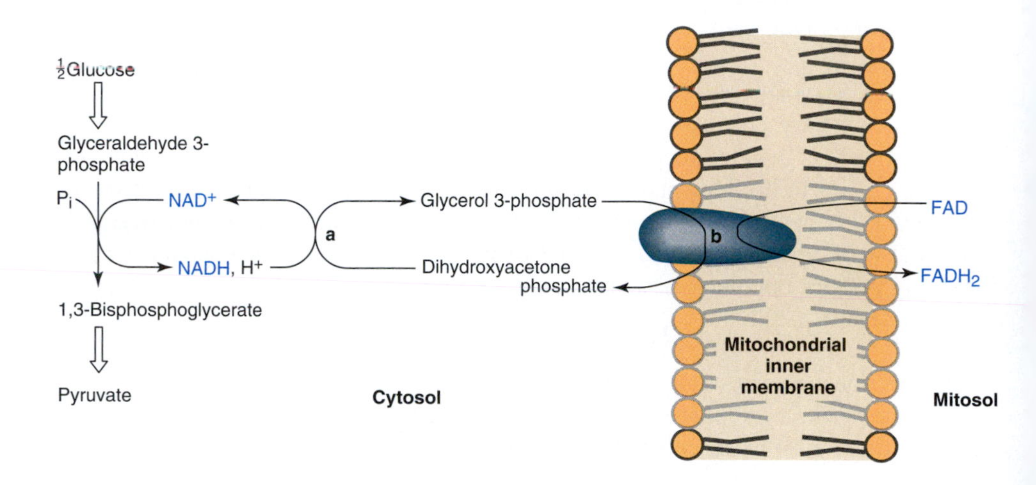

(a)

(b)

FIGURE 14.10
Shuttles for the transport of reducing equivalents from the cytosol to the mitochondrial electron transfer chain.
(*a*) Glycerol phosphate shuttle: **a**, cytosolic glycerol 3-phosphate dehydrogenase oxidizes NADH; **b**, mitochondrial glycerol 3-phosphate dehydrogenase of the outer surface of the inner membrane reduces FAD. (*b*) Malate–aspartate shuttle: **a**, cytosolic malate dehydrogenase reduces oxaloacetate (OAA) to malate; **b**, dicarboxylic acid antiport of the mitochondrial inner membrane catalyzes electrically neutral exchange of malate for α-ketoglutarate (α-KG); **c**, mitochondrial malate dehydrogenase produces intramitochondrial NADH; **d**, mitochondrial aspartate aminotransferase transaminates glutamate and oxaloacetate; **e**, glutamate–aspartate antiport of the mitochondrial inner membrane catalyzes electrogenic exchange of glutamate for aspartate; **f**, cytosolic aspartate aminotransferase transaminates aspartate and α-ketoglutarate.

tosol to the mitosol. The relative proportion of the activities of the two shuttles varies from tissue to tissue, with liver making greater use of the malate–aspartate shuttle, whereas some muscle cells may be more dependent on the glycerol phosphate shuttle. The shuttle systems are irreversible; that is, they represent mechanisms for moving reducing equivalents into the mitosol, but not mechanisms for moving mitochondrial reducing equivalents into the cytosol.

The mitochondrial inner membrane has a large number of transport systems (see p. 583) but lacks one that is effective for oxaloacetate. For this reason oxaloacetate transaminates with glutamate to produce aspartate, which then exits from the mitochondrion in exchange for glutamate. This is the irreversible step of the shuttle. The aspartate entering the cytosol transaminates with α-ketoglutarate to give oxaloacetate and glutamate. The oxaloacetate accepts the reducing equivalents of NADH and becomes malate. Malate then penetrates the mitochondrial inner membrane, where it is oxidized by the mitochondrial **malate dehydrogenase.** This produces NADH within the mitosol and regenerates oxaloacetate to complete the cycle. The overall balanced equation for the sum of all the reactions of the malate–aspartate shuttle is simply

$$\text{NADH}_{\text{cytosol}} + \text{H}^+{}_{\text{cytosol}} + \text{NAD}^+{}_{\text{mitosol}} \longrightarrow \text{NAD}^+{}_{\text{cytosol}} + \text{NADH}_{\text{mitosol}} + \text{H}^+{}_{\text{mitosol}}$$

The glycerol phosphate shuttle is simpler, in the sense that fewer reactions are involved, but $FADH_2$ is generated within the mitochondrial inner membrane rather than NADH within the mitosolic compartment. The irreversible step of the shuttle is catalyzed by the mitochondrial **glycerol 3-phosphate dehydrogenase.** The active site of this enzyme is exposed on the cytosolic surface of the mitochondrial inner membrane, making it unnecessary for glycerol 3-phosphate to penetrate into the mitosol for oxidation. The overall balanced equation for the sum of the reactions of the glycerol phosphate shuttle is

$$\text{NADH}_{\text{cytosol}} + \text{H}^+{}_{\text{cytosol}} + \text{FAD}_{\text{inner membrane}} \longrightarrow \text{NAD}^+{}_{\text{cytosol}} + \text{FADH}_{2\ \text{inner membrane}}$$

Shuttles Are Important in Other Oxidoreductive Pathways

Alcohol Oxidation

The first step of alcohol (i.e., **ethanol**) metabolism is its oxidation to **acetaldehyde** with production of NADH by **alcohol dehydrogenase.**

$$\underset{\text{Ethanol}}{\text{CH}_3\text{CH}_2\text{OH}} + \text{NAD}^+ \longrightarrow \underset{\text{Acetaldehyde}}{\text{CH}_3\overset{\overset{\text{O}}{\|}}{\text{C}}\text{H}} + \text{NADH} + \text{H}^+$$

This enzyme is located almost exclusively in the cytosol of liver parenchymal cells. The acetaldehyde generated traverses the mitochondrial inner membrane for oxidation by a mitosolic **aldehyde dehydrogenase.**

$$\underset{\text{Acetaldehyde}}{\text{CH}_3\overset{\overset{\text{O}}{\|}}{\text{C}}\text{H}} + \text{NAD}^+ + \text{H}_2\text{O} \longrightarrow \underset{\text{Acetate}}{\text{CH}_3\overset{\overset{\text{O}}{\|}}{\text{C}}\text{O}^-} + \text{NADH} + 2\text{H}^+$$

The NADH generated by the last step can be used directly by the mitochondrial electron transfer chain. However, NADH generated by cytosolic alcohol dehydrogenase must be oxidized back to NAD^+ by one of the shuttles. Thus the capacity of human beings to oxidize alcohol is dependent on the ability of their liver to transport reducing equivalents from the cytosol to the mitosol by these shuttle systems.

Glucuronide Formation

The shuttles play an important role in the formation of water-soluble **glucuronides** of **bilirubin** and various **drugs** (see p. 477) so that these compounds can be elim-

CLINICAL CORRELATION 14.1

Alcohol and Barbiturates

Acute alcohol intoxication causes increased sensitivity of an individual to the general depressant effects of barbiturates. Barbiturates and alcohol both interact with the γ-aminobutyrate (GABA)-activated chloride channel. Activation of the chloride channel inhibits neuronal firing, which may explain the depressant effects of both compounds. This drug combination is very dangerous and normal prescription doses of barbiturates have potentially lethal consequences in the presence of ethanol. In addition to the depressant effects of both ethanol and barbiturates on the central nervous system (CNS), ethanol inhibits the metabolism of barbiturates, thereby prolonging the time barbiturates remain effective in the body. Hydroxylation of barbiturates by the endoplasmic reticulum of the liver is inhibited by ethanol. This reaction, catalyzed by the NADPH-dependent cytochrome system, forms water-soluble derivatives of the barbiturates that are eliminated readily from the circulation by the kidneys. Blood levels of barbiturates remain high when ethanol is present, causing increased CNS depression.

Surprisingly, the alcoholic when sober is less sensitive to barbiturates. Chronic ethanol consumption apparently causes adaptive changes in the sensitivity of the CNS to barbiturates (cross-tolerance). It also results in the induction of the enzymes of liver endoplasmic reticulum involved in drug hydroxylation reactions. Consequently, the sober alcoholic is able to metabolize barbiturates more rapidly. This sets up the following scenario. A sober alcoholic has trouble falling asleep, even after taking several sleeping pills, because his/her liver has increased capacity to hydroxylate the barbiturate contained in the pills. In frustration he/she consumes more pills and then alcohol. Sleep results, but may be followed by respiratory depression and death because the alcoholic, although less sensitive to barbiturates when sober, remains sensitive to the synergistic effect of alcohol.

Misra, P. S., Lefevre, A., Ishii, H., Rubin, E., and Lieber, C. S. Increase of ethanol, meprobamate and pentobarbital metabolism after chronic ethanol administration in man and in rats. *Am. J. Med.* 51:346, 1971.

inated from the body in the urine and bile. In this process UDP-glucose (structure on p. 673) is oxidized to **UDP-glucuronic** acid (structure on p. 674).

$$\text{UDP-D-glucose} + 2\ \text{NAD}^+ + \text{H}_2\text{O} \rightarrow \text{UDP-D-glucuronic acid} + 2\ \text{NADH} + 2\ \text{H}^+$$

In a reaction that occurs primarily in the liver, the "activated" glucuronic acid molecule is then transferred to a nonpolar, acceptor molecule, such as **bilirubin** or a compound foreign to the body:

$$\text{UDP-D-glucuronic acid} + \text{R—OH} \rightarrow \text{R—O—glucuronic acid} + \text{UDP}$$

Excess NADH generated by the first reaction has to be reoxidized by the shuttles for this process to continue. Since ethanol oxidation and drug conjugation are properties of the liver, the two of them occurring together may overwhelm the combined capacity of the shuttles. A good thing to tell patients is not to mix the intake of pharmacologically active compounds and alcohol (see Clin. Corr. 14.1).

Two Shuttle Pathways Yield Different Amounts of ATP

The mitosolic NADH formed by the malate–aspartate shuttle activity can be used by the mitochondrial respiratory chain for the production of 2.5 molecules of ATP by oxidative phosphorylation:

$$\text{NADH}_{\text{mitosol}} + \text{H}^+ + \tfrac{1}{2}\,\text{O}_2 + 2\tfrac{1}{2}\,\text{ADP} + 2\tfrac{1}{2}\,\text{P}_\text{i} \rightarrow \text{NAD}^+{}_{\text{mitosol}} + 2\tfrac{1}{2}\,\text{ATP} + \text{H}_2\text{O}$$

In contrast, the $FADH_2$ obtained by the glycerol phosphate shuttle yields only 1.5 molecules of ATP:

$$\text{FADH}_{2\ \text{inner membrane}} + \tfrac{1}{2}\,\text{O}_2 + 1\tfrac{1}{2}\,\text{ADP} + 1\tfrac{1}{2}\,\text{P}_\text{i} \rightarrow \text{FAD}_{\text{inner membrane}} + 1\tfrac{1}{2}\,\text{ATP} + \text{H}_2\text{O}$$

Thus, with the availability of oxygen and mitochondria, the ATP yield from the oxidation of the NADH generated by glyceraldehyde 3-phosphate dehydrogenase in the conversion of one molecule of glucose to two molecules of pyruvate is either 3 or 5, depending on the shuttle system.

Glycolysis Is Inhibited by Sulfhydryl Reagents and Fluoride

Sulfhydryl reagents and **fluoride** are among the best-known inhibitors of the glycolytic pathway. **Glyceraldehyde 3-phosphate dehydrogenase** is vulnerable to inhibition by sulfhydryl reagents because of a catalytically important cysteine residue at its active site. During a catalytic cycle the sulfhydryl group of this cysteine residue normally reacts with glyceraldehyde 3-phosphate to form a thiohemiacetal (Figure 14.8). Sulfhydryl reagents, which are usually **mercury-containing compounds or alkylating compounds such as iodoacetate**, prevent formation of the thiohemiacetal by forming a covalent bond with the cysteine sulfhydryl group (Figure 14.11).

Fluoride is a potent inhibitor of enolase. Mg^{2+} and P_i form an ionic complex with fluoride ion, which is responsible for inhibition of enolase by interfering with binding of its substrate (Mg^{2+} 2-phosphoglycerate^{2-}).

Arsenate Prevents ATP Synthesis but Does Not Inhibit Glycolysis

Pentavalent arsenic or **arsenate** is special with respect to its effects on glycolysis. It is not an inhibitor of the process, and under most conditions actually stimulates glycolytic flux. Arsenate prevents net synthesis of ATP by causing arsenolysis in the glyceraldehyde 3-phosphate dehydrogenase reaction. Arsenate looks like P_i and is able to substitute for P_i in enzyme-catalyzed reactions. The result is the formation of a mixed anhydride of arsenic acid and the carboxyl group of 3-phosphoglycerate during the reaction catalyzed by glyceraldehyde 3-phosphate dehydrogenase (Figure 14.12). 1-Arsenato 3-phosphoglycerate is unstable, undergoing spontaneous hydrolysis to give 3-phosphoglycerate and inorganic arsenate. As a consequence, glycolysis continues unabated in the presence of arsenate, but 1,3-bisphosphoglycerate is not formed, resulting in the loss of the capacity to synthesize ATP at the step catalyzed by phosphoglycerate kinase. Thus net ATP synthesis does not occur when glycolysis is carried out in the presence of arsenate, the ATP invested in the priming stage being balanced by the ATP generated in the pyruvate kinase step. This, along with the fact that arsenolysis also in-

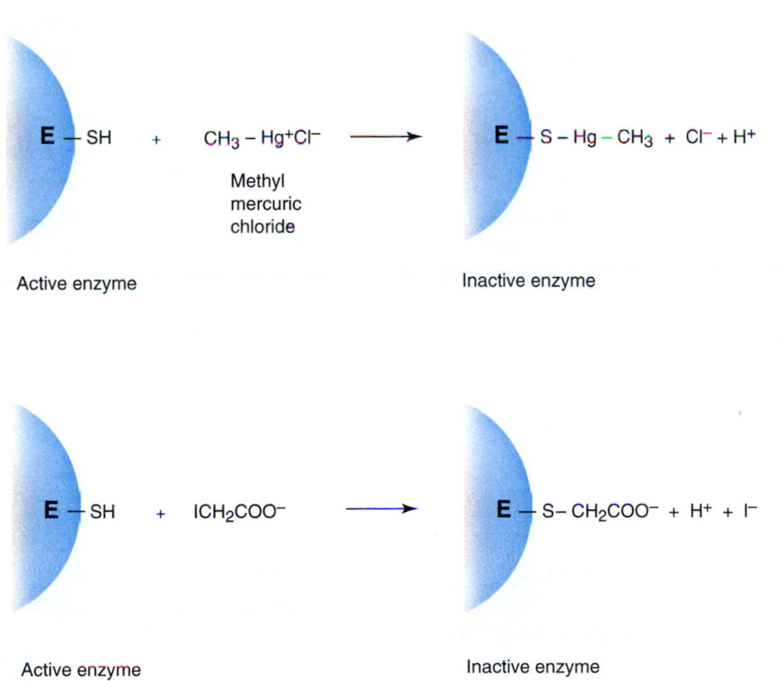

FIGURE 14.11
Mechanism for inactivation of glyceraldehyde 3-phosphate dehydrogenase by sulfhydryl reagents.

O ‖ CH — HCOH — $CH_2OPO_3^{2-}$
D-Glyceraldehyde 3-phosphate
$HAsO_4^{2-}$; NAD^+ → NADH + H^+
O ‖ $COAsO_3^{2-}$ — HCOH — $CH_2OPO_3^{2-}$
1-Arsenato-3-phospho-D-glycerate
spontaneous; H_2O → $HAsO_4^{2-}$ + H^+
O ‖ CO^- — HCOH — $CH_2OPO_3^{2-}$
3-Phospho-D-glycerate

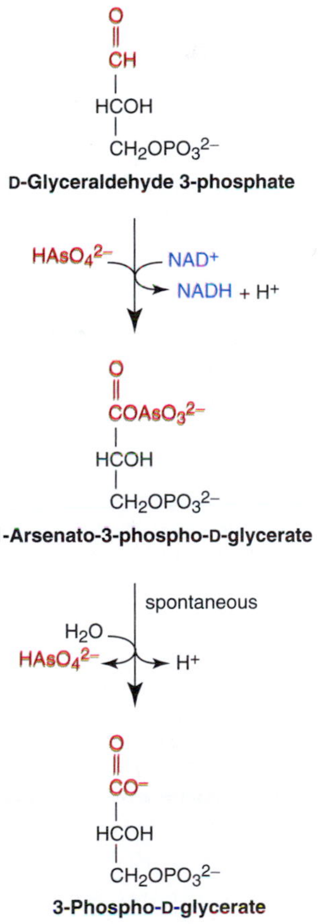

FIGURE 14.12
Arsenate uncouples oxidation from phosphorylation at the step catalyzed by glyceraldehyde 3-phosphate dehydrogenase.

CLINICAL CORRELATION 14.2
Arsenic Poisoning

Most forms of arsenic are toxic, but the trivalent form (arsenite as AsO_2^-) is much more toxic than the pentavalent form (arsenate or $HAsO_4^{2-}$). Less ATP is produced whenever arsenate substitutes for P_i in biological reactions. Arsenate competes for P_i-binding sites on enzymes, resulting in the formation of arsenate esters that are unstable. Arsenite works by a completely different mechanism, involving the formation of a stable complex with enzyme-bound lipoic acid:

For the most part arsenic poisoning is explained by inhibition of those enzymes that require lipoic acid as a coenzyme. These include pyruvate dehydrogenase, α-ketoglutarate dehydrogenase, and branched-chain α-keto acid dehydrogenase. Chronic arsenic poisoning from well water contaminated with arsenical pesticides or through the efforts of a murderer is best diagnosed by determining the concentration of arsenic in the hair or fingernails of the victim. About 0.5 mg of arsenic would be found in a kilogram of hair from a normal individual. The hair of a person chronically exposed to arsenic could have 100 times as much.

Hindmarsh, J. T. and McCurdy, R. F. Clinical and environmental aspects of arsenic toxicity. *CRC Crit. Rev. Clin. Lab. Sci.* 23:315, 1986; and Mudur, G. Half of Bangladesh population at risk of arsenic poisoning. *Br. Med. J.* 320:822, 2000.

terferes with ATP formation by oxidative phosphorylation, makes arsenate a toxic compound (see Clin. Corr. 14.2).

14.4 | REGULATION OF THE GLYCOLYTIC PATHWAY

Like other complex pathways involving multiple steps, regulation of flux through the glycolytic pathway is determined by the activities of several enzymes rather than a single "rate-limiting" enzyme. Furthermore, the relative contribution of an individual enzyme to glycolytic flux varies from tissue to tissue and with changes in nutritional and hormonal state. Quantitative information with respect to the relative contribution of a particular enzyme to flux through a pathway under a particular set of conditions is best obtained by determination of **"control strength"** for the enzyme. Making this measurement is not trivial since it depends on being able to determine the effect that a small inhibition of the activity of one particular enzyme has upon the flux through a pathway. For the purpose of illustration, let's assume inhibition of the activity of a particular enzyme by 10% has no effect upon flux through a pathway. The control strength of the enzyme is then zero (0 divided by 10), which means that the activity of this enzyme does not determine flux under the selected conditions. Now assume that inhibition of an enzyme by 10% results in 10% inhibition of flux through a pathway. In this circumstance, the control strength of the enzyme would be 1.0 (10 divided by 10), which means that flux is entirely dependent on the activity of the enzyme under these conditions; that is, the activity of the enzyme is totally rate limiting for this pathway. Now assume that inhibition of an enzyme by 10% results in only 5% inhibition of flux through the pathway. The control strength of the enzyme is 0.5 (5 divided by 10), which would mean that half of the control of flux through the pathway is determined by this enzyme. The rest of the control is presumably spread over one or more other steps, since the control strengths of steps of a linear pathway must sum to 1.0 by definition.

With the caveat that regulation of flux through glycolysis is dependent on the tissue under consideration and the nutritional and hormonal state of the tissue, the enzymes most generally believed to play the most important regulatory roles in the glycolytic pathway, that is, with the greatest control strength, are **hexokinase, 6-phosphofructo-1-kinase,** and **pyruvate kinase.** A summary of the important regulatory features of these enzymes is presented in Figure 14.13. These enzymes are viewed as regulatory based on their evolved sensitivity to regulation by allosteric effectors and/or covalent modification. An enzyme that is not subject to regulation will most likely catalyze a **"near-equilibrium reaction,"** whereas a regulatory enzyme will most likely catalyze a **"nonequilibrium reaction"** under intracellular conditions. This makes sense because flux through the regulated enzyme is restricted by controls imposed on that enzyme. A nonregulatory enzyme is so active it readily brings its substrates and products to equilibrium concentrations. Whether an enzyme-catalyzed reaction is near equilibrium or nonequilibrium can be determined by comparing the established equilibrium constant for the reaction with the mass–action ratio as it exists within a cell. The equilibrium constant for the reaction A + B → C + D is defined as

$$K_{eq} = \frac{[C][D]}{[A][B]}$$

where the brackets indicate the concentrations at equilibrium. The **mass–action ratio** is calculated in a similar manner, except that the steady-state (ss) concentrations of reactants and products within the cell are used in the equation:

$$\text{mass–action ratio} = \frac{[C]_{ss}[D]_{ss}}{[A]_{ss}[B]_{ss}}$$

If the mass–action ratio is approximately equal to the K_{eq}, the enzyme is said to be active enough to catalyze a near-equilibrium reaction and the enzyme is not

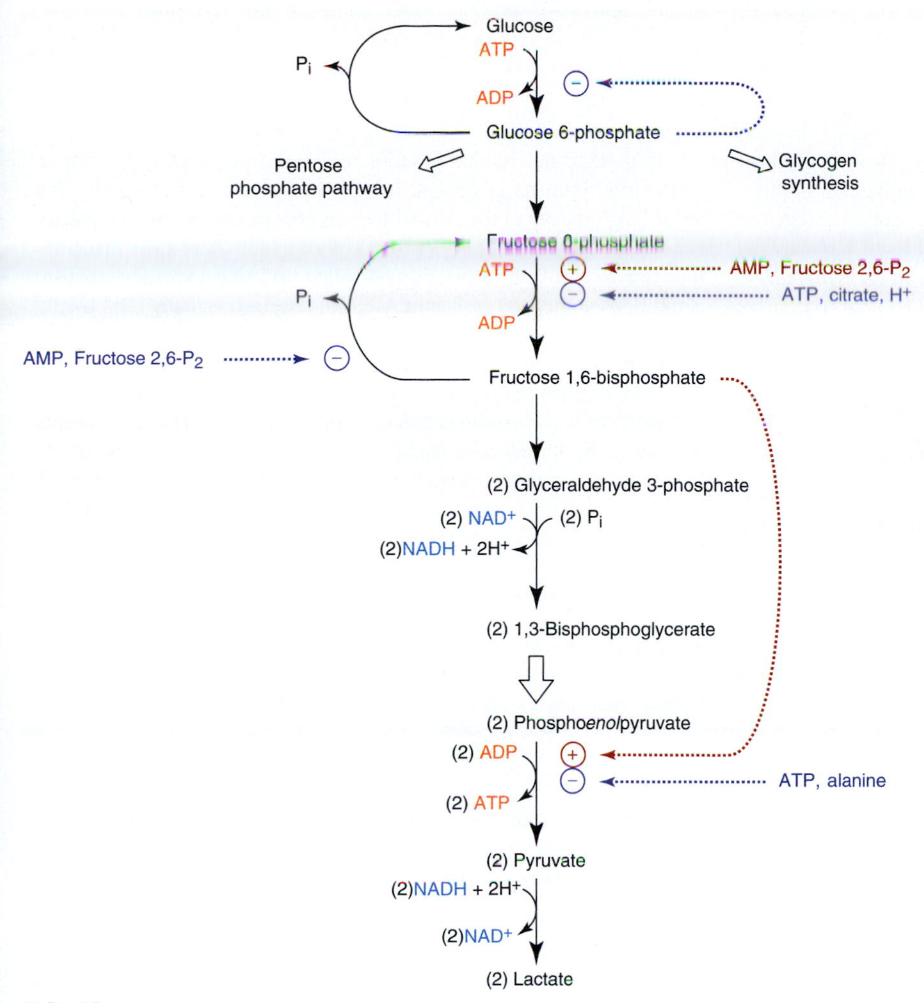

FIGURE 14.13
Important regulatory features of the glycolytic pathway.
Because of tissue differences in isoenzyme expression, not all tissues of the body have all of the regulatory mechanisms shown here.

considered likely to be subject to regulation. When the mass–action ratio is considerably different from the K_{eq}, the enzyme is said to catalyze a nonequilibrium reaction and usually is subject to regulation by one or more mechanisms. Comparison of mass–action ratios and equilibrium constants for reactions catalyzed by the glycolytic enzymes of liver indicate that many of them catalyze equilibrium reactions. The reactions catalyzed by glucokinase (liver isoenzyme of hexokinase), 6-phosphofructo-1-kinase, and pyruvate kinase in the intact liver are far from equilibrium, suggesting that these enzymes are likely sites for regulation of glycolysis in this tissue.

Hexokinase and Glucokinase Have Different Properties

Different isoenzymes of **hexokinase** are expressed in different tissues of the body. The hexokinase isoenzymes found in most tissues have a low K_m for glucose (<0.1 mM) relative to its concentration in blood (~5 mM) and are strongly inhibited by the product of the reaction, glucose 6-phosphate (G6P). The latter is an important regulatory feature because it prevents hexokinase from tying up all of the inorganic phosphate of a cell in the form of phosphorylated hexoses (see Clin. Corr. 14.3). Thus the reaction catalyzed by hexokinase may not be at equilibrium within cells that contain this enzyme because of the inhibition imposed by G6P. Liver

CLINICAL CORRELATION 14.3
Fructose Intolerance

Patients with hereditary fructose intolerance are deficient in the liver aldolase responsible for splitting fructose 1-phosphate into dihydroxyacetone phosphate and glyceraldehyde. Consumption of fructose by these patients results in the accumulation of fructose 1-phosphate and depletion of P_i and ATP in the liver. The reactions involved are those catalyzed by fructokinase and the enzymes of oxidative phosphorylation:

$$\text{fructose} + \text{ATP} \rightarrow \text{fructose 1-phosphate} + \text{ADP}$$
$$\text{ADP} + P_i + \text{“energy provided by electron transport chain”} \rightarrow \text{ATP}$$
$$\text{Net:} \quad P_i + \text{fructose} \rightarrow \text{fructose 1-phosphate}$$

Tying up P_i in the form of fructose 1-phosphate makes it impossible for liver mitochondria to generate ATP by oxidative phosphorylation. The ATP levels fall precipitously, making it also impossible for the liver to carry out its normal work functions. Damage results to the cells in large part because they are unable to maintain normal ion gradients by means of the ATP-dependent cation pumps. The cells swell and eventually lose their internal contents by osmotic lysis.

Although patients with fructose intolerance are particularly sensitive to fructose, humans in general have a limited capacity to handle this sugar. The capacity of the normal liver to phosphorylate fructose greatly exceeds its capacity to split fructose 1-phosphate. This means that fructose use by the liver is poorly controlled and that excessive fructose could deplete the liver of P_i and ATP. Fructose was actually tried briefly in hospitals as a substitute for glucose in patients being maintained by parenteral nutrition. The rationale was that fructose would be a better source of calories than glucose because fructose utilization is relatively independent of the insulin status of a patient. Delivery of large amounts of fructose by intravenous feeding was soon found to result in severe liver damage. Similar attempts have been made to substitute sorbitol and xylitol for glucose. These sugars also tend to deplete the liver of ATP and, like fructose, should not be used for parenteral nutrition.

Gitzelmann, R., Steinmann, B., and Van den Berghe, G. Disorders of fructose metabolism. In: C. R. Scriver, A. L. Beaudet, W. S. Sly, and D. Valle (Eds.), *The Metabolic and Molecular Bases of Inherited Disease*, 7th ed. New York: McGraw-Hill, 1995, pp. 905–934; and Ali, M., Rellos, P., and Cox T. M. Hereditary fructose intolerance. *J. Med.* Genet. 35:353, 1998.

parenchymal cells are unique in that they contain **glucokinase,** an isoenzyme of hexokinase with strikingly different kinetic properties from the other hexokinases. This isoenzyme catalyzes an ATP-dependent phosphorylation of glucose like the other hexokinases, but its $S_{0.5}$ (substrate concentration that gives enzyme activity of one-half maximum velocity) for glucose is considerably higher than the K_m for glucose of the other hexokinases (Figure 14.14). Furthermore, glucokinase is not subject to product inhibition by G6P. The glucose saturation curve for glucokinase is **sigmoidal** (Figure 14.4); that is, **cooperativity** is observed as a function of glucose concentration and therefore does not follow simple Michael–Menten kinetics (see p. 422). The kinetics of glucokinase are therefore described in terms of an $S_{0.5}$ value rather than a K_m value for glucose. Although glucokinase is not inhibited by G6P like other hexokinases, its activity is sensitive to indirect inhibition by fructose 6-phosphate, which is just one step removed and in equilibrium with G6P. A special glucokinase regulatory protein, which surprisingly is located in the nucleus of liver cells, is responsible for this effect (Figure 14.15). It works by sequestering glucokinase in an inactive complex in the nucleus. Fructose 6-phosphate promotes binding of glucokinase to the inhibitory protein by an allosteric mechanism, thereby inhibiting glucokinase. Fructose 6-phosphate in effect promotes translocation of glucokinase from the cytosol to the nucleus where glucokinase is completely inhibited by the regulatory protein (Figure 14.15). This inhibitory effect of fructose 6-phosphate on glucokinase is opposed by glucose and can be completely overcome by a large enough increase in glucose concentration. Again by an allosteric mechanism, glucose triggers dissociation of glucokinase from the regulatory protein, thereby promoting translocation of glucokinase from the nucleus to the cytosol where its catalytic activity can be expressed (Figure 14.15). All of these special regulatory features of glucokinase (high $S_{0.5}$ for glucose, cooperativity with respect to glucose concentration, and glucose stimulated translocation from the nucleus to the cytosol) contribute to the capacity of the liver to “buffer” blood glucose levels. Because GLUT-2, the glucose transporter present in liver cells, brings about rapid equilibration of glucose across the plasma membrane, the concentration of glucose within liver cells is the same as the concentration of glucose in the blood. Since the $S_{0.5}$ of glucokinase for glucose (~7 mM) is considerably greater than normal blood

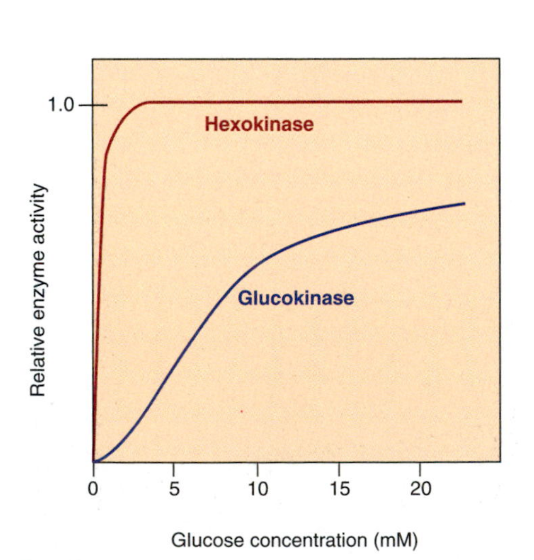

FIGURE 14.14
Comparison of the substrate saturation curves for hexokinase and glucokinase.

glucose concentrations (~5 mM) and glucose promotes translocation of glucokinase from the nucleus to the cytosol, any increase in portal blood glucose above normal leads to a dramatic increase in the rate of glucose phosphorylation by glucokinase in the liver (Figures 14.14 and 14.15). Likewise, any decrease in glucose concentration leads to a marked decrease in the rate of glucose phosphorylation. Thus liver uses glucose at a significant rate only when blood glucose levels are elevated and shuts down its use of glucose when blood glucose levels are low. This buffering effect of liver glucokinase on blood glucose levels would not occur if glucokinase had the low K_m for glucose characteristic of other hexokinases and was, therefore, completely saturated at physiological concentrations of glucose (Figure 14.14). On the other hand, a low K_m form of hexokinase is a good choice for tissues such as the brain in that it allows phosphorylation of glucose even when blood and tissue glucose concentrations are dangerously low.

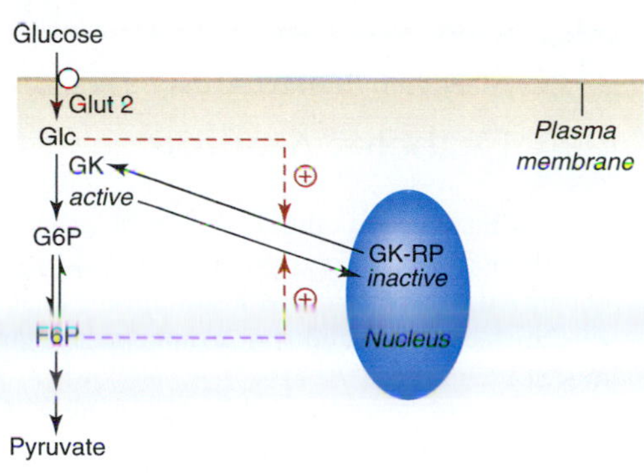

FIGURE 14.15
Glucokinase activity is regulated by translocation of the enzyme between the cytoplasm and the nucleus.
Glucose increases glucokinase (GK) activity by promoting translocation to the cytoplasm. Fructose 6-phosphate decreases GK by stimulating translocation into the nucleus. Binding to its regulatory protein (RP) in the nucleus completely inhibits GK activity.

The reaction catalyzed by glucokinase is not at equilibrium under the intracellular conditions of liver cells. The explanation lies in the rate restriction imposed by the high $S_{0.5}$ of glucokinase for glucose and inhibition of glucokinase activity by the regulatory protein. Yet another important factor is opposition to the activity of glucokinase provided by the activity of **glucose 6-phosphatase.** Like glucokinase, this enzyme has a high K_m (3 mM) with respect to the normal intracellular concentration (~0.2 mM) of its primary substrate, glucose 6-phosphate. Thus flux through this step is almost directly proportional to the intracellular concentration of glucose 6-phosphate. As shown in Figure 14.16, the combined action of glucokinase and glucose 6-phosphatase constitutes a futile cycle; that is, the sum of their reactions is hydrolysis of ATP to give ADP and P_i without the performance of any work. When blood glucose concentrations are about 5 mM, the activity of glucokinase is almost exactly balanced by the opposing activity of glucose 6-phosphatase. The result is that no net flux occurs in either direction. This futile cycling between glucose and glucose 6-phosphate is wasteful of ATP but, combined with the process of gluconeogenesis, contributes significantly to the "buffering" action of the liver on blood glucose levels. Furthermore, it provides a mechanism for preventing glucokinase from tying up all of the P_i of the liver (see Clin. Corr. 14.3).

Fructose, a component of many vegetables, fruits, and sweeteners, promotes hepatic glucose utilization by an indirect mechanism. It is converted in liver to **fructose 1-phosphate** (see Clin. Corr. 14.3), which activates glucokinase activity by having the opposite effect of fructose 6-phosphate on binding of glucokinase to its regulatory protein; that is, fructose 1-phosphate promotes dissociation and therefore translocation out of the nucleus and activation of glucokinase. This may be a factor in the adverse effects, for example, **hypertriacylglycerolemia,** sometimes associated with excessive dietary fructose consumption.

Glucokinase is also an **inducible enzyme.** Under various physiological conditions the amount of the enzyme protein increases or decreases. Induction of synthesis and repression of synthesis of an enzyme are relatively slow processes, usually requiring several hours before significant changes occur. **Insulin** increases the amount of glucokinase by promoting transcription of the glucokinase gene. An increase in blood glucose levels signals an increase in insulin release from the β cells of the pancreas. This results in an increase in blood insulin levels, which promotes

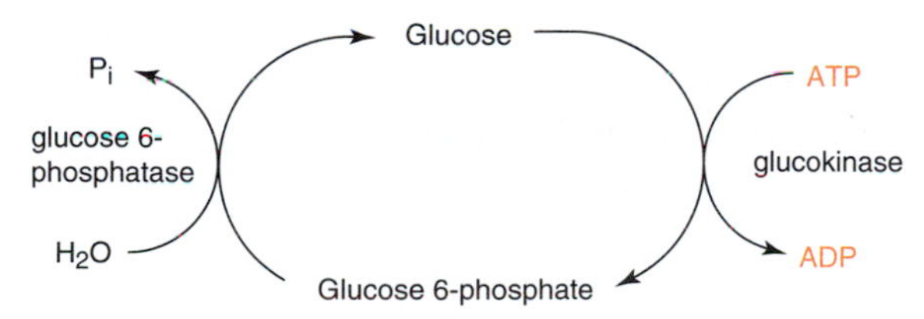

FIGURE 14.16
Phosphorylation of glucose followed by dephosphorylation constitutes a futile cycle in parenchymal cells of the liver.

CLINICAL CORRELATION 14.4

Diabetes Mellitus

Diabetes mellitus is a chronic disease characterized by derangements in carbohydrate, fat, and protein metabolism. Two major types are recognized clinically—type 1 (insulin-dependent diabetes mellitus; see Clin. Corr. 20.8) and type 2 (noninsulin-dependent diabetes mellitus; see Clin. Corr. 20.7).

In patients who do not have fasting hyperglycemia, the oral glucose tolerance test can be used for the diagnosis of diabetes. It consists of determining the blood glucose level in the fasting state and at intervals of 30–60 min for 2 h or more after consuming a 100-g carbohydrate meal. In a normal individual blood glucose returns to normal levels within 2 h after ingestion of the carbohydrate meal. In the diabetic patient, blood glucose will reach a higher level and remain elevated for longer periods of time, depending on the severity of the disease. However, many factors may contribute to an abnormal glucose tolerance test. The patient must have consumed a high-carbohydrate diet for the preceding 3 days, presumably to allow for induction of enzymes of glucose-utilizing pathways, for example, glucokinase, fatty acid synthase, and acetyl-CoA carboxylase. In addition, almost any infection (even a cold) and less well-defined "stress" (presumably by effects on the sympathetic nervous system) can result in (transient) abnormalities of the glucose tolerance test. Because of problems with the glucose tolerance test, elevation of the fasting glucose level should probably be the *sine qua non* for the diagnosis of diabetes. Glucose uptake by cells of insulin-sensitive tissues, that is, muscle and adipose, is decreased in the diabetic state. Insulin is required for glucose uptake by these tissues, and the diabetic patient either lacks insulin or has developed "insulin resistance" in these tissues. Resistance to insulin is an abnormality of the insulin receptor or in subsequent steps mediating the metabolic effects of insulin. Parenchymal cells of the liver do not require insulin for glucose uptake. Without insulin, however, the liver has diminished enzymatic capacity to remove glucose from the blood. This is explained in part by decreased glucokinase activity plus the loss of insulin's action on key enzymes of glycogenesis and the glycolytic pathway.

Taylor, S. I. Diabetes mellitus. In: C. R. Scriver, A. L. Beaudet, W. S. Sly, and D. Valle (Eds.), *The Metabolic and Molecular Bases of Inherited Disease*, 7th ed. New York: McGraw-Hill, 1995, pp. 843–896.

transcription of the glucokinase gene and increases the amount of liver glucokinase enzyme protein. Thus the amount of glucokinase in liver reflects how much glucose is being delivered to the liver via the portal vein. In other words, a person consuming large meals rich in carbohydrate will have greater amounts of glucokinase in the liver than one who is not. The liver in which glucokinase has been induced can make a greater contribution to the lowering of elevated blood glucose levels. The absence of insulin makes the liver of the diabetic patient deficient in glucokinase, in spite of high blood glucose levels, and this is one of the reasons why the liver of the diabetic has less blood glucose "buffering" action (see Clin. Corr. 14.4). Defects in the gene encoding glucokinase cause *m*aturity-*o*nset *d*iabetes of the *y*oung (MODY), a form of type 2 diabetes mellitus.

6-Phosphofructo-1-kinase Is an Important Regulatory Site for Glycolysis

Evidence suggests that **6-phosphofructo-1-kinase** is a very important regulatory site of glycolysis in most tissues. Usually we think of the first step of a pathway as the most logical choice for the most important regulatory step. However, the step that commits carbon to a pathway is an appropriate site for considerable control, and 6-phosphofructo-1-kinase catalyzes the first committed step of the glycolytic pathway. This is because the phosphoglucose isomerase catalyzed reaction is reversible, and every cell uses glucose 6-phosphate in the pentose phosphate pathway and many use it for glycogen synthesis. The reaction catalyzed by 6-phosphofructo-1-kinase commits the cell to the metabolism of glucose by glycolysis and is, therefore, a logical site for regulation by allosteric effectors. **Citrate, ATP,** and **hydrogen ions** (low pH) are the most important negative allosteric effectors, whereas **AMP** and **fructose 2,6-bisphosphate** are the most important positive allosteric effectors (Figure 14.13). Through their actions as strong inhibitors or activators of 6-phosphofructo-1-kinase, these compounds signal different rates of glycolysis in response to changes in (1) energy state of the cell (ATP and AMP), (2) internal environment of the cell (hydrogen ions), (3) availability of alternate fuels such as fatty acids and ketone bodies (citrate), and (4) insulin/glucagon ratio in the blood (fructose 2,6-bisphosphate).

Regulation of 6-Phosphofructo-1-kinase by ATP and AMP

The **Pasteur effect** refers to the inhibition of glucose utilization and lactate accumulation that occurs when respiration (oxygen consumption) is initiated in a suspension of cells or by a tissue. This phenomenon is readily understandable on a thermodynamic basis, the complete oxidation of glucose to CO_2 and H_2O yielding much more ATP than anaerobic glycolysis.

Glycolysis: $\text{D-glucose} + 2\ ADP^{3-} + 2\ P_i^{2-} \rightarrow 2\ \text{L-lactate}^- + 2\ ATP^{4-}$

Complete oxidation: $\text{D-glucose} + 6\ O_2 + 32\ ADP^{3-} + 32\ P_i^{2-} + 32\ H^+ \rightarrow 6\ CO_2 + 6\ H_2O + 32\ ATP^{4-}$

Cells use ATP to meet their metabolic demand, that is, to provide the energy necessary for their inherent work processes. Since so much more ATP is produced from glucose in the presence of oxygen, much less glucose has to be consumed to meet the metabolic demand of the cell. This occurs in part by ATP inhibition of glycolysis at the level of 6-phosphofructo-1-kinase. This can readily be rationalized since ATP is a well-recognized inhibitor of 6-phosphofructo-1-kinase, and more ATP is generated in the presence than in the absence of oxygen. However, ATP levels often do not change significantly between two conditions in which glycolytic flux is greatly altered. Since 6-phosphofructo-1-kinase is severely inhibited at concentrations of ATP (2.5–6 mM) normally present in cells, small changes in ATP concentration cannot account for large changes in flux through 6-phosphofructo-1-kinase. However, much greater changes, percentage wise, occur in the concentrations of AMP, a positive allosteric effector of 6-phosphofructo-1-kinase. The change that occurs in steady-state concentrations of AMP when oxygen is introduced into a system is exactly what might have been predicted; that is, the level goes down dramatically. This change results in less 6-phosphofructo-1-kinase activity, which greatly suppresses glycolysis and accounts in part for the Pasteur effect. Levels of AMP automatically go down in a cell when ATP levels increase. The reason is that the sum of the adenine nucleotides in a cell, that is, ATP + ADP + AMP, is nearly constant under most physiological conditions, but the relative concentrations are such that the amount of ATP is always much greater than the amount of AMP. Furthermore, adenine nucleotides are maintained in equilibrium in the cytosol through action of **adenylate kinase** (also referred to as **myokinase**), which catalyzes the reaction 2ADP → ATP + AMP. The equilibrium constant (K'_{eq}) for this reaction is given by

$$K'_{eq} = \frac{[ATP][AMP]}{[ADP]^2}$$

Since this reaction is "near equilibrium" under intracellular conditions, the concentration of AMP is given by

$$[AMP] = \frac{K'_{eq}[ADP]^2}{[ATP]}$$

Because intracellular [ATP] >> [ADP] >> [AMP], a small decrease in [ATP] causes a substantially greater percentage increase in [ADP], and, since [AMP] is related to the square of [ADP], an even greater percentage increase in [AMP]. Because of this relationship, a small decrease in ATP concentration leads to a greater percent increase in [AMP] than in the percent decrease in [ATP]. This makes the [AMP] an excellent signal of the **energy status** of the cell and allows it to function as an important allosteric effector of 6-phosphofructo-1-kinase activity. Furthermore, [AMP] influences in yet another way the effectiveness of 6-phosphofructo-1-kinase. An enzyme called **fructose 1,6-bisphosphatase** catalyzes an irreversible reaction that opposes the reaction catalyzed by 6-phosphofructo-1-kinase:

$$\text{fructose 1,6-bisphosphate} + H_2O \rightarrow \text{fructose 6-phosphate} + P_i$$

This enzyme sits "cheek by jowl" with 6-phosphofructo-1-kinase in the cytosol of many cells. Together they catalyze a futile cycle ($ATP + H_2O \rightarrow ADP + P_i +$

"heat"), and, at the very least, they decrease "effectiveness" of one another. AMP concentration is a perfect signal of the energy status of the cell—not only because AMP activates 6-phosphofructo-1-kinase but also because AMP inhibits fructose 1,6-bisphosphatase. Thus a small decrease in ATP concentration triggers, via an increase in AMP concentration that percentage wise is much greater, a large increase in net conversion of fructose 6-phosphate into fructose 1,6-bisphosphate. This increases glycolytic flux by increasing the amount of substrate available for the splitting stage. In cells containing hexokinase, there is also greater phosphorylation of glucose because a decrease in fructose 6-phosphate automatically causes a decrease in glucose 6-phosphate, which, in turn, results in less inhibition of hexokinase.

The decrease in lactate production in response to onset of respiration is another feature of the Pasteur effect that can readily be explained. The most important factor is decreased glycolytic flux caused by oxygen. Other factors include competition between lactate dehydrogenase and the pyruvate dehydrogenase complex for pyruvate, as well as competition between lactate dehydrogenase and shuttle systems for NADH. For the most part, lactate dehydrogenase loses the competition in the presence of oxygen.

Intracellular pH Regulates 6-Phosphofructo-1-kinase

It would make sense that lactate, as the end product of glycolysis, should inhibit a regulatory enzyme of glycolysis. It does not. However, **hydrogen ions,** also end products of glycolysis, do inhibit 6-phosphofructo-1-kinase. As shown in Figure 14.17, glycolysis in effect generates **lactic acid,** and the cell must dispose of it as such. This explains why excessive glycolysis in the body lowers blood pH and leads to an emergency medical situation termed **lactic acidosis** (see Clin. Corr. 14.5). Plasma membranes of cells contain a symport for lactate and hydrogen ions that allows release of lactic acid into the bloodstream. This is a defense mechanism, preventing pH from getting so low that everything becomes pickled (see Clin. Corr.

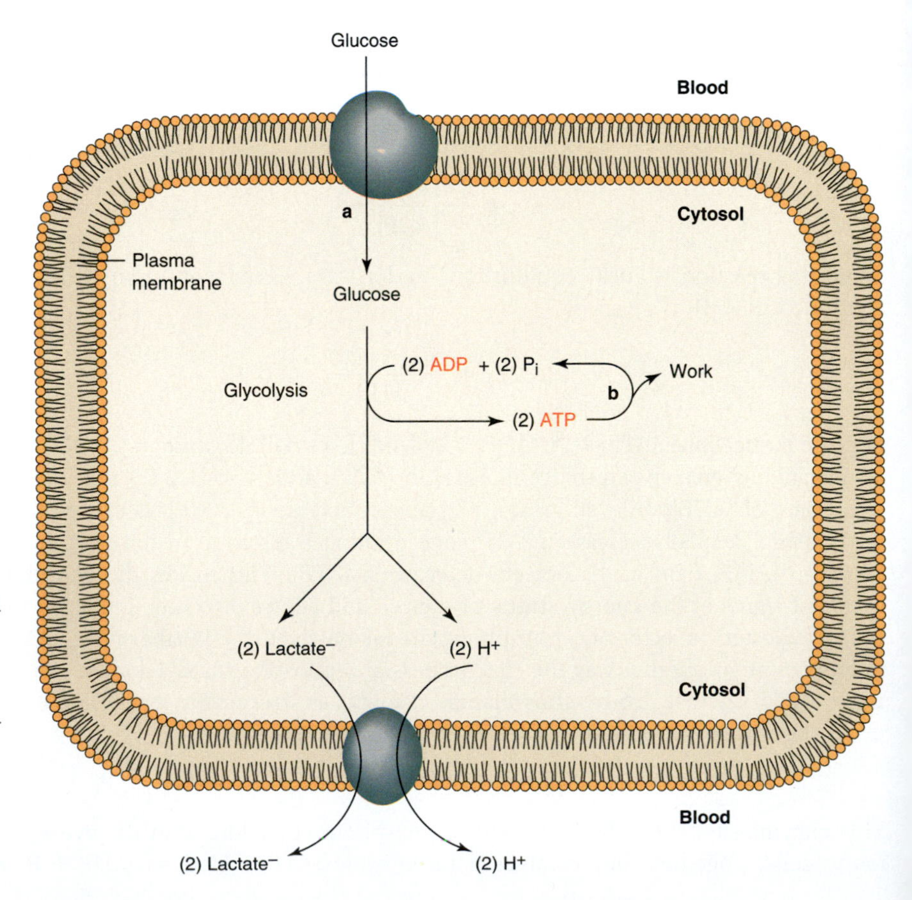

FIGURE 14.17
Unless lactate formed by glycolysis is transported out of the cell, the intracellular pH will decrease by the accumulation of intracellular lactic acid. The low pH decreases 6-phosphofructo-1-kinase activity so that further lactic acid production by glycolysis is shut off. (a) Glucose transport into the cell; (b) all work performances that convert ATP back to ADP and P_i; (c) lactate–H^+ symport (actual stoichiometry of one lactate$^-$ and one H^+ transported by the symport).

CLINICAL CORRELATION 14.5
Lactic Acidosis

This problem is characterized by elevated blood lactate levels, usually greater than 5 mM, along with decreased blood pH and bicarbonate concentrations. Lactic acidosis is the most commonly encountered form of metabolic acidosis and can be the consequence of overproduction of lactate, underutilization of lactate, or both. Lactate production is normally balanced by lactate utilization, with the result that lactate is usually not present in the blood at concentrations greater than 1.2 mM. All tissues of the body have the capacity to produce lactate by anaerobic glycolysis, but most tissues do not produce large quantities because much more ATP can be gained by the complete oxidation of the pyruvate produced by glycolysis. However, all tissues respond with an increase in lactate generation when oxygenation is inadequate. A decrease in ATP resulting from reduced oxidative phosphorylation allows the activity of 6-phosphofructo-1-kinase to increase. These tissues have to rely on anaerobic glycolysis for ATP production under such conditions and this results in lactic acid production. A good example is muscle exercise, which can deplete the tissue of oxygen and cause an overproduction of lactic acid. Tissue hypoxia occurs, however, in all forms of shock, during convulsions, and in diseases involving circulatory and pulmonary failure.

The major fate of lactate in the body is either complete combustion to CO_2 and H_2O or conversion back to glucose by the process of gluconeogenesis. Both require oxygen. Decreased oxygen availability, therefore, increases lactate production and decreases lactate utilization. The latter can also be decreased by liver diseases, ethanol, and a number of other drugs. Phenformin, a drug that was once used to treat the hyperglycemia of insulin-independent diabetes, was well-documented to induce lactic acidosis in certain patients.

Bicarbonate is usually administered in an attempt to control the acidosis associated with lactic acid accumulation. The key to successful treatment, however, is to find and eliminate the cause of the overproduction and/or underutilization of lactic acid and most often involves the restoration of circulation of oxygenated blood.

Newsholme, E. A. and Leech, A. R. *Biochemistry for the Medical Sciences.* New York: Wiley, 1983; and Kruse, J. A. and Carlson, R. W. Lactate metabolism. *Crit. Care Clin.* 3:725, 1985.

14.6). The sensitivity of 6-phosphofructo-1-kinase to hydrogen ions is also part of this mechanism. Hydrogen ions are able to shut off glycolysis, the process responsible for decreasing pH. Transport of lactic acid out of a cell requires that blood be available to carry it away. When blood flow is inadequate, for example, in heavy exercise of a skeletal muscle or an attack of **angina pectoris** in the case of the heart, hydrogen ions cannot escape from cells fast enough. Yet, the need for ATP within such cells may partially override inhibition of 6-phosphofructo-1-kinase by hydro-

CLINICAL CORRELATION 14.6
Pickled Pigs and Malignant Hyperthermia

In patients with malignant hyperthermia, a variety of agents, especially the widely used general anesthetic halothane, will produce a dramatic rise in body temperature, metabolic and respiratory acidosis, hyperkalemia, and muscle rigidity. This genetic abnormality occurs in about 1 in 15,000 children and 1 in 50,000–100,000 adults. It is dominantly inherited. Death may result the first time a susceptible person is anesthetized. Onset occurs within minutes of drug exposure and the hyperthermia must be recognized immediately. Packing the patient in ice is effective and should be accompanied by measures to combat acidosis. The drug dantrolene is also effective.

A phenomenon similar, if not identical, to malignant hyperthermia is known to occur in pigs. Pigs with this problem, called porcine stress syndrome, respond poorly to stress. This genetic disease usually manifests itself as the pig is being shipped to market. Pigs with the syndrome can be identified by exposure to halothane, which triggers the same response seen in patients with malignant hyperthermia. The meat of pigs that have died as a result of the syndrome is pale, watery, and of very low pH (i.e., nearly pickled).

Muscle is the site of the primary lesion in both malignant hyperthermia and porcine stress syndrome. In response to halothane the skeletal muscles become rigid and generate heat and lactic acid. The sarcoplasmic reticulum of such pigs and patients have a genetic abnormality in the ryanodine receptor, a Ca^{2+} release channel, that plays an important function in excitation–contraction coupling in muscle. Because of a defect in this protein, the anesthetic triggers inappropriate release of Ca^{2+} from the sarcoplasmic reticulum. This results in uncontrolled stimulation of a number of heat-producing processes, including myosin ATPase, glycogenolysis, glycolysis, and cyclic uptake and release of Ca^{2+} by mitochondria and sarcoplasmic reticulum.

Muscle cells become irreversibly damaged as a consequence of excessive heat production, lactic acidosis, and ATP loss.

Kalow, W. and Grant, D. M. Pharmacogenetics. In: C. R. Scriver, A. L. Beaudet, W. S. Sly, and D. Valle (Eds.), *The Metabolic and Molecular Bases of Inherited Disease,* 7th ed. New York: McGraw-Hill, 1995, pp. 293–326; and McCarthy T. V., Quane, K. A., and Lynch P. J. Ryanodine receptor mutations in malignant hyperthermia and central core disease. *Hum. Mutat.* 15:410, 2000.

CLINICAL CORRELATION 14.7
Angina Pectoris and Myocardial Infarction

Chest pain associated with reversible myocardial ischemia is termed angina pectoris (literally, strangling pain in the chest). The pain is the result of an imbalance between demand for and supply of blood flow to cardiac muscles and is most commonly caused by narrowing of the coronary arteries. The patient experiences a heavy squeezing pressure or ache substernally, often radiating to either the shoulder and arm or occasionally to the jaw or neck. Attacks occur with exertion, last from 1 to 15 min, and are relieved by rest. The coronary arteries involved are obstructed by atherosclerosis (i.e., lined with characteristic fatty deposits) or less commonly narrowed by spasm. Myocardial infarction occurs if the ischemia persists long enough to cause severe damage (necrosis) to the heart muscle. Commonly, a blood clot forms at the site of narrowing and completely obstructs the vessel. In myocardial infarction, tissue death occurs and the characteristic pain is longer lasting, and often more severe.

Nitroglycerin and other nitrates are frequently prescribed to relieve the pain caused by the myocardial ischemia of angina pectoris. These drugs can be used prophylactically, enabling patients to participate in activities that would otherwise precipitate an attack of angina. Nitroglycerin may work in part by causing dilation of the coronary arteries, improving oxygen delivery to the heart, and washing out lactic acid. Probably more important is the effect of nitroglycerin on the peripheral circulation. Breakdown of nitroglycerin produces nitric oxide (NO), a compound that relaxes smooth muscle, causing venodilation throughout the body. This reduces arterial pressure and allows blood to accumulate in the veins. The result is decreased return of blood to the heart, and a reduced volume of blood the heart has to pump, which reduces the energy requirement of the heart. In addition, the heart empties itself against less pressure, which also spares energy. The overall effect is a lowering of the oxygen requirement of the heart, bringing it in line with the oxygen supply via the diseased coronary arteries. Other useful agents are calcium channel blockers, which are coronary vasodilators, and β-adrenergic blockers. The β-blockers prevent the increase in myocardial oxygen consumption induced by sympathetic nervous system stimulation of the heart, as occurs with physical exertion.

The coronary artery bypass operation is used in severe cases of angina that cannot be controlled by medication. In this operation veins are removed from the leg and interposed between the aorta and coronary arteries of the heart. The purpose is to bypass the portion of the artery diseased by atherosclerosis and provide the affected tissue with a greater blood supply. Remarkable relief from angina can be achieved by this operation, with the patient being able to return to normal productive life in some cases.

Hugenholtz, P. G. Calcium antagonists for angina pectoris. *Ann. N.Y. Acad. Sci.* 522:565, 1988; Feelishch, M. and Noack, E.A. Correlation between nitric oxide formation during degradation of organic nitrates and activation of guanylate cyclase. *Eur. J. Pharmacol.* 139: 19, 1987; and Ignarro, L. J. Biological actions and properties of endothelium-derived nitric oxide formed and released from artery and vein. *Circ. Res.* 65: 1, 1989.

gen ions. Unabated accumulation of hydrogen ions then results in pain, which, in the case of skeletal muscle, can be relieved by simply terminating the exercise. In the case of the heart, rest or pharmacologic agents that increase blood flow or decrease the need for ATP within myocytes may be effective (see Clin. Corr. 14.7).

Intracellular Citrate Levels Regulate 6-Phosphofructo-1-kinase

Many tissues prefer to use fatty acids and ketone bodies as oxidizable fuels in place of glucose. Most of these tissues can use glucose but actually prefer to oxidize fatty acids and ketone bodies. This helps preserve glucose for tissues, such as brain, that are absolutely dependent on glucose as an energy source. Oxidation of both fatty acids and ketone bodies elevates levels of cytosolic **citrate,** which inhibits 6-phosphofructo-1-kinase. The result is decreased glucose utilization by the tissue when fatty acids or ketone bodies are available.

Hormonal Control of 6-Phosphofructo-1-kinase by cAMP and Fructose 2,6-bisphosphate

Fructose 2,6-bisphosphate (Figure 14.18), like AMP, functions as a positive allosteric effector of 6-phosphofructo-1-kinase and as a negative allosteric effector of fructose 1,6-bisphosphatase. Indeed, without the presence of this compound, glycolysis could not occur in liver because 6-phosphofructo-1-kinase would have insufficient activity and fructose 1,6-bisphosphatase would have too much activity for net conversion of fructose 6-phosphate to fructose 1,6-bisphosphate.

Figure 14.19 gives a brief overview of the role of fructose-2,6-bisphosphate in hormonal control of hepatic glycolysis. Understanding this mechanism requires an appreciation of the role of **cAMP** (Figure 14.20) as one of the **"second messengers"** of hormone action. As discussed in more detail in Chapter 20, glucagon is

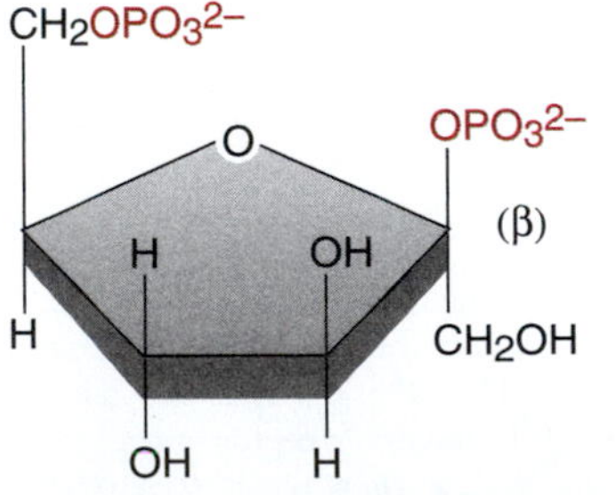

FIGURE 14.18
Structure of fructose 2,6-bisphosphate.

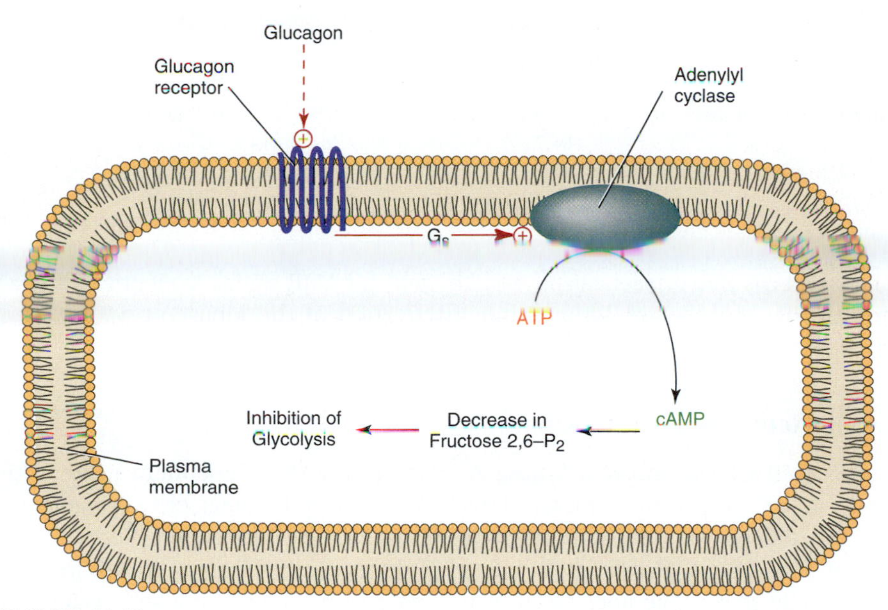

FIGURE 14.19
Overview of mechanism for glucagon inhibition of hepatic glycolysis.
Binding of glucagon to its receptor (a protein that spans the membrane seven times) activates adenylyl cyclase (an intrinsic membrane protein) activity through the action of a stimulatory G protein (G_s; also an intrinsic membrane protein). The (+) symbol indicates activation.

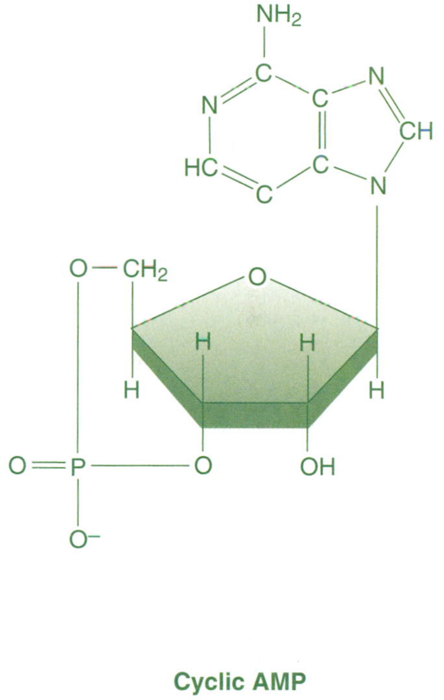

released from α cells of pancreas and circulates in blood until it comes in contact with glucagon receptors located on the outer surface of liver plasma membrane (Figure 14.19). Binding of glucagon to these receptors is sensed by adenylyl cyclase, an enzyme located on the inner surface of the plasma membrane, stimulating it to convert ATP into cAMP. Cyclic AMP triggers a series of intracellular events that result ultimately in a decrease in fructose 2,6-bisphosphate levels. A decrease in this compound makes 6-phosphofructo-1-kinase less effective but makes fructose 1,6-bisphosphatase more effective, thereby severely restricting flux from fructose 6-phosphate to fructose 1,6-bisphosphate in glycolysis.

Fructose 2,6-bisphosphate is a side product, rather than an intermediate, of glycolysis that is produced from F6P by the enzyme **6-phosphofructo-2-kinase** (Figure 14.21). We now have two "phosphofructokinases" to contend with: one produces an intermediate (fructose 1,6-bisphosphate) of glycolysis and the other produces a positive allosteric effector (fructose 2,6-bisphosphate) of the first enzyme. Fructose 2,6-bisphosphatase destroys fructose 2,6-bisphosphate by converting it back to F6P by simple hydrolysis (Figure 14.21). No ATP or ADP is involved in this reaction. A **bifunctional enzyme** in which both catalytic activities are part of the same protein is responsible for both synthesis and degradation of fructose 2,6-bisphosphate. Thus the enzyme is named 6-phosphofructo-2-kinase/fructose 2,6-bisphosphatase to emphasize its ability to synthesize and hydrolyze fructose 2,6-bisphosphate.

cAMP regulates fructose 2,6-bisphosphate levels in liver. How is this possible when the same enzyme carries out both synthesis and degradation of the molecule?

Cyclic AMP

FIGURE 14.20
Structure of cAMP.

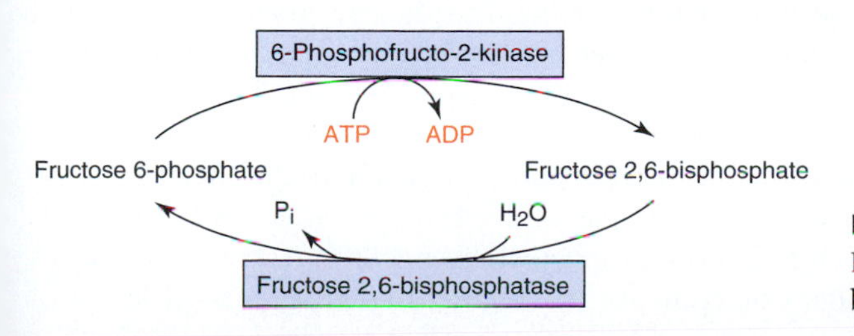

FIGURE 14.21
Reactions involved in the formation and degradation of fructose 2,6-bisphosphate.

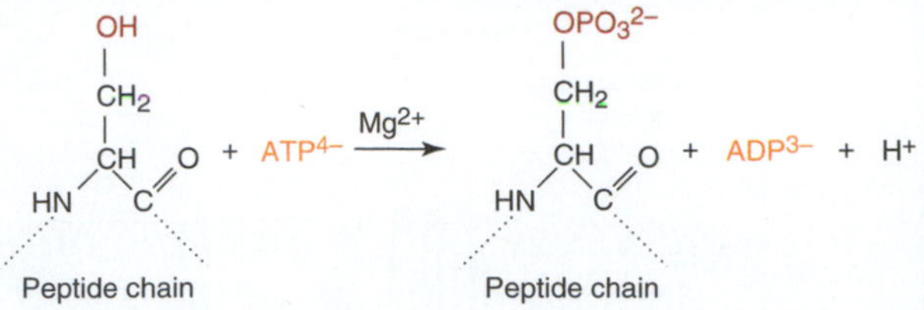

$$\text{Peptide chain–HN–CH(CH}_2\text{OH)–C(=O)} + ATP^{4-} \xrightarrow{Mg^{2+}} \text{Peptide chain–HN–CH(CH}_2\text{OPO}_3^{2-}\text{)–C(=O)} + ADP^{3-} + H^+$$

FIGURE 14.22
Enzymes subject to covalent modification are usually phosphorylated on specific serine residues.
Tyrosine and threonine residues are also important sites of covalent modification by phosphorylation.

The answer is that a mechanism exists by which cAMP signals concurrent activation of the phosphatase activity and inactivation of the kinase activity of this bifunctional enzyme.

cAMP Activates Protein Kinase A

Cyclic AMP activates **protein kinase A** (also called **cAMP-dependent protein kinase**). In its inactive state, protein kinase A is a heterotetramer of two regulatory and two catalytic subunits. Binding of cAMP to the regulatory subunits causes conformational changes that result in release of the catalytic subunits, which are then catalytically active. The liberated protein kinase then catalyzes phosphorylation of specific serine residues present in many different enzymes (Figure 14.22).

Phosphorylation of an enzyme can conveniently be abbreviated as

$$⊡ + \text{ATP} \rightarrow ⊙\text{-P} + \text{ADP}$$

where ⊡ and ⊙-P are used to indicate dephosphorylated and phosphorylated enzymes, respectively. Circle and square symbols are used because phosphorylation causes a change in conformation that affects the active site of the enzyme. The catalytic activity of some enzymes is increased whereas the catalytic activity of others is decreased by this alteration in the active site. The direction of the activity change depends on the enzyme involved. Many enzymes are subject to this type of regulation, an important type of **covalent modification** that can be reversed by simply removing the phosphate. Regardless of whether phosphorylation or dephosphorylation activates the enzyme, the active form of the enzyme is called the *a* form and the inactive form the *b* form. Likewise, regardless of the effect of phosphorylation on catalytic activity, the action of a protein kinase is always opposed by that of a phosphoprotein phosphatase, which catalyzes the reaction of

$$⊙\text{-P} + H_2O \rightarrow ⊡ + P_i$$

Putting these together creates a **cyclic control system** (Figure 14.23), such that the ratio of phosphorylated enzyme to dephosphorylated enzyme is a function of the relative activities of protein kinase and phosphoprotein phosphatase. If kinase activity is increased relative to phosphatase activity, more enzyme will be in the phosphorylated mode, and vice versa. Since activity is determined by whether the enzyme is phosphorylated or dephosphorylated, the relative activities of kinase and phosphatase determine the relative amount of a particular enzyme that is in the catalytically active state.

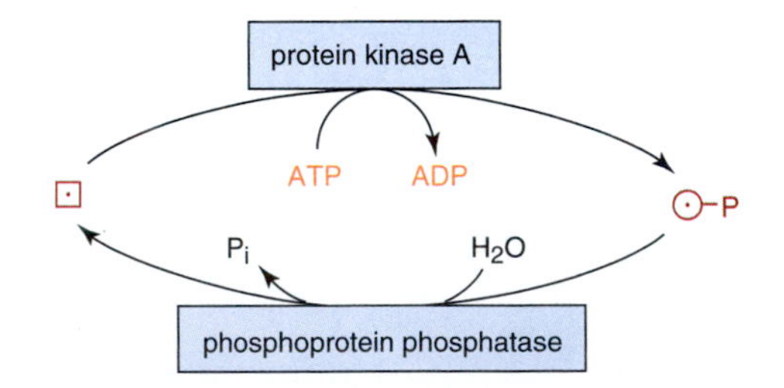

FIGURE 14.23
General model for the mechanism for regulation of enzymes by phosphorylation–dephosphorylation.
The symbols ⊡ and ⊙-P indicate that different conformational and activity states of the enzyme are produced as a result of phosphorylation–dephosphorylation.

6-Phosphofructo-2-kinase/Fructose 2,6-bisphosphatase Is Regulated by Phosphorylation–Dephosphorylation

Most enzymes are either turned on or turned off by phosphorylation. However, with **6-phosphofructo-2-kinase** and **fructose 2,6-bisphosphatase** advantage is taken of the bifunctional nature of the enzyme. Specifically with the liver isoenzyme, phosphorylation causes inactivation of the active site responsible for synthesis of fructose 2,6-bisphosphate but activation of the active site responsible for hydrolysis of fructose 2,6-bisphosphate (Figure 14.24). Dephosphorylation of the enzyme has the opposite effects. A sensitive mechanism has evolved, therefore, to set the intracellular concentration of fructose 2,6-bisphosphate in liver cells in response to changes in blood levels of glucagon or epinephrine (Figure 14.25). Increased levels of

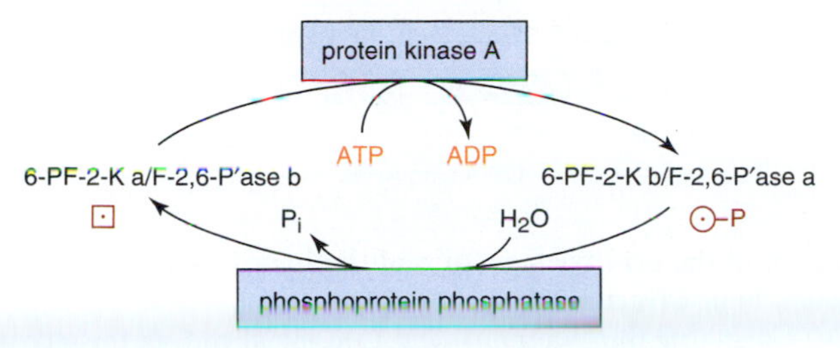

FIGURE 14.24
Mechanism responsible for covalent modification of the bifunctional enzyme 6-phosphofructo-2-kinase/fructose-2,6-bisphosphatase.
Name of the enzyme is abbreviated as 6-PF-2-K/F-2,6-P'ase. Letters a and b indicate the active and inactive forms of the enzymes, respectively.

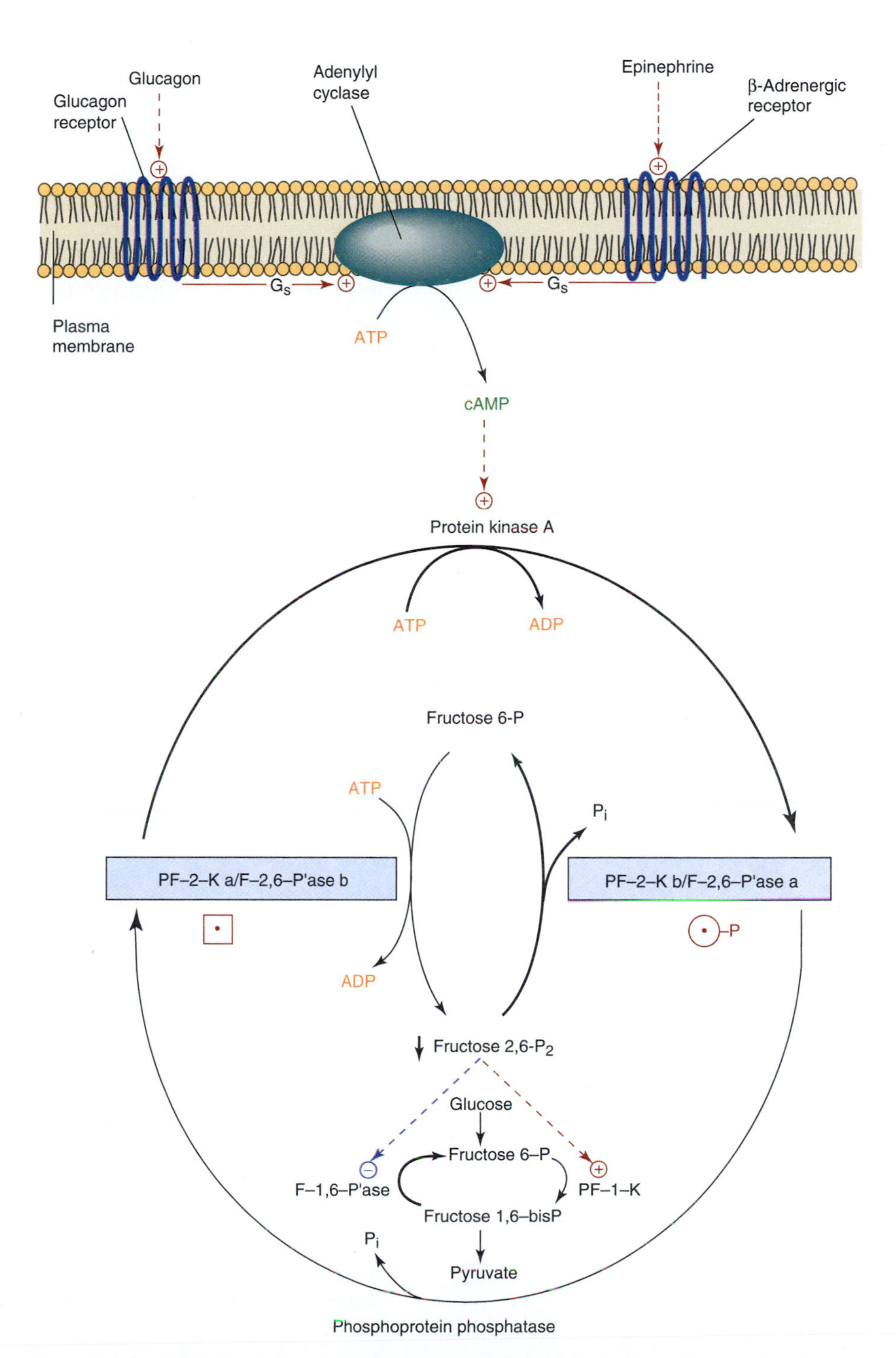

FIGURE 14.25
Mechanism of glucagon and epinephrine inhibition of hepatic glycolysis via cAMP-mediated decrease in fructose 2,6-bisphosphate concentration.
See legend for Figure 14.19. Heavy arrows indicate the reactions that predominate in the presence of glucagon. Small arrow before fructose 2,6-bisphosphate indicates a decrease in concentration of this compound.

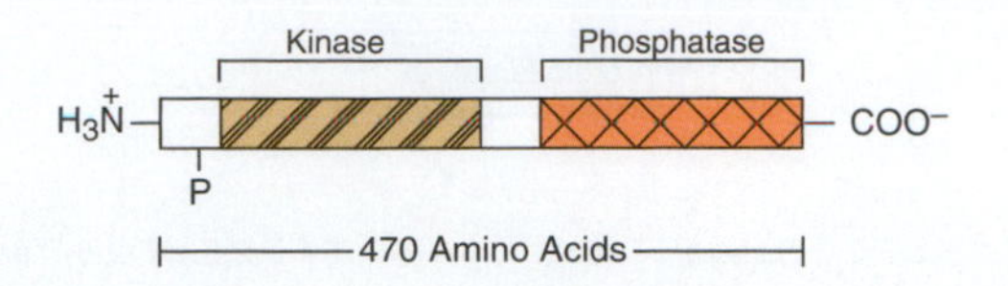

FIGURE 14.26
Schematic diagram of the primary structure of the liver isoenzyme of 6-phosphofructo-2-kinase/fructose-2,6-bisphosphatase.
$N_3\overset{+}{H}$- and -COO^- designate the N-terminal and C-terminal ends of the enzyme, respectively. Domain with kinase activity is located in the N-terminal half of the enzyme; domain with phosphatase activity in the C-terminal half of the enzyme. The letter P indicates the serine 32 phosphorylated by protein kinase A.

glucagon or **epinephrine,** acting through plasma membrane **glucagon receptors** and **β-adrenergic receptors,** respectively, have the common effect of inducing an increase in intracellular levels of cAMP. This second messenger activates protein kinase A, which phosphorylates a single serine residue of 6-phosphofructo-2-kinase/fructose 2,6-bisphosphatase (Figure 14.26). This inhibits 6-phosphofructo-2-kinase and activates fructose 2,6-bisphosphatase. The resulting decrease in fructose 2,6-bisphosphate makes 6-phosphofructo-1-kinase less effective and fructose 1,6-bisphosphatase more effective. The result is inhibition of glycolysis at the level of the conversion of fructose 6-phosphate to fructose 1,6-bisphosphate. Decreased levels of either glucagon or epinephrine in blood result in less cAMP in liver because adenylyl cyclase is less active and cAMP that had accumulated is converted to AMP by the action of cAMP phosphodiesterase. Loss of the cAMP signal results in inactivation of protein kinase A and a corresponding decrease in phosphorylation of 6-phosphofructo-2-kinase/fructose 2,6-bisphosphatase by protein kinase A. A **phosphoprotein phosphatase** removes phosphate from the bifunctional enzyme to produce active 6-phosphofructo-2-kinase and inactive fructose 2,6-bisphosphatase. Fructose 2,6-bisphosphate can now accumulate to a higher steady-state concentration and, by activating 6-phosphofructo-1-kinase and inhibiting fructose 1,6-bisphosphatase, greatly increases the rate of glycolysis. Thus glucagon and epinephrine are extracellular signals that stop liver from using glucose, whereas fructose 2,6-bisphosphate is an intracellular signal that promotes glucose utilization by this tissue.

Insulin opposes the actions of glucagon and epinephrine by means of a signaling cascade that begins with activation of the tyrosine kinase activity of its receptor (see Chapter 21). By mechanisms that are still not entirely defined, insulin brings about activation of cAMP phosphodiesterase (lowers cAMP levels), inhibition of protein kinase A, and activation of phosphoprotein phosphatase, all of which oppose the effects of glucagon and epinephrine (Figure 14.27). Insulin therefore acts in the opposite direction from that of glucagon and epinephrine in determining the levels of fructose 2,6-bisphosphate in liver cells and, therefore, acts to stimulate the rate of glycolysis.

Heart Contains a Different Isoenzyme of the Bifunctional Enzyme

An increase in blood level of epinephrine has a markedly different effect on glycolysis in **heart** from that in liver. Glycolysis is inhibited in liver to conserve glucose for use by other tissues. Epinephrine stimulates glycolysis in heart as part of a mechanism to meet the increased demand for ATP caused by an epinephrine-signaled increase in work load. As in liver, epinephrine acts on the heart by way of a β-adrenergic receptor on the plasma membrane, promoting formation of cAMP by adenylyl cyclase (Figure 14.28). This results in the activation of protein kinase A that in turn phosphorylates 6-phosphofructo-2-kinase/fructose 2,6-bisphosphatase. However, in contrast to what happens in liver, phosphorylation of

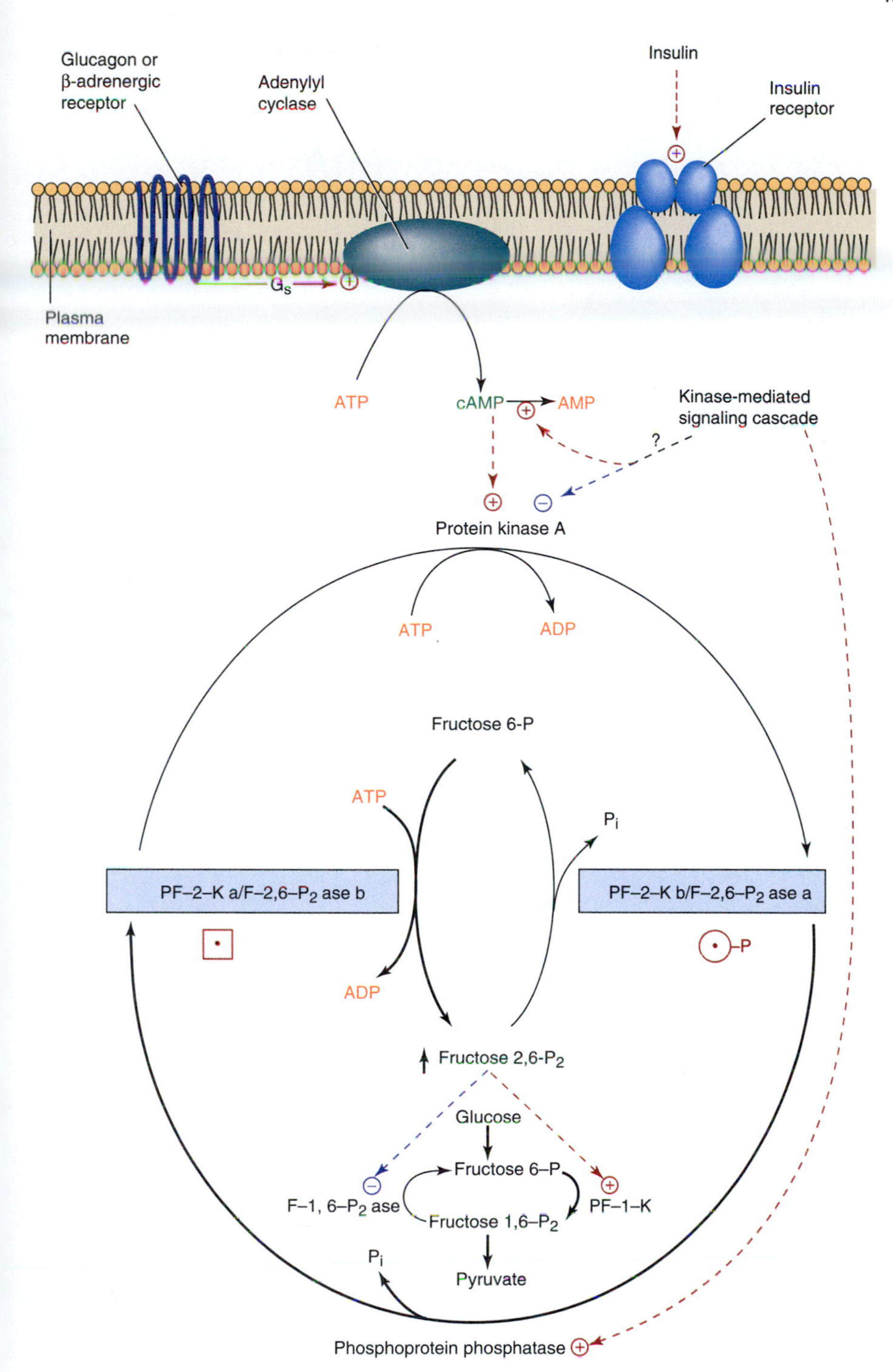

FIGURE 14.27
Mechanism responsible for accelerated rates of hepatic glycolysis when the concentration of glucagon and epinephrine are low and that of insulin is high in the blood.
See legends to Figures 14.19 and 14.25. The insulin receptor is an intrinsic component of the plasma membrane. Small arrow before fructose 2,6-bisphosphate indicates an increase in concentration.

the bifunctional enzyme in heart produces an increase rather than a decrease in fructose 2,6-bisphosphate levels. This is made possible by expression of a different isoenzyme of the bifunctional enzyme in heart. Although it still carries out the same reactions as the liver enzyme, the amino acid sequence of the heart isoenzyme is different, and phosphorylation by protein kinase A occurs at a site that activates rather than inhibits 6-phosphofructo-2-kinase (Figure 14.29). Increased fructose 2,6-bisphosphate concentrations in response to epinephrine result in increased 6-phosphofructo-1-kinase activity and increased glycolytic flux in the heart.

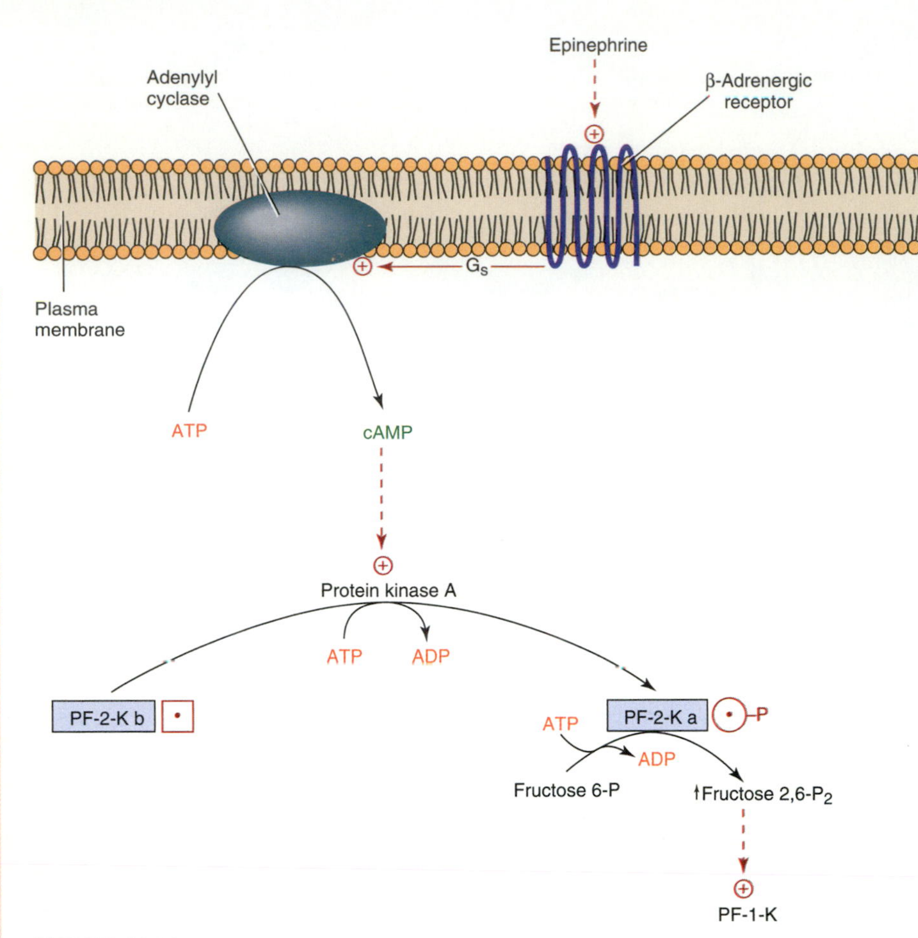

FIGURE 14.28
Mechanism for accelerated rates of glycolysis in the heart in response to epinephrine. See legends to Figures 14.19 and 14.27.

CLINICAL CORRELATION 14.8
Pyruvate Kinase Deficiency and Hemolytic Anemia

Mature erythrocytes are absolutely dependent on glycolytic activity for ATP production. ATP is needed for the ion pumps, especially the Na^+, K^+–ATPase, which maintain the biconcave disk shape of erythrocytes, a characteristic that helps erythrocytes slip through the capillaries as they deliver oxygen to the tissues. Without ATP the cells swell and lyse. Anemia due to excessive erythrocyte destruction is referred to as hemolytic anemia. Pyruvate kinase deficiency is rare but is by far the most common genetic defect of the glycolytic pathway known to cause hemolytic anemia. Most pyruvate kinase-deficient patients have 5–25% of normal red blood cell pyruvate kinase levels and flux through the glycolytic pathway is restricted severely, resulting in markedly lower ATP concentrations. The expected crossover of the glycolytic intermediates is observed; that is, those intermediates proximal to the pyruvate kinase-catalyzed step accumulate, whereas pyruvate and lactate concentrations decrease. Normal ATP levels are observed in reticulocytes of patients with this disease. Although deficient in pyruvate kinase, these "immature" red blood cells have mitochondria and can generate ATP by oxidative phosphorylation. Maturation of reticulocytes into red blood cells results in the loss of mitochondria and complete dependence on glycolysis for ATP production. Since glycolysis is defective, the mature cells are lost rapidly from the circulation. Anemia results because the cells cannot be replaced rapidly enough by erythropoiesis.

Valentine, W. N. The Stratton lecture: hemolytic anemia and inborn errors of metabolism. *Blood* 54:549, 1979; and Tanaka, K. R. and Paglia, D. E. Pyruvate kinase and other enzymopathies of the erythrocyte. In: C. R. Scriver, A. L. Beaudet, W. S. Sly, and D. Valle (Eds.), *The Metabolic and Molecular Bases of Inherited Disease,* 7th ed. New York: McGraw-Hill, 1995, p. 3485.

Pyruvate Kinase Is a Regulated Enzyme of Glycolysis

Pyruvate kinase is another regulatory enzyme of glycolysis (see Clin. Corr. 14.8). This enzyme is drastically inhibited by physiological concentrations of ATP, so much so that its potential activity is probably never fully realized under physiological conditions. The isoenzyme found in liver is greatly activated by fructose 1,6-bisphosphate (FBP), thereby linking regulation of pyruvate kinase to what is happening to 6-phosphofructo-1-kinase. Thus if conditions favor increased flux through 6-phosphofructo-1-kinase, the level of FBP increases and acts as a feed-forward activator of pyruvate kinase. The liver enzyme is also subject to covalent modification by protein kinase A, being active in the dephosphorylated state and inactive in the phosphorylated state (Figure 14.30). Concurrent inhibition of hepatic glycolysis and stimulation of hepatic gluconeogenesis by glucagon can be explained in part

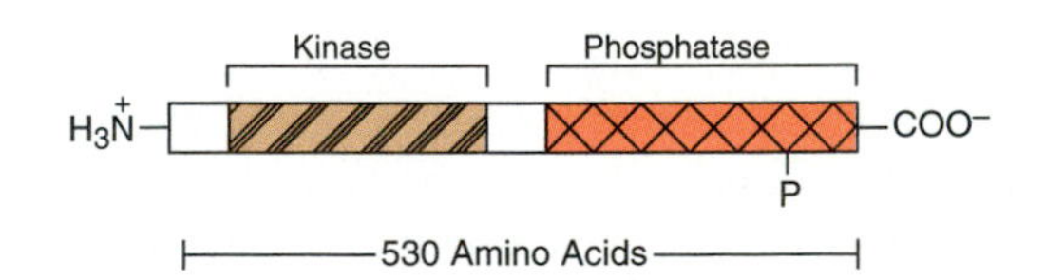

FIGURE 14.29
Schematic diagram of the primary structure of the heart isoenzyme of 6-phosphofructo-2-kinase/fructose-2,6-bisphosphatase.
See legend to Figure 14.26. The letter P indicates the site (serine 466) phosphorylated by protein kinase A.

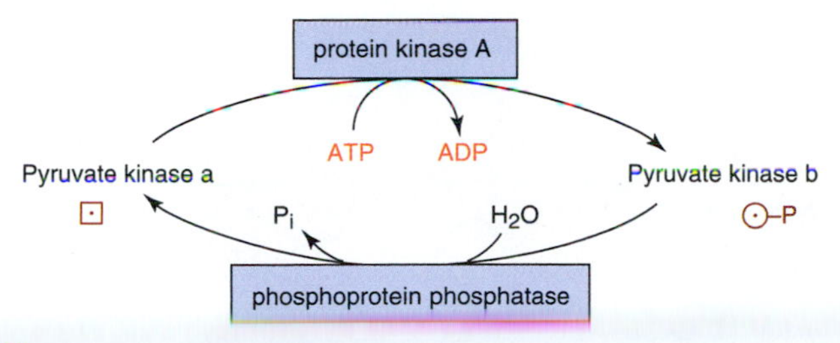

FIGURE 14.30
Glucagon acts via cAMP to cause the phosphorylation and inactivation of hepatic pyruvate kinase.

by inhibition of pyruvate kinase caused by activation of protein kinase A by cAMP. This aspect is explored more thoroughly in Section 14.5 (see p. 641) in the discussion of gluconeogenesis.

Pyruvate kinase, like glucokinase, is induced to higher steady-state concentrations in liver by combination of high carbohydrate intake and high insulin levels. This increase in enzyme concentration is one reason why liver of the well-fed individual has much greater capacity for utilizing carbohydrate than a fasting or diabetic person (see Clin. Corr. 14.4).

14.5 | GLUCONEOGENESIS

Glucose Synthesis Is Required for Survival

Net synthesis or formation of glucose from noncarbohydrate substrates is termed **gluconeogenesis.** This includes use of various amino acids, lactate, pyruvate, propionate, and glycerol as sources of carbon for the pathway (see Figure 14.31). Glucose can also be produced from galactose and fructose. **Glycogenolysis**, that is, formation of glucose or glucose 6-phosphate from glycogen, should be differentiated from gluconeogenesis; glycogenolysis refers to

$$\text{glycogen or (glucose)}_n \rightarrow n \text{ molecules of glucose}$$

and thus does not correspond to *de novo* synthesis of glucose, the hallmark of the process of gluconeogenesis.

The capacity to synthesize glucose is crucial for survival of humans and other animals. Blood glucose levels have to be maintained to support metabolism of tissues that use glucose as their primary substrate (see Clin. Corr. 14.9). These include brain, red blood cells, kidney medulla, lens, cornea, testis, and a number of other tissues. Gluconeogenesis enables the maintenance of blood glucose levels long after all dietary glucose has been absorbed and completely oxidized.

Cori and Alanine Cycles

Two important cycles between tissues are recognized that involve gluconeogenesis. The **Cori cycle** and the **alanine cycle** (Figure 14.32) depend on gluconeogenesis in liver followed by delivery of glucose and its use in a peripheral tissue. Both cycles

CLINICAL CORRELATION 14.9

Hypoglycemia and Premature Infants

Premature and small-for-gestational-age neonates have a greater susceptibility to hypoglycemia than full-term, appropriate-for-gestational-age infants. Several factors appear to be involved. Children in general are more susceptible than adults to hypoglycemia, simply because they have larger brain/body weight ratios and the brain utilizes disproportionately greater amounts of glucose than the rest of the body. Newborn infants have a limited capacity for ketogenesis, apparently because the transport of long-chain fatty acids into liver mitochondria of the neonate is poorly developed. Since ketone body use by the brain is directly proportional to the circulating ketone body concentration, the neonate is unable to spare glucose to any significant extent by using ketone bodies. The consequence is that the neonate's brain is almost completely dependent on glucose obtained from liver glycogenolysis and gluconeogenesis.

The capacity for hepatic glucose synthesis from lactate and alanine is also limited in newborn infants. This is because the rate-limiting enzyme phospho*enol*pyruvate carboxykinase is present in very low amounts during the first few hours after birth. Induction of this enzyme to the level required to prevent hypoglycemia during the stress of fasting requires several hours. Premature and small-for-gestational-age infants are believed to be more susceptible to hypoglycemia than normal infants because of smaller stores of liver glycogen. Fasting depletes their glycogen stores more rapidly, making these neonates more dependent on gluconeogenesis than normal infants.

Ballard, F. J. The development of gluconeogenesis in rat liver: controlling factors in the newborn. *Biochem. J.* 124:265, 1971; and Newsholme, E. A. and Leech, A. R. *Biochemistry for the Medical Sciences.* New York: Wiley, 1983.

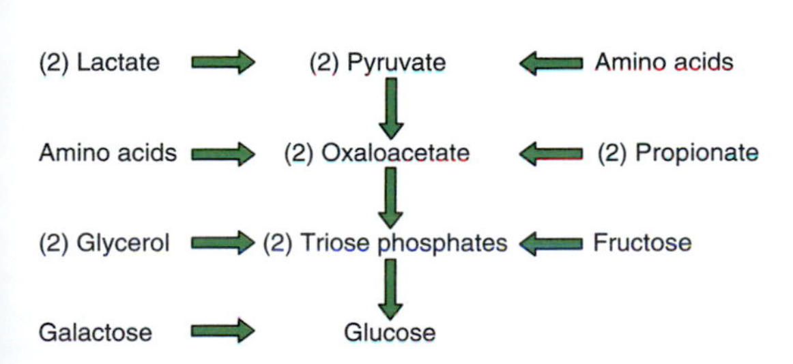

FIGURE 14.31
Abbreviated pathways of gluconeogenesis, illustrating the major substrate precursors for the process.

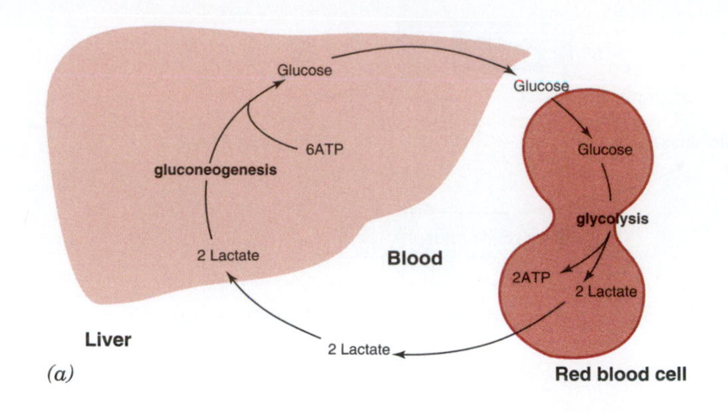

(a)

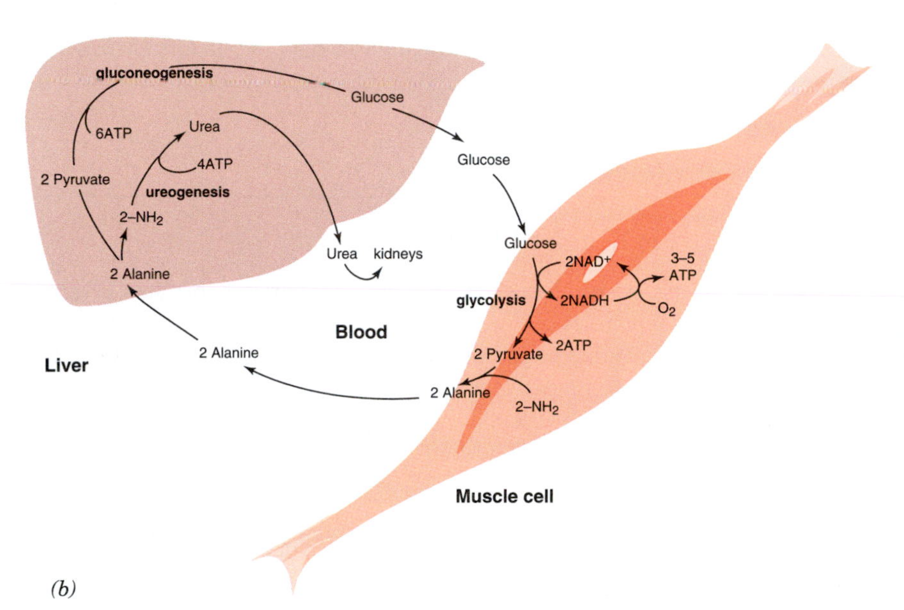

(b)

FIGURE 14.32
Relationship between gluconeogenesis in the liver and glycolysis in the rest of the body.
(*a*) Cori cycle. (*b*) Alanine cycle.

provide a mechanism for continuously supplying tissues that require glucose as their primary energy source. The cycles are only functional between liver and tissues that do not completely oxidize glucose to CO_2 and H_2O. In order to participate in these cycles, peripheral tissues must release either alanine or lactate as the end product of glucose metabolism. The type of recycled three-carbon intermediate is the major difference between the Cori cycle (Figure 14.32*a*) and the alanine cycle (Figure 14.32*b*), carbon returning to liver as lactate in the Cori cycle but as alanine in the alanine cycle. Another difference is that NADH generated by glycolysis in the alanine cycle is not used to reduce pyruvate to lactate, otherwise pyruvate would not be available for conversion to alanine by transamination with glutamate. In tissues that have mitochondria, electrons of NADH can be transported into the mitochondria by the malate–aspartate shuttle or the glycerol phosphate shuttle for the synthesis of ATP by oxidative phosphorylation:

$$\text{NADH} + \text{H}^+ + \tfrac{1}{2}\,\text{O}_2 + 2\tfrac{1}{2}\,\text{ADP} + 2\tfrac{1}{2}\,\text{P}_i \rightarrow \text{NAD}^+ + 2\tfrac{1}{2}\,\text{ATP}$$

or

$$FADH_2 + \frac{1}{2} O_2 + 1\frac{1}{2} ADP + 1\frac{1}{2} P_i \rightarrow FAD + 1\frac{1}{2} ATP$$

The consequence is that five to seven molecules of ATP can be formed per glucose molecule in peripheral tissues that participate in the alanine cycle. This stands in contrast to the Cori cycle in which only two molecules of ATP per molecule of glucose are produced. Overall stoichiometry for the Cori cycle is

$$6\ ATP_{liver} + 2\ (ADP + P_i)_{\text{red blood cells}} \rightarrow 6\ (ADP + P_i)_{liver} + 2\ ATP_{\text{red blood cells}}$$

Six molecules of ATP are needed in liver to provide the energy necessary for glucose synthesis. The alanine cycle also transfers the energy from liver to peripheral tissues and, because of the five to seven molecules of ATP produced per molecule of glucose, is an energetically more efficient cycle. However, participation of alanine in the cycle presents liver with amino nitrogen that must be disposed of as urea (Figure 14.32*b* and p. 787). This is expensive since four ATP molecules are consumed for every urea molecule produced by the urea cycle. This need for urea synthesis increases the amount of ATP required in the liver to 10 molecules per glucose molecule during the alanine cycle:

$$10\ ATP_{liver} + 5\text{–}7\ (ADP + P_i)_{muscle} + O_{2\ muscle} \rightarrow 10\ (ADP + P_i)_{liver} + 5\text{–}7\ ATP_{muscle}$$

This and the requirement in peripheral tissue for oxygen and mitochondria distinguish the alanine cycle from the Cori cycle.

Glucose Synthesis from Lactate

Gluconeogenesis from lactate is an ATP-requiring process with the overall equation of

$$2\ \text{L-lactate}^- + 6\ ATP^{4-} + 6\ H_2O \rightarrow \text{glucose} + 6\ ADP^{3-} + 6\ P_i^{2-} + 4\ H^+$$

Many enzymes of glycolysis are common to the gluconeogenic pathway. Additional reactions have to be involved because glycolysis produces 2 ATP and gluconeogenesis requires 6 ATP per molecule of glucose. Also, certain steps of glycolysis are irreversible under intracellular conditions and are replaced by irreversible steps of the gluconeogenic pathway. The reactions of gluconeogenesis from lactate are given in Figure 14.33. The initial step is conversion of lactate to pyruvate by lactate dehydrogenase. The NADH generated in this step is needed for a subsequent step in the pathway. Pyruvate cannot be converted to phospho*enol*pyruvate (PEP) by pyruvate kinase because the reaction is irreversible under intracellular conditions. Pyruvate is converted into the high-energy phosphate compound PEP by coupling of two reactions requiring high-energy phosphate compounds (an ATP and a GTP). **Pyruvate carboxylase** catalyzes the first reaction, **PEP carboxykinase** the second reaction (see Figure 14.34). **GTP**, required for the PEP carboxykinase, is equivalent to an ATP through the action of nucleoside diphosphate kinase (GDP + ATP $\rightleftharpoons$ GTP + ADP). The CO_2 generated by PEP carboxykinase and the HCO_3^- required by pyruvate carboxylase are linked by the reaction catalyzed by **carbonic anhydrase** ($CO_2 + H_2O \rightleftharpoons H_2CO_3 \rightleftharpoons H^+ + HCO_3^-$). Summing these reactions with the reactions of Figure 14.34 yields

$$\text{pyruvate}^- + 2\ ATP^{4-} \rightarrow \text{phospho}\textit{enol}\text{pyruvate}^{3-} + 2\ ADP^{3-} + Pi^{2-} + 2\ H^+$$

Thus conversion of pyruvate into PEP during gluconeogenesis costs the cell two molecules of ATP. This stands in contrast to the conversion of PEP to pyruvate during glycolysis, which yields the cell only one molecule of ATP.

The intracellular location of pyruvate carboxylase makes the mitochondrion mandatory for conversion of cytosolic pyruvate into cytosolic PEP (Figure 14.33). Because PEP carboxykinase is present in both the cytosolic and mitosolic com-

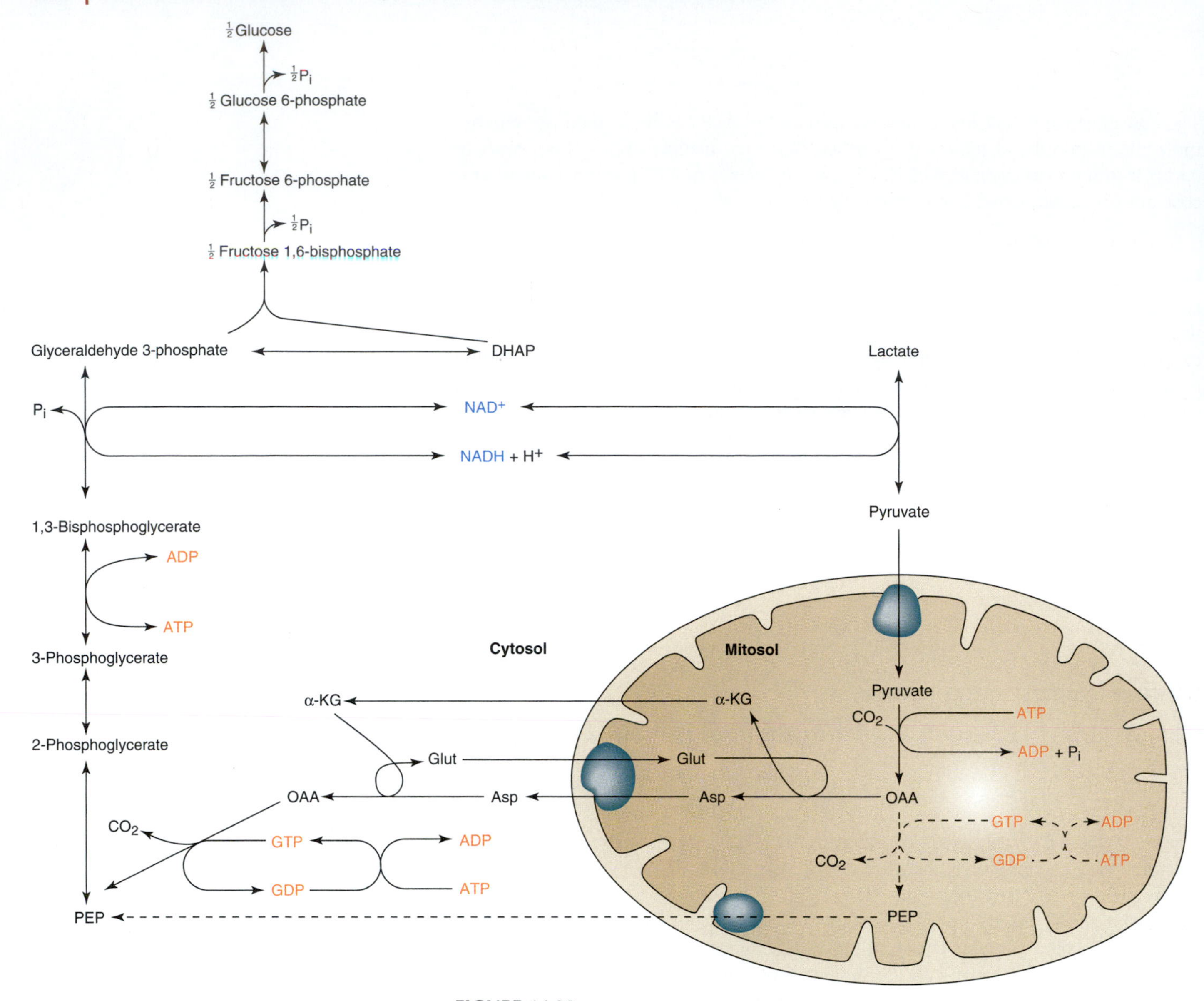

FIGURE 14.33
Pathway of gluconeogenesis from lactate.
The involvement of the mitochondrion in the process is indicated. Dashed arrows refer to an alternate route, which employs mitosolic PEP carboxykinase rather than the cytosolic isoenzyme. Abbreviations: OAA, oxaloacetate; α-KG, α-ketoglutarate; PEP, phospho*enol*pyruvate; and DHAP, dihydroxyacetone phosphate.

partments, there are two routes that oxaloacetate takes to glucose. For the pathway involving the mitochondrial PEP carboxykinase, oxaloacetate is converted within the mitochondrion into PEP that then traverses the mitochondrial inner membrane. The second pathway would be just as simple if oxaloacetate could traverse the mitochondrial inner membrane, but this cannot occur for want of a transporter for oxaloacetate (Figure 14.10*b*). Thus oxaloacetate is converted into aspartate, which exits into the cytosol by way of the **glutamate–aspartate antiport.** Aspartate transaminates with α-ketoglutarate in the cytosol to produce oxaloacetate, which is used by cytosolic PEP carboxykinase for the synthesis of PEP.

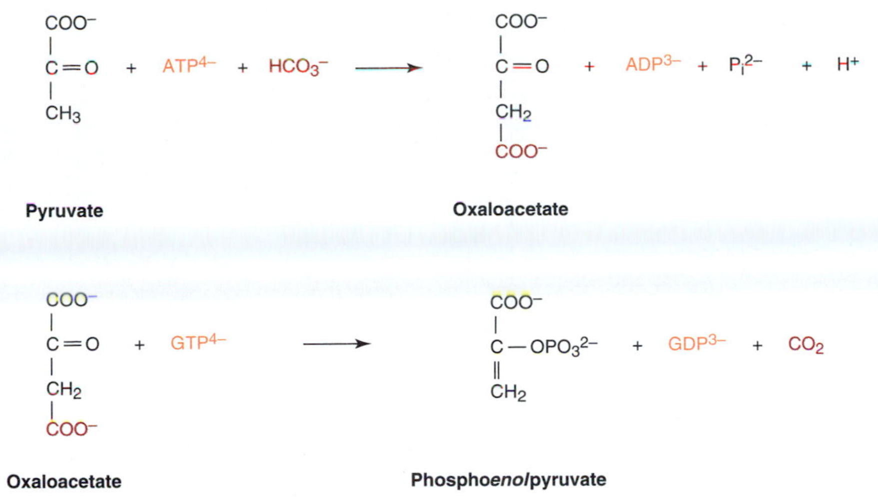

$$\text{Pyruvate } (CH_3{-}CO{-}COO^-) + ATP^{4-} + HCO_3^- \longrightarrow \text{Oxaloacetate } (^-OOC{-}CH_2{-}CO{-}COO^-) + ADP^{3-} + P_i^{2-} + H^+$$

$$\text{Oxaloacetate } (^-OOC{-}CH_2{-}CO{-}COO^-) + GTP^{4-} \longrightarrow \text{Phospho}\textit{enol}\text{pyruvate } (CH_2{=}C(OPO_3^{2-}){-}COO^-) + GDP^{3-} + CO_2$$

FIGURE 14.34
Energy-requiring steps involved in phospho*enol*pyruvate formation from pyruvate.
Reactions are catalyzed by pyruvate carboxylase and PEP carboxykinase, respectively.

Gluconeogenesis Uses Many Glycolytic Enzymes but in the Reverse Direction

Enzymes of the glycolytic pathway operate in reverse to convert PEP to fructose 1,6-bisphosphate during gluconeogenesis. NADH generated by lactate dehydrogenase is utilized by glyceraldehyde 3-phosphate dehydrogenase, establishing an equal balance of generation and utilization of reducing equivalents.

6-Phosphofructo-1-kinase catalyzes an irreversible step in glycolysis and cannot be used for conversion of FBP to fructose 6-phosphate. A way around this step is provided by **fructose 1,6-bisphosphatase**, which catalyzes irreversible hydrolysis of fructose 1,6-bisphosphate to F6P (Figure 14.35).

Phosphoglucose isomerase is freely reversible and functions in both glycolytic and gluconeogenic pathways. **Glucose 6-phosphatase**, which has to be used instead of glucokinase for the last step of gluconeogenesis, catalyzes an irreversible hydrolytic reaction under intracellular conditions (Figure 14.36). Nucleotides are not involved in this reaction; the function of this enzyme is to generate glucose, not to convert glucose into glucose 6-phosphate. Glucose 6-phosphatase is a membrane-bound enzyme, located within the endoplasmic reticulum, with its active site available for G6P hydrolysis on the cisternal surface of the tubules (see Figure 14.37). A translocase for G6P is required to move G6P from the cytosol to its site of hydrolysis within the endoplasmic reticulum. A genetic defect in either the translocase or the phosphatase interferes with gluconeogenesis and results in accumulation of glycogen in liver.

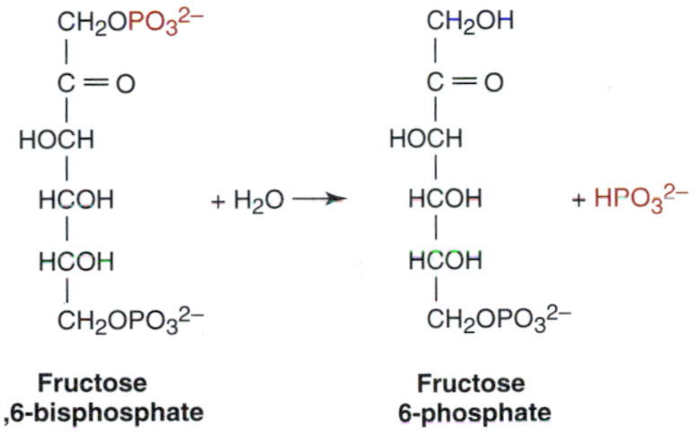

$$\text{Fructose 1,6-bisphosphate} + H_2O \longrightarrow \text{Fructose 6-phosphate} + HPO_3^{2-}$$

FIGURE 14.35
Reaction catalyzed by fructose 1,6-bisphosphatase.

$$\alpha\text{-D-Glucose 6-phosphate} + H_2O \longrightarrow \alpha\text{-D-Glucose} + HPO_3^{2-}$$

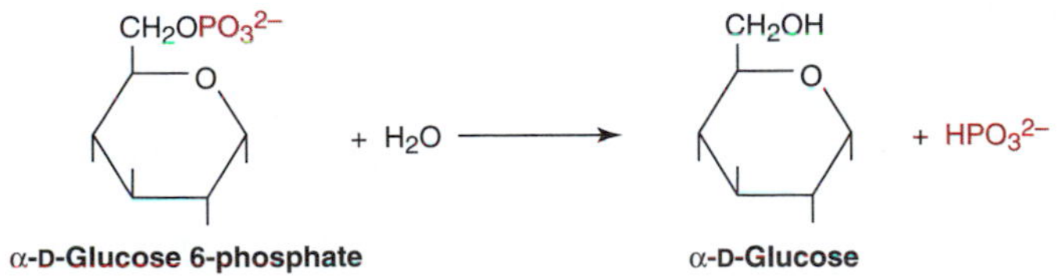

FIGURE 14.36
Reaction catalyzed by glucose 6-phosphatase.

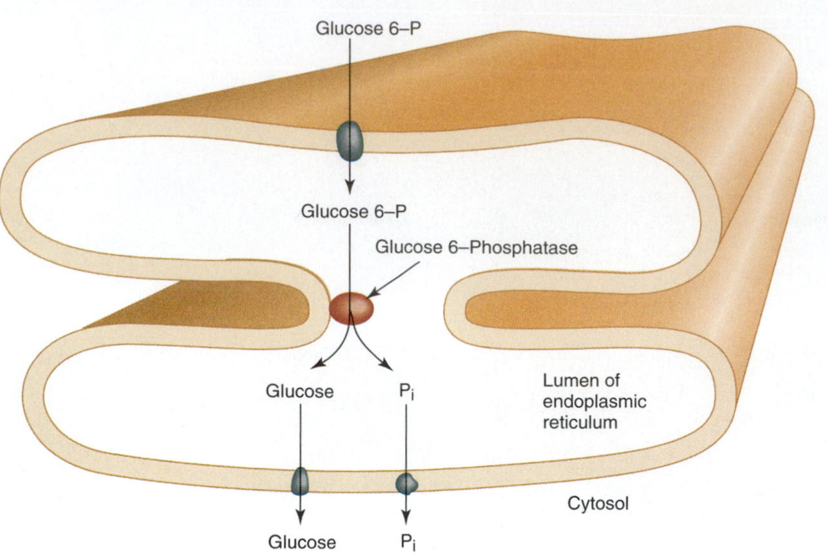

FIGURE 14.37
Glucose 6-phosphate is hydrolyzed by glucose 6-phosphatase located on the cisternal surface of the endoplasmic reticulum.
Three transporters are involved: one moves glucose 6-phosphate into the lumen, a second moves P_i back to the cytosol, and a third moves glucose back into the cytosol.

Glucose Is Synthesized from Carbon Chains of Some Amino Acids

All amino acids except **leucine** and **lysine** can supply carbon for net synthesis of glucose by gluconeogenesis. If catabolism of an amino acid can result in either pyruvate or oxaloacetate formation, then net glucose synthesis can occur from that amino acid. Oxaloacetate is an intermediate in gluconeogenesis and pyruvate is readily converted to oxaloacetate by action of pyruvate carboxylase (Figure 14.33). The abbreviated pathway given in Figure 14.31 shows where amino acid catabolism fits with the process of gluconeogenesis. Catabolism of amino acids feeds carbon into the tricarboxylic cycle at more than one point. As long as net synthesis of a TCA cycle intermediate occurs as a consequence of catabolism of a particular amino acid, net synthesis of oxaloacetate will follow. Reactions that lead to net synthesis of TCA cycle intermediates are called **anaplerotic reactions (anaplerosis)** and support gluconeogenesis because they provide for net synthesis of oxaloacetate. Reactions catalyzed by pyruvate carboxylase and glutamate dehydrogenase are good examples of anaplerotic reactions:

$$\text{pyruvate}^- + ATP^{4-} + HCO_3^- \rightarrow \text{oxaloacetate}^{2-} + ADP^{3-} + P_i^{2-} + H^+$$
$$\text{glutamate}^- + NAD(P)^+ \rightarrow \alpha\text{-ketoglutarate}^{2-} + NAD(P)H + NH_4^+ + H^+$$

On the other hand, the reaction catalyzed by glutamate–oxaloacetate transaminase (α-ketoglutarate + aspartate $\rightarrow$ glutamate + oxaloacetate) is not anaplerotic because net synthesis of a tricarboxylic acid (TCA) cycle intermediate is not accomplished; that is, generation of a TCA cycle intermediate (oxaloacetate) from aspartate is counterbalanced by conversion of a TCA cycle intermediate (α-ketoglutarate) to glutamate.

Since gluconeogenesis from amino acids imposes a nitrogen load on liver, a close relationship exists between urea synthesis and glucose synthesis from amino acids. This relationship is illustrated in Figure 14.38 for alanine. Two alanine molecules are shown being metabolized, one yielding NH_4^+ and the other aspartate, the primary substrates for the urea cycle. Aspartate leaves the mitochondrion and becomes part of the urea cycle after reacting with citrulline. The carbon of aspartate is released from the urea cycle as fumarate, which is converted to malate by cy-

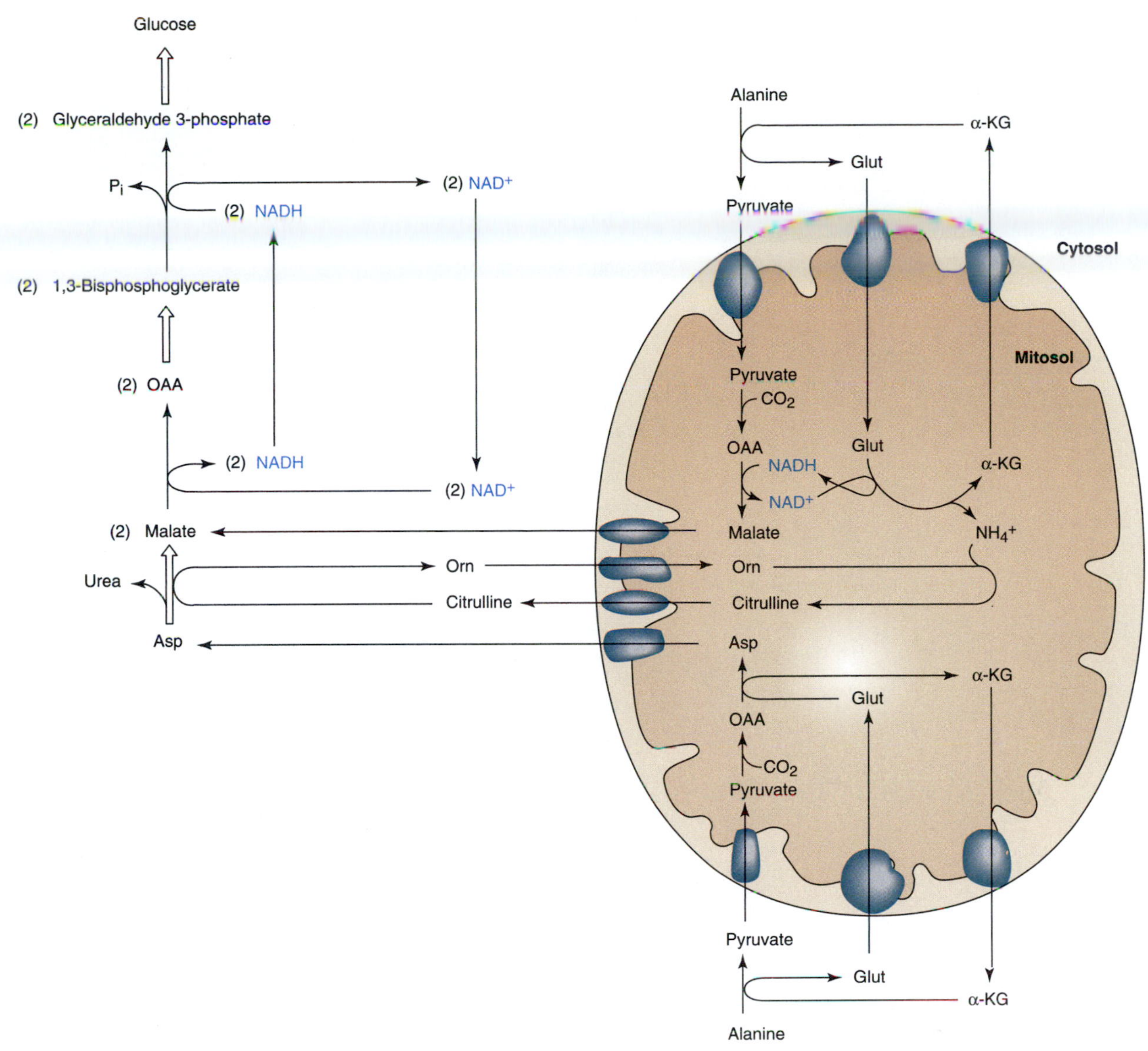

FIGURE 14.38
Pathway of gluconeogenesis from alanine and its relationship to urea synthesis.

tosolic fumarase. This malate along with another malate from the mitochondria are converted to glucose by cytosolic enzymes of gluconeogenesis. A balance is achieved between reducing equivalents (NADH) generated and those required in the cytosol and the mitosol. Summation of the reactions given in Figure 14.38 yields simply

$$2 \text{ alanine} + 10 \text{ ATP} + CO_2 \rightarrow \text{glucose} + \text{urea} + 10 \text{ ADP} + 10 \text{ P}_i$$

Leucine and **lysine** are the only amino acids that cannot function as carbon sources for net synthesis of glucose. These amino acids are **ketogenic** but not **glucogenic.** All other amino acids are classified as glucogenic or both glucogenic and ketogenic (Table 14.1). **Glucogenic amino acids** give rise to net synthesis of either pyruvate or oxaloacetate, whereas amino acids that are both glucogenic and ketogenic also yield the ketone body acetoacetate, or acetyl CoA, which is readily converted into ketone bodies. Acetyl CoA is the end product of lysine metabolism, and acetoacetate and acetyl CoA are end products of leucine metabolism. No pathway exists for converting acetoacetate or acetyl CoA into pyruvate or oxaloacetate in humans and other animals. Acetyl CoA cannot be used for net synthesis of glucose

TABLE 14.1 Glucogenic and Ketogenic Amino Acids

Glucogenic	*Ketogenic*	*Both*
Glycine	Leucine	Threonine
Serine	Lysine	Isoleucine
Valine		Phenylalanine
Histidine		Tyrosine
Arginine		Tryptophan
Cysteine		
Proline		
Hydroxyproline		
Alanine		
Glutamate		
Glutamine		
Aspartate		
Asparagine		
Methionine		

because the reaction catalyzed by the **pyruvate dehydrogenase** complex is irreversible:

$$\text{pyruvate} + \text{NAD}^+ + \text{CoASH} \rightarrow \text{acetyl CoA} + \text{NADH} + \text{CO}_2$$

It might be argued that oxaloacetate is generated from acetyl CoA by the TCA cycle through this series of reactions:

$$\text{acetyl CoA} \rightarrow \text{citrate} \rightarrow \rightarrow 2\ \text{CO}_2 + \text{oxaloacetate}$$

However, this pathway is wrong because it fails to consider the oxaloacetate required for the formation of citrate from acetyl CoA:

$$\text{acetyl CoA} + \text{oxaloacetate} \rightarrow \text{citrate} + \text{CoA}$$

Decarboxylation of citrate by the TCA cycle regenerates the oxaloacetate:

$$\text{citrate} \rightarrow \rightarrow 2\ \text{CO}_2 + \text{oxaloacetate}$$

Summing these reactions gives the sum reaction for one turn of the TCA cycle:

$$\text{acetyl CoA} \rightarrow 2\ \text{CO}_2 + \text{CoA}$$

Thus net synthesis of a TCA cycle intermediate does not occur during complete oxidation of acetyl CoA by the TCA cycle. It is therefore impossible for animals to synthesize glucose from acetyl CoA.

Glucose Can Be Synthesized from Odd-Chain Fatty Acids

Lack of an anaplerotic pathway from acetyl CoA also means that in general it is impossible to synthesize glucose from fatty acids. Most fatty acids found in humans have straight chains with an even number of carbon atoms. Their catabolism by fatty acid oxidation followed by ketogenesis or complete oxidation to CO_2 can be abbreviated as shown in Figure 14.39. Since acetyl CoA and other intermediates of even numbered fatty acid oxidation cannot be converted to oxaloacetate or any other intermediate of gluconeogenesis, it is impossible to synthesize glucose from fatty acids. An exception to this general rule applies to fatty acids with **methyl branches** (e.g., **phytanic acid,** a breakdown product of chlorophyll; see discussion of **Refsum's disease,** Clin. Corr. 16.6) and fatty acids with an odd number of carbon atoms. Catabolism of such compounds yields **propionyl CoA:**

$$\text{fatty acid with an odd number } (n) \text{ of carbon atoms} \rightarrow \frac{(n-3)}{2}\ \text{acetyl CoA} + 1\ \text{propionyl CoA}$$

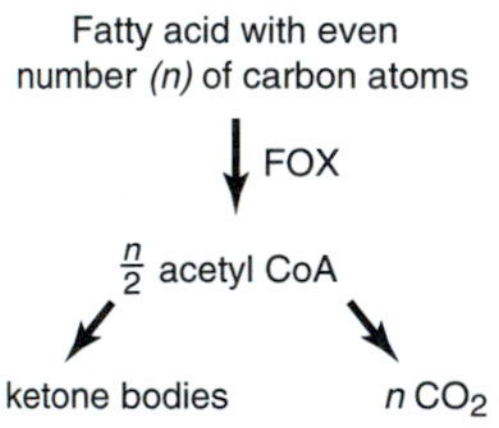

Fatty acid with even number (n) of carbon atoms

↓ FOX

$\frac{n}{2}$ acetyl CoA

↙ ↘

$\frac{n}{4}$ ketone bodies $n\ CO_2$

FFIGURE 14.39
Overview of the catabolism of fatty acids to ketone bodies and CO_2.

Propionate is a good precursor for gluconeogenesis, generating oxaloacetate by the **anaplerotic pathway** shown in Figure 14.40. The coenzyme A ester of propionate is also produced during the catabolism of valine and isoleucine and in the conversion of cholesterol into bile acids.

It is sometimes loosely stated that fat *cannot* be converted into carbohydrate (glucose) by liver. This is certainly true for fatty acids with an even number of carbon atoms. However, the term "fat" usually refers to triacylglycerols, which are composed of three O-acyl groups combined with one glycerol molecule. Hydrolysis of a triacylglycerol yields three fatty acids and glycerol, the latter compound being an excellent substrate for gluconeogenesis (Figure 14.41). Phosphorylation of glycerol by glycerol kinase produces glycerol 3-phosphate, which can be converted by glycerol 3-phosphate dehydrogenase into dihydroxyacetone phosphate, an intermediate of the gluconeogenic pathway (see Figure 14.33). The last stage of glycolysis can compete with the gluconeogenic pathway and convert dihydroxyacetone phosphate into lactate (or into pyruvate for subsequent complete oxidation to CO_2 and H_2O).

Glucose Is Synthesized from Other Sugars

Fructose

Humans consume considerable quantities of **fructose** in the form of **sucrose,** which yields glucose and fructose upon hydrolysis by sucrase in the small bowel. In the liver, fructose is phosphorylated by a special ATP-linked kinase (Figure 14.42), yielding **fructose 1-phosphate** (see Clin. Corr. 14.3). A special aldolase then cleaves fructose 1-phosphate to yield one molecule of dihydroxyacetone phosphate and one of glyceraldehyde. The latter is probably reduced to glycerol and used by the same pathway given in the previous figure. Two molecules of dihydroxyacetone phosphate obtainable from one molecule of fructose can be converted to glucose by enzymes of gluconeogenesis or, alternatively, into pyruvate or lactate by the last stage of glycolysis. In analogy to glycolysis, conversion of fructose into lactate is called **fructolysis.**

The major energy source of **spermatozoa** is fructose, formed from glucose by cells of seminal vesicles as shown in Figure 14.43. An NADPH-dependent reduction of glucose to sorbitol is followed by an NAD^+-dependent oxidation of sorbitol to fructose. Fructose is secreted from seminal vesicles in a fluid that becomes part of semen. Although the fructose concentration in human semen can exceed 10 mM,

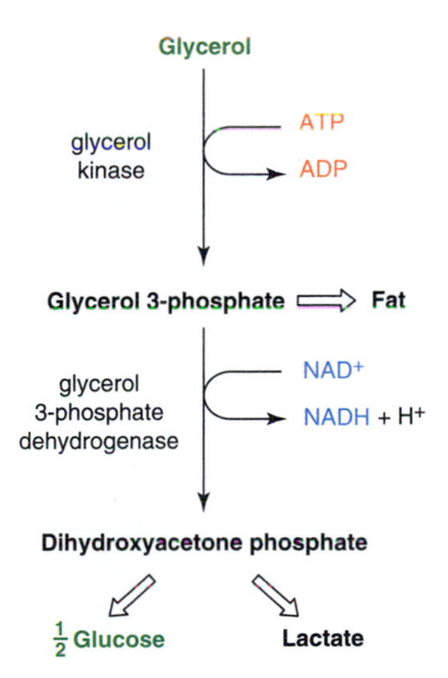

FIGURE 14.41
Pathway of gluconeogenesis from glycerol, along with competing pathways. Large arrows indicate steps of the glycolytic and gluconeogenic pathways that have been given in detail in Figures 14.6 and 14.33, respectively. The large arrow pointing to fat refers to the synthesis of triacylglycerols and glycerophospholipids.

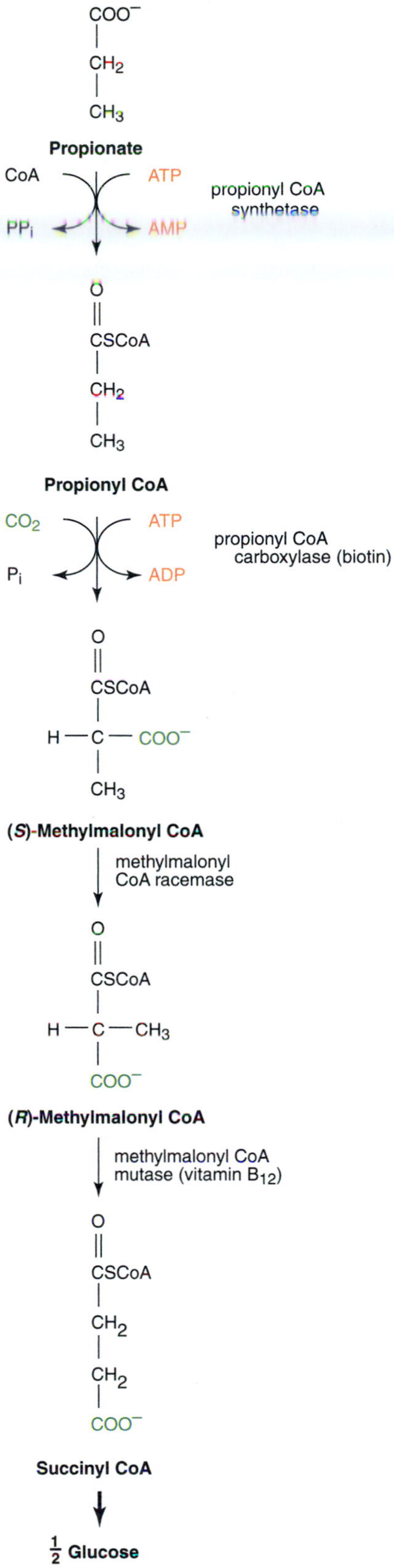

COO^-–CH_2–CH_3
Propionate
CoA, ATP → PP_i, AMP (propionyl CoA synthetase)
CH_3–CH_2–C(=O)SCoA
Propionyl CoA
CO_2, ATP → P_i, ADP (propionyl CoA carboxylase (biotin))
H–C(COO^-)(CH_3)–C(=O)SCoA
(*S*)-Methylmalonyl CoA
methylmalonyl CoA racemase
H–C(CH_3)(COO^-)–C(=O)SCoA
(*R*)-Methylmalonyl CoA
methylmalonyl CoA mutase (vitamin B_{12})
COO^-–CH_2–CH_2–C(=O)SCoA
Succinyl CoA
↓
$\frac{1}{2}$ Glucose

FIGURE 14.40
Pathway of gluconeogenesis from propionate. The large arrow refers to steps of the tricarboxylic acid cycle plus steps of lactate gluconeogenesis (see Figure 14.33).

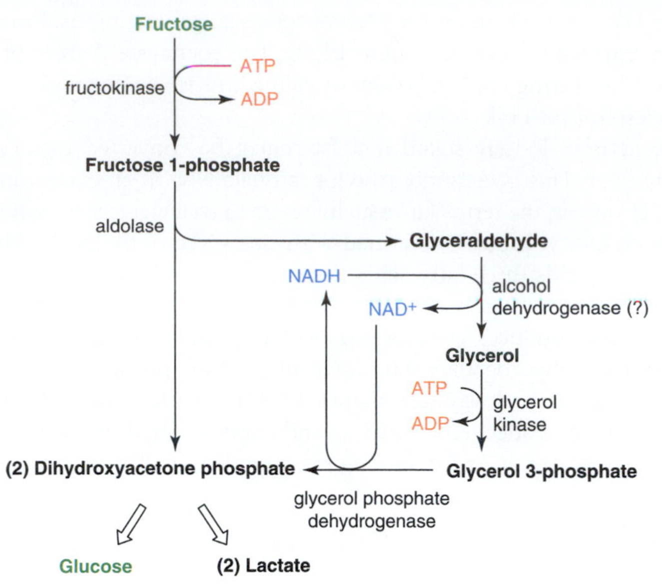

FIGURE 14.42
Pathway of glucose formation from fructose, along with the competing pathway of fructolysis.
Large arrows indicate steps of the glycolytic and gluconeogenic pathways that have been given in detail in Figures 14.6 and 14.33, respectively.

tissues that come in contact with semen utilize fructose poorly, allowing this substrate to be conserved to meet the energy demands of spermatozoa in their search for ova. Spermatozoa contain mitochondria and thus can metabolize fructose completely to CO_2 and H_2O by the combination of fructolysis and TCA cycle activity.

Galactose

Milk sugar or **lactose** is an important source of **galactose** in the human diet. Glucose formation from galactose follows the pathway shown in Figure 14.44. **UDP-glucose** serves as a recycling intermediate in the overall process of converting **galactose** into glucose. Absence of the enzyme **galactose 1-phosphate uridylyltransferase** accounts for most cases of **galactosemia** (see Clin. Corr. 15.5).

Gluconeogenesis Requires Expenditure of ATP

Synthesis of glucose is costly in terms of ATP. Six molecules are required for synthesis of one molecule of glucose from lactate; ten for synthesis of glucose from alanine. ATP needed by liver cells for glucose synthesis is provided in large part by

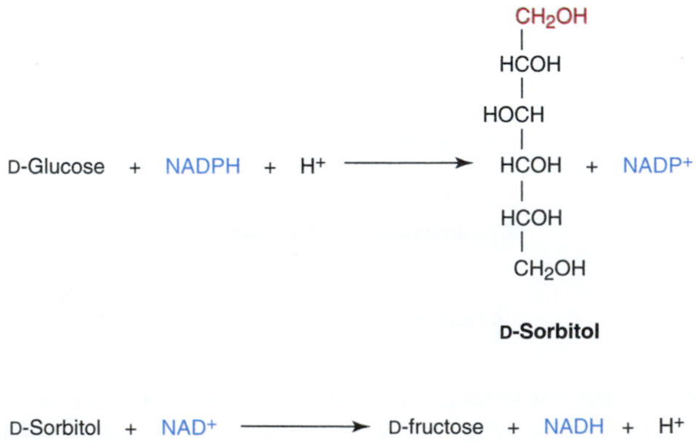

D-Glucose + NADPH + H^+ ⟶ D-Sorbitol (CH_2OH–HCOH–HOCH–HCOH–HCOH–CH_2OH) + $NADP^+$

D-Sorbitol + NAD^+ ⟶ D-fructose + NADH + H^+

FIGURE 14.43
Pathway responsible for the formation of sorbitol and fructose from glucose.

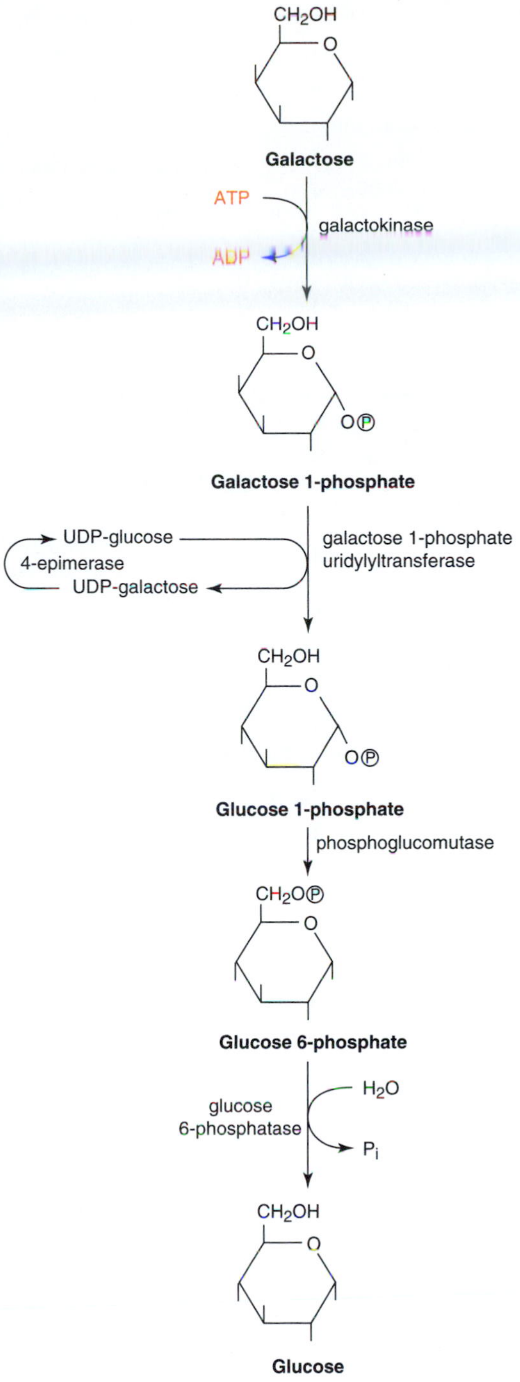

FIGURE 14.44
Pathway of glucose formation from galactose.

fatty acid oxidation. Metabolic conditions under which liver is required to synthesize glucose generally favor increased availability of fatty acids in blood. These fatty acids are oxidized by liver mitochondria to ketone bodies with concurrent production of large amounts of ATP. This ATP is used to support the energy requirements of gluconeogenesis, regardless of the substrate being used as the carbon source for the process.

Gluconeogenesis Has Several Sites of Regulation

Regulation of gluconeogenesis occurs at multiple sites (Figure 14.45). Those enzymes that are used to "go around" the irreversible steps of glycolysis are primarily involved in regulation of the pathway, that is, pyruvate carboxylase, PEP carboxyki-

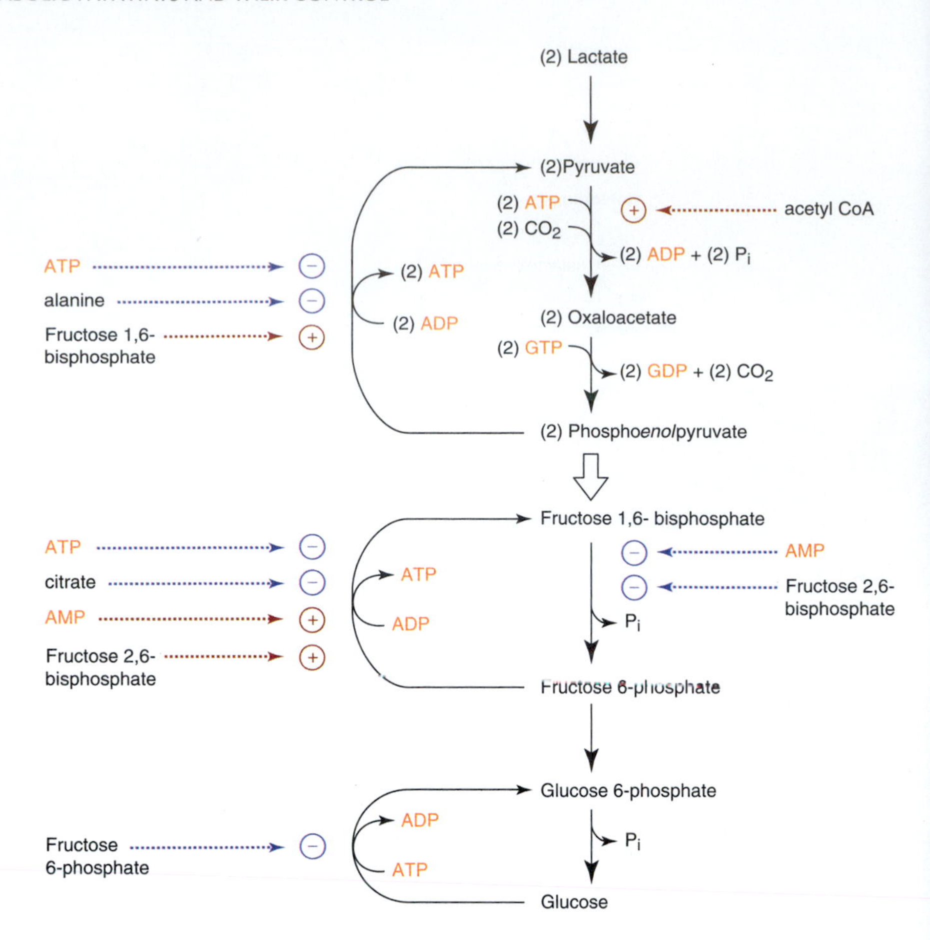

FIGURE 14.45
Important allosteric regulatory features of the gluconeogenic pathway.

nase, fructose 1,6-bisphosphatase, and glucose 6-phosphatase. Regulation of hepatic gluconeogenesis is almost the same as regulation of hepatic glycolysis. Inhibition of glycolysis at its chief regulatory sites, or repressing synthesis of enzymes involved at these sites (glucokinase, 6-phosphofructo-1-kinase, and pyruvate kinase), greatly increases effectiveness of opposing gluconeogenic enzymes. Turning on gluconeogenesis is therefore accomplished in large part by shutting off glycolysis.

Fatty acid oxidation invariably occurs hand in hand with gluconeogenesis. It promotes glucose synthesis, in part by supplying the ATP needed for the process and in part by increasing the steady-state concentrations of mitochondrial acetyl CoA and NADH. Both of these products of fatty acid oxidation are potent activators of **pyruvate dehydrogenase kinase,** which phosphorylates and inactivates the pyruvate dehydrogenase complex. This prevents pyruvate from being converted into acetyl CoA, thereby conserving pyruvate for glucose synthesis. Acetyl CoA is also a positive allosteric effector of pyruvate carboxylase, which, in combination with inhibition of the **pyruvate dehydrogenase complex,** serves to direct the carbon of pyruvate into oxaloacetate for the synthesis of glucose. The increase in oxaloaceteate due to increased pyruvate carboxylase activity along with the increase in acetyl CoA from fatty acid oxidation also promotes greater synthesis of citrate, a negative allosteric effector of 6-phosphofructo-1-kinase. A secondary effect of inhibition of 6-phosphofructo-1-kinase is a decrease in fructose 1,6-bisphosphate concentration, an activator of pyruvate kinase. This decreases the flux of PEP to pyruvate by pyruvate kinase and increases the effectiveness of the combined efforts of pyruvate carboxylase and PEP carboxykinase in conversion of pyruvate to PEP. An increase in ATP levels with the consequential decrease in AMP levels favors gluconeogenesis by way of inhibition of 6-phosphofructo-1-kinase and pyruvate kinase and activation of fructose 1,6-bisphosphatase (see Figure 14.45 and the discussion of regulation of glycolysis, p. 614). A shortage of oxygen for respiration, a shortage of fatty acids for oxidation, or any inhibition or uncoupling of oxidative phosphorylation would cause the liver to turn from gluconeogenesis to glycolysis.

Hormonal Control of Gluconeogenesis Is Critical for Homeostasis

Hormonal control of gluconeogenesis is a matter of regulating the supply of fatty acids to liver and the activities of enzymes of both the glycolytic and gluconeogenic pathways. Glucagon increases plasma fatty acids by promoting lipolysis in adipose tissue, an action opposed by insulin. The greater availability of fatty acids results in more fatty acid oxidation by liver, which promotes glucose synthesis. Insulin has the opposite effect by way of its ability to inhibit lipolysis in the adipose tissue. Glucagon and insulin also regulate gluconeogenesis directly in the liver by influencing the state of phosphorylation of the hepatic enzymes subject to covalent modification. As discussed previously (Figure 14.30), pyruvate kinase is active in the dephosphorylated mode and inactive in the phosphorylated mode. Glucagon activates adenylyl cyclase to produce cAMP, which activates protein kinase A, which, in turn, phosphorylates and inactivates pyruvate kinase. Inactivation of this glycolytic enzyme stimulates the opposing pathway gluconeogenesis by blocking the futile conversion of PEP to pyruvate. Glucagon also stimulates gluconeogenesis at the conversion of fructose 1,6-bisphosphate to fructose 6-phosphate by decreasing the concentration of **fructose 2,6-bisphosphate** in liver. Fructose 2,6-bisphosphate is an allosteric activator of 6-phosphofructo-1-kinase and an allosteric inhibitor of fructose 1,6-bisphosphatase. Glucagon via its second messenger cAMP lowers fructose 2,6-bisphosphate levels by stimulating the phosphorylation of the bifunctional enzyme 6-phosphofructo-2-kinase/fructose 2,6-bisphosphatase. Phosphorylation of this enzyme inactivates the kinase activity that makes fructose 2,6-bisphosphate from F6P but activates the phosphatase activity that hydrolyzes fructose 2,6-bisphosphate back to F6P. The consequence is a glucagon-induced fall in fructose 2,6-bisphosphate levels, leading to a decrease in activity of 6-phosphofructo-1-kinase while fructose 1,6-bisphosphatase becomes more active (Figure 14.45). The overall effect is an increased conversion of FBP to F6P and a corresponding increase in the rate of gluconeogenesis. A resulting increase in fructose 6-phosphate also favors gluconeogenesis by inhibition of glucokinase via its regulatory protein (see discussion of the regulation of glycolysis, p. 615). Insulin has effects opposite to those of glucagon by signaling activation of cAMP phosphodiesterase, inhibition of protein kinase A, and activation of phosphoprotein phosphatase (see discussion of the regulatory effects of insulin, p. 622).

Glucagon and insulin also have long-term effects on hepatic glycolysis and gluconeogenesis by **induction** and **repression** of synthesis of key enzymes of the pathways. A high glucagon/insulin ratio in blood increases the enzymatic capacity for gluconeogenesis and decreases enzymatic capacity for glycolysis in liver. A low glucagon/insulin ratio has the opposite effects. The glucagon/insulin ratio increases when gluconeogenesis is needed and decreases when glucose is plentiful from the gastrointestinal track. Glucagon signals induction of synthesis of greater quantities of PEP carboxykinase, fructose-1,6-bisphosphatase, glucose 6-phosphatase, and various aminotransferases. A model for how this occurs is given in Figure 14.46. Binding of glucagon to its plasma membrane receptor increases cAMP, which activates protein kinase A. Protein kinase A then phosphorylates a protein called the **cAMP-response element binding protein (CREB),** a trans-acting factor that in its phosphorylated form can bind to a **cAMP-response element (CRE),** a cis-acting element within the regulatory region of genes that respond to cAMP. This promotes transcription of genes encoding key gluconeogenic enzymes such as PEP carboxykinase (Figure 14.46). By a similar mechanism, but one that causes repression of gene transcription, glucagon acts to decrease the amounts of glucokinase, 6-phosphofructo-1-kinase, and pyruvate kinase. Insulin opposes the action of glucagon (Figure 14.46), acting through a signal cascade that results in activation of an **insulin-response element binding protein (IREB),** which inhibits transcription of

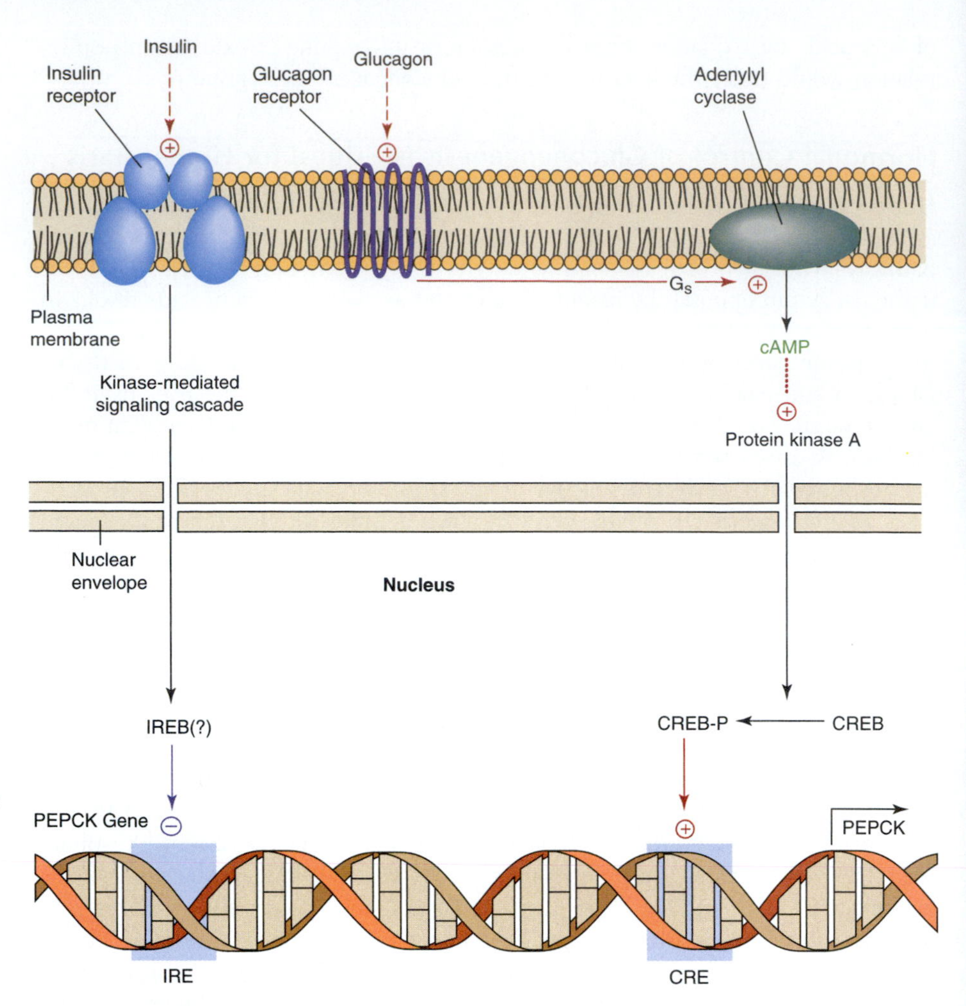

FIGURE 14.46
Glucagon promotes transcription of the gene that encodes PEP carboxykinase.
Abbreviations: PEPCK, PEP carboxykinase; CRE, cAMP-response element; CREB, cAMP-response element binding protein; IRE, insulin-response element; IREB, insulin-response element binding protein.

genes encoding key gluconeogenic enzymes by binding to an **insulin-response element (IRE)** in the regulatory region of such genes. When glucose synthesis is not needed, synthesis of key gluconeogenic enzymes is turned off and synthesis of key glycolytic enzymes is turned on as a consequence of a decrease in the blood glucagon/insulin ratio.

Ethanol Ingestion Inhibits Gluconeogenesis

Ethanol inhibits gluconeogenesis by liver (see Clin. Corr. 14.10). It is oxidized primarily in liver with production of a large load of reducing equivalents that must be transported into the mitochondria by the **malate–aspartate shuttle.** This excess NADH in the cytosol creates problems for liver gluconeogenesis because it forces the equilibrium of the lactate dehydrogenase- and malate dehydrogenase-catalyzed reactions in the directions of lactate and malate formation, respectively:

$$\underset{\text{Ethanol}}{CH_3CH_2OH} + NAD^+ \rightarrow \underset{\text{Acetaldehyde}}{CH_3\overset{\overset{\displaystyle O}{\|}}{C}H} + NADH + H^+$$

$$\text{pyruvate} + NADH + H^+ \rightarrow \text{lactate} + NAD^+$$

$$\text{Sum:}\quad \text{ethanol} + \text{pyruvate} \rightarrow \text{acetaldehyde} + \text{lactate}$$

or

$$\text{oxaloacetate} + NADH + H^+ \rightarrow \text{malate} + NAD^+$$

$$\text{Sum:}\quad \text{ethanol} + \text{oxaloacetate} \rightarrow \text{acetaldehyde} + \text{malate}$$

This inhibits glucose synthesis by limiting the amounts of pyruvate and oxaloacetate available for the reactions catalyzed by pyruvate carboxylase and PEP carboxykinase, respectively.

14.6 | GLYCOGENOLYSIS AND GLYCOGENESIS

Glycogen, a Storage Form of Glucose, Serves as a Ready Source of Energy

Glycogenolysis refers to breakdown of glycogen to glucose or glucose-6-phosphate; and glycogenesis to synthesis of glycogen. These processes occur in almost every tissue but especially in muscle and liver because of the greater importance of glycogen as a stored fuel in these tissues. The **liver** has tremendous capacity for storing glycogen. In the well-fed human, **liver glycogen** content can account for as much as 10% of wet weight of this organ. **Muscle** stores less when expressed on the same basis—a maximum of only 1–2% of its wet weight. However, the average person has more muscle than liver, adding up to about twice as much total muscle glycogen as liver glycogen.

Glycogen is stored in muscle and liver for quite different reasons. **Muscle glycogen** serves as a fuel reserve for the synthesis of ATP within that tissue, whereas liver glycogen functions as a glucose reserve for the maintenance of blood glucose concentrations. Liver glycogen levels vary greatly in response to the intake of food, accumulating to high levels shortly after a meal and then decreasing slowly as it is mobilized to help maintain a nearly constant blood glucose level (see Figure 14.47). Liver glycogen is called into play between meals and to a greater extent during the nocturnal fast. In both humans and the rat, the store of liver glycogen lasts somewhere between 12 and 24 h during fasting, depending greatly, of course, on whether the individual under consideration is caged or running wild.

Muscle glycogen is a source of ATP for increased muscular activity. Most of the glucose of glycogen stored in this tissue is consumed within muscle cells without formation of free glucose as an intermediate. However, because of a special feature of glycogen catabolism to be discussed below, about 8% of muscle glycogen is converted into free glucose within the tissue. Some of this glucose may be released into the bloodstream, but most is metabolized by glycolysis in muscle. Since muscle lacks glucose 6-phosphatase, and most free glucose formed during glycogen breakdown is further catabolized, muscle glycogen is not of quantitative importance in maintenance of blood glucose levels in the fasting state. In contrast, conversion of liver glycogen to glucose by glycogenolysis and glucose 6-phosphatase is of critical importance as a source of blood glucose in the fasting state. Conversion of glucose

CLINICAL CORRELATION 14.10

Hypoglycemia and Alcohol Intoxication

Consumption of alcohol, especially by an undernourished person, can cause hypoglycemia. The same effect can result from drinking alcohol after strenuous exercise. In both cases the hypoglycemia results from the inhibitory effects of alcohol on hepatic gluconeogenesis and thus occurs under circumstances of hepatic glycogen depletion. The problem is caused by the NADH produced during the metabolism of alcohol. The liver simply cannot handle the reducing equivalents provided by ethanol oxidation fast enough to prevent metabolic derangements. The extra reducing equivalents block the conversion of lactate to glucose and promote the conversion of alanine into lactate, resulting in considerable lactate accumulation in the blood. Since lactate has no place to go, lactic acidosis (see Clin. Corr. 14.5) can develop, although it is usually mild.

Low doses of alcohol cause impaired motor and intellectual performance; high doses have a depressant effect that can lead to stupor and anesthesia. Low blood sugar can contribute to these undesirable effects of alcohol. What is more, a patient may be thought to be inebriated when in fact the patient is suffering from hypoglycemia that may lead to irreversible damage to the central nervous system. Children are highly dependent on gluconeogenesis while fasting, and accidental ingestion of alcohol by a child can produce severe hypoglycemia (see Clin. Corr. 14.9).

Krebs, H. A., Freedland, R. A., Hems, R., and Stubbs, M. Inhibition of hepatic gluconeogenesis by ethanol. *Biochem. J.* 112:117, 1969; and Service, F. J. Hypoglycemia. *Med. Clin. North Am.* 79:1, 1995.

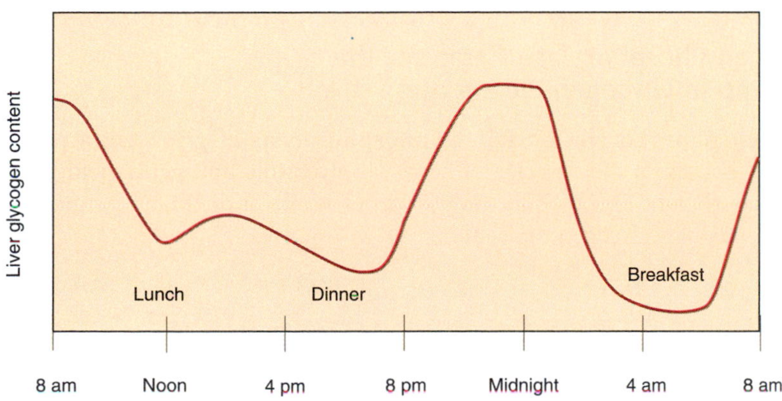

FIGURE 14.47
Variation of liver glycogen levels between meals and during the nocturnal fast.

to glycogen in muscle plays an important role in lowering blood glucose levels elevated by a high-carbohydrate meal. Glycogenesis in liver also contributes to the lowering of blood glucose but may be less important in this regard than glycogen synthesis in muscle.

Exercise of a muscle triggers mobilization of muscle glycogen for formation of ATP. The yield of ATP and the fate of the carbon of glycogen depend on whether the muscle fiber is "white" or "red." **Red muscle fibers** are supplied with a rich blood flow, contain large amounts of myoglobin, and are packed with mitochondria. Glycogen mobilized within these cells is converted into pyruvate, which, because of the availability of O_2 and mitochondria, can be converted into CO_2 and H_2O. In contrast, **white muscle fibers** have less myoglobin and fewer mitochondria. Glycogenolysis within this tissue supplies substrate for glycolysis, with the end product being primarily lactate. White muscle fibers have enormous capacity for glycogenolysis and glycolysis, much more than red muscle fibers. Since their glycogen stores are limited, however, white muscle fibers can only function at full capacity for relatively short periods of time. Breast muscle and the heart of chicken are good examples of white and red muscles, respectively. The heart has to beat continuously and has many mitochondria and a rich supply of blood via the coronary arteries. The heart stores glycogen to be used when a greater work load is imposed. Breast muscle of chicken is not continuously carrying out work. Its important function is to enable the chicken to fly rapidly for short distances, as in fleeing from predators (or amorous roosters). Because glycogen can be mobilized so rapidly, breast muscle is designed for maximal activity for a relatively short period of time. Although white and red muscles are readily recognizable in the chicken, most skeletal muscles of the human body are composed of a mixture of red and white fibers, which provide the muscle with both rapid and sustained capacity for contraction. The distribution of white and red muscle fibers in cross sections of a human skeletal muscle is demonstrated by special staining procedures in Figure 14.48.

Glycogen granules are abundant in liver of well-fed animals but are virtually absent from liver of the 24-h-fasted animal (Figure 14.49). Heavy exercise causes the same loss of glycogen granules in muscle fibers. These granules of glycogen correspond to clusters of glycogen molecules, the molecular weights of which can approach 2×10^7 Da. Glycogen is composed of glucosyl residues, the majority of which are linked together by α-1,4-glycosidic linkages (Figure 14.50). Branches also occur in the glycogen molecule as a result of frequent α-1,6-glycosidic linkages (Figure 14.50). A limb of the glycogen "tree" (see Figure 14.51) is characterized by branches at every fourth glucosyl residue within the more central core of the molecule. These branches occur much less frequently in outer regions of the molecule. An interesting question, which we shall attempt to answer below, is why this polymer is constructed with so many intricate branches and loose ends. Glycogen certainly stands in contrast to proteins and nucleic acids in this regard but, of course, it is a storage form of fuel and never has to catalyze a reaction nor convey information within a cell.

Glycogen Phosphorylase Catalyzes the First Step in Glycogen Degradation

Glycogen phosphorylase catalyzes **phosphorolysis** of glycogen, a reaction in which P_i is used in the cleavage of an α-1,4-glycosidic linkage to yield glucose 1-phosphate (Figure 14.52). This always occurs at one of the many terminal, nonreducing ends of a glycogen molecule:

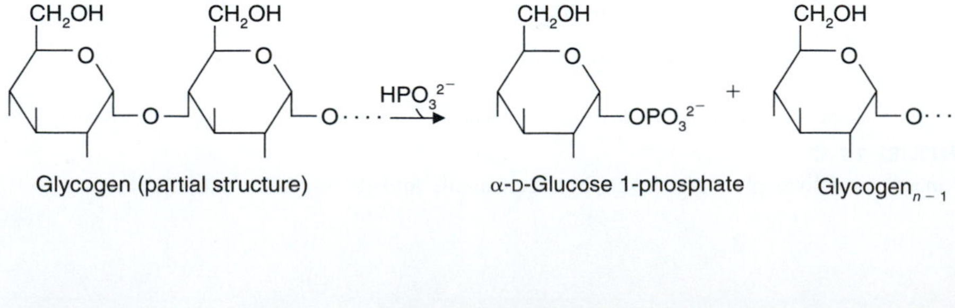

$$\text{Glycogen (partial structure)} \xrightarrow{HPO_3^{2-}} \alpha\text{-D-Glucose 1-phosphate} + \text{Glycogen}_{n-1}$$

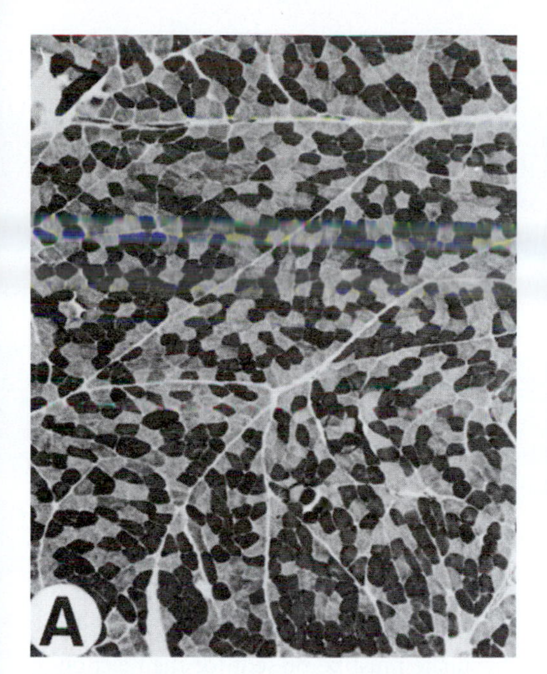

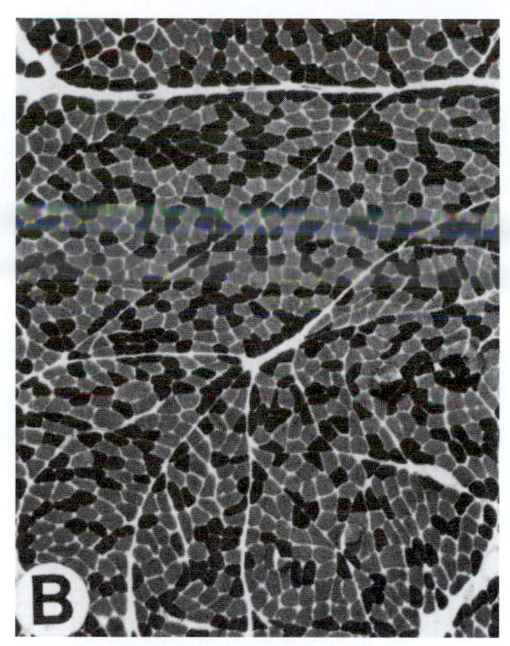

FIGURE 14.48
Cross section of human skeletal muscle showing red and white muscle fibers. Sections were stained for NADH diaphorase activity in (A); for ATPase activity in (B). Red fibers are dark and white fibers are light in (A); vice versa in (B).
Pictures generously provided by Dr. Michael H. Brooke of the Jerry Lewis Neuromuscular Research Center, St. Louis, Missouri.

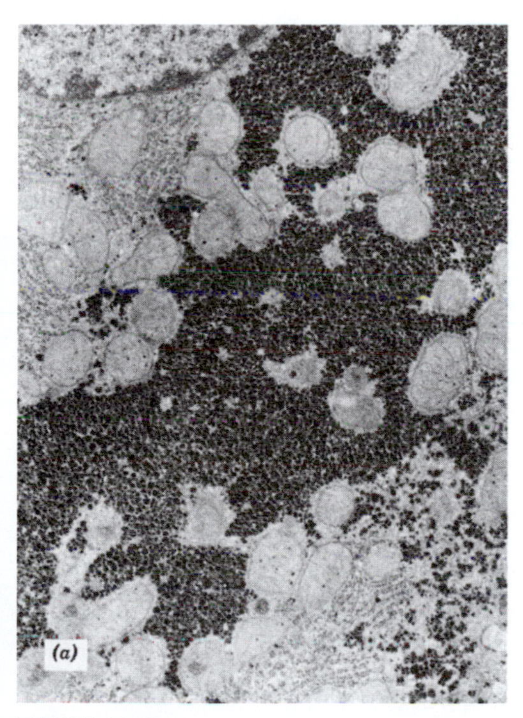

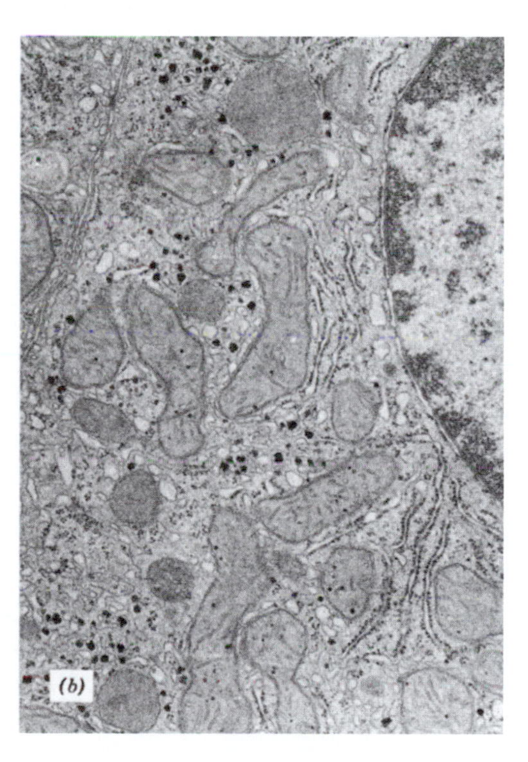

FIGURE 14.49
Electron micrographs showing glycogen granules (darkly stained material) in the liver of a well-fed rat (*a*) and the relative absence of such granules in the liver of a rat starved for 24 h (*b*).
Micrographs generously provided by Dr. Robert R. Cardell of the Department of Anatomy at the University of Cincinnati.

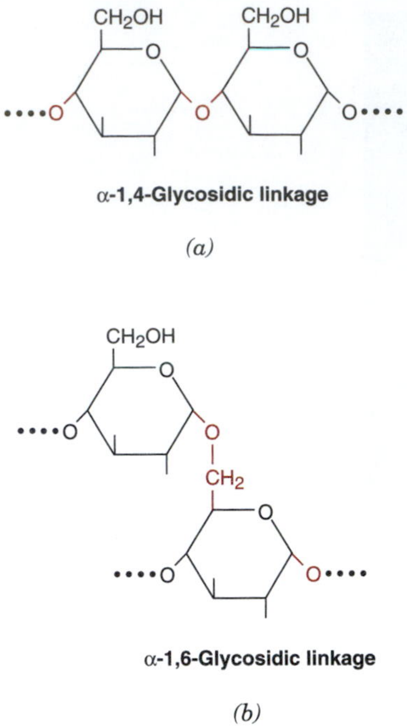

The reaction catalyzed by glycogen phosphorylase should be distinguished from that catalyzed by α-amylase, which degrades glycogen and starch in the gut (see p. 1101). α-Amylase acts by simple hydrolysis, using water rather than inorganic phosphate to cleave α-1,4-glycosidic bonds. Although glycogen may contain up to 100,000 glucose residues, its structure can be abbreviated simply as $(\text{glucose})_n$, where *n* refers to the number of glucosyl residues in the molecule. The reaction catalyzed by glycogen phosphorylase can then be written as

$$(\text{glucose})_n + \text{P}_i^{2-} \rightarrow (\text{glucose})_{n-1} + \alpha\text{-D-glucose 1-phosphate}^{2-}$$

The next step of glycogen degradation is catalyzed by **phosphoglucomutase:**

$$\text{glucose 1-phosphate} \rightarrow \text{glucose 6-phosphate}$$

This is a near-equilibrium reaction under intracellular conditions, allowing it to function in both glycogen degradation and synthesis. Like phosphoglycerate mutase (p. XXX), a bisphosphate compound is an obligatory intermediate:

$$\text{E-P} + \text{glucose 1-phosphate} \rightarrow \text{E} + \text{glucose 1,6-bisphosphate}$$
$$\text{E} + \text{glucose 1,6-bisphosphate} \rightarrow \text{E-P} + \text{glucose 6-phosphate}$$
$$\text{Sum:} \quad \text{glucose 1-phosphate} \rightarrow \text{glucose 6-phosphate}$$

A catalytic amount of glucose 1,6-bisphosphate must be present for the reaction to occur. It is produced in small quantities for this specific purpose by an enzyme called **phosphoglucokinase:**

$$\text{glucose 6-phosphate} + \text{ATP} \rightarrow \text{glucose 1,6-bisphosphate} + \text{ADP}$$

The next enzyme involved in glycogenolysis depends on the tissue under consideration (Figure 14.52). In liver, glucose 6-phosphate produced by glycogenolysis is hydrolyzed by **glucose 6-phosphatase** to give free glucose:

$$\text{glucose 6-phosphate}^{2-} + \text{H}_2\text{O} \rightarrow \text{glucose} + \text{P}_i^{2-}$$

α-1,4-Glycosidic linkage

(a)

α-1,6-Glycosidic linkage

(b)

FIGURE 14.50
Two types of linkage between glucose molecules are present in glycogen.

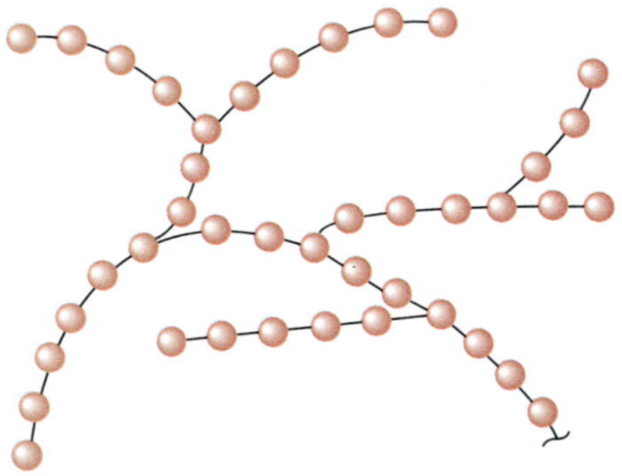

FIGURE 14.51
The branched structure of glycogen.

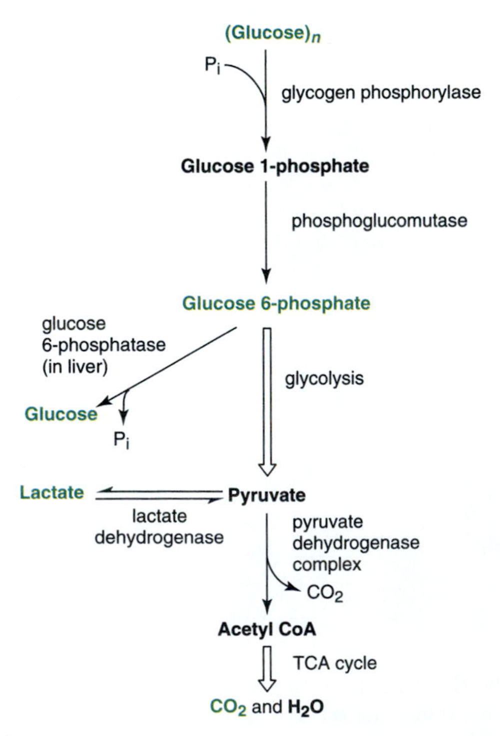

FIGURE 14.52
Glycogenolysis and the fate of glycogen degraded in liver versus its fate in peripheral tissues.

Lack of this enzyme or of the translocase that transports G6P into the endoplasmic reticulum (Figure 14.37) results in type I glycogen storage disease (see Clin. Corr. 14.11). The overall balanced equation for removal of one glucosyl residue from glycogen in liver by glycogenolysis is then

$$(\text{glucose})_n + H_2O \rightarrow (\text{glucose})_{n-1} + \text{glucose}$$

In other words, glycogenolysis in liver involves phosphorolysis but, because the phosphate ester is cleaved by a phosphatase, the overall reaction adds up to hydrolysis of glycogen. ATP is neither used nor produced in the process.

In peripheral tissues the G6P generated by glycogenolysis is used by glycolysis, leading primarily to the generation of lactate in white muscle fibers and primarily to complete oxidation to CO_2 in red muscle fibers. Since no ATP had to be in-

CLINICAL CORRELATION 14.11

Glycogen Storage Diseases

There are a number of well-characterized glycogen storage diseases, all due to inherited defects of one or more of the enzymes involved in the synthesis and degradation of glycogen. The liver is usually the tissue most affected, but heart and muscle glycogen metabolism can also be defective.

Chen, Y.-T. and Burchell, A. Glycogen storage diseases. In: C. R. Scriver, A. L. Beaudet, W. S. Sly, and D. Valle (Eds.), *The Metabolic and Molecular Basis of Inherited Disease,* 7th ed. New York: McGraw-Hill, 1995, p. 935.

Von Gierke's Disease

The most common glycogen storage disease, referred to as type I or von Gierke's disease, is caused by a deficiency of liver, intestinal mucosa, and kidney glucose 6-phosphatase. Thus diagnosis by small bowel biopsy is possible. Patients with this disease can be further subclassified into those lacking the glucose 6-phosphatase enzyme per se (type Ia) and those lacking the glucose 6-phosphate translocase (type Ib) (see Figure 14.37). A genetic abnormality in glucose 6-phosphate hydrolysis occurs in only about 1 person in 200,000 and is transmitted as an autosomal recessive trait. Clinical manifestations include fasting hypoglycemia, lactic acidemia, hyperlipidemia, and hyperuricemia with gouty arthritis. The fasting hypoglycemia is readily explained as a consequence of the glucose 6-phosphatase deficiency, the enzyme required to obtain glucose from liver glycogen and gluconeogenesis. The liver of these patients does release some glucose by the action of the glycogen debrancher enzyme. The lactic acidemia occurs because the liver cannot use lactate effectively for glucose synthesis. In addition, the liver inappropriately produces lactic acid in response to glucagon. This hormone should trigger glucose release without lactate production; however, the opposite occurs because of the lack of glucose 6-phosphatase. Hyperuricemia results from increased purine degradation in the liver; hyperlipidemia because of increased availability of lactic acid for lipogenesis and lipid mobilization from the adipose tissue caused by high glucagon levels in response to hypoglycemia. The manifestations of von Gierke's disease can greatly be diminished by providing carbohydrate throughout the day to prevent hypoglycemia. During sleep this can be done by infusion of carbohydrate into the gut by a nasogastric tube.

Cori, G. T. and Cori, C. F. Glucose-6-phosphatase of the liver in glycogen storage disease. *J. Biol. Chem.* 199:661, 1952.

Pompe's Disease

Type II glycogen storage disease or Pompe's disease is caused by the absence of α-1,4-glucosidase (or acid maltase), an enzyme normally found in lysosomes. The absence of this enzyme leads to the accumulation of glycogen in virtually every tissue. This is somewhat surprising, but lysosomes take up glycogen granules and become defective with respect to other functions if they lack the capacity to destroy the granules. Because other synthetic and degradative pathways of glycogen metabolism are intact, metabolic derangements such as those in von Gierke's disease are not seen. The reason for extralysosomal glycogen accumulation is unknown. Massive cardiomegaly occurs and death results at an early age from heart failure.

Hers, H. G. α-Glucosidase deficiency in generalized glycogen storage disease (Pompe's disease). *Biochem. J.* 86:11, 1963.

Cori's Disease

Also called type III glycogen storage disease, Cori's disease is caused by a deficiency of the glycogen debrancher enzyme. Glycogen accumulates because only the outer branches can be removed from the molecule by phosphorylase. Hepatomegaly occurs, but diminishes with age. The clinical manifestations are similar to but much milder than those seen in von Gierke's disease, because gluconeogenesis is unaffected, and hypoglycemia and its complications are less severe.

Van Hoff, F. and Hers, H. G. The subgroups of type III glycogenosis. *Eur. J. Biochem.* 2:265, 1967.

McArdle's Disease

Also called the type V glycogen storage disease, McArdle's disease is caused by an absence of muscle phosphorylase. Patients suffer from painful muscle cramps and are unable to perform strenuous exercise, presumably because muscle glycogen stores are not available to the exercising muscle. Thus the normal increase in plasma lactate (released from the muscle) following exercise is absent. The muscles are probably damaged because of inadequate energy supply and glycogen accumulation. Release of muscle enzymes creatine phosphokinase and aldolase and of myoglobin is common; elevated levels of these substances in the blood suggest a muscle disorder.

McArdle, B. Myopathy due to a defect in muscle glycogen breakdown. *Clin. Sci.* 10:13, 1951.

vested to produce G6P obtained from glycogen, the overall equation for glycogenolysis followed by glycolysis is

$$(\text{glucose})_n + 3\ \text{ADP}^{3-} + 3\ \text{P}_i^{2-} + \text{H}^+ \rightarrow (\text{glucose})_{n-1} + 2\ \text{lactate}^- + 3\ \text{ATP}^{4-} + 2\ \text{H}_2\text{O}$$

Debranching Enzyme Is Required for Degradation of Glycogen

Glycogen phosphorylase, the first enzyme involved in glycogen degradation, is specific for α-1,4-glycosidic linkages. However, it stops attacking α-1,4-glucosidic linkages four glucosyl residues from an α-1,6-branch point. A glycogen molecule that has been degraded by phosphorylase to the limit caused by the branches is called phosphorylase-limit dextrin. The action of a **debranching enzyme** is what allows glycogen phosphorylase to continue to degrade glycogen. Debranching enzyme is a bifunctional enzyme that catalyzes two reactions necessary for debranching of glycogen. The first is a **4-*α*-D-glucanotransferase** activity in which a strand of three glucosyl residues is removed from a four glucosyl residue branch of the glycogen molecule (Figure 14.53). The strand remains covalently attached to the enzyme until it can be transferred to a free 4-hydroxyl of a glucosyl residue at the end of the same or an adjacent glycogen molecule. The result is a longer amylose chain with only one glucosyl residue remaining in α-1,6-linkage. This linkage is broken hy-

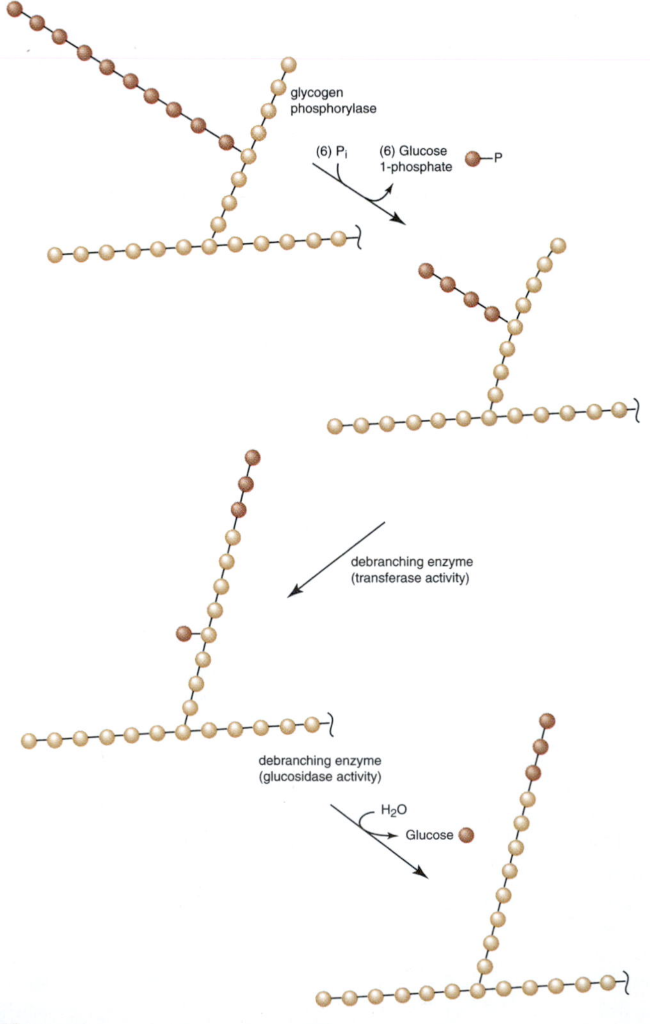

FIGURE 14.53
Action of the glycogen debranching enzyme.

drolytically by the second enzyme activity of debranching enzyme, which is its **amylo-α-1,6-glucosidase** activity:

Glycogen $\xrightarrow{H_2O}$ α-D-Glucose + Glycogen

The cooperative and repetitive action of phosphorylase and debranching enzyme results in complete phosphorolysis and hydrolysis of glycogen. **Glycogen storage diseases** result when either of these enzymes is defective. The average molecule of glycogen yields about 12 molecules of glucose 1-phosphate by action of phosphorylase for every molecule of free glucose produced by the action of debranching enzyme.

There is another, albeit quantitatively less important, pathway for glycogen degradation. A defect in this minor pathway, however, creates a major problem. As pointed out in Clinical Correlation 14.11, a **glucosidase** present in **lysosomes** degrades glycogen that enters these organelles during normal turnover of intracellular components.

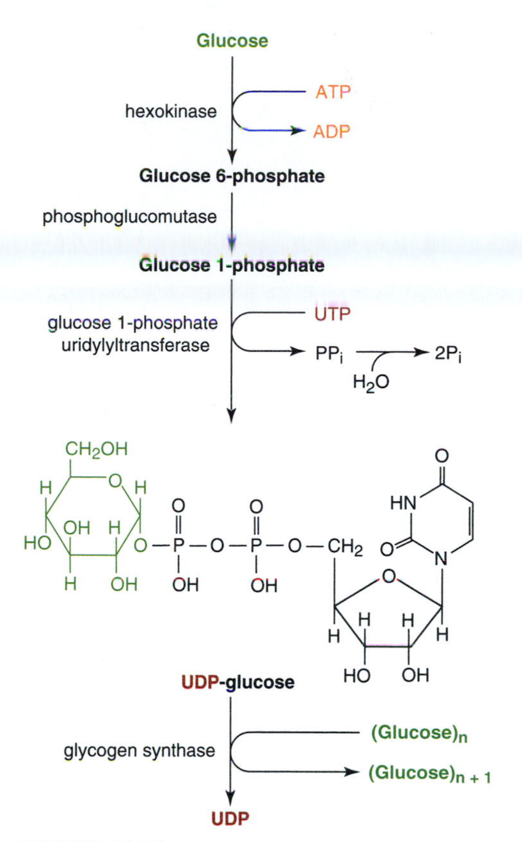

FIGURE 14.54
Pathway of glycogen synthesis.

Synthesis of Glycogen Requires Unique Enzymes

The first reaction involved in glycogen synthesis (Figure 14.54) is already familiar, being catalyzed by glucokinase in hepatic tissue and hexokinase in peripheral tissues:

$$\text{glucose} + \text{ATP} \rightarrow \text{glucose 6-phosphate} + \text{ADP}$$

Phosphoglucomutase, discussed in relation to glycogen degradation, catalyzes a readily reversible reaction as follows:

$$\text{glucose 6-phosphate} \rightarrow \text{glucose 1-phosphate}$$

A unique reaction found at the next step involves formation of UDP-glucose by action of **glucose 1-phosphate uridylyltransferase:**

$$\text{glucose 1-phosphate} + \text{UTP} \rightarrow \text{UDP-glucose} + PP_i$$

This reaction generates UDP-glucose, an "activated glucose" molecule, which is used to build the glycogen molecule. Formation of UDP-glucose is made energetically favorable and irreversible by hydrolysis of pyrophosphate by **pyrophosphatase:**

$$PP_i^{4-} + H_2O \rightarrow 2\,P_i^{2-}$$

Glycogen Synthase

Glycogen synthase catalyzes transfer of the activated glucosyl moiety of UDP-glucose to a glycogen molecule to form a new glycosidic bond between the hydroxyl group of C-1 of the activated sugar and C-4 of a glucosyl residue of the growing glycogen chain. The reducing end of glucose (C-1 of glucose is an aldehyde that can reduce other compounds during its oxidation to a carboxylic acid) is always added to a nonreducing end of the glycogen chain. The glycogen molecule, regardless of its size, theoretically has only one free reducing end tucked away within its core. UDP formed as a product of glycogen synthase is converted back to UTP by action of **nucleoside diphosphate kinase:**

$$\text{UDP} + \text{ATP} \rightleftharpoons \text{UTP} + \text{ADP}$$

Glycogen synthase creates chains of glucose molecules with α-1,4-glycosidic linkages but cannot form the α-1,6-glycosidic branches. Its action alone would only produce amylose, a straight-chain polymer of glucose with α-1,4-glycosidic linkages. Once an amylose chain of at least 11 residues has been formed, a **"branching" enzyme** comes into play. Its name is **1,4-α-glucan branching enzyme** because it removes a block of about seven glucosyl residues from a growing chain and transfers it to another chain to produce an α-1,6-linkage (see Figure 14.55). The new branch has to be introduced at least four glucosyl residues from an adjacent branch point. Thus the creation of the highly branched structure of glycogen requires the concerted efforts of glycogen synthase and branching enzyme. The overall balanced equation for glycogen synthesis by the pathway just outlined is

$$(\text{glucose})_n + \text{glucose} + 2\ \text{ATP}^{4-} + \text{H}_2\text{O} \rightarrow (\text{glucose})_{n+1} + 2\ \text{ADP}^{3-} + 2\ \text{P}_i^{2-} + 2\ \text{H}^+$$

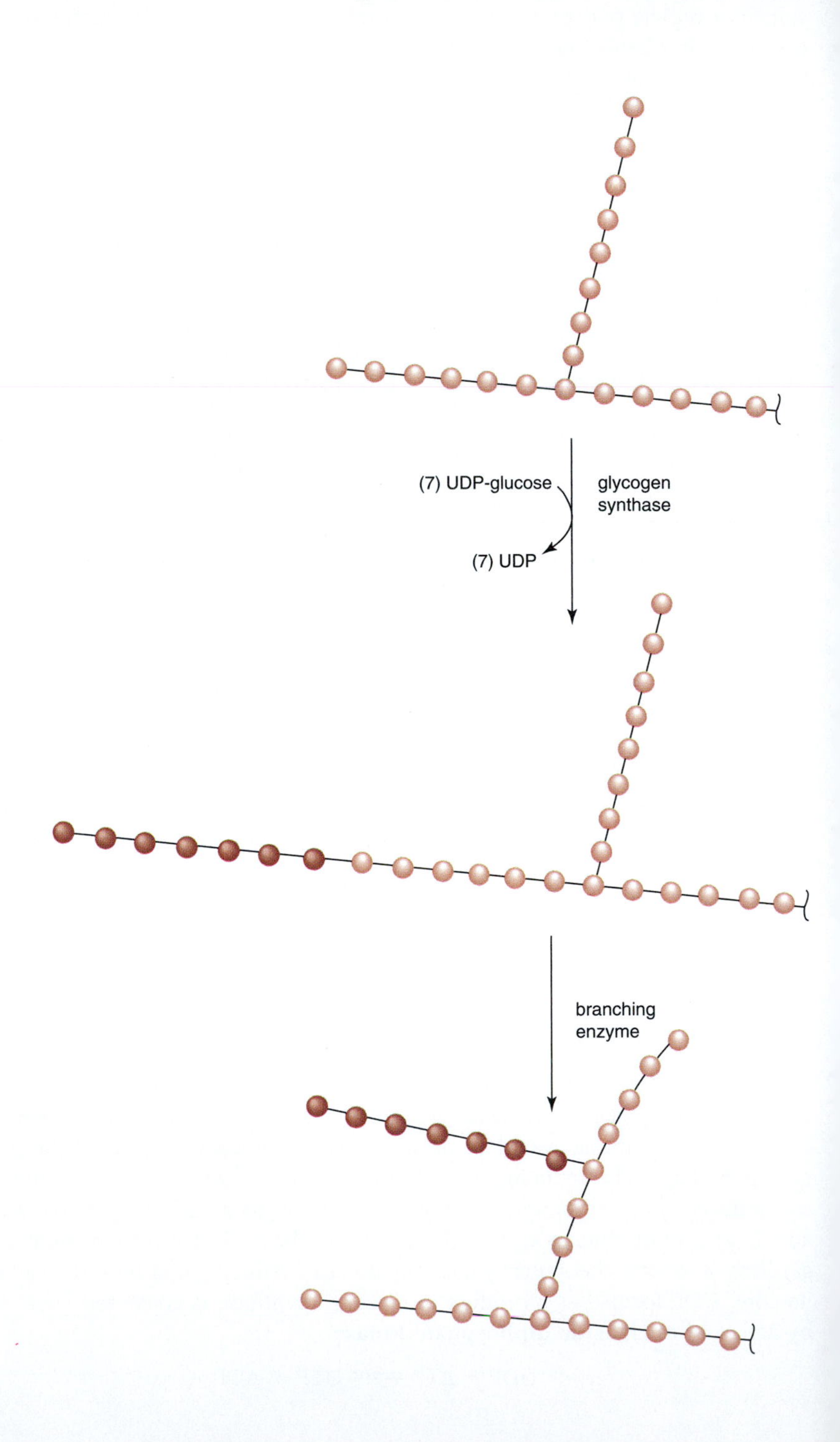

FIGURE 14.55
Action of the glycogen branching enzyme.

As noted above, combination of glycogenolysis and glycolysis yields only three molecules of ATP per glucosyl residue:

$$(\text{glucose})_n + 3\ \text{ADP}^{3-} + 3\ \text{P}_i^{2-} + \text{H}^+ \rightarrow (\text{glucose})_{n-1} + 2\ \text{lactate}^- + 3\ \text{ATP}^{4-} + 2\ \text{H}_2\text{O}$$

Thus the combination of glycogen synthesis plus glycogen degradation to lactate actually yields only one ATP:

$$\text{glucose} + \text{ADP}^{3-} + \text{P}_i^{2-} \rightarrow 2\ \text{lactate}^- + \text{ATP}^{4-} + \text{H}_2\text{O} + \text{H}^+$$

However, glycogen synthesis and degradation are normally carried out at different times in a cell. For example, white muscle fibers synthesize glycogen at rest when glucose is plentiful and less ATP is needed for muscle contraction. Glycogen is then used during periods of exertion. Although in such terms glycogen storage is not a very efficient process, it provides cells with a fuel reserve that can be very quickly and efficiently mobilized.

Special Features of Glycogen Degradation and Synthesis

Why Store Glucose as Glycogen?

The fact that glycogen is such a good fuel reserve makes it obvious why we synthesize and store glycogen in liver and muscle. But why not store our excess glucose calories entirely as fat instead of glycogen? The answer is at least threefold: (1) we do store fat, but fat cannot be mobilized nearly as rapidly as glycogen; (2) fat cannot be used as a source of energy in the absence of oxygen; and (3) fat cannot be converted to glucose to maintain blood glucose levels required by the brain. Why not just pump glucose into cells and store it as free glucose until needed? Why waste so much ATP making a polymer out of glucose? The problem is that glucose is osmotically active. It would cost ATP to "pump" glucose into a cell against a concentration gradient, and glucose would have to reach concentrations of the order of 400 mM in liver cells to match the "glucose reserve" provided by the usual liver glycogen levels. Unless balanced by outward movement of some other osmotically active compound, accumulation of such concentrations of glucose would cause uptake of considerable water and osmotic lysis of the cell. Assuming the molecular mass of a glycogen molecule is of the order of 10^7 Da, 400 mM of glucose is in effect stored at an intracellular glycogen concentration of only 0.01 μM. Storage of glucose as glycogen, therefore, creates no osmotic pressure problem for the cell.

Glycogenin Is Required as a Primer for Glycogen Synthesis

Like DNA synthesis, a **primer** is needed for glycogen synthesis. No template, however, is required. Glycogen itself is the usual primer, in that glycogen synthesis can take place by addition of glucosyl units to glycogen "core" molecules, which are almost invariably present in the cell. The outer regions of the glycogen molecule are removed and resynthesized more rapidly than the inner core. Glycogen within a cell is frequently sheared by the combined actions of glycogen phosphorylase and debranching enzyme but is seldom obliterated before glycogen synthase and branching enzyme rebuild the molecule. This begs the question why glycogen is a branched molecule with only one real beginning (the reducing end) and many branches terminating with nonreducing glucosyl units. The answer is that this gives numerous sites of attack for glycogen phosphorylase on a mature glycogen molecule and the same number of sites that function as primers for the addition of glucosyl units by glycogen synthase. If cells synthesized α-amylose, that is, an unbranched glucose polymer, there would only be one nonreducing end per molecule. This would surely make glycogen degradation and synthesis much slower. As it is, glycogen phosphorylase and glycogen synthase are found in tight association with glycogen granules in a cell. By taking up residence in the branches of the glycogen tree, both enzymes have ready access to a multitude of nonreducing sugars at the ends of the limbs. Why is a primer needed for glycogen synthesis? Because it is impossible to

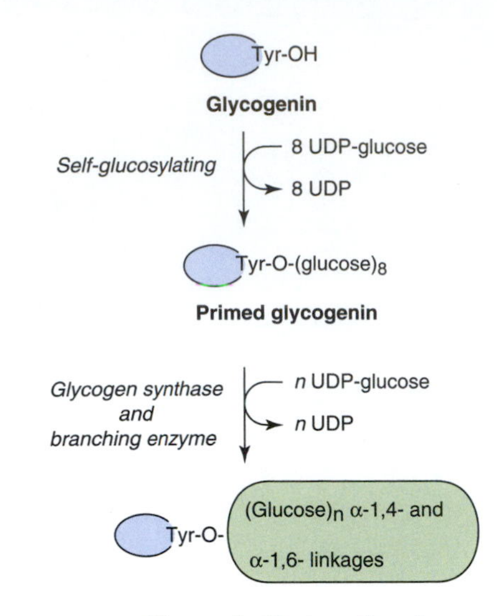

FIGURE 14.56
Glycogenin provides a primer for glycogen synthesis by glycogen synthase.
Y designates a tyrosine residue of glycogenin.

initiate glycogen synthesis with simply a glucose molecule as the acceptor of an activated glucosyl residue from UDP-glucose. Glycogen synthase has a very low K_m for large glycogen molecules, and therefore readily adds glucosyl residues to make even larger glycogen molecules. However, the K_m gets larger and larger as the glycogen molecule gets smaller and smaller. This phenomenon is so pronounced that glucose, at its physiological concentration, cannot function as a primer. This led for some time to the notion that glycogen might be immortal; that is, some glycogen might have to be handed down from one cell generation to the next in order for glycogen to be synthesized. However, it is now known that a polypeptide of 332 amino acids called **glycogenin** functions as a primer for glycogen synthesis. Glycogenin is a self-glucosylating enzyme that uses UDP-glucose to link glucose to one of its own tyrosine residues (Figure 14.56). Glycosylated glycogenin then serves as a primer for synthesis of glycogen. Alas, glycogen is not immortal.

Glycogen Limits Its Own Synthesis

If glycogen synthase becomes more efficient as the glycogen molecule gets bigger, how is synthesis of this ball of sugar curtailed? Adipocytes have an almost unlimited capacity to pack away fat—but then adipocytes have nothing else to do. Muscle cells participate in mechanical activity and liver cells carry out many processes other than glycogen synthesis. Even in the face of excess glucose, there has to be a way to limit the intracellular accumulation of glycogen. Glycogen itself inhibits glycogen synthase by a mechanism discussed later (p. 654).

Glycogen Synthesis and Degradation Are Highly Regulated Pathways

Glycogen synthase and glycogen phosphorylase are regulatory enzymes of glycogen synthesis and degradation, respectively. Both catalyze nonequilibrium reactions, and both are subject to control by allosteric effectors and covalent modification.

Regulation of Glycogen Phosphorylase

Glycogen phosphorylase is subject to allosteric activation by **AMP** and allosteric inhibition by **glucose** and **ATP** (Figure 14.57). Control by these effectors is integrated with a very elaborate control by covalent modification. Phosphorylase exists in an *a* form, which is active, and a *b* form, which has very little activity. These forms are interconverted by the actions of **phosphorylase kinase** and **phosphoprotein phosphatase** (Figure 14.57). A conformational change caused by phosphorylation transforms the enzyme into a more active catalytic state. Although phosphorylase *b* has low catalytic activity, it can greatly be stimulated by AMP. This allosteric effector has little activating effect, however, on the already active phosphorylase *a*. Thus regulation by covalent modification can be bypassed by an AMP-mediated allosteric mechanism and vice versa.

Phosphorylase kinase is responsible for phosphorylation and activation of phosphorylase (Figure 14.57). Moreover, phosphorylase kinase itself is also subject to regulation by a cyclic phosphorylation–dephosphorylation mechanism. Protein kinase A phosphorylates and activates phosphorylase kinase; phosphoprotein phosphatase in turn dephosphorylates and inactivates phosphorylase kinase. Phosphorylase kinase is a large enzyme complex (1.3×10^6 Da) composed of four subunits with four molecules of each subunit in the complex ($\alpha_4\beta_4\gamma_4\delta_4$). Catalytic activity resides with the γ subunit; α, β, and δ subunits exert regulatory control. The α and β subunits are phosphorylated in the transition from the inactive *b* form to the active *a* form of the enzyme. Protein kinase A can only exert an effect on phosphorylase via its ability to phosphorylate and activate phosphorylase kinase. Thus a bicyclic system is required for activation of phosphorylase in response to cAMP-mediated signals (Figure 14.57).

The δ subunit of phosphorylase kinase also plays a regulatory role. It corresponds to a Ca^{2+}-binding regulatory protein, called **calmodulin.** Not unique to

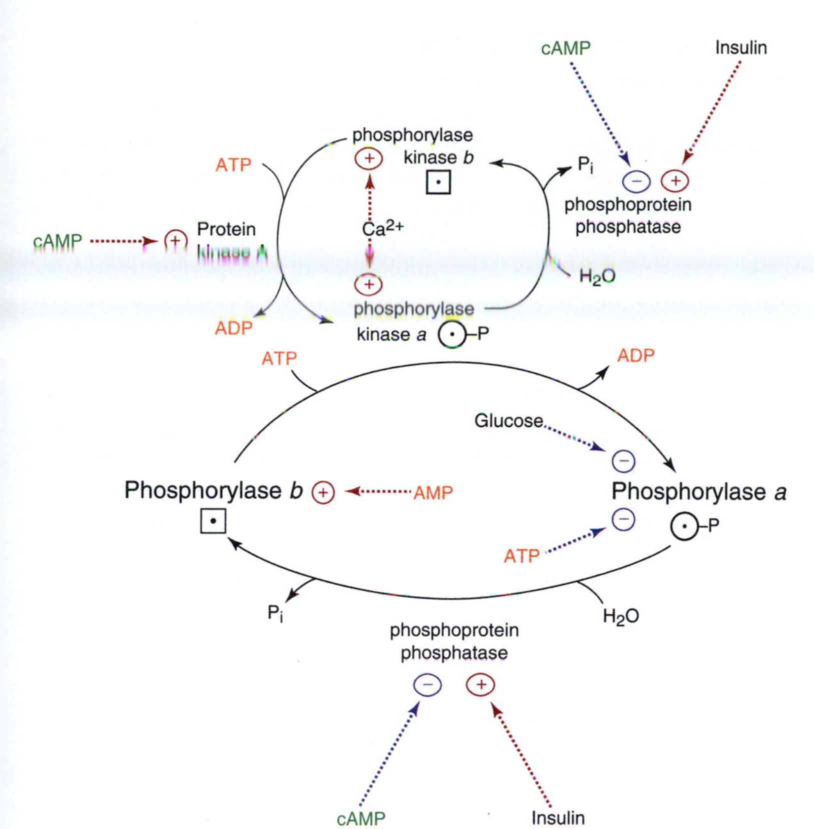

FIGURE 14.57
Regulation of glycogen phosphorylase by covalent modification.
Phosphorylation converts glycogen phosphorylase and phosphorylase kinase from their inactive *b* forms to their active *a* forms.

phosphorylase kinase, calmodulin is found in cells as the free molecule and also bound to other enzyme complexes. It functions as a Ca^{2+} receptor in the cell, responding to changes in intracellular Ca^{2+} concentration and affecting the relative activities of a number of enzyme systems. Binding of Ca^{2+} to the calmodulin subunit of phosphorylase kinase changes the conformation of the complex, making the enzyme more active with respect to the phosphorylation of phosphorylase. As shown in Figure 14.57, Ca^{2+} is an activator of both phosphorylase kinase *a* and phosphorylase kinase *b*. Maximum activation of phosphorylase kinase requires both phosphorylation of specific serine residues of the enzyme and interaction of Ca^{2+} with the calmodulin subunit of the enzyme. This is one mechanism by which Ca^{2+} functions as an important "second messenger" of hormone action, as will be discussed below.

Activation of phosphorylase kinase by phosphorylation and Ca^{2+} will have a substantial effect on the activity of glycogen phosphorylase. It is equally obvious, however, that turning off the phosphoprotein phosphatase that modulates the phosphorylation states of both phosphorylase kinase and glycogen phosphorylase (Figure 14.57) could achieve the same effect. Ultimate control of glycogen phosphorylase would involve the reciprocal regulation of **phosphoprotein phosphatase** and phosphorylase kinase activities. Although numerous details remain to be understood, there is evidence that activities of phosphoprotein phosphatase and phosphorylase kinase are controlled in a reciprocal manner. Regulation of phosphoprotein phosphatase activity may be linked to cAMP (see p. 655). The important point in Figure 14.57 is that hormones that increase cAMP levels, such as glucagon and epinephrine, promote activation of glycogen phosphorylase by signaling activation of phosphorylase kinase and inactivation of phosphoprotein phosphatase. On the other hand, insulin, though a kinase-mediated signal cascade, exerts the opposite effect on phosphorylase by promoting activation of phosphoprotein phosphatase activity.

The Cascade that Regulates Glycogen Phosphorylase Amplifies a Small Signal into a Very Large Effect

There is a good reason for the existence of the bicyclic control system for phosphorylation of glycogen phosphorylase. It provides a tremendous **amplification** mechanism of a very small initial signal. Activation of **adenylyl cyclase** by one molecule of epinephrine causes formation of many molecules of cAMP. Each cAMP molecule activates a protein kinase A molecule, which in turn activates many molecules of phosphorylase kinase and inhibits many molecules of phosphoprotein phosphatase. In turn phosphorylase kinase phosphorylates many molecules of glycogen phosphorylase, which in turn catalyzes phosphorolysis of many glycosidic bonds of glycogen. An elaborate amplification system is provided, therefore, in which the signal provided by just a few molecules of hormone is amplified into production of an enormous number of glucose 1-phosphate molecules. If each step represents, for argument's sake, an amplification factor of 100, then a total of four steps would result in an amplification of 100 million! This system is so rapid, in large part because of this amplification mechanism, that all of the stored glycogen of white muscle fibers could be mobilized completely within just a few seconds.

Regulation of Glycogen Synthase

Glycogen synthase has to be active for glycogen synthesis and inactive for glycogen degradation. The combination of the reactions catalyzed by glycogen synthase, glycogen phosphorylase, glucose 1-phosphate uridylyltransferase, and nucleoside diphosphate kinase adds up to a futile cycle with the overall equation ATP + H_2O $\rightarrow$ ADP + P_i. Hence glycogen synthase needs to be turned off when glycogen phosphorylase is turned on, and vice versa.

Activation of glycogen synthase by glucose 6-phosphate, an allosteric effector, is probably of physiological significance under some circumstances (Figure 14.58). However, as with glycogen phosphorylase, this mode of control is integrated with regulation by covalent modification (Figure 14.58). Glycogen synthase exists in two forms. One is designated the **D form** because it is dependent on the presence of G6P for activity. The other is designated the **I form** because its activity is independent of the presence of G6P. The D form corresponds to the *b* or inactive form of the enzyme, the I form to the *a* or active form of the enzyme. Phosphorylation of glycogen synthase is catalyzed by several different kinases, which in turn are regulated by second messengers of hormone action, including cAMP, Ca^{2+}, diacylglycerol, and probably yet to be identified signaling mechanisms. Each of the protein kinases shown in Figure 14.58 is capable of catalyzing the phosphorylation and contributing to inactivation of glycogen synthase. Although glycogen synthase is a simple tetramer (α_4) of only one subunit type (MW 85,000 Da), it can be phosphorylated on at least nine different serine residues. Eleven different protein kinases have been identified that can phosphorylate glycogen synthase. This stands in striking contrast to glycogen phosphorylase, which is regulated by phosphorylation of one site by one specific kinase.

Cyclic AMP is an extremely important intracellular signal for reciprocally controlling glycogen synthase (Figure 14.58) and glycogen phosphorylase (Figure 14.57). An increase in cAMP signals activation of glycogen phosphorylase and inactivation of glycogen synthase via activation of protein kinase A and inhibition of phosphoprotein phosphatase. Ca^{2+} likewise can influence the phosphorylation state of both enzymes and reciprocally regulate their activity via its effects on phosphorylase kinase. Two cAMP-independent, Ca^{2+}-activated protein kinases have been identified that also may have physiological significance. One of these is a **calmodulin-dependent protein kinase** and the other a Ca^{2+}- and phospholipid-dependent protein kinase **(protein kinase C).** Both enzymes phosphorylate glycogen synthase, but neither can phosphorylate glycogen phosphorylase. Protein kinase C requires phospholipid, **diacylglycerol,** and Ca^{2+} for full activity. There is considerable interest in protein kinase C because tumor-promoting agents called **phorbol esters** have been found to mimic diacylglycerol as activators of this en-

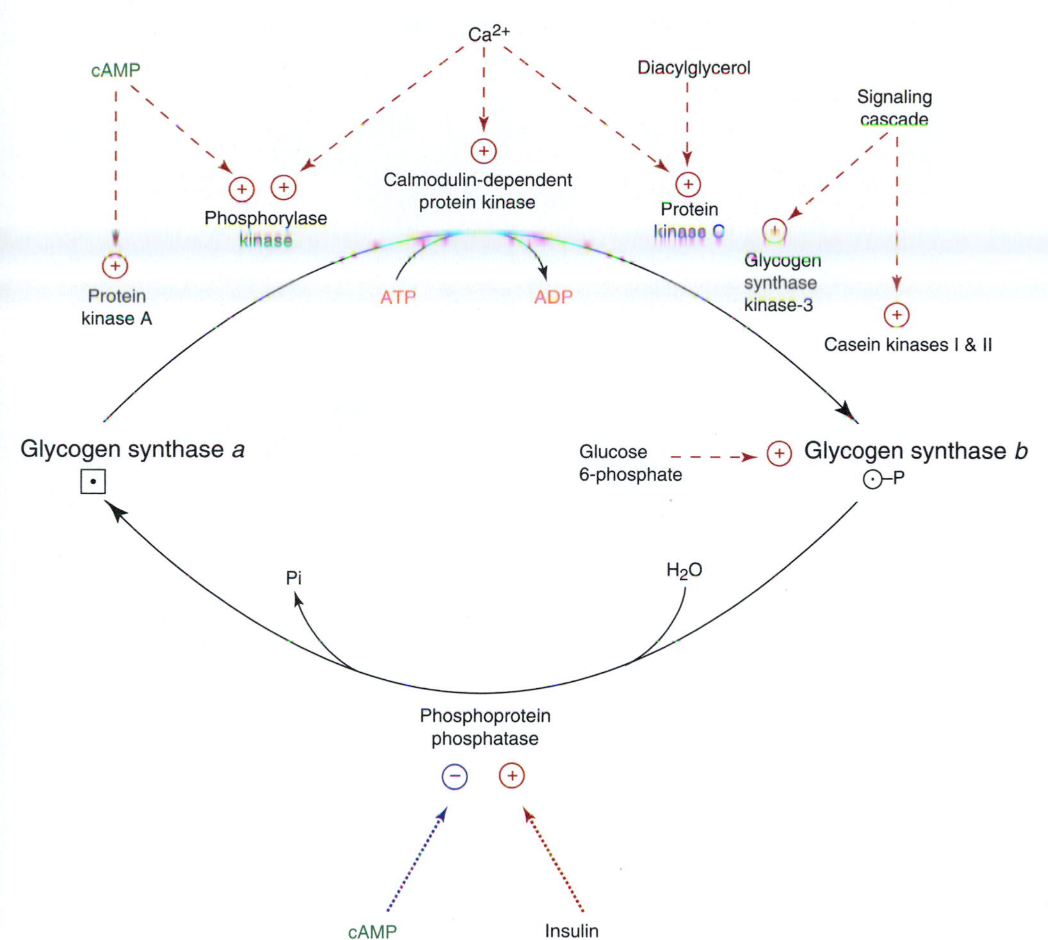

FIGURE 14.58
Regulation of glycogen synthase by covalent modification.
Phosphorylation converts glycogen synthase from its active *a* form to its inactive *b* form.

zyme. Diacylglycerol is considered an important "second messenger" of hormone action, acting via protein kinase C to regulate numerous cellular processes (see p. 930).

Glycogen synthase is also phosphorylated by **glycogen synthase kinase-3**, casein kinase I, and casein kinase II. These kinases are not subject to regulation by cAMP or Ca^{2+}. It is likely, however, that special regulatory mechanisms exist to regulate these kinases. Herein may lie solutions to unsolved problems such as the mechanism of action of **insulin** and other hormones. *Indeed, recent evidence suggests insulin may signal inactivation of glycogen synthase kinase-3, an action that would allow activation of glycogen synthase to occur via dephosphorylation by phosphoprotein phosphatase.*

The phosphoprotein phosphatase that converts glycogen synthase *b* back to glycogen synthase *a* (Figure 14.58) is regulated in a manner analogous to that described in the discussion of glycogen phosphorylase regulation (Figure 14.57). Cyclic AMP promotes inactivation whereas insulin promotes activation of glycogen synthase through opposite effects on phosphoprotein phosphatase activity.

Regulation of Phosphoprotein Phosphatases

Several **phosphoprotein phosphatases** with specificity for removal of phosphate from serine residues of proteins are currently being studied. In general, phosphoprotein phosphatases occur as catalytic subunits associated with a number of different regulatory subunits that control the activity of the catalytic subunit, determine which substrate(s) the catalytic subunit can interact with and dephosphorylate, and target the association of a catalytic subunit with a specific structure or component within a cell. One such regulatory protein important for glycogen metabolism has been given the name **G subunit**, denoting a **glycogen-binding**

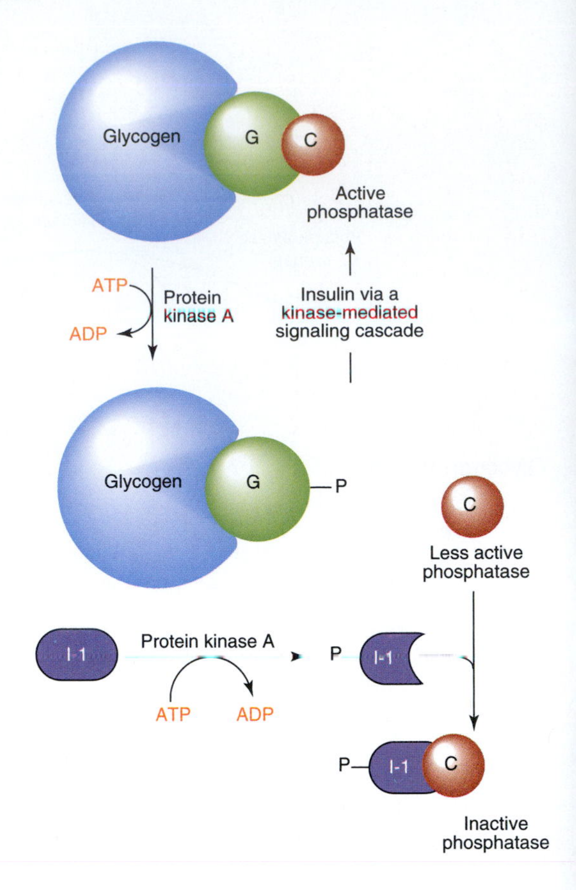

FIGURE 14.59
Mechanism for regulation of a phosphatase that binds to glycogen.
The glycogen binding subunit G binds directly to glycogen; the phosphoprotein phosphatase catalytic subunit C binds to glycogen via the G subunit; and the phosphorylated inhibitor 1 (I-1) binds the free catalytic subunit.

protein. G subunit binds both glycogen and a phosphatase catalytic subunit (Figure 14.59). This association makes the phosphatase ten times more active toward glycogen synthase and glycogen phosphorylase, and thereby greatly promotes their dephosphorylation. However, phosphorylation of the G subunit by protein kinase A results in release of the phosphatase catalytic subunit, which is then less active. Interaction of the free catalytic subunit with yet another regulatory protein (called inhibitor 1) then causes further inhibition of phosphatase activity. Effective inhibition of the residual phosphatase activity of the catalytic subunit requires phosphorylation of inhibitor 1 by protein kinase A, thereby creating yet another link to hormones that increase cAMP levels. Insulin has effects opposite to those of cAMP; that is, insulin promotes activation of the catalytic subunit of phosphoprotein phosphatase. This presumably involves reversal of the steps promoted by cAMP, but details of how this is accomplished remain to be established.

Effector Control of Glycogen Metabolism

Certain muscles are known to mobilize their glycogen stores rapidly in response to anaerobic conditions without marked conversion of phosphorylase *b* to phosphorylase *a* or glycogen synthase *a* to glycogen synthase *b*. Presumably this is accomplished by effector control in which ATP levels decrease, causing less inhibition of phosphorylase; glucose 6-phosphate levels decrease, causing less activation of glycogen synthase; and AMP levels increase, causing activation of phosphorylase. This enables muscle to keep working, for at least a short period of time, by using ATP produced by glycolysis of glucose 6-phosphate obtained from glycogen.

Proof that effector control can operate has also been obtained in studies of a special strain of mice that is deficient in muscle phosphorylase kinase. Phosphorylase *b* in muscle of such mice cannot be converted into phosphorylase *a*. Nevertheless, heavy exercise of these mice results in depletion of muscle glycogen, presumably because of stimulation of phosphorylase *b* by effectors.

Negative Feedback Control of Glycogen Synthesis by Glycogen

Glycogen exerts feedback control over its own formation. The portion of glycogen synthase in the active *a* form decreases as glycogen accumulates in a particular tissue. The mechanism is not well understood, but glycogen may make the *a* form a better substrate for one of the protein kinases, or, alternatively, glycogen may inhibit dephosphorylation of glycogen synthase *b* by phosphoprotein phosphatase. Either mechanism would account for the shift in the steady state in favor of glycogen synthase *b* that occurs in response to glycogen accumulation.

Phosphorylase a *Functions as a "Glucose Receptor" in the Liver*

Consumption of a carbohydrate-containing meal results in an increase in blood and liver glucose, which signals an increase in glycogen synthesis in the latter tissue. The mechanism involves glucose stimulation of insulin release from the pancreas and its effects on hepatic glycogen phosphorylase and glycogen synthase. However, hormone-independent mechanisms also appear to be important in liver (Figure 14.60). Direct inhibition of phosphorylase *a* by glucose is probably of importance. Binding of glucose to phosphorylase makes the *a* form of phosphorylase a better substrate for dephosphorylation by phosphoprotein phosphatase. Therefore phosphorylase *a* functions as a **glucose receptor** in liver. Binding of glucose to phosphorylase *a* promotes inactivation of phosphorylase *a*, with the overall result being inhibition of glycogen degradation by glucose. This "negative feedback" control of glycogenolysis by glucose would not necessarily promote glycogen synthesis. However, there also is evidence that phosphorylase *a* is an inhibitor of dephosphorylation of glycogen synthase *b* by phosphoprotein phosphatase. This inhibition is lost once phosphorylase *a* has been converted to phosphorylase *b* (Figure 14.60). In other words, phosphoprotein phosphatase can turn its attention to glycogen synthase *b* only following dephosphorylation of phosphorylase *a*. Thus, as a result of interaction of glucose with phosphorylase *a*, phosphorylase becomes inactivated, glycogen synthase becomes activated, and glycogen is synthesized rather than degraded in liver. Phosphorylase *a* can serve this function of "glucose receptor" in liver because the concentration of glucose in liver always reflects the blood concentration of glucose. This is not true for extrahepatic tissues. Liver cells have a very high-capacity transport system for glucose (GLUT-2) and a high-K_m enzyme for glucose

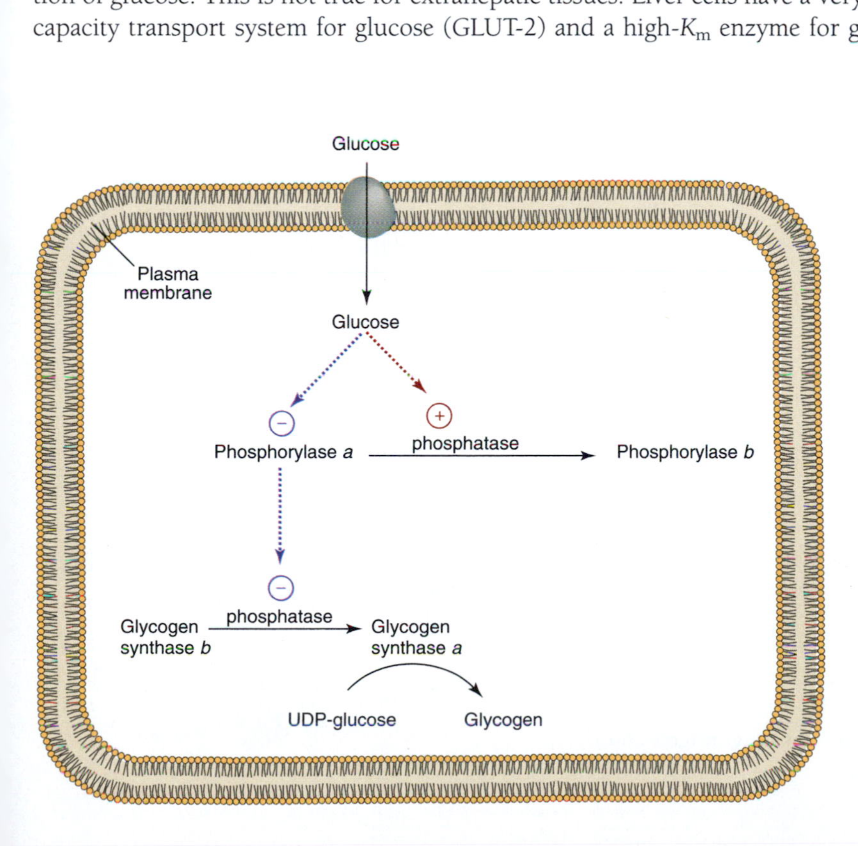

FIGURE 14.60
Overview of the mechanism for glucose stimulation of glycogen synthesis in the liver.

phosphorylation (glucokinase). Cells of extrahepatic tissues as a general rule have glucose transport and phosphorylation systems that maintain intracellular glucose at concentrations too low for phosphorylase *a* to function as a "glucose receptor."

Glucagon Stimulates Glycogen Degradation in the Liver

Glucagon is released from α cells of pancreas in response to low blood glucose levels. One of glucagon's primary jobs during periods of low food intake (fasting or starvation) is to mobilize liver glycogen, that is, stimulate glycogenolysis, in order to ensure that adequate blood glucose is available to meet the needs of glucose-dependent tissues. Glucagon circulates in blood until it interacts with glucagon receptors such as those located on the plasma membrane of liver cells (Figure 14.61). Binding of glucagon to these receptors activates adenylyl cyclase and triggers the cascades that result in activation of glycogen phosphorylase and inactivation of glycogen synthase (Figures 14.57 and 14.58, respectively). Glucagon also inhibits glycolysis at the level of 6-phosphofructo-1-kinase and pyruvate kinase as shown in Figures 14.27 and 14.30, respectively. The net result of these effects of glucagon, all mediated by the second messenger cAMP and covalent modification, is a very rapid increase in blood glucose levels. Hyperglycemia might be expected but does not occur because less glucagon is released from the pancreas as blood glucose levels increase.

Epinephrine Stimulates Glycogen Degradation in the Liver

Epinephrine is released into blood from **chromaffin cells** of the adrenal medulla in response to stress. This hormone is our "fright, flight, or fight" hormone, preparing the body for either combat or escape.

Binding of epinephrine with β-adrenergic receptors in the plasma membrane of liver cells causes activation of adenylyl cyclase (Figure 14.61). The resulting increase in cAMP has the same effect as that caused by glucagon, that is, activation of glycogenolysis and inhibition of glycogenesis and glycolysis to maximize the release of glucose from liver. In addition, binding of epinephrine to **α-adrenergic receptor** in the plasma membrane of liver cells signals formation of **inositol 1,4,5-**

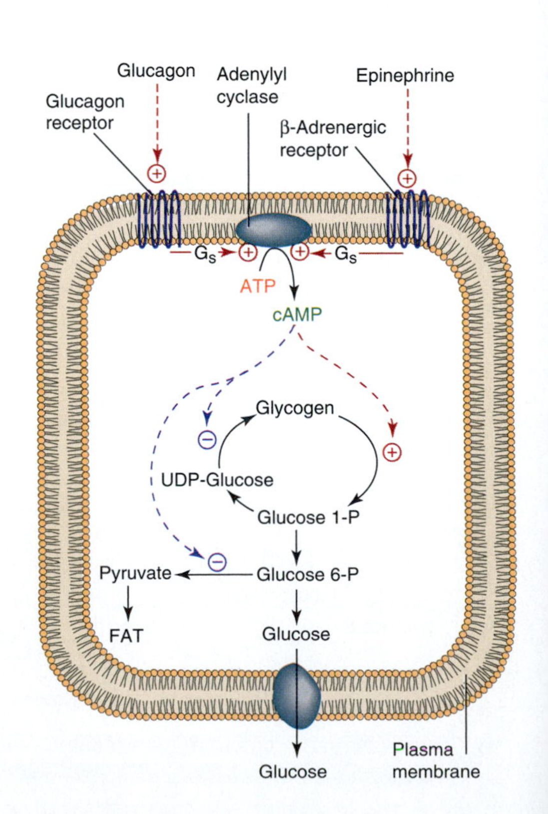

FIGURE 14.61
Cyclic AMP mediates the stimulation of glycogenolysis in liver by glucagon and β agonists (epinephrine).
See legends to Figures 14.19 and 14.25.

trisphosphate (IP_3) and **diacylglycerol** (Figure 14.62 and see p. 930). These compounds are second messengers, produced in the plasma membrane by the action of a phospholipase C on phosphatidylinositol 4,5-bisphosphate (Figure 14.63). Inositol 1,4,5-trisphosphate stimulates the release of Ca^{2+} from the endoplasmic reticulum (Figure 14.62). As previously discussed (Figure 14.57), the increase in Ca^{2+} activates phosphorylase kinase, which in turn activates glycogen phosphorylase. Likewise (Figure 14.58), Ca^{2+}-mediated activation of phosphorylase kinase, calmodulin-dependent protein kinase, and protein kinase C, as well as diacylglycerol-mediated activation of protein kinase C, may all be important for inactivation of glycogen synthase.

An increased rate of glucose release into the blood is a major consequence of epinephrine action on the liver. This makes more blood glucose available to tissues that are called on to meet the challenge of the stressful situation that triggered the release of epinephrine from adrenal medulla.

Epinephrine Stimulates Glycogen Degradation in Heart and Skeletal Muscle

Epinephrine also stimulates glycogen degradation in heart and skeletal muscle. Cyclic AMP, produced in response to epinephrine stimulation of adenylyl cyclase via **β-adrenergic receptors** (Figure 14.64), signals concurrent activation of glycogen phosphorylase and inactivation of glycogen synthase by mechanisms given previously in Figures 14.57 and 14.58, respectively. This does not lead, however, to glucose release into blood from these tissues. Heart and skeletal muscle lack glucose 6-phosphatase, and in these tissues cAMP stimulates glycolysis (see Figure 14.28). Thus the role of epinephrine on glycogen metabolism in heart and skeletal muscle is to make more glucose 6-phosphate available for glycolysis. ATP generated by glycolysis can then be used to meet the metabolic demand imposed on these muscles by the stress that triggered epinephrine release.

Neural Control of Glycogen Degradation in Skeletal Muscle

Nervous excitation of muscle activity is mediated via changes in **intracellular Ca^{2+}** concentrations (Figure 14.65). A nerve impulse causes **membrane depolarization,** which in turn causes Ca^{2+} release from the sarcoplasmic reticulum into the sarcoplasm of muscle cells. This release of Ca^{2+} triggers muscle contraction, whereas reaccumulation of Ca^{2+} by the sarcoplasmic reticulum causes relaxation. The same change in Ca^{2+} concentration effective in causing muscle contraction (from 10^{-8}

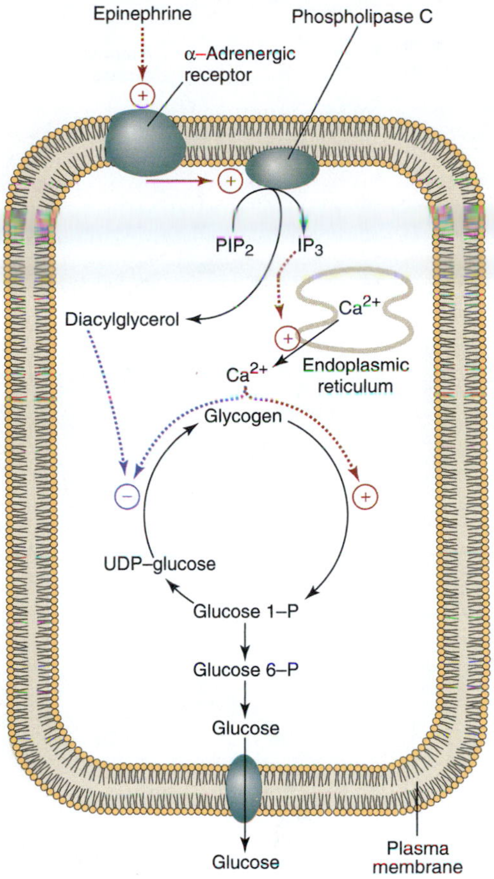

FIGURE 14.62
Inositol trisphosphate (IP_3) and Ca^{2+} mediate the stimulation of glycogenolysis in liver by α agonists.
The α-adrenergic receptor and glucose transporter are intrinsic components of the plasma membrane. Although not indicated, phosphatidylinositol 4,5-bisphosphate (PIP_2) is also a component of the plasma membrane.

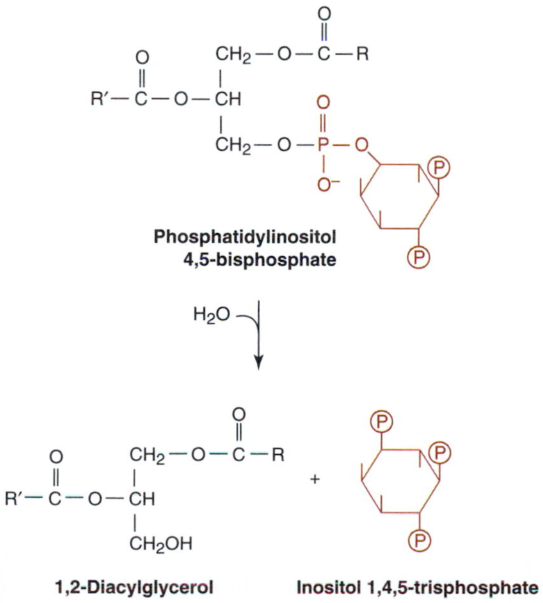

FIGURE 14.63
Phospholipase C cleaves phosphatidylinositol 4,5-bisphosphate to produce 1,2-diacylglycerol and inositol 1,4,5-trisphosphate.

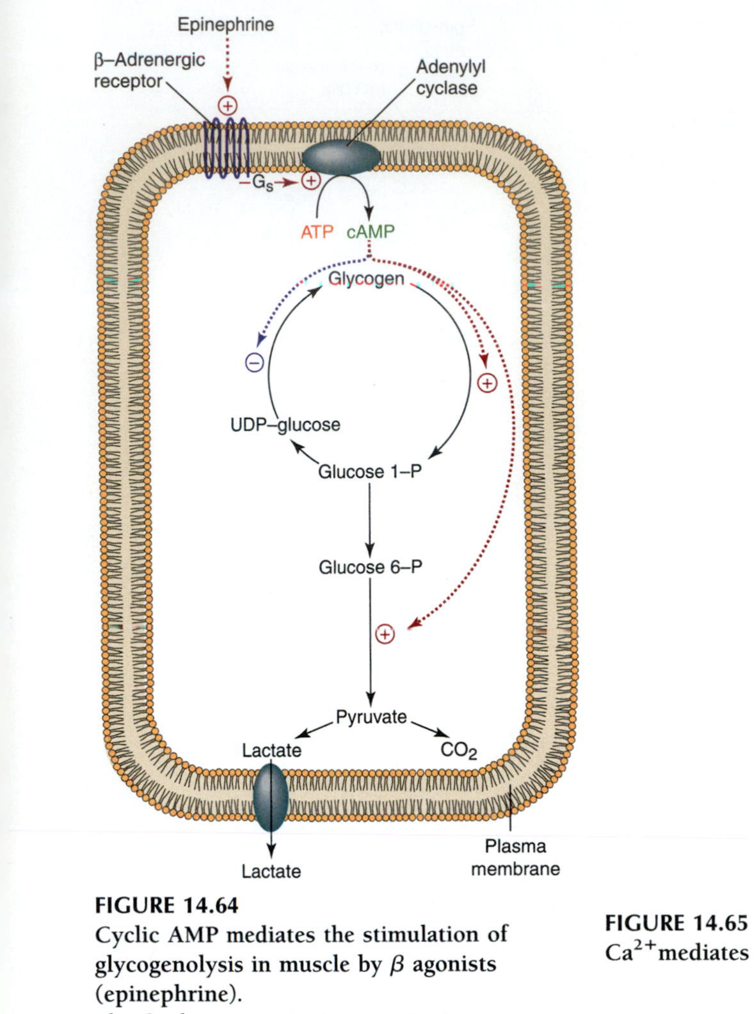

FIGURE 14.64
Cyclic AMP mediates the stimulation of glycogenolysis in muscle by β agonists (epinephrine).
The β-adrenergic receptor is an intrinsic component of the plasma membrane that acts to stimulate adenylyl cyclase via a stimulatory G protein (G_s).

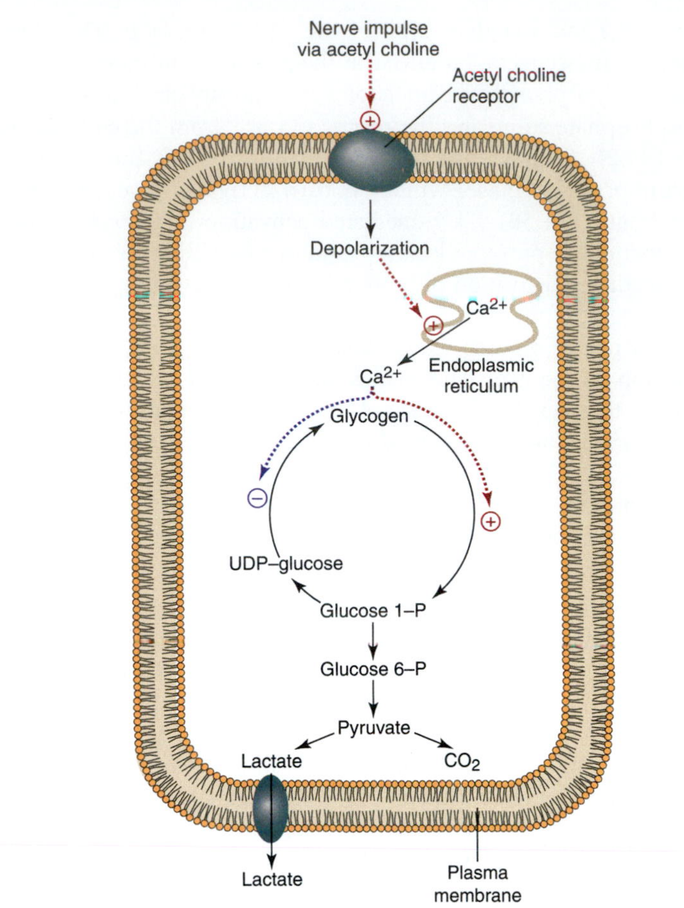

FIGURE 14.65
Ca^{2+} mediates the stimulation of glycogenolysis in muscle by nervous excitation.

to 10^{-6} M) also greatly affects the activity of phosphorylase kinase. As Ca^{2+} concentrations increase there is more muscle activity and a greater need for ATP. Activation of phosphorylase kinase by Ca^{2+} leads to the subsequent activation of glycogen phosphorylase and perhaps the inactivation of glycogen synthase. The result is that more glycogen is converted to glucose 6-phosphate so that more ATP can be produced to meet the greater energy demand of muscle contraction.

Insulin Stimulates Glycogen Synthesis in Muscle and Liver

An increase in blood glucose signals release of insulin from β cells of the pancreas. Insulin receptors, located on the plasma membranes of insulin-responsive cells, respond to insulin binding by inducing a signaling cascade that promotes glucose use within these tissues (Figures 14.66 and 14.67). The pancreas responds to a decrease in blood glucose with less release of insulin but greater release of glucagon. These hormones have opposite effects on glucose utilization by liver, thereby establishing the pancreas as a fine-tuning device that prevents dangerous fluctuations in blood glucose levels.

Insulin increases glucose utilization in part by promoting glycogenesis and inhibiting glycogenolysis in muscle (Figure 14.66) and liver (Figure 14.67). Insulin stimulation of glucose transport at the plasma membrane is essential for these effects in muscle but not liver. Hepatocytes have a high-capacity, **insulin-insensitive glucose transport system (GLUT-2),** whereas muscle cells and adipocytes are equipped with a **glucose transport system (GLUT-4)** that requires insulin for maximum rates of glucose uptake (see p. 520). Insulin stimulates muscle and adipose tissue glucose transport by signaling an increase in the number of functional GLUT-4 proteins associated with the plasma membrane. This is accomplished by promoting

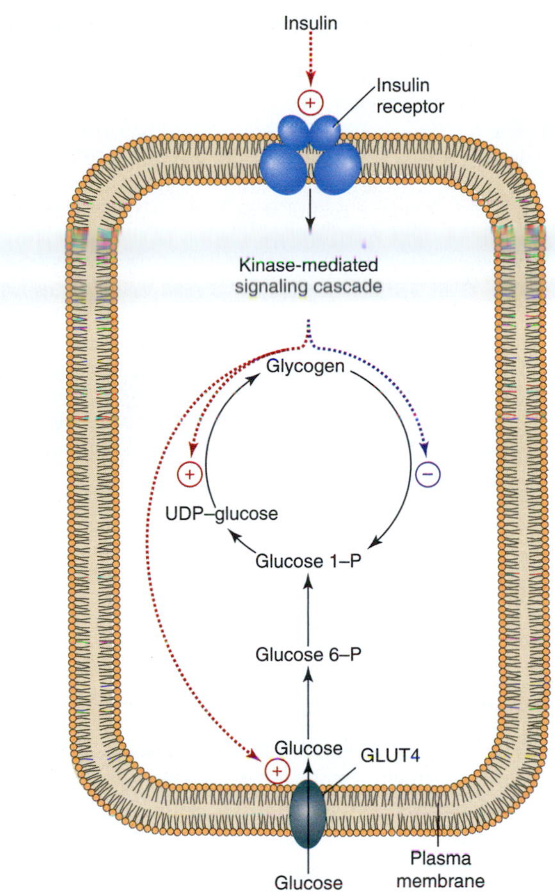

FIGURE 14.66
Insulin acts via a plasma membrane receptor to promote glycogen synthesis in muscle.

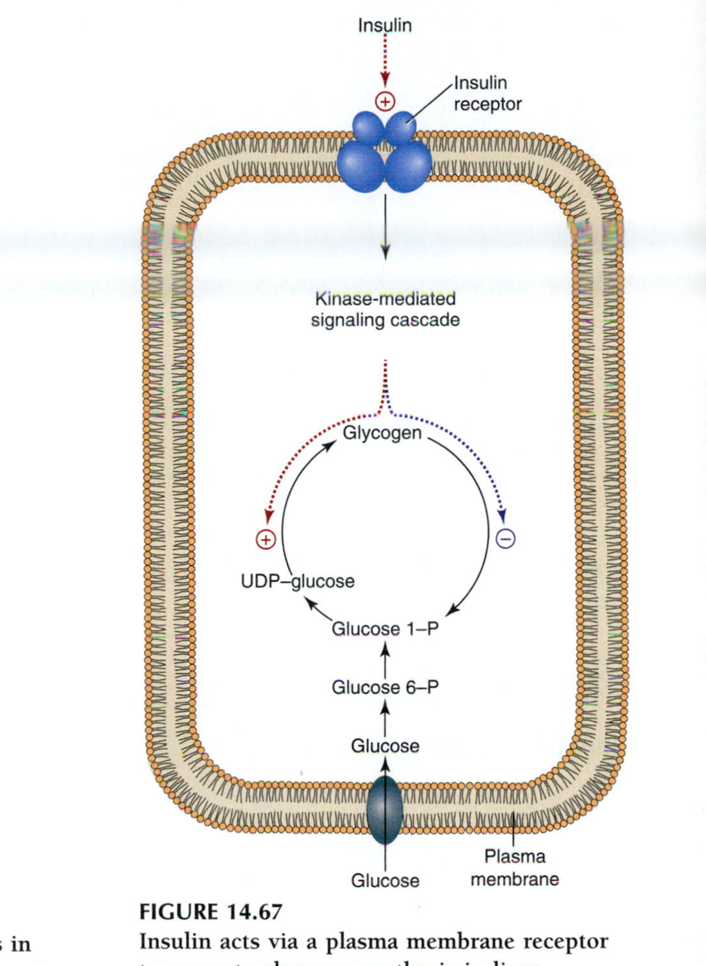

FIGURE 14.67
Insulin acts via a plasma membrane receptor to promote glycogen synthesis in liver.

translocation of GLUT-4 from an intracellular pool to the plasma membrane (see Figure 14.6). Insulin further promotes glycogen accumulation in both tissues by activating glycogen synthase and inhibiting glycogen phosphorylase as discussed previously (Figures 14.57–14.59).

BIBLIOGRAPHY

Arion, W. J., Lange, A. J., Walls, H. E., and Ballas, L. M. Evidence for the participation of independent translocases for phosphate and glucose 6-phosphate in the microsomal glucose 6-phosphatase system. *J. Biol. Chem.* 255:10396, 1980.

Berridge, M. J. Review article: inositol trisphosphate and diacylglycerol as second messengers. *Biochem. J.* 220:345, 1984.

Brooke, M. H. and Kaiser, K. K. The use and abuse of muscle histochemistry. *Ann. N.Y. Acad. Sci.* 228:121, 1974.

Cheatham, B. and Kahn, C. R. Insulin action and the insulin signaling network. *Endocr. Rev.* 16:117, 1995.

Cohen, P. Dissection of the protein phosphorylation cascades involved in insulin and growth factor action. *Biochem. Soc. Trans.* 21:555, 1993.

DeFronzo, R. A. and Ferrannini, E. Regulation of hepatic glucose metabolism in humans. Diabetes Metab. Rev. 3:415, 1987.

DePaoli-Roach, A. A., Park, I.-K., Cerovsky, V., Csortos, C., Durbin, S. D., Kuntz, M. J., Sitikov, A., Tang, P. M., Verin, A., and Zolnierowicz, S. Serine/threonine protein phosphatases in the control of cell function. *Adv. Enzyme Regul.* 34:199, 1994.

Depre, C., Rider, M. H., and Hue L. Mechanisms of control of heart glycolysis. *Eur. J. Biochem.* 258:277, 1998.

Exton, J. H. Mechanisms of hormonal regulation of hepatic glucose metabolism. *Diabetes /Metab. Rev.* 3:163, 1987.

Fell, D. *Understanding the Control of Metabolism.* London: Portland Press, 1997.

Garcia, C. K., Goldstein, J. L., Pathak, R. K., Anderson, R. G. W., and Brown, M. S. Molecular characterization of a membrane transporter for lactate, pyruvate, and other monocarboxylates: implications for the Cori cycle. *Cell* 76:865, 1994.

Geelen, M. J. H., Harris, R. A., Beynen, A. C., and McCune, S. A. Short-term hormonal control of hepatic lipogenesis. *Diabetes* 29:1006, 1980.

Gitzelmann, R., Steinmann, B., and Van den Berghe, G. Disorders of fructose metabolism. In: C. R. Scriver, A. L. Beaudet, W. S. Sly, and D. Valle (Eds.), *The Metabolic and Molecular Bases of Inherited Disease,* 7th ed. New York: McGraw-Hill, 1995, p. 905.

Gould, G. W. and Holman, G. D. The glucose transporter family: structure, function and tissue-specific expression. *Biochem. J.* 295:329, 1993.

Greene, H. L., Slonin, A. E., and Burr, I. M. Type I glycogen storage disease: a metabolic basis for advances in treatment. In: L. A. Barness (Ed.), *Advances in Pediatrics,* Vol. 26. Chicago: Year Book Publishers, 1979, p. 63.

Gurney, A. L., Park, E. A., Liu, J., Giralt, M., McGrane, M. M., Patel, Y. M., Crawford, D. R., Nizielski, S. E., Savon, S., and Hanson, R. W. Metabolic regulation of gene transcription. *J. Nutr.* 124:1533S, 1994.

Hallfrisch, J. Metabolic effects of dietary fructose. *FASEB J.* 4:2652, 1990.

Hanson, R. W. and Mehlman, M. A. (Eds.). *Gluconeogenesis, Its Regulation in Mammalian Species.* New York: Wiley, 1976.

Hunter, T. Protein kinases and phosphatases: the yin and yang of protein phosphorylation and signaling. *Cell* 80:225, 1995.

Isselbacher, K. J., Adams, R. D., Braundwald, E., Petersdorf, R. B., and Wilson, J. D. (Eds.). *Harrison's Principles of Internal Medicine,* 9th ed. New York: McGraw-Hill, 1980.

Lalli, E. and Sassone-Corsi, P. Signal transduction and gene regulation: the nuclear response to cAMP. *J. Biol. Chem.* 269:17359, 1994.

Lieber, C. S. The metabolism of alcohol. *Sci. Am.* 234:25, 1976.

Lomako, J., Lomako, W. M., and Whelan, W. J. A self-glucosylating protein is the primer for rabbit muscle glycogen biosynthesis. *FASEB J.* 2:3097, 1988.

Metzler, D. E. *Biochemistry, the Chemical Reactions of Living Cells.* New York: Academic Press, 1977.

Newsholme, E. A. and Leech, A. R. *Biochemistry for the Medical Sciences.* New York: Wiley, 1983.

Newsholme, E. A. and Start, C. *Regulation in Metabolism.* New York: Wiley, 1973.

Pilkis, S. J., Claus, T. H., Kurland, I. J., and Lange, A. J. 6-Phosphofructo-2-kinase/fructose-2,6-bisphosphatase: a metabolic signaling enzyme. *Annu. Rev. Biochem.* 64:799, 1995.

Pilkis, S. J. and El-Maghrabi, M. R. Hormonal regulation of hepatic gluconeogenesis and glycolysis. *Annu. Rev. Biochem.* 57:755, 1988.

Price, T. B., Rothman, D. L., and Shulman R. G. NMR of glycogen in exercise. *Proc. Nutr. Soc.* 58:851, 1999.

Roach, P. J., Skurat, A. V., and Harris, R. A. Regulation of glycogen metabolism. In L. S. Jefferson and A. D. Charrington (Eds.), *The Endocrine Pancreas and Regulation of Metabolism: Handbook of Physiology.* New York: Oxford, 2001.

Roach, P. J. Hormonal control of glycogen metabolism. In: H. Rupp (Ed.), *Regulation of Heart Function: Basic Concepts and Clinical Applications.* New York: Thieme-Stratton, 1985.

Scriver, C. R., Beaudet, A. L., Sly, W. S., and Valle, D. (Eds.). *The Metabolic and Molecular Bases of Inherited Disease,* 7th ed. New York: McGraw-Hill, 1995.

Stanley, C. A., Anday, E. K., Baker, L., and Delivoria-Papadopolous, M. Metabolic fuel and hormone responses to fasting in newborn infants. *Pediatrics* 64:613, 1979.

Taylor, S. I. Diabetes mellitus. In: C. R. Scriver, A. L. Beaudet, W. S. Sly, and D. Valle (Eds.), *The Metabolic and Molecular Bases of Inherited Disease,* 7th ed. New York: McGraw-Hill, 1995, p. 843.

Van Schaftingen, E., Vandercammen, A., Detheux, M., and Davies, D. R. The regulatory protein of liver glucokinase. *Adv. Enzyme Regul.* 32:133, 1992.

Vaulont, S. and Kahn, A. Transcriptional control of metabolic regulation genes by carbohydrates. *FASEB J.* 8:28, 1994.

QUESTIONS | C. N. ANGSTADT

Multiple Choice Questions

1. In glycolysis ATP synthesis is catalyzed by:
   A. hexokinase.
   B. 6-phosphofructo-1-kinase.
   C. glyceraldehyde 3-phosphate dehydrogenase.
   D. phosphoglycerate kinase.
   E. none of the above.

2. The irreversible reactions of glycolysis include that catalyzed by:
   A. phosphoglucose isomerase.
   B. 6-phosphofructo-1-kinase.
   C. fructose bisphosphate aldolase.
   D. glyceraldehyde 3-phosphate dehydrogenase.
   E. phosphoglycerate kinase.

3. NAD^+ can be regenerated in the cytoplasm if NADH reacts with any of the following EXCEPT:
   A. pyruvate.
   B. dihydroxyacetone phosphate.
   C. oxaloacetate.
   D. the flavin bound to NADH dehydrogenase.

4. Glucokinase:
   A. has a K_m considerably greater than the normal blood glucose concentration.
   B. is found in muscle.
   C. is inhibited by glucose 6-phosphate.
   D. is also known as the GLUT-2 protein.
   E. has glucose 6-phosphatase activity as well as kinase activity.

5. 6-Phosphofructo-1-kinase activity can be decreased by all of the following EXCEPT:
   A. ATP at high concentrations.
   B. citrate.
   C. AMP.
   D. low pH.
   E. decreased concentration of fructose 2,6-bisphosphate.

6. Which of the following supports gluconeogenesis?
   A. α-ketoglutarate + aspartate $\rightleftharpoons$ glutamate + oxaloacetate
   B. pyruvate + ATP + HCO_3 $\rightleftharpoons$ oxaloacetate + ADP + P_i + H^+
   C. acetyl CoA + oxaloacetate + H_2O $\rightleftharpoons$ citrate + CoA
   D. leucine degradation
   E. lysine degradation

7. In the Cori cycle:
   A. only tissues with aerobic metabolism (i.e., mitochondria and O_2) are involved.
   B. a three-carbon compound arising from glycolysis is converted to glucose at the expense of energy from fatty acid oxidation.
   C. glucose is converted to pyruvate in anaerobic tissues, and this pyruvate returns to the liver, where it is converted to glucose.
   D. the same amount of ATP is used in the liver to synthesize glucose as it is released during glycolysis, leading to no net effect on whole-body energy balance.
   E. nitrogen from alanine must be converted to urea, increasing the amount of energy required to drive the process.

8. Gluconeogenic enzymes include all of the following EXCEPT:
   A. fructose 1,6-bisphosphatase.
   B. glucose 6-phosphatase.
   C. phospho*enol*pyruvate carboxykinase.
   D. phosphoglucomutase.
   E. pyruvate carboxylase.

9. When blood glucagon rises, which of the following hepatic enzyme activities FALLS?
   A. adenyl cyclase
   B. protein kinase
   C. 6-phosphofructo-2-kinase
   D. fructose 1,6-bisphosphatase
   E. hexokinase

10. Phospho–dephospho regulation of 6-phosphofructo-1-kinase, 6-phosphofructo-2-kinase, and pyruvate kinase is an important regulatory mechanism in:
    A. brain.
    B. erythrocytes.
    C. intestine.
    D. liver.
    E. skeletal muscle.

Questions 11 and 12: Patients with McArdle's disease suffer from painful muscle cramps and are unable to perform strenuous exercise. This disease, also called type V glycogen storage disease, is caused by the absence of muscle phosphorylase. Unlike patients with Von Gierke's disease (type I glycogen storage disease caused by a deficiency of glucose 6-phosphatase), McArdle's patients do not typically suffer hypoglycemia on fasting. Von Gierke's patients show lactic acidemia but McArdle's patients fail to demonstrate lactic acidemia following exercise.

11. McArdle's patients:
    A. fail to synthesize glycogen appropriately.
    B. do not respond to glucagon in a normal fashion.
    C. show a reduced state of glycolysis during exercise.
    D. have an impaired tricarboxylic acid cycle.
    E. have reduced lactic acid because all of the lactate is converted to pyruvate.

12. Glucose 6-phosphatase, which is deficient in Von Gierke's disease, is necessary for the production of blood glucose from:
    A. liver glycogen.
    B. fructose.
    C. amino acid carbon chains.
    D. galactose.
    E. all of the above.

Questions 13 and 14: Malignant hyperthermia is a genetic abnormality in which exposure to certain agents, especially the widely used general anesthetic halothane, produces a dramatic rise in body temperature, acidosis, hyperkalemia, and muscle rigidity. Death is rapid if the condition is untreated and may occur the first time a susceptible person is anesthetized. The defect causes an inappropriate release of Ca^{2+} from the sarcoplasmic reticulum of muscle. Many heat-producing processes are stimulated in an uncontrolled fashion by the release of Ca^{2+}, including glycolysis and glycogenolysis.

13. Ca^{2+} increases glycogenolysis by:
    A. activating phosphorylase kinase *b*, even in the absence of cAMP.
    B. binding to phosphorylase *b*.
    C. activating phosphoprotein phosphatase.
    D. inhibiting phosphoprotein phosphatase.
    E. protecting cAMP from degradation.

14. Phosphorylation–dephosphorylation and allosteric activation of enzymes play roles in stimulating glycogen degradation. All of the following result in enzyme activation EXCEPT:
    A. phosphorylation of phosphorylase kinase.
    B. binding of AMP to phosphorylase *b*.
    C. phosphorylation of phosphorylase.
    D. phosphorylation of protein kinase A.
    E. dephosphorylation of glycogen synthase.

Problems

15. If a cell is forced to metabolize glucose anaerobically, how much faster would glycolysis have to proceed to generate the same amount of ATP as it would get if it metabolized glucose aerobically?

16. Consumption of alcohol by an undernourished individual or by an individual following strenuous exercise can lead to hypoglycemia. What relevant reactions are inhibited by the consumption of alcohol?

ANSWERS

1. **D** Phosphoglycerate kinase synthesizes ATP in the forward direction. A and B use ATP. C synthesizes 1,3-bisphosphoglycerate.
2. **B** 6-Phosphofructo-1-kinase forms a phosphate ester that has a low energy bond. A, C, and D catalyze reversible reactions. E: The phosphoglycerate kinase reaction is reversible because the product contains a high-energy carboxylic acid–phosphoric acid anhydride link.
3. **D** The flavin is mitochondrial. A may be converted to lactate. B and C are the cytoplasmic acceptors for shuttle systems.
4. **A** Blood glucose is ~5 mM. K_m of glucokinase is ~10 mM. B: Glucokinase is hepatic, and, unlike the muscle hexokinase, it is not inhibited by glucose 6-phosphate.
5. **C** AMP is an allosteric regulator that relieves inhibition by ATP. B and D are probably important physiological regulators in muscle, and E is critical in liver.
6. **B** This reaction is on the direct route of conversion of pyruvate to glucose. A: α-Ketoglutarate and oxaloacetate both give rise to glucose; interconversion of one to the other accomplishes nothing. C: Citrate ultimately gives rise to oxaloacetate, losing two carbon atoms in the process; again nothing is gained. D and E involve the two amino acids that are strictly ketogenic.
7. **B** The liver derives the energy required for gluconeogenesis from aerobic oxidation of fatty acids. A: The liver is an essential organ in the Cori cycle; it is aerobic. C: In anaerobic tissues the end product of glycolysis is lactate; in aerobic tissues it is pyruvate, but there the pyruvate would likely be oxidized aerobically. D: Gluconeogenesis requires 6 ATP per glucose synthesized; glycolysis yields 2 ATP per glucose metabolized. E: Alanine is not part of the Cori cycle.
8. **D** Phosphoglucomutase is on the pathway of glycogen metabolism. A–C and E are the so-called gluconeogenic enzymes; they get around the irreversible steps of glycolysis.
9. **C** As blood glucagon rises, A is activated, producing cAMP; cAMP activates B, and B inactivates C. Low levels of fructose 2,6-bisphosphate increase the activity of D. E is not an important hepatic enzyme; its role is filled in liver by glucokinase.
10. **D** Regulation of these enzymes by hormonally controlled phosphorylation and dephosphorylation is of central importance in liver. 6-Phosphofructo-2-kinase is present in other tissues but does not appear to change its activity in response to hormones except in liver. Other enzymes in extrahepatic tissues, such as those of glycogen metabolism in muscle, are under phospho–dephospho regulation.
11. **C** Muscles use glucose from glycogen as the primary substrate for glycolysis during exercise. Lack of phosphorylase prevents the degradation of glycogen. A: Phosphorylase is not necessary for the synthesis of glycogen, only its degradation. B: Muscles do not respond to glucagon; liver does but liver phosphorylase is not affected in this disease. D: Muscle uses fatty acids and amino acid carbon chains via the tricarboxylic acid cycle. E:

Lactate formation is reduced because glycolysis lacks sufficient substrate.

12. **E** To get into the blood, glucose must be free, not phosphorylated. A: Glycogen is degraded to glucose 1-phosphate, which is converted to glucose 6-phosphate. B: Fructose is metabolized to dihydroxyacetone phosphate, which can either continue through glycolysis or reverse to glucose 6-phosphate, depending on the state of the cell. C: Amino acid carbon chains are substrates for gluconeogenesis. D: Galactose is ultimately converted to glucose 1-phosphate via UDP-galactose.
13. **A** The γ subunit of phosphorylase kinase is a calmodulin-type protein. Both *a* and *b* forms of the enzyme are activated by Ca^{2+}. B, E: These do not happen. C, D: Ca^{2+} does not affect the phosphatase.
14. **D** Protein kinase A catalyzes phosphorylations but is activated by binding cAMP. A, C: Both of these enzymes are phosphorylated in their *a* forms. B: Phosphorylase *b* is allosterically activated by binding AMP. E: The *a* form of the synthase is the nonphosphorylated form.
15. Anaerobically there is a net of 2 mol ATP/mol glucose. Aerobically the same net of 2 ATP is obtained plus 2 NADH because pyruvate is the product. Let us assume the cell uses the malate–aspartate shuttle where each NADH yields $2\frac{1}{2}$ ATP. Thus there is a net of 7 mol ATP/mol glucose. Each pyruvate is converted to AcCoA and the AcCoA is oxidized by the tricarboxylic acid cycle. Each mole of pyruvate then yields 12.5 mol ATP or 25 mol ATP for the 2 pyruvates. This gives a total of 32 mol ATP/mol glucose aerobically (see Chapter 13). Therefore glycolysis must proceed 16 times as rapidly under anaerobic conditions to generate the same amount of ATP as occurs aerobically.
16. In both of these cases, hepatic glycogen is depleted and gluconeogenesis is the primary source of blood glucose. Metabolism of alcohol by the liver produces large amounts of NADH. This shifts the equilibrium of pyruvate–lactate toward lactate and oxaloacetate–malate toward malate. To be used for gluconeogenesis, lactate must first be converted to pyruvate. Oxaloacetate from pyruvate or amino acid carbon chains must be converted to phospho*enol*pyruvate in order to synthesize glucose.

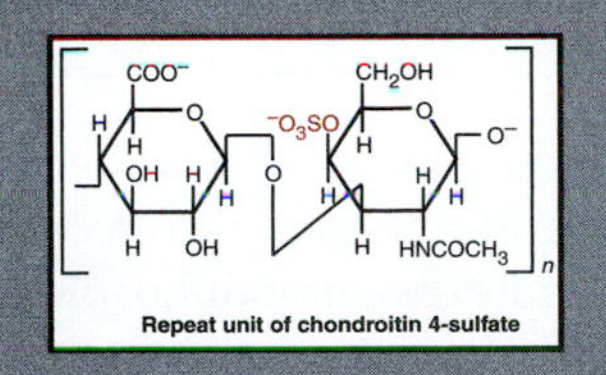

15

CARBOHYDRATE METABOLISM II: SPECIAL PATHWAYS AND GLYCOCONJUGATES

Nancy B. Schwartz

15.1 | OVERVIEW

In addition to catabolism of glucose for the specific purpose of energy production in the form of ATP, several other pathways involving sugar metabolism exist in cells. One, the **pentose phosphate pathway**, known also as the **hexose monophosphate shunt** or the **6-phosphogluconate pathway,** is particularly important in animal cells. It functions side by side with glycolysis and the tricarboxylic acid cycle for production of reducing power in the form of NADPH and pentose intermediates. It has previously been mentioned that NADPH serves as a hydrogen and electron donor in reductive biosynthetic reactions, while in most biochemical reactions NADH is oxidized by the respiratory chain to produce ATP (see Chapter 13). The enzymes involved in this pathway are located in the cytosol, indicating that the oxidation that occurs is not dependent on mitochondria or the tricarboxylic acid cycle. Another important function is to convert hexoses into pentoses, particularly ribose 5-phosphate. This C_5 sugar or its derivatives are components of ATP, CoA, NAD, FAD, RNA, and DNA. The pentose phosphate pathway also catalyzes the **interconversion** of C_3, C_4, C_6, and C_7 sugars, some of which can enter glycolysis. There are also specific pathways for synthesis and degradation of monosaccharides, oligosaccharides, and polysaccharides (other than glycogen), and a profusion of chemical interconversions, whereby one sugar can be changed into another. All monosaccharides, and most oligo- and polysaccharides synthesized from the monosaccharides, can originate from glucose. The interconversion reactions by which one sugar is changed into another can occur directly or at the level of nucleotide-linked sugars. In addition to their important role in sugar transformation, nucleotide-linked sugars are the obligatory activated form for polysaccharide synthesis. Monosaccharides also occur as components of more complex macromolecules like oligo- and polysaccharides, glycoproteins, glycolipids, and proteoglycans. In higher animals these complex carbohydrate molecules are predominantly found in the extracellular space in tissues and associated with cell membranes. As well, they often perform more dynamic functions such as recognition markers and determinants of biological specificity. The discussion of complex carbohydrates in this chapter is limited to the chemistry and biology of those complex carbohydrates found in animal tissues and fluids. The Appendix (see p. 1172) presents the nomenclature and chemistry of the carbohydrates.

15.2 | PENTOSE PHOSPHATE PATHWAY

The Pentose Phosphate Pathway Has Two Phases

The oxidative pentose phosphate pathway provides a means for cutting the carbon chain of a sugar molecule one carbon at a time. However, in contrast to the tricarboxylic acid cycle, the operation of this pathway does not occur as a consecutive set of reactions leading directly from glucose 6-phosphate (G6P) to six molecules of CO_2. The pathway can be visualized as occurring in two stages. In the first stage, hexose is decarboxylated to pentose via two oxidation reactions that lead to formation of NADPH. The pathway then continues and by a series of transformations, six molecules of pentose undergo rearrangements to yield five molecules of hexose.

Glucose 6-Phosphate Is Oxidized and Decarboxylated to a Pentose Phosphate

The first reaction of the pentose phosphate pathway (Figure 15.1) is **dehydrogenation** of G6P at C-1 to form **6-phosphoglucono-δ-lactone** and **NADPH,** catalyzed by glucose 6-phosphate dehydrogenase, a major regulatory site for this pathway. Special interest in this enzyme stems from the severe anemia that may result from the absence of **G6P dehydrogenase** in erythrocytes or from the presence of

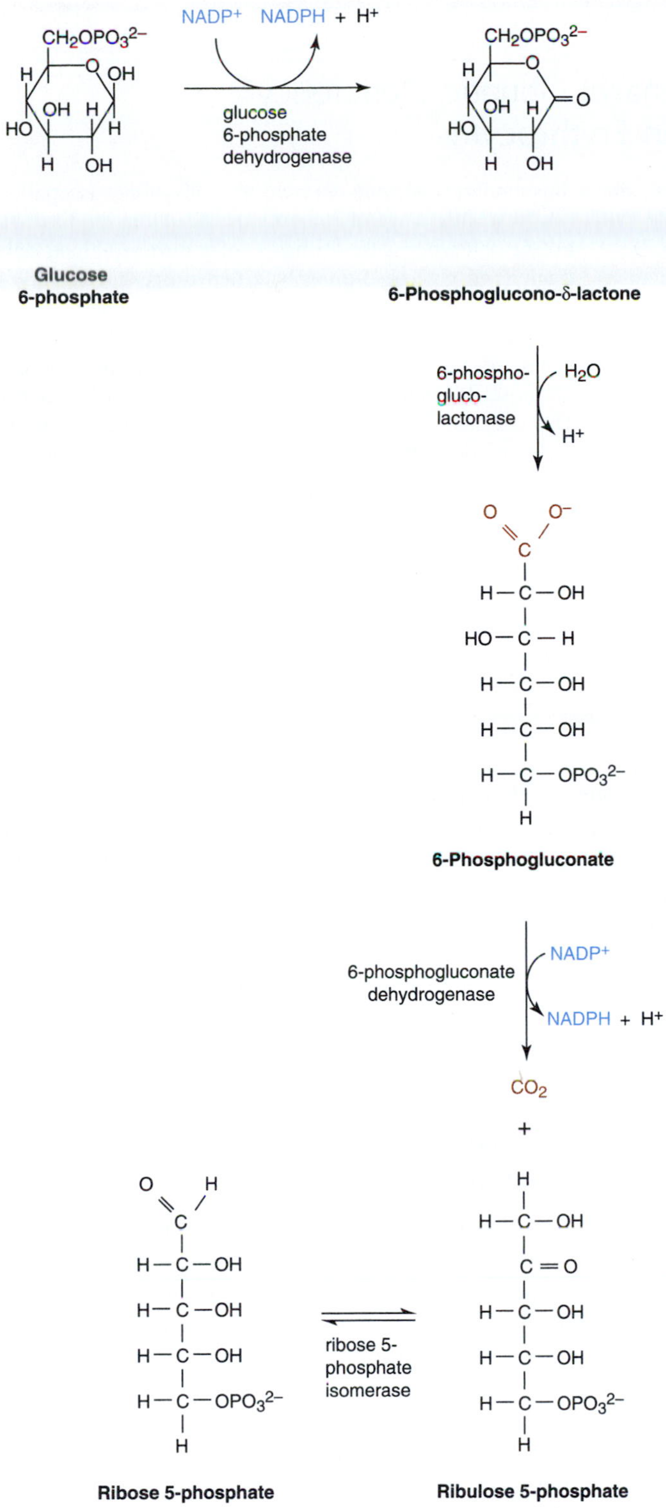

FIGURE 15.1
Oxidative phase of the pentose phosphate pathway: formation of pentose phosphate and NADPH.

one of several genetic variants of the enzyme (see Clin. Corr. 15.1). The intermediate product of this reaction, a lactone, is a substrate for gluconolactonase, which ensures that the reaction goes to completion. The overall equilibrium of these two reactions lies far in the direction of NADPH, maintaining a high NADPH/NADP$^+$ ratio within cells. A second dehydrogenation and decarboxylation is catalyzed by

CLINICAL CORRELATION 15.1

Glucose 6-Phosphate Dehydrogenase: Genetic Deficiency or Presence of Genetic Variants in Erythrocytes

When certain seemingly harmless drugs, such as antimalarials, antipyretics, or sulfa antibiotics, are administered to susceptible patients, an acute hemolytic anemia may result in 48–96 hours. Susceptibility to drug-induced hemolytic disease may be due to a deficiency of G6P dehydrogenase activity in erythrocytes and was one of the early indications that X-linked genetic deficiencies of this enzyme exist. This enzyme, which catalyzes the oxidation of G6P to 6-phosphogluconate and the reduction of $NADP^+$, is particularly important, since the pentose phosphate pathway is the major pathway of NADPH production in the red cell. For example, red cells with the relatively mild A-type of glucose 6-phosphate dehydrogenase deficiency can oxidize glucose at a normal rate when the demand for NADPH is normal. However, if the rate of NADPH oxidation is increased, the enzyme-deficient cells cannot increase the rate of glucose oxidation and carbon dioxide production adequately. In addition, cells lacking glucose 6-phosphate dehydrogenase do not reduce enough NADP to maintain glutathione in its reduced state. Reduced glutathione is necessary for the integrity of the erythrocyte membrane, thus rendering enzyme-deficient red cells more susceptible to hemolysis by a wide range of compounds. Therefore the basic abnormality in G6P deficiency is the formation of mature red blood cells that have diminished glucose 6-phosphate dehydrogenase activity. Young red blood cells may have significantly higher enzyme activity than older cells because of an unstable enzyme variant; following an episode of hemolysis, young red cells predominate and it may not be possible to diagnose this genetic deficiency until the red cell population ages. This enzymatic deficiency, which is usually undetected until administration of certain drugs, illustrates the interplay of heredity and environment on the production of disease. Enzyme defects are only one of several abnormalities that can affect enzyme activity, and others have been detected independent of drug administration. There are more than 300 known genetic variants of this enzyme that contains 516 amino acids, accounting for a wide range of symptoms. These variants can be distinguished from one another by clinical, biochemical, and molecular differences.

6-phosphogluconate dehydrogenase and produces the pentose phosphate, **ribulose 5-phosphate,** and a second molecule of NADPH. The final step in synthesis of ribose 5-phosphate is the **isomerization,** through an enediol intermediate, of ribulose 5-phosphate by **ribose isomerase.** Under certain metabolic conditions, the pentose phosphate pathway can end at this point, with utilization of NADPH for reductive biosynthetic reactions and ribose 5-phosphate as a precursor for nucleotide synthesis. The overall equation may be written as

$$\text{glucose 6-phosphate} + 2\ \text{NADP}^+ + \text{H}_2\text{O} \rightleftharpoons \text{ribose 5-phosphate} + 2\ \text{NADPH} + 2\ \text{H}^+ + \text{CO}_2$$

Interconversions of Pentose Phosphates Lead to Glycolytic Intermediates

In certain cells more NADPH is needed for **reductive biosynthesis** than ribose 5-phosphate for incorporation into nucleotides. A **sugar rearrangement** system (Figure 15.2) forms triose, tetrose, hexose, and heptose sugars from the pentoses, thus creating a **disposal mechanism** for ribose 5-phosphate, and providing a reversible link between the pentose phosphate pathway and glycolysis via intermediates common to both pathways. For the interconversions, another pentose phosphate, **xylulose 5-phosphate,** is formed through isomerization of **ribulose 5-phosphate** by the action of **phosphopentose epimerase.** As a consequence, ribulose 5-phosphate, ribose 5-phosphate, and xylulose 5-phosphate exist as an equilibrium mixture and can then undergo further transformations catalyzed by transketolase and transaldolase.

Transketolase requires thiamine pyrophosphate (TPP) and Mg^{2+}, transfers a C_2 unit designated active "glycolaldehyde" from xylulose 5-phosphate to ribose 5-phosphate, and produces the C_7 sugar sedoheptulose and glyceraldehyde 3-phosphate, an intermediate of glycolysis. Alterations in transketolase can lead to Wernicke–Korsakoff syndrome (see Clin. Corr. 15.2). In a second transfer reaction, catalyzed by **transaldolase**, a C_3 unit (dihydroxyacetone) from sedoheptulose 7-phosphate is transferred to glyceraldehyde 3-phosphate, forming the tetrose, erythrose 4-phosphate, and fructose 6-phosphate, another intermediate of glycolysis. In a third reaction transketolase catalyzes the synthesis of fructose 6-phosphate and

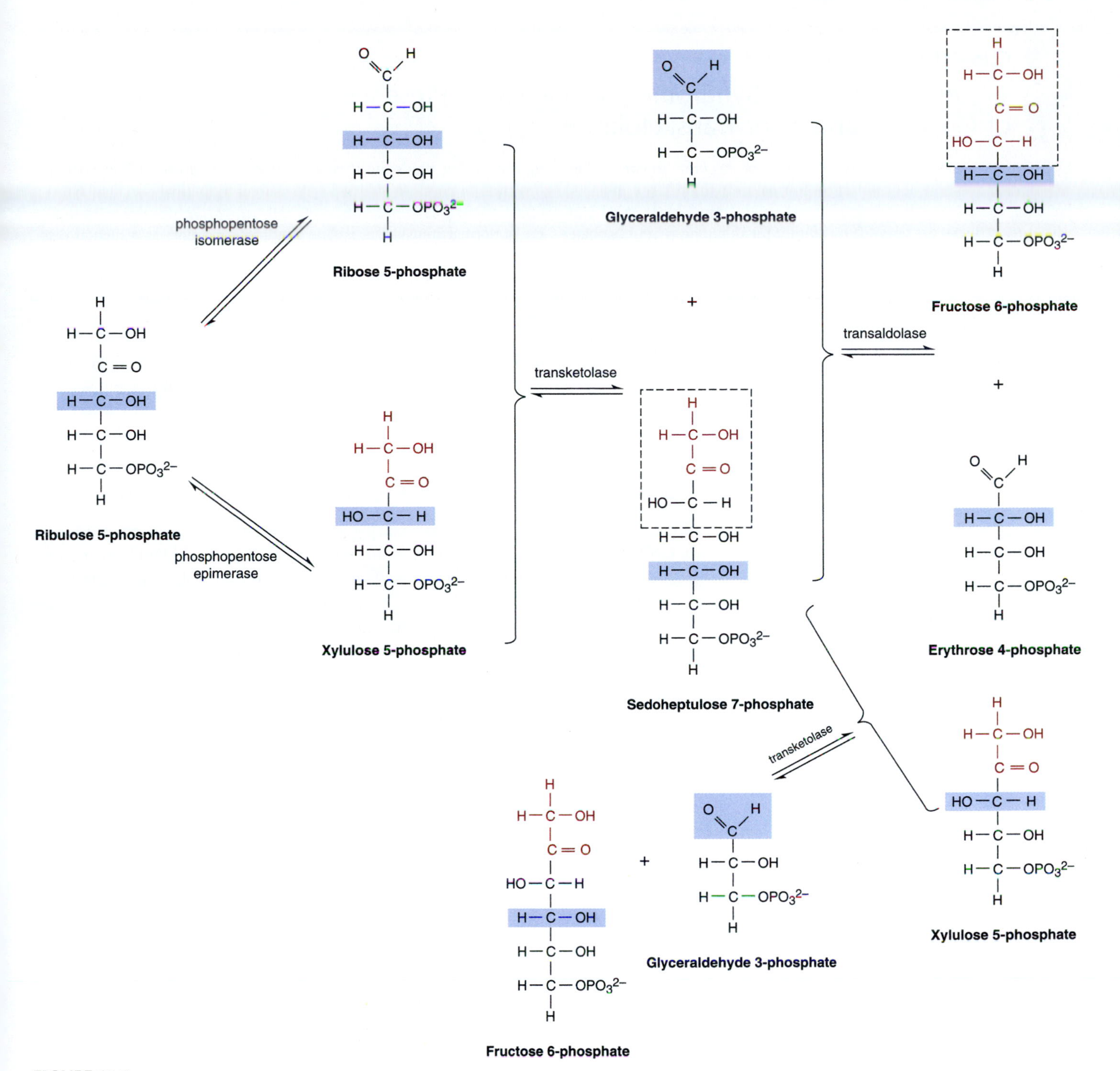

FIGURE 15.2
Nonoxidative reactions of the pentose phosphate pathway: interconversions of pentose phosphates.

glyceraldehyde 3-phosphate, forming two glycolytic intermediates from erythrose 4-phosphate and a second molecule of xylulose 5-phosphate. The sum of these reactions is

$$\text{2 xylulose 5-phosphate} + \text{ribose 5-phosphate} \rightleftharpoons \text{2 fructose 6-phosphate} + \text{glyceraldehyde 3-phosphate}$$

Since xylulose 5-phosphate is derived from ribose 5-phosphate, the net reaction starting from ribose 5-phosphate is

$$\text{3 ribose 5-phosphate} \rightleftharpoons \text{2 fructose 6-phosphate} + \text{glyceraldehyde 3-phosphate}$$

CLINICAL CORRELATION 15.2

Wernicke–Korsakoff Syndrome: Deficiency or Presence of Genetic Variants of Transketolase

Symptoms of Wernicke–Korsakoff syndrome become apparent after moderate stress that does not affect normal individuals. The molecular basis for this syndrome is an alteration in transketolase that significantly reduces (> tenfold) its affinity for thiamine pyrophosphate (TPP). Fluctuations in TPP concentration are common and do not affect normal individuals whose transketolase binds TPP sufficiently tightly so that moderate thiamine deficiencies do not affect enzyme function. In contrast, Wernicke–Korsakoff individuals become symptomatic when TPP concentrations fall below that needed to saturate the abnormal transketolase. The syndrome presents as a mental disorder, with accompanying memory loss and partial paralysis, and can become manifest in alcoholics, whose diets may be vitamin deficient.

Therefore excess ribose 5-phosphate, whether it arises from the first stage of the pentose phosphate pathway or from the degradative metabolism of nucleic acids, is effectively scavenged by conversion to intermediates that can enter the carbon flow of glycolysis.

Glucose 6-Phosphate Can Be Completely Oxidized to CO_2

In certain tissues, like lactating mammary gland, a pathway for complete **oxidation** of **G6P** to CO_2, with concomitant reduction of $NADP^+$ to NADPH, also exists (Figure 15.3). By a complex sequence of reactions, ribulose 5-phosphate produced by the pentose phosphate pathway is recycled into G6P by transketolase, transaldolase, and certain enzymes of the gluconeogenic pathway. Hexose continually en-

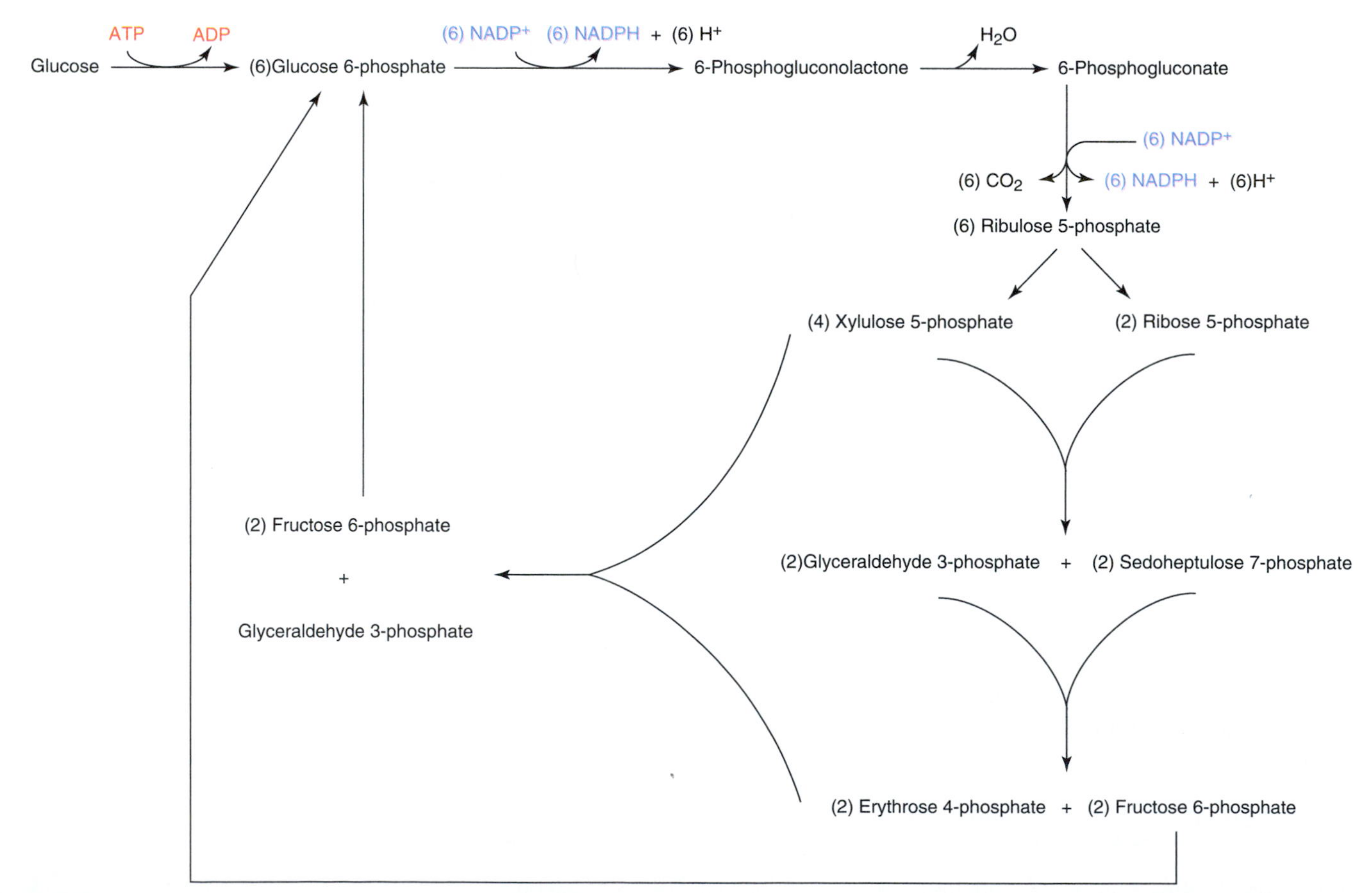

FIGURE 15.3
Pentose phosphate pathway.

ters this system, and CO_2 evolves as the only one carbon compound. A balanced equation for this process would involve the oxidation of six molecules of G6P to six molecules of ribulose 5-phosphate and six molecules of CO_2. This represents essentially the first part of the pentose phosphate pathway and results in transfer of 12 pairs of electrons to $NADP^+$, the requisite amount for total oxidation of one glucose to six CO_2. The remaining six molecules of ribulose 5-phosphate are then rearranged by the pathway described above to regenerate five molecules of G6P. The overall equation can be written as

$$\text{6 glucose 6-phosphate} + 12\ NADP^+ + 7\ H_2O \rightleftharpoons \text{5 glucose 6-phosphate} + 6\ CO_2 + 12\ NADPH + 12\ H^+ + P_i$$

The net reaction is therefore

$$\text{glucose 6-phosphate} + 12\ NADP^+ + 7\ H_2O \rightleftharpoons 6\ CO_2 + 12\ NADPH + 12\ H^+ + P_i$$

Pentose Phosphate Pathway Produces NADPH

The pentose phosphate pathway serves several purposes, including synthesis and degradation of sugars other than hexoses, particularly pentoses necessary for synthesis of nucleotides and other glycolytic intermediates. Most important is the ability to synthesize NADPH, which has a unique role in biosynthetic reactions. The direction of flow and path taken by G6P after entry into the pathway is determined largely by the needs of the cell for NADPH or sugar intermediates. When more NADPH than ribose 5-phosphate is required, the pathway leads to complete oxidation of G6P to CO_2 and resynthesis of G6P from ribulose 5-phosphate. Alternatively, if more ribose 5-phosphate than NADPH is required, G6P is converted to fructose 6-phosphate and glyceraldehyde 3-phosphate by the glycolytic pathway.

The distribution of the pentose phosphate pathway in tissues is consistent with its functions. It is present in erythrocytes for production of NADPH, required to generate reduced glutathione, which is essential for maintenance of normal red cell structure. It is also active in liver, mammary gland, testis, and adrenal cortex, sites of fatty acid or steroid synthesis that also require the reducing power of NADPH. In contrast, in mammalian striated muscle, which exhibits little fatty acid or steroid synthesis, all catabolism proceeds via glycolysis and the TCA cycle and no direct oxidation of glucose 6-phosphate occurs through the pentose phosphate pathway. In some other tissues like liver, 20–30% of the CO_2 produced may arise from the pentose phosphate pathway, and the balance between glycolysis and the pentose phosphate pathway depends on the metabolic requirements of the cell.

15.3 | SUGAR INTERCONVERSIONS AND NUCLEOTIDE-LINKED SUGAR FORMATION

In considering the general principles of carbohydrate metabolism, certain aspects of the origin and fate of other monosaccharides other than glucose and ribose, oligosaccharides, and polysaccharides should be included. Most monosaccharides found in biological compounds derive from glucose. The most common reactions for sugar transformations in mammalian systems are summarized in Figure 15.4.

Isomerization and Phosphorylation Are Common Reactions for Interconverting Carbohydrates

Formation of some sugars can occur directly, starting from glucose via modification reactions, such as the conversion of G6P to fructose 6-phosphate by phosphoglucose isomerase in glycolysis. A similar **aldose–ketose isomerization** catalyzed by **phosphomannose isomerase** results in synthesis of mannose 6-phosphate. Deficiencies

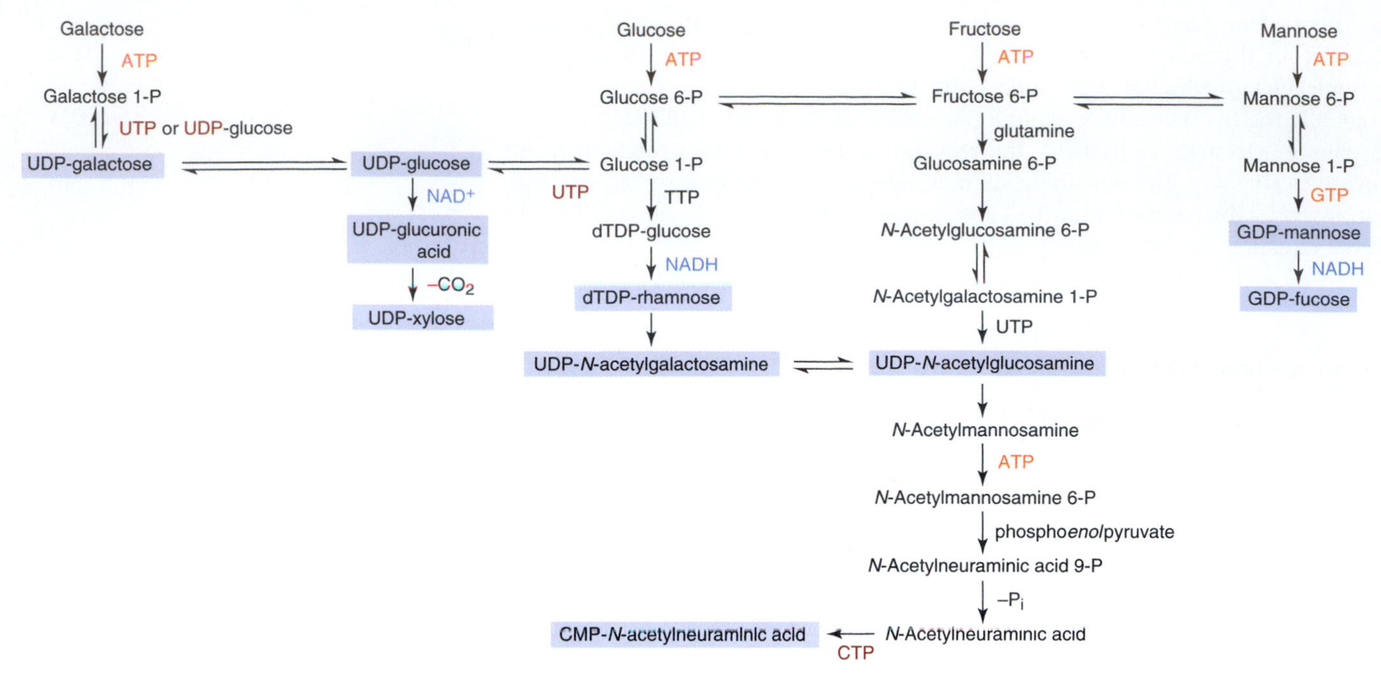

FIGURE 15.4
Pathways of formation of nucleotide-linked sugars and interconversions of some hexoses.

in this enzyme lead to one form of carbohydrate-deficient glycoprotein syndrome (CDGS) (see Clin. Corr. 15.3).

Internal transfer of a phosphate group on the same sugar molecule from one hydroxyl group to another is also a common modification. Glucose 1-phosphate, resulting from enzymatic phosphorolysis of glycogen, is converted to G6P by phosphoglucomutase. Galactose can be phosphorylated directly to galactose 1-phosphate by galactokinase and mannose to mannose 6-phosphate by mannokinase. Free fructose, an important dietary constituent, can be phosphorylated in the liver to fructose 1-phosphate by a special fructokinase. However, no mutase exists to interconvert fructose 1-phosphate and fructose 6-phosphate, nor can phosphofructokinase synthesize fructose 1,6-bisphosphate from fructose 1-phosphate. Rather, a fructose 1-phosphate aldolase cleaves fructose 1-phosphate to dihydroxyacetone

CLINICAL CORRELATION 15.3

Carbohydrate-Deficient Glycoprotein Syndromes (CDGSs)

The CDGSs are an expanding group of clinically heterogeneous autosomal recessive glycosylation disorders. Using the glycosylation state of endogenous serum transferrin as a sensitive indicator, four types of CDGS (I to IV) have been identified thus far. Type I is the most common and is now subdivided into three distinct forms, two of which are in the early mannose biosynthetic pathway (Figure 15.4). Type Ib, which presents in the first year of life with hypoglycemia, vomiting, diarrhea, protein-loss enteropathy, hepatic fibrosis, and susceptibility to recurring thrombosis due to low levels of antithrombin III, is due to a deficiency of phosphomannose isomerase activity. Defects in one form of the next pathway enzyme, phosphomannomutase 2, are responsible for CDGS type Ia and account for approximately 70% of all CDGS type I patients. Clinical symptoms include hypotonia, dysmorphism, liver dysfunction, developmental delays in motor and language development, and failure to thrive; about 20% of patients with this genetic disorder die within the first few years of life. A third form, CDGS type Ic, exhibits a similar transferrin glycosylation phenotype but is due to defects late in the glycoprotein biosynthetic pathway, for example, glucosyltransferase that adds the first glucosyl residue to the $Man_9GlcNac_2$-PP-Dol precursor. CDGS type II is also due to deficiencies in a later step in the pathway in N-acetylglucosminyltransferase II, which initiates the second N-linked antenna. The molecular etiology of CDGS types III and IV have not been identified. Thus there appears to be a constellation of syndromes caused by defects in monosaccharide interconversion or glycoprotein biosynthetic pathways that are characterized by neonatal presentation of severe neurological and metabolic dysfunction.

phosphate (DHAP), which enters the glycolytic pathway directly, and glyceraldehyde, which must first be reduced to glycerol, phosphorylated, and then reoxidized to DHAP. Lack of this aldolase leads to fructose intolerance (see Clin. Corr. 15.4).

CLINICAL CORRELATION 15.4

Essential Fructosuria and Fructose Intolerance: Deficiency of Fructokinase and Fructose 1-Phosphate Aldolase

Fructose may account for 30–60% of the total carbohydrate intake of mammals and is predominantly metabolized by a fructose-specific pathway. The first enzyme in this pathway, fructokinase, is deficient in essential fructosuria. This disorder is a benign asymptomatic metabolic anomaly that appears to be inherited as an autosomal recessive. Following intake of fructose, blood and urinary fructose concentration of affected individuals are unusually high; however, 90% of their fructose intake is eventually metabolized. In contrast, hereditary fructose intolerance is characterized by severe hypoglycemia after ingestion of fructose and prolonged ingestion by affected young children may lead to death. In this disorder fructose 1-phosphate aldolase is deficient, and fructose 1-phosphate accumulates intracellularly (see Clin. Corr. 14.3).

Nucleotide-Linked Sugars Are Intermediates in Many Sugar Transformations

Most other sugar transformation reactions require prior conversion into **nucleotide-linked sugars.** Formation of nucleoside diphosphate (NDP)-sugar involves the reaction of hexose 1-phosphate and nucleoside triphosphate (NTP), catalyzed by a pyrophosphorylase. While these reactions are readily reversible, *in vivo* pyrophosphate is rapidly hydrolyzed by pyrophosphatase, thereby driving the synthesis of nucleotide-linked sugars. These reactions are summarized as follows:

$$\text{NTP} + \text{sugar 1-phosphate} + H_2O \rightleftharpoons \text{NDP-sugar} + PP_i$$
$$PP_i + H_2O \rightleftharpoons 2\ P_i$$
$$\text{NTP} + \text{sugar 1-phosphate} + H_2O \rightleftharpoons \text{NDP-sugar} + 2\ P_i$$

UDP-glucose is a common nucleotide-linked sugar involved in synthesis of glycogen and glycoproteins and is synthesized from glucose 1-phosphate and UTP in a reaction catalyzed by UDP-glucose pyrophosphorylase.

Glucose + UTP ⇌ UDP-Glucose + PP_i

Nucleoside diphosphate-sugars contain two phosphoryl bonds, each with a large negative ΔG of hydrolysis, that contribute to the energized character of these compounds as glycosyl donors in further transformation and transfer reactions, as well as conferring substrate specificity for the enzymes catalyzing these reactions. For instance, uridine diphosphate usually serves as the glycosyl carrier, while ADP, GDP, and CMP act as carriers in other reactions. Many sugar transformation reactions, including epimerization, oxidation, decarboxylation, reduction, and rearrangement, occur only at the level of nucleotide-linked sugars.

Epimerization Interconverts Glucose and Galactose

Epimerization is a common type of reaction in carbohydrate metabolism. Reversible conversion of glucose to galactose in animals occurs by epimerization of UDP-glucose to UDP-galactose, catalyzed by UDP-glucose-4-epimerase (Figure 15.5). UDP-galactose is also an important intermediate in metabolism of free galactose, derived from the hydrolysis of lactose in the intestinal tract. Galactose is phosphorylated by galactokinase and ATP to yield galactose 1-phosphate. Then galactose 1-phosphate uridylyltransferase transforms galactose 1-phosphate into

UDP-glucose ⇌ (UDP-glucose-4-epimerase) UDP-Galactose

FIGURE 15.5
Conversion of glucose into galactose.

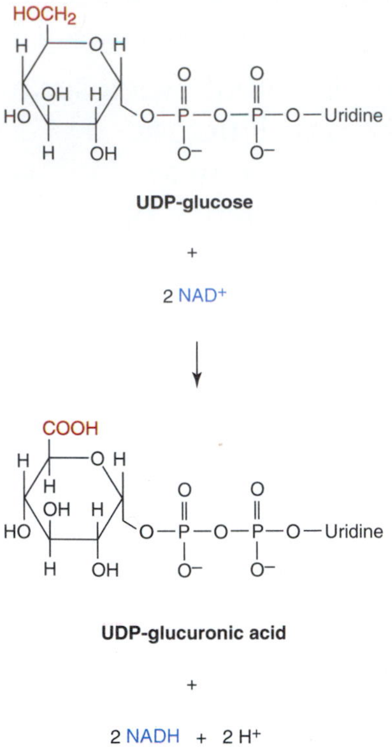

UDP-glucose

+

2 NAD^+

↓

UDP-glucuronic acid

+

2 NADH + 2 H^+

FIGURE 15.6
Formation of UDP-glucuronic acid from UDP-glucose.

UDP-galactose by displacing glucose 1-phosphate from UDP-glucose. These reactions are summarized as follows:

$$\text{galactose} + \text{ATP} \rightleftharpoons \text{galactose 1-phosphate} + \text{ADP}$$

$$\text{UDP-glucose} + \text{galactose 1-phosphate} \rightleftharpoons \text{UDP-galactose} + \text{glucose 1-phosphate}$$

A combination of these reactions allows the efficient transformation of dietary galactose into glucose 1-phosphate, which can then be further metabolized by previously described pathways. Alternatively, the 4-epimerase can operate in the reverse direction when UDP-galactose is needed for biosynthesis. A hereditary disorder, galactosemia, results from the absence of the uridylyltransferase and is characterized by decreased synthesis and availability of UDP-galactose (see Clin. Corr. 15.5).

Epimerization is one of three separate reactions involved in synthesis of GDP-fucose (Figure 15.4). This set begins with the conversion of GDP-mannose to GDP-4-keto-6-deoxymannose catalyzed by GDP-mannose-4,6-dehydratase. GDP-fucose is then produced from GDP-4-keto-6-deoxymannose by a two-step reaction involving 3,5-epimerase-dependent conversion to GDP-4-keto-6-deoxy-L-galactose followed by the 4-reductase-dependent formation of GDP-fucose. These latter two reactions are catalyzed by a single polypeptide, GDP-4-keto-6-deoxymannose 3,5-epimerase-4-reductase; also known as the FX protein, an abundant red cell protein.

Epimerization reactions are not exclusively restricted to nucleotide-linked sugars but also occur at the polymer level; D-glucuronic acid is epimerized to L-iduronic acid after incorporation into heparin and dermatan sulfate (see Section 15.6, p. 684).

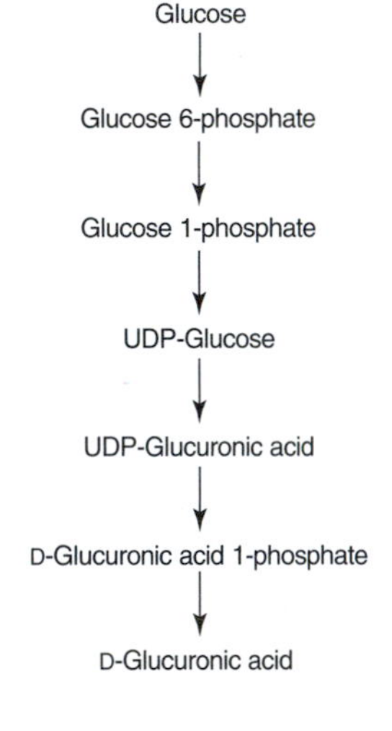

FIGURE 15.7
Biosynthesis of D-glucuronic acid from glucose.

Glucuronic Acid Is Formed by Oxidation of UDP-Glucose

Glucuronic acid is formed by oxidation of UDP-glucose catalyzed by **UDP-glucose dehydrogenase** (Figure 15.6) and most likely follows the path outlined in Figure 15.7. In humans glucuronic acid is converted to L-xylulose, the ketopentose excreted in essential pentosuria (see Clin. Corr. 15.6), and participates in detoxification by formation of glucuronide conjugates (see Clin. Corr. 15.7).

Glucuronic acid is a precursor of **L-ascorbic acid** in those animals that synthesize vitamin C. Free glucuronic acid can be metabolized by reduction with NADPH to L-gulonic acid (Figure 15.8), which is then converted by a two-step process through L-gulonolactone to L-ascorbic acid (vitamin C) in plants and most higher animals. Humans, other primates, and guinea pigs lack the enzyme that converts L-gulonolactone to L-ascorbic acid and therefore must satisfy their needs for

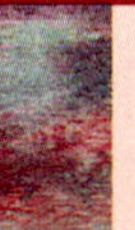

CLINICAL CORRELATION 15.5
Galactosemia: Inability to Transform Galactose into Glucose

Reactions of galactose are of particular interest because in humans they are subject to genetic defects resulting in the hereditary disorder galactosemia. When a defect is present, individuals are unable to metabolize the galactose derived from lactose (milk sugar) to glucose metabolites, often with resultant cataract formation, growth failure, mental retardation, or eventual death from liver damage. The genetic disturbance is expressed as a cellular deficiency of either galactokinase, causing a relatively mild disorder characterized by early cataract formation, or of galactose 1-phosphate uridylyltransferase, resulting in severe disease.

Galactose is reduced to galactitol in a reaction similar to the reduction of glucose to sorbitol. Galactitol is the initiator of cataract formation in the lens and may play a role in the central nervous system damage. Accumulation of galactose 1-phosphate is responsible for liver failure; the toxic effects of galactose metabolites disappear when galactose is removed from the diet.

CLINICAL CORRELATION 15.6

Pentosuria: Deficiency of Xylitol Dehydrogenase

The glucuronic acid oxidation pathway is presumably not essential for human carbohydrate metabolism, since individuals in whom the pathway is blocked suffer no ill effects. A metabolic variation, called idiopathic pentosuria, results from reduced activity of NADP-linked L-xylulose reductase, the enzyme that catalyzes the reduction of xylulose to xylitol. Hence affected individuals excrete large amounts of pentose into the urine, especially following intake of glucuronic acid.

ascorbic acid by its ingestion. Gulonic acid can also be oxidized to 3-ketogulonic acid and decarboxylated to L-xylulose. L-Xylulose is reduced to xylitol, reoxidized to D-xylulose, and phosphorylated with ATP and an appropriate kinase to xylulose 5-phosphate. The latter compound can then reenter the pentose phosphate pathway described previously. The glucuronic acid pathway represents another pathway for oxidation of glucose. This pathway operates in adipose tissue, and its activity is usually increased in tissue from starved or diabetic animals.

Decarboxylation, Oxidoreduction, and Transamidation of Sugars Yield Necessary Products

Although decarboxylation has previously been encountered in the major metabolic pathways, the only known decarboxylation of a nucleotide-linked sugar is the conversion of UDP-glucuronic acid to UDP-xylose. UDP-xylose is necessary for synthesis of proteoglycans (Section 15.6, see p. 683) and is a potent inhibitor of UDP-glucose dehydrogenase, the enzyme that oxidizes UDP-glucose to UDP-glucuronic acid (Figure 15.6). Thus the level of these nucleotide-linked sugar precursors is regulated by this sensitive feedback mechanism.

Deoxyhexoses and **dideoxyhexoses** are also synthesized while the precursor sugars are attached to nucleoside diphosphates, by a multistep process. For example, L-rhamnose is synthesized from glucose by a series of oxidation–reduction reactions starting with dTDP-glucose and yielding dTDP-rhamnose, catalyzed by oxidoreductases. A similar reaction accounts for synthesis of GDP-fucose from GDP-mannose (see previous section) and for various dideoxyhexoses.

Formation of amino sugars, major components of human oligo- and polysaccharides and constituents of antibiotics, occurs by **transamidation.** For example,

CLINICAL CORRELATION 15.7

Glucuronic Acid: Physiological Significance of Glucuronide Formation

The biological significance of glucuronic acid extends to its ability to be conjugated with certain endogenous and exogenous substances, forming a group of compounds collectively termed glucuronides in a reaction catalyzed by UDP-glucuronyltransferase. Conjugation of a compound with glucuronic acid produces a strongly acidic compound that is more water soluble at physiological pH than its precursor and therefore may alter the metabolism, transport, or excretion properties. Glucuronide formation is important in drug detoxification, steroid excretion, and bilirubin metabolism. Bilirubin is the major metabolic breakdown product of heme, the prosthetic group of hemoglobin. The central step in excretion of bilirubin is conjugation with glucuronic acid by UDP-glucuronyltransferase. Development of this conjugating mechanism occurs gradually and may take several days to two weeks after birth to become fully active in humans. "Physiological jaundice of the newborn" results in most cases from the inability of the neonatal liver to form bilirubin glucuronide at a rate comparable to that of bilirubin production. A defect in glucuronide synthesis has been found in a mutant strain of Wistar ("Gunn") rats, is due to a deficiency of UDP-glucuronyltransferase, and results in hereditary hyperbilirubinemia. In humans, a similar defect is found in congenital familial nonhemolytic jaundice (Crigler–Najjar syndrome). Patients with this condition are unable to conjugate foreign compounds efficiently with glucuronic acid.

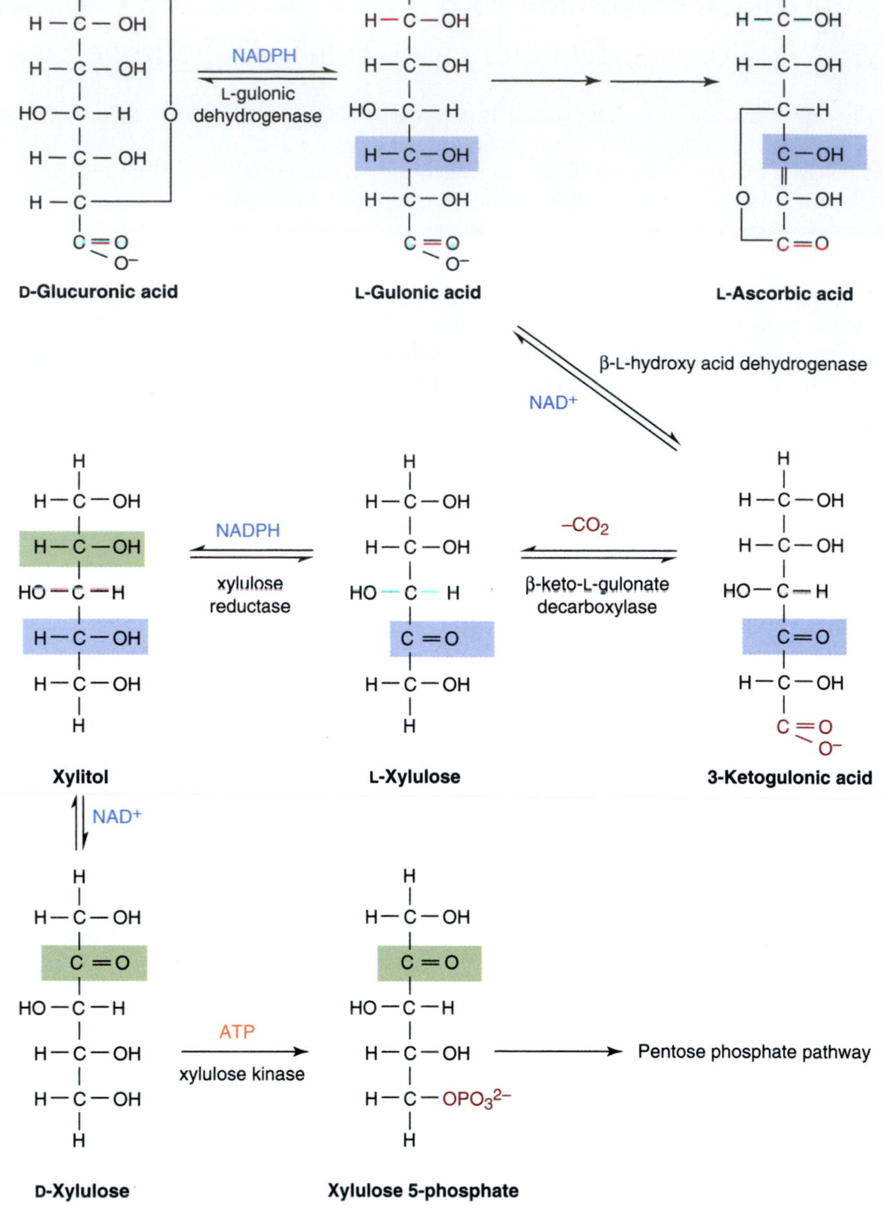

FIGURE 15.8
Glucuronic acid oxidation pathway.

synthesis of glucosamine 6-phosphate occurs by the reaction of fructose 6-phosphate with glutamine.

$$NH_2-\overset{O}{\overset{\|}{C}}(CH_2)_2-\underset{\overset{|}{\overset{+}{N}H_3}}{CH}-COO^-$$

Glutamine

transamidase

$^-OOC(CH_2)_2-\underset{\overset{|}{\overset{+}{N}H_3}}{CH}-COO^-$

Fructose 6-phosphate

Glutamate

Glucosamine 6-phosphate

Glucosamine 6-phosphate can be *N*-acetylated, forming *N*-acetylglucosamine 6-phosphate, followed by isomerization to *N*-acetylglucosamine 1-phosphate. This latter sugar is converted to UDP-*N*-acetylglucosamine by reactions similar to those of UDP-glucose synthesis. UDP-*N*-acetylglucosamine, a precursor of glycoprotein synthesis, can be epimerized to UDP-*N*-acetylgalactosamine, necessary for proteoglycan synthesis. The fructose 6-phosphate–glutamine transamidase reaction is under negative feedback control by UDP-*N*-acetylglucosamine; thus synthesis of both nucleotide-linked sugars is regulated (Figure 15.4). This regulation is meaningful in certain tissues such as skin, in which this pathway can involve up to 20% of glucose flux.

Sialic Acids Are Derived from *N*-Acetylglucosamine

Another product derived from UDP-*N*-acetylglucosamine is **N-acetylneuraminic acid,** one of a family of C_9 sugars, called **sialic acids** (Figure 15.9). The first

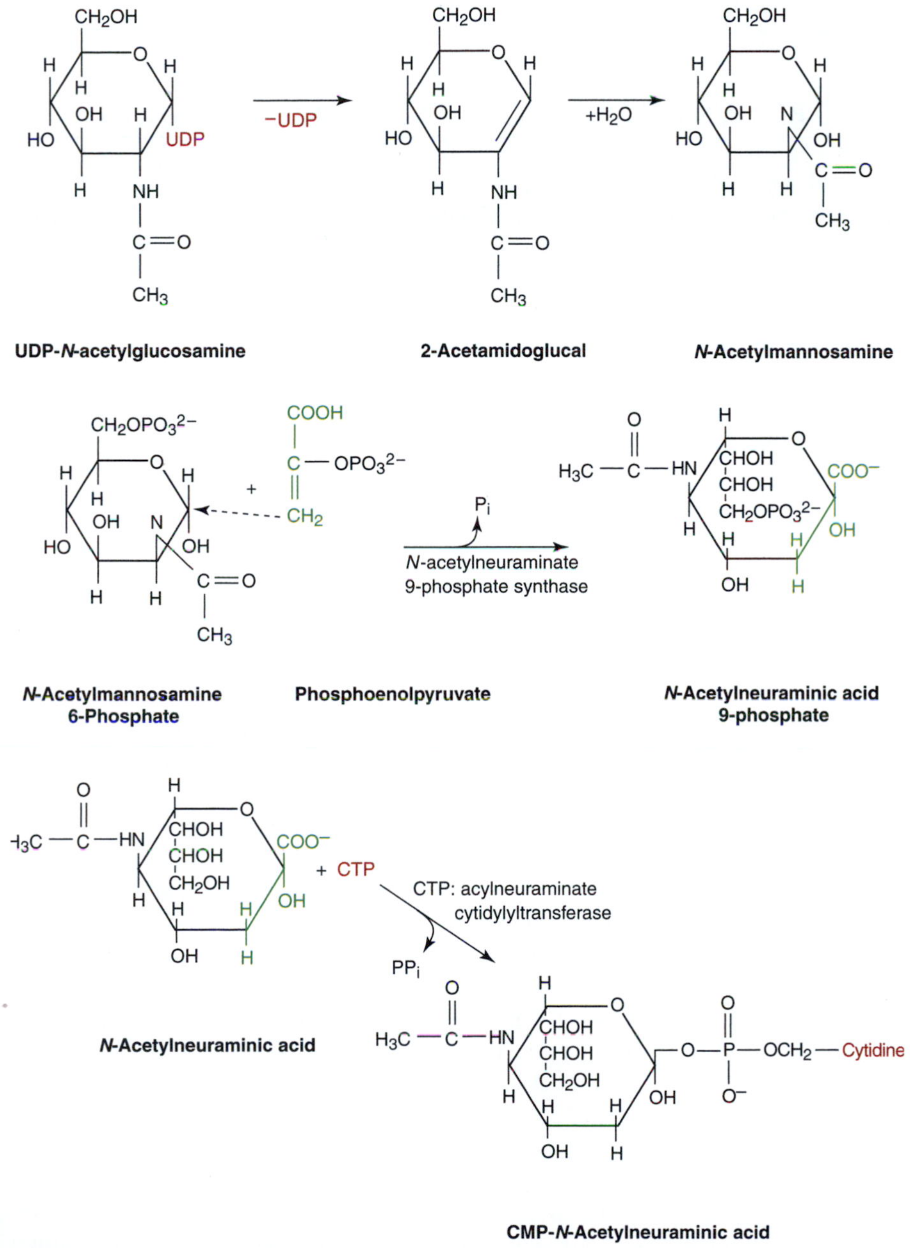

FIGURE 15.9
Biosynthesis of CMP-*N*-acetylneuraminic acid.

reaction involves epimerization of UDP-*N*-acetylglucosamine by a 2-epimerase to *N*-acetylmannosamine, concomitant with elimination of UDP. Since the monosaccharide product is no longer bound to nucleotide, this epimerization is clearly different from those previously encountered. Most likely, this 2-epimerase reaction proceeds by a trans elimination of UDP, with formation of the unsaturated intermediate, 2-acetamidoglucal. In mammalian tissues *N*-acetylmannosamine is presumably phosphorylated by the same enzyme acting as a kinase using ATP to form *N*-acetylmannosamine 6-phosphate, which then condenses with phospho*enol*pyruvate to form *N*-acetylneuraminic acid 9-phosphate. This product is cleaved by a phosphatase to remove the phosphate and activated by CTP to form the CMP derivative, CMP-*N*-acetylneuraminic acid. This is an unusual nucleotide-linked sugar containing only one phosphate group and is formed by an irreversible reaction. All of these reactions occur in the cytosol, with the exception of the last step, which occurs in the nucleus with subsequent export of the activated nucleotide-linked sugar to the cytoplasm. *N*-Acetylneuraminic acid is also a precursor of other sialic acid derivatives, some of which evolve by modification of *N*-acetyl group to *N*-glycolyl or *O*-acetyl after incorporation into glycoprotein.

15.4 | BIOSYNTHESIS OF COMPLEX CARBOHYDRATES

In complex carbohydrate-containing molecules, sugars are linked to other sugars by glycosidic bonds formed by specific **glycosyltransferases.** Energy is required for synthesis of a glycosidic bond and is derived from the nucleotide-linked sugar donor substrate. A glycosyltransferase reaction proceeds by donation of the glycosyl unit from the nucleotide derivative to the nonreducing end of an acceptor sugar. The nature of the bond formed is determined by the specificity of the glycosyltransferase, which is unique for the sugar acceptor, the sugar transferred, and the linkage formed; however, exceptions to the enzyme-one linkage paradigm have emerged. Still, heteropolysaccharide synthesis is controlled by a nontemplate mechanism directed by specific glycosyltransferases. A glycosyltransferase reaction is summarized as follows:

$$\underset{\text{(donor)}}{\text{nucleoside diphosphate-glycose}} + \underset{\text{(acceptor)}}{\text{glycose}} \xrightleftharpoons{\text{glycosyltransferase}} \underset{\text{(glycoside)}}{\text{glycosyl}_1\text{-}O\text{-glycose}_2} + \text{nucleoside diphosphate}$$

More than 40 different glycosidic bonds have been identified in mammalian oligosaccharides and about 15 more in connective tissue polysaccharides. The number of possible linkages is even greater and arises both from the diversity of monosaccharides covalently bonded and from the formation of both α and β linkage with each of the available hydroxyl groups on the acceptor saccharide. The large and diverse number of molecules that can be generated suggests that oligosaccharides have the potential for great informational content. In fact, it is known that the specificity of many biological molecules is determined by the nature of the composite sugar residues. For example, the specificity of the major blood types is determined by the sugars they contain (see Clin. Corr. 15.8). *N*-Acetylgalactosamine is the immunodeterminant of blood type A and galactose of blood type B. Removal of *N*-acetylgalactosamine from type A erythrocytes, or of galactose from type B erythrocytes, converts them to type O erythrocytes. Increasingly, other examples of sugars as determinants of specificity for cell surface receptor and lectin interactions, targeting of cells to certain tissues, and survival or clearance from the circulation of certain molecules are being recognized. All glycosidic bonds identified in biological compounds are degraded by specific hydrolytic enzymes, glycosidases. In addition to being valuable tools for the structural elucidation of oligosaccharides, interest in

CLINICAL CORRELATION 15.8
Blood Group Substances

The surface of human erythrocytes is covered with a complex mosaic of specific antigenic determinants, many of which are polysaccharides. There are about 100 blood group determinants, belonging to 21 independent human blood group systems. The most widely studied are the antigenic determinants of the ABO blood group system and the closely related Lewis system. From the study of these systems, a definite correlation was established between gene activity as it relates to specific glycosyltransferase synthesis and oligosaccharide structure. The genetic variation is achieved through specific glycosyltransferases responsible for synthesis of the heterosaccharide determinants. For example, the *H* gene codes for a fucosyltransferase, which adds fucose to a peripheral galactose in the heterosaccharide precursor. The ABO locus is located on chromosome 9. The *A* allele encodes an *N*-acetylgalactosamine glycosyltransferase, the *B* allele encodes a galactosyltransferase, and the *O* allele encodes an inactive protein. The sugars transferred by the A and B enzymes are added to the H specific oligosaccharide. The Lewis (*Le*) gene codes for another fucosyltransferase, which adds fucose to a peripheral *N*-acetylglucosamine residue in the precursor. Absence of the *H* gene product gives rise to the Le^a-specific determinant, whereas in the absence of both the H and Le enzymes, the interaction product responsible for the Le^b specificity is found. The elucidation of the structures of these oligosaccharide determinants represents a milestone in carbohydrate chemistry. This knowledge is essential to blood transfusion practices and for legal and historical purposes. For example, tissue dust containing complex carbohydrates has been used in serological analysis to establish the blood group of Tutankhamen and his probable ancestral background.

this class of enzymes is based on the many genetic diseases of complex carbohydrate metabolism that result from defects in glycosidases (see Clin. Corr. 15.10 and 15.11).

15.5 | GLYCOPROTEINS

Glycoproteins have restrictively been defined as conjugated proteins that contain, as a prosthetic group, one or more saccharides lacking a serial repeat unit and bound covalently to a peptide chain. This definition excludes proteoglycans, which are discussed in Section 15.6 (see p. 683).

The functions of glycoproteins in the human are of great interest. Glycoproteins in cell membranes may have an important role in the group behavior of cells and other important biological functions of the membrane. Glycoproteins form a major part of the mucus that is secreted by epithelial cells, where they perform an important role in lubrication and in the protection of tissues lining the respiratory and gastrointestinal systems. Many other proteins secreted from cells into extracellular fluids are glycoproteins. These proteins include hormones found in blood, such as follicle-stimulating hormone, luteinizing hormone, and chorionic gonadotropin; and plasma proteins such as the orosomucoids, ceruloplasmin, plasminogen, prothrombin, and immunoglobulins (see Clin. Corr. 9.1).

Glycoproteins Contain Variable Amounts of Carbohydrate

The amount of carbohydrate in glycoproteins is highly variable. Some glycoproteins such as IgG contain low amounts (4%) of carbohydrate by weight, while glycophorin, the human red cell membrane glycoprotein, contains 60% carbohydrate. Human ovarian cyst glycoprotein contains 70% carbohydrate, and human gastric glycoprotein is 82% carbohydrate. The carbohydrate can be distributed fairly evenly along the polypeptide chain or concentrated in defined regions. For example, in human glycophorin A the carbohydrate is found in the NH_2-terminal half of the polypeptide chain that lies on the outside of the cellular membrane.

The carbohydrate attached at one or at multiple points along a polypeptide chain usually contains less than 12–15 sugar residues. In some cases the carbohydrate component consists of a single sugar moiety, as in the submaxillary gland glycoprotein (single *N*-acetyl-α-D-galactosaminyl residue) and in some types of mammalian collagens (single α-D-galactosyl residue). In general, glycoproteins contain sugar residues in the D form, except for L-fucose, L- arabinose, and L-iduronic acid. A glycoprotein from different animal species often has an identical primary structure in the protein component, but a variable carbohydrate component. This heterogeneity of a given protein may even be true within a single organism. For example, pancreatic ribonuclease A and B forms have identical amino acid sequences and a similar kinetic specificity toward substrates but differ significantly in their carbohydrate compositions.

Carbohydrates Are Covalently Linked to Glycoproteins by–or *O*-Glycosyl Bonds

At present, the structures of a limited number of oligosaccharide components have been completely elucidated. Microheterogeneity of glycoproteins, arising from incomplete synthesis or partial degradation, makes structural analyses extremely difficult. However, certain generalities about the structure of glycoproteins have emerged. Covalent linkage of sugars to the peptide chain is a central part of glycoprotein structure, and only a limited number of bond types are found (see Chapter 3). The three major types of **glycopeptide bonds,** as shown in Figures 15.10

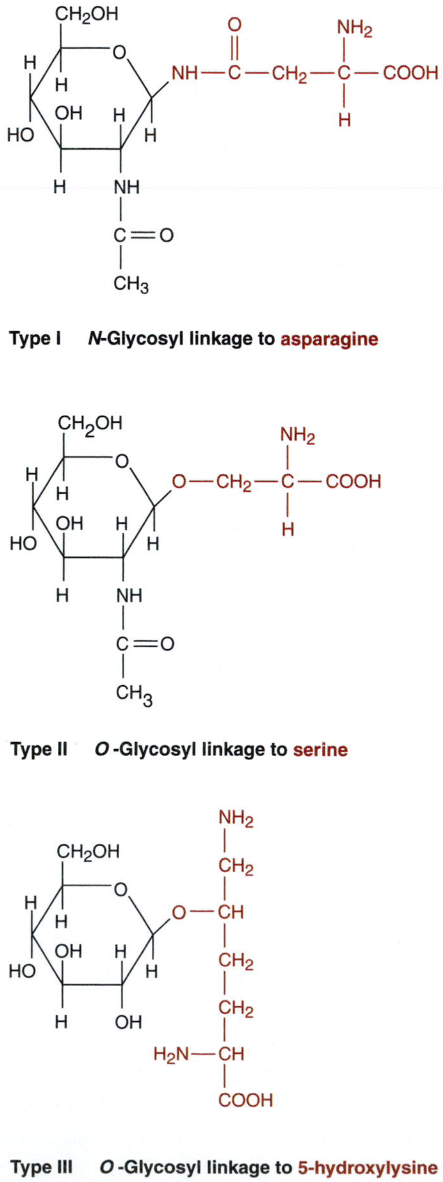

Type I ***N*-Glycosyl linkage to asparagine**

Type II ***O*-Glycosyl linkage to serine**

Type III ***O*-Glycosyl linkage to 5-hydroxylysine**

FIGURE 15.10
Structure of three major types of glycopeptide bond.

and 3.45 (see p. 132), are *N*-glycosyl to **asparagine** (Asn), *O*-glycosyl to **serine** (Ser) or **threonine** (Thr), and *O*-glycosyl to **5-hydroxylysine.** The latter linkage, found for side chains of either a single galactose or the disaccharide glucosylgalactose covalently bonded to hydroxylysine, is generally confined to the collagens. The other two linkages occur in a wide variety of glycoproteins. Of the three major types, only the *O*-glycosidic linkage to serine or threonine is labile to alkali cleavage. By this procedure two types of oligosaccharides (simple and complex) are released. Examination of the simple class from porcine submaxillary mucins reveals some general structural features. A core structure exists, consisting of galactose (Gal) linked β $(1 \rightarrow 3)$ to *N*-acetylgalactosamine (GalNAc) *O*-glycosidically linked to serine or threonine residues. Residues of L-fucose (Fuc), sialic acid (NeuAc), and another *N*-acetylgalactosamine are found at the nonreducing end of this class of glycopeptides. The general structure of this type of glycopeptide is as follows:

$$\begin{array}{ccccccc} \text{GalNAc} & \xrightarrow{1.3} & \text{Gal} & \xrightarrow{1.3} & \text{GalNAc} & \longrightarrow & O\text{-Ser/Thr} \\ & & \uparrow{\scriptstyle 1.2} & & \uparrow{\scriptstyle 2.6} & & \\ & & \text{Fuc} & & \text{NeuAc} & & \end{array}$$

More complex heterosaccharides are also linked to peptides via serine or threonine residues and are exemplified by the blood group substances. Study of these determinants has shown how complex and variable these structures are, as well as how the oligosaccharides of cell surfaces are assembled and how that assembly pattern is genetically determined. An example of how oligosaccharide structures on the surface of red blood cells determine blood group specificity is presented in Clinical Correlation 15.8. Certain common structural features of the oligosaccharide *N*-glycosidically linked to asparagine have also emerged. These glycoproteins commonly contain a core structure consisting of mannose (Man) residues linked to *N*-acetylglucosamine (GlcNAc) in the following structure:

$$(\text{Man})_n \xrightarrow[1.4]{} \text{Man} \xrightarrow[1.4]{} \text{GlcNAc} \xrightarrow[1.4]{} \text{GlcNAc} \longrightarrow \text{Asn}$$

The structural diversity of *N*-linked glycoproteins arises from a large repertoire of high-mannose, hybrid, and complex *N*-glycan subtypes.

Synthesis of *N*-Linked Glycoproteins Involves Dolichol Phosphate

While synthesis of *O*-glycosidically linked glycoproteins involves the sequential action of a series of glycosyltransferases, synthesis of *N*-glycosidically linked glycoproteins involves a somewhat different and more complex mechanism (Figure 15.11). A common core is preassembled as a **lipid-linked oligosaccharide** on the cytosolic side of the ER and then "flipped" across the bilayer prior to incorporation into the polypeptide. During synthesis, the oligosaccharide intermediates are bound to derivatives of **dolichol phosphate.**

$$(CH_2{=}\overset{\overset{\Large CH_3}{|}}{C}{-}CH{=}CH)_n{-}CH_2{-}\overset{\overset{\Large CH_3}{|}}{CH}{-}CH_2{-}CH_2O{-}PO_3H_2$$

Dolichol phosphate

Dolichols are polyprenols (C_{80}–C_{100}) that contain 16–20 isoprene units, in which the final isoprene unit is saturated. These lipids participate in two types of reactions in core oligosaccharide synthesis. The first reaction involves formation of *N*-acetylglucosaminyl pyrophosphoryldolichol with release of UMP from the respective nucleotide-linked sugar. The second involves *N*-acetylglucosamine and the mannose transferase reactions proceed by sugar transfer from the nucleotide without formation of intermediates. Subsequent addition of mannose units occurs via a dolichol-linked mechanism. In the final step, the oligosaccharide is transferred from the dolichol pyrophosphate to an asparagine residue in the polypeptide chain.

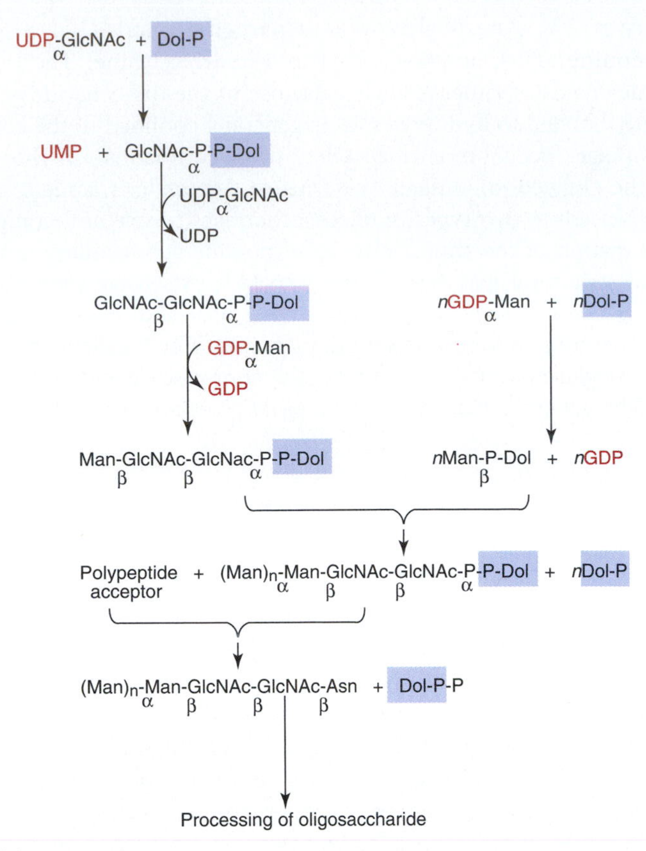

FIGURE 15.11
Biosynthesis of the oligosaccharide core in asparagine-*N*-acetylgalactosamine-linked glycoproteins.
Dol, dolichol.

CLINICAL CORRELATION 15.9

Common Carbohydrate Marker of Lysosomal Targeting and I-Cell Disease

I-cell disease, so-called because of the large inclusion bodies observed in cells cultured from patients, is characterized by severe clinical and radiological features including congenital dislocations, thoracic deformities, hernia, restricted joint mobility, and retarded psychomotor development. This rare and fatal congenital disorder provided the earliest connection between glycoprotein biosynthesis and human disease. In this condition lysosomal function is defective because the acid hydrolase content is low due to a defect in processing the glycoprotein acid hydrolases, causing failure to generate the Man-6-P residue. Without this specific marker, acid hydrolases are not targeted to their ultimate destination, the lysosomes, and are instead secreted into the extracellular milieu. In most patients the GlcNAc phosphotransferase that attaches the mannose-6-phosphate residue has been found to be defective, graphically illustrating the power of glycoprotein processing to direct intracellular protein trafficking.

After synthesis of the specific core region, the oligosaccharide chains are completed by action of glycosyltransferases without further participation of lipid intermediates. A series of early processing reactions, which are highly conserved among vertebrate species and cell types, occur largely in the ER and appear to be coupled with proper folding of the protein to which the glycan chain is covalently attached. Following the initial trimming and release from the ER, *N*-glycans undergo further glycosidase and glycosyltransferase modifications, mostly in the Golgi. Several avenues exist in the processing pathway that determine the final diversity of glycan structure (i.e., high-mannose, hybrid, or complex subtypes) as well as the trafficking fate of glycoproteins. For instance, on certain glycoproteins, the Man_8 $GlcNAc_2$-Asn *N*-glycan is modified by addition of a GlcNAc residue catalyzed by a GlcNAc-phosphotransferase, and subsequent removal by GlcNAc phosphodiester glycosidase, exposing a Man-6-P residue. This key modification is the structural determinant that targets some glycoproteins to the lysosomal compartment and forms the basis of the lysosomal storage disease known as I-cell disease (see Clin. Corr. 15.9).

Just as synthesis of heterooligosaccharides requires specific glycosyltransferases, degradation requires specific glycosidases. Exoglycosidases remove sugars sequentially from the nonreducing end, exposing the substrate for the subsequent glycosidase. The absence of a particular glycosidase prevents the action of the next enzyme, resulting in cessation of catabolism and accumulation of the product (see Clin. Corr. 15.10). Endoglycosidases with broader specificity also exist, and the action of endo- and exoglycosidases results in catabolism of glycoproteins. Many of the same *N*- or *O*-linked glycan chains are found on both glycoproteins and glycolipids; hence certain enzyme defects may affect degradation of both types of glyconjugates, while others appear to be glycoconjugate-specific (see Clin. Corr. 15.11).

CLINICAL CORRELATION 15.10

Aspartylglycosylaminuria: Absence of 4-L-Aspartylglycosamine Amidohydrolase

A group of human inborn errors of metabolism exists that involve storage of glycolipids, glycopeptides, mucopolysaccharides, and oligosaccharides. These diseases are caused by defects in lysosomal glycosidase activity, which prevent the catabolism of oligosaccharides. The disorders involve gradual accumulation in tissues and urine of compounds derived from incomplete degradation of the oligosaccharides and may be accompanied by skeletal abnormalities, hepatosplenomegaly, cataracts, or mental retardation. One disorder resulting from a defect in catabolism of asparagine-*N*-acetylglucosamine-linked oligosaccharides is aspartylglycosylaminuria in which a deficiency of the enzyme 4-L-aspartylglycosylamine amidohydrolase allows accumulation of aspartylglucosamine-linked structures (see accompanying table). Other disorders have been described involving accumulation of oligosaccharides derived from both glycoproteins and glycolipids, which may share common oligosaccharide structures (see table and Clin. Corr. 15.11).

Enzymic Defects in Degradation of Asn-GlcNAc Type Glycoproteins[a]

Disease	*Deficient Enzyme*[b]
Aspartylglycosylaminuria	4-L-Aspartylglycosylamine amidohydrolase (2)
β-Mannosidosis	β-Mannosidosis (7)
α-Mannosidosis	α-Mannosidosis (3)
GM_2 gangliosidosis variant O (Sandhoff–Jatzkewitz disease)	β-*N*-Acetylhexosaminidases (A and B) (4)
GM_1 gangliosidosis	β-Galactosidase (5)
Mucolipidosis I (sialidosis)	Sialidase (6)
Fucosidosis	α-Fucosidase (8)

[a]A typical Asn-GlcNAc oligosaccharide structure.

NeuAc —(6) α— Gal —(5) β— GlcNAc —(4) β— Man
NeuAc —α— Gal —β— GlcNAc —β— Man
Man —(3) α— Man —(7) β— GlcNAc —(2) β— GlcNAc —(1)— Asn
GlcNAc (bisecting, on Man); Fuc —(8)— GlcNAc
NeuAc —α— Gal —β— GlcNAc —β— Man α
NeuAc —(6) α— Gal —(5) β— GlcNAc —(4) β— Man

[b]The numbers in parentheses refer to the enzymes that hydrolyze those bonds.

CLINICAL CORRELATION 15.11

Glycolipid Disorders

A host of human genetic diseases arise from deficiencies in hydrolases, which act predominantly on glycolipid substrates, resulting in accumulation of glycolipid and ganglioside products. The clinical symptoms of diseases associated with each of the glycoconjugates may vary greatly. However, because of the preponderance of lipids in the nervous system, disorders that affect glycolipid degradation often have associated neurodegeneration and severe mental and motor deterioration.

Enzymic Defects in Degradation of Glycolipids

Disease	*Deficiency Enzyme*
Tay–Sachs	β-Hexosaminidase A
Sandhoff's	β-Hexosaminidases A and B
GM_1 gangliosidosis	β-Galactosidase
Sialidosis	Sialidase
Fabry's	α-Galactosidase
Gaucher's	β-Glucoceramidase
Krabbe's	β-Galactoceramidase
Metachromatic leukodystrophy	Arylsulfatase A (cerebroside sulfatase)

15.6 | PROTEOGLYCANS

In addition to glycoproteins, which usually contain proportionally less carbohydrate than protein by weight, there is another class of complex macromolecules that can contain as much as 95% or more carbohydrate, with properties that may resemble those of polysaccharides more than proteins. To distinguish these compounds from other glycoproteins, they are referred to as **proteoglycans.** Their carbohydrate chains are called **glycosaminoglycans** or by their older name, **mucopolysaccharides,** especially in reference to the group of storage diseases, mucopolysaccharidoses, which result from an inability to degrade these molecules (see Clin. Corr. 15.13).

Proteoglycans are high molecular weight polyanionic compounds consisting of many different glycosaminoglycan chains linked covalently to a protein core. Although six distinct classes of glycosaminoglycans are now recognized, certain features are common to all classes. The long unbranched heteropolysaccharide chains are made up largely of disaccharide repeating units, in which one sugar is a hexosamine and the other a uronic acid. Other common constituents of glycosaminoglycans are sulfate groups, linked by ester bonds to certain monosaccharides or by amide bonds to the amino group of glucosamine. However, hyaluronate is not sulfated and is not covalently attached to protein. The carboxyls of uronic acids and sulfate groups contribute to the highly charged polyanionic nature of glycosaminoglycans. Both their electrical charge and macromolecular structure aid in their biological role as lubricants and support elements in connective tissue. Glycosaminoglycans are predominantly components of the extracellular matrices and cell surfaces, and increasingly more dynamic roles in cell adhesion and signaling, as well as a host of other biological activities, have been elucidated for the proteoglycans.

Hyaluronate Is a Copolymer of *N*-Acetylglucosamine and Glucuronic Acid

Hyaluronate differs from the other five types of glycosaminoglycans. It is unsulfated, not covalently linked with protein, and the only glycosaminoglycan not limited to animal tissue; it is also produced by bacteria. It is classified as a glycosaminoglycan because of its structural similarity to these other polymers, since it consists solely of repeating disaccharide units of *N*-acetylglucosamine and glucuronic acid (Figure 15.11). Although hyaluronate has the least complex chemical structure of all the glycosaminoglycans, the chains can reach molecular weights of 10^5–10^7. The large molecular weight, polyelectrolyte character, and large volume it occupies in solution all contribute to the properties of hyaluronate as a lubricant and shock absorbent. Hence it is found predominantly in synovial fluid, vitreous humor, and umbilical cord.

Chondroitin Sulfates Are the Most Abundant Glycosaminoglycans

The most abundant glycosaminoglycans in the body, the **chondroitin sulfates,** have their polysaccharide chains attached to specific serine residues in a protein core of variable molecular weight through a tetrasaccharide linkage region:

$$\text{GlcUA} \xrightarrow[1.3]{} \text{Gal} \xrightarrow[1.3]{} \text{Gal} \xrightarrow[1.4]{} \text{Xyl} \longrightarrow O\text{-Ser}$$

The characteristic repeating disaccharide units of *N*-acetylgalactosamine and glucuronic acid are covalently attached to this linkage region (Figure 15.12). The disaccharides can be sulfated at either the 4 or 6 position of *N*-acetylgalactosamine. Each polysaccharide chain contains between 30 and 50 such disaccharide units, corresponding to molecular weights of 15,000–25,000 Da. An average chondroitin sulfate proteoglycan molecule has approximately 100 chondroitin sulfate chains attached to the protein core, giving a molecular weight of $1.5–2 \times 10^6$ Da. Proteoglycan preparations are, however, extremely heterogeneous, differing in length of protein core, degree of substitution, distribution of polysaccharide chains, length of chondroitin sulfate chains, and degree of sulfation. Chondroitin sulfate proteoglycans also aggregate noncovalently with hyaluronate, forming much larger structures. Proteoglycans are prominent components of cartilage, tendons, ligaments, and aorta and have also been isolated from brain, kidney, and lung.

Dermatan Sulfate Contains L-Iduronic Acid

Dermatan sulfate differs from chondroitin 4- and 6-sulfates in that its predominant uronic acid is L-iduronic acid, although D-glucuronic acid is also present in

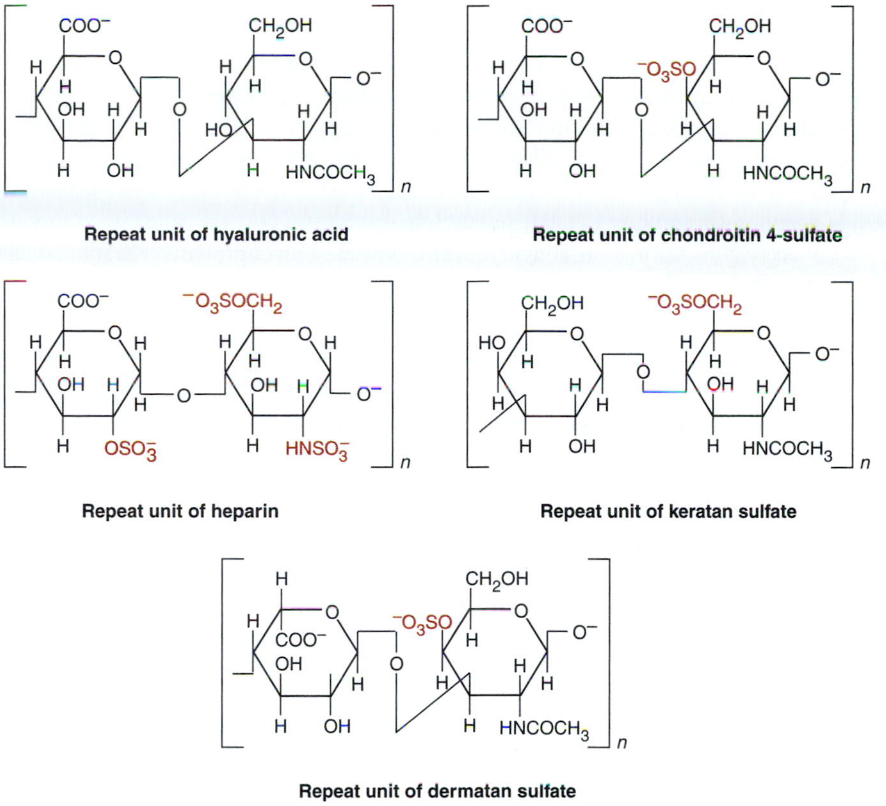

Repeat unit of hyaluronic acid

Repeat unit of chondroitin 4-sulfate

Repeat unit of heparin

Repeat unit of keratan sulfate

Repeat unit of dermatan sulfate

FIGURE 15.12
Major repeat units of glycosaminoglycan chains.

variable amounts. The glycosidic linkages have the same position and configuration as in chondroitin sulfates, with average polysaccharide chains of molecular weights of $2\text{–}5 \times 10^4$ Da. Unlike the chondroitin sulfates, dermatan sulfate is antithrombic like heparin, but in contrast to heparin, it shows only minimal whole blood anticoagulant and blood lipid-clearing activities. As a connective tissue macromolecule, dermatan sulfate is found in skin, blood vessels, and heart valves.

Heparin and Heparan Sulfate Differ from Other Glycosaminoglycans

In **heparin,** glucosamine and D-glucuronic acid or L-iduronic acid form the characteristic disaccharide repeat unit, as in dermatan sulfate (Figure 15.12). In contrast to most other glycosaminoglycans, heparin contains α-glycosidic linkages. Almost all glucosamine residues contain sulfamide linkages, while a small number of glucosamine residues are *N*-acetylated. The sulfate content of heparin, although variable, approaches 2.5 sulfate residues per disaccharide unit in preparations with the highest biological activity. In addition to *N*-sulfate and *O*-sulfate on C-6 of glucosamine, heparin may contain sulfate on C-3 of the hexosamine and C-2 of the uronic acid. Unlike other glycosaminoglycans, heparin is an intracellular component of mast cells and functions predominantly as an anticoagulant and lipid-clearing agent (see Clin. Corr. 15.12).

Heparan sulfate contains a similar disaccharide repeat unit as heparin but has more *N*-acetyl groups, fewer *N*-sulfate groups, and a lower degree of *O*-sulfate groups. Heparan sulfate may be extracellular or an integral and ubiquitous component of the cell surface in many tissues including blood vessel walls, amyloid, and brain.

CLINICAL CORRELATION 15.12

Heparin Is an Anticoagulant

Heparin is a naturally occurring sulfated polysaccharide that is used to reduce the clotting tendency of patients. Both *in vivo* and *in vitro,* heparin prevents the activation of clotting factors but does not act directly on the clotting factors. Rather, the anticoagulant activity of heparin is brought about by the binding interaction of heparin with an inhibitor of the coagulation process. Presumably, heparin binding induces a conformational change in the inhibitor that generates a complementary interaction between the inhibitor and the activated coagulation factor, thereby preventing the factor from participating in the coagulation process. The inhibitor that interacts with heparin is antithrombin III, a plasma protein inhibitor of serine proteases. In the absence of heparin, antithrombin III slowly (10–30 min) combines with several clotting factors, yielding complexes devoid of proteolytic activity; in the presence of heparin, inactive complexes are formed within a few seconds. Antithrombin III contains an arginine residue that combines with the active site serine of factors Xa and IXa; thus the inhibition is stoichiometric. Heterozygous antithrombin III deficiency results in an increased risk of thrombosis in the veins and resistance to the action of heparin.

Keratan Sulfate Exists in Two Forms

Keratan sulfate is composed principally of the repeating disaccharide unit of *N*-acetylglucosamine and galactose, with no uronic acid in the molecule (Figure 15.12). Sulfate content is variable, with ester sulfate present on C-6 of both galactose and hexosamine. Two types of keratan sulfate differ in overall carbohydrate content and tissue distribution. Both contain, as additional monosaccharides, mannose, fucose, sialic acid, and *N*-acetylgalactosamine. Keratan sulfate I, isolated from cornea, is linked to protein by an *N*-acetylglucosamine–asparaginyl bond, typical of glycoproteins. Keratan sulfate II, isolated from cartilage, is attached to protein through *N*-acetylgalactosamine in *O*-glycosidic linkage to either serine or threonine. Skeletal keratan sulfates are often found covalently attached to the same core protein as are the chondroitin sulfate chains.

Biosynthesis of Chondroitin Sulfate Is Typical of Glycosaminoglycan Formation

The heteropolysaccharide chains of proteoglycans are assembled by sequential action of a series of glycosyltransferases, which catalyze the transfer of a monosaccharide from a nucleotide-linked sugar to an appropriate acceptor, either the nonreducing end of another sugar or a polypeptide. Since the biosynthesis of the **chondroitin sulfates** is most thoroughly understood, this pathway will be discussed as the prototype for glycosaminoglycan formation (Figure 15.13).

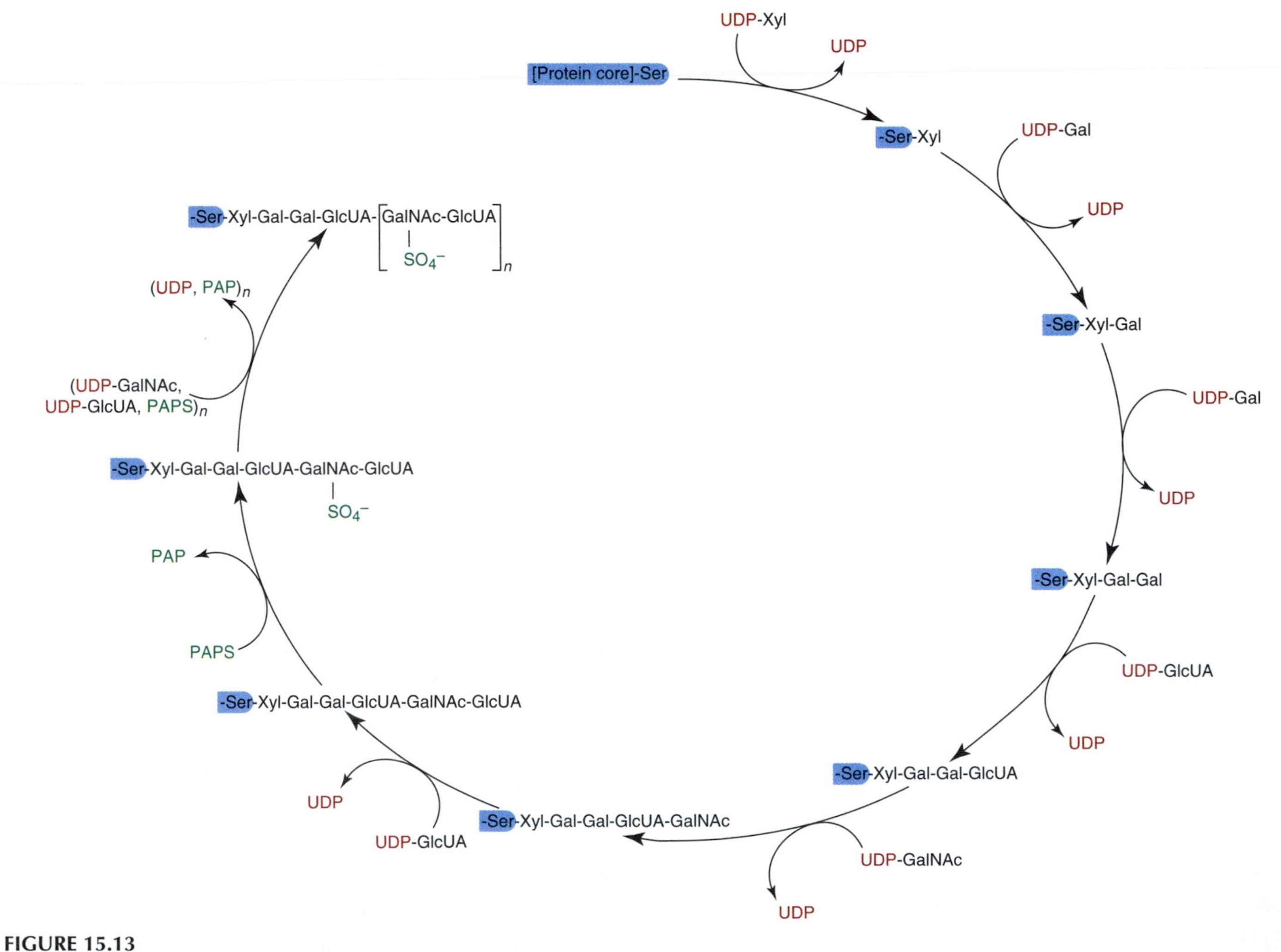

FIGURE 15.13
Synthesis of chondroitin sulfate proteoglycan.
Xyl, xylose; Gal, galactose; GlcUA, glucuronic acid; GalNAc, *N*-acetylgalactosamine; PAPS, 3′-phosphoadenosine 5′-phosphosulfate.

Formation of the core protein of the chondroitin sulfate proteoglycan is the first step in this process, followed by assembly of the polysaccharide chains catalyzed by six different glycosyltransferases. Strict substrate specificity is required for completion of the unique tetrasaccharide linkage region. Polymerization then results from the concerted action of two glycosyltransferases, an *N*-acetylgalactosaminyltransferase and a glucuronosyltransferase, which alternately add the two monosaccharides, forming the characteristic repeating disaccharide units. Sulfation of *N*-acetylgalactosamine at either the 4 or 6 position occurs concomitantly with chain elongation. The sulfate donor, as in other biological systems, is 3′-phosphoadenosine 5′-phosphosulfate (PAPS), which is formed from ATP and sulfate in two steps (Figure 15.14). The importance of sulfation to the function of proteoglycans is highlighted by the preponderance of chondrodystrophic conditions in animals and humans as a result of deficiencies in the sulfation process (see Clin. Corr. 15.13).

Synthesis of other glycosaminoglycans requires additional transferases specific for the appropriate sugars and linkages. Completion often involves modifications in addition to *O*-sulfation, including epimerization, acetylation, and *N*-sulfation. Interestingly, the epimerization of D-glucuronic acid to L-iduronic acid occurs after incorporation into the polymer chain and is coupled with the process of sulfation.

Synthesis of both proteoglycans and glycoproteins is regulated by the same mechanism at the level of hexosamine synthesis. The fructose 6-phosphate–glutamine transamidase reaction (Figure 15.4) is subject to feedback inhibition by UDP-*N*-acetylglucosamine, which is in equilibrium with UDP-*N*-acetylgalactosamine. More specific to proteoglycan synthesis, the concentrations of UDP-xylose and UDP-glucuronic acid are stringently controlled by the inhibition by UDP-xylose of the UDP-glucose dehydrogenase conversion of UDP-glucose to UDP-glucuronic acid (Figure 15.4). Since xylose is the first sugar added during synthesis of chondroitin sulfate, dermatan sulfate, heparin, and heparan sulfate, the earliest effect of decreased core protein synthesis would be accumulation of UDP-xylose, which aids in maintaining a balance between synthesis of protein and polysaccharide moieties of these complex macromolecules.

Proteoglycans, like glycoproteins, are degraded by the sequential action of proteases and **glycosidases**, as well as **deacetylases** and **sulfatases.** Much of the information about metabolism and degradation of proteoglycans has been derived from the study of the **mucopolysaccharidoses** (see Clin. Corr. 15.14). This group of human genetic disorders is characterized by accumulation in tissues and excretion in urine of oligosaccharide products derived from incomplete breakdown of

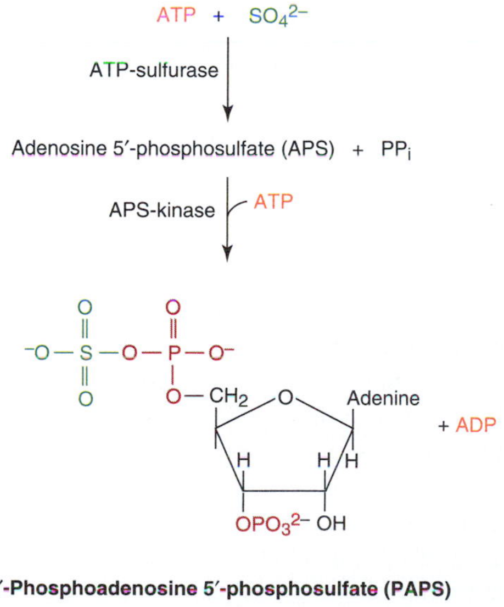

$ATP + SO_4^{2-}$

ATP-sulfurase ↓

Adenosine 5′-phosphosulfate (APS) + PP_i

APS-kinase ↓ ← ATP

$^-O-S(=O)_2-O-P(=O)(O^-)-O-CH_2$ (ribose with Adenine, H, H, H, OPO_3^{2-}, OH) + ADP

3′-Phosphoadenosine 5′-phosphosulfate (PAPS)

FIGURE 15.14
Biosynthesis of 3′-phosphoadenosine 5′-phosphosulfate (PAPS).

CLINICAL CORRELATION 15.13
Chondrodystrophies Due to Sulfation Defects

Sulfation is an essential modification of glycosaminoglycans, the repeating dissaccharides linked to protein cores in the various proteoglycan families. The overall sulfation process is complex, involving multiple components and multiple cellular compartments: transport of inorganic sulfate into the cell via plasma membrane transporters, transformation into phosphoadenosylphosphosulfate (PAPS) via a two-step process catalyzed by PAPS synthetase in the cytosol, then either direct utilization by cytosolic sulfotransferases or transport of PAPS from the cytosol to the Golgi complex for utilization by a host of luminal sulfotransferases. Three autosomal recessive disorders, diastrophic dysplasia (DTD), atelosteogenesis type II (AOII), and achondrogenesis type 1B (ACG-1B), are known to result from different mutations in the *DTDST* gene that encodes a sulfate transporter. Patients with DTD exhibit disproportionate short stature and generalized joint dysplasia, but usually survive a normal lifespan; AOII presents as a perinatally lethal chondrodysplasia; ACG-1B is characterized by extremely short extremities and trunk.

Genetic disorders resulting from defects in the next step in the pathway, synthesis of PAPS by the bifunctional sulfurylase/kinase (PAPS synthetase), have been identified in both animals (i.e., the brachymorphic mouse that exhibits a severe growth disorder resulting in extremely short trunk and limbs and a small skull) and humans (i.e., spondyloepimetaphyseal dysplasia (Pakistani type) characterized by short and bowed lower limbs, enlarged knee joints, and early onset of degenerative joint disease). These identified mutations in the various components of the sulfation process clearly highlight the importance of this post-translational modification to the functioning of proteoglycans, especially in development and maintenance of the skeletal system.

CLINICAL CORRELATION 15.14
Mucopolysaccharidoses

A group of human genetic disorders characterized by excessive accumulation and excretion of the oligosaccharides of proteoglycans exists, collectively called mucopolysaccharidoses. These disorders result from a deficiency of one or more lysosomal hydrolases responsible for the degradation of dermatan and/or heparan sulfate. The enzymes lacking in some specific mucopolysaccharidoses that have been identified are presented in the accompanying table.

Although the chemical basis for this group of disorders is similar, their mode of inheritance as well as clinical manifestations may vary. Hurler's syndrome and Sanfilippo's syndrome are transmitted as autosomal recessives, whereas Hunter's syndrome is X-linked. Both Hurler's syndrome and Hunter's syndrome are characterized by skeletal abnormalities and mental retardation, which in severe cases may result in early death. In contrast, in Sanfilippo's syndrome the physical defects are relatively mild, while the mental retardation is severe. Collectively, the incidence for all mucopolysaccharidoses is 1 per 30,000 births.

In addition to those listed in the table, some others exist. Multiple sulfatase deficiency (MSD) is characterized by decreased activity of all known sulfatases. Recent evidence suggests that a co- or post-translational modification of a cysteine to a 2-amino 3-oxopropionic acid is required for active sulfatases and that a lack of this modification results in MSD. These disorders are amenable to prenatal diagnosis, since the pattern of metabolism by affected cells obtained from amniotic fluid is strikingly different from normal.

Enzyme Defects in the Mucopolysaccharidoses

Syndrome	*Accumulated Products*[a]	*Deficient Enzyme*[b]
Hunter's	Heparan sulfate Dermatan sulfate	Iduronate sulfatase (1)
Hurler–Scheie	Heparan sulfate Dermatan sulfate	α-L-Iduronidase (2)
Maroteaux–Lamy	Dermatan sulfate	*N*-Acetylgalactosamine sulfatase (3)
Mucolipidosis VII	Heparan sulfate Dermatan sulfate	β-Glucuronidase (5)
Sanfilippo's, type A	Heparan sulfate	Heparan sulfamidase (6)
Sanfilippo's, type B	Heparan sulfate	*N*-Acetylglucosaminidase (9)
Sanfilippo's, type C	Heparan sulfate	Acetyl CoA: α-glucosaminide acetyltransferase
Sanfilippo's, type D	Heparan sulfate	*N*-Acetylglucosamine 6-sulfatase (8)
Morquio's, type A	Keratan/chondroitin sulfate	Galactose-6-sulfatase
Morquio's, type B	Keratan sulfate	β-Galactosidase

[a]Structures of dermatan sulfate and heparan sulfate.

Dermatan sulfate —IdUA $\xrightarrow[\alpha]{(2)}$ GalNAc $\xrightarrow[\beta]{(4)}$ GlcUA $\xrightarrow[\beta]{(4)}$ GalNAc $\xrightarrow[\beta]{}$
(IdUA: (1) OSO_3H; first GalNAc: (3) OSO_3H; second GalNAc: OSO_3H)

Heparan sulfate —IdUA $\xrightarrow[\alpha]{(2)}$ GlcN $\xrightarrow[\alpha]{(7)}$ GlcUA $\xrightarrow[\beta]{(5)}$ GlcNAc $\xrightarrow[\alpha]{(9)}$
(IdUA: (1) OSO_3H; GlcN: (6) OSO_3H; GlcNAc: (8) OSO_3H)

[b]The numbers in parentheses refer to the enzymes that hydrolyze those bonds.

the proteoglycans, due to a deficiency of one or more lysosomal hydrolases. In the diseases for which the biochemical defect has been identified, a product accumulates possessing a nonreducing terminus that would have been the substrate for the deficient enzyme.

Although proteoglycans continue to be defined on the basis of the glycosaminoglycan chain they contain, new ones are increasingly being described based largely on functional properties or location. **Aggrecan** and **versican** are the predominant extracellular species; syndecan, CD44, and thrombomodulin are integral membrane proteins; **neurocan, brevican, cerebrocan,** and **phosphacan** are largely restricted to the nervous system. Many proteoglycans (i.e., aggrecan, syndecan, and betaglycan) carry two types of glycosaminoglycan chains, whose size and relative amounts may change with development, age, or disease. Increasingly, the genes encoding the proteoglycan core proteins and biosynthetic enzymes are being cloned, revealing that the relevant proteins belong to families of related origin and possible function. Thus it appears that the versatile structure of these abundant carbohydrate-containing molecules is well exploited by cells in many as yet undiscovered ways.

BIBLIOGRAPHY

Dutton, G. J. (Ed.). *Glucuronic Acid, Free and Combined.* New York: Academic Press, 1966.

Ginsburg, V. and Robbins, P. (Eds.). *Biology of Carbohydrates.* New York: Wiley, 1984.

Horecker, B. L. *Pentose Metabolism in Bacteria.* New York: Wiley, 1962.

Hughes, R. C. *Glycoproteins.* London: Chapman and Hall, 1983.

Kornfeld, R. and Kornfeld, S. Assembly of Asn-linked oligosaccharides. *Annu. Rev. Biochem.* 54:631, 1985.

Lennarz, W. J. (Ed.). *The Biochemistry of Glycoproteins.* New York: Plenum Press, 1980.

Menkes, J. A. Metabolic diseases of the nervous system. In: *Textbook of Child Neurology,* 4th ed. Philadelphia: Lea & Febiger, 1990, p. 28.

Nyhan, W. L. and Sakati, N. A. Mucopolysaccharidoses and related disorders. In: *Diagnostic Recognition of Genetic Disease.* Philadelphia: Lea & Febiger, 1987, p. 371.

Schwartz, N. B., Lyle, S., Ozeran, J. D., Li, H., Deyrup, A., and Westley, J. Sulfate activation and transport in mammals: system components and mechanisms. *Chem. Biol. Interact.* 109:143, 1998.

Schwartz, N. B. and Domowicz, M. Proteoglycan gene mutations and impaired skeletal development. In *Skeletal Morphogenesis and Growth.* AAOS Publications, 1998, p. 413.

Schwartz, N. B., Pirok, E. W., Mensch, J. R., and Domowicz, M. S. Domain organization, genomic structure, evolution and regulation of expression of the aggrecan gene family. *Prog. Nucleic Acid Res. Mol. Biol.* 62:177, 1999.

Schwartz, N. B. Proteoglycans. In *Embryonic Encyclopedia of Life Sciences.* London: Nature Publishing Group, 2000. www.els.net

Schwartz, N. B. Biosynthesis and regulation of expression of proteoglycans. *Front. Biosci.* 5:649, 2000.

Scriver, C. R., Beaudet, A. L., Sly, W. S., and Valle, D. (Eds.). *The Molecular and Metabolic Bases of Inherited Disease,* 7th ed. New York: McGraw-Hill, 1995.

QUESTIONS C. N. ANGSTADT

Multiple Choice Questions

1. All of the following interconversions of monosaccharides (or derivatives) require a nucleotide-linked sugar intermediate EXCEPT:
   A. galactose 1-phosphate to glucose 1-phosphate.
   B. glucose 6-phosphate to mannose 6-phosphate.
   C. glucose to glucuronic acid.
   D. glucuronic acid to xylose.
   E. glucosamine 6-phosphate to *N*-acetylneuraminic acid (a sialic acid).

2. The severe form of galactosemia:
   A. is a genetic deficiency of a uridylyltransferase that exchanges galactose 1-phosphate for glucose on UDP-glucose.
   B. results from a deficiency of an epimerase.
   C. is insignificant in infants but a major problem in later life.
   D. is an inability to form galactose 1-phosphate.
   E. would be expected to interfere with the use of fructose as well as galactose because the deficient enzyme is common to the metabolism of both sugars.

3. All of the following are true about glucuronic acid EXCEPT:
   A. it is a charged molecule at physiological pH.
   B. as a UDP derivative, it can be decarboxylated to a component used in proteoglycan synthesis.
   C. it is a precursor of ascorbic acid in humans.
   D. its formation from glucose is under feedback control by a UDP-linked intermediate.
   E. it can ultimately be converted to xylulose 5-phosphate and thus enter the pentose phosphate pathway.

4. The conversion of fructose 6-phosphate to glucosamine 6-phosphate:
   A. is a transamination reaction with glutamate as the nitrogen donor.
   B. is stimulated by UDP-*N*-acetylglucosamine.
   C. requires that fructose 6-phosphate first be linked to a nucleotide.
   D. is a first step in the formation of *N*-acetylated amine sugars.
   E. occurs only in the liver.

5. Roles for the complex carbohydrate moiety of glycoproteins include all of the following EXCEPT:
   A. determinant of blood type.
   B. template for the synthesis of glycosaminoglycans.
   C. cell surface receptor specificity.
   D. determinant of the rate of clearance from the circulation of certain molecules.
   E. targeting of cells to certain tissues.

6. Fucose and sialic acid:
   A. are both derivatives of UDP-*N*-acetylglucosamine.
   B. are the parts of the carbohydrate chain that are covalently linked to the protein.
   C. can be found in the core structure of certain *O*-linked glycoproteins.
   D. are transferred to a carbohydrate chain when it is attached to dolichol phosphate.
   E. are the repeating unit of proteoglycans.

7. Glycosaminoglycans:
   A. are the carbohydrate portion of glycoproteins.
   B. contain large segments of a repeating unit typically consisting of a hexosamine and a uronic acid.
   C. always contain sulfate.
   D. exist in only two forms.
   E. are bound to protein by ionic interaction.

8. All of the following are true of proteoglycans EXCEPT:
   A. specificity is determined, in part, by the action of glycosyltransferases.
   B. synthesis is regulated, in part, by UDP-xylose inhibition of the conversion of UDP-glucose to UDP-glucuronic acid.
   C. synthesis involves sulfation of carbohydrate residues by PAPS.
   D. synthesis of core protein is balanced with synthesis of the polysaccharide moieties.
   E. degradation is catalyzed in the cytosol by nonspecific glycosidases.

Questions 9 and 10: A patient was admitted to the hospital with symptoms of severe anemia. The onset of symptoms followed two days of therapy with a sulfa drug. Laboratory studies indicated a hemoglobin of 8.5 g/dL and an abnormally low [NADPH]/[$NADP^+$]. Further investigation led to a diagnosis of a defective enzyme of the pentose phosphate pathway.

9. [NADPH]/[$NADP^+$] is maintained at a high level in cells primarily by:
   A. lactate dehydrogenase.
   B. the combined actions of glucose 6-phosphate dehydrogenase and gluconolactonase.
   C. the action of the electron transport chain.
   D. shuttle mechanisms such as the α-glycerophosphate dehydrogenase shuttle.
   E. the combined actions of transketolase and transaldolase.

10. If a cell requires more NADPH than ribose 5-phosphate:
    A. only the first phase of the pentose phosphate pathway would occur.
    B. glycolytic intermediates would flow into the reversible phase of the pentose phosphate pathway.
    C. there would be sugar interconversions but no net release of carbons from glucose 6-phosphate.
    D. the equivalent of the carbon atoms of glucose 6-phosphate would be released as 6 CO_2.
    E. only part of this need could be met by the pentose pathway, and the rest would have to be supplied by another pathway.

Questions 11 and 12: A six-month old infant presented with hypoglycemia, vomiting, diarrhea, protein-loss enteropathy, and hepatic fibrosis. Measurement of the glycosylation state of endogenous serum transferrin revealed type Ib CDGS (carbohydrate-deficient glycoprotein syndrome), which is a defect in phosphomannose isomerase activity.

11. The role of phosphomannose isomerase is the interconversion of:
    A. mannose 6-P and mannose 1-P.
    B. glucose 6-P and mannose 6-P.
    C. fructose 6-P and mannose 6-P.
    D. fructose 1-P and mannose 1-P.
    E. glucose 6-P and mannose 1-P.

12. In type Ic CDGS, a defect in an enzyme transferring a glucosyl residue to a high-mannose dolichol pyrophosphate precursor, the carbohydrate structure would be part of a(n):
    A. *N*-linked glycoprotein.
    B. *O*-linked glycoprotein.
    C. proteoglycan.
    D. glycosaminoglycan.
    E. complex lipid.

Questions 13 and 14: An infant presented with multiple skeletal abnormalities, with the most prominent ones a short trunk and short limbs. Urine was free of partially degraded oligosaccharides or oligosaccharides of proteoglycans. The blood sulfate concentration and extracellular acid hydrolase were normal.

13. The disorder described would most likely be caused by a defect in:
    A. the ability to generate mannose 6-phosphate.
    B. lysosomal glycosidase activity.
    C. the gene that codes for a sulfate transporter.
    D. PAPS synthetase.
    E. a sulfatase.

14. Sulfation is an important component of the synthesis of:
    A. most proteoglycans.
    B. hyaluronate.
    C. most glycoproteins.
    D. conjugated bilirubin.
    E. all of the above.

Problems

15. How is excess ribose 5-phosphate converted to intermediates of glycolysis? Be sure to include any enzymes involved.

16. Essential fructosuria is a defect in fructokinase while fructose intolerance is a defect in fructose 1-phosphate aldolase. Which of these two diseases leads to severe hypoglycemia after ingestion of fructose and why?

ANSWERS

1. **B** The glucose and mannose phosphates are both in equilibrium with fructose 6-phosphate by phosphohexose isomerases. A: This occurs via an epimerase at the UDP-galactose level. C, D: This oxidation of glucose is catalyzed by UDP-glucose dehydrogenase and the product can be decarboxylated to UDP-xylose. E: Again, an epimerization occurs on the nucleotide intermediate.
2. **A** B: The epimerase is normal. C: Galactose is an important sugar for infants. D: In this disease, the galactokinase is normal but is deficient in the mild form of the disease. E: Fructose metabolism does not use the uridylyltransferase that is deficient in galactosemia.
3. **C** Humans do not make ascorbic acid. A: The charged acid group enhances water solubility, which is a major physiological role for glucuronic acid, for example, bilirubin metabolism. B, D: Decarboxylation of UDP-glucuronic acid gives UDP-xylose, which is a potent inhibitor of the oxidation of UDP-glucose to the acid. E: The reduction of D-glucuronic acid to L-gulonic acid leads to ascorbate as well as xylulose 5-phosphate for the pentose phosphate pathway.
4. **D** Glucosamine 6-phosphate is acetylated. UDP-*N*-acetylglucosamine is formed, and the UDP derivative can be epimerized to the galactose derivative. A, C: This conversion is a transamidation of the amide nitrogen of glutamine and does not involve nucleotide intermediates. B: This is a feedback inhibitor of the transamidase reaction, thus controlling formation of the nucleotide sugars. E: May account for 20% of the glucose flux in skin.
5. **B** Synthesis of complex carbohydrates is not template directed but determined by the specificity of individual enzymes. A, C–E: Because of the diversity possible with oligosaccharides, they play a significant role in determining the specificity of many biological molecules.
6. **C** Core structure also contains galactose and *N*-acetylgalactosamine. A: Only sialic acid does. Fucose comes from CDP-mannose. B: Usually found at the periphery of the carbohy-

drate. D: Core structure of *N*-linked carbohydrates contains mannose and *N*-acetylglucosamine. E: Repeating unit is hexosamine and uronic acids.

7. **B** This is a major distinction from glycoproteins, which, by definition, do not have a serial repeating unit. A: These are the carbohydrate of proteoglycans. C: Most do but hyaluronate does not. D: There are at least six different classes. E: Carbohydrates are bound by covalent links.

8. **E** Degradation is lysosomal; deficiencies of one or more lysosomal hydrolases lead to accumulation of proteoglycans in the mucopolysaccharidoses. A: Strict substrate specificity of the enzymes is important in determining the type and quantity of proteoglycans synthesized. Formation of specific protein acceptors for the carbohydrate is also important. B, D: Both xylose and glucuronic acid levels are controlled by this; xylose is the first sugar added in the synthesis of four of the six types and would accumulate if core protein synthesis is decreased. C: This is necessary for the formation of most proteoglycans.

9. **B** Although the glucose 6-phosphate dehydrogenase reaction, specific for NADP, is reversible, hydrolysis of the lactone assures that the overall equilibrium lies far in the direction of NADPH. A, C, D: These all use NAD, not NADP. E: These enzymes are part of the pentose phosphate pathway but catalyze freely reversible reactions that do not involve NADP.

10. **D** A, C–E: Glucose 6-phosphate yields ribose 5-phosphate + CO_2 in the oxidative phase. If this is multiplied by six, the six ribose 5-phosphates can be rearranged to five glucose 6-phosphates by the second, reversible phase. B: If more ribose 5-phosphate than NADPH were required, the flow would be in this direction to supply the needed pentoses.

11. **C** This is similar to the glucose 6-P–fructose 6-P interconversion. A: Mutases catalyze a 1–6 phosphate shift. B, D, E: These conversions do not occur directly.

12. **A** A carbohydrate chain assembled on dolichol phosphate is characteristic of *N*-linked glycoproteins. B, C, D: Synthesis of *O*-linked glycoproteins involves the sequential addition to the *N*-acetylgalactosamine linked to serine or threonine. E: Dolichol phosphate is a lipid but is eliminated when the carbohydrate chain is added to the protein.

13. **D** These structural abnormalities are associated with defects in sulfation of the proteoglycans. A: This would lead to an inability to target acid hydrolase to the lysosome and thus increase the extracellular concentration. B, E: These are lysosomal enzymes whose deficiency would lead to partially degraded oligosaccharides in urine. C: Lack of a transporter would lead to elevated serum levels of sulfate.

14. **A** B: Hyaluronate is one glycosaminoglycan that is not sulfated. C: Glycoproteins don't contain sulfate. D: Bilirubin is conjugated with glucuronic acid.

15. Three pentose phosphates (ribose 5-P + 2 xylulose 5-P) are converted to 2 fructose 6-P and 2 glyceraldehyde 3-P by transaldolase and transketolase via a series of two-carbon and three-carbon transfers. These reactions are reversible.

16. Lack of fructokinase causes accumulation of fructose, which is excreted. Lack of the aldolase results in increased fructose 1-phosphate, which sequesters cellular inorganic phosphate, inhibiting the cell's ability to generate ATP.

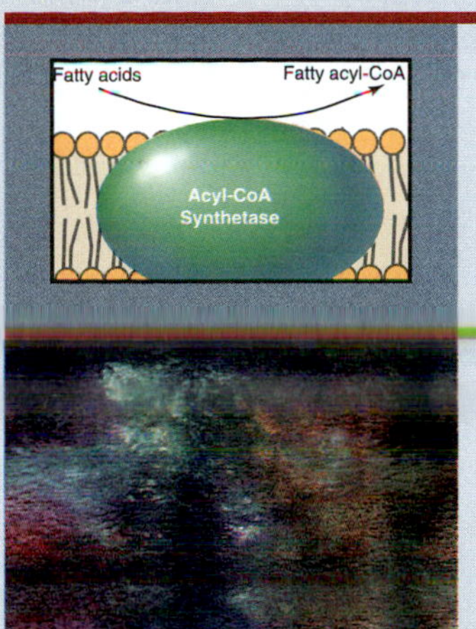

16

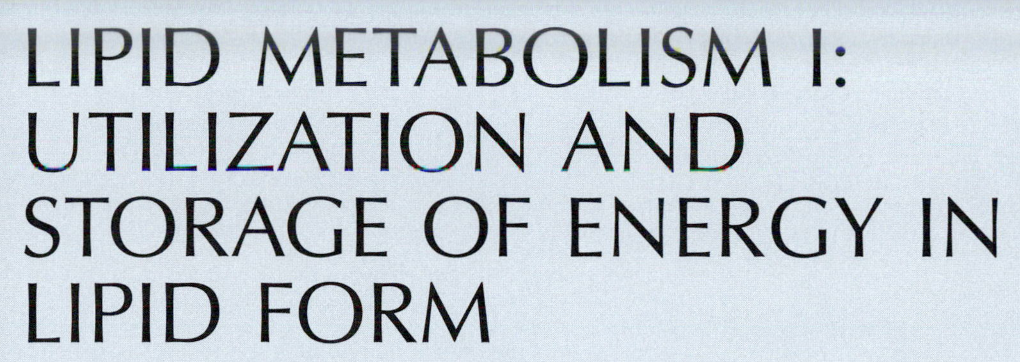

LIPID METABOLISM I: UTILIZATION AND STORAGE OF ENERGY IN LIPID FORM

J. Denis McGarry

16.1 | OVERVIEW

As the human body builds and renews its structures, obtains and stores energy, and performs its various functions, there are many circumstances in which it is essential to use molecules or parts of molecules that do not associate with water. This property of being **nonpolar** and **hydrophobic** is the defining characteristic of substances classed as **lipids.** Most of these are molecules that contain or are derived from **fatty acids.** In the early stages of biochemical research lipids were not investigated as intensively as other body constituents, largely because techniques for studying aqueous systems were easier to develop. This benign neglect led to assumptions that lipids were relatively inert and their metabolism was of lesser importance than that of carbohydrates, for instance.

As the methodology for analyzing lipid metabolism developed, it became evident that fatty acids and their derivatives had at least two major roles in the body. Oxidation of fatty acids was shown to be a major means of metabolic energy production. Their storage in the form of **triacylglycerols** was more efficient and quantitatively more important than storage of carbohydrates as glycogen. As details of the chemistry of biological structures were defined, hydrophobic structures were found to be composed largely of fatty acids and their derivatives. Thus the major separation of cells and subcellular structures into separate aqueous compartments is accomplished with membranes whose hydrophobic characteristics are largely supplied by the fatty acid moieties of complex lipids. These latter compounds contain constituents other than fatty acids and glycerol. They frequently have significant covalently bound hydrophilic moieties, notably carbohydrates in the glycolipids and organic phosphate esters in phospholipids.

Lipids have several other quantitatively less important roles, which are nonetheless of great functional significance. These include the use of surface active properties of some complex lipids for specific functions, such as maintenance of lung alveolar integrity and solubilization of nonpolar substances in body fluids. In addition, several classes of lipids, for example, steroid hormones and prostaglandins, have highly potent and specific physiological roles in control of metabolic processes. The interrelationships of some processes involved in lipid metabolism are outlined in Figure 16.1.

The metabolism of fatty acids and triacylglycerols is so crucial to proper functioning of the human body that imbalances and deficiencies in these processes can have serious pathological consequences. Disease states related to fatty acid and triacylglycerol metabolism include obesity, diabetes, ketoacidosis, and abnormalities in transport of lipids in blood. In addition some unique deficiencies have been found, such as Refsum's disease and familial hypercholesterolemia, which have helped to elucidate some pathways in lipid metabolism.

This chapter is concerned primarily with the structure and metabolism of fatty acids and of their major storage form, triacylglycerols. After discussion of structures

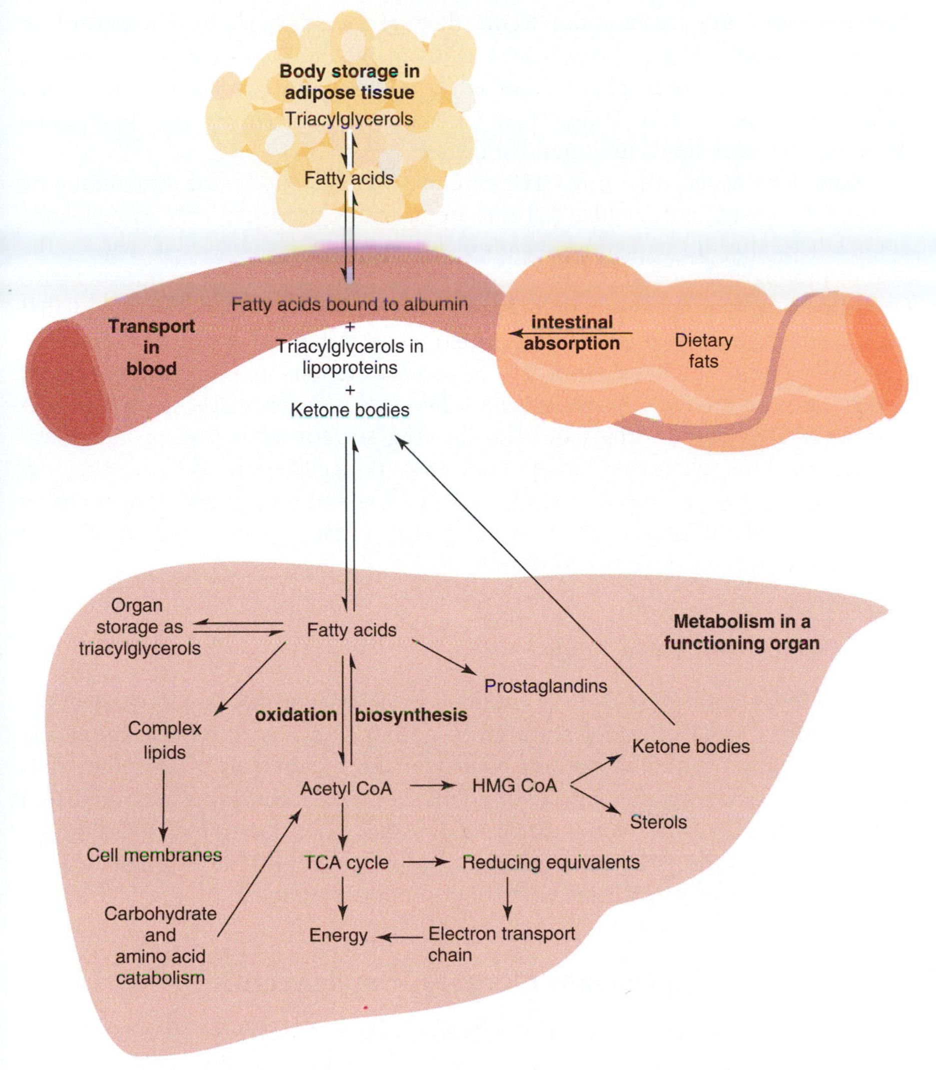

FIGURE 16.1
Metabolic interrelationships of fatty acids in the human.

of the more important fatty acids found in humans, how they are supplied from the diet or by biosynthesis is described. The mechanism for storage as triacylglycerols and how fatty acids are mobilized and transported throughout the body to sites where they are needed is discussed. The central process of energy production from fatty acids is then examined, and finally the mechanisms by which the ketone bodies are synthesized and used are presented. The Appendix (p. 1175) includes the nomenclature and chemistry of lipids and in Chapter 25 (p. 1104) there is a discussion of digestion and absorption of lipids.

16.2 | CHEMICAL NATURE OF FATTY ACIDS AND ACYLGLYCEROLS

Fatty Acids Are Alkyl Chains Terminating in a Carboxyl Group

Fatty acids consist of an **alkyl chain** with a terminal carboxyl group, the basic formula of completely saturated species being $CH_3—(CH_2)_n—COOH$. The important fatty acids for humans have relatively simple structures, although in some organisms they may be quite complex, containing cyclopropane rings or extensive branching. Unsaturated fatty acids occur commonly in humans, with up to six double bonds per chain, these being almost always of the cis configuration. If there is

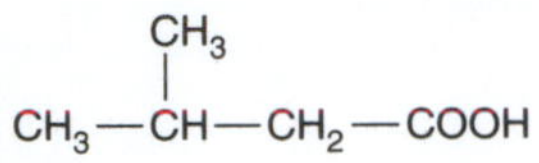

$$CH_3-CH(CH_3)-CH_2-COOH$$

FIGURE 16.2
Isovaleric acid.

more than one double bond per molecule, they are always separated by a **methylene** **($-CH_2-$) group.** Most common fatty acids in biological systems have an even number of carbon atoms, although some organisms synthesize those with an odd number of carbon atoms. Humans can use the latter for energy but incorporate them into complex lipids to a minimal degree.

Some fatty acids with an α-OH group are produced and used structurally by humans. However, more oxidized forms are normally produced only as metabolic intermediates during energy production or for specific physiological activity in the case of prostaglandins and thromboxanes. Some animals, including humans, also produce relatively simple **branched-chain acids,** branching being limited to methyl groups along the chain at one or more positions. These are apparently produced to contribute specific physical properties to some secretions and structures. For instance, large amounts of branched-chain fatty acids, particularly isovaleric acid (Figure 16.2), occur in lipids of echo-locating structures in marine mammals. Elucidation of the role of these compounds in sound focusing should be fascinating.

Most fatty acids in humans are C_{16}, C_{18}, or C_{20}, but several with longer chains occur principally in lipids of the nervous system. These include nervonic acid and a C_{22} acid with six double bonds (Figure 16.3).

$$CH_3-(CH_2)_7-CH{=}CH-(CH_2)_{13}-COOH$$

Nervonic acid

$$CH_3-(CH_2-CH{=}CH)_6-(CH_2)_2-COOH$$

All-*cis*-4,7,10,13,16,19,-docosahexaenoic acid

FIGURE 16.3
Long-chain fatty acids.

Nomenclature of Fatty Acids

The most abundant fatty acids have common names that are accepted for use in the official nomenclature. Examples are given in Table 16.1 with official systematic names. The approved abbreviations consist of the number of carbon atoms followed, after a colon, by the number of double bonds. Carbon atoms are numbered with the carboxyl carbon as number 1, and double bond locations are designated by the number of the carbon atom on the carboxyl side of it. These designations of double bonds are in parentheses after the rest of the symbol.

Most Fatty Acids in Humans Occur as Triacylglycerols

Fatty acids occur primarily as esters of glycerol, as shown in Figure 16.4, when they are stored for future use. Compounds with one **(monoacylglycerols)** or two **(diacylglycerols)** acids esterified occur only in relatively minor amounts and largely as metabolic intermediates in biosynthesis and degradation of glycerol-containing lipids. Most fatty acids in humans exist as **triacylglycerols,** in which all three hydroxyl groups on glycerol are esterified with a fatty acid. These compounds have been called "neutral fats or triglycerides." There are other types of "neutral fats" in the body, and the terms "monoglyceride," "diglyceride," and "triglyceride" are chemically incorrect and should not be used.

The distribution of different fatty acids in the three positions of the glycerol moiety of triacylglycerols in the body at any given time is influenced by many factors, including diet and anatomical location of the triacylglycerols. Compounds

TABLE 16.1 Fatty Acids of Importance to Humans

Numerical Symbol	*Structure*	*Trivial Name*	*Systematic Name*
16:0	$CH_3-(CH_2)_{14}-COOH$	Palmitic	Hexadecanoic
16:1(9)	$CH_3-(CH_2)_5-CH{=}CH-(CH_2)_7-COOH$	Palmitoleic	*cis*-9-Hexadecenoic
18:0	$CH_3-(CH_2)_{16}-COOH$	Stearic	Octadecanoic
18:1(9)	$CH_3-(CH_2)_7-CH{=}CH-(CH_2)_7-COOH$	Oleic	*cis*-9-Octadecenoic
18:2(9,12)	$CH_3-(CH_2)_3-(CH_2-CH{=}CH)_2-(CH_2)_7-COOH$	Linoleic	*cis,cis*-9,12-Octadecadienoic
18:3(9,12,15)	$CH_3-(CH_2-CH{=}CH)_3-(CH_2)_7-COOH$	Linolenic	*cis,cis,cis*-9,12,15-Octadecatrienoic
20:4(5,8,11,14)	$CH_3-(CH_2)_3-(CH_2-CH{=}CH)_4-(CH_2)_3-COOH$	Arachidonic	*cis,cis,cis,cis*-5,8,11,14-Eicosatetraenoic

with the same fatty acid in all three positions of glycerol are rare; the usual case is for a complex mixture.

The Hydrophobic Nature of Lipids Is Important for Their Biological Function

One significant property of fatty acids and triacylglycerols is their lack of affinity for water. Long hydrocarbon chains have negligible possibility for hydrogen bonding. Fatty acids, whether unesterified or in a complex lipid, have a much greater tendency to associate with each other or other hydrophobic structures, such as sterols and hydrophobic side chains of amino acids, than they do with water or polar organic compounds. This hydrophobic character is essential for construction of complex biological structures such as membranes.

The **hydrophobic nature** of triacylglycerols and their highly reduced state make triacylglycerols efficient compounds in comparison to glycogen for storing energy. Three points deserve emphasis. First, on a weight basis pure triacylglycerol yields nearly two and one-half times the amount of ATP on complete oxidation than does pure glycogen. Second, triacylglycerols can be stored without associated water, whereas glycogen is very hydrophilic and binds about twice its weight of water when stored in tissues. Thus the equivalent amount of metabolically recoverable energy stored as hydrated glycogen would weigh about four times as much as if it were stored as triacylglycerol. Third, the average 70-kg person stores about 350 g of carbohydrate as liver and muscle glycogen. This represents about 1400 kcal of available energy, barely enough to sustain bodily functions for 24 hours of fasting. By contrast, a normal complement of fat stores provide sufficient energy for several weeks of survival during total food deprivation.

In humans most of the fatty acids are either saturated or contain only one double bond. Although they are readily catabolized by appropriate enzymes and cofactors, they are fairly inert chemically. The highly unsaturated fatty acids in tissues are much more susceptible to oxidation.

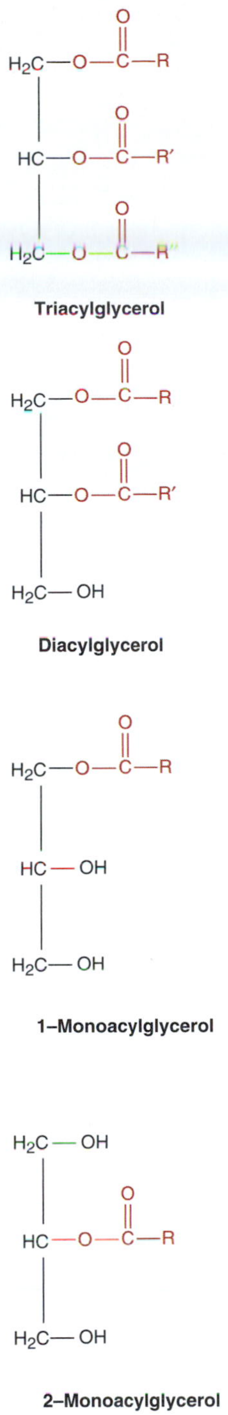

Triacylglycerol

Diacylglycerol

1–Monoacylglycerol

2–Monoacylglycerol

FIGURE 16.4 Acylglycerols.

16.3 | SOURCES OF FATTY ACIDS

Both diet and biosynthesis supply the fatty acids needed by the human body for energy and for construction of hydrophobic parts of biomolecules. Excess dietary protein and carbohydrate are readily converted to fatty acids and stored as triacylglycerols.

Most Fatty Acids Are Supplied in the Diet

Various animal and vegetable lipids are ingested, hydrolyzed at least partially by digestive enzymes, and absorbed through the intestinal mucosa to be distributed through the body, first in the lymphatic system and then in the bloodstream. These processes are discussed in Chapter 25 (see p. 1104). To a large extent dietary supply governs the fatty acid composition of body lipids. Metabolic processes in various tissues modify both dietary and *de novo* synthesized fatty acids to produce nearly all the required structures. With one exception, the actual composition of fatty acids supplied in the diet is relatively unimportant. This exception involves the need for appropriate proportions of relatively highly unsaturated fatty acids because many higher mammals, including humans, are unable to synthesize fatty acids with double bonds near the methyl end of the molecule. Certain **polyunsaturated fatty acids** (PUFAs) with double bonds within the last seven linkages toward the methyl end (linoleic acid) are essential for specific functions. Although all the reasons for this need are not yet explained, one is that some of these acids are precursors of prostaglandins, very active oxidation products (see p. 766).

$$CH_3-(CH_2)_3-(CH_3-CH=CH)_n-(CH_2)_m-COOH$$

Basic formula of the linoleic acid series

$$CH_3-(CH_2-CH=CH)_n-(CH_2)_m-COOH$$

Basic formula of the linolenic acid series

FIGURE 16.5
Linoleic and linolenic acid series.

In humans a dietary precursor is essential for two series of fatty acids. These are the linoleic series and the linolenic series (Figure 16.5), sometimes referred to as n-6 (or ω-6) and n-3 (or ω-3) PUFAs, respectively, to denote the presence of a double bond after the sixth and third carbon atom counting from the terminal methyl (ω) carbon atom.

Palmitate Is Synthesized from Acetyl CoA

The second major source of fatty acids for humans is their biosynthesis from small-molecule intermediates derived from metabolic breakdown of sugars, some amino acids, and other fatty acids. The saturated, straight-chain C16 acid, **palmitic acid,** is first synthesized, and all other fatty acids are made by its modification. Acetyl CoA provides all carbon atoms for this synthesis. Fatty acids are synthesized by sequential addition of two-carbon units to the activated carboxyl end of a growing chain. In mammalian systems the sequence of reactions is carried out by **fatty acid synthase.**

Fatty acid synthase is a fascinating enzyme complex that is still studied intensely. In bacteria it is a complex of several proteins, whereas in mammalian cells it is a single multifunctional protein. Either acetyl CoA or butyryl CoA is the priming unit for fatty acid synthesis, and the methyl end of these primers becomes the methyl end of palmitate. Addition of the rest of the two-carbon units requires activation of the methyl carbon of acetyl CoA by carboxylation to malonyl CoA. However, CO_2 added in this process is lost when condensation of malonyl CoA to the growing chain occurs, so carbon atoms in the palmitate chain originate only from acetyl CoA.

Formation of Malonyl CoA Is the Commitment Step of Fatty Acid Synthesis

The reaction that commits acetyl CoA to fatty acid synthesis is its carboxylation to **malonyl CoA** by the enzyme **acetyl-CoA carboxylase** (Figure 16.6). This reaction is similar in many ways to carboxylation of pyruvate, which is catalyzed by pyruvate carboxylase and starts the process of gluconeogenesis. The reaction requires ATP and HCO_3^- as the source of CO_2. As with pyruvate carboxylase, the first step is formation of activated CO_2 on the biotin moiety of acetyl-CoA carboxylase using energy from ATP. This is then transferred to acetyl CoA.

Acetyl-CoA carboxylase, a key control point in the overall synthesis of fatty acids, can be isolated in a protomeric state that is inactive. The protomers aggregate to form enzymatically active polymers upon addition of citrate *in vitro.* Palmitoyl CoA *in vitro*

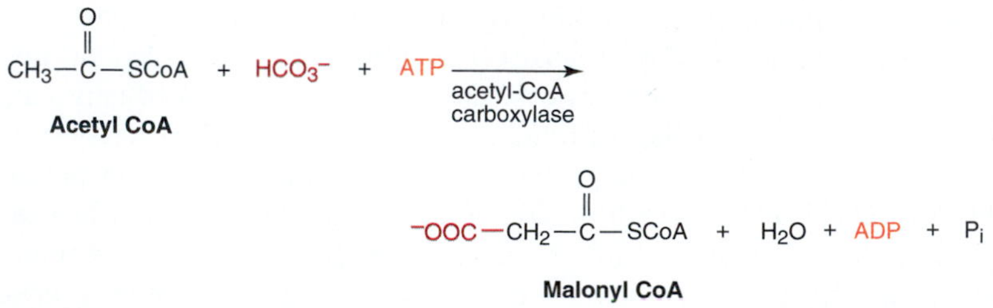

$$\underset{\textbf{Acetyl CoA}}{CH_3-\overset{O}{\overset{\|}{C}}-SCoA} + HCO_3^- + ATP \xrightarrow[\text{carboxylase}]{\text{acetyl-CoA}} \underset{\textbf{Malonyl CoA}}{^-OOC-CH_2-\overset{O}{\overset{\|}{C}}-SCoA} + H_2O + ADP + P_i$$

FIGURE 16.6
Acetyl-CoA carboxylase reaction.

inhibits the enzyme. The action of these two effectors is very logical. Increased synthesis of fatty acids to store energy is desirable when citrate is in high concentration, and decreased synthesis is necessary if high levels of product accumulate. However, the degree to which these regulatory mechanisms operate *in vivo* is still unclear.

Acetyl-CoA carboxylase is also controlled by a cAMP-mediated phosphorylation–dephosphorylation mechanism in which the phosphorylated enzyme is less active than the dephosphorylated one. There is evidence that phosphorylation is promoted by glucagon (via cAMP) as well as by AMP (via an AMP-activated kinase) and that the active form is fostered by insulin. These effects of hormone-mediated phosphorylation are separate from the allosteric effects of citrate and palmitoyl CoA (see Table 16.2).

TABLE 16.2 Regulation of Fatty Acid Synthesis

Enzyme		*Regulatory Agent*	*Effect*
		Palmitate Biosynthesis	
Acetyl-CoA carboxylase	Short term	Citrate	Allosteric activation
		C16–C18 acyl CoAs	Allosteric inhibition
		Insulin	Stimulation
		Glucagon	Inhibition
		cAMP and AMP-mediated phosphorylation	Inhibition
		Dephosphorylation	Stimulation
	Long term	High-carbohydrate diet	Stimulation by increased enzyme synthesis
		Fat-free diet	Stimulation by increased enzyme synthesis
		High-fat diet/PUFAs	Inhibition by decreased enzyme synthesis
		Fasting	Inhibition by decreased enzyme synthesis
		Glucagon	Inhibition by decreased enzyme synthesis
Fatty acid synthase		Phosphorylated sugars	Allosteric activation
		High-carbohydrate diet	Stimulation by increased enzyme synthesis
		Fat-free diet	Stimulation by increased enzyme synthesis
		High-fat diet/PUFAs	Inhibition by decreased enzyme synthesis
		Fasting	Inhibition by decreased enzyme synthesis
		Glucagon	Inhibition by decreased enzyme synthesis
		Biosynthesis of Fatty Acids Other than Palmitate	
Fatty acid synthase		High ratio of methylmalonyl CoA/malonyl CoA	Increased synthesis of methylated fatty acids
		Thioesterase cofactor	Termination of synthesis with short-chain product
Stearyl-CoA desaturase		Various hormones	Stimulation of unsaturated fatty acid synthesis by increased enzyme synthesis
		Dietary polyunsaturated fatty acids (PUFAs)	Decreased activity

Synthesis of acetyl-CoA carboxylase is also regulated, more enzyme being produced by animals on high-carbohydrate or fat-free diets and less on fasting or high-fat diets.

Reaction Sequence for the Synthesis of Palmitic Acid

The first step catalyzed by fatty acid synthase in bacteria is transacylation of the primer molecule, either acetyl CoA or butyryl CoA, to a 4′-phosphopantetheine moiety on a protein constituent of the multienzyme complex. This is the **acyl carrier protein (ACP)** and its phosphopantetheine unit is identical with that in CoA. The mammalian enzyme also contains a phosphopantetheine. Six or seven two-carbon units are then added sequentially to the enzyme complex until the palmitate molecule is completed. After each addition of a two-carbon unit a series of reductive steps takes place. The reaction sequence starting with an acetyl CoA primer and leading to butyryl-ACP is as presented in Figure 16.7.

The next round of synthesis is initiated by transfer of the newly formed fatty acid chain from 4′-phosphopantetheine moiety of ACP to a functional —SH group of ***β*-ketoacyl-ACP synthase** (analogous to Reaction 3a). This liberates the —SH group of ACP for acceptance of a second malonyl unit from malonyl CoA (Reac-

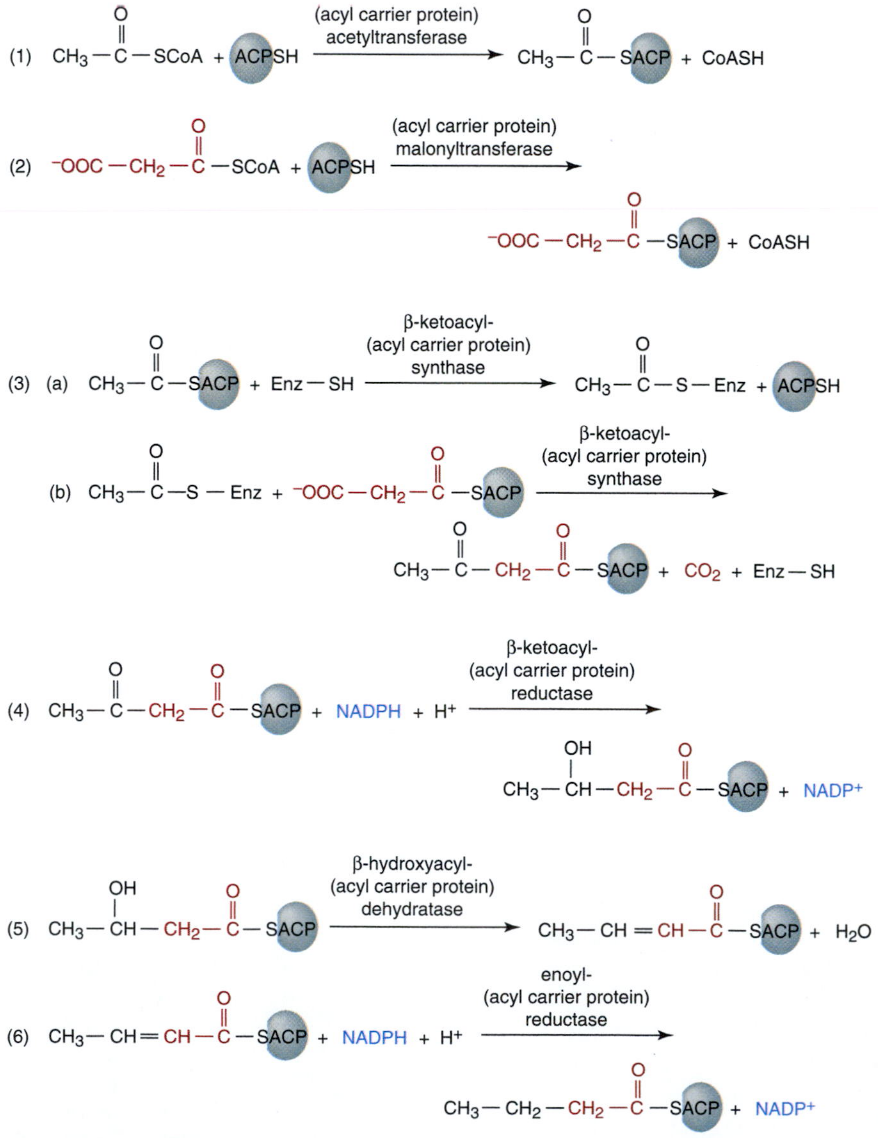

(1) $CH_3-CO-SCoA + ACPSH \xrightarrow{\text{(acyl carrier protein) acetyltransferase}} CH_3-CO-SACP + CoASH$

(2) $^{-}OOC-CH_2-CO-SCoA + ACPSH \xrightarrow{\text{(acyl carrier protein) malonyltransferase}} {}^{-}OOC-CH_2-CO-SACP + CoASH$

(3) (a) $CH_3-CO-SACP + Enz-SH \xrightarrow{\beta\text{-ketoacyl-(acyl carrier protein) synthase}} CH_3-CO-S-Enz + ACPSH$

(b) $CH_3-CO-S-Enz + {}^{-}OOC-CH_2-CO-SACP \xrightarrow{\beta\text{-ketoacyl-(acyl carrier protein) synthase}} CH_3-CO-CH_2-CO-SACP + CO_2 + Enz-SH$

(4) $CH_3-CO-CH_2-CO-SACP + NADPH + H^+ \xrightarrow{\beta\text{-ketoacyl-(acyl carrier protein) reductase}} CH_3-CH(OH)-CH_2-CO-SACP + NADP^+$

(5) $CH_3-CH(OH)-CH_2-CO-SACP \xrightarrow{\beta\text{-hydroxyacyl-(acyl carrier protein) dehydratase}} CH_3-CH=CH-CO-SACP + H_2O$

(6) $CH_3-CH=CH-CO-SACP + NADPH + H^+ \xrightarrow{\text{enoyl-(acyl carrier protein) reductase}} CH_3-CH_2-CH_2-CO-SACP + NADP^+$

FIGURE 16.7
Reactions catalyzed by fatty acid synthase.

tion 2) and allows Reactions 3b to 6 to generate hexanoyl-ACP. The process is repeated five more times at which point palmitoyl-ACP is acted upon by a **thioesterase** with production of free palmitic acid (Figure 16.8). Note that at this stage the sulfhydryl groups of ACP and β-ketoacyl-ACP synthase are both free so that another cycle of fatty acid synthesis can begin.

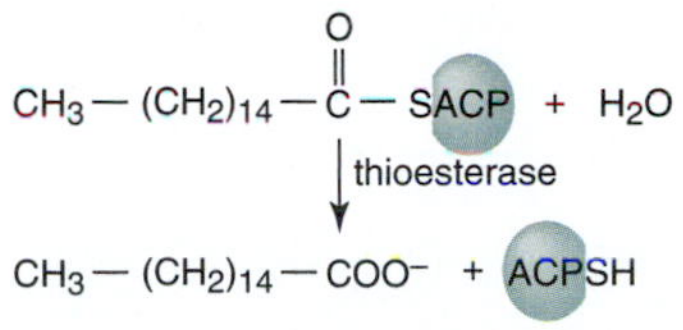

$$CH_3-(CH_2)_{14}-\overset{O}{\overset{\|}{C}}-SACP + H_2O \xrightarrow{\text{thioesterase}} CH_3-(CH_2)_{14}-COO^- + ACPSH$$

FIGURE 16.8
Release of palmitic acid from fatty acid synthase.

Mammalian Fatty Acid Synthase Is a Multifunctional Polypeptide

The reaction sequence given above is the basic pattern for fatty acid synthesis in living systems. Details of the reaction mechanisms are still unclear and may vary between species. The enzyme complex termed fatty acid synthase catalyzes all these reactions, but its structure and properties vary considerably. The individual enzymes in *Escherichia coli* are dissociable. By contrast, **mammalian fatty acid synthase** is composed of two possibly identical subunits, each of which is a multienzyme polypeptide containing all of the necessary catalytic activities in a linear array. There are variations in the enzyme between mammalian species and tissues.

It appears that the growing fatty acid chain is continually bound to the multifunctional protein and is sequentially transferred between the 4′-phosphopantetheine group of ACP, a domain on the protein, and the sulfhydryl group of a cysteine residue on β-ketoacyl-ACP synthase during the condensation reaction (Reaction 3, Figure 16.7) (see also Figure 16.9). An intermediate acylation to a serine residue probably takes place when acyl CoA units add to enzyme-bound ACP in the transacylase reactions.

Regulation of palmitate synthesis probably occurs primarily by control of the rate of synthesis and degradation of the fatty acid synthase. The agents and conditions that do this are given in Table 16.2. They are logical in terms of balancing an efficient utilization of the various biological energy substrates.

Stoichiometry of Fatty Acid Biosynthesis

For acetyl CoA as the primer for palmitate biosynthesis, the overall reaction is

$$CH_3-\overset{O}{\overset{\|}{C}}-SCoA + 7\ {}^-OOC-CH_2-\overset{O}{\overset{\|}{C}}-SCoA + 14\ NADPH + 14\ H^+ \longrightarrow CH_3-(CH_2)_{14}-COO^- + 7\ CO_2 + 14\ NADP^+ + 8\ CoASH + 6\ H_2O$$

To calculate the energy needed for the overall conversion of acetyl CoA to palmitate, we must add the ATP used in formation of malonyl CoA:

$$7\ CH_3-\overset{O}{\overset{\|}{C}}-SCoA + 7\ CO_2 + 7\ ATP \longrightarrow 7\ {}^-OOC-CH_2-\overset{O}{\overset{\|}{C}}-SCoA + 7\ ADP + 7\ P_i$$

Then the stoichiometry for conversion of acetyl CoA to palmitate is

$$8\ CH_3-\overset{O}{\overset{\|}{C}}-SCoA + 7\ ATP + 14\ NADPH + 14\ H^+ \longrightarrow CH_3-(CH_2)_{14}-\overset{O}{\overset{\|}{C}}-O^- + 8\ CoASH + 7\ ADP + 7\ P_i + 6\ H_2O + 14\ NADP^+$$

Acetyl CoA Is Transported from Mitochondria to Cytosol for Palmitate Synthesis

Fatty acid synthase and acetyl-CoA carboxylase are found in the cytosol where biosynthesis of palmitate occurs. Mammalian tissues must use special processes to ensure an adequate supply of acetyl CoA and NADPH for this synthesis in the cytosol. The major source of acetyl CoA is the pyruvate dehydrogenase reaction in the matrix of mitochondria. Since the mitochondrial inner membrane is not readily permeable to acetyl CoA, a process involving citrate moves a C_2 unit to the cytosol for palmitate biosynthesis. This mechanism (Figure 16.10) takes advantage of the facts that citrate exchanges freely from mitochondria to cytosol via a tricarboxylate

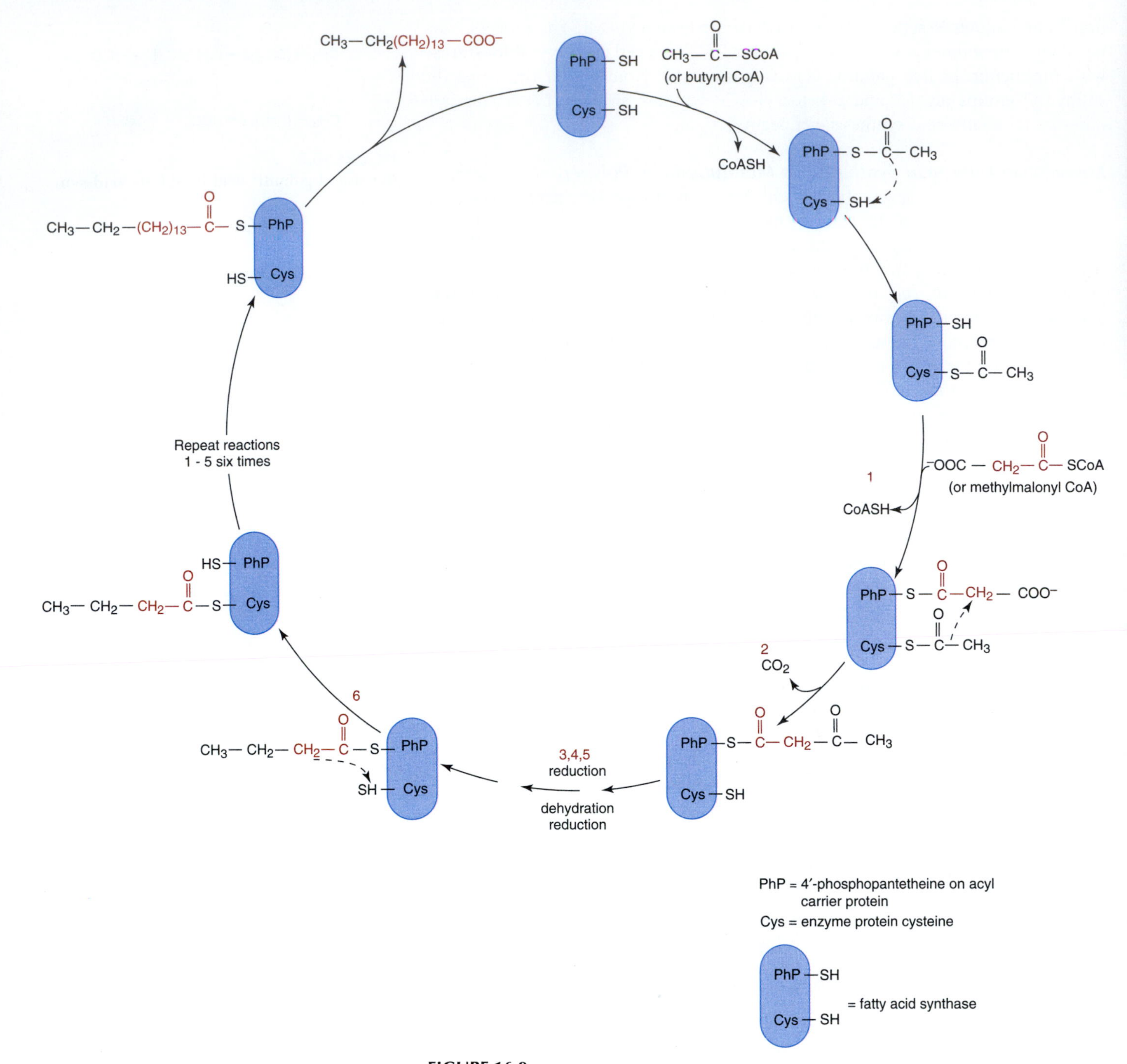

FIGURE 16.9
Proposed mechanism of elongation reactions taking place on mammalian fatty acid synthase.

transporter (see p. 583) and that an enzyme exists in cytosol to convert citrate to acetyl CoA and oxaloacetate. When there is an excess of citrate from the TCA cycle, this intermediate will pass into the cytosol and supply acetyl CoA for fatty acid biosynthesis. The cleavage reaction, which is energy requiring, is catalyzed by **ATP-citrate lyase:**

$$\text{citrate} + \text{ATP} + \text{CoA} \rightarrow \text{acetyl CoA} + \text{ADP} + \text{P}_i + \text{oxaloacetate}$$

This mechanism has other advantages because CO_2 and NADPH for synthesis of palmitate can be produced from excess cytosolic oxaloacetate. As shown in Figure 16.10, NADH reduces oxalacetate to malate via **malate dehydrogenase,** and

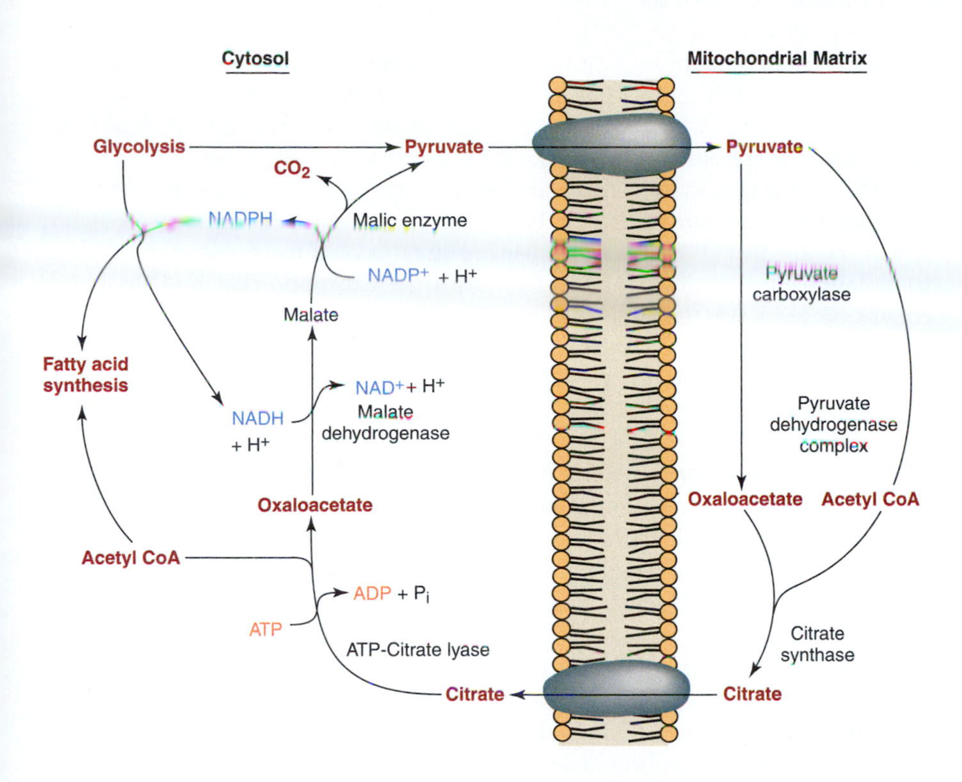

FIGURE 16.10
Mechanism for transfer of acetyl CoA from mitochondria to cytosol for fatty acid biosynthesis.

malate is then decarboxylated by **NADP-linked malic enzyme** (malate:$NADP^+$ oxidoreductase-decarboxylating) to produce NADPH, pyruvate, and CO_2. Thus NADPH is produced from NADH generated in glycolysis. The cycle is completed by return of pyruvate to mitochondria where it can be carboxylated to regenerate oxaloacetate, as described in the process of gluconeogenesis (see p. 633).

In summary, 1 NADH is converted to NADPH for each acetyl CoA transferred from mitochondria to cytosol, each transfer requiring 1 ATP. The transfer of the 8 acetyl CoA used for each molecule of palmitate supplies 8 NADPH. Since the biosynthesis of each palmitate requires 14 NADPH, the other 6 NADPH must be supplied by the pentose phosphate pathway also present in the cytosol. This stoichiometry is, of course, hypothetical. The *in vivo* relationships are complicated because transport of citrate and other di- and tricarboxylic acids across the inner mitochondrial membrane occurs by one-for-one exchanges. The actual flow rates are probably controlled by a composite of the concentration gradients of several of these exchange systems.

Palmitate Is the Precursor of Other Fatty Acids

Humans can synthesize all of the fatty acids they need from palmitate except the essential, polyunsaturated fatty acids (see p. 697). These syntheses involve a variety of enzyme systems in a number of locations in cells. Palmitate produced by fatty acid synthase is modified by three processes: elongation, desaturation, and hydroxylation.

Elongation Reactions

In mammals **elongation of fatty acids** occurs in either the endoplasmic reticulum or mitochondria; the processes are slightly different in these two loci. In the endoplasmic reticulum the sequence of reactions is similar to that occurring in the cytosolic fatty acid synthase with malonyl CoA as the source of two-carbon units and NADPH providing the reducing power. The preferred substrate for elongation is palmitoyl CoA. In contrast to palmitate synthesis, intermediates in subsequent reactions are CoA esters rather than attached to a protein, suggesting that the process is carried out by separate enzymes rather than a complex like fatty acid synthase.

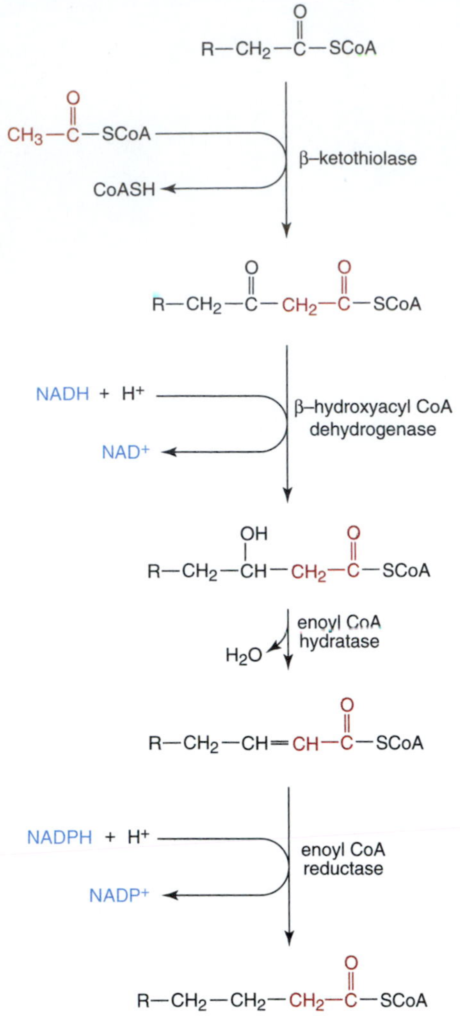

FIGURE 16.11
Mitochondrial elongation of fatty acids.

In most tissues this elongation system in the endoplasmic reticulum converts palmitate to stearate almost exclusively. Brain, however, contains one or more additional elongation systems, which synthesize longer chain acids (up to C_{24}) needed for brain lipids. These other systems also use malonyl CoA as substrate.

The elongation system in mitochondria uses acetyl CoA as the source of the added two-carbon units and both NADH and NADPH serve as reducing agents (Figure 16.11). This system operates by reversal of the pathway of fatty acid β-oxidation (see Section 16.6, p. 713) with the exception that NADPH-linked enoyl-CoA reductase (last step of elongation) replaces FAD-linked acyl-CoA dehydrogenase (first step in β-oxidation). The process has little activity with acyl-CoA substrates of C_{16} or longer, suggesting that it serves primarily in elongation of shorter chain species.

Formation of Monoenoic Acids by Stearoyl-CoA Desaturase

In higher animals **desaturation of fatty acids** occurs in the endoplasmic reticulum, and the oxidizing system used to introduce cis double bonds is significantly different from the main fatty acid oxidation process in mitochondria. The systems in endoplasmic reticulum have sometimes been termed **"mixed function oxidases"** because the enzymes simultaneously oxidize two substrates. In fatty acid desaturation one of these substrates is NADPH and the other is the fatty acid. Electrons from NADPH are transferred through a specific flavoprotein reductase and a cytochrome to **"active oxygen"** so that it will then oxidize the fatty acid. Although the complete mechanism has not been determined, this latter step may involve a hydroxylation. The three components of the system are the **desaturase enzyme, cytochrome *b*$_5$,** and **NADPH-cytochrome *b*$_5$ reductase.** The overall reaction is

$$\text{R—CH}_2\text{—CH}_2\text{—(CH}_2)_7\text{—COOH} + \text{NADPH} + \text{H}^+ + \text{O}_2 \rightarrow$$
$$\text{R—CH=CH—(CH}_2)_7\text{—COOH} + \text{NADP}^+ + 2\ \text{H}_2\text{O}$$

The enzyme specificity is such that the R group must contain at least six carbon atoms.

Mechanisms that regulate conversion of palmitate to unsaturated fatty acids are complex. An important consideration is the control of the amounts synthesized of different unsaturated fatty acids for proper balanced maintenance of the physical state of stored triacylglycerols and membrane phospholipids. A committed step in the formation of unsaturated fatty acids from palmitate or stearate is introduction of the first double bond between C-9 and C-10 atoms by **stearoyl-CoA desaturase** to produce palmitoleic or oleic acid, respectively. The activity of this enzyme and its synthesis are controlled by both dietary and hormonal mechanisms. Increasing the amounts of dietary polyunsaturated fatty acids in experimental animals decreases the activity of stearoyl-CoA desaturase in liver, while insulin, triiodothyronine, and hydrocortisone cause its induction.

Formation and Modification of Polyunsaturated Fatty Acids

A variety of **polyunsaturated fatty acids** are synthesized by humans through a combination of elongation and desaturation reactions. Once the initial double bond has been placed between carbons 9 and 10 by stearoyl-CoA desaturase, additional double bonds can be introduced just beyond C-4, C-5, or C-6 atoms. Desaturation at C-8 probably occurs also in some tissues. The positions of these desaturations are shown in Figure 16.12. The relative specificities of the various enzymes are

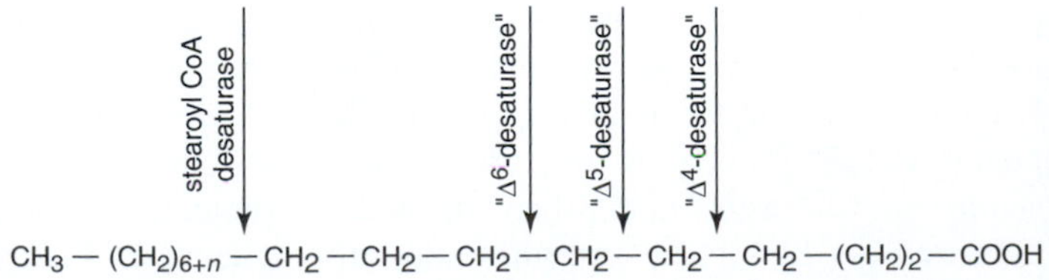

FIGURE 16.12
Positions in the fatty acid chain where desaturation can occur in the human.
There must always be at least six single bonds in the chain toward the methyl end of the molecule just beyond the bond being desaturated.

still to be determined completely, but it seems that elongation and desaturation can occur in either order. Conversion of linolenic acid to all *cis*-4, 7, 10, 13, 16, 19-docosahexaenoic acid in brain is a specific example of such a sequence.

$$CH_3{-}(CH_2{-}CH{=}CH)_3{-}CH_2{-}CH_2{-}CH_2{-}CH_2{-}CH_2{-}CH_2{-}CH_2{-}COOH$$

Linolenic acid

"Δ⁶-desaturase"

$$CH_3{-}(CH_2{-}CH{=}CH)_3{-}CH_2{-}CH{=}CH{-}CH_2{-}CH_2{-}CH_2{-}CH_2{-}COOH$$

elongation

$$CH_3{-}(CH_2{-}CH{=}CH)_3{-}CH_2{-}CH{=}CH{-}CH_2{-}CH_2{-}CH_2{-}CH_2{-}CH_2{-}CH_2{-}COOH$$

"Δ⁵-desaturase"

$$CH_3{-}(CH_2{-}CH{=}CH)_3{-}CH_2{-}CH{=}CH{-}CH_2{-}CH{=}CH{-}CH_2{-}CH_2{-}CH_2{-}COOH$$

elongation

$$CH_3{-}(CH_2{-}CH{=}CH)_3{-}CH_2{-}CH{=}CH{-}CH_2{-}CH{=}CH{-}CH_2{-}CH_2{-}CH_2{-}CH_2{-}CH_2{-}COOH$$

"Δ⁴-desaturase"

$$CH_3{-}(CH_2{-}CH{=}CH)_3{-}CH_2{-}CH{=}CH{-}CH_2{-}CH{=}CH{-}CH_2{-}CH{=}CH{-}CH_2{-}CH_2{-}COOH$$

All-*cis*-4,7,10,13,16,19-docosahexaenoic acid

Polyunsaturated fatty acids, particularly arachidonic acid, are precursors of the highly active prostaglandins, thromboxanes, and leukotrienes. Different classes of prostaglandins are formed depending on the precursor fatty acid and the sequence of oxidations that convert the acids to active compounds. A detailed discussion of these substances and their formation is found in Chapter 17 (see p. 766). Polyunsaturated fatty acids in living systems have a significant potential for auto-oxidation, a process that may have important physiological and/or pathological consequences. Auto-oxidation reactions cause rancidity in fats and curing of linseed oil in paints.

Formation of Hydroxy Fatty Acids in Nerve Tissue

There are apparently two different processes that produce **α-hydroxy fatty acids** in higher animals. One occurs in the mitochondria of many tissues and acts on relatively short-chain fatty acids (see Section 16.6, p. 713). The other has been demonstrated only in tissues of the nervous system where it produces long-chain fatty acids with a hydroxyl group on C-2. These are needed for the formation of some myelin lipids. The specific case of α-hydroxylation of lignoceric acid to cerebronic acid has been studied. These enzymes preferentially use C_{22} and C_{24} fatty acids and show characteristics of the "mixed function oxidase" systems, requiring molecular oxygen and NADH or NADPH. This synthesis may be closely coordinated with biosynthesis of sphingolipids that contain hydroxylated fatty acids.

Fatty Acid Synthase Can Produce Fatty Acids Other Than Palmitate

Synthesis and modification of palmitate account for the great bulk of fatty acid synthesis in the human body, particularly that involved in energy storage. However, smaller amounts of different fatty acids are needed for specific structural or functional purposes. These acids are produced by modification of the process carried out by fatty acid synthase. Two examples are production of fatty acids shorter than palmitate in mammary glands and synthesis of branched-chain fatty acids in certain secretory glands.

Milk produced by many animals contains varying amounts of fatty acids with shorter chain lengths than palmitate. The amounts produced by **mammary gland**

$$CH_3—(CH_2)_n—CH_2OH$$

FIGURE 16.13
Fatty alcohol.

vary with species and especially with the physiological state of the animal. This is probably true of humans, although most investigations have been carried out with rats, rabbits, and various ruminants. The same fatty acid synthase that produces palmitate synthesizes shorter chain acids when the linkage of the growing chain with acyl carrier protein is split before the full C16 chain is completed. This hydrolysis is caused by soluble **thioesterases** whose activity is under hormonal control.

There are relatively few branched-chain fatty acids in higher animals. Until recently, their metabolism has been studied mostly in primitive species such as mycobacteria, where they are present in greater variety and amount. Simple branched-chain fatty acids are synthesized by tissues of higher animals for specific purposes, such as the production of waxes in sebaceous glands and avian preen glands and the elaboration of structures in echo-locating systems of porpoises.

The majority of branched-chain fatty acids in higher animals are synthesized by fatty acid synthase and are methylated derivatives of saturated, straight-chain acids. When **methylmalonyl CoA** is used as a substrate instead of malonyl CoA, a methyl side chain is inserted in the fatty acid, and the reaction is as follows:

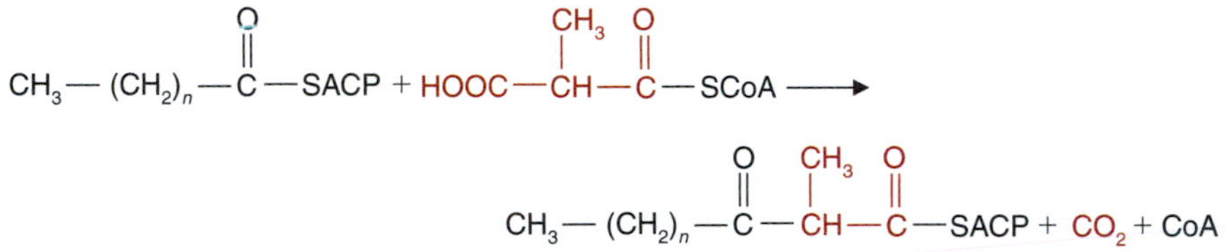

$$CH_3—(CH_2)_n—\overset{\displaystyle O}{\overset{\|}{C}}—SACP + HOOC—\overset{\displaystyle CH_3}{\overset{|}{CH}}—\overset{\displaystyle O}{\overset{\|}{C}}—SCoA \longrightarrow$$

$$CH_3—(CH_2)_n—\overset{\displaystyle O}{\overset{\|}{C}}—\overset{\displaystyle CH_3}{\overset{|}{CH}}—\overset{\displaystyle O}{\overset{\|}{C}}—SACP + CO_2 + CoA$$

Regular reduction steps then follow. These reactions occur in many tissues normally at a rate several orders of magnitude lower than the utilization of malonyl CoA to produce palmitate. The proportion of branched-chain fatty acids synthesized is largely governed by the relative availability of the two precursors. An increase in branching can occur by decreasing the ratio of malonyl CoA to methylmalonyl CoA. A malonyl-CoA decarboxylase capable of causing this decrease occurs in many tissues. It has been suggested that increased levels of methylmalonyl CoA in pathological situations, such as vitamin B_{12} deficiency, can lead to excessive production of branched-chain fatty acids.

Fatty Acyl CoAs May Be Reduced to Fatty Alcohols

As discussed in Chapter 17 (see p. 740), many phospholipids contain fatty acid chains in ether linkage rather than ester linkage. The biosynthetic precursors of these ether-linked chains are **fatty alcohols** (Figure 16.13) rather than fatty acids. These alcohols are formed in higher animals by a two-step, NADPH-linked reduction of fatty acyl CoAs in the endoplasmic reticulum. In tissues that produce relatively large amounts of ether-containing lipids, the concurrent production of fatty acids and fatty alcohols is probably closely coordinated.

16.4 | STORAGE OF FATTY ACIDS AS TRIACYLGLYCEROL

Most body tissues convert fatty acids to triacylglycerol by a common sequence of reactions, but liver and adipose tissue carry out this process to the greatest extent. Adipose tissue is a specialized connective tissue designed for synthesis, storage, and hydrolysis of triacylglycerols. This is the main system for long-term energy storage in humans. We are concerned here with white adipose tissue as opposed to brown adipose tissue, which occurs in much lesser amounts and has other specialized functions. Triacylglycerols are stored as lipid droplets in the cytoplasm, but this is not “dead storage” since they turn over with an average half-life of only a few days. Thus, in a homeostatic situation, there is continuous synthesis and breakdown of triacylglycerols in adipose tissue. Some storage also occurs in skeletal and cardiac muscle, but only for local consumption.

Triacylglycerol synthesis in liver is primarily for production of blood lipoproteins, although the products can serve as energy sources for the liver. Required fatty acids may come from the diet, from adipose tissue via blood transport, or from *de novo* biosynthesis. Acetyl CoA for biosynthesis is derived principally from glucose catabolism.

Triacylglycerols Are Synthesized from Fatty Acyl CoAs and Glycerol 3-Phosphate in Most Tissues

Triacylglycerols are synthesized in most tissues from activated fatty acids and a product of glucose catabolism, which can be either **glycerol 3-phosphate** or **dihydroxyacetone phosphate** (see Figure 16.14). Glycerol 3-phosphate is formed either by reduction of dihydroxyacetone phosphate produced in glycolysis or by phosphorylation of glycerol. White adipose tissue contains little or no glycerol kinase, so it derives glycerol phosphate from glycolytic intermediates. Fatty acids are activated by conversion to their CoA esters in the following reaction:

$$R-\overset{\displaystyle O}{\overset{\|}{C}}-O^- + ATP + CoASH \xrightarrow[\text{synthetase}]{\text{acyl-CoA}} R-\overset{\displaystyle O}{\overset{\|}{C}}-SCoA + AMP + PP_i + H_2O$$

This two-step reaction has an acyl adenylate as intermediate and is driven by hydrolysis of pyrophosphate to P_i.

Synthesis of triacylglycerols involves formation of **phosphatidic acid**, which is a key intermediate in synthesis of other lipids as well (see Chapter 17, p. 734). This may be formed by two sequential acylations of glycerol 3-phosphate, as shown in Figure 16.15. Alternatively, dihydroxyacetone phosphate may be acylated

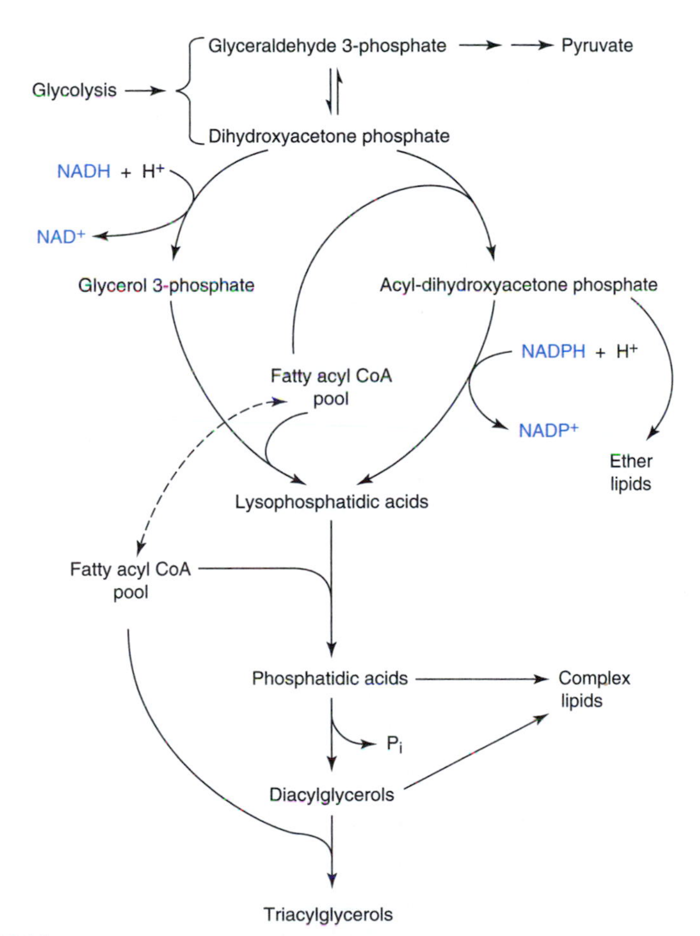

FIGURE 16.14
Alternative pathways for biosynthesis of triacylglycerol from dihydroxyacetone phosphate.

CH_2OH
$HO-CH$
$CH_2-O-PO_3^{2-}$

Glycerol 3-phosphate

acyltransferase: $R-\overset{O}{\overset{\|}{C}}-SCoA$ → CoASH

$CH_2-O-\overset{O}{\overset{\|}{C}}-R$
$HO-CH$
$CH_2-O-PO_3^{2-}$

Lysophosphatidic acid

acyltransferase: $R'-\overset{O}{\overset{\|}{C}}-SCoA$ → CoASH

$CH_2-O-\overset{O}{\overset{\|}{C}}-R$
$R'-\overset{O}{\overset{\|}{C}}-O-CH$
$CH_2-O-PO_3^{2-}$

Phosphatidic acid

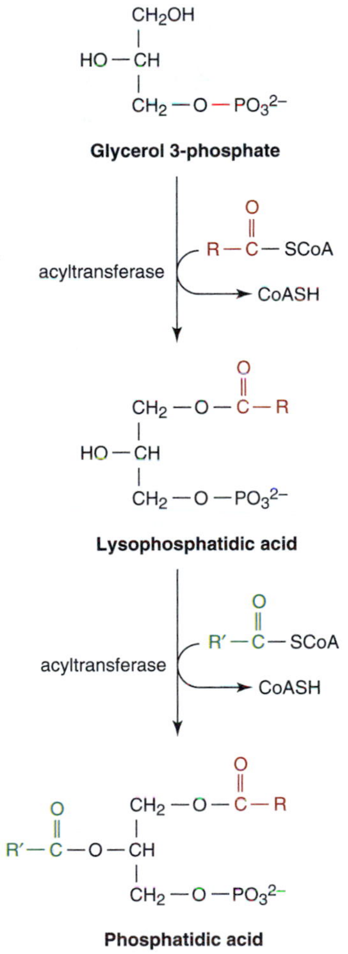

FIGURE 16.15
Synthesis of phosphatidic acid from glycerol 3-phosphate.

directly at C-1 followed by reduction at C-2. The resultant lysophosphatidic acid can then be further esterified, as illustrated in Figure 16.16. If phosphatidic acid from either of these routes is used for synthesis of triacylglycerol, the phosphate group is next hydrolyzed by **phosphatidate phosphatase** to yield diacylglycerol, which is then acylated to triacylglycerol (Figure 16.17).

In intestinal mucosa cells synthesis of triacylglycerol does not require formation of phosphatidic acid as described above. A major product of intestinal digestion of lipids is 2-monoacylglycerols, which are absorbed as such into mucosa cells. An enzyme in these cells catalyzes acylation of these monoacylglycerols with acyl CoA to form 1,2-diacylglycerols, which then can be further acylated.

The specificity of the acylation reactions in all these reactions is still quite controversial. Analysis of **fatty acid patterns** in triacylglycerols from various human tissues shows that the distribution of different acids on the three positions of glycerol is neither random nor absolutely specific. Patterns in various tissues are different. Palmitic acid tends to be concentrated in position 1 and oleic acid in positions 2 and 3 of human adipose tissue triacylglycerols. Two main factors that determine localization of a particular fatty acid to a given position on glycerol are the specificity of acyltransferase involved and relative availability of different fatty acids in the fatty acyl CoA pool. Other factors are probably involved but their relative importance has not been determined.

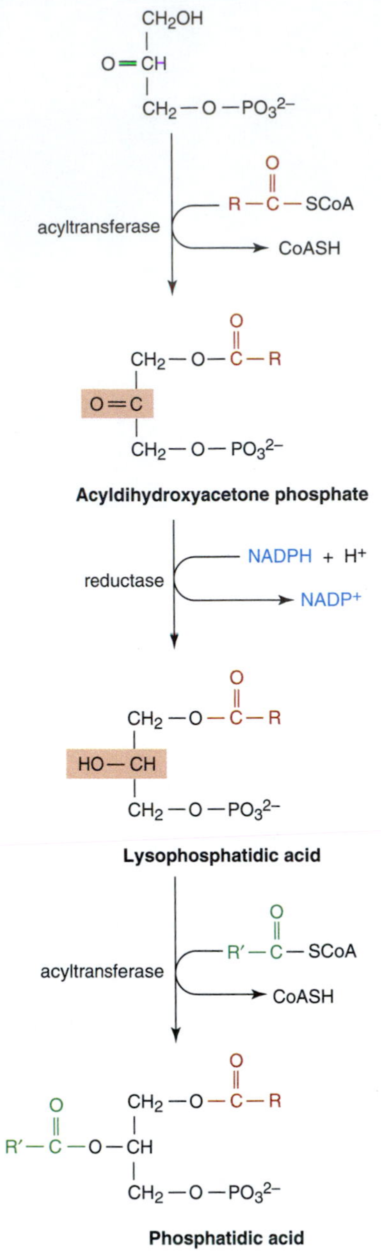

CH_2OH | $O{=}CH$ | $CH_2{-}O{-}PO_3^{2-}$

acyltransferase: $R{-}CO{-}SCoA$ → CoASH

$CH_2{-}O{-}CO{-}R$ | $O{=}C$ | $CH_2{-}O{-}PO_3^{2-}$

Acyldihydroxyacetone phosphate

reductase: NADPH + H^+ → $NADP^+$

$CH_2{-}O{-}CO{-}R$ | $HO{-}CH$ | $CH_2{-}O{-}PO_3^{2-}$

Lysophosphatidic acid

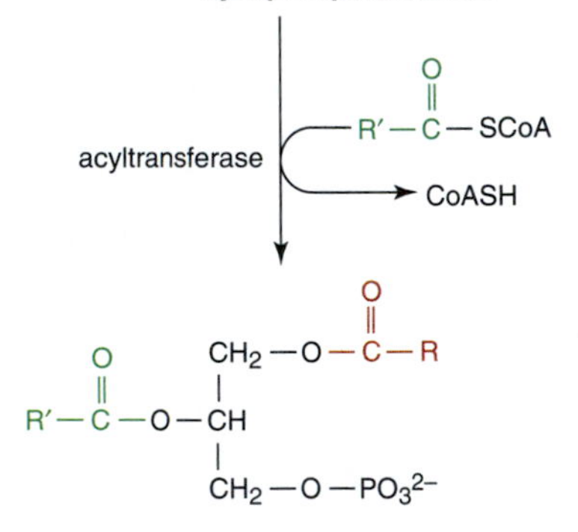

FIGURE 16.16
Synthesis of phosphatidic acid from dihydroxyacetone phosphate.

Phosphatidic acid

phosphatidate phosphatase

Diacylglycerol + P_i

acyltransferase: $R''{-}CO{-}SCoA$ → CoASH

Triacylglycerol

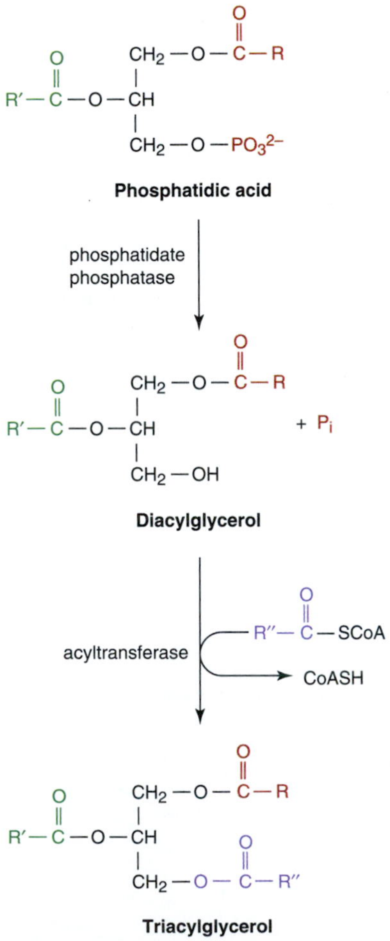

FIGURE 16.17
Synthesis of triacylglycerol from phosphatidic acid.

Mobilization of Triacylglycerols Requires Hydrolysis

The first step in recovering stored fatty acids for energy production is hydrolysis of triacylglycerols. A variety of lipases catalyze this reaction, the sequence of hydrolysis from the three positions on glycerol depending on the specificities of the particular lipases involved.

$$\text{R}'\text{—CO—CH}(\text{CH}_2\text{—O—CO—R})(\text{CH}_2\text{—O—CO—R}'') + 3\,\text{H}_2\text{O} \xrightarrow{\text{lipases}} \text{HO—CH}(\text{CH}_2\text{—OH})_2 + \text{R—COOH} + \text{R}'\text{—COOH} + \text{R}''\text{—COOH}$$

Lipases in adipose tissue are key enzymes for release of the major energy stores and the lipase that removes the first fatty acid is a controlled enzyme, a **hormone-sensitive lipase.** This control of triacylglycerol hydrolysis must be balanced with the process of triacylglycerol synthesis to assure adequate energy stores and avoid obesity (see Clin. Corr. 16.1 and 16.2). Fatty acids and glycerol produced by adipose tissue lipases are released to circulating blood, where fatty acids are bound by serum albumin and transported to tissues for use. Glycerol returns to liver, where it is converted to dihydroxyacetone phosphate and enters glycolytic or gluconeogenic pathways.

CLINICAL CORRELATION 16.1
Obesity

The terms obesity and overweight refer to excess in body weight relative to height. Their definitions are arbitrary and are based on actuarial estimates of ideal body weight (IBW), that is, body weight associated with the lowest morbidity and mortality. Relative weight is body weight relative to IBW: overweight is defined as relative weight up to 20% above normal and obesity is relative weight over 20% above IBW. The body mass index (BMI) is correlated well with measures of body fat and is defined as weight (kg) divided by height^2 (m^2). Overweight is defined as a BMI of 25–30 kg per m^2 and obesity as a BMI $>$ 30 kg per m^2. Skinfold thickness is also a measure of body fat stores.

The cause of most cases of obesity is not known. Endocrine diseases such as hypothyroidism or Cushing's disease (overproduction of corticosteroids) are rare causes. So are mutations in the genes encoding the newly discovered hormone, leptin, or its receptor (see Clin. Corr. 16.2). Genetic factors interact with environmental factors: 80% of children of two obese parents will be obese, while only 14% of children of normal weight parents are obese. The major mechanism of weight gain is consumption of calories in excess of daily energy requirements, but the normal processes controlling food intake and energy dissipation are not very well understood. Rarely, tumors of the hypothalamus result in pathological overeating (hyperphagia). However, a specific defect in most cases of human obesity has not been demonstrated.

Treatment of obesity, a problem of epidemic proportions in developed countries, revolves about dietary restriction, increased physical activity, and behavior modification. The real problem is to modify the patients' eating patterns long-term, and even in those who lose weight, regain of the weight is very common. Currently no pharmacological agents are effective in promoting long-term weight control. Surgery to limit the size of the gastric reservoir can be considered for patients over 100% above IBW. Medical complications of obesity include a two- to threefold increase in hypertension, gallstones, and diabetes and fivefold increase in risk of endometrial carcinoma. Obese patients have decreased plasma antithrombin III levels, which predisposes them to venous thrombosis (see Clin. Corr. 15.12).

Bray, G. A. Complications of obesity. *Ann. Intern. Med.* 103:1052, 1985; and Bray, G. A. The syndromes of obesity: an endocrine approach. In: L. J. DeGroot (Ed.), *Endocrinology*, Vol. 3, 3rd ed. Philadelphia: Saunders, 1995, p. 2624.

CLINICAL CORRELATION 16.2

Leptin and Obesity

In 1994 the *OB* gene of mice, its protein product, and their human homologues were identified. The human gene encodes a polypeptide of 166 amino acids that is expressed in adipose tissue in proportion to the severity of the obesity. The secreted protein, called leptin, contains 146 amino acids, can be measured by immunoassay, and is highly homologous to the murine protein.

Mice of the ob/ob strain that inherit a nonsense mutation in the leptin gene, leading to a truncated protein of 104 amino acids that is not secreted, are obese and diabetic and exhibit reduced activity, metabolism, and body temperature. Injection of recombinant leptin into mice homozygous for this mutation lowered their food intake, body weight, percentage of body fat, and serum glucose and insulin concentrations and increased their metabolic rate, body temperature, and activity level.

A phenotype similar to that of the ob/ob mouse is also found in the db/db mouse and the fa/fa Zucker rat, both of which carry mutations in the leptin receptor gene causing loss of function of this protein. As expected, their massive obesity and hypometabolism are unresponsive to leptin administration.

Inherited defects in the genes for leptin and/or its receptor have recently been found to be the cause of extreme overeating and massive obesity in a few human subjects. One of these was a young girl weighing ~200 lb at age 8 who was found to be homozygous for a mutation in the gene for leptin that prevented the production of a functional leptin molecule. Treatment of this patient with normal leptin has resulted in a dramatic reversal of her hyperphagia and weight gain. However, since most obese individuals produce normal leptin and leptin receptors their tendency to overeat and/or gain excessive weight must stem from some other abnormality. One possibility is that they are resistant to leptin signaling in the hypothalamus at a point distal to the leptin–leptin receptor interaction.

Considine, R. V., Sinha, M. K., Heiman, M. L., et al. Serum immunoreactive-leptin concentrations in normal-weight and obese humans. *N. Engl. J. Med.* 334:292, 1996; Lee, G. H., Proenca, R., Montez, J. M., et al. Abnormal splicing of the leptin receptor in diabetic mice. *Nature* 379:632, 1996; and Farooqi, I. S., Jebb, S. A., Langmack, G., et al. Effects of recombinant leptin therapy in a child with congenital leptin deficiency. *N. Engl. J. Med.* 341:879, 1999.

16.5 | METHODS OF INTERORGAN TRANSPORT OF FATTY ACIDS AND THEIR PRIMARY PRODUCTS

Lipid-Based Energy Is Transported in Blood in Different Forms

The energy available in fatty acids needs to be distributed throughout the body from the site of fatty acid absorption, biosynthesis, or storage to functioning tissues that consume them. This transport is closely integrated with that of other lipids, especially cholesterol, and is intimately involved in pathological processes leading to atherosclerosis.

In humans, three types of substances are used as vehicles to transport lipid-based energy: (1) chylomicrons and other plasma lipoproteins in which triacylglycerols are carried in protein-coated lipid droplets, both of which contain other lipids; (2) fatty acids bound to serum albumin; and (3) so-called ketone bodies, acetoacetate and β-hydroxybutyrate. These three vehicles are used in varying proportions to carry energy in the bloodstream via three routes. One is transport of dietary triacylglycerols as chylomicrons throughout the body from the intestine after absorption. Another is transport of lipid-based energy processed by or synthesized in liver and distributed either to adipose tissue for storage or to other tissues for use; this includes "ketone bodies'" and plasma lipoproteins other than chylomicrons. Finally, there is transport of energy released from storage in adipose tissue to the rest of the body in the form of fatty acids that are bound to serum albumin.

The proportion of energy being transported in any one of the modes outlined above varies considerably with metabolic and physiological state. At any time, the largest amount of lipid in blood is in the form of triacylglycerol in various lipoproteins. Fatty acids bound to albumin, however, are utilized and replaced very rapidly so total energy transport for a given period by this mode may be very significant.

Plasma Lipoproteins Carry Triacylglycerols and Other Lipids

Plasma lipoproteins are synthesized in both intestine and liver and are a heterogeneous group of lipid–protein complexes composed of various types of lipids and apoproteins (see p. 127 or a detailed discussion of structure). The two most important vehicles for delivery of lipid-based energy are **chylomicrons** and **very low density lipoprotein (VLDL),** because they contain relatively large amounts of triacylglycerol. Chylomicrons are formed in the intestine and function in absorption and transport of dietary triacylglycerol, cholesterol, and fat-soluble vitamins. The exact precursor product relationships between the other types of plasma lipoproteins have yet to be completely defined, as do the roles of various protein components. Liver synthesizes VLDL. Fatty acids from triacylglycerols in VLDL are taken up by adipose and other tissues. In the process VLDLs are converted to **low density lipoproteins (LDLs).** The role of **high density lipoprotein (HDL)** in transport of lipid-based energy is yet to be clarified. All of these lipoproteins are integrally involved in transport of other lipids, especially cholesterol. Lipid components can interchange to some extent between different classes of lipoprotein, and some apoproteins probably have functional roles in modifying enzyme activity during exchange of lipids between plasma lipoproteins and tissues. Other apoproteins serve as specific recognition sites for cell surface receptors. Such interaction constitutes the first step in receptor-mediated endocytosis of certain lipoproteins. A detailed discussion of the function of plasma lipoproteins is presented on page 745. Studies of rare genetic abnormalities have been helpful in explaining the roles of some of these apoproteins (see Clin. Corr. 16.3).

Fatty Acids Are Bound to Serum Albumin

Serum albumin acts as a carrier for a number of substances in blood, some of the most important being fatty acids. These acids are water insoluble in themselves, but when they are released into plasma during triacylglycerol hydrolysis they are quickly bound to albumin. Albumin has a number of binding sites for fatty acid, some of them having very high affinity. Saturation of albumin actually with fatty acids is far from maximal, but the turnover of these fatty acids is high, so binding by this mechanism constitutes a major route of energy transfer.

CLINICAL CORRELATION 16.3

Genetic Abnormalities in Lipid-Energy Transport

Diseases that affect the transport of lipid-based energy frequently result in abnormally high plasma triacylglycerol, cholesterol, or both. They are classified as hyperlipidemias. Some of them are genetically transmitted, and presumably they result from the alteration or lack of one or more proteins involved in the production or processing of plasma lipids. The nature and function of all of these proteins is yet to be determined, so the elucidation of exact causes of the pathology in most of these diseases is still in the early stages. However, in several cases a specific protein abnormality has been associated with altered lipid transport in the patient's plasma.

In the extremely rare disease, analbuminemia, there is an almost complete lack of serum albumin. In a rat strain with analbuminemia, a 7-base-pair deletion in an intron of the albumin gene results in the inability to process the nuclear mRNA for albumin. Despite the many functions of this protein, the symptoms of the disease are surprisingly mild. Lack of serum albumin effectively eliminates the transport of fatty acids unless they are esterified in acylglycerols or complex lipids. However, since patients with analbuminemia do have elevated plasma triacylglycerol levels, presumably the deficiency in lipid-based energy transport caused by the absence of albumin to carry fatty acids is filled by increased use of plasma lipoproteins to carry triacylglycerols.

A more serious genetic defect is the absence of lipoprotein lipase. The major problem here is the inability to process chylomicrons after a fatty meal. Pathological fat deposits occur in the skin (eruptive xanthomas) and the patients typically suffer from pancreatitis. If patients are put on a low-fat diet they respond reasonably well.

Another rare but more severe disease, abetalipoproteinemia, is caused by defective synthesis of apoprotein B, an essential component in the formation of chylomicrons and VLDL. Under these circumstances the major pathway for transporting lipid-based energy from the diet to the body is unavailable. Chylomicrons, VLDL, and LDL are absent from the plasma and fat absorption is deficient or nonexistent. There are other serious symptoms, including neuropathy and red cell deformities whose etiology is less clear.

Havel, R. J. and Kane, J. P. Structure and metabolism of plasma lipoproteins. In: C. R. Scriver, A. L. Beaudet, W. S. Sly, and D. Valle (Eds.), *The Metabolic and Molecular Bases of Inherited Disease* Vol. II, 7th ed. New York: McGraw-Hill, 1995, p. 1841; and Brunzell, J. D. Familial lipoprotein lipase deficiency and other causes of the chylomicronemia syndrome. In: C. R. Scriver, A. L. Beudet, W. S. Sly, and D. Valle (Eds.), *The Metabolic and Molecular Bases of Inherited Disease,* Vol. II, 7th ed. New York: McGraw-Hill, 1995, p. 1913.

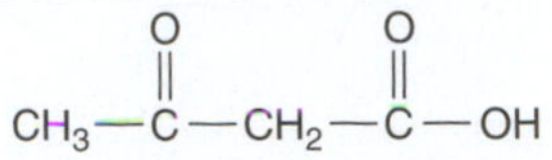

$CH_3-\overset{O}{\overset{\|}{C}}-CH_2-\overset{O}{\overset{\|}{C}}-OH$

Acetoacetic acid

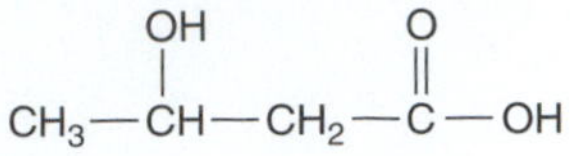

$CH_3-\overset{OH}{\overset{|}{C}H}-CH_2-\overset{O}{\overset{\|}{C}}-OH$

β-Hydroxybutyric acid

FIGURE 16.18
Structures of ketone bodies.

Ketone Bodies Are a Lipid-Based Energy Supply

The third mode of transport of lipid-based energy-yielding molecules is in the form of small water-soluble molecules, **acetoacetate** and **β-hydroxybutyrate** (Figure 16.18), produced primarily by liver during oxidation of fatty acids. The reactions involved in their formation and utilization will be discussed later. Under certain conditions, these substances can reach excessive concentrations in blood, leading to ketosis and acidosis. Spontaneous decarboxylation of acetoacetate to **acetone** also occurs, which is detectable as the smell of acetone in the breath when acetoacetate concentrations are high. This led early investigators to call the group of soluble products "ketone bodies." In fact, β-hydroxybutyrate and acetoacetate are continually produced by liver and, to a lesser extent, by kidney. Skeletal and cardiac muscle utilize them to produce ATP. Nervous tissue, which normally obtains almost all of its energy from glucose, is unable to take up and use fatty acids bound to albumin for energy production. However, it can use ketone bodies when glucose supplies are insufficient such as in prolonged fasting or starvation.

Lipases Hydrolyze Blood Triacylglycerols for Their Fatty Acids to Become Available to Tissues

Fatty acids bound to albumin and ketone bodies are readily taken up by various tissues for oxidation and production of ATP. The energy in fatty acids stored or circulated as triacylglycerols, however, is not directly available, but rather triacylglycerols must be enzymatically hydrolyzed to release the fatty acids and glycerol. **Lipoprotein lipase** hydrolyzes triacylglycerols in plasma lipoproteins. Lipoprotein lipase is located on the surface of endothelial cells of capillaries and possibly of adjoining tissue cells. It hydrolyzes fatty acids from the 1 and/or 3 position of tri- and diacylglycerols present in VLDL or chylomicrons. One of the lipoprotein apoproteins (ApoC-II) must be present to activate the process. Fatty acids released are either bound to serum albumin or taken up by the tissue. Monoacylglycerol products may either pass into the cells or be further hydrolyzed by serum **monoacylglycerol hydrolase.**

A completely distinct type of lipase controls mobilization of fatty acids from triacylglycerols stored in adipose tissue. One of them, the **hormone-sensitive triacylglycerol lipase,** is controlled by a cAMP-mediated mechanism. There are other lipases in the tissue, but the enzyme attacking triacylglycerols initiates the process. Two other lipases then rapidly complete the hydrolysis of mono- and diacylglycerols, releasing fatty acids to plasma where they are bound to serum albumin. Triacylglycerol metabolism is tightly controlled by both hormones and required cofactors. Some of the key regulatory factors are presented in Table 16.3.

TABLE 16.3 Regulation of Triacylglycerol Metabolism

Enzyme	*Regulatory Agent*	*Effect*
	Triacylglycerol Mobilization	
"Hormone-sensitive" lipase	"Lipolytic hormones" (e.g., epinephrine, glucagon, and ACTH)	Stimulation by cAMP-mediated phosphorylation of relatively inactive enzyme
	Insulin	Inhibition
	Prostaglandins	Inhibition
Lipoprotein lipase	Apolipoprotein C-II	Activation
	Insulin	Activation
	Triacylglycerol Biosynthesis	
Phosphatidate phosphatase	Steroid hormones	Stimulation by increased enzyme synthesis

16.6 | UTILIZATION OF FATTY ACIDS FOR ENERGY PRODUCTION

Fatty acids that arrive at the surface of cells are taken up and used for energy production primarily in mitochondria in a process intimately integrated with energy generation from other sources. Energy-rich intermediates produced from fatty acids are the same as those obtained from sugars, that is, NADH and $FADH_2$. The final stages of the oxidation process are exactly the same as for carbohydrates, that is, the metabolism of acetyl CoA by the TCA cycle and production of ATP in the mitochondrial electron transport system.

Utilization of fatty acids for energy production varies considerably from tissue to tissue and depends to a significant extent on the metabolic status of the body, whether it is fed or fasted, exercising or at rest. For instance, nervous tissue oxidizes fatty acids to a minimal degree if at all, but cardiac and skeletal muscle depend heavily on fatty acids as a major energy source. During prolonged fasting most tissues can use fatty acids or ketone bodies for their energy requirements.

β-Oxidation of Straight-Chain Fatty Acids Is the Major Energy-Producing Process

For the most part, fatty acids are oxidized by a mechanism that is similar to, but not identical with, a reversal of the process of palmitate synthesis. That is, two-carbon fragments are removed sequentially from the carboxyl end of the acid after steps of **dehydrogenation, hydration,** and **oxidation** to form a β-keto acid, which is split by **thiolysis.** These processes take place while the acid is activated in a thioester linkage to the 4′-phosphopantetheine of CoA.

Fatty Acids Are Activated by Conversion to Fatty Acyl CoA

The first step in oxidation of a fatty acid is its activation to a fatty acyl CoA. This is the same reaction described for synthesis of triacylglycerol in Section 16.4 (see p. 707) and occurs in the endoplasmic reticulum or the outer mitochondrial membrane.

Carnitine Carries Acyl Groups Across the Inner Mitochondrial Membrane

Whereas most of fatty acyl CoAs are formed outside mitochondria, the oxidizing machinery is inside the inner membrane, which is impermeable to CoA and its derivatives. The cell overcomes this problem by using **carnitine (4-trimethylamino-3-hydroxybutyrate)** as the carrier of acyl groups across the membrane. The steps involved are outlined in Figure 16.19. Enzymes on both sides of the membrane transfer fatty acyl groups between CoA and carnitine.

$$CH_3{-}(CH_2)_n{-}\overset{O}{\overset{\|}{C}}{-}SCoA + HO{-}\underset{}{CH}(CH_2{-}\overset{+}{N}{-}(CH_3)_3){-}CH_2{-}COOH \underset{\text{carnitine palmitoyltransferase}}{\rightleftharpoons} CH_3{-}(CH_2)_n{-}\overset{O}{\overset{\|}{C}}{-}O{-}CH(CH_2{-}\overset{+}{N}{-}(CH_3)_3){-}CH_2{-}COOH + CoASH$$

On the outer mitochondrial membrane the acyl group is transferred to carnitine catalyzed by **carnitine palmitoyltransferase I (CPT I).** Acyl carnitine then exchanges across the inner mitochondrial membrane with free carnitine by a carnitine–acyl carnitine antiporter translocase. Finally, the fatty acyl group is transferred back to CoA by **carnitine palmitoyltransferase II (CPT II)** located on the matrix side of the inner membrane. This process functions primarily in mitochondrial transport of fatty acyl CoAs with chain lengths of C12–C18. Genetic

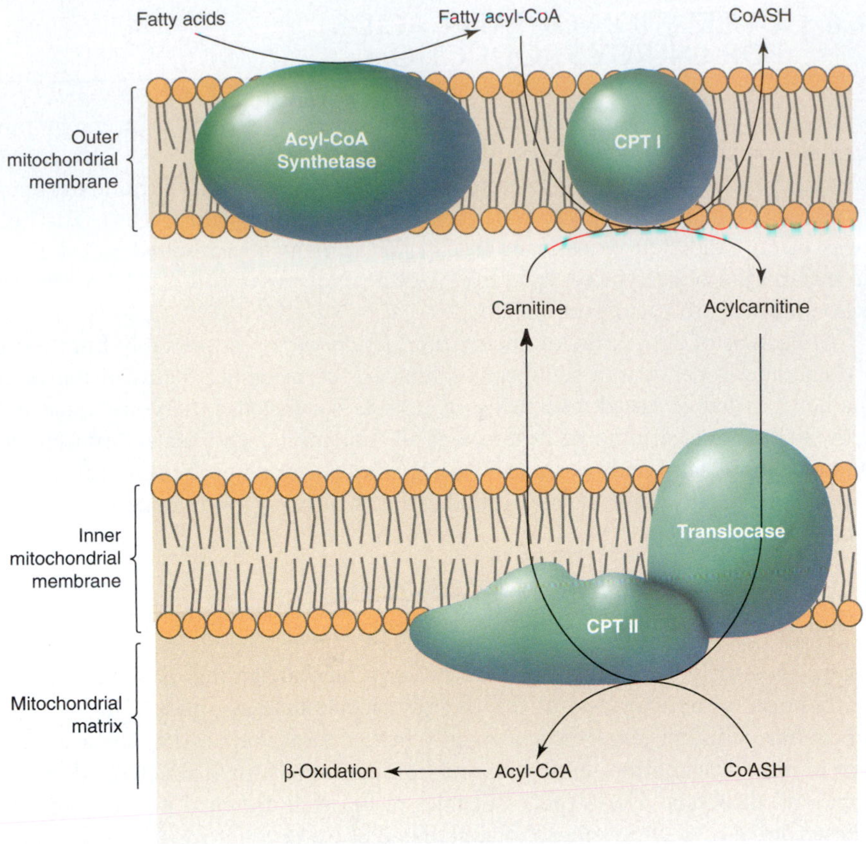

FIGURE 16.19
Mechanism for transfer of fatty acids from the cytosol through the inner mitochondrial membrane for oxidation.

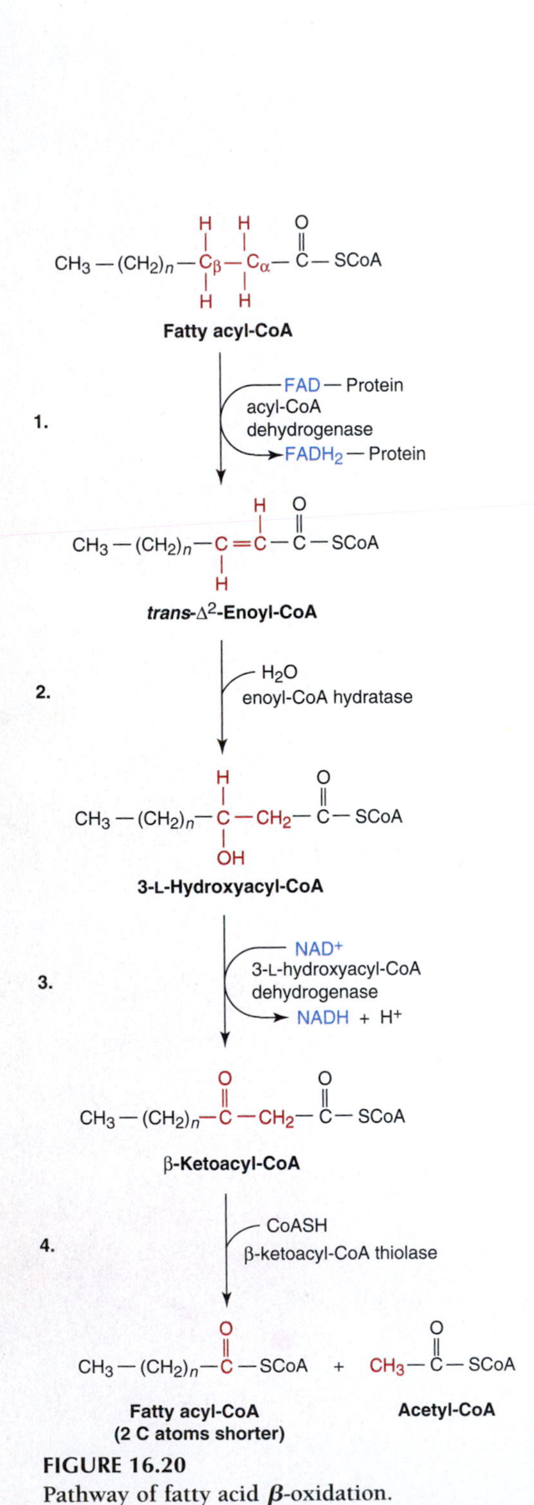

FIGURE 16.20
Pathway of fatty acid β-oxidation.

abnormalities in the system lead to serious pathology (see Clin. Corr. 16.4). By contrast, entry of shorter chain fatty acids is independent of carnitine because they cross the inner mitochondrial membrane and become activated to their CoA derivatives in the matrix.

β-Oxidation Is a Sequence of Four Reactions

The four reactions of β-oxidation are presented in Figure 16.20. Once fatty acyl CoA is formed at the inner surface of the inner mitochondrial membrane it can be oxidized by acyl-CoA dehydrogenase, a flavoprotein that uses FAD as the electron acceptor (Reaction 1). The products are enoyl CoA with a trans double bond between the C-2 and C-3 atoms and enzyme-bound $FADH_2$. As in the TCA cycle, the $FADH_2$ transfers its electrons through the electron transport chain regenerating FAD.

The second step in β-oxidation is hydration of the trans double bond to a **3-L-hydroxyacyl CoA.** This reaction is stereospecific, in that the L isomer is the product when the trans double bond is hydrated. The stereospecificity of the oxidative pathway is governed by the enzyme catalyzing the third reaction, which is specific for the L isomer as its substrate. The final step is the cleavage of the two-carbon fragment by a thiolase. In the overall process then, an acetyl CoA is produced and the acyl CoA product is ready for the next round of oxidation starting with acyl-CoA dehydrogenase.

Each of the four reactions shown in Figure 16.20 is catalyzed by several different enzymes with varying specificities for chain length of the acyl CoA. For example, at least four enzymes are involved in the first dehydrogenation step. These are referred to as very-long-chain, long-chain, medium-chain, and short-chain acyl-CoA dehydrogenases **(VLCAD, LCAD, MCAD,** and **SCAD).** VLCAD, which is thought to handle straight-chain acyl CoAs ranging from C-12 to C-24, differs from

CLINICAL CORRELATION 16.4

Genetic Deficiencies in Carnitine Transport or Carnitine Palmitoyltransferase

A number of diseases result from genetic abnormalities in the transport of long-chain fatty acids across the inner mitochondrial membrane. They stem from deficiencies either in the level of carnitine or in the functioning of the carnitine palmitoyltransferase (CPT) enzyme system.

The clinical symptoms of carnitine deficiency range from mild, recurrent muscle cramping to severe weakness and death. Two categories of the disorder, primary and secondary, are now recognized. Primary carnitine deficiency is caused by a defect in the high-affinity plasma membrane carnitine transporter in tissues such as muscle, kidney, heart, and fibroblasts (but apparently not in liver where a different transporter is operative). It results in extremely low levels of carnitine in affected tissues and also in plasma (because of failure of the kidneys to reabsorb carnitine). The very low carnitine level in heart and skeletal muscle seriously compromises long-chain fatty acid oxidation. Dietary carnitine therapy, by raising the plasma concentration of carnitine and forcing its entry into tissues in a nonspecific manner, is frequently beneficial. Secondary carnitine deficiency is often associated with inherited defects in the β-oxidation pathway that give rise to the accumulation of acyl CoAs and, in turn, acylcarnitines. The latter compounds can be excreted in the urine (see Clin. Corr. 16.5), thus draining the body's carnitine pool; in addition, they are thought to impair the tissue uptake of free carnitine.

CPT deficiency also presents as distinct clinical entities. The most common deficiency results from mutations in the CPT II gene that give rise to a partial loss of enzyme activity. The patient generally experiences muscle weakness during prolonged exercise when muscle relies heavily on fatty acids as an energy source. Myoglobinuria, due to breakdown of muscle tissue, is a frequent accompaniment. The disorder is usually referred to as the "muscular" form of CPT II activity deficiency. Mutations causing more severe (90% or greater) loss of CPT II activity can have serious consequences in early infancy. These are usually precipitated by periods of fasting and include hypoketotic hypoglycemia, hyperammonemia, cardiac malfunction, and sometimes death. Similar morbidity and mortality are associated with mutations in the gene for liver CPT I. To date only a few patients with hepatic CPT I deficiency have been reported, the small number possibly indicating that the disease is frequently lethal and has gone undiagnosed. Muscle CPT I is now known to be a different isoform from its liver counterpart, but no inherited defects at this locus have yet been reported.

The first patient with carnitine–acylcarnitine translocase deficiency was described as recently as 1992. Clinical features included intermittent hypoglycemic coma, hyperammonemia, muscle weakness, and cardiomyopathy. The condition proved fatal at age 3 years. Several additional cases with similar symptomatology have since been reported.

The hallmark of treatment for all inherited disorders of the carnitine transport/CPT system is avoidance of starvation and a diet low in long-chain fatty acids. Supplementary dietary medium-chain triacylglycerols, the fatty acids of which are oxidized by a carnitine-independent mechanism, has proved beneficial.

Stanley, C. A., Hale, D. E., Berry, G. T., Deleeno, S., Boxer, J., and Bonnefont, J. P. A deficiency of carnitine–acylcarnitine translocase in the inner mitochondrial membrane. *N. Engl. J. Med.* 327:19, 1992; Roe, C. R. and Coates, P. M. Mitochondrial fatty acid oxidation disorders. In: C. R. Scriver, A. L. Beaudet, W. S. Sly, and D. Valle (Eds.), *The Metabolic and Molecular Bases of Inherited Disease,* Vol. II, 7th ed. New York: McGraw-Hill, 1995, p. 1501; and Bonnefont, J.-P., Demaugre, F., Prip-Buus, C., Saudubray, J.-M., Brivet, M., Abadi, N., and Thuillier, L. Carnitine palmitoyltransferase deficiencies. *Mol. Gen. Metab.* 68:424, 1999.

the other family members in that it is membrane associated. MCAD has broad chain length specificity but is probably most active toward C-6 and C-8 substrates, while the order of preference for SCAD is $C_4 > C_6 > C_8$. Current thinking is that LCAD is primarily involved in initiating the oxidation of branched-chain fatty acids, for example, 2-methylpalmitoyl CoA.

Another feature unique to the oxidation of long-chain fatty acids is that the enoyl-CoA hydratase, 3-hydroxyacyl-CoA dehydrogenase, and β-ketothiolase steps are all catalyzed by the membrane-bound **trifunctional protein.** This is distinct from the enzymes catalyzing the equivalent reactions in the oxidation of medium- and short-chain acyl CoAs, all of which are soluble and occur in the matrix. Clinical Correlation 16.5 describes genetic deficiencies of acyl-CoA dehydrogenases.

It has not been shown conclusively that any of the enzymes in the β-oxidation scheme are control points, although under rather rigid *in vitro* conditions some apparently have slower maximum rates of reaction than others. It is assumed that control is exerted by availability of substrates and cofactors and by the rate of processing of acetyl CoA by the TCA cycle. One way in which substrate availability is controlled is by regulation of the carnitine shuttle mechanism that transports fatty acids into mitochondria. Thus, under conditions of carbohydrate abundance (high insulin levels), tissue levels of **malonyl CoA,** the product of the acetyl-CoA carboxylase reaction (see p. 698), are elevated, causing suppression of mitochondrial CPT I activity. Conversely, when insulin levels fall (starvation or uncontrolled

CLINICAL CORRELATION 16.5

Genetic Deficiencies in the Acyl-CoA Dehydrogenases

The acyl-CoA dehydrogenase deficiencies represent a recently discovered group of inherited defects that impair the β-oxidation of fatty acids at different stages of the chain shortening process. The affected enzyme may be the very long-chain acyl-CoA dehydrogenase (VLCAD), the long-chain acyl-CoA dehydrogenase (LCAD), the medium-chain acyl-CoA dehydrogenase (MCAD), or the short-chain acyl-CoA dehydrogenase (SCAD) whose substrate specificities vary (see text). These conditions are inherited in autosomal recessive fashion and share many of the same clinical features. The best characterized is MCAD deficiency, which, though first recognized as late as 1982, is now thought to be one of the most common of all inborn errors of metabolism.

Medium-chain acyl-CoA dehydrogenase deficiency usually manifests itself within the first 2 years of life after a fasting period of 12 h or more. Typical symptoms include vomiting, lethargy, and frequently coma, accompanied by hypoketotic hypoglycemia and dicarboxylic aciduria. The absence of starvation ketosis is accounted for by the block in hepatic fatty acid oxidation, which also causes a slowdown of gluconeogenesis. This, coupled with impaired fatty acid oxidation in muscle, which promotes glucose utilization, leads to profound hypoglycemia. Accumulation of medium-chain acyl CoAs in tissues forces their metabolism through alternative pathways including ω-oxidation and transesterification to glycine or carnitine. Excessive urinary excretion of the reaction products (medium-chain dicarboxylic acids together with medium-chain esters of glycine and carnitine) provide diagnostic clues. Most patients with this disorder do well simply by avoiding prolonged periods of starvation, which is consistent with the fact that the metabolic complications of MCAD deficiency are seen only when body tissues become heavily dependent on fatty acids as a source of energy (e.g., with carbohydrate deprivation). In retrospect it now seems likely that many cases previously diagnosed loosely as "Reye-like syndrome" or "sudden infant death syndrome" were in fact due to MCAD deficiency.

Coates, P. M. and Tanaka, K. Molecular basis of mitochondrial fatty acid oxidation defects. *J. Lipid Res.* 33:1099, 1992; and Wanders, R. J. A., Vreken, P., den Boer, M. E. J., et al. Disorders of mitochondrial fatty acyl-CoA β-oxidation. *J. Inher. Metab. Dis.* 22:442, 1999.

diabetes), the tissue malonyl CoA content is reduced, allowing increased fatty acid flux through CPT I and the β-oxidation pathway. This phenomenon is of central importance in the regulation of hepatic ketone body production (see p. 720).

Energy Yield from β-Oxidation of Fatty Acids

Each set of oxidations results in production of one acetyl CoA, one reduced flavoprotein, and one NADH. In the oxidation of palmitoyl CoA seven cleavages take place, and in the last cleavage two acetyl CoA molecules are formed. Thus the products of β-oxidation of palmitate are eight acetyl CoAs, seven reduced flavoproteins, and seven NADH. Based on current knowledge of the yields of ATP in oxidative phosphorylation (see p. 578) each of the reduced flavoproteins can yield 1.5 ATP and each NADH can yield 2.5 ATP when oxidized by the electron transport chain. As described in Chapter 13 (see p. 553) oxidation of each acetyl CoA through the TCA cycle yields 10 ATP, so the eight two-carbon fragments from a palmitate molecule produce 80 ATP. However, 2 ATP equivalents (1 ATP going to 1 AMP) were used to activate palmitate to palmitoyl CoA. Therefore each palmitic acid entering the cell can yield 106 ATP mol^{-1} by complete oxidation. The significance of the role of fatty acids in supplying the energy needs in humans is discussed on p. 872.

Comparison of the β-Oxidation Scheme with Palmitate Biosynthesis

In metabolism of substances, reactions in a catabolic pathway are sometimes quite similar to those in an anabolic pathway, but mechanisms are present for separate control of the two schemes. This is true in the case of fatty acid biosynthesis and β-oxidation. The critical differences between the two pathways are outlined in Table 16.4. They include separation by subcellular compartmentation (β-oxidation occurs in the mitochondria and palmitate biosynthesis in the cytosol), and use of different cofactors (NADPH in biosynthesis; FAD and NAD^+ in oxidation).

Some Fatty Acids Require Modification of β-Oxidation for Metabolism

The β-oxidation scheme accounts for the bulk of energy production from fatty acids in the human. These reactions, however, must be supplemented by other mecha-

TABLE 16.4 Comparison of Schemes for Biosynthesis and β-Oxidation of Palmitate

Parameter	*Biosynthesis*	*β-Oxidation*
Subcellular localization	Primarily cytosolic	Primarily mitochondrial
Phosphopantetheine-containing active carrier	Acyl carrier protein	Coenzyme A
Nature of small carbon fragment added or removed	C-1 and C-2 atoms of malonyl CoA after initial priming	Acetyl CoA
Nature of oxidation–reduction coenzyme	NADPH	FAD when saturated chain dehydrogenated, NAD^+ when hydroxy acid dehydrogenated
Stereochemical configuration of β-hydroxy intermediates	D-β-Hydroxy	L-β-Hydroxy
Energy equivalents yielded or utilized in interconversion of palmitate and acetyl CoA	7 ATP + 14 NADPH = 49 ATP equiv	7 $FADH_2$ + 7 NADH − 2 ATP = 28 ATP equiv

nisms so that ingested odd-chain and unsaturated fatty acids can be oxidized. In addition, reactions catalyze α- and ω-oxidation of fatty acids. **α-Oxidation** occurs at C-2 instead of C-3 as occurs in the β-oxidation scheme. **ω-Oxidation** occurs at the methyl end of the fatty acid molecule. Partial oxidation of fatty acids with cyclopropane ring structures probably occurs in humans, but the mechanisms are not worked out.

Propionyl CoA Is Produced by Oxidation of Odd-Chain Fatty Acids

Oxidation of fatty acids with an odd number of carbon atoms proceeds as described above, but the final product is a molecule of propionyl CoA (Figure 16.21). For this compound to be further oxidized, it undergoes carboxylation to methylmalonyl CoA, molecular rearrangement, and conversion to succinyl CoA. These reactions are identical with those described on page 812 for the metabolism of **propionyl CoA** formed in the metabolic breakdown of some amino acids.

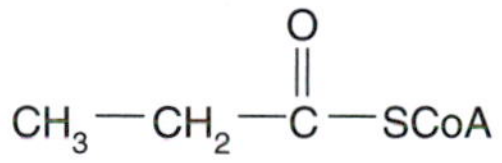

$$CH_3-CH_2-\overset{\overset{\displaystyle O}{\|}}{C}-SCoA$$

FIGURE 16.21
Propionyl CoA.

Oxidation of Unsaturated Fatty Acids Requires Additional Enzymes

Many unsaturated fatty acids in the diet are available for production of energy by humans. Oxidation of linoleoyl CoA, outlined in Figure 16.22, illustrates two special reactions required for oxidation of unsaturated fatty acids. One problem is that in β-oxidation of unsaturated fatty acids the sequential excision of C_2 fragments can generate an acyl CoA intermediate with a double bond between C-3 and C-4 atoms instead of between C-2 and C-3 atoms as required for the enoyl-CoA hydratase reaction. If so, the cis bond between C-3 and C-4 atoms is isomerized into a trans bond between C-2 and C-3 atoms by an auxiliary enzyme, enoyl-CoA isomerase. The regular process can then proceed.

A second problem occurs if the cis double bond of the acyl CoA intermediate resides between C-4 and C-5 atoms. In this case the action of acyl-CoA dehydrogenase gives rise to a *trans*-2, *cis*-4-enoyl CoA. This is acted on by 2,4-dienoyl CoA reductase that, using reducing equivalents from NADPH, produces a *trans*-3-enoyl CoA. This will serve as a substrate for enoyl-CoA isomerase producing *trans*-2-enoyl CoA needed for the next round of β-oxidation.

Some Fatty Acids Undergo α-Oxidation

As noted earlier, several mechanisms for **hydroxylation of fatty acids** exist. Hydroxylation of long-chain acids needed for synthesis of sphingolipids has been

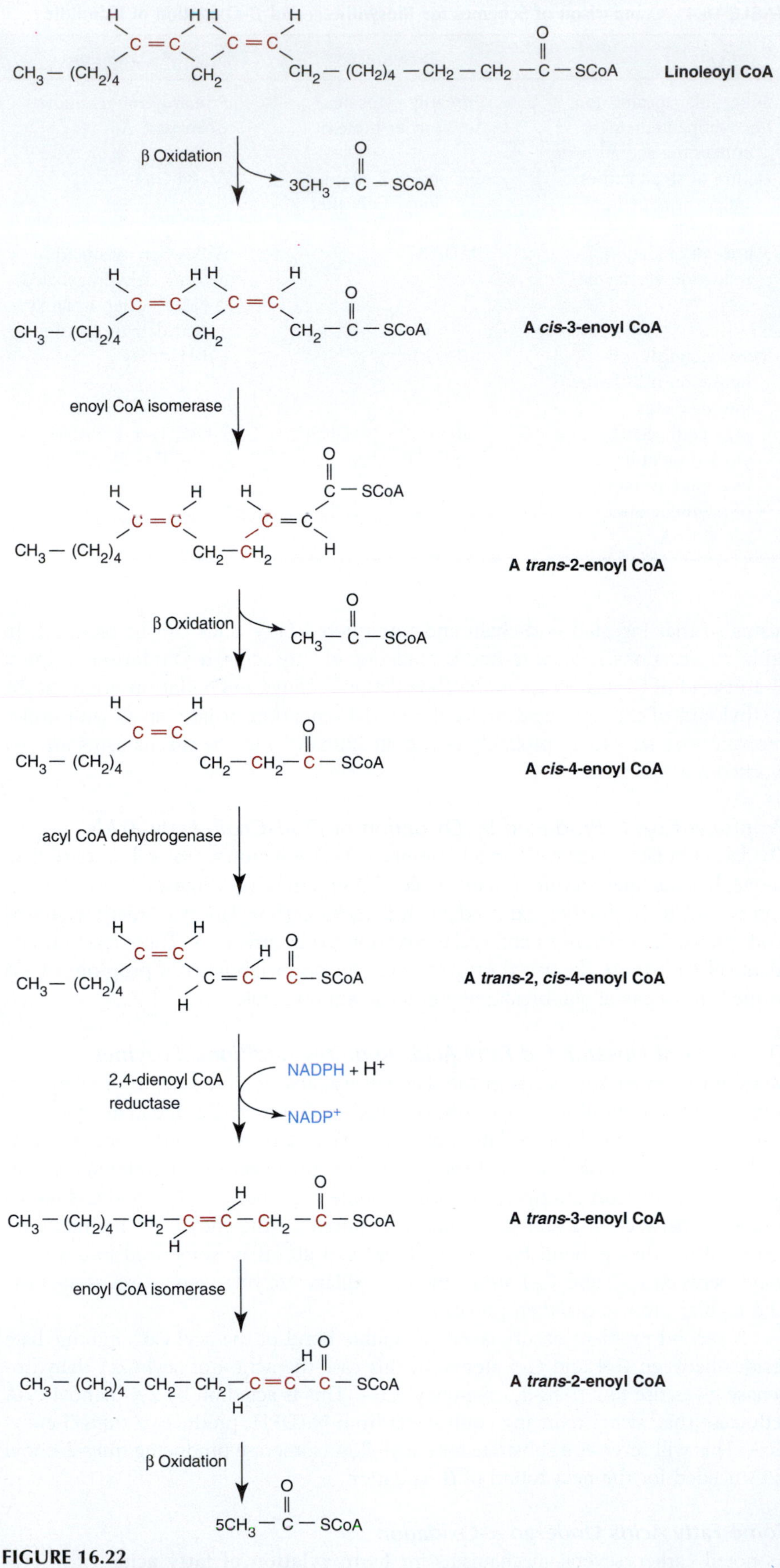

$CH_3-(CH_2)_4-CH=CH-CH_2-CH=CH-CH_2-(CH_2)_4-CH_2-CH_2-C(=O)-SCoA$ **Linoleoyl CoA**

β Oxidation → $3CH_3-C(=O)-SCoA$

$CH_3-(CH_2)_4-CH=CH-CH_2-CH=CH-CH_2-C(=O)-SCoA$ **A *cis*-3-enoyl CoA**

enoyl CoA isomerase

$CH_3-(CH_2)_4-CH=CH-CH_2-CH_2-CH=CH-C(=O)-SCoA$ **A *trans*-2-enoyl CoA**

β Oxidation → $CH_3-C(=O)-SCoA$

$CH_3-(CH_2)_4-CH=CH-CH_2-CH_2-C(=O)-SCoA$ **A *cis*-4-enoyl CoA**

acyl CoA dehydrogenase

$CH_3-(CH_2)_4-CH=CH-CH=CH-C(=O)-SCoA$ **A *trans*-2, *cis*-4-enoyl CoA**

2,4-dienoyl CoA reductase ($NADPH + H^+$ → $NADP^+$)

$CH_3-(CH_2)_4-CH_2-CH=CH-CH_2-C(=O)-SCoA$ **A *trans*-3-enoyl CoA**

enoyl CoA isomerase

$CH_3-(CH_2)_4-CH_2-CH_2-CH=CH-C(=O)-SCoA$ **A *trans*-2-enoyl CoA**

β Oxidation

$5CH_3-C(=O)-SCoA$

FIGURE 16.22
Oxidation of linoleoyl CoA.

described. In addition, there are systems in other tissues that hydroxylate the α carbon of shorter chain acids in order to start their oxidation. The sequence is as follows:

$$CH_3-(CH_2)_n-CH_2-\overset{O}{\overset{\|}{C}}-OH \longrightarrow CH_3-(CH_2)_n-\overset{OH}{\overset{|}{CH}}-\overset{O}{\overset{\|}{C}}-OH \longrightarrow$$

$$CH_3-(CH_2)_n-\overset{O}{\overset{\|}{C}}-\overset{O}{\overset{\|}{C}}-OH \longrightarrow CH_3-(CH_2)_n-\overset{O}{\overset{\|}{C}}-OH + CO_2$$

Some of these hydroxylations occur in the endoplasmic reticulum and mitochondria and involve the "mixed function oxidase" type of mechanism because they require molecular oxygen, reduced nicotinamide nucleotides, and specific cytochromes. Another site of α-hydroxylation, which is particularly important for the metabolism of branched-chain fatty acids (see Clin. Corr. 16.6), is the peroxisome. Here, the acyl CoA (such as phytanoyl CoA) is acted on by a hydroxylase, involving 2-oxoglutarate, Fe^{2+}, and ascorbate with the generation of 2-hydroxyphytanoyl CoA and formyl CoA, the latter being subsequently metabolized to CO_2 via formic acid. The 2-hydroxyphytanoyl CoA is then further metabolized to pristanic acid which can now undergo β-oxidation.

ω-Oxidation Gives Rise to a Dicarboxylic Acid

Another minor pathway for fatty acid oxidation occurs in the endoplasmic reticulum of many tissues. In this case hydroxylation takes place on the methyl carbon at the other end of the molecule from the carboxyl group or on the carbon next to the methyl end. It uses the "mixed function oxidase" type of reaction requiring cytochrome P450, O_2, and NADPH, as well as the necessary enzymes (see Chapter 11). Hydroxylated fatty acid can be further oxidized to a **dicarboxylic acid** via

CLINICAL CORRELATION 16.6
Refsum's Disease

Although the use of the α-oxidation scheme is a relatively minor one in terms of total energy production, it is significant in the metabolism of dietary fatty acids that are methylated. A principal example of these is phytanic acid,

$$\begin{array}{l} CH_3 \\ | \\ CH-CH_3 \\ | \\ (CH_2)_3 \\ | \\ CH-CH_3 \\ | \\ (CH_2)_3 \\ | \\ CH-CH_3 \\ | \\ (CH_2)_3 \\ | \\ CH-CH_3 \\ | \\ CH_2 \\ | \\ COOH \end{array}$$

Phytanic acid

a metabolic product of phytol, which occurs as a constituent of chlorophyll. Phytanic acid is a significant constituent of milk lipids and animal fats, and normally it is metabolized by an initial α-hydroxylation followed by dehydrogenation and decarboxylation. β-Oxidation cannot occur initially because of the presence of the 3-methyl group, but it can proceed after the decarboxylation. The whole reaction produces three molecules of propionyl CoA, three molecules of acetyl CoA, and one molecule of isobutyryl CoA.

In a rare genetic disease called Refsum's disease, the patients lack the peroxisomal α-hydroxylating enzyme and accumulate large quantities of phytanic acid in their tissues and sera. This leads to serious neurological problems such as retinitis pigmentosa, peripheral neuropathy, cerebellar ataxia, and nerve deafness. The restriction of dietary dairy products and meat products from ruminants results in lowering of plasma phytanic acid and regression of neurologic symptoms.

Wanders, R. J. A., van Grunsven, E. G., and Jansen, G. A. Lipid metabolism in peroxisomes: enzymology, functions and dysfunctions of the fatty acid α- and β-oxidation systems in humans. *Biochem. Soc. Trans.* 28:141, 2000.

sequential action of cytosolic **alcohol** and **aldehyde dehydrogenases.** The process occurs primarily with medium chain fatty acids. The overall reactions are

$$CH_3-(CH_2)_n-\overset{O}{\overset{\|}{C}}-OH \longrightarrow HO-CH_2-(CH_2)_n-\overset{O}{\overset{\|}{C}}-OH \longrightarrow \longrightarrow HO-\overset{O}{\overset{\|}{C}}-(CH_2)_n-\overset{O}{\overset{\|}{C}}-OH$$

The dicarboxylic acid so formed can be activated at either end of the molecule to form a CoA ester, which in turn can undergo β-oxidation to produce shorter chain dicarboxylic acids such as adipic (C_6) and succinic (C_4) acids. This process also occurs primarily in peroxisomes (see p. 22)

Ketone Bodies Are Formed from Acetyl CoA

Ketone bodies are water-soluble forms of lipid-based energy and consist of acetoacetic acid and its reduction product β-hydroxybutyric acid. β-Hydroxybutyryl CoA and acetoacetyl CoA are intermediates near the end of the β-oxidation sequence, and it was initially presumed that enzymatic removal of CoA from these compounds was the main route for production of the free acids. However, β-oxidation proceeds completely to acetyl CoA production without accumulation of any intermediates, and acetoacetate and β-hydroxybutyrate are formed subsequently from acetyl CoA by a separate mechanism.

HMG CoA Is an Intermediate in the Synthesis of Acetoacetate from Acetyl CoA

The primary site for formation of ketone bodies is liver, with lesser activity in kidney. The process occurs in the mitochondrial matrix and begins with condensation of two acetyl CoA molecules to form acetoacetyl CoA (Figure 16.23). The enzyme involved, **β-ketothiolase,** is probably an isozyme of that which catalyzes the reverse reaction as the last step of β-oxidation. Acetoacetyl CoA then condenses with another molecule of acetyl CoA to form **β-hydroxy-β-methylglutaryl coenzyme A (HMG CoA).** Cleavage of HMG CoA then yields acetoacetic acid and acetyl CoA.

Acetoacetate Forms Both D-β-Hydroxybutyrate and Acetone

In mitochondria some of the acetoacetate is reduced to D-β-hydroxybutyrate depending on the intramitochondrial [NADH]/[NAD$^+$] ratio. Note that the product of this reaction is D-β-hydroxybutyrate, whereas β-hydroxybutyryl CoA formed during β-oxidation is of the L configuration. **β-Hydroxybutyrate dehydrogenase** is tightly associated with the inner mitochondrial membrane and, because of its high activity in liver, the concentrations of substrates and products of the reaction are maintained close to equilibrium. Thus the ratio of β-hydroxybutyrate to acetoacetate in blood leaving liver can be taken as a reflection of the mitochondrial [NADH]/[NAD$^+$] ratio.

Some acetoacetate continually undergoes slow, spontaneous nonenzymatic decarboxylation to acetone:

$$CH_3-\overset{O}{\overset{\|}{C}}-CH_2-\overset{O}{\overset{\|}{C}}-O^- + H^+ \longrightarrow CH_3-\overset{O}{\overset{\|}{C}}-CH_3 + CO_2$$

Under normal conditions acetone formation is negligible, but when pathological accumulations of acetoacetate occur, as in severe diabetic ketoacidosis (see Clin. Corr. 16.7), the amount of acetone in blood can be sufficient to cause it to be detectable in a patient's breath.

As seen from Figure 16.24, the pathway from acetyl CoA to HMG CoA also operates in the cytosol of liver (indeed, this applies to essentially all tissues of the body). However, HMG CoA lyase is absent in the cytosol and HMG CoA formed

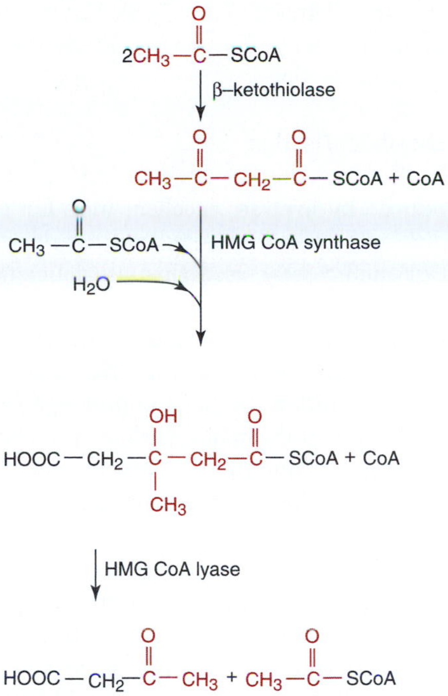

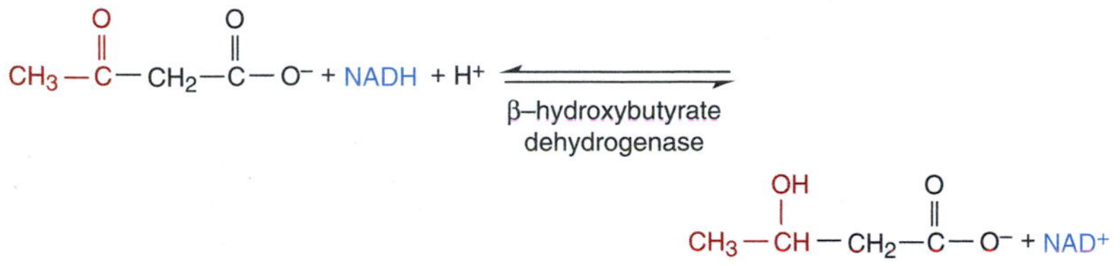

$$2CH_3-\overset{O}{\overset{\|}{C}}-SCoA \xrightarrow{\beta\text{-ketothiolase}} CH_3-\overset{O}{\overset{\|}{C}}-CH_2-\overset{O}{\overset{\|}{C}}-SCoA + CoA$$

$$\xrightarrow[H_2O]{CH_3-\overset{O}{\overset{\|}{C}}-SCoA,\ \text{HMG CoA synthase}} HOOC-CH_2-\underset{CH_3}{\underset{|}{\overset{OH}{\overset{|}{C}}}}-CH_2-\overset{O}{\overset{\|}{C}}-SCoA + CoA$$

$$\xrightarrow{\text{HMG CoA lyase}} HOOC-CH_2-\overset{O}{\overset{\|}{C}}-CH_3 + CH_3-\overset{O}{\overset{\|}{C}}-SCoA$$

$$CH_3-\overset{O}{\overset{\|}{C}}-CH_2-\overset{O}{\overset{\|}{C}}-O^- + NADH + H^+ \underset{\beta\text{-hydroxybutyrate dehydrogenase}}{\rightleftharpoons} CH_3-\overset{OH}{\overset{|}{CH}}-CH_2-\overset{O}{\overset{\|}{C}}-O^- + NAD^+$$

FIGURE 16.23
Pathway of acetoacetate formation.

CLINICAL CORRELATION 16.7
Diabetic Ketoacidosis

Diabetic ketoacidosis (DKA) is a common illness among patients with insulin-dependent diabetes mellitus. Although mortality rates have declined, they are still in the range of 6–10%. The condition is triggered by severe insulin deficiency coupled with glucagon excess and is frequently accompanied by concomitant elevation of other stress hormones, such as epinephrine, norepinephrine, cortisol, and growth hormone. The major metabolic derangements are marked hyperglycemia, excessive ketonemia, and ketonuria. Blood concentrations of acetoacetic plus β-hydroxybutyric acids as high as 20 mM are not uncommon. Because these are relatively strong acids ($pK \sim 3.5$), the situation results in life-threatening metabolic acidosis.

The massive accumulation of ketone bodies in the blood in DKA stems from a greatly accelerated hepatic production rate such that the capacity of nonhepatic tissues to use them is exceeded. In biochemical terms the initiating events are identical with those operative in the development of starvation ketosis; that is, increased glucagon/insulin ratio $\rightarrow$ elevation of liver cAMP $\rightarrow$ decreased malonyl CoA $\rightarrow$ deinhibition of CPT I $\rightarrow$ activation of fatty acid oxidation and ketone production (see text for details). However, in contrast to physiological ketosis, where insulin secretion from the pancreatic β-cells limits free fatty acid (FFA) availability to the liver, this restraining mechanism is absent in the diabetic individual. As a result, plasma FFA concentrations can reach levels as high as 3–4 mM, which drive hepatic ketone production at maximal rates.

Correction of DKA requires rapid treatment that will be dictated by the severity of the metabolic abnormalities and the associated tissue water and electrolyte imbalance. Insulin is essential. It lowers the plasma glucagon level, antagonizes the catabolic effects of glucagon on the liver, inhibits the flow of ketogenic and gluconeogenic substrates (FFA and amino acids) from the periphery, and stimulates glucose uptake in target tissues.

Foster, D. W. and McGarry, J. D. Metabolic derangements and treatment of diabetic ketoacidosis. *N. Engl. J. Med.* 309:159, 1983; and Foster, D. W. and McGarry, J. D. Acute complications of diabetes: ketoacidosis, hyperosmolar coma, lactic acidosis. In: L. J. DeGroot (Ed.), *Endocrinology*, Vol. 2, 3rd ed. Philadelphia: Saunders, 1995, p. 1506.

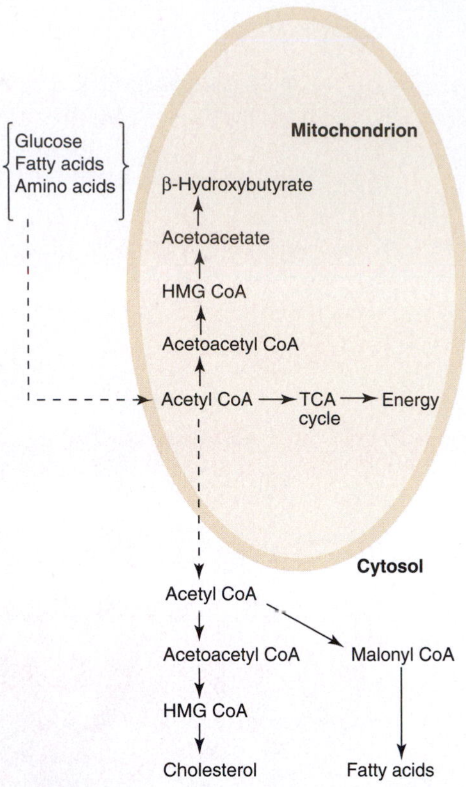

FIGURE 16.24
Interrelationships of ketone bodies with lipid, carbohydrate, and amino acid metabolism in liver.

there is used for cholesterol biosynthesis (see Chapter 17, p. 743). What distinguishes liver from nonhepatic tissues is its high activity of intramitochondrial **HMG CoA synthase** that makes it the primary tissue in ketone body production.

Utilization of Ketone Bodies by Nonhepatic Tissues Requires Formation of Acetoacetyl CoA

Acetoacetate and β-hydroxybutyrate produced by liver are excellent fuels for a variety of nonhepatic tissues, such as cardiac and skeletal muscle, particularly when glucose is in short supply (starvation) or inefficiently used (insulin deficiency). Under these conditions the same tissues can readily use free fatty acids, whose blood concentration rises as insulin levels fall, as a source of energy. A question for many years was why liver should produce ketone bodies in the first place. The answer emerged in the late 1960s with the recognition that during prolonged starvation in humans ketone bodies replace glucose as the major fuel for the central nervous system. Also noteworthy is the fact that during the neonatal period of development, acetoacetate and β-hydroxybutyrate serve as important precursors for cerebral lipid synthesis.

Use of ketone bodies requires that acetoacetate first be reactivated to its CoA derivative. This is accomplished by a mitochondrial enzyme, acetoacetate:succinyl-CoA CoA transferase, present in most nonhepatic tissues but absent from liver. Succinyl CoA serves as the source of the coenzyme. The reaction is depicted in Figure 16.25. Through the action of β-ketothiolase, acetoacetyl CoA is then converted into acetyl CoA, which in turn enters the TCA cycle with production of energy. Mitochondrial β-hydroxybutyrate dehydrogenase in nonhepatic tissues reconverts β-hydroxybutyrate into acetoacetate as the concentration of the latter is decreased.

Starvation and Other Pathological Conditions Lead to Ketosis

Under normal dietary conditions, hepatic production of acetoacetate and β-hydroxybutyrate is minimal and the concentration of these compounds in the blood is very low ($\leq$ 0.2 mM). However, with food deprivation ketone body synthesis is greatly accelerated, and the circulating concentration of acetoacetate plus β-hydroxybutyrate may rise to the region of 3–5 mM. This is a normal response of the body to a shortage of carbohydrate. In the early stages of fasting, use of ketone bodies by heart and skeletal muscle conserves glucose for support of the central nervous system and erythrocytes. With more prolonged starvation, increased blood concentrations of acetoacetate and β-hydroxybutyrate ensures their efficient uptake by brain, thereby further sparing glucose consumption.

In contrast to the **physiological ketosis of starvation,** some pathological conditions, most notably **diabetic ketoacidosis** (see Clin. Corr. 16.7), are characterized by excessive accumulation of ketone bodies in the blood (up to 20 mM). Hormonal and biochemical factors operative in the overall control of hepatic ketone body production are discussed in detail in Chapter 20.

Peroxisomal Oxidation of Fatty Acids Serves Many Functions

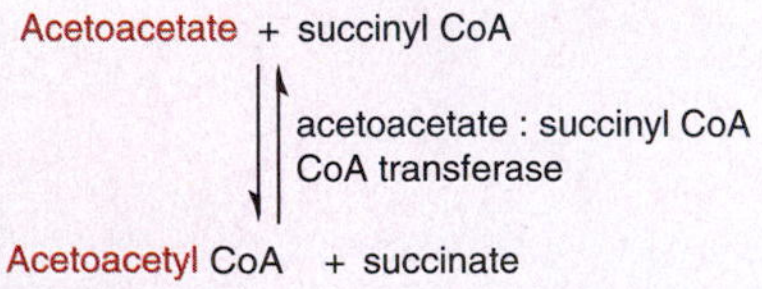

Acetoacetate + succinyl CoA

⇅ acetoacetate : succinyl CoA CoA transferase

Acetoacetyl CoA + succinate

FIGURE 16.25
Initial step in utilization of acetoacetate by nonhepatic tissues.

Although the bulk of cellular fatty acid oxidation occurs in mitochondria, a significant fraction also takes place in **peroxisomes** of liver, kidney, and other tissues. Peroxisomes are a class of subcellular organelles with distinctive morphological and chemical characteristics (see p. 22). Peroxisomes of liver oxidize fatty acids and contain the enzymes needed for the β-oxidation process. The mammalian **peroxisomal fatty acid oxidation** scheme is similar to that in plant glyoxysomes but differs from the mitochondrial β-oxidation system in three important respects. First, the initial dehydrogenation is accomplished by a cyanide-insensitive oxidase system, as shown in Figure 16.26. H_2O_2 formed is eliminated by **catalase.** The remaining steps are the same as in the mitochondrial system. Second, there is evidence that the peroxisomal and mitochondrial enzymes are slightly different and that the specificity in peroxisomes is for somewhat longer chain length. Third, although rat liver mitochondria will oxidize a molecule of palmitoyl CoA to eight molecules of acetyl CoA, the β-oxidation system in liver peroxisomes will not proceed beyond the stage of octanoyl CoA (C8). The possibility is thus raised that one function of peroxisomes is to shorten the chains of relatively long-chain fatty acids to a point at which β-oxidation can be completed in mitochondria. It is interesting to note that a number of drugs used clinically to decrease triacylglycerol levels in patients cause a marked increase in peroxisomes.

Other peroxisomal reactions include chain shortening of dicarboxylic acids, conversion of cholesterol into bile acids, and formation of ether lipids. Given these diverse metabolic roles it is not surprising that the congenital absence of functional peroxisomes, an inherited defect known as Zellweger syndrome, has such devastating effects (see Clin. Corr. 1.5).

$CH_3-(CH_2)_n-CH_2-CH_2-C(=O)-SCoA$

flavoprotein → H_2O_2

flavoprotein-H_2 ← O_2

$CH_3-(CH_2)_n-CH=CH-C(=O)-SCoA$

FIGURE 16.26
Initial step in peroxisomal fatty acid oxidation.

BIBLIOGRAPHY

Bonnefont, J. P., Demaugre, F., Prip-Buus, C., Saudubray, J. M., Brivet, M., Abadi, N., and Thuillier, L. Carnitine palmitoyltransferase deficiencies. *Mol. Gen. Metab.* 68:424, 1999.

Foster, D. W. and McGarry, J. D. Acute complications of diabetes: ketogenesis, hyperosmolar coma, lactic acidosis. In: L. J. DeGroot (Ed.), *Endocrinology,* Vol. 2, 3rd ed. Philadelphia: Saunders, 1995, p. 1506.

Goldstein, J. L., Hobbs, H. H., and Brown, M. S. Familial hypercholesterolemia. In: C. R. Scriver, A. L. Beaudet, W. S. Sly, and D. Valle (Eds.), *The Metabolic and Molecular Bases of Inherited Disease,* Vol. II, 7th ed. New York: McGraw-Hill, 1995, p. 1981.

Gurr, M. I. and James, A. T. *Lipid Biochemistry, An Introduction,* 3rd ed. London: Chapman and Hall, 1980.

Hashimoto, T. Peroxisomal β-oxidation enzymes. *Neurochem. Res.* 24:551, 1999.

Kane, J. P. and Havel, R. J. Disorders of the biogenesis and secretion of lipoproteins containing the B apolipoproteins. In: C. R. Scriver, A. L. Beaudet, W. S. Sly, and D. Valle (Eds.), *The Metabolic and Molecular Bases of Inherited Disease,* Vol. II, 7th ed. New York: McGraw-Hill, 1995, p. 1853.

McGarry, J. D. and Foster, D. W. Regulation of hepatic fatty acid oxidation and ketone body production. *Annu. Rev. Biochem.* 49:395, 1980.

McGarry, J. D. and Brown, N. F. The mitochondrial carnitine palmitoyltransferase system. From concept to molecular analysis. *Eur. J. Biochem.* 244:1, 1997.

Nilsson-Ehle, P., Garfinkel, A. S., and Schotz, M. C. Lipolytic enzymes and plasma lipoprotein metabolism. *Annu. Rev. Biochem.* 49:667, 1980.

Robinson, A. M. and Williamson, D. H. Physiological roles of ketone bodies as substrates and signals in mammalian tissues. *Physiol. Rev.* 60:143, 1980.

Ruderman, N. B., Saha, A. K., Vavvas, D., and Witters, L. A. Malonyl-CoA, fuel sensing, and insulin resistance. *Am. J. Physiol.* 276:E1, 1999.

Stanley, C. A. Plasma and mitochondrial membrane carnitine transport defects. In: P. M. Coates and K. Tanaka (Eds.), *New Developments in Fatty Acid Oxidation.* New York: Wiley-Liss, 1992, p. 289.

Wakil, S. J., Stoops, J. K., and Joshi, V. C. Fatty acid synthesis and its regulation. *Annu. Rev. Biochem.* 52:537, 1983.

Wanders, R. J. A., Vreken, P., Den Boer, M. E. J., Wijburg, F. A., Van Gennip, A. H., and Ijist, L. Disorders of mitochondrial fatty acyl-CoA β-oxidation. *J. Inher. Metab. Dis.* 22:442, 1999.

QUESTIONS | C. N. ANGSTADT

1. Triacylglycerols:
   A. would be expected to be good emulsifying agents.
   B. yield about the same amount of ATP on complete oxidation as would an equivalent weight of glycogen.
   C. are stored as hydrated molecules.
   D. in the average individual, represent sufficient energy to sustain life for several weeks.
   E. are generally negatively charged molecules at physiological pH.

2. In humans, fatty acids:
   A. can be synthesized from excess dietary carbohydrate or protein.
   B. are not required at all in the diet.
   C. containing double bonds cannot be synthesized.
   D. must be supplied entirely by the diet.
   E. other than palmitate, must be supplied in the diet.

3. All of the following statements about acetyl-CoA carboxylase are correct EXCEPT:
   A. it undergoes protomer–polymer interconversion during its physiological regulation.
   B. it requires biotin.
   C. it is inhibited by cAMP-mediated phosphorylation.
   D. it is activated by both palmitoyl CoA and citrate.
   E. its content in a cell responds to changes in fat content in the diet.

4. During the synthesis of palmitate in liver cells:
   A. the addition of malonyl CoA to fatty acid synthase elongates the growing chain by three carbon atoms.
   B. a β-keto residue on the 4′-phosphopantetheine moiety is reduced to a saturated residue by NADPH.

C. palmitoyl CoA is released from the synthase.
D. transfer of the growing chain from ACP to another —SH occurs after addition of the next malonyl CoA.
E. the first compound to add to fatty acid synthase is malonyl CoA.

5. Citrate stimulates fatty acid synthesis by all of the following EXCEPT:
   A. allosterically activating acetyl-CoA carboxylase.
   B. providing a mechanism to transport acetyl CoA from the mitochondria to the cytosol.
   C. participating in a pathway that ultimately produces CO_2 and NADPH in the cytosol.
   D. participating in the production of ATP.

6. Fatty acid synthase:
   A. synthesizes only palmitate.
   B. yields an unsaturated fatty acid by skipping a reductive step.
   C. produces hydroxy fatty acids in nerve tissue.
   D. can stop with the release of a fatty alcohol instead of an acid.
   E. can produce a branched-chain fatty acid if methylmalonyl CoA is used as a substrate.

7. In humans, desaturation of fatty acids:
   A. occurs primarily in mitochondria.
   B. is catalyzed by an enzyme system that uses NADPH and a cytochrome.
   C. introduces double bonds primarily of trans configuration.
   D. can occur only after palmitate has been elongated to stearic acid.
   E. introduces the first double bond at the methyl end of the molecule.

8. All of the following events are usually involved in the synthesis of triacylglycerols in adipose tissue EXCEPT:
   A. addition of a fatty acyl CoA to a diacylglycerol.
   B. addition of a fatty acyl CoA to a lysophosphatide.
   C. a reaction catalyzed by glycerol kinase.
   D. hydrolysis of phosphatidic acid by a phosphatase.
   E. reduction of dihydroxyacetone phosphate.

9. Plasma lipoproteins:
   A. are the only carriers of lipid-based energy in the blood.
   B. usually have a nonpolar core containing triacylglycerols and cholesterol esters.
   C. are composed primarily of free (unesterified) fatty acids.
   D. include chylomicrons generated in the liver.
   E. include high density lipoproteins (HDL) as the major carrier of lipid-based energy.

10. Lipoprotein lipase:
    A. is an intracellular enzyme.
    B. is stimulated by cAMP-mediated phosphorylation.
    C. functions to mobilize stored triacylglycerols from adipose tissue.
    D. is stimulated by one of the apoproteins present in VLDL.
    E. readily hydrolyzes three fatty acids from a triacylglycerol.

11. The high glucagon/insulin ratio seen in starvation:
    A. promotes mobilization of fatty acids from adipose stores.
    B. stimulates β-oxidation by inhibiting the production of malonyl CoA.
    C. leads to increased concentrations of ketone bodies in the blood.
    D. all of the above.
    E. none of the above.

Questions 12 and 13: Following a severe cold that caused a loss of appetite, a one-year-old boy was hospitalized with hypoglycemia, hyperammonemia, muscle weakness, and cardiac irregularities. These symptoms were consistent with a defect in the carnitine transport system. Dietary carnitine therapy was tried unsuccessfully but a diet low in long-chain fatty acids and supplemented with medium-chain triacylglycerols was beneficial.

12. A deficiency of carnitine might be expected to interfere with:
    A. β-oxidation.
    B. ketone body formation from acetyl CoA.
    C. palmitate synthesis.
    D. mobilization of stored triacylglycerols from adipose tissue.
    E. uptake of fatty acids into cells from the blood.

13. The child was diagnosed with carnitine–acylcarnitine translocase deficiency. The dietary treatment was beneficial because:
    A. the child could get all required energy from carbohydrate.
    B. the deficiency was in the peroxisomal oxidation system so carnitine would not be helpful.
    C. medium-chain fatty acids (8–10 carbons) enter the mitochondria before being converted to their CoA derivatives.
    D. medium-chain triacylglycerol contains mostly hydroxylated fatty acids.
    E. medium-chain fatty acids such as C_8 and C_{10} are readily converted into glucose by the liver.

Questions 14 and 15: Medium-chain acyl-CoA dehydrogenase deficiency (MCAD), a defect in β-oxidation, usually produces symptoms within the first two years of life after a period of fasting. Typical symptoms include vomiting, lethargy, and hypoketotic hypoglycemia. Excessive urinary secretion of medium-chain dicarboxylic acids and medium-chain esters of glycine and carnitine help to establish the diagnosis.

14. β-Oxidation of fatty acids:
    A. generates ATP only if acetyl CoA is subsequently oxidized.
    B. is usually suppressed during starvation.
    C. uses only even-chain, saturated fatty acids as substrates.
    D. uses $NADP^+$.
    E. occurs by a repeated sequence of four reactions.

15. The lack of ketone bodies in the presence of low blood glucose in this case is unusual since ketone body concentrations usually rise with fasting-induced hypoglycemia. Ketone bodies:
    A. are formed by removal of CoA from the corresponding intermediate of β-oxidation.
    B. are synthesized from cytoplasmic β-hydroxy-β-methylglutaryl coenzyme A (HMG CoA).
    C. are synthesized primarily in muscle tissue.
    D. include both β-hydroxybutyrate and acetoacetate, the ratio reflecting the intramitochondrial $[NADH]/[NAD^+]$ ratio in liver.
    E. form when β-oxidation is interrupted.

Problems

16. What is the approximate maximum net synthesis of ATP-associated high-energy phosphate bonds (~P) for complete oxidation of 1 mol of oleic acid in a muscle cell?
17. How does oxidation of a 17-carbon fatty acid (from plants) lead to the production of propionyl CoA? Be specific in your answer.

ANSWERS

1. **D** A, C, E. Triacylglycerols are neutral hydrophobic molecules with no hydrophilic portion and, therefore, are not emulsifying agents and are stored anhydrously. B: Their more reduced state, compared to carbohydrates, makes them more energy-rich.
2. **A** It is important to realize that triacylglycerol is the ultimate storage form of excess dietary intake. B–E: We can synthesize most fatty acids, including those with double bonds, except for the essential fatty acids, linoleic and linolenic.
3. **D** A: Acetyl-CoA carboxylase shifts between its protomeric (inactive) and polymeric (active) forms under the influence of a variety of regulatory factors. C: Since cAMP increases at times when energy is needed, it is consistent that a process that uses energy would be inhibited. E: Long-term control is related to enzyme synthesis and responds appropriately to dietary changes.
4. **B** A: Splitting CO_2 from malonyl CoA is the driving force for the condensation reaction so the chain grows two carbon atoms at a time. C: In mammals, palmitate is released as the free acid; the conversion to the CoA ester is by a different enzyme. D: It is important to realize that only ACP binds the incoming malonyl CoA so it must be freed before another addition can be made. E: Acetyl CoA adds first to form the foundation for the rest of the chain.
5. **D** Citrate consumes ATP when acted on by citrate cleavage enzyme. A: Table 16.2. B: Acetyl CoA is generated primarily in mitochondria but does not cross the membrane readily. C: Oxaloacetate generated by citrate cleavage enzyme, when converted to pyruvate, yields CO_2 and NADPH by the malic enzyme.
6. **E** This is much slower than reaction with malonyl CoA, but it is significant. A: In certain tissues, for example, mammary glands, shorter-chain products are formed. B–D: These products are all formed by other processes. Reactions proceeding on a multienzyme complex generally do not "stop" at intermediate steps.
7. **B** A: Desaturation occurs in the endoplasmic reticulum. C: Naturally occurring fatty acids are cis. D: Elongation and unsaturation can occur in any order. E: If this were true we could make linoleic acid.
8. **C** This does not occur to any significant extent in adipose tissue. A, B, D: The sequential addition of fatty acyl CoAs to glycerol 3-phosphate forms lysophosphatidic acid, then phosphatidic acid whose phosphate is removed before the addition of the third fatty acyl residue. E: This is the formation of α-glycerol phosphate in adipose.
9. **B** All lipoproteins have this same general structure, a nonpolar core surrounded by a more polar shell. A, C: Fatty acids bound to serum albumin and ketone bodies are other sources. D: Chylomicrons carry dietary lipid from the intestine. E: HDL function is to carry cholesterol away from tissues.
10. **D** A–C: These are characteristics of hormone-sensitive lipase. E: It generally requires more than one lipase to hydrolyze all of the fatty acids.
11. **D** A high glucagon/insulin ratio results in cAMP-mediated phosphorylations that activate hormone-sensitive lipase in adipose tissue and inhibit acetyl-CoA carboxylase in liver, leading to mobilizaton of fatty acids from adipose tissue and decreased malonyl CoA in liver. The fall in malonyl CoA causes derepression of CPT I activity, allowing accelerated production of acetyl CoA in the mitochondria, and ultimately ketone bodies, from the mobilized fatty acids.
12. **A** Carnitine functions in transport of fatty acyl-CoA esters formed in cytosol into the mitochondria. B: Acetyl CoA for ketone bodies comes from sources in addition to fatty acids. C: Fatty acid synthesis is a cytosolic process. D: Mobilization is under hormonal control. E: Fatty acids cross the plasma membrane.
13. **C** Because medium-chain fatty acids cross the mitochondrial membrane directly, they do not require the carnitine system. A: This is never true. B: Peroxisomal oxidation does not require carnitine but this is not a peroxisomal system. D: Hydroxylated fatty acids are not a common constituent of triacylglycerols. E: Liver cannot synthesize glucose from fatty acids with an even number of carbon atoms.
14. **E** A, D: It is important to realize that β-oxidation, itself, generates $FADH_2$ and NADH, which can be reoxidized to generate ATP. B: Fatty acid oxidation is usually enhanced during fasting. C: β-Oxidation is a general process requiring only minor modifications to oxidize nearly any fatty acid in the cell.
15. **D** A, E: β-Oxidation proceeds to completion; ketone bodies are formed by a separate process. B, C: Ketone bodies are formed, but not used, in liver mitochondria; cytosolic HMG CoA is a precursor of cholesterol. Ketone bodies are not readily synthesized by muscle.
16. Oleic acid has 18 carbons and 1 double bond. Nine Acetyl CoA $\times$ 10 = 90 ATP. Seven reduced flavoproteins $\times$ 1.5 = 10.5 ATP. (*Note*: In one step the flavoprotein reduction is not necessary because the double bond is not present. The isomerase to convert the *cis*-3-enoyl bond to the *trans*-2-enoyl bond does not involve ATP.) 8 NADH $\times$ 2.5 = 20 ATP. Total = 120.5 ATP − 2 ~ P (for activation) = 118.5 ~ P/oleic acid.
17. β-Oxidation proceeds normally but the final thiolase cleavage yields acetyl CoA and propionyl CoA. Propionyl CoA is *not* a substrate for SCAD (short-chain acyl-CoA dehydrogenase) so β-oxidation terminates.

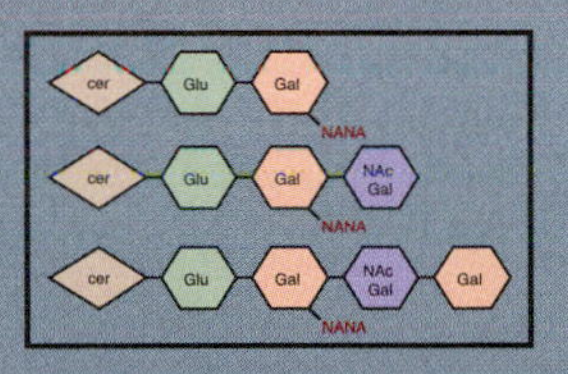

17

LIPID METABOLISM II: PATHWAYS OF METABOLISM OF SPECIAL LIPIDS

Robert H. Glew

17.1 | OVERVIEW

Lipid is a general term that describes substances that are relatively water insoluble and extractable by nonpolar solvents. Complex lipids of humans fall into one of two broad categories: nonpolar lipids, such as triacylglycerols and cholesterol esters, and polar lipids, which are amphipathic in that they contain both a hydrophobic domain and a hydrophilic region in the same molecule. This chapter discusses polar lipids including phospholipids, sphingolipids, and eicosanoids. The hydrophobic and hydrophilic regions are bridged by a glycerol moiety in glycerophospholipids and by sphingosine in sphingomyelin and glycosphingolipids. Triacylglycerols are confined largely to storage sites in adipose tissue, whereas polar lipids occur primarily in cellular membranes. Oils of soybean, palm, rapeseed, sunflower, cottonseed, and peanut account for 80% of worldwide plant oil production and consist mainly of palmitic, stearic, oleic, linoleic, and linolenic acids. Membranes generally contain 40% of their dry weight as lipid and 60% as protein.

Cell–cell recognition, phagocytosis, contact inhibition, and rejection of transplanted tissues and organs are all phenomena of medical significance that involve highly specific recognition sites on the surface of plasma membranes. Synthesis of these complex glycosphingolipids that appear to play a role in these important biological events will be described. Glycolipids are worthy of study because ABO antigenic determinants of blood groups are primarily glycolipid in nature. In addition, various sphingolipids are the storage substances that accumulate in liver, spleen, kidney, or nervous tissue in certain genetic disorders called sphingolipidoses. In order to understand the basis of these enzyme-deficiency states, a knowledge of the relevant chemical structures involved is required.

A very important lipid is cholesterol. This chapter describes the pathway of cholesterol biosynthesis and its regulation, and shows how cholesterol functions as a precursor to bile salts and steroid hormones. Also described is the role of high density lipoprotein (HDL) and lecithin:cholesterol acyltransferase (LCAT) in management of plasma cholesterol.

Finally, the metabolism and function of two pharmacologically powerful classes of hormones derived from arachidonic acid, namely, prostaglandins and leukotrienes, will be discussed. See the Appendix (p. 1175) for a discussion of nomenclature and chemistry of lipids.

17.2 | PHOSPHOLIPIDS

Two major classes of **acylglycerolipids** are **triacylglycerols** and **glycerophospholipids.** They are referred to as glycerolipids because the core of these compounds is provided by the three-carbon polyol, glycerol. Two primary alcohol groups of glycerol are not stereochemically identical and, in the case of phospholipids, it is usually the same hydroxyl group that is esterified to the phosphate residue. The stereospecific numbering system is the best way to designate different hydroxyl groups. In this system, when the structure of glycerol is drawn in the Fischer projection with the C-2 hydroxyl group projecting to the left of the page, the carbon atoms are numbered as shown in Figure 17.1. When the stereospecific numbering

	Carbon number
CH_2OH	1
HO—C—H	2
CH_2OH	3

FIGURE 17.1
Stereospecific numbering of glycerol.

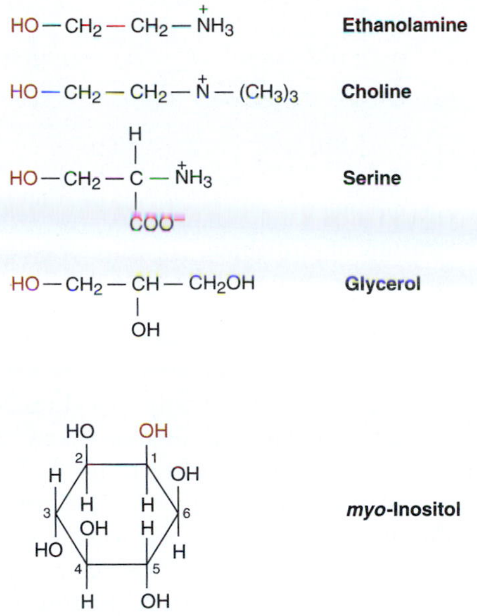

Ethanolamine

Choline

Serine

Glycerol

myo-Inositol

FIGURE 17.2
Structures of some common polar groups of phospholipids.

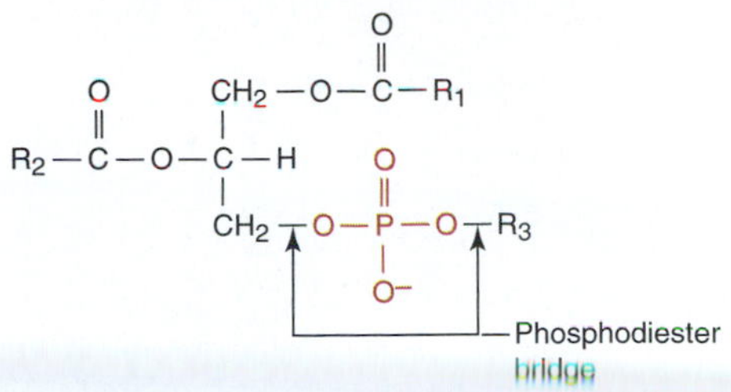

Phosphodiester bridge

FIGURE 17.3
Generalized structure of a phospholipid where R_1 and R_2 represent the aliphatic chains of fatty acids, and R_3 represents a polar group.

(*sn*) system is employed, the prefix *sn*- is used before the name of the compound. Glycerophospholipids usually contain an *sn*-glycerol 3-phosphate moiety. Although each contains the glycerol moiety as a fundamental structural element, neutral triacylglycerols and charged, ionic phospholipids have very different physical properties and functions.

Phospholipids Contain 1,2-Diacylglycerol and a Base Connected by a Phosphodiester Bridge

Phospholipids are polar, ionic lipids composed of 1,2-diacylglycerol and a phosphodiester bridge that links the glycerol backbone to some base, usually a nitrogenous one, such as choline, serine, or ethanolamine (Figures 17.2 and 17.3). The most abundant phospholipids in human tissues are **phosphatidylcholine** (also called lecithin), **phosphatidylethanolamine**, and **phosphatidylserine** (Figure 17.4). At physiologic pH, phosphatidylcholine and phosphatidylethanolamine have no net charge and exist as dipolar zwitterions, whereas phosphatidylserine has a net charge of −1, causing it to be an acidic phospholipid. Phosphatidylethanolamine (PE) is related to phosphatidylcholine in that trimethylation of PE produces lecithin. Most phospholipids contain more than one kind of fatty acid per molecule, so that a given class of phospholipids from any tissue actually represents a family of molecular species. Phosphatidylcholine (PC) contains mostly palmitic acid (16:0) or stearic acid (18:0) in the *sn*-1 position and primarily unsaturated 18-carbon fatty acids oleic, linoleic, or linolenic in the *sn*-2 position. Phosphatidylethanolamine has the same saturated fatty acids as PC at the *sn*-1 position but contains more of the long-chain polyunsaturated fatty acids—namely, 18:2, 20:4, and 22:6—at the *sn*-2 position.

Phosphatidylinositol, an acidic phospholipid that occurs in mammalian membranes (Figure 17.5), is rather unusual because it often contains almost exclusively stearic acid in the *sn*-1 position and arachidonic acid (20:4) in the *sn*-2 position.

Another phospholipid comprised of a polyol polar head group is **phosphatidylglycerol** (Figure 17.5), which occurs in relatively large amounts in mitochondrial membranes and pulmonary surfactant and is a precursor of cardiolipin. Phosphatidylglycerol and phosphatidylinositol both carry a formal charge of −1 at neutral pH and are therefore acidic lipids. **Cardiolipin**, a very acidic (charge, −2) phospholipid, is composed of two molecules of phosphatidic acid linked together

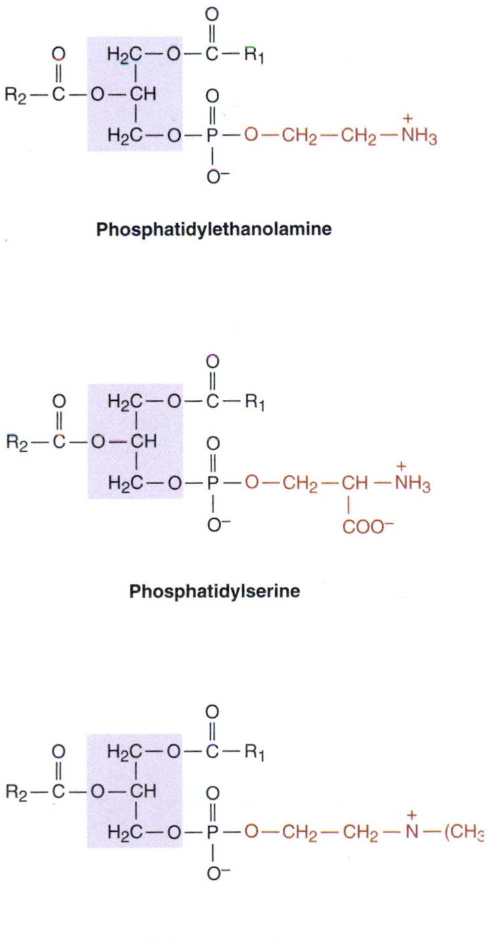

Phosphatidylethanolamine

Phosphatidylserine

Phosphatidylcholine (lecithin)

FIGURE 17.4
Structures of some common phospholipids.

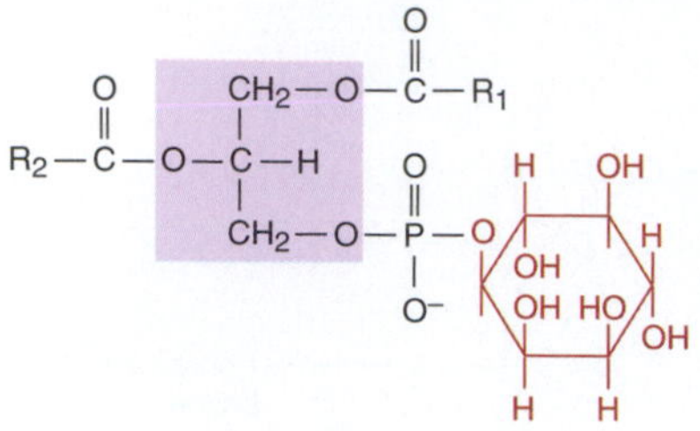

Phosphatidylinositol

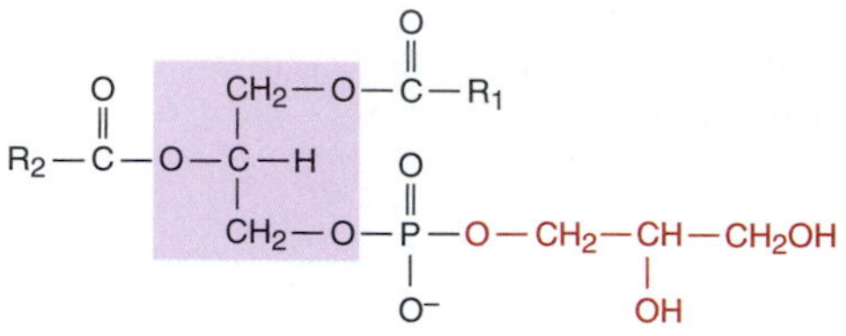

Phosphatidylglycerol

FIGURE 17.5
Structures of phosphatidylglycerol and phosphatidylinositol.

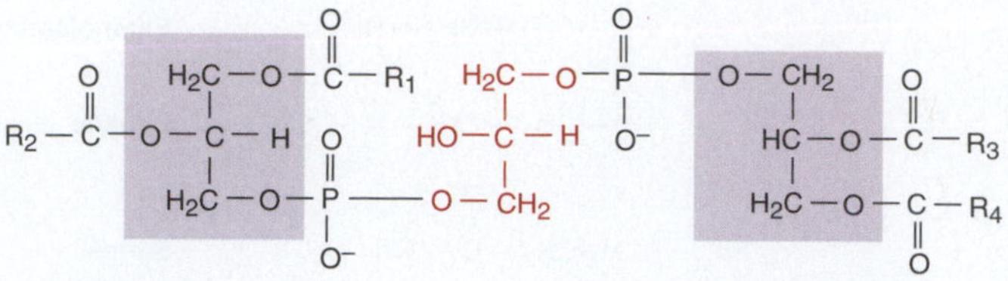

FIGURE 17.6
Structure of cardiolipin.

covalently through a molecule of glycerol (Figure 17.6). It occurs primarily in the inner membrane of mitochondria and in bacterial membranes.

Phospholipids mentioned so far contain only *O*-acyl residues attached to glycerol. *O*-(1-alkenyl) substituents occur at C-1 of the *sn*-glycerol in phosphoglycerides in combination with an *O*-acyl residue esterified to the C-2 position; compounds in this class are known as **plasmalogens** (Figure 17.7). Relatively large amounts of **ethanolamine plasmalogen** (also called plasmenylethanolamine) occur in myelin with lesser amounts in heart muscle where choline plasmalogen is abundant.

An unusual phospholipid called **"platelet-activating factor"** (PAF) (Figure 17.8) is a major mediator of hypersensitivity, acute inflammatory reactions, allergic responses, and anaphylactic shock. In hypersensitive individuals, cells of the polymorphonuclear (PMN) leukocyte family (basophils, neutrophils, and eosinophils), macrophages, and monocytes are coated with IgE molecules that are specific for a particular antigen (e.g., ragweed pollen and bee venom). Subsequent reexposure to the antigen and formation of antigen–IgE complexes on the surface of the aforementioned inflammatory cells provokes synthesis and release of PAF. Platelet-activating factor contains an *O*-alkyl moiety at *sn*-1 and an acetyl residue instead of a long-chain fatty acid in position 2 of the glycerol moiety. PAF is not stored; it is synthesized and released when PMN cells are stimulated. Platelet aggregation, cardiovascular and pulmonary changes, edema, hypotension, and PMN cell chemotaxis are affected by PAF.

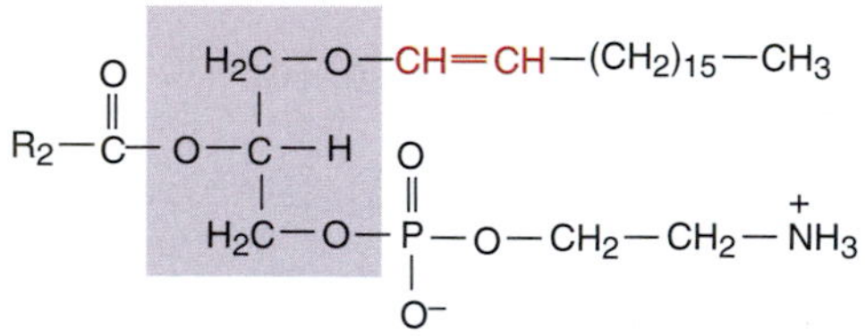

FIGURE 17.7
Structure of ethanolamine plasmalogen.

Phospholipids in Membranes Serve a Variety of Roles

Although present in body fluids such as plasma and bile, phospholipids are found in highest concentration in cellular membranes where they serve as structural and functional components. Nearly one-half the mass of the liver plasma membrane is comprised of various glycerophospholipids (see p. 501). Phospholipids also activate certain enzymes. β-Hydroxybutyrate dehydrogenase, an enzyme imbedded in the inner membrane of mitochondria (see p. 563), has an absolute requirement for phosphatidylcholine; phosphatidylserine and phosphatidylethanolamine cannot substitute.

Dipalmitoyllecithin Is Necessary for Normal Lung Function

Normal lung function depends on a constant supply of **dipalmitoyllecithin** in which the lecithin molecule contains palmitic acid (16:0) residues in both the *sn*-1 and *sn*-2 positions. More than 80% of the phospholipid in the extracellular fluid layer that lines alveoli of normal lungs is dipalmitoyllecithin. This particular phospholipid, called **surfactant,** is produced by type II epithelial cells and prevents atelectasis at the end of the expiration phase of breathing (Figure 17.9). This lipid decreases surface tension of the fluid layer of the lung. Lecithin molecules that do not contain two residues of palmitic acid are not effective in lowering surface tension of the fluid layer lining alveoli. Surfactant also contains phosphatidylglycerol, phosphatidylinositol, and 18- and 36-kDa proteins (designated surfactant proteins), which contribute significantly to the surface tension-lowering property of pulmonary surfactant.

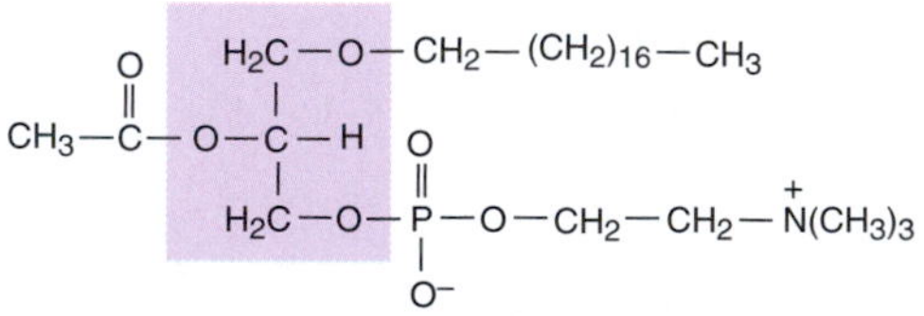

FIGURE 17.8
Structure of platelet-activating factor (PAF).

During the third trimester—before the 28th week of gestation—fetal lung synthesizes primarily sphingomyelin. Normally, at this time, glycogen that has been

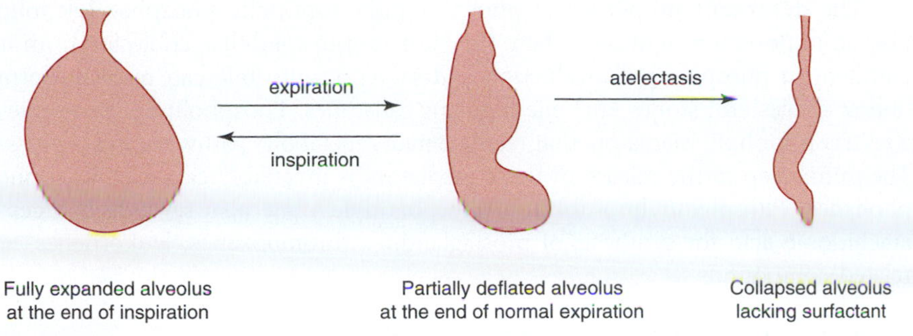

FIGURE 17.9
Role of surfactant in preventing atelectasis.

stored in epithelial type II cells is converted to fatty acids and then to dipalmitoyllecithin. During lung maturation there is a good correlation between increase in lamellar inclusion bodies that represent the intracellular pulmonary surfactant (phosphatidylcholine) storage organelles, called lamellar bodies, and the simultaneous decrease in glycogen content of type II pneumocytes. At the 24th week of gestation the type II granular pneumocytes appear in the alveolar epithelium, and within a few days they produce their typical osmiophilic lamellar inclusion bodies. The number of type II cells increases until the 32nd week, at which time surface active agent appears in the lung and amniotic fluid. Surface tension decreases when inclusion bodies increase in the type II cells. In the few weeks before term, screening tests can be performed on amniotic fluid to detect newborns that are at risk for respiratory distress syndrome (RDS) (see Clin. Corr. 17.1). These tests are useful in timing elective deliveries, in applying vigorous preventive therapy to the newborn infant, and to determine if the mother should be treated with a glucocorticoid drug to accelerate maturation of the fetal lung. Dexamethasone therapy has also been used in neonates with chronic lung disease (bronchopulmonary dysplasia); however, while such corticosteroid therapy may be effective in some cases in improving lung function, in others it causes periventricular abnormalities in the brain.

Respiratory failure due to an insufficiency in surfactant can also occur in adults whose type II cells or surfactant-producing pneumocytes have been destroyed as an adverse side effect of the use of immunosuppressive medications or chemotherapeutic drugs.

CLINICAL CORRELATION 17.1
Respiratory Distress Syndrome

Respiratory distress syndrome (RDS) is a major cause of neonatal morbidity and mortality in many countries. It accounts for approximately 15–20% of all neonatal deaths in Western countries and somewhat less in developing countries. The disease affects only premature babies and its incidence varies directly with the degree of prematurity. Premature babies develop RDS because of immaturity of their lungs, resulting from a deficiency of pulmonary surfactant. The maturity of the fetal lung can be predicted antenatally by measuring the lecithin/sphingomyelin (L/S) ratio in the amniotic fluid. The mean L/S ratio in normal pregnancies increases gradually with gestation until about 31 or 32 weeks when the slope rises sharply. The ratio of 2.0 that is characteristic of the term infant at birth is achieved at the gestational age of about 34 weeks. For predicting pulmonary maturity, the critical L/S ratio is 2.0 or greater. The risk of developing RDS when the L/S ratio is 1.5–1.9 is approximately 40%, and for a ratio less than 1.5 about 75%. Although the L/S ratio in amniotic fluid is still widely used to predict the risk of RDS, the results are unreliable if the amniotic fluid specimen has been contaminated by blood or meconium obtained during a complicated pregnancy.

In recent years determinations of saturated palmitoylphosphatidylcholine (SPC), phosphatidylglycerol, and phosphatidylinositol have been found to be additional predictors of the risk of RDS. Exogenous surfactant replacement therapy using surfactant from human and animal lungs is effective in the prevention and treatment of RDS.

Merritt, T. A., Hallman, M., Bloom, B. T., et al. Prophylactic treatment of very premature infants with human surfactant. *N. Engl. J. Med.* 315:785, 1986; and Simon, N. V., Williams, G. H., Fairbrother, P. F., Elser, R. C., and Perkins, R. P. Prediction of fetal lung maturity by amniotic fluid fluorescence polarization, L/S ratio, and phosphatidylglycerol. *Obstet. Gynecol.* 57:295, 1981.

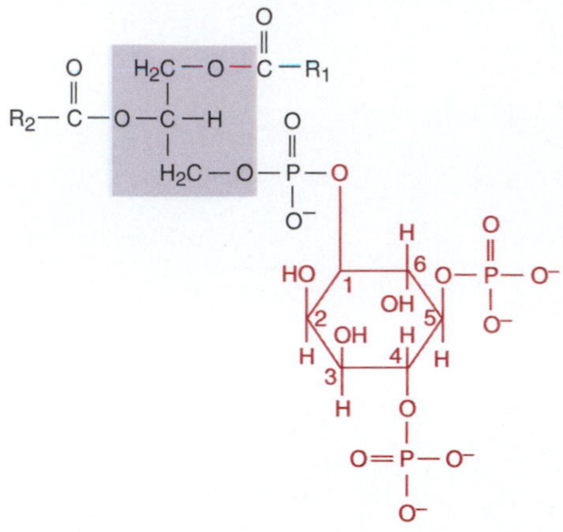

FIGURE 17.10
Structure of phosphatidylinositol 4,5-bisphosphate (PIP_2 or PtdIns $(4,5)P_2$).

The **detergent** properties of phospholipids, especially phosphatidylcholine, play an important role in bile where they function to solubilize cholesterol. An impairment in phospholipid production and secretion into bile can result in formation of cholesterol stones and bile pigment gallstones. Phospholipids also serve as a reservoir for lipid mediators that regulate many metabolic pathways and processes. The initial step in the release of these mediators is invariably catalyzed by a phospholipase. Phosphatidylinositol and phosphatidylcholine also serve as sources of arachidonic acid for synthesis of prostaglandins, thromboxanes, leukotrienes, and related compounds.

Inositides Have a Role in Signal Transduction

Inositol-containing phospholipids (inositides) play a central role in signal transduction systems; the most important is **phosphatidylinositol 4,5-bisphosphate** (PIP_2) (Figure 17.10). When certain ligands bind to their respective receptors on the plasma membrane of mammalian cells (see Chapter 21, p. 928), PIP_2 localized to the inner leaflet of the membrane becomes a substrate for a receptor-dependent phosphoinositidase C (PIC), which hydrolyzes it into two intracellular signals (Figure 17.11): **inositol 1,4,5-trisphosphate** (IP_3), which triggers release of Ca^{2+} from special vesicles of the endoplasmic reticulum, and **1,2-diacylglycerol,** which stimulates activity of protein kinase C (PKC). Removal of the 5-phosphate of IP_3 abolishes the signal and the intracellular Ca^{2+} concentration declines. The 1,2-diacylglycerol signal is removed by conversion to phosphatidic acid in a reaction catalyzed by diacylglycerol kinase, which adds a phosphate group to diacylglycerol (Figure 17.12). Regulatory functions of these products of the PIC reaction are discussed in Chapter 21 (see p. 930). Phosphatidic acid, a product of phospholipase D action on phospholipids, has been implicated as a second messenger.

The complex pathways of inositol phosphate metabolism serve three roles: (1) removal and inactivation of the potent intracellular signal, IP_3, (2) conservation of inositol, and (3) synthesis of polyphosphates such as inositol pentakisphosphate ($InsP_5$) and inositol hexakisphosphate ($InsP_6$), whose functions have not been determined. Inositol 1,4,5-trisphosphate is metabolized by two enzymes: first, a 5-phosphomonoesterase that converts IP_3 to inositol 1,4-bisphosphate and second, a 3-kinase that forms inositol 1,3,4,5-tetrakisphosphate. A family of phosphatases

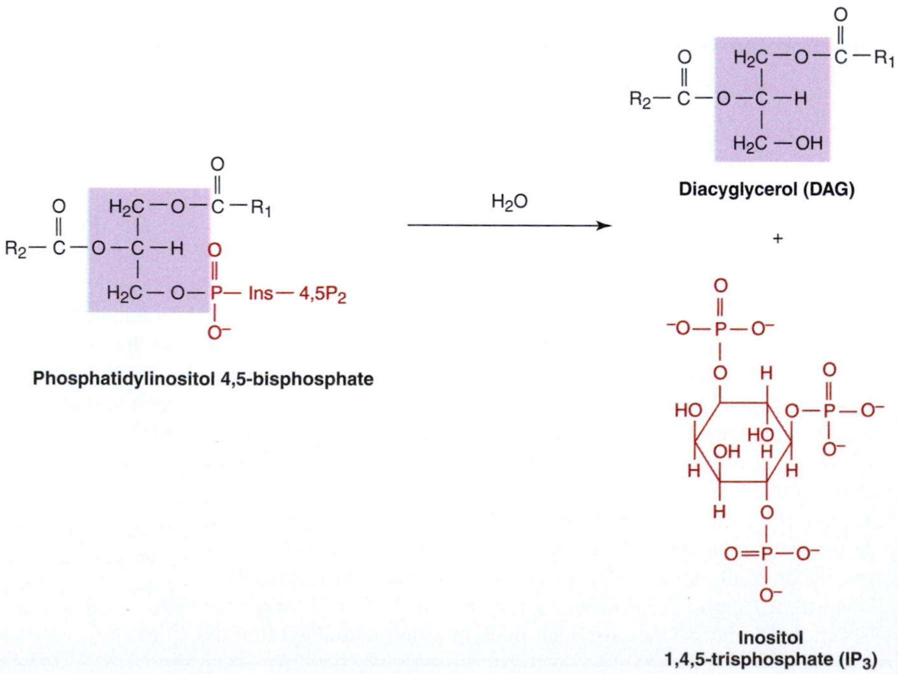

Phosphatidylinositol 4,5-bisphosphate

H_2O

Diacylglycerol (DAG)

+

Inositol 1,4,5-trisphosphate (IP_3)

FIGURE 17.11
Generation of 1,2-diacylglycerol and inositol 1,4,5-trisphosphate by action of phospholipase C on phosphatidylinositol 4,5-bisphosphate.

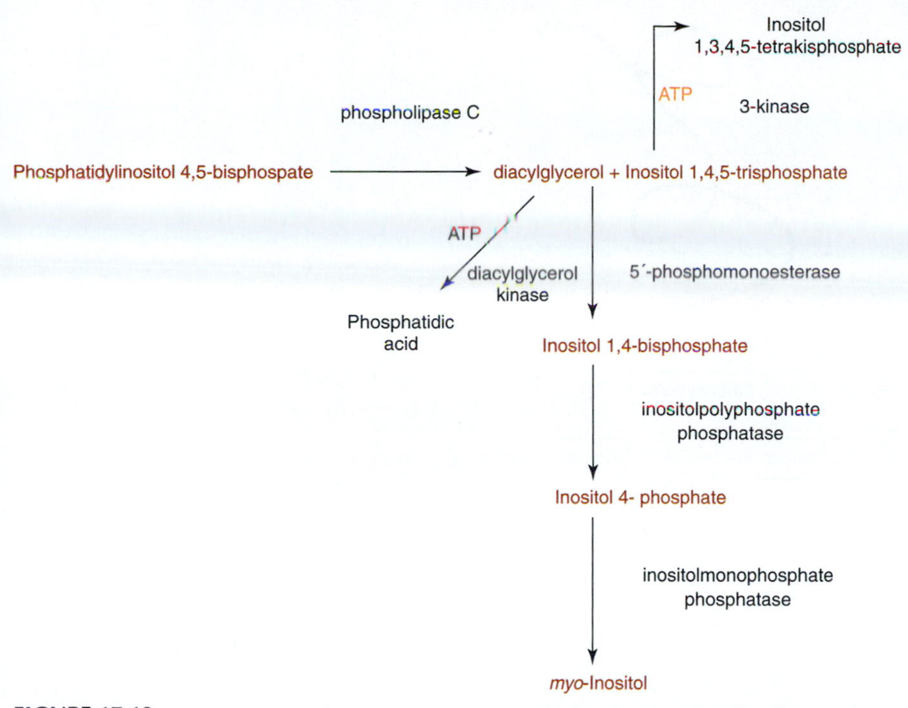

FIGURE 17.12
Pathways for the removal of intracellular inositol 1,4,5-trisphosphate and diacylglycerol.

in turn convert Ins(1,4)P_2 to *myo*-inositol (Figure 17.12). Inositol is eventually reincorporated into the phospholipid pool.

Phosphatidylinositol Serves to Anchor Glycoproteins to the Plasma Membrane

In addition to its role as a structural component of membranes and source of arachidonic acid for prostaglandin and leukotriene synthesis (see p. 772), phosphatidylinositol serves as an anchor to tether certain glycoproteins to the external surface of the plasma membranes. In trypanosomal parasites (e.g., *Trypanosoma brucei*, which causes sleeping sickness), the external surface of the plasma membrane is coated with a protein called **variable surface glycoprotein** (VSG) linked to the membrane through a glycophospholipid anchor, specifically phosphatidylinositol (Figure 17.13). The salient structural features of the protein–lipid linkage region of the **glycosylphosphatidylinositol** (GPI) **anchor** are: (1) the diacylglycerol (DAG) moiety of phosphatidylinositol is integrated into the outer leaflet of the lipid bilayer of the plasma membrane; (2) the inositol residue is linked to DAG through a phosphodiester bond; (3) inositol is linked to glucosamine, which contains a free, unacetylated amino group; (4) there is a mannose-rich glycan domain; and (5) a phosphoethanolamine residue is linked to the protein carboxy terminus. Depending on the protein to which it is attached and the tissue or organism in which it is expressed, the GPI core may be decorated with additional carbohydrates, including mannose, glucose, galactose, *N*-acetylgalactose, *N*-acetylneuraminic acid, and *N*-acetylgalactosamine, and phosphatidylethanolamines that extend from the core mannoses. Other proteins that are attached to the external surface of the plasma membrane include acetylcholinesterase, alkaline phosphatase, and 5′-nucleotidase.

The GPI anchor serves several functions. First, it confers upon the attached protein an unrestricted and lateral mobility within the lipid bilayer, thereby allowing the protein to move about rapidly on the surface of the plasma membrane. Second, the presence of phospholipase C-type activity on the cell surface permits the shedding of the phosphatidylinositol-anchored protein. As an example, this provides trypanosomes with a means for discarding surface antigens, thus changing

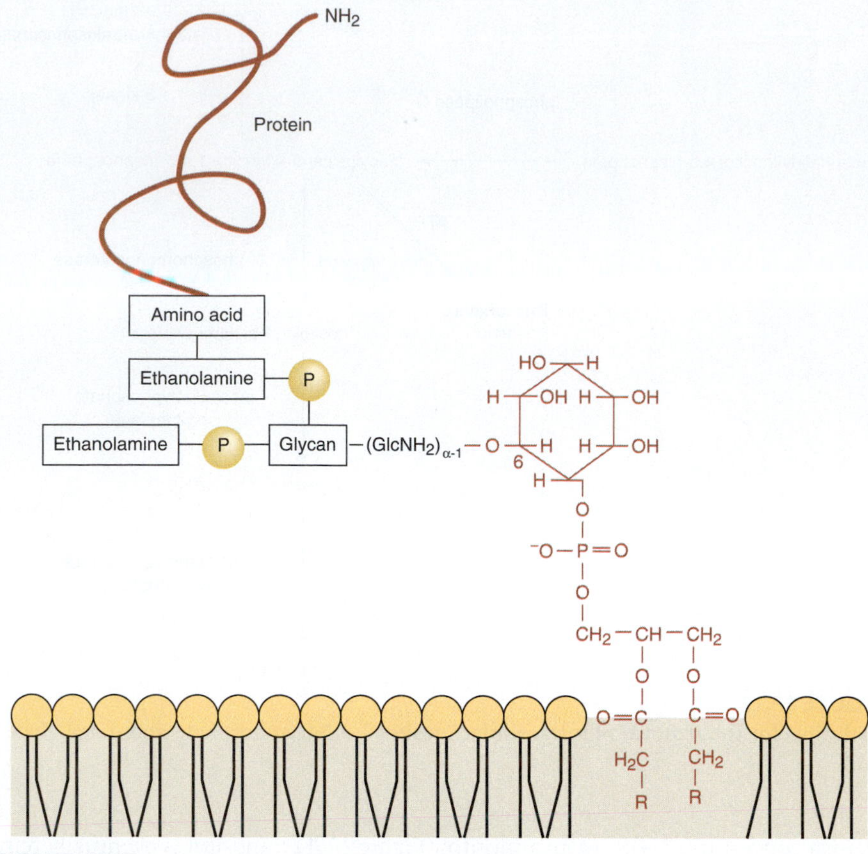

FIGURE 17.13
Structure of a typical phosphatidylinositol membrane protein anchor; $GlcNH_2$, glucosamine.

their coat and escaping antibodies of the host's immune system. Third, phospholipase C action on the phosphatidylinositol anchor releases diacylglyceride, a second messenger that can activate protein kinase C (see p. 930). Biosynthesis of GPI anchors has been characterized extensively.

Other types of protein lipidation (co- or posttranslational modification of proteins by specific lipids) include *N*-myristoylation at the amino terminus of proteins; *S*-palmitoylation at internal cysteines; and *S*-prenylation by farnesyl or geranylgeranyl residues at cysteines at the carboxyl terminus of proteins.

Biosynthesis of Phospholipids

Phosphatidic Acid Is Synthesized from α-Glycerophosphate and Fatty Acyl CoA

l-α-Phosphatidic acid (commonly called phosphatidic acid) and 1,2-diacyl-*sn*glycerol are common intermediates in phospholipid and triacylglycerol biosynthesis (Figure 17.14) and both pathways share some of the same enzymes (see Chapter 16, p. 707). Essentially all cells are capable of synthesizing phospholipids to some degree (except mature erythrocytes), whereas triacylglycerol biosynthesis occurs only in liver, adipose tissue, and intestine. In most tissues, the pathway for phosphatidic acid synthesis begins with α-glycerol 3-phosphate (*sn*-glycerol 3-phosphate). The most general source of α-glycerol 3-phosphate, particularly in adipose tissue, is from reduction of the glycolytic intermediate, dihydroxyacetone phosphate, in the reaction catalyzed by α-glycerol 3-phosphate dehydrogenase:

$$\text{dihydroxyacetone phosphate} + \text{NADH} + \text{H}^+ \rightleftharpoons \alpha\text{-glycerol 3-phosphate} + \text{NAD}^+$$

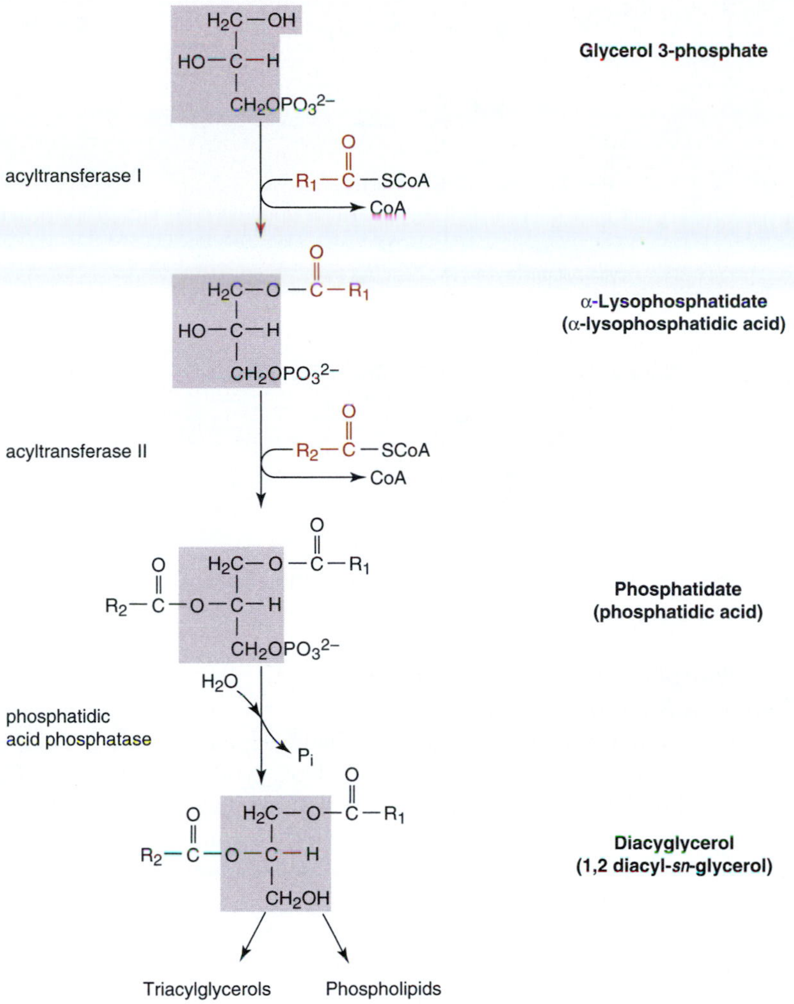

Glycerol 3-phosphate

acyltransferase I

α-Lysophosphatidate
(α-lysophosphatidic acid)

acyltransferase II

Phosphatidate
(phosphatidic acid)

phosphatidic
acid phosphatase

Diacyglycerol
(1,2 diacyl-*sn*-glycerol)

Triacylglycerols Phospholipids

FIGURE 17.14
Phosphatidic acid biosynthesis from glycerol 3-phosphate and the role of phosphatidic acid phosphatase in synthesis of phospholipids and triacylglycerols.

A few specialized tissues, including liver and kidney, derive α-glycerol 3-phosphate by means of the glycerol kinase reaction:

$$\text{glycerol} + \text{ATP} \xrightleftharpoons{\text{Mg}^{2+}} \alpha\text{-glycerol 3-phosphate} + \text{ADP}$$

The next two steps in phosphatidic acid biosynthesis involve stepwise transfer of long-chain fatty acyl groups from fatty acyl CoA. The first acyltransferase (I) is called **glycerol phosphate:acyltransferase** and predominantly attaches saturated fatty acids and oleic acid to the *sn*-1 to produce 1-acylglycerol phosphate or α-lysophosphatidic acid. The second enzyme (II), **1-acylglycerol phosphate:acyltransferase,** acylates the *sn*-2 position, usually with an unsaturated fatty acid (Figure 17.14). In both cases the donor of acyl groups is the CoA thioester derivative of the appropriate long-chain fatty acids.

The specificity of the two acyltransferases does not always match the fatty acid asymmetry found in the phospholipids of a particular cell. Remodeling reactions, discussed below, modify the fatty acid composition at C-1 and C-2 of the glycerol phosphate backbone. Cytosolic phosphatidic acid phosphatase (also called phosphatidic acid phosphohydrolase) hydrolyzes phosphatidic acid (1,2-diacylglycerophosphate) that is generated on the endoplasmic reticulum, thereby yielding 1,2-diacyl-*sn*-glycerol that serves as the branch point in triacylglycerol and phospholipid synthesis (Figure 17.14). Phosphatidic acid can also be formed by a sec-

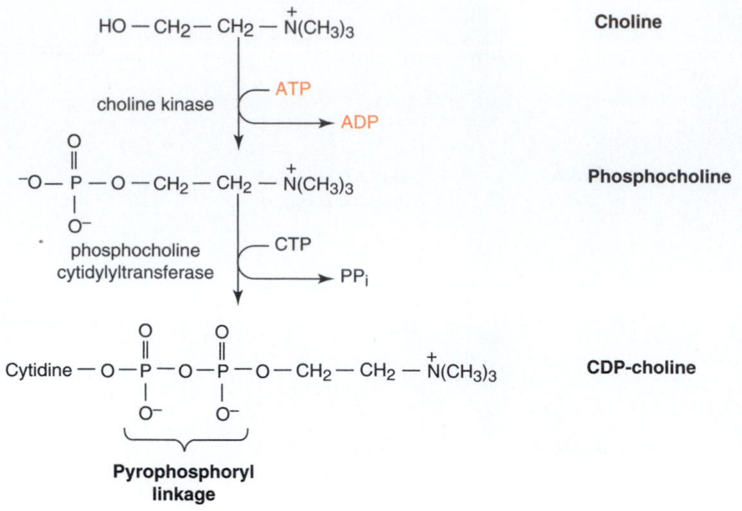

FIGURE 17.15
Biosynthesis of CDP-choline from choline.

ond pathway that begins with dihydroxyacetone phosphate (DHAP). This is usually an alternative supportive route used by some tissues to produce phosphatidic acid.

Specific Phospholipids Are Synthesized by Addition of a Base to Phosphatidic Acid

The major pathway for biosynthesis of phosphatidylcholine (lecithin) involves sequential conversion of choline to phosphocholine, cytidine diphosphate (CDP)-choline, and phosphatidylcholine. In this pathway, the phosphocholine polar head group is activated using cytidine triphosphate (CTP), according to the following reactions. Free choline, a dietary requirement for most mammals including humans, is first phosphorylated by ATP by choline kinase (Figure 17.15). Phosphocholine is converted to CDP-choline at the expense of CTP in the reaction catalyzed by **phosphocholine cytidylyltransferase.** Note inorganic pyrophosphate (PP_i) is a product of this reaction. The high-energy pyrophosphoryl bond in CDP-choline is very unstable and reactive so that the phosphocholine moiety can be transferred readily to the nucleophilic center provided by the OH group at position 3 of 1,2-diacylglycerol by choline phosphotransferase (Figure 17.16). This is the principal pathway for the synthesis of dipalmitoyllecithin in lung.

The rate-limiting step for phosphatidylcholine biosynthesis is the cytidylyltransferase reaction that forms CDP-choline (Figure 17.15). This enzyme is regulated by a novel mechanism involving exchange of enzyme between cytosol and endoplasmic reticulum. The cytosolic form of cytidylyltransferase is inactive and appears to function as a reservoir of enzyme; binding of the enzyme to the membrane results

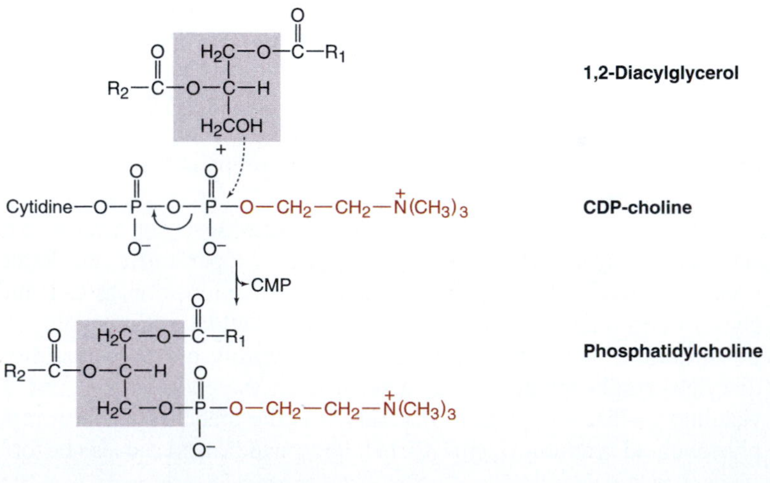

FIGURE 17.16
Choline phosphotransferase reaction.

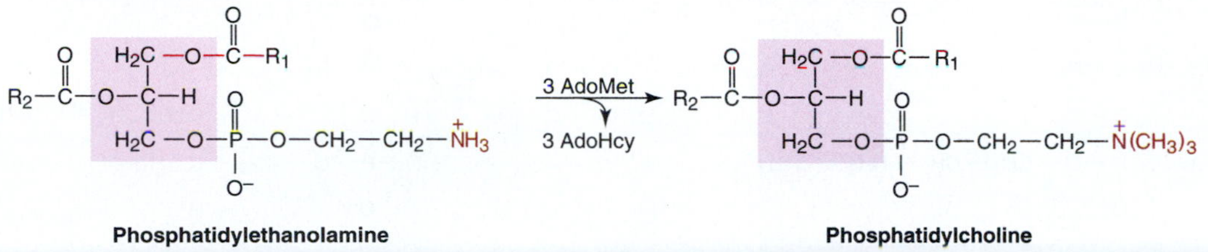

3 AdoMet → 3 AdoHcy

Phosphatidylethanolamine → Phosphatidylcholine

FIGURE 17.17
Biosynthesis of phosphatidylcholine from phosphatidylethanolamine and *S*-adenosylmethionine (AdoMet) and *S*-adenosylhomocysteine (AdoHcy).

in activation. Translocation of cytidylyltransferase from the cytosol to the endoplasmic reticulum is regulated by cAMP and fatty acyl CoA. Reversible phosphorylation of the enzyme by a cAMP-dependent kinase causes it to be released from the membrane, rendering it inactive. Subsequent dephosphorylation will cause cytidylyltransferase to rebind to the membrane and become active. Fatty acyl CoAs activate the enzyme by promoting its binding to the endoplasmic reticulum. In liver only, phosphatidylcholine is formed by repeated methylation of phosphatidylethanolamine. **Phosphatidylethanolamine *N*-methyltransferase** of the endoplasmic reticulum catalyzes transfer of methyl groups one at a time from ***S*-adenosylmethionine (AdoMet)** to phosphatidylethanolamine to produce phosphatidylcholine (Figure 17.17). It is not known if one or more enzymes are involved in these methyl transfers.

The primary pathway for phosphatidylethanolamine synthesis in liver and brain involves **ethanolamine phosphotransferase** of the endoplasmic reticulum that catalyzes the reaction shown in Figure 17.18. This enzyme is particularly abundant in liver. CDP-ethanolamine is formed by **ethanolamine kinase:**

$$\text{ethanolamine} + \text{ATP} \xrightleftharpoons{Mg^{2+}} \text{phosphoethanolamine} + \text{ADP}$$

and **phosphoethanolamine cytidylyltransferase:**

$$\text{phosphoethanolamine} + \text{CTP} \xrightleftharpoons{Mg^{2+}} \text{CDP-ethanolamine} + \text{PP}_i$$

Liver mitochondria also generate phosphatidylethanolamine by decarboxylation of phosphatidylserine; however, this is thought to represent a minor pathway (Figure 17.19).

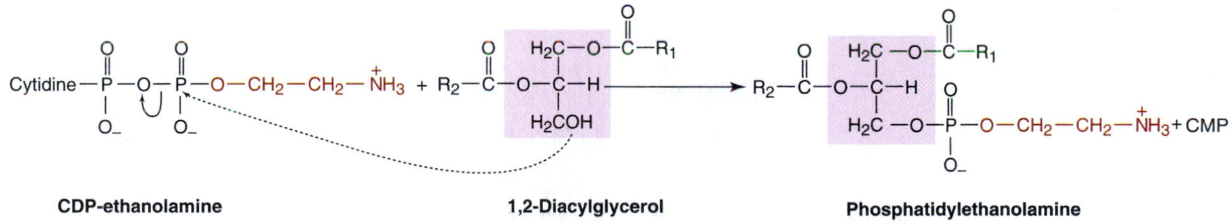

CDP-ethanolamine + 1,2-Diacylglycerol → Phosphatidylethanolamine + CMP

FIGURE 17.18
Biosynthesis of phosphatidylethanolamine from CDP-ethanolamine and diacylglycerol; the reaction is catalyzed by ethanolamine phosphotransferase.

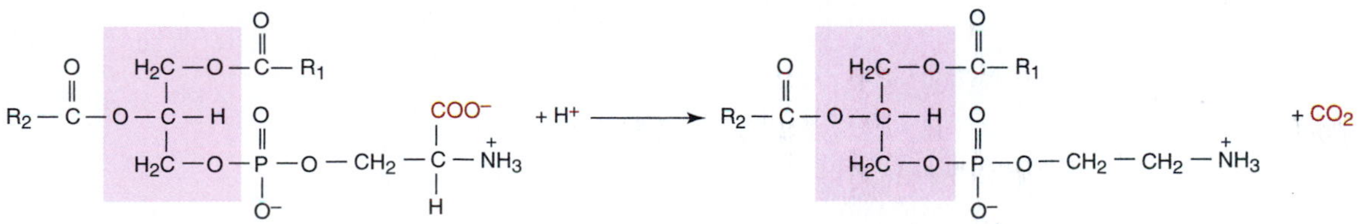

FIGURE 17.19
Formation of phosphatidylethanolamine by the decarboxylation of phosphatidylserine.

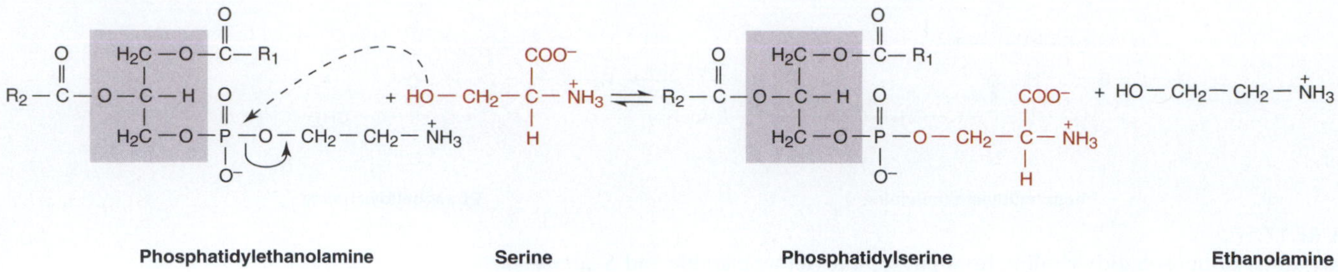

Phosphatidylethanolamine + Serine ⇌ Phosphatidylserine + Ethanolamine

FIGURE 17.20
Biosynthesis of phosphatidylserine from serine and phosphatidylethanolamine by "base exchange."

The major source of phosphatidylserine in mammalian tissues is the "base-exchange" reaction (Figure 17.20) in which the polar head group of phosphatidylethanolamine is exchanged for serine. Since there is no net change in the number or kind of bonds, this reaction is reversible and has no requirement for ATP or any other high-energy compound. The reaction is initiated by attack on the phosphodiester bond of phosphatidylethanolamine by the hydroxyl group of serine.

Phosphatidylinositol is made via CDP-diacylglycerol and free *myo*-inositol (Figure 17.21) catalyzed by **phosphatidylinositol synthase,** another enzyme of the endoplasmic reticulum.

Asymmetric Distribution of Fatty Acids in Phospholipids Is Due to Remodeling Reactions

Two phospholipases, phospholipase A_1 and phospholipase A_2, occur in many tissues and play a role in the formation of specific phospholipid structures containing appropriate fatty acids in the *sn*-1 and *sn*-2 positions. Most fatty acyl CoA transferases and enzymes of phospholipid synthesis discussed above lack the specificity required to account for the asymmetric position or distribution of fatty acids found in many tissue phospholipids. The fatty acids found in the *sn*-1 and *sn*-2 positions of the various phospholipids are often not the same ones transferred to the glycerol backbone in the initial acyltransferase reactions of phospholipid biosynthesis. **Phospholipases A_1** and **A_2** catalyze reactions indicated in Figure 17.22, where X represents the polar head group of a phospholipid. The products of the action of phospholipases A_1 and A_2 are called lysophosphatides.

If it becomes necessary for a cell to remove some undesired fatty acid, such as stearic acid from the *sn*-2 position of phosphatidylcholine, and replace it by a more unsaturated one like arachidonic acid, then this can be accomplished by the action of phospholipase A_2 followed by a reacylation reaction. Insertion of arachidonic acid into the 2 position of *sn*-2-lysophosphatidylcholine can be accomplished either

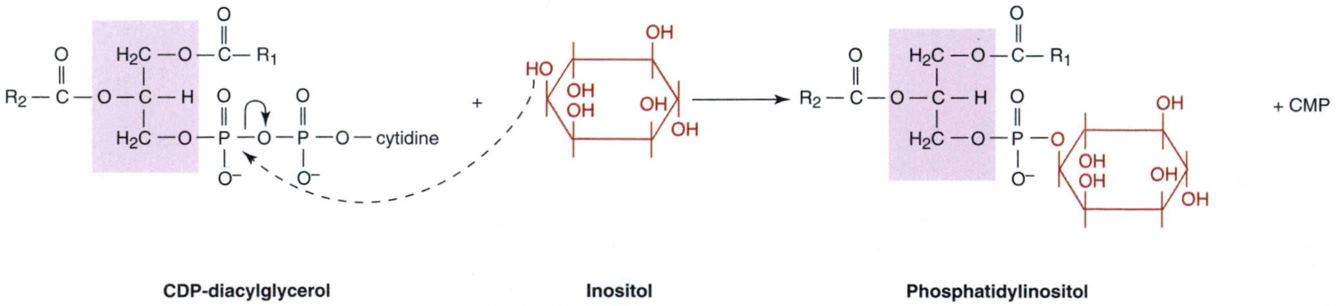

CDP-diacylglycerol + Inositol → Phosphatidylinositol + CMP

FIGURE 17.21
Biosynthesis of phosphatidylinositol.

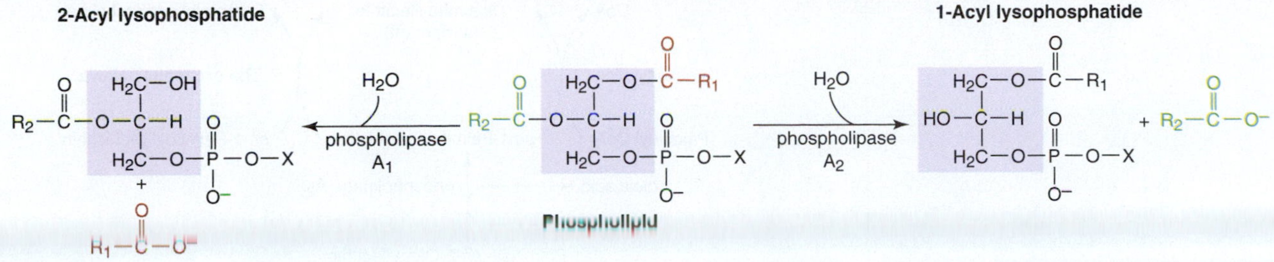

FIGURE 17.22
Reactions catalyzed by phospholipase A_1 and phospholipase A_2.

by direct acylation from arachidonyl CoA involving **arachidonyl CoA transacylase** (Figure 17.23) or from some other arachidonic acid-containing phospholipid by an exchange-type reaction (Figure 17.24) catalyzed by **lysolecithin:lecithin acyltransferase** (LLAT) (Figure 17.24). Since there is no change in either number or nature of the bonds involved in products and reactants, ATP is not required. Reacylation of lysophosphatidylcholine from acyl CoA is the major route for remodeling of phosphatidylcholine.

Lysophospholipids, particularly *sn*-1-lysophosphatidylcholine, can also serve as sources of fatty acid in remodeling reactions. Those involved in synthesis of dipalmitoyllecithin (surfactant) from 1-palmitoyl-2-oleoylphosphatidylcholine are presented in Figure 17.25. Note *sn*-1-palmitoyllysolecithin is the source of palmitic acid in the acyltransferase exchange reaction.

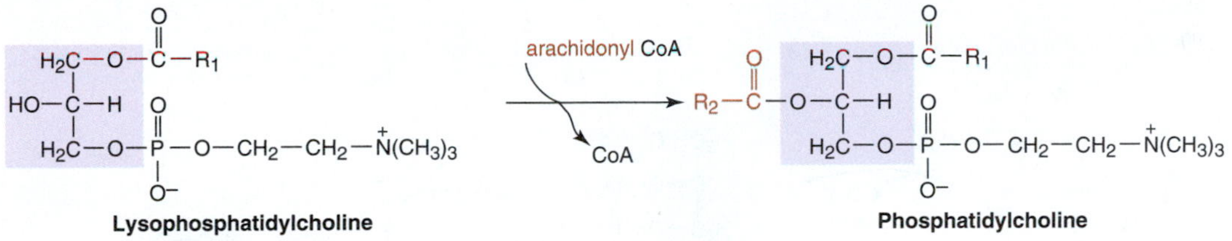

FIGURE 17.23
Synthesis of phosphatidylcholine by reacylation of lysophosphatidylcholine where $R_2-C(=O)-O-$ represents arachidonic acid.
This reaction is catalyzed by acyl-CoA:1-acylglycerol-3-phosphocholine *O*-acyltransferase.

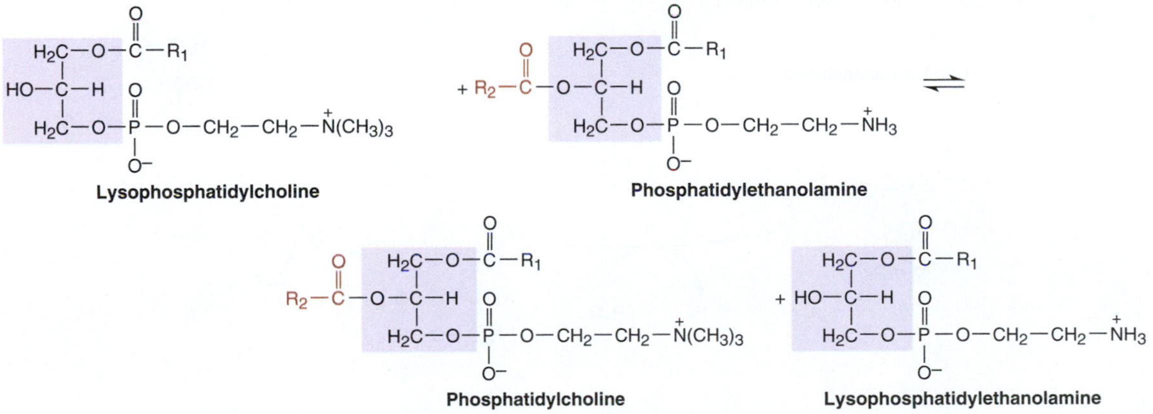

FIGURE 17.24
Formation of phosphatidylcholine by lysolecithin exchange, where $R_2-C(=O)-O-$ represents arachidonic acid.

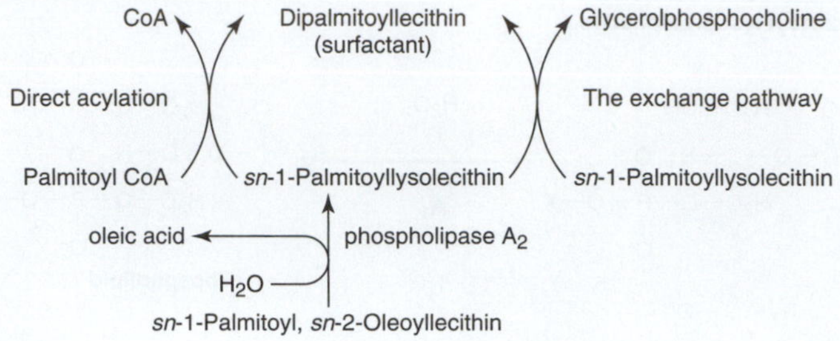

FIGURE 17.25
Two pathways for biosynthesis of dipalmitoyllecithin from *sn*-1 palmitoyllysolecithin.

Plasmalogens Are Synthesized from Fatty Alcohols

Ether glycerolipids are synthesized from DHAP, long-chain fatty acids, and long-chain fatty alcohols; the reactions are summarized in Figure 17.26. Acyldihydroxyacetone phosphate is formed by **acyl CoA: dihydroxyacetone phosphate acyltransferase** (enzyme I) acting on dihydroxyacetone phosphate and long-chain fatty acyl CoA. The ether bond is introduced by **alkyldihydroxyacetone phosphate synthase** (Figure 17.26, enzyme II), which exchanges the 1-*O*-acyl group of acyldihydroxyacetone phosphate with a long chain fatty alcohol. The synthase occurs in peroxisomes. Plasmalogen synthesis is completed by transfer of a long-chain fatty acid from its CoA donor to the *sn*-2 position of 1-alkyl-2-lyso-*sn*-glycero-3-phosphate (Figure 17.26, reaction 4). Patients with Zellweger syndrome lack peroxisomes and cannot synthesize adequate amounts of plasmalogen.

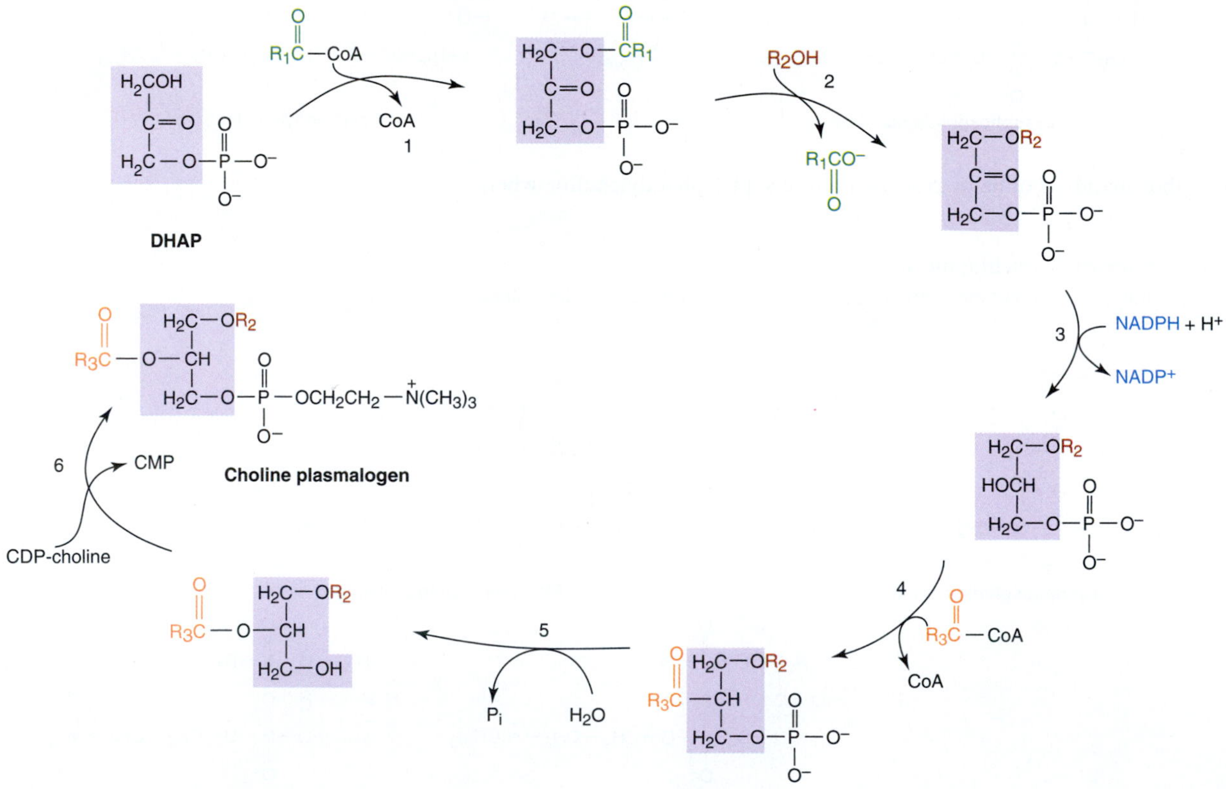

FIGURE 17.26
Pathway of choline plasmalogen biosynthesis from DHAP.
1, Acyl CoA: dihydroxyacetone phosphate acyltransferase; 2, alkyldihydroxyacetone phosphate synthase; 3, NADPH:alkyldihydroxyacetone phosphate oxidoreductase; 4, acyl CoA:1-alkyl-2-lyso-*sn*-glycero-3-phosphate acyltransferase; 5, 1-alkyl-2-acyl-*sn*-glycerol-3-phosphate phosphohydrolase; 6, CDP-choline:1-alkyl-2-acyl-*sn*-glycerol choline phosphotransferase.

17.3 | CHOLESTEROL

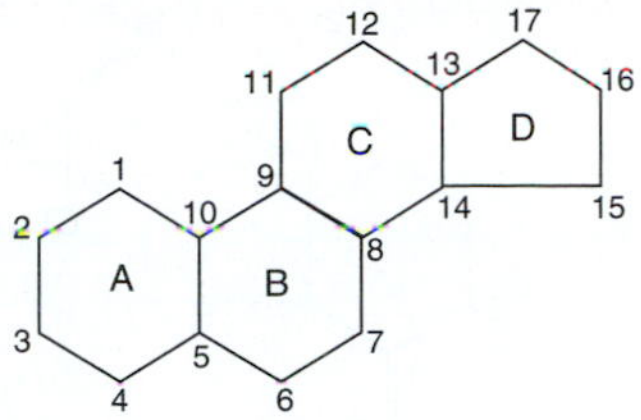

Cholesterol, an Alicyclic Compound, Is Widely Distributed in Free and Esterified Forms

Cholesterol is an alicyclic compound whose structure includes (1) the perhydrocyclopentanophenanthrene nucleus with its four fused rings, (2) a single hydroxyl group at C-3, (3) an unsaturated center between C-5 and C-6 atoms, (4) an eight-membered branched hydrocarbon chain attached to the D ring at position 17, and (5) a methyl group (designated C-19) attached at position 10 and another methyl group (designated C-18) attached at position 13 (see Figures 17.27 and 17.28).

Cholesterol is a lipid with very low solubility in water; at 25°C, the limit of solubility is approximately 0.2 mg/100 mL, or 4.7 mM. The actual concentration of cholesterol in plasma of healthy people is usually 150–200 mg /100 mL. This value is almost twice the normal concentration of blood glucose. This high solubility of cholesterol in blood is due to plasma lipoproteins (mainly LDL and VLDL) that have the ability to bind and thereby solubilize large amounts of cholesterol (see p. 745). Actually, only about 30% of the total plasma cholesterol occurs free (unesterified); approximately 70% of the cholesterol in plasma lipoproteins exists in the form of **cholesterol esters** where some long-chain fatty acid, usually linoleic acid, is attached by an ester bond to the OH group on C-3 of the A ring. The long-chain fatty acid residue enhances the hydrophobicity of cholesterol (Figure 17.29).

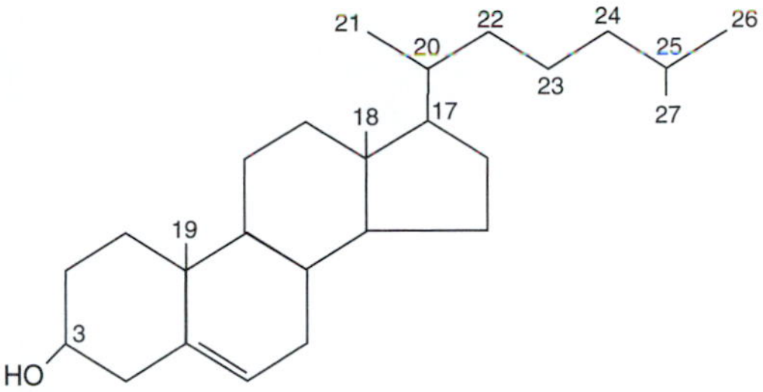

Cholesterol is a ubiquitous and essential component of mammalian cell membranes. It is also abundant in bile where the normal concentration is 390 mg/100 mL. Only 4% of cholesterol in bile is esterified to a long-chain fatty acid. Bile does not contain appreciable amounts of lipoproteins and solubilization of free cholesterol is achieved in part by the detergent property of phospholipids present in bile that are produced in liver (see p. 1105). A chronic disturbance in phospholipid metabolism in liver can result in deposition of cholesterol-rich gallstones. Bile salts, which are metabolites of cholesterol, also aid in solubilizing cholesterol in bile. Cholesterol also appears to protect membranes of the gallbladder from potentially irritating or harmful effects of bile salts.

In the clinical laboratory total cholesterol is estimated by the Liebermann–Burchard reaction. The ratio of free and esterified cholesterol can be determined by gas–liquid chromatography or reverse phase high-pressure liquid chromatography (HPLC).

FIGURE 17.27
The cyclopentanophenanthrene ring.

FIGURE 17.28
Structure of cholesterol (5-cholesten-3β-ol).

Cholesterol Is a Membrane Component and Precursor of Bile Salts and Steroid Hormones

Cholesterol, derived from the diet or synthesized *de novo* in virtually all the cells of humans, has several important roles. It is the major sterol in humans and a component of virtually all plasma and intracellular membranes. Cholesterol is especially abundant in myelinated structures of brain and central nervous system but is present in small amounts in the inner membrane of mitochondria (see p. 501). In

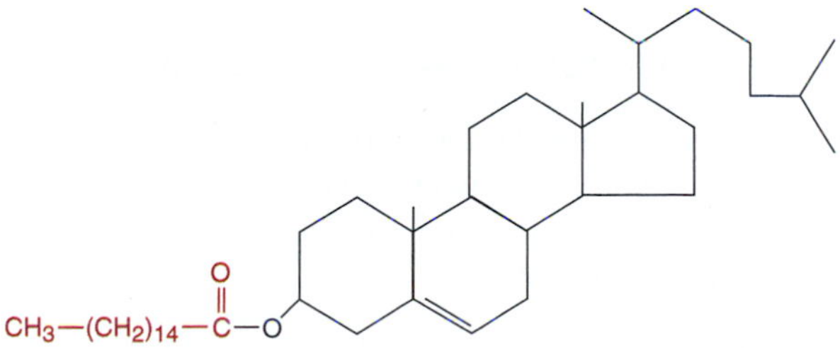

FIGURE 17.29
Structure of cholesterol (palmitoyl) ester.

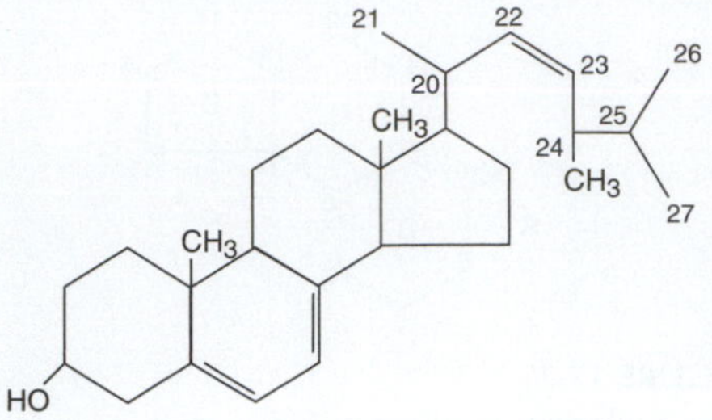

FIGURE 17.30
Structure of ergosterol.

contrast to plasma, most cholesterol in cellular membranes occurs in the free, unesterified form.

Cholesterol is the immediate precursor of **bile acids** that are synthesized in liver and that function to facilitate absorption of dietary triacylglycerols and fat-soluble vitamins (see p. 1105). The ring structure of cholesterol cannot be metabolized to CO_2 and water in humans. Excretion of cholesterol is by way of the liver and gallbladder through the intestine in the form of bile acids.

Another physiological role of cholesterol is as the precursor of various **steroid hormones** (see p. 963). Progesterone is the 21-carbon keto steroid sex hormone secreted by the corpus luteum of the ovary and by placenta. The metabolically powerful corticosteroids of adrenal cortex are derived from cholesterol; they include deoxycorticosterone, corticosterone, cortisol, and cortisone. The mineralocorticoid aldosterone is derived from cholesterol in the zona glomerulu of the cortex of adrenal glands. Cholesterol is also the precursor of female steroid hormones, estrogens (e.g., estradiol) in the ovary, and of male steroids (e.g., testosterone) in the testes. Although all steroid hormones are structurally related to and biochemically derived from cholesterol, they have widely different physiological properties that relate to the menstrual cycle, spermatogenesis, pregnancy, lactation and parturition, mineral balance, and energy (amino acids, carbohydrate, and fat) metabolism.

The hydrocarbon skeleton of cholesterol also occurs in plant sterols, for example, ergosterol, a precursor of vitamin D. **Ergosterol** (Figure 17.30) is converted in skin by ultraviolet irradiation to vitamin D_2, which is involved in calcium and phosphorus metabolism (see p. 1141).

Cholesterol Is Synthesized from Acetyl CoA

Although *de novo* synthesis of cholesterol occurs in virtually all cells, the capacity is greatest in liver, intestine, adrenal cortex, and reproductive tissues, including ovaries, testes, and placenta. From inspection of its structure it is apparent that cholesterol biosynthesis should require a source of carbon atoms and considerable reducing power to generate the numerous carbon–carbon and carbon–hydrogen bonds. All carbon atoms of cholesterol are derived from acetate. Reducing power in the form of NADPH is provided mainly by glucose 6-phosphate dehydrogenase and 6-phosphogluconate dehydrogenase of the pentose phosphate pathway (p. 666). The pathway of cholesterol synthesis occurs in the cytosol and endoplasmic reticulum and is driven in large part by hydrolysis of high-energy thioester bonds of acetyl CoA and phosphoanhydride bonds of ATP.

Mevalonic Acid Is a Key Intermediate

The first compound unique to cholesterol biosynthesis is mevalonic acid derived from acetyl CoA. Acetyl CoA can be obtained from several sources: (1) the β-oxidation of fatty acids (see p. 714), (2) the oxidation of ketogenic amino acids such as leucine and isoleucine (see p. 811), and (3) the pyruvate dehydrogenase reaction. Free acetate can be activated to its thioester derivative at the expense of ATP by **acetokinase,** also referred to as **acetate thiokinase:**

$$\text{ATP} + CH_3COO^- + \text{CoASH} \longrightarrow CH_3\text{—}\overset{\overset{\displaystyle O}{\|}}{C}\text{—SCoA} + \text{AMP} + \text{PP}_i$$

The first two reactions in cholesterol biosynthesis are shared by the pathway that produces ketone bodies (see p. 720). Two molecules of acetyl CoA condense to form acetoacetyl CoA in a reaction catalyzed by **acetoacetyl CoA thiolase (acetyl CoA:acetyl CoA acetyltransferase):**

$$CH_3\text{—}\overset{\overset{\displaystyle O}{\|}}{C}\text{—SCoA} + CH_3\text{—}\overset{\overset{\displaystyle O}{\|}}{C}\text{—SCoA} \longrightarrow CH_3\text{—}\overset{\overset{\displaystyle O}{\|}}{C}\text{—}CH_2\text{—}\overset{\overset{\displaystyle O}{\|}}{C}\text{—SCoA} + \text{CoA—SH}$$

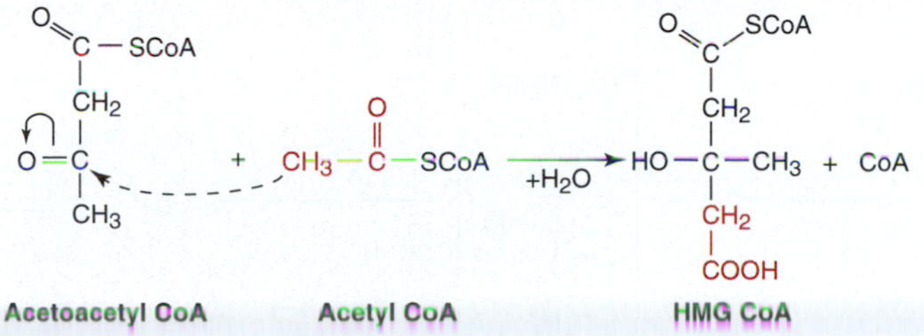

Acetoacetyl CoA + Acetyl CoA $\xrightarrow{+H_2O}$ HMG CoA + CoA

FIGURE 17.31
HMG CoA synthase reaction.

Formation of the carbon–carbon bond in acetoacetyl CoA is favored energetically by cleavage of a thioester bond and generation of free coenzyme A.

The next step introduces a third molecule of acetyl CoA and forms the branched-chain compound **3-hydroxy-3-methylglutaryl CoA** (HMG CoA) (Figure 17.31). This condensation reaction is catalyzed by **HMG CoA synthase (3-hydroxy-3-methylglutaryl CoA:acetoacetyl CoA lyase).** Liver parenchymal cells contain two isoenzyme forms of HMG CoA synthase: the one in the cytosol is involved in cholesterol synthesis, while the other has a mitochondrial location and functions in synthesis of ketone bodies (see p. 720). In the HMG CoA synthase reaction, an aldol condensation occurs between the methyl carbon of acetyl CoA and the β-carbonyl group of acetoacetyl CoA with the simultaneous hydrolysis of the thioester bond of acetyl CoA. The thioester bond in the original acetoacetyl CoA remains intact. HMG CoA can also be formed from oxidative degradation of the branched-chain amino acid leucine, through the intermediates 3-methylcrotonyl CoA and 3-methylglutaconyl CoA (see p. 813).

The step that produces the unique compound mevalonic acid from HMG CoA is catalyzed by the important endoplasmic reticulum enzyme **HMG CoA reductase** (mevalonate:$NADP^+$ oxidoreductase) that has an absolute requirement for NADPH as reductant (Figure 17.32). This reductive step consumes two molecules of NADPH, results in hydrolysis of the thioester bond of HMG CoA, and generates a primary alcohol group in mevalonate. This reduction reaction is irreversible and produces (*R*)-(+) mevalonate, which contains six carbon atoms. HMG CoA reductase catalyzes the rate-limiting reaction in cholesterol biosynthesis. HMG CoA reductase is an intrinsic membrane protein of the endoplasmic reticulum whose carboxyl terminus extends into the cytosol and contains the enzyme's active site. Phosphorylation regulates HMG CoA reductase activity of the cell by diminishing its catalytic activity (V_{max}) and enhancing the rate of its degradation by increasing its susceptibility to proteolytic attack. Increased intracellular cholesterol stimulates phosphorylation of HMG CoA reductase.

The central role of HMG CoA reductase in cholesterol homeostasis is evidenced by the effectiveness of a family of drugs called statins to lower plasma cholesterol levels. Statins (e.g., lovastatin, pravastatin, fluvastatin, cerivastatin, and atorvastatin) inhibit HMG CoA reductase activity, particularly in liver, and commonly lower total plasma cholesterol by as much as 50%.

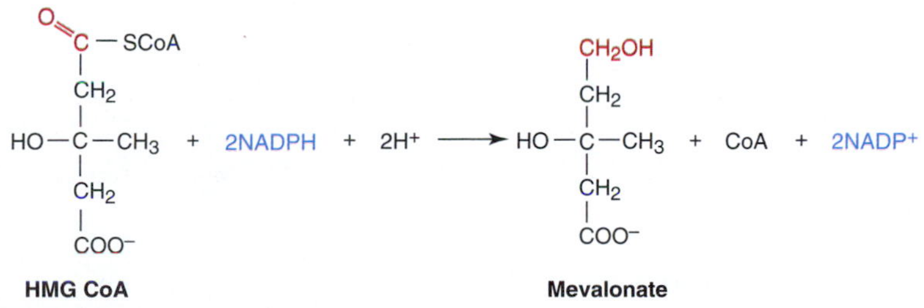

HMG CoA + 2NADPH + $2H^+$ $\longrightarrow$ Mevalonate + CoA + $2NADP^+$

FIGURE 17.32
HMG CoA reductase reaction.

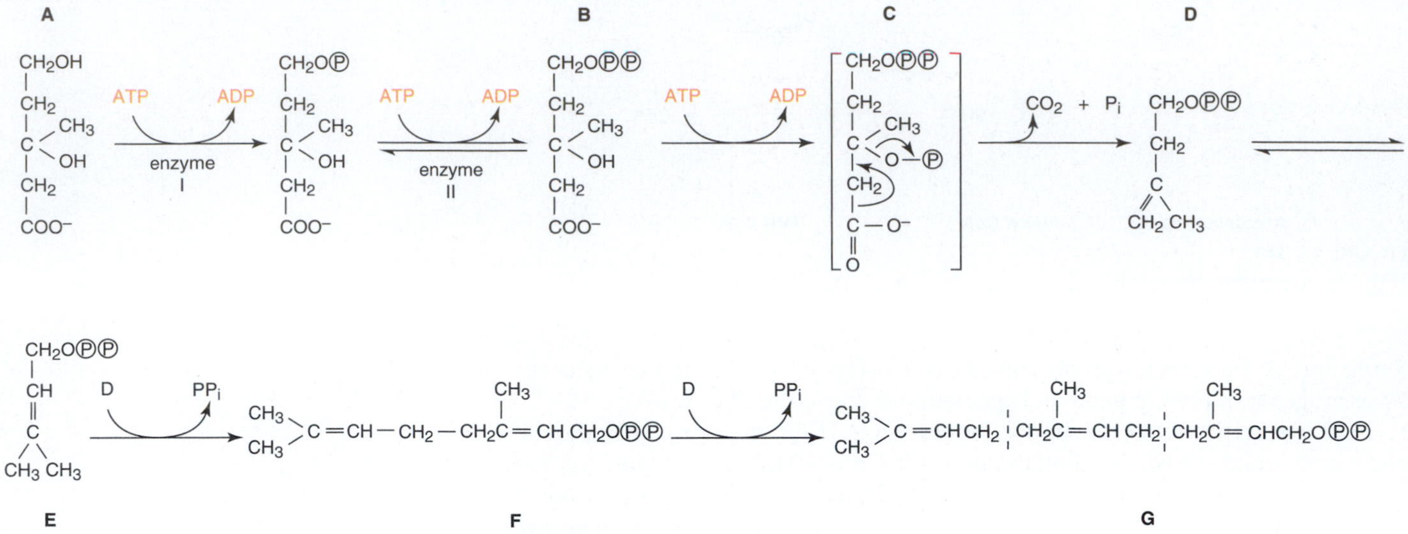

FIGURE 17.33
Formation of farnesyl-PP (F) from mevalonate (A).
Dotted lines divide molecules into isoprenoid-derived units. D is 3-isopentenyl pyrophosphate.

Mevalonic Acid Is a Precursor of Farnesyl Pyrophosphate

Reactions that convert mevalonate to **farnesyl pyrophosphate** are summarized in Figure 17.33. The stepwise transfer of the terminal phosphate group from two molecules of ATP to mevalonate (A) to form 5-pyrophosphomevalonate (B) are catalyzed by **mevalonate kinase** (enzyme I) and **phosphomevalonate kinase** (enzyme II). The next step affects decarboxylation of 5-pyrophosphomevalonate by **pyrophosphomevalonate decarboxylase** and generates Δ^3-isopentenyl pyrophosphate (D). In this ATP-dependent reaction in which ADP, P_i, and CO_2 are produced, it is thought that decarboxylation–dehydration proceeds by way of the triphosphate intermediate, 3-phosphomevalonate 5-pyrophosphate (C). Isopentenyl pyrophos-

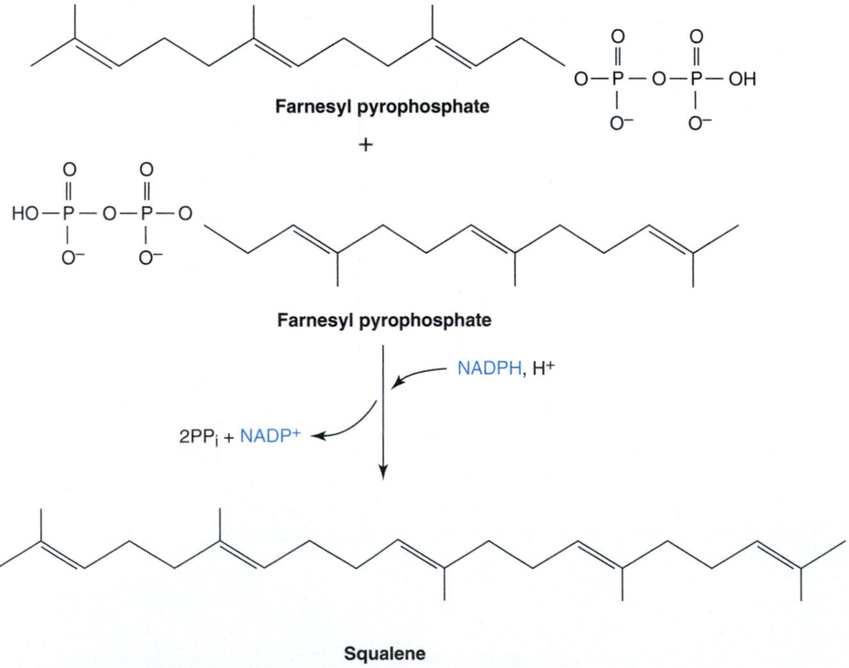

FIGURE 17.34
Formation of squalene from two molecules of farnesyl pyrophosphate.

phate is converted to its allylic isomer 3,3-dimethylallyl pyrophosphate (E) catalyzed by **isopentenyl pyrophosphate isomerase** in a reversible reaction . The condensation of 3,3-dimethylallyl pyrophosphate (E) and 3-isopentenyl pyrophosphate (D) generates geranyl pyrophosphate (F).

The stepwise condensation of three 5-carbon isopentenyl units to form the 15-carbon unit farnesyl pyrophosphate (G) is catalyzed by one enzyme, a cytosolic prenyltransferase called geranyltransferase.

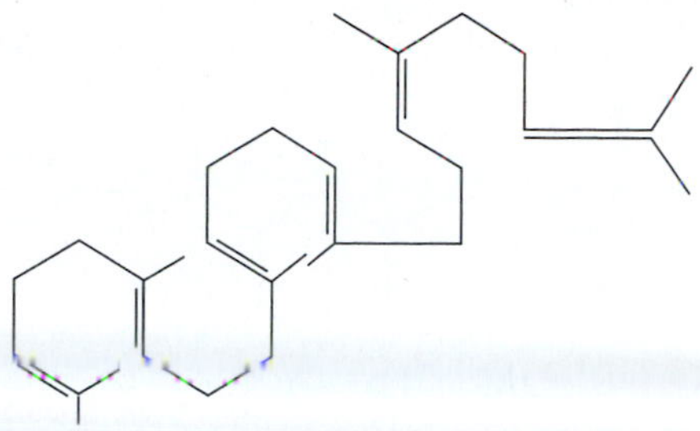

FIGURE 17.35
Structure of squalene, C_{30}.

Cholesterol Is Formed from Farnesyl Pyrophosphate via Squalene

The last steps in cholesterol biosynthesis involve "head-to-head" fusion of two molecules of farnesyl pyrophosphate to form **squalene** and finally cyclization of squalene to yield cholesterol. The reaction producing the 30-carbon molecule of squalene from two 15-carbon farnesyl pyrophosphate moieties (Figure 17.34) is unlike the previous carbon–carbon bond-forming reactions in the pathway (Figure 17.33). **Squalene synthase** present in the endoplasmic reticulum releases two pyrophosphate groups, with loss of a hydrogen from one molecule of farnesyl pyrophosphate and replacement by a hydrogen from NADPH. Several different intermediates probably occur between farnesyl pyrophosphate and squalene. By rotation about carbon–carbon single bonds, the conformation of squalene indicated in Figure 17.35 can be obtained. Note the similarity of the overall shape of squalene to cholesterol and that squalene is devoid of oxygen atoms.

Cholesterol biosynthesis from squalene proceeds through the intermediate **lanosterol,** which contains the fused tetracyclic ring system and an eight-carbon side chain:

$$\text{squalene} \longrightarrow \text{squalene 2,3-epoxide} \longrightarrow \text{lanosterol}$$

The many carbon–carbon bonds formed during cyclization of squalene are generated in a concerted fashion as indicated in Figure 17.36. The OH group of lanosterol projects above the plane of the A ring; this is referred to as the β orientation. Groups that extend down below the ring in a trans relationship to the OH group are designated as α by a dotted line. During this reaction sequence an OH group is added to C-3, two methyl groups undergo shifts, and a proton is eliminated. The oxygen atom is derived from molecular oxygen. The reaction is catalyzed by an endoplasmic reticulum enzyme, **squalene oxidocyclase,** that is composed of at least two activities, squalene epoxidase or monooxygenase and a cyclase (lanosterol cyclase).

The cyclization process is initiated by epoxide formation between what will become C-2 and C-3 of cholesterol, the epoxide being formed at the expense of NADPH:

$$\text{squalene} + O_2 + \text{NADPH} + H^+ \longrightarrow \text{squalene 2,3-epoxide} + H_2O + \text{NADP}^+$$

This reaction is catalyzed by the monooxygenase or epoxidase component. Hydroxylation at C-3 by way of the epoxide intermediate triggers the cyclization of squalene to form lanosterol (Figure 17.36). In the cyclization, two hydrogen atoms and two methyl groups migrate to neighboring positions.

Transformation of lanosterol to cholesterol involves many poorly understood steps and a number of different enzymes. These steps include (1) removal of the methyl group at C-14, (2) removal of two methyl groups at C-4, (3) migration of the double bond from C-8 to C-5, and (4) reduction of the double bond between C-24 and C-25 in the side chain (see Figure 17.37).

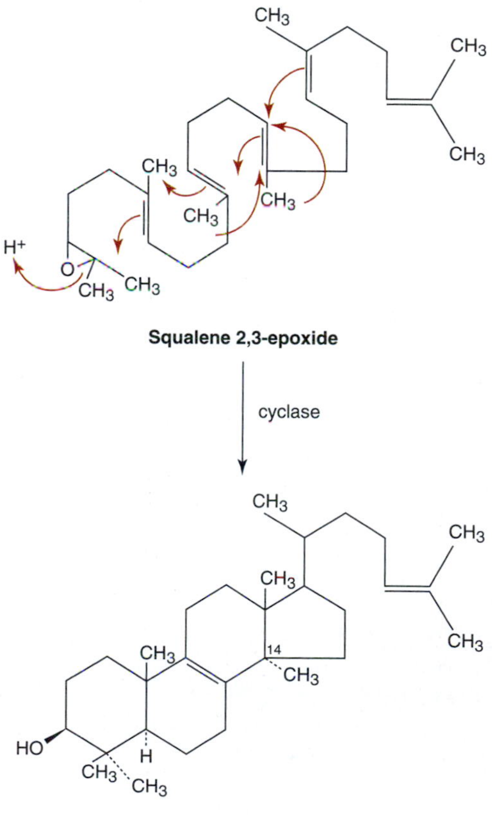

Squalene 2,3-epoxide

cyclase

Lanosterol

FIGURE 17.36
Conversion of squalene 2,3-epoxide to lanosterol.

Plasma Lipoproteins

Blood contains triacylglycerols, cholesterol, and cholesterol esters at concentrations that far exceed their solubilities in water. Blood lipids are kept in solution, or at least thoroughly dispersed in the circulation, by virtue of their being incorporated into macromolecular structures called lipoproteins. The **plasma lipoproteins** facil-

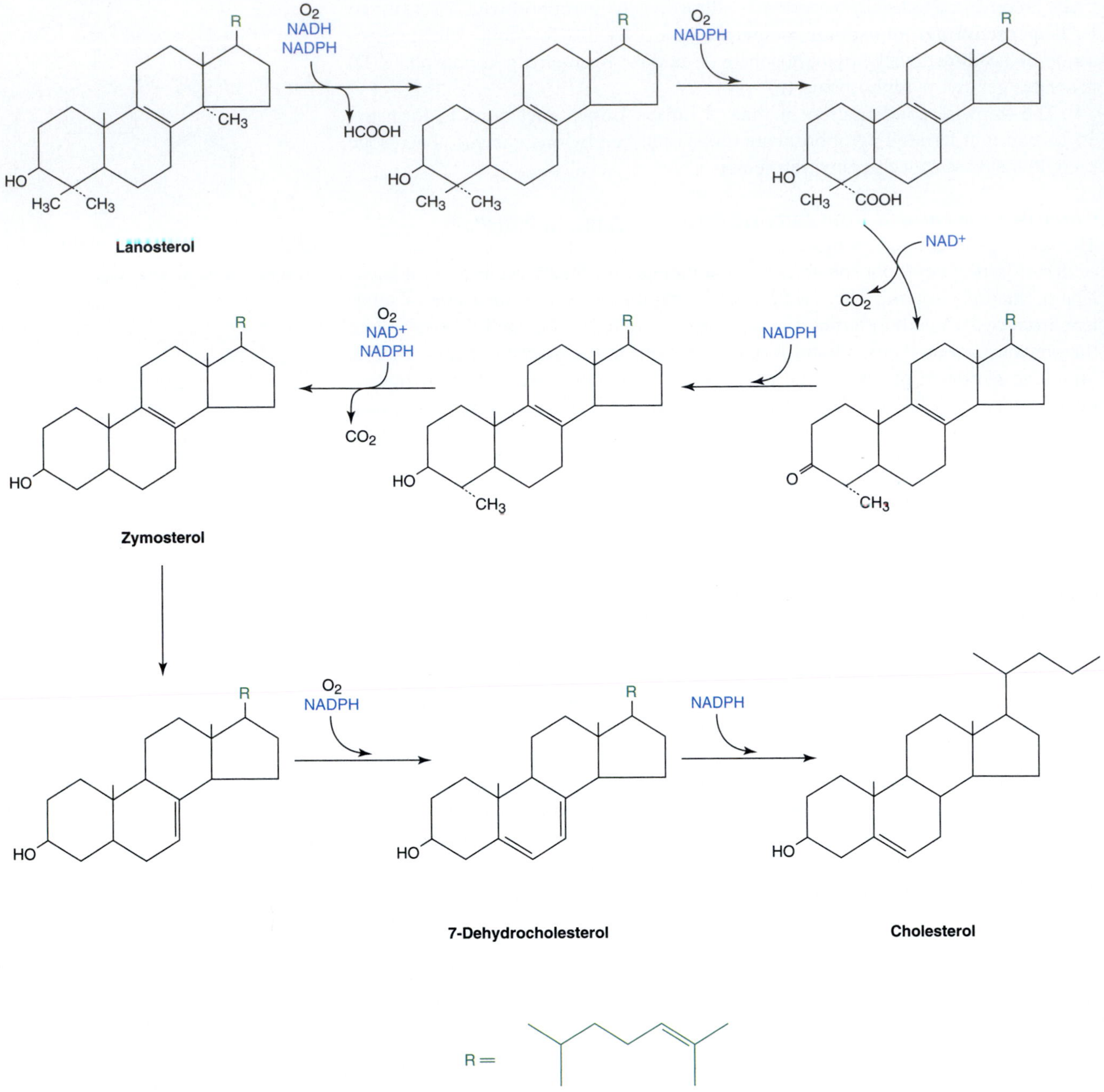

FIGURE 17.37
Conversion of lanosterol to cholesterol.

itate lipid metabolism and the transfer of lipids between tissues. There are four major lipoprotein classes—**high density lipoprotein (HDL)**(1.063–1.210 g/mL), **low density lipoprotein (LDL)**(1.019–1.063 g/mL), **very low density lipoprotein (VLDL)**(0.95–1.006 g/mL), and **chylomicrons** (<0.95 g/mL)—and their lipid and protein compositions are summarized in Table 17.1. All of the plasma lipoproteins contain phospholipids and one or more proteins called apoproteins. There are ten common apoproteins; their designations and the lipoproteins with which they are associated are indicated in Table 17.2. There are subclasses of apoproteins (e.g., apoC-I, apoC-II, apoC-III) and lipoproteins (e.g., HDL_2, HDL_3).

The lipoproteins are spherical particles in which the most hydrophobic lipids such as cholesterol esters and triacylglycerols are located in the core of the structure, sequestered away from water, whereas free (nonesterified) cholesterol, phos-

TABLE 17.1 Characteristics and Functions of the Major Lipoproteins

Characteristics	*Classes of Plasma Lipoproteins*					
	Chylomicrons	*VLDLs*	*IDLs*	*LDLs*	*HDL_2*	*HDL_3*
Density (g mL^{-1})	<0.95	0.95–1.006	1.006–1.019	1.019–1.063	1.063–1.12	1.12–1.21
Diameter (nm)	75–1200	30–80	15–35	18–25	10–20	7.5–10
Molecular weight	400×10^6	$(10–80) \times 10^6$	$(5–10) \times 10^6$	2.3×10^6	$(1.7–3.6) \times 10^6$	
Origin	Intestine	Liver	VLDLs	VLDLs/IDLs/liver	Intestine and liver	
Electrophoretic mobility	Origin	Pre-β	Pre-β to β	β	α	
Physiological role	Exogenous fat transport	Endogenous fat transport	LDL precursor	Cholesterol transport	Reverse cholesterol transport	
Composition (%)						
Triacylglycerol	86	52	28	10	10	5
Cholesteryl ester	3	14	30	38	21	14
Free cholesterol	1	7	8	8	7	3
Phospholipid	8	18	23	22	29	19
Protein	2	8	11	21	33	57

pholipids, and the protein(s) are arrayed on the surface. It is mainly the amphipatic phospholipids and proteins that keep the otherwise highly insoluble lipids like cholesterol and triacylglycerols in solution. The apoproteins on the surface of the particles also serve as structural components, ligands for cell receptors, and cofactors for enzymes involved in lipoprotein metabolism.

The plasma lipoproteins are the vehicles by which cholesterol, cholesterol esters, and triacylglycerols are transported from one tissue or organ to another in the body. For example, LDL delivers cholesterol to various tissues that require cholesterol for membrane structure or steroid hormone synthesis. Furthermore, the cholesterol that LDL carries to the liver from peripheral tissues regulates the rate of hepatic cholesterol synthesis (see Figure 17.38). In contrast, HDL, which is rich in cholesterol and has a low triacylglycerol content (Table 17.1), is the main vehicle for carrying excess cholesterol from the periphery to the liver, where it can be excreted in bile either directly in the form of cholesterol or after conversion to bile salts. Lipoproteins, primarily VLDL and chylomicrons, transport triacylglycerols to sites where they are metabolized for energy (e.g., muscle) or stored (e.g., fat cells).

A second function of plasma lipoproteins is as substrates for lipid-metabolizing enzymes in blood. The enzyme lipoprotein lipase is attached to the vascular en-

TABLE 17.2 Apoproteins of Human Plasma Lipoproteins

Apolipoprotein	*Molecular Weight*	*Plasma Concentration (g L^{-1})*	*Lipoprotein Distribution*
ApoA-I	28,000	1.0–1.2	Chylomicrons, HDL
ApoA-II	17,000	0.3–0.5	Chylomicrons, HDL
ApoA-IV	46,000	0.15–0.16	Chylomicrons, HDL
ApoB-48	264,000	0.03–0.05	Chylomicrons
ApoB-100	512,000	0.7–1.0	VLDL, IDL, LDL
ApoC-I	7000	0.04–0.06	Chylomicrons, VLDL, HDL
ApoC-II	9000	0.03–0.05	Chylomicrons, VLDL, HDL
ApoC-III	9000	0.12–0.14	Chylomicrons, VLDL, HDL
ApoD	33,000	0.06–0.07	HDL
ApoE	38,000	0.03–0.05	Chylomicrons, VLDL, IDL, HDL

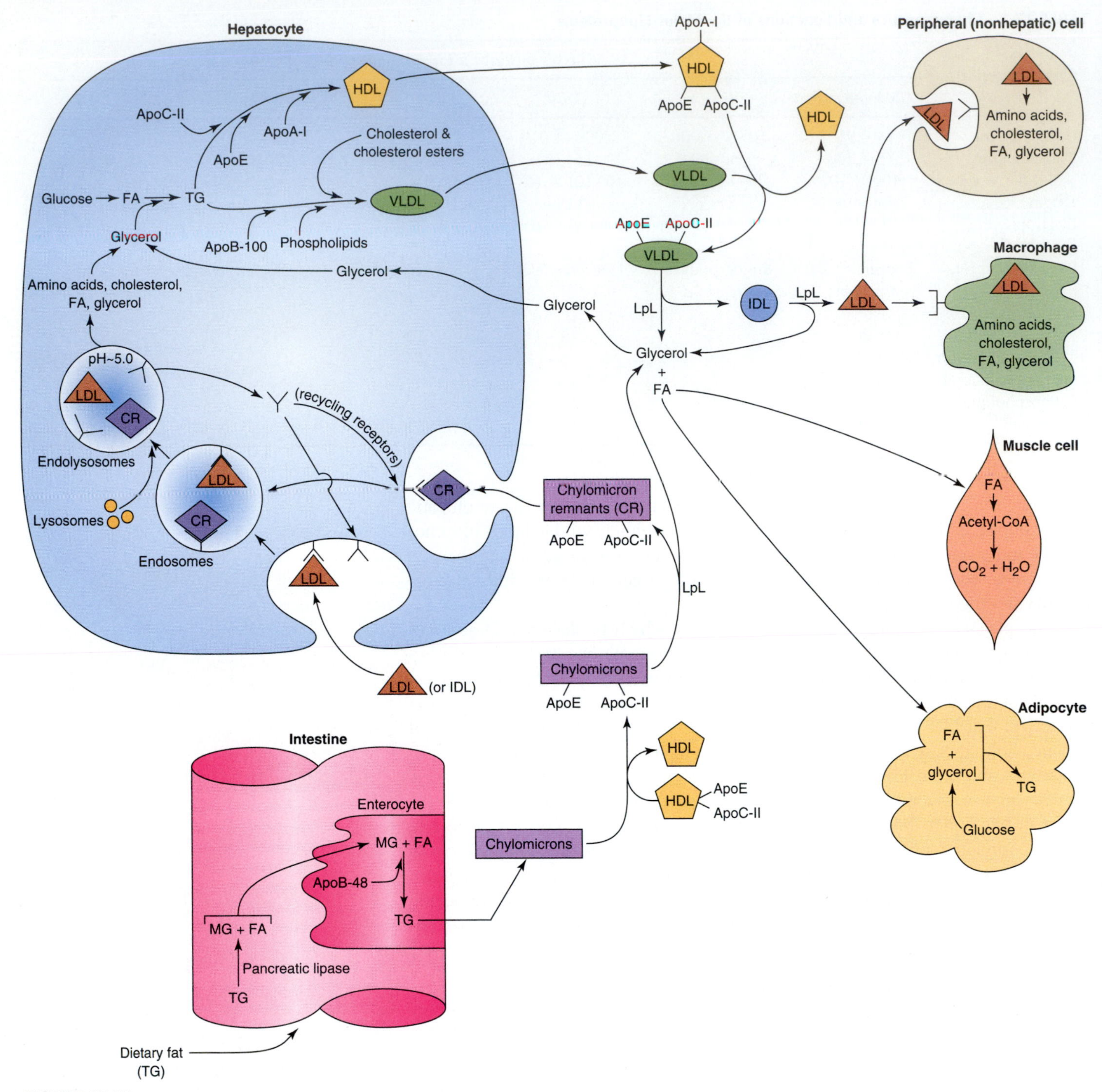

FIGURE 17.38
Organs and pathways involved in plasma lipoprotein metabolism.
FA, fatty acid; TG, triacylglycerol (triglyceride); HDL, high density lipoprotein; LDL, low density lipoprotein; IDL, intermediate density lipoprotein; VLDL, very low density lipoprotein; LpL, lipoprotein lipase; apo-, apoprotein; Y, LDL receptor.

dothelium where, when activated, it can catalyze the hydrolysis of triacylglycerols contained in various lipoprotein particles, VLDL and chylomicrons in particular, assuming the activators of lipoprotein lipase are present. Lipoprotein lipase is activated by **apolipoprotein C-II (apoC-II)** and by heparin released from mast cells and other sites within the macrophage/reticuloendothelial system. The fatty acids released by the esterolytic action of lipoprotein lipase on triacylglycerols can be taken up by cells and oxidized by the β-oxidation pathway to provide energy, incorporated into phospholipids that cells require for membrane assembly, or, in the case of adipocytes, metabolized to and stored in the form of triacylglycerols. As chylomicrons lose their triacylglycerol core they become smaller; triacylglycerol-depleted chylomicrons are termed chylomicron remnants, are rich in cholesterol esters, and are removed from the circulation by means of specific receptors on hepatic membranes and catabolized in the liver (Figure 17.38). Clearly, from these various examples of lipoprotein function one can appreciate the dynamic character of the structure and composition of the plasma lipoproteins.

The main sites of synthesis of the apoprotein components of the plasma lipoproteins are the liver and small intestine. For example, **apoB-48** is produced in the intestine whereas apoB-100 is synthesized mainly in hepatocytes. Thus apoB-48 is synthesized during the course of fat digestion in the small intestine and enters into the circulation when chylomicrons bearing triacylglycerols and cholesterol esters are secreted into the blood following consumption of a fat-containing meal. Conversely, apoB-100 is synthesized when the liver is producing VLDL for secretion into the circulation. The fatty acids contained in the triacylglycerols of newly produced VLDL particles are synthesized *de novo* from sugars, glucose in particular, after the later have been oxidized to acetyl CoA, or from the acetate that results from the oxidation of ethanol. Circulating VLDL acquires cholesterol esters during synthesis and transport in the liver, and from plasma HDL. The transfer of cholesterol esters from HDL to VLDL is facilitated by **cholesterol ester transfer protein (CETP),** which also transfers cholesterol esters to LDL.

Whereas VLDL and chylomicrons are produced by and secreted from specific cells, the other major lipoproteins arise by metabolism of these two lipoproteins in the circulation. The action of lipoprotein lipase on VLDL converts the latter first into **intermediate density lipoprotein (IDL),** and then, as additional triacylglycerol is subjected to lipolysis, into LDL (Figure 17.38).

HDL is synthesized mainly in liver and, to a lesser extent, in the intestine and has the unique function of serving as a reservoir for **apoE** and **apoC-II.** HDL regulates the exchange of proteins and lipids between various lipoproteins in the blood. Initially, HDL particles donate apoE and apoC-II to chylomicrons and VLDL, thereby providing activators for lipoprotein lipase. Once the triacylglycerols in chylomicrons and VLDL have been extensively degraded and these lipoproteins have been transformed into LDL and chylomicron remnants, respectively, the apoE and apoC-II proteins are returned to HDL (Figure 17.38).

Another critical function of HDL is participation in the removal of excess cholesterol from cells and transport of that potentially harmful cholesterol to the liver for elimination from the body by way of the biliary system and feces as cholesterol and bile salts. This process is termed **"reverse cholesterol transport."** From 70% to 75% of the plasma cholesterol contained in lipoproteins is esterified to long-chain fatty acids. It is the free, unesterified form of cholesterol that exchanges readily between different lipoproteins and the plasma membrane of cells. Cholesterol in the plasma membrane of cells and in plasma lipoproteins that is destined for transport to the liver is esterified to a fatty acid in a reaction catalyzed by the enzyme **lecithin:cholesterol acyltransferase (LCAT).** LCAT catalyzes the freely reversible reaction (Figure 17.39), which transfers the fatty acid in the *sn*-2 position of phosphatidylcholine to the 3-hydroxyl of cholesterol and vice versa. LCAT is produced

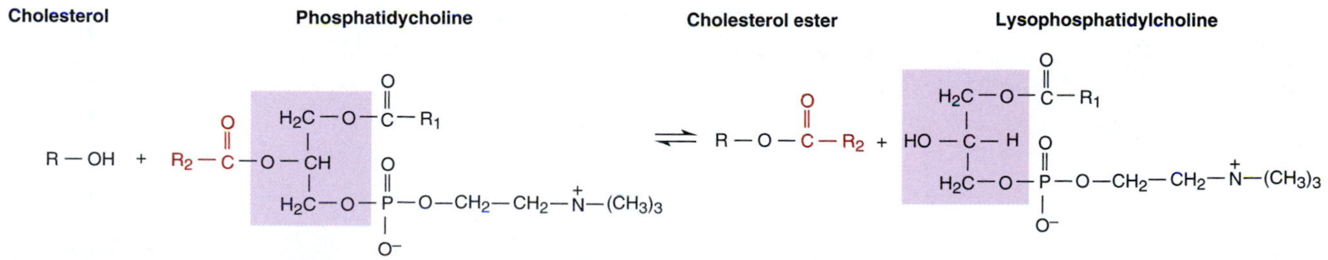

Cholesterol | Phosphatidycholine | Cholesterol ester | Lysophosphatidylcholine

$$\mathrm{R{-}OH} + \mathrm{R_2{-}C({=}O){-}O{-}CH(CH_2{-}O{-}C({=}O){-}R_1)(CH_2{-}O{-}P(O)(O^-){-}O{-}CH_2{-}CH_2{-}\overset{+}{N}{-}(CH_3)_3)} \rightleftharpoons \mathrm{R{-}O{-}C({=}O){-}R_2} + \mathrm{HO{-}C(H)(H_2C{-}O{-}C({=}O){-}R_1)(H_2C{-}O{-}P(O)(O^-){-}O{-}CH_2{-}CH_2{-}\overset{+}{N}{-}(CH_3)_3)}$$

FIGURE 17.39
Lecithin:cholesterol acyltransferase (LCAT) reaction, where R—OH indicates cholesterol.

mainly by the liver. LCAT is bound to HDL in plasma and is activated by apoA-I, which is also a component of HDL. The CETP that is associated with the HDL particle then transfers the cholesterol ester to VLDL or LDL. Cholesterol ester generated in the LCAT reaction diffuses into the core of the HDL particle where it is then transported from the tissues and plasma to the liver, the only organ capable of metabolizing and excreting cholesterol. There is an inverse relationship between plasma HDL concentrations and the incidence of coronary artery disease, and a positive relationship between plasma cholesterol levels and coronary heart disease.

Liver cells metabolize chylomicron remnants by a similar mechanism; however, macrophages and many other kinds of cells possess specific receptors that recognize chylomicron remnants and internalize and degrade them for cellular fuel and structural components. It is the apoE component of the chylomicron remnants that is recognized by these particular receptors. Some LDL is taken up via nonspecific scavenger receptors on certain cells, macrophages in particular.

Cholesterol Biosynthesis Is Carefully Regulated

Elevated plasma cholesterol predisposes people to atherosclerotic vascular disease. In healthy individuals, plasma cholesterol levels are maintained within a relatively narrow concentration range. The liver plays the major role in regulating cholesterol levels: (1) it expresses the majority of the body's LDL receptors; (2) it is the major site for conversion of cholesterol to bile acids; and (3) it has the highest level of HMG CoA reductase activity. The cholesterol pool of the body is derived from absorption of dietary cholesterol and *de novo* synthesis primarily in liver and intestine. When dietary cholesterol is reduced, cholesterol synthesis is increased in liver and intestine to satisfy the needs of other tissues and organs. Cholesterol synthesized *de novo* is transported from liver and intestine to peripheral tissues by VLDLs and chylomicrons, respectively. Liver and intestine are the only tissues that manufacture **apolipoprotein B-100** and **B-48.** Most apolipoprotein B-100 is secreted into the circulation as VLDL, which is converted into LDL by removal of triacylglycerol and apolipoprotein C components, probably in peripheral tissues and liver. When the quantity of dietary cholesterol increases, cholesterol synthesis in liver and intestine is almost totally suppressed. Thus the rate of *de novo* cholesterol synthesis is inversely related to the amount of dietary cholesterol taken up by the body.

The primary site for control of cholesterol biosynthesis is HMG CoA reductase, which catalyzes the step that produces mevalonic acid. This is the committed step and the rate-limiting reaction in the pathway of cholesterol biosynthesis (Figure 17.40). Cholesterol affects **feedback inhibition** of its own synthesis by inhibiting

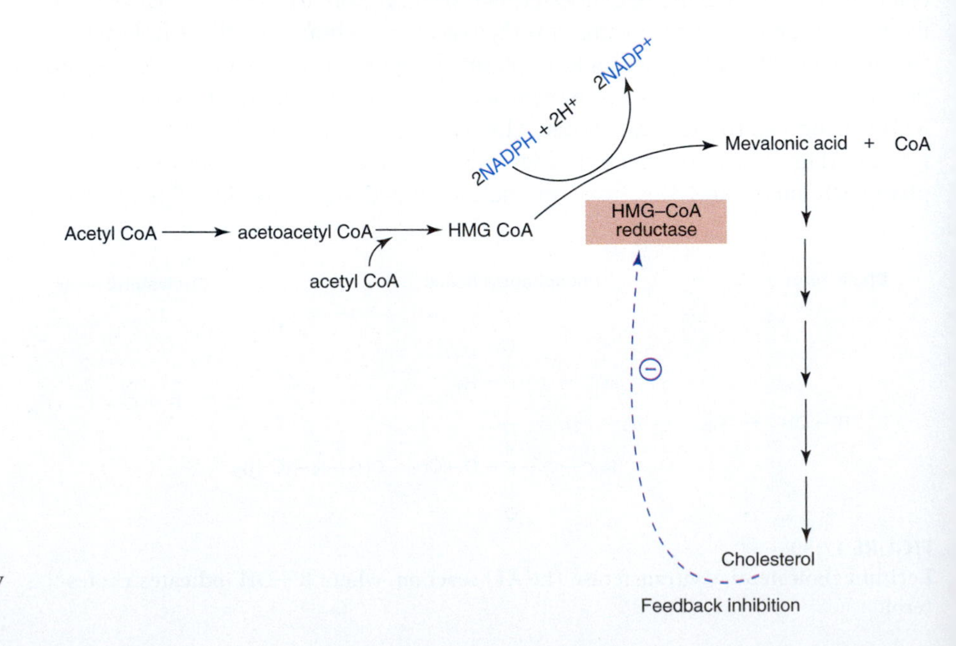

FIGURE 17.40
Summary of cholesterol synthesis indicating feedback inhibition of HMG CoA reductase by cholesterol.

the activity of preexisting HMG CoA reductase and also by promoting rapid inactivation of the enzyme by mechanisms that remain to be elucidated.

In a normal healthy adult on a low-cholesterol diet about 1300 mg of cholesterol is returned to the liver each day for disposal. This comes from cholesterol reabsorbed from the gut by means of the enterohepatic circulation and HDL, which carries cholesterol to the liver from peripheral tissues. Liver disposes of cholesterol by: (1) excretion in bile as free cholesterol and after conversion to bile salts; each day, about 250 mg of bile salts and 550 mg of cholesterol are lost from the enterohepatic circulation; (2) esterification and storage in liver as cholesterol esters; and (3) incorporation into lipoproteins (VLDL and LDL) and secretion into the circulation. On a low-cholesterol diet, liver synthesizes ~800 mg of cholesterol per day to replace bile salts and cholesterol lost from the enterohepatic circulation in the feces.

Suppression of cholesterol biosynthesis by LDL-bound cholesterol involves specific **LDL receptors** that project from the plasma membrane. About 75% of LDL catabolism occurs in liver via this LDL receptor-mediated process. The first step of the regulatory mechanism involves binding of lipoprotein LDL to these LDL receptors, thereby extracting the LDL particles from the plasma. The binding reaction is characterized by its saturability, high affinity, and high degree of specificity. The receptor recognizes only LDL and VLDL, the two plasma lipoproteins that contain **apolipoprotein E (apoE)** and apolipoprotein B-100 (apoB-100). Binding to receptors occurs at sites on the plasma membrane that contain pits coated with a protein called **clathrin,** the cholesterol-charged lipoprotein is endocytosed in the form of clathrin-coated vesicles. This process is called **receptor-mediated endocytosis.** Intracellularly, coated vesicles lose their clathrin and become endosomes (Figure 17.38). The next step involves the fusion of the endosome with a lysosome that contains numerous hydrolytic enzymes, including proteases and cholesterol esterase. In the relatively acid (approx. pH 5.0) environment produced by fusion of the endosome with a lysosome, the LDL receptor separates from LDL and returns to the cell surface. Inside the lysosome the cholesterol esters of LDL are hydrolyzed by cholesterol esterase to produce free cholesterol and a long-chain fatty acid. The protein components of the residual LDL particle undergo proteolysis by lysosomal proteases, whereupon the free amino acids enter the cell's amino acid pool. The free cholesterol then diffuses into the cytosol where, by some unknown mechanism, it inhibits the activity of HMG CoA reductase and suppresses the synthesis of HMG CoA reductase enzyme. There is evidence that cholesterol acts at the level of DNA and protein synthesis to decrease the rate of synthesis of HMG CoA reductase. At the same time, **fatty acyl CoA:cholesterol acyltransferase (ACAT)** in the endoplasmic reticulum is activated by cholesterol, promoting the formation of cholesterol esters, principally cholesterol oleate. Accumulation of intracellular cholesterol inhibits the replenishment of LDL receptors on the cell surface, a phenomenon called down-regulation, thereby blocking further uptake and accumulation of cholesterol.

The LDL receptor is a single-chain glycoprotein; numerous mutations in its gene are associated with familial hypercholesterolemia. The receptor spans the plasma membrane once with the carboxyl terminus on the cytoplasmic face and the amino terminus, which contains the LDL-binding site, extending into the extracellular space. Apoprotein B-100 and apoprotein E, which is present in IDL (intermediate density lipoprotein) and some forms of HDL, are the two proteins through which particular lipoproteins bind to the LDL receptor.

The correlation between high concentrations of plasma cholesterol, particularly LDL cholesterol, and heart attacks and strokes has led to the development of dietary and therapeutic approaches to lower blood cholesterol (see Clin. Corr. 17.2). Patients with familial (genetic) hypercholesterolemia suffer from accelerated atherosclerosis (see Clin. Corr. 17.3). In most cases, there is a lack of functional LDL receptors on the cell surface because the mutant alleles produce little or no LDL receptor protein; these patients are referred to as receptor negative. In others, the LDL receptor is synthesized and transported normally to the cell surface; an amino acid substitution or other alteration in the protein's primary structure, however, adversely affects the LDL-binding region of the receptor. As a result, there is

CLINICAL CORRELATION 17.2
Treatment of Hypercholesterolemia

Many authorities recommend screening asymptomatic individuals by measuring plasma cholesterol. A level less than 200 mg% is considered desirable, and a level over 240 mg% requires lipoprotein analysis, especially determination of LDL cholesterol. Reduction of LDL cholesterol depends on dietary restriction of cholesterol to less than 300 mg day^{-1}, of calories to attain ideal body weight, and of total fat intake to less than 30% of total calories. Approximately two-thirds of the fat should be mono- or polyunsaturated. The second line of therapy is with drugs. Cholestyramine and colestipol are bile salt-binding drugs that promote excretion of bile salts in the stool. This in turn increases the rate of hepatic bile salt synthesis and of LDL uptake by the liver. Lovastatin is an inhibitor of HMG CoA reductase. Since this enzyme is limiting for cholesterol synthesis, lovastatin decreases endogenous synthesis of cholesterol and stimulates uptake of LDL via the LDL receptor. The combination of lovastatin and cholestyramine is sometimes used for severe hyperlipidemia.

Expert Panel. Evaluation and treatment of high blood cholesterol in adults. *Arch. Intern. Med.* 148:36, 1988.

CLINICAL CORRELATION 17.3

Atherosclerosis

Atherosclerosis is the leading cause of death in Western industrialized countries. The risk of developing it is directly related to the plasma concentration of LDL cholesterol and inversely related to that of HDL cholesterol. This explains why the former is frequently called "bad" cholesterol and the latter "good" cholesterol, though chemically there is only one cholesterol. Atherosclerosis is a disorder of the arterial wall characterized by accumulation of cholesteryl esters in cells derived from the monocyte–macrophage line, smooth muscle cell proliferation, and fibrosis. The earliest abnormality is migration of blood monocytes to the subendothelium of the artery. Once there, they differentiate into macrophages. These cells accumulate cholesteryl esters derived from plasma LDL. Why these cells do not regulate cellular cholesterol stores normally is not completely understood. Some of the LDL may be taken up via pathways distinct from the classical LDL receptor pathway. For instance, receptors that mediate uptake of acetylated LDL or LDL complexed with dextran sulfate have been described and these are not regulated by cellular cholesterol content. Distortion of the subendothelium leads to platelet aggregation on the endothelial surface and release of platelet-derived mitogens such as platelet-derived growth factor (PDGF). This is thought to stimulate smooth muscle cell growth. Death of the foam cells results in the accumulation of a cellular lipid that can stimulate fibrosis. The resulting atherosclerotic plaque narrows the blood vessel and serves as the site of thrombus formation, which precipitates myocardial infarction (heart attack).

Ross, R. The pathogenesis of atherosclerosis—an update. *N. Engl. J. Med.* 314:488, 1986.

little or no binding of LDL to the cell, cholesterol is not transferred into the cell, cholesterol synthesis is not inhibited, and the blood cholesterol level increases. Another LDL-deficient group of hypercholesterolemic patients is able to synthesize the LDL receptor but has a defect in the transport mechanism that delivers the glycoprotein to its proper location on the plasma membrane. And finally, there is another subclass of genetically determined hypercholesterolemics whose LDL receptors have a defect in the cytoplasmic carboxyl terminus; they populate their cell surfaces with LDL receptors normally but are unable to internalize the LDL–LDL receptor complex due to an inability to cluster this complex in coated pits.

In specialized tissues such as the adrenal glands and ovaries, the cholesterol derived from LDL serves as a precursor to the steroid hormones, such as cortisol and estradiol, respectively. In liver, cholesterol extracted from LDL and HDL is converted into bile salts that function in intestinal fat digestion.

Cholesterol Is Excreted Primarily as Bile Acids

Bile acids are the end products of cholesterol metabolism. Primary bile acids are synthesized in hepatocytes directly from cholesterol. The most abundant bile acids in humans are derivatives of cholanic acid (Figure 17.41), that is, cholic acid and chenodeoxycholic acid (Figure 17.42). The primary bile acids are 24-carbon compounds containing two or three OH groups, and a side chain that ends in a carboxyl group that is ionized at pH 7.0 (hence the name bile salt). The carboxyl group of the bile acids is often conjugated via an amide bond to either glycine (NH_2—CH_2—COOH) or taurine (NH_2—CH_2—CH_2—SO_3H) to form glycocholic or taurocholic acid, respectively. The structure of glycocholic acid is shown in Figure 17.43.

When the primary bile acids undergo chemical reactions by microorganisms in the gut, they produce secondary bile acids that are also 24C. Examples of secondary bile acids are deoxycholic acid and lithocholic acid, which are derived from

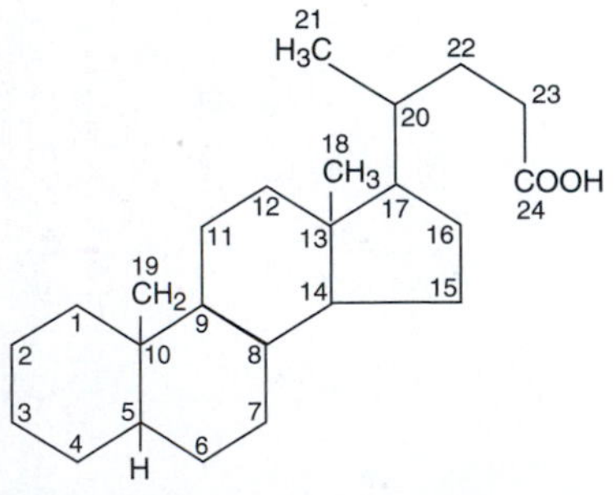

Cholic acid

Deoxycholic acid

FIGURE 17.41
Structure of cholanic acid.

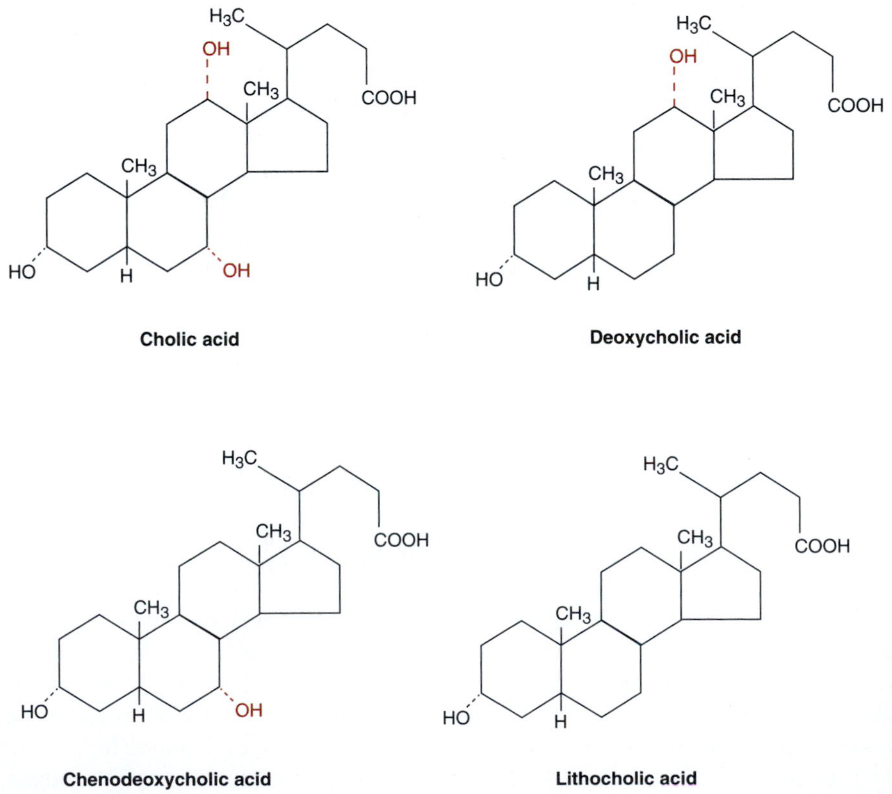

Chenodeoxycholic acid

Lithocholic acid

FIGURE 17.42
Structures of some common bile acids.

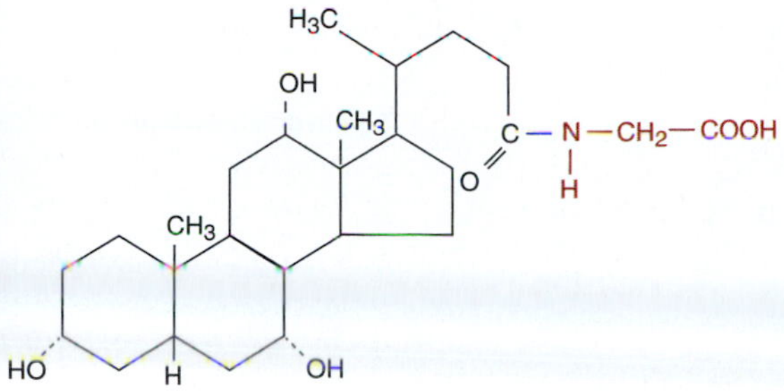

FIGURE 17.43
Structure of glycocholic acid, a conjugated bile acid.

cholic acid and chenodeoxycholic acid, respectively, by the removal of one OH group (Figure 17.42). Transformation of cholesterol to bile acids requires (1) epimerization of the 3β-OH group, (2) reduction of the C-5 double bond, (3) introduction of OH groups at C-7 **(chenodeoxycholic acid)** or at C-7 and C-12 **(cholic acid)**, and (4) conversion of the C-27 side chain into a C-24 carboxylic acid by elimination of a propyl equivalent.

Bile acids are secreted into bile canaliculi, specialized channels formed by adjacent hepatocytes. Bile canaliculi unite with bile ductules, which in turn come together to form bile ducts. The bile acids are carried to the gallbladder for storage and ultimately to the small intestine where they are excreted. The capacity of liver to produce bile acids is insufficient to meet the physiological demands, so the body relies on an efficient **enterohepatic circulation** that carries the bile acids from the intestine back to the liver several times each day. The primary bile acids, after removal of the glycine or taurine residue in the gut, are reabsorbed by an active transport process from the intestine, primarily in the ileum, and returned to the liver by way of the portal vein. Bile acids that are not reabsorbed are acted on by bacteria in the gut and converted into secondary bile acids; a portion of secondary bile acids, primarily deoxycholic acid and lithocholic acid, are reabsorbed passively in the colon and returned to the liver where they are secreted into the gallbladder. Hepatic synthesis normally produces 0.2–0.6 g of bile acids per day to replace those lost in the feces. The gallbladder pool of bile acids is 2–4 g. Because the enterohepatic circulation recycles 6–12 times each day, the total amount of bile acids absorbed per day from the intestine corresponds to 12–32 g.

Bile acids are significant in medicine for several reasons. They represent the only significant way in which cholesterol can be excreted; the carbon skeleton of cholesterol is not oxidized to CO_2 and H_2O in humans but is excreted in bile as free cholesterol and as bile acids. Bile acids prevent the precipitation of cholesterol from solution in the gallbladder. Bile acids and phospholipids function to solubilize cholesterol in bile and act as emulsifying agents to prepare dietary triacylglycerols for hydrolysis by pancreatic lipase. Bile acids play a direct role in the control of pancreatic lipase (see p. 1105) and facilitate the absorption of fat-soluble vitamins, particularly vitamin D, from the intestine.

Vitamin D Is Synthesized from an Intermediate of Cholesterol Biosynthesis

Cholesterol biosynthesis provides substrate for the photochemical production of vitamin D_3 in the skin. Metabolism and function of vitamin D_3 are discussed on page 1141. **Vitamin D_3** is a secosteroid in which the 9,10 carbon bond of the B ring of cholesterol has undergone fission (Figure 17.44). The most important supply of vitamin D_3 is that manufactured in the skin. **7-Dehydrocholesterol** is an intermediate in the pathway of cholesterol biosynthesis and is converted in the skin to provitamin D_3 by irradiation with UV rays of the sun (285–310 nm). Provitamin D_3 is biologically inert and labile and converted thermally and slowly (~36 h) to the double bond isomer by a nonenzymatic reaction to the biologically active vitamin, **cholecalciferol** (vitamin D_3). As little as 10-min exposure to sunlight each day of

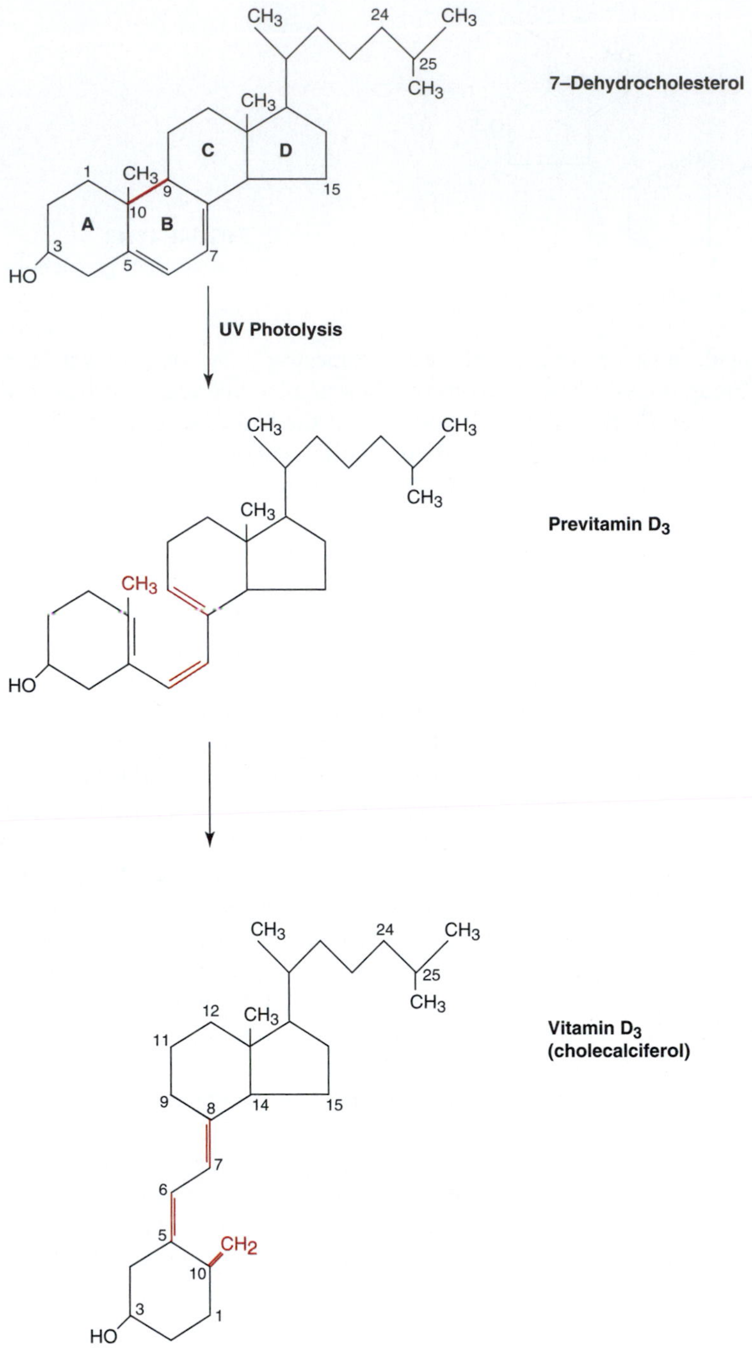

7-Dehydrocholesterol

UV Photolysis

Previtamin D_3

Vitamin D_3 (cholecalciferol)

FIGURE 17.44
Photochemical conversion of 7-dehydrocholesterol to vitamin D_3 (cholecalciferol).

the hands and face will satisfy the body's need for vitamin D. Photochemical action on the plant sterol ergosterol also provides a dietary precursor to **vitamin D_2 (calciferol)** that can satisfy the vitamin D requirement.

17.4 | SPHINGOLIPIDS

Biosynthesis of Sphingosine

Glycerol

Sphingosine

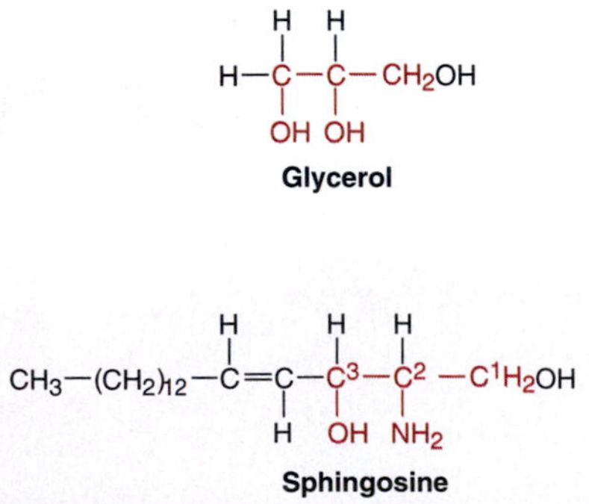

FIGURE 17.45
Comparison of structures of glycerol and sphingosine (*trans*-1,3,dihydroxy-2-amino-4-octadecene).

Sphingolipids are complex lipids whose core structure is provided by the long-chain amino alcohol **sphingosine** (Figure 17.45) (4-sphingenine or *trans*-1,3-dihydroxy-2-amino-4-octadecene). Sphingosine has two asymmetric carbon atoms (C-2 and

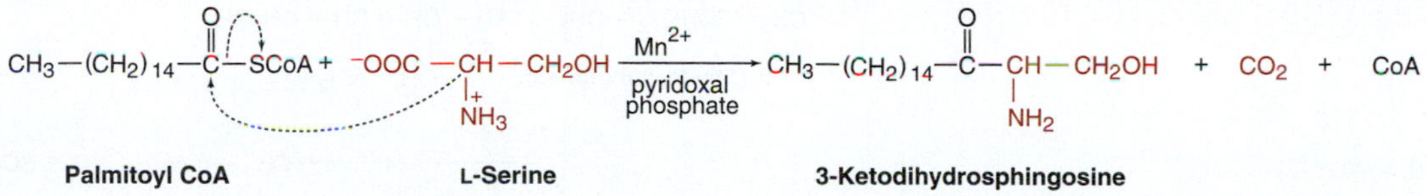

$$CH_3-(CH_2)_{14}-\overset{O}{\overset{\|}{C}}-SCoA + {}^{-}OOC-\underset{NH_3^+}{\underset{|}{CH}}-CH_2OH \xrightarrow[\text{pyridoxal phosphate}]{Mn^{2+}} CH_3-(CH_2)_{14}-\overset{O}{\overset{\|}{C}}-\underset{NH_2}{\underset{|}{CH}}-CH_2OH + CO_2 + CoA$$

Palmitoyl CoA — L-Serine — 3-Ketodihydrosphingosine

FIGURE 17.46
Formation of 3-ketodihydrosphingosine from serine and palmitoyl CoA.

$$CH_3-(CH_2)_{12}-CH_2-CH_2-\overset{O}{\overset{\|}{C}}-\underset{NH_2}{\underset{|}{CH}}-CH_2OH \xrightarrow[H^+ + NADPH \;\; NADP^+]{} CH_3-(CH_2)_{12}-CH_2-CH_2-\underset{OH}{\underset{|}{CH}}-\underset{NH_2}{\underset{|}{CH}}-CH_2OH$$

3-Ketodihydrosphingosine — Sphinganine (dihydrosphingosine)

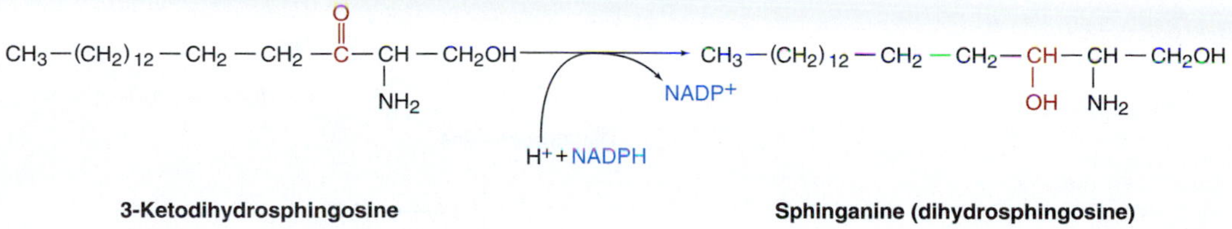

FIGURE 17.47
Conversion of 3-ketodihydrosphingosine to sphinganine.

C-3); of the four possible optical isomers, naturally occurring sphingosine is of the D-erythro form. The double bond of sphingosine has the trans configuration. The primary alcohol group at C-1 is a nucleophilic center that forms covalent bonds with sugars to form glycosphingolipids. When forming a ceramide the amino group at C-2 always bears a long-chain fatty acid (usually C_{20}–C_{26}) in amide linkage. The secondary alcohol at C-3 is always free. It is useful to appreciate the structural similarity of a part of the sphingosine molecule to the glycerol moiety of the acylglycerols (Figure 17.45).

Sphingolipids are present in blood and nearly all body tissues. The highest concentrations are found in the white matter of the central nervous system. Various sphingolipids are components of the plasma membrane of practically all cells.

Sphingosine is synthesized by way of **sphinganine (dihydrosphingosine)** from the precursors serine and palmitoyl CoA. Serine is the source of C-1, C-2, and the amino group of sphingosine, while palmitic acid provides the remaining carbon atoms. Condensation of serine and palmitoyl CoA is catalyzed by **serine palmitoyltransferase**, a pyridoxal phosphate-dependent enzyme. The driving force for the reaction is provided by both cleavage of the reactive, high-energy, thioester bond of palmitoyl CoA and the release of CO_2 from serine (Figure 17.46). Reduction of the carbonyl group in 3-ketodihydrosphingosine to produce sphinganine (Figure 17.47) requires NADPH. The insertion of the double bond into sphinganine to produce sphingosine occurs at the level of ceramide (see below).

Ceramides Are Fatty Acid Amide Derivatives of Sphingosine

Sphingosine does not occur naturally. The core structure of sphingolipids is **ceramide,** a long-chain fatty acid amide derivative of sphingosine. The fatty acid is attached to the 2-amino group of sphingosine through an amide bond (Figure 17.48). Most often the acyl group is **behenic acid,** a saturated C_{22} fatty acid, but other long-chain acyl groups can be used. There are two long-chain hydrocarbon domains in the ceramide molecule; these hydrophobic regions are responsible for the lipid character of sphingolipids.

Ceramide is synthesized from dihydrosphingosine and a long-chain fatty acyl CoA by an endoplasmic reticulum enzyme with dihydroceramide as an intermediate that is then oxidized by dehydrogenation at C-4 and C-5 (Figure 17.49). Ceramide is not a component of membrane lipids but rather is an intermediate in synthesis and catabolism of glycosphingolipids and sphingomyelin. Structures of prominent sphingolipids of humans are presented in Figure 17.50 in diagrammatic form.

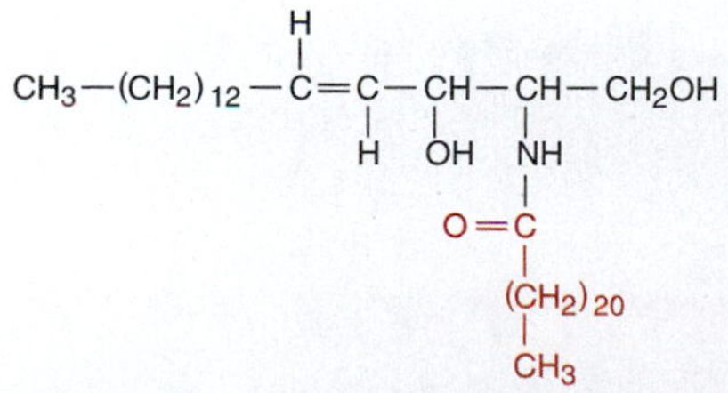

$$CH_3-(CH_2)_{12}-\overset{H}{\overset{|}{C}}=\underset{H}{\underset{|}{C}}-\underset{OH}{\underset{|}{CH}}-\underset{NH-C(=O)-(CH_2)_{20}-CH_3}{\underset{|}{CH}}-CH_2OH$$

FIGURE 17.48
Structure of a ceramide (*N*-acylsphingosine).

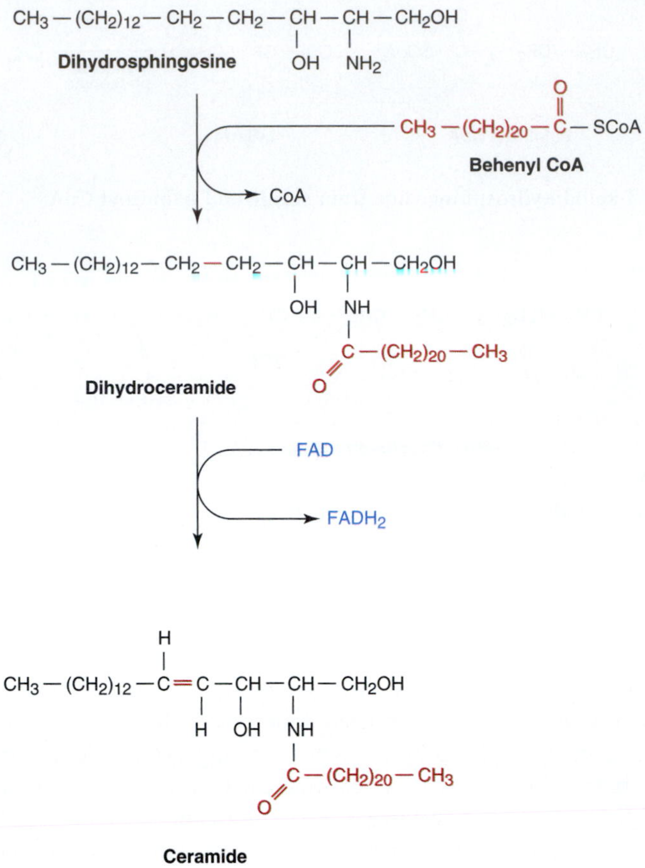

$CH_3-(CH_2)_{12}-CH_2-CH_2-CH(OH)-CH(NH_2)-CH_2OH$

Dihydrosphingosine

$CH_3-(CH_2)_{20}-C(=O)-SCoA$ **Behenyl CoA** → CoA

$CH_3-(CH_2)_{12}-CH_2-CH_2-CH(OH)-CH(NH-C(=O)-(CH_2)_{20}-CH_3)-CH_2OH$

Dihydroceramide

FAD → $FADH_2$

$CH_3-(CH_2)_{12}-CH=CH-CH(OH)-CH(NH-C(=O)-(CH_2)_{20}-CH_3)-CH_2OH$

Ceramide

FIGURE 17.49
Formation of ceramide from dihydrosphingosine.

Sphingomyelin Is the Only Sphingolipid Containing Phosphorus

Sphingomyelin, a major component of membranes of nervous tissue, is the only sphingolipid that is a phospholipid. In sphingomyelin the primary alcohol at C-1 of sphingosine is esterified to choline, through a phosphodiester bridge of the kind that occurs in the acyl glycerophospholipids, and the amino group of sphingosine is attached to a long-chain fatty acid by an amide bond. Sphingomyelin is therefore a ceramide phosphocholine. It contains one negative and one positive charge so that it is neutral at physiological pH (Figure 17.51). The most common fatty acids in sphingomyelin are palmitic (16:0), stearic (18:0), lignoceric (24:0), and nervonic acid (24:1). The sphingomyelin of myelin contains predominantly longer chain fatty acids, mainly lignoceric and nervonic, whereas that of gray matter contains largely stearic acid. Excessive accumulation of sphingomyelin occurs in Niemann–Pick disease.

Sphingomyelin Is Synthesized from a Ceramide and Phosphatidylcholine

Conversion of ceramide to sphingomyelin by **sphingomyelin synthase** involves transfer of a phosphocholine moiety from phosphatidylcholine (lecithin) (Figure 17.52).

Glycosphingolipids Usually Contain a Galactose or Glucose Unit

The principal glycosphingolipid classes are cerebrosides, sulfatides, globosides, and gangliosides. In glycolipids the polar head group is attached to sphingosine via the glycosidic linkage with a sugar molecule rather than a phosphate ester bond, as in phospholipids.

Sphingomyelin

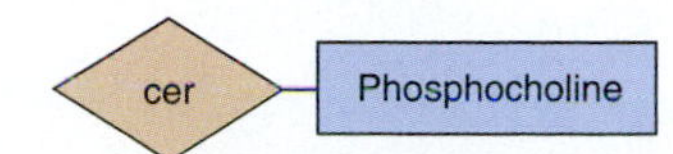

Neutral sphingolipids

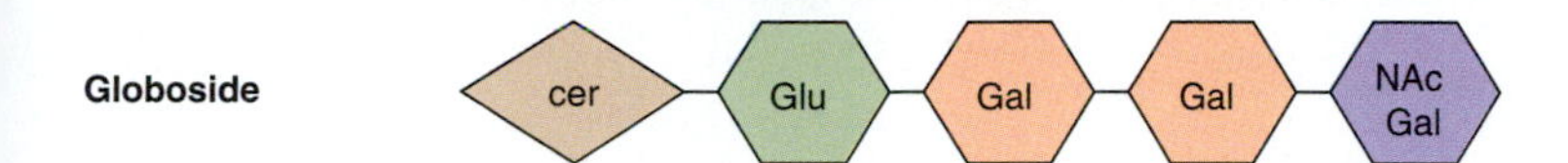

FIGURE 17.50
Structures of some common sphingolipids in diagrammatic form.
Cer, ceramide; Glu, glucose; Gal, galactose; NAcGal, *N*-acetylgalactosamine; and NANA, *N*-acetylneuraminic acid (sialic acid).

$$CH_3-(CH_2)_{12}-CH{=}CH-CH(OH)-CH(NH-C(=O)-(CH_2)_{16}-CH_3)-CH_2-O-P(=O)(O^-)-O-CH_2-CH_2-\overset{+}{N}(CH_3)_3$$

FIGURE 17.51
Structure of sphingomyelin.

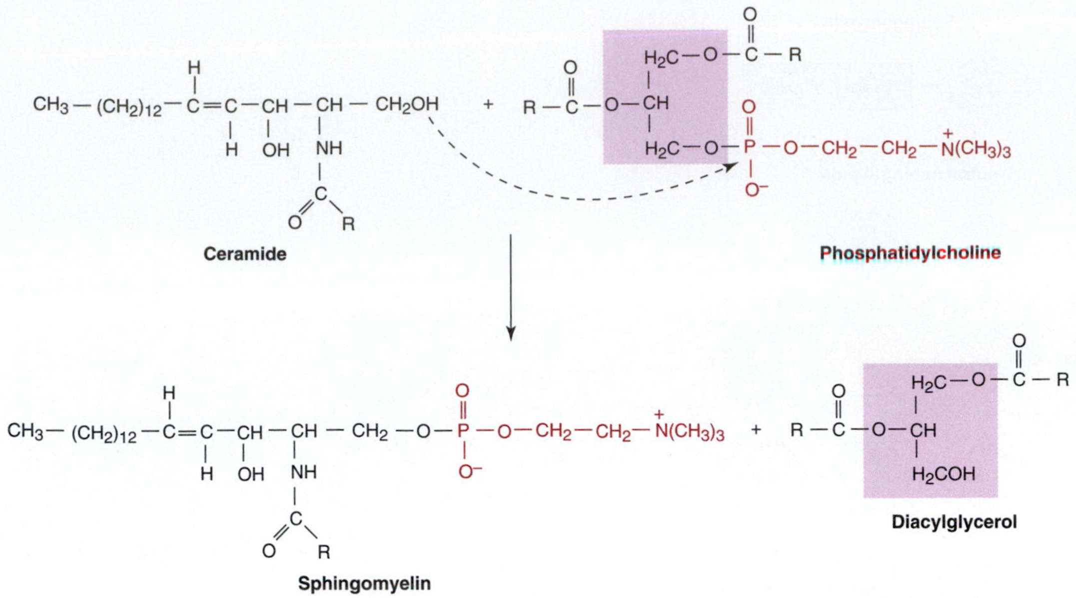

Ceramide + Phosphatidylcholine → Sphingomyelin + Diacylglycerol

FIGURE 17.52
Sphingomyelin synthesis from ceramide and phosphatidylcholine.

Cerebrosides Are Glycosylceramides

Cerebrosides are ceramide monohexosides; the most common are **galactocerebroside** and **glucocerebroside.** Unless specified otherwise, the term cerebroside usually refers to galactocerebroside, also called **"galactolipid."** In Figure 17.53 note that the monosaccharide units are attached at C-1 of the sugar moiety to the C-1 position of ceramide, and the anomeric configuration of the glycosidic bond between ceramide and hexose in both galactocerebroside and glucocerebroside is β. Most galactocerebroside in healthy individuals is found in the brain. Moderately increased amounts of galactocerebroside accumulate in the white matter in Krabbe's disease, also called globoid leukodystrophy, a deficiency in the lysosomal enzyme galactocerebrosidase.

Glucocerebroside (glucosylceramide) is not normally a component of membranes but is an intermediate in the synthesis and degradation of more complex glycosphingolipids (see Figure 17.54). However, 100-fold increases or more in the glucocerebroside content of spleen and liver occur in the genetic lipid storage

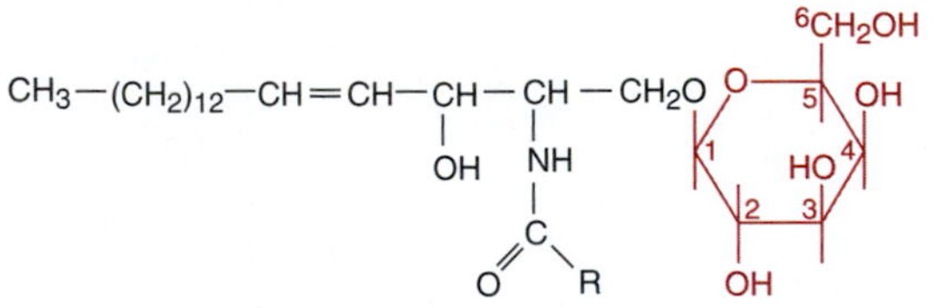

FIGURE 17.53
Structure of galactocerebroside (galactolipid).

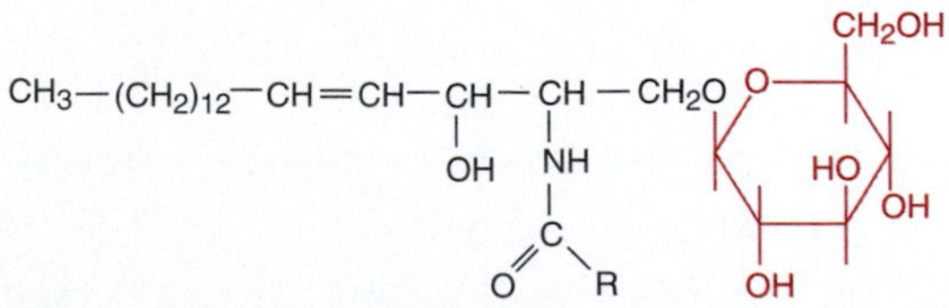

FIGURE 17.54
Structure of glucocerebroside.

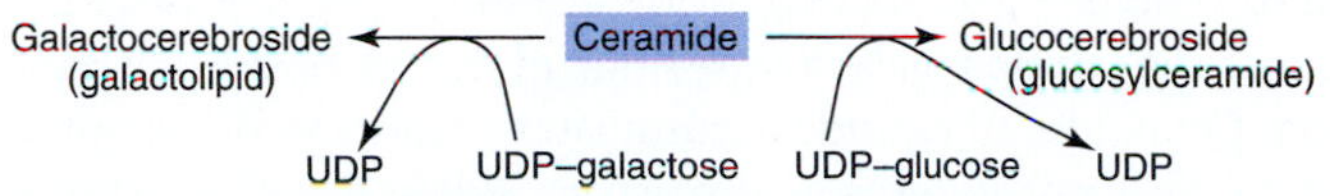

FIGURE 17.55
Synthesis of galacto- and glucocerebrosides.

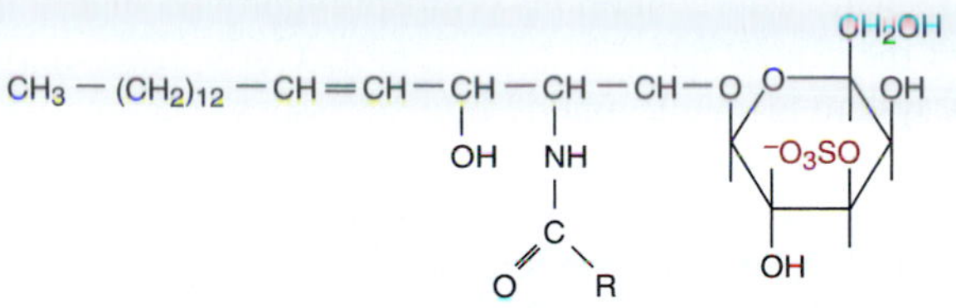

FIGURE 17.56
Structure of galactocerebroside sulfate (sulfolipid).

disorder called Gaucher's disease, which results from a deficiency of lysosomal **glucocerebrosidase.**

Galactocerebroside and glucocerebroside are synthesized from ceramide and the activated nucleotide sugars UDP-galactose and UDP-glucose, respectively. The enzymes that catalyze these reactions, **glucosyltransferase**, and **galactosyltransferase,** are associated with the endoplasmic reticulum (Figure 17.55). In some tissues, synthesis of glucocerebroside (glucosylceramide) proceeds by glucosylation of sphingosine catalyzed by glucosyltransferase:

$$\text{sphingosine} + \text{UDP-glucose} \longrightarrow \text{glucosylsphingosine} + \text{UDP}$$

followed by fatty acylation:

$$\text{glucosylsphingosine} + \text{stearoyl CoA} \longrightarrow \text{glucocerebroside} + \text{CoASH}$$

Sulfatide Is a Sulfuric Acid Ester of Galactocerebroside

Sulfatide, or **sulfogalactocerebroside,** is a sulfuric acid ester of galactocerebroside. Galactocerebroside 3-sulfate is the major sulfolipid in brain, accounting for approximately 15% of the lipids of white matter (see Figure 17.56). Galactocerebroside sulfate is synthesized from galactocerebroside and 3′-phosphoadenosine 5′-phosphosulfate (PAPS) catalyzed by sulfotransferase:

$$\text{galactocerebroside} + \text{PAPS} \longrightarrow \text{PAP} + \text{galactocerebroside 3-sulfate}$$

The structure of PAPS, sometimes referred to as "activated sulfate," is indicated in Figure 17.57. Large quantities of sulfatide accumulate in the central nervous system in metachromatic leukodystrophy due to a deficiency of lysosomal sulfatase.

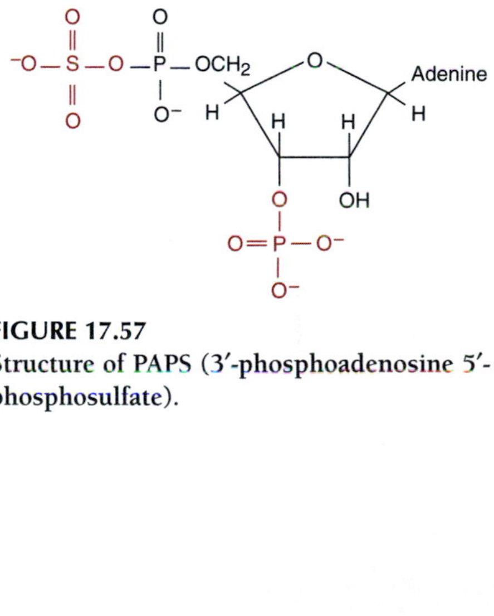

FIGURE 17.57
Structure of PAPS (3′-phosphoadenosine 5′-phosphosulfate).

Globosides are Ceramide Oligosaccharides

Globosides are cerebrosides that contain two or more sugar residues, usually galactose, glucose, or *N*-acetylgalactosamine. The ceramide oligosaccharides are neutral compounds and contain no free amino groups. Lactosylceramide is a component of the erythrocyte membrane (Figure 17.58). Another prominent globoside is **ceramide**

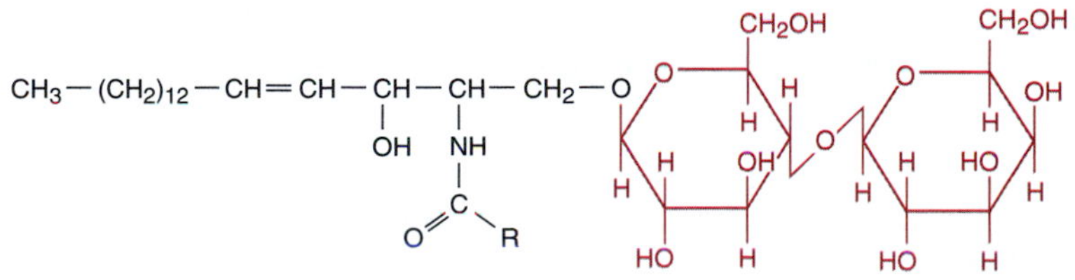

FIGURE 17.58
Structure of ceramide-β-glc-(4 $\longrightarrow$ 1)-β-gal (lactosylceramide).

trihexoside or ceramide galactosyllactoside: ceramide-β-glc($4 \rightarrow 1$)-β-gal($4 \rightarrow 1$)-α-gal. Note that the terminal galactose residue of this globoside has the α-anomeric configuration. Ceramide trihexoside accumulates in kidneys of patients with Fabry's disease who are deficient in lysosomal α-galactosidase A.

Gangliosides Contain Sialic Acid

Gangliosides are sialic acid-containing glycosphingolipids highly concentrated in ganglion cells of the central nervous system, particularly in the nerve endings. The central nervous system is unique among human tissues because more than one-half of the sialic acid is in ceramide–lipid bound form, with the remainder of the sialic acid occurring in the oligosaccharides of glycoproteins. Lesser amounts of gangliosides are present in the plasma membranes of cells of most extraneural tissues where they account for less than 10% of the total sialic acid.

Neuraminic acid (abbreviated Neu) is present in gangliosides, glycoproteins, and mucins. The amino group of neuraminic acid occurs most often as the *N*-acetyl derivative, and the resulting structure is called **N-acetylneuraminic acid** or **sialic acid,** commonly abbreviated **NANA** (see Figure 17.59). The OH group on C-2 occurs most often in the α-anomeric configuration and the linkage between NANA and the oligosaccharide ceramide always involves the OH group on position 2 of *N*-acetylneuraminic acid. Structures of some common gangliosides are indicated in Table 17.3. The principal gangliosides in brain are G_{M1}, G_{D1a}, G_{D1b}, and G_{T1b}. Nearly all of the gangliosides of the body are derived from the family of compounds originating with glucosylceramide. In the nomenclature of the sialoglycosphingolipids, the letter G refers to the name ganglioside. The subscripts M, D, T, and Q indicate mono-, di-, tri-, and quatra(tetra)-sialic acid-containing gangliosides and subscripts 1, 2, and 3 designate the carbohydrate sequence that is attached to ceramide as indicated as follows: 1, Gal-GalNAc-Gal-Glc-ceramide; 2, GalNAc-Gal-Glc-ceramide; and 3, Gal-Glc-ceramide. Consider the nomenclature of the Tay–Sachs ganglioside; the designation G_{M2} denotes the ganglioside structure shown in Table 17.3.

A specific ganglioside on intestinal mucosal cells binds **cholera toxin,** a protein of MW 84,000, secreted by the pathogen *Vibrio cholerae.* The toxin stimulates the secretion of chloride ions into the gut lumen, resulting in the severe diarrhea

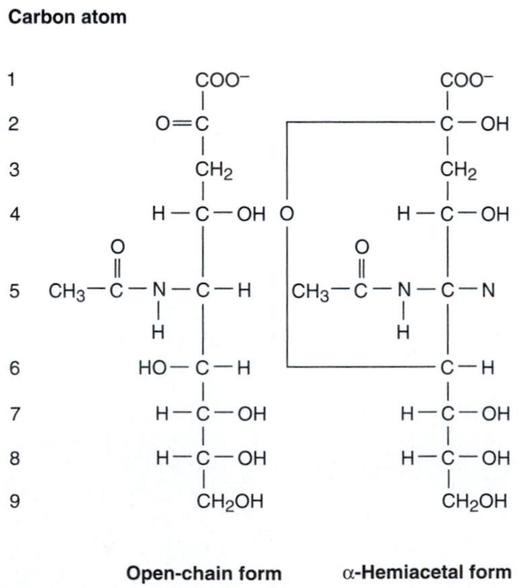

Carbon atom	Open-chain form	α-Hemiacetal form
1	COO^-	COO^-
2	O=C	C—OH
3	CH_2	CH_2
4	H—C—OH	H—C—OH
5	CH_3—C(=O)—N(H)—C—H	CH_3—C(=O)—N(H)—C—N
6	HO—C—H	C—H
7	H—C—OH	H—C—OH
8	H—C—OH	H—C—OH
9	CH_2OH	CH_2OH

FIGURE 17.59
Structure of *N*-acetylneuraminic acid (NANA).

TABLE 17.3 Structures of Some Common Gangliosides

Code Name	*Chemical Structure*
G_{M3}	Galβ → 4Glcβ → Cer 3 (on Galβ) ↑ αNANA
G_{M2}	GalNAcβ → 4Galβ → 4Glcβ → Cer 3 (on Galβ) ↑ αNANA
G_{M1}	Galβ → 3GalNAcβ → 4Galβ → 4Glcβ → Cer 3 (on Galβ) ↑ αNANA
G_{D1a}	Galβ → 3GalNAcβ → 4Galβ → 4Glcβ → Cer 3 (on terminal Galβ) 3 (on internal Galβ) ↑ ↑ αNANA αNANA
G_{D1b}	Galβ → 3GalNAcβ → 4Galβ → 4Glcβ → Cer 3 (on internal Galβ) ↑ αNANA8 ← αNANA
G_{T1a}	Galβ → 3GalNAcβ → 4Galβ → 4Glcβ → Cer 3 (on terminal Galβ) 3 (on internal Galβ) ↑ ↑ αNANA8 ← αNANA αNANA
G_{T1b}	Galβ → 3GalNAcβ → 4Galβ → 4Glcβ → Cer 3 (on terminal Galβ) 3 (on internal Galβ) ↑ ↑ αNANA αNANA8 ← αNANA
G_{Q1b}	Galβ → 3GalNAcβ → 4Galβ → 4Glcβ → Cer 3 (on terminal Galβ) 3 (on internal Galβ) ↑ ↑ αNANA8 ← αNANA αNANA8 ← αNANA

characteristic of cholera. Two kinds of subunits, A and B, comprise the cholera toxin; there is one A subunit (28,000 Da) and five B subunits (~11,000 Da each). After binding to the cell surface membrane through a domain on a B subunit, the active subunit A passes into the cell. There it acts as an **ADP-ribosyltransferase** and transfers ADP-ribose of NAD$^+$ onto the $G_{\alpha s}$ subunit of a G-protein on the cytoplasmic side of the cell membrane (see p. 1094). This leads to activation of adenylate cyclase. The cAMP generated stimulates chloride ion transport and produces diarrhea. The choleragenoid domain, as the B subunits are called, binds to the ganglioside G_{M1} that has the structure shown in Table 17.3.

Gangliosides may also be receptors for other toxins, such as **tetanus toxin,** and certain viruses, such as the influenza viruses. It is also speculated that gangliosides play an informational role in cell–cell interactions by providing specific recognition determinants on the surface of cells.

Several lipid storage disorders involve accumulation of sialic acid-containing glycosphingolipids. The two most common **gangliosidoses** involve the storage of the gangliosides G_{M1} (G_{M1} gangliosidosis) and G_{M2} (Tay–Sachs disease). The G_{M1} gangliosidosis is an autosomal recessive metabolic disease characterized by impaired psychomotor function, mental retardation, hepatosplenomegaly, and death within the first few years of life. The massive cerebral and visceral accumulation of G_{M1} ganglioside is due to a marked deficiency of β-galactosidase.

Sphingolipidoses Are Lysosomal Storage Diseases with Defects in the Catabolic Pathway for Sphingolipids

Sphingolipids are normally degraded within lysosomes of phagocytic cells, particularly the histiocytes or macrophages of the reticuloendothelial system located primarily in liver, spleen, and bone marrow. Degradation of the sphingolipids by visceral organs begins with the engulfment of the membranes of white cells and erythrocytes that are rich in lactosylceramide (Cer-Glc-Gal) and hematoside (Cer-Glc-Gal-NANA). In the brain, the majority of the cerebroside-type lipids are gangliosides. Particularly during the neonatal period, ganglioside turnover in the central nervous system is extensive so that glycosphingolipids are rapidly broken down and resynthesized. Sphingolipid catabolism is summarized in Figure 17.60. Note that among the sphingolipids that comprise this pathway, there occurs a sulfate ester (in sulfolipid or sulfogalactolipid); *N*-acetylneuraminic acid groups (in the gangliosides); an α-linked galactose residue (in ceramide trihexoside); several β-galactosides (in galactocerebroside and G_{M1}); the ganglioside G_{M2}, which terminates

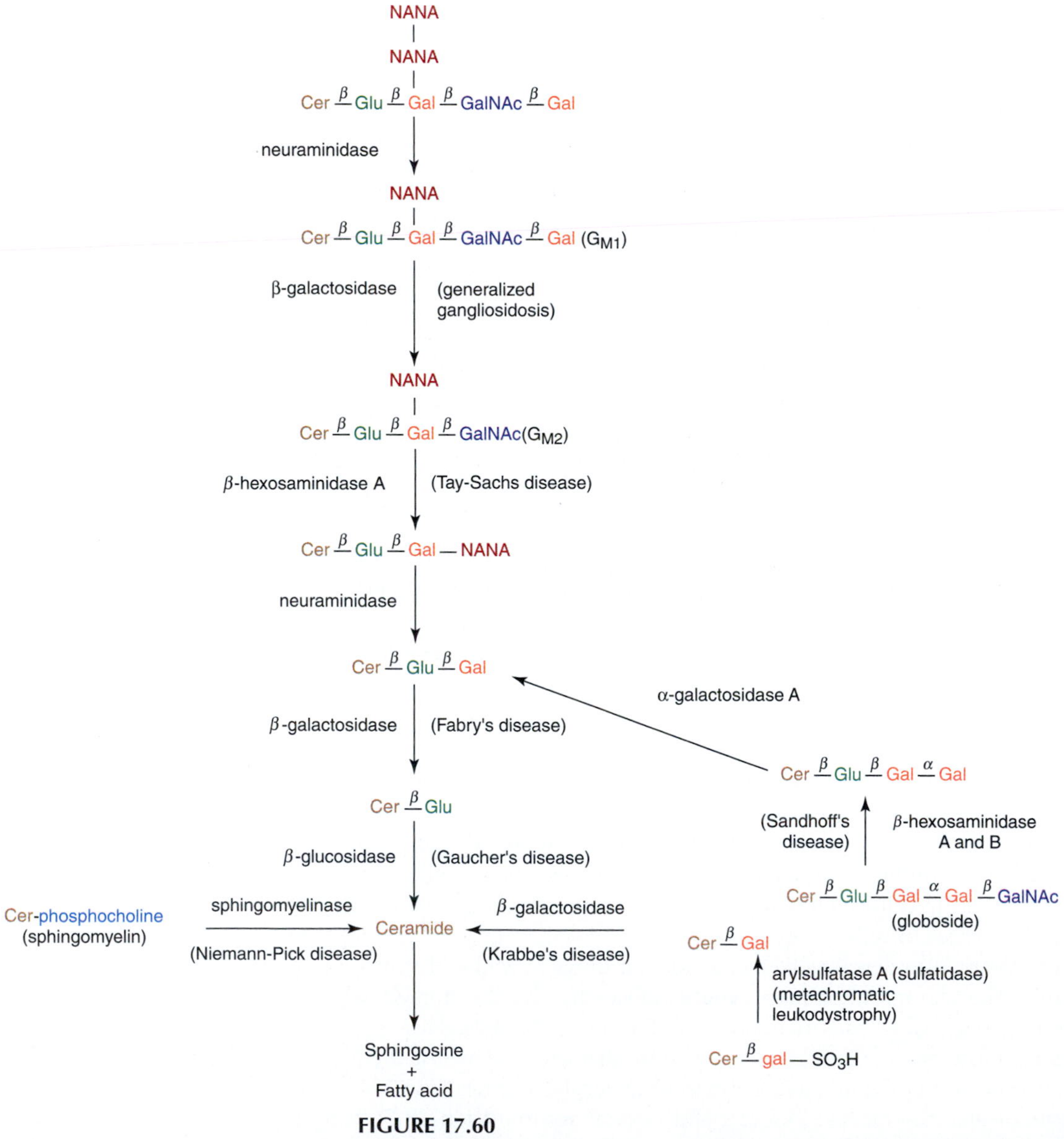

FIGURE 17.60
Summary of the pathways for catabolism of sphingolipids by lysosomal enzymes. The genetically determined enzyme deficiency diseases are indicated in parentheses.

in a β-linked *N*-acetylgalactosamine unit; and glucocerebroside, which is composed of a single glucose residue attached to ceramide through a β linkage. The phosphodiester bond in sphingomyelin is broken to produce ceramide, which is turn is converted to sphingosine by the cleavage of an amide bond to a long-chain fatty acid. This overall pathway requires a series of enzymes that cleave specific bonds in the compounds including α- and β-galactosidases, a β-glucosidase, a neuraminidase, hexosaminidase, sphingomyelin-specific phosphodiesterase (sphingomyelinase), a sulfate esterase (sulfatase), and a ceramide-specific amidase (ceramidase). There are six important features of the sphingolipid catabolic pathway. (1) All the reactions occur within lysosomes; that is, the enzymes of the pathway are contained in lysosomes. (2) The enzymes are hydrolases; therefore one of the substrates in each reaction is water. (3) The pH optimum of each of the hydrolases is in the acid range, pH 3.5–5.5. (4) Most of the enzymes are relatively stable and occur as isoenzymes. For example, hexosaminidase occurs in two forms: hexosaminidase A (HexA) and hexosaminidase B (HexB). (5) The hydrolases of the sphingolipid pathway are glycoprotein in character and often occur firmly bound to the lysosomal membrane. (6) The pathway is composed of intermediates that differ by only one sugar molecule, a sulfate group, or a fatty acid residue. The substrates are converted to products by the sequential, stepwise removal of constituents such as sugars and sulfate, by hydrolytic, irreversible reactions.

In most cases, sphingolipid catabolism functions smoothly, and all of the various complex glycosphingolipids and sphingomyelin are degraded to the level of their basic building blocks, namely, sugars, sulfate, fatty acid, phosphocholine, and sphingosine. However, when the activity of one of the enzymes is markedly reduced due to a genetic error, then the substrate for the defective or missing enzyme accumulates and is deposited within the lysosomes of the tissue responsible for the catabolism of that sphingolipid. For most reactions in Figure 17.60, patients have been identified who lack the enzyme that normally catalyzes that reaction. These disorders, called **sphingolipidoses,** are summarized in Table 17.4.

TABLE 17.4 Sphingolipid Storage Diseases of Humans

Disorder	*Principal Signs and Symptoms*	*Principal Storage Substance*	*Enzyme Deficiency*
1. Tay–Sachs disease	Mental retardation, blindness, cherry red spot on macula, death between second and third year	Ganglioside G_{M2}	Hexosaminidase A
2. Gaucher's disease	Liver and spleen enlargement, erosion of long bones and pelvis, mental retardation in infantile form only	Glucocerebroside	Glucocerebrosidase
3. Fabry's disease	Skin rash, kidney failure, pain in lower extremities	Ceramide trihexoside	α-Galactosidase A
4. Niemann–Pick disease	Liver and spleen enlargement, mental retardation	Sphingomyelin	Sphingomyelinase
5. Globoid leukodystrophy (Krabbe's disease)	Mental retardation, absence of myelin	Galactocerebroside	Galactocerebrosidase
6. Metachromatic leukodystrophy	Mental retardation, nerves stain yellowish brown with cresyl violet dye (metachromasi)	Sulfatide	Arylsulfatase A
7. Generalized gangliosidosis	Mental retardation, liver enlargement, skeletal involvement	Ganglioside G_{M1}	G_{M1} ganglioside: β-galactosidase
8. Sandhoff–Jatzkewitz disease	Same as 1; disease has more rapidly progressing course	G_{M2} ganglioside, globoside	Hexosaminidase A and B
9. Fucosidosis	Cerebral degeneration, muscle spasticity, thick skin	Pentahexosylfucoglycolipid	α-L-Fucosidase

Common features of **lipid storage diseases** are: (1) usually only a single sphingolipid accumulates in the involved organs; (2) the ceramide portion is shared by the various storage lipids; (3) the rate of biosynthesis of the accumulating lipid is normal; (4) a catabolic enzyme is missing in each of these disorders; and (5) the enzyme deficiency occurs in all tissues.

Diagnostic Enzyme Assays for Sphingolipidoses

Diagnosis of a given sphingolipidosis can be made from biopsy of the involved organ, usually bone marrow, liver, or brain, on morphologic grounds on the basis of the highly characteristic appearance of the storage lipid within lysosomes. Assay of enzyme activity confirms the diagnosis of a particular disease. For most of the diseases, peripheral leukocytes, cultured skin fibroblasts, and chorionic villi express the relevant enzyme deficiency and can be used as a source of enzyme for diagnostic purposes. In some cases (e.g., Tay–Sachs disease) serum and even tears are a source of enzyme for the diagnosis of a lipid storage disorder. Sphingolipidoses, for the most part, are autosomal recessive, with the disease occurring only in homozygotes with a defect in both alleles. Enzyme assays can identify carriers or heterozygotes.

In **Niemann–Pick disease,** the deficient enzyme is **sphingomyelinase,** which normally catalyzes the reaction shown in Figure 17.61. Sphingomyelin, radiolabeled in the methyl groups of choline with carbon-14, provides a useful substrate for determining sphingomyelinase activity. Extracts of white blood cells from healthy controls will hydrolyze the labeled substrate and produce the water-soluble product, phosphocholine. Extraction of the final incubation medium with an organic solvent such as chloroform will result in radioactivity in the upper, aqueous phase; the unused, lipid-like substrate sphingomyelin will be found in the chloroform phase. On the other hand, if the white blood cells were derived from a patient with Niemann–Pick disease, then after incubation with labeled substrate and extraction with chloroform, little or no radioactivity (i.e., phosphocholine) would be found in the aqueous phase and the diagnosis would be confirmed.

Another disease that can be diagnosed by use of an artificial substrate is **Tay–Sachs disease,** the most common form of **G_{M2} gangliosidosis.** In this fatal disorder the ganglion cells of the cerebral cortex are swollen and lysosomes are engorged with the acidic lipid, G_{M2} ganglioside. This results in a loss of ganglion cells, proliferation of glial cells, and demyelination of peripheral nerves. The pathognomonic finding is a cherry red spot on the macula caused by swelling and necrosis of ganglion cells in the eye. In Tay–Sachs disease, the artificial substrate 4-

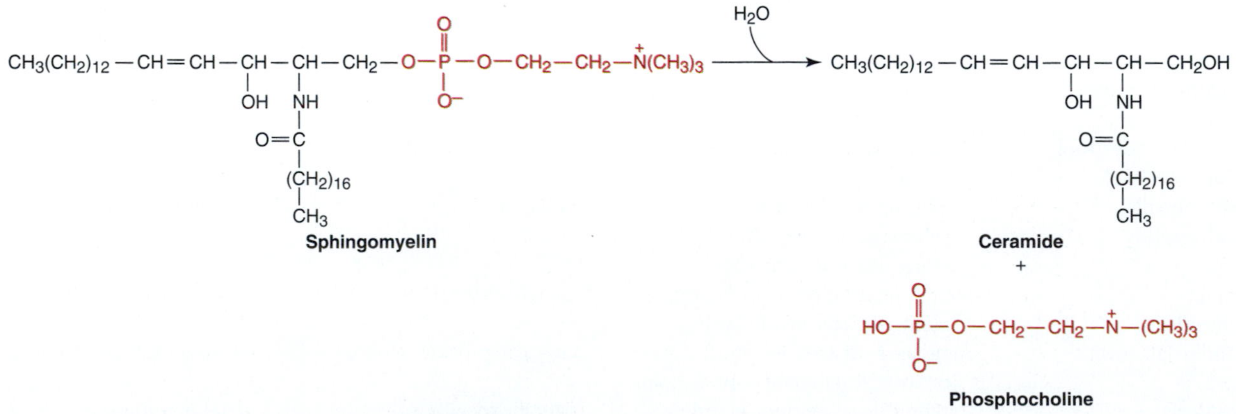

$$CH_3(CH_2)_{12}-CH{=}CH-\underset{OH}{CH}-\underset{NH-C(=O)-(CH_2)_{16}-CH_3}{CH}-CH_2-O-\overset{O}{\overset{\|}{P}}(O^-)-O-CH_2-CH_2-\overset{+}{N}(CH_3)_3 \xrightarrow{H_2O} CH_3(CH_2)_{12}-CH{=}CH-\underset{OH}{CH}-\underset{NH-C(=O)-(CH_2)_{16}-CH_3}{CH}-CH_2OH + HO-\overset{O}{\overset{\|}{P}}(O^-)-O-CH_2-CH_2-\overset{+}{N}-(CH_3)_3$$

Sphingomyelin → Ceramide + Phosphocholine

FIGURE 17.61
Sphingomyelinase reaction.

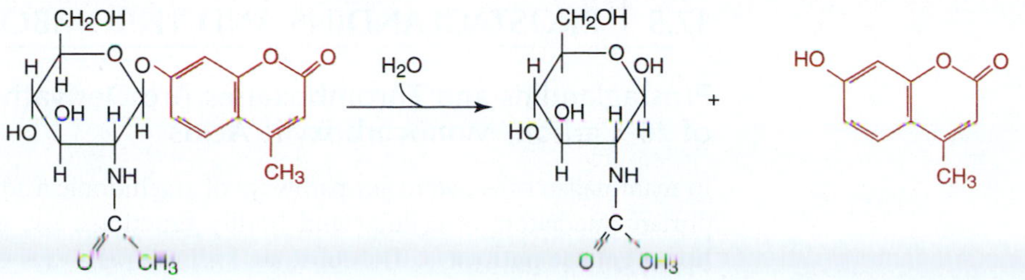

4-Methylumbelliferyl-β-D-*N*-acetylglucosamine → (H_2O) → *N*-Acetylglucosamine + 4-Methylumbelliferone (fluorescent in alkaline medium)

FIGURE 17.62
β-Hexosaminidase reaction.

methylumbelliferyl-β-*N*-acetylglucosamine is used to confirm the diagnosis. The compound is hydrolyzed by hexosaminidase A, the deficient lysosomal hydrolase, to produce the intensely fluorescent product 4-methylumbelliferone (Figure 17.62). Unfortunately, the diagnosis may be confused by the presence of hexosaminidase B in tissue extracts and body fluids. This enzyme is not deficient in the Tay–Sachs patient and will hydrolyze the test substrate. The problem is usually resolved by taking advantage of the relative heat lability of hexosaminidase A and heat stability of hexosaminidase B. The tissue extract or serum specimen to be tested is first heated at 55°C for 1 h and then assayed for hexosaminidase activity. The amount of heat-labile activity is a measure of hexosaminidase A, and this value is used in making the diagnosis.

Enzyme assays of serum or extracts of tissues, peripheral leukocytes, and fibroblasts are useful in heterozygote detection. Once carriers of a lipid storage disease have been identified, or if there has been a previously affected child in a family, the pregnancies at risk for these diseases can be monitored. All nine of the lipid storage disorders are transmitted as recessive genetic abnormalities. In all but one the allele is carried on an autosomal chromosome. The exception is **Fabry's disease,** which is X-chromosome linked. In all of these conditions statistically one of four fetuses will be homozygous (or hemizygous in Fabry's disease), two fetuses will be carriers, and one will be completely normal. The enzyme assays have been used to detect affected fetuses and carriers *in utero,* using cultured fibroblasts obtained by amniocentesis as a source of enzyme.

Except for **Gaucher's disease,** there is no therapy for the sphingolipidoses; the role of medicine at present is prevention through genetic counseling based on enzyme assays of the type discussed above. A discussion of the diagnosis and therapy of Gaucher's disease is presented in Clinical Correlation 17.4.

CLINICAL CORRELATION 17.4
Diagnosis of Gaucher's Disease in an Adult

Gaucher's disease is an inherited disease of lipid catabolism that results in deposition of glucocerebroside in macrophages of the reticuloendothelial system. Because of the large numbers of macrophages in the spleen, bone marrow, and liver, hepatomegaly, splenomegaly and its sequelae (thrombocytopenia or anemia), and bone pain are the most common signs and symptoms of the disease.

Gaucher's disease results from a deficiency of glucocerebrosidase. Although this enzyme deficiency is inherited, different clinical patterns are observed. Some patients suffer severe neurologic deficits as infants, while others do not exhibit symptoms until adulthood. The diagnosis can be made by assaying leukocytes or fibroblasts for their ability to hydrolyze the β-glycosidic bond of artificial substrates (β-glucosidase activity) or of glucocerebroside (glucocerebrosidase activity). Gaucher's disease has been treated with regular infusions of purified glucocerebrosidase.

Friedman, B. A., Vaddi, K., Preston, E. M., Cataldo, J. R., and McPherson, J. M. A comparison of the pharmacological properties of carbohydrate remodeled recombinant and placental-derived β-glucocerebrosidase: implications for clinical efficacy in treatment of Gaucher disease. *Blood* 93:2807, 1999.

17.5 | PROSTAGLANDINS AND THROMBOXANES

Prostaglandins and Thromboxanes Are Derivatives of 20-Carbon, Monocarboxylic Acids

In mammalian cells two major pathways of arachidonic acid metabolism produce important mediators of cellular and bodily functions: the **cyclooxygenase** and the **lipoxygenase pathways.** The substrate for both pathways is unesterified arachidonic acid. The cyclooxygenase pathway leads to a series of compounds including prostaglandins and thromboxanes. Prostaglandins were discovered through their ability to promote contraction of intestinal and uterine muscle and the lowering of blood pressure. Although the complexity of their structures and the diversity of their sometimes conflicting functions often create a sense of frustration, the potent pharmacological effects of the prostaglandins have made them important in human biology and medicine. With the exception of the red blood cell, the prostaglandins are produced and released by nearly all mammalian cells and tissues. Unlike most hormones, prostaglandins are not stored in cells but are synthesized and released immediately.

The three major classes of **prostaglandins** are the **A, E,** and **F** series. Structures of the more common prostaglandins A, E, and F are shown in Figure 17.63. All are related to the prostanoic acid (Figure 17.64). Note that the prostaglandins contain a multiplicity of functional groups; for example, PGE_2 contains a carboxyl group, a β-hydroxyketone, a secondary alkylic alcohol, and two carbon–carbon double bonds. The three classes are distinguished on the basis of the functional groups about the cyclopentane ring: the E series contains a β-hydroxyketone, the F series are 1,3-diols, and those in the A series are α,β-unsaturated ketones. The subscript numerals 1, 2, and 3 refer to the number of double bonds in the side chains. The subscript α refers to the configuration of the C-9 OH group: an α-hydroxyl group projects "down" from the plane of the ring.

The most important dietary precursor of the prostaglandins is linoleic acid (18:2), an essential fatty acid. In adults about 10 g of linoleic acid is ingested daily.

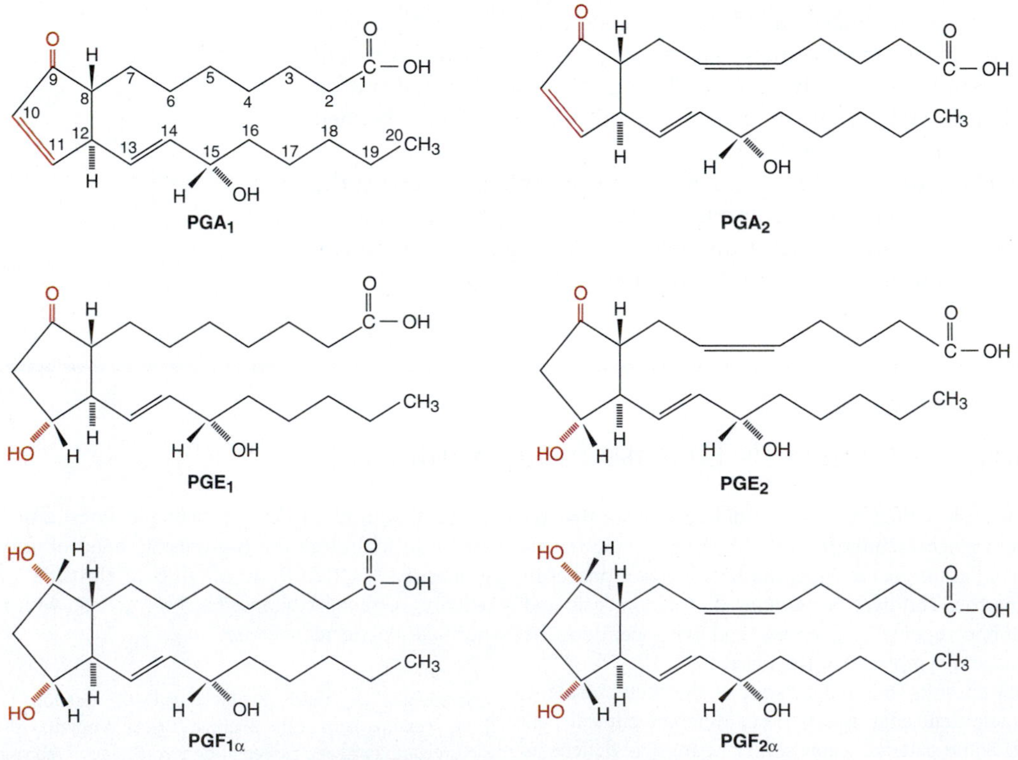

PGA_1 PGA_2

PGE_1 PGE_2

$PGF_{1\alpha}$ $PGF_{2\alpha}$

FIGURE 17.63
Structures of the major prostaglandins.

Only a very small part of this total intake is converted by elongation and desaturation in liver to arachidonic acid (eicosatetraenoic acid) and to some extent also to dihomo-γ-linoleic acid. Since the total daily excretion of prostaglandins and their metabolites is only about 1 mg, it is clear that the formation of prostaglandins is a quantitatively unimportant pathway in the overall metabolism of fatty acids. However, the metabolism of prostaglandins is completely dependent on a regular and constant supply of linoleic acid.

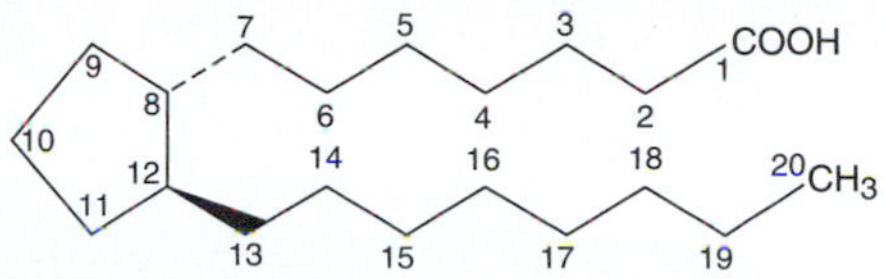

FIGURE 17.64
Structure of prostanoic acid.

Synthesis of Prostaglandins Involves a Cyclooxygenase

The immediate precursors to the prostaglandins are 20-carbon polyunsaturated fatty acids containing 3, 4, and 5 carbon–carbon double bonds. Since **arachidonic**

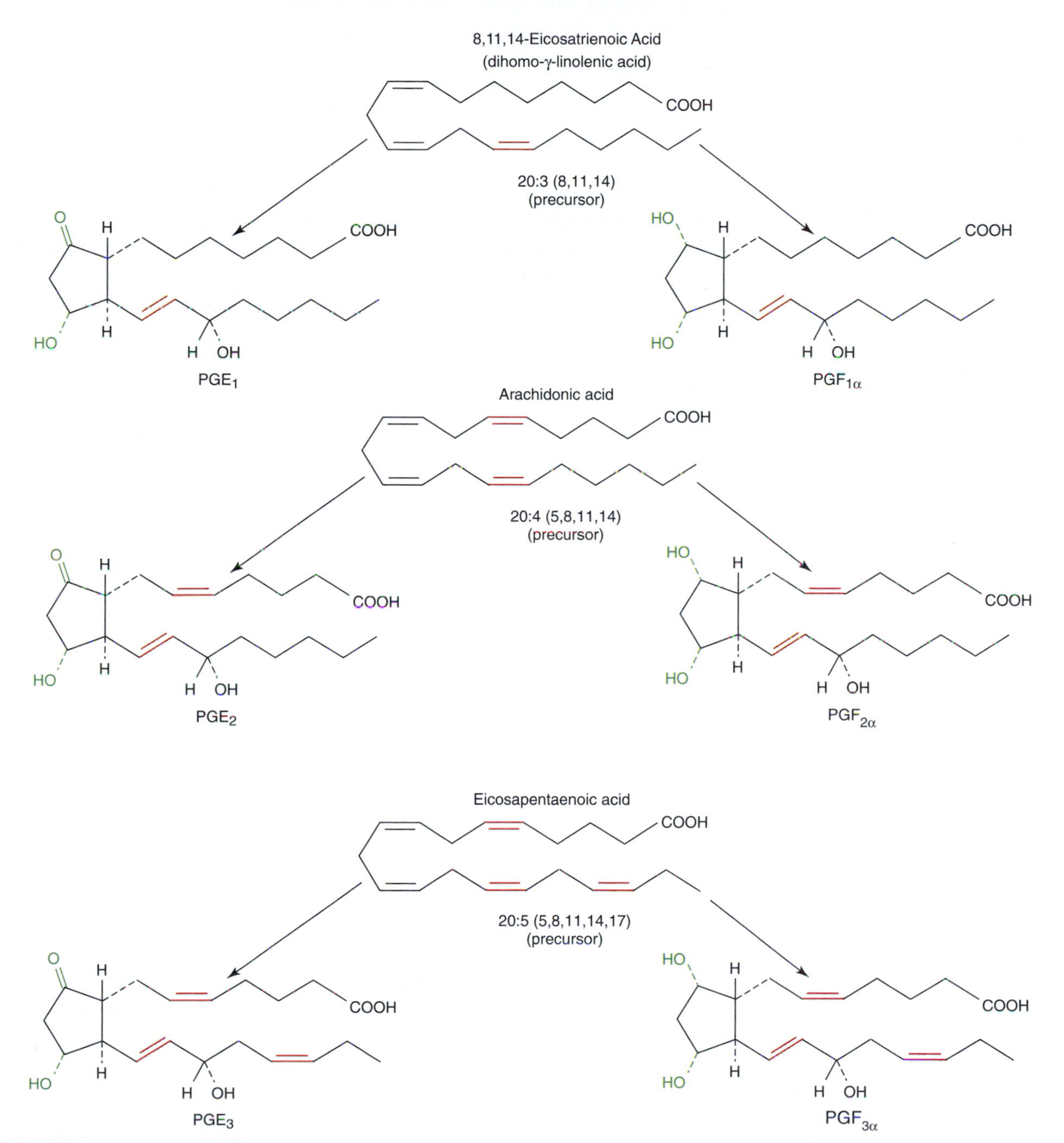

FIGURE 17.65
Synthesis of E and F prostaglandins from fatty acid precursors.

acid and most of its metabolites contain 20 carbons, they are referred to as **eicosanoids.** During their transformation into prostaglandins the fatty acids are cyclized and take up oxygen. Dihomo-γ-linoleic acid (20:3(8,11,14)) is the precursor to PGE_1 and $PGF_{1\alpha}$; arachidonic acid (20:4(5,8,11,14)) to PGE_2 and $PGF_{2\alpha}$; and eicosapentaenoic acid (20:5(5,8,11,14,17)) to PGE_3 and $PGF_{3\alpha}$ (see Figure 17.65).

Compounds of the 2-series derived from arachidonic acid are the principal prostaglandins in humans and are of the greatest significance biologically. The central enzyme system in prostaglandin biosynthesis is the bifunctional **prostaglandin G/H synthase (PGS),** which catalyzes oxidative cyclization of polyunsaturated fatty acids. Arachidonic acid is derived from membrane phospholipids by the action of the hydrolase **phospholipase A_2.** This cleavage step is the rate-limiting step in prostaglandin synthesis and some agents that stimulate prostaglandin production act by stimulating phospholipase A_2. Cholesterol esters containing arachidonic acid also serve as a source of arachidonic acid substrate.

The **cyclooxygenase (COX)** component of the prostaglandin synthase complex catalyzes the cyclization of C-8–C-12 of arachidonic acid to form the cyclic 9, 11-endoperoxide 15-hydroperoxide, PGG_2. The reaction requires two molecules of molecular oxygen (Figure 17.66). The mechanism involves stereospecific removal of the 13-pro-*S*-hydrogen of arachidonic acid. PGG_2 is then converted to prostaglandin H_2 (PGH_2) by a reduced **glutathione-dependent peroxidase (PG hydroperoxidase)** (Figure 17.67). Details of the additional steps leading to individual prostaglandins remain to be elucidated. Reactions that cyclize polyunsaturated fatty acids are found in the membranes of the endoplasmic reticulum. Major pathways of prostaglandin biosynthesis are summarized in Figure 17.68. Formation of primary prostaglandins of the D, E, and F series and of thromboxanes or prostacyclin (PGI_2) is mediated by different specific enzymes, whose presence varies depending on the cell type and tissue. This results in a degree of tissue specificity as to the type and quantity of prostaglandin produced. In kidney and spleen PGE_2 and $PGF_{2\alpha}$ are the major prostaglandins formed. In contrast, blood vessels produce mostly PGI_2 and $PGF_{2\alpha}$. In the heart PGE_2, $PGF_{2\alpha}$, and PGI_2 are formed in about equal amounts. Thromboxane A_2 (TXA_2) is the main prostaglandin endoperoxide formed in platelets.

There are two forms of cyclooxygenase (COX) or prostaglandin G/H synthase (PGS). **COX-1,** or **PGS-1,** is a constitutive enzyme found in gastric mucosa, platelets, vascular endothelium, and kidney. **COX-2,** or **PGS-2,** is inducible and is generated in response to inflammation. COX-2 is mainly expressed in activated macrophages and monocytes when stimulated by **platelet-activating factor** (PAF), **interleukin-1,** or bacterial lipopolysaccharide (LPS) and in smooth muscle cells, epithelial and

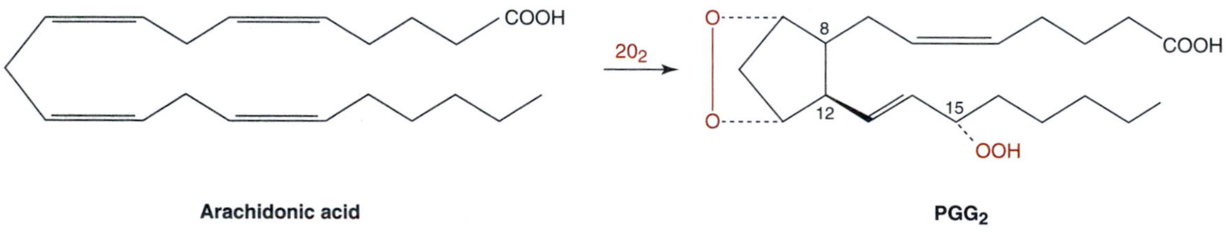

Arachidonic acid → ($2O_2$) → PGG_2

FIGURE 17.66
Cyclooxygenase reaction.

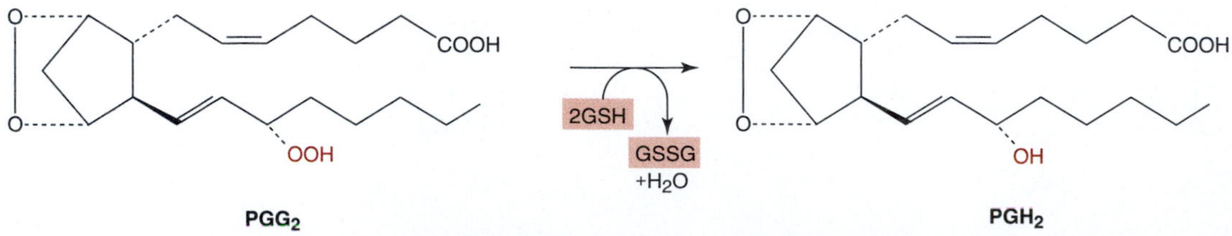

PGG_2 → (2GSH → GSSG + H_2O) → PGH_2

FIGURE 17.67
Conversion of PGG_2 to PGH_2; PG hydroperoxidase (PGH synthase) reaction.

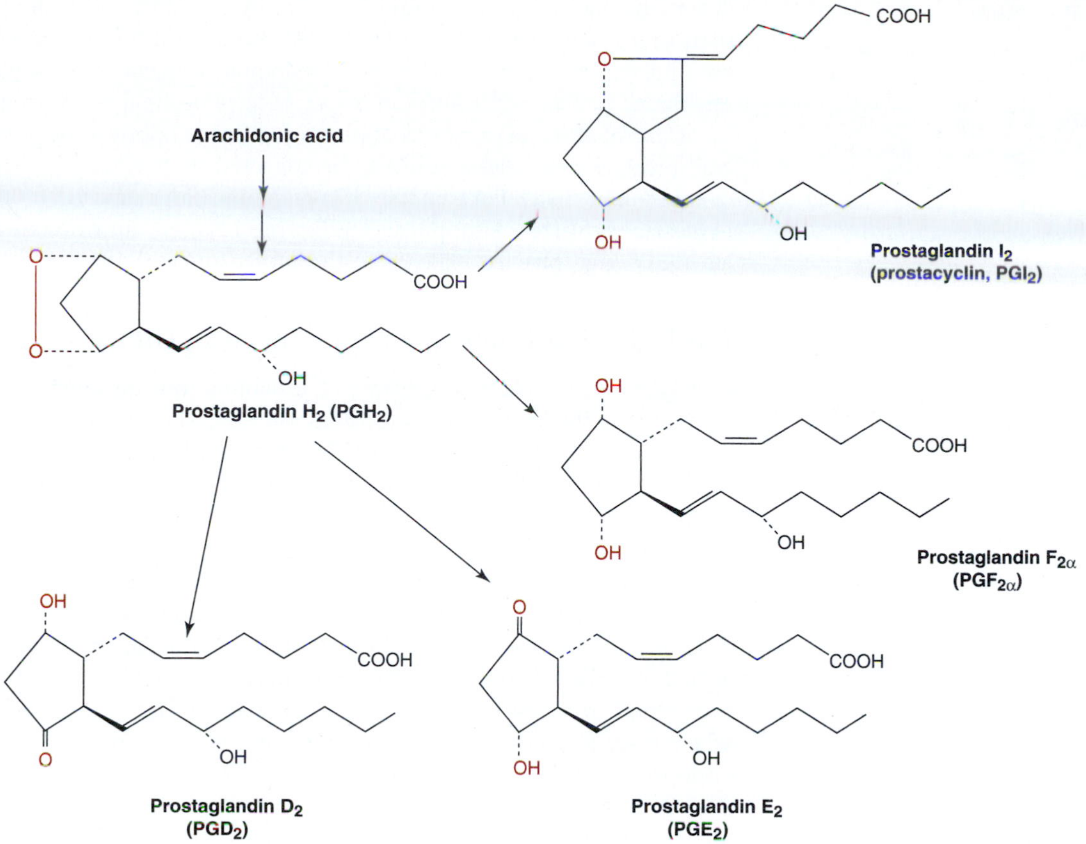

FIGURE 17.68
Major routes of prostaglandin biosynthesis.

endothelial cells, and neurons. PGS-2 induction is inhibited by glucocorticoids. The two forms of PGS catalyze both oxygenation of arachidonic acid to PGG_2 and the reduction of PGG_2 to PGH_2.

Prostaglandins have a very short half-life. Soon after release they are rapidly taken up by cells and inactivated either by oxidation of the 15-hydroxy group or by β-oxidation from the carboxy end of the chain. The lungs have an important role in inactivating prostaglandins. Thromboxanes are highly active metabolites of the PGG_2- and PGH_2-type prostaglandin endoperoxide in which the cyclopentane ring is replaced by a six-membered oxygen-containing (oxane) ring. The term thromboxane is derived from the fact that these compounds have a thrombus-forming potential. **Thromboxane A_2 synthase,** present in the endoplasmic reticulum, is abundant in lung and platelets and catalyzes the conversion of endoperoxide PGH_2 to TXA_2. The half-life of TXA_2 is very short in water ($t_{1/2} \approx 1$ min) as the compound is transformed rapidly into inactive thromboxane B_2 (TXB_2) by the reaction shown in Figure 17.69.

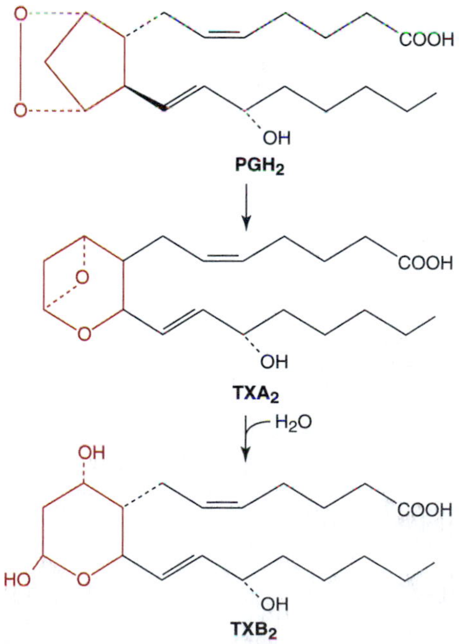

FIGURE 17.69
Synthesis of TXB_2 from PGH_2.

Prostaglandin Production Is Inhibited by Steroidal and Nonsteroidal Anti-inflammatory Agents

The **nonsteroidal, anti-inflammatory drugs (NSAIDs),** such as aspirin (acetylsalicylic acid), indomethacin, and phenylbutazone, block prostaglandin production by irreversibly inhibiting cyclooxygenase. In the case of aspirin, inhibition occurs by acetylation of the enzyme. Other NSAIDs inhibit cyclooxygenase but do so by binding noncovalently to the enzyme instead of acetylating it; they are called "non-aspirin NSAIDS." Aspirin is the only COX inhibitor that covalently modifies the enzyme and it is more potent against COX-1 than COX-2. Most NSAIDs inhibit COX-1 more than COX-2. These drugs have undesirable side effects; aplastic anemia can result from phenylbutazone therapy. COX is an important target for pharmacological agents. **Steroidal anti-inflammatory drugs** like hydrocortisone, prednisone, and

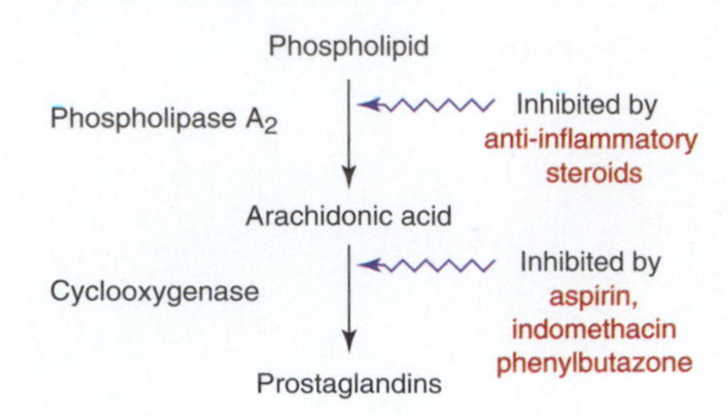

FIGURE 17.70
Sites of action of inhibitors of prostaglandin synthesis.

betamethasone block prostaglandin release by inhibiting phospholipase A_2 activity so as to interfere with mobilization of arachidonic acid (see Figure 17.70). The rate-limiting step in the synthesis of prostaglandins is release of arachidonic acid from membrane phospholipid stores in response to phospholipase A_2 activation.

Control of the biosynthesis of prostaglandins is poorly understood, but, in general, prostaglandin release seems to be triggered by hormonal or neural excitation, or muscular activity. For example, histamine increases prostaglandin concentration in gastric perfusates. Also, prostaglandins are released during labor and after cellular injury (e.g., platelets exposed to thrombin, lungs irritated by dust).

Prostaglandins Exhibit Many Physiological Effects

Prostaglandins are natural mediators of **inflammation.** Inflammatory reactions most often involve the joints (e.g., rheumatoid arthritis), skin (e.g., psoriasis), and eyes; inflammation is treated frequently with corticosteroids that inhibit prostaglandin synthesis. Administration of PGE_2 and PGE_1 induces the signs of inflammation that include redness and heat (due to arteriolar vasodilation), and swelling and edema resulting from increased capillary permeability. PGE_2 generated in immune tissues (e.g., macrophages, mast cells, B cells) evokes chemokinesis of T cells. PGE_2 in amounts that alone do not cause **pain,** prior to administration of the autocoids histamine and bradykinin, enhances both the intensity and duration of pain caused by these two agents. It is thought that pyrogens (**fever**-inducing agents) activate the prostaglandin biosynthetic pathway resulting in release of PGE_2 in the region of the hypothalamus where body temperature is regulated. Aspirin, which is an antipyretic drug, inhibits cyclooxygenase. Prostaglandins have been used extensively as drugs in **reproduction.** Both PGE_2 and PGF_2 have been used to induce parturition and for the termination of an unwanted pregnancy, specifically in the second trimester. There is also evidence that the PGE series of prostaglandins may play some role in infertility in males.

Synthetic prostaglandins have proved to be very effective in inhibiting **gastric acid secretion** in patients with **peptic ulcers.** PGE compounds appear to inhibit cAMP formation in gastric mucosal cells. Prostaglandins also accelerate the healing of gastric ulcers. Prostaglandins play an important role in controlling blood vessel tone and arterial **blood pressure.** The vasodilator prostaglandins, PGE, PGA, and PGI_2, lower systemic arterial pressure, thereby increasing local blood flow and decreasing peripheral resistance. TXA_2 causes contraction of vascular smooth muscle and glomerular mesangium. There is hope that the prostaglandins may eventually prove useful in the treatment of hypertension. PGE_2 functions in the fetus to maintain the patency of the **ductus arteriosus** prior to birth. If the ductus remains open after birth, closure can be hastened by administration of the cyclooxygenase inhibitor indomethacin. In other situations it may be desirable to keep the ductus open. For example, in infants born with congenital abnormalities where the defect can be corrected surgically, infusion of prostaglandins will maintain blood flow through the ductus over this interim period.

Certain prostaglandins, especially PGI_2, inhibit **platelet aggregation,** whereas PGE_2 and TXA_2 promote this clotting process. TXA_2 is produced by platelets and accounts for their spontaneous aggregation when in contact with some foreign surface, collagen, or thrombin. Endothelial cells lining blood vessels release PGI_2 and may account for the lack of adherence of platelets to the healthy blood vessel wall. PGE_2 and PGD_2 dilate renal blood vessels and increase blood flow through the kidney. They also regulate sodium excretion and glomerular filtration rate.

17.6 | LIPOXYGENASE AND OXYEICOSATETRAENOIC ACIDS

Cyclooxygenase directs polyunsaturated fatty acids into the prostaglandin pathway. Another important arachidonic acid oxygenating enzyme, called **lipoxygenase,** is a dioxygenase. There is a family of lipoxygenases that differ in the position of the double bond on the arachidonic acid molecule at which oxygen attack initially occurs (e.g., positions 5, 12, or 15). In humans the most important leukotrienes are the 5-lipoxygenase products that are involved in the mediation of inflammatory disorders. Lipoxygenases occur widely in plants and fungi, as well as in animals, but are absent from yeasts and most prokaryotes. Lipoxygenases contain a single atom of nonheme iron and are active when the iron is in the ferric state.

Monohydroperoxyeicosatetraenoic Acids Are Products of Lipoxygenase Action

Lipoxygenase adds hydroperoxy groups to arachidonic acid to produce **monohydroperoxyeicosatetraenoic acids (HPETEs).** Figure 17.71 shows the conversion of arachidonic acid to the three major HPETEs. In contrast to the cyclooxygenase of prostaglandin endoperoxide synthase, which catalyzes the bis-dioxygenation of unsaturated fatty acids to endoperoxide, lipoxygenases catalyze the monodioxygenation of unsaturated fatty acids to allylic hydroperoxides. Hydroperoxy substitution of arachidonic acid by lipoxygenases may occur at position 5, 12, or 15. A 15-lipoxygenase (15-LOX) oxygenates arachidonic acid at carbon-15. 5-HPETE is the major lipoxygenase product in basophils, polymorphonuclear (PMN) leukocytes, macrophages, mast cells, and any organ undergoing an inflammatory response; 12-HPETE predominates in platelets, pancreatic islet cells, vascular smooth muscle, and glomerular cells; 15-HPETE is the principal lipoxygenase product in reticulocytes, eosinophils, T lymphocytes, and tracheal epithelial cells. The 5-, 12-, and 15-lipoxygenases occur mainly in the cytosol. Specific stimuli or signals determine which type of lipoxygenase product a given type of cell produces. The oxygenated

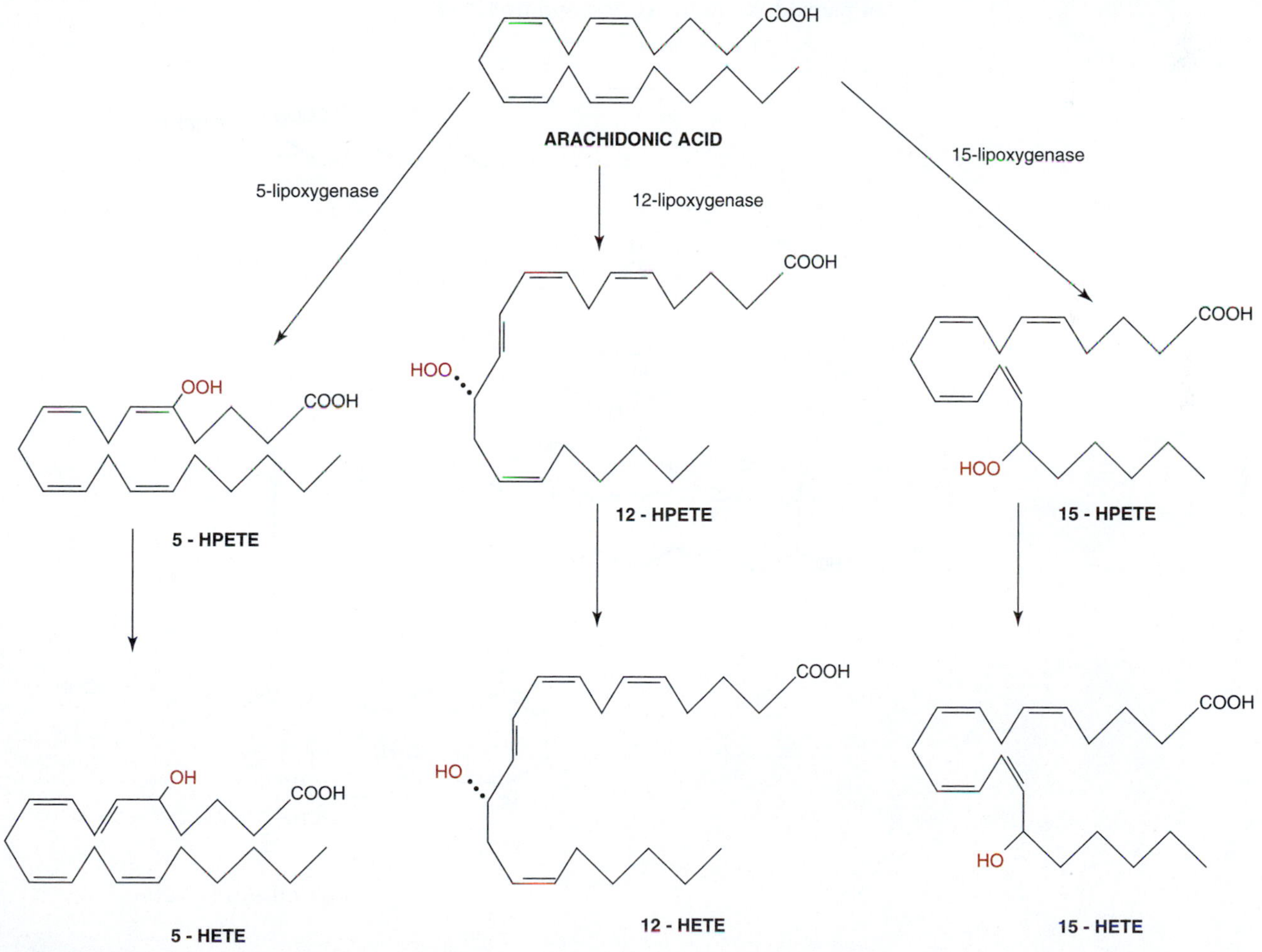

FIGURE 17.71
Lipoxygenase reaction and role of 5-hydroperoxyeicosatetraenoic acids (HPETEs) as precursors of hydroxyeicosatetraenoic acids (HETEs).

carbon atom in HPETEs is asymmetric and there are two possible stereoisomers of the hydroperoxy acid, (*R*) or (*S*). For example, the stereoconfiguration is specified 12*R*-LOX or 12*S*-LOX. All three major HPETEs are of the (*S*) configuration. 5-Lipoxygenase (5-LOX) exhibits both a dioxygenase activity that converts arachidonic acid to 5-HPETE and a dehydrase activity that transforms 5-HPETE to LTA_4. 5-LOX activity is restricted to a few cell types, including B lymphocytes but not T lymphocytes. It is activated by an accessory protein called 5-lipoxygenase-activating protein (FLAP). In human leukocytes, FLAP acts as an arachidonic acid transfer protein that presents the fatty acid substrate to the 5-LOX located on the nuclear membrane.

Leukotrienes and Hydroxyeicosatetraenoic Acids Are Hormones Derived from HPETEs

HPETE hydroperoxides are not hormones but are highly reactive, unstable intermediates that are converted either to the analogous alcohol (hydroxy fatty acid) by reduction of the peroxide moiety or to leukotrienes. **Leukotrienes** are lipoxygenase products containing at least three conjugated double bonds. Figure 17.72 shows how 5-HPETE rearranges to the epoxide leukotriene A_4 (LTA_4), which is then converted to LTB_4 or LTC_4, emphasizing that 5-HPETE is an important branch point in the lipoxygenase pathway.

Peroxidative reduction of 5-HPETE to the stable **5-hydroxyeicosatetraenoic acid (5-HETE)** is illustrated in Figure 17.71. The double bonds in 5-HETE are at positions 6, 8, 11, and 14 and are unconjugated, and the geometry of the double bonds is trans, cis, cis, and cis, respectively. Two other common forms of HETE are 12- and 15-HETE. The HPETEs are reduced either spontaneously or by the action of peroxidases to the corresponding HETEs.

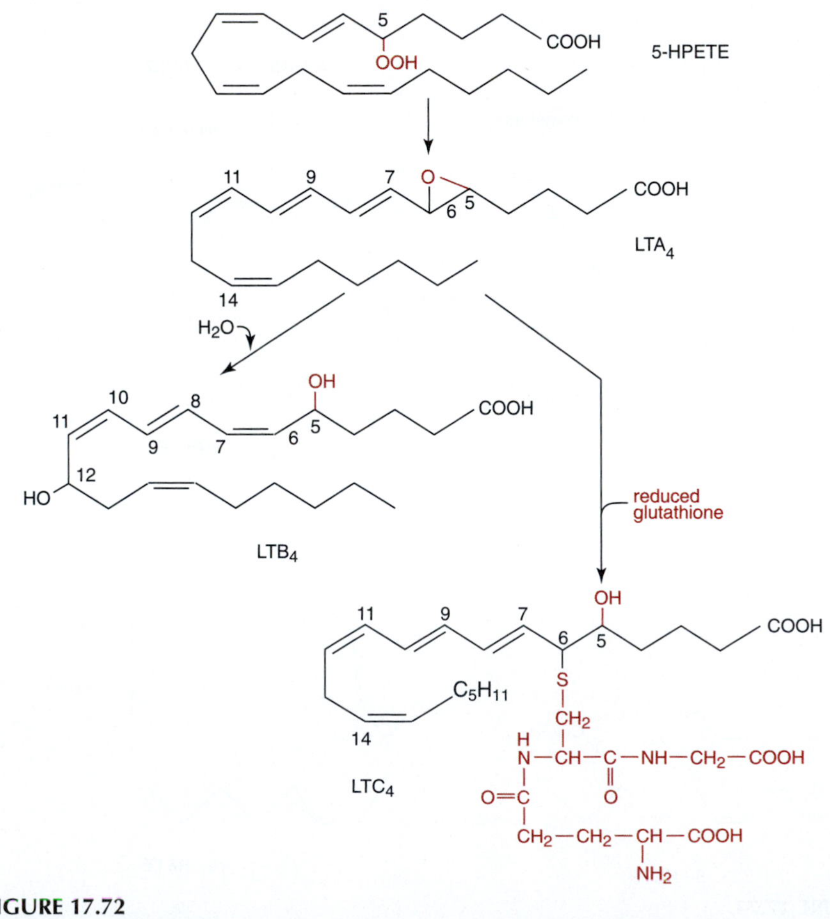

FIGURE 17.72
Conversion of 5-HPETE to LTB_4 and LTC_4 through LTA_4 as intermediate.

Leukotrienes are derived from the unstable precursor 5-HPETE by a reaction catalyzed by **LTA_4 synthase** that generates an epoxide called LTA_4. In the leukotriene series, the subscript indicates the number of double bonds. Thus, while double-bond rearrangement may occur, the number of double bonds in the leukotriene product is the same as in the original arachidonic acid. LTA_4 occurs at a branch point (Figure 17.72) and can be converted either to 5,12-dihydroxyeicosatetraenoic acid (designated leukotriene B_4 or LTB_4) or to LTC_4 and LTD_4.

Conversion of 5-HPETE to the diol LTB_4 (Figure 17.72) occurs by a cytosolic enzyme, **LTB_4 synthase** (LTA_4 hydratase), which adds water to the double bond between C-11 and C-12. The diversion of LTA_4 to leukotrienes LTC_4, LTD_4, and LTE_4 requires the participation of reduced glutathione that opens the epoxide ring in LTA_4 to produce LTC_4 (Figure 17.72). Sequential removal of glutamic acid and glycine residues by specific dipeptidases yields the leukotrienes LTD_4 and LTE_4 (Figure 17.73). The subscript 4 denotes the total number of double bonds.

Leukotrienes and HETEs Affect Several Physiological Processes

Leukotrienes persist for up to 4 h in the body. Stepwise ω-oxidation of the methyl end and β-oxidation of the resulting COOH-terminated fatty acid chain are responsible for the inactivation and degradation of LTB_4 and LTE_4. These reactions

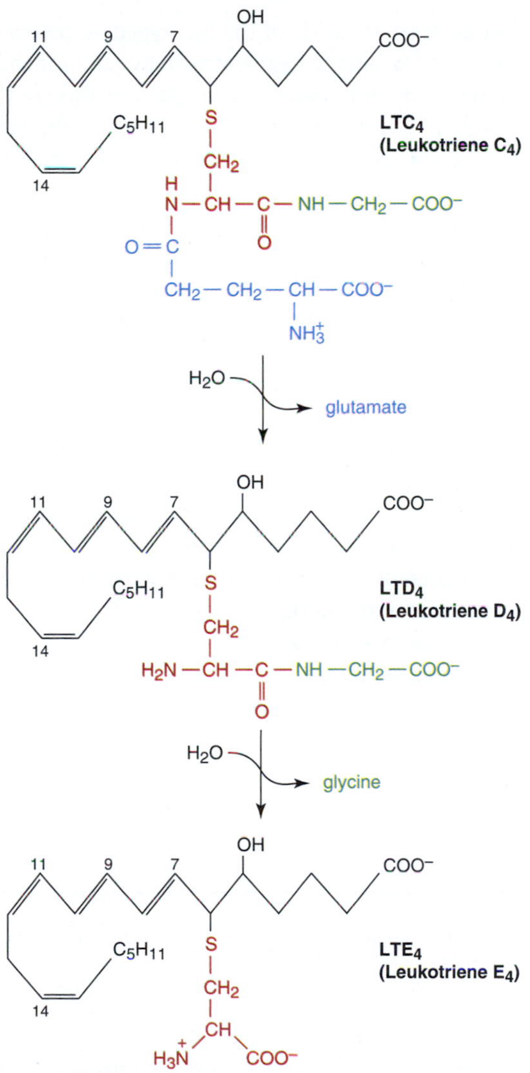

FIGURE 17.73
Conversion of LTC_4 to LTD_4 and LTE_4.

occur in mitochondria and peroxisomes. The actions of the thionyl peptides LTC_4, LTD_4, and LTE_4 comprise the **slow-reacting substance of anaphylaxis** (SRS-A). They cause slowly evolving but protracted contraction of smooth muscles in the airways and gastrointestinal tract. The LTC_4 is rapidly converted to LTD_4 and then slowly converted to LTE_4. These conversions are catalyzed by enzymes in plasma. LTB_4 and the sulfidopeptides LTC_4, LTD_4, and LTE_4 exert their biological actions through specific ligand–receptor interactions. In humans, activation of the 5-LOX of leukocytes produces leukotrienes and dihydroxyeicosanoids (e.g., LTB_4) that provoke bronchoconstriction and inflammation. Current drugs for asthma include 5-LOX inhibitors and leukotriene receptor antagonists.

In general, HETEs (especially 5-HETE) and LTB_4 are involved mainly in regulating neutrophil and eosinophil function: they mediate chemotaxis, stimulate adenylate cyclase, and induce PMN leukocytes to degranulate and release lysosomal hydrolytic enzymes. In contrast, LTC_4 and LTD_4 are humoral agents that promote smooth muscle contraction, constriction of pulmonary airways, trachea, and intestine, and increases in capillary permeability (edema). The HETEs appear to exert their effects by being incorporated into the phospholipids of membranes of target cells. The presence of fatty acyl chains containing a polar OH group may disturb packing of lipids and thus the structure and function of the membrane. LTB_4 has immunosuppressive activity exerted through inhibition of $CD4^+$ cells and proliferation of suppressor $CD8^+$ cells. LTB_4 also promotes neutrophil–endothelial cell adhesion.

Monohydroxyeicosatetraenoic acids of the lipoxygenase pathway are potent mediators of processes involved in allergy (hypersensitivity) and inflammation, secretion (e.g., insulin), cell movement, cell growth, and calcium fluxes. The initial allergic event, namely, the binding of IgE antibody to receptors on the surface of the mast cell, causes the release of substances, including leukotrienes, referred to as mediators of immediate hypersensitivity. Lipoxygenase products are usually produced within minutes after the stimulus. The leukotrienes LTC_4, LTD_4, and LTE_4 are much more potent than histamine in contracting nonvascular smooth muscles of bronchi and intestine. LTD_4 increases the permeability of the microvasculature. Mono-HETEs and LTB_4 stimulate migration (chemotaxis) of eosinophils and neutrophils, making them the principal mediators of PMN leukocyte infiltration in inflammatory reactions.

Eicosatrienoic acids (e.g., dihomo-γ-linolenic acid) and **eicosapentaenoic acid** (Figure 17.65) also serve as lipoxygenase substrates. The amount of these 20-carbon polyunsaturated fatty acids with three and five double bonds in tissues is less than that of arachidonic acid, but special diets can increase their levels. The lipoxygenase products of these tri- and pentaeicosanoids are usually less active than LTA_4 or LTB_4. It remains to be determined if fish oil diets rich in eicosapentaenoic acid are useful in the treatment of allergic and autoimmune diseases.

Pharmaceutical research into therapeutic uses of lipoxygenase and cyclooxygenase inhibitors and inhibitors and agonists of leukotrienes in treatment of inflammatory diseases such as asthma, psoriasis, rheumatoid arthritis, and ulcerative colitis is very active.

BIBLIOGRAPHY

Phospholipid Metabolism

Johnson, D. R., Bhatnager, R. S., Knoll, L. J., and Gordon, J. I. Genetic and biochemical studies of protein N-myristoylation. *Annu. Rev. Biochem.* 63:869, 1994.

Kazzi, S. N., Schurch, S., McLaughlin, K. L., Romero, R., and Janisse, J. Surfactant phospholipids and surface activity among preterm infants with respiratory distress syndrome who develop bronchopulmonary dysplasia. *Acta Paediatr.* 89:1218, 2000.

Kent, C., Carman, G. M., Spence, W., and Dowhan, W. Regulation of eukaryotic phospholipid metabolism. *FASEB J.* 5:2258, 1991.

McConville, M. J. and Ferguson, M. A. J. The structure, biosynthesis and function of glycosylated phosphatidylinositols in the parasitic protozoa and higher eukaryotes. *Biochem. J.* 294:305, 1993.

Nishizuka, Y. Protein kinase C and lipid signaling for sustained cellular responses. *FASEB J.* 9:484, 1995.

Rameh, L. E. and Cantley, L. C. The role of phosphoinositide 3-kinase lipid products in cell function. *J. Biol. Chem.* 274:8347, 1999.

Snyder, F. Platelet-activating factor and its analogs: metabolic pathways and related intracellular processes. *Biochem. Biophys. Acta* 1254:231, 1995.

Stevens, V. L. Biosynthesis of glycosylphosphatidylinositol membrane anchors. *Biochem. J.* 310:361, 1995.

Tjoelker, L. W. and Stafforini, D. M. Platelet-activating factor acetylhydrolases in health and disease. *Biochem. Biophys. Acta* 1488:102, 2000.

Topham, M. K. and Prescott, S. M. Mammalian diacylglycerol kinases, a family of lipid kinases with signaling functions. *J. Biol. Chem.* 274:11447, 1999.

Zwaal, R. F. A., Comfurius, P., and Bevers, E. M. Lipid–protein interactions in blood coagulation. *Biochem. Biophys. Acta* 1376:433, 1998.

Cholesterol Synthesis

Goldstein, J. L. and Brown, M. S. Regulation of the mevalonate pathway. *Nature* 343:425, 1990.

Gordon, D. J. and Rifkind, B. M. High-density lipoprotein: the clinical implications of recent studies. *N. Engl. J. Med.* 321:1311, 1989.

Ness, G. N. and Chambers, C. M. Feedback and hormonal regulation of hepatic 3-hydroxy-3-methylglutaryl coenzyme A reductase: the concept of cholesterol buffering capacity. *Proc. Soc. Exp. Bol. Med.* 164:8, 1999.

Sviridov, D. Intracellular cholesterol trafficking. *Histol. Histopathol* 14:305, 1999.

Lipoproteins

McNamara, D. J. Dietary fatty acids, lipoproteins, and cardiovascular disease. *Adv. Food Nutr. Res.* 36:253, 1999.

White, D. A., Bennett, A. J., Billett, M. A., and Salter, A. M. The assembly of triacylglycerol-rich lipoproteins: an essential role for the microsomal triacylglycerol transfer protein. *Br. J. Nutr.* 80:219, 1998.

Sphingolipids and the Sphingolipidoses

Grabowski, G. A., Gatt, S., and Horowitze, M. Acid β-glucosidase: enzymology and molecular biology of Gaucher disease. *Crit. Rev. Biochem. Mol. Biol.* 25:385, 1990.

Friedman, B. A., Vaddi, K., Preston, E. M., Cataldo, J. R., and McPherson, J. M. A comparison of the pharmacological properties of carbohydrate remodeled recombinant and placental-derived β-glucocerebrosidase: implications for clinical efficacy in treatment of Gaucher disease. *Blood* 93:2807, 1999.

Linke, T., Wilkening, G., Sadeghlar, F., Moczall, H., Bernardo, K., Schuchman, E., and Sandhoff, K. Interfacial regulation of acid ceramidase activity: stimulation of ceramide degradation by lysosomal lipids and sphingolipid activator proteins. *J. Biol. Chem.*, 276:5760, 2001

Lung Surfactant

Ainsworth, S. B., Beresford, M. W., Milligan, D. W. A., Shaw, N. J., Matthews, J. N. S., Fenton, A. C., and Ward-Platt, M. P. Pumactant and poractant alfa for treatment of respiratory distress syndrome in neonates born 25–29 weeks' gestation: a randomized trial. *Lancet* 355:1387, 2000.

Caminici, S. P. and Young, S. The pulmonary surfactant system. *Hosp. Pract.* 26:87, 1991.

Pryhuber, G. S. Regulation of pulmonary surfactant protein B. *Mol. Gen. Metab.* 64:217, 1998.

Prostaglandins, Thromboxanes, and Leukotrienes

Brash, A. R. Lipoxygenases: occurrence, functions, catalysis, and acquisition of substrate. *J. Biol. Chem.* 274:23679, 1999.

Breyer, M. D. and Breyer, R. M. Prostaglandin E receptors and the kidney. *Am. J. Physiol. Renal Physiol.* 279:F12, 2000.

Goetzl, E. J., An, S., and Smith, W. L. Specificity of expression and effects of eicosanoid mediators in normal physiology and human diseases. *FASEB J.* 9:1051, 1995.

Gravito, R. M. and DeWitt, D. L. The cyclooxygenase isoforms: structural insights into the conversion of arachidonic acid to prostaglandins. *Biochem. Biophys. Acta* 1441:278, 1999.

Henderson, W. R. Jr. The role of leukotrienes in inflammation. *Ann. Intern. Med.* 121:684, 1994.

Herschman, H. R. Function and regulation of prostaglandin synthase 2. *Adv. Exp. Med. Biol.* 469:3, 1999.

Kam, P. C. and See, A. U. Cyclooxygenase isoenzymes: physiological and pharmacological role. *Anaesthesia* 55:442, 2000.

Marnett, L. J., Rowlinson, S. W., Goodwin, D. C., Kalgutkar, A. S., and Lanzo, S. A. Arachidonic acid oxygenation by COX-1 and COX-2. *J. Biol. Chem.* 274:22903, 1999.

Mayatepek, E. and Hoffmann, G. Leukotrienes: biosynthesis, metabolism and pathophysiologic significance. *Pediatr. Res.* 37:1, 1995.

Sardesai, V. M. Biochemical and nutritional aspects of eicosanoids. *J. Nutr. Biochem.* 3:562, 1992.

Whittle, B. J. R. COX-1 and COX-2 products in the gut: therapeutic impact of COX-2 inhibitors. *GUT* 47:320, 2000.

Bile Acids

Agellon, L. B. and Torchia E. C. Intracellular transport of bile acids. *Biochem. Biophys. Acta* 1486:198, 2000.

Gilat, T., Somjen, G. L., Mazur, Y., Leikin-Frenkel, A., Rosenberg, R., Halpern, Z., and Konikoff, F. Fatty acid bile acid conjugates (FABACs)—new molecules for the prevention of cholesterol crystallisation in bile. *Gut* 48:75, 2001.

Leonard, M. R., Bogle, M. A., Carey, M. C., and Donovan, J. M. Spread monomolecular films of monohydroxy bile acids and their salts: influence of hydroxyl position, bulk pH, and association with phosphatidylcholine. *Biochemistry* 39:16064, 2000.

QUESTIONS | C. N. Angstadt

Multiple Choice Questions

1. Roles of various phospholipids include all of the following EXCEPT:
   A. cell–cell recognition.
   B. a surfactant function in lung.
   C. activation of certain membrane enzymes.
   D. signal transduction.
   E. mediator of hypersensitivity and acute inflammatory reactions.

2. Which of the following represents a correct group of enzymes involved in phosphatidylcholine synthesis in adipose tissue?
   A. choline phosphotransferase, glycerol kinase, phosphatidic acid phosphatase
   B. choline phosphotransferase, glycerol phosphate:acyltransferase, phosphatidylethanolamine *N*-methyltransferase
   C. glycerol phosphate:acyltransferase, α-glycerolphosphate dehydrogenase, phosphatidic acid phosphatase
   D. glycerol phosphate:acyltransferase, α-glycerolphosphate dehydrogenase, glycerol kinase
   E. α-glycerolphosphate dehydrogenase, glycerol kinase, phosphatidic acid phosphatase

3. CDP-X (where X is the appropriate alcohol) reacts with 1,2-diacylglycerol in the primary synthetic pathway for:
   A. phosphatidylcholine.
   B. phosphatidylinositol.

C. phosphatidylserine.
D. all of the above.
E. none of the above.

4. Phospholipases A_1 and A_2:
   A. have no role in phospholipid synthesis.
   B. are responsible for initial insertion of fatty acids in *sn*-1 and *sn*-2 positions during synthesis.
   C. are responsible for base exchange in the interconversion of phosphatidylethanolamine and phosphatidylserine.
   D. hydrolyze phosphatidic acid to a diglyceride.
   E. remove a fatty acid in an *sn*-1 or *sn*-2 position so it can be replaced by another in phospholipid synthesis.

5. Primary bile acids:
   A. are any bile acids that are found in the intestinal tract.
   B. are any bile acids reabsorbed from the intestinal tract.
   C. are synthesized in the intestinal tract by bacteria.
   D. are synthesized in hepatocytes directly from cholesterol.
   E. are converted to secondary bile acids by conjugation with glycine or taurine.

6. A ganglioside may contain all of the following EXCEPT:
   A. a ceramide structure.
   B. glucose or galactose.
   C. phosphate.
   D. one or more sialic acids.
   E. sphingosine.

7. Structural features that are common to all prostaglandins include:
   A. 20-carbon atoms.
   B. an oxygen-containing internal heterocyclic ring.
   C. a peroxide group at C-15.
   D. two double bonds.
   E. a ketone group.

8. Prostaglandin synthase complex:
   A. catalyzes the rate-limiting step of prostaglandin synthesis.
   B. is inhibited by anti-inflammatory steroids.
   C. contains both a cyclooxygenase and a peroxidase component.
   D. produces PGG_2 as the end product.
   E. uses as substrate the pool of free arachidonic acid in the cell.

9. Thromboxane A_2:
   A. is a long-lived prostaglandin.
   B. is an inactive metabolite of PGE_2.
   C. is the major prostaglandin produced in all cells.
   D. does not contain a ring structure.
   E. is synthesized from the intermediate PGH_2.

10. Hydroperoxyeicosatetraenoic acids (HPETEs):
   A. are derived from arachidonic acid by a peroxidase reaction.
   B. are mediators of hypersensitivity reactions.
   C. are intermediates in formation of leukotrienes.
   D. are relatively stable compounds (persist for as long as 4 h).
   E. are inactivated forms of leukotrienes.

Questions 11 and 12: Hypercholesterolemia is one of the risk factors for cardiovascular disease. Total cholesterol in blood is not as important in considering risk as distribution of cholesterol between LDL ("bad") and HDL ("good") particles. Efforts to reduce serum cholesterol are two-pronged: dietary and drugs if reducing dietary cholesterol and saturated fat is not sufficient. The drugs of choice are statins, inhibitors of the rate-limiting enzyme of cholesterol biosynthesis. If necessary, bile acid-binding resins can be added.

11. In biosynthesis of cholesterol:
   A. 3-hydroxy-3-methyl glutaryl CoA (HMG CoA) is synthesized by mitochondrial HMG CoA synthase.
   B. HMG CoA reductase catalyzes the rate-limiting step.
   C. the conversion of mevalonic acid to farnesyl pyrophosphate proceeds via condensation of 3 molecules of mevalonic acid.
   D. condensation of 2 farnesyl pyrophosphates to form squalene is a freely reversible reaction.
   E. conversion of squalene to lanosterol is initiated by formation of the fused ring system, followed by addition of oxygen.

12. Cholesterol present in LDL (low density lipoprotein):
   A. binds to a cell receptor and diffuses across the cell membrane.
   B. when it enters a cell, suppresses activity of ACAT (acyl CoA: cholesterol acyltransferase).
   C. once in the cell is converted to cholesteryl esters by LCAT (lecithin: cholesterol acyltransferase).
   D. once it has accumulated in the cell, inhibits replenishment of LDL receptors.
   E. represents primarily cholesterol that is being removed from peripheral cells.

Questions 13 and 14: Sphingolipidoses (lipid storage diseases) are a group of diseases characterized by defects in lysosomal enzymes. Rate of synthesis of sphingolipids is normal but undegraded material accumulates in lysosomes. The severity of the particular disease depends, in part, on what tissues are most affected. Those affecting primarily brain and nervous tissue, like Tay–Sachs disease, are lethal within a few years of birth. Since the extent of enzyme deficiency is the same in all cells, easily accessible tissues like skin fibroblasts can be used to assay for enzyme deficiency. There are no treatments for most of the diseases so genetic counseling seems to be the only helpful approach. However, for Gaucher's disease enzyme (glucocerebrosidase) replacement therapy is effective.

13. All of the following are true about degradation of sphingolipids EXCEPT:
   A. it occurs by hydrolytic enzymes contained in lysosomes.
   B. it terminates at the level of ceramides.
   C. it is a sequential, stepwise removal of constituents.
   D. it may involve a sulfatase or a neuraminidase.
   E. it is catalyzed by enzymes that are specific for a type of linkage rather than for a particular compound.

14. In Niemann–Pick disease, the deficient enzyme is sphingomyelinase. Sphingomyelins differ from other sphingolipids in that they are:
   A. not based on a ceramide core.
   B. acidic rather than neutral at physiological pH.
   C. the only types containing *N*-acetylneuraminic acid.
   D. the only types that are phospholipids.
   E. not amphipathic.

Problems

15. Cells from a patient with familial hypercholesterolemia (FH) and cells from an individual without that disease were incubated with LDL particles containing radioactively labeled cholesterol. After incubation, the incubation medium was removed

and the radioactivity of the cells measured. The cells were treated to remove any bound material, lysed, and internal cholesterol content measured. Results are given below. What mutation of the gene for the LDL receptor protein could account for the results?

Cell Type	*Radioactivity of Cell*	*Cholesterol Content*
Normal	3000 cpm/mg cells	Low
FH	3000 cpm/mg cells	High

16. The combination of bile salt-binding resin and an HMG CoA reductase inhibitor is very effective in reducing serum cholesterol for most patients with high cholesterol. Why is this treatment much less effective for patients with familial hypercholesterolemia?

ANSWERS

1. **A** This function appears to be associated with complex glycosphingolipids. B: Especially dipalmitoyllecithin. C: For example, β-hydroxybutyrate dehydrogenase. D: Especially phosphatidylinositols. E: Platelet-activating factor (PAF) does this.
2. **C** A, D, E: Glycerol kinase is not present in adipose tissue, which must rely on α-glycerolphosphate dehydrogenase. B: This is a liver process only.
3. **A** This is the main pathway for choline. B: Phosphatidylinositol is formed from CDP-diglyceride reacting with *myo*-inositol. C: This is formed by "base exchange."
4. **E** Phospholipases A_1 and A_2, as their names imply, hydrolyze a fatty acid from a phospholipid and so are part of phospholipid degradation. They are also important in synthesis, however, in assuring the asymmetric distribution of fatty acids that occurs in phospholipids.
5. **D** They are cholic acid and chenodeoxycholic acid. A, B: The intestinal tract contains a mixture of primary and secondary bile acids, both of which can be reabsorbed. C: Secondary bile acids are formed by bacteria in the intestine by chemical reactions, such as removal of the C-7 OH group. E: Conjugated cholic acid is still a primary bile acid.
6. **C** Glycosphingolipids do not contain phosphate. A, E: Ceramide, which is formed from sphingosine, is the base structure from which glycosphingolipids are formed. B: Glucose is usually the first sugar attached to the ceramide. D: By definition, gangliosides contain sialic acid.
7. **A** Prostaglandins are eicosanoids. B: This is true of thromboxanes but the prostaglandin ring contains only carbons. C: True only of the intermediate of synthesis, PGG_2. D: Number of double bonds is variable. E: True of the A and E series but not of the F series.
8. **C** Cyclooxygenase oxidizes arachidonic acid and peroxidase converts PGG_2 to PGH_2. A, B: The release of the precursor fatty acid by phospholipase A_2 is the rate-limiting step and the one inhibited by anti-inflammatory steroids. D: The peroxidase component converts PGG_2 to PGH_2. E: Arachidonic acid is not free in the cell but is part of the membrane phospholipids or sometimes a cholesteryl ester.
9. **E** PGH_2 is the branch point in the synthesis of the prostaglandins and thromboxanes. A–D: TXA_2 is very active, has a very short half-life (on the order of seconds), contains a six-membered ring, and is the main prostaglandin produced by platelets but not all tissues.
10. **C** These are formed by addition of oxygen to arachidonic acid. A: The enzyme is a lipoxygenase. B–E: HPETEs, themselves, are not hormones but highly unstable intermediates that are converted to either HETEs (mediators of hypersensitivity) or leukotrienes.
11. **B** This enzyme is inhibited by cholesterol. A: Remember that cholesterol biosynthesis is cytosolic; mitochondrial biosynthesis of HMG CoA leads to ketone body formation. C: Mevalonic acid is decarboxylated (among other things) to produce the isoprene pyrophosphates, which are the condensing units. D: Pyrophosphate is hydrolyzed, which prevents reversal. E: The process is initiated by epoxide formation.
12. **D** This is one of the ways to prevent overload in the cell. A: LDL binds to the cell receptor and is endocytosed and then degraded in lysosomes to release cholesterol. B: ACAT is activated to facilitate storage. C: LCAT is a plasma enzyme. E: The primary role of LDL is to deliver cholesterol to peripheral tissues; HDL removes cholesterol from peripheral tissues.
13. **B** Ceramides are hydrolyzed to sphingosine and fatty acid. D: Sulfogalactocerebroside contains sulfate and gangliosides contain one or more *N*-acetylneuraminic acids. E: Many sphingolipids share the same types of bonds, for example, a β-galactosidic bond, and one enzyme, for example, β-galactosidase, will hydrolyze it whenever it occurs.
14. **D** Sphingomyelins are not glycosphingolipids but phosphosphingolipids. A, B, E: They are formed from ceramides, are amphipathic, and are neutral. C is the definition of gangliosides.
15. FH cells have LDL receptors with normal binding properties as indicated by the bound radioactivity being the same as that of normal cells. They are unable to internalize the receptor–LDL complex so cholesterol synthesis is not inhibited as it is with normal cells. The mutation is most likely on the carboxy terminus of the protein, which is involved in internalization. Make sure you understand why mutation in other regions would not lead to the observed results.
16. By binding bile salts, forcing increased excretion, liver has to convert more cholesterol to bile salts. If liver synthesis of cholesterol is inhibited, liver synthesizes more LDL receptors, removing increased LDL particles (and thus cholesterol) from blood. Patients with FH either have no receptors or have nonfunctioning LDL receptors so liver cannot increase uptake of LDL from blood. Synthesis is reduced but this does not have as dramatic an effect on blood cholesterol.

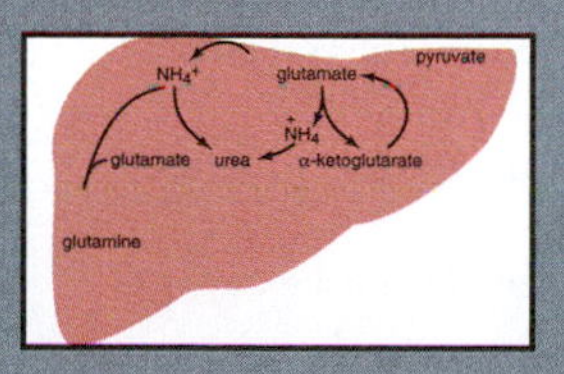

18

AMINO ACID METABOLISM

Marguerite W. Coomes

18.1 | OVERVIEW

Amino acids and the relationship between their structure and the structure and function of proteins were presented in Chapter 3. This chapter describes the metabolism of amino acids, emphasizing the importance of dietary protein as the major source of amino acids for humans.

Molecular nitrogen, N_2, exists in the atmosphere in great abundance. Before it can be utilized by animals it must be "fixed," that is, reduced from N_2 to NH_3 by microorganisms, plants, and electrical discharge from lightning. Ammonia is then incorporated into amino acids and proteins, and these become part of the food chain (Figure 18.1). Humans can synthesize only 11 of the 20 amino acids needed for

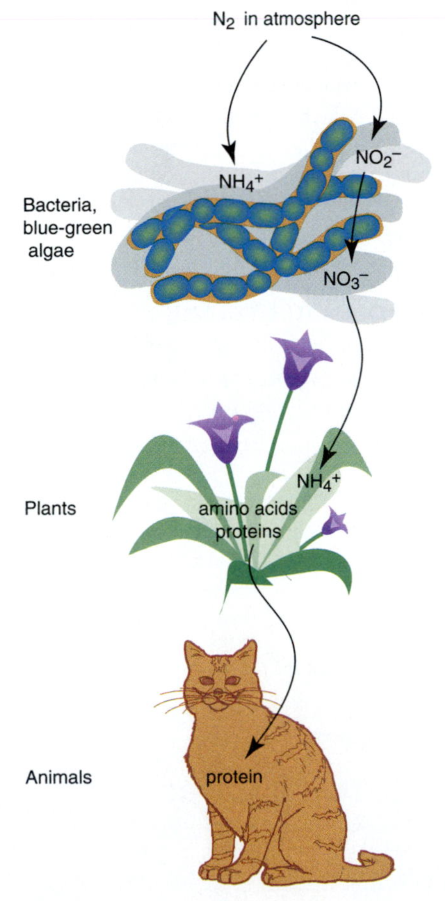

FIGURE 18.1
Outline of entry of atmospheric nitrogen into the animal diet.
This occurs initially by reduction of nitrogen to ammonia by enzymes in microorganisms and plants.

TABLE 18.1

Essential	*Nonessential*
Arginine[a]	Alanine
Histidine	Aspartate
Isoleucine	Asparagine
Leucine	Cysteine
Lysine	Glutamate
Methionine[b]	Glutamine
Phenylalanine[c]	Glycine
Threonine	Proline
Tryptophan	Serine
Valine	Tyrosine

[a]Arginine is synthesized by mammalian tissues, but the rate is insufficient to meet the need during growth.

[b]Methionine is required in large amounts to produce cysteine if the latter is not supplied adequately by the diet.

[c]Phenylalanine is needed in larger amounts to form tyrosine if the latter is not supplied adequately by the diet.

protein synthesis. Those that cannot be synthesized *de novo* are termed **"essential"** because they must be obtained from dietary foodstuffs that contain them (Table 18.1).

This chapter includes discussion of interconversions of amino acids, removal and excretion of ammonia, and synthesis of "nonessential" amino acids by the body. As part of ammonia metabolism, synthesis and degradation of glutamate, glutamine, aspartate, asparagine, alanine, and arginine are discussed. Synthesis and degradation of other nonessential amino acids are then described, as well as the degradation of the essential amino acids. Synthetic pathways of amino acid derivatives and some diseases of amino acid metabolism are also presented.

Carbons from amino acids enter intermediary metabolism at one of seven points. Glucogenic amino acids are metabolized to pyruvate, 3-phosphoglycerate, α-ketoglutarate, oxaloacetate, fumarate, or succinyl CoA. Ketogenic amino acids produce acetyl CoA or acetoacetate. Metabolism of some amino acids results in more than one of the above and they are therefore both glucogenic and ketogenic (Figure 18.2). Products of amino acid metabolism can be used to provide energy. Additional energy-generating compounds, usually NADH, are also produced during degradation of some of the amino acids.

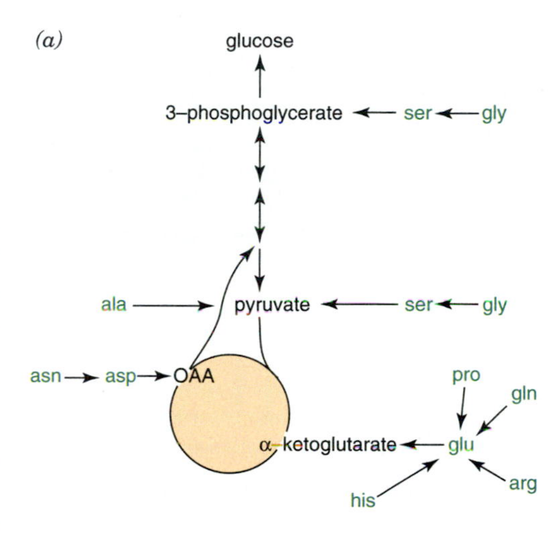

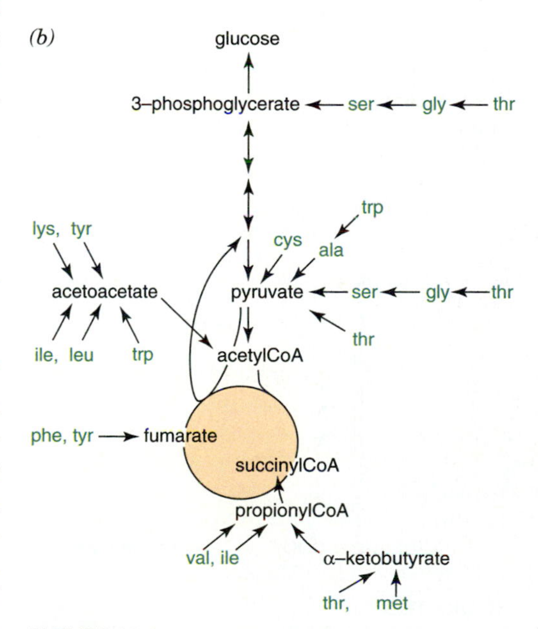

FIGURE 18.2
Metabolic fate of (*a*) nonessential amino acids and (*b*) essential amino acids plus cysteine and tyrosine.

18.2 | INCORPORATION OF NITROGEN INTO AMINO ACIDS

Most Amino Acids Are Obtained from the Diet

A healthy adult eating a varied and plentiful diet is generally in **"nitrogen balance,"** a state where the amount of nitrogen ingested each day is balanced by the amount excreted, resulting in no net change in the amount of body nitrogen. In the well-fed condition, excreted nitrogen comes mostly from digestion of excess protein or from normal turnover. Protein turnover is defined as the synthesis and degradation of protein. Under some conditions the body is either in negative or positive nitrogen balance. In **negative nitrogen** balance more nitrogen is excreted than ingested. This occurs in starvation and certain diseases. During starvation carbon chains of amino acids from proteins are needed for gluconeogenesis; ammonia released from amino acids is excreted mostly as urea and is not reincorporated into protein. A diet deficient in an essential amino acid also leads to a negative nitrogen balance, since body proteins are degraded to provide the deficient essential amino acid, and the other amino acids liberated are metabolized. Negative nitrogen balance may also exist in senescence. **Positive nitrogen balance** occurs in growing children, who are increasing their body weight and incorporating more amino acids into proteins than they break down. Cysteine and arginine are essential in children but not essential in adults because they are synthesized from methionine and ornithine. These amino acids are readily available in adults but limited in children because of their greater use of all amino acids. Positive nitrogen balance also occurs in pregnancy and during refeeding after starvation.

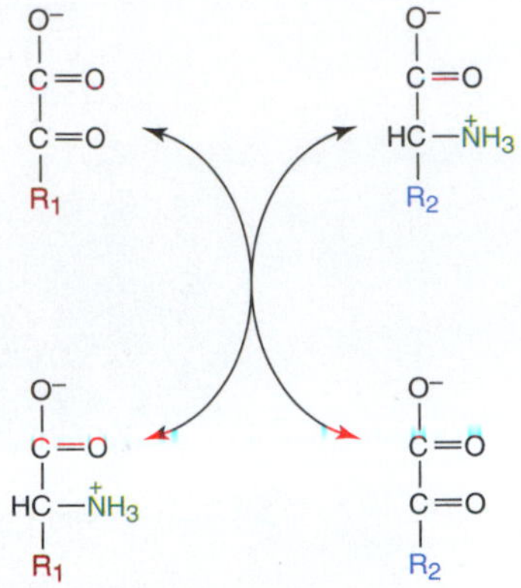

FIGURE 18.3
Aminotransferase reaction.

Amino Groups Are Transferred from One Amino Acid to Form Another

Most amino acids used by the body to synthesize protein or as precursors for amino acid derivatives are obtained from the diet or from protein turnover. When necessary, nonessential amino acids are synthesized from α-keto acid precursors via transfer of a preexisting amino group from another amino acid by **aminotransferases,** also called **transaminases** (Figure 18.3). Transfer of amino groups also occurs during degradation of amino acids. Figure 18.4 shows how the amino group of alanine is transferred to α-ketoglutarate to form glutamate. The pyruvate produced then provides carbons for gluconeogenesis or for energy production via the TCA cycle. This reaction is necessary since ammonia cannot enter the urea cycle directly from alanine but the amino group of glutamate can be utilized. The opposite reaction would occur if there were a need for alanine for protein synthesis that was not being met by dietary intake or protein turnover. Transamination involving essential amino acids is normally unidirectional since the body cannot synthesize the equivalent α-keto acid. Figure 18.5 shows transamination of valine, an essential amino acid. The resulting α-ketoisovalerate is further metabolized to succinyl CoA as discussed on page 814. Transamination is the most common reaction involving free amino acids, and only threonine and lysine do not participate in an aminotransferase reaction. An obligate amino and α-keto acid pair in all of these reactions is glutamate and α-ketoglutarate. This means that amino group transfer between alanine and aspartate would have to occur via coupled reactions, with a glutamate intermediate (Figure 18.6). The equilibrium constant for aminotransferases is close to one so that the reactions are freely reversible. When nitrogen excretion is impaired and **hyperammonemia** occurs, as in liver failure, amino acids, including the essential amino acids, can be replaced in the diet by α-keto acid analogs, with the exception of threonine and lysine as mentioned above. The α-keto acids are transaminated by aminotransferases to produce the different amino acids. Figure18.5 shows valine formation after administration of α-ketoisovalerate as therapy for hyperammonemia.

Tissue distribution of some of the aminotransferase family is used diagnostically by measuring the release of a specific enzyme during tissue damage; for instance, the presence of glutamate aspartate aminotransferase in plasma is a sign of liver damage (see p. 452).

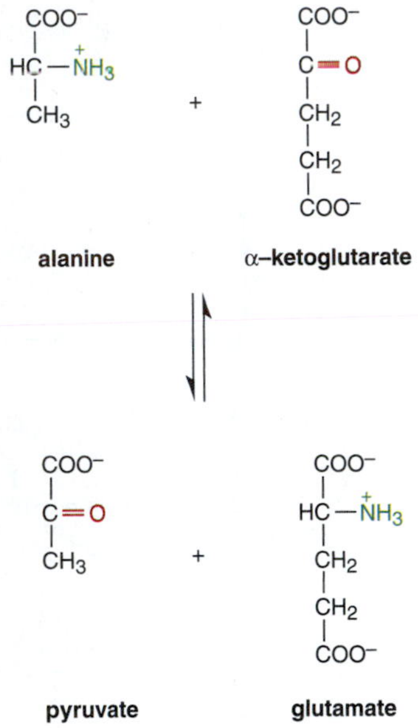

FIGURE 18.4
Glutamate–pyruvate aminotransferase reaction.

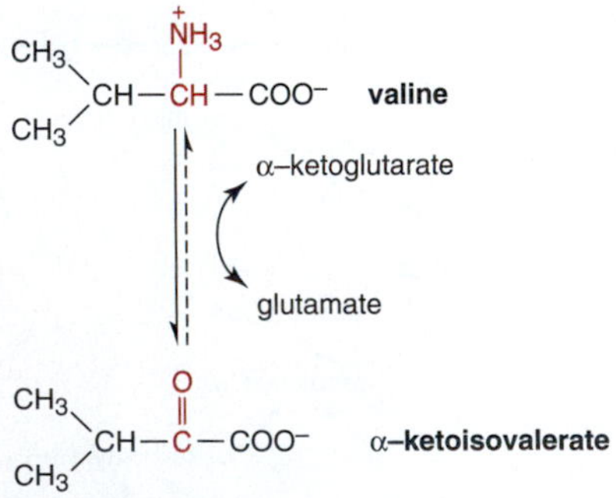

FIGURE 18.5
Transamination of valine.
Valine can be formed from α-ketoisovalerate only when this compound is administered therapeutically.

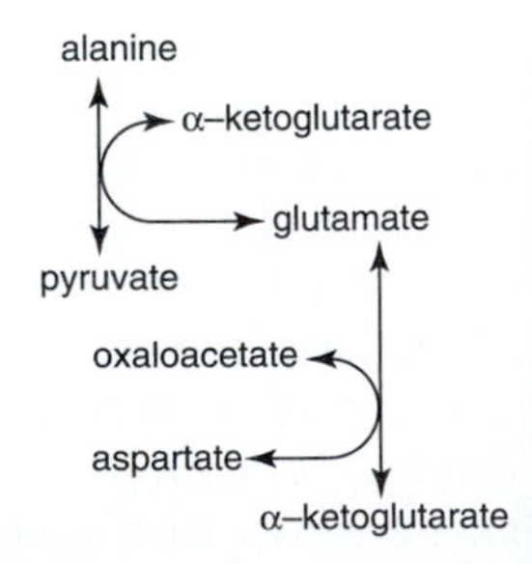

FIGURE 18.6
Coupled transamination reaction.

Pyridoxal Phosphate Is Cofactor for Aminotransferases

Transfer of amino groups occurs via enzyme-associated intermediates derived from **pyridoxal phosphate,** the functional form of vitamin B_6 (Figure 18.7). The active site of the "resting" aminotransferase contains pyridoxal phosphate covalently attached to an ε-amino group of a lysine residue that forms part of the primary structure of the transferase (Figure 18.8). The complex is further stabilized by ionic and hydrophobic interactions. The linkage, —CH=N—, is called a **Schiff base.** The carbon originates in the aldehyde group of pyridoxal phosphate, and the nitrogen is donated by the lysine residue. When a substrate amino acid, ready to be metabolized, approaches the active site, its amino group displaces the lysine ε-amino group and a Schiff base linkage is formed with the amino group of the amino acid substrate (Figure 18.9). At this point the pyridoxal phosphate-derived molecule is no longer covalently attached to the enzyme but is held in the active site only by ionic and hydrophobic interactions between it and the protein. The Schiff base linkage involving the amino acid substrate is in tautomeric equilibrium between an aldimine, —CH=N—CHR$_2$, and a ketimine, —CH$_2$—N=CR$_2$. Hydrolysis of the ketimine liberates an α-keto acid, leaving the amino group as part of the pyridoxamine structure. A reversal of the process is now possible; an α-keto acid reacts with the amine group, the double bond is shifted, and then hydrolysis liberates an amino acid. Pyridoxal phosphate now reforms its Schiff base with the "resting" enzyme (Figure 18.8). Many pyridoxal phosphate-requiring reactions involve transamination, but the ability of the Schiff base to transfer electrons between different atoms allows this cofactor to participate when other groups, such as carboxyls, are to be eliminated. Figure 18.10 shows the reaction of a **pyridoxal phosphate-dependent decarboxylase** and an **α-,β-elimination.** The effective concentration of

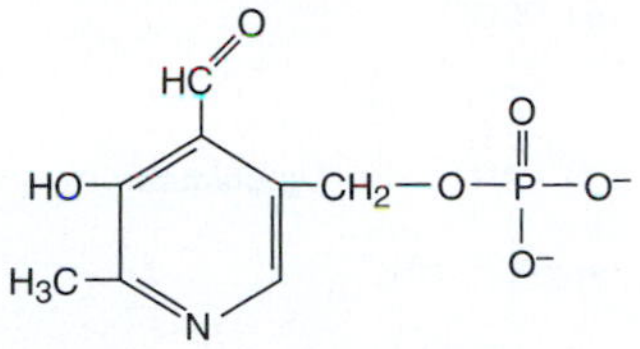

FIGURE 18.7
Pyridoxal phosphate.

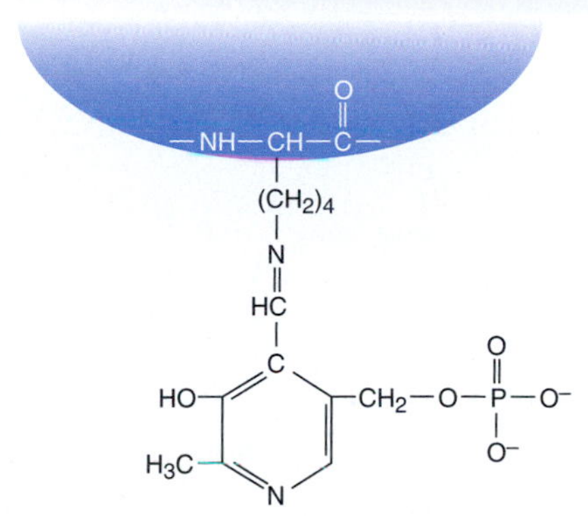

FIGURE 18.8
Pyridoxal phosphate in aldimine linkage to protein lysine residue.

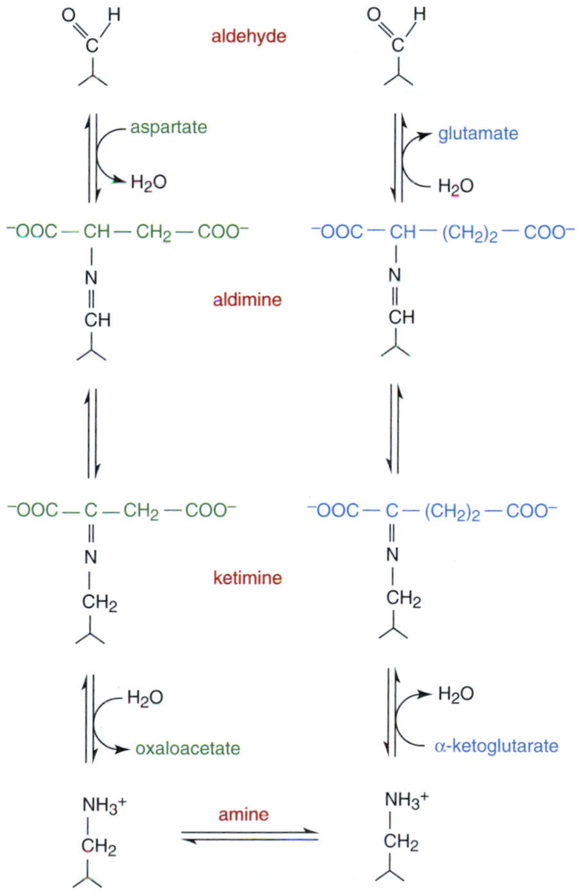

FIGURE 18.9
Different forms of pyridoxal phosphate during a transamination reaction.

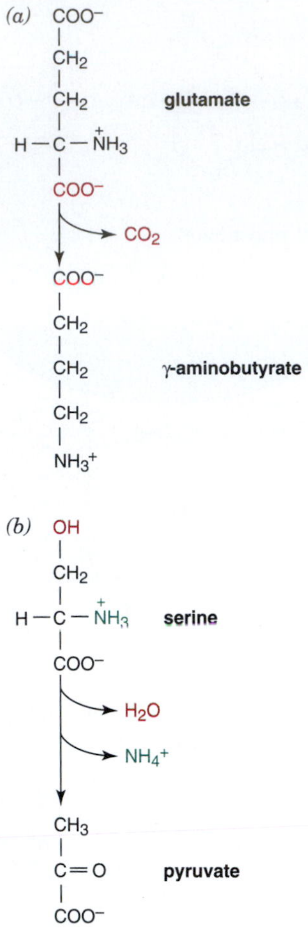

(a) COO^-–CH_2–CH_2–$H-C-\overset{+}{N}H_3$–COO^- glutamate

→ CO_2

COO^-–CH_2–CH_2–CH_2–NH_3^+ γ-aminobutyrate

(b) OH–CH_2–$H-C-\overset{+}{N}H_3$–COO^- serine

→ H_2O

→ NH_4^+

CH_3–C=O–COO^- pyruvate

FIGURE 18.10
Glutamate decarboxylase and serine dehydratase are pyridoxal phosphate-dependent reactions.

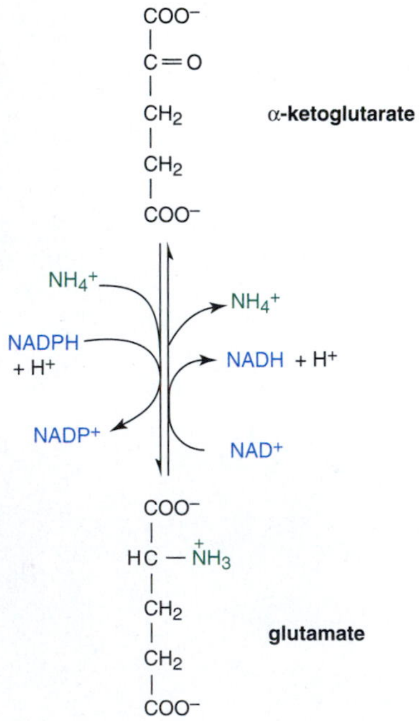

COO^-–C=O–CH_2–CH_2–COO^- α-ketoglutarate

NH_4^+, NADPH + H^+ → $NADP^+$

NAD^+ → NH_4^+, NADH + H^+

COO^-–$HC-\overset{+}{N}H_3$–CH_2–CH_2–COO^- glutamate

FFIGURE 18.11
Glutamate dehydrogenase reaction.

vitamin B_6 in the body may be decreased by administration of certain drugs, such as the antituberculosis drug isoniazid, which forms a Schiff base with pyridoxal, thus making it unavailable for catalysis.

Glutamate Dehydrogenase Incorporates and Produces Ammonia

In liver ammonia is incorporated into glutamate by **glutamate dehydrogenase** (Figure 18.11). The enzyme also catalyzes the reverse reaction. Glutamate always serves as one of the amino acids in transaminations and is thus the "gateway" between amino groups of most amino acids and free ammonia (Figure 18.12). NADPH is used in the synthetic reaction, whereas NAD^+ is used in liberation of ammonia, a degradative reaction. The enzyme produces ammonia from amino acids when these are needed as glucose precursors or for energy. Formation of NADH during the oxidative deamination reaction is a welcome bonus, since it can be reoxidized by the respiratory chain with formation of ATP. The reaction as shown is readily reversible in the test tube but it is likely that *in vivo* it occurs more frequently in the direction of ammonia formation. The concentration of ammonia needed for the reaction to produce glutamate is toxic and under normal conditions would rarely be attained except in the perivenous region of the liver. A major source of ammonia is **bacterial metabolism** in the intestinal lumen, the released ammonia being absorbed and transported to the liver. Glutamate dehydrogenase incorporates this ammonia, as well as that produced locally, into glutamate. The enzyme's dominant role in ammonia removal is emphasized by its location in liver mitochondria, where the initial reactions of the urea cycle occur.

Glutamate dehydrogenase is regulated allosterically by purine nucleotides. When there is need for oxidation of amino acids for energy, the activity is increased in the direction of glutamate degradation by ADP and GDP, which are indicative of a low cellular energy level. GTP and ATP, indicative of an ample energy level, are allosteric activators in the direction of glutamate synthesis (Figure 18.13).

Free Ammonia Is Incorporated into and Produced from Glutamine

Free ammonia is toxic and is preferentially transported in the blood in the form of amino or amide groups. Fifty percent of circulating amino acid molecules are **glutamine**, an ammonia transporter. The amide group of glutamine is a nitrogen donor for several classes of molecules, including purine bases, and the amino group of cytosine. Glutamate and ammonia are substrates for **glutamine synthetase** (Figure 18.14). ATP is needed for activation of the α-carboxyl group to make the reaction energetically favorable.

Removal of the amide group is catalyzed by **glutaminase** (Figure 18.15). There are tissue-specific isozymes. Mitochondrial glutaminase I of kidney and liver requires phosphate for activity. Liver contains glutamine synthetase and glutaminase but is neither a net consumer nor a net producer of glutamine. The two enzymes

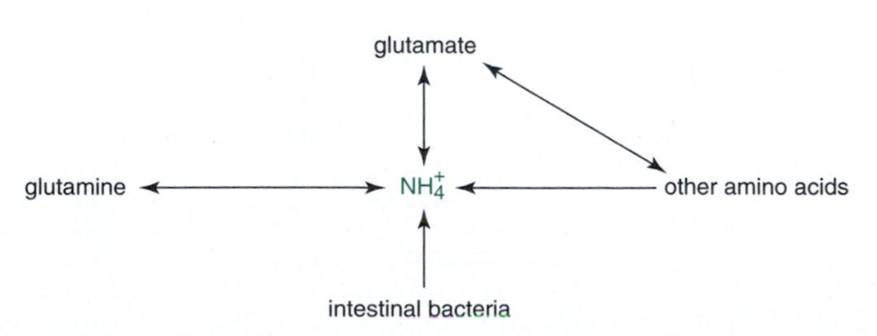

FIGURE 18.12
Role of glutamate in amino acid synthesis, degradation, and interconversion.

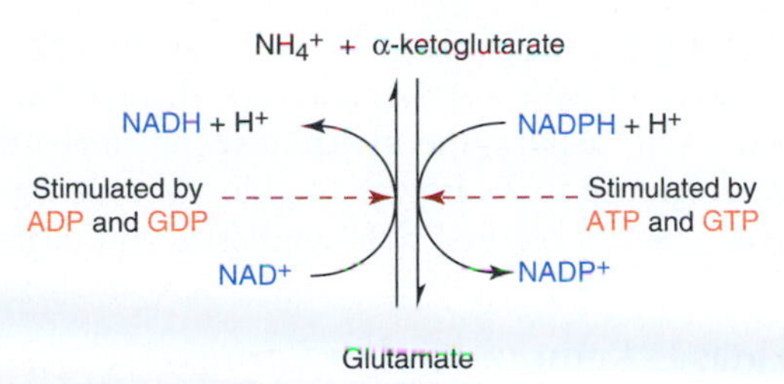

FIGURE 18.13
Allosteric regulation of glutamate dehydrogenase.

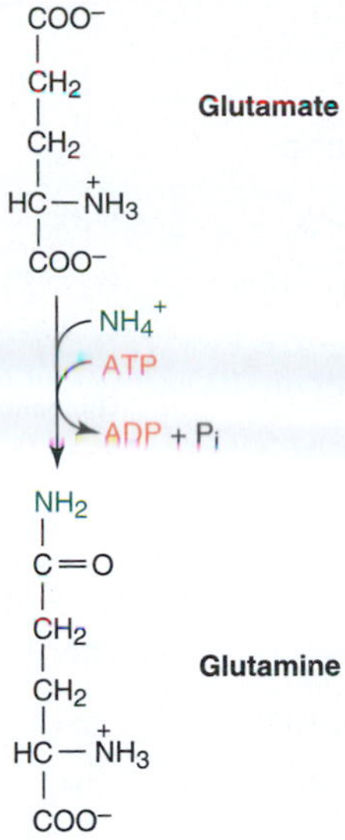

COO⁻–CH₂–CH₂–HC–$\overset{+}{N}H_3$–COO⁻ **Glutamate**

NH_4^+, ATP → ADP + P_i

NH_2–C=O–CH₂–CH₂–HC–$\overset{+}{N}H_3$–COO⁻ **Glutamine**

FIGURE 18.14
Reaction catalyzed by glutamine synthetase.

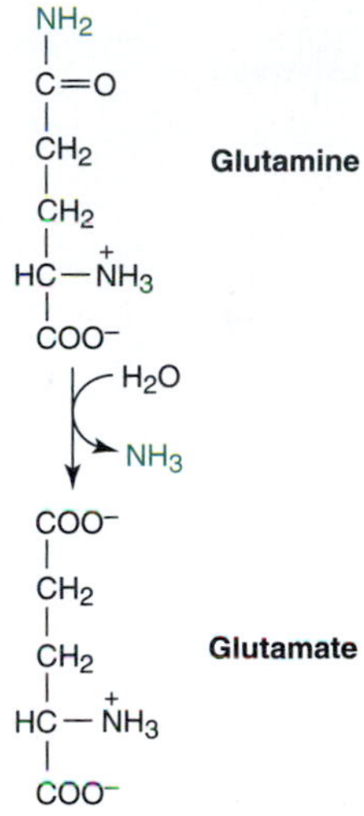

NH_2–C=O–CH₂–CH₂–HC–$\overset{+}{N}H_3$–COO⁻ **Glutamine**

H_2O → NH_3

COO⁻–CH₂–CH₂–HC–$\overset{+}{N}H_3$–COO⁻ **Glutamate**

FIGURE 18.15
Reaction catalyzed by glutaminase.

are confined to parenchymal cells in different segments of the liver. The periportal region contains glutaminase (and the urea cycle enzymes) and is in contact with blood that comes from skeletal muscle. The perivenous area represents 5% of parenchymal cells; blood from it flows to the kidney and cells in this area contain glutamine synthetase. This **"intercellular glutamine cycle"** (Figure 18.16) may be a mechanism for scavenging ammonia that has not been incorporated into urea. The enzymes of urea synthesis are found in the same periportal cells as glutaminase, whereas the uptake of glutamate and α-ketoglutarate for glutamine synthesis predominates in the perivenous region. The glutamine cycle makes it possible to control flux of ammonia either to urea or to glutamine and thence to excretion of ammonia by the kidney under different pH conditions.

Amide Group of Asparagine Is Derived from Glutamine

The amide group of **asparagine** comes from that of glutamine (Figure 18.17), and not from free ammonia, as in the synthesis of glutamine. ATP is needed to activate

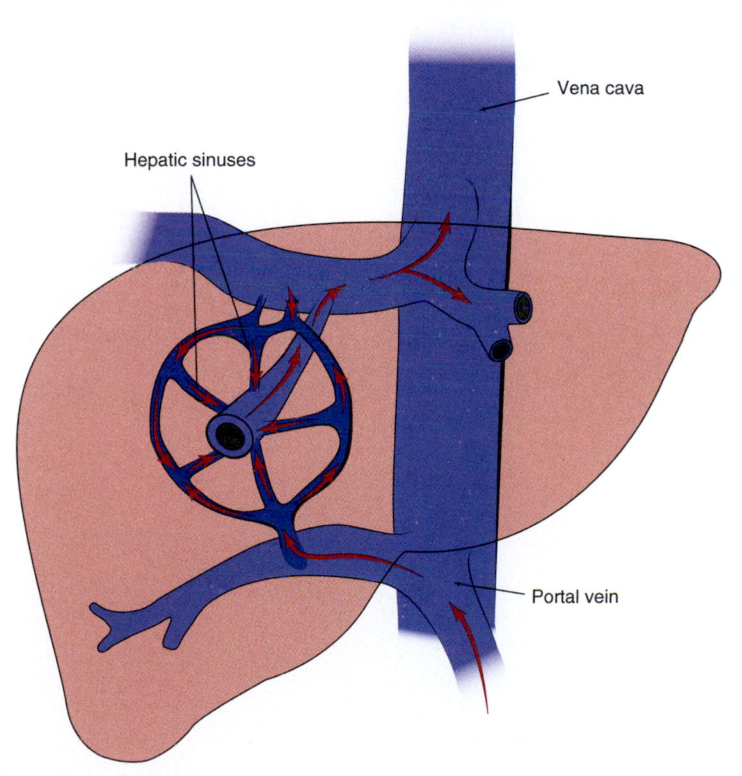

FIGURE 18.16
Intercellular glutamine cycle.
Periportal cells surround incoming blood vessels, and perivenous cells surround outgoing blood vessels.

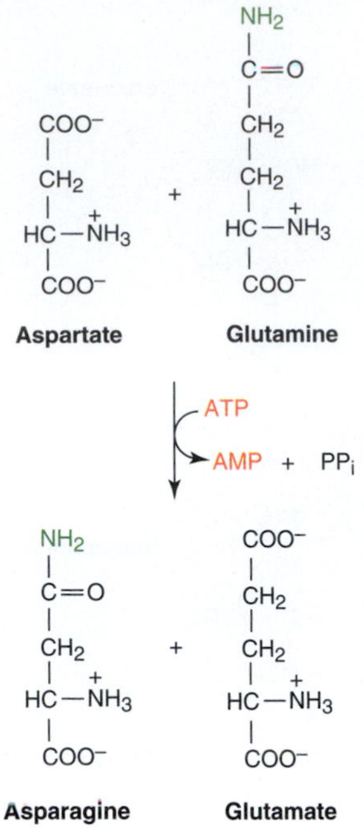

Aspartate + Glutamine → (ATP → AMP + PP_i) → Asparagine + Glutamate

FIGURE 18.17
Synthesis of asparagine.

the receptor α-carboxyl group. Asparagine is readily synthesized in most cells, but some leukemic cells seem to have lost this ability. A therapeutic approach that has been tried in patients with **asparagine synthetase**-deficient tumors is treatment with exogenous **asparaginase** to hydrolyze the blood-borne asparagine on which these cells rely (Figure 18.18). Normal cells synthesize and degrade asparagine.

Amino Acid Oxidases Remove Amino Groups

Many amino acids are substrates for L-amino acid oxidase (Figure 18.19). The significance of this reaction in metabolism of amino acids is uncertain but appears to be small. The enzyme contains flavin mononucleotide (FMN) and produces hydrogen peroxide. After the hydrogen peroxide is reduced to water, the final products are an α-keto acid, ammonia, and water, the same products as for the glutamate dehydrogenase reaction. In the amino acid oxidase reaction, unlike the reaction catalyzed by glutamate dehydrogenase, there is no concomitant production of NADH, and therefore no production of ATP.

A **D-amino acid oxidase** also occurs in human cells. Very little of the D-amino acid isomer is found in humans and the role of D-amino acid oxidase may be in degradation of D-amino acids derived from intestinal bacteria.

18.3 | TRANSPORT OF NITROGEN TO LIVER AND KIDNEY

Protein Is Constantly Being Degraded

Cells die on a regular and programmed basis, a process termed **apoptosis,** and their component molecules are metabolized. Individual proteins also undergo regular turnover under normal conditions. Even though the reactions involved in intracellular protein degradation have been identified, an understanding of the regulation of protein degradation is in its infancy. The half-life of a protein can be an hour or less, such as for ornithine decarboxylase, phosphokinase C, and insulin, several months for hemoglobin and histones, or the life of the organism for the crystallins of the lens. The majority, however, turn over every few days. Selection of a partic-

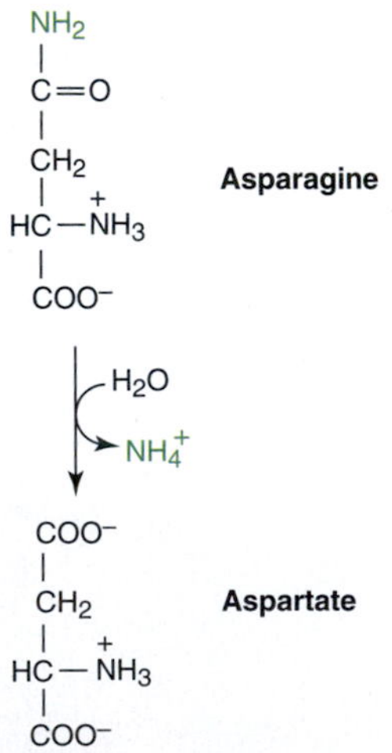

Asparagine → (H_2O → NH_4^+) → Aspartate

FIGURE 18.18
Reaction catalyzed by asparaginase.

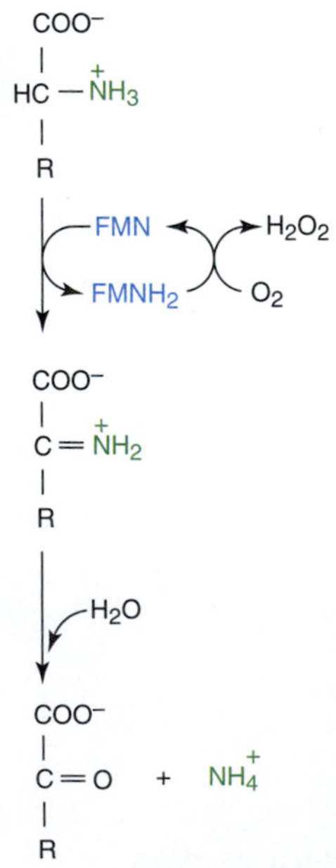

$HC(COO^-)(R)-NH_3^+$ → (FMN → $FMNH_2$; O_2 → H_2O_2) → $C(COO^-)(R)=NH_2^+$ → (H_2O) → $C(COO^-)(R)=O$ + NH_4^+

FIGURE 18.19
Reaction of L-amino acid oxidase, a flavoprotein.

ular protein molecule for degradation is not well understood but may, in many cases, occur by "marking" with covalently bound molecules of a small protein, termed ubiquitin (see p. 272). Ubiquitin contains 76 amino acid residues and is attached via its C-terminal glycine residue to the terminal amino group and to lysine residues in the protein to be marked for degradation. This is a nonlysosomal, ATP-dependent process and requires a complex of three enzymes known as **ubiquitin** protein ligase. Ubiquitination and protein degradation participate in regulating the cell cycle by influencing the availability of proteins required in the S and G1 phases. Other protein degradation occurs in the lysosomes, or extralysosomally by calcium-dependent enzymes.

Amino Acids Are Transported from Muscle After Proteolysis

The majority of body protein, and consequently of amino acids, is in skeletal muscle. Under conditions of energy need, this protein is degraded and amino groups from the amino acids are transferred to produce glutamine and alanine and transported to liver or kidney. Urea is produced in liver and ammonia (from glutamine) in kidneys (Figure 18.20). Carbon skeletons are either used for energy or transported to the liver for gluconeogenesis. Muscle protein responds to conditions such as starvation, trauma, burns, and septicemia, by undergoing massive degradation. Of the amino acids released, most important as a source of fuel are **branched-chain amino acids** (valine, leucine, and isoleucine). The first step in their degradation is transamination, which occurs almost exclusively in muscle. Protein, of course, is degraded throughout the body, but muscle is by far the greatest source of free amino acids for metabolism.

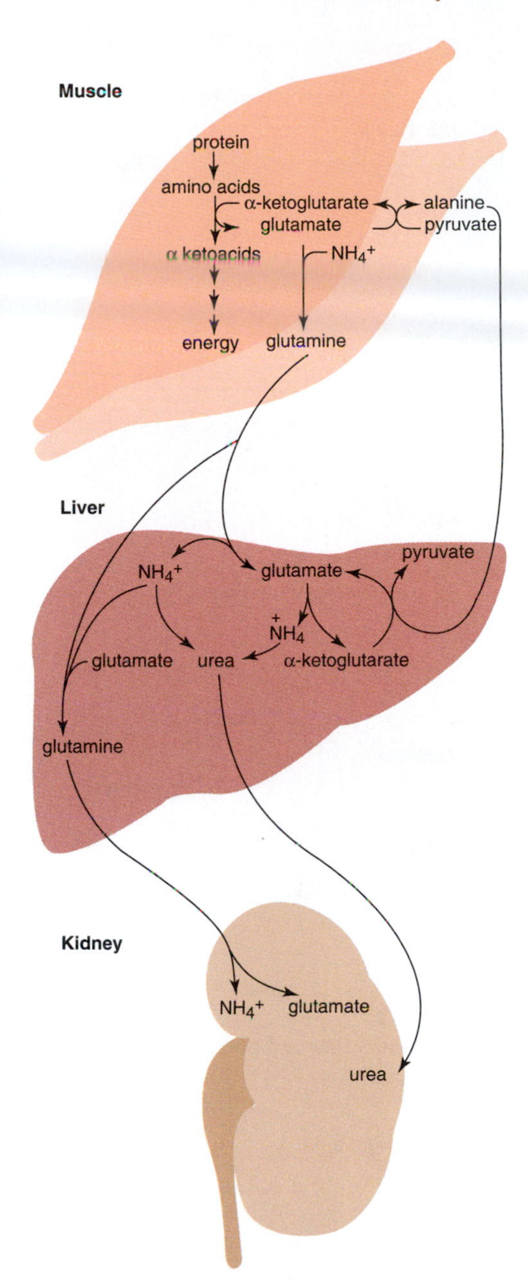

FIGURE 18.20
Major pathways of interorgan nitrogen transport following muscle proteolysis.

Ammonia Is Released in Liver and Kidney

The main destination of glutamine and alanine in the blood is the liver (see Figure 18.20), where ammonia is released by alanine aminotransferase, glutaminase, and glutamate dehydrogenase. Glutamate dehydrogenase not only releases ammonia but also produces NADH and α-ketoglutarate, a glucogenic intermediate. Under conditions of energy need these products are very beneficial. Many malignancies and chronic conditions produce a condition called **cachexia,** characterized by wasting of muscle. This is caused not at the level of regulation of the rate of muscle protein breakdown, but rather by an increase in the rate at which liver removes amino acids from plasma, which, in turn, has a potentiating effect on muscle proteolysis. When circulating glucagon concentration is high (a signal that carbon is required by the liver for gluconeogenesis), it also potentiates amino acid metabolism by stimulating amino acid uptake by the liver.

Some glutamine and alanine is taken up by the kidneys. Ammonia is released by the same enzymes that are active in liver, protonated to ammonium ion and excreted. When acidosis occurs the body shunts glutamine from liver to kidneys to conserve bicarbonate, since formation of urea, the major mechanism for removal of NH_4^+, requires bicarbonate. To avoid excessive use and excretion of this anion as urea during acidosis, uptake of glutamine by liver is suppressed, and more is transported to kidneys for excretion as ammonium ion.

18.4 | UREA CYCLE

Nitrogen Atoms of Urea Come from Ammonia and Aspartate

The **urea cycle** and the tricarboxylic acid (TCA) cycle were discovered by Sir Hans Krebs and co-workers; the urea cycle being described before the TCA cycle.

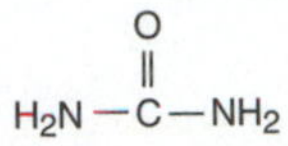

FIGURE 18.21
Urea.

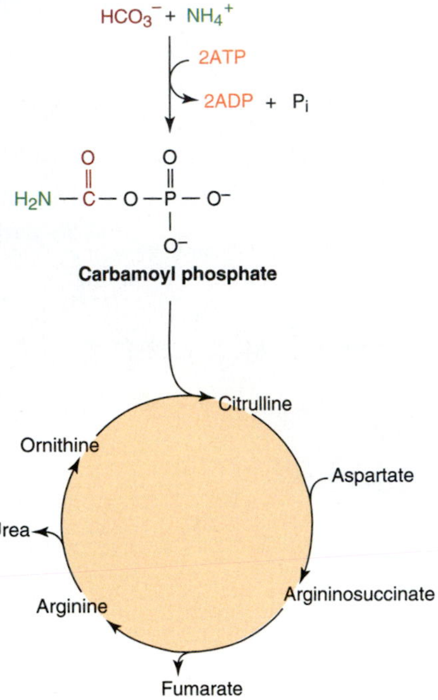

FIGURE 18.22
Synthesis of carbamoyl phosphate and entry into urea cycle.

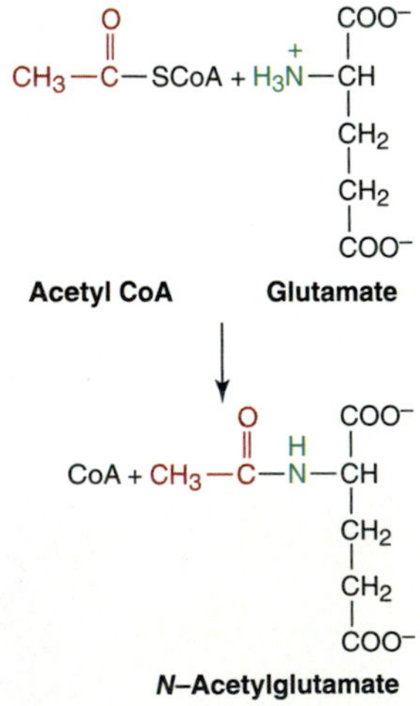

FIGURE 18.23
Reaction catalyzed by *N*-acetylglutamate synthetase.

In land-dwelling mammals, the urea cycle is the mechanism of choice for nitrogen excretion. The two nitrogens in each urea molecule (Figure 18.21) are derived from two sources, free ammonia and the amino group of aspartate. The cycle starts and finishes with **ornithine.** Unlike the TCA cycle, where carbons of oxaloacetate at the start are different from those at the end, the carbons in the final ornithine are the same carbons with which the molecule started.

Ammonia (first nitrogen for urea) enters the cycle after condensation with bicarbonate to form **carbamoyl phosphate** (Figure 18.22), which reacts with ornithine to form **citrulline.** Aspartate (donor of the second urea nitrogen) and citrulline react to form **argininosuccinate,** which is then cleaved to arginine and fumarate. Arginine is hydrolyzed to urea and ornithine is regenerated. Urea is then transported to the kidney and excreted in urine. The cycle requires 4 ATP for each urea molecule produced and excreted. It is therefore more energy efficient to incorporate ammonia into amino acids than to excrete it. The major regulatory step is the initial synthesis of carbamoyl phosphate, and the cycle is also regulated by induction of the enzymes involved.

Synthesis of Urea Requires Five Enzymes

Carbamoyl phosphate synthetase I (CPSI) is technically not a part of the urea cycle, although it is essential for urea synthesis. Free ammonium ion and bicarbonate are condensed, at the expense of 2 ATP, to form carbamoyl phosphate. One ATP activates bicarbonate, and the other donates the phosphate group of carbamoyl phosphate. Carbamoyl phosphate synthetase I occurs in the mitochondrial matrix, uses ammonia as nitrogen donor, and is absolutely dependent on **N-acetylglutamate** for activity (Figure 18.23). Another enzyme with similar activity, **carbamoyl phosphate synthase II (CPSII),** is cytosolic, uses the amide group of glutamine, and is not affected by *N*-acetylglutamate. It participates in pyrimidine biosynthesis (see p. 840).

Formation of citrulline is catalyzed by **ornithine transcarbamoylase** (Figure 18.24) in the mitochondrial matrix. Citrulline is transported out of the mitochondria into the cytosol where the other reactions of the urea cycle occur. Argininosuccinate production by **argininosuccinate synthetase** requires hydrolysis of ATP to AMP and PP_i, the equivalent of hydrolysis of two molecules of ATP. Cleavage of argininosuccinate by **argininosuccinate lyase** produces fumarate and arginine. Arginine is cleaved by **arginase** to ornithine and urea. Ornithine reenters the mitochondrial matrix for another turn of the cycle. The inner mitochondrial membrane contains a **citrulline/ornithine exchange transporter.**

Synthesis of additional ornithine from glutamate for the cycle will be described later. Since arginine is produced from carbons and nitrogens of ornithine, ammonia, and aspartate, it is a nonessential amino acid. In growing children, however, where there is positive nitrogen balance and net incorporation of nitrogen into body constituents, *de novo* synthesis of arginine is inadequate and the amino acid becomes essential.

Carbons from aspartate, released as fumarate, may enter mitochondria and be metabolized to oxaloacetate by fumarase and malate dehydrogenase of the TCA cycle, transaminated, and then can enter another turn of the urea cycle as aspartate. Most oxaloacetate (about two-thirds) derived from fumarate is metabolized via phospho*enol*pyruvate to glucose (Figure 18.25). The amount of fumarate used to form ATP is approximately equal to that required for the urea cycle and gluconeogenesis, meaning that the liver itself gains no net energy in the process of amino acid metabolism.

Since humans cannot metabolize urea it is transported to the kidneys for filtration and excretion. Any urea that enters the intestinal tract is cleaved by **urease**-containing bacteria in the intestinal lumen, the resulting ammonia being absorbed and used by the liver.

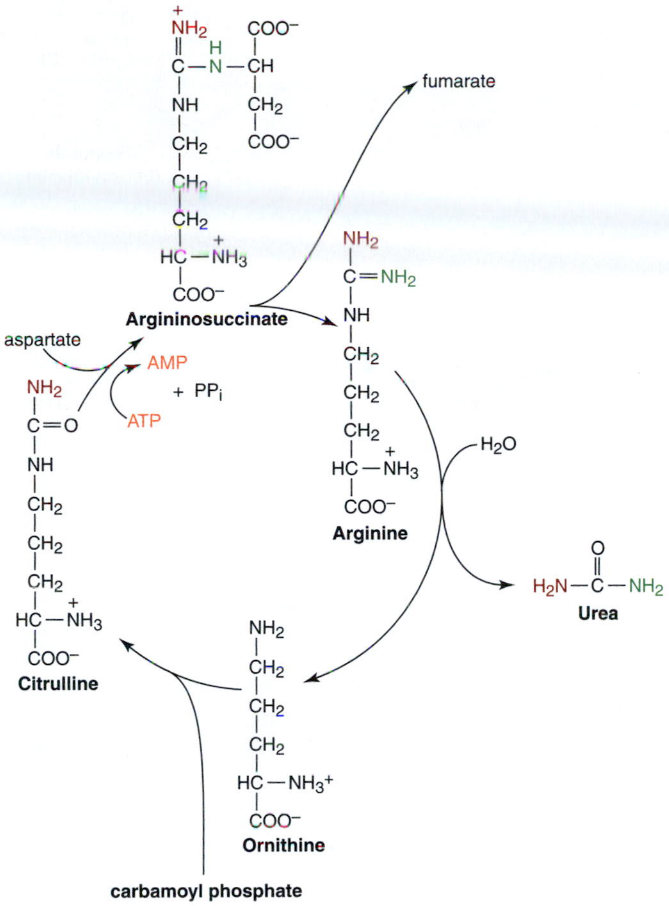

FIGURE 18.24
Urea cycle.

Urea Synthesis Is Regulated by an Allosteric Effector and Enzyme Induction

Carbamoyl phosphate synthetase I requires the allosteric activator **N-acetylglutamate** (see Figure 18.23). This compound is synthesized from glutamate and acetyl CoA by **N-acetylglutamate synthetase,** which is activated by arginine. Acetyl CoA, glutamate, and arginine are needed to supply intermediates or energy for the urea cycle, and the presence of *N*-Acetylglutamate indicates that they are available. Tight regulation is desirable for a pathway that controls the plasma level of potentially toxic ammonia and that is also highly energy dependent.

Induction of urea cycle enzymes occurs (10- to 20-fold) when delivery of ammonia or amino acids to liver rises. Concentration of cycle intermediates also plays a role in its regulation through mass action. A high-protein diet (net excess amino acids) and starvation (need to metabolize body protein in order to provide carbons for energy production) result in induction of urea cycle enzymes.

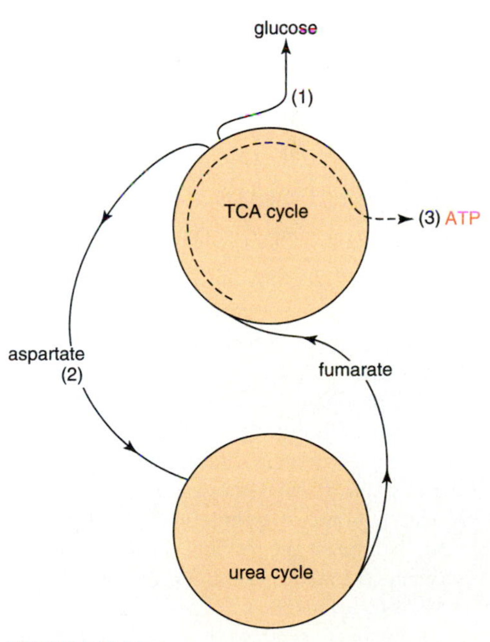

FIGURE 18.25
Fumarate from the urea cycle is a source of glucose (1), aspartate (2), or energy (3).

Metabolic Disorders of Urea Synthesis Have Serious Results

The urea cycle is the major mechanism for utilization and elimination of ammonia, a very toxic substance. Metabolic disorders that arise from abnormal function of enzymes of urea synthesis are potentially fatal and cause coma when ammonia concentrations become high. Loss of consciousness may be a consequence of ATP depletion. The major source of ATP is oxidative phosphorylation, which is linked to

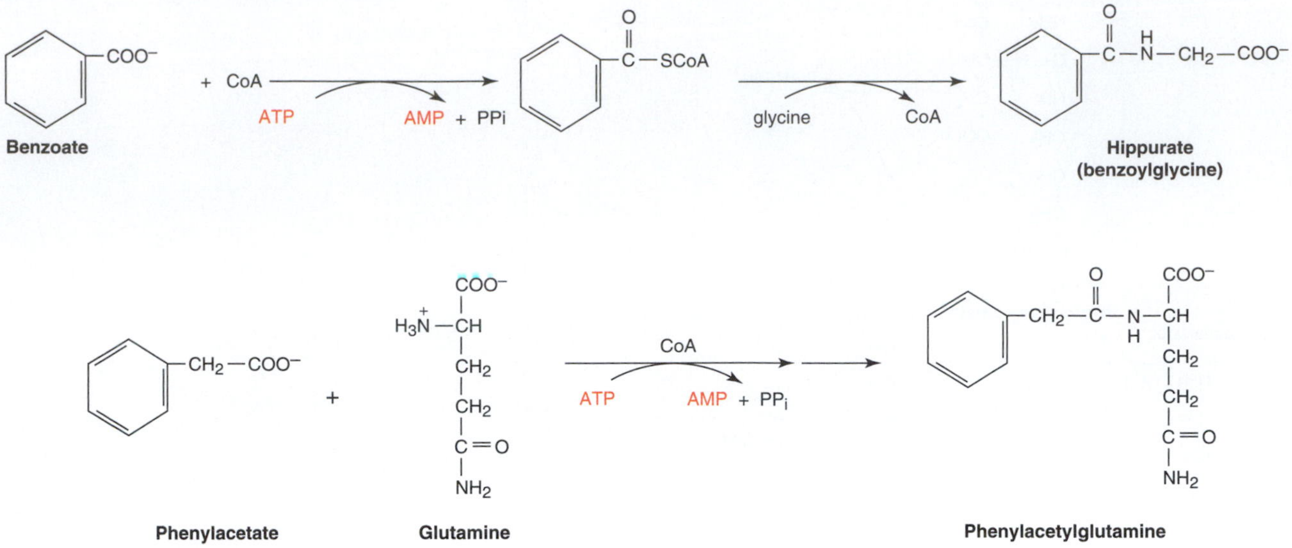

FIGURE 18.26
Detoxification reactions as alternatives to the urea cycle.

transfer of electrons from the TCA cycle down the electron transport chain (see p. 552). A high concentration of ammonia sequesters α-ketoglutarate in the form of glutamate, thus depleting the TCA cycle of important intermediates and reducing ATP production.

Patients with a deficiency in each of the urea cycle enzymes have been found. Therapy for these deficiencies has a threefold basis: (1) to limit protein intake and potential buildup of ammonia, (2) to remove excess ammonia, and (3) to replace any intermediates missing from the urea cycle. The first is accomplished by limiting ingestion of amino acids, replacing them if necessary with the equivalent α-keto acids to be transaminated *in vivo*. The bacterial source of ammonia in the intestines can be decreased by a compound that acidifies the colon, such as levulose, a poorly absorbed synthetic disaccharide that is metabolized by colonic bacteria to acidic products. This promotes the excretion of ammonia in feces as protonated ammonium ions. Antibiotics can also be administered to kill ammonia-producing bacteria. The second is achieved by compounds that bind covalently to amino acids and produce nitrogen-containing molecules that are excreted in urine. Figure 18.26 shows condensation of benzoate and glycine to form **hippurate,** and of phenylacetate and glutamine to form **phenylacetylglutamine.** Phenylacetate is extremely unpalatable and is given as the precursor sodium phenylbutyrate. Both reactions require energy for activation of the carboxyl groups by addition of CoA.

Clinical Correlations 18.1 and 18.2 give examples of therapy for specific enzyme deficiencies, which often includes administration of urea cycle intermediates.

18.5 | SYNTHESIS AND DEGRADATION OF INDIVIDUAL AMINO ACIDS

Other aspects of metabolism of glutamate, glutamine, aspartate, asparagine, alanine, and arginine will now be discussed. Synthesis of other nonessential amino acids and degradation of all the amino acids will be covered, as well as synthesis of physiologically important amino acid derivatives.

Glutamate Is a Precursor of Glutathione and γ-Aminobutyrate

Glutamate is a component of **glutathione** (see p. 818) and a precursor for **γ-aminobutyric acid (GABA),** a neurotransmitter (see p. 1001) (Figure 18.27), and of proline and ornithine (see p. 792).

CLINICAL CORRELATION 18.1

Carbamoyl Phosphate Synthetase and *N*-Acetylglutamate Synthetase Deficiencies

Hyperammonemia has been observed in infants with 0–50% of the normal level of carbamoyl phosphate synthetase activity in their livers. In addition to the treatments described in the text, these infants have been treated with arginine, on the hypothesis that activation of *N*-acetylglutamate synthetase by arginine would stimulate the residual carbamoyl phosphate synthetase. This enzyme deficiency generally leads to mental retardation. A case of *N*-acetylglutamate synthetase deficiency has been described and treated successfully by administering carbamoyl glutamate, an analog of *N*-acetylglutamate, that is also able to activate carbamoyl phosphate synthetase.

CLINICAL CORRELATION 18.2
Deficiencies of Urea Cycle Enzymes

Ornithine Transcarbamoylase Deficiency

The most common deficiency involving urea cycle enzymes is lack of ornithine transcarbamoylase. Mental retardation and death often result, but the occasional finding of normal development in treated patients suggests that the usually associated mental retardation is caused by excess ammonia before adequate therapy. The gene for ornithine transcarbamoylase is on the X chromosome, and males are generally more seriously affected than heterozygotic females. In addition to ammonia and amino acids appearing in the blood in increased amounts, orotic acid also increases, presumably because carbamoyl phosphate that cannot be used to form citrulline diffuses into the cytosol, where it condenses with aspartate, ultimately forming orotate (see p. 840).

Argininosuccinate Synthetase and Lyase Deficiency

The inability to condense citrulline with aspartate results in accumulation of citrulline in blood and its excretion in urine (citrullinemia). Therapy for this normally benign disease requires specific supplementation with arginine for protein synthesis and for formation of creatine. Impaired ability to split argininosuccinate to form arginine resembles argininosuccinate synthetase deficiency in that the substrate, in this case argininosuccinate, is excreted in large amounts. Severity of symptoms in this disease varies greatly so that it is hard to evaluate the effect of therapy, which includes dietary supplementation with arginine.

Arginase Deficiency

Arginase deficiency is rare but causes many abnormalities in development and function of the central nervous system. Arginine accumulates and is excreted. Precursors of arginine and products of arginine metabolism may also be excreted. Unexpectedly, some urea is also excreted; this has been attributed to a second type of arginase found in the kidney. A diet including essential amino acids but excluding arginine has been used effectively.

Brusilow, S. W., Danney, M., Waber, L. J., Batshaw, M. Treatment of episodic hyperammonemia in children with inborn errors of urea synthesis. *N. Engl. J. Med.* 310:1630, 1984.

Arginine Is Also Synthesized in Intestines

Production of arginine for protein synthesis, rather than as an intermediate in the urea cycle, occurs in kidneys, which lack arginase. The major site of synthesis of citrulline to be used as an arginine precursor is intestinal mucosa, which has all necessary enzymes to convert glutamate (via ornithine as described below) to citrulline, which is then transported to the kidneys to produce arginine. Arginine is a precursor for **nitric oxide** (see p. 486); in brain, **agmatine**, a compound that may have antihypertensive properties, is an arginine derivative (Figure 18.28).

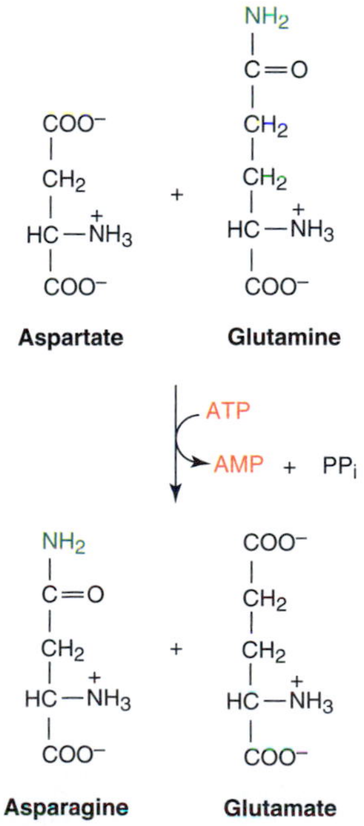

$$\underset{\text{Aspartate}}{^-OOC{-}CH_2{-}CH(\overset{+}{N}H_3){-}COO^-} + \underset{\text{Glutamine}}{H_2N{-}C({=}O){-}CH_2{-}CH_2{-}CH(\overset{+}{N}H_3){-}COO^-} \xrightarrow{ATP \rightarrow AMP + PP_i} \underset{\text{Asparagine}}{H_2N{-}C({=}O){-}CH_2{-}CH(\overset{+}{N}H_3){-}COO^-} + \underset{\text{Glutamate}}{^-OOC{-}CH_2{-}CH_2{-}CH(\overset{+}{N}H_3){-}COO^-}$$

FIGURE 18.27
Synthesis of γ-aminobutyric acid (GABA).

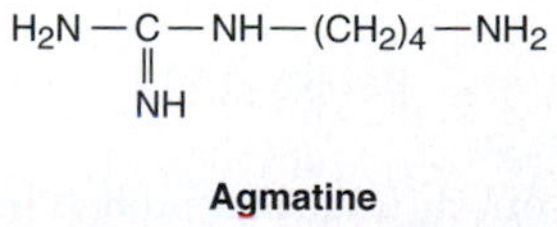

$$H_2N{-}C({=}NH){-}NH{-}(CH_2)_4{-}NH_2$$

Agmatine

FIGURE 18.28
Agmatine.

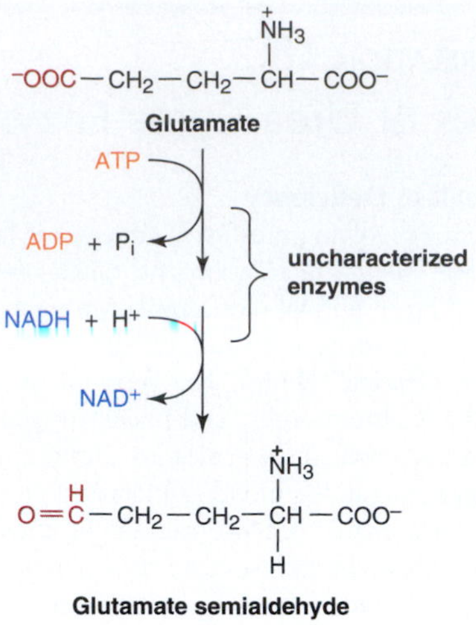

FIGURE 18.29
Synthesis of glutamic semialdehyde.

Ornithine and Proline

Ornithine, the precursor of citrulline and arginine, and **proline** are synthesized from glutamate and degraded, by a slightly different pathway, to glutamate. Synthesis of these two nonessential amino acids starts from α-ketoglutarate with a shared reaction that uses ATP and NADH (Figure 18.29) and forms **glutamic semialdehyde.** This spontaneously will cyclize to form a Schiff base between the aldehyde and amino groups, which is then reduced by NADPH to proline. Glutamic semialdehyde can undergo transamination of the aldehyde group, which prevents cyclization and produces ornithine (Figure 18.30).

Proline is converted back to the Schiff base intermediate, Δ^1-pyrroline 5-carboxylate, which is in equilibrium with glutamic semialdehyde. The transaminase reaction in the ornithine synthetic pathway is freely reversible and forms glutamic semialdehyde from ornithine (Figure 18.30). Proline residues can be hydroxylated after incorporation into a protein to form **3-** or **4-hydroxyproline** (Figure 18.31). When these are released by protein degradation and metabolized they produce glyoxalate and pyruvate, and 4-hydroxy-2-ketoglutarate, respectively.

Ornithine is a precursor of putrescine, the foundation molecule of the polyamines, highly cationic molecules that interact with DNA. Ornithine decarboxylase catalyzes this reaction (Figure 18.32). It is regulated by phosphorylation at several sites, presumably in response to specific hormones, growth factors, or cell cycle regulatory signals. It can also be induced, often the first easily measurable sign that cell division is imminent, since polyamines must be synthesized before mitosis can occur. Other common **polyamines** are **spermidine** and **spermine** (see Figure 18.59), which are synthesized from putrescine by addition of propylamine, a product of methionine metabolism (see p. 806).

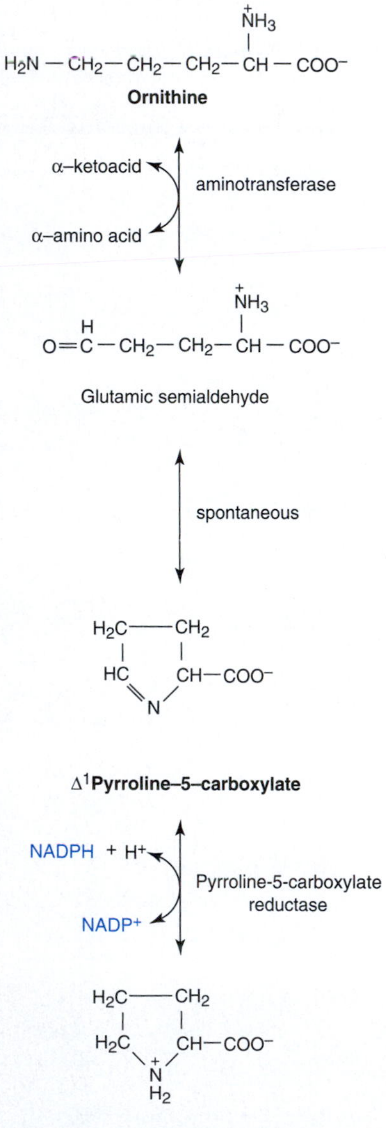

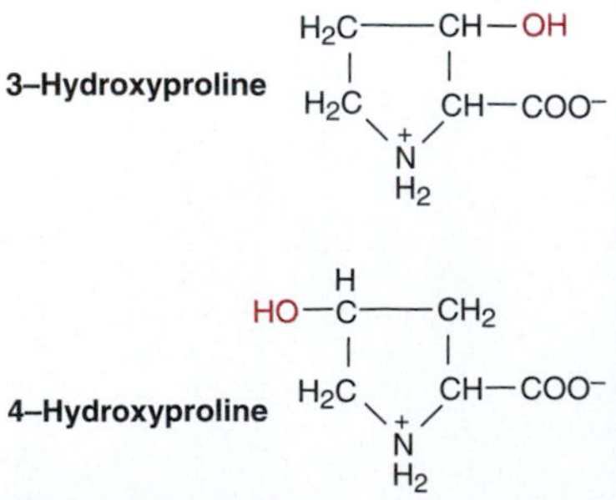

FIGURE 18.30
Synthesis of ornithine and proline from glutamic semialdehyde, a shared intermediate.

FIGURE 18.31
Hydroxyprolines.

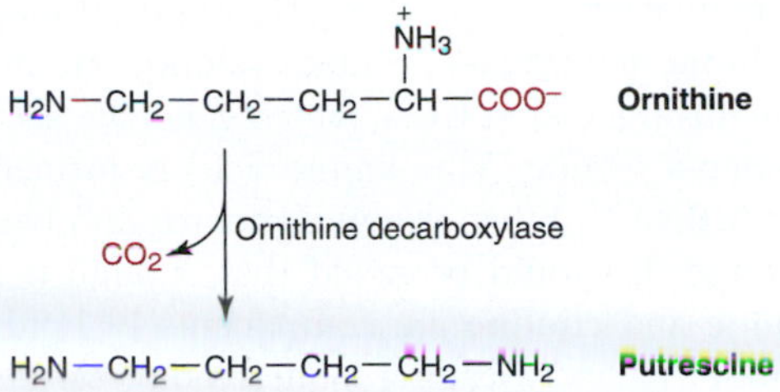

$$H_2N-CH_2-CH_2-CH_2-\overset{\overset{+}{N}H_3}{\overset{|}{C}H}-COO^-$$ **Ornithine**

Ornithine decarboxylase (CO_2 released)

$$H_2N-CH_2-CH_2-CH_2-CH_2-NH_2$$ **Putrescine**

FIGURE 18.32
Decarboxylation of ornithine to putrescine.
Structures of spermidine and spermine are shown in Figure 18.59.

Serine and Glycine

Serine is synthesized *de novo* from 3-phosphoglycerate from the glycolytic pathway. When serine provides gluconeogenic intermediates, 3-phosphoglycerate is an intermediate in its degradation, although the enzymes and intermediates in the two pathways are different. Synthesis of serine uses phosphorylated intermediates starting from 3-phosphoglycerate (Figure 18.33*a*), loss of the phosphate being the last step. From serine to 3-phosphoglycerate the intermediates are unphosphorylated, with addition of a phosphate being the last step. The enzymes of these two pathways are not the same (Figure 18.33*b*). Another reaction for entry of serine into intermediary metabolism is via **serine dehydratase,** which forms pyruvate with loss of the amino group as NH_4^+ (Figure 18.34). The same enzyme catalyzes a similar reaction with threonine (see p. 797).

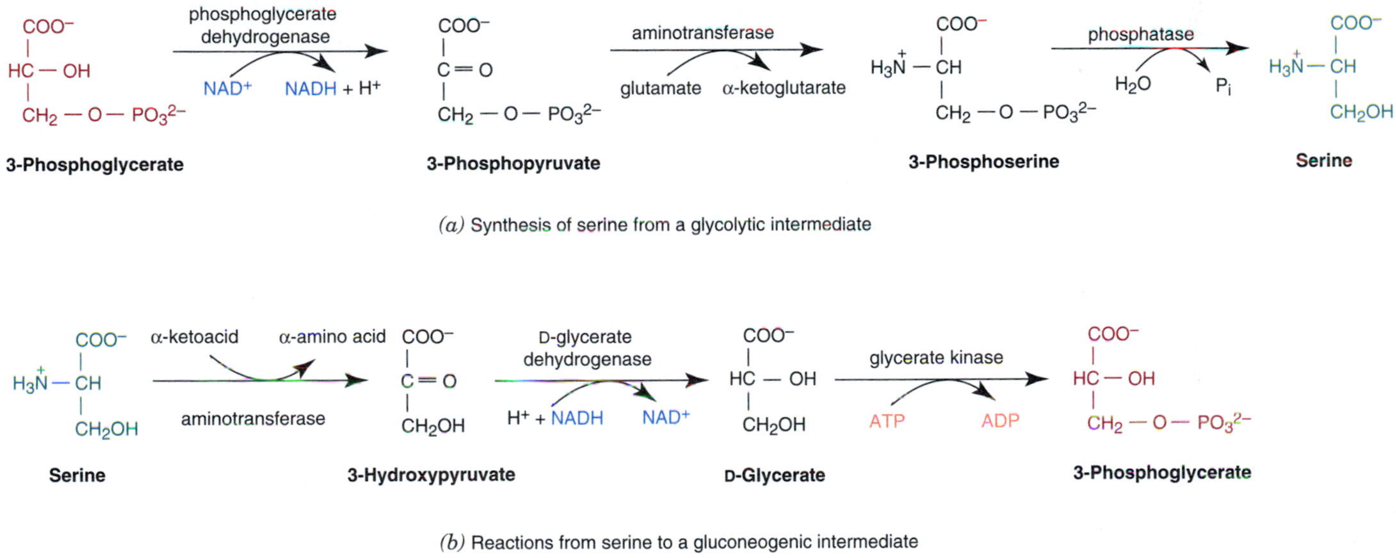

3-Phosphoglycerate (COO^- / $HC-OH$ / $CH_2-O-PO_3^{2-}$) → phosphoglycerate dehydrogenase (NAD^+ → $NADH + H^+$) → **3-Phosphopyruvate** (COO^- / $C=O$ / $CH_2-O-PO_3^{2-}$) → aminotransferase (glutamate → α-ketoglutarate) → **3-Phosphoserine** (COO^- / $H_3\overset{+}{N}-CH$ / $CH_2-O-PO_3^{2-}$) → phosphatase (H_2O → P_i) → **Serine** (COO^- / $H_3\overset{+}{N}-CH$ / CH_2OH)

(a) Synthesis of serine from a glycolytic intermediate

Serine (COO^- / $H_3\overset{+}{N}-CH$ / CH_2OH) → aminotransferase (α-ketoacid → α-amino acid) → **3-Hydroxypyruvate** (COO^- / $C=O$ / CH_2OH) → D-glycerate dehydrogenase ($H^+ + NADH$ → NAD^+) → **D-Glycerate** (COO^- / $HC-OH$ / CH_2OH) → glycerate kinase (ATP → ADP) → **3-Phosphoglycerate** (COO^- / $HC-OH$ / $CH_2-O-PO_3^{2-}$)

(b) Reactions from serine to a gluconeogenic intermediate

FIGURE 18.33
Pathways for (*a*) synthesis of serine and (*b*) metabolism of serine for gluconeogenesis.

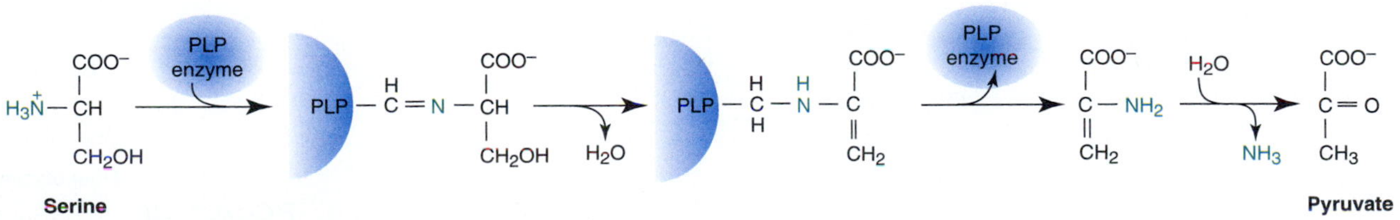

Serine (COO^- / $H_3\overset{+}{N}-CH$ / CH_2OH) → PLP enzyme → $PLP-\overset{H}{C}=N-CH(COO^-)(CH_2OH)$ → H_2O → $PLP-CH_2-NH-C(COO^-)=CH_2$ → PLP enzyme → COO^- / $C-NH_2$ / $\|CH_2$ → H_2O in, NH_3 out → **Pyruvate** (COO^- / $C=O$ / CH_3)

FIGURE 18.34
Reaction of serine dehydratase requires pyridoxal phosphate.

Serine is precursor of an unusual but important amino acid. Certain proteins, notably **glutathione peroxidase,** contain **selenocysteine** (Figure 18.35). In mRNA for selenoproteins the codon UGA, which generally serves as a termination codon, codes for selenocysteine. This amino acid is formed from serine after formation of the seryl-$tRNA^{Ser}$. Ethanolamine, choline, and betaine (Figure 18.36) are derivatives of serine. It should be noted that choline is now classified as a vitamin. **Ethanolamine** and **choline** are components of lipids, and **betaine** is a methyl donor in a minor pathway leading to methionine salvage (see p. 806). Serine is also a sulfhydryl group acceptor from homocysteine in cysteine synthesis (see p. 804).

In some enzymes a serine residue is modified to form a prosthetic group. In humans the only example described so far is ***S*-adenosylmethionine decarboxylase** (discussed below in relation to polyamine formation; see p. 807). The prosthetic group formed is pyruvate. *S*-adenosylmethionine decarboxylase is synthesized in precursor form that is then cleaved autocatalytically between a glutamate and a serine residue to form two polypeptides. During cleavage other reactions convert the new N-terminal serine of one of the resulting peptides into a pyruvate (Figure 18.37). The pyruvate functions in decarboxylation by forming a Schiff base with the amino group of *S*-adenosylmethionine.

Serine is converted reversibly to **glycine** in a reaction that requires pyridoxal phosphate and tetrahydrofolate. **N^5,N^{10}-methylenetetrahydrofolate** (N^5,N^{10}-THF or N^5,N^{10}-H_4 folate) is produced (Figure 18.38). The demand for serine or glycine and the amount of N^5,N^{10}-THF available determine the direction of this reaction. Glycine is degraded to CO_2 and ammonia by a **glycine cleavage complex** (Figure 18.39; see Clin. Corr. 18.3). This reaction is reversible *in vitro,* but not *in vivo,* as the K_m values for ammonia and N^5,N^{10}-THF are much higher than their respective physiological concentrations.

Glycine is the precursor of **glyoxalate,** which can be transaminated back to glycine or oxidized to **oxalate** (Figure 18.40). Excessive production of oxalate forms the insoluble calcium oxalate salt, which may lead to kidney stones. The role of glycine as a neurotransmitter is described on page 993.

Tetrahydrofolate Is a Cofactor in Some Reactions of Amino Acids

Tetrahydrofolate is the reduced form of folic acid, one of the B vitamins, and often occurs as a poly-γ-glutamyl derivative (Figure 18.41). Tetrahydrofolate is a one-**carbon carrier** that facilitates interconversion of methenyl, formyl, formimino, methylene, and methyl groups (Figure 18.42). This occurs at the expense of pyridine nucleotide reduction or oxidation and occurs while the carbon moiety is attached to H_4 folate (Figure 18.43). The most oxidized forms, formyl and methenyl, are

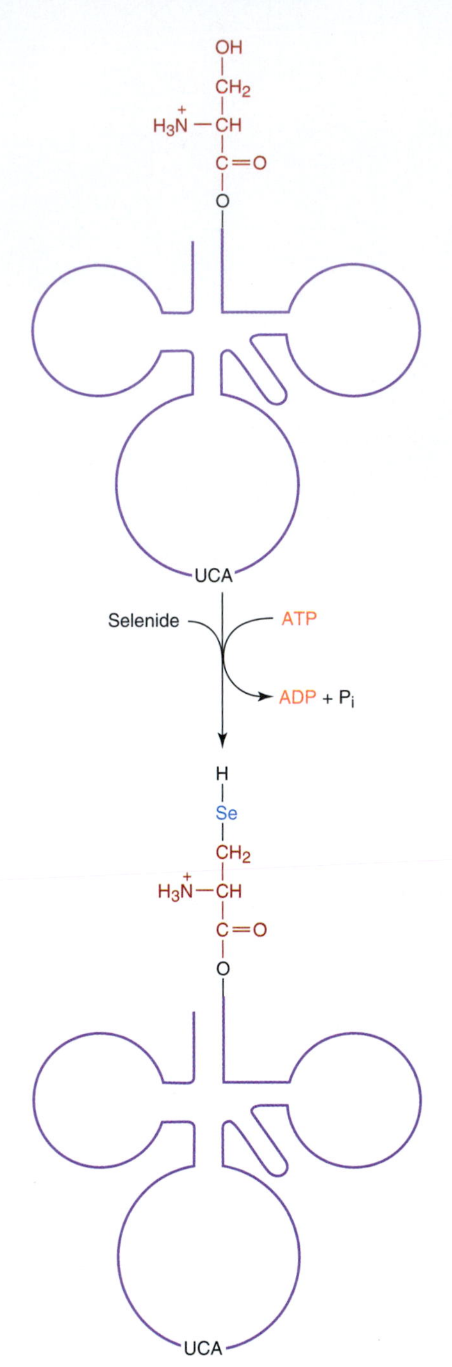

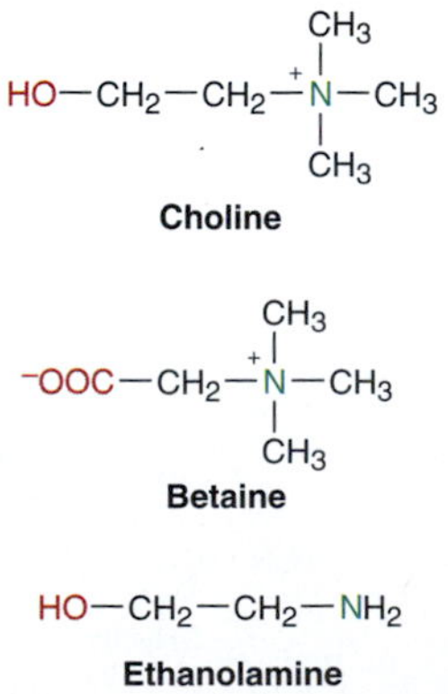

FIGURE 18.35
Formation of selenocysteinyl tRNA from seryl tRNA is via a phosphoseryl tRNA intermediate.

$HO-CH_2-CH_2-\overset{+}{N}(CH_3)_3$

Choline

$^{-}OOC-CH_2-\overset{+}{N}(CH_3)_3$

Betaine

$HO-CH_2-CH_2-NH_2$

Ethanolamine

FIGURE 18.36
Choline and related compounds.

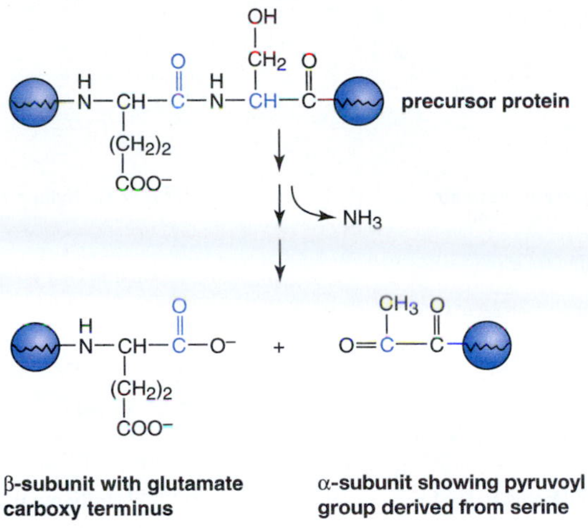

FIGURE 18.37
Formation of enzyme with covalently bound pyruvoyl prosthetic group.

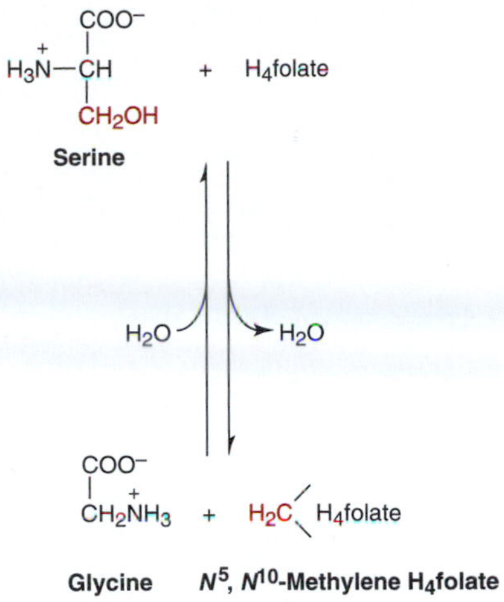

FIGURE 18.38
Serine hydroxymethyltransferase.

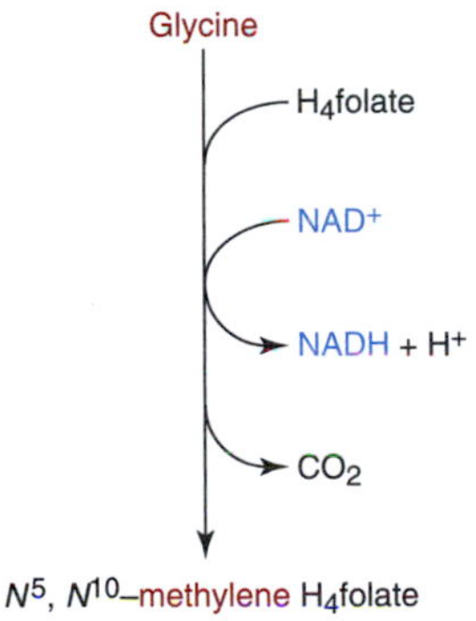

FIGURE 18.39
Glycine cleavage is pyridoxal phosphate dependent.

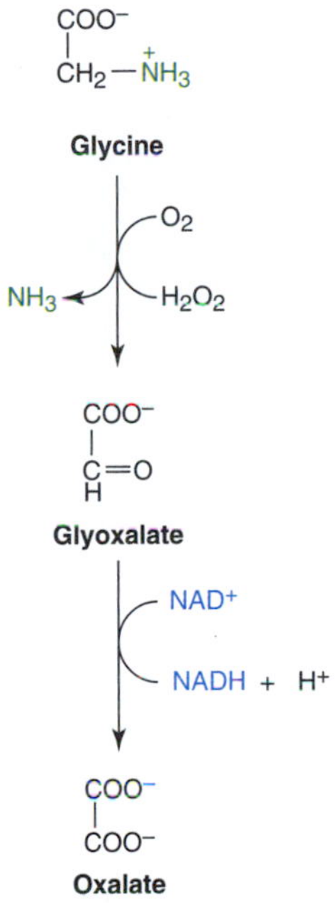

FIGURE 18.40
Oxidation of glycine.

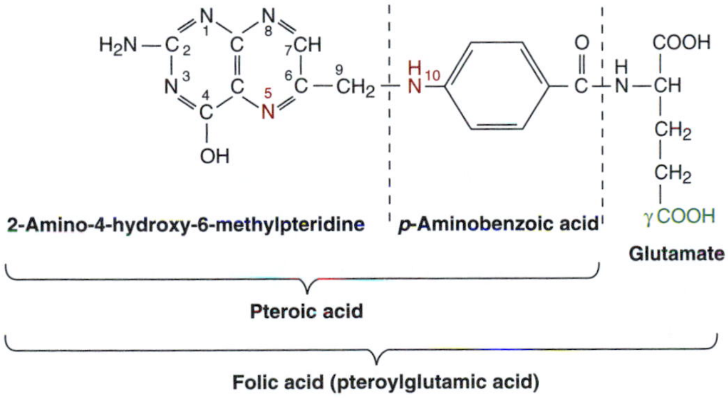

FIGURE 18.41
Components of folate.
Poly-γ-glutamate can be added to the γ-carboxyl group.

CLINICAL CORRELATION 18.3
Nonketotic Hyperglycinemia

Nonketotic hyperglycinemia is characterized by severe mental deficiency. Many patients do not survive infancy. The name of this very serious disease is meant to distinguish it from ketoacidosis in abnormalities of branched-chain amino acid metabolism in which, for unknown reasons, the glycine level in the blood is also elevated. Deficiency of glycine cleavage complex has been demonstrated in homogenates of liver from several patients, and isotopic studies *in vivo* have confirmed that this enzyme is not active in these patients. The glycine cleavage complex consists of four different protein subunits. Inherited abnormalities have been found in three of them. The severity of this disease suggests that glycine cleavage is of major importance in the catabolism of glycine. Glycine is a major inhibitory neurotransmitter, which probably explains some neurological complications of the disease. Vigorous measures to reduce the glycine levels fail to alter the course of the disease.

Nyhan, W. L. Metabolism of glycine in the normal individual and in patients with non-ketotic hyperglycinemia. *J. Inherit. Metab. Dis.* 5:105, 1982.

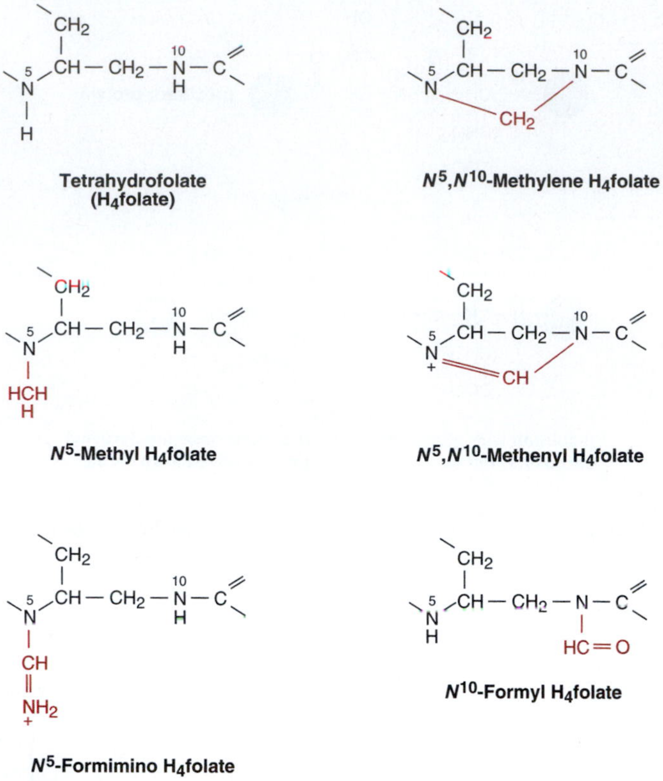

Tetrahydrofolate (H4folate)

N^5,N^{10}-Methylene H4folate

N^5-Methyl H4folate

N^5,N^{10}-Methenyl H4folate

N^5-Formimino H4folate

N^{10}-Formyl H4folate

FIGURE 18.42
Active center of H_4 folate.
N^5 is the site of attachment of methyl groups; N^{10} is the site for formyl and formimino; methylene and methenyl groups form bridges between N^5 and N^{10}.

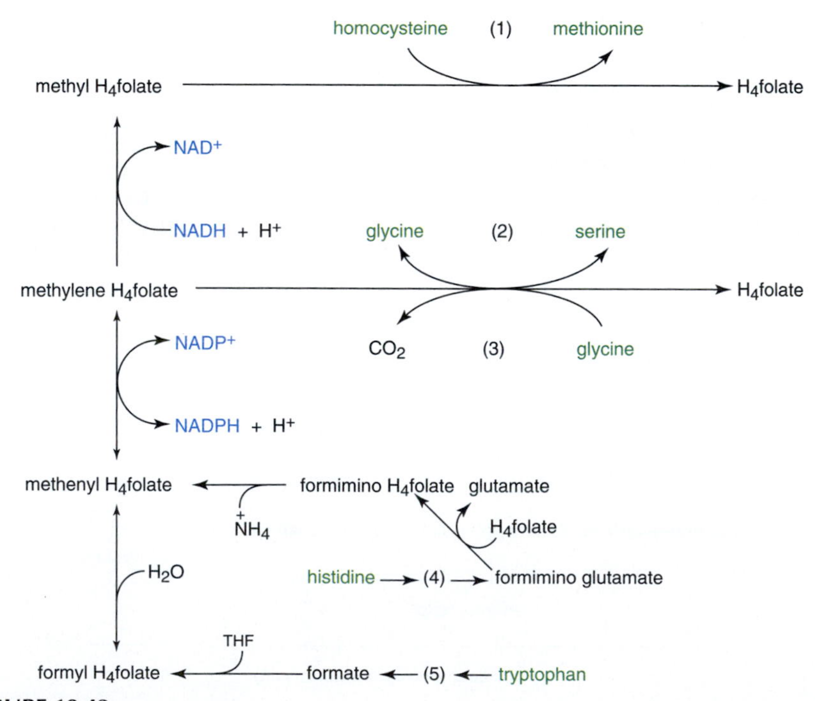

FIGURE 18.43
Interconversion of derivatized H_4 folate and roles in amino acid metabolism.
(1) Methionine salvage, (2) serine hydroxymethyltransferase, (3) glycine cleavage complex, (4) histidine degradation, and (5) tryptophan metabolism.

bound to N^{10} of the pteridine ring, methylene forms a bridge between N^5 and N^{10}, and methyl is bound to N^5. The interconversions permit use of a carbon that is removed from a molecule in one oxidation state for addition in a different oxidation state to a different molecule (Figure 18.42).

In reduction of the N^5,N^{10}-methylene bridge of tetrahydrofolate to a methyl group for transfer to the pyrimidine ring (Figure 18.44), a reaction found in **thymidylate synthesis** (see p. 845), the reducing power comes not from pyridine nucleotide but from the pteridine ring itself. The resulting dihydrofolate has no physiological role and is reduced back to tetrahydrofolate by NADPH-dependent dihydrofolate reductase (see Clin. Corr. 18.4). The net result of the two reactions is oxidation of NADPH and reduction of the methylene bridge to a methyl group, analogous to the one-step reactions shown in Figure 18.43.

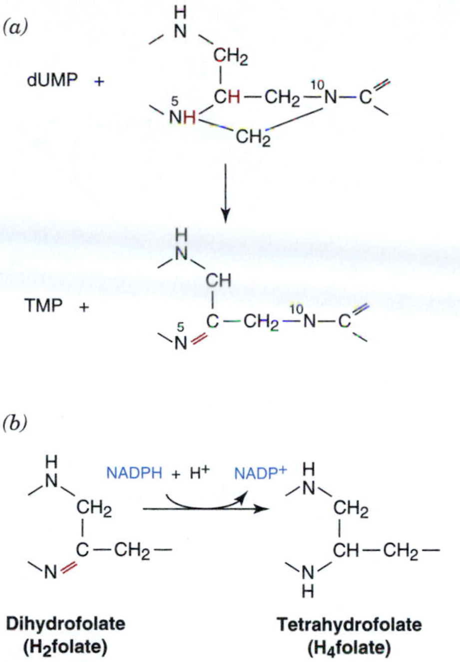

(a) dUMP + [H-N, CH_2, 5 NH, CH–CH_2–10 N–C, CH_2] → TMP + [H-N, CH, 5 N, C–CH_2–10 N–C]

(b) Dihydrofolate (H_2folate) —NADPH + H^+ → $NADP^+$→ Tetrahydrofolate (H_4folate)

FIGURE 18.44
Reduction reactions involving H_4 folate.
(*a*) Reduction of methylene group on H_4 folate to a methyl group and transfer to dUMP to form TMP. (*b*) Reduction of resulting dihydrofolate to tetrahydrofolate.

Threonine

Threonine is usually metabolized to pyruvate (Figure 18.45), but an intermediate in this pathway can undergo thiolysis with CoA to acetyl CoA and glycine. Thus the α-carbon atom of threonine can contribute to the one-carbon pool. In an alternative, but less common pathway, **serine dehydratase** (see p. 793) converts threonine to α-ketobutyrate. A complex similar to pyruvate dehydrogenase metabolizes this to propionyl CoA.

Phenylalanine and Tyrosine

Tyrosine and **phenylalanine** are discussed together, since tyrosine results from hydroxylation of phenylalanine and is the first product in phenylalanine degradation. Because of this, tyrosine is not usually considered to be essential, whereas phenylalanine is. Three-quarters of ingested phenylalanine is hydroxylated to tyrosine. This is catalyzed by phenylalanine hydroxylase (Figures 18.46 and 18.47; see Clin. Corr. 18.5), which is tetrahydrobiopterin dependent (Figure 18.48). The reaction occurs only in the direction of tyrosine formation, and phenylalanine cannot be synthesized from tyrosine. Biopterin resembles folic acid in containing

CLINICAL CORRELATION 18.4
Folic Acid Deficiency

The 100–200 mg of folic acid required daily by an average adult can theoretically be obtained easily from conventional Western diets. Deficiency of folic acid, however, is not uncommon. It may result from limited diets, especially when food is cooked at high temperatures for long periods, which destroys the vitamin. Intestinal diseases, notably celiac disease, are often characterized by folic acid deficiency caused by malabsorption. Inability to absorb folate is rare. Folate deficiency is usually seen only in newborns and produces symptoms of megaloblastic anemia. Of the few cases studied, some were responsive to large doses of oral folate but one required parenteral administration, suggesting a carrier-mediated process for absorption. Besides the anemia, mental and other central nervous system symptoms are seen in patients with folate deficiency, and all respond to continuous therapy although permanent damage appears to be caused by delayed or inadequate treatment. A classical experiment was carried out by a physician, apparently serving as his own experimental subject, to study the human requirements for folic acid. His diet consisted only of foods (boiled repeatedly to extract the water-soluble vitamins) to which vitamins (and minerals) were added but not folic acid. Symptoms attributable to folate deficiency did not appear for seven weeks, altered appearance of blood cells and formiminoglutamate excretion were seen only at 13 weeks, and serious symptoms (irritability, forgetfulness, and macrocytic anemia) appeared only after four months. Neurological symptoms were alleviated within two days after folic acid was added to the diet; the blood picture became normal more slowly. The occurrence of folic acid in essentially all natural foods makes deficiency difficult, and apparently a normal person accumulates more than adequate reserves of this vitamin. For pregnant women the situation is very different. Needs of the fetus for normal growth and development include constant, uninterrupted supplies of coenzymes (in addition to amino acids and other cell constituents). Recently, folate deficiency has been implicated in spina bifida.

Herbert, V. Experimental nutritional folate deficiency in man. *Trans. Assoc. Am. Physicians* 75:307, 1962.

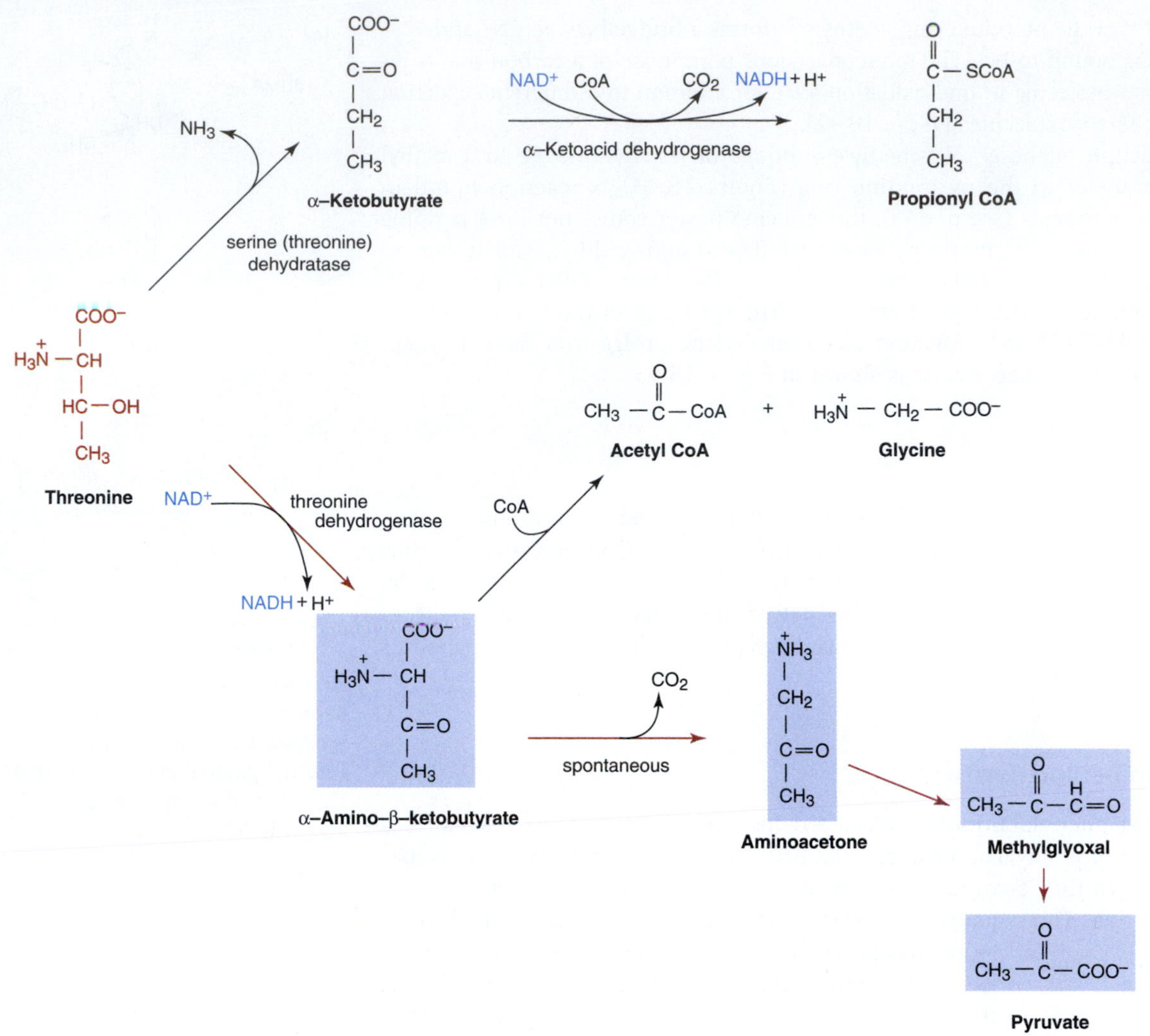

FIGURE 18.45
Outline of threonine metabolism.
Major pathway is in color.

a pteridine ring but is not a vitamin. It is synthesized from GTP (see Clin. Corr. 18.5).

Tyrosine Metabolism Produces Fumarate and Acetoacetate

The first step in metabolism of tyrosine is transamination by **tyrosine aminotransferase** to *p*-hydroxyphenylpyruvate (Figure 18.49). The enzyme is inducible, its synthesis being increased by glucocorticoids and dietary tyrosine. ***p*-Hydroxyphenylpyruvate oxidase** produces **homogentisic acid.** This complex reaction involves decarboxylation, oxidation, migration of the carbon side chain, and hydroxylation. Ascorbic acid is required for at least one of these activities, but

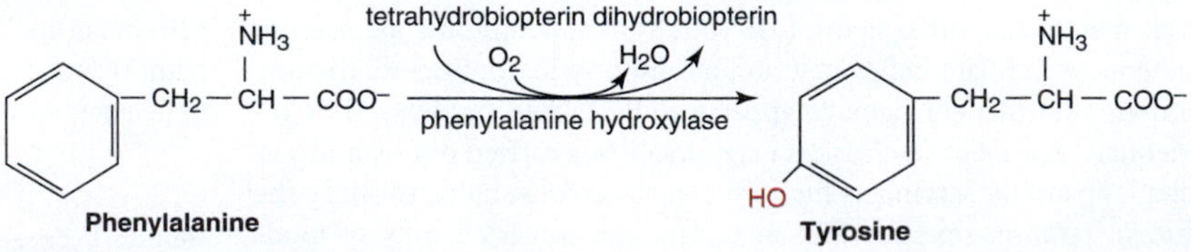

tetrahydrobiopterin dihydrobiopterin
O_2 H_2O
phenylalanine hydroxylase
Phenylalanine
Tyrosine

FIGURE 18.46
Phenylalanine hydroxylase.

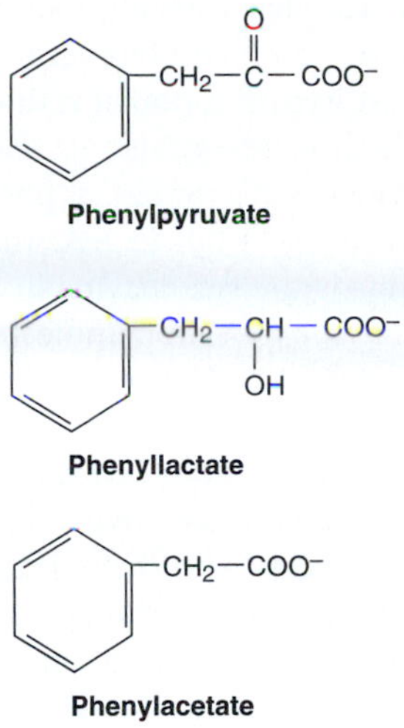

Phenylpyruvate

Phenyllactate

Phenylacetate

FIGURE 18.47
Minor products of phenylalanine metabolism.

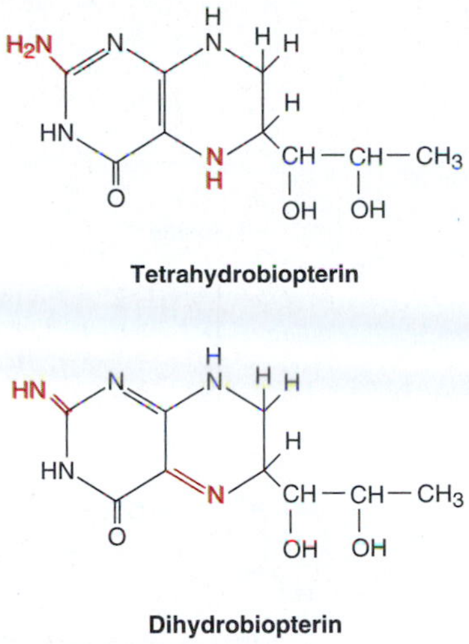

Tetrahydrobiopterin

Dihydrobiopterin

FIGURE 18.48
Biopterin.
Dihydro (quinonoid) form is produced during oxidation of aromatic amino acids and is then reduced to the tetrahydro form by a dehydrogenase using NADH and H^+.

all four are catalyzed by the one enzyme. The aromatic ring is next cleaved by the iron-containing homogentisate oxidase to maleylacetoacetate. This will isomerize from cis to trans to give fumarylacetoacetate, in a reaction catalyzed by maleylacetoacetate isomerase, an enzyme that seems to require glutathione for activity. Fumarylacetoacetate is then cleaved to fumarate and acetoacetate. Fumarate can be further utilized in the TCA cycle for energy or for gluconeogenesis. Acetoacetate can be used, as acetyl CoA, for lipid synthesis or energy (see Clin. Corr. 18.6).

Dopamine, Epinephrine, and Norepinephrine Are Derivatives of Tyrosine

Most tyrosine not incorporated into proteins is metabolized to acetoacetate and fumarate. Some is used as precursor of **catecholamines.** The eventual metabolic fate of tyrosine carbons is determined by the first step in each pathway. Catecholamine

CLINICAL CORRELATION 18.5

Phenylketonuria

Phenylketonuria (PKU) is the most common disease caused by a deficiency of an enzyme of amino acid metabolism. The name comes from the excretion of phenylpyruvic acid, a phenylketone, in the urine. Phenyllactate is also excreted (Figure 18.47), as is an oxidation product of phenylpyruvate, phenylacetate, which gives the urine a "mousey" odor. These three metabolites are found only in trace amounts in urine in the healthy person. The symptoms of mental retardation associated with this disease can be prevented by a diet low in phenylalanine. Routine screening is required by governments in many parts of the world. Classical PKU is an autosomal recessive deficiency of phenylalanine hydroxylase. Over 170 mutations in the gene have been reported. In some cases there are severe neurological symptoms and very low IQ. These are generally attributed to toxic effects of phenylalanine, possibly because of reduced transport and metabolism of other aromatic amino acids in the brain due to competition from the high phenylalanine concentration. Another characteristic is light color of skin and eyes, due to underpigmentation because of tyrosine deficiency. Conventional treatment is to feed affected infants a synthetic diet low in phenylalanine, but including tyrosine, for about four to five years and impose dietary protein restriction for several more years or for life.

About 3% of infants with high levels of phenylalanine have normal hydroxylase but are defective in either synthesis or reduction of biopterin. Biopterin deficiency can be treated by addition of this compound to the diet. Deficiency in dihydrobiopterin reductase is more serious. Since biopterin is also necessary for the synthesis of catecholamines and serotonin, which function as neurotransmitters, central nervous system functions are more seriously affected and treatment includes administration of precursors of serotonin and catecholamines.

Brewster, T. G., Moskowitz, M. A., and Kaufman, S. Dihydrobiopterin reductase deficiency associated with severe neurologic disease and mild hyperphenylalanemia. *Pediatrics* 63:94, 1979; Kaufman, S. Regulation of the activity of hepatic phenylalanine hydroxylase. *Adv. Enzyme Regul.* 25:37, 1986; Scriver, C. R. and Clow, L. L. Phenylketonuria: epitome of human biochemical genetics. *N. Engl. J. Med.* 303:1336,1980; and Woo, S. L. C. Molecular basis and population genetics of phenylketonuria. *Biochemistry* 28:1, 1989.

Tyrosine

α-ketoglutarate

glutamate

p-Hydroxyphenylpyruvate

O_2

CO_2

p-hydroxyphenylpyruvate oxidase

Homogentisate

O_2

homogentisate oxidase

Maleylacetoacetate

maleylacetoacetate isomerase

Fumarylacetoacetate

fumarylacetoacetate hydrolase

Fumarate + Acetoacetate

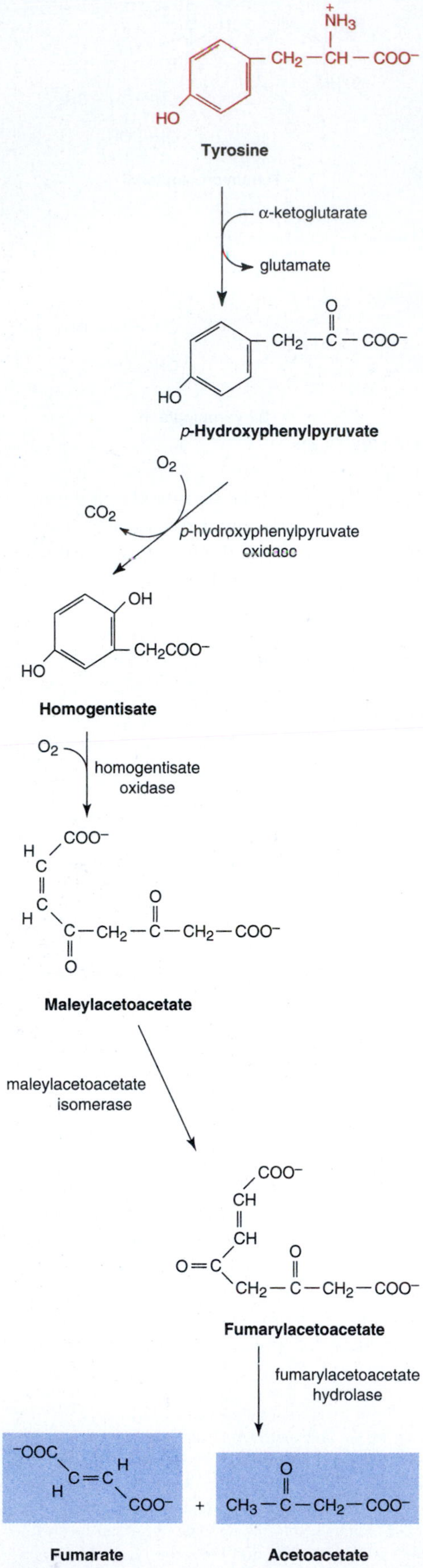

FIGURE 18.49
Degradation of tyrosine.

synthesis (Figure 18.50) starts with **tyrosine hydroxylase,** which, like phenylalanine and tryptophan hydroxylase, is dependent on tetrahydrobiopterin. All three are affected by **biopterin** deficiency or a defect in **dihydrobiopterin reductase** (see Figure 18.48). Tyrosine hydroxylase produces dihydroxyphenylalanine, also known as DOPA, dioxophenylalanine. **DOPA decarboxylase,** with pyridoxal phosphate as cofactor, forms **dopamine,** the active neurotransmitter, from DOPA. In the substantia nigra and some other parts of the brain, this is the last enzyme in this pathway (see Clin. Corr. 18.7). The adrenal medulla converts dopamine to **norepinephrine** and **epinephrine** (also called **adrenaline**). The methyl group of epinephrine is derived from *S*-adenosylmethionine (see p. 803).

Brain plasma tyrosine regulates norepinephrine formation. Estrogens decrease tyrosine concentration and increase tyrosine aminotransferase activity, diverting tyrosine into the catabolic pathway. Estrogen sulfate competes for the pyridoxal phosphate site on DOPA decarboxylase. These three effects combined may help explain mood variations during the menstrual cycle. Tyrosine is therapeutic in some cases of depression and stress. Its transport appears to be reduced in skin fibroblasts from schizophrenic patients, indicating other roles for tyrosine derivatives in mental disorders.

Catecholamines are metabolized by **monoamine oxidase** and **catecholamine *O*-methyltransferase**. Major metabolites are shown in Figure 18.51. Absence of

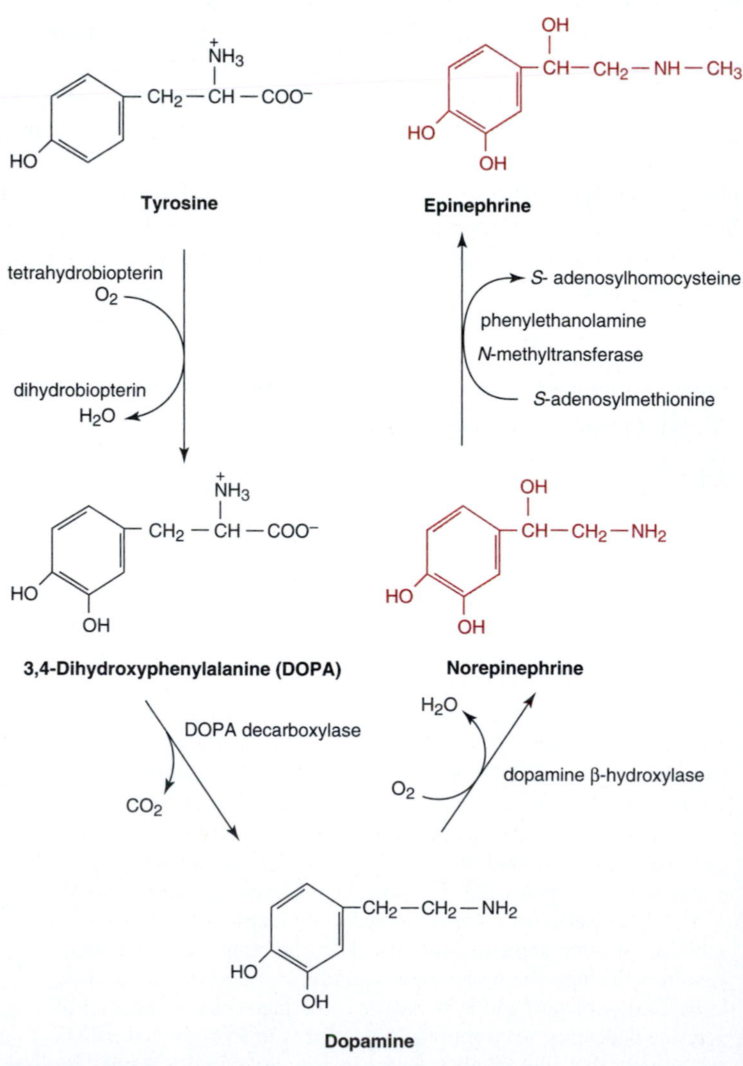

FIGURE 18.50
Synthesis of catecholamines.

CLINICAL CORRELATION 18.6

Disorders of Tyrosine Metabolism

Tyrosinemias

The absence or deficiency of tyrosine aminotransferase produces accumulation and excretion of tyrosine and metabolites. The disease, oculocutaneous or type II tyrosinemia, results in eye and skin lesions and mental retardation. Type I, hepatorenal tyrosinemia, is more serious, involving liver failure, renal tubular dysfunction, rickets, and polyneuropathy, caused by a deficiency of fumarylacetoacetate hydrolase. Accumulation of fumarylacetoacetate and maleylacetate, both of which are alkylating agents, can lead to DNA alkylation and tumorigenesis. Both diseases are autosomal recessive and rare.

Alcaptonuria

The first condition identified as an "inborn error of metabolism" was alcaptonuria. Individuals deficient in homogentisate oxidase excrete almost all ingested tyrosine as the colorless homogentisic acid in their urine. This autooxidizes to the corresponding quinone, which polymerizes to form an intensely dark color. Concern about the dark urine is the only consequence of this condition early in life. Homogentisate is slowly oxidized to pigments that are deposited in bones, connective tissue, and other organs, a condition called ochronosis because of the ochre color of the deposits. This is thought to be responsible for the associated arthritis, especially in males. The study of alcaptonuria by Archibald Garrod, who first indicated its autosomal recessive genetic basis, includes an unusual historical description of the iatrogenic suffering of the first patient treated for the condition, which is frequently benign.

Albinism

Skin and hair color are controlled by an undetermined number of genetic loci in humans and exist in infinite variation; in mice 147 genes have been identified in color determination. Many conditions have been described in which the skin has little or no pigment. The chemical basis is not established for any except classical albinism, which results from a lack of tyrosinase. Lack of pigment in the skin makes albinos sensitive to sunlight, increasing the incidence of skin cancer in addition to burns; lack of pigment in the eyes may contribute to photophobia.

Fellman, J. H., Vanbellinghan, P. J., Jones, R. T., and Koler, R. D. Soluble and mitochondrial tyrosine aminotransferase. Relationship to human tyrosinemia. *Biochemistry* 8:615, 1969; and Kvittingen, E. A. Hereditary tyrosinemia type I. An overview. *Scand. J. Clin. Lab. Invest.* 46:27, 1986.

these metabolites in urine is diagnostic of a deficiency in synthesis of catecholamines. Lack of synthesis of serotonin (see p. 802) is indicated by lack of 5-hydroxyindole-3-acetic acid, shown in the same figure.

Tyrosine Is Required for Synthesis of Melanin, Thyroid Hormone, and Quinoproteins

Conversion of tyrosine to melanin requires **tyrosinase**, a copper-containing protein (Figure18.52*a*). The two-step reaction uses DOPA as a cofactor internal to the

CLINICAL CORRELATION 18.7

Parkinson's Disease

Usually in people over the age of 60 years but occasionally earlier, tremors may develop that gradually interfere with motor function of various muscle groups. This condition is named for the physician who described "shaking palsy" in 1817. The primary cause is unknown, and there may be more than one etiological agent. The defect is caused by degeneration of cells in certain small nuclei of the brain called substantia nigra and locus caeruleus. Their cells normally produce dopamine as a neurotransmitter, the amount released being proportional to the number of surviving cells. A dramatic outbreak of parkinsonism occurred in young adult drug addicts using a derivative of pyridine (methylphenyltetrahydropyridine, MPTP). It (or a contaminant produced during its manufacture) appears to be directly toxic to dopamine-producing cells of substantia nigra. Symptomatic relief, often dramatic, is obtained by administering DOPA, the precursor of dopamine. Clinical problems developed when DOPA (L-DOPA, levo-DOPA) was used for treatment of many people who have Parkinson's disease. Side effects included nausea, vomiting, hypotension, cardiac arrhythmias, and various central nervous system symptoms. These were explained as effects of dopamine produced outside the central nervous system. Administration of DOPA analogs that inhibit DOPA decarboxylase but are unable to cross the blood–brain barrier has been effective in decreasing side effects and increasing effectiveness of the DOPA. The interactions of the many brain neurotransmitters are very complex, cell degeneration continues after treatment, and elucidation of the major biochemical abnormality has not yet led to complete control of the disease. Recently, attempts have been made at treatment by transplantation of fetal adrenal medullary tissue into the brain. The adrenal tissue synthesizes dopamine and improves the movement disorder.

Calne, D. B. and Langston, J. W. Aetiology of Parkinson's disease. *Lancet* 2:1457, 1983; and Cell and tissue transplantation into the adult brain. *Ann. N.Y. Acad. Sci.* 495, 1987.

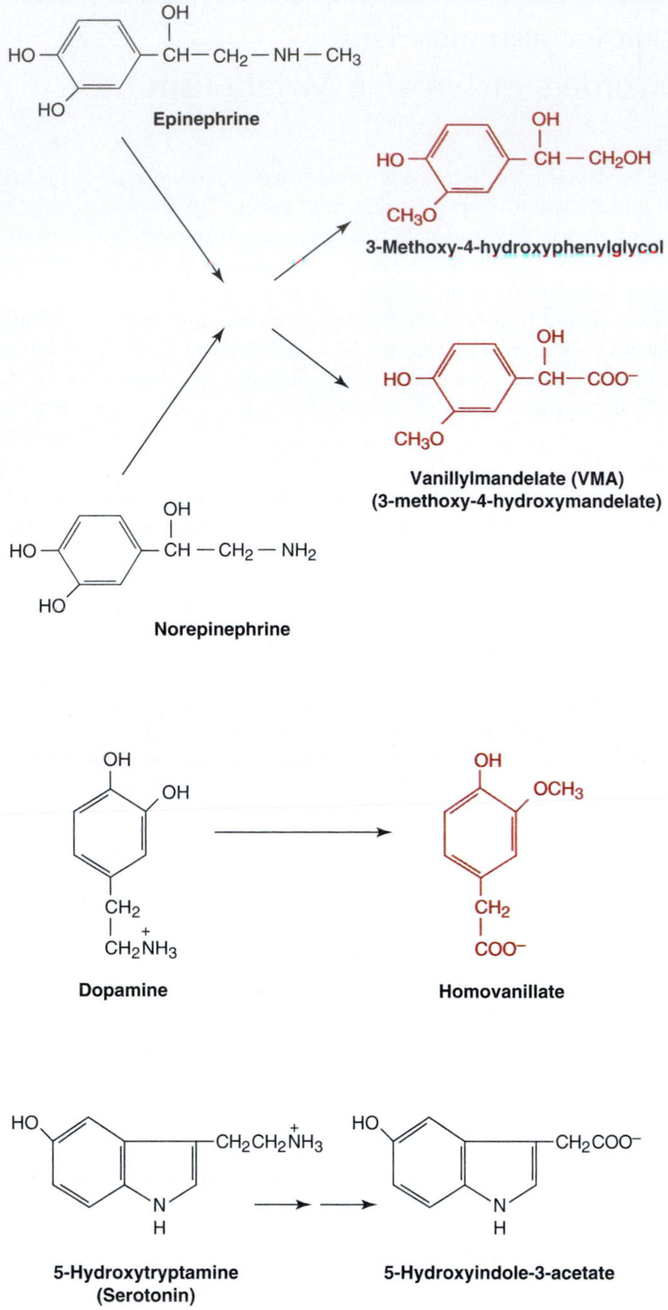

FIGURE 18.51
Major urinary excretion products of dopamine, epinephrine, norepinephrine, and serotonin.

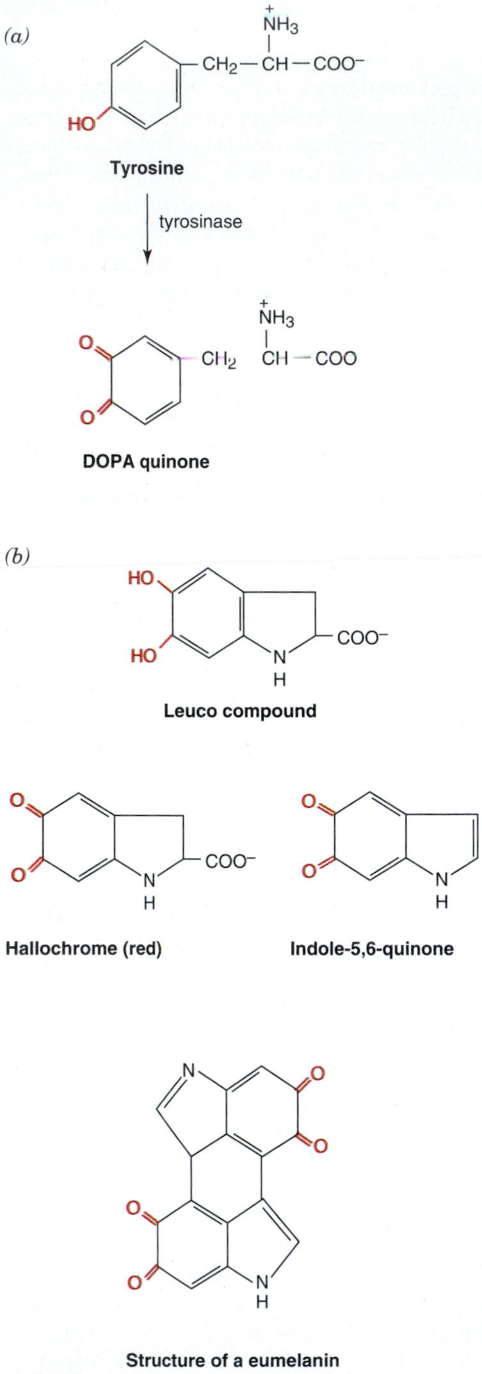

FIGURE 18.52
(*a*) Tyrosinase uses DOPA as a cofactor/intermediate; (*b*) some intermediates in melanin synthesis and an example of the family of black eumelanins.

reaction and produces **dopaquinone.** During melanogenesis, following exposure to UVB light, tyrosinase and a protein called tyrosinase-related protein, which may function in posttranslational modification of tyrosinase, are induced. A lack of tyrosinase activity produces **albinism.** There are various types of **melanin** (Figure 18.52*b*). All are aromatic quinones and the conjugated bond system gives rise to color. The dark pigment that is usually associated with melanin is eumelanin, from the Greek for "good melanin." Other melanins are yellow or colorless. The role of tyrosine residues of thyroglobulin in thyroid hormone synthesis is presented in the chapter on hormones (see p. 919).

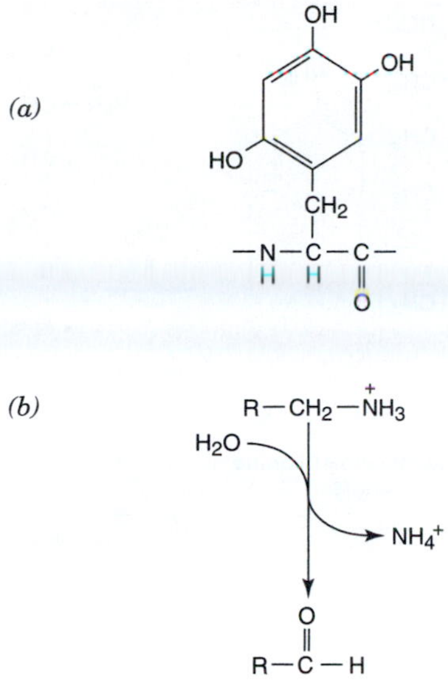

FIGURE 18.53
(*a*) Topaquinone and (*b*) amine oxidase reaction.

Some proteins use a modified tyrosine residue as a prosthetic group in oxidation–reduction reactions. The only example reported in humans is **topaquinone** (trihydroxyphenylalanylquinone),which is present in some plasma amine oxidases (Figure 18.53).

Methionine and Cysteine

De novo synthesis of **methionine** does not occur and methionine is essential. **Cysteine,** however, is synthesized by transfer of the sulfur atom derived from methionine to the hydroxyl group of serine. As long as the supply of methionine is adequate, cysteine is nonessential. The disposition of individual atoms of methionine and cysteine is a prime example of how cells regulate pathways to fit their immediate needs for energy or for other purposes. Conditions under which various pathways are given preference will be described below.

Methionine First Reacts with Adenosine Triphosphate

When excess methionine is present its carbons can be used for energy or for gluconeogenesis, and the sulfur retained as the sulfhydryl of cysteine. Figure 18.54 shows the first step, catalyzed by **methionine adenosyltransferase.** All phosphates of ATP are lost, and the product is ***S*-adenosylmethionine** (abbreviated **AdoMet,** or **SAM** in older references). The sulfonium ion is highly reactive, and the methyl is a good leaving group. AdoMet as a methyl group donor will be described below. After a methyltransferase removes the methyl group, the resulting ***S*-adenosylhomocysteine** is cleaved by **adenosylhomocysteinase** (Figure 18.55). Note that homocysteine is one carbon longer than cysteine. Although the carbons are destined for intermediary metabolism, the sulfur, a more specialized atom, will be conserved through transfer to serine to form cysteine. This requires the pyridoxal phosphate-dependent **cystathionine synthase** and **cystathionase** (Figure 18.55; see Clin. Corr. 18.8). Since the bond to form cystathionine is made on one side of the sulfur, and that cleaved is on the other side, the result is a **transsulfuration** (see Clin. Corr. 18.9). Homocysteine produces α-ketobutyrate and ammonia. α-Ketobutyrate is decarboxylated by a multienzyme complex resembling pyruvate

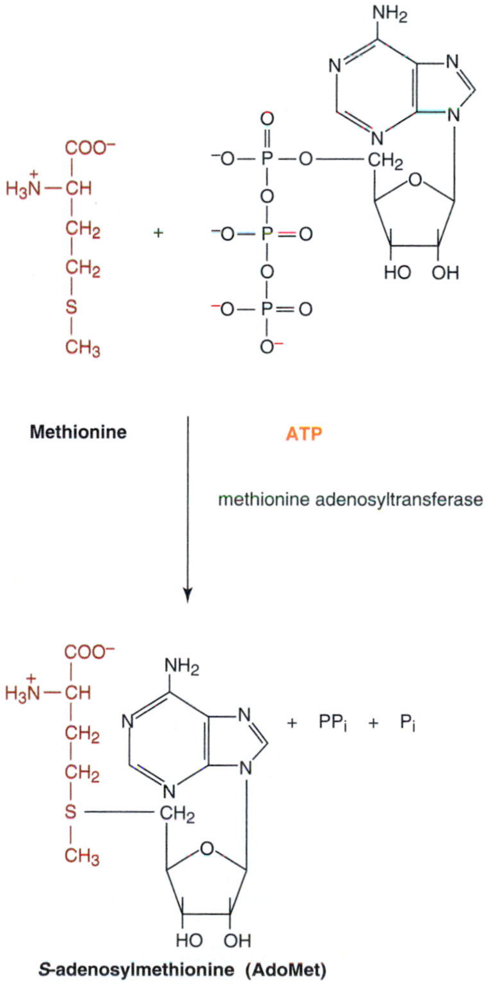

FIGURE 18.54
Synthesis of *S*-adenosylmethionine.

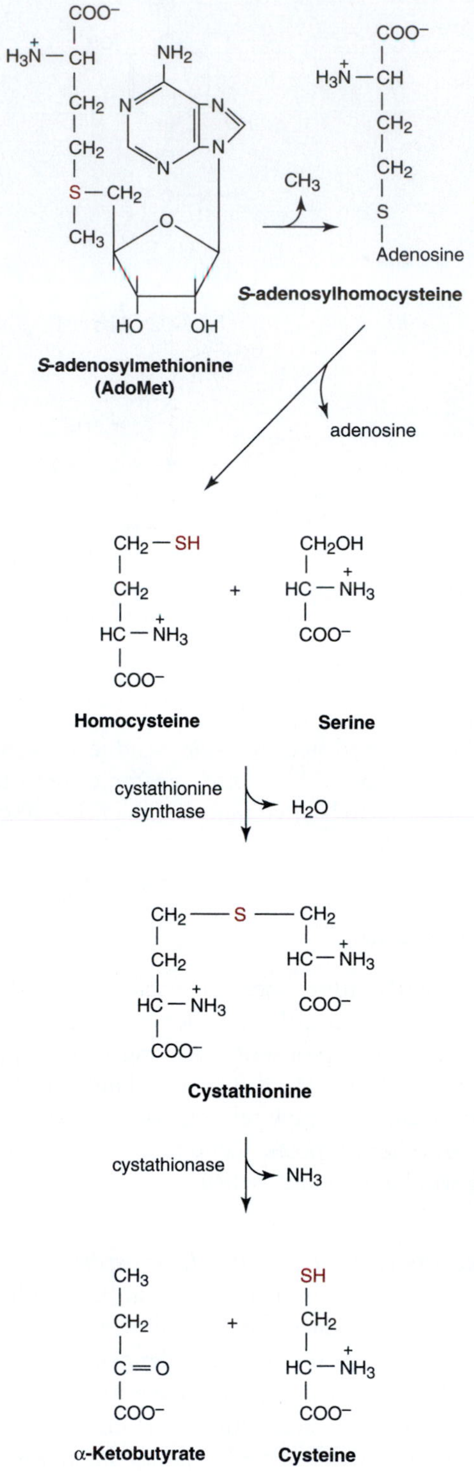

FIGURE 18.55
Synthesis of cysteine from *S*-adenosylmethionine.

dehydrogenase to yield propionyl CoA, which is then converted to succinyl CoA as described on page 814.

When the need is for energy, and not for cysteine, homocysteine produced in the above pathway is metabolized by **homocysteine desulfhydrase** to α-ketobutyrate, NH_3, and H_2S (Figure 18.56).

CLINICAL CORRELATION 18.8

Hyperhomocysteinemia and Atherogenesis

Deficiency of cystathionine synthase causes homocysteine to accumulate, and remethylation leads to high levels of methionine. Many minor products of these amino acids are formed and excreted. No mechanism has been established to explain why accumulation of homocysteine should lead to some of the pathological changes. Homocysteine may react with and block lysyl aldehyde groups on collagen. The lens of the eye is frequently dislocated some time after the age of three years, and other ocular abnormalities often occur. Osteoporosis develops during childhood. Mental retardation is frequently the first indication of this deficiency. Attempts at treatment include restriction of methionine intake and feeding of betaine (or its precursor, choline). In some cases significant improvement has been obtained by feeding pyridoxine (vitamin B_6), suggesting that the deficiency may be caused by more than one type of gene mutation; one type may affect the K_m for pyridoxal phosphate and others may alter the K_m for other substrates, V_{max}, or the amount of enzyme. A theory relating hyperhomocysteinemia to atherogenesis has been proposed. Excess homocysteine can form homocysteine thiolactone, a highly reactive intermediate that thiolates free amino groups in low density lipoproteins (LDLs) and causes them to aggregate and be endocytosed by macrophages. The lipid deposits form atheromas. Homocysteine can have other effects, including lipid oxidation and platelet aggregation, which in turn lead to fibrosis and calcification of atherosclerotic plaques. About one-quarter of patients with atherosclerosis who exhibit none of the other risk factors (such as smoking or oral contraceptive therapy) have been found to be deficient in cystathionine synthase activity.

Kaiser-Kupfer, M. I., Fujikawa, L., Kuwabara, T., et al. Removal of corneal crystals by topical cysteamine in nephrotic cystinosis. *N. Engl. J. Med.* 316:775, 1987; and McCully, K. S. Chemical pathology of homocysteine I. Atherogenesis. *Ann. Clin. Lab. Sci.* 23:477, 1993.

S-Adenosylmethionine Is a Methyl Group Donor

The role of tetrahydrofolate as a one-carbon group donor has been described (see p. 794). Although this cofactor could in theory serve as a source of methyl groups, the vast majority of methyltransferase reactions utilize **S-adenosylmethionine.** Methyl group transfer from AdoMet to a methyl acceptor is irreversible. An example is shown in Figure 18.57. *S*-adenosylhomocysteine left after methyl group transfer can be metabolized to cysteine, α-ketobutyrate, and ammonia. When cells need to resynthesize methionine, since the methyltransferase reaction is irreversible, another enzyme is required (Figure 18.58). **Homocysteine methyltransferase** is one of two enzymes known to require a vitamin B_{12} cofactor (the other is described on p. 814). The methyl group comes from N^5-methyltetrahydrofolate. This is the only reaction known that uses this form of tetrahydrofolate as a methyl donor. The net

CLINICAL CORRELATION 18.9

Other Diseases of Sulfur Amino Acids

Congenital deficiency of any of the enzymes involved in transsulfuration results in accumulation of sulfur-containing amino acids. Hypermethioninemia has been attributed to deficiency of methionine adenosyltransferase, probably caused by a K_m mutant that requires higher than normal concentrations of methionine for saturation but functions normally in methylation reactions. Lack of cystathionase does not seem to cause any clinical abnormalities other than cystathioninuria. The first reported case of this deficiency was about a mentally retarded patient and the retardation was attributed to the deficiency. Apparently the mental retardation was coincidental, the condition being benign. The amount of cysteine synthesized in these deficiencies is unknown, but treatment with a low methionine diet for hypermethioninemia is unnecessary.

Diseases Involving Cystine

Cystinuria is a defect of membrane transport of cystine and basic amino acids (lysine, arginine, and ornithine) that results in their increased renal excretion. Extracellular sulfhydryl compounds are quickly oxidized to disulfides. Low solubility of cystine results in crystals and the formation of calculi, a serious feature of this disease. Treatment is limited to removal of stones, prevention of precipitation by drinking large amounts of water or alkalinizing the urine to solubilize cystine, or formation of soluble derivatives by conjugation with drugs. Much more serious is cystinosis in which cystine accumulates in lysosomes. The stored cystine forms crystals in many cells, with a serious loss of function of the kidneys, usually causing renal failure within ten years. The defect is believed to be in the cystine transporter of lysosomal membranes.

Seashore, M. R., Durant, J. L., and Rosenberg, L. E. Studies on the mechanisms of pyridoxine responsive homocystinuria. *Pediatr. Res.* 6:187, 1972; Mudd, S. H. The natural history of homocystinuria due to cystathione β-synthase deficiency. *Am. J. Hum. Genet.* 37:1, 1985; and Frimpter, G. W. Cystathionuria: nature of the defect. *Science* 149:1095, 1965.

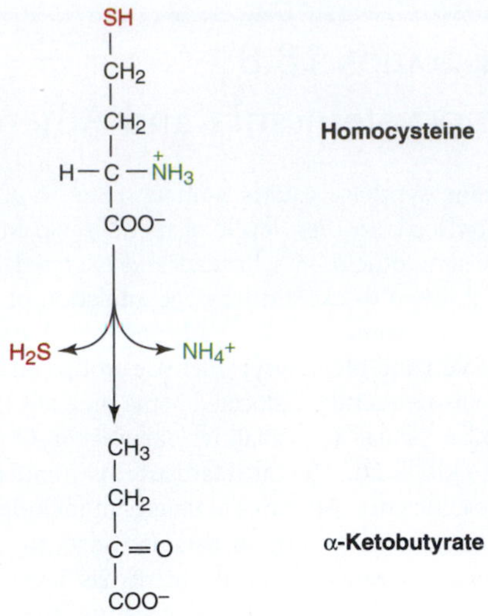

Homocysteine

H_2S NH_4^+

α-Ketobutyrate

FIGURE 18.56
Homocysteine desulfhydrase.

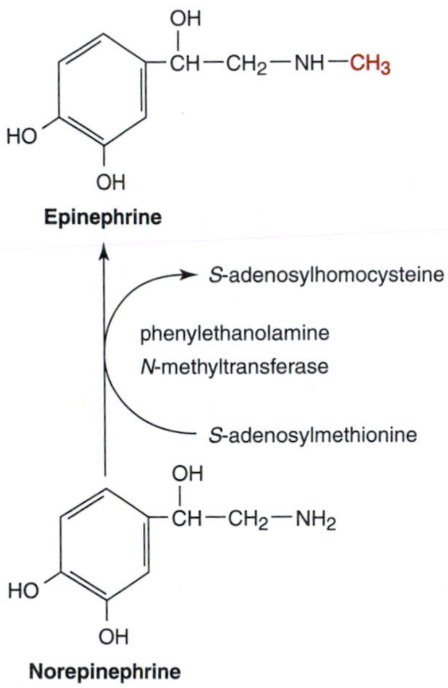

Epinephrine

S-adenosylhomocysteine

phenylethanolamine *N*-methyltransferase

S-adenosylmethionine

Norepinephrine

FIGURE 18.57
***S*-adenosylmethyltransferase reaction.**

result of reactions in Figures 18.57 and 18.58 is donation of a methyl group and regeneration of methionine under methionine-sparing conditions. A minor salvage pathway uses a methyl group from **betaine** instead of N^5-methyltetrahydrofolate.

S-Adenosylmethionine Is the Precursor of Spermidine and Spermine

Propylamine added to **putrescine** (see p. 793) to form **spermidine** and **spermine** is also derived from AdoMet, leaving methylthioadenosine. Putrescine is formed by decarboxylation of ornithine (see p. 793), and with propylamine forms spermidine. Addition of another propylamine gives spermine (Figure 18.59). The methylthioadenosine that remains can be used to resynthesize methionine. Much of the polyamine needed by the body is provided by microflora in the gut or from the diet and is carried by the enterohepatic circulation. Meat has a high content of putrescine, but other foods contain more spermidine and spermine.

The butylamino group of spermidine is used for posttranslational modification of a specific lysine residue in eIF-4D, an initiation factor for eukaryotic protein synthesis. The group is then hydroxylated, and the modified residue that results is called **hypusine** (Figure 18.60).

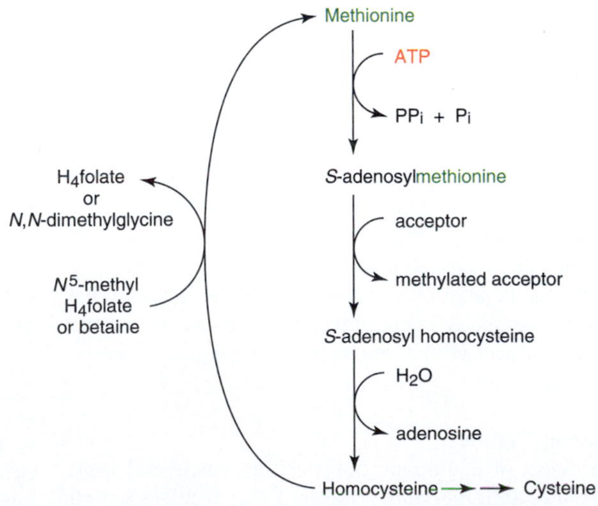

FIGURE 18.58
Resynthesis of methionine, a methylcobalamin-dependent reaction.

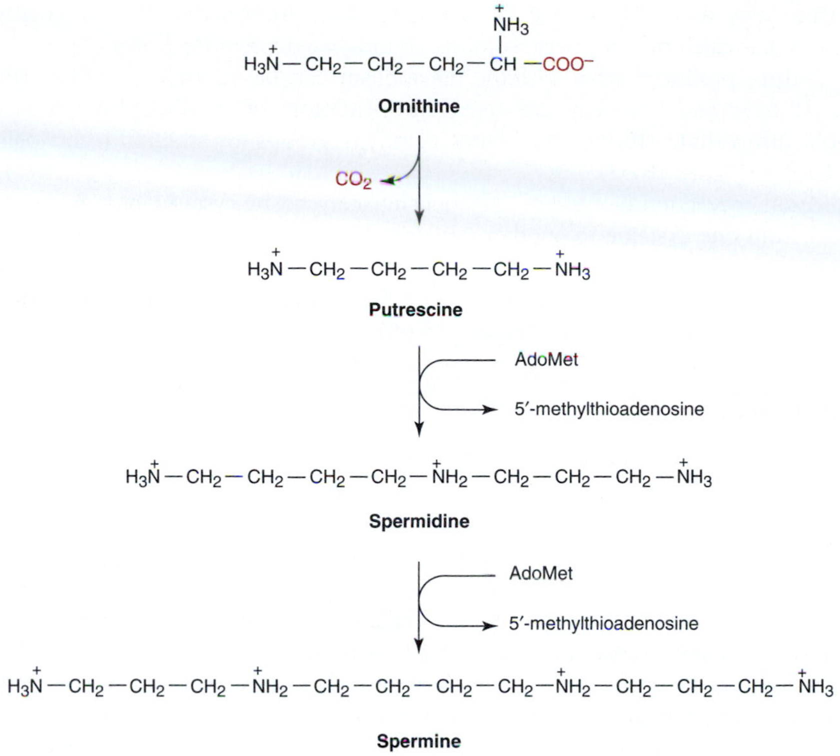

$H_3\overset{+}{N}-CH_2-CH_2-CH_2-CH(\overset{+}{N}H_3)-COO^-$

Ornithine

CO_2

$H_3\overset{+}{N}-CH_2-CH_2-CH_2-CH_2-\overset{+}{N}H_3$

Putrescine

AdoMet

5′-methylthioadenosine

$H_3\overset{+}{N}-CH_2-CH_2-CH_2-CH_2-\overset{+}{N}H_2-CH_2-CH_2-CH_2-\overset{+}{N}H_3$

Spermidine

AdoMet

5′-methylthioadenosine

$H_3\overset{+}{N}-CH_2-CH_2-CH_2-\overset{+}{N}H_2-CH_2-CH_2-CH_2-CH_2-\overset{+}{N}H_2-CH_2-CH_2-CH_2-\overset{+}{N}H_3$

Spermine

FIGURE 18.59
Polyamine synthesis.

Metabolism of Cysteine Produces Sulfur-Containing Compounds

Cysteine, derived from the sulfur of homocysteine and a molecule of serine, is metabolized in several ways. The pathway chosen is determined by the needs of the cell. The major metabolite is **cysteinesulfinate** (Figure 18.61). This is further metabolized to sulfite and pyruvate, or to hypotaurine and taurine. Taurine is an abundant intracellular free amino acid, but its exact role is unknown. It appears to play a necessary role in brain development. It forms conjugates with bile acids (see p. 1107) and may enhance bile flow and increase cholesterol clearance by the liver.

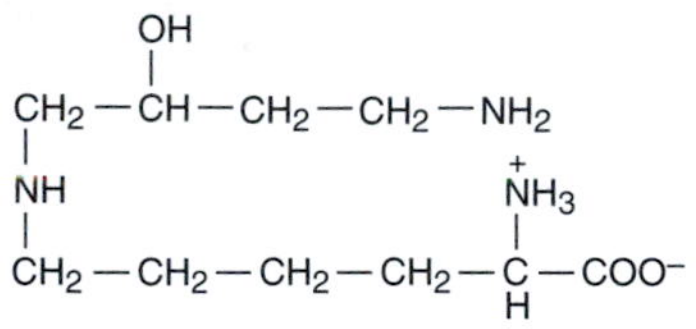

OH

$CH_2-CH(OH)-CH_2-CH_2-NH_2$

NH

$\overset{+}{N}H_3$

$CH_2-CH_2-CH_2-CH_2-C(H)-COO^-$

FIGURE 18.60
Hypusine.

Cysteine: $SH-CH_2-HC(\overset{+}{N}H_3)-COO^-$ → Cysteinesulfinate: $SO_2^--CH_2-HC(\overset{+}{N}H_3)-COO^-$ → Hypotaurine: $SO_2^--CH_2-CH_2-\overset{+}{N}H_3$

Cysteine **Cysteinesulfinate** **Hypotaurine**

Taurine: $SO_3^--CH_2-CH_2-\overset{+}{N}H_3$

pyruvate

HSO_3^-
Bisulfite

glutamate αKg

Taurine

$$HSO_3^- + O_2 + H_2O \xrightarrow{\text{sulfite oxidase}} SO_4^{2-} + H_2O_2 + H^+$$

Bisulfite

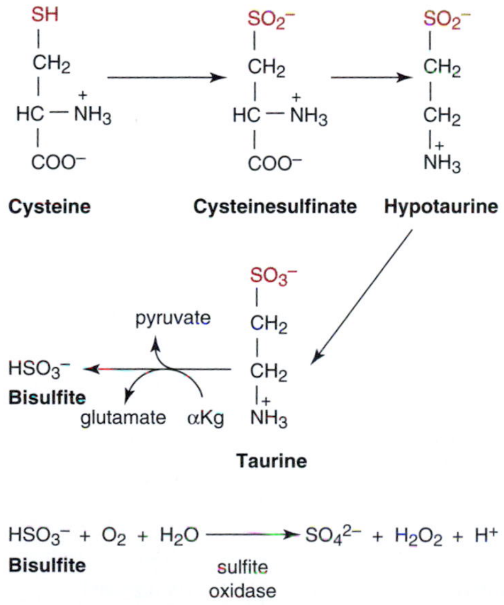

FIGURE 18.61
Formation of taurine and sulfate from cysteine.

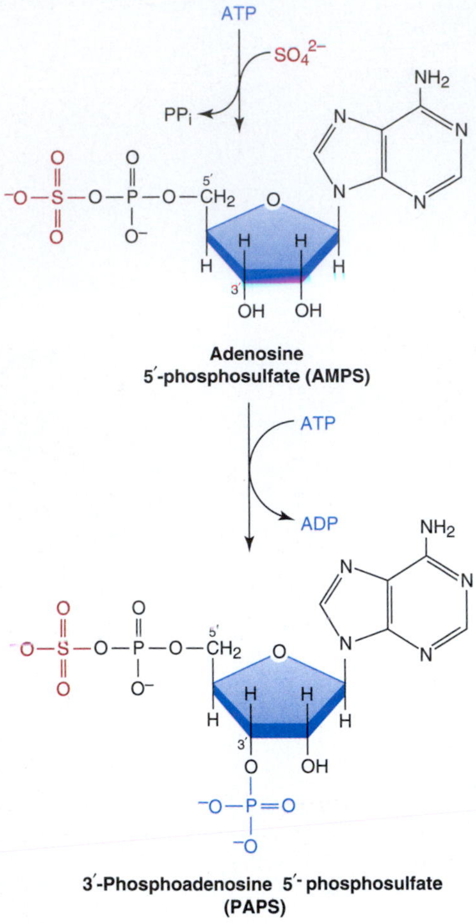

Taurine may also play a role in salvaging toxic intermediates, in regulating intracellular calcium, and, because of its abundance, in osmoregulation.

Sulfite produced from cysteine metabolism can be oxidized to sulfate (Figure 18.61), and this can be used in formation of **3′-phosphoadenosine 5′-phosphosulfate (PAPS),** the source of sulfate groups for addition to biological molecules (Figure 18.62).

Another reaction of cysteine metabolism catalyzed by cystathionase moves the sulfur from one cysteine to another cysteine (Figure 18.63) to form **thiocysteine.** Thiosulfate is formed from cysteine as shown in Figure 18.64. An enzyme called **rhodanese** can incorporate a sulfur from thiosulfate or thiocysteine into other molecules such as cyanide ion (Figure 18.65).

FIGURE 18.62
Synthesis of PAPS.

Tryptophan

Metabolism of **tryptophan** has many branch points. The dominant or oxidative pathway of tryptophan in the human (Figure 18.66, in color) starts with oxidation of tryptophan to *N*-formylkynurenine by a heme-containing enzyme, **tryptophan dioxygenase**, also called **tryptophan pyrrolase** or **tryptophan oxygenase**, because the pyrrole ring is cleaved in the reaction. Tryptophan dioxygenase is induced by glucocorticoids and glucagon. It is found in liver; other tissues contain a similar enzyme called indolamine dioxygenase, which is less substrate specific. Formamidase then hydrolyzes formylkynurenine to formate and kynurenine. At this point the pathway begins to branch. In the dominant pathway, reactions lead to 3-hydroxykynurenine, 3-hydroxyanthranilic acid and alanine, amino-carboxymuconic semialdehyde, and, by decarboxylation, to aminomuconic semialdehyde. This can be further metabolized in several steps to glutarate and eventually acetoacetyl CoA, or recyclized nonenzymatically to **picolinic acid**, which is excreted in the urine.

Tryptophan Is a Precursor of NAD

Tryptophan is the precursor of approximately 50% of the body's pyridine nucleotides. The rest is obtained from the diet. The branch point leading to nicotinate mononucleotide can be seen in Figure 18.66 at the stage of amino-carboxymuconic semialdehyde. Picolinate carboxylase forms 2-aminomuconic semialdehyde; this compound has a low K_m and is easily saturated with substrate. Since

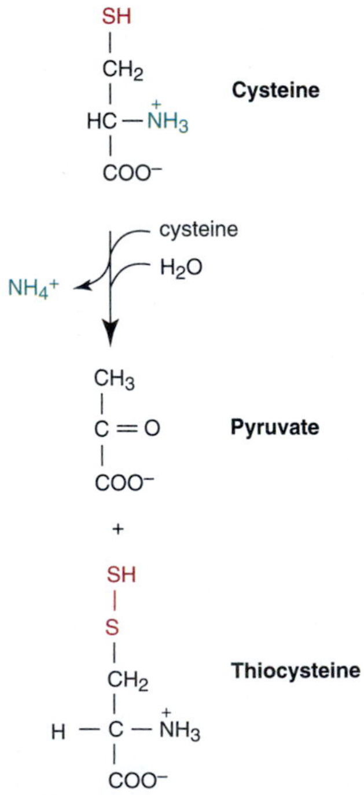

FIGURE 18.63
Synthesis of thiocysteine.

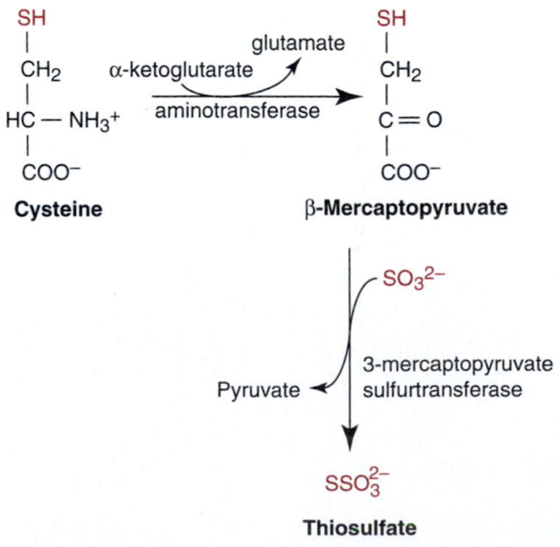

FIGURE 18.64
Formation of thiosulfate.

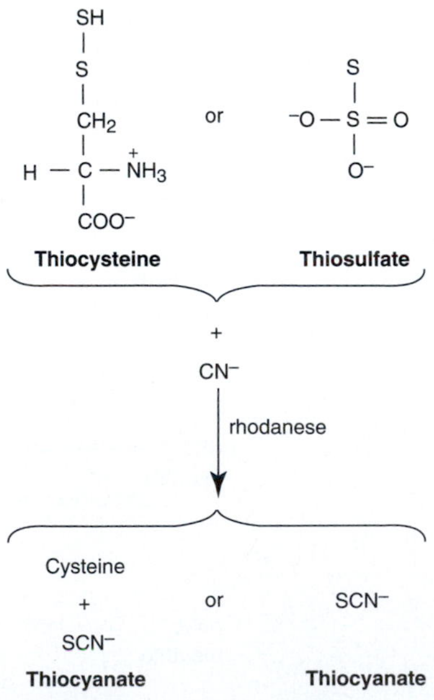

FIGURE 18.65
Detoxification of cyanide by products of cysteine metabolism.

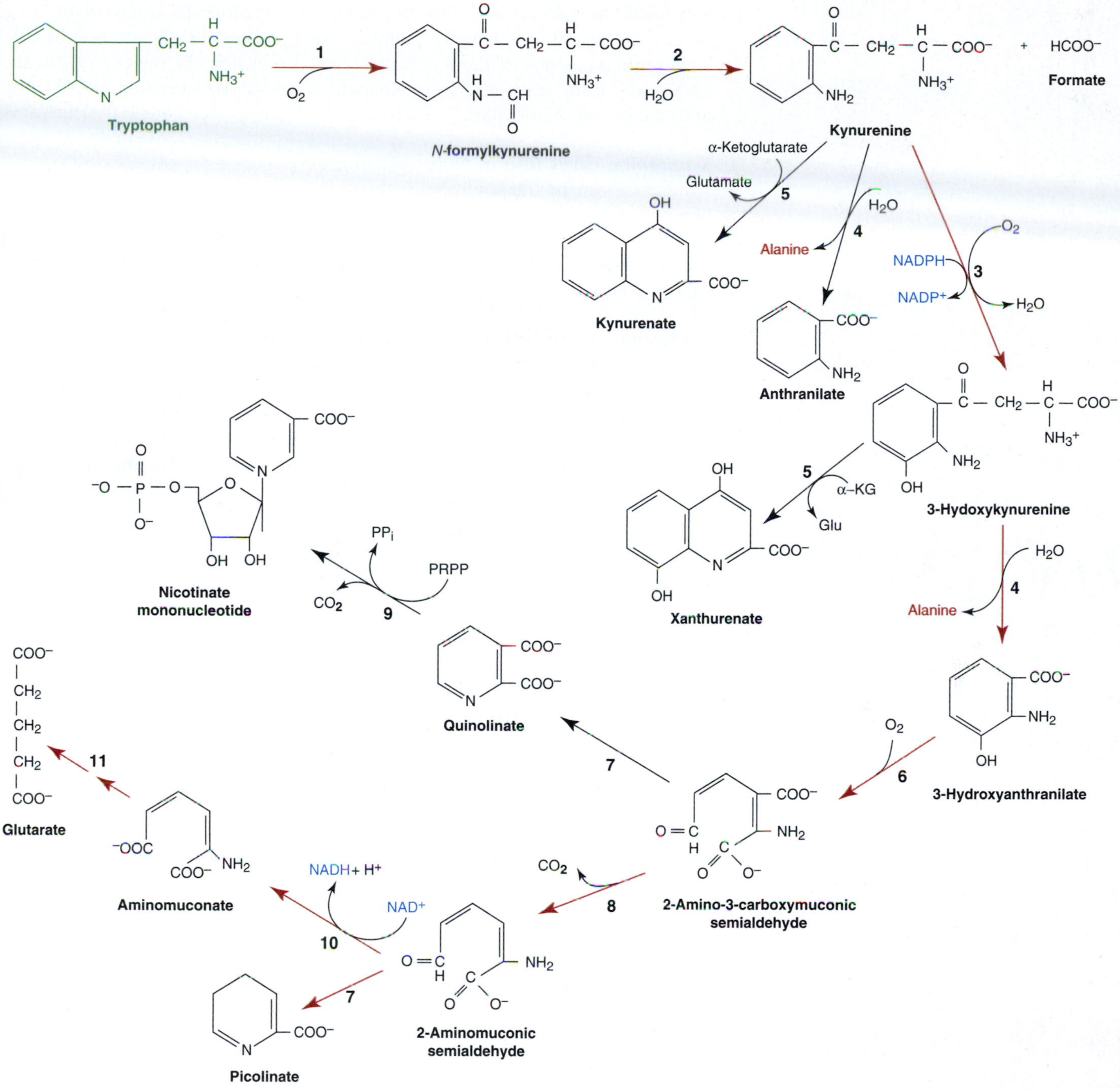

FIGURE 18.66
Metabolism of tryptophan.
Major pathway is shown in red. Enzymes indicated by number are (1) tryptophan oxygenase, (2) kynurenine formamidase, (3) kynurenine hydroxylase, (4) kynureninase, (5) aminotransferase, (6) 3-hydroxyanthranilate oxidase, (7) spontaneous nonenzymatic reaction, (8) picolinate carboxylase, (9) quinolinate phosphoribosyltransferase, (10) aldehyde dehydrogenase, and (11) complex series of reactions.

picolinate carboxylase has low activity in liver, some amino-carboxymuconic semialdehyde is cyclized in a nonenzymatic reaction to quinolinic acid. Phosphoribosylpyrophosphate provides a ribonucleotide moiety and the final step is a decarboxylation leading to nicotinate mononucleotide. Note that the nicotinic acid ring is synthesized as a part of a nucleotide. Because **kynurenine hydroxylase** is inhibited by estrogen, women are more susceptible to **pellagra,** the disease produced by niacin deficiency (from the Italian *pelle,* skin, and *agra,* rough).

Pyridoxal Phosphate Has a Prominent Role in Tryptophan Metabolism

Many enzymes in this lengthy pathway are pyridoxal phosphate dependent. **Kynureninase** is one of them and is affected by vitamin B_6 deficiency (Figure 18.66), resulting in excess kynurenine and xanthurenate excretion, which give urine a greenish-yellow color. This is a diagnostic symptom of vitamin B_6 deficiency.

Kynurenine Gives Rise to Neurotransmitters

Kynurenine can be transaminated with condensation of the side chain to form a two-ring compound, **kynurenic acid.** This reaction is also depicted in Figure 18.66. Kynurenic acid, its decarboxylated metabolite **kynuramine,** and **quinolinate** have all been shown to act as tryptophan-derived neurotransmitters, possibly as antiexcitotoxics and anticonvulsives.

Serotonin and Melatonin Are Tryptophan Derivatives

Serotonin (5-hydroxytryptamine) results from hydroxylation of tryptophan by a tetrahydrobiopterin-dependent enzyme and decarboxylation by a pyridoxal phosphate-containing enzyme (Figure 18.67*a*). It is a neurotransmitter in brain and causes contraction of smooth muscle of arterioles and bronchioles. It is found widely in the body and may have other physiological roles. **Melatonin,** a sleep-

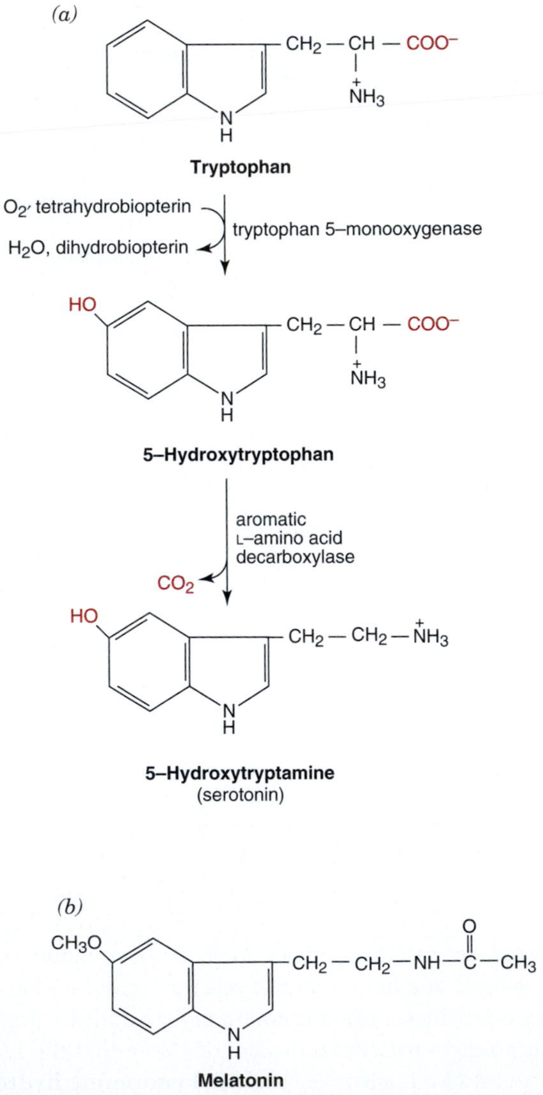

(*a*)

$CH_2-CH-COO^-$ / $\overset{+}{N}H_3$ / N H

Tryptophan

O_2, tetrahydrobiopterin → H_2O, dihydrobiopterin; tryptophan 5–monooxygenase

HO / $CH_2-CH-COO^-$ / $\overset{+}{N}H_3$ / N H

5–Hydroxytryptophan

aromatic L–amino acid decarboxylase; CO_2

HO / $CH_2-CH_2-\overset{+}{N}H_3$ / N H

5–Hydroxytryptamine
(serotonin)

(*b*)

CH_3O / $CH_2-CH_2-NH-\overset{O}{\overset{\|}{C}}-CH_3$ / N H

Melatonin

FIGURE 18.67
(*a*) Synthesis of serotonin (5-hydroxytryptamine) and (*b*) structure of melatonin.

inducing molecule, is **N-acetyl-5-methoxytryptamine** (Figure 18.67*b*). The acetyltransferase needed for its synthesis is present in pineal gland and retina. Melatonin is involved in regulation of circadian rhythm, being synthesized mostly at night. It appears to function by inhibiting synthesis and secretion of other neurotransmitters such as dopamine and GABA (see p. 1000).

Tryptophan Induces Sleep

Ingestion of foods rich in tryptophan leads to sleepiness because the resulting serotonin is also sleep-inducing. Reducing availability of tryptophan in the brain can interfere with sleep. Tryptophan availability is reduced when other amino acids compete with it for transport through the blood–brain barrier. Elevated plasma concentrations of other amino acids, after a high-protein meal, diminish transport of tryptophan and induce wakefulness. The sleep-inducing effect of carbohydrates is due to decreased plasma amino acid levels, since carbohydrate stimulates release of insulin, and insulin causes removal of amino acids from plasma and uptake into muscle. This alleviates competition and increases the amount of tryptophan that can enter the brain. Strangely, extra serotonin appears to lead to sleepiness in females, but only sedation in males.

Branched-Chain Amino Acids

Metabolism of **branched-chain amino acids (BCAAs)—valine, isoleucine**, and **leucine**—is unusual, being initiated in muscle. NADH is formed during their metabolism, making them an excellent source of energy. Branched-chain amino acid aminotransferase is present at a much higher concentration in muscle than liver. Although these three amino acids produce different products, the first steps in their metabolism are similar.

Initial Reactions of BCAA Metabolism Are Shared

BCAA aminotransferase exists in three isozymes distributed differently between tissues, sometimes found in cytosol and sometimes in mitochondria (Figure 18.68). Two handle all three BCAAs, and one is specific for leucine. Starvation induces the muscle aminotransferases but does not affect the enzymes in liver. The resulting α-keto branched-chain acids are oxidatively decarboxylated by an inner mitochondrial membrane enzyme complex similar to the pyruvate dehydrogenase complex, which produces NADH and CO_2. When phosphorylated the dehydrogenase component of the complex has some activity, but this is greatly increased by dephosphorylation. All three α-keto branched-chain acids appear to be metabolized by the same enzyme. The more active form is found in liver in the fed state, and in muscle during starvation, reflecting the metabolism of dietary BCAAs by liver, and of muscle BCAAs to provide energy during fasting. The resulting CoA compounds are one carbon shorter than the original amino acids and are next acted on by an enzyme that resembles the first dehydrogenase of fatty acid β-oxidation.

Pathways of Valine and Isoleucine Metabolism Are Similar

Valine and isoleucine continue down a common pathway, with addition of water across the double bond to form a hydroxylated intermediate (Figure 18.69). The hydroxyl group on the isoleucine derivative is oxidized by NAD^+ followed by thiolysis to give acetyl CoA and propionyl CoA. The valine derivative loses CoA and is oxidized by NAD^+ to methylmalonate semialdehyde, which is then converted to propionyl CoA.

The Leucine Pathway Differs from Those of the Other Two Branched-Chain Amino Acids

The position of the methyl side chain in leucine prohibits the oxidation step of the metabolism of the other BCAAs (Figure 18.70). The double-bond-containing derivative is carboxylated, hydroxylated, and cleaved to acetoacetate and acetyl CoA.

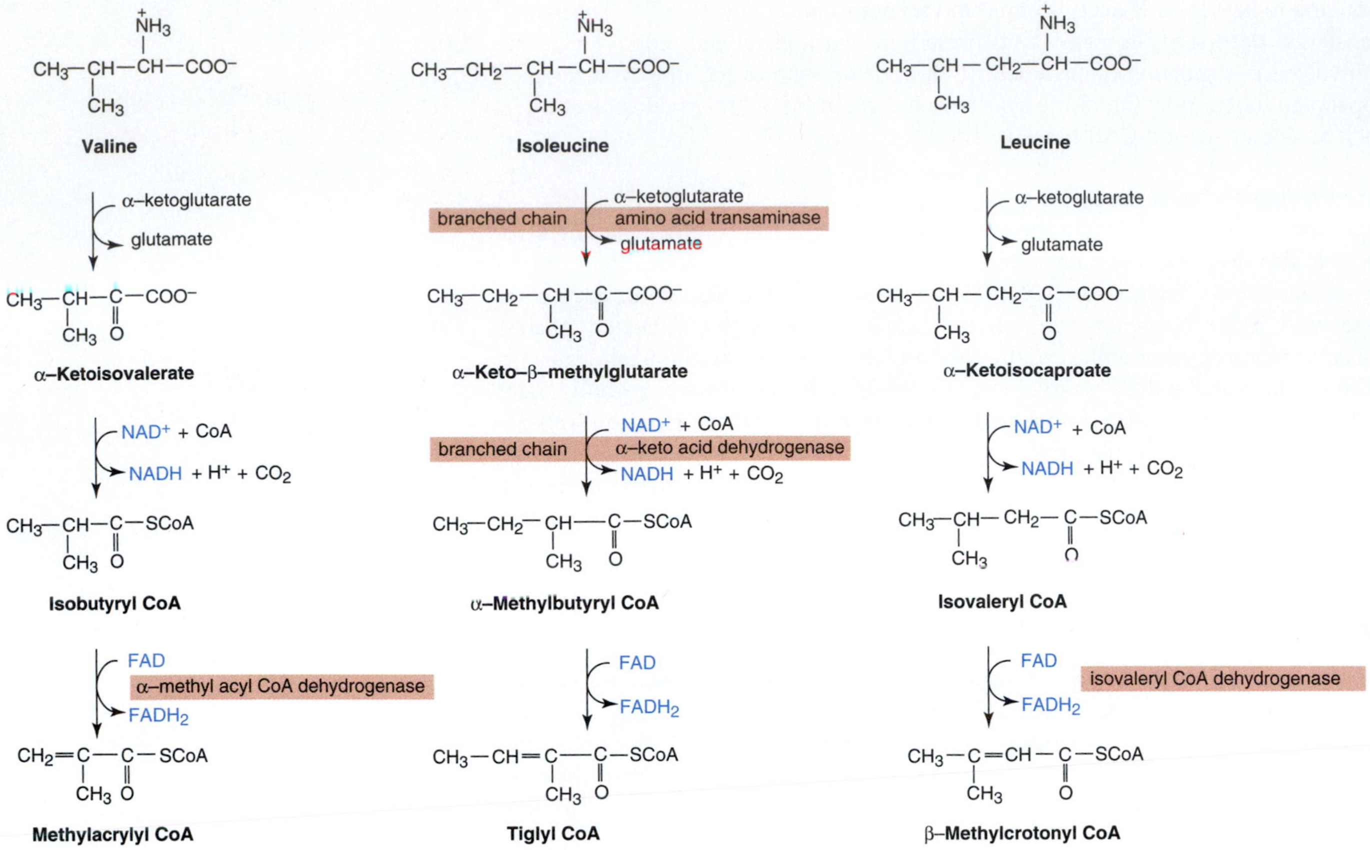

Valine | Isoleucine | Leucine

α–ketoglutarate → glutamate (branched chain amino acid transaminase)

α–Ketoisovalerate | α–Keto–β–methylglutarate | α–Ketoisocaproate

NAD^+ + CoA → NADH + H^+ + CO_2 (branched chain α–keto acid dehydrogenase)

Isobutyryl CoA | α–Methylbutyryl CoA | Isovaleryl CoA

FAD → $FADH_2$ (α–methyl acyl CoA dehydrogenase; isovaleryl CoA dehydrogenase)

Methylacrylyl CoA | Tiglyl CoA | β–Methylcrotonyl CoA

FIGURE 18.68
Common reactions in degradation of branched-chain amino acids.

One intermediate is ***β*-hydroxy-*β*-methylglutaryl CoA**, an intermediate in cytosolic sterol synthesis (see p. 743). Since BCAA degradation occurs in mitochondria the two pools do not mix. Leucine also has a minor alternative pathway (not shown), which results in excretion of 3-hydroxyvaleric acid, and can be utilized in the case of blockage in the leucine degradative pathway (see Clin. Corr. 18.10).

Propionyl CoA Is Metabolized to Succinyl CoA

Propionyl CoA is an end product of isoleucine, valine, and methionine metabolism, odd-chain fatty acid oxidation, and degradation of the side chain of cholesterol. The first step in the conversion of the three-carbon propionyl CoA to the four-carbon succinyl CoA is initiated by **propionyl CoA carboxylase**, which is biotin dependent (Figure 18.71; see Clin. Corr. 18.11), yielding D-methylmalonyl CoA. A racemase converts the product to a mixture of D- and L-methymalonyl CoA. **Methylmalonyl mutase**, which requires 5'-deoxyadenosylcobalamin (a derivative of vitamin B_{12}), converts the L-isomer to succinyl CoA. This is the second enzyme known to be dependent on vitamin B_{12} (see p. 805). The reaction is very unusual, removing a methyl side chain and inserting it as a methylene group into the backbone of the compound.

Lysine

Lysine like leucine is entirely ketogenic. The carbons enter intermediary metabolism as acetoacetyl CoA. Lysine has an ε- and an α-amino group. The initial transamination of the ε-group requires α-ketoglutarate as acceptor and cosubstrate

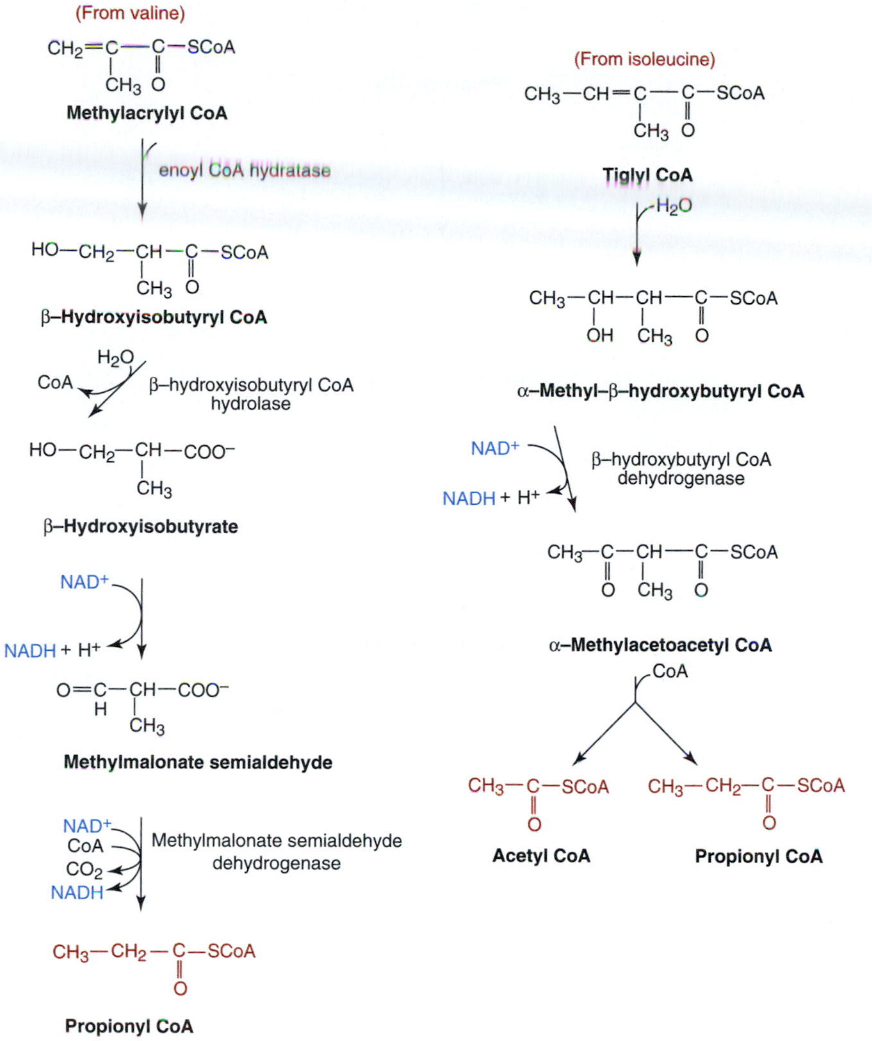

FIGURE 18.69
Terminal reactions in degradation of valine and isoleucine.

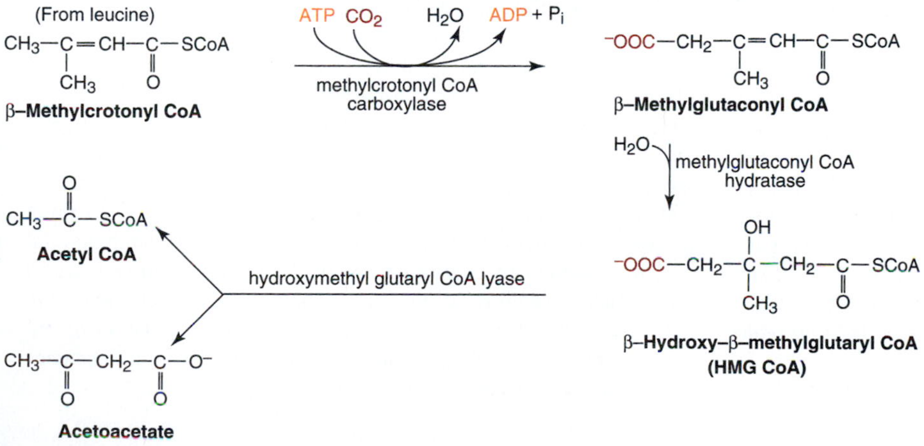

FIGURE 18.70
Terminal reactions of leucine degradation.

CLINICAL CORRELATION 18.10
Diseases of Metabolism of Branched-Chain Amino Acids

Enzyme deficiencies in catabolism of branched-chain amino acids are not common. In general they produce acidosis in newborns or young children. Very rare instances have been reported of hypervalinemia and hyperleucine-isoleucinemia. It has been suggested that the two conditions indicate existence of specific aminotransferases for valine and for leucine and isoleucine. Alternatively, mutation could alter the specificity of a single enzyme. The most common abnormality is deficiency of branched-chain keto acid dehydrogenase complex activity. There are several variations, but all patients excrete the branched-chain α-keto acids and corresponding hydroxy acids and other side products; an unidentified product imparts characteristic odor as indicated by the name maple syrup urine disease. Some cases respond to high doses of thiamine. A large percentage show serious mental retardation, ketoacidosis, and short life span. Dietary treatment to reduce the branched-chain ketoacidemia is effective in some cases. Some cases have been reported of deficiency of enzymes in later reactions of branched-chain amino acids. These include a blockage of oxidation of isovaleryl CoA with accumulation of isovalerate (which gives urine a smell of sweaty feet), β-methylcrotonyl CoA carboxylase deficiency (in which urine smells like that of a cat), deficiency of β-hydroxy-β-methylglutaryl CoA lyase, and deficiency of β-ketothiolase that splits α-methylacetoacetyl CoA (with no defect in acetoacetate cleavage). In the latter condition, development is normal and symptoms appear to be related only to episodes of ketoacidosis.

Zhang, B., Edenberg, H. J., Crabb, D. W., and Harris, R. A. Evidence for both a regulatory and structural mutation in a family with maple syrup urine disease. *J. Clin. Invest.* 83:1425, 1989.

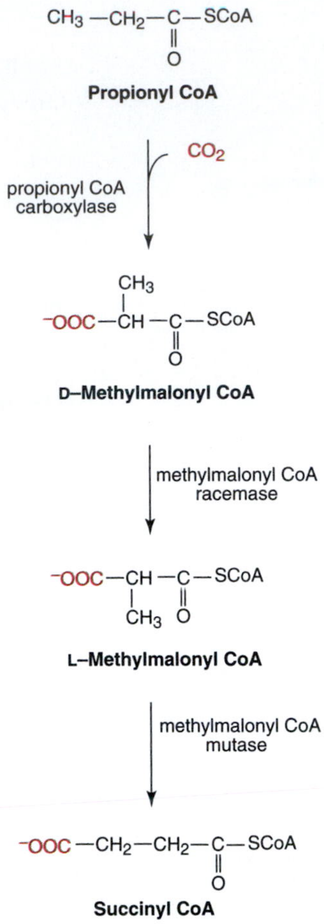

$CH_3-CH_2-C(=O)-SCoA$

Propionyl CoA

↓ propionyl CoA carboxylase (CO_2)

$^{-}OOC-CH(CH_3)-C(=O)-SCoA$

D-Methylmalonyl CoA

↓ methylmalonyl CoA racemase

$^{-}OOC-CH(CH_3)-C(=O)-SCoA$

L-Methylmalonyl CoA

↓ methylmalonyl CoA mutase

$^{-}OOC-CH_2-CH_2-C(=O)-SCoA$

Succinyl CoA

FIGURE 18.71
Interconversion of propionyl CoA, methylmalonyl CoA, and succinyl CoA. The mutase requires 5′-deoxyadenosylcobalamin for activity.

CLINICAL CORRELATION 18.11

Diseases of Propionate and Methylmalonate Metabolism

Deficiencies of the three enzymes shown in Figure 18.71 contribute to ketoacidosis. Propionate is formed in the degradation of valine, isoleucine, methionine, threonine, the side chain of cholesterol, and odd-chain fatty acids. The amino acids appear to be the main precursors since decreasing or eliminating dietary protein immediately minimizes acidosis. A defect in propionyl-CoA carboxylase results in accumulation of propionate, which is diverted to alternative pathways, including incorporation into fatty acids for an acetyl group forming odd-chain fatty acids. The extent of these reactions is very limited. In one case large amounts of biotin were reported to produce beneficial effects, suggesting that more than one defect decreases propionyl-CoA carboxylase activity. Possibilities are a lack of intestinal biotinidase that liberates biotin from ingested food for absorption or a lack of biotin holocarboxylase that incorporates biotin into biotin-dependent enzymes. Children have been found with acidosis caused by high levels of methylmalonate, which is normally undetectable in blood. Enzymes analyzed from liver taken at autopsy or from cultured fibroblasts have shown that some cases were due to deficiency of methylmalonyl-CoA mutase. One group was unable to convert methylmalonyl CoA to succinyl CoA under any conditions, but another group carried out the conversion when 5′-adenosylcobalamin was added. Clearly, those with an active site defect in the enzyme cannot metabolize methylmalonate, but those with defects in handling vitamin B_{12} respond to massive doses of the vitamin. Other cases of methylmalonic aciduria suffer from a more fundamental inability to use vitamin B_{12} that leads to deficiency in methylcobalamin (coenzyme of methionine salvage) and in 5′-adenosylcobalamin deficiency (coenzyme of methylmalonyl CoA isomerization).

Mahoney, M. J. and Bick, D. Recent advances in the inherited methylmalonic acidemias. *Acta Paediatr. Scand.* 76:689, 1987.

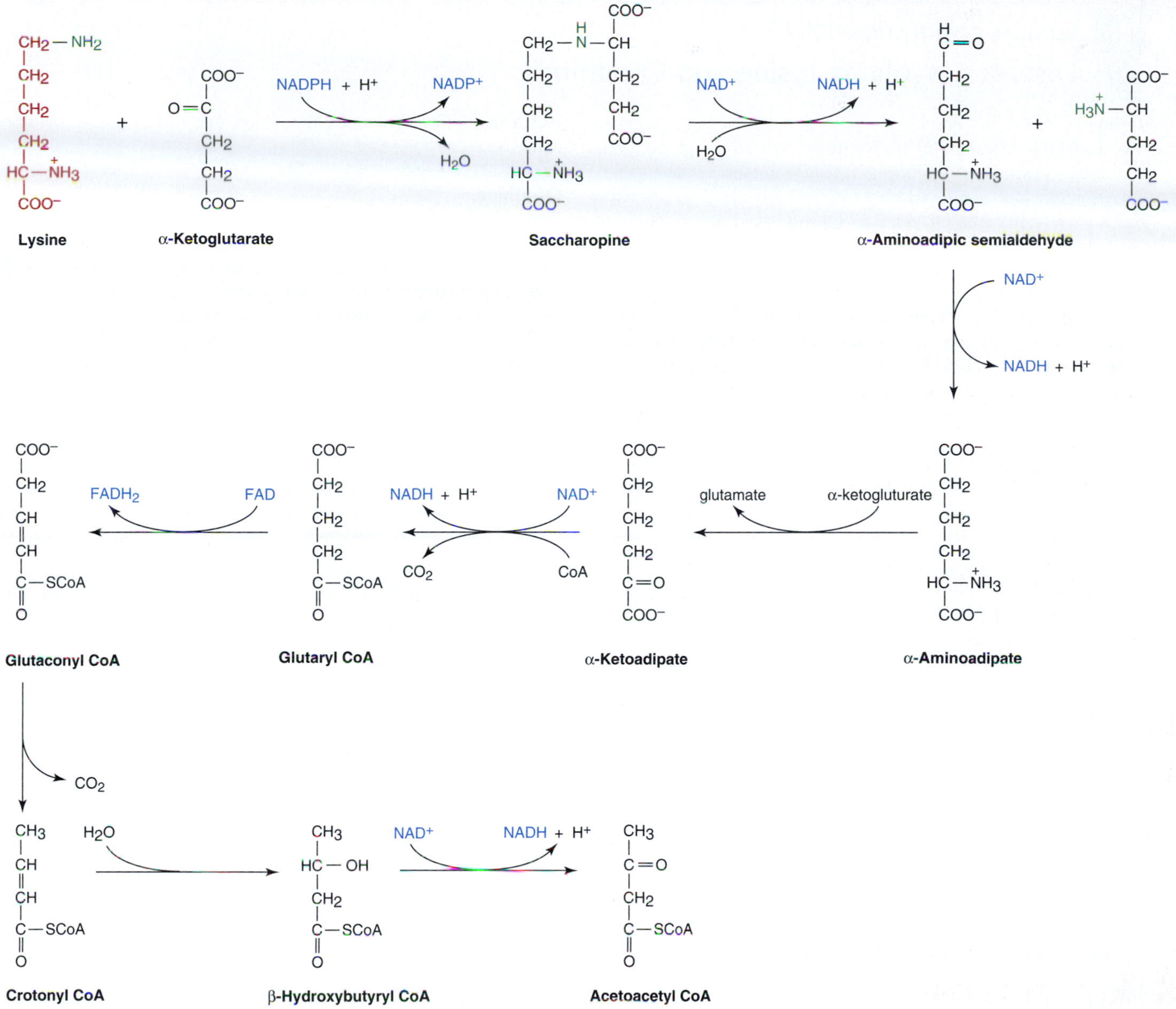

FIGURE 18.72
Principal pathway of lysine degradation.

(Figure 18.72). Instead of the pyridoxal phosphate–Schiff base mechanism, an intermediate called **saccharopine** is formed, which is then cleaved to glutamate and a semialdehyde compound. The usual Schiff base electronic rearrangement mechanism is replaced by an oxidation and a reduction, but the products are effectively the same. The semialdehyde is then oxidized to a dicarboxylic amino acid, and a transamination of the α-amino group occurs in a pyridoxal-dependent manner. Further reactions lead to acetoacetyl CoA. A minor pathway starts with removal of the α-amino group and goes via the cyclic compound **pipecolate** (Figure 18.73) to join the major pathway at the level of the semialdehyde intermediate. This does not replace the major pathway even in a deficiency of enzymes in the early part of the pathway (see Clin. Corr. 18.12).

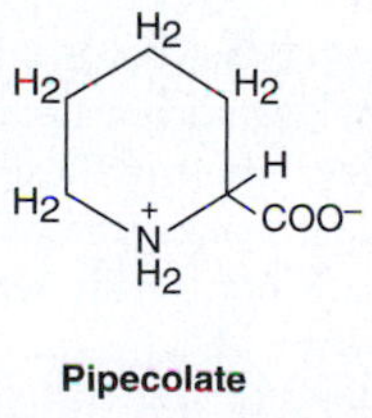

Carnitine Is Derived from Lysine

Medium- and long-chain fatty acids are transported into mitochondria for β-oxidation as **carnitine** conjugates (see p. 713). Carnitine is synthesized not from free lysine but rather from lysine residues in certain proteins. The first step is trimethylation of the ε-amino group of the lysine side chain, with AdoMet as the

Pipecolate

FIGURE 18.73
Minor product of lysine metabolism.

CLINICAL CORRELATION 18.12
Diseases Involving Lysine and Ornithine

Lysine

Two metabolic disorders of lysine are recognized. α-Amino adipic semialdehyde synthase is deficient in a small number of patients who excrete lysine and smaller amounts of saccharopine. This has led to the discovery that the enzyme has both lysine-α-ketoglutarate reductase and saccharopine dehydrogenase activities. Single proteins with multiple enzymatic activities are also found in pyrimidine synthesis and fatty acid synthesis. It is thought that hyperlysinemia is benign. More serious is familial lysinuric protein intolerance due to failure to transport dibasic amino acids across intestinal mucosa and renal tubular epithelium. Plasma lysine, arginine, and ornithine are decreased to one-third or one-half of normal. Patients develop marked hyperammonemia after a meal containing protein. This is thought to arise from deficiency of urea cycle intermediates ornithine and arginine in liver, limiting the capacity of the cycle. Consistent with this view, oral supplementation with citrulline prevents hyperammonemia. Other features are thin hair, muscle wasting, and osteoporosis, which may reflect protein malnutrition due to lysine and arginine deficiency.

Ornithine

Elevated ornithine levels are generally due to deficiency of ornithine δ-aminotransferase. A well-defined clinical entity, gyrate atrophy of the choroid and retina, characterized by progressive loss of vision leading to blindness by the fourth decade, is caused by deficiency of this mitochondrial enzyme. The mechanism of changes in the eye is unknown. Progression of the disease may be slowed by dietary restriction in arginine and/or pyridoxine therapy, which reduces ornithine in body fluids.

O'Donnell, J. J., Sandman, R. P., and Martin, S. R. Gyrate atrophy of the retina: inborn error of L-ornithine:2-oxoacid aminotransferase. *Science* 200.200, 1978; and Rajantie, J., Simell, O., and Perheentupa, J. Lysinuric protein intolerance. Basolateral transport defect in renal tubuli. *J. Clin. Invest.* 67:1078, 1981.

methyl donor (Figure 18.74). Free trimethyllysine is released by hydrolysis of these proteins and is metabolized in four steps to carnitine.

Histidine

Histidase (see Clin. Corr. 18.13) releases free ammonia from histidine and leaves a compound with a double bond called urocanate (Figure 18.75). Two other reactions

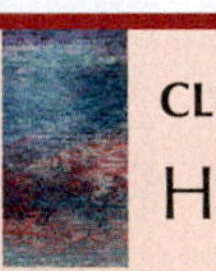

CLINICAL CORRELATION 18.13
Histidinemia

Histidinemia is due to histidase deficiency. A convenient assay for this enzyme uses skin, which produces urocanate as a constituent of sweat; urocanase and other enzymes of histidine catabolism found in liver do not occur in skin. A finding that urocanate is absent in sweat can only be interpreted as a lack of synthesis, and not as accelerated disappearance by further metabolism. Histidase deficiency can be confirmed by enzyme assay in skin biopsies. Incidence of the disorder is high, about 1 in 10,000 newborns screened. Most reported cases of histidinemia have shown normal mental development. Restriction of dietary histidine normalizes the biochemical abnormalities but is not usually required.

Scriver, C. R. and Levy, H. L. Histidinemia: reconciling retrospective and prospective findings. *J. Inherit. Metab. Dis.* 6:51, 1983.

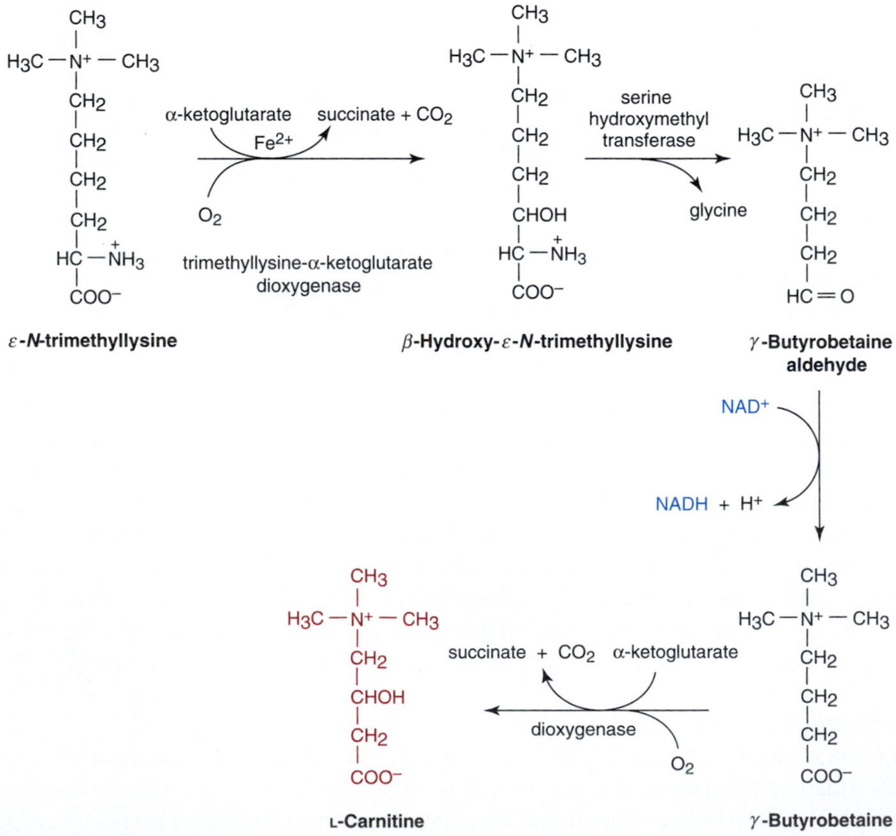

FIGURE 18.74
Biosynthesis of carnitine.

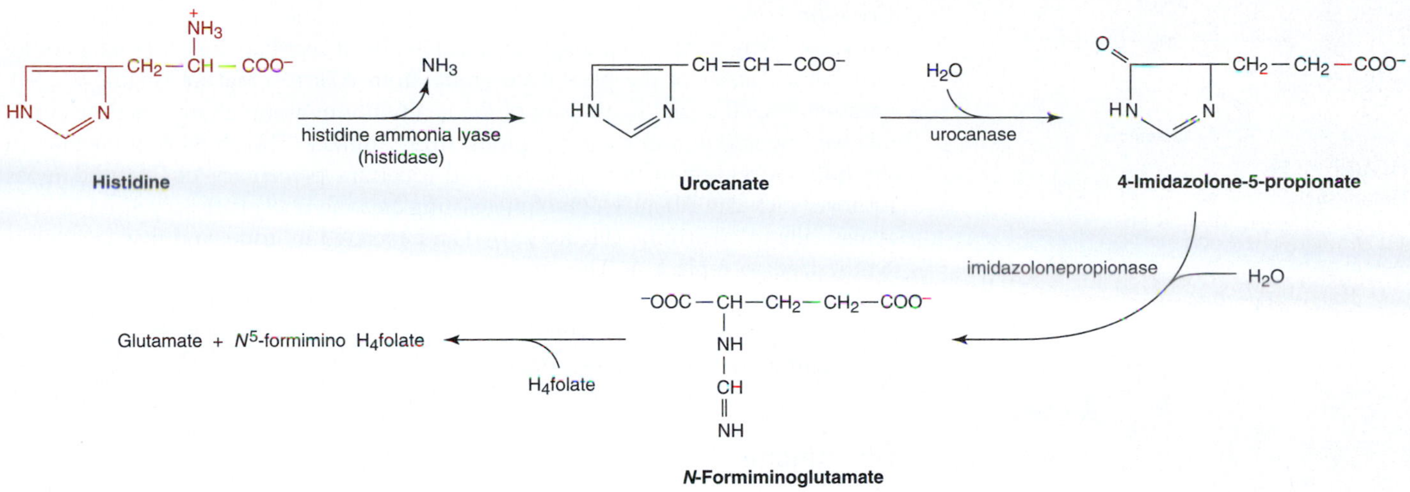

FIGURE 18.75
Degradation of histidine.

lead to **formiminoglutamate (FIGLU).** The formimino group is then transferred to tetrahydrofolate.

Urinary Formiminoglutamate Is Increased in Folate Deficiency

The formimino group of formiminoglutamate must be transferred to tetrahydrofolate before the final product, glutamate, can be produced. When there is insufficient tetrahydrofolate available, this reaction decreases and FIGLU is excreted in urine. This is a diagnostic sign of folate deficiency if it occurs after a test dose of histidine is ingested (see Clin. Corr. 18.14).

Histamine, Carnosine, and Anserine Are Produced from Histidine

Histamine (Figure 18.76), released from cells as part of an allergic response, is produced from histidine by **histidine decarboxylase.** Histamine has many physiological roles, including dilation and constriction of certain blood vessels. An overreaction to histamine can lead to asthma and other allergic reactions. **Carnosine** (β-alanylhistidine) and **anserine** (β-alanylmethylhistidine) are dipeptides (Figure 18.77) found in muscle. Their function is unknown.

CLINICAL CORRELATION 18.14

Diseases of Folate Metabolism

A significant fraction of absorbed folic acid must be reduced to function as a coenzyme. Symptoms of folate deficiency may be due to deficiency of dihydrofolate reductase. Parenteral administration of N^5-formyltetrahydrofolate, the most stable of the reduced folates, is effective in these cases. In some cases of central nervous system abnormality attributed to deficiency of methylene folate reductase there is homocystinuria. Decreased enzyme activity lowers the N^5-methyltetrahydrofolate formed so that the source of methyl groups for the salvage of homocysteine is limiting. Large amounts of folic acid, betaine, and methionine reversed the biochemical abnormalities and, in at least one case, the neurological disorder. Patients with widely divergent presentations have shown deficiencies in transfer of the formimino group from formiminoglutamate to tetrahydrofolate. They excreted varying amounts of FIGLU; some responded to large doses of folate, but others did not. The mechanism whereby a deficiency of formiminotransferase produces pathological changes is unclear. It is not sure whether this deficiency causes a disease state. One patient showed symptoms of folate deficiency and had tetrahydrofolate methyltransferase deficiency. The associated anemia did not respond to vitamin B_{12} but showed some improvement with folate. It was suggested that the patient formed inadequate N^5-methyltetrahydrofolate to promote remethylation of homocysteine. This left the coenzyme "trapped" in the methylated form and unavailable for use in other reactions.

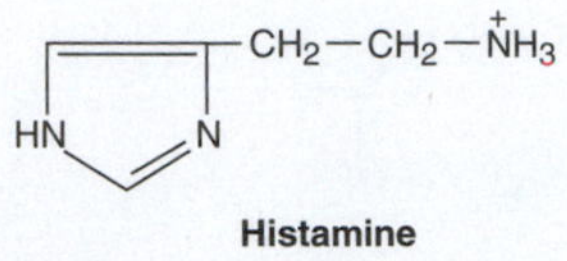

Histamine

FIGURE 18.76
Histamine.

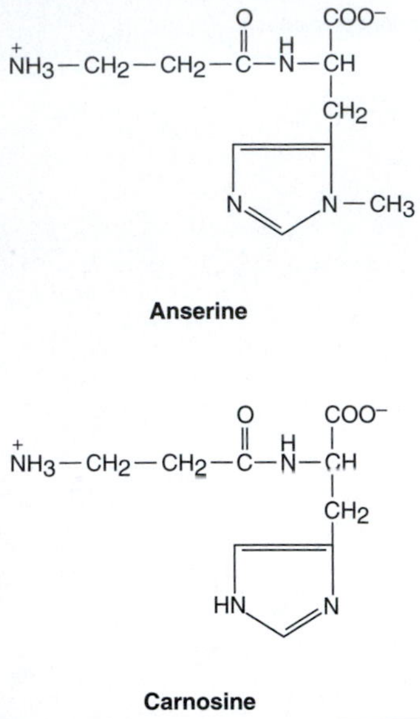

Anserine

Carnosine

FIGURE 18.77
Anserine and carnosine.

Creatine

Storage of "high-energy" phosphate, particularly in cardiac and skeletal muscle, occurs by transfer of the phosphate group from ATP to **creatine** (Figure 18.78). Creatine is synthesized by transfer of the guanidinium group of arginine to glycine, followed by addition of a methyl group from AdoMet. The amount of creatine in the body is related to muscle mass, and a certain percentage of this undergoes turnover each day. About 1–2% of preexisting creatine phosphate is cyclized nonenzymatically to **creatinine** (Figure 18.79) and excreted in urine, and new creatine is synthesized to replace it. The amount of creatinine excreted by an individual is therefore constant from day to day. When a 24-h urine sample is requested, the amount of creatinine in the sample can be used to determine whether the sample truly represents a whole day's urinary output.

Glutathione

Glutathione, the tripeptide **γ-glutamylcysteinylglycine,** has several important functions. It is a reductant, conjugated to drugs to make them more water soluble, involved in transport of amino acids across cell membranes, part of some leukotriene structures (see p. 772), a cofactor for some enzymatic reactions, and an aid in the rearrangement of protein disulfide bonds. Glutathione as reductant is very important in maintaining stability of erythrocyte membranes. Its sulfhydryl group can be used to reduce peroxides formed during oxygen transport (see p. 592). The resulting oxidized form of GSH consists of two molecules joined by a disulfide bond. This is reduced to two molecules of GSH at the expense of NADPH (Figure 18.80).

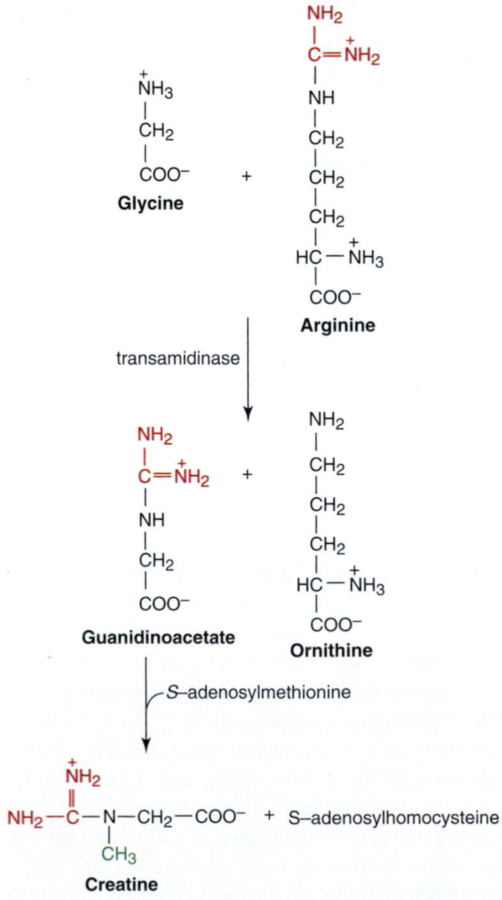

Glycine + Arginine

transamidinase

Guanidinoacetate + Ornithine

S-adenosylmethionine

Creatine + S-adenosylhomocysteine

FIGURE 18.78
Synthesis of creatine.

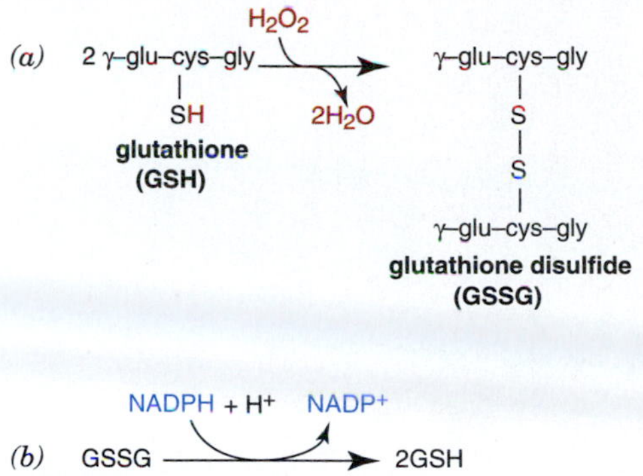

FIGURE 18.80
(*a*) Scavenging of peroxide by glutathione peroxidase and (*b*) regeneration of reduced glutathione by glutathione reductase.

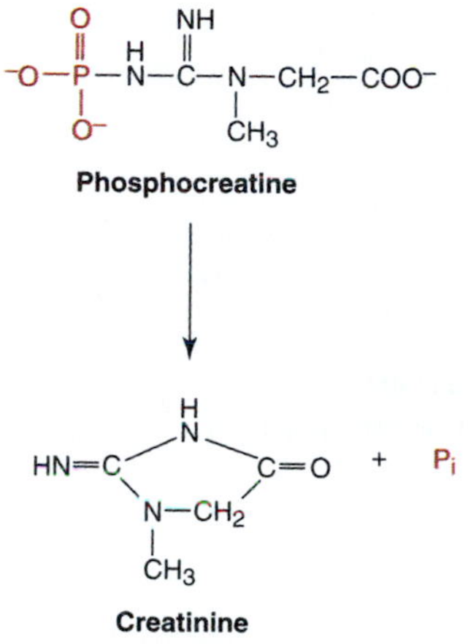

FIGURE 18.79
Spontaneous reaction forming creatinine.

The usual steady-state ratio of GSH to GSSG in erythrocytes is 100:1. Conjugation of drugs by glutathione, often after a preliminary reaction catalyzed by cytochrome P450 (see Chapter 11), renders them more polar for excretion (Figure 18.81).

Glutathione Is Synthesized from Three Amino Acids

Glutathione is synthesized by formation of the dipeptide γ-glutamylcysteine and the subsequent addition of glycine. Both reactions require activation of carboxyl groups by ATP (Figure 18.82). Synthesis of glutathione is largely regulated by cysteine availability.

The γ-Glutamyl Cycle Transports Amino Acids

There are several mechanisms for transport of amino acids across cell membranes. Many are symport or antiport mechanisms (see p. 526) and are coupled to sodium transport. The **γ-glutamyl cycle** is an example of "group transfer" transport. It is more energy-requiring than other mechanisms, but is rapid and has high capacity, and functions in kidneys and some other tissues. It is particularly important in renal epithelial cells. **γ-Glutamyl transpeptidase** is located in the cell membrane. It shuttles GSH to the cell surface to interact with an amino acid. γ-Glutamyl amino acid is transported into the cell, and the complex is hydrolyzed to liberate the amino acid (Figure 18.83). Glutamate is released as **5-oxoproline,** and cysteinylglycine is cleaved to its component amino acids. To regenerate GSH glutamate is reformed from oxoproline in an ATP-requiring reaction, and GSH is resynthesized from its three component parts. Three ATP molecules are used in the regeneration of glutathione, one in formation of glutamate from oxoproline and two in formation of the peptide bonds.

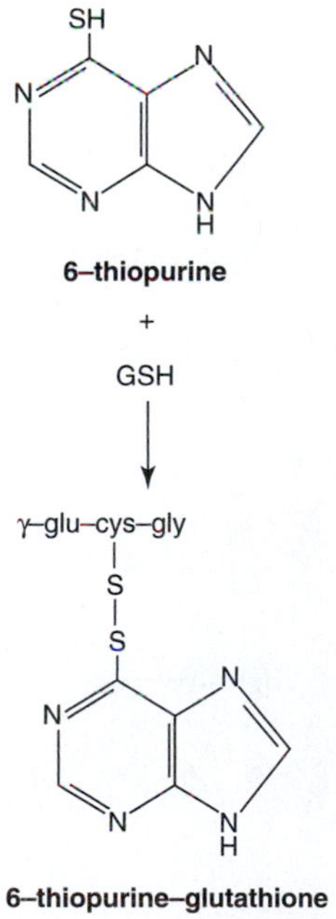

FIGURE 18.81
Conjugation of a drug by glutathione transferase.

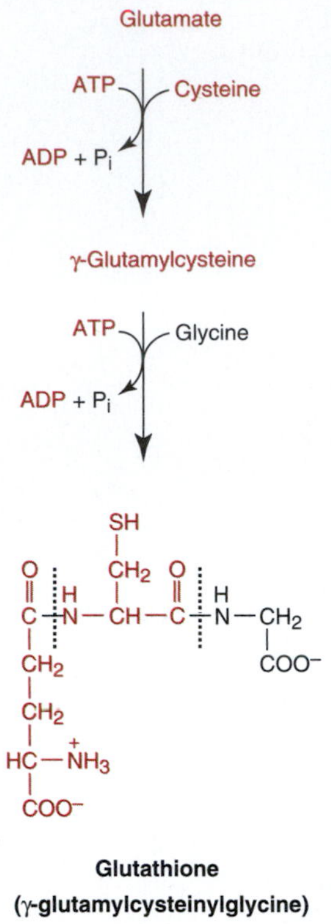

FIGURE 18.82
Synthesis of glutathione.

Glutathione Concentration Affects the Response to Toxins

When the body encounters toxic conditions such as peroxide formation, ionizing radiation, alkylating agents, or other reactive intermediates, it is beneficial to increase the level of GSH. Cysteine and methionine have been administered as GSH precursors, but they have the disadvantage of being precursors of an energy-expensive pathway to GSH. A more promising approach is administration of a soluble diester of GSH, such as **γ-(α-ethyl)glutamylcysteinylethylglycinate**.

Very premature infants have a very low concentration of cysteine because of low cystathionase activity in liver. This keeps the GSH concentration low and makes them more susceptible to oxidative damage, especially from hydroperoxides formed in the eye after hyperbaric oxygen treatment. Under certain circumstances, such as rendering tumor cells more sensitive to radiation or parasites more sensitive to drugs, it is desirable to lower GSH levels. This can be achieved by administration of the glutamate analog **buthionine sulfoximine** (Figure 18.84) as a competitive inhibitor of GSH synthesis.

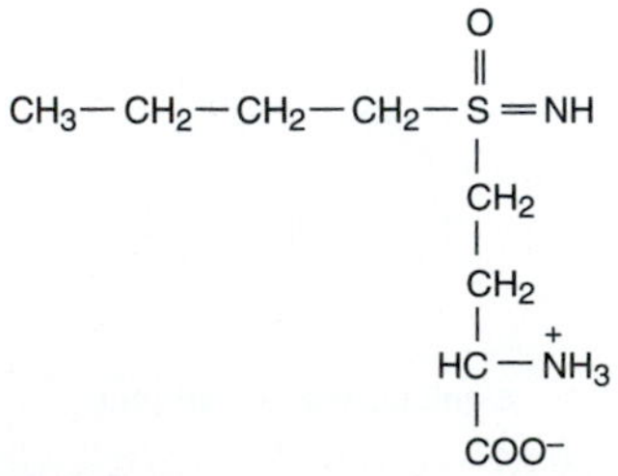

FIGURE 18.84
Buthionine sulfoximine.

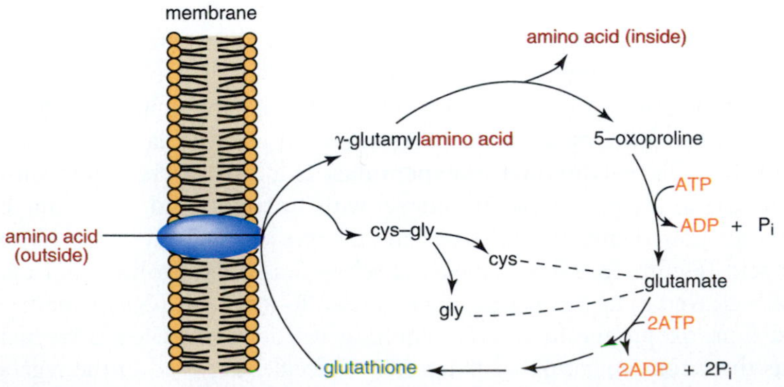

FIGURE 18.83
γ-Glutamyl cycle for transporting amino acids.

BIBLIOGRAPHY

General

Meister, A. *Biochemistry of the Amino Acids,* 2nd ed. New York: Academic Press, 1965.

Pyridoxal Phosphate

Dolphin, P., Poulson, R., and Avramovic, O. (Eds.). *Vitamin B_6 Pyridoxal Phosphate.* New York: Wiley, 1986.

Glutamate and Glutamine

Bode, B. L., Kaminski, D. L., Souba, W. W., and Li, A. P. Glutamine transport in human hepatocytes and transformed liver cells. *Hepatology* 21:511, 1995.

Fisher, H. F. Glutamate dehydrogenase. *Methods Enzymol.* 113:16, 1985.

Haussinger, D. Nitrogen metabolism in liver: structural and functional organization and physiological relevance. *Biochem. J.* 267:281, 1990.

Urea Cycle

Holmes, F. L. Hans Krebs and the discovery of the ornithine cycle. *Fed. Proc.* 39:216, 1980.

Jungas, R. L., Halperin, M. L., and Brosnan, J. T. Quantitative analysis of amino acid oxidation and related gluconeogenesis in humans. *Physiol. Rev.* 72:419, 1992.

Branched-Chain Amino Acids

Shander, P., Wahren, J., Paoletti, R., Bernardi, R., and Rinetti, M. *Branched Chain Amino Acids.* New York: Raven Press, 1992.

Serine

Snell, K. The duality of pathways for serine biosynthesis is a fallacy. *Trends Biochem. Sci.*11:241, 1986.

Arginine

Reyes, A. A., Karl, I. E., and Klahr, S. Editorial review: role of arginine in health and renal disease. *Am. J. Physiol.* 267:F331, 1994.

Sulfur Amino Acids

Lee, B. J., Worland, P. J., Davis, J. N., Stadtman, T. C., and Hatfield, D. L. Identification of a selenocysteyl-tRNASer in mammalian cells that recognizes the nonsense codon, UGA. *J. Biol. Chem.* 264:9724, 1989.

Stepanuk, M. H. Metabolism of sulfur-containing amino acids. *Annu. Rev. Nutr.* 6:179, 1986.

Wright, C. E., Tallan, H. H., Lin, Y. Y., and Gaull, G. E. Taurine: biological update. *Annu. Rev. Biochem.* 55:427, 1986.

Polyamines

Perin, A., Scalabrino, G., Sessa, A., and Ferioloini, M. E. *Perspectives in Polyamine Research.* Milan: Watchdog Editors, 1988.

Tabor, C. W. and Tabor, H. Polyamines. *Annu. Rev. Biochem.* 53:749, 1984.

Folates and Pterins

Blakley, R. L. and Benkovic, S. J. *Folate and Pterins.* New York: Wiley, Vol. 1: 1984, Vol. 2, 1985.

Quinoproteins

Davidson, V. L. (Ed.). *Principles and Applications of Quinoproteins.* New York: Marcel Dekker, 1993.

Carnitine

Bieber, L. L. Carnitine. *Annu. Rev. Biochem.* 57:261, 1988.

Glutathione

Taniguchi, N., Higashi, T., Sakamoto, Y., and Meister, A. *Glutathione Centennial: Molecular Perspectives and Clinical Implications.* New York: Academic Press, 1989.

Tryptophan

Stone, T. W. *Quinolinic Acid and the Kynurenines.* Boca Raton, FL: CRC Press, 1989.

Schwarcz, R. Metabolism and function of brain kynurenines. *Biochem. Soc. Trans.* 21:77, 1993.

Disorders of Amino Acid Metabolism

Rosenberg, L. E., and Scriver, C. R. Disorders of amino acid metabolism. In: P. K. Bondy and L. E. Rosenberg (Eds.), *Metabolic Control and Disease,* 8th ed. Philadelphia: Saunders, 1980.

Scriver, C. R., Beaudet, A. L., Sly, W. S., and Valle, D. (Eds.). *The Metabolic and Molecular Bases of Inherited Disease,* 7th ed. New York: McGraw-Hill, 1995.

Wellner, D. and Meister, A. A survey of inborn errors of amino acid metabolism and transport. *Annu. Rev. Biochem.* 50:911, 1980.

QUESTIONS | C. N. ANGSTADT

Multiple Choice Questions

1. Amino acids considered nonessential for humans are:
   A. those not incorporated into protein.
   B. not necessary in the diet if sufficient amounts of precursors are present.
   C. the same for adults as for children.
   D. the ones made in specific proteins by posttranslational modifications.
   E. generally not provided by the ordinary diet.

2. Aminotransferases:
   A. usually require α-ketoglutaramate or glutamine as one of the reacting pair.
   B. catalyze reactions that result in a net use or production of amino acids.
   C. catalyze irreversible reactions.
   D. require pyridoxal phosphate as an essential cofactor for the reaction.
   E. are not able to catalyze transamination reactions with essential amino acids.

3. The production of ammonia in the reaction catalyzed by glutamate dehydrogenase:
   A. requires the participation of NADH or NADPH.
   B. proceeds through a Schiff base intermediate.
   C. may be reversed to consume ammonia if it is present in excess.
   D. is favored by high levels of ATP or GTP.
   E. would be inhibited when gluconeogenesis is active.

4. The amide nitrogen of glutamine:
   A. represents a nontoxic transport form of ammonia.
   B. is a major source of ammonia for urinary excretion.
   C. is used in the synthesis of asparagine, purines, and pyrimidines.

D. can be recovered as ammonia by the action of glutaminase.
E. all of the above are correct.

5. All of the following are correct about ornithine EXCEPT:
   A. it may be formed from or converted to glutamic semialdehyde.
   B. it can be converted to proline.
   C. it plays a major role in the urea cycle.
   D. it is a precursor of putrescine, a polyamine.
   E. it is in equilibrium with spermidine.

6. *S*-Adenosylmethionine:
   A. contains a positively charged sulfur (sulfonium) that facilitates the transfer of substituents to suitable acceptors.
   B. yields α-ketobutyrate in the reaction in which the methyl is transferred.
   C. donates a methyl group in a freely reversible reaction.
   D. generates H_2S by transsulfuration.
   E. provides the carbons for the formation of cysteine.

7. In humans, sulfur of cysteine may participate in all of the following EXCEPT:
   A. the conversion of cyanide to less toxic thiocyanate.
   B. the formation of thiosulfate.
   C. the formation of taurine.
   D. the donation of the sulfur for methionine formation.
   E. the formation of PAPS.

8. Lysine as a nutrient:
   A. may be replaced by its α-keto acid analog.
   B. produces pyruvate and acetoacetyl CoA in its catabolic pathway.
   C. is methylated by *S*-adenosylmethionine.
   D. is the only one of the common amino acids that is a precursor of carnitine.
   E. all of the above are correct.

9. Histidine:
   A. unlike most amino acids, is not converted to an α-keto acid when the amino group is removed.
   B. is a contributor to the tetrahydrofolate one-carbon pool.
   C. decarboxylation produces a physiologically active amine.
   D. forms a peptide with β-alanine.
   E. all of the above are correct.

10. Glutathione does all of the following EXCEPT:
    A. participate in the transport of amino acids across some cell membranes.
    B. scavenge peroxides and free radicals.
    C. form sulfur conjugates for detoxication of compounds.
    D. convert hemoglobin to methemoglobin.
    E. act as a cofactor for some enzymes.

Questions 11 and 12: Defects in the metabolism of the branched-chain amino acids are rare but serious. The most common one is called maple syrup urine disease (named from the smell of the urine), which is a deficiency of branched-chain keto acid dehydrogenase complex. The disease is characterized by mental retardation, ketoacidosis, and a short life span.

11. All of the following are true about the branched-chain amino acids EXCEPT:
    A. they are essential in the diet.
    B. they differ in that one is glucogenic, one is ketogenic, and one is classified as both.
    C. they are catabolized in a manner that bears a resemblance to β-oxidation of fatty acids.
    D. they are oxidized by a dehydrogenase complex to branched-chain acyl CoAs one carbon shorter than the parent compound.
    E. they are metabolized initially in the liver.

12. Valine and isoleucine give rise to propionyl CoA, a precursor of succinyl CoA. A disease related to a defect in this conversion is methylmalonic aciduria. Some patients respond to megadoses of vitamin B_{12}. Which of the following statements about the conversion of propionyl CoA to succinyl CoA is/are correct?
    A. The first step in the conversion is a biotin-dependent carboxylation.
    B. Some methylmalonic aciduria patients respond to B_{12} because the defect in the mutase converting malonyl CoA to succinyl CoA is poor binding of the cofactor.
    C. The same pathway of propionyl CoA to succinyl CoA is part of the metabolism of odd-chain fatty acids.
    D. All of the above are correct.
    E. None of the above is correct.

Questions 13 and 14: Hyperammonemia caused by deficiencies of the enzymes involved in carbamoyl phosphate synthesis or any of the enzymes of the urea cycle is a serious condition. Untreated, the result is early death or, if the patient survives, mental retardation and other developmental abnormalities. Ornithine transcarbamoylase deficiency is the most common error in the cycle. A major focus of treatment is measures to relieve the hyperammonemia and sometimes supplementation with arginine.

13. In the formation of urea from ammonia, all of the following are correct EXCEPT:
    A. aspartate supplies one of the nitrogens found in urea.
    B. part of the large negative free energy change of the process may be attributed to the hydrolysis of pyrophosphate.
    C. the rate of the cycle fluctuates with the diet.
    D. fumarate is produced.
    E. ornithine transcarbamoylase catalyzes the rate-limiting step.

14. Carbamoyl phosphate synthetase I:
    A. is a flavoprotein.
    B. is controlled primarily by feedback inhibition.
    C. is unresponsive to changes in arginine.
    D. requires *N*-acetylglutamate as an allosteric effector.
    E. requires ATP as an allosteric effector.

Problems

15. An inability to generate tetrahydrobiopterin would have what specific effects on the metabolism of phenylalanine, tyrosine, and tryptophan?

16. Untreated phenylketonuria patients, in addition to mental retardation, have diminished production of catecholamines and light skin and hair. If the defect is in phenylalanine hydroxylase itself, a diet lacking phenylalanine but including tyrosine alleviates these conditions. If the defect is in the ability to produce tetrahydrobiopterin, the same dietary treatment may alleviate the mental retardation and light hair but not the diminished catecholamine production. What is the rationale explaining these findings?

17. If serine labeled with ^{14}C in the carbon bearing the hydroxyl group serves as the source of a one-carbon group, how does this labeled carbon become the labeled methyl group of epinephrine? Be specific.

ANSWERS

1. **B** B, E: Although most of our supply of nonessential amino acids comes from the diet, we can make them if necessary, given the precursors. A: All of the 20 common amino acids are incorporated into protein. C: Arginine is not believed to be required for adults. D: These are the amino acids for which there are no codons.
2. **D** The mechanism of action begins with the formation of a Schiff base with pyridoxal phosphate. A: Most mammalian aminotransferases use glutamate or α-ketoglutarate. B: One amino acid is converted into another amino acid; there is neither net gain nor net loss. C: The reactions are freely reversible. E: Only lysine and threonine do not have aminotransferases.
3. **C** This is an important mechanism for reducing toxic ammonia concentrations. A: This would favor ammonia consumption. B: The cofactor is a pyridine nucleotide not pyridoxal phosphate. D: These are inhibitory. E: Since part of the role is to provide amino acid carbon chains for gluconeogenesis, this would be active.
4. **E** It is in the form of the amide nitrogen of glutamine that much of amino acid nitrogen is made available in a nontoxic form. D: Glutaminase releases the ammonia in the liver and kidney.
5. **E** Spermidine is formed by adding propylamine to putrescine. A, B: Both amino acids give rise to glutamic semialdehyde and are formed from it. C: It is both a substrate and product of the cycle. D: This is a decarboxylation.
6. **A** The reactive, positively charged sulfur reverts to a neutral thioether when the methyl group is transferred to an acceptor. B: The product, *S*-adenosylhomocysteine, is hydrolyzed to homocysteine. C: Transmethylations from AdoMet are irreversible. D: Transsulfuration refers to the combined action of cystathionine synthase and cystathionase transferring methionine's sulfur to serine to yield cysteine. E: Methionine provides only the sulfur; carbons are from serine.
7. **D** Methionine is the source of sulfur for cysteine (via homocysteine), but the reverse is not true in humans. A, B: Transamination to β-mercaptopyruvate with subsequent formation of thiosulfate and/or conversion of cystine to thiocysteine allows transfer of the sulfur to detoxify cyanide. C: Taurine is deaminated cysteine. E: SO_4^{2-}, the most oxidized form of sulfur found physiologically, is either excreted or activated as PAPS for use in detoxifying phenolic compounds or in biosynthesis.
8. **D** Free lysine is not methylated, but lysyl residues in a protein are methylated in a posttranslational modification. Intermediates of carnitine synthesis are derived from trimethyllysine liberated by proteolysis. A: Lysine does not participate in transamination probably in part because the α-keto acid exists as a cyclic Schiff base. B: This is one of two purely ketogenic amino acids. C: See above.
9. **E** A: Elimination of ammonia from histidine leaves a double bond (urocanate) unlike both transamination and oxidative deamination reactions. B: A portion of the ring is released as formimino THF. C, D: Carnosine is β-alanylhistidine and anserine is β-alanylmethylhistidine.
10. **D** Most of the functions of glutathione listed are dependent on the sulfhydryl group (—SH). A major role of glutathione in red blood cells is reduction of methemoglobin. Glutathione reductase helps to maintain the ratio of GSH/GSSG at about 100:1.
11. **E** A: BCAA aminotransferase, the first enzyme, is much higher in muscle than in liver. B–D: Although their catabolism is similar, the end products are different because of the differences in the branching. After transamination, the α-keto acids are oxidized by a dehydrogenase complex in a fashion similar to pyruvate dehydrogenase. The similarity to β-oxidation comes in steps like oxidation to an α,β-unsaturated CoA, hydration of the double bond, and oxidation of a hydroxyl to a carbonyl.
12. **D** A: This is a typical carboxylation reaction like pyruvate carboxylase. B: Megadoses of a cofactor if the defect is in binding has been used in a number of diseases. This reaction is one of two reactions that use B_{12}. C: The final cleavage of odd-chain fatty acids produces propionyl CoA.
13. **E** Carbamoyl phosphate synthetase I catalyzes the rate-limiting step. A, B, D: One of the nitrogen atoms is supplied as aspartate, with its carbon atoms being released as fumarate. This reaction is physiologically irreversible because of the hydrolysis of pyrophosphate. C: The level of CPSI and the synthesis of *N*-acetylglutamate increase as protein in the diet increases.
14. **D** The primary control is by the allosteric effector, *N*-acetylglutamate. B: This is an activation, not an inhibition. C: Synthesis of the effector, and therefore activity of CPSI, is increased in the presence of arginine. E: ATP is a substrate.
15. Tetrahydrobiopterin is a necessary component of phenylalanine, tyrosine, and tryptophan hydroxylases. Its deficiency would inhibit normal degradation of both phenylalanine and tyrosine because their degradative pathways begin with the respective hydroxylases. Catecholamine formation (norepinephrine and epinephrine) begin with the formation of DOPA from tyrosine via tyrosine hydroxylase so catecholamine synthesis would be inhibited. The initial step in the conversion of tryptophan to serotonin is catalyzed by tryptophan hydroxylase.
16. Mental retardation seems to be caused by the elevated levels of phenylalanine in the blood so removing phenylalanine from the diet is beneficial. Light skin and hair is a secondary effect of phenylalanine competing with tyrosine for tyrosinase, not an enzyme defect, which is necessary to form melanins. Low tyrosine because of the inability to convert phenylalanine to tyrosine could also be a factor. Since tyrosine is low, the production of catecholamines is affected. If the defect is in the production of tetrahydrobiopterin, diet is less effective because tyrosine hydroxylase leading to DOPA (the first step in catecholamine formation) is tetrahydrobiopterin dependent.
17. Serine donates the labeled carbon to tetrahydrofolate to become labeled N^5, N^{10}-methylenetetrahydrofolate. This gets reduced to labeled N^5-methyltetrahydrofolate which reacts with homocysteine, in the presence of vitamin B_{12}, to produce methyl-labeled methionine. After activation to AldoMet, the methionine donates this labeled methyl to norepinephrine to produce epinephrine.

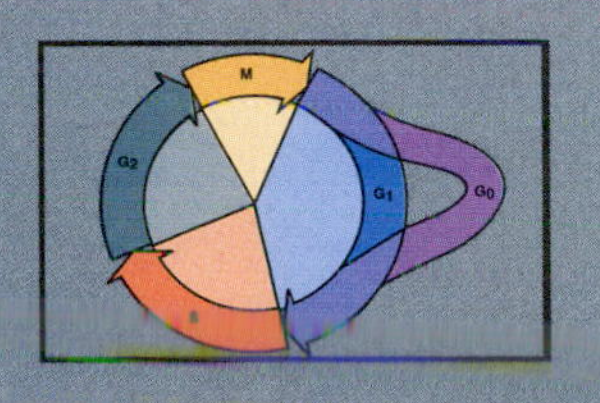

19

PURINE AND PYRIMIDINE NUCLEOTIDE METABOLISM

Joseph G. Cory

19.1 | OVERVIEW

The material in this chapter is limited to mammalian cells and where possible to nucleotide metabolism in humans. There are major differences between nucleotide metabolism in bacteria and mammalian cells and even some differences between humans and animals. Purine and pyrimidine nucleotides participate in many critical cellular functions. The metabolic roles of nucleotides very widely from serving as monomeric precursors of RNA and DNA to serving as second messengers. The sources of the purine and pyrimidine nucleotides are via *de novo* synthetic pathways and salvage of exogenous and endogenous nucleobases and nucleosides. Amino acids, CO_2, and ribose-5-phosphate (from the pentose phosphate pathway) serve as sources for carbon, nitrogen, and oxygen atoms of purines and pyrimidine nucleotides.

The intracellular concentrations of nucleotides are finely controlled by allosterically regulated enzymes in the pathways in which nucleotide end-products regulate key steps in the pathways. 2′-Deoxyribonucleotides required for DNA replication are generated directly from ribonucleotides and their production is also carefully regulated by nucleotides acting as positive and negative effectors. In addition to the regulation of nucleotide metabolism via allosteric regulation, concentrations of key enzymes in their metabolic pathways are altered during the cell cycle with many of the increases in enzyme activity occurring during late G1/early S phase just preceding DNA replication.

Defects in the metabolic pathways for *de novo* synthesis or salvage of nucleotides result in clinical diseases or syndromes. Furthermore, defects in degradation of nucleotides also lead to clinical problems. These include gout (defect in *de novo* purine nucleotide synthesis), Lesch–Nyhan syndrome (defect in purine nucleobase salvage), orotic aciduria (defect in *de novo* pyrimidine nucleotide synthesis), and immunodeficiency diseases (defects in purine nucleoside degradation). Because nucleotide synthesis is required for DNA replication and RNA synthesis in dividing cells, drugs that block *de novo* pathways of nucleotide synthesis have been successfully used as antitumor and antiviral agents.

19.2 | METABOLIC FUNCTIONS OF NUCLEOTIDES

Nucleotides and their derivatives play critical and diverse roles in cellular metabolism. Many different nucleotides are present in mammalian cells. Some, such as ATP, are present in the millimolar range while others, such as cyclic AMP, are orders of magnitude lower in concentration. The functions of nucleotides include the following:

1. Role in **Energy Metabolism:** As seen in earlier chapters, ATP is the principal form of chemical energy available to cells. ATP is generated in cells via either oxidative or substrate-level phosphorylation. ATP drives reactions as a phosphorylating agent and is involved in muscle contraction, active transport, and maintenance of ion gradients. ATP also serves as phosphate donor for generation of other nucleoside 5′-triphosphates.
2. **Monomeric Units of Nucleic Acids:** RNA and DNA consist of sequences of nucleotides. Nucleoside 5′-triphosphates are substrates for reactions catalyzed by RNA and DNA polymerases.
3. **Physiological Mediators:** Nucleosides and nucleotides serve as physiological mediators of key metabolic processes. Adenosine is important in control of

coronary blood flow; ADP is critical in platelet aggregation and hence blood coagulation; cAMP and cGMP act as second messengers; and GTP is required for capping of mRNA, signal transduction through GTP-binding proteins, and in microtubule formation.

4. **Precursor Function:** GTP is the precursor for formation of the cofactor, tetrahydrobiopterin, required for some hydroxylation reactions and nitric oxide generation.
5. **Components of Coenzymes:** The coenzymes NAD^+, $NADP^+$, and FAD and their reduced forms and coenzyme A all contain as part of their structure a 5′-AMP moiety.
6. **Activated Intermediates:** Nucleotides also serve as carriers of "activated" intermediates required for a variety of reactions. UDP-glucose is a key intermediate in synthesis of glycogen and glycoproteins. GDP-mannose, GDP-fucose, UDP-galactose, and CMP-sialic acid are all intermediates in reactions in which sugar moieties are transferred for synthesis of glycoproteins. CTP is utilized to generate CDP-choline, CDP-ethanolamine, and CDP-diacylglycerols, which are involved in phospholipid metabolism. Other activated intermediates include *S*-adenosylmethionine (SAM) and 3′-phosphoadenosine 5′-phosphosulfate (PAPS). *S*-adenosylmethionine is a methyl donor in reactions involving methylation of sugar and base moieties of RNA and DNA and in formation of compounds such as phosphatidylcholine from phosphatidylethanolamine and carnitine from lysine. *S*-adenosylmethionine also provides aminopropyl groups for synthesis of spermine from ornithine. PAPS is used as the sulfate donor to generate sulfated biomolecules such as proteoglycans and sulfatides.
7. **Allosteric Effectors:** Many of the regulated steps of metabolic pathways are controlled by intracellular concentrations of nucleotides. Many examples have already been discussed in previous chapters, and the roles of nucleotides in regulation of mammalian nucleotide metabolism will be discussed in this chapter.

Distributions of Nucleotides Vary with Cell Type

The principal purine and pyrimidine compounds found in cells are the 5′-nucleotide derivatives. ATP is the nucleotide found in the highest concentration. The distributions of nucleotides in cells vary with cell type. In red blood cells, adenine nucleotides far exceed the concentrations of guanine, cytosine, and uracil nucleotides; in other tissues, such as liver, there is a complete spectrum of nucleotides and their derivatives, which include NAD^+, NADH, UDP-glucose, and UDP-glucuronic acid. In normally functioning cells, nucleoside 5′-triphosphates predominate, whereas in hypoxic cells the concentrations of nucleoside 5′-monophosphates and nucleoside 5′-diphosphates are greatly increased. Free nucleobases, nucleosides, nucleoside 2′- and 3′-monophosphates, and "modified" bases represent degradation products of endogenous or exogenous nucleotides or nucleic acids.

The concentrations of **ribonucleotides** in cells are in great excess over the concentrations of **2′-deoxyribonucleotides.** For example, the concentration of ATP in Ehrlich tumor cells is 3600 pmol per 10^6 cells compared to dATP concentrations of 4 pmol per 10^6 cells. However, at the time of DNA replication the concentrations of dATP and other deoxyribonucleoside 5′-triphosphates are markedly increased to meet the substrate requirements for DNA synthesis.

In normal cells, the total concentrations of nucleotides are essentially constant. Thus the total concentration of AMP plus ADP plus ATP remains constant, but there can be major changes in the individual concentration such that the ratio of ATP/(ATP + ADP + AMP) is altered depending on the energy state of the cell. The same is true for NAD^+ and NADH. The total concentration of NAD^+ plus NADH is normally fixed within rather narrow concentration limits. Consequently, when it is stated that the NADH level is increased, it follows that the concentration of NAD^+ is correspondingly decreased in that cell. The basis for this "fixed" concentration of nucleotides is that *de novo* synthesis and salvage pathways for nucleotides, nucleosides, and nucleobases are very rigidly controlled under normal conditions.

Adenine

Guanine

Hypoxanthine

Xanthine

FIGURE 19.1
Purine bases.

19.3 | CHEMISTRY OF NUCLEOTIDES

The major purine derivatives in cells are **adenine** and **guanine.** Other purine bases encountered are hypoxanthine and xanthine (Figure 19.1). Nucleoside derivatives of these molecules contain either ribose or 2-deoxyribose linked to the purine ring through a ***β*-*N*-glycosidic bond** at N-9. Ribonucleosides contain ribose, while deoxyribonucleosides contain deoxyribose as the sugar moiety (Figure 19.2). Nucleotides are **phosphate esters** of purine nucleosides (Figure 19.3). 3′-Nucleotides such as adenosine 3′-monophosphate (3′-AMP) may occur in cells but only as a result of nucleic acid degradation. In normally functioning cells, tri- and diphosphates are found to a greater extent than monophosphates, nucleosides, or free bases.

The pyrimidine nucleotides found in highest concentrations in cells are those containing **uracil** and **cytosine.** The structures of the bases are shown in Figure 19.4. Uracil and cytosine nucleotides are the major pyrimidine components of RNA. As with purine derivatives, the pyrimidine nucleosides or nucleotides contain either ribose or 2-deoxyribose. The sugar moiety is linked to the pyrimidine in a *β*-*N*-glycosidic bond at N-1. Nucleosides of pyrimidines are uridine, cytidine, and deoxythymidine (Figure 19.5). Phosphate esters of pyrimidine nucleosides are UMP, CMP, and dTMP. In cells the major pyrimidine derivatives are tri- and diphosphates (Figure 19.6).

See the Appendix (p. 1179) for a summary of the nomenclature and chemistry of the purines and pyrimidines.

Properties of Nucleotides

Cellular components containing either purine or pyrimidine bases can easily be detected because of their strong absorption of UV light. Purine bases, nucleosides, and

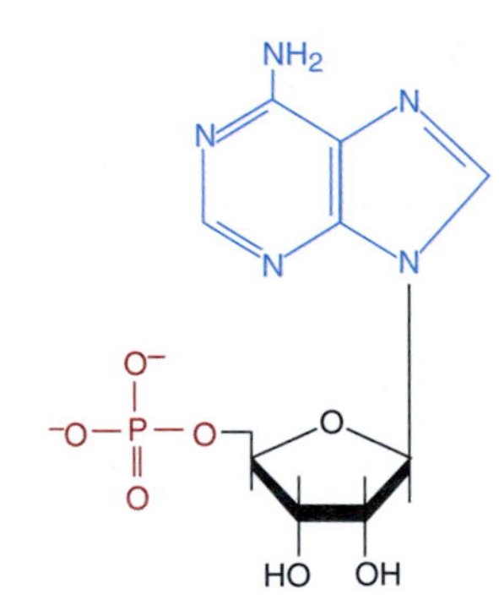
Adenosine 5′-monophosphate (AMP)

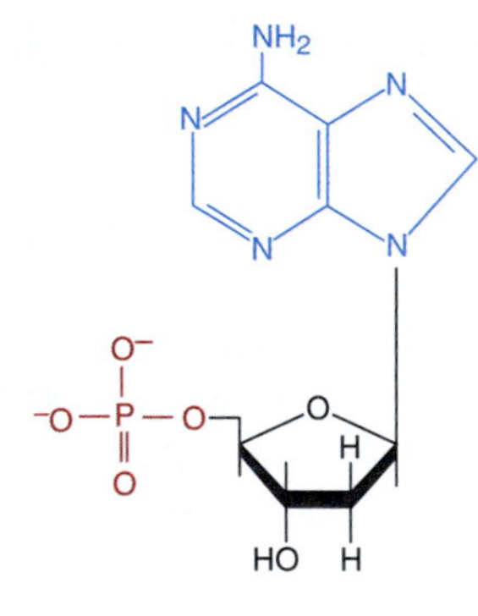
Deoxyadenosine 5′-monophosphate (dAMP)

Adenosine

Deoxyadenosine

FIGURE 19.2
Adenosine and deoxyadenosine.

Adenosine 5′-diphosphate (ADP)

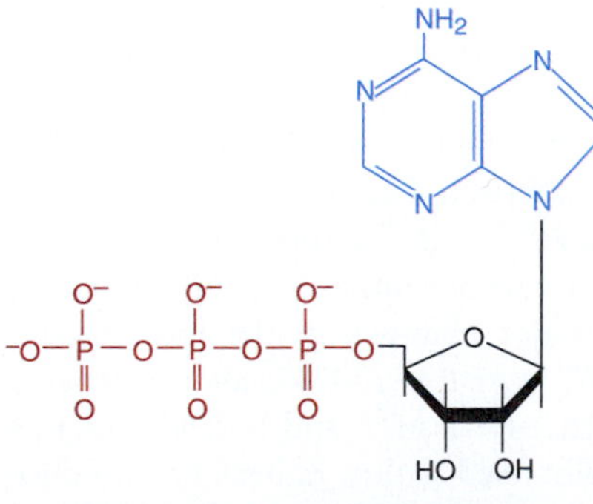
Adenosine 5′-triphosphate (ATP)

FIGURE 19.3
Adenine nucleotides.

Uracil

Cytosine

Thymine

FIGURE 19.4
Pyrimidine bases.

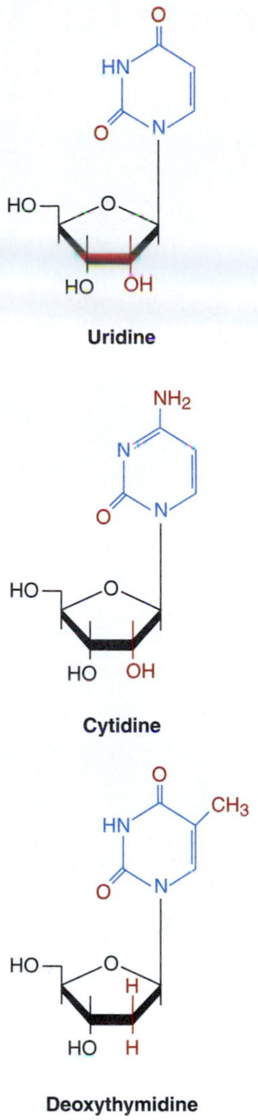

Uridine

Cytidine

Deoxythymidine

FIGURE 19.5
Pyrimidine nucleosides.

nucleotides have stronger absorptions than pyrimidines and their derivatives. The wavelength at which maximum absorption occurs varies with the particular base component, but in most cases the UV maximum is close to 260 nm. The UV spectrum for each derivative responds differently to changes in pH. The UV absorptions provide the basis for sensitive methods in assaying these compounds. For example, deamination of adenine nucleosides or nucleotides to the corresponding hypoxanthine derivatives causes a marked shift in absorption maximum from 265 to 250 nm, which is easily determined. Because of the high molar extinction coefficients of purine and pyrimidine bases and their high concentrations in nucleic acids, the absorbance at 260 nm is used to quantitate the amount of nucleic acid in RNA and DNA preparations.

The *N*-glycosidic bond of nucleosides and nucleotides is stable to alkali. However, stability of this bond to acid hydrolysis differs markedly. The *N*-glycosidic bond of purine nucleosides and nucleotides is easily hydrolyzed by dilute acid at elevated temperatures (e.g., 60°C) to yield free purine base and sugar or sugar phosphate. On the other hand, the *N*-glycosidic bond of uracil, cytosine, and thymine nucleosides and nucleotides is very stable to acid treatment. Strong conditions, such as perchloric acid (60%) and 100°C, release free pyrimidine but with complete destruction of the sugar. The *N*-glycosidic bond of dihydrouracil nucleoside and dihydrouracil nucleotide is labile in mild acid.

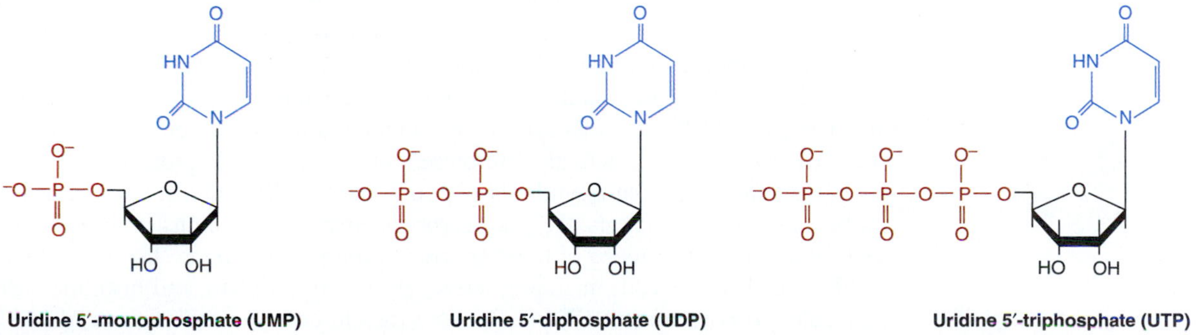

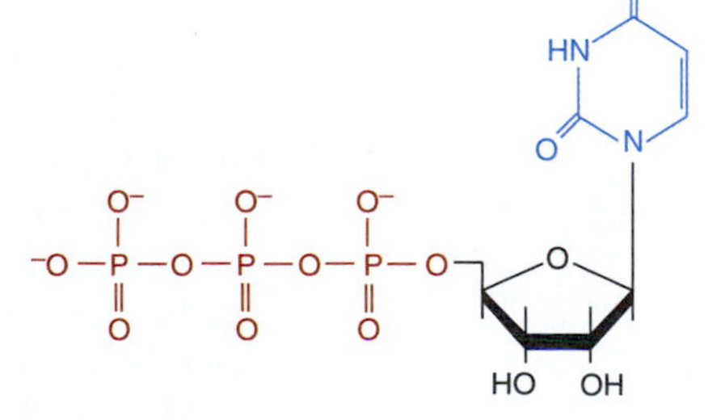

Uridine 5′-monophosphate (UMP)

Uridine 5′-diphosphate (UDP)

Uridine 5′-triphosphate (UTP)

FIGURE 19.6
Uracil nucleotides.

Because of the highly polar phosphate group, purine and pyrimidine nucleotides are much more soluble in aqueous solutions than are their nucleosides and free bases. In general, nucleosides are more soluble than free bases.

Purine and pyrimidine bases and their nucleoside and nucleotide derivatives can easily be separated by a variety of techniques. These methods include paper chromatography; thin-layer chromatography (TLC), utilizing plates with cellulose or ion-exchange resins; electrophoresis; and ion-exchange column chromatography. With high-performance liquid chromatography (HPLC), nanomole quantities of these components can easily and quickly be separated, detected, and quantitated.

19.4 | METABOLISM OF PURINE NUCLEOTIDES

The purine ring is synthesized *de novo* in mammalian cells utilizing amino acids as carbon and nitrogen donors and also CO_2 as a carbon donor. The ***de novo* pathway** for purine nucleotide synthesis leading to **inosine 5′-monophosphate (IMP)** consists of ten metabolic steps. Hydrolysis of ATP is required to drive several reactions in this pathway. Overall, the *de novo* pathway for purine nucleotide synthesis is expensive in terms of moles of ATP utilized per mole of IMP synthesized.

Purine Nucleotides Are Synthesized by a Stepwise Buildup of the Ring to Form IMP

All enzymes involved in synthesis of purine nucleotides are found in the cytosol. However, not all cells (e.g., red cells) are capable of *de novo* purine nucleotide synthesis. In the *de novo* pathway, a stepwise series of reactions leads to synthesis of IMP, which is the precursor for both **adenosine 5′-monophosphate (AMP)** and **guanosine 5′-monophosphate (GMP).** This pathway is highly regulated by AMP and GMP; IMP is not normally found to any extent in cells.

Formation of IMP is shown in Figure 19.7. Several points should be emphasized about this pathway: 5′-phosphoribosyl-1-pyrophosphate (PRPP) is synthesized from ribose 5-phosphate generated by the pentose phosphate pathway; the equivalent of 6 mol of ATP is utilized per mole of IMP synthesized; formation of **5-phosphoribosylamine** (the first step) is the committed and regulated step. In formation of 5-phosphoribosylamine, the N—C bond is formed that will ultimately be the *N*-glycosidic bond of the purine nucleotide; there are no known regulated steps between 5-phosphoribosylamine and IMP; and **tetrahydrofolate** serves as a "C_1" carrier (N^{10}-formyl H_4folate, Figure 19.8) in this pathway. It is important to point out that in reactions 3 and 9 (Fig. 19.7) tetrahydrofolate is regenerated. **Phosphoribosyl-5-aminoimidazole carboxylase,** which catalyzes the reaction in which CO_2 is used to introduce carbon-6 of the ring, is not a biotin-dependent carboxylase.

The enzyme activities catalyzing several steps in the pathway reside on separate domains of **multifunctional proteins.** The activities of 5′-phosphoribosylglycinamide synthetase, 5′-phosphoribosylglycinamide transformylase, and 5′-phosphoribosylaminoimidazole synthetase form part of a trifunctional protein. 5′-Phosphoribosylaminoimidazole carboxylase and 5′-phosphoribosyl-4-(*N*-succinocarboxamide)-5-aminoimidazole synthetase activities are on the same bifunctional protein. 5′-Phosphoribosyl-4-carboxamide-5-aminoimidazole transformylase and IMP cyclohydrolase activities are present on a bifunctional protein.

To summarize, *de novo* synthesis of purine nucleotides requires amino acids as carbon and nitrogen donors, CO_2 as a carbon source, and "C_1-units" transferred via H_4folate. The contributions of these sources to the purine ring is shown in Figure 19.9. Several amino acids including serine, glycine, tryptophan, and histidine yield "C_1-units" to H_4folate (see Chapter 18) and therefore they contribute to carbons-2 and-8 of the ring.

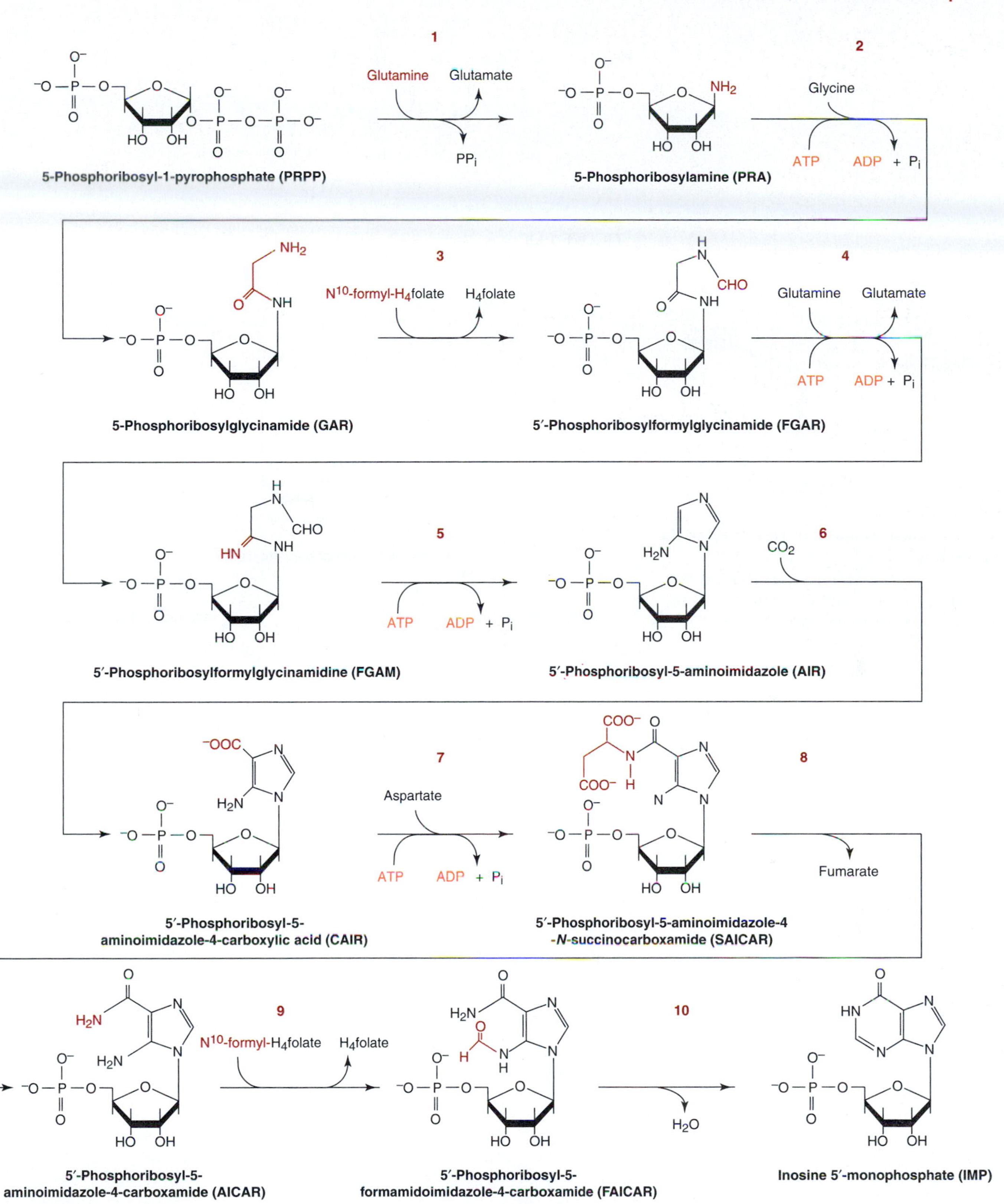

FIGURE 19.7
***De novo* synthesis of purine ribonucleotides.**
The enzymes catalyzing the reactions are (1) glutamine PRPP amidotransferase, (2) GAR synthetase, (3) GAR transformylase, (4) FGAM synthetase, (5) AIR synthetase, (6) AIR carboxylase, (7) SAICAR synthetase, (8) adenylosuccinate lyase, (9) AICAR transformylase, and (10) IMP cyclohydrolase.

H
H2N
N
N
H
H
H
H
N
N
OH
N
H
CH2
N
C
H
O
O
H
COO−
C
N
C
H
CH2
CH2
COO−

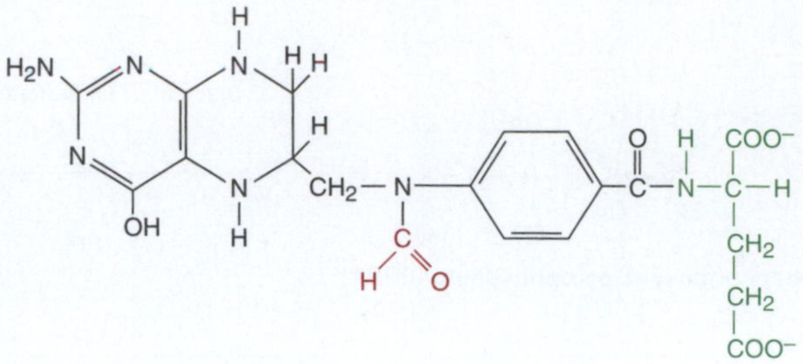

FIGURE 19.8
Structure of N^{10}-formyl H_4folate.

HCO3−
Glycine
Aspartate amine
"C1"–H4folate
"C1"–H4folate
Glutamine amide
1
2
3
4
5
6
7
8
9

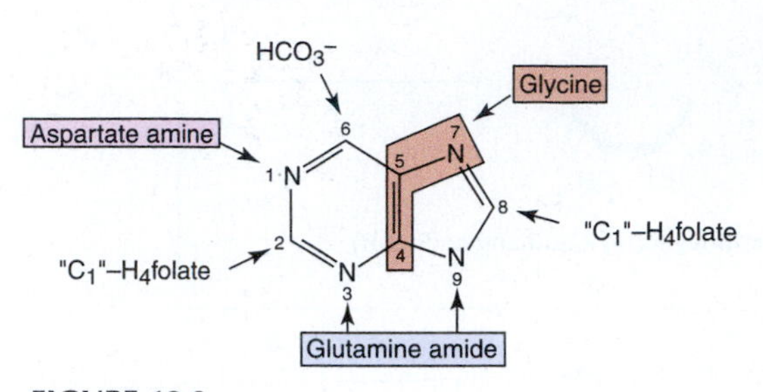

FIGURE 19.9
Sources of carbon and nitrogen atoms in the purine ring.
C-4, C-5, and N-7 from glycine; N-3 and N-9 from glutamine; C-2 and C-8 from "C_1"-H_4folate; N-1 from aspartate; and C-6 from CO_2.

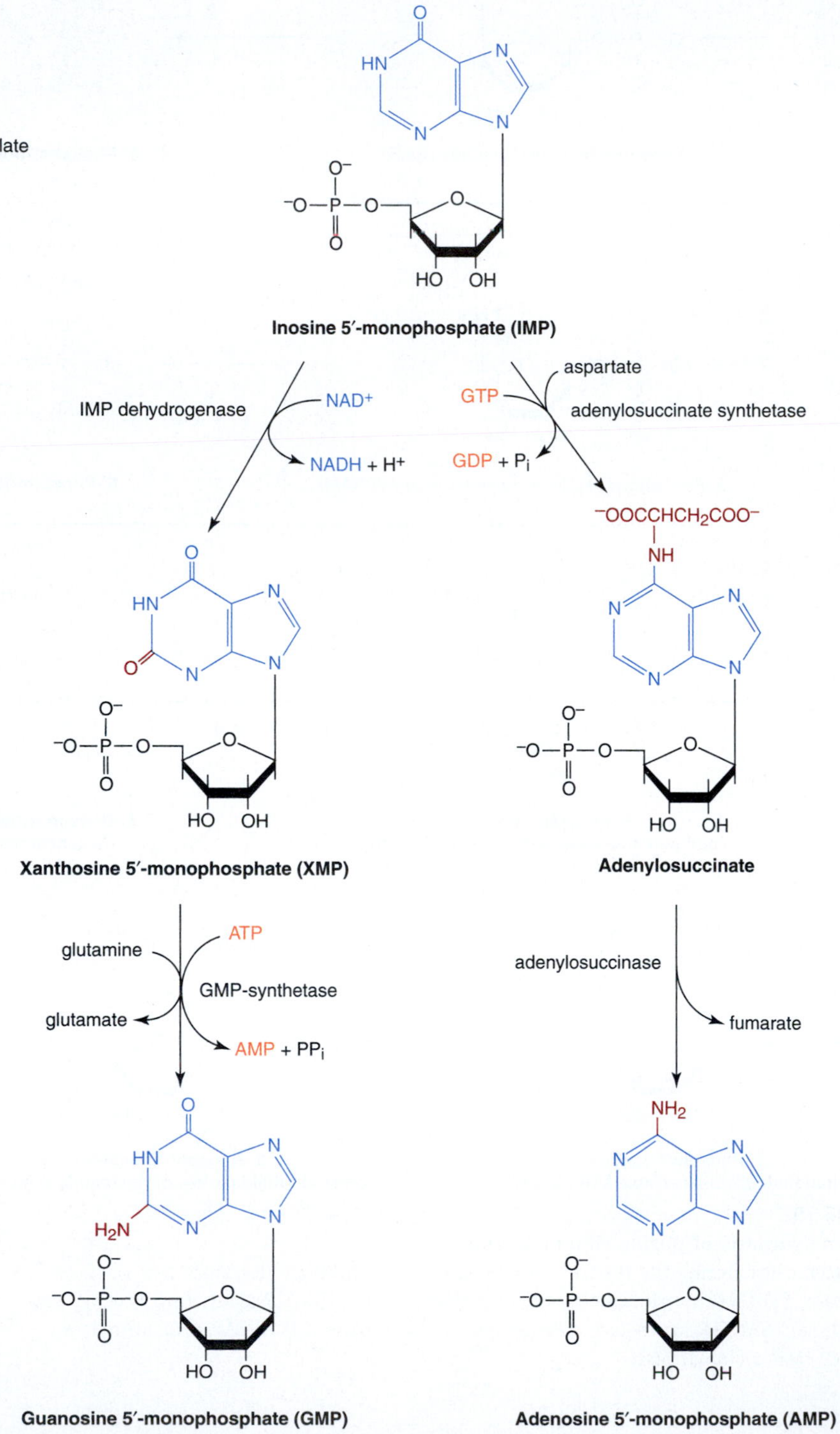

Inosine 5′-monophosphate (IMP)
IMP dehydrogenase
NAD+
NADH + H+
GTP
aspartate
adenylosuccinate synthetase
GDP + Pi
−OOCCHCH2COO−
NH
Xanthosine 5′-monophosphate (XMP)
Adenylosuccinate
glutamine
ATP
GMP-synthetase
glutamate
AMP + PPi
adenylosuccinase
fumarate
NH2
H2N
Guanosine 5′-monophosphate (GMP)
Adenosine 5′-monophosphate (AMP)

FIGURE 19.10
Formation of AMP and GMP from IMP branch point.

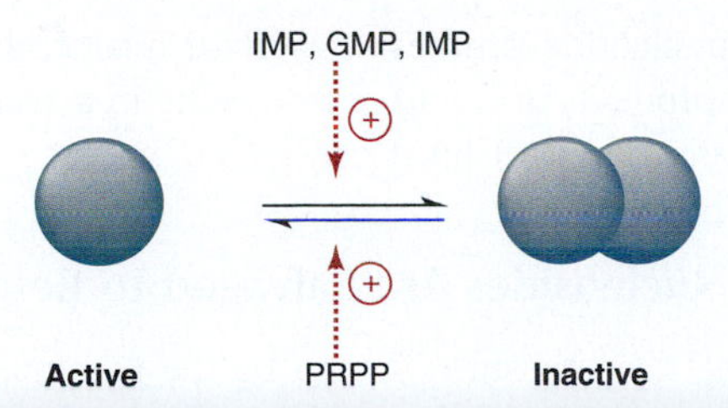

FIGURE 19.11
Effects of allosteric modulators on molecular forms of glutamine PRPP amidotransferase.

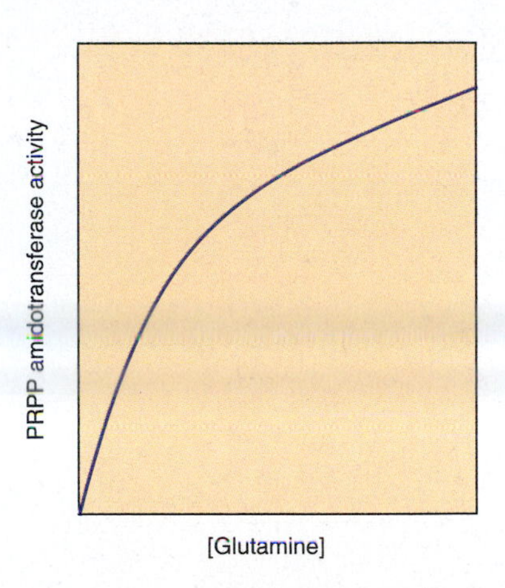

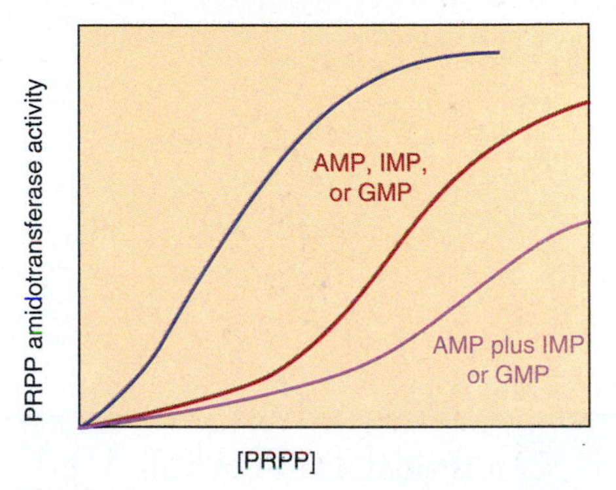

FIGURE 19.12
Glutamine PRPP amidotransferase activity as a function of glutamine or PRPP concentrations.

IMP Is Common Precursor for AMP and GMP

IMP, the first ribonucleotide formed in the *de novo* pathway, is the common precursor for AMP and GMP synthesis (Figure 19.10). AMP and GMP are converted to ATP and GTP, respectively, utilizing **nucleoside 5′-monophosphate kinases** and **nucleoside 5′-diphosphate kinases.** Conversion of IMP to AMP and GMP does not occur randomly. Formation of GMP from IMP requires ATP as the energy source, whereas formation of AMP from IMP requires GTP as the energy source. This is thought of as a reciprocal relationship: that is, when there is sufficient ATP in the cell, GMP will be synthesized from IMP and when there is sufficient GTP, AMP will be synthesized from IMP.

Purine Nucleotide Synthesis Is Highly Regulated

The committed step of a metabolic pathway is generally the site of metabolic regulation. In *de novo* purine nucleotide synthesis, formation of **5-phosphoribosylamine** from glutamine and 5-phosphoribosyl-1-pyrophosphate (PRPP) is the committed step in IMP formation. The enzyme catalyzing this reaction, **glutamine PRPP amidotransferase,** is rate-limiting and is regulated allosterically by the end products of the pathway—IMP, GMP, and AMP. These nucleotides serve as negative effectors. On the other hand, PRPP is a positive effector. Glutamine PRPP amidotransferase is a monomer of 135 kDa. In the presence of IMP, AMP, or GMP, the enzyme forms a dimer that is much less active. The presence of PRPP favors the active monomeric form of the enzyme (Figure 19.11).

The enzyme from human tissues has distinct nucleotide-binding sites. One site specifically binds oxypurine nucleotides (IMP and GMP) while the other site specifically binds aminopurine nucleotides (AMP). When AMP and GMP or IMP are present simultaneously, the enzyme activity is inhibited synergistically. Glutamine PRPP amidotransferase displays hyperbolic kinetics with respect to glutamine as the substrate and sigmoidal kinetics with respect to PRPP (Figure 19.12). Since the intracellular concentration of glutamine varies relatively little and is close to the K_m of the enzyme, the glutamine concentration has little effect in regulating IMP synthesis. The intracellular concentration of PRPP, however, varies widely and can be 10 to 100 times less than the K_m for PRPP. As a result, the concentration of PRPP plays an important role in regulating synthesis of purine nucleotides.

Between formation of 5-phosphoribosylamine and IMP, there are no known regulated steps. However, there is regulation at the branch point of IMP to AMP and IMP to GMP. From IMP to GMP, **IMP dehydrogenase** is the rate-limiting enzyme and it is regulated by GMP acting as a competitive inhibitor of IMP dehydrogenase. **Adenylosuccinate synthetase** is rate-limiting in conversion of IMP to AMP with AMP acting as a competitive inhibitor.

There must be other mechanisms that regulate the ATP/GTP ratio within relatively narrow limits. In most cells the total cellular concentration of adenine nucleotides (ATP plus ADP plus AMP) is four to six times that of guanine nucleotides (GTP plus GDP plus GMP). The overall regulation of purine nucleotide synthesis is summarized in Figure 19.13. Defects in the metabolic pathway that lead to loss of

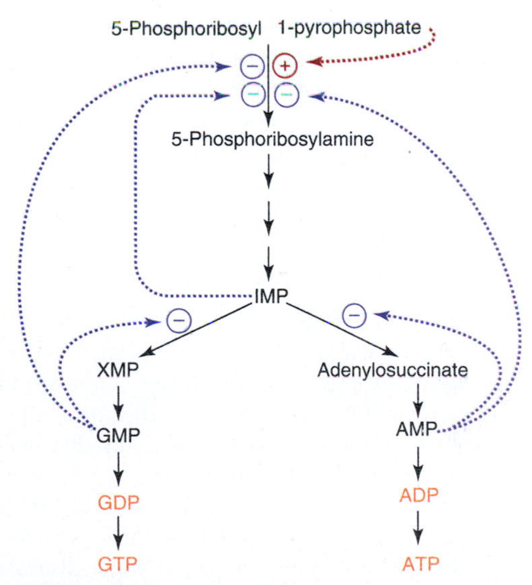

FIGURE 19.13
Regulation of purine nucleotide synthesis.

regulation of purine nucleotide synthesis result in overproduction of purine nucleotides and the end product, uric acid. This results in a relatively common clinical condition known as gout (see Clin. Corr. 19.1).

Purine Bases and Nucleosides Are Salvaged to Reform Nucleotides

The efficiency of normal metabolism is shown by the presence of two distinct **"salvage pathways."** One pathway utilizes the bases—hypoxanthine, guanine, and adenine—as substrates while the other pathway utilizes preformed nucleosides as the substrates. Each pathway is specific with respect to the base or nucleoside being "salvaged." The "salvage" of bases requires the activity of **phosphoribosyltransferases** that utilize PRPP as the ribose phosphate donor. There are two distinct phosphoribosyltransferases. **Hypoxanthine–guanine phosphoribosyltransferase (HGPRTase)** catalyzes the reactions

$$\text{hypoxanthine} + \text{PRPP} \longrightarrow \text{IMP} + \text{PP}_i$$

and

$$\text{guanine} + \text{PRPP} \longrightarrow \text{GMP} + \text{PP}_i$$

and **adenine phosphoribosyltransferase (APRTase)** catalyzes

$$\text{adenine} + \text{PRPP} \longrightarrow \text{AMP} + \text{PP}_i$$

CLINICAL CORRELATION 19.1

Gout

Gout is characterized by elevated uric acid concentrations in blood and urine due to a variety of metabolic abnormalities that lead to overproduction of purine nucleotides via the *de novo* pathway. Many, if not all, of the clinical symptoms associated with elevated concentrations of uric acid arise because of the very poor solubility of uric acid in the aqueous environment. Sodium urate crystals deposit in joints of the extremities and in renal interstitial tissue and these events tend to trigger the sequelae. Hyperuricemia from overproduction of uric acid via the *de novo* pathway can be distinguished from hyperuricemia that results from kidney disease or excessive cell death (e.g., increased degradation of nucleic acids from radiation therapy). Feeding of [^{15}N] glycine to a patient who is an overproducer will result in uric acid excreted in urine that is enriched in ^{15}N at the N-7 of uric acid while in a patient who is not an overproducer, there will be no enrichment of ^{15}N in uric acid from these patients.

Studies of "gouty" patients have shown that multiple and heterogeneous defects are the cause of overproduction of uric acid. In some cases, biochemical defects have not been defined. Examples of biochemical defects that result in increased purine nucleotide synthesis include the following:

1. Increased PRPP synthetase activity: Increased PRPP synthetase activity results in increased intracellular concentrations of PRPP. As discussed in the section on regulation of purine nucleotide synthesis, PRPP acts as a positive effector of glutamine–PRPP amidotransferase, leading to increased flux through the *de novo* pathway since activity of the rate-limiting step is markedly increased.
2. Partial HGPRTase activity: Partial decrease in HGPRTase activity has two fallouts with respect to the *de novo* pathway for purine nucleotide synthesis. First, since there is decreased salvage of hypoxanthine and guanine, PRPP is not consumed by the HGPRTase reaction and PRPP can activate glutamine–PRPP amidotransferase activity. Second, with decreased salvage of hypoxanthine and guanine, IMP and GMP are not formed via this pathway so that regulation of the PRPP amidotransferase step by IMP and GMP as negative effectors is compromised.
3. Glucose 6-phosphatase deficiency: In patients who have glucose 6-phosphatase deficiency (von Gierke's disease, type I glycogen storage disease) there is frequently hyperuricemia and gout as well. Loss of glucose 6-phosphatase activity results in more glucose 6-phosphate being shunted to the hexose monophosphate shunt. As a result of increased hexose monophosphate shunt activity, more ribose 5-phosphate is generated and the intracellular level of PRPP is increased. PRPP is a positive effector of PRPP amidotransferase.

These examples show that factors that increase the rate-limiting step in *de novo* synthesis of purine nucleotide synthesis lead to increased synthesis and degradation to uric acid. There are different approaches to the treatment of gout that include colchicine, antihyperuricemic drugs, and allopurinol. Allopurinol and its metabolite, alloxanthine, are effective inhibitors of xanthine oxidase and will cause a decrease in uric acid levels. In overproducers who do not have a severe deficiency of HGPRTase activity, allopurinol treatment inhibits xanthine oxidase, thereby increasing the concentrations of hypoxanthine and xanthine. These purine bases are then salvaged via HGPRTase to form IMP and XMP. These reactions consume PRPP and generate purine nucleotides that inhibit PRPP amidotransferase. The overall effect is that allopurinol treatment decreases both uric acid formation and *de novo* synthesis of purine nucleotides.

Becker, M.A. and Roessler, B.J. Hyperuricemia and Gout in The Metabolic and Molecular Bases of Inherited Disease. Scriver, C.R., Beaudet, A. L. Sly, W. S. and Valle, D (Eds.) Vol II, Chapter 49, 1655, 1995.

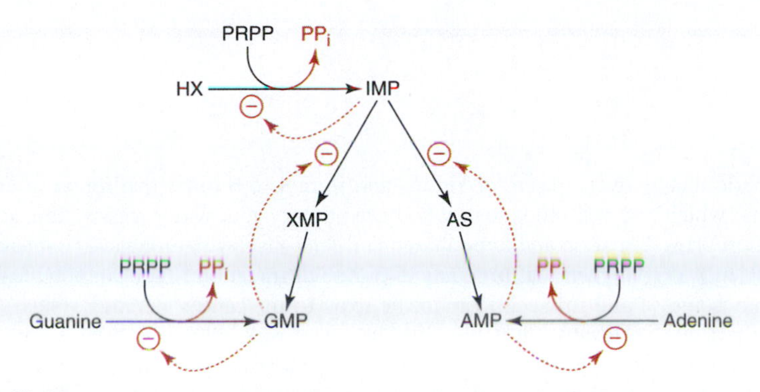

FIGURE 19.14
Salvage of purine nucleobases via phosphoribosyltransferases.
Effects of products on AMP and GMP synthesis from IMP. The dashed lines represent sites of regulation; ⊖ inhibition.

These two enzymes do not overlap in substrate specificity. The reactions are regulated by their end products. IMP and GMP are competitive inhibitors, with respect to PRPP, of HGPRTase while AMP is a competitive inhibitor, with respect to PRPP, of APRTase. In this way, salvage of purine bases is regulated.

Hypoxanthine and guanine for salvage arise from degradation of endogenous or exogenous purine nucleotides. Adenine utilized in the APRTase reaction appears to be mainly from synthesis of polyamines (see p. 807). For each molecule of **spermine** synthesized, two molecules of **5′-methylthioadenosine** are generated that is degraded to **5-methylthioribose-1-phosphate** and adenine by the **5′-methylthioadenosine phosphorylase**-catalyzed reaction. The adenine base is salvaged through the APRTase reaction.

Generation of AMP and GMP through these phosphoribosyltransferase reactions effectively inhibits the *de novo* pathway at the PRPP amidotransferase step. First, PRPP is consumed, decreasing the rate of formation of 5-phosphoribosylamine; and second, AMP and GMP serve as feedback inhibitors at this step (Figure 19.14).

HGPRTase activity is markedly depressed in the **Lesch–Nyhan syndrome** (see Clin. Corr. 19.2), which is characterized clinically by hyperuricemia, mental retardation, and self-mutilation.

Nucleosides such as adenosine are "salvaged" by adenosine kinase, a 5′-phosphotransferase that utilizes ATP as the phosphate donor. The substrate specificity of the 5′-phosphotransferases vary with the particular nucleoside kinase.

Overall, these salvage reactions conserve energy and permit cells to form nucleotides from the free bases. Erythrocytes, for example, do not have glutamine PRPP amidotransferase and hence cannot synthesize 5-phosphoribosylamine, the first unique metabolite in the pathway of purine nucleotide synthesis. As a consequence, they must depend on purine phosphoribosyltransferases and 5′-phosphotransferase (adenosine kinase) to replenish their nucleotide pools.

Purine Nucleotides Are Interconverted to Maintain the Appropriate Balance of Cellular Adenine and Guanine Nucleotides

De novo synthesis of purine nucleotides is under very fine control, executed at the committed step catalyzed by **glutamine PRPP amidotransferase** and at the branch point, IMP to AMP and IMP to GMP. Additional enzymes present in mammalian cells allow for interconversions of adenine and guanine nucleotides to maintain the appropriate balance of cellular concentrations of these purine nucleotides. These interconversions occur by indirect steps. There is no direct one-step pathway for conversion of GMP to AMP or AMP to GMP. In each case, AMP or GMP is metabolized to IMP (Figure 19.15). These reactions are catalyzed by separate enzymes, each of which is under separate controls. Reductive deamination of GMP to IMP is catalyzed by **GMP reductase.** GTP activates this step while xanthosine monophosphate

CLINICAL CORRELATION 19.2
Lesch–Nyhan Syndrome

Lesch–Nyhan syndrome is characterized clinically by hyperuricemia, excessive uric acid production, and neurological problems, which may include spasticity, mental retardation, and self-mutilation. This disorder is associated with a very severe or complete deficiency of HGPRTase activity. The gene for HGPRTase is on the X chromosome, hence the deficiency is virtually limited to males. In a study of the available patients, it was observed that if HGPRTase activity was less than 2% of normal, mental retardation was present, and if the activity was less than 0.2% of normal, the self-mutilation aspect was expressed. This defect also leads to excretion of hypoxanthine and xanthine.

There are more than a hundred disease-related mutations defined in the HGPRTase gene from Lesch–Nyhan patients. These have led to the loss of HGPRTase protein, loss of HGPRTase activity, "K_m mutants," HGPRTase protein with a short half-life, and so on.

The role of HGPRTase is to catalyze reactions in which hypoxanthine and guanine are converted to nucleotides. The hyperuricemia and excessive uric acid production that occur in patients with Lesch–Nyhan syndrome are easily explained by the lack of HGPRTase activity. Hypoxanthine and guanine are not salvaged, leading to increased intracellular pools of PRPP and decreased levels of IMP or GMP. Both of these factors promote *de novo* synthesis of purine nucleotides without regard for proper regulation of this pathway at PRPP amidotransferase.

It is not understood why a severe defect in this salvage pathway leads to neurological problems. Adenine phosphoribosyltransferase activity in these patients is normal or elevated. With this salvage enzyme, presumably the cellular needs for purine nucleotides could be met by conversion of AMP to IMP followed by the conversion of IMP to GMP if the cell's *de novo* pathway were not functioning. The normal tissue distribution of HGPRTase activity perhaps could explain the neurological symptoms. The brain (frontal lob, basal ganglia, and cerebellum) has 10–20 times the level of enzyme activity found in liver, spleen, or kidney and from 4 to 8 times that found in erythrocytes. Individuals who have primary gout with excessive uric acid formation and hyperuricemia do not display neurological problems. It is argued that products of purine degradation (hypoxanthine, xanthine, and uric acid) cannot be toxic to the central nervous system (CNS). However, it is possible that these metabolites are toxic to the developing CNS or that lack of enzyme leads to an imbalance in the concentrations of purine nucleotides at critical times during development.

If IMP dehydrogenase activity in brain were extremely low, lack of HGPRTase could lead to decreased amounts of intracellular GTP due to decreased salvage of guanine. Since GTP is a precursor of tetrahydrobiopterin, a required cofactor in the biosynthesis of neurotransmitters, and is required in other functions such as signal transduction via G-proteins and protein synthesis, low concentrations of GTP during development could be the triggering factor in the observed neurological manifestations.

Treatment of Lesch–Nyhan patients with allopurinol will decrease the amount of uric acid formed, relieving some of the problems caused by sodium urate deposits. However, since the Lesch–Nyhan patient has a marked reduction in HGPRTase activity, hypoxanthine and guanine are not salvaged, PRPP is not consumed, and consequently *de novo* synthesis of purine nucleotides is not shut down. There is no treatment for the neurological problems. These patients usually die from kidney failure, resulting from high sodium urate deposits.

Rossiter, B. J. F. and Caskey, C. T. Hypoxanthine–guanine phosphoribosyltransferase deficiency: Lesch–Nyhan syndrome and gout. In: C. R. Scriver, A. L. Beaudet, W. S. Sly, and D. Valle (Eds.), *The Metabolic and Molecular Bases of Inherited Disease,* 7th ed., Vol. II, Chap. 50. New York: McGraw-Hill, 1995, p. 1679; and Nyhan, W. L. The recognition of Lesch–Nyhan syndrome as an inborn error of purine metabolism. *J. Inherit. Metab. Dis.* 20:171,1997.

(XMP) is a strong competitive inhibitor of the reaction. GTP, while not required by the enzyme, increases enzyme activity by lowering the K_m with respect to GMP and by increasing V_{max}.

AMP deaminase (5′-AMP aminohydrolase) catalyzes deamination of AMP to IMP and is activated by K^+ and ATP and inhibited by P_i, GDP, and GTP. In the absence of K^+, the velocity versus AMP concentration curve is sigmoidal. The presence of K^+ is not required for maximum activity; rather K^+ is a positive allosteric effector reducing the apparent K_m for AMP.

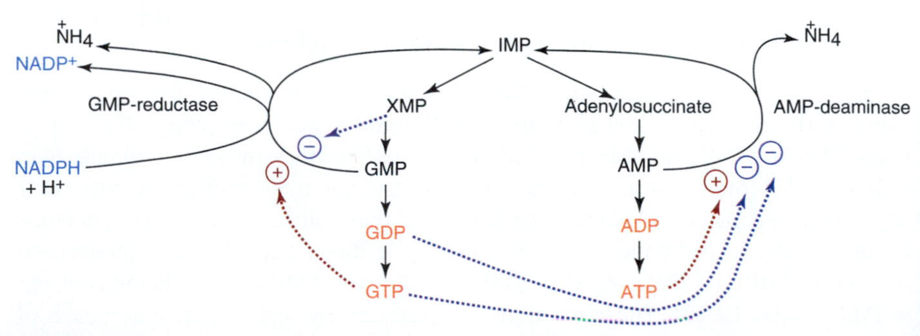

FIGURE 19.15
Interconversions of purine nucleotides.

The net effect of these reactions is that cells interconvert adenine and guanine nucleotides to meet cellular needs, while maintaining control over these reactions.

GTP Is Precursor of Tetrahydrobiopterin

GTP is the direct precursor for **tetrahydrobiopterin** synthesis (Figure 19.16). Reactions from GTP to tetrahydrobiopterin are catalyzed by **GTP cyclohydrolase I, 6-pyruvoyl-tetrahydropterin synthase,** and **sepiapterin reductase,** with GTP cyclohydrolase I being rate-limiting. Many cell types synthesize tetrahydrobiopterin. Tetrahydrobiopterin is a required cofactor in hydroxylation reactions involving phenylalanine, tyrosine, and tryptophan (see p. 797) and is involved in the generation of nitric oxide. Inhibitors of IMP dehydrogenase cause a marked reduction in cellular levels of tetrahydrobiopterin, demonstrating the importance of GTP as the precursor of tetrahydrobiopterin and of IMP dehydrogenase as the rate-limiting enzyme in GTP formation.

End Product of Purine Degradation in Humans Is Uric Acid

The degradation of purine nucleotides, nucleosides, and bases follow a common pathway leading to formation of uric acid (Figure 19.17). The enzymes involved in

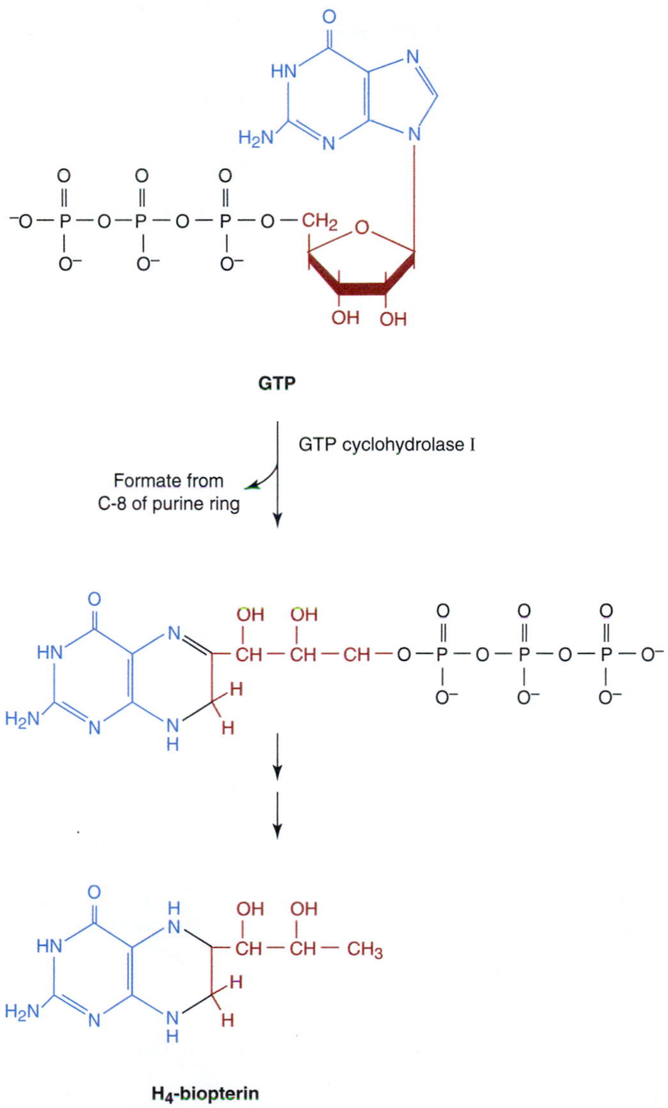

GTP

GTP cyclohydrolase I

Formate from C-8 of purine ring

H_4-biopterin

FIGURE 19.16
Synthesis of tetrahydrobiopterin from GTP.

degradation of nucleic acids and nucleotides and nucleosides vary in specificity. **Nucleases** are specific toward either RNA or DNA and also toward the bases and position of cleavage site at the 3′,5′-phosphodiester bonds. **Nucleotidases** range from those with relatively high specificity, such as 5′-AMP nucleotidase, to those with broad specificity, such as the acid and alkaline phosphatases, which will hydrolyze any of the 3′- or 5′-nucleotides (see Clin. Corr. 19.3). AMP deaminase is specific for AMP. **Adenosine deaminase** is less specific, since not only adenosine, but also 2′-deoxyadenosine and many other 6-aminopurine nucleosides are deaminated by this enzyme.

Purine nucleoside phosphorylase catalyzes the reversible reactions

$$\text{inosine} + P_i \leftrightarrow \text{hypoxanthine} + \text{ribose 1-P}$$

or

$$\text{guanosine} + P_i \leftrightarrow \text{guanine} + \text{ribose 1-P}$$

or

$$\text{xanthosine} + P_i \leftrightarrow \text{xanthine} + \text{ribose 1-P}$$

Deoxyinosine and deoxyguanosine are also natural substrates for purine nucleoside phosphorylase. This is important since removal of deoxyguanosine

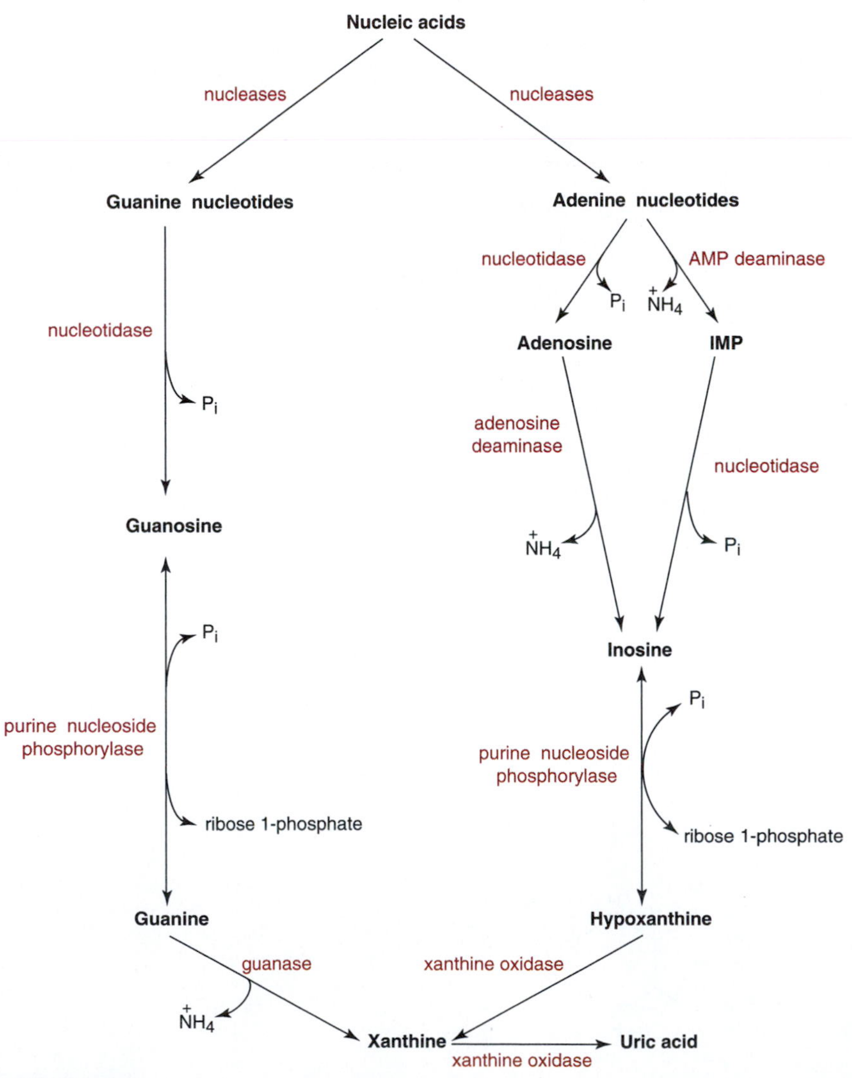

FIGURE 19.17
Degradation of purine nucleotides.

CLINICAL CORRELATION 19.3
Increased Cytosolic 5′-Nucleotidase Activity

A recent report described four patients in whom 5′-nucleotidase activity in fibroblast lysates as measured using either 5′-AMP or 5′-UMP as substrates was increased to six- to tenfold in these patients compared to the controls. These four unrelated children had problems associated with "developmental delay, seizures, ataxia, infections, severe language deficit, and an unusual behavioral phenotype characterized by hyperactivity, short attention span, and poor social interaction." Since it was possible that the increased cellular 5′-nucleotidase activity was causing decreased nucleotide pools, the patients were treated with oral uridine. Remarkably, all four patients treated with uridine improved dramatically in all aspects of the physical and medical behaviors. These findings again point to the fact that either overproduction or blockage of nucleotide synthesis, and decreased degradation or increased degradation of nucleotides or nucleosides can have major consequences on the well-being of the individual.

Page, T., Yu, A., Fontanesi, J., and Nyhan, W. L. Developmental disorder associated with increased cellular nucleotidase activity. *Proc. Natl. Acad. Sci. USA* 94:11601, 1997.

prevents uncontrolled accumulation of dGTP, which is toxic to cells at high concentrations. While the equilibrium constants for reactions catalyzed by purine nucleoside phosphorylase favor the direction of nucleoside synthesis, cellular concentrations of free purine base and ribose-1-phosphate are too low to support nucleoside synthesis under normal conditions. The main function of the enzyme is the degradative rather than synthetic pathway. Adenosine deaminase deficiency is associated with **severe combined immunodeficiency,** while purine nucleoside phosphorylase deficiency leads to a defective **T-cell immunity** but normal B-cell immunity (see Clin. Corr. 19.4).

CLINICAL CORRELATION 19.4
Immunodeficiency Diseases Associated with Defects in Purine Nucleoside Degradation

Two distinct immunodeficiency diseases are associated with defects in adenosine deaminase (ADA) and purine nucleoside phosphorylase (PNP), respectively. These enzymes are involved in the degradative pathways leading to formation of uric acid. Natural substrates for adenosine deaminase are adenosine and deoxyadenosine while natural substrates for purine nucleoside phosphorylase are inosine, guanosine, deoxyinosine, and deoxyguanosine. A deficiency in ADA is associated with a severe combined immunodeficiency involving both T-cell and B-cell functions. PNP deficiency is associated with an immunodeficiency involving T-cell functions with the sparing of effects on B-cell function. In neither case is the mechanism(s) by which the lack of these enzymes leads to immune dysfunction known. However, in ADA-deficient patients, intracellular concentrations of dATP and S-adenosylhomocysteine are greatly increased. Several hypotheses have been put forth to explain the biochemical consequences of a lack of ADA: (1) high concentrations of dATP inhibit ribonucleotide reductase activity and as a consequence inhibit DNA synthesis; (2) deoxyadenosine inactivates *S*-adenosylhomocysteine hydrolase leading to decreased *S*-adenosylmethionine required for methylation of bases in RNA and DNA; and (3) increased concentrations of adenosine result in increased cAMP levels. It is possible that each of these mechanisms contributes to the overall effect of immune dysfunction. There is not, however, a suitable explanation for the specificity of the effects on only T cells and B cells.

Treatments of children with ADA deficiency have included blood transfusions, bone marrow transplantation, enzyme replacement therapy with ADA–polyethylene glycol conjugate (ADA–PEG), and most recently gene therapy. Each of these treatments has disadvantages. Blood transfusions produce problems of "iron overload" and safety of the source. Bone marrow transplantation, while curative, requires a suitably matched donor. Enzyme replacement therapy with ADA–PEG has been the most successful to date, but the treatment requires constant monitoring of ADA levels and frequent injections of very costly ADA–PEG. Gene therapy presents the hope for the future. There are strong indications in gene therapy trials that the ADA gene has been successfully transfected into stem cells of ADA-deficient children.

Hershfield, M. S. and Mitchell, B. S. Immunodeficiency diseases caused by adenosine deaminase deficiency and purine nucleoside phosphorylase deficiency. In: C. R. Scriver, A. L. Beaudet, W. S. Sly, and D. Valle (Eds.), *The Metabolic and Molecular Bases of Inherited Disease,* 7th ed., Vol. II, Chap. 52, New York McGraw-Hill, 1995, p. 1725; Hoogerbrugge, P. M., von Beusechem, V. W., Kaptein, L. C., Einerhard, M. P., and Valerio, D. Gene therapy for adenosine deaminase deficiency. *Br. Med. Bull.* 51:72, 1995; Hershfield, M. S. Adenosine deaminase deficiency: clinical expression, molecular basis, and therapy. *Semin. Hematol.* 35:291, 1998; Russell, C. S. and Clarke, L. A. Recombinant proteins for genetic disease. *Clin. Genet.* 55:389, 1999; Onodera, M., Nelson, D. M., Sakiyama, Y., Candotti, F., and Blaese, R. M. Gene therapy for severe combined immunodeficiency caused by adenosine deaminase deficiency: improved retroviral vectors for clinical trials. *Acta Haematol.* 101:89, 1999; and Carpenter, P. A., Ziegler, J. B., and Vowels, M. R. Late diagnosis and correction of purine nucleoside phosphorylase deficiency with allogenic bone marrow transplantation. *Bone Marrow Transplant.* 17:121, 1996.

CLINICAL CORRELATION 19.5

Cancer Patients Undergoing Radiation or Chemotherapy Treatment

Cancer patients with large tumor burdens who undergo radiation therapy or chemotherapy treatments also show increased serum and urine concentrations of uric acid. The source of this increased uric acid is not due to increased purine nucleotide synthesis but rather from the destruction of the tumor cells that in turn release degraded nucleic acids and cellular nucleotides that are further metabolized to uric acid. Many of the cancer treatment protocols include allopurinol as one of the drugs for the sole purpose of limiting the buildup of uric acid in the patients.

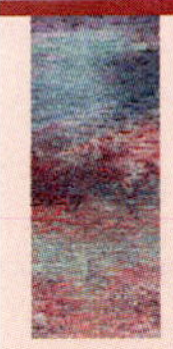

CLINICAL CORRELATION 19.6

Subclass of Patients with Autism

Recently a subclass of children with infantile autism was shown to excrete uric acid that is greater than two standard deviations above the normal mean. This group of children represents approximately 20% of the autistic population. With cultured fibroblasts from these autistic children, it was found that *de novo* synthesis of purine nucleotides as measured by [^{14}C]formate incorporation into purine nucleotides was increased fourfold over that seen in the normal control fibroblasts. As of this time the molecular basis for the increased *de novo* synthesis of purine nucleotides in these autistic children is not known. An unusual aspect of this increased purine nucleotide synthesis in these children is that the excretion of uric acid is elevated but the serum concentrations of uric acid is in the normal range.

Page, T. and Coleman, M. Purine metabolism abnormalities in a hyperuricosuric subclass of autism. *Biochim. Biophys. Acta* 1500:291, 2000.

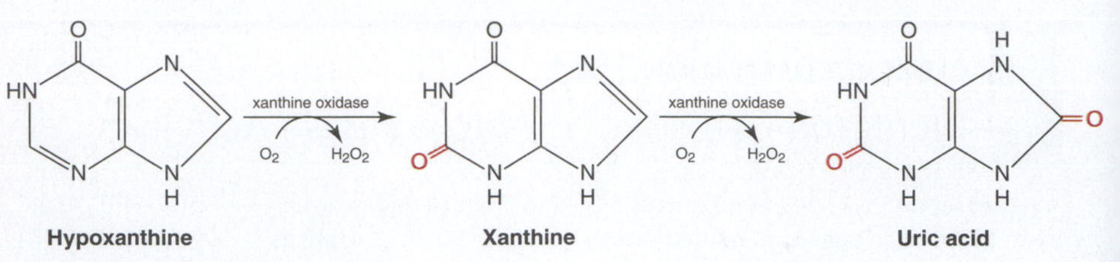

FIGURE 19.18
Reactions catalyzed by xanthine oxidase.

Formation of Uric Acid

As shown in Figure 19.17, adenine nucleotides end up as hypoxanthine while guanine nucleotides are metabolized to xanthine. These purines are metabolized by **xanthine oxidase** to form **uric acid,** a unique end product of purine nucleotide degradation in humans. The reactions are shown in Figure 19.18.

Xanthine oxidase contains FAD, Fe, and Mo and requires molecular oxygen as a substrate. Since uric acid is not very soluble in aqueous medium, there are clinical conditions in which elevated levels of uric acid results in deposition of sodium urate crystals primarily in joints. Hyperuricemia is a clinical condition characterized by excess levels of uric acid in the blood and generally increased levels of uric acid excretion in the urine (hyperuricuria). Since uric acid is the unique end product of purine degradation in humans, excess levels of uric acid indicate some metabolic situation that may or may not be serious. There are several instances in which the cause of the hyperuricemia/hyperuricuria can be defined as a metabolic alteration and other situations in which there is not a defined metabolic alteration (see Clin. Corr. 19.1, 19.2, 19.5, and 19.6).

19.5 | METABOLISM OF PYRIMIDINE NUCLEOTIDES

The pyrimidine ring is synthesized *de novo* in mammalian cells utilizing amino acids as carbon and nitrogen donors and CO_2 as a carbon donor. *De novo* synthesis of pyrimidine nucleotide leads to **uridine 5′-monophosphate** (UMP) in six metabolic steps. ATP hydrolysis (or equivalent) is required to drive several steps in the pathway.

Pyrimidine Nucleotides Are Synthesized by a Stepwise Series of Reactions to Form UMP

In contrast to *de novo* purine nucleotide synthesis, not all enzymes for *de novo* synthesis of pyrimidine nucleotides are in the cell cytosol. Reactions leading to formation of UMP are shown in Figure 19.19. The following important aspects of the pathway should be noted. The pyrimidine ring is formed first and then ribose 5-phosphate is added via PRPP. The enzyme catalyzing formation of carbamoyl phosphate, **carbamoyl phosphate synthetase II,** is cytosolic and is distinctly different from **carbamoyl phosphate synthetase I** found in the mitochondria as part of the urea cycle. Synthesis of *N*-carbamoylaspartate is the committed step in pyrimidine nucleotide synthesis but formation of cytosolic carbamoyl phosphate is the regulated step. Formation of **orotate** from dihydroorotate is catalyzed by the mitochondrial **dihydroorotate dehydrogenase.** The other enzymes of the pathway are found in the cytosol on multifunctional proteins. The activities of carbamoyl phosphate synthetase II, **aspartate carbamoyltransferase,** and **dihydroorotase** are found on a trifunctional protein (CAD), and **orotate phosphoribosyltransferase** and **OMP decarboxylase** activities are found on a bifunctional protein, defined as UMP synthase. A defect in this bifunctional protein that affects either phosphoribosyltransferase activity or decarboxylase activity leads to a rare clinical condition

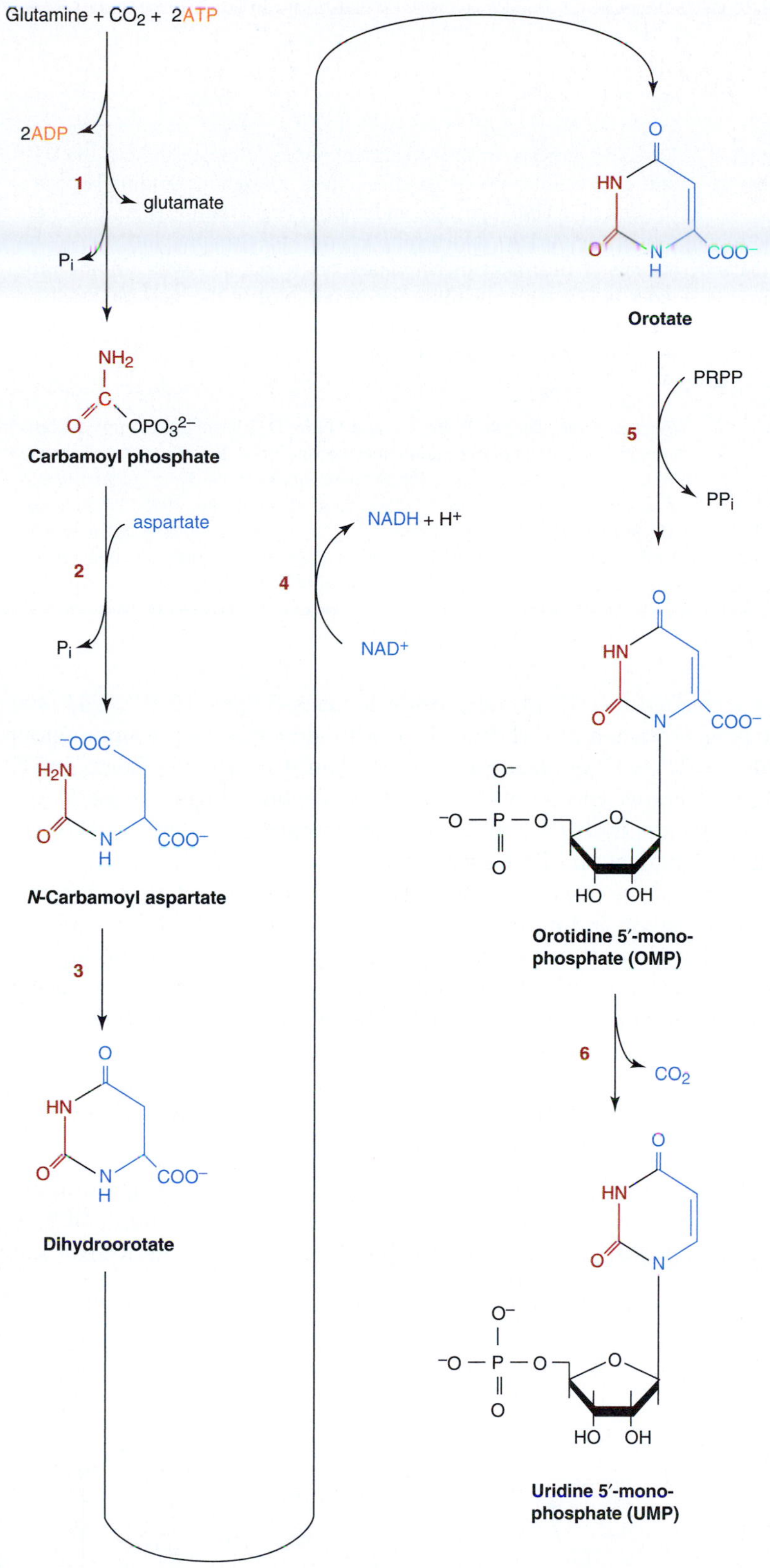

FIGURE 19.19
De novo **synthesis of pyrimidine nucleotides.** Enzyme activities catalyzing the reactions are (1) carbamoyl phosphate synthetase II, (2) aspartate carbamoyltransferase, (3) dihydroorotase, (4) dihydroorotate dehydrogenase, (5) orotate phosphoribosyltransferase, and (6) OMP decarboxylase. The activities of 1, 2, and 3 are on a trifunctional protein (CAD); the activities of 5 and 6 are on a bifunctional protein (UMP synthase).

known as hereditary orotic aciduria (see Clin. Corr. 19.7). Recent studies have shown that an immunosuppressive drug, leflunomide, which is used in the treatment of rheumatoid arthritis, inhibits *de novo* synthesis of pyrimidine nucleotides specifically at dihydroorotate dehydrogenase, the mitochondrial-localized enzyme of *de novo* pyrimidine nucleotide synthesis.

CLINICAL CORRELATION 19.7

Hereditary Orotic Aciduria

Hereditary orotic aciduria results from a defect in *de novo* synthesis of pyrimidine nucleotides. This genetic disease is characterized by severe anemia, growth retardation, and high levels of orotic acid excretion. The biochemical basis for orotic aciduria is a defect in one or both of the activities (orotate phosphoribosyltransferase or orotidine decarboxylase) associated with UMP synthase, the bifunctional protein . It is a very rare disease but the understanding of the metabolic basis for this disease has led to successful treatment of the disorder. Patients are fed uridine, which leads not only to reversal of the hematologic problem but also to decreased formation of orotic acid. Uridine is taken up by cells and converted by uridine phosphotransferase to UMP that is sequentially converted to UDP and then to UTP. UTP formed from exogenous uridine, in turn, inhibits carbamoyl phosphate synthetase II, the major regulated step in the *de novo* pathway. As a result, orotic acid synthesis via the *de novo* pathway is markedly decreased to essentially normal levels. Since UTP is also the substrate for CTP synthesis, uridine treatment serves to replenish both the UTP and CTP cellular pools. In effect, then, exogenous uridine bypasses the defective UMP synthase and supplies cells with UTP and CTP required for nucleic acid synthesis and other cellular functions. The success of treatment of hereditary orotic aciduria with uridine provides *in vivo* data regarding the importance of the step catalyzed by carbamoyl phosphate synthase II as the site of regulation of pyrimidine nucleotide synthesis in man.

Webster, D. R., Becroft, D. M. O., and Suttle, D. P. Hereditary orotic aciduria and other disorders of pyrimidine metabolism. In: C. R. Scriver, A. L. Beaudet, W. S. Sly, and D. Valle (Eds.), *The Metabolic and Molecular Bases of Inherited Disease*, 7th ed., Vol. II, Chap. 55. New York: McGraw-Hill, 1995, 1799; Suchi, M., Mizuno, H., Kawai, Y., Tonboi, T., et al. Molecular cloning of the human UMP synthase gene and characterization of a point mutation in two hereditary orotic aciduria families. *Am. J. Hum. Genet.* 60:525, 1997.

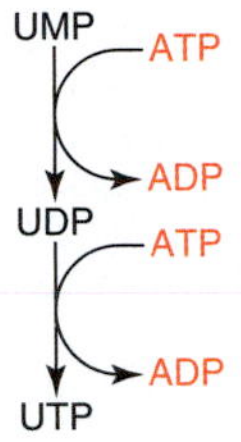

UMP — ATP → ADP
UDP — ATP → ADP
UTP

FIGURE 19.20
Formation of UTP from UMP.

UMP is converted to UTP by **nucleotide kinases** (Figure 19.20). **CTP synthetase** catalyzes formation of CTP from UTP with glutamine as the amino group donor (Figure 19.21). CTP synthetase displays homotropic sigmoidal kinetics; CTP, the product, is a negative effector of the reaction as shown in Figure 19.22. By regulating CTP synthetase in this way cells maintain an appropriate ratio of UTP and CTP for cellular functions and RNA synthesis.

To summarize, *de novo* synthesis of pyrimidine nucleotides requires aspartate and glutamine as carbon and nitrogen donors and CO_2 as a carbon donor (Figure 19.23). Five of the six reactions in the pathway take place in the cytosol of the cell, while the other reaction occurs in the mitochondria. The cytosolic enzyme activities reside on multifunctional proteins. UTP is the direct precursor of CTP.

Pyrimidine Nucleotide Synthesis in Humans Is Regulated at the Level of Carbamoyl Phosphate Synthetase II

Regulation of pyrimidine nucleotide synthesis in mammalian cells occurs at the carbamoyl phosphate synthetase II step. Carbamoyl phosphate synthetase II is a cytosolic enzyme and distinct from the mitochondrial carbamoyl phosphate syn-

Uridine 5′-triphosphate (UTP) — Glutamine, ATP → Glutamate, ADP + P_i → Cytidine 5′-triphosphate (CTP)

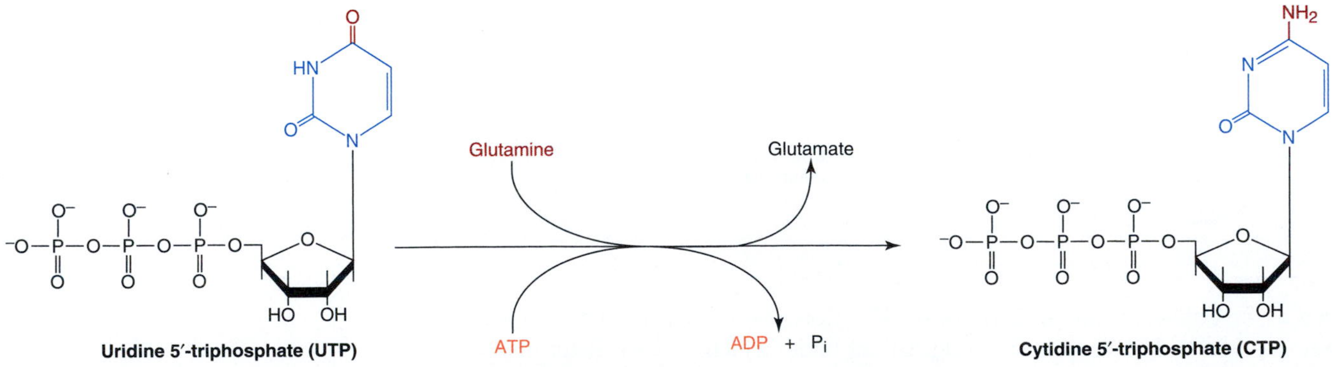

FIGURE 19.21
Formation of CTP from UTP.

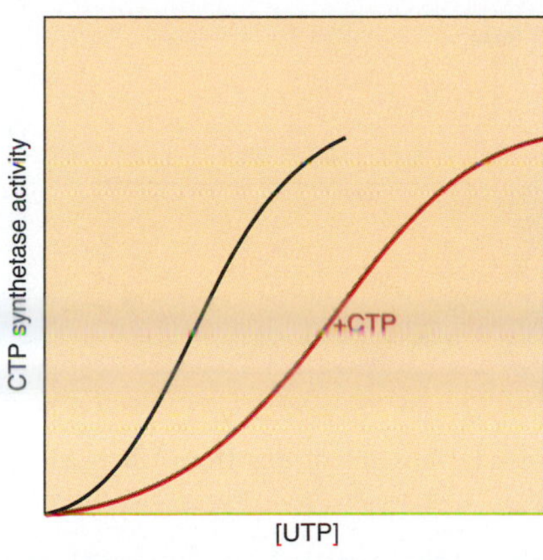

FIGURE 19.22
Regulation of CTP synthetase.

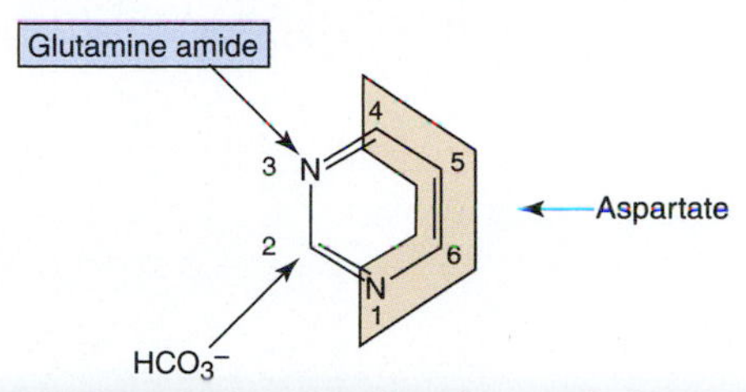

FIGURE 19.23 Sources of carbon and nitrogen atoms in pyrimidines.
C-4, C-5, and C-6 and N-3 from aspartate; N-1 from glutamine and C-2 from CO_2.

thetase I, which utilizes ammonia as the amino donor and is activated by *N*-acetylglutamate. Carbamoyl phosphate synthetase II is inhibited by UTP, an end product of the pathway, and is activated by PRPP. The K_i for UTP and the K_a for PRPP are in the range of values that would allow intracellular levels of UTP and PRPP to have an effect on the control of pyrimidine nucleotide synthesis. Carbamoyl phosphate synthetase II is the only source of carbamoyl phosphate in extrahepatic tissues. However, in liver, under stressed conditions in which there is excess ammonia, carbamoyl phosphate synthetase I generates carbamoyl phosphate in mitochondria, which ends up in the cytosol and serves as a substrate for pyrimidine nucleotide synthesis. This pathway serves to detoxify excess ammonia. Elevated levels of orotic acid are excreted as a result of ammonia toxicity in humans. This points to carbamoyl phosphate synthetase II as being the major regulated activity in pyrimidine nucleotide metabolism.

UMP does not inhibit carbamoyl phosphate synthetase II but competes with OMP to inhibit **OMP decarboxylase** (Figure 19.24). As discussed earlier, conversion of UTP to CTP is also regulated so that cells maintain a balance between uridine and cytidine nucleotides.

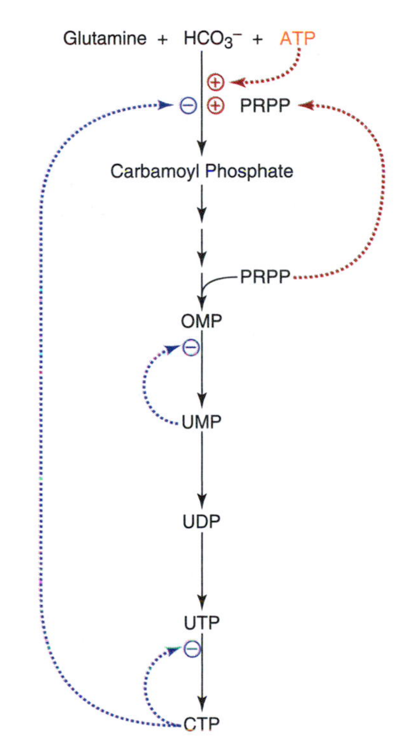

FIGURE 19.24
Regulation of pyrimidine nucleotide synthesis.
Solid arrows represent enzyme-catalyzed reactions and dashed arrows inhibition by products of the reactions.

Pyrimidine Bases Are Salvaged to Reform Nucleotides

Pyrimidines are "salvaged" by conversion to nucleotides by **pyrimidine phosphoribosyltransferase.** The general reaction is

$$\text{pyrimidine} + \text{PRPP} \rightarrow \text{pyrimidine nucleoside 5}'\text{-monophosphate} + \text{PP}_i$$

The enzyme from human erythrocytes utilizes orotate, uracil, and thymine as substrates but not cytosine. These salvage reactions divert the pyrimidine bases from the degradative pathway to the nucleotide level. As a pyrimidine base becomes available to cells, there are competing reactions that will either result in degradation and excretion or reutilization of the bases. For example, when normal liver is presented with uracil, it is rapidly degraded to β-alanine, whereas in proliferating tumor cells uracil is converted to UMP. This is the result of the availability of PRPP, enzyme levels, and metabolic state of the cells.

19.6 | DEOXYRIBONUCLEOTIDE FORMATION

The concentrations of **deoxyribonucleotides** are extremely low in nonproliferating cells. Only at the time of DNA replication (S phase) does the deoxyribonucleotide pool increase to support the required DNA synthesis.

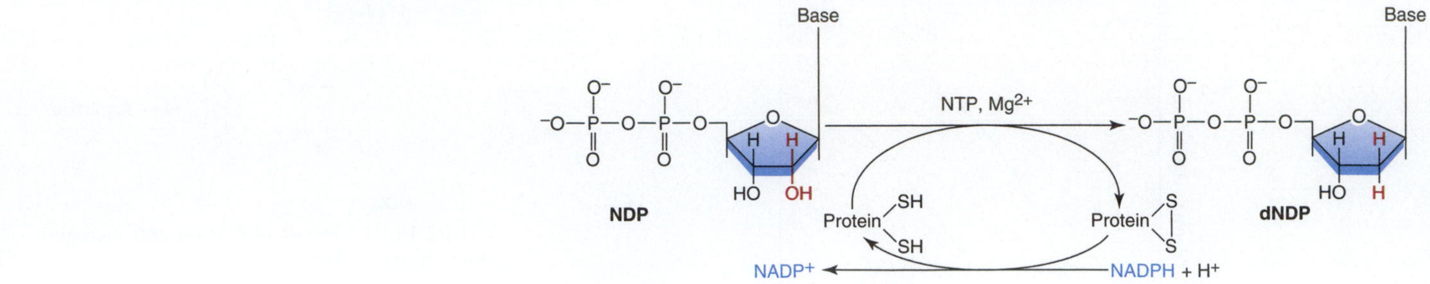

FIGURE 19.25
De novo **synthesis of 2′-deoxyribonucleotides from ribonucleotides.**

Deoxyribonucleotides Are Formed by Reduction of Ribonucleoside Diphosphates

Nucleoside 5′-diphosphate reductase (ribonucleotide reductase) catalyzes the conversion of ribonucleoside 5′-diphosphate to 2′-deoxyribonucleotide. The reaction is controlled by the amount of enzyme present in cells and by a very finely regulated allosteric control mechanism. The reaction is summarized in Figure 19.25. Reduction of a particular substrate requires a specific nucleoside 5′-triphosphate as a positive effector. For example, reduction of CDP or UDP requires ATP as the positive effector, while reduction of ADP and GDP requires the presence of dGTP and dTTP, respectively (Table 19.1). A small molecular weight protein, **thioredoxin** or **glutaredoxin,** is involved in reduction at the 2′ position through oxidation of its sulfhydryl groups. To complete the catalytic cycle, NADPH is used to regenerate free sulfhydryl groups on the protein. **Thioredoxin reductase**, a flavoprotein, is required if thioredoxin is involved; glutathione and **glutathione reductase** are involved if glutaredoxin is the protein.

Mammalian ribonucleotide reductase consists of two nonidentical protein subunits (heterodimer), neither of which alone has enzymatic activity. The larger subunit has at least two different effector-binding sites and the smaller subunit contains a nonheme iron and a stable tyrosyl free radical. The two subunits make up the active site of the enzyme. The two subunits are encoded by different genes on separate chromosomes. The mRNAs for these subunits, and consequently the proteins, are differentially expressed as cells transit the cell cycle.

The activity of ribonucleotide reductase is under allosteric control. While reduction of each substrate requires the presence of a specific positive effector, the products serve as potent negative effectors of the enzyme. The effects of nucleoside 5′-triphosphates as regulators of ribonucleotide reductase activity are summarized in Table 19.1. DeoxyATP is a potent inhibitor of the reduction of all four substrates, CDP, UDP, GDP, and ADP; dGTP inhibits reduction of CDP, UDP, and GDP; dTTP inhibits reduction of CDP, UDP, and ADP. Thus dGTP and dTTP serve as either positive or negative effectors of ribonucleotide reductase. This means that while dGTP is the required positive activator for ADP reduction, it also serves as an effective inhibitor of CDP and UDP reduction; dTTP is the positive effector of GDP reduction

TABLE 19.1 Nucleoside 5′-Triphosphates as Regulators of Ribonucleotide Reductase Activity

Substrate	*Major Positive Effector*	*Major Negative Effector*
CDP	ATP	dATP, dGTP, dTTP[a]
UDP	ATP	dATP, dGTP, dTTP[a]
ADP	dGTP	dATP
GDP	dTTP	dATP

[a]In decreasing order of effectiveness.

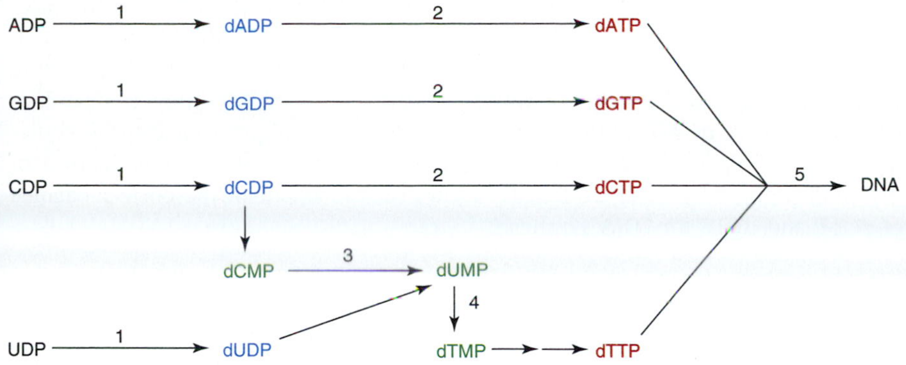

FIGURE 19.26
Role of ribonucleotide reductase in DNA synthesis.
The enzymes catalyzing the reactions are (1) ribonucleotide reductase, (2) nucleoside 5′-diphosphate kinase, (3) deoxycytidylate deaminase, (4) thymidylate synthase, and (5) DNA polymerase.

and serves as an inhibitor of CDP and UDP. Effective inhibition of ribonucleotide reductase by dATP, dGTP, or dTTP explains why deoxyadenosine, deoxyguanosine, and deoxythymidine are toxic to a variety of mammalian cells.

Ribonucleotide reductase is uniquely responsible for catalyzing the rate-limiting reactions in which 2′-deoxyribonucleoside 5′-triphosphates are synthesized *de novo* for DNA replication as summarized in Figure 19.26. Inhibitors of ribonucleotide reductase are potent inhibitors of DNA synthesis and, hence, of cell replication.

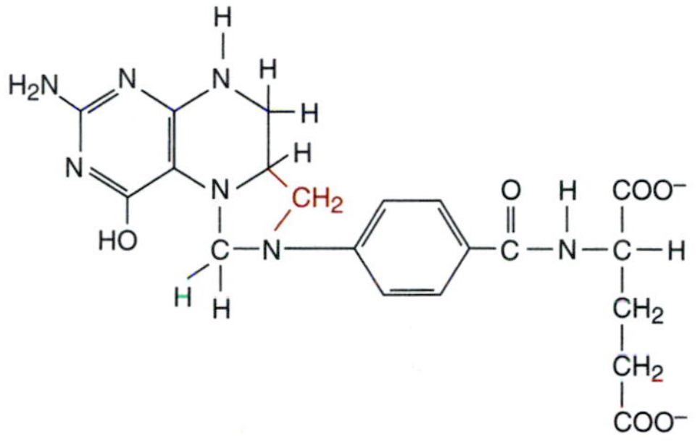

FIGURE 19.27
Structure of N^5, N^{10}-methylene H_4folate.

Deoxythymidylate Synthesis Requires N5, N10-Methylene H4Folate

Deoxythymidylate (dTMP) is formed from 2′-deoxyuridine 5′-monophosphate (dUMP) in a unique reaction. **Thymidylate synthase** catalyzes the transfer of a one-carbon unit from N^5, N^{10}-methylene H_4folate (Figure 19.27), which is transferred to dUMP and simultaneously reduced to a methyl group. The reaction is presented in Figure 19.28. In this reaction, N^5, N^{10}-methylene H_4folate serves a one-carbon donor and as a reducing agent. This is the only reaction in which H_4folate, acting as a one-carbon carrier, is oxidized to H_2folate. There are no known regulatory mechanisms for this reaction.

The substrate for this reaction can come from two different pathways as shown below:

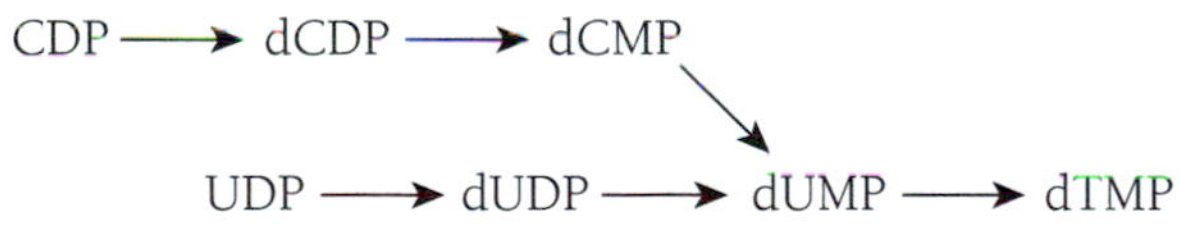

CDP ⟶ dCDP ⟶ dCMP ↘
UDP ⟶ dUDP ⟶ dUMP ⟶ dTMP

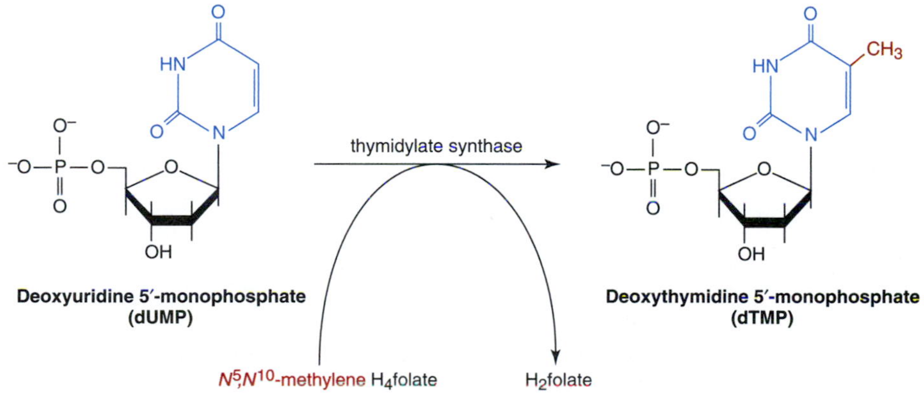

FIGURE 19.28
Synthesis of deoxythymidine nucleotide.

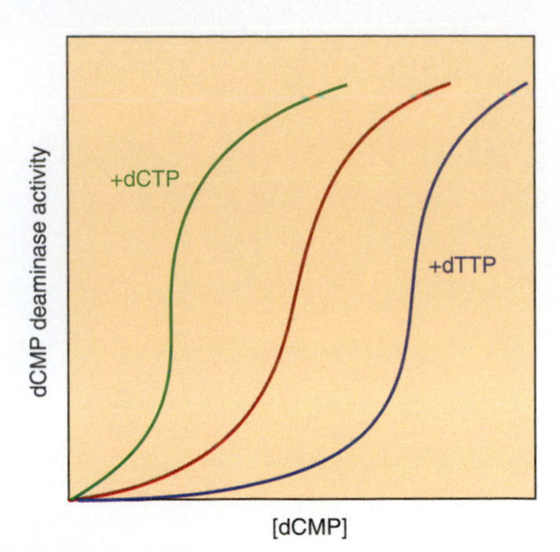

FIGURE 19.29
Regulation of dCMP deaminase.

In both pathways deoxyribonucleotides, dCDP or dUDP, are generated in the reactions catalyzed by ribonucleotide reductase. In one pathway, dUMP is generated from dUDP while in the other pathway, dCMP is deaminated to dUMP. Labeling studies indicate that the major pathway for formation of dUMP involves deamination of dCMP by **dCMP deaminase,** an enzyme that is subject to allosteric regulation by dCTP (positive) and dTTP (negative) (Figure 19.29). This regulation of dCMP deaminase by dCTP and dTTP allows cells to maintain the correct balance of dCTP and dTTP for DNA synthesis.

Pyrimidine Interconversions with Emphasis on Deoxyribopyrimidine Nucleosides and Nucleotides

As shown in Section 19.4, metabolic pathways exist for interconversion of purine nucleotides and are regulated to maintain an appropriate balance of adenine and guanine nucleotides. Pathways also exist for interconversion of pyrimidine nucleotides and are of particular importance for pyrimidine deoxyribonucleosides and deoxyribonucleotides as summarized in Figure 19.30. Note that dCTP and dTTP are major positive and negative effectors of the interconversions and salvage of deoxyribonucleosides.

Pyrimidine Nucleotides Are Degraded to β-Amino Acids

Turnover of nucleic acids results in release of pyrimidine and purine nucleotides. Degradation of pyrimidine nucleotides follows the pathways shown in Figure 19.31. Pyrimidine nucleotides are converted to nucleosides by nonspecific phosphatases. Cytidine and deoxycytidine are deaminated to uridine and deoxyuridine, respectively, by **pyrimidine nucleoside deaminase. Uridine phosphorylase** catalyzes phosphorolysis of uridine, deoxyuridine, and deoxythymidine to uracil and thymine.

Uracil and thymine are degraded further by analogous reactions, although the final products are different, as shown in Figure 19.32. Uracil is degraded to β-alanine, NH_4^+, and CO_2. None of these products is unique to uracil degradation, and consequently the turnover of cytosine or uracil nucleotides cannot be estimated from the end products of this pathway. Thymine degradation proceeds to

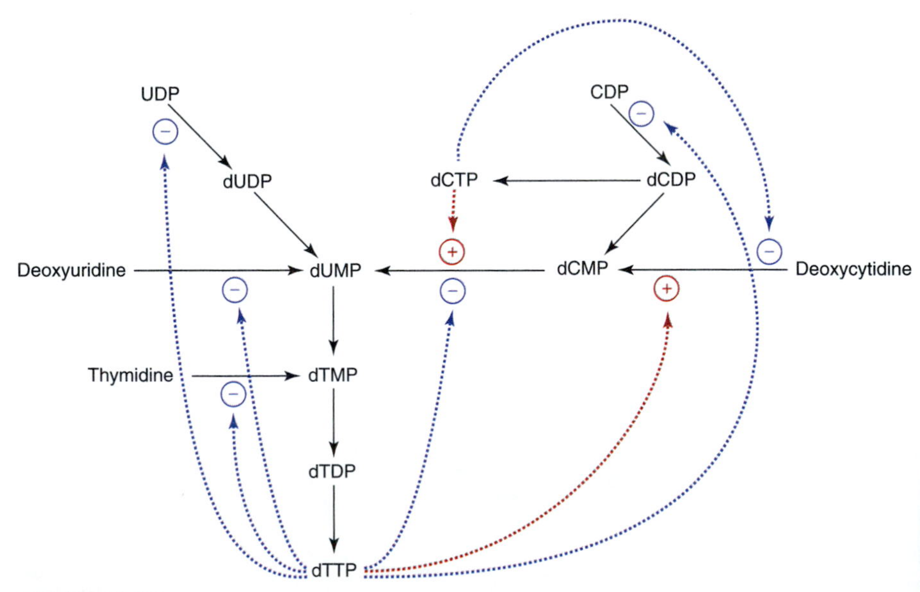

FIGURE 19.30
Interconversions of pyrimidine nucleotides with emphasis on deoxyribonucleotide metabolism.
The solid arrows indicate enzyme-catalyzed reactions; the dashed lines represent positions of negative control points.

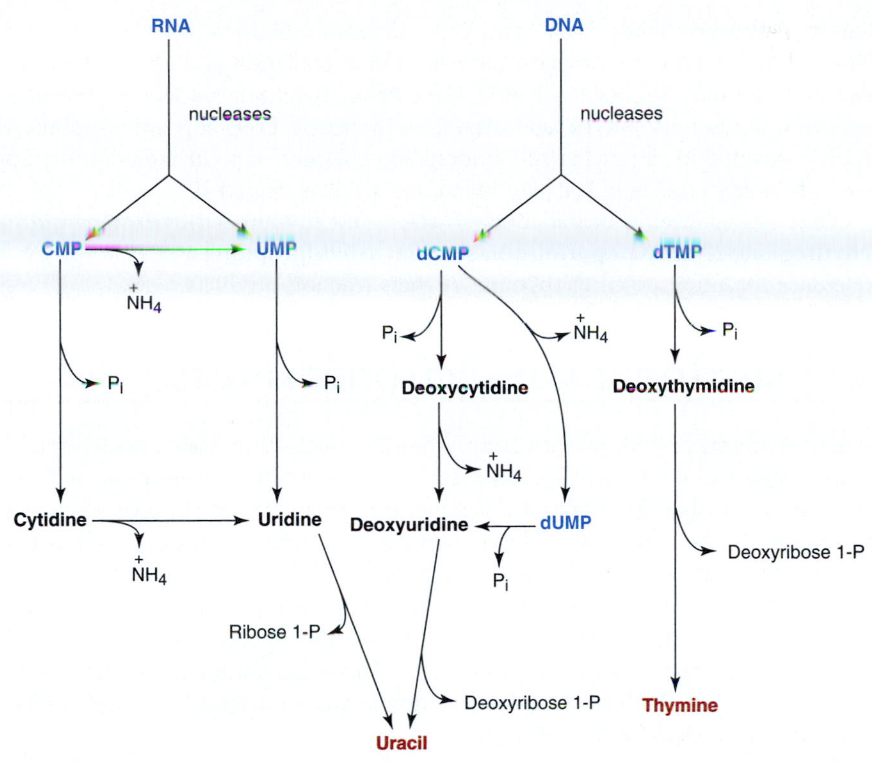

FIGURE 19.31
Pathways for degradation of pyrimidine nucleotides.

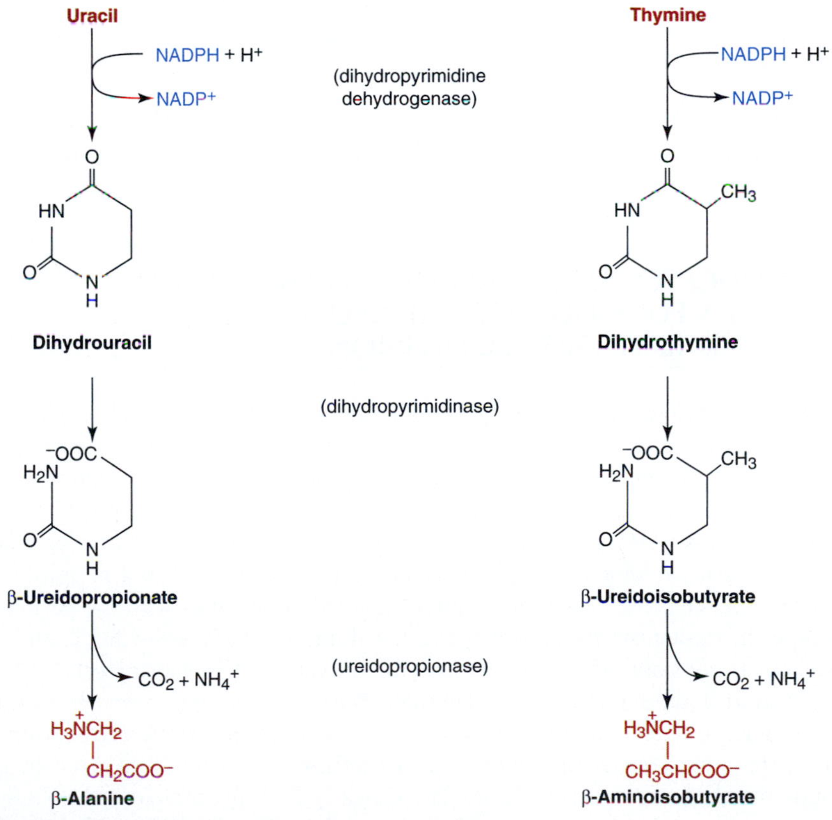

FIGURE 19.32
Degradation of uracil and thymine to end products.

β-aminoisobutyric acid, NH_4^+, and CO_2. β-Aminoisobutyric acid is excreted in urine of humans and originates exclusively from degradation of thymine. Thus it is possible to estimate the turnover of DNA or deoxythymidine nucleotides by measurement of β-aminoisobutyric acid excretion. Increased levels of β-aminoisobutyric acid are excreted in cancer patients undergoing chemotherapy or radiation therapy in which large numbers of cells are killed and DNA is degraded.

Enzymes catalyzing degradation of uracil and thymine **(dihydropyrimidine dehydrogenase, dihydropyrimidinase,** and **uriedopropionase)** do not show a preference for either uracil or thymine or their reaction products.

19.7 | NUCLEOSIDE AND NUCLEOTIDE KINASES

De novo synthesis of both purines and pyrimidine nucleotides yields nucleoside 5′-monophosphates (see Figures 19.7 and 19.19). Likewise, the salvage of nucleobases via the phosphoribosyltransferases or nucleosides via nucleoside kinases also yields nucleoside 5′-monophosphate. This is particularly important in cells such as erythrocytes that cannot form nucleotides *de novo*.

In addition to nucleoside kinases, there are nucleotide kinases that convert a nucleoside 5′-monophosphate to nucleoside 5′-diphosphate and nucleoside 5′-diphosphates to nucleoside 5′-triphosphates. These are important reactions since most reactions in which nucleotides function require nucleoside 5′-triphosphate (primarily) or nucleoside 5′-diphosphate.

Nucleoside kinases show a high degree of specificity with respect to the base and sugar moieties. There is also substrate specificity in nucleotide kinases. Mammalian cells also contain, in high concentration, nucleoside diphosphate kinase that is relatively nonspecific for either phosphate donor or phosphate acceptor in terms of purine or pyrimidine base or the sugar. This reaction is as follows:

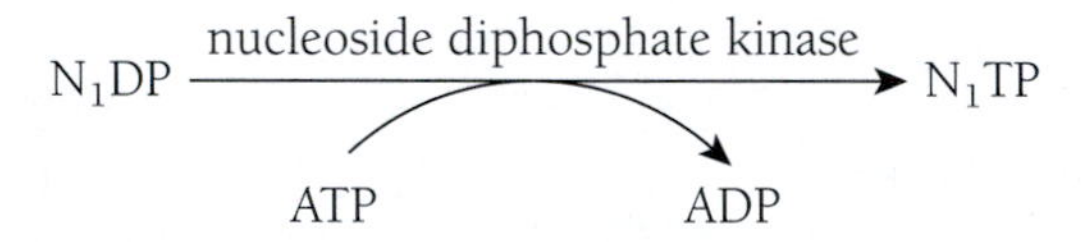

Since ATP is present in the highest concentration and most readily regenerated on a net basis via glycolysis or oxidative phosphorylation, it is probably the major donor for these reactions.

19.8 | NUCLEOTIDE-METABOLIZING ENZYMES AS A FUNCTION OF THE CELL CYCLE AND RATE OF CELL DIVISION

For cell division to occur, essentially all components of cells must double. The term **cell cycle** describes the events that lead from formation of a daughter cell as a result of mitosis to completion of the processes needed for its own division into two daughter cells. The cell cycle is represented in Figure 19.33 (see also p. 178). The phases are mitosis (M), gap 1 (G1), synthesis (S), and gap 2 (G2). Some cells will enter G0, a state in which cells are viable and functional but are in a nonproliferative or quiescent phase. The total length of the cell cycle varies with the particular cell type. In most mammalian cell types, the duration of phases of M, S, and G2 are relatively constant, while time periods for G1 vary widely, causing cells to have long or short doubling times. There are many "factors" that will cause cells to leave the G0 state and reenter the cell cycle. In preparation for **DNA replication** (S phase), there are considerable increases in synthesis of enzymes involved in nucleotide metabolism, especially during late G1/early S. While protein and RNA synthesis occur throughout G1, S, and G2 phases of the cell cycle, DNA replication occurs only during S phase.

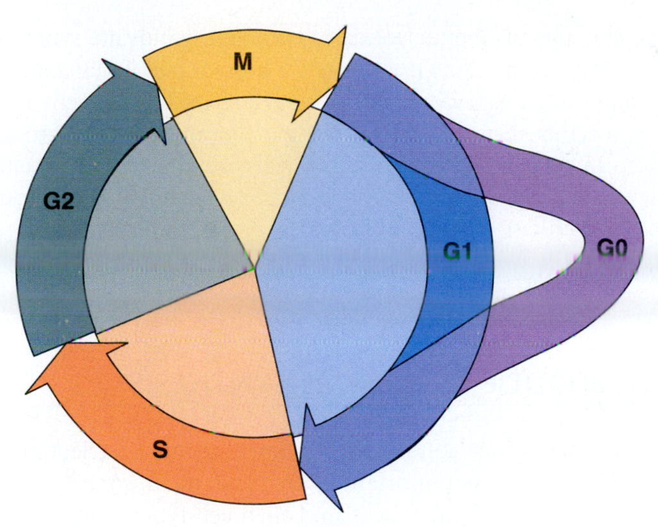

FIGURE 19.33
Diagrammatic representation of the cell cycle.
For a mammalian cell with a doubling time of 24 h, G1 would last ~12 h; S, 7 h; G2, 4 h; and M, 1 h. Cells would enter the G0 state if they became quiescent or nonproliferative.

Enzymes of Purine and Pyrimidine Nucleotide Synthesis Are Elevated During S Phase

Strict regulation of nucleotide synthesis requires that mechanisms must be available to the cell to meet the requirements for ribonucleotide and deoxyribonucleotide precursors at the time of increased RNA synthesis and DNA replication. To meet these needs, cells increase levels of specific enzymes involved with nucleotide formation during very specific periods of the cell cycle.

Enzymes involved in purine nucleotide synthesis and interconversions that are elevated during the S phase of the cell cycle are PRPP amidotransferase and IMP dehydrogenase. Adenylosuccinate synthetase and adenylosuccinase do not increase. Enzymes involved in pyrimidine nucleotide synthesis that are elevated during the S phase include aspartate carbamoyltransferase, dihydroorotase, dihydroorotate dehydrogenase, orotate phosphoribosyltransferase, and CTP synthetase. Many enzymes involved in synthesis and interconversions of deoxyribonucleotides are also elevated during the S phase. Included in these enzymes are ribonucleotide reductase, thymidine kinase, dCMP deaminase, thymidylate synthase, and TMP kinase.

As discussed previously, the deoxyribonucleotide pool is extremely small in "resting" cells (less than 1 μM). As a result of the increase in ribonucleotide reductase, concentrations of deoxyribonucleotides reach 10–20 μM during DNA synthesis. However, this concentration would sustain DNA synthesis for only minutes, while complete DNA replication requires hours. Consequently, levels of ribonucleotide reductase activity not only must increase but also must be sustained during S phase in order to provide the necessary substrates for DNA synthesis.

Growing tissues such as regenerating liver, embryonic tissues, intestinal mucosal cells, and erythropoietic cells are geared toward DNA replication and RNA synthesis. These tissues will show elevated levels of the key enzymes involved with purine and pyrimidine nucleotide synthesis and interconversion, and complementary decreases in amount of enzymes that catalyze reactions in which these precursors are degraded. These changes reflect the proportion of the cells in that tissue that are in S phase.

An ordered pattern of biochemical changes occurs in tumor cells. Utilizing a series of liver, colon, and kidney tumors of varying growth rates, these biochemical changes have been defined as (1) **transformation-linked** (meaning that all tumors regardless of growth rate show certain increased and certain decreased amounts of enzyme), (2) **progression-linked** (alterations that correlate with growth rate of tumors), and (3) **coincidental alterations** (not connected to the malignant state).

As examples, amounts of ribonucleotide reductase, thymidylate synthase, and IMP dehydrogenase increase as a function of tumor growth rate. PRPP amidotransferase, UDP kinase, and uridine kinase are examples of enzymes whose activity is increased in all tumors, whether they are slow-growing or the most rapidly growing.

Alterations in gene expression in tumor cells are not only quantitative changes in enzyme levels but also qualitative changes (isozyme shifts). While some enzymes are increased in both fast-growing normal tissue (e.g., embryonic and regenerating liver) and tumors, the overall quantitative and qualitative patterns for normal and tumor tissue can be distinguished.

19.9 | NUCLEOTIDE COENZYME SYNTHESIS

Nicotinamide adenine dinucleotide (NAD), flavin adenine nucleotide (FAD), and coenzyme A (CoA) serve as important coenzymes in intermediary metabolism. These are synthesized by a variety of mammalian cell types. Figures 19.34, 19.35,

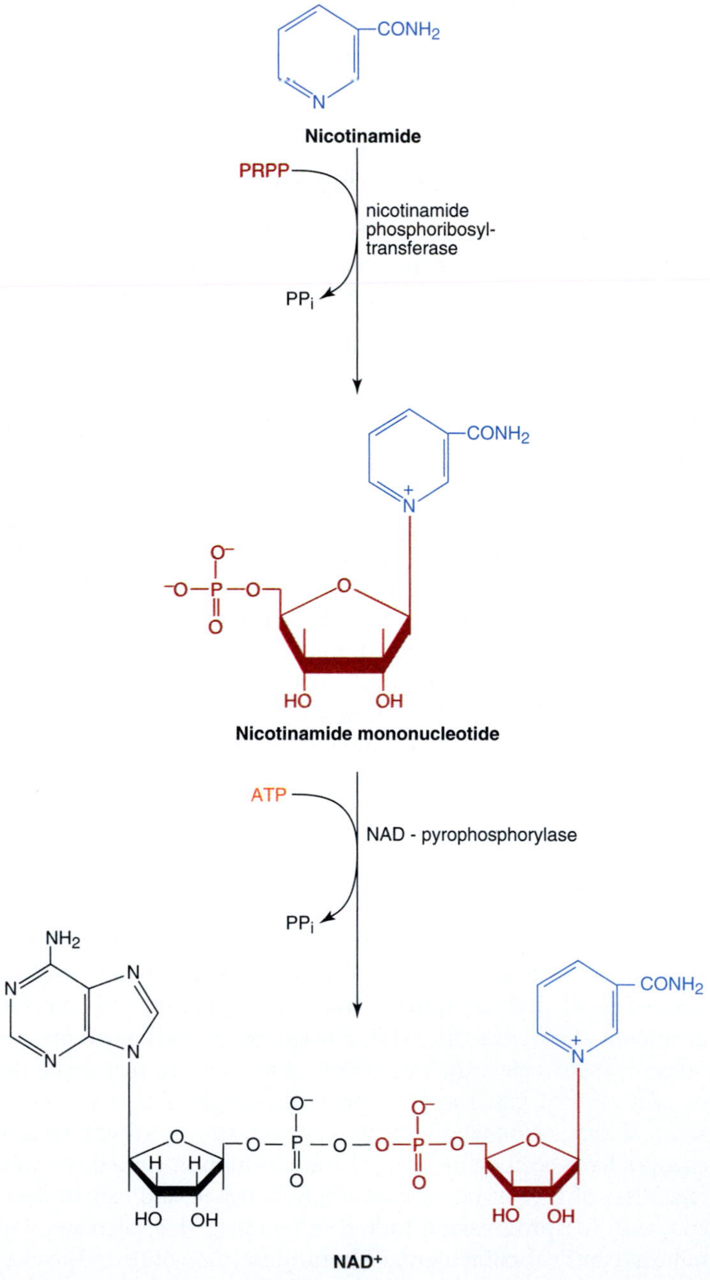

FIGURE 19.34
Pathway for NAD^+ synthesis.

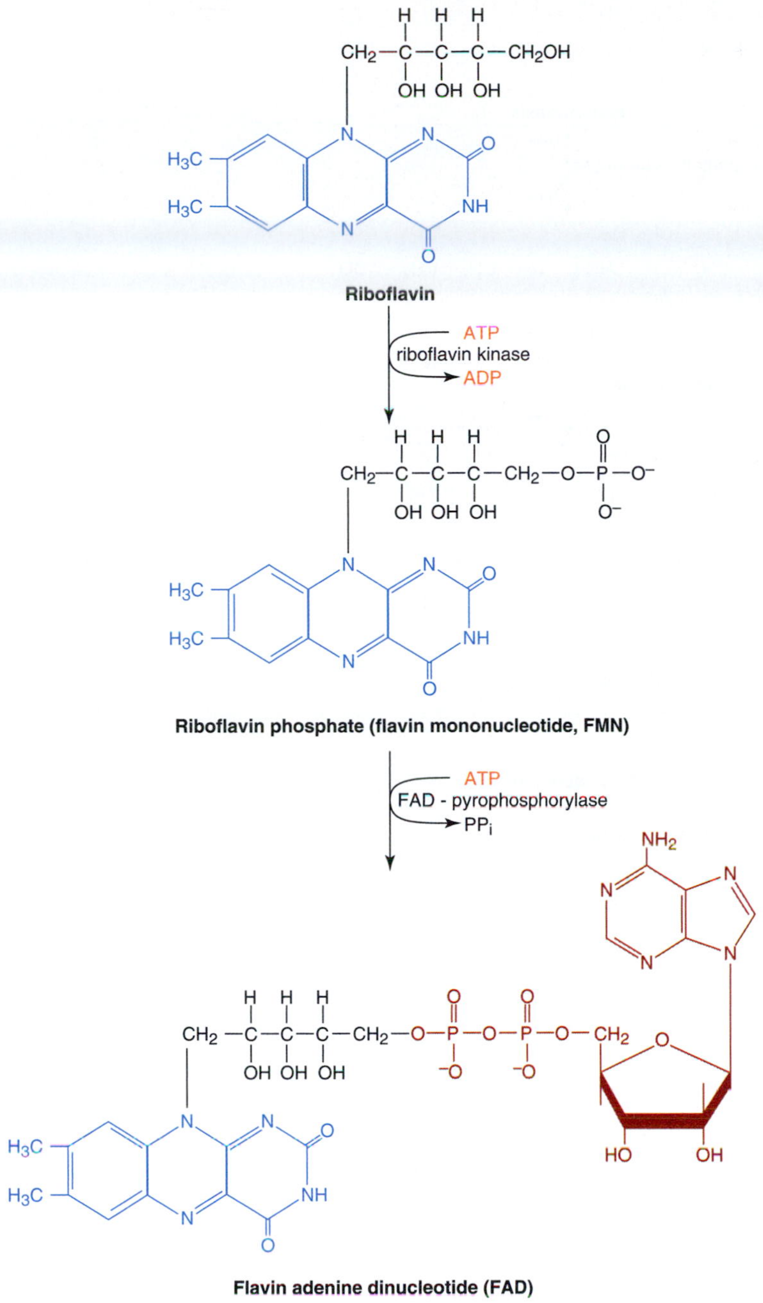

Riboflavin

ATP
riboflavin kinase
ADP

Riboflavin phosphate (flavin mononucleotide, FMN)

ATP
FAD - pyrophosphorylase
PP_i

Flavin adenine dinucleotide (FAD)

FIGURE 19.35
Synthesis of FAD.

and 19.36 present the biosynthetic pathways for each. NAD synthesis requires niacin, FAD synthesis requires riboflavin, and CoA requires pantothenic acid. NAD is synthesized by three different pathways starting from tryptophan (see p. 808), nicotinate, or nicotinamide. When tryptophan is in excess of the amount needed for protein synthesis and serotonin synthesis (see p. 810), it is used for NAD synthesis. This situation is not likely in most normal diets and, consequently, niacin is required in the diet.

Each of these coenzymes contains an AMP moiety, which is not directly involved in the functional part of the molecule since electron transfer in NAD or FAD occurs via the niacin or riboflavin rings, respectively, and activation of acyl groups occurs through the —SH group of CoA. Synthesis of NAD^+ by any of the three pathways requires utilization of PRPP as the ribose 5-phosphate donor.

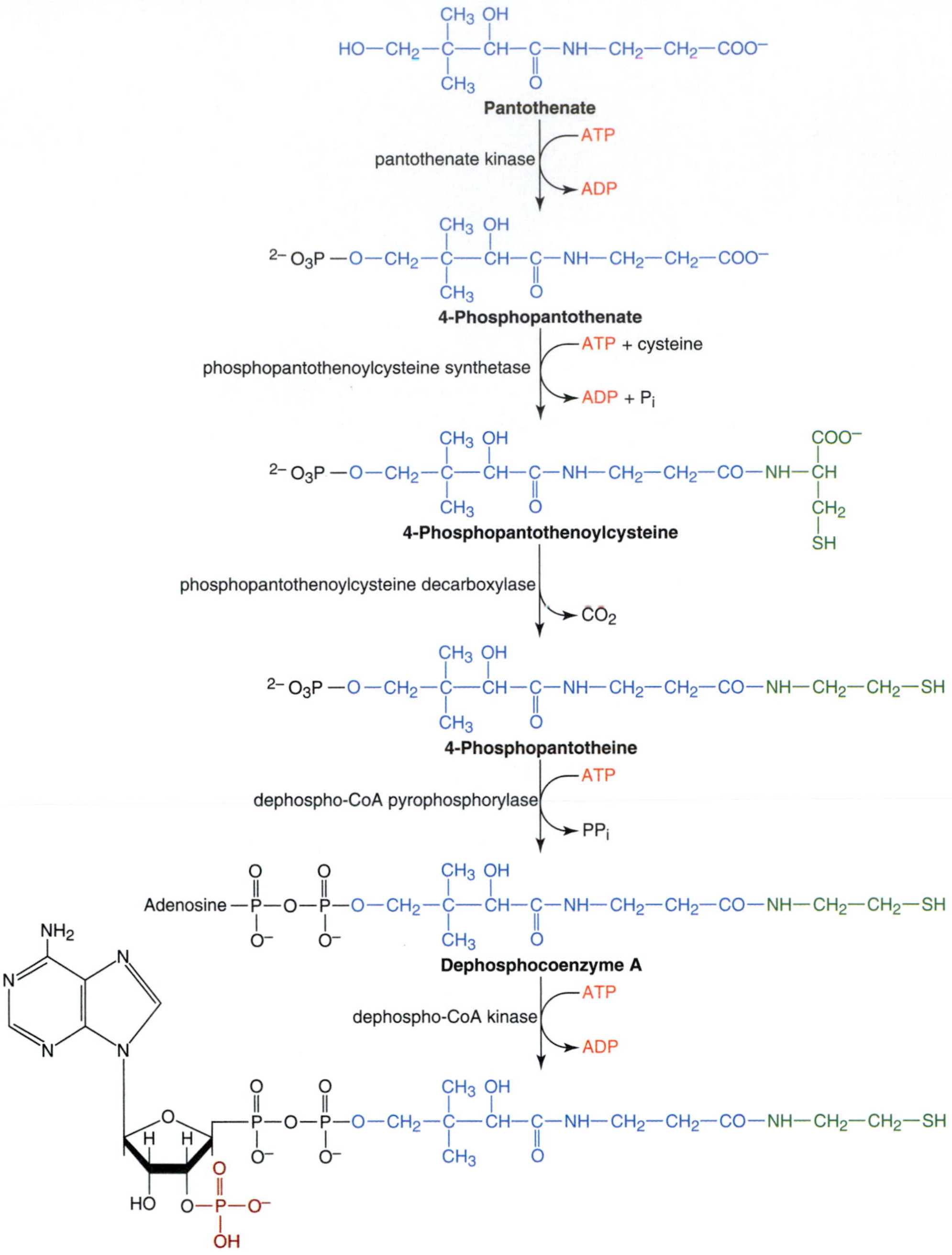

FIGURE 19.36
Synthesis of CoA.

Nicotinamide adenine dinucleotide phosphate ($NADP^+$) is derived by phosphory
NAD^+. NAD^+ is used not only as a cofactor in oxidation–reduction reactions but also
strate in ADP-ribosylation reactions (e.g., DNA repair and pertussis toxin poisoning; se
These reactions lead to the turnover of NAD^+. The end product of NAD^+ degradation
done-5-carboxamide, which is excreted in urine.

Synthesis of nucleotide coenzymes is regulated such that there are essentially "fi
centrations of these coenzymes in the cell. When the statement is made that a certain
condition is favored when the concentration of NAD^+ is low, the concentration of NAI
respondingly high. As an example, for glycolysis to continue under anaerobic conditio
must be regenerated constantly by reduction of pyruvate to lactate by lactate dehydrog
p. 609).

19.10 | SYNTHESIS AND UTILIZATION OF 5-PHOSPHORIBOSYL-1-PYROPHOSPHATE

5-Phosphoribosyl-1-pyrophosphate (PRPP) is a key molecule in *de novo* synthesis of purine and pyrimidine nucleotides, salvage of purine and pyrimidine bases, and synthesis of NAD^+. PRPP synthetase catalyzes the reaction presented in Figure 19.37. Ribose 5-phosphate used in this reaction is generated from glucose 6-phosphate metabolism by the **pentose phosphate pathway** or from ribose 1-phosphate generated by phosphorolysis of nucleosides via nucleoside phosphorylase.

PRPP synthetase has an absolute requirement for inorganic phosphate and is strongly regulated. The velocity versus P_i concentration curve for PRPP synthetase activity is sigmoidal rather than hyperbolic, meaning that at the normal cellular concentration of P_i, the enzyme activity is depressed. The activity is further regulated by ADP, 2,3-bisphosphoglycerate, and other nucleotides. ADP is a competitive inhibitor of PRPP synthetase with respect to ATP; 2,3-bisphosphoglycerate is a competitive inhibitor with respect to ribose 5-phosphate; and nucleotides serve as noncompetitive inhibitors with respect to both substrates. 2,3-Bisphosphoglycerate may be important in regulating PRPP synthetase activity in red cells.

Concentrations of PRPP are low in "resting" or confluent cells but increase rapidly at the time of rapid cell division. Increased flux of glucose 6-phosphate through the pentose phosphate pathway results in increased cellular levels of PRPP and increased production of purine and pyrimidine nucleotides. PRPP is important not only because it serves as a substrate in the glutamine PRPP amidotransferase and the phosphoribosyltransferase reactions, but also because it serves as a positive effector of the major regulated steps in purine and pyrimidine nucleotide synthesis, namely, PRPP amidotransferase and carbamoyl phosphate synthetase II.

Reactions and pathways in which PRPP is required are as follows:

1. *De novo* purine nucleotide synthesis
   a. PRPP + glutamine → 5-phosphoribosylamine + glutamate + PP_i
2. Salvage of purine bases
   a. PRPP + hypoxanthine (guanine) → IMP (GMP) + PP_i
   b. PRPP + adenine → AMP + PP_i
3. *De novo* pyrimidine nucleotide synthesis
   a. PRPP + orotate → OMP + PP_i
4. Salvage of pyrimidine bases
   a. PRPP + uracil → UMP + PP_i
5. NAD^+ synthesis
   a. PRPP + nicotinate → nicotinate mononucleotide + PP_i
   b. PRPP + nicotinamide → nicotinamide mononucleotide + PP_i
   c. PRPP + quinolinate → nicotinate mononucleotide + PP_i

Ribose 5-phosphate — ATP → AMP, Mg^{2+} → 5-Phosphoribosyl-1-pyrophosphate (PRPP)

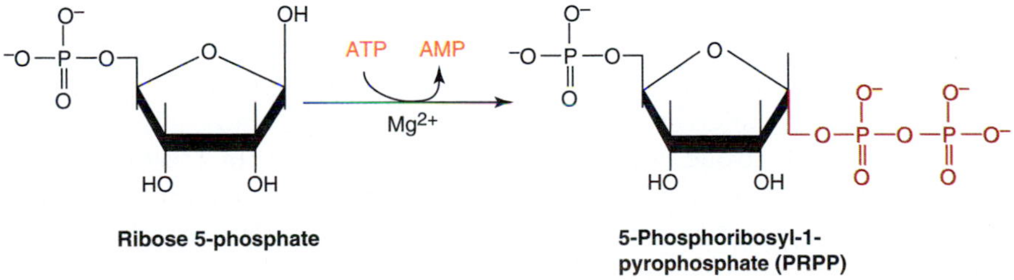

FIGURE 19.37
Synthesis of PRPP.

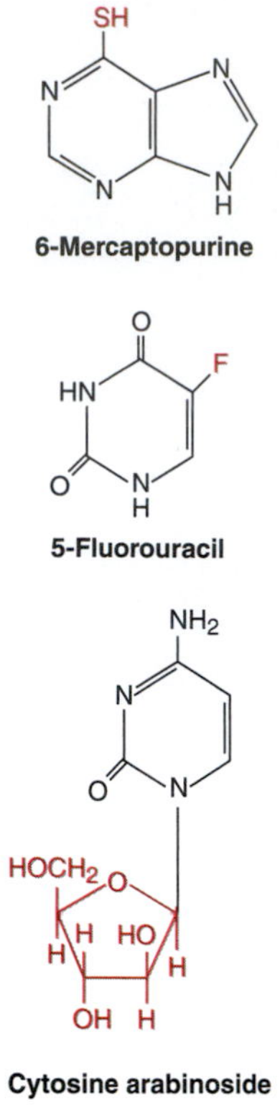

19.11 | COMPOUNDS THAT INTERFERE WITH CELLULAR PURINE AND PYRIMIDINE NUCLEOTIDE METABOLISM: CHEMOTHERAPEUTIC AGENTS

De novo synthesis of purine and pyrimidine nucleotides is critical for normal cell replication, maintenance, and function. Regulation of these pathways is important since disease states arise from defects in the regulatory enzymes. Many compounds, both synthetic and natural products of plants, bacteria, or fungi, are structural analogs of the bases or nucleosides used in metabolic reactions. These compounds are relatively specific inhibitors of enzymes involved in nucleotide synthesis or interconversion and have proved to be useful in the therapy of diverse clinical problems. They are generally classified as antimetabolites, antifolates, glutamine antagonists, and other agents.

6-Mercaptopurine

5-Fluorouracil

Cytosine arabinoside

FIGURE 19.38
Structures of 6-mercaptopurine, 5-fluorouracil, and cytosine arabinoside.

Antimetabolites Are Structural Analogs of Bases or Nucleosides

Antimetabolites are usually structural analogs of purine and pyrimidine bases or nucleosides that interfere with very specific metabolic reactions. They include **6-mercaptopurine** and **6-thioguanine** for treatment of acute leukemia; **azathioprine** for immunosuppression in patients with organ transplants; **allopurinol** for treatment of gout and hyperuricemia; and acyclovir for treatment of herpes virus infection. The detailed understanding of purine nucleotide metabolism aided in the development of these drugs. Conversely, study of the mechanism of action of these drugs has led to a better understanding of normal nucleotide metabolism in humans.

Three antimetabolites will be discussed to show: (1) the importance of *de novo* synthetic pathways in normal cell metabolism; (2) that regulation of these pathways occurs *in vivo;* (3) the concept of the requirement for metabolic activation of the drugs; and (4) that inactivation of these compounds greatly influences their usefulness.

6-Mercaptopurine (6-MP) (Figure 19.38) is a useful antitumor drug in humans. Its cytotoxic activity is related to formation of 6-mercaptopurine ribonucleotide by the tumor cell. Utilizing PRPP and HGPRTase, 6-mercaptopurine ribonucleoside 5′-monophosphate accumulates in cells and is a negative effector of PRPP amidotransferase, the committed step in the *de novo* pathway. This nucleotide also acts as an inhibitor of the conversion of IMP to GMP at the IMP dehydrogenase step and IMP to AMP at the adenylosuccinate synthetase step. Since 6-mercaptopurine is a substrate for xanthine oxidase and is oxidized to 6-thiouric acid, allopurinol is generally administered to inhibit degradation of 6-MP and to potentiate the antitumor properties of 6-MP.

5-Fluorouracil (Fura) (Figure 19.38) is an analog of uracil. 5-Fluorouracil is not the active species. It must be converted by cellular enzymes to the active metabolites 5-fluorouridine 5′-triphosphate (FUTP) and 5-fluoro-2′-deoxyuridine 5′-monophosphate (FdUMP). FUTP is efficiently incorporated into RNA and once incorporated inhibits maturation of 45S precursor rRNA into the 28S and 18S species and alters splicing of pre-mRNA into functional mRNA. FdUMP is a potent and specific inhibitor of thymidylate synthase. In the presence of H_4folate, FdUMP, and thymidylate synthase, a ternary complex is formed that results in covalent binding of FdUMP to thymidylate synthase. This inhibits dTMP synthesis and leads a "thymineless death" for cells.

Cytosine arabinoside (araC) (Figure 19.38) is used in the treatment of several forms of human cancer. It must be metabolized by cellular enzymes to cytosine arabinoside 5′-triphosphate (araCTP) to exert its cytotoxic effects. AraCTP competes with dCTP in the DNA polymerase reaction and araCMP is incorporated into DNA. This inhibits synthesis of the growing DNA strand. Clinically, the efficacy of araC as an antileukemic drug correlates with the concentration of araCTP that is

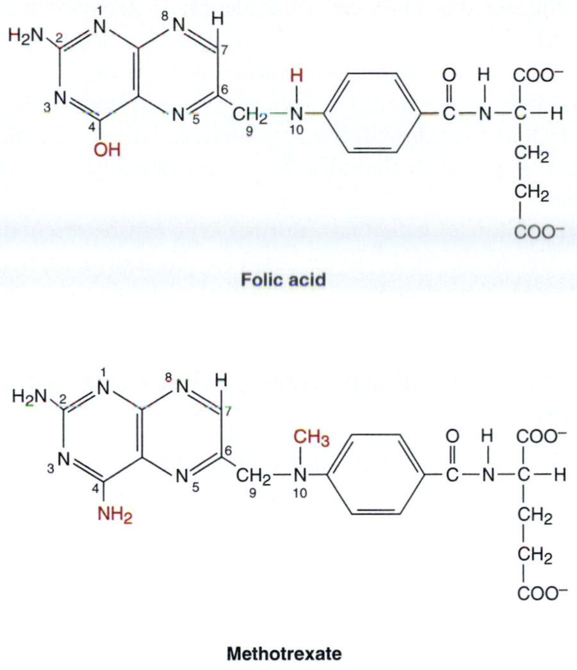

Folic acid

Methotrexate

FIGURE 19.39
Comparison of the structures of folic acid and methotrexate.

achieved in the tumor cell, which in turn determines the amount of araCMP incorporated into DNA. Formation of araCMP via deoxycytidine kinase appears to be the rate-limiting step in activation to araCTP.

Antifolates Inhibit Formation of Tetrahydrofolate

Antifolates interfere with formation of H_4folate from H_2folate or H_2folate from folate by inhibition of H_2folate reductase. **Methotrexate (MTX), a** close structural analog of folic acid, is used as an antitumor agent in the treatment of human cancers. The two structures are presented Figure 19.39. MTX and folate differ at C-4, where an amino group replaces a hydroxyl group, and at N-10, where a methyl group replaces a hydrogen atom. MTX specifically inhibits H_2folate reductase with a K_i in the range of 0.1 nM. The reactions inhibited are shown in Figure 19.40.

MTX at very low concentrations is cytotoxic to mammalian cells in culture. The inhibition of dihydrofolate reductase by MTX results in the lowering of both ribonucleoside 5′-triphosphate and 2′-deoxyribonucleoside 5′-triphosphate intracellular pools. The effects can be prevented by addition of deoxythymidine and hypoxanthine to the culture medium. Reversal of the MTX effects by thymidine and

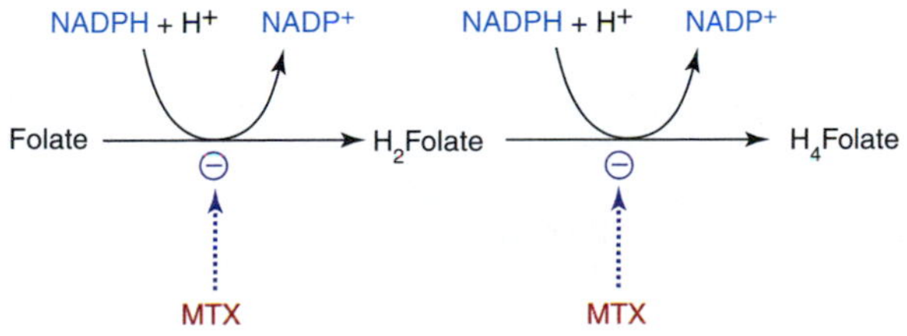

FIGURE 19.40
Sites of inhibition by methotrexate.

hypoxanthine indicates that MTX causes depletion of deoxythymidine and purine nucleotides in cells. Figure 19.41 shows the relationship between H_4folate, *de novo* purine nucleotide synthesis, and dTMP formation. It is important to note that in the thymidylate synthase reaction, H_2folate is generated and unless it is readily reduced back to H_4folate via dihydrofolate reductase, cells would not be capable of *de novo* synthesis of purine nucleotides or thymidylate synthesis due to depletion of H_4folate pools.

In treatment of human leukemias, normal cells can be rescued from the toxic effects of "high-dose MTX" by N^5-formyl-H_4folate **(leucovorin).** This increases the clinical efficacy of MTX treatment.

Glutamine Antagonists Inhibit Enzymes that Utilize Glutamine as Nitrogen Donors

Many reactions in mammalian cells utilize glutamine, which serves as the amino group donor. By contrast, bacteria primarily utilize ammonia as the amino donor in a similar reaction. These amidation reactions are critical in *de novo* synthesis of purine nucleotide (N-3 and N-9), synthesis of GMP from IMP, formation of cytosolic carbamoyl phosphate, synthesis of CTP from UTP, and synthesis of NAD^+.

Compounds that inhibit these reactions are referred to as glutamine antagonists. **Azaserine** (*O*-diazoacetyl-L-serine) and **6-diazo-5-oxo-l-norleucine (DON)** (Figure 19.42), which were first isolated from cultures of *Streptomyces,* are very effective inhibitors of enzymes that utilize glutamine as the amino donor. Since azaserine and DON inactivate the enzymes involved, addition of glutamine alone will not reverse the effects of either of these two drugs. It would necessitate that many metabolites such as guanine, cytosine, hypoxanthine (or adenine), and nicotinamide be provided to bypass the many sites blocked by these glutamine antagonists. As expected from the fact that so many key steps are inhibited by DON and azaserine, these agents are extremely toxic and not of clinical use.

Other Agents Inhibit Cell Growth by Interfering with Nucleotide Metabolism

Tumor cells treated with **hydroxyurea** (Figure 19.43) show a specific inhibition of DNA synthesis with little or no inhibition of RNA or protein synthesis. Hydroxyurea

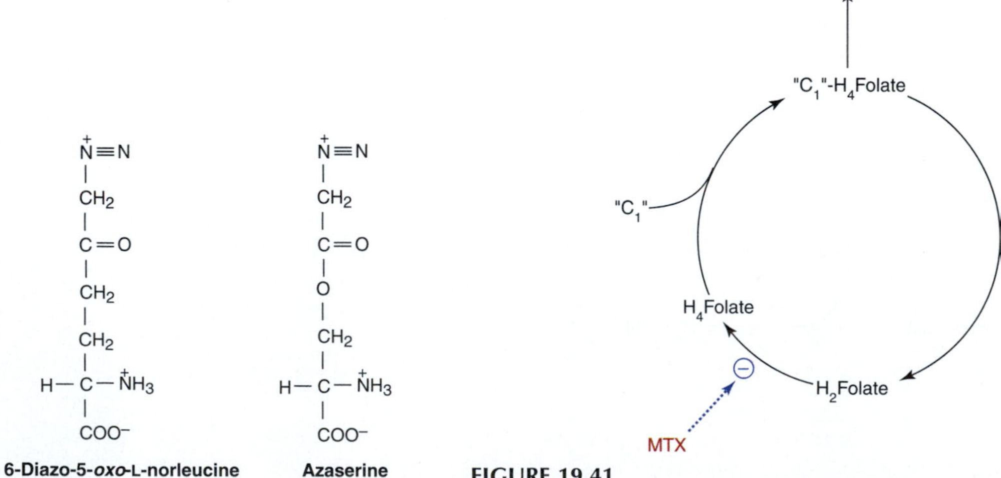

FIGURE 19.42
Structure of glutamine antagonists.

FIGURE 19.41
Relationship between H_4folate, *de novo* purine nucleotide synthesis, and dTMP synthesis.

inhibits ribonucleotide reductase to block reduction of CDP, UDP, GDP, and ADP to the corresponding 2′-deoxyribonucleoside 5′-diphosphates. Toxicity results from depletion of 2′-deoxyribonucleoside 5′-triphosphates required for DNA replication. Clinical use of hydroxyurea as an antitumor agent is limited because of its rapid rate of clearance and the high drug concentration required for effective inhibition. However, hydroxyurea has recently been utilized in the treatment of sickle cell anemia. By a mechanism not understood, hydroxyurea treatment of sickle cell patients results in the "turn-on" of the fetal (γ) hemoglobin gene resulting in the increased expression of fetal hemoglobin. The increased expression of fetal hemoglobin results in the decreased frequency of sickle cell crises in the patients. It does not appear that the effects of hydroxyurea in sickle cell erythrocytes are directly related to inhibition of ribonucleotide reductase.

Tiazofurin (Figure 19.43) is converted by cellular enzymes to the NAD^+ analog, **tiazofurin adenine dinucleotide** (TAD). TAD inhibits IMP dehydrogenase, the rate-limiting enzyme in GTP synthesis, with a K_1 of 0.1 μM. As a result of IMP dehydrogenase inhibition, the concentration of GTP is markedly depressed. It is interesting that, although there are many dehydrogenases in the cells that utilize NAD^+ as a substrate, IMP dehydrogenase is the enzyme that is most affected. This could be because IMP dehydrogenase catalyzes a rate-limiting step in a pathway and is quantitatively limiting in concentration.

These clinically useful drugs serve as examples in which the knowledge of basic biochemical pathways and mechanisms leads to generation of effective drugs.

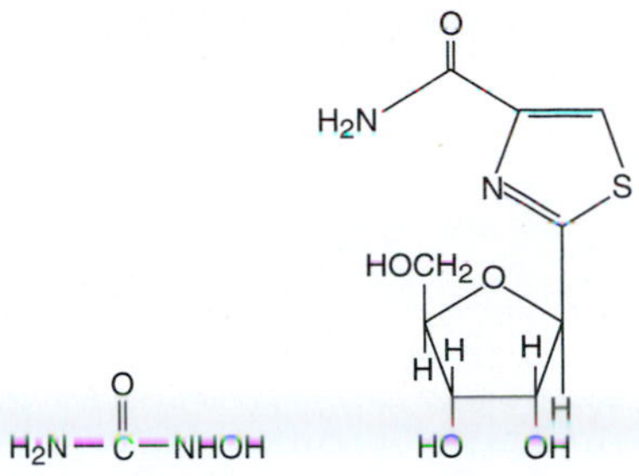

Hydroxyurea Tiazofurin

FIGURE 19.43
Structure of hydroxyurea and tiazofurin.

Purine and Pyrimidine Analogs as Antiviral Agents

Herpes virus (HSV) and **human immunodeficiency virus** (HIV), the causative agent of AIDS infections, present major clinical problems. Two antimetabolites are used in the control-treatment (but not cure) of HSV and HIV infections. These drugs , acyclovir (acycloguanosine), a purine analog, and **3′-azido-3′-deoxythymidine (AZT),** a pyrimidine analog (Figure 19.44), require metabolism to phosphorylated compounds to yield the active drug. Acycloguanosine is activated to the monophosphate by a specific HSV–thymidine kinase, encoded by the HSV genome, which catalyzes phosphorylation of acycloguanosine. The host cellular thymidine kinase cannot utilize acyclovir as a substrate. Acycloguanosine monophosphate is then phosphorylated by the cellular enzymes to the di- and triphosphate forms. Acycloguanosine triphosphate serves as a substrate for the HSV-specific DNA polymerase and is incorporated into the growing viral DNA chain, causing chain termination. The specificity of acycloguanosine and its high therapeutic index reside, therefore, in the fact that only HSV-infected cells can form the acycloguanosine monophosphate.

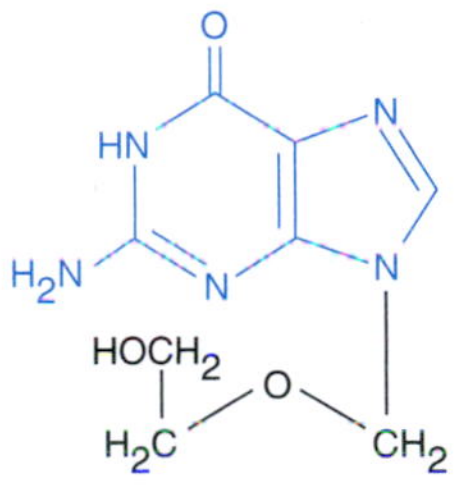

AZT is phosphorylated by cellular kinases to AZT triphosphate, which blocks HIV replication by inhibiting HIV–DNA polymerase (an RNA-dependent polymerase). The selectivity of AZT for HIV-infected versus uninfected cells occurs because DNA polymerase from HIV is at least 100-fold more sensitive to AZT triphosphate than is host cell DNA-dependent DNA polymerase.

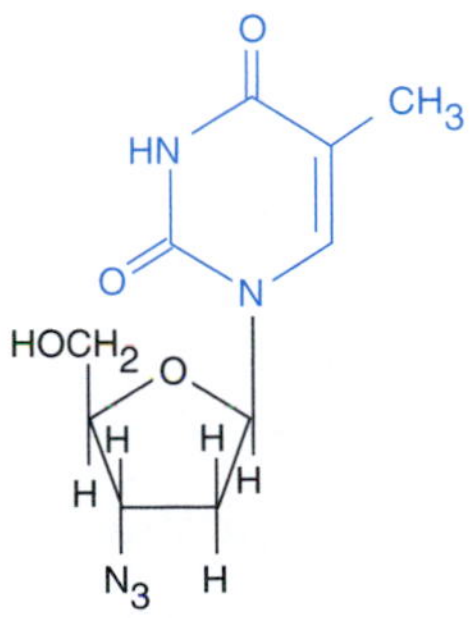

These two antiviral agents demonstrate the diversity of responses required for selectivity. In one case, enzyme activity encoded by the viral genome is mandatory for activation of the drug (acycloguanosine); in the second example, although cellular enzymes activate AZT, the viral gene product (HIV–DNA polymerase) is the selective target.

Acycloguanosine

AZT

FIGURE 19.44
Structure of the antiviral agents acyclovir and AZT.

Biochemical Basis for Development of Drug Resistance

Failure of chemotherapy in treatment of human cancer is often related to development of tumor cell populations that are resistant to the cytotoxic effects of the particular drug. Tumors represent a very heterogeneous population of cells and in many instances drug-resistant cells are present. Upon therapy, drug-sensitive cells are killed off and a resistant cell population becomes enriched. In some cases, drug treatment causes genetic alterations that result in the drug-resistant phenotype. Resistance to drugs can be categorized as "specific drug resistance" or **"multidrug resistance."**

Biochemical and molecular mechanisms that account for drug resistance have been determined for many drugs. For example, resistance to methotrexate develops as a result of several different alterations. These include: a defect or loss of the transporter for N^5-formyl-H_4folate and N^5-methyl-H_4folate, which results in decreased cellular uptake of MTX; amplification of the dihydrofolate reductase gene, which results in a marked increase in cellular dihydrofolate reductase, the target enzyme; alterations in the dihydrofolate reductase gene, which result in a "mutant" dihydrofolate reductase that is less sensitive to inhibition by MTX; and decreased levels of folylpolyglutamate synthetase, which results in lower levels of polyglutamylated MTX, the "trapped" form of MTX. A MTX-resistant population could have any one or a combination of these alterations. The net result of any of these resistance mechanisms is to decrease the ability of MTX to inhibit dihydrofolate reductase at clinically achievable MTX concentrations. Other specific drug resistance mechanisms could be described for compounds such as cytosine arabinoside, 5-fluorouracil, and hydroxyurea.

In multiple drug resistance, the drug-resistant population is cross-resistant to a series of seemingly unrelated antitumor agents. These compounds include drugs such as the vinca alkaloids, adriamycin, actinomycin D, and etoposide. All of these drugs are natural products or derived from natural products and they are not chemically related in structure. They have different mechanisms of action as antitumor agents but appear to act on some nuclear event.

Multidrug-resistant tumor cells express high levels (compared to the drug-sensitive tumor cell phenotype) of a protein called **MDR1** (P-glycoprotein) or another protein called **MRP (multidrug resistance-associated protein).** See p. 525 for a discussion of these transporters. These proteins are membrane bound and have a mass around 170 kDa but are distinctly separate proteins. These proteins function as **ATP-dependent "pumps"** to efflux drugs from cells. As a result, cellular concentration of a drug decreases below its cytotoxic concentration.

Development of drug-resistant tumor cells presents major clinical problems. Study of the mechanisms of drug resistance has greatly aided in our understanding of cancer cells.

BIBLIOGRAPHY

Arner, E. S. J. and Eriksson, S. Mammalian deoxyribonucleoside kinases. *Pharmacol. Ther.* 67:155, 1995.

Cory, J. G. Role of ribonucleotide reductase in cell division. In: J. G. Cory and A. H. Cory (Eds.), *Inhibitors of Ribonucleoside Diphosphate Reductase Activity, International Encyclopedia of Pharmacology and Therapeutics.* New York: Pergamon Press, 1989, p. 1.

Elion, G. B. The purine path to chemotherapy. *Science* 244:41, 1989.

Scriver, C. R., Beaudet, A. L., Sly, W. S., and Valle, D. (Eds.). *The Metabolic and Molecular Bases of Inherited Disease,* 7th ed. Vol. II, Chaps. 49–55. New York: McGraw-Hill, 1995.

Weber, G. Biochemical strategy of cancer cells and the design of chemotherapy: *G.H.A. Clowes Mem. Lecture Cancer Res.* 43:3466, 1983.

Zalkin, H. and Dixon, J. E. *De novo* purine nucleotide biosynthesis. *Prog. Nucleic Acid Res.* 42:259, 1992.

QUESTIONS | C. N. ANGSTADT

Multiple Choice Questions

1. Nucleotides serve all of the following roles EXCEPT:
   A. monomeric units of nucleic acids.
   B. physiological mediators.
   C. sources of chemical energy.
   D. structural components of membranes.
   E. structural components of coenzymes.

2. The amide nitrogen of glutamine is a source of nitrogen for the:
   A. *de novo* synthesis of purine nucleotides.
   B. *de novo* synthesis of pyrimidine nucleotides.

C. synthesis of GMP from IMP.
D. all of the above.
E. none of the above.

3. The two purine nucleotides found in RNA:
A. are formed in a branched pathway from a common intermediate.
B. are formed in a sequential pathway.
C. must come from exogenous sources.
D. are formed by oxidation of the deoxy forms.
E. are synthesized from nonpurine precursors by totally separate pathways.

4. The type of enzyme known as a phosphoribosyltransferase is involved in all of the following EXCEPT:
A. salvage of pyrimidine bases.
B. the *de novo* synthesis of pyrimidine nucleotides.
C. the *de novo* synthesis of purine nucleotides.
D. salvage of purine bases.

5. Uric acid is:
A. formed from xanthine in the presence of O_2.
B. a degradation product of cytidine.
C. deficient in the condition known as gout.
D. a competitive inhibitor of xanthine oxidase.
E. oxidized, in humans, before it is excreted in urine.

6. In nucleic acid degradation, all of the following are correct EXCEPT:
A. there are nucleases that are specific for either DNA or RNA.
B. nucleotidases convert nucleotides to nucleosides.
C. the conversion of a nucleoside to a free base is an example of a hydrolysis.
D. because of the presence of deaminases, hypoxanthine rather than adenine is formed.
E. a deficiency of adenosine deaminase leads to an immunodeficiency.

7. Deoxyribonucleotides:
A. cannot be synthesized so they must be supplied preformed in the diet.
B. are synthesized *de novo* using dPRPP.
C. are synthesized from ribonucleotides by an enzyme system involving thioredoxin.
D. are synthesized from ribonucleotides by nucleotide kinases.
E. can be formed only by salvaging free bases.

8. The conversion of nucleoside 5′-monophosphates to nucleoside 5′-triphosphates:
A. is catalyzed by nucleoside kinases.
B. is a direct equilibrium reaction.
C. utilizes a relatively specific nucleotide kinase and a relatively nonspecific nucleoside diphosphate kinase.
D. generally uses GTP as a phosphate donor.
E. occurs only during the S phase of the cell cycle.

9. If a cell were unable to synthesize PRPP, which of the following processes would likely be directly impaired?
A. FAD synthesis
B. NAD synthesis
C. coenzyme A synthesis
D. ribose 5-phosphate synthesis
E. dTMP synthesis

10. Which of the following chemotherapeutic agents works by impairing *de novo* purine synthesis?
A. acyclovir (acycloguanosine)
B. 5-fluorouracil (antimetabolite)
C. methotrexate (antifolate)
D. hydroxyurea
E. AZT (3′-azido-3′-deoxythymidine)

Questions 11 and 12: Hereditary orotic aciduria is characterized by severe anemia, growth retardation, and high levels of orotic acid excretion. The defect may be in either orotate phosphoribosyltransferase, orotidine decarboxylase, or both. The preferred treatment for this disease is dietary uridine, which reverses the anemia and decreases the formation of orotic acid.

11. Elements involved in the effectiveness of the dietary treatment include:
A. conversion of exogenous uridine to UMP by uridine phosphotransferase.
B. UTP from exogenous uridine providing substrate for synthesis of CTP.
C. inhibition of carbamoyl phosphate synthetase II by UTP.
D. all of the above.
E. none of the above.

12. In the *de novo* synthesis of pyrimidine nucleotides:
A. reactions take place exclusively in the cytosol.
B. a free base is formed as an intermediate.
C. PRPP is required in the rate-limiting step.
D. UMP and CMP are formed from a common intermediate.
E. UMP inhibition of OMP-decarboxylase is the major control of the process.

Questions 13 and 14: Gout is a disease characterized by hyperuricemia from an overproduction of purine nucleotides via the *de novo* pathway. In Lesch–Nyhan syndrome, the specific cause is a severe deficiency of HGPRTase. Allopurinol is used in the treatment of gout to reduce the production of uric acid. In Lesch–Nyhan syndrome, the decrease in uric acid is balanced by an increase in xanthine plus hypoxanthine in blood. In the other forms of gout, the decrease in uric acid is greater than the increase in xanthine plus hypoxanthine.

13. The explanation for this difference in the two forms of gout is:
A. it is an experimental artifact and the decrease in uric acid and increase in xanthine plus hypoxanthine in non-Lesch–Nyhan gout is the same.
B. allopurinol is less effective in non-Lesch–Nyhan gout.
C. there is an increased excretion of xanthine and hypoxanthine in non-Lesch–Nyhan gout.
D. PRPP levels are reduced in Lesch–Nyhan.
E. in non-Lesch–Nyhan gout hypoxanthine and xanthine are salvaged to IMP and XMP and inhibit PRPP amidotransferase.

14. Which of the following is/are aspects of the overall regulation of *de novo* purine nucleotide synthesis?
A. AMP, GMP, and IMP cause a shift of PRPP amidotransferase from a small form to a large form.
B. PRPP levels in the cell can be severalfold less than the K_m of PRPP amidotransferase for PRPP.
C. GMP is a competitive inhibitor of IMP dehydrogenase.
D. All of the above are correct.
E. None of the above is correct.

Question 15: Rheumatoid arthritis is an autoimmune disease that can lead to severe problems from inflammation. The first line of treatment is anti-inflammatory drugs, which may control the symptoms but do not halt the progress of the disease. A more aggressive treatment is low does of methotrexate, a dihydrofolate reductase inhibitor. A new immunosuppressive drug, Arava, an inhibitor of dihydroorotate dehydrogenase, has the same ultimate effect as methotrexate but with fewer side effects.

15. Which of the following statements is/are correct?
    A. Methotrexate inhibits the *de novo* synthesis of UMP.
    B. Arava inhibits the *de novo* synthesis of pyrimidine nucleotides.
    C. Arava inhibits the conversion of dUMP to dTMP.
    D. All of the above.
    E. None of the above.

Problems

16. If a cell capable of *de novo* synthesis of purine nucleotides has adequate AMP but is deficient in GMP, how would the cell regulate synthesis to increase [GMP]? If both AMP and GMP were present in appropriate concentrations, what would happen?
17. How would you measure the turnover of DNA?

ANSWERS

1. **D** Both cAMP and cGMP are physiological mediators. NAD, FAD, and CoA all contain AMP as part of their structures.
2. **D** Nitrogen atoms 3 and 9 of purine nucleotides and N-3 of pyrimidine nucleotides are supplied by glutamine in *de novo* synthesis. The 2-amino group of GMP also comes from this source.
3. **A** GMP and AMP are both formed from the first purine nucleotide, IMP, in a branched pathway. B: The pyrimidine nucleotides UMP and CTP are formed in a sequential pathway from orotic acid. C: Humans are capable of synthesizing purine nucleotides. D: Deoxy forms are formed by reduction of the ribose forms. E: IMP is the common precursor.
4. **C** In purine nucleotide synthesis, the purine ring is built up stepwise on ribose-5-phosphate and not transferred to it. A, B, D: Phosphoribosyltransferases are important salvage enzymes for both purines and pyrimidines and are also part of the synthesis of pyrimidines since OPRT catalyzes the conversion of orotate to OMP.
5. **A** The xanthine oxidase reaction produces uric acid. B, E: Uric acid is an end product of purines, not pyrimidines. C: Gout is characterized by excess uric acid.
6. **C** The product is ribose-1-phosphate rather than the free sugar, a phosphorolysis. A: They can also show specificity toward the bases and positions of cleavage. B: A straight hydrolysis. D: AMP deaminase and adenosine deaminase remove the 6-NH_2 as NH_3. The IMP or inosine formed is eventually converted to hypoxanthine. E: This is called severe combined immunodeficiency.
7. **C** Deoxyribonucleotides are synthesized from the ribonucleoside diphosphates by nucleoside diphosphate reductase that uses thioredoxin as the direct hydrogen-electron donor. A, B, E: There is a synthetic mechanism as just described but it is not a *de novo* pathway. D: Nucleotide kinases are enzymes that add phosphate to a base or nucleotide.
8. **C** These two enzymes are important in interconverting the nucleotide forms. A: These convert nucleosides to nucleoside monophosphates. B: Two steps are required. D: ATP is present in highest concentration and is the phosphate donor. E: Occurs during the S phase but this is a general reaction for the cell.
9. **D** PRPP is formed from ribose 5-phosphate in an irreversible reaction. A, B, C: All of these contain nucleotides and require PRPP at some point. E: dTMP is formed directly from dUMP, which doesn't have to be made *de novo*.
10. **C** Antifolates reduce the concentration of THF compounds that are necessary for two steps of purine synthesis. A, E: These are antiviral agents that inhibit DNA synthesis. B: 5-Fluorouracil is a pyrimidine analog not a purine analog. D: Hydroxyurea inhibits the reduction of ribonucleotides to deoxyribonucleotides so is not involved in *de novo* purine synthesis. E: Allopurinol potentiates the effect of 6-mercaptopurine but is not an inhibitor of purine synthesis.
11. **D** It is common for an exogenous agent to require conversion to an active form; in this case the uridine is "salvaged" to the monophosphate and ultimately to the triphosphate. The cell is deficient in UTP and CTP because the conversion of orotic acid is blocked so the exogenous uridine provides a bypass around the block. Orotic acid formation is decreased since UTP inhibits carbamoyl phosphate synthetase II, the control enzyme.
12. **B** This is in contrast to purine *de novo* synthesis. A: One enzyme is mitochondrial. C: PRPP is required to convert orotate to OMP but this is not rate-limiting. D: OMP to UMP to CTP is a sequential process. E: This does occur but the rate-limiting step is that catalyzed by CPS II.
13. **E** Not only is uric acid production directly inhibited, *de novo* synthesis is as well, thus reducing production of xanthine and hypoxanthine. In Lesch–Nyhan, the only effect is the direct inhibition of xanthine oxidase. A: It is a real effect. B: Actually, it is more effective because of the dual roles. C: This does not happen. D: PRPP levels are very high because of the lack of the salvage of bases, leading to improper or lack of regulation of pathways.
14. **D** A is the mechanism of inhibition since the large form of the enzyme is inactive. B: PRPP amidotransferase shows sigmoidal kinetics with respect to PRPP so large shifts in concentration of PRPP have the potential for altering velocity. C plays a major role in controlling the branched pathway of IMP to GMP or AMP.
15. **B** The conversion of dihydroorotate to orotic acid is a key step in *de novo* synthesis of the pyrimidine nucleotides. A: Methotrexate reduces the tetrahydrofolate pool but tetrahydrofolate is not involved in the *de novo* synthesis of pyrimidines. C: This is the step that methotrexate inhibits by lowering the tetrahydrofolate pool.
16. AMP would partially inhibit *de novo* synthesis of IMP by its allosteric inhibition of PRPP amidotransferase. The IMP formed would be directed toward GMP because AMP is an inhibitor of its own synthesis from IMP. If both AMP and GMP are adequate, the synergistic effect of the two on PRPP amidotransferase would severely inhibit *de novo* synthesis.
17. Thymine is derived almost exclusively from DNA. The degradation of thymine leads to the unique product β-aminoisobutyrate, which is excreted. This can be measured in urine.

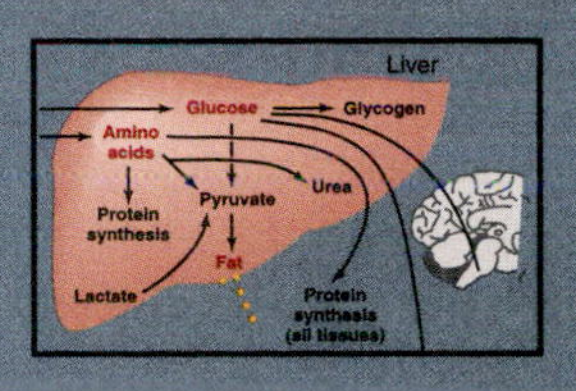

20

METABOLIC INTERRELATIONSHIPS

Robert A. Harris and David W. Crabb

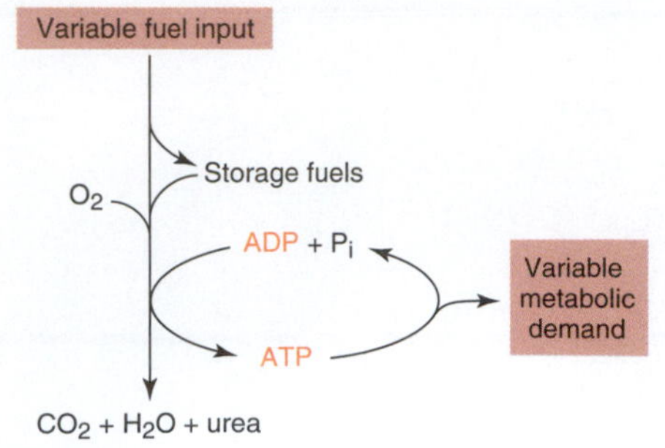

FIGURE 20.1
Humans can use a variable fuel input to meet a variable metabolic demand.

20.1 | OVERVIEW

In this chapter the interdependence of metabolic processes of the major tissues of the body will be stressed. Not all of the major metabolic pathways operate in every tissue at any given time. Given the nutritional and hormonal status of a patient, we need to know qualitatively which major metabolic pathways of the body are functional and how these pathways relate to one another.

The metabolic processes with which we are concerned are glycogenesis, glycogenolysis, gluconeogenesis, glycolysis, fatty acid synthesis, lipogenesis, lipolysis, fatty acid oxidation, glutaminolysis, tricarboxylic acid (TCA) cycle activity, ketogenesis, amino acid oxidation, protein synthesis, proteolysis, and urea synthesis. It is important to know (1) which tissues are most active in these various processes, (2) when these processes are most or least active, and (3) how these processes are controlled and coordinated in different metabolic states.

The best way to gain an understanding of the relationships of the pathways to one another is to become familiar with the changes in metabolism that occur during the **starve–feed cycle.** As shown in Figure 20.1, the starve–feed cycle allows a variable fuel and nitrogen consumption to meet a variable metabolic and anabolic demand. Feed refers to the intake of meals (the variable fuel input) after which we store the fuel (in the form of glycogen and triacylglycerol) to be used to meet our metabolic demand while we fast. Note the participation of an **ATP cycle** within the starve–feed cycle (Figure 20.1). Adenosine triphosphate is the agent that transfers energy in the starve–feed cycle, being like money to the cell.

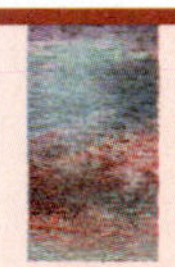

CLINICAL CORRELATION 20.1
Obesity

Obesity is the most common nutritional problem in the United States. It can reduce life span because it is a risk factor in development of diabetes mellitus, hypertension, endometrial carcinoma, osteoarthritis, cirrhosis, gallstones, and cardiovascular diseases. In fact, the quartet of obesity, insulin resistance, dyslipidemia, and hypertension is called either syndrome X or the metabolic syndrome and contributes greatly to the high rate of cardiovascular death in Western countries. Obesity is easy to explain—an obese person has eaten more calories than he/she expended. The accumulation of massive amounts of body fat is not otherwise possible. For unknown reasons, the neural control of caloric intake to balance energy expenditure is abnormal. Rarely, obesity is secondary to a correctable disorder, such as hypothyroidism or Cushing's syndrome. The latter is the result of increased secretion of glucocorticoids, which cause fat deposition in the face and trunk, with wasting of the limbs, and glucose intolerance. These effects are due to increased protein breakdown in muscle and conversion of the amino acids to glucose and fat. Less commonly, tumors, vascular accidents, or maldevelopment of the nervous system hunger control centers in the hypothalamus cause obesity.

Genetic models of obesity in rodents have led to breakthroughs in our understanding of the control of body mass. The obese (*ob/ob*) mouse was discovered in the 1950s, and the defective gene cloned in 1994. This *ob* gene encodes a 146 amino acid secreted protein (called leptin for its slimming effect) that is produced in adipocytes and detectable in blood. The *ob/ob* mice have a nonsense mutation in the gene and produce no leptin. Injection of leptin into *ob/ob* mice causes increased energy expenditure and reduced eating, with marked weight loss. This effect on appetite is mimicked by intracerebroventricular injection. Leptin also reduced appetite and weight of normal mice. Obese humans do not generally have defective *ob* genes and in fact tend to have high blood levels of leptin. This suggests that their nervous systems are insensitive to leptin, analogous to the insulin resistance seen in many diabetic patients.

In the most common type of obesity, the number of adipocytes of the body does not increase, they just get large as they become engorged with triacylglycerols. If obesity develops before puberty, however, an increase in the number of adipocytes can also occur. In the latter case, both hyperplasia (increase in cell number) and hypertrophy (increase in cell size) are contributing factors to the magnitude of the obesity. Obesity in men tends to be centered on the abdomen and mesenteric fat, while in women it is more likely to be on the hips. The male pattern, characterized by a high waist to hip circumference ratio, is more predictive of premature coronary heart disease.

The only effective treatment of obesity is reduction in the ingestion or increase in the use of calories. Practically speaking, this means dieting, since even vigorous exercise such as running only consumes 10 kcal/min of exercise. Thus an hour long run (perhaps 5–6 miles) uses the energy present in about two candy bars. However, exercise programs can be useful to help motivate individuals to remain on their diets. Unfortunately, the body compensates for decreased energy intake with reduced formation of triiodothyronine and a corresponding decrease in the basal metabolic rate. Thus there is a biochemical basis for the universal complaint that it is far easier to gain than to lose weight. Furthermore, about 95% of people who are able to lose a significant amount of weight regain it within one year.

Bray, G. D. Effect of caloric restriction on energy expenditure in obese patients. *Lancet* 2:397, 1969; Baringer, M. Obese protein slims mice. *Science* 269:475, 1995; and Ahima, R. S. and Flier, J. S. Leptin. *Annu. Rev. Physiol.* 62:413, 2000.

CLINICAL CORRELATION 20.2

Protein Malnutrition

Protein malnutrition is the most important and widespread nutritional problem among young children in the world today. The clinical syndrome, called kwashiorkor, occurs mainly in children 1–3 years of age and is precipitated by weaning an infant from breast milk onto a starchy, protein-poor diet. The name originated in Ghana, meaning "the sickness of the older child when the next baby is born." It is a consequence of feeding the child a diet adequate in calories but deficient in protein. It may become clinically manifest when protein requirements are increased by infection, for example, malaria, helminth infestation, or gastroenteritis. The syndrome is characterized by poor growth, low plasma protein and amino acid levels, muscle wasting, edema, diarrhea, and increased susceptibility to infection. The presence of subcutaneous fat clearly differentiates it from simple starvation. The maintenance of fat stores is due to the high carbohydrate intake and resulting high insulin levels. In fact, the high insulin level interferes with the adaptations described for starvation. Fat is not mobilized as an energy source, ketogenesis does not take place, and there is no transfer of amino acids from the skeletal muscle to the visceral organs, that is, the liver, kidneys, heart, and immune cells. The lack of dietary amino acids results in diminished protein synthesis in all tissues. The liver becomes enlarged and infiltrated with fat, reflecting the need for hepatic protein synthesis for the formation and release of lipoproteins. In addition, protein malnutrition impairs the function of the gut, resulting in malabsorption of calories, protein, and vitamins, which accelerates the disease. The consequences of the disease depend somewhat on when in development the deficiency occurs. Children with low weight for height are called "wasted" but can make a good recovery when properly fed. Those with low height for weight are called "stunted" and never regain full height or cognitive potential.

Protein–calorie malnutrition is also a problem for the elderly when they become sick. Both the energy requirements and food intake of well elderly decline with age. On a lower calorie diet, there is the risk that insufficient intake of protein and of certain nutrients such as iron, calcium, and vitamins will be lower than needed. Deficiencies in these nutrients may accelerate loss of lean body mass and strength (leading to falls), anemia, loss of bone strength, and, rarely, vitamin deficiency states. Furthermore, the chronic illnesses that are more common in the elderly often impair appetite, food intake, or nutrient assimilation. As a result, elderly patients are more frequently found to suffer from protein–calorie malnutrition than younger adults.

Chase, H. P., Kumar, V., Caldwell, R. T., and O'Brien, D. Kwashiorkor in the United States. *Pediatrics* 66:972, 1980; Schlienger, J. L., Pradignac, A., and Grunenberger, F. Nutrition of the elderly: a challenge between facts and needs. *Horm. Res.* 43:46, 1995; Omran, M. L. and Morley, J. E. Assesment of protein energy malnutrition in older persons, part I: history, examination, body composition, and screening tools. *Nutrition* 16:50, 2000; Omran, M. L. and Morley, J. E. Assessment of protein energy malnutrition in older persons, part II: laboratory evaluation. *Nutrition* 16:131, 2000; and Corish, C. A. and Kennedy, N. P. Protein–energy undernutrition in hospital in-patients. *Br. J. Nutr.* 83:575, 2000.

Humans have the capacity to consume food at a rate far greater than their basal caloric requirements, which allows them to survive from meal to meal. We thus store calories as glycogen and triacylglycerol and utilize them as needed. Unfortunately, an almost unlimited capacity to consume food is matched by an almost unlimited capacity to store it as triacylglycerol. **Obesity** is the consequence of excess food consumption and is the commonest form of malnutrition in affluent countries (see Clin. Corr. 20.1), whereas other forms of **malnutrition** are more prevalent in developing countries (see Clin. Corr. 20.2 and 20.3). The regulation of food

CLINICAL CORRELATION 20.3

Starvation

Starvation leads to the development of a syndrome known as marasmus. Marasmus is a word of Greek origin meaning "to waste." Although not restricted to any age group, it is most common in children under 1 year of age. In developing countries early weaning of infants from breast milk is a common cause of marasmus. This may result from pregnancies in rapid succession, the desire of the mother to return to work, or the use of overdiluted artificial formulas (to make the expensive formulas last longer). This practice leads to insufficient intake of calories. Likewise, diarrhea and malabsorption can develop if safe water and sterile procedures are not used.

In contrast to kwashiorkor (see Clin. Corr. 20.2), subcutaneous fat, hepatomegaly, and fatty liver are absent in marasmus because fat is mobilized as an energy source and muscle temporarily provides amino acids to the liver for the synthesis of glucose and hepatic proteins. Low insulin levels allow the liver to oxidize fatty acids and to produce ketone bodies for other tissues. Ultimately, energy and protein reserves are exhausted, and the child starves to death. The immediate cause of death is often pneumonia, which occurs because the child is too weak to cough. Adults can suffer from marasmus as a result of diseases that prevent swallowing (cancer of the throat or esophagus) or interfere with access to food (strokes or dementia). A related and very common disorder is cancer cachexia, which in part is due to anorexia (loss of appetite and, therefore, starvation) and in part to body wasting. The latter differs from simple starvation in that skeletal muscle is not spared and both muscle and fat are used to supply energy needs (see Clin. Corr. 20.10).

Waterlow, J. C. Childhood malnutrition—the global problem. *Proc. Nutr. Soc.* 38:1, 1979; Uvin, P. The state of world hunger. *Nutr. Rev.* 52:151, 1994; and Body, J. J. The syndrome of anorexia–cachexia. *Curr. Opin. Oncol.* 11:255, 1999.

consumption is complex and not well understood. One clearly important factor is leptin, a protein synthesized and secreted into the blood by adipocytes that regulates energy expenditure and appetite through effects it exerts on the hypothalamus (see Clin. Corr. 20.1). The tight control needed is indicated by the calculation that eating two extra pats of butter (~100 cal) per day over caloric expenditures results in a 10-lb weight gain per year. A weight gain of 10 lb may not sound excessive, but multiplied by 10 years it equals obesity!

20.2 | STARVE–FEED CYCLE

In the Well-Fed State the Diet Supplies the Energy Requirements

Figure 20.2 shows the fate of glucose, amino acids, and fat obtained from food. Glucose passes from the intestinal epithelial cells to the liver by way of the portal vein. Amino acids are partially metabolized in the gut before being released into portal blood. Triacylglycerols, contained in **chylomicrons,** is secreted by the intestinal epithelial cells into lymphatics, which drain the intestine. The lymphatics lead to the thoracic duct, which, by way of the subclavian vein, delivers chylomicrons to the blood at a site of rapid blood flow. This rapidly distributes the chylomicrons and prevents their coalescence.

Liver has a major role in handling dietary glucose. It can be converted into glycogen by glycogenesis or into pyruvate and lactate by glycolysis, or it can be used

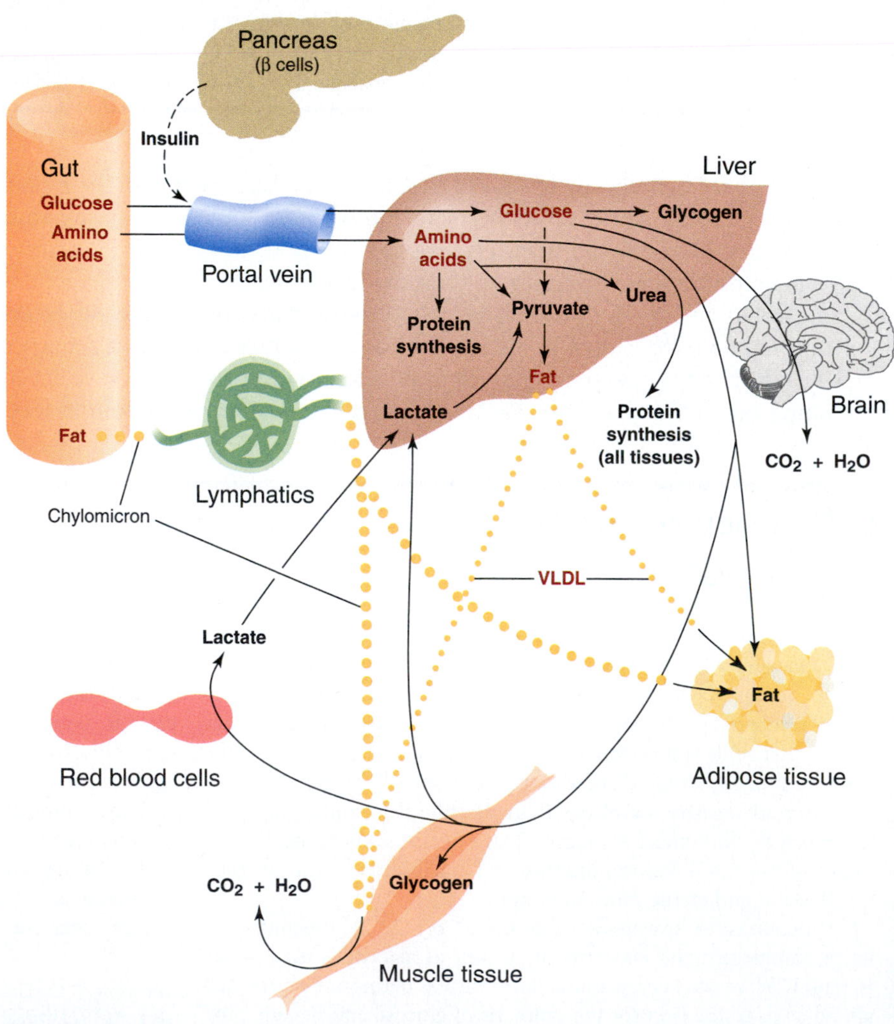

FIGURE 20.2
Disposition of glucose, amino acids, and fat by various tissues in the well-fed state.

in the pentose phosphate pathway for the generation of NADPH for synthetic processes. Pyruvate can be oxidized to acetyl CoA, which, in turn, can be converted into triacylglycerols or oxidized to CO_2 and water by the TCA cycle. Much of the glucose coming from the intestine passes through the liver to reach other organs, including brain and testis, which are almost solely dependent on glucose for the production of ATP, red blood cells and renal medulla, which can only convert glucose to lactate and pyruvate, and adipose tissue, which converts it into triacylglycerols. Muscle also has good capacity to use glucose, converting it to glycogen or using it in the glycolytic and the TCA cycle pathways. A number of tissues produce lactate and pyruvate from circulating glucose, which are taken up by the liver and converted to triacylglycerols. In the very well-fed state, the liver uses glucose and does not engage in gluconeogenesis. Thus the **Cori cycle** (the conversion of glucose to lactate in the peripheral tissues followed by conversion of lactate back to glucose in liver) is interrupted in the well-fed state.

Dietary protein is hydrolyzed in the intestine, the cells of which use some amino acids as an energy source. Most dietary amino acids are transported into the portal blood, but the intestinal cells metabolize aspartate, asparagine, glutamate, and glutamine and release alanine, lactate, citrulline, and proline into portal blood. Liver then has the opportunity to remove absorbed amino acids from the blood (Figure 20.2) but lets most of each of them pass through, unless the concentration of the amino acid is unusually high. This is especially important for the essential amino acids, needed by all tissues of the body for protein synthesis. Liver catabolizes amino acids, but the K_m values for amino acids of many of the enzymes involved is high, allowing the amino acids to be present in excess before significant catabolism can occur. In contrast, the tRNA-charging enzymes that generate **aminoacyl-tRNAs** have much lower K_m values for amino acids. This ensures that as long as all the amino acids are present, protein synthesis occurs as needed for growth and protein turnover. Excess amino acids can be oxidized completely to CO_2, urea, and water, or the intermediates generated can be used as substrates for lipogenesis. Amino acids that escape the liver are used for protein synthesis or energy in other tissues.

Glucose, lactate, pyruvate, and amino acids can support hepatic lipogenesis (Figure 20.2). Fat formed from these substrates is released from the liver in the form of **very low density lipoproteins (VLDLs)**, whereas dietary triacylglycerol is delivered to the bloodstream in the form of **chylomicrons.** Both VLDLs and chylomicrons circulate in the blood until acted on by an extracellular enzyme **lipoprotein lipase** attached to the surface of endothelial cells in the lumen of the capillaries of a number of tissues, but particularly adipose tissue.

Lipoprotein lipase liberates fatty acids by hydrolysis of the triacylglycerols present in VLDLs and chylomicrons. The fatty acids are then taken up by the adipocytes, reesterified with glycerol 3-phosphate to form triacylglycerols, and stored as fat droplets. Glycerol 3-phosphate is generated from glucose, using the first half of the glycolytic pathway to generate dihydroxyacetone phosphate, which is reduced to glycerol 3-phosphate by glycerol 3-phosphate dehydrogenase. Most of the fatty acids present in the triacylglycerols of human adipose tissue originate from fatty acids delivered to adipocytes in the form of triacylglycerols of VLDLs and chylomicrons rather than from *de novo* fatty acid synthesis within adipocytes.

The β cells of the pancreas are very responsive to the influx of glucose and amino acids in the fed state. The β cells release insulin during and after eating, which is essential for the metabolism of these nutrients by liver, muscle, and adipose tissue. The role of insulin in the starve–feed cycle is discussed in more detail in Section 20.3.

In the Early Fasting State Hepatic Glycogenolysis Is an Important Source of Blood Glucose

Hepatic glycogenolysis is very important for maintenance of blood glucose during early fasting (Figure 20.3). Lipogenesis is curtailed, and lactate, pyruvate, and amino acids used by that pathway are diverted into formation of glucose, complet-

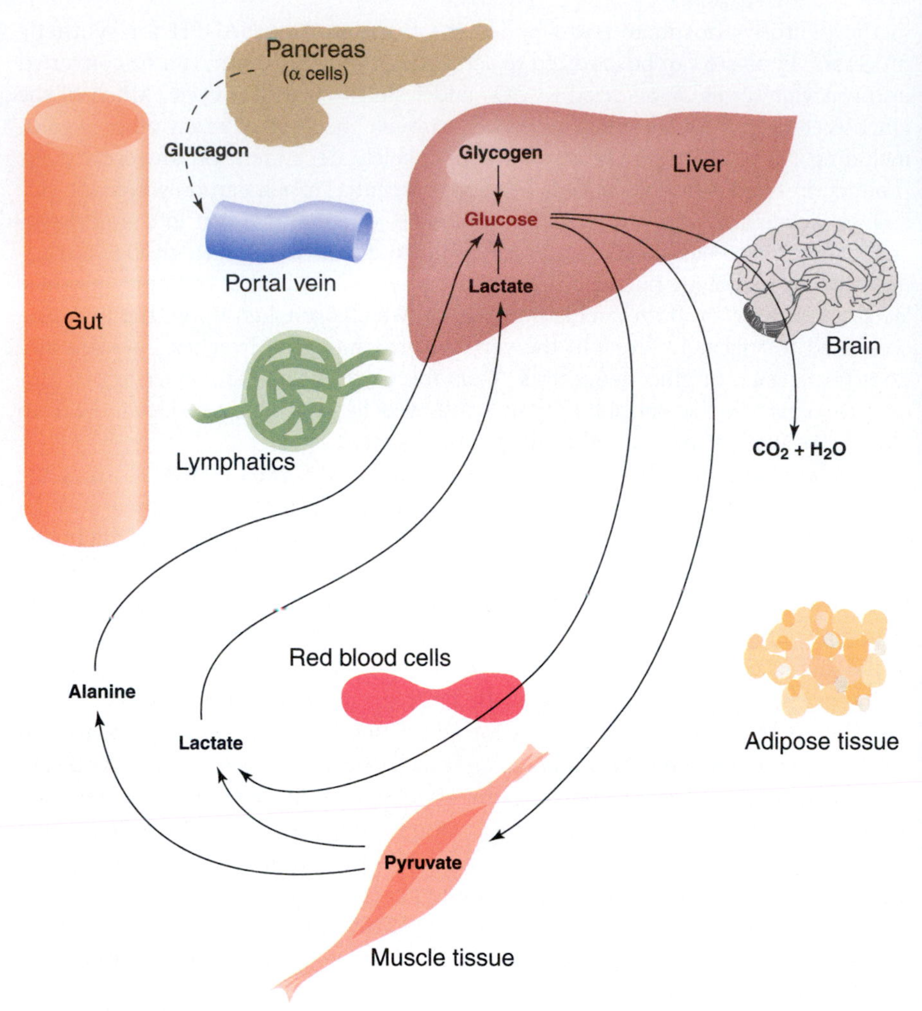

FIGURE 20.3
Metabolic interrelationships of major tissues in early fasting state.

ing the Cori cycle. The **alanine cycle,** in which carbon and nitrogen return to the liver in the form of alanine, also becomes important. Catabolism of amino acids for energy is greatly diminished in early fasting because less is available.

Fasting State Requires Gluconeogenesis from Amino Acids and Glycerol

Since no fuel enters from the gut and little glycogen is left in the liver in the fasting state (Figure 20.4), the body is dependent on hepatic **gluconeogenesis,** primarily from lactate, glycerol, and alanine. The **Cori** and **alanine cycles** play important roles but do not provide carbon for net synthesis of glucose. This is because glucose formed from lactate and alanine by the liver merely replaces that which was converted to lactate and alanine by peripheral tissues. In effect, these cycles transfer energy from fatty acid oxidation in the liver to peripheral tissues that cannot oxidize triacylglycerol. The brain oxidizes glucose completely to CO_2 and water. Hence net glucose synthesis from some other source of carbon is mandatory in fasting. Fatty acids cannot be used for the synthesis of glucose, because no pathway exists by which the acetyl CoA produced by fatty acid oxidation can be converted to glucose. Glycerol, a by-product of lipolysis in adipose tissue, is an important substrate for glucose synthesis. However, protein, especially from skeletal muscle, supplies most of the carbon needed for net glucose synthesis. Proteins are hydrolyzed within muscle cells and most amino acids are partially metabolized. **Alanine** and

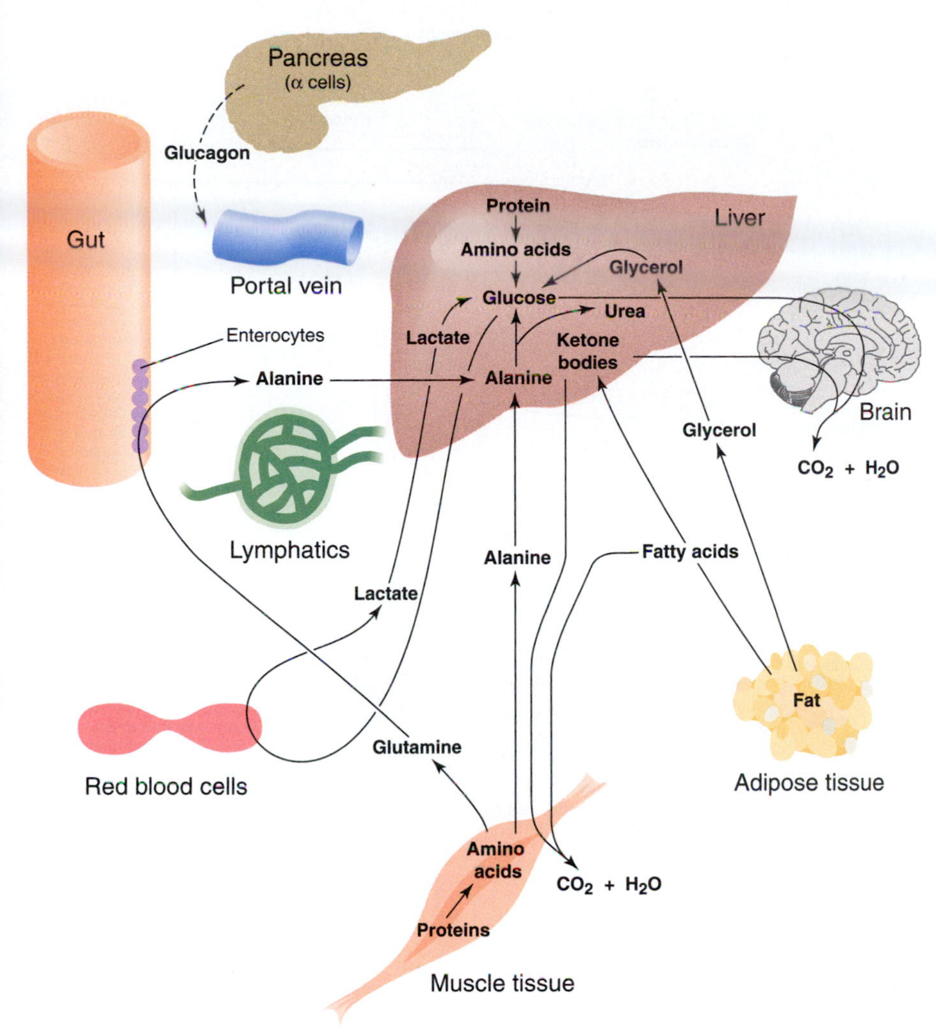

FIGURE 20.4
Metabolic interrelationships of major tissues in fasting state.

glutamine are the amino acids released from muscle in the largest amounts. The other amino acids are for the most part metabolized to give intermediates (pyruvate and α-ketoglutarate), which can yield alanine and glutamine. Branched-chain amino acids are a major source of nitrogen for the production of alanine and glutamine in muscle. Branched-chain α-keto acids produced from the branched-chain amino acids by transamination are partially released into the blood for uptake by the liver, which synthesizes glucose from the α-keto acid of valine, ketone bodies from the α-keto acid of leucine, and both glucose and ketone bodies from the α-keto acid of isoleucine.

Part of the glutamine released from muscle is converted into alanine by the intestinal epithelium. Glutamine is partially oxidized in enterocytes to supply energy and precursor molecules for synthesis of pyrimidines and purines, with the carbon and amino groups left over being released back into the bloodstream in part as alanine and NH_4^+. This pathway, sometimes called **glutaminolysis** because glutamine is only partially oxidized, involves formation of malate from glutamine via the TCA cycle and the conversion of malate to pyruvate by malic enzyme (Figure 20.5*a*). Pyruvate then transaminates with glutamate to give alanine, which is released from the cells.

Glutaminolysis is also used by cells of the immune system (lymphocytes and macrophages) to meet a large portion of their energy needs (Figure 20.5*b*). Aspartate rather than alanine is the major end product of glutaminolysis in lymphocytes. Enterocytes and lymphocytes use glutamine as their major fuel source as a way to

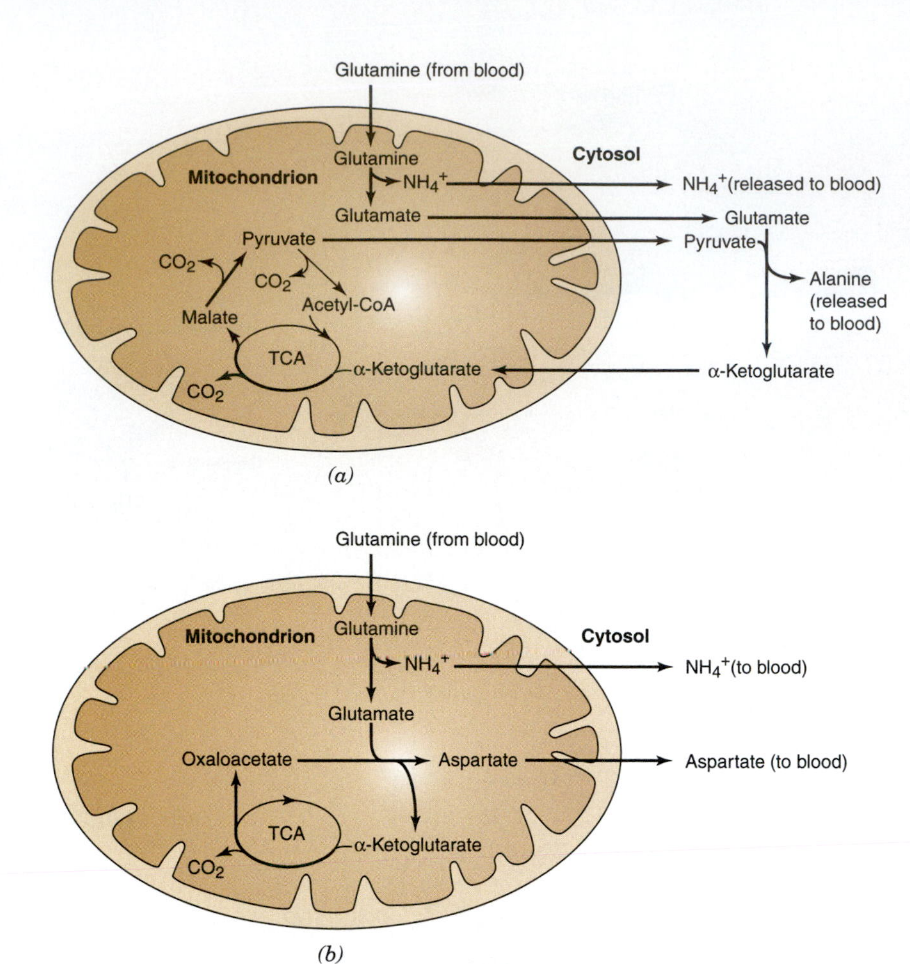

FIGURE 20.5
Glutamine catabolism by rapidly dividing cells.
(*a*) Enterocytes and (*b*) lymphocytes.
Part (a) redrawn from Duée, P.-H., Darcy-Vrillon, B., Blachier, F., and Morel, M.-T. Proc. Nutr. Soc. *54:83, 1995.*

ensure a continuous supply of the precursor molecules (glutamine and aspartate) required for synthesis of purines and pyrimidines that these rapidly dividing cells need for the synthesis of RNA and DNA. Unfortunately, glutamine is not a component of parenteral nutrition solutions.

Synthesis of glucose in the liver during fasting is closely linked to synthesis of urea. Most amino acids can give up their amino nitrogen by transamination with α-ketoglutarate, forming glutamate and a new α-keto acid, which often can be utilized for glucose synthesis. Glutamate provides both nitrogenous compounds required for urea synthesis: ammonia from oxidative deamination by glutamate dehydrogenase, and aspartate from transamination of oxaloacetate by aspartate aminotransferase. An additional important source of ammonia and precursors of ornithine such as citrulline is the gut mucosa (described in more detail in Section 20.4).

Adipose tissue is also very important in the fasting state. Because of low blood insulin levels during fasting, **lipolysis** is greatly activated in this tissue. This raises the blood level of fatty acids, which are used in preference to glucose by many tissues. In heart and muscle, the oxidation of fatty acids inhibits glycolysis and pyruvate oxidation. In liver, fatty acid oxidation provides most of the ATP needed for gluconeogenesis. Very little acetyl CoA generated by fatty acid oxidation in liver is oxidized completely. The acetyl CoA is converted instead into ketone bodies by liver mitochondria. **Ketone bodies** (acetoacetate and β-hydroxybutyrate) are released into the blood and are a source of energy for many tissues. Like fatty acids, ketone bodies are preferred by many tissues over glucose. Fatty acids are not oxidized by the brain because they cannot cross the blood–brain barrier. Once their blood concentration is high enough, ketone bodies can enter the brain and serve as an alternative fuel. They are unable, however, to completely replace the brain's need for glucose. Ketone bodies may also suppress proteolysis and branched-chain amino acid oxidation in muscle, and decrease alanine release. This both decreases muscle wasting and reduces the amount of glucose synthesized in liver. As long as ketone body levels are maintained at a high level by hepatic fatty acid oxidation, there is less need for glucose, less need for gluconeogenic amino acids, and less need for breaking down precious muscle tissue.

The interrelationships among liver, muscle, and adipose tissue in supplying glucose for the brain are shown in Figure 20.4. Liver synthesizes the glucose, muscle and gut supply the substrate (alanine), and adipose tissue supplies the ATP (via fatty acid oxidation in the liver) needed for hepatic gluconeogenesis. These relationships are disrupted in **Reye's syndrome** (see Clin. Corr. 20.4) and by alcohol (see Clin. Corr. 14.10). This tissue cooperation is dependent on the appropriate blood hormone levels. Glucose levels are lower in fasting, reducing the secretion of insulin but favoring release of glucagon from the pancreas and **epinephrine** from the adrenal medulla. In addition, fasting reduces formation of **triiodothyronine,** the active form of thyroid hormone, from **thyroxine.** This reduces the daily basal energy requirements by as much as 25%. This response is useful for survival but makes weight loss more difficult than weight gain (see Clin. Corr. 20.1).

In the Early Refed State, Fat Is Metabolized Normally and Normal Glucose Metabolism Is Slowly Reestablished

Figure 20.6 shows what happens soon after fuel is absorbed from the gut. Triacylglycerol is metabolized as described above for the well-fed state. In contrast, the liver extracts glucose poorly and, in fact, the liver remains in the gluconeogenic mode for a few hours after feeding. Rather than providing blood glucose, however, hepatic gluconeogenesis provides glucose 6-phosphate for glycogenesis. This means

CLINICAL CORRELATION 20.4
Reye's Syndrome

Reye's syndrome is a devastating but now rare illness of children that is characterized by evidence of brain dysfunction and edema (irritability, lethargy, and coma) and liver dysfunction (elevated plasma free fatty acids, fatty liver, hypoglycemia, hyperammonemia, and accumulation of short-chain organic acids). It appears that hepatic mitochondria are specifically damaged, which impairs fatty acid oxidation and synthesis of carbamoyl phosphate and ornithine (for ammonia detoxification) and oxaloacetate (for gluconeogenesis). On the other hand, the accumulation of organic acids has suggested that the oxidation of these compounds is defective and that the CoA esters of some of these acids may inhibit specific enzymes, such as carbamoyl phosphate synthetase I, pyruvate dehydrogenase, pyruvate carboxylase, and the adenine nucleotide transporter, all present in mitochondria. The issue has not yet been resolved. The use of aspirin by children with varicella was linked to the development of Reye's syndrome, and parents have been urged not to give aspirin to children with viral infections. The incidence of the syndrome subsequently decreased. The therapy for Reye's syndrome consists of measures to reduce brain edema and the provision of glucose intravenously. Glucose administration prevents hypoglycemia and elicits a rise in insulin levels that may (1) inhibit lipolysis in adipose cells and (2) reduce proteolysis in muscles and the release of amino acids, which (3) reduces the deamination of amino acids to ammonia. Although Reye's syndrome has become rare, other syndromes related to injury to the hepatocyte mitochondria have been recognized in recent years. These include toxicity of an experimental hepatitis drug (fialuridine) and of several antiretroviral nucleoside analogs. Injury to the mitochondria results in lactic acidosis, hypoglycemia, and fat accumulation in the liver, resulting in some cases in liver failure. These agents appear to interfere with mitochondrial DNA synthesis, resulting in accumulation of mutations in key mitochondrially encoded proteins and RNAs.

Reye, R. D. K., Morgan, G., and Baval, J. Encephalopathy and fatty degeneration of the viscera, a disease entity in childhood. *Lancet* 2:749, 1963; Treem, W. R. Inherited and acquired syndromes of hyperammonemia and encephalopathy in children. *Semin. Liver Dis.* 14:236, 1994; and Semino-Mora, C., Leon-Monzon, M., and Dalakas, M. C. Mitochondrial and cellular toxicity induced by fialuridine in human muscle *in vitro*. *Lab. Invest.* 76:487, 1997.

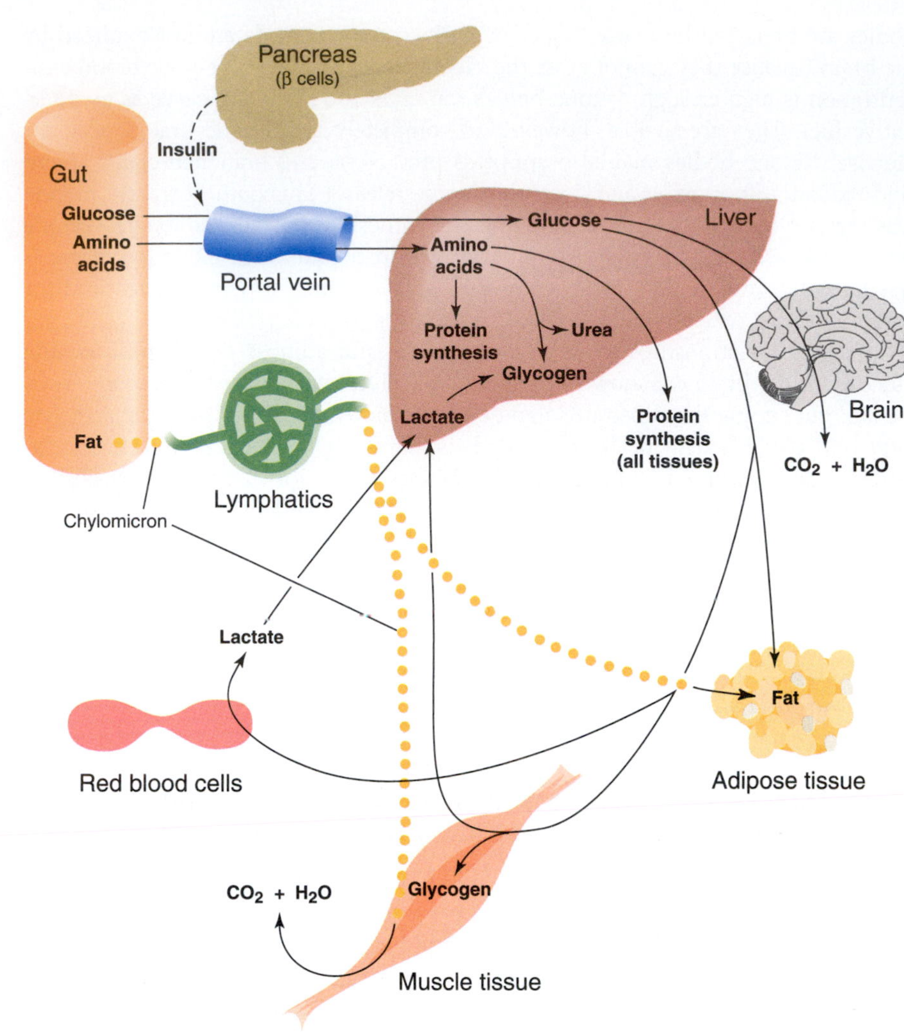

FIGURE 20.6
Metabolic interrelationships of major tissues in early refed state.

that liver glycogen is not repleted after a fast by direct synthesis from blood glucose. Rather, glucose is catabolized in peripheral tissues to lactate, which is converted in the liver to glycogen by the *indirect pathway of glycogen synthesis* (i.e., gluconeogenesis):

$$\begin{array}{c} \text{Glycogen} \\ \uparrow \\ \text{glucose} \xrightarrow{\textit{direct}} \text{glucose 6-phosphate} \xleftarrow{\textit{indirect}} \text{lactate} \end{array}$$

Gluconeogenesis from specific amino acids entering from the gut also plays an important role in reestablishing normal liver glycogen levels by the indirect pathway. After the rate of gluconeogenesis declines, glycolysis becomes the predominant means of glucose disposal in the liver, and liver glycogen is sustained by the *direct pathway of synthesis* from blood glucose.

Other Important Interorgan Metabolic Interactions

An important pathway exists in the intestinal epithelium for the conversion of glutamine to **citrulline** (Figure 20.7). One of the enzymes (ATP-dependent glutamate reductase) necessary for this conversion is expressed only in enterocytes. Citrulline produced in the gut is metabolized by the kidney to arginine, which can be con-

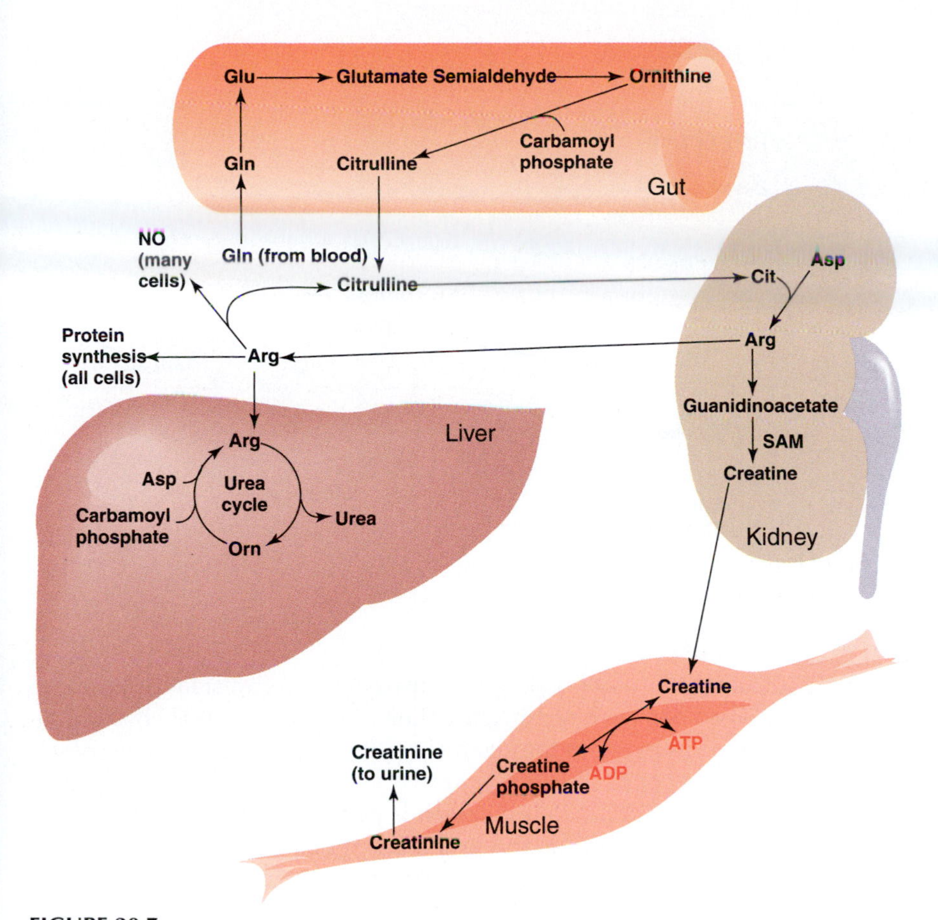

FIGURE 20.7
Gut and kidney function together in synthesis of arginine from glutamine.
Abbreviations: Cit, citrulline; Arg, arginine; Asp, aspartate; Gln, glutamine; Glu, glutamate; NO, nitric oxide; Orn, ornithine; SAM, *S*-adenosylmethionine.

verted to creatine or released into the blood. The liver uses blood arginine to generate ornithine, which expands the capacity of the **urea cycle** during periods of increased protein intake. Although perhaps not immediately obvious, this pathway is of great importance for urea cycle activity in the liver.

The liver contains an enzyme system that irreversibly converts ornithine into glutamate:

$$\text{ornithine} \longrightarrow \text{glutamate semialdehyde} \longrightarrow \text{glutamate}$$

Depletion of ornithine by these reactions inhibits urea synthesis in the liver for want of ornithine, the intermediate of the urea cycle that must recycle. Replenishment of ornithine is necessary and completely dependent on a source of blood arginine. Thus urea synthesis in the liver is dependent on citrulline synthesis by the gut and arginine synthesis by the kidney. Arginine is also used by many cells for the production of **nitric oxide (NO)** (Figure 20.7), an activator of guanylyl cyclase, which produces cGMP, an important second messenger (see p. 950).

Citrulline participates in another interesting interorgan shuttle. The arginine generated from citrulline in the kidney can be metabolized further to **creatine** (Figure 20.7). The first enzyme in this pathway is glycine transamidinase (GTA), which generates guanidinoacetate from arginine and glycine (see p. 818). GTA is found predominantly in renal cortex, pancreas, and liver. After methylation in a reaction that requires **S-adenosylmethionine (SAM)**, creatine is formed. This is quantitatively the most important use of SAM in the body. One to two grams of creatine is synthesized per day. Creatine then circulates to other tissues and accumulates in muscle, where it serves as a high-energy reservoir when phosphorylated to **creatine**

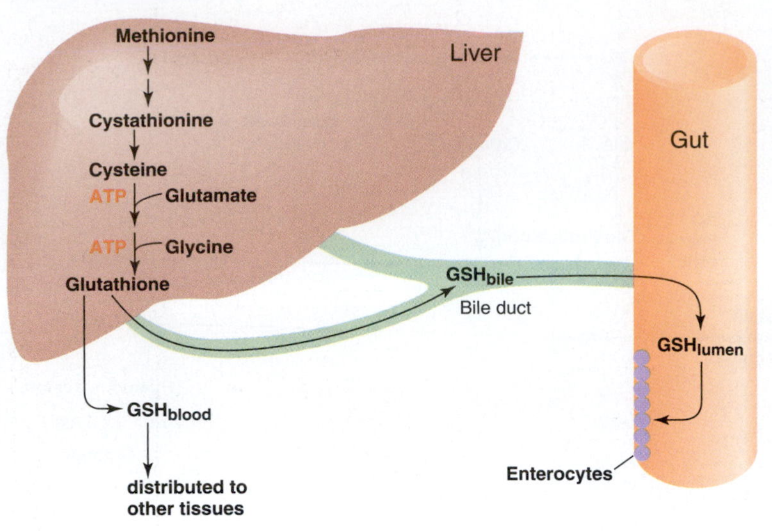

FIGURE 20.8
Liver provides glutathione for other tissues.

phosphate. Creatine phosphate undergoes nonenzymatic conversion to creatinine. Creatinine is released into the bloodstream and removed from the body by renal filtration. Excretion of creatinine is thus used clinically, both as a measure of muscle mass and of renal function.

Two other compounds related to amino acids participate in interorgan shuttles. **Glutathione (GSH)** is a tripeptide that is important in detoxification of endogenously generated peroxides and exogenous chemical compounds (see p. 818). Liver plays a major role in the synthesis of GSH from glutamate, cysteine, and glycine (Figure 20.8). Synthesis is limited by the availability of cysteine. Cysteine present in plasma is not taken up well by liver, which utilizes dietary methionine to form cysteine via the cystathionine pathway (see p. 803). Hepatic GSH is released both to the bloodstream and to the bile. Kidney removes a substantial amount of plasma GSH. Enterocytes may be able to take up biliary-excreted GSH from the intestinal lumen. Release to plasma is the same in fed and fasting states, providing a stable source of this compound and its constituent amino acids, especially cysteine, for most tissues of the body.

Carnitine is derived from lysyl residues on various proteins, which are *N*-methylated utilizing SAM to form trimethyllysyl residues (Figure 20.9). Free **trimethyllysine** is released when the proteins are degraded. It is hydroxylated and then cleaved, releasing glycine and γ-butyrobetaine aldehyde. The latter is oxidized to γ-butyrobetaine and then hydroxylated to form carnitine. Both hydroxylation steps require **vitamin C** as a cofactor. Kidney and to a lesser extent liver are the only tissues that can carry out the complete pathway, and thus they supply other tissues, especially muscle and heart, with the carnitine needed for fatty acid oxidation. Skeletal muscle can form γ-butyrobetaine but must release it for its final conversion to carnitine by liver or kidney.

Energy Requirements, Reserves, and Caloric Homeostasis

The average person leading a sedentary life consumes 180–280 g of carbohydrate, 70–100 g of protein, and 70–100 g of fat daily. This meets a daily energy requirement of 1600–2400 kcal. As shown in Table 20.1, the **energy reserves** of an average-sized person are considerable. These reserves are called on between meals and overnight to maintain blood glucose. Although the ability to mobilize glycogen rapidly is indeed very important, our glycogen reserves are minuscule with respect to our fat reserves (Table 20.1). Fat stores are only called on during more prolonged

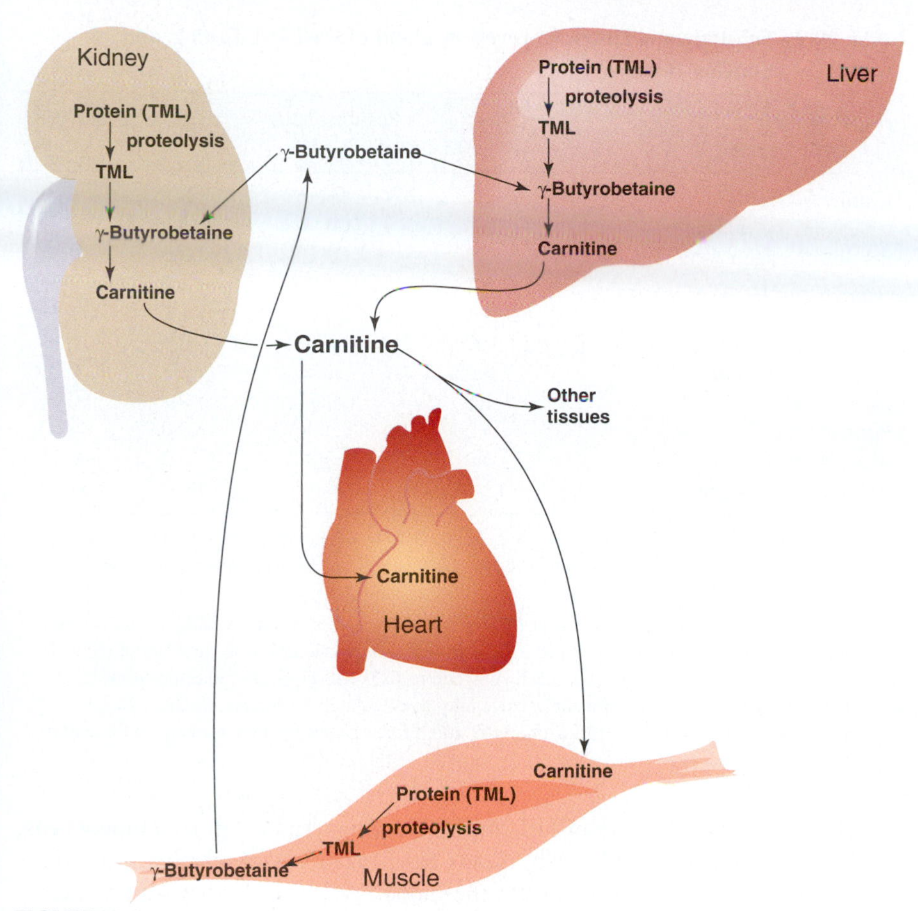

FIGURE 20.9
Kidney and liver provide carnitine for other tissues.
Abbreviations: (TML), trimethyllysyl residues in protein molecules; TML, free trimethyllysine.

fasting. The fat stores of obese subjects can weigh as much as 80 kg, adding another 585,000 kcal to their energy reserves. Protein is listed in Table 20.1 as an energy reserve because it can be used to provide amino acids for oxidation. On the other hand, protein is not inert like stored triacylglycerol and glycogen. Proteins make up the muscles that allow us to move and breathe and the enzymes that carry out metabolism. Hence protein is not dispensable like triacylglycerol and glycogen and is given up by the body more reluctantly.

TABLE 20.1 The Energy Reserves of Humans[a]

		Fuel Reserves	
Stored Fuel	*Tissue*	*(g)*	*(kcal)*
Glycogen	Liver	70	280
Glycogen	Muscle	120	480
Glucose	Body fluids	20	80
Fat	Adipose	15,000	135,000
Protein	Muscle	6,000	24,000

[a]Data are for a normal subject weighing 70 kg. Carbohydrate supplies 4 kcal g^{-1}; fat, 9 kcal g^{-1}; protein, 4 kcal g^{-1}.

TABLE 20.2 Substrate and Hormone Levels in Blood of Well-Fed, Fasting, and Starving Humans[a]

Hormone or Substrate (units)	*Very Well Fed*	*Postabsorptive 12 h*	*Fasted 3 days*	*Starved 5 weeks*
Insulin (μU mL^{-1})	40	15	8	6
Glucagon (pg mL^{-1})	80	100	150	120
Insulin/glucagon ratio (μU pg^{-1})	0.50	0.15	0.05	0.05
Glucose (mM)	6.1	4.8	3.8	3.6
Fatty acids (mM)	0.14	0.6	1.2	1.4
Acetoacetate (mM)	0.04	0.05	0.4	1.3
β-Hydroxybutyrate (mM)	0.03	0.10	1.4	6.0
Lactate (mM)	2.5	0.7	0.7	0.6
Pyruvate (mM)	0.25	0.06	0.04	0.03
Alanine (mM)	0.8	0.3	0.3	0.1
ATP equivalents (mM)	313	290	380	537

Source: From Ruderman, N. B., Aoki, T. T., and Cahill, G. F. Jr. Gluconeogenesis and its disorders in man. In: R. W. Hanson and M. A. Mehlman (Eds.), *Gluconeogenesis, Its Regulation in Mammalian Species*. New York: Wiley, 1976, p. 515.

[a]Data are for normal-weight subjects except for the 5-week starvation values, which are from obese subjects undergoing therapeutic starvation. ATP equivalents were calculated on the basis of the ATP yield expected on complete oxidation of each substrate to CO_2 and H_2O: 38 molecules of ATP for each molecule of glucose; 144 for the average fatty acid (oleate); 23 for acetoacetate; 26 for β-hydroxybutyrate; 18 for lactate; 15 for pyruvate; and 13 (corrected for urea formation) for alanine.

The constant availability of fuels in the blood is termed **caloric homeostasis,** which, as illustrated in Table 20.2, means that, regardless of whether a person is well-fed, fasting, or starving to death, the blood level of fuels that supply a comparable amount of ATP when metabolized does not fall below certain limits. Note that blood glucose concentrations are controlled within very tight limits, whereas fatty acid and ketone body concentrations in the blood can vary by one or two orders of magnitude, respectively. Glucose is carefully regulated because of the absolute need of the brain for this substrate. If the blood glucose level falls too low ($<$1.5 mM), coma and death will follow shortly unless the glucose concentration is restored. On the other hand, **hyperglycemia** must be avoided because glucose will be lost in the urine, resulting in dehydration and sometimes hyperosmolar, hyperglycemic coma (see Clin. Corr. 20.5). Chronic hyperglycemia results in glycation of

CLINICAL CORRELATION 20.5

Hyperglycemic, Hyperosmolar Coma

Type II diabetic patients sometimes develop a condition called hyperglycemic, hyperosmolar coma. This is particularly common in the elderly and can even occur in individuals under severe metabolic stress who were not recognized as having diabetes beforehand. Hyperglycemia, perhaps worsened by failure to take insulin or hypoglycemic drugs, by an infection, or by a coincidental medical problem such as a heart attack, leads to urinary losses of water, glucose, and electrolytes (sodium, chloride, and potassium). This osmotic diuresis reduces the circulating blood volume, a stress that results in the release of hormones that worsen insulin resistance and hyperglycemia. In addition, elderly patients may be less able to sense thirst or to obtain fluids. Over the course of several days these patients can become extremely hyperglycemic (glucose $>$ 1000 mg dL^{-1}), dehydrated, and comatose. Ketoacidosis does not develop in these patients possibly because free fatty acids are not always elevated or because adequate insulin concentrations exist in the portal blood to inhibit ketogenesis (although it is not high enough to inhibit gluconeogenesis). Therapy is aimed at restoring water and electrolyte balance and correcting the hyperglycemia with insulin. The mortality of this syndrome is considerably higher than that of diabetic ketoacidosis.

Arieff, A. I. and Carroll, H. J. Nonketotic hyperosmolar coma with hyperglycemia. Clinical features, pathophysiology, renal function, acid–base balance, plasma–cerebrospinal fluid equilibria, and the effects of therapy in 37 cases. *Medicine* 51:73, 1972; Genuth, S. M. Diabetic ketoacidosis and hyperglycemic hyperosmolar coma. *Curr. Ther. Endocrinol. Metab.* 6:438, 1997; and Lorber, D. Nonketotic hypertonicity in diabetes mellitus. *Med. Clin. North Am.* 79:39, 1995.

CLINICAL CORRELATION 20.6
Hyperglycemia and Protein Glycation

Glycation of enzymes is known to cause changes in their activity, solubility, and susceptibility to degradation. In the case of hemoglobin A, glycation occurs by a nonenzymatic reaction between glucose and the amino-terminal valine of the β chain. A Schiff base forms between glucose and valine, followed by a rearrangement of the molecule to give a 1-deoxyfructose molecule attached to the valine. The reaction is favored by high glucose levels and the resulting protein, called hemoglobin A_{1c}, is a good index of how high a person's average blood glucose concentration has been over the previous several weeks. The concentration of this modified protein increases in an uncontrolled diabetic and is low in patients who control their glucose level closely.

It has been proposed that glycation of proteins may contribute to the medical complications caused by diabetes, for example, coronary heart disease, retinopathy, nephropathy, cataracts, and neuropathy. Increased glycation of lens proteins may contribute to the development of diabetic cataracts. Collagen, laminin, vitronectin, and other matrix proteins can become glycated and undergo alterations in biological properties, such as self-assembly and binding of other matrix molecules. Glycated proteins and lipoproteins can also be recognized by receptors present on macrophages, which are intimately involved in the formation of atherosclerotic plaques. It is likely that these phenomena favor the accelerated atherosclerosis that occurs in diabetics. The compound aminoguanidine inhibits the formation of the glycation products and is being tested for its ability to prevent diabetic complications.

Brownlee, M. Glycation products and the pathogenesis of diabetic complications. *Diabetes Care* 15:1835, 1992; Ceriello, A. Hyperglycemia: the bridge between non-enzymatic glycation and oxidative stress in the pathogenesis of diabetic complications. *Diabetes Nutr. Metab.* 12:42, 1999; and Schmidt, A. M., Yan, S. D., Wautier, J. L., and Stern, D. Activation of receptor for advanced glycation end products: a mechanism for chronic vascular dysfunction in diabetic vasculopathy and atherosclerosis. *Circ. Res.* 84:489, 1999.

a number of proteins and endothelial dysfunction, which contributes to the complications of **diabetes** (see Clin. Corr. 20.6). The changes in **insulin/glucagon ratio** shown in Table 20.2 are crucial to the maintenance of caloric homeostasis. Simply stated, well-fed individuals have high insulin/glucagon ratios that favor storage of glycogen and triacylglycerol, while starving individuals have low insulin/glucagon ratios that stimulate glycogenolysis, lipolysis, ketogenesis, proteolysis, and gluconeogenesis.

Glucose Homeostasis Has Five Phases

Figure 20.10 shows the work of Cahill and his colleagues with obese patients undergoing long-term starvation for weight loss. It illustrates the effects of starvation on those processes that are used to maintain **glucose homeostasis** and is divided arbitrarily into five phases. Phase I is the well-fed state, in which glucose is provided by dietary carbohydrate. Once this supply is exhausted, hepatic glycogenolysis maintains blood glucose levels during phase II. As this supply of glucose starts to dwindle, hepatic gluconeogenesis from lactate, glycerol, and alanine becomes increasingly important until, in phase III, gluconeogenesis is the major source of blood glucose. These changes occur within 20 hours or so of fasting, depending on how well-fed the individual was prior to the fast, how much hepatic glycogen was present, and the sort of physical activity occurring during the fast. Several days of fasting move one into phase IV, where the dependence on gluconeogenesis actually decreases. As discussed above, ketone bodies have accumulated to high enough concentrations for them to enter the brain and meet some of its energy needs. Renal gluconeogenesis also becomes significant in this phase. Phase V occurs after very prolonged starvation of extremely obese individuals and is characterized by even less dependence on gluconeogenesis. The energy needs of almost every tissue are met to a large extent by either fatty acid or ketone body oxidation in this phase.

As long as ketone body concentrations are high, proteolysis will be somewhat restricted, and conservation of muscle proteins and enzymes will occur. This continues until practically all of the fat is gone as a consequence of starvation. After all of it is gone, the body has to use muscle protein. Before it's gone—you're gone (see Clin. Corr. 20.3).

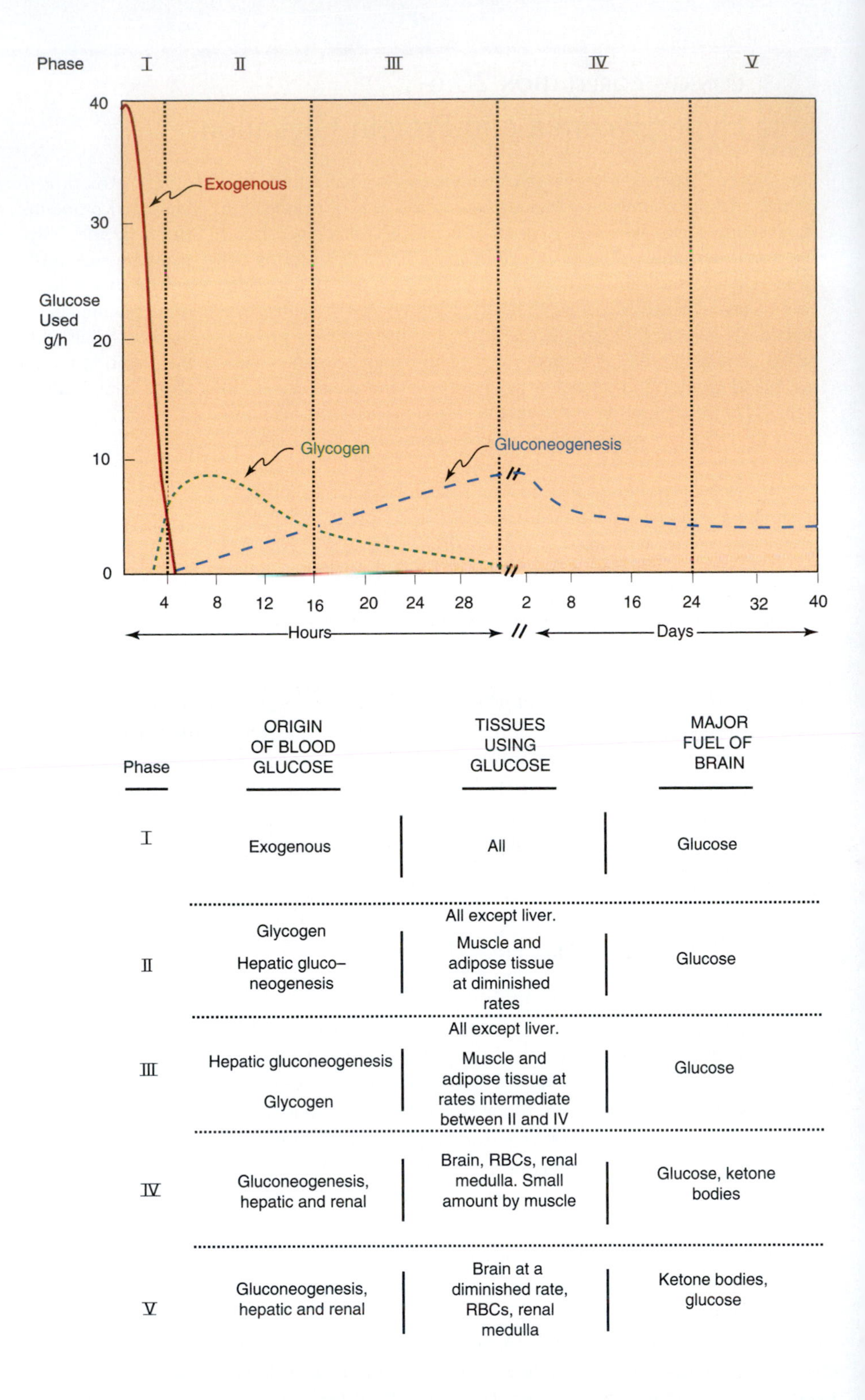

Phase	ORIGIN OF BLOOD GLUCOSE	TISSUES USING GLUCOSE	MAJOR FUEL OF BRAIN
I	Exogenous	All	Glucose
II	Glycogen Hepatic gluco–neogenesis	All except liver. Muscle and adipose tissue at diminished rates	Glucose
III	Hepatic gluconeogenesis Glycogen	All except liver. Muscle and adipose tissue at rates intermediate between II and IV	Glucose
IV	Gluconeogenesis, hepatic and renal	Brain, RBCs, renal medulla. Small amount by muscle	Glucose, ketone bodies
V	Gluconeogenesis, hepatic and renal	Brain at a diminished rate, RBCs, renal medulla	Ketone bodies, glucose

FIGURE 20.10
The five phases of glucose homeostasis in humans.
Reprinted with permission from Ruderman, N. B., Aoki, T. T., and Cahill, G. F. Jr. In: R. W. Hanson, and M. A. Mehlman (Eds.), Gluconeogenesis, Its Regulation in Mammalian Species. *New York: Wiley, 1976, p. 515.*

20.3 | MECHANISMS INVOLVED IN SWITCHING LIVER METABOLISM BETWEEN THE WELL-FED AND STARVED STATES

The liver of a well-fed person is actively engaged in processes that favor the synthesis of glycogen and triacylglycerol; such a liver is **glycogenic, glycolytic,** and **lipogenic.** The liver of the fasting person is quite a different organ; it is **glycogenolytic, gluconeogenic, ketogenic,** and **proteolytic.** The strategy is to store calories when food is available, but then to mobilize these stores when the rest of the body is in need. The liver is switched between these metabolic extremes by a variety of regulatory mechanisms: substrate supply, allosteric effectors, covalent modification, and induction–repression of enzymes.

Substrate Availability Controls Many Metabolic Pathways

Because of other, more sophisticated levels of control, the importance of **substrate supply** is often ignored. However, the concentration of fatty acids in blood entering the liver is clearly a major determinant of the rate of ketogenesis. Excess triacylglycerol is not synthesized unless one consumes excessive amounts of substrates that can be used for lipogenesis. Glucose synthesis by the liver is also restricted by the rate at which gluconeogenic substrates flow to the liver. Delivery of excess amino acids to the liver of the diabetic, because of accelerated and uncontrolled proteolysis, increases the rate of gluconeogenesis and exacerbates the hyperglycemia characteristic of diabetes. In addition, high glucose levels increase the rate of synthesis of sorbitol, which may contribute to diabetic complications. On the other hand, failure to supply the liver adequately with glucogenic substrate (mainly alanine) explains some types of hypoglycemia, such as that observed during pregnancy or advanced starvation.

Another pathway regulated by substrate supply is **urea synthesis.** Amino acid metabolism in the intestine provides a substantial fraction of the ammonia used by the liver for urea production. As discussed above, the intestine also releases citrulline, the metabolic precursor of ornithine. A larger ornithine pool permits increased urea synthesis after a high protein meal. In protein deficiency, the rate of urea formation declines but does not cease.

We can conclude that substrate supply is a major determinant of the rate at which virtually every metabolic process of the body operates. However, variations in substrate supply are not sufficient to account for the marked changes in metabolism that must occur in the starve–feed cycle. Finer tuning of the pathways is required.

Negative and Positive Allosteric Effectors Regulate Key Enzymes

Figures 20.11 and 20.12 summarize the effects of negative and positive **allosteric effectors** important in the liver in the well-fed and starved states, respectively. As shown in Figure 20.11, glucose activates glucokinase (indirectly by promoting its translocation from the nucleus to the cytoplasm; see p. 615), thereby promoting phosphorylation of glucose; glucose also inactivates glycogen phosphorylase and activates glycogen synthase (indirectly; see p. 657), thereby preventing degradation and promoting synthesis of glycogen; fructose 2,6-bisphosphate stimulates 6-phosphofructo-1-kinase and inhibits fructose 1,6-bisphosphatase, thereby stimulating glycolysis and inhibiting gluconeogenesis; fructose 1,6-bisphosphate activates pyruvate kinase, thereby stimulating glycolysis; pyruvate activates the pyruvate dehydrogenase complex (indirectly by inhibition of pyruvate dehydrogenase kinase; see p. 639); citrate activates acetyl-CoA carboxylase, thereby stimulating fatty acid synthesis; and malonyl CoA inhibits carnitine palmitoyltransferase I, thereby inhibiting fatty acid oxidation.

As shown in Figure 20.12, acetyl CoA stimulates gluconeogenesis in the fasted state by activating pyruvate carboxylase and inhibiting the pyruvate dehydrogenase complex (directly by an inhibitory allosteric effect and indirectly through stimulation of pyruvate dehydrogenase kinase; see p. 639); long-chain acyl-CoA esters inhibit acetyl-CoA carboxylase, which lowers the level of malonyl CoA and permits greater carnitine palmitoyltransferase I activity and fatty acid oxidation rates; fructose 6-phosphate inhibits glucokinase (indirectly by promoting its translocation from the cytoplasm to the nucleus); citrate, which is increased in concentration as a consequence of greater fatty acid oxidation, inhibits 6-phosphofructo-1-kinase as

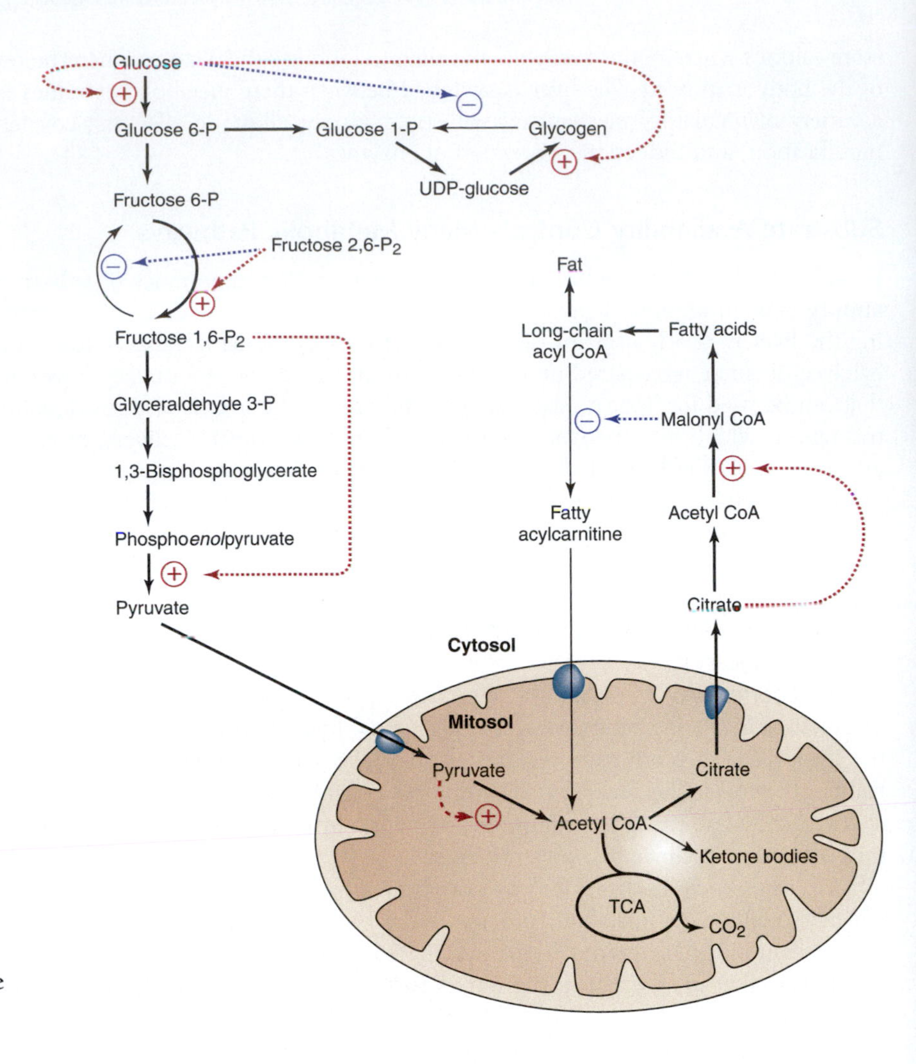

FIGURE 20.11
Control of hepatic metabolism in the well-fed state by allosteric effectors.

well as 6-phosphofructo-2-kinase (not shown); and NADH produced by fatty acid oxidation inhibits TCA cycle activity.

Allosteric effectors also regulate flux through metabolic pathways in nonhepatic tissues. **Citrate,** for example, probably serves as a *sensor of excess fuel availability* in a number of tissues. Through its ability to act as a negative allosteric effector for 6-phosphofructo-1-kinase and a positive allosteric effector for acetyl-CoA carboxylase, citrate regulates flux through both glycolysis and fatty acid oxidation. The latter effect is indirect, involving citrate activation of acetyl-CoA carboxylase, which increases the level of **malonyl CoA,** a negative allosteric effector of carnitine palmitoyltransferase I. Since glucose and fatty acid catabolism can both increase citrate levels and both can be inhibited by citrate, cells are able to sense the amount of fuel available for catabolism by their citrate level.

The malonyl CoA that is synthesized in the liver serves as an intermediate in fatty acid synthesis and also as a regulator of fatty acid oxidation via its negative allosteric effect on carnitine palmitoyltransferase I. Malonyl CoA is also synthesized in a number of other tissues, for example, skeletal muscle and heart, but in these tissues the only purpose for its synthesis is regulation of fatty acid oxidation at the level of carnitine palmitoyltransferase I. Steady-state levels of malonyl CoA are set in nonhepatic tissues by the relative rates of acetyl-CoA carboxylase, a highly regulated enzyme, and malonyl-CoA decarboxylase, a newly discovered enzyme that will probably also be found subject to tight regulation:

$$\text{malonyl CoA} \rightarrow \text{acetyl CoA} + CO_2$$

Although not shown in Figure 20.12, the concentration of cAMP, an important allosteric effector in its own right, is also increased in concentration in liver during

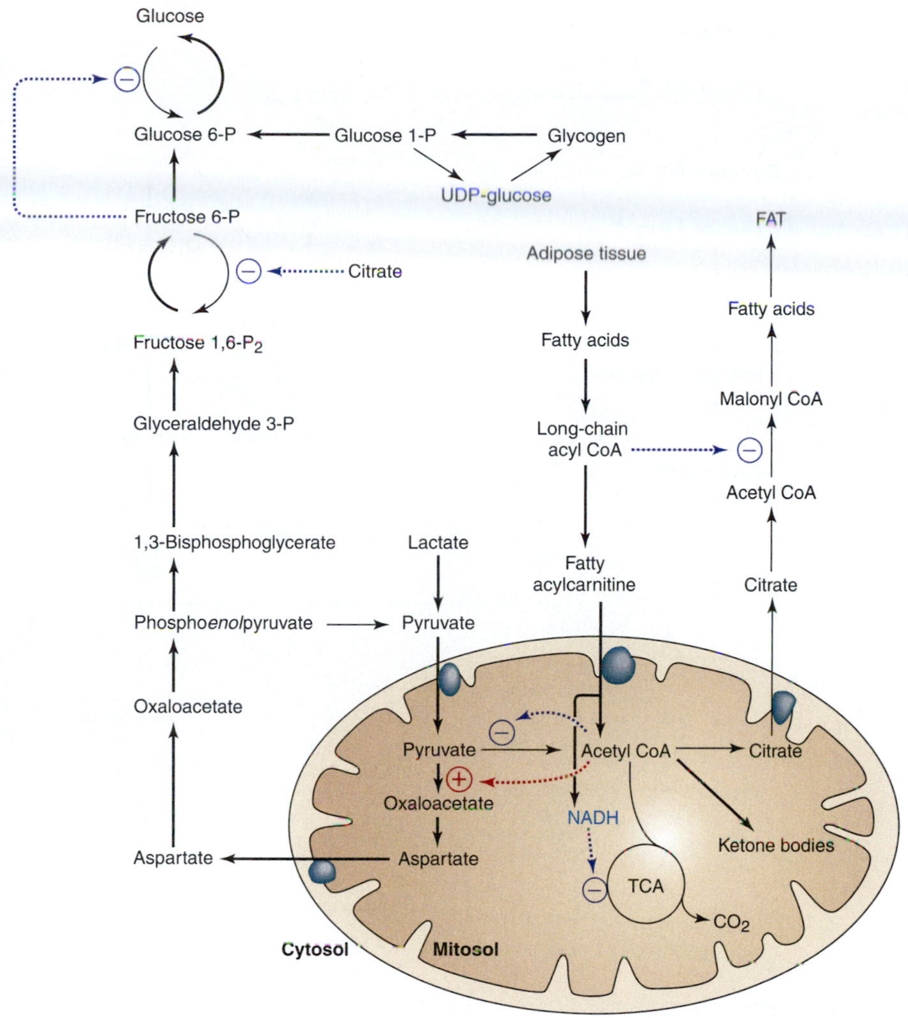

FIGURE 20.12
Control of hepatic metabolism in the fasting state by allosteric effectors.

starvation. It functions as a positive effector of protein kinase A (also called cAMP-dependent protein kinase), which, in turn, is responsible for changing the kinetic properties of several important regulatory enzymes by covalent modification, as summarized next.

Covalent Modification Regulates Key Enzymes

Figures 20.13 and 20.14 point out the enzymes subject to **covalent modification** that play important roles in switching the liver between the well-fed and starved states. The regulation of enzymes by covalent modification has been discussed in Chapters 10 and 14. Recall that ⊡ and ⊙-P represent the nonphosphorylated and phosphorylated states of an enzyme, respectively.

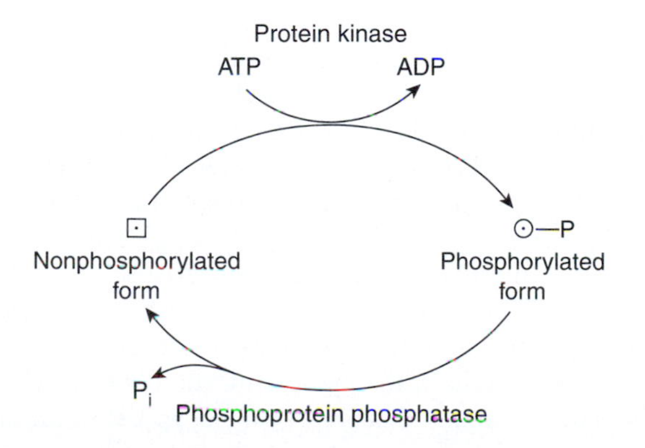

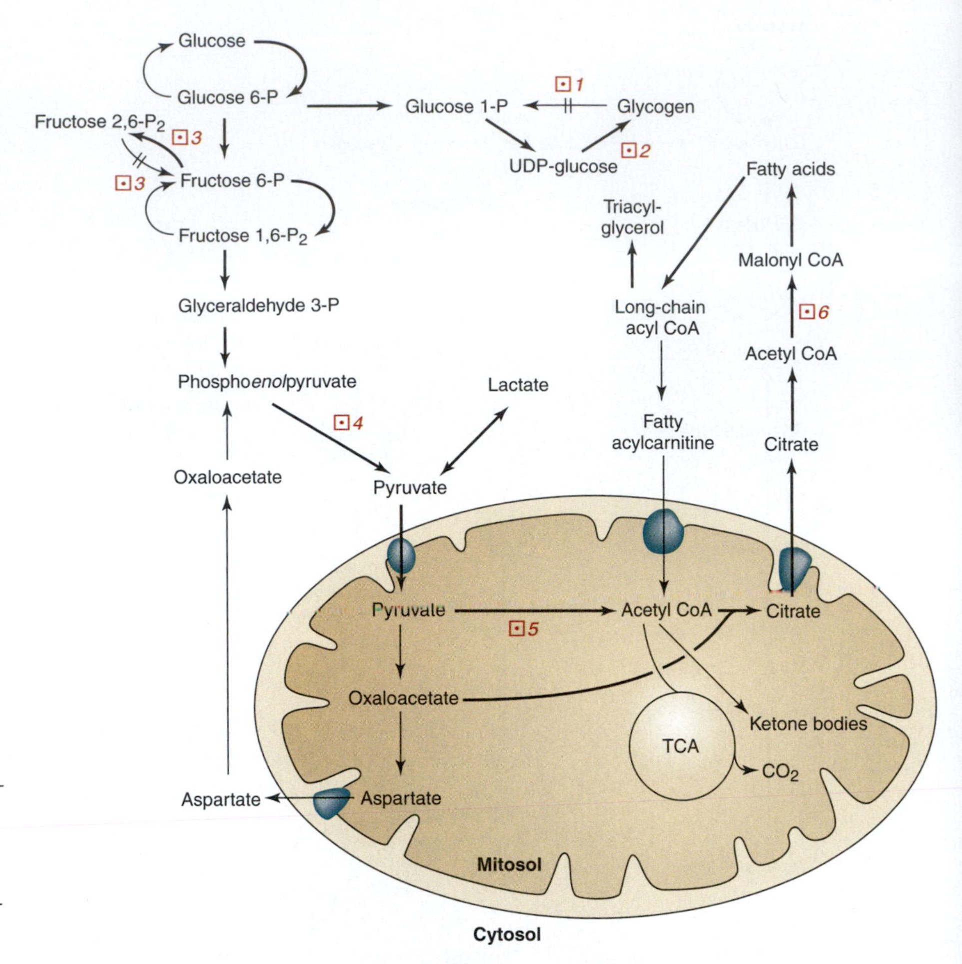

FIGURE 20.13
Activity and state of phosphorylation of enzymes subject to covalent modification in the lipogenic liver.
Dephosphorylated mode is indicated by ⊡. The interconvertible enzymes numbered are: *1*, glycogen phosphorylase; *2*, glycogen synthase; *3*, 6-phosphofructo-2-kinase/fructose- 2,6-bisphosphatase (bifunctional enzyme); *4*, pyruvate kinase; *5*, pyruvate dehydrogenase; and *6*, acetyl-CoA carboxylase.

The important points are as follows: (1) enzymes subject to covalent modification undergo phosphorylation on one or more serine residues by a protein kinase; (2) the phosphorylated enzyme can be returned to the dephosphorylated state by phosphoprotein phosphatase; (3) phosphorylation of the enzyme changes its conformation and its catalytic activity; (4) some enzymes are active only in the dephosphorylated state, others only in the phosphorylated state; (5) cAMP is the messenger that signals the phosphorylation of many, but not all, of the enzymes subject to covalent modification; (6) **cAMP** acts by activating **protein kinase A;** (7) cAMP also indirectly promotes phosphorylation of enzymes subject to covalent modification by signaling inactivation of phosphoprotein phosphatase; (8) glucagon and α-adrenergic agonists (epinephrine) increase cAMP levels by activating adenylyl cyclase; (9) insulin (see p. 944) opposes the action of glucagon and epinephrine, in part by lowering cAMP and in part by mechanisms independent of cAMP; and (10) the action of insulin in general promotes dephosphorylation of enzymes subject to covalent modification.

Hepatic enzymes subject to covalent modification are dephosphorylated in well-fed animals (Figure 20.13). Although not shown in this figure, phosphorylase kinase is also dephosphorylated in this state. Insulin/glucagon ratios are high in blood, and cAMP levels are low in liver. This results in low activity of protein kinase A and high activity of **phosphoprotein phosphatase.** Protein kinase A is responsible for phosphorylation of glycogen synthase, glycogen phosphorylase (via phosphorylase kinase), **6-phosphofructo-2-kinase/fructose 2,6-bisphosphatase** (bifunctional enzyme), pyruvate kinase, and acetyl-CoA carboxylase. Although the mitochondrial pyruvate dehydrogenase complex is not regulated by protein kinase A, which is located exclusively in the cytoplasm, the phosphorylation state of the pyruvate dehydrogenase complex changes in parallel with the enzymes listed in Fig-

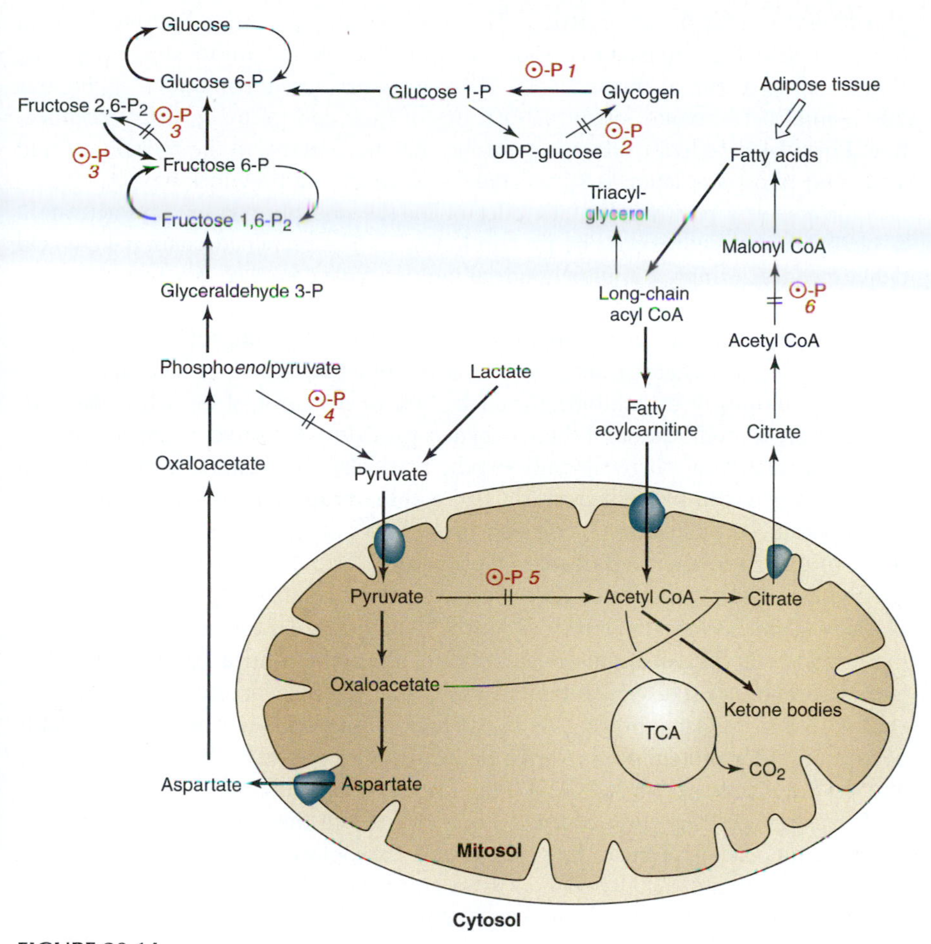

FIGURE 20.14
Activity and state of phosphorylation of enzymes subject to covalent modification in the glucogenic liver.
Phosphorylated mode is indicated by ⊙-P. Numbers refer to the same enzymes as in Figure 20.13.

ure 20.13. Since protein kinase A and **pyruvate dehydrogenase kinase** are relatively inactive in the liver of the well-fed animal, the dephosphorylated states of all of these enzymes prevail in the well-fed state. Three of the enzymes—**glycogen phosphorylase, phosphorylase kinase,** and the **fructose-2,6-bisphosphatase** activity of the bifunctional enzyme—are inactive when dephosphorylated. The others are all active. As a consequence of the dephosphorylated state of these enzymes, glycogenesis, glycolysis, and lipogenesis are greatly favored in the liver of the well-fed animal. On the other hand, the opposing pathways—glycogenolysis, gluconeogenesis, and ketogenesis—are inhibited.

As shown in Figure 20.14, the hepatic enzymes subject to covalent modification are in the phosphorylated mode in the liver of the fasting animal. Insulin is low but glucagon is high in the blood, resulting in an increase in hepatic cAMP levels. This activates protein kinase A and inactivates phosphoprotein phosphatase. The net effect is a greater degree of phosphorylation of the regulatory enzymes than in the well-fed state. In the starved state, three enzymes—glycogen phosphorylase, phosphorylase kinase, and the fructose-2,6-bisphosphatase of the bifunctional enzyme—are in the active catalytic state. All the other enzymes subject to covalent modification are inactive in the phosphorylated mode. As a result, glycogenesis, glycolysis, and lipogenesis are shut down almost completely, and glycogenolysis, gluconeogenesis, and ketogenesis dominate.

Two additional hepatic enzymes, **phenylalanine hydroxylase** and the **branched-chain α-keto acid dehydrogenase complex,** are also controlled by phosphorylation/dephosphorylation. These enzymes catalyze rate-limiting steps in the disposal of phenylalanine and the branched-chain amino acids (leucine, isoleucine, and valine), respectively. These enzymes are not included in Figures 20.13 and 20.14 because of special features of their control by covalent modification. Phenylalanine hydroxylase, a cytosolic enzyme, is active in the phosphorylated state, and phosphorylation is stimulated by glucagon via protein kinase A. Phenylalanine acts as a positive allosteric effector for the phosphorylation and activation of phenylalanine hydroxylase by protein kinase A. The branched-chain α-keto acid dehydrogenase complex, a mitochondrial multienzyme complex, is active in the dephosphorylated state, with its activity determined by the relative activities of a branched-chain α-keto acid dehydrogenase kinase and a phosphoprotein phosphatase. Branched-chain α-keto acids activate branched-chain α-keto acid dehydrogenase indirectly by inhibiting branched-chain α-keto acid dehydrogenase kinase. Covalent modification of these enzymes provides a sensitive means for control of the degradation of phenylalanine and the branched-chain amino acids. The clinical experience with **phenylketonuria** and **maple syrup urine disease** emphasizes the importance of regulating blood and tissue levels of these amino acids. Of note, the artificial sweetener **aspartame** (Nutrasweet™) is ***N*-aspartylphenylalanine methyl ester.** The amount in a liter of sweetened drinks may approach the amount of phenylalanine normally obtained from the daily diet. This is of no harm to normal individuals but aspartame is a threat to **phenylketonuria** patients on a low phenylalanine diet. Phenylalanine and the branched-chain amino acids cannot be synthesized in humans, making them essential amino acids that must be available continuously for protein synthesis. Thus the activities of phenylalanine hydroxylase and the branched-chain α-keto acid dehydrogenase complex must carefully be controlled to prevent depletion of body stores. Therefore the tissue requirements for these amino acids supersede the phase of the starve–feed cycle in establishing the phosphorylation and activity state of these enzymes.

Adipose tissue responds almost as dramatically as liver to the starve–feed cycle because it also contains enzymes subject to covalent modification. Pyruvate kinase, the pyruvate dehydrogenase complex, acetyl-CoA carboxylase, and hormone-sensitive lipase (not found in liver) are all in the dephosphorylated mode in the adipose tissue of the well-fed person. As in liver, the first three enzymes are active when dephosphorylated. **Hormone-sensitive lipase** is inactive when dephosphorylated. A high insulin level in the blood and a low cAMP concentration in adipose tissue are important determinants of the phosphorylation state of these enzymes, which favors lipogenesis in the well-fed state. During fasting, as a consequence of the decrease in the insulin level and an increase in epinephrine, adipocytes quickly shut down lipogenesis and activate lipolysis. This is accomplished in large part by the phosphorylation of the enzymes described above. In this manner, adipose tissue is transformed from a fat storage tissue into a source of fatty acids for oxidation in other tissues and glycerol for gluconeogenesis in the liver.

Conservation of glucose as well as three-carbon compounds that can readily be converted to glucose (lactate, alanine, pyruvate) by the liver is crucial for survival in the starved state. Certain cells, particularly those of the central nervous system, are absolutely dependent on a continuous supply of glucose. Tissues that can use alternative fuels invariably shut down their use of glucose and three-carbon precursors. This is referred to as the glucose–fatty acid cycle in recognition that increased availability of fatty acids for oxidation spares glucose in the starved state. Inactivation of the pyruvate dehydrogenase complex by phosphorylation is an important feature of the **glucose–fatty acid cycle.** This occurs in skeletal muscle, heart, kidney, and liver, but not in the central nervous system, when the alternative fuels (fatty acids and ketone bodies) of the starved state become abundant. Activation of pyruvate dehydrogenase kinase by products of the catabolism of the alternative fuels (acetyl CoA and NADH) is responsible for the greater degree of phosphorylation and therefore lower activity of the pyruvate dehydrogenase complex.

Covalent modification, like allosteric effectors and substrate supply, is a short-term regulatory mechanism, operating on a minute-to-minute basis. On a longer time scale, enzyme activities are controlled at the level of expression.

Changes in Levels of Key Enzymes Are a Longer Term Adaptive Mechanism

The **adaptive change in enzyme levels** is a mechanism of regulation involving changes in the rate of synthesis or degradation of key enzymes. Whereas allosteric effectors and covalent modification affect either the K_m or V_{max} of an enzyme, this mode of regulation involves the actual quantity of an enzyme in a tissue. Because of the influence of hormonal and nutritional factors, there are more or fewer enzyme molecules present in the tissue. For example, when a person is maintained in a well-fed or overfed condition, the liver improves its capacity to synthesize triacylglycerol. This can be explained in part by increased substrate supply, appropriate changes in allosteric effectors (Figure 20.11), and the conversion of the interconvertible enzymes into the dephosphorylated form (Figure 20.13). This is not the entire story, however, because the liver also has more of those enzyme molecules that play a key role in triacylglycerol synthesis (see Figure 20.15). A whole battery of enzymes is induced, including glucokinase, 6-phospho-1-fructokinase, and pyruvate kinase for faster rates of glycolysis; glucose 6-phosphate dehydrogenase, 6-phosphogluconate dehydrogenase, and malic enzyme to provide greater quantities of NADPH for reductive synthesis; and citrate cleavage enzyme, acetyl-CoA carboxylase, fatty acid synthase, and Δ^9-desaturase for more rapid rates of fatty acid synthesis. All of these enzymes are present at higher levels in the well-fed state because of an increase in the blood of the insulin/glucagon ratio and glucose. While these enzymes are induced, there is a decrease in the enzymes that

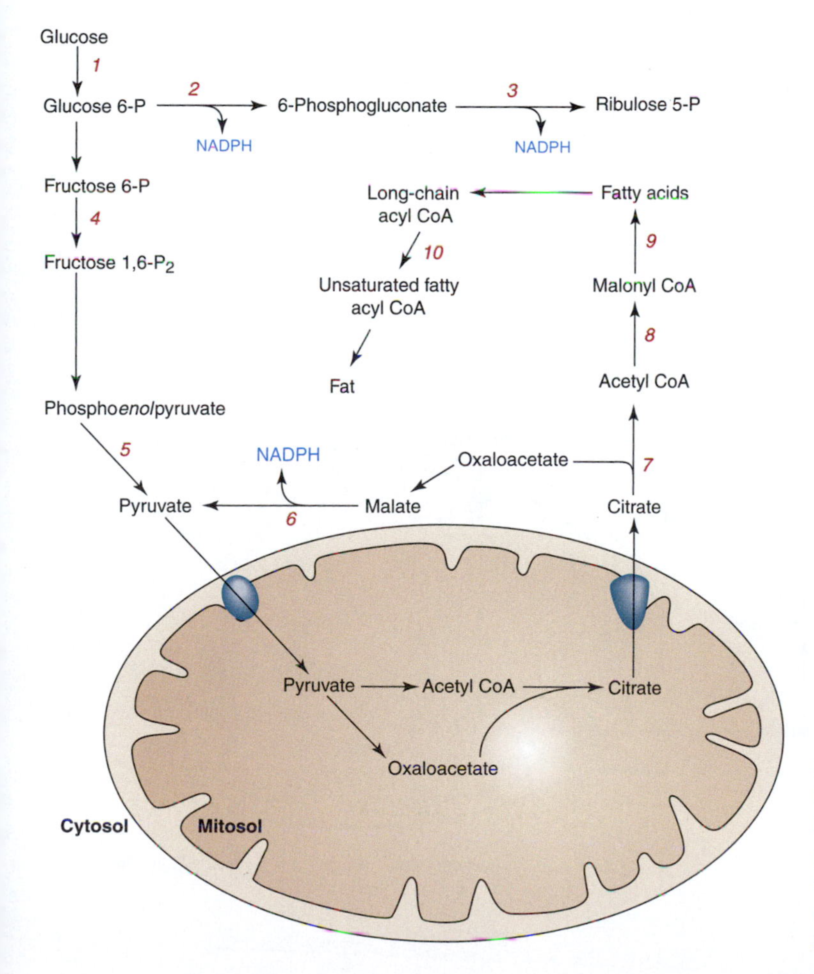

FIGURE 20.15
Enzymes induced in the liver of the well-fed individual. Inducible enzymes numbered are: *1*, glucokinase; *2*, glucose 6-phosphate dehydrogenase; *3*, 6-phosphogluconate dehydrogenase; *4*, 6-phosphofructo-1-kinase; *5*, pyruvate kinase; *6*, malic enzyme; *7*, citrate cleavage enzyme; *8*, acetyl-CoA carboxylase; *9*, fatty acid synthase; and *10*, Δ^9-desaturase.

favor glucose synthesis. **Phospho*enol*pyruvate carboxykinase,** pyruvate dehydrogenase kinase, **pyruvate carboxylase,** fructose 1,6-bisphosphatase, **glucose 6-phosphatase,** and some aminotransferases are decreased in amount; that is, their synthesis is reduced or degradation is increased in response to increased circulating glucose and insulin.

In fasting, the enzyme pattern of the liver changes dramatically (Figure 20.16). The enzymes involved in lipogenesis decrease in quantity, possibly because their synthesis is decreased or degradation of these proteins is increased. At the same time a number of enzymes (glucose 6-phosphatase, fructose 1,6-bisphosphatase, phospho*enol*pyruvate carboxykinase, pyruvate carboxylase, and various aminotransferases) favoring gluconeogenesis are induced, making the liver much more effective in synthesizing glucose. Starvation also induces pyruvate dehydrogenase kinase (not shown in Figure 20.16), the enzyme responsible for phosphorylation and inactivation of the pyruvate dehydrogenase complex, which prevents conversion of pyruvate to acetyl CoA, thereby conserving lactate, pyruvate, and amino acid carbon for glucose synthesis. Induction of carnitine palmitoyltransferase I and 3-hydroxy-3-methylglutaryl-CoA synthase likewise increases the capacity of the liver for fatty acid oxidation and ketogenesis, the former serving as the primary source of the ATP needed for hepatic glucose synthesis. In addition, the enzymes of the urea cycle and other amino acid metabolizing enzymes such as liver **glutaminase,** tyrosine aminotransferase, serine dehydratase, proline oxidase, and histidase are induced, possibly by the presence of higher blood glucagon levels. This permits the disposal of nitrogen, as urea, from the amino acids used in gluconeogenesis.

The switches that control these long-term processes are becoming better understood. A member of the nuclear receptor family, called the **peroxisome proliferator-activated receptor α(PPARα)** is likely to serve as a receptor for fatty acids, particularly in the liver, kidney, and heart. When the receptor binds fatty acids, it is able to activate transcription of a number of genes involved in disposal

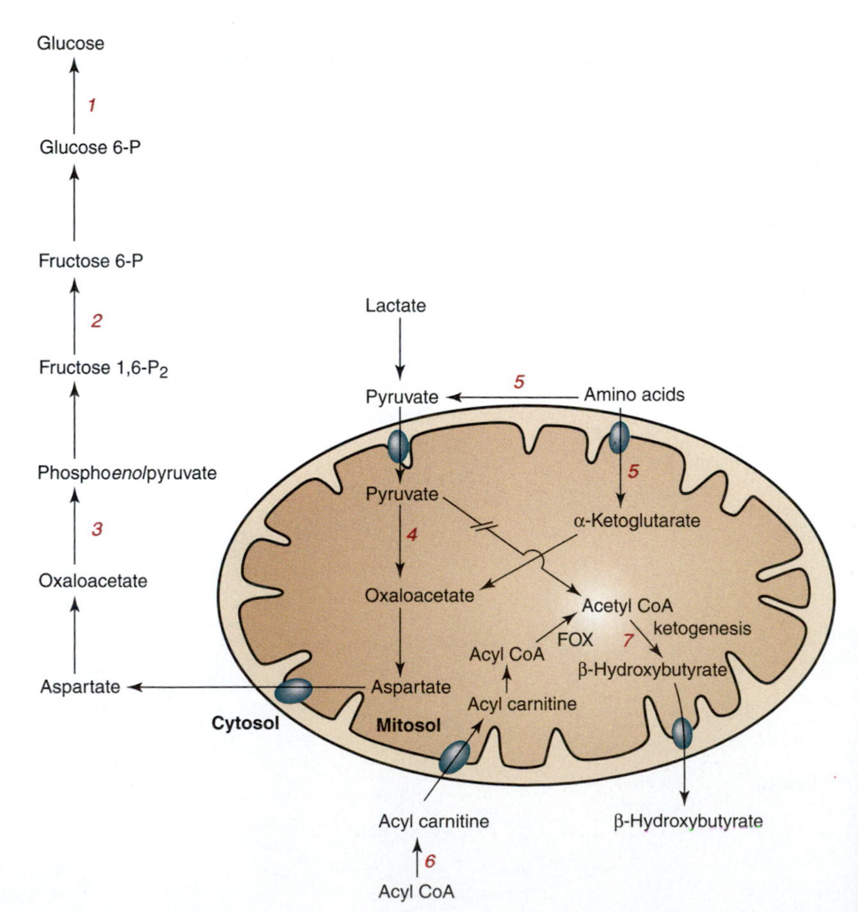

FIGURE 20.16
Enzymes induced in the liver of an individual during fasting.
The inducible enzymes numbered are: *1*, glucose 6-phosphatase; *2*, fructose 1,6-bisphosphatase; *3*, phospho*enol*pyruvate carboxykinase; *4*, pyruvate carboxylase; *5*, various aminotransferases; *6*, carnitine palmitoyltransferase I; and *7*, 3-hydroxy-3-methylglutaryl-CoA synthase. Parallel lines intersecting arrow from pyruvate to acetyl CoA denote inhibition (due to phosphorylation) of the pyruvate dehydrogenase complex due to induction of pyruvate dehydrogenase kinase. Abbreviation: FOX, fatty acid oxidation.

of fatty acids, including enzymes of the peroxisomal, microsomal, and mitochondrial fatty acid oxidation systems, apolipoprotein genes needed for export of triacylglycerol from the liver as VLDL, and enzymes of ketogenesis. In adipose tissue, another isoform of PPAR, PPARγ is differentially expressed. When activated (perhaps by various fatty acid derivatives such as prostaglandins), it orchestrates the differentiation of pre-adipocytes to adipocytes. This increases the ability of the adipocytes to store fatty acids. Another transcription factor involved in this level of control is the **steroid regulatory element binding factor (SREBP).** When levels of cholesterol fall, SREBP, which is anchored in the endoplasmic reticulum, is cleaved by a protease, releasing the N-terminal fragment of the protein that acts as a transcriptional activator of enzymes involved in sterol and fatty acid biosynthesis such as ATP citrate lyase, fatty acid synthase, 3-hydroxy-3-methylglutaryl-CoA synthase, 3-hydroxy-3-methylglutaryl-CoA reductase, squalene synthase, and the LDL receptor. The level of SREBP is also modulated by insulin and carbohydrate intake and probably controls the increase in fatty acid synthesis that accompanies long-term increased carbohydrate intake.

These adaptive changes are clearly important in the starve–feed cycle, greatly affecting the capacity of the liver for its various metabolic processes. The adaptive changes also influence the effectiveness of the short-term regulatory mechanisms. For example, long-term starvation or uncontrolled diabetes decreases the level of acetyl-CoA carboxylase. Taking away long-chain acyl CoA esters that inhibit this enzyme, increasing the level of citrate that activates this enzyme, or creating conditions that activate this enzyme by dephosphorylation will not have any effect when the enzyme is virtually absent. Another example is afforded by the **glucose intolerance** of starvation. A chronically starved person cannot effectively utilize a load of glucose because of the absence of the key enzymes needed for glucose metabolism. A glucose load, however, will set into motion the induction of the required enzymes and the reestablishment of short-term regulatory mechanisms.

20.4 | METABOLIC INTERRELATIONSHIPS OF TISSUES IN VARIOUS NUTRITIONAL AND HORMONAL STATES

Many changes that occur in various nutritional and hormonal states are variations on the starve–feed cycle and are completely predictable from what we have learned about the cycle. Some examples are given in Figure 20.17. Others are so obvious that a diagram is not necessary; for example, in rapid growth of a child, amino acids are directed away from catabolism and into protein synthesis. However, the changes that occur in some physiologically important situations are rather subtle and poorly understood. An example of the latter is aging, which seems to lead to a decreased sensitivity of the major tissues of the body to hormones. The important consequence is a decreased ability of the tissues to respond normally during the starve–feed cycle. Whether this is a contributing factor to or a consequence of the aging process is not known.

Staying in the Well-Fed State Results in Obesity and Insulin Resistance

Figure 20.17*a* illustrates the metabolic interrelationships prevailing in an obese person. Most of the body fat of the human is either provided by the diet or synthesized in the liver and transported to the adipose tissue for storage. Obesity is caused by a person staying in such a well-fed state that stored fat does not get used up during the fasting phase of the cycle. The body then has no option other than to accumulate fat (see Clin. Corr. 20.1).

Obesity almost always causes some degree of **insulin resistance.** Insulin resistance is a poorly understood phenomenon in which the tissues fail to respond to insulin. The number or affinity of insulin receptors is reduced in some patients;

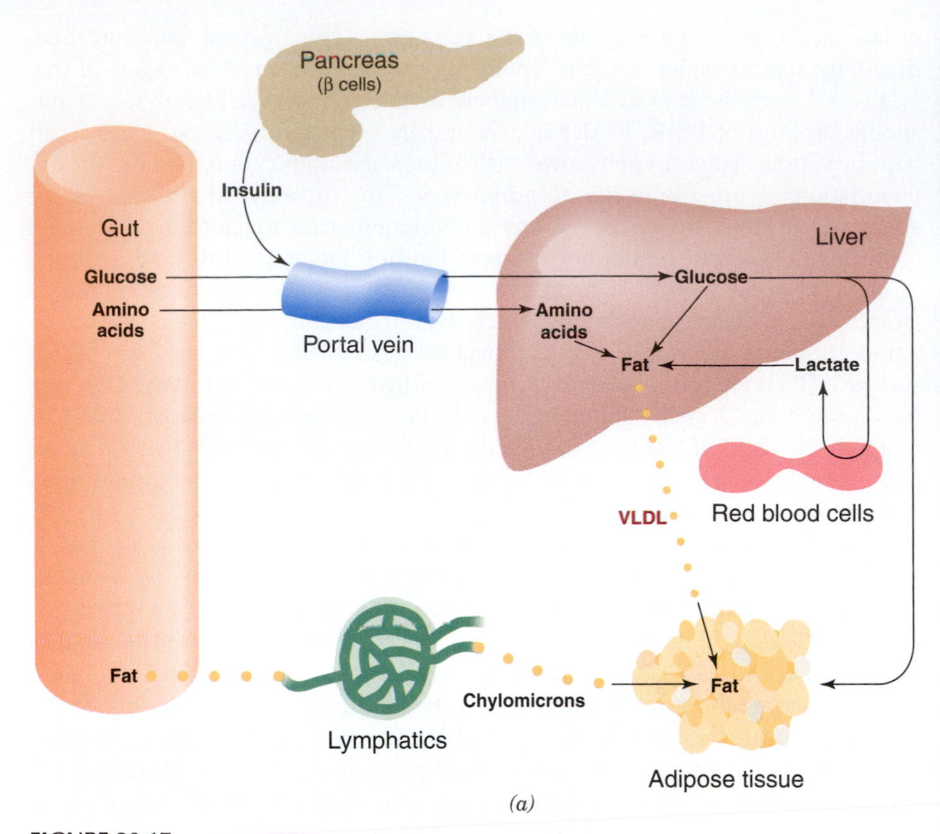

FIGURE 20.17
Metabolic interrelationships of tissues in various nutritional, hormonal, and disease states.
(*a*) Obesity.

others have normal insulin binding, but abnormal postreceptor responses, such as the activation of glucose transport. As a general rule, the greater the quantity of body fat, the greater the resistance of normally insulin-sensitive cells to the action of insulin. Fat cells produce two hormones, tumor necrosis factor α (TNF-α) and a recently discovered protein called resistin, that appear responsible for induction of insulin resistance. As a consequence of insulin resistance, plasma insulin levels are greatly elevated in the blood of an obese individual. As long as the β cells of the pancreas produce enough insulin to overcome the insulin resistance, an obese individual will have relatively normal blood levels of glucose and lipoproteins. The insulin resistance of obesity can lead, however, to the development of type 2 diabetes, as discussed next.

Type 2 Diabetes Mellitus

Figure 20.17*b* shows the metabolic interrelationships characteristic of a person with **type 2 diabetes,** which is also called noninsulin-dependent diabetes. In contrast to type 1 or insulin-dependent diabetes, insulin is present in type 2 diabetes (see Clin. Corr. 20.7), with the problem being resistance to the action of insulin along with insufficient production of insulin by β cells to overcome the resistance. The majority of patients with type 2 are obese, and although their insulin levels often are high, they are not as high as those of a nondiabetic but similarly obese person. The pancreases of these diabetic patients do not produce enough insulin to overcome their insulin resistance. Hence this form of diabetes is also a form of β-cell failure, and exogenous insulin will reduce the hyperglycemia, and very often must be administered to control blood glucose levels of noninsulin-dependent diabetic patients. Hyperglycemia results mainly because of poor uptake of glucose by peripheral tissues, particularly muscle. Whereas insulin normally stimulates translocation of intracellular vesicles bearing **GLUT-4 (*glucose transporter isoform 4*)** to the plasma mem-

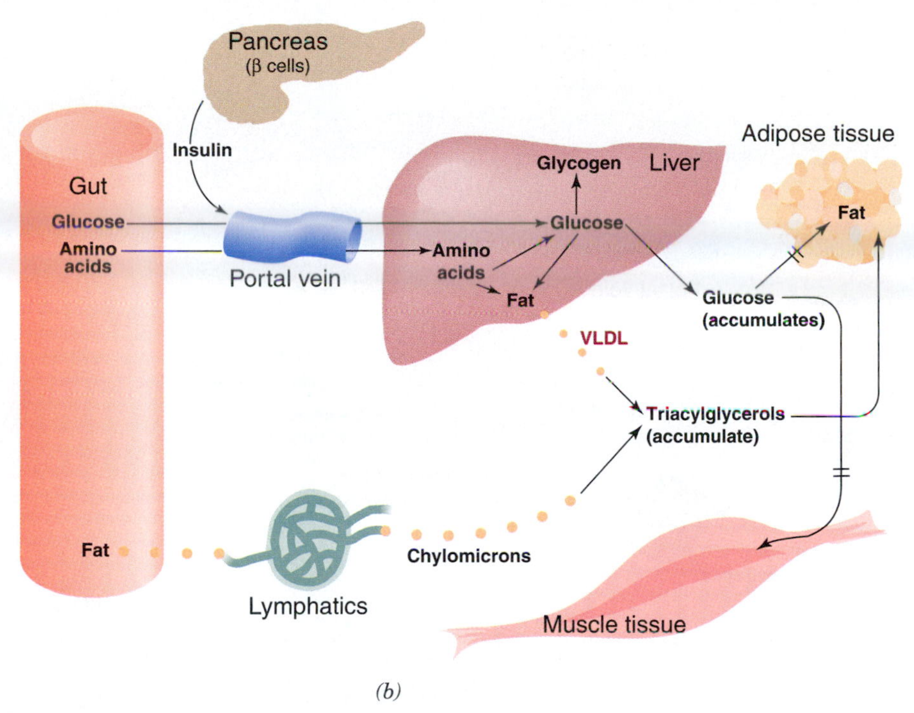

FIGURE 20.17 (continued)
(*b*) Noninsulin-dependent diabetes mellitus.
(

brane, this is largely prevented in type 2 diabetes by resistance to insulin. Ketoacidosis rarely develops in type 2 diabetes, which stands in contrast to the clinical experience with type 1 diabetes, perhaps because enough insulin is present in type 2 diabetes to prevent uncontrolled release of fatty acids by lipolysis in adipocytes. Hypertriacylglycerolemia is characteristic of type 2 diabetes but usually results from

CLINICAL CORRELATION 20.7

Type 2 (Noninsulin-Dependent) Diabetes Mellitus

Type 2 diabetes mellitus accounts for 80–90% of the diagnosed cases of diabetes. It also is called noninsulin-dependent diabetes mellitus (NIDDM) to differentiate it from insulin-dependent diabetes mellitus (IDDM). It usually occurs in middle-aged to older obese people. Noninsulin-dependent diabetes is characterized by hyperglycemia, often with hypertriglyceridemia and other features of syndrome X (see Clin. Corr. 20.1). The ketoacidosis characteristic of the insulin-dependent disease is not observed. Increased levels of VLDL are probably the result of increased hepatic triacylglycerol synthesis stimulated by hyperglycemia and hyperinsulinemia. Insulin is present at normal to elevated levels in this form of the disease. Obesity often precedes the development of insulin-independent diabetes and appears to be the major contributing factor. Obese patients are usually hyperinsulinemic. Recent data implicate increased levels of expression of tumor necrosis factor α (TNF-α) and a new protein called resistin by adipocytes of obese individuals as a cause of the resistance. The greater the adipose tissue mass, the greater the production of TNF-α and resistin, which act to impair insulin receptor function. An inverse relationship between insulin levels and the number of insulin receptors has been established. The higher the basal level of insulin, the fewer receptors present on the plasma membranes. In addition, there are defects within insulin-responsive cells at sites beyond the receptor. An example is the ability of insulin to recruit glucose transporters (GLUT-4) from intracellular sites to the plasma membrane. As a consequence, insulin levels remain high, but glucose levels are poorly controlled because of the lack of normal responsiveness to insulin. Although the insulin level is high, it is not as high as in a person who is obese but not diabetic. In other words, there is a relative deficiency in the insulin supply from the β cells. This disease is caused, therefore, not only by insulin resistance but also by impaired β-cell function, resulting in relative insulin deficiency. Diet alone can control the disease in the obese diabetic. If the patient can be motivated to lose weight, insulin receptors will increase in number, and the postreceptor abnormalities will improve, which will increase both tissue sensitivity to insulin and glucose tolerance. The type 2 diabetic tends not to develop ketoacidosis but nevertheless develops many of the same complications as the insulin-dependent diabetic, that is, nerve, eye, kidney, and coronary artery disease.

Kahn, B. B. and Flier, J. S. Obesity and insulin resistance. *J. Clin. Invest.* 106:473, 2000; Cavaghan, M. K., Ehrmann, D. A., and Polonsky, K. S. Interactions between insulin resistance and insulin secretion in the development of glucose intolerance. *J. Clin. Invest.* 106:329, 2000; and Steppan, C. M., Bailey, S. T., Bhat, S., Brown, E. J., Banerjee, R. R., Wright, C. M., Patel, H. R., Shima, R. S., and Lazar, M. A. The hormone resistin links obesity to diabetes. *Nature* 409:307, 2001.

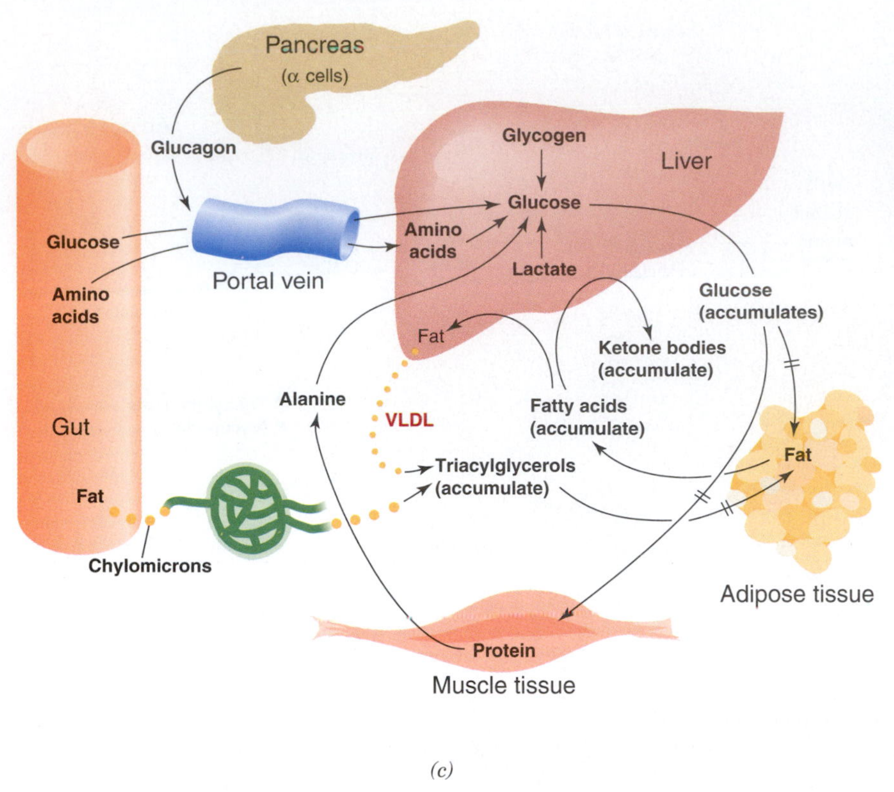

FIGURE 20.17 (continued)
(*c*) Insulin-dependent diabetes mellitus.

an increase in VLDL without hyperchylomicronemia. This is most likely explained by rapid rates of *de novo* hepatic synthesis of fatty acids and VLDL rather than increased delivery of fatty acids from the adipose tissue.

Type 1 Diabetes Mellitus

Figure 20.17*c* shows the metabolic interrelationships that exist in **type 1 or insulin-dependent diabetes mellitus** (see Clin. Corr. 20.8 and 20.9). In contrast to type

CLINICAL CORRELATION 20.8
Type 1 (Insulin-Dependent) Diabetes Mellitus

Type 1 or insulin-dependent diabetes mellitus (IDDM) was once called juvenile-onset diabetes because it usually appears in childhood or in the teens, but it is not limited to these patients. Insulin is either absent or nearly absent in this disease because of defective or absent β cells in the pancreas. The β cells are destroyed by an autoimmune process. Untreated, type 1 diabetes is characterized by hyperglycemia, hyperlipoproteinemia (chylomicrons and VLDL), and episodes of severe ketoacidosis. Far from being a disease of defects in carbohydrate metabolism alone, diabetes causes abnormalities in fat and protein metabolism in such patients as well. The hyperglycemia results in part from the inability of the insulin-dependent tissues to take up plasma glucose and in part by accelerated hepatic gluconeogenesis from amino acids derived from muscle protein. The ketoacidosis results from increased lipolysis in the adipose tissue and accelerated fatty acid oxidation in the liver. Hyperchylomicronemia is the result of low lipoprotein lipase activity in adipose tissue capillaries, an enzyme dependent on insulin for its synthesis.

Although insulin does not cure the diabetes, its use markedly alters the clinical course of the disease. The injected insulin promotes glucose uptake by tissues and inhibits gluconeogenesis, lipolysis, and proteolysis. The patient has the difficult job of trying to adjust the insulin dose to a variable dietary intake and variable physical activity, the other major determinant of glucose disposal by muscle. Tight control demands the use of several injections of insulin per day and close blood sugar monitoring by the patient. Tight control of blood sugar has now been proved to reduce the microvascular complications of diabetes (renal and retinal disease).

Atkinson, M. A. and Maclaren, N. K. The pathogenesis of insulin dependent diabetes mellitus. *N. Engl. J. Med.* 331:1428, 1994; Clark, C. M. and Lee, D. A. Prevention and treatment of the complications of diabetes mellitus. *N. Engl. J. Med.* 332:1210, 1994; Luppi, P. and Trucco, M. Immunolgical models of type 1 diabetes. *Horm. Res.* 52:1, 1999; and Kukreja, A. and Maclaren, N. K. Autoimmunity and diabetes. *J. Clin. Endocrinol. Metab.* 84:4371, 1999.

CLINICAL CORRELATION 20.9

Complications of Diabetes and the Polyol Pathway

Diabetes is complicated by several disorders that may share a common pathogenesis. The lens, peripheral nerve, renal papillae, Schwann cells, glomerulus, and possibly retinal capillaries contain two enzymes that constitute the polyol pathway (the term polyol refers to polyhydroxy sugars). The first is aldose reductase, an NADPH-requiring enzyme. It reduces glucose to form sorbitol. Sorbitol is further metabolized by sorbitol dehydrogenase, an NAD^+-requiring enzyme that oxidizes sorbitol to fructose. Aldose reductase has a high K_m for glucose; therefore this pathway is only quantitatively important during hyperglycemia. It is known that in diabetic animals the sorbitol content of lens, nerve, and glomerulus is elevated. Sorbitol accumulation may damage these tissues by causing them to swell. There are now inhibitors of the reductase that prevent the accumulation of sorbitol in these tissues and thus retard the onset of these complications. This is a very controversial area because differences in potency of the inhibitors, experimental designs, and length of trials have resulted in different studies reaching different conclusions. A recent metanalysis concluded that these drugs slow the progression of damage to motor nerves, but they have not been shown to ameliorate the damage to sensory nerves. The latter is important because of the condition known as painful diabetic neuropathy involving these nerves.

Frank, R. N. The aldose reductase controversy. *Diabetes* 43:169, 1994; King, G. L. and Brownlee, M. The cellular and molecular mechanisms of diabetic complications. *Endocrinol. Metab. Clin. North Am.* 25:255, 1996; Clark, C. M. and Lee, D. A. Prevention and treatment of the complications of diabetes mellitus. *N. Engl. J. Med.* 332:1210, 1994; and Airey, M., Bennett, C., Nicolucci, A., and Williams, R. Aldose reductase inhibitors for the prevention and treatment of diabetic peripheral neuropathy. *Cochrane Database Syst. Rev.* CD002:182, 2000.

2 diabetes, there is a complete absence of insulin production by the pancreas in this disease. Because of defective β-cell production of insulin, blood levels of insulin do not increase in response to elevated blood glucose levels. Even when dietary glucose is being delivered from the gut, the insulin/glucagon ratio cannot increase, and the liver remains gluconeogenic and ketogenic. Since it is impossible to switch to the processes of glycolysis, glycogenesis, and lipogenesis, the liver cannot properly buffer blood glucose levels. Indeed, since hepatic gluconeogenesis is continuous, the liver contributes to hyperglycemia in the well-fed state. Failure of some tissues, especially muscle and adipose tissue, to take up glucose in the absence of insulin because **GLUT-4** remains sequestered within cells, contributes greatly to the hyperglycemia. Accelerated gluconeogenesis, fueled by substrate made available by tissue protein degradation, maintains the hyperglycemia even in the starved state.

The absence of insulin in patients with type 1 diabetes mellitus results in uncontrolled rates of lipolysis in adipose tissue. This increases blood levels of fatty acids and results in accelerated ketone body production by the liver. If ketone bodies are not used as rapidly as they are formed, diabetic ketoacidosis develops due to accumulation of ketone bodies and hydrogen ions. The pathway of fatty acid oxidation and ketogenesis cannot handle the quantity of fatty acids taken up by liver. The excess is esterified and directed into VLDL synthesis. **Hypertriacylglycerolemia** results because VLDL is synthesized and released by the liver more rapidly than these particles can be cleared from the blood by lipoprotein lipase. The expression of this enzyme is dependent on insulin, and as a result, untreated type 1 diabetics often have hyperchylomicronemia. In summary, in diabetes every tissue continues to play the catabolic role that it was designed to play in starvation, in spite of delivery of adequate or even excess fuel from the gut. This results in a gross elevation of all fuels in the blood with severe wasting of body tissues and ultimately in death unless insulin is administered, hence the term "insulin-dependent": these patients will die if not given insulin.

Aerobic and Anaerobic Exercise Require Different Fuels

It is important to differentiate between two distinct types of **exercise—aerobic and anaerobic.** Aerobic exercise is exemplified by long-distance running, anaerobic exercise by sprinting or weight lifting. During anaerobic exercise there is really very little interorgan cooperation. The blood vessels within the muscles are compressed during peak contraction; thus their cells are isolated from the rest of the body. Muscle largely relies on its own stored glycogen and phosphocreatine. **Phosphocreatine** serves as a source of high-energy phosphate for ATP synthesis (Figure 20.7) until glycogenolysis and glycolysis are stimulated. Glycolysis becomes the primary source of ATP for want of oxygen. Aerobic exercise is metabolically more interesting (Figure 20.17*d*). For moderate exercise, much of the energy is derived from glycolysis of muscle glycogen. This biochemical fact is the basis for **carbohydrate loading.** Muscle glycogen content can be increased by exhaustive exercise that depletes glycogen, followed by rest and a high-carbohydrate diet. Glucose uptake from the blood is also increased as a consequence of translocation of GLUT-4 to the plasma membrane, a process that occurs independently of insulin during exercise. There is also stimulation of branched-chain amino acid oxidation, ammonium production, and alanine release from the exercising muscle. However, a well-fed individual does not store enough glycogen and cannot take up enough glucose to provide the energy needed for running long distances. The **respiratory quotient,** the ratio of carbon dioxide exhaled to oxygen consumed, falls during distance running. This indicates the progressive switch from glycogen to fatty acid oxidation during a race. Lipolysis gradually increases as glucose stores are exhausted, and, as in the fasted state, muscles oxidize fatty acids in preference to glucose as the former become available. The rate of fatty acid oxidation also increases as a consequence of a decrease in malonyl CoA concentration, which allows greater carnitine palmitoyltransferase I activity. A decrease in ATP due to the demand of muscle contraction results in an increase in AMP, a positive allosteric effector for AMP-activated protein kinase, which in turn phosphorylates and inactivates acetyl-CoA carboxylase. This, along with an increase in long-chain acyl CoA esters, which are negative allosteric effectors of acetyl-CoA carboxylase, result in a lower steady-state concentration of malonyl CoA and therefore greater carnitine palmitoyltransferase I activity and fatty acid oxidation. Unlike fasting, there is little increase in blood ketone body concen-

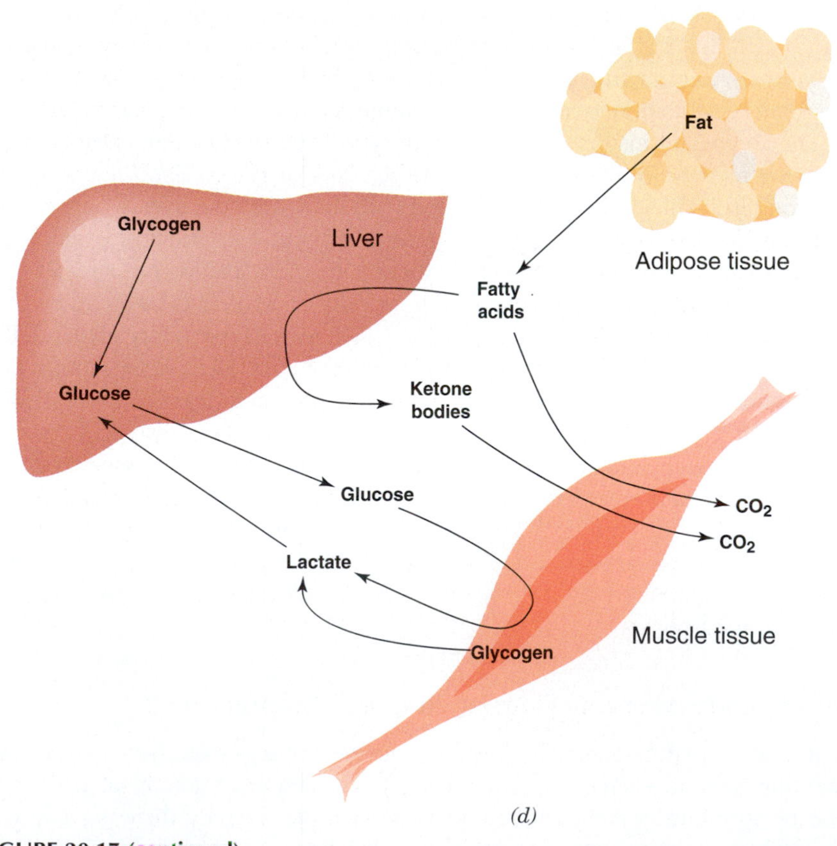

FIGURE 20.17 (continued)
(*d*) Exercise.

tration during exercise. This may reflect a balance between hepatic ketone body synthesis and muscle ketone body oxidation.

Changes in Pregnancy Are Related to Fetal Requirements and Hormonal Changes

The fetus can be considered as another nutrient-requiring tissue (Figure 20.17*e*). It mainly uses glucose for energy but may also use amino acids, lactate, fatty acids, and ketone bodies. Lactate produced in the **placenta** by glycolysis has two fates: part is directed to the fetus to be used as a fuel and the rest returns to the maternal circulation to establish a Cori cycle with the liver. Maternal LDL cholesterol is an important precursor of placental steroids (estradiol and progesterone). During **pregnancy,** the starve–feed cycle is perturbed. The placenta secretes a polypeptide hormone, **placental lactogen,** and two steroid hormones, estradiol and progesterone. Placental lactogen stimulates lipolysis in adipose tissue, and the steroid hormones induce an insulin-resistant state. After meals, pregnant women enter the starved state more rapidly than do nonpregnant women as a result of increased consumption of glucose and amino acids by the fetus. Plasma glucose, amino acids, and insulin levels fall rapidly, and glucagon and placental lactogen levels rise and stimulate lipolysis and ketogenesis. The consumption of glucose and amino acids by the fetus may be great enough to cause maternal hypoglycemia. On the other hand, in the fed state pregnant women have increased levels of insulin and glucose and demonstrate resistance to exogenous insulin. These swings of plasma hormones and fuels are even more exaggerated in pregnant diabetic women and make control of their blood glucose difficult. This is important, as hyperglycemia adversely affects fetal development.

Lactation Requires Synthesis of Lactose, Triacylglycerols, and Protein

In late pregnancy placental hormones induce lipoprotein lipase in the mammary gland and promote the development of milk-secreting cells and ducts. During

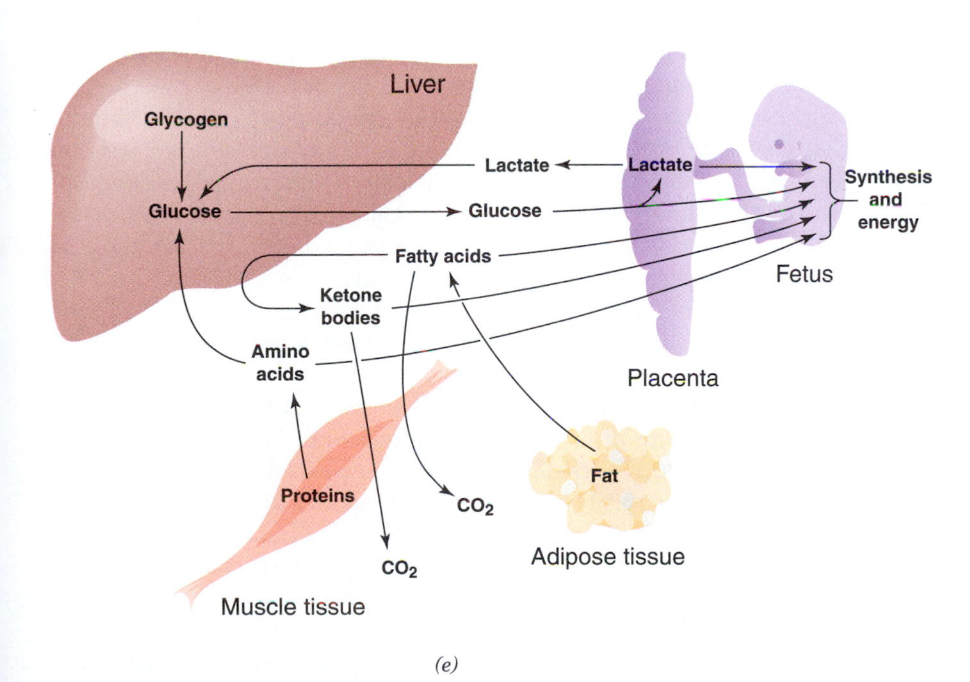

FIGURE 20.17 (continued)
(*e*) Pregnancy.

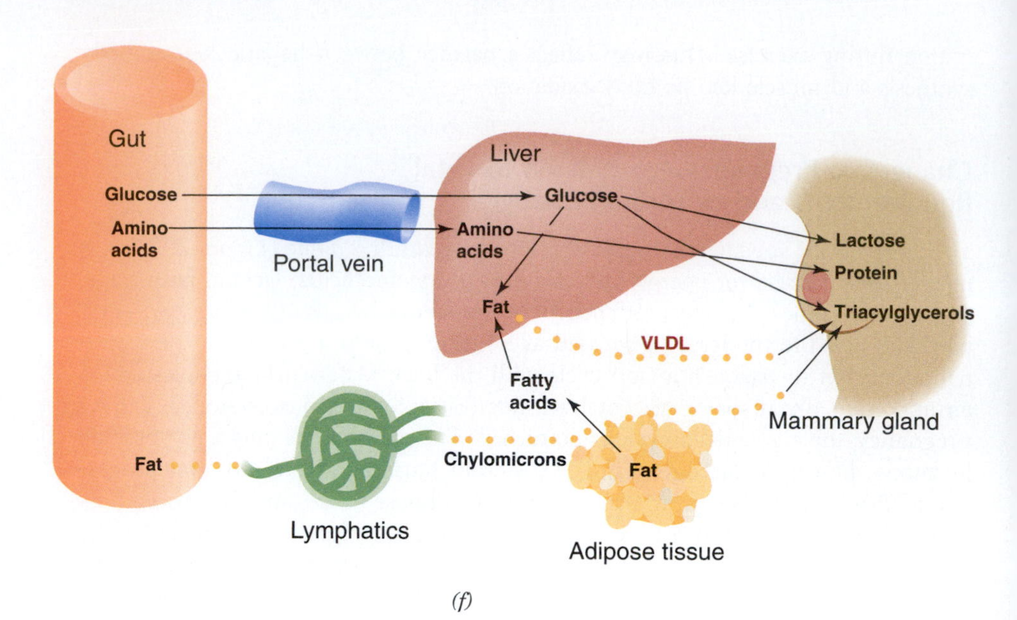

FIGURE 20.17 (continued)
(*f*) Lactation.

lactation (see Figure 20.17*f*) the breast utilizes glucose for **lactose** and triacylglycerol synthesis, as well as its major energy source. Amino acids are taken up for protein synthesis, and chylomicrons and VLDL are utilized as sources of fatty acids for triacylglycerol synthesis. If these compounds are not supplied by the diet, proteolysis, gluconeogenesis, and lipolysis must supply them, resulting eventually in maternal malnutrition and poor quality milk. The lactating breast also secretes a hormone with some similarity to parathyroid hormone (see p. 915). This hormone probably is important for the absorption of calcium and phosphorus from the gut and bone.

Stress and Injury Lead to Metabolic Changes

Physiological stresses include **injury, surgery, renal failure, burns,** and **infections** (Figure 20.17*g*). Characteristically, blood cortisol, glucagon, **catecholamines,** and **growth hormone** levels increase. The patient is resistant to insulin. Basal metabolic rate and blood glucose and free fatty acid levels are elevated. However, ketogenesis is not accelerated as in fasting. For incompletely understood reasons, the intracellular muscle glutamine pool is reduced, resulting in reduced protein synthesis and increased protein breakdown. It can be very difficult to reverse this protein breakdown, although now it is common to replace amino acids, glucose, and triacylglycerol by infusing solutions of these nutrients intravenously. However, these solutions lack glutamine, tyrosine, and cysteine because of stability and solubility constraints. Supplementation of these amino acids, perhaps by the use of more stable dipeptides, may help to reverse the catabolic state better than can be accomplished at present.

Negative nitrogen balance of injured or infected patients is mediated by monocyte and lymphocyte proteins, such as **interleukin-1, interleukin-6,** and **TNF-α** (see Clin. Corr. 20.10). These cytokines are responsible for causing fever as well as a number of other metabolic changes. Interleukin-1 activates proteolysis in skeletal muscle. Interleukin-6 stimulates the synthesis of a number of hepatic proteins called acute phase reactants by the liver. **Acute phase reactants** include fibrinogen, complement proteins, some clotting factors, and α_2-macroglobulin, which are presumed to play a role in defense against injury and infection. TNF-α suppresses adipocyte triacylglycerol synthesis, prevents uptake of circulating triacylglycerol by inhibiting lipoprotein lipase, stimulates lipolysis, inhibits release of insulin, and promotes insulin resistance. These cytokines appear responsible for much of the wasting seen in chronic infections.

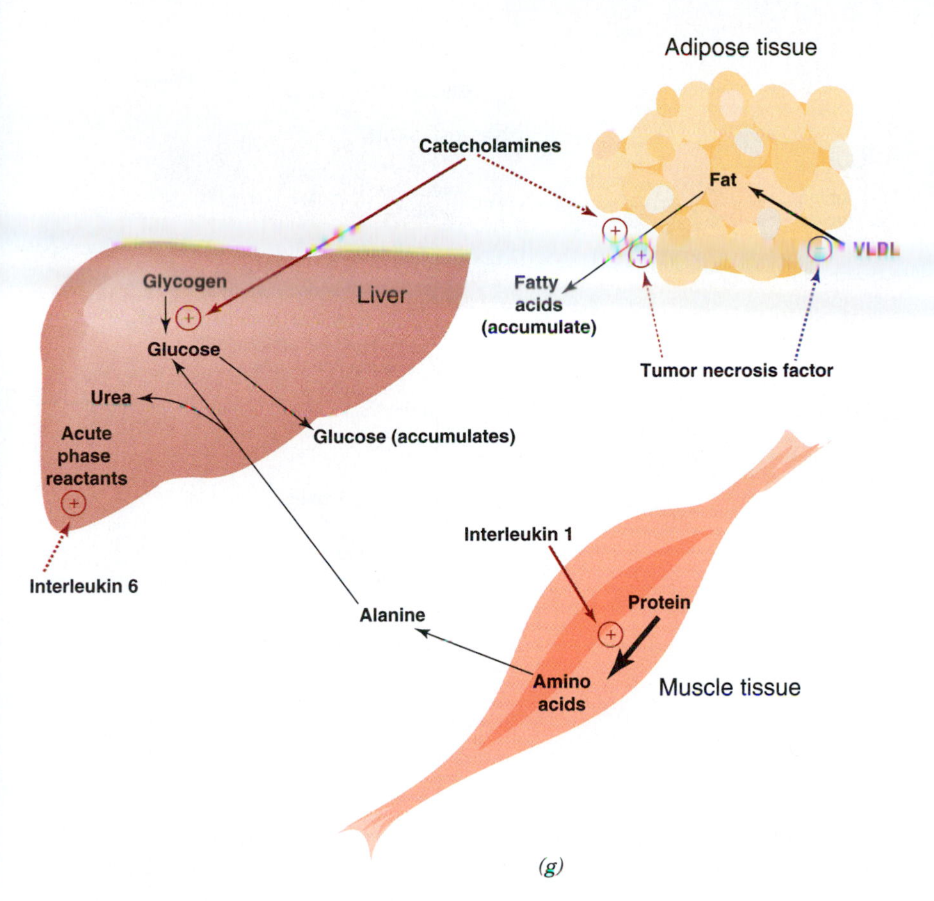

FIGURE 20.17 (continued)
(*g*) Stress.

Liver Disease Causes Major Metabolic Derangements

Since the liver is central to the body's metabolic interrelationships, advanced **liver disease** can be associated with major metabolic derangements (Figure 20.17*h*). The most important abnormalities are those in the metabolism of amino acids. The liver is the only organ capable of **urea synthesis.** In patients with **cirrhosis,** the liver is

CLINICAL CORRELATION 20.10
Cancer Cachexia

Unexplained weight loss may be a sign of malignancy, and weight loss is common in advanced cancer. Decreased appetite and food intake contribute to but do not entirely account for the weight loss. The weight loss is largely from skeletal muscle and adipose tissue, with relative sparing of visceral protein (i.e., liver, kidney, and heart). Although tumors commonly exhibit high rates of glycolysis and release lactate, the energy requirement of the tumor probably does not explain weight loss because weight loss can occur with even small tumors. In addition, the presence of another energy-requiring growth, the fetus in a pregnant woman, does not normally lead to weight loss. Several endocrine abnormalities have been recognized in cancer patients. They tend to be insulin-resistant, have higher cortisol levels, and have a higher basal metabolic rate compared with controls matched for weight loss. Two other phenomena may contribute to the metabolic disturbances. Some tumors synthesize and secrete biologically active peptides such as ACTH, nerve growth factor, and insulin-like growth factors, which could modify the endocrine regulation of energy metabolism. It is also possible that the host response to a tumor, by analogy to chronic infection, includes release of interleukin-1 (IL-1), interleukin 6 (IL-6), and tumor necrosis factor α (TNF-α) by cells of the immune system. TNF-α is also called cachexin because it produces wasting. TNF-α and IL-1 may act in a paracrine fashion, as plasma levels are not elevated. They do induce the synthesis of IL-6, which has been detected in cachectic patients' blood at increased levels. These cytokines stimulate fever, proteolysis, lipolysis, and the synthesis of acute phase reactants by the liver. It seems likely that additional, as yet uncharacterized, cytokines play a role in cancer cachexia, and that the mechanism for this phenomenon may vary with different tumor types.

Beutler, B. and Cerami, A. Tumor necrosis, cachexia, shock, and inflammation: a common mediator. *Annu. Rev. Biochem.* 57:1505, 1988; Tracey, K. J. and Cerami, A. Tumor necrosis factor: a pleiotropic cytokine and therapeutic target. *Annu. Rev. Med.* 45:491, 1994; Nitenberg, G. and Raynard, B. Nutritional support of the cancer patient: issues and dilemmas. *Crit. Rev. Oncol. Hematol.* 34:137, 2000; Tisdale, M. J. Cancer cachexia: metabolic alterations and clinical manifestations. *Nutrition* 13:1, 1997; and Tisdale, M. J. New cachexic factors. *Curr. Opin. Clin. Metab. Care* 1:253, 1998.

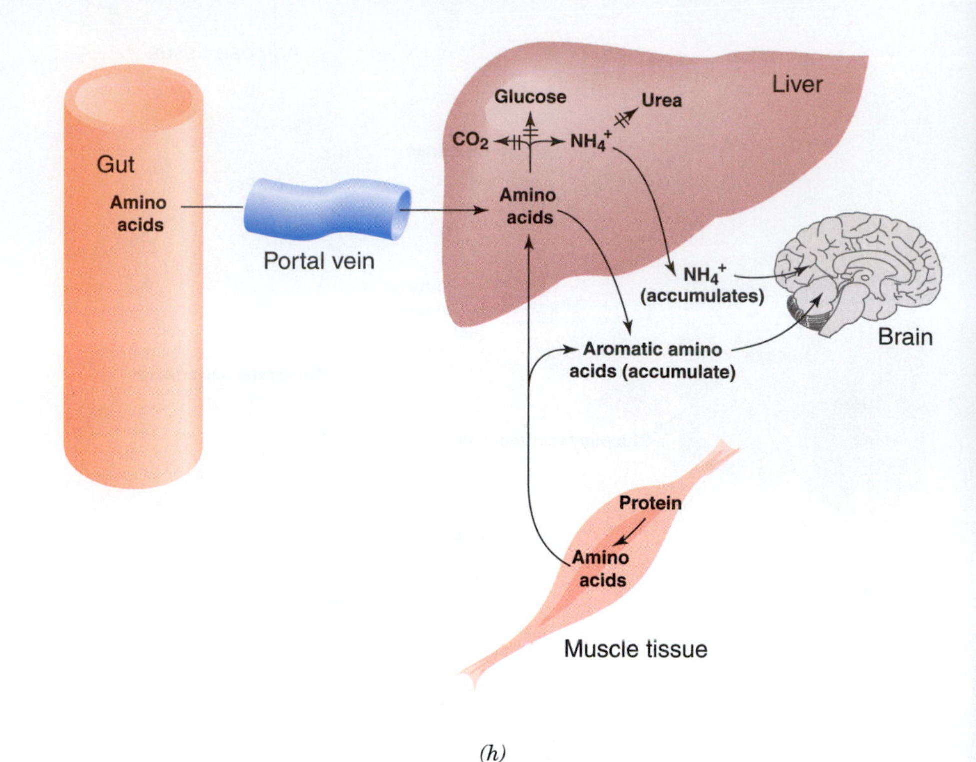

FIGURE 20.17 (continued)
(*h*) Liver disease.

unable to convert ammonia into urea and glutamine rapidly enough, and the blood ammonia level rises. Part of this problem is due to abnormalities of blood flow in the cirrhotic liver, which interferes with the intercellular **glutamine cycle** (see p. 898). Ammonia arises from several enzyme reactions, such as glutaminase, glutamate dehydrogenase, and adenosine deaminase, during metabolism in intestine and liver, and from intestinal lumen, where bacteria split urea into ammonia and carbon dioxide. Ammonia is very toxic to the central nervous system and is a major reason for the coma that sometimes occurs in patients in liver failure.

In advanced liver disease, **aromatic amino acids** accumulate in the blood to higher levels than branched-chain amino acids, apparently because of defective hepatic catabolism of the aromatic amino acids. This is important because aromatic amino acids and branched-chain amino acids are transported into the brain by the same carrier system. An elevated ratio of aromatic amino acids to branched-chain amino acids in liver disease results in increased brain uptake of aromatic amino acids. This may increase synthesis of **neurotransmitters** such as **serotonin,** which may be responsible for some of the neurological abnormalities characteristic of liver disease. The liver is also a major source of insulin-like growth factor I (IGF-I). Cirrhotics suffer muscle wasting because of deficient IGF-I synthesis in response to growth hormone. They also often demonstrate insulin resistance and may exhibit diabetes. Finally, in outright liver failure, patients sometimes die of hypoglycemia because the liver is unable to maintain the blood glucose level by gluconeogenesis.

In Renal Disease Nitrogenous Wastes Accumulate

In chronic renal disease, there are many abnormalities of nitrogen metabolism. Levels of amino acids normally metabolized by kidney (glutamine, glycine, proline, and citrulline) increase. Nitrogen end products, for example, urea, uric acid, and creatinine, also accumulate (Figure 20.17*i*). This accumulation is worsened by high dietary protein intake or accelerated proteolysis. The fact that gut bacteria can split urea into ammonia and that liver uses ammonia and α-keto acids to form nonessential amino acids has been used to control the level of nitrogenous wastes in renal

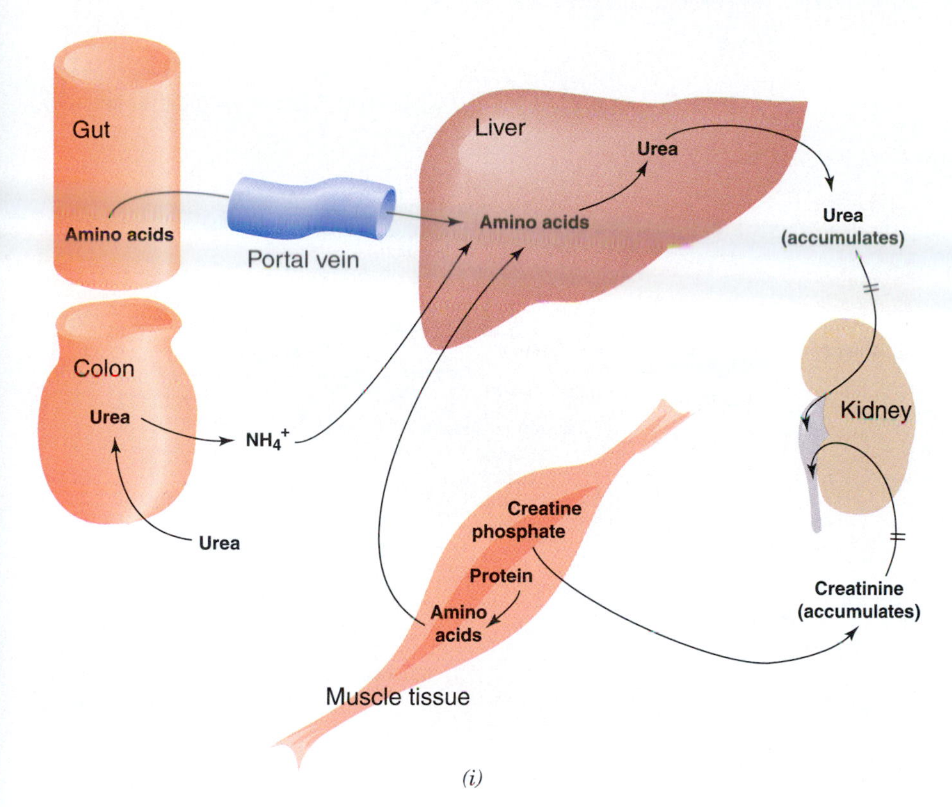

FIGURE 20.17 (continued)
(i) Kidney failure.

patients. Patients are given a diet high in carbohydrate, and the amino acid intake is limited as much as possible to essential amino acids. Under these circumstances, the liver synthesizes nonessential amino acids from TCA cycle intermediates. This type of diet therapy may extend the time before the patient requires dialysis.

Oxidation of Ethanol in Liver Alters the NAD^+/NADH Ratio

The liver is primarily responsible for the first two steps of the **ethanol** catabolism:

$$\text{ethanol } (CH_3CH_2OH) + NAD^+ \rightarrow \text{acetaldehyde } (CH_3CHO) + NADH + H^+$$

$$\text{acetaldehyde } (CH_3CHO) + NAD^+ + H_2O \rightarrow \text{acetate } (CH_3COO^-) + NADH + H^+$$

The first step, catalyzed by **alcohol dehydrogenases** in the cytosol, generates NADH; the second step, catalyzed by **aldehyde dehydrogenase,** also generates NADH but occurs largely in the mitochondrial matrix space. Liver disposes of NADH generated by these reactions by the only pathway it has available—the mitochondrial electron-transport chain. Intake of even moderate amounts of ethanol generates too much NADH. Many enzymes, for example, several involved in gluconeogenesis and fatty acid oxidation, are sensitive to product inhibition by NADH. Thus, during alcohol metabolism, these pathways are inhibited (Figure 20.17*j*), and fasting hypoglycemia and the accumulation of hepatic triacylglycerols (fatty liver) are consequences of alcohol ingestion. Lactate can accumulate as a consequence of inhibition of lactate gluconeogenesis but rarely causes overt metabolic acidosis.

Liver mitochondria have a limited capacity to oxidize acetate to CO_2, because the activation of acetate to acetyl CoA requires GTP, a product of the succinyl-CoA synthetase reaction. The TCA cycle and, therefore, GTP synthesis are inhibited by high NADH levels during ethanol oxidation. Much of the acetate made from ethanol escapes the liver to the blood. Virtually every other cell with mitochondria can oxidize it to CO_2 by way of the TCA cycle. Acetaldehyde, the intermediate in the formation of acetate from ethanol, can also escape from the liver. Acetaldehyde is a reactive compound that readily forms covalent bonds with functional groups of biologically important compounds. Formation of acetaldehyde adducts with proteins

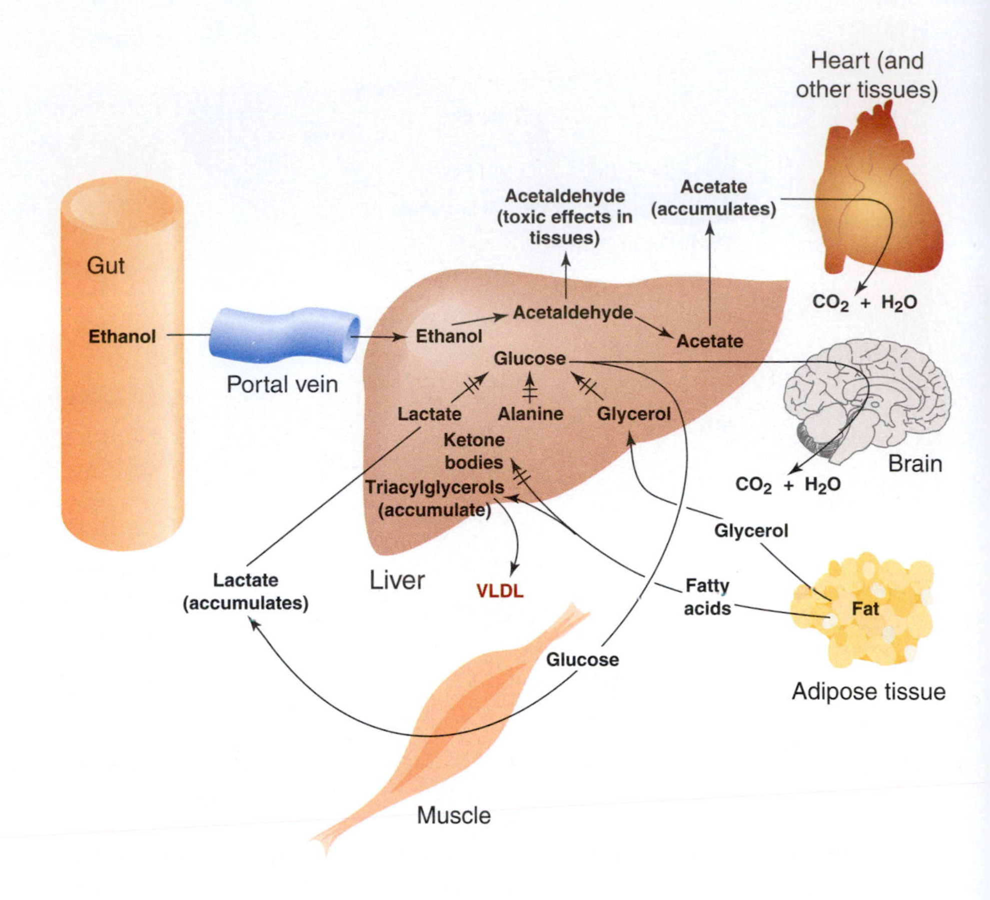

FIGURE 20.17 (continued)
(*j*) Ethanol ingestion.

in liver and blood of animals and humans drinking alcohol has been demonstrated. Such adducts may provide a marker for past drinking activity of an individual, just as hemoglobin A_{1c} has proved useful as an index of blood glucose control in diabetic patients.

In Acid–Base Regulation, Glutamine Plays a Pivotal Role

Regulation of **acid–base balance**, like that of nitrogen excretion, is shared by the liver and kidney. Although complete catabolism of most of the amino acids derived from proteins yields neutral products (CO_2, H_2O, and urea), this is not the case for positively charged amino acids (**arginine**, **lysine**, and **histidine**) and sulfur-containing amino acids (**methionine** and **cysteine**), the metabolism of which result in net formation of protons (acid). Although catabolism of negatively charged amino acids (**glutamate**, **aspartate**) results in net utilization of protons, this consumes and therefore balances only part of the protons produced by catabolism of positively charged and sulfur-containing amino acids. Therefore, in order for the body to remain in acid–base balance, the excess protons produced each day by catabolism of proteins must be matched by an equivalent amount of base. The role of the kidney in achieving this balance is critically important. Glutamine, readily taken up by kidney cells, is deaminated by glutaminase to give glutamate, which is oxidatively deaminated by glutamate dehydrogenase to give α-ketoglutarate, which is converted by enzymes of the TCA cycle to malate, which is converted by enzymes of gluconeogenesis to glucose. Summation of all steps reveals net production of glucose and, more importantly, ammonium and bicarbonate ions:

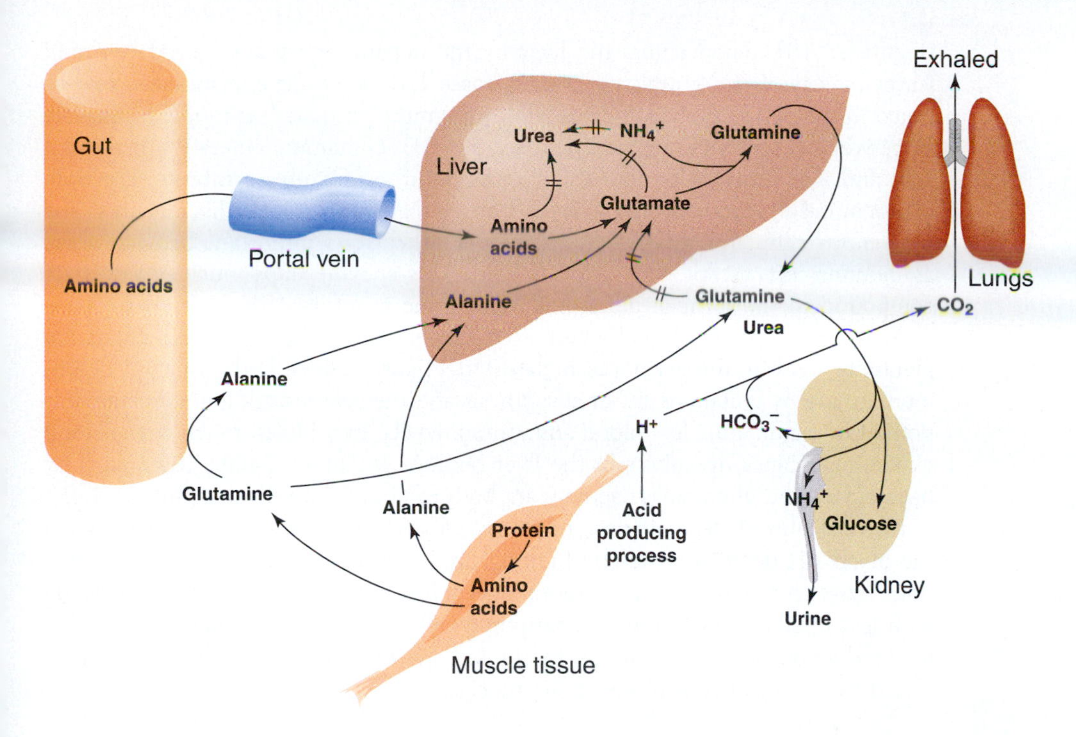

FIGURE 20.17 (continued)
(*k*) Acidosis.

$$\text{glutamine} + 1\tfrac{1}{2}\,O_2 + 3H_2O \rightarrow \tfrac{1}{2}\,\text{Glucose} + 2HCO_3^- + 2NH_4^+$$

The ammonium ions are transported into the glomerular filtrate while the bicarbonate ions are transported into the blood to neutralize the protons produced by amino acid catabolism:

$$HCO_3^- + H^+ \rightleftharpoons H_2CO_3 \rightleftharpoons CO_2 + H_2O$$

CO_2 produced by this reaction is readily blown off in the lungs, thereby effectively eliminating excess acid produced by amino acid oxidation. This mechanism assumes even greater importance in **metabolic acidosis** (see Figure 20.17*k*), a condition in which much more acid is being produced in the body because some metabolic process is out of control, for example, lactic acid production by anaerobic glycolysis or β-hydroxybutyric acid production by ketogenesis. In metabolic acidosis the activities of renal glutaminase, glutamate dehydrogenase, phospho*enol*pyruvate carboxykinase, and the mitochondrial glutamine transporter increase and correlate with increased renal gluconeogenesis from glutamine. This is because adaptive changes occur in the amounts of these enzymes in the kidney that greatly increase the kidney's capacity for glucose synthesis from glutamine by the pathway outlined above. The consequence is increased urinary excretion of ammonium ions and greater generation of bicarbonate ion for neutralization of the acid that is being produced. The liver does its part in this situation by synthesizing less urea, thereby making more glutamine available for the kidney. The opposite occurs in **alkalosis;** that is, urea synthesis increases in the liver, and gluconeogenesis and ammonium ion excretion (and bicarbonate generation) by the kidney decrease.

An intercellular glutamine cycle is an important feature of liver metabolism that enables this tissue to play a central role in the regulation of blood pH. The liver is composed of two types of hepatocytes, both of which are uniquely involved in glutamine metabolism: **periportal hepatocytes** near the hepatic arteriole and portal venule and **perivenous** scavenger **hepatocytes** located near the central venule

(Figure 20.18). Blood enters the liver by the hepatic artery and portal vein and leaves by way of the central vein. Glutaminase and urea cycle enzymes are concentrated in the periportal hepatocytes while glutamine synthetase is found exclusively in perivenous scavenger hepatocytes (see p. 784). Glutamine enters the periportal cells and is hydrolyzed to contribute ammonium ion for urea synthesis. The bulk of glutamine and ammonium nitrogen entering the liver leaves the liver as urea. The perivenous cellular location of glutamine synthetase is important because some ammonium ions escape conversion to urea. This enzyme traps much of this toxic compound in the form of glutamine. Glutamine is released from the perivenous scavenger cells into the blood where it circulates back to the liver and reenters the glutamine cycle in the periportal hepatocytes. Thus, in liver, both donation of ammonium ion by glutamine for urea synthesis and the synthesis of glutamine are important in maintaining low blood ammonium levels. In acidosis, much of the blood glutamine escapes hydrolysis in the liver because glutamine uptake (transport) by hepatocytes and glutaminase activity are both inhibited by a decrease in blood pH. Carbamoyl phosphate synthetase of periportal hepatocytes is also less active when the blood pH decreases, thereby limiting urea synthesis. This permits the perivenous scavenger cells to convert more ammonium ion to glutamine, which, along with less uptake of glutamine by periportal cells, makes more glutamine available for production of bicarbonate ion by the kidneys to neutralize the acid being produced by an out of control metabolic process.

The Colon Salvages Energy from the Diet

Unlike the small intestine, which uses glutamine for its major energy source, the **colon** utilizes **short-chain fatty acids: butyrate, propionate, isobutyrate,** and **acetate** (Figure 20.19). It obtains most of these fatty acids from the lumen of the colon, where bacteria produce them by fermentation of unabsorbed dietary components. These short-chain fatty acids would otherwise be lost in stool, so their use by cells of the colon (colonocytes) represents a way of gaining as much energy from

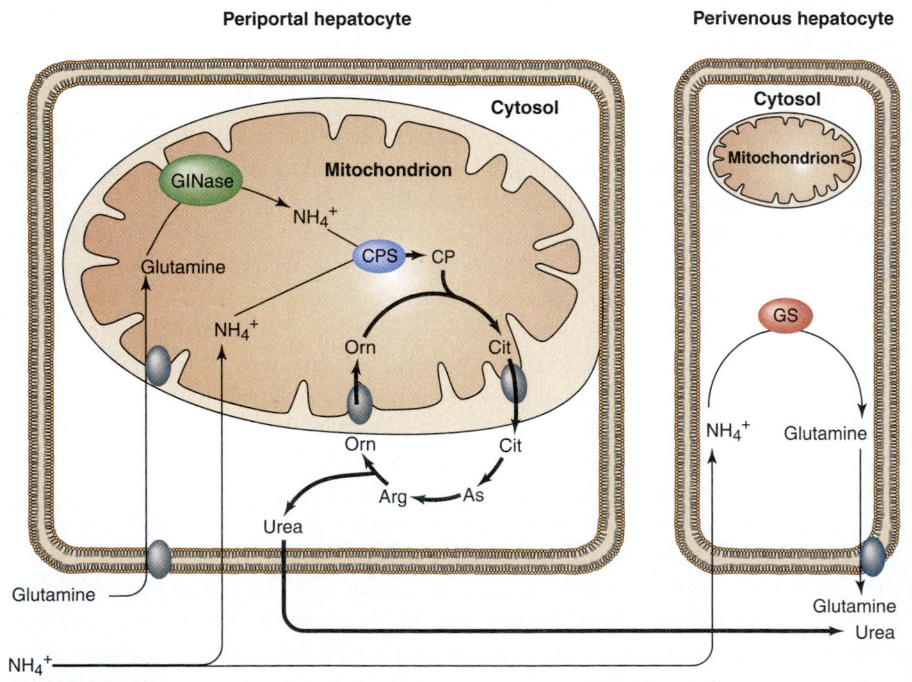

FIGURE 20.18
Intercellular glutamine cycle of the liver.
Abbreviations: GlNase, glutaminase; GS, glutamine synthetase; CPS, carbamoyl phosphate synthetase I; CP, carbamoyl phosphate; Cit, citrulline; AS, argininosuccinate; Arg, arginine; Orn, ornithine.
Redrawn from Häussinger, D. Adv. Enzymol. Relat. Areas Mol. Biol. 72:43, 1998.

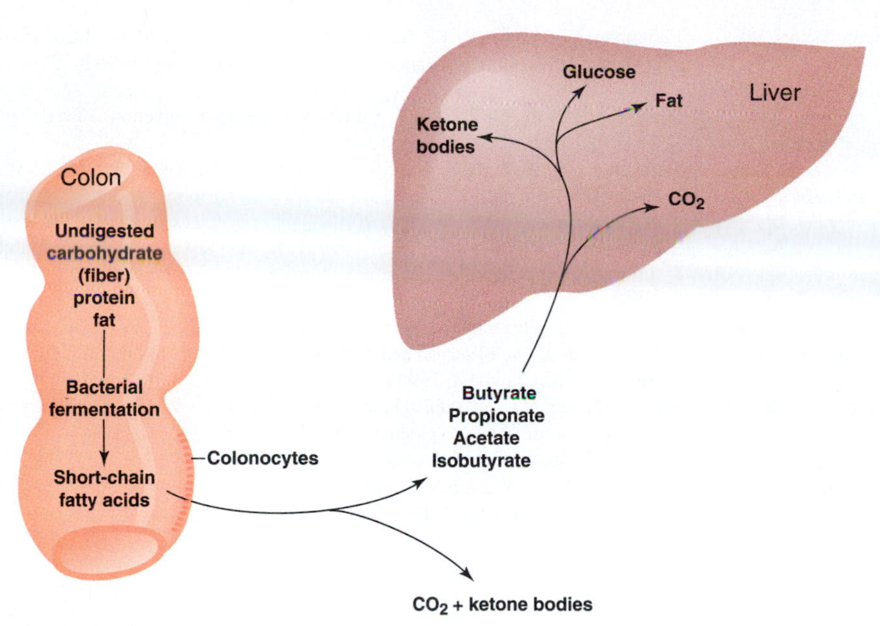

FIGURE 20.19
Bacterial fermentation generates fuel for colonocytes.

dietary sources as possible. If produced in excess of the needs of the colon, short-chain fatty acids pass into the portal blood for use by the liver. Interestingly, colonocytes can produce ketone bodies from butyrate, presumably releasing them into the portal blood for use as fuel by extrahepatic tissues. When surgery is performed that bypasses the colon (e.g., an ileostomy), some patients develop a form of **colitis** called diversion colitis. In some cases, providing enemas containing the short-chain fatty acids has healed the colitis.

BIBLIOGRAPHY

Amiel, S. A. Organ fuel selection: brain. *Proc. Nutr. Soc.* 54:151, 1995.

Brownlee, M., Vlassara, H., and Cerami, A. Nonenzymatic glycosylation and the pathogenesis of diabetes complications. *Ann. Intern. Med.* 101:527, 1984.

Calder, P. C. Fuel utilization by cells of the immune system. *Proc. Nutr. Soc.* 54:65, 1995.

Cerami, A. and Koenig, R. H. Hemoglobin A_{1c} as a model for the development of the sequelae of diabetes mellitus. *Trends Biochem. Sci.* 3:73, 1978.

Cohen, P. and Cohen, P. T. W. Protein phosphatases come of age. A review. *J. Biol. Chem.* 264:21435, 1989.

Crabb, D. W. and Lumeng, L. Metabolism of alcohol and the pathophysiology of alcoholic liver disease. In: G. Gitnick (Ed.), *Principles and Practice of Gastroenterology and Hepatology.* New York: Elsevier, 1988, p. 1163.

Curthoys, N. P. and Watford, M. Regulation of glutaminase activity and glutamine metabolism. *Annu. Rev. Nutr.* 15:133, 1995.

DeLeve, L. D. and Kaplowitz, N. Glutathione metabolism and its role in hepatotoxicity. *Pharmacol. Ther.* 52:287, 1991.

Duée, P.-H., Darcy-Vrillon, B., Blachier, F., and Morel, M.-T. Fuel selection in intestinal cells. *Proc. Nutr. Soc.* 54:83, 1995.

Elia, M. General integration and regulation of metabolism at the organ level. *Proc. Nutr. Soc.* 54:213, 1995.

Fulop, M. Alcoholism, ketoacidosis, and lactic acidosis. *Diabetes Metab. Rev.* 5:365, 1989.

Frayn, K. N., Humphreys, S. M., and Coppack, S. W. Fuel selection in white adipose tissue. *Proc. Nutr. Soc.* 54:177, 1995.

Geelen, M. J. H., Harris, R. A., Beynen, A. C., and McCune, S. A. Short-term hormonal control of hepatic lipogenesis. *Diabetes* 29:1006, 1980.

Kersten, S., Seydoux, J., Peters, J. M., Gonzalez, F. J., Desvergne, B., and Wahli, W. Peroxisome proliferator-activated receptor alpha mediates the adaptive response to fasting. *J. Clin. Invest.* 103:1489, 1999.

Harris, R. A., Huang, B., and Wu, P. Control of pyruvate dehydrogenase kinase gene expression. *Adv. Enzyme. Regul.* 41:269, 2001.

Häussinger, D. Hepatic glutamine transport and metabolism. *Adv. Enzymol. Relat. Areas Mol. Biol.* 72:43, 1998.

Henriksson, J. Muscle fuel selection: effect of exercise and training. *Proc. Nutr. Soc.* 54:125, 1995.

Hers, H. G. and Hue, L. Gluconeogenesis and related aspects of glycolysis. *Annu. Rev. Biochem.* 52:617, 1983.

Horton, J. D., Bashmakov, Y., Shimomura, I., and Shimano, H. Regulation of sterol regulatory element binding proteins in livers of fasted and refed mice. *Proc. Natl. Acad. Sci. USA* 95:5987, 1998.

Katz, J. and McGarry, J. D. The glucose paradox. Is glucose a substrate for liver metabolism? *J. Clin. Invest.* 74:1901, 1984.

Krebs, H. A. Some aspects of the regulation of fuel supply in omnivorous animals. *Adv. Enzyme Regul.* 10:387, 1972.

Krebs, H. A., Williamson, D. H., Bates, M. W., Page, M. A., and Hawkins, R. A. The role of ketone bodies in caloric homeostasis. *Adv. Enzyme Regul.* 9:387, 1971.

Kurkland, I. J. and Pilkis, S. J. Indirect and direct routes of hepatic glycogen synthesis. *FASEB J.* 3:2277, 1989.

Lecker, S. H., Solomon, V., Mitch, W. E., and Goldberg, A. L. Muscle protein breakdown and the critical role of the ubiquitin–proteasome pathway in normal and disease states. *J. Nutr.* 129:227S, 1999.

MacDonald, I. A. and Webber, J. Feeding, fasting and starvation: factors affecting fuel utilization. *Proc. Nutr. Soc.* 54:267, 1995.
McGarry, J. D. and Brown, N. F. The mitochondrial carnitine-palmitoyl-CoA transferase system. *Eur. J. Biochem.* 244:1, 1997.
Newsholme, E. A. and Leech, A. R. *Biochemistry for the Medical Sciences.* New York: Wiley, 1983.
Nosadini, R., Avogaro, A., Doria, A., Fioretto, P., Trevisan, R., and Morocutti, A. Ketone body metabolism: a physiological and clinical overview. *Diabetes Metab. Rev.* 5:299, 1989.
Ookhtens, M. and Kaplowitz, N. Role of the liver in interorgan homeostasis of glutathione and cyst(e)ine. *Semin. Liver Dis.* 18:313, 1998.
Pedersen, O. The impact of obesity on the pathogenesis of non-insulin-dependent diabetes mellitus: a review of current hypotheses. *Diabetes Metab. Rev.* 5:495, 1989.
Pedersen, O. and Beck-Nielsen, H. Insulin resistance and insulin-dependent diabetes mellitus. *Diabetes Care* 10:516, 1987.
Pilkis, S. J., Claus, T. H., Kurland, I. J., and Lange, A. J. 6-Phosphofructo-2-kinase/fructose-2,6-bisphosphatase: a metabolic signaling enzyme. *Annu. Rev. Biochem.* 64:799, 1995.
Randle, P. J. Metabolic fuel selection: general integration at the whole-body level. *Proc. Nutr. Soc.* 54:317, 1995.
Ruderman, N. B., Saha, A. K., Vavvas, D., and Witters, L. A. Malonyl CoA, fuel sensing, and insulin resistance. *Am. J. Physiol. Endocrinol. Metab.* 276:E1, 1999.
Shulman, G. I. and Landau, B. R. Pathways of glycogen repletion. *Physiol. Rev.* 72:1019, 1992.
Steppan, C. M., Bailey, S. T., Bhat, S., Brown, E. J., Banerjee, R., Wright, C. M., Patel, H. R., Ahima, R. S., and Lazar, M. A. The hormone resistin links obesity to diabetes. *Nature* 409:307, 2001.
Sugden, M. C., Holness, M. J., and Palmer, T. N. Fuel selection and carbon flux during the starved-to-fed transition. A review article. *Biochem. J.* 263:313, 1989.
Taylor, S. I. Diabetes mellitus. In: C. R. Scriver, A. L. Beaudet, W. S. Sly, and D. Vallee, (Eds.), *The Metabolic and Molecular Basis of Inherited Disease.* 7th ed. New York: McGraw-Hill, 1995.
Toth, B., Bollen, M., and Stalmans, W. Acute regulation of hepatic phosphatases by glucagon, insulin, and glucose. *J. Biol. Chem.* 263:14061, 1988.
Ulwin, P. The state of world hunger. *Nutr. Rev.* 52:151, 1994.
Williamson, D. H. and Lund, P. Substrate selection and oxygen uptake by the lactating mammary gland. *Proc. Nutr. Soc.* 54:165, 1995.

QUESTIONS | C. N. ANGSTADT

Multiple Choice Questions

1. The fact that the K_m of aminotransferases for amino acids is much higher than that of aminoacyl-tRNA synthetases means that:
   A. at low amino acid concentrations, protein synthesis will take precedence over amino acid catabolism.
   B. the liver cannot accumulate amino acids.
   C. amino acids will undergo transamination as rapidly as they are delivered to the liver.
   D. any amino acids in excess of immediate needs for energy must be converted to protein.
   E. amino acids can be catabolized only if they are present in the diet.

2. In the early refed state:
   A. the fatty acid concentration of blood rises.
   B. liver glycogenolysis continues to maintain blood glucose levels.
   C. liver replenishes its glycogen by synthesis of glucose 6-phosphate from lactate.
   D. glucose being absorbed from the gut is used primarily for fatty acid synthesis by the liver.
   E. amino acids cannot be used for protein synthesis.

3. Carnitine:
   A. is formed in all cells for their own use.
   B. is synthesized directly from free lysine.
   C. formation requires that lysyl residues in protein be methylated by *S*-adenosylmethionine.
   D. formation is inhibited by vitamin C.
   E. is cleaved to γ-butyrobetaine.

4. The largest energy reserve (in terms of kilocalories) in humans is:
   A. blood glucose.
   B. liver glycogen.
   C. muscle glycogen.
   D. adipose tissue triacylglycerol.
   E. muscle protein.

5. All of the following represent control of a metabolic process by substrate availability EXCEPT:
   A. increased urea synthesis after a high-protein meal.
   B. rate of ketogenesis.
   C. hypoglycemia of advanced starvation.
   D. response of glycolysis to fructose 2,6-bisphosphate.
   E. sorbitol synthesis.

6. Conversion of metabolically important hepatic enzymes from their nonphosphorylated form to their phosphorylated form:
   A. always activates the enzyme.
   B. is always catalyzed by a cAMP-dependent protein kinase.
   C. is signaled by insulin.
   D. is more likely to occur in the fasted than in the well-fed state.
   E. usually occurs at threonine residues of the protein.

7. Changing the level of enzyme activity by changing the number of enzyme molecules:
   A. is considerably slower than allosteric or covalent modification methods.
   B. may involve enzyme induction.
   C. may override the effectiveness of allosteric control.
   D. may be caused by hormonal influences or by changing the nutritional state.
   E. all of the above are correct.

8. Muscle metabolism during exercise:
   A. is the same in both aerobic and anaerobic exercise.
   B. shifts from primarily glucose to primarily fatty acids as fuel during aerobic exercise.
   C. uses largely glycogen and phosphocreatine in the aerobic state.
   D. causes a sharp rise in blood ketone body concentration.
   E. uses only phosphocreatine in the anaerobic state.

9. In type 2 (noninsulin-dependent) diabetes mellitus:
   A. hypertriglyceridemia does not occur.
   B. ketoacidosis in the untreated state is always present.
   C. β cells of the pancreas are no longer able to make any insulin.
   D. may be accompanied by high levels of insulin in the blood.
   E. severe weight loss always occurs.

10. Glutaminase:
    A. in renal cells is unaffected by blood pH.

B. in liver is confined to perivenous hepatocytes.
C. requires ATP for the reaction it catalyzes.
D. is more active in liver in acidosis.
E. in renal cells increases in acidosis.

Questions 11 and 12: Protein and calorie malnutrition are two important nutritional problems, especially among children. Both often occur in developing countries when the child is weaned from breast milk. Protein malnutrition, kwashiorkor, occurs when the child is fed a diet adequate in calories (mostly carbohydrate) but deficient in protein. Inadequate caloric intake is called marasmus. In kwashiorkor, insulin levels are high and there is subcutaneous fat. Children with low weight for height can make a good recovery when properly fed although those with the reverse situation do not. Children with marasmus lack subcutaneous fat.

11. Adipose tissue responds to low insulin/glucagon ratio by:
    A. dephosphorylating the interconvertible enzymes.
    B. stimulating the deposition of fat.
    C. increasing the amount of pyruvate kinase.
    D. activation of hormone-sensitive lipase.
    E. stimulating phenylalanine hydroxylase.

12. Which of the following would favor gluconeogenesis in the fasted state?
    A. fructose 1,6-bisphosphate stimulation of pyruvate kinase
    B. acetyl CoA activation of pyruvate carboxylase
    C. citrate activation of acetyl-CoA carboxylase
    D. malonyl CoA inhibition of carnitine palmitoyltransferase I
    E. fructose 2,6-bisphosphate stimulation of 6-phosphofructo-1-kinase

Questions 13 and 14: Reye's syndrome is a serious, but now rare, illness of children. The use of aspirin by children with chickenpox was linked to the development of the disease and parents are now warned not to give aspirin to children with viral infections. Reye's syndrome, which damages hepatic mitochondria, is characterized by brain dysfunction (irritability, lethargy, and coma) and liver dysfunction (elevated plasma free fatty acids, fatty liver, hypoglycemia, hyperammonemia, and accumulation of short-chain organic acids). Impairing mitochondria could impair fatty acid oxidation, synthesis of carbamoyl phosphate and ornithine for ammonia detoxication, and synthesis of oxaloacetate for gluconeogenesis. Part of the therapy is administration of intravenous glucose, which, among other things, reduces proteolysis in muscle and therefore produces fewer amino acids that must be eventually deaminated.

13. All of the following statements about interorgan interactions are correct EXCEPT:
    A. ornithine for the urea cycle is synthesized from glutamate in the kidney.
    B. citrulline is a precursor for arginine synthesis by the kidney.
    C. kidney uses arginine in the synthesis of creatine for distribution to muscle.
    D. arginine synthesized by the kidney is the source of nitric oxide for many cells.
    E. creatinine cleared by the kidney is generated from creatine phosphate in muscle.

14. Muscle proteolysis releases branched-chain amino acids, among others. Branched-chain amino acids:
    A. can also be synthesized from alanine.
    B. can be catabolized by muscle but not liver.
    C. are the main dietary amino acids metabolized by intestine.
    D. promote phosphorylation of the branched-chain α-ketoacid dehydrogenase complex.
    E. are a major source of nitrogen for alanine and glutamine produced in muscle.

Problems

15.

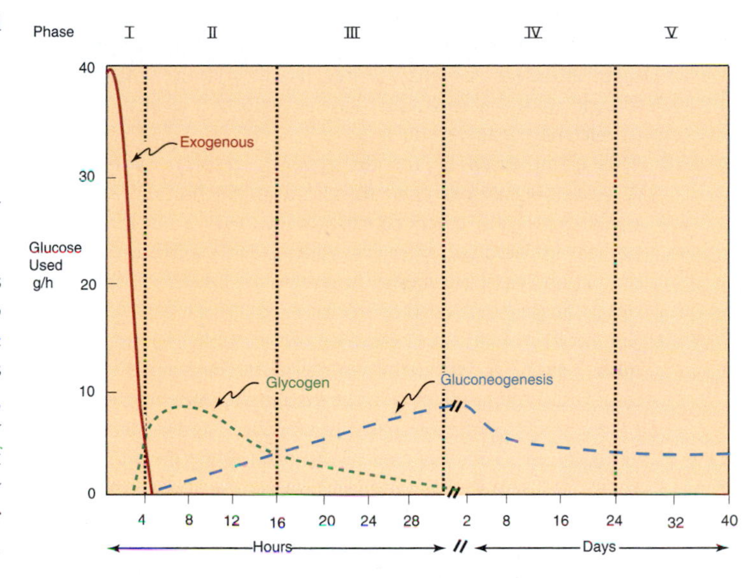

What metabolic and hormonal changes have occurred that account for decreased gluconeogenesis in phase IV in the above diagram of the five phases of glucose homeostasis in humans?

16. Why is the measurement of hemoglobin A_{1c} a better measure of a diabetic's blood glucose status than a daily blood glucose measurement?

ANSWERS

1. **A** A high K_m means that a reaction will proceed slowly at low concentration, whereas a low K_m means the reaction can be rapid under the same circumstances. Protein synthesis requires only that all amino acids be present. Unless amino acids are in high enough concentration, the liver does not catabolize them. C: A high K_m requires a higher concentration of substrate for rapid activity.
2. **C** This is the indirect pathway. A: If any change, the concentration would go down because of less lipolysis in the adipose tissue. B: Glucose coming in from the gut provides blood glucose. D: Enzymatic capacity for fatty acid synthesis develops relatively slowly after refeeding. E: Amino acids can be used for protein synthesis in this state.
3. **C** These trimethyllysines are released when protein is hydrolyzed. A: Only liver and kidney have the complete synthetic pathway. B: Lysine must first be present in cellular protein. D: There are two hydroxylations that require this vitamin. E: This is a precursor.

4. **D** The caloric content of adipose tissue fat is more than 5 times as great as that of muscle protein and almost 200 times as great as that of the combined carbohydrates. A: Blood glucose must be maintained but is a relatively minor reserve. B, C: Glycogen is a rapidly mobilizable reserve of energy but not a large one. E: Protein can be used for energy, but that is not its primary role.
5. **D** Fructose 2,6-bisphosphate is an allosteric effector (activates the kinase and inhibits the phosphatase) of the enzyme controlling glycolysis. A: After a high-protein meal, the intestine produces ammonia and precursors of ornithine for urea synthesis. B: Ketogenesis is dependent on the availability of fatty acids. C: This represents lack of gluconeogenic substrates. E: This leads to complications in diabetes.
6. **D** In the well-fed state, the insulin/glucagon ratio is high and cAMP levels are low. A: Some enzymes are active when phosphorylated; for others the reverse is true. B: This is the most common, though not only, mechanism of phosphorylation. C: Glucagon signals the phosphorylation of hepatic enzymes by elevating cAMP. E: The most common site for phosphorylation in metabolic enzymes is serine.
7. **E** A: Adaptive changes are examples of long-term control. B, D: Both hormonal and nutritional effects are involved in inducing certain enzymes and/or altering their rate of degradation. C: If there is little or no enzyme because of adaptive changes, allosteric control is irrelevant. This is important to keep in mind in refeeding a starved person.
8. **B** This is indicated by the drop in the respiratory quotient. A: Anaerobically exercised muscle uses glucose almost exclusively; aerobically exercised muscle uses fatty acids and ketone bodies. D: Ketone bodies are good aerobic substrates so the blood concentration does not increase greatly. E: Phosphocreatine is only a short-term source of ATP.
9. **D** The problem is insulin resistance, not complete failure to produce insulin. A: Hypertriglyceridemia is characteristic. B: Ketoacidosis is common only in the insulin-dependent type. C: See correct answer. E: Most patients are obese and remain obese.
10. **E** Glutaminase activity is elevated in the kidney during acidosis as the first step in generation of bicarbonate for neutralization of excess acid. A: See previous comment. B: This is the site of glutamine synthesis. C: Glutamine synthetase requires ATP; glutaminase catalyzes simple hydrolysis of glutamine to glutamate and ammonia. D: Less flux through liver glutaminase during acidosis permits glutamine to escape liver for use by the kidney.
11. **D** A: Low insulin/glucagon ratio means high cAMP and, thus, high activity of cAMP-dependent protein kinase and protein phosphorylation. B, D: Phosphorylation activates hormone-sensitive lipase to mobilize fat. C: cAMP works by stimulating covalent modification of enzymes. E: This is a liver enzyme.
12. **B** Pyruvate carboxylase is a key gluconeogenic enzyme. A, E: Stimulation of these enzymes stimulates glycolysis, opposing gluconeogenesis. C, D: Malonyl CoA inhibits transport of fatty acids into mitochondria for β oxidation, a necessary source of energy for gluconeogenesis.
13. **A** Kidney lacks the enzyme needed to convert glutamate to glutamate semialdehyde. B: This is true in both kidney and liver. C: The reaction requires *S*-adenosylmethionine. E: Creatinine is thus a measure of both muscle mass and renal function.
14. **E** When branched-chain amino acids are derived from muscle protein, transamination transfers the nitrogen to alanine or glutamine, which is transported to the liver and kidney. A: Branched-chain amino acids are essential amino acids that have to be supplied by the diet and cannot be synthesized from other amino acids. B: Muscle has high levels of the aminotransferases for branched-chain amino acids, whereas liver has high levels of enzymes for the catabolism of the branched-chain α-ketoacids. C: Intestine metabolizes several dietary amino acids but not these. D: Branched-chain ketoacids derived by transamination of the branched-chain amino acids inhibit the branched-chain α-ketoacid dehydrogenase kinase and therefore prevent phosphorylation of the branched-chain α-ketoacid dehydrogenase complex.
15. In phase IV, most tissues are using primarily fatty acids and ketone bodies for fuel. The ketone body concentration is now sufficiently high that they can enter the brain and supply a significant amount of energy. This reduces the requirement for glucose. The low insulin/glucagon ratio stimulates lipolysis and gluconeogenesis. Increased fatty acid oxidation in liver increases the level of acetyl CoA and NADH. High NADH allosterically inhibits the tricarboxylic acid cycle and the accumulating acetyl CoA is converted to ketone bodies. The low insulin/glucagon ratio also decreases catabolism of muscle protein, which is the major source of substrate for gluconeogenesis. The decrease in gluconeogenesis conserves protein for a longer period of time.
16. A single measurement of glucose, even if taken daily, measures blood glucose concentration only at that particular time. Blood glucose levels vary throughout the day with such things as food intake, exercise, and emotional state. Hemoglobin A_{1c} results from the nonenzymatic reaction between glucose and the amino-terminal valine of the β-chain of hemoglobin. The concentration of hemoglobin A_{1c} is correlated with the average blood glucose concentration over a period of several weeks (three months, which is the lifetime of a red blood cell).

PART V

PHYSIOLOGICAL PROCESSES

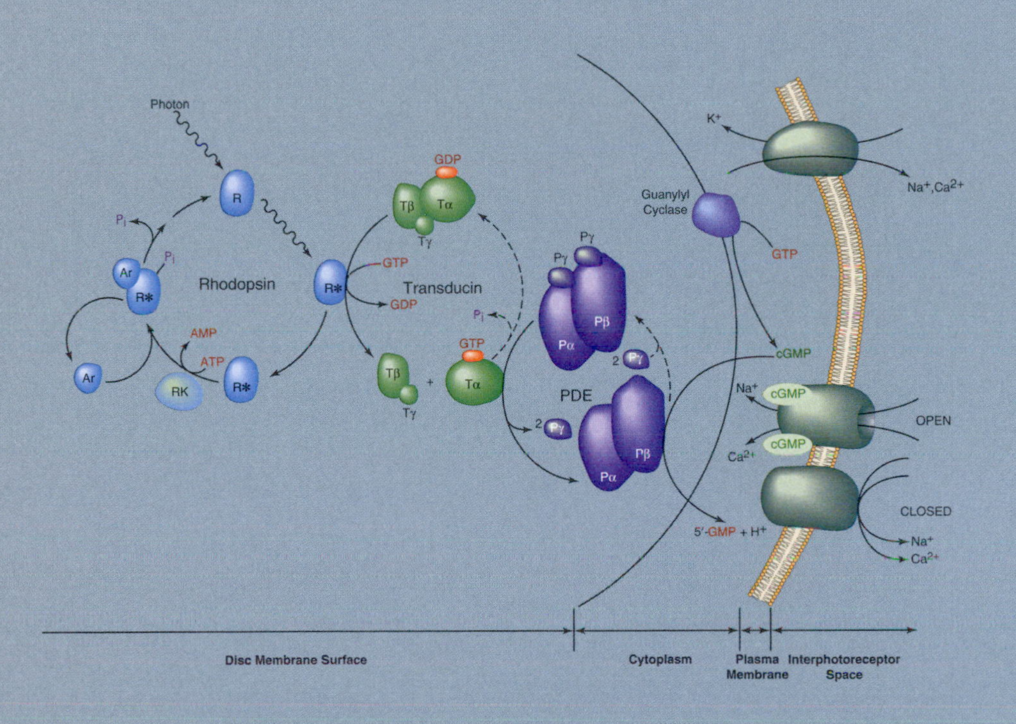

Many of the biochemical reactions discussed in Part IV occur in both prokaryotes and eukaryotes. Higher organisms, including humans, have developed a variety of specialized functions, ranging from conduction of electrical impulses along a nerve fiber to specialized processes for digestion and absorption of food. In animals these are often referred to as physiological processes. The figure above represents one such process, the biochemistry of vision; details are presented in Figure 23.25. Physiological processes require coordination because they often involve different cells and organs. Hormones serve as the messengers relaying instructions between cells and tissues. In this Part, the major classes of hormones are described first, followed by presentations of important physiological processes. Part V concludes with chapters on our nutritional requirements.

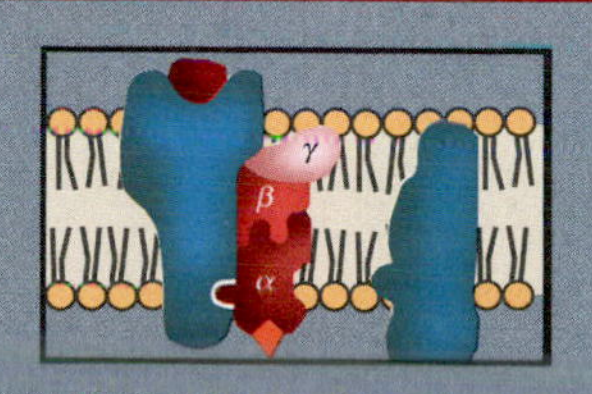

21

BIOCHEMISTRY OF HORMONES I: POLYPEPTIDE HORMONES

Gerald Litwack and
Thomas J. Schmidt

21.1 | OVERVIEW

Cells are regulated by many hormones, growth factors, neurotransmitters, and certain toxins through interactions of these diverse ligands with their **cognate receptors** located at the cell surface. This collection of receptors is the major mechanism through which **peptide hormones** and **amino acid-derived hormones** exert their effects at the cellular level. Another mechanism involves permeation of the cell membrane by **steroid hormones** that subsequently interact with their intracellular cognate receptors (see Chapter 22). These two sites, the plasma membrane and the intracellular milieu, represent the principal locations of the initial interaction between ligands and cellular receptors and are diagrammed in Figure 21.1. Polypeptide hormones and several amino acid-derived hormones bind to cognate receptors in the plasma membrane. One exception is thyroid hormone, which binds to a receptor that resides in the nucleus much like certain steroid hormone receptors.

The **hormonal cascade system** is applicable to many, but not all, hormones. It begins with signals in the central nervous system (CNS), followed by hormone secretion by the hypothalamus and pituitary, and, finally, interaction with an end target organ. In this chapter major polypeptide hormones are summarized and the synthesis of specific hormones is described. Synthesis of the amino acid-derived hormones, epinephrine and triiodo-L-thyronine, is also outlined. Examples of hormone inactivation and degradation are presented. The remainder of this chapter

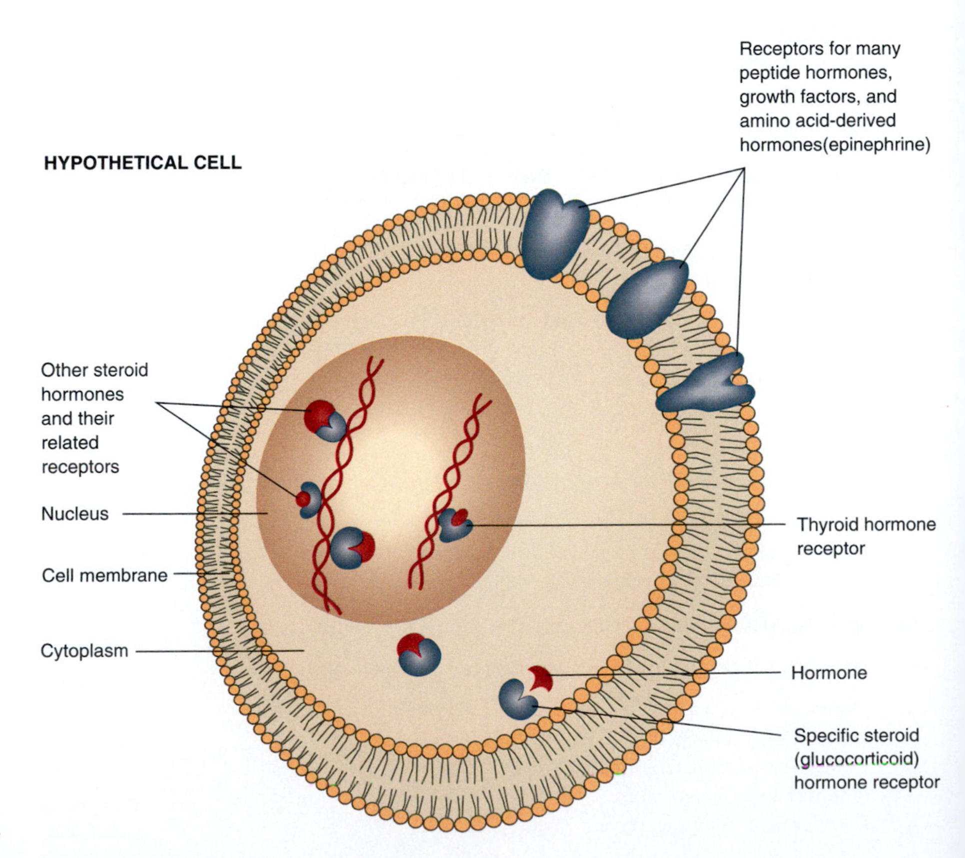

FIGURE 21.1
Diagram showing the different locations of classes of receptors expressed by a target cell.

focuses on receptors, **signal transduction,** and **second messenger pathways.** Receptor internalization is described and examples of cyclic hormonal cascade systems are introduced. Finally, a discussion of oncogenes and receptor function is presented.

In terms of receptor mechanisms, aspects of hormone–receptor interactions are presented with a brief mathematical analysis. Signal transduction is considered, especially in reference to GTP-binding proteins. Second messenger systems discussed include cAMP and the protein kinase A pathway, inositol trisphosphate–diacylglycerol and the Ca^{2+}–protein kinase C pathway, and cGMP and the protein kinase G pathway. These pathways are discussed in the context of representative hormone action. Newly identified components of these signal transduction pathways are defined in terms of the kinase system(s) involved. In addition, the insulin receptor and its tyrosine kinase and second messenger pathways are considered.

21.2 | HORMONES AND THE HORMONAL CASCADE SYSTEM

The definition of a hormone has been expanded over the last several decades. Hormones secreted by endocrine glands were originally considered to represent all of the physiologically relevant hormones. Today, the term **hormone** refers to any substance in an organism that carries a signal to generate some sort of alteration at the cellular level. Thus **endocrine hormones** represent a class of hormones that arise in one tissue, or "gland," and travel a considerable distance through the circulation to reach a target cell expressing cognate receptors. **Paracrine hormones** arise from a cell and travel a relatively small distance to interact with their cognate receptors on another neighboring cell. **Autocrine hormones** are produced by the same cell that functions as the target for that hormone (neighboring cells may also be targets). Thus we can classify hormones based on their radii of action. Often, endocrine hormones that travel long distances to their target cells may be more stable than autocrine hormones that exert their effects over very short distances.

Cascade System Amplifies a Specific Signal

For many hormonal systems in higher animals, the signal pathway originates in the brain and culminates in the ultimate target cell. Figure 21.2 outlines the sequence

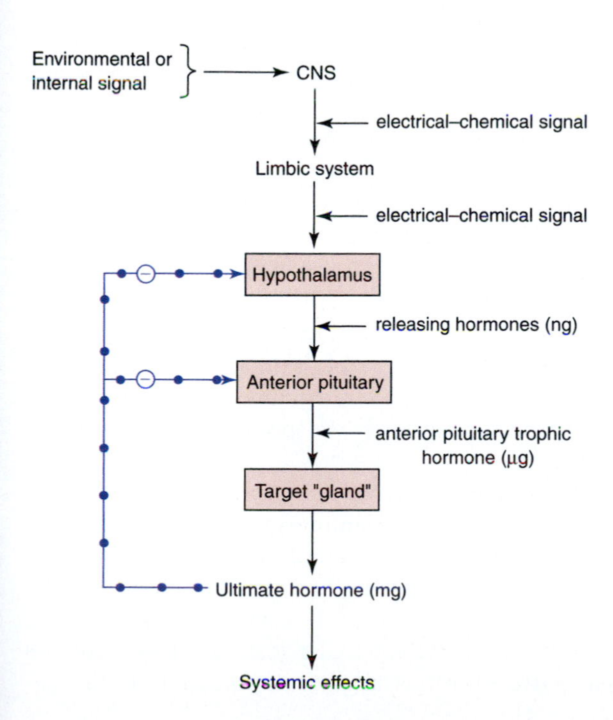

FIGURE 21.2
Hormonal cascade of signals from CNS to ultimate hormone.
The target "gland" refers to the last hormone-producing tissue in the cascade, which is stimulated by an appropriate anterior pituitary hormone. Examples would be thyroid gland, adrenal cortex, ovary, and testes. Ultimate hormone feeds back negatively on sites producing intermediate hormones in the cascade. Amounts [nanogram (ng), microgram (μg), and milligram (mg)] represent approximate quantities of hormone released.
Redrawn from Norman, A. W., and Litwack, G. Hormones. *New York: Academic Press, 1987, p. 38.*

of events in this cascade. A stimulus may originate in the external environment or within the organism in this cascade. This signal may be transmitted as an electrical pulse (action potential) or as a chemical signal or both. In many cases, but not all, such signals are forwarded to the limbic system and subsequently to the hypothalamus, the pituitary, and the target gland that secretes the final hormone. This hormone then affects various target cells to a degree that is frequently proportional to the number of cognate receptors expressed by that cell. This may be a true **cascade** in the sense that increasing amounts of hormone are generated at successive levels (hypothalamus, pituitary, and target gland) and also because the half-lives of these blood-borne hormones tend to become longer in progression from the hypothalamic hormone to the ultimate hormone. In the case of environmental stress, for example, there is a single stressor (change in temperature, noise, trauma, etc.). This stressor results in a signal to the hippocampal structure in the limbic system that signals the hypothalamus to release a hypothalamic releasing hormone, corticotropin-releasing hormone (CRH), which is usually secreted in nanogram amounts and may have a $t_{1/2}$ in the bloodstream of several minutes. CRH travels down a closed portal system to gain access to the **anterior pituitary,** where it binds to its cognate receptor in the cell membrane of corticotropic cells and initiates a set of metabolic changes resulting in the release of adrenocorticotropic hormone (ACTH) as well as β-lipotropin. ACTH, which is released in microgram amounts and has a longer $t_{1/2}$ than CRH, circulates in the bloodstream until it binds to its cognate receptors expressed in the membranes of cells located in the inner layer of the cortex of the adrenal gland (target gland). Here it affects metabolic changes leading to the synthesis and release in 24 h of the ultimate hormone, cortisol, in multimilligram amounts and this active glucocorticoid hormone has a substantial $t_{1/2}$ in blood. Cortisol is taken up by a wide variety of cells that express varying amounts of the intracellular glucocorticoid receptor. The ultimate hormone, in this case cortisol, feeds back negatively on cells of the anterior pituitary, hypothalamus, and perhaps higher levels to shut down the overall pathway in a process that is also mediated by the glucocorticoid receptor. At the **target cell** cortisol–receptor complexes mediate specific transcriptional responses and the individual hormonal effects summate to produce the systemic effects of cortisol. The cascade is represented in this example by a single environmental stimulus generating a series of hormones in progressively larger amounts and with increasing stabilities, and by the ultimate hormone that affects most of the cells in the body. Many other systems operate similarly, there being different specific releasing hormones, anterior pituitary tropic hormones, and ultimate hormones involved in the process. Clearly, the number of target cells affected may be large or small depending on the distribution of receptors for each ultimate hormone.

A related system involves the **posterior pituitary hormones,** oxytocin and vasopressin (antidiuretic hormone), which are stored in the posterior pituitary but are synthesized in neuronal cell bodies in the hypothalamus. This system is represented in Figure 21.3; elements of Figure 21.2 appear in the central vertical pathway. The posterior pituitary system branches to the right from the hypothalamus. Oxytocin and vasopressin are synthesized in separate cell bodies of hypothalamic neurons. More cell bodies dedicated to synthesis of vasopressin are located in the supraoptic nucleus and more cell bodies dedicated to synthesis of oxytocin are located in the paraventricular nucleus. Their release from the posterior pituitary gland along with **neurophysin,** a stabilizing protein, occurs separately via specific stimuli impinging on each of these types of neuronal cells.

Highly specific signals dictate release of polypeptide hormones along the cascade of this system. Thus there are a variety of **aminergic neurons** (secreting amine-containing substances like dopamine and serotonin) that connect to neurons involved in the synthesis and release of the **releasing hormones** of the hypothalamus. Releasing hormones are summarized in Table 21.1. These aminergic neurons fire depending on various types of internal or external signals and their activities account for **pulsatile release patterns** of certain hormones, such as the go-

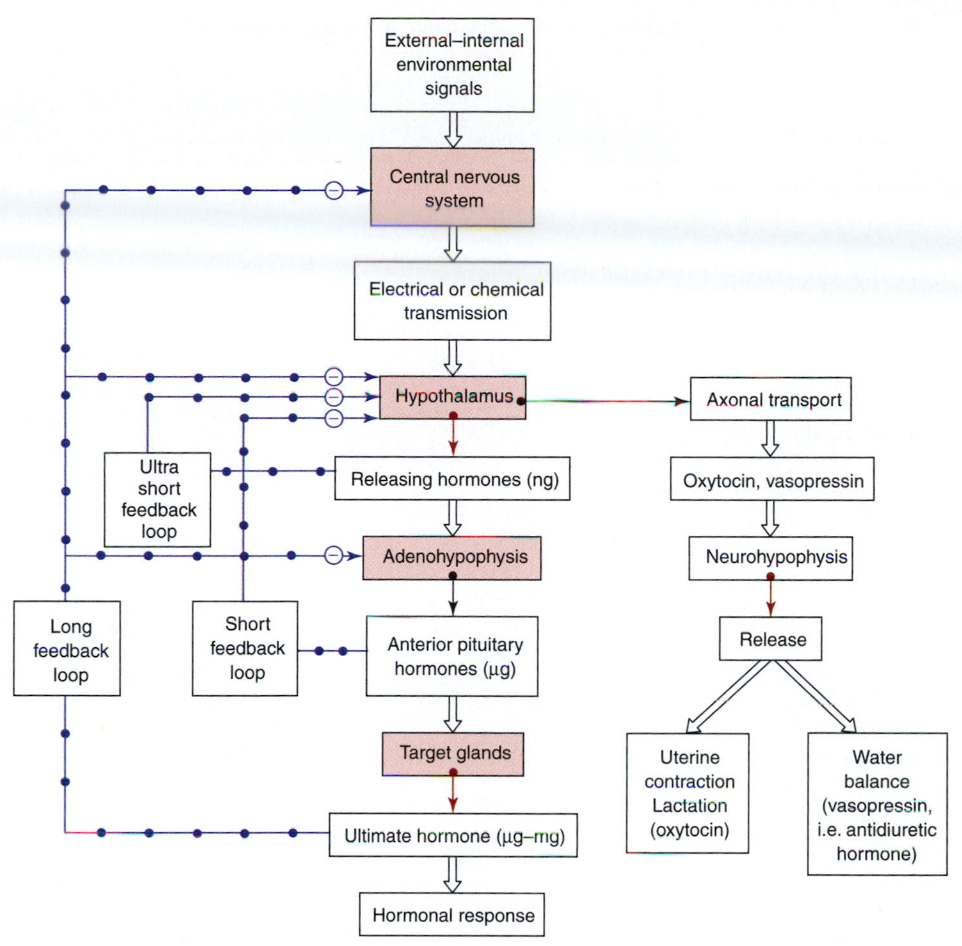

FIGURE 21.3

Many hormonal systems involve the hypothalamus.

Cascade of hormonal signals starting with an external or internal environmental signal. This is transmitted first to the CNS and may involve components of the limbic system, such as the hippocampus and amygdala. These structures innervate the hypothalamus in a specific region, which responds with secretion of a specific releasing hormone, usually in nanogram amounts. Releasing hormones are transported down a closed portal system connecting the hypothalamus and anterior pituitary, bind to cell membrane receptors, and cause the secretion of specific anterior pituitary hormones, usually in microgram amounts. These access the general circulation through fenestrated local capillaries and bind to specific target gland receptors. The interactions trigger release of an ultimate hormone in microgram to milligram daily amounts, which generates the hormonal response by binding to receptors in several target tissues. In effect, this overall system is an amplifying cascade. Releasing hormones are secreted in nanogram amounts and they have short half-lives on the order of a few minutes. Anterior pituitary hormones are produced often in microgram amounts and have longer half-lives than releasing hormones. Ultimate hormones can be produced in daily milligram amounts with much longer half-lives. Thus the products of mass × half-life constitute an amplifying cascade mechanism. With respect to differences in mass of hormones produced from hypothalamus to target gland, the range is nanograms to milligrams, or as much as one million-fold. When the ultimate hormone has receptors in nearly every cell type, it is possible to affect the body chemistry of virtually every cell by a single environmental signal. Consequently, the organism is in intimate association with the external environment, a fact that we tend to underemphasize. Solid arrows indicate a secretory process. Long arrows studded with open or closed circles indicate negative feedback pathways (ultra-short, short, and long feedback loops).

Redrawn from Norman, A. W. and Litwack, G. Hormones. *New York: Academic Press, 1987, p. 102.*

nadotropin-releasing hormone (GnRH), and the **rhythmic cyclic release** of other hormones like cortisol.

Another prominent feature of the hormonal cascade (Figure 21.3) is the **negative feedback** system operating when sufficiently high levels of the ultimate hormone have been secreted into the circulation. Generally, there are three feedback loops—the **long feedback**, the **short feedback**, and the **ultra-short feedback loops.** In the long feedback loop, the final hormone binds a cognate receptor in/on cells of the anterior pituitary, hypothalamus, and CNS to prevent further elaboration of hormones from those cells that are involved in the cascade. The short

TABLE 21.1 Hypothalamic Releasing Hormones[a]

Releasing Hormone	*Number of Amino Acids in Structure*	*Anterior Pituitary Hormone Released or Inhibited*
Thyrotropin-releasing hormone (TRH)	3	Thyrotropin (TSH); can also release prolactin (PRL) experimentally
Gonadotropin-releasing hormone (GnRH)	10	Luteinizing and follicle-stimulating hormones (LH and FSH) from the same cell type; leukotriene C_4 (LTC_4) can also release LH and FSH by a different mechanism
Gonadotropin release-inhibiting factor (GnRIF)	12.2 kDa molecular weight	LH and FSH release inhibited
Corticotropin-releasing hormone (CRH)	41	ACTH, β-lipotropin (β-LPH), and some β-endorphin
Arginine vasopressin (AVP)	9	Stimulates CRH action in ACTH release
Angiotensin II (AII)	8	Stimulates CRH action in ACTH release; releases ACTH weakly
Somatocrinin (GRH)	44	Growth hormone (GH) release
Somatostatin (GIH)	14	GH release inhibited
Hypothalamic gastrin-releasing peptide		Inhibits release of GH and PRL
Prolactin-releasing factor (PRF)		Releases prolactin (PRL)
Prolactin release-inhibiting factor (PIF)		Evidence that a new peptide may inhibit PRL release; dopamine also inhibits PRL release and was thought to be PIF for some time; dopamine may be a secondary PIF: oxytocin may inhibit PRL release

[a]Melanocyte-stimulating hormone (MSH) is a major product of the pars intermedia (Figure 21.5) in the rat and is under the control of aminergic neurons. Humans may also secrete α-MSH from pars intermedia-like cells although this structure is anatomically indistinct in the human.

CLINICAL CORRELATION 21.1

Testing Activity of the Anterior Pituitary

Releasing hormones and chemical analogs, particularly of the smaller peptides, are now routinely synthesized. The gonadotropin-releasing hormone, a decapeptide, is available for use in assessing the function of the anterior pituitary. This is of importance when a disease situation may involve either the hypothalamus, the anterior pituitary, or the end organ. Infertility is an example of such a situation. What needs to be assessed is which organ is at fault in the hormonal cascade. Initially, the end organ, in this case the gonads, must be considered. This can be accomplished by injecting the anterior pituitary hormone LH or FSH. If sex hormone secretion is elicited, then the ultimate gland would appear to be functioning properly. Next, the anterior pituitary would need to be analyzed. This can be done by i.v. administration of synthetic GnRH; by this route GnRH can gain access to the gonadotropic cells of the anterior pituitary and elicit secretion of LH and FSH. Routinely, LH levels are measured in the blood as a function of time after the injection. These levels are measured by radioimmunoassay (RIA) in which radioactive LH or hCG is displaced from binding to an LH-binding protein by LH in the serum sample. The extent of the competition is proportional to the amount of LH in the serum. In this way a progress of response is measured that will be within normal limits or clearly deficient. If the response is deficient, the anterior pituitary cells are not functioning normally and are the cause of the syndrome. On the other hand, normal pituitary response to GnRH would indicate that the hypothalamus was nonfunctional. Such a finding would prompt examination of the hypothalamus for conditions leading to insufficient availability/production of releasing hormones. Obviously, the knowledge of hormone structure and the ability to synthesize specific hormones permit the diagnosis of these disease states.

Marshall, J. C. and Barkan, A. L. Disorders of the hypothalamus and anterior pituitary. In: W. Kelley (Ed.), *Internal Medicine*. New York: Lippincott, 1989, p. 2159; and Conn, P. M. The molecular basis of gonadotropin-releasing hormone action. *Endocr. Rev.* 7:3, 1986.

feedback loop is exemplified by the pituitary hormone that feeds back negatively on the hypothalamus operating through a cognate receptor. In ultra-short feedback loops the hypothalamic releasing factor feeds back at the level of the hypothalamus to inhibit further secretion of this releasing factor. These mechanisms provide tight controls on the operation of the cascade, responding to stimulating signals as well as negative feedback, and render this system highly responsive to the hormonal milieu. Clinical Correlation 21.1 describes approaches for testing the responsiveness of the anterior pituitary gland.

Polypeptide Hormones of the Anterior Pituitary

Polypeptide hormones of the anterior pituitary are shown in Figure 21.4 with their controlling hormones from the hypothalamus. The major hormones of the anterior pituitary are growth hormone (GH), thyrotropin or thyroid-stimulating hormone (TSH), adrenocorticotropic hormone (ACTH), β-lipotropin (β-LTH),

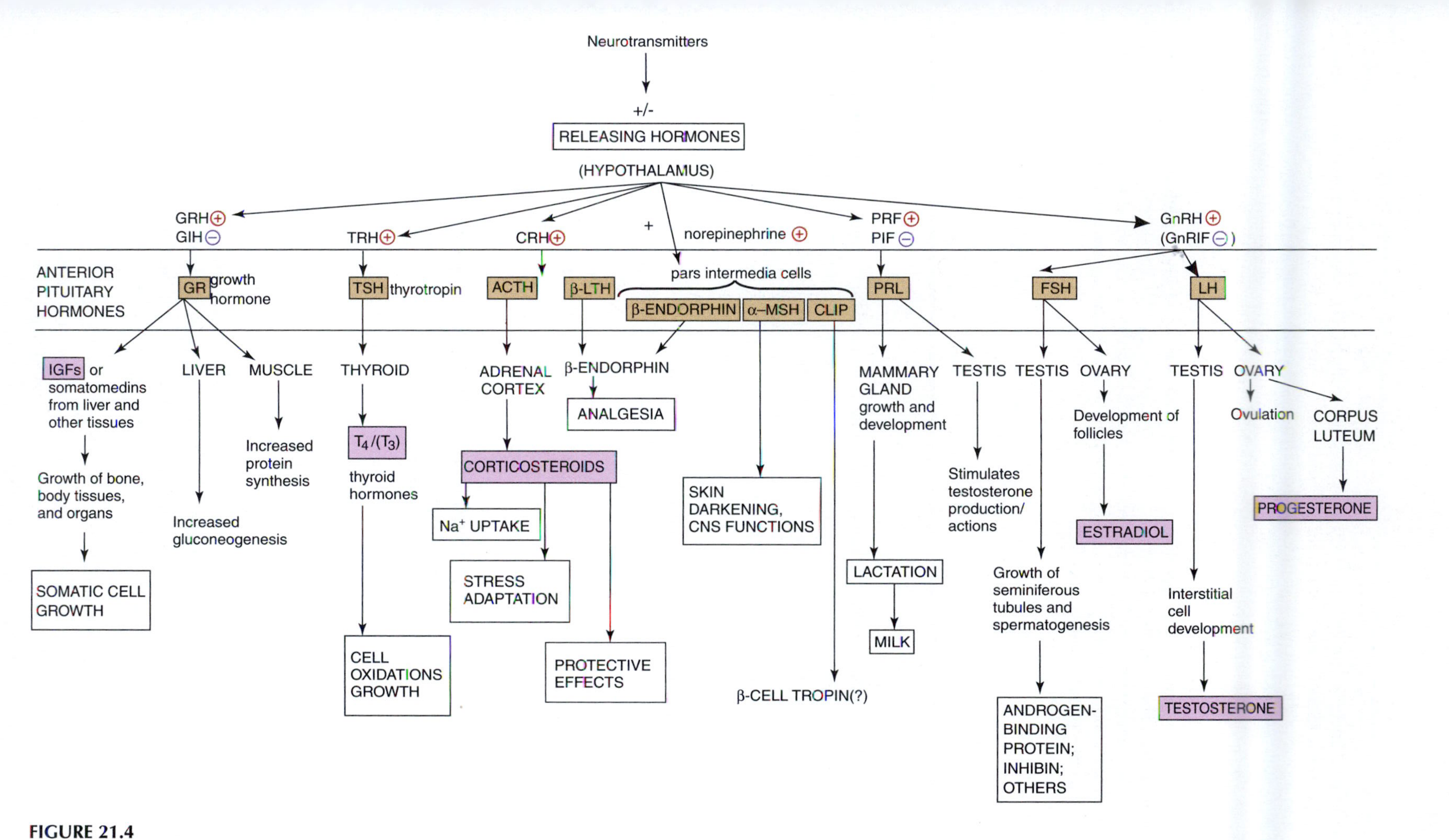

FIGURE 21.4
Overview of anterior pituitary hormones with hypothalamic releasing hormones and their actions.

CLINICAL CORRELATION 21.2

Hypopituitarism

The hypothalamus is connected to the anterior pituitary by a delicate stalk that contains the portal system through which releasing hormones, secreted from the hypothalamus, gain access to the anterior pituitary cells. In the cell membranes of these cells are specific receptors for releasing hormones. In most cases, different cells express different releasing hormone receptors. The connection between the hypothalamus and anterior pituitary can be disrupted by trauma or tumors. Trauma can occur in automobile accidents or other local damaging events that may result in severing of the stalk and preventing the releasing hormones from reaching their target anterior pituitary cells. When this happens, the anterior pituitary cells no longer have their signaling mechanism for the release of anterior pituitary hormones. In the case of tumors of the pituitary gland, all of the anterior pituitary hormones may not be shut off to the same degree or the secretion of some may disappear sooner than others. In any case, if hypopituitarism occurs this condition may result in a life-threatening situation in which the clinician must determine the extent of loss of pituitary hormones, especially ACTH. Posterior pituitary hormones—oxytocin and vasopressin—may also be lost, precipitating a problem of excessive urination (vasopressin deficiency) that must be addressed. The usual therapy involves administration of the end organ hormones, such as thyroid hormone, cortisol, sex hormones, and progestin; with female patients it is also necessary to maintain the ovarian cycle. These hormones can easily be administered in oral form. Growth hormone deficiency is not a problem in the adult but would be an important problem in a growing child. The patient must learn to anticipate needed increases of cortisol in the face of stressful situations. Fortunately, these patients are usually maintained in reasonably good condition.

Marshall, J. C. and Barkan, A. L. Disorders of the hypothalamus and anterior pituitary. In: W. N. Kelley (Ed.), *Internal Medicine*. New York: Lippincott, 1989, p. 2159; and Robinson, A. G. Disorders of the posterior pituitary. In: W. N. Kelley (Ed.), *Internal Medicine*. New York: Lippincott, 1989, p. 2172.

β-endorphin (from pars intermedia-like cells), α-MSH (from pars intermedia-like cells), β-MSH (from pars intermedia-like cells), corticotropin-like intermediary peptide (CLIP; from pars intermedia-like cells), prolactin (PRL), follicle-stimulating hormone (FSH), and luteinizing hormone (LH). All are single polypeptide chains, except TSH, FSH, and LH, which are dimers that share a similar or identical α subunit. Since the intermediate lobe in humans is rudimentary, the circulating levels of free α- and β-MSH are relatively low. It is of interest, particularly in the human, that MSH receptors recognize and are activated by ACTH, since the first 13 amino acids of ACTH contain the α-MSH sequence. For this reason, ACTH may be an important contributing factor to skin pigmentation and may exceed the importance of MSH, especially in conditions where the circulating level of ACTH is high. The clinical consequences of hypopituitarism are presented in Clinical Correlation 21.2.

21.3 | MAJOR POLYPEPTIDE HORMONES AND THEIR ACTIONS

Since cellular communication is so specific, it is not surprising that there are a large number of hormones in the body and that new hormones continue to be discovered. Limitations of space permit a summary of only a few of the well-characterized hormones. Table 21.2 presents some major polypeptide hormones and their actions. Inspection of Table 21.2 shows that many hormones cause release of other substances, some of which may themselves be hormones. This is particularly the case for hormonal systems that are included in cascades like that presented in Figures 21.2 and 21.3. Other activities of receptor–hormone complexes located in cell membranes are to increase the flux of ions into cells, particularly calcium ions, and to activate or suppress activities of enzymes in contact with the receptor or a transducing protein with which the receptor interacts. Examples of these kinds of activities are discussed later in this chapter. In the functioning of most membrane–receptor complexes, stimulation of enzymes or flux of ions is followed by a chain of events, which may be described as intracellular cascades, during which a high degree of amplification is obtained.

TABLE 21.2 Important Polypeptide Hormones in the Body and Their Actions[a]

Source	*Hormone*	*Action*
Hypothalamus	Thyrotropin-releasing hormone (TRH)	Acts on thyrotrope to release TSH
	Gonadotropin-releasing hormone (GnRH)	Acts on gonadotrope to release LH and FSH from the same cell
	Growth hormone-releasing hormone or somatocrinin (GRH)	Acts on somatotrope to release GH
	Growth hormone release inhibiting hormone or somatostatin (GIH)	Acts on somatotrope to prevent release of GH
	Corticotropin-releasing hormone (CRH) Vasopressin is a helper hormone to CRH in releasing ACTH; angiotensin II also stimulates CRH action in releasing ACTH	Acts on corticotrope to release ACTH and β-liptropin
	Prolactin-releasing factor (PRF) (not well established)	Acts on lactotrope to release PRL
	Prolactin release inhibiting factor (PIF) (not well established may be a peptide hormone under control of dopamine or may be dopamine itself)	Acts on lactotrope to inhibit release of PRL)
Anterior pituitary	Thyrotropin (TSH)	Acts on thyroid follicle cells to bring about release of $T_4(T_3)$
	Luteinizing hormone (LH) (human chorionic gonadotropin, hCG, is a similar hormone from the placenta)	Acts on Leydig cells of testis to increase testosterone synthesis and release; acts on corpus luteum of ovary to increase progesterone production and release
	Follicle-stimulating hormone (FSH)	Acts on Sertoli cells of seminiferous tubule to increase proteins in sperm and other proteins; acts on ovarian follicles to stimulate maturation of ovum and production of estradiol
	Growth hormone (GH)	Acts on a variety of cells to produce IGFs (or somatomedins), cell growth, and bone sulfation
	Adrenocorticotropic hormone (ACTH)	Acts on cells in the adrenal gland to increase cortisol production and secretion
	β-Endorphin	Acts on cells and neurons to produce analgesic and other effects
	Prolactin (PRL)	Acts on mammary gland to cause differentiation of secretory cells (with other hormones) and to stimulate synthesis of components of milk
	Melanocyte-stimulating hormone (MSH)	Acts on skin cells to cause the dispersion of melanin (skin darkening)
Ultimate gland hormones	Insulin-like growth factors (IGF)	Respond to GH and produce growth effects by stimulating cell mitosis
	Thyroid hormone (T_4/T_3) (amino acid-derived hormone)	Responds to TSH and stimulates oxidation in many cells
	Opioid peptides	May derive as breakdown products of γ-lipotropin or β-endorphin or from specific gene products; can respond to CRH or dopamine and may produce analgesia and other effects
	Inhibin	Responds to FSH in ovary and in Sertoli cell; regulates secretion of FSH from anterior pituitary. Second form of inhibin (activin) may stimulate FSH secretion
	Corticotropin-like intermediary peptide (CLIP)	Derives from intermediate pituitary by degradation of ACTH; contains β-cell tropin activity, which stimulates insulin release from β cells in presence of glucose

(continued)

TABLE 21.2 *(continued)*

Source	*Hormone*	*Action*
Peptide hormones responding to other signals than anterior pituitary hormones	Arginine vasopressin (AVP; antidiuretic hormone, ADH)	Responds to increased activity in osmoreceptor, which senses extracellular [Na^+]; increases water reabsorption from distal kidney tubule
	Oxytocin	Responds to suckling reflex and estradiol; causes milk "let down" or ejection in lactating female, involved in uterine contractions of labor; luteolytic factor produced by corpus luteum; decreases steroid synthesis in testis
β Cells of pancreas respond to glucose and other blood constituents to release insulin	Insulin	Increases tissue utilization of glucose
α Cells of pancreas respond to low levels of glucose and falling serum calcium	Glucagon	Decreases tissue utilization of glucose to elevate blood glucose
Derived from protein by actions of renin and converting enzyme	Angiotensin II and III (AII and AIII)	Renin initially responds to decreased blood volume or decreased [Na^+] in the macula densa of the kidney. AII/AIII stimulate outer layer of adrenal cortex to synthesize and release aldosterone
Released from heart atria in response to hypovolemia; regulated by other hormones	Atrial natriuretic factor (ANF) or atriopeptin	Acts on outer adrenal cells to decrease aldosterone release; has other effects also
Generated from plasma, gut, or other tissues	Bradykinin	Modulates extensive vasodilation resulting in hypotension
Hypothalamus and intestinal mucosa	Neurotensin	Effects on gut; may have neurotransmitter actions
Hypothalamus, CNS, and intestine	Substance P	Pain transmitter, increases smooth muscle contractions of the GI tract
Nerves and endocrine cells of gut; hypothermic hormone	Bombesin	Increases gastric acid secretion
	Cholecystokinin (CCK)	Stimulates gallbladder contraction and bile flow; increases secretion of pancreatic enzymes
Stomach antrum	Gastrin	Increases secretion of gastric acid and pepsin
Duodenum at pH values below 4.5	Secretin	Stimulates pancreatic acinar cells to release bicarbonate and water to elevate duodenal pH
Hypothalamus and GI tract	Vasointestinal peptide (VIP)	Acts as a neurotransmitter in peripheral autonomic nervous system; relaxes vascular smooth muscles; increases secretion of water and electrolytes from pancreas and gut
Kidney	Erythropoietin	Acts on bone marrow for terminal differentiation and initiates hemoglobin synthesis
Ovarian corpus luteum	Relaxin	Inhibits myometrial contractions; its secretion increases during gestation
	Human placental lactogen (hPL)	Acts like PRL and GH because of large amount of hPL produced
Salivary gland	Epidermal growth factor	Stimulates proliferations of cells derived from ectoderm and mesoderm together with serum; inhibits gastric secretion
Thymus	Thymopoietin (α-thymosin)	Stimulates phagocytes; stimulates differentiation of precursors into immune competent T cells

(continued)

TABLE 21.2 *(continued)*

Source	*Hormone*	*Action*
Parafollicular C cells of thyroid gland	Calcitonin (CT)	Lowers serum calcium
Parathyroid glands	Parathyroid hormone (PTH)	Stimulates bone resorption; stimulates phosphate excretion by kidney; raises serum calcium levels
Endothelial cells of blood vessels	Endothelin	Vasoconstriction

Source: Part of this table is reproduced from Norman, A. W. and Litwack, G. *Hormones*. Orlando, FL: Academic Press, 1987.

[a]This is only a partial list of polypeptide hormones in humans. TSH, thyroid-stimulating hormone or thyrotropin; LH, luteinizing hormone; FSH, follicle-stimulating hormone; GH, growth hormone; ACTH, adrenocorticotropic hormone; PRL, prolactin; T_4, thyroid hormone (also T_3); IGF, insulin-like growth factor. For the releasing hormones and for some hormones in other categories, the abbreviation may contain "H" at the end when the hormone has been well characterized, and "F" in place of H to refer to "Factor" when the hormone has not been well characterized. Names of hormones may contain "tropic" or "trophic" endings; tropic is mainly used here. Tropic refers to a hormone generating a change, whereas trophic refers to growth promotion. Both terms can refer to the same hormone at different stages of development. Many of these hormones have effects in addition to those listed here.

21.4 | GENES AND FORMATION OF POLYPEPTIDE HORMONES

Genes for polypeptide hormones contain the coding sequence for the hormone and the control elements upstream of the structural gene. In some cases, more than one hormone is encoded in a gene. One example is **proopiomelanocortin,** a protein hormone precursor that encodes the following hormones: ACTH, β-lipotropin, and other hormones like γ-lipotropin, γ-MSH, α-MSH, CLIP, β-endorphin, and potentially β-MSH and enkephalins. Oxytocin and vasopressin are each encoded on separate genes together with information for each respective **neurophysin,** a protein that binds to the completed hormone and stabilizes it.

Proopiomelanocortin Is a Precursor Polypeptide for Eight Hormones

Proopiomelanocortin, as schematized in Figure 21.5, generates at least eight hormones from a single gene product. All products are not expressed simultaneously in a single cell type, but occur in separate cells based on their content of specific proteases required to cleave the propeptide, specific metabolic controls, and the presence of different positive regulators. Thus, while proopiomelanocortin is expressed in both the corticotropic cell of the anterior pituitary and the pars intermedia cell, the stimuli and products are different as summarized in Table 21.3. The pars intermedia is a discrete anatomical structure located between the anterior and posterior pituitary in the rat (Figure 21.6). In the human, however, the pars intermedia is not a discrete anatomical structure, although the cell type may be present in the equivalent location.

Many Polypeptide Hormones Are Encoded by a Single Gene

Other genes encoding more than one peptide are the genes for vasopressin and oxytocin and their accompanying neurophysins, products that are released from the posterior pituitary upon specific stimulation. In much the same manner that ACTH and β-lipotropin (β-LPH) are split out of the proopiomelanocortin precursor peptide, so are the products vasopressin, neurophysin II, and a glycoprotein of as yet unknown function split out of the vasopressin precursor. A similar situation exists for oxytocin and neurophysin I (Figure 21.7).

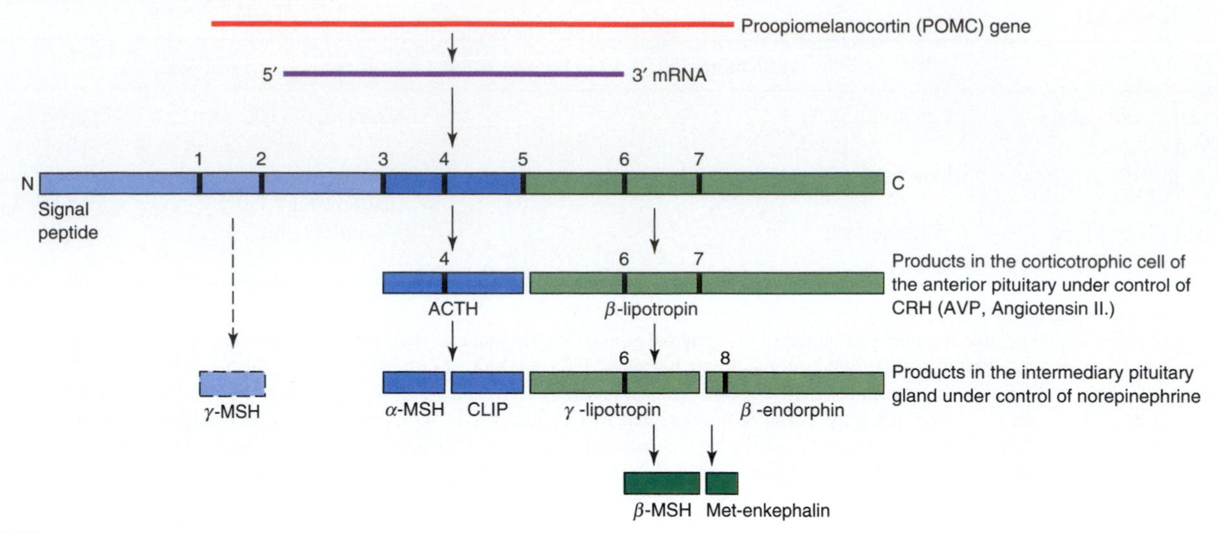

FIGURE 21.5

Proopiomelanocortin is a polypeptide product encoded by a single gene. The dark vertical bars represent proteolytic cleavage sites for specific enzymes. The cleavage sites are Arg-Lys, Lys-Arg, or Lys-Lys. Some specificity also may be conferred by neighboring amino acid residues. In the corticotrophic cell of the anterior pituitary, enzymes are present that cleave at sites 3 and 5, releasing the major products, ACTH and β-lipotropin, into the general circulation. In the pars intermedia, especially in vertebrates below humans, these products are further cleaved at major sites 4, 6, and 7 to release α-MSH, CLIP, γ-lipotropin, and β-endorphin into the general circulation. Some β-lipotropin arising in the corticotroph may be further degraded to form β-endorphin. These two cell types appear to be under separate controls. The corticotrophic cell of the anterior pituitary is under the positive control of the CRH and its auxiliary helpers, arginine vasopressin (AVP) and angiotensin II. AVP by itself does not release ACTH but enhances the action of CRH in this process. The products of the intermediary pituitary, α-MSH, CLIP (corticotropin-like intermediary peptide), γ-lipotropin, and β-endorphin, are under the positive control of norepinephrine, rather than CRH, for release. Obviously there must exist different proteases in these different cell types in order to generate a specific array of hormonal products. β-Endorphin also contains a pentapeptide, enkephalin, which potentially could be released at some point (hydrolysis at 8).

Vasopressin and **neurophysin II** are released by the activity of baroreceptors and osmoreceptors, which sense a fall in blood pressure or a rise in extracellular sodium ion concentration, respectively. Generally, **oxytocin** and **neurophysin I** are released from the posterior pituitary by the suckling response in lactating females or by other stimuli mediated by a specific cholinergic mechanism. Oxytocin–neurophysin I release is triggered by injection of estradiol. Release of vasopressin–neurophysin II is stimulated by administration of nicotine. The two separate and specific releasing agents, estradiol and nicotine, prove that oxytocin and vasopressin, together with their respective neurophysins, are synthesized and released

TABLE 21.3 Summary of Stimuli and Products of Proopiomelanocortin[a]

Cell type	Corticotroph	Pars intermedia
Stimulus	CRH (+) (Cortisol (−))	Dopamine (−) Norepinephrine (+)
Auxiliary stimulus	AVP, AII	
Major products	ACTH, β-lipotropin (β-endorphin)	α-MSH, CLIP, γ-lipotropin, β-endorphin

[a]CRH, corticotropin-releasing hormone; AVP, arginine vasopressin; AII, angiotensin II; ACTH, adrenocorticotropin; α-MSH, α melanocyte-stimulating hormone; CLIP, corticotropin-like intermediary peptide.

Note: Although there are *pars intermedia cells* in the human pituitary gland, they do not represent a distinct lobe.

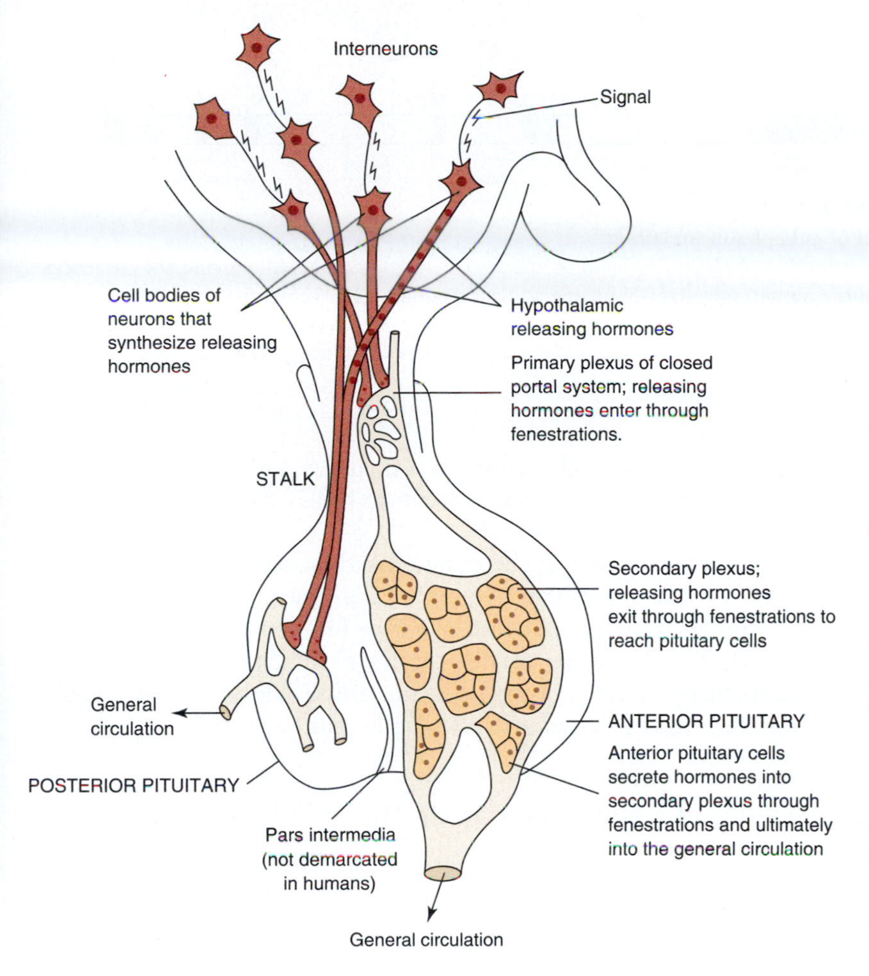

FIGURE 21.6
The hypothalamus with nuclei in various locations in which the hypothalamic releasing hormones are synthesized.
Shown is the major vascular network consisting of a primary plexus where releasing hormones enter its circulation through fenestrations and the secondary plexus in the anterior pituitary where the releasing hormones are transported out of the circulation, again through fenestrations, to the anterior pituitary target cells. Also shown are the effects of the hypothalamic releasing hormones causing the secretion into the general circulation of the anterior pituitary hormones.
Adapted from Norman, A. W. and Litwack, G. Hormones. *New York: Academic Press, 1987, p. 104.*

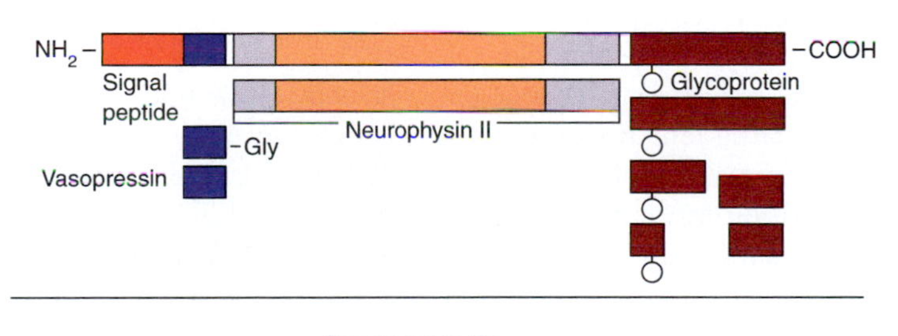

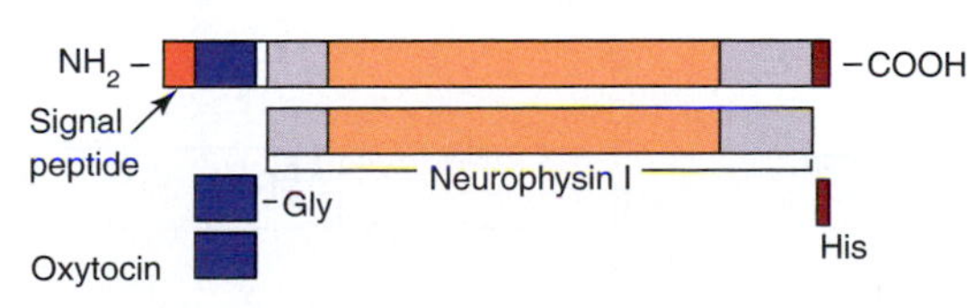

FIGURE 21.7
Prepro-vasopressin and prepro-oxytocin.
Proteolytic maturation proceeds from top to bottom for each precursor. The organization of the gene translation products is similar in either case except that a glycopeptide is included on the proprotein of vasopressin in the C-terminal region. Orange bars of the neurophysin represent conserved amino acid regions; gray bars represent variable C and N termini.
Redrawn with permission from Richter, D. Trends Biochem. Sci. *8:278, 1983.*

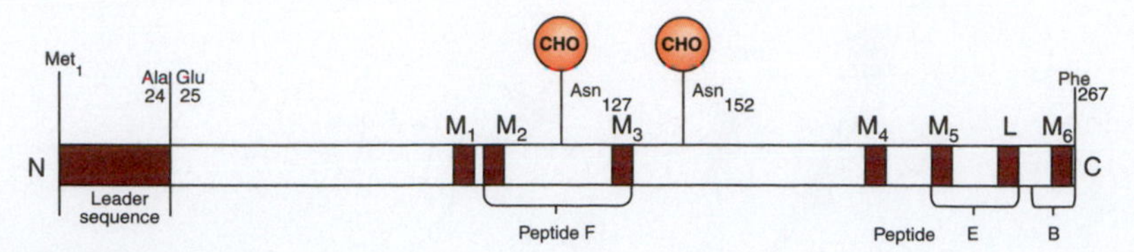

FIGURE 21.8
Model of enkephalin precursor.
The distribution of Met-enkephalin sequences (M_1–M_6) and Leu-enkephalin (L) sequences within the protein precursor of bovine adrenal medulla. CHO, potential carbohydrate attachment sites.
Redrawn from Comb, M., Seeburg, P. H., Adelman, J., Eiden, L., and Herbert, E. Nature *295:663, 1982.*

from different cell types. Although oxytocin is well known for its milk-releasing action in the lactating female, in the male it seems to be associated with an increase in testosterone synthesis in the testes.

Other polypeptide hormones are being discovered that are co-encoded together by a single gene. An example is the discovery of the gene encoding GnRH, a decapeptide that appears to reside to the left of a gene for the GnRH-associated peptide (GAP), which, like dopamine, may be capable of inhibiting prolactin release. Thus both hormones—GnRH and the prolactin release inhibiting factor GAP—appear to be co-secreted by the same hypothalamic cells.

Multiple Copies of a Hormone Can Be Encoded by a Single Gene

An example of multiple copies of a single hormone encoded on a single gene is the enkephalins secreted by chromaffin cells of the adrenal medulla. **Enkephalins** are pentapeptides with opioid activity; methionine-enkephalin (Met-ENK) and leucine-enkephalin (Leu-ENK) have the structures

Tyr-Gly-Gly-Phe-Met (Met-ENK)

Tyr-Gly-Gly-Phe-Leu (Leu-ENK)

A model of the enkephalin precursor is presented in Figure 21.8, which encodes several Met-ENK (M) molecules and a molecule of Leu-ENK (L). Again, the processing sites to release enkephalin molecules from the protein precursor involve Lys-Arg, Arg-Arg, and Lys-Lys bonds.

Many genes for hormones encode only one copy of the hormone and this may be the general situation. An example of a single hormone gene is shown in Figure 21.9. Information for the hormone CRH is contained in the second exon and the information in the first exon is not expressed. Having cDNAs for use as probes that contain the information for expression of CRH allows for the localiza-

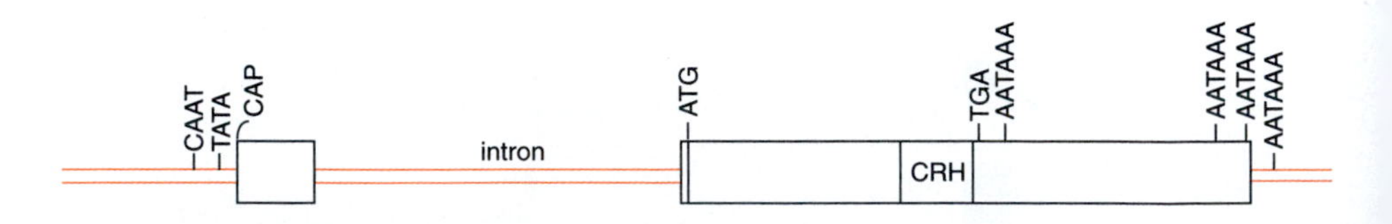

FIGURE 21.9
Nucleic acid sequence of rat *proCRH* genes.
A schematic representation of the rat *proCRH* gene. Exons are shown as blocks and the intron by a double red line. The TATA and CAAT sequence, putative cap site, translation initiation ATG, translation terminator TGA, and poly(A) addition signals (AATAAA) are indicated. The location of the CRH peptide is indicated by CRH.
Redrawn from Thompson, R. D., Seasholz, A. F., and Herbert, E. Mol. Endocrinol. *1:363, 1987.*

tion of the hormone in tissues. RNA extracts from different tissues probed with this DNA reveal the location of CRH mRNA in testis, brain stem, and adrenal gland in addition to pituitary and hypothalamus. The presence of the hormone in extrahypothalamic–pituitary axis tissues and its functions there are subjects of active investigation.

21.5 | SYNTHESIS OF AMINO ACID-DERIVED HORMONES

Many hormones and neurotransmitters are derived from amino acids, principally from tyrosine and phenylalanine. Glutamate, aspartate, and other compounds are important neurotransmitter substances as well. Although there may be some confusion about which compounds are neurotransmitters and which are hormones, it is clear that epinephrine from the adrenal medulla is a hormone, whereas norepinephrine is a neurotransmitter. This section considers epinephrine and thyroxine or triiodothyronine. The biogenic amines, such as dopamine, which are considered to be neurotransmitters, are discussed in Chapter 22.

Epinephrine Is Synthesized from Phenylalanine/Tyrosine

Synthesis of epinephrine occurs in the adrenal medulla. A number of steroid hormones, including aldosterone, cortisol, and dehydroepiandrosterone, are produced in the adrenal cortex and are discussed in Chapter 22. The biochemical reactions that form **epinephrine** from tyrosine or phenylalanine are presented in Figure 21.10. Epinephrine is a principal hormone secreted from the adrenal medulla chromaffin cell along with some norepinephrine, enkephalins, and some of the enzyme dopamine-β-hydroxylase. Secretion of epinephrine is signaled by the neural response to stress, which is transmitted to the adrenal medulla by way of a preganglionic acetylcholinergic neuron (Figure 21.11). Release of acetylcholine by the neuron increases the availability of intracellular calcium ion, which stimulates exocytosis and release of the material stored in the **chromaffin granules** (Figure 21.11*b*). This overall system of epinephrine synthesis, storage, and release from the adrenal medulla is regulated by neuronal controls and also by glucocorticoid hormones synthesized in and secreted from the adrenal cortex in response to stress. Since the products of the adrenal cortex are transported through the adrenal medulla on their way out to the general circulation, cortisol becomes elevated in the medulla and induces **phenylethanolamine N-methyltransferase (PNMT)**, a key enzyme catalyzing the conversion of norepinephrine to epinephrine. Thus, in biochemical terms, the stress response at the level of the adrenal cortex ensures the production of epinephrine from the adrenal medulla (Figure 21.12). Presumably, epinephrine once secreted into the bloodstream not only affects α receptors of hepatocytes to ultimately increase blood glucose levels as indicated, but also interacts with α receptors on vascular smooth muscle cells and on pericytes to cause cellular contraction and increase blood pressure.

Synthesis of Thyroid Hormone Requires Incorporation of Iodine into a Tyrosine of Thyroglobulin

An outline of the biosynthesis and secretion of thyroid hormone, **tetraiodo-L-thyronine (T_4)**, also called **thyroxine**, and its active cellular counterpart, **triiodo-L-thyronine (T_3)** (structures presented in Figure 21.13), is presented in Figure 21.14. The thyroid gland is differentiated to concentrate iodide from the blood and through the reactions shown in Figures 21.13 and 21.14, monoiodotyrosine (MIT), diiodotyrosine (DIT), T_4, and T_3 are produced within **thyroglobulin** (TG). Thus the iodinated amino acids and thyronines are stored in the thyroid follicle as part of thyroglobulin. There are hot spots (regions for very active iodination) in the

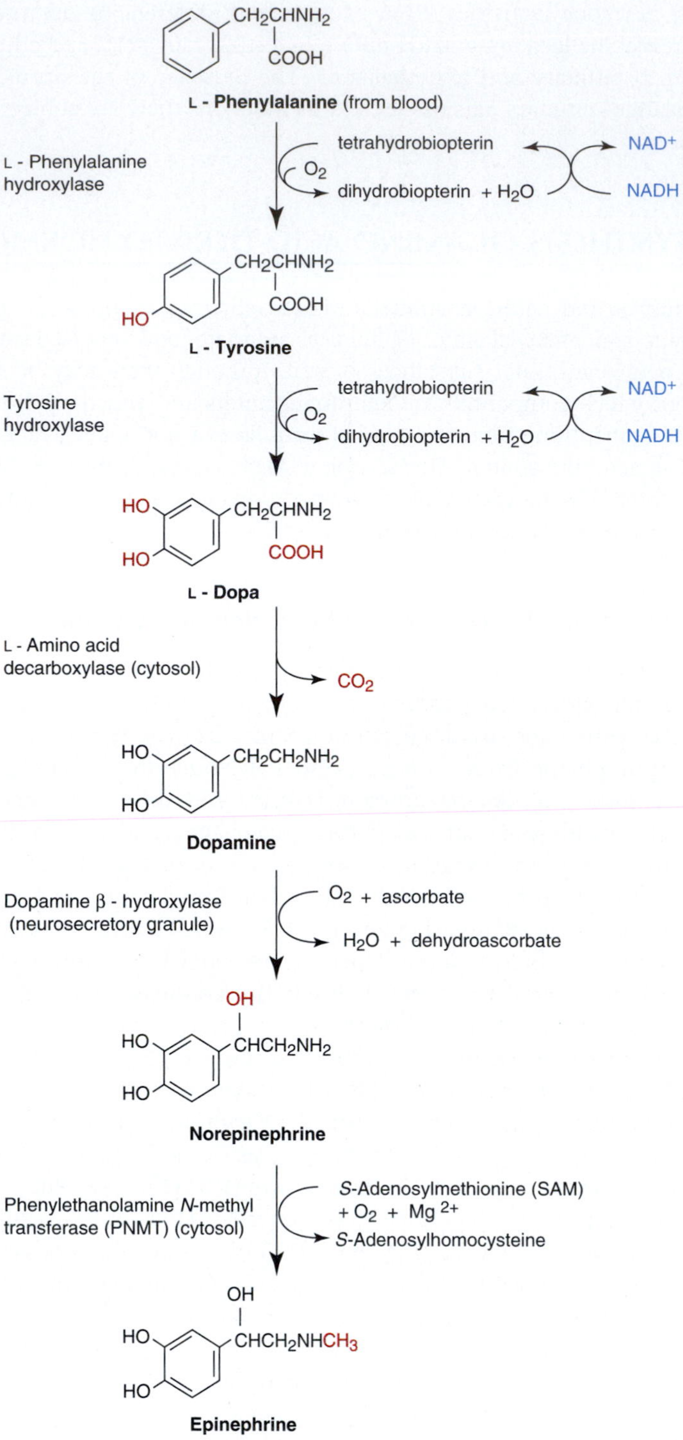

L - Phenylalanine (from blood)

L - Phenylalanine hydroxylase: tetrahydrobiopterin, O_2 → dihydrobiopterin + H_2O; NAD^+ / NADH

L - Tyrosine

Tyrosine hydroxylase: tetrahydrobiopterin, O_2 → dihydrobiopterin + H_2O; NAD^+ / NADH

L - Dopa

L - Amino acid decarboxylase (cytosol): → CO_2

Dopamine

Dopamine β - hydroxylase (neurosecretory granule): O_2 + ascorbate → H_2O + dehydroascorbate

Norepinephrine

Phenylethanolamine *N*-methyl transferase (PNMT) (cytosol): *S*-Adenosylmethionine (SAM) + O_2 + Mg^{2+} → *S*-Adenosylhomocysteine

Epinephrine

FIGURE 21.10
Synthesis of epinephrine by the chromaffin cell of the adrenal medulla.

thyroglobulin sequence for the incorporation of iodine. Sequences around iodotyrosyls occur in three consensus groups: Glu/Asp-Tyr, associated with the synthesis of thyroxine or iodotyrosines; Ser/Thr-Tyr-Ser, associated with the synthesis of iodothyronine and iodotyrosine; and Glu-X-Tyr, associated with the remaining iodotyrosyls in the sequence. Secretion of T_3 and T_4 into the bloodstream requires endocytosis of the thyroglobulin stored in the follicle and its proteolysis within the epithelial cell. The released DIT and MIT are then deiodinated and the released iodide ions are recycled and reutilized for hormone synthesis.

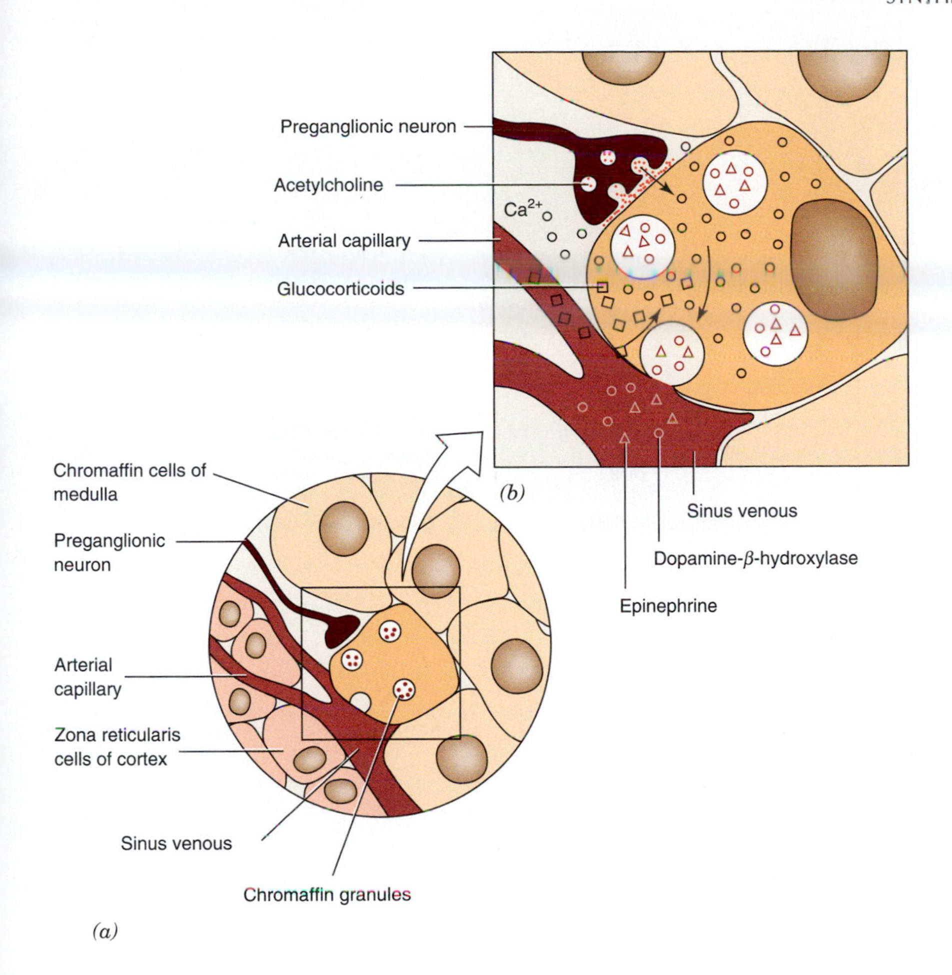

FIGURE 21.11
Relationship of adrenal medulla chromaffin cells to preganglionic neuron innervation and the structural elements involved in the synthesis of epinephrine and the discharge of catecholamines in response to acetylcholine.
(*a*) Functional relationship between cortex and medulla for control of synthesis of adrenal catecholamines. Glucocorticoids that stimulate enzymes catalyzing the conversion of norepinephrine to epinephrine reach the chromaffin cells from capillaries shown in (*b*). (*b*) Discharge of catecholamines from storage granules in chromaffin cells after nerve fiber stimulation resulting in the release of acetylcholine. Calcium enters the cells as a result, causing the fusion of granular membranes with the plasma membrane and exocytosis of the contents.
Redrawn from Krieger, D. T. and Hughes, J. C. (Eds.). Neuroendocrinology. *Sunderland, MA: Sinauer Associates, 1980.*

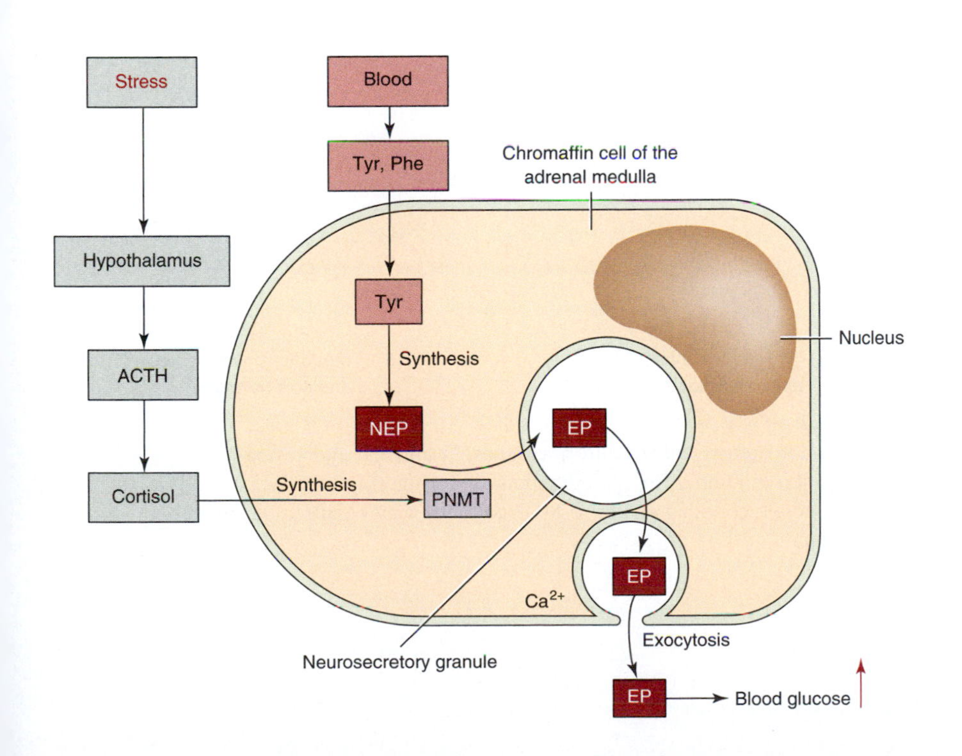

FIGURE 21.12
Biosynthesis, packaging, and release of epinephrine in the adrenal medulla chromaffin cell.
PNMT, phenylethanolamine *N*-methyltransferase; EP, epinephrine; NEP, norepinephrine. Neurosecretory granules contain epinephrine, dopamine β-hydroxylase, ATP, Met- or Leu-enkephalin, as well as larger enkephalin-containing peptides or norepinephrine in place of epinephrine. Epinephrine and norepinephrine are stored in different cells. Enkephalins could also be contained in separate cells, although that is not completely clear.
Adapted from Norman, A. W. and Litwack, G. Hormones. *New York: Academic Press, 1987, p. 464.*

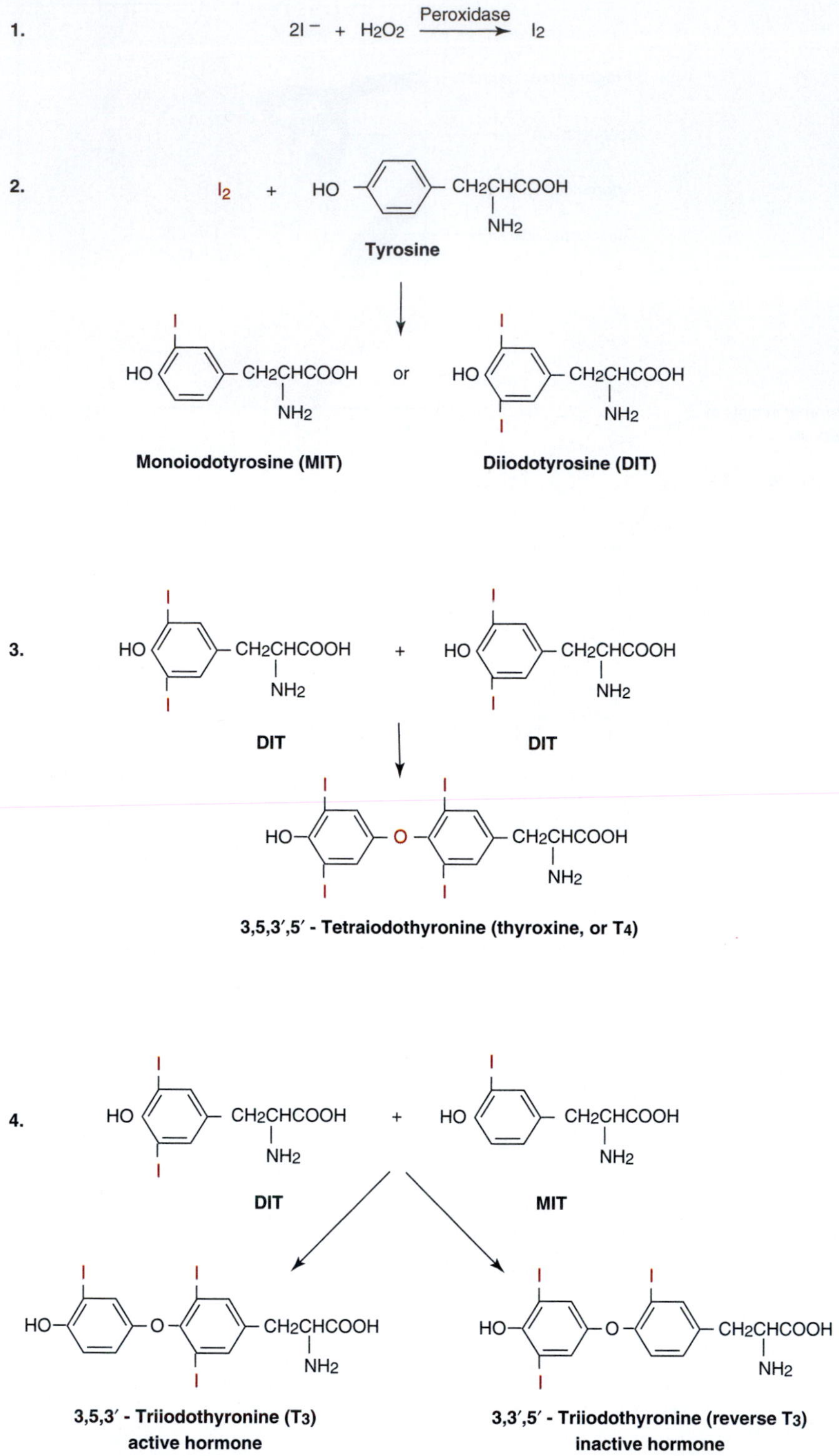

1. $2I^- + H_2O_2 \xrightarrow{\text{Peroxidase}} I_2$

2. I_2 + HO–C_6H_4–$CH_2CH(NH_2)COOH$

Tyrosine

HO–C_6H_3(I)–$CH_2CH(NH_2)COOH$ or HO–C_6H_2(I_2)–$CH_2CH(NH_2)COOH$

Monoiodotyrosine (MIT) | Diiodotyrosine (DIT)

3. DIT + DIT

3,5,3′,5′ - Tetraiodothyronine (thyroxine, or T_4)

4. DIT + MIT

3,5,3′ - Triiodothyronine (T_3) active hormone

3,3′,5′ - Triiodothyronine (reverse T_3) inactive hormone

FIGURE 21.13
Synthesis and structures of thyroid hormones T_4 and T_3, and reverse T_3.
Step 1, oxidation of iodide; Step 2, iodination of tyrosine residues; Step 3, coupling of DIT to DIT; Step 4, coupling of DIT to MIT (coupling may be intramolecular or intermolecular).

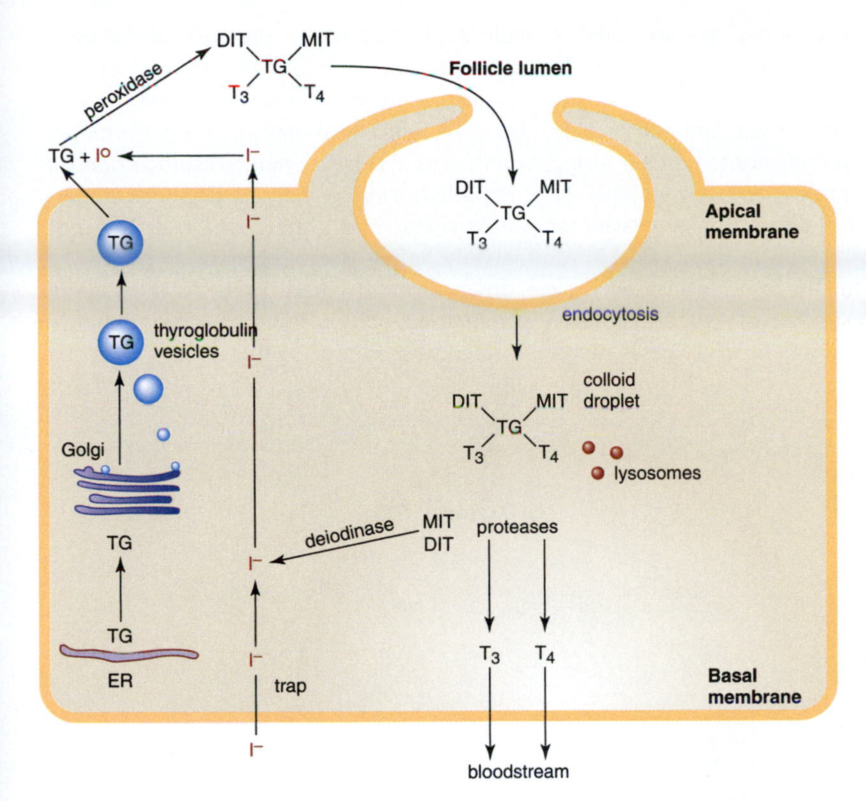

FIGURE 21.14
Cellular mechanisms for T_3 and T_4 release into the bloodstream.
Iodide trapping by basal membrane of the thyroid epithelium concentrates iodide approximately 30-fold. Secretion of T_3 and T_4 into bloodstream requires endocytosis of thyroglobulin and subsequent proteolysis. DIT and MIT are deiodinated and the released iodide ions are reutilized for hormone synthesis.
Redrawn from Berne, R. M. and Levy, M. L. (Eds.) Physiology, *2nd ed. St. Louis, MO: C.V. Mosby, 1988, p. 938.*

21.6 | INACTIVATION AND DEGRADATION OF HORMONES

Most polypeptide hormones are degraded to amino acids by proteinases presumably in **lysosomes.** Certain hormones, however, contain modified amino acids; for example, among the hypothalamic releasing hormones, the N-terminal amino acid can be **cycloglutamic acid** (or pyroglutamic acid) (Table 21.4) and a C-terminal amino acid amide. Some of the releasing hormones that have either or both of these

TABLE 21.4 Hypothalamic Releasing Hormones Containing an N-Terminal Pyroglutamate,[a] a C-Terminal Amino Acid Amide, or Both

Hormone	*Sequence*
Thyrotropin-releasing hormone (TRH)	*pGlu*-H-*Pro-NH*$_2$
Gonadotropin-releasing hormone (GnRH)	*pGlu*-HWSYGLRP-*Gly-NH*$_2$
Corticotropin-releasing hormone (CRH)	SQEPPISLDLTFHLLREVLEMTKADQLAQQAHSNRKLLDI-*Ala-NH*$_2$
Somatocrinin (GRH)	YADAIFTNSYRKVLGQLSARKLLQDIMSRQQGESNQERG-ARAR-*Leu-NH*$_2$

[a]The pyroglutamate structure is

H H
H– –H
–H
O= N C- - - -›Peptide
H ‖
O

[b]Single-letter abbreviations used for amino acids: Ala, A; Arg, R; Asn, N; Asp, D; Cys, C; Glu, E; Gln, Q; Gly, G; His, H; Ile, I; Leu, L; Lys, K; Met, M; Phe, F; Pro, P; Ser, S; Thr, T; Trp, W; Tyr, Y; Val, V.

TABLE 21.5 Examples of Hormones Containing a Cystine Disulfide Bridge Structure

Hormone	*Sequence*[a]
Somatostatin (GH)	FFNKCGA[1] / | W S | | K S \ | TFTSC[14]
Oxytocin	YC[1] / | I S | | E S \ | NCPLG—NH_2
Arginine vasopressin	YC / | F S | | Q S \ | NCPRG—NH_2

[a]Letters refer to single-letter amino acid abbreviations (see Table 21.4).

amino acid derivatives are listed in Table 21.4. Apparently, breakage of the cyclic glutamate ring or cleavage of the C-terminal amide leads to inactivation of many of these hormones and such enzymic activities have been reported in blood. This activity probably accounts, in part, for the short half-life of many of these hormones.

Some hormones contain a ring structure joined by a cystine disulfide bond. A few examples are given in Table 21.5. Peptide hormones may be degraded initially by the random action of **cystine aminopeptidase** and **glutathione transhydrogenase** as shown in Figure 21.15. Alternatively, as has been suggested in the case of oxytocin, the peptide may be broken down through partial proteolysis to shorter peptides, some of which may have hormonal actions on their own. Maturation of **prohormones** in many cases involves proteolysis (Figure 21.5).

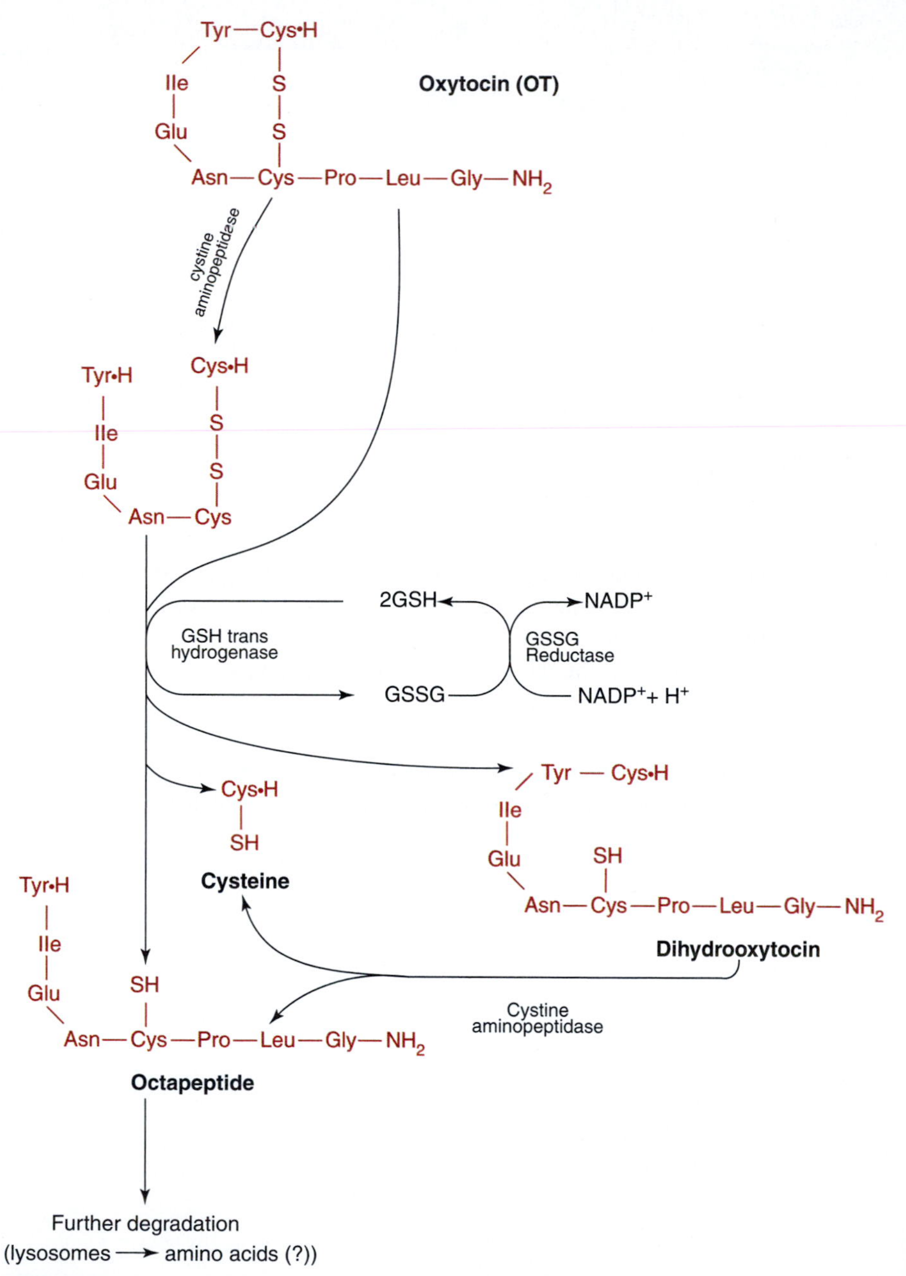

FIGURE 21.15
Degradation of posterior pituitary hormones.
Oxytocin transhydrogenase is similar to degrading enzymes for insulin; presumably, these enzymes also degrade vasopressin.
Redrawn from Norman, A. W. and Litwack, G. Hormones. *New York: Academic Press, 1987, p. 167.*

21.7 | CELL REGULATION AND HORMONE SECRETION

Hormonal secretion is under specific control. In the cascade system displayed in Figures 21.2 and 21.3, hormones must emanate from one source, cause hormonal release from the next cell type in line, and so on, down the cascade system. The correct responses must follow from a specific stimulus. The precision of these signals is defined by the hormone and the receptor as well as by the activities of the CNS, which precedes the first hormonal response in many cases. Certain generalizations can be made. Polypeptide hormones generally bind to their cognate receptors located in cell membranes. The receptor recognizes structural features of the hormone to generate a high degree of specificity and affinity. The affinity constants for these interactions are in the range of 10^9–10^{11} M^{-1}, representing tight binding. This interaction usually activates or complexes with a transducing protein in the membrane, such as a G-protein (GTP-binding protein), or other transducer and causes an activation of some enzymatic function on the cytoplasmic side of the membrane. In some cases receptors undergo **internalization** to the cell interior; these receptors may or may not (e.g., the insulin receptor) be coupled to transducing proteins in the cell membrane. A discussion of internalization of receptors is presented in Section 21.11. The "activated" receptor complex could physically open a membrane ion channel or have other profound impacts on membrane structure and function. For example, binding of the hormone to the receptor may cause conformational changes in the receptor molecule, enabling it to associate with transducer in which further conformational changes may occur to permit interaction with an enzyme on the cytoplasmic side of the plasma membrane. This interaction may cause conformational changes in an enzyme so that its catalytic site becomes active.

G-Proteins Serve as Cellular Transducers of Hormone Signals

Most transducers of receptors in the plasma membrane are GTP-binding proteins and are referred to as G-proteins. **G-Proteins** consist of three types of **subunits—α, β, and γ.** The **α subunit** is the guanine nucleotide-binding component and is thought to interact with the receptor indirectly through the β and γ subunits and then directly with an enzyme, such as adenylate cyclase, resulting in enzyme activation. Actually there are two forms of the α subunit, designated α_s for a stimulatory subunit and α_i for an inhibitory subunit. Two types of receptors, and thus hormones, control the adenylate cyclase reaction: hormone receptors that lead to stimulation of the adenylate cyclase and those that lead to inhibition of the cyclase. This is depicted in Figure 21.16 with an indication of the role of α_s and α_i and some of the hormones that interact with the stimulatory and inhibitory receptors.

The events that occur when hormone and receptor interact are presented in Figure 21.17. The sequence is as follows: receptor binds hormone in the membrane (Step 1); which produces a conformational change in receptor to expose a site for G-protein (β, γ dimer) attachment (Step 2); G-protein can be either stimulatory, G_s, or inhibitory, G_i, referring to the ultimate effects on the activity of adenylate cyclase; the receptor interacts with β, γ dimer of G-protein, enabling the α subunit to exchange GTP for bound GDP (Step 3); dissociation of GDP causes separation of G-protein α subunit from β, γ dimer and the α-binding site for interaction with adenylate cyclase appears on the surface of the G-protein α subunit (Step 4); α subunit binds to adenylate cyclase and activates the catalytic center, so that ATP is converted to cAMP (Step 5); GTP is hydrolyzed to GDP by the GTPase activity of the α subunit, returning it to its original conformation and allowing its interaction with β, γ subunit once again (Step 6); GDP associates with the α subunit and the system is returned to the unstimulated state awaiting another cycle of activity. It is important to note that there is also evidence suggesting that β, γ dimers may have an important role in regulating certain effectors including adenylate cyclase.

Where an inhibitory G-protein is coupled to the receptor, the events are similar but inhibition of adenylate cyclase activity may arise by direct interaction of the

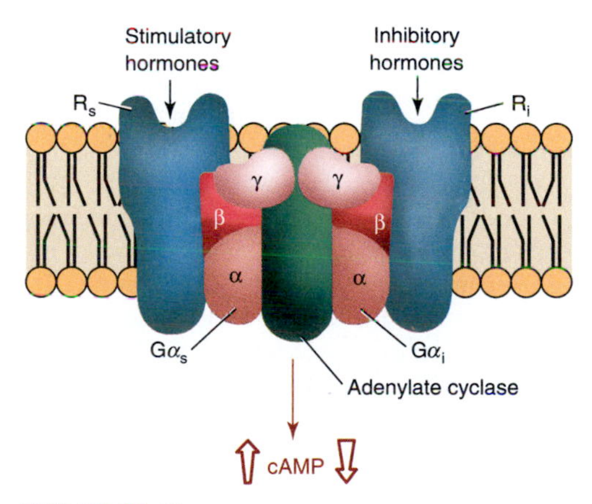

FIGURE 21.16
Components of a hormone-sensitive adenylate cyclase system.
Adenylate cyclase converts of ATP to cAMP. The occupancy of R_s by stimulatory hormones stimulates adenylate cyclase via formation of an active dissociated $G\alpha_s$ subunit. The occupancy of R_i by inhibitory hormones forms an "active" $G\alpha_i$ complex and concomitant reduction in cycling activity. The fate of β and γ subunits in these dissociation reactions is not yet known. R_s, stimulatory hormone receptor; R_i, inhibitory hormone receptor.

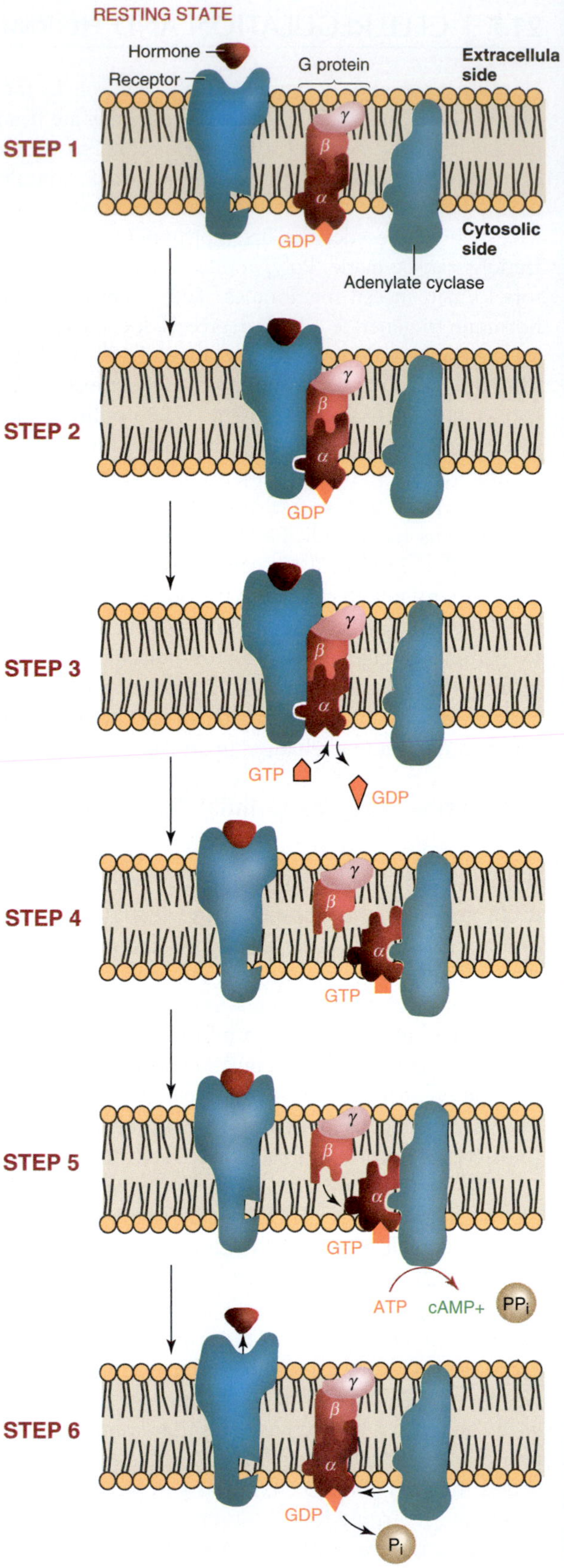

FIGURE 21.17
Activation of adenylate cyclase by binding of a hormone to its receptor. The cell membrane is depicted, which contains on its outer surface a receptor protein for a hormone. On the inside surface of the membrane is adenylate cyclase protein and the transducer protein G. In the resting state GDP is bound to the α subunit of the G-protein. When a hormone binds to the receptor, a conformational change occurs (Step 1). The activated receptor binds to the G-protein (Step 2), which activates the latter so that it releases GDP and binds GTP (Step 3), causing the α and the complex of β and γ subunits to dissociate (Step 4). Free Gα subunit binds to the adenylate cyclase and activates it so that it catalyzes the synthesis of cAMP from ATP (Step 5); this step may involve a conformational change in Gα. In some cases the β, γ complex may play an important role in regulation of certain effectors including adenylate cyclase. When GTP is hydrolyzed to GDP, a reaction most likely catalyzed by Gα itself, Gα is no longer able to activate adenylate cyclase (Step 6), and Gα and Gβ, γ reassociate. The hormone dissociates from the receptor and the system returns to its resting state.
Redrawn from Darnell, J., Lodish, H., and Baltimore, D. Molecular Cell Biology. *New York: Scientific American Books, 1986, p. 682.*

TABLE 21.6 Activities Transduced by G-Protein Subfamilies

α Subunit	*Expression*	*Effector*
G_s	Ubiquitous	↑ Adenylate cyclase, Ca^{2+} channel
G_{olf}	Olfactory	↑ Adenylate cyclase
G_{+1} (transducin)	Rod photoreceptors	↑ cGMP-phosphodiesterase
G_{+2} (transducin)	Cone photoreceptors	↑ cGMP-phosphodiesterase
G_{i1}	Neural > other tissues	↓ Adenylate cyclase
G_{i2}	Ubiquitous	
G_{i3}	Other tissues > neural	
G_o	Neural, endocrine	↓ Ca^{2+} channel
G_q	Ubiquitous	↑ Phospholipase C
G_{11}	Ubiquitous	
G_{14}	Liver, lung, kidney	
$G_{15/16}$	Blood cells	

Source: Adapted from Spiegel, A.M., Shenker, A., and Weinstein, L. S. *Endocr. Rev.* 13:536, 1992.

inhibitory α subunit with adenylate cyclase or, alternatively, the inhibitory α subunit may interact directly with the stimulatory α subunit on the other side and prevent the stimulation of adenylate cyclase activity indirectly. Immunochemical evidence suggests multiple G_i subtypes and molecular cloning of cDNAs encoding putative α subunits has provided evidence for multiple α_i subtypes.

Purification and biochemical characterization of G-proteins (G_s as well as G_i) have revealed somewhat unanticipated diversity in this subfamily. Polymerase chain reaction-based cloning has now brought the number of distinct genes encoding mammalian α subunits to at least 15. With regard to α subunits, further diversity is achieved by alternative splicing of the α_s (four forms) gene. There also appears to be diversity among the mammalian β and γ subunits. At least four distinct β subunit cDNAs and probably as many γ subunits have been described. What is not clear is how these complexes combine to form distinct β,γ complexes. Some data suggest that different β,γ complexes may have distinct properties with respect to α subunit and receptor interactions, but additional research will be required to fully describe these unique interactions. Table 21.6 lists some activities transduced by G-protein subfamilies.

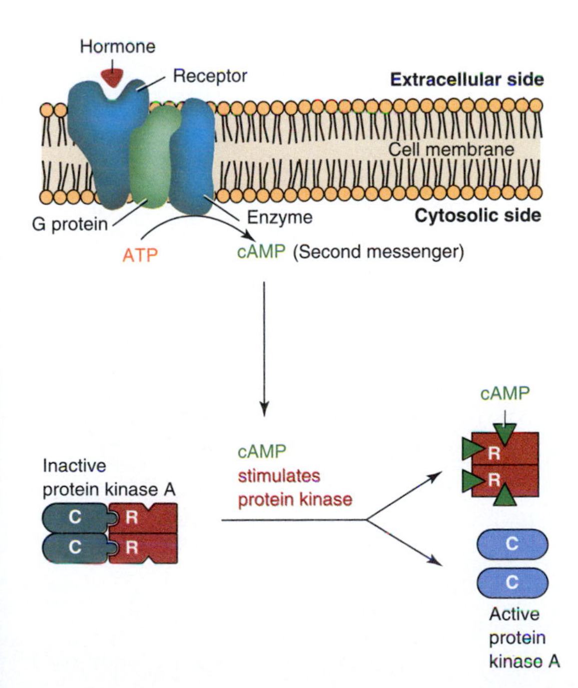

FIGURE 21.18
Activation of protein kinase A.
Hormone–receptor-mediated stimulation of adenylate cyclase and subsequent activation of protein kinase A. C, catalytic subunit; R, regulatory subunit.

Cyclic AMP Activates Protein Kinase A Pathway

The generation of cAMP usually activates protein kinase A, referred to as the **protein kinase A pathway.** The overall pathway is presented in Figure 21.18. Four cAMP molecules are used in the reaction to complex two regulatory subunits (R) and liberate two protein kinase catalytic subunits (C). The liberated catalytic subunits are able to phosphorylate proteins to produce a cellular effect. In many cases the cellular effect leads to the release of preformed hormones. For example, ACTH binds to membrane receptors, elevates intracellular **cAMP** levels, and releases cortisol from the zona fasciculata cells of the adrenal gland by this general mechanism. Part of the mechanism of release of thyroid hormones from the thyroid gland involves the cAMP pathway as outlined in Figure 21.19. TSH has been shown to stimulate numerous key steps in this secretory process, including iodide uptake and endocytosis of thyroglobulin (Figure 21.14). The protein kinase A pathway is also responsible for the release of testosterone by testicular Leydig cells as presented in Figure 21.20. There are many other examples of hormonal actions mediated by cAMP and the protein kinase A pathway.

Inositol Trisphosphate Formation Leads to Release of Ca^{2+} from Intracellular Stores

Uptake of calcium from the cell exterior through calcium channels may be affected directly by hormone–receptor interaction at the cell membrane. In some cases, ligand–receptor interaction is thought to open calcium channels directly in the cell

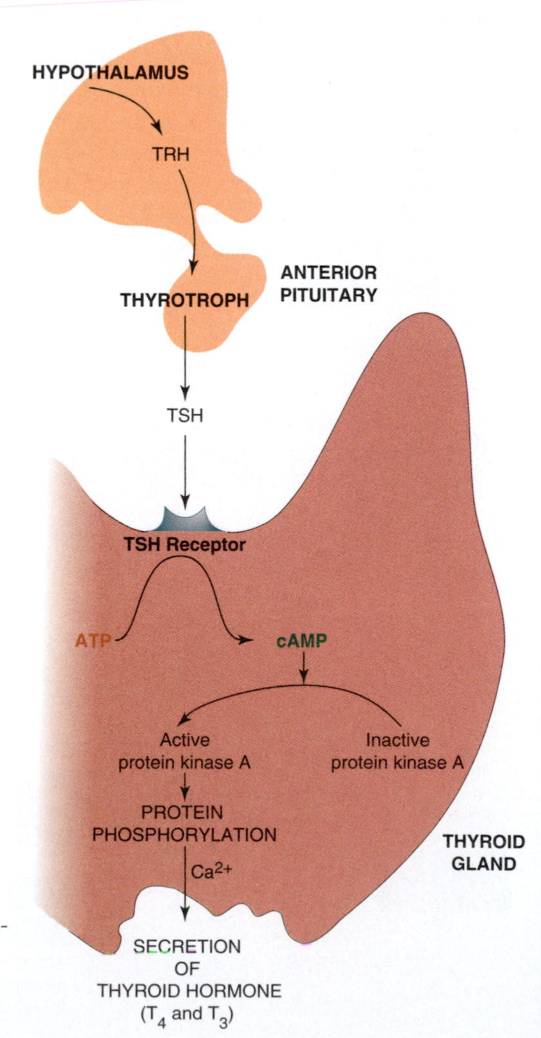

FIGURE 21.19
Effect of TSH on secretion of thyroid hormone.
TSH stimulates all steps in the synthesis and secretion of T_3 and T_4. These effects are mediated by binding of hormone to TSH receptors located on basal membrane of thyroid epithelial cells, elevation of cAMP levels, and subsequent cascade of phosphorylation reactions.

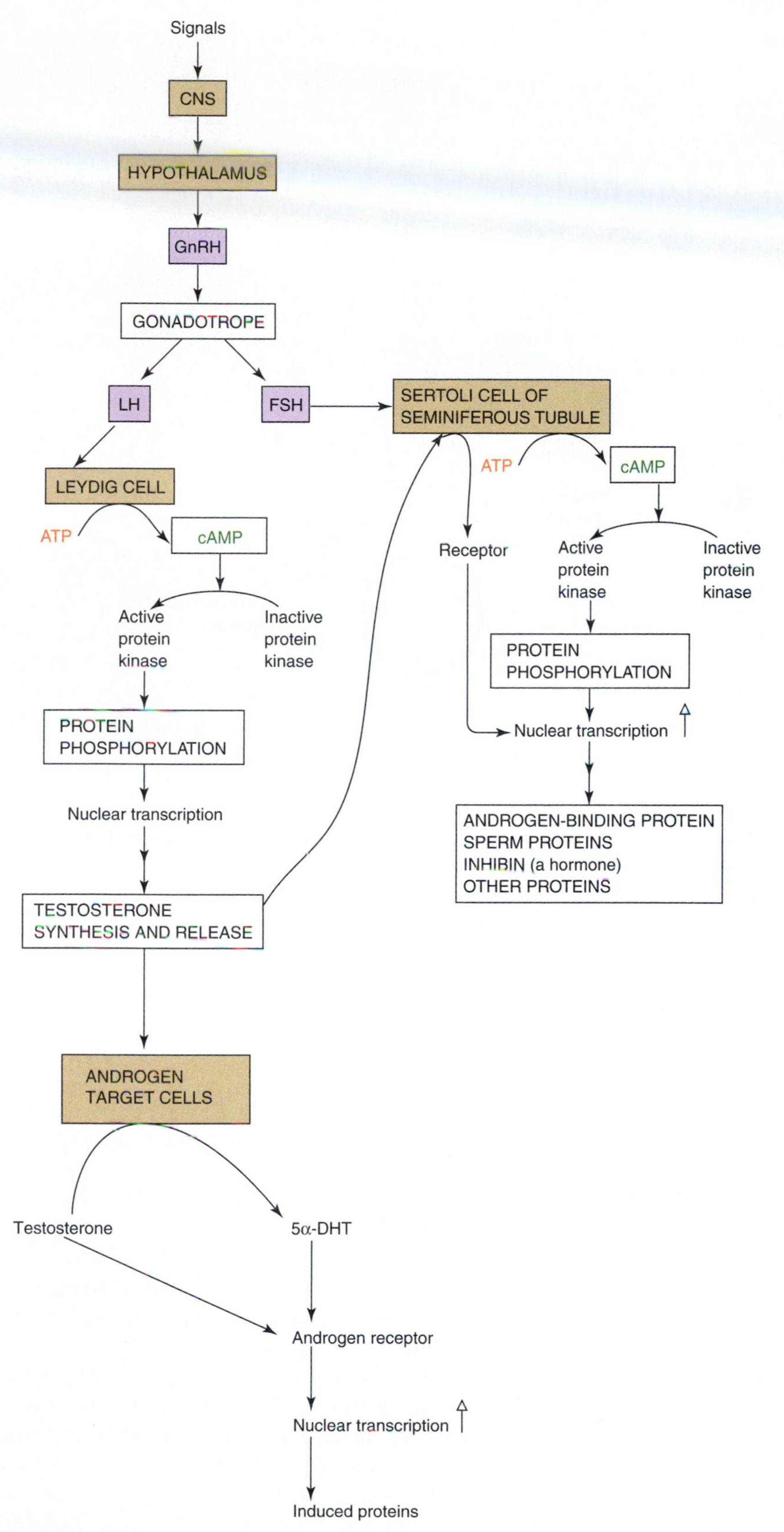

FIGURE 21.20
Overview of the secretion controls, some general actions of the gonadotropes, and testosterone release in males.
In some, but not all, androgen target cells, testosterone is reduced to the more potent androgen, 5α-dihydrotestosterone (5α-DHT).

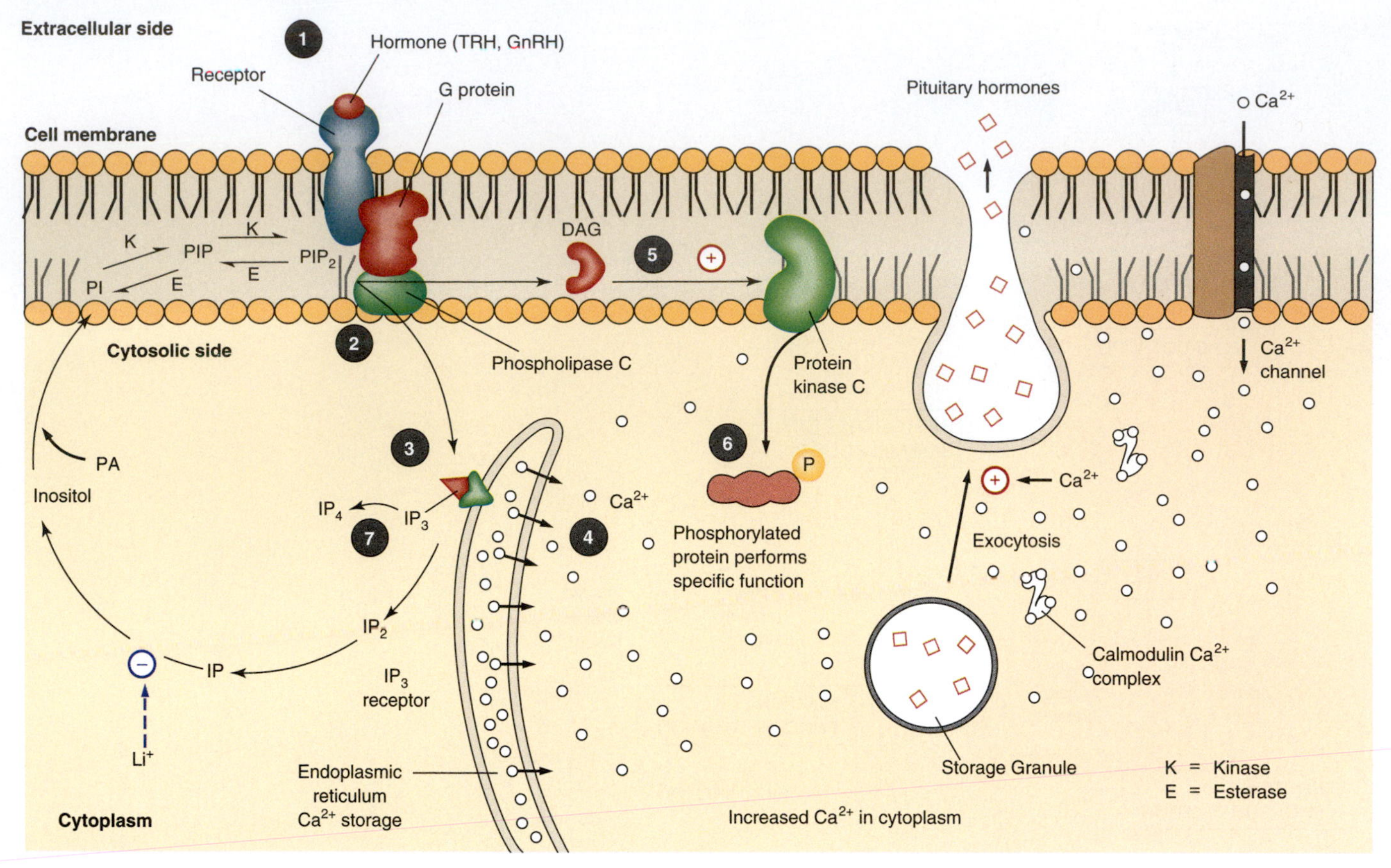

FIGURE 21.21

Overview of hormonal signaling through the phosphatidylinositol system generating the second messengers, inositol-1,4,5-trisphosphate (IP_3) and diacylglycerol (DAG).

The action of IP_3 is to increase cytoplasmic Ca^{2+} levels by a receptor-mediated event in the cellular calcium store. Steps in pathway: (1) binding of hormone to cell membrane receptor; (2) production of IP_3 from PIP_2; (3) binding of IP_3 to receptor on calcium storage site; (4) release of free calcium to the cytosol; (5) release of DAG and subsequent binding to protein kinase C; (6) phosphorylation of protein substrates by protein kinase C activated by DAG and Ca^{2+}; and (7) phosphorylation of IP_3 to yield IP_4. DAG, diacylglycerol; PA, phosphatidic acid; IP, inositol phosphate; IP_2, inositol bisphosphate; IP_3, inositol 1,4,5-trisphosphate; IP_4, inositol 1,3,4,5-tetrakisphosphate; PIP, phosphatidylinositol phosphate; PIP_2, phosphatidylinositol 4,5-bisphosphate; K, Kinase; E, esterase.

membrane (see p. 516). Another system to increase intracellular Ca^{2+} concentration derives from hormone–receptor activation of **phospholipase C** activity transduced by a G-protein (Figure 21.21).

A hormone operating through this system binds to a specific cell membrane receptor, which interacts with a G-protein in a mechanism similar to that of the protein kinase A pathway and transduces the signal, resulting in stimulation of phospholipase C. This enzyme catalyzes the hydrolysis of **phosphatidylinositol-4,5-bisphosphate** (PIP_2) to form two **second messengers, diacylglycerol (DAG)** and **inositol 1,4,5-triphosphate (IP_3).**

IP_3 diffuses to the cytosol and binds to an IP_3 receptor on the membrane of a particulate **calcium** store, either separate from or part of the endoplasmic reticulum. IP_3 binding results in the release of calcium ions contributing to the large increase in cytosolic Ca^{2+} levels. Calcium ions are important in exocytosis by facilitating fusion of secretory granules to the cytoplasmic side of the plasma membrane, in microtubular aggregation, or in the function of contractile proteins, which may be part of the structure of the exocytosis, or all of these.

The IP_3 is metabolized by stepwise removal of phosphate groups (Figure 21.21) to form inositol. This combines with phosphatidic acid (PA) to form phosphatidylinositol (PI) in the cell membrane. PI is phosphorylated twice by a kinase to form PIP_2, which is ready to undergo another round of hydrolysis and formation

CLINICAL CORRELATION 21.3

Lithium Treatment of Manic–Depressive Illness: The Phosphatidylinositol Cycle

Lithium has been used for years in the treatment of manic–depression. Our newer knowledge suggests that lithium therapy involves the phosphatidylinositol (PI) pathway. This pathway generates the second messengers inositol 1,4,5-trisphosphate (IP_3) and diacylglycerol following the hormone/neurotransmitter–membrane receptor interaction and involves the G-protein complex and activation of phospholipase C. IP_3 and its many phosphorylated derivatives are ultimately dephosphorylated in a stepwise fashion to generate free inositol. Inositol is then used for the synthesis of phosphatidylinositol monophosphate. The phosphatase that dephosphorylates IP to inositol is inhibited by Li^+. In addition, Li^+ may also interfere directly with G-protein function. The result of Li^+ inhibition is that the PI cycle is greatly slowed even in the face of continued hormonal/neurotransmitter stimulation and the cell becomes less sensitive to these stimuli. Manic–depressive illness may occur through the overactivity of certain CNS cells, perhaps as a result of abnormally high levels of hormones or neurotransmitters whose actions are to stimulate the PI cycle. The chemotherapeutic effect of the Li^+ could be to decrease the cellular responsiveness to elevated levels of agents that might promote high levels of PI cycle and precipitate manic–depressive illness.

Avissar, S. and Schreiber, G. Muscarinic receptor subclassification and G-proteins: significance for lithium action in affective disorders and for the treatment of the extrapyramidal side effects of neuroleptics. *Biol. Psychiatry* 26:113, 1989; Hallcher, L. M. and Sherman, W. R. The effects of lithium ion and other agents on the activity of myoinositol 1-phosphatase from bovine brain. *J. Biol. Chem.* 255:896, 1980; and Pollack, S. J., Atack, J., Knowles, M. R., McAllister, G., Ragan, C. I., Baker, R., Fletcher, S. R., Iversen, L. L., and Broughton, H. B. Mechanism of inositol monophosphatase, the putative target of lithium therapy. *Proc. Natl. Acad. Sci. USA* 91:5766, 1994.

of second messengers (DAG and IP_3) upon hormonal stimulation. If the receptor is still occupied by hormone, several rounds of the cycle could occur before the hormone–receptor complex dissociates or some other feature of the cycle becomes limiting. It is interesting that the conversion of inositol phosphate to inositol is inhibited by **lithium ion** (Li^+) (Figure 21.21). This could be the metabolic basis for the beneficial effects of Li^+ in manic–depressive illness (see Clin. Corr. 21.3). Finally, it is important to note that not all of the generated IP_3 is dephosphorylated during hormonal stimulation. Some of the IP_3 is phosphorylated via IP_3 kinase to yield inositol 1,3,4,5-tetrakisphosphate (IP_4), which may mediate some of the slower or more prolonged hormonal responses or facilitate replenishment of intracellular Ca^{2+} stores from the extracellular fluid, or both.

Diacylglycerol Activates Protein Kinase C Pathway

At the same time that IP_3 produced by hydrolysis of PIP_2 is increasing the concentration of Ca^{2+} in the cytosol, the other cleavage product, DAG, mediates different effects. Importantly, DAG activates a crucial serine/threonine protein kinase called protein kinase C because it is Ca^{2+} dependent (details of protein kinase C are discussed on p. 948). The initial rise in cytosolic Ca^{2+} induced by IP_3 is believed to somehow alter kinase C so that it translocates from the cytosol to the cytoplasmic face of the plasma membrane. Once translocated, it is activated by a combination of Ca^{2+}, DAG, and the negatively charged membrane phospholipid, phosphatidylserine. Once activated, protein kinase C then phosphorylates specific proteins in the cytosol or, in some cases, in the plasma membrane. These phosphorylated proteins can be either activated or inactivated by phosphorylation. For example, a phosphorylated protein could potentially migrate to the nucleus and stimulate mitosis and growth. It is also possible that a phosphorylated protein could play a role in the secretion of preformed hormones.

21.8 | CYCLIC HORMONAL CASCADE SYSTEMS

Hormonal cascade systems are generated by external signals as well as by internal signals. Examples of this are the **diurnal variations** in levels of cortisol secretion from the adrenal cortex probably initiated by serotonin and vasopressin, in the secretion of melatonin from the pineal gland, and the internal regulation of the ovarian cycle. Some of these biorhythms operate on a cyclic basis, often dictated by daylight and darkness, and are referred to as **chronotropic control** of hormone secretion.

Melatonin and Serotonin Syntheses Are Controlled by Light/ Dark Cycles

The release of melatonin from the pineal gland, presented in Figure 21.22, is an example of a biorhythm. Here, as in other such systems, the internal signal is provided by a neurotransmitter, in this case norepinephrine produced by an adrenergic neuron. In this system, control is exerted by light entering the eyes and is transmitted to the pineal gland by way of the CNS. The adrenergic neuron innervating the pinealocyte is inhibited by light transmitted through the eyes. Norepinephrine released as a neurotransmitter in the dark stimulates cAMP formation through a β receptor in the pinealocyte cell membrane, which leads to the enhanced synthesis of **N-acetyltransferase.** The increased activity of this enzyme causes the conversion of **serotonin** to **N-acetylserotonin,** and **hydroxyindole-*O*-methyltransferase (HIOMT)** then catalyzes the conversion of *N*-acetylserotonin to **melatonin,** which is secreted in the dark hours but not during light hours. Melatonin-sensitive cells contain receptors that generate effects on reproductive and other functions. For example, melatonin exerts an antigonadotropic effect, although the physiological significance of this effect is unclear.

Ovarian Cycle Is Controlled by Gonadotropin-Releasing Hormone

An example of a pulsatile release mechanism is regulation of the periodic release of GnRH. A periodic control regulates the release of GnRH at definitive periods (of about 1 h in higher animals) and is controlled by aminergic neurons, which may be adrenergic (norepinephrine secreting) in nature. The initiation of this function occurs at puberty and is important in both the male and female. While the male system functions continually, the female system is periodic and known as the **ovarian cycle.** This system is presented in Figure 21.23. In the male, the cycling center in the CNS does not develop because its development is blocked by androgens before birth. In the female, a complicated set of signals needs to be organized

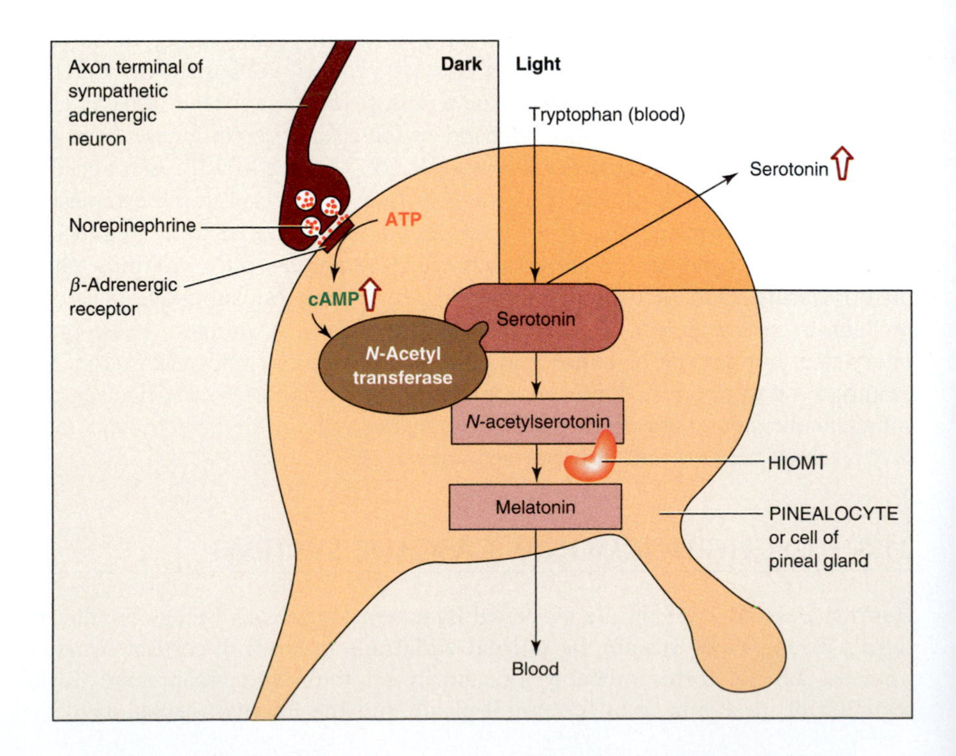

FIGURE 21.22
Biosynthesis of melatonin in pinealocytes. HIOMT, hydroxyindole-*O*-methyltransferase. *Redrawn from Norman, A. W. and Litwack, G.* Hormones. *New York: Academic Press, 1987, p. 710.*

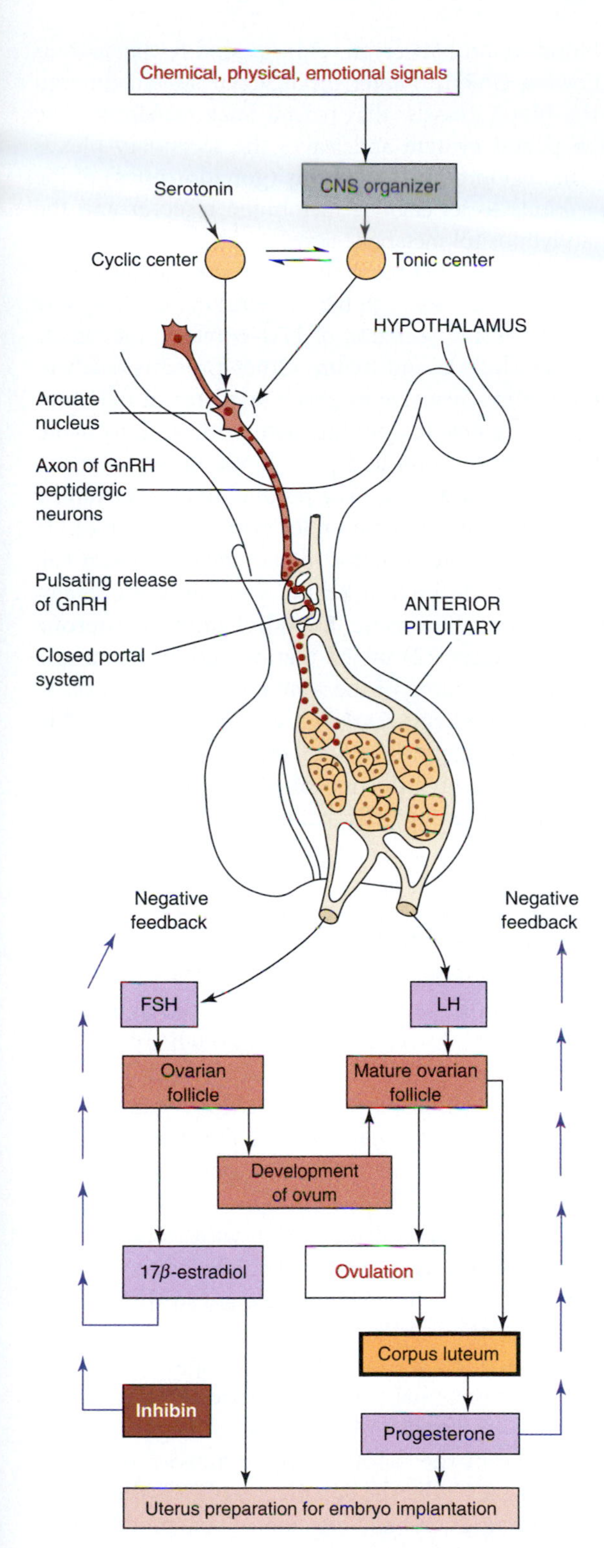

FIGURE 21.23
Ovarian cycle in terms of generation of hypothalamic hormone, pituitary gonadotropic hormones, and sex hormones.
To begin the cycle at puberty, several centers in the CNS coordinate with the hypothalamus so that GnRH is released in a pulsatile fashion. This causes the release of LH and FSH, which in turn affect the ovarian follicle, ovulation, and the corpus luteum. The hormone inhibin selectively inhibits FSH secretion. Products of the follicle and corpus luteum, respectively, are β-estradiol and progesterone. GnRH, gonadotropin-releasing hormone; FSH, follicle-stimulating hormone; LH, luteinizing hormone.

in the CNS before the initial secretion of GnRH occurs at puberty. The higher centers (CNS organizer) must harmonize with the tonic and cycling centers and these interact with each other to prime the hypothalamus. The pulsatile system, which innervates the arcuate nucleus of the hypothalamus, must also function for GnRH to be released, and this system apparently must be functional throughout life for these cycles to be maintained. Release of GnRH from the axon terminals of the cells that synthesize this hormone is followed by entry of the hormone into the primary plexus of the closed portal system connecting the hypothalamus and the anterior pituitary (Figure 21.23). The blood–brain barrier preventing peptide transport is overcome in this process by allowing GnRH to enter the vascular system through fenestrations, or openings in the blood vessels, that permit such transport. The GnRH is then carried down the **portal system** and leaves the secondary plexus through fenestrations, again, in the region of the target cells **(gonadotropes)** of the anterior pituitary. The hormone binds to its cognate membrane receptor and the signal, mediated by the phosphatidylinositol metabolic system, causes the release of both FSH and LH from the same cell. The **FSH** binds to its cognate membrane receptor on the ovarian follicle and, operating through the protein kinase A pathway via cAMP elevation, stimulates synthesis and secretion of 17β-estradiol, the female sex hormone, and maturation of the follicle and ovum. Other proteins, such as **inhibin,** are also synthesized. Inhibin is a negative feedback regulator of FSH production in the gonadotrope. When the follicle reaches full maturation and the ovum also is matured, LH binds to its cognate receptor and plays a role in ovulation together with other factors, such as prostaglandin $F_{2\alpha}$. The residual follicle remaining after ovulation becomes the functional corpus luteum under primary control of LH (Figure 21.23). The **LH** binds to its cognate receptor in the corpus luteum cell membrane and, through stimulation of the protein kinase A pathway, stimulates synthesis of progesterone, the progestational hormone. **Estradiol** and **progesterone** bind to intracellular receptors (see Chapter 22) in the uterine endometrium and cause major changes resulting in the thickening of the wall and vascularization in preparation for implantation of the fertilized egg. Estradiol, which is synthesized in large amount prior to production of progesterone, induces the progesterone receptor as one of its inducible phenotypes. This induction of progesterone receptors primes the uterus for subsequent stimulation by progesterone.

Absence of Fertilization

If fertilization of the ovum does not occur, the corpus luteum involutes as a consequence of diminished LH supply. Progesterone levels fall sharply in the blood with the regression of the corpus luteum. Estradiol levels also fall due to the cessation of its production by the corpus luteum. Thus the stimuli for a thickened and vascularized uterine endometrial wall are lost. Menstruation occurs through a process of apoptosis (programmed cell death) of the uterine endometrial cells until the endometrium reaches its unstimulated state. Ultimately, the fall in blood steroid levels releases the negative feedback inhibition on the gonadotropes and hypothalamus and the cycle starts again with release of FSH and LH by the gonadotropes in response to GnRH.

The course of the ovarian cycle is shown in Figure 21.24 with respect to the relative blood concentration of hormones released from the hypothalamus, anterior pituitary, ovarian follicle, and corpus luteum. In addition, changes in the maturation of the follicle and ovum as well as the uterine endometrium are shown. The steroid hormones, estradiol and progesterone, are discussed in Chapter 22.

The cycle first begins at puberty when GnRH is released, corresponding to day 1 in Figure 21.24. GnRH is released in a pulsatile fashion, causing the gonadotrope to release FSH and LH; there is a rise in the blood concentrations of these gonadotropic hormones in subsequent days. Under the stimulation of FSH the follicle begins to mature (lower section of Figure 21.24) and estradiol (E_2) is produced. In response to estradiol the uterine endometrium begins to thicken (there would have been no prior menstruation in the very first cycle). Eventually, under the continued action of FSH, the follicle matures with the maturing ovum, and extraordinarily high concentrations of estradiol are produced (around day 13 of the cycle). These levels of estradiol, instead of causing feedback inhibition, now generate, through **feedback stimulation,** a huge release of LH and to a lesser extent FSH from the gonadotrope. The FSH responds to a smaller extent due to the ovarian production of the hormone inhibin under the influence of FSH. Inhibin is a specific negative feedback inhibitor of FSH, but not of LH, and probably suppresses the synthesis of the β subunit of FSH. The high midcycle peak of LH is referred to as the "LH spike." Ovulation then

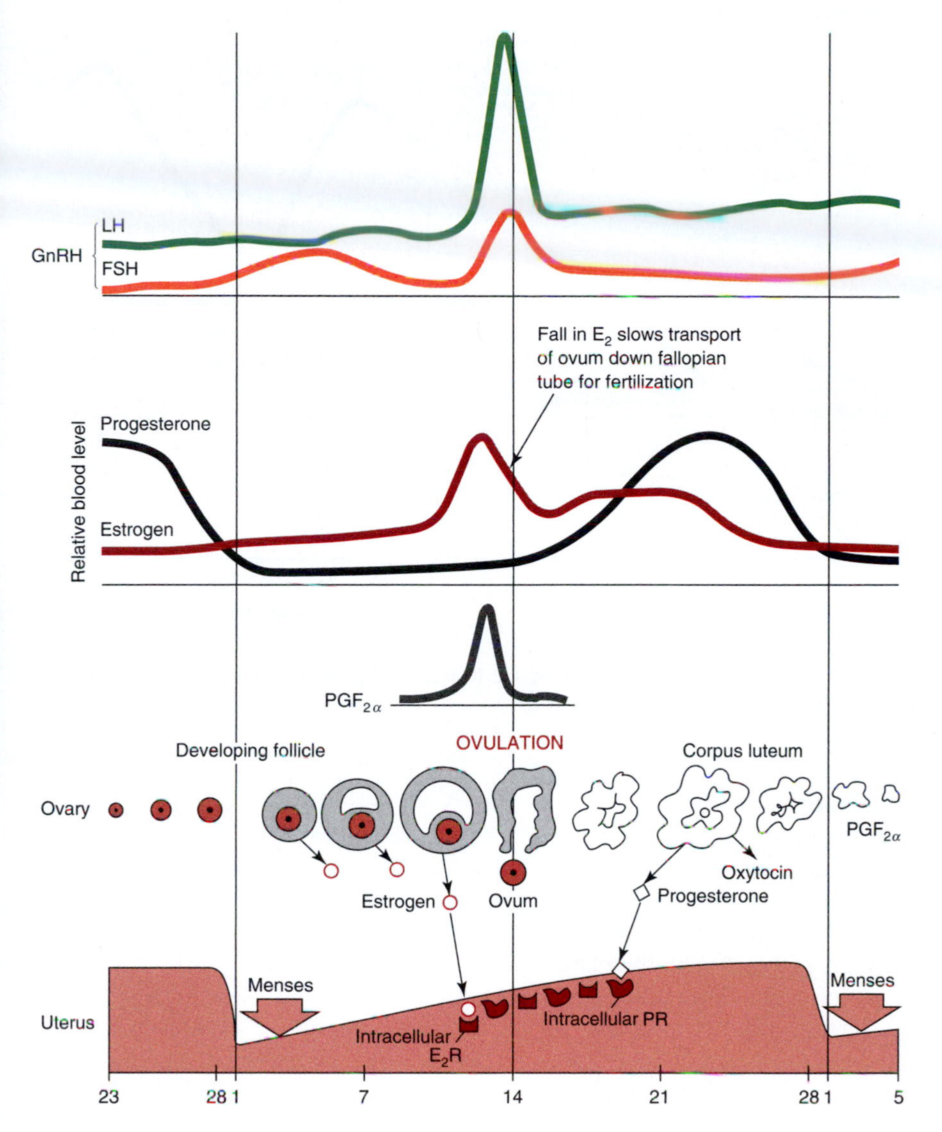

FIGURE 21.24
The ovarian cycle.
In the upper diagram, relative blood levels of GnRH, LH, FSH, progesterone, estrogen, and $PGF_{2\alpha}$ are shown. In the lower diagram, events in the ovarian follicle, corpus luteum, and uterine endometrium are diagrammed. GnRH, gonadotropin-releasing hormone; LH, luteinizing hormone; FSH, follicle-stimulating hormone; $PGF_{2\alpha}$, prostaglandin $F_{2\alpha}$; E_2, estradiol; E_2R, intracellular estrogen receptor; and PR, intracellular progesterone receptor.

occurs at about day 14 (midcycle) through the effects of high LH concentration together with other factors, such as $PGF_{2\alpha}$. Both LH and $PGF_{2\alpha}$ act on cell membrane receptors. After ovulation, the function of the follicle declines as reflected by the fall in blood estrogen concentration. The spent follicle now differentiates into the functional corpus luteum driven by the still high levels of blood LH (Figure 21.24, top). Under the influence of prior high levels of estradiol (estrogen) and the high levels of progesterone produced by the now functional corpus luteum, the uterine endometrial wall reaches its greatest development in preparation for implantation of the fertilized egg, should fertilization occur. Note that the previous availability of estradiol in combination with the estrogen receptor (E_2R) produces elevated levels of progesterone receptor (PR) within the cells of the uterine wall. The blood levels of estrogen fall with the loss of function of the follicle but some estrogen is produced by the corpus luteum in addition to the much greater levels of progesterone. In the absence of fertilization the corpus luteum continues to function for about 2 weeks, then involutes because of the loss of high levels of LH. The production of oxytocin by the corpus luteum itself and the production or availability of $PGF_{2\alpha}$ cause inhibition of progesterone synthesis and enhance luteolysis by a process of programmed cell death (see p. 985). With the death of the corpus luteum there is a profound decline in blood levels of estradiol and progesterone so that the thickened endometrial wall can no longer be maintained and menstruation occurs, followed by the start of another cycle with a new developing follicle.

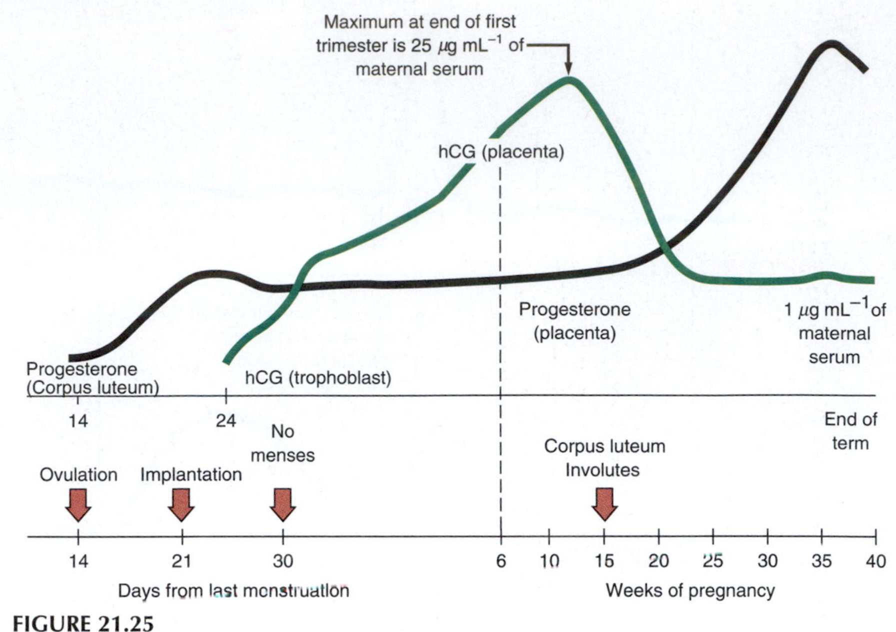

FIGURE 21.25
Effect of fertilization on the ovarian cycle in terms of progesterone and secretion of human chorionic gonadotropin (hCG).

Fertilization

The situation changes if fertilization occurs as shown in Figure 21.25. The corpus luteum, which would have ceased function by 28 days, remains viable due to the production of **chorionic gonadotropin** (CG), which resembles and acts like LH, from the trophoblast. Eventually, the production of **human chorionic gonadotropin** (hCG) is taken over by the placenta, which continues to produce the hormone at very high levels throughout most of the gestational period. Nevertheless, the corpus luteum, referred to as the "corpus luteum of pregnancy," eventually dies and, by about 12 weeks of pregnancy, the placenta has taken over the production of progesterone, which is secreted at high levels throughout pregnancy. Although both progesterone and estrogen are secreted in progressively greater quantities throughout pregnancy, from the seventh month onward estrogen secretion continues to increase while progesterone secretion remains constant or may even decrease slightly (Figure 21.25). The increased production of a progesterone-binding protein may also serve to lower the effective concentration of free progesterone in the myometrium. Thus the estrogen/progesterone ratio increases toward the end of pregnancy and may be partly responsible for the increased uterine contractions. Oxytocin secreted by the posterior pituitary also contributes to these uterine contractions. The fetal membranes also release prostaglandins ($PGF_{2\alpha}$) at the time of parturition and they also increase the intensity of uterine contractions. Finally, the fetal adrenal glands secrete cortisol, which not only stimulates fetal lung maturation by inducing surfactant but may also stimulate uterine contractions.

As mentioned before, the system in the male is similar, but less complex in that cycling is not involved, and it progresses much as outlined in Figure 21.25.

21.9 | HORMONE–RECEPTOR INTERACTIONS

Receptors are proteins and differ by their specificity for ligands and by their location in the cell (see Figure 21.1). The interaction of ligand with receptor essentially resembles a semienzymatic reaction:

$$\text{hormone} + \text{receptor} \rightleftharpoons \text{hormone–receptor complex}$$

The **hormone–receptor complex** usually undergoes conformational changes resulting from interaction with the hormonal ligand. These changes allow for a subsequent interaction with a transducing protein (G-protein) in the membrane or for activation to a new state in which active domains become available on the surface of the receptor. The mathematical treatment of the interaction of hormone and receptor is a function of the concentrations of the reactants, hormone [H] and receptor [R], in the formation of the hormone–receptor complex [RH], and the rates of formation and reversal of the reaction:

$$[\mathrm{H}] + [\mathrm{R}] \underset{k_{-1}}{\overset{k_{+1}}{\rightleftharpoons}} [\mathrm{RH}]$$

The reaction can be studied under conditions, such as low temperature, that will further reduce reactions involving the hormone–receptor complex. The equilibrium can thus be expressed in terms of the association constant, K_a, which is equal to the inverse of the dissociation constant, K_d:

$$K_a = \frac{[\mathrm{RH}]}{[\mathrm{H}][\mathrm{R}]} = \frac{k_{+1}}{k_{-1}} = \frac{1}{K_d}$$

The concentrations are equilibrium concentrations that can be restated in terms of the forward and reverse velocity constants, k_{+1} being the on-rate and k_{-1} being the off-rate (**on** refers to hormone association with the receptor and **off** refers to hormone dissociation). Experimentally, equilibrium under given conditions is determined by a progress curve of binding that reaches saturation. A saturating amount of hormone is determined using variable amounts of free hormone and measuring the amount bound with some convenient assay. The half-maximal value of a plot of receptor-bound hormone (ordinate) versus total free-hormone concentration (abscissa) approximates the dissociation constant, which will have a specific hormone concentration in molarity as its value. Hormone bound to receptor is corrected for nonspecific binding of the hormone to the membrane or other nonreceptor intracellular proteins. This can be measured conveniently if the hormone is radiolabeled, by measuring receptor binding using labeled hormone ("hot" or "uncompeted") and receptor binding using labeled hormone after the addition of an excess (100–1000 times) of unlabeled hormone ("hot" + "cold" or competed). The excess of unlabeled hormone will displace the high-affinity hormone-binding sites but not the low-affinity nonspecific binding sites. Thus when the "competed" curve is subtracted from the "uncompeted" curve, as seen in Figure 21.26, an intermediate curve will represent specific binding of labeled hormone to receptor. This is of critical importance when receptor is measured in a system containing other proteins. As an approximation, 20 times the K_d value of hormone is usually enough to saturate the receptor.

Scatchard Analysis Permits Determination of the Number of Receptor-Binding Sites and Association Constant for Ligand

Most measurements of K_d are made using **Scatchard analysis,** which is a manipulation of the equilibrium equation. The equation can be developed by a number of routes but can be envisioned from mass action analysis of the equation presented above. At equilibrium the total possible number of binding sites (B_{max}) equals the unbound plus the bound sites, so that $B_{\mathrm{max}} = \mathrm{R} + \mathrm{RH}$, and the unbound sites (R) will be equal to $\mathrm{R} = B_{\mathrm{max}} - \mathrm{RH}$. To consider the sites left unbound in the reaction the equilibrium equation becomes

$$K_a = \frac{[\mathrm{RH}]}{[\mathrm{H}](B_{\mathrm{max}} - [\mathrm{RH}])}$$

Thus

$$\frac{\text{bound}}{\text{free}} = \frac{[\mathrm{RH}]}{[\mathrm{H}]} = K_a(B_{\mathrm{max}} - [\mathrm{RH}]) = \frac{1}{K_d}(B_{\mathrm{max}} - [\mathrm{RH}])$$

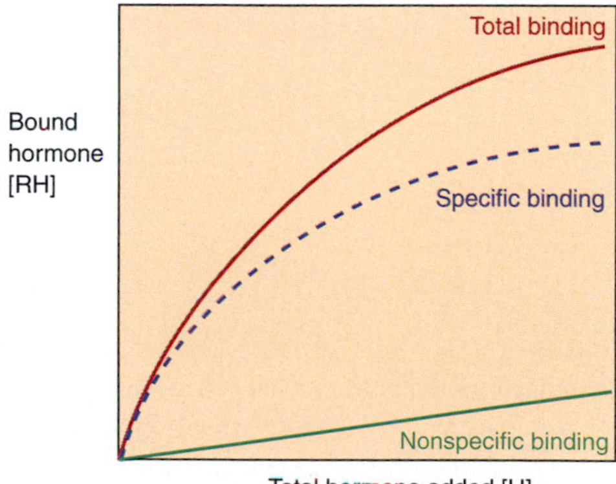

FIGURE 21.26
Typical plot showing specific hormone binding.

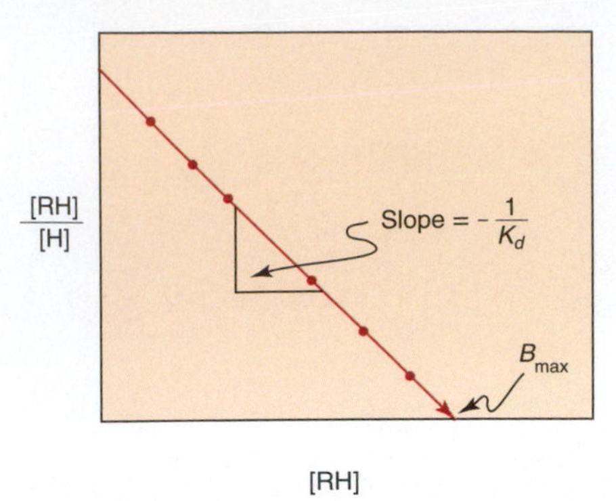

FIGURE 21.27
Typical plot of Scatchard analysis of specific binding of ligand to receptor.

The Scatchard plot of bound/free = [RH]/[H] on the ordinate versus bound on the abscissa yields a straight line, as shown in Figure 21.27. When the line is extrapolated to the abscissa, the intercept gives the value of $B_{\max}$ (the total number of specific receptor-binding sites). The slope of the negative straight line is $-K_a$ or $-1/K_d$.

These analyses are sufficient for most systems but become more complex when there are two components in the Scatchard plot. In this case the straight line usually curves as it approaches the abscissa and a second phase is observed somewhat asymptotic to the abscissa while still retaining a negative slope (Figure 21.28*a*). In order to obtain the true value of K_d for the steeper, higher-affinity sites, the low-affinity curve must be subtracted from the first set, which also corrects the extrapolated value of $B_{\max}$. From these analyses information is obtained on K_d, the number of classes of binding sites (usually one or two), and the maximal number of high-affinity receptor sites (receptor number) in the system (see Figure 21.28*b*). These curvilinear Scatchard plots result not only from the existence of more than one distinct binding component but also as a consequence of what is referred to as negative cooperativity. This term refers to the fact that in some systems the affinity of the receptor for its ligand is gradually decreased as more and more ligand binds. From application to a wide variety of systems it appears that K_d values for many

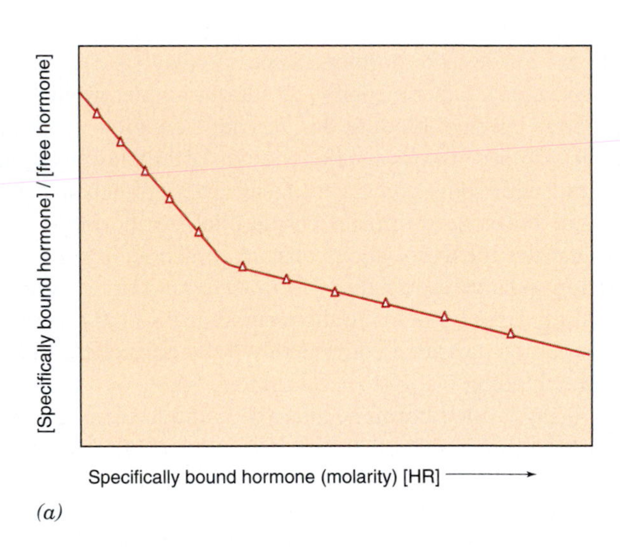

(a)

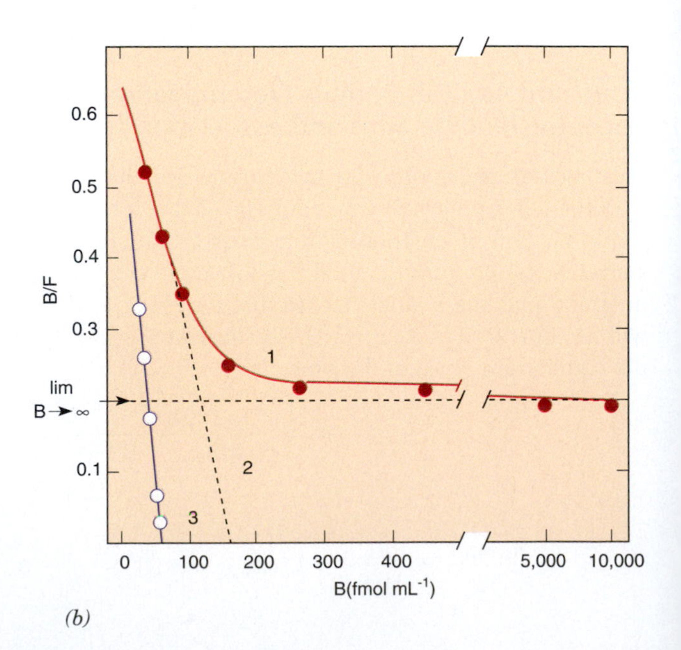

(b)

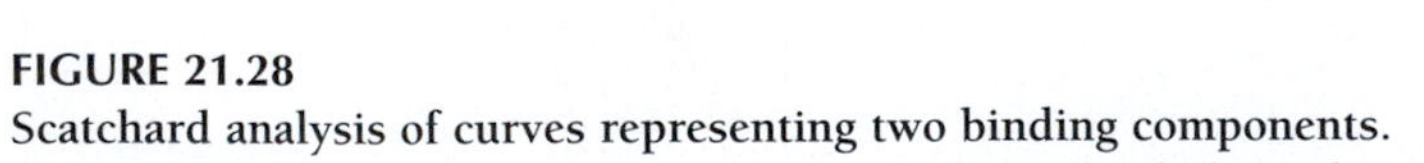

FIGURE 21.28
Scatchard analysis of curves representing two binding components.
(*a*) Scatchard curve showing two components. (*b*) Scatchard plot with correction of high-affinity component by subtraction of nonspecific binding attributable to the low-affinity component. Curve 1: Total binding. Curve 2: Linear extrapolation of high-affinity component that includes contribution from low-affinity component. Curve 3: Specific binding of high-affinity component after removal of nonspecific component.
Redrawn from Chamness, G. C. and McGuire, W. L. Steroids 26:538, 1975.

hormone receptors range from 10^{-9} to 10^{-11} M, indicating very tight binding. These interactions are generally marked by a high degree of specificity so that both parameters describe interactions of a high order, indicating the uniqueness of receptors and the selectivity of signal reception.

Some Hormone–Receptor Interactions Involve Multiple Hormone Subunits

The anterior pituitary hormones **thyrotropin (TSH), luteinizing hormone (LH),** and **follicle-stimulating hormone (FSH)** each contain an α and a β subunit. The α subunit for all three hormones is nearly identical and the α subunit of any of the three can substitute for the other two. Consequently, the α subunit performs some function in common to all three hormones in their interaction with receptor but is obviously not responsible for the specificity required for each cognate receptor. The hormones cannot replace each other in binding to their specific receptor. Thus the specificity of receptor recognition is imparted by the β subunit, whose structure is unique for the three hormones.

On the basis of topological studies with monoclonal antibodies, a picture of the interaction of LH with its receptor has been suggested as shown in Figure 21.29. In this model, the receptor recognizes both subunits of the hormonal ligand, but the β subunit is specifically recognized by the receptor to lead to a response. With the TSH–receptor complex there may be more than one second messenger generated. In addition to the stimulation of adenylate cyclase and the increased intracellular level of cAMP, the phosphatidylinositol pathway (Figure 21.21) is also turned on. The preferred model is one in which there is a single receptor whose interaction with hormone activates both the adenylate cyclase and the phospholipid second messenger systems, as shown in Figure 21.30.

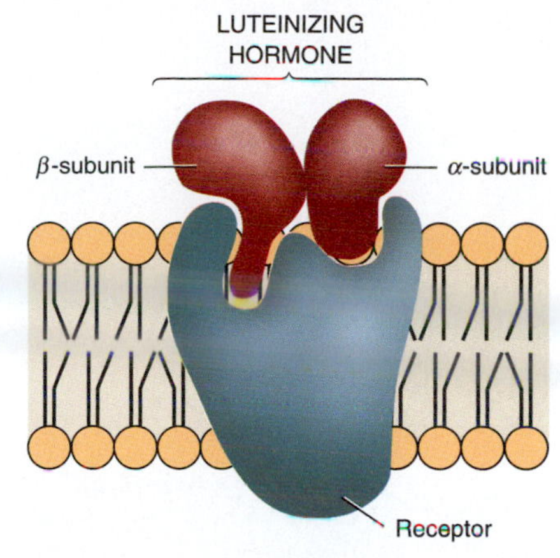

FIGURE 21.29
Interaction of the α and β subunits of LH with the LH receptor of rat Leydig cells.
The interaction was determined by topological analysis with monoclonal antibodies directed against epitopes on the α and β subunits of the hormone. Both α and β subunits participate in LH receptor binding.
Adapted from Alonoso-Whipple, C., Couet, M. L., Doss, R., Koziarz, J., Ogunro, E. A., and Crowley, W. E. Jr. Endocrinology *123:1854, 1988.*

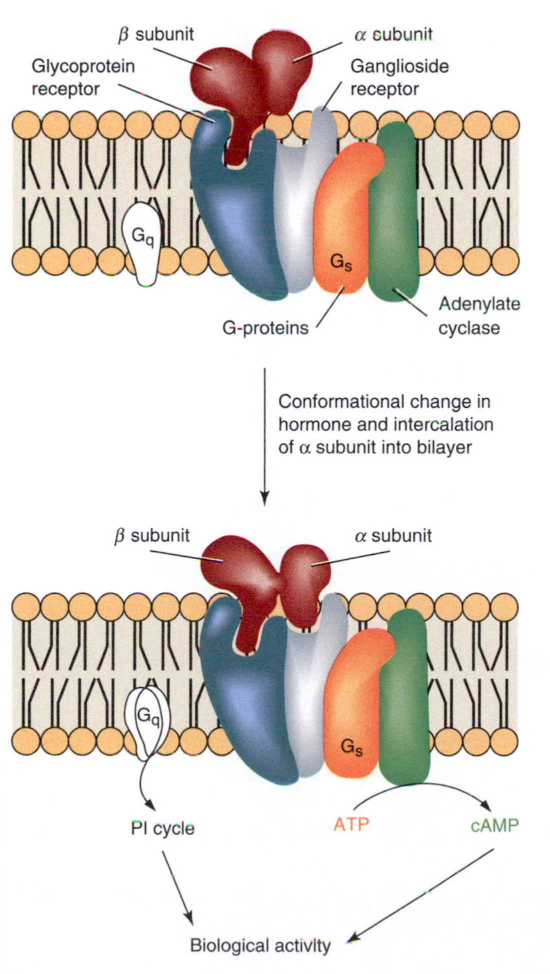

FIGURE 21.30
Model of TSH receptor.
The receptor is composed of glycoprotein and ganglioside components. After the TSH β subunit interacts with the receptor, the hormone changes its conformation and the α subunit is brought into the bilayer, where it interacts with other membrane components. The β subunit of TSH may carry the primary determinants recognized by the glycoprotein receptor component. It is suggested that the TSH signal to adenylate cyclase is via the ganglioside; the glycoprotein component appears more directly linked to phospholipid signal system. PI, phosphatidylinositol; G_s, G-protein linked to activation of adenyl cyclase; G_q, G-protein linked to PI cycle.
Adapted with modifications from Kohn, L. D., et al. In: G. Litwack (Ed.), Biochemical Actions of Hormones, *Vol. 12. New York: Academic Press, 1985, p. 466.*

21.10 | STRUCTURE OF RECEPTORS: β-ADRENERGIC RECEPTOR

Structures of receptors are conveniently discussed in terms of functional domains. Consequently, for membrane receptors there will be functional **ligand-binding domains** and **transmembrane domains,** which for many membrane receptors involve protein kinase activities. In addition, specific **immunological domains** contain primary epitopes of antigenic regions. Several membrane receptors have been cloned and studied with regard to structure and function, including the β-adrenergic receptors (β_1 and β_2), which recognize catecholamines, principally norepinephrine, and stimulate adenylate cyclase. The β_1- and β_2-adrenergic receptors are subtypes that differ in affinities for norepinephrine and for synthetic antagonists. Thus β_1-adrenergic receptor binds norepinephrine with a higher affinity than epinephrine, whereas the order of affinities is reversed for the β_2-adrenergic receptor. The drug isoproterenol has a greater affinity for both receptors than the two hormones. In Figure 21.31 the amino

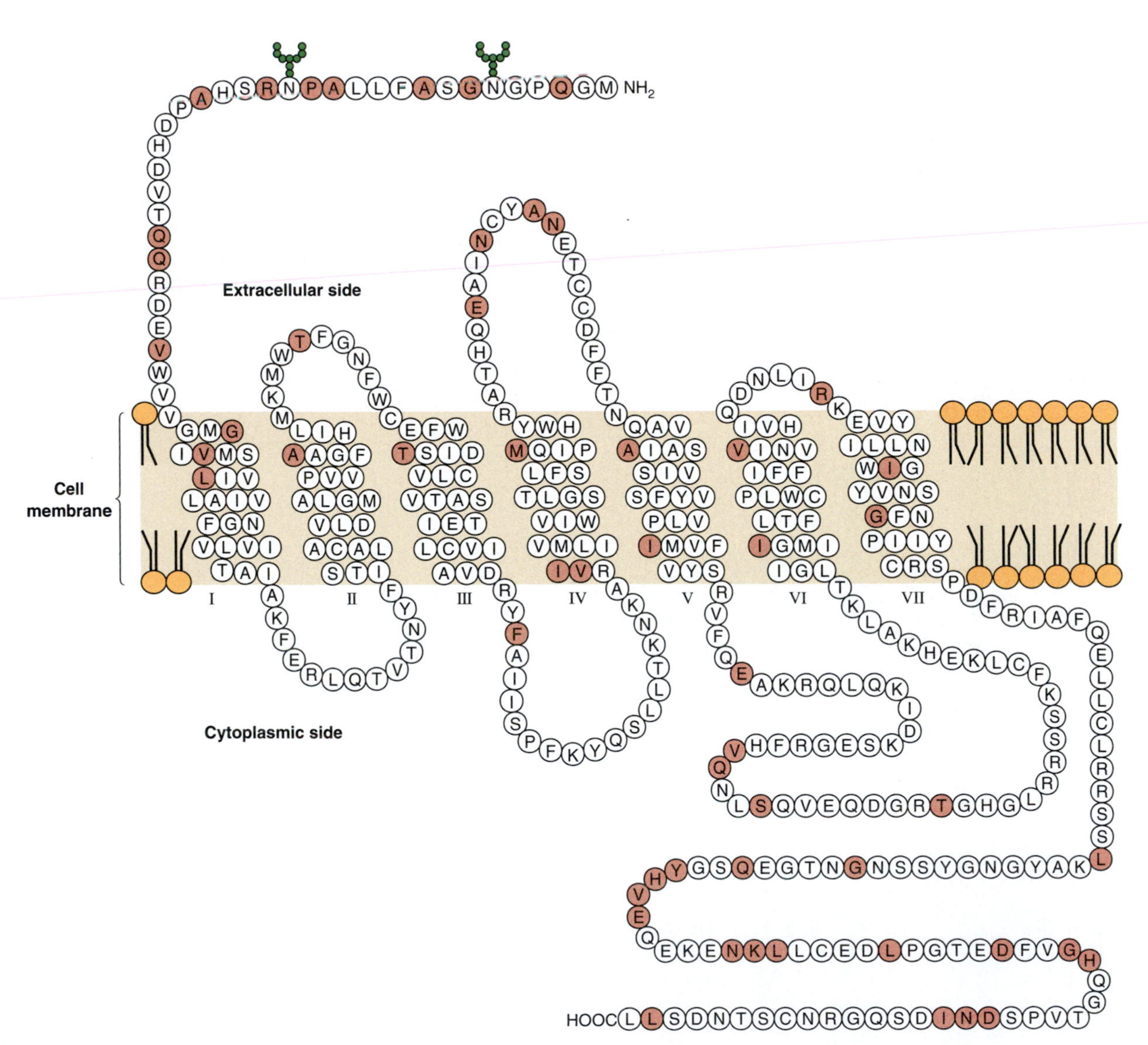

FIGURE 21.31
Proposed model for insertion of the β_2-adrenergic receptor (AR) in the cell membrane.
The model is based on hydropathicity analysis of the human β_2-AR. Standard one-letter codes for amino acid residues are used. Hydrophobic domains are represented as transmembrane helices. Pink circles with black letters indicate residues in the human sequence that differ from those in hamster. Also noted are the potential sites of N-linked glycosylation.
Redrawn from Kobilka, B. K., Dixon, R. A., Frielle, T., Doblman, H. G., et al. Proc. Natl. Acad. Sci. USA *84*:46, *1987*.

acid sequence is shown (with single-letter abbreviations for amino acids; see Table 21.4 for list) for the β_2-adrenergic receptor. A polypeptide stretch extending from α helix I extends to the extracellular space. There are seven membrane-spanning domains and these appear also in the β_1 receptor, where there is extensive homology with the β_2 receptor. Cytosolic peptide regions extend to form loops between I and II, III and IV, and V and VI and an extended chain from VII. The long extended chain from VII may contain phosphorylation sites (serine and threonine residues) of the receptor, which are important in terms of the receptor regulatory process involving receptor desensitization. Phosphorylation of these residues within the cytoplasmic tail of the receptor results in the binding of an inhibitory protein, called β arrestin, which blocks the receptor's ability to activate G_s. Cell exterior peptide loops extend from II to III, IV to V, and VI to VII, but mutational analysis suggests that the external loops do not take part in ligand binding. It appears that ligand binding may occur in a pocket arranged by the location of the membrane-spanning cylinders I–VII, which for the β_1 receptor appear to form a ligand pocket, as shown from a top view in Figure 21.32. Recently reported work suggests that transmembrane domain VI may play a role in the stimulation of adenylate cyclase activity. By substitution of a specific cysteine residue in this transmembrane domain, a mutant was generated that displays normal ligand-binding properties but a decreased ability to stimulate the adenylate cyclase.

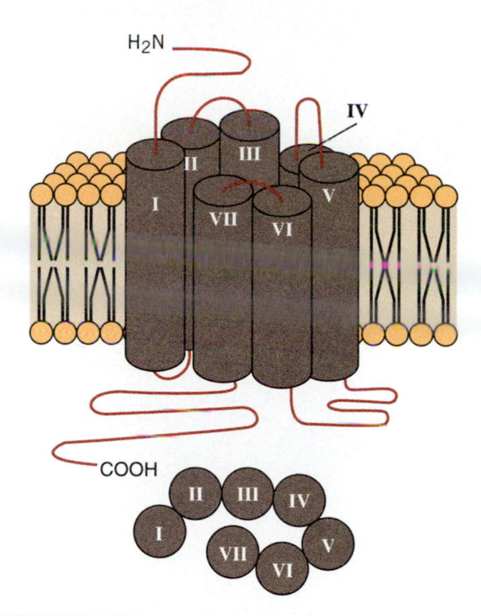

FIGURE 21.32
Proposed arrangement of the β-adrenergic receptor helices in the membrane.
Lower portion of figure is a view from above the plane of the plasma membrane. It is proposed that helices IV, VI, and VII are arranged in the membrane in such a way as to delineate a ligand-binding pocket, with helix VII centrally located.
Adapted from Frielle, T., Daniel, K. W., Caron, M. G., and Lefkowitz, R. J. Proc. Natl. Acad. Sci. USA 85:9494, *1988.*

21.11 | INTERNALIZATION OF RECEPTORS

Up to now we have described receptor systems that transduce signals through other membrane proteins, such as G-proteins, which move about in the fluid cell membrane. However, many types of cell membrane hormone–receptor complexes are internalized, that is, moved from the cell membrane to the cell interior by a process called **endocytosis.** This would represent the opposite of exocytosis in which components within the cell are moved to the cell exterior. The process of endocytosis as presented in Figure 21.33 involves the polypeptide–receptor complex bound in **coated pits**, which are indentations in the plasma membrane that invaginate into the cytosol and pinch off from the membrane to form **coated vesicles.** The vesicles shed their coats, fuse with each other, and form vesicles called receptosomes. The receptors and ligands on the inside of these **receptosomes** have different fates. Receptors are recycled to the cell surface following fusion with the Golgi apparatus. Alternatively, the vesicles fuse with lysosomes for degradation of both the receptor and hormone. In addition, some hormone–receptor complexes are dissociated in the lysosome and only the hormone is degraded, while the receptor is returned intact to the membrane. In some systems, the receptor may also be concentrated in coated pits in the absence of exogenous ligand and cycle in and out of the cell in a constitutive, nonligand-dependent manner.

Clathrin Forms a Lattice Structure to Direct Internalization of Hormone–Receptor Complexes from the Plasma Membrane

The major protein component of the coated vesicle is **clathrin**, a nonglycosylated protein of MW 180,000 whose amino acid sequence is highly conserved. The coated vesicle contains 70% clathrin, 5% polypeptides of about 35 kDa, and 25% polypeptides of 50–100 kDa. Aspects of the structure of a coated vesicle are shown in Figure 21.34. Coated vesicles have a lattice-like surface structure comprised of hexagons and pentagons. Three clathrin molecules generate each polyhedral vertex and two clathrin molecules contribute to each edge. The smallest such structure would contain 12 pentagons with 4–8 hexagons and 84 or 108 clathrin molecules. A 200-nm diameter coated vesicle contains about 1000 clathrin molecules. Clathrin forms flexible lattice structures that act as scaffolds for vesicular budding.

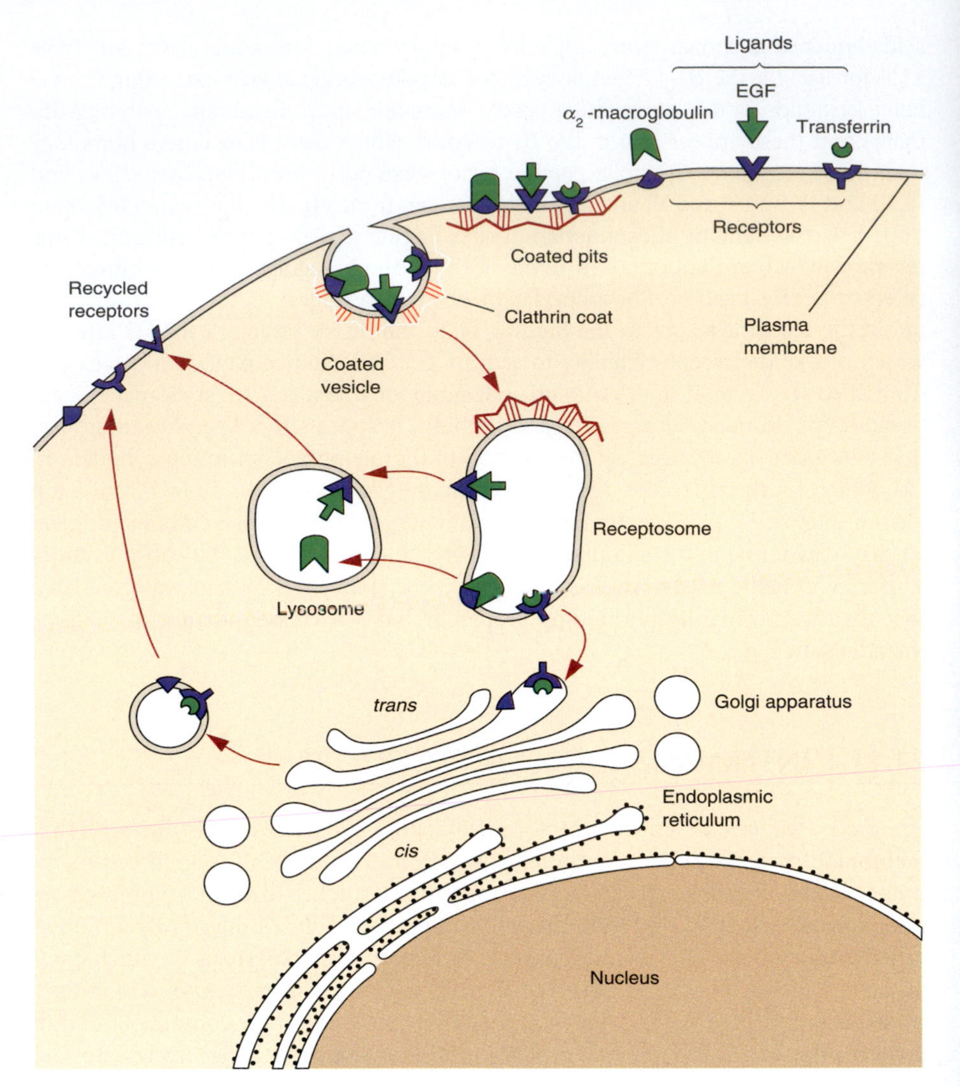

FIGURE 21.33
Diagrammatic summary of the morphological pathway of endocytosis in cells.
The morphological elements of the pathway of endocytosis are not drawn to scale. The ligands shown as examples are EGF, transferrin, and α_2-macroglobulin. EGF is an example of a receptor system in which both ligand and receptor are delivered to lysosomes; transferrin is shown as an example of a system in which both the ligand and receptor recycle to the surface; α_2-macroglobulin is shown as an example of a system in which the ligand is delivered to lysosomes but the receptor recycles efficiently back to the cell surface via the Golgi apparatus.
Adapted from Pastan, I. and Willingham, M. C. (Eds.). Endocytosis. *New York: Plenum Press, 1985, p. 3.*

Completion of the budding process results in the mature vesicle being able to enter the cycle.

The events following endocytosis are not always clear with respect to a specific membrane receptor system. This process is a means to introduce the intact receptor or ligand to the cell interior in cases where the nucleus is thought to contain a receptor or ligand-binding site. Consider, for example, growth factors that are known to bind to a cell membrane receptor but trigger events leading to mitosis. Signal transmission may occur by the alteration of a specific cytosolic protein, perhaps by membrane growth factor receptor-associated protein kinase activity, resulting in the nuclear translocation of the covalently modified cytosolic protein. In the case of internalization, delivery of an intact ligand (or portion of the ligand) could interact with a nuclear receptor. Such mechanisms are speculative. Nevertheless, these ideas could constitute a rationale for the participation of endocytosis in signal transmission to intracellular components.

Endocytosis renders a cell less responsive to hormone. Removal of the receptor to the interior, or cycling of membrane components, alters responsiveness or metabolism (e.g., glucose receptors are shuttled between the cell interior and the cell membrane under the control of hormones in certain cells). In another type of downregulation, a hormone–receptor complex translocated to the nucleus represses its own receptor mRNA levels by interacting with a specific DNA sequence. More about this form of receptor downregulation is mentioned in Chapter 22.

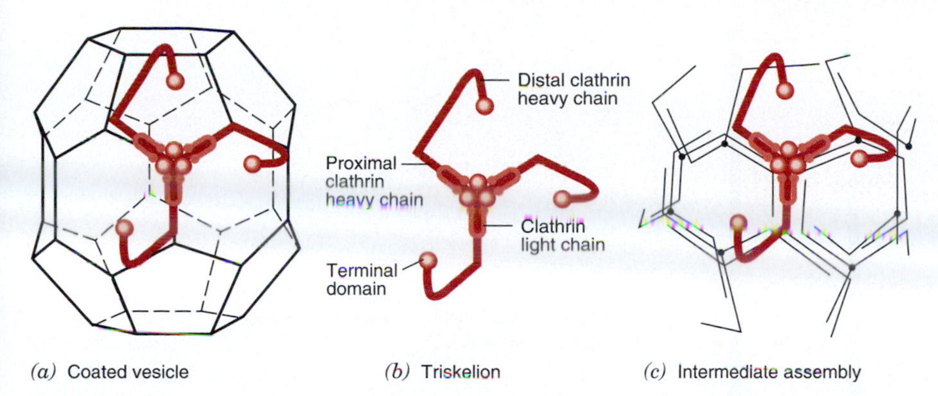

FIGURE 21.34
Structure and assembly of a coated vesicle.
(*a*) A typical coated vesicle contains a membrane vesicle about 40 nm in diameter surrounded by a fibrous network of 12 pentagons and 8 hexagons. The fibrous coat is constructed of 36 clathrin triskelions. One clathrin triskelion is centered on each of the 36 vertices of the coat. Coated vesicles having other sizes and shapes are believed to be constructed similarly: each vesicle contains 12 pentagons but a variable number of hexagons. (*b*) Detail of a clathrin triskelion. Each of three clathrin heavy chains is bent into a proximal arm and a distal arm. A clathrin light chain is attached to each heavy chain, most likely near the center. (*c*) An intermediate in the assembly of a coated vesicle, containing 10 of the final 36 triskelions, illustrates the packing of the clathrin triskelions. Each of the 54 edges of a coated vesicle is constructed of two proximal and two distal arms intertwined. The 36 triskelions contain $36 \times 3 = 108$ proximal and 108 distal arms, and the coated vesicle has precisely 54 edges.
See Crowther, R. A. and Pearse, B. M. F. J. Cell Biol. *91:790, 1981. Redrawn from Nathke, I. S., Heuser, J., Lupas, A., Stock, J., Turck, C. W., and Brodsky, E. M.* Cell *68:899, 1992; and from Darnell, J., Lodish, H., and Baltimore, D.* Molecular Cell Biology. *New York: Scientific American Books, 1986, p. 647.*

21.12 | INTRACELLULAR ACTION: PROTEIN KINASES

Many amino acid-derived hormones or polypeptides bind to cell membrane receptors (except for thyroid hormone) and transmit their signal by (1) elevation of cAMP and transmission through the **protein kinase A pathway;** (2) triggering of the hydrolysis of phosphatidylinositol 4,5-bisphosphate and stimulation of the **protein kinase C** and IP_3–Ca^{2+} pathways; or (3) stimulation of intracellular levels of cGMP and activation of the **protein kinase G pathway.** There are also other less prevalent systems for signal transfer, which, for example, affect molecules in the membrane like phosphatidylcholine. As previously discussed in the case of TSH–receptor signaling, it may be possible that two of these pathways are activated.

The cAMP system operating through protein kinase A activation has been described. Specific proteins are expected to be phosphorylated by this kinase compared to other protein kinases, such as protein kinase C. Both protein kinase A and C phosphorylate proteins on **serine** or **threonine** residues. An additional protein kinase system involves phosphorylation of **tyrosine,** which occurs in cytoplasmic domains of some membrane receptors especially growth factor receptors. This system is important for the insulin receptor, IGF receptor, and certain oncogenes discussed below. The cellular location of these protein kinases is presented in Figure 21.35.

The catalytic domain in the protein kinases is similar in amino acid sequence, suggesting that they have all evolved from a common primordial kinase. The three **tyrosine-specific kinases** shown in Figure 21.35 are transmembrane receptor proteins that, when activated by the binding of specific extracellular ligands, phosphorylate proteins (including themselves) on tyrosine residues inside the cell. Both chains of the insulin receptor are encoded by a single gene, which produces a precursor protein that is cleaved into the two disulfide-linked chains. The extracellular domain of the PDGF receptor is thought to be folded into five

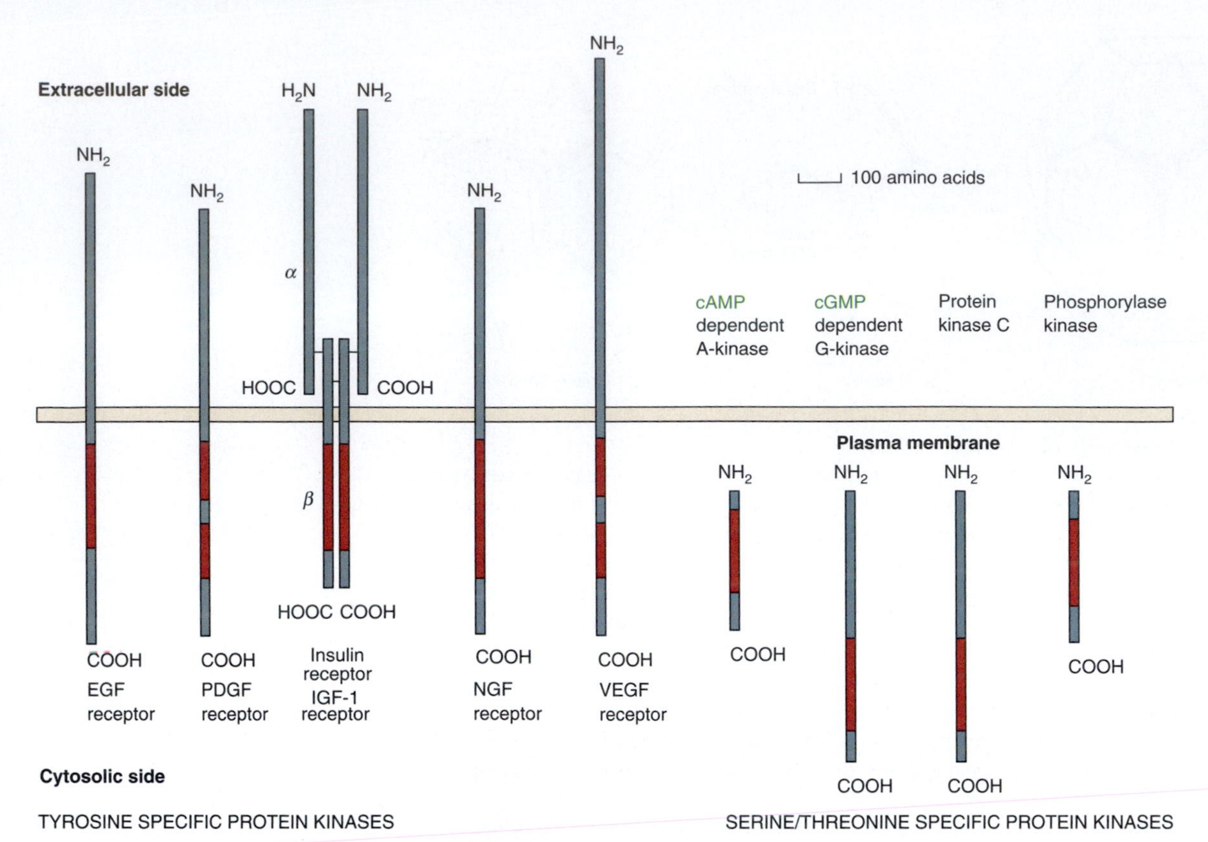

FIGURE 21.35
Protein kinases showing size and location of their catalytic domain.
In each case the catalytic domain (red region) is about 250 amino acid residues long. The regulatory subunits normally associated with A kinase and with phosphorylase kinase are not shown. EGF, epidermal growth factor; NGF, nerve growth factor; VEGF, vascular endothelial growth factor.
Redrawn from Alberts, B., Bray, D., Lewis, J., Raff, M., Roberts, K., and Watson, J. D. Molecular Biology of the Cell, *3rd ed. New York: Garland Publishing, 1994, p. 760.*

immunoglobulin (Ig)-like domains, suggesting that this protein belongs to the Ig superfamily.

Proteins that are regulated by phosphorylation–dephosphorylation have multiple phosphorylation sites and may be phosphorylated by more than one class of protein kinase.

Insulin Receptor: Transduction Through Tyrosine Kinase

From Figure 21.35 it is seen that the α subunits of the **insulin receptor** are located outside the cell membrane and serve as the insulin-binding sites. Following insulin binding the receptor complex undergoes an activation sequence, involving conformational changes and phosphorylation **(autophosphorylation)** of tyrosine residues located within the cytoplasmic portions (β subunits) of the receptor. This autophosphorylation results in activation of the tyrosine kinase activity located in the β subunit, which is now able to phosphorylate cytosolic proteins, such as the protein termed **insulin receptor substrate-1 (IRS-1)**, that acts as an intracellular docking protein for proteins mediating insulin action. Although it is not totally clear whether protein phosphorylation is the primary mechanism for all of insulin's actions, the net results of these initial phosphorylation events include a series of short-term metabolic effects, such as a rapid increase in the uptake of glucose, as well as longer-term effects of insulin on cellular differentiation and growth. Although, as already indicated, the insulin receptor itself functions as an **autotyro-**

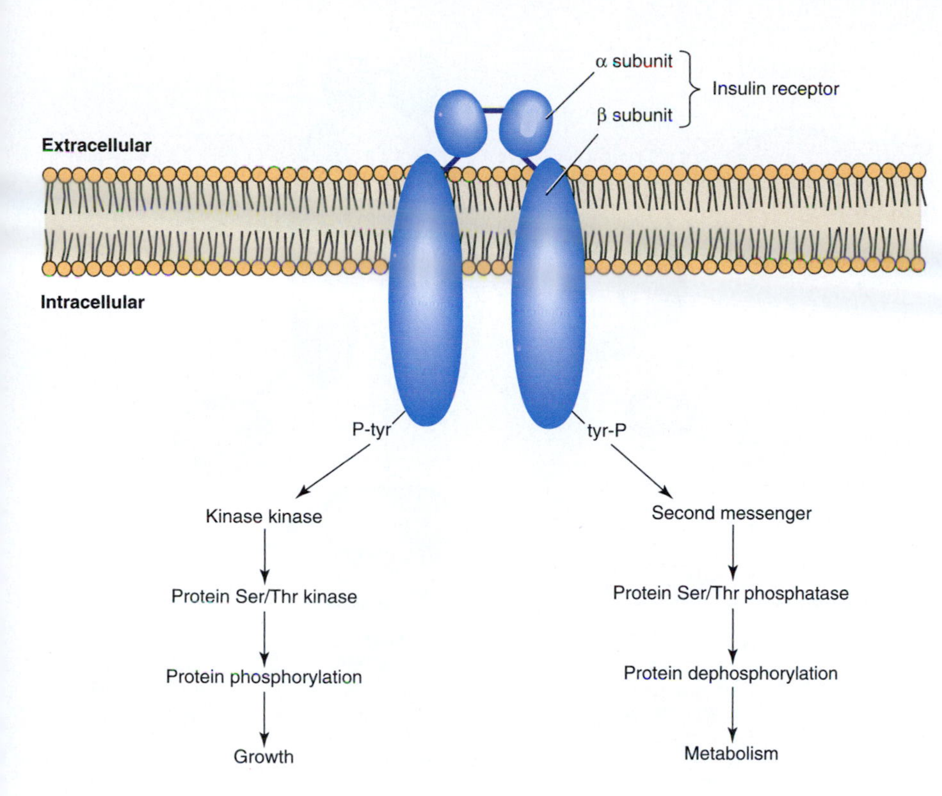

FIGURE 21.36
Hypothetical model depicting two separate biochemical pathways to explain paradoxical effects of insulin on protein phosphorylation. Insulin simultaneously produces increases in the serine/threonine phosphorylation of some proteins and decreases in others. This paradoxical effect may result from the activation of both kinases and phosphatases. Model explains (1) the generation of a soluble second messenger that directly or indirectly activates serine/threonine phosphatase, and (2) the stimulation of a cascade of protein kinases, resulting in the phosphorylation of cellular proteins.
Redrawn from Saltiel, A. R. FASEB J. 8:1034, 1994.

sine kinase that is activated by hormone binding, the subsequent changes in protein phosphorylation occur predominantly on serine and threonine residues, as indicated in Figure 21.36. As also shown, insulin simultaneously stimulates the phosphorylation of some proteins and the dephosphorylation of other proteins. Either of these biochemical events leads to activation or inhibition of specific enzymes involved in mediating the effects of insulin. These opposite effects (phosphorylation and dephosphorylation) mediated by insulin suggest that perhaps separate signal transduction pathways may originate from the insulin receptor to produce these pleiotropic actions. A hypothetical scheme for this bifurcation of signals in insulin's action is presented in Figure 21.37. The substrates of the insulin receptor tyrosine kinase are an important current research effort since phosphorylated proteins could produce the long-term effects of insulin. On the other hand, there is evidence that an insulin second messenger may be developed at the cell membrane to account for the short-term metabolic effects of insulin. The substance released as a result of insulin–insulin receptor interaction may be a glycoinositol derivative that, when released from the membrane into the cytosol, could be a stimulator of phosphoprotein phosphatase. This activity would dephosphorylate a variety of enzymes, either activating or inhibiting them, and produce effects already known to be associated with the action of insulin. In addition, this second messenger, or the direct phosphorylating activity of the receptor tyrosine kinase, might explain the movement of glucose receptors (transporters) from the cell interior to the surface to account for enhanced cellular glucose utilization in cells that utilize this mechanism to control glucose uptake. These possibilities are reviewed in Figure 21.37. Activation of the enzymes indicated in this figure leads to increased metabolism of glucose while inhibition of the enzymes indicated leads to decreased breakdown of glucose or fatty acid stores (see Clin. Corr. 21.4).

Activity of Vasopressin: Protein Kinase A

An example of the activation of the **protein kinase A** pathway by a hormone is the activity of arginine vasopressin (AVP) on the distal kidney cell. Here the action of **vasopressin (VP)**, also called the antidiuretic hormone (Table 21.5), is to cause in-

CLINICAL CORRELATION 21.4
Decreased Insulin Receptor Kinase Activity in Gestational Diabetes Mellitus

During pregnancy an important maternal metabolic adaptation is a decrease in insulin sensitivity. This adaptation helps provide adequate glucose for the developing fetus. However, in 3–5% of pregnant women glucose intolerance develops. Gestational diabetes mellitus (GDM) is characterized by an additional decrease in insulin sensitivity and an inability to compensate with increased insulin secretion. Although both pregnancy-induced insulin resistance and GDM are generally reversible after pregnancy, approximately 30–50% of women with a history of GDM go on to develop type 2 diabetes later in life, particularly if they are obese. Although the cellular mechanisms responsible for the insulin resistance in GDM are not fully understood, the resistance to insulin-mediated glucose transport appears to be greater in skeletal muscle from GDM subjects than in women who are pregnant but do not have GDM. Recent data indicate that defects in insulin action, rather than a decrease in insulin receptor-binding affinity, may contribute to the pathogenesis of GDM. More specifically, skeletal muscle cells of GDM subjects appear to overexpress plasma cell membrane glycoprotein-1 (PC-1), which has been reported to inhibit the tyrosine kinase activity of the insulin receptor by directly interacting with α subunits and blocking the insulin-induced conformational change. Additionally, excessive phosphorylation of serine/threonine residues located within muscle insulin receptors appears to downregulate tyrosine kinase activity in GDM. Thus an overexpression of PC-1 and a decrease in receptor kinase activity, coupled with a decreased expression and phosphorylation (tyrosine residues) of the insulin receptor substrate-1 (IRS-1; see Figure 21.37), may underlie the insulin resistance in GDM.

Shao, J., Catalono, P. M., Hiroshi, Y., Ruyter, I., Smith, S., Youngren, J., and Friedman, J. E. Decreased insulin receptor tyrosine kinase activity and plasma cell membrane glycoprotein-1 overexpression in skeletal muscle from obese women with gestational diabetes mellitus (GDM). *Diabetes* 49(4): 603, 2000; and Maddux, B. A. and Goldfine, I. D. Membrane glycoprotein PC-1 inhibition of insulin receptor function occurs via direct interaction with the receptor alpha-subunit. *Diabetes* 49(1): 13, 2000.

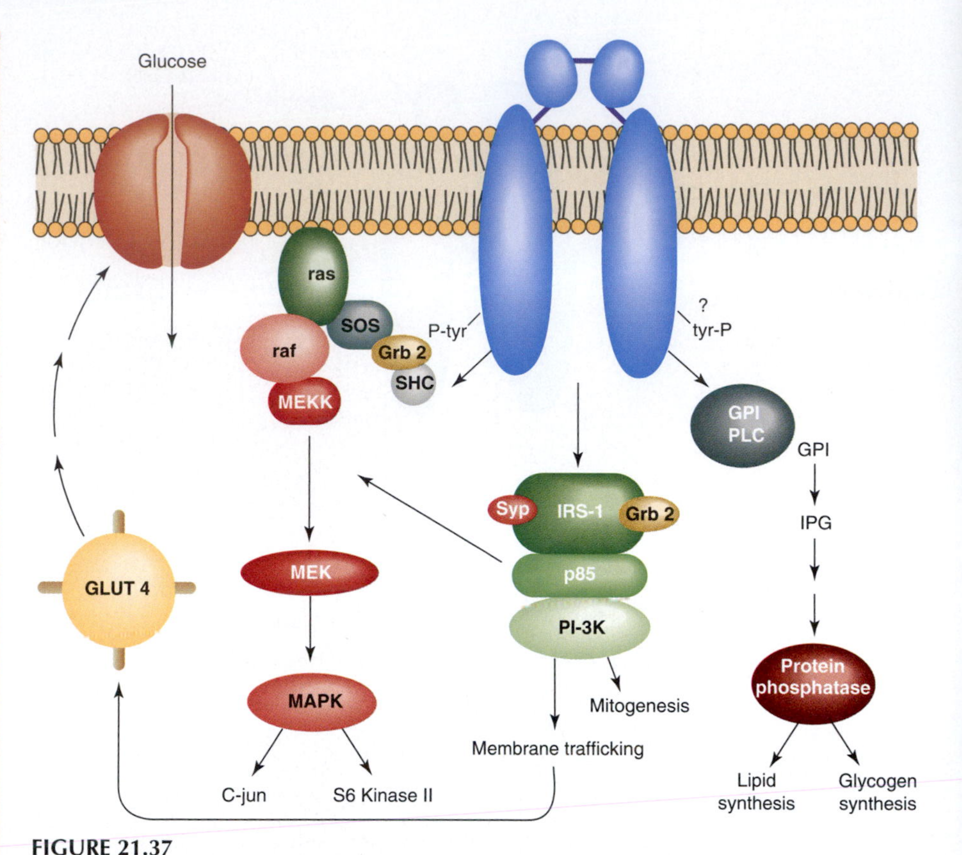

FIGURE 21.37

Hypothetical scheme for signal transduction in insulin action.

The insulin receptor undergoes tyrosine autophosphorylation and subsequent kinase activation upon hormone binding. The receptor phosphorylates intracellular substrates including IRS-1 and Shc proteins, which associate with SH2-containing proteins like p85, SYP, or Grb2 upon phosphorylation. Formation of the IRS-1–p85 complex activates PI 3-kinase; the IRS-1/SYP complex activates SYP, leading to MEK activation. Formation of the Shc–Grb2 complex mediates the stimulation of $P21^{ras}$ GTP binding, leading to a cascade of phosphorylations. These phosphorylations probably occur sequentially and involve *raf* protooncogene, MEK, MAP kinase, and S6 kinase II. The receptor is probably separately coupled to the activation of a specific phospholipase C that catalyzes the hydrolysis of the glycosyl-PI molecules in the plasma membrane. A product of this reaction, inositol phosphate glycan (IPG), may act as a second messenger, especially with regard to activation of serine/threonine phosphatases and the subsequent regulation of lipid and glucose metabolism. Abbreviations: IRS-1, insulin receptor substrate-1; SH, *src* homology; MAP kinase, mitogen-activated protein kinase; MEK, MAP kinase kinase; GPI, glycosylphosphatidylinositol; PLC, phospholipase; SOS, son of sevenless.

Redrawn from Saltiel, A. R. FASEB J. 8:1034, 1994.

creased water reabsorption from the urine in the distal kidney. A mechanism for this system is shown in Figure 21.38. Neurons synthesizing AVP (vasopressinergic neurons) are signaled to release AVP from their nerve endings by interneuronal firing from a **baroreceptor** responding to a fall in blood pressure or from an **osmoreceptor** (probably an interneuron), which responds to an increase in extracellular salt concentration. The high extracellular salt concentration apparently causes shrinkage of the osmoreceptor cell and generates an electrical signal transmitted down the axon of the osmoreceptor to the cell body of the VP neuron generating an action potential. This signal is then transmitted down the long axon from the VP cell body to its nerve ending where, by depolarization, the VP–neurophysin II complex is released into the extracellular space. The complex enters local capillaries through fenestrations and progresses to the general circulation. The complex dissociates and free VP is able to bind to its cognate membrane receptors in the dis-

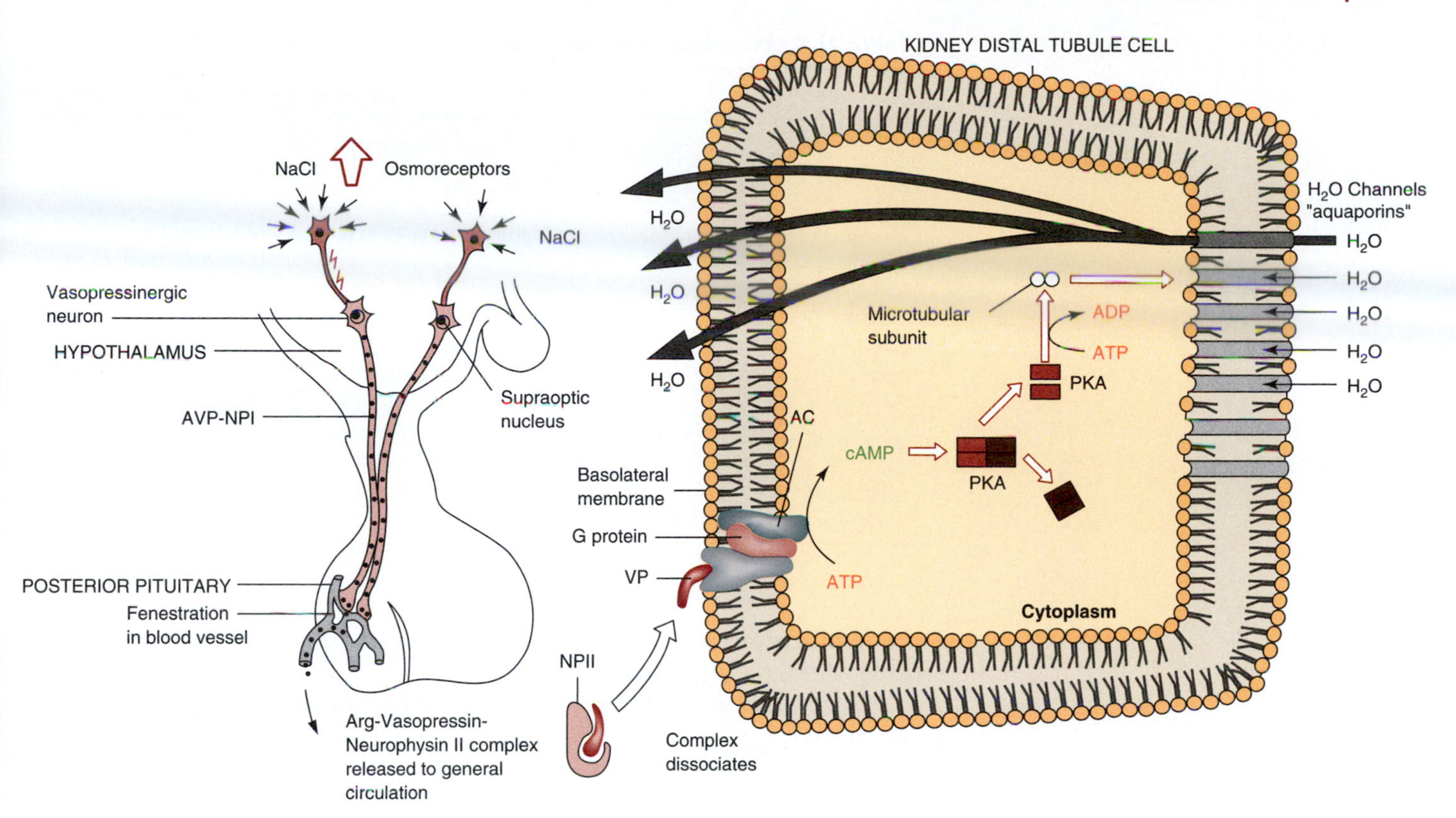

FIGURE 21.38
Secretion and action of arginine vasopressin in the distal kidney tubules.
The release of arginine vasopressin (AVP or VP) from the posterior pituitary begins with a signal from the osmoreceptor, or baroreceptor (not shown), in the upper right-hand corner of figure. The signal can be an increase in the extracellular concentration of sodium chloride, which causes the osmoreceptor neuron to shrink and send an electrical message down its axon, which interfaces with the vasopressinergic cell body. This signal is transmitted down the long axon of the vasopressinergic neuron and depolarizes the nerve endings causing the release, by exocytosis, of the VP–neurophysin complex stored there. They enter the local circulation through fenestrations in the vessels and perfuse the general circulation. Soon after release, neurophysin dissociates from VP and VP binds to its cognate receptor in the cell membrane of the kidney distal tubule cell (other VP receptors are located on the corticotrope of the anterior pituitary and on the hepatocytes and their mechanisms in these other cells are different from the one for the kidney tubule cell). NPII, neurophysin II; VP, vasopressin; R, receptor; AC, adenylate cyclase; MF, myofibril; GP, glycogen phosphorylase; PK_i, inactive protein kinase; PK_a, active protein kinase; R-Ca, regulatory subunit–cyclic AMP complex; TJ, tight junction; PD, phosphodiesterase. Vasopressin–neurophysin complex dissociates at some point and free VP binds to its cell membrane receptor in the plasma membrane surface. Through a G-protein, adenylate cyclase is stimulated on the cytoplasmic side of the cell membrane, generating increased levels of cAMP from ATP. Cyclic AMP-dependent protein kinases are stimulated and phosphorylate various proteins (perhaps including microtubular subunits), which, through aggregation, insert as water channels (aquaporins) in the luminal plasma membrane, thus increasing the reabsorption of water by free diffusion.
Redrawn in part from Dousa, T. P. and Valtin, H. Kidney Int. *10:45, 1975.*

tal kidney, anterior pituitary, hepatocyte, and perhaps other cell types. After binding to the kidney receptor, VP causes stimulation of adenylate cyclase through the stimulatory G-protein and activates protein kinase A. The protein kinase phosphorylates protein subunits that aggregate to form specific water channels, referred to as aquaporins (see Figure 21.38), which are inserted into the luminal membrane for admission of larger volumes of water than would occur by free diffusion. Water is transported across the kidney cell to the basolateral side and to the general circulation, causing a dilution of the original high salt concentration (signal) and an increase in blood pressure. These aquaporins, which are a family of integral membrane proteins that function as selective water channels, consist of six transmembrane α-helical domains. Although aquaporin monomers function as water channels or pores, their stability and proper functioning may require a tetrameric assembly. Specific mutations in the amino acid sequences of the intracellular and extracellular loops of these proteins result in nonfunctional aquaporins and the de-

TABLE 21.7 Examples of Hormones that Operate Through the Protein Kinase A Pathway

Hormone	*Location of Action*
CRH	Corticotrope of anterior pituitary
TSH	Thyroid follicle
LH	Leydig cell of testis Mature follicle at ovulation and corpus luteum
FSH	Sertoli cell of seminiferous tubule ovarian follicle
ACTH	Inner layers of cells of adrenal cortex
Opioid peptides	Some in CNS function on inhibitory pathway through G_i
AVP	Kidney distal tubular cell (the AVP hepatocyte receptor causes phospholipid turnover and calcium ion uptake; the AVP receptor in anterior pituitary causes phospholipid turnover)
PGI_2 (prostacyclin)	Blood platelet membrane
Norepinephrine/epinephrine	β-Receptor

velopment of diabetes insipidus, which is characterized by increased thirst and production of a large volume of urine.

Some hormones that operate through the protein kinase A pathway are listed in Table 21.7.

Gonadotropin-Releasing Hormone (GnRH): Protein Kinase C

Table 21.8 presents examples of polypeptide hormones that stimulate the phosphatidylinositol pathway. An example of a system operating through stimulation of the phosphatidylinositol pathway and subsequent activation of the **protein kinase C** system is **GnRH** action, shown in Figure 21.39. Probably under aminergic interneuronal controls, a signal is generated to stimulate the cell body of the GnRH-ergic neuron where GnRH is synthesized. The signal is transmitted down the long axon to the nerve ending where the hormone is stored. The hormone is released from the nerve ending by exocytosis resulting from depolarization caused by signal transmission. The GnRH enters the primary plexus of the closed portal system connecting the hypothalamus and anterior pituitary through fenestrations. Then GnRH

TABLE 21.8 Examples of Polypeptide Hormones that Stimulate the Phosphatidylinositol Pathway

Hormone	*Location of Action*
TRH	Thyrotrope of the anterior pituitary releasing TSH
GnRH	Gonadotrope of the anterior pituitary releasing LH and FSH
AVP	Corticotrope of the anterior pituitary; assists CRH in releasing ACTH; hepatocyte: causes increase in cellular Ca^{2+}
TSH	Thyroid follicle: releasing thyroid hormones causes increase in phosphatidylinositol cycle as well as increase in protein kinase A pathway
Angiotensin II/III	Zona glomerulosa cell of adrenal cortex: releases aldosterone
Epinephrine (thrombin)	Platelet: releasing ADP/serotonin; hepatocyte via α receptor: releasing intracellular Ca^{2+}

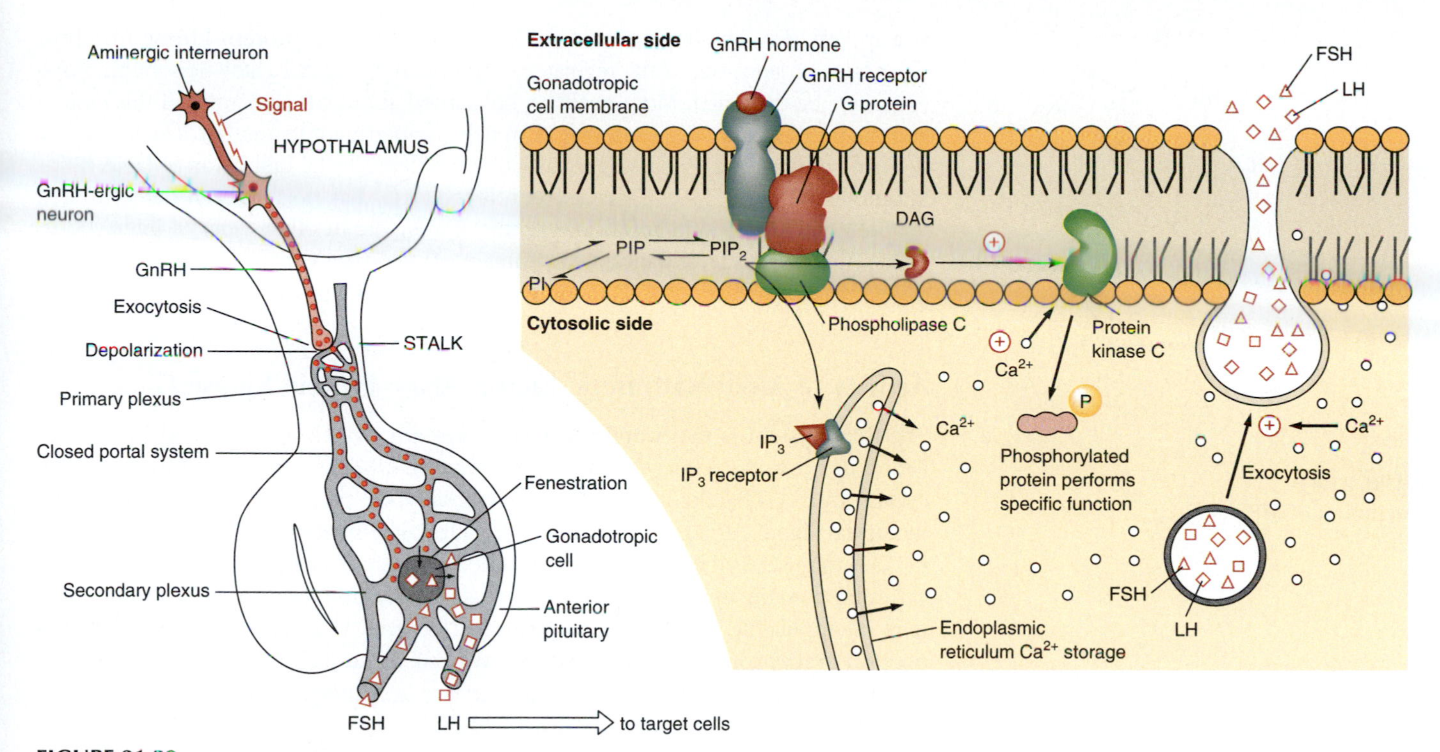

FIGURE 21.39
Regulation of secretion of LH and FSH.
A general mode of action of GnRH to release the gonadotropes from the gonadotropic cell of the anterior pituitary is presented. GnRH, gonadotropin-releasing hormone; FSH, follicle-stimulating hormone; LH, luteinizing hormone; DAG, diacylglycerol.

exits the closed portal system through fenestrations in the secondary plexus and binds to cognate receptors in the cell membrane of the gonadotrope (see enlarged view in Figure 21.39). The signal from the hormone–receptor complex is transduced (through a G-protein) and phospholipase C is activated. This enzyme catalyzes the hydrolysis of PIP_2 to form DAG and IP_3. Diacylglycerol activates protein kinase C, which phosphorylates specific proteins, some of which may participate in the resulting secretory process to transport LH and FSH to the cell exterior. The product IP_3, which binds to a receptor on the membrane of the calcium storage particle, probably located near the cell membrane, stimulates the release of calcium ion. Elevated cytosolic Ca^{2+} causes increased stimulation of protein kinase C and participates in the exocytosis of LH and FSH from the cell.

Protein kinase C is a family of kinases (Figure 21.40). The enzyme consists of two domains, a regulatory and a catalytic domain, which can be separated by proteolysis

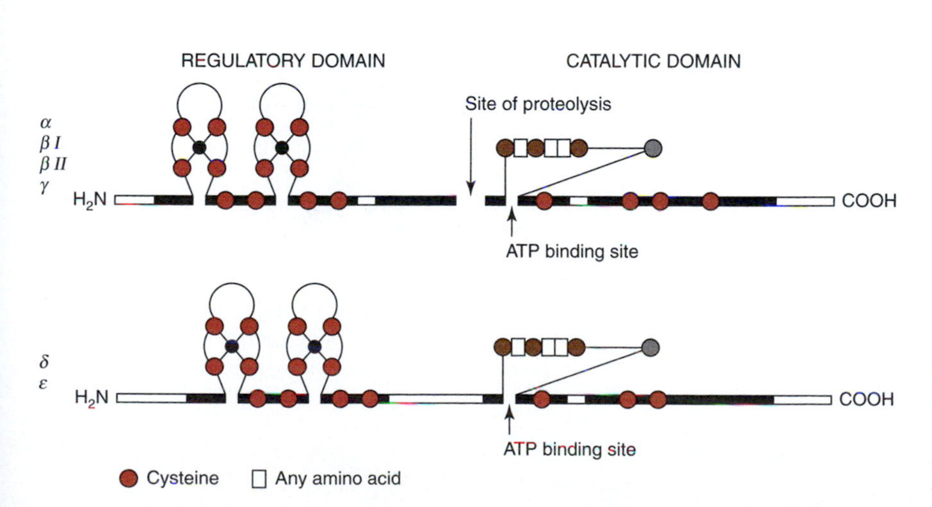

FIGURE 21.40
Common structure of protein kinase C subspecies.
Modified from Kikkawa, U., Kishimoto, A., and Nishizuka, Y. Annu. Rev. Biochem. *58:31, 1989.*

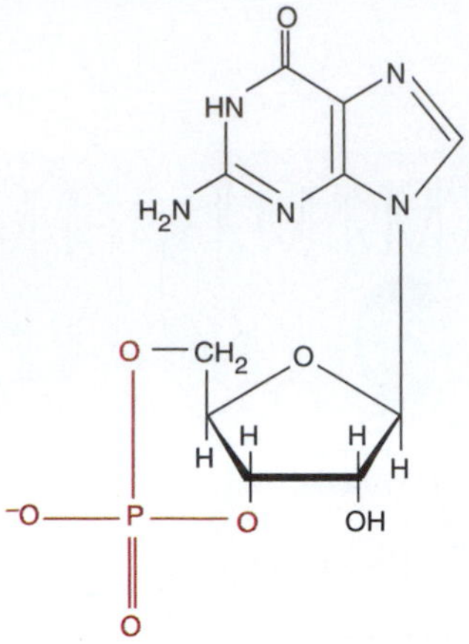

FIGURE 21.41
Structure of cGMP.

at a specific site. The free catalytic domain, formerly called **protein kinase M,** phosphorylates proteins free of the regulatory components. The free catalytic subunit, however, may be degraded. More needs to be learned about the dynamics of this system and the translocation of the enzyme from one compartment to another. The regulatory domain contains two Zn^{2+} fingers usually considered to be hallmarks of DNA-binding proteins (see p. 388). This DNA-binding activity has not yet been demonstrated for protein kinase C and metal fingers may participate in other types of interactions. The ATP-binding site in the catalytic domain contains the G box, GXGXXG, which is a consensus sequence for ATP binding with a downstream lysine residue.

Activity of Atrial Natriuretic Factor (ANF): Protein Kinase G

The third system is the **protein kinase G system,** which is stimulated by the elevation of cytosolic cGMP (Figure 21.41). **Cyclic GMP** is synthesized by guanylate cyclase from GTP. Like adenylate cyclase, guanylate cyclase is linked to a specific biological signal through a membrane receptor. The guanylate cyclase extracellular domain may serve the role of the hormone receptor. This is directly coupled to the cytosolic domain through one membrane-spanning domain (Figure 21.42), which may be applicable to the **atrial natriuretic factor** (ANF; also referred to as atriopeptin) **receptor–guanylate cyclase system.** Thus the hormone-binding site, transmembrane domain, and guanylate cyclase activities are all served by a single polypeptide chain.

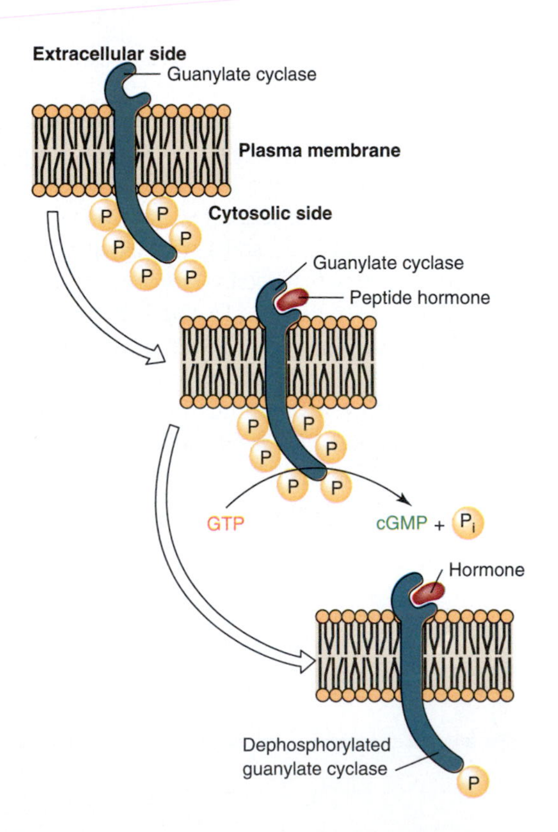

FIGURE 21.42
Model for the regulation of guanylate cyclase activity after peptide hormone binding. The enzyme exists in a highly phosphorylated state under normal conditions. Binding of hormone markedly enhances enzyme activity, followed by a rapid dephosphorylation of guanylate cyclase and a return of activity to basal state despite continued presence of hormonal peptide.
Redrawn from Schultz, S., Chinkers, M., and Garbers, D. L. FASEB J. *3:2026, 1989.*

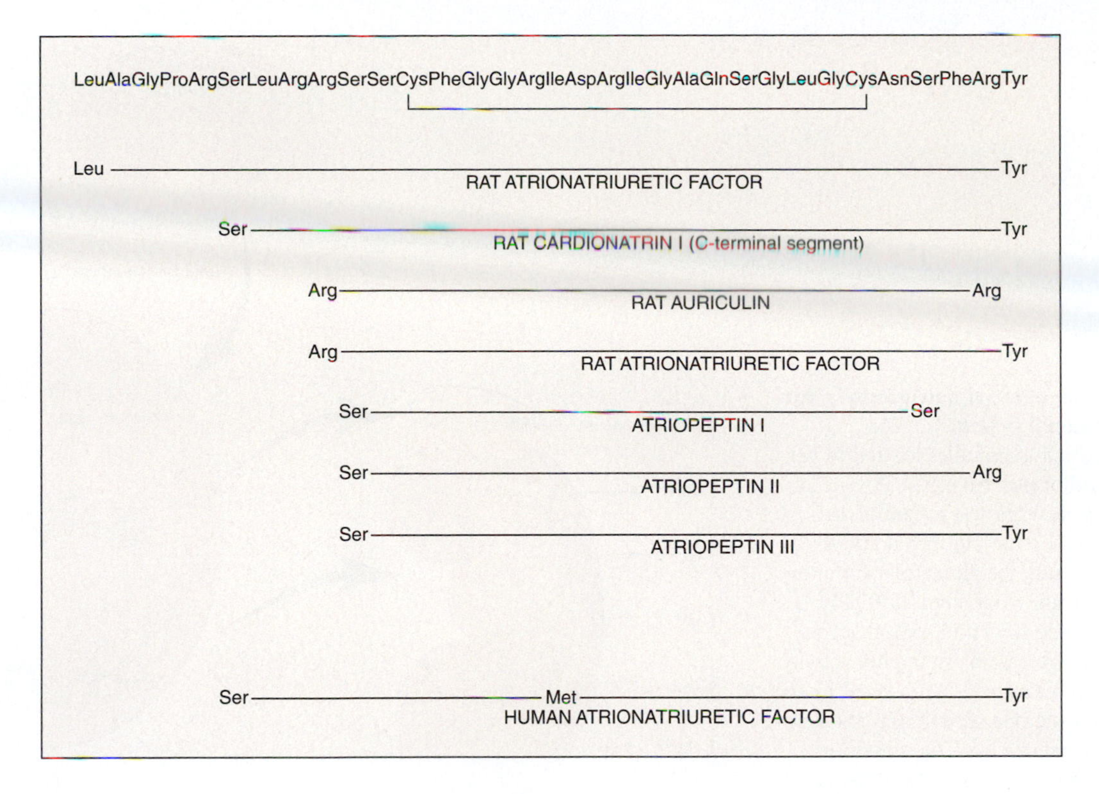

FIGURE 21.43
Atrial natriuretic peptides.
These peptides relax vascular smooth muscle and produce vasodilation and natriuresis. Other effects are discussed in the text.
Adapted from Carlin, M. and Genest, J. Endocr. Rev. *6:107, 1985.*

ANF is one of a family of peptides, as shown in Figure 21.43; a sequence of human ANF is shown at the bottom. The functional domains of the ANF receptor are illustrated in Figure 21.44. Atrial natriuretic factor is released from cardiac atrial cells under control of several hormones. Data from atrial cell culture suggest that ANF secretion is stimulated by activators of cardiac protein kinase C and decreased by activators of protein kinase A. These opposing actions may be mediated by the actions of α- and β-adrenergic receptors, respectively. An overview of the secretion of ANF and its general effects is shown in Figure 21.45. ANF is released by a number of signals, such as blood volume expansion, elevated blood pressure directly induced by vasoconstrictors, high salt intake, and increased heart pumping rate. ANF is secreted as a dimer that is inactive for receptor interaction and is converted in plasma to a monomer capable of interacting with receptor. The actions of ANF (Figure 21.45) are to increase the glomerular filtration rate, leading to

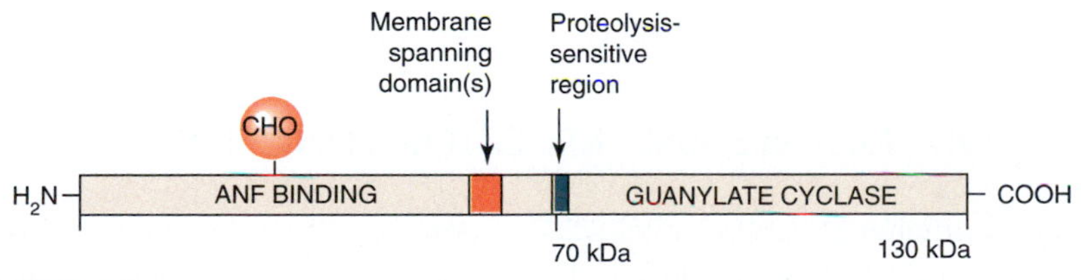

FIGURE 21.44
Functional domains of ANF-R_1 receptor.
Hypothetical model shows the sequence of an ANF-binding domain, a membrane spanning domain(s), a proteolysis-sensitive region, a guanylate cyclase catalytic domain, glucosylation sites (CHO), and amino and carboxyl terminals of the receptor.
Redrawn from Liu, B., Meloche, S., McNicoll, N., Lord, C., and DeLéan, A. Biochemistry *28:5599, 1989.*

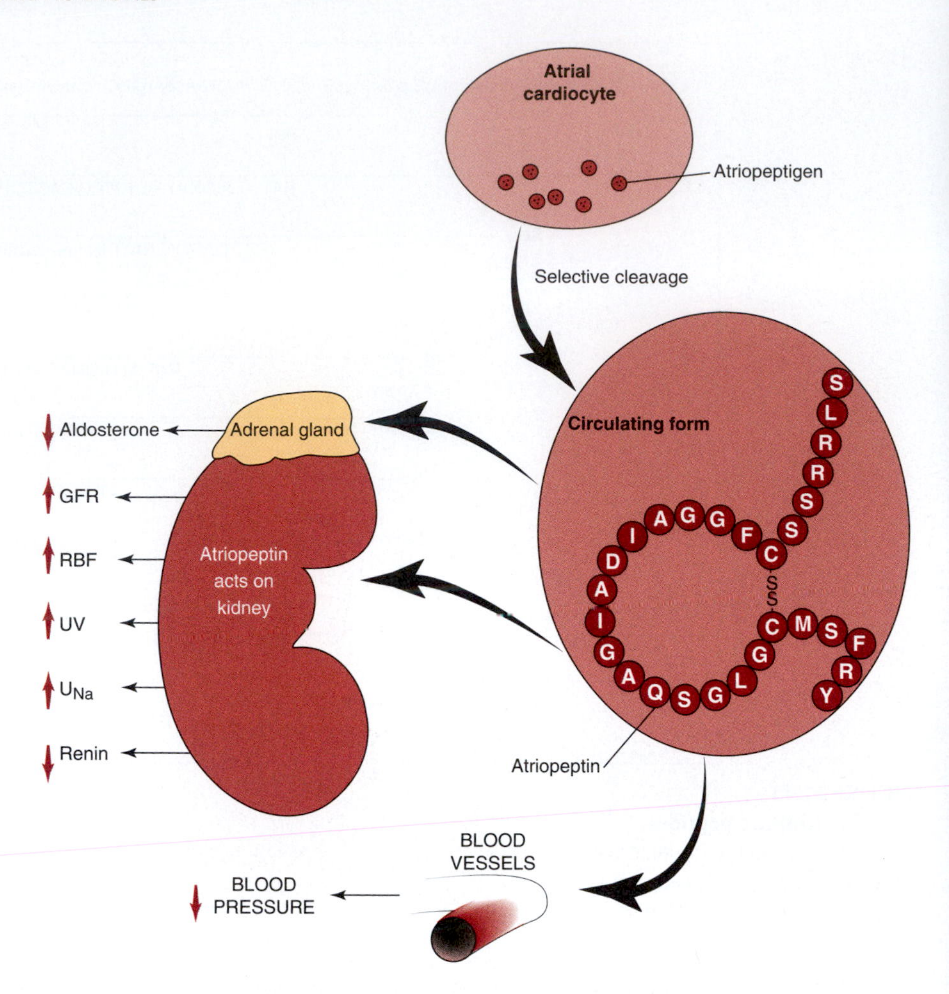

FIGURE 21.45
Schematic diagram of atrial natriuretic factor–atriopeptin hormonal system.
Prohormone is stored in granules located in perinuclear atrial cardiocytes. An elevated vascular volume results in cleavage and release of atriopeptin, which acts on the kidney (glomeruli and papilla) to increase the glomerular filtration rate (GFR), to increase renal blood flow (RBF), to increase urine volume (UV) and sodium excretion (U_{Na}), and to decrease plasma renin activity. Natriuresis and diuresis are also enhanced by the suppression of aldosterone secretion by the adrenal cortex and the release from the posterior pituitary of arginine vasopressin. Vasolidation of blood vessels also results in a lowering of blood pressure (BP). Diminution of vascular volume provides a negative feedback signal that suppresses circulating levels of atriopeptin.
Redrawn from Needleman, P. and Greenwald, J. E. N. Engl. J. Med. *314:828, 1986.*

increased urine volume and excretion of sodium ion. Renin secretion is also reduced and aldosterone secretion by the adrenal cortex is lowered. This action reduces aldosterone-mediated sodium reabsorption. ANF inhibits the vasoconstriction produced by angiotensin II and relaxes the constriction of the renal vessels, other vascular beds, and large arteries. ANF operates through its membrane receptor, which appears to be the extracellular domain of guanylate cyclase. The cGMP produced activates protein kinase G, which further phosphorylates cellular proteins to express many of the actions of this pathway. More needs to be learned about protein kinase G. Using analogs of ANF it has been shown that the majority of receptors expressed in the kidney are biologically silent, since they fail to elicit a physiological response. This new class of receptors may serve as specific peripheral storage–clearance binding sites and as such act as a hormonal buffer system to modulate plasma levels of ANF.

21.13 | ONCOGENES AND RECEPTOR FUNCTIONS

Normal, nonmalignant cells contain DNA sequences homologous to retroviral **oncogenes**, and these sequences are designated as **protooncogenes.** These protooncogenes are grouped according to their function or location in the cell: growth factors, growth factor receptors, nuclear proteins, and membrane proteins. The subcellular locations and functions of these proteins suggest that in normal cells they play important roles in terms of growth, development, and differentiation. Mutations, deletions, or insertions in protooncogene sequences activate these sequences into transforming oncogenes. A cancer cell may express few or many

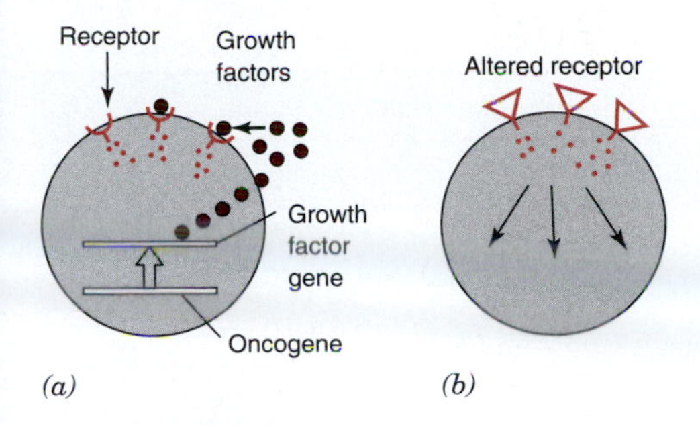

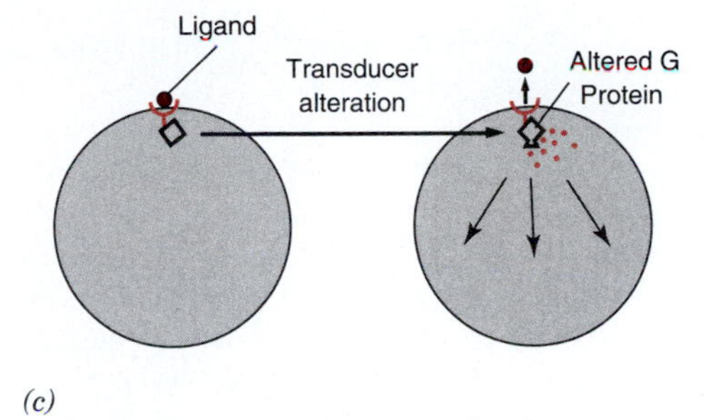

FIGURE 21.46
Mechanisms by which oncogenes can allow a cell to escape dependence on exogenous growth factors.
(*a*) By autocrine mechanism, where the cytoplasmic oncogene indirectly stimulates expression of growth factor gene and oversecretion of growth factors, which then overstimulates receptors on same cell; (*b*) by receptor alteration, so that receptor is "permanently turned on" without a requirement for growth factor binding; and (*c*) by transducer alteration, where the intermediate between the receptor and its resultant activity, that is, the GTP stimulatory protein, is permanently activated, uncoupling the normal requirement of ligand–receptor binding.

oncogenes that dictate the aberrant uncontrolled behavior of the cell. There are three mechanisms by which oncogenes allow a cell to escape dependence on exogenous growth factors; these are presented in Figure 21.46. Some oncogenes are genes for parts of receptors, most often related to growth factor hormone receptors, which function in the absence of the hormonal ligand. Thus an oncogene may represent a truncated gene where the ligand-binding domain is missing. This would result in production of the receptor protein, insertion into the cell membrane, and continuous constitutive function in the absence or presence of ligand (Figure 21.46*b*,*c*). In this situation the second messengers would be produced constitutively at a high rate, instead of being regulated by ligand, and the result would be uncontrolled growth of the cell. Some oncogenes may have tyrosine protein kinase activity and therefore function like tyrosine kinase normally related to certain cell membrane receptors. Other oncogenes relate to thyroid and steroid hormone receptors (see p. 974) while still others are DNA-binding proteins, some of which may be transactivating factors or related to such factors. Oncogene-encoded proteins that bind to DNA may be identical with or related to transactivating factors. The oncogene *Jun*, for example, is a component of activator protein 1 (AP1), a transactivating factor that regulates transcription. The ***HER-2/neu*** (also know as ***c-erbB-2***) oncogene encodes for a 185-kDa transmembrane receptor tyrosine kinase with homology to members of the epidermal growth factor (EGF) receptors. Unlike other EGF receptors, which transduce the EGF signal and thus facilitate hormone-mediated proliferation of mammary epithelial cells, HER-2/neu has intrinsic tyrosine kinase activity that activates receptor-mediated signal transduction in the absence of ligand. The *HER-2/neu* protein is overexpressed, mostly as a result of gene amplification, in 20–30% of human breast cancers. Overexpression of this oncogene in breast as well as ovarian cancer has been linked to poor prognosis, which includes shorter disease-free interval and shorter survival time, and has also been shown to have prognostic and predictive value for treatment with chemotherapy. *HER-2/neu* overexpression has been shown to induce resistance to **tumor necrosis factor** (TNF), which causes cancer cells to escape from host immune defenses. Recent data suggest that *HER-2/neu* constitutively activates the **Akt/NF-κB** anti-apoptotic cascade to confer this TNF resistance. Monoclonal antibodies directed against *HER-2/neu* have been developed and are being used in clinical practice. Table 21.9 reviews some oncogenes, or cancer-causing genes, and the functions of their protooncogenes (normal gene) (see Clin. Corr. 21.5).

TABLE 21.9 Known Oncogenes, Their Products and Functions[a]

Name of Oncogene	*Retrovirus*	*Tumor*	*Oncogenic Protein*	
			Cellular Location	*Protooncogene Function*
src	Chicken sarcoma	Chicken sarcoma	Plasma membrane	Tyrosine-specific protein kinase
yes	Chicken sarcoma		Plasma membrane (?)	
fgr	Cat sarcoma		(?)	
abl	Mouse leukemia	Human leukemia	Plasma membrane	Tyrosine-specific protein kinase
fps	Chicken sarcoma		Cytoplasm (plasma membrane?)	
fes	Cat sarcoma	Sarcoma	Cytoplasm (cytoskeleton?)	Tyrosine-specific protein kinase
ros	Chicken sarcoma		(?)	
erb-B	Chicken leukemia	Erythroleukemia, fibrosarcoma	Plasma and cytoplasmic membranes	EGF receptor's cytoplasmic tyrosine-specific protein kinase domain
fms	Cat sarcoma	Sarcoma	Plasma and cytoplasmic membranes	Tyrosine-specific protein kinase; macrophage colony-stimulating factor receptor
mil	Chicken carcinoma		Cytoplasm	(?)
raf	Mouse sarcoma	Sarcoma	Cytoplasm	Protein kinase (serine threonine) activated by Ras
mos	Mouse sarcoma	Mouse leukemia	Cytoplasm	(?)
sis	Monkey sarcoma	Monkey sarcoma	Secreted	PDGF-like growth factor, β-chain
Ha-ras	Rat sarcoma	Human carcinoma, rat carcinoma	Plasma membrane	GTP-binding protein
Ki-ras	Rat sarcoma	Human carcinoma, leukemia, and sarcoma	Plasma membrane	GTP-binding protein
N-ras	—	Human leukemia and carcinoma	Plasma membrane	
myc	Chicken leukemia	Human lymphoma	Nucleus	DNA binding related to cell proliferation; transcriptional control
myb	Chicken leukemia	Human leukemia	Nucleus	(?)
B-lym	—	Chicken lymphoma, human lymphoma	Nucleus (?)	(?)
ski	Chicken sarcoma		Nucleus (?)	(?)
rel	Turkey leukemia	Reticuloendotheliosis	(?)	(?)
erb-A	Chicken leukemia		(?)	Thyroid hormone receptor (c-erb-Aαl); related to steroid hormone receptors. retinoic acid receptor, and vitamin D_3 receptor
ets	Chicken leukemia		(?)	DNA binding
elk (*ets*-like)				DNA-binding protein
jun		Osteosarcoma		Products associate to form AP1 gene transcription factor
fos		Fibrosarcoma		Products associate to form AP1 gene transcription factor

Source: Adapted from Hunter, T. The proteins of oncogenes, *Sci. Am.* 251:70, 1984.

[a]The second column gives the source from which each viral oncogene was first isolated and the cancer induced by the oncogene. Some names, such as *fps* and *fes,* may be equivalent genes in birds and mammals. the third column lists human and animal tumors caused by agents other than viruses in which the *ras* oncogene or an inappropriately expressed protooncogene has been identified.

CLINICAL CORRELATION 21.5

Advances in Breast Cancer Chemotherapy

Breast carcinoma is moderately sensitive to multiple antitumor agents and combinations of drugs produce higher response rates and longer durations of response and survival than single-agent therapy. Within the past decade a number of new agents have been developed and proved effective in the treatment of breast cancer. The most prominent are the taxanes (plant alkaloids), including paclitaxel and docetaxel. These taxanes are currently being used in combination with other active drugs including the anthracyclines (doxorubicin), cisplatin and its analogs (carobplatin), alkylating agents (cyclophosphamide), antimetabolites (5-fluorouracil), vinca alkaloids (predominantly the norvinblastine analog vinorelbine), and anti-estrogens (tamoxifen). Recent advances in our understanding of the basic biology of breast carcinoma, including the internal and external stimuli that result in malignant transformation, progression, transformation, and metastasis, have provided additional potential targets for therapeutic intervention. More specifically, growth factor receptors, including the EGF receptor (HER-1), and *HER-2/neu* (c-erbB-2) have been targeted with a variety of treatment strategies. Monoclonal antibodies specific for each of these receptors have been developed and the first to undergo complete preclinical and clinical evaluation was the anti-HER-2 monoclonal antibody, trasluzumab (Herceptin). This antibody binds with high affinity to the extracellular domain of HER-2 and inhibits transmission of the growth-stimulatory signal. Other approaches to target HER-2 have included the development of tyrosine kinase inhibitors as well as other signal transduction inhibitors (vaccines, gene therapy, antisense therapy). The tyrosine kinase inhibitor emodin (1,3,8-trihydroxy-6-methylanthraquinone) has been shown to suppress growth of *HER-2/neu*-overexpressing breast cancer cells in athymic nude mice and to actually sensitize these cells to paclitaxel. Another class of compounds, the 4-anilinoquinazolines, has been shown to selectively target and irreversibly inactivate the EGF receptor tyrosine kinase through specific covalent modification of a cysteine residue present in the ATP-binding pocket of this enzyme. These new cytotoxic agents and antitumor strategies thus offer promising treatment options for patients with metastatic breast carcinoma.

Hortobagyi, G. N. Developments in chemotherapy of breast cancer. *Cancer* 88:3073, 2000; Zhang, L., Lau, Y.-K., Xia, W., Hortobagyi, G. N., and Hung, M.-C. Tyrosine kinase inhibitor emodin suppresses growth of HER-2/neu-overexpressing breast cancer cells in athymic mice and sensitizes these cells to the inhibitory effect of paclitaxel. *Clin. Cancer Res.* 5:343, 1999; and Fry, D. W., Bridges, A. J., Denny, W. A., Doherty, A., Greis, K. D., Hicks, J. L., Hook, K. E., Keller, P. R., Leopold, W. R., Loo, J. A., McNamara, D. J., Nelson, J. M., Sherwood, V., Smaill, J., Trump-Kallmeyer, S., and Dobrusin, E. M. Specific, irreversible inactivation of the epidermal growth factor receptor and erbB2 by a new class of tyrosine kinase inhibitor. *Proc. Natl. Acad. Sci. USA* 95:12022, 1998.

BIBLIOGRAPHY

Alberts, B., Bray, D., Lewis, J., Raff, R., Roberts, K., and Watson, J. D. *Molecular Biology of the Cell,* 3rd ed. New York: Garland Publishing, 1994.

Cuatrecasas, P. Hormone receptors, membrane phospholipids, and protein kinases. *The Harvey Lectures Ser.* 80:89, 1986.

DeGroot, L. J. (Ed.). *Endocrinology.* Philadelphia: Saunders, 1995.

Hunter, T. The proteins of oncogenes. *Sci. Am.* 251:70, 1984.

Krieger, D. T. and Hughes, J. C. (Eds.). *Neuroendocrinology.* Sunderland, MA: Sinauer Associates, 1980.

Litwack, G. (Ed.). *Biochemical Actions of Hormones,* Vols. 1–14. New York: Academic Press, 1973–1987.

Litwack, G. (Ed. in Chief). *Vitamins and Hormones,* Vol. 50. Orlando, FL: Academic Press, 1995.

Norman, A. W. and Litwack, G. *Hormones,* 2nd ed. Orlando, FL: Academic Press, 1997.

Richter, D. Molecular events in expression of vasopressin and oxytocin and their cognate receptors. *Am. J. Physiol.* 255:F207, 1988.

Ryan, R. J., Charlesworth, M. C., McCormick, D. J., Milius, R. P., and Keutmann, H. T. *FASEB J.* 2:2661, 1988.

Saltiel, A. R. The paradoxical regulation of protein phosphorylation in insulin action. *FASEB J.* 8:1034, 1994.

Spiegel, A. M., Shenker, A., and Weinstein, L. S. Receptor–effector coupling by G proteins: implication for normal and abnormal signal-transduction. *Endocr. Rev.* 13:536, 1992.

Struthers, A. D. (Ed.) *Atrial Natriuretic Factor.* Boston: Blackwell Scientific Publications, 1990.

Weinberg, R. A. The action of oncogenes in the cytoplasm and nucleus. *Science* 230:770, 1985.

QUESTIONS | C. N. ANGSTADT

Multiple Choice Questions

1. In a cascade of hormones (e.g., hormones from hypothalamus to pituitary to target tissue), at each successive level:
   A. quantity of hormone released and its half-life can be expected to increase.
   B. quantity of hormone released increases, but its half-life does not change.
   C. quantity of hormone released and its half-life are approximately constant.
   D. quantity of hormone released decreases, but its half-life does not change.
   E. quantity of hormone released and its half-life can both be expected to decrease.

2. If a single gene contains information for synthesis of more than one hormone molecule:
   A. all the hormones are produced by any tissue that expresses the gene.
   B. all hormone molecules are identical.

C. cleavage sites in the gene product are typically pairs of basic amino acids.
D. all peptides of the gene product have well-defined biological activity.
E. hormones all have similar function.

3. The direct effect of cAMP in the protein kinase A pathway is to:
A. activate adenylate cyclase.
B. dissociate regulatory subunits from protein kinase.
C. phosphorylate certain cellular proteins.
D. phosphorylate protein kinase A.
E. release hormones from a target tissue.

4. With anterior pituitary hormones, TSH, LH, and FSH:
A. α subunits are all different.
B. β subunits are specifically recognized by the receptor.
C. β subunit alone can bind to the receptor.
D. hormonal activity is expressed through activation of protein kinase B.
E. intracellular receptors bind these hormones.

5. In the interaction of a hormone with its receptor all of the following are true EXCEPT:
A. more than one polypeptide chain of the hormone may be necessary.
B. more than one second messenger may be generated.
C. an array of transmembrane helices may form the binding site for the hormone.
D. receptors have a greater affinity for hormones than for synthetic agonists or antagonists.
E. hormones released from their receptor after endocytosis could theoretically interact with a nuclear receptor.

6. Some hormone–receptor complexes are internalized by endocytosis. This process may involve:
A. binding of hormone–receptor complex to a clathrin coated pit.
B. recycling of receptor to cell surface.
C. degradation of receptor and hormone in lysosomes.
D. formation of a receptosome.
E. all of the above.

In the following questions, match the numbered hormone with the lettered kinase it stimulates.

A. protein kinase A
B. tyrosine kinase
C. protein kinase C
D. protein kinase G

7. Atrial natriuretic factor.
8. Gonadotrophin-releasing hormone.
9. Insulin.
10. Vasopressin.

Questions 11 and 12: Hypopituitarism may result from trauma, such as an automobile accident severing the stalk connecting the hypothalamus and anterior pituitary, or from tumors of the pituitary gland. In trauma, usually all of the releasing hormones from hypothalamus fail to reach the anterior pituitary. With a tumor of the gland, some or all of the pituitary hormones may be shut off. Posterior pituitary hormones may also be lost. Hypopituitarism can be life-threatening. Usual therapy is administration of end organ hormones in oral form.

11. If the stalk between the hypothalamus and anterior pituitary is severed, the pituitary would fail to cause the ultimate release of all of the following hormones EXCEPT:
A. ACTH.
B. estradiol.
C. oxytocin.
D. testosterone.
E. thyroxine.

12. In hypopituitarism it is necessary to maintain the ovarian cycle in female patients. In the ovarian cycle:
A. GnRH enters the vascular system via transport by a specific membrane carrier.
B. corpus luteum rapidly involutes only if fertilization does not occur.
C. inhibin works by inhibiting synthesis of α subunit of FSH.
D. FSH activates a protein kinase A pathway.
E. LH is taken up by corpus luteum and binds to cytoplasmic receptors.

Questions 13 and 14: Manic–depression may be caused by overactivity of certain central nervous system cells, perhaps caused by abnormally high levels of hormones or neurotransmitters, which stimulate the phosphatidylinositol (PI) cycle. Lithium has been used for many years to treat manic–depression. In the presence of Li^+, the PI cycle is slowed despite continued stimulation and cells become less sensitive to these stimuli. Li^+ may have two functions—inhibition of the phosphatase that dephosphorylates inositol trisphosphate and direct interference with the function of G-proteins.

13. PI cycle begins with activation of phospholipase C, which initiates a sequence of events including all of the following EXCEPT:
A. activation of IP_3 by action of a phosphatase.
B. increase in intracellular Ca^{2+} concentration.
C. release of diacylglycerol (DAG) from a phospholipid.
D. activation of protein kinase C.
E. phosphorylation of certain cytoplasmic proteins.

14. Which of the following statements concerning G-proteins is correct?
A. G-proteins bind the appropriate hormone at the cell surface.
B. GTP is bound to G-protein in the resting state.
C. α Subunit may be either stimulatory or inhibitory because it has two forms.
D. Adenylate cyclase can be activated only if α, β, and γ subunits of G-protein are associated with each other.
E. Hydrolysis of GTP is necessary for G-protein subunits to separate.

Problems

15. Using the Scatchard plot shown, calculate the maximum number of binding sites of the receptor for its hormone and the dissociation constant for binding.

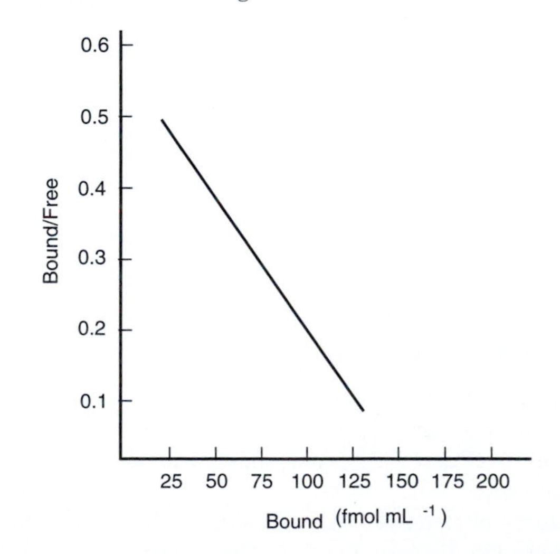

16. How would you determine whether an inability to produce cortisol in response to stress was caused by a problem in the hypothalamus, the anterior pituitary, or the adrenal cortex?

ANSWERS

1. **A** Each successive step typically releases a larger amount of a longer lived hormone.
2. **C** One or more trypsin-like proteases catalyze the reaction. A: The POMC gene product is cleaved differently in different parts of the anterior pituitary. B: Multiple copies of a single hormone may occur, but not necessarily. D: Some fragments have no known function. E: ACTH and β-endorphin, for example, hardly have similar functions.
3. **B** cAMP binding causes a conformational change in regulatory subunits, resulting in release of active protein kinase A. A: Adenylate cyclase produces cAMP. C: This is what protein kinase does.
4. **B** These are different in each hormone. A: α Subunits are identical or nearly so. C: Although specificity is conferred by β subunits, binding to the receptor requires both subunits. D: It is protein kinase A, and perhaps also protein kinase C in the case of TSH. E: These large glycoprotein hormones do not penetrate the cell membrane; they bind to receptors on the cell surface.
5. **D** β Receptors bind isoproterenol more tightly than their hormones. A, B: These are true of glycoprotein hormones. C: This appears to be true for β_1 receptor. E: This is possible, but entirely speculative; there are currently no known examples.
6. **E** A and D always happen. B or C happen sometimes.
7. **D** This is activated by cGMP.
8. **C** This hormone stimulates phospholipase C.
9. **B** The β subunit of the receptor has tyrosine kinase activity. The action of insulin is complex.
10. **A** Vasopressin is only one of many hormones that stimulate protein kinase A.
11. **C** Oxytocin is released from posterior pituitary. A, B, D, and E all require releasing hormones from hypothalamus for anterior pituitary to release them.
12. **D** A: GnRH enters the vascular system through fenestrations. B. The corpus luteum is replaced by placenta if fertilization occurs. C: Glycoprotein hormones share a common α subunit. Specific control of them would not involve a subunit they share. E: LH interacts with receptors on the cell membrane.
13. **A** IP_3 is the active form. Action of phosphatase on it renders it inactive as it is converted to inositol. B: IP_3 causes release of calcium. C, D: This is the other second messenger and it activates protein kinase C, E: This is what protein kinase C does.
14. **C** Which form is released depends on the specific hormone and receptor that have interacted. A: Receptor binds the hormone and the complex interacts with G-protein. B, E: GDP is bound to the resting enzyme. Its replacement by GTP causes α subunit to dissociate. Actually hydrolysis of GTP allows subunits to re-associate. D: α Subunit, when dissociated from the other two, interacts with the enzyme to either activate or inhibit it depending on whether it is an α_s or an α_i .
15. From the plot below, the intercept on the abscissa is 150 fmol mL^{-1}, which is the maximum number of binding sites. The slope of the line is -0.004 so K_d is 250 fmol mL^{-1}.

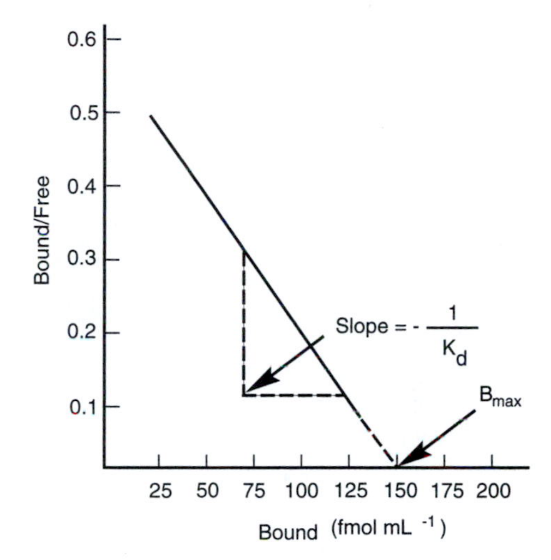

16. The end organ is adrenal cortex. If cortisol is increased in response to an injection of ACTH, the adrenal cortex is functioning properly. An intravenous infusion of CRH (corticotropic-releasing hormone) should be able to reach the anterior pituitary. If ACTH increases in response to this, the pituitary is functioning so the problem is probably with the hypothalamus. If ACTH does not increase in response to CRH, the anterior pituitary is not functioning.

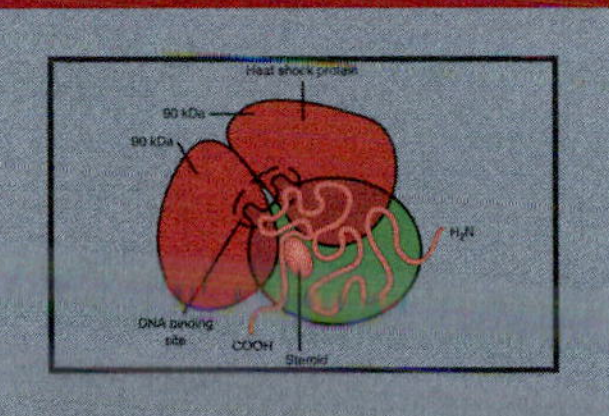

22

BIOCHEMISTRY OF HORMONES II: STEROID HORMONES

Gerald Litwack and
Thomas J. Schmidt

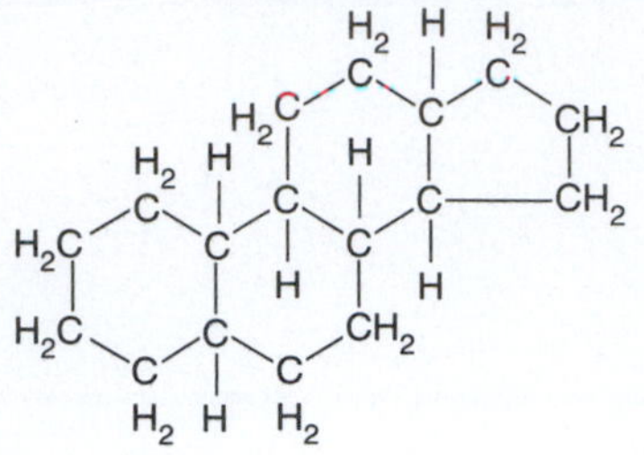

Cyclopentanoperhydrophenanthrene nucleus

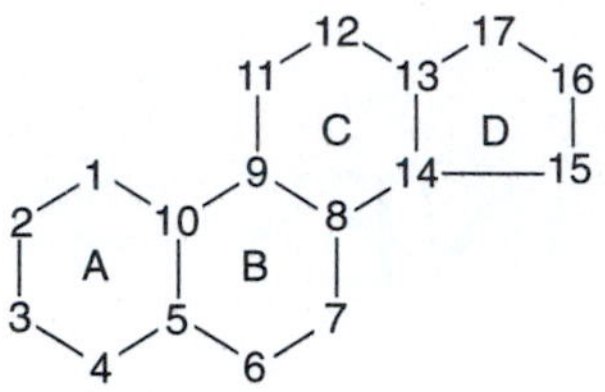

Numbering system of carbons

FIGURE 22.1
The steroid nucleus.

22.1 | OVERVIEW

Steroid hormones in the human include cortisol as the major glucocorticoid or anti-stress hormone, aldosterone as an important regulator of Na^+ uptake, and the sex and progestational hormones. Sex hormones are 17β-estradiol in females and testosterone in males. Progesterone is the major progestational hormone. Testosterone is reduced in some target tissues to dihydrotestosterone, a higher affinity ligand for the androgen receptor. Vitamin D_3 is converted to the secosteroid hormone, dihydroxy vitamin D_3. Genes in the steroid receptor supergene family include retinoic acid receptors and thyroid hormone receptor, although the ligands for these additional receptors are not derivatives of cholesterol. Retinoic acid and thyroid hormone, however, have six-membered ring structures that could be considered to resemble the A ring of a steroid.

Steroid hormone structure will be reviewed, as will synthesis and inactivation of steroid hormones. Regulation of synthesis of steroid hormones is reviewed with respect to the renin–angiotensin system for aldosterone, the gonadotropes, especially follicle-stimulating hormone for 17β-estradiol, and the vitamin D_3 mechanism. Steroid hormone transport by transporting proteins in blood is reviewed. A general model for steroid hormone action at the cellular level is presented with information on receptor activation and regulation of receptor levels. Specific examples of steroid hormone action for programmed cell death and for stress are presented. Finally, the roles of steroid hormone receptors as transcriptional transactivators and repressors are reviewed.

22.2 | STRUCTURES OF STEROID HORMONES

Steroid hormones are derived in specific tissues in the body and are divided into two classes: the **sex** and **progestational hormones**, and the **adrenal hormones.** They are synthesized from cholesterol and all of these hormones pass through the required intermediate, Δ^5-pregnenolone. The structure of steroid hormones is related to the **cyclopentanoperhydrophenanthrene** nucleus. The numbering of the cyclopentanoperhydrophenanthrene ring system and the lettering of the rings are presented in Figure 22.1. The ring system of the steroid hormones is stable and not catabolized by mammalian cells. Conversion of active hormones to less active or inactive forms involves alteration of ring substituents rather than the ring structure itself. The parental precursor of the steroids is **cholesterol,** shown in Figure 22.2. The biosynthesis of cholesterol is given on page 742.

The major steroid hormones and their actions in humans are presented in Table 22.1. Many are similar in gross structure, although the specific receptor for each hormone is able to distinguish the cognate ligand. For cortisol and aldos-

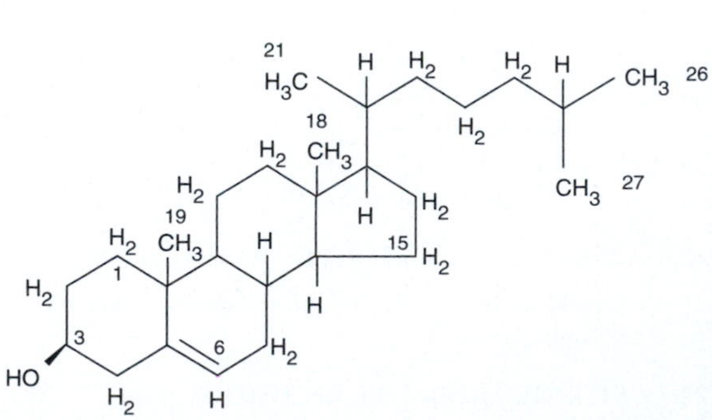

Cholesterol

FIGURE 22.2
Structure of cholesterol.

TABLE 22.1 Major Steroid Hormones of Humans

Hormone	*Structure*	*Secretion from*	*Secretion Signal*	*Functions*
Progesterone	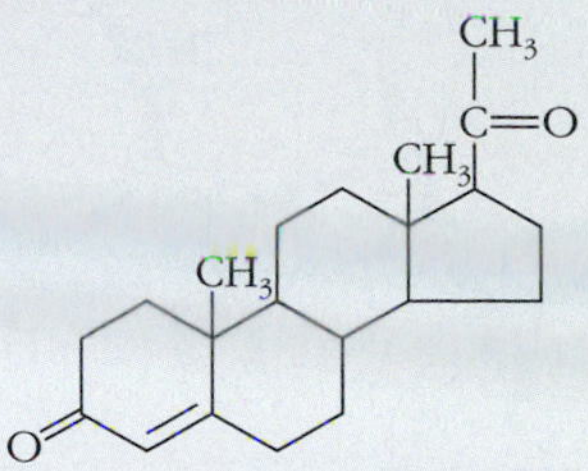	Corpus luteum	LH	Maintains (with estradiol) the uterine endometrium for implantation; differentiation factor for mammary glands
17β-Estradiol	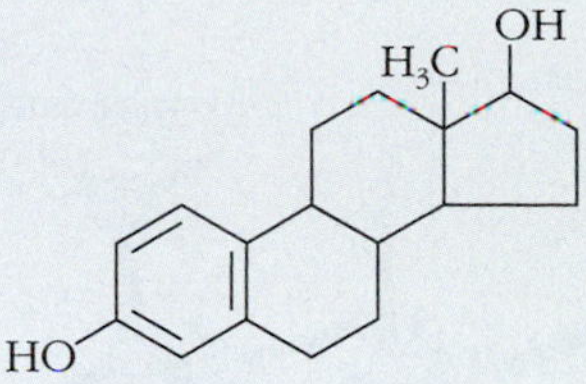	Ovarian follicle; corpus luteum; (Sertoli cell)	FSH	Female: regulates gonadotropin secretion in ovarian cycle (see Chapter 21); maintains (with progesterone) uterine endometrium; differentiation of mammary gland. Male: negative feedback inhibitor of Leydig cell synthesis of testosterone
Testosterone	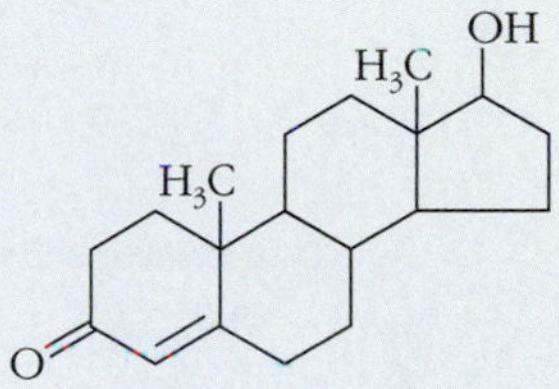	Leydig cells of testis; (adrenal gland); ovary	LH	Male: after conversion to dihydrotestosterone, production of sperm proteins in Sertoli cells; secondary sex characteristics (in some tissues testosterone is active hormone)
Dehydroepiandrosterone	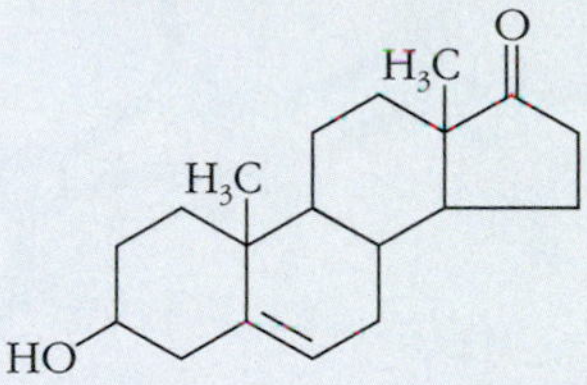	Reticularis cells	ACTH	Various protective effects; weak androgen; can be converted to estrogen; no receptor yet found; inhibitor of G6-PDH: regulates NAD^+ coenzymes
Cortisol	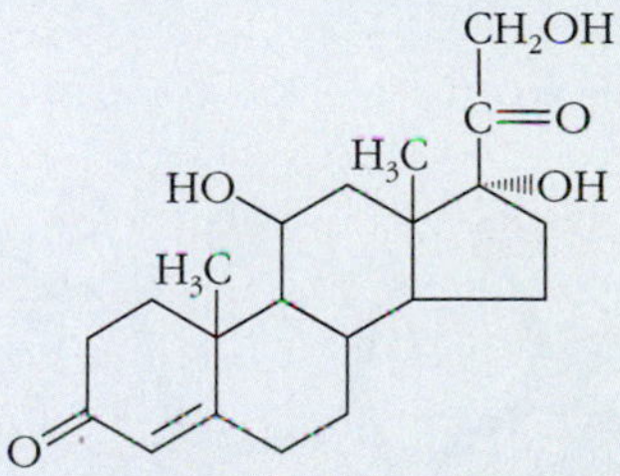	Fasciculata cells	ACTH	Stress adaptation through various cellular phenotypic expressions; slight elevation of liver glycogen; killing effect on certain T cells in high doses; elevates blood pressure; sodium uptake in luminal epithelia
Aldosterone	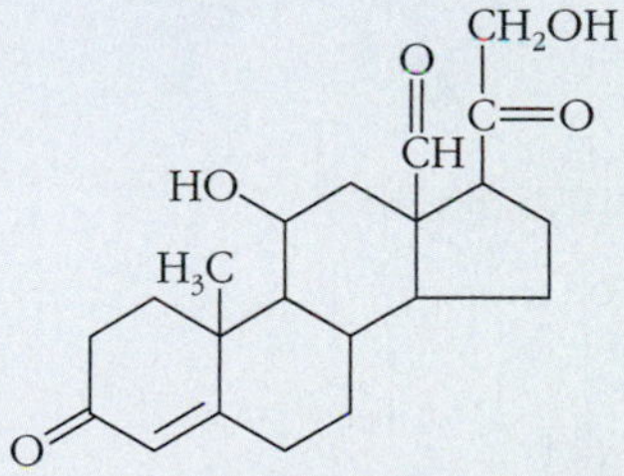	Glomerulosa cells of adrenal cortex	Angiotensin II/III	Causes sodium ion uptake via conductance channel; occurs in high levels during stress; raises blood pressure; fluid volume increased
1,25-Dihydroxyvitamin D_3	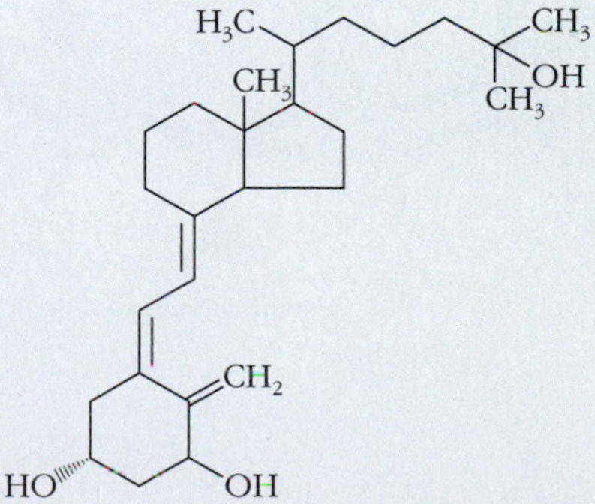	Vitamin D arises in skin cells after irradiation and then successive hydroxylations occur in liver and kidney to yield active form of hormone	PTH (stimulates kidney proximal tubule hydroxylation system)	Facilitates Ca^{2+} and phosphate absorption by intestinal epithelial cells; induces intracellular calcium-binding protein

[a]LH, luteinizing hormone; FSH, follicle-stimulating hormone; ACTH, adrenocorticotropic hormone; PTH, parathyroid hormone.

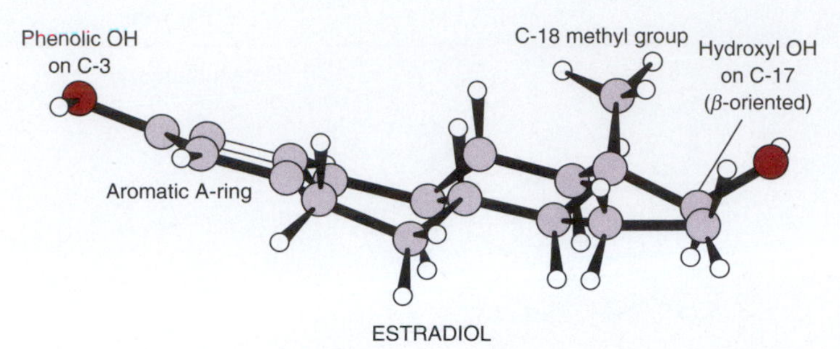

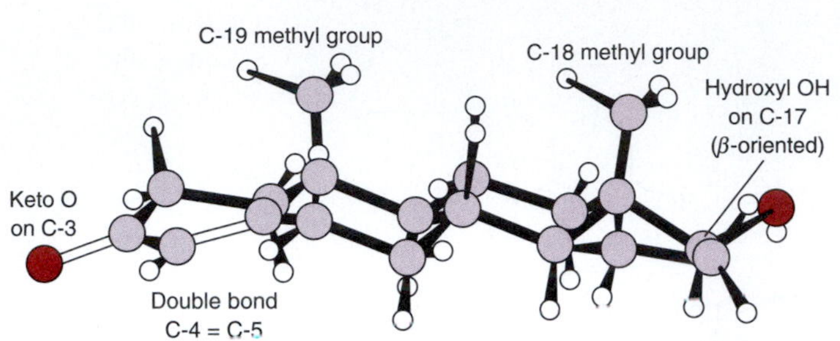

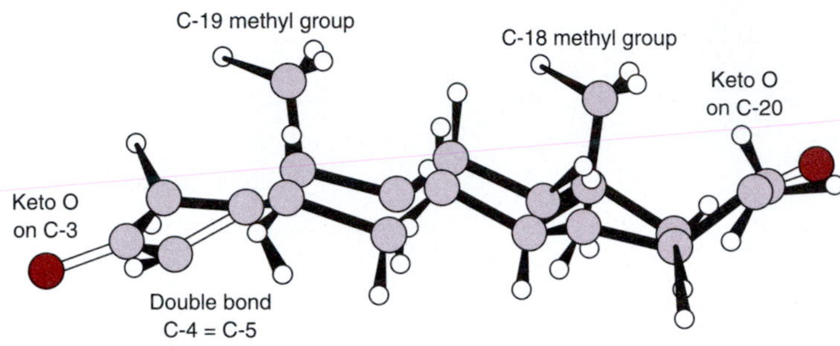

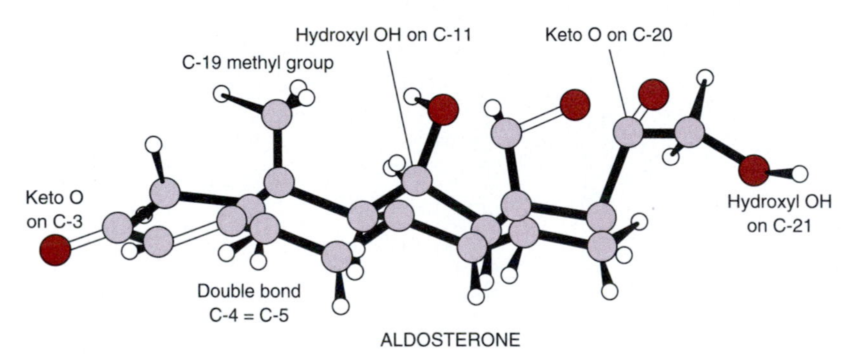

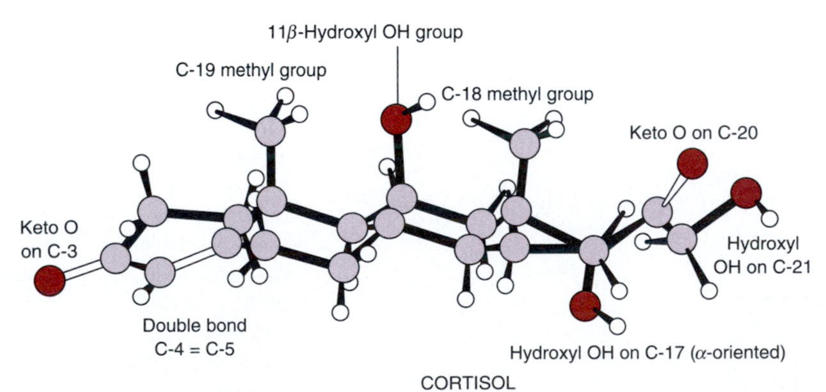

FIGURE 22.3
"Ball-and-stick" representations of some steroid hormones determined by X-ray crystallographic methods.
Details of each structure are labeled. In aldosterone the acetal grouping is $R{-}CH\langle^{OR_1}_{OR_2}$ and the hemiketal grouping is $\,^{R_1}_{R_2}\rangle C \langle^{OR_3}_{OH}$, where R_1, R_2 and R_3 refer to different substituents.
Reprinted with permission from Glusker, J. P. In: G. Litwack (Ed.), Biochemical Actions of Hormones, *Vol. 6. New York: Academic Press, 1979, pp. 121–204.*

terone, however, there is overlap in the ability of each specific receptor to bind both ligands. Thus the availability and concentrations of each receptor and the relative amounts of each hormone in a given cell become paramount considerations. The steroid hormones listed in Table 22.1 are described as classes based on the number of carbons in their structures. Thus a C-27 secosteroid is $1,25(OH)_2D_3$; C-21 steroids are progesterone, cortisol, and aldosterone; C-19 steroids are testosterone and dehydroepiandrosterone; and a C-18 steroid is **17β-estradiol.** Classes, such as sex hormones, can be distinguished easily by the carbon number, C-19 being androgens, C-18 being estrogens, and C-21 being progestational or adrenal steroids. Aside from the number of carbon atoms in a class structure, certain substituents in the ring system are characteristic. For example, glucocorticoids and mineralocorticoids (typically aldosterone) possess a C-11 OH or oxygen function. In rare exceptions, certain synthetic compounds elicit a response without a C-11 OH group but they require a new functional group in proximity within the A-B ring system. Estrogens do not have a C-19 methyl group and their A ring contains three double bonds. Many receptors recognize the ligand A ring primarily; the estrogen receptor can distinguish the A ring of estradiol stretched out of the plane of the B-C-D rings compared to other steroids in which the A ring is coplanar with the B-C-D rings. These relationships are shown in Figure 22.3.

22.3 | BIOSYNTHESIS OF STEROID HORMONES

Steroid Hormones Are Synthesized from Cholesterol

Hormonal regulation of steroid hormone biosynthesis is mediated by an elevation of intracellular **cAMP** and $\mathbf{Ca^{2+}}$, although generation of **inositol trisphosphate** may also be involved, as shown in Figure 22.4. The stimulatory response of cAMP

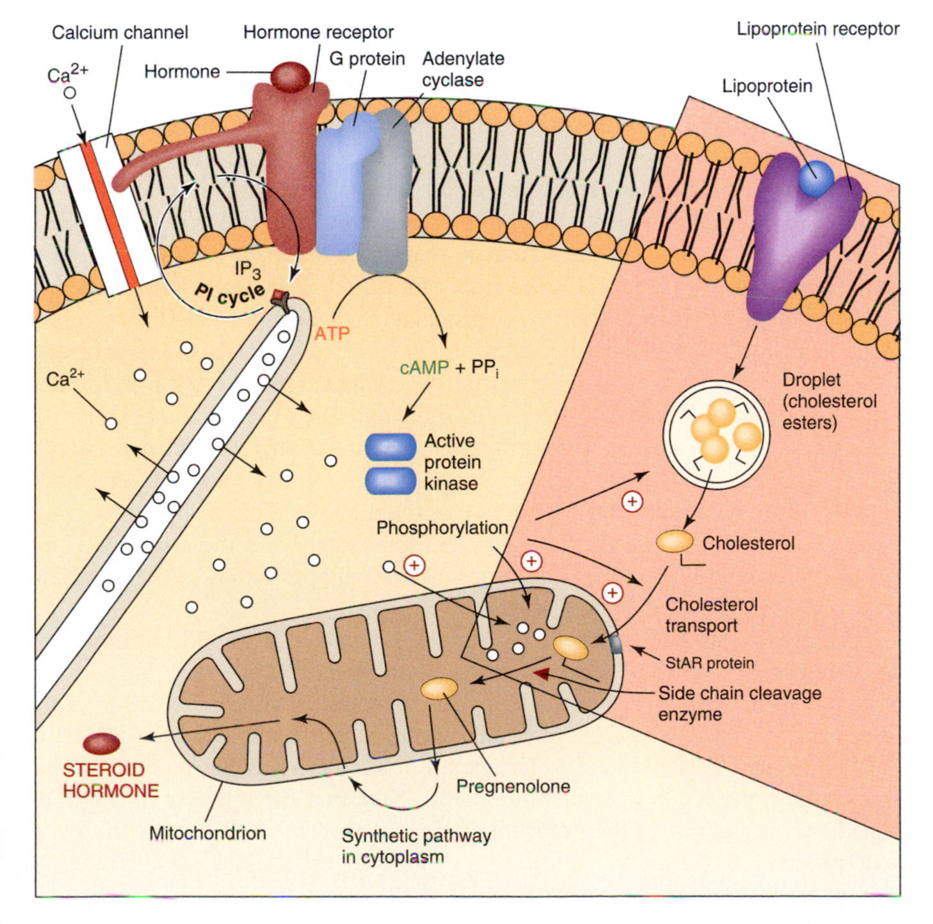

FIGURE 22.4
Overview of hormonal stimulation of steroid hormone biosynthesis.
Nature of the hormone (top of figure) depends on the cell type and receptor (ACTH for cortisol synthesis; FSH for estradiol synthesis; LH for testosterone synthesis, etc., as given in Table 22.1). It binds to cell membrane receptor and activates adenylate cyclase mediated by a stimulatory G-protein. Receptor, activated by hormone, may directly stimulate a calcium channel or indirectly stimulate it by activating the phosphatidylinositol cycle (PI cycle) as shown in Figure 21.21. If the PI cycle is concurrently stimulated, IP_3 could augment cytosol Ca^{2+} levels from the intracellular calcium store. The increase in cAMP activates protein kinase A (Figure 22.18) whose phosphorylations cause increased hydrolysis of cholesterol esters from the droplet to free cholesterol and increase cholesterol transport into the mitochondrion. The combination of elevated Ca^{2+} levels and protein phosphorylation, as well as induction of the StAR protein, result in increased side chain cleavage and steroid biosynthesis. These combined reactions overcome the rate-limiting steps in steroid biosynthesis and more steroid is produced, which is secreted into the extracellular space and circulated to the target tissues in the bloodstream.

is mediated via acute (occurring within seconds to minutes) and chronic (requiring hours) effects on steroid synthesis. The acute effect is to mobilize and deliver cholesterol, the precursor for all steroid hormones, to the mitochondrial inner membrane, where it is metabolized to pregnenolone by the cytochrome P450 cholesterol side chain cleavage enzyme (see Chapter 11 for discussion of P450 enzymes). In contrast, the chronic effects of cAMP are mediated via increased transcription of the genes that encode the steroidogenic enzymes and are thus responsible for maintaining optimal long-term steroid production. Apparently a protein is induced and this newly synthesized regulatory protein actually facilitates the translocation of cholesterol from outer to inner mitochondrial membrane where the P450 enzyme is located. This 30-kDa phosphoprotein is designated as the **steroidogenic acute regulatory (StAR)** protein. In humans, StAR mRNA has been shown to be specifically expressed in testis and ovary, known sites of steroidogenesis. Patients with lipoid congenital adrenal hyperplasia (LCAH), an inherited disease in which both adrenal and gonadal steroidogenesis is significantly impaired and lipoidal deposits occur in these tissues, express truncated and nonfunctional StAR proteins. These biochemical and genetic data strongly suggest that StAR protein is the hormone-induced protein factor that mediates acute regulation of steroid hormone biosynthesis.

Pathways for conversion of cholesterol to the adrenal cortical steroid hormones are presented in Figure 22.5. Cholesterol undergoes side chain cleavage to form Δ^5-pregnenolone releasing a C_6 aldehyde, isocaproaldehyde. **Δ^5-Pregnenolone** is mandatory in the synthesis of all steroid hormones. Pregnenolone is converted directly to progesterone by two cytoplasmic enzymes, **3β-ol dehydrogenase** and **$\Delta^{4,5}$-isomerase.** The dehydrogenase converts the 3-OH group of pregnenolone to a 3-keto group and the isomerase moves the double bond from the B ring to the A ring to produce progesterone. In the corpus luteum the bulk of steroid synthesis stops at this point. Progesterone is further converted to aldosterone or cortisol. Conversion of pregnenolone to **aldosterone** in the adrenal zona glomerulosa cells requires endoplasmic reticulum 21-hydroxylase, and mitochondrial 11β-hydroxylase and 18-hydroxylase. To form cortisol, primarily in adrenal zona fasciculata cells, endoplasmic reticulum **17-hydroxylase** and **21-hydroxylase** are required together with mitochondrial **11β-hydroxylase.** The endoplasmic reticulum (ER) hydroxylases are all cytochrome P450 enzymes (see Chapter 11). Δ^5-Pregnenolone is converted to **dehydroepiandrosterone** in the adrenal zona reticularis cells by the action of 17α-hydroxylase of the endoplasmic reticulum to form 17α-hydroxypregnenolone and then by the action of a carbon side chain cleavage system to form dehydroepiandrosterone. Cholesterol is converted to the sex hormones by way of Δ^5-pregnenolone (Figure 22.6) and **progesterone.** This is converted to testosterone by the action of cytoplasmic enzymes and 17-dehydrogenase. **Testosterone,** so formed, is a major secretory product in the Leydig cells of the testis and undergoes conversion to dihydrotestosterone in some androgen target cells before binding to the androgen receptor. This conversion requires the activity of **5α-reductase** located in the ER and nuclear fractions. Pregnenolone can enter an alternative pathway to form dehydroepiandrosterone as described above. This compound is converted to 17β-estradiol via the aromatase enzyme system and the action of 17-reductase. Also, estradiol is formed from testosterone by the action of the aromatase system.

The hydroxylases of endoplasmic reticulum involved in steroid hormone synthesis are cytochrome P450 enzymes. Molecular oxygen (O_2) is a substrate with one oxygen atom incorporated into the steroidal substrate (as an OH) and the second atom incorporated into a water molecule. Electrons are generated from NADH or NADPH through a flavoprotein to ferredoxin or similar nonheme protein. Various agents induce the levels of cytochrome P450. Note that there is movement of intermediates in and out of the mitochondrial compartment during the steroid synthetic process.

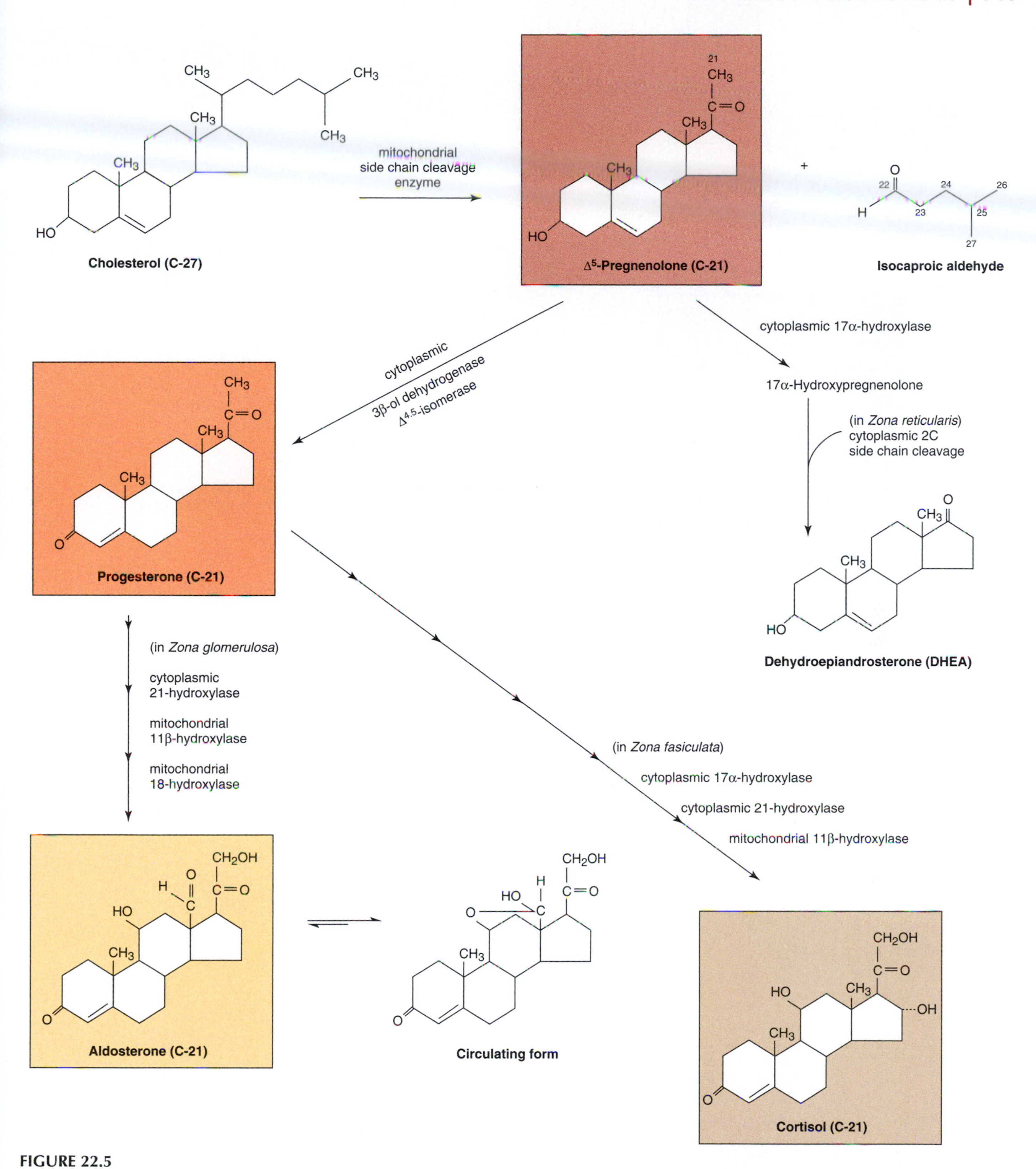

FIGURE 22.5
Conversion of cholesterol to adrenal cortical hormones.

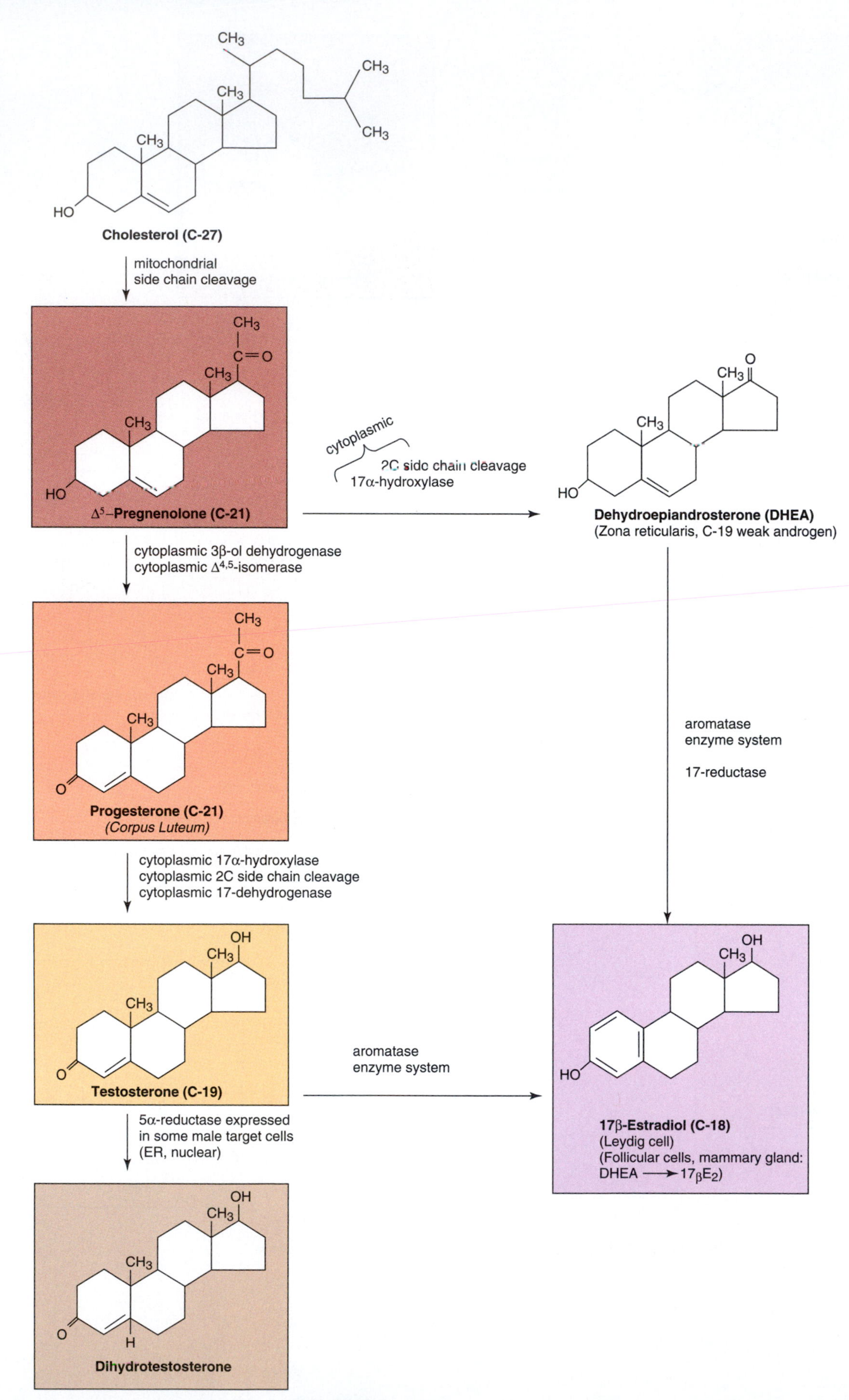

FIGURE 22.6
Conversion of cholesterol to sex hormones.

22.4 | METABOLIC INACTIVATION OF STEROID HORMONES

The steroid ring system is very stable. For the most part, inactivation of steroid hormones involves reduction. Testosterone is initially reduced to a more active form by the enzyme 5α-reductase to form **dihydrotestosterone,** the preferred ligand for the androgen receptor. However, further reduction similar to the other steroid hormones results in inactivation. The inactivation reactions predominate in liver and generally render the steroids more water soluble, as marked by subsequent conjugation with glucuronides or sulfates that are excreted in the urine. Table 22.2 summarizes reactions leading to inactivation and excretory forms of the steroid hormones.

22.5 | CELL–CELL COMMUNICATION AND CONTROL OF SYNTHESIS AND RELEASE OF STEROID HORMONES

Secretion of steroid hormones from cells where they are synthesized is elicited by other hormones. Many, but not all, such systems are described in Chapter 21 (Figures 21.2, p. 907 and 21.3, p. 909). The hormones that directly stimulate the biosynthesis and secretion of the steroid hormones are summarized in Table 22.3. The signals for stimulation of biosynthesis and secretion of steroid hormones are polypeptide hormones operating through cognate cell membrane receptors. In some systems where both cAMP and the phosphatidylinositol (PI) cycle are involved, it is not clear whether one second messenger predominates. In many cases, for example, aldosterone synthesis and secretion, probably several components (i.e., acetylcholine muscarinic receptor, atriopeptin receptor, and their second messengers) are involved in addition to the signal listed in Table 22.3.

Steroid Hormone Synthesis Is Controlled by Specific Hormones

The general mechanism for hormonal stimulation of steroid hormone synthesis is presented in Figure 22.4. Figure 22.7 presents the system for stimulation of cortisol biosynthesis and release. The role of Ca^{2+} in steroid synthesis and/or secretion is unclear. Rate-limiting steps in the biosynthetic process involve the availability of cholesterol from cholesterol esters in the droplet, the transport of cholesterol to the inner mitochondrial membrane (StAR protein), and the upregulation of the otherwise rate-limiting side chain cleavage reaction.

Aldosterone

Figure 22.8 shows the reactions leading to the secretion of aldosterone in the adrenal zona glomerulosa cell. The regulatory control on aldosterone synthesis and secretion is complicated. The main driving force is **angiotensin II** generated from the signaling to the **renin–angiotensin system** shown in Figure 22.9. Essentially, the signal is generated under conditions when blood [Na^+] and blood pressure (blood volume) are required to be increased. The N-terminal decapeptide of circulating **α_2-globulin (angiotensinogen)** is cleaved by **renin**, a protease. This decapeptide is the hormonally inactive precursor, angiotensin I. It is converted to the octapeptide hormone, angiotensin II, by the action of converting enzyme. Angiotensin II is converted to the heptapeptide, angiotensin III, by an aminopeptidase. Both angiotensins II and III bind to the angiotensin receptor (Figure 22.8), which activates the phosphatidylinositol cycle to generate IP_3 and DAG. The IP_3 stimulates release of calcium ions from the intracellular calcium storage vesicles. In addition, the activity of the Ca^{2+} channel is stimulated by the angiotensin–receptor complex. K^+ ions are also required to stimulate the Ca^{2+} channel and these events lead to a greatly increased level of cytoplasmic Ca^{2+}. The enhanced cytoplasmic Ca^{2+} has a role in aldosterone secretion and together with diacylglycerol stimulates protein ki-

TABLE 22.2 Excretion Pathways for Steroid Hormones

Steroid Class	*Starting Steroid*	*Inactivation Steps*	*A:B Ring Junction*	*Steroid Structure Representations of Excreted Product*	*Principal Conjugate Present*[a]
Progestins	Progesterone	1. Reduction of C-20 2. Reduction of 4-ene-3-one	(cis)	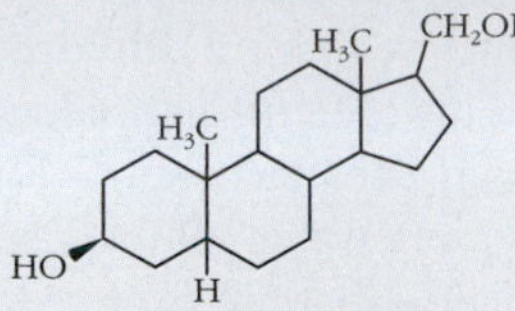 Pregnanediol (5β-pregnane-3$^{\alpha}$, 2$^{\alpha}$-diol)	G
Estrogens	Estradiol	1. Oxidation of 17β-OH 2. Hydroxylation at C-2 with subsequent methylation 3. Further hydroxylation or ketone formation at a variety of positions (e.g., C-6, C-7, C-14, C-15, C-16, C-18)		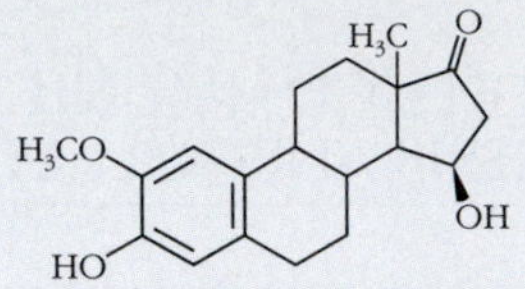 One of many possible compounds	G
Androgens	Testosterone	1. Reduction of 4-ene-3-one 2. Oxidation of C-17 hydroxyl	(cis and trans)	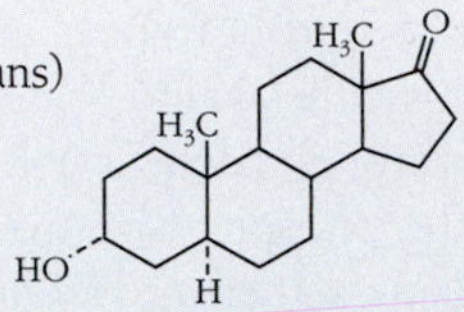 Androsterone 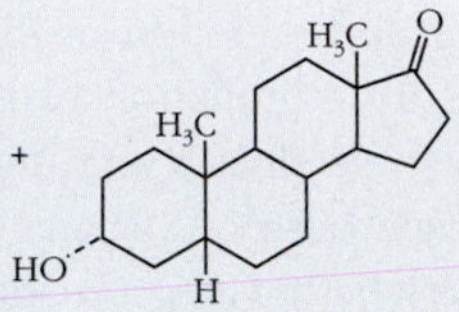 + Etiocholanolone	G, S
Glucocorticoids	Cortisol	1. Reduction of 4-ene-3-one 2. Reduction of 20-oxo group 3. Side chain cleavage	(trans)	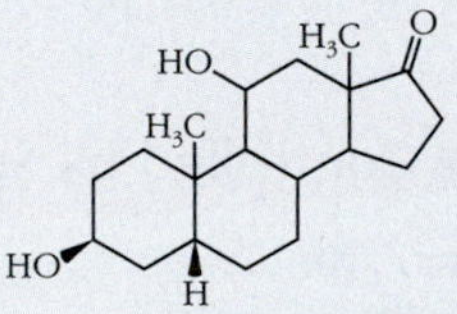 11β-OH-androsterone 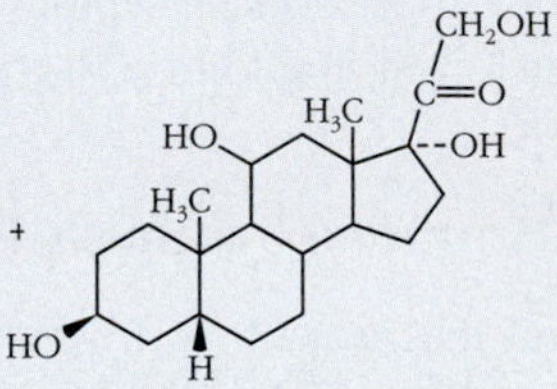 + Allo tetrahydrocortisone	G
Mineralocorticoids	Aldosterone	1. Reduction of 4-ene-3-one	(trans)	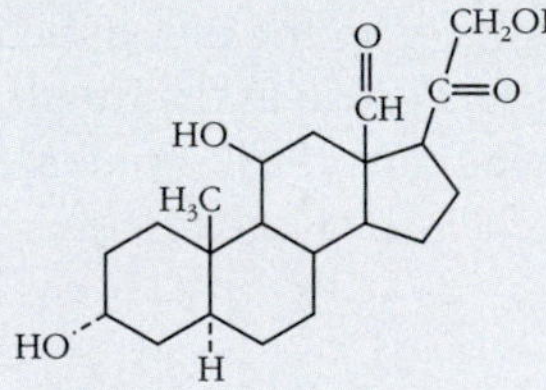 3α, 11β, 21-$(OH)_3$-20-oxo-5β-pregnane-18-al	G
Vitamin D metabolites	$1,25(OH)_2D_3$	1. Side chain cleavage between C-23 and C-24		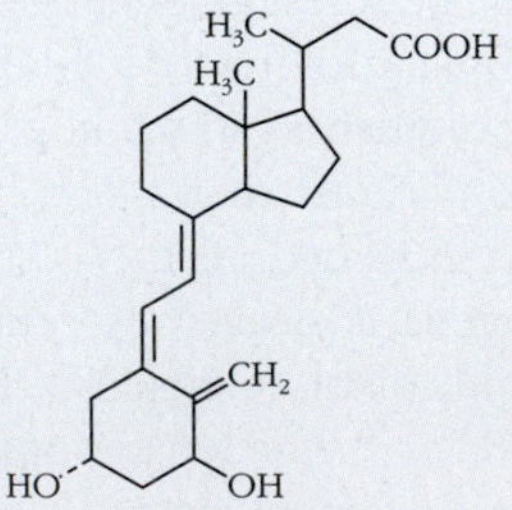 Calcitroic acid	?

Source: From Norman, A. W. and Litwack, G. *Hormones*. Orlando, FL: Academic Press, 1987.

[a]G, Glucuronide; S, sulfate.

TABLE 22.3 Hormones that Directly Stimulate Synthesis and Release of Steroid Hormones

Steroid Hormone	*Steroid-Producing Cell or Structure*	*Signal*[a]	*Second Messenger*	*Signal System*
Cortisol	Adrenal zona fasciculata	ACTH	cAMP, PI cycle, Ca^{2+}	Hypothalamic–pituitary cascade
Aldosterone	Adrenal zona glomerulosa	Angiotensin II/III	PI cycle, Ca^{2+}	Renin–angiotensin system
Testosterone	Leydig cell	LH	cAMP	Hypothalamic–pituitary cascade
17β-Estradiol	Ovarian follicle	FSH	cAMP	Hypothalamic–pituitary–ovarian cycle
Progesterone	Corpus luteum	LH	cAMP	Hypothalamic–pituitary–ovarian cycle
1,25$(OH)_2$ D_3	Kidney	PTH	cAMP	Sunlight, parathyroid glands, plasma Ca^{2+} level

[a]ACTH, adrenocorticotropic hormone; LH, luteinizing hormone; FSH, follicle-stimulating hormone; PI, phosphatidylinositol; PTH, parathyroid hormone.

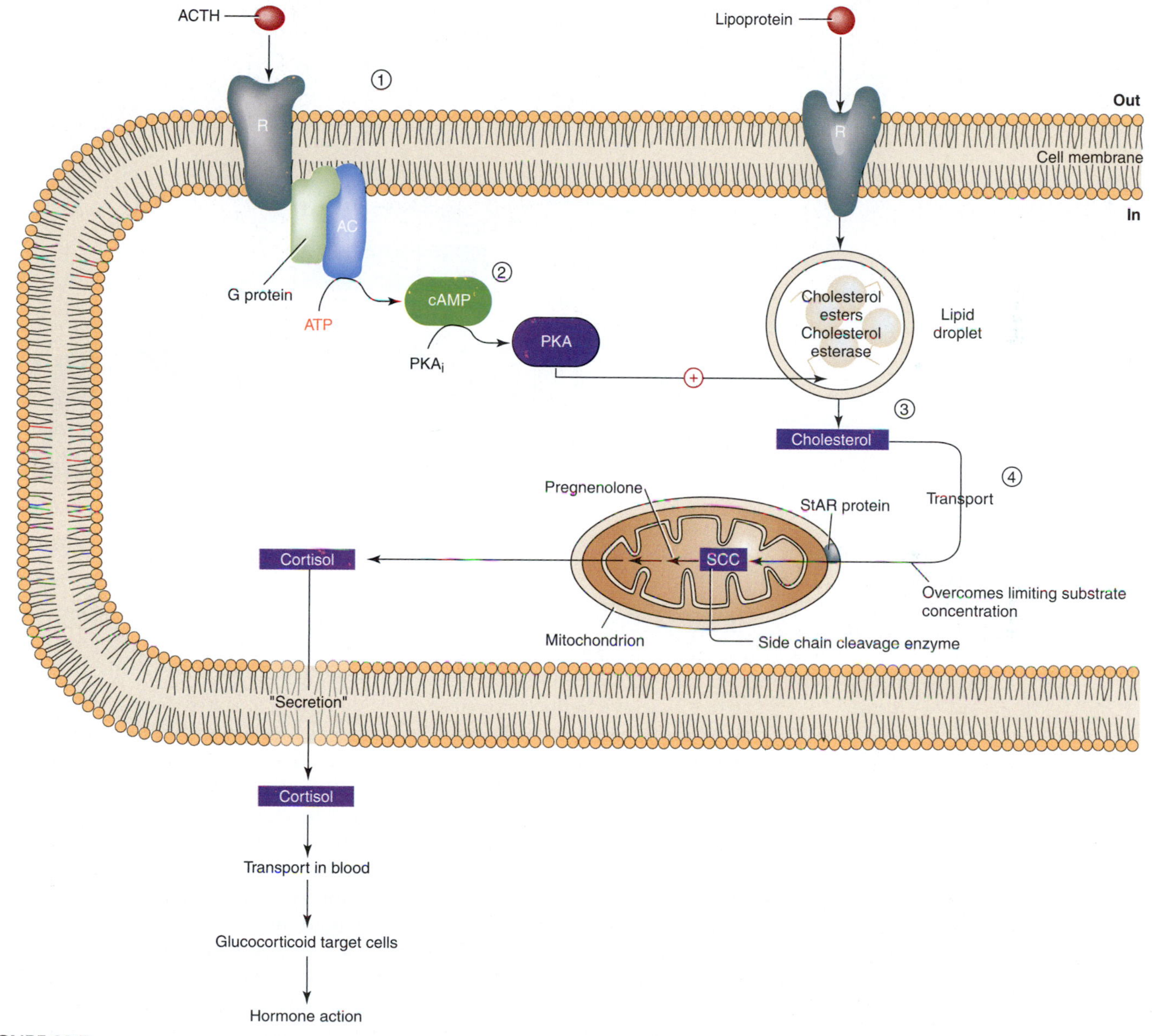

FIGURE 22.7
Action of ACTH on adrenal fasciculata cells to enhance production and secretion of cortisol.
1. Binding of ACTH to G-protein coupled receptor (R). 2. Activation of adenylate cyclase and generation of cAMP, which activates protein kinase A. 3. Hydrolysis of cholesterol esters by cholesterol esterase to generate cholesterol. 4. Transport of cholesterol into mitochondria by steroidogenic acute regulatory protein. AC, adenylate cyclase; cAMP, cyclic AMP; PKA, protein kinase A; SCC, side chain cleavage system of enzymes. StAR (steroidogenic acute regulatory) protein is a cholesterol transporter functioning between the outer and inner mitochondrial membranes.

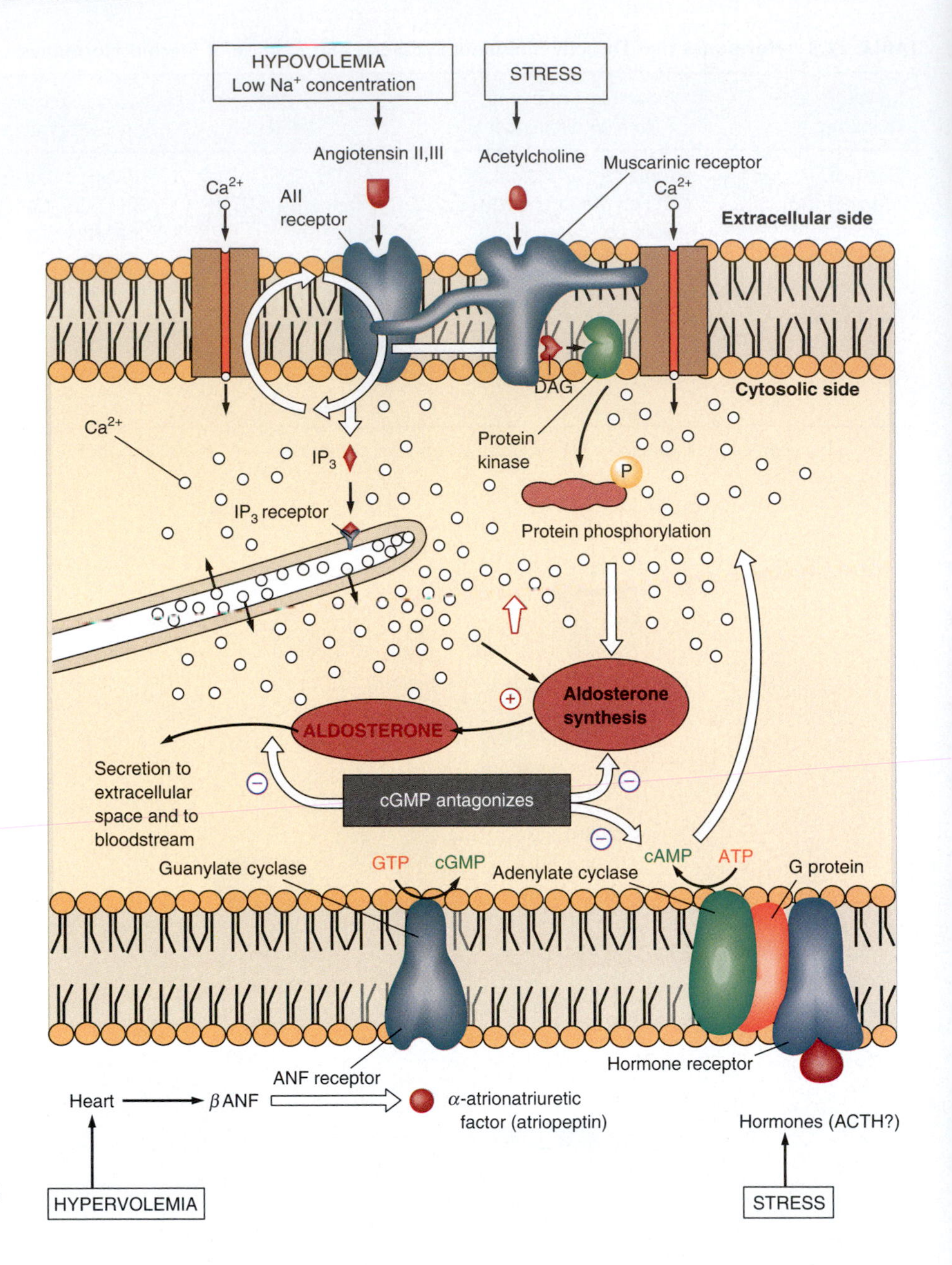

FIGURE 22.8
Reactions leading to the secretion of aldosterone in the adrenal zona glomerulosa cell. cGMP, cyclic GMP; ANF, atrial natriuretic factor; see Figure 22.7 for additional abbreviations.

nase C. **Acetylcholine** released through the neuronal stress signals has similar effects mediated by the muscarinic acetylcholine receptor to further reinforce Ca^{2+} uptake by the cell and stimulation of protein kinase C. Enhanced protein kinase C activity leads to protein phosphorylations that stimulate the rate-limiting steps of aldosterone synthesis leading to elevated levels of aldosterone, which are then secreted into the extracellular space and finally into the blood. Once in the blood, aldosterone enters the distal kidney cell, binds to its receptor, which initially may be cytoplasmic, and ultimately stimulates expression of proteins that increase the transport of Na^+ from the glomerular filtrate to the blood.

Signals opposite to those that activate the formation of angiotensin generate the **atrial natriuretic factor** (ANF) or atriopeptin from the heart atria (Figure 22.8; see also Figure 21.45). ANF binds to a specific zona glomerulosa cell membrane receptor and activates guanylate cyclase, which is part of the same receptor polypeptide so that the cytosolic level of cGMP increases. Cyclic GMP antagonizes the synthesis and secretion of aldosterone as well as the formation of cAMP by adenylate cyclase.

Aldosterone should be regarded as a stress hormone since its presence in elevated levels in blood occurs as a result of stressful situations. In contrast, cortisol,

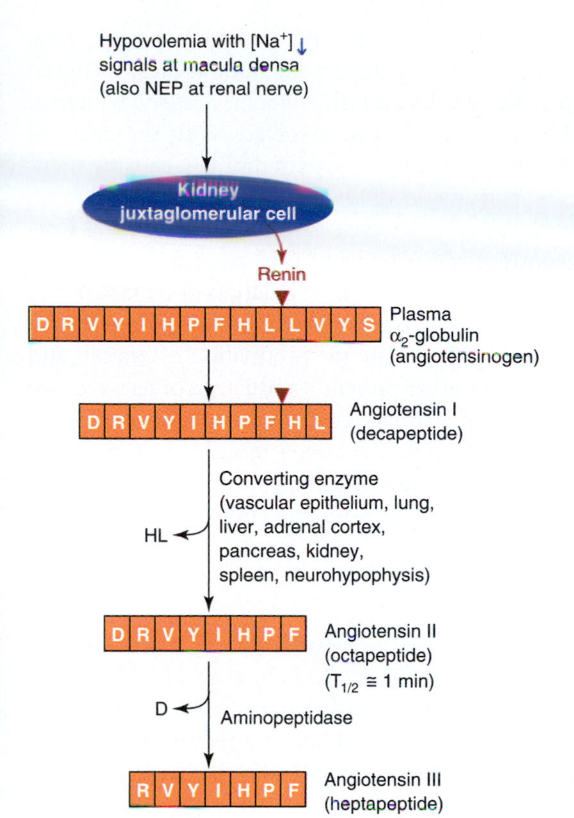

FIGURE 22.9
Renin–angiotensin system.
Amino acid abbreviations are found on page 97. NEP, norepinephrine.

also released in stress, has an additional biorhythmic release (possibly under control of serotonin and vasopressin), which accounts for a substantial reabsorption of Na^+ probably through glucocorticoid stimulation of the Na^+/H^+ antiport in luminal epithelial cells in addition to the many other activities of cortisol (e.g., anti-inflammatory action, control of T-cell growth factors, synthesis of glycogen, and effects on carbohydrate metabolism).

Estradiol

Control of formation and secretion of **17β-estradiol,** the female sex hormone, is shown in Figure 22.10. During development, control centers for the steady-state and cycling levels arise in the CNS. Their functions are required to initiate the ovarian cycle at puberty. These centers must harmonize with the firing of other neurons, such as those producing a clock-like mechanism via release of catecholamines or other amines to generate the pulsatile release of gonadotropic-releasing hormone (GnRH), probably at hourly intervals. Details of these reactions are presented on page 933. The FSH circulates and binds to its cognate receptor on the cell membrane of the ovarian follicle cell and through its second messengers, primarily cAMP and the activation of cAMP-dependent protein kinase, there is stimulation of the synthesis and secretion of the female sex hormone, 17β-estradiol. At normal stimulated levels of 17β-estradiol, there is a negative feedback on the **gonadotrope** (anterior pituitary), suppressing further secretion of FSH. Near ovarian midcycle, however, there is a superstimulated level of 17β-estradiol produced that has a positive, rather than a negative, feedback effect on the gonadotrope. This causes very high levels of LH to be released, referred to as the LH spike, and elevated levels of FSH. The level of FSH released is substantially lower than LH because the follicle produces **inhibin,** a polypeptide hormone that specifically inhibits FSH release without affecting LH release. The elevation of LH in the LH spike participates in the process of ovulation. After ovulation, the remnant of the follicle is differentiated into the functional corpus luteum, which now synthesizes progesterone (and also some estradiol), under the influence of elevated LH levels. Progesterone, however, is a

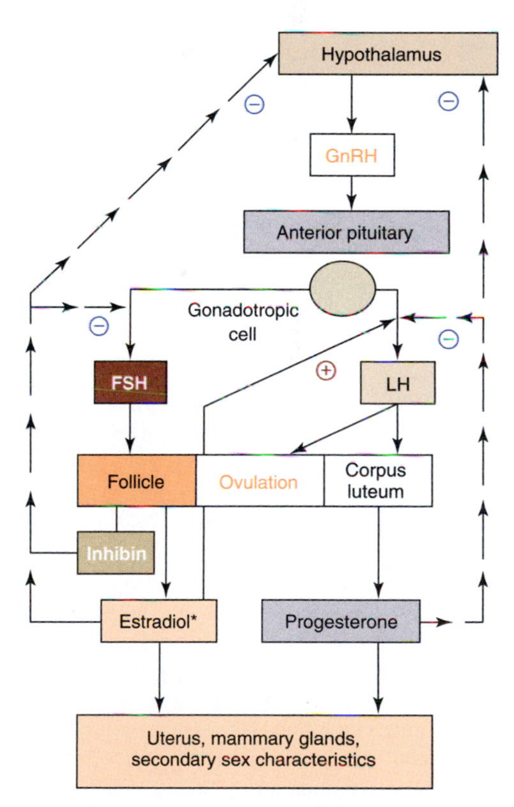

FIGURE 22.10
Formation and secretion of 17β-estradiol and progesterone.

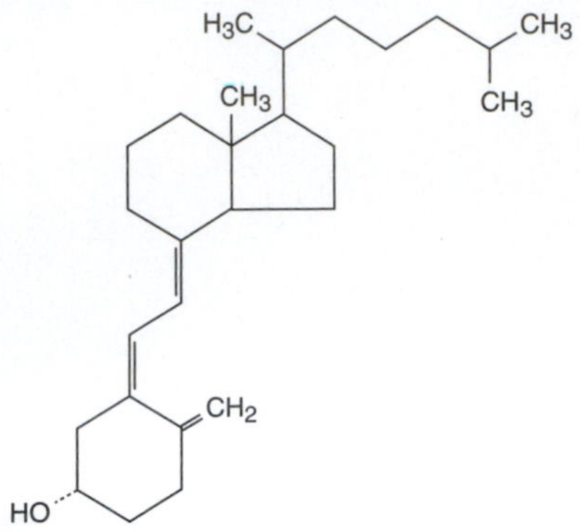

FIGURE 22.11
Structure of 1α,25-dihydroxycholecalciferol or calcitriol, the biologically active form of vitamin D.

feedback inhibitor of LH synthesis and release (operating through a progesterone receptor in the gonadotropic cell) and eventually the corpus luteum dies, owing to a fall in the level of available LH and the production of oxytocin, a luteolytic agent, by the corpus luteum. Prostaglandin $F_{2\alpha}$ may also be involved. With the death of the corpus luteum, the blood levels of progesterone and estradiol fall, causing menstruation as well as a decline in the negative feedback effects of these steroids on the anterior pituitary and hypothalamus and the cycle begins again. Clinical Correlation 22.1 describes how oral contraceptives interrupt this sequence.

The situation is similar in males with respect to the regulation of gonadotropin secretion, but LH acts principally on Leydig cells to stimulate production of testosterone, and FSH acts on the Sertoli cells to stimulate the production of inhibin and other proteins that include an androgen binding protein. Production of testosterone is subject to the negative feedback effect of 17β-estradiol synthesized in the Sertoli cell. The 17β-estradiol operates through a nuclear estrogen receptor in the Leydig cell to inhibit testosterone synthesis at the transcriptional level. In all cases of steroid hormone production, the synthetic system resembles that shown in Figure 22.4.

CLINICAL CORRELATION 22.1

Oral Contraception

Oral contraceptives usually contain an estrogen and a progestin. Taken orally, the levels of these steroids increase in blood to a level where secretion of FSH and LH is repressed. Consequently, the gonadotropic hormone levels in blood fall and there is insufficient FSH to drive the development of the ovarian follicle. As a result, the follicle does not mature and ovulation cannot occur. In addition any corpora lutea cannot survive because of low LH levels. In sum, the ovarian cycle ceases. The uterine endometrium thickens and remains in this state, however, because of elevated levels of estrogen and progestin. Pills without the steroids (placebos) are usually inserted in the regimen at about the 28th day and, as a result, blood levels of the steroids fall dramatically and menstruation occurs. When oral contraceptive steroids are resumed, the blood levels of estrogen and progestin increase again and the uterine endometrium thickens. This sequence creates a false "cycling" because of the occurrence of menstruation at the expected time in the cycle. The ovarian cycle and ovulation are suppressed by the oral contraceptive based on the negative feedback effects of estrogen and progestin on the secretion of the anterior pituitary gonadotropes. It is also possible to provide contraception by implanting in the skin silicone tubes containing progestins. The steroid is slowly released, providing contraception for up to 3–5 years.

Zatuchni, G. I. Female contraception. In: K. L. Becker (Ed.), *Principles and Practice of Endocrinology and Metabolism.* New York: Lippincott, 1990, p. 861; and Shoupe, D. and Mishell, D. R. Norplant: subdermal implant system for long term contraception. *Am. J. Obstet. Gynecol.* 160:1286, 1988.

Vitamin D_3

The active form of vitamin D, frequently referred to as **calcitriol,** is a **secosteroid,** a steroid in which one of the rings has been opened (see Figure 22.11). The active form of vitamin D stimulates intestinal absorption of dietary calcium and phosphorus, the mineralization of bone matrix, bone resorption, and reabsorption of calcium and phosphate in the renal tubule. The **vitamin D endocrine system** is diagramed in Figure 22.12. 7-Dehydrocholesterol is activated in the skin by sunlight to form **vitamin D_3 (cholecalciferol).** This form is hydroxylated first in the liver to **25-hydroxy vitamin D_3 (25-hydroxycholecalciferol)** and subsequently in the kidney to form **1α,25-vitamin D_3 (1,25$(OH)_2D_3$)(1α,25-dihydroxycholecalciferol).** The hormone binds to nuclear 1,25$(OH)_2D_3$ receptors in intestine, bone, and kidney and then activates genes encoding calcium-binding proteins whose actions

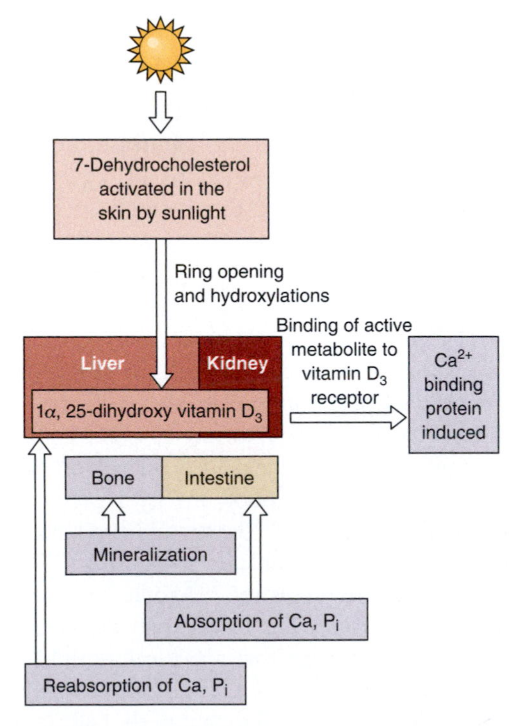

FIGURE 22.12
The vitamin D endocrine system.
P_i, inorganic phosphate.
Adapted from Norman, A. W. and Litwack, G. Hormones, 2nd ed. New York: Academic Press, 1997, p. 379.

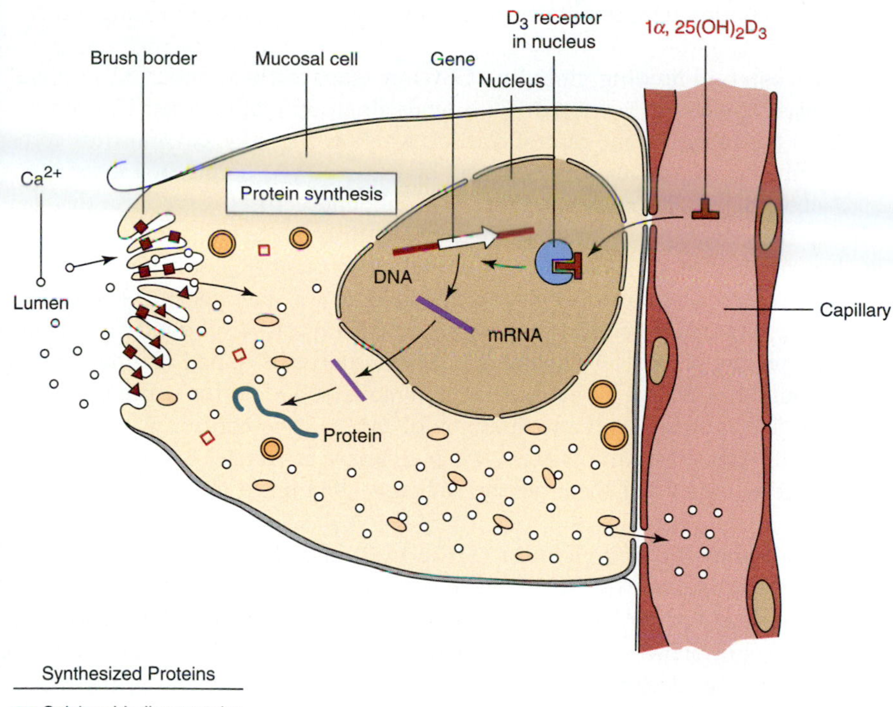

FIGURE 22.13
Schematic model of the action of 1,25$(OH)_2D_3$ in the intestine in stimulating intestinal calcium transport.
Redrawn from Nemere, I. and Norman, A. W. Biochim. Biophys. Acta *694:307, 1982.*

may lead to the absorption and reabsorption of Ca^{2+} (as well as phosphorus). The subcellular mode of action is presented in Figure 22.13. In this scheme calcitriol enters the intestinal cell from the blood side and migrates to the nucleus, where its unoccupied receptor is associated with the chromatin fraction. Once the active form of the hormone binds to this nuclear receptor, the complex then associates with a specific DNA sequence, referred to as the vitamin D_3-responsive element, and by doing so selectively regulates the rate of transcription of responsive genes. Messenger RNA is produced and translated in the cytoplasm; these RNAs encode Ca^{2+}-binding proteins, Ca^{2+}-ATPase, other ATPases, membrane components, and facilitators of vesicle formation. Increased levels of Ca^{2+}-binding proteins may cause increased uptake of Ca^{2+} from the intestine or may simply buffer the cytosol against high Ca^{2+} levels.

With each of the steroid-producing systems discussed, feedback controls are operative whereby sufficient amounts of the circulating steroid hormone inhibit the further production and release of intermediate hormones in the pathway at the levels of the pituitary and hypothalamus, as viewed in Figure 21.3 (see p. 909). In the case of the vitamin D systems, the controls are different since the steroid production is not stimulated by the cascade process applicable to estradiol. When the circulating levels of the active form of vitamin D (1,25$(OH)_2D_3$) are high, hydroxylations at the 24 and 25 positions are favored and the inactive 24,25$(OH)_2$-vitamin D_3 compound is generated.

22.6 | TRANSPORT OF STEROID HORMONES IN BLOOD

Steroid Hormones Are Bound to Specific Plasma Proteins or Albumin

Four major plasma proteins account for binding of the steroid hormones in blood. They assist in maintaining a level of these hormones in the circulation and protect the hormone from metabolism and inactivation. The proteins are corticosteroid-binding globulin protein, sex hormone-binding protein, androgen-binding protein, and albumin.

Corticosteroid-binding globulin (CBG) or **transcortin** is about 52 kDa and human plasma contains 3–4 mg dL^{-1}. It binds about 80% of the total 17-hydroxysteroids in the blood. In the case of cortisol, which is the principal antistress corticosteroid in humans, about 75% is bound by CBG, 22% is bound in a loose manner to albumin, and 8% is in free form. The unbound cortisol is the form that can permeate cells and bind to intracellular receptors to produce biological effects. The CBG has a high affinity for cortisol with a binding constant (K_a) of 2.4×10^7 M^{-1}). Critical structural determinants for steroid binding to CBG are the Δ^4-3-ketone and 20-ketone groups. Aldosterone binds weakly to CBG but is also bound by albumin and other plasma proteins. Normally 60% of aldosterone is bound to albumin and 10% is bound to CBG. In human serum, albumin is 1000-fold the concentration of CBG and binds cortisol with an affinity of 10^3 M^{-1}. Thus cortisol will always fill CBG-binding sites first. During stress, when secretion of cortisol is very high, CBG sites will be saturated but there will be sufficient albumin to accommodate excess cortisol.

Sex hormone-binding globulin (SHBG) (40 kDa) binds androgens with an affinity constant of about 10^9 M^{-1}. One to three percent of testosterone is unbound in the circulation and 10% is bound to SHBG, with the remainder bound to albumin. The level of SHBG is probably important in controlling the balance between circulating androgens and estrogens along with the actual amounts of these hormones produced in given situations. About 97–99% of bound testosterone is bound reversibly to SHBG but much less estrogen is bound to this protein in the female. The plasma concentration of SHBG before puberty is about the same in males and females, but, at puberty, when the functioning of the sex hormones becomes important, there is a small decrease in the level of circulating SHBG in females and a larger decrease in males, ensuring a relatively greater amount of the unbound (active) testosterone and 17β-estradiol. In adults, males have about one-half as much circulating SHBG as females, so that the unbound testosterone in males is about 20 times greater than in females. In addition, the total (bound plus unbound) concentration of testosterone is about 40 times greater in males. Testosterone itself lowers SHBG levels in blood, whereas 17β-estradiol raises SHBG levels in blood. These effects have important ramifications in pregnancy and in other conditions.

Androgen-binding protein (ABP) is produced by the Sertoli cells in response to testosterone and FSH, both of which stimulate protein synthesis in these cells. Androgen-binding protein is doubtless not of great importance in the entire blood circulation but is important because it maintains a ready supply of testosterone for the production of protein constituents of spermatozoa. Its role may be to maintain a high local concentration of testosterone in the vicinity of the developing germ cells within the tubules.

22.7 | STEROID HORMONE RECEPTORS

Steroid Hormones Bind to Specific Intracellular Protein Receptors

A general model for steroid hormone action presented in Figure 22.14 takes into account the differences among steroid receptors in terms of their location within the cell. In contrast to polypeptide hormone receptors that are generally located on/in the cell surface, steroid hormone receptors, as well as other related receptors for nonsteroids (i.e., thyroid hormone, retinoic acid, vitamin D_3) are located in the cell interior. Among the steroid receptors there appear to be some differences as to the subcellular location of the **non-DNA-binding forms** of these **receptors.** The glucocorticoid receptor and possibly the aldosterone receptor appear to reside in the cytoplasm, whereas the other receptors, for which suitable data have been collected, may be located within the nucleus, possibly in association with DNA, although not necessarily at productive acceptor sites on the DNA. Starting at the top of Figure

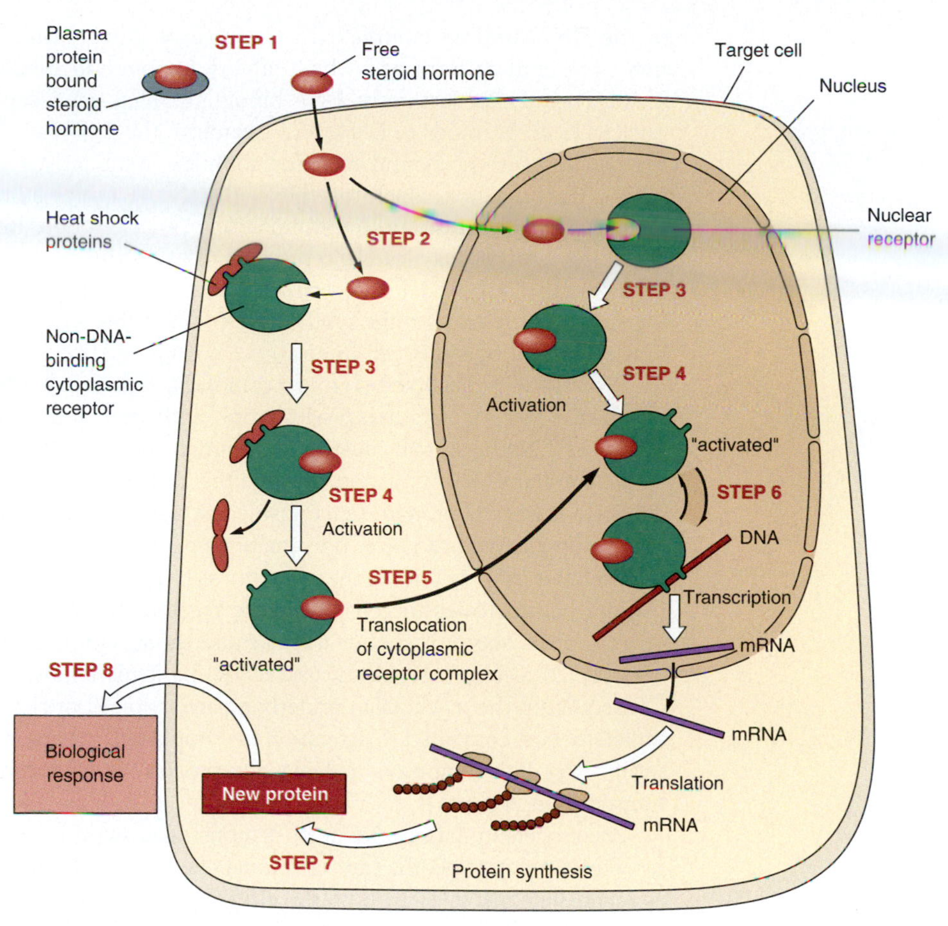

FIGURE 22.14
Model of steroid hormone action.
Step 1—Dissociation of free hormone (biologically active) from circulating transport protein; Step 2—diffusion of free ligand into cytosol or nucleus; Step 3—binding of ligand to unactivated cytoplasmic or nuclear receptor; Step 4—activation of cytosolic or nuclear hormone–receptor complex to activated, DNA-binding form; Step 5—translocation of activated cytosolic hormone–receptor complex into nucleus; Step 6—binding of activated hormone–receptor complexes to specific response elements within the DNA; Step 7—synthesis of new proteins encoded by hormone-responsive genes; and Step 8—alteration in phenotype or metabolic activity of target cell mediated by specifically induced proteins.

22.14, Step 1 shows a bound and a free form of a steroid hormone(s). The free form may enter the cell by a process of diffusion. In the case of glucocorticoids, like cortisol, the steroid would bind to an unactivated receptor with an open ligand-binding site (Step 3). The binding constant for this reaction is on the order of 10^9 M^{-1}, compared to about 10^7 M^{-1} for the binding to CBG (see above). The non-DNA-binding form also referred to as the unactivated or nontransformed receptor has a mass of about 300 kDa, because other proteins may be associated in the complex. A dimer of a 90-kDa protein, which is a **heat shock protein** that is induced when cells are stressed, is associated with the receptor in this form and occludes its DNA-binding domain, accounting for its non-DNA-binding activity. Associated with this dimer of hsp90 is another heat shock protein designated as hsp56, which interestingly also functions as an immunophilin and, as such, binds to a number of potent immunosuppressive drugs. The dimer of the 90-kDa heat shock protein is depicted by the pair of red ovals attached to the cytoplasmic receptor that block the DNA-binding domain pictured as a pair of “fingers” in the subsequently activated form. Activation or transformation to the DNA-binding form is accomplished by release of the 90-kDa heat shock proteins (Step 4). It is not clear what actually drives the activation step(s). Clearly, the binding of the steroidal ligand is important but other factors may be involved. A low molecular weight component has been proposed to be part of the cross-linking between the nonhomologous proteins and the receptor in the DNA-binding complex. In the case of glucocorticoid receptor, only the non-DNA-binding form has a high affinity for binding steroidal ligand. Following activation and exposure of the DNA-binding domain, the receptor translocates to the nucleus (Step 5), binds to DNA, and "searches" the DNA for a high-affinity acceptor site. At this site the bound receptor complex, frequently a homodimer, acts as a transactivation factor, which together with other transactivators allows for the starting of RNA polymerase and the stimulation of transcription. In some cases the binding of the receptor may lead to repression of transcription and this effect is less well understood. New mRNAs are translocated to the cytoplasm and assembled into translation complexes for the synthesis of proteins (Step 7) that alter metabolism and functioning of the target cell (Step 8).

When the unoccupied steroid hormone receptor is located in the nucleus, as may be the case for estradiol, progesterone, androgen, and vitamin D_3 receptors (see Figure 22.13), the steroid must travel through the cytoplasm and cross the perinuclear membrane. This transport through the cytoplasm (aqueous environment) may require a transport protein. Once inside the nucleus the steroid binds to the high-affinity, unoccupied receptor, presumably already on DNA, and causes it to be "activated" to a form bound to the acceptor site. The ligand might promote a conformation that decreases the off-rate of the receptor from its acceptor, if it is located on or near its acceptor site, or might cause the receptor to initiate searching if the unoccupied receptor associates with DNA at a locus remote from the acceptor site. Consequently, the mechanism underlying activation of nuclear receptors is less well understood as compared to activation of cytoplasmic receptors. After binding of activated receptor complexes to DNA acceptor sites, enhancement or repression of transcription occurs.

Heterogeneity has clearly been demonstrated within a given subtype of steroid receptor. For example, the transcriptional effects of 17β-estradiol in humans as well as rodents are mediated by both the **alpha** and **beta** forms of the **estrogen receptor** (**ERα** and **ERβ**, respectively). Both ERα and ERβ bind 17β-estradiol with the same high affinity and both activated complexes bind to estrogen response elements within the DNA. ERα and ERβ share a high degree of amino acid homology. However, there are specific differences in regions of these estrogen receptors that would be predicted to influence transcriptional activity. ERα contains two distinct activation domains, AF-1 and AF-2, whose transcriptional activity is influenced by cell promoter context. Although ERβ contains an AF-2 domain, it does not contain a strong AF-1 domain within its amino terminus, but rather contains a repressor domain. Consistent with this structural difference are the observations that ERβ functions as a transdominant inhibitor of ERα transcriptional activity at subsaturating hormone levels and that ERβ decreases overall cellular sensitivity to 17β-estradiol. One explanation for these inhibitory effects of ERβ on ERα function is that ERβ can form **heterodimers** with ERα, which in turn regulate ER functions. Analyses of the specific effects mediated by 17β-estradiol in ERα or ERβ knockout mice, coupled with the observation that these two forms of the ER exhibit distinct and only partially overlapping patterns of expression, suggest that these two receptors play different biological roles. In fact, when ERα and ERβ are coexpressed in neurons, they trigger different intracellular signals, leading to distinct metabolic responses. From a physiological perspective it has been suggested the ERβ might play a more relevant role for neural cell differentiation, whereas activation of ERα might be implicated in synaptic plasticity, both during development and in mature neurons. The uterus is a target tissue in which endogenously expressed ERα and ERβ appear to play different roles. The uterus is composed of heterogeneous cell types that undergo continuous synchronized changes in proliferation rates and differentiation in response to changes in the circulating levels of estrogens and progesterone. ERα and ERβ levels in uterine tissue also appear to vary during the menstrual cycle, with the highest level of both present during the proliferative phase. 17β-Estradiol plays an important role in the differentiation of the uterus by regulating the expression of target genes, including that for the **progesterone receptor (PR).** Induction of PR by 17β-estradiol in both uterine and breast tissue is important physiologically, since it prepares these target tissues for subsequent stimulation by progesterone. Recent data demonstrate that ERα appears to be responsible for this induction of PR in the uterine stroma and glandular epithelial cells. In contrast, ERβ appears to be responsible for downregulation of PR in the luminal epithelium. Thus the differential expression of ERα and ERβ within specific cell types dictates what type of response will be elicited by 17β-estradiol.

Consensus DNA sequences defining specific **hormone response elements (HREs)** for the binding of various activated steroid hormone–receptor complexes are summarized in Table 22.4. Receptors for glucocorticoids, mineralocorticoids, progesterone, and androgen all bind to the same HRE on the DNA. Thus, in a given cell type, the extent and type of receptor expressed will determine the hormone sensitivity. For example, sex hormone receptors are expressed in only a few cell types and the progesterone receptor is likewise restricted to certain cells, whereas the glucocorticoid receptor is expressed in a wide variety of cell types. In cases where aldosterone and cortisol receptors are coexpressed, only one form may predominate depending on the cell type. Some tissues, such as the kidney and colon, are known targets for aldosterone and express relatively high levels of mineralocorticoid receptors as well as glucocorticoid receptors. These mineralocorticoid target tissues express the enzyme **11β-hydroxysteroid dehydrogenase** (see Clin. Corr. 22.2). This enzyme converts cortisol and corticosterone, both of which bind to the mineralocorticoid receptor with high affinity, to their 11-keto analogs, which bind poorly to the mineralocorticoid receptor. This inactivation of corticosterone and cortisol, which circulate at much higher concentrations than aldosterone, facilitates binding of aldosterone to mineralocorticoid receptors in these tissues. In tissues that express mineralocorticoid receptors but are not considered target tissues, this enzyme may not be expressed, and the mineralocorticoid receptors may simply function as pseudoglucocorticoid receptors and mediate the effects of low circulating levels of cortisol (predominant glucocorticoid in humans). Thus the mineralocorticoid and glucocorticoid receptors may regulate the expression of an overlapping gene network in various target tissues. As indicated in Table 22.4, the estrogen–receptor complex recognizes a unique response element. All of the response elements listed at the top of Table 22.4 function as positive elements, since binding of the receptors results in an increase in transcription of the associated gene.

Glucocorticoid hormones also repress transcription of specific genes. For example, glucocorticoids are known to repress transcription of the **proopiome-**

TABLE 22.4 Steroid Hormone Receptor Responsive DNA Elements: Consensus Acceptor Sites

Element	*DNA Sequence*[a]
POSITIVE	
Glucocorticoid responsive element (GRE)	5′-GGTACAnnnTGTTCT-3′
Mineralocorticoid responsive element (MRE)	5′-GGTACAnnnTGTTCT-3′
Progesterone responsive element (PRE)	5′-GGTACAnnnTGTTCT-3′
Androgen responsive element (ARE)	5′-GGTACAnnnTGTTCT-3′
Estrogen responsive element (ERE)	5′-AGGTCAnnnTCACT-3′
NEGATIVE	
Glucocorticoid responsive element	5′-ATYACNnnnTGATCW-3′

Source: Data are summarized from work of Beato, M. *Cell* 56:355, 1989.

[a]n, any nucleotide; Y, a purine; W, a pyrimidine.

CLINICAL CORRELATION 22.2

Apparent Mineralocorticoid Excess Syndrome

Some patients (usually children) exhibit symptoms, including hypertension, hypokalemia, and suppression of the renin–angiotensin–aldosterone system, that would be expected if they were hypersecreting aldosterone. Since bioassays of plasma and urine sometimes fail to identify any excess of mineralocorticoids, these patients are said to suffer from the apparent mineralocorticoid excess (AME) syndrome. This syndrome is a consequence of the failure of cortisol inactivation by the 11β-hydroxysteroid dehydrogenase. This gives cortisol direct access to the renal mineralocorticoid receptor. Since cortisol circulates at much higher concentrations than aldosterone, this glucocorticoid saturates these mineralocorticoid receptors and functions as an agonist, causing sodium retention and suppression of the renin–angiotensin–aldosterone axis. Although this AME syndrome can result from a congenital defect in the distal nephron 11β-hydroxysteroid dehydrogenase isoform, which renders the enzyme incapable of converting cortisol to cortisone (binds poorly to mineralocorticoid receptors), it can also be acquired by ingesting excessive amounts of licorice. The major component of licorice is glycyrrhizic acid and its hydrolytic product, glycyrrhetinic acid (GE). This active ingredient (GE) acts as a potent inhibitor of 11β-hydroxysteroid dehydrogenase. By blocking the activity of this inactivating enzyme, GE facilitates the binding of cortisol to renal mineralocorticoid receptors and hence induces AME syndrome.

Edwards, C. R. W. Primary mineralocorticoid excess syndromes. In: L. J. DeGroot (Ed.), *Endocrinology*. Philadelphia: Saunders, 1995, p. 1775; and Shackleton, C. H. L. and Stewart, P. M. The hypertension of apparent mineralocorticoid excess syndrome. In: E. G. Biglieri and J. C. Melby (Eds.), *Endocrine Hypertension*. New York: Raven Press, 1990, pp. 155–173.

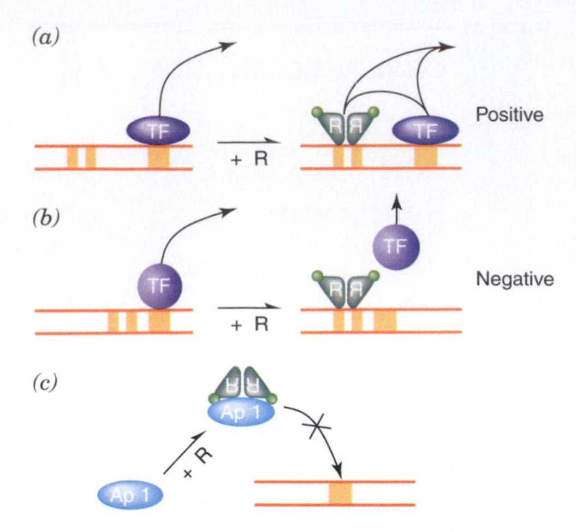

FIGURE 22.15
Positive and negative transcriptional effects of steroid receptors.
TF, transcription factor; R, receptor. (*a*) Binding of receptor dimer immediately adjacent to a transcription factor leads to synergistic activation of transcription. (*b*) Binding of receptor dimer to a negative hormone response element (HRE) may displace a positive transcription factor. (*c*) Protein–protein interaction between receptor dimer and positive transcription factor such as Ap 1 may block Ap 1 DNA binding and repress transcriptional response.
Redrawn from Renkawitz, R. Ann. N. Y. Acad. Sci. *684:1, 1993.*

lanocortin (POMC) gene, which contains the ACTH sequences. Glucocorticoid-mediated repression of *POMC* gene expression thus plays a key role in the negative feedback loop regulating the rate of secretion of ACTH and ultimately cortisol. Negative glucocorticoid response elements (nGREs) mediate this repression of the *POMC* gene as well as other important genes. A general model of positive as well as negative transcriptional effects mediated by steroid receptors is shown in Figure 22.15: In (*a*) binding of a steroid receptor (R) homodimer to its response element allows it to interact synergistically with a positive transcription factor (TF) and hence induce gene transcription; in (*b*) binding of a receptor dimer to its response element displaces a positive transcription factor (TF) but has no or weak transactivation potential because no synergizing factor is nearby; and in (*c*) the DNA-AP 1 (positive factor) may interact in a protein–protein fashion in such a way that the transactivating functions of both proteins are inhibited and gene transcription is repressed.

Endogenous as well as exogenous glucocorticoids exert major anti-inflammatory actions and these steroids have been utilized for decades for this very purpose. These endogenous and synthetic hormones appear to suppress the immune system via several biochemical mechanisms. For example, glucocorticoids are known to inhibit immune responses by inhibiting **prostaglandin** production. Nuclear glucocorticoid–receptor complexes (see Figure 22.16) have been shown to induce **annexin-I** (also referred to as **lipocortin**), a protein of about 40,000 molecular weight. This polypeptide functions as an inhibitor of membrane phospholipase A_2, whose activity is responsible for the release of fatty acid precursors, like arachidonic acid, for prostaglandin synthesis. Glucocorticoids are also known to inhibit expression of **cyclooxygenase** (see p. 767), which catalyzes the generation of prostaglandins and other related compounds which are potent inflammatory agents. Two forms of cyclooxygenase exist: **COX1** and **COX2.** COX1 appears to be a constitutively expressed enzyme that may function in vascular responses and may catalyze the generation of required hormones under noninflammatory conditions. On the other hand, the COX2 enzyme (prostaglandin G/H synthase-2) can be induced in inflammatory cells as part of the inflammatory response. This endoplasmic reticulum enzyme catalyzes the oxidation of arachidonic acid to the peroxy form (cyclic endoperoxide). From this intermediate a variety of prostaglandins and thromboxanes (mediators of inflammatory response) and prostacyclins can be synthesized. This suppression of COX2 synthesis by glucocorticoids accounts for a major part of their anti-inflammatory effects.

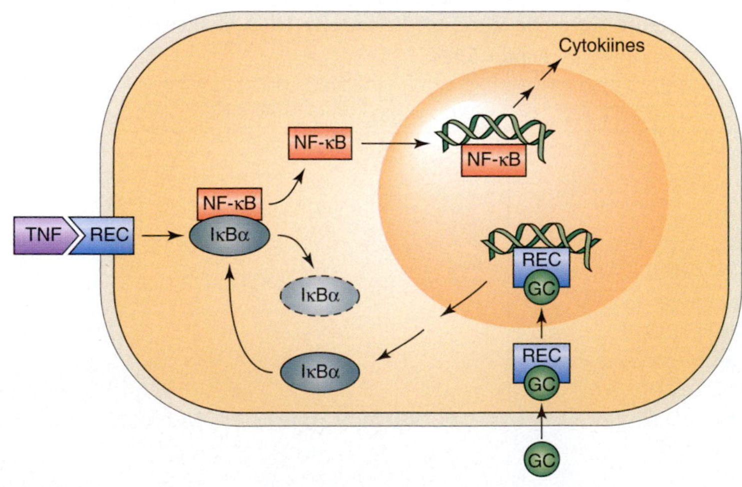

FIGURE 22.16
Action of glucocorticoids in suppressing immune and inflammatory responses mediated by cytokines.
REC, receptor; GC, glucocorticoid hormones; TNF, tumor necrosis factor.
Redrawn from Marx, J. Science *270:232, 1995.*

Glucocorticoids are anti-inflammatory by another mechanism that involves inhibition of a key transcription factor. In unstimulated immune cells, the transcription factor **NF-κB** is retained in the cytoplasm in its inactive form complexed with another protein, either **IκBα** or the structurally related protein **IκBβ.** Stimulation of these cells by any one of a variety of immune signals, such as tumor necrosis factor, results in phosphorylation of the **IκBα** on serines 32 and 36. This posttranslational modification, which leads to ubiquitination and subsequent degradation of **IκBα** via a proteosome-mediated mechanism, triggers the release of NF-κB from the cytoplasmic complex. The released (activated) NF-κB then migrates into the nucleus, where it activates cytokine (mediators of inflammation) and other genes. Glucocorticoids effectively suppress this immune cell activation by two mechanisms (see Figure 22.16). These steroids, via binding to intracellular receptors, induce IκBα transcription and this additional IκBα holds NF-κB in the cytoplasm in its inactive form, even under conditions where it would normally be released to move into the nucleus. Additionally, potential binding of activated nuclear glucocorticoid–receptor complexes to nuclear NF-κB may prevent the latter from binding to appropriate DNA response elements and contribute to steroid-mediated immunosuppression.

Some members of this receptor supergene family mediate gene silencing. Silencer elements in DNA, by analogy to enhancer elements, function independently of their position and orientation. The silencer for a particular gene consists of modules that independently repress gene activity. In the absence of their specific ligands- the **thyroid hormone receptor** (T_3R) and **retinoic acid receptor** (RAR) appear to bind to specific silencer elements and repress gene transcription. This silencing activity may occur via destabilization of the transcription–initiation complex or via direct or indirect effects on the carboxy-terminal domain of RNA polymerase II. After binding of their respective ligands, these two receptors lose this silencing activity and are converted into transactivators of gene transcription.

As indicated in Figure 22.15, dimerization of receptor monomers is a prerequisite for efficient DNA binding and transcriptional activation by most steroid receptors. Strong interactions between these monomers are mediated by the ligand-binding domains of several steroid receptors. The dimerization domain of the ligand-binding domain has been proposed to form a leucine zipper-like structure or a helix–turn–zipper motif (see p. 387), which is known to be necessary for the dimerization of other transcription factors. Although the majority of receptors in this superfamily form homodimers, heterodimers have also been detected. More specifically, a class of retinoic acid receptors, classified as retinoid X receptors (RXRs), regulate gene expression via heterodimerization with the other form of the retinoic acid receptor (RAR), the thyroid hormone receptor, and other members of this receptor superfamily. A model for the stabilization of the transcriptional preinitiation complex by an RXR/RAR heterodimer is presented in Figure 22.17.

Thus the changes produced in different cells by the activation of steroid hormone receptors may differ. The whole process is triggered by the entry of the steroidal ligand in amounts that supersede the dissociation constant of the receptor. The phenotypic changes in different cell types in response to a specific hormone then summate to give the systemic or organismic response to the hormone.

Some Steroid Receptors Are Part of the c-*Erb*A Family of Protooncogenes or Are Orphan Receptors

The glucocorticoid receptor has three major **functional domains** (Figure 22.18). Starting at the C terminus the steroid-binding domain is indicated and has 30–60% homology with the **ligand-binding domains** of other receptors in the steroid receptor family. The more alike two steroids that bind different receptors are, the greater the extent of homology to be anticipated in this domain. The steroid-binding domain contains a sequence that may be involved in the binding of molybdate and a dimer of the 90-kDa heat shock protein whose function would theoretically

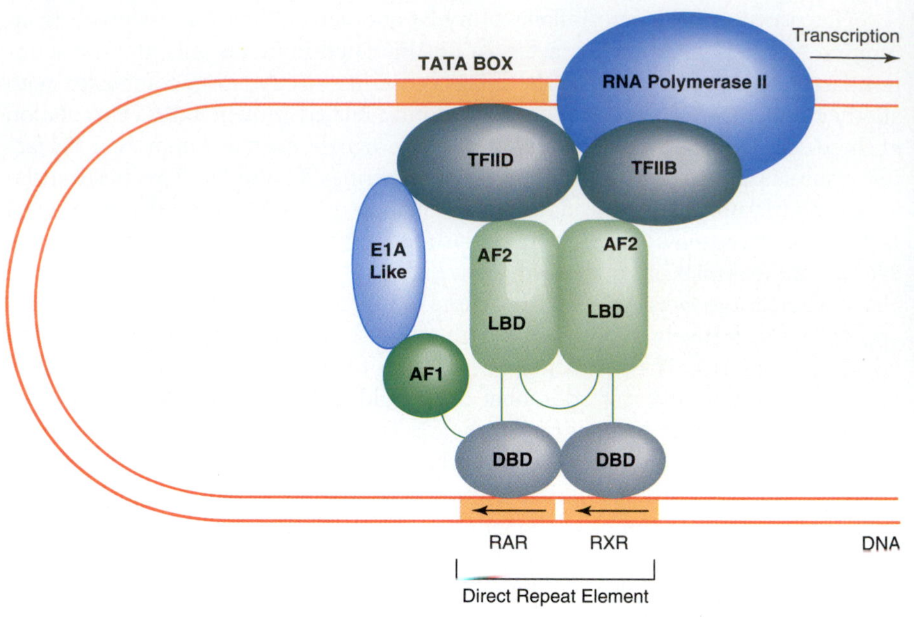

FIGURE 22.17
Model for the stabilization of the preinitiation complex by an RXR/RAR heterodimer. TF, transcription factor; LBD, ligand-binding domain; DBD, DNA-binding domain; AF1, activation function located in amino-terminal region of receptor, which may provide contact with cell-specific proteins; AF2, activation function located within ligand-binding domain, which interacts directly with transcriptional machinery.

result in the assembly of the high molecular weight unactivated–nontransformed steroid–receptor complex. To the left of that domain is a region that modifies transcription. In the center of the molecule is the **DNA–binding domain.** Among the steroid receptors there is 60–95% homology in this domain. Two zinc fingers (see p. 388) interact with DNA. The structure of the zinc finger DNA-binding motif is shown in Figure 22.19. The N-terminal domain contains the principal **antigenic domains** and a site that modulates transcriptional activation. The amino acid sequences in this site are highly variable among the steroid receptors. These features are common to all steroid receptors. The family of steroid receptors is diagrammed in Figure 22.20. The ancestor to which these receptor genes are related is v-*erbA* or c-*erbA*. v-*ErbA* is an oncogene that binds to DNA but has no ligand-binding domain. In some cases the DNA-binding domains are homologous enough that more than one receptor will bind to a common responsive element (consensus sequence on DNA) as shown in Table 22.4. In addition to those genes pictured in Figure 22.20, the **aryl hydrocarbon receptor** (Ah) may also be a member of this family.

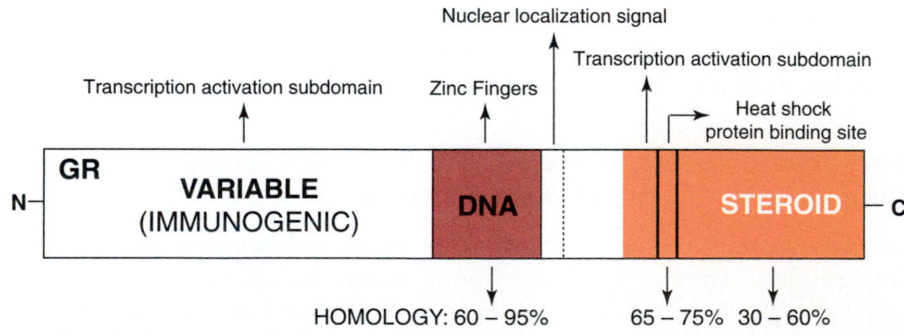

FIGURE 22.18
Model of a typical steroid hormone receptor.
The results are derived from studies on cDNA in various laboratories, especially those of R. Evans and K. Yamamoto.

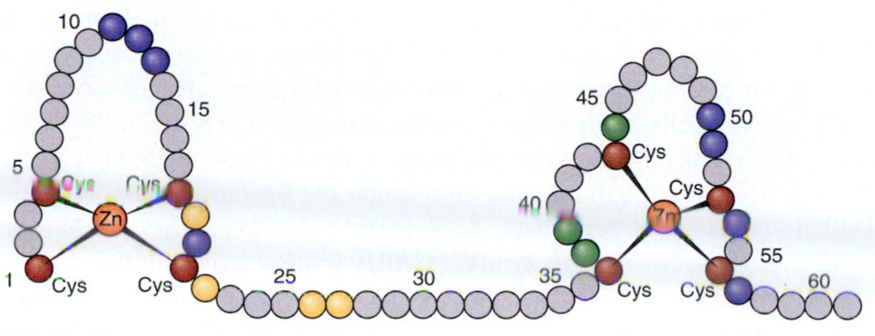

FIGURE 22.19
Structure of the zinc finger located within the glucocorticoid receptor DNA-binding domain as determined by X-ray crystallography.
Yellow circles indicate amino acid residues (located in GR monomer) that interact with base pairs. Blue circles are those making phosphate backbone contacts. Green circles are those participating in dimerization.
Redrawn from Luisi, B. F., Schwabe, J. W. R., and Freedman, L. P. In: G. Litwack (Ed.), Vitamins and Hormones, *Vol. 49. San Diego: Academic Press, 1994, pp. 1–47.*

The Ah receptor binds carcinogens with increasing affinity paralleling increasing carcinogenic potency and translocates the carcinogen to the cellular nucleus unless the receptor is already located in the nucleus. The N-terminal portions of the receptors usually contain major antigenic sites and may also contain a site that is active in modulating binding of the receptor to DNA.

Thyroid hormone and retinoic acid receptors are also members of the same superfamily of receptors although their ligands are not steroids. They do contain six-membered rings as shown in Figure 22.21. For some steroid receptors the A ring is the prominent site of recognition by the receptor, presenting the likelihood that the A ring inserts into the binding pocket of the receptor. In some cases, derivatives of the structures with a six-membered ring bind to the estradiol and glucocorticoid receptors. Thus the ring structures of thyroid hormone and retinoic acid have structural similarities not unlike many of the steroidal ligands involved in binding.

Thus all the receptors of this large gene family, acting as either homodimers or heterodimers, function as ligand-activated transcription factors and as such modulate (induce or repress) the expression of specific genes.

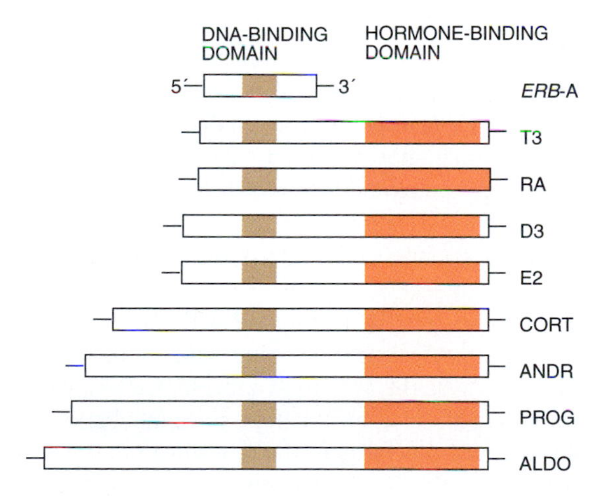

FIGURE 22.20
Steroid receptor gene superfamily.
T3, triiodothyronine; RA, retinoic acid; D3, dihydroxy vitamin D_3; E2, estradiol; CORT, cortisol; ANDR, androgen; PROG, progesterone; ALDO, aldosterone. Figure shows roughly the relative sizes of the genes for these receptors. The information derives from the laboratories of R. Evans, K. Yamamoto, P. Chambon, and others. In some cases there is high homology in DNA-binding domains and lower homologies in ligand-binding domains.

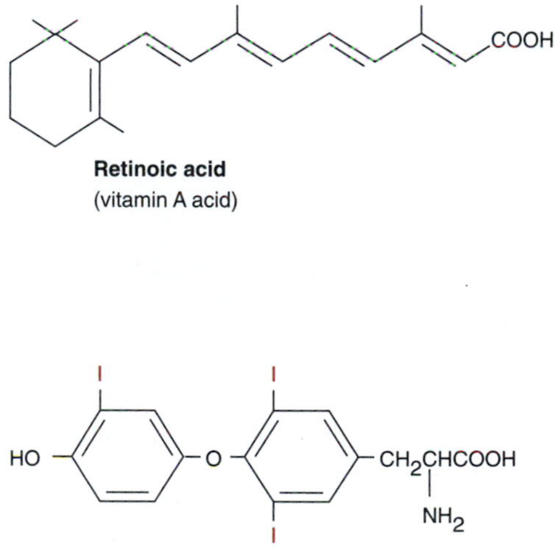

COOH
Retinoic acid
(vitamin A acid)

I I HO O $CH_2CHCOOH$ NH_2 I

FIGURE 22.21
Structures of retinoic acid (vitamin A acid) and 3,5,3′-triiodothyronine.

A large number of related receptors are referred to as **orphan receptors.** Although these nuclear receptors are structurally related to known receptors in this superfamily, no physiological ligands or activators for these orphan receptors were originally known. Collectively these receptors represent a diverse and ancient component of the nuclear receptor superfamily and are found in almost all animal species. Examples of orphan receptors for which the ligand has been identified include **BXR (*b*enzoate *X* *r*eceptor), RXR (*r*etinoid *X* *r*eceptor), PPAR (*p*eroxisome *p*roliferator-*a*ctivated *r*eceptor), CARβ (*c*onstitutive *a*ndrostane *r*eceptor), PXR (*p*regnane *X* *r*eceptor),** SXR **(*s*teroid and *x*enobiotic *r*eceptor),** and FXR **(*f*arnesoid *X* *r*eceptor).** PXR, SXR, and CARβ are highly expressed in the liver and respond to specific steroidal ligands. These hormones are actually the first new steroidal ligands to be described since aldosterone was discovered in 1952. These three ligand-responsive orphan receptors are known to require heterodimerization with RXR for DNA binding. These recently discovered orphan receptors and their respective ligands may clearly be important physiologically and have an impact on specific human diseases. For example, the human SXR can be activated by a diverse group of steroid agonists and antagonists. Activation of this broad-specificity sensing receptor appears to facilitate the detoxification and removal of various endogenous hormones, dietary steroids, drugs, and xenobiotic compounds with biological activity. The ligand-activated SXR complex appears to directly regulate the activity of inducible cytochrome P450 enzymes in response to the presence of its substrates. Thus the activated SXR increases steroid catabolism by inducing transcription of several genes encoding key metabolizing enzymes. Hence in patients receiving steroid replacement therapy or women taking oral contraceptives, certain drugs (i.e., rifampicin) that bind to SXR will cause rapid depletion of the administered steroids due to this increased steroid catabolism.

22.8 | RECEPTOR ACTIVATION: UPREGULATION AND DOWNREGULATION

Little is known about activation of steroid receptors. Activation converts a non-DNA-binding form (unactivated–nontransformed) of the receptor to a form (activated–transformed) that is able to bind nonspecific DNA or specific DNA (hormone-responsive element). The likelihood that certain receptors are cytoplasmic (glucocorticoid receptor and possibly the mineralocorticoid receptor) while others are nuclear (progesterone, estradiol, vitamin D_3, and androgen receptors) may have a bearing on the significance of the activation phenomenon. Most information is available for cytoplasmic receptors. The current view is that the non-DNA-binding form is a heteromeric trimer consisting of one molecule of receptor and a dimer of 90-kDa heat shock protein, as shown in Figure 22.22. The DNA-binding site of the receptor is blocked by the heteromeric proteins or by some other factor or by a combination of both. Upon activation–transformation a stepwise disaggregation of this complex could occur, leading to the activated receptor having its DNA-binding site fully exposed.

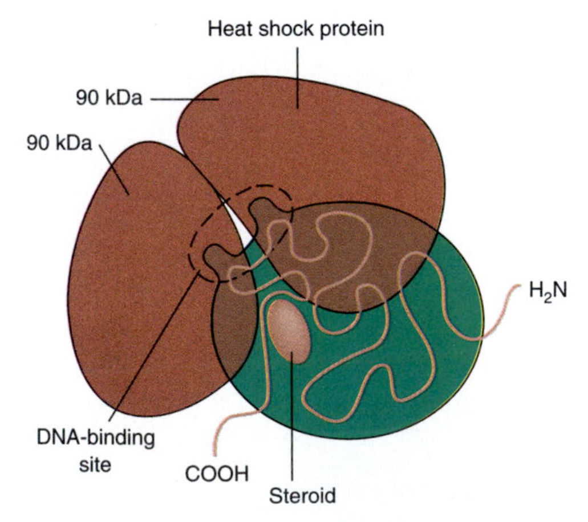

FIGURE 22.22
Hypothetical model of a non-DNA-binding form of a steroid receptor.
This form of the receptor cannot bind to DNA because the DNA-binding site is blocked by the 90-kDa hsp proteins or by some other constituent. Mass of this complex is approximately 300 kDa.

Although the conditions required to induce activation *in vitro* are well known, the primary signal within the cell is not. Many believe that the binding of ligand alone is not sufficient to cause the activation process. Clearly, elevated temperature is a requirement for this conformational change, since incubation of target cells with appropriate steroids at low temperatures fails to result in *in vivo* activation and subsequent translocation. Once the liberated receptor is free in the cytoplasm, it crosses the perinuclear membrane, perhaps through a nucleopore, to enter the nucleus. It binds nonspecifically and specifically to chromatin, probably as a dimer, presumably in search of the specific response element (Table 22.4). Thus these receptors are transactivating factors and may act in concert with other transacting factors to provide the appropriate structure to initiate transcription. Most steroid receptors have in their DNA-binding domains an SV40-like sequence (i.e., Pro-Lys-Lys-Lys-Arg-

Lys-Val) known to code for nuclear translocation. Steroid receptors have variants of this sequence; some degeneracy is permitted but probably a specific lysine residue cannot be altered. This signal may provide recognition for the nucleopore.

Steroid Receptors Can Be Upregulated or Downregulated Depending on Exposure to the Hormone

Many membrane or intracellular receptors are downregulated when the cell has been exposed to a certain amount of the hormonal ligand. In some cases, the downregulation is called "desensitization." **Downregulation** takes many forms. For membrane receptors the mechanism may be internalization by endocytosis of the receptors after exposure to hormone. Internalization reduces the number of receptors on the cell surface and renders the cell less responsive to hormone, that is, desensitizes the cell. In the case of intracellular steroid receptors, downregulation generally takes the form of reducing the level of receptor mRNA, which decreases the concentration of receptor molecules. The receptor gene may have a specific responsive element on its promoter whose action results in an inhibition of transcription of receptor mRNA or the receptor may stimulate transcription of a gene that codes for a protein that degrades the mRNA of the receptor. Sequences are now being recognized on receptor gene promoters that may bind activated steroid–receptor complexes and result in inhibition of transcription (Table 22.4). Downregulation of receptors by their own ligands plays an important physiological role because it prevents overstimulation of target cells when circulating hormone levels are elevated.

Although downregulation of steroid receptor levels by their cognate hormones appears to be the most frequent form of autoregulation, it is not common to all target cells. In fact, glucocorticoid-mediated upregulation has been reported in a number of responsive cells. Since all of these cells are growth inhibited by these hormones, it was initially suggested that hormone-mediated upregulation may be required for subsequent growth inhibition. However, the fact that glucocorticoid-mediated upregulation also occurs in human lymphoid cells that express glucocorticoid receptors but are not growth inhibited by these steroids, demonstrates that this positive **autoregulation** is neither the result nor cause of hormone-mediated growth arrest (see Clin. Corr. 22.3).

Nuclear Hormone Receptors and Coactivators and Corepressors

Several levels of control exerted on hormone action include availability of the ligand to its receptor, extent of expression of the specific receptor in the cell, and cell-specific factors. Proteins that regulate the transactivational functions of nuclear receptors are referred to as **coactivators** and **corepressors.** Steroid receptor coactivators are proteins that bind to the receptor while it is in contact with DNA and structurally facilitate subsequent transcription, a state of the complex known as *activation.* In the inactive state, the receptor, while bound to DNA at a specific gene promoter governed by the hormone response element, is contacted by another set of proteins, corepressors, that structurally interfere with the activated transcriptional state of the complex. Well-known corepressors for steroid hormone nuclear receptors are SMRT and NCoR (GRIP1 is also in this category resembling NCoR). These corepressors affect mSin3 and histone deacetylases (HDAC) to cause deacetylation of histones producing condensation of chromatin resulting in transcriptional repression. Binding of cognate hormonal ligand to the receptor complex induces a conformational alteration of the receptor resulting in the dissociation of corepressors and recruitment of coactivators. These coactivators then catalyze the acetylation of histones and lead to transcriptional activation (see Figure 22.23). Coactivators are proteins in the SRC family designated SRC-1, SRC-2, and SRC-3. SRCs interact with receptors through conserved motifs with the amino acid sequence LXXLL, where L is leucine and X is any amino acid.

CLINICAL CORRELATION 22.3

Mineralocorticoid Receptor Mutation Results in Hypertension and Toxemia of Pregnancy

A very recent finding provides a molecular explanation for an inherited form of hypertension and a possible molecular basis for the toxemia of pregnancy, also known as eclampsia. The causes of hypertension, especially the hypertension associated with eclampsia, are only conjectured about and specific molecular endpoints that have been implicated involve the renin–angiotensin system and various factors related to this system, including the angiotensinogen-converting enzyme and the mineralocorticoid receptor. This steroid receptor is normally activated by aldosterone, which promotes sodium reabsorption in the distal nephron of the kidney. However, a recent report has focused on the hypertension that sometimes occurs during pregnancy and has described a new mutation in the mineralocorticoid receptor. In this mutated receptor the serine residue at position 810 has been replaced with a leucine residue (referred to as S810L mutation). This mutated receptor appears to be responsible for the hypertension that appears early in pregnancy and is magnified during pregnancy (approximately 6% of pregnancies are complicated by development of hypertension). The amino acid serine 810 is located in the hormone-binding domain of the receptor and is conserved in all mineralocorticoid receptors across many species. As a consequence of this specific mutation in the hormone-binding domain, the mineralocorticoid receptor binds progesterone with the same high affinity with which it binds aldosterone. Although progesterone binds to the normal wild-type mineralocorticoid receptor, it does so with low affinity and functions as an antagonist. However, progesterone bound to the mutated form of the receptor functions as an agonist, thus activating the receptor and inducing a physiological response (reabsorption of sodium in kidney). Given that during pregnancy plasma progesterone levels increase significantly (see Figure 20.35), this mutated form of the receptor is continuously saturated with this steroid. As a consequence of progesterone functioning as an agonist, the blood pressure of the pregnant woman expressing this mutated receptor may reach very high and dangerous levels. For example, in subjects under age 35 carrying the mutated form of the receptor, the systolic to diastolic blood pressure ratio was reported to be 167/110, whereas the ratio was 126/78 (normal range) in noncarriers. Many other steroids fail to bind to either the normal (wild-type) or mutated form of this receptor. However, spironolactone, which usually functions as an aldosterone antagonist when bound to the normal receptor, also functions as an agonist when bound to the mutated receptor. Therefore spironolactone probably should not be used to treat hypertensive patients bearing the S810L mutation in their mineralocorticoid receptor. Since the early development of heart failure occurs in the seriously hypertensive group of patients, it seems possible that this mutated mineralocorticoid receptor could be an important contributing factor.

Geller, D. S., Farhi, A., Pinkerton, N., Fradley, M., Moritz, M., Spitzer, A., Meinke, G., Tsai, F. T. F., Sigler, P. B., and Lifton, R. P. Activating mineralocorticoid receptor mutation in hypertension exacerbated by pregnancy. *Science* 289:119, 2000.

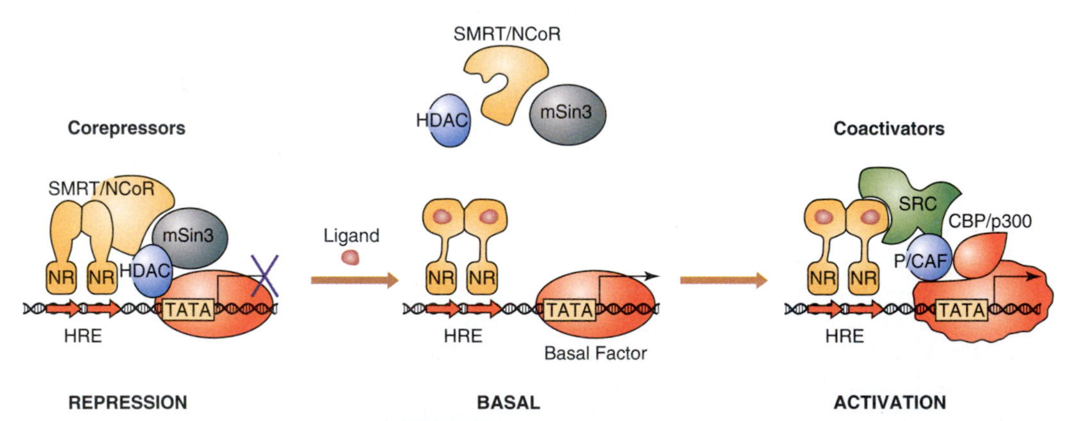

FIGURE 22.23
Role of corepressors and coactivators in transcriptional regulation by steroid/nuclear hormone receptors.
HRE, hormone response element; NR, nuclear receptor; SMRT, NCoR, corepressors; HDAC, histone deacetylase; P/CAF, histone acetyltransferase; SRC, coactivator; mSin3, transcriptional repressor that binds to repressive transcription factors and recruits HDAC into repressor complex; CBP/p300, general transcriptional activator. This model only holds for nuclear receptors that bind to DNA in the absence of a cognate ligand.

22.9 | SPECIFIC EXAMPLES OF STEROID HORMONE ACTION AT THE CELLULAR LEVEL: PROGRAMMED DEATH

Programmed cell death or **apoptosis** is a process by which cells die according to a program that may be beneficial for the organism. It results from the rise or fall in the level of a specific hormone(s). Uterine endometrial cells at the beginning of menstruation are an example where programmed cell death is initiated by the fall in the levels of progesterone and estradiol in the blood (see Clin. Corr. 22.4). Another case is apoptosis of thymus cells during development when the adrenal cortex becomes functional and begins to synthesize and secrete relatively large amounts of cortisol. A newborn has a large thymus but when cortisol is synthesized and released the thymus cortical cells begin to die until a resistant core of cells is reached and the gland achieves its adult size. Thus programmed cell death is a mechanism used in development for the maturation of certain organs as well as in cyclic systems where cells proliferate and then regress until another cycle is initiated to begin the proliferation all over again, as is the case with the ovarian cycle.

Glucocorticoid-induced apoptosis in thymocytes is mediated by the intracellular glucocorticoid receptor. There are two phases to this complex process: inhibition of cell proliferation (cytostatic phase) followed by a cytolytic phase characterized by internucleosomal DNA cleavage and ultimate cell death (cytolytic phase). These two phases are not necessarily linked, since some cells are growth inhibited, but not lysed, by glucocorticoid hormones. The precise mechanism by which glucocorticoid–receptor complexes induce cell death is not fully understood. Exposure to hormone may result in a conformational change in chromatin with the unmasking of internucleosomal linker DNA regions, which are substrates for a nuclease. Treatment of thymocytes with glucocorticoids results in the activation of a constitutive, endogenous Ca^{2+}/Mg^{2+}-dependent endonuclease, while similar treatment of human leukemic T cells results in the activation of Ca^{2+}/Mg^{2+}-independent nuclease. Recent studies have demonstrated that the Ca^{2+}/Mg^{2+}-dependent nuclease that is activated by glucocorticoids in rat thymocytes is homologous with a cyclophilin. These proteins are high-affinity binding proteins for the immunosuppressive drug

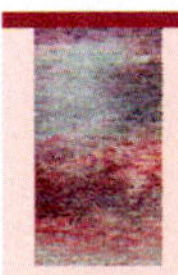

CLINICAL CORRELATION 22.4
Programmed Cell Death in the Ovarian Cycle

During the ovarian cycle, the ovarian follicle expels the mature ovum at day 14 and the remaining cells of the follicle are differentiated into a functional corpus luteum. The corpus luteum produces some estradiol to partially replace that provided earlier by the maturing follicle. However, its principal product is progesterone. Estradiol and progesterone are the main stimulators of uterine endometrial wall thickening in preparation for implantation. One of the proteins induced by estradiol action in the endometrium is the progesterone receptor. Thus the uterine endometrial cells become exquisitely sensitive to estradiol as well as progesterone. The corpus luteum supplies the latter, but in the absence of fertilization and development of an embryo, the corpus luteum lives only for a short while and then atrophies because of lack of LH or chorionic gonadotropin, a hormone produced by the early embryo. The production of oxytocin and $PGF_{2\alpha}$ in the ovary may bring about the destruction of the corpus luteum (luteolysis). Blood levels of estradiol and progesterone fall dramatically after luteolysis and the stimulators of uterine endometrial cells disappear, causing degeneration of this thickened, vascularized layer of tissue and precipitating menstruation. These cells die by programmed cell death (apoptosis) due to the withdrawal of steroids. The hallmark of programmed cell death is internucleosomal cleavage of DNA. Thus programmed cell death appears to play specific roles in development and in tissue cycling either due to a specific hormonal stimulus or to withdrawal of hormone(s).

Erickson, G. F. and Schreiber, J. R. Morphology and physiology of the ovary. In: K. L. Becker (Ed.), *Principles and Practice of Endocrinology and Metabolism.* New York: Lippincott, 1990, p. 776; Rebar, R. W., Kenigsberg, D., and Hogden, G. D. The normal menstrual cycle and the control of ovulation. In: K Becker (Ed.), *Principles and Practice of Endocrinology and Metabolism.* New York: Lippincott, 1990, p. 788; and Hamburger, L., Hahlin, M., Hillensjo, T., Johanson, C., and Sjogren, A. Luteotropic and luteolytic factors regulating human corpus luteum function. *Ann. N. Y. Acad. Sci.* 541:485, 1988.

cyclosporin A and have Ca^{2+}/Mg^{2+}-dependent nuclease activity. The mechanism(s) by which glucocorticoid hormones induce lysis of thymocytes versus leukemic T cells appears to differ in several other respects. Treatment of sensitive T cells with these hormones results in upregulation of glucocorticoid receptor mRNA levels, while identical treatment of thymocytes appears to result in downregulation of mRNA levels. Also, the mRNA levels for an important growth factor, c-*myc*, are repressed in glucocorticoid-treated T cells, but induced in thymocytes. Thus the cytostatic and cytolytic phases of apoptosis may be mediated by slightly different pathways in these two different cell types.

BIBLIOGRAPHY

Argentin, S., Sun, Y. L., Lihrmann, I., Schmidt, T. J., Drouin, J., and Nemer, M. Distal cis-acting promoter sequences mediate glucocorticoid stimulation of cardiac ANF gene transcription. *J. Biol. Chem.* 266:23315, 1991.

Baulieu, E. E. Steroid hormone antagonists at the receptor level: a role for heat-shock protein MW 90,000 (hsp 90). *J. Cell. Biochem.* 35:161, 1987.

Beato, M. Gene regulation by steroid hormones. *Cell* 56:335, 1989.

Carson-Jurica, M. A., Schrader, W. T., and O'Malley, B. W. Steroid receptor family: structure and functions. *Endocr. Rev.* 11:201, 1990.

Blumberg, B. and Evans, R. M. Orphan nuclear receptors—new ligands and new possibilities. *Genes Dev.* 12:3149, 1998.

Chen, J. D. Steroid/nuclear receptor coactivators. In: G. Litwack (Ed.), *Vitamins and Hormones,* Vol. 28. San Diego: Academic Press, 2000, p. 391.

Chrousos, G. P., Loriaux, D. L., and Lipsett, M. B. (Eds.). *Steroid Hormone Resistance.* New York: Plenum Press, 1986.

Drouin, J., Sun, Y. L. Tramblay, S., Schmidt, T. J., deLean A., and Nemer, M. Homodimer formation is rate-limiting for high affinity DNA binding by glucocorticoid receptor. *Mol. Endocrinol.* 6:1299, 1992.

Evans, R. M. The steroid and thyroid hormone receptor superfamily. *Science* 240:889, 1988.

Giguere, V., Hollenberg, S. M., Rosenfeld, M. G., and Evans, R. M. Functional domains of the human glucocorticoid receptor. *Cell* 46:645, 1986.

Green, S., Kumar, V., Theulaz, I., Wahli, W., and Chambon, P. The N-terminal DNA-binding "zinc-finger" of the estrogen and glucocorticoid receptors determines target gene specificity. *EMBO J.* 7:3037, 1988.

Gustafsson, J. A., et al. Biochemistry, molecular biology, and physiology of the glucocorticoid receptor. *Endocr. Rev.* 8:185, 1987.

Huft, R. W. and Pauerstein, C. J. *Human Reproduction: Physiology and Pathophysiology.* New York: Wiley, 1979.

Litwack, G. (Ed.). *Biochemical Actions of Hormones,* Vols. 1–14. New York: Academic Press, 1973–1987.

Mester, J. and Baulieu, E. E. Nuclear receptor superfamily. In: L. J. DeGroot et al. (Eds), *Endocrinology,* 3rd ed., Philadelphia: Saunders, 1995, p. 93.

Norman, A. W. and Litwack, G. *Hormones,* 2nd ed. Orlando, FL: Academic Press, 1997.

O'Malley, B. W., Tsai, S. Y., Bagchi, M., Weigel, N. L., Schrader, W. T., and Tsai, M. J. Molecular mechanism of action of a steroid hormone receptor. *Recent Prog. Horm. Res.* 47:1, 1991.

Renkawitz, R. Repression mechanisms of v-erbA and other members of the steroid receptor superfamily. *Ann. N.Y. Acad. Sci.* 684:1, 1993.

Rusconi, S. and Yamamoto, K. R. Functional dissection of the hormone and DNA binding activities of the glucocorticoid receptor. *EMBO J.* 6:1309, 1987.

Schmidt, T. J. and Meyer, A. S. Autoregulation of corticosteroid receptors. How, when, where and why? *Receptor* 4:229, 1994.

Schwabe, J. W. R. and Rhodes, D. Beyond zinc fingers: steroid hormone receptors have a novel structural motif for DNA recognition. *Trends Biochem. Sci.* 16:291, 1991.

Wahli, W. and Martinez, E. Superfamily of steroid nuclear receptors—positive and negative regulators of gene expression. *FASEB J.* 5:2243, 1991.

QUESTIONS | C. N. ANGSTADT

Multiple Choice Questions

1. The C-21 steroid hormones include:
   A. aldosterone.
   B. dehydroepiandrosterone.
   C. estradiol.
   D. testosterone.
   E. vitamin D_3.

2. Side chain cleavage enzyme complex activity may be stimulated by all of the following EXCEPT:
   A. cAMP.
   B. Ca^{2+} released via stimulation of the IP_3 pathway.
   C. Ca^{2+} entering the cell through a channel.
   D. 5′-AMP.
   E. induction of the StAR protein.

3. Δ^5-Pregnenolone is a precursor of all of the following EXCEPT:
   A. aldosterone.
   B. cortisol.
   C. 17β-estradiol.
   D. progesterone.
   E. vitamin D_3.

4. Retinoic acid and its derivatives:
   A. may activate gene transcription by eliminating the silencing activity of receptor proteins.
   B. bind to homodimeric proteins, which in turn bind to DNA.
   C. bind directly to DNA via leucine zipper motifs.
   D. are vitamin derivatives and hence have no effect on regulation of gene expression.
   E. may substitute for thyroid hormones in binding to the thyroid hormone receptor.

Refer to the following for Questions 5 to 8:

A. corticosteroid-binding globulin
B. serum albumin
C. sex hormone-binding globulin

D. androgen-binding protein
E. transferrin

5. Major aldosterone carrier in blood.

6. Supplies testosterone to spermatozoa.

7. Binds 75–80% of 17-hydroxysteroids in the plasma.

8. At puberty decreases more in males than in females.

9. The receptor for which of the following recognizes a unique response element in DNA?
A. estrogen
B. glucocorticoid
C. mineralocorticoid
D. progesterone

10. All of the following receptors may belong to the steroid receptor gene superfamily EXCEPT:
A. aryl hydrocarbon receptor.
B. *erb*A protein.
C. retinoic acid receptor.
D. thyroid hormone receptor.
E. α-tocopherol receptor.

Questions 11 and 12: Aldosterone bound to its receptor promotes sodium reabsorption in the distal nephron of the kidney. Elevated sodium in the blood leads to hypertension, which can be a serious problem in pregnancy. The causes of hypertension in pregnancy are not yet known although defects in the system leading to aldosterone release may be involved. In some cases, hypertension that appears in early pregnancy and increases with time has been shown to be caused by a mutation in the mineralocorticoid receptor. The mutation allows progesterone to bind with the same affinity as aldosterone and thus act as an agonist. Because of the high levels of progesterone during pregnancy, the mutated receptor remains saturated and blood pressure can become dangerously high. Spironolactone, which acts as an antagonist of aldosterone with a normal receptor, acts as an agonist with the mutated receptor and should not be used to treat this kind of hypertension.

11. All of the following are normal events leading to or following secretion of aldosterone from the adrenal gland EXCEPT:
A. renin is released by the kidney in hypovolemia.
B. angiotensinogen binds to membrane receptors.
C. the PI cycle is activated producing IP_3 and DAG.
D. Ca^{2+} levels in the cell rise.
E. aldosterone is secreted into the blood.

12. Once ovulation occurs, the pathway followed differs when the egg is fertilized and when it is not. Which of the following statements about this process is/are correct?
A. FSH, via cAMP as a second messenger, stimulates the follicle to release 17β-estradiol.
B. Blood levels of progesterone fall as pregnancy progresses because the corpus luteum dies.
C. Inhibin produced by the follicle prevents release of LH.
D. The primary influence for the corpus luteum to produce progesterone and estradiol is FSH.
E. All of the above are correct.

Questions 13 and 14: A newborn has a large thymus but, when the adrenal cortex becomes functional, the thymus cells begin to die until the gland reaches adult size. This is an example of programmed cell death called apoptosis. Apoptosis can result from either the rise or fall of specific hormones. In this case, apoptosis arises as the adrenal cortex synthesizes and secretes relatively large amounts of cortisol. The cortisol–receptor complex is responsible for the cell death. Although the precise mechanism is not clear, it has been shown that rat thymocytes treated with glucocorticoids show activation of a Ca^{2+}/Mg^{2+}-dependent nuclease that may cleave DNA.

13. Glucocorticoid receptors are in the cytoplasm. All of the following statements about the process by which the hormone influences transcription are correct EXCEPT:
A. the hormone must be in the free state to cross the cell membrane.
B. cytoplasmic receptors may be associated with heat shock proteins.
C. the receptor–hormone complex is not activated/transformed until it is translocated to the nucleus.
D. in the nucleus, the activated/transformed receptor–hormone complex searches for specific sequences on DNA called HREs (hormone response elements).
E. the activated receptor–hormone complex may either activate or repress transcription of specific genes (only one activity per gene).

14. Another instance of apoptosis caused by changes in hormone levels is seen in the ovarian cycle. When there is no fertilization of the ovum, the endometrial cells die because:
A. LH levels rise after ovulation.
B. estradiol levels are not involved in the LH surge phenomenon.
C. estradiol inhibits the induction of the progesterone receptor in the endometrium.
D. oxytocin and $PGF_{2\alpha}$ directly destroy the endometrium.
E. the involution of the corpus luteum causes estradiol and progesterone levels to fall dramatically.

Problems

15. Steroidal anti-inflammatory drugs inhibit prostaglandin synthesis in at least two ways—inhibition of phospholipase A_2 and inhibition of cyclooxygenase. Why are the new generation of anti-inflammatory drugs called COX2 inhibitors (Celebrex, Vioxx) better tolerated than the older drugs?
16. What is the relationship between 7-dehydrocholesterol and 1α, 25-dihydroxycholecalciferol?

ANSWERS

1. **A** This is a mineralocorticoid. B, D: These are C-19 androgens. C: Estradiol is a C-18 estrogen. E: Vitamin D_3 is a C-27 compound.
2. **D** This is the inactivated form of cAMP. A–C: The hormone–receptor complex activates both adenylate cyclase through the G-protein and the calcium channel. E: StAR protein facilitates the transfer of cholesterol to the inner mitochondrial membrane where the P450 is located.
3. **E** The synthesis of vitamin D_3 occurs from 7-dehydrocholesterol. A–D: Pregnenolone is an obligatory intermediate in the synthesis of all the steroid hormones.

4. **A** The retinoic acid receptor (RAR) binds to specific silencer elements in the absence of the ligand, retinoic acid. When retinoic acid is bound, the receptor loses silencing activity and activates gene transcription. B: In addition, there are retinoid X receptors (RXR), which also affect gene expression, via heterodimerization with RAR. C: Retinoic acid binds to its receptor. The receptor protein may bind to DNA via a leucine zipper. E: The thyroid hormone receptor is a different protein than RAR.
5. **B** Sixty percent is transported this way.
6. **D** This protein is secreted by Sertoli cells in response to testosterone and FSH.
7. **B** Corticosteroid-binding globulin has a high affinity for cortisol.
8. **C** As a result, there is more unbound testosterone circulating in the blood of adult males.
9. **A** The *positive* glucocorticoid response element is the same as the mineralocorticoid response element and the progesterone response element. The estrogen response element differs.
10. **E** B: Note that c-*erb*A is a protooncogene.
11. **B** Angiotensinogen is cleaved by renin to angiotensin I, which must further be cleaved by converting enzyme to active angiotensin II. A: This is a major signal. C, D: These lead to increased Ca^{2+} and activation of protein kinase C.
12. **A** Activation of cAMP-dependent protein kinase stimulates the synthesis and secretion of estradiol. B: This is what happens in the absence of fertilization. In pregnancy, the corpus luteum eventually dies but the placenta produces high levels of progesterone. C: Inhibin controls FSH release. D: LH controls the corpus luteum.
13. **C** Dissociation of the heat shock protein from the receptor–hormone complex in the cytosol activates the complex. A: Steroid hormones travel bound to plasma proteins but some is always free. D: These are consensus sequences in DNA. E: Activation is more common but glucocorticoids repress transcription of the proopiomelanocortin gene.
14. **E** Estradiol and progesterone are the main stimulators of the endometrium thickening in preparation for implantation of the fertilized ovum. A: LH rises prior to ovulation. B: Estrogen rises midcycle, progesterone a little later, but these would tend to maintain the endometrium. C: Estradiol induces the progesterone receptor. D: Oxytocin and $PGF_{2\alpha}$ are involved in luteolysis (degradation of the corpus luteum) along with reduction of LH levels.
15. There are two cyclooxygenases. COXl is a constitutively expressed enzyme that catalyzes the synthesis of required prostaglandins, for example, protection of the stomach lining and blood coagulation processes. COX2, however, is induced as part of the inflammatory response. NSAIDs (nonsteroidal anti-inflammatory drugs like aspirin) inhibit both COX1 and COX2, so stomach irritation is common. COX2 drugs inhibit production of only the inflammatory prostaglandins.
16. 1α,25-Dihydroxycholecalciferol (1,25 $(OH)_2D_3$) is the active form of vitamin D. Ultraviolet light acting on the skin converts 7-dehydrocholesterol to cholecalciferol. This compound must be hydroxylated in liver to 25-hydroxycholecalciferol and subsequently in kidney to yield the active 1,25$(OH)_2D_3$.

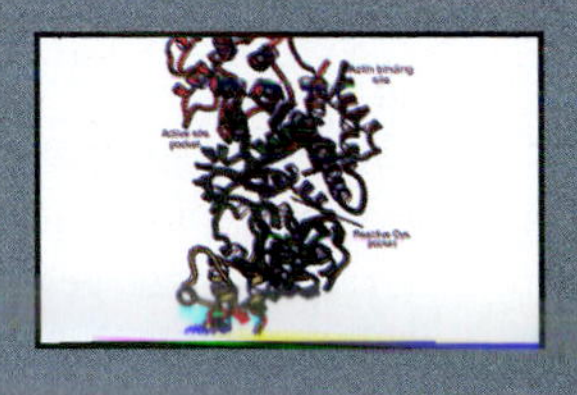

23

MOLECULAR CELL BIOLOGY

Thomas E. Smith

23.1 | OVERVIEW

Animals sense their environment through the responses of specific organs to stimuli: touch, pain, heat, cold, intensity (light or noise), color, shape, position, pitch, quality, acid, sweet, bitter, salt, alkaline, fragrance, and so on. Externally, these generally reflect responses of the skin, eye, ear, tongue, and nose to stimuli. Some of these signals are localized to the point at which they occur; others—sound and sight—are projected in space, that is, the environment outside and distant to the animal.

Discrimination of these signals occurs at the point of reception, but acknowledgment of what they are occurs as a result of secondary stimulation of the nervous system and transmission of the signals to the brain. In many instances, a physical response is indicated and results in voluntary or involuntary muscular activity. Common to these events is electrical activity associated with signal transmission along neurons, and chemical activity associated with signal transmission across synaptic junctions. In all cases, stimuli received from the environment in the form of pressure (skin, feeling), light (sight, eye), noise (ear, hearing), taste (tongue), or smell (nose) are converted (transduced) into electrical impulses and to some other form of energy in order to effect the desired terminal response dictated by the brain. A biochemical component is associated with each of these events.

General biochemical mechanisms of signal transduction and amplification will be discussed as they relate to biochemical events involved in nerve transmission, vision, and muscular contraction. Finally, a specialized case of biochemical signal amplification will be discussed, namely, blood coagulation. The latter topic is in effect hemostasis with emphasis on the enzymes and ancillary proteins involved in phases of the process ranging from coagulation to fibrinolysis. A common theme of all topics discussed in this chapter is that processes are initiated by some stimuli, propagated and/or amplified by some biochemical processes, and terminated and repositioned for the next event. Energy and intermediary metabolites necessary for many of these events may be mentioned briefly in this chapter since they are covered in sufficient details elsewhere in this textbook.

23.2 | NERVOUS TISSUE: METABOLISM AND FUNCTION

Knowledge of the chemical composition of the brain began with the work of J. L. W. Thudichum in 1884 and publication of his monogram entitled "A Treatise on the Chemical Composition of the Brain, Based Throughout on Original Research" (cited in West and Todd, *Textbook of Biochemistry,* MacMillan, 1957). Thudichum's research was supplemented by the work of others during those earlier years. There have been almost explosive advances during recent years, through the use of molecular biological techniques, not only in knowledge of the composition of the brain but also of molecular mechanisms involved in many brain/neuronal functions.

About 2.4% of an adult's body weight is nervous tissue, of which approximately 83% is the brain. The nervous system provides the communications network between the senses, the environment, and all parts of the body. The brain is the command center. This system is always functioning and requires a large amount of energy to keep it operational. Under normal conditions, the brain derives its energy from glucose metabolism. Ketone bodies can cross the blood–brain barrier and be metabolized by brain tissue. Their metabolism becomes more prominent during **starvation,** but even then they cannot totally replace the need for glucose. The human brain uses approximately 103 g of glucose per day. For a 1.4-kg brain, this corresponds to a rate of utilization of approximately 0.3 μmol min^{-1} g^{-1} of tissue. This rate of glucose utilization represents a capacity for ATP production through the tricarboxylic acid (TCA) cycle alone of approximately 6.8 μmol min^{-1} g^{-1} of tissue. Of course, the TCA cycle is not 100% efficient for ATP production nor is all of the glucose metabolized through it. Most of the ATP used by the brain and other nervous tissue is generated aerobically through the TCA cycle that functions at near maximum capacity. Glycolysis functions at approximately 20% capacity. Much of the energy used by the brain is to maintain ionic gradients across the plasma membranes, to effect various storage and transport processes, and for the synthesis of neurotransmitters and other cellular components.

Two features of brain composition are worth noting. It contains specialized and complex lipids, which appear to function to maintain membrane integrity rather than in metabolic roles. There is generally a rapid **turnover rate** of brain proteins relative to other body proteins in spite of the fact that neuronal cells generally do not divide after they have differentiated.

Cells of the nervous system responsible for collecting and transmitting messages are the **neurons.** They are very highly specialized (Figure 23.1). Each neuron consists of a cell body, **dendrites** that are short antenna-like protrusions that receive signals from other cells, and an **axon** that extends from the cell body and transmits signals to other cells. The central nervous system (CNS) is a highly integrated system where individual neurons can receive signals from a variety of different sources, including inhibitory and excitatory stimuli.

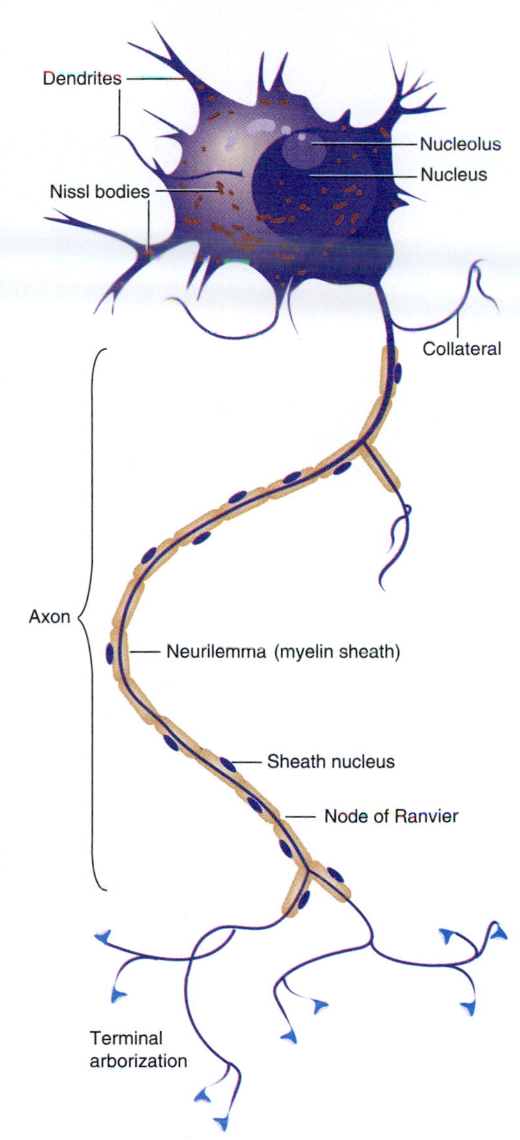

FIGURE 23.1
A motor nerve cell and investing membranes.

Cells other than neurons exist in the CNS. The brain has about ten times more glial cells than neurons. Glial cells occupy spaces between neurons and provide some electrical insulation. They are generally not electrically active, but they are capable of division. There are basically five types of glial cells: Schwann cells, oligodendrocytes, microglia, ependymal cells, and astrocytes. Each type of glial cell has a specialized function, but only astrocytes appear to be directly associated with a biochemical function related to neuronal activity. One of their functions is metabolic (see discussion below on GABA, p. 1001), the other is anatomical.

Astrocytes send out processes at the external surfaces of the CNS. These processes are linked to form anatomical complexes that provide sealed barriers and isolate the CNS from the external environment. **Astrocytes** also send out similar processes to the circulatory system inducing the endothelial cells of the capillaries to become sealed by forming tight junctions that prevent the passive entry into the brain of water-soluble molecules. These tight junctions form what is commonly known as the **blood–brain barrier.** Generally water-soluble compounds enter the brain only if there are specific membrane transport systems for them.

A normal adult has between 10^{11} and 10^{13} neurons, and communication between them is by electrical and chemical signals. Electrical signals transmit nerve impulses down the axon and chemicals transmit signals across the gap between cells. Some of the biochemical events that give the cell its electrical properties and are involved in propagation of impulses will be discussed.

ATP and Transmembrane Electrical Potential in Neurons

Adenosine triphosphate generated from the metabolism of glucose is used to help maintain an equilibrium electrical potential across the membrane of the neuron of approximately −70 mV, with the inside being more negative than the outside. This potential is maintained by the **Na^+/K^+ exchanging ATPase** (see p. 522), the energy for which is derived from the hydrolysis of ATP to ADP and inorganic phosphate. This system pumps Na^+ out of the cell by an antiport mechanism, where K^+ is moved into the cell. The channels through which Na^+ enters the cell are voltage gated; that is, the proteins of the channel undergo a charge-dependent conformational change that opens the channel when the electrical potential across the membrane decreases (specifically, becomes less negative) by a value greater than some threshold value. When the membrane becomes depolarized, Na^+, whose concentration is higher outside the cell than inside, flows into the cell and K^+, whose concentration is greater inside the cell, flows out of the cell, both moving down their respective concentration gradients. The channels are open in a particular region of the cell membrane for fractions of milliseconds (Figure 23.2). The localized depolarization (voltage change) causes a conformational change in the neighboring

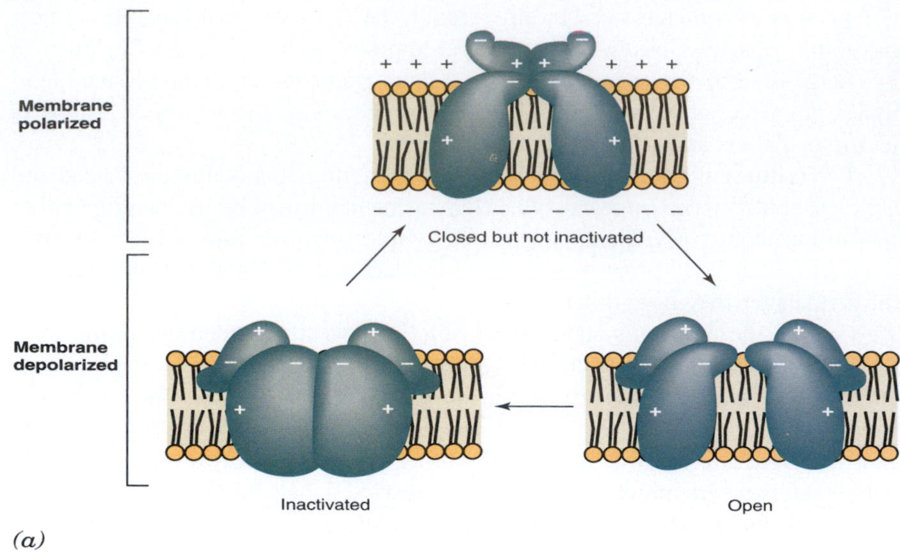

(a)

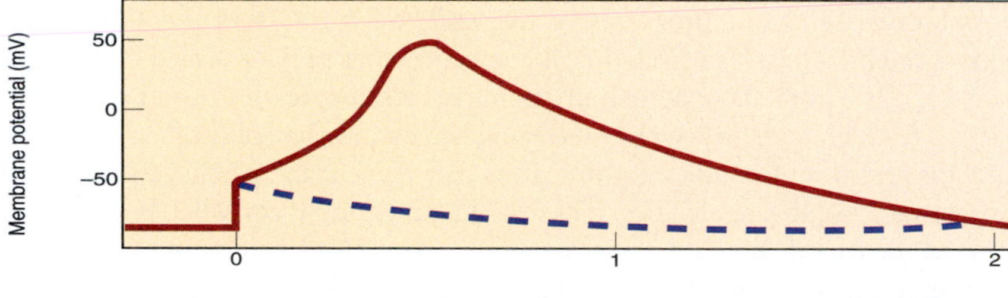

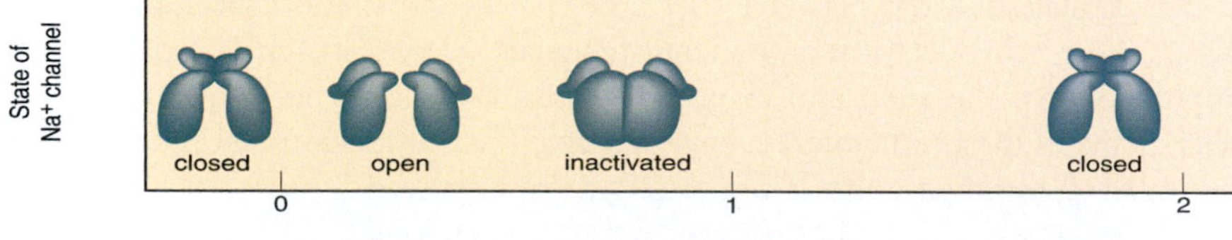

(b)

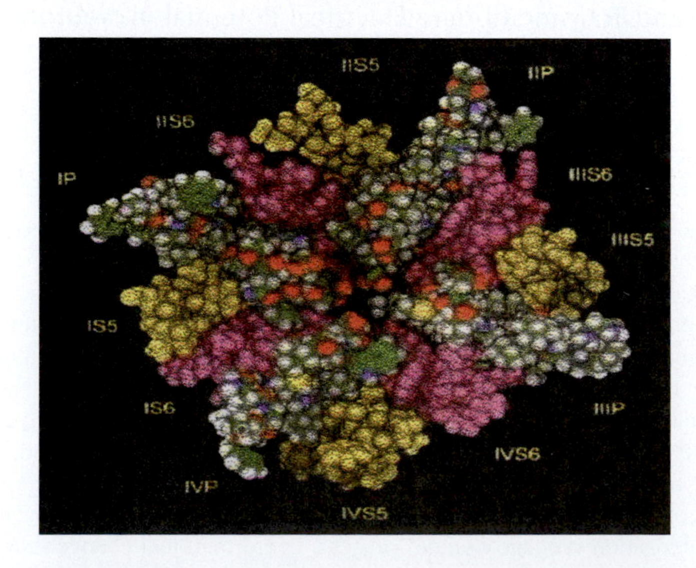

(c)

FIGURE 23.2
Na^+ channels.
Schematics (*a*, *b*) of Na^+ channels opening and closing during nerve impulse transmission. (*c*) Molecular model of a Na^+ channel pore. This view is down the funnel section of the pore. Notations ending in S5 and S6 denote helical structures and those ending in P denote loops. *Parts (a) and (b) redrawn from Alberts, B., Bray, D., Lewis, J., Raff, M., Roberts, K., and Watson, J.* Molecular Biology of the Cell, *2d ed. New York: Garland Publishing, 1989, p. 1071; part (c) reproduced with permission from Lipkin, G. M. and Fozzard, H. A.* Biochemistry *39:8161, 2000.* Copyright (2000) American Chemical Society.

proteins that make up the **"voltage-gated" ion channels.** These channels open momentarily to allow more ions in and, thus, by affecting adjacent channel proteins, allow the process to continue down the axon. There is a finite recovery time, during which channel proteins cannot reopen. Thus, charge propagation proceeds in one direction. It is the progressive depolarization and repolarization along the length of the axon that allows electrical impulses to be propagated undiminished in amplitude. Electrical impulse transmission is a continuous process in nervous tissue, and it is the ATP generated primarily from the metabolism of glucose that keeps the system operational.

Up to this date, only the 3-D structure for a potassium voltage-gated channel from a bacterial species has been determined. Using that information and the fact that Na^+, K^+, and Ca^{2+} channels bear striking similarities, a model for a Na^+ channel has been developed and is shown in Figure 23.2*c*. Mutations at critical points within these structures can lead to disease states.

TABLE 23.1 Some of the Neurotransmitters Found in Nervous Tissue

EXCITATORY
Acetylcholine
Aspartate
Dopamine
Histamine
Norepinephirine
Epinephrine
Glutamate
5-Hydroxytryptamine
INHIBITORY
4-Aminobutyrate
Glycine
Taurine

Neuron-Neuron Interaction Occurs Through Synapses

There are two mechanisms for neuron–neuron interaction: through **electrical synapses** or through **chemical synapses.** Electrical synapses permit the more rapid transfer of signals from cell to cell. Chemical synapses allow for various levels of versatility in cell–cell communication. T. R. Elliot, in 1904, was one of the first scientists to clearly express the idea that signaling between nerves could be chemical (cited in Fried, 1995). Considerably more information is now available about this mode of neuron–neuron communication. The 2000 Nobel Prize in Physiology and Medicine was awarded for research in this and related areas. Chemical synapses are of two types: those in which the neurotransmitter binds directly to an ion channel and causes it to open or to close, and those in which the neurotransmitter binds to a receptor that releases or leads to generation of a second messenger that reacts with the ion channel to cause it to open or to close. Primary emphasis here is on chemical synapses.

Chemical neurotransmitters have the following properties: (1) they are found in the presynaptic axon terminal; (2) enzymes necessary for their syntheses are present in the presynaptic neuron; (3) stimulation under physiological conditions results in their release; (4) mechanisms exist (within the synaptic junction) for rapid termination of their action; (5) their direct application to the postsynaptic terminal mimics the action of nervous stimulation; and (6) drugs that modify the metabolism of the neurotransmitter should have predictable physiological effects *in vivo,* assuming that the drug is transported to the site where the neurotransmitter acts.

Chemical neurotransmitters may be excitatory or inhibitory. **Excitatory neurotransmitters** include acetylcholine and the catecholamines. **Inhibitory neurotransmitters** include **γ-aminobutyric acid** (**GABA** or 4-aminobutyric acid), glycine, and taurine (Table 23.1). The two major inhibitory neurotransmitters in the central nervous system are glycine and GABA. Glycine acts predominantly in the spinal cord and the brain stem; GABA acts predominantly in all other parts of the brain. **Strychnine** (Figure 23.3), a highly poisonous alkaloid obtained from *Nux vomica* and related plants of the genus *Strychnos,* binds to **glycine receptors** of the CNS. It has been used in very small doses as a CNS stimulant. Can you propose how it works? The **GABA receptor** also reacts with a variety of pharmacologically significant agents such as **benzodiazepines** (Figure 23.4) and barbiturates. As with strychnine and glycine, there is little structural similarity between GABA and benzodiazepines.

The genes for the **nicotinic acetylcholine receptor** (see p. 518), the glycine receptor, and the GABA receptor have been cloned and their amino acid sequences inferred. There is a relatively high degree of homology in their primary amino acid sequences.

A model of one-half of the GABA receptor is shown in Figure 23.5. This receptor has an $\alpha_2\beta_2$ composition. The polypeptides are synthesized with "signal peptides" that direct their transport to the membrane. The α subunit has 456 amino

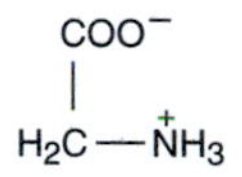

COO^- / $H_2C—\overset{+}{N}H_3$

Glycine

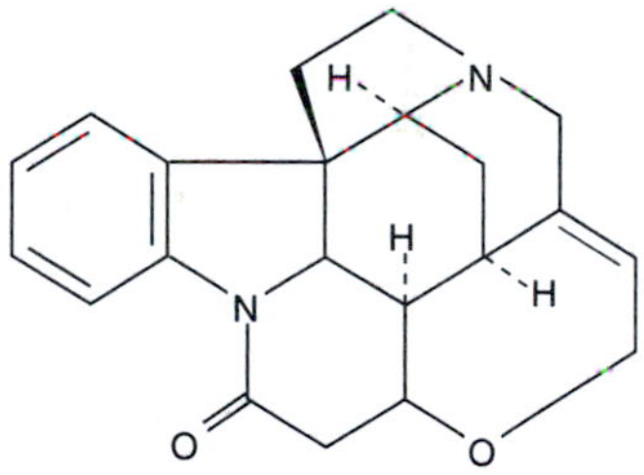

Strychnine

FIGURE 23.3
Structures of glycine and strychnine.

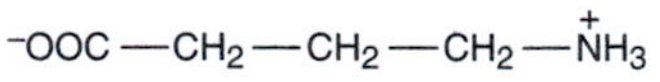

$^-OOC—CH_2—CH_2—CH_2—\overset{+}{N}H_3$

GABA

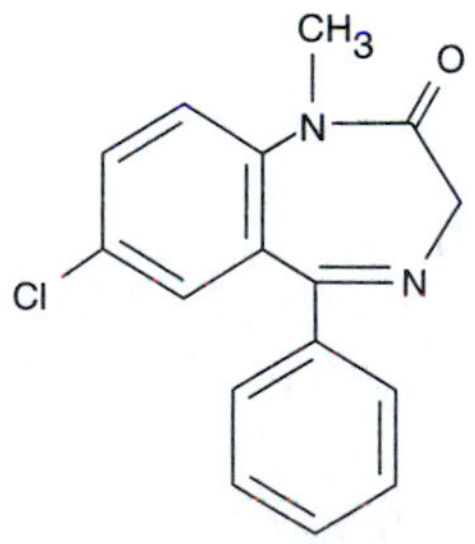

Diazepam

FIGURE 23.4
Structures of GABA and diazepam.

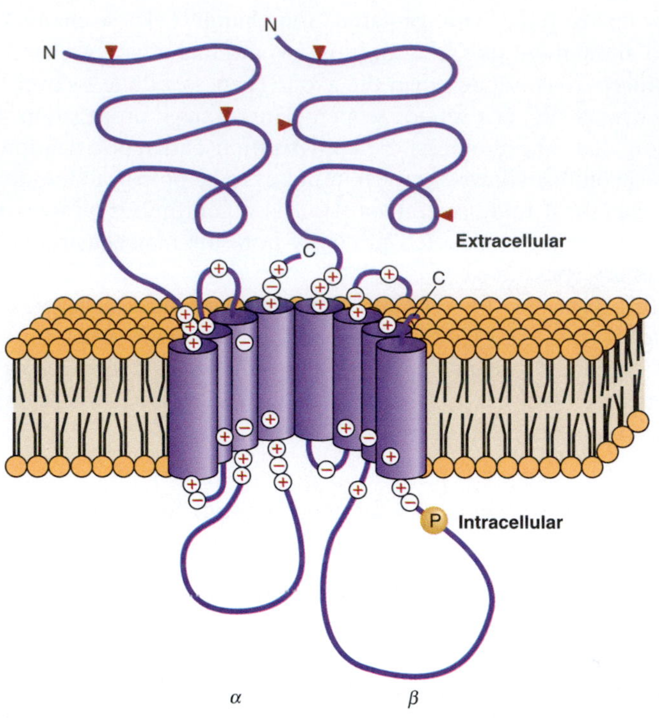

FIGURE 23.5
Schematic model of one-half of the GABA receptor embedded in the cell membrane. The complete receptor has an $\alpha_2\beta_2$ structure and forms an ion channel. The site labeled P is a serine residue that may be phosphorylated by a cAMP-dependent protein kinase.
Redrawn from Schofield, P. R., Darlison, M. G., Fujita, N., et al. Nature *328:221, 1987.*

acid residues and the β subunit has 474. The signal peptides are cleaved, leaving α and β subunits of 429 and 449 amino acid residues, respectively. Interestingly, pharmaceutical agents bind to the α subunit, whereas GABA, the natural inhibitory neurotransmitter, binds to the β subunit. The protrusion of an extended length of the amino-terminal end of each polypeptide to the extracellular side of the membrane suggests that the residues to which the channel regulators bind are at the N terminal. A smaller C-terminal segment is also on the extracellular side of the membrane. The four subunits of the receptor form a channel through which small negative ions (Cl^-) can flow, depending on what is bound to the receptor end of the molecule.

All neurotransmitters are synthesized and stored in **presynaptic neurons.** They are released after stimulation of the neuron, traverse the synapse, and bind to specific receptors on the postsynaptic junction to elicit a response in the next cell. If the neurotransmitter is an excitatory one, it causes depolarization of the membrane as described above. If it is an inhibitory neurotransmitter, it binds to a channel-linked receptor, causing a conformational change that opens the pore and permits small negatively charged ions, specifically Cl^-, to enter. The net effect is to increase the chloride conductance of the postsynaptic membrane, making it more difficult to become depolarized, that is, effectively causing **hyperpolarization.**

Synthesis, Storage, and Release of Neurotransmitters

Nonpeptide neurotransmitters may be synthesized in almost any part of the neuron, in the cytoplasm near the nucleus, or in the axon. Most nonpeptide neurotransmitters are amino acids or derivatives of amino acids.

Neurotransmitters travel rapidly across the **synaptic junction** (which is about 20 nm across), bind to receptors on the postsynaptic side, induce conformational changes in receptors and/or the membrane of the presynaptic neuron, and initiate the process of electrical impulse propagation. Storage and release of neurotransmit-

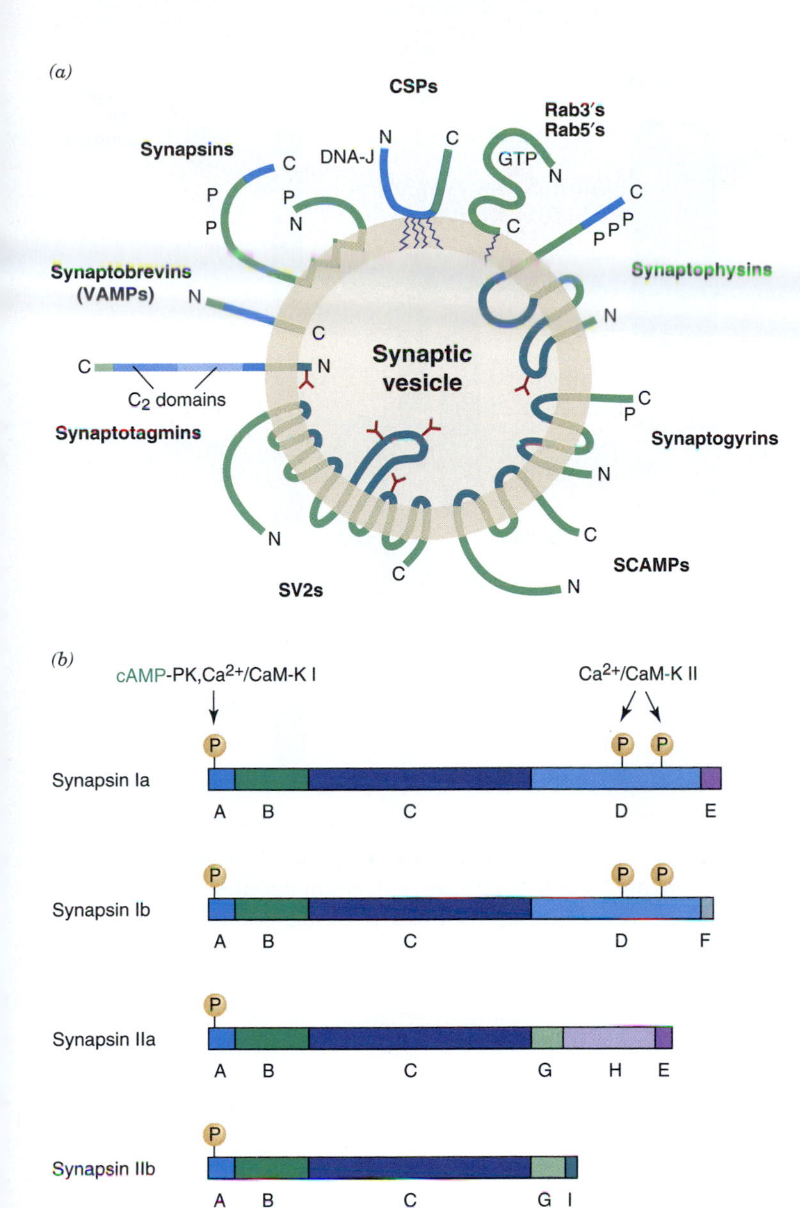

FIGURE 23.6
Synaptic vesicle and synapsin family of proteins.
(*a*) Schematic drawing of the relative arrangement of proteins of the synaptic vesicle (SV). Rab proteins are attached by isoprenyl groups and cysteine string proteins by palmitoyl chains to SVs. The N and C termini of proteins are marked by N and C, respectively. Phosphorylation sites are indicated by P. (*b*) Structural arrangement of the synapsin family of proteins.
Part (a) redrawn from Sudhof, T. C. Nature *375:645, 1995; part (b) redrawn from the work of Chilcote, T. J., Siow, Y. L., Scaeffer, E., et al.* J. Neurochem. *63:1568, 1994.*

ters are intricate processes, but many details of these processes have begun to unfold. It has been shown by conventional techniques that some neurons contain more than one chemical type of neurotransmitter. The significance of this observation is not clear. Release of neurotransmitter is a quantal event. A nerve impulse reaching the presynaptic terminal results in the release of transmitters from a fixed number of **synaptic vesicles.** Release of neurotransmitters involves attachment of the synaptic vesicle to the membrane and **exocytosis** of their content into the synaptic cleft.

Storage of neurotransmitters occurs in large or small vesicles in the presynaptic terminal. Small vesicles predominate and exist in two pools, free and attached to cytoskeletal proteins, mainly actin. Small vesicles contain only "classical" small molecule type transmitters. A schematic diagram of a small synaptic vesicle is shown in Figure 23.6*a*. Large vesicles may contain "classical" small molecule neurotransmitters and neuropeptides. Some may also contain enzymes for synthesis of norepinephrine from dopamine. A list of some of the proteins in large vesicles is presented in Table 23.2. Synapsin (Figure 23.6*b*) has a major role in regulating activity of these vesicles. Figure 23.7 shows schematically how some of the proteins may be arranged on the synaptic vesicle and how they may interact with the plasma membrane of the presynaptic neuron.

TABLE 23.2 List of Synaptic Vesicle Proteins

Synapsin	Ia
	Ib
	IIa
	IIb
Synaptophysin	
Synaptotagmin	
Syntaxin	
Synaptobrevin/VAMP	
Rab3 and rabphilin	
SV-2	
Vacuolar proton pump	

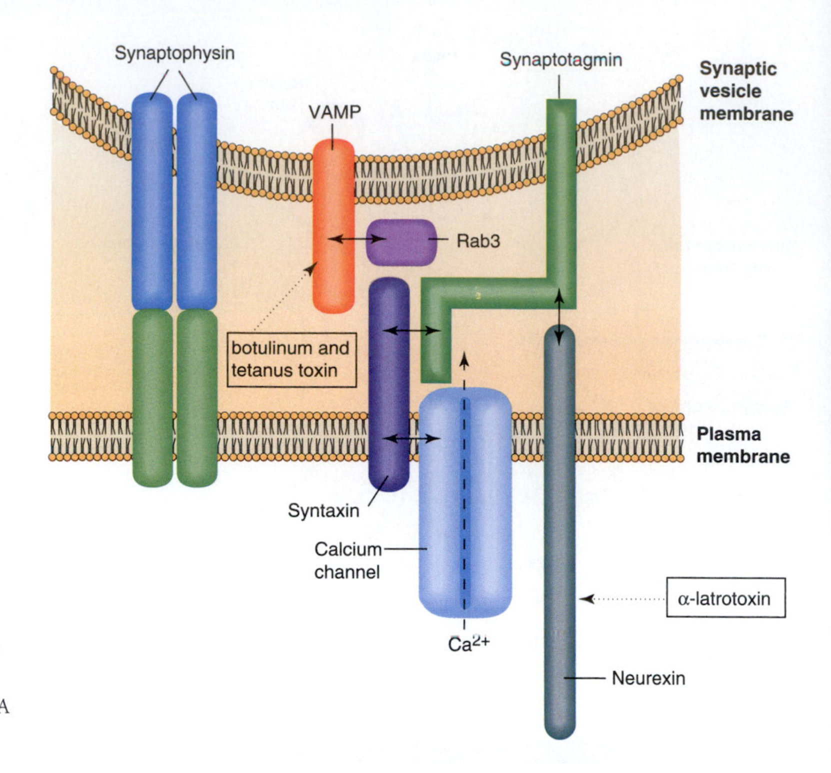

FIGURE 23.7
Schematic diagram showing how some of the synaptic vesicle proteins may interact with plasma membrane proteins.
Redrawn from Bennett, M. K. and Scheller, R. H. Proc. Natl. Acad. Sci. USA *90:2559, 1993.*

1. **Synapsin** is a family of proteins, encoded by two genes, that differ primarily in the C-terminal end (Figure 23.6*b*). Synapsins constitute about 9% of the total protein of the synaptic vesicle membrane. All can be phosphorylated near their N termini by either **cAMP-dependent protein kinase** and/or **calcium–calmodulin (CaM) kinase I.** Synapsins Ia and Ib can also be phosphorylated by **CaM kinase II** near their C termini, a region that is missing in synapsins IIa and IIb.

Synapsin has a major role in determining whether the synaptic vesicles are free and available for binding to the presynaptic membrane (Figure 23.8). Synaptic vesicles exist in the presynaptic neuron in either a free state or bound to cytoskeletal proteins. Nerve stimulation leads to the entry of Ca^{2+} into the presynaptic neuron

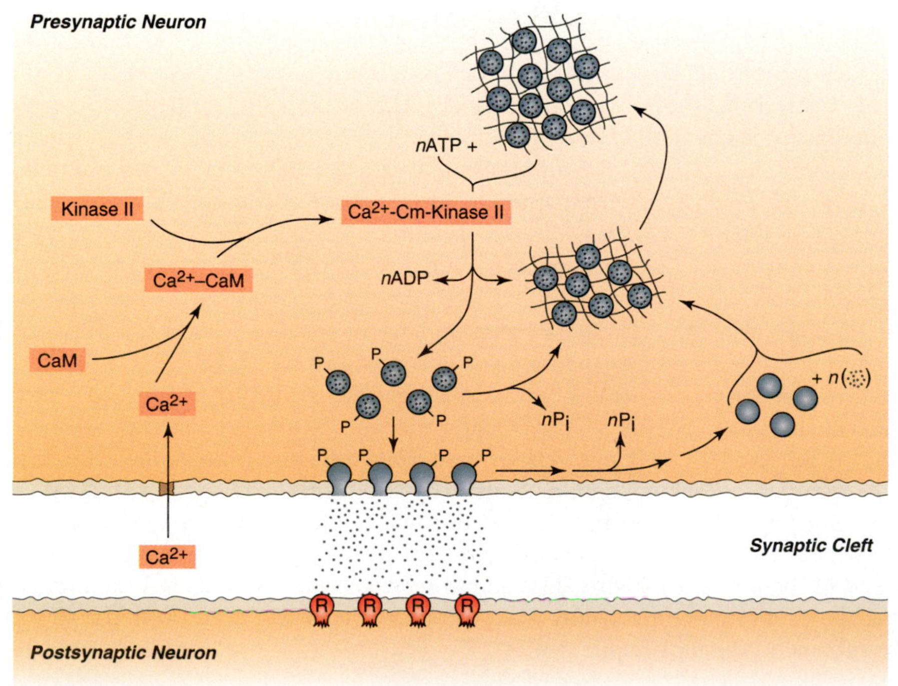

FIGURE 23.8
Model of the mechanism by which calcium ions and calmodulin kinase II regulate the function of synaptic vesicles.
Green circles within the skeletal mesh represent bound, nonphosphorylated synaptic vesicles. Synapsin on those vesicles with P has been phosphorylated and they are in the free pool. Some of them can attach to the presynaptic membrane and release neurotransmitters into the synaptic cleft. Released neurotransmitters can react with receptors (red) on the postsynaptic membrane. The process of recovery of synaptic vesicles and repackaging is also schematically illustrated.

(see Clin. Corr. 23.1). Ca^{2+}–CaM kinases I and II are activated and phosphorylate synapsin. This either prevents binding of synaptic vesicles to the cytoskeletal proteins or releases them from those binding sites. The result is an increase in the free pool of synaptic vesicles. **Calcium–calmodulin** (see p. 524) can bind synapsin and competitively block its interaction with actin and presumably other cytoskeletal proteins. Calcium–calmodulin therefore regulates the number of synaptic vesicles that are free or bound.

2. **Synaptophysin** is an integral membrane protein of synaptic vesicles that is structurally similar to gap junction proteins. It may be involved in formation of a channel from the synaptic vesicle through the presynaptic membrane to permit passage of neurotransmitters into the synaptic cleft.

3. **Synaptotagmin** is also an integral membrane protein of synaptic vesicles that interacts in a Ca^{2+}-dependent manner with specific proteins localized on the presynaptic plasma membrane. It is probably involved in the process of docking of synaptic vesicles to the membrane.

4. **Syntaxin** is an integral membrane protein of the plasma membrane of the presynaptic neuron. Syntaxin binds synaptotagmin and mediates its interaction with Ca^{2+} channels at the site of release of the neurotransmitters. It also appears to have a role in exocytosis.

5. **Synaptobrevin/VAMP** (or vesicle-associated membrane protein) is a family of two small proteins of 18 and 17 kDa that are anchored in the cytoplasmic side of the membrane through a single C-terminal domain and appear to be involved in **vesicle transport** and/or exocytosis. VAMPs appear to be involved in the release of synaptic vesicles from the plasma membrane of the presynaptic neuron. **Tetanus** and **botulinum toxins** bind VAMPs to cause a slow and irreversible inhibition of transmitter release.

6. **Rab3** belongs to the large rab family of **GTP-binding proteins.** Rab3 is specific for synaptic vesicles and is involved in the docking and fusion process of exocytosis. Rab3 is anchored to the membrane through a polyprenyl side chain near its C-terminal end. Elimination by genetic engineering of the polyprenyl side chain-binding site did not alter its function *in vitro,* but it is not clear whether this is also true *in vivo.*

CLINICAL CORRELATION 23.1
Lambert–Eaton Myasthenic Syndrome

Lambert–Eaton myasthenic syndrome (LEMS) is an autoimmune disease in which the body raises antibodies against voltage-gated calcium channels (VGCC) located on presynaptic nerve termini. Upon depolarization of presynaptic neurons, calcium channels at presynaptic nerve termini open, permitting the influx of Ca^{2+}. This increase in Ca^{2+} concentration initiates events of the synapsin cycle and leads to release of neurotransmitters into synaptic junctions. When autoantibodies against VGCC react with neurons at neuromuscular junctions, Ca^{2+} cannot enter and the amount of acetylcholine released into synaptic junctions is diminished. Since action potentials to muscles may not be induced, the effect mimics that of classic myasthenia gravis.

LEMS has been observed in conjunction with other conditions such as small cell lung cancer. Some patients have shown a neurological disorder manifesting itself as subacute cerebellar degeneration (SCD). Plasma exchange (removal of antibodies) and immunosuppressive treatments have been effective for LEMS, but the latter treatment is less effective on SCD.

Diagnostic assays for LEMS depend on the detection of antibodies against VGCC in patients' sera. There are at least four subtypes of VGCC: T, L, N, and P. It has been found that the P subtype may be the one responsible for initiating neurotransmitter release at the neuromuscular junction in mammals. A peptide toxin produced by a cone snail (*Conus magnus*) binds to P-type VGCC in cerebella extracts. This small peptide has been labeled with ^{125}I and bound to VGCC in cerebella extracts, and the radiolabeled complex has been precipitated by sera of patients that have been clinically and electrophysiologically defined as LEMS positive. This assay may prove useful not only in detecting LEMS but as a means of finding out more about the antigenicity of the area(s) on the VGCCs to which antibodies are raised.

Goldstein, J. M., Waxman, S. G., Vollmer, T. L., et al. Subacute cerebellar degeneration and Lambert–Eaton myasthenic syndrome associated with antibodies to voltage-gated calcium channels: differential effect of immunosuppressive therapy on central and peripheral defects. *J. Neurol. Neurosurg. Psychiatry* 57:1138, 1994; and Motomura, M., Johnston, I., Lang, B., et al. An improved diagnostic assay for Lambert–Eaton myasthenic syndrome. *J. Neurol. Neurosurg. Psychiatry* 58:85, 1995.

7. **SV-2** is a large glycoprotein with 12 transmembrane domains. Its function is unclear.

8. **Vacuolar proton pump** is a membrane ATPase that is responsible for the transport of neurotransmitters into the synaptic vesicle.

Termination of Signals at Synaptic Junctions

Neurotransmitter action may be terminated by metabolism, reuptake, and/or diffusion into other cell types. Neurotransmitters responsible for fast responses are generally inactivated by one or both of the first two mechanisms. The following sections outline some of the pathways involved in the synthesis and the degradation of representative fast-acting neurotransmitters, specifically, acetylcholine, catecholamines, 5-hydroxytryptamine, and γ-aminobutyrate (GABA).

Acetylcholine

Reactions involving **acetylcholine** at the synapse are summarized in Figure 23.9. Acetylcholine is synthesized by the condensation of choline and acetyl CoA catalyzed by **choline acetyltransferase** in the cytosol of the neuron. The reaction is

$$\underset{\text{Choline}}{(CH_3)_3\overset{+}{N}CH_2CH_2OH} + CH_3CO{-}SCoA \longrightarrow \underset{\text{Acetylcholine}}{(CH_3)_3\overset{+}{N}CH_2CH_2OCOCH_3} + CoASH$$

Choline is derived mainly from the diet; however, some may come from reabsorption from the synaptic junction or other metabolic sources (see p. 794). The major source of acetyl CoA is decarboxylation of pyruvate by pyruvate dehydrogenase complex in mitochondria. Since choline acetyltransferase is present in the cytosol, acetyl coenzyme A must get into the cytosol for the reaction to occur. The same mechanism discussed previously (see p. 703) for getting acetyl CoA across the inner mitochondrial membrane (as citrate) operates in presynaptic neurons.

Acetylcholine is released and reacts with the nicotinic–acetylcholine receptor located in the postsynaptic membrane (see Clin. Corr. 23.2). The action of acetyl-

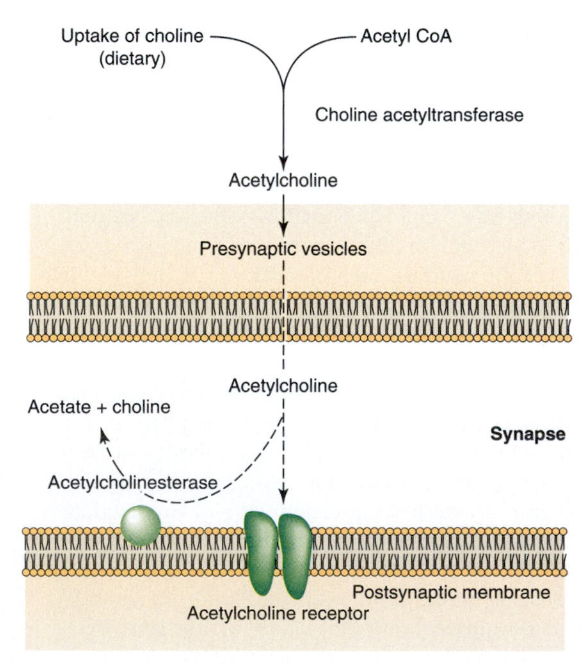

FIGURE 23.9
Summary of the reactions of acetylcholine at the synapse.
Acetyl CoA, acetyl coenzyme A.

CLINICAL CORRELATION 23.2

Myasthenia Gravis: A Neuromuscular Disorder

Myasthenia gravis is an acquired autoimmune disease characterized by muscle weakness due to decreased neuromuscular signal transmission. The neurotransmitter involved is acetylcholine. The sera of more than 90% of patients with myasthenia gravis have antibodies to the nicotinic–acetylcholine receptor (AChR) located on the postsynaptic membrane of the neuromuscular junction. Antibodies against the AChR interact with it and inhibit its function, either its ability to bind acetylcholine or its ability to undergo conformational changes necessary to effect ion transport. Evidence in support of myasthenia gravis as an autoimmune disease affecting the AChR is the finding that the number of AChRs is reduced in patients with the disease, and experimental models of myasthenia gravis have been generated by either immunizing animals with the AChR or by injecting them with antibodies against it.

It is not known what events trigger the onset of the disease. There are a number of environmental antigens that have epitopes resembling those on the AChR. A rat monoclonal antibody of the IgM type prepared against AChRs reacts with two proteins obtained from the intestinal bacterium *E. coli.* Both of the proteins are membrane proteins of 38 and 55 kDa, the smaller of which is located in the outer membrane. This does not suggest that exposure to *E. coli* proteins is likely to trigger the disease. The sera of both normal individuals and myasthenia gravis patients have antibodies against a large number of *E. coli* proteins. Some environmental antigens from other sources also react with antibodies against AChRs.

The thymus gland, which is involved in antibody production, is also implicated in this disease. Antibodies have been found in thymus glands of myasthenia gravis patients that react with AChRs and with environmental antigens. The relationship between environmental antigens, thymus antibodies against AChRs, and onset of myasthenia gravis is not clear.

Myasthenia gravis patients may receive one or a combination of several therapies. Pyridostigmine bromide, a reversible inhibitor of acetylcholine esterase (AChE) that does not cross the blood–brain barrier, has been used. The inhibition of AChE within the synapse by drugs of this type increases the half-time for acetylcholine hydrolysis. This leads to an increase in the concentration of acetylcholine, stimulation of more AChR, and increased signal transmission. Other treatments include use of immunosuppressant drugs, steroids, and surgical removal of the thymus gland to decrease the rate of production of antibodies. Future treatment may include the use of anti-idiotype antibodies to the AChR antibodies, and/or the use of small nonantigenic peptides that compete with AChR epitopes for binding to the AChR antibodies.

Stefansson, K., Dieperink, M. E., Richman, D. P., Gomez, C. M., and Marton, L. S. *N. Engl. J. Med.* 312:221, 1985; Drachman, D. B. (Ed.). Myasthenia gravis: biology and treatment. *Ann. N.Y. Acad. Sci.* 505:1, 1987; and Steinman, L. and Mantegazza, R. *FASEB J.* 4:2726, 1990.

choline at the postsynaptic membrane is terminated by the action of the enzyme **acetylcholinesterase**, which hydrolyzes the acetylcholine to acetate and choline.

$$\text{Acetylcholine} + H_2O \rightleftharpoons \text{acetate} + \text{choline}$$

Choline is largely taken up by the presynaptic membrane and reutilized for synthesis of more acetylcholine. Acetate probably gets reabsorbed into the blood and is metabolized by tissues other than nervous tissue.

An X-ray crystallographic structure of acetylcholine esterase is shown in Figure 23.10. Its mechanism of action is similar to that of serine proteases (see p. 377). It has a **catalytic triad,** but the amino acids in that triad, from N to C termini, are in reverse order to those of the serine proteases and contain glutamate instead of aspartate.

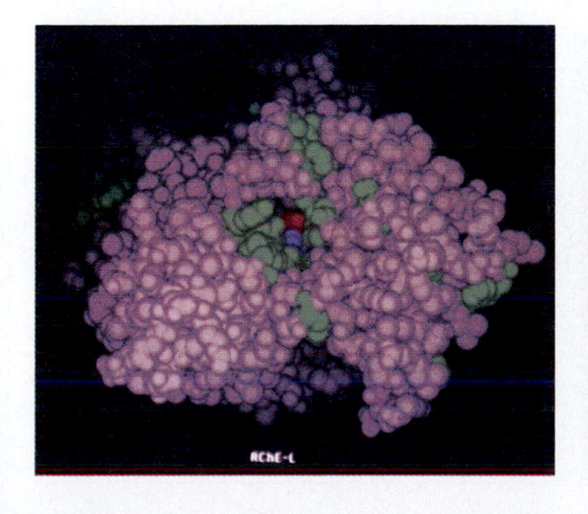

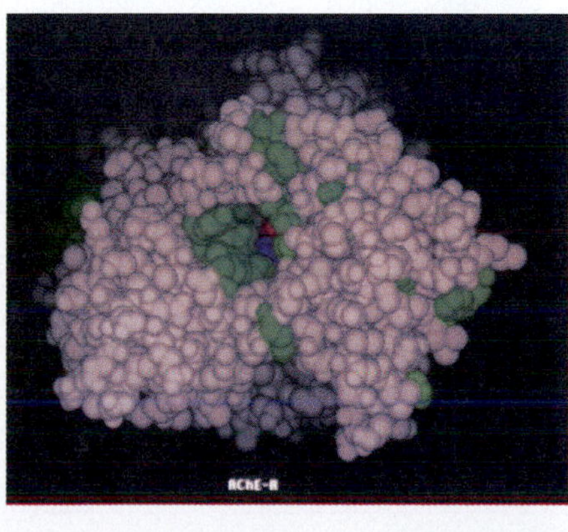

FIGURE 23.10
Space-filling stereo view of acetylcholinesterase looking down into the active site.
Aromatic residues are in green, Ser^{200} is red, Glu^{199} is cyan, and other residues are gray.
Reprinted with permission from Sussman, J. L., Harel, M., Frolow, F., et al. Science 253:872, 1991. Copyright © 1984 by American Association for the Advancement of Science.

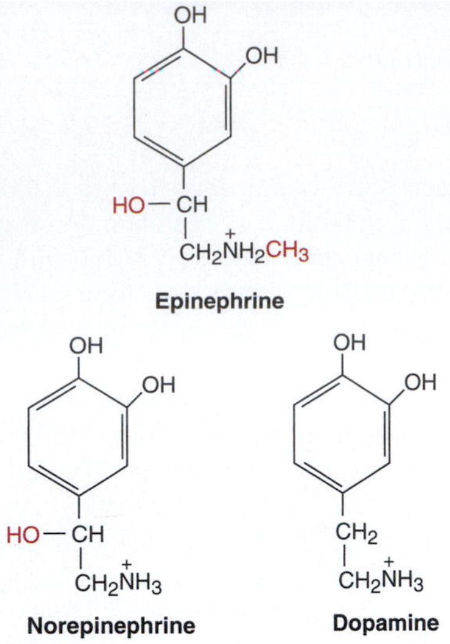

Epinephrine

Norepinephrine

Dopamine

FIGURE 23.11
Catecholamine neurotransmitters.

Catecholamines

The **catecholamine neurotransmitters** are **dopamine** (3,4-dihydroxyphenylethylamine), **norepinephrine,** and **epinephrine** (Figure 23.11). Their biosynthesis has been discussed (see p. 799). The action of catecholamine neurotransmitters is terminated by reuptake into the presynaptic neuron by specific transporter proteins. Cocaine, for example, binds to the **dopamine transporter** and blocks its reuptake. Dopamine remains within the synapse for a prolonged period of time and continues to stimulate the receptors of the postsynaptic neuron. After reuptake, these neurotransmitters may be either repackaged into synaptic vesicles or metabolized. **Catechol-*O*-methyltransferase** and **monoamine oxidase** are involved in their metabolism (Figure 23.12). Catechol-*O*-methyltransferase catalyzes the transfer of a methyl group from *S*-adenosylmethionine to one of the phenolic OH groups. Monoamine oxidase catalyzes the oxidative deamination of these amines to aldehydes and ammonium ions. Monoamine oxidase can use them as substrates whether or not they have been altered by the methyltransferase. The end product of dopamine metabolism is homovanillic acid, and that of epinephrine and norepinephrine is 3-methoxy-4-hydroxymandelic acid.

MHMA

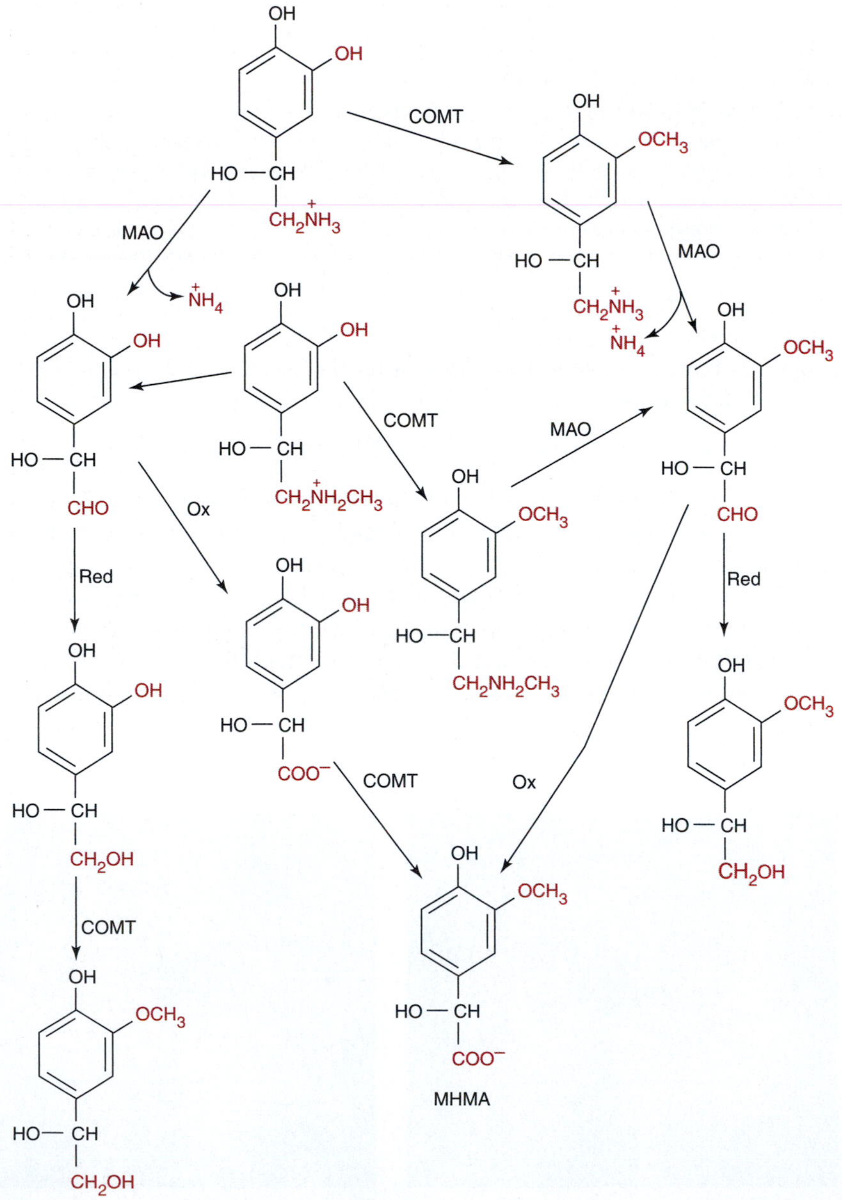

FIGURE 23.12
Pathways of catecholamine degradation.
COMT, catechol-*O*-methyltransferase (requires *S*-adenosylmethionine); MAO, monoamine oxidase; Ox, oxidation; Red, reduction. The major end product of epinephrine and norepinephrine metabolism is 3-methoxy-4-hydroxymandelic acid (MHMA).

5-Hydroxytryptamine (Serotonin)

Serotonin, 5-hydroxytryptamine, is derived from tryptophan (see p. 810). Like dopamine, its action is terminated by **reuptake** by a specific transporter. Some types of depression are associated with low brain levels of serotonin. The action of antidepressants such as Paxil (paroxethine hydrochloride), Prozac (fluoxetine hydrochloride), and Zoloft (sertraline hydrochloride) is linked to their ability to inhibit serotonin reuptake. Once inside the presynaptic neuron, serotonin may be either repackaged in synaptic vesicles or metabolized. The primary route for its degradation is oxidative deamination to the corresponding acetaldehyde in a reaction catalyzed by the enzyme monoamine oxidase (Figure 23.13). The aldehyde is further oxidized to 5-hydroxyindole-3-acetate by an aldehyde dehydrogenase.

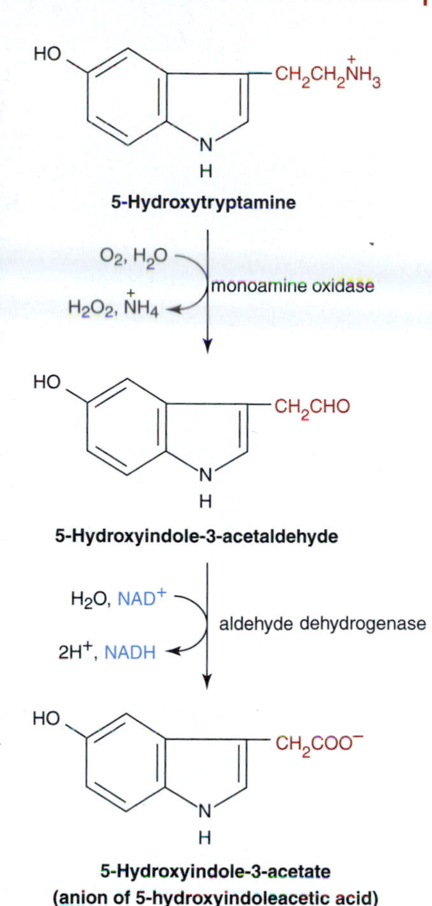

FIGURE 23.13
Degradation of 5-hydroxytryptamine (serotonin).

γ-Aminobutyrate

γ-Aminobutyrate (GABA), an inhibitory neurotransmitter, is synthesized and degraded through reactions commonly known as the **GABA shunt.** In brain tissue, GABA and glutamate, an excitatory neurotransmitter, may share some common routes of metabolism in astrocytes (Figure 23.14). Both are taken up by astrocytes and converted to glutamine, which is then transported back into presynaptic neurons. In excitatory neurons, glutamine is converted to glutamate and repackaged in synaptic vesicles. In inhibitory neurons, glutamine is converted to glutamate and then to GABA, which is repackaged in synaptic vesicles.

It has been suggested that brain levels of GABA in some epileptic patients may be low. **Valproic acid** (2-propylpentanoic acid) apparently increases brain levels of GABA. The mechanism by which it does so is not clear. Valproic acid is metabolized primarily in the liver by glucuronidation and urinary excretion of the glucuronides, or by mitochondrial β oxidation and oxidation by endoplasmic reticulum enzymes.

Neuropeptides Are Derived from Precursor Proteins

Peptide neurotransmitters are generally synthesized as larger proteins that are cleaved by proteolysis to produce the neuropeptide molecules. Their synthesis requires the same biochemical machinery as any protein synthesis and takes place in

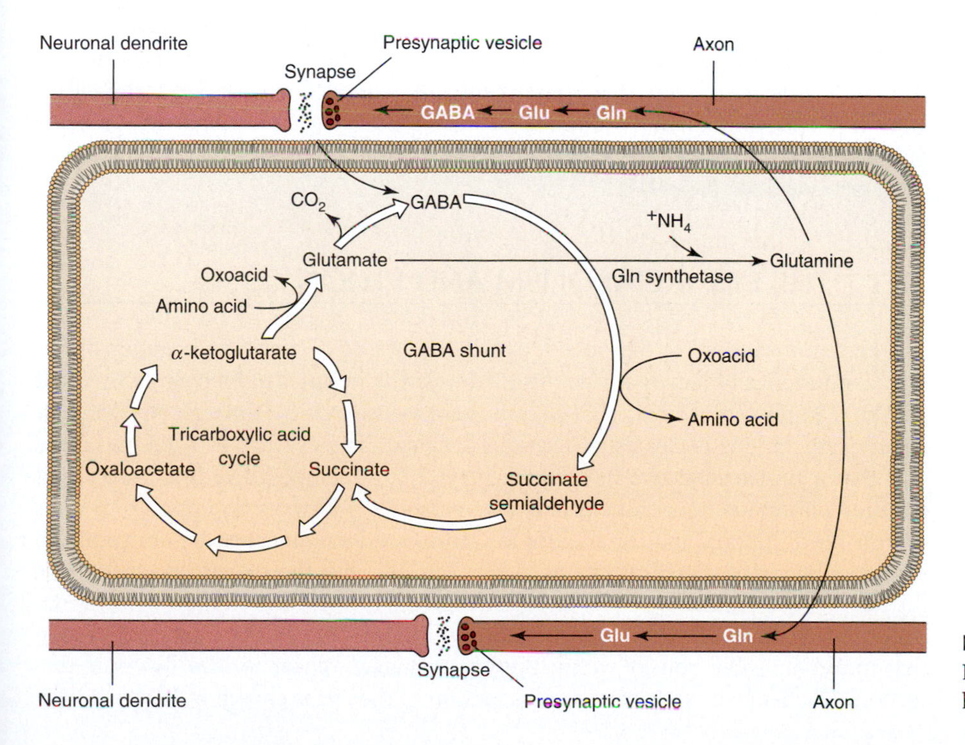

FIGURE 23.14
Involvement of the astrocytes in the metabolism of GABA and glutamate.

TABLE 23.3 Peptides Found in Brain Tissue[a]

Peptide	*Structure*
β-Endorphin	Y G G F M T S E K S Q T P L V T L F K N A I I K N A Y K K G E
Met-enkephalin	Y G G F M
Leu-enkephalin	Y G G F L

Somatostatin	A G C K N F F W
	        | |
	        C S T F T K

Luteinizing hormone–releasing hormone	p-E H W S Y G L R P G NH_2
Thyrotropin-releasing hormone	p-E H P-NH_2
Substance P	R P K P E E F F G L M-NH_2
Neurotensin	p-E L Y E N K P R R P Y I L
Angiotensin I	D R V Y I H P F H L
Angiotensin II	D R V Y I H P F
Vasoactive intestinal peptide	H S D A V F T D N Y T R L R- K E M A V K K Y L N S I L N-NH_2

[a]Peptides with "p" preceding the structure indicate that the N terminal is pyroglutamate. Those with NH_2 at the end indicate that the C terminal is an amide.

the cell body, not the axon. They travel down the axon to the presynaptic region by one of two generic mechanisms: **fast axonal transport** at a rate of about 400 mm day^{-1} and slow axonal transport at a rate of 1–5 mm day^{-1}. Since axons may vary in length from 1 mm to 1 m in length, theoretically the total transit time could vary from 150 ms to 200 days. It is highly unlikely that the latter transit time occurs under normal physiological conditions, and the upper limit is probably hours rather than days. There is some experimental evidence to suggest that the faster transit times prevail.

Neuropeptides mediate sensory and emotional responses such as those associated with hunger, thirst, sex, pleasure, and pain. Included in this category are **enkephalins, endorphins**, and **substance P**. Substance P is an excitatory neurotransmitter that has a role in pain transmission, whereas endorphins have roles in eliminating the sensation of pain. Some of the peptides found in brain tissue are shown in Table 23.3. Note that Met-enkephalin is derived from the N-terminal region of β-endorphin. The N-terminal or both the N- and C-terminal amino acids of many of the neuropeptide transmitters are modified. For a further discussion of these peptides, see page 918.

23.3 | THE EYE: METABOLISM AND VISION

The eye, our window to the outside world, allows us to view the beauties of nature, the beauties of life, and, *vide* this textbook, the beauties of biochemistry. What are the features of this organ that permit these views? A view through any window or any camera lens is clearest when unobstructed. The eye has evolved in such a way that a similar objective has been achieved. It is composed of live tissues that require continuous nourishment and use metabolic pathways appropriate to their unique needs. Energy and metabolites for growth and maintenance are derived from nutrients by conventional biochemical mechanisms, but the structures responsible for these processes are arranged and distributed so as not to interfere with the visual process. Also, the brain has devised an enormously efficient filtering system that makes invisible objects within the eye that may appear to lead to visual distortion. A schematic diagram of a cross section of the eye is shown in Figure 23.15.

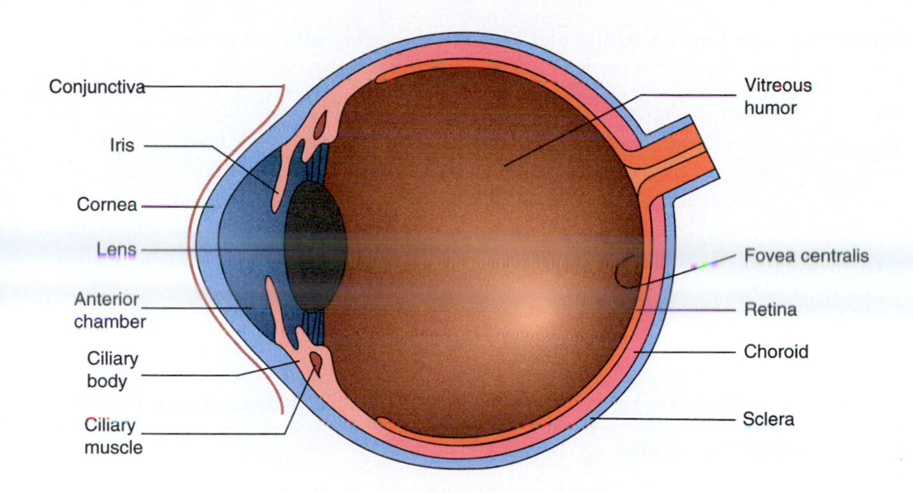

FIGURE 23.15
Schematic of a horizontal section of the left eye.

Light entering the eye passes progressively through the **cornea,** the anterior chamber that consists of aqueous humor, the lens, the vitreous body that consists of vitreous humor, and finally focuses on the **retina** that contains the visual sensing apparatus. Tears bath the exterior of the cornea while the interior is bathed by the aqueous humor, an isoosmotic fluid containing salts, albumin, globulin, glucose, and other constituents. The aqueous humor brings nutrients to the cornea and to the lens, and it removes end products of metabolism from them. The vitreous humor is a collagenous or gelatinous mass that helps maintain the shape of the eye while allowing it to remain somewhat pliable.

Cornea Derives ATP from Aerobic Metabolism

The eye is an extension of the nervous system and, like other tissues of the central nervous system, its major metabolic fuel is glucose. The cornea is not a homogeneous tissue and obtains a relatively large percentage of its ATP from aerobic metabolism. About 30% of the glucose used by the cornea is metabolized by glycolysis and about 65% by the pentose phosphate pathway (see p. 666). On a relative weight basis, the cornea has the highest activity of the pentose phosphate pathway of any other mammalian tissue. It also has a high activity of **glutathione reductase,** an enzyme that requires NADPH, a product of the pentose phosphate pathway. Corneal epithelium is permeable to atmospheric oxygen that is necessary for various oxidative reactions. The reactions of oxygen can result in the formation of various **active oxygen species** that are harmful to the tissues, perhaps in some cases by oxidizing protein sulfhydryl groups to disulfides. Reduced glutathione (GSH) is used to reduce those disulfide bonds back to their original native states while GSH itself is converted to oxidized glutathione (GSSG). Furthermore, GSSG may also be formed by autooxidation. Glutathione reductase uses NADPH to reduce GSSG to 2 GSH.

$$\text{GSSG} + \text{NADPH} + \text{H}^+ \xrightleftharpoons{\text{GSH reductase}} 2\ \text{GSH} + \text{NADP}^+$$

Activities of the pentose phosphate pathway and of glutathione reductase maintain this tissue in an appropriately reduced state by effectively neutralizing the active oxygen species.

Lens Consists Mostly of Water and Protein

The **lens** is bathed on one side by aqueous humor and supported on the other side by vitreous humor. It has no blood supply but is metabolically active. It gets nutrients from the aqueous humor and eliminates waste into the aqueous humor. The lens is mostly water and proteins. The majority of the proteins are the $\boldsymbol{\alpha}$, $\boldsymbol{\beta}$,

TABLE 23.4 Eye Lens Crystallins and Their Relationships with Other Proteins

Crystallin	*Distribution*	[Related] *or Identical*
α	All vertebrates	**Small heat shock proteins** (αB) [*Schistosoma mansoni* antigen]
β	All vertebrates	[*Myxococcus xanthus* protein S]
γ	(embryonic γ not in birds)	[*Physarum polycephatum* spherulin 3a]
Taxon-specific enzyme crystallins		
δ	Most birds, reptiles	**Argininosuccinate lyase** (δ2)
ε	Crocodiles, some birds	**Lactate dehydrogenase B**
ξ	Guinea pig, camel, llama	**NADPH: quinone oxidoreductase**
η	Elephant shrew	**Aldehyde dehydrogenase I**

Source: Wistow, G. *Trends Biochem. Sci.* 18:301, 1993.

and γ-crystallins. There are also albuminoids, enzymes, and membrane proteins that are synthesized in an epithelial layer around the edge of the lens. Some other types of proteins that are found in lens, including the lens of species other than vertebrates, are shown in Table 23.4. Lens proteins may have different genetic origins and different functions in other tissues. The most important physical requirement of these proteins is that they maintain a clear crystalline state. The central area of the lens, the core, consists of the lens cells that were present at birth. The lens grows from the periphery (Figure 23.16). The human lens increases in weight and thickness with age and becomes less elastic. This is accompanied by a loss of near vision (Table 23.5), a condition referred to as **presbyopia.** On average the lens may increase threefold in size and approximately $1\frac{1}{2}$-fold in thickness from birth to about age 80.

Lens proteins must be maintained in a native unaggregated state. They are sensitive to various insults such as changes in oxidation–reduction state, osmolarity, excessively increased concentrations of metabolites, and UV irradiation. Structural integrity of the lens is maintained by the Na^+/K^+ exchanging ATPase (see p. 522) for osmotic balance, glutathione reductase for redox state balance, and protein synthesis for growth and maintenance. Energy for these processes comes from the metabolism of glucose. About 85% of the glucose metabolized by the lens is by glycolysis, 10% by the hexose monophosphate pathway, and 3% by the tricarboxylic acid cycle, presumably by the cells located at the periphery.

Cataract is the only known disease of the lens. Cataracts are opacities of lenses brought about by a loss of osmolarity and a change in solubility of some of the pro-

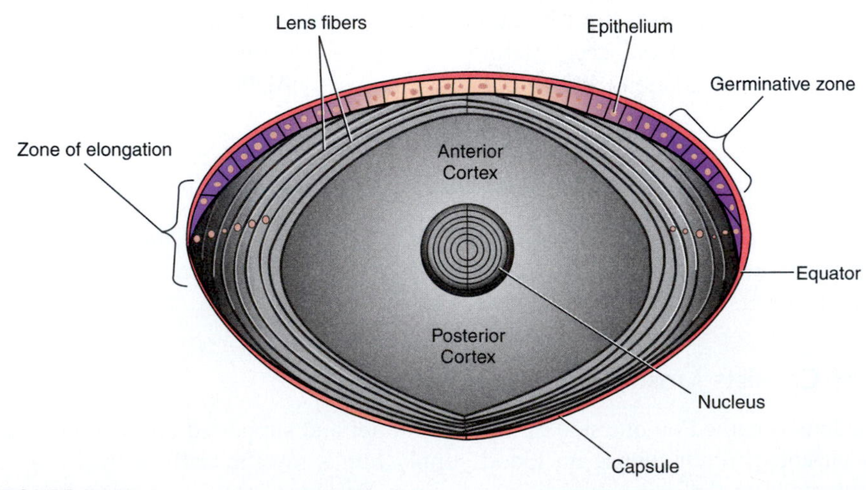

FIGURE 23.16
Schematic representation of a meridional section of a mammalian lens.

teins, resulting in regions of high light scatter. Cataracts affect about one million people per year in the United States, and there are no known cures or preventative measures. The remedy is lens replacement, a very common operation in the United States. There are basically two types of cataracts: **senile cataracts** and **diabetic cataracts.** Both are the result of changes in the solubility and aggregation state of the lens crystallins. In senile cataracts, changes in the architectural arrangement of the lens crystallins are age-related and due to such changes as breakdown of the protein molecules starting at the C-terminal ends, deamidation, and racemization of aspartyl residues. Diabetic cataracts result from increased osmolarity of the lens due to the activity of **aldose reductase** and **polyol (aldose) dehydrogenase** of the polyol metabolic pathway. When the glucose concentration in the lens is high, aldose reductase reduces some of it to **sorbitol** (Figure 23.17) that may be converted to **fructose** by polyol dehydrogenase. In human lens, the ratio of activities of these two enzymes favors sorbitol accumulation, especially since sorbitol is not used otherwise, and it diffuses out of the lens rather slowly. Accumulation of sorbitol in the lens increases osmolarity of the lens, affects the structural organization of the crystalline proteins within the lens, and enhances the rate of protein aggregation and denaturation. Areas where this occurs will have increased light scattering properties—which is the definition of cataracts. Normally, sorbitol formation is not a problem because the K_m of aldose reductase for glucose is about 200 mM and very little sorbitol would be formed. In diabetics, where the circulating concentration of glucose is high, activity of this enzyme can be significant.

TABLE 23.5 Changes in Focal Distance with Age

Age (years)	*Focal Distance (in.)*
10	2.8
20	4.4
35	9.8
45	26.2
70	240.0

Source: Adapted from Koretz, J. F. and Handelman, G. H. *Sci. Am.*, 92, July 1988.

Retina Derives ATP from Anaerobic Glycolysis

The **retina,** like the lens, depends heavily on anaerobic glycolysis for ATP production. Unlike the lens, the retina is a vascular tissue, but there are essentially no blood vessels in the area where visual acuity is greatest, the **fovea centralis** (see Clin. Corr. 23.3). Mitochondria are present in retinal cells, rods, and cones but no mitochondria are in the outer segments of the rods and cones where the visual pigments are located.

NADH produced during glycolysis reduces pyruvate to lactate. Lactate dehydrogenase of the retina uses either NADH or NADPH. It is not clear whether lac-

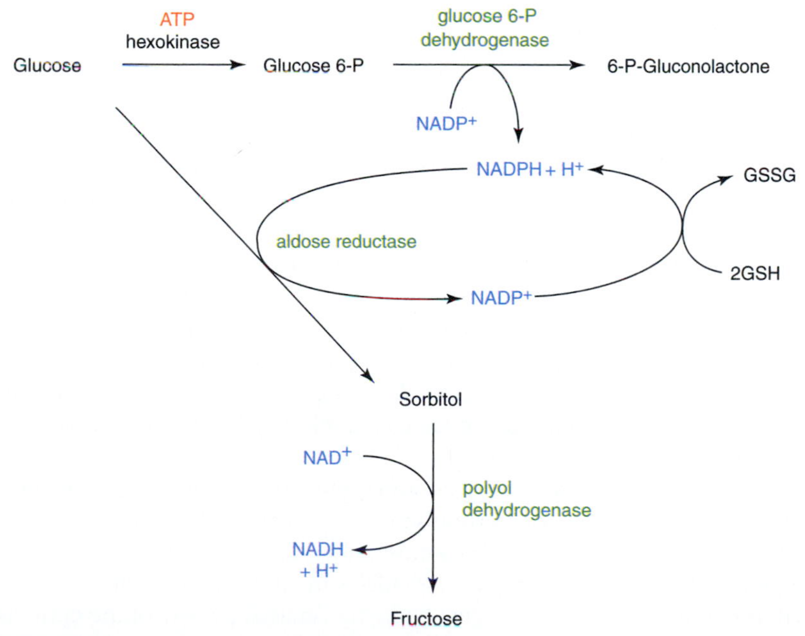

FIGURE 23.17
Metabolic interrelationships of lens metabolism.

CLINICAL CORRELATION 23.3

Macula Degeneration and Other Causes of Loss of Vision

Many diseases of the eye affect vision, not all of which have clear, direct biochemical origins. The most serious eye diseases are those that result in blindness. Glaucoma is the most common and has a direct causal relationship with diabetes, the biochemistry of which is fairly well known. Glaucoma can be treated and blindness does not have to be a result. Macula degeneration leads to blindness and there is no cure. The macula is a circular area of the retina, the center of which is the fovea centralis, the area containing the greater concentration of cones and the one of greatest visual acuity. Macula degeneration may be among the leading causes of blindness of people over the age of 50. Macula degeneration is of two types: dry and wet. The dry form develops gradually over time, whereas the wet form develops rapidly and can lead to blindness within days. Macula degeneration occurs when blood vessels rupture under the macula, leading to a loss of the nutrient supply and a rapid loss of vision. Experimental procedures are in progress to remove surgically scar tissue that develops and to transplant tissue from the rear of the eye to restore nourishment to the photoreceptor cells.

Rupture of blood vessels that obscure macula details and result in rapid onset of blindness may be temporary in some cases. Six cases of sudden visual loss associated with sexual activity have been reported that were not associated with a sexually transmitted disease. Vision was lost in one eye apparently during, but most often was reported a few days after engaging in, "highly stimulatory" sexual activity. Blindness was due to rupture of blood vessels in the macula area. When patients did see an ophthalmologist, most were reluctant to discuss what they were doing when sight lost was first observed. Four of the patients recovered with restoration of vision upon reabsorption of blood. In one case, where blood was trapped between the vitreous humor and the retinal surface directly in front of the fovea, the hemorrhage cleared only slightly during the next month, but visual acuity did not improve. The patient did not return for a follow-up examination, but there was no indication that the condition was permanent. Since most of the persons affected by this phenomenon were over the age of 39, it may be a worry more to professors than to students. It also may give a new meaning to the phrase, "love is blind."

Friberg, T. R., Braunstein, R. A., and Bressler, N. M. *Arch. Ophthalmol.* 113:738, 1995.

tate dehydrogenase of the retina plays a role in regulation of glucose metabolism through the glycolytic or hexose monophosphate pathways by its selective use of NADH or NADPH.

Visual Transduction Involves Photochemical, Biochemical, and Electrical Events

Figure 23.18 shows an electron micrograph and schematic of the retinal membrane. Light entering the eye through the lens passes the optic nerve fibers, the ganglion neurons, the bipolar neurons, and nuclei of rods and cones before it reaches the outer segment of the rods and cones where the signal transduction process begins. The **pigmented epithelial** layer, the choroid, lies behind the retina, absorbs the excess light, and prevents reflections back into the rods and cones where it may cause distortion or blurring of the image (see Clin. Corr. 23.4).

The eye may be compared to a video camera, which collects images, converts them into electrical pulses, records them on magnetic tape, and allows their visualization by decoding the taped information. The eye focuses on an image by projecting that image onto the retina. A series of events begins, the first of which is photochemical, followed by biochemical events that amplify the signal, and finally electrical impulses are sent to the brain where the image is reconstructed in "the mind's eye." During this process, the initial event has been transformed from a physical event to a chemical event through a series of biochemical reactions, to an electrical event, to a conscious acknowledgment of the presence of an object in the environment outside the body.

Photons of light are absorbed by photoreceptors in the **outer segments** of **rods** or **cones**, where they cause isomerization of the visual pigment, **retinal,** from the 11-*cis*-form to the all-*trans*-form. This isomerization causes a conformational change in the protein moiety of the complex and affects the resting membrane potential of the cell, resulting in an electrical signal being transmitted by way of the optic nerve to the brain. These processes will be discussed later in more detail.

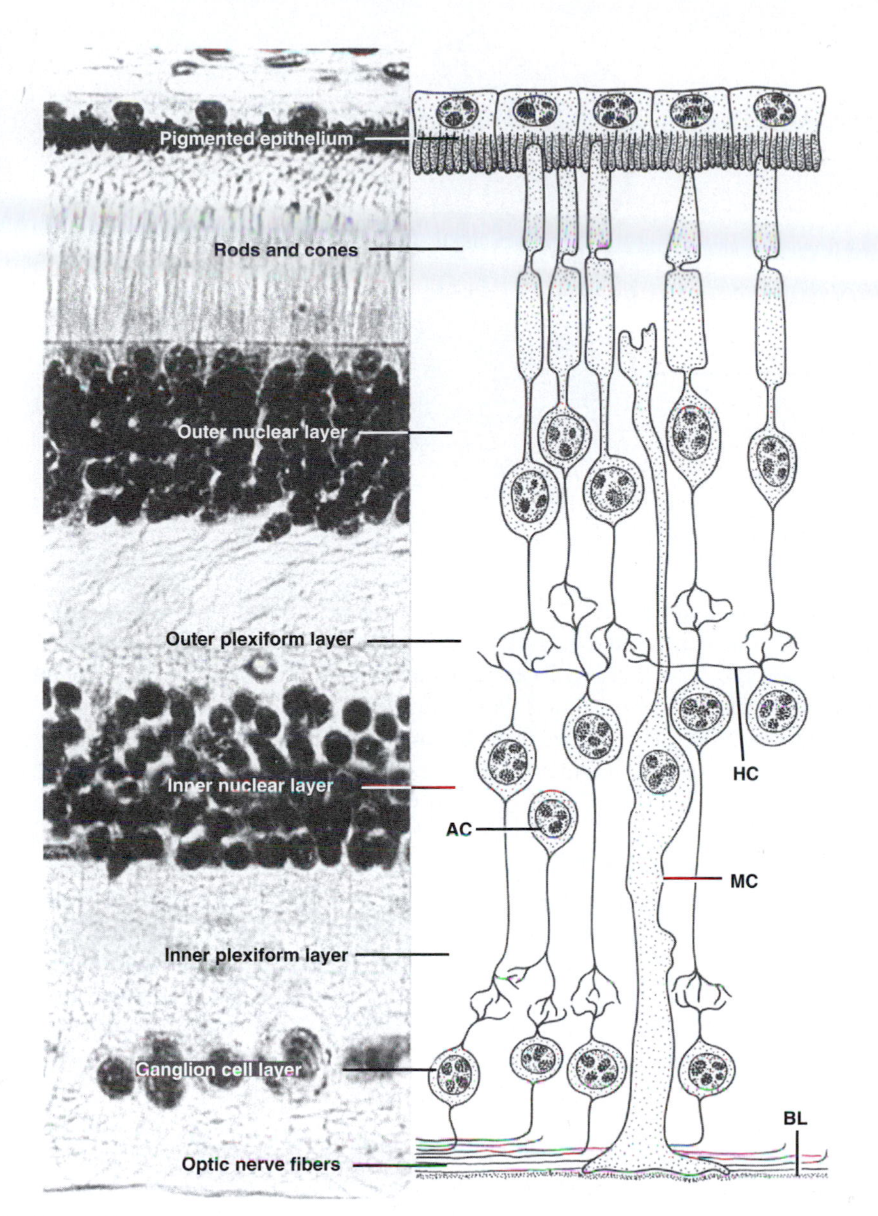

FIGURE 23.18
Electron micrograph and schematic representation of cells of the human retina.
Tips of rods and cones are buried in the pigmented epithelium of the outermost layer. Rods and cones form synaptic junctions with many bipolar neurons, which in turn form synapses with cells in the ganglion layer that send axons through the optic nerve to the brain. The synapse of a rod or cone with many cells is important for the integration of information. HC, horizontal cells; AC, amacrine cell; MC, Muller cell; BL, basal lamina.
Reprinted with permission from Kessel, R. G. and Kardon, R. H. Tissues and Organs: A Text-Atlas of Scanning Electron Microscopy. *New York: W. H. Freeman, 1979, p. 87. Copyright by R. ©. Kessel and R. H. Kardon, all rights reserved.*

CLINICAL CORRELATION 23.4
Niemann–Pick Disease and Retinitis Pigmentosa

There are central nervous system disorders associated with the Niemann–Pick group of diseases that can become evident by ocular changes. Some of these are observed as abnormal macula with gray discoloration and granular pigmentation or granule opacities about the fovea.

Acute type I Niemann–Pick disease, lipidosis with sphingomyelinase deficiency and primary sphingomyelin storage, may show a cherry red spot in the retina in as many as 50% of patients. Macula halo syndrome applies to the crystalloid opacities seen in some patients with subacute type I disease. They form a halo approximately one-half the disk diameter at their outer edge and are scattered throughout the various layers of the retina. They do not interfere with vision.

In an 11-year-old girl who had type II disease, more extensive ocular involvement was observed. There was sphingomyelin storage in the keratocytes of the cornea, the lens, the retinal ganglion cells, the pigmented epithelium, the corneal tract, and the fibrous astrocytes of the optic nerve.

Retinitis pigmentosa is a secondary effect of the abnormal biochemistry associated with Niemann–Pick disease.

Spence, M. W. and Callahan, J. W. Sphingomyelin-cholesterol lipidoses: the Niemann–Pick group of diseases. In: C. R. Schriver, A. L. Beaudet, W. Sly, and D. Valle (Eds.), *The Metabolic Basis of Inherited Disease.* New York: McGraw-Hill, 1989, p. 1656.

Rods and Cones Are Photoreceptor Cells

The **photoreceptor cells** of the eye are the rods and the cones (Figure 23.18). Each type has flattened disks that contain a photoreceptor pigment. This pigment is **rhodopsin** in the rod cells, and red, green, or blue pigment in the cone cells. Rhodopsin is a transmembrane protein to which is bound a prosthetic group, **11-*cis*-retinal.** Rhodopsin minus its prosthetic group is **opsin.** The three proteins that form the red, green, and blue pigments of cone cells are different from each other and from opsin.

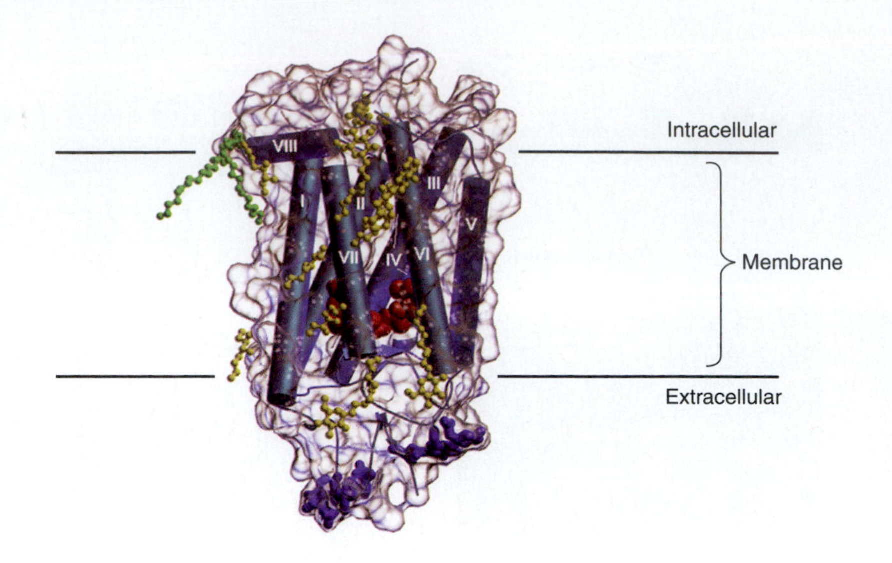

FIGURE 23.19
Crystal structure of bovine rhodopsin at 2.8-Å resolution.
Rhodopsin is a transmembrane protein. The width of the membrane into which rhodopsin is imbedded is approximately equivalent to the length of the helices (blue rods). The intracellular side of the membrane approximately transects helix VIII. β-Strands are shown as blue arrows. Structures in green on the intracellular side are two palmitoyl groups oriented such that the hydrophobic groups can interact with hydrophobic regions of the membrane. The blue ball and stick structures at the bottom (extracellular side) of the molecule are carbohydrates. The yellow structures located near the hydrophobic surface of the protein are nonylglucoside and heptaneol molecules.
Reprinted with permission from Teller, D. C., Okada, T., Behnke, C. A., Palczewski, K., and Stenkamp, R.. E. Biochemistry 40: 7761, 2001. Copyright, 2001, American Chemical Society. Figure generously supplied by Dr. K. Palczewski and Dr. C. A. Behnke.

Rhodopsin, about 40 kDa, contains seven transmembrane α-helices. An 11-*cis*-retinal is attached through a protonated Schiff base to the ε-amino group of Lys^{296} on the seventh helix that lies about midway between the two faces of the membrane (Figure 23.19). See also Clinical Correlation 23.5.

Formation of 11-*cis*-retinal from **β-carotene** and rhodopsin from opsin and 11-*cis*-retinal are shown in Figure 23.20. The 11-*cis*-retinal is derived from **vitamin A** and/or β-carotene of the diet. These are transported to specific sites in the body while attached to specific carrier proteins. Cleavage of β-carotene yields two molecules of **all-*trans*-retinol.** There is an enzyme in the pigmented epithelial cell layer of the retina that catalyzes the isomerization of all-*trans*-retinol to **11-*cis*-retinol.** Oxidation of the 11-*cis*-retinol to 11-*cis*-retinal and its binding to opsin occur in the rod outer segment.

The absorption spectra of 11-*cis*-retinal and the four visual pigments are shown in Figure 23.21. There is a shift in the wavelength of maximum absorption of 11-*cis*-retinal upon binding to opsin and the protein components of the other visual pigments. Absorption bands for the pigments are coincident with their light sensitivity.

The magnitude of change in the electrical potential of photoreceptor cells following exposure to a light pulse differs in magnitude from that of neurons during depolarization. The **resting potential** of a rod cell membrane is approximately −30 mV compared to −70 mV for neurons. Excitation of rod cells causes **hyperpolarization** of the membrane, from about −30 mV to about −35 mV (Figure 23.22). It takes hundreds of milliseconds for the potential to reach its maximum state of hyperpolarization. A number of biochemical events occur during this time interval and before the potential returns to its resting state.

The initial events, absorption of photons and the subsequent isomerization of 11-*cis*-retinal, are rapid, requiring only picoseconds. Following this, changes occur in rhodopsin, leading to various short-lived conformational states (Figure 23.23), each with its specific absorption characteristics. The exact structures for these intermediates are not known. It is clear that they result from the conformational changes the protein moiety of rhodopsin undergoes when 11-*cis*-retinal is isomerized to all-*trans*-retinal. Figure 23.24, a schematic from the 3-D structure of rhodopsin, shows the amino acid side chains surrounding 11-*cis*-retinal as viewed from the cytoplasmic side. When 11-*cis*-retinal undergoes isomerization to the all-*trans*-retinal, the β-ionone ring of all-*trans*-retinal reaches Ala^{169} in helix IV. The protein must undergo several conformational changes for this to occur. The intermediate species listed in Figure 23.23 represent some of those conformations. Finally, rhodopsin dissociates yielding opsin and all-*trans*-retinal.

CLINICAL CORRELATION 23.5

Retinitis Pigmentosa from a Mutation of the Gene for Peripherin

A group of heterogeneous diseases of variable clinical and genetic origins have been placed under the category of retinitis pigmentosa (RP). Several have origins in abnormal lipid metabolism. Approximately 1.5 million people throughout the world are affected by this disease. It is a slowly progressive condition associated with loss of night and peripheral vision. It can be inherited through an autosomal dominant, recessive, or X-linked mode. RP has been associated with mutations in the protein moiety of rhodopsin and in a related protein, peripherin/RDS, both of which are integral membrane proteins. Peripherin is a 344 amino acid residue protein located in the rim region of the disk membrane. Structural models of these two proteins are shown in the figure below. Filled circles and other notations in the figure mark residues or regions that have been correlated with RP or other retinal degenerations.

A case has been described where a *de novo* mutation in exon 1 of the gene coding for peripherin resulted in the onset of RP. Using molecular biological techniques, Lam et al. (1995) found the specific change in peripherin to be a C-to-T transition in the first nucleotide of codon 46. This resulted in changing an arginine to a stop codon (R46X). The pedigree of this family is shown in the figure below. Neither parent had the mutation and genetic typing analysis (20 different short tandem repeat polymorphisms) showed that the probability that the proband's parents are not his actual biological parents is less than 1 in 10 billion. This establishes with near certainty that the mutation is *de novo*.

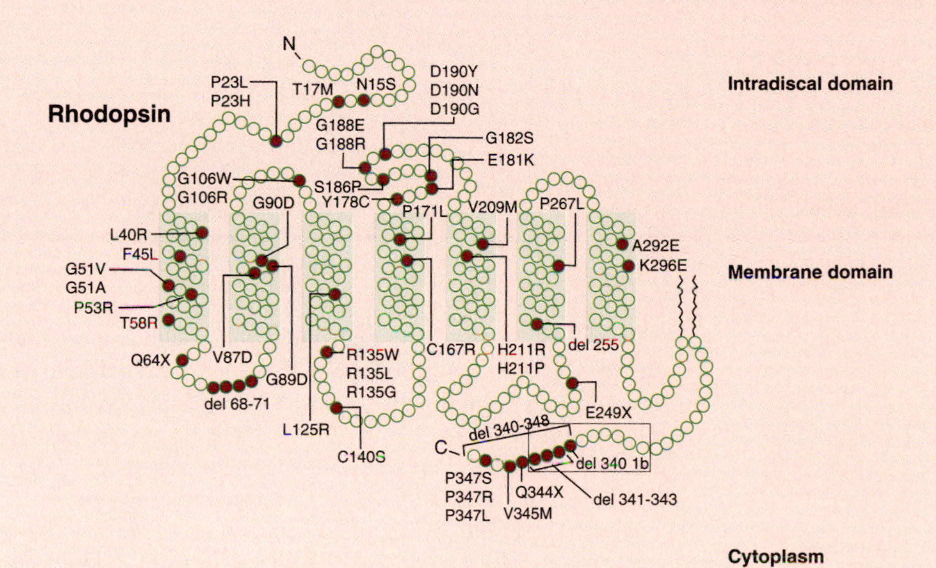

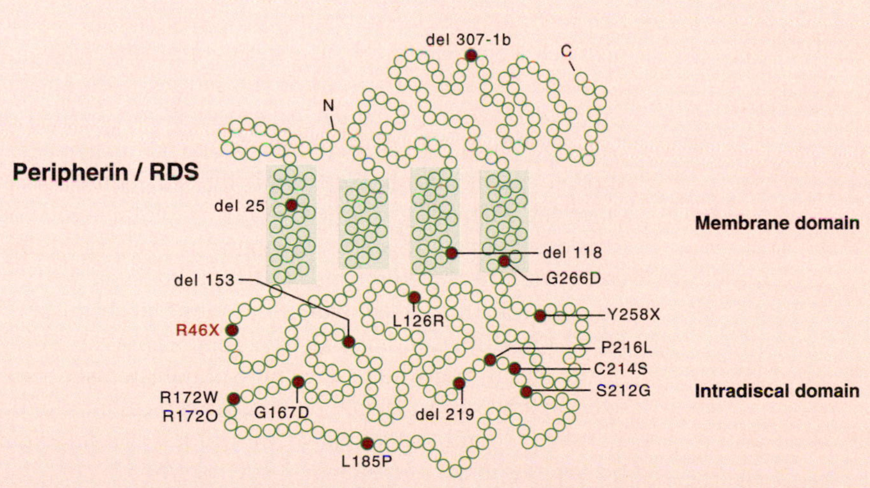

Schematic representation of structural models for rhodopsin (top) and peripherin/RDS (bottom).
The location of mutations in amino acid residues that segregate with RP or other retinal degenerations are shown as solid circles.
From Lam et al. (1995).

Continued on page 1010.

Clinical Correlation 23.5 (continued)

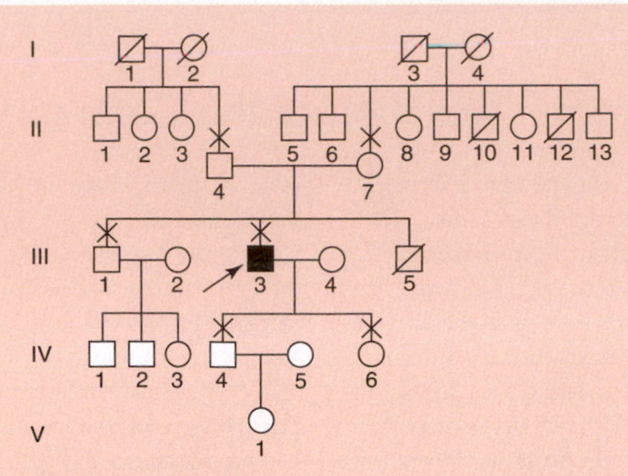

Pedigree of family.
Males are squares, females are circles. Solid square indicates the proband. A slash through symbol indicates deceased.
From Lam et al. (1995).

This R46X mutation has been observed in another unrelated patient. These observations demonstrate the importance of the use of DNA analysis to establish the genetic basis for RP, especially considering that RP symptoms have been associated with a variety of other diseases, such as those related to abnormal lipid metabolism.

Shastry B. S. Retinitis pigmentosa and related disorders: phenotypes of rhodopsin and peripherin/RDS mutations. *Am. J. Med. Genet.* 52:467, 1994: and Lam, B. L., Vandenburgh, K., Sheffield, V. C., and Stone, E. M. Retinitis pigmentosa associated with a dominant mutation in codon 46 of the peripherin/RDS gene (Arg^{46}stop). *Am. J Ophthalmol.* 119:65, 1995.

At 37°C, activated rhodopsin decays in slightly more than 1 ms through several intermediates to **metarhodopsin II,** which has a half-life of approximately 1 min. It is the **active rhodopsin** species, R*, that is involved in the biochemical reactions of interest. Metarhodopsin II will have begun to form within hundredths of microseconds of the initial event. All of the first series of reactions shown in Figure 23.23 occur in the disk of the rod outer segment. Upon dissociation of metarhodopsin into opsin and all-*trans*-retinal, the all-*trans*-retinal is enzymatically converted to all-*trans*-retinol by **all-*trans*-retinol dehydrogenase** that is located in the rod outer segment. All-*trans*-retinol is transported (or diffuses) into the pigmented epithelium, where a specific isomerase converts it to 11-*cis*-retinol. The 11-*cis*-retinol is then transported (or diffuses) back into the rod outer segment and is reoxidized to 11-*cis*-retinal. Since the all-*trans*-retinol dehydrogenase appears to have only about 6% as much activity with 11-*cis*-retinal as substrate, it appears that another enzyme may be responsible for oxidation of 11-*cis*-retinol to 11-*cis*-retinal. Once the aldehyde is formed, it can recombine with opsin to form rhodopsin. Rhodopsin is now in a state to begin the cycle again. Similar events occur in the cones with the three proteins of the red, green, and blue pigments.

There are three interconnecting "mini" biochemical cycles in the conversion of light energy to nerve impulses (Figure 23.25). These cycles describe the reactions of rhodopsin, **transducin,** and **phosphodiesterase,** respectively. The net result is to cause a hyperpolarization of the plasma membrane of the rod (or cone) cells, that is, from −30 mV to approximately −35 mV. It is important to understand how the plasma membrane is maintained at −30 mV.

Rod cells of a fully dark-adapted human can detect a flash of light that emits as few as five photons. The rod is a specialized type of neuron in that the signal generated does not depend on an all-or-none event. The signal may be graded in intensity and reflect the extent that the millivolt potential changes from its steady-state value of −30 mV. This **steady-state potential** is maintained at a more positive value because **Na^+ channels** of the photoreceptor cells are **ligand-gated** and are maintained in a partially opened state. The ligand responsible for keeping some of the Na^+ channels open is **cyclic GMP (cGMP).** cGMP binds to them in a

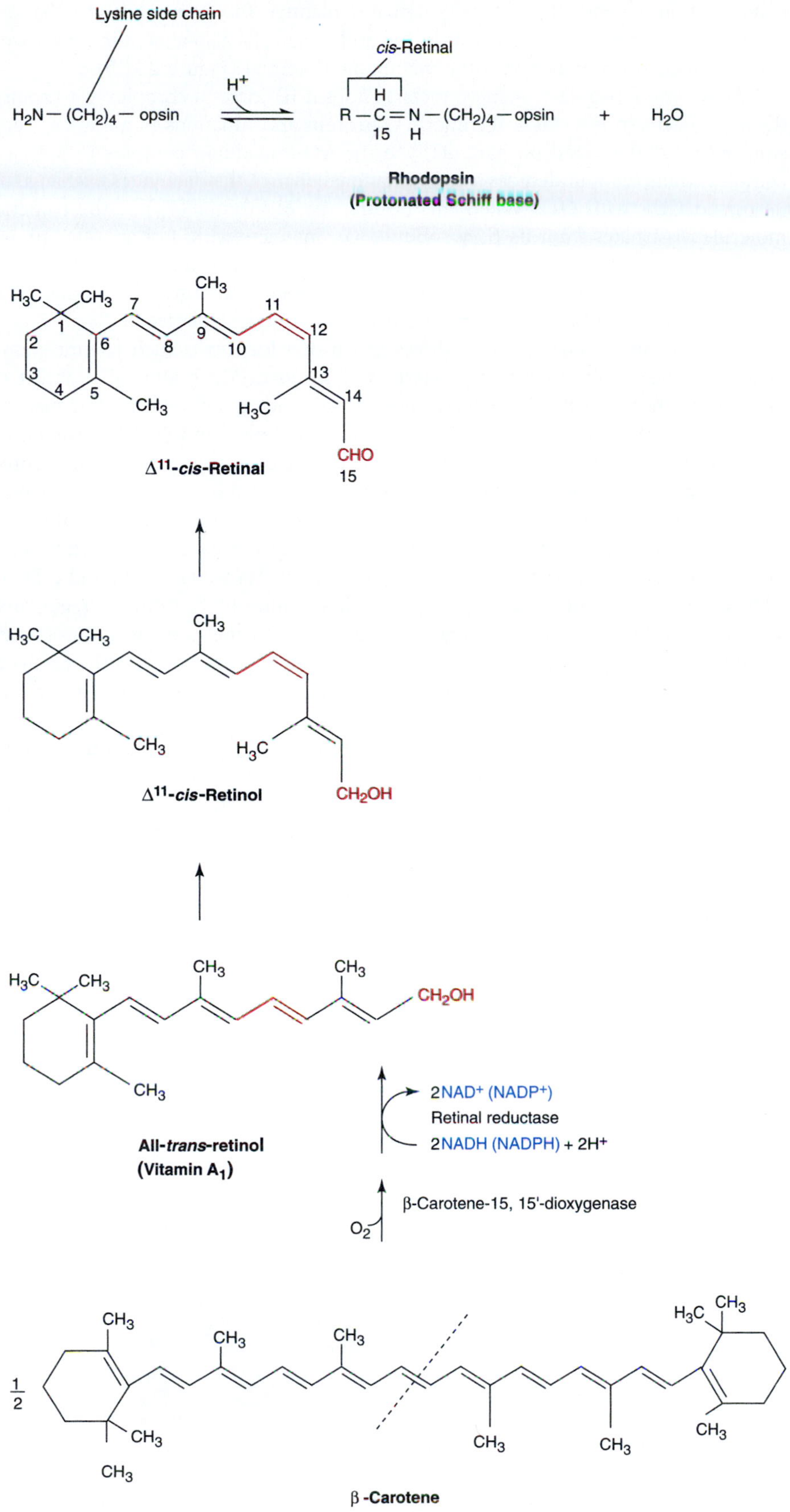

FIGURE 23.20
Formation of 11-*cis*-retinal and rhodopsin from β-carotene.

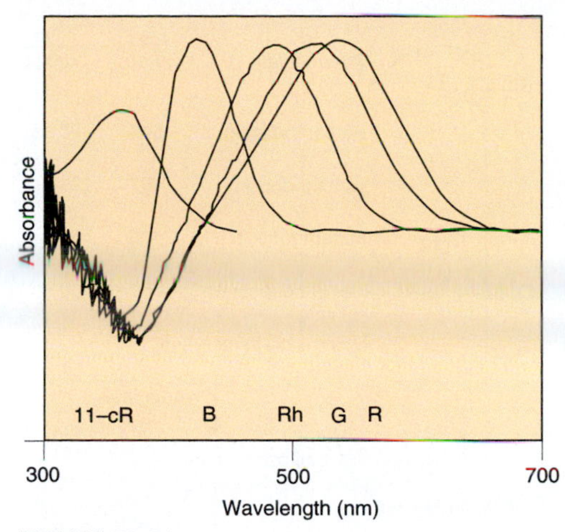

FIGURE 23.21
Absorption spectra of 11-*cis*-retinal and the four visual pigments.
Absorbance is relative and was obtained for pigments as difference spectra from reconstituted recombinant apoproteins. The spectrum for 11-*cis*-retinal (11-cR) is in the absence of protein. Other abbreviations: B, blue pigment; Rh, rhodopsin; G, green; R, red.
Modified from Nathans, J. Cell *78:357, 1994.*

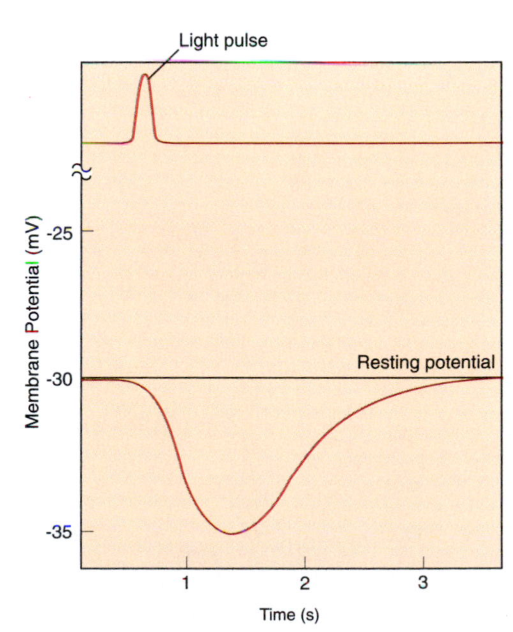

FIGURE 23.22
Changes in the potential of a rod cell membrane after a light pulse.
Redrawn from Darnell, J., Lodish, H., and Baltimore, D. Molecular Cell Biology. *New York: Scientific American Books, 1986, p. 763.*

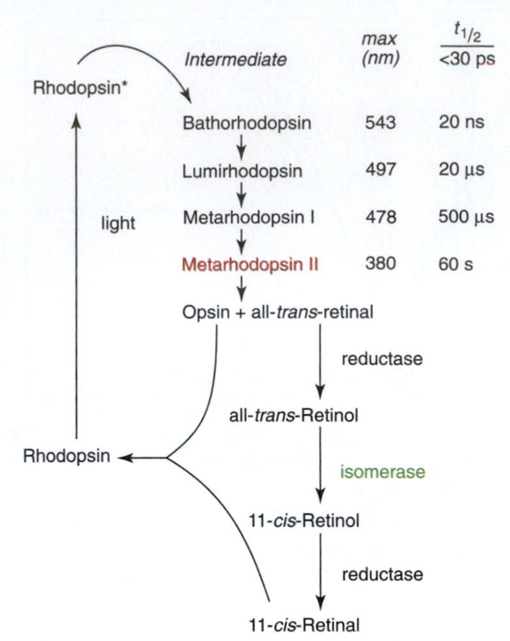

FIGURE 23.23
Light activation of rhodopsin.

concentration-dependent, kinetically dynamic manner. Biochemical events that affect the concentration of cGMP within rod and cone cells also affect the number of Na^+ channels open and, hence, the membrane potential (Figure 23.25).

Active rhodopsin (R*, namely, metarhodopsin II) forms a complex with transducin. Transducin is a classical trimeric **G-protein** and functions in a manner very similar to that described on page 925. In the R*–transducin complex ($R^*\text{–}T_{\alpha,\beta,\gamma}$ complex) transducin undergoes a conformational change that facilitates exchange of its bound GDP with GTP. When this occurs, the α subunit (T_α) of the trimeric molecule dissociates from its β, γ subunits. T_α interacts with and activates **phosphodiesterase (PDE),** which hydrolyzes cGMP to 5′-GMP, resulting in a decreased concentration of cGMP and a decrease in the number of channels held open. The membrane potential becomes more negative, that is, hyperpolarized.

The diagram of Figure 23.25 shows in cartoon form two such channels embedded in the plasma membrane, one of which has cGMP bound to it and it is open. The other does not have cGMP bound to it and it is closed. By this mechanism, the concentration of Na^+ in the cell is directly linked to the concentration of cGMP and, thus, also the membrane potential. PDE in rod cells is a **heterotetrameric protein** consisting of one each α and β catalytic subunits and two γ regulatory subunits. T_α-GTP forms a complex with the γ subunits of PDE resulting in their dissociation from the catalytic subunits, freeing the catalytically active α,β-dimeric PDE subunit complex. T_α has GTPase activity. Hydrolysis of bound GTP to GDP and inorganic phosphate (P_i) results in dissociation of T_α from the regulatory γ subunits of PDE, permitting them to reassociate with the catalytic subunits and to inhibit the PDE activity. The same reactions occur in cone cell, but the catalytic subunit of cone cell PDE is composed of two α catalytic subunits instead of α,β subunits as are present in rod cells.

cGMP concentration is regulated by intracellular Ca^{2+} concentration. Calcium enters rod cells in the dark through sodium channels, increasing its concentration to the 500-nM range. At these concentrations, activity of **guanylate cyclase** is low. When sodium channels are closed, Ca^{2+} entry is inhibited, but efflux mediated by the sodium/calcium–potassium exchanger is unchanged (top complex of the plasma membrane in Figure 23.25). This results in a decrease in the intracellular Ca^{2+} concentration and leads to activation of guanylate cyclase and increased production of cGMP from GTP.

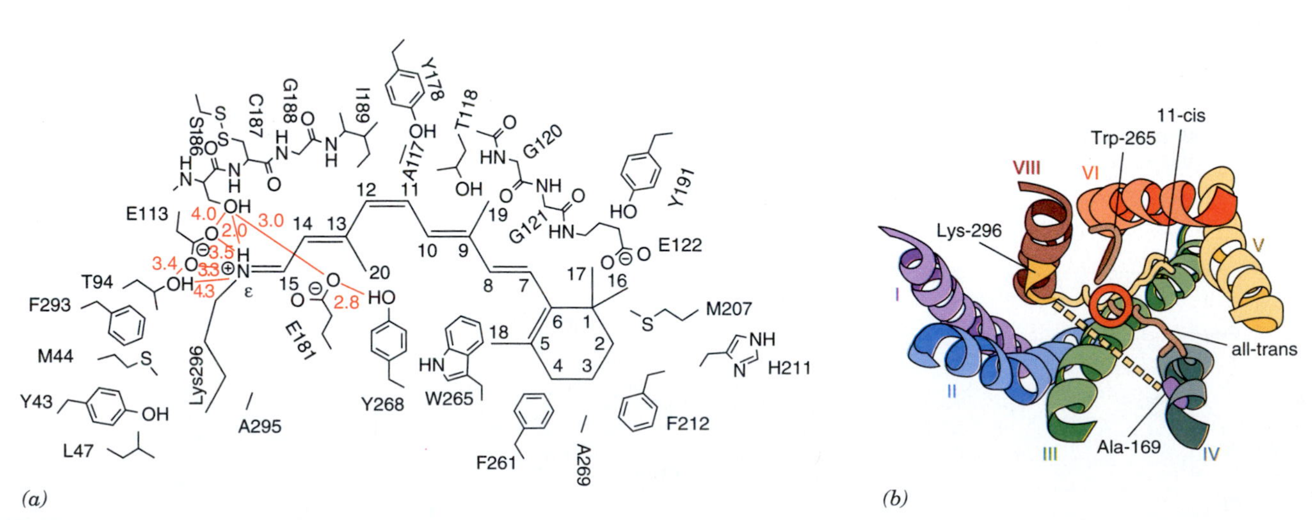

FIGURE 23.24
Schematics of 11-*cis*-retinal and its all-*trans* form.
(*a*) The amino acid side chains surrounding 11-*cis*-retinal. When 11-*cis*-retinal isomerizes to the all-*trans* form (*b*), the β-ionone ring of all-*trans*-retinal interacts with Ala[169] of helix IV.

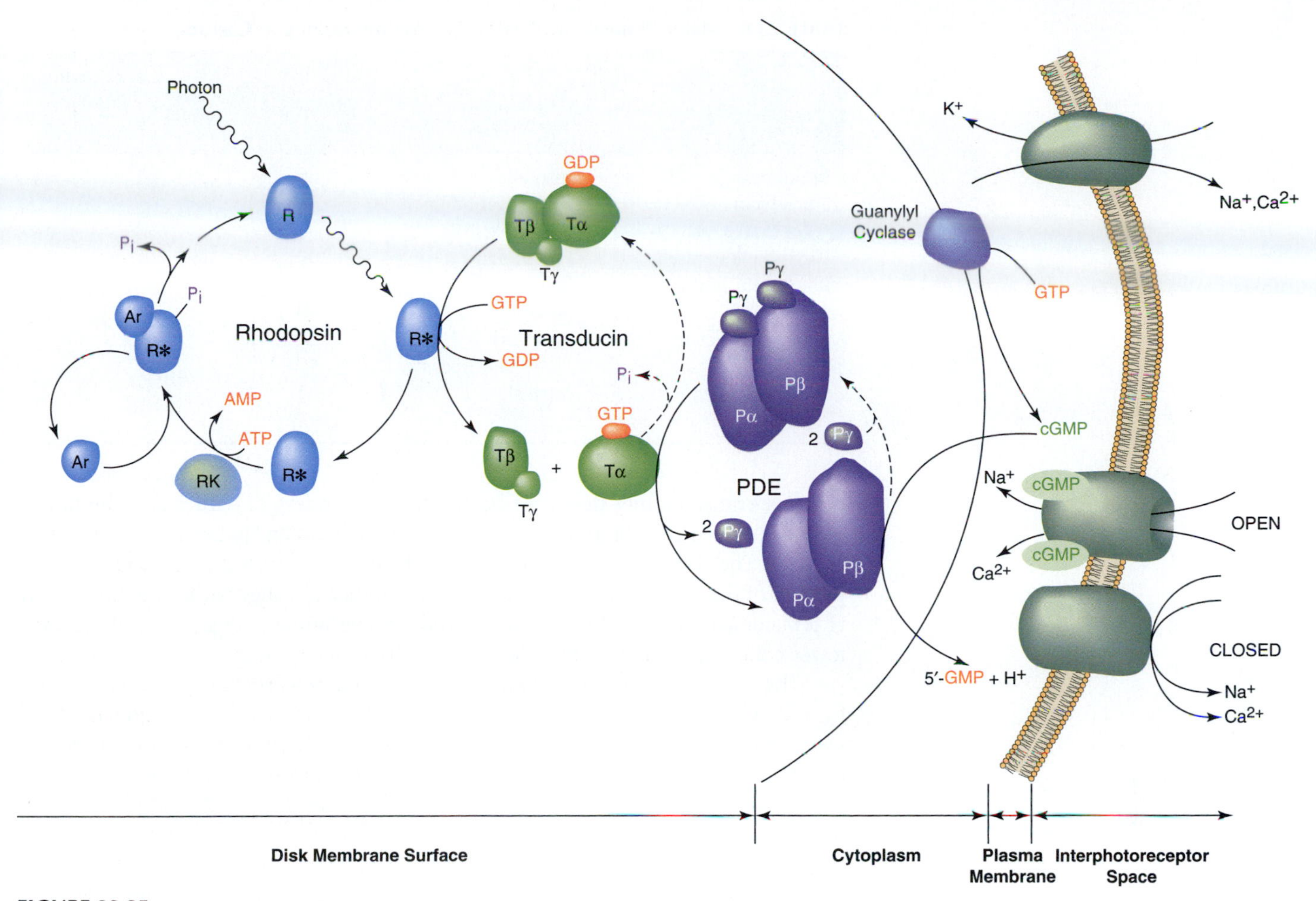

FIGURE 23.25
Cascade of biochemical reactions involved in the visual cycle.
Redrawn from Farber, D. B. Invest. Ophthalmol. Vis. Sci. *36:263, 1995.*

Resynthesis of cGMP and the hydrolysis of T_α-GTP play important roles in stopping the reactions of the visual cycle. Inactivation of activated rhodopsin, R*, is also very important. Activated rhodopsin, R*, is phosphorylated by **rhodopsin kinase** in the presence of ATP (Figure 23.25). The R*–P_i has high binding affinity for the cytosolic protein **arrestin.** The arrestin–R*–P_i complex is no longer capable of interacting with transducin. The kinetics of arrestin binding to the activated-phosphorylated rhodopsin is sufficiently rapid *in vivo* to stop the cascade of reactions.

Rhodopsin is regenerated through another series of reactions and the cycle can be initiated again by photons of light. Figure 23.23 shows that the series of reactions leading to the regeneration of rhodopsin includes the dissociation of all-*trans*-retinal from metarhodopsin. Regeneration of 11-*cis*-retinal from all-*trans*-retinal occurs by reactions previously described and occurs before it is used again to form rhodopsin.

Some of the major proteins involved in the visual cycle are listed in Table 23.6. Mutations can occur in virtually any one of them and result in visual impairment. In fact, a recently observed recessive retinitis pigmentosa appears to be associated with a human gene that encodes an also recently observed δ subunit of cGMP. The disease has been characterized by an abnormal retinal cGMP metabolism.

Color Vision Originates in the Cones

Even though photographic artists, such as the late Ansel Adams, make the world look beautiful in black and white, the intervention of colors in the spectrum of life's

TABLE 23.6 Major Proteins Involved in the Phototransduction Cascade

Protein	Relation to Membrane	Molecular Mass (kDa)	Concentration in Cytoplasm (μM)
Rhodopsin	Intrinsic	39	—
Transducin ($\alpha + \beta + \gamma$)	Peripheral or soluble	80	500
Phosphodiesterase	Peripheral	200	150
Rhodopsin kinase	Soluble	65	5
Arrestin	Soluble	48	500
Guanylate cyclase	Attached to cytoskeleton	?	?
cGMP-activated channel	Intrinsic	66	?

pictures brings another degree of beauty to the wonders of nature and the beauty of life . . . even the ability to make a distinction between tissues from histological staining. The ability of humans to distinguish colors resides within a relatively small portion of the visual system, the cones. The number of cones within the human eye is few compared with the number of rods. Some animals (e.g., dogs) have even fewer cones, and other animals (e.g., birds) have many more.

The mechanism by which light stimulates cone cells is exactly the same as it is for rod cells. There are three types of cone cells, defined by the blue, green, or red visual pigments they contain. They are also referred to as short (S), medium (M), or long (L) wavelength-specific opsins, respectively. Normally, only one type of visual pigment occurs per cell. During fetal development, however, multiple pigments may occur in the same cell. Under those conditions, the blue pigment seems to appear first followed by both green and red pigments. The number of cones containing these multiple pigments decreases after birth and can hardly be found at all in adults. The blue pigment has optimum absorbance at 420 nm, green pigment at 535 nm, and red pigment at 565 nm (Figure 23.21). Each of these pigments has 11-*cis*-retinal as the prosthetic group, and, when activated by light, 11-*cis*-retinal isomerizes to all-*trans*-retinal as in rod cells. Colors other than those of the visual pigments are distinguished by graded stimulation of the different cones and comparative analysis by the brain. Color vision is **trichromatic.**

Color discrimination by cone cells is an inherent property of the proteins of the visual pigments to which the 11-*cis*-retinal is attached. The 11-*cis*-retinal is attached to each of the proteins through a protonated Schiff base. The absorption spectrum produced by the conjugated double-bond system of 11-*cis*-retinal is influenced by its environment (Figure 23.21). When 11-*cis*-retinal is bound to different visual proteins, amino acid residues in the local areas around the protonated base and the conjugated π-bond system influence the energy level and give different absorption spectra with absorption maxima that are different for the different color pigments.

The genes for the visual pigments have been cloned and their amino acid sequences inferred from the gene sequences. A comparison of these sequences is presented in Figure 23.26. Open circles represent amino acids that are the same, and closed circles represent amino acids that are different. A string of closed circles at either end may represent an extension of the chain of one protein relative to the other. Red and green pigments show the greatest degree of homology, about 96% identity, whereas the degree of homology between different pairs of the others is between 40% and 45%.

Genes encoding the visual pigments have been mapped to specific chromosomes (see Clin. Corr. 23.6). The rhodopsin gene resides on **chromosome 3,** the gene encoding the blue pigment resides on **chromosome 7,** and the two genes for the red and green pigments are on the **X chromosome.** In spite of their great similarity, the red and green pigments are distinctly different proteins. Individuals have

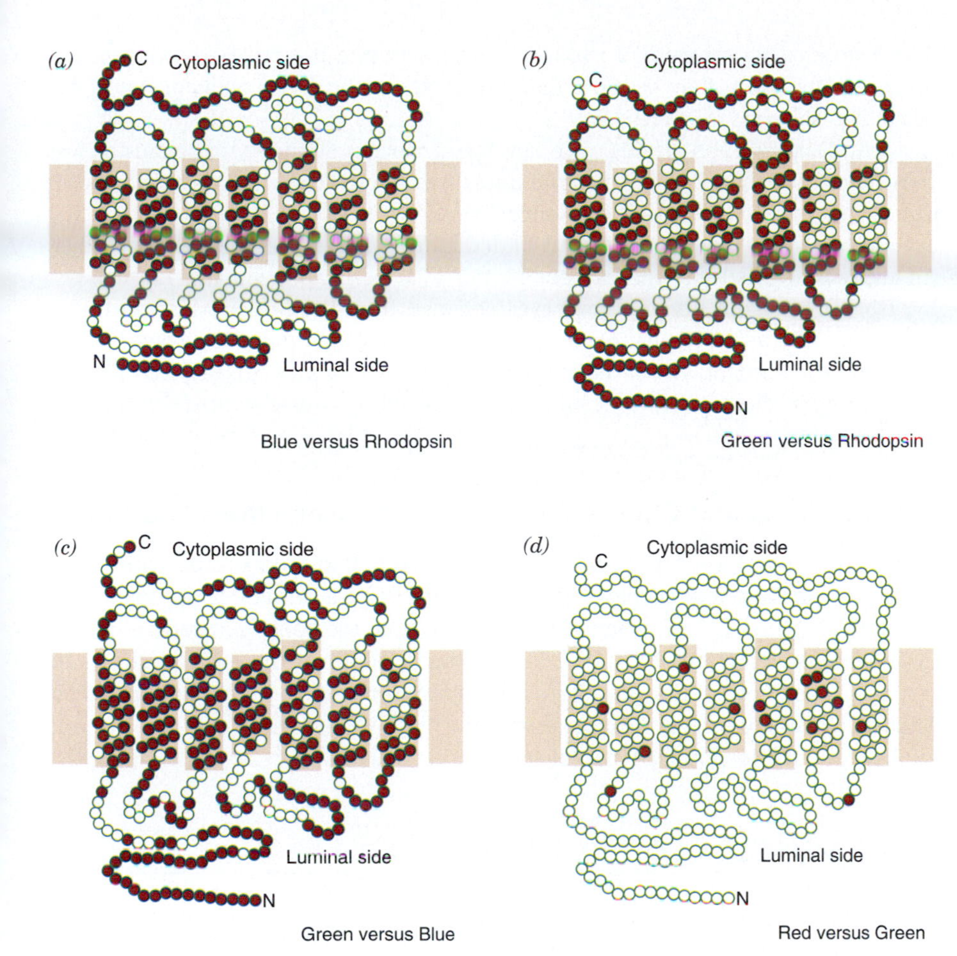

FIGURE 23.26
Comparisons of the amino acid sequences of the human visual pigments. Each black dot indicates an amino acid difference.
Adapted from Nathans, J. Annu. Rev. Neurosci. *10:163, 1987.*

CLINICAL CORRELATION 23.6
Chromosomal Location of Genes for Vision

The chromosomal arrangement of genes for vision precludes inheritance of a single defective gene from one parent that would render recipients sightless. Genes that code for visual pigments occur on chromosomes that exist in pairs except in males where there is a single X chromosome containing the genes for red and green pigments. In females, there is a pair of X chromosomes and, therefore, color vision abnormalities in females are rare, affecting only about 0.5% of the population. By contrast, about 8% of males have abnormal color vision that affects red or green perception and, on rare occasions, both. For the sake of simplicity, the proteins coded for by the different genes will be referred to as pigments in spite of the fact that they become visual pigments only when they form complexes with 11-*cis*-retinal.

The gene that codes for the protein moiety of rhodopsin, the rod pigment, is located on the third chromosome. Genes that code for the three pigment proteins of cone cells are located on two different chromosomes. The gene for the blue pigment is on the seventh chromosome. The genes for the red and green pigments are tightly linked and are on the X chromosome, which normally contains one gene for the red pigment and from one to three genes for the green pigment. In a given set of cones, only one of these gene types is expressed, either the gene for the red pigment or one of the genes for the green pigment. Genetic mutations may cause structural abnormalities in the proteins that influence the binding of retinal or the environment in which retinal resides. In addition, the gene for the protein of a specific pigment may not be expressed. If 11-*cis*-retinal does not bind or one of the proteins is not expressed, the individual will have dichromatic color vision and be color blind for the color of the missing pigment. If the mutation changes the environment around the 11-*cis*-retinal, shifting the absorption spectrum of the pigment, the individual will have abnormal trichromatic color vision; that is, the degree of stimulation of one or more of the three cone pigments will be abnormal. This will result in a different integration of the signal and hence a different interpretation of color.

Vollrath, D., Nathans, J., and Davis, R. W. Tandem array of human visual pigment genes at X. 28. *Science* 240:1669, 1988; and Nathans, J. In the eye of the beholder: visual pigments and inherited variations in human vision. *Cell* 78:357, 1994.

been identified with inherited variations that affect one but not both pigments simultaneously. There may be more than one gene for the green pigment, but it appears that only one is expressed.

John Dalton (1766–1844), who developed the atomic theory of chemistry, was color blind. He thought his color blindness was due to the vitreous humor being tinted blue, selectively absorbing longer wavelengths of light. He instructed that after his death his eyes be examined to determine whether his theory was correct. An autopsy revealed that the vitreous humor was "perfectly pellucid," normal. Using DNA analysis on his preserved eyes obtained from the British Museum, it has now been demonstrated that Dalton was missing the blue pigment. Thus, instead of having trichromatic vision, he was dichromatic with a vision type referred to as **deuteranopia.** The type of color blindness of one who is missing the green pigment is called **protanopia.**

Other Physical and Chemical Differences between Rods and Cones

Sensitivity and response time of rods differ from that of cones. Absorption of a single photon by photoreceptors in rod cells generates a current of approximately 1–3 pA ($1–3 \times 10^{-12}$ A), whereas the same event in the cones generates a current of approximately 10 fA (10×10^{-15} A), about 1/100th of the rod response. The response time of cone cells, however, is about four times faster than that of rod cells. Thus the cones are better suited for discerning rapidly changing events and the rods are better suited for low-light visual sensitivity.

23.4 | MUSCLE CONTRACTION

The **sliding filament model** for muscle contraction was proposed on the basis of an extensive evaluation of electron micrographs of skeletal muscle tissue. This simple but eloquent model has weathered the test of time. Genes for many of the proteins found in muscle tissue have been cloned, and the amino acid sequences of the proteins they encode inferred from their cDNA sequences. Three-dimensional structures of several of these proteins have also been published. Although the detailed picture of muscle contraction has not been completed, a clearer understanding of the process is emerging. In this section, some biochemical aspects of the mechanism of muscle contraction will be discussed. Primary emphasis will be on skeletal muscle rather than cardiac and smooth muscles. Some aspects of cardiac muscle function will be addressed in clinical correlations.

Skeletal Muscle Contraction Follows an Electrical to Chemical to Mechanical Path

The signal for skeletal muscle contraction begins with an electrical impulse from a nerve. This is followed by a chemical change occurring within the unit cell of the muscle and is followed by contraction, a mechanical process. Thus the **signal transduction** process goes from electrical to chemical to mechanical.

Figure 23.27 is a schematic diagram of the structural organization of skeletal muscle. Muscle consists of bundles of fibers (diagram *c*). Each bundle is called a fasciculus (diagram *b*). The fibers are made up of myofibrils (diagram *d*), and each myofibril is a continuous series of muscle cells or units called **sarcomeres.** The muscle cell is multinucleated and is no longer capable of division. Most muscle cells survive for the life of the animal, but they can be replaced when lost or lengthened by fusion of **myoblast cells.**

A muscle cell is shown diagrammatically in Figure 23.28. Note that the myofibrils are surrounded by a membranous structure called the **sarcoplasmic reticulum.** At discrete intervals along the fasciculi and connected to the terminal cisterna of the sarcoplasmic reticulum are transverse tubules. These are connected to the ex-

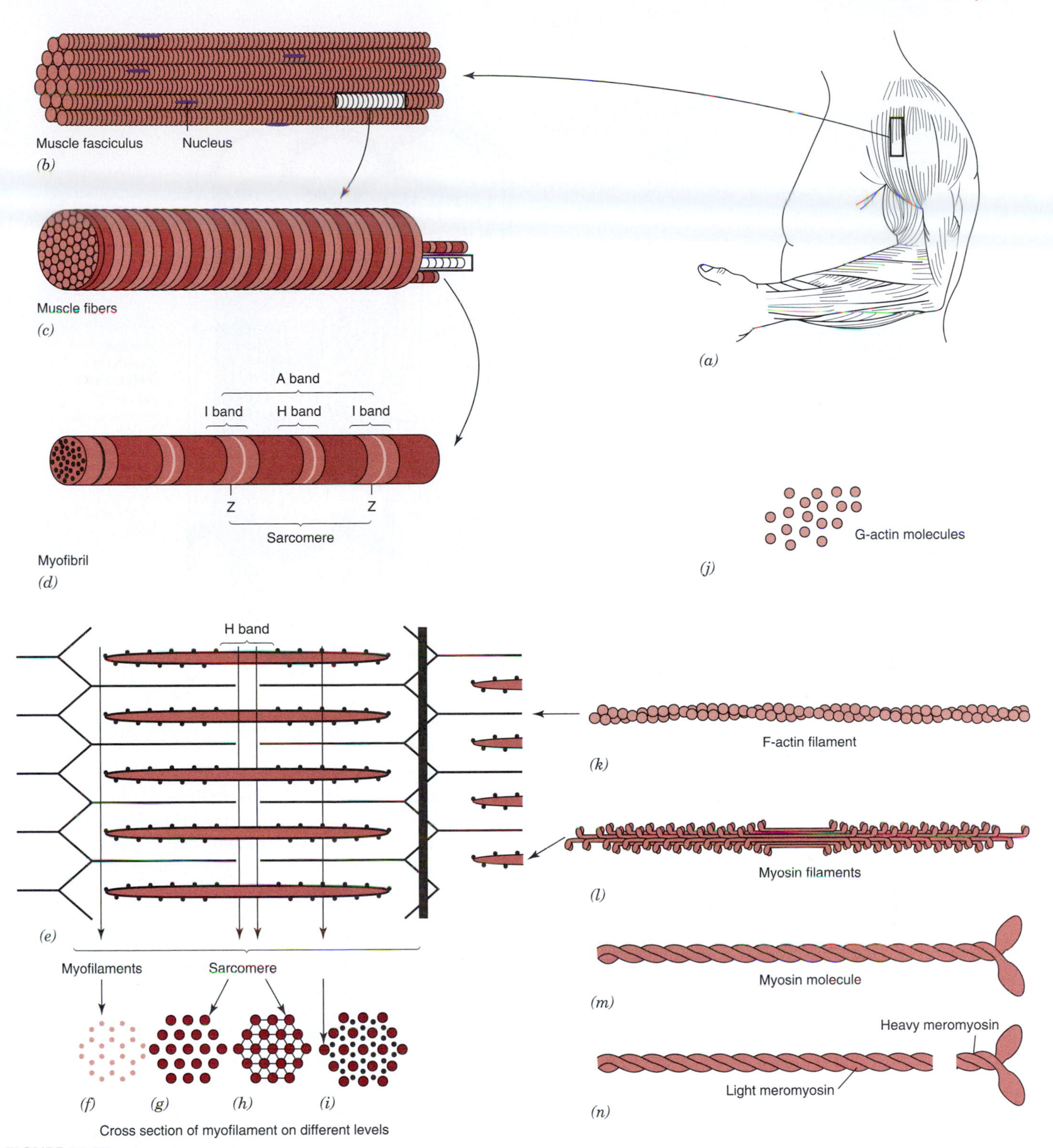

FIGURE 23.27
Structural organization of skeletal muscle.
Redrawn from Bloom, W. D. and Fawcett, D. W. Textbook of Histology, 10th ed. Philadelphia: Saunders, 1975.

ternal plasma membrane that surrounds the entire structure. The nuclei and the mitochondria lie just inside the plasma membrane.

The single contractile unit, the sarcomere, extends from Z line to Z line (Figures 23.27*d* and 23.28). Bands seen in the sarcomere are due to the arrangement of specific proteins (Figure 23.27*e*). Two types of fibers are apparent: long thick fibers with protrusions on both ends lie near the center of the sarcomere, and long thin

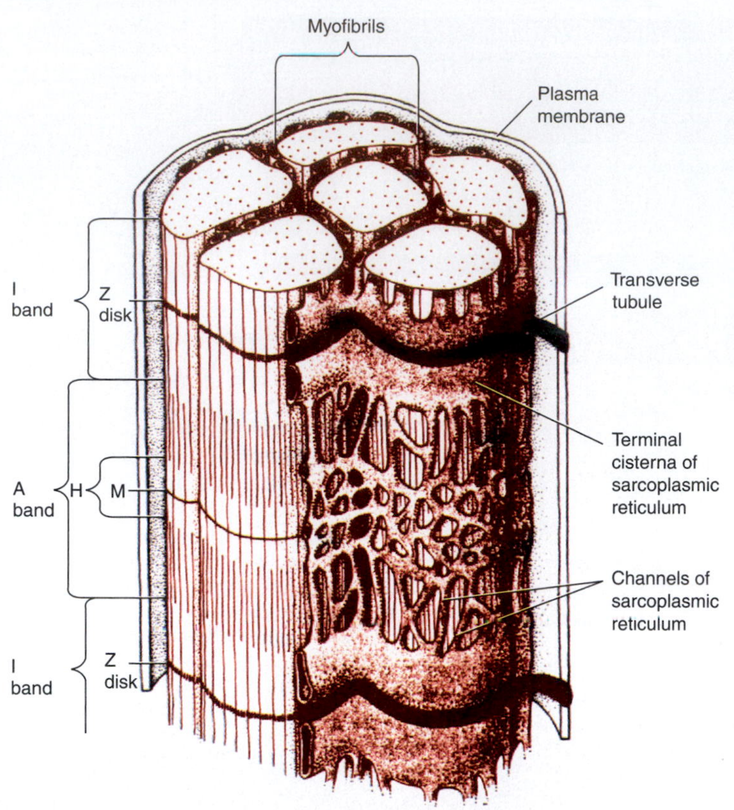

FIGURE 23.28
Schematic representation of a bundle of six myofibrils.
The lumen of the transverse tubules connects with the extracellular medium and enters the fibers at the Z disk.
Reprinted with permission from Darnell, J., Lodish, H., and Baltimore, D. Molecular Cell Biology. *New York: Scientific American Books, 1986, p. 827.*

fibers are attached to the **Z line.** The **I band (isotropic)** extends for a short distance on both sides of the Z line. This region contains only **thin filaments** that are attached to a protein band within the Z line. The **H band** is in the center of the sarcomere. There are no thin filaments within this region. In the middle of the H band, there is a somewhat diffuse band due to the presence of other proteins that assist in cross-linking the fibers of the **heavy filaments** (Figure 23.27, pattern *h*). The **A band (anisotropic)** is located between the inner edges of the I bands. When muscle contracts, the H and I bands shorten, but the distance between the Z line and the near edge of the H band remains constant. The distance between the innermost edges of the I bands on both ends of the sarcomere also remains constant. This occurs because the lengths of thin filaments and of thick filaments do not change during contraction. Contraction, therefore, results when these filaments "slide" past each other.

The contractile elements, sarcomeres, consist of many different proteins, eight of which are listed in Table 23.7. The two most abundant are myosin and actin. About 60–70% of the muscle protein is myosin and about 20–25% is actin. Thick filaments are mostly myosin and the thin filament is mostly actin. Three of the other proteins listed in Table 23.7 are associated with thin filaments, and two are associated with thick filaments. (See also Clinical Correlation 23.7 for reference to cardiac-related proteins.)

Myosin Forms the Thick Filaments of Muscle

The schematic drawing of a myosin molecule in Figure 23.29*b* is a representation of the electron micrographs in Figure 23.29*a*. Myosin, a long fibrous molecule with two globular heads on one end, is composed of two heavy chains of about 230 kDa each. Bound to each heavy chain in the vicinity of the head group is a dissimilar pair of light chains, each of which is approximately 20 kDa. The light chains are "calmodulin-like" proteins that bind calcium. *In vitro,* one of these light chains can be removed from each myosin molecule without affecting myosin's ability to function in the contraction process.

The carboxyl end of each myosin chain is located in the tail section. The tail sections of two heavy chains are coiled around each other in a coil–coil arrangement (Figure 23.29*b*). **Trypsin** cleaves the tail section at about one-third of its length from the head to produce **heavy meromyosin** (the head group and a short tail) and **light meromyosin** (the remainder of the tail section). Only light meromyosin has the ability to aggregate under simulated physiological conditions, suggesting that aggregation is one of its roles in heavy chain formation *in vivo.* The head section can be separated from the remainder of the tail section by treatment with **papain.** The myosin head group resulting from this cleavage is referred to as **subfragment 1** or **S-1.** Action of these proteases also demonstrates that the molecule has at least two hinge points in the vicinity of the head–tail junction (Figure 23.29*b*).

cDNAs for myosin from many different species and from different types of muscle have been cloned and amino acid sequences from them inferred. Myosin has evolved very slowly. There is a very high degree of homology among myosins obtained from different sources, particularly within the head, or globular, region. There is somewhat less sequence homology within the tail region, but functional homology exists to an extraordinarily high degree regardless of length, which ranges from about 86 to about 150 nm for different species. The myosin head group contains nearly one-half (about 839 to about 850) of the total number of amino acid residues in the entire molecule in mammals.

Myosin forms a **symmetrical tail-to-tail aggregate** around the M line of the H zone in the sarcomeres. Its tail sections are aligned in a parallel manner on both sides of the M line with the head groups pointing toward the Z line. Each thick filament contains about 400 molecules of myosin. The C protein (Table 23.7) is involved in their assembly. The M protein is also involved, presumably to hold the tail sections together as well as to anchor them to the M line of the H zone.

The globular head section of myosin contains the **ATPase** activity that provides energy for contraction and the **actin-binding site.** The S-1 fragment also contains the binding sites for the **essential light chain** and the **regulatory light chain.** A space-filling model of the three-dimensional structure of the myosin S-1 fragment is shown in Figure 23.30. The actin-binding region is located at the lower right-hand corner and the cleft, which is visible in that region of the mol-

TABLE 23.7 Approximate Molecular Weights of Skeletal Muscle Contractile Proteins

Myosin	500,000
Heavy chain	200,000
Light chain	20,000
Actin monomer (G-actin)	42,000
Tropomyosin	70,000
Troponin	76,000
Tn-C subunit	18,000
Tn-I subunit	23,000
Tn-T subunit	37,000
α-Actinin	200,000
C-protein	150,000
β-Actinin	60,000
M-protein	100,000

CLINICAL CORRELATION 23.7

Familial Hypertropic Cardiomyopathies and Mutations in Muscle Proteins

Hypertropic cardiomyopathies are characterized by enlargement/thickening of the left and/or right ventricle. Arrhythmias and premature death can result from these conditions. Familial hypertrophic cardiomyopathy has been shown, so far, to result from mutations involving seven different genes coding for some of the proteins listed in Table 23.7. Those proteins are β-myosin heavy chain, ventricular myosin regulatory light chain, cardiac troponin T, cardiac troponin I, α-tropomyosin, and cardiac myosin binding protein C. Some of these genes express different isoforms of their proteins that may be expressed only in cardiac tissue. Thus some of the cardiac effects noted are not generally applicable to other types of muscle. Mutations in genes coding for each of these proteins could alter structure and function of the sarcomere and lead to an alteration in cardiac function. Alteration in cardiac function may include decrease in force generated from myosin–actin interaction, faulty anchoring of myosin (protein C) within the sarcomere, and/or interference with any of the normal functions of tropomyosin and troponin subunits. These genetic abnormalities lead to production of abnormal proteins and abnormal myocyte structure and function. Schematic representations of genomic and protein structures of these mutants are given in the reference cited below.

Bonne, C., Carrier, L., Pascal, R., Hainque, B., and Schwartz, K. Familial hypertropic cardiomyopathy: from mutations to functional defects. *Circ. Res.* 83:580, 1998.

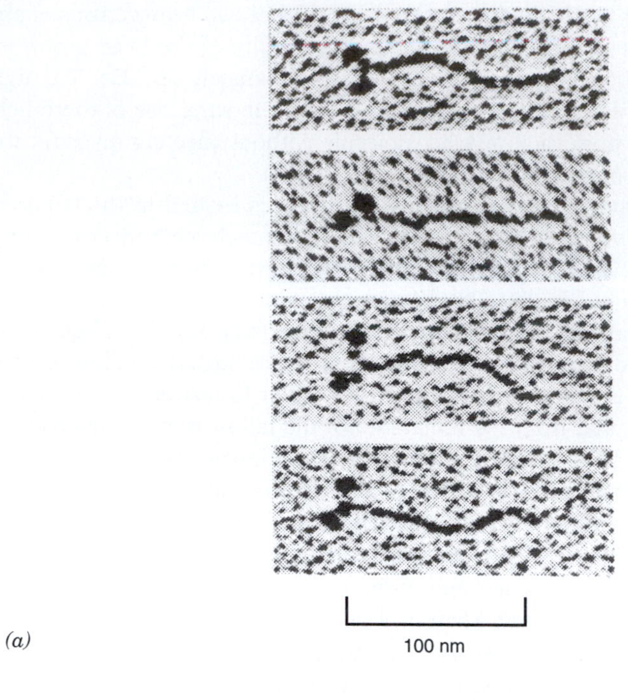

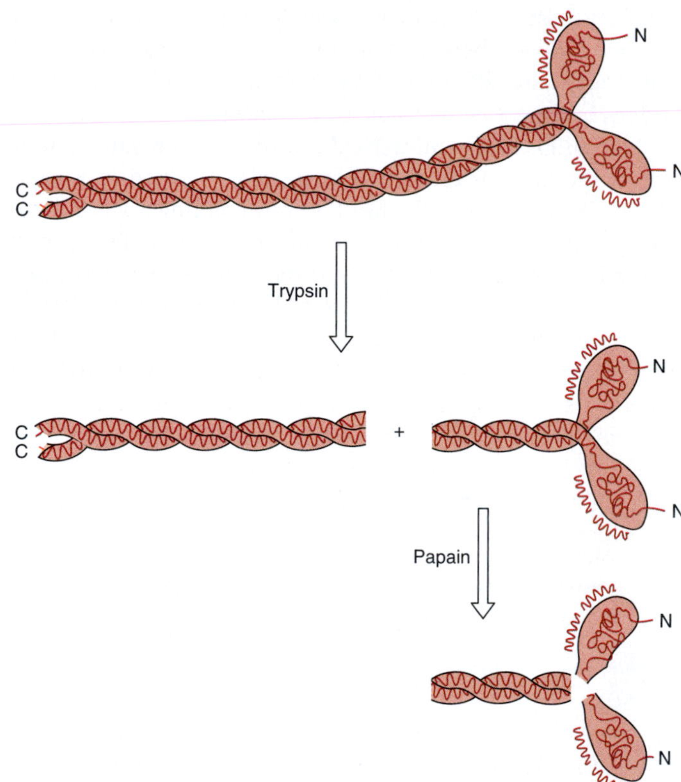

FIGURE 23.29
(*a*) Electron micrographs of the myosin molecule and (*b*) schematic drawing of a myosin molecule.
Diagram shows the two heavy chains and the two light chains of myosin. Also shown are the approximate positions of cleavage by trypsin and papain.
Reprinted with permission from Alberts, B., Bray, D., Lewis, J., Raff, M., Roberts, K., and Watson, J. Molecular Biology of the Cell, *2nd ed. New York: Garland Publishing, 1983.*

ecule, points toward the site where ATP binds. The 25-, 50-, and 20-kDa domains of the heavy chain are colored green, red, and blue, respectively. The essential light chain (ELC) and the regulatory light chain (RLC) are shown in yellow and magenta, respectively.

The ATP-binding site is also an open cleft about 13 Å deep and 13 Å wide. It is separated from the actin-binding site by approximately 35 Å. Myosin binds to actin in a stereospecific manner. The ELC and RLC associate with a single long helix that connects the head region with the tail section. There is room for flexibility between the ELC and the connecting single helix. The conformation of myosin that

FIGURE 23.30
Space-filling model of the amino acid residues in myosin S-1 fragment.
The 25-, 50-, and 20-kDa domains of the heavy chain are green, red, and blue, respectively. The essential and regulatory light chains are yellow and magenta, respectively.
Reprinted with permission from Rayment, I., Rypniewski, W. R., Schmidt-Base, K., Smith, R., et. al. Science 261:50, 1993. Copyright © 1993 by American Association for the Advancement of Science.

has ATP bound to it has an affinity for actin that is 1/10,000th that of the conformation of myosin that does not have ATP bound to it! Thus the transduction of chemical energy to mechanical work depends on the primary event of protein conformational changes that occur upon binding of ATP, its hydrolysis, and product dissociation.

Actin, Tropomyosin, and Troponin Are Thin Filament Proteins

Actin is a major protein of thin filaments and makes up about 20–25% of muscle protein. It is synthesized as a 42-kDa globular protein. Among a variety of species, better than 90% of its amino acid sequence is conserved. This is shown in Table 23.8 for skeletal muscle, smooth muscle, and cardiac muscle actin in three species. Differences are observed at most in about seven positions. Sequences of more than 30 actin isotypes, the longest containing 375 amino acid residues, reveal that a maximum of only 32 residues in any of them have been substituted (see Clin. Corr. 23.8). A significant number of these substitutions occur at the N terminal, which may be predicted considering that all actin molecules are posttranslationally modified at the N terminal. The N-terminal methionine is acetylated and removed, and the next amino acid is acetylated. The process may end at this stage or it may be repeated one or two times. In all cases, the N-terminal amino acid is acetylated.

Actin is first synthesized in a globular form called G-actin. A representation of its structure is shown in Figure 23.31. Actin has two distinct domains of approximately equal size that, historically, have been designated as large (left) and small (right) domains. Each consists of two subdomains. Both the N-terminal and C-terminal amino acid residues are located within subdomain 1 of the small domain. The molecule has polarity, and when it aggregates to form **F-actin,** or fibrous actin, it does so with a specific directionality. This is important for the "stick and pull" processes involved in sarcomere shortening during muscular contraction.

TABLE 23.8 Summary of the Amino Acid Differences Among Chicken Gizzard Smooth Muscle Actin, Skeletal Muscle Actin, and Bovine Cardiac Actin

	Residue Number						
Actin Type	*1*	*2*	*3*	*17*	*89*	*298*	*357*
Skeletal muscle[a]	Asp	Glu	Asp	Val	Thr	Met	Thr
Cardiac muscle[b]		Asp	Glu			Leu	Ser
Smooth muscle[c]	Absent		Glu	Cys	Ser	Leu	Ser

Source: Adapted from Vandekerckhove, J. and Weber, K. *FEBS Lett.* 102:219, 1979.
[a]From rabbit, bovine, and chicken skeletal muscle.
[b]From bovine heart.
[c]From chicken gizzard.

CLINICAL CORRELATION 23.8
Mutations in Actin and Dilated Cardiomyopathy

Dilated cardiomyopathy is characterized by thin walls of the heart with an inability to pump blood out effectively. A genetic cause of this condition can be defective actin. The amino acid sequence of actin is highly conserved. There are several isoforms of actin resulting from expression of different genes; five of six known ones are expressed in skeletal and cardiac myocytes. In adult cardiac myocytes, approximately 80% of actin results from expression of specific cardiac isoforms. Mutations in the invariant regions of these cardiac isoforms can result in serious health effects. It has been demonstrated that an inherited single amino acid substitution, A312H in one patient and Q361G in another patient, resulted in dilated cardiomyopathy, a condition in which cardiac transplant is the only definitive treatment for end-stage disease. The positions of these mutations in the tertiary structure of G-actin are shown in the figure below.

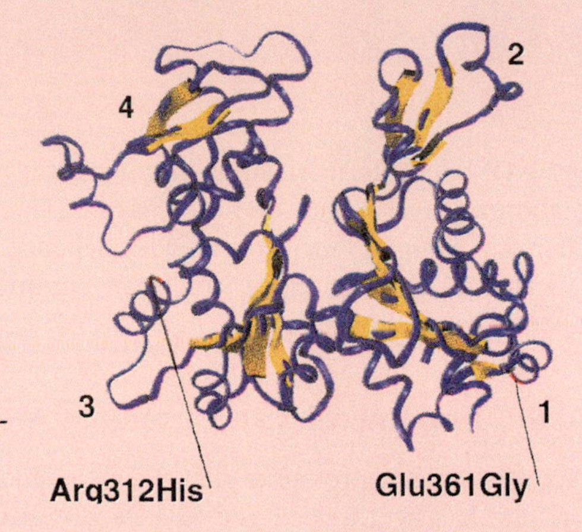

Schematic representation of the cardiac actin monomer and location of idiopathic dilated cardiomyopathic mutations.
From Olson et al. (1998).

Comparison of the positions of those substitutions with the structure of actin shown in Figure 23.31 shows that they occur in two different functional regions of actin.

Olson, T. M., Michels, V. V., Thibodeau, S. N., Tai, Y-S., and Keating, M. T. Actin mutations in dilated cardiomyopathy, a heritable form of heart failure. *Science* 280:750, 1998. Copyright © 1998 by American Association for the Advancement of Science.

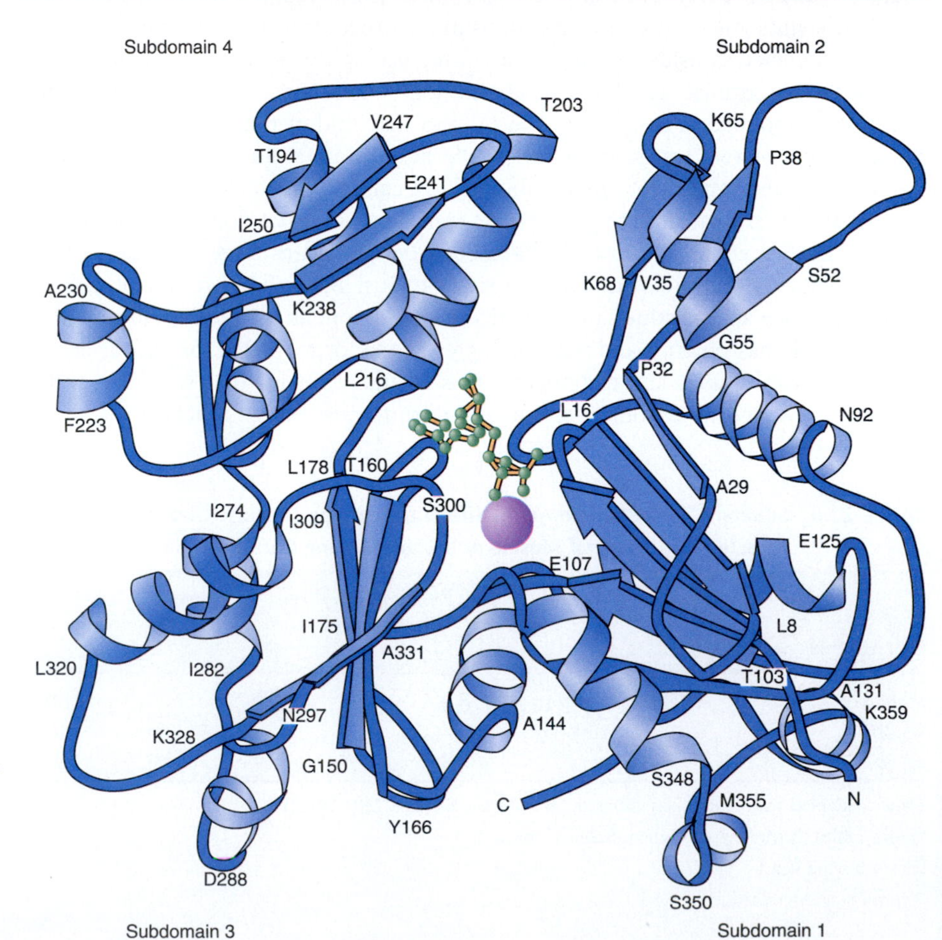

FIGURE 23.31
Secondary structural elements of G-actin crystal structure.
ADP and the metal ion are shown in the cleft between the two large domains.
Reproduced with permission from Lorenz, M., Popp, D., and Holmes, K. C. J. Mol. Biol. 234:826, 1993 by permission of the publisher Academic Press Limited, London.

G-actin contains a specific binding site, between the two major domains, for ATP and a divalent metal ion. Mg^{2+} is most likely the physiologically important cation, but Ca^{2+} also binds tightly and competes with Mg^{2+} for the same tight binding site. The **G-actin–ATP–Mg^{2+} complex** aggregates to form the **F-actin polymer** (see Figure 23.35*a*). Aggregation can occur from either direction, but kinetic data indicate that the preferred direction of aggregation is by extension from the large end of the molecule where the rate is diffusion controlled. ATP hydrolysis is orders of magnitude faster in the aggregated actin than it is in the monomer. G-actin–ADP–Mg^{2+} also aggregates to form F-actin but at a slower rate. Orientation of G-actin molecules in F-actin is such that subdomains 1 and 2 are to the outside where myosin-binding sites are located. F-actin may be viewed as either (1) a single-start, left-handed single-stranded helix with rotation of the monomers through an approximate 166° with a rise of 27.5 Å or (2) a two-start, right-handed double-stranded helix with a half pitch of 350–380 Å.

A number of cytosolic proteins bind actin. **β-Actinin** binds to F-actin and plays a major role in limiting the length of thin filaments. **α-Actinin,** a homodimeric protein with a subunit molecular mass of 90–110 kDa, binds adjacent actin monomers of F-actin at positions 86–117 and 350–375 and strengthens the fiber. It also helps to anchor the actin filament to the Z line of the sarcomere. Major proteins associated with the thin filament are tropomyosin and troponin.

Tropomyosin is a rod-shaped protein consisting of two dissimilar subunits, each of about 35 kDa. It forms aggregates in a head-to-tail configuration. This polymerized protein interacts in a flexible manner with the thin filament throughout its entire length. It fits within the groove of the helical assembly of the actin monomers of F-actin. Each of the single tropomyosin molecules interacts with about seven monomers of actin. The site on actin to which tropomyosin interacts is between subdomain 1 and 3. Tropomyosin helps to stabilize the thin filament and to transmit signals for conformational change to other components of the thin filament upon Ca^{2+} binding. Bound to each individual tropomyosin molecule is one molecule of troponin.

Troponin consists of three dissimilar subunits designated Tn-C, Tn-I, and Tn-T with molecular masses of about 18 kDa, 21 kDa, and 37 kDa, respectively. The Tn-T subunit binds to tropomyosin. The Tn-I subunit is involved in the inhibition of the binding of actin to myosin in the absence of Ca^{2+}. The Tn-C subunit, a calmodulin-like protein, binds Ca^{2+} and induces a conformational change in Tn-I and tropomyosin, resulting in exposure of the actin–myosin-binding sites.

A three-dimensional structure of Tn-C shows it to be dumbbell-shaped with much similarity to calmodulin. A structural model of the calcium saturated Tn-C–Tn-I complex is shown in Figure 23.32. The Tn-I subunit fits around the central region of Tn-C in a helical coil conformation and forms caps over it at each end. The cap regions of Tn-I are in close contact with Tn-C when Tn-C is fully saturated with Ca^{2+}. Tn-C

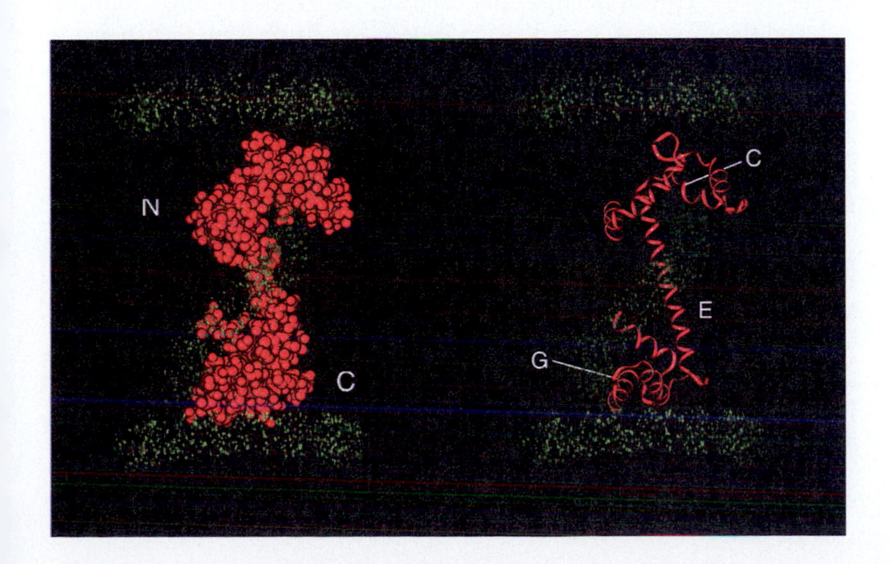

FIGURE 23.32
Best fit model for the $4Ca^{2+}$–Tn-C–Tn-I complex.
A model for the complex of $4Ca^{2+}$–Tn-C–Tn-I based on neutron scattering studies with deuterium labeling and contrast variation (Olah, C. C. and Trewhella, J., *Biochemistry* 33:12800, 1994). (Right) A view showing the spiral path of Tn-I (green crosses) winding around the $4Ca^{2+}$–Tn-C that is represented by an α-carbon backbone trace (red ribbon) with the C, E, and G helices labeled. (Left) The same view with $4Ca^{2+}$–Tn-C represented as a CPK model.
Photograph generously supplied by Dr. J. Trewhella. The publisher recognizes that the U. S. Government retains a nonexclusive, royalty-free license to publish or reproduce the published form of this contribution or to allow others to do so, for U. S. Government purposes.

CLINICAL CORRELATION 23.9
Troponin Subunits as Markers for Myocardial Infarction

Troponin contains three subunits (Tn-T, Tn-I, and Tn-C), each of which is expressed by more than one gene. Two genes code for skeletal muscle Tn-I, one in fast and one in slow skeletal muscle; and one gene codes for cardiac muscle Tn-I. The genes that code for Tn-T have the same distribution pattern. They differ in that the slow skeletal muscle gene for Tn-I is also expressed in the fetal heart tissue. The gene for the cardiac form of Tn-I appears to be specific for heart tissue. Tn-C is encoded by two genes, but neither gene appears to be expressed only in cardiac tissue.

The cardiac form of Tn-I in humans is about 31 amino acids longer than the skeletal muscle form, which makes it easy to differentiate from others. Serum levels of Tn-I increase within 4 h of an acute myocardial infarction and remain high for about 7 days in about 68% of patients tested. Almost 25% of one group of patients tested also showed a slight increase in the cardiac form of Tn-I after acute skeletal muscle injury. This would be a good but not a very sensitive test for myocardial infarction.

Two isoforms of cardiac Tn-T, Tn-T_1 and Tn-T_2, are present in adult human cardiac tissue. Two additional isoforms are also present in fetal heart tissue. Speculation is that the isoforms are the result of alternative splicing of mRNA. Serum levels of Tn-T_2 increase within 4 h of acute myocardial infarction and remain high for up to 14 days. The appearance of Tn-T_2 in serum is 100% sensitive and 95% specific for detection of myocardial infarction. In the United States, the Food and Drug Administration has given approval for marketing of the first Tn-T assay for acute myocardial infarction. Myocardial infarcts are either undiagnosed or misdiagnosed in hospital patients admitted for other causes, or in 5 million or more people who go to doctors for episodes of chest pain. It is believed that this test will be sufficiently specific to diagnose myocardial incidents and to help direct doctors to proper treatment of these individuals.

Anderson, P. A. W., Malouf, N. N., Oakeley, A. E., Pagani, E.D., and Allen, P.D. Troponin T isoform expression in humans. *Circ. Res.* 69:1226, 1991; and Ottlinger, M. E. and Sacks, D. B. *Clin. Lab. News*, 33, 1994.

has four divalent metal ion-binding sites. Two of these sites are in the C-terminal region and have high affinity for calcium ions (K_D of about 10^{-7} M). These two sites are presumed to be always occupied by divalent metal ions (Ca^{2+} or Mg^{2+}) since the concentrations of these ions in resting cells are within the same order of magnitude as the K_D. Under these conditions, Tn-I has a conformation that permits its interaction with binding sites on actin, inhibits myosin binding, and prevents contraction. Upon excitation, the calcium ion concentration increases to about 10^{-5} M, high enough to effect calcium binding to sites within the N-terminal region of Tn-C. Tn-I now binds preferentially to Tn-C in a capped structural conformation as shown in Figure 23.32. Myosin-binding sites on actin are now exposed. The relatively loose interaction of tropomyosin with actin permits it the flexibility to alter its conformation as a function of calcium ion concentration and to assist in blockage of the myosin-binding sites on actin. (See Clinical Correlation 23.9 for additional information about troponin.)

Figure 23.27*i* shows a schematic cross section of a sarcomere and the relative arrangement of thin and thick filaments. Six thin filaments surround each thick filament. The arrangement and flexibility of myosin head groups make it possible for each thick filament to interact with multiple thin filaments. When cross-bridges form between thick and thin filaments, they do so in patterns consistent with that shown in

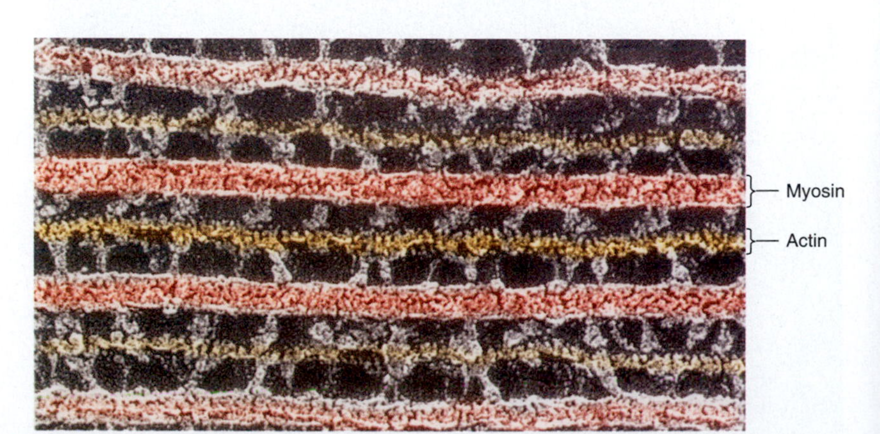

FIGURE 23.33
Colorized electron micrograph of actin–myosin cross-bridges in a striated insect flight muscle.
Reproduced with permission from Darnell, J., Lodish, H., and Baltimore, D. Molecular Cell Biology. *New York: Scientific American Books, 1986.*

the colorized electron micrograph of Figure 23.33. This shows a two-dimensional view of a thick filament interacting with thin filaments lying on either side of it. Similar interactions of myosin occur with actin of the other four thin filaments that surround it.

Muscle Contraction Requires Ca^{2+}

Contraction of skeletal muscle is initiated by transmission of nerve impulses across the **neuromuscular junction** mediated by release into the synaptic cleft of the neurotransmitter **acetylcholine. Acetylcholine receptors** are associated with the plasma membrane and are ligand-gated. Binding of acetylcholine causes them to open and to permit Ca^{2+}/Na^{+} to enter the sarcomere. The electron micrograph and accompanying diagrams of Figure 23.34 provide a picture of the anatomical relationship between the presynaptic nerve and sarcomere.

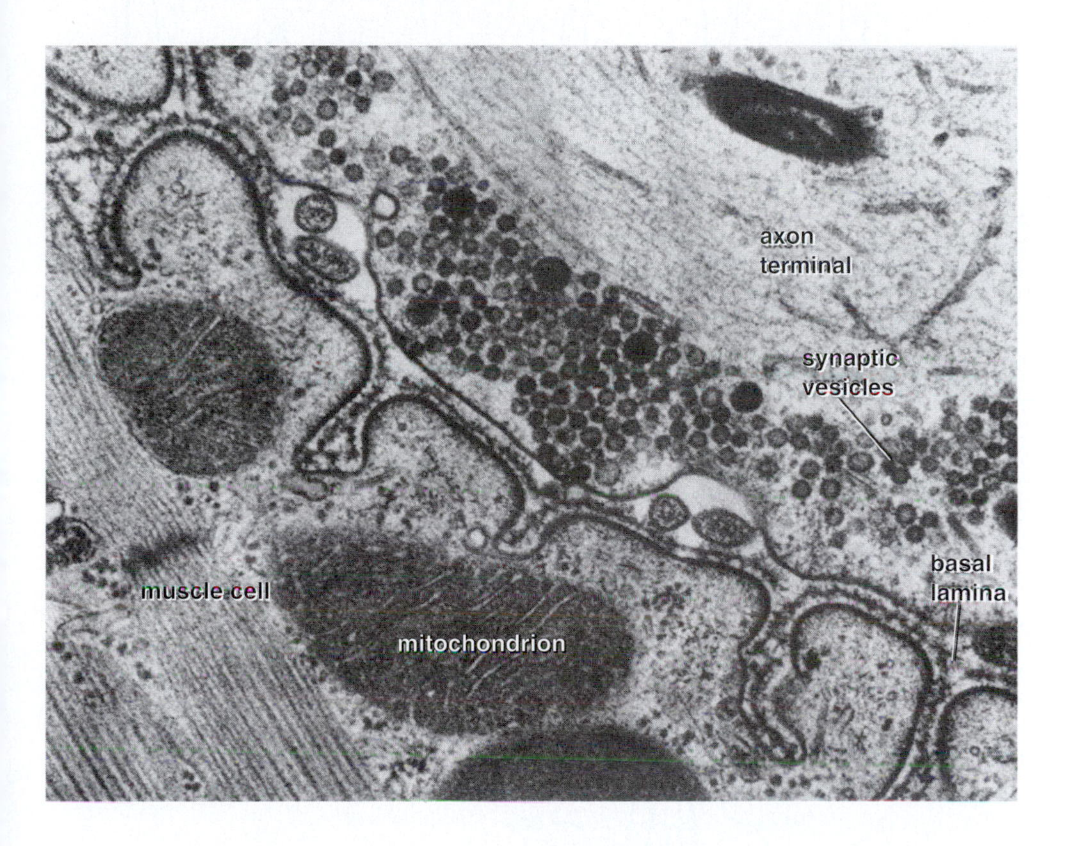

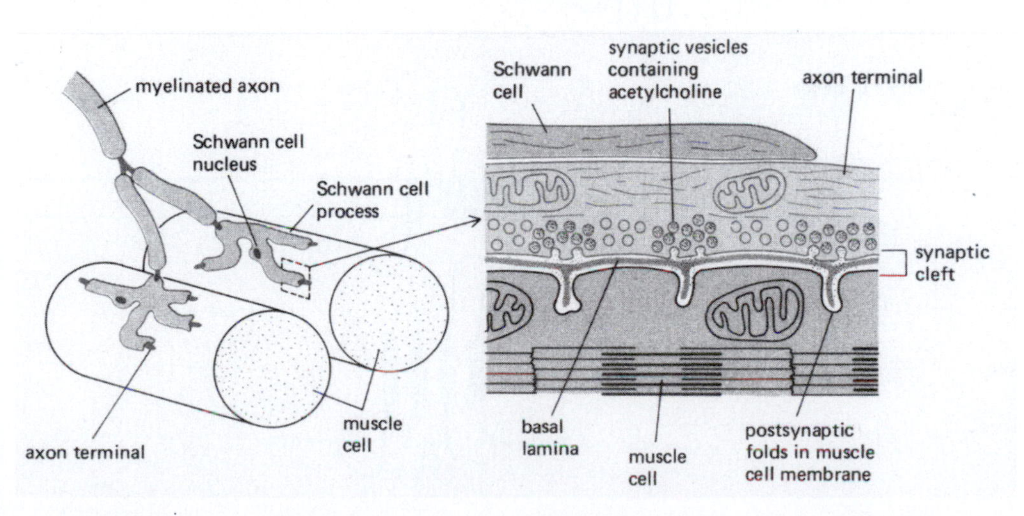

FIGURE 23.34
(*a*) Electron micrograph of a neuromuscular junction and (*b*) schematic diagram of the neuromuscular junction shown in (*a*).
Reproduced with permission from Alberts, B., Bray, D., Lewis, J., Raff, M., Roberts, K., and Watson, J. Molecular Biology of the Cell. *New York: Garland Publishing, 1983.*

CLINICAL CORRELATION 23.10
Voltage-Gated Ion Channelopathies

Action potentials in nerve and muscle are propagated by the operation of voltage-gated ion channels. Generally, there are three recognized types of voltage-gated cation channels: Na^+, Ca^{2+}, and K^+. Each of these has been cloned, the primary sequence has been inferred from the DNA sequence, models have been constructed of how each may be assembled in the membrane, and a 3-D structure of a K^+ channel has been determined. Each is a heterogeneous protein consisting of various numbers of α and β subunits. A linear model of the arrangement of each of these is shown in the figure below. In actual fact, they are arranged in a more-or-less circular manner with a funnel-like channel formed through the middle of α subunits. Roles of β subunits are still being elucidated, but they appear to help stabilize and/or regulate activity of α subunits.

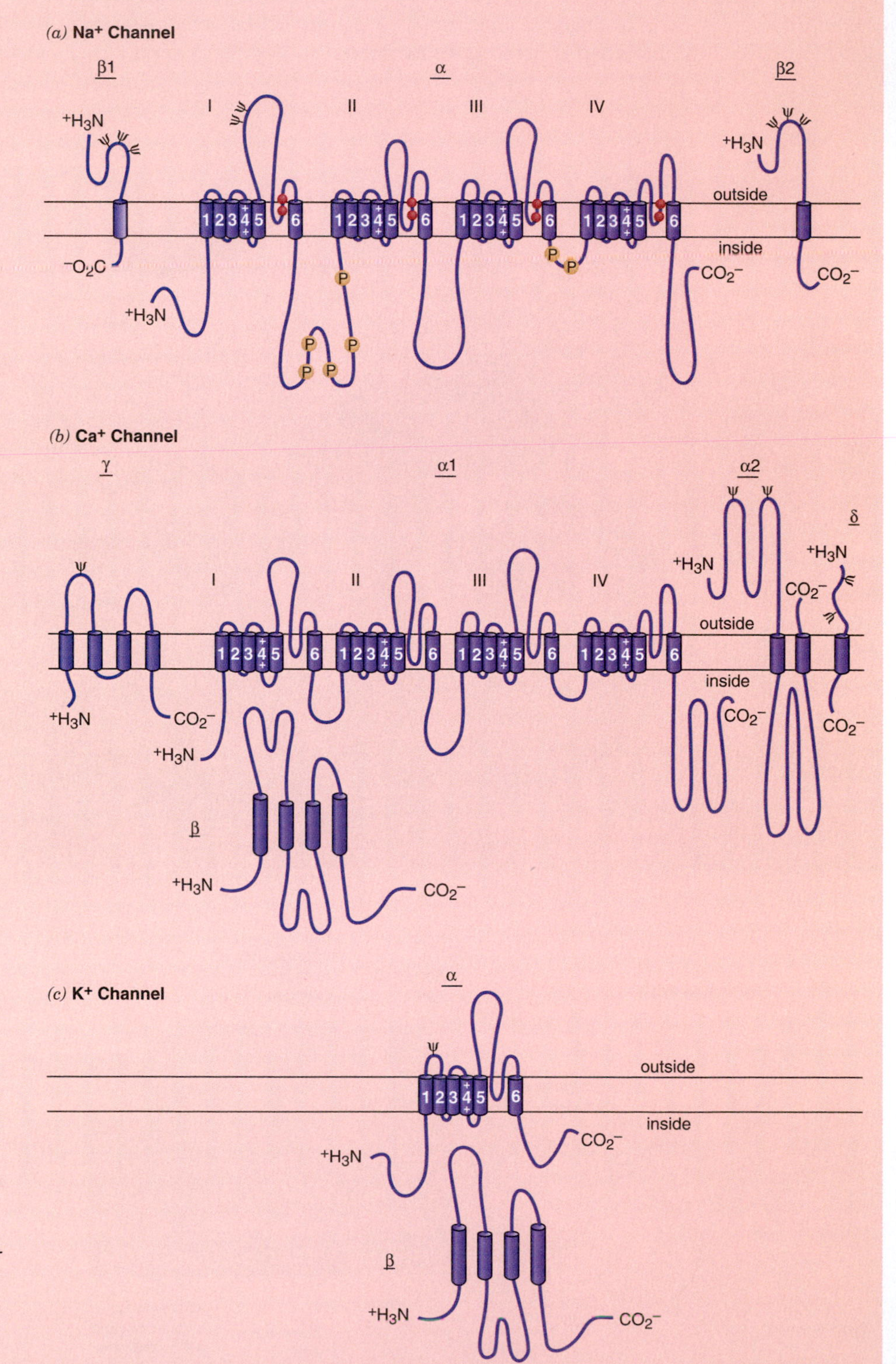

Transmembrane organization of ion channel subunits.
Glycosylation and phosphorylation sites are marked.
From Catterall (1995).

Continued on page 1027.

Clinical Correlation 23.10 (continued)

Toxins and mutagenesis are being used to study subunit function. Tetrodotoxin and saxitoxin block Na^+ channel pores of the α subunit. Scorpion toxins also bind to the α subunit and appear to affect activation and inactivation gating. Experiments of this type suggest that the α subunit is involved in both conductance and gating.

Voltage-gated channels from nerve and muscle tissue show high homology in many of the transmembrane domains but are less conserved in the intracellular connecting loops. A 3-D structure of a K^+ channel has been determined and other voltage-gated channels have been modeled after it. Some effects of channel mutations are beginning to be understood better. A common effect of mutations in Na^+ channels is muscle weakness or paralysis. Some inherited Na^+ voltage-gated ion channelopathies are listed below. Each of these is reported to result from a single amino acid change in the α subunit. The inheritance pattern generally is dominant.

Disorder	*Unique Clinical Feature*
Hyperkalemic periodic paralysis	Induced by rest after exercise, or the intake of K^+
Paramyotonia congenita	Cold-induced myotonia
Sodium channel myotonia	Constant myotonia

It has been surmised that if the membrane potential is slightly more positive (i.e., changes from −70 to −60 mV) the myofiber can reach the threshold more easily and the muscle becomes hyperexcitable. If the membrane potential becomes even more positive (i.e., up to −40 mV) the fiber cannot fire an action potential. This inability to generate an action potential is synonymous with paralysis. The fundamental biochemical defect in each case is a mutation in the channel protein.

Catterall, W. A. Structure and function of voltage-gated ion channels. *Annu. Rev. Biochem.* 64:493, 1995; Hoffmann, E. P. Voltage-gated ion channelopathies: inherited disorders caused by abnormal sodium, chloride, and calcium regulation in skeletal muscle. *Annu. Rev. Med.* 46:431, 1995; and Abraham, M. R., Jahangir, A., Alekseev, A. E., and Terzic, A. Channelopathies of inwardly rectifying potassium channels. *FASEB J.* 13:1901, 1999.

Transverse tubules along the membrane in the vicinity of Z lines are connected to terminal cisternae of the sarcoplasmic reticulum. Nerve impulses lead to depolarization of the plasma membrane and of the transverse tubules. This depolarization causes an influx of Ca^{2+} into the sarcomere. As indicated above, Ca^{2+} concentration increases about 100-fold, permitting it to bind to the low-affinity sites of Tn-C and to initiate the contraction process. (See Clinical Correlation 23.10 and Clinical Correlation 23.11 for cardiac muscle events.)

Energy for Muscle Contraction Is Supplied by ATP Hydrolysis

ATP is an absolute requirement for muscular contraction. ATP hydrolysis by **myosin–ATPase** to give the myosin–ADP complex and inorganic phosphate leads to a myosin conformation that has an increased binding affinity for actin. Dissociation of ADP and binding of additional ATP is required for the dissociation of the myosin–actin complex.

In normal muscle tissue, the concentration of ATP in sarcomeres remains fairly constant even during strenuous muscle activity. This is because of increased meta-

CLINICAL CORRELATION 23.11

Ion Channels and Cardiac Muscle Disease

Voltage-gated ion channels in cardiac muscles, like those of other tissues, require a finite recovery time after excitation. The heart contracts and relaxes on a continuous basis and in a rhythmic manner that cannot be altered significantly without causing serious problems such as arrhythmias, fibrillation, and possibly death. The recovery time between contractions of cardiac muscle is measured on electrocardiograms as the QT interval. Initiating the excitation phase is the opening of Na^+ channels for a finite period of time. Na^+ moves into the cell causing depolarization. K^+ channels then open to permit K^+ to move out of the cell—both flowing down their respective chemical concentration gradients. Opening of the K^+ channels is key to shutting off the action potential. There are inherited conditions known generically as long QT syndrome or LQTS. Inherited LQTS has been linked to genes that affect K^+ channels (*KVLQT1, HERG, mink*) and Na^+ channels (*SCN5A*). Defects in these channels in cardiac muscle can cause sudden death, particularly in young people who are physically active, have never had an electrocardiogram, and have no knowledge that they have LQTS. Prevalence for this condition is approximately 1 in 10,000.

Balser, J. R. Structure and function of the cardiac sodium channels. *Cardiovasc. Res.* 42:327, 1999; Ackerman, M. J., Schroeder, J. J., Berry, R., Schaid, D. J., Porter, C. J., Michels, V. V., and Thibodeau, S. N. A novel mutation in KVLQT1 is the molecular basis of inherited long QT syndrome in a near-drowning patient's family. *Pediatr. Res.* 44:148, 1998; and Barinaga, M. Tracking down mutations that can stop the heart. *Science* 281:32, 1998.

bolic activity and the action of two enzymes: **creatine phosphokinase** and **adenylate kinase.** Creatine phosphokinase catalyzes transfer of phosphate from phosphocreatine to ADP in an energetically favored manner:

$$\text{phosphocreatine} + \text{ADP} \leftrightarrow \text{ATP} + \text{creatine}$$

If the metabolic process is insufficient to keep up with the energy demand, the creatine phosphokinase system serves as a "buffer" to maintain cellular levels of ATP. The second enzyme is adenylate kinase that catalyzes the reaction

$$2\ \text{ADP} \leftrightarrow \text{ATP} + \text{AMP}$$

ATP depletion has serious consequences for muscle cells. When the ATP supply is exhausted and the intracellular Ca^{2+} concentration is no longer controlled, myosin will exist exclusively bound to actin, a condition called **rigor mortis.** A major function of ATP in muscular contraction is to promote dissociation of the actin–myosin complex, not to promote its formation.

Model for Skeletal Muscle Contraction

A model of the myosin–actin complex is shown in Figure 23.35. Myosin heads undergo conformational changes upon binding of ATP, hydrolysis of ATP, and release

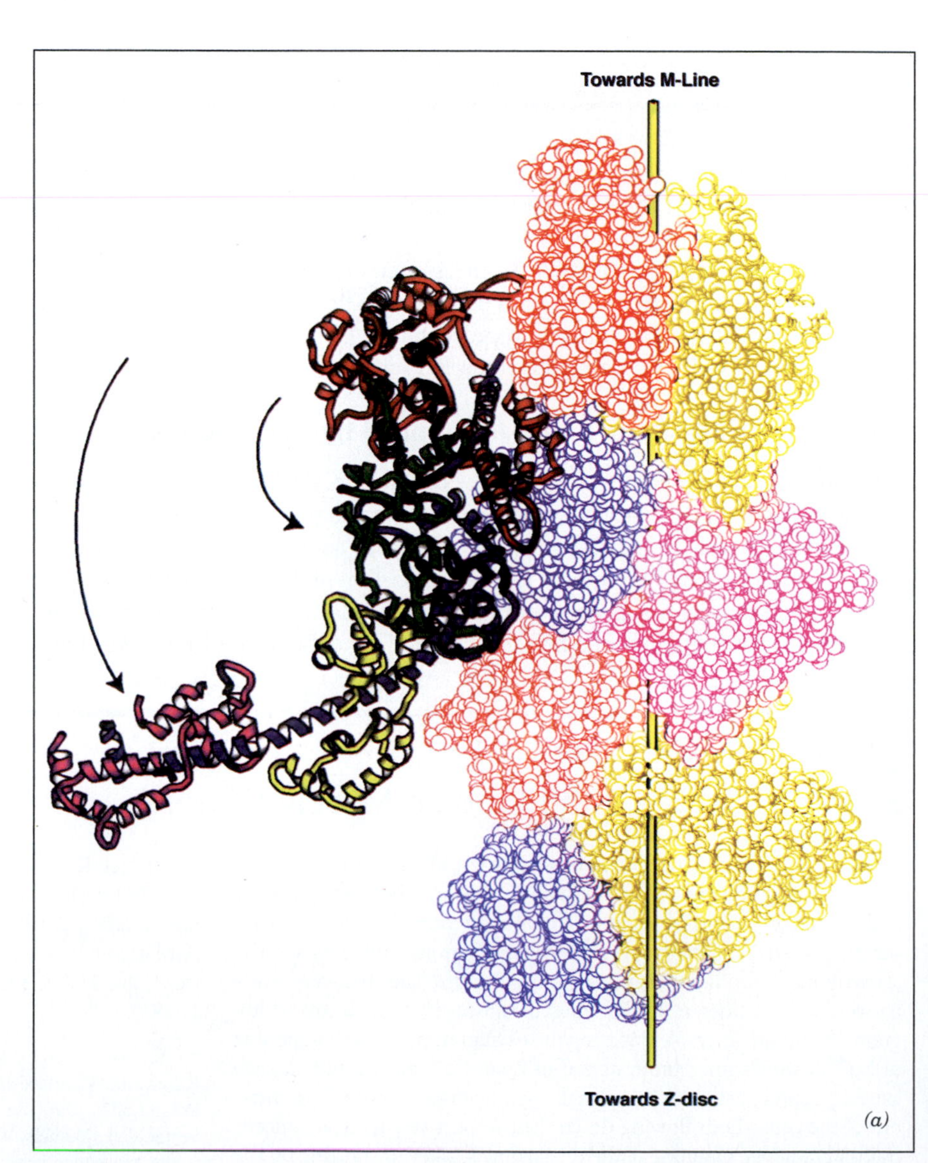

FIGURE 23.35
Model of actin–myosin interaction.
(*a*) Myosin is shown as a ribbon structure and actin as space-filling structure. Each G-actin monomer is represented by different colors. (*b*) Stereo view of myosin showing the pocket that contains the mobile "reactive" cysteine residues. (*c*) Schematic diagram showing how conformation changes in the myosin head group and arm can pull actin filaments in a direction away from the Z disk during hydrolysis of ATP and release of inorganic phosphate.
Part (a) and (b) reprinted from Rayment, I. and Holden, H. M. The three dimensional structure of a molecular motor. Trends Biol. Sci. 19:129, 1994, with permission from Elsevier Science.

Continued on page 1029.

of products. ATP binding leads to closure of the active site cleft and opening of the cleft in the region of the actin-binding site. Hydrolysis of ATP results in closure of the cleft in the actin-binding region, which opens again upon or for release of inorganic phosphate. The conformational change that occurs is evident by the movement of two cysteine-containing helices. The distance between Cys^{697} and Cys^{707} changes from about 19 Å to about 2 Å. If further conformational change is prevented by cross-linking of these two cysteines, ADP becomes trapped within its binding site. A stereo view of myosin showing the reactive cysteine pocket is shown in Figure 23.35*b*. Conformational changes in myosin that occur in conjunction with closing and opening of the cleft in the actin-binding region during ATP hydrolysis are associated with the so-called **power stroke.**

Many of the major proteins involved in muscle contraction have been crystallized and their 3-D structures are known; major efforts are underway to determine exactly how they work together to effect contraction. Details of the kinetics and additional structural information will refine knowledge of this process. A logical sequence of events based on current knowledge may be summarized as follows. A myosin–ATP complex exists in the resting state. Also in the resting state, the troponin complex of the thin filament is in a conformation with actin that prevents myosin–actin interaction. Upon depolarization, Ca^{2+} enters the sarcomere, binds to Tn-C of the troponin complex, and effects a conformational change in Tn-I that is propagated through tropomyosin to actin, resulting in exposure of the myosin-binding sites on actin. The initial interaction between myosin and actin is weak, but

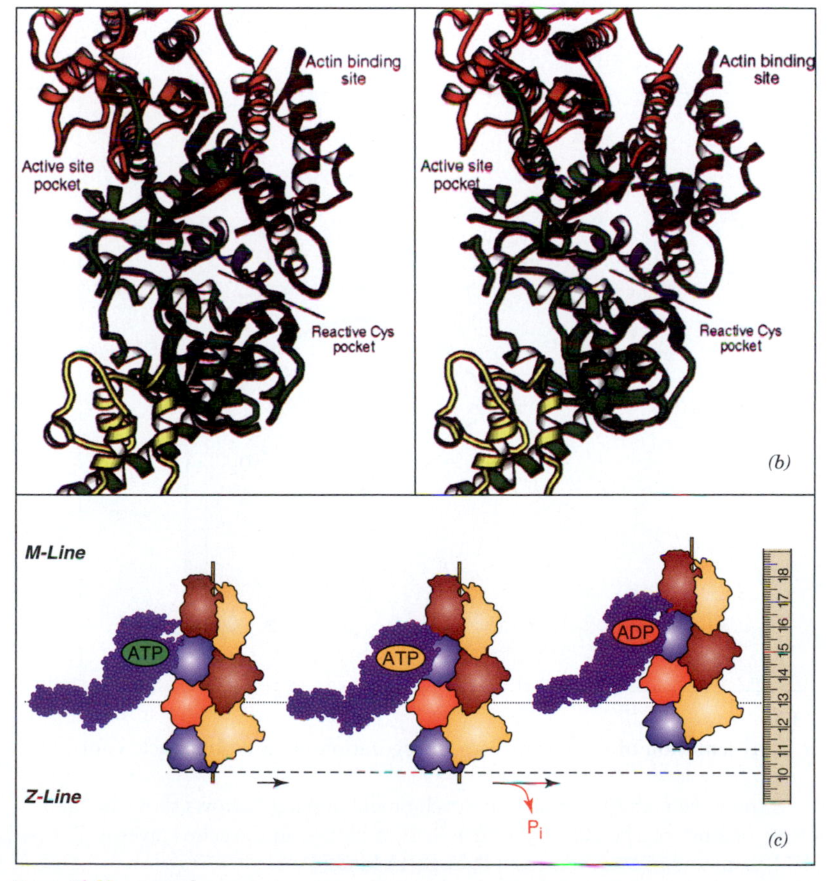

Figure 23.35 continued

is propagated through tropomyosin to actin, resulting in exposure of the myosin-binding sites on actin. The initial interaction between myosin and actin is weak, but it stimulates myosin's ATPase activity. ATP is hydrolyzed to ADP and P_i, leading to a myosin conformation that has a stronger affinity for actin, approximately 10,000-fold stronger. Myosin then undergoes a major conformational change from a closed to an open state with release of inorganic phosphate. This conformational change leads to or is accompanied by movement of the myosin head group and arm region as shown in Figure 23.35*a* with concomitant shortening of the sarcomere, the power stroke, bringing the Z lines closer together. Actin filament can be moved as much as 10 nm during this process. The open conformation of myosin is now in a position to bind ATP and start the cycle over again. A schematic of how this process might occur is shown in Figure 23.35*c*.

The thick filament is anchored in the center of the sarcomere and the myosin head groups of this filament are polarized in opposite directions on each side of the M line. Each thick filament contains hundreds of S-1 or myosin head units surrounded by six actin-containing thin filaments. Individual myosin units functioning in an asynchronous manner—possibly like changes in the position of hands on a rope in the game of tug-of-war—can maintain force or create maximum force in this manner. Thus, when some myosin head groups bind with high affinity, others have low affinity and this binding, pulling (conformational change), release, binding action of myosin shortens the sarcomere in an appropriately forceful manner.

Calcium Regulates Smooth Muscle Contraction

Calcium ions play an important role in smooth muscle contraction also, but there are some important differences in the mechanism by which it acts. A mechanism for calcium regulation of smooth muscle contraction is shown in Figure 23.36. Key

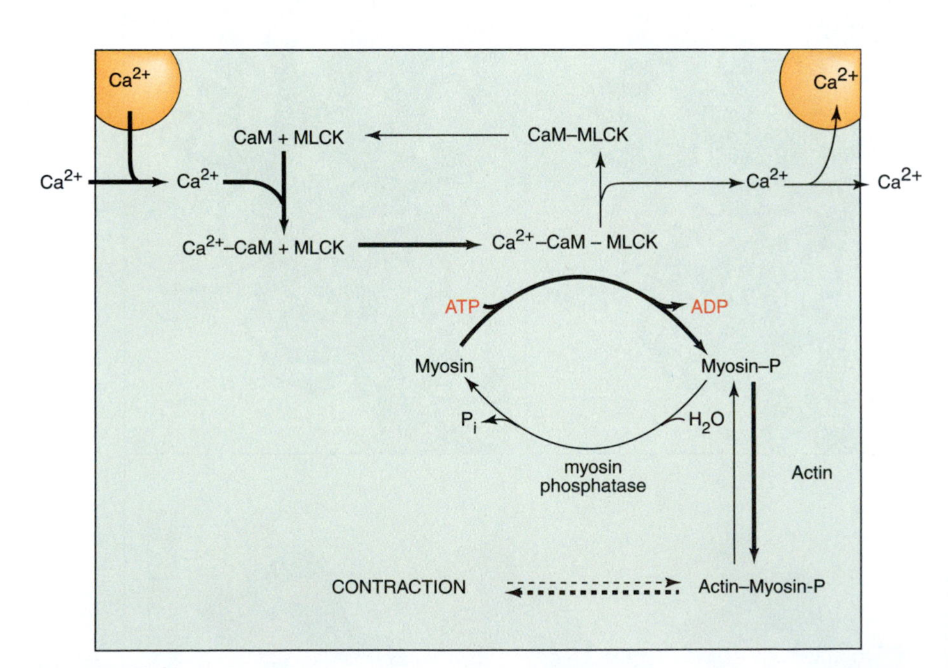

FIGURE 23.36
Schematic representation of the mechanism of regulation of smooth muscle contraction.
Heavy arrows show the pathway for tension development and light arrows show the pathway for release of tension. The Mg^{2+}-ATPase activity is highest in the actin–myosin–P complex. CM, calmodulin; MLCK, myosin light chain kinase.
Adapted from Kramm, K. E. and Stull, J. T. Annu. Rev. Pharmacol. Toxicol. *25:593, 1985.*

elements of this mechanism are as follows (1) A phosphorylated myosin light chain stimulates myosin Mg-ATPase, which supplies energy for the contractile process. (2) Myosin light chain is phosphorylated by a **myosin light chain kinase (MLCK).** (3) A Ca^{2+}–calmodulin (CaM) complex activates MLCK. (4) Formation of the Ca^{2+}–CM complex is dependent on the concentration of intracellular Ca^{2+}. Release of Ca^{2+} from its intracellular stores or an increase in its flux across the plasma membrane is important for control. (5) Contraction is stopped or decreased by the action of a **myosin phosphatase** or the transport of Ca^{2+} out of the cell. It is apparent that, in smooth muscle, many more biochemical steps are involved in the regulation of contraction, steps that can be regulated by hormones and other agents. They serve the function of smooth muscles well, giving them the ability to develop various degrees of tension and to retain it for prolonged periods of time.

23.5 | MECHANISM OF BLOOD COAGULATION

Circulation of blood is essential for life, and the integrity of the process must be maintained. The importance of blood circulation in the transport of oxygen and nutrients to cells, in the transport of carbon dioxide and waste products from cells, and in the maintenance of pH are well known. Blood circulation occurs in a very specialized type of closed system in which the volume of circulating fluid is maintained fairly constant. This system is one in which the transfer of solutes across its boundaries is a necessary function. As in any system of pipes and tubes, leaks can occur as a result of various types of insults and they must be repaired in order to maintain a state of hemostasis, that is, no bleeding. This section concentrates on some of the biochemical processes involved in hemostasis, particularly blood clotting.

Biochemical Processes of Hemostasis

Hemostasis implies that the process of clot formation **(procoagulation)** is in balance with the processes of stopping clot formation **(anticoagulation)** and of clot dissolution **(fibrinolosis).** The process of procoagulation leads to production of fibrin from fibrinogen and aggregation of fibrin into an insoluble network, or clot, which covers the ruptured area and prevents further loss of blood. During the same time period, aggregation of blood platelets occurs at the site of injury. Platelet aggregation forms a physical plug to help stop the leak. The morphological changes platelets undergo cause the release of chemicals they contain that aid in other aspects of the overall process. Some of these released chemicals enhance platelet aggregation in an autofacilitative manner, some cause vasoconstriction to reduce blood flow to the area, and some are enzymes that aid directly in clot formation. The steps involved in the initial phase of this process are shown schematically in the following diagram.

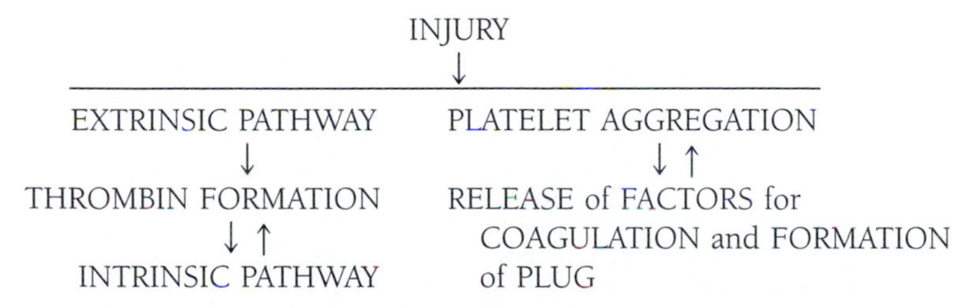

The dual arrows are not intended to imply that those steps are directly reversible, but to indicate that the processes they join are mutually facilitative.

The next phase of hemostasis, the anticoagulation phase, begins early in the process, as soon as it is kinetically feasible, and prevents excessive clot formation. Finally, the process of clot dissolution or fibrinolosis occurs when the injured vessel itself is repaired sufficiently well enough to prevent further bleeding. Many of the proteins involved in blood coagulation contain **epidermal growth factor**

(EGF)-like domains that may act directly or indirectly to facilitate regrowth of damaged areas in blood vessels. All of these processes are dynamic and most are driven by kinetic and mass action processes. So one phase does not end before another phase begins. Blood clotting is not a process of signal transduction in the same sense as are other topics in this chapter. Instead, it is a dynamic process of signal amplification and modulation.

Some of the major proteins involved in this process are listed in Table 23.9, not necessarily in order of appearance. All are important and others are sure to be added. It is important to emphasize that structural anomalies, primarily mutations in critical regions of any of these proteins, may affect adversely the clotting and clot dissolution processes and lead to disease states.

Clot formation has been discussed as a process involving two separate pathways: **intrinsic** or **contact factor pathway** and **extrinsic** or **tissue factor pathway**, both of which lead to the activation of factor X. From this point on, there is a single pathway for clot formation. Historically, the term intrinsic pathway came from the observation that blood clotting could occur spontaneously when blood was placed in clean glass test tubes, leading to the idea that all components of the clotting process were intrinsic to circulating blood. Glass contains **anionic surfaces** that form nucleation points that initiate the process. In mammals, anionic surfaces are exposed upon rupture of the **endothelial lining** of blood vessels and are binding and activation sites for specific factors that initiate clotting in the intrinsic pathway. Similarly, extrinsic came from the observation that there was a factor extrinsic to circulating blood that facilitates blood clotting. This factor was identified as

TABLE 23.9 Some of the Factors Involved in Blood Coagulation, Control, and Clot Dissolution

Factor	*Name*	*Pathway*	*Characteristic*	*Concentration*[a]
I	Fibrinogen	Both		9.1
II	Prothrombin	Both	Contains N-terminal Gla residues	1.4
III	Tissue factor	Extrinsic	Transmembrane protein	—
IV	Calcium ions	Both		—
V	Proaccelerin	Both	Protein cofactor	0.03[b]
VII	Proconvertin	Extrinsic	Endopeptidase with Gla residues	0.01[c]
VIII	Antihemophilic	Intrinsic	Protein cofactor	0.0003[b]
IX	Christmas factor	Intrinsic	Endopeptidase with Gla residues	0.089
X	Stuart factor	Both	Endopeptidase with Gla residues	0.136
XI	Thromboplastin antecedent	Intrinsic	Endopeptidase	0.031
XII	Hageman factor	Intrinsic	Endopeptidase	0.375
XIII	Proglutamidase	Both	Transpeptidase	0.031[b]
	α_2-Antiplasmin		Plasmin inhibitor	0.953
	Antithrombin III	Both	Thrombin inhibitor	3.0
	Heparin Co-II	Both	Thrombin inhibitor	1.364
	HMWK[d]	Intrinsic	Receptor protein	0.636
	α_2-Macroglobulin		Proteinase inhibitor	2.9
	Plasminogen		Zymogen/clot dissolution	2.4
	Prekallikrein	Intrinsic	Zymogen/activator of factor XII	0.581
	Protein C	(Both)	Endopeptidase with Gla residues	0.065
	Protein C inhibitor		Protein C inhibitor	0.070
	Protein S	(Both)	Cofactor with Gla residues	0.030
	Protein Z	Both	Gla-containing cofactor for protein Z inhibitor	0.08
	Protein Z inhibitor	Both	Inhibitor of factor Xa	0.018
	TFPI[e]		Tissue factor pathway inhibitor	0.003

[a]Concentrations are approximate and shown as micromolar.

[b]These values approximate solution concentrations since some are complexed with other proteins in platelets.

[c]This factor probably circulates as both FVII and FVIIa.

[d]HMWK is high molecular weight kininogen.

[e]TFPI is tissue factor pathway inhibitor, formely known as liporotein-associated coagulation factor (LAC1).

factor III, or tissue factor (see Figure 23.38*a*), an integral membrane protein that also becomes exposed upon rupture of blood vessels. Whether intrinsic or extrinsic, the process of blood coagulation is initiated on the membrane and is continued on the membrane surface at the site of injury. It is becoming more apparent that there is a great degree of interdependence on both traditionally designated pathways for effecting the coagulation process.

Throughout this section, coagulation factors in their nonactivated form are referred to by the letter "F" followed by the traditional Roman number designation for that factor, for example, **FVII** for factor VII. The letter "a" will indicate activated forms, for example, FVIIa for the activated form of factor VII. Other abbreviations will be defined as required.

Procoagulation Phase of Hemostasis (Phase 1)

Kinetic analyses of the appearance and concentrations of various proteins and zymogens involved in the blood coagulation cascade demonstrate that the extrinsic pathway initiates formation of a blood clot. This is a one-step process that activates the key enzyme, FX to FXa, that is responsible for prothrombin activation. Thrombin, which results from prothrombin during the activation process, is a key enzyme for clot formation and for the initiation of many other reactions involved in the overall process of hemostasis. A schematic of the reactions involved in the procoagulation phase, including both extrinsic and intrinsic pathways, is shown in Figure 23.37.

Extrinsic Pathway and Initiation of Coagulation

The membrane receptor that initiates the procoagulation phase is FIII or **tissue factor (TF),** a transmembrane protein of 263 amino acids. Residues 1–219 are on

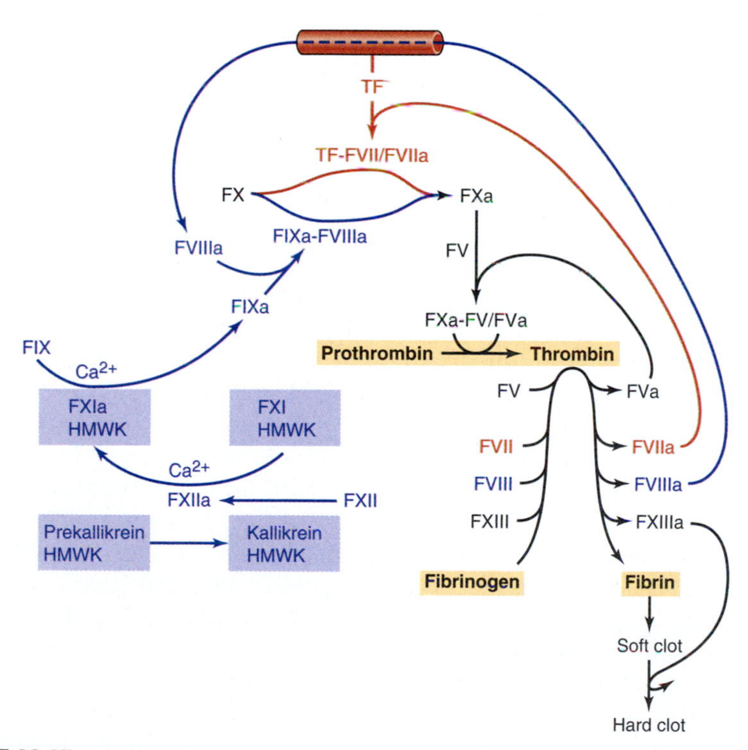

FIGURE 23.37
Diagrammatic representation of phase 1 of the blood coagulation scheme. Reactions of the extrinsic pathway are shown in red, those of the intrinsic pathway are shown in blue, and those common to both pathways are shown in black. Major end products of the scheme are highlighted in yellow. HMWK, high molecular weight kininogen. Activated factors are designated with an "a."

the extracellular side of the membrane, are exposed after injury, and form the receptor to which FVII binds. FVII is a **γ-carboxyglutamyl** (Gla)-containing protein that binds to tissue factor only in the presence of Ca^{2+}. The resulting complex (FVII–Ca^{2+}–TF) is the initial enzyme complex of the extrinsic pathway, and its action initiates the blood clotting process. TF and FVII are unique to the extrinsic pathway and are essentially all of its major components. The zymogen form of FVII is initially activated through protein–protein interaction as a result of its binding to TF. Additional FVII is activated by thrombin through proteolytic cleavage. A small amount of FVIIa normally exists. Unlike other proteinases of the blood coagulation scheme, FVIIa has a long half-life in circulating blood. Once dissociated from TF, FVIIa is not catalytically active, and its presence in blood is harmless. Formation of the initial complex with TF could involve some of the already preformed FVIIa. Thus it is not possible to state with absolute certainty whether it is the zymogen form of FVII in complex with TF or the activated form, FVIIa, that is responsible for the initial activation of FX.

Thrombin Formation

FXa and FV form a complex referred to as prothrombinase, which catalyzes formation of thrombin from prothrombin through a proteolytic cleavage reaction. The relatively small amount of thrombin formed in this initiation phase (Figure 23.37) catalyzes the activation of FV, FVII, FVIII, and FXIII in addition to the formation of fibrin from fibrinogen.

Reactions described so far are those involved in initiating the process but are insufficient to form fibrin fast enough to form an effective clot. Formation of thrombin is key to sustaining and accelerating the extrinsic pathway as well as activating one of the major protein cofactors, FVIII, that is required for the intrinsic pathway. The components of the system are now activated and poised for rapid clot formation.

Reactions of the Intrinsic Pathway

Reactions of the intrinsic pathway are also shown in Figure 23.37. Injury to the endothelial lining of blood vessels also exposes anionic membrane surfaces. The proteinase zymogen, **FXII,** binds directly to some of these anionic surfaces and undergoes a conformational change that increases its catalytic activity 10^4- to 10^5-fold. **Prekallikrein** and **FXI,** also zymogens, circulate in blood as separate complexes with **high molecular weight kininogen (HMWK)**—either as a FXI–HMWK complex or as a prekallikrein–HMWK complex. FXI and prekallikrein are attached to anionic sites of exposed membrane surfaces through their interactions with HMWK. This interaction brings those zymogens to the site of injury and in direct proximity to FXII. The membrane-bound "activated" form of FXII activates prekallikrein to yield **kallikrein.** Kallikrein further activates FXII to give FXIIa. FXI is membrane bound through its noncovalent attachment to HMWK and is activated by FXIIa through proteolytic cleavage to give FXIa. FXIa activates **FIX** to FIXa (see Clin. Corr. 23.12). FIXa in the presence of **FVIIIa** also activates **FX** to FXa. This is essentially a four-step cascade started by the "contact" activation of FXII and the autocatalytic action between FXII and kallikrein to give FXIIa (step 1). FXIIa activates FXI (step 2); FXIa activates FIX (step 3); and FIXa, in the presence of FVIIIa, activates FX (step 4). If each enzyme molecule activated also catalyzed the formation of 100 others before it is inactivated, the amplification factor would be 1×10^6 from this part of the pathway alone. As shown in the diagram, however, several feedback loops accelerate the overall process to produce a fibrin clot in a rapid and efficient manner. During this time, FXIII, a transglutaminase that has been activated also by thrombin, is actively forming a hard clot by catalyzing the formation of cross-links between fibrin monomers of the soft clot. This is the overall process of clot formation. Now it will be instructive to discuss some properties of the proteins involved in this scheme.

Tissue factor (TF), (FIII): TF (Figure 23.38*a*) is a transmembrane protein of 263 amino acids. Residues 243–263 are located on the cytosolic side of the membrane sequence. Residues 1–219 are on the outside of the membrane, are exposed after injury, and form the receptor for formation of the initial complex of the extrinsic pathway. This domain is glycosylated and contains four cysteine residues. A stereo representation of a section of it highlighting some of the amino acid residues involved in **FVII** binding is shown in Figure 23.38*b*.

Factor VII: A 3-D ribbon structural representation of FVIIa is shown in Figure 23.39. The region for TF interaction, Ca^{2+} binding, and the substrate-binding pocket are highlighted.

Factor X: A stereo view of FXa is shown in Figure 23.40. Both the extrinsic and the intrinsic pathways form FXa by cleavage of FX at positions 145 and 151 with elimination of a six amino acid peptide.

Thrombin: Thrombin circulates in plasma as prothrombin. Prothrombin is a 72-kDa protein (Figure 23.41) that contains ten **γ-carboxyglutamate (Gla) residues** in its N-terminal region. Binding of Ca^{2+} to these residues facilitates binding of prothrombin to membrane surfaces and to the FXa–FVa complex at the site of injury. The prothrombinase complex (FXa–FVa) activates prothrombin by making two proteolytic cleavages on the carboxyl side of arginine residues, first at position 320 and then at position 284. The active thrombin molecule (α-thrombin) consists of two chains, 6 kDa and 31 kDa, that are covalently linked by a disulfide bond. A stereo view of the active α-thrombin molecule is shown in Figure 23.42. Regions involved in some of its functions are highlighted. The substrate for thrombin is fibrinogen.

Fibrinogen/Fibrin: Fibrinogen is a large molecule of approximately 340 kDa consisting of two tripeptide units with α, β, γ structure (Figure 23.43). The subunits are linked together at their N-terminal regions by disulfide bonds. Fibrinogen has three globular domains, one on each end and one in the middle where the chains are joined. The globular domains are separated by rod-like domains. A short segment of the free N-terminal regions project out from the central globular domain. The N-terminal regions of the $\alpha\alpha'$ and the $\beta\beta'$ subunits, through charge–charge repulsion, prevent aggregation of fibrinogen. Thrombin cleaves these N-terminal peptides and allows the resulting fibrin molecules to aggregate and to form a **"soft" clot.** The soft clot is stabilized and strengthened by the action of **FXIIIa**, a transglutaminase. This enzyme catalyzes formation of an isopeptide linkage by re-

CLINICAL CORRELATION 23.12
Intrinsic Pathway Defects: Prekallikrein Deficiency

Components of the intrinsic pathway include factor XII (Hageman factor), factor XI, prekallikrein (Fletcher factor), and high molecular weight kininogen. Clinical disorders have been associated with defects in each of these components. Inherited disorders in each appear to be autosomal recessive. Each appears to be associated with an increase in activated partial thromboplastin time (APTT). The only one of these components directly associated with a clinical bleeding disorder is factor XI deficiency.

In some cases where there is a prekallikrein (Fletcher factor) deficiency, autocorrection after prolongation of the preincubation phase of the APTT test occurs. This phenomenon is explained by the ability of factor XII to be activated by an autocatalytic mechanism. The reaction is very slow in prekallikrein deficiency since the rapid reciprocal autoactivation between factor XII and prekallikrein cannot take place. Prekallikrein deficiency may be due to a decrease in the amount of the protein synthesized, to a genetic alteration in the protein itself that interferes with its ability to be activated, or its ability to activate factor XII. A lack of knowledge of the structure of the gene for prekallikrein precludes definitive explanations of the mechanisms operational in patients with prekallikrein deficiency. Specific deficiencies of the intrinsic pathway, however, can be localized to a specific factor if the appropriate number of tests are performed. These may include a direct measurement of the amount of each of the factors present in the patients plasma in addition to APTT test performed with and without prolonged preincubation time. Use of these direct measurements helped diagnose a prekallikrein deficiency in a 9-year-old girl who had a prolonged APTT. The functional level of prekallikrein in this patient was less than 1/50th of the minimum normal value. An immunological test (ELISA, see p. 456) showed an antigen level of 20–25%, suggesting that she was synthesizing a dysfunctional molecule.

Coleman, R. W., Rao, A. K., and Rubin, R. N. Fletcher factor deficiency in a 9-year-old girl: mechanisms of the contact pathway of blood coagulation. *Am. J. Hematol.* 48:273, 1995.

(a)

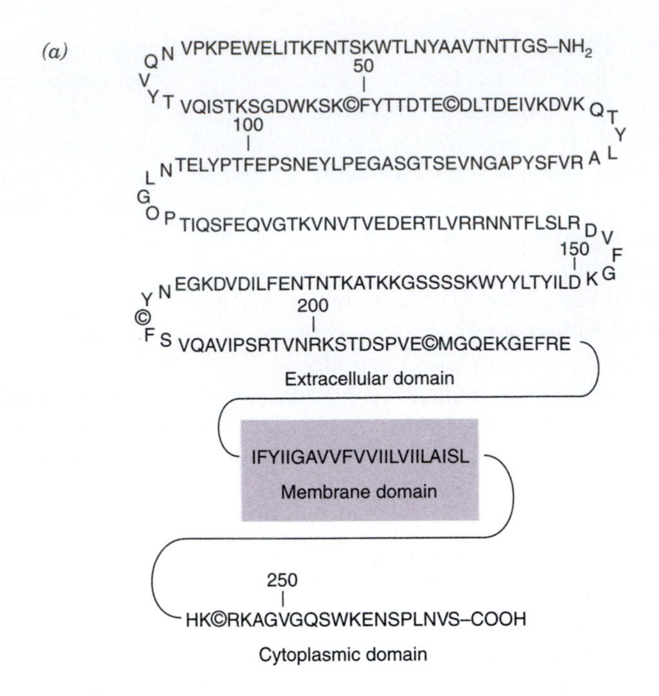

(b)

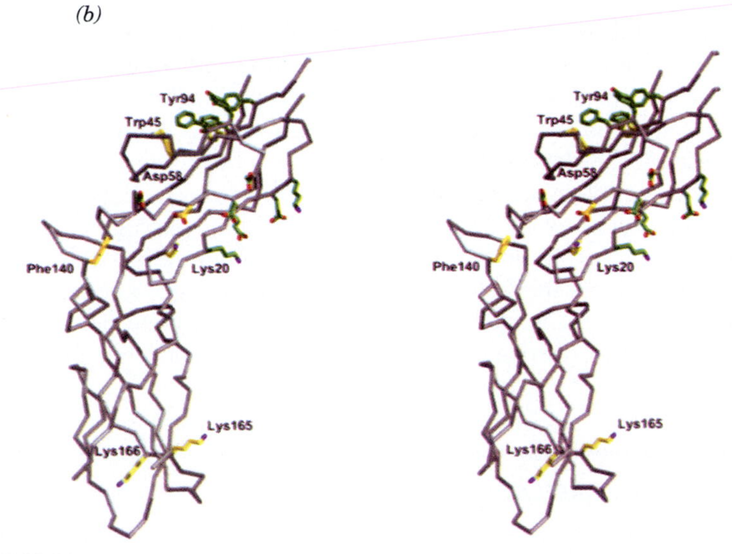

FIGURE 23.38
Tissue factor.
(*a*) Amino acid sequence of human tissue factor derived from its cDNA sequence. (*b*) A stereo representation of the carbon chain of the extracellular domain of tissue factor. Residues important for binding of factor VII are shown in yellow. Clusters of aromatic and charged residues are shown in light blue.
Part (a) redrawn from Spicer, E. K., et al. Proc. Natl. Acad. Sci. USA *84:5148, 1987; part (b) reproduced with permission from Muller, Y. A., Ultsch, M. H., Kelley, R. F., and deVos, A. M.* Biochemistry *33:10864, 1994. Copyright, 1994, American Chemical Society.*

placing the δ-amide group of glutamine residues of one chain with the ε-amino group of lysine residues of another chain (Figure 23.44) with the release of ammonia. This cross-linking of fibrin completes the steps involved in formation of the hard clot.

Factor V: α-Thrombin activates the protein cofactors FV and FVIII and is involved in platelet aggregation. FV is a 330-kDa protein. Activation of FV by thrombin occurs through proteolytic cleavage at Arg^{709} and Arg^{1545}. FVa is a heterodimer consisting of an N-terminal domain (105 kDa) and a C-terminal domain (74 kDa) held together noncovalently by Ca^{2+} (Figure 23.45).

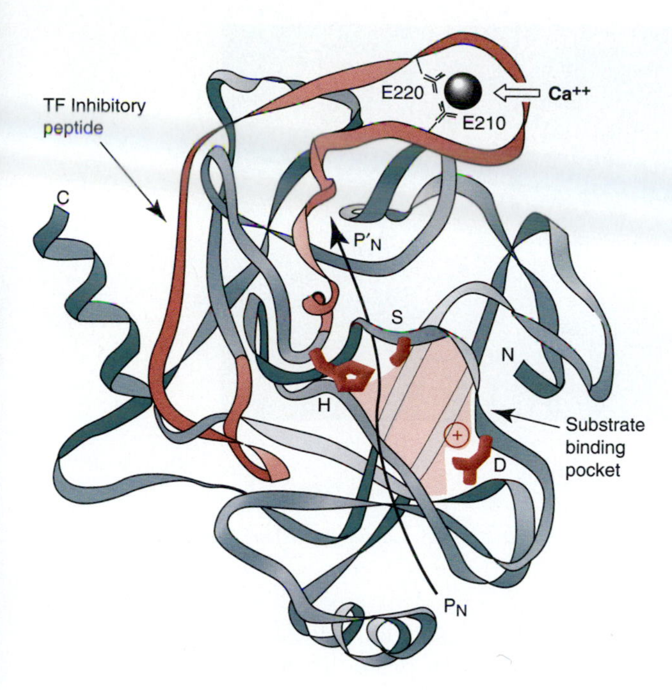

FIGURE 23.39
Ribbon structural representation of the protease domain of factor VIIa.
The dark ribbon labeled "TF inhibitory peptide" represents a section involved in binding to tissue factor. The catalytic triad is shown in the substrate binding pocket as H, S, and D for His 193, Ser 344, and Asp 338, respectively. The arrow labeled P_N–P'_N lies in the putative extended substrate-binding region.
Reproduced with permission from Sabharwal, A. K., Birktoft, J. J., Gorka, J., et al. J. Biol. Chem. *270:15523, 1995.*

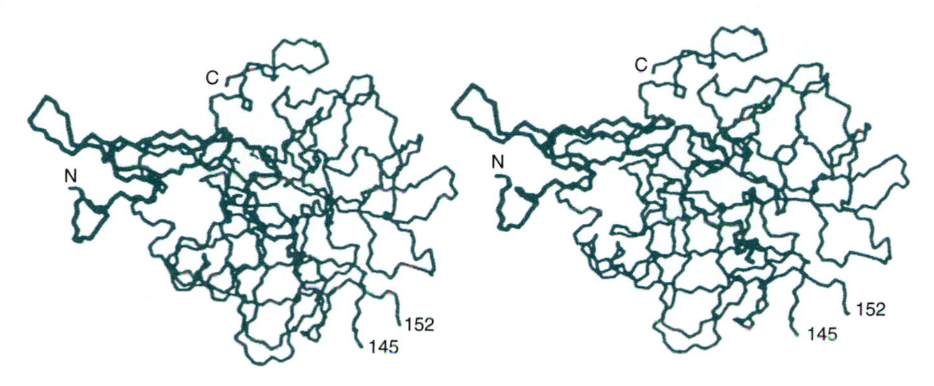

FIGURE 23.40
Stereo view of the CN backbone structure of factor Xa.
The EGF-like domain is in bold.
Reproduced from the work of Padmanabhan, K., Padmanabhan, K. P., Tulinsky, A., et al. J. Mol. Biol. *232:947, 1993.*

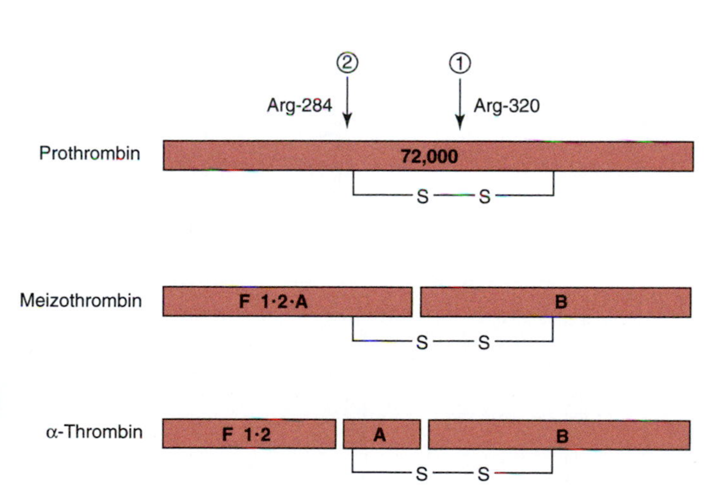

FIGURE 23.41
Schematic diagram of prothrombin activation.

Factor VIII: FVIII circulates in plasma attached to another protein, **von Willebrand's factor (vWF).** FVIII is a 285-kDa protein that is activated by thrombin cleavage at Arg^{372}, Arg^{740}, Arg^{1648}, and Arg^{1689}. This last cleavage releases FVIIIa from vWF. FVIIIa is a heterotrimer (Figure 23.45) composed of N-terminal peptides of 40 kDa (A_2) and 50 kDa (A_1), and a C-terminal peptide of 74 kDa (A_3). FVIIIa also contains a Ca^{2+} bridge between the N- and C-terminal domains. Classic he-

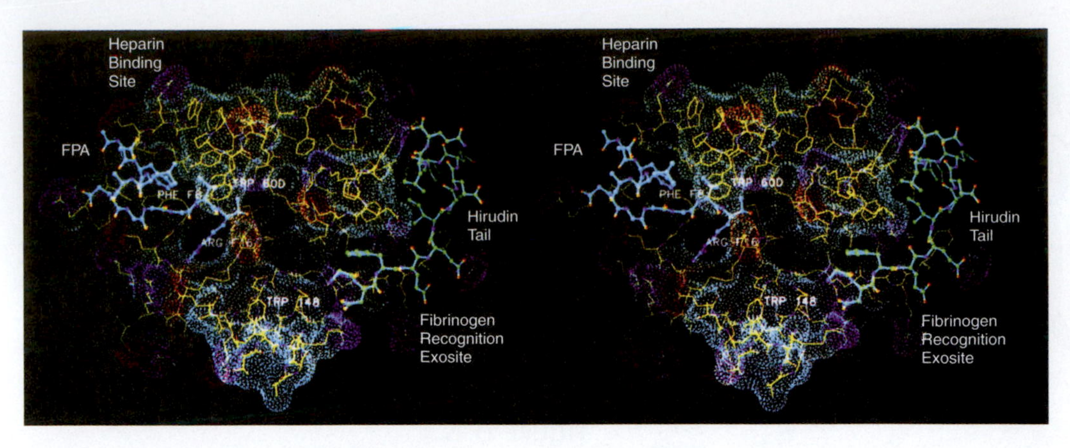

FIGURE 23.42
Stereo view of the active site cleft of human α-thrombin.
Dark blue, basic amino acids; red, acid; light blue, neutral. The active site goes from left to right.
Reprinted from Stubbs, M. T. and Bode. W. The clot thickens: clues provided by thrombin structure. Trends Biol. Sci. 20; 23, 1995, with permission from Elsevier Science.

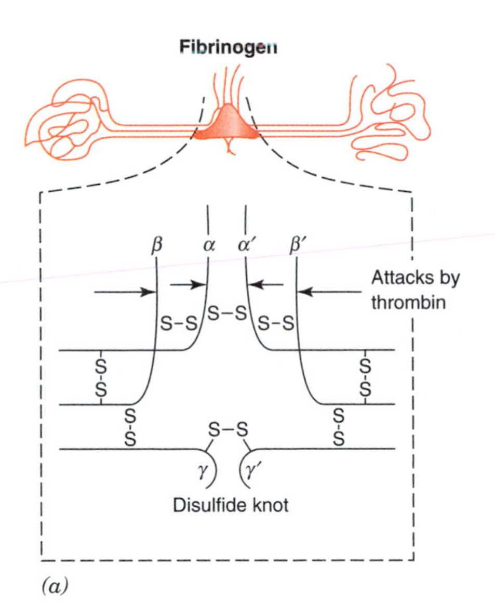

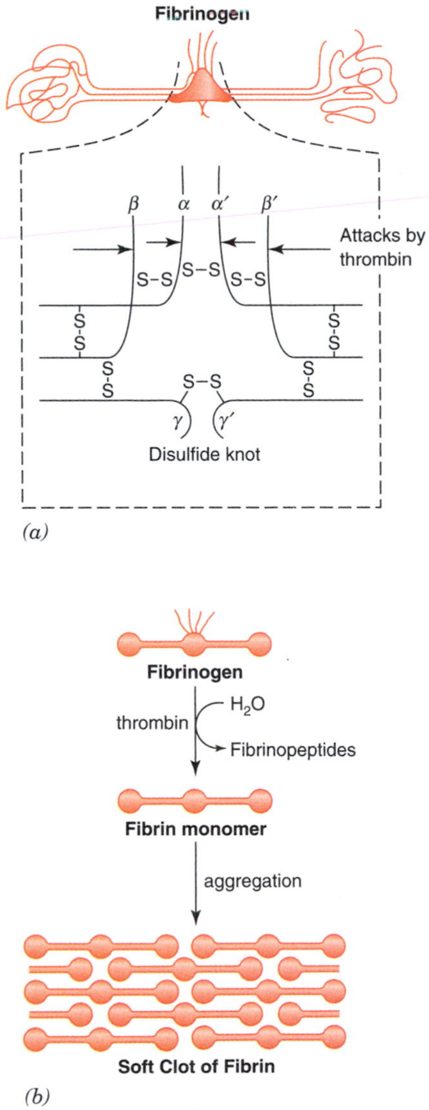

mophilia results from a deficiency in FVIII (see Clin. Corr. 23.13). It should be apparent from description of the procoagulation phase that reactions of the intrinsic pathway are necessary for rapid formation of a clot at the site of injury.

Factor XIII: Thrombin also activates FXIII, transglutaminase (Figure 23.46). **Protransglutaminase** exists in both plasma and platelets. The structural form of the platelet enzyme is α_2, whereas that of the plasma form is $\alpha_2\beta_2$. Thrombin activates FXIII by specific cleavage of a peptide bond in the α subunit of platelet and plasma forms of transglutaminase. Cleavage in the α subunit of the plasma form of the enzyme leads to dissociation of the β subunit, which is not catalytically active. The platelet form of the enzyme is released at the site of fibrin aggregation and is activated just by a bond cleavage within the α subunit.

High Molecular Weight Kininogen (HMWK): Two of the proteins involved in the intrinsic pathway circulate in blood as separate complexes with HMWK, prekallikrein and FXI. A molecule of HMWK binds to one or the other of these pro-

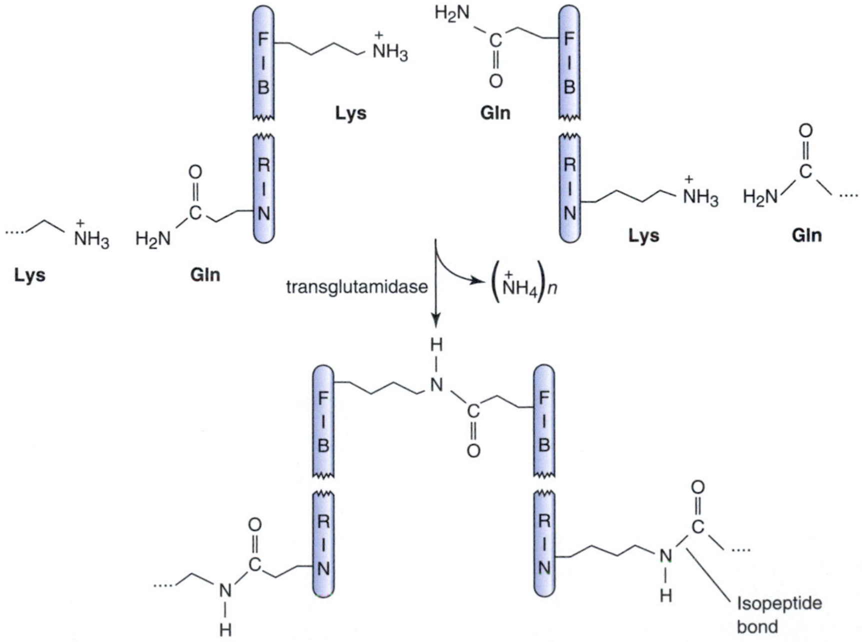

Fibrinogen
thrombin — H_2O → Fibrinopeptides
Fibrin monomer
aggregation
Soft Clot of Fibrin
(b)

FIGURE 23.43
Diagrammatic representation of the fibrinogen molecule and its conversion to the soft clot of fibrin.

Lys, Gln, FIB, RIN, $\overset{+}{N}H_3$, H_2N, transglutamidase, $(\overset{+}{N}H_4)_n$, Isopeptide bond

FIGURE 23.44
Reactions catalyzed by transglutaminase.

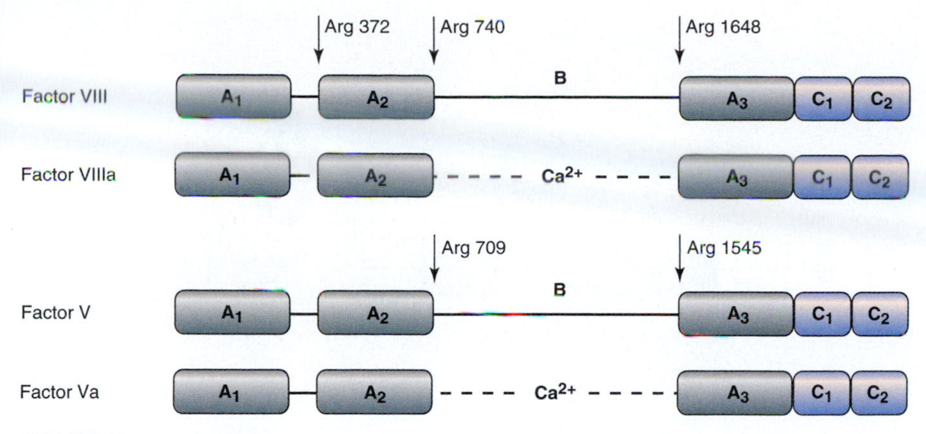

FIGURE 23.45
Organizational structure of cofactor proteins, factors VIII and V.
Positions for thrombin cleavage are shown. A's and C's represent structural domains.
Redrawn from Kalafatis, M., Swords, N. A., Rand, M. D., and Mann, K. G. Biochim. Biophys. Acta *1227:113, 1994.*

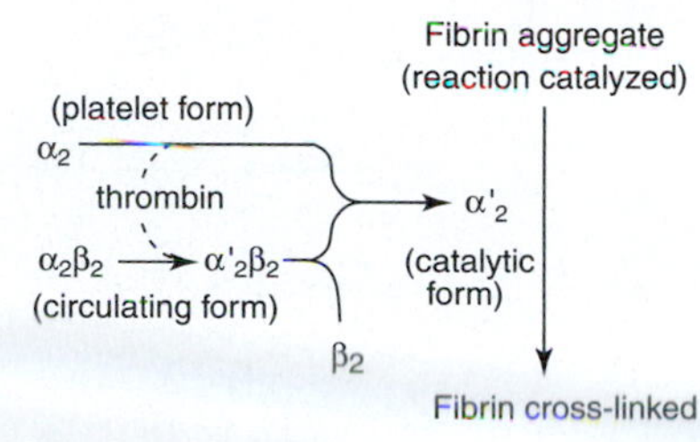

FIGURE 23.46
Activation of transglutaminase by thrombin.

or a prekallikrein–HMWK complex. Figure 23.47 shows a schematic diagram of the functional regions of HMWK. The binding site on HMWK for prekallikrein consists of approximately 31 amino acid residues. FXI binds to approximately 58 amino acid residues that include the 31 to which prekallikrein binds. Prekallikrein is a 619 amino acid protein. FXIIa converts prekallikrein to kallikrein by cleavage of the peptide bond between Arg^{371} and Ile^{372}. Kallikrein contains two chains covalently linked by a single disulfide bond. The C-terminal domain of 248 amino acids contains the catalytic site.

Formation of Platelet Plug

Clumping of platelets at the site of injury is mediated by thrombin. There is a **thrombin receptor,** a member of the seven-transmembrane-domain family of receptors, on endothelial cells. This receptor is exposed upon injury and is activated by α-thrombin. Aggregation of platelets is facilitated by their initial binding to this activated receptor. In addition to formation of a physical plug, platelets undergo a morphological change and release chemicals such as ADP, serotonin, some types of

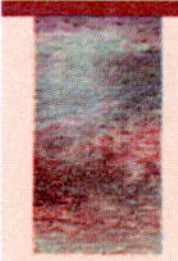

CLINICAL CORRELATION 23.13

Classic Hemophilia

Hemophilia is an inherited disorder characterized by a permanent tendency for hemorrhages, spontaneous or traumatic, due to a defective blood clotting system. Classic hemophilia, hemophilia A, is an X-linked recessive disorder characterized by a deficiency of factor VIII. About 1 in 10,000 males is born with a deficiency of factor VIII. Of the approximate 25,000 hemophiliacs in the United States, more than 80% are of the A type. Hemophilia B is due to a dysfunction in factor IX.

Some hemophilia A patients may have a normal prothrombin time if the concentration of tissue factor is high. One possible explanation for this is that factor V in human plasma is much lower in concentration than factor X. Activation of an amount of factor X to Xa in excess of that required to bind all of factor Va would initiate blood clotting by the extrinsic pathway and give a normal prothrombin time. The intrinsic pathway would not function normally due to the deficiency in factor VIII. Without the two pathways operating in concert, the overall process of blood clotting would be impaired. Both factor Xa and thrombin activate factor V and are involved in a number of other reactions. If the overall process is not accelerated at its onset by intervention of the intrinsic pathway, due to kinetics of the interaction of thrombin and factor Xa with the normally low concentration of factor V, the clotting disorder is expressed. The blood levels of factor VIII in severe hemophilia A patients are less than 5% of normal. These patients have generally been treated by blood transfusion with its associated dangers: the possibility of contraction of hepatitis or HIV, and the 6% possibility of patients making autoantibodies. Treatment of hemophiliacs has been made much safer as a result of cloning and expression of the gene for factor VIII. The pure protein can be administered to patients with none of the associated dangers mentioned above.

Nemerson, Y. *Tissue factor and hemostasis.* Blood 71:1, 1988.

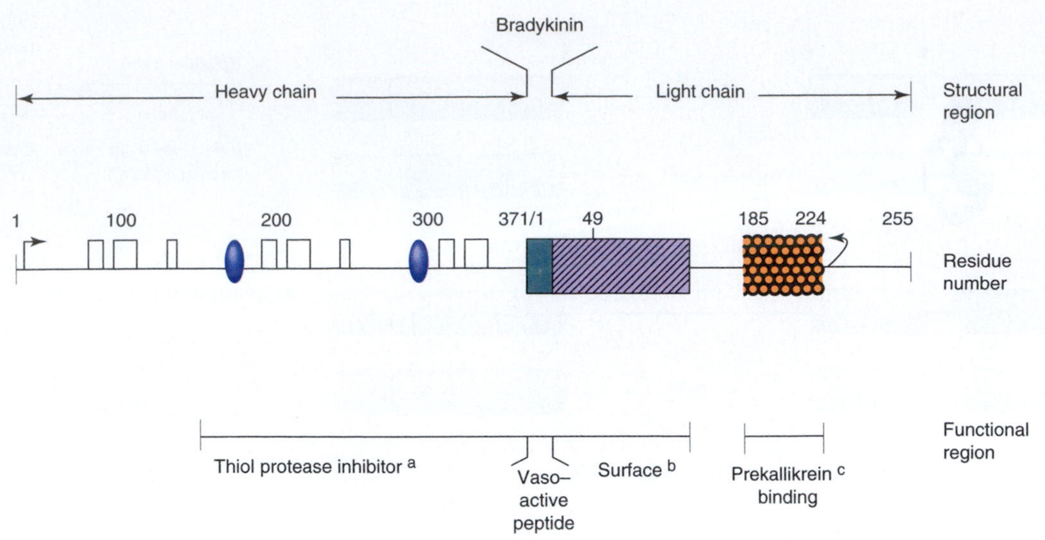

FIGURE 23.47
Schematic diagram of the functional regions of human high molecular weight kininogen (HMWK).
Bradykinin is derived from near the middle of HMWK by proteolysis. The resulting two chains are held together by disulfide bonds, horizontal arrows.
Redrawn from Tait, J. F. and Fujikawa, K. J. Biol. Chem. 261:15396, 1986.

morphological change and release chemicals such as ADP, serotonin, some types of phospholipids, and proteins that aid in coagulation and tissue repair (Figure 23.48). A glycoprotein, von Willebrand's factor (vWF), is released, concentrates in the area of the injury, and forms a link between the exposed receptor and the platelets. von Willebrand's factor also serves as a carrier for FVIII.

Platelet aggregation becomes autocatalytic with the release of ADP and **thromboxane A_2.** Platelet FIV, a **heparin-binding protein,** prevents premature formation of heparin–antithrombin III complexes and it attracts cells with anti-inflammatory

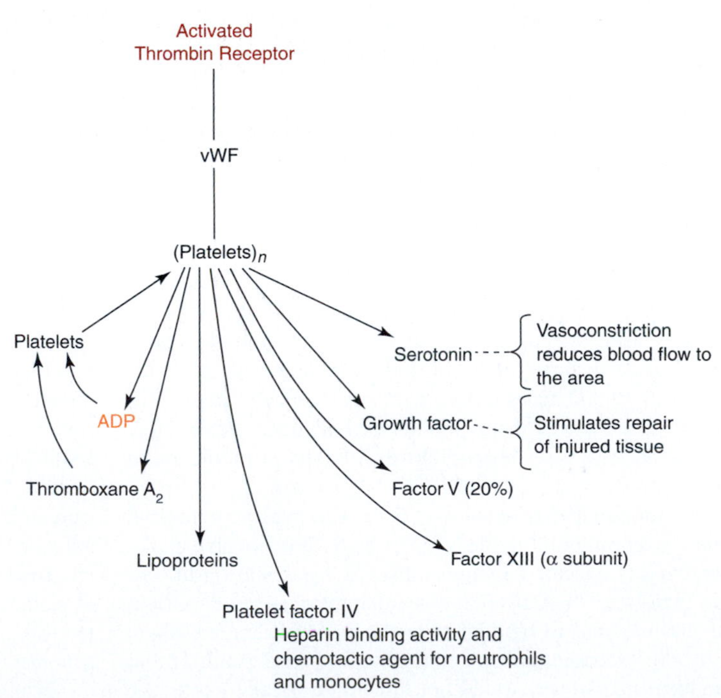

FIGURE 23.48
Action of platelets in blood coagulation.

activity to the site of injury. About 20% of FV and one form of FXIII, the transglutaminase, exist in platelets. Intact normal vascular endothelium does not normally initiate platelet aggregation since receptors and other elements are not exposed and activators such as ADP are rapidly degraded or are not in blood in sufficient concentration to be effective. The endothelium also secretes **prostacyclin (PGI_2)**, a potent inhibitor of platelet aggregation.

Anticoagulation Phase of Hemostasis (Phase 2)

Inhibition of proteinases is a kinetic process that begins almost as soon as coagulation itself. Initially, formation of inhibitor complexes is slow because concentrations of the enzymes with which the inhibitors interact are low. As activation of the enzymes proceeds, inhibition increases and become more prominent. These reactions, and destruction of protein cofactors, eventually stop the coagulation process completely. In general, **proteinase–inhibitor complexes** do not dissociate readily and are removed intact from blood by the liver.

Inhibition of Extrinsic Pathway

Inhibition of the extrinsic pathway, that is, the TF–FVIIa–Ca^{2+}–FXa complex, is unique and involves specific interaction with **tissue factor pathway inhibitor (TFPI),** also known as **lipoprotein-associated coagulation inhibitor (LACI)** and as **anticonvertin.** TFPI is a 32-kDa protein that contains three tandem domains (Figure 23.49). Each domain is a functionally homologous protease inhibitor that resembles other individual protease inhibitors such as the bovine **pancreatic trypsin inhibitor.** TFPI inhibits the extrinsic pathway by interacting specifically with the TF–FVIIa–Ca^{2+}–FXa complex. Domain 1 binds to FXa and domain 2 binds to FVIIa of the complex. Binding of TFPI to FVIIa does not occur unless FXa is present. Two aspects of TFPI action are worthy of additional emphasis. The first is that TFPI is a multienzyme inhibitor in which each of its separate domains inhibits the action of one of the enzymes of the multienzyme complex of the extrinsic pathway. The second is that the TFPI–FXa complex mediates internalization of FVIIa by an endocytosis mechanism. The C terminal (third domain) of TFPI appears to be necessary for this internalization process to occur. Most of FVIIa is degraded within the cells but a small amount returns to the surface as intact protein and is apparently one of the sources of circulating FVIIa. As stated above, FVIIa has no detrimental effects since it is active as a protease only when in a complex with TF.

Other proteinase inhibitors in blood interact with other enzymes of the blood coagulation system. Most of those inhibitors fit into the ***serine* proteinase *inhibitor* (serpin)** family of proteins. There is a similarity in tertiary structure among them and a common core domain of about 350 amino acids.

Antithrombin III (AT3) is a major serpin and inhibits several serine proteinases of the coagulation system, most specifically thrombin and FXa. AT3 inhibits thrombin and FXa by different mechanisms, both of which involve the interaction of heparin. Heparin is a highly sulfated oligosaccharide of the glucosaminoglycan type. It usually exists as a mixture of oligosaccharides spanning a range of molecular sizes. Inhibition of thrombin by AT3 occurs in a ternary complex of AT3, heparin, and thrombin. Heparin interaction in this complex is size-dependent. At least 18 saccharide units are required for the effective formation of this inhibited thrombin complex:

$$\text{thrombin} + \text{heparin}_{(>18)} + \text{AT3} \longrightarrow \text{AT3:heparin}_{(>18)}\text{:thrombin}_{(\text{inactive})}$$

AT3 also forms a complex with a specific pentasaccharide form of heparin. The structure of this pentasaccharide is shown in Figure 23.50. Shown in Figure 23.51*a* and 23.51*b* are structural models of AT3 alone and bound to this heparin pentasaccharide, respectively. This latter structure is sufficient to inhibit FXa.

$$\text{AT3} + \text{heparin}_{(5)} \longrightarrow \text{AT3:heparin}_{(5)} \xrightarrow{\text{FXa}} \text{AT3:heparin}_{(5)}\text{:FX}_{\text{inh}}$$

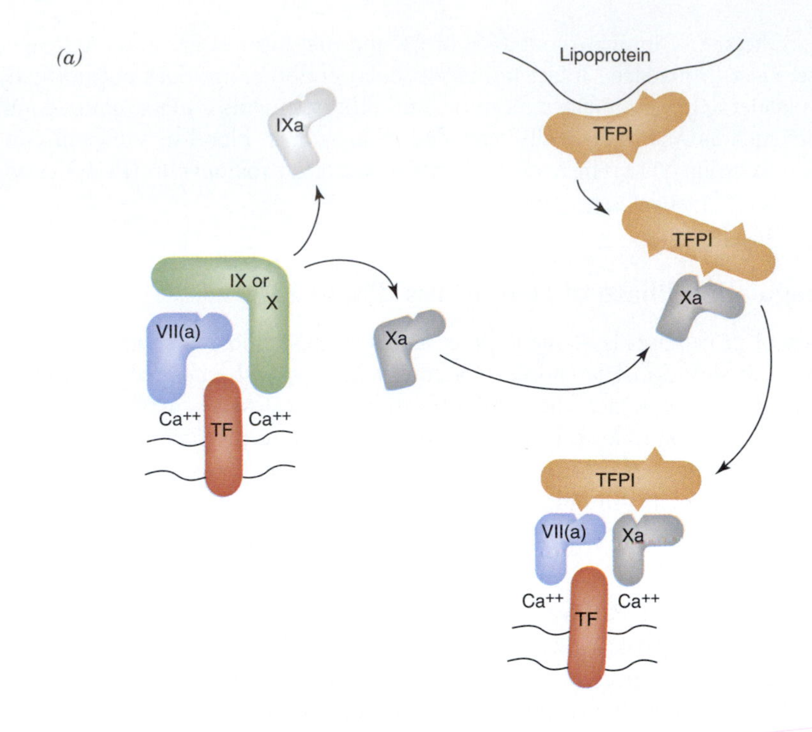

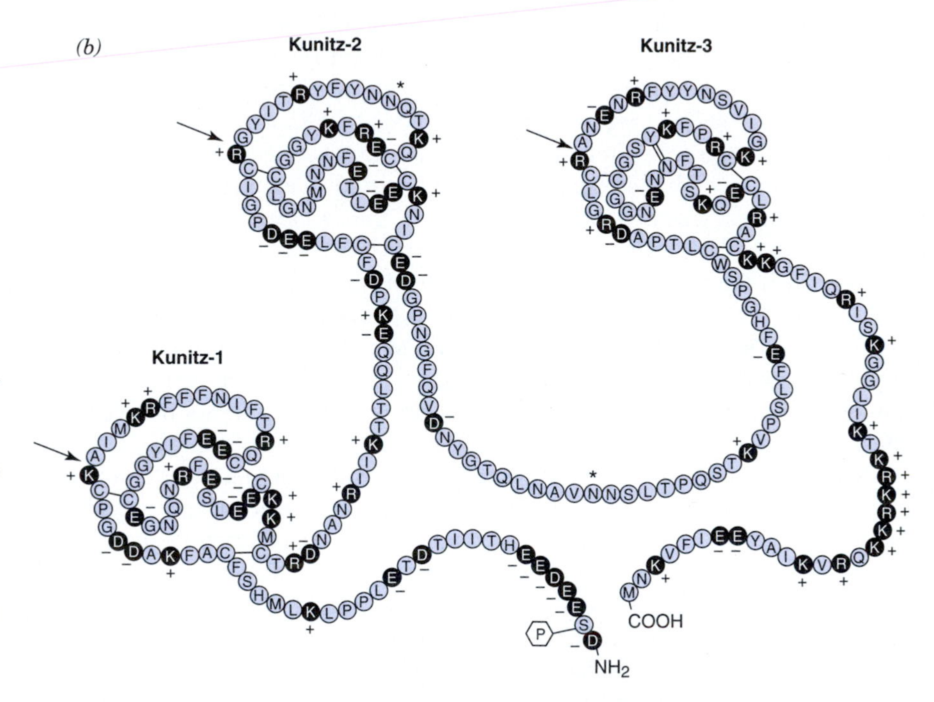

FIGURE 23.49
Proposed mechanism of inhibition of the extrinsic pathway.
TFPI is tissue factor pathway inhibitor whose structure is shown in (*b*). Kunitz domain 1 inhibits factor VIIa and Kunitz domain 2 inhibits factor Xa. Domain 3 is necessary for endocytosis of the complex. Arrows indicate the presumed location of the active-site inhibitor region for each domain.
Reproduced from Brooze, G. J., Girard, T. J., and Novotny, W. F. Biochemistry 29:7539, 1990. Copyright (1990) American Chemical Society.

Although AT3 can inhibit thrombin and FXa in the absence of heparin, heparin enhances inhibition of thrombin by a factor of approximately 9000 and inhibition of FXa by a factor of approximately 17,000.

There is another pathway for inhibition of FXa. Blood contains a 62-kDa glycoprotein, designated **protein Z (PZ),** that contains Gla residues in a structure similar to other Gla-containing proteases. Protein Z, however, does not have protease activity. In the presence of Ca^{2+} it can interact with membrane and form a complex with another plasma protein, **protein Z-dependent protease inhibitor (ZPI),** that inhibits FXa.

$$\text{FXa} + \text{Ca}^{2+} + \text{PZ} + \text{ZPI} \rightarrow (\text{Ca}^{2+}\text{:PZ})\text{:ZPI:FX}_{\text{inh}}$$

Protein Z-dependent protease inhibitor, 72 kDa, has a high affinity for FXa, but not for other proteases, including thrombin, FVIIa, FIXa, and protein C.

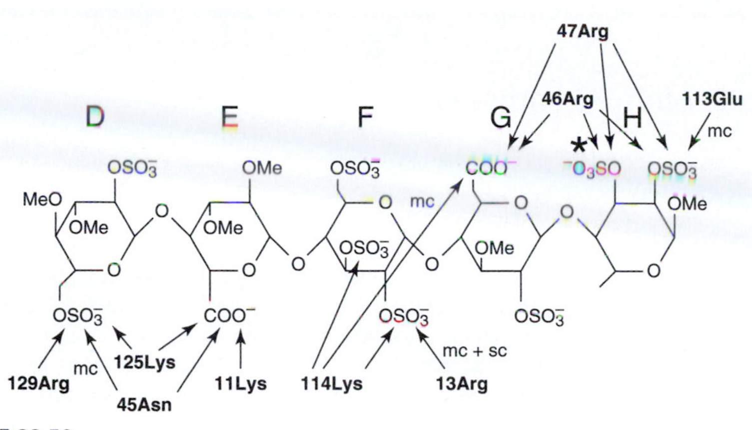

FIGURE 23.50
Chemical structure of the heparin pentasaccharide and its interactions with antithrombin.
Redrawn with permission from Whisstock, J. C., Pike, R. N., Jin, L., et al. J. Mol. Biol. *301:1287, 2000.*

Inactivation of FVa and FVIIIa

Another protein involved in the anticoagulation process is **protein C (PC).** Protein C, a Gla-containing protein, is activated in a membrane-bound complex of thrombin, **thrombomodulin,** and calcium ions. Protein C, in the presence of another protein cofactor, **protein S (PS),** inhibits coagulation by inactivating factors Va and VIIIa. Protein S is a 75-kDa Gla-containing protein. Deficiency and/or mutations in

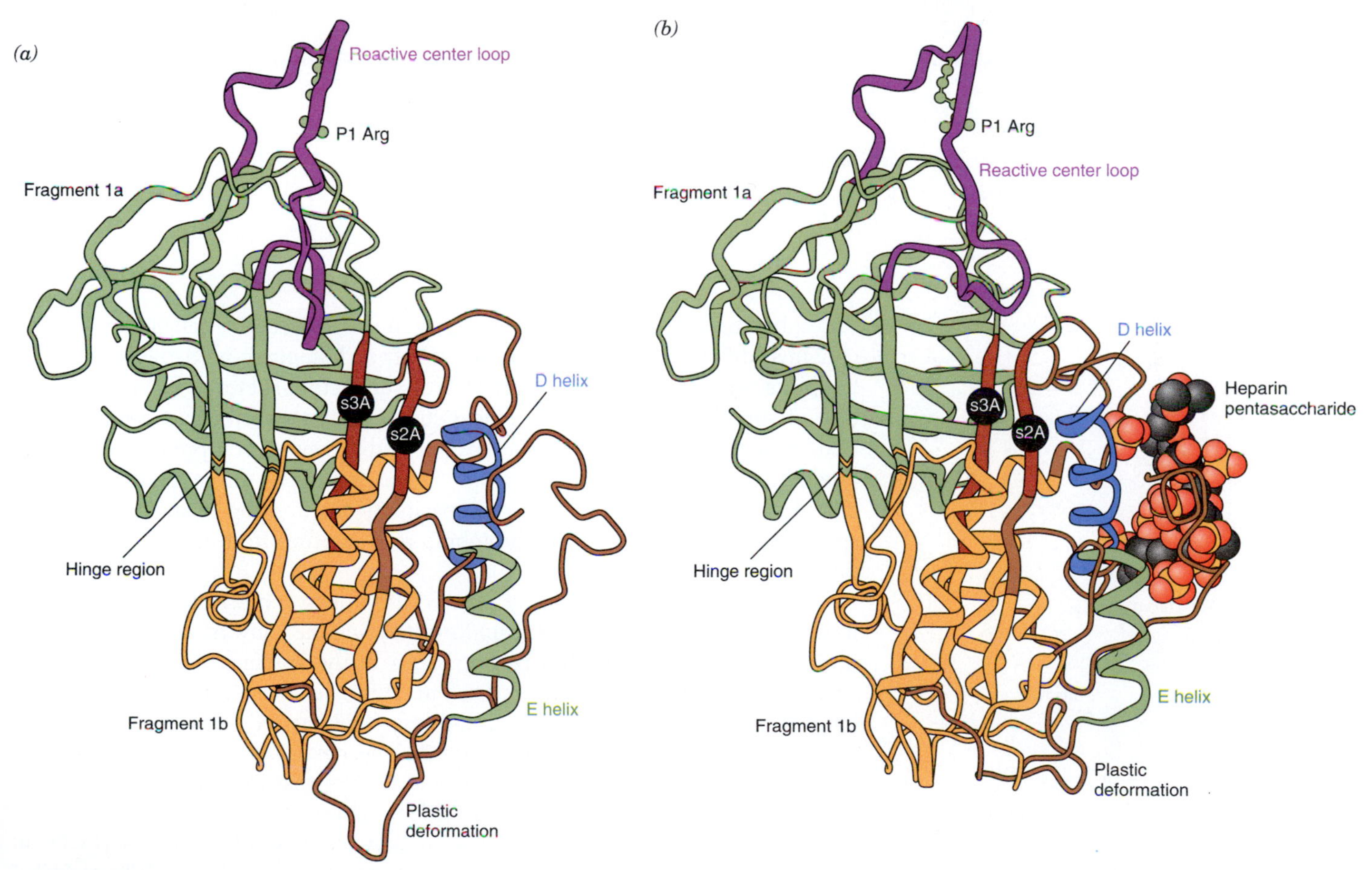

FIGURE 23.51
Fragments of antithrombin to which the heparin pentasaccharide binds.
Fragment (*a*) in the absence of the heparin pentasaccharide and (*b*) in the presence of the heparin pentasaccharide.

Reproduced with permission from Whisstock, J. C., Pike, R. N., Jin, L., et al. J. Mol. Biol. *301:1287, 2000.*

CLINICAL CORRELATION 23.14

Thrombosis and Defects of the Protein C Pathway

Four major proteins are involved in the action of protein C in regulating blood coagulation: protein C itself; protein S, a cofactor for protein C action; factor Va; and factor VIIIa. The latter two are substrates for catalytic action of the proteins C–S complex. Mutations, generally inherited, in any of them can result in venous thrombosis with various degrees of severity.

De novo mutations have also been identified in patients showing type I protein C deficiency. One was the result of a missense mutation, a transition of T to C resulting in the change of a codon for amino acid residue 270 from TCG to CCG. This gave Pro instead of Ser at that position, resulting in a conformational change that affected activity. The gene for protein C is on chromosome 2 and has 9 exons and 8 introns. In another patient, a *de novo* mutation located at the exon VI–intron f junction was detected. A 5-bp deletion (underlined below) occurred, resulting in a "read through" of sections of the intron.

Exon VI <> Intron f

Normal sequence: **CAC CCC GCA G**<>GTGAGAAGCCCCCAATAT---
His Pro Ala

Mutated sequence: **CAC CCC GCA** GGA GCC CCC AAT AT---
His Pro Ala Gly Ala Pro Asn -----

The sequence normally translated is in bold type. The degree of severity of thrombotic events depends on the extent to which the gene inherited from the other parent is normal and the extent to which it is expressed.

Resistance to the action of activated protein C as a result of single point mutations in its substrates, factor Va and factor VIIIa, can occur. This prevents or retards their inactivation through the proteolytic action of protein C. The most commonly identified cause of inherited resistance to the action of activated protein C is a single point mutation in the gene for factor V, R506Q also known as FV-Leiden.

A third cause of protein c-related thrombosis is a defect in protein S. Fewer specific details are available that permit a definition of the mechanism of the interaction between protein C and protein S, and likewise of the mutations that affect its function. It is quite clear, however, that protein S deficiency leads to thrombotic events. Venous thrombosis occurs in almost one-half of patients at some stage of their lives if they have deficiencies in functional amounts of protein S.

Gandrille, S., Jude, B., Alhenc-gelas, H., et al. First *de novo* mutations in the protein C gene of two patients with type I deficiency: a missense mutation and a splice site deletion. *Blood* 84:2566, 1994; Zoller, B., Bernts dotter, A., Garcia de Frutos, P., et al. Resistance to activated protein C as an additional genetic risk factor in hereditary deficiency of protein S. *Blood* 85:3518, 1995; Reitsma, P. H., Bernardi., F., Deig, R. G., et al. Protein C deficiency: a database of mutations, 1995 update. *Thromb. Haemost.* 73:876, 1995; and Irani-Hakime, N., Tamim, H., Kreidy, R., and Almawi, W. Y. The prevalence of factor V R506Q mutation-Leiden among apparently healthy Lebanese. *Am. J. Hematol.* 65:45, 2000.

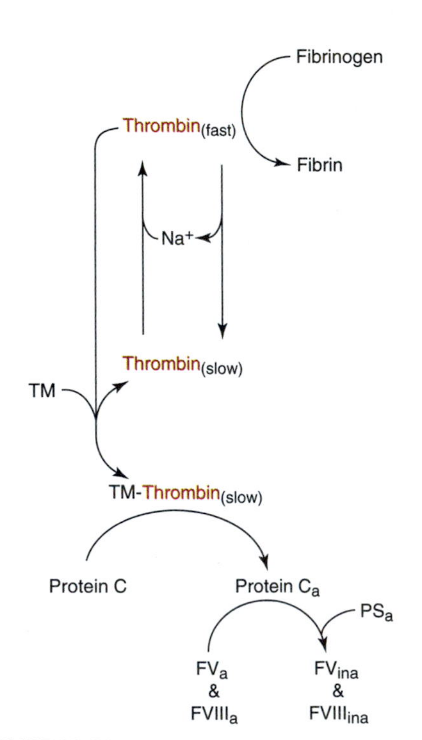

FIGURE 23.52
Allosteric reactions of thrombin and its actions on fibrinogen and protein C.

protein S or protein C lead to **thrombotic diseases** (see Clin. Corr. 23.14). Protein C inactivation of FVa and FVIIIa occurs by cleavage of peptide bonds at specific arginine residues:

$$\text{Va and VIIIa} \xrightarrow{\text{PC:PS}} \text{V}_{\text{inh}} \text{ and VIII}_{\text{inh}}$$

Thrombomodulin is an integral glycoprotein of endothelial cell membranes. It contains 560 amino acids and shows sequence homology with the low-density lipoprotein receptor, but very little with tissue factor. There is, however, a great deal of similarity in functional domains between tissue factor and thrombomodulin, each of which functions as a receptor and activator for a proteinase. Thrombomodulin is a receptor for thrombin. In the thrombin–thrombomodulin–Ca^{2+} complex, thrombin has a decreased specificity for fibrinogen and an increased specificity for protein C. Thus thrombin's role switches from that of procoagulation to anticoagulation.

Thrombin (Figure 23.52) may exist *in vitro* in two conformations: one is stabilized by Na^+ and has high specificity for catalyzing the conversion of fibrinogen to fibrin; the other conformation predominates in the absence of sodium, has low specificity for fibrinogen conversion, but high specificity for thrombomodulin binding and activity on protein C. These forms are referred to as "fast" and "slow," respectively. This type of dynamic "feedback" mechanism is important for stopping the clotting process at its point of origin. It is clear that thrombin plays this dual role, but the role of Na^+ in this process *in vivo* is not clear.

There is also a specific inhibitor for protein C. **Protein C inhibitor (PCI)** has been found in plasma, platelets, and megakaryocytic cells. About 30% of PCI is released from platelets upon stimulation by ADP, epinephrine, thrombin, and other molecules that stimulate platelet activity. Inactivation of activated protein C (APC) by PCI occurs on membrane surfaces like most of the other reactions discussed.

Fibrinolysis Phase of Hemostasis (Phase 3)

Fibrinolysis Requires Plasminogen and Tissue Plasminogen Activator (t-PA) to Produce Plasmin

Reactions of fibrinolysis are shown in Figure 23.53. Lysis of a fibrin clot occurs through action of the enzyme plasmin that is formed from plasminogen through the action of **tissue plasminogen activator (t-PA** or **TPA).** Plasminogen has high affinity for fibrin clots and forms complexes with fibrin throughout various regions of the fibrin network. t-PA also binds to fibrin clots and activates plasminogen to plasmin by specific bond cleavage. The clot is then solubilized by the action of plasmin.

t-PA is a 72-kDa protein with several functional domains. It has a growth factor domain near its N terminus, two adjacent **Kringle domains** that interact with fibrin, and a domain with protease activity that is close to its C terminus. Kringle domains are conserved sequences that fold into large loops stabilized by disulfide bonds. These domains are important structural features for protein–protein interactions that occur with several blood coagulation factors. t-PA is activated by cleavage between an Arg–Ile bond resulting in a molecule with a heavy and a light chain. The serine protease activity is located within the light chain.

Protein inhibitors regulate activity of t-PA. Four immunologically distinct types of inhibitors have been identified, two of which are of greater physiological significance because they react rapidly with t-PA and are specific for it. They are **plasminogen activator–inhibitor type 1 (PAI-1)** and **plasminogen activator–inhibitor type 2 (PAI-2).** Human PAI-2 is a 415 amino acid protein.

Starting and stopping blood coagulation follow essentially the same type of process—interactions of proteins, formation of multienzyme complexes, and proteolysis. Both are one-direction processes, and the only mechanism for replenishing the proteins once they have been used is by resynthesis.

Role of Gla Residues in Blood Coagulation Factors

Posttranslational modification of many of the proteins involved in blood coagulation that result in formation of γ-carboxyglutamyl (Gla) residues transforms them into excellent chelators of calcium ions. Ca^{2+} have at least two important functions in blood coagulation. They form complexes with Gla residues of those factors and induce conformational and electronic states that facilitate their interaction with

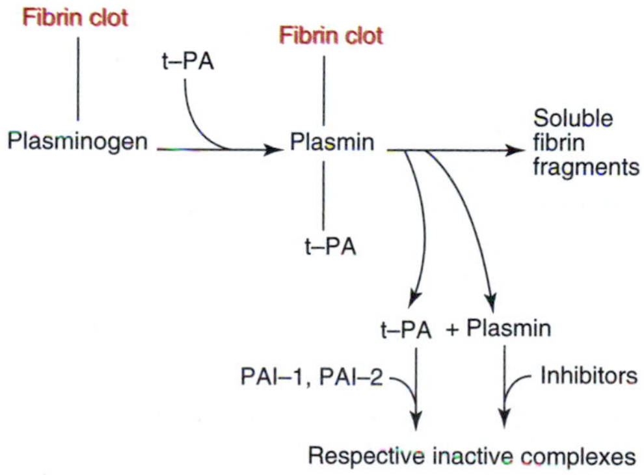

Fibrin clot
t–PA
Fibrin clot
Plasminogen
Plasmin
Soluble fibrin fragments
t–PA
t–PA + Plasmin
PAI–1, PAI–2
Inhibitors
Respective inactive complexes

FIGURE 23.53
Reactions involved in clot dissolution.

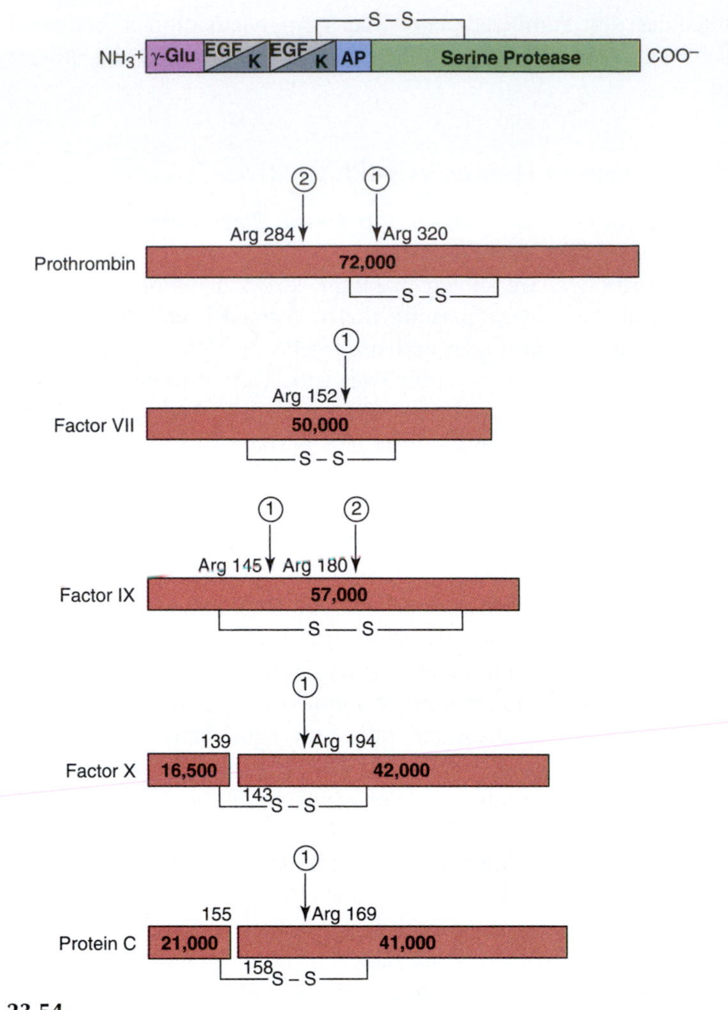

FIGURE 23.54
(*a*) General structure of the γ-carboxyglutamyl-containing proteins and (*b*) structural organization of the zymogens and their cleavage sites for activation.

membrane "receptors." Ca^{2+} also bind at sites other than Gla residues, producing protein conformational changes that enhance catalytic activity. Evidence for this second role for Ca^{2+} comes from the observation that activation of at least one of the enzymes leads to both the cleavage and elimination of the N-terminal region containing the Gla residues, but calcium ions are still required for its effective participation in blood coagulation.

Schematic representations of the structure of five of the **Gla-containing proteins** listed in Table 23.9 are shown in Figure 23.54. Gla-containing residues are located in the N-terminal region of the molecules followed by a structural component that resembles epidermal growth factor. The position of proteolytic cleavage by activation proteinases is generally at an amino acid residue located between cysteine residues that form a disulfide bond. Activation may or may not result in loss of a small peptide. Prothrombin is the only one whose activation occurs by cleavage outside the bridging disulfide bond and results in elimination of the Gla peptide (Figure 23.41). Factor VII is activated by cleavage of a single Arg^{152}–Ile^{153} bond; factor IX by cleavages at Arg^{145} and Arg^{180} with the release of an approximate 11-kDa peptide. Factor X consists of two chains connected by a disulfide bridge. It is activated by cleavage of its heavy chain at Arg^{194}–Ile^{195}. The Gla residues are located in the light chain. Protein C also consists of a heavy and a light chain connected by a disulfide bond. Cleavage of an Arg–Ile bond at position 169 results in its activation.

Role of Vitamin K in Protein Carboxylase Reactions

Modification of prothrombin, **protein C, protein S, protein Z,** and factors VII, IX, and X, to form Gla residues, occurs during synthesis by a carboxylase located on the luminal side of the rough endoplasmic reticulum. **Vitamin K** (phytonadione, the "koagulation" vitamin) is an essential cofactor for this carboxylase. During the reaction, the dihydroquinone or reduced form of vitamin K (Figure 23.55), vitamin K(H_2), is oxidized to the epoxide form, vitamin K(O), by O_2. A plausible mechanism involves the addition of molecular oxygen to the C-1 position of dihydrovitamin K and its subsequent rearrangement to an alkoxide with a pK_a of ~20. This intermediate serves as a strong base and abstracts a proton from the γ-methylene carbon of glutamate yielding a carbanion to which CO_2 can add by a nucleophilic mechanism (Figure 23.55). The **vitamin K epoxide** formed is converted back to the **dihydroquinone** by enzymes that require dithiols like **thioredoxin** as cofactors. Analogs of vitamin K inhibit dithiol-requiring vitamin K reductases and result in

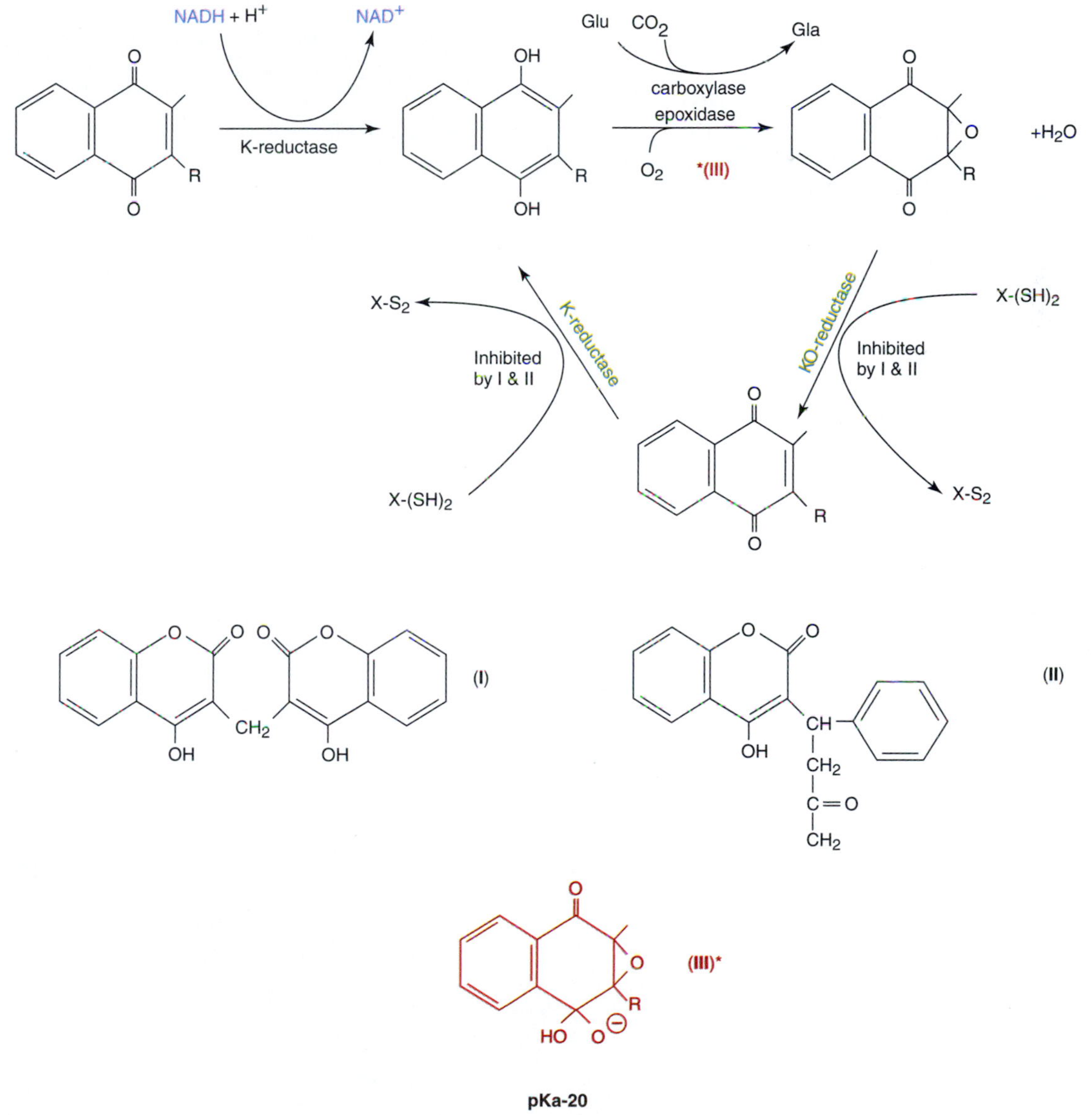

NADH + H+ NAD+
K-reductase
Glu CO_2 Gla
carboxylase
epoxidase
O_2 *(III)
+H_2O
X-S_2
X-$(SH)_2$
Inhibited by I & II
K-reductase
KO-reductase
(I)
(II)
(III)*
pKa-20

FIGURE 23.55
The vitamin K cycle as it functions in protein glutamyl carboxylation reaction.
X-$(SH)_2$ and X-S_2 represent the reduced and oxidized forms, respectively, of a thioredoxin. The NADH-dependent and the dithiol-dependent vitamin K reductases are different enzymes. The dithiol-dependent K and KO reductases are inhibited by dicoumarol (I) and warfarin (II). *Possible alkoxide intermediate (III).
Redrawn and modified from Vermeer, C. Biochem. J. 266:625, 1990.

conversion of all available vitamin K to the epoxide form that is not functional in this reaction. The overall carboxylation reaction is

$$\mathrm{HOOC{-}CH_2{-}CH_2{-}CH(NH{-}R_1){-}C(=O){-}R_2} \xrightarrow[\text{Protein carboxylase}]{\text{vitamin K}(H_2) + O_2 + CO_2 \;\rightarrow\; \text{vitamin K(O)} + H_3O^+} \mathrm{(HOOC)_2CH{-}CH_2{-}CH(NH{-}R_1){-}C(=O){-}R_2}$$

The structures of two analogs, **dicoumarol** and **warfarin**, that interfere with the action of vitamin K are shown in Figure 23.55. In animals treated with these compounds, prothrombin, protein C, protein S, factors VII, IX, X, and other Gla-

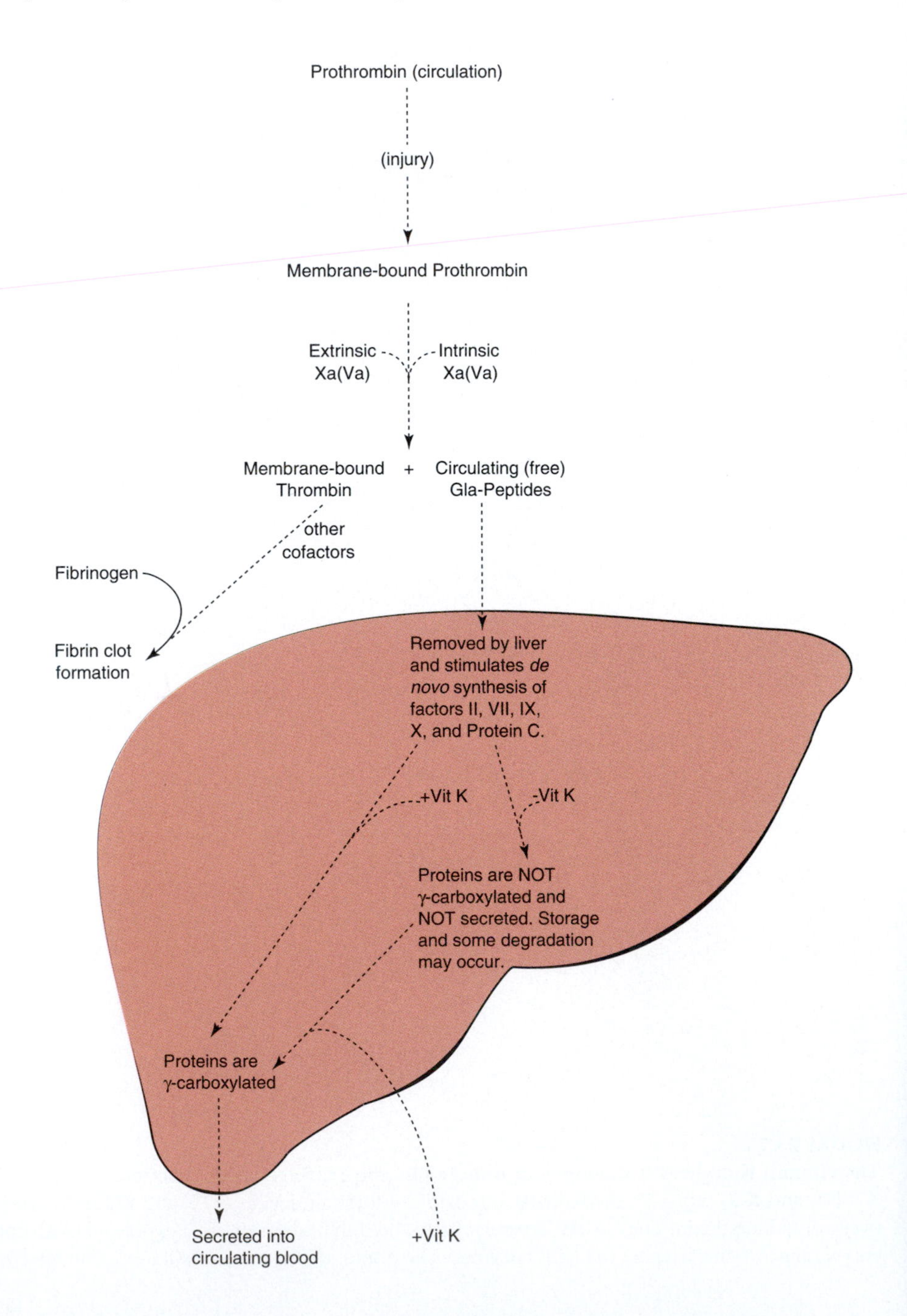

FIGURE 23.56
Role of Gla-peptides in the regulation of *de novo* synthesis of coagulation factors. Schematic representation of structural models for rhodopsin (top) and peripherin/RDS (bottom).

containing proteins are not posttranslationally modified, are ineffective in binding Ca^{2+}, and cannot participate in blood coagulation. Dicoumarol and warfarin have no effect on blood coagulation in the test tube.

Control of the Synthesis of Gla-Proteins

Gla-peptides that are released from prothrombin upon activation are removed from circulation by the liver. These N-terminal Gla-containing peptides stimulate the *de novo* synthesis of Gla-requiring proteins of the blood coagulation scheme (Figure 23.56). The proteins are synthesized even in the absence of vitamin K or in the presence of antagonists of vitamin K. They, however, are not secreted into the circulation. When vitamin K is restored, or is added in high enough concentrations to overcome the effects of antagonists, the preformed proteins are carboxylated and secreted into the circulation.

Activation of blood coagulation is a one-way process. The use of the activation peptides released from prothrombin to signal the liver to synthesize more of these proteins is an efficient mechanism for maintaining their concentrations in blood at effective levels. Monitoring of patients on long-term therapy with vitamin K antagonists is necessary to assure that posttranslational modification to produce the Gla-containing proteins is not shut down completely. When all of these processes function properly, hemostasis is achieved.

BIBLIOGRAPHY

General

Alberts, B., Bray, D., Lewis, J., Raff, M., Roberts, K., and Watson, J. *Molecular Biology of the Cell,* 3rd ed. New York: Garland Publishing, 1994.

Barany, M. (Ed.). *Biochemistry of Smooth Muscle Contraction.* New York: Academic Press, 1996.

Nerve

Bennett, M. K. and Scheller, R. H. The molecular machinery for secretion is conserved from yeast to neurons. *Proc. Natl. Acad. Sci. USA* 90:2559, 1993.

Fried, G. Synaptic vesicles and release of transmitters: new insights at the molecular level. *Acta Physiol. Scand.* 154:1, 1995.

Greengard, P., Valtorta, F., Czernik, A. J., and Benfenati, F. Synaptic vesicle phosphoproteins and regulation of synaptic function. *Science* 259:780, 1993.

Grenningloh, G., et al. The strychnine-binding subunit of the glycine receptor shows homology with nicotinic acetylcholine receptors. *Nature* 328:215, 1987.

Goodman, S. R., Zimmer, W. E., Clark, M. B., Zegon, I. S., Barker, J. E., and Bloom, M. L. Brain spectrin: of mice and men. *Brain Res. Bull.* 36:593, 1995.

Pleribone, V. A., Shupllakov, O., Brodin, L., Hilfiker-Rothenfluh, S., Czernik, A. J., and Greengard, P. Distinct pools of synaptic vesicles in neurotransmitter release. *Nature* 375:493, 1995.

Sudhof, T. The synaptic vesicle cycle: a cascade of protein–protein interactions. *Nature* 375:645, 1995.

Taylor, P. The cholinesterases. *J. Biol. Chem.* 266:4025, 1991.

Vision

Abrahamson, E. W. and Ostroy, S. E. (Eds.). *Molecular Processes in Vision,* Benchmark Papers in Biochemistry/3. Stroudsburg, PA: Hutchinson Ross Publishing Company, 1981.

Farber, D. B. From mice to men: the cyclic GMP phosphodiesterase gene in vision and disease. *Invest. Ophthamol. Vis. Sci.* 36:263, 1995.

Nathans, J. Molecular biology of visual pigments. *Annu. Rev. Neurosci.* 10:163, 1987.

Nathans, J., et al. Molecular genetics of human blue cone monochromacy. *Science* 245:831, 1989.

Palczewski, K. Is vertebrate phototransduction solved? New insights into the molecular mechanism of phototransduction. *Invest. Ophthamol. Vis. Sci.* 35:3577, 1994.

Stryer, L. Visual excitation and recovery. *J. Biol. Chem.* 266:10711, 1991.

Zigler, J. S. Jr. and Goosey, J. Aging of protein molecules: lens crystallins as a model system. *Trends Biochem. Sci.* 7:133, 1981.

Muscle

Anderson, P. A. W., Malouf, N. N., Oakley, A. E., Pagani, E. D., and Allen, P. D. Troponin T isoform expression in humans: a comparison among normal and failing adult heart, fetal heart, and adult and fetal skeletal muscle. *Circ. Res.* 69:1226, 1991.

Carlier, M.-F. Actin: protein structure and filament dynamics. *J. Biol. Chem.* 266:1, 1991.

da Silva, A. C. R. and Reinach, F. C. Calcium binding induces conformational changes in muscle regulatory proteins. *Trends Biochem. Sci.* 16:53, 1991.

dos Remedios, C. G. and Moens, P. D. J. Actin and the actomyosin interface: a review. *Biochim. Biophys. Acta* 1228:99, 1995.

Ebashi, S. Excitation–contraction coupling and the mechanism of muscle contraction. *Annu. Rev. Physiol.* 53:1, 1991.

Geeves, M. A. and Holmes, K. C. Structural mechanism of muscle contraction. *Annu. Rev. Biochem.* 68:687, 1999.

Gerisch, G., Noegel, A. A., and Schleicher, M. Genetic alteration of proteins in actin-based motility systems. *Annu. Rev. Psychol.* 53:607, 1991.

Hirose, K., Franzini-Armstrong, C., Goldman, Y. E., and Murray, J. M. Structural changes in muscle crossbridges accompanying force generation. *J. Cell Biol.* 127:763, 1994.

Huxley, H. E. The mechanism of muscular contraction. *Science* 164:1356, 1969.

Lorenz, M., Popp, D., and Holmes, K. C. Refinement of the F-actin model against X-ray fiber diffraction data by the use of a directed mutation algorithm. *J. Mol. Biol.* 234:826, 1993.

Blood

Antalis, T. M., et al. Cloning and expressing of a cDNA coding for a human monocyte-derived plasminogen activator inhibitor. *Proc. Natl. Acad. Sci. USA* 85:985, 1988.

Brummel, K. E., Butenas, S., and Mann, K. G. An integrated study of fibrinogen during blood coagulation. *J. Biol. Chem.* 274:22862, 1999.

Butenas, S., van't Veer, C., and Mann, K. G. Evaluation of the initiation phase of blood coagulation using ultrasensitive assays for serine proteases. *J. Biol. Chem.* 272:21527, 1997.

Colombatti, A. and Bonaldo, P. The superfamily of proteins with von Willebrand factor type A-like domains: one theme common to components of extracellular matrix, hemostasis, cellular adhesion, and defense mechanisms. *Blood* 77:2305, 1991.

Cooper, D. N. The molecular genetics of familial venous thrombosis. *Blood Rev.* 5:55, 1991.

Dowd, P., Hershline, R., Ham, S. W., and Naganathan, S. Vitamin K and energy transduction: a base strength amplification mechanism. *Science* 269:1684, 1995.

Fujimaki, K., Yamazaki, T., Taniwaki, M., and Ichinose, A. The gene for human protein Z is localized to chromosome 13 at band q34 and is coded by eight regular exons and one alternative exon. *Biochemistry* 37:6838, 1998.

Han, X., Fiehler, R., and Broze, G. J. Jr. Isolation of a protein Z-dependent plasma protease inhibitor. *Proc. Natl. Acad. Sci. USA* 95:9250, 1998.

Han, X., Huang, Z. F., Fiehler, R., and Broze, G. J. Jr. The protein Z-dependent protease inhibitor is a serpin. *Biochemistry* 38:11073, 1999.

Iakhiaev, A., Pendurthi, U. R., Voigt, J., Ezban, M., and Rao, L. V. M. Catabolism of factor VIIa bound to tissue factor fibroblasts in the presence and absence of tissue factor pathway inhibitor. *J. Biol. Chem.* 274:36995, 1999

Kalafatis, M., Sworde, N. A., Rand, M. D., and Mann, K. G. Membrane-dependent reactions in blood coagulation: role of the vitamin K-dependent enzyme complexes. *Biochim. Biophys. Acta* 1227:113, 1994.

Kuliopulus, A., Hubbard, B. R., Lam, Z., Koski, I. J., Furie, B., Furie, B. C., and Walsh, C. T. Dioxygen transfer during vitamin K-dependent carboxylase catalysis. *Biochemistry* 31:7722, 1992.

McClure, D. B., Walls, J. D., and Grinnell, B. W. Post-translational processing events in the secretion pathway of human protein C, a complex vitamin K-dependent antithrombotic factor. *J. Biol. Chem.* 267:19710, 1992.

Nishioka, J., Ning, M., Hayashi, T., and Suzuki, K. Protein C inhibitor secreted from activated platelets efficiently inhibits activated protein C on phosphotidylethanolamine of platelet membrane and microvesicles. *J. Biol. Chem.* 273:11281, 1998.

Ny, T., Elgh, F., and Lund, B. The structure of the human tissue-type plasminogen activator gene: correlation of introns and exon structures to functional and structural domains. *Proc. Natl. Acad. Sci. USA* 81:5355, 1984.

Palston, P. A. and Gettings, P. G. W. A database of recombinant wild-type and mutant serpins. *Thromb. Haemost.* 72:166, 1994.

Reitsma, P. H., et al. Protein C deficiency: a database of mutations, 1995 update. *Thromb. Haemost.* 73:876, 1995.

Shriver, Z., Sundaram, M., Venkataraman, G., Fareed, J., Linhardt, R., Biemann, K., and Sasisekharan, R. Cleavage of the antithrombin III binding site in heparin by heparinases and its implication in the generation of low molecular weight heparin. *Proc. Natl. Acad. Sci. USA* 97:10365, 2000.

Usalan, C., Erdem, Y., Altum, B., Arici, M., Haznedaroglu, I. C., Yasavul, U., Turgan, C., and Caglar, S. Protein Z levels in haemodialysis patients. *Int. Urol. Nephrol.* 31:541, 1999.

Van't Veer, C., Golden, N. J., Kalafatis, M., and Mann, K. G. Inhibitory mechanism of the protein C pathway on tissue factor-induced thrombin generation: synergistic effect in combination with tissue factor inhibitor. *J. Biol. Chem.* 272:7983, 1997.

Vermeer, C. γ-Carboxyglutamate-containing proteins and the vitamin K-dependent carboxylase. *Biochem. J.* 266:625, 1990.

Yin, Z.-F., Huang, Z.-F., Cui, J., Fiehler, R., Lasky, N., Ginsburg, D., and Broze, G. J. Jr. Prothrombin phenotype of protein Z deficiency. *Proc. Natl. Acad. Sci. USA* 97:6734, 2000.

Zeheb, R. and Gelehrter, T. D. Cloning and sequencing of cDNA for the rat plasminogen activator inhibitor-1. *Gene* 73:459, 1988.

QUESTIONS | C. N. ANGSTADT

Multiple Choice Questions

1. In the propagation of a nerve impulse by an electrical signal:
   A. the electrical potential across the membrane maintained by the ATP-driven Na^+, K^+ ion pump becomes more negative.
   B. local depolarization of the membrane causes protein conformational changes in ion channels that allow Na^+ and K^+ to move down their concentration gradients.
   C. charge propagation is bidirectional along the axon.
   D. "voltage-gated" ion channels have a finite recovery time so the amplitude of the impulse changes as it moves along the axon.
   E. astrocytes function as antenna-like protrusions and receive signals from other cells.

2. All of the following are characteristics of nonpeptide neurotransmitters EXCEPT:
   A. they transmit the signal across the synapse between cells.
   B. they must be made in the cell body and then travel down the axon to the presynaptic terminal.
   C. electrical stimulation increasing Ca^{2+} in the presynaptic terminal fosters their release from storage vesicles.
   D. binding to receptors on the postsynaptic terminal induces a conformational change in proteins of that membrane.
   E. their actions are terminated by specific mechanisms within the synaptic junction.

3. Which of the following statements about biochemical events occurring in the eye is true?
   A. Glucose in the lens is metabolized primarily by the TCA cycle in order to provide sufficient ATP for the Na^+/K^+–ATPase.
   B. Controlling the blood glucose level at or below 110 mg dL^{-1} might reduce the incidence of diabetic cataracts by allowing the production of sorbitol.
   C. The high rate of the hexose monophosphate pathway in the cornea is necessary to provide NADPH as a substrate for glutathione reductase.
   D. The retina contains mitochondria so it depends on the TCA cycle for its production of ATP.
   E. Cataracts are the result of increasing blood flow in the lens leading to disaggregation of lens proteins.

4. All of the following statements about the transduction of the light signal by rhodopsin are true EXCEPT:
   A. absorption of photons of light initiates the conversion of rhodopsin to activated rhodopsin.
   B. it involves the G protein, transducin.
   C. cGMP concentration is increased in the presence of an activated rhodopsin–transducin–GTP complex.
   D. the signal is turned off, in part, by the GTPase activity of the α subunit of transducin.
   E. both guanylate cyclase and phosphodiesterase are regulated by calcium concentration.

5. All of the following statements about actin and myosin are true EXCEPT:

A. the globular head section of myosin has domains for binding ATP and actin.
B. actin is the major protein of the thick filament.
C. the binding of ATP to the actin–myosin complex promotes dissociation of actin and myosin.
D. F-actin, formed by aggregation of G-actin–ATP–Mg^{2+} complex, is stabilized when tropomyosin is bound to it.
E. binding of calcium to the calmodulin-like subunit of troponin induces conformational changes that permit myosin to bind to actin.

6. ATP concentration is maintained relatively constant during muscle contraction by:
   A. increasing the metabolic activity.
   B. the action of adenylate kinase.
   C. the action of creatine phosphokinase.
   D. all of the above.
   E. none of the above.

7. Platelet aggregation:
   A. is initiated at the site of an injury by conversion of fibrinogen to fibrin.
   B. is inhibited in uninjured blood vessels by the secretion of prostacyclin by intact vascular endothelium.
   C. causes morphological changes and a release of the vasodilator serotonin.
   D. is inhibited by the release of ADP and thromboxane A_2.
   E. is inhibited by von Willebrand's factor (vWF).

8. In the formation of a blood clot:
   A. proteolysis of γ-carboxyglutamate residues from fibrinogen to form fibrin is required.
   B. the clot is stabilized by the cross-linking of fibrin molecules by the action of factor XIII, transglutaminase.
   C. thrombin's only role is in activation of factor VII.
   D. tissue factor, factor III or TF, must be inactivated for the clotting process to begin.
   E. the role of calcium is primarily to bind fibrin molecules together to form the clot.

9. Factor Xa, necessary for conversion of prothrombin to thrombin, is formed by the action of the TF–FVII–Ca^{2+} complex on factor X:
   A. only in the extrinsic pathway for blood clotting.
   B. only in the intrinsic pathway for blood clotting.
   C. as part of both the extrinsic and intrinsic pathways.
   D. only if the normal blood clotting cascade is inhibited.

10. Lysis of a fibrin clot:
   A. is in equilibrium with formation of the clot.
   B. begins when plasmin binds to the clot.
   C. requires the hydrolysis of plasminogen into heavy and light chains.
   D. is regulated by the action of protein inhibitors on plasminogen.
   E. requires the conversion of plasminogen to plasmin by t-PA (tissue plasminogen activator).

Questions 11 and 12: Genes that code for the proteins of the visual pigments are located on different chromosomes. The rhodopsin gene is on the third chromosome; the blue pigment gene is on the seventh chromosome; and the genes for the red and green pigments are on the X chromosome. Females have two X chromosomes so color vision abnormalities are rare, but males have only one X chromosome. About 8% of males have abnormal color vision that affects either red or green perception or, rarely, both red and green color perception. At least one individual is known who has blue color blindness.

11. The cones of the retina:
   A. are responsible for color vision.
   B. are much more numerous than the rods.
   C. have red, blue, and green light-sensitive pigments that differ because of small differences in the retinal prosthetic group.
   D. do not use transducin in signal transduction.
   E. are better suited for discerning rapidly changing visual events because a single photon of light generates a stronger current than it does in the rods.

12. Which of the following statements about rhodopsin is true?
   A. Rhodopsin is the primary photoreceptor of both rods and cones.
   B. The prosthetic group of rhodopsin is all-*trans*-retinol derived from cleavage of β-carotene.
   C. Conversion of rhodopsin to activated rhodopsin, R*, by a light pulse requires depolarization of the cell.
   D. Rhodopsin is located in the cytosol of the cell.
   E. Absorption of a photon of light by rhodopsin causes an isomerization of 11-*cis*-retinal to all-*trans*-retinal.

Questions 13 and 14: Ion channels in cardiac muscle, as in other tissues, require a finite recovery time after excitation. The recovery time between cardiac contractions is measured on an electrocardiogram as the QT interval. Initiating the excitation phase is the opening of Na^+ channels, causing depolarization. Then K^+ channels open to permit K^+ to move out of the cell, which is a key element in shutting off the action potential. There are inherited defects in these channels leading to a condition known as long QT syndrome (LQTS). LQTS can cause sudden death, especially in physically active young people, who may not even know that they have LQTS because they have not previously had any symptoms of cardiac irregularities.

13. The nerve impulse that initiates muscular contraction:
   A. begins with the binding of acetylcholine to receptors in the sarcoplasmic reticulum.
   B. causes both the plasma membrane and the transverse tubules to undergo hyperpolarization.
   C. causes opening of calcium channels, which leads to an increase in calcium concentration within the sarcomere.
   D. prevents Na^+ from entering the sarcomere.
   E. prevents Ca^{2+} from binding to troponin C.

14. When a muscle contracts, the:
   A. transverse tubules shorten, drawing the myofibrils and sarcoplasmic reticulum closer together.
   B. thin filaments and the thick filaments of the sarcomere shorten.
   C. light chains dissociate from the heavy chains of myosin.
   D. H bands and I bands of the sarcomere shorten because the thin filaments and thick filaments slide past each other.
   E. cross-linking of proteins in the heavy filaments increases.

Problems

15. In the presence of warfarin, an analog of vitamin K, several proteins of the blood coagulation pathway are ineffective because they cannot bind Ca^{2+} efficiently. Why?
16. Organophosphate compounds are irreversible inhibitors of acetylcholinesterase. What effect does an organophosphate inhibitor have on the transmission of nerve impulses?

ANSWERS

1. **B** This is the mechanism for impulse propagation. A: The potential becomes less negative. C: It is unidirectional. D: Voltage-gated channels do have a finite recovery time so the amplitude remains constant. E: This describes dendrites. Astrocytes are glial cells that are involved in processes that insulate neurons from their external environment.
2. **B** This is true for neuropeptides, but many nonpeptide neurotransmitters are synthesized in the presynaptic terminal. A: This is a difference between electrical and chemical signals. C: Synapsin plays a role in this process. E: Make sure you know the three types of processes involved.
3. **C** Make sure you understand the role of glutathione in protecting against harmful by-products from atmospheric oxygen. A: Most of the ATP (85%) in the lens is generated by glycolysis. B: Controlling glucose reduces sorbitol formation. D: Retina metabolism is similar to that of other eye tissues directly involved in the visual process. Thus its major source of energy is from glycolysis. E: Lens has no blood supply. In diabetic cataracts there is increased aggregation of lens proteins because of increasing sorbitol concentrations and osmolarity.
4. **C** The transducin complex activates the phosphodiesterase, thus lowering [cGMP]. A: This process is very rapid. B, D: Transducin meets the criteria for a typical G-protein. E: The enzymes are regulated in opposite directions by Ca^{2+}, thus controlling [cGMP].
5. **B** Myosin forms the thick filament. Actin is in the thin filament. A: These are both important in myosin's role. C: Note that the role of ATP in contraction is to favor dissociation, not formation, of the actin–myosin complex. D, E: Tropomyosin, tropinin, and actin are the three major proteins of the filament. Their actions are closely interconnected.
6. **D** Make sure you know the reactions catalyzed by these two enzymes.
7. **B** The "ying–yang" nature of PGI_2 and TXA_2 help to control platelet aggregation until there is a need for it. A: Initiation is by contact with an activated receptor at the site of injury. Clot formation requires activation of various enzymes. C: Serotonin is a vasoconstrictor. Vasodilation would be contraindicated in this situation. D: TXA_2 facilitates aggregation. E: vWF forms a link between the receptor and platelets, promoting aggregation.
8. **B** The cross-linking occurs between a glutamine and a lysine. A, E: γ-Carboxyglutamate residues are on various enzymes; they bind calcium and facilitate the interaction of these proteins with membranes that form the sites for initiation of reaction. C: Thrombin activates factors V, VII, VIII, and XIII and fibrinogen. D: TF, factor III, is the primary receptor for initiation of the clotting process.
9. **A** Tissue factor and factor VII are unique to the extrinsic pathway. B, C: The membrane interaction with the intrinsic pathway is with high molecular weight kininogen and prekallikrein.
10. **E** The clot is solubilized by plasmin. A: Both formation and lysis of clots are unidirectional. B: Both plasminogen and t-PA bind to the clot. C, D: Both of these refer to t-PA.
11. **A** Rods are responsible for low-light vision. C: All three pigments have 11-*cis*-retinal; the proteins differ and are responsible for the different spectra. D: The biochemical events are believed to be the same in rods and cones. E: Cones are better suited for rapid events because their response rate is about four times faster than rods, even though their sensitivity to light is much less.
12. **E** This causes the conformational change of the protein that affects the resting membrane potential and initiates the rest of the events. A: Cones have the same prosthetic group but different proteins, so rhodopsin is in rods only. B: This is the precursor of the prosthetic group 11-*cis*-retinal. C: Isomerization of the prosthetic group leads to hyperpolarization. D: Rhodopsin is a transmembrane protein.
13. **C** Ca^{2+} enters with Na^+. A: Acetylcholine receptors are on the plasma membrane. B: The impulse results in depolarization of both of these structures. D: Both Ca^{2+} and Na^{2+} enter the sarcomere when the channels open. E: Binding of Ca^{2+} to Tn-C initiates contraction.
14. **D** This occurs because of association–dissociation of actin and myosin. A: Depolarization in the transverse tubules may be involved in transmission of the signal but not directly in the contractile process. B: The filaments do not change in length, but slide past each other. C: This is not physiological. E: Cross-linking occurs in the H band of the sarcomere but does not change during the contractile process.
15. The proteins affected are those that have γ-carboxyglutamyl (Gla) residues, which are excellent calcium chelators. The formation of Gla is a posttranslational modification catalyzed by a carboxylase whose essential cofactor is vitamin K. Warfarin prevents the reduction of vitamin K epoxide formed during the carboxylation back to the dihydroquinone form, which is necessary for the reaction.
16. Acetylcholine is an excitatory neurotransmitter, causing channels to open, Na^+ to enter, and a depolarization of the membrane. Hydrolysis of acetylcholine allows repolarization. The presence of the inhibitor keeps the channel open and prevents repolarization, and thus continued transmission of nerve impulses.

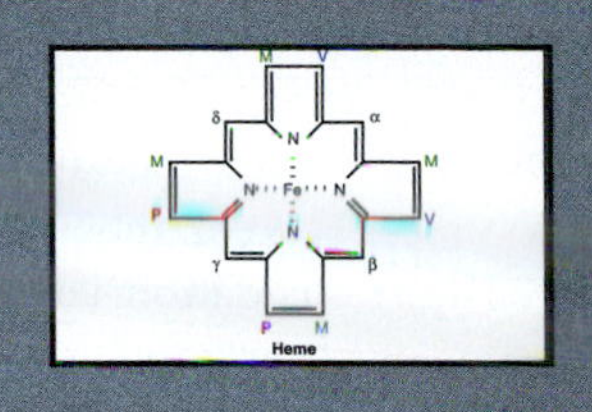

24

IRON AND HEME METABOLISM

William M. Awad, Jr.

24.1 | IRON METABOLISM: OVERVIEW

Iron is closely involved in the metabolism of oxygen, permitting the transportation and participation of oxygen in a variety of biochemical processes. The common oxidation states are either ferrous (Fe^{2+}) or ferric (Fe^{3+}). Higher oxidation levels occur as short-lived intermediates in certain redox processes; as an example, during the reaction catalyzed by cytochrome $P450_{cam}$ (see Chapter 11) from *Pseudomonas putida,* the iron of one of the intermediates must be in the ferryl (Fe^{4+}) form. Iron has an affinity for electronegative atoms such as oxygen, nitrogen, and sulfur, which provide the electrons that form the bonds with iron. These can be of very high affinity when favorably oriented on macromolecules. In forming complexes, no bonding electrons are derived from iron. There is an added complexity to the structure of iron: the nonbonding electrons in the outer shell of the metal (the incompletely filled 3*d* orbitals) can exist in two states. Where bonding interactions with iron are weak, the outer nonbonding electrons will avoid pairing and distribute throughout the 3*d* orbitals. Where bonding electrons interact strongly with iron, however, there is pairing of the outer nonbonding electrons, favoring lower energy 3*d* orbitals. These two different distributions for each oxidation state of iron can be determined by electron spin resonance measurements. Dispersion of 3*d* electrons to all orbitals leads to the high-spin state, whereas restriction of 3*d* electrons to lower energy orbitals, because of electron pairing, leads to a low-spin state. Some iron–protein complexes reveal changes in spin state without changes in oxidation during chemical events (e.g., binding and release of oxygen by hemoglobin).

At neutral and alkaline pH ranges, the redox potential for iron in aqueous solutions favors the Fe^{3+} state; at acid pH values, the equilibrium favors the Fe^{2+} state. In the Fe^{3+} state iron slowly forms large polynuclear complexes with hydroxide ion, water, and other anions that may be present. These complexes can become so large as to exceed their solubility products, leading to their aggregation and precipitation with pathological consequences.

Iron can bind to and influence the structure and function of various macromolecules, with deleterious results to the organism. To protect against such reactions, several iron-binding proteins function specifically to store and transport iron. These proteins have a very high affinity for the metal and, in the normal physiological state, also have incompletely filled iron-binding sites. The interaction of iron with its ligands has been well characterized in some proteins (e.g., hemoglobin, myoglobin, and transferrin).

24.2 | IRON-CONTAINING PROTEINS

Iron binds to proteins either by incorporation into a **protoporphyrin IX** ring (see below) or by interaction with other protein ligands. Ferrous–and ferric–protoporphyrin IX complexes are designated **heme** and **hematin,** respectively. Heme-containing proteins include those that transport (e.g., hemoglobin) and store (e.g., myoglobin) oxygen; and certain enzymes that contain heme as part of their prosthetic groups (e.g., catalase, peroxidases, tryptophan pyrrolase, prostaglandin synthase, guanylate cyclase, NO synthase, and the microsomal and mitochondrial cytochromes). Discussions on structure–function relationships of heme proteins are presented in Chapters 9 and 13.

Nonheme proteins include transferrin, ferritin, a variety of redox enzymes that contain iron at the active site, and iron–sulfur proteins. A significant body of information has been acquired that relates to the structure–function relationships of some of these molecules.

Transferrin Transports Iron in Serum

The protein in serum involved in the transport of iron is **transferrin,** a β1-glycoprotein synthesized in liver, that consists of a single polypeptide of 78,000 Da with two noncooperative iron-binding sites. The protein is a product of gene duplication derived from a putative ancestral gene coding for a protein binding only one atom of iron. Several metals bind to transferrin; the highest affinity is for Fe^{3+}; Fe^{2+} ion is not bound. The binding of each Fe^{3+} ion is absolutely dependent on the coordinate binding of an anion, which in the physiological state is carbonate, as indicated below:

$$\text{transferrin} + Fe^{3+} + CO_3^{2-} \rightarrow \text{transferrin} \cdot Fe^{3} \cdot CO_3^{2-}$$

$$\text{transferrin} \cdot Fe^{3+} \cdot CO_3^{2-} \rightarrow \text{transferrin} \cdot 2\ (Fe^{3+}CO_3^{2-})$$

Estimates of the association constants for the binding of Fe^{3+} to transferrins from different species range from 10^{19} to 10^{31} M^{-1}, indicating that for practical purposes wherever there is excess transferrin free ferric ions will not be found. In the normal physiological state, approximately one-ninth of all transferrin molecules are saturated with iron at both sites; four-ninths of transferrin molecules have iron at either site; and four-ninths of circulating transferrin are free of iron. Unsaturated transferrin protects against infections (see Clin. Corr. 24.1). The two iron-binding sites show differences in sequence and in affinity for other metals. Transferrin binds to specific cell surface receptors that mediate the internalization of the protein.

The **transferrin receptor** is a transmembrane protein consisting of two subunits of 90,000 Da each, joined by a disulfide bond. Each subunit contains one transmembrane segment and about 670 residues that are extracellular and bind a transferrin molecule, favoring the diferric form. Internalization of the receptor–transferrin complex is dependent on receptor phosphorylation by a Ca^{2+}–calmodulin–protein kinase C complex. Release of the iron atoms occurs within the acidic milieu of lysosomes after which the receptor–apotransferrin complex returns to the cell surface where the apotransferrin is released to be reutilized in the plasma.

Lactoferrin Binds Iron in Milk

Milk contains iron that is bound almost exclusively to a glycoprotein, **lactoferrin,** closely homologous to transferrin, with two sites binding the metal. The iron content of the protein varies, but it is never saturated. Studies on the function of lactoferrin have been directed toward its antimicrobial effect, protecting the newborn from gastrointestinal infections. Microorganisms require iron for replication and function. Presence of incompletely saturated lactoferrin results in the rapid binding of any free iron, leading to the inhibition of microbial growth by preventing a sufficient amount of iron from entering these microorganisms. Other microbes, such as *Escherichia coli,* which release competitive iron chelators, are able to proliferate despite the presence of lactoferrin, since the chelators transfer the iron specifically to the microorganism. Lactoferrin is present in granulocytes being released during bacterial infections; it is also present in mucous secretions. Besides its bacteriostatic function it is believed to facilitate iron transport and storage in milk. Lactoferrin has been found in the urine of premature infants fed human milk.

CLINICAL CORRELATION 24.1
Iron Overload and Infection

If an individual is overloaded with iron by any of several causes, the serum transferrin value may be close to saturation with iron, making small amounts of free serum iron available. Microorganisms that are usually nonpathogenic, because they are iron dependent and cannot compete against partially saturated transferrin in the normal individual, can now become pathogenic under these circumstances. For example, *Vibrio vulnificus,* a marine halophile, is found in a small percentage of oysters and commercial shellfish. Individuals who are iron overloaded can develop a rapidly progressive infection, with death ensuing within 24 h after ingestion of the offending meal, whereas normal individuals consuming the same food are entirely free of symptoms.

Muench, K. H. Hemochromatosis and infection: alcohol and iron, oysters and sepsis. *Am. J. Med.* 87:3, 1989.

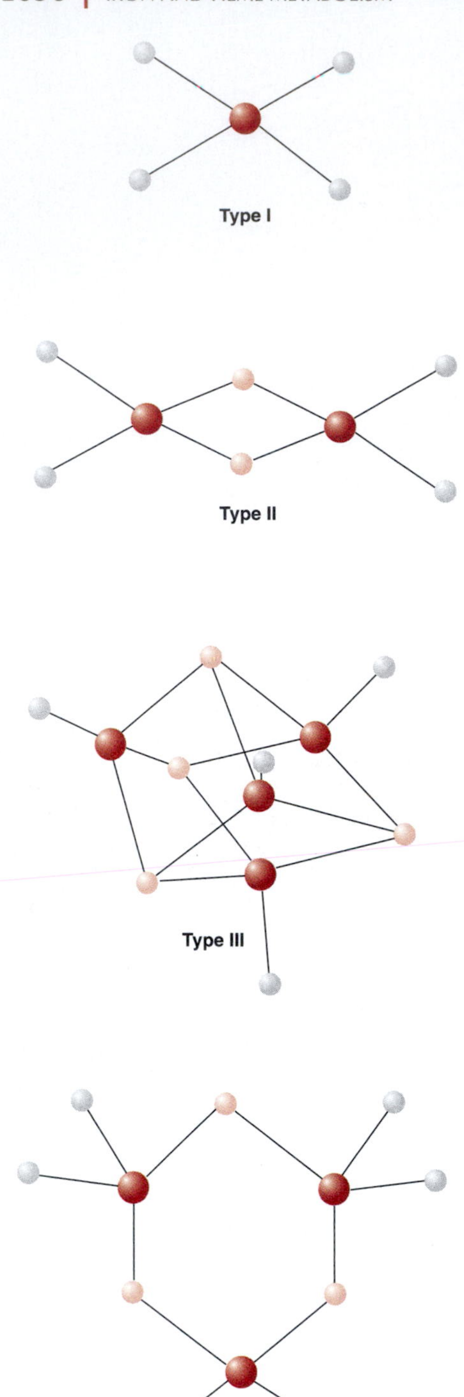

FIGURE 24.1
Structure of iron and sulfur atoms in ferrodoxins.
Dark colored circles represent iron atoms; light colored circles represent the inorganic sulfur atoms; and small stippled circles represent cysteinyl sulfur atoms derived from the polypeptide chain. Variation in type IV ferredoxins can occur where one of the cysteinyl residues can be substituted by a solvent oxygen atom of an OH group.

Ferritin Is a Protein Involved in Storage of Iron

Ferritin is the major protein involved in storage of iron. It consists of an outer polypeptide shell 130 Å in diameter with a central ferric-hydroxide-phosphate core 60 Å across. The apoprotein, **apoferritin,** consists of 24 subunits of a varying mixture of H subunits (178 amino acids) and L subunits (171 amino acids) that provide various isoprotein forms. H subunits predominate in nucleated blood cells and heart, L subunits in liver and spleen. Synthesis of the subunits is regulated mainly by the concentration of free intracellular iron. The bulk of iron storage occurs in hepatocytes, reticuloendothelial cells, and skeletal muscle. The ratio of iron to polypeptide is not constant, since the protein has the ability to gain and release iron according to physiological needs. With a capacity of 4500 iron atoms, the molecule contains usually less than 3000. Channels from the surface permit the accumulation and release of iron. When iron is in excess, the storage capacity of newly synthesized apoferritin may be exceeded. This leads to iron deposition adjacent to ferritin spheres. Histologically such amorphous iron deposition is called **hemosiderin.** The H chains of ferritin oxidize ferrous ions to the ferric state. Ferritins derived from different tissues of the same species differ in electrophoretic mobility in a fashion analogous to the differences noted with isoenzymes. In some tissues ferritin spheres form lattice-like arrays, which are identifiable by electron microscopy.

Plasma ferritin (low in iron, rich in L subunits) has a half-life of 50 h and is cleared by reticuloendothelial cells and hepatocytes; its concentration, although very low, correlates closely to the size of the body iron stores.

Other Nonheme Iron-Containing Proteins Are Involved in Enzymatic Processes

Many iron-containing proteins are involved in enzymatic processes, most of which are related to oxidation mechanisms. The structural features of the ligands binding the iron are not well known, except for a few components involved in mitochondrial electron transport. These latter proteins, termed **ferredoxins,** are characterized by iron being bonded, with one exception, only to sulfur atoms. Several types of iron–sulfur clusters are known (see Figure 24.1). The smallest, type I (e.g., nebredoxin), found only in microorganisms, consists of a small polypeptide with a mass of about 6000 and contains one iron atom bound to four cysteine residues. Type II consists of ferredoxins found in plants and animal tissues where two iron atoms are found, each liganding to two separate cysteine residues and sharing two sulfide anions. The most complicated of the iron–sulfur proteins are the bacterial ferredoxins, type III, which contain four atoms of iron, each of which is linked to a single separate cysteine residue but also shares three sulfide anions with neighboring iron atoms to form a cube-like structure. In some anaerobic bacteria a family of ferredoxins may contain two type III iron–sulfur groups per macromolecule. Type IV ferredoxins contain structures with three atoms of iron each linked to two cysteine residues and each sharing two sulfide anions, to form a planar ring. In one example of this ferredoxin type, an exception of iron atoms being liganded only to sulfur atoms was found where the sulfur of a cysteinyl residue was substituted by a solvent oxygen atom. The redox potential afforded by these different ferredoxins varies widely and is in part dependent on the environment of the surrounding polypeptide that envelops these iron–sulfur groups. In nebredoxin the iron undergoes ferric–ferrous conversion during electron transport. With the plant and animal ferredoxins (type II iron–sulfur proteins) both irons are in the Fe^{3+} form in the oxidized state; upon reduction only one goes to Fe^{2+}. In the bacterial ferredoxin (type III iron–sulfur protein) the oxidized state can be either $2\ Fe^{3+} \cdot 2\ Fe^{2+}$ or $3\ Fe^{3+} \cdot Fe^{2+}$, with corresponding reduced forms of $Fe^{3+} \cdot 3\ Fe^{2+}$ or $2\ Fe^{3+} \cdot 2\ Fe^{2+}$. Recently, several more complicated iron–sulfur clusters have been characterized, revealing an expanding role for the clusters in the redox mechanism (see Clin. Corr. 24.2 and 24.3).

CLINICAL CORRELATION 24.2

Iron–Sulfur Cluster Synthesis and Human Disease

Iron–sulfur clusters are synthesized in the mitochondrion and exported to the cytoplasm, a poorly understood process involving mitochondrial chaperones and other proteins, some with as yet unknown function. A mutation in the human *ABC7* transporter gene leads to one form of X-linked sideroblastic anemia seen in patients with an abnormal gait (cerebellar ataxia) and associated with impaired development of cytosolic iron–sulfur proteins.

Bekri, S., Kispal, G., Lane, H., et al. Human ABC7 transporter gene structure and mutation causing X-linked sideroblastic anemia with ataxia with disruption of cytosolic iron–sulfur protein maturation. *Blood* 96:3256, 2000.

CLINICAL CORRELATION 24.3

Friedreich Ataxia

Frataxin is a mitochondrial protein of unknown function, the mutations of which lead to the disease Friedreich ataxia, associated with a marked reduction in this protein's concentration. The disease is characterized by marked abnormalities in gait, impaired heart muscle, and diabetes. There is marked increase in mitochondrial iron accumulation. In this condition a GAA triplet expansion to sometimes over 1000-fold in intron 1 of frataxin is the cause of the condition in about 98% of cases. No anemia is seen with this disease.

Delatycki, M. B., Williamson, R., and Forrest, S. M. Friedreich ataxia: an overview: *J. Med. Genet.* 37:1, 2000.

24.3 | INTESTINAL ABSORPTION OF IRON

The high affinity of iron for both specific and nonspecific macromolecules leads to little significant formation of free iron salts, and thus this metal is not lost via usual excretory routes. Rather, iron excretion occurs only through normal sloughing of tissues that are not reutilized (e.g., epidermis and gastrointestinal mucosal cells). In the healthy adult male the loss is about 1 mg day^{-1}. In premenopausal women, the normal physiological events of menses and parturition substantially augment iron loss. A wide variation of such loss exists, depending on the amounts of menstrual flow and the multiplicity of births. In the extremes of the latter settings, a premenopausal woman may require an amount of iron that is four to five times that needed by an adult male for prolonged periods of time. The postmenopausal woman who is not iron-deficient has an iron requirement similar to that of the adult male. Children and patients with blood loss naturally have increased iron requirements.

Cooking of food facilitates breakdown of ligands attached to iron, increasing the availability of the metal in the gut. The low pH of stomach contents permits the reduction of Fe^{3+} to Fe^{2+}, facilitating dissociation from ligands. The latter requires the presence of an accompanying reductant, which is usually achieved by adding ascorbate to the diet. The absence of a normally functioning stomach reduces substantially the amount of iron that is absorbed. Some iron-containing compounds bind the metal so tightly that it is not available for assimilation. Contrary to popular belief, spinach is a poor source of iron because of an earlier erroneous record of the iron content and because some of the iron is bound to phytate (inositol hexaphosphate), which is resistant to the chemical actions of the gastrointestinal tract. Specific protein cofactors derived from the stomach or pancreas have been suggested as being facilitators of iron absorption in the small intestine.

The major site of absorption of iron is in the small intestine, with the largest amount being absorbed in the duodenum and a gradient of lesser absorption occurring in the more distal portions of the small intestine. The metal enters the mucosal cell either as the free ion or as heme; in the latter case the metal is released

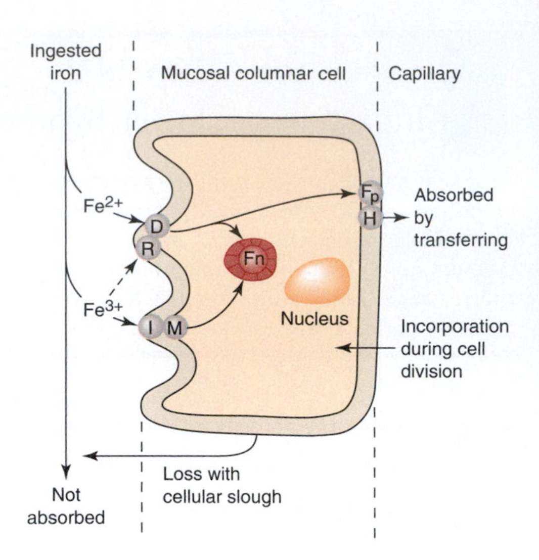

FIGURE 24.2
Intestinal mucosal regulation of iron absorption.
The flux of iron in the duodenal mucosal cell is indicated. A fraction of the iron that is potentially acceptable is transferred from the intestinal lumen into the epithelial cell. A large portion of ingested iron is not absorbed, in part because it is not presented in a readily acceptable form. Some iron is retained within the cell, bound by apoferritin to form ferritin. This iron is sloughed into the intestinal lumen with the normal turnover of cells. A portion of the iron within the mucosal cell is absorbed and transferred to the capillary bed to be incorporated into transferrin. During cell division, which occurs at the bases of the intestinal crypts, iron is incorporated for cellular requirements. These fluxes change dramatically in iron-depleted or iron-excess states. The following structures are indicated: D, DMT 1; I, integrin; M, mobilferrin; Fn, ferritin; Fp, ferroportin 1; H, hephaesin; and R, ferrireductase, presumably acting with DMT 1. (see Clin. Corr. 24.4).

from the porphyrin ring in the mucosal cytoplasm. The large amount of bicarbonate secreted by the pancreas neutralizes the acidic material delivered by the stomach and thus favors the oxidation of Fe^{2+} to Fe^{3+}. Whatever the requirements of the host are, in the face of an adequate delivery of iron to the lumen, a substantial amount of iron enters mucosal cells. Regulation of iron transfer occurs between the mucosal cell and the capillary bed (see Figure 24.2 and Clin. Corr. 24.4). In the

CLINICAL CORRELATION 24.4

Duodenal Iron Absorption

Iron is probably absorbed in three forms from the intestinal lumen into the duodenal mucosa: as ferrous ion, as ferric ion, and as heme. The mechanism of absorption of heme is unknown. Ferrous ion binds to a transmembrane protein, divalent metal transporter 1 (DMT 1), known also as divalent cation transporter 1 (DCT 1) or as natural resistance associated macrophage protein 2 (Nramp 2). Transfer of iron is dependent on a proton-coupled mechanism. This protein can transport other metals such as zinc, manganese, cobalt, cadmium, and lead. The original member of this family of proteins was found in macrophages and is called Nramp 1. Recent evidence suggests that the latter protein's function is to present iron to the macrophage allowing the cell to undergo important redox functions in host resistance. The gene for DMT 1 contains 17 exons. The mRNA can be spliced alternatively at the 3′ end including different 3′ untranslated regions. One form contains a single 3′ IRE indicating posttranscriptional control is possible for that protein's expression. A ferrireductase has been described in the microvillus of intestinal mucosa and may be closely associated with DMT 1.

Ferric iron absorption can follow a different course. Mucin in the duodenal lumen helps to solubilize ferric ions with presentation of the metal to an integrin, a transmembrane protein consisting of a heterodimer of 230 kDa. The cytosolic surface of the integrin interacts with a soluble 56-kDa protein known as mobilferrin. The integrin transfers the iron from the luminal to cytoplasmic surface of the cell, where it is bound by a soluble protein, mobilferrin, which acts as a cytosolic shuttle transferring iron either to cytosolic ferritin or to the opposite pole of the cell. A transmembrane protein ferroportin 1 on the basolateral surface of the mucosal cell transfers iron from the cytosol through the endothelium to transferrin and thus acts as an iron exporter. The mRNA for this protein has an IRE in the 5′ untranslated region. A high level of expression of this protein is also seen in Kupffer cells and macrophages. Ferroportin 1 transfers ferrous ions. It is closely associated with a second protein, hephaestin, which is homologous to ceruloplasmin; the assumption is that hephaestin acts as a ferroxidase, converting the ferrous ion into the ferric form and allowing its incorporation into transferrin. The different locations of the IREs in DMT 1 and ferroportin 1 would appear to be counterproductive, since it would be expected ordinarily that both would be upregulated or downregulated in concert. The apparent function of both proteins is to control the presentation of iron to transferrin. However, in any iron state, one messenger RNA would be stabilized whereas the other would be inhibited. This issue remains to be resolved. In the iron-deprived state DMT 1 concentration rises. Mobilferrin responds to iron deprivation not be increasing in concentration but by moving from the cytoplasm to the microvillus, thereby being closer to the site of iron influx.

Donovan, A., Brownlie, A., Zhou, Y., et al. Positional cloning of zebrafish ferroportin 1 identifies a conserved vertebrate iron exporter. *Nature* 403:776, 2000; Conrad, M. E, Umbreit, J. N., Moore, E. G., et al. Separate pathways for cellular uptake of ferric and ferrous iron. *Am. J. Physiol. Gastrointest. Liver Physiol.* 279:G767,2000; Umbreit, J. M., Conrad, M. E., Simovich, M. J., et. al. Identification and localization of iron transport protein in normal and iron deficient cells. *Blood* 96:227a, abstract #968, 2000; and Riedel, H. D., Remus, A. J., Fitscher, B. A., et al. Characterization and partial purification of a ferrireductase from human duodenal microvillus membranes. *Biochem. J.* 309:745, 1995.

normal state, certain processes define the amount of iron that will be transferred. Where there is iron deficiency, the amount of transfer increases; where there is iron overload in the host, the amount transferred is curtailed substantially. One mechanism that has been demonstrated to regulate this transfer of iron across the mucosal–capillary interface is the synthesis of apoferritin by the mucosal cell. In situations in which little iron is required by the host, a large amount of apoferritin is synthesized to trap the iron within mucosal cells and prevent transfer to the capillary bed. As the cells turn over (within a week), their contents are extruded into the intestinal lumen without absorption occurring. In situations in which there is iron deficiency, virtually no apoferritin is synthesized so as not to compete against the transfer of iron to the deficient host. Figure 24.2 and Clinical Correlation 24.4 provide details on the molecular aspects of iron absorption.

24.4 | MOLECULAR REGULATION OF IRON UTILIZATION

Cytosol contains at least two proteins that respond to changes in iron concentration. They act as effector molecules controlling the translation of mRNAs, which are important in iron metabolism. These **iron regulatory proteins (IRPs)** bind to specific **stem–loop structures** on certain mRNAs. IRP-1 is the best defined of these proteins. It contains an Fe_4S_4 cubane group when the cellular concentration of iron is high. This prosthetic group endows IRP-1 so that it possesses an **aconitase** (see p. 554) activity. However, since neither citrate nor isocitrate is present in significant amounts in the cytosol, the activity is only a potential one. At low iron concentrations, the cubane structure collapses, dissociating from the protein and leaving an apoenzyme without catalytic activity. However, it can now bind to specific mRNA stem–loop structures, known as **iron responsive elements (IREs)** (see Figure 24.3). Seven mRNAs, encoding proteins with defined functions in iron metabolism, are known to contain IREs (see Table 24.1 for details). However, a search of human data bank sequences has revealed about 70 genes which contain IREs, emphasizing the important role of iron in many as yet to be identified metabolic processes. This is a field of rapidly expanding knowledge. Five of these mRNA have single 5′ stem–loop structures; two mRNAs have 3′ IREs. Transferrin receptor has five 3′ stem–loop structures whereas **divalent metal transporter 1** (DMT 1) has only one. The binding of the 5′ and 3′ flanking IREs leads to different translational effects. In the iron-deprived state, binding to the 3′ IRE of transferrin receptor (see Figure 24.4) leads to stabilization of the mRNA with reduced turnover and, therefore, an increased number of receptor-specific RNA

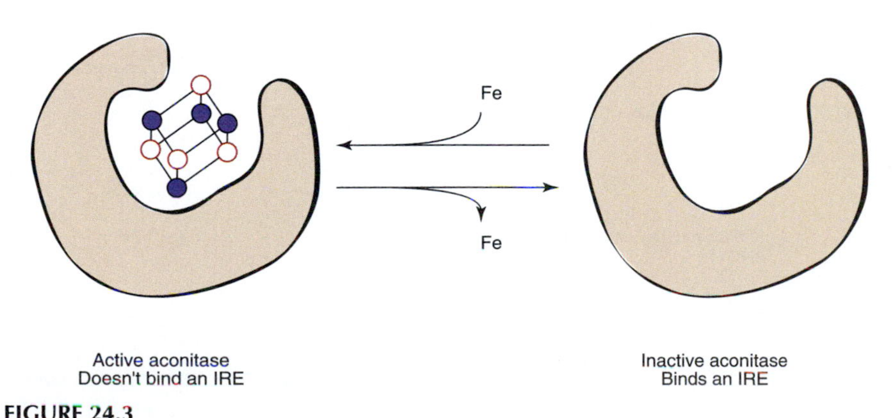

FIGURE 24.3
Ironresponsive protein 1.
Dark colored circles represent iron atoms and open circles inorganic sulfur atoms.

CLINICAL CORRELATION 24.5
Mutant Iron-Responsive Elements

Single mutations have been described in the loop segment of the iron-responsive element of ferritin light chain mRNA with an increased amount of apoferritin being synthesized but without an increase in total body iron. This mutation leads to a 28-fold lower affinity for IRP-1 in one case. The L-ferritin content of the lens can increase ninefold compared to the content in the lens from control subjects. As a consequence, crystals of pure ferritin light chains appear in the lens, leading to cataracts. The greatly increased synthesis of ferritin in the lens may lead to an increased amount of iron-catalyzed reactions with well-described oxidative lenticular damage.

Very recently a mutation in a Japanese kindred was described in the IRE of ferritin H chain mRNA with a marked increase in IRP affinity. The trapping of IRPs leads to decreased H chain synthesis and increased L chain synthesis. This condition is associated with a significant increase in body iron content, but apparently without evidence of cataracts.

Girelli, D., Corrocher, R., Bisceglia, L., et al. Molecular basis for the recently described hereditary hyperferritinemia–cataract syndrome: a mutation in the iron-responsive elements of ferritin L-subunit gene (the "Verona mutation"). *Blood* 86:4050, 1995; Beaumont, C., Leneuve, P., Devaux, I., et al. Mutation in the iron responsive element of the L ferritin mRNA in a family with dominant hyperferritinaemia and cataract. *Nature Genet.* 11:444, 1995; Mumford, A. D., Cree I. A., Arnold, J. D., et al. The lens in hereditary hyperferritinemia cataract syndrome contains crystalline deposits of L-ferritin. *Br. J. Ophthalmol.* 84:697, 2000; and Kato, J., Fujikawa, K., Kanda, M., et al. Mutation in the iron-responsive element of H-ferritin subunit in a family associated with hepatic siderosis. *Blood* 96:484a, abstract #2081, 2000.

TABLE 24.1 IRE-Containing mRNAs

mRNA	*Site*	*Number*	*Compartment*
L-ferritin	5′	Single	Cytosol
H-ferritin	5′	Single	Cytosol
Erythrocyte ALA synthase	5′	Single	Mitochondrion
Mitochondrial aconitase	5′	Single	Mitochondrion
Ferroportin 1	5′	Single	Cytosol
Transferrin receptor	3′	Multiple	Cytosol
Divalent metal transporter 1	3′	Single	Cytosol

molecules, thereby leading to the increased synthesis of receptor protein. The single 5′ stem–loop of ferritin mRNA (see Figure 24.5) is homologous to the 3′ stem–loops of the transferrin receptor mRNA. However, in the former case, binding of the IRP leads to a decreased rate of translation of the mRNA and, thereby, to a decreased concentration of ferritin molecules. Note that the molecular events that are controlled are different in the synthesis of transferrin receptor and apoferritin (see Clin. Corr. 24.5).

In summary, low iron concentrations lead to activation of an IRP that binds to mRNAs for transferrin receptor and ferritin. In the former case, more receptor is synthesized while in the latter case, less apoferritin is synthesized. The net effect is utilization of iron by proliferating cells. In contrast, high iron concentrations lead to loss of binding by the IRPs to IREs with a shift of iron from uptake by proliferating cells to storage in liver.

IRP-1 is regulated by its change from active to inactive states in mRNA-binding properties as noted above. **IRP-2,** a second regulatory protein, also responds to varying concentrations of iron but is regulated by increased synthesis at low iron concentrations and increased degradation by proteasomes at high iron concentrations. In addition to the effects of change in iron concentration, increased production of NO (see p. 486) also regulates IRPs. Further activation can occur by phosphorylation.

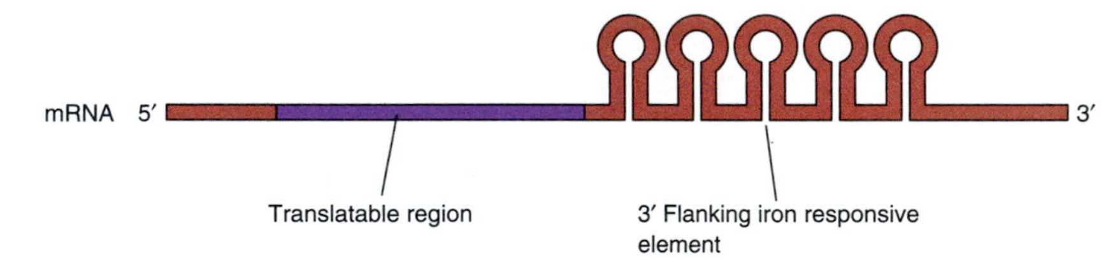

FIGURE 24.4
Structure of transferrin receptor mRNA.

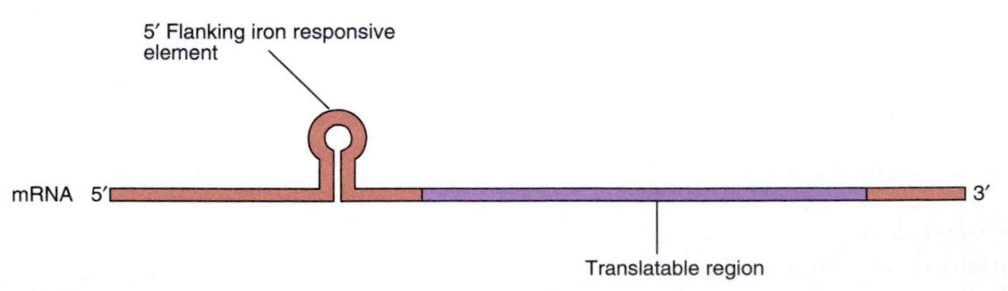

FIGURE 24.5
Structure of apoferritin H subunit mRNA.

24.5 | IRON DISTRIBUTION AND KINETICS

A normal 70-kg male contains 3–4 g of iron, of which only 0.1% (3.5 mg) is in the plasma. Approximately 2.5 g is in hemoglobin. Table 24.2 lists the distribution of iron in humans. Normally about 33% of the sites on transferrin contain iron. Iron absorbed from the intestine is delivered primarily to the bone marrow for incorporation into the hemoglobin of red blood cells. Mobilization of iron from intestinal mucosa and from storage sites involves in part the reduction of iron to the ferrous state and its reoxidation to the ferric form. The reduction mechanisms have not been well described. On the other hand, conversion of the Fe^{2+} back to the Fe^{3+} state is regulated by serum enzymes called ferroxidases as indicated below:

$$Fe^{2+} + \text{ferroxidase} \rightarrow Fe^{3+} + \text{reduced ferroxidase}$$

Ferroxidase I is also known as **ceruloplasmin** (see Clin. Corr. 24.6). Another serum protein, **ferroxidase II,** appears to be the major serum component that oxidizes ferrous ions. In any disease process in which iron loss exceeds iron repletion, a sequence of physiological responses occurs. The initial events produce no symptoms to the subject and involve depletion of iron stores without compromise of any physiological function. This depletion is manifested by reduction or absence of iron stores in the liver and in the bone marrow and also by a decrease in the content of the very small amount of ferritin that is normally present in plasma. Serum ferritin levels reflect slow release from storage sites during the normal cellular turnover that occurs in the liver; measurements are made by radioimmune assays. Serum ferritin is mostly apoferritin in form, containing very little iron. During this early phase, the level and percentage saturation of serum transferrin is not distinctly abnormal. As the iron deficiency progresses, the level of hemoglobin begins to fall and morphological changes appear in the red blood cells. Concurrently, the serum iron falls with a rise in the level of total serum transferrin, the latter reflecting a physiological adaptation in an attempt to absorb more iron from the gastrointestinal tract. At this state of iron depletion a very sensitive index is the percentage saturation of serum transferrin with iron (normal range 21–50%). At this point the patient usually comes to medical attention, and the diagnosis of iron deficiency is made. In countries where iron deficiency is severe without corrective medical measures being available, a third and severe stage of iron deficiency can occur, where there begins a depletion of iron-containing enzymes leading to very pronounced metabolic effects (see Clin. Corr. 24.7).

Iron overload can occur in patients so that the iron content of the body can be elevated to values as high as 100 g. This may happen for a variety of reasons. Some

TABLE 24.2 Approximate Iron Distribution: 70-kg Man

	g	%
Hemoglobin	2.5	68
Myoglobin	0.15	4
Transferrin	0.003	0.1
Ferritin, tissue	1.0	27
Ferritin, serum	0.0001	0.004
Enzymes	0.02	0.6
Total	3.7	100

CLINICAL CORRELATION 24.6

Ceruloplasmin Deficiency

A deficiency, but not the absence, of ceruloplasmin, a copper-containing protein, is associated with Wilson's disease in which there is progressive hepatic failure and degeneration of the basal ganglia, associated with a characteristic copper deposition in the cornea (Kayser–Fleischer rings). Because there was no evidence for significant impairment of mobilization of iron in Wilson's disease, it was originally thought that the ferroxidase activity of ceruloplasmin was not physiologically important. However, a recently discovered very rare genetic defect in ceruloplasmin biosynthesis, where the protein was virtually absent in serum, leads to a marked elevation of liver iron content and serum ferritin levels. These patients develop diabetes mellitus, retinal degeneration, and central nervous system findings. The diabetes and central nervous system findings are associated with increased iron in the pancreas and brain, respectively. In iron deficiency the transcription of the ceruloplasmin gene increases fourfold. Thus, in contrast to earlier considerations, it appears that ceruloplasmin has a significant role in iron metabolism.

Harris, E. D. The iron–copper connection: the link to ceruloplasmin grows stronger. *Nutr. Rev.* 53:226, 1995; Mukhopadhyay, C. K., Mazumder, B., and Fox, P. L. Role of hypoxia-inducible factor-1 in transcriptional activation of ceruloplasmin by iron deficiency. *J. Biol. Chem.* 275:21048, 2000; and Van Eden, M. E. and Aust, S. D. Intact human ceruloplasmin is required for the incorporation of iron into human ferritin. *Arch. Biochem. Biophys.* 381:119, 2000.

CLINICAL CORRELATION 24.7
Iron-Deficiency Anemia

Microscopic examination of blood smears from patients with iron-deficiency anemia usually reveals the characteristic findings of microcytic (small in size) and hypochromic (underpigmented) red blood cells. These changes in the red cell result from decreased rates of globin synthesis when heme is not available. A bone marrow aspiration will reveal no storage iron to be present and serum ferritin values are virtually zero. The serum transferrin value (expressed as the total iron-binding capacity) will be elevated (upper limits of normal: 410 mg dL^{-1}) with a serum iron saturation of less than 16%. Common causes for iron deficiency include excessive menstrual flow, multiple births, and gastrointestinal bleeding that may be occult. The common causes of gastrointestinal bleeding include medications that can cause ulcers or erosion of the gastric mucosa (especially aspirin or cortisone-like drugs), hiatal hernia, peptic ulcer disease, gastritis associated with chronic alcoholism, and gastrointestinal tumor. Management of such patients must include both a careful examination for the cause and source of bleeding and supplementation with iron. The latter is usually provided in the form of oral ferrous sulfate tablets; occasionally, intravenous iron therapy may be required. Where the iron deficiency is severe, transfusion with packed red blood cells may also be indicated. The weakness and fatigue associated with iron deficiency may be due to the inhibition of translation of the mitochondrial aconitase mRNA through binding of IRP to its IRE.

Hentze, M. W. and Kuhn, L. C. Molecular control of vertebrate iron metabolism: mRNA based circuits operated by iron, nitric oxide, and oxidative stress. *Proc. Natl. Acad. Sci. USA* 93:8175, 1996; and Eisenstein, R. S. and Blemings, K. P. Iron regulatory proteins, iron responsive elements and iron homeostasis. *J. Nutr.* 128:2295, 1998.

patients have a recessive heritable disorder associated with a marked inappropriate increase in iron absorption. In such cases the serum transferrin can be almost completely saturated with iron. This state, which is known as **idiopathic hemochromatosis,** is more commonly seen in men because women with the abnormal gene are protected somewhat by menstrual and childbearing events. Accumulation of iron in the liver, pancreas, and heart can lead to cirrhosis and liver tumors, diabetes mellitus, and cardiac failure, respectively. Treatment consists of periodic withdrawal

CLINICAL CORRELATION 24.8
Hemochromatosis Type I: Molecular Genetics and the Issue of Iron-Fortified Diets

The hemochromatosis gene is heterozygous in about 9% of the U.S. population. The disease is expressed primarily in the homozygous state; about 0.25% of all individuals are at risk. Normal individuals have a major histocompatibility complex class-1 gene (*HFE*) that encodes for the α chain, containing three immunoglobulin-like domains. The associated β chain is β_2-microglobulin. The normal gene product has a structure that cannot present an antigen. Most individuals with hemochromatosis are homozygous for a Cys 282 $\rightarrow$ Tyr mutation, which prevents formation of the normal conformation of an immunoglobulin domain. Other individuals can have a single Cys 282 $\rightarrow$ Tyr mutation on one allele with the other allele containing a mutation at another site. Normal HFE complexes with the transferrin receptor dimer to form a heterotetramer that contains two receptor monomers and two HFEs. Structure–function studies show that the binding sites for transferrin and HFE overlap on the receptor, leading to the suggestion that this binding of HFE serves to control the rate of iron transfer to specific cells. Where there is a mutation in the HFE that prevents binding to the receptor, transferrin binds to a greater extent, leading to the increased amount of iron stores in certain cells. However, this is an incomplete story. It is not yet explained why there is such an increase in iron absorption in the intestinal mucosa in this disease.

A controversy has developed as to whether food should be fortified with iron because of the prevalence of iron-deficiency anemia, especially among premenopausal women. It was suggested that dietary iron deficiency would be reduced if at least 50 mg of iron was incorporated per pound of enriched flour. Others suggested that the risk of toxicity from excess iron absorption through iron fortification was too great. Sweden has mandated iron fortification for 50 years and about 42% of the average daily intake of iron is derived from these sources. However, 5% of males had elevation of serum iron values, with 2% having iron stores consonant with the distribution found in early stages of hemochromatosis, pointing out the danger of iron-fortified diets. In countries where iron deficiency is widespread, however, fortification may still be the most appropriate measure.

McLaren, C. E., Gorddeuk, V. R., Looker, A. C., et al. Prevalence of heterozygotes for hemochromatosis in the white population of the United States. *Blood* 86:2021, 1995; Feder, J. N., Gnirki, A., Thomas, W., et al. A novel MHC class 1-like gene is mutated in patients with hereditary haemochromatosis. *Nature Genet.* 13:399, 1996; Olsson, K. S., Heedman, P. A., and Staugard, F. Preclinical hemochromatosis in a population on a high-iron-fortified diet. *J. Am. Med. Assoc.* 239:1999, 1978; Olsson, K. S., Marsell, R., Ritter, B., Olander, B., et al. Iron deficiency and iron overload in Swedish male adolescents. *J. Intern. Med.* 237:187, 1995; Bennett, M. J., Lebron, J. A., and Bjorkman, P. J. Crystal structure of the hereditary haemochromatosis protein HFE complexed with transferrin receptor. *Nature* 403:46, 2000; and Steinberg, K. K., Cogswell, M. E., Chang, J. C., et al. Prevalence of C282Y and H63D Mutations in the Hemochromatosis (HFE) Gene in the United States. *J. Am. Med. Assoc.* 283:2216, 2001.

CLINICAL CORRELATION 24.9

Hemochromatosis Type II

There are other causes for pathologically increased iron accumulation, most of which remain to be characterized. Recently, a protein homologous with transferrin receptor has been characterized and associated with mucosal iron absorption. Mutations in this protein transferrin receptor 2 have been found to be associated with hemochromatosis despite normal HFE genes. This receptor, in contrast to transferrin receptor 1, neither has a stem–loop in its mRNA nor does it bind HFE.

West, A. P. Jr., Bennett, M. J., Sellers, V. M. D., et al. Comparison of the interactions of transferrin receptor and transferrin receptor 2 with transferrin and the hereditary hemochromatosis protein HFE. *J. Biol. Chem.* 275:38135, 2000; Camaschella, C., Roetto, A., Cali, A., et al. The gene *TRF 2* is mutated in a new type of haemochromatosis mapping to *7q22*. *Nat. Genet.* 25:14, 2000; and Kawabati, H., Yang, R., Hirawa, T., et al. Molecular cloning of transferrin receptor 2. A new member of the transferrin receptor-like family. *J. Biol. Chem.* 274: 20826, 1999.

of large amounts of blood where the iron is contained in the hemoglobin. Another group of patients have severe anemias, among the most common of which are the thalassemias, a group of hereditary **hemolytic anemias.** In these cases the subjects require transfusions throughout their lives, leading to the accumulation of large amounts of iron derived from the transfused blood. Clearly bleeding would be an inappropriate measure in these cases; rather, the patients are treated by the administration of iron chelators, such as desferrioxamine, which leads to the excretion of large amounts of complexed iron in the urine. A rare third group of patients acquire excess iron because they ingest large amounts of both iron and ethanol, the latter promoting iron absorption. In these cases excess stored iron can be removed by bleeding (see Clin. Corr. 24.8 and 24.9).

24.6 | HEME BIOSYNTHESIS

Heme is produced in virtually all mammalian tissues. Synthesis is most pronounced in the bone marrow and liver because of the requirements for incorporation into hemoglobin and the cytochromes, respectively. As depicted in Figure 24.6, heme is a largely planar molecule. It consists of one ferrous ion and a tetrapyrrole ring, **protoporphyrin IX.** The diameter of the iron atom is a little too large to be accommodated within the plane of the porphyrin ring, and thus the metal puckers out to one side as it coordinates with the apical nitrogen atoms of the four pyrrole groups. Heme is one of the most stable of compounds, reflecting its strong resonance features.

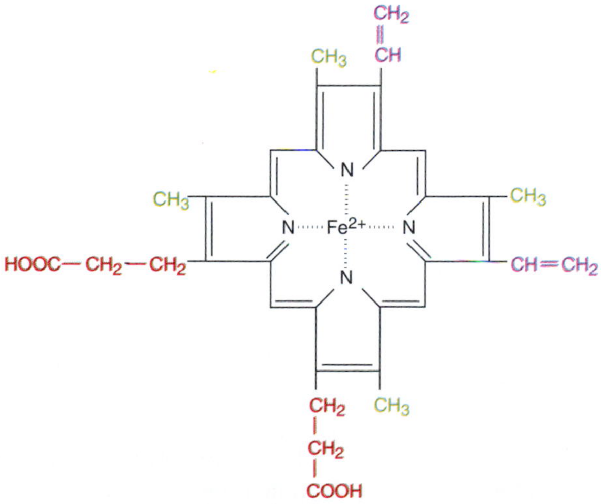

FIGURE 24.6
Structure of heme.

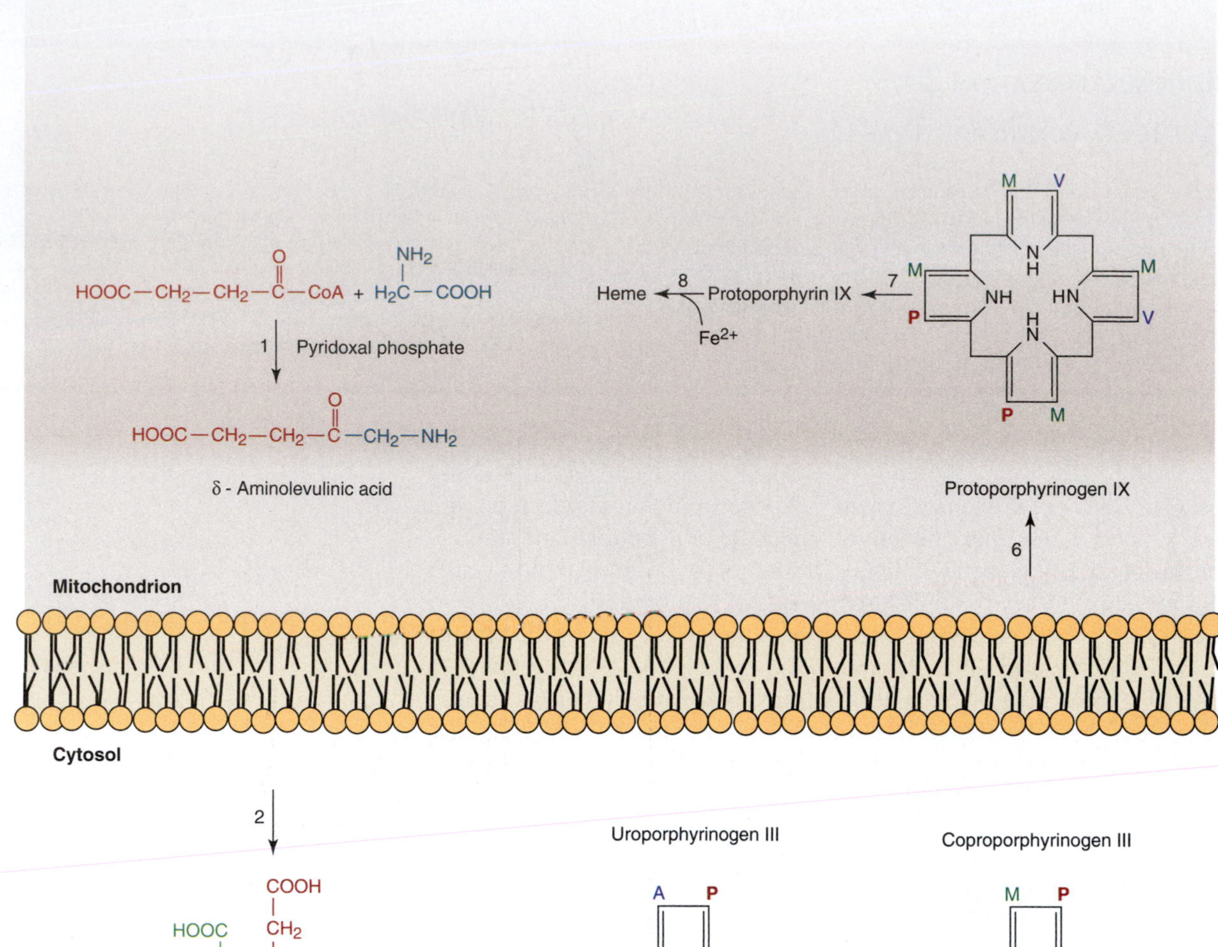

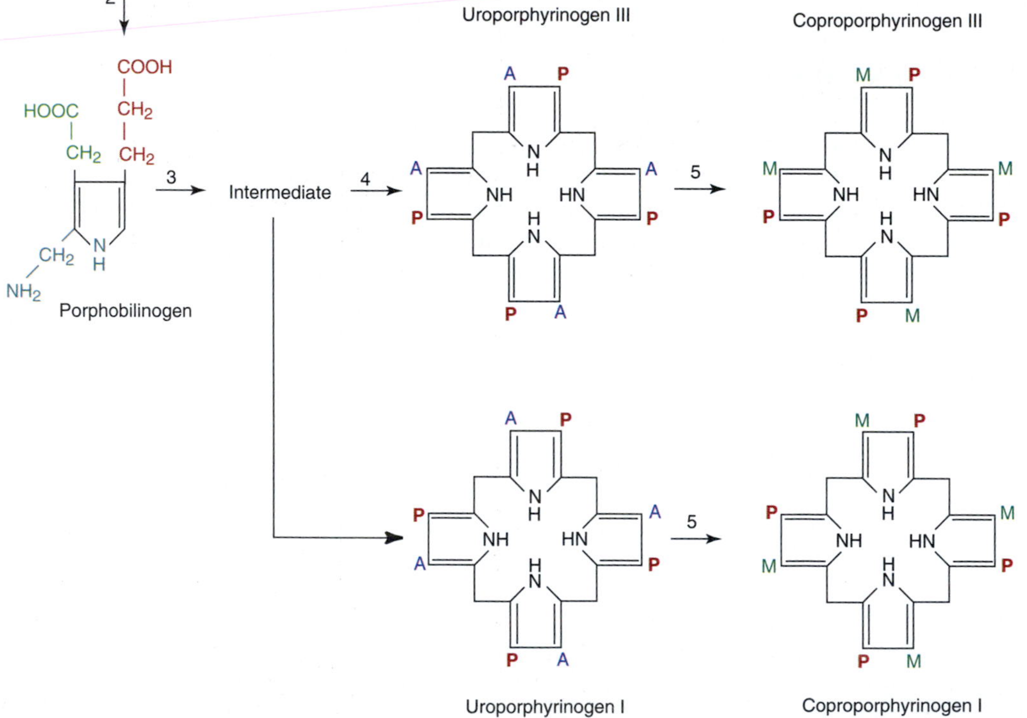

FIGURE 24.7
Pathway for heme biosynthesis.
Numbers indicate enzymes involved in each step as follows: 1, Ala synthase; 2, Ala dehydratase; 3, porphobilinogen deaminase; 4, uroporphyrinogen III cosynthase; 5, uroporphyrinogen decarboxylase; 6, coproporphyrinogen III oxidase; 7, protoporphyrinogen IX oxidase; 8, ferrochelatase. Pyrrole ligands are indicated as follows: P, propionic; A, acetic; M, methyl; V, vinyl.

Figure 24.7 depicts the pathway for heme biosynthesis. The following are the important aspects to be noted. First, the initial and last three enzymatic steps are catalyzed by enzymes that are in the mitochondrion, whereas the intermediate steps take place in the cytosol. This is important in considering the regulation by heme of the first biosynthetic step; this aspect is discussed below. Second, the organic portion of heme is derived totally from eight residues each of glycine and succinyl CoA. Third, the reactions occurring on the side groups attached to the tetrapyrrole ring involve the colorless intermediates known as **porphyrinogens.** The latter compounds, though exhibiting resonance features within each pyrrole ring, do not demonstrate resonance between the pyrrole groups. As a consequence, the porphyrinogens are unstable and can readily be oxidized, especially in the presence of light, by nonenzymatic means to their stable **porphyrin** products. In the latter cases resonance between pyrrole groups is established by oxidation of the four methylene bridges. Figure 24.8 depicts the enzymatic conversion of protoporphyrinogen to protoporphyrin by this oxidation mechanism. This is the only enzymatic porphyrinogen oxidation that is enzyme regulated in humans; all other porphyrinogen–porphyrin conversions are nonenzymatic and are catalyzed by light rather than by specific enzymes. Fourth, once the tetrapyrrole ring is formed, the order of the R groups as one goes clockwise around the tetrapyrrole ring defines which of the four possible types of **uro-** or **coproporphyrinogens** are being synthesized. These latter compounds have two different substituents, one each for every pyrrole group. Going clockwise around the ring, the substituents can be arranged as ABABABAB (where A and B are different substituents), forming a type I porphyrinogen, or the arrangement can be ABABABBA, forming a type III porphyrinogen. In principle, two other arrangements can occur to form porphyrinogens II and IV, and these can be synthesized chemically; however, they do not occur naturally. In protoporphyrinogen and protoporphyrin there are three types of substituents, and the classification becomes more complicated; type IX is the only form that is synthesized naturally.

Derangements of porphyrin metabolism are known clinically as the **porphyrias.** This family of diseases is of great interest because it has revealed that the regulation of heme biosynthesis is complicated. The clinical presentations of the different porphyrias provide a fascinating exposition of biochemical regulatory abnormalities and their relationship to pathophysiological processes. Table 24.3 lists the details of the different porphyrias (see Clin. Corr. 24.10).

Enzymes in Heme Biosynthesis Occur in Both Mitochondria and Cytosol

δ-Aminolevulinic Acid Synthase

δ-Aminolevulinic acid (ALA) synthase controls the rate-limiting step of heme synthesis in all tissues studied. The synthesis of the enzyme occurs in the cytosol, being directed by mRNA derived from the nucleus. The enzyme is incorporated into the matrix of the mitochondrion. Succinyl CoA, an intermediate of the tricarboxylic acid cycle, is one of the substrates and occurs only in mitochondria. This pro-

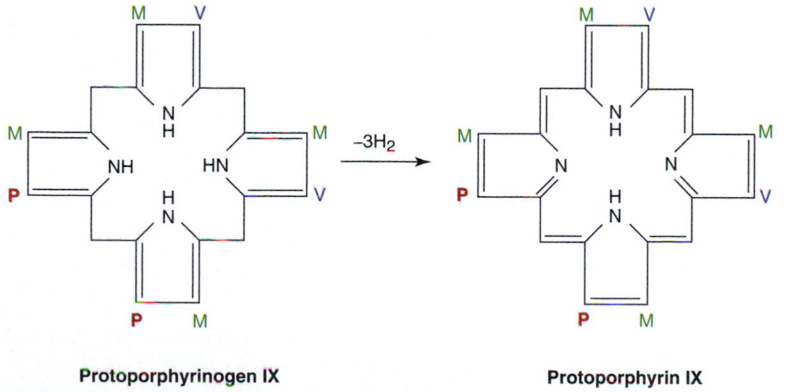

M V M P NH HN M V P M $-3H_2$ Protoporphyrinogen IX Protoporphyrin IX

FIGURE 24.8
Action of protoporphyrinogen IX oxidase, an example of the conversion of a porphyrinogen to a porphyrin.

TABLE 24.3 Derangements in Porphyrin Metabolism

Disease State	*Genetics*	*Tissue*	*Enzyme*	*Activity*	*Organ Pathology*
Acute intermittent porphyria	Dominant	Liver	1. ALA synthase 2. Porphobilinogen deaminase 3. Δ-5α-Reductase	Increase Decrease Decrease	Nervous system
Hereditary coproporphyria	Dominant	Liver	1. ALA synthase 2. Coproporphyrinogen oxidase	Increase Decrease	Nervous system; skin
Variegate porphyria	Dominant	Liver	1. ALA synthase 2. Protoporphyrinogen oxidase	Increase Decrease	Nervous system; skin
Porphyria cutanea tarda	Dominant	Liver	1. Uroporphyrinogen decarboxylase	Decrease	Skin, induced by liver disease
Hereditary protoporphyria	Dominant	Marrow	1. Ferrochelatase	Decrease	Gallstones, liver disease, skin
Erythropoietic porphyria	Recessive	Marrow	1. Uroporphyrinogen III cosynthase	Decrease	Skin and appendages; reticuloendothelial system
Lead poisoning	None	All tissues	1. ALA dehydrase 2. Ferrochelatase	Decrease Decrease	Nervous system; blood; others

CLINICAL CORRELATION 24.10

Acute Intermittent Porphyria

A 40-year-old woman appears in the emergency room in an agitated state, weeping and complaining of severe abdominal pain. She has been constipated for several days and has noted marked weakness in the arms and legs and that "things do not appear to be quite right." Physical examination reveals a slightly rapid heart rate (100/min) and moderate hypertension (blood pressure of 160/110 mmHg). There have been earlier episodes of severe abdominal pain; operations undertaken on two occasions revealed no abnormalities. The usual laboratory tests are normal. The neurological complaints are not localized to an anatomical focus. A decision is made that the present symptoms are largely psychiatric in origin and have a functional rather than an organic basis. The patient is sedated with 60 mg of phenobarbital; a consultant psychiatrist agrees by telephone to see the patient in about 4 h. The staff notices a marked deterioration; generalized weakness rapidly appears, progressing to a compromise of respiratory function. This ominous development leads to immediate incorporation of a ventilatory assistance regimen, with transfer to intensive care for physiological monitoring. Her condition deteriorates and she dies 48 h later. A urine sample of the patient is reported later to have a markedly elevated level of porphobilinogen. This patient had acute intermittent porphyria, a disease of incompletely understood derangement of heme biosynthesis. There is a dominant pattern of inheritance associated with an overproduction of the porphyrin precursors, ALA and porphobilinogen. Three enzyme abnormalities have been noted in cases that have been studied carefully. These include (1) a marked increase in ALA synthase, (2) a reduction by one-half of activity of porphobilinogen deaminase, and (3) a reduction by one-half of the activity of steroid D4-5α-reductase. The change in content of the second enzyme is consonant with a dominant expression. The change in content of the third enzyme is acquired and not apparently a heritable expression of the disease. It is believed that a decrease in porphobilinogen deaminase leads to a minor decrease in content of heme in liver. The lower concentration of heme leads to a failure both to repress the synthesis and to inhibit the activity of ALA synthase. Almost never manifested before puberty, the disease is thought to appear only with the induction of D4-5β-reductase at adolescence. Without a sufficient amount of D4-5α-reductase, the observed increase in the 5β steroids is due to a shunting of D4 steroids into the 5β-reductase pathway. The importance of abnormalities of this last metabolic pathway in the pathogenesis of porphyria is controversial.

Pathophysiologically, the disease poses a great riddle: the derangement of porphyrin metabolism is confined to the liver, which anatomically appears normal, whereas the pathological findings are restricted to the nervous system. In the present case, involvement of (1) the brain led to the agitated and confused state and the respiratory collapse, (2) the autonomic system led to the hypertension, increased heart rate, constipation, and abdominal pain, and (3) the peripheral nervous system and spinal cord led to the weakness and sensory disturbances. Experimentally, no known metabolic intermediate of heme biosynthesis can cause the pathology noted in acute intermittent porphyria. There should have been a greater suspicion of the possibility of porphyria early in the patient's presentation. The analysis of porphobilinogen in the urine is a relatively simple test. The treatment would have been glucose infusion, the exclusion of any drugs that could cause elevation of ALA synthase (e.g., barbiturates), and, if her disease failed to respond satisfactorily despite these measures, the administration of intravenous hematin to inhibit the synthesis and activity of ALA synthase. Acute hepatic porphyria is of historic political interest. The disease has been diagnosed in two descendants of King George III, suggesting that the latter's deranged personality preceding and during the American Revolution could possibly be ascribed to porphyria.

Meyer, U. A., Strand, L. J., Dos, M., et al. Intermittent acute porphyria: demonstration of a genetic defect in porphobilinogen metabolism. *N. Engl. J. Med.* 286:1277, 1972; and Stein, J. A. and Tschudy, D. D. Acute intermittent porphyria: a clinical and biochemical study of 46 patients. *Medicine (Baltimore)* 49:1, 1970.

tein has been purified to homogeneity from rat liver mitochondria. The cytosolic protein is a homodimer of a 71,000-Da subunit, containing a basic N-terminal signaling sequence that directs the enzyme into the mitochondrion. An ATP-dependent 70,000-Da cytosolic component, known as a chaperone protein, maintains ALA synthase in the unfolded extended state, the only form that can pass through the mitochondrial membrane. Thereafter, the N-terminal signaling sequence is cleaved by a metal-dependent protease in the mitochondrial matrix, to yield ALA synthase subunits of 65,000 Da each. Within the matrix another oligomeric chaperon protein, of 14 subunits of 60,000 Da each, catalyzes the correct folding of the protein in a second ATP-dependent process (see Figure 24.9). The ALA synthase has a short biological half-life (~60 min). Both the synthesis and the activity of the enzyme are subject to regulation by a variety of substances; 50% inhibition of activity occurs in the presence of 5-mM hemin, and virtually complete inhibition is noted at a 20-mM concentration. The enzymatic reaction involves the condensation of a **glycine** residue with a residue of **succinyl CoA** to produce δ-aminolevulinic acid. The reaction has an absolute requirement for **pyridoxal phosphate.** Two isoenzymes exist for ALA synthase; only the mRNA of the erythrocytic form contains an IRE. Mutations in the erythrocytic form cause a second type of X-linked sideroblastic anemia; in this case no ataxia is seen.

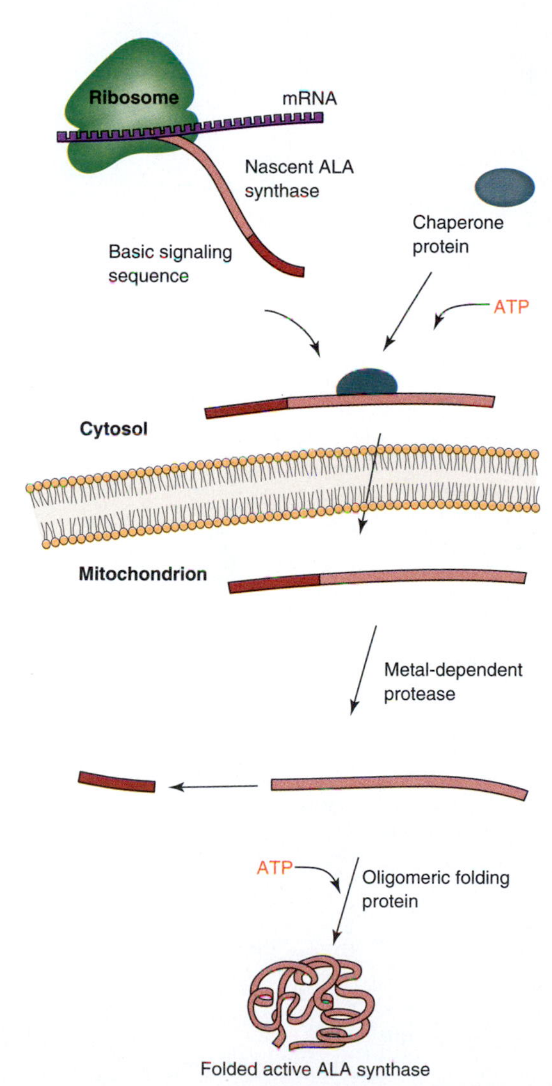

FIGURE 24.9
Synthesis of δ-aminolevulinic acid (ALA) synthase.

Aminolevulinic Acid Dehydratase

Aminolevulinic acid dehydratase (280 Da) is cytosol and consists of eight subunits, of which only four interact with the substrate. This protein interacts with the substrate to form a Schiff base, but in this case the ε-amino group of a lysine residue binds to the ketonic carbon of the substrate molecule (Figure 24.10). Two molecules of ALA condense asymmetrically to form **porphobilinogen.** The ALA dehydrase is a zinc-containing enzyme and is very sensitive to inhibition by heavy metals, particularly lead. A characteristic finding of **lead poisoning** is the elevation of ALA in the absence of an elevation of porphobilinogen.

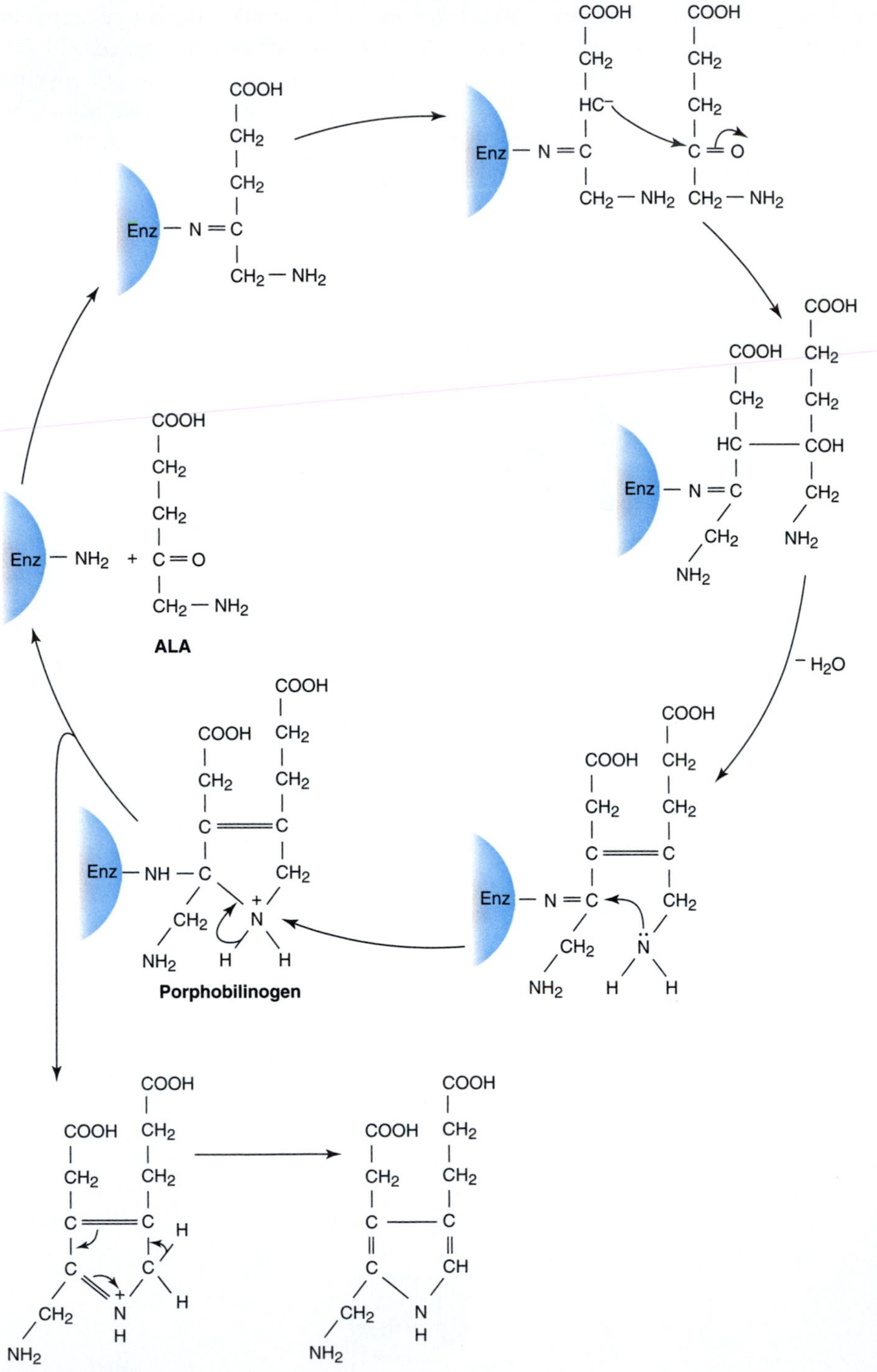

FIGURE 24.10
Synthesis of porphobilinogen.

Porphobilinogen Deaminase and Uroporphyrinogen III Synthase

The synthesis of the porphyrin ring is a complicated process that has recently been defined. A sulfhydryl group on **porphobilinogen deaminase** forms a thioether bond with a porphobilinogen residue through a deamination reaction. Thereafter, five additional porphobilinogen residues are deaminated successively to form a linear hexapyrrole adduct with the enzyme. The adduct is cleaved hydrolytically to form both an enzyme–dipyrrolomethane complex and the linear tetrapyrrole, hydroxymethylbilane. The enzyme–dipyrrolomethane complex is then ready for another cycle of addition of four porphobilinogen residues to generate another tetrapyrrole. Thus dipyrrolomethane is the covalently attached novel cofactor for the enzyme. Porphobilinogen deaminase has no ring-closing function; hydroxymethylbilane closes in an enzyme-independent step to form uroporphyrinogen I if no additional factors are present. However, the deaminase is closely associated with a second protein, **uroporphyrinogen III synthase,** which directs the synthesis of the III isomer. The formation of the latter involves a spiro intermediate generated from hydroxymethylbilane; this allows the inversion of one of the pyrrole groups (see Figure 24.11). In

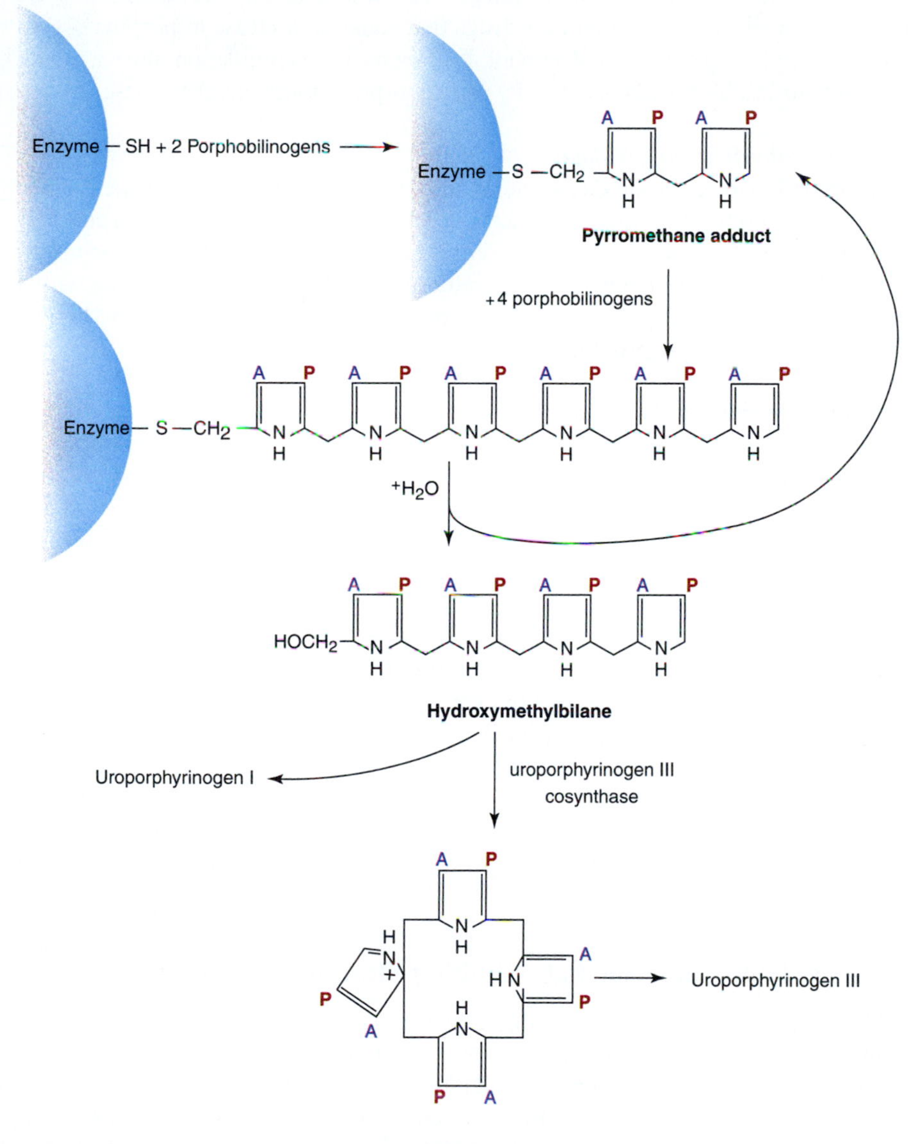

FIGURE 24.11
Synthesis of uroporphyrinogens I and III.
Enzyme in blue is uroporphyrinogen I synthase.

the absence of the uroporphyrinogen III synthase, uroporphyrinogen I is synthesized slowly; in its presence, the III isomer is synthesized rapidly. A rare recessively inherited disease, erythropoietic porphyria, associated with marked cutaneous light sensitization, is due to an abnormality of reticulocyte uroporphyrinogen III synthase. Here large amounts of the type I isomers of uroporphyrinogen and coproporphyrinogen are synthesized in the bone marrow. Two isoenzymes exist for porphobilinogen deaminase due to alternative splicing of exon 1 or exon 2 to the rest of the mRNA.

Uroporphyrinogen Decarboxylase

Uroporphyrinogen decarboxylase acts on the side chains of uroporphyrinogens to form coproporphyrinogens. The protein catalyzes the conversion of both I and III isomers of uroporphyrinogen to the respective coproporphyrinogen isomers. Uroporphyrinogen decarboxylase is inhibited by iron salts. Clinically the most common cause of porphyrin derangement is associated with patients who have a single gene abnormality for uroporphyrinogen decarboxylase, leading to 50% depression of the enzyme's activity. This disease, which shows cutaneous manifestations primarily with sensitivity to light, is known as **porphyria cutanea tarda.** The condition is not expressed unless patients either take drugs that cause an increase in porphyrin synthesis or drink large amounts of alcohol, leading to the accumulation of iron, which then acts to inhibit further the activity of uroporphyrinogen decarboxylase.

Coproporphyrinogen Oxidase

Coproporphyrinogen oxidase, a mitochondrial enzyme, is specific for the type III isomer of coproporphyrinogen, not acting on the type I isomer. Coproporphyrinogen III enters mitochondria and is converted to protoporphyrinogen IX. The mechanism of action is not understood. A dominant hereditary disease associated with a deficiency of this enzyme leads to a form of hereditary hepatic porphyria, known as **hereditary coproporphyria.**

Protoporphyrinogen Oxidase

Protoporphyrinogen oxidase, another mitochondrial enzyme, generates protoporphyrin IX, which, in contrast to the other heme precursors, is very water-insoluble. Excess amounts of protoporphyrin IX that are not converted to heme are excreted by the biliary system into the intestinal tract. A dominant disease, **variegate porphyria,** is due to a deficiency of protoporphyrinogen oxidase.

Ferrochelatase

Ferrochelatase inserts ferrous iron into protoporphyrin IX in the final step of heme synthesis. Reducing substances are required for its activity. The protein is sensitive to the effects of heavy metals (especially lead) and, of course, to iron deprivation. In these latter instances, zinc instead of iron is incorporated to form a zinc–protoporphyrin IX complex. In contrast to heme, the zinc–protoporphyrin IX complex is brilliantly fluorescent and easily detectable in small amounts. Prokaryotic ferrochelatase contains no prosthetic groups; however, the mammalian enzyme contains an Fe_2S_2 group.

ALA Synthase Catalyzes Rate-Limiting Step of Heme Biosynthesis

ALA synthase controls the rate-limiting step of heme synthesis in all tissues. Succinyl CoA and glycine are substrates for a variety of reactions. The modulation of the activity of ALA synthase determines the quantity of the substrates that will be shunted into heme biosynthesis. Heme (and also hematin) acts both as a repressor of the synthesis of ALA synthase and as an inhibitor of its activity. Since heme resembles neither the substrates nor the product of the enzyme's action, it is probable that the latter inhibition occurs at an allosteric site. Almost 100 different drugs and metabolites can cause induction of ALA synthase; for example, a 40-fold increase is noted in the rat after treatment with 3,5-dicarbethoxy-1,4-dihydrocollidine. The effect of pharmacological agents has led to the important clinical feature where some patients with certain kinds of porphyria have had exacerbations of their condition following the inappropriate administration of certain drugs (e.g., barbiturates). ALA dehydratase is also inhibited by heme; but this is of little physiological consequence, since the activity of ALA dehydrase is about 80-fold greater than that of ALA synthase, and thus heme-inhibitory effects are reflected first in the activity of ALA synthase.

Glucose or a proximal metabolite serves to inhibit heme biosynthesis in a mechanism that is not yet defined. This is of clinical relevance, since some patients manifest their porphyric state for the first time when placed on a very low caloric (and therefore glucose) intake. Other regulators of porphyrin metabolism include certain steroids. Steroid hormones (e.g., oral contraceptive pills) with a double bond in ring A between C-4 and C-5 atoms can be reduced by two different reductases. The product of 5α-reduction has little effect on heme biosynthesis; however, the product of 5β-reduction serves as a stimulus for the synthesis of ALA synthase.

24.7 | HEME CATABOLISM

Catabolism of heme-containing proteins presents two requirements to the mammalian host: (1) development of a means of processing the hydrophobic products of porphyrin ring cleavage and (2) retention and mobilization of the contained iron so that it may be reutilized. Red blood cells have a life span of approximately 120 days. Senescent cells are recognized by their membrane changes and removed and engulfed by the reticuloendothelial system at extravascular sites. The globin chains denature, releasing heme into the cytoplasm. The globin is degraded to its constituent amino acids, which are reutilized for general metabolic needs.

Figure 24.12 depicts the events of heme catabolism. Heme is degraded primarily by a microsomal enzyme system in reticuloendothelial cells that requires molecular oxygen and NADPH. **Heme oxygenase** is substrate inducible and catalyzes the cleavage of the α-methene bridge, which joins the two pyrrole residues containing the vinyl substituents. The α-methene carbon is converted quantitatively to carbon monoxide. The only endogenous source of **carbon monoxide** in humans is the α-methene carbon. A fraction of the carbon monoxide is released via the respiratory tract. Thus the measurement of carbon monoxide in an exhaled breath provides an index to the quantity of heme that is degraded in an individual. The oxygen present in the carbon monoxide and in the newly derivatized lactam rings is generated entirely from molecular oxygen. The stoichiometry of the reaction requires 3 mol of oxygen for each ring cleavage. Heme oxygenase will only use heme as a substrate, with the iron possibly participating in the cleavage mechanism. Thus free protoporphyrin IX is not a substrate. The linear tetrapyrrole **biliverdin IX** is the product formed by the action of heme oxygenase. Biliverdin IX is reduced by **biliverdin reductase** to **bilirubin IX.**

Bilirubin Is Conjugated to Form Bilirubin Diglucuronide in Liver

Bilirubin is derived not only from senescent red cells but also from turnover of other heme-containing proteins, such as the cytochromes. Studies with labeled glycine as a precursor have revealed that an early-labeled bilirubin, with a peak within 1–3 h, appears a very short time after a pulsed administration of the labeled precursor. A larger amount of bilirubin appears much later at about 120 days, reflecting the turnover of heme in red blood cells. Early-labeled bilirubin can be divided into two parts: an early–early part, which reflects the turnover of heme proteins in the liver, and a late–early part, which consists of both the turnover of heme-containing hepatic proteins and the turnover of bone marrow heme, which

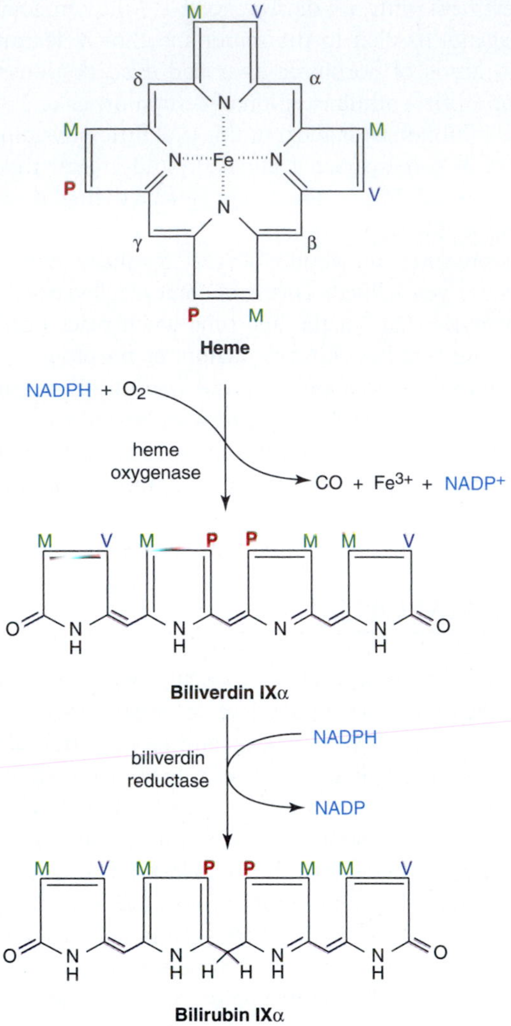

FIGURE 24.12
Formation of bilirubin from heme.
Greek letters indicate labeling of methene carbon atoms in heme.

is either poorly incorporated or easily released from red blood cells. The latter is a measurement of ineffective erythropoiesis and can be very pronounced in disease states such as pernicious anemia and the thalassemias.

Bilirubin is poorly soluble in aqueous solutions at physiological pH values. When transported in plasma, it is bound to serum albumin with an association constant greater than $10^6\ M^{-1}$. Albumin contains one such high-affinity site and another with a lesser affinity. At the normal albumin concentration of 4 g dL^{-1}, about 70 mg of bilirubin per dL of plasma can be bound on the two sites. However, bilirubin toxicity **(kernicterus)**, which is manifested by the transfer of bilirubin to membrane lipids, commonly occurs at concentrations greater than 25 mg dL^{-1}. This suggests that the weak affinity of the second site does not allow it to serve effectively in the transport of bilirubin. Bilirubin on serum albumin is rapidly cleared by the liver, where there is a free bidirectional flux of the tetrapyrrole across the sinusoidal–hepatocyte interface. Once in the hepatocyte, bilirubin is bound to several cytosolic proteins, of which only one has been well characterized. The latter component, **ligandin,** is a small basic component making up to 6% of the total cytosolic protein of rat liver. Ligandin has been purified to homogeneity from rat liver and characterized as having two subunits of molecular mass of 22 kDa and 27 kDa. Each subunit contains glutathione *S*-epoxidetransferase activity, a function important in detoxification mechanisms of aryl groups. The stoichiometry of binding is one bilirubin molecule per complete ligandin molecule.

Once in the hepatocyte the propionyl side chains of bilirubin are conjugated to form a diglucuronide (see Figure 24.13). The reaction utilizes uridine diphosphoglucuronate derived from the oxidation of uridine diphosphoglucose. The former serves as a glucuronate donor to bilirubin. In normal bile the diglucuronide is the major form of excreted bilirubin, with only small amounts of the monoglucuronide or other glycosidic adducts. **Bilirubin diglucuronide** is much more water-soluble than free bilirubin, and thus the transferase facilitates the excretion of the bilirubin into bile. Bilirubin diglucuronide is poorly absorbed by the intestinal mucosa. The glucuronide residues are released in the terminal ileum and large intestine by bacterial hydrolases; the released bilirubin is reduced to the colorless linear tetrapyrroles known as **urobilinogens.** Urobilinogens can be oxidized to colored products known as **urobilins,** which are excreted in the feces. A small fraction of urobilinogen can be reabsorbed by the terminal ileum and large intestine to be removed by hepatic cells and resecreted in bile. When urobilinogen is reabsorbed in large amounts in certain disease states, the kidney serves as a major excretory site.

In the normal state plasma bilirubin concentrations are 0.3–1 mg dL^{-1}, and this is almost all in the unconjugated state. In the clinical setting conjugated bilirubin is expressed as **direct bilirubin** because it can be coupled readily with diazonium salts to yield azo dyes; this is the direct **van den Bergh reaction.** Unconjugated bilirubin is bound noncovalently to albumin and will not react until it is released by the addition of an organic solvent such as ethanol. The reaction with diazonium salts yielding the azo dye after the addition of ethanol is the indirect van den Bergh reaction, and this measures the **indirect bilirubin** or the unconjugated bilirubin. Unconjugated bilirubin binds so tightly to serum albumin and lipid that it does not diffuse freely in plasma and therefore does not lead to an elevation of bilirubin in the urine. Unconjugated bilirubin has a high affinity for membrane

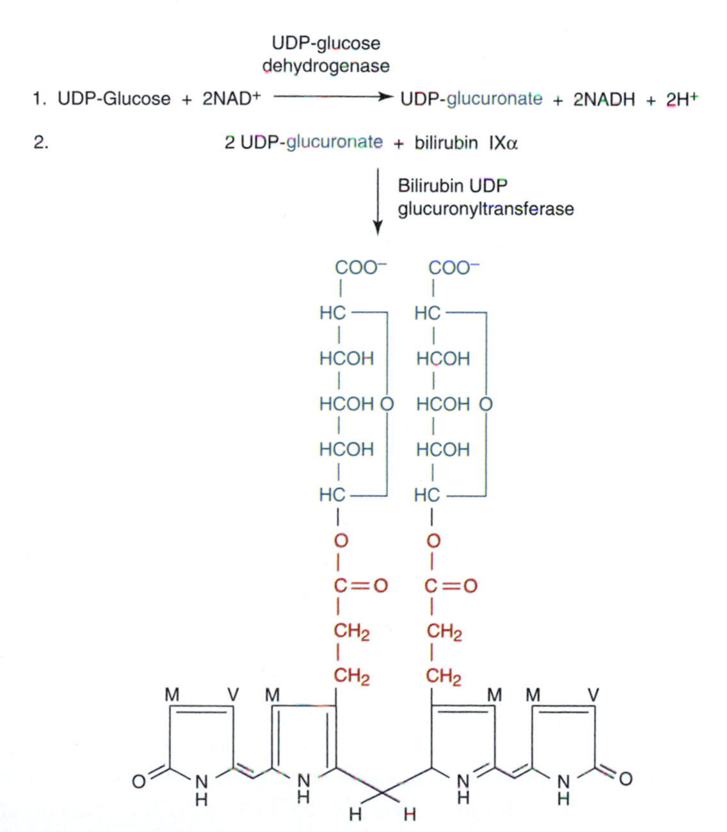

FIGURE 24.13
Biosynthesis of bilirubin diglucuronide.

CLINICAL CORRELATION 24.11
Neonatal Isoimmune Hemolysis

Rh-negative women pregnant with Rh-positive fetuses develop antibodies to Rh factors. These antibodies cross the placenta to hemolyze fetal red blood cells. Usually this is not of clinical relevance until about the third Rh-positive pregnancy, in which the mother has had antigenic challenges from earlier babies. Antenatal studies reveal rising maternal levels of IgG antibodies against Rh-positive red blood cells, indicating that the fetus is Rh-positive. Before birth, placental transfer of fetal bilirubin occurs with excretion through the maternal liver. Because hepatic enzymes of bilirubin metabolism are poorly expressed in the newborn, infants may not be able to excrete the large amounts of bilirubin that can be generated from red cell breakdown. At birth these infants usually appear normal; however, the unconjugated bilirubin in the umbilical cord blood is elevated up to 4 mg dL^{-1}, due to the hemolysis initiated by maternal antibodies. During the next 2 days the serum bilirubin rises, reflecting continuing isoimmune hemolysis, leading to jaundice, hepatosplenomegaly, ascites, and edema. If untreated, signs of central nervous system damage can occur, with the appearance of lethargy, hypotonia, spasticity, and respiratory difficulty, constituting the syndrome of kernicterus. Treatment involves exchange transfusion with whole blood, which is serologically compatible with both the infant's blood and maternal serum. The latter requirement is necessary to prevent hemolysis of the transfused cells. Additional treatment includes external phototherapy, which facilitates the breakdown of bilirubin. The entire problem can be prevented by treating Rh-negative mothers with anti-Rh globulin. These antibodies recognize the fetal red cells, block the Rh antigens, and cause them to be destroyed without stimulating an immune response in the mothers.

Maur, H. M., Shumway, C. N., Draper, D. A., and Hossaini, A. Controlled trial comparing agar, intermittent phototherapy, and continuous phototherapy for reducing neonatal hyperbilirubinemia. *J. Pediatr.* 82:73, 1973; Bowman, J. J. Management of Rh-isoimmunization. *Obstet. Gynecol.* 52:1, 1978; and Dennery, P. A., Seidman, D. S. and Stevenson, D. K. Neonatal hyperbilirubinemia. *N. Engl. J,. Med.* 344:581, 2001.

lipids, which leads to the impairment of cell membrane function, especially in the nervous system. In contrast, conjugated bilirubin is relatively water-soluble, and elevations of conjugated bilirubin lead to high urinary concentrations with the characteristic deep yellow-brown color. The deposition of conjugated and unconjugated bilirubin in skin and the sclera gives the yellow to yellow-green color seen in patients with jaundice.

A third form of plasma bilirubin occurs only with hepatocellular disease in which a fraction of the bilirubin binds so tightly that it is not released from serum albumin by the usual techniques and is thought to be linked covalently to the protein. In some cases up to 90% of total bilirubin can be in this covalently bound form.

The normal liver has a very large capacity to conjugate and mobilize the bilirubin that is delivered. As a consequence, hyperbilirubinemia due to excess heme destruction, as in hemolytic diseases, rarely leads to bilirubin levels that exceed 5 mg dL^{-1}, except in situations in which functional derangement of the liver is present (see Clin. Corr. 24.11). Thus marked elevation of unconjugated bilirubin reflects primarily a variety of hepatic diseases, including those that are heritable and those that are acquired (see Clin. Corr. 24.12).

Elevations of conjugated bilirubin level in plasma are attributable to liver and/or biliary tract disease. In simple uncomplicated biliary tract obstruction, the major component of the elevated serum bilirubin is the diglucuronide form, which is released by the liver into the vascular compartment. Biliary tract disease may be extrahepatic or intrahepatic, the latter involving the canaliculi and biliary ductules (see Clin. Corr. 24.13).

Intravascular Hemolysis Requires Scavenging of Iron

In certain diseases destruction of red blood cells occurs in the intravascular compartment rather than in the extravascular reticuloendothelial cells. In the former case the appearance of free hemoglobin and heme in plasma could lead potentially to the excretion of these substances through the kidney with a substantial loss of iron. To prevent this occurrence, specific plasma proteins are involved in scavenging mechanisms. Transferrin binds free iron and thus permits its reutilization. Free

CLINICAL CORRELATION 24.12
Bilirubin UDP-Glucuronosyltransferase Deficiency

Bilirubin UDP-glucuronosyltransferase has two isoenzyme forms, derived from alternative mRNA splicing between variable forms of exon 1 and common exons 2, 3, 4, and 5. The latter exons define the part of the protein that binds the UDP-glucuronate, whereas the various exons 1 have defined specificities for either bilirubin or other acceptors, such as phenol. Two exons have bilirubin specificity leading to two forms of bilirubin UDP-glucuronosyltransferase forms. Two major families of diseases are seen with deficiencies of the enzyme. Crigler–Najjar syndrome occurs in infants and is associated with extraordinarily high serum unconjugated bilirubin due to an autosomal recessive inheritance of mutations on both alleles in exons 2, 3, 4, or 5. Gilbert's syndrome is also associated with a deficiency of the enzyme's activity, but only to about 25% of normal. The patients appear jaundiced but without other clinical symptoms. The major complication is an exhaustive search by the physician looking for some serious liver disease and failing to recognize the benign condition. Two different findings that may be restricted to different populations account for the condition. In Japan a dominant pattern of inheritance is noted with a mutation on only one allele. The 75% reduction of activity is ascribed to the fact that the enzyme exists as an oligomer, where mutant and normal monomers might associate to form heterooligomers. The explanation is that not only is the mutant monomer inactive, but it forces conformational effects on the normal subunit, reducing its activity substantially. In contrast, in the Western world the condition is due largely to a homozygous expansion of the bases in the promoter region with less efficient transcription of the gene.

Aono, S., Adachi, Y., Uyama, S., et al. Analysis of genes for bilirubin UDP-glucuronosyltransferase in Gilbert's syndrome. *Lancet* 345:958, 1995; Bosma, P. J., Chowdhury, J. R., Bakker, C., et al. The genetic basis of the reduced expression of bilirubin UDP-glucuronosyltransferase 1 in Gilbert's syndrome. *N. Engl. J. Med.* 333:1171, 1995; and Burchell, B. and Hume, R. Molecular genetic basis of Gilbert's syndrome. *J. Gastroenterol. Hepatol.* 14: 960, 1999.

hemoglobin, after oxygenation in the pulmonary capillaries, dissociates into α,β dimers, which are bound to a family of circulating plasma proteins, the **haptoglobins,** which have a high affinity for the oxyhemoglobin dimer. Since deoxyhemoglobin does not dissociate into dimers in physiological settings, it is not bound by haptoglobin. The stoichiometry of binding is two α,β-oxyhemoglobin dimers per haptoglobin molecule. Interesting studies have been made with rabbit antihuman–hemoglobin antibodies on the haptoglobin–hemoglobin interaction. Human haptoglobin interacts with a variety of hemoglobins from different species. The binding of human haptoglobin with human hemoglobin is not affected by the binding

CLINICAL CORRELATION 24.13
Elevation of Serum Conjugated Bilirubin

Elevations of serum conjugated bilirubin are attributable to liver and/or biliary tract disease. In simple uncomplicated biliary tract obstruction, the major component of the elevated serum bilirubin is the diglucuronide form, which is released by the liver into the vascular compartment. Biliary tract disease may be extrahepatic or intrahepatic, the latter involving the canaliculi and biliary ductules.

Hydrophobic compounds have an affinity for membranes and thus enter cells. To remove these compounds ATP-dependent membrane channels pump them into the circulation, where they can be bound by serum albumin and transferred to the liver. These pumps were initially described in the setting of cancer chemotherapy where the agents are commonly hydrophobic and used in high concentrations. One kind of resistance to therapy was found to be due to the increased activity of the pumps in tumor cells, reducing intracellular drug concentrations. Accordingly, these pumps have been called MRPs (multidrug resistance proteins). At least six kinds are known (see p. 525). The property of hydrophobicity reduces the need of a specific pump for each compound since capture from a dilute aqueous medium is a minor issue, in contrast to hydrophilic compounds where channels have to be very specific. The MRPs serve also to transfer physiological hydrophobic compounds such as steroids from the adrenal gland and bilirubin from hepatocytes into bile. Dubin–Johnson syndrome is an autosomal recessive disease involving a defect in the biliary secretory mechanism of the liver. Excretion from the hepatocyte to the canaliculi relies on MRP 2. In Dubin–Johnson syndrome mutations occur in this protein (known also as cMOAT, *c*analicular *m*ultispecific *o*rganic *a*nion *t*ransporter). Excretion through the biliary tract of a variety (but not all) of organic anions is affected. Retention of melanin-like pigment in the liver in this disorder leads to a characteristic gray-black color of this organ. A second heritable disorder associated with elevated levels of serum conjugated bilirubin is Rotor's syndrome. In this poorly defined disease no hepatic pigmentation occurs.

Iyanagi, E. Y. and Accoucheur, S. Biochemical and molecular disorders of bilirubin metabolism. *Biochim. Biophys. Acta* 1407:173, 1998; *Hepatology* 15:1154,1992; and Tsugi, H., Konig, J., Rost, D., et al. Exon–intron organization of the human multidrugresistance protein 2 (MRP 2) gene mutated in Dubin–Johnson syndrome. *Gastroenterology* 117:653, 1999.

of rabbit antihuman-hemoglobin antibody. These studies suggest that haptoglobin binds to sites on hemoglobin that are highly conserved in evolution and therefore are not sufficiently antigenic to generate antibodies. The most likely site for the molecular interaction of hemoglobin and haptoglobin is the interface of the α and β globins of the tetramer that dissociates to yield α,β dimers. Sequence determinations have indicated that these contact regions are highly conserved in evolution.

Haptoglobins are α_2-globulins. Synthesized in liver, they consist of two pairs of polypeptide chains (α being the lighter and β the heavier). The α and β chains are derived from a single mRNA, generating a single polypeptide chain that is cleaved to form the two different chains. The β chains are glycopeptides of 39 kDa and are invariant in structure; α chains are of several kinds. The haptoglobin chains are joined by disulfide bonds between the α and β chains and between the two α chains.

Interaction of haptoglobin with hemoglobin forms a complex that is too large to be filtered through the renal glomerulus. Free hemoglobin (appearing in renal tubules and in urine) will occur during intravascular hemolysis only when the binding capacity of circulating haptoglobin has been exceeded. Haptoglobin delivers hemoglobin to the reticuloendothelial cells. The heme in free hemoglobin is relatively resistant to the action of heme oxygenase, whereas the heme residues in an α,β dimer of hemoglobin bound to haptoglobin are very susceptible.

Measurement of serum haptoglobin is used clinically as an indication of the degree of intravascular hemolysis. Patients who have significant intravascular hemolysis have little or no plasma haptoglobin because of the removal of haptoglobin–hemoglobin complexes by the reticuloendothelial system. Haptoglobin levels can also be low in severe extravascular hemolysis, in which the large load of hemoglobin in the reticuloendothelial system leads to the transfer of free hemoglobin into plasma.

Free heme and hematin appearing in plasma are bound by a β-globulin, **hemopexin** (57 kDa). One heme residue binds per hemopexin molecule. Hemopexin transfers heme to liver, where further metabolism by heme oxygenase occurs. Normal plasma hemopexin contains very little bound heme, whereas in intravascular hemolysis, the hemopexin is almost completely saturated by heme and is cleared with a half-life of about 7 h. In the latter, excess heme binds to albumin, with newly synthesized hemopexin serving as a mediator for the transfer of the heme from albumin to the liver. Hemopexin also binds free protoporphyrin.

BIBLIOGRAPHY

Anderson, K. E., Sassa, S., Bishop, D. F., and Desnick, R. J. Disorders of Heme Biosynthesis: X-Linked Sideroblastic Anemia and the Porphyrias. In: C. R. Scriver, A. L. Beaudet, W. S. Sly, and D. Valle (Eds.), The Metabolic and Molecular Bases of Inherited Disease, 8th ed., Vol II. New York: McGraw-Hill, 2001, p. 2991.

Andrews, N. C. Medical Progress. Disorders of iron metabolism. *N. Engl. J. Med.* 341:1986,1999.

Battersby, A. R. The Bakerian Lecture, 1984. Biosynthesis of the pigments of life. *Proc. R. Soc. London B* 225:1, 1985.

Beutler, E., Bothwell, T. M., Charlton, R. W., and Motulsky, A. G. Hereditary Hemochromatosis. In: C. R. Scriver, A. L. Beaudet, W. S. Sly, and D. Valle (Eds.), The Metabolic and Molecular Bases of Inherited Disease, 8th ed., Vol II, New York: McGraw-Hill, 2001, p. 3127.

Braig, K., Otwinowski, Z., Hegde, R., Boisvert, D. C., Joachimiak, A., Horwich, A. L., and Sigler, P. B. The crystal structure of the bacterial chaperonin GroEL at 2.8 Å. *Nature* 371:578, 1994.

Casey, J. L., Hentze, M. W., Koeller, D. H., Caughman, S., Rouault, T. A., Klausner, R. D., and Harford, J. B. Iron-responsive elements: regulatory RNA sequences that control mRNA levels and translation. *Science* 240:924, 1988.

Chowdhury, J. R., Wolkoff, A. W., Chowdhury, N. R., and Irias, I. M. Hereditary Jaundice and Disorders of Bilirubin Metabolism. In: C. R. Scriver, A. L. Beaudet, W. S. Sly, and D. Valle (Eds.), The Metabolic and Molecular Bases of Inherited Disease, 8th ed., Vol. II, New York: McGraw-Hill, 2001, p. 3063.

Eisenstein, R. S. Iron regulatory proteins and the molecular control of mammalian iron metabolism. *Annu. Rev. Nutr.* 20:627, 2000.

Fenton, W. A., Kashi, Y., Furtak, K., and Horwich, A. L. Residues in chaperonin GroEL required for polypeptide binding and release. *Nature* 371:614, 1994.

Ferreira, G. C. Ferrochelatase binds the iron-responsive element present in the erythroid δ-aminolevulinate synthase mRNA. *Biochem. Biophys. Res. Commun.* 214:875, 1995.

Fleet, J. C. A new role for lactoferrin: DNA binding and transcription activation. *Nutr. Rev.* 53:226, 1995.

Guo, B., Phillips, J. D., Yu, Y., and Leibold, E. A. Iron regulates the intracelluar degradation of iron regulatory protein 2 by the proteasome. *J. Biol. Chem.* 270:21645, 1995.

Hartl, F. U. Secrets of a double-doughnut. *Nature* 371:557, 1994.

Huebers, H. A. and Finch, C. A. Transferrin: physiologic behavior and clinical implications. *Blood* 64:763, 1984.

Lieu, P. T., Heiskala, M., Peterson, P. A., et al. The role of iron in health and disease. *Mol. Aspects Med.* 22:1, 2001.

Lustbader, J. W., Arcoleo, J. P., Birken, S., and Greer, J. Hemoglobin-binding site on haptoglobin probed by selective proteolysis. *J. Biol. Chem.* 258:1227, 1983.

Maeda, N., Yang, F., Barnett, D. R., Bowman, B. H., and Smithies, O. Duplication within the haptoglobin *Hp2* gene. *Nature* 309:131, 1984.

Mascotti, D. P., Rup, D., and Thach, R. E. Regulation of iron metabolism: translational effects mediated by iron, heme, and cytokines. *Annu. Rev. Nutr.* 15:239, 1995.

May, W. S., Sahyoun, N., Jacobs, S., Wolf, M., and Cuatrecasas, P. Mechanism of phorbol-diester-induced regulation of surface transferrin receptor involves the action of activated protein kinase C and an intact cytoskeleton. *J. Biol. Chem.* 260:9419, 1985.

Melefors, O. and Hentze, M. W. Iron regulatory factor—the conductor of cellular iron regulation. *Blood Rev.* 7:251, 1993.

Osterman, J., Horwich, A. L., Neupert, W., and Hartl, F.-U. Protein folding in mitochondria requires complex formation with hsp60 and ATP hydrolysis. *Nature* 341:125, 1989.

Pantopoulos, K., Gray, N. K., and Hentze, M. W. Differential regulation of two related RNA-binding proteins, iron regulatory protein (IRP) and IRP_B. *RNA* 1:155, 1995.

Pantopoulos, K. and Hentze, M. W. Nitric oxide signalling to iron-regulatory protein: direct control of ferritin mRNA translation and transferrin receptor mRNA stability in transfected fibroblasts. *Proc. Natl. Acad. Sci. USA* 92:1267, 1995.

Theil, E. L. and Eisenstein, R. S. Combinatorial mRNA regulation: iron regulatory proteins and iso-iron responsive elements (isoIREs). *J. Biol. Chem.* 275, 40659 (2000).

Weiss, J. S., et al. The clinical importance of a protein-bound fraction of serum bilirubin in patients with hyperbilirubinemia. *N. Engl. J. Med.* 309:147, 1983.

Yamashiro, D. J., Tycko, B., Fluss, S. R., and Maxfield, F. R. Segregation of transferrin to a mildly acidic (pH 6.5) para-Golgi compartment in the recycling pathway. *Cell* 37:789, 1984.

QUESTIONS | C. N. ANGSTADT

Multiple Choice Questions

Refer to the following for Questions 1–3.

A. ferritin
B. ferredoxin
C. hemosiderin
D. lactoferrin
E. transferrin

1. A type of protein in which iron is specifically bound to sulfur.
2. Major protein responsible for the storage of iron.
3. Delivers iron to tissues by binding to specific cell surface receptors.

4. In the intestinal absorption of iron:
   A. the presence of a reductant like ascorbate enhances the availability of the iron.
   B. the regulation of uptake occurs between the lumen and the mucosal cells.
   C. the amount of apoferritin synthesized in the mucosal cell is directly related to the need for iron by the host.
   D. iron bound tightly to a ligand, such as phytate, is more readily absorbed than free iron.
   E. low pH in the stomach inhibits absorption by favoring Fe^{3+}.

5. Which of the following statements about iron distribution is correct?
   A. Iron overload cannot occur because very efficient excretory mechanisms are available.
   B. Cells cannot regulate their uptake of iron with changing iron content.
   C. Transferrin decreases in iron deficiency to facilitate storage of iron.
   D. Iron homeostasis is maintained in part by iron regulatory proteins binding to iron responsive elements in mRNA.
   E. In the early stages of iron depletion, serum ferritin levels rise rapidly as iron is released from storage forms.

6. The biosynthesis of heme requires all of the following EXCEPT:
   A. propionic acid.
   B. succinyl CoA.
   C. glycine.
   D. ferrous ion.

7. Lead poisoning would be expected to result in an elevated level of:
   A. aminolevulinic acid.
   B. porphobilinogen.
   C. protoporphyrin I.
   D. heme.
   E. bilirubin.

8. Ferrochelatase:
   A. is an iron-chelating compound.
   B. releases iron from heme in the degradation of hemoglobin.
   C. binds iron to sulfide ions and cysteine residues.
   D. is inhibited by heavy metals.
   E. is involved in the cytoplasmic portion of heme synthesis.

9. Heme oxygenase:
   A. can oxidize the methene bridge between any two pyrrole rings of heme.
   B. requires molecular oxygen.
   C. produces bilirubin.
   D. produces carbon dioxide.
   E. can use either heme or protoporphyrin IX as substrate.

10. Haptoglobin:
    A. helps prevent loss of iron following intravascular red blood cell destruction.
    B. levels in serum are elevated in severe intravascular hemolysis.
    C. inhibits the action of heme oxygenase.
    D. binds heme and hematin as well as hemoglobin.
    E. binds α,β-deoxyhemoglobin dimers.

Questions 11 and 12: A woman appears in an emergency room in an agitated state with severe abdominal pain and marked weakness in all her limbs. Usual laboratory tests are normal and no specific abnormality can be found. She is sedated with phenobarbital (a barbiturate) and observed over the next few hours. Her condition rapidly deteriorates and further study finds an elevated level of porphobilinogen in her urine. A diagnosis of acute intermittent porphyria is made. Enzyme abnormalities in this disease are a marked increase in ALA synthetase and a 50% reduction in porphobilinogen deaminase.

11. Aminolevulinic acid synthase:
    A. requires NAD for activity.
    B. is allosterically activated by heme.
    C. synthesis is inhibited by steroids.
    D. is synthesized in mitochondria.
    E. synthesis can be induced by a variety of drugs.

12. Normally, porphobilinogen is an intermediate in the pathway to heme biosynthesis. There are many other intermediates. Uroporphyrinogen III:
   A. is synthesized rapidly from porphobilinogen in the presence of a cosynthase.
   B. does not contain a tetrapyrrole ring.
   C. is formed from coproporphyrinogen III by decarboxylation.
   D. is converted directly to protoporphyrinogen IX.
   E. formation is the primary control step in heme synthesis.

Questions 13 and 14: If an Rh-negative woman becomes pregnant with Rh-positive fetuses, she will develop antibodies to Rh factors. About the third Rh-positive pregnancy, the antibodies will cross the placenta and hemolyze fetal red blood cells. Because hepatic enzymes of bilirubin metabolism are not fully developed in the newborn, within a few days after birth the infant will develop jaundice, hepatosplenamegaly, ascites, and edema. If untreated, central nervous system damage can occur.

13. The substance deposited in skin and sclera in jaundice is:
   A. biliverdin.
   B. only unconjugated bilirubin.
   C. only direct bilirubin.
   D. both bilirubin and bilirubin diglucuronide.
   E. hematin.

14. Conjugated bilirubin is:
   A. transported in blood bound to serum albumin.
   B. deficient in Crigler–Najjar syndrome, a deficiency of a UDP-glucuronosyltransferase.
   C. reduced in serum in biliary tract obstruction.
   D. the form of bilirubin most elevated in hepatic (liver) disease.
   E. less soluble in aqueous solution than the unconjugated form.

Problems

15. In the early stages of iron deficiency, physiological changes occur although there are no clinical symptoms. The level and percent saturation of serum transferrin are relatively normal. As the iron deficiency progresses, serum iron falls and the level of total serum transferrin increases but its percent saturation falls. How is serum iron level maintained in the early stages? What is the purpose of the increased transferrin in the later stages?

16. Microorganisms require iron for replication and function. Infants are protected against gastrointestinal infections because of the way iron is handled in human milk. In individuals with an iron overload, normally nonpathogenic microorganisms can become pathogenic because of a breakdown in the normal handling of iron. What is the mechanism of iron handling that accounts for these two conditions?

ANSWERS

1. **B** Animal ferredoxins, also known as nonheme iron-containing proteins, have two irons bound to two cysteine residues and sharing two sulfide ions.
2. **A** Hemosiderin is an amorphous deposit of iron around ferritin when iron is in excess.
3. **E** Internalization of the receptor–transferrin complex is mediated by a Ca^{2+}–calmodulin–protein kinase C complex. Internalization is followed by release of the iron and recycling of the apotransferrin to the plasma.
4. **A** Ascorbate facilitates reduction to the ferrous state and, therefore, dissociation from ligands and absorption. B: Substantial iron enters the mucosal cell regardless of need, but the amount transferred to the capillary beds is controlled. C: Iron bound to apoferritin is trapped in mucosal cells and not transferred to the host. D: Iron must dissociate from ligands for absorption. This is why the iron in spinach is not a good source of iron. E: Oxidation to Fe^{3+} is favored by higher pH.
5. **D** D, B: In the presence of low iron this mechanism leads to increased synthesis of transferrin receptor and decreased synthesis of apoferritin. A: The high affinity of many macromolecules for iron prevents efficient excretion. C: Transferrin increases in iron deficiency to improve absorption. E: Serum ferritin is normally small and decreases.
6. **A** The organic portion of heme comes totally from glycine and succinyl CoA; the propionic acid side chain comes from the succinate. D: The final step of heme synthesis is the insertion of the ferrous ion.
7. **A** Lead inhibits ALA dehydratase so it results in accumulation of ALA. B–D: Synthesis of porphobilinogen and subsequent compounds is inhibited. Heme certainly would not be elevated, because lead also inhibits ferrochelatase. E: Bilirubin is a breakdown product of heme, not an intermediate in synthesis.
8. **D** This enzyme catalyzes the last step of heme synthesis, the insertion of Fe^{2+}, and is sensitive to the effects of heavy metals. A, B: It is an enzyme of synthesis. C: This describes an iron–sulfur protein. E: The last step of heme synthesis is a mitochondrial process.
9. **B** Oxygenases usually use O_2. A: The enzyme is specific for the methene between the two rings containing the vinyl groups (α-methene bridge). C, D: The products are biliverdin and CO; the measurement of CO in the breath is an index of heme degradation. E: Iron is necessary for activity.
10. **A** Haptoglobin is part of the scavenging mechanism to prevent urinary loss of heme and hemoglobin from intravascular degradation of red blood cells. B: Since the scavenged complex is taken up by the reticuloendothelial system, the haptoglobin levels in serum are low. C: Heme residues in the dimers bound to haptoglobin are more susceptible than free heme to oxidation by heme oxygenase. D: Heme and hematin are bound by a β-globulin, while haptoglobin is an α-globulin. E: Deoxyhemoglobin does not dissociate to dimers physiologically.
11. **E** The enzyme is induced in response to need as well as by drugs and metabolites. Giving the porphyria patient phenobarbital increased her already high ALA synthase. A: The mechanism involves a Schiff base with glycine, therefore the coenzyme is pyridoxal phosphate. B: Heme both allosterically inhibits and suppresses synthesis of the enzyme. C: One reduction product of catabolic steroids stimulates the synthesis. D: The gene for this enzyme is on nuclear DNA.
12. **A** In the absence of cosynthase, uroporphyrinogen I is synthesized slowly. B: The tetrahydropyrrole ring has formed by this point. C, D: The decarboxylation goes from uroporphyrinogen to coproporphyrinogen, which is then converted to protopor-

phyrinogen. E: The synthesis of aminolevulinic acid is the rate-limiting step.

13. **D** Both conjugated (direct) and unconjugated (indirect) bilirubin are deposited. They are coming from the breakdown of the red cells.

14. **B** Conjugated bilirubin is bilirubin diglucuronide. A: Unconjugated bilirubin is transported bound to albumin. C: Liver conjugates bilirubin but if the bile duct is obstructed it won't be excreted. D: Liver is responsible for the conjugation, which would be impaired in hepatic damage. E: The purpose of conjugation is to increase water solubility.

15. In the absence of iron, the storage forms of iron in liver and bone marrow will be depleted to maintain serum iron. The normal small amount of serum ferritin decreases and what is there is mostly apoferritin with little iron bound. In the later stages, transferrin increases in an attempt to absorb more iron from the gastrointestinal tract.

16. Normally there is very little free iron present because iron-binding proteins are less than completely saturated. In milk, the protein is lactoferrin, which rapidly binds free iron. In serum the protein is transferrin. Normally there is excess iron-binding capacity but in iron overload, transferrin may be almost completely saturated. This makes small amounts of free iron available.

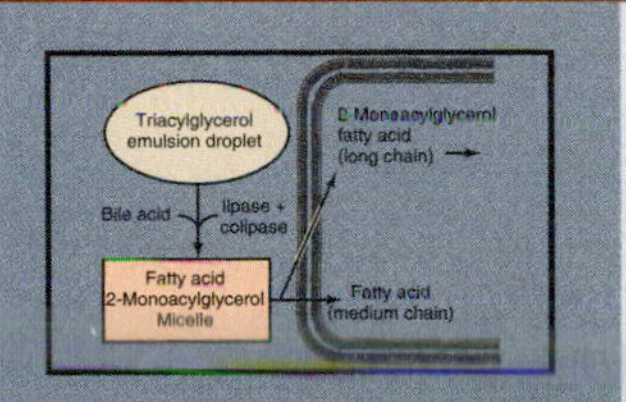

25

DIGESTION AND ABSORPTION OF BASIC NUTRITIONAL CONSTITUENTS

Ulrich Hopfer

25.1 | OVERVIEW

Secretion of digestive fluids and digestion of food were among the earliest biochemical events to be investigated at the beginning of the era of modern science. Major milestones were the discovery of hydrochloric acid secretion by the stomach and enzymatic hydrolysis of protein and starch by gastric juice and saliva, respectively. The discovery of gastric HCl production goes back to the American physician William Beaumont (1785–1853). In 1822 he treated a patient with a stomach wound. The patient recovered from the wound, but retained a gastric fistula (abnormal opening through the skin). Beaumont seized the opportunity to obtain and study gastric juice at different times during and after meals. Chemical analysis revealed, to the surprise of chemists and biologists, the presence of the inorganic acid HCl. This discovery established the principle of unique secretions into the gastrointestinal tract, which are elaborated by specialized glands.

Soon thereafter, the principle of enzymatic breakdown of food was recognized. In 1836 Theodor Schwann (1810–1882), a German anatomist and physiologist, noticed that gastric juice degraded albumin in the presence of dilute acid. He recognized that a new principle was involved and coined the word pepsin from the Greek *pepsis*, meaning digestion. Today the process of secretion of digestive fluids, digestion of food, and absorption of nutrients and of electrolytes can be described in considerable detail.

The major nutrients are proteins, carbohydrates, and fats. Many different types of food can satisfy the nutritional needs of humans, even though they differ in the ratio of protein to carbohydrate and to fat and in the ratio of digestible to nondigestible material. Unprocessed plant products are especially rich in **fibrous material** that cannot be digested by human enzymes or easily degraded by intestinal bacteria. The fibers are mostly carbohydrates, such as **cellulose** (β-1,4-glucan) or **pectins** (mixtures of methyl esters of polygalacturonic acid, polygalactose, and polyarabinose). High-fiber diets enjoy a certain popularity nowadays because of added bulk to stool and a postulated preventive effect on development of colonic cancer.

Table 25.1 describes average contributions of different food classes to the diet of North Americans. Intake by individuals may substantially deviate from the average, as food consumption depends mainly on availability and individual tastes. The ability to utilize a wide variety of food is possible because of the great adaptability and digestive reserve capacity of the gastrointestinal tract.

Knowledge of the nature of proteins and carbohydrates in the diet is important from a clinical point of view. Certain proteins and carbohydrates, although good nutrients for most humans, cannot be properly digested by some individuals and produce gastrointestinal disease. In this case, omission of the offending food constituent eliminates the problems. Examples of nutrients that can be the cause of gastrointestinal disorders in some people are **gluten**, one of the protein fractions of wheat, and **lactose**, the disaccharide in milk.

Multiple Gastrointestinal Organs Contribute to Food Digestion

The bulk of ingested nutrients consists of large polymers that have to be broken down to monomers before they can be absorbed and made available to all cells of

TABLE 25.1 Contribution of Major Food Groups to Daily Nutrient Supplies in the United States

Type of Nutrient	*Total Daily Consumption (g)*	*Dairy Products, Except Butter (%)*	*Meat, Poultry, Fish (%)*	*Eggs (%)*	*Fruits, Nuts, Vegetables (%)*	*Flour, Cereal (%)*	*Sugar, Sweeteners (%)*	*Fats, Oils (%)*
Protein	100	22	42	6	12	18	0	0
Carbohydrate	381	7	0.1	0.1	19	36	37	0
Fat	155	13	35	3	4	1	0	42

the body. The complete process from food intake to absorption of nutrients into the blood consists of a complicated sequence of events, which at the minimum includes (Figure 25.1):

1. Mechanical homogenization of food and mixing of ingested solids with fluids secreted by the glands of the gastrointestinal tract.

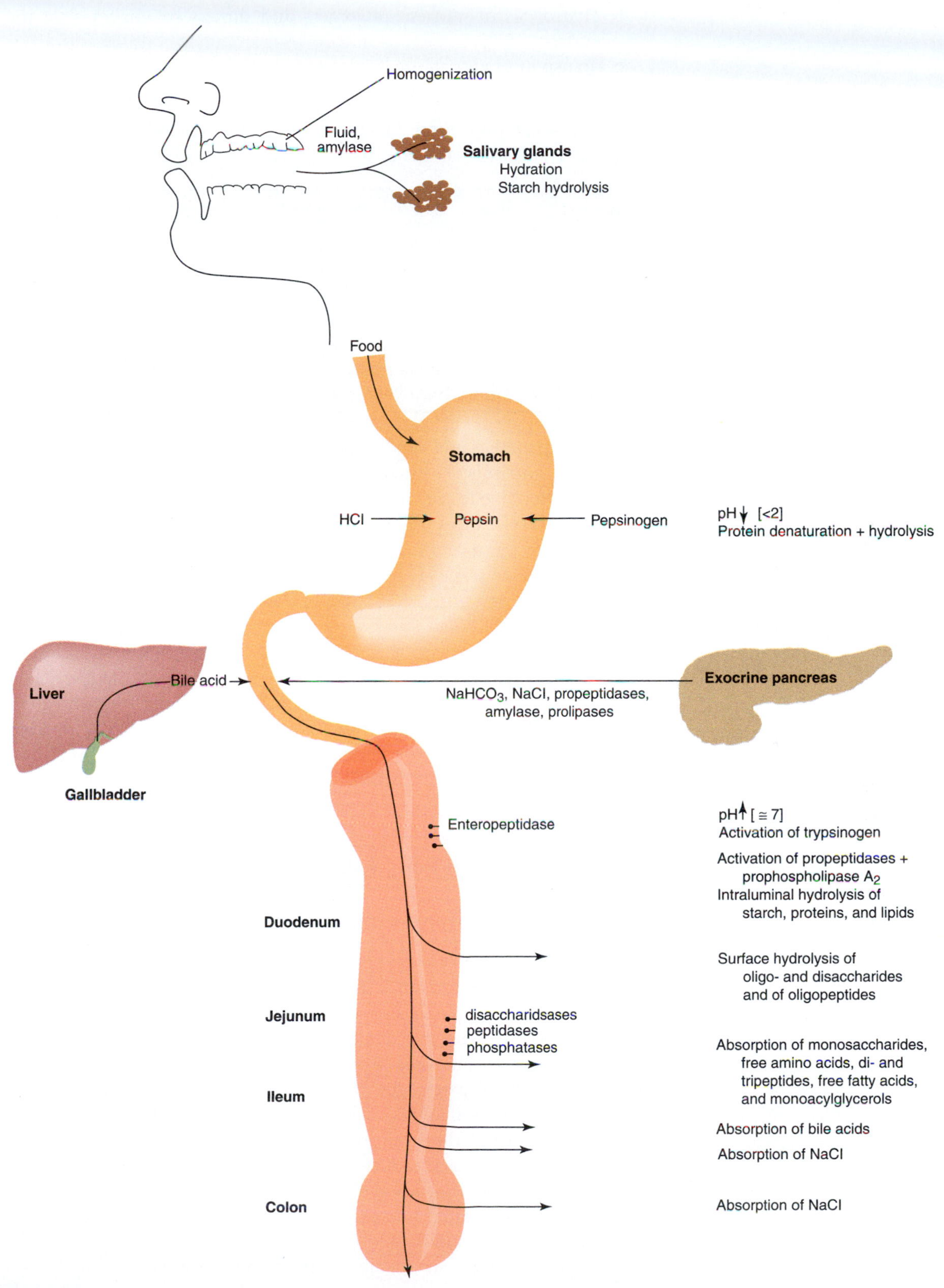

FIGURE 25.1
Gastrointestinal organs and their functions.

2. Secretion of digestive enzymes that hydrolyze macromolecules to oligomers, dimers, or monomers.
3. Secretion of electrolytes, acid, or base to provide an appropriate environment for optimal enzymatic digestion.
4. Secretion of bile acids as detergents to solubilize lipids and facilitate their absorption.
5. Hydrolysis of nutrient oligomers and dimers by enzymes on the intestinal surface.
6. Transport of nutrient molecules and electrolytes from the intestinal lumen across the epithelial cells into blood or lymph.

To accomplish these functions, the gastrointestinal tract contains specialized glands and surface epithelia:

Organ	*Major Function in Digestion and Absorption*
Salivary glands	Elaboration of fluid and digestive enzymes
Stomach	Elaboration of HCl and digestive enzymes
Pancreas	Elaboration of $NaHCO_3$ and enzymes for intraluminal digestion
Liver	Elaboration of bile acids
Gallbladder	Storage and concentration of bile
Small intestine	Terminal digestion of food, absorption of nutrients and electrolytes
Large intestine	Absorption of electrolytes

The **pancreas** and **small intestine** are essential for digestion and absorption of all basic nutrients. Fortunately, both organs have large reserve capacities. Thus maldigestion due to pancreatic failure becomes a problem only when the pancreatic secretion rate of digestive enzymes drops below one-tenth of the normal rate. The secretion of the liver **(bile)** is important for efficient lipid absorption, which depends on the presence of bile acids. In contrast, gastric digestion of food is nonessential for adequate nutrition, and loss of this function can be compensated for by the pancreas and the small intestine. Yet normal gastric digestion greatly increases the smoothness and efficiency of the total digestive process. The stomach aids in the digestion through its reservoir function, its churning ability, and initiation of protein and lipid hydrolysis, which, although small, is important for stimulation of pancreatic and gallbladder output. Peptides, amino acids, and fatty acids liberated in the stomach stimulate the coordinated release of pancreatic juice and bile into the lumen of the small intestine, thereby ensuring efficient digestion of food.

25.2 | DIGESTION: GENERAL CONSIDERATIONS

Different Sites of Intestinal Digestion

Most breakdown of food is catalyzed by soluble enzymes and occurs within the lumen of the stomach or small intestine. The **pancreas,** not the stomach, is the major organ that synthesizes and secretes the large amounts of enzymes needed for digestion. Secreted enzymes amount to at least 30 g of protein per day in a healthy adult. **Pancreatic enzymes** together with bile are poured into the lumen of the second (descending) part of the duodenum, so that the bulk of the **intraluminal digestion** occurs distal to this site in the small intestine. However, pancreatic enzymes cannot completely digest all nutrients to absorbable forms. Even after exhaustive contact with pancreatic enzymes, a substantial portion of carbohydrates and amino acids is present as dimers and oligomers that depend for final digestion on enzymes present on the luminal surface or within the chief epithelial cells that line the lumen of the small intestine **(enterocytes).**

The luminal plasma membrane of enterocytes is enlarged by a regular array of projections, termed microvilli, which give it the appearance of a brush and have led to the name **brush border** for the luminal pole of enterocytes. This membrane contains on its external surface many **di-** and **oligosaccharidases, amino-** and **dipeptidases,** as well as esterases (Table 25.2). Many of these enzymes protrude up to 100 Å into the lumen, attached to the plasma membrane by an anchoring polypeptide that itself has no role in the hydrolytic activity. The substrates for these enzymes are the oligomers and dimers that result from pancreatic digestion. The surface enzymes are glycoproteins that are relatively stable against digestion by pancreatic proteases or the effects of detergents.

A third site of digestion is the cytoplasm of enterocytes. **Intracellular digestion** is important for hydrolysis of di- and tripeptides, which can be absorbed across the luminal plasma membrane.

Digestive Enzymes Are Secreted as Proenzymes

Salivary glands, gastric mucosa, and pancreas contain specialized cells that synthesize and store digestive enzymes until the enzymes are needed during a meal. The enzymes are then released into the lumen of the gastrointestinal tract (Figure 25.2). This secretion is termed **exocrine** because of its direction toward the lumen. Proteins destined for secretion are synthesized on polysomes of the rough endoplasmic reticulum (see p. 257 for synthesis and glycosylation of membrane and secreted proteins) and transported via the Golgi complex to storage vesicles in the apical cytoplasm. The storage vesicles **(zymogen granules)** have a diameter of about 1 μm and are bounded by a typical cellular membrane. Most digestive enzymes are produced and stored as inactive **proenzymes (zymogens)** (see p. 265). When an appropriate stimulus for secretion is received by the cell, the granules move closer to the luminal plasma membrane, where their membranes fuse with the plasma membrane and release their contents into the lumen **(exocytosis).** Activation of proenzymes occurs only after they are released from the cells.

Secretion Is Regulated by Many Secretagogues

Secretion of enzymes and electrolytes is regulated and coordinated. Elaboration of electrolytes and fluids simultaneously with that of enzymes is required to flush any discharged digestive enzymes out of the gland into the gastrointestinal lumen. Regulation of secretion occurs through **secretagogues** that interact with receptors on the **exocrine cell** surface (Table 25.3). Neurotransmitters, hormones, pharmacological agents, and certain bacterial toxins can be secretagogues. Different exocrine cells, for example, in different glands, usually possess different sets of receptors.

TABLE 25.2 Digestive Enzymes of the Small Intestinal Surface

Enzyme (Common Name)	*Substrate*
Maltase	Maltose
Sucrase/isomaltase	Sucrose/α-limit dextrin
Glucoamylase	Amylose
Trehalase	Trehalose
β-Glucosidase	Glucosylceramide
Lactase	Lactose
Endopeptidase	Protein (cleavage at internal hydrophobic amino acids)
Aminopeptidase A	Oligopeptide with acidic NH_2 terminus
Aminopeptidase N	Oligopeptide with neutral NH_2 terminus
Dipeptidyl aminopeptidase IV	Oligopeptide with X-Pro or X-Ala at NH_2 terminus
Leucine aminopeptidase	Peptides with neutral amino acid at NH_2 terminus
γ-Glutamyltransferase	Glutathione + amino acid
Enteropeptidase (enterokinase)	Trypsinogen
Alkaline phosphatase	Organic phosphates

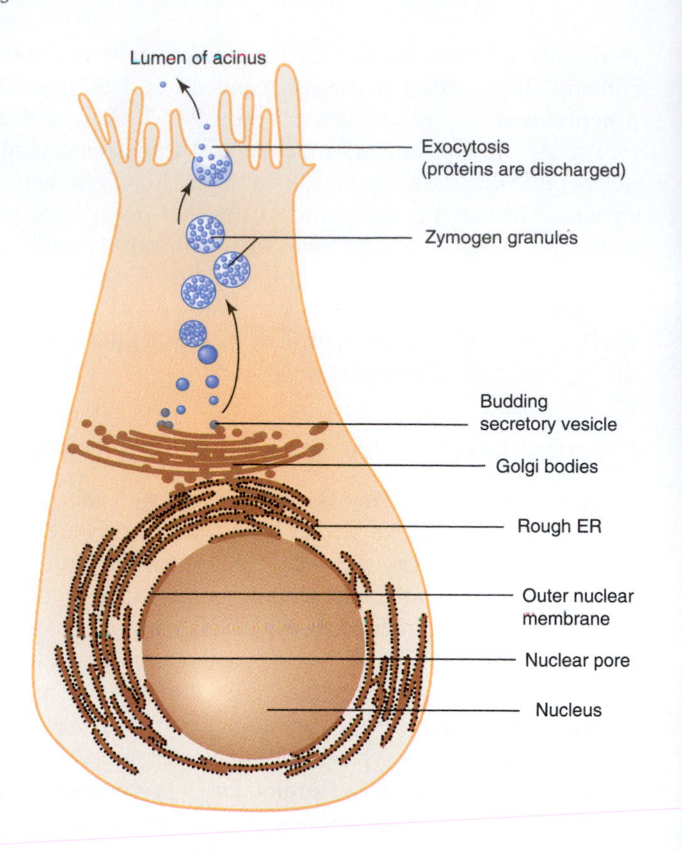

FIGURE 25.2
Exocrine secretion of digestive enzymes.
Redrawn with permission from Jamieson, J. D. In: G. Weissmann and R. Claiborne (Eds.), Cell Membranes: Biochemistry, Cell Biology and Pathology. *New York: HP Publishing, 1975. Figure by B. Tagawa.*

Binding of secretagogues to receptors sets off a chain of signaling events that ends with fusion of zymogen granules with the plasma membrane. Two major signaling pathways (Figure 25.3) are (1) activation of phosphatidylinositol-specific **phospholipase C** with liberation of **inositol 1,4,5-trisphosphate** and **diacylglycerol** (see p. 930), in turn, triggering Ca^{2+} release from the endoplasmic reticulum into the cytosol and activation of protein kinase C, respectively; and (2) activation of **adenylate** or **guanylate cyclase,** resulting in elevated cAMP or cGMP levels, respectively (see p. 925). Secretion can be stimulated through either pathway.

Acetylcholine (Figure 25.4) elicits salivary, gastric, and pancreatic enzyme and electrolyte secretion. It is the major neurotransmitter for stimulating secretion, with input from the central nervous system in salivary and gastric glands, or via local reflexes in gastric glands and the pancreas. The acetylcholine receptor of exocrine cells is of the muscarinic type; that is, it can be blocked by atropine (Figure 25.5). Most people have experienced the effect of atropine because it is used by dentists to "dry up" the mouth for dental work.

The biogenic amines, histamine and **5-hydroxytryptamine (serotonin),** are also secretagogues. Histamine (Figure 25.6) is a potent stimulator of HCl secretion. It interacts with a gastric-specific histamine receptor, also referred to as the **H_2 receptor,** on the contraluminal plasma membrane of gastric parietal cells. Histamine is normally secreted by specialized regulatory cells in the stomach wall **(entero-**

TABLE 25.3 Physiological Secretagogues

Organ	*Secretion*	*Secretagogue*
Salivary gland	NaCl, amylase	Acetylcholine, (catecholamines?)
Stomach	HCl, pepsinogen	Acetylcholine, histamine, gastrin
Pancreas-acini	NaCl, digestive enzymes	Acetylcholine, cholecystokinin (secretin)
Pancreas-duct	$NaHCO_3$, NaCl	Secretin
Small intestine	NaCl	Acetylcholine, serotonin, vasoactive intestinal peptide (VIP), guanylin

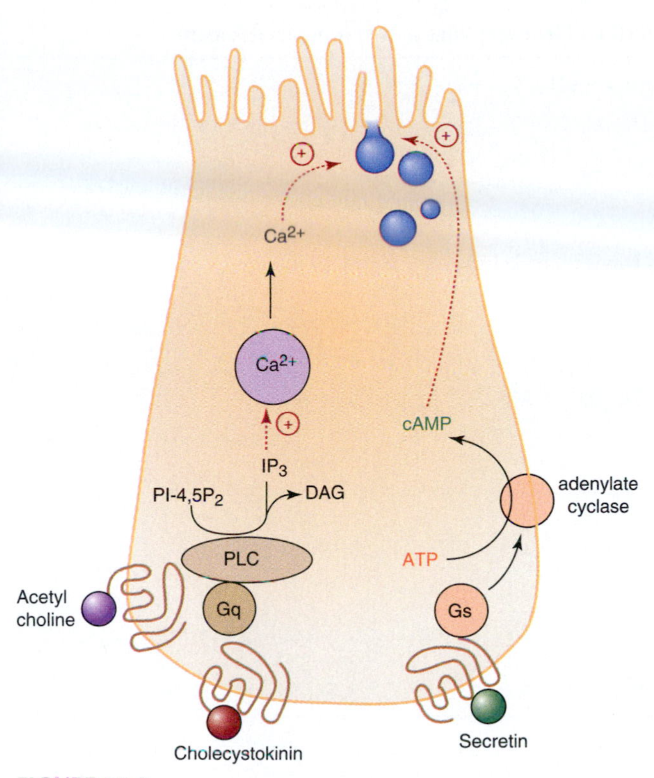

FIGURE 25.3
Cellular regulation of exocrine secretion in the pancreas.
Abbreviations: PI-4,5P$_2$, phosphatidylinositol-4,5-bisphosphate; DAG, diacylglycerol; IP$_3$, inositol-1,4,5-trisphosphate; PLC, phospholipase C.
Adapted from Gardner, J. D. Annu. Rev. Physiol. *41:63, 1979. Copyright © 1979 by Annual Reviews, Inc.*

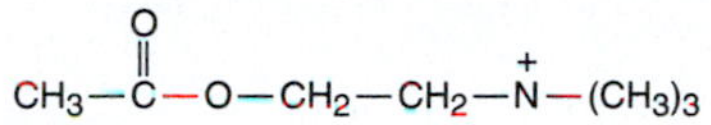

$CH_3-\overset{O}{\overset{\|}{C}}-O-CH_2-CH_2-\overset{+}{N}-(CH_3)_3$

FIGURE 25.4
Acetylcholine.

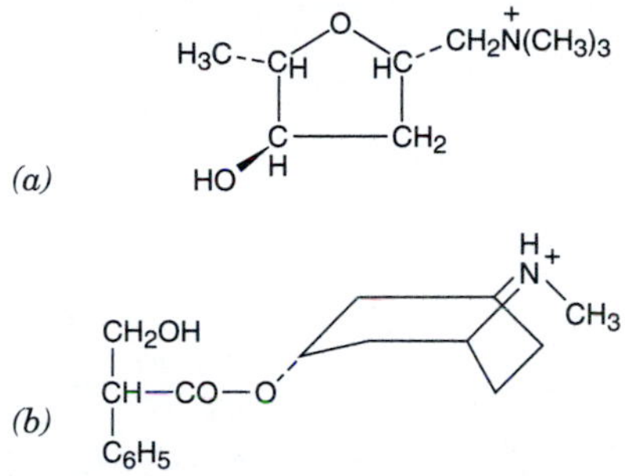

FIGURE 25.5
(*a*) L(+)-Muscarine and (*b*) atropine.

FIGURE 25.6
Histamine.

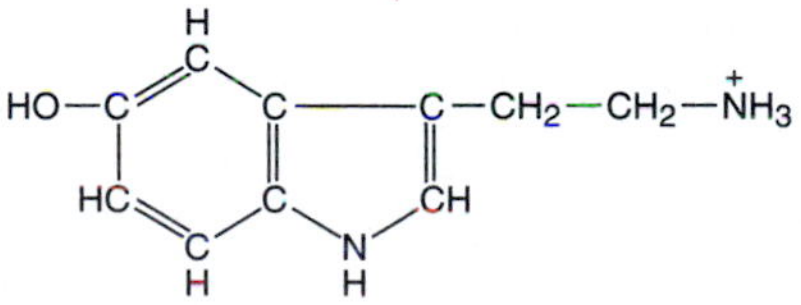

FIGURE 25.7
5-OH-tryptamine (serotonin).

chromaffin-like or **ECL cells).** Histamine analogs that are antagonists at the H_2 receptor are used medically to decrease HCl output during treatment for peptic ulcers. 5-Hydroxytryptamine is present in relatively high amounts in the gastrointestinal tract (Figure 25.7). It serves as a neurotransmitter but also directly stimulates secretion of NaCl by the small intestinal mucosa.

A third class of secretagogues consists of peptide neurotransmitters and hormones (Table 25.4). The intestinal nerve cells are rich in peptide neurotransmitters that stimulate NaCl secretion. **Vasoactive intestinal peptide (VIP)** is a particularly potent one in this respect in the intestines and pancreas. The gastrointestinal tract contains many specialized epithelial cells that produce biologically active amines and peptides. The peptides are stored in granules, usually close to the contraluminal pole of these cells, and are released into the blood. Hence these cells are classified as **epithelial endocrine cells.** Of particular importance are the peptides gastrin, cholecystokinin (pancreozymin), and secretin. The peptide **guanylin** is in part released into the lumen and stimulates NaCl secretion by binding to a brush border receptor that activates guanylate cyclase and thus elevates cytosolic cGMP levels.

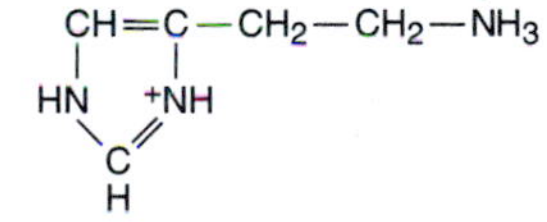

Gastrin occurs as either a peptide of 34 amino acids (G-34) or of 17 amino acids (G-17) from the COOH terminus of G-34. The functional portion of gastrin resides mainly in the last five amino acids of the COOH terminus. Thus pentagastrin, a synthetic pentapeptide containing only the last five amino acids, can be used specifically to stimulate gastric HCl and pepsin secretion. Gastrin and cholecystokinin have an interesting chemical feature, a **sulfated tyrosine,** which considerably enhances the potency of both hormones.

Cholecystokinin and **pancreozymin** denote the same peptide. The different names refer to the different functions elicited by the peptide and were coined before it was purified. The peptide stimulates gallbladder contraction (hence

TABLE 25.4 Secretory Intestinal Neuropeptides and Hormones (Human)

Vasoactive intestinal peptide (VIP)	
His-Ser-Asp-Ala-Val-Phe-Thr-Asp-Asn-Tyr-Thr-Arg-Leu-Arg-Lys-Gln-Met-Ala-Val-Lys	
|	
Asn-Leu-Ile-Ser-Asn-Leu-Tyr-Lys	
Secretin	
His-Ser-Asp-Gly-Thr-Phe-Thr-Ser-Glu-Leu-Ser-Arg-Leu-Arg-Glu-Gly-Ala-Arg-Leu-Gln	
|	
[c]NH_2-Val-Leu-Gly-Gln-Leu-Leu-Arg	
Guanylin	
Pro-Gly-Thr-Cys-Glu-Ile-Cys-Ala-Tyr-Ala-Ala-Cys-Thr-Gly-Cys	
Gastrin G-34-II[a]	G-17-II
[b]Glp-Leu-Gly-Pro-Gln-Gly-Pro-Pro-His-Leu-Val-Ala-Asp-Pro-Ser-Lys-Lys-Gln	
|	
[c]NH_2-Phe-Asp-Met-Trp-Gly-Tyr(SO_3H)-Ala-$(Glu)_5$-Leu-Trp-Pro-Gly	
Cholecystokinin	
Lys-Ala-Pro-Ser-Gly-Arg-Met-Ser-Ile-Val-Lys-Asn-Leu-Gln-Asn-Leu-Asp-Pro	
|	
[c]NH_2-Phe-Asp-Met-Trp-Gly-Met-Tyr(SO_3H)-Asp-Arg-Asp-Ser-Ile-Arg-His-Ser	

Source: Yanaihara, C. In: B. B. Rauner, G. M. Makhlouf, and S. G. Schultz (Eds.), *Handbook of Physiology: Section 6: Alimentary Canal Vol II: Neural and Endocrine Biology*. Bethesda. MD: American Physiological Society, 1989, pp. 95–62.

[a]Gastrin I is not sufated.

[b]Glp = pyrrolidino carboxylic acid, derived from Glu through internal amide formation.

[c]NH_2 = amide of carboxy-terminal amino acid.

cholecystokinin) as well as secretion of pancreatic enzymes (hence pancreozymin). It is secreted by epithelial endocrine cells of the small intestine, particularly in the duodenum, and this secretion is stimulated by luminal amino acids and peptides, usually derived from gastric proteolysis, by fatty acids, and by an acid pH. Cholecystokinin and gastrin are thought to be related in an evolutionary sense, as both share an identical amino acid sequence at the COOH terminus.

Secretin, a polypeptide of 27 amino acids, is secreted by yet another type of endocrine cell in the small intestine. Its secretion is stimulated particularly by luminal pH less than 5. The major biological activity of secretin is stimulation of secretion of pancreatic juice rich in $NaHCO_3$. Pancreatic $NaHCO_3$ is essential for neutralization of gastric HCl in the duodenum. Secretin also enhances pancreatic enzyme release, acting synergistically with cholecystokinin.

25.3 | EPITHELIAL TRANSPORT

Solute Transport May Be Transcellular or Paracellular

Solute movement across an epithelial cell layer is determined by the properties of epithelial cells, particularly their plasma membranes, and by the intercellular tight junctional complexes (Figure 25.8). The **tight junctions** extend in a belt-like manner around the perimeter of each epithelial cell and connect neighboring cells. They constitute part of the barrier between the two extracellular spaces on either side of the epithelium, that is, the gastrointestinal lumen and the intercellular (interstitial) space on the other (blood or serosal) side. The tight junction marks the boundary between the **luminal** and **contraluminal** region of the **plasma membrane** of epithelial cells.

Two potentially parallel pathways for **solute transport** across epithelial cell layers can be distinguished: that through the cells **(transcellular)** and that through the tight junctions between cells **(paracellular)** (Figure 25.8). The transcellular route

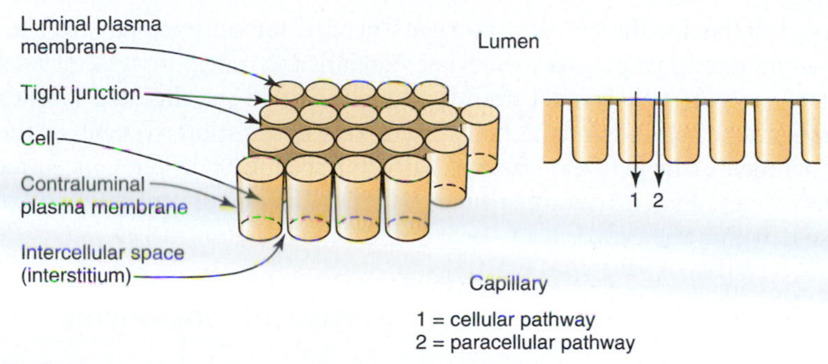

FIGURE 25.8
Pathways for transport across epithelia.

in turn consists mainly of two barriers in series, formed by the luminal and contraluminal plasma membranes. Because of this combination of different barriers in parallel (cellular and paracellular pathways) and in series (luminal and contraluminal plasma membranes), information on all three barriers as well as their mutual influence is required for understanding the overall transport properties of the epithelium.

A major function of gastrointestinal epithelial cells is active transport of nutrients, electrolytes, and vitamins. The cellular basis for this **vectorial solute movement** lies in the properties of different transporters in the luminal and contraluminal regions of the plasma membrane. The small intestinal cells provide an example of the differentiation and specialization of the two types of membrane. The luminal and contraluminal plasma membranes differ in morphological appearance, enzymatic composition, chemical composition, and transport functions (Table 25.5). The luminal membrane is in contact with the nutrients in the chyme (the semifluid mass of partially digested food) and is specialized for terminal digestion of nutrients through hydrolytic enzymes on its external surface and for nutrient absorption through transport systems that accomplish concentrative uptake. Transport systems are present for monosaccharides, amino acids, peptides, and electrolytes. In contrast, the contraluminal plasma membrane, which is in contact with the intercellular fluid, capillaries, and lymph, has properties similar to the plasma membrane of most cells. It possesses receptors for hormonal or neuronal regulation of cellular functions, a Na^+/K^+-exchanging ATPase for removal of Na^+ from the cell, and

TABLE 25.5 Characteristics Differences Between Luminal and Contraluminal Plasma Membrane of Small Intestinal Epithelial Cells

Parameter	*Luminal*	*Contraluminal*
Morphological appearance	Microvilli in ordered arrangement (brush border)	Few microvilli
Enzymes	Di- and oligosaccharidases Aminopeptidase Dipeptidases γ-Glutamyltransferase Alkaline phosphatase Guanylate cyclase	Na^+/K^+ ATPase Adenylate cyclase
Transport systems	Na^+–monosaccharide contransport (SGLT-1)	Facilitated monosaccharide transprot (GLUT-2)
	Facilitated fructose transport (GLUT-5) Na^+–neutral amino acid cotransport Na^+–bile acid cotransport (ASBT), H^+–peptide cotransport (PepT1)	Facilitated neutral amino acid transport

Abbreviations: SGLT, sodium glucose transporter; GLUT, glucose transporter; and ASBT, apical sodium bile acid transporter.

transport systems for the entry of nutrients needed for cell maintenance. In addition, the contraluminal plasma membrane contains the transport systems necessary for exit of the nutrients derived from the lumen so that the digested food can become available to all cells of the body. Some of the transport systems in the contraluminal membrane may catalyze exit when intracellular nutrient concentration is high after a meal and entry at other times when the blood concentrations are higher than those within the cell.

NaCl Absorption Has Both Active and Passive Components

Transport of Na^+ plays a crucial role not only for epithelial NaCl absorption or secretion, but also in the energization of nutrient uptake. The **Na^+/K^+-exchanging ATPase** provides the dominant mechanism for transduction of chemical energy in the form of ATP into osmotic energy of a concentration (chemical) or a combined concentration and electrical (electrochemical) ion gradient across the plasma membrane. In epithelial cells this enzyme is located exclusively in the contraluminal plasma membrane (Figure 25.9). The stoichiometry of the Na^+/K^+-

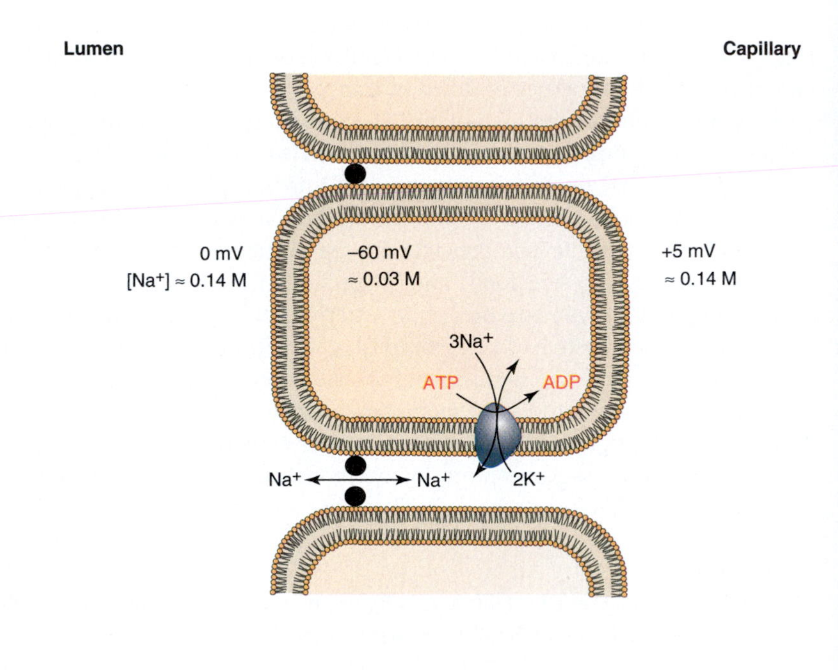

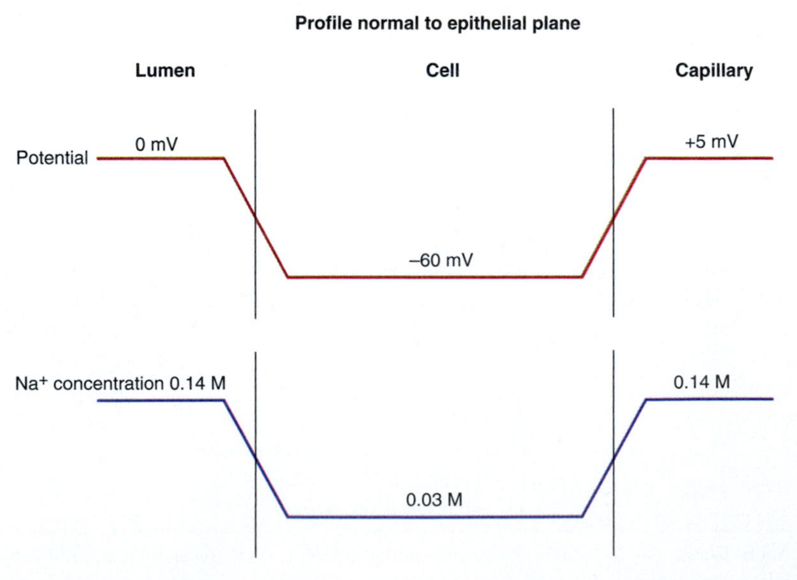

FIGURE 25.9
Na^+ concentrations and electrical potentials in enterocytes.

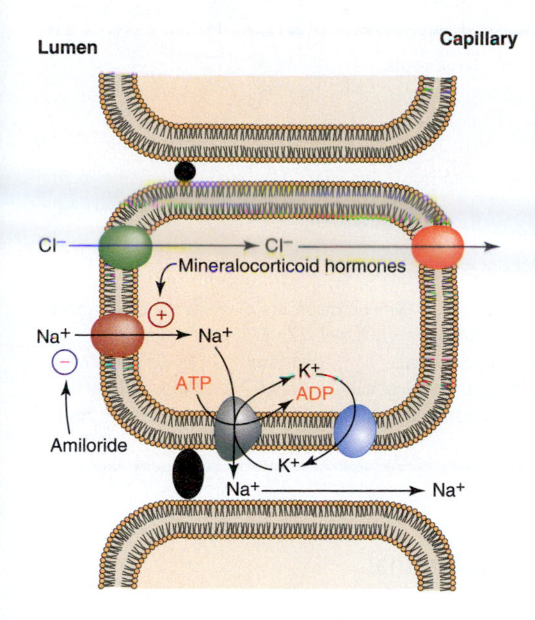

FIGURE 25.10
Model for electrogenic NaCl absorption in the lower large intestine.
The mechanisms of the Na^+ transporters are well established but the nature of the Cl^- transporter is not known.

exchanging ATPase reaction is 1 mol of ATP coupled to the outward pumping of 3 mol of Na^+ and the simultaneous inward pumping of 2 mol of K^+. The Na^+/K^+-exchanging ATPase maintains the high K^+ and low Na^+ concentrations in the cytosol and is directly or indirectly responsible for an electrical potential of about −60 mV of the cytosol relative to the extracellular solution. The direct contribution comes from the charge movement when 3 Na^+ ions are replaced by 2 K^+; the indirect contribution is by way of the K^+ gradient, which becomes the dominant force for establishing the electrical potential by an outward movement of K^+ through **K^+ channels.**

Transepithelial NaCl movements are produced by the combined actions of the Na^+/K^+-exchanging ATPase and additional "passive" transport systems in the plasma membrane, which allow the entry of Na^+ or Cl^- into the cell. NaCl absorption results from Na^+ entry into the cell across the luminal plasma membrane and its extrusion by the Na^+/K^+-exchanging ATPase across the contraluminal membrane. Epithelial cells of the lower portion of the large intestine possess a luminal Na^+ channel **(epithelial Na^+ channel** or **ENaC)** that allows the uncoupled entry of Na^+ down its electrochemical gradient (Figure 25.10). This Na^+ flux is **electrogenic;** that is, it is associated with an electrical current, and it can be inhibited by the diuretic drug amiloride at micromolar concentrations (Figure 25.11). This transport system, and hence NaCl absorption, is regulated by mineralocorticoid hormones of the adrenal cortex.

Epithelial cells of the small intestine have a transport system in their brush border membrane, which catalyzes an electrically neutral Na^+/H^+ exchange (**Na^+/H^+ exchanger** or **NHE,** with NHE3 the predominant isotype in the intestine) (Figure 25.12). The exchange is not affected by low concentrations of amiloride and not regulated by mineralocorticoids. The Na^+/H^+ exchange sets up a H^+-gradient that secondarily drives Cl^- absorption through a specific **Cl^-/HCO_3^- exchanger** in the luminal plasma membrane, as illustrated in Figure 25.12. The intestinal Cl^-/HCO_3^- exchanger is coded for by the **"downregulated in adenoma"** or ***DRA*** gene (see Clin. Corr. 25.1). The need for two types of NaCl absorption may arise from the different functions of upper and lower intestine, which require different regulation. The upper intestine absorbs the bulk of NaCl from the diet and from secretions of the exocrine glands after each meal, while the lower intestine participates in the fine regulation of NaCl retention, depending on the overall electrolyte balance of the body.

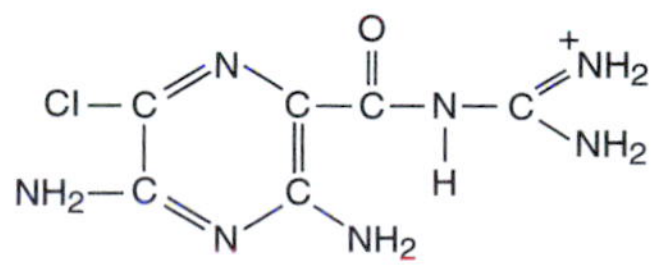

FIGURE 25.11
Amiloride.

CLINICAL CORRELATION 25.1

Familial Chloridorrhea Causes Metabolic Alkalosis

Mutations in the human *DRA* (downregulated in adenoma) gene cause familial chloridorrhea. Patients with this disease have moderate diarrhea, generate stool with an acidic pH, and hence suffer from a metabolic alkalosis. The normal *DRA* gene product confers Na^+-independent Cl^-/HCO_3^- exchange activity on epithelial cells of the lower ileum and the colon. This transporter serves to recover Cl^- from stool and to secrete bicarbonate to neutralize normal proton secretion (via sodium/proton exchange (NHE3)). Loss of Cl^-/HCO_3^- exchange activity explains the lack of Cl^- absorption and excess loss of HCl in the feces. In turn, chronic HCl loss with the stool is responsible for chronic metabolic alkalosis. The extra electrolytes in stool are responsible for its greater fluid contents, that is, diarrhea, due to osmotic effects.

Kere, J., Lohi, H., and Hoglund, P. Genetic disorders of membrane transport III. Congenital chloride diarrhea. *Am. J. Physiol.* 276:G7, 1999; and Melvin, J. E., Park, K., Richardson, L., Schultheis, P. J., and Shull, G. E. Mouse down-regulated in adenoma (DRA) is an intestinal Cl^-/HCO_3^- exchanger and is up-regulated in colon of mice lacking the NHE3 Na^+/H^+ exchanger. *J. Biol. Chem.* 274:22855, 1999.

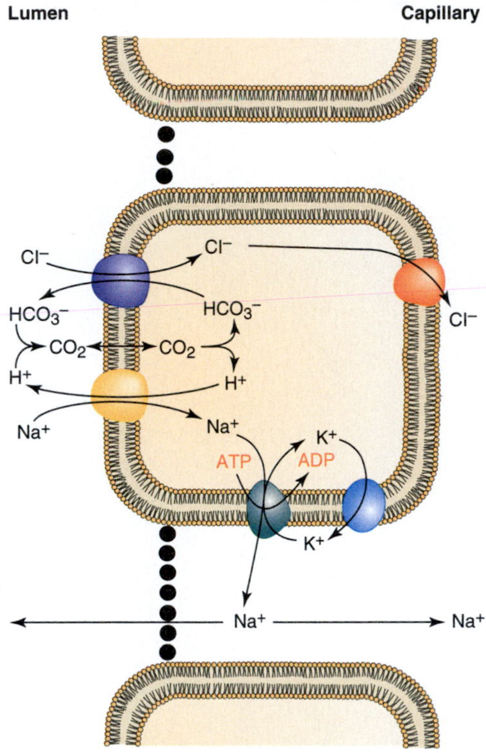

FIGURE 25.12
Model for electrically neutral NaCl absorption in the small intestine.

NaCl Secretion Depends on Contraluminal Na^+/K^+-Exchanging ATPase

Epithelial cells of most regions of the gastrointestinal tract have the potential for electrolyte and fluid secretions. The major secreted ions are Na^+ and Cl^-. Water follows passively because of the osmotic forces exerted by any secreted solute. Thus NaCl secretion secondarily results in fluid secretion. The fluid may be either hypertonic or isotonic, depending on its contact time with the epithelium and the tissue permeability to water. The longer the contact and the greater the water permeability, the closer the secreted fluid gets to osmotic equilibrium, that is, isotonicity. Ionic compositions of gastrointestinal secretions are presented in Figure 25.13.

NaCl secretion involves the Na^+/K^+-exchanging ATPase located in the contraluminal plasma membrane of epithelial cells (Figure 25.14). The enzyme is implicated because cardiac glycosides, inhibitors of this enzyme, abolish salt secretion.

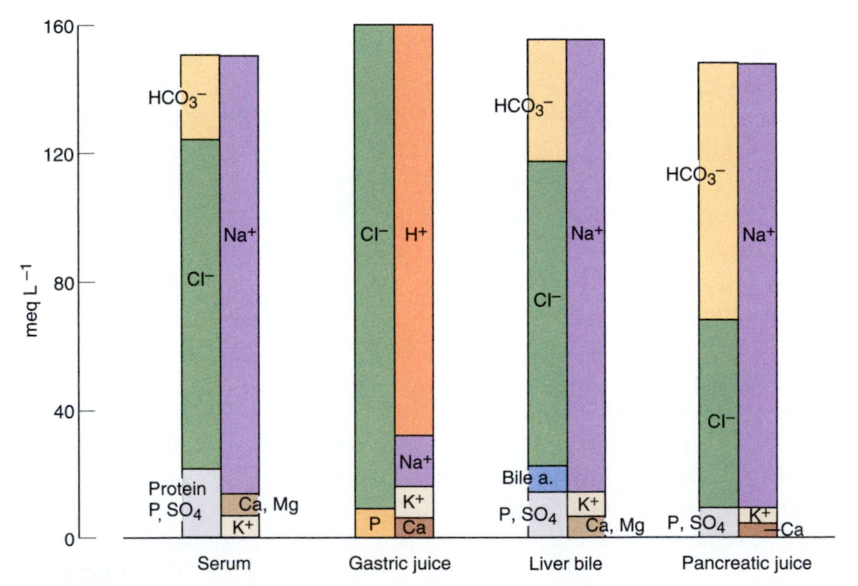

FIGURE 25.13
Ionic composition of secretions of the gastrointestinal tract.
Serum included for comparisons. Note the high H^+ concentration in gastric juice (pH = 1) and the high HCO_3^- concentration in pancreatic juice. P, organic and inorganic phosphate; SO_4, inorganic and organic sulfate; Ca, calcium; Mg, magnesium; bile a., bile acids. *Adapted from* Biological Handbooks. Blood and Other Body Fluids. *Federation of American Societies for Experimental Biology, 1961.*

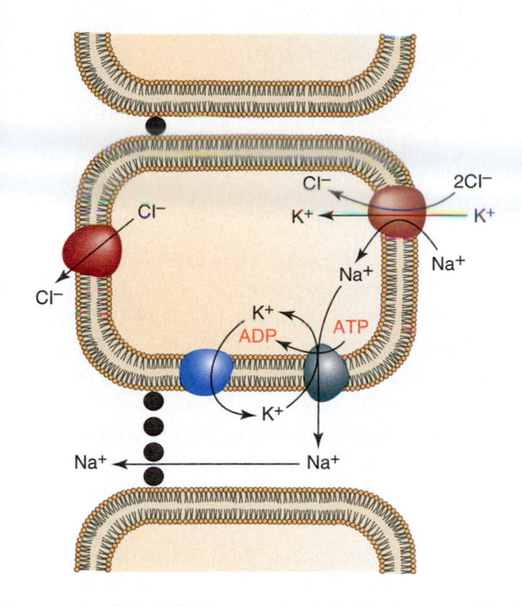

FIGURE 25.14
Model for epithelial NaCl secretion.

The involvement of Na^+/K^+-exchanging ATPase does not provide a straightforward explanation for NaCl movement from the capillary side to the lumen because the enzyme extrudes Na^+ from the cell toward the capillary side. Thus the active step of Na^+ transport across one of the plasma membranes has a direction opposite to that of overall transepithelial NaCl movements. This paradox is resolved by an electrical coupling of Cl^- secretion across the luminal plasma membrane and Na^+ movements via the paracellular route, illustrated in Figure 25.14. The Cl^- secretion depends on coupled uptake of 2 Cl^- ions with Na^+ and K^+ via a specific cotransporter in the contraluminal plasma membrane and specific luminal Cl^- channels (predominantly **cystic fibrosis transmembrane regulatory** or **CFTR** protein). The **$Na^+/K^+/2Cl^-$-cotransporter (NKCC1)**, which can be identified by specific inhibitors such as the common diuretic **furosemide** (Figure 25.15), utilizes energy of the Na^+ gradient to accumulate Cl^- within the cytoplasm above its electrochemical equilibrium concentration. Subsequent opening of luminal Cl^- channels allows efflux of Cl^- together with a negative charge (see Clin. Corr. 25.2 and 25.3).

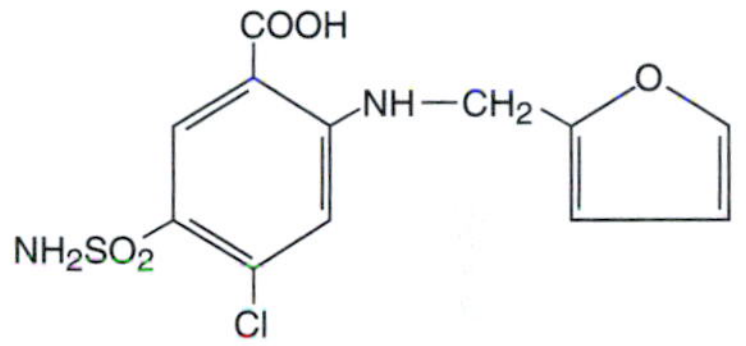

In the pancreas acinar cells secret a fluid rich in Na^+ and Cl^-, which provides the vehicle for the movement of digestive enzymes from the acini, where they are released, to the lumen of the duodenum. The fluid is modified in the ducts by the

FIGURE 25.15
Furosemide.

CLINICAL CORRELATION 25.2
Cystic Fibrosis

Cystic fibrosis is an autosomal recessive inherited disease due to a mutation in the cystic fibrosis transmembrane regulatory (CFTR) protein. This protein contains 1480 amino acids organized into two membrane-spanning portions, which contain six transmembrane regions each, two ATP-binding domains, and a regulatory domain that undergoes phosphorylation by cAMP-dependent protein kinase. Over 400 mutations have been discovered since the gene was cloned in 1989.

The normal form of this protein is the predominant Cl^- channel in the luminal plasma membrane of epithelial cells in many tissues. The channel is normally closed but opens when phosphorylated by protein kinase A, thus providing regulated Cl^- and fluid secretion. The most common and severe mutation lacks one phenylalanine (ΔF508 CFTR), which prevents the protein from maturing properly and reaching the plasma membrane. Individuals who inherit this mutant CFTR protein from both parents lack Cl^- and fluid secretion in tissues that depend on CFTR protein for this function. Failure to secrete fluid, in turn, can lead to gross organ impairment due to partial or total blockage of passageways, for example, the ducts in the pancreas, the lumen of the intestine, or airways. (See Clin. Corr. 25.3 for activation of the CFTR Cl^- channel.)

CLINICAL CORRELATION 25.3

Bacterial Toxigenic Diarrheas and Electrolyte Replacement Therapy

Voluminous, life-threatening intestinal electrolyte and fluid secretion (diarrhea) occurs in patients with cholera, an intestinal infection by *Vibrio cholerae*. Certain strains of *E. coli* also cause (traveler's!) diarrhea that can be serious in infants. The secretory state is a result of enterotoxins produced by the bacteria. The mechanisms of action of some of these enterotoxins are well understood at the biochemical level. Cholera toxin activates adenylate cyclase by causing ADP-ribosylation of the $G_{\alpha s}$-protein, which stimulates the cyclase (see p. 926). Elevated cAMP levels in turn activate protein kinase A, which opens the luminal CFTR Cl^- channel and inhibits the Na^+/H^+ exchanger by protein phosphorylation. The net result is gross NaCl secretion. *Escherichia coli* produces a heat-stable toxin that binds to the receptor for the physiological peptide "guanylin," namely, the brush border guanylate cyclase. When the receptor is occupied on the luminal side by either guanylin or the heat-stable *E. coli* toxin, the guanylate cyclase domain of the protein on the cytosolic side is activated and cGMP levels rise. Elevated cGMP levels have the same effect on Cl^- secretion as elevated cAMP, except that a cGMP-activated protein kinase is involved in protein phosphorylation.

Modern, oral treatment of cholera takes advantage of the presence of Na^+–glucose cotransport in the intestine, which is not downregulated by cAMP and remains fully active in this disease. In this case, the presence of glucose allows uptake of Na^+ to replenish body NaCl. Composition of solution for oral treatment of cholera patients is glucose 110 mM, Na^+ 99 mM, Cl^- 74 mM, HCO_3^- 29 mM, and K^+ 4 mM. The major advantages of this form of therapy are its low cost and ease of administration when compared with intravenous fluid therapy.

The composition of sport drinks for electrolyte replacement is based on the same principle, namely, more rapid sodium absorption in the presence of glucose.

Carpenter, C. C. J. In: M. Field, J. S. Fordtran, and S. G. Schultz (Eds.), *Secretory Diarrhea*. Bethesda, MD: American Physiological Society, 1980, p. 67.

additional secretion of **$NaHCO_3$** (Figure 25.16). The HCO_3^- concentration in the final pancreatic juice can reach concentrations of up to 120 mM.

Permeability of **tight junctions** to H_2O, Na^+, or other ions modifies active transepithelial solute movements. High permeability is necessary to allow Na^+ to equilibrate between extracellular solutions of the intercellular and luminal compartments during NaCl or $NaHCO_3$ secretion. Different regions of the gastrointestinal tract differ with respect to the transport systems that determine the passive entry (see above for amiloride-sensitive Na^+ channel and Na^+/H^+ exchange) and with respect to the permeability characteristics of the tight junction. The distal portion (colon) is much tighter so as to prevent leakage of Na^+ from blood to lumen, in accordance with its function of scavenging of NaCl from the lumen.

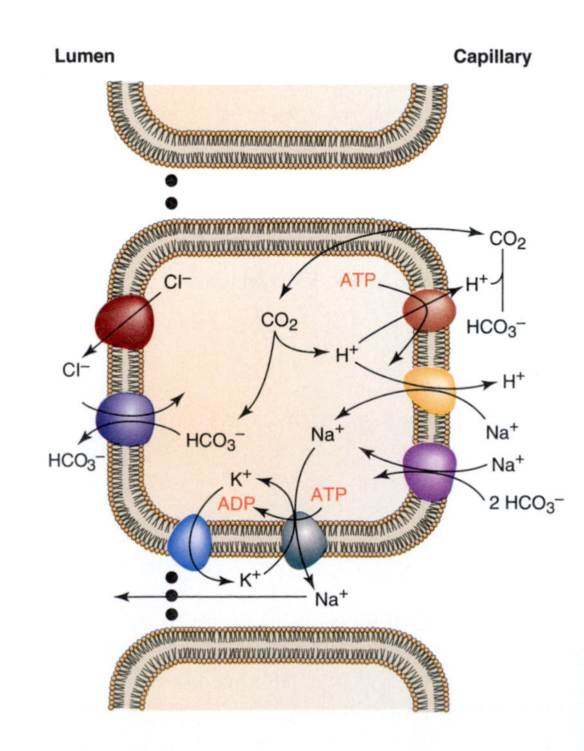

FIGURE 25.16
Model for $NaHCO_3$ secretion by pancreatic duct cells.
Note that three different mechanisms exist for bicarbonate influx into the cell (or its equivalent proton secretion) at the contraluminal plasma membrane: (1) Na^+/H^+ exchange, (2) H^+-ATPase, and (3) Na^+–$2HCO_3^-$ cotransport.

Concentration Gradients or Electrical Potentials Drive Transport of Nutrients

Many solutes are absorbed across the intestinal epithelium against a concentration gradient. Energy for this "active" transport is directly derived from the Na^+ concentration gradient or the electrical potential across the luminal plasma membrane, rather than from the chemical energy of a covalent bond change, such as ATP hydrolysis. Glucose transport is an example of uphill solute transport driven directly by the electrochemical Na^+ gradient and only indirectly by ATP (Figure 25.17).

Glucose is absorbed from the intestinal lumen into the blood against a concentration gradient. This vectorial transport is the combined result of several separate membrane events (Figure 25.18): (1) ATP-dependent Na^+ transport efflux at the contraluminal pole establishes an electrochemical Na^+ gradient across the plasma membrane; (2) K^+ channels that convert a K^+ gradient into a membrane potential; (3) different transport systems for glucose in the luminal and contraluminal plasma membranes; and (4) coupling of Na^+ and glucose transport across the luminal membrane.

The transport system in the luminal plasma membrane facilitates a tightly coupled movement of Na^+ and D-glucose or structurally similar sugars **(sodium glucose transporter** or **SGLT).** The most common intestinal sodium–glucose cotransporter is **SGLT-1** and it couples the movement of 2 Na^+ ions with that of 1 glucose molecule. It mediates glucose and Na^+ transport equally well in both directions. However, because of the higher Na^+ concentration in the lumen and the negative potential within the cell, the physiological direction of glucose movement is from lumen to cell, even if the cellular glucose concentration is higher than the luminal one. In other words, downhill Na^+ movement normally supports concentrative glucose transport. Concentration ratios of up to 20-fold between intracellular and extracellular glucose have been observed *in vitro* under conditions of blocked efflux of cellular glucose. In some situations Na^+ uptake via this route is physiologically more important than glucose uptake (see Clin. Corr. 25.3).

The contraluminal plasma membrane contains a member of the ***glucose transporter*** (or **GLUT**) family, which facilitates glucose exit and entry. The intestine contains the **GLUT-2** transporter, which accepts many monosaccharides, including glucose. The direction of net flux is determined by the sugar concentration gradient. The two glucose transport systems SGLT-1 and GLUT-2 in the luminal and contraluminal plasma membranes, respectively, share glucose as substrate, but otherwise differ considerably in terms of amino acid sequence, secondary protein structure,

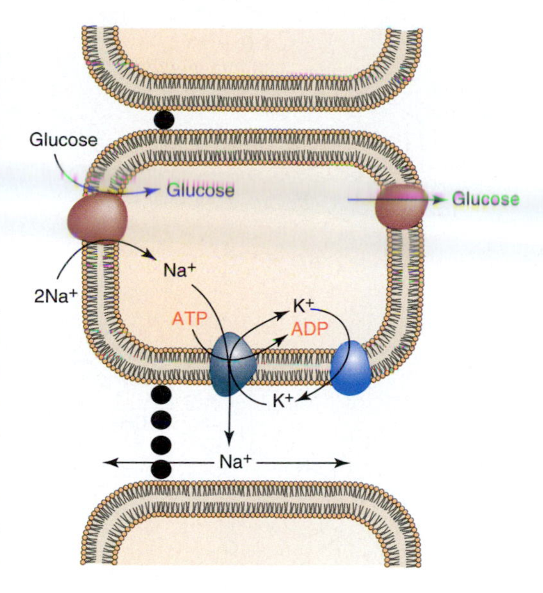

Glucose concentration profile normal to epithelial plane

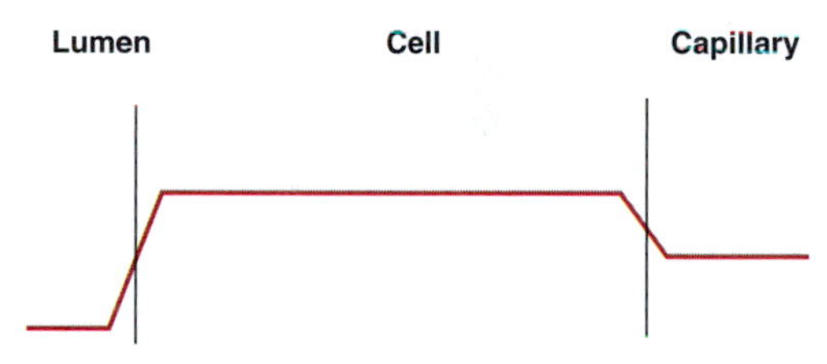

FIGURE 25.17
Model for epithelial glucose absorption.

$$2X \quad 3Na^+_{cell} + 2K^+_{interstitium} + ATP_{cell} \xrightarrow{Na^+,\,K^+\text{–ATPase}} 3Na^+_{interstitium} + 2K^+_{cell} + ADP_{cell} + P_{cell}$$

$$4X \quad K^+_{cell} \xrightleftharpoons{K^+\text{ channel}} K^+_{interstitium}$$

$$3X \quad 2Na^+_{lumen} + Glc_{lumen} \xrightleftharpoons{\text{SGLT-1}} 2Na^+_{cell} + Glc_{cell}$$

$$3X \quad Glc_{cell} \xrightleftharpoons{\text{GLUT-2}} Glc_{interstitium}$$

$$6X \quad Na^+_{interstitium} \xrightleftharpoons{\text{Tight junction}} Na^+_{lumen}$$

$$\text{Sum} \quad 3Glc_{lumen} + 2ATP_{cell} \longrightarrow 3Glc_{interstitium} + 2ADP_{cell} + 2P_{cell}$$

FIGURE 25.18
Transepithelial glucose transport as translocation reactions across the plasma membranes and the tight junction.
SGLT-1 (sodium glucose transporter 1) and GLUT-2 (glucose transporter 2) are specific intestinal gene products mediating Na^+–glucose cotransport and facilitated glucose transport, respectively. Numbers in the left column indicate the minimal turnover of individual reactions to balance the overall reaction.

Na^+ as cosubstrate, specificity for other sugars, sensitivity to inhibitors, or biological regulation. Since both SGLT and GLUT are not inherently directional, "active" transepithelial glucose transport can be maintained under steady-state conditions only if the Na^+/K^+-exchanging ATPase continues to move Na^+ out of the cell. Thus the active glucose transport is indirectly dependent on a supply of ATP and an active Na^+/K^+-exchanging ATPase.

The advantage of an electrochemical Na^+ gradient serving as intermediate is that the Na^+/K^+-exchanging ATPase can energize transport of many different nutrients. The only requirement is presence of a system catalyzing cotransport of the nutrient with Na^+.

Gastric Parietal Cells Secrete HCl

The parietal (oxyntic) cells of gastric glands are capable of secreting HCl into the gastric lumen. Luminal H^+ concentrations of up to 0.14 M (pH 0.8) have been observed (see Figure 25.13). As the plasma pH = 7.4, the parietal cell transports protons against a concentration gradient of $10^{6.6}$. The free energy required for **HCl secretion** under these conditions is minimally 9.1 kcal/mol of HCl (= 38 J/mol of HCl), as calculated from

$$\Delta G' = RT(2.3 \log 10^{6.6}) \qquad RT = 0.6 \text{ kcal mol}^{-1} \text{ at } 37^\circ\text{C}$$

A **H^+/K^+ exchanging-ATPase** (or gastric proton pump) is intimately involved in the mechanism of active HCl secretion. This enzyme is unique to the parietal cell and is found only in the luminal region of the plasma membrane. It couples the hydrolysis of ATP to an electrically neutral obligatory exchange of K^+ for H^+, secreting H^+ and moving K^+ into the cell. The stoichiometry appears to be 1 mol of transported H^+ and K^+ for each mole of ATP.

$$\text{ATP}_{\text{cell}} + \text{H}^+_{\text{cell}} + \text{K}^+_{\text{lumen}} \rightleftharpoons \text{ADP}_{\text{cell}} + \text{P}_{\text{i,cell}} + \text{H}^+_{\text{lumen}} + \text{K}^+_{\text{cell}}$$

As the H^+/K^+-exchanging ATPase generates a very acidic solution, protein reagents that are activated by acid can become specific inhibitors of this enzyme. Figure 25.19 shows an example of a **proton pump inhibitor** used to treat peptic ulcers. In the steady state, HCl is elaborated by H^+/K^+-exchanging ATPase only if the luminal membrane is permeable to K^+ and Cl^- and the contraluminal plasma membrane catalyzes an exchange of Cl^- for HCO_3^- (Figure 25.20). The exchange of Cl^- for HCO_3^- is essential to resupply the cell with Cl^- and to prevent accumulation of base within the cell. Thus, under steady-state conditions, secretion of HCl into the gastric lumen is coupled to movement of HCO_3^- into the plasma.

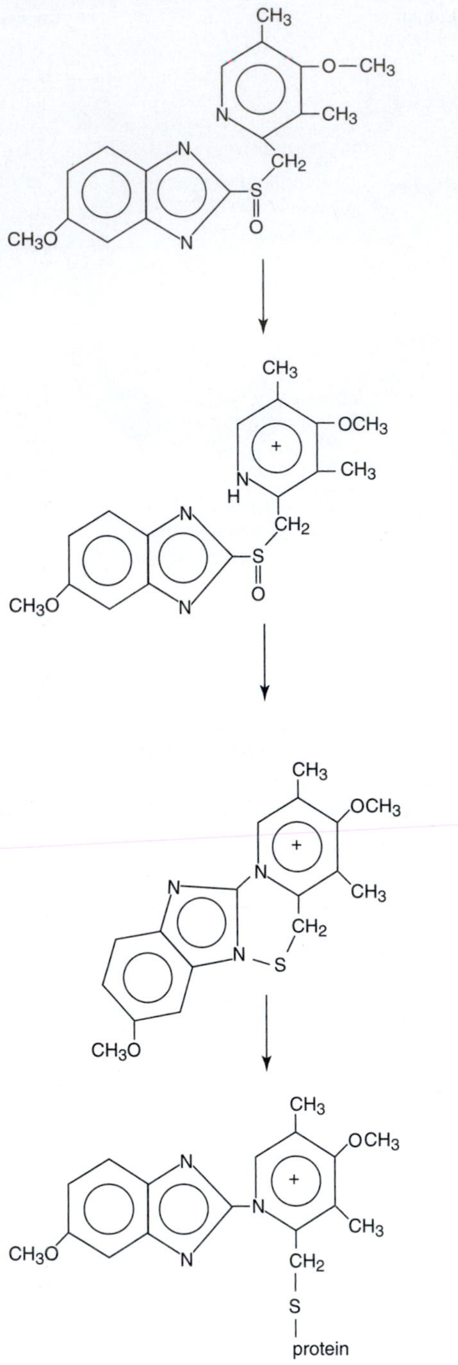

FIGURE 25.19
Omeprazole, an inhibitor of H^+/K^+ ATPase. This drug accumulates in an acidic compartment ($pK_a \sim 4$) and is converted to a reactive sulfenamide, which reacts with cysteine SH groups.
From Sachs, G. In: L. R. Johnson (Ed.), Physiology of the Gastrointestinal Tract. *New York: Raven Press, 1994, p. 1133.*

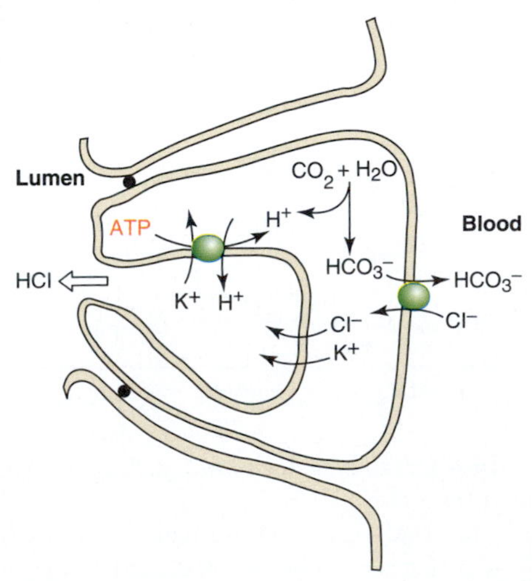

FIGURE 25.20
Model for secretion of hydrochloric acid.

25.4 | DIGESTION AND ABSORPTION OF PROTEINS

Peptidases Assure Efficient Protein Digestion

The total daily protein load to be digested consists of about 70–100 g of dietary proteins and 35–200 g of endogenous proteins from digestive enzymes and sloughed-off cells. Digestion and absorption of proteins are very efficient processes in healthy humans, since only about 1–2 g of nitrogen are lost through feces each day, which is equivalent to 6–12 g of protein.

Except for a short period after birth, oligo- and polypeptides are not absorbed intact in appreciable quantities by the intestine. Proteins are hydrolyzed by peptidases specific for peptide bonds. This class of enzymes is divided into **endopeptidases** (proteases), which attack internal bonds and liberate large peptide fragments, and **exopeptidases,** which cleave off one amino acid at a time from either the COOH **(carboxypeptidases)** or the NH_2 terminus **(aminopeptidases).** Endopeptidases are important for an initial breakdown of long polypeptides into smaller products, which can then be attacked more efficiently by exopeptidases. The final products are free amino acids and di- and tripeptides. Both types of end products are absorbed by epithelial cells (Figure 25.21).

Protein digestion can be divided into a gastric, a pancreatic, and an intestinal phase, depending on the source of peptidases.

Pepsins Catalyze Gastric Digestion of Protein

Gastric juice contains HCl, a low pH of less than 2, and proteases of the pepsin family. The acid serves to kill off microorganisms and to **denature proteins.** Denaturation makes proteins more susceptible to hydrolysis by proteases. **Pepsins** are unusual enzymes in that they are acid stable; in fact, they are active at acid but not at neutral pH. The catalytic mechanism that is effective for peptide hydrolysis at the acid pH depends on two carboxylic groups at the active site of the enzymes. Pepsin A, the major gastric protease, prefers peptide bonds formed by the amino group of aromatic acids (Phe, Tyr) (Table 25.6).

Active pepsin is generated from the proenzyme **pepsinogen** by removal of 44 amino acids from the NH_2 terminus (pig enzyme). Cleavage between residues 44 and 45 occurs as either an intramolecular reaction **(autoactivation)** below pH 5 or by active pepsin (autocatalysis). The liberated peptide from the NH_2 terminus

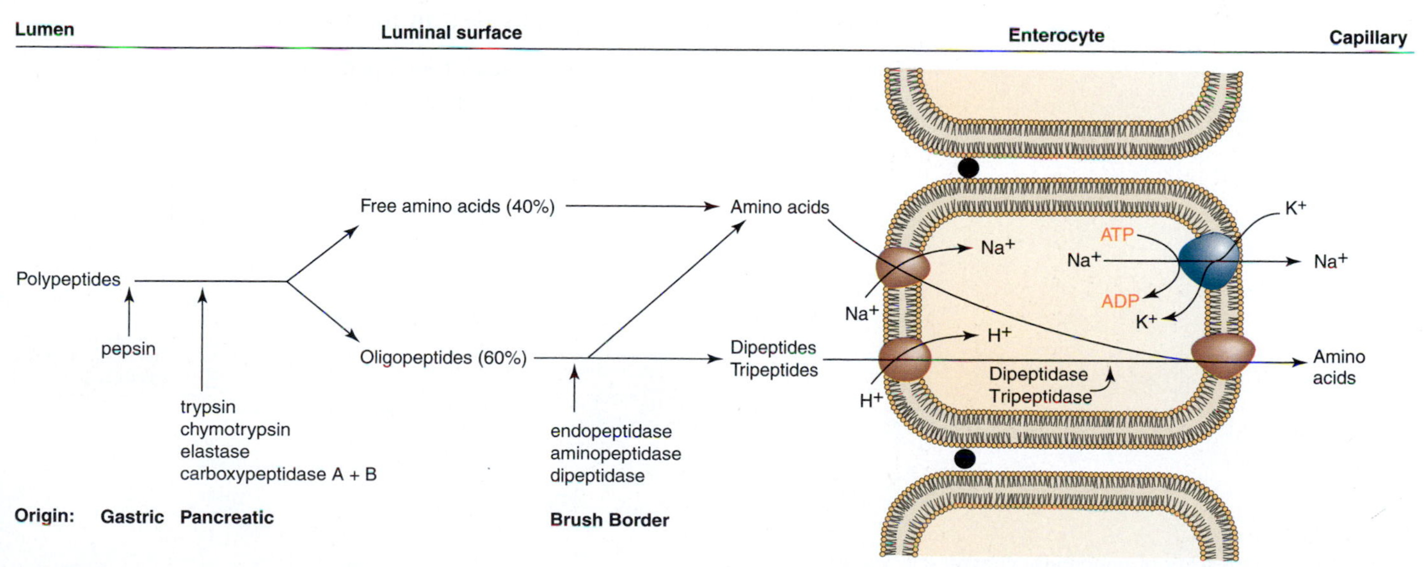

FIGURE 25.21
Digestion and absorption of proteins.

TABLE 25.6 Gastric and Pancreatic Peptidases

Enzyme	*Proenzyme*	*Activator*	*Cleavage Point*	*R*
CARBOXYL PROTEASES				
Pepsin A	Pepsinogen A	Autoactivation, pepsin	—CO↓—NHCH(R)CO↓—NHCH(R′)CO—	Tyr, Phe, Leu
SERINE PROTEASES				
Trypsin	Trypsinogen	Enteropeptidase, trypsin	—CO—NHCH(R)CO↓—NHCH(R′)CO—	Arg, Lys
Chymotrypsin	Chymotrypsinogen	Trypsin	—CO—NHCH(R)CO↓—NHCH(R′)CO—	Tyr, Trp, Phe, Met, Leu
Elastase	Proelastase	Trypsin	—CO—NHCH(R)CO↓—NHCH(R′)CO—	Ala, Gly, Ser
ZINC PEPTIDASES				
Carboxypeptidase A	Procarboxypeptidase A	Trypsin	—CO↓—NHCH(R)COO⁻	Val, Leu, Ile, Ala
Carboxypeptidase B	Procarboxypeptidase B	Trypsin	—CO↓—NHCH(R)COO⁻	Arg, Lys

remains bound to pepsin and acts as "pepsin inhibitor" above pH 2. This inhibition is released either by a drop of the pH below 2 or further degradation of the peptide by pepsin. Thus pepsinogen is converted to pepsin by autoactivation and subsequent autocatalysis at an exponential rate.

The major products of pepsin action are large peptide fragments and some free amino acids. The importance of gastric protein digestion does not lie so much in its contribution to the breakdown of ingested macromolecules, but rather in the generation of peptides and amino acids that act as stimulants for **cholecystokinin** release in the duodenum. The gastric peptides are instrumental in the initiation of the pancreatic phase of protein digestion.

Pancreatic Zymogens Are Activated in Small Intestine

Pancreatic juice is rich in **proenzymes** of endopeptidases and carboxypeptidases (Figure 25.22), which are activated after they reach the lumen of the small intestine.

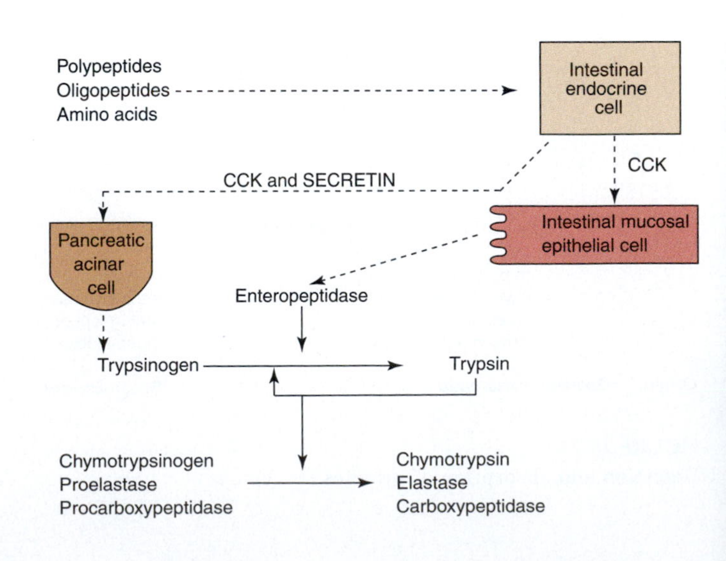

FIGURE 25.22
Secretion and activation of pancreatic enzymes.
Abbreviation: CCK, cholecystokinin.
Reproduced with permission from Freeman, H. J. and Kim, Y. S. Annu. Rev. Med. 29:102, 1978. Copyright © 1978 by Annual Reviews, Inc.

Enteropeptidase (or **enterokinase,** the older name), a protease produced by duodenal epithelial cells, activates pancreatic **trypsinogen** to **trypsin** by scission of a hexapeptide from the NH_2 terminus. Trypsin autocatalytically activates more trypsinogen to trypsin and also acts on the other proenzymes, thus liberating the endopeptidases **chymotrypsin** and **elastase** and the **carboxypeptidases A** and **B.** Pancreatic juice normally contains a small-molecular-weight peptide that acts as a **trypsin inhibitor** and neutralizes any trypsin formed prematurely within the pancreatic cells or pancreatic ducts.

The substrate specificities of trypsin, chymotrypsin, and elastase are presented in Table 25.6. The enzymes are active only at neutral pH and depend on pancreatic $NaHCO_3$ for neutralization of gastric HCl. Their mechanism of catalysis involves an **essential serine** residue (see p. 385) and is thus similar to serine esterases, such as acetylcholine esterase. Reagents that interact with serine and modify it, inactivate serine esterases and peptidases. An example is the highly toxic diisopropylphosphofluoridate, which was developed originally for chemical warfare (neurotoxic because of inhibition of acetylcholine esterase).

Peptides generated from ingested proteins are degraded within the small intestinal lumen by carboxypeptidases A and B, which are **Zn^{2+} metalloenzymes** and possess a different catalytic mechanism than carboxyl or serine peptidases. The combined action of pancreatic peptidases results in the formation of free amino acids and small peptides of 2–8 residues. Peptides account for about 60% of the amino nitrogen at this point.

Brush Border and Cytoplasmic Peptidases Digest Small Peptides

Since pancreatic juice does not contain appreciable aminopeptidase activity, final digestion of di- and oligopeptides depends on enzymes of the small intestine. The luminal surface of epithelial cells is particularly rich in endopeptidase and aminopeptidase activity, but also contains dipeptidases (Table 25.2). This digestion at the cell surface produces free amino acids and di- and tripeptides, which are absorbed via specific **amino acid** or **peptide transport systems.** Transported di- and tripeptides are generally hydrolyzed in the cytoplasm before they leave the cell. Cytoplasmic dipeptidases explain why practically only free amino acids are found in the portal blood after a meal. The virtual absence of peptides used to be taken as evidence that luminal protein digestion had to proceed all the way to free amino acids before absorption could occur. However, it is now established that a large portion of dietary amino nitrogen is absorbed in the form of small peptides with subsequent intracellular hydrolysis. Di- and tripeptides containing proline and hydroxyproline or unusual amino acids, such as β-alanine in carnosine (β-alanylhistidine) or anserine (β-alanyl 1-methylhistidine), are absorbed without intracellular hydrolysis because they are not good substrates for the intestinal cytoplasmic dipeptidases. β-Alanine is present in chicken meat.

Amino Acids and Dipeptides Are Absorbed by Carrier-Mediated Transport

The small intestine has a high capacity to absorb free amino acids and small peptides. Most L-amino acids can be transported across the epithelium against a concentration gradient, although the need for concentrative transport *in vivo* is not obvious, since luminal concentrations are usually higher than the plasma levels of 0.1–0.2 mM. Amino acid and peptide transport in the small intestine has all the characteristics of carrier-mediated transport, such as discrimination between D- and L-amino acids and energy and temperature dependence. Genetic defects are known to occur in humans (see Clin. Corr. 25.4.)

On the basis of genetics, transport experiments, and expression cloning, at least seven brush border transport systems for the uptake of L-amino acids or small peptides in the luminal membrane can be distinguished (transporter name and typical substrates in parentheses): (1) for neutral amino acids with short or polar side

CLINICAL CORRELATION 25.4

Neutral Amino Aciduria: Hartnup Disease

Transport functions, like enzyme functions, are subject to modification by mutations. An example of a genetic defect in epithelial amino acid transport is Hartnup disease, named after the family in which the disease entity resulting from the defect was first recognized. The disease is characterized by the inability of renal and intestinal epithelial cells to absorb neutral amino acids from the lumen. In the kidney, in which plasma amino acids reach the lumen of the proximal tubule through the glomerular ultrafiltrate, the inability to reabsorb amino acids manifests itself as excretion of amino acids in the urine (amino aciduria). The intestinal defect results in malabsorption of amino acids from the diet. Therefore the clinical symptoms are mainly those due to essential amino acid and nicotinamide deficiencies. The pellagra-like features (see p. 1151) are explained by a deficiency of tryptophan, which serves as precursor for nicotinamide. Investigations of patients with Hartnup disease revealed the existence of intestinal transport systems for di- or tripeptides, which are different from the ones for free amino acids. The genetic lesion does not affect transport of peptides via PepT1, which remains as a pathway for absorption of small peptide products of digestion.

Silk, D. B. A. Disorders of nitrogen absorption. In: J. T. Harries (Ed.), *Clinics in Gastroenterology: Familial Inherited Abnormalities,* Vol. 11: London: Saunders, 1982, p. 47.

chains (ASCT-1 for Ala, Ser, Thr); (2) for neutral amino acids with aromatic or hydrophobic side chains (Phe, Tyr, Met, Val, Leu, Ile); (3) for imino acids (Pro, Hyp); (4) for β-amino acids (Beta/Taut for β-Ala, taurine); (5) for basic amino acids and cystine (Lys, Arg, Cys-Cys); (6) for acidic amino acids (EAAT-3 for Asp, Glu); and (7) for di- and tripeptides (PepT1 for Gly-sarcosine).

The concentration mechanisms for neutral L-amino acids appear to be similar to those discussed for D-glucose (see Figure 25.17). Na^+-dependent transport systems have been identified in the luminal (brush border) membrane and Na^+-independent transporters in the contraluminal plasma membrane of small intestinal epithelial cells. Similarly, as for active glucose transport, the energy for concentrative amino acid transport is derived directly from the electrochemical Na^+ gradient and only indirectly from ATP. Amino acids are not chemically modified during membrane transport, although they may be metabolized within the cytoplasmic compartment. The brush border transport for the other amino acids is energized in more complicated ways. For example, EAAT-3 mediates cotransport of the amino acid with 2 Na^+ ions and countertransport with 1 K^+ ion.

Neutral dipeptides are cotransported with a H^+ and thus are energized through the proton electrochemical gradient across this membrane. However, because of Na^+/H^+ exchange, the H^+ gradient is similar to the Na^+ gradient established by Na^+/K^+-exchanging ATPase. The dipeptide transporter also accepts β-lactam antibiotics (aminopenicillins) and is important for absorption of orally administered antibiotics of this class.

25.5 | DIGESTION AND ABSORPTION OF CARBOHYDRATES

Di- and Polysaccharides Require Hydrolysis

Dietary carbohydrates provide a major portion of the daily caloric requirement. They consist of mono-, di-, and polysaccharides (Table 25.7). The major ones in Western diets are **sucrose** (table sugar), starch, and lactose. Monosaccharides are absorbed directly. Disaccharides require the small intestinal surface enzymes for hydrolysis into monosaccharides, while polysaccharides depend on pancreatic amylase and surface enzymes for this process (Figure 25.23).

Starch, a major nutrient, is a plant polysaccharide with a molecular mass of more than 100 kDa. It consists of a mixture of linear chains of glucose molecules linked by α-1,4-glucosidic bonds **(amylose)** and of branched chains with branch points made up by α-1,6 linkages **(amylopectin).** The ratio of 1,4- to 1,6-glucosidic bonds is about 20:1. **Glycogen** is an animal polysaccharide similar in structure to amylopectin. The two compounds differ in terms of the number of branch points, which occur more frequently in glycogen.

TABLE 25.7 Dietary Carbohydrates

Carbohydrate	*Typical Source*		*Structure*
Amylopectin	Potatoes, rice, corn, bread	α-Glc(1 → 4)$_n$Glc with α-Glc(1 → 6) branches	
Amylose	Potatoes, rice, corn, bread	α-Glc(1 → 4)$_n$ Glc	
Sucrose	Table sugar, desserts	α-Glc(1 → 2)β-Fru	
Trehalose	Young mushrooms	α-Glc(1 → 1)α-Glc	
Lactose	Milk, milk products	β-Gal(1 → 4)Glc	
Fructose	Fruit, honey	Fru	
Glucose	Fruit, honey, grape	Glc	
Raffinose	Leguminous seeds	α-Gal(1 → 6)α-Glc (1 → 2)β-Fru	

Hydrated starch and glycogen are attacked by the endosaccharidase **α-amylase** present in saliva and pancreatic juice (Figure 25.24). Hydration of the polysaccharides occurs during heating and is essential for efficient digestion. Amylase is specific for internal α-1,4-glucosidic bonds; α-1,6 bonds are not attacked, nor are α-1,4 bonds of glucose units that serve as branch points. The pancreatic amylase is secreted in large excess relative to starch intake and is more important than the salivary enzyme from a digestive point of view. The products of the digestion by

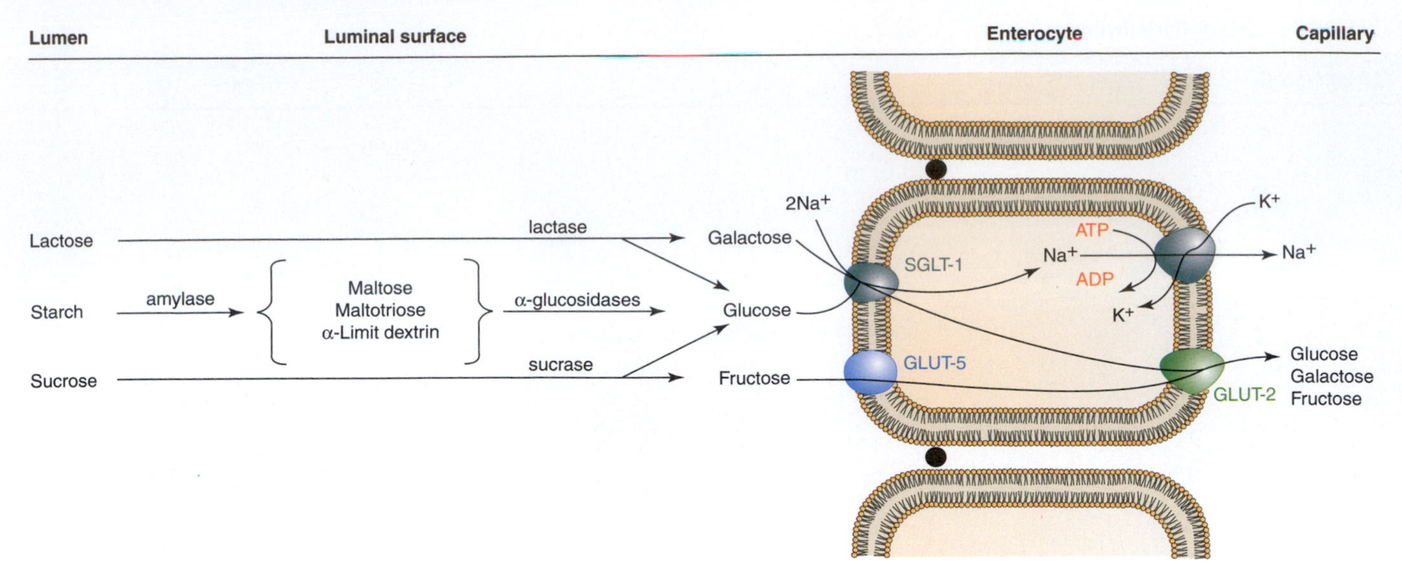

FIGURE 25.23
Digestion and absorption of carbohydrates.

α-amylase are mainly the disaccharide **maltose,** the trisaccharide **maltotriose,** and so-called **α-limit dextrins** containing on the average eight glucose units with one or more α-1,6-glucosidic bonds.

Final hydrolysis of di- and oligosaccharides to monosaccharides is carried out by surface enzymes of the small intestinal epithelial cells (Table 25.8). The surface oligosaccharidases are exoenzymes that cleave off one monosaccharide at a time from the nonreducing end. The capacity of **α-glucosidases** is normally much greater than that needed for completion of the digestion of starch. Similarly, there is usually excess capacity for sucrose hydrolysis relative to dietary intake. In contrast, **β-galactosidase (lactase)** for hydrolysis and utilization of lactose, the major milk carbohydrate, can be rate-limiting in humans (see Clin. Corr. 25.5).

Di-, oligo-, and polysaccharides not hydrolyzed by α-amylase and/or intestinal surface enzymes cannot be absorbed; therefore they reach the lower tract of the intestine, which, from the lower ileum on, contains bacteria. Bacteria can utilize many of the remaining carbohydrates because they possess many more types of saccharidases than humans. Monosaccharides that are released as a result of **bacterial enzymes** are predominantly metabolized anaerobically by the bacteria themselves, resulting in degradation products such as short-chain fatty acids, lactate, hydrogen gas (H_2), methane (CH_4), and carbon dioxide (CO_2). In excess, these compounds can cause fluid secretion, increased intestinal motility, and cramps, either because of increased intraluminal osmotic pressure and distension of the gut, or a direct irritant effect of the bacterial degradation products on the intestinal mucosa.

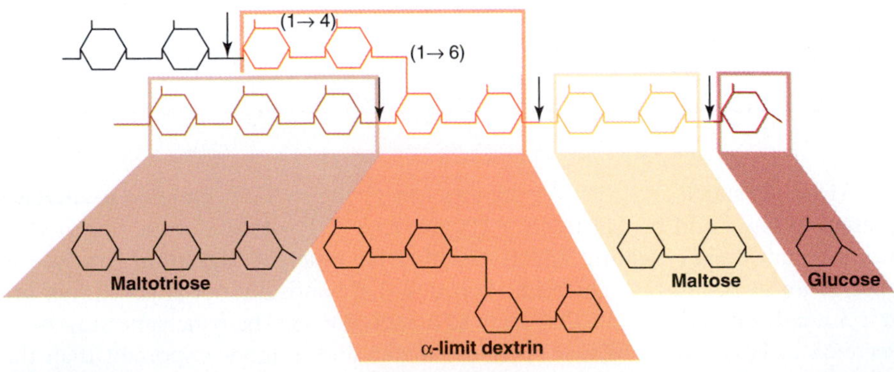

FIGURE 25.24
Digestion of amylopectin by salivary and pancreatic α-amylase.

TABLE 25.8 Saccharidases of the Surface Membrane of the Small Intestine

Enzyme	*Specificity*	*Natural Substrate*	*Product*
exo-1, 4-α-Glucosidase (glucoamylase)	α-(1 → 4)Glucose	Amylose	Glucose
Oligo-1, 6-glucosidase (isomaltase)	α-(1 → 6)Glucose	Isomaltose, α-dextrin	Glucose
α-Glucosidase (maltase)	α-(1 → 4)Glucose	Maltose, maltotriose	Glucose
Surcose-α-glucosidase (sucrase)	α-Glucose	Sucrose	Glucose fructose
α,α-Trehalase	α-(1 → 1)Glucose	Trehalose	Glucose
β-Glucosidase	β-Glucose	Glucosylceramide	Glucose, ceramide
β-Galactosidase (lactase)	β-Galactose	Lactose	Glucose, galactose

The well-known problem of flatulence after ingestion of leguminous seeds (beans, peas, and soya) is caused by oligosaccharides that cannot be hydrolyzed by human intestinal enzymes. The seeds contain modified sucrose to which one or more galactose moieties are linked. The glycosidic bonds of galactose are in the α configuration, which can only be split by bacterial enzymes. The simplest sugar of this family is **raffinose** (see Table 25.7). **Trehalose**, a disaccharide that occurs in young mushrooms, requires a special disaccharidase, trehalase.

Monosaccharides Are Absorbed by Carrier-Mediated Transport

The major monosaccharides that result from digestion of di- and polysaccharide are **D-glucose**, **D-galactose**, and **D-fructose.** Absorption of these and other minor monosaccharides is by carrier-mediated processes that exhibit substrate specificity, stereospecificity, saturation kinetics, and inhibition by specific inhibitors.

At least two types of monosaccharide transporters catalyze monosaccharide uptake from the lumen into the cell: (1) a **Na^+-monosaccharide cotransporter (SGLT)**, existing probably as a tetramer of 75-kDa peptides, has high specificity for D-glucose and D-galactose and catalyzes "active" sugar absorption; and (2) a **Na^+-independent, facilitated-diffusion** type of **monosaccharide transport system** with specificity for D-fructose **(GLUT-5).** In addition, a **Na^+-independent monosaccharide transporter (GLUT-2)**, which accepts all three monosaccharides, is present in the contraluminal plasma membrane. GLUT-2 also occurs in the liver and kidney, and other members of the GLUT family of transporters are found in all cells. All GLUT transporters mediate uncoupled D-glucose flux down its concentration gradient. GLUT-2 of gut, liver, and kidney moves D-glucose out of the cell into the blood under physiological conditions, while in other tissues GLUT-1 (in erythrocytes and brain) or the insulin-sensitive GLUT-4 (in adipose and muscle tissue) are mainly involved in D-glucose uptake. Properties of intestinal SGLT-1 and of GLUT-2 are compared in Table 25.9, and their role in transepithelial glucose absorption is illustrated in Figure 25.23.

TABLE 25.9 Characteristics of Glucose Transport Systems in the Plasma Membranes of Enterocytes

Characteristic	*Luminal*	*Contraluminal*
Designation	SGLT-1	GLUT-2
Subunit molecular weight (kDa)	75	57
Effect of Na^+	Cotransport with Na^+	None
Good substrates	D-Glc, D-Gal, α-methyl-D-Glc	D-Glc, D-Gal, D-Man, 2-deoxy-D-Glc

CLINICAL CORRELATION 25.5
Disaccharidase Deficiency

Intestinal disaccharidase deficiencies are encountered relatively frequently in humans. Deficiency can be present in one enzyme or several enzymes for a variety of reasons (genetic defect, physiological decline with age, or the result of "injuries" to the mucosa). Of the disaccharidases, lactase is the most common enzyme with an absolute or relative deficiency, which is experienced as milk intolerance. The consequences of an inability to hydrolyze lactose in the upper small intestine are inability to absorb lactose and bacterial fermentation of ingested lactose in the lower small intestine. Bacterial fermentation results in the production of gas (distension of gut and flatulence) and osmotically active solutes that draw water into the intestinal lumen (diarrhea). The lactose in yogurt has already been partially hydrolyzed during the fermentation process of making yogurt. Thus individuals with lactase deficiency can often tolerate yogurt better than unfermented dairy products. Lactase is commercially available to pretreat milk so that the lactose is hydrolyzed.

Buller, H. A. and Grant, R. G. Lactose intolerance. *Annu. Rev. Med.* 41:141, 1990.

25.6 | DIGESTION AND ABSORPTION OF LIPIDS

Digestion of Lipids Requires Overcoming Their Limited Water Solubility

An adult human ingests about 60–150 g of lipid per day. **Triacylglycerols** constitute more than 90% of the dietary fat. The rest is made up of phospholipids, cholesterol, cholesterol esters, and free fatty acids. In addition, 1–2 g of cholesterol and 7–22 g of phosphatidylcholine (lecithin) are secreted into the small intestinal lumen as constituents of bile.

Lipids are defined by their good solubility in organic solvents and poor or nonexisting solubility in aqueous solutions. The poor water solubility presents problems for digestion because the substrates are not easily accessible to the digestive enzymes in the aqueous phase. In addition, even if ingested lipids are hydrolyzed into simple constituents, the products tend to aggregate to larger complexes that make poor contact with the cell surface and therefore are not easily absorbed. These problems are overcome by (1) increases in the interfacial area between the aqueous and lipid phase and (2) "solubilization" of lipids with **detergents.** Thus changes in the physical state of lipids are intimately connected to chemical changes during digestion and absorption.

At least five different phases can be distinguished (Figure 25.25): (1) hydrolysis of triacylglycerols to free fatty acids and monoacylglycerols; (2) solubilization of lipids by detergents (bile acids) and transport from the intestinal lumen toward the cell surface; (3) uptake of free fatty acids and monoacylglycerols into the mucosal cell and resynthesis to triacylglycerols; (4) packaging of newly synthesized triacylglycerols into special lipid-rich globules, called chylomicrons; and (5) exocytosis of chylomicrons from mucosal cells into lymph.

Lipids Are Digested by Gastric and Pancreatic Lipases

Hydrolysis of triacylglycerols is initiated in the stomach by **lingual** and **gastric lipases.** Gastric digestion can account for up to 30% of total triacylglycerol hydrolysis. However, the rate of hydrolysis is slow because the ingested triacylglycerols form a separate lipid phase with a limited water–lipid interface. Lipases adsorb to that interface and convert triacylglycerols into fatty acids and diacylglycerols (Figure 25.26). The importance of the initial hydrolysis is that some of the water-immiscible triacylglycerols are converted to products that possess both polar and nonpolar groups. Such surfactive products spontaneously adsorb to water–lipid interfaces and confer a hydrophilic surface to lipid droplets, thereby providing additional stable interface with the aqueous environment. At constant volume of the lipid phase, any increase in interfacial area produces dispersion of the lipid phase

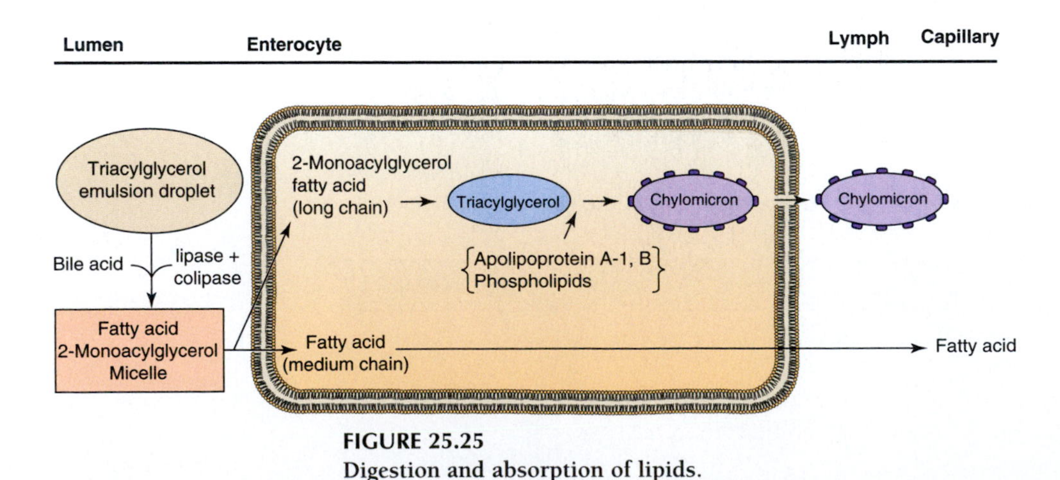

FIGURE 25.25
Digestion and absorption of lipids.

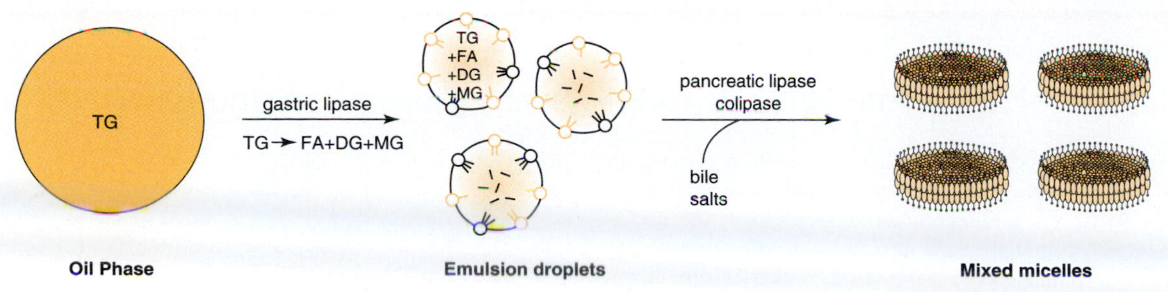

FIGURE 25.26
Changes in physical state during triacylglycerol digestion.
Abbreviations: TG, triacylglycerol; DG, diacylglycerol; MG, monoacylglycerol; FA, fatty acid.

into smaller droplets **(emulsification)** and provides more sites for adsorption of more lipase molecules.

Pancreatic lipase is the major enzyme for triacylglycerol hydrolysis (Figure 25.27). This enzyme is specific for esters in the α-position of glycerol and prefers long-chain fatty acids (longer than ten carbon atoms). Hydrolysis by the pancreatic enzyme also occurs at the water–lipid interface of emulsion droplets. The products are **free fatty acids** and **β-monoacylglycerols.** The purified form of the enzyme is strongly inhibited by the bile acids that normally are present in the small intestine during lipid digestion. The problem of inhibition is overcome by **colipase,** a small protein (12 kDa) that binds to both the water–lipid interface and to lipase, thereby anchoring and activating the enzyme. It is secreted by the pancreas as procolipase and depends on tryptic removal of a NH_2-terminal decapeptide for full activity. Clinical Correlation 25.6 describes two commercial strategies to reduce lipid absorption as a means to decrease obesity.

Pancreatic juice also contains an **unspecific lipid esterase,** which acts on cholesterol esters, monoacylglycerols, or other lipid esters, such as esters of vitamin A with carboxylic acids. In contrast to triacylglycerol lipase, this lipid esterase requires bile acids for activity.

Phospholipids are hydrolyzed by specific phospholipases. Pancreatic secretions are especially rich in **prophospholipase A_2** (Figure 25.28). As other pancreatic proenzymes, this one is also activated by trypsin. Phospholipase A_2 requires bile acids for activity.

Triacylglycerol

Fatty acids and monoacylglycerol

R = hydrocarbon chain

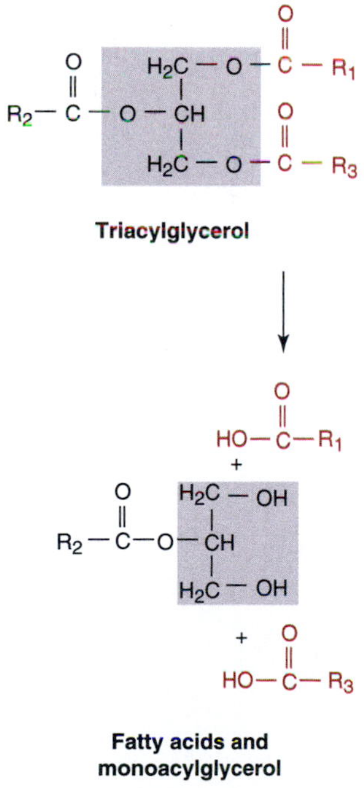

FIGURE 25.27
Mechanism of action of lipase.

Bile Acid Micelles Solubilize Lipids During Digestion

Bile acids are biological detergents synthesized by liver and secreted as conjugates of glycine or taurine with the bile into the duodenum (Figure 25.29). At physiological pH, they are anions and possess detergent properties. The terms bile acids and **bile salts** are often used interchangeably. Bile acids at pH values above the pK (Table 25.10) reversibly form aggregates at concentrations above 2–5 mM. These aggregates are called **micelles** (see p. 502), and the minimal concentration necessary for micelle formation is the **critical micellar concentration** (Figure 25.30). Bile acid molecules in micelles are in equilibrium with those free in solution. Thus micelles, in contrast to emulsified lipids, are equilibrium structures with well-defined sizes that are much smaller than emulsion droplets. Micelle sizes typically range between 4 and 60 nm depending on bile acid concentration and the ratio of bile acids to lipids.

In micelles bile acids have their hydrophobic portions removed from contact with water, while hydrophilic groups remain exposed to the water. The hydrophobic region of bile acids is formed by one surface of the fused ring system, while the carboxylate or sulfonate ion and the hydroxyl groups on the other side of the ring system are hydrophilic. Major driving forces for micelle formation are the removal of apolar, hydrophobic groups from and the interaction of polar

CLINICAL CORRELATION 25.6
Pharmacological Interventions to Prevent Fat Absorption and Obesity

Obesity is a major problem in modern society as food is generally very abundant. Therefore interest in weight reduction methods is widespread. Two recent commercial strategies exploit the understanding of intestinal lipid absorption for this purpose.

Olestra® is a commercial lipid produced by esterification of natural fatty acids with sucrose instead of glycerol (Figure (*a*)). Six to eight fatty acids are covalently coupled to sucrose. These hexa-, hepta-, or octa-acyl sucrose compounds taste like natural lipids; however, they cannot be hydrolyzed to absorbable constituents and are therefore excreted unchanged.

Pancreatic lipase is the major enzyme that breaks down dietary triacylglycerols to absorbable fatty acids and glycerol. Orlistat® (Figure (*b*)) is a powerful inhibitor of pancreatic lipase and hence lipid digestion and absorption. Ingestion of Orlistat® blocks lipid absorption and results in lipid excretion, with the added benefit of a slower intestinal transit time due to feedback from lipids reaching the terminal small intestine and colon. Orlistat® is a nonhydrolysable analog of a triacylglycerol.

Olestra is a registered trademark of Proctor and Gamble and Orlistat of Roche, Basel, Switzerland.

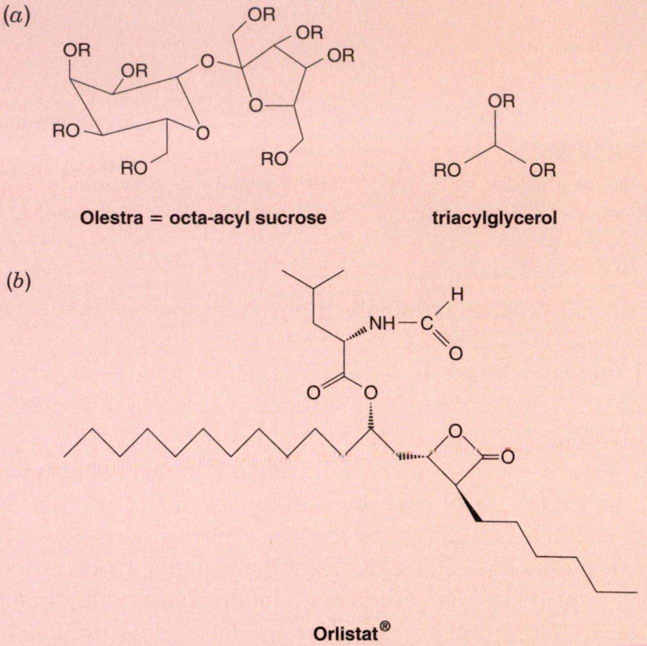

(*a*)

Olestra = octa-acyl sucrose **triacylglycerol**

(*b*)

Orlistat®

Thompson, A. B. R., et al. In: A. B. Christophe, and S. DeVriese, (Eds), *Fat Digestion and Absorption*. Champaign, IL: AOCS Press, 2000, p. 383; and Golay, A. In: A. B. Christophe, and S. DeVriese, (Eds.), *Fat Digestion and Absorption*. Champaign, IL: AOCS Press, 2000, p. 420.

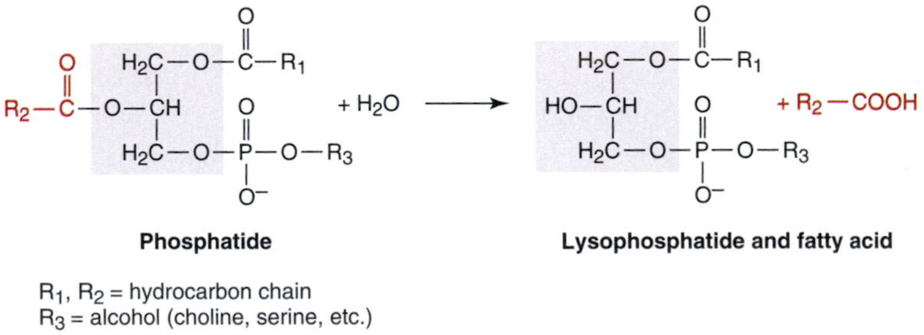

Phosphatide $+ H_2O \longrightarrow$ **Lysophosphatide and fatty acid** $+ R_2-COOH$

R_1, R_2 = hydrocarbon chain
R_3 = alcohol (choline, serine, etc.)

FIGURE 25.28
Mechanism of action of phospholipase A_2.

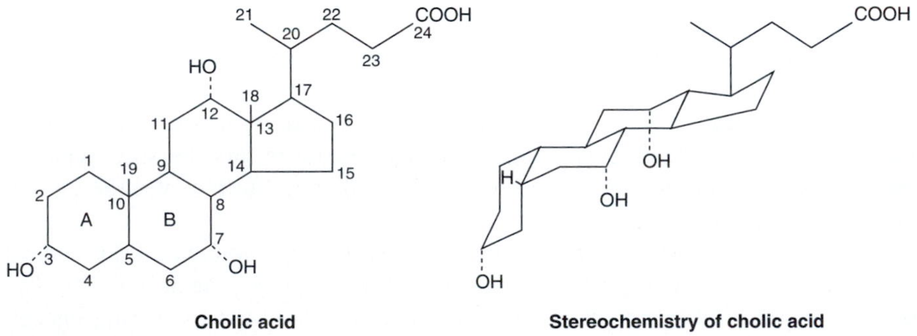

Cholic acid **Stereochemistry of cholic acid**

FIGURE 25.29
Cholic acid, a bile acid.

TABLE 25.10 Effect of Conjugation on the Acidity of Cholic, Deoxycholic, and Chenodeoxycholic Acids

Bile Acid	*Ionized Group*	pK_a
Unconjugated bile acids	$—COO^-$ of cholestanoic acid	≃ 5
Glycoconjugates	$—COO^-$ of glycine	≃ 3.7
Tauroconjugates	$—SO_3^-$ of taurine	≃ 1.5

Primary: Cholic (OH, COOH, HO, OH); Chenic (COOH, HO, OH)

Secondary: Deoxycholic (OH, COOH, HO)

+ glycine $NH_3^+CH_2COO^-$ → $R—C(=O)—N(H)—CH_2COO^-$: Cholylglycine, Chenylglycine, Deoxycholylglycine

+ taurine $NH_3^+(CH_2)_2SO_3^-$ → $R—C(=O)—N(H)—(CH_2)_2SO_3^-$: Cholyltaurine, Chenyltaurine, Deoxycholyltaurine

Source: Reproduced with permission from Hofmann, A. F. *Handbook of Physiology* 5:2508, 1968.

groups with water molecules; the distribution of polar and apolar regions places constraints on the stereochemical arrangements of bile acid molecules within a micelle. Four bile acid molecules are sufficient to form a very simple micelle as shown in Figure 25.31. Bile acid micelles can solubilize other lipids, such as phospholipids and fatty acids. These **mixed micelles** have disk-like shapes, whereby the phospholipids and fatty acids form a bilayer and the bile acids occupy the edge positions, rendering the edge of the disk hydrophilic (Figure 25.32). Within the mixed phospholipid–bile acid micelles, other water-insoluble lipids, such as cholesterol, can be accommodated and thereby "solubilized" (for potential problems see Clin. Corr. 25.7).

During triacylglycerol digestion, free fatty acids and monoacylglycerols are released at the surface of fat emulsion droplets. In contrast to triacylglycerols, which are water-insoluble, free fatty acids and monoacylglycerols are slightly water-soluble, and molecules at the surface equilibrate with those in solution and in bile acid micelles. Thus the products of triacylglycerol hydrolysis are continuously transferred from emulsion droplets to the micelles (see Figure 25.26).

Micelles are the major vehicle for moving lipids from the lumen to the mucosal surface where absorption occurs. Because the fluid layer next to the cell surface is poorly mixed, the major transport mechanism for solute flux across this **"unstirred" fluid layer** is diffusion down the concentration gradient. With this type of transport mechanism, the delivery rate of nutrients at the cell surface is proportional to their concentration difference between luminal bulk phase and cell surface. Obviously, the unstirred fluid layer presents problems for sparingly soluble or insoluble nutrients, in that reasonable concentration gradients and delivery rates cannot be achieved. Bile acid micelles overcome this problem for lipids by increasing their effective concentration in the unstirred layer. The increase in transport rate is nearly proportional to the increase in effective concentration and can be 1000-fold over that of individually solubilized fatty acids, in accordance with the different water solubility of fatty acids as micelles or as individual molecules. This relationship between flux and effective concentration holds because the diffusion constant, another parameter that determines the flux, is only slightly

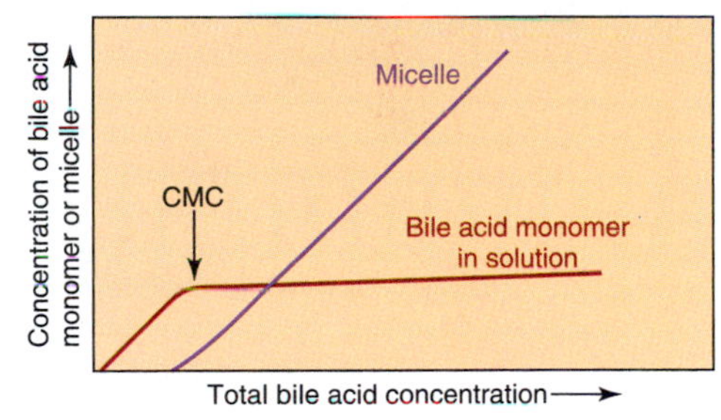

FIGURE 25.30
Solubility properties of bile acids in aqueous solutions.
Abbreviation: CMC, critical micellar concentration.

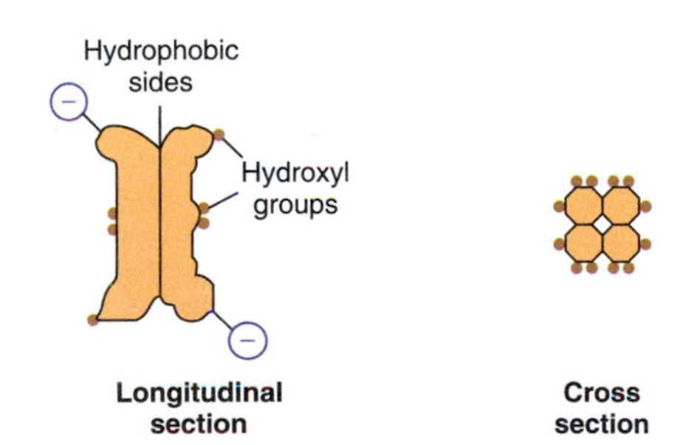

FIGURE 25.31
Diagrammatic representation of a Na^+ cholate micelle.
Adapted from Small, D. M. Biochim. Biophys. Acta *176:178, 1969.*

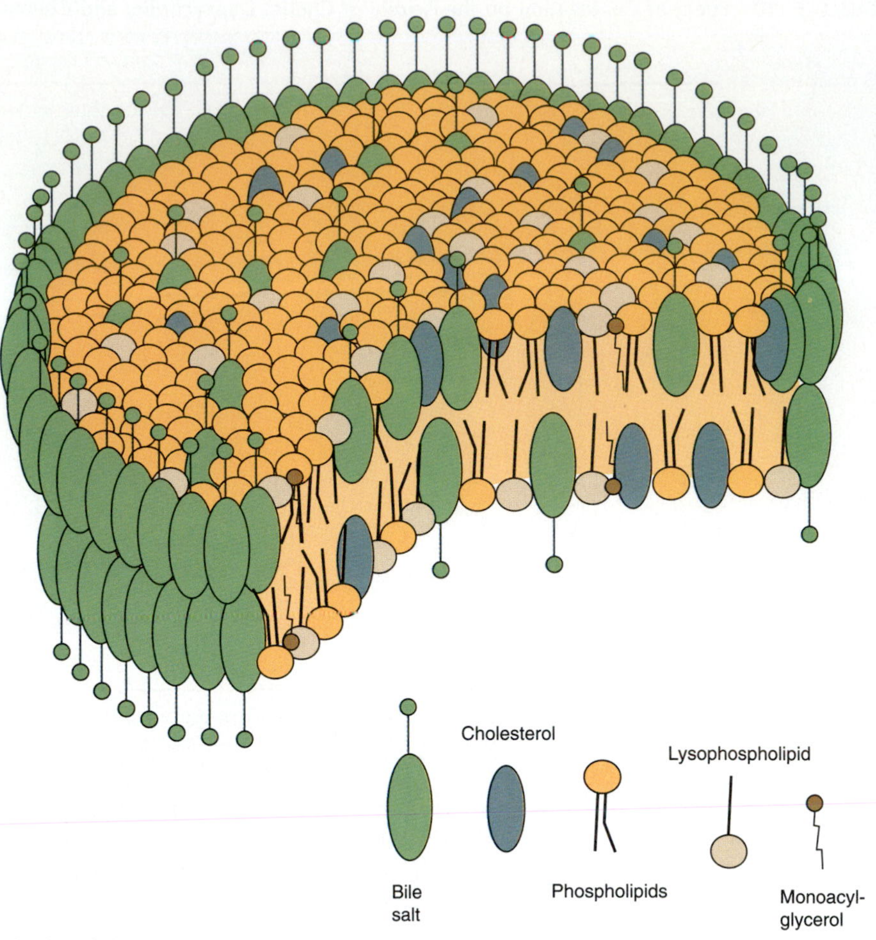

FIGURE 25.32
Proposed structure of the intestinal mixed micelle.
The bilayer disk has a band of bile salt at its periphery and other, more hydrophobic components (fatty acids, monoacylglycerol, phospholipids, and cholesterol) within its interior. *Redrawn based on figure from Carey, M. C. In: A. M. Arias, H. Popper, D. Schachter, et al. (Eds.),* The Liver: Biology and Pathology. *New York: Raven Press, 1982.*

smaller for the mixed micelles as compared to lipid molecules free in solution. Thus efficient lipid absorption depends on the presence of sufficient bile acids to "solubilize" the ingested and hydrolyzed lipids in micelles. In the absence of bile acids, the absorption of triacylglycerols does not stop completely, although the efficiency is reduced drastically. The residual absorption depends on the slight water solubility of the free fatty acids and monoacylglycerols. Unabsorbed lipids reach the lower intestine where a small part can be metabolized by bacteria. The bulk of unabsorbed lipids, however, is excreted with the stool (this is called **steatorrhea**).

Micelles also transport cholesterol and the lipid-soluble **vitamins A, D, E,** and **K** through the unstirred fluid layers. Bile acid secretion is absolutely essential for their absorption.

Most Absorbed Lipids Are Incorporated into Chylomicrons

Uptake of lipids by the epithelial cells occurs by diffusion through the plasma membrane. Absorption is virtually complete for fatty acids and monoacylglycerols, which are slightly water-soluble. It is less efficient for water-insoluble lipids. For example, only 30–40% of the dietary cholesterol is absorbed.

Within the intestinal cells, the fate of absorbed fatty acids depends on chain length. **Fatty acids** of **short** and **medium chain** length ($\leq$ 10 carbon atoms) pass through the cell into the portal blood without modification. **Long-chain fatty acids**

CLINICAL CORRELATION 25.7
Cholesterol Stones

Liver secretes phospholipids and cholesterol together with bile acids into the bile. Because of the limited solubility of cholesterol, its secretion in bile can result in cholesterol stone formation in the gallbladder. Stone formation is a relatively frequent complication; up to 20% of North Americans will develop stones during their lifetime.

Cholesterol is practically insoluble in aqueous solutions. However, it can be incorporated into mixed phospholipid–bile acid micelles up to a mole ratio of 1:1 for cholesterol/phospholipids and thereby "solubilized" (see accompanying figure). The liver can produce supersaturated bile with a higher ratio than 1:1 of cholesterol/phospholipid. This excess cholesterol has a tendency to come out of solution and to crystallize. Such bile with excess cholesterol is considered lithogenic, that is, stoneforming. Crystal formation usually occurs in the gallbladder, rather than the hepatic bile ducts, because contact times between bile and any crystallization nuclei are greater in the gallbladder. In addition, the gallbladder concentrates bile by absorption of electrolytes and water. The bile salts chenodeoxycholate (Figure 25.33) and its stereoisomer ursodeoxycholate (7-hydroxy group in β-position) are now available for oral use to dissolve gallstones. Ingestion of these bile salts reduces cholesterol excretion into the bile and allows cholesterol in stones to be solubilized.

The tendency to secrete bile supersaturated with respect to cholesterol is inherited and found more frequently in females than in males, often associated with obesity. Supersaturation also appears to be a function of the size and nature of the bile acid pool as well as the secretion rate.

Schoenfield, L. J. and Lachin, J. M. Chenodiol (chenodeoxycholic acid) for dissolution of gallstones: The National Cooperative Gallstone Study. A controlled trial of safety and efficacy. *Ann. Intern. Med.* 95:257, 1981; and Carey, M. C. and Small, D. M. The physical chemistry of cholesterol solubility in bile. *J. Clin. Invest.* 61:998, 1978.

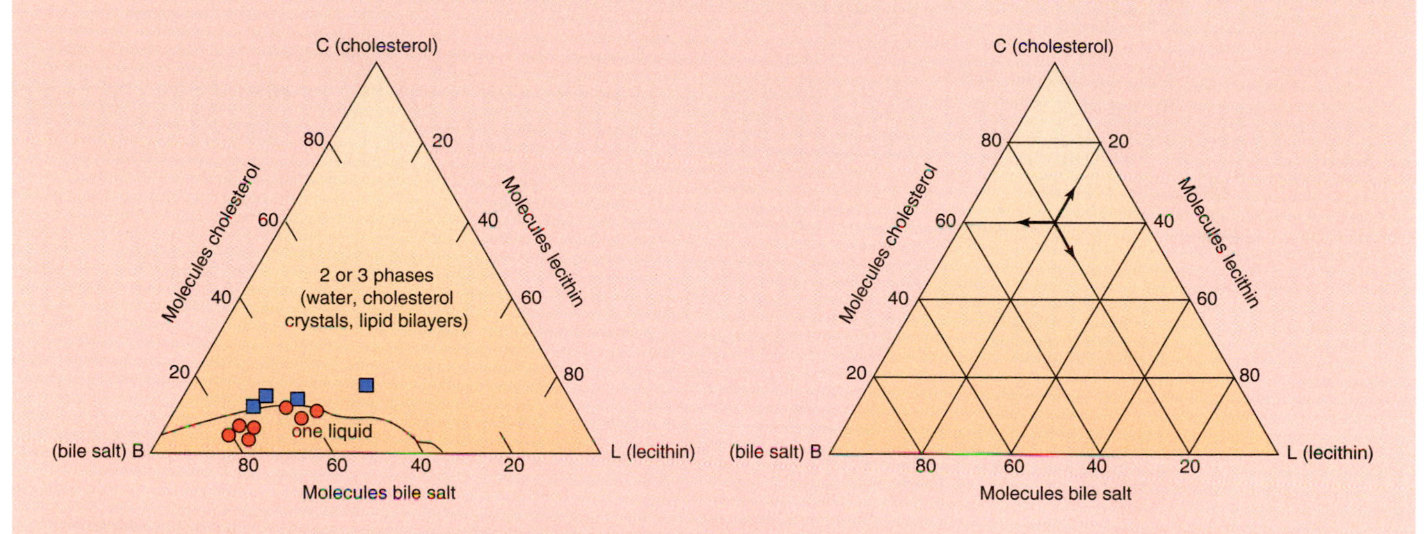

Diagram of the physical states of mixtures of 90% water and 10% lipid.
The 10% lipid is made up of bile acids, phosphatidylcholine (lecithin), and cholesterol, and the triangle represents all possible ratios of the three lipid constituents. Each point within the triangle corresponds to a particular composition of the three components, which can be read off the graph as indicated; each point on one of the sides corresponds to a particular composition of just two components. The left triangle contains the composition of gallbladder bile samples from patients without stones (red circles) and with cholesterol stones (blue squares). Lithogenic bile has a composition that falls outside the "one liquid" area in the lower left corner.
Redrawn from Hofmann, A. F. and Small, D. M. Annu. Rev. Med. *18:362, 1967. Copyright © 1967 by Annual Reviews, Inc.*

(>12 carbon atoms) become bound to a cytoplasmic fatty acid-binding protein (intestinal FABP or I-FABP), and are transported to the endoplasmic reticulum, where they are resynthesized into triacylglycerols. Glycerol for this process is derived from the absorbed 2-monoacylglycerols and, to a minor degree, from glucose. The resynthesized triacylglycerols form lipid globules to which surface-active phospholipids and special proteins, termed **apolipoproteins,** adsorb. The lipid globules migrate within membrane-bounded vesicles through the Golgi to the contraluminal plasma membrane. They are finally released into the intercellular space by fusion of the vesicles with the contraluminal plasma membrane and travel from there through intestinal lymph vessels (or lacteals) and the thoracic duct to the large systemic veins. The lipid globules are called **chylomicrons** because they can be several

CLINICAL CORRELATION 25.8
A-β-Lipoproteinemia

A-β-lipoproteinemia is an autosomal recessive disorder characterized by the absence of all lipoproteins containing apo-β-lipoprotein B, that is, chylomicrons, very low density lipoproteins (VLDLs), and low density lipoproteins (LDLs) (see p. 745). Serum cholesterol is extremely low. This defect is associated with severe malabsorption of triacylglycerol and lipid-soluble vitamins (especially tocopherol and vitamin E) and accumulation of apolipoprotein B in enterocytes and hepatocytes. The defect does not appear to involve the gene for apolipoprotein B, but rather one of several proteins involved in processing of apolipoprotein B in liver and intestinal mucosa, or in assembly and secretion of triacylglycerol-rich lipoproteins, that is, chylomicrons and VLDLs from these tissues, respectively.

Kane, J. P. Apolipoprotein B: structural and metabolic heterogeneity. *Annu. Rev. Physiol.* 45:673, 1983; and Kane, J. P. and Havel, R. J. In: C. R. Scriver, A. L. Beaudet, W. S. Sly, and D. Valle (Eds.), *The Metabolic and Molecular Bases of Inherited Disease*, Vol. 1, 7th ed. New York: McGraw-Hill, 1995, p. 1853.

micrometers in diameter and leave through lymph vessels (chyle = milky lymph derived from the Greek *chylos*, which means juice). The intestinal apolipoproteins are distinctly different from those of the liver (see p. 127) and are designated A-1 and B. **Apolipoprotein B** is essential for chylomicron release from enterocytes (see Clin. Corr. 25.8).

While dietary medium-chain fatty acids reach the liver directly with the portal blood, the long-chain fatty acids first reach adipose tissue and muscle via the systemic circulation before coming into contact with the liver. Fat and muscle cells take up large amounts of dietary lipids for storage or metabolism. The bypass of the liver may have evolved to protect this organ from a lipid overload after a meal.

The differential handling of medium- and long-chain fatty acids by intestinal cells can be specifically exploited to provide the liver with high-caloric nutrients in the form of fatty acids. Short- and medium-chain fatty acids are not very palatable; however, triacylglycerols synthesized from these fatty acids are quite palatable and can be used as part of the diet. Short-chain fatty acids are produced physiologically, particularly in the colon, by bacteria from residual carbohydrates. These fatty acids are absorbed into portal blood.

25.7 | BILE ACID METABOLISM

Bile acids are synthesized within the liver from cholesterol but are modified by bacterial enzymes in the intestinal lumen. **Primary bile acids** synthesized by the liver

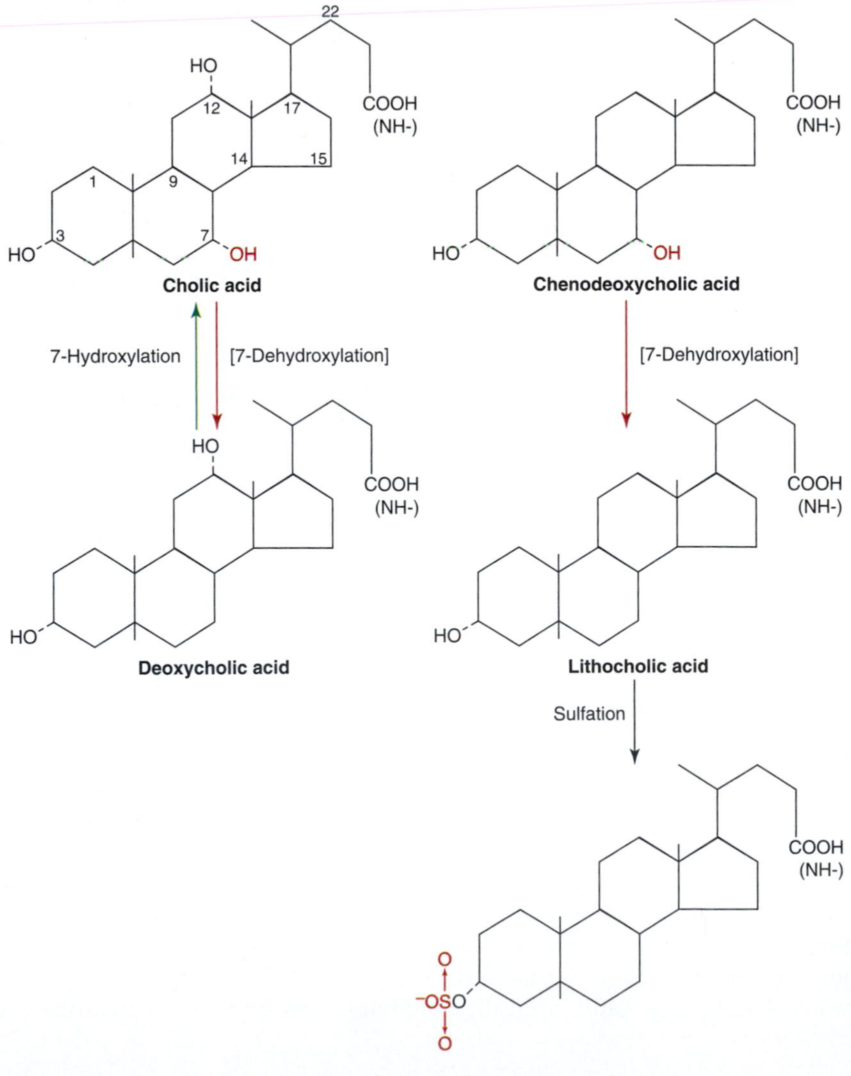

FIGURE 25.33
Bile acid metabolism in the rat.
Green and black arrows indicate reactions catalyzed by liver enzymes; red arrows indicate those of bacterial enzymes within the intestinal lumen. (NH—), glycine or taurine conjugate of the bile acids.

are **cholic** and **chenodeoxycholic** (or chenic) **acid. Secondary bile acids** are derived from the primary bile acids by bacterial reduction in position 7 of the ring structure, resulting in **deoxycholate** and **lithocholate**, respectively (Figure 25.33).

Primary and secondary bile acids are reabsorbed by the intestine into the portal blood, taken up by the liver, and then resecreted into bile. Within the liver, primary as well as secondary bile acids are linked to either glycine or taurine via an isopeptide bond. These derivatives are called **glyco-** and **tauroconjugates**, respectively, and constitute the forms that are secreted into bile. With the conjugation, the carboxyl group of the unconjugated acid is replaced by an even more polar group. The p*K* values of the carboxyl group of glycine and of the sulfonyl group of taurine are lower than that of unconjugated bile acids, so that conjugated bile acids remain ionized over a wider pH range. The conjugation is partially reversed within the intestinal lumen by hydrolysis of the isopeptide bond.

The total amount of conjugated and unconjugated bile acids secreted per day by the liver is 20–30 g for an adult. However, the body maintains only a very small pool of 3–5 g. A small pool is advantageous because bile acids become toxic at high concentrations due to their detergent properties, for example, through their ability to lyse cells. Therefore to achieve the observed secretion rates given above, bile acids are actively reabsorbed in the lower ileum, recirculated to the liver, and resecreted 4–10 times per day. In other words, during a meal, bile acids from the liver and gallbladder are released into the lumen of the upper small intestine, pass with chyme down the small intestinal lumen, are reabsorbed into the portal blood by ileal enterocytes, and are then extracted from blood by the liver parenchymal cells. The process of secretion and reuptake is referred to as the **enterohepatic circulation** (Figure 25.34). Reabsorption of bile acids by the intestine is quite efficient as only about 0.8 g of bile acids escapes reuptake each day and is secreted with the feces. Serum levels of bile acids normally vary with the rate of reabsorption and therefore are highest during a meal.

Cholate, deoxycholate, chenodeoxycholate, and their conjugates continuously participate in the enterohepatic circulation. In contrast, most of the lithocholic acid that is produced by bacterial enzymes is sulfated during the next passage through

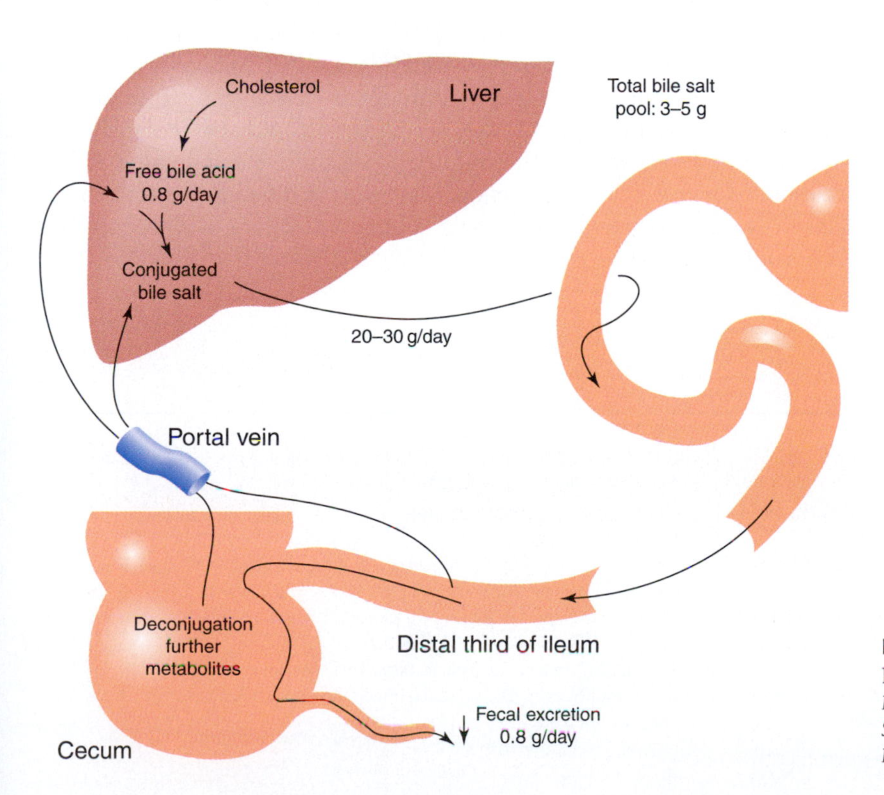

FIGURE 25.34
Enterohepatic circulation of bile acids.
Redrawn from Clark, M. L. and Harries, J. T. In: I. McColl and G. E. Sladen (Eds.), Intestinal Absorption in Man. *New York: Academic Press, 1975, p. 195.*

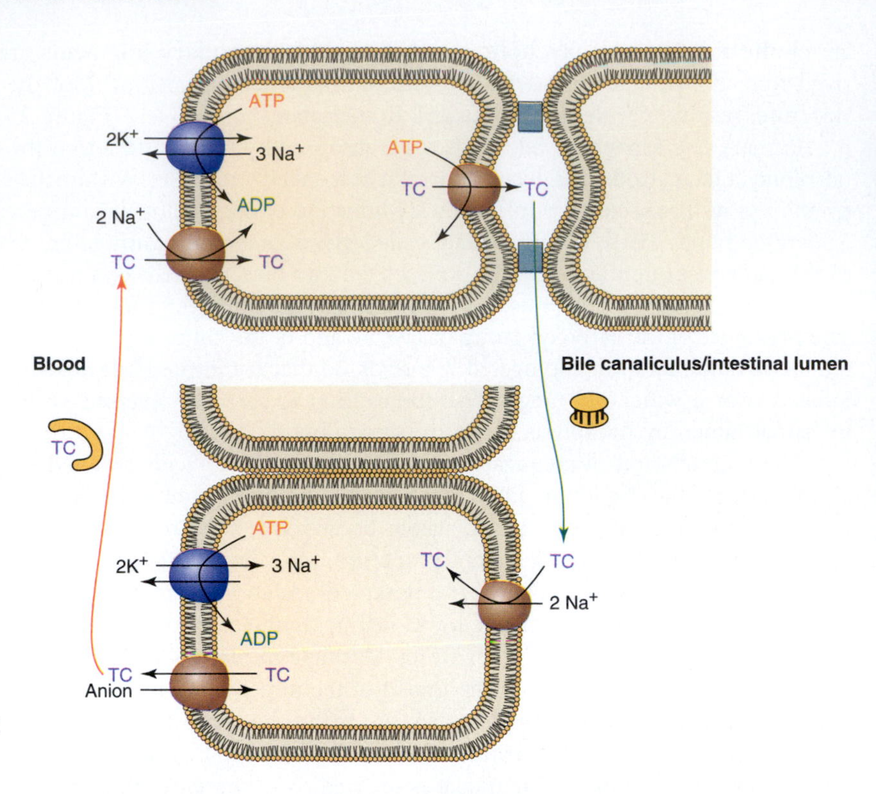

FIGURE 25.35
Molecular mechanisms of taurocholate (TC) transport during enterohepatic circulation.

the liver. The **sulfate ester** of **lithocholic acid** is not a substrate for the bile acid transport system in the ileum and therefore is excreted in the feces.

The molecular mechanisms for active bile acid transport by epithelial cells in the lower ileum and liver cells have been determined (Figure 25.35). Ileal absorption is mediated by secondary active transport via a luminal Na^+–bile acid cotransport system **(apical sodium-dependent bile acid transporter** or **ASBT)** with a stoichiometry of 2:1 for Na^+/bile acid. Bile acids exit the cells on the blood side in exchange for other types of anions, for example, bicarbonate. The uptake of bile acids from blood by liver cells is also by secondary active Na^+–bile acid cotransport, however, by a different transporter **(Na^+ taurocholate cotransporting polypeptide** or **NTCP).** In contrast, secretion of bile acids by liver cells across the canalicular plasma membrane into bile is by primary active transport (**bile salt export pump** or **BSEP**, an **ABC-type ATPase**, see p. 525). To achieve reasonable total bile acid concentrations within cells and in plasma while maintaining very low free concentrations without detergent action, bile acids are bound to specific binding proteins in ileal enterocytes and liver cells. Albumin functions as a binding protein for bile acids in plasma.

BIBLIOGRAPHY

Bahar, R. J. and Stolz, A. Bile acid transport. *Gastroenterol. Clin. North Am.* 28:27, 1999.

Chang, E. B., Sitrin, M. D., and Black, D. D. *Gastrointestinal, Hepatobiliary, and Nutritional Physiology.* Philadelphia: Lippincott-Raven Publ., 1996.

Christophe, A. B. and DeVriese, S. (Eds.). *Fat Digestion and Absorption.* Champaign, IL: AOCS Press, 2000.

Cristofaro, E., Mottu, F., and Wuhrmann, J. J. Involvement of the raffinose family of oligosaccharides in flatulence. In: H. L. Sipple and K. W. McNutt (Eds.), *Sugars in Nutrition.* New York: Academic Press, 1974, p. 314.

Field, M. and Semrad, C. E. Toxigenic diarrheas, congenital diarrheas, and cystic fibrosis. *Annu. Rev. Physiol.* 55:631, 1993.

Hediger, M. A. and Rhoads, D. B. Molecular physiology of sodium–glucose cotransporters. *Physiol. Rev.* 74:993, 1994.

http://gastroenterology.medscape.com

http://www.lactose.net

Johnson, L. R. (Ed.-in-chief). *Physiology of the Gastrointestinal Tract,* Vols. 1 and 2, 2nd and 3rd eds. New York: Raven Press, 1987, 1994.

Kere, J. and Hoglund, P. Inherited disorders of ion transport in the intestine. *Curr. Opin. Genet. Dev.* 10:306, 2000.

Mathews, D. M. *Protein Absorption.* New York: Wiley-Liss, 1991.

Meier, P. J. and Stieger, B. Molecular mechanisms of bile formation. *News Physiol. Sci.* 15:89, 2000.

Pandol, S. and Raybauld, H. E. Integrated response to a meal. *The Undergraduate Teaching Project in Gastroenterology and Liver Disease,* Unit #29. American Gastroenterological Assoc. Timonium, MD: Milner-Fenwick, Inc., 1995.

Porter, R. and Collins, G. M. *Brush Border Membranes,* Vol. 95, Ciba Foundation Symposium. London: Pitman, 1983.

Schultz, S. G. (section Ed.). *Handbook of Physiology. Section 6: The Gastrointestinal System, Vol. IV. Intestinal Absorption and Secretion.* (M. Field, and R. A. Frizzell, Eds.). Bethesda, MD: American Physiological Society, 1991.

Sleisenger, M. H. (Ed.). Malabsorption and nutritional support. *Clin. Gastroenterol.* 12:323, 1983.

Soleimani, M. and Ulrich, C. D. How cystic fibrosis affects pancreatic ductal bicarbonate secretion. *Med. Clin. North Am.* 84:641, 2000.

Thomson, A. B. R., Schoeller, C., Keelan, M., Smith, L., and Clandinin, M.T. Lipid absorption: passing through the unstirred layers near the brush border membrane, and beyond. *Can. J. Physiol. Pharmacol.* 71:531, 1993.

Van-Loo, J., Cummings, J., Delzenne, N., Englyst, H., Franck, A., Hopkins, M., Kok, N., Macfarlane, G., Newton, D., Quigley, M., Roberfroid, M., van Vliet, T., and van den Heuvel, E. Functional food properties of non-digestible oligosaccharides: a consensus report from the ENDO project (DGXII AIRII-CT94-1095). *Br. J. Nutr.* 81:121, 1999.

QUESTIONS | C. N. ANGSTADT

Multiple Choice Questions

1. Active forms of most enzymes that digest food may normally be found in all of the following EXCEPT:
   A. in soluble form in the lumen of the stomach.
   B. in the saliva.
   C. attached to the luminal surface of the plasma membrane of intestinal epithelial cells.
   D. dissolved in the cytoplasm of intestinal epithelial cells.
   E. in zymogen granules of pancreatic exocrine cells.

2. Histamine is a potent secretagogue of:
   A. amylase by the salivary glands.
   B. HCl by the stomach.
   C. gastrin by the stomach.
   D. hydrolytic enzymes by the pancreas.
   E. $NaHCO_3$ by the pancreas.

3. The contraluminal plasma membranes of small intestinal epithelial cells contain:
   A. aminopeptidases.
   B. Na^+/K^+ ATPase.
   C. disaccharidases.
   D. GLUT-5.
   E. Na^+–monosaccharide transport (SGLT-1).

4. Starch digestion is more efficient after heating the starch with water because heating:
   A. hydrates the starch granules, making them more susceptible to pancreatic amylase.
   B. converts α-1,4 links to β-1,4 links, which are more susceptible to attack by mammalian amylases.
   C. partly hydrolyses α-1,6 links.
   D. converts the linear amylose to branched amylopectin, which resembles glycogen.
   E. inactivates amylase inhibitors, which are common in the tissues of starchy plants.

5. Micelles:
   A. are the same as emulsion droplets.
   B. form from bile acids at all bile acid concentrations.
   C. although they are formed during lipid digestion, do not significantly enhance utilization of dietary lipid.
   D. always consist of only a single lipid species.
   E. are essential for the absorption of vitamins A and K.

6. Epithelial cells of the lower ileum express a Cl^-/HCO_3^- exchange coded for by the DRA gene. These cells:
   A. mediate an electrogenic exchange of 2 luminal Na^+ for 1 cytosolic H^+.
   B. absorb Cl^- into the cell in exchange for HCO_3^- moving into the lumen.
   C. prevent a metabolic acidosis due to loss of HCl.
   D. mediate Na^+ movement out of the cell as Cl^- moves into the cell.
   E. mediate Na^+ movement into the cell via a Na^+ channel.

7. Certain tissues effect Cl^- secretion via a Cl^- channel (CFTR protein—cystic fibrosis transmembrane regulatory protein). Cholera toxin abnormally opens the channel leading to a loss of NaCl. Treatment for cholera and sports drinks for electrolyte replacement are fluids high in Na^+ and glucose. The presence of glucose enhances NaCl replenishment because:
   A. absorbing any nutrient causes Na^+ uptake.
   B. glucose prevents Na^+ excretion.
   C. Na^+ and glucose are transported in opposite directions.
   D. glucose is absorbed across intestinal epithelial cells via a Na^+-dependent cotransporter.
   E. glucose inhibits the Na^+/K^+ ATPase.

Questions 8 and 9: A young woman finds that every time she eats dairy products she feels highly uncomfortable. Her stomach becomes distended, and she has gas and, frequently, diarrhea. A friend suggested that she try yogurt to get calcium and she is able to tolerate that. These symptoms do not appear when she eats food other than dairy products. Like many adults, she is deficient in an enzyme required for carbohydrate digestion.

8. The most likely enzyme in which she is deficient is:
   A. α-amylase.
   B. β-galactosidase (lactase).
   C. α-glucosidase (maltase).
   D. sucrose-α-glucosidase (sucrase).
   E. α, α-trehalase.

9. Monosaccharides are absorbed from the intestine:
   A. by a Na^+-dependent cotransporter for glucose and galactose.
   B. by a Na^+-independent facilitated transport for fructose.
   C. by a Na^+-independent transporter (GLUT-2) across the contraluminal membrane.

D. against a concentration gradient if the transporter is Na^+ dependent.
E. all of the above.

Questions 10 and 11: A woman comes to the emergency room with severe abdominal pain in the right upper quadrant as well as severe pain in her back. The pain began several hours after she consumed a meal of fried chicken and cheese-coated french fries. The symptoms indicated "gallstones" and this was confirmed by ultrasound. While surgery might be necessary in the future, conservative treatment was tried first. She was instructed to limit fried foods and high-fat dairy products. She was also given chenodeoxycholate to take orally to try to dissolve the gallstones.

10. Cholesterol "stones":
 A. usually form during passage of bile through the hepatic bile duct.
 B. occur when the mixed phospholipid–bile acid micelles are very high in phospholipid.
 C. can be dissolved by excess bile acids because the bile acids help to solubilize in micelles the water-insoluble cholesterol.
 D. rarely occur because cholesterol is not a normal part of bile.
 E. are a necessary part of lipid digestion.

11. In the metabolism of bile acids:
 A. the liver synthesizes the cholic and deoxycholic acids, which are therefore considered primary bile acids.
 B. secondary bile acids are produced by conjugation of primary bile acids to glycine or taurine.
 C. 7-dehydroxylation of bile acids by intestinal bacteria produces secondary bile acids, which have similar detergent and physiological properties as primary bile acids.
 D. daily bile acid secretion by the liver is approximately equal to daily bile acid synthesis.
 E. conjugation reduces the polarity of bile acids, enhancing their ability to interact with lipids.

Questions 12 and 13: Hartnup disease is a genetic defect in an amino acid transport system. The specific defect is in the neutral amino acid transporter in both intestinal and renal epithelial cells. Clinical symptoms of the disease result from deficiencies of essential amino acids and nicotinamide (because of a deficiency specifically of tryptophan).

12. In addition to tryptophan, which of the following amino acids is likely to be deficient in Hartnup disease?
 A. aspartate
 B. leucine
 C. lysine
 D. proline
 E. all of the above

13. Hartnup disease patients are able to get some of the benefit of the protein they consume because:
 A. only the neutral amino acid carrier is defective.
 B. di- and tripeptides, normal products of protein digestion, are absorbed by a different carrier (PepT1).
 C. their endo- and exopeptidases are normal.
 D. all of the above.
 E. none of the above.

Problems

14. Using known endo- and exopeptidases, suggest a pathway for the complete degradation of the following peptide:

 His-Ser-Lys-Ala-Trp-Ile-Asp-Cys-Pro-Arg-His-His-Ala

15. At what point would the lack of coplipase inhibit the digestion of fat?

ANSWERS

1. **E** Zymogen granules contain inactive proenzymes or zymogens, which are not activated until after release from the cell (amylase from the pancreas and salivary glands is an exception).
2. **B** Stimulation of H_2 receptors of the stomach causes HCl secretion. A, D: Acetylcholine is the secretagogue. C: Gastrin is a secretagogue. E. This is stimulated by secretin.
3. **B** Only the contraluminal surface contains the Na^+/K^+ ATPase. All other activities are associated with the luminal surface.
4. **A** α-Amylase attacks hydrated starch more readily than unhydrated; heating hydrates the starch granules. B: Humans don't hydrolyze β-1,4 links. C–E: None of these could be accomplished by heat.
5. **E** The lipid-soluble vitamins must be dissolved in mixed micelles as a prerequisite for absorption. A: Micelles are of molecular dimensions and are equilibrium structures; emulsion droplets are much larger metastable structures. B: Micelle formation occurs only above the critical micellar concentration (CMC); below that concentration the components are in simple solution. C: Micelles are necessary so that the enzyme has access to the triacylglycerols. D: Micelles may consist of only one component, or they may be mixed.
6. **B** The proton gradient generated by an electrically neutral Na^+/H^+ exchange drives this. A: The Na^+/H^+ exchange via the expressed NHE3 transporter is electrically neutral. C: Constant loss of HCl leads to a metabolic alkalosis. D: The direction of Na^+ movement is into the cell with subsequent removal by the Na^+/K^+ ATPase. E: This occurs in the large intestine, not here.
7. **D** This is a cheaper and easier way to increase body electrolytes than intravenous fluid therapy. A: Only certain nutrients are absorbed by Na^+-dependent cotransporters. B: These are not related. C: Na^+-dependent cotransporters transfer the two substances in the same direction. E: Glucose has no effect on this enzyme.
8. **B** Dairy products contain lactose. Undigested lactose is fermented and produces the symptoms. A, C, D: Deficiency of any of these would cause problems with most carbohydrates. E: Trehalose is found in mushrooms, not dairy products.
9. **E** All of these play an important role in absorbing the monosaccharides from digestion. D: This is especially important for the uptake of most of dietary glucose, the most abundant monosaccharide.

10. **C** Stones occur when bile is supersaturated (> 1:1 ratio of cholesterol/phospholipid). The ingested bile salts increase the solubilization in micelles. A: Stones usually form in the gallbladder. B: The problem is too little phospholipid relative to cholesterol. D: Actually stones are relatively common. Cholesterol is a normal component of bile. E: Bile salts are necessary for lipid digestion but stones are not.
11. **C** The primary bile acids (cholic and chenodeoxycholic acids) are synthesized in the liver. In the intestine they may be reduced by bacteria to form the secondary bile acids, deoxycholate and lithocholate. D: Only a small fraction of the bile acid escapes reuptake; this must be replaced by synthesis. Both primary and secondary bile acids are reabsorbed and recirculated (enterohepatic circulation). E: Both are conjugated to glycine or taurine, increasing their polarity.
12. **B** Trp shares a carrier with tyr, phe, val, leu, ile, and met. A: Asp is acidic. C: Lysine is basic. D: Proline is an imino acid. All of these use separate carriers.
13. **D** The body has at least seven transporters for the various classes of amino acids (see Answer 12). It also has a carrier for di- and tripeptides. C: Since we don't ingest intact protein, the presence of the endo- and exopeptidases is essential.
14. Trypsin cleavage gives (a) His-Ser-Lys + (b) Ala-Trp-Ile-Met-Cys-Gly-Pro-Arg + (c) His-His-Ala. Further degradation of (a) is accomplished by elastase and dipeptidase. Further degradation of (b) would start with chymotrypsin and also use dipeptidases, tripeptidase, and carboxypeptidase B. To degrade (c) carboxypeptidase A and dipeptidase would be enough. The point is that several peptidases with varying specificities are required.
15. There would be some hydrolysis of the triacylglycerols in the stomach by gastric lipase. In the small intestine, the emulsion droplet would be stabilized by bile salts but pancreatic lipase, which is responsible for the bulk of the hydrolysis, is inactive in the absence of colipase. Handling of the other lipid material would not be affected.

26

PRINCIPLES OF NUTRITION I: MACRONUTRIENTS

Stephen G. Chaney

26.1 | OVERVIEW

Nutrition is best defined as the composition and quantity of food intake and the utilization of the food intake by living organisms. Since the process of food utilization is biochemical, the major thrust of the next two chapters is a discussion of basic nutritional concepts in biochemical terms. Simply understanding basic nutritional concepts is no longer sufficient. Nutrition attracts more than its share of controversy in our society, and a thorough understanding of nutrition almost demands an understanding of the issues behind these controversies. These chapters will explore the biochemical basis for some of the most important nutritional controversies.

Study of human nutrition can be divided into three areas: undernutrition, overnutrition, and ideal nutrition. The primary concern in this country is not with **undernutrition** because nutritional deficiency diseases are now quite rare. **Overnutrition** is a particularly serious problem in developed countries. Current estimates suggest that between 15% and 30% of the U. S. population is obese, and obesity is known to have a number of serious health consequences. Finally, there is increasing interest in the concept of ideal, or **optimal, nutrition.** This is a concept that has meaning only in an affluent society. Only when food supply becomes abundant enough so that deficiency diseases are a rarity does it become possible to consider long-range effects of nutrients on health. This is probably the most exciting area of nutrition today.

26.2 | ENERGY METABOLISM

Energy Content of Food Is Measured in Kilocalories

You should be well acquainted with the energy requirements of the body. Much of the food we eat is converted to ATP and other high-energy compounds, which are utilized to drive biosynthetic pathways, generate nerve impulses, and power muscle contraction. Energy content of foods is generally described in terms of **calories.** Technically speaking, this refers to **kilocalories of** heat energy released by combustion of that food in the body. Some nutritionists prefer the term **kilojoule** (a measure of mechanical energy), but since the American public is likely to be counting calories rather than joules in the foreseeable future, we will restrict ourselves to that term. Caloric values of protein, fat, carbohydrate, and alcohol are roughly 4, 9, 4, and 7 kcal g^{-1}, respectively. Given these data and the amount and composition of the food, it is simple to calculate the caloric content (input) of the foods we eat. Calculating caloric content of foods does not appear to be a major problem in this country. Millions of Americans are able to do it with ease. The problem lies in balancing caloric input with caloric output. Where do these calories go?

Energy Expenditure Is Influenced by Four Factors

Four principal factors affect individual energy expenditure: surface area (which is related to height and weight), age, sex, and activity level. (1) The effects of surface area are thought to be simply related to the rate of heat loss by the body—the greater the surface area, the greater the rate of heat loss. While it may seem surprising, a lean individual actually has a greater surface area, and thus a greater energy requirement, than an obese individual of the same weight. (2) Age may reflect two factors: growth and lean muscle mass. In infants and children more energy expenditure is required for rapid growth, which is reflected in a higher basal metabolic rate (rate of energy utilization in resting state). In adults (even lean adults), muscle tissue is gradually replaced by fat and water during the aging process, resulting in a 2% decrease in **basal metabolic rate (BMR)** per decade of adult life. (3) Women tend to have a lower BMR than men due to a smaller percentage of lean muscle mass and the effects of female hormones on metabolism. (4) The effect of activity levels on energy requirements is obvious. However, most overemphasize the immediate, as opposed to the long-term, effects of exercise. For example, one would need to jog for over an hour to burn up the calories present in one piece of apple pie.

The effect of a regular **exercise** program on energy expenditure can be quite beneficial. Regular exercise increases basal metabolic rate, allowing calories to burn up more rapidly 24 hours a day. A regular exercise program should be designed to increase lean muscle mass and should be repeated 3 to 5 days a week but need not be aerobic exercise to have an effect on basal metabolic rate. For an elderly or infirm individual, even daily walking may help to increase basal metabolic rate slightly.

Hormone levels are important also, since thyroxine, sex hormones, growth hormone, and, to a lesser extent, epinephrine and cortisol increase BMR. The effects of epinephrine and cortisol probably explain in part why severe stress and major trauma significantly increase energy requirements. Finally, energy intake itself has an inverse relationship to expenditure in that during periods of **starvation** or semi-starvation BMR can decrease up to 50%. This is of great survival value in cases of genuine starvation, but not much help to the person who wishes to lose weight on a calorie-restricted diet.

26.3 | PROTEIN METABOLISM

Dietary Protein Serves Many Roles Including Energy Production

Protein carries a certain mystique as a "body-building" food. While protein is an essential structural component of all cells, it is equally important for maintaining essential secretions such as digestive enzymes and peptide or protein hormones. Protein is also needed for synthesis of plasma proteins, which are essential for maintaining osmotic balance, transporting substances through the blood, and maintaining immunity. However, the average North American adult consumes far more protein than needed to carry out these essential functions. Excess protein is treated as a source of energy, with the glucogenic amino acids being converted to glucose and the ketogenic amino acids to fatty acids and keto acids. Both kinds of **amino acids** are eventually converted to triacylglycerol in adipose tissue if fat and carbohydrate supplies are already adequate to meet energy requirements. Thus for most of us the only body-building obtained from high-protein diets is in adipose tissue.

It has been popular to say that the body has no storage depot for protein, and thus adequate dietary protein must be supplied with every meal. However, in actuality, this is not quite accurate. While there is no separate class of "storage" protein, there is a certain percentage of body protein that undergoes a constant process of breakdown and resynthesis. In the fasting state the breakdown of this store of body protein is enhanced, and the resulting amino acids are utilized for glucose production, for the synthesis of nonprotein nitrogenous compounds, and for the synthesis of the essential secretory and plasma proteins mentioned above. Even in the fed state, some of these amino acids are utilized for energy production and as biosynthetic precursors. Thus the turnover of body protein is a normal process—and an essential feature of what is called nitrogen balance.

Nitrogen Balance Relates Intake to Excretion of Nitrogen

Nitrogen balance (Figure 26.1) is the relationship between **intake** of nitrogen (chiefly in the form of protein) and **excretion** of nitrogen (chiefly in the form of undigested protein in the feces and urea and ammonia in urine). A normal adult is in nitrogen equilibrium, with losses just balanced by intake. Negative nitrogen balance results from inadequate dietary intake of protein, since amino acids utilized for energy and biosynthetic reactions are not replaced. It also occurs in injury when there is net destruction of tissue and in major trauma or illness when the body's *(a)* Positive nitrogen balance (growth, pregnancy, lactation and recovery from metabolic stress).

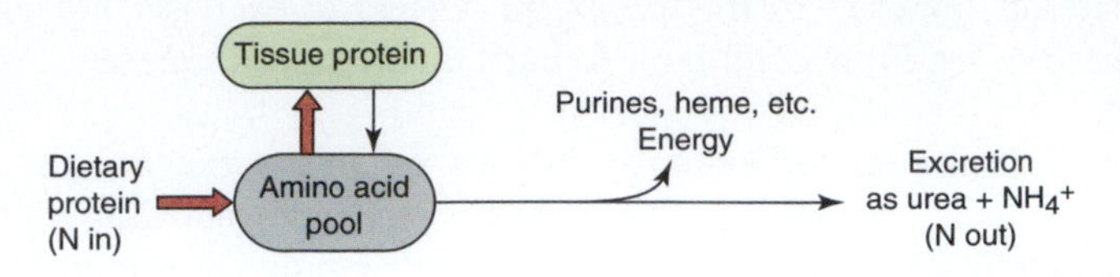

(b) Negative nitrogen balance (metabolic stress).

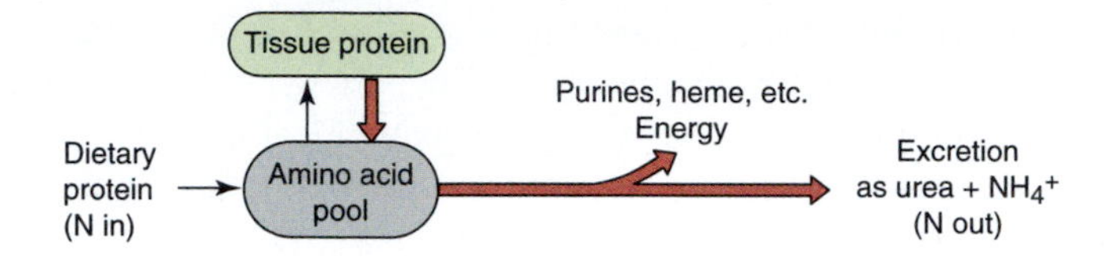

(c) Negative nitrogen balance (inadequate dietary protein).

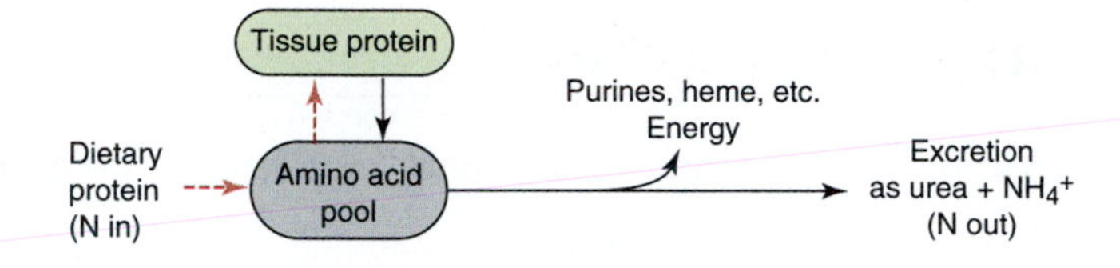

(d) Negative nitrogen balance (lack of an essential amino acid).

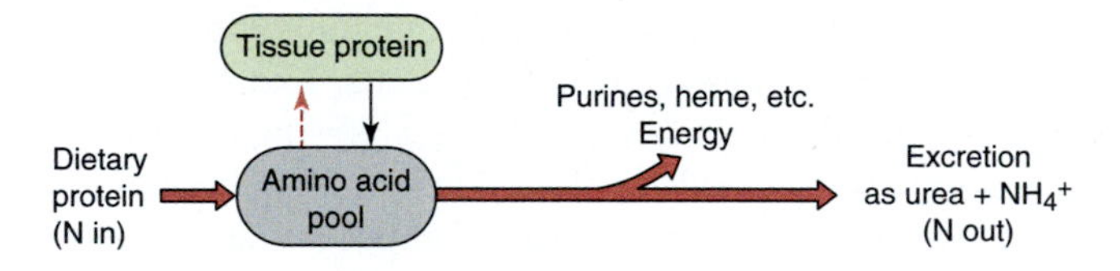

FIGURE 26.1
Factors affecting nitrogen balance.
Schematic representations of the metabolic interrelationship involved in determining nitrogen balance. Each figure represents the nitrogen balance resulting from a particular set of metabolic conditions. The dominant pathways in each situation are indicated by heavy red arrows.

adaptive response causes increased catabolism of body protein stores. Positive nitrogen balance occurs when there is a net increase in body protein stores, such as in growing children, pregnant women, or convalescing adults.

Essential Amino Acids Must Be Present in the Diet

In addition to the amount of protein in the diet, several other factors must be considered. One is the complement of **essential amino acids** present in the diet. Essential amino acids are amino acids that cannot be synthesized by the body (see p. 781). If just one of these essential amino acids is missing from the diet, the body cannot synthesize new protein to replace the protein lost due to normal turnover, and a negative nitrogen balance results (Figure 26.1). Obviously, the complement of essential amino acids in dietary protein determines how well it can be used by the body.

Most animal proteins generally contain all essential amino acids in about the quantities needed by the human body. Vegetable proteins, on the other hand, often lack one or more essential amino acids and may, in some cases, be more difficult to digest. Even so, vegetarian diets can provide adequate protein provided enough extra protein is consumed to provide sufficient quantities of the essential amino acids and/or two or more different proteins are consumed together, which complement each other in amino acid content. For example, if corn (which is deficient in lysine) is combined with legumes (deficient in methionine but rich in lysine), the efficiency

CLINICAL CORRELATION 26.1

Vegetarian Diets and Protein–Energy Requirements

One of the most important problems of a purely vegetarian diet (as opposed to a lacto-ovo vegetarian diet) is the difficulty in obtaining sufficient calories and protein. Potential caloric deficit results since the caloric densities of fruits and vegetables are much less than the meats they replace (30–50 cal per 100 g versus 150–300 cal per 100 g). The protein problem is threefold: (1) most plant products contain much less protein (1–2 g of protein per 100 g versus 15–20 g per 100 g); (2) most plant protein is of low biological value; and (3) some plant proteins are not completely digested. Actually, well-designed vegetarian diets usually provide enough calories and protein for the average adult. In fact, the reduced caloric intake may well be of benefit because strict vegetarians do tend to be lighter than their nonvegetarian counterparts.

However, whereas an adult male may require about 0.8 g of protein and 40 cal kg^{-1} of body weight, a young child may require 2–3 times that amount. Similarly, a pregnant woman needs an additional 10 g of protein and 300 cal day^{-1} and a lactating woman an extra 15 g of protein and 500 cal. Thus both young children and pregnant and lactating women run a risk of protein–energy malnutrition. Children of vegetarian mothers generally have a lower birth weight than children of mothers consuming a mixed diet. Similarly, vegetarian children generally have a slower rate of growth through the first 5 years, but generally catch up by age of 10.

It is possible to provide sufficient calories and protein even for these high-risk groups provided the diet is adequately planned. Three principles should be followed to design a calorie–protein-sufficient vegetarian diet for young children: (1) whenever possible, include eggs and milk in the diet; they are both excellent sources of calories and high-quality protein; (2) include liberal amounts of those vegetable foods with high-caloric density in the diet, including nuts, grains, dried beans, and dried fruits; and (3) include liberal amounts of high-protein vegetable foods that have complementary amino acid compositions. It used to be thought that these complementary proteins must be present in the same meal. Recent animal studies, however, suggest that a meal low in (but not devoid of) an essential amino acid may be supplemented by adding the limiting amino acid at a subsequent meal.

First International Congress on Vegetarian Nutrition. *Proc. Am. J. Clin. Nutr.* 48 (Suppl. 1):707, 1988; and Saunders, T. A. B. Vegetarian diets and children. *Pediatr. Nutr.* 42:955, 1995.

of utilization for the combination of the two vegetable proteins approaches that of animal protein. The adequacy of vegetarian diets with respect to protein and calories is discussed more fully in Clinical Correlation 26.1, and the need for high-quality protein in the low-protein diets used for treatment of renal disease is discussed in Clinical Correlation 26.2.

CLINICAL CORRELATION 26.2

Low-Protein Diets and Renal Disease

Chronic renal failure is characterized by accumulation of the end products of protein catabolism, mainly urea. Some degree of dietary protein restriction is usually necessary because these toxic end products are responsible for many of the symptoms associated with renal failure. The amount of protein restriction is dependent on the severity of the disease. It is easy to maintain patients in nitrogen balance for prolonged periods on diets containing as little as 40 g protein day^{-1} if the diet is sufficient in calories. Diets containing less than 40 g day^{-1} pose problems. Protein turnover continues and a balance must be found between providing enough protein to avoid negative nitrogen balance and little enough to avoid buildup of waste products.

The strategy employed in such diets is twofold: (1) provide a minimum of protein, primarily of high biological value, and (2) provide the bulk of the daily calories as carbohydrates and fats. The goal is to provide just enough essential amino acids to maintain positive nitrogen balance. In turn, the body should be able to synthesize the nonessential amino acids from other nitrogen-containing metabolites. Enough carbohydrate and fat is provided so that essentially all dietary protein can be spared from energy metabolism. With this type of diet, it is possible to maintain a patient on 20 g of protein per day for considerable periods. Because of the difficulty in maintaining nitrogen equilibrium at such low-protein intakes, the patient's protein status should be monitored. This can be done by measuring parameters such as serum albumin and transferrin.

Unfortunately, such diets are extremely monotonous and difficult to follow. A typical 20-g protein diet consists of: (1) one egg plus $\frac{3}{4}$ cup milk or 1 additional egg or 1 oz of meat; (2) one-half pound of deglutenized (low protein) wheat bread; all other breads and cereals must be avoided—this includes almost all baked goods; (3) a limited amount of low-protein, low-potassium fruits and vegetables; (4) sugars and fats to make up the rest of the needed calories; however, cakes, pies, and cookies should be avoided.

The palatability of these diets can be improved considerably by starting with a vegan diet and supplementing it with a mixture of essential amino acids and keto acid analogs of the essential amino acids. This technique will help preserve renal function and allow a somewhat greater variety of foods.

Goodship, T. H. J. and Mitch, W. E. Nutritional approaches to preserving renal function. *Adv. Intern. Med.* 33:377, 1988; Dwyer, J. Vegetarian diets for treating nephrotic syndrome. *Nutr. Rev.* 51:44, 1993; and Barsotti, G., Morrell, E., Cupisti, A., Bertoncini, P., and Giovannetti, S. A special supplemented "vegan" diet for nephrotic patients. *Am. J Nephrol.* 11:380, 1991.

Protein Sparing Is Related to Dietary Content of Carbohydrate and Fat

Another factor that determines protein requirement is dietary intake of fat and carbohydrate. If these components are present in insufficient quantities, some dietary protein must be used for energy generation and becomes unavailable for building and replacing tissue. Thus, as energy (calorie) content of the diet from carbohydrate and fat increases, the need for protein decreases. This is referred to as **protein sparing.** Carbohydrate is somewhat more efficient at protein sparing than fat—presumably because carbohydrate can be used as an energy source by almost all tissues, whereas fat cannot.

Normal Adult Protein Requirements Depend on Diet

Assuming adequate calorie intake and a 75% efficiency of utilization, which is typical of mixed protein in the average American diet, the **recommended protein intake** is 0.8 g/kg body wt day^{-1}. This amounts to about 58 g protein day^{-1} for a 72-kg (160-lb) man and about 44 g day^{-1} for a 55-kg (120-lb) woman. These recommendations would need to be increased on a vegetarian diet if overall efficiency of utilization were less than 75%.

Protein Requirements Are Increased During Growth and Recovery Following an Illness

Because dietary protein is essential for synthesis of new body tissue, as well as for maintenance and repair, the need for protein increases markedly during periods of rapid growth. Such growth occurs during pregnancy, infancy, childhood, and adolescence.

Once growth requirements have been considered, age does not seem to have much effect on protein requirements. If anything, the protein requirement may decrease slightly with age. However, older people need and generally consume fewer calories, so high-quality protein should provide a larger percentage of their total calories. Furthermore, some older people may have special protein requirements due to malabsorption problems.

Illness, major trauma, and surgery all cause a major **catabolic response.** Both energy and protein needs are very large, and the body responds by increasing production of glucagon, glucocorticoids, and epinephrine. In these situations breakdown of body protein is greatly accelerated and a negative nitrogen balance results unless protein intake is increased (Figure 26.1). Although the increased protein requirement is of little significance in short-term illness, it can be vitally important in the recovery of hospitalized patients as discussed in the next section (see also Clin. Corr. 26.3).

26.4 | PROTEIN–ENERGY MALNUTRITION

The most common form of malnutrition in the world is **protein–energy malnutrition** (PEM). In developing countries inadequate intake of protein and energy is all too common, especially in infants and young children. While the symptoms of protein–energy insufficiency vary widely from case to case, it is common to classify most cases as either **marasmus** or **kwashiorkor.** Marasmus is defined as inadequate intake of both protein and energy. Kwashiorkor is defined as inadequate intake of protein with adequate energy intake. Often the diets associated with marasmus and kwashiorkor are similar, with the kwashiorkor being precipitated by conditions of increased protein demand such as infection. The marasmic infant will have a thin, wasted appearance and will be small for his/her age. If PEM continues long enough the child will be permanently stunted in both physical and mental development. In kwashiorkor the patient will often have a deceptively plump appearance due to edema. Other telltale symptoms associated with kwashiorkor are

CLINICAL CORRELATION 26.3

Providing Adequate Protein and Calories for Hospitalized Patients

The normal metabolic response to infection, trauma, and surgery is a complex and carefully balanced catabolic state. As discussed in the text, epinephrine, glucagon, and cortisol are released, greatly accelerating the rates of lipolysis, proteolysis, and gluconeogenesis. The net result is an increased supply of fatty acids and glucose to meet the increased energy demands of such major stress. The high serum glucose results in elevation of circulating insulin levels, which is more than counterbalanced by increased levels of epinephrine and other hormones. Skeletal muscle uses very little of the serum glucose, but relies on free fatty acids and its own catabolized protein as a primary source of energy. It also continues to export amino acids, primarily alanine, for use elsewhere in the body, resulting in a very rapid depletion of body protein stores.

A highly catabolic hospitalized patient may require 35–45 kcal kg^{-1} day^{-1} and 2–3 g protein kg^{-1} day^{-1}. A patient with severe burns may require even more. A physician has a number of options available to provide this postoperative patient with sufficient calories and protein to ensure optimal recovery. When the patient is unable to ingest enough food, it may be adequate to supplement the diet with high-calorie–high-protein preparations, which are usually mixtures of homogenized cornstarch, egg, milk protein, and flavorings. When the patient is unable to ingest solid food or to digest complex mixtures of foods adequately, elemental diets are usually administered via a nasogastric tube. Elemental diets consist of small peptides or purified amino acids, glucose and dextrins, some fat, vitamins, and electrolytes. These diets are sometimes sufficient to meet most of the short-term caloric and protein needs of a moderately catabolic patient.

When a patient is severely catabolic or unable to digest and absorb foods normally, parenteral (intravenous) nutrition is necessary. The least invasive method is to use a peripheral, slow-flow vein in a manner similar to any other i.v. infusion. The main limitation of this method is hypertonicity. However, a solution of 5% glucose and 4.25% purified amino acids can be used safely. This solution will usually provide enough protein to maintain positive nitrogen balance, but will rarely provide enough calories for long-term maintenance of a catabolic patient.

The most aggressive nutritional therapy is total parenteral nutrition. Usually an indwelling catheter is inserted into a large fast-flow vessel such as the superior vena cava, so that the very hypertonic infusion fluid can rapidly be diluted. This allows solutions of up to 60% glucose and 4.25% amino acids to be used, providing sufficient protein and most of the calories for long-term maintenance. Intravenous lipid infusion is often added to boost calories and provide essential fatty acids. All of these methods can prevent or minimize the negative nitrogen balance associated with surgery and trauma. The actual choice of method depends on the patient's condition. As a general rule it is preferable to use the least invasive technique.

Streat, S. J. and Hill, G. L. Nutritional support in the management of critically ill patients in surgical intensive care. *World J. Surg.* 11:194, 1987; and The Veterans Affairs Total Parenteral Nutrition Cooperative Study Group. Perioperative total parenteral nutrition in surgical patients. *N. Engl. J. Med.* 325:25, 1991.

dry, brittle hair, diarrhea, dermatitis of various forms, and retarded growth. The most devastating result of both marasmus and kwashiorkor is reduced ability of the afflicted individuals to fight off infection. They have a reduced number of T lymphocytes (and thus diminished cell-mediated immune response) as well as defects in the generation of phagocytic cells and production of immunoglobulins, interferon, and other components of the immune system. Many of these individuals die from secondary infections, rather than from starvation.

The most common form of PEM seen in the United States occurs in the hospital setting. A typical course of events is as follows. The patient has not been eating well for several weeks or months prior to entering the hospital due to chronic or debilitating illness. He/she enters the hospital with major trauma, severe infection, or for major surgery, all of which cause a large negative nitrogen balance. This is often compounded by difficulties in feeding the patient or by the necessity of fasting in preparation for surgery or diagnostic tests. The net result is PEM as reflected by low levels of serum albumin and other serum proteins or by decreased cellular immunity tests. Hospitalized patients with demonstrable PEM have delayed wound healing, decreased resistance to infection, increased mortality, and increased length of hospitalization. Most major hospitals have programs that monitor the nutritional status of their patients and intervene where necessary to maintain a positive nitrogen and energy balance (see Clin. Corr. 26.3).

26.5 | EXCESS PROTEIN–ENERGY INTAKE

Much has been said in recent years about the large amount of protein that the average American consumes. Certainly most consume far more than needed to maintain nitrogen balance. An average American currently consumes about 99 g of protein, 68% from animal sources. However, a healthy adult can consume this amount of protein with no apparent harm. Concern has been raised about possible effects of high-protein intake on calcium requirements. Some studies suggest that high-protein intake increases urinary loss of calcium and may accelerate bone demineralization associated with aging. However, this issue is far from settled.

Obesity Has Dietary and Genetic Components

Perhaps the more serious and frequent nutritional problem in this country is excessive energy consumption. In fact, **obesity** is the most frequent nutritional disorder in the United States. It would, however, be unfair to label obesity as simply a problem of excess consumption. Overeating plays an important role in many individuals, as does inadequate exercise, but there is also a strong genetic component as well. While the biochemical mechanisms for this genetic predisposition are unclear, an obesity gene in mice and humans that appears to regulate obesity through effects on both appetite and deposition of fat has been identified. This gene produces a protein called leptin. Initially, it was thought that leptin might be a useful therapy for obesity. However, it is now apparent that most obese individuals already overproduce leptin and that it has deleterious as well as beneficial effects (see below). Detailed characterization of this and other genes that predispose to obesity in animals may yield valuable clues to the causes and treatment of obesity in humans.

Metabolic Consequences of Obesity Have Significant Health Implications

A full discussion of the treatment of obesity is clearly beyond the scope of this chapter, but it is worthwhile to consider some of the metabolic consequences of obesity. One striking clinical feature of overweight individuals is a marked elevation of serum free fatty acids, cholesterol, and triacylglycerols irrespective of the dietary intake of fat. Why is this? Obesity is obviously associated with an increased number and/or size of **adipose tissue cells.** These cells overproduce factors such as **leptin** and **tumor necrosis factor α (TNF-α)**, which cause cellular resistance to **insulin** by interfering with **autophosphorylation** of the **insulin receptor (IR)** and the subsequent phosphorylation of the **insulin receptor substrate 1 (IRS-1)** (see p. 944). Initially, the pancreas maintains glycemic control by overproducing insulin. Thus many obese individuals with apparently normal blood glucose control have a syndrome characterized by insulin resistance of the peripheral tissue and high concentrations of insulin in the circulation. This hyperinsulinemia appears to stimulate the sympathetic nervous system, leading to sodium and water retention and vasoconstriction, which increase blood pressure.

The insulin resistance in adipose tissue results in increased activity of the hormone-sensitive lipase, which along with the increased mass of adipose tissue is probably sufficient to explain the increase in circulating free fatty acids. These excess fatty acids are carried to the liver and converted to triacylglycerol and cholesterol. Excess triacylglycerol and cholesterol are released as very low density lipoprotein particles, leading to higher circulating levels of both triacylglycerol and cholesterol. For reasons that are not entirely clear this also results in a decrease in high density lipoprotein particles.

Eventually, the capacity of the pancreas to overproduce insulin declines, which leads to higher fasting blood sugar levels and decreased glucose tolerance. Fully 80% of adult-onset (type II) diabetics are overweight. Because of these metabolic changes, obesity is a primary risk factor in coronary heart disease, hypertension, and diabetes mellitus. This is nutritionally significant because all of these metabolic changes are reversible. Quite often reduction to ideal body weight is the single most important aim of nutritional therapy. Furthermore, when the individual is at ideal body weight, the composition of the diet becomes a less important consideration in maintaining normal serum lipid and glucose concentrations.

Discussion of weight reduction regimens should include mention of one other metabolic consequence of obesity. As discussed above, obesity can lead to increased retention of both sodium and water. As the fat stores are metabolized, they produce water (which is denser than the fat), and the water may largely be retained. In fact, some individuals may actually observe short-term weight gain on certain diets, even though the diet is working perfectly well in terms of breaking down their adipose tissue. This metabolic fact of life can be psychologically devastating to dieters, who expect quick results for all their sacrifice.

26.6 | CARBOHYDRATES

The chief metabolic role of dietary carbohydrates is for energy production. Any carbohydrate in excess of that needed for energy is converted to glycogen and triacylglycerol for long-term storage. The body can adapt to a wide range of dietary carbohydrate levels. Diets high in carbohydrate result in higher steady-state levels of glucokinase and some of the enzymes involved in the pentose phosphate pathway and triacylglycerol synthesis. Diets low in carbohydrate result in higher steady-state levels of some of the enzymes involved in gluconeogenesis, fatty acid oxidation, and amino acid catabolism. **Glycogen stores** are also affected by the carbohydrate content of the diet (see Clin. Corr. 26.4).

The most common nutritional problems involving carbohydrates are those of carbohydrate intolerance. The most common form of **carbohydrate intolerance** is diabetes mellitus, caused either by subnormal insulin production or lack of insulin receptors. This causes an intolerance to glucose and sugars that are readily converted to glucose. Dietary treatment of diabetes is discussed in Clinical Correlation 26.5.

CLINICAL CORRELATION 26.4
Carbohydrate Loading and Athletic Endurance

The practice of carbohydrate loading dates back to observations made in the early 1960s that endurance during vigorous exercise was limited primarily by muscle glycogen stores. Of course, glycogen stores are not the sole energy source for muscle. Free fatty acids are present in the blood during vigorous exercise and are utilized by muscle along with the muscle glycogen stores. Once the glycogen stores have been exhausted, however, muscle cannot rely entirely on free fatty acids without tiring rapidly. This is probably related to the fact that muscle becomes increasingly hypoxic during vigorous exercise. While glycogen stores are utilized equally well aerobically or anaerobically, fatty acids can only be utilized aerobically. Under anaerobic conditions, fatty acids cannot provide ATP rapidly enough to serve as the sole energy source.

Thus the practice of carbohydrate loading to increase glycogen stores was devised for track and other endurance athletes. Originally, it was thought that it would be necessary to trick the body into increasing glycogen stores. The original carbohydrate loading regimen consisted of a 3–4-day period of heavy exercise while on a low-carbohydrate diet, followed by 1–2 days of light exercise while on a high-carbohydrate diet. The initial low-carbohydrate–high-energy demand period caused a depletion of muscle glycogen stores. Apparently, the subsequent change to a high-carbohydrate diet resulted in a slight rebound effect, with the production of higher than normal levels of insulin and growth hormone. Under these conditions glycogen storage was favored and glycogen stores reached almost twice the normal amounts. This practice did increase endurance significantly. In one study, test subjects on a high-fat and high-protein diet had less than 1.6 g of glycogen per 100 g of muscle and could perform a standardized workload for only 60 min. When the same subjects then consumed a high-carbohydrate diet for 3 days, their glycogen stores increased to 4 g per 100 g of muscle and the same workload could be performed for up to 4 h.

While the technique clearly worked, the athletes often felt lethargic and irritable during the low-carbohydrate phase of the regimen, and the high-fat diet ran counter to current health recommendations. Fortunately, recent studies indicate that regular consumption of a high-complex-carbohydrate low-fat diet during training increases glycogen stores without the need for tricking the body with sudden dietary changes. Current recommendations are for endurance athletes to consume a high-carbohydrate diet (with emphasis on complex carbohydrates) during training. Then carbohydrate intake is increased further (to 70% of calories) and exercise tapered off during the 2–3 days just prior to an athletic event. This increases muscle glycogen stores to levels comparable to the previously described carbohydrate loading regimen.

Conlee, R. K. Muscle glycogen and exercise endurance: a twenty-year perspective. *Exer. Sports Sci. Rev.* 15:1, 1987; Ivey, J. L., Katz, A. L., Cutler, C. L., Sherman, W. M., and Cayle, E. F. Muscle glycogen synthesis after exercise: effect of time of carbohydrate ingestion. *J. Appl. Physiol.* 64:1480, 1988; and Probart, C. K., Bird, P. J., and Parker, K. A. Diet and athletic performance. *Med. Clin. North Am.* 77:757, 1993.

CLINICAL CORRELATION 26.5

High-Carbohydrate Versus High-Fat Diet for Diabetics

For years the American Diabetes Association has recommended diets that were low in fat and high in complex carbohydrates and fiber for diabetics. The logic of such a recommendation seemed to be inescapable. Diabetics are prone to hyperlipidemia with attendant risk of heart disease, and low-fat diets seemed likely to reduce risk of hyperlipidemia and heart disease. In addition, numerous clinical studies had suggested that the high-fiber content of these diets improved control of blood sugar. This recommendation has proved to be controversial. An understanding of the controversies involved illustrates the difficulties in making dietary recommendations for population groups rather than individuals. In the first place, it is very difficult to make any major changes in dietary composition without changing other components of the diet. In fact, most of the clinical trials of the high-carbohydrate–high-fiber diets have resulted in significant weight reduction, either by design or because of the lower caloric density of the diet. Since weight reduction improves diabetic control, it is not entirely clear whether the improvements seen in the treated group were due to the change in diet composition *per se* or because of the weight loss. Second, there is significant individual variation in how diabetics respond to these diets. Many diabetic patients appear to show poorer control (as evidenced by higher blood glucose levels, elevated VLDL and/or LDL levels, and reduced HDL levels) on the high-carbohydrate–high-fiber diets than they do on diets high in monounsaturated fatty acids. However, diets high in monounsaturated fatty acids tend to have higher caloric density and are inappropriate for overweight individuals with type 2 (non-insulin-dependent) diabetes. Thus a single diet may not be equally appropriate for all diabetics. Even the "glycemic index" concept (Table 26.2) may also turn out to be difficult to apply to the diabetic population as a whole, because of individual variation. In 1994 the American Diabetes Association abandoned the concept of a single diabetic diet. Instead, their recommendations focus on achievement of glucose, lipid, and blood pressure goals, with weight reduction and dietary recommendations based on individual preferences and what works best to achieve metabolic control in that individual.

Jenkins, D. J. A., Wolener, T. M. S., Jenkins, A. L., and Taylor, R. H. Dietary fiber, carbohydrate metabolism and diabetes. *Mol. Aspects Med.* 9:97, 1987; Garg, A., Grundy, S. M., and Unger, R. H. Comparison of the effects of high and low carbohydrate diets on plasma lipoproteins and insulin sensitivity in patients with mild NIDDM. *Diabetes* 41;1278, 1992; American Diabetes Association. Nutritional recommendations and principles for people with diabetes. *Diabetes Care* 17:519, 1994; and Nuttall, F. Q. and Chasuk, R. M. Nutrition and the management of type 2 diabetes. *J. Fam. Pract.* 47:S45, 1998.

Lactase insufficiency (see p. 1103) is also a common disorder of carbohydrate metabolism, affecting over 30 million people in the United States alone. It is most prevalent among blacks, Asians, and Hispanics. Without the enzyme lactase, dietary lactose is not significantly hydrolyzed or absorbed. It remains in the intestine where it acts osmotically to draw water into the gut and serves as a substrate for conversion to lactic acid and CO_2, by intestinal bacteria. The end result is bloating, flatulence, and diarrhea—all of which can be avoided simply by eliminating milk and milk products from the diet.

26.7 | FATS

Triacylglycerols, or fats, are directly utilized by many tissues of the body as an energy source and, as phospholipids, are an important constituent of membranes. Excess fat in the diet can only be stored as triacylglycerol. As with carbohydrate, the body adapts to a wide range of fat intakes. However, problems develop at the extremes (either high or low) of fat consumption. At the low end, **essential fatty acid (EFA) deficiency** may become a problem. The fatty acids linoleic and linolenic cannot be made by the body and thus are essential components of the diet. These EFAs are needed for maintaining the function and integrity of membrane structure, for fat metabolism and transport, and for synthesis of **prostaglandins** and related compounds. The most characteristic symptom of essential fatty acid deficiency is a scaly dermatitis. EFA deficiency is very rare in the United States, occurring primarily in low-birth-weight infants fed artificial formulas lacking EFAs and in hospitalized patients maintained on total parenteral nutrition for long periods of time. At the high end of the scale, there is concern that excess dietary fat causes elevation of serum lipids and thus an increased risk of heart disease. Recent studies suggest that high-fat intakes are associated with increased risk of colon, breast, and prostate cancer, but it is not yet certain whether the cancer risk is associated with fat intake *per se,* or with the excess calories associated with a high-fat diet. Animal studies suggest that polyunsaturated fatty acids of the ω-6 series may be more tumorigenic than other unsaturated fatty acids. The reason for this is not known, but it has been suggested that prostaglandins derived from the ω-6 fatty acids may stimulate tumor progression.

26.8 | FIBER

Dietary fiber comprises those components of food that cannot be broken down by human digestive enzymes. It is incorrect, however, to assume that fiber is indigestible since some fibers are, in fact, at least partially broken down by intestinal bacteria. Knowledge of the role of fiber in human metabolism has expanded significantly in the past decade. Our current understanding of the metabolic roles of dietary fiber is based on three important observations: (1) there are several different types of dietary fiber, (2) they each have different chemical and physical properties, and (3) they each have different effects on human metabolism, which can be understood, in part, from their unique properties.

The major types of fiber and their properties are summarized in Table 26.1. **Cellulose** and most **hemicelluloses** increase stool bulk and decrease transit time and are associated with the effects of fiber on regularity. They decrease intracolonic pressure and appear to play a beneficial role with respect to diverticular diseases. By diluting out potential carcinogens and speeding their transit through the colon, they may also play a role in reducing the risk of colon cancer. **Lignins** have a slightly different role. In addition to their bulk-enhancing properties, they adsorb organic substances such as cholesterol so as to lower plasma cholesterol concentration. **Mucilaginous fibers,** such as **pectin** and **gums,** tend to form viscous gels in the stomach and intestine and slow the rate of gastric emptying, thus slowing the rate of absorption of many nutrients. The most important clinical role of these fibers is to slow the rate at which carbohydrates are digested and absorbed. Thus both the rise in blood sugar and the subsequent rise in insulin levels are significantly decreased if these fibers are ingested along with carbohydrate-containing foods. **Water-soluble fibers** (pectins, gums, some hemicelluloses, and storage polysaccharides) also help to lower serum cholesterol levels in most people. Whether this is due to their effect on insulin levels (insulin stimulates cholesterol synthesis and export) or to other metabolic effects (perhaps caused by end products of partial bacterial digestion) is not known. Vegetables, wheat, and most grain fibers are the best sources of the water-insoluble cellulose, hemicellulose, and lignin. Fruits, oats, and

TABLE 26.1 Major Types of Fiber and Their Properties

Type of Fiber	*Major Source in Diet*	*Chemical Properties*	*Physiological Effects*
Cellulose	Unrefined cereals	Nondigestible	Increases stool bulk
	Bran	Water insoluble	Decreases intestinal transit time
	Whole wheat	Absorbs water	Decreases intracolonic pressure
Hemicellulose	Unrefined cereals	Partially digestible	Increases stool bulk
	Some fruits and vegetables	Usually water insoluble	Decreases intestinal transit time
	Whole wheat	Absorbs water	Decreases intracolonic pressure
Lignin	Woody parts of vegetables	Nondigestible	Increases stool bulk
		Water insoluble	Bind cholesterol
		Absorbs organic substances	Bind carcinogens
Pectin	Fruits	Digestible	Decreases rate of gastric emptying
		Water soluble	Decreases rate of sugar uptake
		Mucilaginous	Decreases serum cholesterol
Gums	Dried beans	Digestible	Decreases rate of gastric emptying
	Oats	Water soluble	Decreases rate of sugar uptake
		Mucilaginous	Decreases serum cholesterol

legumes are the best source of the water-soluble fibers. Obviously, a balanced diet should include food sources of both soluble and insoluble fiber.

26.9 | COMPOSITION OF MACRONUTRIENTS IN THE DIET

From the foregoing discussion it is apparent that there are relatively few instances of macronutrient deficiencies in the American diet. Thus much of the interest in recent years has focused on whether there is an ideal diet composition consistent with good health. It would be easy to pass off such discussions as purely academic, yet our understanding of these issues could well be vital. Heart disease, stroke, and cancer kill many Americans each year, and if some experts are even partially correct, many of these deaths could be preventable with prudent diet.

Composition of the Diet Affects Serum Cholesterol

With respect to heart disease, the current discussion centers around two key issues: (1) Can serum cholesterol and triacylglycerol concentration be controlled by diet? (2) Does lowering serum cholesterol and triacylglycerol levels protect against heart disease? The controversies centered around dietary control of cholesterol levels illustrate perfectly the trap one falls into by trying to look too closely at each individual component of the diet instead of the diet as a whole. For example, there are at least four dietary components that can be identified as having an effect on serum cholesterol: cholesterol itself, **polyunsaturated fatty acids (PUFAs)**, **saturated fatty acids (SFAs)**, and **fiber.** It would seem that the more cholesterol one eats, the higher the serum cholesterol should be. However, cholesterol synthesis is tightly regulated and decreases in dietary cholesterol have relatively little effect on serum cholesterol levels (see p. 750). One can obtain a more significant reduction in cholesterol and triacylglycerol levels by increasing the ratio of PUFA/SFA in the diet. Finally, some plant fibers, especially the water-soluble fibers, appear to decrease cholesterol levels significantly.

While the effects of various lipids in the diet can be dramatic, the biochemistry of their action is still uncertain. Saturated fats inhibit receptor-mediated uptake of LDLs, but the mechanism is complex. Palmitic acid (saturated, C16) raises serum cholesterol levels while stearic acid (saturated, C18) is neutral. Polyunsaturated fatty acids lower both LDL and HDL cholesterol levels, while oleic acid (monounsaturated, C18) appears to lower LDL without affecting HDL levels. Furthermore, the ω-3 and ω-6 polyunsaturated fatty acids have slightly different effects on lipid profiles (see Clin. Corr. 26.6). However, these mechanistic complexities do not significantly affect dietary recommendations. Most foods high in saturated fats contain both palmitic and stearic acid and are atherogenic. Since oleic acid lowers LDL levels, olive oil and, possibly, peanut oil may be considered as beneficial as polyunsaturated oils.

There is very little disagreement with respect to these data. The question is: What can be done with the information? Much of the disagreement arises from the tendency to look at each dietary factor in isolation. For example, it is debatable whether it is worthwhile placing a patient on a highly restrictive 300-mg cholesterol diet (1 egg contains about 213 mg of cholesterol) if his serum cholesterol is lowered by only 5–10%. Likewise, changing the **PUFA/SFA ratio** from 0.3 (the current value) to 1.0 would either require a radical change in the diet by elimination of foods containing saturated fat (largely meats and fats) or an addition of large amounts of rather unpalatable polyunsaturated fats to the diet. For many Americans this would be unrealistic. Fiber is another good example. One could expect, at the most, a 5% decrease in serum cholesterol by adding a reasonable amount of fiber to the diet. (Very few people would eat the 10 apples per day needed to lower serum cholesterol by 15%.) Are we to conclude then that any di-

CLINICAL CORRELATION 26.6

Polyunsaturated Fatty Acids and Risk Factors for Heart Disease

Recent studies confirm that reduction of elevated serum cholesterol levels can reduce the risk of heart disease. They have rekindled interest in the effects of diet on serum cholesterol levels and other risk factors for heart disease. One of the most important dietary factors regulating serum cholesterol levels is the ratio of polyunsaturated fats (PUFAs) to saturated fats (SFAs) in the diet. One of the most interesting recent developments is the discovery that different types of polyunsaturated fatty acids have different effects on lipid metabolism and on other risk factors for heart disease. As discussed on page 698, there are two families of polyunsaturated essential fatty acids—the ω-6, or linoleic family, and the ω-3, or linolenic family. Recent clinical studies have shown that the ω-6 PUFAs (chief dietary source is linoleic acid from plants and vegetable oils) primarily decrease serum cholesterol levels, with only modest effects on serum triacylglycerol levels. The ω-3 PUFAs (chief dietary source is eicosapentaenoic acid from certain ocean fish and fish oils) cause modest increases in serum cholesterol levels and significantly lower serum triacylglycerol levels. The mechanisms behind these effects on serum lipid levels are unknown.

The ω-3 PUFAs have yet another unique physiological effect that may decrease the risk of heart disease: they decrease platelet aggregation. The mechanism of this effect is a little clearer. Arachidonic acid (ω-6 family) is known to be a precursor of thromboxane A_2 (TXA_2), which is a potent proaggregating agent, and prostaglandin I_2 (PGI_2), which is a weak antiaggregating agent (see p. 766). The ω-3 PUFAs are thought to act by one of two mechanisms. (1) Eicosapentaenoic acid (ω-3 family) may be converted to thromboxane A_3 (TXA_3), which is only weakly proaggregating, and prostaglandin I_3 (PGI_3), which is strongly antiaggregating. Thus the balance between proaggregation and antiaggregation would be shifted toward a more antiaggregating condition as the ω-3 PUFAs displace ω-6 PUFAs as a source of precursors to the thromboxanes and prostaglandins. (2) The ω-3 PUFAs may also act by simply inhibiting the conversion of arachidonic acid to TXA_2. The unique potential of eicosapentaenoic acid and other ω-3 PUFAs in reducing the risk of heart disease is being tested in numerous clinical trials.

Holub, B. J. Dietary fish oils containing eicosapentaenoic acid and the prevention of atherosclerosis and thrombosis. *Can. Med. Assoc. J.* 139:377, 1988; Simopoulos, A. P. Omega-3 fatty acids in health and disease and in growth and development. *Am. J. Clin. Nutr.* 54:438, 1991; and Gapinski, J. P., Van Ruiswyk, J. V., Heudebert, G. R., and Schectman, G. S. Preventing restenosis with fish-oils following coronary angioplasty. A meta analysis. *Arch. Intern. Med.* 153:1595, 1993.

etary means of controlling serum cholesterol levels is useless? Only if each element of the diet is examined in isolation. For example, recent studies have shown that vegetarians, who have lower cholesterol intakes plus higher PUFA/SFA ratios and higher fiber intakes, may average 25–30% lower cholesterol levels than their nonvegetarian counterparts. Perhaps, more to the point, diet modifications of the type acceptable to the average American have been shown to cause a 10–15% decrease in cholesterol levels in long-term studies. A 7-year clinical trial sponsored by the National Institutes of Health has proved conclusively that lowering serum cholesterol levels reduces the risk of heart disease in men. While a consensus has been reached that lowering serum cholesterol levels reduces the risk of heart disease, it is important to remember that serum cholesterol is just one of many cardiovascular risk factors.

Effects of Refined Carbohydrate in the Diet Are Complex

Much of the nutritional dispute in the area of carbohydrate intake centers around the amount of refined carbohydrate in the diet. Until recently, simple sugars (primarily sucrose) were blamed for almost every ill from tooth decay to heart disease and diabetes. In the case of tooth decay, these assertions were clearly correct. In the case of heart disease, however, the linkage is more obscure (see Clin. Corr. 26.7). The situation with respect to diabetes is probably even less direct. Whereas restriction of simple sugars is often desirable in patients who already have diabetes, recent studies show less than expected correlation between the type of carbohydrate ingested and the subsequent rise in serum glucose levels (Table 26.2). Ice cream, for example, causes a much smaller increase in serum glucose levels than either potatoes or whole wheat bread. This is because other components of food—such as protein, fat, and the soluble fibers—are much more important than the type of dietary carbohydrate in determining how rapidly glucose will enter the bloodstream.

CLINICAL CORRELATION 26.7

Metabolic Adaptation: Relationship Between Carbohydrate Intake and Serum Triacylglycerols

In evaluating the nutrition literature, it is important to be aware that most clinical trials are of rather short duration (2–6 weeks), while some metabolic adaptations may take considerably longer. Thus even apparently well-designed clinical studies may lead to erroneous conclusions that will be repeated in the popular literature for years to come. For example, several studies conducted in the 1960s and 1970s tried to assess the effects of carbohydrate intake on serum triacylglycerol levels. Typically, young college-age males were given a diet in which up to 50% of their fat calories were replaced with sucrose or other simple sugars for a period of 2–3 weeks. In most cases serum triacylglycerol levels increased markedly (up to 50%). This led to the tentative conclusion that high intake of simple sugars, particularly sucrose, might increase the risk of heart disease, a notion that was popularized by nutritional best sellers such as "Sugar Blues" and "Sweet and Dangerous." Unfortunately, while the original conclusions were promoted in the lay press, the experiments themselves were questioned. Subsequent studies showed that if these trials were continued for longer periods of time (3–6 months), the triacylglycerol levels usually normalized. The nature of this slow metabolic adaptation is unknown.

While the interpretation of the original clinical trials may have been faulty, the ensuing dietary recommendations may not have been entirely incorrect. Many snack and convenience foods in the American diet are high in sugar, fat, and caloric density. Thus removing some of these foods from the diet can aid in weight control, and overweight is known to contribute to hypertriacylglycerolemia. Some individuals exhibit carbohydrate-induced hypertriacylglycerolemia. Triacylglycerol levels in these individuals respond dramatically to diets that substitute foods containing complex carbohydrates and fiber for those foods containing primarily simple sugars as a carbohydrate source.

Parks, E. J. and Hellerstein, M. K. Carbohydrate-induced hypertriacylglycerolemia: historical perspective and review of biological mechanisms. *Am. J. Clin. Nutr.* 71:412, 2000.

Mixing Vegetable and Animal Proteins Meets Nutritional Protein Requirements

Concern has been voiced recently about the type of protein in the American diet. Epidemiologic data and animal studies suggest that consumption of animal protein is associated with increased incidence of heart disease and various forms of cancer. One could assume that it is probably not the animal protein itself that is involved, but the associated fat and cholesterol. What sort of protein should we consume? Although the present diet may not be optimal, a strictly vegetarian diet may not be

TABLE 26.2 Glycemic Index[a] of Some Selected Foods

Food	Glycemic index	Food	Glycemic index
Grain and cereal products		Root vegetables	
Bread (white)	69 ± 5	Beets	64 ± 16
Bread (whole wheat)	72 ± 6	Carrots	92 ± 20
Rice (white)	72 ± 9	Potato (white)	70 ± 8
Sponge cake	46 ± 6	Potato (sweet)	48 ± 6
Breakfast cereals		Dried legumes	
All bran	51 ± 5	Beans (kidney)	29 ± 8
Cornflakes	80 ± 6	Beans (soy)	15 ± 5
Oatmeal	49 ± 8	Peas (blackeye)	33 ± 4
Shredded wheat	67 ± 10	Fruits	
Vegetables		Apple (Golden Delicious)	39 ± 3
Sweet corn	59 ± 11	Banana	62 ± 9
Frozen peas	51 ± 6	Oranges	40 ± 3
Dairy products		Sugars	
Ice cream	36 ± 8	Fructose	20 ± 5
Milk (whole)	34 ± 6	Glucose	100
Yogurt	36 ± 4	Honey	87 ± 8
		Sucrose	59 ± 10

Source: Data from Jenkins, D. A., et al. Glycemic index of foods: a physiological basis for carbohydrate exchange. *Am. J. Clin. Nutr.* 34:362, 1981.

[a]Glycemic index is defined as the area under the blood glucose response curve for each food expressed as a percentage of the area after taking the same amount of carbohydrate as glucose (mean: 5–10 individuals).

acceptable to many Americans. Perhaps a middle road is best. Clearly there are no known health dangers associated with a mixed diet that is lower in animal protein than the current American standard.

An Increase in Fiber from Varied Sources Is Desirable

Because of current knowledge about effects of fiber on human metabolism, most suggestions for a prudent diet recommend an increase in dietary fiber. The main question is: "How much is enough?" The current fiber content of the American diet is about 14–15 g per day. Most experts feel that an increase to at least 25–30 g would be safe and beneficial. Since different types of fibers have different physiological roles, the increase in fiber intake should come from a wide variety of fiber sources—including fresh fruits, vegetables, and legumes as well as the more popular cereal sources of fiber (which are primarily cellulose and hemicellulose).

Current Recommendations Are for a "Prudent Diet"

Several private and government groups have made specific recommendations with respect to the ideal dietary composition for the American public in recent years. This movement was spearheaded by the Senate Select Committee on Human Nutrition, which first published its *Dietary Goals for the United States* in 1977. This Committee recommended that the American public reduce consumption of total calories, total fat, saturated fat, cholesterol, simple sugars, and salt to "ideal" goals more compatible with good health (Figure 26.2). In recent years the USDA, the American Heart Association, the American Diabetes Association, the National Research Council, and the Surgeon General all have published similar recommendations, and the USDA has used these recommendations to design revised recommendations for a **balanced** or **"prudent diet"** (Figure 26.3). How valid is the

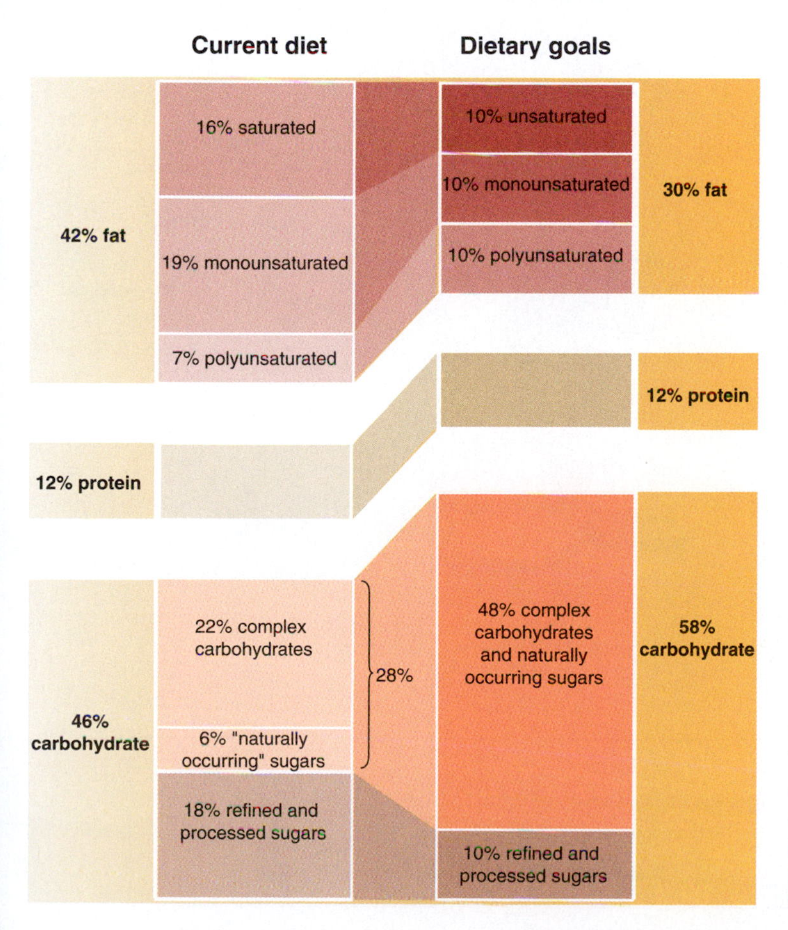

FIGURE 26.2
United States dietary goals.
Graphical comparison of the composition of the current U.S. diet and the dietary goals for the U.S. population suggested by the Senate Select Committee on Human Nutrition.
From Dietary Goals for the United States, *2nd ed. Washington, DC: U.S. Government Printing Office, 1977.*

scientific basis of the recommendations for a prudent diet? Is there evidence that a prudent diet will improve the health of the general public? These remain controversial questions.

An important argument against such recommendations is that we do not have enough information to set concrete goals. We might be creating some problems while solving others. For example, the goals of reducing total and saturated fat in the diet are best met by replacing animal protein with vegetable protein. This might reduce the amount of available iron and vitamin B_{12} in the diet. It is also quite clear that the same set of guidelines do not apply for every individual. For example, exercise is known to raise serum HDL cholesterol and obesity is known to elevate serum cholesterol and triacylglycerols and reduce glucose tolerance. Thus a very active individual who maintains ideal body weight can likely tolerate higher fat and sugar intakes than an obese individual.

On the "pro" side, however, it clearly can be argued that all of the dietary recommendations are aimed at reducing nutritional risk factors in the general population. Besides, similar diets have been consumed by our ancestors and by people in other countries with no apparent harm. Whatever the outcome of this debate, it will undoubtedly shape many of our ideas concerning the role of nutrition in medicine.

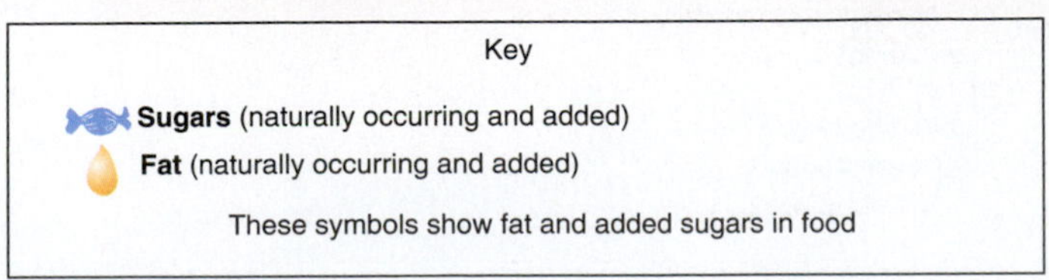

FIGURE 26.3
USDA food pyramid.
Graphical representation of USDA recommendations for a balanced diet.
HG Bulletin #252. *Washington, DC: U.S. Government Printing Office, 1992.*

BIBLIOGRAPHY

Protein Energy Malnutrition in Hospitalized Patients

The Veterans Affairs Total Parenteral Nutrition Cooperative Study Group. Perioperative total parenteral nutrition in surgical patients, *N. Engl. J. Med.* 325:525, 1991.

Metabolic Consequences of Obesity

Maxwell, M. H. and Waks, A. U. Obesity and hypertension. *Bibl. Cardiol.* 41:29, 1987.

Simopoulos, A. P. Obesity and carcinogenesis: historical perspective. *Am. J. Clin. Nutr.* 45:271, 1987.

Pi-Sunyer, F. X. Health implications of obesity *Am. J. Clin. Nutr.* 53:15955, 1991

Boden, G. Free fatty acids, insulin resistance, and type 2 diabetes mellitus. *Proc. Assoc. Am. Physicians* 111, 241, 1999.

Hotamisligii, G. S. The role of TNFα and TNF receptors in obesity and insulin resistance. *J. Intern. Med.* 245:621, 1999.

Metabolic Predisposition to Obesity

Bjorntorp, P. Fat cell distribution and metabolism. *Ann. N. Y. Acad. Sci.* 499:66, 1987.

Bray, G. A. Obesity—a disease of nutrient or energy balance? *Nutr. Rev.* 45:33, 1987.

Dulloo, A. G. and Miller, D. S. Obesity: a disorder of the sympathetic nervous system. *World Rev. Nutr. Diet.* 50:1, 1987.

Halaas, J. L., Gajiwala, K. S., Maffei, M., Cohen, S. L., Chait, B. T., Rabinowitz, D., Lallone, R. L., Burley, S. K., and Friedman, J. M. Weight-reducing effects of the plasma protein encoded by the obese gene. *Science* 269:543, 1995.

Campfield, L. A., Smith, F. J., Guisez, Y., Devos, R., and Burn, P. Recombinant mouse OB protein: evidence for a peripheral signal linking adiposity and central neural networks. *Science* 269:546, 1995.

Complex Carbohydrates and Fiber

Anderson, J. W. Fiber and health: an overview. *Am. J. Gastroenterol.* 81:892, 1986.

Eastwood, M. Dietary fiber and the risk of cancer. *Nutr. Rev.* 45:193, 1987.

Shankar, S. and Lanza, E. Dietary fiber and cancer prevention. *Hematol. Oncol. Clin. North Am.* 5:25, 1991.

Wolever, T. M. S., Jenkins, D. J. A., Jenkins, A. L., and Josse, R. G. The glycemic index: methodology and clinical implications. *Am. J. Clin. Nutr.* 54:846, 1991.

Miller, J. C. B. Importance of glycemic index in diabetes. *Am. J. Clin. Nutr.* 59(suppl.):747S, 1994.

Macronutrient Composition and Health

Gorlin, R. The biological actions and potential clinical significance of dietary ω-3 fatty acids. *Arch. Intern. Med.* 148:2043, 1988.

Grundy, S. M. Monounsaturated fatty acids, plasma cholesterol, and coronary heart disease. *Am. J. Clin. Nutr.* 45:1168, 1987.

Kisselbah, A. and Schetman, G. Polyunsaturated and saturated fat, cholesterol, and fatty acid supplementation. *Diabetes Care* 11:129, 1988.

Kritchevsky, D. and Klurfeld, D. M. Caloric effects in experimental mammary tumorigenesis. *Am. J. Clin. Nutr.* 45:236, 1987.

Welsh, C. W. Enhancement of mammary tumorigenesis by dietary fat: review of potential mechanisms. *Am. J. Clin. Nutr.* 45:191, 1987.

Rasmussen, O. W., Thomsen, C., Hansen, K. W., Vesterland, M., Winther, E., and Hermansen, K. Effects on blood pressure, glucose, and lipid levels of a high monounsaturated fat diet compared with a high-carbohydrate diet in NIDDM subjects. *Diabetes Care* 16:156S, 1993.

Gardener, C. D. and Kraemer, H. C. Monounsaturated versus polyunsaturated dietary fat and serum lipids. *Arterioscler. Thromb.Vasc. Biol.* 15:1917, 1995.

Simopoulus, A. P. Omega-3 fatty acids in health and disease and in growth and development. *Am. J. Clin. Nutr.* 54:438, 1991.

Berry, E. M. Dietary fatty acids in the management of diabetes mellitus. *Am. J. Clin. Nutr.* 66:991S, 1997.

Nicholson, A. S., Sklar, M., Barnard, N. D., Gore, S., Sullivan, R., and Browning, S. Toward improved management of NIDDM: a randomized, controlled, pilot intervention using a lowfat, vegetarian diet. *Prev. Med.* 29:87, 1999.

Dietary Recommendations

American Heart Association. *Recommendations for Treatment of Hyperlipidemia in Adults.* Dallas: American Heart Association. 1984.

Food and Nutrition Board of the National Academy of Sciences. *Towards Healthful Diets.* Washington, DC: U.S. Government Printing Office, 1980.

National Research Council. *Diet, Nutrition and Cancer.* Washington: National Academy Press, 1982.

Senate Select Committee on Human Nutrition. *Dietary Goals for the United States,* 2nd ed, Stock No. 052-070-04376-8. Washington, DC: U.S. Government Printing Office, 1977.

Truswell, A. S. Evolution of dietary recommendations, goals, and guidelines. *Am. J. Clin. Nutr.* 45:1060, 1987.

U.S. Department of Agriculture. *Nutrition and Your Health, Dietary Guidelines for Americans.* Stock No. 017-001-00416-2. Washington, DC: U.S. Government Printing Office, 1980.

U.S. Department of Health and Human Services. *The Surgeon General's Report on Nutrition and Health.* Stock No. 017-001-00465-1. Washington, DC: U.S. Government Printing Office, 1988.

U.S. Department of Agriculture. *The Food Guide Pyramid.* Stock No. HSG-252. Hyatsville, MD: Human Nutrition Information Service, 1992.

QUESTIONS | C. N. ANGSTADT

Multiple Choice Questions

1. Of two people with approximately the same weight, the one with the higher basal energy requirement would most likely be:
   A. taller.
   B. female if the other were male.
   C. older.
   D. under less stress.
   E. all of the above.

2. Basal metabolic rate:
   A. is not influenced by energy intake.
   B. increases in response to starvation.
   C. may decrease up to 50% during periods of starvation.
   D. increases in direct proportion to daily energy expenditure.
   E. is not responsive to changes in hormone levels.

3. The primary effect of the consumption of excess protein beyond the body's immediate needs will be:
   A. excretion of the excess as protein in the urine.
   B. an increase in the "storage pool" of protein.
   C. an increased synthesis of muscle protein.
   D. an enhancement in the amount of circulating plasma proteins.
   E. an increase in the amount of adipose tissue.

4. Which of the following individuals would most likely be in nitrogen equilibrium?
   A. A normal, adult male
   B. A normal, pregnant female
   C. A growing child
   D. An adult male recovering from surgery
   E. A normal female on a very low protein diet

5. Kwashiorkor is:
   A. the most common form of protein–calorie malnutrition in the United States.
   B. characterized by a thin, wasted appearance.
   C. an inadequate intake of food of any kind.
   D. an adequate intake of total calories but a specific deficiency of protein.
   E. an adequate intake of total protein but a deficiency of the essential amino acids.

6. An excessive intake of calories:
   A. usually does not have adverse metabolic consequences.
   B. leads to metabolic changes that are usually irreversible.
   C. frequently leads to elevated serum levels of free fatty acids, cholesterol, and triacylglycerols.
   D. is frequently associated with an increased number of insulin receptors.
   E. is the only component of obesity.

7. A complete replacement of animal protein in the diet by vegetable protein:
   A. would be expected to have no effect at all on the overall diet.
   B. would reduce the total amount of food consumed for the same number of calories.
   C. might reduce the total amount of iron and vitamin B_{12} available.
   D. would be satisfactory regardless of the nature of the vegetable protein used.
   E. could not satisfy protein requirements.

8. Dietary fat:
   A. is usually present, although there is no specific need for it.
   B. if present in excess, can be stored as either glycogen or adipose tissue triacylglycerol.
   C. should include linoleic and linolenic acids.
   D. should increase on an endurance training program in order to increase the body's energy stores.
   E. if present in excess, does not usually lead to health problems.

9. Which of the following statements about dietary fiber is/are correct?
   A. Water-soluble fiber helps to lower serum cholesterol in most people.
   B. Mucilaginous fiber slows the rate of digestion and absorption of carbohydrates.
   C. Insoluble fiber increases stool bulk and decreases transit time.
   D. All of the above are correct.
   E. None of the above is correct.

10. Which one of the following dietary regimens would be *most* effective in lowering serum cholesterol?
    A. restrict dietary cholesterol
    B. increase the ratio of polyunsaturated to saturated fatty acids
    C. increase fiber content
    D. restrict cholesterol and increase fiber
    E. restrict cholesterol, increase PUFA/SFA, increase fiber

Questions 11 and 12: A young man suffered third degree burns over much of his body and is hospitalized in a severe catabolic state. An individual in this state requires about 40 kcal kg^{-1} day^{-1} and 2 g protein kg^{-1} day^{-1} to be in positive caloric and nitrogen balance. This young man weighs 140 lb (64 kg). Total parenteral nutrition (TPN) is started with a solution containing 20% glucose and 4.25% amino acids (the form in which protein is supplied).

11. If 3000 g of solution is infused per day:
    A. the patient would not be getting sufficient protein.
    B. the calories supplied would be inadequate.
    C. both protein and calories would be adequate to meet requirements.
    D. this is too much protein being infused.

12. Sometimes a lipid solution is also infused in a patient on TPN. In the case of this young man, the purpose of the lipid solution would be to:
    A. supply additional calories to meet caloric needs.
    B. supply essential fatty acids.
    C. improve the palatability of the mixture.
    D. provide fiber.
    E. assure an adequate supply of cholesterol for membrane building.

Questions 13 and 14: For many years, the American Diabetic Association recommended a diet high in complex carbohydrates and fiber and low in fat for diabetics. It was later found that some individuals did not do as well on such a diet as on one high in monounsaturated fatty acids. Since 1994, the ADA has abandoned the concept of a single diabetic diet and now recommends a focus on achieving glucose, lipid, and blood pressure goals with weight reduction if necessary.

13. Which of the following statements is/are correct?
    A. A high-carbohydrate–high-fiber diet often results in significant weight reduction because it has a lower caloric density than a diet high in fat.
    B. A diet high in monounsaturated fatty acids would be most appropriate for an overweight diabetic.
    C. The goal for lipids is to reduce all lipoprotein levels in the blood.
    D. Obesity aggravates diabetes because it inhibits the production of insulin by the pancreas.
    E. All of the above are correct.

14. For diabetics:
    A. the only carbohydrate that must be eliminated in the diet is sucrose.
    B. fiber increases the rate at which carbohydrate is digested and absorbed.
    C. not all carbohydrate foods raise blood glucose levels at the same rate because the glycemic index of all foods is not the same.
    D. who are normally in good control, stress will have no effect on their blood sugar levels.
    E. a vegetarian diet is the only appropriate choice.

Problems

15. Calculate the number of grams each of carbohydrate, lipid, and protein a person on a 2300-kcal diet should consume to meet the guidelines established by the Senate Select Committee on Human Nutrition. Assuming the individual weighs 180 lb and the protein is from mixed animal/vegetable sources with a 75% efficiency of utilization, does the amount of protein you calculated above meet the recommended amount of protein?
16. A 120-lb woman is consuming a diet with adequate total calories and 44 g of protein per day. The protein is exclusively from vegetable sources, primarily corn based. What would be her state of nitrogen balance?

ANSWERS

1. **A** A taller person with the same weight would have a greater surface area. B: Males have higher energy requirements than females. C: Energy requirements decrease with age. D: Stress, probably because of the effects of epinephrine and cortisol, increase energy requirements.
2. **C** This is part of the survival mechanism in starvation. A, B: BMR decreases when energy intake decreases. D: BMR as defined is independent of energy expenditure. Only when the exercise is repeated on a daily basis so that lean muscle mass is increased does BMR also increase. E: Many hormones increase BMR.
3. **E** Excess protein is treated like any other excess energy source and stored (minus the nitrogen) eventually as adipose tissue lipid. A: Protein is not found in normal urine except in very small amounts. The excess nitrogen is excreted as NH_4^+ and urea, whereas the excess carbon skeletons of the amino acids are used as energy sources. B–D: There is no discrete storage form of protein, and although some muscle and structural protein is expendable, there is no evidence that increased intake leads to generalized increased protein synthesis.
4. **A** B–D: Although normal, pregnancy is also a period of growth, requiring positive balance as does a period of convalescence. E: Inadequate protein intake leads to negative balance.
5. **D** This is the definition. A: The most common protein–calorie malnutrition occurs in severely ill, hospitalized patients who would be more likely to have generalized malnutrition. B, C: These are the characteristics of marasmus. E: This would lead to negative nitrogen balance but does not have a specific name.
6. **C** Probably because an increased number and/or size of adipose cells will contain fewer or insensitive insulin receptors. A: Excess caloric intake will lead to obesity if continued long enough. B: Fortunately most of the changes accompanying obesity can be reversed if weight is lost. D: Many of the adverse effects of obesity are associated with an increased number of adipocytes that are deficient in or have insensitive insulin receptors. E: Inadequate exercise and a genetic component also play roles in obesity.
7. **C** A, C: This would reduce the amount of fat, especially saturated fat, but could also reduce the amount of necessary nutrients that come primarily from animal sources. B: The protein content of vegetables is quite low, so much larger amounts of vegetables would have to be consumed. D, E: It is possible to satisfy requirements for all of the essential amino acids completely if vegetables with complementary amino acid patterns, in proper amounts, are consumed.
8. **C** A, C: Linoleic and linolenic acids are essential fatty acids and so must be present in the diet. B, D: Excess carbohydrate can be stored as fat but the reverse is not true. D: Carbohydrate loading has been shown to increase endurance. E: High-fat diets are associated with many health risks.
9. **D** These each illustrate the different properties and roles of the common kinds of fiber.
10. **E** Any of the measures alone would decrease serum cholesterol slightly, but to achieve a reduction of more than 15% requires all three.
11. **C** This amount of solution would supply 128 g protein and 2900 kcal (2400 from glucose and 512 from amino acids), both enough to meet the stated requirements.
12. **B** Patients on TPN need to have essential fatty acids supplied. A: The original solution supplies adequate calories although additional calories shouldn't hurt this young man. C: The patient isn't tasting this mixture so palatability is irrelevant. D: Fiber is supplied by complex carbohydrate sources. E: The body can make its own cholesterol.
13. **A** Since carbohydrate has less than half the caloric density of fat and fiber provides no calories, such diets have low caloric density. B: High fat would tend to increase weight. C: The goal is to reduce the LDL and VLDL levels but not HDL level. D: The problem is that factors released by adipose cells invoke insulin resistance.
14. **C** Foods vary widely in their glycemic index. Bread and rice raise blood glucose levels more rapidly than does sucrose. A: See C. B: One of the benefits of fiber is that it decreases the rate of carbohydrate absorption. D: Stress raises blood glucose for everyone because of release of epinephrine and glucocorticoids. E: An appropriately designed vegetarian diet is perfectly acceptable but is certainly not the only choice.
15. According to the SSCHN guidelines a 2300-kcal diet should consist of 333.5 g of carbohydrate (no more than 57.5 g as simple sugar), 69 g of protein, and 76.7 g of fat (no more than 25.6 g of saturated fat). A 180-lb man weighs 81.8 kg times 0.8 g kg^{-1} = 65.5 g of protein per day to meet requirements. Twelve percent of 2300 kcal supplies sufficient protein.
16. 44 g protein/120 lb is 0.8 g protein kg^{-1} day^{-1}. However, this is inadequate because pure vegetable protein is less than 75% efficient in utilization. Also, heavy reliance on one protein (corn is deficient in lysine) would likely lead to a deficiency of one or more essential amino acids. Therefore this woman is in negative nitrogen balance.

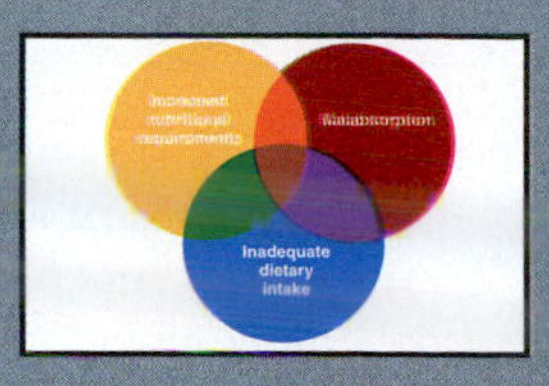

27

PRINCIPLES OF NUTRITION II: MICRONUTRIENTS

Stephen G. Chaney

27.1 | OVERVIEW

Micronutrients play a vital role in human metabolism, being involved in almost every known biochemical reaction and pathway. However, their biochemistry is of little interest unless we also know if dietary deficiencies are likely. The American diet is undoubtedly the best it has ever been. Our current food supply provides an abundant variety of foods all year long and deficiency diseases have become medical curiosities. However, our diet is far from optimal. The old adage that we get everything we need from a balanced diet is true only if we eat a balanced diet. Unfortunately, many Americans do not consume a balanced diet. Foods of high caloric density and low nutrient density (often referred to as empty calories or junk food) are an abundant and popular part of the American diet, and our nutritional status suffers because of these food choices. Obviously then, neither alarm nor complacency is justified. We need to know how to evaluate the adequacy of our diet.

27.2 | ASSESSMENT OF MALNUTRITION

There are three increasingly stringent criteria for measuring **malnutrition.**

1. **Dietary intake studies,** which are usually based on a 24-hour recall, are the least stringent. The 24-hour recalls tend to overestimate the number of people with deficient diets. Also, poor dietary intake alone is not a problem in this country unless the situation is compounded by increased need.
2. **Biochemical assays,** either direct or indirect, are a more useful indicator of the nutritional status of an individual. At their best, they indicate **subclinical nutritional deficiencies,** which can be treated before deficiency diseases develop. However, all biochemical assays are not equally valid—an unfortunate fact that is not sufficiently recognized. Changes in biochemical parameters due to stress need to be interpreted with caution. The distribution of many nutrients in the body changes dramatically in a stress situation such as illness, injury, and pregnancy. A drop in level of a nutrient in one tissue compartment (usually blood) need not signal a deficiency or an increased requirement. It could simply reflect a normal metabolic adjustment to stress.
3. The most stringent criterion is the appearance of **clinical symptoms.** However, it is desirable to intervene long before symptoms became apparent.

The question remains: When should dietary surveys or biochemical assays be interpreted to indicate the need for nutritional intervention? The following general guidelines are useful. Dietary surveys are seldom a valid indication of general malnutrition unless the average intake for a population group falls significantly below the Estimated Average Requirement (EAR) for one or more nutrients. However, by looking at the percentage of people within a population group who have suboptimal intake, it is possible to identify high-risk population groups that should be monitored more closely. Biochemical assays can definitely identify subclinical cases of malnutrition where nutritional intervention is desirable provided (a) the assay has been shown to be reliable, (b) the deficiency can be verified by a second assay, and (c) there is no unusual stress situation that may alter micronutrient distribution. In assessing nutritional status, it is important for the clinician to be aware of those population groups at risk, the most reliable biochemical assays for monitoring nutritional status, and the symptoms of deficiencies if they should occur.

27.3 | DIETARY REFERENCE INTAKES

Dietary Reference Intakes (DRIs) are quantitative estimates of nutrient intakes to be used for planning and assessing diets for healthy people. In assessing quantitative standards for nutrient intake the Food and Nutrition Board of the National Research Council considers the amount of nutrients and food components required for preventing deficiency diseases and, where the data are definitive, for promoting optimal health. The first step in this process is determining the **Estimated Average Requirement (EAR),** the amount of nutrient estimated to meet the nutrient requirement of half of the healthy individuals in a life-style and gender group. The **Recommended Dietary Allowance (RDA)** is normally set at 2 standard deviations above the EAR and is assumed to be the dietary intake amount that is sufficient to meet the nutrient requirement of nearly all (97–98%) of healthy individuals in a group. If a nutrient is considered essential but the experimental data are considered inadequate for determining an EAR, an **adequate intake (AI)** is set rather than an RDA. The AI is based on observed or experimentally determined approximations of nutrient intake by a group of individuals. For example, the AI for young infants is based on the daily mean nutrient intake supplied by human milk for healthy, full-term infants who are exclusively breast-fed. Finally, for most nutrients the Food and Nutrition Board sets a Tolerable Upper Intake Level (UL). The UL is defined as the highest level of daily nutrient intake that is likely to pose no risks of adverse health effects to almost all individuals in the general population. The RDAs, AIs, and ULs are designed to be of use in planning and evaluating diets for individuals. The EAR is designed to be used in setting goals for nutrient intake and assessing the prevalence of inadequate intake in a population group.

These determinations are relatively easy to make for those nutrients associated with dramatic deficiency diseases, for example, vitamin C and scurvy. In other instances more indirect measures must be used, such as tissue saturation or extrapolation from animal studies. There are no criteria that can be used for all micronutrients, and there are always some uncertainty and debate as to the correct criteria. The criteria are changed by new research. The Food and Nutrition Board normally meets every 6–10 years to consider currently available information and update its recommendations.

RDAs serve as a useful general guide in evaluating adequacy of individual diets. However, the RDAs have several limitations that should be kept in mind. Important limitations are as follows:

1. RDAs are designed to meet the needs of healthy people and do not take into account special needs arising from infections, metabolic disorders, or chronic diseases.
2. Since present knowledge of nutritional needs is incomplete, there may be unrecognized nutritional needs. To provide for these needs, the RDAs should be met from as varied a selection of foods as possible. No single food can be considered complete, even if it meets the RDA for all known nutrients. This is important, especially in light of the current practice of fortifying foods of otherwise low nutritional value.
3. As currently formulated, RDAs may not define the "optimal" level of any nutrient, since optimal amounts are difficult to define. Because of information suggesting that optimal intake of certain micronutrients (e.g., vitamins A, C, and E) may reduce heart disease and cancer risk, the RDAs for these nutrients have recently been increased slightly; however, some experts feel that the current RDAs may not be sufficient to promote optimal health.

27.4 | FAT-SOLUBLE VITAMINS

Vitamin A Is Derived from Plant Carotenoids

The active forms of **vitamin A** are **retinol, retinal (retinaldehyde),** and **retinoic acid.** These precursors are synthesized by plants as the **carotenoids** (Figure 27.1), which are cleaved to retinol by most animals and stored in the liver as retinol palmitate. Liver, egg yolk, butter, and whole milk are good sources of the preformed retinol. Dark green

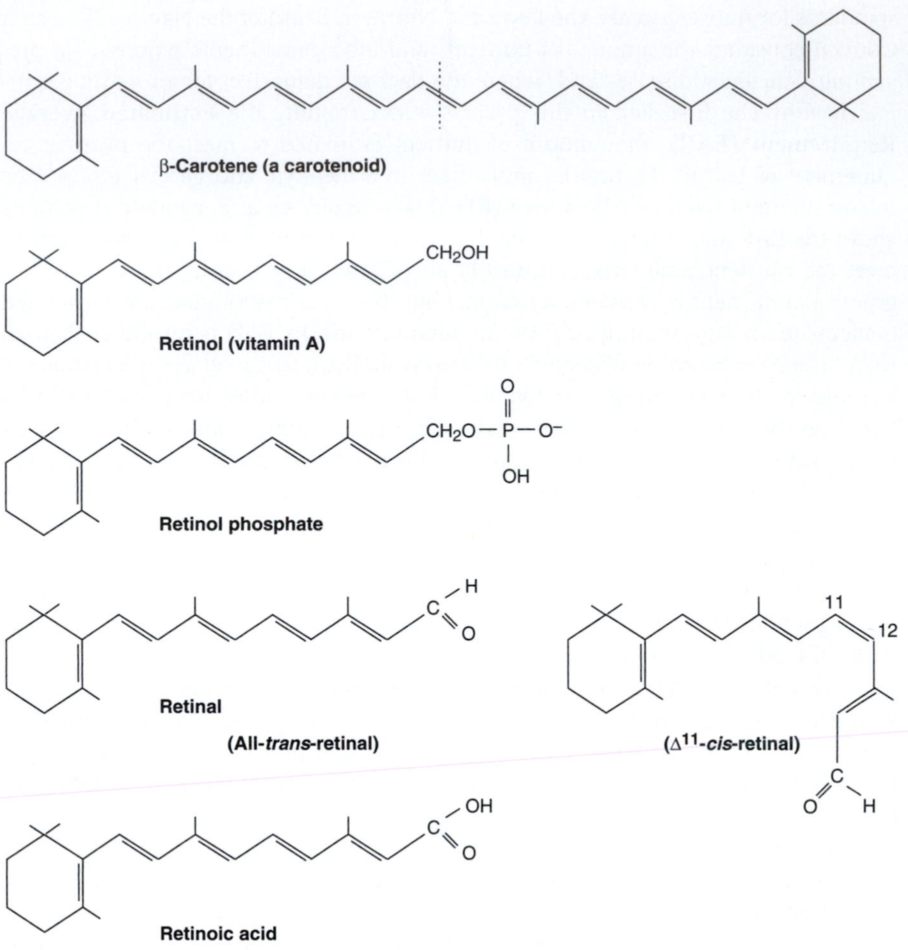

β-Carotene (a carotenoid)

CH_2OH

Retinol (vitamin A)

$CH_2O-P(=O)(OH)-O^-$

Retinol phosphate

Retinal

(All-*trans*-retinal)

(Δ^{11}-*cis*-retinal)

Retinoic acid

FIGURE 27.1
Structures of vitamin A and related compounds.

and yellow vegetables are generally good sources of the carotenoids. Conversion of carotenoids to retinol is rarely 100%, so that the vitamin A potency of various foods is expressed in terms of retinol equivalents (1 RE is equal to 1 mg retinol, 6 mg ***β*-carotene**, and 12 mg of other carotenoids). *β*-Carotene and other carotenoids are major sources of vitamin A in the American diet. These carotenoids are first cleaved to retinol and converted to other vitamin A metabolites in the body (Figure 27.1).

Vitamin A serves a number of functions in the body. Only in recent years has its biochemistry become well understood (Figure 27.2). *β*-Carotene and some other carotenoids have an important role as **antioxidants.** At the low oxygen tensions prevalent in the body, *β*-carotene is a very effective antioxidant and may reduce the risk of those cancers initiated by free radicals and other strong oxidants. Several epidemiologic studies suggested that adequate dietary *β*-carotene may be important in reducing the risk of lung cancer, especially in people who smoke. However, supplemental *β*-carotene alone did not provide any detectable benefit and may have actually increased cancer risk for smokers in two multicenter prospective studies. This illustrates the danger of making dietary recommendations on epidemiologic studies alone.

Retinol is converted to **retinyl phosphate** in the body. The retinyl phosphate appears to serve as a **glycosyl donor** in the synthesis of some glycoproteins and mucopolysaccharides in much the same manner as dolichol phosphate (see p. 259). Retinyl phosphate is essential for the synthesis of certain glycoproteins needed for normal growth regulation and for mucus secretion. Both retinol and retinoic acid bind to specific intracellular receptors, which then bind to DNA and affect the synthesis of proteins involved in the regulation of cell growth and differentiation. Thus

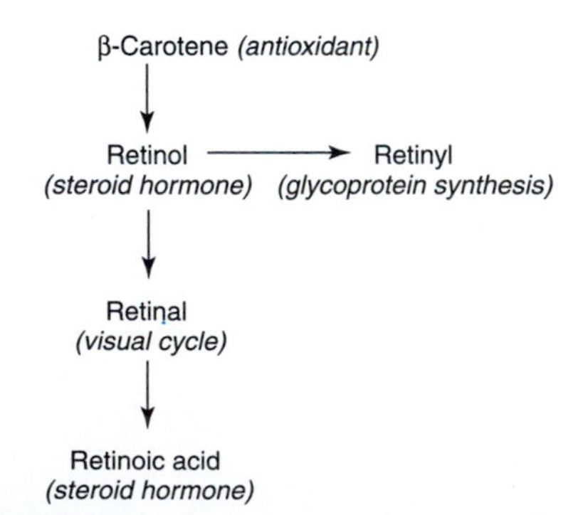

FIGURE 27.2
Vitamin A metabolism and function.

both retinol and retinoic acid can be considered to act like **steroid hormones** in regulating growth and differentiation.

Finally, in the Δ^{11}-*cis*-retinal form, vitamin A becomes reversibly associated with the **visual proteins** (the opsins). When light strikes the retina, a number of complex biochemical changes take place, resulting in the generation of a nerve impulse, conversion of the retinal to the all-trans form, and its dissociation from the visual protein (see p. 1012). Regeneration of functional visual pigments requires isomerization back to the Δ^{11}-*cis* form (Figure 27.3). In addition to the direct role of vitamin A in the visual cycle, clinical studies suggest that the carotenoids lutein and zeaxanthin reduce the risk of macular degeneration.

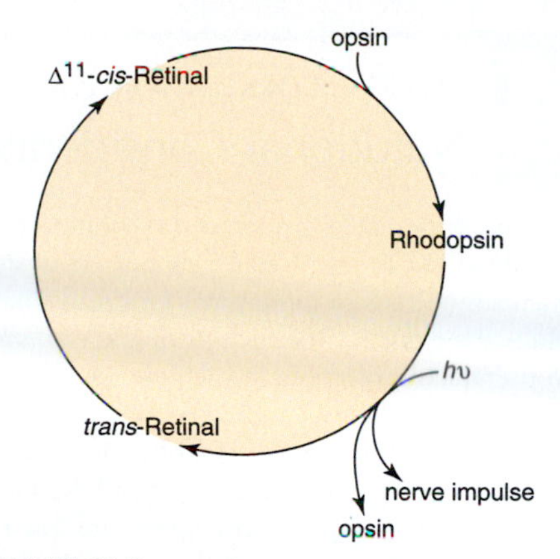

FIGURE 27.3
Role of vitamin A in vision.

Based on what is known about the biochemical mechanisms of vitamin A action, its biological effects are easier to understand. Vitamin A is required for the maintenance of healthy epithelial tissue. Retinol and/or retinoic acid are required to prevent the synthesis of high molecular weight forms of **keratin** and retinyl phosphate is required for the synthesis of glycoproteins (an important component of the **mucus** secreted by many epithelial tissues). The lack of mucus secretion leads to a drying of these cells, and the excess keratin synthesis leaves a horny keratinized surface in place of the normal moist and pliable epithelium. Vitamin A deficiency can lead to anemia caused by impaired mobilization of iron from the liver because retinol and/or retinoic acid are required for the synthesis of the iron transport protein transferrin.

Vitamin A-deficient animals are more susceptible to both infections and cancer. Decreased resistance to infections may be due to keratinization of mucosal cells lining the respiratory, gastrointestinal, and genitourinary tracts. Under these conditions fissures readily develop in the mucosal membranes, allowing microorganisms to enter. Vitamin A deficiency may also impair the immune system. The protective effect of vitamin A against many forms of cancer may result from the antioxidant potential of carotenoids and the effects of retinol and retinoic acid in regulating cell growth.

Since vitamin A is stored in the liver, deficiency can develop only over prolonged periods of inadequate intake. Mild **vitamin A deficiencies** are characterized by **follicular hyperkeratosis** (rough keratinized skin resembling "goosebumps"), **anemia** (biochemically equivalent to iron deficiency anemia, but in the presence of adequate iron intake), and increased susceptibility to infection and cancer. **Night blindness** is an early symptom of vitamin A deficiency. Severe vitamin A deficiency leads to a progressive keratinization of the cornea of the eye known as **xerophthalmia** in its most advanced stages. Infection usually sets in, with resulting hemorrhaging of the eye and permanent loss of vision.

For most people (unless they happen to eat liver) the dark green and yellow vegetables are the most important dietary source of vitamin A. Unfortunately, these foods are most often missing from the American diet. Dietary surveys indicate that 40–60% of the population consumes less than two-thirds of the RDA for vitamin A. Clinical symptoms of vitamin A deficiency are rare in the general population, but vitamin A deficiency is a fairly common consequence of severe liver damage or diseases that cause fat malabsorption (see Clin. Corr. 27.1).

Vitamin A accumulates in the liver. Intake of large amounts over prolonged periods can be toxic. Doses of 25,000–50,000 RE per day over months or years are toxic for many children and adults. The usual symptoms include bone pain, scaly dermatitis, enlargement of liver and spleen, nausea, and diarrhea. It is virtually impossible to ingest toxic amounts of vitamin A from normal foods unless one eats polar bear liver (6000 RE g^{-1}) regularly. Most instances of **vitamin A toxicity** are due to the use of massive doses of vitamin A supplements. Fortunately, this practice is relatively rare because of increased public awareness of vitamin A toxicity.

Vitamin D Synthesis in the Body Requires Sunlight

Technically, **vitamin D** should be considered a hormone rather than a vitamin. **Cholecalciferol (D_3)** is produced in skin by UV irradiation of 7-dehydrocholesterol

CLINICAL CORRELATION 27.1

Nutritional Considerations in Cystic Fibrosis

Patients with malabsorption diseases often develop malnutrition. Cystic fibrosis (CF) involves a generalized dysfunction of the exocrine glands that leads to formation of a viscid mucus, which progressively plugs their ducts. Obstruction of the bronchi and bronchioles leads to pulmonary infections, which are usually the direct cause of death. In many cases, however, the exocrine cells of the pancreas are also affected, leading to a deficiency of pancreatic enzymes and sometimes a partial obstruction of the common bile duct.

Deficiency of pancreatic lipase and bile salts leads to severe malabsorption of fat and fat-soluble vitamins. Calcium tends to form insoluble salts with the long-chain fatty acids, which accumulate in the intestine. While these are the most severe problems, some starches and proteins are also trapped in the fatty bolus of partially digested foods. This physical entrapment, along with the deficiencies of pancreatic amylase and pancreatic proteases, can lead to severe protein–calorie malnutrition as well. Excessive mucus secretion on the luminal surface of the intestine may also interfere with the absorption of several nutrients, including iron.

Fortunately, microsphere preparations of pancreatic enzymes are now available that can greatly alleviate many of these malabsorption problems. With these preparations, protein and carbohydrate absorption are returned to near normal. Fat absorption is improved greatly but not normalized, since deficiencies of bile salts and excess mucus secretion persist. Because dietary fat is a major source of calories, these patients have difficulty obtaining sufficient calories from a normal diet. This is complicated by increased protein and energy needs resulting from the chronic infections often seen in these patients. Thus many experts recommend energy intakes ranging from 120–150% of the RDA to combat the poor growth and increased susceptibility to infection. The current recommendations are for high-energy–high-protein diets without any restriction of dietary fat (50% carbohydrate, 15% protein, and 35% fat). If caloric intake from the normal diet is inadequate, dietary supplements or enteral feedings may be used. The dietary supplements most often contain easily digested carbohydrates and milk protein mixtures. Medium-chain triglycerides are sometimes used as a partial fat replacement since they can be absorbed directly through the intestinal mucosa in the absence of bile salts and pancreatic lipase.

Since some fat malabsorption is present, deficiencies of the fat-soluble vitamins often occur. Children aged 2–8 years need a standard adult multivitamin preparation containing 400 IU of vitamin D and 5000 IU of vitamin A per day. Older children, adolescents, and adults need a standard multivitamin at a dose of 1–2 per day. If serum vitamin A levels become low, water-miscible vitamin A preparations should be used. For vitamin E the recommendations are: ages 0–6 mo, 25 IU day^{-1}; 6–12 mo, 50 IU day^{-1}; 1–4 years, 100 IU day^{-1}; 4–10 years. 100–200 IU day^{-1}; and >10 years, 200–400 IU day^{-1}; in water-soluble form. Vitamin K deficiency has not been studied adequately, but the current recommendations are: ages 0–12 mo, 2.5 mg $week^{-1}$ or 2.5 mg twice a week if on antibiotics; ages >1 year, 5.0 mg twice weekly when on antibiotics or it cholestatic liver disease is present. Iron deficiency is fairly common but iron supplementation is not usually recommended because of concern that higher iron levels in the blood might encourage systemic bacterial infections. Calcium levels in the blood are usually normal. However, since calcium absorption is probably suboptimal, it is important to make certain that the diet provides at least RDA levels of calcium.

Littlewood, J. M. and MacDonald, A. Rationale of modern dietary recommendations in cystic fibrosis. *J. R. Soc. Med.* 80(Suppl. 15):16, 1987; and Ramsey, B. W., Farrell, P. M., and Pencharz, P. Nutritional assessment and management in cystic fibrosis; a consensus report. *Am J. Clin. Nutr.* 55:108, 1992.

(see Figure 27.4). Thus, as long as the body is exposed to adequate sunlight, there is little or no dietary requirement for vitamin D. The best dietary sources of vitamin D_3 are saltwater fish (especially salmon, sardines, and herring), liver, and egg yolk. Milk, butter, and other foods are routinely fortified with **ergocalciferol (D_2)** prepared by irradiating ergosterol from yeast. Vitamin D potency is measured in terms of milligrams cholecalciferol (1 mg cholecalciferol or ergocalciferol = 40 IU).

Both cholecalciferol and ergocalciferol are metabolized identically. They are carried to the liver where the 25-hydroxy derivative is formed. **25-Hydroxycholecalciferol** [25-(OH)D] is the major circulating derivative of vitamin D, and it is in turn converted into the biologically active **1-α,25-dihydroxycholecalciferol** (also called **calcitriol**) in the proximal convoluted tubules of kidney (see Clin. Corr. 27.2).

1,25-$(OH)_2$D acts in concert with **parathyroid hormone (PTH),** which is also produced in response to low serum calcium. Parathyroid hormone plays a major role in regulating the activation of vitamin D. High PTH amounts stimulate the production of 1,25-$(OH)_2$D, while low PTH amounts induce formation of an inactive 24,25-$(OH)_2$D. Once formed, the 1,25-$(OH)_2$D acts as a typical steroid hormone in intestinal mucosal cells, where it induces synthesis of a protein, calbinden, required for calcium transport. In bone 1,25-$(OH)_2$D and PTH act synergistically to promote bone resorption (demineralization) by stimulating osteoblast formation and activity. Finally, PTH and 1,25-$(OH)_2$D inhibit calcium excretion in the kidney by stimulating calcium reabsorption in the distal renal tubules. Calcitonin is produced when serum

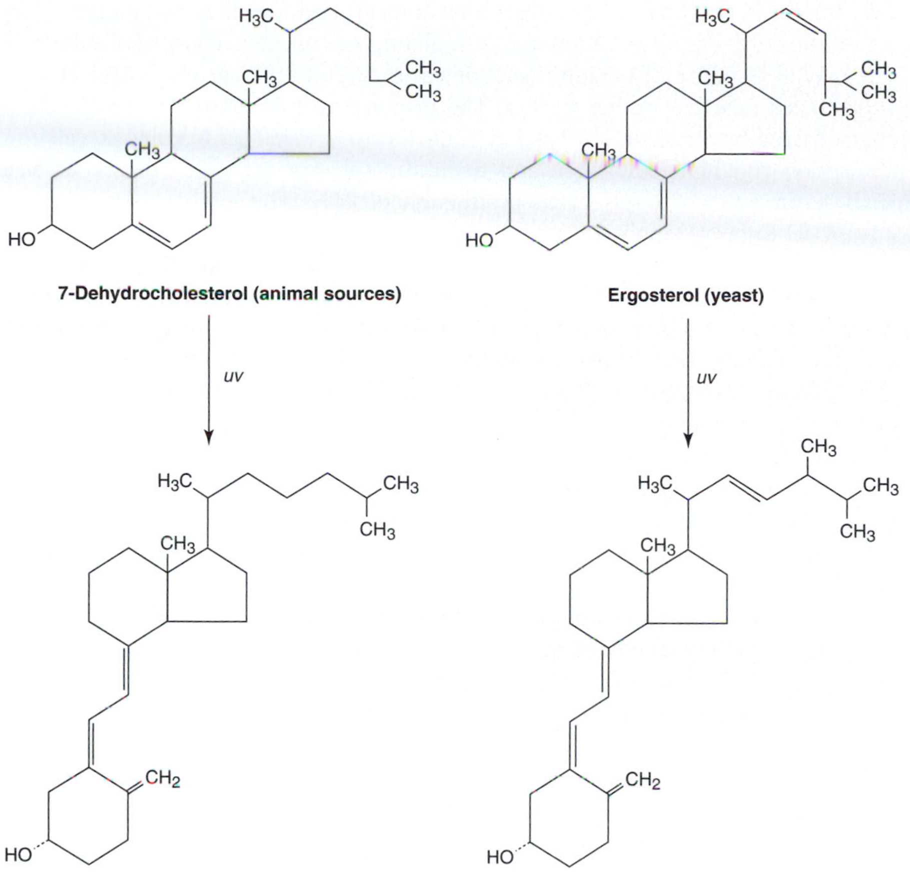

7-Dehydrocholesterol (animal sources)

Ergosterol (yeast)

uv

uv

FIGURE 27.4
Structures of vitamin D and related compounds.

CLINICAL CORRELATION 27.2
Renal Osteodystrophy

In chronic renal failure, a complicated chain of events leads to a condition known as renal osteodystrophy. The renal failure results in an inability to produce 1,25-$(OH)_2$D, and thus bone calcium becomes the only important source of serum calcium. In the later stages, the situation is complicated further by increased renal retention of phosphate and resulting hyperphosphatemia. The serum phosphate levels are often high enough to cause metastatic calcification (i.e., calcification of soft tissue), which tends to lower serum calcium levels further (the solubility product of calcium phosphate in the serum is very low and a high serum level of one component necessarily causes a decreased concentration of the other). The hyperphosphatemia and hypocalcemia stimulate parathyroid hormone secretion, and the resulting hyperparathyroidism further accelerates the rate of bone loss. The result is bone loss and metastatic calcification. In this case, administration of high doses of vitamin D or its active metabolites would not be sufficient since the combination of hyperphosphatemia and hypercalcemia would only lead to more extensive metastatic calcification. The readjustment of serum calcium levels by high calcium diets and/or vitamin D supplementation must be accompanied by phosphate reduction therapies. The most common technique is to use phosphate-binding antacids that make phosphate unavailable for absorption. Orally administered 1,25-$(OH)_2$D is effective at stimulating calcium absorption in the mucosa but does not enter the peripheral circulation in significant amounts. Thus, in severe hyperparathyroidism, intravenous 1,25-$(OH)_2$D may be necessary. There is also some promising research in progress with calciumimetic agents that bind to a calcium sensor located on the extracellular membrane of the parathyroid gland and decrease parathyroid hormone production and release.

Delmez, J. M. and Siatopolsky, E. Hyperphosphatemia: its consequences and treatment in patients with chronic renal disease. *Am J. Kidney Dis.* 19:303,1992; Ritz, E., Schomig, M., and Bommer, J. Osteodystrophy in the millennium. *Kidney Int.* 56(Suppl.73): 594, 1999; and Hoyland, J. A. and Picton, M. L. Cellular mechanisms of osteodystrophy. *Kidney Int.* 56(Suppl. 73): 508, 1999.

calcium levels are high (usually right after a meal) and lowers serum calcium levels by blocking bone resorption and stimulating calcium excretion by the kidney. The overall response of calcium metabolism to several different physiological situations is summarized in Figure 27.5. The response to low serum calcium levels is characterized by elevation of PTH and 1,25-$(OH)_2D$, which act to enhance calcium absorption and bone resorption and to inhibit calcium excretion (Figure 27.5*a*). High serum calcium levels block production of PTH. The low PTH levels allow 25-(OH)D to be metabolized to 24,25-$(OH)_2D$ instead of 1,25-$(OH)_2D$. In the absence of PTH and 1,25-$(OH)_2D$ bone resorption is inhibited and calcium excretion is enhanced. The high serum calcium levels also stimulate production of calcitonin, which contributes to the inhibition of bone resorption and the increase in calcium excretion. Finally, the high levels of both serum calcium and phosphate increase the rate of bone mineralization (Figure 27.5*b*). Thus bone is a very important reservoir of the calcium and phosphate needed to maintain homeostasis of serum levels. When dietary vitamin D and calcium are adequate, no net loss of bone calcium occurs. However, when dietary calcium is low, PTH and 1,25-$(OH)_2D$ will cause net demineralization of bone to maintain normal serum calcium levels. Vitamin D deficiency also causes net demineralization of bone due to elevation of PTH (Figure 27.5*c*).

The most common symptoms of **vitamin D deficiency** are **rickets** in young children and **osteomalacia** in adults. Rickets is characterized by continued formation of osteoid matrix and cartilage that are improperly mineralized, resulting in soft, pliable bones. In adults demineralization of preexisting bone causes bone to become softer and more susceptible to fracture. Osteomalacia is easily distinguishable from the more common osteoporosis, by the fact that the osteoid matrix remains intact in the former, but not in the latter. Vitamin D may be involved in more than regulation of calcium homeostasis. Receptors for 1,25-$(OH)_2D$ occur in many tissues including parathyroid gland, islet cells of pancreas, keratinocytes of skin, and myeloid stem cells in bone marrow.

Because of fortification of dairy products with vitamin D, dietary deficiencies are very rare and are most often seen in low-income groups, the elderly (who often also have minimal exposure to sunlight), strict vegetarians (especially if their diet is also low in calcium and high in fiber), and chronic alcoholics. Most cases of vitamin D deficiency are a result of diseases causing **fat malabsorption** or severe liver and kidney disease (see Clin. Corr. 27.1 and 27.2). Certain drugs also interfere with vitamin D metabolism. For example, corticosteroids stimulate the conversion of vitamin D to inactive metabolites and cause bone demineralization when used for long periods of time.

Vitamin D can also be toxic in high doses. The Tolerable Upper Intake Level (UL) for adults is 2000 IU day^{-1}. The mechanism of **vitamin D toxicity** is summarized in Figure 27.5*d*. Enhanced calcium absorption and bone resorption cause hypercalcemia, which can lead to metastatic calcification. The enhanced bone resorption causes demineralization similar to that seen in vitamin D deficiency. Finally, the high serum calcium leads directly to hypercalciuria, which predisposes the patient to formation of renal stones.

Vitamin E Is a Mixture of Tocopherols

For many years **vitamin E** was described as the "vitamin in search of a disease." While deficiency diseases are still virtually unknown, its metabolic role has become better understood in recent years. Vitamin E occurs in the diet as a mixture of several closely related compounds, called tocopherols. **α-Tocopherol** is the most potent form of vitamin E.

First and foremost, the tocopherols are important naturally occurring **antioxidants.** Due to their lipophilic character they accumulate in circulating lipoproteins, cellular membranes, and fat deposits, where they react very rapidly with molecular oxygen and free radicals. They act as scavengers for these compounds, protecting

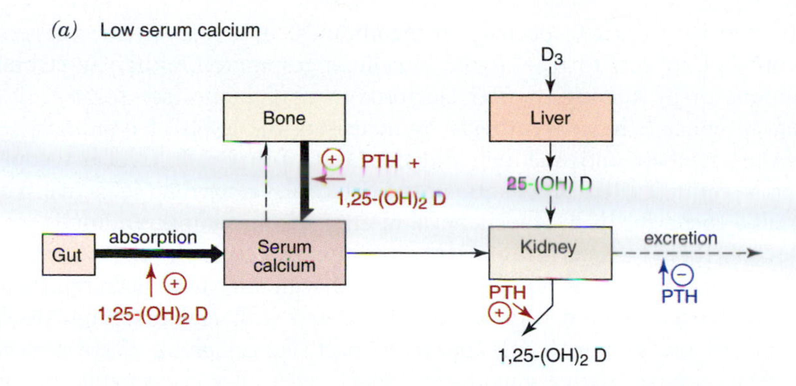

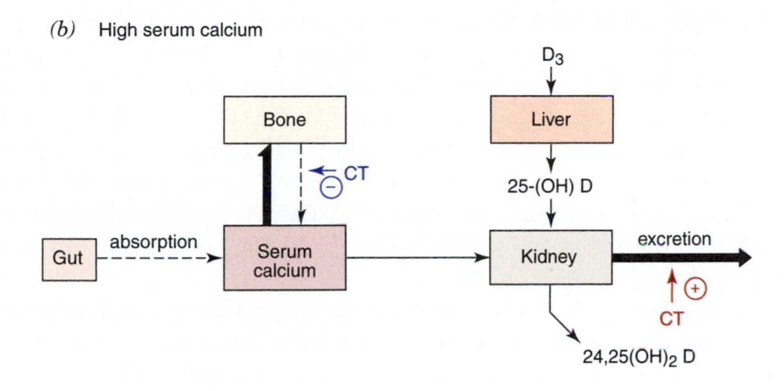

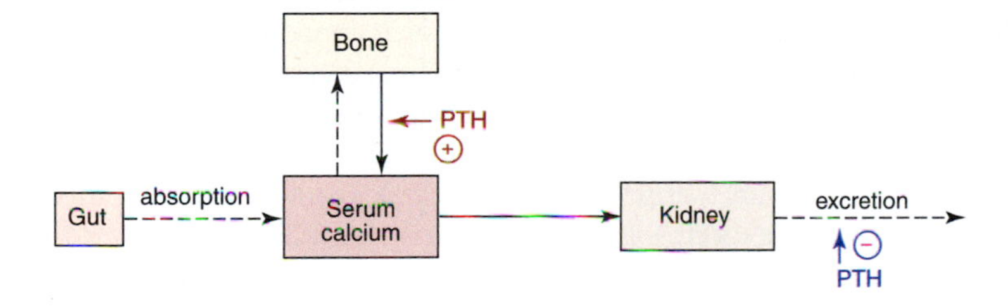

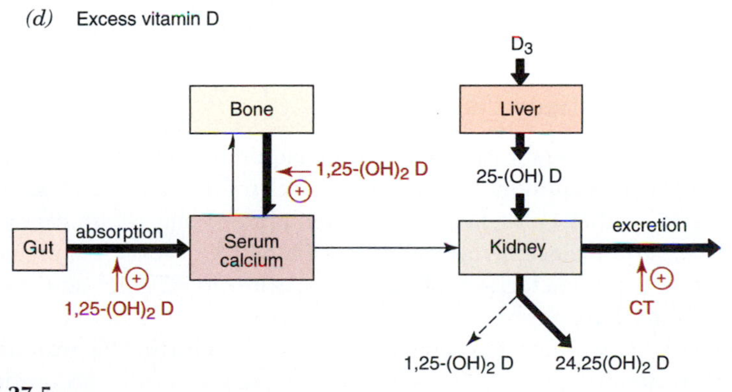

FIGURE 27.5
Vitamin D and calcium homeostasis.
The dominant pathways of calcium metabolism under each set of metabolic conditions are shown with heavy arrows. Effect of various hormones is shown by red arrows for stimulation or blue arrows for repression. PTH, parathyroid hormone; CT, calcitonin; D, cholecalciferol; 25-(OH)D, 25-hydroxycholecalciferol; 1,25-$(OH)_2$D, 1-α,25-dihydroxycholecalciferol.

unsaturated fatty acids (especially in membranes) from peroxidation reactions. α-Tocopherol appears to play a role in cellular respiration, either by stabilizing ubiquinone or by helping transfer electrons to ubiquinone (see p. 567). It also appears to enhance **heme synthesis** by increasing the levels of δ-aminolevulinic acid (ALA) synthase and ALA dehydratase. Most of these effects of α-tocopherol may be an indirect effect of its antioxidant potential, rather than its actual participation as a coenzyme in any biochemical reactions. For example, an important role of α-tocopherol in humans is to prevent **oxidation of LDL,** since the oxidized form of LDL may be atherogenic. Finally, neurological symptoms have been reported following prolonged vitamin E deficiency associated with malabsorption diseases. γ-Tocopherol, on the other hand, appears to inactivate fat-soluble electrophilic mutagens, thus complementing glutathione, which inactivates electrophilic mutagens in the aqueous compartments of the cell.

Setting the recommended intake levels of vitamin E has been hampered by the difficulty of producing severe vitamin E deficiency in humans. It is generally assumed that the vitamin E levels in the American diet are sufficient, since no major vitamin E deficiency diseases have been found. However, vitamin E requirements increase as intake of polyunsaturated fatty acids (PUFAs) increases. While the recent emphasis on high PUFA diets to reduce serum cholesterol may be of benefit in controlling heart disease, the propensity of PUFA to form free radicals on exposure to oxygen may lead to an increased cancer risk. Thus it is prudent to increase vitamin E intake in high PUFA diets.

Premature infants fed formulas low in vitamin E may develop a form of hemolytic anemia that can be corrected by vitamin E supplementation. Adults suffering from fat malabsorption have a decreased red blood cell survival time. Hence supplementation may be necessary with premature infants and in cases of fat malabsorption. Recent studies have suggested that supplementation with at least 100 mg day^{-1} of vitamin E may decrease the risk of heart disease. However, these studies are not definitive, so the Food and Nutrition Board has recommended an increase in the RDA for vitamin E to 15 mg day^{-1}, which corresponds to approximately 22 International Units (IU) of α-tocopherol per day. This is more than the amount of vitamin E found in the typical American diet, but far less than the levels used in clinical intervention studies evaluating the effect of vitamin E on the heart disease risk.

Interestingly, vitamin E appears to be one case where the natural form of the vitamin is clearly more effective than the synthetic form. Because the α-tocopherol transfer protein in the liver specifically binds to the natural *RRR*-α-tocopherol, the *RRR* form of α-tocopherol is retained by the body four to six times longer than the synthetic *all rac* or *d,l* form of α-tocopherol. Vitamin E appears to be the least toxic of the fat-soluble vitamins. The UL for vitamin E has been set at 1000 mg day^{-1} (1500 IU of *RRR*-α-tocopherol) primarily because high levels of vitamin could potentiate the effects of blood thinning medications such as dicumarol.

Vitamin K Is a Quinone Derivative

Vitamin K is found naturally as K_1 **(phytylmenaquinone)** in green vegetables and K_2 **(multiprenylmenaquinone),** which is synthesized by intestinal bacteria. The body converts synthetical menaquinone **(menadione)** and a number of water-soluble analogs to a biologically active form of vitamin K . Dietary requirements are measured in terms of micrograms of vitamin K_1 with the RDA for adults being in the range of 60–80 μg day^{-1}.

Vitamin K_1 is required for conversion of several **clotting factors** and **prothrombin precursors** to the active state. The mechanism of this action has been most clearly delineated for prothrombin (see p. 1047). Prothrombin is synthesized as an inactive precursor called preprothrombin. Conversion to the active form requires a vitamin K-dependent **carboxylation** of specific glutamic acid residues to **γ-carboxyglutamic** acid (Figure 27.6). The γ-carboxyglutamic acid residues are good chelators and allow prothrombin to bind Ca^{2+}. The prothrombin–Ca^{2+} com-

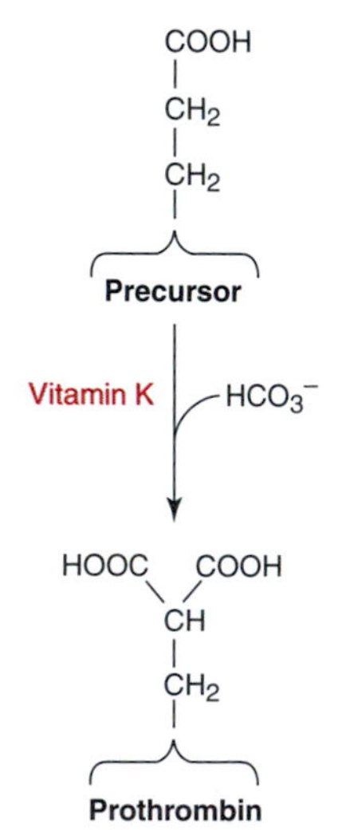

FIGURE 27.6
Function of vitamin K.

plex in turn binds to the phospholipid membrane, where proteolytic conversion to thrombin can occur *in vivo*. The mechanism of the carboxylation reaction has not been fully clarified but appears to involve the intermediate formation of a 2,3-epoxide derivative of vitamin K. **Dicumarol,** a naturally occurring anticoagulant, inhibits the reductase, which converts the epoxide back to the active vitamin.

Vitamin K is also essential for the synthesis of γ-carboxyglutamic acid residues in the protein **osteocalcin,** which accounts for 15–20% of the noncollagen protein in the bone of most vertebrates. These residues are responsible for most of the calcium-binding properties of osteocalcin. Because osteocalcin synthesis is controlled by vitamin D and osteocalcin may play an important role in bone remodeling, vitamin K may be important for bone formation.

The only readily detectable symptom in humans of **vitamin K deficiency** is increased coagulation time. Some studies have indicated that vitamin K deficiency may be a factor in **osteoporosis** as well. Since vitamin K is synthesized by intestinal bacteria, deficiencies have long been assumed to be rare. However, intestinally synthesized vitamin K may not be efficiently absorbed and marginal vitamin K deficiencies may be common. The most common deficiency occurs in newborn infants (see Clin. Corr. 27.3), especially those of mothers on anticonvulsant therapy (see Clin. Corr. 27.4). Vitamin K deficiency also occurs in patients with **obstructive jaundice** and other diseases leading to severe **fat malabsorption** (see Clin. Corr. 27.1) and patients on long-term antibiotic therapy (which may destroy vitamin K-synthesizing organisms in the intestine). Deficiency is sometimes seen in the elderly, who are prone to poor liver function (reducing preprothrombin synthesis) and fat malabsorption. Clearly vitamin K deficiency should be suspected in patients demonstrating easy bruising and prolonged clotting time.

CLINICAL CORRELATION 27.3
Nutritional Considerations in the Newborn

Newborn infants are at special nutritional risk because of very rapid growth and needs for many nutrients are high. Some micronutrients (such as vitamins E and K) do not cross the placental membrane well and tissue stores are low in the newborn. The gastrointestinal tract may not be fully developed, leading to malabsorption problems (particularly with respect to the fat-soluble vitamins). The gastrointestinal tract is also sterile at birth and the intestinal flora that normally provide significant amounts of certain vitamins (especially vitamin K) take several days to become established. If the infant is born prematurely, the nutritional risk is slightly greater, since the gastrointestinal tract will be less well developed and the tissue stores will be less.

The most serious nutritional complications of newborns appear to be hemorrhagic disease. Newborns, especially premature infants, have low tissue stores of vitamin K and lack the intestinal flora necessary to synthesize the vitamin. Breast milk is a relatively poor source of vitamin K. Approximately 1 out of 400 live births shows some signs of hemorrhagic disease, which can be prevented by 1 mg of the vitamin given at birth.

Most newborn infants have sufficient reserves of iron to last 3–4 months (although premature infants have smaller reserves). Since iron is present in low amounts in both cow's milk and breast milk, iron supplementation is usually initiated at a relatively early age by the introduction of iron-fortified cereal. Vitamin D levels are also somewhat low in breast milk and supplementation with vitamin D is usually recommended. However, recent studies have suggested that iron in breast milk is present in a form that is particularly well utilized by infants and that earlier studies probably underestimated the amount of vitamin D available in breast milk. Other vitamins and minerals are present in adequate amounts in breast milk if the mother is getting a good diet. When infants must be maintained on assisted ventilation with high oxygen concentrations, supplemental vitamin E may reduce the risk of bronchopulmonary dysplasia and retrolental fibroplasia, potential complications of oxygen therapy. The anemia of prematurity may respond to supplemental folate and vitamin B_{12}.

In summary, supplemental vitamin K is given at birth to prevent hemorrhagic disease. Breast-fed infants are usually provided with supplemental vitamin D, with iron being introduced along with solid foods. Bottle-fed infants are provided with supplemental iron. If infants must be maintained on oxygen, supplemental vitamin E may be beneficial.

Barness, L. A. Pediatrics. In: H. Schneider, C. E. Anderson, and D. B. Coursin (Eds.), *Nutritional Support of Medical Practice,* 2nd ed. New York: Harper & Row, 1983, p. 541; Huysman, M. W. and Sauer, P. J. The vitamin K controversy. *Curr. Opin. Pediatr.* 6:129,1994; Worthington-White, D. A., Behnke, M., and Gross, S. Premature infants require additional folate and vitamin B_{12} to reduce the severity of anemia of prematurity. *Am. J. Clin. Nutr.* 60:930, 1994, and Mueller, D. P. R. Vitamin E therapy in retinopathy of prematurity. *Eye* 6:221, 1992.

CLINICAL CORRELATION 27.4
Anticonvulsant Drugs and Vitamin Requirements

Anticonvulsant drugs such as phenobarbital or diphenylhydantoin (DPH) present an excellent example of the type of drug–nutrient interactions that are of concern to the physician. Metabolic bone disease appears to be the most significant side effect of prolonged anticonvulsant therapy. Whereas children and adults who take these drugs seldom develop rickets or severe osteomalacia, as many as 65% of those on long-term therapy will have abnormally low serum calcium and phosphorus and abnormally high serum alkaline phosphatase. Some bone loss is usually observed in these cases. While the hypocalcemia and bone loss are thought to be an effect of the anticonvulsant drugs on vitamin D metabolism, not all studies have shown decreased levels of 25-(OH)D and 1,25-$(OH)_2D$ in patients on these drugs. However, supplemental vitamin D in the range of 2000–10,000 units day^{-1} appears to correct both the hypocalcemia and osteopenia. Anticonvulsants also tend to increase needs for vitamin K, leading to an increased incidence of hemorrhagic disease in infants born to mothers on anticonvulsants. Anticonvulsants appear to increase the need for folic acid and B_6. Low serum folate levels occur in 75% of patients on anticonvulsants and megaloblastic anemia may occur in as many as 50% without supplementation. By biochemical parameters, 30–60% of the children on anticonvulsants exhibit some form of B_6 deficiency. Clinical symptoms of B_6 deficiency are rarely seen, however. From 1 to 5 mg of folic acid and 10 mg of vitamin B_6 appear to be sufficient for most patients on anticonvulsants. Since folates may speed up the metabolism of some anticonvulsants, it is important that excess folic acid not be given.

Moslet, U. and Hansen, E. S. A review of vitamin K, epilepsy and pregnancy. *Acta Neurol. Scand.* 85:39, 1992: Rivery, M. D. and Schottelius, D. D. Phenytoin-folic acid: a review. *Drug Intell. Clin. Pharmacy* 18:292,1984; and Tjellesen, L. Metabolism and action of vitamin D in epileptic patients on anticonvulsant treatment and healthy adults. *Dan. Med. Bull.* 41:139,1994.

27.5 | WATER-SOLUBLE VITAMINS

Water-soluble vitamins differ from fat-soluble vitamins in several respects. Most are readily excreted once their concentration surpasses the renal threshold. Thus toxicities are rare. Deficiencies of these vitamins occur relatively quickly on an inadequate diet. Their metabolic stores are labile and depletion can often occur in a matter of weeks or months. Since the water-soluble vitamins are coenzymes for many common biochemical reactions, it is often possible to assay vitamin status by measuring one or more enzyme activities in isolated red blood cells. These assays are especially useful if one measures the endogenous activity and the stimulated activity following addition of the active coenzyme derived from that vitamin.

Most water-soluble vitamins are converted to coenzymes, used in the pathways for energy generation or hematopoiesis. Deficiencies of the energy-releasing vitamins produce a number of overlapping symptoms. In many cases the vitamins participate in so many biochemical reactions that is impossible to pinpoint the exact biochemical cause of any given symptom. However, because of the central role these vitamins play in energy metabolism, deficiencies show up first in rapidly growing tissues. Typical symptoms include **dermatitis, glossitis** (swelling and reddening of the tongue), **cheilitis** at the corners of the lips, and **diarrhea.** In many cases nervous tissue is also involved due to its high-energy demand or specific effects of the vitamin. Some of the common neurological symptoms include **peripheral neuropathy** (tingling of nerves at the extremities), depression, mental confusion, lack of motor coordination, and **malaise.** In some cases demyelination and degeneration of nervous tissues also occurs. These deficiency symptoms are so common and overlapping that they can be considered as properties of the energy-releasing vitamins as a class, rather than being specific for any one.

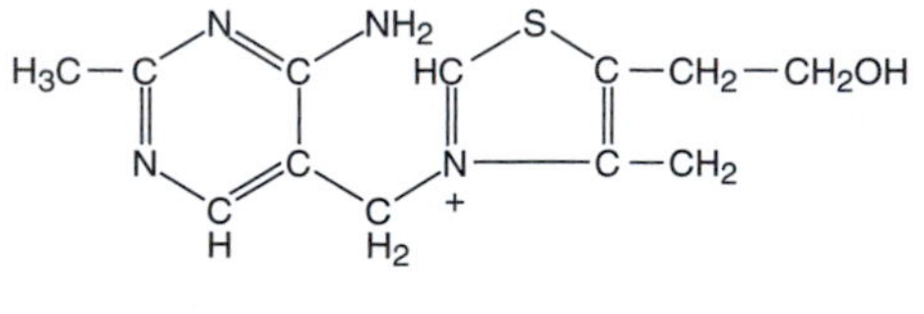

Thiamine

FIGURE 27.7
Structure of thiamine.

27.6 | ENERGY-RELEASING WATER-SOLUBLE VITAMINS

Thiamine (Vitamin B_1) Forms the Coenzyme Thiamine Pyrophosphate (TPP)

Thiamine (Figure 27.7) is rapidly converted to the coenzyme **thiamine pyrophosphate (TPP),** which is required for the key reactions catalyzed by pyruvate dehydrogenase complex and α-ketoglutarate dehydrogenase complex (see Figure 27.8). Cellular energy generation is severely compromised in thiamine deficiency. TPP is also required for the transketolase reactions of the pentose phosphate pathway. The pentose phosphate pathway is the sole biosynthetic source of ribose for the synthesis of nucleic acid precursors and the major source of NADPH for fatty acid biosynthesis and other biosynthetic pathways. Red blood cell transketolase is also the enzyme most commonly used for measuring thiamine status in the body. TPP appears to function in transmission of nerve impulses. TPP (or a related metabolite, thiamine triphosphate) is localized in peripheral nerve membranes. It appears to be required for acetylcholine synthesis and may also be required for ion translocation reactions in stimulated neural tissue.

Although the reactions involving TPP are fairly well characterized, it is not clear how they result in symptoms of **thiamine deficiency.** Pyruvate dehydrogenase and transketolase reactions are the most sensitive to thiamine levels. Thiamine deficiency appears to selectively inhibit carbohydrate metabolism, causing an accumulation of pyruvate. Cells may directly be affected by lack of available energy and NAPDH or may be poisoned by the accumulated pyruvate. Other symptoms of thiamine deficiency involve the neural tissue and may result from the direct role of TTP in nerve transmission.

Loss of appetite, constipation, and nausea are among the earliest symptoms of thiamine deficiency. Mental depression, peripheral neuropathy, irritability, and fatigue are other early symptoms. These symptoms of thiamine deficiency are most often seen in the elderly and low-income groups on restricted diets. Symptoms of moderately severe thiamine deficiency include **mental confusion, ataxia** (unsteady gait while walking and general inability to achieve fine control of motor functions), and **ophthalmoplegia** (loss of eye coordination). This set of symptoms is usually referred to as **Wernicke–Korsakoff syndrome** and is most commonly seen in chronic **alcoholics** (see Clin. Corr. 27.5). Severe thiamine deficiency is known as beriberi. Dry beriberi is characterized primarily by advanced neuromuscular symptoms,

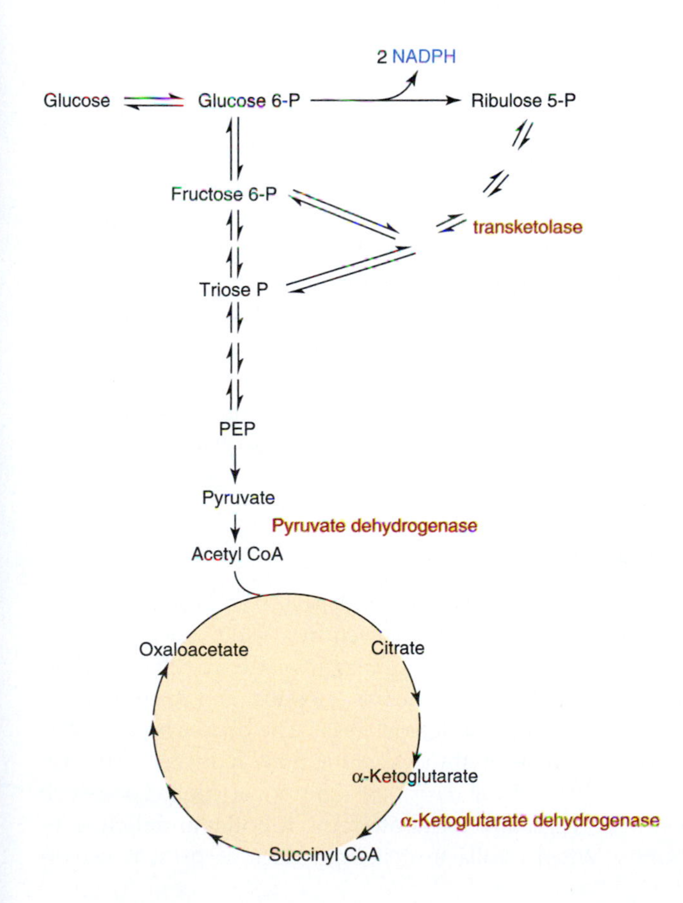

FIGURE 27.8
Summary of important reactions involving thiamine pyrophosphate. The reactions involving thiamine pyrophosphate are indicated in red.

CLINICAL CORRELATION 27.5
Nutritional Considerations in Alcoholics

Chronic alcoholics run considerable risk of nutritional deficiencies. The most common problems are neurologic symptoms associated with thiamine or pyridoxine deficiencies and hematological problems associated with folate or pyridoxine deficiencies. The deficiencies seen with alcoholics are not necessarily due to poor diet alone, although it is often a strong contributing factor. Alcohol causes pathological alterations of the gastrointestinal tract that often directly interfere with absorption of certain nutrients. The severe liver damage associated with chronic alcoholism appears to interfere directly with storage and activation of nutrients and vitamins.

Up to 40% of hospitalized alcoholics have megaloblastic erythropoiesis due to folate deficiency. Alcohol interferes with folate absorption and alcoholic cirrhosis impairs storage of this nutrient. Another 30% of hospitalized alcoholics have sideroblastic anemia or identifiable sideroblasts in erythroid marrow cells characteristic of pyridoxine deficiency. Some alcoholics also develop a peripheral neuropathy that responds to pyridoxine supplementation. This problem appears to result from impaired activation and increased degradation of pyridoxine. In particular, acetaldehyde (an end product of alcohol metabolism) displaces pyridoxal phosphate from its carrier protein in the plasma. Pyridoxal phosphate then rapidly degraded to inactive compounds and excreted.

The most dramatic nutritionally related neurological disorder is the Wernicke–Korsakoff syndrome. The symptoms include mental disturbances, ataxia (unsteady gait and lack of fine motor coordination), and uncoordinated eye movements. Congestive heart failure similar to that seen with beriberi may also occur. While this syndrome may only account for 1–3% of alcohol-related neurologic disorders, the response to supplemental thiamine is so dramatic that it is usually worth consideration. The thiamine deficiency probably arises from impaired absorption, although alcoholic cirrhosis may also affect the storage of thiamine in the liver.

Deficiencies of almost any of the water-soluble vitamins can occur and cases of alcoholic scurvy and pellagra are occasionally reported. Chronic ethanol consumption causes an interesting redistribution of vitamin A stores in the body. Vitamin A stores in the liver are rapidly depleted while levels of vitamin A in the serum and other tissues may be normal or slightly elevated. Apparently, ethanol causes both increased mobilization of vitamin A from the liver and increased catabolism of liver vitamin A to inactive metabolites by the hepatic cytochrome P450 system. Alcoholic patients have decreased bone density and an increased incidence of osteoporosis. This probably relates to a defect in the 25-hydroxylation step in the liver as well as an increased rate of metabolism of vitamin D to inactive products by an activated cytochrome P450 system. Alcoholics generally have decreased serum levels of zinc, calcium, and magnesium due to poor dietary intake and increased urinary losses. Iron deficiency anemia is very rare unless there is gastrointestinal bleeding or chronic infection. In fact, excess iron is a more common problem with alcoholics. Many alcoholic beverages contain relatively high iron levels, and alcohol may enhance iron absorption.

Hayumpa, A. M. Mechanisms of vitamin deficiencies in alcoholism. *Alcohol. Clin. Exp. Res.* 10:573, 1986; and Lieber, C. S. Alcohol, liver and nutrition. *J. Am. Coll. Nutr.* 10:602,1991.

including muscular atrophy and weakness. When these symptoms are coupled with edema, the disease is referred to as wet beriberi. Both forms of beriberi can be associated with an unusual type of heart failure characterized by high cardiac output. Beriberi occurs primarily in populations relying exclusively on polished rice for food, although cardiac failure is sometimes seen in alcoholics as well.

The thiamine requirement is proportional to caloric content of the diet being in the range of 1.0–1.2 mg day^{-1} for the normal adult. This should be increased somewhat if carbohydrate intake is excessive or if the metabolic rate is elevated (due to fever, trauma, pregnancy, or lactation). Coffee and tea contain substances that destroy thiamine, but this is not a problem with normal consumption. Routine enrichment of cereals has assured that most Americans have an adequate intake of thiamine on a mixed diet.

Riboflavin Forms FAD and FMN

Riboflavin is the precursor of flavin adenine dinucleotide (FAD) and flavin mononucleotide (FMN) coenzymes, both of which are involved in a wide variety of redox reactions essential for energy production and cellular respiration. Characteristic symptoms of riboflavin deficiency are angular cheilitis, glossitis, and scaly dermatitis (especially around the nasolabial folds and scrotal areas). The best flavin-requiring enzyme for assaying riboflavin status is erythrocyte glutathione reductase. The recommended riboflavin intake is 1.0–1.3 mg day^{-1} for the normal adult. Foods rich in riboflavin include milk, meat, eggs, and cereal products. **Riboflavin deficiencies** are quite rare in this country. When riboflavin deficiency does occur, it is usually seen in chronic alcoholics. Hypothyroidism slows the conversion of riboflavin to FMN and FAD. It is not known whether this affects riboflavin requirements.

Niacin Forms the Coenzymes NAD and NADP

Niacin is not a vitamin in the strictest sense of the word, since some niacin can be synthesized from tryptophan. However, conversion of tryptophan to niacin is relatively inefficient (60 mg of tryptophan is required for the production of 1 mg of niacin) and occurs only after all of the body requirements for tryptophan (protein synthesis and energy production) have been met. Since synthesis of niacin requires thiamine, pyridoxine, and riboflavin, it is also very inefficient on a marginal diet. Thus most people require dietary sources of both tryptophan and niacin. Niacin (nicotinic acid) and niacinamide (nicotinamide) are both converted to the ubiquitous oxidation–reduction coenzymes NAD^+ and $NADP^+$ in the body.

Borderline **niacin deficiencies** result in a glossitis (redness) of the tongue, somewhat similar to riboflavin deficiency. Pronounced deficiencies lead to **pellagra,** which is characterized by the three D's: **dermatitis, diarrhea,** and **dementia.** The dermatitis is characteristic in that it is usually seen only in skin areas exposed to sunlight and is symmetric. The neurologic symptoms are associated with actual degeneration of nervous tissue. Because of food fortification, pellagra is a medical curiosity in the developed world. Today it is primarily seen in **alcoholics,** patients with severe **malabsorption** problems, and **elderly** on very restricted diets. Pregnancy, lactation, and chronic illness lead to increased needs for niacin, but a varied diet will usually provide sufficient amounts.

Since tryptophan can be converted to niacin, and niacin can exist in a free or bound form, the calculation of available niacin for any given food is not a simple matter. For this reason, niacin requirements are expressed in terms of niacin equivalents (1 niacin equiv = 1 mg free niacin). The current recommendation of the Food and Nutrition Board for a normal adult is 12–16 niacin equivalents (NE) per day. The richest food sources of niacin are meats, peanuts and other legumes, and enriched cereals.

Pyridoxine (Vitamin B_6) Forms the Coenzyme Pyridoxal Phosphate

Pyridoxine, pyridoxamine, and **pyridoxal** are naturally occurring forms of vitamin B_6 (see Figure 27.9). They are efficiently converted by the body to **pyridoxal phosphate,** which is required for the synthesis, catabolism, and interconversion of amino acids. The role of pyridoxal phosphate in amino acid metabolism has been discussed (see p. 783). While pyridoxal phosphate-dependent reactions are numerous, there are a few instances in which the biochemical lesion seems to be directly associated with the symptoms of B_6 deficiency (Figure 27.10). Pyridoxal phosphate is essential for energy production from amino acids and can be considered an energy-releasing vitamin. Thus some symptoms of severe B_6 deficiency are similar to those of the other energy-releasing vitamins. Pyridoxal phosphate is also required for the synthesis of the neurotransmitters serotonin and norepinephrine and for synthesis of the sphingolipids necessary for myelin formation. This may explain the irritability, nervousness, and depression seen with mild deficiencies and the peripheral neuropathy and convulsions observed with severe deficiencies. Pyridoxal phosphate is required for the synthesis of δ-aminolevulinic acid, a precursor of heme. **B_6 deficiencies** occasionally cause sideroblastic microcytic anemia in the presence of high serum iron. Pyridoxal phosphate is covalently linked to a lysine residue of glycogen phosphorylase and stabilizes the enzyme. This may explain the decreased glucose tolerance associated with deficiency, although B_6 appears to have some direct effects on the glucocorticoid receptor as well. Vitamin B_6 is also required for the conversion of homocysteine to cysteine. This is important because **hyperhomocysteinemia** appears to be a risk factor for cardiovascular disease. Finally, pyridoxal phosphate is one of the cofactors required for the conversion of tryptophan to NAD. While this

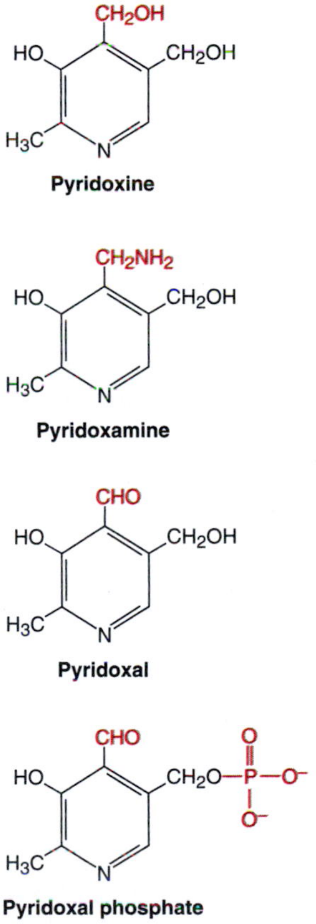

Pyridoxine

Pyridoxamine

Pyridoxal

Pyridoxal phosphate

FIGURE 27.9
Structures of vitamin B_6.

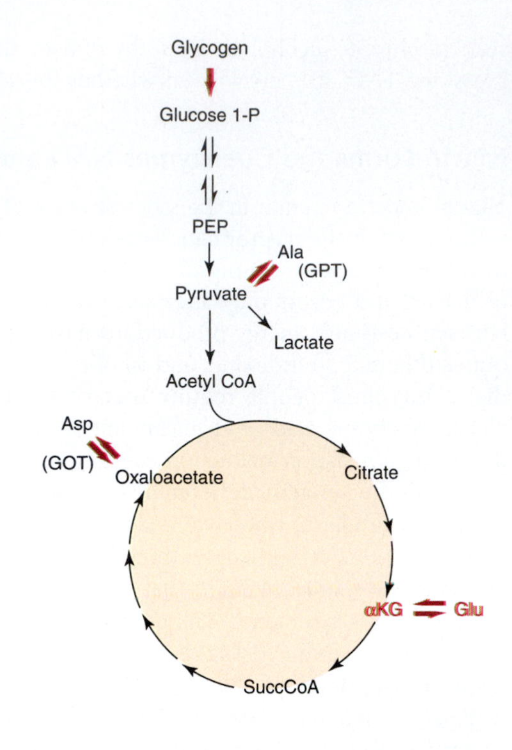

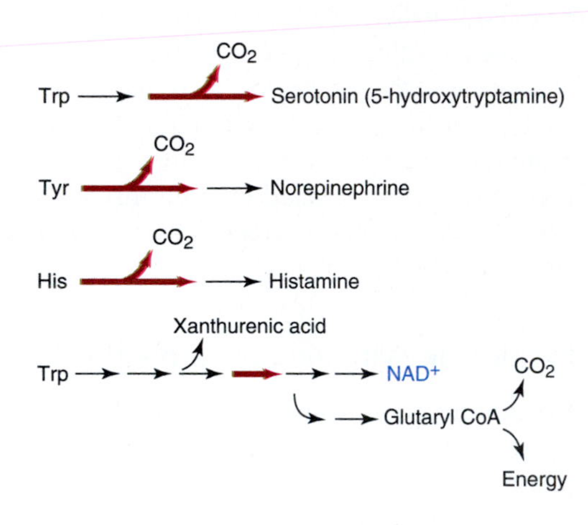

FIGURE 27.10
Some important metabolic roles of pyridoxal phosphate.
Reactions requiring pyridoxal phosphate are indicated with red arrows. ALA, δ-aminolevulinic acid; αKG, α-ketoglutarate; GPT, glutamate pyruvate aminotransferase; and GOT, glutamate oxaloacetate aminotransferase.

may not be directly related to the symptomatology of B_6 deficiency, a tryptophan load test is a sensitive indicator of vitamin B_6 status (see Clin. Corr. 27.6).

The dietary requirement for B_6 is roughly proportional to the protein content of the diet. Assuming that the average American consumes close to 100 g of protein per day, the RDA for vitamin B_6 is 1.3–1.7 mg day^{-1} for normal adults. This is increased during pregnancy and lactation and may increase somewhat with age. Vitamin B_6 is fairly widespread in foods, but meat, vegetables, whole-grain cereals, and egg yolks are among the richest sources.

Evaluation of B_6 nutritional status has become a controversial topic in recent years as discussed in Clinical. Correlation. 27.6. It has usually been assumed that the average American diet is adequate in B_6 and it is not routinely added to flour and other fortified foods. However, recent nutritional surveys have cast doubt on this assumption. A significant fraction of the survey population was found to consume less than the EAR for B_6.

CLINICAL CORRELATION 27.6

Vitamin B_6 Requirements in Users of Oral Contraceptives

The controversy over B_6 requirements for users of oral contraceptives best illustrates the potential problems associated with biochemical assays. For years, one of the most common assays for vitamin B_6 status had been the tryptophan load assay. This assay is based on the observation that when tissue pyridoxal phosphate levels are low, the normal catabolism of tryptophan is impaired and most of the tryptophan is catabolized by a minor pathway leading to synthesis of xanthurenic acid. Under many conditions, the amount of xanthurenic acid recovered in a 24-h urine sample following ingestion of a fixed amount of tryptophan is a valid indicator of vitamin B_6 status. When the tryptophan load test was used to assess the vitamin B_6 status of users of oral contraceptives, however, alarming reports started appearing in the literature. Not only did oral contraceptive use increase the excretion of xanthurenic acid considerably but the amount of pyridoxine needed to return xanthurenic acid excretion to normal was 10 times the RDA and almost 20 times the level required to maintain normal B_6 status in control groups. As might be expected, this observation received much popular attention in spite of the fact that most classical symptoms of vitamin B_6 deficiency were not observed in oral contraceptive users.

More recent studies using other measures of vitamin B_6 have painted a different picture. For example, erythrocyte glutamate pyruvate aminotransferase and erythrocyte glutamate oxaloacetate aminotransferase are both pyridoxal phosphate-containing enzymes. One can also assess vitamin B_6 status by measuring the endogenous activity of these enzymes and the degree of stimulation by added pyridoxal phosphate. These types of assays show a much smaller difference between nonusers and users of oral contraceptives. The minimum level of pyridoxine needed to maintain normal vitamin B_6 status as measured by these assays was only 2.0 mg day^{-1}, which is 1.5-fold greater than the RDA and about twice that needed by nonusers.

Why the large discrepancy? For one thing, it must be kept in mind that enzyme activity can be affected by hormones as well as vitamin cofactors. Kynureninase is the key pyridoxal phosphate-containing enzyme of the tryptophan catabolic pathway and is regulated both by pyridoxal phosphate availability and by estrogen metabolites. Even with normal vitamin B_6 status most of the enzyme exists in the inactive apoenzyme form. However, this does not affect tryptophan metabolism because tryptophan oxygenase, the first enzyme of the pathway, is rate limiting. Thus the small amount of active holoenzyme is more than sufficient to handle the metabolites produced by the first part of the pathway. However, kynureninase is inhibited by estrogen metabolites. Thus with oral contraceptive use its activity is reduced to a level where it becomes rate limiting and excess tryptophan metabolites are shunted to xanthurenic acid. Higher than normal levels of vitamin B_6 overcome this problem by converting more apoenzyme to holoenzyme, thus increasing the total amount of enzyme. Since the estrogen was having a specific effect on the enzyme used to measure vitamin B_6 states in this assay, it did not necessarily mean that pyridoxine requirements were altered for other metabolic processes in the body.

Does this mean that vitamin B_6 status is of no concern to users of oral contraceptives? Oral contraceptives do appear to increase vitamin B_6 requirements slightly. Several dietary surveys have shown that a significant percentage of women in the 18–24-year age group consume diets containing less than the recommended 1.3 mg pyridoxine day^{-1}. If these women are also using oral contraceptives, they are at some increased risk for developing a borderline deficiency. Thus, while the tryptophan load test was clearly misleading in a quantitative sense, it did alert the medical community to a previously unsuspected nutritional risk.

Bender, D. A. Oestrogens and vitamin B_6: actions and interactions. *World Rev. Nutr. Diet.* 51:140, 1987; and Kirksey, A., Keaton, K., Abernathy, R. P., and Grager, J. L. Vitamin B_6 nutritional status of a group of female adolescents. *Am. J. Clin. Nutr.* 31:946, 1978.

Pantothenic Acid and Biotin Form Coenzymes Involved in Energy Metabolism

Pantothenic acid is a component of **coenzyme A** (CoA) and the phosphopantetheine moiety of fatty acid synthase and thus is required for the metabolism of all fat, protein, and carbohydrate via the citric acid cycle and for fatty acid and cholesterol synthesis. More than 70 enzymes have been described to date that utilize CoA or derivatives. In view of the importance of these reactions, one would expect pantothenic acid deficiencies to be a serious concern in humans. This is not the case for two reasons: (1) pantothenic acid is very widespread in natural foods, probably reflecting its widespread metabolic role, and (2) most symptoms of pantothenic acid deficiency are vague and mimic those of other B vitamin deficiencies.

Biotin is the covalently bound prosthetic group of pyruvate carboxylase (needed for synthesis of oxaloacetate for gluconeogenesis and replenishment of the citric acid cycle), acetyl-CoA carboxylase (fatty acid biosynthesis) and propionyl-CoA carboxylase (methionine, leucine, and valine metabolism). Biotin occurs in peanuts, chocolate, and eggs and is synthesized by intestinal bacteria. Recent studies suggest that the biotin synthesized by intestinal bacteria may not be present in a location or a form that can contribute significantly to absorbed biotin.

27.7 | HEMATOPOIETIC WATER-SOLUBLE VITAMINS

Folic Acid (Folacin) Functions as Tetrahydrofolate in One-Carbon Metabolism

The simplest form of **folic acid** is pteroylmonoglutamic acid. However, folic acid usually occurs as **polyglutamate** derivatives with from 2 to 7 glutamic acid residues (Figure 27.11). These compounds are taken up by intestinal mucosal cells and the extra glutamate residues removed by **conjugase,** a lysosomal enzyme. The free folic acid is then reduced to **tetrahydrofolate** by the enzyme **dihydrofolate reductase** and circulated in the plasma primarily as the free N^5-methyl derivative of tetrahydrofolate (Figure 27.11). Inside cells tetrahydrofolates are found primarily as polyglutamate derivatives, and these appear to be the biologically most potent forms. Folic acid is also stored as tetrahydrofolate polyglutamate in liver.

Various one-carbon tetrahydrofolate derivatives are used in biosynthetic reactions (see Figure 27.12) such as the synthesis of choline, serine, glycine, purines, and dTMP. In addition, tetrahydrofolate, vitamin B_{12}, and pyridoxal phosphate are required for the conversion of homocysteine to methionine. Methionine, of course, is also converted to *S*-adenosylmethionine, which is used in many methylation reactions.

The most pronounced effect of **folate deficiency** is inhibition of DNA synthesis due to decreased availability of purines and dTMP. This leads to arrest of cells in S phase and a characteristic "megaloblastic" change in size and shape of nuclei of rapidly dividing cells. The block in DNA synthesis slows down maturation of red blood cells, causing production of abnormally large "macrocytic" red blood cells with fragile membranes. Thus a **macrocytic anemia** associated with megaloblastic changes in the bone marrow is characteristic of folate deficiency. In addition, **hyperhomocysteinemia** is fairly common in the elderly population and appears to be due to inadequate intake and/or decreased utilization of folate, vitamin B_6, and vitamin B_{12}. As mentioned earlier, hyperhomocysteinemia appears to be a risk factor for cardiovascular disease. Elevated homocysteine levels usually respond to supplementation with folic acid, vitamin B_6, and vitamin B_{12}.

Causes of folate deficiency include inadequate intake, increased need, impaired absorption, increased demand, and impaired metabolism. Some dietary surveys suggest that inadequate intake may be more common than previously supposed. However, as with most other vitamins, inadequate intake is probably not sufficient

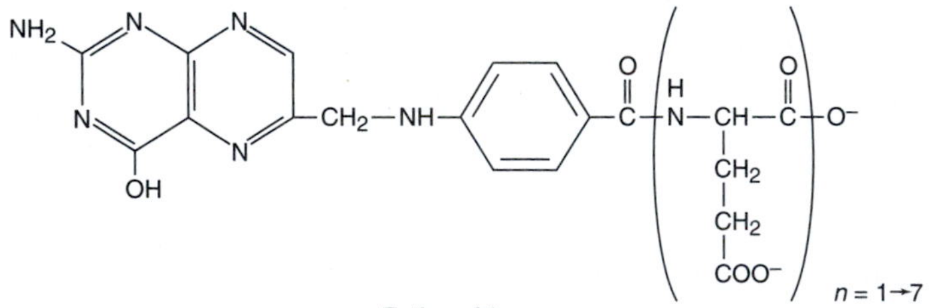

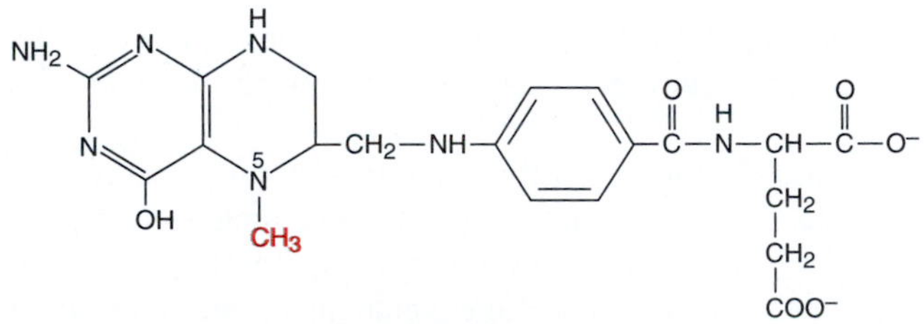

Folic acid ($n = 1 \rightarrow 7$)

N^5-Methyltetrahydrofolate

FIGURE 27.11
Structure of folic acid and N^5-methyltetrahydrofolate.

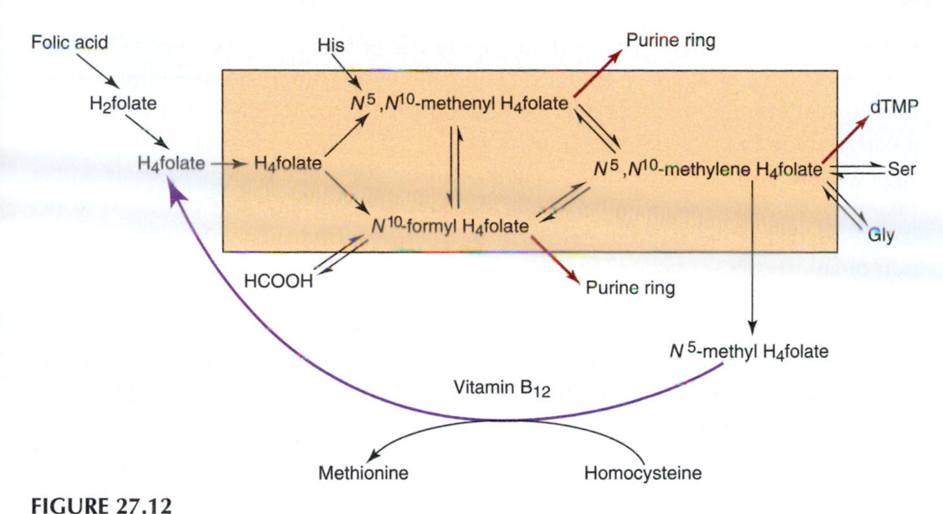

FIGURE 27.12
Metabolic roles of folic acid and vitamin B_{12} in one-carbon metabolism.
Metabolic interconversions of folic acid and its derivatives are indicated with black arrows. Pathways relying exclusively on folate are shown with red arrows. The important B_{12}-dependent reaction converting N^5-methyl H_4folate back to H_4folate is shown with a blue arrow. The box encloses the "pool" of C_1 derivatives of H_4folate.

to trigger symptoms of folate deficiency in the absence of increased requirements or decreased utilization. Recent studies suggest that gene polymorphisms that increased the need for folate may be more common than previously thought (see Clin. Corr. 27.7). However, the most common example of increased need occurs during **pregnancy** and **lactation.** As the blood volume and the number of rapidly dividing cells in the body increase, the need for folic acid increases. By the third trimester the folic acid requirement has almost doubled. In the United States almost 20–25% of otherwise normal pregnancies are associated with low serum folate levels, but actual megaloblastic anemia is rare and is usually seen only after multiple pregnancies. However, recent studies have shown that inadequate folate levels during the early stages of pregnancy increase the risk for **neural tube defects,** a type

CLINICAL CORRELATION 27.7
Gene Polymorphisms and Folic Acid Requirements

Folic acid supplementation lowers the risk of neural tube defects and decreases serum homocysteine levels, which may lower the risk of heart disease. This information has led to an increase in the RDA for folic acid and to fortification of grain products with folic acid. Yet it is clear that even on a marginal diet not all adults have elevated homocysteine levels and not all mothers give birth to babies with neural tube defects. What determines these individual responses to inadequate folate intake? Gene polymorphisms may be part of the answer. A common polymorphism results from a C $\rightarrow$ T substitution in the gene for 5,10-methylenetetrahydrofolate reductase (MTHFR) that produces the 5-methyltetrahydrofolate required for the conversion of homocysteine to methionine. This C $\rightarrow$ T substitution at bp 677 results in a substitution of valine for alanine and causes the enzyme to have lower specific activity and reduced stability. Approximately 12% of the Caucasians and Asians are homozygous (T/T) and 50% are heterozygous (C/T) for this polymorphism. Plasma folate concentrations are significantly lower and plasma homocysteine levels are significantly higher in individuals with the T/T genotype consuming normal diets. In addition, the T/T genotype is a significant risk factor for neural tube defects, perhaps accounting for up to 15% of the cases. An active investigation of genetic polymorphisms in the other genes that code for enzymes involved in folate metabolism is underway. Polymorphisms have been described for the genes coding for methionine synthetase and folate receptor α, the cellular receptor required for 5-methyltetrahydrofolate uptake. Both appear to be benign. However, recent studies suggest that the absorption of folate by the intestine may be lower in mothers with a history of neural tube defect pregnancies than in control mothers. The genetics of this effect has not yet been studied.

Bailey, L. B. and Gregory, J. F. Polymorphisms of methylenetetrahydrofolate reductase and other enzymes: metabolic significance, risks, and impact on folate requirement. *J. Nutr.* 129:919, 1999; Barber, R. C., Lammer, E .J., Shaw, G. M., Greer, K. A., and Finnell, R. H. The role of folate transport and metabolism in neural tube defect risk. *Mol. Genet. Metab.* 66:1, 1999; and Bodie, A. M., Dedlow, E. R., Nackoski, J. A., Opalko, F. J., Kauwell, G. P.A., Gregory, J. F., and Bailey, L. B. Folate absorption in women with a history of neural tube defect-affected pregnancy. *Am. J. Clin. Nutr.* 72:154, 2000.

of birth defect. Normal diets seldom supply the 600 mg of folate needed during pregnancy. Thus, in January 1998, the Food and Drug Administration implemented new legislation requiring that enriched grain products be fortified with folic acid at a concentration of 1.4 mg per gram of product. This level of fortification was designed to raise the average intake of folic acid by 100 mg day^{-1}. This decision has proved to be somewhat controversial. An extra 100 mg day^{-1} is projected to decrease the risk of neural tube defects and hyperhomocysteinemia by ≥20%. Addition of an extra 200 mg day^{-1} to the American diet would offer much greater protection against both neural tube defects and hyperhomocysteinemia, but supplementation with this level of folic acid could mask the symptoms of vitamin B_{12} deficiency as described below. Thus most physicians routinely recommend supplementation for women during the child-bearing years and for the elderly. Folate deficiency is common in alcoholics (see Clin. Corr. 27.5). Folate deficiencies are also seen in a number of malabsorption diseases and are occasionally seen in the elderly, due to a combination of poor dietary habits and poor absorption.

Anticonvulsants and **oral contraceptives** may interfere with folate absorption and anticonvulsants appear to increase catabolism of folates (see Clin. Corr. 27.4). Oral contraceptives and estrogens interfere with folate metabolism in their target tissue. Long-term use of any of these drugs can lead to folate deficiencies unless adequate supplementation is provided. For example, 20% of patients using oral contraceptives develop megaloblastic changes in the cervicovaginal epithelium, and 20–30% show low serum folate.

Vitamin B_{12} (Cobalamine) Contains Cobalt in a Tetrapyrrole Ring

Pernicious anemia, a megaloblastic anemia associated with neurological deterioration, was invariably fatal until 1926 when liver extracts were shown to be curative. Subsequent work showed the need for both an extrinsic factor present in liver and an **intrinsic factor** produced by the body; **vitamin B_{12}** was the extrinsic factor. Vitamin B_{12} consists of **cobalt** in a coordination state of six: coordinated in four positions by a tetrapyrrole (a corrin) ring, in one position by a benzimidazole nitrogen, and in the sixth position by one of several different ligands (Figure 27.13). The crystalline forms of B_{12} used in supplementation are usually hydroxycobalamine or cyanocobalamine. In foods B_{12} usually occurs bound to protein in the methyl or 5′-deoxyadenosyl forms. To be utilized the B_{12} must first be removed from the protein by acid hydrolysis in the stomach or trypsin digestion in the intestine. It must then combine with intrinsic factor, a protein secreted by the stomach, which carries it to the ileum for absorption.

In humans there are two major symptoms of **B_{12} deficiency** (hematopoietic and neurological), and only two biochemical reactions in which B_{12} is known to participate (Figure 27.14). Thus it is very tempting to speculate on exact cause and effect mechanisms. The methyl derivative of B_{12} is required for the methionine synthesis reaction, which converts homocysteine to methionine, and the 5-deoxyadenosyl derivative is required for the methylmalonyl CoA mutase reaction (methylmalonyl CoA → succinyl CoA), a key reaction in the catabolism of some branched-chain amino acids. The neurologic disorders seen in B_{12} deficiency are due to progressive demyelination of nervous tissue. It has been proposed that the methylmalonyl CoA that accumulates interferes with myelin sheath formation in two ways. (1) Methylmalonyl CoA is a competitive inhibitor of malonyl CoA in fatty acid biosynthesis. Because the myelin sheath is continually turning over, any severe inhibition of fatty acid biosynthesis will lead to its degeneration. (2) In fatty acid synthesis, methylmalonyl CoA can substitute for malonyl CoA, leading to synthesis of branched-chain fatty acids, which might disrupt membrane structure. There is some evidence supporting both mechanisms.

The megaloblastic anemia associated with B_{12} deficiency is thought to reflect the effect of B_{12} on folate metabolism. The B_{12}-dependent methionine synthesis reaction (homocysteine $+N^5$-methyl THF → methionine + THF) appears to be the only major pathway by which N^5-methyltetrahydrofolate can return to the tetrahydrofolate

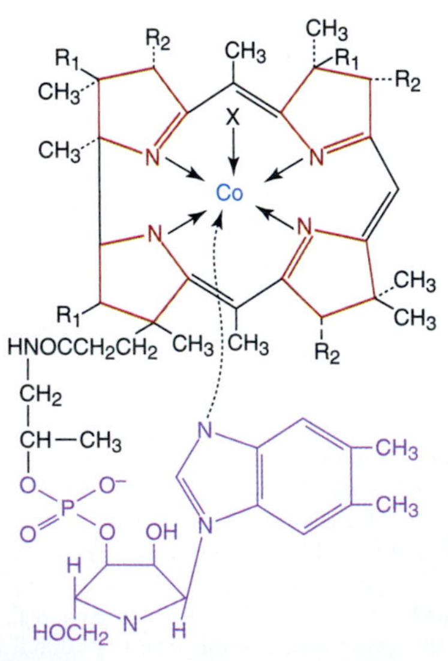

FIGURE 27.13
Structure of vitamin B_{12} (cobalamine).

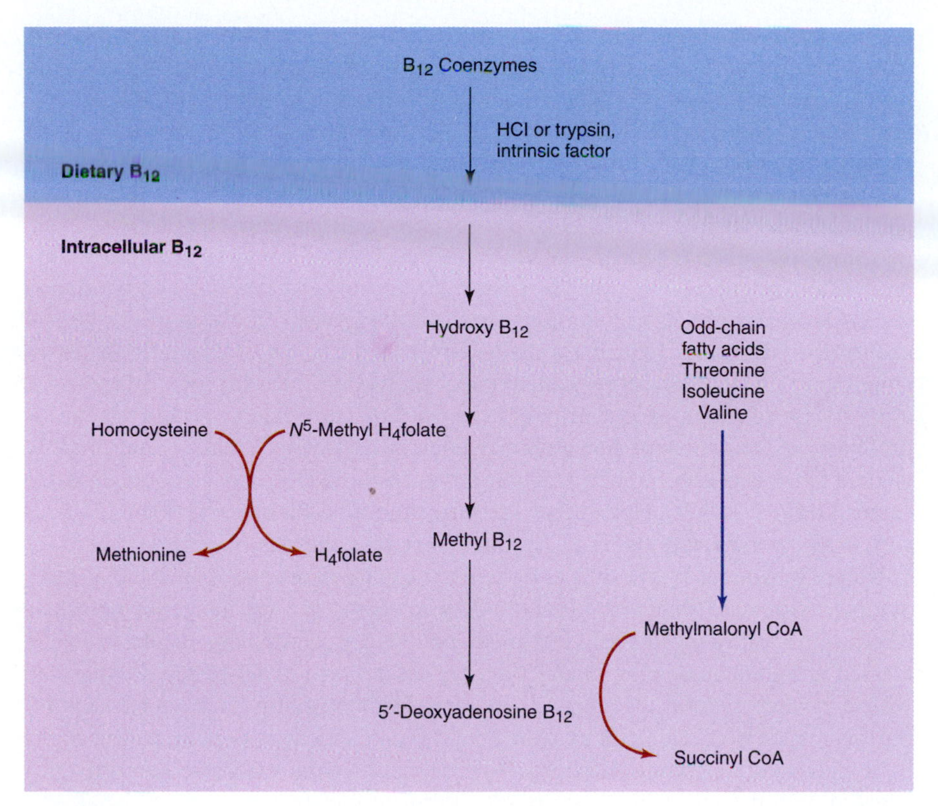

FIGURE 27.14
Metabolism of vitamin B_{12}.
The metabolic interconversions of B_{12} are indicated with black arrows, and B_{12}-requiring reactions are indicated with red arrows. Other related pathways are indicated with blue arrows.

pool (Figure 27.14). Thus in B_{12} deficiency there is a buildup of N^5-methyltetrahydrofolate and a deficiency of the tetrahydrofolate derivatives needed for purine and dTMP biosynthesis. Essentially all of the folate becomes "trapped" as the N^5-methyl derivative. Inhibition of the conversion of N^5-methyltetrahydrofolate to tetrahydrofolate also affects the ability of the cell to convert tetrahydrofolate to its polyglutaminated form, since N^5-methyltetrahydrofolate is a poor substrate for the polyglutamination reaction. This also increases the requirement for folic acid because it is the polyglutaminated form of tetrahydrofolate that is retained in cells. Large amounts of supplemental folate can overcome the megaloblastic anemia associated with B_{12} deficiencies, but not the neurological problems. This is the crux of the current debate on the optimal levels for folate fortification of foods. It is the megaloblastic anemia that usually brings the patient into the doctor's office. Thus, by masking the anemia, routine fortification of foods with high levels of folate could prevent the detection of B_{12} deficiency until the neurological damage had become irreversible.

Vitamin B_{12} is widespread in foods of animal origin, especially meats, and the liver stores up to a 6-year supply of vitamin B_{12}. Thus deficiencies of B_{12} are rare. They are most often seen in older people due to insufficient production of intrinsic factor and/or HCl in the stomach. B_{12} deficiency can also be seen in patients with severe malabsorption diseases and in long-term vegetarians.

27.8 | OTHER WATER-SOLUBLE VITAMINS

Ascorbic Acid Functions in Reduction and Hydroxylation Reactions

Vitamin C or **ascorbic acid** is a six-carbon compound closely related to glucose. Its main biological role is as a reducing agent for the hydroxylation of lysine and proline in protocollagen. Without this hydroxylation protocollagen cannot properly cross-link into normal collagen fibrils. Thus vitamin C is obviously important for maintenance of normal connective tissue and for wound healing, since the connective tissue is laid down first. Vitamin C is also necessary for bone formation, since bone tissue has an organic matrix containing collagen as well as the inorganic, calcified portion. Collagen is a component of the ground substance surrounding capillary walls, so vitamin C deficiency is associated with **capillary fragility.**

Since vitamin C is concentrated in the adrenal gland it may be required for hydroxylation reactions in synthesis of some corticosteroids especially in periods of stress Ascorbic acid acts as a nonenzymatic reducing agent. For example, it aids in **absorption of iron** by reducing it to the ferrous state in the stomach. It spares vitamin A, vitamin E, and some B vitamins by protecting them from oxidation. Also, it enhances utilization of folic acid, either by aiding conversion of folate to tetrahydrofolate or formation of polyglutamate derivatives of tetrahydrofolate. Finally, vitamin C appears to be a biologically important antioxidant. The National Research Council has recently concluded that adequate amounts (RDA levels) of antioxidants such as β-carotene and vitamin C in the diet reduce the risk of cancer.

Most symptoms of **vitamin C deficiency** can be related to its metabolic roles. Symptoms of mild deficiency include easy bruising and formation of petechiae (small, pinpoint hemorrhages in skin) due to increased capillary fragility and decreased immunocompetence. **Scurvy,** a more severe form of deficiency, is associated with decreased wound healing, osteoporosis, hemorrhaging, and anemia. Osteoporosis results from the inability to maintain the collagenous organic matrix of the bone, followed by demineralization. Anemia results from extensive hemorrhaging coupled with defects in iron absorption and folate metabolism.

Vitamin C is readily absorbed so that deficiencies almost invariably result from poor diet and/or increased need. There is uncertainty over the need in periods of stress. In severe stress or trauma there is a rapid drop in serum vitamin C levels. In these situations most of the body's supply of vitamin C is mobilized to the adrenals and/or the area of the wound. Does this represent an increased demand for vitamin C, or merely a normal redistribution to those areas where it is needed most? Do the lowered serum levels of vitamin C impair its functions in other tissues in the body? The current consensus seems to be that the lowered serum vitamin C levels indicate an increased demand, but there is little agreement as to how much.

Smoking causes lower serum levels of vitamin C. In fact, the 2000 RDAs recommend that smokers consume 110–125 mg of vitamin C per day instead of the 75–90 mg day^{-1} needed by nonsmoking adults. **Aspirin** appears to block uptake of vitamin C by white blood cells. **Oral contraceptives** and **corticosteroids** also lower serum levels of vitamin C. While there is no universal agreement as to the seriousness of these effects, the possibility of marginal vitamin C deficiencies should be considered with any patient using these drugs over a long period of time, especially if dietary intake is less than optimal.

Very controversial is the use of megadoses of vitamin C to prevent and cure the **common cold.** Ever since this use of vitamin C was first popularized by Linus Pauling in 1970, the issue has generated considerable controversy. However, some double-blind studies suggest that while vitamin C supplementation does not appear to be useful in preventing the common cold, it may moderate the symptoms or shorten the duration. The mechanism by which vitamin C ameliorates the symptoms of the common cold is not known. It has been suggested that vitamin C is required for normal leukocyte function or for synthesis and release of histamine during stress situations.

While megadoses of vitamin C are probably no more harmful than the widely used over-the-counter cold medications, some potential side effects of high vitamin C intake should be considered. For example, oxalate is a major metabolite of ascorbic acid. Thus high ascorbate intakes could theoretically lead to the formation of oxalate kidney stones in predisposed individuals. However, most studies have shown that excess vitamin C is primarily excreted as ascorbate rather than oxalate. Pregnant mothers taking megadoses of vitamin C may give birth to infants with abnormally high vitamin C requirements. Earlier suggestions that megadoses of vitamin C interfered with B_{12} metabolism have proved to be incorrect. The UL for vitamin C has been set at 2000 mg day^{-1} because higher levels can cause diarrhea in some individuals.

Choline Performs Many Functions in the Body

The vitamin **choline** is required for synthesis and release of acetycholine, an important **neurotransmitter** involved in memory storage, motor control, and other functions. It is also a precursor for synthesis of the phospholipids phosphatidylcholine (lecithin) and sphingomyelin, which are important for membrane function, intracellular signaling, and hepatic export of very low density lipoproteins. Phosphatidylcholine also plays a role in the removal of cholesterol from tissues due to the importance of the lecithin–cholesterol acyltransferase reaction in reverse cholesterol transport (see p. 749). Finally, choline is a precursor for the formation of the methyl donor betaine.

The dietary requirement for choline is not firmly established. Studies in rodents indicate that **choline deficiency** increases the risk of liver cancer and memory deficits in aged animals, but these effects have not been demonstrated in humans. Lecithin supplementation has a modest serum cholesterol lowering effect and both choline and betaine supplementation appear to lower serum homocysteine levels in humans. However, current data are insufficient to draw any firm conclusions about whether choline and/or lecithin supplementation have any effects on cardiovascular risk. There is an endogenous pathway for *de novo* biosynthesis of choline via the sequential methylation of phosphatidylethanolamine using *S*-adenosylmethionine as a methyl donor. Thus the demand for dietary choline is modified by the dietary intake of methionine, folate, and vitamin B_{12}. In one published study on the effects of inadequate dietary choline in healthy men with adequate dietary intakes of methionine, folate, and vitamin B_{12}, at the end of 3 weeks, the subjects had decreased choline stores and elevated alanine aminotransferase, which indicates possible liver damage. Individuals fed with total parenteral nutrition solutions devoid of choline but adequate for methionine and folate also developed fatty livers and elevated serum alanine aminotransferase. On the basis of these and other studies, AIs of 425 mg day^{-1} for women and 550 mg day^{-1} for men have been set by the Food and Nutrition Board. Choline is widely distributed in foods. It is estimated that the average American consumes 730–1040 mg day^{-1}, so dietary deficiencies of choline are unlikely.

27.9 | MACROMINERALS

Calcium Has Many Physiological Roles

Calcium is the most abundant mineral in the body. Most is in bone, but the small amount of Ca^{2+} outside bone functions in a number of essential processes. It is required for many enzymes, mediates some hormonal responses, and is essential for blood coagulation. It is also essential for muscle contractility and normal neuromuscular irritability. In fact only a relatively narrow range of serum Ca^{2+} levels is compatible with life. Since maintenance of constant serum levels is so vital, an elaborate homeostatic control system has evolved (see p. 1145). Low serum Ca^{2+} stimulates formation of 1,25-dihydroxycholecalciferol, which enhances intestinal absorption. If dietary Ca^{2+} intake is insufficient to maintain serum calcium, 1,25-dihydroxycholecalciferol and parathyroid hormone stimulate bone resorption. Long-term dietary Ca^{2+} insufficiency, therefore, almost always results in net loss of calcium from the bones.

Dietary Ca^{2+} requirements, however, vary considerably from individual to individual due to the existence of other factors that affect availability of Ca^{2+}. For example, vitamin D is required for optimal utilization of calcium. Excess dietary protein may upset calcium balance by causing more rapid excretion of Ca^{2+}. Exercise increases the efficiency of calcium utilization for bone formation. Calcium balance studies carried out on Peruvian Indians, who have extensive exposure to sunlight, get extensive exercise, and subsist on low-protein vegetarian diets, indicate a need for only 300–400 mg Ca^{2+} per day. However, calcium balance studies carried out in this country consistently show higher requirements and the RDA has been set at 1000–1300 mg day^{-1}.

The chief symptoms of **Ca^{2+} deficiency** are similar to those of vitamin D deficiency, but other symptoms such as muscle cramps are possible with marginal deficiencies. A significant portion of low-income children and adult females in this country do not have adequate Ca^{2+}. This is of particular concern because these are the population groups with particularly high needs for Ca^{2+}. For this reason, the U.S. Congress has established the WIC (Women and Infant Children) program to assure adequate protein, Ca^{2+}, and iron for indigent families with pregnan/lactating mothers or young infants.

CLINICAL CORRELATION 27.8
Diet and Osteoporosis

The controversies raging over the relationships between calcium intake and osteoporosis illustrate the difficulties we face in making simple dietary recommendations for complex biological problems. Based on the TV ads and wide variety of calcium-fortified foods on the market, it would be easy to assume that all an older woman needs to prevent osteoporosis is a diet rich in calcium. However, that may be like closing the barn door after the horse has left. There is strong consensus that the years from age 10 to 35, when the bone density is reaching its maximum, are the most important for reducing the risk of osteoporosis. The maximum bone density obtained during these years is clearly dependent on both calcium intake and exercise and dense bones are less likely to become seriously depleted of calcium following menopause. Unfortunately, most American women are consuming far too little calcium during these years. The RDA for calcium is 1300 mg day^{-1} (4 or more glasses of milk day^{-1}) for women from age 11 to 18, 1000 mg day^{-1} (3 or more glasses of milk day^{-1}) for women from 19 to 50, and 1200 mg day^{-1} (4 glasses of milk day^{-1}) for women over 50. The median calcium intake for women in this age range is only about 500 mg day^{-1}. Thus it is clear that increased calcium intake should be encouraged in this group.

But what about postmenopausal women? After all, many of the advertisements seem to be targeted at this group. Do they really need more calcium? The 1994 NIH consensus panel on osteoporosis recommended that postmenopausal women consume up to 1500 mg calcium day^{-1}, but this recommendation has been vigorously disputed by other experts in the field. Let's examine the evidence. Calcium balance studies have shown that many postmenopausal women need 1200–1500 mg calcium day^{-1} to maintain a positive calcium balance (more calcium coming in than is lost in the urine), but that does not necessarily mean that the additional calcium will be stored in their bones. In fact, some recent studies have failed to find a correlation between calcium intake and loss of bone density in postmenopausal women while others have reported a protective effect. All of those studies have been complicated by the discovery that calcium intake may have different effects on different types of bones. Calcium intakes in the range of 1000–1500 mg day^{-1} appear to slow the decrease in density of cortical bone, such as that found in the hip, hand, and some parts of the forearm. Similar doses, however, appear to have little or no effect on loss of density from the trabecular bone found in the spine, wrist, and other parts of the forearm. At least some of the confusion in the earlier studies appears to have resulted from differences in the site used for measurement of bone density. Thus the effect of high calcium intakes alone on slowing bone loss in postmenopausal women remains controversial at present. It is clear that elderly women should be getting the RDA for calcium in their diet. With the recent concern about the fat content of dairy products, calcium intakes in this group appear to be decreasing rather than increasing. Furthermore, even with estrogen replacement therapy, calcium intake should not be ignored. Recent studies have shown that with calcium intakes in the range of 1000–1500 mg day^{-1}, the effective dose of estrogen can be reduced significantly.

While advertisements and the popular literature focus on calcium intake, we also need to remember that bones are not made of calcium alone. If the diet is deficient in other nutrients, the utilization of calcium for bone formation will be impaired. Vitamin C is needed to form the bone matrix and magnesium and phosphorus are an important part of bone structure. Vitamin K and a variety of trace minerals, including copper, zinc, manganese, and boron are important for bone formation. Thus calcium supplements may not be optimally utilized if the overall diet is inadequate. Vitamin D is required for absorption and utilization of calcium. It deserves special mention since it may be a particular problem for the elderly (see Clin. Corr. 27.9). Finally, an adequate exercise program is just as important as estrogen replacement therapy and an adequate diet for preventing the loss of bone density.

Schaafsma, G., Van Berensteyn, E. C. H., Raymakers, J. A., and Dursma, S. A. Nutritional aspects of osteoporosis. *World Rev. Nutr. Diet.* 49:121, 1987; Heaney, R. P. Calcium in the prevention and treatment of osteoporosis. *J. Intern. Med.* 231:169, 1992: and National Institutes of Health. *Optimal Calcium Intake.* NIH Consensus Statement, 12 (Nov. 4), 1994.

Dietary surveys show that 34–47% of the population over 60 years of age consumes less than the EAR for Ca^{2+}. This is the group most at risk of developing osteoporosis, characterized by loss of bone organic matrix as well as progressive demineralization. Causes of **osteoporosis** are multifactorial and largely unknown, but it appears likely that part of the problem has to do with Ca^{2+} metabolism (see Clin. Corr. 27.8). Recent studies suggest that inadequate intake of Ca^{2+} may result in elevated blood pressure. Although this hypothesis has not been conclusively demonstrated, it is of great concern because most low-sodium diets (which are recommended for patients with high blood pressure) severely limit dairy products, the main source of Ca^{2+} for Americans.

Magnesium Is Required by Many Enzymes

Magnesium is required for many enzyme activities, particularly those involving ATP forming an ATP–Mg^{2+} complex, and for neuromuscular transmission. Amounts of Mg^{2+} are significantly reduced during processing of food items, and recent dietary surveys have shown that the average Mg^{2+} intake in Western countries is often below the EAR. Deficiency is observed in conditions of alcoholism, use of certain diuretics, and metabolic acidosis. The main symptoms of **Mg^{2+} deficiency** are weakness, tremors, and cardiac arrhythmia. There is some evidence that supplemental Mg^{2+} may help prevent the formation of calcium oxalate stones in the kidney. Mg^{2+} supplementation has also been shown to lower blood pressure in several, but not in all, clinical studies, and the recent Health Professional Follow-Up Study demonstrated an inverse effect between dietary Mg^{2+} intake and the total risk of stroke.

27.10 | TRACE MINERALS

Iron Is Efficiently Reutilized

Iron metabolism is unique in that it operates largely as a closed system, with iron stores being efficiently reutilized by the body. Iron losses are minimal (<1 mg day^{-1}), but iron absorption is also minimal under the best of conditions. Iron usually occurs in foods in the ferric form bound to protein or organic acids. Before absorption can occur, the iron must be split from these carriers (a process that is facilitated by the acid secretions of the stomach) and reduced to the ferrous form (a process that is enhanced by ascorbic acid). Only 10% of the iron in an average mixed diet is usually absorbed, but the efficiency of absorption can be increased to 30% by severe iron deficiency. Iron absorption and metabolism have been discussed in Chapter 24 and are summarized in Figure 27.15.

As a component of hemoglobin and myoglobin, iron is required for O_2 and CO_2 transport. As a component of cytochromes and nonheme iron proteins, it is required for oxidative phosphorylation. As a component of the essential lysosomal enzyme myeloperoxidase, it is required for proper phagocytosis and killing of bacteria by neutrophils. The best-known symptom of iron deficiency is a microcytic hypochromic anemia. **Iron deficiency** is also associated with decreased immunocompetence.

Assuming a 10–15% efficiency of absorption, the Food and Nutrition Board has set a recommended dietary allowance of 10 mg day^{-1} for normal adult males and 15 mg day^{-1} for menstruating females. For pregnant females this allowance is raised to 30 mg day^{-1}. While 10 mg day^{-1} of iron can easily be obtained from a normal diet, 15 mg is marginal at best and 30 mg can almost never be obtained. The best dietary sources are meats, dried legumes, dried fruits, and enriched cereal products.

Iron-deficiency anemia is the most prevalent nutritional disorder in the United States. Young children and pregnant females need enough iron for a continuing

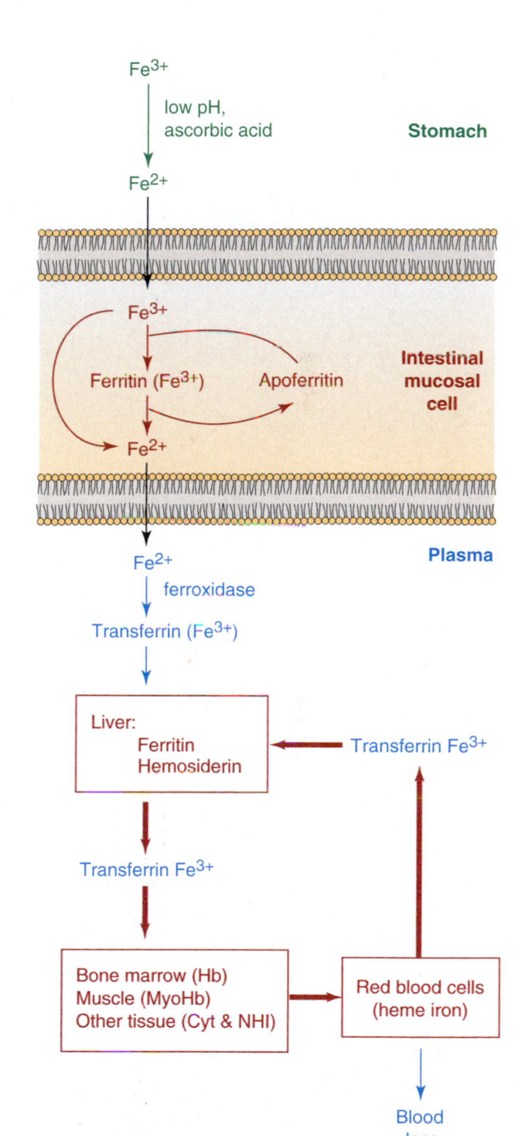

FIGURE 27.15
Overview of iron metabolism.
This figure reviews some of the features of iron metabolism discussed in Chapter 24. Red arrows indicate that most of the body's iron is efficiently reutilized by the pathway shown. Hb, hemoglobin; MyoHb, myoglobin; Cyt, cytochromes; and NHI, nonheme iron.

increase in blood volume. Menstruating females lose iron through blood loss and lactating females through production of lactoferrin (see p. 1055). Thus iron deficiency anemia is primarily a problem for these population groups. Dietary surveys indicate that 95% or more of children and menstruating females are not obtaining adequate iron in their diet. Biochemical measurements reveal a 10–25% incidence of iron deficiency anemia in this same group. Iron deficiency anemia is also a problem with the elderly due to poor dietary intake and increased frequency of achlorhydria.

Because iron deficiency anemia is widespread, government programs of nutritional intervention such as the WIC program have emphasized iron-rich foods. There is concern among some nutritionists that iron deficiency has been overemphasized. Some recent studies have suggested that excess iron intake may increase the risk of cardiovascular disease. Thus iron supplementation and the consumption of iron-fortified foods may be inappropriate for adult men and post menopausal women. **Excess iron** can also lead to a rare condition called **hemochromatosis** in which iron deposits are found in abnormally high levels in many tissues. This can lead to liver, pancreatic, and cardiac dysfunction as well as pigmentation of the skin. Hemochromatosis is also occasionally seen in hemolytic anemias and liver disease.

Iodine Is Incorporated into Thyroid Hormones

Dietary **iodine** is efficiently absorbed and transported to the **thyroid** gland, where it is stored and used for the synthesis of the thyroid hormones triiodothyronine and thyroxine. These hormones function in regulating the **basal metabolic rate** of adults and the growth and development of children. Saltwater fish are the best natural food sources of iodine and in earlier years population groups living in inland areas suffered from the endemic deficiency disease **goiter.** The most characteristic symptom of goiter is the enlargement (sometimes massive) of the thyroid gland. Since iodine has been routinely added to table salt, goiter has become relatively rare. However, in some inland areas, mild forms of goiter may still be seen in up to 5% of the population.

Zinc Is a Cofactor for Many Enzymes

Zinc absorption appears to be proportional to **metallothionein** levels in intestinal mucosa cells. Metallothionein in zinc transport is uncertain; it may serve as a buffer for zinc ions as the metal transverses the intestinal cells. Over 300 zinc metalloenzymes have been described to date, including a number of regulatory proteins and both RNA and DNA polymerases. Zinc deficiencies in children are usually marked by poor growth and impairment of sexual development. In both children and adults zinc deficiencies result in poor wound healing. Zinc is also present in gustin, a salivary polypeptide that appears to be necessary for normal development of taste buds. Thus zinc deficiencies also lead to decreased taste acuity. Zinc is required for optional cytokine production by monocytes and T cells. Thus zinc deficiencies are also associated with impaired immune function.

The few dietary surveys indicate that zinc intake may be marginal for many individuals, and zinc supplementation has been shown to improve immune status in the elderly. Severe **zinc deficiency** is seen primarily in alcoholics (especially if they have cirrhosis), patients with chronic renal disease or severe malabsorption diseases, and occasionally in people after long-term parenteral nutrition (TPN). The most characteristic early symptom of zinc-deficient patients on TPN is dermatitis. Zinc is occasionally used therapeutically to promote wound healing and may be of some use in treating gastric ulcers.

Copper Is a Cofactor for Important Enzymes

Copper absorption may depend on metallothionein, since excess intake of either copper or zinc interferes with the absorption of the other. Copper is present in a number of important metalloenzymes, including cytochrome *c* oxidase, dopamine β-hydroxylase, superoxide dismutase, lysyl oxidase, and C_{18},Δ^9-desaturase. The C_{18},Δ^9-desaturase is responsible for converting stearic acid (a C_{18} saturated fatty acid) to oleic acid (a C_{18} monounsaturated fatty acid). This may be responsible for the fact that dietary stearic acid does not have the cholesterol-raising property of the other saturated fatty acids. Lysyl oxidase is necessary for the conversion of certain lysine residues in collagen and elastin to allysine, which is needed for cross-linking. Symptoms of **copper deficiency** include **hypercholesterolemia,** demineralization of bones, leukopenia, anemia, fragility of large arteries, and demyelination of neural tissue. Anemia appears to be due to a defect in iron metabolism because the copper-containing enzyme ferroxidase is necessary for conversion of iron from the Fe^{2+} state (in which form it is absorbed) to the Fe^{3+} state (in which form it can bind to the plasma protein transferrin). Bone demineralization and blood vessel fragility can be traced directly to defects in collagen and elastin formation. Hypercholesterolemia may be related to increases in the ratio of saturated to monounsaturated fatty acids of the C_{18} series due to reduced activity of the C_{18}, Δ^9-desaturase.

Balance studies in human volunteers indicate a minimum copper requirement of 1.0–2.6 mg day^{-1}. The RDA is 1.5–3 mg day^{-1}. Most dietary surveys find the average American diet provides only 1 mg at < 2000 cal day^{-1}. This is puzzling since few symptoms of copper deficiency have been identified in the general public. It is not known whether there exist widespread marginal copper deficiencies, or whether the copper balance studies are inaccurate. Recognizable symptoms of copper deficiency are usually seen only as a result of excess zinc intake and in Menkes' syndrome, a relatively rare X-linked hereditary disease associated with a defect in copper transport. Wilson's disease, an autosomal recessive disease, is associated with abnormal accumulation of copper in various tissues and can be treated with the naturally occurring copper chelating agent penicillamine.

Chromium Is a Component of Chromodulin

Chromium is a component of the low molecular weight protein **chromodulin,** which potentiates the effects of insulin, presumably by facilitating insulin binding to cell receptor sites. The chief symptom of **chromium deficiency** is impaired **glucose tolerance,** a result of decreased insulin effectiveness.

The frequency of chromium deficiency is unknown. The RDA for chromium has been set at 50–200 mg day^{-1} for a normal adult. The best current estimate is that the average consumption of chromium is around 30 mg day^{-1} in the United States. Unfortunately, the range of intakes is very wide (5–100 mg day^{-1}) even for individuals otherwise consuming balanced diets. Those most likely to have marginal or low intakes of chromium are individuals on low caloric intakes or consuming large amounts of processed foods. Some concern has been voiced that many Americans may be marginally deficient in chromium.

Selenium Is a Scavenger of Peroxides

Selenium occurs in selenoproteins such as **glutathione peroxidase** and **thioredoxin reductase.** These proteins contain one or more selenocysteine residues, which are incorporated through a unique cotranslational process. Glutathione peroxidase destroys peroxides in the cytosol. Since the effect of vitamin E on peroxide formation is limited primarily to the membrane, both glutathione peroxidase and vitamin E appear to be necessary for efficient scavenging of peroxides. Selenium is one of the few nutrients not removed by the milling of flour and is usually thought to be present in adequate amounts in the diet. The selenium levels are very low in the soil in certain parts of the country, however, and foods raised in these regions will be low in selenium. Fortunately, this effect is minimized by the current food distribution system, which assures that the foods marketed in any one area are derived from a number of different geographical regions. A recent clinical study suggested that supplementation with greater than RDA levels of selenium may reduce the risk of certain cancers. Other clinical studies are now in progress to evaluate this hypothesis.

Manganese, Molybdenum, Fluoride, and Boron Are Also Important

Manganese is a component of pyruvate carboxylase and probably other metalloenzymes as well. **Molybdenum** is a component of xanthine oxidase (see p. 840). **Fluoride** is known to strengthen bones and teeth and is usually added to drinking water. **Boron** may also play an important role in bone formation.

27.11 | THE AMERICAN DIET: FACT AND FALLACY

Much has been said about the supposed deterioration of the American diet. How serious a problem is this? Clearly Americans are eating much more processed food than their ancestors. These foods differ from simpler foods in that they have a higher caloric density and a lower nutrient density than the foods they replace. However, these foods are almost uniformly enriched with iron, thiamine, riboflavin, niacin, and low levels of folic acid. In many cases they are even fortified (usually as much for sales promotion as for nutritional reasons) with as many as 11–15 vitamins and minerals. Unfortunately, it is not practical to replace all of the nutrients lost, especially the trace minerals and phytonutrients such as the carotenoids. Imitation foods present a special problem in that they are usually incomplete in more subtle ways. For example, imitation cheese and imitation milkshakes that are widely sold in this country usually contain the protein and calcium one would expect of the food they replace, but often do not contain the riboflavin, which one would also obtain from these items. Fast food meals tend to be high in calories and fat, and low in certain vitamins and trace minerals. For example, the standard fast food meal provides over 50% of the calories the average adult needs for the entire day, while providing $< 5\%$ of the vitamin A and $< 30\%$ of biotin, folic acid, and pantothenic acid. Unfortunately, much of the controversy in recent years has centered around whether these trends are "good" or "bad." This simply obscures the issue at hand. Clearly it is possible to obtain a balanced diet that includes some processed, imitation, and fast foods if one compensates by selecting foods for the other meals that are low in caloric density and rich in nutrients. Without such compensation the "balanced diet" becomes a myth.

27.12 | ASSESSMENT OF NUTRITIONAL STATUS IN CLINICAL PRACTICE

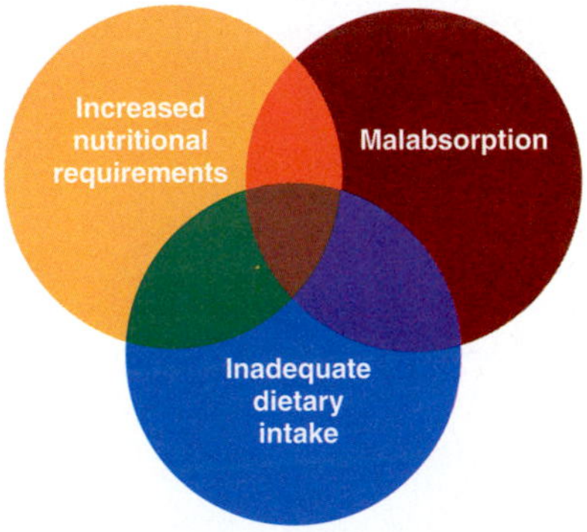

Having surveyed the major micronutrients and their biochemical roles, it might seem that the process of evaluating the **nutritional status** of an individual patient would be an overwhelming task. It is perhaps best to recognize that there are three factors that can add to nutritional deficiencies: poor diet, malabsorption, and increased nutrient need. Only when two or three components overlap in the same person (Figure 27.16) do the risks of symptomatic deficiencies become significant. For example, infants and young children have increased needs for iron, calcium, and protein. Dietary surveys show that many of them consume diets inadequate in iron and some consume diets that are low in calcium. Protein is seldom a problem unless the children are being raised as strict vegetarians. Thus the chief nutritional concerns for most children are iron and calcium. Teenagers tend to consume diets low in calcium, magnesium, vitamin A, vitamin B_6, and vitamin C. Of all these nutrients, their needs are particularly high for calcium and magnesium during the teenage years, so these are the nutrients of greatest concern. Young women are likely to consume diets low in iron, calcium, magnesium, vitamin B_6, folic acid, and

FIGURE 27.16
Factors affecting individual nutritional status. Schematic representation of three important risk factors in determining nutritional status. A person in the periphery would have very low risk of any nutritional deficiency, whereas those in the green, orange, purple, or center areas would be much more likely to experience some symptoms of nutritional deficiencies.

zinc—and all of these nutrients are needed in greater amounts during pregnancy and lactation. Adult women often consume diets low in calcium, yet they may have a particularly high need for calcium to prevent rapid bone loss. Finally, the elderly have unique nutritional needs (see Clin. Corr. 27.9) and tend to have poor nutrient intake due to restricted income, loss of appetite, and loss of the ability to prepare a wide variety of foods. They are also more prone to suffer from malabsorption problems and to use multiple prescription drugs that increase nutrient needs (Table 27.1).

Illness and **metabolic stress** often cause increased demand or decreased utilization of certain nutrients. For example, diseases leading to fat malabsorption cause a particular problem with absorption of calcium and the fat-soluble vitamins. Other malabsorption diseases can result in deficiencies of many nutrients depending on the particular malabsorption disease. Liver and kidney disease can prevent activation of vitamin D and storage or utilization of many other nutrients including vitamin A, vitamin B_{12}, and folic acid. Severe illness or trauma increase the need for calories, protein, and possibly some micronutrients such as vitamin C and certain B vitamins. Long-term use of many drugs in the treatment of chronic disease states can affect the need for certain micronutrients. Some of these are summarized in Table 27.1.

Who then is at a nutritional risk? Obviously, the answer depends on many factors. Nutritional counseling is an important part of treatment for infants, young children, and pregnant/lactating women. A brief analysis of a dietary history and further nutritional counseling are important when dealing with high-risk patients.

TABLE 27.1 Drug–Nutrient Interactions

Drug	*Potential Nutrient Deficiencies*
Alcohol	Thiamine Folic acid Vitamin B_6
Anticonvulsants	Vitamin D Folic acid Vitamin K
Cholestyramine	Fat-soluble vitamins Iron
Corticosteroids	Vitamin D and calcium Zinc Potassium
Diuretics	Potassium Zinc
Isoniazid	Vitamin B_6
Oral contraceptives and estrogens	Vitamin B_6 Folic acid and B_{12}

CLINICAL CORRELATION 27.9

Nutritional Needs of Elderly Persons

If current trends continue, one of five Americans will be over the age of 65 by the year 2030. With this projected aging of the American population, there has been increased interest in defining the nutritional needs of the elderly. Recent research shows altered needs of elderly persons for several essential nutrients. For example, the absorption and utilization of vitamin B_6 has been shown to decrease with age. Dietary surveys have consistently shown that B_6 is a problem nutrient for many Americans and the elderly appear to be no exception. Many older Americans get less than 50% of the RDA for B_6 from their diet. Vitamin B_{12} deficiency is also more prevalent in the elderly. Many older adults develop atrophic gastritis, which results in decreased acid production in the stomach. That along with a tendency toward decreased production of intrinsic factor leads to poor absorption of B_{12}. Recent research has suggested that elevated blood levels of the amino acid homocysteine may be a risk factor for atherosclerosis. Homocysteine is normally metabolized to methionine and cysteine in reactions requiring folic acid, B_{12}, and B_6. Vitamin D can be a problem as well. Many elderly do not spend much time in the sunlight and, to make matters worse, the conversion of both 7-dehydrocholesterol to vitamin D in the skin and 25-(OH)D to 1,25-$(OH)_2$D in the kidney decrease with age. These factors often combine to produce significant deficiencies of 1,25-$(OH)_2$D in the elderly, which can in turn lead to negative calcium balance. These changes do not appear to be the primary cause of osteoporosis but they certainly may contribute to it.

There is some evidence for increased need for chromium and zinc as well. Chromium is not particularly abundant in the American diet and many elderly appear to have difficulty converting dietary chromium to the biologically active chromodulin. The clinical relevance of these observations is not clear but chromium deficiency could contribute to adult-onset diabetes. Similarly, dietary surveys show that most elderly consume between one-half and two-thirds the RDA for zinc. Conditions such as atrophic gastritis can also interfere with zinc absorption. Symptoms of zinc deficiency include loss of taste acuity, dermatitis, and a weakened immune system. All of these symptoms are common in the elderly population and it has been suggested that zinc deficiency might contribute.

Not all of the news is bad, however. Vitamin A absorption increases with age and the ability of the liver to clear vitamin A from the blood decreases, so it remains in the circulation for a longer time. In fact, not only does the need for vitamin A decrease as we age, but the elderly also need to be particularly careful to avoid vitamin A toxicity. While this does not restrict their choice of foods or multivitamin supplements, they should generally avoid separate vitamin A supplements.

Munro, H. N., Suter, P. M., and Rusell, R. M. Nutritional requirements of the elderly. *Annu. Rev. Nutr.* 7:23, 1987; Russell, R. M. and Suter, P. M. Vitamin requirements of elderly people: an update. *Am. J. Clin. Nutr.* 58:4, 1993; Ubbink, J. B., Vermoak, W. J., van der Merne, A., and Becker, P. J. Vitamin B_{12}, vitamin B_6 and folate nutritional status in men with hyperhomocysteinemia. *Am. J. Clin. Nutr.* 57:47, 1993; Joosten, E., van der Berg, A., Riezler, R., Neurath, H. J., Linderbaum, J., Stabler, S. P., and Allen, R. H. Metabolic evidence that deficiencies of vitamin B_{12}, folate and vitamin B_6 occur commonly in elderly people. *Am. J. Clin. Nutr.* 58:468,1993.

BIBLIOGRAPHY

Recommended Dietary Allowances

Food and Nutrition Board of the National Academy of Sciences. *Recommended Daily Allowances,* 10th ed. Washington, DC: National Academy of Sciences, 1989.

Food and Nutrition Board of the National Academy of Sciences. Dietary reference intakes for thiamin, riboflavin, niacin, vitamin B_6, folate, vitamin B_{12}, pantothenic acid, biotin, and choline. www.nap.edu. 1999.

Food and Nutrition Board of the National Academy of Sciences. Dietary reference intakes for vitamin C, vitamin E, selenium, and carotenoids. www.nap.edu. 2000.

Yates, A. A., Schlicker, S. A., and Suitor, C. W. Dietary reference intakes: the basis for the recommendation for calcium and related nutrients, B vitamins, and choline. *J. Am. Diet. Assoc.* 98:699, 1998.

Vitamin A

Goodman, D. S. Vitamin A and retinoids in health and disease. *N. Engl. J. Med.* 310:1023, 1984.

Thurnham, D. I. and Northrop-Clews, C. A. Optimal nutrition: vitamin A and the carotenoids. *Proc. Nutr. Soc.* 58:449, 1999.

van Poppel, G. and Goldbohm, R. A. Epidemiolgic evidence for β-carotene and cancer prevention. *Am. J. Clin. Nutr.* 62:1393S,1995.

Vitamin D

Darwish, H. and DeLuca, H. F. Vitamin D-regulated gene expression. *Crit. Rev. Eukaryot. Gene Expr.* 3:89,1993.

DeLuca, H. F. The vitamin D story. A collaborative effort of basic science and clinical medicine. *FASEB J.* 2:224, 1988.

Langman, C. B. New developments in calcium and vitamin D metabolism. *Curr. Opin. Pediatr.* 12:135, 2000.

Vitamin E

Bieri, J. G., Coresh, L., and Hubbard, V. S. Medical uses of vitamin E. *N. Engl. J. Med.* 308:1063, 1983.

Brizeluis-Flohe, R., and Traber, M.G. Vitamin E: function and metabolism. *FASEB J.* 13:1145, 1999.

Das, S. Vitamin E in the genesis and prevention of cancer. A review. *Acta Oncol.* 33:615, 1994.

Meydani, S. N., Wu, D., Santos, M. S., and Hayek, M. C. Antioxidants and immune response in aged persons: overview of present evidence. *Am. J. Clin. Nutr.* 62:1462S, 1995.

Stampfer, M. J. and Rimm, E. B. Epidemiologic evidence for vitamin E in prevention of cardiovascular disease. *Am. J. Clin. Nutr.* 62:1365S,1995.

Tribble, D. L. Antioxidant consumption and the risk of coronary heart disease: emphasis on vitamin C, vitamin E, and β-carotene. A statement for healthcare professionals from the American Heart Association. *Circulation* 99:591, 1999.

Vitamin K

Binkley, N. C. and Suttie, J. W. Vitamin K nutrition and osteoporosis. *J. Nutr.* 125:1812,1995.

Lipsky, J. J. Nutritional sources of vitamin K. *Mayo Clin. Proc.* 69:462,1994.

Nelsestuen, G. L., Shah, A. M., and Harvey, S. B. Vitamin K—dependent proteins. *Vitam. Horm.* 58:355, 2000.

Vitamin B_6

Merril, A. H. Jr. and Henderson, J. M. Diseases associated with defects in vitamin B_6 metabolism or utilization. *Annu. Rev. Nutr.* 7:137, 1987.

Tully, D. B., Allgood, V. E., and Cidlowski, J. A. Modulation of steroid receptor-mediated gene expression by vitamin B_6. *FASEB J.* 8:343,1994.

Folate

Landgren, F., Israelsson, B., Lindgren, A., Hultsberg, B., Anderson, A., and Brettstrom, L. Plasma homocysteine in acute myocardial infarction: homocysteine-lowering effects of folic acid. *J. Intern. Med.* 237:381, 1995.

McNulty, H., Cuskelly, G. J., and Wood, M. Response of red blood cell folate to intervention: implications for folate recommendations for the prevention of neural tube defects. *Am. J. Clin. Nutr.* 71(suppl):13085, 2000.

Milunsky, A., Jick, H., Jick, S. S., Bruell, C. L., MacLaugin, D. S., Rothman, K. J., and Willet, W. Multivitamin/folic acid supplementation in early pregnancy reduces the prevalence of neural tube defects. *JAMA* 262:2847,1989.

Vitamin B_{12}

Seethoram, B., Bose, S., and Li, N. Cellular import of cobalamine (vitamin B_{12}). *J. Nutr.* 129:1761, 1999.

Weir, D. G. and Scott, J. M. Brain function in the elderly: role of vitamin B_{12} and folate. *Br. Med. Bull.* 55:669, 1999.

Vitamin C

Simon, J. A. Vitamin C and cardiovascular disease: a review. *J. Am. Coll. Nutr.* 11:107, 1992.

Hemila, H. and Douglas, R. M. Vitamin C and acute respiratory infections. *Int. J. Tuberc. Lung Dis.* 3:756, 1999.

Third International Conference on Vitamin C. *Annu. N. Y. Acad. Sci.* 498:1, 1987.

Simon, J. A. Vitamin C and cardiovascular disease: a review. *J. Am. Coll. Nutr.* 11:107,1992.

Choline

Food and Nutrition Board, *Dietary Reference Intakes for Thiamin, Riboflavin, Niacin, Vitamin B_6, Folate, Vitamin B_{12}, Pantothenic Acid, Biotin, and Choline.* Washington, DC: National Academy of Science Press, 1999.

Calcium

Hatton, D. C. and McCarron, D. A. Dietary calcium and blood pressure in experimental models of hypertension. *Hypertension*, 23:513,1994.

National Institutes of Health. *Optimal Calcium Intake.* NIH Consensus Statement, Vol. 12, No. 4, 1994.

Iron

Dollman, P. R. Biochemical basis for the manifestations of iron deficiency. *Annu. Rev. Nutr.* 6:13,1986.

Chromium

Mertz W. Chromium in human nutrition: a review. *J. Nutr.* 123:626,1993.

Vincent, J. B. The biochemistry of chromium. *J. Nutr.* 130:715, 2000.

Selenium

Ganther, H. E. Selenium metabolism, selenoproteins and mechanisms of cancer prevention: complexities with thioredoxin reductase. *Carcinogenesis.* 20:1657, 1999.

Other Trace Minerals

Symposium on metal metabolism and disease. *Clin. Physiol. Biochem.* 4:1, 1986.

Dietary Surveys

Block, G. Dietary guidelines and the results of food consumption surveys. *Am. J. Clin. Nutr.* 53:3565,1991.

Kritchevsy, D. Dietary guidelines. The rationale for intervention. *Cancer* 72:1011,1993.

Pao, E. M. and Mickle, S. J. Problem nutrients in the United States. *Food Technol.* 35:58, 1981.

QUESTIONS | C. N. ANGSTADT

Multiple Choice Questions

1. Recommended dietary allowances (RDAs):
   A. are standards for all individuals.
   B. meet special dietary needs arising from chronic diseases.
   C. include all nutritional needs.
   D. define optimal levels of nutrients.
   E. are useful only as general guides in evaluating the adequacy of diets.

2. The Estimated Average Requirement (EAR) of a nutrient is:
   A. the same as the RDA of that nutrient.
   B. an amount that should meet the requirement of half of the healthy individuals of a particular group.
   C. based on the observed nutrient intake of a particular group.
   D. the highest level of nutrient deemed to pose no risk or adverse health effects on the particular population.
   E. two standard deviations higher than the RDA.

3. The effects of vitamin A may include all of the following EXCEPT:
   A. prevention of anemia.
   B. serving as an antioxidant.
   C. cell differentiation.
   D. the visual cycle.
   E. induction of certain cancers.

4. Ascorbic acid may be associated with all of the following EXCEPT:
   A. iron absorption.
   B. bone formation.
   C. acute renal disease when taken in high doses.
   D. wound healing.
   E. participation in hydroxylation reactions.

5. In assessing the adequacy of a person's diet:
   A. age of the individual usually has little relevance.
   B. trauma decreases activity, and hence need for calories and possibly some micronutrients.
   C. a 24-hour dietary intake history provides an adequate basis for making a judgment.
   D. currently administered medications must be considered.
   E. intestine is the only organ whose health has substantial bearing on nutritional status.

Refer to the following for Questions 6–10.

A. calcium
B. iron
C. iodine
D. copper
E. selenium

6. Absorption is inhibited by excess dietary zinc.
7. Excess dietary protein causes rapid excretion.
8. Risk of nutritional deficiency is high in young children.
9. Unsupplemented diets of populations living in inland areas may be deficient.
10. Essential component of glutathione peroxidase.

Questions 11 and 12: Cystic fibrosis is a generalized dysfunction of the exocrine glands leading to a viscid mucus that plugs various ducts. Pulmonary infections are common and are usually the direct cause of death. Cystic fibrosis patients, however, also have severe malabsorption problems because pancreatic enzymes are deficient and there may also be a partial obstruction of the common bile duct. Malabsorption of fat, fat-soluble vitamins, and calcium is the most common, but not only, problem. Patients have increased protein and energy needs because of chronic infections.

11. Serum calcium levels are usually normal in spite of suboptimal calcium absorption and vitamin D deficiency. Serum calcium is being maintained:
   A. by low parathyroid hormone (PTH) levels inhibiting calcium excretion.
   B. by an increase in calcitonin.
   C. by increased bone resorption stimulated by elevated PTH.
   D. because the kidney reabsorbs calcium from the urine.
   E. lack of 1,25-dihydroxy vitamin D, which prevents bone from taking calcium from blood.

12. Cystic fibrosis patients are frequently on antibiotics for infections. Antibiotics exacerbate the fat malabsorption problem for obtaining:
   A. vitamin A.
   B. vitamin C.
   C. vitamin D.
   D. vitamin E.
   E. vitamin K.

Questions 13 and 14: Chronic alcoholics are at risk of multiple nutritional abnormalities because of poor diet, pathological changes in the gastrointestinal tract leading to malabsorption problems, and severe liver damage that interferes with activation and storage of certain nutrients. Neurological symptoms associated with thiamine and pyridoxine deficiencies and hematological problems associated with folate or pyridoxine deficiencies are common.

13. Cellular energy generation is severely compromised in thiamine deficiency because TPP is:
   A. a major cofactor for oxidation–reduction reactions.
   B. an essential cofactor for pyruvate and α-ketoglutarate dehydrogenases.
   C. a cofactor for transketolase.
   D. necessary for utilizing amino acids for energy via transamination.
   E. necessary for oxidation of alcohol.

14. Alcohol impairs both absorption and storage of folate. Megaloblastic erythropoiesis occurs because cells are arrested in the S phase since DNA synthesis is inhibited. DNA synthesis is inhibited in folate deficiency because tetrahydrofolate is required:
   A. in the synthesis of purine nucleotides and dTMP.
   B. in the conversion of homocysteine to methionine.
   C. for the utilization of vitamin B_{12}.
   D. because of all of the above.
   E. because of none of the above.

Problems

15. The neurological disorders seen in vitamin B_{12} deficiency are caused by progressive demyelination of nervous tissue. How does lack of B_{12} interfere with formation of the myelin sheath?

16. What is the chemical reaction in which vitamin K participates. How is this reaction involved in blood coagulation and bone formation?

ANSWERS

1. **E** RDAs should meet the needs of about 97% of the population but do not take into account special requirements. A: RDAs are designed for most individuals; exceptions occur. B: Diseases often change dietary requirements. C: Some nutritional needs may be unknown; the requirements for all known nutrients are not even clear. D: Optimal levels of nutrients are hard to define; it depends on the criterion for optimal.
2. **B** This is the first step in defining nutritional guidelines. A, E: The RDA is set at two standard deviations above the EAR. C: This is called the Adequate Intake (AI). D: This is the Total Upper Intake Level (UL).
3. **E** May have protective effects against some cancers. A: Retinyl phosphate serves as a glycosyl donor in the synthesis of certain glycoproteins, including transferrin. B: Various carotenoids are antioxidants. C: Retinol and retinoic acid may function like steroid hormones. D: Retinol cycles between the Δ^{11}-cis and all-trans forms.
4. **C** There has been speculation, *not borne out* by studies designed to shed light on the issue, that high levels of ascorbic acid could lead to oxalate kidney stones. A: Ascorbic acid aids in iron absorption by reducing iron. B: Ascorbic acid is essential for collagen synthesis, which is critical in bone formation. D, E: Ascorbic acid is required for the hydroxylation of lysine and proline residues in protocollagen and, therefore, is required for wound healing.
5. **D** Corticosteroids stimulate vitamin D inactivation. Isoniazid affects B_6, as do oral contraceptives. A: Dramatic differences may occur at different ages. B: Trauma increases caloric requirements and probably requirements for specific micronutrients. C: You cannot be sure that any 24-hour diet history is either accurate or representative of the individual's typical diet. E: While the intestine must function well enough to absorb nutrients, further metabolic changes are typically required. The metabolism of vitamin D by the liver and kidney and the conversion of β-carotene to vitamin A in the liver exemplify these interorgan interrelations.
6. **D** These two probably share the same protein, metallothionein.
7. **A** This is a concern about a diet that is very much higher than required in protein.
8. **B** Rapid growth in children causes high demands for iron.
9. **C** The problem is rare in the United States due to the common use of iodized salt.
10. **E** The selenium is bound to a cysteine residue.
11. **C** PTH promotes bone resorption and inhibits calcium excretion. A: PTH is elevated when vitamin D is low. B: Elevated calcitonin is a response to high serum calcium. D: Kidney does not do this. E: 1,25-Dihydroxy vitamin D promotes bone resorption but lack does not necessarily affect bone's ability to take up calcium from blood.
12. **E** Some of our vitamin K is obtained from bacterial synthesis in the intestine, which is wiped out by antibiotics. A, C, D: All of these require adequate fat absorption but come from foods not intestinal bacteria. B: Vitamin C is water-soluble.
13. **B** Pyruvate dehydrogenase is very sensitive to thiamine deficiency and this is on the main route to energy for carbohydrate and several amino acids. A, E: Oxidation–reduction reactions use cofactors derived from niacin or riboflavin. C: TPP is a cofactor for transketolase but this enzyme is in the hexose monophosphate pathway, which is not a major energy pathway. D: This cofactor is derived from pyridoxine.
14. **A** *De novo* synthesis of purine nucleotides and dUMP to dTMP conversion require tetrahydrofolate. These components are required for DNA synthesis. B: Tetrahydrofolate is required in this process but the process is not involved in DNA synthesis. C: Vitamin B_{12} is required for the release of tetrahydrofolate from N^5-methyltetrahydrofolate but this is not part of DNA synthesis.
15. One of the reactions in which vitamin B_{12} participates is conversion of methylmalonyl CoA to succinyl CoA (a step in the catabolism of valine and isoleucine). Methylmalonyl CoA is a competitive inhibitor of malonyl CoA in fatty acid biosynthesis, necessary for the maintenance of the myelin sheath. Secondly, methylmalonyl CoA can be used in fatty acid synthesis leading to formation of branched-chain fatty acids, which might disrupt normal membrane structure.
16. Vitamin K is necessary for carboxylation of specific glutamic acid residues in certain proteins to form γ-carboxyglutamic acid residues. In blood coagulation, this step is required for the conversion of preprothrombin to prothrombin. In bone formation, this is required to form the calcium-binding residues of the protein osteocalcin.

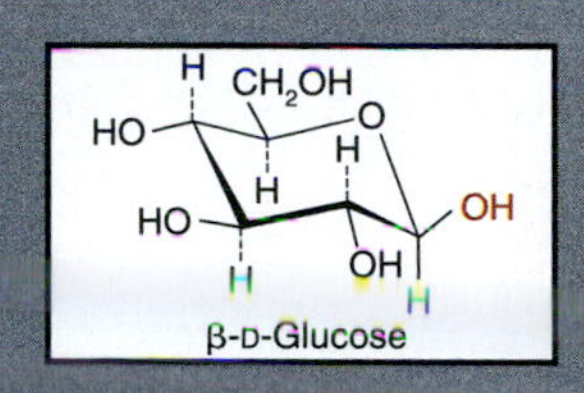

APPENDIX

REVIEW OF ORGANIC CHEMISTRY

Carol N. Angstadt

FUNCTIONAL GROUPS

Alcohols

The general formula of **alcohols** is R—OH, where R equals an alkyl or aryl group. They are classified as *primary, secondary,* or *tertiary,* according to whether the hydroxyl (OH)-bearing carbon is bonded to no carbon or one, two, or three other carbon atoms:

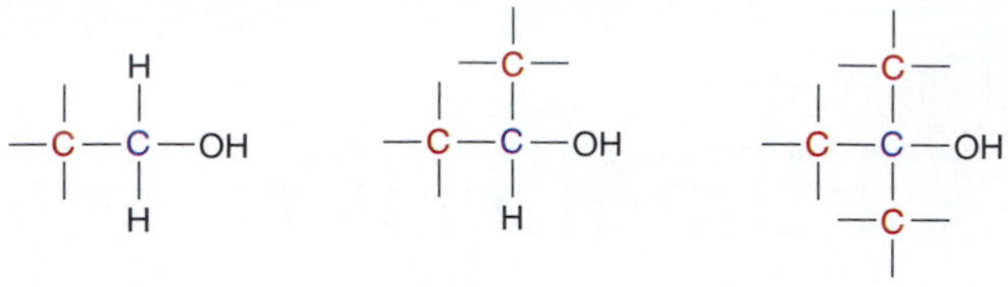

$$\text{—C—CH}_2\text{—OH} \qquad \text{(—C—)}_2\text{CH—OH} \qquad \text{(—C—)}_3\text{C—OH}$$

Primary   Secondary   Tertiary

Aldehydes and Ketones

Aldehydes and **ketones** contain a carbonyl group:

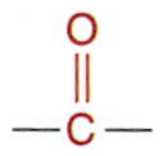

$$\text{—C(=O)—}$$

Aldehydes are

$$\text{R—C(=O)—H}$$

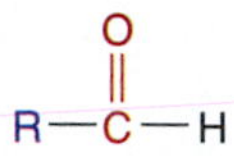

and a *ketone* has two groups (alkyl and/or aryl) at the carbonyl group

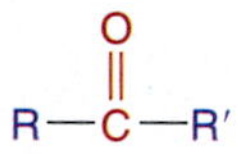

$$\text{R—C(=O)—R}'$$

Acids and Acid Anhydrides

Carboxylic acids contain the functional group

$$\text{—C(=O)—OH}$$

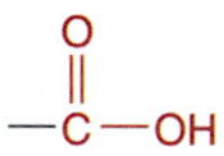

(—COOH). Dicarboxylic and tricarboxylic acids contain two or three carboxyl groups. A carboxylic acid dissociates in water to a negatively charged carboxylate ion:

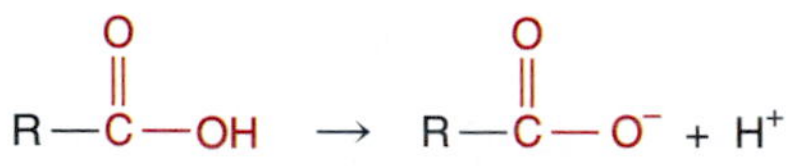

$$\text{R—C(=O)—OH} \rightarrow \text{R—C(=O)—O}^- + \text{H}^+$$

Carboxylic acid   Carboxylate ion

Names of carboxylic acids usually end in -ic and the carboxylate ion in -ate. **Acid anhydrides** are formed when two molecules of acid react with loss of a molecule of water. An acid anhydride may form between two organic acids, two inorganic acids, or an organic and an inorganic acid:

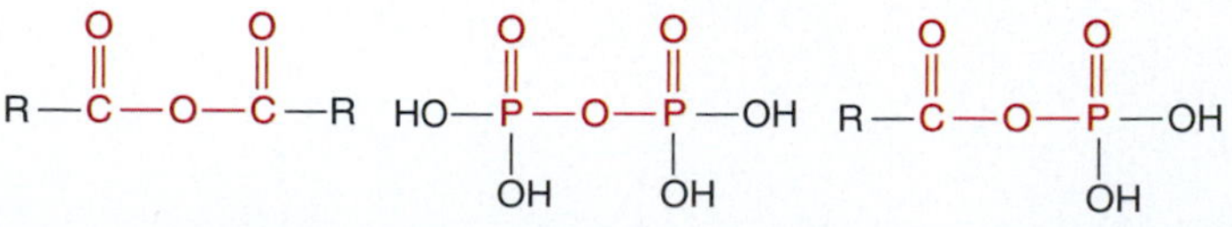

$$\text{R—C(=O)—O—C(=O)—R} \qquad \text{HO—P(=O)(OH)—O—P(=O)(OH)—OH} \qquad \text{R—C(=O)—O—P(=O)(OH)—OH}$$

Organic anhydride   Inorganic anhydride   Organic–inorganic anhydride

Esters

Esters form in the reaction between a carboxylic acid and an alcohol:

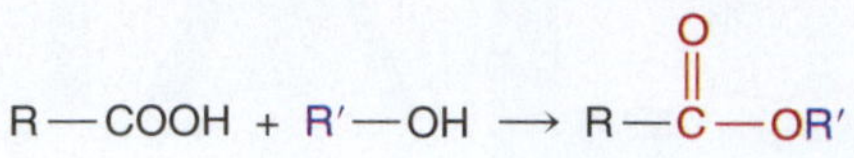

$$\text{R—COOH} + \text{R}'\text{—OH} \longrightarrow \text{R—C(=O)—OR}'$$

Esters may form between an inorganic acid and an organic alcohol, for example, glucose 6-phosphate.

Hemiacetals, Acetals, and Lactones

A reaction between an aldehyde and an alcohol gives a **hemiacetal,** which may react with another molecule of alcohol to form an **acetal:**

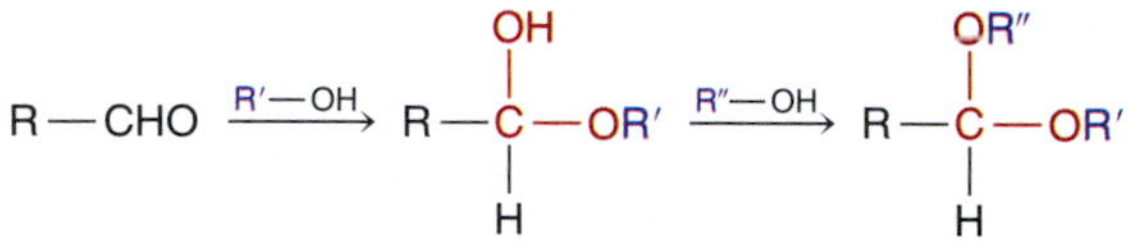

$$\text{R—CHO} \xrightarrow{\text{R}'\text{—OH}} \text{R—CH(OH)—OR}' \xrightarrow{\text{R}''\text{—OH}} \text{R—CH(OR}'')\text{—OR}'$$

Hemiacetal   Acetal

Lactones are cyclic esters formed when an acid and an alcohol group on the same molecule react and usually require that a five- or six-membered ring be formed.

Unsaturated Compounds

Unsaturated compounds are those containing one or more carbon–carbon multiple bonds, for example, a double bond:—C=C—.

Amines and Amides

Amines, R—NH_2, are organic derivatives of NH_3 and are classified as *primary, secondary,* or *tertiary,* depending on the number of alkyl groups (R) bonded to the nitrogen. When a fourth substituent is bonded to the nitrogen, the species is positively charged and called a *quaternary ammonium ion*:

$$\text{R—NH}_2 \qquad \text{R—NH—R}' \qquad \text{R—N(R}'')\text{—R}' \qquad \text{R—N}^+\text{(H(or R}'''))(\text{R}'')\text{—R}'$$

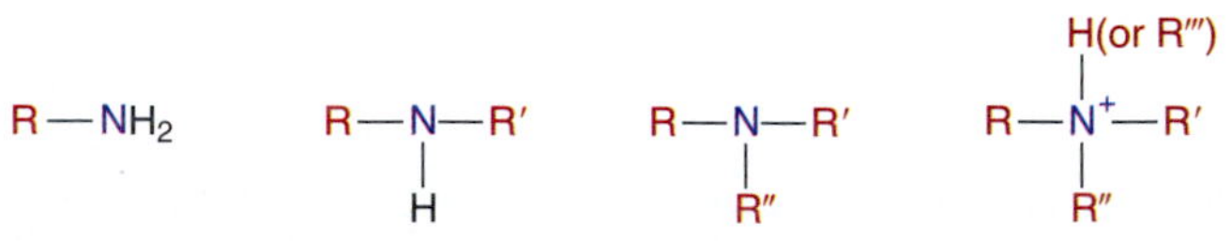

Primary amine   Secondary amine   Tertiary amine   Quaternary ammonium ion

Amides contain the functional group

$$\text{—C(=O)—N(H)—X}$$

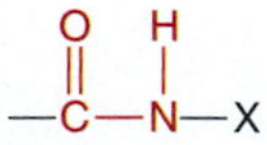

where X can be H (simple) or R (*N* substituted). The carbonyl group is from an acid, and the *N* is from an amine. If both functional groups are from amino acids, the amide bond is referred to as a **peptide bond.**

TYPES OF REACTIONS

Nucleophilic Substitutions at an Acyl Carbon

If the acyl carbon is on a carboxylic group, the leaving group is water. Nucleophilic substitution on carboxylic acids usually requires a catalyst or conversion to a more reactive intermediate; biologically this occurs via enzyme catalysis. X—H may be an alcohol (R—OH), ammonia, amine (R—NH_2), or another acyl compound. Types of nucleophilic substitutions include *esterification, peptide bond* formation, and *acid anhydride* formation.

$$\mathrm{R(L)C{=}\ddot{O}{:}} + \mathrm{{:}X{-}H} \rightleftharpoons \mathrm{R{-}C(XH^{+})(L){-}\ddot{O}{:}^{-}} \rightleftharpoons \mathrm{R{-}C(X)(LH^{+}){-}\ddot{O}{:}^{-}} \longrightarrow \mathrm{X(R)C{=}O} + \mathrm{LH}$$

New compound　Leaving group

Hydrolysis and Phosphorolysis Reactions

Hydrolysis is the cleavage of a bond by water:

$$\mathrm{R{-}C({=}O){-}OR'} + \mathrm{H_2O} \longrightarrow \mathrm{R{-}C({=}O){-}OH} + \mathrm{R'{-}OH}$$

Hydrolysis is often catalyzed by either acid or base. *Phosphorolysis* is the cleavage of a bond by inorganic phosphate:

$$\text{glucose–glucose} + \mathrm{HO{-}P({=}O)(O^{-}){-}O^{-}} \longrightarrow \text{glucose 1-phosphate} + \text{glucose}$$

Oxidation–Reduction Reactions

Oxidation is the loss of electrons; **reduction** is the gain of electrons. Examples of oxidation are as follows:

1. Fe^{2+} + acceptor → Fe^{3+} + acceptor. · e^-
2. S(ubstrate) + O_2 + DH_2 → S—OH + H_2O + D
3. S—H_2 + acceptor → S + acceptor · H_2

Some of the group changes that occur on oxidation–reduction are:

1. $\mathrm{{-}CH_2OH} \rightleftharpoons \mathrm{{-}C(H){=}O}$
2. $\mathrm{{>}C{-}OH} \rightleftharpoons \mathrm{{>}C{=}O}$
3. $\mathrm{{-}C(H){=}O} \rightleftharpoons \mathrm{{-}C({=}O){-}OH}$
4. $\mathrm{{-}CH_2NH_2} \rightleftharpoons \mathrm{{-}C(H){=}O} + \mathrm{NH_3}$
5. $\mathrm{{-}CH_2{-}CH_2{-}} \rightleftharpoons \mathrm{{-}CH{=}CH{-}}$

STEREOCHEMISTRY

Stereoisomers are compounds with the same molecular formulas and order of attachment of constituent atoms but with different arrangements of these atoms in space.

Enantiomers are stereoisomers in which one isomer is the mirror image of the other and requires the presence of a chiral atom. A chiral carbon (also called an asymmetric carbon) is one that is attached to four different groups:

$$\mathrm{C(B)(E)(D)(A)} \qquad \mathrm{C(B)(E)(A)(D)}$$

Enantiomers will be distinguished from each other by the designations *R* and *S* or D and L. The maximum number of stereoisomers possible is 2^n, where n is the number of chiral carbon atoms. A molecule with more than one chiral center will be an achiral molecule if it has a point or plane of symmetry.

Diastereomers are stereoisomers that are not mirror images of each other and need not contain chiral atoms. **Epimers** are diastereomers that contain more than one chiral carbon and differ in configuration about *only one* asymmetric carbon.

Anomers are a special form of carbohydrate epimers in which the difference is specifically about the anomeric carbon (see p. 1173). Diastereomers can also occur with molecules in which there is restricted rotation about carbon–carbon bonds. Double bonds exhibit **cis–trans isomerism.** The double bond is in the cis configuration if the two end groups of the longest contiguous chain containing the double bond are on the same side and is trans if the two ends of the longest chain are on opposite sides. Fused ring systems, such as those found in steroids (see p. 1177), also exhibit cis–trans isomerism.

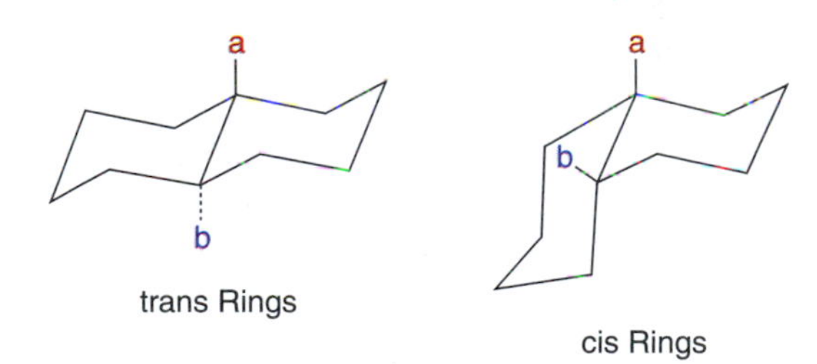

trans Rings　　cis Rings

TYPES OF FORCES INVOLVED IN MACROMOLECULAR STRUCTURES

A **hydrogen** bond is a dipole–dipole attraction between a hydrogen atom attached to an electronegative atom and a nonbonding electron pair on another electronegative atom:

$$\mathrm{{:}\ddot{X}^{\delta^-}{-}H^{\delta^+} \cdots\cdots {:}\ddot{X}^{\delta^-}{-}H^{\delta^+}}$$

Hydrogen bonds of importance in macromolecular structures occur between two nitrogen atoms, two oxygen atoms, or an oxygen and a nitrogen atom.

A **hydrophobic interaction** is the association of nonpolar groups in a polar medium. *Vander Waals* forces consist of dipole and induced dipole interactions between two nonpolar groups. A nonpolar residue dissolved in water induces a highly ordered, thermodynamically unfavorable, solvation shell. Interaction of nonpolar residues with each other, with the exclusion of water, increases the entropy of the system and is thermodynamically favorable.

Ionic (electrostatic) interactions between charged groups can be attractive if the charges are of opposite signs or repulsive if they are of the same sign. The strength of an electrostatic interaction in the interior of a protein molecule may be high. Most charged groups on the surface of a protein molecule interact with water rather than with each other. A **disulfide bond** (S—S) is a covalent bond formed by the oxidation of two sulfhydryl (SH) groups.

CARBOHYDRATES

Carbohydrates are polyhydroxy aldehydes or ketones or their derivatives. **Monosaccharides** (simple sugars) are those carbohydrates that cannot be hydrolyzed into simpler compounds. The generic name of a monosaccharide includes the type of function, a Greek prefix indicating the number of carbon atoms, and the ending -ose; for example, *aldohexose* is a six-carbon aldehyde and *ketopentose* a five-carbon ketone. Monosaccharides may react with each other to form larger molecules. With fewer than eight monosaccharides, either a Greek prefix indicating the number or the general term **oligosaccharide** may be used. **Polysaccharide** refers to a polymer with more than eight monosaccharides. Oligo- and polysaccharides may be either homologous or mixed.

Most *monosaccharides* are asymmetric, an important consideration since enzymes usually work on only one isomeric form. The simplest carbohydrates are glyceraldehyde and dihydroxyacetone whose structures, shown as Fischer projections, are as follows:

```
 H—C=O          H—C=O         CH2OH
   |              |             |
 H—C—OH        HO—C—H          C=O
   |              |             |
  CH2OH          CH2OH         CH2OH
```

D-Glyceraldehyde | L-Glyceraldehyde | Dihydroxyacetone

D-Glyceraldehyde may also be written as follows:

```
      H
     /
    C=O
    |
H ◄—C—► OH
    |
   CH2OH
```

In the Cahn–Ingold–Prelog system, the designations are (*R*) (*rectus;* right) and (*S*) (*sinister;* left).

The configuration of monosaccharides is determined by the stereochemistry at the asymmetric carbon furthest from the carbonyl carbon (number 1 for an aldehyde; lowest possible number for a ketone). Based on the *position* of the OH on the highest number asymmetric carbon, a monosaccharide is D if the OH projects to the *right* and L if it projects to the *left.* The D and L monosaccharides with the same name are **enantiomers,** and the substituents on all asymmetric carbon atoms are reversed as in

```
  H—C=O          H—C=O
    |              |
  H—C—OH        HO—C—H
    |              |
 HO—C—H          H—C—OH
    |              |
  H—C—OH        HO—C—H
    |              |
  H—C—OH        HO—C—H
    |              |
   CH2OH          CH2OH
```

D-Glucose | L-Glucose

Epimers (e.g., glucose and mannose) are stereoisomers that differ in the configuration about *only one* asymmetric carbon. The relationship of OH groups to *each other* determines the specific monosaccharide. Three aldohexoses and three pentoses of importance are

```
  H—C=O          H—C=O          H—C=O
    |              |              |
  H—C—OH        HO—C—H          H—C—OH
    |              |              |
 HO—C—H         HO—C—H         HO—C—H
    |              |              |
  H—C—OH         H—C—OH        HO—C—H
    |              |              |
  H—C—OH         H—C—OH         H—C—OH
    |              |              |
   CH2OH          CH2OH          CH2OH
```

D-Glucose | D-Mannose | D-Galactose

```
  H—C=O           CH2OH          CH2OH
    |              |              |
  H—C—OH           C=O            C=O
    |              |              |
  H—C—OH         H—C—OH        HO—C—H
    |              |              |
  H—C—OH         H—C—OH         H—C—OH
    |              |              |
   CH2OH          CH2OH          CH2OH
```

D-Ribose | D-Ribulose | D-Xylulose

Fructose, a ketohexose, differs from glucose only on carbon atoms 1 and 2:

```
     1CH2OH
       |
     2C=O
       |
  HO—3C—H
       |
   H—4C—OH
       |
   H—5C—OH
       |
     6CH2OH
```

Five- and six-carbon monosaccharides form **cyclic hemiacetals** or **hemiketals** in solution. A new asymmetric carbon is generated so two isomeric forms are possible:

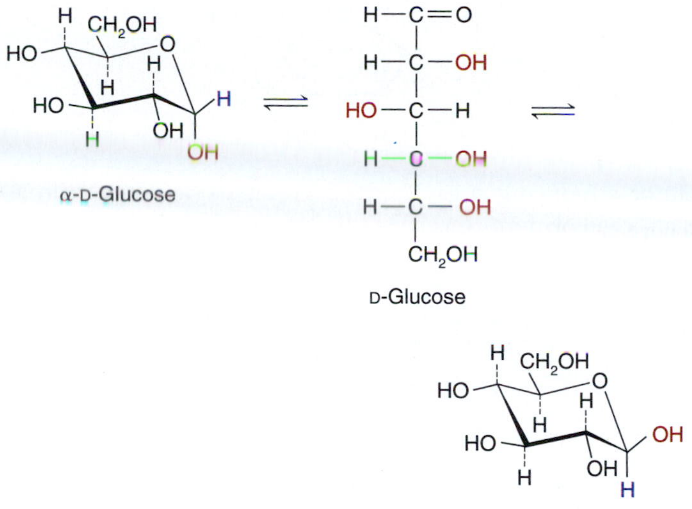

α-D-Glucose

D-Glucose

β-D-Glucose

Both five-membered (furanose) and six-membered (pyranose) ring structures are possible, although pyranose rings are more common. A furanose ring is written as follows:

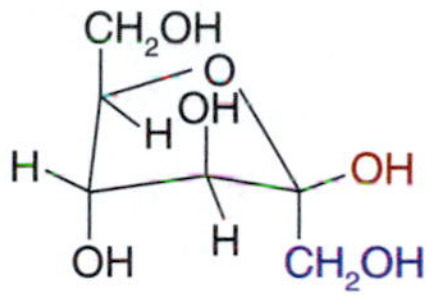

β-D-Fructose

The isomer is designated α if the OH group and the CH_2OH group on the two carbon atoms linked by the oxygen are trans to each other and β if they are cis. The hemiacetal or hemiketal forms may also be written as modified *Fischer projection formulas*: α if OH on the acetal or ketal carbon projects to the same side as the ring and β if on the opposite side:

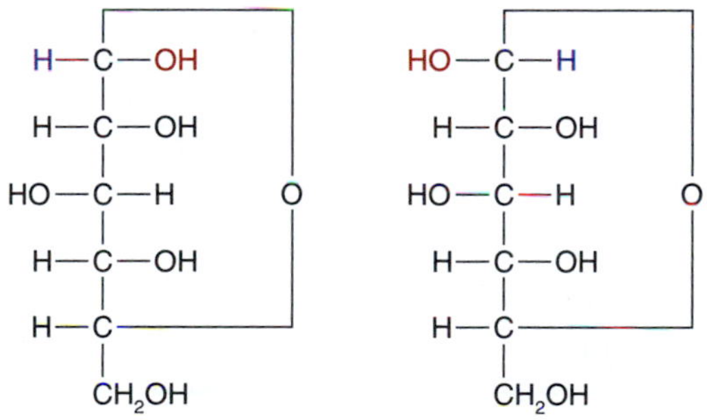

α-D-Glucose β-D-Glucose

Haworth formulas are used most commonly:

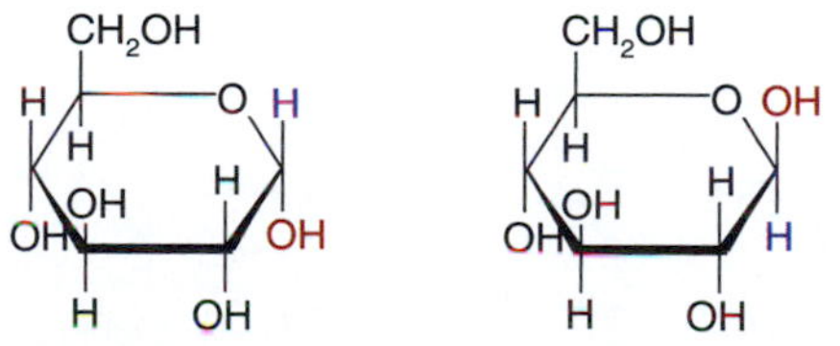

α-D-Glucose β-D-Glucose

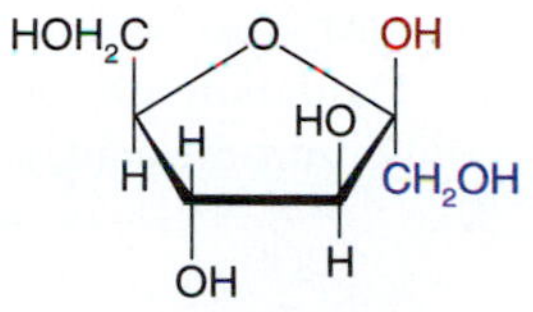

β-D-Fructose

The ring is perpendicular to the plane of the paper with the oxygen written to the back (upper) right, C-1 to the right, and substituents above or below the plane of the ring. The OH at the acetal or ketal carbon is below in the α isomer and above in the β. Anything written to the right in the Fischer projection is written down in the Haworth formula.

The α and β forms of the same monosaccharide are special forms of epimers called *anomers*, differing only in the configuration about the anomeric (acetal or ketal) carbon. Monosaccharides exist in solution primarily as a mixture of the hemiacetals (or hemiketals) but react chemically as aldehydes or ketones. **Mutarotation** is the equilibration of α and β forms through the free aldehyde or ketone. Substitution of the H of the anomeric OH prevents mutarotation and fixes the configuration in either the α or β form.

Monosaccharide Derivatives

A **deoxymonosaccharide** is one in which an OH has been replaced by H. In biological systems, this occurs at C-2 unless otherwise indicated. An **amino monosaccharide** is one in which an OH has been replaced by NH_2, again at C-2 unless otherwise specified. The amino group of an amino sugar may be *acetylated*:

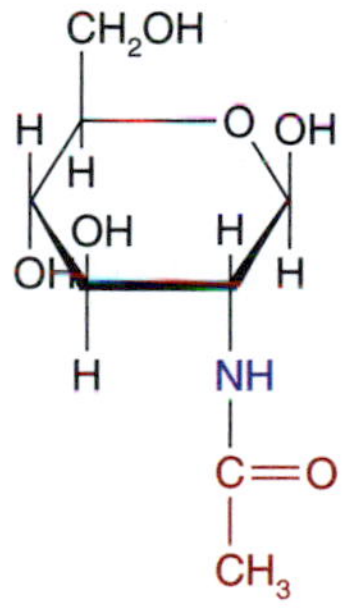

β-*N*-Acetylglucosamine

An aldehyde is reduced to a primary and a ketone to a secondary **monosaccharide alcohol (alditol).** Alcohols are named with the base name of the sugar plus the ending *-itol* or with a trivial name (glucitol = sorbitol). Monosaccharides that differ around only two of the first three carbon atoms yield the same alditol. D-Glyceraldehyde and dihydroxyacetone give glycerol:

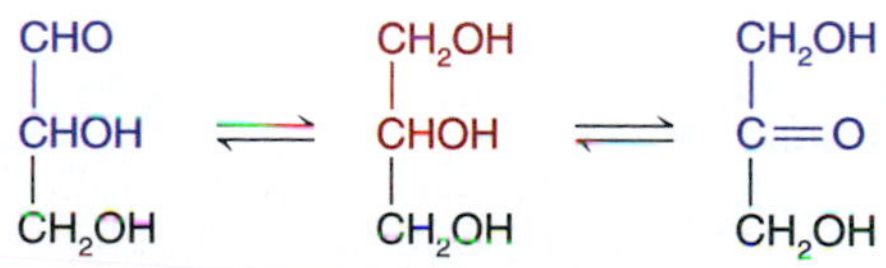

$$CHO{-}CHOH{-}CH_2OH \rightleftharpoons CH_2OH{-}CHOH{-}CH_2OH \rightleftharpoons CH_2OH{-}C{=}O{-}CH_2OH$$

D-Glucose and D-fructose give D-sorbitol; D-fructose and D-mannose give D-mannitol. Oxidation of the terminal CH_2OH, but not of the CHO, yields a **-uronic acid,** a *monosaccharide acid*:

D-Glucuronic acid

Oxidation of the CHO, but not the CH_2OH, gives an **-onic acid:**

D-Gluconic acid

D-Glyceric acid

Oxidation of both the CHO and CH_2OH gives an **-aric acid:**

D-Glucaric acid

Ketones do not form acids. Both -onic and -uronic acids can react with an OH in the same molecule to form a **lactone** (see p. 1170):

D-Glucono-5-lactone

L-Ascorbic acid (derivative of L-gulose)

Reactions of Monosaccharides

The most common *esters* of monosaccharides are phosphate esters at carbon atoms 1 and/or 6:

To be a **reducing sugar,** mutarotation must be possible. In alkali, enediols form that may migrate to 2,3 and 3,4 positions:

Enediols may be oxidized by O_2, Cu^{2+}, Ag^+, and Hg^{2+}. Reducing ability is more important in the laboratory than physiologically. A hemiacetal or hemiketal may react with the OH of another monosaccharide to form a disaccharide (*acetal: glycoside*) (see below):

(May be either α or β)

α-1,4-Glycosidic linkage

One monosaccharide still has a free anomeric carbon and can react further. Reaction of the anomeric OH may be with any OH on the other monosaccharide, including the anomeric one. The anomeric OH that has reacted is fixed as either α or β and cannot mutarotate or reduce. If the glycosidic bond is not between two anomeric carbon atoms, one of the units will still be free to mutarotate and reduce.

Oligo- and Polysaccharides

Disaccharides have two monosaccharides, either the same or different, in glycosidic linkage. If the glycosidic linkage is between the two anomeric carbon atoms, the disaccharide is nonreducing:

Maltose

Isomaltose

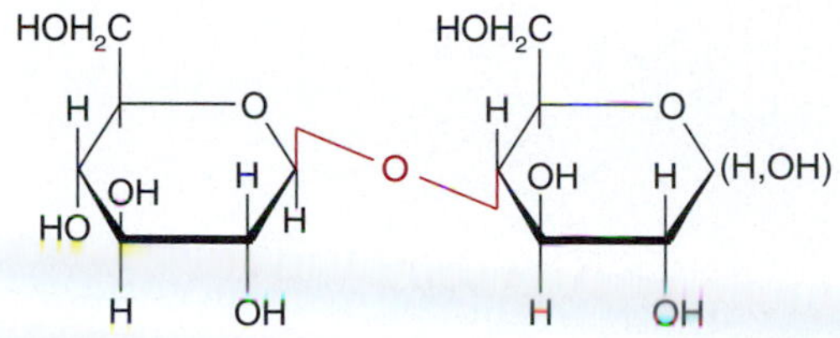

Cellobiose

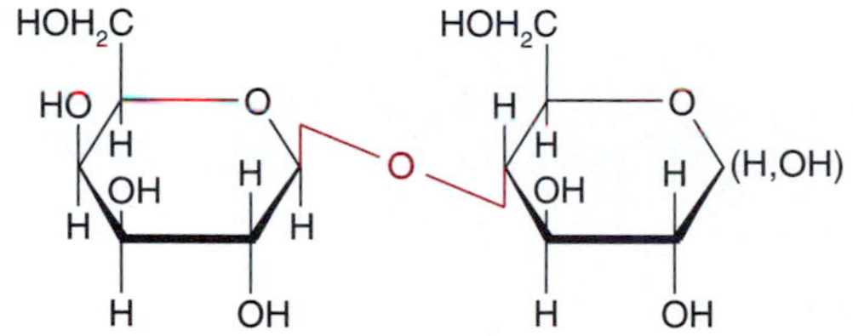

Lactose

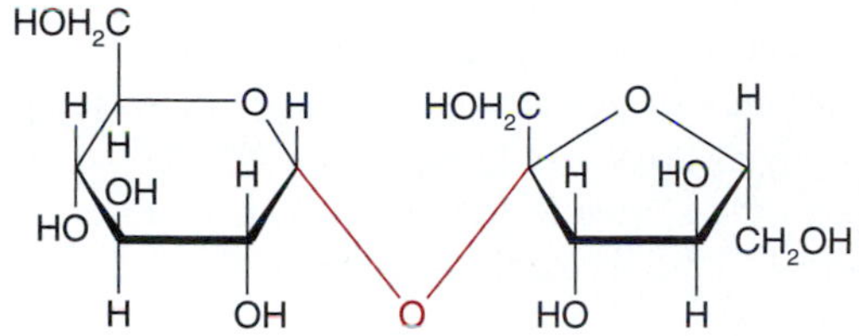

Sucrose

Maltose = 4-*O*-(α-D-glucopyranosyl)D-glucopyranose; reducing

Isomaltose = 6-*O*-(α-D-glucopyranosyl)D-glucopyranose; reducing

Cellobiose = 4-*O*-(β-D-glucopyranosyl)D-glucopyranose; reducing

Lactose = 4-*O*-(β-D-galactopyranosyl)D-glucopyranose; reducing

Sucrose = α-D-glucopyranosyl-β-D-fructofuranoside; nonreducing

As many as thousands of monosaccharides, either the same or different, may be joined by glycosidic bonds to form *polysaccharides.* The anomeric carbon of one unit is usually joined to C-4 or C-6 of the next unit. The ends of a polysaccharide are not identical (reducing end = free anomeric carbon; nonreducing = anomeric carbon linked to next unit; branched polysaccharide = more than one nonreducing end). The most common carbohydrates are homopolymers of glucose; for example, starch, glycogen, and cellulose. Plant starch is a mixture of **amylose,** a linear polymer of maltose units, and **amylopectin,** branches of repeating maltose units (glucose–glucose in α-1,4 linkages) joined via isomaltose linkages. **Glycogen,** the storage form of carbohydrate in animals, is similar to amylopectin, but the branches are shorter and occur more frequently. **Cellulose,** in plant cell walls, is a linear polymer of repeating cellobioses (glucose–glucose in β-1,4 linkages).

Mucopolysaccharides contain amino sugars, free and acetylated, uronic acids, sulfate esters, and sialic acids in addition to the simple monosaccharides. **N-Acetylneuraminic acid,** a sialic acid, is

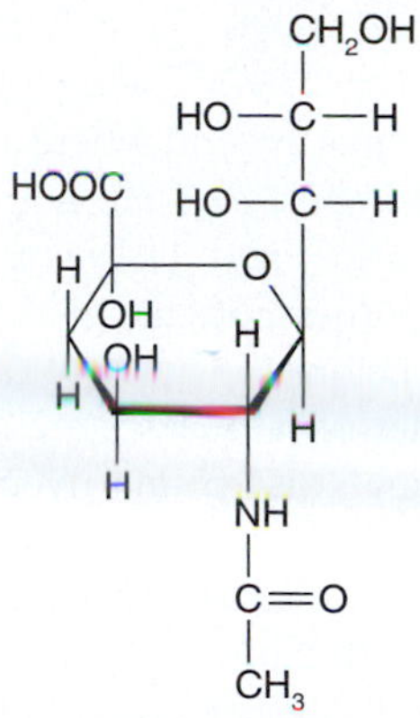

LIPIDS

Lipids are a diverse group of chemicals related primarily because they are insoluble in water, soluble in nonpolar solvents, and found in animal and plant tissues.

Saponifiable lipids yield salts of fatty acids upon alkaline hydrolysis. *Acylglycerols* = glycerol + fatty acid(s); *phosphoacylglycerols* = glycerol + fatty acids + HPO_4^{2-} + alcohol; *sphingolipids* = sphingosine + fatty acid + polar group (phosphorylalcohol or carbohydrate); *waxes* = long-chain alcohol + fatty acid. *Nonsaponifiable lipids* (*terpenes, steroids, prostaglandins,* and related compounds) are not usually subject to hydrolysis. *Amphipathic* lipids have both a polar "head" group and a nonpolar "tail." Amphipathic molecules can stabilize emulsions and are responsible for the lipid bilayer structure of membranes.

Fatty acids are monocarboxylic acids with a short (<6 carbon atoms), medium (8–14 carbon atoms), or long (>14 carbon atoms) aliphatic chain. Biologically important ones are usually linear molecules with an even number of carbon atoms (16–20). Fatty acids are numbered using either arabic numbers (COOH is 1) or the Greek alphabet (COOH is not given a symbol; adjacent carbon atoms are α, β, γ, etc.). **Saturated fatty acids** have the general formula $CH_3(CH_2)_nCOOH$. (*Palmitic acid* = C_{16}; *stearic acid* = C_{18}.) They tend to be extended chains and solid at room temperature unless the chain is short. Both trivial and systematic (prefix indicating number of carbon atoms + *anoic acid*) names are used. $CH_3(CH_2)_{14}COOH$ = palmitic acid or hexadecanoic acid.

Unsaturated fatty acids have one or more double bonds. Most naturally occurring fatty acids have cis double bonds and are usually liquid at room temperature. Fatty acids with trans double bonds tend to have higher melting points. A double bond is indicated by Δ^n, where n is the number of the first carbon of the bond. *Palmitoleic* = Δ^9-hexadecenoic acid; *oleic* = Δ^9-octadecenoic acid; *linoleic* = $\Delta^{9,12}$-octadecadienoic acid; *linolenic* = $\Delta^{9,12,15}$-octadecatrienoic acid; and *arachidonic* = $\Delta^{5,8,11,14}$-eicosatetraenoic acid. Since fatty acids are elongated *in vivo* from the carboxyl end, biochemists use alternate terminology to assign these fatty acids to families: omega (ω) minus x (or $n - x$), where x is the number of carbon atoms from the methyl end where a double bond is first encountered. *Palmitoleic* and *oleic* are $\omega - 9$ acids, *linoleic* and *arachidonic* are $\omega - 6$ acids, and *linolenic* is an $\omega - 3$ acid. Addition of carbon atoms does not change the family to which an unsaturated fatty acid belongs.

Since the pK values of fatty acids are about 4–5, in physiological solutions, they exist primarily in the ionized form, called salts or "soaps." Long-chain fatty acids are insoluble in water, but soaps form micelles. Fatty acids form esters with alcohols and thioesters with CoA.

Biochemically significant reactions of unsaturated fatty acids are:

1. *Reduction* —CH=CH— + XH_2 → —CH_2CH_2— + X
2. *Addition of water* —CH=CH— + H_2O → —CH(OH)—CH_2—
3. *Oxidation* R—CH=CH—R′ → R—CHO + R′—CHO

Prostaglandins, thromboxanes, and *leukotrienes* are derivatives of C_{20}, polyunsaturated fatty acids, especially arachidonic acid. **Prostaglandins** have the general structure:

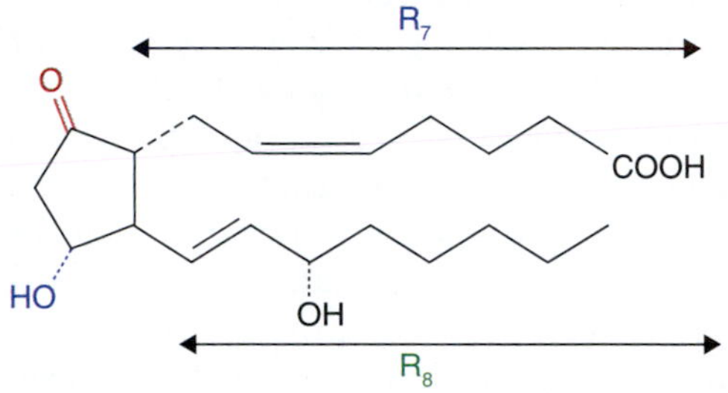

PGE_2

The series differ from each other in the substituents on the ring and whether C-15 contains an OH or O · OH group. The subscript indicates the number of double bonds in the side chains. Substituents indicated by —(β) are above the plane of the ring; ··· (α) below:

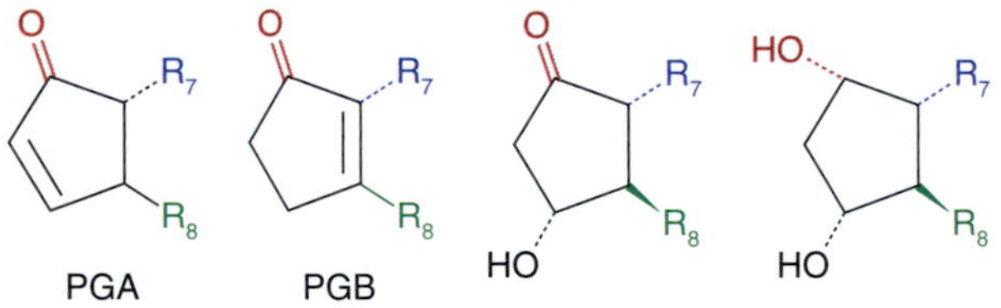

PGA PGB PGE PGF

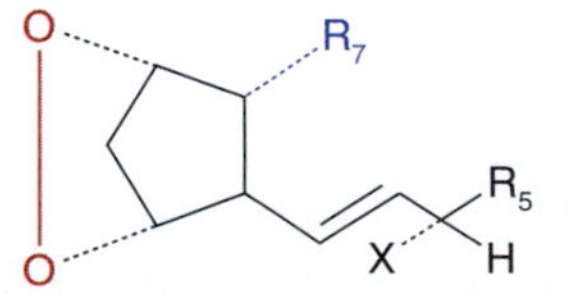

PGG(X=OH); PGH(X=OOH) 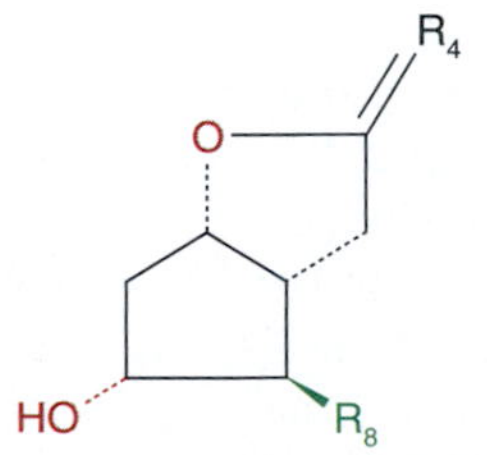 PGI

Thromboxanes have an oxygen incorporated to form a six-membered ring:

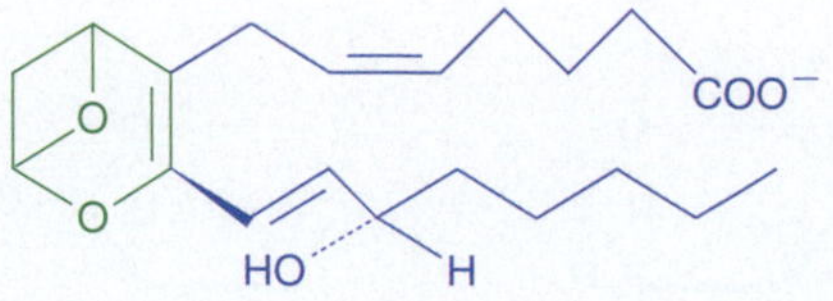

TXA_2

Leukotrienes are substituted derivatives of arachidonic acid in which no internal ring has formed; R is variable:

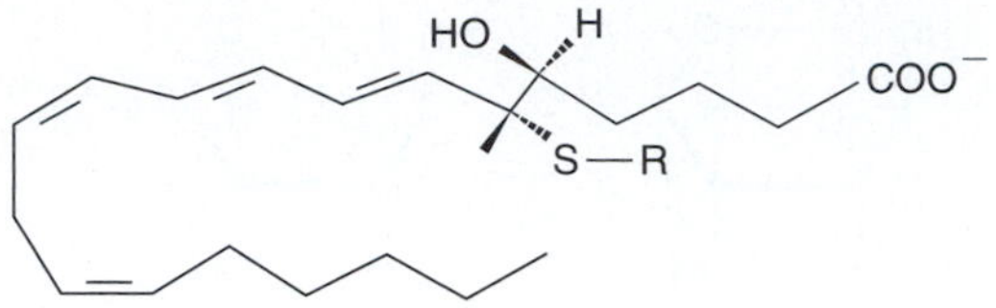

Leukotriene C, D, or E

Acylglycerols are compounds in which one or more of the three OH groups of glycerol is esterified. In **triacylglycerols** (triglycerides) all three OH groups are esterified to fatty acids. At least two of the three R groups are usually different. If R_1 is not equal to R_3, the molecule is asymmetric and of the L configuration:

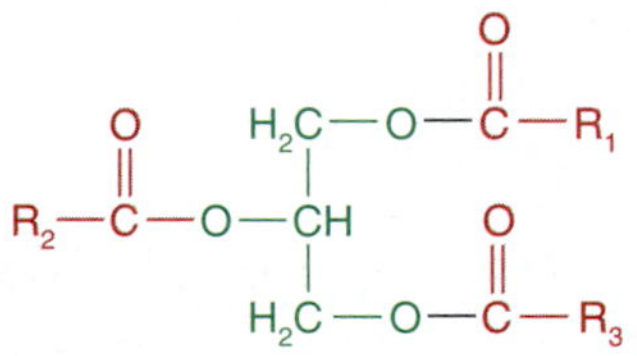

The properties of the triacylglycerols are determined by those of the fatty acids they contain: with *oils,* liquids at room temperature (preponderance of short-chain and/or cis-unsaturated fatty acids), and *fats,* solid (preponderance of long-chain, saturated, and/or trans-unsaturated).

Triacylglycerols are hydrophobic and do not form stable micelles. They may be hydrolyzed to glycerol and three fatty acids by strong alkali or enzymes (lipases). *Mono-* [usually with the fatty acid in the β(2) position] and *diacylglycerols* also exist in small amounts as metabolic intermediates. Mono- and diacylglycerols are slightly more polar than triacylglycerols. *Phosphoacylglycerols* are derivatives of L-α-glycerolphosphate (L-glycerol 3-phosphate):

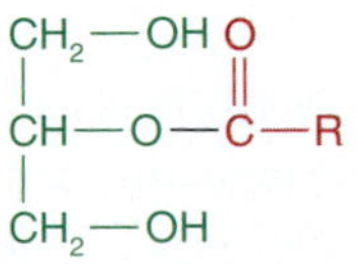

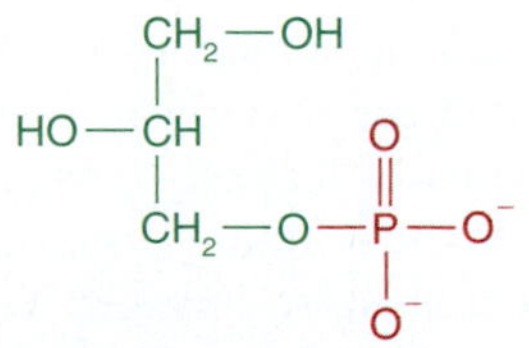

The parent compound, **phosphatidic acid** (two OH groups of L-α-glycerolphosphate esterified to fatty acids), has its phosphate esterified to an alcohol (XOH) to form several series of phosphoacylglycerols. These are amphipathic molecules, but the net charge at pH 7.4 depends on the nature of X—OH.

X—OH	*Phosphoacylglycerol*
$HO—CH_2—CH_2—\overset{+}{N}—(CH_3)_3$	Phosphatidylcholines (lecithins)
$HO—CH_2—CH_2—\overset{+}{N}H_3$	Phosphatidylethanolamines (cephalins)
$HO—CH_2—CH(\overset{+}{N}H_3)—COO^-$	Phosphatidylserines
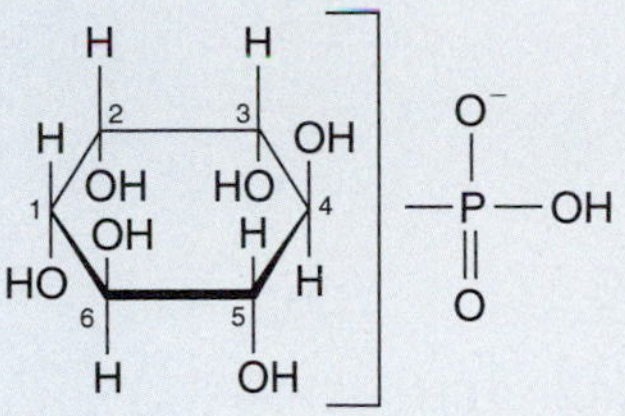	Phosphatidylinositols phospate on 4, or 4 and 5

In **plasmalogens**, the OH on C-1 is in *ether,* rather than ester, linkage to an alkyl group. If *one* fatty acid (usually β) has been hydrolyzed from a phosphoacylglycerol, the compound is a *lyso*-compound; for example, lysophosphatidylcholine (lysolecithin):

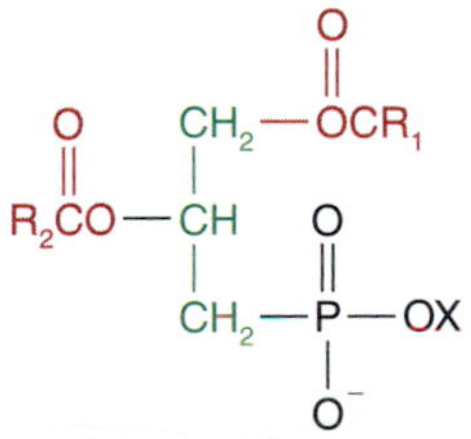

A phosphoacylglycerol

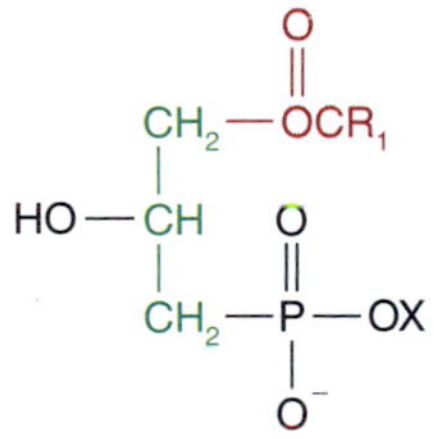

A lyso-compound

Sphingolipids are complex lipids based on the C-18, unsaturated alcohol, sphingosine. In *ceramides,* a long-chain fatty acid is in amide linkage to sphingosine:

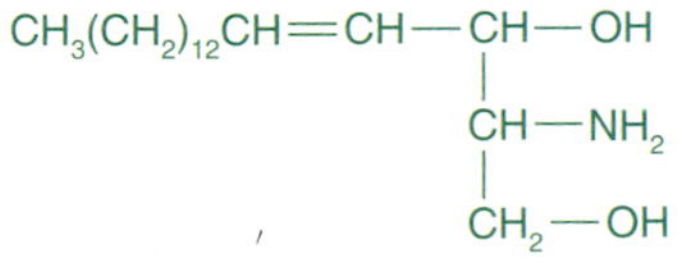

Sphingosine

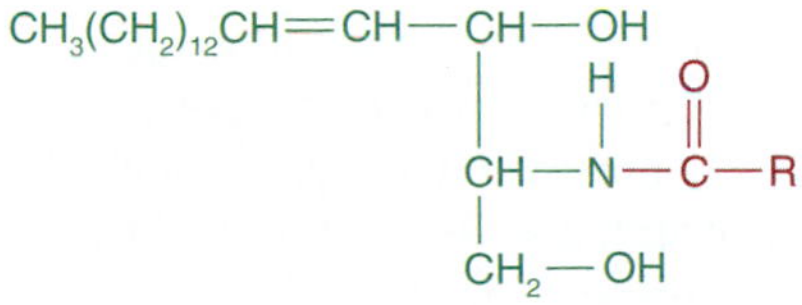

A ceramide

Sphingomyelins, the most common sphingolipids, are a family of compounds in which the primary OH group of a ceramide is esterified to phosphorylcholine (phosphorylethanolamine):

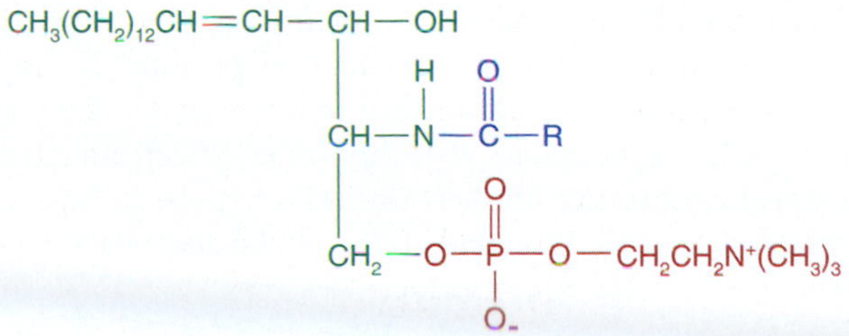

They are amphipathic molecules, existing as zwitterions at pH 7.4 and the only sphingolipids that contain phosphorus. *Glycosphingolipids* do not contain phosphorus but contain carbohydrate in glycosidic linkage to the primary alcohol of a ceramide. They are amphipathic and either neutral or acidic if the carbohydrate moiety contains an acidic group. **Cerebrosides** have a single glucose or galactose linked to a ceramide. *Sulfatides* are galactosylceramides esterified with sulfate at C-3 of the galactose:

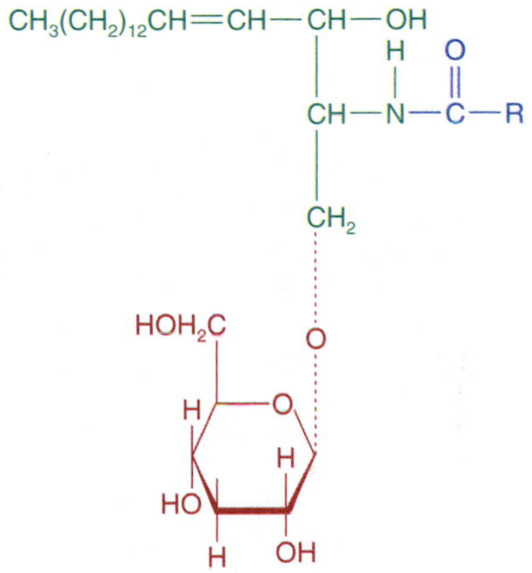

Glucosylceramide (glucocerebroside)

Globosides (*ceramide oligosaccharides*) are ceramides with two or more neutral monosaccharides, whereas **gangliosides** are an oligosaccharide containing one or more sialic acids.

Steroids are derivatives of cyclopentanoperhydrophenanthrene. The steroid nucleus is a rather rigid, essentially planar structure with substituents above or below the plane of the rings designated β (solid line) and those below called α (dotted line):

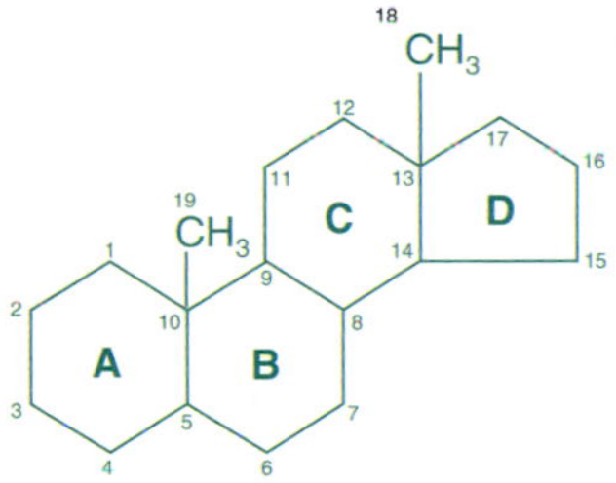

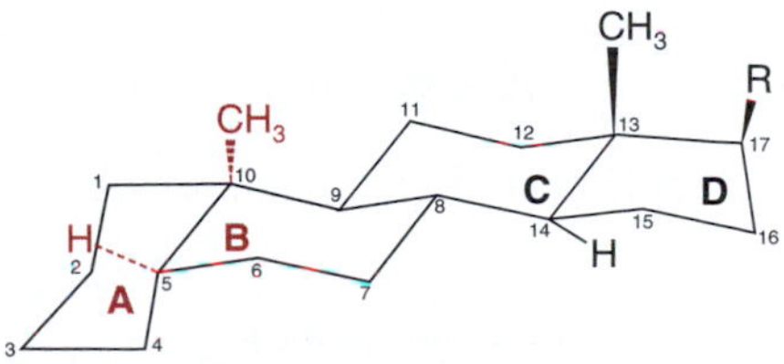

A and B rings—cis;
the others—trans

Most steroids in humans have methyl groups at positions 10 and 13 and frequently a side chain at position 17. *Sterols* contain one or more OH groups, free or esterified to a fatty acid. Most steroids are nonpolar. In a liposome or cell membrane, **cholesterol** orients with the OH toward any polar groups; cholesterol esters do not. **Bile acids** (e.g., cholic acid) have a polar side chain and so are amphipathic:

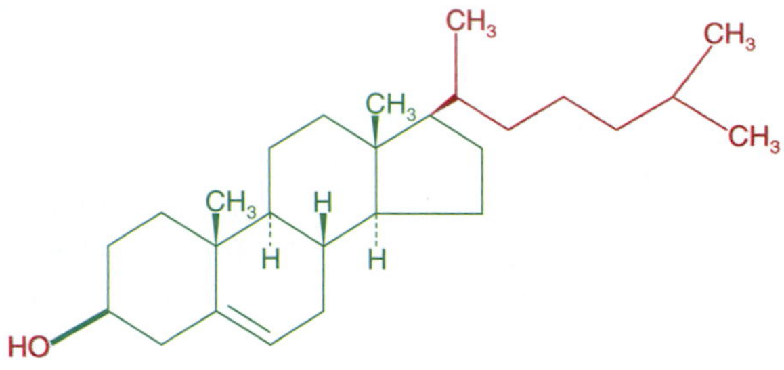

Cholesterol

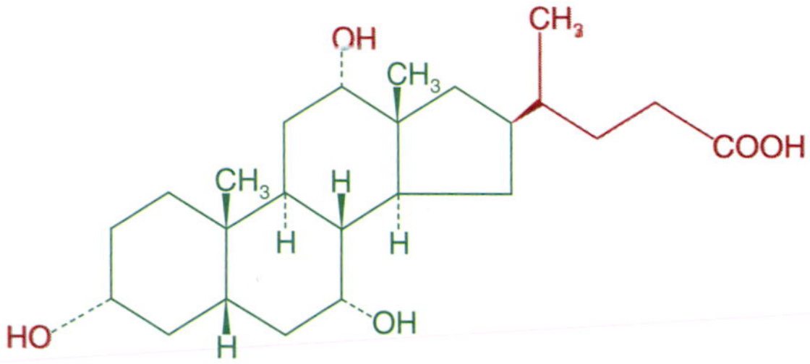

Cholic acid

Steroid hormones are oxygenated steroids of 18–21 carbon atoms. *Estrogens* have 18 carbon atoms, an aromatic ring A, and no methyl at C-10. *Androgens* have 19 carbon atoms and no side chain at C-17. *Glucocorticoids* and *mineralocorticoids* have 21 carbon atoms including a C_2, oxygenated side chain at C-17. *Vitamin D_3 (cholecalciferol)* is not a sterol but is derived from 7-dehydrocholesterol in humans:

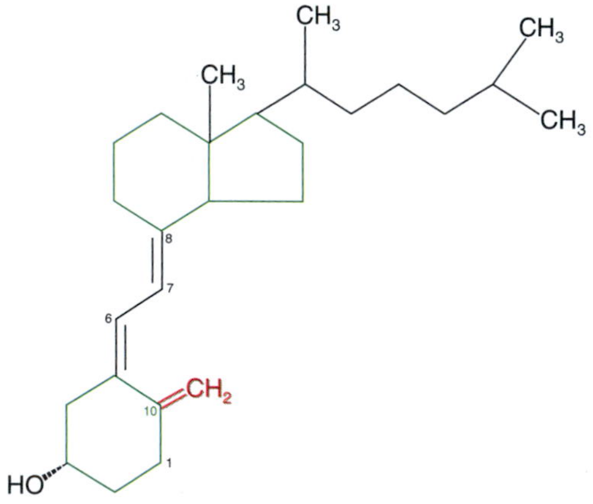

Cholecalciferol

Terpenes are polymers of two or more isoprene units. **Isoprene** is

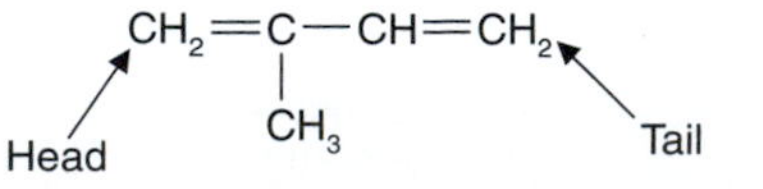

$$CH_2{=}C(CH_3){-}CH{=}CH_2$$

Head   Tail

Terpenes may be linear or cyclic, with the isoprenes usually linked head to tail and most double bonds trans (but may be cis as in vitamin A). *Squalene,* the precursor of cholesterol, is a linear terpene of six isoprene units. Fat-soluble *vitamins* (A, D, E, and K) contain isoprene units:

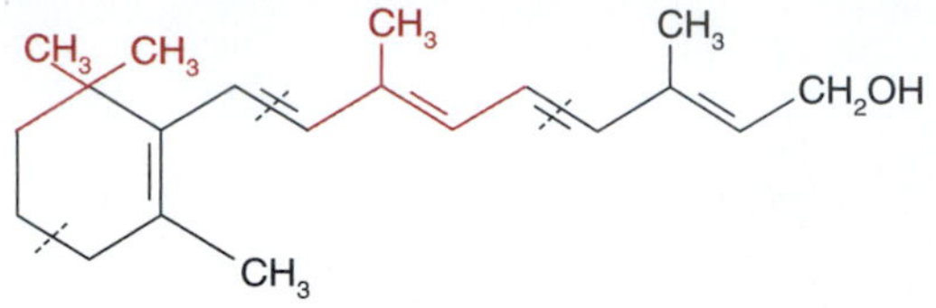

Vitamin A

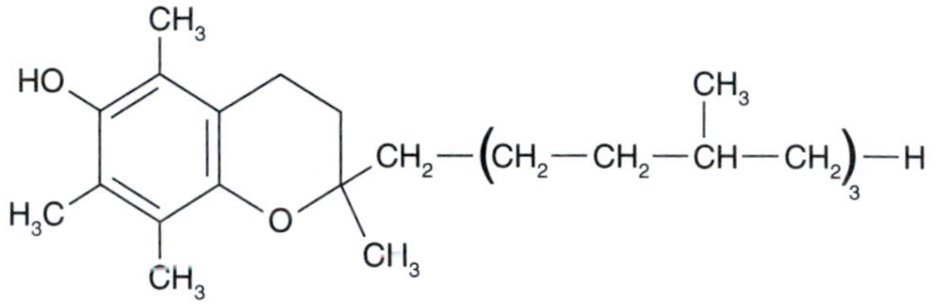

Vitamin E (α-tocopherol)

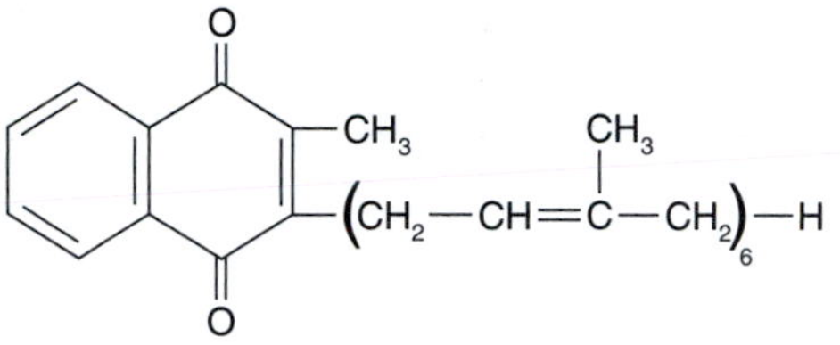

Vitamin K_2

AMINO ACIDS

Amino acids contain both an *amino* (NH_2) and a *carboxylic acid* (COOH) group. Biologically important amino acids are usually α-amino acids with the formula

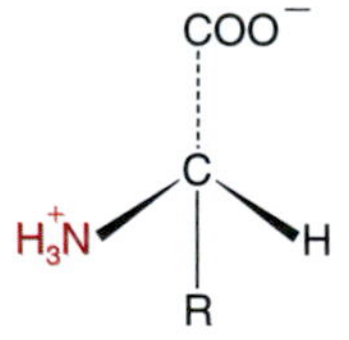

L-α-Amino acid

The amino group, with an unshared pair of electrons, is basic, with a pK_a of about 9.5, and exists primarily as—NH_3^+ at pH values near neutrality. The carboxylic acid group (p$K \sim 2.3$) exists primarily as a carboxylate ion. If R is anything but H, the molecule is asymmetric with most naturally occurring ones of the L configuration (same relative configuration as L-glyceraldehyde: see p. 1172).

The *polarity* of amino acids is influenced by their side chains (R groups) (see p. 96 for complete structures). *Nonpolar* amino acids include those with large, aliphatic, aromatic, or undissociated sulfur groups (aliphatic = Ala, Ile, Leu, Val; aromatic = Phe, Trp; sulfur = Cys, Met). *Intermediate* polarity amino acids include Gly, Pro, Ser, Thr, and Tyr (undissociated).

Amino acids with ionizable side chains are *polar.* The pK values of the side groups of arginine, lysine, glutamate, and aspartate are such that these are nearly always charged at physiological pH, whereas the side groups of histidine (pK = 6.0) and cysteine (pK = 8.3) exist as both charged and uncharged species at pH 7.4 (acidic = Glu, Asp, Cys; basic = Lys, Arg, His). Although undissociated cysteine is nonpolar, cysteine in dissociated form is polar.

All amino acids are at least *dibasic acids* because of the presence of both the α-amino and α-carboxyl groups, the ionic state being a function of pH. The presence of another ionizable group will give a tribasic acid as shown for cysteine.

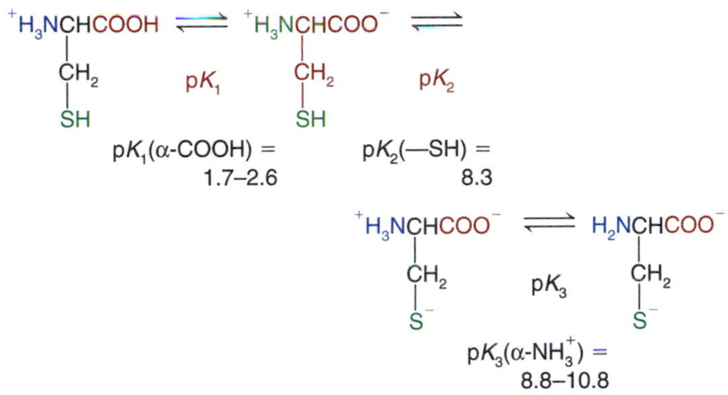

$^{+}H_3NCHCOOH$ ⇌ (pK_1) $^{+}H_3NCHCOO^-$ ⇌ (pK_2) $^{+}H_3NCHCOO^-$ ⇌ (pK_3) $H_2NCHCOO^-$
(side chains: CH_2SH, CH_2SH, CH_2S^-, CH_2S^-)

pK_1(α-COOH) = 1.7–2.6
pK_2(—SH) = 8.3
pK_3(α-NH_3^+) = 8.8–10.8

The **zwitterionic form** is the form in which the *net* charge is zero. The *isoelectric point* is the average of the two pK values involved in the formation of the zwitterionic form. In the above example this would be the average of $pK_1 + pK_2$.

PURINES AND PYRIMIDINES

Purines and pyrimidines, often called *bases*, are nitrogen-containing heterocyclic compounds with the structures

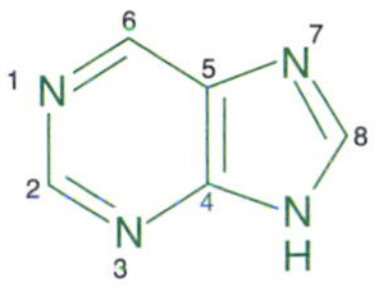

Purine

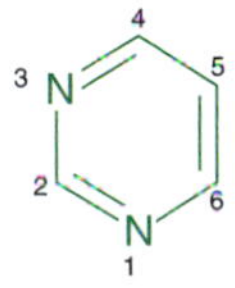

Pyrimidine

Major bases found in nucleic acids and as cellular nucleotides are the following:

Purines	*Pyrimidines*
Adenine: 6-amino	Cytosine: 2-oxy, 4-amino
Guanine: 2-amino, 6-oxy	Uracil: 2,4-dioxy
	Thymine: 2,4-dioxy, 5-methyl
Other important bases found primarily as intermediates of synthesis and/or degradation are	
Hypoxanthine: 6-oxy	Orotic acid: 2,4-dioxy, 6-carboxy
Xanthine: 2,6-dioxy	

Oxygenated purines and pyrimidines exist as *tautomeric* structures with the keto form predominating and involved in hydrogen bonding between bases in nucleic acids:

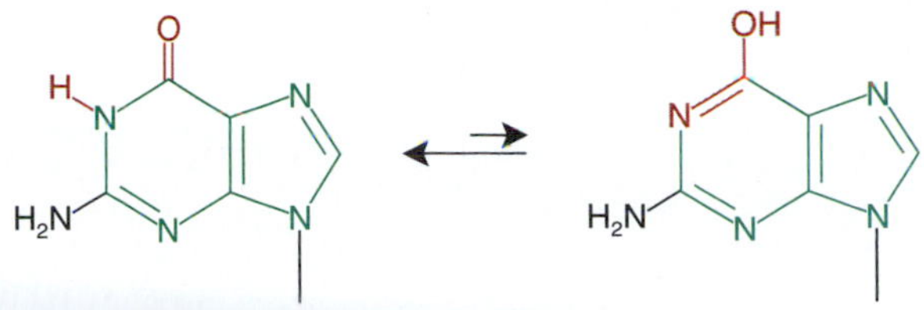

Keto ⟷ Enol

Nucleosides have either β-D-ribose or β-D-2-deoxyribose in an *N*-glycosidic linkage between C-1 of the sugar and N-9 (purine) or N-1 (pyrimidine).

Nucleotides have one or more phosphate groups esterified to the sugar. Phosphates, if more than one is present, are usually attached to each other via phosphoanhydride bonds. Monophosphates may be designated as either the base monophosphate or as an *-ylic acid* (AMP: adenylic acid):

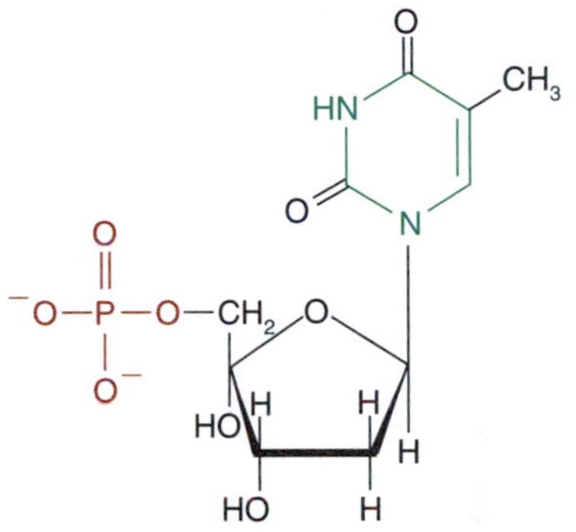

By conventional rules of *nomenclature*, the atoms of the base are numbered 1–9 in purines or 1–6 in pyrimidines and the carbon atoms of the sugar 1′–5′. A nucleoside with an unmodified name indicates that the sugar is ribose and the phosphate(s) is/are attached at C-5′ of the sugar. Deoxy forms are indicated by the prefix d (dAMP = deoxyadenylic acid). If the phosphate is esterified at any position other than 5′, it must be so designated [3′-AMP; 3′-5′-AMP (cyclic AMP = cAMP)]. The nucleosides and nucleotides (ribose form) are named as follows:

Base	*Nucleoside*	*Nucleotide*
Adenine	Adenosine	AMP, ADP, ATP
Guanine	Guanosine	GMP, GDP, GTP
Hypoxanthine	Inosine	IMP
Xanthine	Xanthosine	XMP
Cytosine	Cytidine	CMP, CDP, CTP
Uracil	Uridine	UMP, UDP, UTP
Thymine	dThymidine	dTMP, dTTP
Orotic acid	Orotidine	OMP

Minor (modified) bases and nucleosides also exist in nucleic acids. *Methylated* bases have a methyl group on an amino group (*N*-methyl guanine), a ring atom (1-methyl adenine), or on an OH group of the sugar (2′-*O*-methyl adenosine). *Dihydrouracil* has the 5–6 double bond saturated. In *pseudouridine,* the ribose is attached to C-5 rather than to N-1.

In **polynucleotides** (*nucleic acids*), the mononucleotides are joined by phosphodiester bonds between the 3′-OH of one sugar (ribose or deoxyribose) and the 5′-OH of the next (see p. 33 for the structure).

INDEX